2022 IEEE 49th Photovoltaics Specialists Conference (PVSC 2022)

Philadelphia, Pennsylvania, USA
5-10 June 2022

Pages 1-683

IEEE Catalog Number: CFP22PSC-POD
ISBN: 978-1-7281-6118-1

**Copyright © 2022 by the Institute of Electrical and Electronics Engineers, Inc.
All Rights Reserved**

Copyright and Reprint Permissions: Abstracting is permitted with credit to the source. Libraries are permitted to photocopy beyond the limit of U.S. copyright law for private use of patrons those articles in this volume that carry a code at the bottom of the first page, provided the per-copy fee indicated in the code is paid through Copyright Clearance Center, 222 Rosewood Drive, Danvers, MA 01923.

For other copying, reprint or republication permission, write to IEEE Copyrights Manager, IEEE Service Center, 445 Hoes Lane, Piscataway, NJ 08854. All rights reserved.

*** This is a print representation of what appears in the IEEE Digital Library. Some format issues inherent in the e-media version may also appear in this print version.

IEEE Catalog Number:	CFP22PSC-POD
ISBN (Print-On-Demand):	978-1-7281-6118-1
ISBN (Online):	978-1-7281-6117-4

Additional Copies of This Publication Are Available From:

Curran Associates, Inc
57 Morehouse Lane
Red Hook, NY 12571 USA
Phone: (845) 758-0400
Fax: (845) 758-2633
E-mail: curran@proceedings.com
Web: www.proceedings.com

2022 IEEE 49th Photovoltaics Specialists Conference (PVSC 2022)

Philadelphia, Pennsylvania, USA
5-10 June 2022

Pages 1-683

IEEE Catalog Number: CFP22PSC-POD
ISBN: 978-1-7281-6118-1

TABLE OF CONTENTS

Investigation of Degradation Kinetics of Perovskite Solar Cells by Accelerated Aging 1
  Dhruba B. Khadka, Yasuhiro Shirai, Masatoshi Yanagida, Kenjiro Miyano

Effect of Phenethylammonium Thiocyanate Additive in Tin Perovskite for Efficient and Stable Pb-Free Perovskite Solar Cells 4
  Dhruba B. Khadka, Yasuhiro Shirai, Masatoshi Yanagida, Kenjiro Miyano

Investigation of the Circular Economy Approach in Asian Solar PV Manufacturing 7
  Viktor Dancza, Kai Cheng

Three General Methods for Predicting Bifacial Photovoltaic Performance Including Spectral Albedo 8
  Erin M. Tonita, Christopher E. Valdivia, Michael Martinez-Szewczyk, Mariana I. Bertoni, Karin Hinzer

Evaluation of Auger Limited Behavior in Thermoradiative Cells 9
  Jamie D. Phillips

Fabrication of a Chemically Exfoliated 2D MoS2 Nanoparticle Based Solar Cell 12
  Wafa Alnaqbi, Ayman Rezk, Aisha Alhammadi, Ammar Nayfeh

Evaluating the Durability of Balance of Systems Components Using Combined-Accelerated Stress Testing 15
  David Miller, Greg Perrin, Kent Terwilliger, Joshua Morse, Chuanxiao Xiao, Bobby To, Chun-Sheng Jiang, Peter Hacke

High Throughput Boron Emitter Formation from Pre-Deposited APCVD BSG Layers for TOPCon Solar Cells 18
  Marius Meßmer, Sattar Bashardoust, Udo Belledin, Sven Seren, Heiko Zunft, Sebastian Mack, Andreas Wolf

Uncertainty in Annual Energy Resulting from Uncertain Irradiance Measurements 22
  Clifford W. Hansen, Aaron Scheiner

Impact of Photovoltaic Plant Tilt on the Need for Storage 27
  Russell K Jones, Sarah Kurtz

Precursor Ink Design for Scalable Fabrication of Perovskite Solar Cells Via High-Speed Flexography 28
  Julia E Huddy, Youxiong Ye, William J Scheideler

Terrain Aware Backtracking Via Forward Ray Tracing 29
  Kurt Rhee

Sustainability of PV Repowering 31
  Ian Marius Peters, Jens Hauch, Christoph Brabec

Correlations in Spatial Variability When Accounting for Cloud Advection 32
  Joseph Ranalli

Realization of Ultrathin GaAs Photonic Power Converters with Rear-Side Metal Grating on Full 4" Wafers 38

  Oliver Höhn, Meike Schauerte, Patrick Schygulla, Hubert Hauser, David Lackner, Benedikt Bläsi, Henning Helmers

Human Health Risk Assessment of Solvents and Lead Toxicity in Emerging Perovskite Solar Cells 43

  Sherif A. Khalifa, Sabrina Spatari, Aaron T. Fafarman, Vasilis M. Fthenakis, Patrick Gurian, Jason B. Baxter

Spectral Shape Changes the Optimal Perovskite Thickness of the 2-Terminal Perovskite/Silicon Tandem Solar Cell............ 44

  Dong C. Nguyen, Yasuaki Ishikawa

Leveraging Undoped CdSeTe for >950 mV............ 47

  Pascal Jundt, James Sites

Low-Temperature PECVD Deposition of Highly Conductive N-Type Microcrystalline Silicon Thin Films for Optoelectronic Applications 52

  Brahim Aissa, Amir A. Abdallah, Juan Lopez Garcia

Analysis for Solar Coverage and CO2 Emission Reduction of Photovoltaic-Powered Vehicles 58

  Masafumi Yamaguchi, Taizo Masuda, Takashi Nakado, Kazumi Yamada, Kenichi Okumura, Akinori Satou, Yasuyuki Ota, Kenji Araki, Kensuke Nishioka

Interrelated Characterizations of 2D/3D Perovskite Solar Cells Aged Under Damp Heat Conditions 59

  Cynthia Farha, Emilie Planès, Lara Perrin, David Martineau, Lionel Flandin

An Intelligent Algorithm for Maximum Power Point Tracking in PV Systems Through Load Management 62

  Kelvin Tan, Joseph A. Azzolini, William J. Parquette, Christian R. Polo, Meng Tao

Large Area Survey Grain Size and Texture Optimization for Thin Film CdTe Solar Cells Using Xenon-Plasma Focused Ion Beam (PFIB)............ 63

  Vladislav Kornienko, Ochai Oklobia, Stuart Irvine, Steve Jones, Giray Kartopu, Ali Abbas, Yau Yau Tse, Jake Bowers, Kurt Barth, Michael Walls

Diffraction-Optimized Surface Structures for Enhanced Light Harvesting in Organic Solar Cells 69

  Milena Merkel, Jörg Imbrock, Cornelia Denz

Validation of Photovoltaic Spectral Effects Derived from Satellite-Based Solar Irradiance Products............ 72

  Sophie Pelland, Christian A. Gueymard

Elimination of the Carbon-Rich Layer in Cu2ZnSn(S, Se)4 Absorbers Prepared from Nanoparticle Inks............ 73

  Stephen Campbell, Martial Duchamp, Neil Beattie, Michael Jones, Guillaume Zoppi, Vincent Barrioz, Yongtao Qu

Light Trapping Characteristics of Photonic Crystal Constructs and Randomly Textured Thin Silicon............ 74

  Sara M. Almenabawy, Yibo Zhang, Rajiv Prinja, Nazir P. Kherani

Gallium-Boron Spin-On Co-Doping for Polycrystalline Silicon Passivating Contacts............ 75

  Thien Truong, Matthew Young, Mowafak Al-Jassim, Daniel Macdonald, Hieu Nguyen, Josua Stuckelberger

Chemical Surface and Interface Structure of Sulfur-Passivated Silicon with a SiNx Capping Layer 76
Amandee Hua, Nan Jiang, Ajay Upadhyaya, Issac Lam, Tasnim K Mouri, Dirk Hauschild, Lothar Weinhardt, Wanli Yang, Ajeet Rohatgi, Ujjwal Das, Clemens Heske

Data Mining of Solar Cells Production Data Using Factorial Analysis 77
Johnson Wong, Dinica Li, Gordon Deans

Statistical and Engineering Process Control of Phosphorus Diffused Solar Wafers Using Contactless Infrared Reflectometry 80
Johnson Wong, Divya Ananthanarayanan, Gordon Deans

Proposal of Connection Assessment Diagrams to Speed Up the Studies of Hosting Capacity of PV Generators in MV Distribution Systems... 83
Pedro A. V. Pato, Fernanda C. L. Trindade, Tiago R. Ricciardi, Paulo Meira, Walmir Freitas

Ray Tracing of Bent Applications of Luminescent Solar Concentrator PV Modules 89
Xitong Zhu, Michael G. Debije, Angèle H. M. E. Reinders

Optical Modeling of Light Trapping Using an ITO-Based Electrodynamic Dust Shield Structure 92
Nicole Swatton, Andrey Semichaevsky

Assessing the Alignment of Solar Facilities with Global Climate Goals 95
Parikhit Sinha, Liv Hammann

Preparation of Plasmonic Ag and Au Nanoparticle Interfaces for Photocurrent Enhancement in Si Solar Cells 98
Brahim Aïssa, Adnan Ali, Rui N. Pereira, Anirban Mitra

The Natural and Accelerated Evolution of EVA Adhesion Through Intermediate Exposures........................ 106
Patrick Thornton, Nick Bosco, Reinhold H Dauskardt

High-Performance O- Band Photonic Power Converters Under Non-Uniform Laser Illumination 107
Meghan N. Beattie, Henning Helmers, Gavin P. Forcade, Christopher E. Valdivia, David Lackner, Oliver Höahn, Karin Hinzer

Post-Annealing Treatment on Hydrothermally Grown Sb2(S, Se)3 Thin Films for Efficient Solar Cells.. 108
Suman Rijal, Zhaoning Song, Deng-Bing Li, Jaehoon Chung, Sandip S Bista, Dipendra Pokhrel, Sabin Neupane, Randy Ellingson, Yanfa Yan

Wafer-Scale Pulsed Laser Deposition of ITO for Silicon Heterojunction Solar Cells: Reduced Damage Vs Interfacial Resistance 109
Yury Smirnov, Pierre-Alexis Repecaud, Leonard Tutsch, Ileana Florea, Pere Roca I Cabarrocas, Martin Bivour, Monica Morales-Masis

Potential Capacity and Targeted Costs for Floating Photovoltaics in North America.................................... 110
Leonardo Micheli

Techno-Economic Analysis of Novel PV Plant Designs for Extreme Cost Reductions 111
Nicholas Pilot, Robin Bedilion, Daniel Fregosi, Sean Hackett, Michael Bolen, Joseph Stekli

Potentiostatic Photoluminescence Imaging of Charge Extraction in Perovskite Solar Cells........................... 114
Lukas Wagner, Patrick Schygulla, Jan Philipp Herterich, Mohamed Elshamy, Dmitry Bogachuk, Salma Zouhair, Simone Mastroianni, Uli Würfel, Yuhang Liu, Shaik M. Zakeeruddin, Michael Grätzel, Andreas Hinsch, Stefan W. Glunz

Flexible GaAs Solar Cell Using Water-Soluble Sacrificial Layer for Epitaxial Lift-Off Process 115
Sahil Sharma, Carlos A Favela, Bo Yu, Eduard Galstyan, Venkat Selvamanickam

An Evaluation of Empirical Models for Use in Normalizing PV Plant Performance Data 116
Daniel Fregosi, Michael Bolen

Performance Investigation of Batteries Supporting Solar Power in U.S. 121
Farzan Zareafifi, Daniel Baerwaldt, Socheata Hour, Yi Hao Xie, Sarah Kurtz

Numerical Modeling of Capacitance Signatures of Perovskite Solar Cells 126
*Rasha Awni, Zhaoning Song, Chongwen Li, Lei Chen, Suman Rijal, Sandip Bista, Tao Zhu,
Xiaoming Wang, Yanfa Yan*

Reverse Energy Injustice on Molokai Island to the Underserved Communities with 100% Energy
from the Sun (Light & Heat) for Energy Cost Savings Equity 127
John O. Borland

Performance Loss Rate Estimation for Systems Affected by Potential Induced Degradation 131
Panagiotis Goumenos, Andreas Livera, Michalis Florides, George E. Georghiou

Assessing and Optimizing Free Space Luminescent Solar Concentrators for Urban Façade
Installation 134
Shweta Pal, Rebecca Saive

Curvilinear Prismatic Window Which Eliminates Glare and Reduces Front-Surface Reflections for
PV Modules and Other Surfaces 137
Mark O'Neill, Chris Youtsey

Monochromatic Light Trapping in Photonic Power Converters 143
*Nicholas P. Irvin, Neda Nouri, Chaomin Zhang, Christopher E. Valdivia, Karin Hinzer,
Richard R. King, Christiana B. Honsberg*

Implications of Agriculturally Co-Located Solar PV Installations on the FEW Nexus in the Central
Valley 144
Jacob T Stid, Siddharth Shukla, Annick Anctil, Anthony D Kendall, David W Hyndman

Photovoltaic Investigation on the Lunar Surface (PILS): Design Considerations and Ground
Testing 145
*Jeremiah S McNatt, Timothy J Peshek, Norman F Prokop, Greeta J Thaikattil, Michael J
Krasowski, Amy R Stalker, Brian J Tomko, Mathew R Deminico*

Reference Cell Performance and Modeling on a One-Axis Tracking Surface 146
*Frank Vignola, Josh Peterson, Rich Kessler, Sean Snider, Peter Gotseff, Manajit Sengupta,
Aron Habte, Afshin Andreas, Fotis Mavromatakis*

Optimization of CdTe Solar Cells Using Co-Sputtered CdSeTe 154
*Deng-Bing Li, Sandip Singh Bista, Neupane Sabin, Xiaomeng Duan, Manoj K Jamarkattel,
Abdul Quader, Adam Phillips, Michael Heben, Randall J Ellingson, Yanfa Yan*

Parametric Study of Building-Integrated Photovoltaic Windows 155
Yuan Gao, Jacob Jonsson, Charlie Curcija

Impact of Daily Irradiance Profiles on Intra-Day Solar Forecasting 156
Javier Lopez-Lorente, Spyros Theocharides, George Makrides, George E. Georghiou

Hyperspectral Imaging of Localized, Optically-Active Defects in GaAs Solar Cells 164
Behrang H. Hamadani, Margaret A. Stevens, Brianna Conrad, Matthew P. Lumb, Eric Armour, Kenneth J. Schmieder

Achieving Global Decarbonization by Photovoltaic Electrification: Impact of Disruptive Technologies 168
Billy J. Stanbery, Jao Van De Lagemaat

Harsh Sequential Stress Tests for Improved PV Durability 169
Jean Patrice Rakotoniaina, Romain Couderc, Eszter Voroshazi, Jérémie Aimé

PV Module Operating Temperature Model Equivalence and Parameter Translation 172
Anton Driesse, Marios Theristis, Joshua S. Stein

Accelerating Simulation for High-Fidelity PV Inverter System Reliability Assessment with High-Performance Computing... 178
Liwei Wang, Ramanathan Thiagarajan, Shuangshuang Jin, Zheyu Zhang

Inverter Reliability Estimation for Advanced Inverter Functionality 183
Jack Flicker, Jay Johnson, Matthew J. Reno, Joseph A. Azzolini, Peter Hacke, Ramanathan Thiagarajan

Grid-Forming and Grid-Following Inverter Comparison of Droop Response 190
Nicholas S. Gurule, Javier Hernandez Alvidrez, Matthew J. Reno, Wei Du, Kevin Schneider

Inferring PV System Specifications from Net Load 197
Upama Nakarmi, Thomas E. Hoff, Marc Perez, Philip Gruenhagen

Planarizing HVPE Growth on GaAs Substrates Produced by Controlled Spalling........................... 198
Anna K Braun, William E McMahon, Allison N Perna, Kevin L Schulte, Corinne E Packard, Aaron J Ptak

Lamination Process Induced Residual Stress in Glass-Glass Vs. Glass-Backsheet Modules........................ 199
Farhan Rahman, Ian M. Slauch, Rico Meier, Jared Tracy, Elizabeth C. Palmiotti, Mariana I. Bertoni, James Y. Hartley

Near-Busbar Degradation of Screen-Printed Metallization in Silicon Photovoltaic Modules........................ 200
Dana B. Sulas-Kern, Helio Moutinho, Tristan Erion-Lorico, Steve Johnston

Improving Behind-The-Meter PV Impact Studies with Data-Driven Modeling and Analysis........................ 204
Joseph A. Azzolini, Samuel Talkington, Matthew J. Reno, Santiago Grijalva, Logan Blakely, David Pinney, Stanley McHann

A Model to Predict Daily Snow Albedo Change Over Time 205
Christopher Pike, Daniel Riley, Laurie Burnham

High-Specific-Power Schottky-Junction Photovoltaics from CVD-Grown MoS2 208
Timothy Ismael, Kazi M. Islam, Muhammad A. Abbas, George B. Ingrish, Claire E. Luthy, Orhan Kizilkaya, Carlos M. Gutierrez, Meghan E. Bush, Jeremiah S. McNatt, Anthony J. Hoffman, Matthew D. Escarra

Highly Stretchable, Durable and Lightweight Lego®-Style 3-Dimensional Photovoltaic............................. 209
Min Ju Yun, Yeon Hyang Sim, Dong Yoon Lee, Seung I. Cha

Optimal Strategy for Using Biomass to Enable California High Penetration Solar 212
Mahmoud Y. Abido, Sarah R. Kurtz

Automatic Crack Segmentation in Electroluminescence Images of Solar Modules and Maximum Inactive Area Prediction 213
Xin Chen, Todd Karin, Anubhav Jain

Experimental Assessment of Temperature Estimation Models of Bifacial Photovoltaic Modules 214
Gaetano Mannino, Giuseppe Marco Tina, Mario Cacciato, Lorenzo Todaro, Fabrizio Bizzarri, Andrea Canino

Seasonal Dependence of Diurnal Efficiency Degradation and Recovery in Perovskite Mini-Modules During Outdoor Testing.. 217
Vasiliki Paraskeva, Maria Hadjipanayi, Matthew Norton, Aranzazu Aguirre, Afshin Hadipour, Rita Ebner, George E. Georghiou

Applying Unsupervised Machine Learning for the Detection of Shading on a Portfolio of Commercial Roof-Top Power Plants in Germany 223
Nicolas Holland, Klaus Kiefer, Christian Reise, E. A. Sarquis Filho, Bernd Kollosch, Björn Müller

Electron Selective TiO_x Contact for Ultrathin Amorphous Germanium Solar Cells.................................... 228
Norbert Osterthun, Hosni Meddeb, Nils Neugebohrn, Kai Gehrke, Martin Vehse

Magnetic Field Imaging (MFI) of Shingle Solar Modules 231
Julian Weber, Stephan Hoffmann, Kai Kaufmann, Angela De Rose

Evaluating Electroluminescence Imaging and Image Processing as a Quantitative Solar Cell Characterization Method 232
Meghan E Bush, Timothy J Peshek, Erica N Montbach

CdTe-Based Photovoltaics Using a CdTe/CdSe/CdTe Absorber Layer Structure 233
Jacob F Leaver, Ken Durose, Jonathan D Major

Effects of Growth Temperature on Electrical Conductivity in Low-Dimensional, Ruddlesden-Popper Perovskite Thin Films Deposited by RIR-MAPLE.. 234
Niara E. Wright, Adrienne D. Stiff-Roberts

$GeCl_4$-Based High Quality Ge Epitaxy on Engineered Ge Substrates for Thin Multi-Junction Solar Cells... 235
Jinyoun Cho, Clément Porret, Valérie Depauw, Guillaume Courtois, Daniel McDermott, Roger Loo, Kristof Dessein, Rufi Kurstjens

Investigation of Degradation Mechanisms in Carbon-Based Perovskite Solar Cells Exposed to Damp-Heat Conditions... 239
Nikoleta Kyranaki, Cynthia Farha, Lara Perrin, Lionel Flandin, Emilie Planès, Lukas Wagner, Karima Saddedine, David Martineau, Stéphane Cros

Selective Etching of 6.1 Å Materials for Transfer-Printed Devices 240
Margaret A. Stevens, Jill A. Nolde, Shawn Mack, Kenneth J. Schmieder

Inorganic Perovskite Solar Cells with Very High Voltage and Excellent Stability Against Thermal and Environmental Degradation.. 244
Saba Sharikadze, Junhao Zhu, Ranjith Kottokkaran, Arkadi Akopian, Vikram Dalal

23.5% Efficiency GaAs Solar Cells Fabricated with Low-Cost, Non-Vacuum Processing 247
Phillip R Jahelka, Harry A Atwater, Aaron Ptak, Christiane Frank-Rotsch, Frank Kiessling, Cora Went, Michael Kelzenberg

Metallic Lead Recovery Via Electrowinning from Lead Acetate for Silicon Solar Module Recycling 248
> Natalie Click, Meng Tao

Rear Junction Bifacial Screen-Printed Double Side Passivated Contact Si Solar Cells 251
> Young-Woo Ok, Vijaykumar D Upadhyaya, Brian Rounsaville, Ajay D Upadhyaya, Wook-Jin Choi, Ajeet Rohatgi, Gabby De Luna, John Derek Arcebal, Pradeep Padhamnath, Shubham Duttagupta

Micro-Scale III-V/Ge Multijunction Solar Cell with Through Cell Via Contacts 254
> Mathieu De Lafontaine, Guillaume Gay, Erwine Pargon, Camille Petit-Etienne, Romain Stricher, Serge Ecoffey, Artur Turala, Maïté Volatier, Abdelatif Jaouad, Simon Fafard, Vincent Aimez, Maxime Darnon

Spatiotemporal Modeling of Real World Backsheets Field Survey Data: Hierarchical (Multilevel) Generalized Additive Models 255
> Raymond J. Wieser, Zelin Zack Li, Stephanie L. Moffitt, Ruben Zabalza, Evan Boucher, Silvana Ayala, Matthew Brown, Xiaohong Gu, Liang Ji, Colleen O'Brien, Adam W. Hauser, Greg S. O'Brien, Xuanji Yu, Roger H. French, Michael D. Kempe, Jared Tracy, Kausik R. Choudhury, William J. Gambogi, Laura S. Bruckman, Kenneth P. Boyce

Mapping of Local Defects and Voltages in Solar Cells Using Non-Contact Electrostatic Voltmeter Method 261
> Hamza Ahmad Raza, Govindasamy Tamizhmani

Tellurium Oxide as a Back-Contact Buffer Layer for CdTe Solar Cells 264
> Camden Kasik, Ramesh Pandey, Akash Shah, James Sites

Intelligent Cloud-Based Monitoring and Control Digital Twin for Photovoltaic Power Plants 267
> Andreas Livera, George Paphitis, Loucas Pikolos, Ioannis Papadopoulos, Jesús Montes-Romero, Javier Lopez-Lorente, George Makrides, Juergen Sutterlueti, George E. Georghiou

Predicting Solar Cell Recombination from C-V-F Fingerprints Using Machine Learning 275
> Isaac K. Lam, Austin G. Kuba, Nathan J. Rollins, William N. Shafarman

Current & Future Photovoltaic System Impacts on City-Wide Grid Performance & Neighborhood Microgrids 276
> C. Birk Jones, William F. Vining, Thad Haines

Passivating Surface Iodide Defects Slows the CsPbI3 Phase Transformation 283
> Jeffrey A Christians, Jonathan Outen, Rory M Campagna, Zachery R Wylie, Peter Ruffolo

Which Potential for Kesterite Absorbers in Tandem Solar Cells: A Quantitative Modelling Approach 284
> Alex Jimenez, Alejandro Navarro, Sergio Giraldo, Kunal Jogendra Tiwari, Marcel Placidi, Lorenzo Calvo-Barrio, Joaquim Puigdollers, Edgardo Saucedo, Zacharie Jehl Li-Kao

Measuring Global, Direct, Diffuse, and Ground-Reflected Irradiance Using a Reference Cell Array 285
> Michael Gostein, Adam Hoffman, Bruce H. King, Audrey Marquis

Validation of In-Situ I-V Measurement Unit for PV System Monitoring Applications 291
> Audrey Marquis, Michael Gostein, Bruce H. King

Effect of Near-Interface Compensation of CdScTe Absorber Layers on Solar Cell Performance 295
> Brian Good, Eric Colegrove, Matthew O. Reese

Millions of Small Pressure Cycles Drive Damage in Cracked Solar Cells.. 298
    Timothy J Silverman, Nick Bosco, Michael Owen-Bellini, Cara Libby, Michael G Deceglie

Quantifying Energy Flows in PV Circularity Processes.. 299
    Heather M. Mirletz, Silvana Ovaitt, Ashley Gaulding, Seetharaman Sridhar, Teresa Barnes

Uncertainty Quantification of Bifacial Performance Modeling.. 302
    Matthew J. Prilliman, Janine M. Freeman Keith

Field Experience Detecting PV Underperformance in Real Time Using Existing Instrumentation................ 307
    Scott Sheppard, Tim Cook, Daniel Fregosi, Christopher Perullo, Michael Bolen

Enhancing Temporal Variability of 5-Minute Satellite-Derived Solar Irradiance Data 314
    Jing Huang, Richard Perez, James Schlemmer, Marc Perez, Akanksha Bhat, Patrick Keelin,
    Alex Kubiniec

The Materials Degradation in Encapsulants for Application in Glass/Glass PV Modules After
Accelerated Aging.. 319
    Sona Ulicna, Archana Sinha, David C. Miller, Laura T. Schelhas, Michael Owen-Bellini

Perovskite PV Design for Stable Space Operation.. 320
    Kaitlyn T. Vansant, Ahmad R. Kirmani, Jay B. Patel, Laura E. Mundt, David P. Ostrowski,
    Brian M. Wieliczka, Gabriella D. Lahti, Michael D. McGehee, Laura T. Schelhas, Joseph M.
    Luther, Timothy J. Peshek, Lyndsey B. McMillon-Brown

Thermoradiative Cell Technology: Analysis and Loss Mechanisms.. 321
    Geoffrey A. Landis

Estimation of Shade Losses in Unlabeled PV Data.. 326
    Bennet E. Meyers, David J. F. Rodriguez

Contribution of Na+ from Glass to PID-S in Solar Modules: Na Migration in EVA.................................... 327
    Jacob A. Clenney, Erick Martinez Loran, Guillaume Von Gastrow, Tanguy Terlier, David P.
    Fenning, Rico Meier, Mariana I. Bertoni

Importance of Ideality Factors in perovskite/Si Tandem Solar Cell Design 328
    Benjamin Williams, Benjamin Daiber, Chris Case

A Deep Learning Approach to Increase Luminescence Image Resolution of Solar Cells 329
    Priya Dwivedi, Robert Lee Chin, Thorsten Trupke, Ziv Hameiri

Agrivoltaics Using Bi-Facial PVs for Permaculture in Utility-Scale Projects 330
    P. M. Jansson, M. G. Newberry, S. M. Myers

Dedicated Cold-Climate Field Laboratory for Photovoltaic System and Component Studies: The
Michigan Regional Test Center as a Case Study.. 333
    Laurie Burnham, Daniel Riley, Bruce H. King, Jennifer Braid, Paul Dice, Ana Dyreson,
    William Snyder, Christopher Pike

Efficient Self-Protected Thin Film c-Si Solar Cell Against Reverse-Biasing Condition: A
Simulation Study 336
    Omar M. Saif, Abdelhalim Zekry, Ahmed Shaker, M. Abouelatta, Ahmed Saeed

Dynamic Simulation of a Load-Matching Photovoltaic System for Green Hydrogen Production................ 339
    Christian R. Polo, William J. Parquette, Kelvin Tan, Meng Tao

Embodied Energy and CO2 from the Manufacture of Cadmium Telluride and Silicon Photovoltaics 344
Hope Wikoff, Samantha B Reese, Matthew O Reese

Collection of Heat Loss in Photovoltaic System by Parallely Connected Thermoelectric Network 345
Joel Erickson, Jing Bai

Ultra-Thin and Lightweight CdS/CdTe Solar Cell Fabricated on Ceramic Substrate for Space
Applications 348
*Manoj K. Jamarkattel, Adam B. Phillips, Geethika K. Liyanage, Fadhil K. Alfadhili, Ebin
Bastola, Victor V. Plotnikov, Alvin D. Compaan, Randy J. Ellingson, Michael J. Heben*

Continuous Flash Sublimation of Inorganic Halide Perovskites: Enabling Industrially Compatible
Deposition Rates 351
*Tobias Abzieher, Christopher P. Muzzillo, Mirzo Mirzokarimov, Ahmad R. Kirmani,
Gabriella Lahti, Wylie Kau, Daniel M. Kroupa, Joseph M. Luther, David T. Moore*

Temperature- And Illumination-Dependent Characterization of Wide Bandgap Sulfide CIGS and
CZTS Solar Cells 352
*Simon M. F. Zhang, Guojun He, Chang Yan, Kaiwen Sun, Xiaojing Hao, Ivan Perez-Wurfl,
Ziv Hameiri*

Fill Factor Prediction of Modern Industrial Cells: Potential Gaps and Improvements 353
Gaia Maria N Javier, Priya Dwivedi, Yoann Buratti, Thorsten Trupke, Ziv Hameiri

Real-Time Prediction Algorithms to Detect Clouds and Forecast Photovoltaic System Performance 354
*Maqsood Ali Mughal, Habeebullah Adua, Muhammad Hammad Uddin, Evan Sauter, Stephen
Natale, Timothy Lewis, Jonathan G. Ferreira*

Surrogate Modeling for Rapid Prediction of Energy Yield from Vehicle-Integrated Photovoltaics 362
Timofey Golubev

Novel Laser Oxidation for Screen-Printed Selective Area Front Poly-Silicon Contacts for TOPCon
Cells 366
*Sagnik Dasgupta, Young-Woo Ok, Vijaykumar D. Upadhyaya, Wook-Jin Choi, Ying-Yuan
Huang, Shubham Duttagupta, Ajeet Rohatgi*

Contactless Determination of Emitter Sheet Resistance for Diffused Silicon Wafers 367
Yan Zhu, Thorsten Trupke, Ziv Hameiri

Photodoping Causes Inconsistencies in the Injection-Dependent Lifetimes of Perovskite Thin Films 368
*Robert A Lee Chin, Arman Soufiani, Jianghui Zheng, Paul Fassl, Anita Ho-Baillie, Ulrich
Paetzold, Thorsten Trupke, Ziv Hameiri*

Investigating the Impurity Gettering Rate in Polycrystalline-Silicon Based Passivating Contacts 369
*Zhongshu Yang, Jan Krügener, Frank Feldmann, Jana-Isabelle Polzin, Bernd Steinhauser,
Tien T. Le, Daniel Macdonald, Anyao Liu*

Differences of CIGS Cell Performance with Zn(O, S)/(Zn, Mg)O Or CdS/i-ZnO Buffers System
Explored by Numerical Simulations 370
Giovanna Sozzi, Dimitrios Hariskos, Wolfram Witte

Fabricating High Aspect Ratio Front Contacts for Solar Cells by String-Printing 374
Mathis Van De Voorde, Rebecca Saive

Benchmarking PV Performance Models with High Quality IEC 61853 Matrix Measurements
(Bilinear Interpolation, SAPM, PVGIS, MLFM and 1-Diode) 375
Steve Ransome

Determining the Decomposition Voltage of $Cu(In_{1-x}Ga_x)Se_2$ 381
Klaas Bakker, Joaquin Coll Matas, Johan Bosman, Nicolas Barreau, Arthur Weeber, Mirjam Theelen

A Combined Shading and Radiation Simulation Tool for Defining Agrivoltaic Systems 384
Haomiao Wang, Henry J. Williams, Xiaotong Bu, K. Max Zhang

Measurement of Band Alignment Between ZnO Based Front Emitters and $CdCl_2$ Treated CdSeTe/CdTe Absorbers 387
Xiaolei Liu, Luke Jones, Luksa Kujovic, Nicholas Hunwick, Luis Infante-Ortega, Michael Walls, Tushar Shimpi, Walajabad Sampath, Kurt Barth, Stephen Jones, Ochai Oklobia, Stuart Irvine

Performance Investigation and Analysis of Anti-Soiling Coatings in Hot Desert Climate 390
Hebatalla Alhamadani, Shaikha Hassan, Gerhard Mathiak, Omar Albadwawi, Vivian Alberts

Light Distribution and Uniformity Evaluation of Cross Compound Parabolic Concentrators 395
Mazin Al-Shidhani, Mohammad Alnajideen, Gao Min

A Comparative Study of 3D Printed Non-Imaging Solar V-Trough and Compound Parabolic Concentrators for Low-Cost, High-Performance CPV Applications 396
Mohammad Alnajideen, Mazin Al-Shidhani, Gao Min

Development of Photovoltaic Inverter Model with Islanding Detection Using the Sandia Frequency Shift Method 398
Nelson E. Saavedra-Peña, Rachid Darbali-Zamora, Edgardo Desarden-Carrero, Erick Aponte-Bezares

Optical Properties of Thin Film Sb2Se3 and Identification of Its Electronic Losses in Photovoltaic Devices 405
Niva K. Jayswal, Suman Rijal, Biwas Subedi, Indra Subedi, Zhaoning Song, Robert W. Collins, Yanfa Yan, Nikolas J. Podraza

Racking Reflection and Shading Effects on Single Axis Tracked Bifacial Photovoltaic Modules 406
Mandy R Lewis, Trevor J Coathup, Annie C J Russell, Javier Guerrero-Perez, Christopher E Valdivia, Karin Hinzer

Spectroscopic Ellipsometry Analysis and Quantum Efficiency Simulation of $CuInSe_2$ Solar Cells 407
Dhurba R. Sapkota, Ambalanath Shan, Balaji Ramanujam, Puja Pradhan, Richard Irving, Adam B. Phillips, Michael J. Heben, Randy J. Ellingson, Sylvain Marsillac, Nikolas J. Podraza, Robert W. Collins

Novel Interconnection Method for Micro-CPV: 132 Solar Cell Prototype 413
Norman Jost, Steve Askins, Richard Dixon, Mathieu Ackermann, Cesar Dominguez, Ignacio Anton

Understanding the Behavior of Fixed Composition $CdSe_xTe_{1-x}$ (CST) Solar Cells 414
Ebin Bastola, Adam B. Phillips, Abasi Abudulimu, Vlad Kornienko, Manoj K. Jamarkattel, Zulkifl H. Rabbani, Jared D. Friedl, Prabodika N. Kaluarachchi, Ali Abbas, Abdul Quader, Xavier Mathew, Michael Walls, Randy J. Ellingson, Michael J. Heben

Effects of Novel In+RbF Post-Deposition Treatment on $Cu(In_xGa_{1-x})Se_2$ Solar Cells 415
Jake Wands, Polyxeni Tsoulka, Thomas Lepetit, Nicolas Barreau, Angus Rockett

Seasonal Dependence of Bifacial Photovoltaic Array Gain Due to Inverter Clipping 418
Thunchanok Kaewnukultorn, Steven Hegedus

Accuracy of Potential High Limit Estimation for Solar Plants in the Southeast US 419
William B. Hobbs, David J. Ault, Vahan Gevorgian, Govind Saraswat

Improved Efficiency of Non-Toxic Cu3BiS3 Thin Film Solar Cell Employing PCBM Electron
Transport Layer 424
Sandip Das

Evaluating Intrinsic Defects Across CIGS Absorber Via X-Ray Absorption Near Edge Structures 425
Srisuda Rojsatien, Tara Nietzold, Niranjana Kumar, Barry Lai, Jeff Bailey, Arun Mannodi-
Kanakkithodi, Maria K. Y. Chan, Mariana Bertoni

Accelerated Durability Evaluation of Emerging Cell Interconnect Technologies................................. 426
Fang Li, Dylan J. Colvin, Kristopher O. Davis, Andrew Gabor, Govindasamy Tamizhmani

Tracking Se Local Structures Across CdSeTe Absorber with X-Ray Microscopy 429
Srisuda Rojsatien, Niranjana Kumar, Trumann Walker, Barry Lai, Dan Mao, Arun Mannodi-
Kanakkithodi, Maria K. Y. Chan, Mariana Bertoni

Epitaxial Growth of Detachable GaAs/Ge Heterostructure on Mesoporous Ge Substrate for Layer
Separation and Substrate Reuse... 430
Nicolas Paupy, Bouraoui Ilahi, Zakaria Oulad Elhmaidi, Valentin Daniel, Tadeáš Hanuš,
Roxana Arvinte, Alexandre Heintz, Alex Brice Poungoué Mbeunmi, Thierno Mamoudou
Diallo, Richard Arès, Abderraouf Boucherif

Analyzing Hosting Capacity Protection Constraints Under Time-Varying PV Inverter Fault
Response.. 431
Joseph A. Azzolini, Nicholas S. Gurule, Rachid Darbali-Zamora, Matthew J. Reno

Tuning Thermal Induced Porous-Ge Reconstruction for Layer Transfer and Substrate Re-Use 439
Ahmed Ayari, Bouraoui Ilahi, Roxana Arvinte, Tadeas Hanus, Laurie Mouchel, Denis
Machon, Abderraouf Boucherif

Simulation-Based Determination of Shockley-Read-Hall Recombination Lifetimes in Group-V
Doped P-N Junction CdTe Devices.. 440
Alexandra M. Bothwell, Darius Kuciauskas

Hardware-In-The-Loop Lab for Testing Grid Supporting Functions of Smart Inverters 441
Thunchanok Kaewnukultorn, Sergio Sepúlveda-Mora, Steven Hegedus

A Deep Learning Approach to Denoise Electroluminescence Images of Solar Cells 442
Grace Liu, Priya Dwivedi, Thorsten Trupke, Ziv Hameiri

Temperature Dependence of Silicon-Dielectric Interface Recombination 443
Anh Huy Tuan Le, Eduardo Prieto Ochoa, Ruy Sebastian Bonilla, Nino Borojevic, Ziv
Hameiri

Optical Simulations of All-Inorganic CsPbBr3 Perovskite Quantum Dot Intermediate Band Solar
Cells (QDIBSCs)... 444
Ola Rashwan, Chase Sasala

Life-Cycle Analysis of crystalline-Si "Direct Wafer" and Tandem Perovskite PV Modules and
Systems.. 447
Enrica Leccisi, Adam Lorenz, Vasilis Fthenakis

Bandgap Model Using Symbolic Regression for Environmentally Compatible Lead-Free Inorganic
Double Perovskites... 452
Ahmer A. B. Baloch, Omar Albadwawi, Badreyya Alshehhi, Vivian Alberts

Enhancement in the Efficiency of Rear Emitter SHJ Solar Cells by Using a CaF2/ITO Double-Layer Anti-Reflective Coating 456

Muhammad Aleem Zahid, Muhammad Quddamah Khokhar, Youngkuk Kim, Junsin Yi

A Thermal Model for Bifacial PV Panels.................... 457

Shahzada Pamir Aly, Jim Joseph John, Gerhard Mathiak, Omar Albadwawi, Luis Pomares, Vivian Alberts

Planar Transparent Conductive Oxide/Ag Rear Contacts for High Efficiency III-V Photovoltaics.................. 460

Christopher T. Gregory, Sean J. Babcock, Richard R. King

Quantitative Measurement of Active Dopant Density Distribution in Black Silicon Solar Cell Using Scanning Nonlinear Dielectric Microscopy.................... 461

Yasuo Cho, Beniamino Iandolo, Ole Hansen

Flexible and Lightweight CdS/CdTe Solar Cells Via a Water-Assisted Lift-Off Process.................... 464

Sandip S Bista, Deng-Bing Li, Suman Rijal, Sabin Neupane, Rasha A Awni, Randy J. Ellingson, Zhaoning Song, Adam Phillips, Michael Heben, Yanfa Yan

Public Road Tests of Toyota Prius Equipped with High Efficiency PV Module with Output Power of 860W.................... 467

Taizo Masuda, Takashi Nakado, Masafumi Yamaguchi, Tatsuya Takamoto, Kensuke Nishioka, Kazumi Yamada

Revealing Sub-Cell Degradation of Multi-Junction Solar Cells by Absolute Electroluminescence Imaging.................... 468

Youyang Wang, Liying Li, Xiaobo Hu, Yun Jia, Guoen Weng, Xianjia Luo, Shaoqiang Chen, Hidefumi Akiyama

Using Machine Learning to Predict the Complete Degradation of Accelerated Damp Heat Testing in Just 10% of Testing Time.................... 472

Zubair Abdullah-Vetter, Priya Dwivedi, Robert Lee Chin, Brendan Wright, Thorsten Trupke, Ziv Hameiri

CuCl Doping Variations in High Efficiency Polycrystalline CdSeTe/CdTe Thin Film Solar Cells 475

Zachary F. Lustig, Tushar M. Shimpi, Akash Shah, Walajabad S. Sampath

Automated Analysis of Internal Quantum Efficiency Using Chain Order Regression.................... 476

Zubair Abdullah-Vetter, Priya Dwivedi, Yoann Buratti, Alfred Krzywicki, Arcot Sowmya, Thorsten Trupke, Ziv Hameiri

Photon Recycling and Luminescent Coupling in All-Perovskite Tandem Solar Cells Quantified by Full Opto-Electronic Device Simulation 479

Urs Aeberhard, Simon J. Zeder, Beat Ruhstaller

The National Solar Radiation Database (NSRDB): Current Status.................... 480

Aron Habte, Manajit Sengupta, Yu Xie, Grant Buster, Michael Rossol, Paul Edwards, Galen Maclaurin, Evan Rosenlieb, Jaemo Yang, Haiku Sky, Mike Bannister, Billy Roberts

FTO Delamination for Photovoltaic Module Separation.................... 481

Jongwon Ko, Soohyun Bae, Yoonmook Kang, Hae-Seok Lee, Donghwan Kim

Investigation of CsF - Treatment Effects on Cu(In,Ga)(S,Se)2 Solar Cells Using Photothermal Atomic Force Microscopy Under Various Photoexcitation Conditions 482

Ayaka Yamada, Takuji Takahashi

Polysilicon Passivating Contact Layer for Crystalline Silicon Solar Cells: A Dopant-Grading Approach 483
Duy Phong Pham, Junsin Yi

Effects of (i)a-Si:H Deposition Temperature on Passivation Quality and Performance of High-Efficiency Silicon Heterojunction Solar Cells.................... 484
Yifeng Zhao, Paul Procel, Arno H. M. Smets, Luana Mazzarella, Can Han, Liqi Cao, Guangtao Yang, Zhirong Yao, Arthur Weeber, Miro Zeman, Olindo Isabella

Investigation of the Crack Propensity of Co-Extruded Polypropylene Based Backsheets.................... 485
Gernot Oreski, Chiara Barretta, Astrid Macher, Gabriele Eder, Lukas Neumaier, Markus Feichtner, Minna Aarnio-Winterhof

Comparative Life Cycle Assessment of Crystalline Silicon Glass-Sheet Based PV Modules and Plastic PV Modules 489
Sakthi Guhan Somasundaram, Xitong Zhu, Angele Reinders

Study of ALD-Grown Tin Oxide as an Electron Selective Layer for NIP Perovskite-Based Solar Cells.................... 497
Félix Gayot, Elise Bruhat, Matthieu Manceau, Eric De Vito, Denis Mariolle, Stéphane Cros

Investigation of Lead-Free 2D/3D Mixed-Dimensional Tin Perovskite Solar Cell Embedded with Plasmonic Metal Nanoparticles 504
Atanu Purkayastha, Manoranjan Minz, Ramesh Kumar Sonkar, Arun Tej Mallajosyula

High-Efficiency Solar Cell by Combining High and Low Thermal Budget for Si Passviting Contacts.................... 507
Muhammad Quddamah Khokhar, Shahzada Qamar Hussin, Muhammad Aleem Zahid, Duy Phong Pham, Eun-Chel Cho, Junsin Yi

The Role of the European Green Deal for the Photovoltaic Market Growth in the European Union.................... 508
Arnulf Jäger-Waldau, Georgia Kakoulaki, Nigel Taylor, Sandor Szábo

Assessment of Mechanical Robustness of Conventional and CFRP-Based Lightweight PV Module Architectures Under Static Loads.................... 512
Umang Desai, Aparna Singh

Exploring the Role of Temperature and Hole Transport Layer on the Ribbon Orientation and Efficiency of Sb2Se3 Cells Deposited Via Thermal Evaporation 516
Ryan Voyce, Stephen Campbell, Oliver S. Hutter, Guillaume Zoppi, Neil S. Beattie, Elizabeth A. Gibson, Vincent Barrioz

Decentralized BESS Control on a Real Low Voltage System with a Large Number of Prosumers.................... 517
Bruno Cortes, Ricardo Torquato, Tiago R. Ricciardi, Fernanda C. L. Trindade, Walmir Freitas, Victor B. Riboldi, Kunlin Wu

Hydrogen Complexes Present After Different Firing Profiles and Their Influence on LeTID Degradation 525
Benjamin Hammann, Nicole Assmann, Philip M. Weiser, Wolfram Kwapil, Tim Niewelt, Florian Schindler, Rune Søndenå, Eduard V. Monakhov, Martin C. Schubert

Superior Performance of Two-Phase Triple Halide Inorganic Perovskites.................... 526
Deniz N. Cakan, Rishi E. Kumar, Connor Dolan, Moses Kodur, Yanqi Luo, Tao Zhou, Zhonghou Cai, Barry Lai, Martin Holt, David P. Fenning

Thermally Evaporated Titanium Dioxide Film as an Electron-Selective Contact for Silicon Solar Cells......527
Changhyun Lee, Soohyun Bae, Hyunju Lee, Yoonmook Kang, Hae-Seok Lee, Donghwan Kim

3 MeV Proton Radiation Tolerance Study of Ultra-Thin Gallium Arsenide Solar Cells for Space Applications......528
Larkin Sayre, Armin Barthel, Andrew Johnson, Louise C Hirst

Monolithic Perovskite/Silicon Tandem Solar Cells on P-Type POLO/PERC Silicon Bottom Cells......529
Silvia Mariotti, Klaus Jäger, Marvin Diederich, Marlene S. Härtel, Bor Li, Kári Sveinbjörnsson, Eike Köhnen, Rolf Brendel, Sarah Kajari-Schröder, Robby Peibst, Steve Albrecht, Lars Korte, Tobias Wietler

Multiple Substrate Reuse: A Straightforward Reconditioning of Ge Wafers After Porous Separation......530
Alexandre Chapotot, Javier Arias-Zapata, Tadeáš Hanuš, Bouraoui Ilahi, Nicolas Paupy, Valentin Daniel, Zakaria Oulad El Hmaidi, Jérémie Chrétien, Gwenaëlle Hamon, Maxime Darnon, Abderraouf Boucherif

Short Drying Processes for Silicon Solar Cells......531
Daniel Ourinson, Michael Linse, Markus Klawitter, Andreas Lorenz

Sn4+-Free, Stable Tin Perovskite Films for Lead-Free Perovskite Solar Cells......532
Ajay Singh, Jeremy Hieulle, Himanshu Phirke, Joana A. F. Machado, Sevan Gharabeiki, Rukhsar Ahmad, Susanne Siebentritt, Alex Redinger

Fill Factor Losses in Cu(In,Ga)Se2 Based Solar Cells Due to Metastabel Defects — the Effect of Ag Addition......533
Thomas P. Weiss, Omar Ramirez, Taowen Wang, Valentina Serrano-Escalante, Stefan Paetel, Wolfram Witte, Jiro Nishinaga, Thomas Feurer, Ayodhya N. Tiwari, Susanne Siebentritt

Analysis of the Soiling Effects on Commissioning of Photovoltaic Systems: Short-Circuit Current Correction......534
Dênio Alves Cassini, Suellen C. Silva Costa, Antonia Sonia A. C. Diniz, Lawrence L. Kazmerski

Optical Absorption of MoS2 Based Ultrathin Solar Cells......538
Carlos Bueno-Blanco, Simon Aurel Svatek, Elisa Antolin

Microinverter Testing Update Using High Power Modules: Efficiency, Yield, and Conformity to a New "Estimation Formula" for Variation of PV Panel Size......539
Stefan Krauter, Jörg Bendfeld, Marius Möller

Glare Potential Evaluation of Structured PV Glass Based on Gonioreflectometry......544
Markus Babin, Sune Thorsteinsson, Adrian A. Santamaria Lancia, Michael L. Jakobsen, Sergiu V. Spataru

Upstream-Downstream Optimization of Volt-Var Control in Smart Grids......545
Laura R. Fardin, Christiano Lyra, Fernanda C. L. Trindade

Growth of GaAs on Ge/Si (001) Nanovoided Virtual Substrate......550
Jonathan Henriques, Alexandre Heintz, Bouraoui Ilahi, Richard Arès, Abderraouf Boucherif

Monte Carlo Evaluation of Multijunction Solar Systems in Tandem and 4-Terminal Configurations......551
Roberto Corso, Marco Leonardi, Andrea Scuto, Salvatore A. Lombardo

Outdoor Energy Performances for Standard and Bi-Facial Modules as Well on the Failure Modes Observed in Outdoor Conditions 554
A. Ottanà, F. Rametta, W. Gangemi, C. Colletti, A. Di Stefano, A. Canino, M. Foti, C. Gerardi, F. Bizzarri

Nanoabsorbers for Semitransparent Photovoltaics 557
Maximilian Götz-Köhler, Hosni Meddeb, Norbert Osterthun, Nils Neugebohrn, Kai Gehrke, Martin Vehse

State of the Art of Modelling Soiling and Snow Losses in PV Systems 562
Sébastien Arbaretaz, Murielle Stepec, Eszter Voroshazi

Paving the Way to Building-Integrated Translucent Tandem Photovoltaics: Process Optimization and Transfer to Perovskite-Perovskite 2-Terminal Tandem Cells 565
David Benedikt Ritzer, Marco Alejandro Ruiz-Preciado, Bahram Abdollahi Nejand, Tobias Abzieher, Ulrich Wilhelm Paetzold

Terawatt-Scale Photovoltaics Enabled by Technological Learning 566
Lukas Wagner, Robert Pietzcker, Lorenz Friedrich, Jan Christoph Goldschmidt

Effects of Solar Spectrum and Albedo on the Performance of Bifacial Si Heterojunction Mini-Modules 567
Marco Leonardi, Roberto Corso, Andrea Scuto, Gabriella Milazzo, Carmelo Connelli, Marina Foti, Cosimo Gerardi, Fabrizio Bizzarri, Stefania M. S. Privitera, Salvatore A. Lombardo

Role of Back-Side Indium Tin Oxide on the Degradation Mechanism of Silicon Heterojunction Solar Cells 570
Gbenga D. Obikoya, Anishkumar Soman, Ujjwal K. Das, Steven S. Hegedus

Vinyl Acetate Content Tailoring in Ethylene Vinyl Acetate Improves the Resilience Against Environmental Stressors 574
Umang Desai, Bhuwanesh Kumar Sharma, Aparna Singh

Poisson Drift Diffusion Modeling of Valley Photovoltaic Devices 575
Daixi Xia, Hassan Allami, Jacob J. Krich

2T Mechanically Stacked Perovskite/Si Tandem Cells Beyond 28%: The Role of 2D Materials in Perovskite Top Cells Coupled with a Commercially Available Bifacial c-Si Heterojunction Cell 576
Antonio Agresti, Sara Pescetelli, Fabio Matteocci, Erica Magliano, Elisa Nonni, Giuseppe Bengasi, Carmelo Connelli, Cosimo Gerardi, Hanna Pazniak, Sebastiano Bellani, Francesco Bonaccorso, Fabrizio Bizzarri, Marina Foti, Aldo Di Carlo

CdTe:In - Post-Growth Doping and Proposals for Photovoltaic Devices 577
Luke Thomas, Theo DC Hobaon, Laurie J Phillips, Kieran J Cheetham, Neil Tarbuck, Mark Isaacs, Huw Sheil, Vin Dhanak, Tim D Veal, Stephen Campbell, Vincent Barrioz, Jon D Major, Ken Durose

Multiple Inverter Microgrid Experimental Fault Testing 578
Nicholas S. Gurule, Javier Hernandez Alvidrez, Matthew J. Reno, Jack Flicker

Reactive Anisotropic Conductive Adhesive Wafer Bonding for Solar Cells 584
Eric M. Rehder, Shoghig Mesropian, Xing-Quan Liu

The Effect of Dust Hygroscopicity on Soiling and Self-Cleaning Processes in a Condensing Environment 588
Jordan Eidlisz, Nadera Sultana, Illya Nayshevsky, Qianfeng Xu, Alan M. Lyons

On the Stability of Indium Tin Oxide with Functional Layers Back Contact Applications in Semitransparent Cu(In,Ga)Se2 Solar Cells 591

Robert Fonoll-Rubio, Marcel Placidi, Torsten Hoelscher, Angelica Thomere, Zacharie Jehl Li-Kao, Maxim Guc, Victor Izquierdo-Roca, Roland Scheer, Alejandro Pérez-Rodríguez

Bulk Lifetime Study of P-Type Czochralski Silicon with Different Processing History Using Quinhydrone-Methanol Surface Passivation 592

Tasnim K. Mouri, Ajay Upadhyaya, Ajeet Rohatgi, William N Shafarman, Ujjwal K. Das

PV Module Degradation Due to Frequent and Prolonged Inverter Clipping: A Preliminary Study 596

Manjunath Matam, Ryan M. Smith, Hubert Seigneur

ETFE and Its Role in the Fabrication of Lightweight c-Si Solar Modules 604

Fabiana Lisco, Farwa Bukhari, Luke Jones, Adam Law, John Michael Walls, Christophe Ballif

Thermal Stability of 2D/3D Halide Perovskites 605

Jeffrey A Christians, Josephine L Surel, Elizabeth V Cutlip

Life Cycle Assessment Analysis of Thin-Film, Flexible Solar Panels Produced in the Netherlands 606

Gianluca Limodio, Seba Makhlouf, Edward Hamers, Arno Smets

Front SiON/TCO Stacks Development for Double Side Poly-Si/SiOX Passivated Contacts Solar Cells 607

Charles Seron, Thibaut Desrues, Christine Denis, Raphaël Cabal, Frédéric Jay, Adeline Lanterne, Quentin Rafhay, Anne Kaminski, Sébastien Dubois

Comparing the Accuracy of Horizon Shade Modelling Based on Digital Surface Models Versus Fisheye Sky Imaging 608

Daniel Alvarez Mira, Martin Bartholomäus, Sebastian Poessl, Peter B. Poulsen, Sergiu V. Spataru

Ongoing Performance Assessment Strategies & Operational Challenges When Managing Hundreds of Distributed Photovoltaic Assets Across Asia 614

André M. Nobre, Anusha Agarwal, Sai Pranav

Development of Highly Uniform and Reproducible DI-O_3 Layers for Photovoltaic Applications and Beyond 620

Munan Gao, Vibhor Kumar, Ngwe Zin

Characterizing the Capacitance of Different c-Si PV Cell Technologies Using Impedance Spectroscopy 623

David A. Van Nijen, Patrizio Manganiello, Mirco Muttillo, Miro Zeman, Olindo Isabella

Demonstration of a Monolithically Integrated Hybrid Electroabsorptive Modulator/Photovoltaic Device for Bidirectional Free Space Optical Communication at 1.55 μm 624

Emily Kessler-Lewis, Stephen J. Polly, Elijah Sacchitella, Seth M. Hubbard, Raymond Hoheisel

Glued III-V on Si Tandem Solar Cells Using Hybrid Transparent Conductive Layers 625

Phuong-Linh Nguyen, Jeronimo Buencuerpo, Philippe Baranek, Oliver Hoehn, David Lackner, Frank Dimroth, Marco Faustini, Stephane Collin, Andrea Cattoni

Flexible All-Perovskite Tandem Solar Cells with High Specific Power 626

Zhaoning Song, Cong Chen, Chongwen Li, Lei Chen, Yanfa Yan

Impact of Thermal Annealing on the Mechanical Properties of Ge Epilayer on Mesoporous Germanium for Layer Separation and Substrate Re-Use 627
Firas Zouaghi, Ahmed Ayari, Bouraoui Ilahi, Jeremie Chretien, Tadeas Hanus, Nicolas Paupy, Nicolas Quaegebeur, Abderraouf Boucherif

Development of a Novel Soiling Chamber for Testing Antisoiling Coatings 628
Matthew T Muller

Progress and Demonstration of Micro-CPV Module with Integrated Planar Tracking and Diffuse Light Collection 629
Steve A Askins, Guido Vallerotto, César Dominguez, Mathilde Duchemin, Gaël Nardin, Mathieu Ackermann, Delphine Petri, Matthieu Despeisse, Jacques Levrat, Xavier Niquille, Christophe Ballif, Juan F Martinez, Marc Steiner, Gerald Siefer, Ignacio Antón

A Deep Learning Approach for PV Failure Mode Detection in Infrared Images: First Insights 630
Daniel Rocha, Miguel Lopes, Jennifer P. Teixeira, Paulo A. Fernandes, Modesto Morais, Pedro M. P. Salome

What Are PVDF-Based Backsheets Made Of? 633
Chiara Barretta, Eric Helfer, Astrid E. Macher, Gernot Oreski

Excess Current Due to Embedded Superlattices in Graphene/Ox/n-GaAs Solar Cells, at 50 Suns and Above 637
AC Varonides

Stability of Silicon Heterojunction Solar Cells Having Hydrogen Plasma Treated Intrinsic Layer 643
Anishkumar Soman, Gbenga Obikoya, Steve Johnston, Steven Harvey, Ujjwal Das, Steven Hegedus

Snow Shedding Properties of Bifacial PV Panels 646
Ajay Singh, Derek Jones

Glass-Glass PV Modules: Characterization of Chemical and Mechanical Degradation 649
Laura Spinella, Sona Ulicna, Archana Sinha, Dana B. Sulas-Kern, Michael Owen-Bellini, Steve Johnston, Laura T. Schelhas

Model of an Automous PV Home Using a Hybrid Storage System Based on Li-Ion Batteries and Hydrogen Storage with Waste Heat Utilization 650
Marius Möller, Stefan Krauter

Towards High Efficiency All-Perovskite Tandem Solar Cell by Preventing Performance Loss Arising from Physically Mixed Interfacial Layers 653
Biwas Subedi, Alex Bordovalos, Lei Chen, Zhaoning Song, Cong Chen, Yanfa Yan, Nikolas J Podraza

Nonparametric Temporal Downscaling of GHI Clear-Sky Indices Using Gaussian Copula 654
Jing Huang, Marc Perez, Richard Perez, Dazhi Yang, Patrick Keelin, Tom Hoff

Ga-Doping of MZO in CdSeTe/CdTe Thin Film Solar Cells 658
Mustafa Togay, Tushar Shimpi, Sampath S. Walajabad, Kurt L. Barth, Eric Don, Gabor Parada, J. Michael Walls, Jake W. Bowers

Firm PV Power Generation in Switzerland 661
Jan Remund, Marc Perez, Richard Perez

Demonstration of Point Contact Geometry for Solar Cells Using Single Walled Carbon Nanotube 667
Fadhil K. Alfadhili, Adam B. Phillips, Manoj K. Jamarkattel, Bhuiyan M. Anwar, Prabodika N. Kaluarachchi, Zahrah S. Almutawah, Abdul Quader, Deng-Bing Li, Yanfa Yan, Randy J. Ellingson, Michael J. Heben

Accelerate Cycles of Learning: Unencapsulated Silicon Photovoltaic Cells to Environmental
Stressors 668
Nafis Iqbal, Nitin K. Chockalingam, Kehley A. Coleman, Jeffrie Fina, Kristopher O. Davis, Laura S. Bruckman, Ina T. Martin

End of Use, Circularity, and Sustainability Considerations in Solar Photovoltaic Module Design
and Product Development and Support 675
Chris Powicki, Wayne Li, Cara Libby

Degradation of Crystalline Silicon Photovoltaic Modules Installed in Different Climates 680
Chiara Barretta, Astrid E. Macher, Julián Ascencio-Vásquez, Marc Köntges, Marko Topic, Gernot Oreski

Comparing Fluorinated and Non-Fluorinated Anti-Soiling Coatings for Solar Panel Cover Glass 683
Luke O. Jones, Adam M. Law, Gary Critchlow, John M. Walls

Extraction of Prevailing Soiling Rates from Soiling Measurement Data 684
Josh Peterson, Julie Chard, Justin Robinson

Mismatch Losses in Simulated Commercial and Utility-Scale PV Arrays Due to Shortened Strings 692
Ryan M. Smith, Manjunath Matam, Hubert Seigneur

Fault Analysis and Relay Assessment on a Substation System with High Penetration of PV
Generation 693
Biqi Wang, Genesis Alvarez, Micah J. Till, Kevin Jones, Mathew Gardner, Rolando Burgos, Bo Wen

Determining Surface Recombination Velocity and Band Bending at the Back Interface of CdTe
Devices Using Back Illuminated Quantum Efficiency 701
Adam B. Phillips, Jared D. Friedl, Zhaoning Song, Ramez Hosseinian Ahangharnejhad, Ebin Bastola, Zulkifl H. Rabbani, Deng-Bing Li, Yanfa Yan, Randy J. Ellingson, Michael J. Heben

Use of a Selenium-Telluride Alloy as a Back Interface for CdTe-Based Cells 702
Daniel Z. Shaw, Camden L. Kasik, Andrew C. Treglia, James R. Sites

Effect of Dilute Acid Exposure on Sol-Gel Porous Silica Anti-Reflection Coatings 705
F. Bukhari, L. Jones, A. Law, A. Abbas, J. M. Walls

Insights into the Stability of Amorphous/Crystalline Silicon Interface Under Light and Temperature 708
Salman Manzoor, Mariana Bertoni

Automated Shift Detection in Sensor-Based PV Power and Irradiance Time Series 709
Kirsten Perry, Matthew Muller

The Effect of Inverter Loading Ratio on Energy Estimate Bias 714
Kevin S. Anderson, William B. Hobbs, William F. Holmgren, Kirsten R. Perry, Mark A. Mikofski, Rounak A. Kharait

Life Cycle Assessment of High-Efficiency Si Solar Modules 721
Estefania Papaioannou, Pritpal Singh, Ross Lee

Alternative Rear Contacts for Ultrathin $CdSe_xTe_{1-x}$ Solar Cells.. 722
Bérengère Frouin, Andrea Cattoni, John Moseley, David Albin, Joel Duenow, Abderrahime Sekkat, David Muñoz-Rojas, Stéphane Collin

Feeder Open-Phase Detection by Smart Inverters... 725
Yiwei Ma, Xiaojie Shi, Aminul Huque, Roland Bründlinger, Ron Ablinger

Design with Integrated PV Technologies in Various Products and Environments 731
Eli Shirazi, Wouter Eggink, Xitong Zhu, Angele Reinders

Electric Field and Its Effect on Hot Carriers in InGaAs Valley Photovoltaic Devices 732
Kyle R. Dorman, Vincent R. Whiteside, David K. Ferry, Tetsuya D. Mishima, Hamidreza Esmaielpour, Michael B. Santos, Ian R. Sellers

Single-Axis Tracker Control Optimization Potential for the Contiguous United States 733
Kevin Anderson, Saurabh Aneja

Outdoor Characterization of Hybrid HCPV- T Module Featuring a Passive Tracking System 739
Guido Vallerotto, Steve Askins, Javier Van Herpt, David Martí, Jaime Caselles, Ignacio Antón

CIGS Degradation Due to Water Ingress: Post-Mortem Analysis of a Field-Exposed PV Module 740
Simona Villa, Remi Anitat, Pelin Yilmaz, Aldo Kingma, Mikolaj Dziechciarz, Joran Van Den Berg, Klaas Bakker, Mirjam Theelen

Evaluation of Cellular Based DER Direct Transfer Trip (DTT) Technologies 741
Yiwei Ma, Aminul Huque, Joseph Estrada, Tim Godfrey, Charles Brewster

Towards a Shade Tolerant Monolithically Interconnected Perovskite Module for Use in Four Terminal Tandem Devices 748
Klaas Bakker, Jacopo Sala, Mehrdad Najafi, Michaël Daenen, Bart Geerligs

Panel Segmentation: A Python Package for Automated Solar Array Metadata Extraction Using Satellite Imagery... 751
Kirsten Perry, Christopher Campos

Optical Determination of Carrier Concentrations in ITO, PEDOT:PSS, and (FASnI3)0.6(MAPbI3)0.4 Within a PV Device 752
Madan K Mainali, Prakash Uprety, Zhaoning Song, Changlei Wang, Indra Subedi, Kiran Ghimire, Maxwell M Junda, Yanfa Yan, Nikolas J Podraza

Strategies to Optimize and Validate Tracking Performance of Single-Axis Trackers on Diffuse Sites 753
Kendra Passow, Kyumin Lee, Sanket Shah, Daniel Fusaro, Jon Sharp

Development of Hierarchical Control for a Lunar Habitat DC Microgrid Model Using Power Hardware-In-The-Loop 754
Andrew R. R. Dow, Rachid Darbali-Zamora, Jack D. Flicker, Felipe Palacios, Jeffrey T. Csank

Influence of Se Grading on the Free Carrier Profile of CdSeTe/CdTe Solar Cells 761
Jared D. Friedl, Ebin Bastola, Rasha A. Awni, Xavier Mathew, Adam B. Phillips, Yanfa Yan, Michael J. Heben

Statistical Performance Analysis on ≈ 320 Perovskite Single- And Two-Junction Solar Cells and Modules from >30 Global Sources 766
Tao Song, Charles Mack, Rafell Williams, Josh Gallon, Allan Anderberg, Larry Ottoson, Daniel J. Friedman, Nikos Kopidakis

GPU-Accelerated Machine Learning for Analysis of Time-Resolved Photoluminescence Data 769
Calvin Fai, Anthony J. C. Ladd, Charles J. Hages

Micro-Fabrication and Transfer of a Detachable Ge Epitaxial Layer Grown on Porous Germanium 770
Valentin Daniel, Jeremie Chretien, Gwenaelle Hamon, Mathieu De Lafontaine, Nicolas Paupy, Zakaria Oulad El Hmaidi, Bouraoui Ilahi, Tadeàš Hanus, Maxime Darnon, Abderraouf Boucherif

Native Oxide Growth on CdSeTe for Improved Back Surface Passivation 773
Adam Danielson, Carey Reich, Mason Mahaffey, Arthur Onno, Zach Holman, Walajabad Sampath

Optical Characterization of Thin Film Cu_xAlO_y in the CdTe Device Configuration 774
Indra Subedi, Kamala Khanal Subedi, Prabin Dulal, Adam B. Phillips, Michael J. Heben, Randy J. Ellingson, Nikolas J. Podraza

A Tool for the Simulation, Evaluation and Teaching the Operation of Low Power Microgrids 778
Johann A. Hernández M, Adolfo A. Jaramillo M, Carlos A. Arredondo-Orozco

Translating Material-Level Characterization of Carbon-Nanotube-Reinforced Composite Gridlines to Module-Level Degradation 783
Andre Chavez, Brian Rummel, April Jeffries, Sang M. Han, Nick Bosco, Brian Rounsaville, Ajeet Rohatgi

Testing the Abrasion Resistance of Porous SiO_2 Anti-Reflection Coatings for Solar Cover Glass 786
Adam M Law, Farwa Bukhari, Luke O Jones, Ali Abbas, John Michael Walls

Hydrothermally Deposited Antimony Sulfide Solar Cells with V_{OC} Approaching 800 mV 792
Dipendra Pokhrel, Nini Rose Mathew, Suman Rijal, Ebin Bastola, Abasi Abudulimu, Tamanna Mariam, Xavier Mathew, Adam B Phillips, Michael J Heben, Zhaoning Song, Yanfa Yan, Randy J Ellingson

Multiresonant Light Trapping in Ultra-Thin Solar Cells with Transparent Quasi-Random Structures 795
Eduardo Camarillo Abad, Hannah J. Joyce, Louise C. Hirst

FAIRification, Quality Assessment, and Missingness Pattern Discovery for Spatiotemporal Photovoltaic Data 796
William C. Oltjen, Yangxin Fan, Jiqi Liu, Liangyi Huang, Xuanji Yu, Mengjie Li, Hubert Seigneur, Xusheng Xiao, Kristopher O. Davis, Laura S. Bruckman, Yinghui Wu, Roger H. French

A Sparse and Low Rank Penalized Signal Decomposition Model with Constraints: Anomaly Detection in PV Systems 802
Wei Yang, Daniel Fregosi, Michael Bolen, Kamran Paynabar

Cloud Segmentation and Motion Tracking in Sky Images 805
Benjamin G Pierce, Joshua S Stein, Jennifer L Braid, Daniel Riley

Towards Standardization of Accelerated Stress Testing Protocols for Metal-Halide Perovskite Photovoltaic Modules 806

Michael Owen-Bellini, Timothy J Silverman, Michael G. Deceglie, Paul Ndione, Nikos Kopidakis, Ingrid Repins, Mickey Wilson, Dana B. Sulas-Kern, Joseph Berry, Laura T. Schelhas, Colin Sillerud, Jinsong Huang, Michael J. Heben, Yanfa Yan, Devin Mackenzie, Joshua S. Stein

The Balance of Thermodynamic Potentials in Solar Cells Investigated by Numerical Device Simulations 807

Felix Komoll, Uwe Rau

Hybrid Functional Calculations for Antimony Doping in CdTe 808

Intuon Chatratin, Shagorika Mukherjee, Anderson Janotti

GaAs-Based Photovoltaic Infrared Energy Harvesting for Microscale Biomedical Implants 811

Y. Sun, J. Letner, J. Lee, N. Ahmed, C. Chestek, D. Blaauw, J. Phillips

Evaluation of an LED Simulator for Single- And Multi-Junction PV Cell Performance Testing 814

Nikos Kopidakis, Tao Song, Charles Mack, Rafell Williams, Hal Friesen, Justin Bertagnolli, John Walmsley

Metastability and Degradation of CdTe Solar Cells Investigated by nm-Scale Electrical Potential Imaging 819

Chun-Sheng Jiang, David Albin, Marco Nardone, Kassidy H. Howard, Adam Danielson, Amit Munshi, Tushar Shimpi, Walajabad Sampath, Chuanxiao Xiao, Helio R. Moutinho, Mowafak M. Al-Jassim, Glenn Teeter

Photovoltaic Thermal Management in Luminescent Solar Concentrators 820

Megan E. Phelan, David R. Needell, Maggie M. Potter, Haley C. Bauser, Catherine N. Ryczek, Ralph G. Nuzzo, Harry A. Atwater

PVRPM in Python: An Overview of New Capabilities 826

Paul Lunis, Brandon Silva, Marios Theristis, Hubert Seigneur

Oxygen and Temperature Effects on NiO Buffer Layers for CdTe Solar Cells 827

Nicholas Hunwick, Xiaolei Liu, Patrick J. M. Isherwood, John. M. Walls

Optimizing $CdCl_2$ Treatment on CdTe Solar Cells Using Spray Deposition Method 828

Prabodika N. Kaluarachchi, Shannon E. Costello, Ryan Madden, Jacob M. Gibbs, Tyler R. Brau, Aesha P. Patel, Manoj K. Jamarkattel, Jared D. Friedl, Kevin G. Schaffer, Kristof J. Nieschwitz, Ebin Bastola, Adam B. Phillips, Randy J. Ellingson, Michael J. Heben

External Quantum Efficiency and Device Reflectance of CIGS PV for Terrestrial and Space Based Applications 833

Bishal Shrestha, Indra Subedi, Robert W. Collins, Nikolas J. Podraza

Radiation Tolerance, High Temperature Stability, and Self-Healing of Triple Halide Perovskite Solar Cells 836

Hadi Afshari, Sergio A Chacon, Brandon K Durant, Rose Crawford, Bibhudutta Rout, Giles E Eperon, Ian R Sellers

Hyperspectral Luminescence Imaging Analysis of Solar Cells with Localized Radiative Defects 837

Brianna Conrad, Behrang H. Hamadani

Impact of In-Situ Cd Saturation MOCVD Grown CdTe Solar Cells on as Doping and VOC 838
Ochai Oklobia, Steve Jones, Giray Kartopu, Dingyuan Lu, Wes Miller, Rajni Mallick,
Xiaoping Li, Gang Xiong, Vladislav Kornienko, Martin Bliss, Ali Abbas, Michael Walls,
Stuart J. C. Irvine

Insertion of Photovoltaic Generation in the Planning of Electricity Distribution Systems Based on
Its Economic Potential.. 839
João Cardoso Das Neves Neto, Miguel Edgar Morales Udaeta, Carlos Frederico Meschini
Almeida, Henrique Fernandes Camilo

Physics-Guided Machine Learning Identifies 5 Optimum Test Locations to Predict Global PV
Energy Yield for Arbitrary Farm Topologies 843
Jabir Bin Jahangir, Muhammed Tahir Patel, Muhammad A. Alam

Analyzing Effects of Solar Variability and System Location on LMP Prices 847
Mesude Bayrakci-Boz, Joseph Ranalli

Demystifying the Effect of Hydrogen Treatment on Silicon Photovoltaics 854
Govind Nanda, Sara Almenabawy, Rajiv Prinja, Geetu Sharma, Nazir P. Kherani

Bill of Materials Variation and Module Degradation in Utility-Scale PV Systems 855
Michael G. Deceglie, E. Ashley Gaulding, John S. Mangum, Timothy J Silverman, Steve W.
Johnston, James A. Rand, Mason J. Reed, Robert Flottemesch, Ingrid L. Repins

The Thermodynamics Behind the Photovoltage Generation and Photocurrent Collection in Solar
Cells... 856
Uwe Rau

Tuning the Band Gap of Magnesium Zinc Oxide to Enhance Band Alignment with CdTe Based
Photovoltaic Devices... 857
Kerrie M Morris, Mustafa Togay, Rachael C Greenhalgh, Jake W Bowers, John M Walls

Sensitivity of Sub-Hourly Modeling Error to Project Size 858
Christopher Hayes, Abhishek Parikh, Mark Mikofski, Rounak Kharait

Energy-Based Soiling Loss Monitoring Approach for Solar PV System 859
Pavan Fuke, Shoubhik De, Narendra Shiradkar, Anil Kottantharayil

Predicting Materials Parameters in Colloidal Quantum Dot Photovoltaic Devices Using Machine
Learning Models Trained on Experimental Data 862
Hoon Jeong Lee, Ariana B. Hofelmann, Yida Lin, Susanna M. Thon

Development of HVPE-Grown III-V Solar Cells Passivated with AlInP.. 867
Jacob T Boyer, Kevin L Schulte, Aaron J Ptak, John Simon

Progress in PV Material Durability Test Methodologies.. 868
William J. Gambogi

Measuring Irradiance with Bifacial Reference Panels.. 871
Nicholas Riedel-Lyngskær, Jan Vedde, Peter B. Poulsen, Sergiu Spataru

Local nm-Scale Imaging of Electrical Contact for Series Resistance Degradation of Silicon Solar
Cells... 872
C.-S. Jiang, S. Johnston, E. A. Gauding, M. G. Deceglie, R. Flottemesch, C. Xiao, R.
Moutinho, D. B. Sulas-Kern, J. Mangum, M. M. Al-Jassim, I. L. Repins

Perimeter Recombination in GaAs Solar Cells with Different Geometries.................... 875
Natasha Gruginskie, Gerard Bauhuis, Peter Mulder, Elias Vlieg, John Schermer

Global Ranking of Losses to Photovoltaic Power 876
A. Kubiniec, K. Seymour, A. Bhat, J. Hazari, T. Haley, Marc Perez

PV Module Toxicity Testing Methods and Results: A Literature Review 879
F. Li, S. Shaw, C. Libby, B. Bicer, G. Tamizhmani

Validation of Novel Bifacial Photovoltaic Performance Model with 3D Shading for Fixed-Tilt and
Single-Axis Tracked Systems.. 883
Annie Russell, Christopher E. Valdivia, Cédric Bohémier, Joan E. Haysom, Karin Hinzer

Characterizing the Back-Contact Interface of Bi-Facial Poly-Crystalline CdTe Devices Using
Transmission Electron Microscopy 884
John Farrell, Ebin Bastola, Manoj Jamarkattel, Michael Heben, Robert F. Klie

Photon Management in CdSeTe Absorber Solar Cells: The Case for Increased Attention to Optical
Cell Design 887
Carey L. Reich, Arthur Onno, Adam Danielson, Zachary C. Holman, Walajabad S. Sampath

Geographic Analysis for Determining the Value of Different Photovoltaic Performance Factors.................. 888
Madhuri Kumari, Marios Theristis, Joshua S. Stein

Computerized Tool for Students Training in Solar Geometry.................................... 893
Johjan Stiven Zea Fernández, Mario Luna-Delrisco, Sebastián Villegas Moncada, Carlos
Ernesto Arrieta González, Johann A. Hernández M, Carlos A Arredondo Orozco

Anisotropy-Induced Fluctuations in Cu(In, Ga)Se2 897
Diego Colombara

Development of Spatial Mapping and Degradation Monitoring for Perovskite Films.................... 898
Emily J Miller, Biwas Subedi, Jaehoon Chung, Chongwen Li, Yanfa Yan, Nikolas J Podraza

PV+ Storage Operation and Maintenance 899
Natalie Gayoso, Nicole D Jackson, Thushara Gunda, Jal Desai, Andy Walker

Effect of Microstructure on the Photoactivity of Thin Film CdSe 900
Rachael Greenhalgh, Kerrie Morris, Vladislav Kornienko, Martin Bliss, Ali Abbas, Jake
Bowers, Michael Walls

AgriPV Citizen Science Lab: A Collaborative Model for Engineers, Youth Scholars and
Communities 904
Stuart Bowden, Jazmine Cordon, Myla Dykes, Michael Hernandez, Michelle Jordan, Alex
Killam, Jasmine Martinez Castillo, Alex Park, Alondra Pita, Maryan Robledo, Steve Zuiker

Behavioral and Population Data-Driven Distribution System Load Modeling 907
Isaac Bromley-Dulfano, Xiangqi Zhu, Barry Mather

Thin-Film Multijunction Inverted Metamorphic Solar Cells with Light Management for Space
Applications... 913
Julia D'Rozario, Steve Polly, Rao Tatavarti, Seth Hubbard

Probing Dynamic Influence of Moisture Ingress on Cell Deflection in Photovoltaic Modules 914
Ian M Slauch, Rishi E Kumar, Tala Sidawi, Jared Tracy, William Gambogi, Rico Meier,
David P Fenning, Mariana I Bertoni

Potovoltaic Module R&D Considerations for Soiling Mitigation 915
    Lin J. Simpson, Matthew Brantl, Ryo Huntamer

Seven-Level Cascaded H-Bridge Multilevel Single-Phase Inverter Implemented with an ATMEGA
Microprocessor 916
    Edgardo Desarden-Carrero, Rachid Darbali Zamora, Erick Aponte-Bezares, Eduardo I.
    Ortiz-Rivera

Annual Energy Production Uncertainty of Bifacial PV Plants Caused by Inaccuracies in Albedo
Data: Case Studies Using SAM 923
    Vicente Lara Fanego

Arsenic Doped CdSeTe Solar Cells: Charge Collection and Defects 924
    Niranjana Mohan Kumar, Srisuda Rojsatien, Trumann Walker, Tara Nietzold, Barry Lai,
    Arun K. M. Kanakkithodi, Maria Chan, Dan Mao, Mariana Bertoni

Observations on a Colorado Electric-Utility Resource Plan for Increasing Renewables from 55% to
80% by 2030 925
    Ronald A. Sinton

Estimation of Soiling Losses in Unlabeled PV Data 930
    Bennet Meyers

Understanding the Solar Cell Contacts with Atmospheric Screen-Printed Copper 937
    Sandra Huneycutt, Abasifreke Ebong, Krishnamraju Ankireddy, Ruvini Dharmadasa, Thad
    Druffel

Na Diffusion and Device Performance of AgBr Treated CuGaSe$_2$ Thin Films 941
    Elizabeth Palmiotti, Polyxeni Tsoulka, Thomas Lepetit, Nicolas Barreau, Angus Rockett

Extensive Evaluation and Uncertainty Estimation of Albedo Data Sources 945
    Vicente Lara-Fanego, Christian A. Gueymard, Jose A. Ruiz-Arias, Tomas Cebecauer, Juraj
    Betak

Sampling Solar Irradiance with Copula 948
    Mesude Bayrakci-Boz

Analysis of Temperature Dependence of Solar Cell Performance Through Light Soaking 953
    Samuel Seibert, Aesha P. Patel, Manoj Rajakaruna, Sandip S. Bista, Lei Chen, Randy J.
    Ellingson, Yanfa Yan, Zhaoning Song

Demonstrating the Thermoradiative Diode: Generating Electrical Power Through Radiative
Emission 959
    Nicholas J Ekins-Daukes, Michael P Nielsen, Andreas Pusch, Muhammad H Sazzad, Phoebe
    M Pearce, Peter J Reece

Vertical Bifacial Solar Panels as a Candidate for Solar Canal Design 960
    Jeremiah B Reagan, Sarah Kurtz

Arrhenius Analysis of the Degradation Modes in Emerging Photovoltaic Backsheets 961
    Naila M. Al Hasan, Rachael Arnold, David C. Miller, Jimmy Newkirk, Emily Rago, Michael
    Thuis, Bruce H. King, Laura T. Schelhas, Archana Sinha, Kent Terwilliger, Sona Ulicná,
    Peter Pasmans, Christopher Thellen

Comparison of Measured and Modeled Snow Losses for Photovoltaic Systems in Colorado 964
    Owen W. Westbrook, Sara M. Macalpine, David A. Bowersox

PV Hosting Capacity Estimation: Experiences with Scalable Framework.. 967
Wenbo Wang, Daniel Thom, Kwami Senam Sedzro, Sherin Ann Abraham, Yiyun Yao, Jianli Gu, Shibani Ghosh

Properties of Co-Sputtered $(In_xGa_{(1-x)})_2O_3$ Layers Used in CdTe Solar Cells 972
Manoj K. Jamarkattel, Adam B. Phillips, Indra Subedi, Abasi Abudulimu, Ebin Bastola, Deng-Bing Li, Zhaoning Song, Xavier Matthew, Yanfa Yan, Randy J. Ellingson, Nikolas J. Podraza, Michael J. Heben

High Efficiency Solar Cells Grown on Spalled Germanium Without Polishing 975
John S. Mangum, Anthony D. Rice, Jie Chen, Jason Chenenko, Evan Wong, Anna K. Braun, Steve Johnston, John F. Geisz, Aaron J. Ptak, Corinne E. Packard

The Effect of $CdSe_xTe_{1-x}$ Thickness on the $CdSe_xTe_{1-x}$/CdTe Solar Cell Performance........................ 976
Md Zahangir Alom, Sheikh Tawsif Elahi, Vasilios Palekis, Wei Wang, Chris Ferekides

Overall Performance Losses and Activated Mechanisms in Double Glass and Glass-Backsheet Photovoltaic Modules with Monofacial and Bifacial PERC Cells, Under Accelerated Exposures.................. 980
Jiqi Liu, Sameera Nalin Venkat, Jennifer L. Braid, Xuanji Yu, Brenton Brownell, Xinjun Li, Jean-Nicolas Jaubert, Kaushik Roy Choudhury, Laura S. Bruckman, Roger H. French

Chlorine Doped n-Type CdTe Solar Cells.. 988
Wei Wang, Vasilios Palekis, Md Zahangir Alom, Sheikh Elahi Tawsif, Chris Ferekides

26.7% AM0, 30.2% AM1.5G Dual Junction Solar Cell with 50x InGaAs Quantum Wells, GaAsP Strain Compensation, and Distributed Bragg Reflector 991
Stephen J Polly, Brandon Bogner, Anastasiia Fedorenko, Subhra Chowdhury, Dhrubes Biswas, Seth M Hubbard

Spectral Rear Irradiance Testing and Modeling for Degradation and Performance of Solar Fields.............. 992
Silvana Ovaitt, Matthew Brown, Chris Deline, Michael D. Kempe

Effects of Satellite Sampling on Subhourly Modeling Errors 995
Mark A. Mikofski, William F. Holmgren, Jeff Newmiller, Rounak Kharait

Room Temperature, Dip Coating Organic Passivation for c-Si Surface 996
Kejun Chen, Abigail. R Meyer, Harvey Guthrey, William Nemeth, San Theingi, Matthew Page, Sumit Agarwal, David. L Young, Paul Stradins

Investigation of Underperformance in Fielded N-Type Monocrystalline Silicon Photovoltaic Modules.. 997
E. Ashley Gaulding, Steve W. Johnston, Dana B. Sulas-Kern, Mason J. Reed, James A. Rand, Robert Flottemesch, Timothy J Silverman, Michael G. Deceglie

Fill Factor Loss in Perovskite Solar Cells Using Fullerene ETLs Caused by Air Exposure 998
Austin G Kuba, Alexander J Harding, Raphael Richardson, Ujjwal K Das, Kevin D Dobson, William N Shafarman

The Effect of Residual PbI2 on 2-Step Vapor-Processed P-I-N and N-I-P MAPbI3 Solar Cells.................... 999
Austin G Kuba, Alexander J Harding, Chaiwarut Santiwipharat, Ujjwal K Das, Kevin D Dobson, William N Shafarman

Characterization of DER Momentary Cessation and Rate-Of-Change-Of-Frequency Response................... 1000
Rasel Mahmud, Li Yu, Andy Hoke

No Time to Waste: Quickly Optimizing Perovskite Composition with Off-The-Shelf Active Learning Methods.. 1003
    Rishi E Kumar, Moses Kodur, Arun Kumar Mannodi Kannakithodi, David P Fenning

Reactive Silver Inks as Front Electrodes for TCO Coated Solar Cells....................................... 1004
    Michael W. Martinez-Szewczyk, Steven Digregorio, Owen Hildreth, Mariana I. Bertoni

Flexible CdTe/MgCdTe Double-Heterostructure Solar Cells Made from Epitaxial Lift-Off Thin Films... 1007
    Xin Qi, Jia Ding, Zheng Ju, Stephen Schaefer, Yong-Hang Zhang

Machine Learning Driven Studies of Performance Degradation in a-Si:H/C-Si Heterojunction Solar Cells.. 1010
    Davis Unruh, Reza Vatan Meidanshahi, Zitong Zhao, Stephen M. Goodnick, Gergely T. Zimanyi

Stability Analysis and Volt-Watt Control Setting Guideline for Distributed Energy Resources 1013
    Wenzong Wang, Wei Ren, Aminul Huque, Devin Van Zandt, Reigh Walling

Moisture Ingress and Distribution in Bifacial Silicon Photovoltaics.. 1019
    Rishi E Kumar, Tala Sidawi, Ian M Slauch, Rico Meier, Mariana I Bertoni, David P Fenning

Evaluation of PV Module Packaging Strategies of Monofacial and Bifacial PERC Using Degradation Pathway Network Modeling 1020
    Sameera Nalin Venkat, Jiqi Liu, Xuanji Yu, Jakob Wegmueller, Kunal Rath, Xinjun Li, Jean-Nicolas Jaubert, Jennifer L. Braid, Roger H. French, Laura S. Bruckman

Material Use and Life Cycle Impact of Crystalline Silicon PV Modules Over Time 1028
    Luyao Yuan, Annick Anctil

Drift-Diffusion Modelling of Four-Junction InGaP/InGaAs/SiGeSn/Ge Solar Cells 1031
    Laurier S. Baribeau, Robert F. H. Hunter, Christopher E. Valdivia, Karin Hinzer

Impact of Humidity, Temperature, and Oxygen on the Stability of $FA_{0.7}MA_{0.3}Sn_{0.5}Pb_{0.5}I3$ Perovskites... 1032
    Alex Bordovalos, Marie S Tumusange, Biwas Subedi, Lei Chen, Zhaoning Song, Yanfa Yan, Nikolas J Podraza

The Influence of Wind and Module Tilt on the Operating Temperature of Single-Axis Trackers................. 1033
    Keith R. McIntosh, Malcolm D. Abbott, Ben A. Sudbury, Saurabh Aneja, Mitch Bowman, Lance Brown, Ben Kahane, Norm Nicholas, Kristian Nolde

Impacts of Nonuniform Soiling on Photovoltaic Production 1037
    Lin J. Simpson, Ian K. Teague, Jody Ford, Nathan Shih, Mahfujur Rahman, Jorge I. T. Marchand, Kirsten Perry, Chris Deline

Operability of a Power System with Synchronous Condensers and Grid-Following Inverters...................... 1038
    Marena Trujillo, Rick Wallace Kenyon, Gemini Yau, Li Yu, Andy Hoke, Bri-Mathias Hodge

Vapor Treatment for Growth of High-Quality Oxide Barriers Within P-I-N Perovskite Solar Cells and Tandems... 1043
    Samuel A. Johnson, Michael D. McGehee, Joseph J. Berry, Axel F. Palmstrom

Long-Term UV Durability of Laminated Glass/Transparent Backsheet Coupons for Bifacial Photovoltaics: Backsheet Side Exposure.. 1044
    Soshana Smith, Stephanie Moffitt, Stefan Mitterhofer, Song-Syun Jhang, Stephanie Watson, Li-Piin Sung, Lakesha Perry, Deborah Jacobs, Xiaohong Gu

Silicon Heterojunction Solar Cells with High Bulk Resistivities Over 1,000 Ω·cm in Relevant Field Conditions of Illumination and Temperature... 1045
  Anh Huy Tuan Le, Apoorva Srinivasa, Stuart G. Bowden, Ziv Hameiri, André Augusto

Electroluminescence Analysis and Grading of Hail Damaged Solar Panels 1046
  Andrew M. Gabor, Phillip J. Knodle, Maurice Covino, Dylan J. Colvin, Kristopher O. Davis

What is the Role of Recycling in the Solar Terawatt Future? 1047
  Pablo R Dias, Moonyong Kim, Alison Lennon, Brett Hallam

Solar Panel Power Simulation for Shade Detection 1048
  David Jose Florez Rodriguez, Bennet E. Meyers

Evaluation and Demonstration of Slot-Die Coating for Perovskite Thin Film Mini-Modules for Space Photovoltaics... 1055
  Manoj Rajakaruna, Amir Hossein Ghahremani, Tao Zhu, Jaehoon Chung, Tamanna Mariam, Tyler Brau, Adam Phillips, Michael J. Heben, Zhaoning Song, Randy J. Ellingson, Yanfa Yan

Deleterious Effect of Light Trapping on the Temperatures of Solar Modules 1059
  Nicholas P. Irvin, D. Martínez Escobar, Aaron Wheeler, Tomas Leijtens, Hyunjong Lee, Annikki Santala, Richard R. King, Christiana B. Honsberg, Sarah R. Kurtz

Evaluating the Environmental Benefit of Residential Photovoltaic Modules Early Retirement in California... 1060
  Mallika Kothari, Annick Anctil

From Femtoseconds to Gigaseconds: The SolDeg Project to Analyze Si Heterojunction Cell Degradation with Machine Learning... 1061
  Gergely Zimanyi, Davis Unruh, Reza Vatan, Zitong Zhao, Andrew Diggs, Stephen Goodnick

Critical Transport Behavior in Quantum Dot Solids 1062
  Michael Kovtun, Zachary Crawford, Adam Goga, Gergely T. Zimanyi

Extended Accelerated Stress Testing (EAST) of Glass/Glass, Glass/Backsheet and Glass/Transparent Backsheet PV Modules: Influence of EVA and POE Encapsulants........................... 1065
  Akash Kumar, Ashwini Pavgi, Peter Hacke, Kaushik Roy Choudhury, Govindasamy Tamizhmani

Development of a Co-Anneal Process for Double-Side TOPCon Precursor Fabricated by Ex-Situ POCl3 and APCVD Boron Diffusion... 1068
  Wook-Jin Choi, Young-Woo Ok, Keeya Madani, Shubham Duttagupta, Ajeet Rohatgi

Complex Refractive Index and Complex Dielectric Function Modeling of Film Stack in Perovskite Solar Cells Using Spectroscopic Ellipsometry 1069
  Maria Fernanda Villa Bracamonte, Jose Raul Montes Bojorquez, Arturo Ayon

Spatially-Resolved X-Ray Excited Optical Luminescence of Metal Halide Perovskites 1070
  Connor Dolan, Deniz N. Cakan, Rishi E. Kumar, Moses Kodur, Yanqi Luo, Barry Lai, David P. Fenning

Exploring the Composition Space of Wide Band-Gap Absorbers for Silicon-Perovskite Tandems 1071
  Moses Kodur, Rishi E. Kumar, Deniz N. Cakan, Connor Dolan, Yanqi Luo, Barry Lai, David P Fenning

A Machine Vision Tool for Facilitating the Optimization of Large-Area Perovskite Photovoltaics............. 1072
  Mathilde Fievez, Nina Taherimakhsousi, Benjamin P. Macleod, Edward P. Booker, Muriel Matheron, Matthieu Manceau, Stéphane Cros, Solenn Berson, Curtis P. Berlinguette

Effective Irradiance Monitoring Using Reference Modules 1073
Jennifer L. Braid, Joshua S. Stein, Bruce H. King, Christopher Raupp, Jaya Mallineni, Justin Robinson, Steve Knapp

Designing a Multi-Quantum-Dot Array for Efficient Light-Harvesting in Solar Cells 1079
Jose Raul Montes-Bojorquez, Maria Fernanda Villa-Bracamonte, Arturo A. Ayon

Chemomechanics of Halide Perovskites: Linking Mechanical Behavior with Reliability 1082
Nicholas J Rolston

Planar and Nanowire InP Thin Solar Cells for Ultralight Space Power Applications 1083
Sara Anjum, Pilar Espinet Gonzalez, Harry A. Atwater

Effective Passivation of CdTe Rear Interface Via Thin Selenium Interface Layer Indicated by
Surface Photovoltage Spectroscopy 1086
Michael A Scarpulla, Nathan D Rock, Amit Munshi

Study of Perovskite Solar Cells Under High-Fluence, Low-Energy Proton Radiation 1087
Michael D Kelzenberg, Ahmad R. Kirmani, Kaitlyn T. Vansant, Joseph M. Luther, Harry A. Atwater

Photophysical Properties of CdSe/CdTe Bilayer Solar Cells: A Confocal Raman and
Photoluminescence Microscopy Study 1088
Abasi Abudulimu, Jaroslav Kulicek, Ebin Bastola, Adam B Phillips, Aesha Patel, Dipendra Pokhrel, Manoj K. Jamarkattel, Michael J Heben, Bohuslav Rezek, Randy J Ellingson

The Capability of a Grid-Forming Inverter to Support Dynamic Microgrids with High Penetrations
of Photovoltaics Systems 1091
Rachid Darbali-Zamora, C. Birk Jones, Matthew S. Lave, Erick E. Aponte-Bezares

Parametric Analysis of Capacitance-Voltage Data for In-Situ Heat and Light Soaking Behavior of
CIGS Solar Cells 1099
Shubhra Bansal, Mohsen Jahandardoost

MoS2 Solar Cell with 120 Nm-Absorber and 3.8% AM1.5G Efficiency 1100
Elisa Antolin, Simon A. Svatek, Carlos Bueno-Blanco, Antonio Marti, Der-Yuh Lin, Micaela Rodriguez-Peña, Monica Luna

Effects of Arsenic Doping on CdSe$_x$Te$_{1-x}$/CdTe Solar Cells 1101
Sheikh Tawsif Elahi, Md Zahangir Alom, Wei Wang, Vasilios Palekis, Chris Ferekides

What is the Optimal Electricity Share for Very Inexpensive Solar PV? 1105
Adam Dvorak, Marta Victoria

Transparent Oxides as a Protective Encapsulant for Perovskite Solar Cells in Low Earth Orbit 1109
Kyle M Crowley, Kaitlyn Vansant, Timothy J Peshek, Lyndsey B McMillon-Brown

Barriers to Solar Photovoltaic (PV) Adoption on a National Scale in the United States 1110
Casey Corrado, Emily Holt, Lauren Schambach

Laser-Weld Qualification Methods for Al Foil Interconnection of Back-Contacted Cells to Predict
Module Reliability 1118
Barry B. Hartweg, Kathryn C. Fisher, Zhengshan J. Yu, Zachary C. Holman

Highly Efficient Perovskite-On-Silicon Tandem Solar Cells on Planar and Textured Silicon...................... 1119
Christian M. Wolff, Xin Yu Chin, Deniz Türkay, Kerem Artuk, Mohammadreza Golobostanfard, Florent Sahli, Daniel Jacobs, Quentin Guesnay, Peter Fiala, Mostafa Othman, Bosky Sharma, Brett Kamino, Aïcha Hessler-Wyser, Mathieu Boccard, Quentin Jeangros, Christophe Ballif

Corrosion Testing of Solar Cells: Insights to Wear-Out Mechanisms... 1120
Andrew Fairbrother, Luca Gnocchi, Christophe Ballif, Alessandro Virtuani

NRG-X-Change and Cooperative Game Strategies as an Alternative to Net-Metering for Solar
Generation 1121
Hector Lopez, Ali Zilouchian

Nanostructured ZnO Electron Transporting Materials for Hysteresis-Free Perovskite Solar Cell 1122
Vilko Mandic, Ivana Panzic, Floren Radovanovic-Peric, Thomas Rath

Variable Renewable Energy Participation in U.S. Ancillary Services Markets: Economic
Evaluation and Key Issues.. 1123
James Hyungkwan Kim, Fredrich Kahrl, Andrew Mills, Ryan Wiser, Cristina Crespo Montañés, Will Gorman

Amorphous Manganese Sulfide Enables Efficient and Stable All-Inorganic Antimony
Selenosulfide Solar Cells... 1126
Chen Qian, Jianjun Li, Kaiwen Sun, Chenhui Jiang, Jialiang Huang, Rongfeng Tang, Martin Green, Bram Hoex, Tao Chen, Xiaojing Hao

A Data-Driven Feeder Selection Method for Distribution System Planning Studies 1127
Alexandre B. Nassif, Fernanda C. L. Trindade

Fireable Passivating Tunnel Oxide Contacts for Crystalline Silicon Solar Cell.......................... 1132
Franz-Josef Haug, Mario Lehmann, Sofia Libraro, Ezgi Genç, Audrey Morisset, Christophe Ballif

Antisolvent Effect on Acetamidinium Substituted 2D Ruddlesden-Popper Perovskite Solar Cells.............. 1133
Vani Pawar, Anuj Kumar Palariya, Nisheka Anadkat, Sandeep Kumar, Sushobhan Avasthi

Performance and Stability of Electrodeposited Mixed Perovskites $MAPbI_{3-x}Cl_x$ and
$MA_{1-y}FA_yPbI_{3-x}Br_x$ 1136
Mirella Al Katrib, Lara Perrin, Emilie Planes

Performance Assessment of a Residential Building Integrated Photovoltaic (BIPV) System in
Dhaka City.. 1139
Md. Mahbub Ali, Nur Jahan Beanta Sorower, Abu Niem Seum, Md. Shifain Mahathir Alvi, Rawnak Reza Raka, Mohaimenul Islam, Md. Mosaddequr Rahman

Elucidating Materials Paradigm of CIGS by Structure--Composition-- Performance Correlations.............. 1145
Niklas Pyrlik, Christina Ossig, Giovanni Fevola, Svenja Patjens, Jan Hense, Catharina Ziska, Martin Seyrich, Frank Seiboth, Andreas Schropp, Jan Garrevoet, Gerald Falkenberg, Christian G. Schroer, Romain Carron, Michael E. Stuckelberger

An Experimental Comparison Between View Factor and Ray Tracing Models for Energy
Estimation of Bifacial Modules.. 1146
Hugo Sánchez, Sebastian Dittmann, Carlos Meza, Ralph Gottschalg

Field Assessment of Transparent Conductive Oxides Stability Under Outdoor Conditions 1151
Brahim Aïssa, Amir A. Abdallah, Juan Lopez Garcia

Influence of Business Models on PV-Battery Dispatch Decisions and Market Value: A Pilot Study of Operating Plants 1158
Joachim Seel, Cody Warner, Andrew Mills

Wide Bandgap AlGaInP-Based Photovoltaic Cell for Indoor Ambient Energy Harvesting 1159
Aditya Prabaswara, Jack Browne, Yongjie Zou, Richard King, Stephen Goodnick, Brian Corbett

A Comparison Study of the Performance of Vertical Vs Single Axis Tracking Bifacial Agrivoltaic Systems in Belgium 1162
Brecht Willockx, Jan Cappelle

Power Factors 2022 PV System Efficiency Benchmarks 1163
Stephen Lightfoote, Samantha Wilson, Steve Voss

Thin, Radiation-Resilient III-V PV Devices Utilizing Quantum Structures and Epitaxial Light Reflectors 1167
Brandon M. Bogner, Stephen J. Polly, Seth M. Hubbard, Roger E. Welser

Rapid Thermal Annealing (RTA) of Hydrogenated Poly-Si Under Air and Nitrogen and Blister Formation 1168
Arpan Sinha, Sagnik Dasgupta, Ajeet Rohatgi, Mool Gupta

Investigation of High Open Circuit Voltage in CdTe-Based Solar Cells Using Oxide Back Buffer Layers 1169
Abdul Quader, Manoj K. Jamarkattel, Ebin Bastola, Kamala Khanal Subedi, Dipendra Pokhrel, Indra Subedi, Adam B. Phillips, Nikolas J. Podraza, Randy J. Ellingson, Michael J. Heben

Solution-Processed Copper Selenium Oxide (CuSeO$_3$) as Hole Transport Layer for CdS/CdTe Solar Cells 1170
Sandip S Bista, Deng-Bing Li, Suman Rijal, Sabin Neupane, Manoj K Jamarkattel, Rasha A Awni, Zhaoning Song, Adam Phillips, Michael Heben, Randy J. Ellingson, Yanfa Yan

Prediction of Electron Band Gap of A$_2$XY$_6$ Perovskite Compounds Using Machine Learning 1173
Jatin Chaudhary, Swastik Bhattacharya, Jukka Heikkonen, Rajeev Kanth

A Silicon Learning Curve and Polysilicon Requirements for Broad-Electrification with Photovoltaics by 2050 1177
Brett Hallam, Moonyong Kim, Robert Underwood, Storm Drury, Li Wang, Pablo R Dias

Photovoltaic Surfaces to Reverse Global Warming 1178
Christiana B. Honsberg, Stuart G. Bowden, Ian R. Sellers, Richard R. King, Stephen M. Goodnick

Strategies for Implementing of Very Large Scale Solar and Wind Power Plants in the Gobi Desert for the Northeast Asia Regional Energy Market 1179
Enebish Namjil, Keiichi Komoto

Interposed Versus Juxtaposed Solar Array Configurations for Agrivoltaics 1182
M. Sojib Ahmed, M. Rezwan Khan, Anisul Haque, Muhammad A. Alam, M. Ryyan Khan

Fabrication of Microscale Back-Contact Arrays for Local Charge Transport Measurements 1185
Kaden M. Powell, Yu-Lin Hsu, Etee Kawna Roy, David J. Magginetti, Heayoung P. Yoon

Agrivoltaic Modules Optimizing Light for Crops in Dryland Regions 1189
Christiana B. Honsberg, Greg Barron-Gafford, Stuart G. Bowden, Robert Sampson

Understanding Device Performance Limiting Factors by Reproducing the Current-Voltage Characteristics from Transient Optoelectrical Measurements 1190
Abasi Abudulimu, Klaus Eckstein, Mirella El Gemayel, Imge Namal, Adam B Phillips, Michael J Heben, Tobias Hertel, Sebastian B Meier, Larry Lüer, Randy J Ellingson

Impact of the 2019-2020 Australian Black Summer Wildfires on Photovoltaic Energy Production 1191
Ethan Ford, Bram Hoex, Ian Marius Peters

Combining Nanoscale 3D Printing with Spark Ablation to Achieve Novel Nanostructured Surfaces for Photovoltaic Applications 1195
Ivana Panzic, Alexander Jelinek, Floren Radovanovic-Peric, Daniel Kiener, Vilko Mandic

Understanding Configuration of Geopolymer Materials for Application in Solar-Cells 1196
Arijeta Bafti, Filip Brlekovic, Vilko Mandic, Luka Pavic, Ivana Panžić, Andraž Krajne

The Potential Use of Spark Ablation in Development of AgNP Decorated Copper Oxide Thin Films for Photodetection Applications 1197
Floren Radovanovic-Peric, Vilko Mandic, Ivana Panžic

Impact of Anti-Soiling Coating on Potential Induced Degradation of Silicon PV Modules 1198
Farrukh Ibne Mahmood, Govindasamy Tamizhmani

Extreme Solar: Towards 24–7 Renewable Energy 1201
Sijo Augustine, Sathishkumar Ranade, Valerio De Angelis, Gabriel Cowles, Jinchao Huang, Olga Lavrova, Stanley Atcitty

Impact of Indium Chloride Treatment on the Properties of $CuInSe_2$ Thin Films 1204
Deewakar Poudel, Adam Masters, Benjamin Belfore, Elizabeth Palmiotti, Angus Rockett, Sylvain Marsillac

On the Effect of Indium Chloride Dose on the Recrystallization of $Cu(In,Ga)Se_2$ Thin Films and Associated Devices 1209
Deewakar Poudel, Adam Masters, Benjamin Belfore, Angus Rockett, Sylvain Marsillac

Effect of Metal Halides Treatment on High Throughput Low Temperature CIGS Solar Cells 1214
Deewakar Poudel, Benjamin Belfore, Adam Masters, Angus Rockett, Sylvain Marsillac

Grain Enhancement in Polycrystalline $CuGaSe_2$ by AgBr Vapor Treatment 1219
Deewakar Poudel, Benjamin Belfore, Adam Masters, Elizabeth Palmiotti, Angus Rockett, Sylvain Marsillac

Post-Deposition Metal Halide Treatment of $CuGaSe_2$ for Photovoltaic Application 1224
Deewakar Poudel, Benjamin Belfore, Adam Masters, Elizabeth Palmiotti, Angus Rockett, Sylvain Marsillac

Current or Power Matching? A Third Option for Monolithic All-Perovskite Tandem Solar Cells 1229
Yuan Gao, Renxing Lin, Ke Xiao, Xin Luo, Jin Wen, Xu Yue, Hairen Tan

Simulation of High Open-Circuit Voltage Perovskite/CIGS-GeTe Tandem Cell 1230
Mohamed Mousa, Mostafa M. Salah, A. Zekry, Mohamed Abouelatta, Ahmed Shaker, Fathy Z. Amer, Roaa I. Mubarak, Ahmed Saeed

Unlocking 1550 nm Laser Power Conversion by InGaAs Single- And Multi-Junction PV Cells 1235
Henning Helmers, Oliver Höhn, Thomas Tibbits, Meike Schauerte, H. M. Noman Amin, David Lackner

Estimation and Degradation Analysis of Physics-Based Circuit Parameters for PV Systems Using Only DC Operation and Weather Data 1236
Baojie Li, Xin Chen, Todd Karin, Anubhav Jain

The Effect of Moisture Ingress on Titania Antireflection Coatings in Field-Aged Photovoltaic Modules 1237
Oscar Kwame Segbefia, Naureen Akhtar, Tor Oskar Sætre

Improved STC and Energy Yield Performance of Bifacial Modules with White-Grid Rear Reflectors 1245
Robert Witteck, Michael Siebert, Tobias Wietler, Marc Köntges, Paulius Laurikenas, Julius Denafas

4D- Printed Shape Memory Polymer Based Solar Tracker 1248
Serhii Tytov, Fhad Al-Modaf, Shicheng Su, Nazek El-Atab

Study of Degradation of $Cu(In,Ga)Se_2$ Solar Cell Parameters Due to Temperature 1252
Rabee B. Alkhayat, Dhurba R. Sapkota

Flight Demonstration Test of State-Of-The-Art Photovoltaic Devices on JAXA's New ISS Transfer Vehicle HTV-X 1257
Mitsuru Imaizumi, Teppei Okumura, Tetsuya Nakamura, Shusaku Kanaya, Taishi Sumita

Mechanical Degradation Studies on Flexible CIGS Cells and Modules for Floating PV 1258
Wim Soppe, Aldo Kingma, Dorrit Roosen

Influence of Temperature and Magnetic Field on the Transient Voltage Decay of a Silicon Solar Cell with Parallel Vertical Junction in Open Circuit 1262
Pape Diop, Papa Touty Traore, Papa Monzon Samake, Babou Dione, Fatimata Ba, Modou Pilor

Progression in Grain Size of Novel Photoferroic Absorber Bournonite (CuPbSbS3) 1263
Oliver M Rigby, Budhika G Mendis, Marek Szablewski

Direct Observation of an Atomic Thin Inversion Layer at the Native Oxide/ n-Si Interface 1264
Yibo Zhang, Joel Y. Y. Loh, Andrew G. Flood, Chengliang Mao, Geetu Sharma, Nazir P. Kherani

Ultrathin III-V Solar Cells with Light-Trapping Structures Fabricated in Situ Using an HVPE Reactor 1265
Allison N. Perna, Anna K. Braun, Kevin L. Schulte, John Simon, Corinne E. Packard, Aaron J. Ptak

New Substituted Small a cation(Acetamidinium) Based Tin Perovskite Solar Cell 1266
Soumen Kundu, Sushobhan Avasthi

Significance of Power and Energy Ratings of Modules in Large-Scale PV Plants 1271
Bijaya Paudyal, Donny Campos Paniagua

FEM Based Thermal Model of an Agrivoltaic System 1275
Karan Rane, Navni Verma, Ardeshir Contractor, Narendra Shiradkar

Towards Understanding of Cementation of Particulate Soils on PV Cover Glass Materials 1279
Mohamed Adawi, Adedoyin Abe, Min Zou, Robert A. Fleming

Thin-Film Solar Cells with MgF$_2$/Ag Back Mirror Patterning for Improved near-IR Reflectance 1280
Lara Barros Rebouças, Gerard J. Bauhuis, Jens Olhmann, Jeroen Maasen, Elias Vlieg, John J. Schermer

InAs Thermophotovoltaic Cells with Low Reverse Saturation Current 1283
Eric J. Tervo, Andrew J. Ferguson, Myles A. Steiner, Ryan M. France

Toxicity Assessment of Lead and Other Metals Used in Perovskite Solar Panels 1284
Gonzalo Rodriguez-Garcia, Jon J. Kellar, Zhengtao Zhu, Ilke Celik

Molecular Beam Epitaxy Growth of CdSe for Si-Based Tandem Cell Application 1288
Stephen Schaefer, Zheng Ju, Allison McMinn, Xin Qi, Yong-Hang Zhang

Fabrication of Ultrathin Ge Template for Growth of Multijunction Solar Cells Based on Wafer-Scale Porous Ge 1291
Tadeáš Hanuš, Javier Arias-Zapata, Bouraoui Ilahi, Philippe-Olivier Provost, Alexandre Chapotot, Abderraouf Boucherif

Prediction of Novel Phosphors Using Machine Learning for Efficiency Enhancement of Silicon Solar Cells 1292
Tae-Gwan Kim, Eun-Gyeong Kim, M. Shaheer Akhtar, O-Bong Yang

Predictive Modeling of Cracks Within Flexible Perovskite Thin Films 1293
Melissa A Davis, Rebekah Sweat, Zhibin Yu

Results of Environmental-Based PV Soiling Models After Extreme Dust Events: The Case of Saharan Dust Intrusions in Southern Spain 1294
João Gabriel Bessa, Álvaro F Solas, Florencia A Cruz, Eduardo F Fernández, Leonardo Micheli

Optical Design Considerations for Thin Photonic Power Converters with Textured Back Reflector 1295
Neda Nouri, Christopher E. Valdivia, Meghan N. Beattie, Jacob J. Krich, Karin Hinzer

Interrelation of CdTe Grain Size, Post-Growth Processing and Window Layer Selection on Solar Cell Performance 1299
Thomas P Shalvey, Heath Bagshaw, Jonathan D Major

DIrect Sunlight into CO Conversion 1300
Thierry De Vrijer, Arno Smets

Novel 1D Van Der Waals SbSeI Micro-Columnar Solar Cells by a Self-Catalyzed High Pressure Process 1304
Ivan Caño-Prades, Alejandro Navarro-Güell, Sergio Giraldo, Joaquim Puigdollers, Marcel Placidi, Edgardo Saucedo

Developement of Phosphors by Magnetron Sputtering for Solar Cells Improvement 1305
Eduardo Salas, Miguel Modesto Tardio, Elisa García-Tabares, Gracia Belén Perea, Rosa De La Cruz, Stavros Athanasopoulos, Clement Kanyinda-Malu, Juan Enrique Muñoz-Santiuste, Beatriz Galiana

High Efficiency Silicon Heterojunction Metal Wrap Through Produced in Industrial Pilot Line 1306
Marina Foti, Nicolas Guillevin, Eric Kossen, Lars Okel, Eelko Hoek, Anna Carr, Bas Van Aken, Petra Manshanden, Francesco Rametta, Marcello Sciuto, Antonio Spampinato, Alfredo Di Matteo, Antonino Ragonesi, Gianluca Coletti, Cosimo Gerardi

DERConnect – a Distributed Energy Resources Testbed for Solar Power Integration 1310
  Jan Kleissl, Adil Khurram, Keaton Chia, Scott Brown, Aditya Mishra, Jorge Cortes, Raymond
  De Callafon, Rajesh Gupta, Sonia Martinez, David Victor

Slot-Die Fabrication of Solution-Processed Kesterite Solar Cells for Product Integrated
Photovoltaics 1311
  Xinya Xu, Matthew C Naylor, Michael Jones, Bethan Ford, Stephen Campbell, Yongtao Qu,
  Vincent Barrioz, Guillaume Zoppi, Neil S Beattie

N-Type CdTe Thin Films Via In-Situ Indium Doping 1312
  Theodore D C Hobson, Luke Thomas, Laurie J Phillips, Leanne A H Jones, Matthew J Smiles,
  Christopher H Don, Pardeep K Thakur, Vinod R Dhanak, Tim D Veal, Jonathan D Major,
  Ken Durose

Tellurium Availability for the PV Industry Using a System Dynamics Approach 1313
  Francis Hanna, Annick Anctil

Progress on Substrate Reuse Using Sonic Lift-Off for GaAs- Based Photovoltaics 1314
  Andrew B. Sindermann, Stephen J. Polly, Pablo Guimera Coll, Elijah J. Sacchitella, Brandon
  M. Bogner, Mariana I. Bertoni, Seth M. Hubbard

Bandgap Dependence of Near-Conduction Band State in $(Ag_yCu_{1-y})(In_xGa_{1-x})Se_2$ Solar Cells 1315
  Michael F. Miller, Alexandra M. Bothwell, Nicholas Valdes, Stefan Paetel, Rouin Farshchi,
  Ana Kanevce, William Shafarman, Darius Kuciauskas, Aaron R. Arehart

Improvement in PV Plant Performance for Convection Heat Transfer Changes from Altered Plant
Layout 1319
  Matthew Prilliman, Sarah Smith, Brooke Stanislawski, Marc Calaf, Raul Bayoan Cal, Tim
  Silverman, Janine M. F. Keith

High Altitude Flight Results Using Selenium, a PV Measurement Ecosystem 1320
  Don Walker, Colin J. Mann, John Nocerino, Kevin Lopez, Alexandra Pettengill, Jonathan
  Ortiz, Katrina Baumgarten, Misha Dowd, Yao Lao, Simon H. Liu

Time-Evolving Electroluminescence Imaging in Perovskite Solar Cells 1321
  Jackson W. Schall, Hsinhan Tsai, Harvey Guthrey, Chun-Sheng Jiang, Steve Johnston, Dana
  Kern, Andrew Norman, Mowafak Al-Jassim

The Nuts and Bolts of PV: Maturing Solar PV Racking and Module Mounting Critical Bolted Joint
Technologies for LCOE Reductions and Increased Reliability 1322
  James Elsworth, Gerald Robinson, Jon Ness, Joe Cain

Atomic Layer Deposited Bilayers and the Influence on Metal-Insulator-Semiconductor Schottky
Barriers 1323
  Benjamin E. Davis, Nicholas C. Strandwitz

Optimizing Perovskite Solar Cells by Understanding the Bulk Properties of Contact Layers 1329
  Mason Mahaffey, Zhengshan Jason Yu, Vidya Krishnan, David Quispe, Arthur Onno,
  Zachary Holman

Development and Qualification of IMMβ and Z4J+, Radiation Hard III-V Solar Cells 1332
  John T Hart, Dan Aiken, Zac Bittner, Ben Cho, Daniel Derkacs, Khalid Emshadi, Andrew
  Espenlaub, Frank Fencl, Jeremy Leshin, Ahmad Mansoori, Nate Miller, Pravin Patel, Albert
  Perry, Janine Walker

The Profound Influence of Substrate Thermal Resistance on the Photovoltaic Properties of Solution-Processed Cu(In,Ga)Se2 1333
Kyle G Weideman, Rakesh Agrawal

Soiling Measurement Based on Checkered Pattern Image Analysis 1334
Bing Guo, Wasim Javed

Modeling Efficiency of Inverters with Multiple Inputs.. 1335
Clifford Hansen, Jay Johnson, Rachid Darbali-Zamora, Nicholas Gurule

Pathways to High Efficiency Perovskite Monolithic Solar Modules 1338
Xuezeng Dai, Shangshang Chen, Yehao Deng, Allen Wood, Guang Yang, Chengbin Fei, Jinsong Huang

Simulation of Hot-Carrier Filtering in InAs-InP Nanowire Heterostructures 1339
Urs Aeberhard

A Simple Approach to Ohmic Contacts for Transition Metal Dichalcogenide Solar Cells............. 1340
Mario Martinez, Simon A. Svatek, Carlos Bueno-Blanco, Der-Yuh Lin, Ines Duran, Antonio Marti, Elisa Antolin

Luminescent Solar Concentrators for Building Integrated Photovoltaic Devices 1341
Liam J. Halloran

LETID in Legacy and Modern PV Modules: Accelerated Testing and Field Deployment............. 1342
Joseph Karas, Ingrid Repins

Measurement of Snow Loading on a Tilted PV Module in Northern Michigan 1343
Daniel Riley, Laurie Burnham, William Snyder, Bruce King, Paul Dice

Probabilistic Assessment of Narrowband Vs Broadband Solar Irradiance Temporal Variability in Ottawa 1346
Nick Anderson, Viktar Tatsiankou, Karin Hinzer, Richard Beal, Henry Schriemer

Evaluation of Solar Capacity Factor of ~2000 Solar Plants Across the United States Using Multilayer Perceptron Regressor Models 1347
Samantha S. Wilson, Stephen Lightfoote, Stephen Voss

Radiant/Non-Radiant Lifetime Switching in Chlorophyll and Application to Energy Storing Photovoltaic Cells.. 1350
Julie B. Liu, Nahian Rahman, Aaron Song, Elizabeth Nazginov, Mia Pancari, Amina Exilhomme, Charles M. Fortmann

Measuring Carrier Concentration on the Back Side of Thin Film Solar Cells 1355
Nathan Rosenblatt, Alex Polizzotti, Sachit Grover, Xiaoping Li, Wyatt K. Metzger

Reproducibility and Photostability of High-Efficiency Perovskite Solar Cells in Scalable Manufacturing 1358
Rohit Prasanna

Optimized Near-Field Thermophotovoltaic Cell Using InAs and InAsSbP................................... 1359
Gavin P Forcade, Christopher E Valdivia, Sean Molesky, Shengyuan Lu, Alejandro W Rodriguez, Jacob J Krich, Raphael St-Gelais, Karin Hinzer

22% Efficiency Module Combining Silicon Heterojunction Solar and Shingle Interconnection 1360
Marina Foti, Marco Galiazzo, Enrico Sovernigo, Nicola Frasson, Cosimo Gerardi, Alfredo Guglielmino, Grazia Litrico, Marcello Sciuto, Antonio Spampinato, Antonino Ragonesi, Francesco Rametta, Andrea Canino, Agata Carbonaro, Fabrizio Coco, Agnese Di Stefano, Fabrizio Bizzarri

Polyethienimine Interface Dipole Tuning for Electron Selective Contacts 1363
Eloi Ros Costals, Thomas Tom, Gerard Masmitjà, Benjamin Pusay, Estefania Almache, Maykel Jimenez, Julià Lopez, Edgardo Saucedo, Pablo Ortega, Joan Bertomeu, Joaquim Puigdollers, Cristobal Voz

Impact of Snow Depth on Single-Axis Tracked Bifacial Photovoltaic System Performance 1366
Annie C. J. Russell, Christopher E. Valdivia, Joan E. Haysom, Karin Hinzer

Author Index

Investigation of Degradation Kinetics of Perovskite Solar Cells by Accelerated Aging

Dhruba B. Khadka[1], Yasuhiro Shirai[1], Masatoshi Yanagida[1] and Kenjiro Miyano[1]

[1] Photovoltaics Materials Group, Global Research Center for Environment and Energy based on Nanomaterials Science (GREEN), National Institute for Materials Science (NIMS), 1-1 Namiki, Tsukuba, Ibaraki 305-0044, Japan.

Abstract—The operational stability of encapsulated perovskite solar cells (PSCs) is imperative for their commercialization. Here, we have investigated the degradation of PSCs with organic (PTAA) and inorganic (NiO$_x$) HTLs under light and thermal stress. The device parameters under 1-sun illumination were monitored over time at different temperatures; 20 to 85°C. The device degradation kinetics has been discussed by adopting the Arrhenius model with temperature and humidity prefactor correction. The temperature-dependent device parameters analysis showed a lower value of degradation activation energy for the device with the PTAA (~ 0.274 eV) than that for the NiO$_x$ device (~ 0.495 eV).

Keywords—Perovskite degradation, Arrhenius model, activation energy, device stability, humidity/thermal stress.

I. INTRODUCTION

Perovskite solar cells (PSCs) have huge potential to be fabricated with cheaper materials and production costs than conventional semiconductor solar cells.[1] Despite the high device efficiency, the commercialization of PSCs is impeded by long-term stability. [2]–[4] The device under operation accelerates the loss of power conversion efficiency (PCE). The cause of operational instability of PSCs with different interface layers also largely remains elusive. Therefore, it is important to dig up the mechanistic parameters to track the degradation propagation.

An accelerating stress testing could provide a real picture of operational stability which is decisive for its commercialization.[1] The mean life time (MLT) of solar cell devices is carried out by accelerated life testing (ALT) under overstress conditions.[5] The most common stress conditions for ALT are:[6] (a) thermal stress followed by high temperature and/or temperature cycling and/or temperature gradients,(b) electrical stress with power cycling/voltage cycling/power extremes/voltage extremes, (c) mechanical stress with mechanical shock/stress/vibrational tests/creep-stress relaxation tests,(d) environmental stress i.e. humidity or radiation. Moreover, the ALT tests are also carried out by the combination of these tests. For example, thermal stress/humidity (typically T-85°C/RH-85%) or illumination/thermal stress conditions are popularly used for the testing of the different kinds of solar cells[6] including HaPCs.[7] Several reports have discussed the degradation of perovskite monitoring the device parameters and discussed the

different characteristics features.[8] However, the analysis of the degradation trend of device parameters has not discussed under ALT test conditions using an analytical model to explore the degradation kinetics.

In this report, we analyzed the degradation trend of the device parameters of PSCs under continuous illumination driven by thermal stress. ALT was performed at various temperatures *(20-85°C)* and under constant illumination of 1 sun. The device parameters monitored with time were fitted with a linear or exponential functional model to evaluate the degradation kinetics. The mathematical analysis revealed that the poor stability of the device with PTAA is attributed to a lower activation energy for degradation compared to the device with NiO$_x$.

II. EXPERIMENTAL

Device fabrication:

The details of the precursor solution and device fabrication can be found in the earlier reports.[9]–[12] In brief, for the fabrication of inverted PSCs, we have used ITO substrate and PTAA (2 mg/ml in CB) deposited by spin coating, NiO$_x$ deposited by sputtering mentioned in our earlier.[10], [11] For the fabrication of perovskite films, we adopted two stept deposition PbI$_2$ deposition followed by dripping of the MAX precursor solution (a mixture of MAI + MACl.[9] For completion of device, PCBM as ETL, AZO layer and Ag. Devices were sealed by encapsulation glass and UV-curable resins (UV-RESIN XNR5516Z) before the subsequent measurement in ambient conditions.

For operational stability testing:

The encapsulated devices (PTAA and NiOx devices) in enclosed system under air ambient. The devices were kept under constant illumination and set temperatures (20, 60 and 85 °C). We have tested HaPSCs with two different HTLs; PTAA and NiO$_x$. During thermal stress, the devices were placed under constant illumination at relative humidity of 35-40% at room temperature in enclosed system.

The *J–V* scans were done every 7 h for the first 7 days and then once every 24 h until over 1000 h.

III. RESULTS AND DISCUSSION

To investigate the degradation kinetics, the PSCs with HTLs; PTAA or NiO$_x$ were monitored under continuous illumination of one sun placing at thermal stress. The *J–V* characteristics of fresh devices are depicted in Fig. 1a. The PSC with the PTAA has a PCE of ~19.32% and the device with sputtered NiO$_x$ has an efficiency of 15.60%. The characteristics

978-1-7281-6118-1/22 $31.00 © 2022 IEEE

insights and have been discussed in our previous reports.[9], [11] The detail of degradation characteristics with the point of view of materials properties and interface quality has been documented in our earlier report.[13] In this report, we discussed the degradation kinetics of these devices under thermal and light stress considering the ALT model.

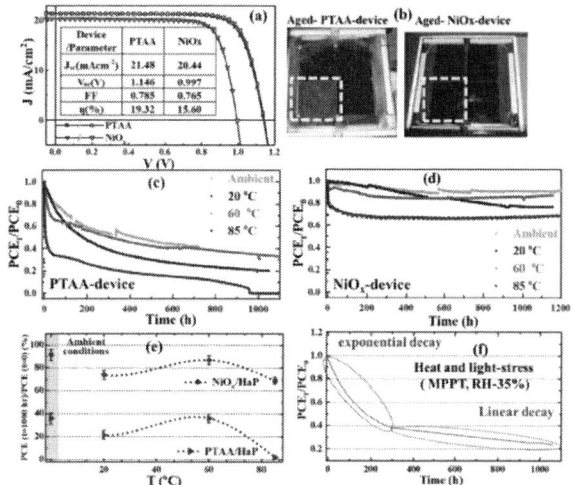

Fig. 1. Current density-voltage (J-V) characteristics of PTAA and NiO$_x$ devices (a). Photographs of aged devices after 1000 h (60°C) (b). The operational stability of the (c) PTAA/HaP and (d) NiO$_x$/HaP devices at different temperatures under maximum power point conditions (1 sun). (e) The relative performance of PTAA or NiO$_x$ devices aged under different thermal stress under continuous illumination. (f) Schematic display of degradation trend of PSCs showing exponential decay at the initial stage followed by linear decay. The shaded region indicates ambient conditions.

Figure 1b displays Photographs of aged devices of PTAA and NiOx devices. Figures 1c,d demosntrate the device degradation trend at 20, 60, 85 °C, and ambient conditions (27 °C<T<35 °C) under continuous illuminations (one sun) at RH of 35-40%for t>1000 h for respective devices. These devices showed a stark difference degradation rate. This is attributed to the difference in the interface chemistries at PTAA/HaP or NiO$_x$/HaP and HaP bulk with external stimuli. As given in Fig. 1e, the aged PSCs with NiO$_x$ retained ~74.8% of PCE$_0$ at 20 °C, 86.7% at 60 °C, 68.7% at 85 °C, and ~90.1% under ambient condition whereas the devices with PTAA device dropped to 22.5%, 35.9%, 2.0%, and 22.6% of PCE$_0$, respectively. The monitored data showed that the device parameters are primarily governed by the J_{sc} trend.

From the device stability data, we can notice that the J_{sc} of the PTAA/HaP device dropped faster than the NiO$_x$/HaP device. Particularly, an initial steep drop in PCE at 85°C was observed for both devices. A loss of fill factor in both devices is also affected by a gradual increase of the series resistance of the devices.

The degradation trend of the PSCs with either of HTLs followed by two characteristics functional regimes (Fig.1f). At the beginning, the degradation trend has the exponential decay which is followed by a linear decay. It is speculated that the rate of exponential or linear decay is influenced by the carrier transport layer used in the device configuration. Indeed, these trends are the consequence of collective effects of the partial

degradation of CTL, defect induced in HaP bulk by crystal phase transformation, and CTL/HaP interface.

A. Accelerated Degradation Testing Model

The degradation kinetics can be analysed by ALT models.[14] The Arrhenius model is a fundamental mathematical model for the analysis of optoelectronic device characteristics. The degradation constant for this model is given by:

$$k = Ae^{\left(-\frac{E_A}{K_B T}\right)} \qquad (1)$$

Here, E$_A$- the activation energy, k$_B$- the Boltzmann constant 8.62×10^{-5} eVK^{-1}, T- the temperature in Kelvin and A - a constant dependent on the degradation mechanisms and the experimental conditions. The Arrhenius model primarily explains the kinetics of the temperature-dependent process.

To get more insights into the degradation kinetics under light and heat stress, the Arrhenius model equation (1) can be modified as:

$$K = \frac{k_{deg}(T_H)}{k_{deg}(T_L)} = \exp\left[\frac{E_A}{K_B}\left(\frac{1}{T_L} - \frac{1}{T_H}\right)\right] \qquad (2)$$

where T$_L$ and T$_H$ correspond to low and high temperatures (reference temperature). In our analysis, we considered T_H=85 °C (standard temperature for reliability test). Note that a higher value of acceleration factor (K) denotes better device stability.

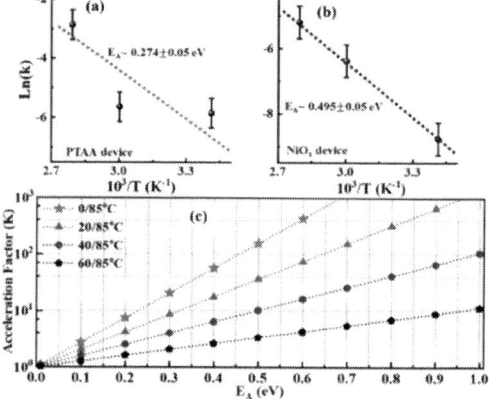

Fig. 2. The degradation activation energy of PSCs with (a) the PTAA and (b) NiO$_x$ estimated by analysing the trend of device parameters (J$_{SC}$). The estimation of acceleration factor of the device considering prefactor with the varying working low temperatures (T$_L$) with reference to the standard test temperature (T$_H$=85 °C) adopting the Arrhenius equation (1).

Since the J_{SC} degradation trend is parallel to the loss of PCE.[13] we analysed the J_{sc} trend at different thermal stress to calculate the degradation activation energy. The Arrhenius equation (1) was applied for the evaluation of the temperature dependence (Fig. 2a,b). It is found that the activation energy of the device with NiO$_x$ for degradation (0.495±0.02 eV) is higher than that of the PTAA device (0.274±0.02 eV). It indicates that the PTAA device, with lower activation energy, degrades faster. We found that the estimated values of E$_A$ for our devices are lower than Si, CIGS, or CdTe- based devices which are well-known for stable device performance with much higher MLT. It implicates that the value of E$_A$ is directly proportional to the MLT of PSCs. We believed that the difference in E$_A$ for

978-1-7281-6118-1/22 $31.00 © 2022 IEEE

the PSCs with PTAA and NiO_x is attributed to the structural defect in perovskite, interfacial deterioration of HTL/HaP, and HTL under heat and light stress. A lower value of E_A for the PSC with PTAA could have stemmed from structural defect induced from thermal stress[15] and the partial deterioration of the PTAA film due to absorption of a high-energy photon under illumination. While higher activation energy for the device with the NiO_x is analogous to the relatively stable device due to the robust nature of NiO_x which results in a comparatively stable interface and immune to ionic diffusion for the perovskite bulk.

Furthermore, we have calculated the acceleration factor of degradation using equation (2) at working temperature (T_L) (T_L= 0 °C (winter), 20 °C (~below room temperature), 32 °C (~ ambient working temperature), and 60 °C (~working at the roof temperature) (Fig. 2c) with respect to 85 °C. This result shows a higher acceleration factor at a lower working temperature. This means the MLT of the device is longer at a lower working temperature (T<85 °C). Unlike the Arrhenius estimation, the experimental device data showed poor stability for T<20 °C. It is to be noted that the environmental stress (temperature and ambient humidity) induces the defects such as structural defect and deterioration of interface or bulk layer used in device structures which could be the main factors for the anomalous trend (Fig. 1c, d).

IV. SUMMARY AND CONCLUSIONS

In summary, we investigated the light-heat stress accelerated degradation of PSCs with different interface layers. The degradation analysis of device parameters showed higher activation energy for PSCs with NiO_x than PTAA indicating the interfacial effect on degradation potentials. The device degradation kinetics of PSCs can be described by the Arrhenius model with the introduction of temperature and humidity prefactor especially at lower working temperature (T<RT) while device stability at higher working temperature (T>30 °C) follows the regular trend with temperature prefactor. This work corroborates that the degradation of PSCs can be slowed down by stabling the structural phase of HaP bulk and engineering the interfacial layer with higher activation energy.

ACKNOWLEDGMENT

This work was supported by Yazaki Memorial Foundation for Science and Technology.

REFERENCES

[1] M. Cai, Y. Wu, H. Chen, X. Yang, Y. Qiang, and L. Han, "Cost-Performance Analysis of Perovskite Solar Modules," *Adv. Sci.*, vol. 4, no. 1, p. 1600269, Jan. 2017, doi: 10.1002/advs.201600269.

[2] W. Tress *et al.*, "Performance of perovskite solar cells under simulated temperature-illumination real-world operating conditions," *Nat. Energy*, vol. 4, no. 7, pp. 568–574, 2019, doi: 10.1038/s41560-019-0400-8.

[3] D. B. Khadka, Y. Shirai, M. Yanagida, and K. Miyano, "Pseudohalide Functional Additives in Tin Halide Perovskite for Efficient and Stable Pb-Free Perovskite Solar Cells," *ACS Appl. Energy Mater.*, vol. 4, no. 11, pp. 12819–12826, Nov. 2021, doi: 10.1021/acsaem.1c02496.

[4] I. Gueye *et al.*, "Chemical and Electronic Investigation of Buried NiO 1−δ, PCBM, and PTAA/MAPbI 3− x Cl x Interfaces Using Hard X-ray Photoelectron Spectroscopy and Transmission Electron Microscopy," *ACS Appl. Mater. Interfaces*, vol. 13, no. 42, pp. 50481–50490, Oct. 2021, doi: 10.1021/acsami.1c11215.

[5] S. Schuller, P. Schilinsky, J. Hauch, and C. J. Brabec, "Determination of the degradation constant of bulk heterojunction solar cells by accelerated lifetime measurements," *Appl. Phys. A*, vol. 79, no. 1, pp. 37–40, Jun. 2004, doi: 10.1007/s00339-003-2499-4.

[6] T. Walter, "Reliability Issues of CIGS-Based Thin Film Solar Cells," in *Semiconductors and Semimetals*, vol. 92, 2015, pp. 111–150.

[7] K. Domanski, E. A. Alharbi, A. Hagfeldt, M. Grätzel, and W. Tress, "Systematic investigation of the impact of operation conditions on the degradation behaviour of perovskite solar cells," *Nat. Energy*, vol. 3, no. 1, pp. 61–67, Jan. 2018, doi: 10.1038/s41560-017-0060-5.

[8] D. B. Khadka, Y. Shirai, M. Yanagida, and K. Miyano, "Degradation of encapsulated perovskite solar cells driven by deep trap states and interfacial deterioration," *J. Mater. Chem. C*, vol. 6, no. 1, pp. 162–170, 2018, doi: 10.1039/C7TC03733C.

[9] D. B. Khadka, Y. Shirai, M. Yanagida, T. Masuda, and K. Miyano, "Enhancement in efficiency and optoelectronic quality of perovskite thin films annealed in MACl vapor," *Sustain. Energy Fuels*, vol. 1, no. 4, pp. 755–766, 2017, doi: 10.1039/C7SE00033B.

[10] D. B. Khadka, Y. Shirai, M. Yanagida, J. W. Ryan, and K. Miyano, "Exploring the effects of interfacial carrier transport layers on device performance and optoelectronic properties of planar perovskite solar cells," *J. Mater. Chem. C*, vol. 5, no. 34, pp. 8819–8827, 2017, doi: 10.1039/C7TC02822A.

[11] D. B. Khadka, Y. Shirai, M. Yanagida, and K. Miyano, "Unraveling the Impacts Induced by Organic and Inorganic Hole Transport Layers in Inverted Halide Perovskite Solar Cells," *ACS Appl. Mater. Interfaces*, vol. 11, no. 7, pp. 7055–7065, Feb. 2019, doi: 10.1021/acsami.8b20924.

[12] D. B. Khadka, Y. Shirai, M. Yanagida, and K. Miyano, "Ammoniated aqueous precursor ink processed copper iodide as hole transport layer for inverted planar perovskite solar cells," *Sol. Energy Mater. Sol. Cells*, vol. 210, p. 110486, 2020, doi: https://doi.org/10.1016/j.solmat.2020.110486.

[13] D. B. Khadka, Y. Shirai, M. Yanagida, and K. Miyano, "Insights into Accelerated Degradation of Perovskite Solar Cells under Continuous Illumination Driven by Thermal Stress and Interfacial Junction," *ACS Appl. Energy Mater.*, vol. 4, no. 10, pp. 11121–11132, Oct. 2021, doi: 10.1021/acsaem.1c02037.

[14] H. Tian, F. Mancilla-david, K. Ellis, P. Jenkins, and E. Muljadi, "A Detailed Performance Model for Photovoltaic Systems Preprint," *Sol. Energy J.*, no. July, 2012.

[15] J. Lim *et al.*, "Kinetics of light-induced degradation in semi-transparent perovskite solar cells," *Sol. Energy Mater. Sol. Cells*, vol. 219, p. 110776, Jan. 2021, doi: 10.1016/j.solmat.2020.110776.

Effect of Phenethylammonium Thiocyanate Additive in Tin Perovskite for Efficient and Stable Pb-free Perovskite Solar Cells

Dhruba B. Khadka[1], Yasuhiro Shirai[1], Masatoshi Yanagida[1] and Kenjiro Miyano[1]

[1] Photovoltaics Materials Group, Global Research Center for Environment and Energy based on Nanomaterials Science (GREEN), National Institute for Materials Science (NIMS), 1-1 Namiki, Tsukuba, Ibaraki 305-0044, Japan.

Abstract— The phenethylammonium thiocyanate (PEASCN) was introduced into the $FASnI_3$ perovskite film as a pseudohalide functional additive. This results in the suppression of Sn-oxidation and compact and larger grain film with better crystallinity. The device with PEASCN additive improved the power conversion efficiency (PCE) from 4.52% (control) to 9.65% (PEASCN). The device analysis revealed that the PEASCN additive has improved the optoelectronic properties coupled with a higher diffusion potential and passivation of defect densities in the Sn-PSCs. This report suggests that the pseudohalide-based functional additive is propitious for film growth, modification of surface chemistry, and defects at interface and bulk.

Keywords—Tin perovskite, Pseudohalide, tin oxidation, additive, stability, defect density.

I. INTRODUCTION

Tin and Bismuth based-perovskite solar cells (Sn/Bi-PSC) are alternative to lead-based halide perovskite devices to solve the toxicity issue of lead.[1]–[4] The intrinsic instability of Sn-HaP due to the facile oxidation of Sn^{2+} to Sn^{4+} leads to the formation of tin vacancy and metal-like behavior that deteriorates the device performance.[5] Several approaches have been introduced that passivate the catastrophic effects in Sn-HaP film quality by structural regulation and additive engineering which results in improving the device performance and its stability.[6] A number of additives have been used as reducing agents to inhibit the oxidation and facilitate the formation of a pinhole-free uniform film.[7] The A-site alloying in $FASnI_3$ crystal lattice has demonstrated improvement in the device performance and stability by regulating 3D structure.[8], [9] Besides this, a thin 2D layer was formed on the top of the 3D layer by post-treatment.[10], [11] Moreover, additives with pseudohalide functional derivatives increase the device performance of HaPSCs and stability coupled with the hydrophobicity and optoelectronic quality of HaP film.[12], [13] The additive with N-, S-, and O-based electron donors conjugate with tin halides by donating a lone pair electron to the divalent tin resulting increase in covalency.[14] Noting that the pseudohalide derivatives is beneficial for stabilizing the halide bonding.[15]–[17]

In this work, the bulky organic cation pseudohalide PEASCN was added in the $FASnI_3$ precursor to control the extent of tin oxidation and film growth. The pseudohalide of the PEASCN additive effectively improves the film quality and optoelectronic properties. The PCE enhanced from 4.52% for pristine to 9.65% for PEASCN added device with better stability. We have explored the characteristics insight into the effect of PEASCN additive by analysing the growth, optophysical, optoelectronic properties.[18] This report has discussed the additive approach to pave the way for addressing the oxidation and instability issue of Sn-HaPSCs.

II. EXPERIMENTAL

A. Device fabrication

For the fabrication of $FASnI_3$; 0.8 M of FAI, SnI_2, and SnF_2 (0.08 M) and for PEASCN incorporated $FASnI_3$, PEASCN: FAI (x, 1-x); 0.8 M), SnI_2 (0.8 M), and SnF_2 (0.08 M) were dissolved in dimethyl sulfoxide (DMSO) solvent for one hour. The Sn-HaP precursor was deposited on the PEDOT:PSS (30 nm)/ITO substrate. These films were annealed on the hot plate at 60 °C for 3min and 100 °C for 10 min. Then, the device is completed depositing PCBM /BCP thin films were spun–coated on top of the Sn-HaP films. The detailed fabrication can be found in our earlier report.[18], [19] Finally, Ag (100 nm) was thermally evaporated and get device of ~0.26 cm^2 area.

B. Materials and device characterizations

The XRD results were collected using Rigaku Smart Lab, CuK_α radiation, λ=1.5405Å. The SEM images were obtained by a high-resolution scanning electron microscope (SEM) at 5 kV accelerating voltage (Hitachi, S-4800). XPS spectra were obtained using a Versa Probe II (ULVAC-PHI, Japan). The current density–voltage (J-V) curves were measured under 1 sun with an AM1.5G spectral filter (100 mW/cm^2) coupled with an MPPT system (Systemhouse Sunrise Corp.). capacitance spectra (C–f) were collected using an LCR meter (IM3536, Hioki) under dark.

III. RESULTS AND DISCUSSION

Figure 1a shows a complete device with the structure of ITO/PEDOT:PSS/FA$_{1-x}$PEASCN$_x$SnI$_3$/PCBM/BCP/Ag. The device parameters with PEASCN additive are given in Fig. 1c and J-V curves (Fig. 1b). The device with PEASCN additive (x≤0.08) exhibited a PCE of ~9.65% (J$_{SC}$~ 22.16 mAcm^{-2},

$V_{OC} \sim 0.667V$, and FF= 65.3) with negligible hysteresis. It showed a significant improvement in PCE of 4.52% for control device. The device with PEASCN additive (Fig.1d) revealed superior device stability compared to the control device.

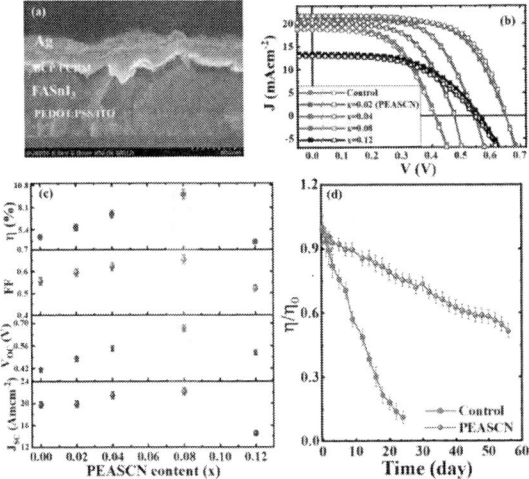

Fig. 1. Cross-sectional image of the device (a). The *J-V* curves of FA$_{1-x}$PEASCN$_x$SnI$_3$ devices (for x=0 - 0.12) (b). Device parameters with varying content of PEASCN (c). x=0.08 is optimal denoted as PEASCN hereafter. normalized efficiency with the stability of pristine and PEASCN incorporated devices (d).

The SEM images (Fig. 1a, b) shows a better film coverage and large grains for the film with PEASCN (x=0.08) compared to the control (x=0). It grows with highly oriented crystallographic planes of (100) and (200) for with PEASCN additive whereas the control film has with multiple crystal orientations i.e. (102), (122), (222), (213), etc. No 2D phase features were observed with PEASCN additive for x≤ 0.08.[18] This supports the improvement in device performance.

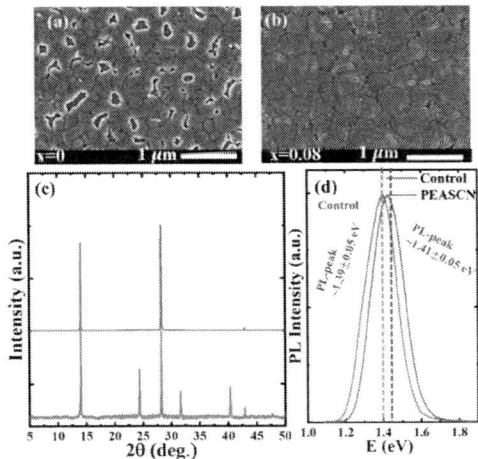

Fig. 2. SEM images (a,b), XRD patterns (c) and PL spectra (d) of the FA$_{1-x}$PEASCN$_x$SnI$_3$ films with contents (x=0, 0.08 (PEASCN)).

The Photoluminescence (PL) spectra of the corresponding films (Fig.2d) reveal a slight blueshift of characteristics peaks with the PEASCN additive. The characteristic PL peaks are centred at 1.40±0.02 eV for x=0 and 1.41±0.02 eV for x=0.08.

This modifies the interface band alignment which increase the carrier transport.

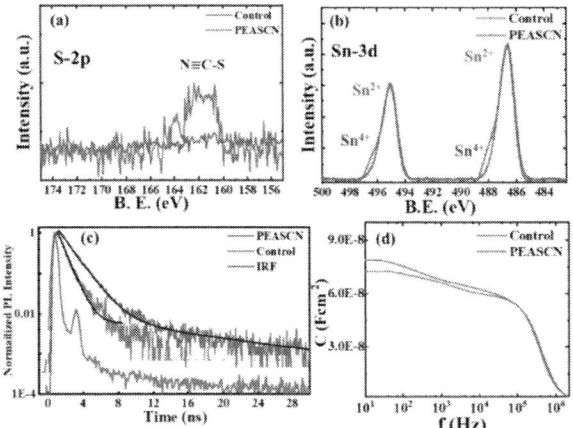

Fig. 3. XPS spectra (S-2p, Sn-3d) of the respective film (a,b). TRPL results of films (c). C-f spectra at room temperature (d).

To get insight into the effect of the PEASCN, the surface chemistry of FASnI$_3$ film was examined by X-ray photoelectron spectroscopy (XPS) measurement. Figure 3a shows the XPS core of S-2p indicating existence of SCN component on surface of film with PEASCN additive. The characteristics peak convoluted from the XPS spectra (Fig. 3b) at ~ 486.7 (495.2) eV and 487.3 (495.7) eV are assigned to the Sn^{2+} and Sn^{4+} species, respectively. We found that the control FASnI$_3$ film has a higher atomic percentage of Sn^{4+} compared with the PEASCN additive film. It corroborates that the PEASCN incorporation in the FASnI$_3$ lattice controls the extent of Sn^{2+} oxidation and hence ameliorates the film quality. This is concurrent with the device results. Figure 3c a depicts time-resolved PL (TRPL) characteristics of Sn-HaP film. It shows a longer carrier lifetime of 1.68 ns for the PEASCN added FASnI$_3$ film compared to the control film (0.64 ns). This implicates a lower defect density in the film with an additive that must be a consequence of the controlled growth of morphology with large grain and better crystallinity of film with additive.

Furthermore, the capacitance-frequency (*C-f*) response (Fig. 3d) for the control device shows a slightly larger value in the range of 1kHz to 50 kHz that arises from the absorber bulk. It is correlated with a higher defect density in the control Sn-HaP film which is prone to inferior device performance. The capacitance at a lower frequency has a lower value for the device with PEASCN additive. It is related to the reduction in ion or charge accumulation at the interfacial layer or electrode. This could result in improvement in device stability with PEASCN additive.

IV. SUMMARY AND CONCLUSIONS

In summary, we achieved Sn-PSCs of PEC ~9.65% using phenethylammonium thiocyanate (PEASCN) additive with improved stability. The Sn-HaP film with the PEASCN additive remarkably improved the film morphology and highly oriented crystal growth with control of Sn^{2+} to Sn^{4+} oxidation. The optophysical properties show that the FASnI$_3$ film with additive increases the carrier lifetime and reduces the band offset. Our report implicates that the pseudohalide functional is beneficial for the improvement in optoelectronic quality of FASnI$_3$ film which collectively results in the performance and stability.

ACKNOWLEDGMENT

This work was supported by JST-Mirai Program Grant Number JPMJMI21E6, Japan.

REFERENCES

[1] G. Nasti and A. Abate, "Tin Halide Perovskite (ASnX3) Solar Cells: A Comprehensive Guide toward the Highest Power Conversion Efficiency," *Adv. Energy Mater.*, vol. 10, no. 13, p. 1902467, Apr. 2020, doi: 10.1002/aenm.201902467.

[2] D. B. Khadka, Y. Shirai, M. Yanagida, and K. Miyano, "Tailoring the film morphology and interface band offset of caesium bismuth iodide-based Pb-free perovskite solar cells," *J. Mater. Chem. C*, vol. 7, no. 27, pp. 8335–8343, 2019, doi: 10.1039/C9TC02181G.

[3] M. G. M. Pandian et al., "Effect of solvent vapour annealing on bismuth triiodide film for photovoltaic applications and its optoelectronic properties," *J. Mater. Chem. C*, vol. 8, no. 35, pp. 12173–12180, 2020, doi: 10.1039/D0TC02455D.

[4] L. Liang and P. Gao, "Lead-Free Hybrid Perovskite Absorbers for Viable Application: Can We Eat the Cake and Have It too?," *Adv. Sci.*, vol. 5, no. 2, p. 1700331, Feb. 2018, doi: 10.1002/advs.201700331.

[5] T. Nakamura et al., "Sn(IV)-free tin perovskite films realized by in situ Sn(0) nanoparticle treatment of the precursor solution," *Nat. Commun.*, vol. 11, no. 1, p. 3008, 2020, doi: 10.1038/s41467-020-16726-3.

[6] Y. Yan, T. Pullerits, K. Zheng, and Z. Liang, "Advancing Tin Halide Perovskites: Strategies toward the ASnX3 Paradigm for Efficient and Durable Optoelectronics," *ACS Energy Lett.*, vol. 5, no. 6, pp. 2052–2086, Jun. 2020, doi: 10.1021/acsenergylett.0c00577.

[7] C. Wang et al., "Illumination Durability and High-Efficiency Sn-Based Perovskite Solar Cell under Coordinated Control of Phenylhydrazine and Halogen Ions," *Matter*, vol. 4, no. 2, 2021, doi: 10.1016/j.matt.2020.11.012.

[8] D. B. Khadka, Y. Shirai, M. Yanagida, and K. Miyano, "Attenuating the defect activities with a rubidium additive for efficient and stable Sn-based halide perovskite solar cells," *J. Mater. Chem. C*, vol. 8, no. 7, pp. 2307–2313, 2020, doi: 10.1039/C9TC06206H.

[9] X. Jiang et al., "One-Step Synthesis of SnI2·(DMSO)x Adducts for High-Performance Tin Perovskite Solar Cells," *J. Am. Chem. Soc.*, vol. 143, no. 29, pp. 10970–10976, Jul. 2021, doi: 10.1021/jacs.1c03032.

[10] M. Liao et al., "Efficient and Stable FASnI3 Perovskite Solar Cells with Effective Interface Modulation by Low-Dimensional Perovskite Layer," *ChemSusChem*, vol. 12, no. 22, pp. 5007–5014, Nov. 2019, doi: 10.1002/cssc.201902000.

[11] T. Wu et al., "Efficient and Stable Tin Perovskite Solar Cells Enabled by Graded Heterostructure of Light-Absorbing Layer," *Sol. RRL*, vol. 4, no. 9, p. 2000240, Sep. 2020, doi: 10.1002/solr.202000240.

[12] Y. Numata, Y. Sanehira, R. Ishikawa, H. Shirai, and T. Miyasaka, "Thiocyanate Containing Two-Dimensional Cesium Lead Iodide Perovskite, Cs 2 PbI 2 (SCN) 2: Characterization, Photovoltaic Application, and Degradation Mechanism," *ACS Appl. Mater. Interfaces*, vol. 10, no. 49, pp. 42363–42371, 2018, doi: 10.1021/acsami.8b15578.

[13] H. Kim, J. W. Lee, G. R. Han, S. K. Kim, and J. H. Oh, "Synergistic Effects of Cation and Anion in an Ionic Imidazolium Tetrafluoroborate Additive for Improving the Efficiency and Stability of Half-Mixed Pb-Sn Perovskite Solar Cells," *Adv. Funct. Mater.*, vol. 31, no. 11, p. 2008801, Mar. 2021, doi: 10.1002/adfm.202008801.

[14] X. Meng et al., "Highly Stable and Efficient FASnI 3 -Based Perovskite Solar Cells by Introducing Hydrogen Bonding," *Adv. Mater.*, vol. 31, no. 42, p. 1903721, Oct. 2019, doi: 10.1002/adma.201903721.

[15] B. Yu et al., "Efficient (>20%) and Stable All-Inorganic Cesium Lead Triiodide Solar Cell Enabled by Thiocyanate Molten Salts," *Angew. Chemie Int. Ed.*, vol. 60, no. 24, pp. 13436–13443, Jun. 2021, doi: 10.1002/anie.202102466.

[16] Y. Yu et al., "Synergistic Effects of Lead Thiocyanate Additive and Solvent Annealing on the Performance of Wide-Bandgap Perovskite Solar Cells," *ACS Energy Lett.*, vol. 2, no. 5, pp. 1177–1182, May 2017, doi: 10.1021/acsenergylett.7b00278.

[17] M. Rameez, S. Shahbazi, P. Raghunath, M. C. Lin, C. H. Hung, and E. W.-G. Diau, "Development of Novel Mixed Halide/Superhalide Tin-Based Perovskites for Mesoscopic Carbon-Based Solar Cells," *J. Phys. Chem. Lett.*, vol. 11, no. 7, pp. 2443–2448, Apr. 2020, doi: 10.1021/acs.jpclett.0c00479.

[18] D. B. Khadka, Y. Shirai, M. Yanagida, and K. Miyano, "Pseudohalide Functional Additives in Tin Halide Perovskite for Efficient and Stable Pb-Free Perovskite Solar Cells," *ACS Appl. Energy Mater.*, vol. 4, no. 11, pp. 12819–12826, Nov. 2021, doi: 10.1021/acsaem.1c02496.

[19] D. B. Khadka, Y. Shirai, M. Yanagida, T. Masuda, and K. Miyano, "Enhancement in efficiency and optoelectronic quality of perovskite thin films annealed in MACl vapor," *Sustain. Energy Fuels*, vol. 1, no. 4, pp. 755–766, 2017, doi: 10.1039/C7SE00033B.

Investigation of the Circular Economy Approach in Asian Solar PV Manufacturing

Viktor Dancza, Kai Cheng

Brunel University London, London, United Kingdom

Blueleaf Energy Asia, Singapore, Singapore

While supporting our fight against climate change, the surge in PV module production over the past 3 decades put a huge strain on finite natural resources and will result in vast amounts of waste in this decade, due to first generation PV modules reaching their 25-to-30-year useful life. Recycling and end-of-life treatment have received attention among researchers, but the transition of manufacturing methods to a circular economy model is a relatively new area in the solar PV industry, due to the longevity of the panels. To identify the key barriers and enablers of this transition, this research took the questions directly to the stakeholders, in the form of open access and targeted surveys. Social media was used to distribute the public survey and Asian PV manufacturers were contacted via email. Analysing the results using descriptive methods, the public and manufacturing industry opinions showed strong correlation with the findings of the literature review. The two sets of answers also aligned in many cases, most likely due to the social media survey predominantly answered by PV professionals and thus may have already included manufacturing opinions. The survey results validated the findings of many publications from the past, shining light on the social, economic, political, and environmental challenges the PV industry is still yet to overcome. At the same time, the PV waste problem was once more identified as a potential foundation for new industries and untapped opportunities promising billions of dollars in revenue.

Three general methods for predicting bifacial photovoltaic performance including spectral albedo

Erin M. Tonita, Christopher E. Valdivia, Michael Martinez-Szewczyk, Mariana I. Bertoni, Karin Hinzer

SUNLAB, University of Ottawa, Ottawa, ON, Canada

DEfECT Lab, Arizona State University, Tempe, AZ, United States

Modelling of bifacial system performance is often limited due to over-estimating rear irradiance, and/or neglecting the impact of spectral albedo on the rear-face. We present three methods for evaluating bifacial performance with spectral albedo, and assess their performance in comparison to the 0-400 W/m2 range that typically occurs on the rear of a panel during outdoor operation. We investigate the impact of spectral albedo: (1) assuming the irradiance contribution of the rear side is reduced only by albedo; (2) scaling all spectral albedos to have a rear incident intensity of 200 W/m2 at AM1.5; and (3) selecting a typical spectral albedo to represent 200 W/m2 with all other albedos scaled proportionally. We evaluate performance with an optoelectronic model of a typical bifacial silicon heterojunction cell, validated with quantum efficiency and Suns-Voc measurements. All spectral albedo results are compared to non-spectral IEC 60904-1-2 bifacial measurement standards, highlighting discrepancies between spectrally-resolved and spectrally-flat albedos. For example, method (3) shows the most significant maximum power (Pmax) differences for green grass, white sand, and snow of +1.7%, +3.1%, and +3.3%, respectively. Efficiency correspondingly increases between 0.4-0.7% abs. for these albedos. Average albedos calculated over the absorption range of the technology will reduce large discrepancies found in Pmax but will not account for efficiency variation which is mainly driven by the spectral shape. Overall, we identify our third spectral bifacial illumination method as most closely emulating field data. The methods presented can be used to improve the accuracy of energy yield predictions.

Evaluation of Auger Limited Behavior in Thermoradiative Cells

Jamie D. Phillips*

Electrical and Computer Engineering Department
University of Delaware, Newark, DE 19716, USA
*jphilli@udel.edu

Abstract—The impact of Auger recombination on thermoradiative cell power density and power conversion efficiency are investigated with simulations based on recombination rates for narrow bandgap semiconductor materials. A crossover point occurs for non-radiatively/radiatively limited cell performance that determines the minimum bandgap energy and corresponding maximum intrinsic carrier density for a given cell temperature. Calculations for HgCdTe as a representative bulk narrow bandgap material spanning the full range of infrared show that the maximum power conversion efficiency and power density for radiatively limited performance are constrained well below the calculated ideal performance limits. This work is expected to serve as a basis to set realistic expectations for thermoradiative cell performance for bulk semiconductor materials, and to highlight the importance of suppressing Auger recombination rates as perhaps the most critical need for improving thermoradiative cells.

Index Terms—Semiconductor physics, energy conversion devices, thermal energy harvesting, optoelectronics, non-radiative recombination

I. INTRODUCTION

THERMORADIATIVE cells are energy conversion devices that generate electrical power through net radiative emission from a hot object staring into a cold scene. This may be considered the inverse process of a photovoltaic cell, based on the classical description of net radiative absorption from a hot object to produce electricity in a cold cell. Theoretical efficiency limits have been established for thermoradiative cells to establish power conversion efficiency limits versus bandgap energy and temperature [1]–[5], analogous to the Shockley-Queisser limiting efficiency [6] established for photovoltaic cells as a function of bandgap and incident solar irradiation. The introduction of non-radiative generation-recombination processes in thermoradiative cells inhibit the net outflow of radiation from the cell, and arguably represents the primary limiting non-ideality for energy conversion efficiency. Realistic thermal energy conversion will require narrow bandgap semiconductor materials, where non-radiative recombination processes will be intrinsically limited by Auger recombination (impact ionization). While prior literature on thermoradiative cell efficiency has identified the importance of non-radiative processes, a more detailed analysis of constraints placed on thermoradiative cells by Auger processes has not been published. This work explores the specific crossover cases to achieve radiatively limited performance in key narrow bandgap semiconductor materials.

II. THERMORADIATIVE ENERGY CONVERSION EFFICIENCY AND POWER DENSITY

The primary objective of thermoradiative cells is the efficient conversion of thermal power to electrical power. The electrical power generated by a thermoradiative cell without non-radiative losses is given by

$$P = IV = q\Delta\mu\left[\phi_c\left(\Delta\mu\right) - \phi_a\right] \quad (1)$$

where q is the charge of an electron, ϕ_c and ϕ_a are each the radiative flux emitted by the cell and absorbed by the cell from the ambient, respectively, and $\Delta\mu$ is the cell electrochemical potential. The thermoradiative power conversion efficiency is the ratio of electric power output to the supplied thermal power, given by [2]

$$\eta = \frac{P}{Q_{in}} = \frac{P}{P + \dot{E}_{rad} - \dot{E}_{abs}} \quad (2)$$

where Q_{in} is the input heat flux for the cell at a given temperature and $\dot{E}_{rad}$ and $\dot{E}_{abs}$ are the radiative energy flux emitted and absorbed by the cell, respectively. The radiative energy flux is given by

$$\frac{2\pi}{h^3 c^2} \int_{E_G}^{\infty} \frac{E^3}{e^{E - \Delta\mu/k_B T} - 1} dE. \quad (3)$$

It should be noted that there is a key distinction for thermoradiative cells in that the temperature of the energy source of heat flux is not always fixed (unlike solar or thermophotovoltaics), such that the maximum power point (MPP) and maximum efficiency point are not necessarily the same. Non-radiative processes impact the net outflow of radiation and ability to produce current flow in the second quadrant of current-voltage characteristics to generate power [7]. Electrical power generation and conversion efficiency ultimately depend on limiting radiative and non-radiative conversion efficiency in the thermoradiative cell.

III. RADIATIVE AND AUGER RECOMBINATION RATES

The intrinsic limit on radiative efficiency is determined by the fraction of radiative electron recombination relative to total electron recombination. The recombination rates may be

978-1-7281-6118-1/22 $31.00 © 2022 IEEE

given by the "ABC model" that representes the carrier density dependence of recombination rates by [8]

$$U = An + Bn^2 + Cn^3 \qquad (4)$$

where U is the net recombination rate, n is the charge carrier density, An is the Shockley-Read-Hall (SRH) recombination rate, Bn^2 is the radiative recombination rate, and Cn^3 is the Auger recombination rate. The non-radiative Shockley-Read-Hall recombination rate will depend on the density of unintentional material defects, with a weak dependence on the nature of the host material electronic bandstructure. For the purpose of this work, SRH recombination will be considered a "solvable problem", under the assumption that sufficiently high material quality may be obtained to reach either radiatively or Auger limited behavior.

At increasing carrier injection, Auger recombination rates will become dominant over radiative recombination, degrading radiative efficiency. This is a commonly attributed to the efficiency droop observed in light emitting diodes at high injection. For thermoradiative cells, the minimum carrier density will be defined by thermally induced generation rates, approximately at the intrinsic carrier concentration at thermal equilibrium. The crossover point to achieve radiatively limited behavior at a given temperature can be approximated in the ABC model as

$$n_i < \frac{B}{C}. \qquad (5)$$

This behavior is shown in Fig. 5 using values for bulk III-V semiconductor materials [9] and HgCdTe alloys [10], [11] at T = 300 K. The crossover point implies that a minimum bandgap energy of approximately 0.3 eV (corresponding to a maximum carrier density of approximately 3×10^{15} cm^{-3}) is required to achieve radiatively limited performance. Below this bandgap energy (and above this intrinsic carrier density), Auger recombination will become dominant, reducing the short circuit current density and ability to generate electrical power.

IV. AUGER LIMITS ON THERMORADIATIVE EFFICIENCY AND POWER DENSITY

The impact of Auger limited behavior on thermoradiative power conversion efficiency and power density can be estimated based on the crossover point defined by Eq. 5. To explore this impact for a broad range of cell temperatures and materials, HgCdTe alloys provide a good benchmark due to well known material models that span bandgap energies across the infrared region of interest. The power conversion efficiency and power density are determined at MPP for a given bandgap energy and cell temperature. An example calculation of thermoradiative cell current and power density characteristics at $T_C = 500K$ is shown in Fig. 2. The example compares an ideal cell at a given temperature and bandgap energy to a HgCdTe cell that accounts for the impact of Auger recombination, illustrating the reduction in short circuit current density and maximum power density. The crossover point at $T_C = 500K$ is determined to be $E_G = 0.4839eV$, where the

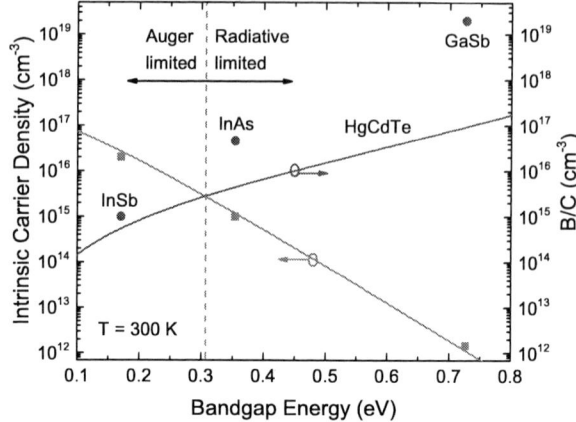

Fig. 1. Determination of the crossover point between radiatively and Auger limited thermoradiative cell performance based on the ABC model relation $n_i < B/C$ and data from select bulk compound semiconductor materials.

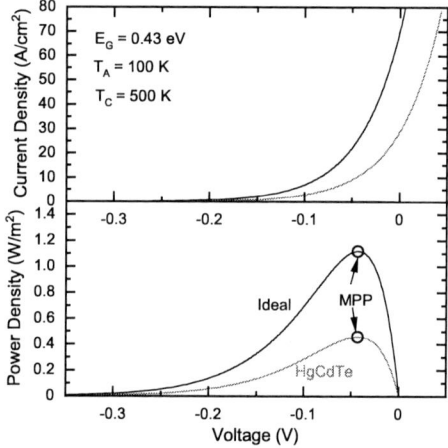

Fig. 2. Comparison of calculated current density versus voltage and power density versus voltage for thermoradiative cells based on an ideal material with 0.43 eV bandgap energy and Hg$_{0.568}$Cd$_{0.432}$Te with the same bandgap energy while also considering Auger recombination.

curves in Fig. 2 correspond to a bandgap energy that is below the crossover point by approximately kT.

The reduction in power density and power conversion efficiency for thermoradiative cells can be calculated by inserting a radiative efficiency term $B/(B + Cn)$ to describe the cell current density. A comparison between ideal and HgCdTe parameters is shown in Fig. 3, where there is a clear reduction in efficiency and saturation in achievable power density in the Auger limited region.

The calculated crossover points for radiatively limited performance in HgCdTe are compared to the theoretical power conversion efficiency limits in Fig. 4. Under the assumptions of this analysis, it is clear that Auger processes will present an intrinsic limit on power conversion efficiency, which is approximately 7 % for HgCdTe. The nearly temperature independent

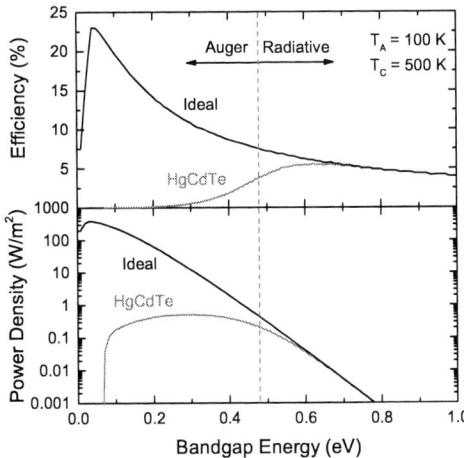

Fig. 3. Comparison of calculated ideal and HgCdTe power conversion and efficiency and power density as a function of bandgap energy at $T_C = 500K$ and $T_A = 100K$.

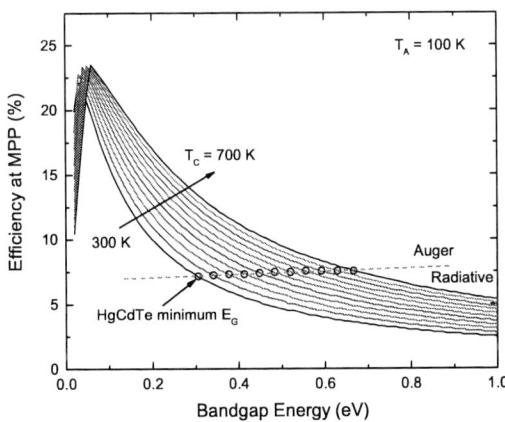

Fig. 4. Thermoradiative cell power conversion efficiency theoretical limits and comparison to estimated radiatively limited behavior for HgCdTe.

limit on power conversion efficiency shown in Fig. 4 may be qualitatively viewed as the counteracting and nearly balanced mechanisms of both increasing radiative emission and Auger recombination with increasing temperature.

The crossover points for radiatively limited performance in HgCdTe are similarly shown on the ideal power density curves in Fig. 5. Similar to power conversion efficiency in Fig. 4, the crossover points lie significantly below the ideal maximum thermoradiative power density values. The power density at the crossover point between Auger and radiatively limited behavior maintains a monotonic increase with cell temperature, primarily due to the higher energy of emitted radiation and corresponding larger electric potential at MPP.

V. CONCLUSION

Auger recombination is expected to play a critical role in determining thermoradiative cell performance. Based on the

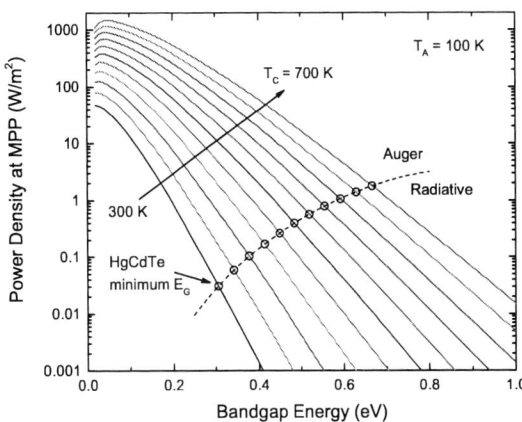

Fig. 5. Thermoradiative cell power density theoretical limits and comparison to estimated radiatively limited behavior for HgCdTe.

common carrier density dependent ABC model for recombination rates, there will be a crossover minimum bandgap energy to achieve radiatively limited performance for a given cell temperature and recombination parameters defined by the material bandstructure. Realistic estimates for Auger processes in HgCdTe project a significant reduction in cell performance below the radiatively limited crossover point. While the modeling used in this work is not comprehensive in defining the intricate details of recombination processes and related non-idealities for a thermoradiative cell, the results provide a reasonable estimation of non-radiative limits and highlights the vital importance of minimizing Auger processes in order for thermoradiative cells to succeed.

REFERENCES

[1] S. J. Byrnes, R. Blanchard, and F. Capasso, "Harvesting renewable energy from earth's mid-infrared emissions," *Proceedings of the National Academy of Sciences*, vol. 111, no. 11, pp. 3927–3932, 2014.
[2] R. Strandberg, "Theoretical efficiency limits for thermoradiative energy conversion," *Journal of Applied Physics*, vol. 117, no. 5, p. 055105, 2015.
[3] W.-C. Hsu, J. K. Tong, B. Liao, Y. Huang, S. V. Boriskina, and G. Chen, "Entropic and near-field improvements of thermoradiative cells," *Scientific Reports*, vol. 6, no. 1, p. 34837, 2016.
[4] X. Zhang, W. Peng, J. Lin, X. Chen, and J. Chen, "Parametric design criteria of an updated thermoradiative cell operating at optimal states," *Journal of Applied Physics*, vol. 122, no. 17, p. 174505, 2017.
[5] T. Deppe and J. N. Munday, "Nighttime photovoltaic cells: Electrical power generation by optically coupling with deep space," *ACS Photonics*, vol. 7, no. 1, pp. 1–9, 2020.
[6] W. Shockley and H. J. Queisser, "Detailed balance limit of efficiency of p-n junction solar cells," *Journal of Applied Physics*, vol. 32, no. 3, pp. 510–519, 1961.
[7] J. D. Phillips, "Thermoradiative Cell Equivalent Circuit Model," *IEEE Transactions on Electron Devices*, vol. 68, no. 2, 2021.
[8] N. F. Mott, "Recombination; a survey," *Solid-State Electronics*, vol. 21, pp. 1275–1280, 1978.
[9] IoffeInstitute, "New semiconductor materials. characteristics and properties." [Online]. Available: http://www.ioffe.ru/SVA/NSM/Semicond/
[10] J. S. Blakemore, *Semiconductor Statistics*. Dover Publications, 1987.
[11] S. E. Schacham and E. Finkman, "Recombination mechanisms in p-type HgCdTe: Freezeout and background flux effects," *Journal of Applied Physics*, vol. 57, no. 6, pp. 2001–2009, 1985.

Fabrication of a Chemically Exfoliated 2D MoS₂ Nanoparticle Based Solar Cell

Wafa Alnaqbi, Ayman Rezk, Aisha Alhammadi, and Ammar Nayfeh

Khalifa University, Abu Dhabi, 127788, UAE

*Corresponding Author: *ammar.nayfeh@ku.ac.ae*

Abstract—In this work, a 2D MoS₂ nanoparticle based solar cells is fabricated. The MoS₂ nanoparticles (NPs) are synthesized through a chemical exfoliation method then drop casted on the Si substrate. Both SEM and AFM are sed to characterize and observe the MoS₂ NPs coverage on the substrate. The MoS₂ solar cell performance is tested under AM1.5G showing a proof of concept. In addition, it an increased in the MoS₂ photodiode's responsivity conmpared to a control Si diode is seen. Moreover, the MoS₂ NPs can absrob significant amount of photons making them suitable for future low cost solar cells

I. INTRODUCTION

Recently interest in finding new materials for solar cells has rapidly increased. This is especially true for thin film solar cells to increase the optical path length and current [1]. One advantage of thin films is they can be deposited on variety of substrates like flexible or rigid, metal or insulator, using variety of physical, chemical, electrochemical, plasma-based deposition techniques, allowing for many different research to be implemented [2]. As the technology in solar cells is moving toward thinner devices, the 2D materials like the graphene, and the other layered semiconducting 2D materials have been investigated for the use in solar cells. These 2D materials have various attractive properties, are ultrathin, flexible, and they have shown excellent mechanical properties, in addition to being highly efficient absorbers, with high conductivity [3, 4].

Molybdenum Disulfide (MoS₂) is one of widely studied layered two-dimensional transition metal dichalcogenides (2-D TMDs) materials, it has unique electronic, optical, and attractive semiconducting properties, in addition, it is abundant in the nature [5,6]. Like the other 2D layered materials, the interlayer force in MoS₂ is the weak van der Waals force, that make it possible to be exfoliated into mono to few layers using different techniques [7]. MoS₂ has a tunable band gap. The bulk MoS₂ has an indirect band gap of 1.2 eV, and due to the quantum confinement effect, this band gap increases, as the number of the MoS₂ layers decreases, where the monolayer of MoS₂ has a direct band gap of 1.9 eV [8]. These attractive properties of MoS₂ have made it suitable to be used in solar cells.

In this paper, MoS₂ nanoparticles (NPs) is synthesized using chemical exfoliation method, and it is characterized, it is used then to fabricate MoS₂ NPs/Si solar cell, the performance of the solar cell is investigated by performing electrical measurements under 1 Sun.

II. METHODOLOGY

MoS₂ NPs/p-Si Solar Cell Fabrication:

MoS₂ nanoparticles synthesized utilizing chemical exfoliation method are used to fabricate MoS₂ NPs/p-Si photodiode, to synthesize the MoS₂ NPs, a 0.5 g of the MoS₂ powder is dispersed in a 50 mL of N-Methyl-2-pyrrolidone (NMP), then it is sonicated in an ice bath using a probe sonicator, followed by two stage centrifugation steps, finally NMP is removed, and the MoS₂ NPs is filtered and re-dispersed in a 50 mL of IPA. The detailed synthetization process of the nanoparticles is presented in a previous work [9]. Figure 1(a) shows the MoS₂ NPs after removing the NMP, and in figure 1 (b) the MoS₂ NPs dispersed in IPA is shown. The fabrication of MoS₂ NPs/p-Si solar cell started with cutting a square piece with an area of 1.5 × 1.5 cm² from a 390 μm thick p-type Si wafer with a resistivity of 1-5 Ω.cm as a starting substrate. The Si substrate is cleaned first by sonicate it for 1 minute in acetone to remove the tiny pieces on the substrate resulted from cutting the wafer, then by rinsing it with acetone, then isopropanol to remove the organic contamination, followed by cleaning with deionized (DI) water, and drying with nitrogen. After that, the substrate is dipped in a solution of 250 ml of DI water to 25 ml Hydrofluoric acid (HF) (10:1) for 30 second to remove the native oxide, then it is rinsed with DI water. After the cleaning process, 100 μL drop of the chemically exfoliated MoS₂ NPs is drop casted on the substrate with a time difference of 1 hour between them for drying. Later Au/Pl is sputtered through a finger mask for the front contact and full layer for the back contact. The fabrication process is summarized in table 1, and a schematic of the solar cell fabrication process is presented in figure 2. A photo of the fabricated solar cell is shown in figure 3 (a), and figure 3 (b) shows the final device.

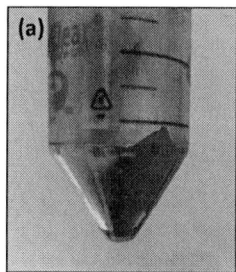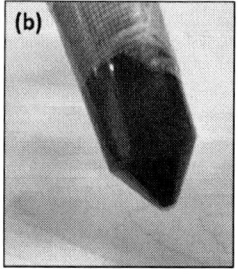

Figure 1: MoS₂ NPs (a) after removing the NMP, (b) dispersed in IPA.

Step Number	Step name	Step details
1	p-Si wafer cutting	Area:1.5 cm x 1.5 cm Doping: Boron Resistivity:1-5 Ω.cm Thickness:390 μm
2	p-Si substrate cleaning	- Sonicate with acetone - Rinse with acetone/IPA/ DI water
3	p-Si substrate native oxide removing	Dip in 10:1 DI water: HF for 30 s
4	MoS₂ NPs Deposition	Drop cast a 100 μm drop
5	Back contact Au/Pl sputtering	Thickness: 50 nm
6	Top layer Au/Pl sputtering	Thickness: 50 nm
7	Front contact Au/Pl sputtering	-Through shadow mask - Thickness: 50 nm

Table 1: MoS₂ NPs/p-Si solar cell fabrication process summary.

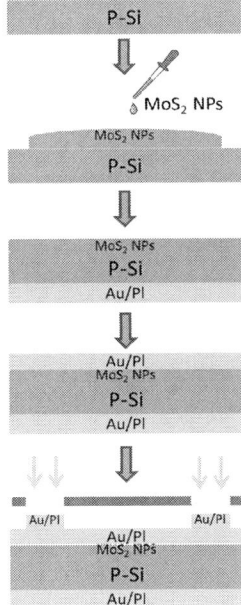

Figure 2: Schematic of the MoS₂ NPs/p-Si solar cell fabrication process.

Figure 3: (a) The fabricated MoS₂ NPs/p-Si Solar Cell. (b) Schematic of the final structure of the fabricated MoS₂ NPs/p-Si Solar Cell.

III. CHARACTERIZATION

To characterize the synthesized MoS₂ NPs, it is drop casted in a Si substrate. The topography of the sample is obtained using AFM that is shown in figure 4. The image shows an abundant of particles with an average height of approximately 1.8 nm.

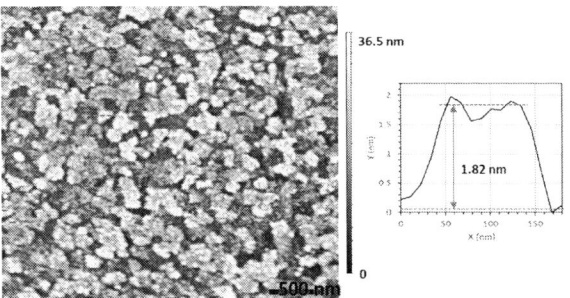

Figure 4: AFM image of the chemically exfoliated MoS₂ NPs on Si.

The morphology of the sample is obtained using SEM, the obtained image is shown in figure 5. it shows that the MoS₂ NPs gives a good coverage of the surface.

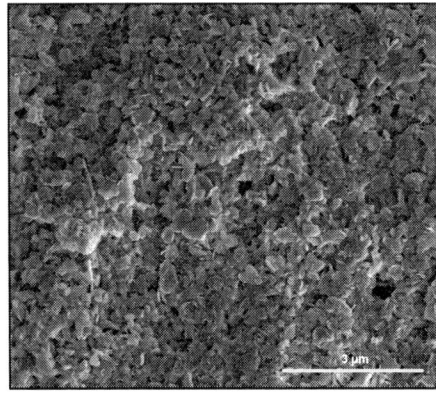

Figure 5: SEM image of the chemically exfoliated MoS₂ NPs on Si.

IV. RESULTS AND DISCUSSION

To test the performance of the fabricated solar cell, a solar simulator (Sol3a 94123A) is used. Figure 6 shows the measured J-V curve characteristics under 1sun, A.M. 1.5G of the solar with open circuit voltage (V_{OC}) is 0.25 V. The obtained J-V curve has shown a solar cell function and proof of concept using MoS2 nano particles to make the junction. The low current is expected due to thin metal and no optimized interface. However, the collected J-V curves shows the behavior of the MoS₂/Si junction.

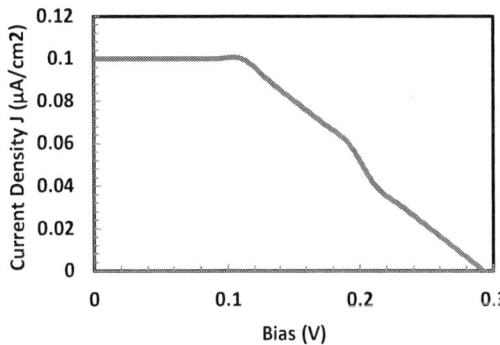

Figure 6: The J-V characteristics of the MoS₂ NPs/p-Si solar cell under 1 Sun, A.M. 1.5G.

978-1-7281-6118-1/22 $31.00 © 2022 IEEE

The solar cell is tested as a photodetector, and the current is measured with and without light in figure 7. The increase is due to the large absorption of the light, and the EHP generation in the MoS_2 layer. A control photodiode with metal on silicon without MoS_2 is also fabricated and its current density is also plotted in Figure 7.

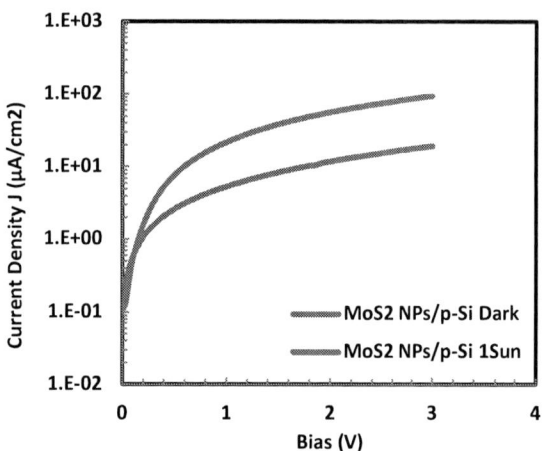

Figure 7: The J-V characteristics of the MoS_2 NPs/p-Si solar cell under 1sun, and in the dark.

Figure 8 shows the responsivity results of both the p-Si photodiode, and the MoS_2 NPs/p-Si solar cell, and it shows that the solar cell has higher responsivity than the p-Si photodiode, indicating a better junction and increased absorption.

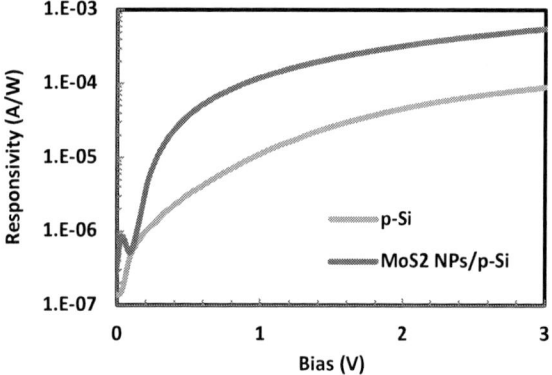

Figure 8: The Responsivity-V characteristics of the MoS2 NPs/p-Si solar cell and the p-Si substrate.

IV. CONCLUSION

To summarize, 2D MoS_2 nanoparticles are synthesized through chemical exfoliation method. The SEM images shows a good coverage of the MoS_2 NPs on the substrate. A MoS_2 nano particle based solar cell is fabricated, and its performance is investigated under 1 Sun. The obtained J-V characteristic confirms the device's performing as a solar cell. The solar cell is also tested as photodiode, the light response is measured, and compared to a control p-Si photodiode. It is observed that the MoS_2 based photodiode has an increased responsivity indicating the higher absorption in the MoS_2 NPs layer, thus its suitability for the photovoltaic applications.

REFERENCES

[1] R. Singh, V. Rangari, Sanagapalli, V. Jayaraman, S. Mahendra, V. Singh, "Nano-structured CdTe, CdS and TiO2 for thin film solar cell applications," *Solar Energy Materials and Solar Cells*, vol. 82, no. 1-2, pp. 315–330, 2004.
[2] K. L. Chopra, P. D. Paulson, and V. Dutta, "Thin-film solar cells: an overview," *Progress in Photovoltaics: Research and Applications*, vol. 12, no. 23, pp. 69–92, 2004.
[3] L. Wang, L. Huang, W. C. Tan, X. Feng, L. Chen, X. Huang, and K.-W. Ang, "2D Photovoltaic Devices: Progress and Prospects," *Small Methods*, vol. 2, no. 3, p. 1700294, 2018.
[4] S. Das, D. Pandey, J. Thomas, and T. Roy, "The Role of Graphene and Other 2D Materials in Solar Photovoltaics," *Advanced Materials*, vol. 31, no. 1, p. 1802722, 2018.
[5] D. Bhattacharya, S. Mukherjee, R. K. Mitra, and S. K. Ray, "Size-dependent optical properties of MoS2 nanoparticles and their photo-catalytic applications," *Nanotechnology*, vol. 31, no. 14, p. 145701, 2020.
[6] A. Rezk, A. Alhammadi, W. Alnaqbi, and A. Nayfeh, "Utilizing trapped charge at Bilayer 2d MoS_2/SiO_2 interface for memory applications," Nanotechnology, vol. 33, no. 27, p. 275201, 2022.
[7] Z. He and W. Que, "Molybdenum disulfide nanomaterials: Structures, properties, synthesis and recent progress on hydrogen evolution reaction," *Applied Materials Today*, vol. 3, pp. 23–56, 2016.
[8] C. P. Veeramalai, F. Li, Y. Liu, Z. Xu, T. Guo, and T. W. Kim, "Enhanced field emission properties of molybdenum disulphide few layer nanosheets synthesized by hydrothermal method," *Applied Surface Science*, vol. 389, pp. 1017–1022, 2016.
[9] W. Alnaqbi, J. M. Ashraf, A. Rezk, S. Abdul Hadi, A. Alhammadi and A. Nayfeh, "Absorption in the UV-Vis Region from Chemically Exfoliated MoS2 Nanoparticles for Solar Applications," 2021 IEEE 48th Photovoltaic Specialists Conference (PVSC), 2021, pp. 0784-0787

Evaluating the Durability of Balance of Systems Components Using Combined-Accelerated Stress Testing

David Miller[1]*, Greg Perrin[1], Kent Terwilliger[1], Joshua Morse[1],

Chuanxiao Xiao[1], Bobby To[1], Chun-Sheng Jiang[1], and Peter Hacke[1]

[1]National Renewable Energy Laboratory (NREL), Golden, CO, 80401, USA

*Corresponding author: David.Miller@nrel.gov

Abstract—The degradation of photovoltaic (PV) balance of systems (BoS) components is not well-studied, but the consequences include offline modules, strings, and inverters; system shutdown; arc faults; and fires. A utility provider experienced a ~30% failure rate in their power transfer chain, originally attributed to branch connectors. Field-failed specimen assemblies were therefore examined, consisting of cable connector, branch connector, and discrete fuse components. In this study, unused field-vintage specimens are examined using combined-accelerated stress testing (C-AST) to clarify the most influential environmental stressors as well as the effect of external mechanical perturbation. A benchtop prototype fixture was used to develop the perturbation capability for the C-AST chamber. The benchtop experiments were also used to develop the in-situ data acquisition of specimen current, voltage, and temperature. A significant increase in operating temperature (~100°C from ~40°C) and a different failure mode were observed promptly once periodic mechanical perturbation was applied. The current at failure was decreased from 35 A (with failure in the fuses) to 15 A (failure at the male/female metal pin connection). After initial examination using X-ray computed tomography, the external plastic was machined away from failed specimens to allow for failure analysis, including the extraction of the internal convolute springs for morphological examination (optical and electron microscopy).

Keywords—balance of systems, branch connector, C-AST, durability, DuraMAT, fuse, reliability, SEM, XCT

I. BACKGROUND

Balance of systems (BoS) components in photovoltaic (PV) systems include: cable connectors, cables, branch connectors, fuses (discrete), and fuse blocks. To date, the durability of BoS components has received limited research. After inverters, PV-specific studies have focused mostly on connectors, e.g., through steady-state or single-factor accelerated testing [1]. The industry generally speaks of a quantifiable replacement rate that may approach the order of ~1% of strings per year, whereas owners and operators may choose to extend system life beyond typical PV modules' 25-year warranties. Consequences of BoS degradation and failure include offline modules, strings, and inverters; system shutdown; arc faults; and fires [2].

Standards for connectors include IEC 62852 [3] and UL 6703 [4]. Focusing primarily on safety, these standards apply a variety of functionality metrics (e.g., current capacity, electrical insulation, and mechanical insertion force) and accelerated test methods (e.g., separate current-, temperature-, and humidity-cycling tests as well as mechanical bending). Component durability is typically examined using tailored methods designed to address historically observed degradation modes. In contrast, combined accelerated stress testing (C-AST) has recently been developed to examine PV modules [5]. In C-AST, multiple stressors (including UV-VIS radiation, temperature, moisture, electrical current, and external mechanical perturbation) are applied simultaneously in a developed time sequence, rather than in steady state. C-AST is intended to screen for degradation and failure modes, based solely on the extremes of the PV application in challenging environments. While C-AST was initially developed for PV modules, it may also be applied to PV materials and components.

This study explores an event where a utility provider experienced a ~30% failure rate in their power transfer chain at multiple PV installations in Arizona and New Jersey, originally attributed to branch connectors. "Failure" refers to overheating, softening, and physical distortion—with overt temperature rises observed in thermographic imaging. The worst consequences included broken circuits, electrical arcs, and local combustion. Here, in collaboration with the utility provider, we sought to further diagnose and advise on the BoS components. This study was also used to develop accelerated BoS test capability.

II. INTRODUCTION

In this study, C-AST was adapted for the examination of PV branch connectors. The PV installations featured branch connectors fixed to the system using long leads (up to ~1 meter) of cable, allowing a variety of inadvertent mechanical motions. To study the mechanical motion possible in the installations, a custom fixture was used to develop the capability to apply periodic external multi-axial mechanical perturbation during weathering. While the chamber version is presently in study, a benchtop prototype was used to compare static and dynamic

978-1-7281-6118-1/22 $31.00 © 2022 IEEE

(externally actuated) specimen assemblies. The key activities of the C-AST BoS project include:

- Development of accelerated testing, in-situ data acquisition, and subsequent read point characterization methods for the use of C-AST with BoS components.

- Demonstration of the ability to distinguish known bad components using C-AST and identify their corresponding degradation modes.

- Identification of the significance of external mechanical perturbation in C-AST, relative to testing with no external actuation.

III. EXPERIMENTAL

A. Specimens

Because the number and contribution of damage-susceptible components was unknown, the adjoining system components were examined together as an assembly consisting of: two cable connectors (specimen ends), two branch connectors, and two fuses (specimen middle). Components of a similar vintage to the utility system components were purchased from an electronics vendor to aid development of the test capability. Unaged spare field specimens were obtained from the utility provider for use in C-AST. Field-failed specimens were obtained from the utility provider to compare to the assemblies subject to accelerated testing. Upon the utility provider's request, component make and models will be kept confidential.

B. Test Fixture and Its Application

A single specimen push-pull test fixture was developed in the preliminary benchtop experiments. During operation, a polymer collar was directed in the opposite direction of a polymer push-rod using a linked mechanism. A DC motor was used to displace the middle of the fuse within the specimen by 3 mm initially, which was later reduced to 1 mm in subsequent experiments. The DC motor was confirmed to operate at 16 rpm, faster than the hydraulic actuators in C-AST, i.e., with on and off durations on the order of seconds. Separate GENESYS+ GH20-75 (TDK-Lambda Corp.) power supplies gave direct current to separate static and dynamic (actuated) assemblies, starting at 20 A and 10 A, respectively, and initially incremented by 1 A each day.

C. Data Acquisition and Subsequent Characterizations

A 9174 chassis, equipped with 9220 analog-in and 9264 analog-out cards (all National Instruments Corp.) was used for current and voltage monitoring. Operating temperature was monitored with a 2700 system (Keithley Instruments LLC) using discrete T-type thermocouples (TCs). A separate TC was placed on each fuse and branch connector in each specimen assembly. Electrical and temperature data were obtained at 100 Hz and 0.5 Hz, respectively, and then analyzed in 1-minute bins. A T420 camera (FLIR Systems Inc.) was used for supplemental thermography imaging. An X3000 system (North Star Imaging Inc.) was used for X-ray computed tomography (XCT) to obtain the three-dimensional structures of the specimen assemblies (unaged and degraded). Brightfield microscopy was performed using a white-balanced VHX-5000 microscope (Keyence Corp).

IV. RESULTS AND DISCUSSION

A. Benchtop Comparison of Static and Dynamic Specimens

Fig. 1 compares assemblies tested simultaneously, including those: with external mechanical perturbation using the benchtop test fixture prototype ("dynamic"); and those without actuation ("static"). The dynamic specimen temperature of ~100°C is greater than the static specimen temperature of ~40°C despite the greater DC current to the static assembly. In successive experiments, dynamic specimens failed at 15 A at the fuse/branch connector connection. Localized arcing is suspected based on observations including smoke, heterogeneous melting of plastic, discolored metal pins, and increased fuse resistance. In successive experiments, static specimens failed at 35 A, with the greatest heating observed at the filament-based cartridge at the center of the discrete fuses. No additional visible indication was observed for failed static specimens, but an open circuit was verified for at least one of the fuses in each failed assembly. Temperature and current behavior like in Fig. 1 were observed in two successive static specimens, then two successive dynamic specimens, then two side-by-side static and dynamic benchtop experiments.

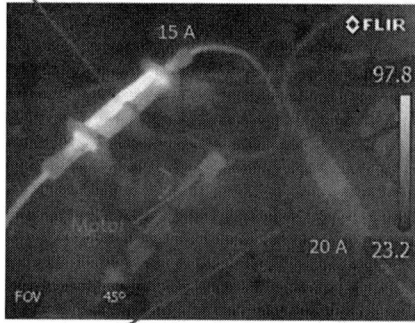

Fig. 1. Infrared image comparing sample assemblies in separate circuits. The maximum temperature of the dynamic sample (~100°C, for $\delta = 1$ mm) is greater than that of the static sample ($\delta = 0$ mm) with no external mechanical actuation (~40°C), despite a greater DC current to the static sample.

The temperature history for the dynamic specimen in Fig. 1 is shown in Fig. 2. The branch connectors (TC 106 and TC 108) and the adjoining discrete fuse (TC 107) both occasionally exceeded 100°C. While the second discrete fuse (TC 105) remained coldest overall, the rank order of the temperature for the other three components varied through the experiment. Supplemental infrared imaging, which included the TCs and the adjacent sample regions, identified localized temperatures exceeding 130°C. The specimen temperature in Fig. 2 is irregular, with spikes and broader peaks that do not always correspond to the electrical current increments. The voltage drop across the current-controlled specimen assembly was also irregular (not shown), with peaks occurring corresponding to the temperature spikes in Fig. 2. The TC data for the static assembly (not shown) is more monotonic, with the greatest temperature approaching 40°C in the fuses and 30°C in the branch connectors.

The specimen temperatures shown in Fig. 1 and Fig. 2 readily identify the significance of external mechanical strain on the durability of branch connector/fuse assemblies. Not only is the temperature greatly increased by mechanical perturbation, but the hottest location also occurs in a different component within the specimen assembly. The measured temperature in Fig. 2 identifies that the specimen temperature loses stability within the first day of the experiment. The data in Fig. 2 is the average temperature through 1 minute; binning and averaging reduce its variation relative to the instantaneous temperature. Temperature spikes are observed even when the current is not incremented, identifying the stochastic effect of mechanical perturbation. The temperatures in Fig. 2 are representative of the specimen temperature, but do not give the maximum specimen temperature because the thermocouples are positioned near but not exactly on the hottest location in each component.

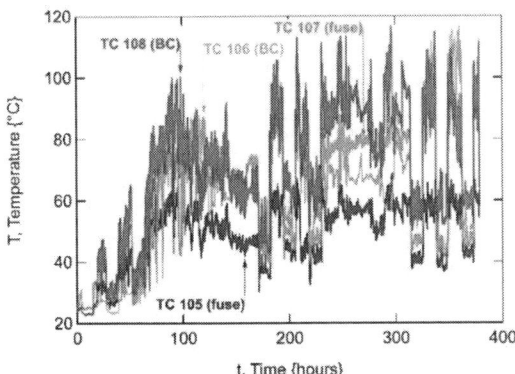

Fig. 2. Measured temperature for the dynamic sample assembly in Fig. 1. The thermocouples at the center of the discrete fuses ("fuse") are compared to those of the branch connectors ("BC"), as the current was initially incremented in steps and modulated occasionally to simulate the effect of light-level variation.

B. Assessing the Failure of Field and Accelerated Specimens

Failed specimens were examined iteratively: after testing, in their failed state; after milling the external plastic to reveal the internal metal components; and after extracting the convolute springs within the female metal pins. Fig. 3 compares specimens, field-failed (left) and after accelerated testing (right). An asymmetry is observed in (a), where the internal connection extends further on the right side (green arrow). Discoloration of the surface of the metal pins (originally silver in color) is seen in Fig. 3 (c) and (d). The Sn surface of the convolute springs is also discolored in Fig. 3 (e) and (f). Corrosion and bending of the members are observed along the length of the field-failed spring in Fig. 3 (e), whereas the spring from the dynamic specimen assembly is circumferentially deformed in Fig. 3 (f).

The failure analysis in Fig. 3 identifies a variety of degradation modes, including: oxidation (with discoloration from heating); inelastic deformation [longitudinally in Fig. 3 (e) and circumferentially in Fig. 3 (f)]; and additional corrosion in Fig. 3 (e). The most discolored metal pins and most affected convolute springs presumably reveal the hottest locations within the assembled components. More corrosion might be observed when the environmental stressors are applied during C-AST, going beyond the standard laboratory environment in the benchtop experiments in Fig. 3 (f).

V. SUMMARY

The C-AST method is being developed for use with BoS components. The use of external mechanical perturbation was

found to quickly and greatly affect the results of benchtop experiments. Failure analysis diagnosed the hottest location, presumably the location of greatest degradation: the connection between male and female metal pins in both the benchtop (with mechanical perturbation) and field-failed specimens. The validation of the accelerated testing relative to the field installations in this study remains an ongoing effort in addition to the present examination of specimens using C-AST.

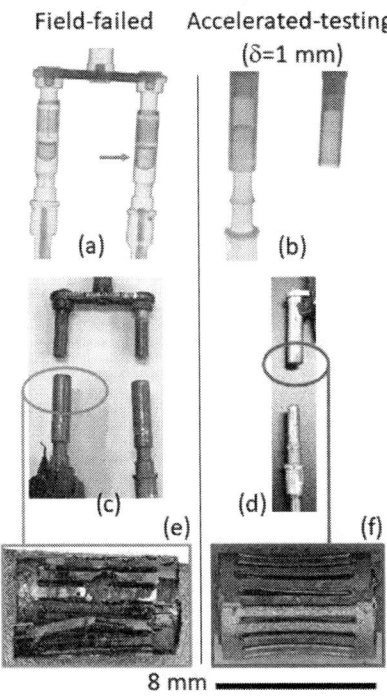

Fig. 3. Comparison of field-failed (left) and benchtop-failed (right, from Fig. 1 and Fig. 2) specimens, including: (a) and (b) X-ray computed tomography prior to removing the exterior plastic; (c) and (d) visual appearance of the interior metal components after removing the exterior plastic; and (e) and (f) optical micrographs of the convolute springs extracted from the failed connections.

ACKNOWLEDGMENTS

Funding was provided as part of the Durable Module Materials Consortium (DuraMAT), an Energy Materials Network Consortium funded under Agreement 32509 by the U.S. Department of Energy (DOE) Office of Energy Efficiency and Renewable Energy Solar Energy Technologies Office. This work was authored by the National Renewable Energy Laboratory, operated by Alliance for Sustainable Energy, LLC, for the U.S. Department of Energy under Contract No. DE-AC36-08GO28308.

REFERENCES

[1] Yang et. al., "Reliability model development for photovoltaic connector lifetime", Proc. IEEE PVSC, 2013, 139-144.

[2] L. Fiorentini, "PV fires experiences in Italy: from forensic activities to fire risk assessment of existing and new PV plants", Proc. PV Rel. Work., 2019, 192-233, https://www.nrel.gov/docs/fy20osti/77361.pdf.

[3] Connectors for DC-application in photovoltaic systems – Safety requirements and tests, IEC 62852, International Electrotechnical Commission, Geneva, Switzerland, 2014.

[4] Connectors for Use in Photovoltaic Systems, UL 6703, UL LLC, Northbrook, USA, 2017.

[5] Spataru et. al., "Combined-accelerated stress testing system for photovoltaic modules", Proc. IEEE PVSC, 2018, 3943-3948.

High Throughput Boron Emitter Formation from Pre-deposited APCVD BSG Layers for TOPCon Solar Cells

Marius Meßmer[1], Sattar Bashardoust[1], Udo Belledin[1], Sven Seren[2], Heiko Zunft[2], Sebastian Mack[1] and Andreas Wolf[1]

[1]Fraunhofer Institute for Solar Energy Systems, Heidenhofstr. 2, 79110 Freiburg, Germany
[2]Gebr. SCHMID GmbH, Robert-Bosch-Str. 32-36, 72250 Freudenstadt, Germany

Abstract—**This work presents an alternative boron diffusion approach for tunnel oxide passivated contact TOPCon solar cells enabling a highly increased throughput compared to the typically used BBr3 diffusion. We use APCVD BSG layers as the boron dopant source and combine them with a subsequent thermal anneal for dopant drive-in in a quartz tube furnace. Here, we use either a conventional single slot quartz boat configuration, or, for highly increased throughput, a vertical wafer stack configuration with the wafer surfaces in direct contact to each other. We investigate the emitter dark saturation current densities j_{0e} as well as the energy conversion efficiency of TOPCon solar cells fabricated for each configuration and compare the results to those of a BBr3 reference process.**

Keywords—*TOPCon, boron diffusion, high throughput, stack diffusion, APCVD*

I. INTRODUCTION

The passivated emitter and rear (PERC) solar cell [1, 2] on p-type monocrystalline-silicon is the working horse in the photovoltaic (PV) industry but passivated contacts mainly on n-type mono-silicon wafers are widely seen as the forthcoming technology and within this decade are expected to gain a considerably higher market share [3]. The PV industry is currently looking for a cost-effective way to transfer the tunnel oxide passivated contact (TOPCon) [4] solar cell concept into mass production. For small area laboratory type solar cells high energy conversion efficiencies of 26% and high open-circuit voltages V_{oc} up to 725 mV are reported [5, 6] and also from industry promising results are published [7–9]. Nevertheless, the TOPCon process still faces high cost, as apart from higher metallization cost, also boron-diffusion requires much higher temperatures and longer process times compared to the POCl3-diffusion used for PERC solar cells. Therefore, this work investigates an alternative high throughput approach for the cost effective boron emitter formation by implementing the stacked diffusion approach [10] enabling an approximately 3 to 6 times higher throughput for the thermal process [11].

II. EXPERIMENTAL

Dopant diffusion from pre-deposited dopant sources [12–17] are well known as well as stack diffusion [10]. We developed a stack diffusion process from pre-deposited atmospheric pressure chemical vapor deposition (APCVD) borosilicate glass (BSG)

layer [11], which we implement in the process sequence for fabrication of TOPCon solar cells (see Fig. 1). For all groups, n-type phosphorus-doped Czochralski-grown silicon (Cz-Si:P) wafers with M2 format and a base resistivity of $\rho_b \approx 1$ Ωcm serve as starting material. After saw damage etching (SDE) and alkaline texturing, tube furnace diffusion using boron tribromide BBr3 as liquid dopant precursor is performed for group 1 serving as the reference group. For group 2 and 3 an APCVD process forms the dopant source on one side of the wafer. Here, we deposit 20 nm BSG layer capped by 20 nm silicon oxide SiO_x layer. For the BSG deposition we use a ratio

$$R_{B2H6} = \phi_{B2H6} / \phi_{SiH4} \qquad (1)$$

for the gas flow of diborane B_2H_6 and silane SiH_4, ϕ_{B2H6} and ϕ_{SiH4}, respectively, of $R_{B2H6} = 0.125$. This results in an expected boron concentration C_B within the BSG layer of $C_B \approx 4$ wt% from comparison to literature data from Kurachi and Yoshioka [12]. The in-diffusion of the boron dopants is shifted to a subsequent thermal tube furnace process. For group 2, the wafers are placed next to each other in dedicated slots in a conventional quartz boat, whereas for group 3, the wafers are stacked vertically in small groups of 4 wafers. In this

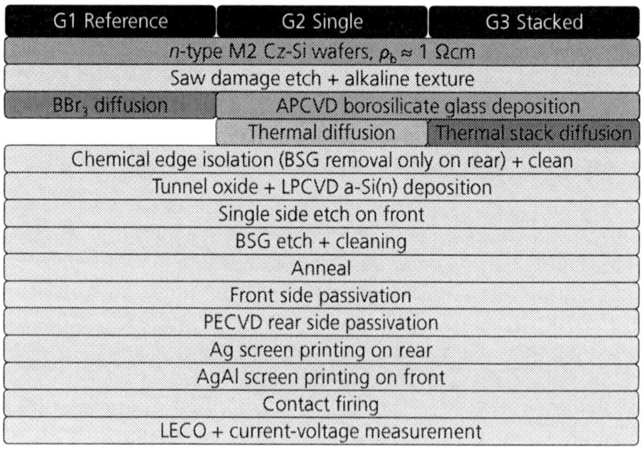

Fig. 1 Process scheme for the fabrication of TOPCon solar cells: Group 1 represents the reference process with boron diffusion from boron tribromide BBr3. Groups 2 and 3 use APCVD borosilicate glass BSG layers as a dopant source with the in-diffusion process in single slot or stack configuration, respectively.

experimental setup, only 4 stacked wafers in the diffusion process are possible but Refs. [18–21] already showed promising results with a higher amount of stacked wafers. After the diffusion processes, a single side etching process removes the BSG and the emitter on the rear side of the wafer, while the BSG layer on the front side is kept on the surface. After a cleaning step, a tube furnace process first grows a thin tunnel oxide followed by a low-pressure chemical vapor deposition (LPCVD) process forming the phosphorus-doped amorphous-silicon (a-Si(n)) layer. A single side etch (SSE) process using atmospheric etching [22] then removes the parasitic a-Si(n) layer on the front side of the wafer. During the SSE, the BSG layer on the front side serves as an etch barrier [22]. Prior to the annealing of the a-Si(n), the BSG layer is removed, and the surfaces are wet-chemically cleaned. Then, the front side is passivated growing a thin low-temperature thermal oxide of 1-2 nm thickness in a tube furnace, followed by plasma enhanced chemical vapor deposition (PECVD) in a tube furnace process of a stack of aluminum oxide (AlO$_x$) and silicon nitride (SiN$_x$). The rear side is also passivated in a PECVD tube furnace process with a SiN$_x$ layer. Afterwards, metallization is performed by screen-printing of a silver (Ag) grid on the rear side and a silver-aluminum (AgAl) grid on the front side. Then, the contacts are formed in an industrial conveyor belt furnace. As a final process step, a laser-enhanced contact optimization (LECO) [23, 24] process is performed prior to the current-voltage measurements which yield the solar cell performance.

Besides the solar cells, we fabricate test samples for characterization purposes. For the determination of the emitter dark saturation current density j_{0e}, symmetrical lifetime samples are used. For this, textured n-type material with a base resistivity of $\rho_b \approx 6$ Ωcm serves as starting material. Then, the boron diffusion processes, BBr$_3$ and APCVD (both side BSG deposition) + thermal diffusion process (single slot and vertically stacked), are performed. After BSG etching and cleaning, the samples are annealed along with the cells and both sides passivated with the front side passivation stack. After firing, QSSPC measurements yield the effective lifetime on 5 spots over the wafer. Then, j_{0e} is obtained by the procedure described in [25].

III. RESULTS AND DISCUSSION

Introducing the APCVD technology into the TOPCon manufacturing sequence starts with the development of a suitable thermal diffusion process for the in-diffusion from the pre-deposited BSG layers. For the single slot and vertically stacked wafers, the same thermal diffusion process is performed. The temperature-time profile for this thermal diffusion process is close to the one from the reference BBr$_3$ diffusion process and aims for an identical sheet resistance and doping profile. Fig. 2 shows the emitter sheet resistance for the reference BBr$_3$ diffusion process as well as for both configurations of the thermal diffusion process with APCVD BSG layers measured by four-point probe (4pp) in a 10 x 10 pattern over the wafer surface. The reference BBr$_3$ process yields an emitter sheet resistance $R_{sh} = (122.0 \pm 5.0)$ Ω/sq. For the APCVD single slot configuration the thermal diffusion processes yields $R_{sh} = (112.3 \pm 6.5)$ Ω/sq, while stacking shows a stronger in-diffusion of the dopants with $R_{sh} = (97.5 \pm 3.6)$ Ω/sq. Details on the diffusion process development and the differences between

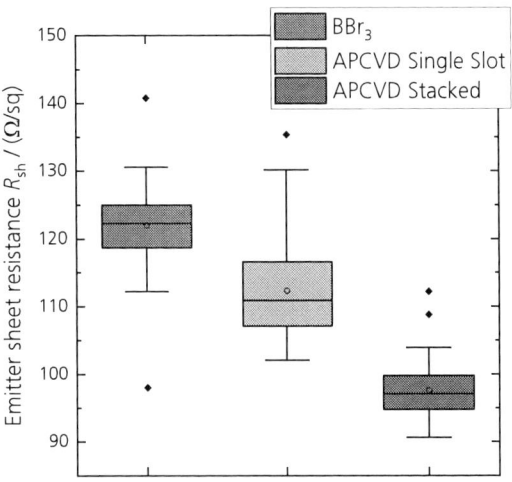

Fig. 2 Emitter sheet resistances R_{sh} after the diffusion process for the different approaches measured by four-point probe (4pp) in a 10 x 10 pattern on one wafer per group.

single slot and stacked configuration will be published in an upcoming journal paper.

Fig. 3 depicts the determined j_{0e} results on symmetrical lifetime samples. The reference BBr$_3$ diffusion reaches $j_{0e} = (26 \pm 3)$ fA/cm^2. With APCVD BSG layers and subsequent thermal diffusion process in single slot configuration, j_{0e} slightly increases to $j_{0e} = (32 \pm 4)$ fA/cm^2, whereas the stack configuration further increases j_{0e} to $j_{0e} = (36 \pm 5)$ fA/cm^2. This increase is at least partly related to the higher doping level and thus, increased Auger recombination for the single slot and stacked configuration (compare Fig. 2).

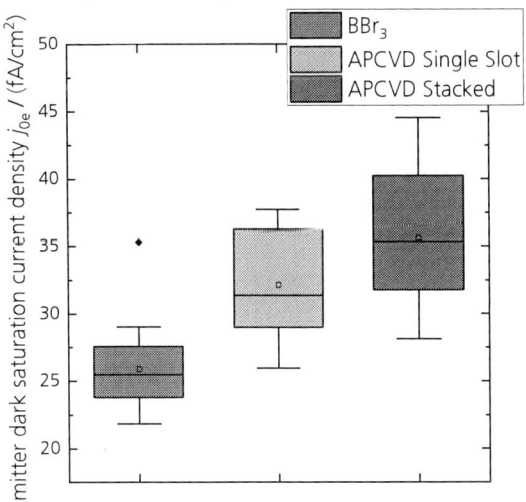

Fig. 3 Emitter dark saturation current density j_{0e} determined from QSSPC measurement (5 points over the wafer surface) for the different emitter formation processes, the reference processing using BBr$_3$ as well as BSG deposition by APCVD with the thermal diffusion process in single slot and stacked configuration.

The implementation of boron diffusion from APCVD BSG layers into solar cell processing in combination with single slot and stacked configuration yields the solar cell results shown in Table 1. The reference sequence using BBr$_3$ diffusion yields a

978-1-7281-6118-1/22 $31.00 © 2022 IEEE

Table 1 IV parameter for the different emitter formations, reference processing using BBr$_3$ as well as BSG deposition by APCVD with the thermal diffusion process in single slot and stacked configuration measured by an industrial cell tester.

	Diffusion Configuration	IV parameter			
		V_{oc} / mV	j_{sc} / mA/cm^2	FF / %	η / %
Median	BBr$_3$ Reference	695 ± 2	40.3 ± 0.1	80.8 ± 0.3	22.6 ± 0.2
	APCVD-BSG Single Slot	693 ± 2	40.2 ± 0.1	80.9 ± 0.2	22.5 ± 0.1
	APCVD-BSG Stacked	683 ± 6	40.1 ± 0.2	80.2 ± 0.5	21.9 ± 0.3
Best	BBr$_3$ Reference	698	40.4	81.1	22.9
	APCVD BSG Single Slot	696	40.2	81.0	22.7
	APCVD BSG Stacked	694	40.2	80.6	22.5

median energy conversion efficiency of $\eta = (22.6 \pm 0.2)$% with a peak energy conversion efficiency of $\eta = 22.9$%. Performing the boron diffusion out of APCVD BSG layers in a subsequent thermal diffusion process in single slot configuration yields a median energy conversion efficiency of $\eta = (22.5 \pm 0.1)$% which is slightly lower than the reference. This results from a slightly decreased open-circuit voltage V_{oc}. Here, also high peak energy conversion efficiency of $\eta = 22.7$% are reached. Using stacked wafers with APCVD BSG layers results in a median energy conversion efficiency of $\eta = (21.9 \pm 0.3)$% with decreased FF and V_{oc} by 10 mV in comparison to the single slot configuration. Nevertheless, a high peak energy conversion efficiency of $\eta = 22.5$% is achieved for this group. Further investigations are necessary to identify the reason for the loss in V_{oc} for the stacked wafers compared to the single slot configuration, as the slight increase in j_{0e} due to stacking shown in Fig. 3 would reduce V_{oc} by only 1 to 2 mV between Groups 2 and 3. This holds for the best cell results but not for the median values. In other applications, results for stacking of wafers during thermal processing from [18–21] showed no significant negative influences on the emitter dark saturation current density.

IV. SUMMARY AND OUTLOOK

In this work, we present an approach for high throughput boron emitter formation with pre-deposited APCVD BSG layers for TOPCon solar cells. Emitter dark saturation current densities of $j_{0e} = (36 \pm 5)$ fA/cm^2 at $R_{sh} = (97.5 \pm 3.6)$ Ω/sq for stacked samples are reached, which shows slightly higher j_{0e} than for unstacked single slot wafers. With this approach, we reach up to $\eta = 22.5$% efficiency for TOPCon solar cells with a highly increased throughput potential. Our results still show a performance gap compared to BBr$_3$-diffused reference solar cells, which requires further investigation. With further optimization of the thermal diffusion process for these layers, further efficiency improvements will be possible.

In an upcoming journal paper, we will present cost of ownership COO calculations for the approach in comparison to state-of-the-art processing using BBr$_3$ diffusion. Further, experimental details on the development of the boron emitter formation with the APCVD BSG layers and the subsequent thermal diffusion process will be shown.

ACKNOWLEDGMENT

The authors would like to thank the APCVD development team at Gebr. SCHMID GmbH and all colleagues at the Fraunhofer ISE PV-TEC. This work was funded by the German Federal Ministry of Economic Affairs and Climate Action within the research project "NextTec" under contract number 03EE1001A.

REFERENCES

[1] A. W. Blakers, A. Wang, A. M. Milne, J. Zhao, and M. A. Green, "22.8% efficient silicon solar cell," *Appl. Phys. Lett.*, vol. 55, no. 13, pp. 1363–1365, 1989.

[2] R. Preu, E. Lohmüller, S. Lohmüller, P. Saint-Cast, and J. M. Greulich, "Passivated emitter and rear cell—Devices, technology, and modeling," *Applied Physics Reviews*, vol. 7, no. 4, p. 41315, 2020.

[3] ITRPV consortium, "International Technology Roadmap for Photovoltaic (ITRPV): Results 2020," 2021.

[4] F. Feldmann, M. Bivour, C. Reichel, M. Hermle, and S. W. Glunz, "Passivated rear contacts for high-efficiency n-type Si solar cells providing high interface passivation quality and excellent transport characteristics," *Solar Energy Materials and Solar Cells*, vol. 120, pp. 270–274, 2014.

[5] A. Richter, J. Benick, F. Feldmann, A. Fell, M. Hermle, and S. W. Glunz, "n-Type Si solar cells with passivating electron contact: Identifying sources for efficiency limitations by wafer thickness and resistivity variation," *Solar Energy Materials and Solar Cells*, vol. 173, pp. 96–105, 2017.

[6] F. Haase, C. Hollemann, S. Schäfer, A. Merkle, M. Rienäcker, and J. Krügener et al., "Laser contact openings for local poly-Si-metal contacts enabling 26.1%-efficient POLO-IBC solar cells," *Solar Energy Materials and Solar Cells*, vol. 186, pp. 184–193, 2018.

[7] D. Chen, Y. Chen, Z. Wang, J. Gong, C. Liu, and Y. Zou et al., "24.58% total area efficiency of screen-printed, large area industrial silicon solar cells with the tunnel oxide passivated contacts (i-TOPCon) design," *Solar Energy Materials and Solar Cells*, vol. 206, p. 110258, 2020.

[8] *Press Release - Trina Solar refreshes i-TOPCon cell efficiency by 25.5%*. Trina Solar, 2022.

[9] *Press Release - JinkoSolar achieves 25.7% efficiency for n-type TOPCon solar cell*. JinkoSolar Holding Co., Ltd., 2022.

[10] J. Horzel, D. Franke, G. Blendin, M. Jahn, and W. Schmidt, "Verfahren zur Ausbildung eines Dotierstoffprofils," WO 2010/066626 A2, Jun 17, 2010.

[11] M. Meßmer, S. Bashardoust, U. Belledin, B. Kafle, B. S. Goraya, and S. Seren et al., "Stack Diffusion Process for

Cost- and Energy-efficient Boron Emitter Formation," to be published.

[12] I. Kurachi and K. Yoshioka, "Investigation of Boron Solid-Phase Diffusion from BSG Film Deposited by AP-CVD for Solar Cell Application," (en), 2012.

[13] R. Cabal, J. Jourdan, B. Grange, Y. Veschetti, and D. Heslinga, "Investigation of the Potential of Boron Doped Oxide Deposited by PECVD - Application to Advanced Solar Cells Fabrication Processes," (eng), 2009.

[14] S. Gloger, A. Herguth, J. Engelhardt, G. Hahn, and B. Terheiden, "A 3-in-1 doping process for interdigitated back contact solar cells exploiting the understanding of co-diffused dopant profiles by use of PECVD borosilicate glass in a phosphorus diffusion," *Prog. Photovolt: Res. Appl.*, vol. 24, no. 7, pp. 955–967, 2016.

[15] F. Book, H. Knauss, C. Demberger, F. Mutter, and G. Hahn, "Phosphorous Doping from APCVD Deposited PSG," (eng), 2016.

[16] Y. Schiele, F. Book, C. Demberger, K. Jiang, and G. Hahn, "Co-Diffused APCVD Boron Rear Emitter with Selectively Etched-Back FSF for Industrial n-Type Si Solar Cells," (eng), 2014.

[17] S. Meier, S. Wiesnet, S. Mack, S. Werner, S. Maier, and S. Unmüßig et al., "Co-Diffusion for p-Type PERT Solar Cells Using APCVD BSG Layers as Boron-Doping Source," (eng), 2016.

[18] M. Meßmer, S. Lohmüller, F. Braun, J. Weber, and A. Wolf, "High Throughput Solar Cell Processing by Oxidation of Wafer Stacks," in *38th EU PVSEC*, 2021.

[19] M. Meßmer, S. Lohmüller, J. Weber, A. Piechulla, S. Nold, and J. Horzel et al., "High Throughput Low Energy Industrial Emitter Diffusion and Oxidation," in *37th EU PVSEC*, online, 2020, pp. 370–377.

[20] M. Meßmer, S. Lohmüller, J. Weber, and A. Wolf, "Industrial High Throughput Emitter Formation and Thermal Oxidation for Silicon Solar Cells by the High Temperature Stack Oxidation Approach," *Phys. Status Solidi A*, 2021.

[21] M. Meßmer, A. Wolf, M. Zimmer, F. Meyer, G. Hoppe, and A. Lorenz et al., "Increasing the Throughput of Industrial PERC Production," *Photovoltaics International*, no. 47, 2021.

[22] B. Kafle, S. Mack, C. Teßmann, S. Bashardoust, L. Clochard, and E. Duffy et al., "Atmospheric Pressure Dry Etching of Polysilicon Layers for Highly Reverse Bias‐Stable TOPCon Solar Cells," *Sol. RRL*, p. 2100481, 2021.

[23] T. Fellmeth, H. Höffler, S. Mack, E. Krassowski, K. Krieg, and B. Kafle et al., "Laser Enhanced Contact Optimization on iTOPCon Solar Cells," *Prog. Photovolt: Res. Appl.*, to be published 2022.

[24] R. Mayberry, K. Myers, V. Chandrasekaran, A. Henning, H. Zhao, and E. Hofmüller, "Laser Enhanced Contact Optimization (LECO) and LECO-Specific Pastes - A Novel Technology for Improved Cell Efficiency," *36th European Photovoltaic Solar Energy Conference and Exhibition*, 2019.

[25] A. Kimmerle, J. Greulich, and A. Wolf, "Carrier-diffusion corrected J0-analysis of charge carrier lifetime measurements for increased consistency," *Solar Energy Materials and Solar Cells*, vol. 142, pp. 116–122, 2015.

Uncertainty in Annual Energy Resulting from Uncertain Irradiance Measurements

Clifford W. Hansen[1], Aaron Scheiner[2]

1 Sandia National Laboratories, Albuquerque, NM 87185-1033, USA.

2 Rutgers University, New Brunswick, NJ 08901, USA

Abstract— We report an analysis quantifying the contribution to uncertainty in annual energy projections from uncertainty in ground-measured irradiance. Uncertainty in measured irradiance is quantified for eight instruments by the difference from a well-maintained, secondary standard pyranometer which is regarded as truthful. We construct a statistical model of irradiance uncertainty and apply the model to generate a sample of 100 annual time series of irradiance for each instrument. The sample is propagated through a common performance model for a reference photovoltaic system to quantify variation in annual energy. Although the measured irradiance varies from the reference by a few percent (standard deviation of 1-2%) the uncertainty in annual energy is on the order of a fraction of one percent. We propose a model for a factor that represents uncertainty in modeled annual energy that arises from uncertainty in ground-measured irradiance.

Keywords— photovoltaic, energy modeling, uncertainty

I. INTRODUCTION

Financing institutions often require engineers to estimate uncertainty in projected annual energy for a proposed photovoltaic (PV) energy system in order to establish the project's risk of loan repayment. Annual energy production is estimated by applying a sequence of models to weather data (e.g., irradiance, air temperature) representative of the proposed system's location and operating condition. Uncertainty in annual energy production arises from uncertainties in the weather data, models or system characteristics.

The uncertainty in annual energy is often estimated by applying multipliers to a base estimate of annual energy (e.g. [1]) although more complex methods have been proposed (e.g. [2]). Here we consider the approach outlined in [1] as a practical approach to quantifying uncertainty in annual yield Y:

$$Y = [\sum_t f(W(t), P)] \times \prod_{i=1}^{M}(1 - \Delta_i) \qquad (1)$$

In Eq. 1, f represents a performance model for the PV system of interest, $W(t)$ is a time-indexed vector of weather inputs representing a typical or base year, P is a vector of parameters for the performance model. The summation of $f(W(t), P)$ provides the annual energy yield for the base year. Uncertainty in this deterministic quantity is described by a set of uncertainty factors, Δ_i, each of which represents the contribution to uncertainty in annual yield of an independent process,

parameter or data source. Each factor Δ_i is quantified as a fraction of annual energy yield. When each Δ_i is regarded as a random variable, Y is also a random variable and the result of Equation 1 is a distribution of annual energy yield.

In this paper, we quantify a factor Δ_i representing uncertainty in projected annual energy arising from uncertainty in ground-measured global horizontal irradiance (GHI).

II. DATA SOURCE AND DATA PREPARATION

We downloaded measured GHI for calendar year 2020 for a total of nine instruments (Table 1) at the National Renewable Energy Laboratory's (NREL) Solar Radiation Research Laboratory in Golden, CO, USA, from the Measurement and Instrumentation Data Center (MIDC [3]. Each GHI instrument is regularly calibrated and maintained and is located in close proximity to all other instruments. We regard these GHI data as among the best achievable measurements for each type of instrument. In addition, we downloaded measured diffuse horizontal irradiance (DHI) measured by a horizontal Kipp and Zonen CMP22 pyranometer with a shade ball, direct normal irradiance (DNI) measured by a Kipp and Zonen DHP1 pyrheliometer, snow depth (cm), precipitation (cm) and air temperature (°C).

One instrument, a Kipp and Zonen CMP22 pyranometer is selected and regarded as the "true" value of GHI. This selection is somewhat arbitrary and not critically important, because our intent is to quantify the effect on annual energy projections of variation of other GHI instruments from the reference. Table I lists the selected irradiance instruments, and the mean and standard deviation of the percent difference in GHI at each time relative to the reference instrument. Instruments are grouped by accuracy class and classes are listed in order of decreasing accuracy. Two other CMP22 instruments (CMP22-1 and CMP22-2) are included, as well as two Licor 200 pyranometers, one of which has a custom temperature adjustment (LI-200R). Nomenclature follows that used by the MIDC to aid in reproducing our analysis. Statistics in Table 1 are not weighted by irradiance and thus a mean difference of e.g. −0.28% for the LI-200 instrument does not imply that the annual insolation would differ by the same percent. Relative differences in GHI are generally more variable for instruments of lower accuracy (Figure 1).

TABLE I.   SUMMARY OF GHI INSTRUMENTS AND RELATIVE DIFFERENCES FROM REFERENCE

Instrument	Mean	St. Dev.	Comments
CMP22	-	-	Reference, sec. standard
CMP22-1	0.335%	1.087%	Sec. standard
CMP22-2	0.191%	1.046%	Sec. standard
PSP	0.208%	1.989%	Sec. standard
SPP	0.136%	1.219%	Sec. standard
CMP11	−0.439%	1.422%	Class A
SPLite2	1.03%	1.930%	Economical
LI-200R	0.533%	1.852%	Economical
LI-200	−0.277%	2.323%	Economical

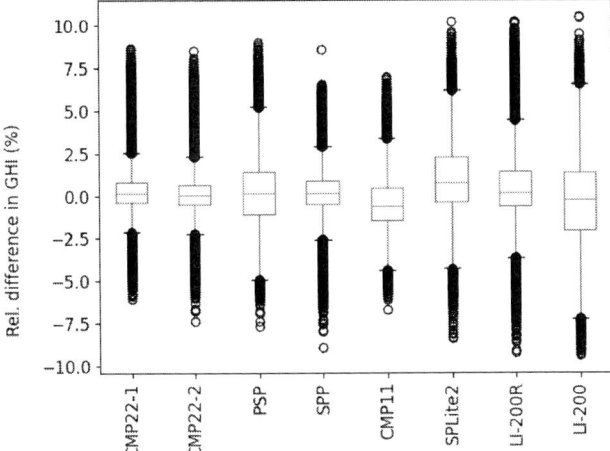

Fig. 1. Relative differences from reference GHI for each instrument.

Data are recorded at 1-minute intervals. For quality checking we also downloaded diffuse horizontal irradiance (DHI for the "Diffuse CMP22-2" instrument), direct normal irradiance (DNI for the "Direct CHP1-1" instrument), solar zenith, precipitation (mm), and snow depth (cm). From the full year of data, we selected the subset where:

- Solar elevation exceeds 5 degrees.

- Recorded GHI is within 5% of GHI estimated as DHI + DNI × cos(Z), where Z is solar zenith.

- Snow depth is 0 cm, and precipitation is less than 2 mm.

These criteria select data where it is unlikely that shadows or other external factors cause differences between instruments.

III. METHODOLOGY

Uncertainty is present in each measurement of GHI in an instrument's time series. Our goal is to understand how the uncertainty in these measurements translates to uncertainty in annual AC energy for a PV system.

To elucidate the uncertainty in annual AC energy, for each instrument we construct many simulated time series of GHI with statistics consistent with the single, measured time series of GHI. We define a reference PV system and a performance model

[1] https://pkg.robjhyndman.com/forecast/

for the system. Keeping all other model inputs (e.g. air temperature) the same for each instrument, the simulated time series of GHI provide an estimate of the distribution of annual AC energy.

To simulate time series of GHI for each instrument, we create a model for the relative difference in GHI between each instrument and the reference GHI. The model ensures that simulated time series have similar distributions of values and similar long-memory autocorrelation that is observed in the measurements. For each instrument k the uncertainty propagation is done as follows:

Step 1: Compute the relative difference $X_k(t)$ in GHI from the reference instrument (subscript R):

$$X_k(t) = (G_k(t) - G_R(t)) / G_R(t) \qquad (1)$$

Step 2: Bin the relative differences $X_k(t)$ by irradiance $G_k(t)$ (in increments of 100 W/m^2), air temperature (in increments of 10°C) and solar zenith (in increments of 10 degrees). Within each bin, fit an empirical cumulative distribution function (ECDF). Record each GHI value's rank (percentile) within the appropriate bin's ECDF. Binning is necessary because the distribution of relative difference depends on GHI, air temperature and zenith (e.g., Figure 2). The rank transformation allows for simulations that transition between bins.

Step 3: Use the logit function to transform each rank from the range [0, 1] to (−∞, ∞) to avoid having constraints on fitting of the time series model.

Step 4: Fit an autoregressive fractionally integrated moving average (ARFIMA) model to the time series of transformed ranks, using the `arfima` function in the R package `forecast` version 8.15[1]. The ARFIMA form is chosen because the time series of relative differences exhibits autocorrelation at long lags (Figure 3). Model order is selected automatically by the `arfima` function.

Step 5: Generate a sample of 100 independent time series of synthetic transformed ranks, and invert the logit and rank transformations and use Eq. 1 to recover 100 independent time series of synthetic GHI, $\tilde{G}_{k,i}(t), i=1,\ldots,100$.

Step 6: Simulate annual energy $E_{AC,k}(t)$ for each synthetic time series $\tilde{G}_{k,i}(t)$ for a reference PV system at Golden, CO, USA. The reference system is set at 35-degree tilt, azimuth 180°, 1 kW$_{DC}$ and 1 kW$_{AC}$ capacities, with a power temperature coefficient of −0.4%/°C. Simulations are done with pvlib-python [4]. DC and AC output are modeled using functions based on the PVWatts v5 model. Cell temperature is modeled using the Sandia Array Performance Model (SAPM) and coefficients representative of a glass-polymer module on open racking.

Step 7: Compute the relative difference between $E_{AC,k}(t)$ and the annual energy simulated for the reference system using the reference GHI:

$$D_k(t) = (E_{AC,k}(t) - E_{AC,R}(t)) \, / \, E_{AC,R}(t) \qquad (1)$$

The mean and standard deviation of $D_k(t)$ describe the uncertainty in annual energy arising from uncertainty in the GHI measurements for instrument k.

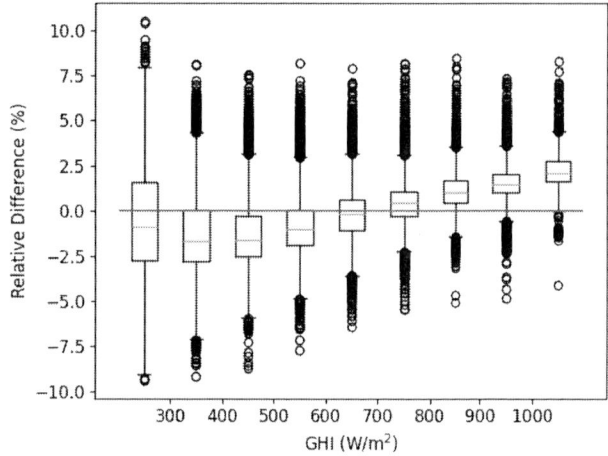

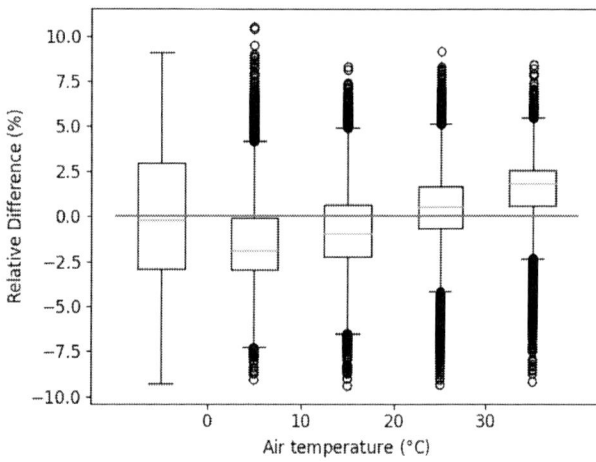

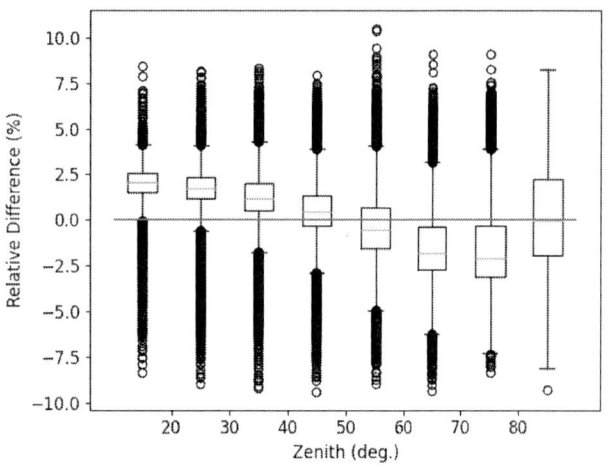

Fig. 2. Relative differences in GHI binned by a) GHI b) air temperature and c) solar zenith for the LI-200 instrument.

IV. VERIFICATION OF GHI SIMULATIONS

We compared statistics for the relative difference in GHI for the simulations to statistics for the relative difference in observed GHI. Figure 3 shows that autocorrelation matches closely to about lag 10; beyond lag 10 the simulated GHI's autocorrelation exceeds that of the observations. Greater autocorrelation at long lags will tend to overestimate somewhat the variance in annual energy, because the simulated GHI will tend to change less rapidly than the observed GHI and thus remain away from central values for longer periods. We compared the distributions of the relative difference in GHI within each irradiance, temperature and zenith bin (e.g., Figure 4). The distributions are similar for the simulated and observed GHI. These comparisons verify that the statistical model produces GHI time series with statistics that are consistent with the measurements.

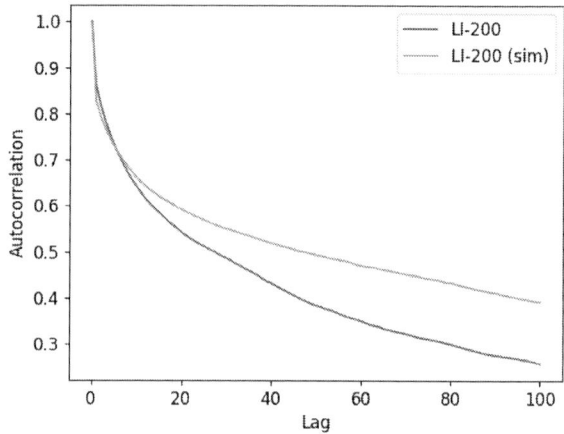

Fig. 3. Autocorrelations for the relative difference in GHI for the LI-200 instrument and for one simulated time series of GHI.

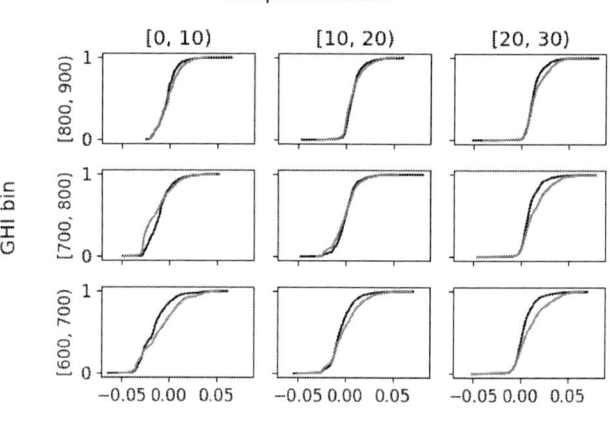

Fig. 4. Distributions of relative difference in GHI: measured (black) and simulated (red) for the LI-200 instrument.

V. RESULTS

Table II summarizes statistics for the distribution of relative difference in AC energy computed for each instrument from the 100 simulated time series of GHI. For each instrument, the distribution of relative difference in AC energy is well-described by a normal distribution (e.g., Figure 5). Confidence intervals indicate that variation among instruments is greater than the uncertainty due to the Monte Carlo generation of simulated GHI time series. Mean relative difference in AC energy is comparable in magnitude to and roughly correlated with the mean relative difference in irradiance (Figure 6a). The correlation is expected because of the strong correlation between insolation (integrated irradiance) and annual AC energy. Interestingly, the mean relative differences for instruments in higher accuracy classes (CMP22-1, CMP22-2, PSP and SPP) are similar to those for economical pyranometers. We hypothesize that this similarity results from careful and accurate calibration of each instrument to a common baseline. The similarity is noteworthy given the conventional wisdom that lower confidence should be assigned to data collected with economical pyranometers, even with proper calibration and maintenance. Even with proper calibration, variation among two pairs of similar instruments (CMP22-1 and CMP22-2, and LI-200 and LI-200R) indicates that instrument calibration or inherent differences can be as great as the variation between instruments of different manufacture.

TABLE II.　SUMMARY OF RELATIVE DIFFERENCES IN AC ENERGY

Instrument	Mean	St. Dev.
CMP22-1	0.211%	0.218%
CMP22-2	-0.007%	0.271%
PSP	-0.283%	0.515%
SPP	0.316%	0.432%
CMP11	-0.476%	0.469%
SPLite2	0.674%	0.618%
LI-200R	0.212%	0.374%
LI-200	-0.220%	0.559%

Except for the CMP22-1 and CMP22-2 instruments, the standard deviation of relative differences in AC energy (Figure 6b) is comparable for all other instruments. The relatively low magnitude and small variability for the CMP22-1 and CMP22-2 instruments is likely due to their similarity with the CMP22 instrument selected as the reference GHI. Variance in the simulations of AC energy arises primarily from variance in the relative differences from the reference GHI of the subject instrument. The standard deviation in AC energy is roughly 30% of the standard deviation in irradiance. The reduction in variance likely results from the weighting inherent in the calculation of AC energy (i.e., higher irradiance values contribute more to annual AC energy than to lower irradiance values) combined with the smaller variance in relative differences in irradiance at higher irradiance (e.g., Figure 2a).

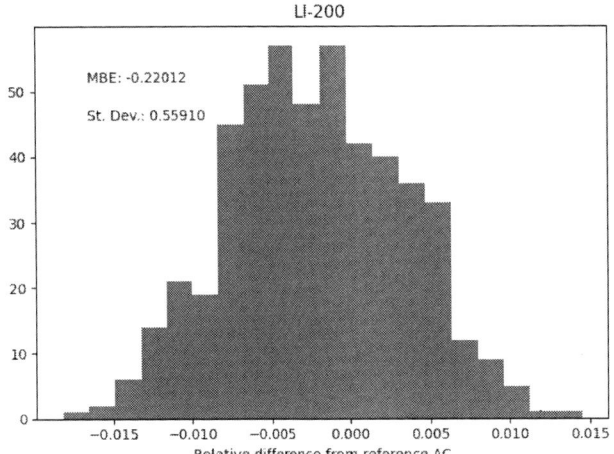

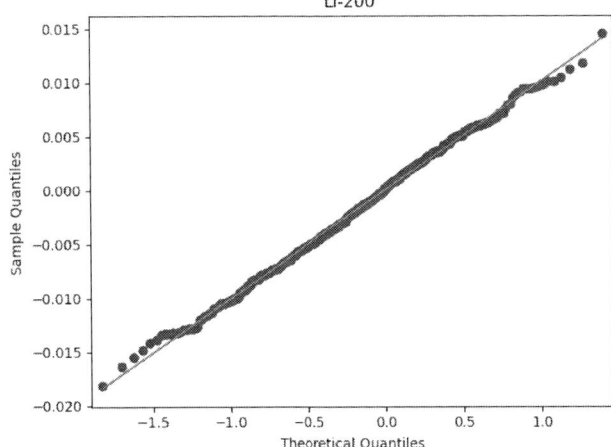

Fig. 5. Distribution of relative difference in AC energy for the LI-200 instrument.

VI. MODEL FOR THE UNCERTAINTY FACTOR

We propose the following two-level model for a factor Δ (see Equation 1) which represents uncertainty in annual AC energy resulting from uncertainty in ground-measured irradiance:

- The uncertainty factor Δ is described by a normal distribution.

- The mean of the normal distribution is sampled from a uniform distribution with a range from −0.4% to 0.4%.

- The standard deviation of the normal distribution is sampled from a uniform distribution [0.35%, 0.6%].

The distributions for the mean and standard deviation are defined by excluding the SPLite2 instrument as an outlier in both mean and standard deviation, and by excluding the CMP22-1 and CMP22-2 instruments from the distribution for standard deviation, due to their similarity to the reference GHI instrument. We regard the remaining instruments as representative of all pyranometers with proper calibration. The results in Figure 6 do not appear to support assigning different uncertainty values for higher or lower accuracy instruments.

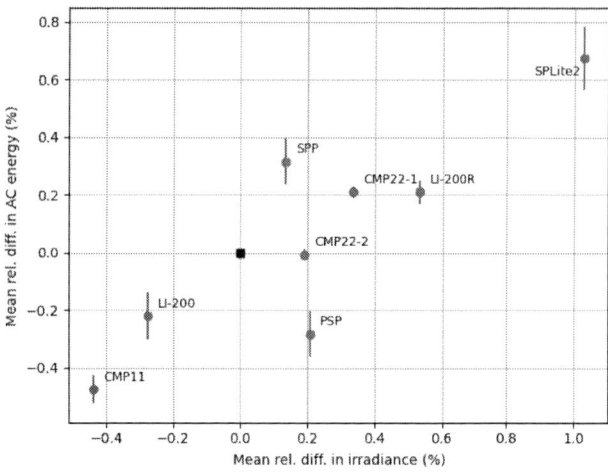

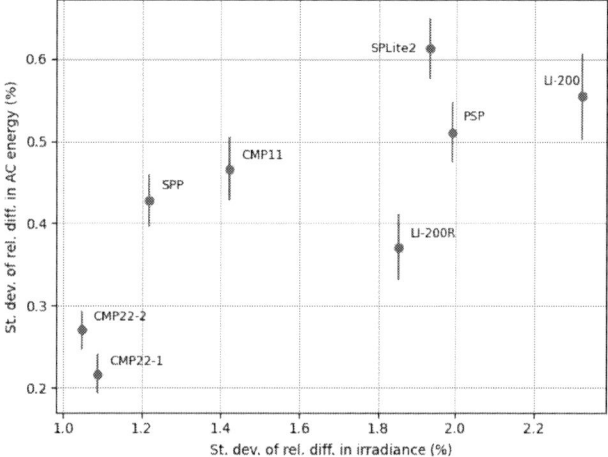

Fig. 6. Mean (a) and standard deviation of relative difference in AC energy compared to relative difference in irradiance. Error bars are confidence intervals (alpha of 0.05) about the overall mean of 5 replicates each of size 100. Error bars indicate uncertainty in overall mean due to sample size.

VII. SUMMARY AND CONCLUSIONS

Our analysis found that uncertainty in annual AC energy arising from uncertainty in measured GHI can be modeled by a normal distribution. Even for well-calibrated instruments, estimated AC energy can be biased relative to a reference calculation. We found that instruments regarded as less precise (e.g., silicon photodiodes similar to the LI-200) did not necessary correspond with greater uncertainty in annual AC energy. The narrowest uncertain ranges are observed for instruments CMP22-1 and CMP22-2 that are most like the reference CMP22 instrument.

When using an annual factor approach to representing the total uncertainty in AC energy (see Equation 1, and [1] Section 3.2), Table II offers distribution parameters (mean and standard deviation) for several classes of irradiance instruments. However, the distribution means in particular could vary significantly from the mean in Table II when uncertainty in instrument calibration, or instrument to instrument variation, is taken into account. Consequently, we recommend a two-level

model for the uncertainty factor, as described in Section VI, that does not distinguish between the type of irradiance instrument.

VIII. ACKNOWLEDGMENT

Sandia National Laboratories is a multimission laboratory managed and operated by National Technology and Engineering Solutions of Sandia, LLC., a wholly owned subsidiary of Honeywell International, Inc., for the U.S. Department of Energy's National Nuclear Security Administration under contract DE-NA-0003525.

IX. REFERENCES

[1] C. Reise, B. Müller. "Uncertainties in PV System Yield Predictions and Assessments," Intl. Enegy Agency Report IEA-PVPS T13-12:2018, 2018. ISBN 978-3-906042-51-0. 2018.

[2] C. W. Hansen, C. E. Martin. "Photovoltaic System Modeling: Uncertainty and Sensitivity Analyses," Sandia National Laboratories Report SAND2015-6700. 2015.

[3] A. Andreas, T. Stoffel. NREL Solar Radiation Research Laboratory (SRRL): Baseline Measurement System (BMS); Golden, Colorado (Data); NREL Report No. DA-5500-56488. 1981. http://dx.doi.org/10.5439/1052221.

[4] W. F. Holmgren, C. W. Hansen, M. A. Mikofski. "pvlib python: a python package for modeling solar energy systems." Journal of Open Source Software, 3(29), 884, (2018). https://doi.org/10.21105/joss.00884.

Impact of Photovoltaic Plant Tilt on the Need for Storage

Russell K Jones, Sarah Kurtz

Jones Solar Engineering, Manhattan Beach, CA, United States

University of California at Merced, Merced, CA, United States

This paper explores the application of optimizing tilt of photovoltaic (PV) plants as a statewide strategy to best match the California statewide load over the year and thus minimize storage requirements for a carbon-free grid. Through a simple cost model and energy balance model examining PV + storage in isolation, we show that, even though horizontal trackers produce the lowest cost electricity when the timing of generation is ignored, high-tilt PV plants have the potential to reduce overall system cost substantially by reducing the required storage capacity and by better utilizing surplus electricity. California should consider tilted PV configurations in capacity expansion planning and consider PV electricity pricing or incentives that encourage new PV installations that better match the seasonal load to reduce storage requirements.

Precursor Ink Design for Scalable Fabrication of Perovskite Solar Cells via High-Speed Flexography

Julia E Huddy, Youxiong Ye, William J Scheideler

Dartmouth College, Hanover, NH, United States

Perovskites could deliver terawatt-scale photovoltaic capacity via low-cost manufacturing, but upscaling demands more reliable, large-area deposition methods for transport layers (CTLs) and absorbers. We present a method for scaling ultrathin NiOx hole transport layers and double cation perovskite absorbers using high-speed flexography for the fastest (60 m/min) reported fabrication of inorganic CTLs and perovskites. By engineering precursor rheology, NiOx HTLs and $MA_{0.6}FA_{0.4}PbI_3$ absorbers were printed with high uniformity and ultralow pinhole densities, improving photovoltaic performance (PCE > 15%) over spin-coated devices. These results guide ink design for scalable solar module manufacturing and reveal opportunities to enhance large area performance.

Terrain Aware Backtracking via Forward Ray Tracing

Kurt Rhee
Nevados Engineering
San Francisco, California, United States
kurt@nevadosengineering.com

Abstract—**Trackers with conventional backtracking strategies can incur significant shading and electrical mismatch losses due to the realities of as-built world terrain, and current best methods for slope awareness only account for single slopes in the cross axis and tracker axis directions. Nevados engineering has developed a new methodology for eliminating terrain shading on horizontal single axis tracking solar sites which can handle arbitrary tracker azimuths as well as multiple slopes in any direction, including installations where the slope changes in both the axis and the cross axis directions.**

Keywords—*horizontal single axis tracker, tracker, backtracking, ray tracing, terrain*

I. INTRODUCTION

Horizontal single axis trackers, which are often used in utility scale photovoltaic power plants, can increase the amount of light captured per module compared to a fixed tilt array and can likewise improve said project's economic return. Two factors that may decrease the overall positive effect of horizontal single axis trackers are the amount of grading required to install a standard single axis tracking system and deficiencies in the tracker rotation strategy used to control how the trackers follow the sun.

Nevados single axis trackers are designed specifically to reduce or eliminate the grading requirement by introducing specialized bearings which can handle post to post net angle changes of up to 15 degrees. Within a single tracker there may be some number of bays, each with it's own distinct normal vector. This design, which improves the grading aspect of project economics, by definition also makes the tracker system more vulnerable to cross axis and intra-tracker axis slopes. This vulnerability is solved for by introducing an iterative forward ray tracing algorithm which tests for module shading.

The most basic tracking algorithm available to system and tracker designers, entitled true-tracking, "minimizes the incidence angle between the panel normal (vector) and incoming beam irradiance from the sun" [1]. "This approach lays modules flat at solar noon and tilts modules at steep angles in early morning and late afternoon to face towards then sun when it is at low elevation. At steep tilt and low sun elevation, inter-(tracker) shading causes cell to cell electrical mismatch in modules with a conventional cell layout, resulting in severe nonlinear power loss" [1].

The next improvement of the tracking algorithm, entitled back-tracking, reduces the module angle during the early morning and late evening hours in order to prevent shade from one tracker falling onto another tracker. This is often called "row to row shading". This algorithm assumes that trackers are entirely within a single plane (flat) and will fail to prevent shading when this condition is not met [1].

The third backtracking strategy, which we will call cross-slope aware back tracking, takes into account the fact that one tracker may lie in a different plane (higher or lower) than the tracker directly to its east or west. The angle between these two planes, called the cross-axis slope, extends infinitely and all trackers that fall within the scope of each algorithm run will inherit the same cross axis slope. This methodology can fail to prevent shading in three ways. One, if the cross axis slope is not constant throughout the whole site and the algorithm fails to take into account the variety of slope changes there may be row to row shading. Two, if the two planes which hold the modules are not parallel with one another, meaning there is a slope in the tracker axis direction, the algorithm will not be able to prevent shade. Three, if slope changes from pile to pile within a tracker, a strategy which only takes into account the tracker as a whole will fail to prevent shading from occurring.

Nevados terrain tolerant trackers have bays which can have both cross axes slopes and slopes in the tracker axis direction, as well as intra-tracker variability in pile top elevations. These facts necessitated the development of a new backtracking algorithm which could account for these additional factors.

II. METHODOLOGY

The relevant information required to perform terrain aware backtracking in all dimensions begins with the elevation encoded locations points within a given project. The points should describe a set of rectangular polygons which enclose each contiguous set of modules with a common normal vector (a bay of modules).

From this point, sun angles can be determined and a schedule of true tracking angles can be created for each tracker. The methodology described by Anderson and Mikofski [1] has proven very useful.

Looking at each time step individually one can set up the backtracking problem as a traditional forward ray tracer with no screen object necessary through which to trace rays. In the simplest incantation of a backwards ray tracing program such as those used in the movie and video game industries, rays are traced from a "camera", through the pixels on a screen and into the scene and then finally towards a light source. Because a realistic image is not necessary to determine if a shadow is cast by the direct light from the sun, we can remove the camera and screen objects from our problem and trace the light from the light source (the sun) instead of towards it.

Within the scene, a set of rays can be spawned for each bay in a tracker. Each ray can be defined by a direction corresponding to the corner of the bay and an origin corresponding to the unit vector coming from the direction of the sun.

Because each bay of trackers can be described by a plane with a point and a normal vector pointing outward from the module faces, an intersection test can then be performed for a given ray and possibly shaded bays. If the rays of the sun passing one bay cause shade on another bay, then both trackers will be backtracked by 1 degree until no shading occurs.

Due to the fact that ray tracing is a time consuming process, a number of methods can be implemented so that the problem solves in a reasonable amount of time. These include reducing the amount of intersection tests required by choosing only reasonable objects that could possibly block direct irradiance, and having the ray tracing processes run in parallel.

In addition to the utility a ray tracing backtracking strategy presents for tracker systems that have both cross-axis slopes and variability in intra-tracker axis angle, it can also provide additional energy benefit to a system that is affected by shade objects. At sites that are affected by external shade objects such as those that are bordered on the sides by forest or buildings, the same ray tracing logic can be applied to those structures, and shading can be avoided.

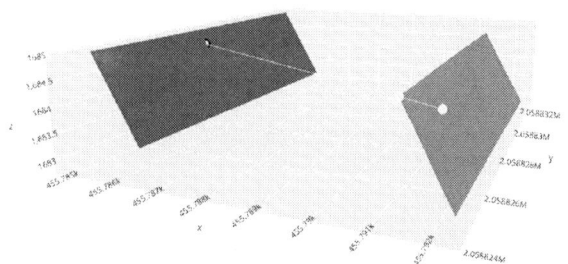

Fig. 1. A ray tracing scene between two bays of modules. In this case the intersection test is positive and backtracking will need to occur.

Fig. 2. Simplified backwards ray tracing schematic [3]

III. Conclusion

Forward ray tracing represents a viable way to avoid shading in cases where cross-axis slope is not consistent throughout the site, where intra-tracker axis angle variations occur, where azimuth is not consistent throughout the site, and where object shading or far shading may occur. This method allows Nevados trackers to remain unshaded

Acknowledgment

A word of thanks to Kevin Anderson at NREL for publishing the original methodologies to which this work seeks to add and for helping the author in finding additional reference material for related work.

References

[1] K. Anderson "Maximizing Yield with Improved Single-Axis Backtracking on Cross-Axis Slopes" National Renewable Energy Laboratory NREL/CP-5K00-76023

[2] K.Anderson and M. Mikofski "Slope-Aware Backtracking for Single-Axis Trackers" National Renewable Energy Laboratory NREL/TP-5K00-76626

[3] Henrik "https://commons.wikimedia.org"

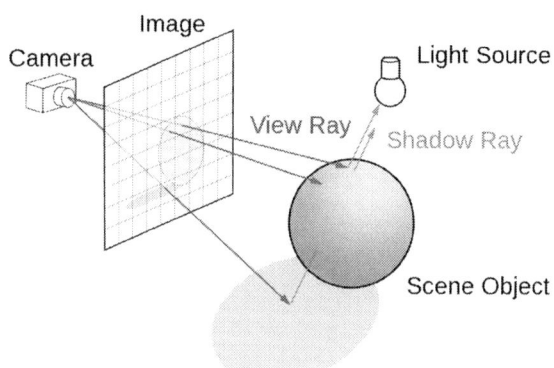

Sustainability of PV Repowering

Ian Marius Peters, Jens Hauch, Christoph Brabec

FZ Jülich, Erlange, Germany

PV-Repowering describes the processes of replacing old PV modules with new and better ones. The dynamics of this process are driven by two factors: degradation and innovation. Faster degradation prepones the ideal moment for module replacement. Innovation has a more complex impact; better technology makes replacements attractive earlier, yet a great rate of progress may make waiting for tomorrow' technology the better choice. While these two factors affect the ideal timing for repowering from an economical as well as from a sustainability point of view, they do so in different ways. Economically, we see advantages for module replacement after 20 to 25 years of operation in the field, yet when considering CO_2 reductions, repowering is opportune only within a limited window. In Germany, this window pertains to PV modules installed before 2015. Yet even then repowering only results in minor improvements of the CO_2 balance at best. In all cases, whether considering economics or sustainability, repowering is inferior to using new PV modules for additional PV-installations. To achieve carbon neutrality, PV capacity expansion at the earliest possible time has to be prioritized. PV-repowering is a second order effect, yet it has its place wherever capacity expansion is not possible, for example for rooftop installations used for electricity self-consumption.

Correlations In Spatial Variability When Accounting For Cloud Advection

Joseph Ranalli
Penn State Hazleton
Hazleton, PA
jar339@psu.edu

Abstract—**Spatiotemporal variability of irradiance has been a topic of interest in the literature. This study attempts to separate advective and uncorrelated portions of the spatially distributed irradiance by comparing cross correlations between site pairs within an irradiance measurement network and accounting for time lag when calculating those cross correlations. Following techniques in the literature, cross correlations are computed as a function of site pair separation distance for multiple wavelet transform timescales. Results show that the well known form of the correlation's decline can be maintained by considering the lagging correlation and the component of the site pair separation distance perpendicular to the overall cloud motion. This may open the door for additional fidelity in modeling of spatiotemporal irradiance that more formally represents how the loss of correlation with distance depends on the cloud motion vector.**

Index Terms—**variability, spatial aggregation, cloud motion vector, wavelet variability model**

I. INTRODUCTION

Spatiotemporal variability of irradiance is a topic of interest for modelers of photovoltaic (PV) power generation. All solar generation is made up of spatially distributed modules that collect the solar resource, converting solar irradiance to electricity. In contrast, irradiance monitoring efforts often take the form of pyranometer stations that provide individual point-based measurements. Even the best measurements consist of a few point-based measurement locations. The output of spatially distributed PV generators is known to exhibit a lower degree of variability than these point measurements. This difference can be attributed to the spatial smoothing effect as a distributed plant aggregates (integrates) the irradiance over its geographic extent. Models of this effect exist, and represent a range of different phenomenological bases.

II. BACKGROUND

Models of spatial aggregation of irradiance attempt to relate the variability of a single point irradiance time series to that of an entire PV plant.

One of the most widely used models of this phenomenon is the Wavelet Variability Model (WVM) [1]. The WVM represents a distributed plant as a set of discrete points and infers a reduction in variability based on the reduction of correlation across this point population. It uses a functional representation of the correlation between spatially distant points i and j that depends on the separation distance, d_{ij},

the cloud motion speed, V_c, and the wavelet mode timescale, $\bar{t}$ [2]:

$$\rho_{ij} = \exp\left(-\frac{d_{ij}}{\frac{V_c}{2}\bar{t}}\right) \qquad (1)$$

In computing the plant's output, the magnitude of each wavelet mode is scaled by a factor that depends on the correlation for all possible site pairs making up the plant. The shortest scales (highest frequencies) experience the greatest reduction, due to the inverse relationship between the correlation and the timescale. Here, the scale of the wavelet mode for the aggregate plant, w_P, is scaled down relative to the reference as the square root of the sum of all site pair correlations as found via Eq. 1.

$$w_P(t) = w_{ref}(t)\sqrt{\frac{\rho_{sum}}{N^2}} \qquad (2)$$

The basis of the scaling, along with the derivation of the best fit to real data, is laid out in the two sources that were cited in the preceding text [1], [2]. The correlations described are instantaneous correlations, i.e. the site pairs are tested only for simultaneous variability, without allowing for any time lag. However, as noted by the original authors [2], *"along-wind sites can become negatively correlated ... if there is frozen cloud field advection."* These negative correlations were also observed in data by other investigators [3], [4]. This occurrence actually implies that analyzing only instantaneous (i.e. zero-lag) correlation can be misleading and does not provide any discrimination of the relationship between the actual variability signal (i.e. cloud-advection-induced variability) and uncorrelated "noise" in the spatial variability of the plant.

The Cloud Advection Model (CAM) [5], [6] was developed as an attempt to describe the fully frozen condition, which is akin to assuming perfect correlation between all site pairs in the direction of the cloud motion, when accounting for time lag. The original reports of the CAM describe improvements in representing the frequency domain characteristics of the aggregate plant response. However, they also describe situations in which the assumption of perfect correlation is too strong. That is, high frequency oscillations are observed to be incoherent, implying a loss of lagging correlation with increasing frequency. As such, some modification is necessary to account for the decorrelation present in the observed data.

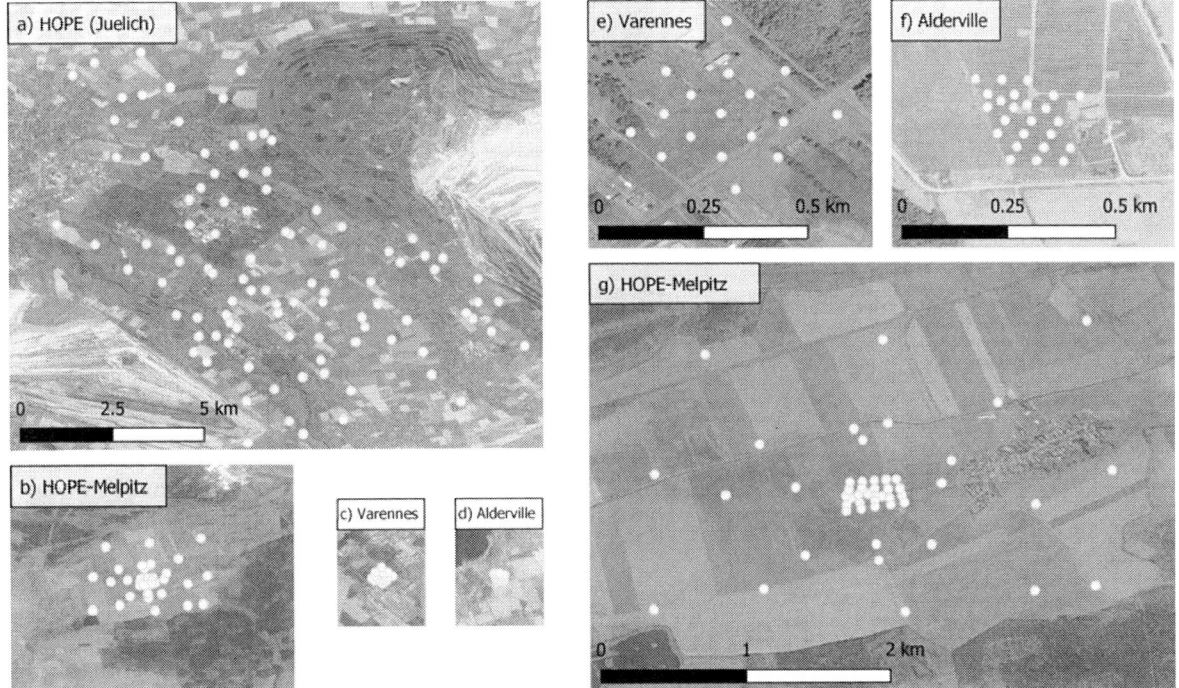

Fig. 1. Maps of the measurement campaign layouts for each of the datasets used. Note that subplots a-d share a scale, while e-g repeat previous figures at a higher scale to show additional detail for these configurations. a) HOPE (Jülich), b) HOPE-Melpitz, c) Varennes, d) Alderville, e) Varennes (detail view), f) Alderville (detail view), g) HOPE-Melpitz (detail view).

The analysis in this paper seeks to describe the middle ground between the WVM (no advective lag) and the CAM (frozen advection) by investigating the relationship between *lagging* correlations between multi-scale wavelet modes and the distance of site-pair separation.

III. DATA

Two different distributed irradiance data sources were used in this study. The HOPE campaign [7] made distributed measurements of irradiance in Germany using a network of point sensors. The original campaign was conducted with 100 sensors near Jülich, Germany, while an second campaign was conducted with 50 of those sensors installed near Melpitz, Germany. The data for both campaigns have a time interval of 1 second. An additional distributed irradiance dataset consists of two distributed measurement sites in Eastern Canada [8], one near Varennes, Quebec and one near Alderville, Ontario. Data from these sites was sampled at a maximum rate of 100 Hz, but data were only recorded for time intervals when the irradiance underwent a significant change. As a result, back-filling was necessary to complete the time series. In order to provide a consistent time step with the HOPE data, the Canadian site measurements have been resampled by averaging to a 1 second sample rate for this study. This has the added benefit of removing the somewhat unrealistic step changes from the irradiance timeseries.

Throughout the remainder of this paper, the four sites will be abbreviated with designations JUL (Jülich), MEL (Melpitz), VAR (Varennes) and ALD (Alderville). An image of the distribution of the sensors for each campaign is given in Fig. 1.

These datasets represent various levels of geographic extent and variable cloud motion conditions. The largest site is JUL covering a total region of roughly 10km x 5km along its diagonal axis, while the two Canadian sites are smallest at approximately 0.25km x 0.25km. MEL spans roughly 3km x 2km, but also contains a relatively dense section of centrally located points. For this study, data used from VAR and ALD was limited to the publicly available, "variable" and "highly variable" days on the Natural Resources Canada (NRCAN) website [1].

A. Calculation Methodology

Time series were obtained for each dataset and all were resampled to a sampling period of 1 second. Irradiance was converted to clearness index using the simplified solis clear sky model as implemented in PVLIB-Python [9]. A time window of interest was chosen for each dataset, during which the cloud motion vector was observed to be approximately constant. For ALD and VAR, this window was from 08:00-16:00 on the days indicated. For MEL, the window chosen was September 8, 2013 from 09:00-13:00 and for JUL, it was May 15, 2013 from 09:00-15:00. Cloud motion vectors were computed using the method described by Jamaly and Kleissl [10], which was

---

[1]https://www.nrcan.gc.ca/energy/renewable-electricity/solar-photovoltaic/18409

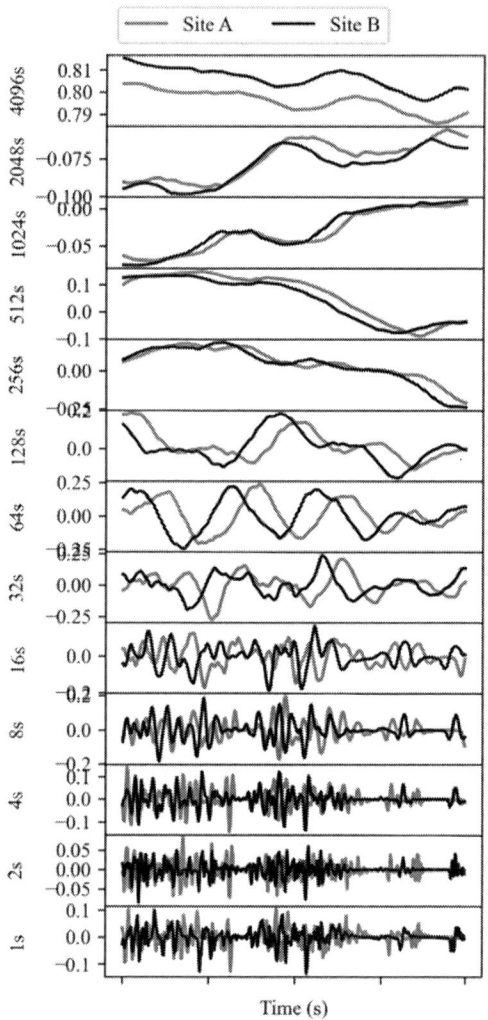

Fig. 2. Comparison of multiple timescales of wavelet modes for a 5 minute period for two separate sites in the ALD dataset.

compared to and yielded similar results to the method of Gagné et al [11], [12].

Wavelet modes were calculated using the implementation of the WVM in PVLIB-Python [9]. A total of 12 modes with timescales from 2s up to 4098s were computed for each time series. We excluded the shortest timescale as its parent wavelet shape was inconsistent with the top-hat wavelet used for all other modes. The cross correlation between every possible site pair, at every wavelet timescale, was computed using the signal processing library within scipy [13]. Scaling was applied to the cross correlation results, such that the value represented the correlation coefficient, ranging from zero to one. Results were computed at all possible lags without any biasing applied, which provides a slight favoritism to smaller lags. An example of the wavelet modes for two separate sites is shown in Fig. 2. The lagged nature of the correlation between these sites is clearly visible, especially for intermediate timescale modes.

IV. RESULTS

Plots similar to those of Lave et al [2] were generated to show the relationship between pairwise correlation, separation distance and wavelet timescale. To understand the results with respect to cloud advection, the site-pair separation distance was further differentiated into parallel and perpendicular directions, relative to the cloud motion vector. The absolute value of the perpendicular distance is used in the colorization of the plot. The results for the ALD very-variable day are shown in Fig 3.

As is evident from the leftmost plot, the proposed correlation of Lave et al. [2] is a good fit for the zero-lag correlation case. However, when computing the the maximum correlation allowing for a temporal offset (lag), we achieve the result in the central plot in Fig 3, which exhibits an upward shift for nearly every point, but especially those with small separation distances perpendicular to the cloud motion. Locations with a larger perpendicular separation are not as significantly affected, indicating that the correlation for these points is already well explained by the zero-lag case. We attempt to correct for this biasing of the correlation by replacing the absolute distance in Eq. 1 with the perpendicular separation distance in computing the x-axis variable (see Eq. 3). As seen in the right panel of Fig. 3, use of this distance helps return the dispersion of the data to the prediction line. The suitability of this match indicates that the uncorrelated variability tends to occur perpendicular to, rather than along, the cloud motion.

$$\rho_{ij,lag} = \exp\left(-\frac{d_{\perp ij}}{\frac{V_c}{2}\bar{t}}\right) \qquad (3)$$

While Fig. 3 maintains continuity with how this data is presented elsewhere in the literature, the x-axis scaling has the undesirable effect of compressing all the short timescale points almost onto the y-axis. As utilizing the peak lagging correlation mostly affects short- and moderate timescales, it would be helpful to utilize an axis scaling that better highlights those timescales. In Fig. 5a, we re-present the data from Fig. 3, but include the cloud speed coefficient in the exponent. This results in a plot that linearly represents the curve fit described by Eq. 3. The same effects described previously are clearly visible in this data.

One exception to the quality of fit is the moderate number of scattered points that appear in the lower right of the rightmost plot. Recoloring these points by their raw absolute distance, as in Fig. 4, demonstrates that though these are predominantly points that, despite having very small perpendicular separation distances, have large separation distances in the direction of cloud motion. The lagged correlation of these points remains much lower than would be expected for frozen advection. This implies that for large spatial distances, the cloud-advection is not truly frozen and highlights the fact that both cloud-advection and uncorrelated noise components play some role in the overall spatiotemporal variability.

Results for multiple test sites are shown in the other parts of Fig. 5. The best performing cases are ALD highly variable,

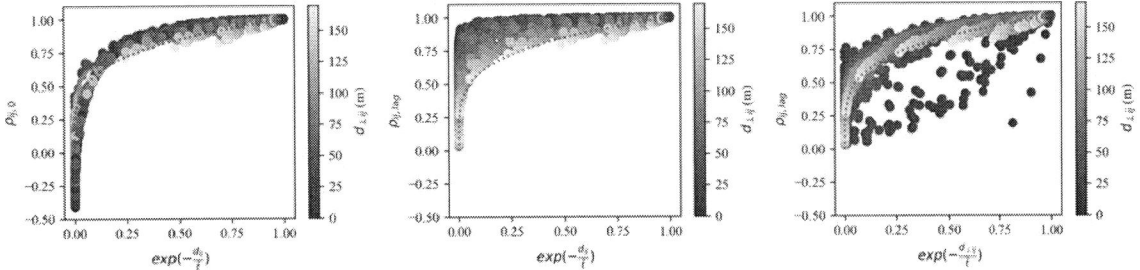

Fig. 3. Dependence of correlation on separation distance and timescale as in [2] for ALD, Aug. 12, 2015 from 08:00 through 16:00. left) correlation at zero lag and absolute distance, center) maximum correlation at any lag and absolute distance, right) maximum correlation at any lag and distance perpendicular to cloud motion. All plots are colored by the distance perpendicular to the cloud motion direction. Dotted line is correlation as proposed by Lave et al.

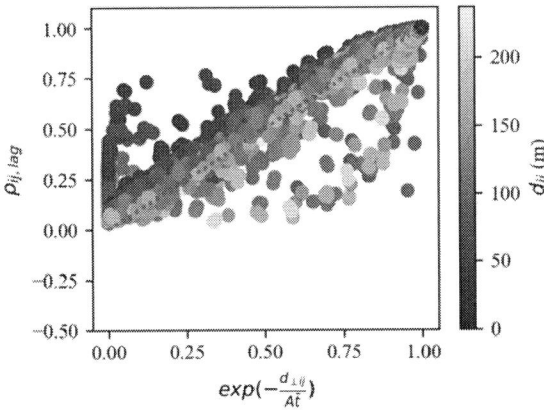

Fig. 4. Data from the rightmost panel of Fig. 5a, colorized to show the absolute pair separation distance.

VAR variable and MEL (a, c and e). All three cases show a clear increase in the correlation between short timescale modes when allowing for lag (as in the center column). The improvement provided by utilizing the model based on Eq. 3 is also evident in the right column of these plots. Less benefit is seen when considering ALD variable, VAR highly variable and JUL cases (b, d and f). For all three of these cases with lesser performance, we do observe that tracking the peak lagging correlation does increase the overall level of correlation, pushing the points toward the upper left of the plot in the central column. The predominant difference seen for these cases is the degree to which points with a large perpendicular separation distance are also shifted to the upper left, implying that the correlation is stronger than expected at greater separation distances perpendicular to the cloud motion. These points experience less shifting when switching to the modified form of the equation, as they correspond to points whose absolute and perpendicular separation distances are similar. Points with short perpendicular separation distances are in fact shifted closer to alignment with the $x = y$ predicted line.

To actually quantify the suitability of the modified model,

TABLE I
QUALITY OF FIT FOR EACH CASE.

Site	Day	Cloud Spd. (m/s)	Base ρ	Peak Lag ρ	Modified Model ρ
ALD	HV	8.5	0.976	0.934	0.964
ALD	V	30.4	0.899	0.870	0.825
VAR	HV	40.7	0.884	0.725	0.750
VAR	V	10	0.981	0.958	0.985
MEL	Sept 8	19.5	0.967	0.928	0.955
JUL	May 15	17.8	0.932	0.887	0.889

the quality of the best-fit line between the model and the data points can be computed for each of the plots. These results are shown in Table I. For each case except ALD variable, switching to the modified distance measurement improves our ability to predict the correlation between site pairs when allowing for lag. As seen in the table, the cases with the best performance corresponded to those with the lowest cloud motion speeds.

V. CONCLUSION

This study demonstrates that when considering separated site-pair correlations, the use of a lagging cross correlation can improve discrimination of the variability induced by advection from that introduced by "noise" in the cloud field. Accounting for the lagging correlation increases the level of measured correlation between wavelet modes, particularly for sites whose time series are closely related by virtue of lying parallel to the cloud motion direction. The analysis demonstrated that the empirical model proposed previously by Lave et al. [2] better fits the lagging correlation when scaling by the perpendicular separation between sites relative to the cloud motion, rather than the absolute distance. This confirms that the uncorrelated "noise" in the wavelet modes is more significant in the direction perpendicular to the cloud motion, and frozen advection is important along the direction of cloud motion. In the test cases considered here, performance of this modified model was better for conditions with slower cloud motion speeds. Further studies are warranted to further investigate the use of this modified distance measurement and to incorporate discrimination of advection- and randomness-

978-1-7281-6118-1/22 $31.00 © 2022 IEEE

induced variability into models of the smoothing of plant power generation time series.

VI. ACKNOWLEDGMENT

The author would like to acknowledge Sophie Pelland of Natural Resources Canada for sharing access to the NRCAN test site data and code for variability analysis, and Matthew Lave of Sandia National Laboratories for sharing WVM-related modeling code.

REFERENCES

[1] M. Lave, J. Kleissl, and J. S. Stein, "A Wavelet-Based Variability Model (WVM) for Solar PV Power Plants," *IEEE Transactions on Sustainable Energy*, vol. 4, no. 2, pp. 501–509, Apr. 2013.

[2] M. Lave and J. Kleissl, "Cloud speed impact on solar variability scaling – Application to the wavelet variability model," *Solar Energy*, vol. 91, pp. 11–21, May 2013. [Online]. Available: http://www.sciencedirect.com/science/article/pii/S0038092X13000406

[3] V. P. A. Lonij, A. E. Brooks, A. D. Cronin, M. Leuthold, and K. Koch, "Intra-hour forecasts of solar power production using measurements from a network of irradiance sensors," *Solar Energy*, vol. 97, pp. 58–66, Nov. 2013. [Online]. Available: http://www.sciencedirect.com/science/article/pii/S0038092X13003125

[4] B. Elsinga, "Chasing the Clouds: Irradiance Variability and Forecasting for Photovoltaics," Dissertation, Utrecht University, Nov. 2017. [Online]. Available: http://dspace.library.uu.nl/handle/1874/356774

[5] J. Ranalli, E. E. Peerlings, and T. Schmidt, "Cloud Advection and Spatial Variability of Solar Irradiance," in *47th IEEE Photovoltaic Specialist Conference (PVSC)*. Virtual: IEEE, Jun. 2020, p. 8.

[6] J. Ranalli and E. E. M. Peerlings, "Cloud advection model of solar irradiance smoothing by spatial aggregation," *Journal of Renewable and Sustainable Energy*, vol. 13, no. 3, p. 033704, May 2021, publisher: American Institute of Physics. [Online]. Available: https://aip.scitation.org/doi/abs/10.1063/5.0050428

[7] A. Macke, P. Seifert, H. Baars, C. Barthlott, C. Beekmans, A. Behrendt, B. Bohn, M. Brueck, J. Bühl, S. Crewell, T. Damian, H. Deneke, S. Düsing, A. Foth, P. D. Girolamo, E. Hammann, R. Heinze, A. Hirsikko, J. Kalisch, N. Kalthoff, S. Kinne, M. Kohler, U. Löhnert, B. L. Madhavan, V. Maurer, S. K. Muppa, J. Schween, I. Serikov, H. Siebert, C. Simmer, F. Späth, S. Steinke, K. Träumner, S. Trömel, B. Wehner, A. Wieser, V. Wulfmeyer, and X. Xie, "The HD(CP)^2 Observational Prototype Experiment (HOPE) – an overview," *Atmospheric Chemistry and Physics*, vol. 17, no. 7, pp. 4887–4914, Apr. 2017. [Online]. Available: https://www.atmos-chem-phys.net/17/4887/2017/acp-17-4887-2017-discussion.html

[8] A. Gagné, D. Turcotte, N. Goswamy, and Y. Poissant, "High resolution characterisation of solar variability for two sites in Eastern Canada," *Solar Energy*, vol. 137, pp. 46–54, Nov. 2016. [Online]. Available: https://www.sciencedirect.com/science/article/pii/S0038092X16303000

[9] W. Holmgren, C. Hansen, and M. Mikofski, "pvlib python: a python package for modeling solar energy systems," *Journal of Open Source Software*, vol. 3, no. 29, p. 884, Sep. 2018. [Online]. Available: https://joss.theoj.org/papers/10.21105/joss.00884

[10] M. Jamaly and J. Kleissl, "Robust cloud motion estimation by spatio-temporal correlation analysis of irradiance data," *Solar Energy*, vol. 159, pp. 306–317, Jan. 2018. [Online]. Available: http://www.sciencedirect.com/science/article/pii/S0038092X17309556

[11] A. Gagné, N. Ninad, J. Adeyemo, D. Turcotte, and S. Wong, "Directional Solar Variability Analysis," in *2018 IEEE Electrical Power and Energy Conference (EPEC)*, Oct. 2018, pp. 1–6, iSSN: 2381-2842.

[12] S. Pelland, A. Gagné, M. A. Allam, D. Turcotte, and N. Ninad, "Spatiotemporal Interpolation of High Frequency Irradiance Data for Inverter Testing," in *2021 IEEE 48th Photovoltaic Specialists Conference (PVSC)*, Jun. 2021, pp. 0211–0218, iSSN: 0160-8371.

[13] P. Virtanen, R. Gommers, T. E. Oliphant, M. Haberland, T. Reddy, D. Cournapeau, E. Burovski, P. Peterson, W. Weckesser, J. Bright, S. J. van der Walt, M. Brett, J. Wilson, K. J. Millman, N. Mayorov, A. R. J. Nelson, E. Jones, R. Kern, E. Larson, C. J. Carey, I. Polat, Y. Feng, E. W. Moore, J. VanderPlas, D. Laxalde, J. Perktold, R. Cimrman, I. Henriksen, E. A. Quintero, C. R. Harris, A. M. Archibald, A. H. Ribeiro, F. Pedregosa, P. van Mulbregt, and SciPy 1.0 Contributors, "SciPy 1.0: fundamental algorithms for scientific computing in Python," *Nature Methods*, vol. 17, no. 3, pp. 261–272, Mar. 2020.

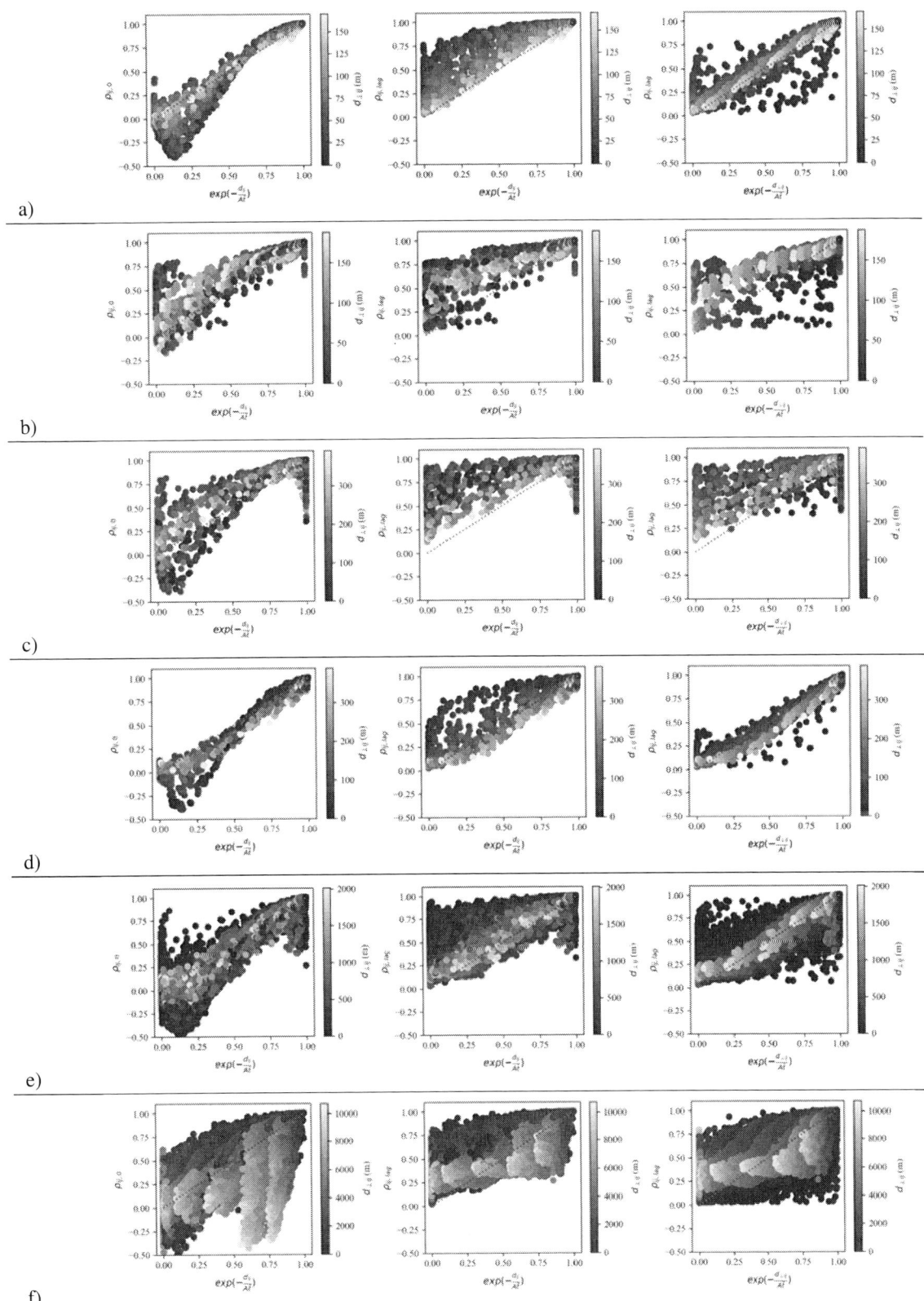

Fig. 5. Similar to Fig. 3, but with linearized scaling on the x-axis. left col) correlation at zero lag and absolute distance, center col) maximum correlation at any lag and absolute distance, right col) maximum correlation at any lag and distance perpendicular to cloud motion. Rows show the different test sites: a) ALD, Highly Variable (same as Fig. 3). b) ALD, Variable. c) VAR, Highly Variable. d) VAR, Variable. e) MEL, Sept 8, 2013 09:00-13:00. f) JUL, May 15, 2013 09:00-15:00

Realization of Ultrathin GaAs Photonic Power Converters with Rear-Side Metal Grating on Full 4" Wafers

Oliver Höhn, Meike Schauerte, Patrick Schygulla, Hubert Hauser, David Lackner, Benedikt Bläsi, and Henning Helmers

Fraunhofer Institute for Solar Energy Systems ISE, Freiburg, 79110, Germany

Abstract— **Photonic power converters (PPCs) attract more and more attention as they enable galvanically isolated optical power transmission. In many power by light applications, conversion efficiency is the key metric for PPCs. A remarkable efficiency of 68.9% was recently achieved by utilizing Fabry-Pérot resonances in a GaAs PPC. In this work, we extend this idea of leveraging optical resonance towards strong grating coupled guided mode resonances. A proof of concept is demonstrated by ultrathin GaAs PPCs with an absorber thickness of 160 nm with nanostructured photonic reflectors. Reference ultrathin devices on substrate and with planar mirror show close-to-bandgap EQE of 13% and 20%, respectively, while guided modes resonances yield an EQE of 45% at 840 nm for devices with photonic mirror. Devices were realized on full 4-inch wafers, using processes compatible with industrial manufacturing.**

Keywords—Ultrathin GaAs, Photonic Power Converter, Photonic Structure

I. INTRODUCTION

In the last years, photonic power converters (PPCs) have received increasing attention [1–11]. as they enable an increasing number of optically powered applications. Recently, we have demonstrated a record efficiency of 68.9% by leveraging photon recycling as well as Fabry-Pérot resonances in a thin film GaAs PPC [12].

The performance of PPCs is principally limited by the tradeoff between transmission and thermalization losses. The cell thickness is finite, and thus photons with an energy close to the bandgap are not fully absorbed. Decreasing the wavelength of light results in increased absorptance, however, then thermalization starts to play a major role [13]. Introduction of a planar back surface reflector allows to change the boundary conditions of this relation. Nearly perfect absorption close to the bandgap was demonstrated; namely an EQE of 94.4% was achieved in a GaAs PPC at a wavelength of 858 nm (thermalization ~1%rel) [12]. However, still a relatively thick absorber (1.75 µm) was needed to achieve this efficiency.

In contrast to this, Chen *et al.* [14] showed an ultrathin GaAs solar cell (absorber thickness <200 nm) with a solar conversion efficiency of 19.9% by using a periodically structured rear-side metal, facilitating multi-resonant light trapping and thus a significant broadband absorption enhancement. This structure was realized on a few square centimeters. As PPCs, in contrast

to solar cells, operate under monochromatic light, one single strong resonance could be used instead of multi-resonant light trapping. Thus, the effect of resonant light trapping is ideally suited for such devices.

In this paper, we show a proof of principle of an ultrathin GaAs PPC (absorber thickness 160 nm) that makes use of one single guided mode resonance close to its band gap to enhance absorption at 840 nm. To this end, a crossed grating was implemented. It is suited for ultrathin GaAs PPCs as well as solar cells; however, the grating dimensions have not yet been optimized for a GaAs PPC. The main aim of this work is to prove that such devices including a photonic structure can be realized not only on a few square centimeters or even only a few square millimeters, but that this process is suitable for full 4" wafers. A single large area nanoimprint step in combination with standard microfabrication processes were applied to demonstrate the suitability for industrial fabrication. Finally, we present a comparison of experimental data with a rigorous coupled-wave analysis (RCWA) model, showing that the position of the resonances can be modeled well, while the overall height of the absorption still requires a more detailed model and refined input data.

II. MODELING AND STRUCTURE DEFINITION

For the optical modeling the Rigorous Coupled Wave Analysis (RCWA) ("reticolo" implementation) as provided by Lalanne, Hugonin et al. [15–17] is used. The method allows for the definition of a complex refractive index profile with a complete structure with periodicity *P* consisting of multiple textures stacked in vertical direction. Within each texture, the refractive index is constant in vertical direction, but can vary laterally.

Fig. 1 AFM micrograph of the realized grating. The surface roughness is caused by the residual layer etching step.

The grating used for the experiments in this paper is a crossed grating with a period of 1 μm, a height of 250 nm with a close to binary height profile (see Fig. 1) and a metal fill factor of 43 %. A sketch of the modeled structure is shown in Fig. 2. Note that the geometric fill factor in this case is defined as the proportion of metal in the structured area.

We modeled such a binary crossed grating with the same height and metal fill factor as the experimental structure. However, the period as the most relevant influence parameter was varied in a wide range.

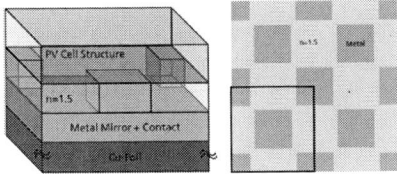

Fig. 2. Sketch of the simulation domain. While a constant height of 250 nm as well as a volume fill factor of the binary metal features of 43 % was assumed in the structured region, the period of the grating was varied. The cell structure above was assumed exactly like the realized structure shown in Fig. 4 (from window to BSF). The metal in the variation was assumed to be silver (Ag) and the surrounding material as a material with a constant refractive index of 1.5.

The modeling result of the period variation assuming Ag as metal is shown in Fig. 3.

Different absorption maxima can be seen: a period-independent maximum (like the one around 675 nm) can be identified as Fabry-Pérot resonance, while the period dependent resonances are guided mode resonances (compare also [14]). The experimentally used grating with 1 μm period shows such guided mode resonances close to the band gap of GaAs (870 nm). While they are not extremely pronounced, they already show a 3-fold increase as compared to a planar device. A considerably more pronounced resonance close to the band gap of GaAs can be found for grating periods around 620-650 nm.

Even though the grating with a period of 1 μm is not optimal, it shows clear resonances and thus is suited for a first proof of principle for realizing such a structure on a full 4" wafer.

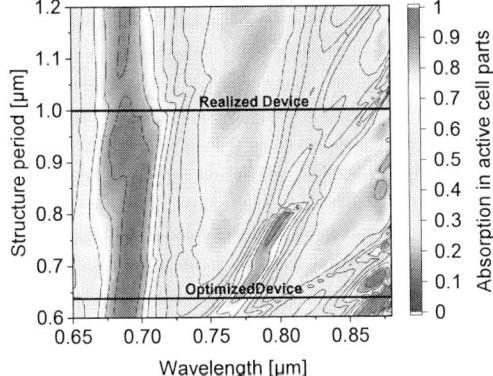

Fig. 3. Modeled absorption of the device depicted in Fig. 2 for varying grating periods. The modeled absorption is the absorption in the active solar cell. Note that the modeled parasitic absorption loss in structured metals severely depends on the exact structure and the metal quality and thus might deviate significantly from an experimental result.

III. TECHNOLOGICAL REALIZATION

A. Metal Organic Vapor Phase Epitaxy of the PPC

The epitaxial layer structure depicted in Fig. 4 was grown by metal organic vapor phase epitaxy (MOVPE) in an AIXTRON AIX2800G4-TM reactor.

Growth was performed on 450 μm thick 4" GaAs wafers with an offcut of 6° towards the (111)-B direction. Arsine (AsH_3) and phosphine (PH_3) were employed as group-V and trimethylgallium (TMGa), -aluminium (TMAl), -indium (TMIn) as group-III precursors. Silane (SiH_4) and dimethylzinc (DMZn) served for n-type and p-type doping, respectively. The solar cell was implemented as a rear-heterojunction with increasing Al-content from the n-GaAs absorber towards the p-$Al_{0.3}Ga_{0.7}As$ base. Further details on the epitaxy can be found elsewhere [12].

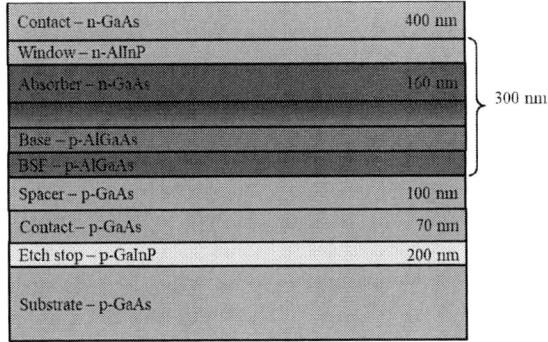

Fig. 4. Nominal layer structure of the realized ultrathin solar cell. The thick contact and spacer layer at the front and rear side are structured during fabrication and are not part of the active cell in the end.

B. Microfabrication

The device front side was fabricated using standard microfabrication processes similar to the procedure described in Ref. [12]. Afterwards the device was temporarily bonded to a sapphire wafer and the substrate was removed to access the rear side.

Fig. 5 shows a sketch and a photograph of a cell with completed front side processing after substrate removal. The device is semitransparent as the GaAs layers are ultrathin.

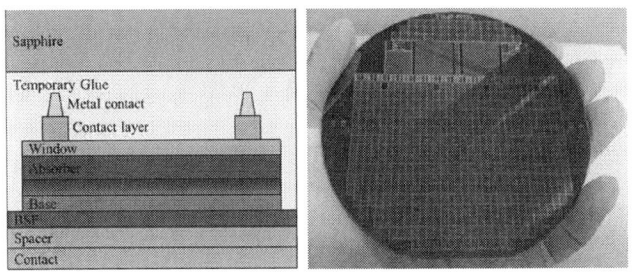

Fig. 5. Left: Sketch of the solar cells on temporary bond after front side processing and substrate removal. Right: Photograph of the cells on the temporary sapphire substrate after GaAs substrate removal and rear side contact layer removal. The solar cell structure is so thin that it appears semitransparent without the rear-side photonic contact.

To realize an electrical contact on the rear side, point contacts were fabricated: Point-shaped metal stacks (Pd/Zn/Pd/Au) with a diameter of 10 μm were processed by photolithography, metal evaporation and lift-off. The point contacts were arranged in a hexagonal pattern with a spacing of 100 μm. To remove the contact layer and the spacer between the contacts citric acid:H_2O_2 (3:1) was used. Afterwards, the photonic structure was realized via roller-UV-nanoimprint-lithography [18] in a negative tone photoresist (SU-8) with the help of a flexible stamp. The residual layer after imprinting was removed using plasma ashing. In the first variation, the Ag reflector (200 nm) and in the second variation the Au reflector (200 nm) was deposited via PVD. Both variations were finally reinforced by a 30 μm-thick electroplated Cu layer for mechanical support. Before characterization, the wafer was released from the sapphire wafer by soaking in a combination of acetone and limone.

The most relevant processing steps are sketched in Fig. 6.

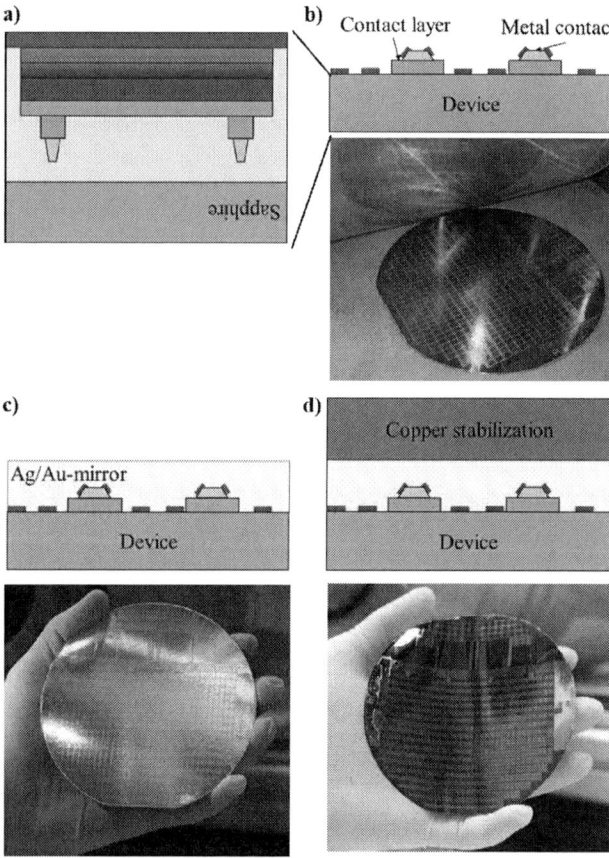

Fig. 6. a) Sketch of the front side processed cell before rear-side processing of the photonic contact. b) Device just after imprinting of the rear side in the roller nanoimprint tool developed at Fraunhofer ISE [18]. c) Device after metal evaporation. d) Device after copper plating for stabilization and removal of the temporary bond. The resulting PV cells are highly flexible and ultralight.

A structural sketch of the resulting PPCs is shown on the left in Fig. 7. On the right in Fig. 7 a face-up microscopy image is shown, which reveals the partial transparency of the ultrathin absorber: The hexagonal point contact pattern on the rear is clearly observable even below the active area.

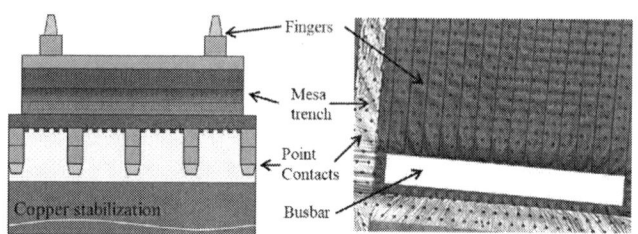

Fig. 7. Structural sketch (left) and microscopy image (right) of the fabricated ultrathin PPCs.

IV. RESULTS

After microfabrication, all PPCs on the wafers were characterized using an automated light *I-V* measurement setup under broad band illumination from a Xenon flash bulb at illumination conditions resulting in a short circuit current density of about 6 A/cm². Example *I-V* curves of devices with photonic reflector supplemented with either Ag or Au reflector are shown in Fig. 8. Box plots of the measured open-circuit voltages (V_{oc}) of all cells on the wafer are shown in Fig. 9.

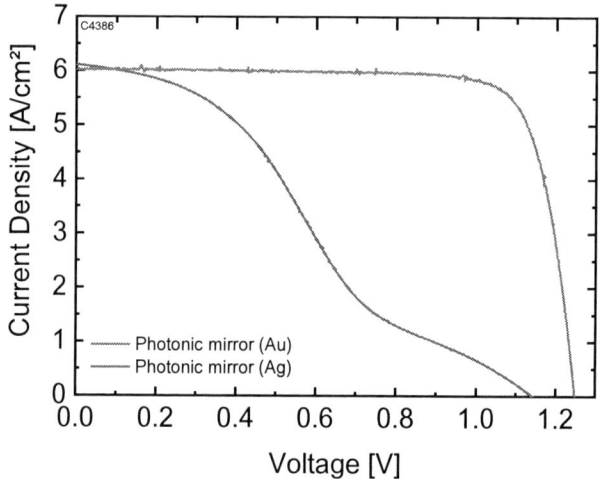

Fig. 8. *I-V* curves of two selected cells. Cells with photonic silver reflector show an s-shape that is presumably caused by imperfect rear side contacting. Cells with photonic gold reflector do not show this issue.

The *I-V* curves of the cells with photonic silver reflector, show an s-shape. This is attributed to non-ohmic contacts at the rear point contacts, which were probably caused by process imperfections during rear-side processing. The cells with photonic gold reflector do not show this issue. Here, gold forms an ohmic contact to the p-GaAs rear contact layer as well as to the p-AlGaAs back surface field layer, resulting in ohmic contacts despite the processing issues at the point contacts. However, a large fraction of PPCs on both samples are operational at similar voltages, which indicates that the microfabrication including the photonic structure succeeded homogeneously across the whole wafer and suggests that photonic structures can be implemented in wafer fabrication of PPCs or solar cells.

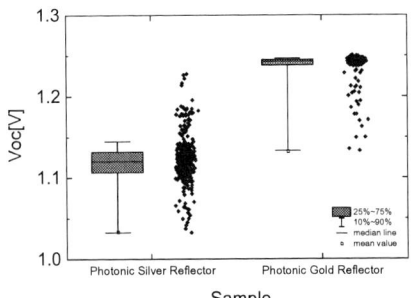

Fig. 9. Open circuit voltage distribution of all PPCs on the wafer. A major part of the cells is operational. The cells with gold mirror show higher voltages due to the above-mentioned s-shape in the *I-V* curves for silver.

In a second step the external quantum efficiencies (EQE) of the PPCs were measured in the relevant spectral range for GaAs, see Fig. 10. Resonances can be seen on the cells with either planar or photonic reflector. The resonances that occur both in the cells with planar reflector and photonic mirror are the above-mentioned Fabry-Pérot resonances. The observed resonances closest to the absorption edge at 870 nm only occur in the PPCs with photonic rear-side mirror. They can be attributed to grating coupled guided mode resonances, similar to the ones used for light trapping in solar cells in the work by Chen *et al.* [14]. For PPCs these resonances close to the bandgap are crucial to reach highest efficiencies as these allow for tuning the optical system to nearly unity absorptance in proximity to the bandgap and, thus, can be used to minimize thermalization losses without compromising for absorptance.

While the positions of the absorptance peaks are reasonably well reproduced by an RCWA modeling of the devices including all III-V layers and the photonic contact, the heights of the absorptance peaks do not yet fit. This can have several reasons: In the model, a perfectly binary grating was assumed, which is currently not the case in the realized cells. The grating is rounded there. Also, we expect that parasitic absorptions within the structured metal were underestimated in the model. Further, parasitic losses at the point contacts were neglected. Note that the used grating was not yet optimized for GaAs PPC operation, but rather should be considered as a proof for large area fabrication on full 4"-wafers.

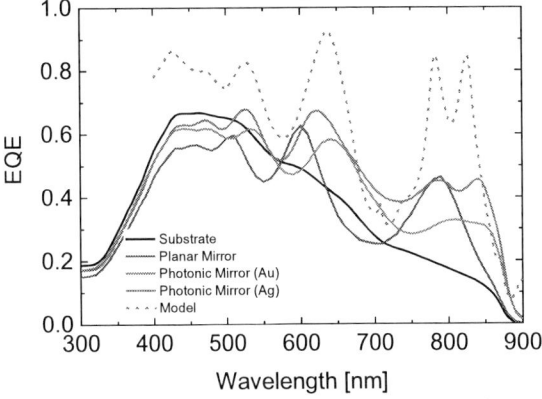

Fig. 10. Measured EQE of PPCs on substrate, with a planar mirror and with photonic gold and silver reflector (continuous lines). For comparison, also the modeled absorptance of a PPC with a photonic silver reflector is shown (dashed line). Note that the cell thickness for this modeling was adjusted to the measured epitxial thickness (approx + 5%) instead of the nominal ones as used above.

In future development, the model will be refined, and to this point neglected details and related losses will be considered. Based on such a refined model an optimized photonic structure for ultrathin PPCs will be determined aiming at an increase in EQE close to the bandgap as compared with a thick counterpart.

V. CONCLUSION

We demonstrated ultrathin GaAs PPCs with photonic reflector on full 4" wafers. The microfabrication is based on standard microfabrication steps. The photonic rear reflector is realized by a full wafer roller-UV-nanoimprint lithography. Hence, the entire process chain is compatible with industrial production. We showed that guided mode resonances (840 nm) close to the bandgap of the ultrathin PPCs lead to EQEs of 45%, whereas a thin film cell with planar mirror and a reference cell on substrate only showed 20% and 13%, respectively. This clearly proves that the grating adds significant useable light trapping as compared to a planar mirror. With an optimized optical system and a fine-tuned grating, close-to-bandgap EQEs above 90% are in reach for ultrathin PPCs, eventually promising to outperform thin film PPCs with planar back reflector and absorber thicknesses in the range of a few micrometers.

ACKNOWLEDGEMENT

This project is supported by the Vector Stiftung within the project "PhotonikPV" and the German Federal Ministry of Education and Research in the project "AlIR-Power" (#01DM21006A). P. Schygulla acknowledges a PhD scholarship from the Heinrich-Böll-Stiftung.

VI. REFERENCES

[1] Y. Zhao, P. Liang, H. Ren, and P. Han, "Enhanced efficiency in 808 nm GaAs laser power converters via gradient doping," *AIP Advances*, vol. 9, no. 10, p. 105206, 2019, doi: 10.1063/1.5109133.

[2] E. Oliva, F. Dimroth, and A. W. Bett, "GaAs Converters for High Power Densities of Laser Illumination," *Prog Photovolt Res Appl*, vol. 4, no. 16, pp. 289–295, 2008, doi: 10.1002/pip.811.

[3] S. Fafard *et al.,* "Ultrahigh efficiencies in vertical epitaxial heterostructure architectures," *Appl. Phys. Lett.*, vol. 108, no. 7, p. 71101, 2016, doi: 10.1063/1.4941240.

[4] O. Höhn, A. W. Walker, A. W. Bett, and H. Helmers, "Optimal laser wavelength for efficient laser power converter operation over temperature," *Appl. Phys. Lett.*, vol. 108, no. 24, p. 241104, 2016, doi: 10.1063/1.4954014.

[5] J. Schubert, E. Oliva, F. Dimroth, W. Guter, R. Löckenhoff, and A. W. Bett, "High-Voltage GaAs Photovoltaic Laser Power Converters," *IEEE Trans. Electron Devices*, vol. 56, no. 2, pp. 170–175, 2009, doi: 10.1109/TED.2008.2010603.

[6] S. K. Reichmuth, H. Helmers, S. P. Philipps, M. Schachtner, G. Siefer, and A. W. Bett, "On the temperature dependence of dual-junction laser power converters," *Prog Photovolt Res Appl*, vol. 25, no. 1, pp. 67–75, 2017, doi: 10.1002/pip.2814.

[7] F. Proulx *et al.,* "Measurement of strong photon recycling in ultra-thin GaAs n/p junctions monolithically integrated in high-photovoltage vertical epitaxial heterostructure architectures with conversion efficiencies exceeding 60%," *Phys. Status Solidi RRL*, vol. 11, no. 2, p. 1600385, 2017, doi: 10.1002/pssr.201600385.

[8] N. A. Kalyuzhnyy *et al.,* "Optimization of photoelectric parameters of InGaAs metamorphic laser (λ=1064 nm) power converters with over 50% efficiency," *Sol. Energy Mater. Sol. Cells*, vol. 217, p. 110710, 2020, doi: 10.1016/j.solmat.2020.110710.

[9] J. Mukherjee, S. Jarvis, M. Perren, and S. J. Sweeney, "Efficiency limits of laser power converters for optical power transfer applications," *J. Phys. D, Appl. Phys.*, vol. 46, no. 26, 264006, 2013, doi: 10.1088/0022-3727/46/26/264006.

[10] R. Kimovec, H. Helmers, A. W. Bett, and M. Topič, "Comprehensive electrical loss analysis of monolithic interconnected multi-segment laser power converters," *Prog Photovolt Res Appl*, vol. 27, no. 3, pp. 199–209, 2019, doi: 10.1002/pip.3075.

[11] E. Lopez *et al.,* "Experimental coupling process efficiency and benefits of back surface reflectors in photovoltaic multi‐junction photonic power converters," *Prog Photovolt Res Appl*, vol. 29, no. 4, pp. 461–470, 2021, doi: 10.1002/pip.3391.

[12] H. Helmers *et al.,* "68.9% Efficient GaAs‐Based Photonic Power Conversion Enabled by Photon Recycling and Optical Resonance," *Phys. Status Solidi RRL*, vol. 15, no. 7, p. 2100113, 2021, doi: 10.1002/pssr.202100113.

[13] H. Helmers *et al.,* "Pushing the boundaries of photovoltaic light to electricity conversion: A GaAs based photonic power converter with 68.9% efficiency," in *48th IEEE Photovoltaic Specialists Conference (PVSC)*, Online, 2021, p. 2100113. Accessed: May 31 2021.

[14] H.-L. Chen *et al.,* "A 19.9%-efficient ultrathin solar cell based on a 205-nm-thick GaAs absorber and a silver nanostructured back mirror," *Nat. Energy.*, vol. 4, pp. 761–767, 2019, doi: 10.1038/s41560-019-0434-y.

[15] J. P. Hugonin and P. Lalanne, *Reticolo software for grating analysis.* Orsay, Frace: Institut d'Optique, 2005.

[16] M. G. Moharam, E. B. Grann, D. A. Pommet, and T. K. Gaylord, "Formulation for stable and efficient implementation of the rigorous coupled-wave analysis of binary gratings," *J. Opt. Soc. Am. A*, vol. 12, no. 5, pp. 1068–1076, 1995.

[17] P. Lalanne and G. M. Morris, "Highly improved convergence of the coupled-wave method for TM polarization," *Journal of the Optical Society of America. A, Optics, image science, and vision*, vol. 13, no. 4, p. 779, 1996, doi: 10.1364/JOSAA.13.000779.

[18] H. Hauser *et al.,* "Honeycomb Texturing of Silicon Via Nanoimprint Lithography for Solar Cell Applications," *IEEE J. Photovolt.*, vol. 2, no. 2, pp. 114–122, 2012, doi: 10.1109/JPHOTOV.2012.2184265.

Human Health Risk Assessment of Solvents and Lead Toxicity in Emerging Perovskite Solar Cells

Sherif A. Khalifa, Sabrina Spatari, Aaron T. Fafarman, Vasilis M. Fthenakis, Patrick Gurian, Jason B. Baxter

Department of Chemical and Biological Engineering, Drexel University, Philadelphia, PA, United States

Faculty of Civil and Environmental Engineering, Technion - Israel Institute of Technology, Haifa, Israel

Center for Life Cycle Analysis, Columbia University, New York, NY, United States

Department of Civil and Environmental Engineering, Drexel University, Philadelphia, PA, United States

Emerging lead halide perovskite photovoltaics (LHP PV) continue to spur research efforts due to their outstanding efficiency, low manufacturing costs, and fabrication versatility, potentially making them competitive with existing PV technologies. While ongoing efforts are focused on tackling stability and scalability of LHPs, the potential toxicity of solvents and lead (Pb) remains a major sustainability challenge to their large-scale commercialization. Immediate risks to human health of local communities must be evaluated for regulatory compliance and appropriate mitigation actions identified. Here, we present a screening-level human health risk assessment of 1) potential accidental solvent releases during manufacturing and 2) potential Pb releases following catastrophic breakage of LHP PV modules. We estimated exposure point concentrations of Pb in air, groundwater, and soil media using US EPA-compliant fate and exposure modeling to develop order-of-magnitude level estimations of potential risk. We found out that inhalation risk of commonly used N,N-Dimethylformamide (DMF) solvent for a resident downwind of a manufacturing site showed no adverse potential health effects under our maximum emissions assumptions. Pb exposure concentrations in air, groundwater, and soil could exceed US EPA maximum permissible limits under some worst-case scenarios upon catastrophic failure at large scales. Our results are sensitive to many assumptions including soil types, PV plant capacity, and leaching potential of the Pb. This study does not substitute the need for site-specific analysis, but it indicates the need for further attention to design fail-safe perovskite modules and safe manufacturing and deployment sites.

Spectral shape changes the optimal perovskite thickness of the 2-terminal perovskite/silicon tandem solar cell

Dong C. Nguyen[*] and Yasuaki Ishikawa[*]

College of Science and Engineering, Aoyama Gakuin University, Sagamihara, Kanagawa 252-5258, Japan
[*]Email: dongnc@ee.aoyama.ac.jp, yishikawa@ee.aoyama.ac.jp

Abstract—In this work, we simulate and analyze the impact of spectral shape on the 2-terminal perovskite/silicon heterojunction tandem solar cell performance. Under the standard test condition (spectrum AM1.5G 100 mW/m^2, 300K), the optimal perovskite thickness of the 2-terminal tandem solar cell to maximize the tandem efficiency is 640 nm. However, we found that this optimal value depends significantly on the actual spectral shape. Specifically, the optimal perovskite thickness obtained is thinner than 640 nm under the blue-rich spectrum while thicker than 640 under the red-rich spectrum. This finding helps design suitable tandem structures in different climatic zones.

Index Terms—Tandem cell, perovskite, Atlas, TCAD, spectral shape, silicon heterojunction.

I. INTRODUCTION

In the future, the tandem solar cell technology might be a potential option to replace the single-junction silicon solar cell technology because its power conversion efficiency (PCE) can reach 44% [1], which overcomes the Shockley–Queisser limit [2]. The 2-terminal perovskite/silicon tandem solar cell is preferred because of its high PCE and low manufacturing costs. However, a considerable challenge for this technology is the current matching condition between the top and bottom sub-cells, which depends on many factors, including the climatic conditions, such as spectral irradiance and shape, incident angle, temperature, etc. So, assessing the actual environmental impacts is critical to designing a suitable tandem structure for each climatic zone.

In this work, we focus on simulation and analysis of the impact of spectral shape (blue-rich and red-rich spectrum) on the 2-terminal perovskite/silicon heterojunction (SHJ) tandem solar cell performance by using a device simulator (Atlas Silvaco).

II. SIMULATION AND ANALYSIS METHODOLOGY

In this work, the physical structure of the 2-terminal perovskite/SHJ tandem solar cell has been coded in the Silvaco Atlas module based on the 2-dimensional (2D) simulation framework. The tandem structure includes the perovskite (top) and SHJ (bottom) sub-cells, which are connected by a tunnel recombination junction (TRJ) layer in series. The top sub-cell consists of a lithium fluoride (LiF) anti-reflective coating (ARC), an indium tin oxide (ITO) front contact, an electron transport titanium oxide (TiO$_2$), a perovskite absorber, and a hole transport contact layer (spiro-OMeTAD). Meanwhile,

the bottom sub-cell is composed of n$^+$ doped and intrinsic (i) hydrogenated amorphous silicon (a-Si:H) layers, the n-type crystalline silicon (c-Si) absorber (substrate), i and p$^+$ doped a-Si:H layers, and a silver (Ag) rear contact. The TRJ layer comprises p-type hydrogenated microcrystalline silicon oxide (p-μc-Si$_{1-x}$O$_x$:H), and a-Si:H(n) [3]. We consider all interfaces to be optically flat.

The data of the complex refractive index (n, k) and the absorption coefficient in the tandem solar cell layers was taken from the available literature: LiF [4], ITO [5], TiO$_2$ [5], 1.67-eV-perovskite [6], Spiro-OMeTAD [5], p-μc-Si$_{1-x}$O$_x$:H [7], p$^+$-a-Si:H [8], a-Si:H(i) [8], c-Si(n) [9], n$^+$-a-Si:H [8]. Input material parameters of the top cell's layers [10], [11], the bottom cell's layers [12], [13], and the TRJ layer [11] are tabulated in Table I. Our simulation also took into account the defect states of the TiO$_2$/Perovskite interface [14], [15] and the a-Si:H/c-Si:H interface [12]. The defect densities of the c-Si(n)/a-Si(i) and TiO$_2$/perovskite interfaces were chosen at a fixed value of 1×10^{11} cm^{-2}eV^{-1}. The surface recombination velocity of a-Si:H layers was fixed $v_{surf,n} = v_{surf,p} = 10$ cm/s. Series resistance (R_s) of this solar cell was set 4.9 Ω, which is referenced from a recent work [16].

The average photon energy (APE) is the characterization of the energetic distribution in an irradiance spectrum, which is calculated by Eq. 1:

$$APE_{350-1200} = \frac{\sum G(\lambda)}{e \sum \Phi(\lambda)} \tag{1}$$

where $G(\lambda)$, $\Phi(\lambda)$, and e is the spectral irradiance (W/m^2), the photon flux density (number of photons/m^2/sec), and electron charge, respectively. The APE value of the spectrum AM1.5G in the $350 - 1200$ nm wavelength range is 1.785 eV. We propose the irradiance spectrum be divided into two spectral types composed of the blue-rich and red-rich spectrum based on the APE value, taking AM1.5G as reference. The blue and red-rich spectrum have more than and less than 1.785 eV, respectively.

In this work, we examine the 2-terminal tandem solar cell performance under the blue-rich spectrum (APE values of 1.795, 1.805, 1.814, and 1.825), the red-rich spectrum (APE values of 1.775, 1.766, 1.756, and 1.748), and the AM1.5G spectrum, as illustrated in Fig. 1. Herein, we assume that the visible and near-infrared (NIR) light regions in a

978-1-7281-6118-1/22 $31.00 © 2022 IEEE

TABLE I: Input electrical parameters of the materials used in the 2-terminal tandem solar cell

Parameters	TiO$_2$	Perovskite	Spiro	SiO$_x$:H	a-Si:H(n)	a-Si:H(i)	c-Si(n)	a-Si:H(p)
Thickness d (nm)	10	Various (400-1200)	20	75	5	5	260000	5
Electron affinity χ (eV)	3.9	3.9	2.45	4.05	3.9	3.9	4.17	3.9
Permittivity ϵ_r	9	6.5	6.5	11.9	11.8	11.8	11.7	11.8
Energy band gap E_g (eV)	3.2	1.67	3	1.124	1.72	1.72	1.12	1.72
Doping concentration N_d (cm^{-3})	1×10^{16}	1×10^{14}	2×10^{16}	1×10^{20}	1×10^{19}	2.2×10^{15}	1×10^{16}	1×10^{19}
Electron mobility μ_n (cm^2.V^{-1}s^{-1})	20	2	2×10^{-4}	10	1	1	1300	1
Hole mobility μ_p (cm^2.V^{-1}s^{-1})	10	2	2×10^{-4}	1	1	1	450	1
Effective DOS in conduction band N_c (cm^{-3})	2.2×10^{18}	2.2×10^{18}	2.2×10^{18}	2.8×10^{19}	2.5×10^{20}	2.5×10^{20}	2.8×10^{19}	2.5×10^{20}
Effective DOS in valence band N_v (cm^{-3})	1.8×10^{19}	1.8×10^{19}	1.8×10^{19}	2.68×10^{19}	2.5×10^{20}	2.5×10^{20}	1×10^{19}	2.5×10^{20}
Electron lifetime τ_n (sec)	1×10^{-7}	5×10^{-7}	1×10^{-7}	1×10^{-7}	1×10^{-7}	1×10^{-7}	1×10^{-2}	1×10^{-7}
Hole lifetime τ_p (sec)	1×10^{-7}	5×10^{-7}	1×10^{-7}	1×10^{-7}	1×10^{-7}	1×10^{-7}	1×10^{-2}	1×10^{-7}

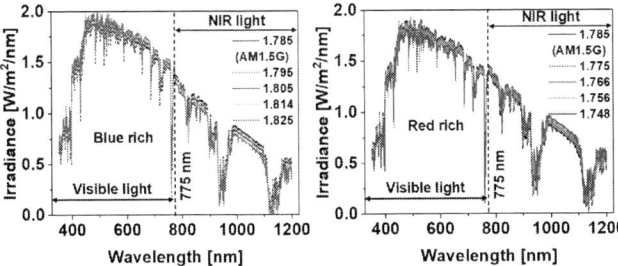

Fig. 1: Spectral shapes: blue-rich spectrum (left), and red-rich spectrum (right).

spectrum are divided at the wavelength of 775 nm. The blue-rich spectrum derived from the spectrum AM1.5G has higher spectral irradiance within the visible light and lower spectral irradiance within the NIR light than the AM1.5G spectrum. Inversely, the red-rich spectrum derived from the AM1.5G spectrum has lower spectral irradiance within the visible light and higher spectral irradiance within the NIR light than the AM1.5G spectrum. The higher the APE value of a spectrum is, the larger within the visible light and the lower within the NIR light the spectral irradiance has. However, their total spectral irradiance in each spectrum is the same as 1000 W/m^2 to avoid the impact of the spectral irradiance on the tandem performance change.

III. RESULTS AND DISCUSSIONS

After examining the perovskite thickness under the standard test condition (STC), we obtained the current matching condition between the top and bottom sub-cells of the 2-terminal tandem solar cell at the optimal perovskite thickness of 660 nm. In that case, the tandem solar cell attains the optimal electrical characteristics composed of PCE of 30.58%, V_{oc} of 1.92 V, J_{sc} of 19.74 mA/cm^2, and FF of 80.73%. However, this matching condition does not get the maximum tandem PCE value. The evidence is that we acquired a tandem PCE value of 30.6% with the values of V_{oc} of 1.92 V, J_{sc} of 19.63 mA/cm^2, and FF of 81.19% at perovskite thickness 640 nm,

which is the maximum PCE value in this work. Indeed, the current matching condition is a preferable condition to optimize the 2-terminal tandem structure [17], but not a sufficient condition to maximize the tandem performance because some experimental reports presented the maximum tandem PCE has not been achieved at the current matching condition [18].

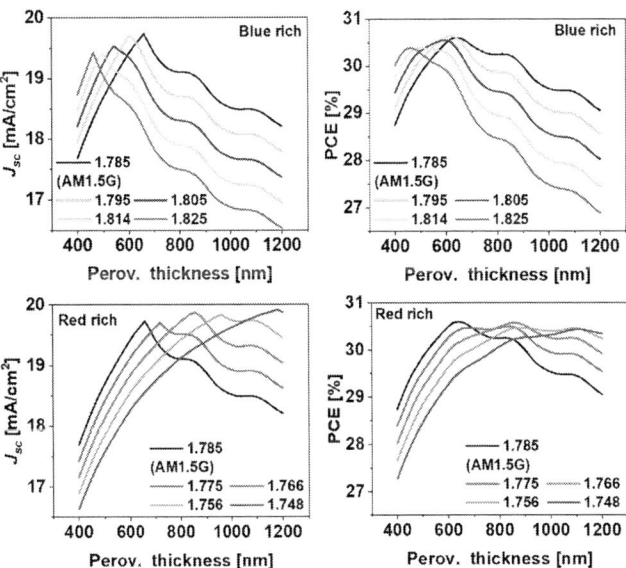

Fig. 2: The tandem J_{sc} and PCE values under the blue-rich spectrum versus the perovskite thickness (upper images), the tandem J_{sc} and PCE values under the red-rich spectrum versus the perovskite thickness (lower images).

Fig. 2 illustrates the tandem J_{sc} and PCE values under the blue-rich and red-rich spectrum versus the perovskite thickness. Table II lists the optimal values of perovskite thickness and the corresponding optimal tandem J_{sc} and PCE values in the current matching condition and the maximum PCE condition. The maximum tandem PCE seems to be attained at the limited top cell J_{sc} under the red-rich spectrum, at the limited bottom cell J_{sc} under the blue-rich spectrum.

TABLE II: Comparison of the optimal tandem performance under different blue-rich and red-rich spectrum, taking spectrum AM1.5G as reference.

APE value	Matching			Maximum			J_{sc} limited by
	Perov. Thickness (nm)	J_{sc} (mA/cm^2)	PCE [%]	Perov. Thickness (nm)	J_{sc} (mA/cm^2)	PCE [%]	
1.748	1180	19.92	30.36	1100	19.81	30.44	Top cell
1.756	960	19.83	30.42	900	19.70	30.47	Top cell
1.766	860	19.87	30.58	860	19.87	30.58	Matching
1.775	720	19.70	30.45	720	19.70	30.45	Matching
1.785	**660**	**19.73**	**30.58**	**640**	**19.63**	**30.60**	**Top cell**
1.795	600	19.70	30.62	620	19.66	30.65	Bottom cell
1.805	540	19.54	30.45	600	19.35	30.55	Bottom cell
1.814	500	19.40	30.39	500	19.40	30.39	Matching
1.825	460	19.43	30.39	460	19.43	30.39	Matching

As shown the results, we observed that the current matching condition is preferable, but not a sufficient condition to maximum the 2-terminal tandem PCE under the blue-rich, red-rich spectrum even AM1.5G spectrum.

It is seen that the spectral shape impacts the STC-based optimal perovskite thickness, leading to the significant change of the maximum tandem PCE. Precisely, the tandem solar cell attains the maximum tandem PCE at the perovskite thickness less than 640 nm under the blue-rich spectrum but at the perovskite thickness more than 640 nm under the red-rich spectrum. In other words, the optimal perovskite thickness is proportional to the APE value of the spectrum. Indeed, the tandem J_{sc} is limited by the bottom cell under the blue-rich spectrum with the higher visible and lower NIR spectral irradiance, so the perovskite thickness needs to be thinner to reduce the exceeded absorption in the top sub-cell,hence enhance the bottom J_{sc}. Inversely, the tandem J_{sc} is limited by the top cell under the red-rich spectrum with the lower visible and higher NIR spectral irradiance, the photon current density absorbed in the top sub-cell needs to be enhanced by increasing the perovskite thickness.

The world's six climatic zones include tropical-humid, subtropical-arid (desert), subtropical coastal, temperate-coastal, high elevation, and temperate continental. The distribution of the APE value is a function of the spectrum for each climatic zone, so the spectral shapes vary in different climatic zones. The blue-rich spectrum is primarily distributed in tropical-humid, while the red-rich spectrum exists in the other climatic zones as a reference in climatic datasets (IEC61853-4). Thus, the optimal structure of the 2-terminal tandem solar cells is suitably designed depending on the intended use in each climatic zone. Understanding the effect of the spectral shape on the tandem cell performance is advantageous for that design.

IV. CONCLUSIONS

By examining the perovskite thickness in simulations under different spectral shapes, we revealed that the current matching condition is preferable, not sufficient to maximize the 2-terminal tandem performance. The maximum tandem PCE seems to be attained at the limited bottom cell J_{sc} under the blue-rich spectrum, at the limited top cell J_{sc} under the red-rich spectrum, but this point needs to be investigated more

clearly. The optimal perovskite thickness is thinner and thicker than the STC-based optimal value under the blue-rich and red-rich spectrum, respectively. This finding is helpful in suitably designing the 2-terminal tandem structure.

ACKNOWLEDGMENT

This work was supported by the New Energy and Industrial Technology Development Organization (NEDO), Japan.

REFERENCES

[1] T. Leijtens, K. A. Bush, R. Prasanna, and M. D. McGehee, "Opportunities and challenges for tandem solar cells using metal halide perovskite semiconductors," Nature Energy, vol. 3. pp. 828–838, 2018.

[2] A. Richter, M. Hermle, and S. W. Glunz, "Reassessment of the limiting efficiency for crystalline silicon solar cells," IEEE Journal of Photovoltaics, vol. 3, pp. 1184–1191, 2013.

[3] I. A. Yunaz, K. Sriprapha, S. Hiza, A. Yamada, and M. Konagai, "Effects of temperature and spectral irradiance on performance of silicon-based thin film multijunction solar cells," Japanese Journal of Applied Physics, Part 1: Regular Papers and Short Notes and Review Papers, vol. 46, pp. 1398–1403, 2007.

[4] H. H. Li, "Refractive index of alkali halides and its wavelength and temperature derivatives," Journal of Physical and Chemical Reference Data, vol. 5, pp. 329–528, 1976.

[5] E: Raoult et al., "Optical characterizations and modelling of semi-transparent perovskite solar cells for tandem applications," in 36th European Photovoltaic Solar Energy Conference and Exhibition, 2019, pp. 757–763.

[6] J. Werner et al., "Complex refractive indices of cesium-formamidinium-based mixed-halide perovskites with optical band gaps from 1.5 to 1.8 eV," ACS Energy Letters, vol. 3, pp. 742–747, 2018.

[7] J. Peter Seif et al., "Amorphous silicon oxide window layers for high-efficiency silicon heterojunction solar cells," Journal of Applied Physics, vol. 115, pp. 024502:1-024502:8, 2014.

[8] Z. C. Holman et al., "Current losses at the front of silicon heterojunction solar cells," IEEE Journal of Photovoltaics, vol. 2, pp. 7–15, 2012.

[9] M. A. Green, "Self-consistent optical parameters of intrinsic silicon at 300 K including temperature coefficients," Solar Energy Materials and Solar Cells, vol. 92, pp. 1305–1310, 2008.

[10] T. Minemoto and M. Murata, "Theoretical analysis on effect of band offsets in perovskite solar cells," Solar Energy Materials and Solar Cells, vol. 133, pp. 8–14, 2015.

[11] A. Nakanishi, Y. Takiguchi, and S. Miyajima, "Device simulation of $CH_3NH_3PbI_3$ perovskite/heterojunction crystalline silicon monolithic tandem solar cells using an n-type a-Si:H/p-type μc-Si$_{1-x}$O$_x$:H tunnel junction," Physica Status Solidi (A) Applications and Materials Science, vol. 213, pp. 1–6, 2016.

[12] M. Lu, U. Das, S. Bowden, S. Hegedus, and R. Birkmire, "Optimization of interdigitated back contact silicon heterojunction solar cells: Tailoring hetero-interface band structures while maintaining surface passivation," Progress in Photovoltaics: Research and Applications, vol. 19, pp. 326–338, 2011.

[13] A. Froitzheim, K. Brendel, L. Elstner, W. Fuhs, K. Kliefoth, and M. Schmidt, "Interface recombination in heterojunctions of amorphous and crystalline silicon," Journal of Non-Crystalline Solids, vol. 299–302, pp. 663–667, 2002.

[14] F. Zhang et al., "Interfacial oxygen vacancies as a potential cause of hysteresis in perovskite solar cells," Chemistry of Materials, vol. 28, pp. 802–812, 2016.

[15] F. Wang, S. Bai, W. Tress, A. Hagfeldt, and F. Gao, "Defects engineering for high-performance perovskite solar cells," npj Flexible Electronics, vol. 2. pp. 1–14, 2018.

[16] A. Al-Ashouri et al., "Monolithic perovskite/silicon tandem solar cell with >29% efficiency by enhanced hole extraction," Science, vol. 370, pp. 1300–1309, 2020.

[17] Y. Hu, L. Song, Y. Chen, and W. Huang, "Two-Terminal Perovskites Tandem Solar Cells: Recent Advances and Perspectives," Solar RRL, vol. 3. pp. 1900080:1–21, 2019. doi: 10.1002/solr.201900080.

[18] M. Boccard and C. Ballif, "Influence of the Subcell Properties on the Fill Factor of Two-Terminal Perovskite-Silicon Tandem Solar Cells," ACS Energy Letters, vol. 5, pp. 1077–1082, 2020.

Leveraging Undoped CdSeTe for >950 mV

Pascal Jundt and James Sites

Colorado State University, Fort Collins, Colorado, 80523, United States

Abstract—**Despite recent measurements indicating excellent quality of CdTe-based absorber materials, specifically the CdSeTe alloy, the technology has yet to significantly close its longstanding voltage deficit in cells employing polycrystalline absorbers. To address this, we propose utilizing CdSeTe as an intrinsic layer in an n-i-p cell structure, preserving its exceptional lifetime and shifting the burden of doping to other materials. We evaluate the feasibility of this design using SCAPS 1D, and use this modeling to identify which material parameters are most important. Based on these results, we believe the n-type and absorber materials which are already available are sufficient, and the remaining challenge is to find a suitable p-type material. Several candidates are identified based on the crucial parameters identified with the modeling, and two are explored experimentally.**

Keywords—CdTe, p-i-n diode, transparent conducting oxide

I. INTRODUCTION

Cadmium telluride's (CdTe) most significant assets are high speed and low cost of manufacture, due to polycrystalline thin-film depositions which are very defect tolerant. Current module and research cell performance is respectable; short-circuit current density and fill factor are both near their respective Shockley-Queisser limits. However, for decades, polycrystalline CdTe has suffered from low open-circuit voltage relative to its band gap [1], typically not exceeding ~860 mV. V_{OC} over 1 V has been demonstrated on a few occasions [1][2], but always with single-crystal CdTe in highly non-traditional cell structures which do not translate into economical large-scale production. Increasing open-circuit voltage while retaining all the benefits of polycrystalline fabrication is the biggest research challenge facing the technology today.

Several pathways for increasing V_{OC} have been identified. Modeling has consistently indicated that simultaneous improvement of CdTe absorber carrier concentration and lifetime is necessary [3]. Unfortunately, both characteristics have proven extremely difficult to achieve at the same time. Copper has long been utilized as a dopant in CdTe, but in addition to achieving only mediocre carrier concentration [4], copper is known to create defects which facilitate recombination and severely reduce carrier lifetime [5], as well as accelerate cell degradation. Although more recent efforts at doping with group-V elements such as arsenic have demonstrated higher carrier concentration along with long lifetime [6], voltages still have not shown significant improvement.

The absorber material is not the only limiting factor. While the front interface of traditional poly-CdTe cells appears to be very well optimized, the back interface remains a challenge. In particular, CdTe's deep valence band maximum forms a Schottky barrier with most common contacting metals,

necessitating additional back layers and treatments to mitigate voltage-reducing downward band bending. Present approaches to fix this problem are only partly successful, and back-interface passivation remains an issue as well.

In current cell structures utilizing p-type CdTe, by far the most common, the demands of the absorber are numerous. In this structure, the CdTe absorbers are responsible for absorbing photons and creating carrier pairs, providing a high-quality medium with minimal recombination in which these carriers move, creating a junction field in conjunction with the n-type emitter, and being robust to multiple post-deposition layers and treatments. Satisfying all of these conditions simultaneously is difficult. With a different approach, the requirements of the absorber may be less onerous. Alloying CdTe with selenium has been shown to have a substantial passivating effect [7], and undoped CdSeTe deposited at Colorado State University (CSU) by close-space-sublimation (CSS) has demonstrated lifetimes of several μs. The ideal ~1.4 eV bandgap, excellent absorption, and very high lifetime of this material are all hallmarks of an exceptional absorber. However, these attributes are heavily compromised if the material is doped p-type, and most attempts to fabricate devices with CdSeTe-only absorbers using conventional cell architectures have yielded poor results.

In this paper, we propose a cell architecture uncommon in CdTe: the n-i-p structure. The aim is to maximally leverage the strength of intrinsic CdSeTe by leaving it undoped, shifting the burden of doping to the p-type contact, where carrier transport and recombination are less important. The CdSeTe absorber is therefore responsible only for creating electron-hole pairs and providing a long-lifetime medium in which carriers travel. Built-in voltage is thus solely the responsibility of the n-type and p-type layers. Ideally, these materials are highly-doped, transparent, and conductive, serving as both the junction and the contacts of the cell. This minimizes the number of layers, reducing complexity, and may allow for bifacial structures.

With this scheme, the research challenge is narrowed. As the front contact and absorber already appear to be sufficiently optimized, the focus is on finding a suitable p-type material(s) for the back contact, and engineering the interface between this material and the CdSeTe absorber. This approach aims to solve several chronic challenges with CdTe simultaneously, including hole selectivity at the back contact, low lifetimes in completed cells, and bifaciality.

In this paper, we use SCAPS 1D to model a CdSeTe n-i-p cell structure to both assess its feasibility and to determine which absorber and contact parameters are most important. We then identify p-contact candidates based upon the criteria identified in the simulations.

978-1-7281-6118-1/22 $31.00 © 2022 IEEE

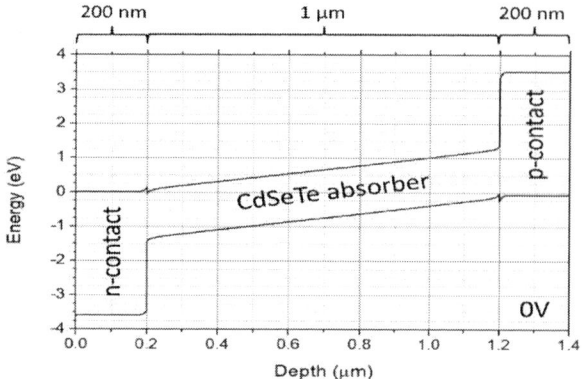

Fig. 1. Band diagram at 0 V of ideal n-i-p structure with a CdSeTe asborber.

II. MODELING

Fig. 1 shows the band diagram of an ideal CdSeTe n-i-p structure. The n-type front contact is an accurate representation of SnO:F currently used in many thin-film solar cells, and the locations of the bands in the absorber is accordant with $CdSe_{0.4}Te_{0.6}$. The p-type back contact is an unknown ideal material, with similar parameters to the n-contact, save for inverted conduction and valence band offsets and p-type doping. This basic structure is explored with simulation to determine what absorber and p-contact parameters are likely to be most impactful on cell performance.

A. Absorber Requirements

As the strength of the field is given by the potential difference between the contacts divided by their separation, the field strength is largest when the contacts are as close together as possible. In addition, a thinner absorber reduces the opportunities for carriers to recombine and has economic advantages for industrial fabrication. However, if the absorber is too thin, incomplete absorption of sunlight leads to current loss. Fig. 2a demonstrates the performance sensitivity to CdSeTe absorber thickness, illustrating this tradeoff. For this particular simulated cell structure, ~1 μm appears to be the ideal CdSeTe thickness. An absorber of this thickness has the additional advantages of reduced materials usage and deposition time. Maintaining good performance at this thickness requires

excellent passivation of *both* contact interfaces, which is a challenge.

As the absorber itself is undoped, the junction fields within the cell are relatively weak at the desired operating voltage. Therefore, carriers will spend a comparatively long time within the absorber, and so excellent absorber lifetime is vital. Fig. 2b shows the effect of absorber lifetime on performance for a 1-μm absorber structure. It is clear that lifetime within the absorber is a crucial parameter, potentially decisive on its own. The influence of lifetime on conversion efficiency is due mostly to changes in open-circuit voltage.

As carriers travel predominantly within the absorber material, carrier mobility of the material is also important. Fig. 2c shows the effect of mobility in the 1-μm-absorber structure assuming a CdSeTe lifetime of 1 μs – the solid line represents varying hole mobility, and the dashed line, electron mobility. Electron mobility appears to have very little effect over the range tested, while the effect of hole mobility is more substantial. This is likely because electrons generated from front-side excitation have on average a very short distance to travel to the front contact, while holes must move through the entire absorber to the back. In this 1-um absorber example, a hole mobility of 1 cm^2/Vs seems sufficient to attain near maximum conversion efficiency, while performance drops off precipitously as mobility decreases. Mobility primarily influences fill factor.

Based on the modeling, an intrinsic material should have lifetime >1 μs, hole mobility > 1 cm^2/Vs, and should be kept relatively thin, around 1 μm. The undoped CdSeTe deposited by CSS at CSU already satisfies the lifetime requirement. Mobility is much harder to discern; CdTe measurements have varied widely between single crystal (~500 cm^2/Vs) [8] and polycrystalline material (~1-10 cm^2/Vs) [9]. CdSeTe is even less studied; to our knowledge, mobility in CdSeTe has only been determined by optical methods, yielding ambipolar mobility of ~50-100 cm^2/Vs in polycrystalline material [10]. Uncertainty is quite high, but as the hole mobility requirements are quite modest, it nevertheless appears likely that mobility in this material is already sufficient.

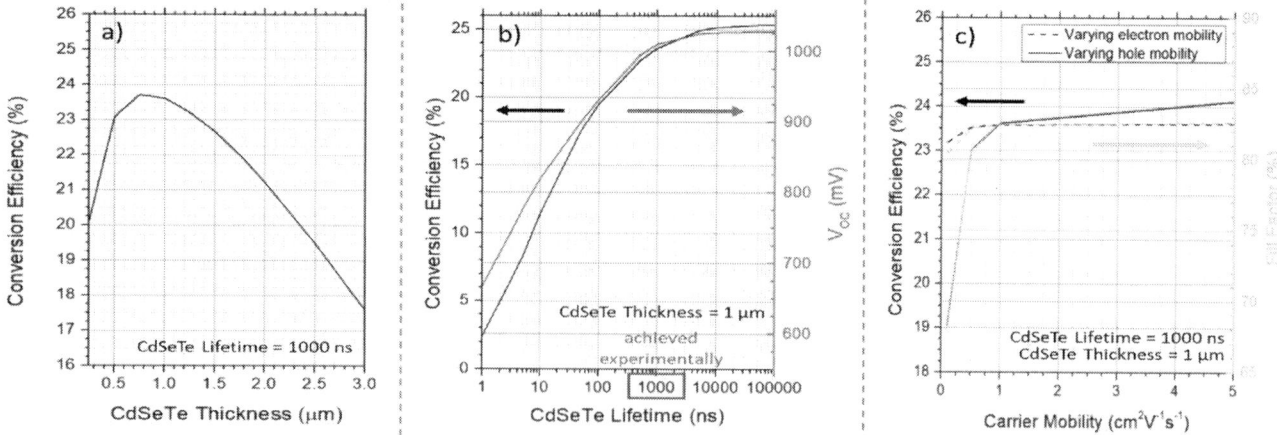

Fig. 2. a) Effect of CdSeTe absorber thickness on conversion efficiency of n-i-p structure. b) Effect of CdSeTe lifetime on efficiency and open-circuit voltage. c) Effect of carrier mobility within CdSeTe on efficiency and fill factor. Note the different conversion efficiency scales on the y-axis of each plot.

B. Back Contact Requirements

We now turn our attention to the back contact. The band diagram in Fig. 1 and results in Fig. 2 assume 10^{18} cm^{-3} carrier concentration in both contact layers. When either of these are reduced, the built-in voltage across the cell drops. At 10^{15} cm^{-3}, extraordinarily long lifetimes would be required to attain even modest performance. Therefore, it is imperative that the p-contact material achieves a high carrier concentration, at least 10^{16} cm^{-3}.

In addition to carrier concentration, the alignment of the valence band between the CdSeTe absorber and the p-contact is critical. The presence of a hole barrier due to misaligned bands which impedes hole travel is a serious issue. To investigate these effects, p-contacts with increasingly deviating valence band positions relative to CdSeTe were simulated at a range of carrier concentrations. The results are summarized in Fig. 3 with a 1-μm CdSeTe absorber with 1-μs lifetime. Valence band offset (VBO) is clearly decisive. While well-aligned bands with 0-eV offset yield very high efficiency, deviation from the ideal results in a dramatic reduction in performance.

It is notable that deviation from 0 eV VBO in *either* direction has substantial negative effects on performance. The negative VBO illustrated in Fig. 3 is defined as the p-contact having a valence band position below that of CdSeTe. This directly creates an abrupt energy step which impedes hole travel, so it is not surprising that performance suffers greatly as this barrier height is increased. If the carrier concentration in the p-contact is low, this barrier is wide and impenetrable. With higher p-contact carrier concentration, the valence band in the bulk of the p-contact is pushed up towards the fermi level, turning the hole barrier into a spike. As the p-contact doping increases, the spike narrows, allowing easier tunneling of carriers. This enables retention of performance for small negative VBO, and higher p-contact carrier concentration allows more leniency. This suggests that p-contact candidates with a small negative VBO may still work if they can be sufficiently doped. However, with a negative VBO greater than a few tenths of an eV, hole extraction inevitably becomes impossible, no matter the p-contact doping.

It is a bit less obvious how a *positive* VBO would also harm performance, as shown in Fig. 3. In this case, the p-contact valence band is above the CdSeTe valence band edge, so there is no direct energy barrier to holes. However, as the VBO increases, the higher valence band energy of the p-contact raises its fermi level by the same amount, reducing the fermi level splitting ΔE_F between the two contacts and thus the cell's built-in voltage, resulting in weaker fields. Furthermore, as ΔE_F between the contacts is reduced in forward bias, the necessity of maintaining the positive VBO will begin to force the bands in the intrinsic CdSeTe to bend downwards at the back of the cell, impeding hole transport, similar to a Schottky barrier with a low work function metal. These two effects both ultimately reduce the performance of the cell, and both are intensified as the VBO becomes more positive. As these effects are not related to the carrier concentration of the p-contact, the performance reduction at positive VBO shown in Fig. 3 is relatively consistent for all simulated doping levels.

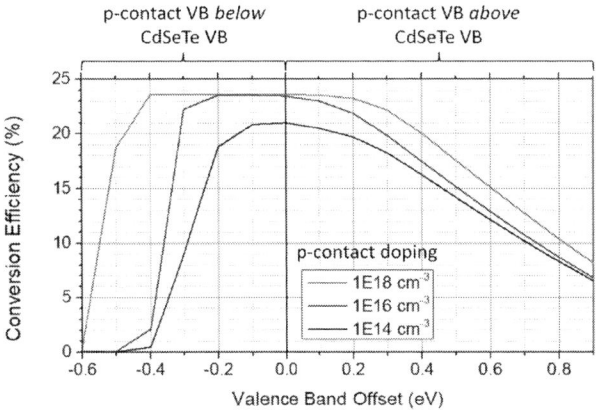

Fig. 3. Effect of CdSeTe/p-contact valence band offset on conversion efficiency. High p-contact doping mitigates losses to a degree, but a large offset always results in a major efficiency reduction.

III. BACK CONTACT CANDIDATES

While n-type transparent conducting oxides (TCOs) abound, there are comparatively very few p-type conductive oxides. Those which are known to exist have less ideal parameters than their n-type counterparts. In addition to the high conductivity and transparency expected of a TCO, the modeling makes clear that correct valence band alignment between the absorber and the back contact is essential. Relatively high doping of the back contact is also very important, both to provide adequate fields and to help mitigate the effects of misaligned valence bands. Other considerations seem to be far less important. Therefore, in evaluating potential p-type TCO candidates, heavy weight is given to dopability, conductivity, and valence band alignment with CdSeTe.

Some p-type TCOs which have demonstrated particularly high conductivity and carrier concentration include LaCuOSe:Mg [11] and CuCrO:Mg [12], but the deposition processes of these materials seem incompatible with CdSeTe absorbers, and valence band alignment with CdSeTe is unknown. Poly[bis(4-phenyl)(2-4-6-trimethylphenyl)amine] (PTAA) is a widely used hole selective layer in perovskites, and this polymer has been demonstrated to be relatively compatible with CdTe [13], though substantial V_{OC} improvements have yet to be realized. We chose instead to investigate more traditional materials which have been known to the community for some time, repurposed in this novel application. Two of these materials were incorporated into CdSeTe cells with mixed results.

A. Nickel Oxide

NiO is perhaps the earliest known p-type TCO. While its incorporation into traditional CdTe cells has led to only very modest performance increases, it has not proven detrimental [14], evidence of a fairly defect-free back interface. Intrinsic NiO is very resistive, but it can be readily doped with copper, which can dramatically increase conductivity and can drastically increase the carrier density to upwards of 10^{20} cm^{-3} [15], sufficient for use as a back TCO.

To investigate this material, NiO films were deposited using RF sputtering from a 4-inch diameter ceramic target at 60 W power in a 15 mTorr Ar ambient. As-deposited films were

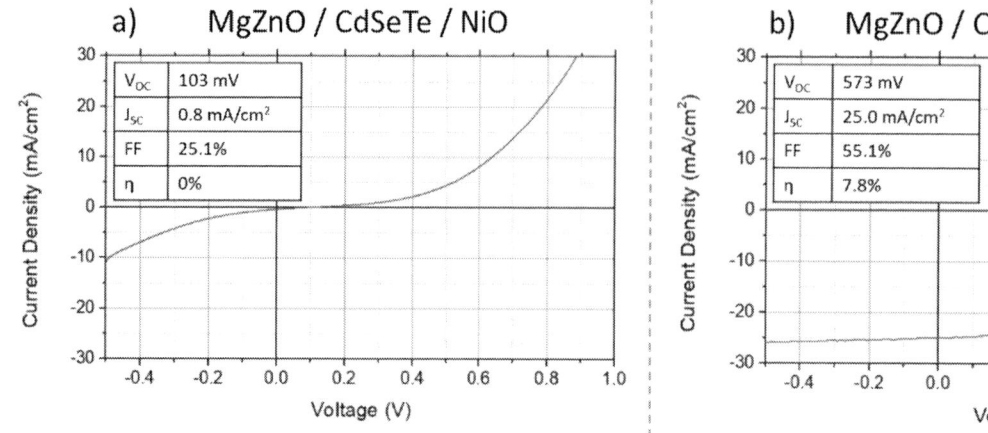

Fig. 4. a) J-V measurement of n-i-p cell with intrinsic CdSeTe absorber and p-type NiO. Strong kinking and absence of current collection are evidence of severely impeded hole transport. b) J-V measurement of n-i-p cell with intrinsic CdSeTe absorber and p-type ZnTe:Cu. Excellent current density and absence of kinking are evidence of good ZnTe conductivity and CdSeTe/ZnTe band alignment. Low voltage needs to be addressed.

transparent with a slight yellowish tint. As expected, the resistivity of these films was too high to measure. Following the procedure outlined by others [15][16], small chips of 99.9% Cu were placed on the target during sputtering in an attempt to create NiO:Cu films. This had a strong visual effect – films deposited in this way were more opaque and had a darker rust color – but resistivity was still too high to measure, indicative of very poor doping. It is unknown why our replication of this doping method did not yield the same results reported in the literature.

Despite the apparent unsuitability of the films, cells using NiO as a back contact were fabricated. These cells had the simple superstrate structure TEC10 / MgZnO (100nm) / CdSeTe (1 μm) / NiO (50 nm) / Ni (~300 nm). CdSeTe was deposited by close-space sublimation [17], MgZnO and NiO by RF sputtering, and the Ni electrode by DC sputtering. Note that the CdSeTe undergoes no intentional doping treatment. Cells with intrinsic NiO and NiO sputtered with different amounts of copper chips were fabricated, but all cells yielded J-V curves similar to Fig. 4a. In all cases, there was a complete absence of current collection and a strongly kinked shape. Both are evidence of a substantial hole barrier, almost certainly at the CdSeTe/NiO interface. Between CdSeTe and NiO is a VBO of approximately +0.5 eV [14], which the modeling in Section II has demonstrated to be quite detrimental even if the NiO were to be appropriately doped. Adjusting the concentration of oxygen in the film may allow downward adjustment of the valence band edge [18], reducing the VBO to an acceptable level, but as this might turn out to be quite difficult, NiO has several other shortcomings, and there appear to be better candidates, we instead focus our efforts on other materials.

B. Zinc Telluride

As our modeling suggests that valence band alignment with CdSeTe appears to be the single most crucial p-contact requirement, we sought a material which fulfills this criterion. ZnTe is a II-VI semiconductor with a VBO with CdTe = 0.0 ± 0.05 eV [19], and its wider bandgap of ~1.7-2.3 eV [20] creates a beneficial electron barrier with CdSeTe. While this material is relatively opaque and unsuitable as a transparent back contact

on its own, the sacrifice of bifaciality is acceptable if a functioning n-i-p cell can be achieved. ZnTe has been known to the CdTe community for decades, but to our knowledge its incorporation with undoped CdSeTe-only absorbers is novel.

ZnTe:Cu films were deposited using RF sputtering from a 4-inch diameter $Zn_{0.47}Te_{0.47}Cu_{0.06}$ ceramic target at 60 W power in an 18 mTorr Ar ambient. As previously reported [21], we also found that deposition at 250°C substrate temperature was necessary to make good quality films, with a subsequent 20-minute anneal at 250°C in Ar. In contrast with NiO, these ZnTe:Cu films were fairly conductive, measuring <1 kΩcm resistivity with a 2-point measurement. Hall measurements will be performed in the future.

Cells with the structure TEC10 / MgZnO (100nm) / CdSeTe (1 μm) / ZnTe:Cu (50 nm) / Ni (~300 nm) were fabricated. All layers were deposited as specified above. As with the bare films, ZnTe:Cu was deposited at a 250°C substrate temperature with a 20-minute anneal. J-V of the highest-performing cell is shown in Fig. 4b, though all cells with this structure performed similarly well. Short-circuit current is very good, indicating that the ZnTe films are both sufficiently conductive and allow unimpeded transmission of holes out of the CdSeTe absorber. The high fill factor (considering the low voltage) and complete lack of any kinking in the J-V curve is further evidence that holes are entirely unobstructed at the back of the cell.

These respectable results achieved with a very preliminary cell are promising. Cells fabricated at CSU with CdSeTe-only absorbers have never attained greater than ~2% conversion efficiency to date, and achieving ~8% with a nominally entirely undoped CdSeTe absorber is a new result in CdTe technology. The challenge is to address the very low open-circuit voltage in this cell. It is possible that doping in the ZnTe is not sufficiently high, limiting the fermi level separation and the built-in voltage. Additionally, based on steady-state photoluminescence measurements, it appears that the lifetime in the CdSeTe material is very low, estimated as <10 ns; this will be confirmed with time-resolved photoluminescence in the future. As Fig. 2b shows, CdSeTe lifetime has a decisive effect on V_{OC} in this structure, so much of the V_{OC} loss can be attributed to CdSeTe

978-1-7281-6118-1/22 $31.00 © 2022 IEEE

lifetime. As Cu is known to dramatically reduce lifetime in CdSeTe [22], it appears that Cu is diffusing into the CdSeTe absorber from the ZnTe:Cu during its high temperature deposition and anneal. In future work, this diffusion may be prevented by depositing a thin layer of undoped ZnTe or perhaps CdZnTe between CdSeTe and ZnTe:Cu. CdZnTe may have the additional benefit of mitigating the lattice mismatch between CdSeTe and ZnTe [23], reducing interface recombination. A second approach is to remove Cu from the cell entirely, and dope ZnTe with something else. Nitrogen has been shown to dope ZnTe effectively [20][24] and would be well-suited for this new application.

IV. CONCLUSION

In conclusion, modeling has demonstrated the feasibility of n-i-p cell structures utilizing an undoped CdSeTe-only absorber and has identified high p-type doping and a ~0 eV valence band offset with CdSeTe as the most important parameters for the p-type contact. While NiO is a widely-known p-type transparent conducting oxide, its valence band position seems incompatible with CdSeTe. On the other hand, ZnTe:Cu has produced a functional n-i-p cell in a preliminary experiment, albeit with low voltage. While changes to the structure are needed to preserve CdSeTe lifetime and improve the back contact, ZnTe appears to be a promising candidate.

ACKNOWLEDGMENT

This material is based upon work supported by the U.S. Department of Energy's Office of Energy Efficiency and Renewable Energy (EERE) under the Solar Energy Technologies Office award number DE-EE0008974. The authors would like to thank Jen Drayton and Camden Kasik for assistance with cell fabrication.

REFERENCES

[1] J. M. Burst et al., "CdTe solar cells with open-circuit voltage breaking the 1 V barrier," in *Nature Energy* 1, 16015, 2016.

[2] Y. Zhao et al., "Monocrystalline CdTe solar cells with open-circuit voltage over 1 V and efficiency of 17%," in *Nature Energy* 1, 16067, 2016.

[3] A. Kanevce et al., "The roles of carrier concentration and interface, bulk, and grain boundary recombination for 25% efficient CdTe solar cells," in *J. Appl. Phys.* **121**, 214506, 2017.

[4] J-H. Yang et al., "Enhanced p-type dopability of P and As in CdTe using non-equilibrium thermal processing," in *J. Appl. Phys.* **118**, 025102, 2015.

[5] D. Krasikov and I. Sankin, "Defect interactions and the role of complexes in the CdTe solar cell absorber," in *J. Mat. Chem. A* **5**, no. 7, pp. 3503-3513, 2017.

[6] D. Kuciauskas et al., "Microsecond carrier lifetimes in polycrystalline CdSeTe heterostructures and in CdSeTe thin film solar cells," in *47th IEEE PVSC*, pp. 0082-0084, 2020.

[7] J. Guo et al., "Effect of selenium and chlorine co-passivation in polycrystalline CdSeTe devices," in *Appl. Phys. Lett.* **115**, 153901, 2019.

[8] P. Ščajev et al., "Excitation-dependent carrier lifetime and diffusion length in bulk CdTe determined by time-resolved optical pump-probe techniques," in *J. Appl. Phys.* **123**, 025704, 2018.

[9] Q. Long et al., "Electron and hole drift mobility measurements on thin film CdTe solar cells," in *Appl. Phys. Lett.* **105**, 042106, 2014.

[10] D. Kuciauskas et al., "Voltage loss comparison in CdSe/CdTe solar cells and polycrystalline CdSeTe heterostructures," in *IEEE J. Photovolt.* **12**, no. 1, pp. 6-10, 2022.

[11] H. Hiramatsu et al., "Heavy hole doping of epitaxial thin films of a wide gap p-type semiconductor, LaCuOSe, and analysis of the effective mass," in *Appl. Phys. Lett.* **91**, 012104, 2007.

[12] R. Nagarajan, A. Draeseke, A. Sleight, and J. Tate, "P-type conductivity in $CuCr_{1-x}Mg_xO_2$ films and powders," in *J. Appl. Phys.* **89**, no. 12, pp. 8022-8025, 2001.

[13] J. Hack, C. Lee, S. Grover, and G. Xiong, "Hole transport material for passivated back contacts on CdTe solar cells," in *48th IEEE PVSC*, pp. 1880-1882, 2021.

[14] D. Xiao et al., "CdTe thin film solar cell with NiO as a back contact buffer layer," in *Sol. Energy Mater. Sol. Cells* **169**, pp. 61-67, 2017.

[15] S. Chen, T. Kuo, Y. Lin, and H. Lin, "Preparation and properties of p-type transparent conductive Cu-doped NiO films," in *Thin Solid Films* **519**, no. 15, pp. 4944-4947, 2011.

[16] E. Hassan, A. Saeed, and A. Elttayef, "Doping and thickness variation influence on the structural and sensing properties of NiO film prepared by RF-magnetron sputtering," in *J Mater Sci: Mater Electron* **27**, pp. 1270-1277, 2016.

[17] D. Swanson et al., "Single vacuum chamber with multiple close space sublimation sources to fabricate CdTe solar cells," in *J. Vac. Sci. Technol. A* **34**, 021202, 2016.

[18] R. Yadav et al., "Tuning the band structure of nickel oxide for efficient hole extraction in perovskite solar cells," in *4th IEEE ICEE*, pp. 1-5, 2018.

[19] D. Rioux, D. Niles, and H. Höchst, "ZnTe: a potential interlayer to form low resistance back contacts in CdS/CdTe solar cells," in *J. Appl. Phys.* **73**, pp. 8381-8385, 1993.

[20] T. Shimpi, J. Drayton, D. Swanson, and W. Sampath, "Properties of nitrogen-doped zinc telluride films for back contact to cadmium telluride photovoltaics," in *J. Electron. Mater.* **46**, pp. 5112-5120, 2017.

[21] A. Kindvall, A. Munshi, T. Shimpi, A. Danielson, and W. Sampath, "Effect of process temperature and copper doping on the performance of ZnTe:Cu back contacts in CdTe photovoltaics," in *46th IEEE PVSC*, pp. 0189-0192, 2019.

[22] A. Onno et al., "Understanding what limits the voltage of polycrystalline CdSeTe solar cells," in *Nature Energy* **7**, pp. 400-408, 2022.

[23] N. Amin, A. Yamada, and M. Konagai, "Effect of ZnTe and CdZnTe alloys at the back contact of 1-μm-thick CdTe thin film solar cells," in *Jpn. J. Appl. Phys.* **41**, pp. 2834-2841, 2002.

[24] A. Rakhshani, "Effect of growth temperature, thermal annealing and nitrogen doping on optoelectronic properties of sputter-deposited ZnTe films," in *Thin Solid Films* **536**, pp. 88-93, 2013.

978-1-7281-6118-1/22 $31.00 © 2022 IEEE

Low-temperature PECVD deposition of highly conductive n-type microcrystalline silicon thin films for optoelectronic applications

Brahim Aïssa, Amir A. Abdallah and Juan Lopez Garcia

Qatar Environment and Energy Research Institute (QEERI), Hamad bin Khalifa University (HBKU), Qatar Foundation, P.O. Box 5825, Doha, Qatar

Abstract—We report on the characterization results of doped n-type microcrystalline hydrogenated-silicon (μc-Si: H) films deposited by a plasma-enhanced chemical vapor deposition in the temperature range between 50 and 200 °C. The interest in these films arises from their ability to combine a high optical absorption of amorphous silicon part with the electronic behavior of the crystalline silicon one, making them interesting for the production of large electronic devices such as solar cells, image sensors, and flat panels. It is demonstrated that n-type μc-Si: H films with high electrical conductivity can be obtained even at low temperature deposition, around 50 °C (σ=12.8 S cm^{-1}). The structural properties of the films have been studied by Raman and infrared spectroscopy that allowed for the determination of the crystalline fraction.

Keywords—Microcrystalline silicon, Raman spectroscopy, FTIR, Hall Effects, Silicon Heterojunction Solar Cell

I. INTRODUCTION

In order to fabricate reliable thin film transistors or solar cells at low cost with acceptable conversion efficiency, several groups around the world directed their efforts to the research of microcrystalline hydrogenated-silicon films (μc-Si:H) developing a new material with high optical and electrical properties [1]. Since then, the applications of this promising material in solar cells or thin film transistors have been achieved with unexpected results. Low-temperature deposition methods are needed for producing large-area electronics on alternative and low-cost substrates such as polymers. It is well known that μc-Si:H films have an amorphous silicon matrix in which silicon crystalline grains are embedded [2]. The electrical conductivity of this inhomogeneous material is a result of the contribution of the different parts of it, i.e. microcrystalline and amorphous ones.

Contacts utilizing microcrystalline silicon (μc-Si:H) doped layers represent an excellent compromise between the passivation quality of all-Si contact stacks and the reduced parasitic absorption of more transparent materials. By leveraging the indirect nature of the crystalline silicon bandgap, μc-Si:H contact layers exhibit improved transparency compared with their amorphous counterparts. Indeed, this strategy has received attention in the literature [3]–[13], but difficulties in depositing highly crystalline layers thin enough to realize a reduction in parasitic absorption have persisted. Publication of an effective pretreatment method for seeding microcrystalline growth on amorphous passivation layers has been a key development in the deposition of thin μc-Si:H films [14], [15].

In particular, μc-Si:H is interesting as an absorber layer for solar cells. The low absorption coefficient for light with $k > 700$ nm ($\alpha = 10^2 - 10^3$ cm^{-1}) of this material requires thick layers (1– 3 μm) to collect the radiation efficiently [3]. Therefore, high deposition rates are necessary to make this material a good candidate for industrial production [4]. In this regard, many studies based on plasma-enhanced chemical vapor deposition (PECVD) have been made to improve the growth rate and the film quality, especially for the crystallinity [5]. The PECVD technique is one of the most acceptable approaches to grow device quality μc-Si thin films within a short fabrication time, owing to efficient gas dissociation and lower ion energy [5]. The thin film solar cells with high efficiencies have already been made using this technique. However, the optimization of the devices requires a deeper investigation of the material properties [6]. In this paper, the correlation between the substrate temperature, structural properties, and electric conductivity has been investigated in n-type μc-Si:H thin films. To extract a correlation between morphological and electrical properties, we combine structural measurement techniques (Raman and Infrared spectroscopy) with electrical measurement techniques (Hall Effect, d.c. conductivities). The effects of experimental parameters on microstructure and optoelectronic properties of the intrinsic μc-Si:H thin films were also studied, and the results on preliminary μc-Si:H thin film solar cells were also investigated.

II. EXPERIMENTAL

Solar cells were prepared on textured, 4", *n*-type float-zone wafers with resistivity of 2–3 Ωcm, and thickness of 180–220 μm. Plasma-enhanced chemical vapor deposition was used to deposit all thin film silicon layers. Doped *p* or *n* type μc-Si:H contact layers were deposited in the same chamber at a frequency of 40.68 MHz at temperatures ranging from 50 to 200 °C.

978-1-7281-6118-1/22 $31.00 © 2022 IEEE

Intrinsic a-Si:H for surface passivation was deposited in a separate chamber at a fixed temperature of 200 °C for all cells. Dopant gases used for the silicon contact layers were tri-methyl boron (TMB) for *p*-type and PH3 for *n*-type. The thickness of the doped μc-Si:H layers was 40 nm on glass, which is known from previous calibration measurements to yield ~27 nm on the textured wafer. A plasma pretreatment to oxidize the surface of the intrinsic a-Si:H was applied prior to the μc-Si:H layer deposition in all cells [15]. Tin-doped indium oxide (ITO) was deposited by sputtering (through a shadow mask on the front and full area on the back) and silver was sputtered (full area) on the back. A silver grid was then screen printed on the front to complete five, 4 cm² solar cells. Wafers were then annealed at 210 °C for 20 min in air. Current–voltage characteristics (*JV*) were collected on finished devices using a Wacom Electric Co. Super solar simulator with AM 1.5G illumination and Keithley source meters. For series resistance extraction, *JV* characteristics for each cell were collected at 5% illumination to obtain a curve in which *V*oc approached *V*mpp for the 1 sun illumination curve. By linear translation of the 5% illumination curve downward to coincide with the curve collected at 1 sun, the difference between *V*5% oc and the corresponding voltage (*V*A) at the same current value on the 1 sun *JV* curve could be determined, and used to calculate *Rs* in (1).

$$Rs = V_{oc(shaded)} - V_A / J_{sc(full)} - J_{sc(shaded)} \qquad (1)$$

EQE was collected using a home-built spectral response measurement system, equipped with xenon arc lamp. Crystalline volume fraction of doped μc-Si:H layers was measured directly on each cell by Raman spectroscopy using 325 nm light to avoid contributions from the wafer. Resulting spectra were fit with three Gaussian curves centered at 480, 510, and 520 cm⁻¹ [16, 17].

Variable-angle spectroscopic ellipsometry was performed on *n*-type μc-Si:H on glass deposited at each temperature over the energy range 0.6–6 eV at angles of 50°, 60°, and 70°. Spectra were collected using a Horiba Jobin Yvon ellipsometer, and modeling was performed in the DeltaPsi2 software. Transmission spectra were collected for the same *n*-type μc-Si:H films on glass as measured by ellipsometry using a Perkin Elmer Lambda 950 UV/VIS/NIR spectrometer over the wavelength range 250–2000 nm.

The electrical conductivity (σ), carrier concentration (*Ne*), and Hall mobility (*μ*Hall) were performed with a HMS-5000 system using the Van Der Pauw method. Glass substrates were also used to focus the study on n-type μc-Si:H especially to point out the effect of the deposition temperature. For this, evaporated aluminum contacts were deposited onto these glass substrates to form the

electrodes. The optical transmittance (*T*) and reflectance (*R*) of the films were measured by an UV-Vis-NIR spectrometer with an integrating sphere, and the absorptance was determined from 100%–*T*–*R*. Depth profiling was carried out using IONTOF TOF-SIMS5 model. Primary analysis was performed by positive Bi⁺ ion beam at 30 keV and ~1.29 pA current over a 100 × 100 μm² analyzed area using sawtooth rastering mode. The depth profile has been conducted in positive polarity, which targets the positive ions with secondary sputtering source.

III. RESULTS

Figure 1 shows the Raman spectra of the deposited films at different temperatures. The existence of an amorphous phase in the matrix is responsible for a distribution of lengths and angles in the silicon-to-silicon bonds.

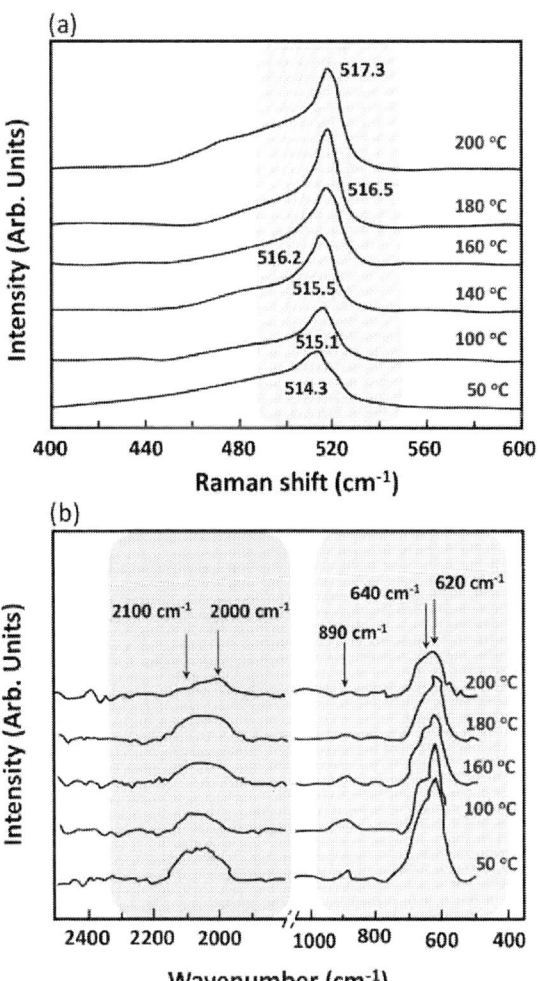

Fig. 1. (a) Raman spectra and (b) Infrared spectra of films deposited in the temperature range within 50 and 200 °C.

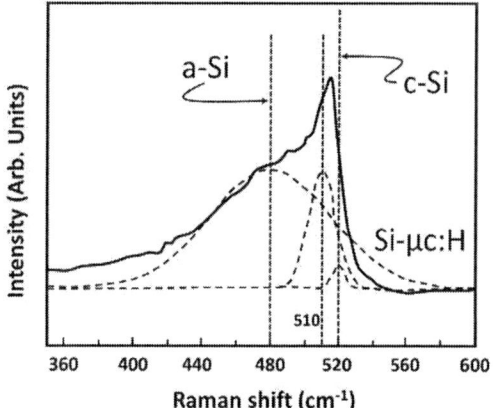

Fig. 2. Raman dispersion of microcrystalline hydrogenated-silicon samples showing the deconvolution of spectrum for crystalline (520 cm⁻¹), intermediate (510 cm⁻¹), and amorphous (480 cm⁻¹) phases.

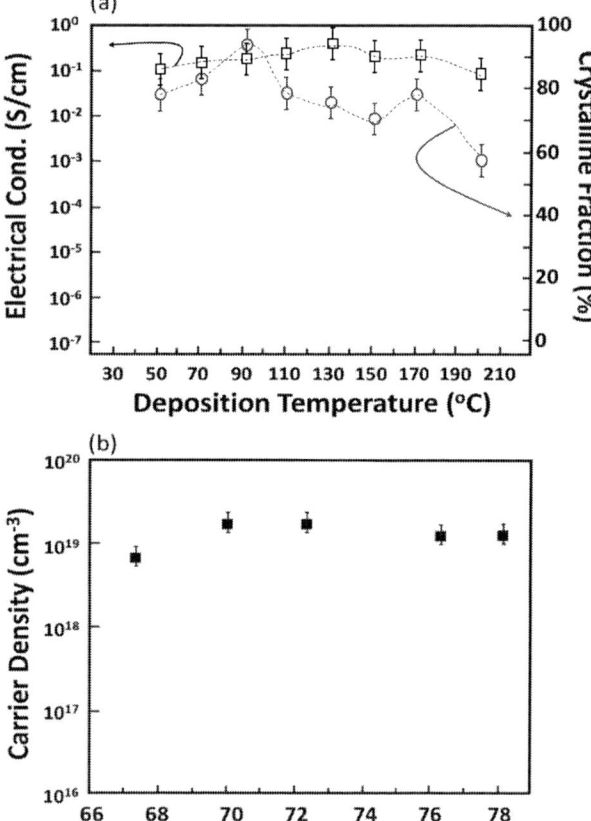

Fig. 3. (a) Conductivity and crystalline fraction as a function of the substrate temperature for representative hydrogenated-silicon films. (b) Carrier density as function of the crystalline volume fraction Xc (%).

Thus, there is a broadening in the Raman spectra as well as a shift to lower energies. We observed that, with decreasing deposition temperature, the crystalline component (520 cm¹) broadens and its maximum shifts to lower frequencies [18-22].

The infrared spectra shown in Fig. 1b indicates that with an increase of the deposition temperature there is a decrease in the stretching absorption modes at 2000 and 2100 cm⁻¹ (Si-H and Si-H₂, respectively) and also at ~ 630 cm⁻¹, characteristic of the bending, rocking, and wagging modes, ascribed to the Si-H bonds. A reduction of the absorption at 890 cm⁻¹ is also observed. The interpretation of the Si-Hx (x=1, 2 or 3 modes) associated with this spectral region is as follows: the absorption at 890 cm⁻¹ (i.e. "scissors" and/or "wagging") and at 2090-2100 cm⁻¹ ("stretching") is ascribed to the presence of [Si-H₂]n type of bonds, whilst the absorption at 880 cm⁻¹ and 907 cm⁻¹ is ascribed to the "bending" modes of hydrogen bonded as Si-H₂ and Si-H₃. A possible interpretation for the observed result is that at higher temperatures, the precursors are more energetic and, therefore, due to their greater mobility, a rearrangement to a more stable configuration in the growing surface occurs. The crystalline fraction Xc obtained from the deconvolution of the Raman spectra for wave numbers between 480 and 520 cm⁻¹ attributed to the vibrational TO modes of the network phonons on the amorphous and crystalline silicon, respectively, has a quasi-linear dependence with the temperature (see Fig. 2).

The crystalline fraction (Xc) was calculated through the empiric expression ($Xc= Ic + Im / Ic + Im + Ia$), where Ic corresponds to the integrated intensity of the 520 cm⁻¹ peak, ascribed to the crystalline contribution Im is the integrated intensity in the Gaussian related to the intermediate peak 510 cm⁻¹, and Ia is the integrated intensity of the 480 cm⁻¹ peak, ascribed to the amorphous sections of the material. Fig. 2 shows the deconvoluted spectrum that allows for the calculations of the crystalline phases [23-27].

In the calculation of Xc the intermediate 510 cm⁻¹ peak was assumed to be related to grain boundaries, where an increase in the bond lengths of the crystalline regions shifts the Raman dispersion spectra to lower energies. With this procedure, results that are more reliable are obtained since all the contributions from the material structure are taken into account in the matrix phonon scattering for the Raman spectra deconvolution. Crystalline fractions up to 30% were obtained at temperatures below 100 ºC. As the deposition temperature increased, the maximum crystalline fraction was approximately 80% in our samples. This crystalline volume fraction value confirms the supposition about the structural relaxation to a more stable configuration (crystalline) with an increase in the mobility of the species adsorbed in the growing surface. It is worth noting that these crystallinity fractions (66% at $Ts=100$ °C for undoped samples) are obtained at temperatures lower than those usually used in the CVD microcrystalline silicon film deposition process. This is an indication that there are factors affecting the crystallization mechanisms other than the substrate deposition temperature. The way the hydrogen is incorporated in the network can be indicative of the film structure. Fig. 3a shows that the electrical conductivity exhibits an increase of five orders of magnitude when the substrate deposition temperature is increased from 50 to 200 °C, with the crystalline fraction showing a similar behavior.

The measured d.c. conductivity values for samples deposited at temperatures below 100 °C are smaller than the typical conductivities of n-type doped amorphous silicon films ($\sim 10^{-2}$ S cm^{-1}). Some hypotheses to explain this fact are presented in the literature: (i) the phosphorus atoms are not activated due to the low substrate deposition temperatures utilized in the preparation of the samples with a consequent reduction in the number of carriers, (ii) the phosphorus atoms in interstitial structure sites contribute to increase the structural defect density, thus reducing the electronic mobility and are segregated to the amorphous matrix region around the crystallites impairing the electric conduction process, and (iii) the properties of the amorphous matrix depend both on the hydrogen content and on the hydrogen-silicon Considering just the samples deposited at temperatures above 100 °C, we observe that the density of carriers, as evaluated by Hall effect measurements, is proportional to the crystalline volume fraction of the film (Fig. 3b).

J-V characteristics comparison between cells with μc-Si:H(n) and a-Si:H(n) is listed in TABLE I. The conversion efficiencies of the cells using μc-Si:H(n) layer, and a-Si:H(n) layer, are 17.61% and 18.79%, respectively. As listed in TABLE I, the application of μc-Si:H(n) layer brings 2.4% gain in Jsc but suffers from lower FF.

Window	Eta %	Voc (mV)	Jsc (mA/cm2)	FF %
μc-Si:H	17.61	718	33.80	72.60
A-Si:H	18.79	715	33.07	79.50

Table I: J/V results with solar cells with different N layers Measurement performed for films grown at 130 °C

As well known, very thin amorphous silicon layer can passivate the silicon dangling bonds at c-Si surface, explaining high Voc of a-Si/c-Si heterojunction solar cells [2,6]. In the following defect-rich doped layers deposition, the passivation of a-Si/c-Si interface could be affected, and the minority carrier lifetime (MCLT) would drop. Therefore, we measured minority carrier lifetimes after doped layers deposition. For the cell with μc-Si:H(n), MCLT measurement gave 1691 μs, while 1395

μs for cell with a-Si:H(n). This MCLT difference between using μc-Si:H(n) and a-Si:H(n) might be from the valence band barrier at the interface between a-Si:H(i) and μc-Si:H(n), which prevents the back diffusion of photo-generated holes and reduces the recombination of photo generated holes. Vac from J-V measurement of the two samples shown in TABLE I were very close, 718 mV and 715 mV correspondingly, slightly in favor of μc-Si:H.

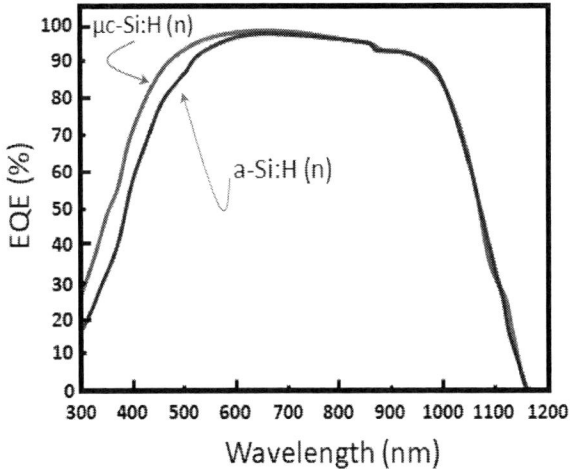

Fig. 4. EQE spectra of a-Si:H/c-Si solar cells with different n layers

EQE spectra for both samples are shown in Fig. 4 in the range of 300-1200 nm [28, 29]. From EQE spectra, the photocurrent gain in the cell with μc-Si:H(n) is mainly from better spectral response in the region between 300-700 nm. We attribute it to superior optical and electrical properties of μc-Si:H(n) film. In order to confirm that, we carried out optical transmittance and reflectance measurements for i/n stack layers (7nm i layer plus 8nm n layer) with different n layers, and calculated the effective transmittance (TTe) by the following formula:

$$TTe = T/1 - R$$

where T and R represent the measured transmittance and reflectance respectively. The effective transmittance spectra of different i/n stack layers. For the layers with μc-Si:H(n), TTe is higher than its counterpart a-Si:H(n) in the range of 300-700 nm, agreeing with the EQE result shown in Fig. 4 [30-33]. The high transmittance in the short wavelength region is resulted from the volume fraction of crystallinity content in the film. At the front surface of SHJ solar cell, since ITO/n/i layers form a grading refractive index structure, it helps to reduce the reflectance and enhance transmittance, agreeing with the slightly EQE improvement in the wavelength region between 700-1000 nm in Fig. 4.

IV. CONCLUSIONS

It has been shown that the substrate temperature deposition is not a *sine-qua-non* condition to produce microcrystalline hydrogenated-silicon films. With the proper choice of deposition parameters, deposition at low temperature of microcrystalline silicon films exhibiting high conductivity is possible in a conventional PECVD reactor. The influence of the temperature on the electrical conductivity of the μc-Si:H films may be related to the phosphorus activation, which increases with increasing deposition temperatures. The increase in the electrical conductivity depends on an increased incorporation of the phosphorus dopant in a P_4 configuration (coordination number 4). If there is insufficient energy in the system to permit the phosphorus atoms to bond in this configuration, they will be interstitially located, increasing the defect density, or they may segregate to the grain boundaries in a P_3 configuration. In both cases, the result is a decrease of the crystallinity and the conductivity of the samples. For the lower substrate deposition temperatures, the precursor adatom species Si-H_3, Si-H_2, Si-H, P-H_2, P-H, H) mobility is smaller than at higher deposition temperatures. Hydrogen bonded to a crystalline grain may act as a barrier, inhibiting grain growth. Also, due to the chemical reactivity of the hydrogen plasma with silicon, the formation of polymerlike chains [Si-H_2]n may occur. This situation leads to a reduction in the electrical conductivity and, to a smaller degree, in the crystalline volume fraction. This will be deeply detailed in the full version of this paper.

ACKNOWLEDGMENT

The authors thank PV-LAB (EPFL) in Neuchatel (Switzerland) for the PECVD deposition and Core labs (QEERI) in Qatar for the materials characterizations.

REFERENCES

[1] T. Jana and R. Goswami "Optoelectronic and Structural Properties of Plasma Deposited Nanocrystalline Hydrogenated Silicon Oxide Thin Films" Nano Vol. 16, No. 10, 2150115 (2021).

[2] Y. Alajlani et al. "Inorganic Thin Film Materials for Solar Cell Applications" Encyclopedia of Smart Materials Vol. 1, 386-399 (2018).

[3] J. Cattin et al. "Optimized Design of Silicon Heterojunction Solar Cells for Field Operating Conditions" IEEE Journal of Photovoltaics 9 (6), 1541-1547 (2019).

[4] M.M. Kivambe et al. "Record-Efficiency n-Type and High-Efficiency pType Monolike Silicon Heterojunction Solar Cells with a HighTemperature Gettering Process" ACS Applied Energy Materials 2 (7), 4900-4906 (2019).

[5] K. Yoshikawa et al. "Silicon heterojunction solar cell with interdigitated back contacts for a photo conversion efficiency over 26%," Nature Energy, vol. 2(5) 17032 (2017).

[6] J. Melskens et al. "Passivating contacts for crystalline silicon solar cells: from concepts and materials to prospects," IEEE Journal of Photovoltaics, vol. 8(2) 373–388, (2018).

[7] K. Masuko, M. Shigematsu, T. Hashiguchi, D. Fujishima, M. Kai, N. Yoshimura, et al. "Achievement of More Than 25% Conversion Efficiency With Crystalline Silicon Heterojunction Solar Cell," IEEE Journal of Photovoltaics, vol. 4, pp. 1433-1435, Nov 2014.

[8] S. De Wolf, A. Descoeudres, Z. C. Holman, and C. Ballif: "High-efficiency Silicon Heterojunction Solar Cells: A Review," green, vol. 2, 2012.

[9] M. Taguchi, A. Yano, S. Tohoda, K. Matsuyama, Y. Nakamura, T. Nishiwaki, et al. , "24.7% Record Efficiency HIT Solar Cell on Thin Silicon Wafer," Photovoltaics, IEEE Journal of Photovoltaics, vol. 4, pp. 96-99, 2014.

[10] J. Ramanujam and A. Verma, "Photovoltaic Properties of a-Si:H Films Grown by Plasma Enhanced Chemical Vapor Deposition: A Review," Materials Express, vol. 2, pp. 177-196, 2012.

[11] Z. C. Holman, A. Descoeudres, L. Barraud, F. Z. Fernandez, J. P. Seif: S. De Wolf, et al. "Current Losses at the Front of Silicon Heterojunction Solar Cells," IEEE Journal of Photovoltaics, vol. 2, pp. 7-15, Jan 2012.

[12] M. Tanaka, M. Taguchi, T. Matsuyama, T. Sawada, S. Tsuda, S. Nakano, et al. , "Development of New a-Si/C-Si Heterojunction Solar-Cells: Acj-Hit (Artificially Constructed Junction-Heterojunction with Intrinsic Thin-Layer)," Japanese Journal of Applied Physics Part I-Regular Papers Short Notes & Review Papers, vol. 31, pp. 3518-3522, Nov 1992.

[13] H. Fujiwara and M. Kondo, "Effects of a-Si:H layer thicknesses on the performance of a-Si:Wc-Si heterojunction solar cells," Journal of Applied Physics, vol. 101, p. 054516, 2007.

[14] N. Jensen, U. Rau, R. M. Hausner, S. Uppal, L. Oberbeck, R. B. Bergmann, et al. , "Recombination mechanisms in amorphous silicon/crystalline silicon heterojunction solar cells," Journal of Applied Physics, vol. 87, pp. 2639-2645, Mar I 2000.

[15] M. R. Page, E. lwaniczko, Y. Q. XU, L. Roybal, F. Hasoon, Q. Wang, et al. , "Amorphous/crystalline silicon heterojunction solar cells with varying i-layer thickness," Thin Solid Films, vol. 519, pp. 4527-4530, 2011.

[16] H. Watanabe, K. Haga, and T. Lohner, "Structure of High-photosensitivity silicon-oxygen alloy films," Journal of Non-Crystalline Solids, vol. 164-166, p. 1088, 1993.

[17] S. De Wolf and G. Beaucarne, "Surface passivation properties of boron-doped plasma-enhanced chemical vapor deposited hydrogenated amorphous silicon films on p-type crystalline Si substrates," Applied Physics Letters, vol. 88, Jan 9 2006.

[18] S. De Wolf and M. Kondo, "Boron-doped a-Si: H/c-Si interface passivation: Degradation mechanism," Applied Physics Letters, vol. 91, Sep 10 2007.

[19] A. Samanta and D. Das, "Structural investigation of

nC-Si/SiOx:H thin films from He diluted (SiH4+C02) plasma at low temperature," Applied Surface Science, vol. 259, pp.477-485. 10/15/ 2012.

[20] P. Cuony, D. T. L. Alexander, L. LOfgren, M. Krumrey, M. Marending, M. Despeisse, et al. , "Mixed phase silicon oxide layers for thin-film silicon solar cells," MRS Proceedings, vol. 1321. 2011.

[21] P. Sichanugrist, T. Sasaki, A. Asano, Y. Ichikawa, and H. Sakai, "Amorphous silicon oxide and its application to metal/n-i-p/ITO type a-Si solar cells," Solar Energy Materials and Solar Cells, vol. 34, pp. 415-422, 1994.

[22] M. Liebhaber, M. Mews, T. F. Schulze, L. Korte, B. Rech, and K. Lips, "Valence band offset in heterojunctions between crystalline silicon and amorphous silicon (sub) oxides (a-SiOx:H, 0< x < 2)," Applied Physics Letters, vol. 106, p. 031601, 2015.

[23] J. M. Essick, Z. Nobel, Y. M. Li, and M. S. Bennett, "Conduction- and valence-band offsets at the hydrogenated amorphous silicon-carbon/crystalline silicon interface via capacitance techniques," Physical Review B, vol. 54, pp. 4885-4890, Aug 15 1996.

[24] H. Mimura and Y. Hatanaka, "Energy-Band Discontinuities in a Heterojunction of Amorphous Hydrogenated Si and Crystalline Si Measured by Internal Photoemission," Applied Physics Letters, vol. 50, pp. 326-328, Feb 9 1987.

[25] R. Carius, "Photoluminescence in the amorphous system SiO," Journal of Applied Physics, vol. 52, p. 4241, 1981.

[26] R. m. Biron, C. Pahud, F. -J. Haug, J. Escarn, K. Soderstrom, and C. Ballif, "Window layer with p doped silicon oxide for high Voe thin-film silicon n-i-p solar cells," Journal of Applied Physics, vol. 110, p. 124511, 2011.

[27] A. Klein, C. Korber, A. Wachau, F. Sauberlich, Y. Gassenbauer, S. P. Harvey, et all. , "Transparent Conducting Oxides for Photovoltaics: Manipulation of Fermi Level, Work Function and Energy Band Alignment," Materials, vol. 3, pp. 4892-4914, 2010.

[28] M. Iqbal, M.M. Nauman, F.U. Khan, P.E. Abas, Q. Cheok, A. Iqbal, B. Aissa, "Vibration-based piezoelectric, electromagnetic, and hybrid energy harvesters for microsystems applications: A contributed review, International journal of energy research 45 (1), 65-102, https://doi.org/10.1002/er.5643

[29]H. Gavi, Balla D. Ngom, A.C. Beye, A.M. Strydom, B. Aissa, V.V. Srinivasu, M. Chaker, N. Manyala, "Low-field microwave absorption in pulse laser deposited FeSi thin film, Journal of Magnetism and Magnetic Materials 324 (6), 1172-1176, https://doi.org/10.1016/j.jmmm.2011.11.003

[30]J. Haschke, R. Lemerle, B. Aïssa, A.A. Abdallah, M.M. Kivambe, M. Boccard, C. Ballif, "Annealing of silicon heterojunction solar cells: Interplay of solar cell and indium tin oxide properties, " IEEE Journal of Photovoltaics 9 (5), 1202-1207, DOI: 10.1109/JPHOTOV.2019.2924389

[31]B. Aïssa, M.A. El Khakani, "The channel length effect on the electrical performance of suspended-single-wall-carbon-nanotube-based field effect transistors,"

Nanotechnology 20 (17), 175203, https://doi.org/10.1088/0957-4484/20/17/175203

[32] A. Bentouaf, R. Mebsout, H. Rached, S. Amari, A.H. Reshak, B. Aïssa "Theoretical investigation of the structural, electronic, magnetic and elastic properties of binary cubic C15-Laves phases TbX2 (X = Co and Fe)," Journal of alloys and compounds 689 (25), 885-893, https://doi.org/10.1016/j.jallcom.2016.08.046

[33] R. Nechache, M. Nicklaus, N. Diffalah, A. Ruediger, F. Rosei, "Pulsed laser deposition growth of rutile TiO2 nanowires on Silicon substrates," Applied Surface Science 313, 48-52

Analysis for Solar Coverage and CO2 Emission Reduction of Photovoltaic-powered Vehicles

Masafumi Yamaguchi, Taizo Masuda, Takashi Nakado, Kazumi Yamada, Kenichi OKumura, Akinori Satou, Yasuyuki Ota, Kenji Araki, Nensuke Nishioka

Toyota Tech. Inst., Nagoya, Japan

Toyota Motor Co., Susono, Japan

Toyota Motor Co., Toyota, Japan

University of Miyazaki, Miyazaki, Japan

Photovoltaic (PV)-powered vehicles are expected to play a critical role in a future carbon neutrality society because it has been reported that the vehicle integrated PVs (VIPVs) have great ability to reduce CO_2 emission from the transport sector. Development of high-efficiency solar cell modules is very important due to the limited installable area of PV on vehicle exterior. This paper presents the importance of developing high-efficiency solar cell modules for PV-powered vehicles and test driving data of the Toyota Prius demonstration car installed with high-efficiency III-V compound triple-junction solar cell module with an efficiency of more than 30%. Average daily driving distance by VIPV of 17.0km under the global horizontal irradiance of 4.0 $kWh/m2/day$ and 62% CO_2 emission reduction from passenger cars in Japan are demonstrated. Analytical results for effectiveness of high-efficiency solar cell modules from the point-views of PV-powered driving distance and reduction in CO_2 emission are also presented. The PV usage is shown to be very effective for heavy vehicles with significant CO_2 emission reduction of 6-10 t-CO2-eq/year in the case of trailer and 1-3 t-CO2-eq/year in the case of truck.

Interrelated Characterizations of 2D/3D Perovskite Solar Cells Aged Under Damp Heat Conditions

Cynthia Farha[1], Emilie Planès*[1], Lara Perrin[1], David Martineau[2], and Lionel Flandin[1]

[1] Univ. Grenoble Alpes, Univ. Savoie Mont Blanc, CNRS, Grenoble INP, LEPMI, 38000 Grenoble, France

[2] Solaronix S.A., Rue de l'Ouriette 129, 1170 Aubonne, Switzerland

Abstract—**Among alternative device structures, carbon-based perovskite solar cells look highly promising due to their inherent high stability. A one step perovskite solution with ammonium valeric acid iodide additive was pipetted to infiltrate the mesoporous layers through the carbon layer. To further investigate their stability, aging campaigns at 85°C/85%RH have been conducted during 1000 h. In this study, matured encapsulated system based on glass and a surlyn gasket was used, enabling the humidity permeation up to solar cells and inducing probably an accelerated degradation of devices. Thanks to dedicated characterization techniques, the local performances have been correlated to the degradation inhomogeneity.**

Keywords—perovskite, solar cell, maturation, pipetting process, aging, characterization, durability

I. INTRODUCTION

Perovskite solar cells (PSCs) have stunned the photovoltaic community with an incredible progress in recent years, showing conversion efficiencies more than 25% [1]. These perovskites have a large absorption coefficient, high carrier mobility, direct band gap, and high stability. Despite its excellent performance, the $MAPbI_3$ perovskite is also a model sample for the instability studies against external environmental stress such as moisture and heat. 5-Ammonium valeric acid iodide (AVAI) doped $MAPbI_3$ was the first reported reduced-dimension perovskite to significantly improve the crystallinity and stability of 3D perovskite [2]. The presence of the AVAI additive in the perovskite precursors solution clearly had an impact on the final material.

In this study, HTM-free 2D/3D-AVAI-$MAPbI_3$-based solar cells were developed substituting the HTM with hydrophobic carbon electrodes (C-PSCs). When measuring the photovoltaic parameters of our cells, we observed that they do not achieve their maximal performance immediately following their completion, but rather require a certain maturation period [3]. Within this configuration, we demonstrate remarkable long-term stability of 1000 h, corresponding to 42 days under 85°C with 85%R.H stress conditions.

II. THE STUDIED SYSTEMS

A. Cell structure

Since the perovskite precursor solution is added after all the cell layers are fabricated, most of them need to be highly porous and no planar configurations can be implemented. The overall structure of a typical High Temperature Carbon Based (HTCB)-cell without a hole-selective layer consists of a front FTO-coated glass, a mesoporous inorganic electron-selective layer, an insulating layer and a mesoporous HTCB-electrode. To improve the efficiency by avoiding non-radiative recombination at the front electrode, a compact hole blocking layer (c-TiO_2) was grown by spray-pyrolysis and has proven to be essential for reaching high PCEs. Mesoporous TiO_2, ZrO_2 and carbon paste layers were added by screen printing. Emerging from the dye-sensitized solar cells, anatase mesoporous TiO_2 layer has been widely used as electron selective layer. Porous dielectric oxide layers of ZrO_2 have been mainly used as insulators, due to their ability to avoid direct shunts and inability to extract charge carriers. After the insulator layer is fabricated, the mesoporous cell stack is completed by depositing HTCB-layer. The perovskite precursor solution was finally dropped in the center of the cell (*fig. 1*) using a micropipette, and then annealed [4][5]. The obtained active area of the full cells was 1.5 cm² (performances were then measured either with or without a mask as indicated). Obtained cells are encapsulated using a glass/glass encapsulation with a Surlyn® gasket.

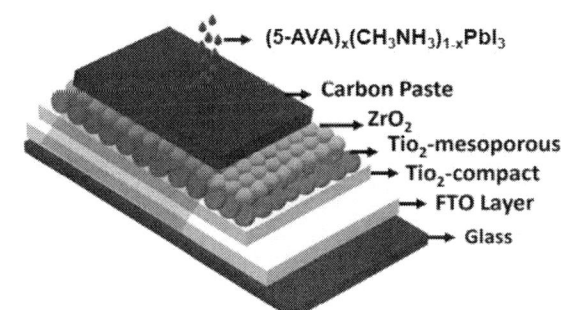

Figure 1: Carbon-based Perovskite Solar Cell architecture

B. Maturation proccess

The devices undergo a post-treatment method, which does not only improve the perovskite and its interfaces with others layers, but also provides a pathway to fabricate high performance and highly stable C-PSCs with low hysteresis. In this work, we observed significant improvements in the photovoltaic parameters when C-PSCs were exposed to high humidity (75%) at a constant temperature (40°C) for up to 100-150 hours in an environmental chamber [6].

978-1-7281-6118-1/22 $31.00 © 2022 IEEE

III. DIFFERENT CHARACTERIZATIONS TECHNIQUES

Different characterization methods have been used to correlate both solar cell performance and perovskite properties in order to evaluate potential degradation mechanisms. UV-Visible spectroscopy was used to study the optical properties of the perovskite layer. This method also allows the detection of the different phases present in the layers of the solar cell. Due to the opacity of the carbon layer, the measurements were performed in diffuse reflection mode. The transition of our perovskite (PK) is equal to approximately 765 nm. By the Tauc Plot method, it is then possible to evaluate the band gap, which is close to 1.59 eV. Thanks to XRD analyzes, it is possible to identify all the crystalline phases present in our perovskite and their main orientations. A relative crystallinity can be evaluated by this way. Furthermore, photoluminescence spectroscopy measurements from 600 nm to 900 nm with an excitation at 520 nm have been performed to identify the defects present in the perovskite layer or at the interfaces with the active layer. The PV performances of solar cells were determined by I-V measurements realized under AM 1.5 conditions (1000 W.m^{-2}). Thanks to the measurements it is possible to extract: the open circuit voltage (Voc), the short-circuit current (Jsc), Fill factor (FF), Series resistance (Rs), Shunt resistance (Rsh) and the Power conversion efficiency (PCE). Finally, the LBIC (Light Beam Induced Current) measurements have performed to study local performances of solar cells. The equipment used was developed in the lab, the cells are excited with a green laser beam (520 nm) with a diameter of 0.7 mm which moves with a step of 1 mm on the surface of the cell. We obtain as a result a LBIC mapping showing the dispersion of the measurement of short-circuit current within the cell.

IV. RESULTS

A. J-V Curves

J-V measurements, were made either by putting a 0.64 cm² mask on the cell to study its center, or without any mask allowing to illuminate the full cell with an active area equal to 1.5 cm². At T0, we notice that the maturated cells have greater performance (PCE) than those without maturation, cells measured with a mask having better efficiency than those without (*fig.2(a)*). The higher PCE values obtained using only the center of the cell is due to the deposition method with drop deposit at the center of the cell, followed by a diffusion step.

Between 0 and 24h, we note a very different behavior between non-matured and matured cells: an important decrease of PCE and Jsc can be observed for matured cells while an opposite behavior is observed for the Jsc of non-matured cells (*fig.2(b)*). Their PCE still decreases due to Voc and FF decrease. We can therefore consider that non-matured cells undergo a competition between maturation and aging during this step. After 24h, the PV parameters are close whatever the cell type. Then, each kind of solar cell undergo its own way, and after 1000h of aging, both cells have similar performances of about 6.5% with a similar Jsc value of about 10mA/cm². This corresponds to a decrease of the initial performances of about 45% for both cells.

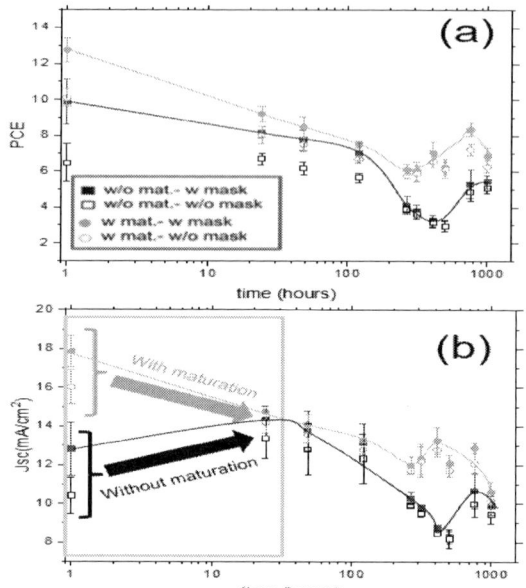

Figure 2: Evolution of the photovoltaic parameters a) PCE; b) Jsc for the two batches of cells with and without maturation (and with and without a measurement mask), as a function of aging time

B. LBIC mapping

LBIC measurements were performed throughout aging on the two cell types. In order to have a global view, a violin plots representation was chosen, as presented in figure 3. As expected by the macroscopic measurement of Jsc using J-V curves, the local Isc measured using LBIC confirms that matured solar cells delivers higher currents than non-matured ones. However, after 1000h of ageing, both types of cells present close LBIC patterns.

For the cells without maturation, the short circuit current globally increases between 0 and 264h indicating the slight initial maturation step, and then stabilizes until the end of aging (*fig. 3*). Concerning the matured cells, we can notice a decrease of Isc up to 264h of aging, probably explained by an aging effect. Between 264 and 500 hours, an increase in Isc is observed, which is probably related to the increase in PCE. Then, the Isc decreases again until the end of aging. We notice that at 1000 hours of aging, the Isc is the same for both types of cells, and this is correlated to the J(V) measurements.

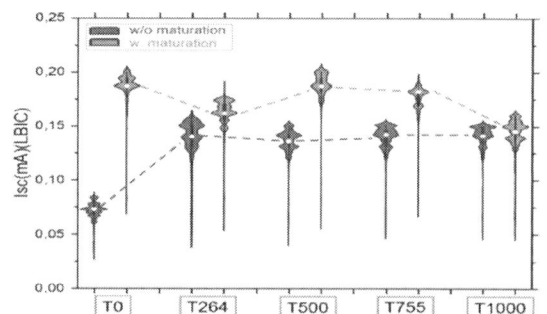

Figure 3: Violin Plots representing the Short Circuit Currents from LBIC mapping of cells with and without maturation over the aging

C. Photoluminescence

Between 0 and 300h, a decrease in PL intensity occurs for non-matured cells, reaching the intensity level of matured cells (*fig. 4*). The maturation phenomenon for non-matured cells seems to be confirmed, while matured cells probably undergo aging. Then, a significant red shift and an increase in PL emission were observed for both cell types, indicating a probable degradation of the PK and/or its interfaces with the other layers. Finally, a small variation in emission wavelengths is observed while a decrease in PL intensities is observed. This correlates with the observed increase in Voc, FF, and thus PCE.

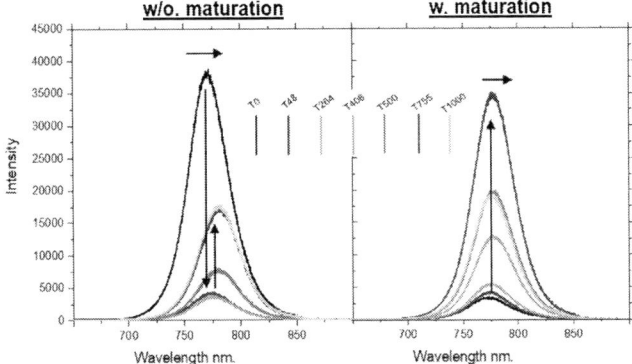

Figure 4: Photoluminescence spectra of both cells (from TiO₂ side) during the aging campaign

D. X-ray Diffraction

The XRD diffractograms of the two types of cells were performed at T0, T500 and T1000h (*fig. 5*). Whatever the type of cell, at 500h a peak appears at 10° (grey arrow), which can be attributed to the perovskite monohydrate phase[7], probably explained by the absorption of moisture by the perovskite. This result is to be correlated with the decrease of PCE and the increase of PL intensity. At 1000h of aging, the characteristic lines of PbI₂ are observable (red arrows) while that of PK monohydrate disappears: the degradation mechanism continues. The degradation mechanism seems to be limited since an increase of the PCE and a decrease of the PL intensity are noted in parallel.

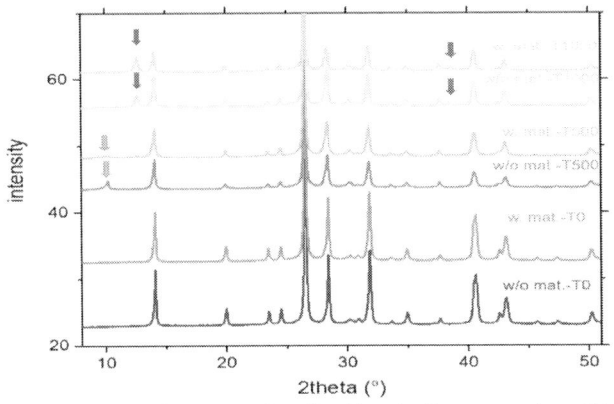

Figure 5: XRD Diffractograms for cells with and without maturation at T0, T500 and T1000h

CONCLUSION

In the initial state, the matured cells have better performance than the non-matured cells. After 24h of aging at 85°C/85%RH, both types of cells present similar performances indicating a limited effect of maturation on the stability of these devices. Thanks to a rather advanced study of the perovskite present within the solar cell, it was possible to highlight several stages of its degradation: formation of a monohydrated perovskite phase, then of PbI₂. According to these analyses, the perovskite shows a rather low level of degradation explaining the complete functionality of the solar cells after 1000h of aging. We can note a PCE decrease of about 45% for the non-matured and matured devices when using the 0.64 cm² measurement mask. Without the latter, a less severe impact is observed with 21% PCE loss for non-matured cells and 38% for matured ones.

ACKNOWLEDGMENT

This work has been partly founded by both: the European "UNIQUE" project, supported under the umbrella of SOLAR-ERA.NET_Cofund by ANR, PtJ, MIUR, MINECO-AEI, SWEA (Cofund ERA-NET Action, N° 691664), and the "PROPER" project supported by "EIG Concert Japan" and financed from the French National Centre for Scientific Research under the funding number "IRUEC 222437". This work has also been supported by the French National Research Agency, through Investments for Future Program (ref. ANR-18-EURE-0016-Solar Academy).

REFERENCES

[1] National Renewable Energy Laboratory (NREL). Best Research-Cell Efficiency Chart https://www.nrel.gov/pv/cell-efficiency.html (accessed Dec. 2021).

[2] Wei, N.; Chen, Y.; Miao, Y.; Zhang, T.; Wang, X.; Wei, H.; Zhao, Y. 5-Ammonium Valeric Acid Iodide to Stabilize MAPbI3 via a Mixed-Cation Perovskite with Reduced Dimension. *J. Phys. Chem. Lett.* 2020, *11* (19), 8170–8176. https://doi.org/10.1021/acs.jpclett.0c02528.

[3] Schneider, A.; Alon, S.; Etgar, L. Evolution of Photovoltaic Performance in Fully Printable Mesoscopic Carbon-Based Perovskite Solar Cells. *Energy Technol.* 2019, 7 (7), 1–9. https://doi.org/10.1002/ente.201900481.

[4] Bogachuk, D.; Zouhair, S.; Wojciechowski, K.; Yang, B.; Babu, V.; Wagner, L.; Xu, B.; Lim, J.; Mastroianni, S.; Pettersson, H.; Hagfeldt, A.; Hinsch, A. Low-Temperature Carbon-Based Electrodes in Perovskite Solar Cells. *Energy Environ. Sci.* 2020, *13* (11), 3880–3916. https://doi.org/10.1039/d0ee02175j.

[5] Bogachuk, D.; Tsuji, R.; Martineau, D.; Narbey, S.; Herterich, J. P.; Wagner, L.; Suginuma, K.; Ito, S.; Hinsch, A. Comparison of Highly Conductive Natural and Synthetic Graphites for Electrodes in Perovskite Solar Cells. *Carbon N. Y.* 2021, *178*, 10–18. https://doi.org/10.1016/j.carbon.2021.01.022.

[6] Hashmi, S. G.; Martineau, D.; Dar, M. I.; Myllymäki, T. T. T.; Sarikka, T.; Ulla, V.; Zakeeruddin, S. M.; Grätzel, M. High Performance Carbon-Based Printed Perovskite Solar Cells with Humidity Assisted Thermal Treatment. *J. Mater. Chem. A* 2017, 5 (24), 12060–12067. https://doi.org/10.1039/c7ta04132b.

[7] Leguy, A. M. A.; Hu, Y.; Campoy-Quiles, M.; Alonso, M. I.; Weber, O. J.; Azarhoosh, P.; Van Schilfgaarde, M.; Weller, M. T.; Bein, T.; Nelson, J.; Docampo, P.; Barnes, P. R. F. Reversible Hydration of CH3NH3PbI3 in Films, Single Crystals, and Solar Cells. *Chem. Mater.* 2015, 27 (9), 3397–3407. https://doi.org/10.1021/acs.chemmater.5b00660.

An Intelligent Algorithm for Maximum Power Point Tracking in PV Systems through Load Management

Kelvin Tan, Joseph A. Azzolini, William J. Parquette, Christian R. Polo, Meng Tao

Arizona State University, Tempe, AZ, United States

Practically all of today' photovoltaic (PV) systems employ a maximum power point tracker (MPPT) to maximize the power output of a PV array under different temperature, weather, and irradiance conditions. We proposed and demonstrated a load-matching PV system which performs maximum power point tracking by varying the number of loads connected to the PV array, without a conventional MPPT. However, the control algorithm in our system makes many unsuccessful switches as it does not know the optimum switch points for the loads. This paper presents an intelligent algorithm that can estimate the optimum switch point before attempting a switch. Simulation and experimental results show that the proposed algorithm is effective in minimizing unsuccessful switches. These results demonstrate an improved algorithm for maximum power point tracking through load management.

Large Area Survey Grain Size and Texture Optimization For Thin Film CdTe Solar Cells Using Xenon-Plasma Focused Ion Beam (PFIB)

Vladislav Kornienko[1,2], Ochai Oklobia[3], Stuart Irvine[3], Steve Jones[3], Giray Kartopu[3],Ali Abbas[1], Yau Yau Tse[2], Jake Bowers[1], Kurt Barth, and Michael Walls[1]

CREST[1], Department of Materials[2], Loughborough University, Loughborough, LE11 3TU, UK,

Centre for Solar Energy Research (CSER)[3], Faculty of Science & Engineering, Swansea University, St Asaph, LL17 0JD, UK

Abstract—Microstructural analysis of high efficiency thin film CdTe solar cells has been obtained over large areas. Analysis regions are device cross-sections approximately 0.325 mm in length. The samples have been prepared using a xenon-plasma focused ion beam (Xe-PFIB). The detailed images of the microstructure were obtained using backscattered electron imaging and electron backscatter diffraction (EBSD). As deposited devices and those with a low level of cadmium chloride treatment both show strong (111) growth texture. A high density of twins is seen in the columnar grains. Three As doped FTO/CdZnS/CdTe with varying process conditions we devices with 13.1%, 16.3% and 17% conversion efficiency were investigated. Lowest efficiency device was $CdCl_2$ treated at 420°C for 10 minutes while the 16.3 and 17% devices were both treated at 440°C for 10 minutes. The large area analysis revealed a partial recrystallisation state in the 16.3% efficient device which was induced by an incomplete chloride activation process. The analysis confirms that the efficiency of the devices tends to correlate with grain size. It also showed that a strong correlation exists between device efficiency and the randomization of the texture away from the (111) grain orientation. EBSD can be used to survey large areas and to mark out features for more detailed analysis using transmission electron microscopy (TEM). As an example, we show how using an EBSD scanned cross-sectional area can identify a partially recrystallized region which is then extracted and analyzed in detail using TEM.

Keywords—CdTe, texture, grain size, twins, EBSD, TEM

I. INTRODUCTION

It is commonly believed that the champion conversion efficiency of thin film CdTe devices can be increased to ~25% [1]. Some of the improvement requires optimization of the device microstructure. The $CdCl_2$ activation treatment has a dramatic effect on grain boundary and other defect passivation, and grain growth[2]. The effect of grain size and texture on conversion efficiency has been reported in relation to the randomization of [111]//GD (growth direction) textures for completely recrystallized absorbers[3]. XRD data is often used to illustrate the texture changes. However, the site-specific nature of the recrystallization process has not been extensively studied using Electron Back Scatter Diffraction (EBSD). This means observing clear defined region where new large grains with random orientation are visible next to small, columnar and highly orientated grains.

Cross-sectional EBSD characterization of the microstructure of thin film CdTe devices is often restricted by the small analysis area due to the slow Ga-FIB sample preparation. Large area texture analysis of CdTe devices has been carried out using standard Ga-FIB where the area approaches $80 \times 80 \mu m$[4][5][6]. However, this area was only analyzed in the surface plane and not in device cross-section. New spin polishing techniques can be used to prepare $100 \times 100 \mu m$ in plane view areas by using a Xe plasma focused ion beam (PFIB)[7][8], but it is difficult to corelate the texture, grain size and grain boundary information as a function of depth within the film. It is challenging to measure accurately exact depth the scan was taken and to ensure homogeneous milling across areas surface.

Here we report on large area EBSD analysis of three thin film As-doped CdTe devices each treated with different $CdCl_2$ activation processes yielding low, medium, and high performance. The results show the performance of the devices can be correlated to microstructural features such as texture, grain size, and grain boundary structure. Cross-sections >300μm in width were prepared and scanned using EBSD using a Helios G4 Xe-plasma focused ion beam (PFIB). Large-area EBSD provides much improved statistics for the microstructural analysis. We have also been able to select a particular region from the large survey cross-section and carry out Transmission Electron Microscopy (TEM) analysis on the same area.

The method used in this paper for large area cross-sectional EBSD analysis allows the back and front interfaces of the device to be imaged without requiring the typical lift-out procedure needed for transmission electron microscopy (TEM) sample preparation[9]. Cross-sections of EBSD scans of a lift out sample has been observed to be approximately 60μm in width when prepared using Ga-FIB [10]. This limitation on

978-1-7281-6118-1/22 $31.00 © 2022 IEEE

width reduces the statistical accuracy due to the small number of grains, typically <200 for CdTe devices depending on grain size. Lift out cross section EBSD analysis is also time consuming.

II. EXPERIMENTAL

The device structure consisted of FTO/CdZnS/CdTe where the CdTe was As doped. The CdTe layers were deposited by metalorganic chemical vapor deposition (MOCVD). Three devices were characterized. The first (device 1) was deposited with a nominally stoichiometric Cd to Te ratio in the CdTe. The CdCl$_2$ was 10 minutes at 420°C. The two other devices (2 and 3) had CdTe absorbers deposited with Cd-rich conditions and CdCl$_2$ treatments at 440°C for 10 minutes.

The samples were cleaved using a handheld diamond glass scriber and applying bending pressure to break the glass and produce a smooth cross-sectional edge with the device intact. An optical microscope was used to confirm a good edge was obtained. Rough or uneven cross sections require excessive PFIB milling. If un-milled glass protruded to far beyond the film, the signals would be blocked. Good alignment reduces expensive ion milling time and is an important consideration if large numbers of samples are to be analyzed. Fig.1 shows an example of an ideal cross-section edge prior to ion milling. The device is flush with the edge of the glass.

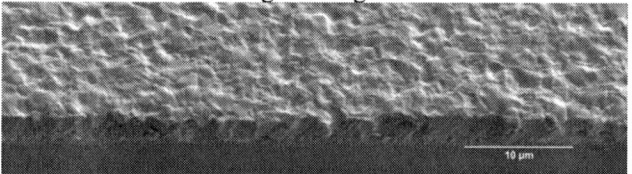

Fig.1. Example of device cross-section edge prior to milling, that is suitable for efficient ion milling.

A Helios G4 PFIB was used to prepare approximately 350μm of a cross-section width and scanned using EBSD and backscatter imaging. The ion beam, EBSD and backscatter detector are all housed in the same vacuum chamber inside the PFIB system. A 36° pre-tilt holder was used to tilt the device to 54° during ion milling and to 70° during EBSD scanning.

The total PFIB processing time was approximately 2 hours per device. The milling rate achieved is at least 10x the milling rate possible with a conventional Gallium ion FIB[11]. A 3μm layer of Pt was deposited at the cross-section edge to protect the device during milling (at 16kV and 30nA current). Milling at the device edge was applied since this helped to avoid the "chunk lift-out" procedure as well for TEM. This would have significantly increased the system time used per device. The ion milling current was set to 60nA (total milling area ~ 1.5 μm x 350 μm or 525μm^2).

The PFIB used is equipped with the latest CMOS symmetry EBSD detector. EBSD scans are optimized for CdTe device analysis to balance speed, resolution, and scan time. The speed setting was set to 600 patterns per second, 2μs beam dwell time, standard scan mode with ~85nm step size, a single 325μm scan in this case takes approximately 20-30 minutes after all the adjustments. Reducing the area of analysis and lowering the step size down to 25 nm enables higher resolution EBSD with a similar scan time to the wider 325μm scan.

Note that the resolution of the EBSD camera is also limited by the glass substrate used for the fabrication of the CdTe device. Regardless of the conductivity of the tin oxide layer, the glass is insulating, and this causes charging problems when attempting to increase dwell time or reduce the step size. This occurs even after careful sample preparation. The problem was mitigated by ensuring the surface was conductive by sputter-coating the device with Au/Pd for 60 seconds before PFIB milling.

After EBSD analysis, small lines could be ion etched to mark the area of interest for subsequent additional detailed analysis. This could include isolating a small section of the sample with the FIB, removing it, and then transferring it to a different transmission electron microscope (TEM). This was achieved using 30pA beam (30kV) at 54° tilt near the edge of the sample. This procedure results in a sample suitable for high resolution TEM imaging and energy dispersive X-ray spectroscopy (EDX) analysis on a known area defined during the large cross-sectional survey using EBSD and backscatter imaging. TEM analysis was carried out using FEI Techai 20 operated at 200kV, and equipped with an Oxford Instruments X-max N8 TLE SSD EDX detector. A site-specific TEM lamella of device 2 was prepared using a Ga-FIB at a partially recrystallized area of the device with medium performance.

EBSD pattern quality depends on milling surface quality which is influenced by the ion beam milling current. PFIB can exceed 2.5μA of current during milling, which is much greater than used in the Ga FIB. Exceeding current above 60nA (~350x1μm area) in the PFIB leads to less control of beam and more likelihood for curtaining effects to appear. This would appear as streaks going down the cross-section and interfere with EBSD scanning if they are severe. The Pt layer improves surface quality milling, but the roughness of the CdTe device surface also matters. The lower the beam current density, the smoother the surface after milling, resulting in sharper the EBSD patterns. Therefore, the advantage of PFIB is that higher current (>2.5μA) can be used for rapid bulk milling, but for cross-section polishing the lower (60nA) current is preferable. There is an optimum combination of milling current and final polishing current to reduce milling time and maintain the high quality of EBSD patterns. For the large cross-sections produced in this study, 60nA was used for the polishing step. However, for a thinner coating or smaller grain size, using 15nA or 4nA is an option for a similar cross-section width. Using a lower current density increases the ion milling time. However, reducing the current density is an option if analysis of the FTO layer or any other film in the region of 200nm thickness is of interest.

III. RESULTS

Table 1 shows the device J-V parameters and the microstructure data obtained using EBSD. Devices are labeled 1-3 from poor to best performance and from undertreated to fully treated. Conversion efficiencies of the devices 1, 2 and 3 were 13.1%, 16.3% and 17% respectively. The increase in grain size and the randomization of the as deposited [111]//growth direction (GD) texture is summarized and linked to the conversion efficiency. A peak texture intensity values (multiples of uniform density-mud) of 1 means the texture is completely random and >5 indicates a strong fiber (mostly in one direction) texture component such as [111]//GD.

Table 1. Device performance and EBSD data.

Device	CdCl$_2$ (°C/min)	η %	Jsc (mA cm^{-2})	Voc (V)	FF (%)	Average grain size (µm)	St dev (σ) Grain size	Texture intensity (mud)
1	420/10	13.1	24.5	0.758	71	0.9	0.5	10.54
2	440/10	16.3	26.5	0.849	72	1.1	0.6	9.06
3	440/10	17.0	26.8	0.868	73	1.5	1.1	2.16

Fig. 2 shows the band contrast (BC) and EBSD from the device 1. The high intensity of (111) grains is shown as blue in the inverse pole figure (IPF) map. IPF maps show the plotted crystal directions that are parallel with the growth direction of the films. Note that due to the large area of analysis, the single long scan (325µm) is divided into 5 segments. These are presented in Fig. 2, 3 and 5 for device 1, 2 and 3, respectively. The colored inverse pole figure maps show the orientations parallel to the growth direction (GD).

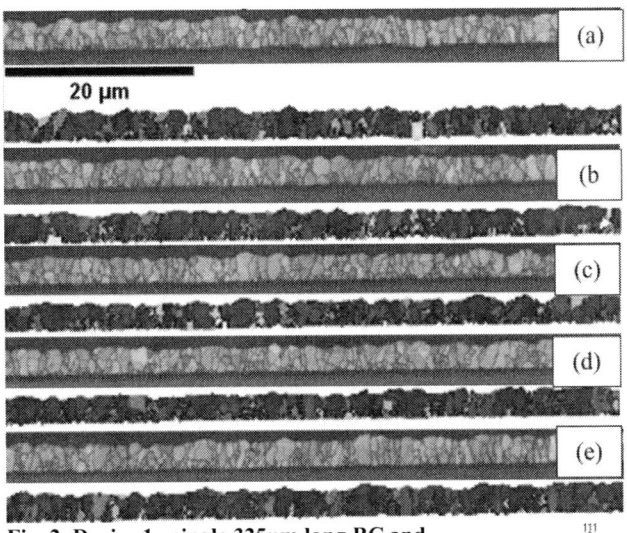
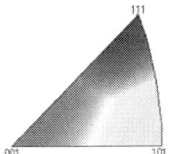

Fig. 2. Device 1 - single 325µm long BC and IPF map split into 5 × 65µm sections from left to right (a-e).

The lowest performing device 1, produced using the lowest CdCl$_2$ treatment temperature, had the smallest average grain size and highest intensity of (111) texture component. Device 2 exhibited a transitionary (partial recrystallized) structure as shown in Fig. 3. Strong (111) texture (blue), similar to device 1, is seen. Additionally, large grains extending through the absorber thickness are seen. They have a more random orientation than the smaller predominantly (111) grains. This suggests that these larger grains with a more random orientation are agglomerated and recrystallized smaller (111) grains. Examples of these are shown under horizonal arrows in Fig.3

(a, b, c and e). The recrystallization is only observed in specific areas, highlighting the importance of large-area analysis.

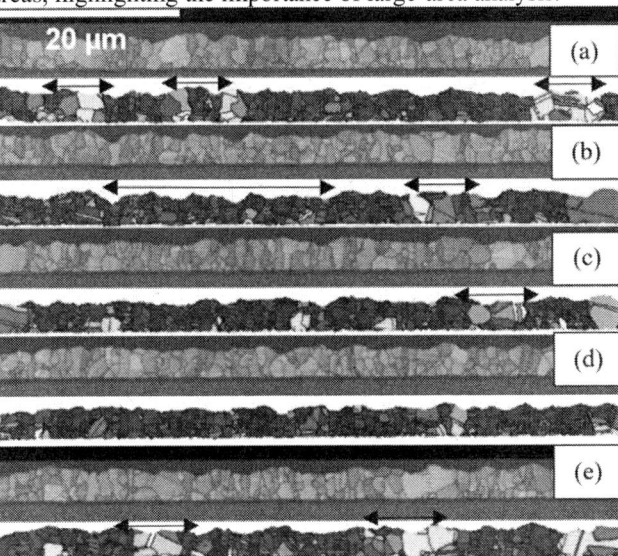

Fig. 3. Device 2 - single 325µm long BC and IPF map split into 5 × 65µm sections from left to right (a-e).

Analysis of device 2 also revealed a high density of twins in the non-recrystallized (111) grains. The larger non-(111) grains contained lower twin density. The high twin density is hard to resolve in Fig.3 (EBSD) and is more easily observed in the higher-resolution backscatter image, as indicated by arrows in Fig. 4.

Fig. 4. Backscatter image from an area selected in device 2 with the high twin density regions highlighted with arrows.

EBSD analysis with a smaller step size (25nm) and reduced area was also carried out on device 2 (see Fig. 5). Note that not all the twins can be identified in the high-density regions (111) orientated grains using the EBSD technique due to the spatial resolution limits. However, the presence of twins is clearer in the backscatter image presented in Fig. 5.

978-1-7281-6118-1/22 $31.00 © 2022 IEEE

Fig. 5. Backscatter and IPF map of device 2 scanned with 25nm step size.

EBSD quality is dependent on the surface roughness and defect density of the material. For example, the unidentified orientation and twins in Fig.5 make the analysis of twin density and texture less accurate.

Fig.6 shows the analysis taken from device 3.. The (111) texture is greatly reduced with an almost a 5× reduction in the peak intensity (mud) compared to device 1. The largest average grain size was also measured in device 3 at ~1.5μm. The EBSD scans for the three devices showed that they contained 852, 869, and 470 grains for device 1, 2, and 3, respectively across the 325μm scans.

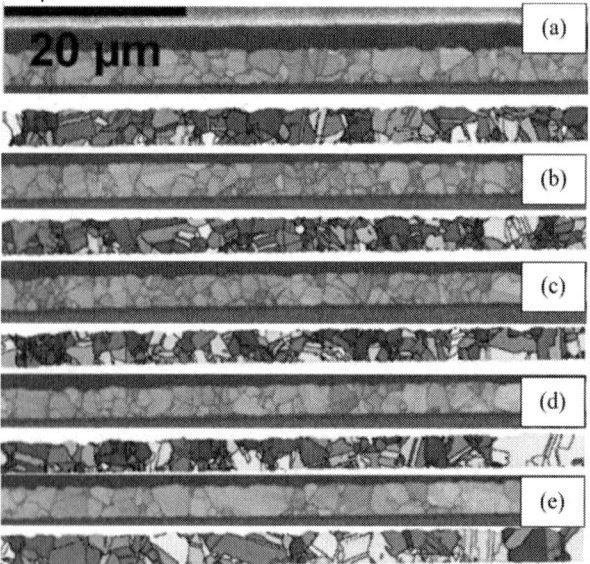

Fig. 6 Device 3 - single 325μm long BC and IPF map split into 5 × 65μm sections from left to right (a-e).

The large area surveying capabilities of the backscatter detector enables observation of a very large cross-sectional area of the devices (>1mm²), while retaining sufficient resolution to observe features such as voids and other defects. This is a useful capability particular at the front interface. By adjusting the contrast in image J software, the grain boundary contrast can be more defined.

The Helios G4 PFIB is equipped with an immersion mode for the backscatter imaging. Immersion mode involves the use of magnetic lens, which narrows the cone of electrons, which

improves spatial resolution. This also lowers the field of view to around 118μm but enables higher resolution images taken at 3kV. The backscatter images at full field of view are much higher resolution than EBSD, hence twin features and interface features are more visible. Figure 7 shows the comparison between the IPF orientation map and backscatter images for device 2. Therefore, backscatter imaging enables us to achieve sufficient resolution without necessarily carrying out TEM analysis with similar magnification. This is crucial particularly when observing the front interface. Voids are commonly observed in CdTe and CdSeTe/CdTe devices after the $CdCl_2$ treatment. Large area analysis $\geq 300\mu m$ facilitates determining the density of voids at the junction interface.

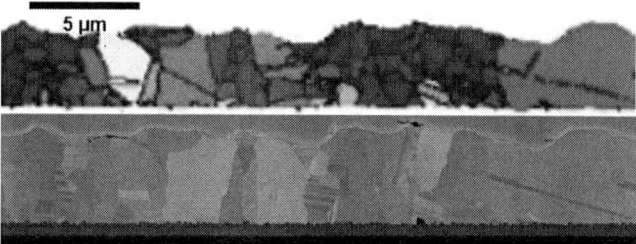

Fig.7. EBSD Coloured IPF map (85nm step size resolution) compared with backscatter image in the same area for device 2.

Large voids can be observed across the front interface in Fig 8. However, there are regions of 30-50μm of cross section where there are few or no voids and small area analysis could be misleading. The large area survey also reveals disparity in grain size across the device shown by Fig.9, which suggests that the $CdCl_2$ treatment is non-uniform. It is also possible to observe that the larger grains often have twin boundaries which are in random positions. The smaller grains in these devices had undertreated (111) orientated twins with much higher twin density than the fully recrystallized devices.

Fig. 8. Cropped backscatter image showing voids near the front interface of device 2.

Fig.9. Backscatter image of device 2 showing a veery large grain next to smaller grains with higher twin density.

Using ImageJ (open source image processing software), it was possible determine the grain size distribution, measure the length of all the potential twin boundaries, and from that obtain boundary length per unit area. For areas of particular interest, the superior resolution of Transmission Electron Microscopy (TEM) is extremely beneficial and complementary to EBSD and backscatter analysis. It is possible to select an area from the backscatter and mark it by ion etching and then extract a lamellae for TEM of that exact point. A selected area from device 2 is shown in Fig.10. The grain boundary per unit length highlighted in Fig.10 (a) was 6 μm^{-1} marked by red line for a recrystallized and (b) 31μm^{-1} for high density and partially recrystallized twinned grain respectively (marked by yellow line. There are significantly more twin boundaries visible in the TEM than indexed in the EBSD scan, or in the backscatter images. Therefore, values such as coincident site lattice boundary length (CSL) can be underestimated in EBSD if the grains are small and under-treated during CdCl$_2$ activation.

Fig. 10. Cross-sectional STEM of device 2, highlighting aa grain of measured twin density.

The EDX elemental maps shown in Fig.11 were carried out on the partially recrystallized section of device 2. From the EBSD data, the grains with high twin density are commonly (111) orientated, which appear as blue on the EBSD maps. There is a slight decrease in chlorine content in the grain interior for the long high density twinned grains, but the device was passivated, and chlorine is visible at the grain boundaries.

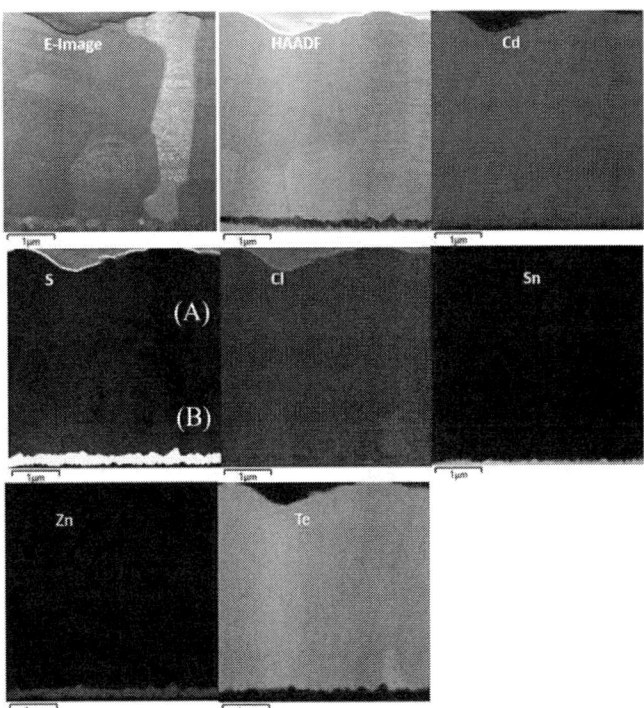

Fig.11. STEM images (bright field and HAADF) with corresponding EDX elemental maps for partially recrystallized device 2.

IV. DISCUSSION

This work has shown that EBSD is a useful technique to study variations in the grain size, texture, and boundary structure of large areas of thin film CdTe PV devices. The Xe-PFIB technique enables large-area sample preparation, high resolution backscatter imaging and EBSD to be achieved in about 2 hours per device.

Large-area analysis provides an improved overview of the uniformity achieved after device processing. For example, we observed areas of partial recrystallization state in device 2 adjacent to larger, through-thickness recrystallized grains. This quality of information highlights the importance of process uniformity to assist device making, considering that device 3 also had the same CdCl$_2$ treatment.

The during the CdCl$_2$ treatment the columnar (111) grains coalesce and recrystallize leading to more rounded randomly orientated grains with fewer defects. The lower temperature CdCl$_2$ treatment carried out at 420°C left a strong (111) texture. Increasing the temperature to 440°C led to a more randomized texture. Although, devices 2 and 3 were fabricated using the same processing conditions, the micro-structural analysis shows that device 2 did not fully recrystallize. This reveals the high sensitivity of the crucial CdCl$_2$ process.

Confirming our previous observations [7], there is a clear correlation observed between the CdTe average grain size and conversion efficiency for these samples. The average grain size in device 3 was 1.5μm and the efficiency was 17%. In contrast, the average grain size for device 1 was 0.9μm and the efficiency was 13.1%.

The results also reveal a correlation between the randomization of the grain orientation with conversion efficiency for these samples. The larger voids observed above the CdZnS buffer layer were between 300-500nm. There were also smaller voids that are resolvable in the backscatter and TEM images, but they are very small (<50nm in size).

This study, in agreement with previous ones conducted on close-space sublimated CdTe devices [11], show that the best performing devices exhibit a more randomized texture. The randomness of the texture can be measured using the multiple of uniform density (mud) value extractable from the EBSD analysis. In this work, the best device had a low mud value of ~2. A mud value of 1 would correspond to a completely randomized texture. The effects of the CdCl$_2$ process on all parts of the device are extremely complex involving grain boundary and other defects, passivation, diffusion, defect removal, recrystallization, and grain growth, etc. There is a clear correlation for randomization of texture and increase in grain size for higher efficiency. However, device interfaces and processing conditions need to be simultaneously optimized, to obtain higher efficiencies.

Stacking fault are removed during chlorine treatment. It is however a byproduct of the treatment and the presence of chlorine itself in the grain boundaries causes electronic defect passivation[2]. The results in Fig. 11 show a very minor change in chlorine concentration between grains with low and high density of twins. Larger recrystallized grains show a reduction in twin boundary density, and twin orientation is no longer primarily just parallel with the substrate. The (111) texture component is reduced after recrystallisation. The more random grains with lower twin density have progressed to a further recrystallized state. This is important to consider when dealing with interdiffusion of elements in devices with Se for example[13]. The Sulphur map in Fig. 11 shows higher counts near the recrystallized grain than the partially recrystallized grain as labeled in B and A respectively. which highlights again that the CdCl$_2$ process is inhomogeneous.

V. CONCLUSION

Three CdTe devices were characterized in a PFIB using EBSD and high-resolution backscatter imaging. The large 0.350 mm long cross-section survey capability of the PFIB enables a more complete analysis of microstructure. Areas of interest were be marked for subsequent complementary analysis using TEM. As an example, a small partially recrystallized region was identified and extracted for TEM analysis. The degree of recrystallisation can be quantified using EBSD by measuring the texture intensity or mud value. The mud of the lowest performing devices was 9.06-10.54 compared with 2.16 for a more fully recrystallized structure. Grains with a predominantly (111) orientation showed a high twin density. Regions with more random orientation also had generally larger grains. There is a correlation between larger, more random grains and device performance. The twin density is underestimated in EBSD due to its resolution limits. TEM example measurements estimated twin density of 31um^{-1} for partial recrystallized grain compared with 6µm^{-1} for a lower twin density grain. EDX data showed an increase in chlorine and sulphur in the recrystallized regions compared to uncrystallized grains with high twin density.

ACKNOWLEDGEMENTS

The authors are grateful to UKRI, EPSRC and Loughborough materials characterization center (LMCC) for funding this project through grants EPW00092X/1, EP/W000555/1 and EP/P030599/1)

REFERENCES

[1] G.M. Wilson *et al.*, "The 2020 Photovoltaic Technologies Roadmap," *J. Phys. D. Appl. Phys.*,vol. 53,no.51, pp. 493001, 2020, doi: 10.1088/1361-6463/ab9c6a

[2] P. Hatton, M. J. Watts, A. Abbas, J. M. Walls, R. Smith, and P. Goddard, "Chlorine activated stacking fault removal mechanism in thin film CdTe solar cells: the missing piece," *Nat. Commun.*, vol. 12, no. 1, p. 4938, 2021, doi: 10.1038/s41467-021-25063-y.

[3] W. J. J. Luschitz, B. Siepchen, J. Schaffner, K. Lakus-Wollny, G. Haindl, A. Klein, "CdTe thin film solar cells: Interrelation of nucleation, structure, and performance," *Thin Solid Films*, vol. 517, pp. 2125–2131, 2008.

[4] D. Mao, C. E. Wickersham, and M. Gloeckler, "Measurement of chlorine concentrations at CdTe grain boundaries," *IEEE J. Photovoltaics*, vol. 4, no. 6, pp. 1655–1658, 2014, doi: 10.1109/JPHOTOV.2014.2357258.

[5] M. M. Nowell, M. A. Scarpulla, N. R. Paudel, K. A. Wieland, A. D. Compaan, and X. Liu, "Characterization of Sputtered CdTe Thin Films with Electron Backscatter Diffraction and Correlation with Device Performance," *Microsc. Microana.*, vol. 21, no. 4, pp. 927–935, 2015, doi: 10.1017/S143192761500077X.

[6] T. Baines, L. Bowen, B. G. Mendis, and J. D. Major, "Microscopic analysis of interdiffusion and void formation in CdTe$_{1-x}$Se$_x$ and CdTe layers," *ACS Appl. Mater. Interfaces*, 2020, doi: 10.1021/acsami.0c09381.

[7] V.Kornienko *et al.*, "Effects of Successive CdCl$_2$ Treatments on the Texture of Thin Film CdTe Solar Cells," *IEEE 47th Photovolt. Spec. Conf.*, pp. 1–3, 2020.

[8] B. E. McCandless *et al.*, "Overcoming Carrier Concentration Limits in Polycrystalline CdTe Thin Films with In Situ Doping," *Sci. Rep.*, vol. 8, no. 1, pp. 1–14, 2018, doi: 10.1038/s41598-018-32746-y.

[9] M. Lekstrom, M. A. McLachlan, S. Husain, D. W. McComb, and B. A. Shollock, "Using the in situ lift-out technique to prepare TEM specimens on a single-beam FIB instrument," *J. Phys. Conf. Ser.*, vol. 126, p. 12028, 2008, doi: 10.1088/1742-6596/126/1/012028.

[10] J. Poplawsky, C. Li, N. Paudel, Y. Yan, and S. Pennycook, "Identifying the Electronic Properties of Grain Boundaries in CdTe Thin-film Solar Cells Using Electron Backscatter Diffraction and Electron Beam Induced Current Techniques," *Microsc. Microanal.*, vol. 19, no. S2, pp. 718–719, 2013, doi: 10.1017/S1431927613005588.

[11] S. Mayr *et al.*, "Xenon plasma focused ion beam milling for obtaining soft x-ray transparent samples," *Crystals*, vol. 11, no. 5, pp. 1–7, 2021, doi: 10.3390/cryst11050546.

[12] T. Fiducia *et al.*, "Selenium passivates grain boundaries in alloyed CdTe solar cells," *Sol. Energy Mater. Sol. Cells*, vol. 238, no. May 2022, p. 111595, 2022, doi: 10.1016/j.solmat.2022.111595.

Diffraction-optimized surface structures for enhanced light harvesting in organic solar cells

Milena Merkel, Jörg Imbrock, and Cornelia Denz

Institute of Applied Physics, University of Muenster, 48149 Muenster, Germany

Abstract—Surface phase gratings for organic solar cells are designed, compared and optimized with respect to their ability to diffract incident light and thus increase its path length in the active layer and finally its absorption probability. The gratings are based on concentric rings, a Fermat's spiral or an Archimedean spiral. They exhibit circular symmetric diffraction patterns and therefore diffract light independent of its azimuthal angle. The grating pillars are arranged on these patterns according to a periodic sequence or the deterministic aperiodic Thue-Morse or Rudin-Shapiro sequences to further tailor the desired diffraction pattern. After additional parameter optimization, one of the most promising patterns, the Thue-Morse sequence on an Archimedean spiral, leads to a current density increase of 5%.

Index Terms—surface structure, organic solar cell, Thue-Morse, Rudin-Shapiro, path length elongation

I. Introduction

Organic solar cells exhibit several important advantages over their inorganic counterparts, such as their flexibility and light weight, but, most importantly, allow for an energy- and cost-effective production [1]. However, due to the low charge carrier mobility in organic active materials, the active layer must be thin for efficient charge carrier extraction, which in turn reduces optical absorption. The final active layer thickness thus always constitutes a compromise between the two antagonistic processes of optical absorption and charge carrier extraction to maximize the final current density [2].

Structured surface layers are used to resolve this contradiction. They diffract the incident light, thereby increasing the active path length of the incident light and enhancing its absorption probability, while maintaining efficient charge carrier extraction [3], [4].

Here, novel surface structures that diffract light independent of its azimuthal angle are designed, evaluated and optimized [5]. They are based on concentric rings, a Fermat's spiral, and an Archimedean spiral, all of which exhibit circular symmetric diffraction patterns. The grating pillars are arranged on the rings and the spirals according to a periodic sequence or the two famous deterministic aperiodic Thue-Morse and Rudin-Shapiro sequences [6], [7]. The diffraction characteristics of the corresponding phase gratings are compared and, assuming a realistic, state-of-the-art organic solar cell, the path length elongation and the resulting current density increase are calculated. Finally, the pillar depth and diameter are adjusted for a further optimization of the surface structure.

DFG within PAK 943 DE 486/22-1

II. Surface Structure Design

Phase gratings are applied on solar cell surfaces to diffract incident light and enhance its active path length. This increases the absorption probability of the light and ultimately the current density. Here, the phase gratings consist of the transparent material polydimethylsiloxane (PDMS). Fig. 1 depicts the diffraction of incident light with wavelength λ by such a phase grating with pillar height Δh, pillar diameter d, grating vectors $|G|$ and refractive index n_{PDMS}. The light is diffracted into angle δ, leading to a path length elongation P in the active layer, compared to the the light path without diffractive surface structure (dashed line). We construct various different surface structures based on concentric rings, a Fermat's, or an Archimedean spiral, all of which boast a circular symmetric diffraction pattern. As a result, they diffract light independent of its azimuthal angle, which simplifies their application. On these general shapes, the grating pillars are arranged according to both a periodic and the two deterministic aperiodic Thue-Morse and Rudin-Shapiro sequences.

Periodic sequences generally exhibit discrete diffraction peaks, whereas the Rudin-Shapiro (RS) sequence has an absolutely continuous diffraction spectrum and thus resembles a random sequence. The Thue-Morse (TM) sequence, in turn, exhibits a singular continuous diffraction spectrum. The arrangement of these different sequences on the circular symmetric shapes therefore results in a variety of diffraction patterns.

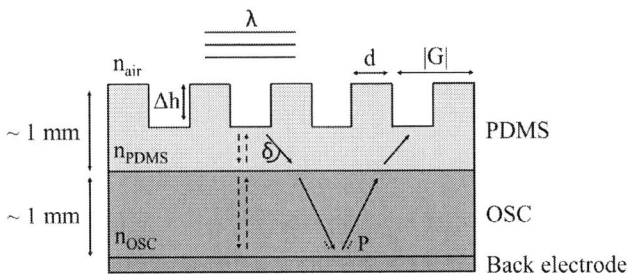

Fig. 1. Scheme of light diffraction from a PDMS phase grating with a relief height Δh, a pillar diameter d and grating vectors $|G|$ on an organic solar cell (OSC) with reflecting back electrode. n is the respective refractive index of the different materials and λ the wavelength of the incident light. The light is diffracted at an angle δ, leading to a path length elongation P in the active layer compared to the case without phase grating (dashed line). For the accurate calculation of the absorption increase, multiple reflections and refractions at layers adjacent to the active layer are taken into account.

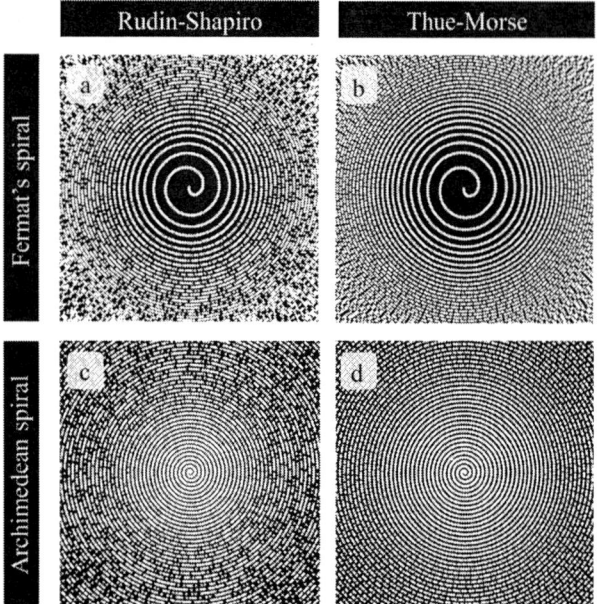

Fig. 2. Selection of structures used as surface phase gratings: Rudin-Shapiro and Thue-Morse sequences arranged on a Fermat's spiral (a and b) and on an Archimedean spiral (c and d), respectively.

In Fig. 2, a selection of the different surface structures studied is shown, namely the Rudin-Shapiro and the Thue-Morse sequences on a Fermat's spiral and on an Archimedean spiral. For better comparability, all of these structures are constructed with a duty cycle of 50 %.

III. PATH LENGTH ELONGATION

For each of the surface structures, the path length elongation in the active layer of a high-absorption, state-of-the-art solar cell with a PBDB-T:ITIC bulk heterojunction active layer is calculated. In Fig. 3 the calculations are visualized exemplarily for the Thue-Morse sequence on an Archimedean spiral.

The phase difference $\Delta\phi(x, y, \lambda)$ between the light incident on the grating at position (x,y) and the part incident on the grating grooves is given by

$$\Delta\phi(x, y, \lambda) = 2\pi(n_{\text{PDMS}} - n_{\text{air}})\frac{h(x, y)}{\lambda}. \quad (1)$$

From this phase difference, the intensity distribution I of the multitude of grating vectors G_x and G_y of the aperiodic phase grating in x- and y- direction is calculated via its Fourier Transform (FT):

$$I(G_x, G_y, \lambda) = |FT(e^{i\Delta\phi(x,y,\lambda)})|^2 \quad (2)$$

It corresponds to the diffraction efficiency of the different grating vectors and is plotted in Fig. 3a for $\lambda = 400\,\text{nm}$. Since this intensity distribution has a strong radial symmetry, it can be determined independent of direction:

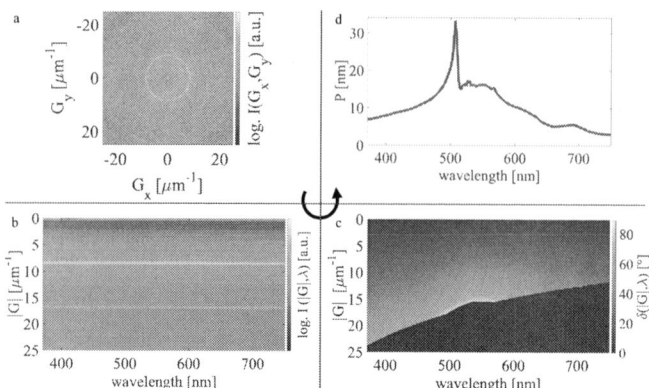

Fig. 3. Calculation steps to determine the path length elongation in the active layer for a Thue-Morse Archimedean spiral surface structure. a) Intensity distribution of the grating vectors G_x and G_y of the phase grating in x- and y-direction. b) Distribution of the absolute value of grating vector, $|G|$, for wavelengths between 370 nm and 750 nm. c) Diffraction angle in the active layer in dependence of the wavelength and the grating vector $|G|$. d) Active path elongation P.

$$I(|G|, \lambda) = \sum_{|G| = \sqrt{G_x^2 + G_y^2}} I(G_x, G_y, \lambda), \quad (3)$$

as shown in Fig. 3b for the whole wavelength range considered (370 nm to 750 nm).

Light with wavevector $k(\lambda)$ incident on this phase grating is consequently diffracted at angles $\delta(|G|, \lambda)$:

$$\delta(|G|, \lambda) = \sin^{-1}\left(\frac{|G|}{k(\lambda)}\right). \quad (4)$$

When propagating through the different layers of the solar cell, the diffraction angle changes depending on the refractive index of the respective materials according to Snell's law. Fig. 3c shows the final diffraction angle in the active layer as a function of $|G|$ and λ. For wave vectors smaller than the grating vector, the incident light is not diffracted, so the diffraction angle is zero, as in the lower right corner of the diagram. In the wavelength range between 480 nm and 580 nm, strong transitions between diffraction angles greater than 80° and 0° occur. This is due to the fact that at these wavelengths the refractive index of the active material is lower than that of the layer above (zinc oxide). Therefore, at certain angles of incidence on this interface, the light is diffracted at steep angles, while at others, total reflection occurs and the light cannot enter the active layer.

The path length elongation $P(\lambda)$ in the active layer with thickness t is finally obtained via

$$P(|G|, \lambda) = \frac{t}{\cos(\delta(|G|, \lambda))} - t \quad (5)$$

This path length elongation $P(|G|, \lambda)$ can be weighted by how often the respective grating vector $|G|$ actually occurs, i.e., by Fig. 3b:

$$P(\lambda) = \frac{\sum_{|G|} P(|G|, \lambda) I(|G|, \lambda)}{\sum_{|G|} I(|G|, \lambda)} \qquad (6)$$

This weighted path length elongation is depicted in Fig. 3d for a standard thickness of $t = 100$ nm. The broader peak corresponds to diffraction into steep angles in Fig. 3c, while the narrow peak on top corresponds to the overlap of the second line of higher intensity in Fig. 3b and the area of the steep diffraction angles in Fig. 3c.

The other surface structures cause a path length elongation with a similar broader, but different additional narrow peaks. However, for a final comparison of the different structures, the increase in current density they induce is calculated in a next step.

IV. CURRENT DENSITY INCREASE

To determine the final current density increase due to the application of the surface phase gratings, both the absorption spectrum of the active material and the solar spectrum must be considered.

The absorption A_{single} after a single pass through the active layer with and without the surface layer is given by:

$$A_{\text{single}}(|G|, \lambda) = T(|G|, \lambda) \left(1 - e^{-\alpha(\lambda)(P(|G|, \lambda) + 100\,\text{nm})} \right), \qquad (7)$$

where α is the absorption coefficient of the active material and $T(|G|, \lambda)$ is the portion of light transmitted through all cell layers above the active layer. In the case of a cell without surface structure, $P(|G|, \lambda)$ is set to zero, otherwise the values calculated in the previous section are used. To determine the total absorption $A_{\text{tot}}(\lambda)$, multiple reflections in the active layer, as well as in adjacent layers and at the back electrode are taken into account and finally a weighting according to the occurrence of the respective grating vector $|G|$ is performed.

Assuming optimal quantum efficiency, the current density J can be calculated via

$$J(\lambda) = A_{\text{tot}}(\lambda) I_{\text{sol}}(\lambda) e \qquad (8)$$

with $I_{\text{sol}}(\lambda)$ being the solar irradiance spectrum and e the elementary charge. The total increase in current density is calculated from the difference in current density with and without surface layer by integration over the entire wavelength range studied. The current density (J) increase is listed in Table I for the different surface structures.

For the selected pillar height and diameter, the surface structures cause J increases between 4.2% and 4.7%. The increase in J strongly depends on the overlap of the maxima of the intensity distribution of the grating vectors with the regions of diffraction into steep angles. Structures based on the Fermat's spiral show the lowest overall J increase, while those based on the Archimedean spiral show the highest. Moreover, fine structuring based on the RS sequence leads to smaller J increases than those based on the periodic and TM sequences.

TABLE I
INCREASE IN CURRENT DENSITY J USING THE DIFFERENT SURFACE STRUCTURES FOR A PILLAR HEIGHT AND DIAMETER OF 500 nm EACH.

Pattern	J increase (%)
RS Fermat	4.2
TM Fermat	4.2
Periodic Fermat	4.2
RS Rings	4.2
TM Rings	4.3
Periodic Rings	4.4
RS Archimedean	4.6
TM Archimedean	4.7
Periodic Archimedean	4.7

One of the two most promising structures, the TM Archimedean spiral, is further optimized in terms of pillar diameter and depth. For a diameter of 460 nm and a depth of 610 nm the strongest current density increase of 5.0% is achieved.

V. CONCLUSION

In summary, we investigated the properties of several intricate, novel surface structures based on periodic and aperiodic sequences arranged on concentric rings, a Fermat's spiral and an Archimedean spiral. All of these structures diffract light independent of its azimuthal angle, simplifying their application.

To identify the optimal surface structure, the respective current density increase was determined, taking into account the calculated path elongation in the active layer as well as the absorption spectrum of the active material and the solar spectrum. For one of the most promising structures, the Thue-Morse sequence on an Archimedean spiral, the pillar depth and diameter were further optimized and a current density enhancement of 5% was achieved.

These newly developed surface structures and the corresponding calculations can easily be adapted to other solar cell types. Moreover, while a strongly absorbing active layer was considered here, the current density can increase even more for thinner or weakly absorbing active layers.

REFERENCES

[1] G. Wang, M. A. Adil, J. Zhang, and Z. Wei, "Large-area organic solar cells: Material requirements, modular designs, and printing methods," Adv. Mater., vol. 31, pp. 1805089, 2019.

[2] D. H. Apaydın, D. E. Yıldız, A. Cirpan, and L. Toppare, "Optimizing the organic solar cell efficiency: role of the active layer thickness," Sol. Energy Mater. Sol. Cells, vol. 113, pp. 100–105, 2013.

[3] J. Cui, Á. Rodríguez-Rodríguez, M. Hernández, M. García-Gutiérrez, A. Nogales, M. Castillejo et al., "Laser-induced periodic surface structures on P3HT and on its photovoltaic blend with PC71BM," ACS Appl. Mater. Interfaces, vol. 8, pp. 31894–31901, 2016.

[4] M. Merkel, T. Schemme, and C. Denz, "Aperiodic biomimetic Vogel spirals as diffractive optical elements for tailored light distribution in funtional polymer layers," J. Opt., vol. 23, pp. 065401, 2021.

[5] M. Merkel, J. Imbrock, and C. Denz, "Diffraction-optimized aperiodic surface structures for enhanced current density in organic solar cells," unpublished.

[6] H. M. Morse, "Recurrent geodesics on a surface of negative curvature," Trans. Am. Math. Soc., vol. 22, pp. 84–100, 1921.

[7] W. Rudin, "Some theorems on Fourier coefficients," Proc. Am. Math. Soc., vol. 10, pp. 855–859, 1959.

Validation of Photovoltaic Spectral Effects Derived From Satellite-Based Solar Irradiance Products

Sophie Pelland, Christian A. Gueymard

CanmetENERGY, Varennes, QC, Canada

Solar Consulting Services, Colebrook , NH, United States

The Satellite Application Facility on Climate Monitoring (CM-SAF) Spectral Resolved Irradiance (SRI) and National Renewable Energy Laboratory (NREL) National Solar Radiation Database Spectral on Demand (NSRDB-S) satellite-based spectral irradiance products are tested here against benchmark data and models at seven ground stations: one in Spain for CM-SAF SRI and six in North America for NSRDB-S. Benchmarks include WISER spectroradiometers, spectra modeled from SolarSIM-G measurements and the SMARTS radiative code with two alternate input sources: AErosol RObotic NETwork (AERONET) and the Modern-Era Retrospective analysis for Research and Applications, Version 2 (MERRA-2) reanalysis. The satellite products are tested in terms of their ability to estimate photovoltaic (PV) spectral effects for six PV module technologies. The spectra are also compared directly under clear-sky conditions. Both CM-SAF SRI and NSRDB-S outperformed the simple benchmark of neglecting spectral effects in terms of predicting instantaneous spectral mismatch factors, but only CM-SAF SRI did better at predicting the long-term spectral derate factors. The clear-sky results revealed systematic differences between NSRDB-S and benchmark spectra, likely due to the NSRDB-S treatment of aerosols. Meanwhile, the mean SMARTS spectra with AERONET and MERRA-2 inputs were in good agreement, showing promise for the use of MERRA-2 as input to clear-sky models.

Elimination of the carbon-rich layer in Cu2ZnSn(S,Se)4 absorbers prepared from nanoparticle inks

Stephen Campbell, Martial Duchamp, Neil Beattie, Michael Jones, Guillaume Zoppi, Vincent Barrioz, Yongtao Qu

Northumbria University, Newcastle upon Tyne, United Kingdom

Nanyang Technological University, Singapore, Singapore

Kesterite Cu2ZnSn(S,Se)4 (CZTSSe) is a promising photovoltaic material attracting significant research interests in recent years. Among the variety of techniques employed for preparation of the absorber thin films, the best results are observed for a hydrazine-based method with efficiency up to 12.6 %. On the other hand, Cu2ZnSnS4 (CZTS) nanoparticle inks annealed in the presence of Se have shown efficiency as high as 9.3 %. Importantly, CZTS nanoparticle inks have the power to be compatible with high volume, high value manufacturing with a variety of substrates including flexible foils, plastics and ultra-thin glass. However, one of the current limitations of the nanoparticle ink technology is the presence of a fine-grain (FG) layer between the CZTSSe large grain (LG) layer and the back contact. The presence of this FG layer is likely to reduce device performance via carrier recombination through traps, interface states and increased grain boundary density. CZTS nanoparticles were synthesized by injection of cold sulphur (25 Â°C) into hot metallic precursors (225 Â°C). The long carbon chain molecule, oleylamine used in the nanoparticle synthesis step is believed to be the direct reason of the FG layer. Herein, a higher soft-baking temperature of 400 Â°C is studied to evaporate the carbon rich solvent efficiently from the nanoparticle precursor thin films before the selenization process. As a result, the absorber is found to be composed of a single LG CZTSSe layer where the carbon-rich FG layer is eliminated.

Light Trapping Characteristics of Photonic Crystal Constructs and Randomly Textured Thin Silicon

Sara M. Almenabawy, Yibo Zhang, Rajiv Prinja, Nazir P. Kherani

Department of Electrical and Computer Engineering, University of Toronto, Toronto, ON, Canada

Department of Materials Science & Engineering, University of Toronto, Toronto, ON, Canada

We present an experimental study investigating the light trapping properties of ultra-thin silicon with inverted pyramidic photonic crystals of nano-scale mesas, as well as random pyramidal textured silicon. Using conventional industry compatible technologies - photolithography and wet alkaline etching - photonic crystals are fabricated with uniformity, homogeneity, high reproducibility wherein the mesas are of nanoscale widths (i.e., below 40 nm). For thin-silicon foils of 20 and 40 µm thicknesses, both structures demonstrate significant enhancement in absorption compared to flat thin silicon, especially after the addition of double-layer anti-reflection coating and a metallic back reflector. For photonic crystals, the enhancement is due to the significant increase in optical path length especially at long wavelengths where effects such as parallel to interface refraction are observed. For random pyramids, the wide range of pyramid sizes obtained here (from 500 nm to 5 µm) show a broadband absorption enhancement where small pyramids suppress the reflection at short wavelengths and large pyramids contribute to trapping the long-wavelength photons. The maximal photocurrent densities for these structures are comparable to that for the Lambertian surface and exceed 40 mA/cm2 for silicon foil thickness down to 20 µm.

Gallium-Boron Spin-on co-Doping for Polycrystalline Silicon Passivating Contacts

Thien Truong, Matthew Young, Mowafak Al-Jassim, Daniel Macdonald, Hieu Nguyen, Josua Stuckelberger

Australian National University, Canberra, Australia

National Renewable Energy Laboratory, Golden, CO, United States

Herein, we present an alternative doping technique for p-type poly-Si/SiOx passivating contacts using a spin-on method for different mixtures of Ga and B glass solutions. Effects of solution mixing ratios on the contact performance, represented by an implied open circuit voltage iVoc and a contact resistivity ϱc, are investigated. For all as-annealed samples at different drive-in temperatures, increasing the percentage of Ga in the solution shows a decrement in iVoc (from ~680 to ~610 mV) and increment in ϱc (from ~3 to ~800 mΩ.cm2). After a hydrogenation treatment by depositing a SiNx/AlOx stack followed by forming gas annealing, all samples show improved iVoc (up to ~700 mV with Ga and B co-doped, and ~720 mV with all Ga). Interestingly, when co-doping Ga with B, even a small amount of B in the mixing solution is likely to have negative effects on the surface passivation. Free carrier density and total dopant density profiles obtained by electrical capacitance voltage and secondary ion mass spectrometry measurements, respectively, reveal a relatively low percentage of electrically active Ga and B after their diffusion into the poly-Si and the Si substrate. These results help to understand the different features of the two dopants: a low contact resistivity with B, a good passivation with Ga, their degree of activation inside the poly-Si and c-Si layers, and the effects of the annealing temperature.

Chemical surface and interface structure of sulfur-passivated silicon with a SiNx capping layer

Amandee Hua, Nan Jiang, Ajay Upadhyaya, Issac Lam, Tasnim K Mouri, Dirk Hauschild, Lothar Weinhardt, Wanli Yang, Ajeet Rohatgi, Ujjwal Das, Clemens Heske

Department of Chemistry and Biochemistry, University of Nevada Las Vegas (UNLV), Las Vegas, NV, United States

School of Electrical and Computer Engineering, Georgia Institute of Technology, Atlanta, GA, United States

Institute of Energy Conversion, University of Delaware, Newark, DE, United States

Institute for Photon Science and Synchrotron Radiation (IPS), Karlsruhe Institute of Technology (KIT), Karlsruhe, Germany

Institute for Chemical Technology and Polymer Chemistry (ITCP), Karlsruhe, Germany

Advanced Light Source (ALS), Lawrence Berkeley National Laboratory, Berkeley, CA, United States

SiO2 passivation is commonly used to improve the efficiency of silicon-based photovoltaics. However, SiO2 passivation requires high processing temperatures, potentially leading to a deterioration of the Si bulk quality. A novel sulfur-based passivation, requiring lower processing temperatures (~550 degree C), has been introduced, which, however, can suffer from degradation during the subsequent manufacturing process. Hence, a SiNx capping layer is required to protect the passivation layer. In this study, we have investigated sulfur-passivated n-n+ diffused silicon wafers with SiNx capping layers of varying thicknesses, as well as the impact of subsequent rapid thermal processing (RTP) on these layers, using x-ray photoelectron spectroscopy (XPS) and x-ray emission spectroscopy (XES). The surface-sensitive XPS data gives detailed insights into the local chemical bonding environments at the surface. In particular, it shows sulfur in a sulfite-like chemical environment and the presence of Si-O bonds on the sulfur-passivated silicon sample. The more bulk-sensitive XES S L2,3 spectra reveal the presence of S-Si bonds, which is maintained upon SiNx layer deposition. Subsequent RTP causes an increase in oxygen and sulfur content at the surface, accompanied with the formation of sulfates. A detailed description of the various chemical structure findings will be discussed in view of their ability to protect and passivate the Si surface.

Data Mining of Solar Cells Production Data Using Factorial Analysis

Johnson Wong, Dinica Li, Gordon Deans

Aurora Solar Technologies Inc, Vancouver, British Columbia, V7P3N4, Canada

Abstract — For a large solar cell factory floor, factorial Analysis is a technique which makes use of the entirety of production data and the path information, to delineate performance differences in the process tools of every process step. This is a data mining approach that has a number of important advantages over single factor experiments using test batches to compare tool performance: 1) there is no experimental overhead, no disruptions brought to the production that could lower overall throughput, 2) being a surveillance technique, it yields simultaneous information about all factors (all process tools of each step) as opposed to one factor at a time, 3) drawing conclusions on 100% of production data, it has inherently much greater statistical resolution and accuracy over sampling experiments. Moreover, factorial analysis is compatible with all production floors that have end-of-line I-V measurements, and does not require midstream process quality control measurement data to be effective, and is therefore positioned as a very cost effective method to monitor the production.

I. INTRODUCTION

A typical solar cell manufacturing environment consists of multiple process steps, with each step comprising of a number of parallel process tools whose combined throughput is commensurate with the production floor's capacity, as shown in Figure 1. For a 3GW factory, each day roughly 1000 batches of solar cells are routed through non-deterministic process paths. Each batch's solar cell mean performance, as characterized by end-of-line I-V measurements, is the combined result of all the process tools which created it.

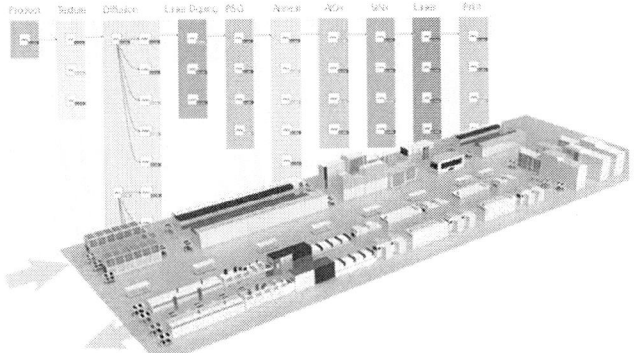

Fig. 1. Representation of the solar cell factory floor and the process steps and process tools.

In order to delineate the different tool's performance, auxiliary midstream measurement data, such as emitter sheet resistance, coating optical properties, or photoluminescence (PL) images may be collated, but most production lines do not have a comprehensive set of midstream inline measurement tools, and also these midstream measured wafer properties may have no direct relationship to the end-of-line solar cell I-V parameters. This paper presents an alternative approach that uses big data analytics to mine readily available I-V tester data and batch process path information across the factory.

II. TOOL DRIFT AND TOOL TO TOOL DISPARITIES

There are generally two kinds of attributable performance losses in a process step: 1) tool to tool disparities, which arises from difference in process tool's performance, and tool drift, which arises from temporal deviations from optimum level in a tool's performance. Figure 2 illustrates these two losses for four rear passivation tools (AlOx1 to 4). The performance is expressed in efficiency terms, generated from simulations of typical time series that may be seen in a solar cell production environment. Associated with these efficiency variations, there are corresponding changes in the main I-V parameters of short circuit current (Isc), open circuit voltage (Voc), fill factor (FF) in modes that are typical of variations in rear passivation quality, derived from simulations in Griddler 2.5 PRO [1].

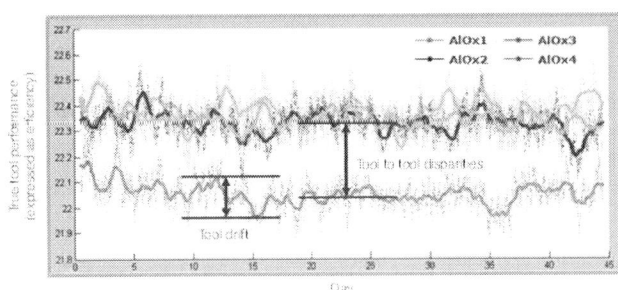

Fig. 2. Tool to tool disparities and tool drift in four rear passivation tools (simulation).

Figure 3 plots the average efficiency loss attributable to tool to tool disparities and tool drift for different processes. In total, tool drift accounts for up to 0.4% absolute loss in average cell efficiency, and tool to tool disparities accounts for up to 0.24% absolute loss. Therefore, there is a lot of value in accurately determining which tools are systemically low performing, and to monitor carefully the temporal drifts of tools and devise optimal maintenance scheduling.

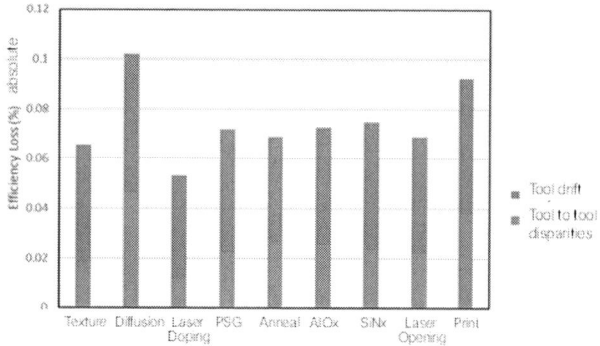

Fig. 3. Average efficiency loss attributable to tool to tool disparities and tool drift for different processes.

III. FACTORIAL ANALYSIS VERSUS HOT RUNS

If only end-of-line I-V data were available, human analysis of low performing tools will involve firstly some guesswork of which tools are underperforming, followed by running experimental batches or hot runs to verify the guess. Hot runs are comparative batches, in which the baseline batches are routed through a "golden path" consisting of the most reliable process tools, and the test batches are routed through an identical path except at the process step of interest, at which it is processed by the suspected underperforming tool. This is essentially a single factor experiment based on very limited samples. Often, to resolve difference in efficiency contributions between the suspected tool and the golden path, only one small batch is used to draw the conclusion. When the batch to batch run variation is greater than 1 bin (0.1%), then these hot runs will not be able to resolve anything but the most obvious poorly performing tools that are 0.5% absolute efficiency points below average.

In contrast, factorial analysis is a data mining technique that simultaneously deduces from thousands of batches routed through a myriad of different process paths, the overall picture of each tool's performance. The difference between factorial analysis and hot runs is diagrammatically sketched in Figure 4.

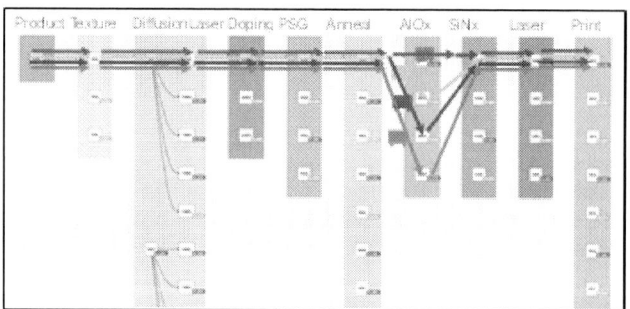

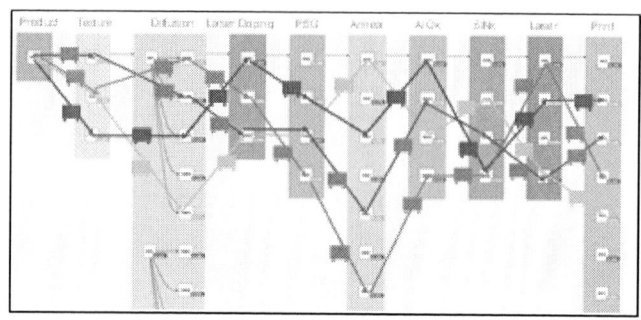

Fig. 4. Upper: Single factor experiments to delineate process tool performance differences; Lower: Factorial analysis to survey entire factory production data to delineate performance differences in all tools

Factorial analysis is a surveillence technique that acts on all batches that run through the production lines. It has the inherent advantages of making full use of all available information, rather than sampling using small batches. Moreover, it can draw conclusions about all tools simultaneously. Factorial analysis is also assoicated with a set of comprehensive statistical tools, which calculate the uncertainty associated with the tool performance differences. Even in a 24 hour period, hundreds of tool differences can be resolved with certainty levels that are equivalent to hot runs that made use of 4-20 full batches. At such a comparatively high certainty level, less than half bin (0.05%) differences can be resolved.

Factorial analysis allows one to uncover the true tool performance time series. Figure 4, upper graph, shows a simple plot of batches efficiency per tool. Each data point represents the daily average in efficiency of all batches which passed through a particular rear passivation tool. Because the batch efficiency is influenced by all process steps, the data point can be biased upwards or downwards depending on the proportion that the batches have been routed through other tools of other process steps. Therefore, this kind of simple plot is not an accurate depiction of true tool performance. In contrast, the middle graph is a plot after correcting for the systematic long term contributions to efficiency by tools of other process steps, accordig to the factorial analysis results. As seen, this middle graph is much closer to the bottom graph, which are the true tool performance timelines. Thus, plotting corrected time series using factorial analysis, allows engineers to draw far more accurate conclusions about the performnace differences between tools, as well as the temporal variation in tool performance.

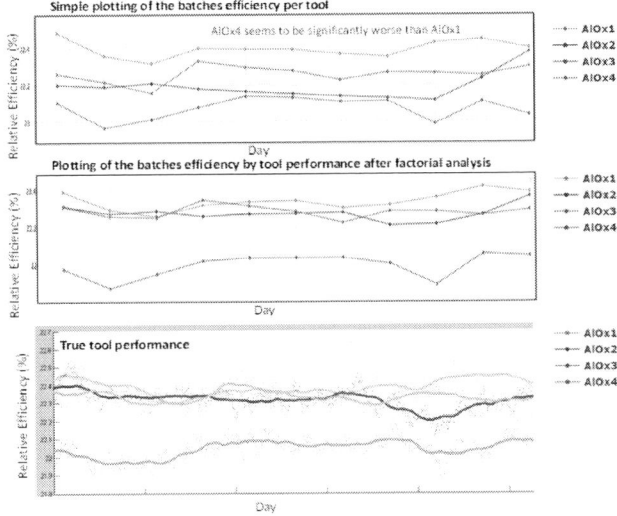

Fig. 5. Upper graph: Simple plotting of batches efficiency per tool (raw data); Middle graph: Plots of batch efficiency after correction by factorial analysis; Lower graph: true tool performance.

CONCLUSIONS

Factorial analysis is a highly practical statistical tool for solar cell plant monitoring that fits the bottom line: finding the fastest, shortest and most cost-effective route to maximizing the watts produced and preserving the profit margin. It can increase the engineering team's productivity by doing the heavy lifting of tracing batch fault occurrences, problematic process tools, and discerning process tool to tool disparities. By means of data mining, the entirety of production data is surveyed with no experimental overhead, allowing the team to see the big picture with accuracy and confidence, and freeing their time to implement systematic ways to raise plant efficiency, whether that be optimizing tool maintenance timing, preferential routing, or process optimization and continuous improvement.

REFERENCES

[1] J. Wong, Griddler: Intelligent computer aided design of complex solar cells, Proc. 40th IEEE PVSC, Tampa 2013, pp. 933-938.

Statistical and Engineering Process Control of Phosphorus Diffused Solar Wafers Using Contactless Infrared Reflectometry

Johnson Wong, Divya Ananthanarayanan, Gordon Deans

Aurora Solar Technologies Inc, Vancouver, British Columbia, V7P3N4, Canada

Abstract — For the silicon solar cell diffusion process, statistical process control (SPC) and engineering process control (EPC) are commonly used to ensure that yield and thoughput targets are met during high-volume industrial production. SPC uses control charts to detect out-of-specification (OOS) wafers that need to be reworked (typically those with emitters that are too lightly diffused to be contactable by the screen printed metal electrode)., EPC involves tuning of diffusion furnace parameters based on feedback from the emitter sheet resistance measurements of prior diffusion batches to reduce process variations and the frequency of OOS events. Both of these process control techniques, SPC and EPC, rely on sampled measurements of diffused wafers over each diffusion batch. This paper focuses on best practices and unique advantages of the use of contactless infrared reflectometry (IR) measurements of phosphorus-diffused wafers to implement SPC and EPC.

I. INTRODUCTION

Phosphorus diffusion is a mature process for the formation of the p-n junction and emitter layer in PERC solar cell manufacturing. Nevertheless, as the industry over time has adopted more lightly doped emitter layers for laser-doped selective emitter PERC cells, these light emitter designs are presenting more challenging conditions for the diffusion furnace equipment in maintaining the process within the allowable condition space or "window". Nowadays, the optimal emitter layer rests on a fine balance between having enough dopant concentration at the surface (after laser doping drive-in) for the screen printed metal electrode contacts to minimize metal induced recombination (which mainly impacts the fill factor FF) [1-2], and keeping the emitter dopant concentration as low as possible to minimize emitter recombination and free carrier absorption losses [3], which impact mainly the open circuit voltage Voc and short circuit current Isc. This balance is often chosen to be near the limit of the process window in which the diffusion furnace can operate with sufficient stability and uniformity. Because the process is therefore not tuned to maximize stability, it becomes important for inline measurement of the emitter properties following the process for both quality assurance and process control purposes.

Generally, there are two complementary uses of inline measurement data for process control: 1) Statistical process control (SPC), which in effect is monitoring, flagging and taking remedial actions for out of specification (OOS) diffused wafers; and 2) Engineering process control (EPC),

which is the procedure of regularly applying adjustments to the process tool operating parameters based on feedback from the measurements, in order to reduce process variance.

In this work, we report on the use of contactless infrared (IR) reflectometry measurements of phosphorus-diffused wafers to implement SPC and EPC. The principles and use of this measurement technique for the characterization of doped emitter properties can be found in Ref. [4].

II. STATISTICAL PROCESS CONTROL

Statistical process control (SPC) involves process monitoring using control charts, to detect incidental outlier wafers that might require rework or out-of-control (OOC) events [5]. This approach of focusing on special causes for remedial action is very practical for the silicon solar cell manufacturing environment and tends to be cost-effective.

One of the advantages of contactless IR reflectometry is its ability to deduce multiple properties of the diffused solar wafer, including emitter surface concentration and texture pyramid size, in addition to the emitter sheet resistance. As a research activity, a study was carried out to examine the effectiveness of creating control charts in terms of these various properties, in live industrial production, inferred by a commercial IR reflectometry tool. Figure 1 shows these graphs over a roughly three-week period, and in each graph the upper control limit (UCL) and lower control limit (LCL) are drawn at three standard deviations away from the mean. Using these charts, periods of large deviations from the everyday pattern can be detected when there is relatively large frequency of points lying above the UCL or below the LCL. Here we have circled in red what appear to be periods of instability leading momentarily to large numbers of outliers, and in blue periods of large drifts away from the mean. These are periods that are deemed out of control by the SPC.

In terms of traditional quality assurance as implemented by monitoring the emitter sheet resistance and flagging outlier wafers, a double sampling scheme was implemented in which IR reflectometry serves as the automated and primary detector of OOS wafers, usinga low alarm threshold, and in cases where that threshold is exceeded, a four-point probe serves as a secondary (offline) tester to confirm whether the wafers in question are indeed out of specification and need to be routed for rework. Use of the four-point probe is minimized by choosing appropriate alarm thresholds based on observed performance. This double-sampling method is an effective means to avoid false positives or negatives. Figure 2 illustrates its principles visualized on a control chart. Wafers that are above a set upper alarm level are deemed to be questionable and are collected for measurement by the four-point probe. If the four-point probe determines that the average sheet resistance of the wafer is above the upper specification limit, then the wafer is deemed to be out of specification, and the group of wafers corresponding to this out of specification wafer will be routed for rework. By optimizing the primary detector's alarm threshold, nearly 100% of out of specification wafers can be detected with only 2/100 wafers being false positives that are extra burdens on the quality assurance operators' time.

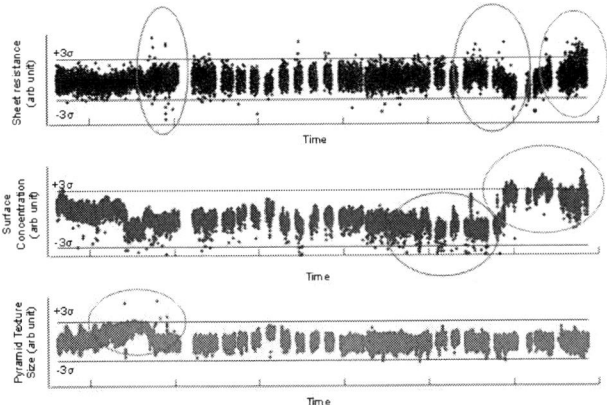

Fig. 1. Shewhart control charts for the sheet resistance, surface concentration, and pyramid texture size of solar wafers in a production line, extracted by the DM110h instrument.

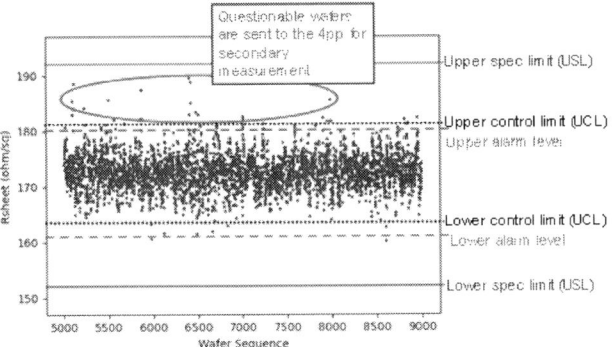

Fig. 2. Doubling sampling scheme to flag wafers that are out of specifications.

III. ENGINEERING PROCESS CONTROL

SPC is essentially process monitoring, which is commonly applied to processes that vary about a fixed mean, and where successive observations are viewed as statistically independent. In contrast, engineering process control (EPC) involves the application of active feedback or feed-forward control, based on measurements taken as the process is performed. It is usually applied to processes in which successive measurements are related over time, and where the mean drifts dynamically [5].

EPC can be done with measurements from the infrared reflectometry measurement alone. The measurement quantity used for feedback can either be the emitter sheet resistance, or it can be weighted more heavily on the emitter surface concentration---both of which can be determined by IR reflectometry. Using surface concentration for feedback has the advantage that it is a more consistent measurement, in the sense that batch-to-batch autocorrelation in surface concentration is often greater than that of sheet resistance. Therefore there is more room to reduce the variance in surface concentration than there is in sheet resistance. Moreover, because surface concentration has a more direct relationship to metallization contact resistance as well as the amount of emitter recombination, it is a parameter that is more critical to quality compared to sheet resistance. In sum, the ability of the IR reflectometry technique to determine multiple characteristics of the emitter simultaneously, offers some unique advantages in the implementation of EPC.

Thanks to maximizing the frequency of sampling (every run), and using rigorous feedback mechanisms like PI control, the variance in emitter sheet resistance produced by the diffusion process can be significantly reduced. In a simulation, with the modelled diffusion process and relative measurement accuracy taken into consideration [6], Figure 3 plots the process variation over 10,000 successive runs, for the cases of no PI control (gray), PI control in the presence of measurement error typical of the Aurora Solar Technologies DM commercial IR reflectometry tool versus a four-point probe reference (blue), and PI control where the measurement is perfectly accurate (green). The cases with PI control applied clearly led to much reduced variance in deviations from the target. With PI control in the presence of measurement error, the standard deviation of the sheet resistance variations is reduced from 5.89 to 2.84 ohm/sq.

Therefore, with effective PI control, the standard deviation in the diffused wafers sheet resistance using feedback control can be typically halved compared to the uncontrolled scenario. More importantly, the occurrence of OOS wafers is reduced by more than one order of magnitude, drastically reducing the frequency of rework wafers. For PI control, the infrared reflectometry technique not only has the advantages that the measurements are fully automated, but it is also not prone to some major drawbacks of the four-point probe, such as the tendency to erroneously read an overly-high sheet resistance emitter as, contrarily, having a low sheet resistance due to leakage current into the base, which can lead to critical instabilities in the control loop.

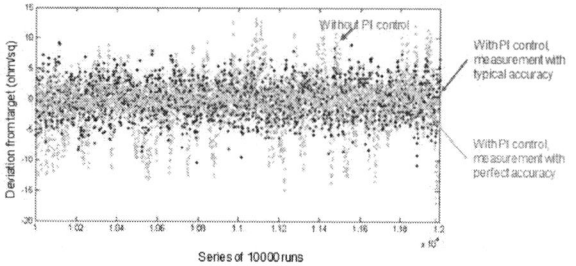

Fig. 3. Time series of et, the deviation from target, with and without PI control

CONCLUSIONS

PERC PV cell efficiency is significantly influenced by conformance of the emitter profile to design specifications. Therefore, it is important to effectively control variations in the emitter fabrication process in industrial production. Effective control is primarily dependent on the relationship of the controlled process output property to finished good yield, on how amenable that property is to feedback control, and on the frequency at which it is measured.

Thanks to the contactless nature and high throughput measurement, IR reflectometry proves to be a solar wafer emitter characterization technique which is highly suited to both SPC and EPC. In SPC, using the IR reflectometry as a primary detector of suspected outlier wafers and verification using four-point probe as the secondary measurement, one can very effectively implement a double sampling scheme to screen high sheet resistance wafers that are out of specification. In EPC, inline IR reflectometry measurements alone are sufficient to implement effective PI control loops that can reduce the standard deviation of sheet resistance variations by half, in a typical manufacturing environment.

REFERENCES

[1] V. Shanmugam, J. Cunnusamy, A. Khanna, P. K. Basu, Y. Zhang, C. Chen, A. F. Stassen, M. B. Boreland, T. Mueller, B. Hoex, and A. G. Aberle, "Electrical and Microstructural Analysis of Contact Formation on Lightly Doped Phosphorus Emitters Using Thick-Film Ag Screen Printing Pastes", IEEE Journal of Photovoltaics, Volume 4, No. 1, January 2014.

[2] T. Dullweber, H. Hannebauer, S. Dorn, S. Schimanke, A. Merkle, C. Hampe and R. Brendel, "Emitter saturation current densities of 22fA/cm2 applied to industrial PERC solar cells approaching 22% conversion efficiency", Prog. Photovolt: Res. Appl. 2017; 25:509–514

[3] P. Jäger, U. Baumann, T. Dullweber, "Impact of the Thermal Budget of the Emitter Formation on the pFF of PERC+ Solar Cells", SiliconPV 2019, the 9th International Conference on Crystalline Silicon Photovoltaics, AIP Conf. Proc. 2147, 140005-1–140005-6

[4] D. Ananthanarayanan, J. Wong, N Balaji, A. G.Aberle, J. W. Ho, "Mid-Infrared Reflectance and Transmittance Characterization of Phosphorus and Boron Diffused Silicon Solar Wafers" Solar Energy Materials and Solar Cells Volume 205, February 2020, 110286

[5] D. C. Montgomery, J. B. Keats, G. C. Runger and W. S. Messina, Integrating Statistical Process Control and Engineering Process Control, Journal of Quality Technology, Volume 26, 1994 - Issue 2.

[6] G. Box and T. Kramer (1992) Statistical Process Monitoring and Feedback Adjustment—A Discussion, Technometrics, 34:3, 251-267

Proposal of Connection Assessment Diagrams to Speed up the Studies of Hosting Capacity of PV Generators in MV Distribution Systems

Pedro A. V. Pato, Fernanda C. L. Trindade, Tiago R. Ricciardi, Paulo Meira, Walmir Freitas

University of Campinas, Campinas, Sao Paulo, 13083-852, Brazil

Abstract— In 2021, an average of 5.5 generators were connected every day in Brazilian MV distribution systems, and 98.9% of these generators are photovoltaic. The increased penetration of MV distributed generation has been accompanied by a rise in the workload of utility planning engineers, who must study and propose solutions to enable the connection of every MV generator. In this context, this work proposes a quick first-assessment approach to identify if the required connection can be approved or if further studies are required. The proposed approach focuses on the most restrictive steady-state technical impacts (overvoltage and overload) related to the increased penetration of MV distributed generators. The proposed approach consists of building connection assessment diagrams that can successfully speed up the required analyses and, consequently, decrease person-hour costs.

Keywords— *Electric power distribution systems, hosting capacity, photovoltaic generators*

I. INTRODUCTION

The connection of photovoltaic (PV) generation at medium voltage (MV) and low voltage (LV) levels of distribution systems is increasing due to technical advancements, price decrease, and the urgent need to intensify the participation of renewable energy sources. In Brazil, the current penetration level of distributed generators (DGs) connected to LV systems is not yet a strong concern to distribution utility engineers. However, the connections at MV levels usually require many studies, overloading the planning engineers.

Several works propose methods to estimate the hosting capacity (HC) of distribution systems to distributed energy resources and active network management actions to increase it. In [1], the PV HC of LV distribution systems is assessed using Monte Carlo simulations, and it shows that the two most restrictive criteria are overvoltage followed by conductor overload. Ref. [2] provides helpful PV HC sensitivity analysis and also uses Monte Carlo simulations in its studies. In [3], the HC of MV systems is evaluated considering the robust operation of on-load tap changers (OLTCs) and static var compensators (SVCs). Ref. [4] assesses the HC of DGs and electric vehicles simultaneously.

This work was supported by Coordenação de Aperfeiçoamento de Pessoal de Nível Superior - Brazil (CAPES), by Conselho Nacional de Desenvolvimento Científico e Tecnológico (CNPq), grants 142275/2020-4 and 304373/2020-6, and by São Paulo Research Foundation (FAPESP), grant 2020/10523-4.

In this work, a simple, fast, and conservative solution that does not require load flow simulations or optimization modeling is proposed to list the MV buses of radial distribution systems that can safely receive DGs of given rated powers without leading to steady-state overcurrent and overvoltage problems. It is useful to avoid unnecessary studies and provides supportive information regarding the connection of the PV generator by quickly showing possible bottlenecks.

This abstract is organized as follows. Section I introduces the work. Section II presents the methodology, followed by a brief description of the case studies in Section III and the main results in Section IV. Finally, Section V presents the main conclusions.

II. METHODOLOGY

This section presents the methodology proposed in this work to provide a quick first assessment of the connection of PV generators at MV level.

The methodology builds connection assessment diagrams indicating buses of the distribution system that can safely host DGs of given rated powers without network reinforcements or steady-state overcurrent and overvoltage. Among the advantages are the simplicity and speed compared with load flow simulations.

Then, anytime planning engineers receive a request for the installation of a new MV PV generator, they can use the diagrams to decide if the connection can be approved without further detailed studies.

A. Required Information

To build the connection assessment diagrams, the proposed methodology needs the MV network topology, including the location of switches, reclosers, capacitor banks, line voltage regulators (LVRs), and transformers; the impedance of lines and transformers; the control settings of LVRs and switchable capacitor banks; typical and/or measured load profiles (active and reactive powers); and a representative PV generation profile with the corresponding rated power of the PV generator(s).

B. Simplified Network Model

The method assumes that the MV network operates radially and balanced, so it can be modeled as depicted in Fig. 1, where

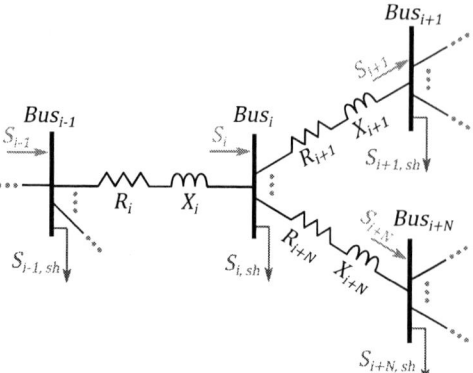

Fig. 1. Generic radial three-phase MV distribution system.

$S_i = P_i + jQ_i$ is the 3-phase complex power injected in bus i through the upstream line and $S_{i,sh} = P_{i,sh} + jQ_{i,sh}$ is the shunt 3-phase complex power drained from bus i; R_i and X_i are the positive sequence resistance and reactance, respectively, from the line that connects bus i to the upstream bus (if the line's 3-phase impedance matrix is not symmetrical, then the positive sequence impedance is assumed in this work to be the difference between the average of the diagonal elements and the average of the off-diagonal elements).

The LV circuits and the 1-phase and 2-phase MV laterals do not need to be modeled in detail, being represented only by the aggregated shunt powers $S_{i,sh}$.

C. Step-by-Step of the Proposed Methodology

First, for each generic bus i, the value of $S_{i,sh}(t)$ is assigned to be the sum, at the time instant t, of all measured/typical complex demands of the MV and LV loads connected to bus i, either directly or through distribution transformers or 1-phase/2-phase laterals. If LV network data is available, the losses can be estimated at rated voltage and added to $S_{i,sh}(t)$, otherwise, they can be simply neglected. DGs connected to bus i add to the real part of $S_{i,sh}(t)$ a negative amount of kW (the imaginary part is neglected, assuming a unity power factor operation). Capacitor banks connected to bus i add to the imaginary part of $S_{i,sh}(t)$ a negative amount of kvar.

Next, assuming that $S_{i,sh}(t)$ has a Δt-minute resolution and duration of ND days, the maximum and minimum daily shunt powers at each generic bus i are obtained using (1) and (2), respectively, where $t \in \{0, \Delta t, 2\Delta t, \dots, 24\,\text{h}\}$ and $\Omega_t = \{t + k \cdot 24\,\text{h} : k = 1, 2, \dots, ND\}$. Fig. 2 illustrates the daily curves and the corresponding limits.

$$\overline{S_{i,sh}}(t) = \max_{\tau \in \Omega_t}\{P_{i,sh}(\tau)\} + j \max_{\tau \in \Omega_t}\{Q_{i,sh}(\tau)\} \quad (1)$$

$$\underline{S_{i,sh}}(t) = \min_{\tau \in \Omega_t}\{P_{i,sh}(\tau)\} + j \min_{\tau \in \Omega_t}\{Q_{i,sh}(\tau)\} \quad (2)$$

Then, starting from the end buses and going towards the substation bus (backward sweep), (3) and (4) are used to calculate the maximum and the minimum daily injected powers, respectively, where Ω_i represents the set of indices of all buses immediately downstream bus i, and $\overline{L_k}(t)$ and $\underline{L_k}(t)$

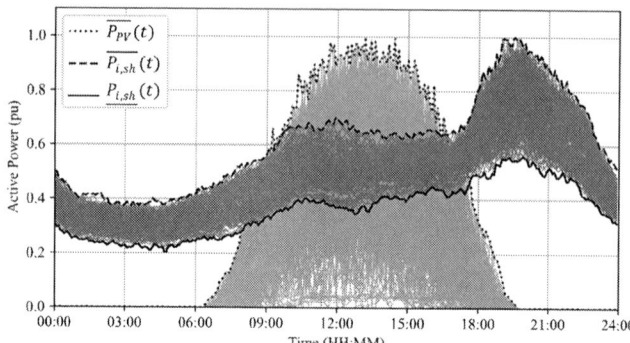

Fig. 2. One-year power measurements of a PV generator (orange) and a typical distribution transformer connected at a generic bus i (blue). The maximum and minimum daily power profiles are indicated by black lines.

are estimations for the maximum and minimum line losses at time instant t and rated voltage V_k^{rated}, which are calculated using (5) to (10).

$$\overline{S_i}(t) = \overline{P_i}(t) + j\overline{Q_i}(t) = \overline{S_{i,sh}}(t) + \sum_{k \in \Omega_i}\left(\overline{S_k}(t) + \overline{L_k}(t)\right) \quad (3)$$

$$\underline{S_i}(t) = \underline{P_i}(t) + j\underline{Q_i}(t) = \underline{S_{i,sh}}(t) + \sum_{k \in \Omega_i}\left(\underline{S_k}(t) + \underline{L_k}(t)\right) \quad (4)$$

$$\overline{L_k}(t) = \frac{R_k + jX_k}{(V_k^{rated})^2} \cdot \left(\overline{P_k^{sqr}}(t) + \overline{Q_k^{sqr}}(t)\right) \quad (5)$$

$$\underline{L_k}(t) = \frac{R_k + jX_k}{(V_k^{rated})^2} \cdot \left(\underline{P_k^{sqr}}(t) + \underline{Q_k^{sqr}}(t)\right) \quad (6)$$

$$\overline{P_k^{sqr}}(t) = \max\left\{\left(\overline{P_k}(t)\right)^2, \left(\underline{P_k}(t)\right)^2\right\} \quad (7)$$

$$\overline{Q_k^{sqr}}(t) = \max\left\{\left(\overline{Q_k}(t)\right)^2, \left(\underline{Q_k}(t)\right)^2\right\} \quad (8)$$

$$\underline{P_k^{sqr}}(t) = \begin{cases} 0, & \text{if } \overline{P_k}(t) \geq 0, \underline{P_k}(t) \leq 0 \\ \min\left\{\left(\overline{P_k}(t)\right)^2, \left(\underline{P_k}(t)\right)^2\right\}, & \text{otherwise} \end{cases} \quad (9)$$

$$\underline{Q_k^{sqr}}(t) = \begin{cases} 0, & \text{if } \overline{Q_k}(t) \geq 0, \underline{Q_k}(t) \leq 0 \\ \min\left\{\left(\overline{Q_k}(t)\right)^2, \left(\underline{Q_k}(t)\right)^2\right\}, & \text{otherwise} \end{cases} \quad (10)$$

Next, starting from the substation bus and going towards the end buses (forward sweep), (11) is used to estimate the magnitude of the maximum possible reverse current that a new PV generator installed at bus i can cause while equation (12) calculates the maximum voltage magnitude, where $\overline{P_{PV}}(t)$ represents the maximum daily power profile of the PV generator at a given rated power, as indicated in Fig. 2. In (12), $\overline{V_{up,i}}(t)$ is the maximum voltage magnitude of the bus immediately upstream bus i, while $V_{reg,i}$ and $B_{reg,i}$ are the voltage reference and band settings [5], respectively, of the LVR connected to bus i (if the bus i is the substation bus, then these are the settings from the substation's OLTC transformer).

$$\bar{I}_i(t) = \frac{\sqrt{\left(\underline{P_i}(t) - \overline{P_{PV}}(t)\right)^2 + \overline{Q_i^{sqr}}(t)}}{\sqrt{3} \cdot V_i^{rated}} \qquad (11)$$

$$\overline{V}_i(t) = \begin{cases} V_{reg,i} + B_{reg,i}/2 \,, & \text{if bus } i \text{ is regulated} \\ \overline{V_{up,i}}(t) - \dfrac{R_i \cdot \left(\underline{P_i}(t) - \overline{P_{PV}}(t)\right) + X_i \cdot \underline{Q_i}(t)}{V_i^{rated}} \,, \text{o/w} \end{cases} \qquad (12)$$

Finally, the connection assessment diagram is built by classifying the buses into two categories: *those that can safely host the new PV generator* and *those that need further assessment*. A generic bus i is assigned to the first category if both conditions (13) and (14) are met for all t, and to the second one, otherwise. In (13), I_i^{max} is the ampacity of the conductor that connects the bus i to its upstream bus. In (14), V_{pu}^{max} is the regulatory limit for steady-state overvoltage given in pu.

$$\overline{I}_i(t) \leq I_i^{max} \qquad (13)$$

$$\overline{V}_i(t) \leq V_{pu}^{max} \cdot V_i^{rated} \qquad (14)$$

D. Additional Comments

- The step-by-step procedure described previously needs to be executed for each possible topology configuration that the opening/closing of the MV switches can generate. The final diagram is then obtained, listing as safe buses only the buses that are safe for all possible operating topologies.
- If a recloser is connected at bus i, then $\underline{S_i}(t)$ is set to be the minimum between zero and the value from the right-hand side of equation (4), since the recloser may open at any time the DG is operating.
- The described methodology assumes the new PV generator operates with a unity power factor (based on Brazilian regulation), if that is not the case, (11) and (12) can easily be adapted to account for the worst possible case of the DG's reactive power.
- Equation (13) assumes that for any directed path p that connects the substation to a generic bus i, the cables ampacity only decreases ($I_k^{max} \geq I_l^{max}$ for $l \in \Omega_k$ and $k, l \in p$). If that is not the case, then condition (13) must consider the minimum ampacity of the path p.
- The set of all buses capable of hosting a DG generator of a given rated power is called a *hosting zone* in this work. Since for any grid's path p the short-circuit level decreases as the buses get farther from the substation, it is natural to imagine that the forward-sweep application of (11) and (12) would form a connected hosting zone starting in the substation and stopping in the first layer of buses that present overvoltage and/or overcurrent when the DG is connected to them. However, (12) shows that a voltage regulator can break the hosting zone connectivity by either regulating the output voltage too low or too high. In the first case, a sufficiently low $V_{reg,i}$ allows the i-th bus' regulator to prevent overvoltage in its downstream buses despite possible overvoltage problems in its upstream buses. In the second case, if $V_{reg,i} + B_{reg,i}/2 > V_{pu}^{max} \cdot V_i^{rated}$, then the hosting

zone has to stop in the i-th bus but might start again in a further downstream bus if the lines' voltage drop brings the voltage back to adequate levels considering the presence of the DG (usually when its rated power is not enough to cause reverse power flow). These phenomena are illustrated in the results section IV.

- If the utility accepts a risk-based approach, then the maximum ($max\{\cdot\}$) and minimum ($min\{\cdot\}$) operators in (1) and (2) can be replaced by sufficiently high and low percentiles operators, respectively. For instance, $max\{\cdot\}$ can be replaced by the 95th percentile operator and $min\{\cdot\}$ for the 5th percentile operator. Clearly, higher risks lead to bigger hosting zones for the same DG rated power.
- Finally, an important observation is that each diagram refers to a given generator rated power. Therefore, it is possible to build diagrams for some strategic power values or one diagram for a conservative high value of rated power.

III. CASE STUDY

The proposed method was programmed in Python language and applied to two real electric circuits (ckt1 and ckt2) to build connection assessment diagrams for PV generators with rated powers of 0.5 MW, 1.0 MW, and 1.5 MW. Table I presents the main features of both circuits located in the Brazilian Southeast.

To validate the diagrams, eight test buses located in the boundaries of the hosting zones are selected for each of the two circuits and each of the three rated powers. Then, for each test bus, a PV generator with the corresponding rated power is installed and a one year-long quasi-static time series simulation [6] with 5 minutes timestep is performed using the DSS Python module [7] to verify if overvoltage and/or overcurrent transgressions occur. The overvoltage transgression is evaluated according to the Brazilian regulation [8] that requires, among other constraints, the voltage at MV consumers to not be above 1.05 pu for more than 0.5% of the time.

TABLE I. MAIN FEATURES OF CKT1 AND CKT2.

Feature		ckt1 – 12.5 MVA 69 kV:13.8 kV		ckt2 – 6.25 MVA 69 kV:11.95 kV	
		Fdr. 1	Fdr. 2	Fdr. 1	Fdr. 2
Farthest bus (km)		27.88	25.41	35.32	48.15
Feeder length (km)	MV	225.3	159.9	252.0	394.3
	LV[a]	34.73	93.49	43.50	74.50
# of consumers	MV	44	9	3	9
	LV	1,906	4,814	2,319	4,083
# of transformers[b]	15 kVA	106	73	189	312
	30 kVA	87	111	68	177
	45 kVA	40	72	36	77
	75 kVA	27	44	19	33
	150 kVA	31	5	0	1
	Others	46	23	33	34
# of LVRs	Open Δ	1	0	1	2
	Closed Δ	1	1	0	0
# of capacitors	150 kvar	0	0	1	1
	300 kvar	2	0	0	0
	600 kvar	0	2	0	1

a. LV grids are 3-phase, 4-wire, 220 V/127 V (line-to line/line-to-neutral).
b. Transformers are 3-phase and delta-grounded wye connected.

The utility responsible for ckt1 and ckt2 provided the networks' necessary information and the demand curves of the smart metered MV consumers. The demand curves of the LV consumers are synthesized using [9]. The orange curve in Fig. 2 is used to model the PV generator's power injection. All LVRs are assumed to operate in cogeneration mode to prevent misoperation in the case of reverse power flow [10].

IV. MAIN RESULTS

The obtained connection assessment diagrams are shown in Fig. 3 and Fig. 4 for the circuits ckt1 and ckt2, respectively. The hosting zone is indicated in green, and the validation test buses in red. As expected, the hosting zone decreases as the DG rated power increases, which can be verified by comparing the 0.5, 1.0, and 1.5 MW hosting zones.

In Fig. 3 (a) is possible to see that the test bus 8 only belongs to the hosting zone because its upstream LVR is regulating the voltage at a sufficiently low value. If there were no LVR between test buses 4 and 8, the hosting zone would end at test bus 4. Similarly, the hosting zone of Fig. 3 (a) and (b) would stop at test bus 5 if there were no LVR between test buses 5 and 6. These examples illustrate how properly adjusted voltage regulators can enable the connection of PV generators by mitigating the overvoltage problem.

On the other hand, the hosting zones of Fig. 4 do not start at the substation because the OLTC located there regulates the output voltage to 1.04 pu $\pm$ 0.018 pu (V_{pu}^{max}= 1.05 pu) which prevents the nearby buses from hosting the DG. Hence, to increase the PV host capacity of circuits, the utilities might have to change the practice of setting the regulators V_{reg} close to V_{pu}^{max}. This historically common strategy minimizes the risk of consumers having undervoltage problems but is based on the premise that distribution systems have a decreasing voltage profile, which is not necessarily true if DGs are connected.

The quasi-static time series simulation results are presented in Fig. 5 and Fig. 6 for the circuits ckt1 and ckt2, respectively. As expected, in no case the line ampacity was violated or the test bus voltage stayed above 1.05 pu for longer than 0.5% of the time. The validation of the proposed method in the test buses also validates it for any generic hosting zone bus i belonging to a path p that connects the substation to a test bus since the closer a bus is to the substation the grater its short circuit level is.

Fig. 5 and Fig. 6 also show that the most restrictive criterion for defining the hosting zone boundaries was overvoltage. For instance, the line current stayed under 50% of its ampacity in almost all test bus.

Table II illustrates how the application of the proposed methodology can speed up the analysis required to allow the connection of a PV generator at MV. The connection

TABLE II. TIME COMPARISON OF THE PROPOSED METHOD AND TRADITIONAL LOAD FLOW STUDIES[a]

	Ckt1	Ckt2
Proposed method[b]	88 s	76 s
Proposed method validation[c]	78,203 s (21.72 h)	75,465 s (21.96 h)

a. All simulations done in an Intel Core i7-8700 CPU, 3.20GHz, 32 GB RAM.
b. Time to build the 3 connection assessment diagrams.
c. Time to run the quasi-static time series simulation for the 24 test buses.

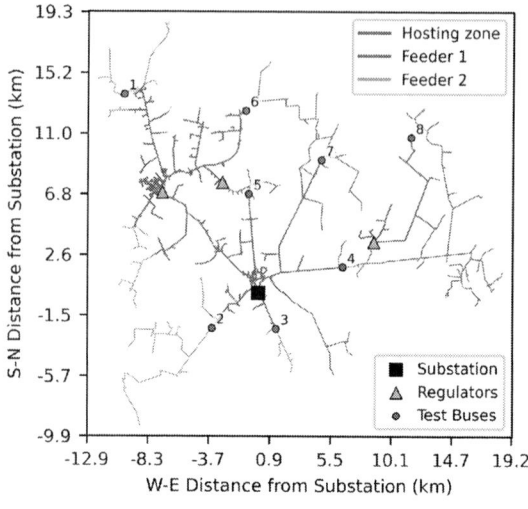

(a)

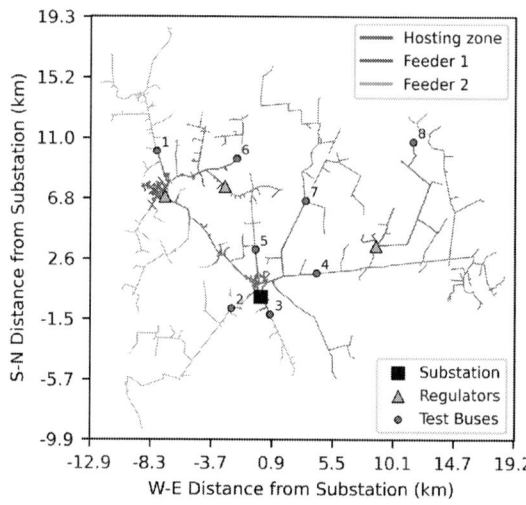

(b)

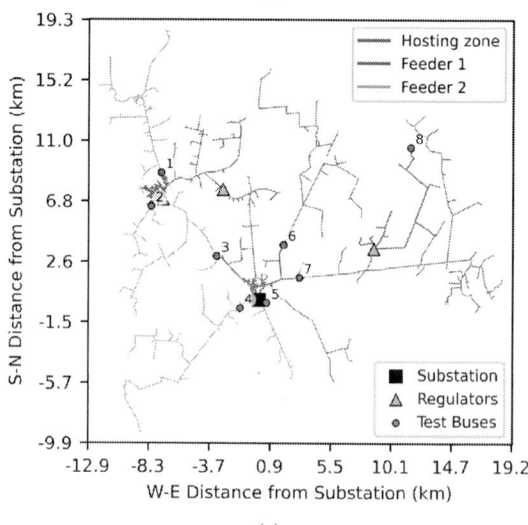

(c)

Fig. 3. Connection assessment diagrams of ckt1 for a (a) 0.5 MW, (b) 1.0 MW and (c) 1.5 MW PV generator. The hosting zones are shown in green and the test buses in red.

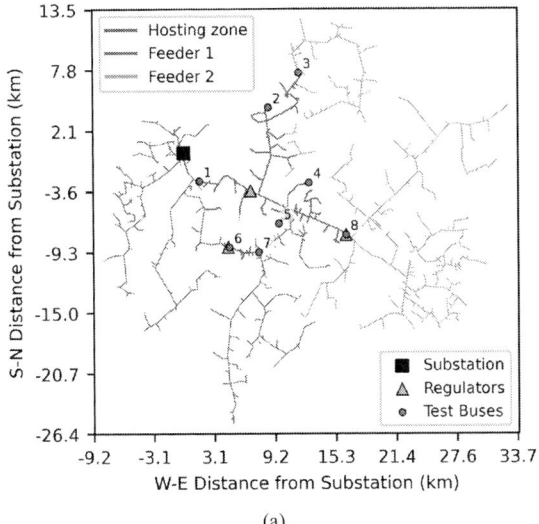

(a)

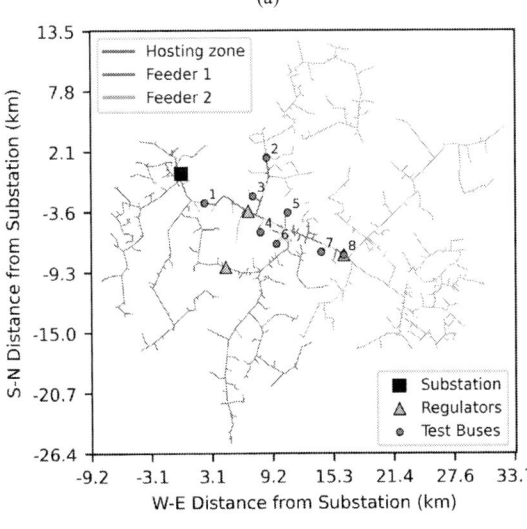

(b)

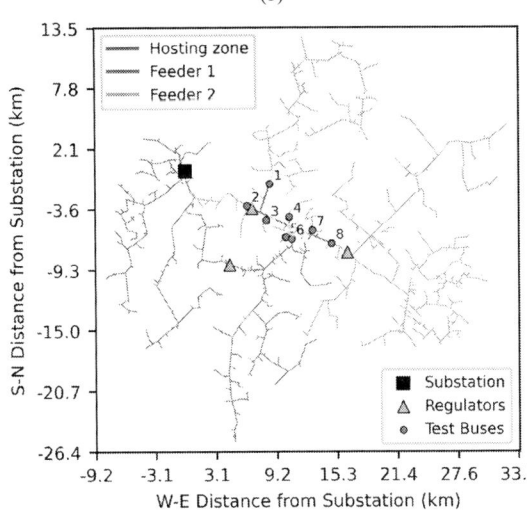

(c)

Fig. 4. Connection assessment diagrams of ckt2 for a (a) 0.5 MW, (b) 1.0 MW and (c) 1.5 MW PV generator. The hosting zones are shown in green and the test buses in red.

assessment diagrams are fast to build (several orders of magnitude faster than traditional load flow studies) so a utility can update the hosting capacity diagrams of all its feeders in a daily basis.

V. CONCLUSIONS

The method proposed in this work builds, for a given MV PV rated power, a connection assessment diagram with the hosting zone, where the generator can be connected without requiring grid reinforcement and without causing overvoltage and overcurrent problems (the most restrictive steady-state technical impacts).

The quasi-static time series simulations performed in strategic chosen test buses showed that the hosting zones from the diagrams are conservative.

The time necessary to build the diagrams is negligible compared to traditional load flow studies; so, if needed, the utilities can update the diagrams of all its circuits in a daily basis for a range of PV rated powers of interest.

The proposed approach decreases the utilities' person-hour costs by quickly identifying if the connection of a PV generator in some MV bus can be approved or if further studies are required.

REFERENCES

[1] R. Torquato, D. Salles, C. O. Pereira, P. C. M. Meira, and W. Freitas, "A comprehensive assessment of PV hosting capacity on low-voltage distribution systems," in *IEEE Trans. Power Del.*, vol. 33, no. 2, pp. 1002-1012, April 2018, DOI: 10.1109/TPWRD.2018.2798707.

[2] A. Dubey and S. Santoso, "On estimation and sensitivity analysis of distribution circuit's photovoltaic hosting capacity," *IEEE Trans. Power Syst.*, vol. 32, no. 4, pp. 2779-2789, Jul. 2017.

[3] S. Wang, S. Chen, L. Ge, and L. Wu, "Distributed generation hosting capacity evaluation for distribution systems considering the robust optimal operation of OLTC and SVC," in *IEEE Trans Sustain Energy*, vol. 7, no. 3, pp. 1111-1123, July 2016, DOI: 10.1109/TSTE.2016.2529627.

[4] E. C. da Silva, O. D. Melgar-Dominguez, and R. Romero, "Simultaneous distributed generation and electric vehicles hosting capacity assessment in electric distribution systems," in *IEEE Access*, vol. 9, pp. 110927-110939, 2021, DOI: 10.1109/ACCESS.2021.3102684.

[5] W. H. Kersting, "The modeling and application of step voltage regulators," in *Proc. IEEE Power Syst. Conf. Expo.*, 2009, pp. 1-8.

[6] M. J. Reno, J. Deboever and B. Mather, "Motivation and requirements for quasi-static time series (QSTS) for distribution system analysis," in *Proc. IEEE Power Energy Soc. Gen. Meet.*, 2017, pp. 1-5.

[7] DSS Extensions, "DSS C-API: An unofficial C API for EPRI's OpenDSS". Available: https://github.com/dss-extension

[8] Agência Nacional de Energia Elétrica (ANEEL), "Procedimentos de distribuição de energia elétrica no sistema elétrico nacional – PRODIST", Módulo 8 – Qualidade da Energia Elétrica, Revisão 10, pp. 1–88, 2018. [Online]. Available at: https://antigo.aneel.gov.br/prodist (in Portuguese).

[9] R. Torquato, Q. Shi, W. Xu and W. Freitas, "A monte carlo simulation platform for studying low voltage residential networks," *IEEE Trans. Smart Grid*, vol. 5, no. 6, pp. 2766-2776, Nov. 2014.

[10] R. A. Walling, R. Saint, R. C. Dugan, J. Burke, and L. A. Kojovic, "Summary of distributed resources impact on power delivery systems," in *IEEE Transactions on Power Delivery*, vol. 23, no. 3, pp. 1636-1644, July 2008.

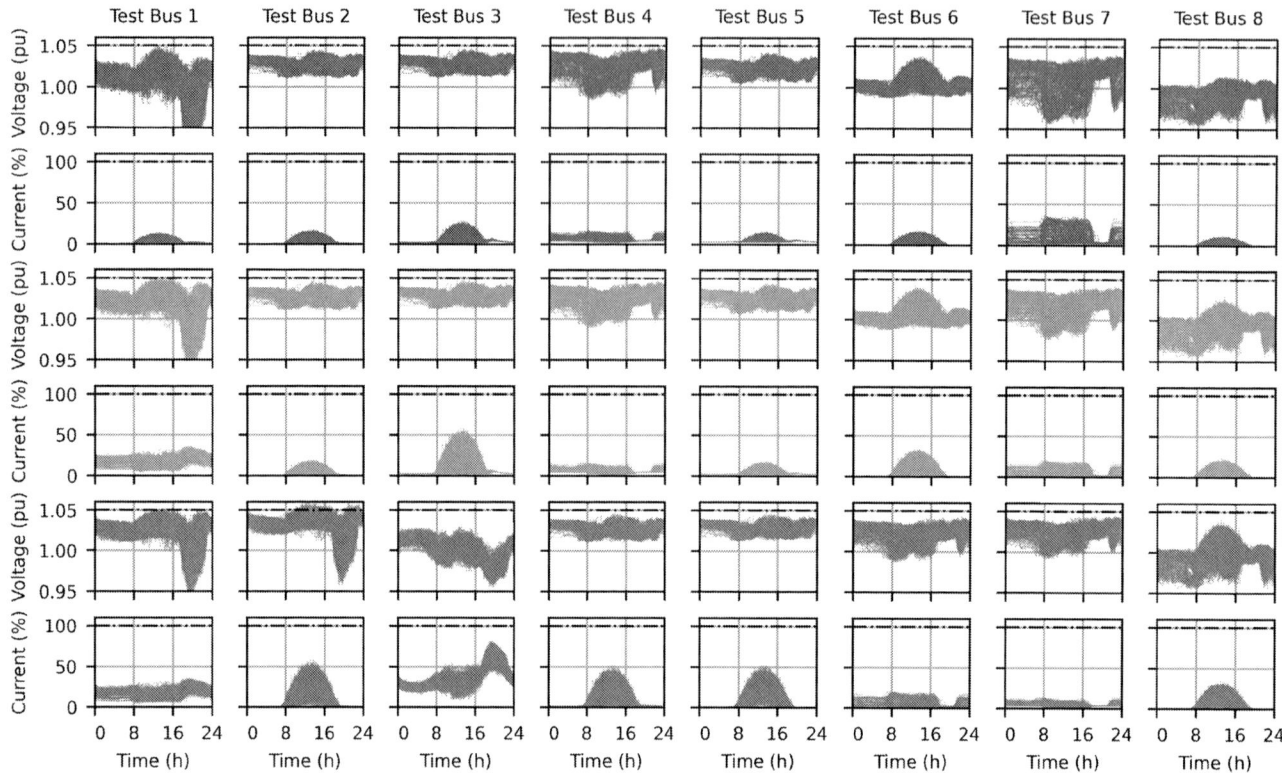

Fig. 5. One-year quasi-static time series simulation results for the voltages and currents of all test buses in ckt1. The curves in blue, orange and green refer to the 0.5 MW, 1.0 MW and 1.5 MW PV generator, respectively. Currents are given in percentage of the cable maximum ampacity.

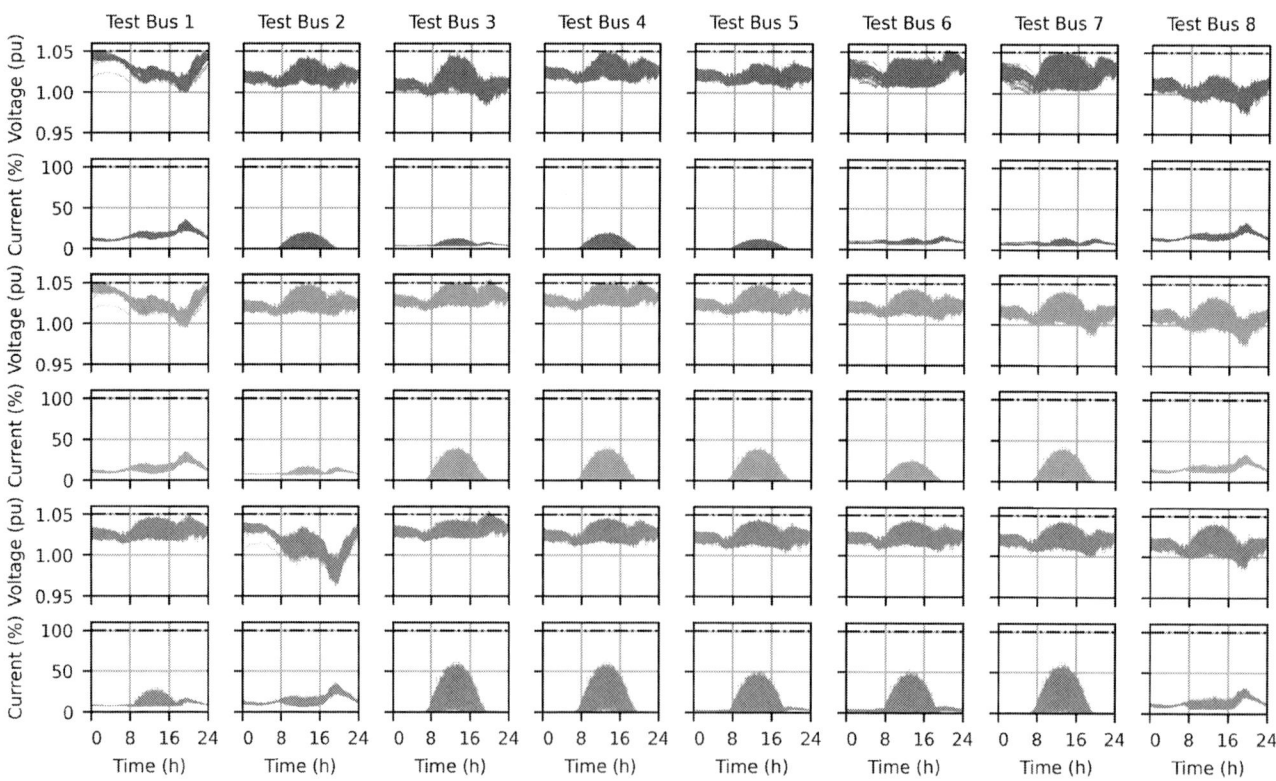

Fig. 6. One-year quasi-static time series simulation results for the voltages and currents of all test buses in ckt2. The curves in blue, orange and green refer to the 0.5 MW, 1.0 MW and 1.5 MW PV generator, respectively. Currents are given in percentage of the cable maximum ampacity.

Ray Tracing of Bent Applications of Luminescent Solar Concentrator PV Modules

Xitong Zhu[1], Michael G. Debije[2], Angèle H.M.E. Reinders[1,3]

1) Energy Technology Group, Department of Mechanical Engineering, Eindhoven University of Technology, 5612 AE Eindhoven, The Netherlands
2) Stimuli-responsive Funct. Materials & Devices Group, Department of Chemical Engineering and Chemistry, Eindhoven University of Technology, 5612 AE Eindhoven, The Netherlands
3) Department of Design, Production and Management, Faculty of Engineering Technology, University of Twente, 7522 NB Enschede, The Netherlands

Abstract— **Flat luminescent solar concentrator PV (LSC-PV) devices have in development for nearly 45 years. However, to enhance their integration potential in buildings and vehicles, this study is focused on bent LSC-PV devices. LSC PV modules with a 20x20 cm^2 top surface area and 6 different curvatures (k, from 1 to 10 m^{-1}) have been designed in Dassault Systèmes Solidworks and simulated by means of ray tracing in Synopsys LightTools. These modules are 20 mm thick, are made of PMMA with 110 parts per million (ppm) of Lumogen Red 305 dye, and silicon solar cells attached only to their rectangular edges with a total PV cell coverage area of 80 cm^2 per module. The simulation results show better optical and electrical performances for the bent LSC-PV modules than for flat planar modules. The performance increases with the increase of curvature; for instance, an LSC-PV module with a curvature of k=10 m^{-1} is 40% more efficient than a flat reference. Furthermore, in this study, the effect of reflection layers and sizing of the front surface of the bent LSC PV modules on their performance is investigated.**

Keywords— Luminescent Solar Concentrator (LSC), Ray Tracing, Simulation, Building Integrated Photovoltaic (BIPV), Vehicle Integrated Photovoltaic (VIPV) (key words)

I. INTRODUCTION

Harvesting solar energy by using luminescent solar concentrator photovoltaic (LSC PV) modules were originally proposed in the 1970s[1]–[4]. The main body of the LSC is a lightguide made of inexpensive plastic or glass embedded with luminescent materials, or with luminescent materials applied in a separate layer on top and/or bottom of the lightguide. The luminescent materials can be organic fluorescent dyes, inorganic phosphors, or quantum dots[5]. Sunlight penetrates the top surface of the lightguide and is absorbed by the luminescent materials, and the light is re-emitted at longer wavelengths. Due to total internal reflection, a fraction of the re-emitted light is guided to small PV cells attached to the edges of the lightguide. As there are many unique advantages: the lightguide elements can be colorful and manufactured in a variety of sizes; the device performs well under both direct and indirect light.[6] Thanks to its interesting design features, such as customized coloring and

formability, LSC PV is an attractive technology for integration in buildings and products such as vehicles[7], [8].

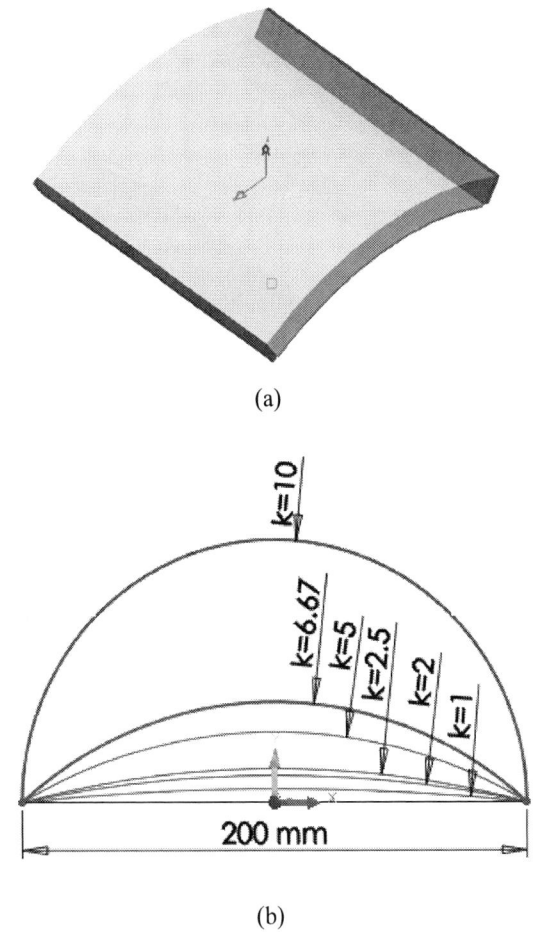

(a)

(b)

Fig. 1. a) The example of the simulated modules (k=5) of this study (the translucent pink area is the lightguide, the violet areas are PV cells, the yellow area is the rear reflection layer) b) The curvature sketch of the top surface of the lightguide in different modules.

However, in the past decades, almost all LSC PV research has been limited to flat devices [9]–[11]. In some literature, some flat LSC-PV devices already have outstanding performance in simulation or practical measurement[9], [12]–[14], such as cube LSC matrix simulated power conversion efficiency reach 18%[15], and a functional prototype LSC-PV module with the geometrical concentration of factor 3.6, measured power conversion efficiency as 5.8%[13]. However, the integration of LSC PV devices in objects will involve different physical forms: some research has already shown the potential application of curved LSC PV devices[6]. Therefore, it is necessary to better understand the effect of the curvature of an LSC on its optical and electrical performances. To explore the suitability of giving the LSC PV devices curvature in various applications, this study evaluates the performance of bent LSC devices by means of ray-tracing simulations.

II. METHOD

This study simulates bent LSC devices with 6 curvatures of 1 to 10 m^{-1}. All the models assume a PMMA lightguide with 110 ppm Lumogen Red 305 dye; Fig. 1 illustrates one of the simulated modules ($k = 5$) with a size of 200x200x20 mm^3 (LxWxH). The LSC PV modules have been designed in Dassault Systèmes Solidworks and simulated by means of ray tracing in Synopsys LightTools. As a reference, a flat LSC PV module with a similar size has been simulated as well. In some simulations, an ideal reflection layer has been integrated on the rear surface of lightguide.

The LSC PV performance has been quantified by the optical collection efficiency (η_{OCE}) and power conversion efficiency (η_{PCE}). The optical collection efficiency is calculated by the followed formula:

$$\eta_{OCE} = \frac{\sum \dot{Q}_{PV}}{\dot{Q}_{in}} \qquad (1)$$

Where:

$\dot{Q}_{in}$ is the radiant flux which falls on the aperture area (W)

$\dot{Q}_{PV}$ is the radiant flux which falls on each cell (W)

The power conversion efficiency which is used to evaluate the electrical performance is calculated by the followed formula:

$$\eta_{PCE} = \frac{\sum P}{\dot{Q}_{in}}$$
$$= \frac{\sum_{i=1}^{n=total\ PV\ cells} \int_{\lambda=0}^{4500\ nm} S_{PV}^{i}(\lambda)\ SR\ (\lambda)\ FF\ Voc\ A_{PV}^{i}\ d\lambda}{\int_{0}^{\infty} Sap(\lambda) A_{ap}\ d\lambda} \qquad (2)$$

Where:

P is the power production of each cells (W)

S_{PV} is the spectral distributed irradiance on each solar cell (W/m^{2}nm)

SR is the spectral response of the spectral distribution (A/W)

FF is the fill factor of the solar cell

Voc is the open-circuit voltage of the solar cell (V)

A_{PV} is the area of the solar cell (m2)

S_{ap} is the spectral distributed irradiance on the aperture area (W/m^{2}nm)

A_{ap} is the area of the aperture area (m^2)

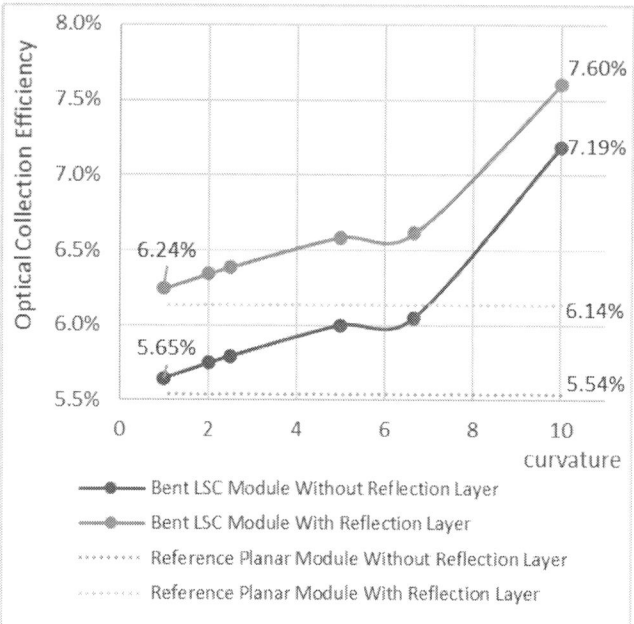

(a)

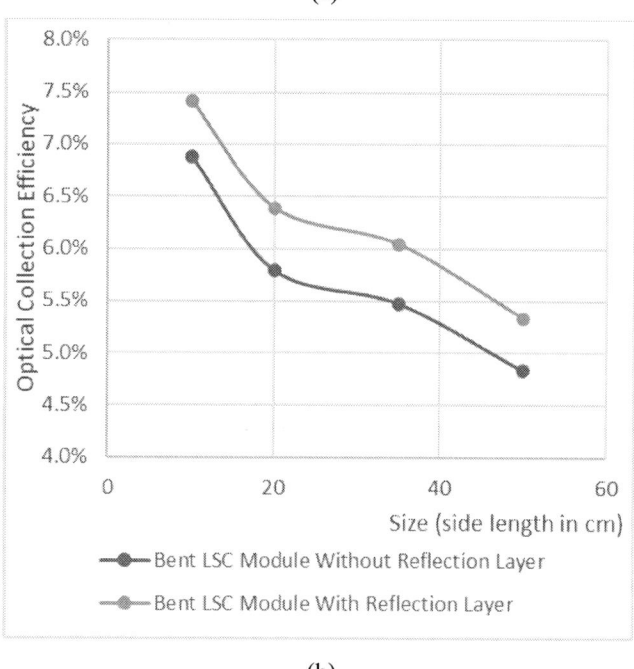

(b)

Fig. 2. a) Optical Collection Efficiency of two rectangle edges of the bent LSC modules without (blue) and with (orange) rear reflection layer. b) Optical Collection Efficiency of two rectangle edges of the bent LSC modules with different edge lengths at k=2.5 m^{-1}. Lines have been added to aid the eye.

III. SIMULATION RESULT

Simulation results are shown in Fig. 2.; in both simulations, bent LSC PV modules show better optical performance than flat LSC PV panels. With the increase of curvature, the optical collection efficiency of two edges would increase from 5.54% to 7.78%. With a reflection layer though, it increases from 6.14% to 8.23%. The reflection layer can hence increase the optical collection efficiency of the flat module by 11.8%, and the boosting effect of the reflection layer decreases with curvature to 5.7% at a curvature of k=10 m^{-1}. With the assumption of 20% efficiency of the solar cells, the flat module has a 1.11% power conversion efficiency, and the bent LSC-PV module with rear reflection layer, at a curvature of k=10 m^{-1} has a 1.52% power conversion efficiency.

For the effect of size on bent LSC PV modules, we selected the curvature of k=2.5 m^{-1}, as figure 2b shows, four different sizes models show the optical collection efficiency decreased with the size increase. This is similar to the flat LSC performance change we researched before.

IV. CONCLUSIONS

This study shows that LSC PV technology has an excellent potential to be applied to curved surfaces in buildings and in solar-powered cars, as examples. Namely, the optical collection efficiency at the edges of bent LSC PV modules is higher than that of a flat reference LSC PV module of the same surface area. With the same 400 cm^2 top view area, compared with 5.54% of two edges optical collection efficiency for the planar module, the module with a curvature of 10 m^{-1} could reach 7.78%. By assuming a flat 20% electrical conversion efficiency of the solar cell, compared with 1.11% of total power conversion efficiency for the flat module, the module with a curvature of 10 could reach 1.44%. Considering the wavelength of received irradiance and spectral response of solar cells, the actual power conversion efficiency of the flat module would be 1.58%. The module with a curvature of 10 could reach 1.92%.

According to the calculation, in the lightguide of convex curved LSC-PV modules, the irradiance emitted by the dyes close to the upper surface has a higher ratio to be conducted to the edge, while the dyes close to the upper surface in the lightguide have a higher probability of absorbing and emitting irradiance. Moreover, in the spectrum distribution at the edge of bent LSC-PV modules, there is some irradiance whose wavelength is out of the emission range of the dye, this is because the bent lightguide could conduct some solar irradiance to the edge. These two reasons may account for the higher performance of bent LSC devices than planar LSC devices.

Moreover, the bent LSC PV modules show similar efficiency attenuation to the flat LSC PV modules with the size increase, which means the application of bent LSC PV modules also should consider the size limitation or matrix small size modules.

REFERENCES

[1] W. H. Weber and J. Lambe, "Luminescent greenhouse collector for solar radiation," *Appl. Opt.*, vol. 15, no. 10, 1976, doi: 10.1364/ao.15.002299.

[2] A. Goetzberger and W. Greube, "Solar energy conversion with fluorescent collectors," *Appl. Phys.*, vol. 14, no. 2, pp. 123–139, Oct. 1977, doi: 10.1007/BF00883080.

[3] A. Goetzberger, "Fluorescent Planar Collector-Concentrators: a Review," *Sol. Cells*, vol. 4, pp. 3–23, 1980.

[4] P. S. Friedman, "Progress on the Development of

Luminescent Solar Concentrators.," *Proc. Soc. Photo-Optical Instrum. Eng.*, vol. 248, no. November 1980, pp. 98–104, 1980, doi: 10.1117/12.970591.

[5] S. J. Gallagher, B. Norton, and P. C. Eames, "Quantum dot solar concentrators: Electrical conversion efficiencies and comparative concentrating factors of fabricated devices," *Sol. Energy*, vol. 81, no. 6, pp. 813–821, 2007, doi: 10.1016/j.solener.2006.09.011.

[6] B. Vishwanathan *et al.*, "A comparison of performance of flat and bent photovoltaic luminescent solar concentrators ScienceDirect A comparison of performance of flat and bent photovoltaic luminescent solar concentrators," *Sol. ENERGY*, vol. 112, no. January 2019, pp. 120–127, 2015, doi: 10.1016/j.solener.2014.12.001.

[7] A. Reinders, R. Kishore, L. Slooff, and W. Eggink, "Luminescent solar concentrator photovoltaic designs," *Jpn. J. Appl. Phys.*, vol. 57, no. 8, 2018, doi: 10.7567/JJAP.57.08RD10.

[8] A. Reinders, *Designing with Photovoltaics*. 2020.

[9] M. G. Debije and P. P. C. Verbunt, "Thirty years of luminescent solar concentrator research: Solar energy for the built environment," *Adv. Energy Mater.*, vol. 2, no. 1, pp. 12–35, 2012, doi: 10.1002/aenm.201100554.

[10] W. G. J. H. M. van Sark *et al.*, "Luminescent Solar Concentrators - A review of recent results," *Opt. Express*, vol. 16, no. 26, p. 21773, 2008, doi: 10.1364/oe.16.021773.

[11] M. Rafiee, S. Chandra, H. Ahmed, and S. J. McCormack, "An overview of various configurations of Luminescent Solar Concentrators for photovoltaic applications," *Opt. Mater. (Amst).*, vol. 91, no. March 2018, pp. 212–227, 2019, doi: 10.1016/j.optmat.2019.01.007.

[12] M. Aghaei, M. Nitti, N. J. Ekins-Daukes, and A. H. M. E. Reinders, "Simulation of a novel configuration for luminescent solar concentrator photovoltaic devices using bifacial silicon solar cells," *Appl. Sci.*, vol. 10, no. 3, pp. 1–10, 2020, doi: 10.3390/app10030871.

[13] A. Reinders, M. G. Debije, and A. Rosemann, "Measured Efficiency of a Luminescent Solar Concentrator PV Module Called Leaf Roof," *IEEE J. Photovoltaics*, vol. 7, no. 6, pp. 1663–1666, 2017, doi: 10.1109/JPHOTOV.2017.2751513.

[14] L. H. Slooff *et al.*, "A Luminescent Solar Concentrator with 7.1% power conversion efficiency," *Phys. Status Solidi - Rapid Res. Lett.*, vol. 2, no. 6, pp. 257–259, 2008, doi: 10.1002/pssr.200802186.

[15] M. Aghaei, X. Zhu, M. Debije, W. Wong, T. Schmidt, and A. Reinders, "Simulations of Luminescent Solar Concentrator Bifacial Photovoltaic Mosaic Devices Containing Four Different Organic Luminophores," *IEEE J. Photovoltaics*, vol. 12, no. 3, pp. 771–777, 2022, doi: 10.1109/JPHOTOV.2022.3144962.

Optical modeling of light trapping using an ITO-based electrodynamic dust shield structure

Nicole Swatton[1], Andrey Semichaevsky[2]

1. Arizona State University, 1151 S. Forest Ave, Tempe, AZ 85281
2. Lincoln University (PA), 1570 Baltimore Pike, Lincoln University, PA 19352

Abstract—**Optimal design of light-trapping coatings for PV cells is essential for enhancing absorption of light in active layers. Some conductive polymer coatings can serve as an electrodynamic dust shield (EDS), e.g., for a lunar PV module. This paper discusses an approach to optical modeling of nanostructured surfaces for light trapping. PV structures include thin EDS and light-trapping layers and much thicker optically transparent spacer and active pn-junction layers. We used a combination of physical optics for the nanosurface and the transfer-matrix method for the whole structure. Examples illustrate how absorption can be improved over a range of incidence angles or wavelengths by nanopatterning an ITO layer.**

Keywords—*light trapping, computational design, metasurfaces.*

I. INTRODUCTION

PV cells for some space and terrestrial applications use multifunctional ultrathin coatings. One of the examples is an electrodynamic dust shield (EDS) [1]. The primary purpose of the EDS is to remove microscopically fine dust particles from PV modules by inducing low-frequency high-strength AC electric fields along the surface. Modern electrically conductive and optically transparent materials, such as thin-film indium tin oxide (ITO) can help design novel structures that serve multiple purposes. It is also important to maximize light absorption in active layers of the PV structure over a broad range of wavelengths and incidence angles. Particularly, a light-trapping structure should improve the solar cell's external efficiency when solar light falls on it obliquely, since mechanical tracking may not be available. Anisotropic (chiral) coatings designed for high transmittance and low emissivity have recently attracted attention for related applications [2]. Light can be trapped in thin PV structures using various gratings [3] and dielectric-dielectric photonic crystals applied on the top surfaces of thin PV structures [4] or by including small scatterers randomly inside organic polymer cells [5].

In this paper we study the light-trapping performance of an ITO-based diffracting nanosurface that also serves as the ground electrode for an EDS. It can be fabricated by lithographically patterning a thin film of deposited ITO and applying a bonding material layer over it. Similar light-trapping gratings nanoimprinted in polymer adhesive layers on the top surface of PV cells were also described in [6]-[7].

The FEM solution is obtained for the Helmholtz equation for our dielectric-dielectric grating. Using a combination of numerical full-wave electromagnetic scattering simulations and the transfer-matrix method [8], we quantify absorption of light

in a GaAs layer at the bottom of a stack over a range of angles and wavelengths. The FEM solution gives us the angular spectra of scattered plane waves that are then propagated numerically through the PV structure using the transfer-matrix method.

II. LIGHT-TRAPPING STRUCTURES

The EDS on the top of the stack is periodic, composed of planar spiral-like ITO structures when viewed from the top. A profile of nanoimprinted EDS coming from Zygo profilometer measurements and the cross-section of the PV structure are presented in Figures 1a and 1b, respectively.

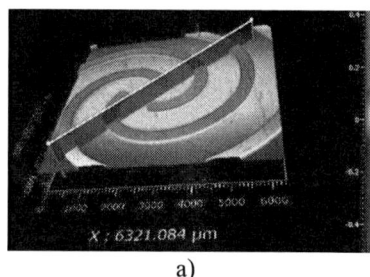

a)

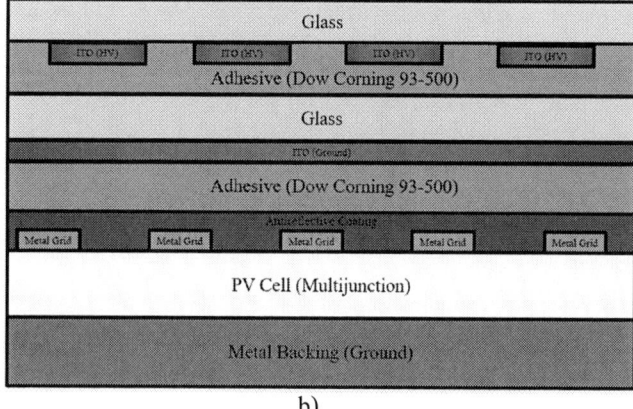

b)

Fig. 1. a) Profile of the EDS and b) Cross-section of the PV structure.

For the structure in Figure 1b, ITO films were applied on the surface of a glass slide through photolithography methods, the ITO is structured to form the EDS features, and then embedded in the Dow Corning 93-500 adhesive layer. The ITO was deposited by Kennedy Space Center to make samples of varying thicknesses between 100 and 1000 nm by MLD and

Abrisa. Such thin layers of ITO absorb only a very small amount of incident light in the visible.

The 'ground electrode' of the original EDS was not patterned. Let us assume that one can also fabricate large-scale structured nanosurfaces using ITO, as in [9]. This modification could be made in a way that does not affect electrode's role of an electric conductor. We assume that the period of the 2D rectangular grating is 300 nm, and the duty cycle is 50%; these values are subject to changes during further optimization. We also assumed that the depth of the etch is 150 nm. The thickness of the adhesive layer under the EDS ground is assumed to be about 100 microns. The dielectric-dielectric grating then originates from the mismatching refractive indices of the ITO and adhesive. The glass layer has a thickness of about 1 mm. We expect that EQE of the PV structure can be enhanced at grazing incidences due to the additional angular spread introduced by the ITO-adhesive nanosurface.

III. MODELS AND CODES

The nanosurface is modeled using finite elements for the Helmholtz equation in COMSOL, assuming wavelength-dependent complex refractive indices of materials. Periodic boundary conditions were applied, as the structure is essentially a very large periodic array of surface scatterers.

The Helmholtz equation (1) is solved assuming incident plane waves, and the scattered electric field E is obtained as an average of those for the two polarizations.

$$\nabla^2 \vec{E} + \left(\frac{\omega}{c}\right)^2 \vec{\varepsilon} \cdot \vec{E} = 0 , \quad (1)$$

where $\vec{\varepsilon}$ is the complex permittivity tensor.

The scattered E-field is transformed using into an angular spectrum using the near-to-far-field transformation (NFFF) [10], (2).

$$\vec{E}(\vec{r}) = A(\vec{r}) \oint_C f(\vec{E}(\vec{r}\,'), \vec{H}(\vec{r}\,'))e^{-j\vec{k}\cdot\vec{r}\,'}dl\,', \quad (2)$$

where f denotes a function, unprimed coordinates are in the far zone, and primed ones are in the near zone, C is a near-zone scattering boundary, A denotes an amplitude in the far zone, k is the wavenumber.

Discrete scattered plane waves with amplitudes computed using the NFFF (2) are propagated through the structure using the transfer-matrix method (TMM). At first, a standard TMM approach for isotropic media was used, and nonpatterned ITO and other layers were assumed in all models. We also plan to apply an anisotropic wave propagation model based on [8]. In both cases, the solutions are computed for TM and TE incidences. Codes were implemented using MATLAB R21. The wavelength- and incidence-dependent reflectance and transmittance of the structure was computed, and the absorption in the active layer was then found.

IV. RESULTS AND DISCUSSION

We modeled light absorption in the GaAs layer of the structure shown in Figure 1 using methods described in section III with the complex refractive indices of the constituent materials extracted from public sources and our own

measurements. First, the ITO EDS ground was assumed to be nonpatterned.

The model was run over 1000 incident angles uniformly distributed in the [0...90] deg. interval and for 100 discrete wavelengths between 400 and 800 nm. In addition, the transmissivity of the stack was measured in a series of spectrophotometry experiments, showing a good agreement with the model. Figure 2 shows fraction of the incident intensity absorbed by the GaAs layer as a function of incident angle and wavelength.

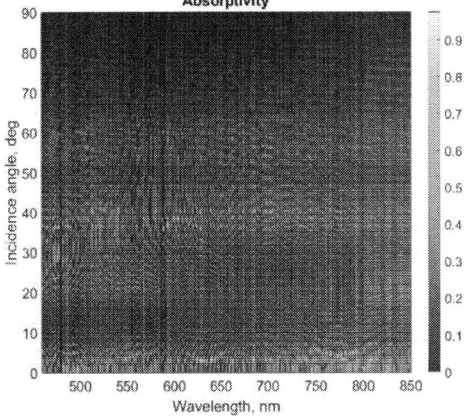

Fig. 2. Absoptivity in the GaAs layer as a function of the incidence angle and free-space wavelength of light.

Simulations for the nonpatterned EDS ground shown in Figure 2 indicate that variations in the absorptance of the active layer with the wavelength are mainly due to Fabry-Perot resonances. There is also a significant reduction in the active layer absorptance with the increasing angle of incidence for all wavelengths.

Next, we assumed that the side of the ITO EDS ground facing the adhesive is patterned into a 2D grating. Figure 3 presents the angular spectrum from the surface nanostructure, as it was computed for the dimensions from section II at the 600 nm free-space wavelength. The scattered plane waves are assumed to propagate in the adhesive and reach the GaAs PV cell.

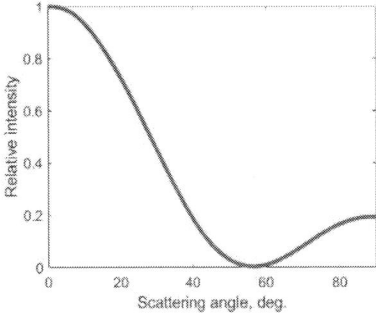

Fig. 3. Diffraction pattern produced by the ITO-adhesive nanosurface.

At normal incidence, the ITO-adhesive 2D grating produces a mainlobe with FWHM of 56 degrees. For oblique incidences, the direction of the maximum is shifted according to the grating equation ($\theta' = \theta_0 + \theta_{inc}$), but the FWHM remains close to that at normal incidence. Using our computed diffracted fields as inputs into the TMM, we quantified the absorption in the GaAs PV cell.

Figure 4 shows the fraction of incident light absorbed, A=1-R-T, in a GaAs solar cell as a function of the incidence angle; the solid line for the stack with a nonpatterned film of ITO EDS ground, and the dashed line shows the same parameter when the EDS ground ITO layer is patterned.

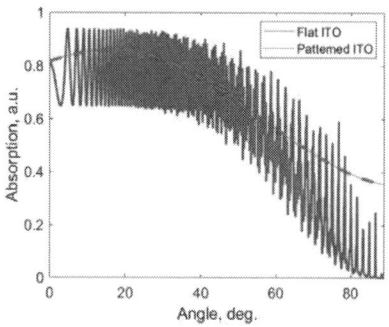

Fig. 4. Light absorption in the GaAs layer as a function of the incidence angle at 600 nm incident wavelength (flat ITO vs. nanostructured ITO layers).

Results shown in Figure 4 indicate that the patterned nanosurface significantly improves the light absorption in the active layer at incidence angles over 45 degrees compared to a thin film of the same thickness. Some improvement is also seen at angles around 20 degrees. The oscillations due to Fabry-Perot resonances are also disappearing when the diffraction grating is applied. This can be explained by the 'convolution-like' effect that the grating has on the optical transfer function of the stack.

V. CONCLUSIONS

A combination of full-wave analysis with the TMM is highly suitable for computational design of structured surfaces for light trapping in PV. Cells with nonpatterned thin planar ITO layers show significant reduction of EQE at grazing incidences. This can be resolved by a diffracting nanopatterned ITO surface serving as the EDS ground.

ACKNOWLEDGMENTS

The authors acknowledge support from NASA Glenn Research Center and useful conversations with Mr. Jeremiah McNatt and other EDS design team members.

REFERENCES

[1] EMC test report. Electrodynamic Dust Shield, EML-0069-REF, Rev. Basic, NASA Kennedy Space Center, 2014.

[2] Z. Gao, G. Xu, R. Zhu, Y. Zhai, J. Wang, S. Qu, and Q. Fan, "Multifunctional anisotropic coding metasurface with low emissivity and high optical transmittance," Infrared Physics and Technology, vol. 117, 2021, 103845.

[3] K. Li, S. Haque, A. Martins, E. Fortunato, R. Martins, M.J. Mendes, and C.S. Schuster, "Light trapping in solar cells: simple design rules to maximize absorption," Optica, vol. 7, 2020, pp. 1377-1384.

[4] K.J. Yu, L. Gao, J. S. Park, Y. R. Lee, C.J. Corcoran, R.G. Nuzzo, D. Chanda, J.A. Rogers, "Light Trapping in Ultrathin Monocrystalline Silicon Solar Cells ," Adv. Energy Mater., vol. 3, 2013, pp. 1401–1406.

[5] L. McMillon, M. Mariano, Y.L. Lin, S.M. Halumi, J. Li, A. Semichaevsky, B. Rand, A. Taylor, "Light trapping in polymer solar cells by processing with nanostructured diatomaceous earth," Organic Electronics, vol. 50, 2017, pp. 7-15.

[6] A. Lin, S.M. Fu, B. Chen, S. Yan, Y.K. Zhong, M.Kao, C. Shen, J. Shieh, and T.Y. Tseng, "The external light trapping for perovskite solar cells using nanoimprinted polymer metamaterial patterns," 2016 IEEE 43rd Photovoltaic Specialists Conference (PVSC).

[7] S. Yoshinaga, Y. Ishikawa, S. Araki, and Y. Uraoka, "Light trapping effect of nanoimprinted-textured crystalline silicon solar cells," 2013 IEEE 39th Photovoltaic Specialists Conference (PVSC).

[8] J. Hao and L. Zhou, "Electromagnetic wave scatterings by anisotropic metamaterals: Generalized 4x4 transfer-matrix method," Phys. Rev. B, vol. 77, 2008, 094201.

[9] H.S. Jeong, H.-J. Jeon, Y. H. Kim, M.B. Oh, P. Kumar, S.-W. Kang, and H.-T. Jung, "Bifunctional ITO layer with a high resolution, surface nano-pattern for alignment and switching of LCs in device applications," NPG Asia Materials, vol. 4, 2012, pp. 1-7.

[10] A. Taflove, S. Hagness, "Computational Electrodynamics: The Finite-Difference Time Domain Method," Artech House, 1995.

Assessing the Alignment of Solar Facilities with Global Climate Goals

Parikhit Sinha[1] and Liv Hammann[2]

[1]First Solar, Tempe, AZ, 85281, USA, [2]right. based on science, Frankfurt, Hesse, 60314, Germany

Abstract — **Climate science-based targets have become the state-of-the-art approach for greenhouse gas goal setting by companies and institutions. As companies try to maximize the climate benefit of their renewable energy investments and lower their Scope 3 emissions, climate science-based target setting can be extended to solar facilities themselves. By evaluating the embodied carbon and economic emissions intensity of a solar facility and globally extrapolating, the solar park's temperature alignment can be calculated with the X-Degree Compatibility Model. A case study of 100 MWdc solar facilities in North Carolina indicates that solar facilities are well aligned with global climate goals for a 1.75°C (i.e. 'well below 2°C') warming scenario. While the analysis shows that both, CdTe and mono-c-Si PV systems, are compatible with the chosen global warming scenario, the CdTe PV system has a lower climate impact, measurable in °C. The most sensitive variables contributing to economic emissions intensity are PPA price, O&M cost, system lifetime, and embodied carbon. Continued progress in lowering the embodied carbon and increasing the lifetime of PV systems is needed to counteract the tendency for increasing economic emissions intensity from declining PPA prices.**

I. INTRODUCTION

Efforts to mitigate global climate change have evolved beyond central government action, with corporate and regional entities establishing individual metrics and targets. The most sophisticated methods for target setting are known as climate science-based targets [1], in which global greenhouse gas (GHG) emissions scenarios (e.g., 1.75°C or 2°C warming scenarios) are used to estimate allowable emissions for a given company or entity.

While solar photovoltaic (PV) facilities have no direct GHG emissions during operation, they have emissions over the system life cycle, particularly in the production of components, as well as during construction and decommissioning. These life cycle emissions have an important role in the net displacement of grid electricity GHG emissions by solar energy projects [2].

As corporate and regional entities make procurement decisions for solar energy, they are beginning to advocate for decarbonizing the solar supply chain [3]. Since the embodied carbon of purchased electricity is part of Scope 3 GHG accounting [4], these efforts would lower a buyer's Scope 3 emissions and maximize the climate benefit of their investments.

This study provides a quantitative method for solar energy buyers to assess the alignment of solar facilities with global climate goals (concretely, with IEA's B2DS scenario which is a 1.75°C warming scenario [5]). The method is applied to ground-mount solar facilities, with identification of key system parameters that influence alignment.

II. METHODS

The analysis is based on the X-Degree Compatibility (XDC) model [6], version 2.0, which utilizes four steps:

1) What quantity of emissions (CO_2-eq) does the facility generate per unit of gross value added (GVA; $) from a base year to 2050?

2) What quantity of emissions (CO_2-eq) would reach the atmosphere if the entire world operated as emission intensively until 2050 as the facility under consideration?

3) What degree of global warming would be expected if that quantity of emissions reached the atmosphere?

4) What is the difference between the solar park's XDC calculated in step 3 and the sector-specific Target XDC? This results in the XDC Gap. A positive XDC Gap indicates misalignment with the 1.75°C scenario while a negative XDC Gap indicates alignment with the 1.75°C scenario.

For step 1, electricity production from a 100MWdc utility-scale 1-axis tracking facility in North Carolina, USA with 30-year system lifetime and 1.25 DC:AC ratio was modeled with PlantPredict software, using both cadmium telluride (CdTe; 1730 MWh/MWdc/yr 1st year specific yield; 0.2%/yr degradation rate) and mono-crystalline silicon (mono-c-Si; 1712 MWh/MWdc/yr 1st year specific yield; 0.5%/yr degradation rate) PV module technology. A 2019 average U.S. power purchase agreement (PPA) price of $24/MWh [7] was used in conjunction with electricity production to estimate annual revenue. GVA is defined as PPA revenue minus operations and maintenance (O&M) expenses ($10/kWdc per yr) [8] minus decommissioning costs ($83/kWac) [9] in a given year. Note all financial values are considered as 2018 constant $.

Also, for step 1, the quantity of life cycle emissions are from the IEA PVPS (2020) life cycle inventory [10] implemented in Simapro 9.2.0.1 software with UVEK DQRv2:2018 background database and IPCC 2013 GWP 100a impact method (705 and 1177 metric tons CO2-eq/MWdc for CdTe and mono-c-Si ground-mount PV systems, respectively; production in USA/Malaysia and China, respectively). Of these life cycle emissions, 4.4% and 2.9% are assumed to occur during decommissioning for CdTe and mono-c-Si ground-mount PV systems, respectively [11], and the remainder are due to the supply chain of components and facility construction. During project operation, minor GHG emissions are estimated with electricity use for 1-axis tracking (2.35 MWh/MWdc per yr) [12] using the life cycle carbon intensity of the regional electricity grid (SERC, 0.621 metric tons CO2-eq/MWh; Ecoinvent 3.6).

For step 2, annual GHG emissions are normalized by annual GVA to obtain annual economic emissions intensity (EEI). The facilities' GHG emissions are not emitted directly by the facility (Scope 1) but are indirect emissions (Scope 2 or 3) [4].

978-1-7281-6118-1/22 $31.00 © 2022 IEEE

In the XDC model, indirect emissions are weighted by 50% by convention [6]; hence, the EEI is weighted by 50%. Because the annual EEI is heterogeneous over the system lifetime with most emissions occurring initially, effective EEI is also calculated as a constant value that leads to the same upscaled cumulative emissions when multiplying with the global GVA. Global emissions are obtained by multiplying the solar facility's effective EEI by global GVA ($73.8 trillion in 2018; 1.93% annual growth rate) [13] through 2050.

For step 3, global emissions estimated in step 2 are used as input to the FaIR climate model [14] to estimate global warming associated with these emissions.

In step 4, the results are compared to the sector-specific Target XDC since the contribution to global warming and the leverage for emissions reductions differs between sectors [6]. The Target XDC is the benchmark which actors in the energy sector (OECD region, NACE 35, electricity, gas, steam, and air conditioning supply) should reach in order to be aligned with the chosen 1.75°C scenario (B2DS scenario).

III. RESULTS AND DISCUSSION

Based on the inputs for step 1 and 2 of the XDC Model, the EEI and effective EEI for CdTe and mono-c-Si ground-mount PV systems are shown in Fig. 1. As explained above, the EEI is highest initially due to the embodied carbon in the PV system components and facility construction activities, and has a smaller peak at project end due to decommissioning. The effective EEI is the constant EEI which is calculated as described above.

When the effective EEI values are globally extrapolated and input to the FaIR climate model (XDC steps 2 and 3 above), they give an XDC of 1.8°C for CdTe and 2.1°C for mono-c-Si PV systems. In both cases, the ground-mount PV facilities are well within the sector-specific benchmark for the NACE 35 sector of 3.2°C, resulting in XDC Gaps of -1.4 and -1.1°C, respectively. Hence the solar facilities are aligned with 1.75°C of global warming. However, the difference between PV systems is also apparent in their XDCs, with the CdTe PV system's climate impact 0.3°C lower than for mono-c-Si PV.

In order to understand the main contributors to the temperature alignment, model parameters were varied (Table 1). Several parameters influence the GVA estimate, including PPA price, O&M and decommissioning cost, energy yield, and

degradation rate. Of these variables, the PPA price and O&M cost are most sensitive. PPA prices have been declining with time and scale of PV deployment [7], resulting in a tendency to increase the climate impact as measured by the XDC Model as less GVA is created. This is partly offset by a trend of decreasing O&M costs.

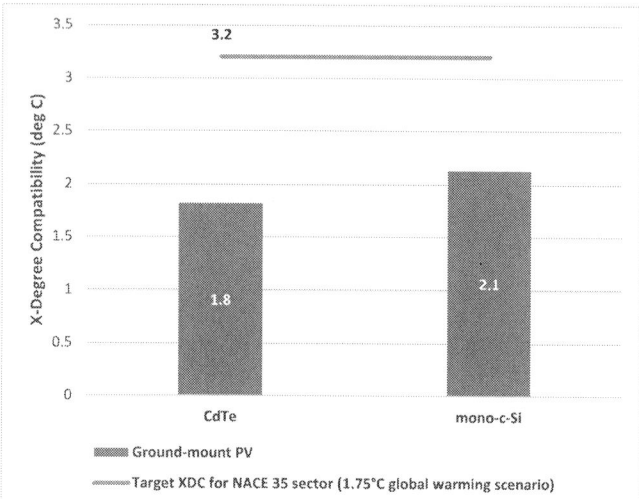

Fig. 2. Alignment of solar facilities with global climate goals. NACE 35 refers to electricity, gas, steam, and air conditioning supply sector in OECD region.

The EEI is estimated by normalizing annual GHG emissions by annual GVA. While the annual GVA is tending to decrease due to lower PPA prices, the annual GHG emissions are also tending to decrease with time and scale of PV deployment [10]. The sensitivity analysis in Table 1 and Figure 3 shows that embodied carbon is a sensitive model parameter along with PPA price and O&M cost.

Lastly, the system lifetime is also a sensitive parameter that influences EEI, with longer lifetimes reducing the effective EEI, since emissions are normalized over a longer duration. Improvements in PV module and system stability [15] can help counteract the tendency toward increasing EEI from lower PPA prices.

With regards to the OECD NACE 35 sector benchmark (3.2°C), the sensitivity analysis shows that ground-mount PV systems are compatible with this sector benchmark under the various model parameter variations. All the cases analyzed in

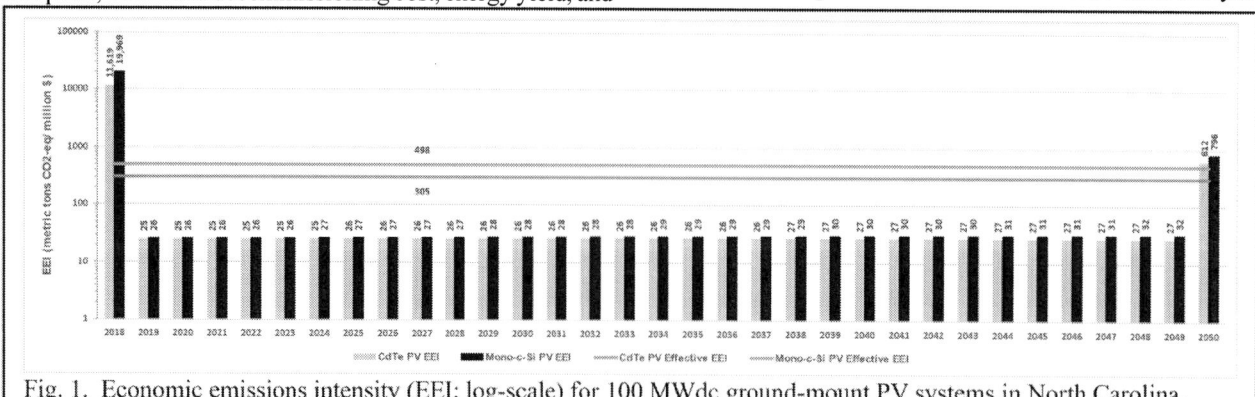

Fig. 1. Economic emissions intensity (EEI; log-scale) for 100 MWdc ground-mount PV systems in North Carolina.

Table 1. Sensitivity analysis of model parameters

	Model parameter		X-Degree Compatibility in °C		X-Degree Compatibility Gap Compared to NACE 35 sector target (3.2°C)	
	Low	High	Low	High	Low	High
O&M ($/kWdc per yr)	4	14	1.7	1.9	-1.3	-1.5
Decommissioning ($/kWac)	40	80	1.7	1.8	-1.5	-1.5
PPA price ($/MWh)	20.00	40.00	1.6	1.9	-1.3	-1.7
Embodied carbon (metric tons CO2e/MWdc)	705	3047	1.8	3.0	-0.2	-1.5
End-of-life fraction of embodied carbon	1.60%	6.60%	1.7	1.8	-1.4	-1.5
Tracking electricity usage (MWh/MWdc per yr)	1.50	3.00	1.7	1.8	-1.4	-1.5
Life cycle grid electricity carbon footprint (SERC; metric tons CO2e/MWh)	0.200	0.700	1.7	1.8	-1.5	-1.5
1st yr specific yield (MWh/MWdc/yr)	1700	1800	1.7	1.8	-1.4	-1.5
Degradation rate (%/yr)	0.1%	0.5%	1.8	1.8	-1.4	-1.5
Lifetime	25	33	1.7	1.9	-1.3	-1.5

Table 1 fall within this benchmark. However, when running the analysis with very high values for embodied carbon, the solar facility's climate impact is only slightly within the sector-specific Target XDC. Therefore, continued progress in lowering the embodied carbon in PV systems is crucial for maintaining compatibility with climate goals. The EPEAT registry for sustainable electronics is developing criteria for low carbon PV modules that can further incentivize reductions in embodied carbon [16].

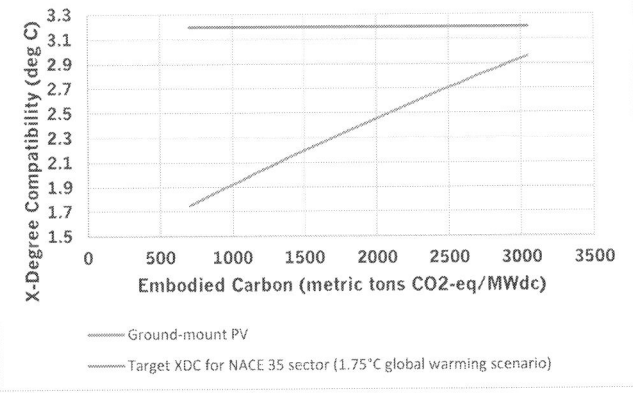

Fig. 3. Sensitivity analysis of X-Degree Compatibility with embodied carbon for 100 MWdc ground-mount PV systems in North Carolina. Lower and upper bound values for embodied carbon are from [10] and [17], respectively.

IV. CONCLUSIONS

Assessment of the economic emissions intensity of ground-mount PV facilities indicates that they are aligned with global climate change and OECD electricity sector-specific goals for a 1.75°C warming scenario. The most sensitive variables contributing to economic emissions intensity are PPA price, O&M cost, system lifetime, and embodied carbon. Continued progress in lowering the embodied carbon and increasing the lifetime of PV systems is needed to counteract the tendency of increasing economic emissions intensity from declining PPA prices.

REFERENCES

[1] SBTi, "Science based targets," available at: https://sciencebasedtargets.org/
[2] H. Richardson, "Lifecycle and avoided emissions of solar technologies," WattTime, Oakland, CA, 2020.
[3] REBA, "Decarbonizing Industrial Supply Chain Energy (DISC-e)," available at: https://cebi.org/programs/disc-e/
[4] WRI/WBCSD, "The greenhouse gas protocol," available at: https://ghgprotocol.org/corporate-standard
[5] IEA, "Energy Technology Perspectives 2017," available at: https://www.iea.org/reports/energy-technology-perspectives-2017.
[6] H. Helmke, H. P. Hafner, F. Gebert, and A. Pankiewicz, "Provision of climate services—the XDC model," in *Handbook of Climate Services*, pp. 223-249, 2020.
[7] LBNL, "Utility-scale solar data update: 2020 edition," available at: https://emp.lbl.gov/utility-scale-solar.
[8] R. Jones-Albertus, D. Feldman, R. Fu, K. Horowitz, and M. Woodhouse, "Technology advances needed for photovoltaics to achieve widespread grid price parity," Prog. Photov., vol. 24(9), pp. 1272-1283, 2016.
[9] EPRI, "PV plant decommissioning salvage value," Report 3002013116, 2018.
[10] IEA PVPS, "Life cycle inventories and life cycle assessment of photovoltaic systems," Report T12-19:2020, 2020.
[11] P. Stolz, R. Frischknecht, F. Wyss, and M. de Wild-Scholten, "PEF screening report of electricity from photovoltaic panels in the context of the EU Product Environmental Footprint Category Rules (PEFCR) Pilots (V.2.0)," 2016.
[12] P. Sinha, M. S. Schneider, S. N. Dailey, C. Jepson, M.J. de Wild-Scholten, "Eco-efficiency of CdTe photovoltaics with tracking systems," in *39th Photovoltaic Specialists Conference* 2013, pp. 3374-3378.
[13] World Bank, "Gross value added at basic prices," available at: https://data.worldbank.org/indicator/NY.GDP.FCST.CD
[14] C. J. Smith, P. M. Forster, M. Allen, N. Leach, R. J. Millar, G. A. Passerello, and L. A. Regayre, "FAIR v1.3: a simple emissions-based impulse response and carbon cycle model," vol. 11(6), pp. 2273-2297, 2018.
[15] I. M. Peters, J. Hauch, C. Brabec, and P. Sinha, "The value of stability in photovoltaics," *Joule*, 2021
[16] GEC, "EPEAT Registry," available at: https://www.epeat.net/
[17] IEA PVPS, "ENVI-PV", available at: http://viewer.webservice-energy.org/envi-pv_v2.0/

Preparation of Plasmonic Ag and Au Nanoparticle Interfaces for Photocurrent Enhancement in Si Solar Cells

Brahim Aïssa[1] (SM IEEE), Adnan Ali[1], Rui N. Pereira[2] and Anirban Mitra[3]

[1]Qatar Environment and Energy Research Institute, Hamad bin Khalifa University, Qatar Foundation, P.O. Box 34110, Doha, Qatar

[2]Department of Physics and i3N - Institute for Nanostructures, Nanomodelling and Nanofabrication, University of Aveiro, 3810-193 Aveiro, Portugal

[3]Department of Physics, Indian Institute of Technology Roorkee, Roorkee- 247667, Uttarakhand, India
*Corresponding author: baissa@hbku.edu.qa; Brahim.aissa@mpbc.ca

Abstract—**Nanoparticle (NP) arrays of noble metals strongly absorb light in the visible to infrared wavelengths through resonant interactions between the incident electromagnetic field and the metal's free electron plasma. Such plasmonic interfaces enhance light absorption and photocurrent in solar cells. We report here broadband plasmonic interfaces consisting of silver nanoparticles (NPs) formed by depositing by e-beam evaporation a thin film of Ag followed by dewetting process under thermal annealing. The NP interface yields a clear photocurrent enhancement (PE) in thin film silicon devices. For coatings produced from Ag NPs, an optimal value of 15% surface coverage (SC) was observed. Scanning electron microscopy of interface morphologies revealed that low SC is resulting in broadband PE; while at higher coverage, strings and clusters are formed and caused red shifting of the PE peak and a narrower spectral response. The multi-physics impacts of size and radius distribution of plasmonic-NPs (including Ag and Au) were also studied. For optimal PV performance, a parametric analysis was performed for system sizes of (3×3, 5×5, 7×7), and radii varying from 10 to 150 nm. Total spectral heat absorbed was also investigated by integrating total spectral heating from 300 nm to 1200 nm. The optimum performance for a Schottky-like device were obtained for a 70 nm radius and a surface coverage of 22.72 NP/μm², revealing a maximum short circuit current gain of about 47 % when compared to bare silicon solar cell.**

Keywords— Plasmonic, Nanoparticle, Multi-physics model; Photocurrent, Surface Coverage, Schottky solar cell

I. INTRODUCTION

The existing global photovoltaic solar cell market is 90% c-Si based solar cells [5–7]. For fulfilling global energy demand from photovoltaics, enhancement in light conversion efficiency and cost reduction are the main research targets. Different techniques have been applied for enhancing light absorption in the active layer [9–12]. Usually, increase in active layer thickness is applied, while an optical absorption enhancement may provide the freedom to decrease the thickness of the active layer, which has direct effect on the cost. Besides this, to

enhance the power conversion efficiency and decrease the cost, nanophotonic approach for light entrapment has been recently explored, and was demonstrated to contribute to photo-stability and long term yielding [13,14]. Plasmonics refers to the generation, detection, and manipulation of signals at optical frequencies along metal-dielectric interfaces in the nanometer scale. It finds applications in sensing, microscopy, optical communications, biophotonics and in light trapping enhancement for solar energy conversion. Although plasmonics has a short history of development, it has made substantial advancement in enhancing the absorption of the solar spectrum as well as in the charge carrier separation efficiency. Recently, huge developments have been made in understanding the basic parameters and mechanisms governing the application of plasmonics including the effects of nanoparticles size, their arrangement, geometry and how all these factors impact the dielectric field in the surrounding medium of the plasmons.

In the specific scheme of plasmon-enhanced solar cells, the metal nanoparticles associated with the local surface plasmon resonance (LSP), aiming at triggering a highly confined electric field and large scattering cross-sections (σ), are associated with parasitic ohmic losses that consequently undergo local temperature rising, thereby altering the photoelectric conversion efficiency and influence the stability of solar cells. The previous studies of utilizing the plasmonic effect of nanoparticles on solar cells mainly focused on optical and electrical aspects [5-8], i.e. maximization of the light absorption and improvement of carrier collection. However, the thermal effects in the PSSCs should be considered in performance optimization, particularly the energy dissipation and stability issues after the incorporation of the metal nanostructure.[9, 10] The resonant interaction of light with the NPs at plasmonic resonance can result in both scattering and absorption of light. The beneficial effect of scattering consists in the extended path of the light inside the photovoltaic material, which increases the probability of scattered photons to be absorbed and generate photo-carriers. On the other hand, the parasitic absorption i.e. absorption occurring inside the NPs, can severely limit the overall photocurrent enhancement that can be

978-1-7281-6118-1/22 $31.00 © 2022 IEEE

produced in a solar cell due to light trapping [16–18]. Since the optical properties of plasmonic NPs depend strongly on their size, shape and local environment [19], the NPs for PV application should be properly designed and engineered in order to maximize the scattering and minimize the parasitic losses over the wavelength range important for light trapping.

The process commonly used for the fabrication of the NPs, known as solid-state dewetting (SSD), consists in thermally-induced transformation of morphology from a thin film into an array of droplets or nanoparticles [20,21]. Silver is the usual material of choice among the materials commonly used in plasmonics, due to its high radiative efficiency and low imaginary permittivity in the visible and near-infrared (NIR) spectrum [22,23]. Initial designs depended on NPs placed on the front surface of the solar cell. Further studies showed that it is more advantageous to incorporate the NPs in the rear side of the cell i.e. in proximity to the rear mirror. In such configuration, known as plasmonic back reflector (PBR), NPs interact only with light which was not absorbed during the first pass through the cell material [24,25] thus decreasing the parasitic loses. A standard configuration PBR consists of a flat silver mirror, a thin aluminum doped zinc oxide (AZO) spacer layer and the NPs deposited on the top. Thereafter, such stack can serve as a substrate for the deposition of any thin-film photovoltaic absorber [26,27]. It has been demonstrated that self-assembled NPs in a PBR configuration can provide light trapping performance comparable to state of the art random textures in thin film Si solar cells [28–43].

Here, we report on the formation of Ag NPs networks of different surface densities by a simple dewetting of Ag thin film deposited by e-beam evaporation followed by annealing process. These Ag network were then used as broadband plasmonic interfaces for c-Si solar cells, thereby yielding to a photocurrent enhancement (PE) by up to 50% as supported by our simulation results. The effects of the surface densities on the PE is also discussed. This extended abstract contains the main finding while detailed results will be presented in the full version of this paper.

II. EXPERIMENTAL AND MODELING

All chemicals were bought from Aldrich and used as received unless otherwise stated. The evaporation was performed using a Denton Vacuum Explorer™ e-beam evaporator under 2×10^{-4} Torr of a pure Ag target (99.999% from Kurt J Lesker). Deposition was performed onto SOI wafers (purchased from University wafers) and a witness soda lime glass (SLG). Films were then thermally annealed at different temperatures to get Ag NPs network of different densities as indicated by the surface coverage. Ocean Optics UV/Vis spectrometer (USB4000- UV-VIS) was used to characterize NP absorption spectra. SOI cells were used to measure the NP effects on photocurrent. SOI cells are simple devices that have an accessible open surface that enabled easy monolayer coverage of NPs close to the c-Si active layer within a few nanometers. The wafer was cleaned using the RCA process and metal finger contacts (40 nm thick) were deposited by thermal evaporation using a SS mask. The wafer was a bonded c-Si n-type wafer, of 1-10 Ω-cm resistivity.

We have developed an opto-thermo-electrical framework on COMSOL for plasmonic Schottky solar cell (PSSC) with monofacial configuration decorated with Au-NPs on the top surface of silicon (Si) absorber, as highlighted in Figure 1a. The multi-physics effect of Au-NP's fractional coverage assessed by system size and radius are investigated. Sensitivity analysis is carried out for system size (3×3, 5×5, 7×7) and NP's radius (10 – 150 nm) as shown in Figure 1b.

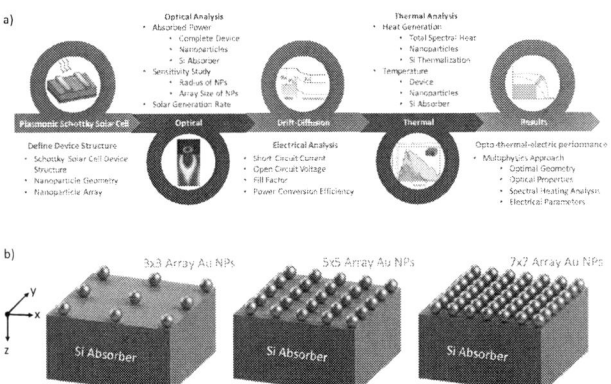

Figure 1: a) Multi-physics approach to assess the optical, thermal and electrical performance of Plasmonic Schottky Solar Cell, and b) Three Au nanoparticles (NP) systems (3×3, 5×5, 7×7) decorated on top of silicon absorber considered for investigation.

The multi-physics model was developed by coupling wave optics, semi-conductor (drift-diffusion), and heat transfer modules. Here, by optimizing the geometrical parameters of NPs for AM1.5G solar radiation and temperature of 300K, we analyzed light absorption leading to maximum short circuit current (J_{sc}). From a thermal perspective, spectral absorption in the device, spectral heating in NPs and thermalization heating in Si absorber were calculated using an energy balance approach. The total spectral heating content was then employed to assess the device and NP temperature. The optimum performance was achieved for a system with a 5×5 NP array of 70 nm radius with a maximum J_{sc} gain of 46.8%. However, this electrical enhancement is accompanied by increased thermal gain in NPs (182.5%). Interestingly, the relation between NP's geometry and multi-physics performance is not direct and an optimal point exists within the design space. World-wide performance gain for annual energy yield in comparison to bare-silicon absorber was also analyzed after optimization to appreciate the global potential of PSSC. Importantly, the work and design guidelines presented here are a step in the right direction for analyzing the holistic performance especially since currently no multi-physics model exists for coupling optical, electrical, and thermal response of PSSC. Furthermore, the developed framework is not only limited to monofacial configurations and Schottky solar cells, but can be extended to other solid-state devices.

The mathematical analysis for plasmonic-based solar cells primarily required three major physical sciences areas: 1) Optics; 2) Electrical; and 3) Thermal, to ensure the proper multi-physics coupling. In particular, commercial software

(COMSOL MULTIPHYSICS[30], MATLAB[31] and SCAPS[32] were used to develop plasmonic/solar cell interaction. For this purpose, the literature review was conducted to understand the limitation and practicalities of each solver available as an open-source or commercial packages. Modeling of plasmonic solar cells (PSC) is vital for assessing the geometrical and operating conditions for optimal optoelectronic performance. In principle, modeling of PSC is a multi-physics problem [33, 34] comprising of optics and electronics as shown in Figure 1a. Optical physics is attributed to the plasmonic structure and light propagation. While electronics is associated with the physics of solar cells which is required for investigating carrier transport and extraction. Electric field intensity, resonance modes, Absorption (A), Reflection (R), Transmission (T), scattering and extinction parameters are the key indicators for optics as highlighted in Fig. 1a, whereas carrier generation, transport, recombination, short circuit current, and quantum efficiency drive the solar cell performance. Device designs using propagation of electromagnetic (EM) waves include surface plasmon polaritons (SPP) in plasmonic waveguides and nanoparticle arrays as shown in Figure 1b[35]. Finite Difference Time Domain (FDTD) and discontinuous Galerkin time-domain (DGTD) methods for solving Maxwell's equations are best suited for simulating propagation-based plasmonic devices[36].

A. OPTICAL MODELING:

COMSOL has been used to measure optical absorption by using the FDTD method, which solves Maxwell equations for electrical and magnetic fields for the plasmonic material and associated solar cell geometry. We measure the absorbed power, $P_{abs}(x, y, z, \omega)$ as a wavelength function and the generation rate $g(x, y, z, \omega)$ depending on the photon energy and wavelength frequency $\left(\omega = \frac{2\pi c}{\lambda}\right)$.

$$P_{abs}(x, y, z, \omega) = \frac{1}{2}\omega|E(x, y, z, \omega)|^2 \, imag\{\varepsilon(x, y, z, \omega)\}$$

$$g(x, y, z, \omega) = \frac{P_{abs}(x, y, z, \omega)}{\hbar\,\omega}$$

The solar spectrum AM1.5G is then used to measure the generation rates to deliver the weighted generation rate as:

$$G(x, y, z, \omega) = g(x, y, z, \omega)\left[\frac{P_{solar}(\omega)}{P_{incident}(\omega)}\right]$$

$P_{solar}(\omega)$ is the AM1.5G solar energy spectrum and $P_{incident}(\omega)$ is the incident power for optical simulations. Finally, the broadband generation rate is calculated by integrating for frequencies.

$$G(x, y, z) = \int G(x, y, z, \omega)d\omega$$

A. ELECTRICAL MODELING:

The diode equations (Poissons, drift/diffusion and current-continuity models for electrons and holes) are used for electrical simulations. The computational method used in solving the non-linear Poisson and drift-diffusion equations is performed using SCAPS[32]. Equations for carrier transport includes Poisson equation, Continuity, Drift-Diffusion and

Recombination currents where $n_{n,p}$ is the electron/hole concentration and $G_{n,p}$ is the generation profile coming from an optical simulation.

$$\text{Poisson Equation: } \epsilon_{rel}\epsilon_0\nabla^2\psi = -q\left(n_p - n_n\right)$$
$$\text{Continuity: } \nabla J_{n,p} = \left(G_{n,p} - R_{n,p}(n_p, n_n)\right)$$
$$\text{Drift-Diffusion: } J_{n,p} = \mu_{n,p}n_{n,p}(-\nabla\psi) \pm D_{n,p}\nabla n_{n,p}$$
$$\text{Recombination: } R_{n,p}(n_p, n_n) = B\left(n_p n_n - n_i^2\right) + \frac{n_n n_p - n_i^2}{\tau(n_n + n_p)}$$

Here, ϵ_{rel} is the dielectric of the charge transport layer, $R_{n,p}$ is the radiative and trap assisted Shockley-Read-Hall (SRH) recombination, $\mu_{n,p}$ is the mobility, $D_{n,p}$ is the diffusion constant, n_i is the inherent charges, B is the direct recombination constant, and τ is carrier lifetime.

The average solar cell photocurrent can be measured as the short circuit current density J_{sc}.

$$\vec{J}_{sc}(\lambda) = \frac{1}{V}\iiint_0^V \left|\vec{J}_n(\vec{x}, \lambda) + \vec{J}_p(\vec{x}, \lambda)\right|dV$$

Where V is the solar cell volume and the integrated short circuit current of the wavelength can be determined as

$$\vec{J}_{sc} = \int \vec{J}_{sc}(\lambda)d\lambda.$$

The solar cell response of the current-voltage (I-V) can be reported as:

$$\vec{J}(V) = \vec{J}_0\left[exp\left(\frac{qV}{k_BT}\right) - 1\right] - \vec{J}_{sc}$$

where $\vec{J}_0$ is the leakage current.

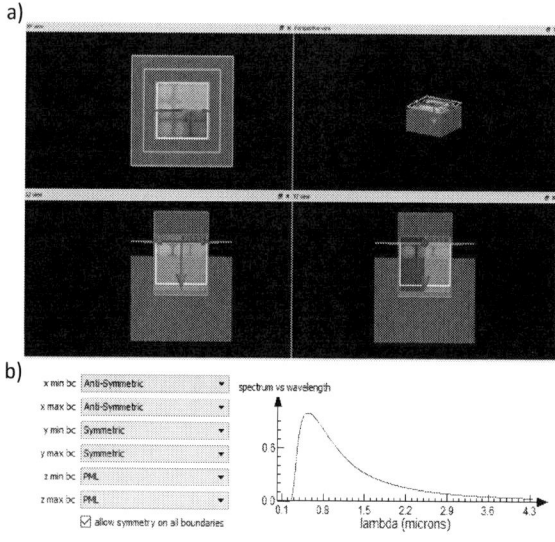

Figure 2: a) Computational domain for PSSC employed for optical analysis. Here, the active top surface has a frontal area of 1.1 µm² and Au nanoparticles (NP) arrays, and b) Boundary conditions on the device with incident solar spectrum. PML implies a perfectly matched layer, which is used for truncating domains in computational models to simulate open boundaries.

The voltage at which the open-circuit voltage (V_{OC}) is located at $\vec{J}(V) = 0$:

978-1-7281-6118-1/22 $31.00 © 2022 IEEE

$$V_{OC} = \frac{k_B T}{q} ln\left(\frac{\vec{J}_{sc}}{\vec{J}_0} + 1\right)$$

The peak power (P_{max}) density is:

$$P_{max} = max\{\vec{J}(V) \cdot V\} = max\left\{\vec{J}_0 \cdot V\left[exp\left(\frac{qV}{k_B T}\right) - 1\right] - \vec{J}_{sc} \cdot V\right\}.$$

The fill factor is:

$$FF = \frac{P_{max}}{J_{sc} \cdot V_{OC}}$$

III. RESULTS

The SOI device was a metal-semiconductor-metal (MSM) photodetector that has a single doped n-type c-Si layer with lateral Schottky barrier contacts. The MSM hotodetector is favorable for photocurrent measurement experiments because the photocurrent generated is linearly proportional to the optical power of the incident light. Photocurrent is a key variable in determining both the open current voltage and the short circuit current of a solar cell. Improving the photocurrent generated would translate into greater solar cell efficiency.

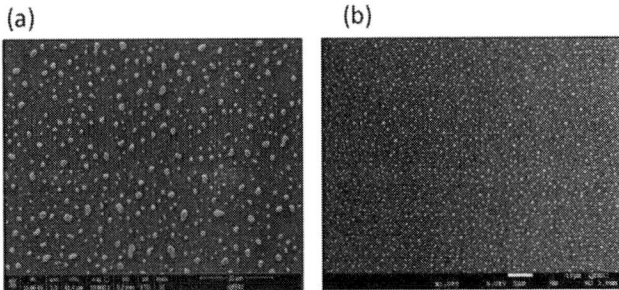

Figure 3: SEM micrograph of the Ag thin-film after annealing process, showing formation of Ag-NPs of an average size of 100 nm and a coverage of (a) 10 % and (b) 30 % of the surface.

Fig. 3 depicts the morphology of Ag NPs fabricated on Si substrates, formed from 12 nm thick Ag precursor films and annealed for 1 h under different temperatures (350 °C and 450 °C, respectively). The size (longitudinal diameter) and surface coverage histograms for samples were calculated. In the low annealing temperature regime, the formation of irregular nanoclusters was observed. Well defined and ellipsoidal NPs were obtained above 350 °C. With a further increase of the annealing temperature, up to 450 °C, an aggregation process of NPs initiates, the size uniformity of the NPs increases and the number of small particles decreases. Furthermore, the mean surface coverage (SC) size of NPs, determined by the maximum of Gaussian single-peak fitting to the surface coverage histograms, increases from 36 ± 0.8 up to 99 ± 0.3 nm, with almost unchanged full width at half maximum and the total SC increases from 10.1% at 350 °C to 30.5% f at 450 °C. Light trapping by scattering from metal nanoparticles at the surface of the solar cell. Light is preferentially scattered and trapped into the semiconductor thin film by multiple and high-angle

scattering, causing an increase in the effective optical path length in the cell.

Figure 4 shows a schematic of the SOI device with nanoparticles. Absorption (Abs) and total reflection (R_{Total}) of the Ag NPs networks deposited on glass substrates. The light source was a halogen lamp under AM 1.5 standard conditions. The beam was directed to an optical microscope and focused on the sample and reference using an internal beam splitter. The induced photocurrent was measured via probes using a pre-amplifier across the contacts. The signal was extracted using a lock in amplifier and sent to the control computer. The short circuit current was recorded as a function of wavelength over the contact area using the analysis computer. The photocurrent response for the SOI device was recorded several times on each sample at different points on the p-n metal contacts and averaged to get the photocurrent response. The PE is defined as the ratio of the difference between the photocurrent generated after and before NP deposition to that before deposition expressed in terms of a percentage. Therefore, the uncoated SOI device was first measured to get the relative photocurrent. As little time as possible was left between the measurements to be sure the collector time variation was small.

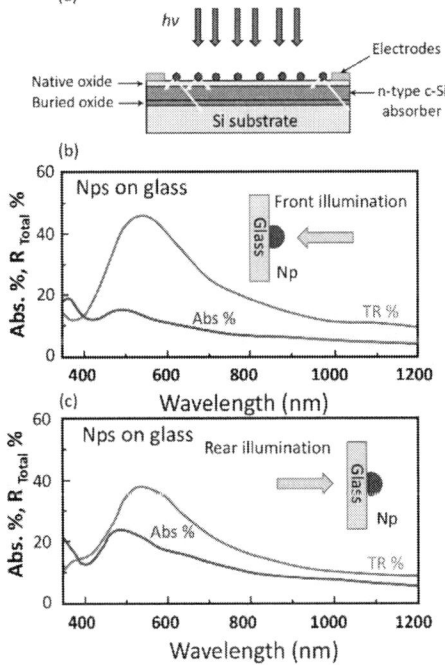

Figure 4: (a) Schematic of the SOI device with nanoparticles. The aluminum contacts were deposited using thermal evaporation. (b-c) Absorption (Abs) and total reflection (R_{Total}) of the Ag NPs deposited on glass substrates from front and rear illumination.

Improving the photocurrent generated would translate into greater solar cell efficiency. Significant differences in the optical properties of the NPs were observed for measurements performed in two distinct configurations: front-side illumination (Fig. 4b), when the incident light impinges the NPs first, and rear side illumination (Fig. 4c), when the incident light impinges the glass substrate first. These two configurations correspond to the way in which NPs interact with the sunlight in substrate and superstrate thin film solar cell structures, respectively. Interestingly, the extinction spectra (sum of all interactions between incident light and the NPs) were found to be independent of illumination configuration. For the rear side illumination we measured a lower R_{Total}, which for constant extinction leads to a notably higher absorption. For application in solar cells, there is a clear trade-off between the beneficial effects of light scattering and the adverse parasitic absorption limiting the achievable enhancements. The values of both scattering and absorption are higher for rear-side illumination, however, a more pronounced parasitic absorption surpasses the beneficial effect of higher scattering observed in the rear side illumination.

Design and simulation of PSSC were performed using COMSOL/MATLAB/SCAPS for multiphysics modeling. The developed model is capable of calculating optical properties such as absorption in different layered mediums in PSSC as shown in spectral absorption curves of Figure 3, regarding the absorption spectrum for variable systems sizes (3×3, 5×5, 7×7) and NP radii. In principle, device absorption can be further segregated into total NP absorption and Si absorber absorption for carrier generation. The spectral analysis supported by broadband integrated analysis of short circuit current (Figures 5a and 5b) showed the system size of 5×5 with NP radius of 70 nm resulting in the best electrical performance with J_{sc} of 11.54 ma/cm² with a gain of 46.8% when compared to bare-Si Schottky solar cell. Therefore, the optimal fractional coverage is 25 NPs/1.1 µm² and in terms of the frontal area of NPs on the top of Si absorber is

$$\frac{25\pi[0.07^2]}{1.1^2} = 34.9\%.$$

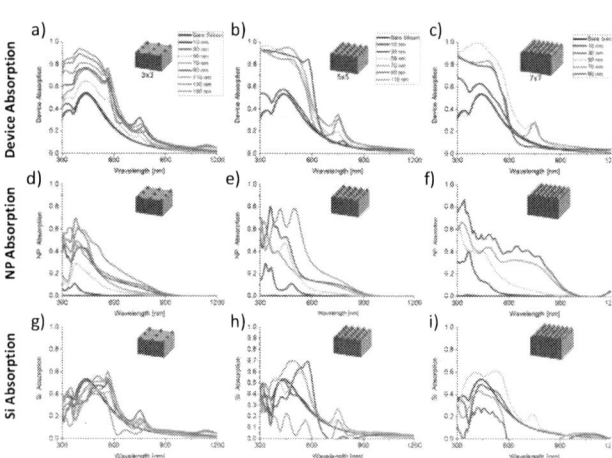

Figure 5: Absorption spectrum for variable systems sizes (3x3, 5x5, 7x7) and NP radii. a-c) Device absorption which can be further segregated into d-f) total NP absorption and g-i) Si absorber absorption for carrier generation.

Figure 5a presents an efficient design of a solar cell based on the combination of the effect of plasmonic resonance and Schottky junction. J_{sc} was estimated using a semiconductor model considering device and absorber geometry, and by integrating the generation profile from optical simulations.

Figure 5h shows higher absorption in the visible range matching the spectral response of silicon-based solar cells. The design of nano-particle plasmon arrays allows the silicon layers, which are as thin as 2 µm, to improve significantly in absorption. It can lead to a higher carrier generation profile inside the absorber layer.

To analyze the thermal functioning of the PSSC, Figure 5a shows the spectral approach to assess different thermal processes in the device for an optimal configuration (70 nm and 5×5), and percentage contribution of thermal processes from integrated values. Expected spectral heating of the cells were calculated by using AM1.5G solar irradiance ($P_{solar}(\lambda)$), spectral absorption in device ($P_{abs}(\lambda)$), spectral electrical output ($P_{elect}(\lambda)$) and total spectral heat absorbed in device ($Q_{total\ heat}(\lambda)$).[42, 43] $Q_{total\ heat}(\lambda)$ was found to be 28.70 %, which can be further divided into absorption in nanoparticles (($P_{abs,NP}(\lambda)$ = 13.80 %) and thermalization heating ($P_{therm,Si}(\lambda)$ = 14.89 %) as shown in Figure 5b. Since the analysis was conducted for λ_1, λ_2 from 300 nm to 1200 nm infra-red parasitic absorption was neglected. This is also assumed due to the thin silicon wafer used for optical analysis.

$$Q_{total\ heat} = \int_{\lambda_1}^{\lambda_2} Q_{total\ heat}(\lambda)$$

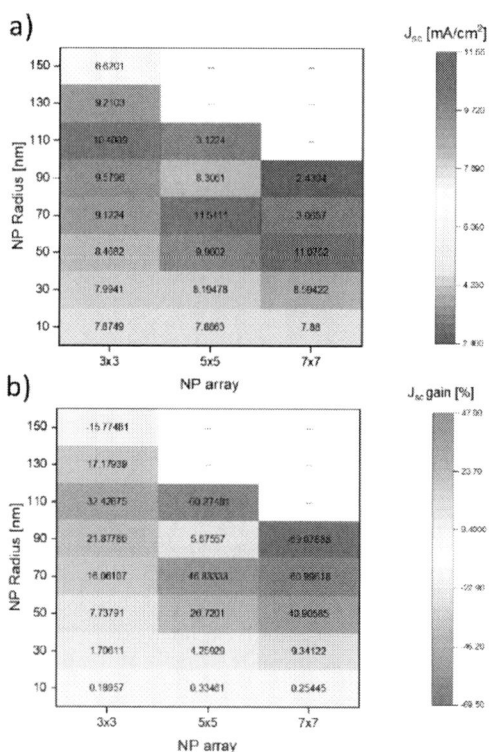

Figure 6: a) Short circuit current (J_{sc}) generated in Si absorber with Au NP as a function of NP radii and array size and b) Short circuit current (J_{sc}) gain percentage compared to Bare-Si Schottky solar cell.

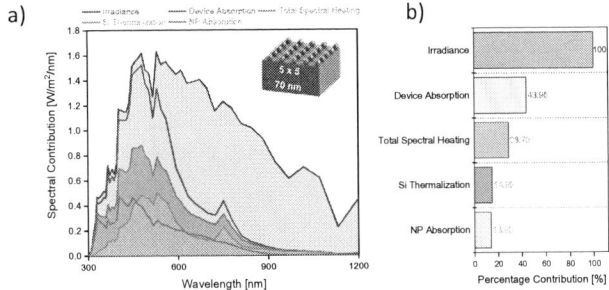

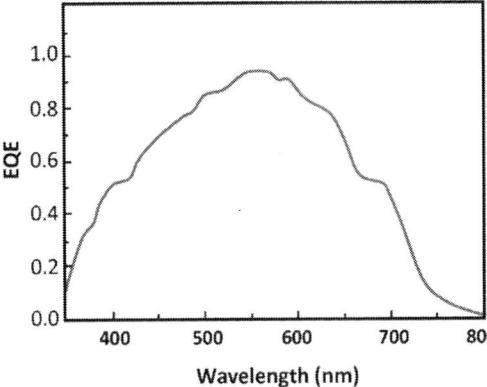

Figure 7: a) Spectral approach to assess different thermal processes in the device for an optimal configuration (70 nm and 5×5), and b) Percentage contribution of thermal processes from integrated values.

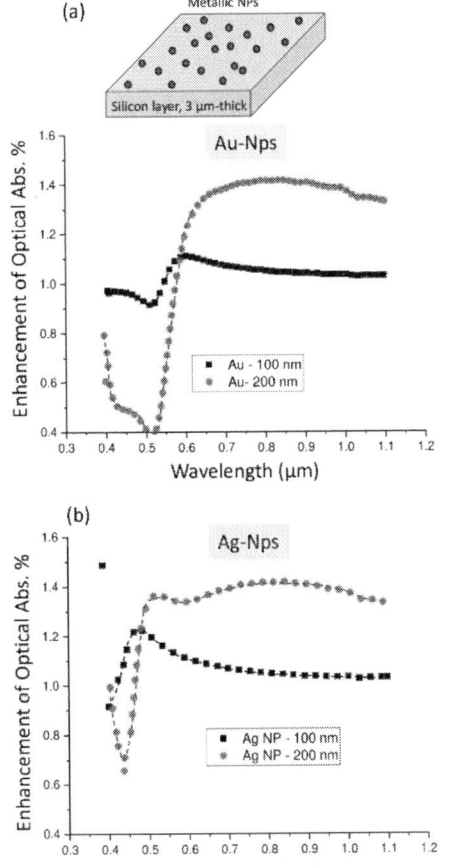

Figure 8: COMSOL simulation of the plasmonics NPs size effect on the optical absorption by using the Finite Difference Time Domain (FSDTD) method (a) network of Au NPs (100 nm and 200 nm), and (b) network of Ag NPs (100 nm and 200 nm).

COMSOL has been used to measure optical absorption by using the Finite Difference Time Domain (FSDTD) method, which solves Maxwell equations for electrical and magnet fields for the plasmonic material and associated solar cell geometry. We measure the absorbed power, as a frequency function and the generation rate $g(x, y, z, w)$ depending on the photon energy and frequency. Figure 8 shows the results obtained by COMSOL simulation of the plasmonics NPs size

effect (both Au and Ag) on the optical absorption by using the Finite Difference Time Domain (FSDTD) method, performed on a 25 NPs /μm^2 density and 100 nm and 200 nm NPs average size. It is suggested that the bigger NPs size led to a higher optical absorption. Also, SC of 16 % was identified as the optimal value.

Figure 9: EQE curve of solar cells formed from 12 nm thick Ag films annealed and 450 °C for 1 h (15 % SC).

Finally, complete thin film photovoltaic devices were fabricated to investigate the benefits of plasmonic light trapping with our best-performing. NPs formed from a 12 nm thick Ag film annealed at 450 °C for 1 h was selected. The EQE characteristics are shown in Fig. 9. Ag NPS network provides a superior cell performance relative to the reference sample, over the entire measured range. In this spectral region all the incident light is absorbed in the Si layers during the first pass without interacting with the NPs. Ag NP provide pronounced absorption enhancement at longer wavelengths [13,15]. Therefore, the main light trapping effect leading to the observed EQE enhancements in the red-NIR region is attributed to plasmon-assisted light scattering.

IV. CONCLUSIONS

Optimized SOI wafers based solar cells containing NPs formed from a 12 nm thick Ag film annealed at 450 °C for 1 h allowed for a pronounced broadband photocurrent enhancement. These results indicate that plasmonic light-trapping is a valid solution for implementation in the thin film PV. For a coupled model, commercial software COMSOL MULTIPHYSICS, MATLAB and SCAPS were used to develop plasmonic/solar cell interaction. To optimize electrical performance, sensitivity analysis was carried out for variable system sizes and radii of gold NPs. Global performance gain for annual energy yield in comparison to bare-silicon absorber was also analyzed after optimization to highlight the world-wide potential of PSSC. The developed framework is not limited to monofacial configurations and Schottky solar cells, but can be extended to other solid-state devices.

ACKNOWLEDGMENT

This work was achieved within the project "Light Management in Solar Cells using Fault-Tolerant Plasmonics and Metamaterials [MetaSol]", under the National Priority Research program (NPRP) NPRP#11S-0117-180330, financed by Qatar National Research Funds (QNRF), a member of Qatar Foundation. Authors would like to thank QEERI Core Labs Team for the materials characterizations.

REFERENCES

[1] IRENA (2019), Future of Solar Photovoltaic: Deployment, investment, technology, grid integration and socio-economic aspects (A Global Energy Transformation: paper), International Renewable Energy Agency, Abu Dhabi, n.d. https://www.irena.org/-/media/Files/IRENA/Agency/Publication/2019/Nov/IRENA_Future_of_Solar_PV_2019.pdf.

[2] A. Ali, H. Park, R. Mall, B. Aïssa, S. Sanvito, H. Bensmail, A. Belaidi, F. El-Mellouhi, Machine Learning Accelerated Recovery of the Cubic Structure in Mixed-Cation Perovskite Thin Films, Chem. Mater. 32 (2020) 2998–3006. https://doi.org/10.1021/acs.chemmater.9b05342.

[3] M. Liang, A. Ali, A. Belaidi, M.I. Hossain, O. Ronan, C. Downing, N. Tabet, S. Sanvito, F. EI-Mellouhi, V. Nicolosi, Improving stability of organometallic-halide perovskite solar cells using exfoliation two-dimensional molybdenum chalcogenides, Npj 2D Mater. Appl. 4 (2020) 40. https://doi.org/10.1038/s41699-020-00173-1.

[4] S.K. Cushing, N. Wu, Plasmon-Enhanced Solar Energy Harvesting, Interface Mag. 22 (2013) 63–67. https://doi.org/10.1149/2.f08132if.

[5] A.P. Amalathas, M.M. Alkaisi, Fabrication and Replication of Periodic Nanopyramid Structures by Laser Interference Lithography and UV Nanoimprint Lithography for Solar Cells Applications, in: 2020. https://doi.org/10.5772/intechopen.72534.

[6] G. Kakavelakis, I. Vangelidis, A. Heuer-Jungemann, A.G. Kanaras, E. Lidorikis, E. Stratakis, E. Kymakis, Plasmonic Backscattering Effect in High-Efficient Organic Photovoltaic Devices, Adv. Energy Mater. 6 (2016) 1501640. https://doi.org/10.1002/aenm.201501640.

[7] C. Sun, Z. Wang, X. Wang, J. Liu, A Surface Design for Enhancement of Light Trapping Efficiencies in Thin Film Silicon Solar Cells, Plasmonics. 11 (2016) 1003–1010. https://doi.org/10.1007/s11468-015-0135-8.

[8] S. In, N. Park, Inverted Ultrathin Organic Solar Cells with a Quasi-Grating Structure for Efficient Carrier Collection and Dip-less Visible Optical Absorption, Sci. Rep. 6 (2016) 21784. https://doi.org/10.1038/srep21784.

[9] B. Paci, G. Kakavelakis, A. Generosi, J. Wright, C. Ferrero, E. Stratakis, E. Kymakis, Improving stability of organic devices: a time/space resolved structural monitoring approach applied to plasmonic photovoltaics, Sol. Energy Mater. Sol. Cells. 159 (2017) 617–624. https://doi.org/https://doi.org/10.1016/j.solmat.2016.01.003.

[10] D. Baran, R.S. Ashraf, D.A. Hanifi, M. Abdelsamie, N. Gasparini, J.A. Röhr, S. Holliday, A. Wadsworth, S. Lockett, M. Neophytou, C.J.M. Emmott, J. Nelson, C.J. Brabec, A. Amassian, A. Salleo, T. Kirchartz, J.R. Durrant, I. McCulloch, Reducing the efficiency–stability–cost gap of organic photovoltaics with highly efficient and stable small molecule acceptor ternary solar cells, Nat. Mater. 16 (2017) 363–369. https://doi.org/10.1038/nmat4797.

[11] M. Iqbal, M.M. Nauman, F.U. Khan, P.E. Abas, Q. Cheok, A. Iqbal, B. Aissa, "Vibration-based piezoelectric, electromagnetic, and hybrid energy harvesters for microsystems applications: A contributed review, International journal of energy research 45 (1), 65-102, https://doi.org/10.1002/er.5643

[12] H. Gavi, Balla D. Ngom, A.C. Beye, A.M. Strydom, B. Aissa, V.V. Srinivasu, M. Chaker, N. Manyala, "Low-field microwave absorption in pulse laser deposited FeSi thin film, Journal of Magnetism and Magnetic Materials 324 (6), 1172-1176, https://doi.org/10.1016/j.jmmm.2011.11.003

[13] J. Haschke, R. Lemerle, B. Aïssa, A.A. Abdallah, M.M. Kivambe, M. Boccard, C. Ballif, "Annealing of silicon heterojunction solar cells: Interplay of solar cell and indium tin oxide properties, " IEEE Journal of Photovoltaics 9 (5), 1202-1207, DOI: 10.1109/JPHOTOV.2019.2924389

[14] B. Aïssa, M.A. El Khakani, "The channel length effect on the electrical performance of suspended-single-wall-carbon-nanotube-based field effect transistors," Nanotechnology 20 (17), 175203, https://doi.org/10.1088/0957-4484/20/17/175203

[15] A. Bentouaf, R. Mebsout, H. Rached, S. Amari, A.H. Reshak, B. Aïssa "Theoretical investigation of the structural, electronic, magnetic and elastic properties of binary cubic C15-Laves phases TbX2 (X = Co and Fe)," Journal of alloys and compounds 689 (25), 885-893, https://doi.org/10.1016/j.jallcom.2016.08.046

[16] R. Nechache, M. Nicklaus, N. Diffalah, A. Ruediger, F. Rosei, "Pulsed laser deposition growth of rutile TiO2 nanowires on Silicon substrates," Applied Surface Science 313, 48-52

[17] A. Shang, X. Zhai, C. Zhang, Y. Zhan, S. Wu, and X. Li, "Nanowire and nanohole silicon solar cells: a thorough optoelectronic evaluation," Progress in Photovoltaics: Research and Applications, https://doi.org/10.1002/pip.2613 vol. 23, no. 12, pp. 1734-1741, 2015/12/01 2015, doi: https://doi.org/10.1002/pip.2613.

[18] M. G. Deceglie, V. E. Ferry, A. P. Alivisatos, and H. A. Atwater, "Design of Nanostructured Solar Cells Using Coupled Optical and Electrical Modeling," Nano Letters, vol. 12, no. 6, pp. 2894-2900, 2012/06/13 2012, doi: 10.1021/nl300483y.

[19] J. Nelson, The Physics of Solar Cells (The Physics of Solar Cells).

[20] T. J. Kucharski, Y. Tian, S. Akbulatov, and R. Boulatov, "Chemical solutions for the closed-cycle storage of solar energy," Energy & Environmental Science, 10.1039/C1EE01861B vol. 4, no. 11, pp. 4449-4472, 2011, doi: 10.1039/C1EE01861B.

[21] S. Chander, A. Purohit, A. Nehra, S. Nehra, and M. S. Dhaka, "A Study on Spectral Response and External Quantum Efficiency of Mono-Crystalline Silicon Solar Cell," INTERNATIONAL JOURNAL of RENEWABLE ENERGY RESEARCH, vol. 5, pp. 41-44, 12/20 2014.

[22] M. Eisapour, A. H. Eisapour, s. M. j. Hosseini, and P. Talebizadeh Sardari, "Exergy and Energy Analysis of Wavy Tubes Photovoltaic-Thermal Systems Using Microencapsulated PCM Nano-Slurry Coolant Fluid," Applied Energy, 03/14 2020, doi: 10.1016/j.apenergy.2020.114849.

[23] M. Hosseinzadeh, A. Salari, M. Sardarabadi, and M. Passandideh-Fard, "Optimization and parametric analysis of a nanofluid based photovoltaic thermal system: 3D numerical model with experimental validation," Energy Conversion and Management, vol. 160, 03/15 2018, doi: 10.1016/j.enconman.2018.01.006.

[24] Comsol, "COMSOL Multiphysics® v. 5.4," ed: Stockholm, Sweden, 2019.

[25] H. Yazdani, Matlab R2019b. 2019.

[26] M. Burgelman, P. Nollet, and S. Degrave, "Modeling polycrystalline semiconductor solar cell," Thin Solid Films, vol. 361, pp. 527-532, 02/01 2000, doi: 10.1016/S0040-6090(99)00825-1.

[27] X. Li, N. P. Hylton, V. Giannini, K.-H. Lee, N. J. Ekins-Daukes, and S. A. Maier, "Bridging electromagnetic and carrier transport calculations for three-dimensional modelling of plasmonic solar cells," Opt. Express, vol. 19, no. S4, pp. A888-A888, 2011, doi: 10.1364/oe.19.00a888.

[28] S. Chaoudhary, A. Dewasi, V. Rastogi, R. N. Pereira, A. Sinopoli, B. Aïssa and A. Mitra "Broadband self-powered photodetection with p-NiO/n-Si heterojunctions enhanced with plasmonic Ag nanoparticles deposited with pulsed laser ablation" J. Mater. Sci: Mater Electron (2022), 1-13, doi: https://doi.org/10.1007/s10854-022-08058-3

[29] X. Guo, Y. Ma, Y. Wang, and L. Tong, "Nanowire plasmonic waveguides, circuits and devices," Laser & Photonics Reviews, vol. 7, 11/01 2013, doi: 10.1002/lpor.201200067.

[30] J. Khurgin, "How to face the loss in plasmonics and metamaterials," Nature Nanotechnology volume 10, pages 2–6 (2015).

[31] S. Chaoudhary, A. Dewasi, V. Rastogi, R. N. Pereira, A. Sinopoli, B. Aïssa and A. Mitra "Laser ablation fabrication of a p-NiO/n-Si heterojunction for broadband and self-powered UV-Visible-NIR photodetection" Nanotechnology (article in press), https://doi.org/10.1088/1361-6528/ac5ca6

[32] A. Ali, F. El-Mellouhi, A. Mitra, B. Aïssa "Research Progress of Plasmonic Nanostructure-Enhanced Photovoltaic Solar Cells" Nanomaterials 12, no. 5: 788. https://doi.org/10.3390/nano12050788

[33] M. Usama Siddiqui, A. F. M. Arif, L. Kelley, and S. Dubowsky, "Three-dimensional thermal modeling of a photovoltaic module under varying conditions," *Solar Energy,* vol. 86, no. 9, pp. 2620-2631, 2012/09/01/ 2012, doi: https://doi.org/10.1016/j.solener.2012.05.034.

[34] W. C. Swinbank, "Long-wave radiation from clear skies," *Quarterly Journal of the Royal Meteorological Society,* vol. 89, no. 381, pp. 339-348, 1963, doi: https://doi.org/10.1002/qj.49708938105.

[35] S. Chaoudhary, A. Dewasi, S. Ghosh, R.J. Choudhary, D.M. Phase, T. Ganguli, V. Rastogi, R.N. Pereira, A. Sinopoli, B. Aïssa, A. Mitra "X-ray photoelectron spectroscopy and spectroscopic ellipsometry analysis of the p-NiO/n-Si heterostructure system grown by pulsed laser deposition" Thin Solid Films 743, 139077, 2022. https://doi.org/10.1016/j.tsf.2021.139077

[36] Y. Liu, G. Sun, D. Wu, and S. Dong, "Investigation on the correlation between solar absorption and the size of non-metallic nanoparticles," *Journal of Nanoparticle Research,* vol. 21, 07/17 2019, doi: 10.1007/s11051-019-4576-4.

[37] V. Pustovalov, "Light-to-heat conversion and heating of single nanoparticles, their assemblies, and the surrounding medium under laser pulses," *RSC Adv.,* vol. 6, pp. 81266-81289, 08/25 2016, doi: 10.1039/C6RA11130K.

[38] A. Ali, A. Mitra, B. Aïssa "Metamaterials and Metasurfaces: A Review from the Perspectives of Materials, Mechanisms and Advanced Metadevices" Nanomaterials 2022, 12(6), 1027; https://doi.org/10.3390/nano12061027

[39] E. Fares, B. Aïssa and R. Isaifan "Inkjet printing of metal oxide coatings for enhanced photovoltaic soiling environmental applications" Global J.

Environ. Sci. Manage. 8(4): 1-18 (2022), http://dx.doi.org/10.22034/GJESM.2022.04.03

[40] H. Sadok Cherif, A. Bentouaf, Z.A. Bouyakoub, H. Rached, and B. Aïssa "Computational determination of structural, electronic, magnetic and thermodynamic properties of Co2HfZ (Z = Al, Ga, Si and Sn) full Heusler compounds for spintronic applications" Journal of Alloys and Compounds, 2021, 162503, https://doi.org/10.1016/j.jallcom.2021.162503.

[41] M.I. Hossain, A. Khandakar, M.E.H. Chowdhury, S. Ahmed, M.M. Nauman and B. Aïssa, "Numerical and Experimental Investigation of Infrared Optical Filter based on Metal Oxides Thin Films for Temperature Mitigation in Photovoltaics" Journal of Electronic Materials 51, 179-189, 2022. https://doi.org/10.1007/s11664-021-09269-w

[42] M. Arunachalam, A. Sinopoli, F. Aidoudi, S. E. Creager, R. Smith, B. Merzougui, and B. Aïssa "High Performance of Anion Exchange Blend Membranes Based on Novel Phosphonium Cation Polymers for All-Vanadium Redox Flow Battery Applications" ACS Applied Materials & Interfaces, 2021, 13, 38, 45935–45943, https://doi.org/10.1021/acsami.1c10872

[43] B. Aïssa, A. Ali, F. El Mellouhi "Oxide and Organic–Inorganic Halide Perovskites with Plasmonics for Optoelectronic and Energy Applications: A Contributive Review" Catalysts 2021, 11(9), 1057; https://doi.org/10.3390/catal11091057

The Natural and Accelerated Evolution of EVA Adhesion through Intermediate Exposures

Patrick Thornton, Nick Bosco, Reinhold H Dauskardt

Stanford University, Stanford, CA, United States

National Renewable Energy Laboratory, Golden, CO, United States

Ethylene vinyl acetate (EVA) encapsulants comprise the majority of the encapsulants currently in use; much work has been done to understand and model the adhesive characteristics of EVA-encapsulated modules, but limited work has provided reliable insight into adhesion during the intermediate stages of exposure, limiting the ability to validate model predictions in this range. Here we provide the critical adhesion energy measurements for EVA adhesion after up to 4 years of field aging in 3 different US locations and 10,000 hours of accelerated aging. Both field and accelerated aging reveal a distinct plateau that emerges during the intermediate exposure periods (after 1 year in the field and after 1,000 hours in a chamber). After 10,000 hours, adhesion strength within accelerated aged mini-modules falls by an order of magnitude and the failure path transitions to the glass/EVA interface - trends generally seen after long-term field exposures (>15 years). Previous adhesive modeling predicted that adhesion would continuously decrease over the lifetime of a module, but these current results uncover an intermediate plateauing trend that is critical to accurately modeling the evolution of adhesion and predicting potential adhesive failure.

High-Performance O-Band Photonic Power Converters Under Non-Uniform Laser Illumination

Meghan N. Beattie, Henning Helmers, Gavin P. Forcade, Christopher E. Valdivia, David Lackner, Oliver Höhn, Karin Hinzer

SUNLAB, Centre for Research in Photonics, University of Ottawa, Ottawa, ON, Canada

Fraunhofer Institute for Solar Energy Systems ISE, Freiburg, Germany

Photonic power converters designed and fabricated at Fraunhofer ISE for operation in the O-band were measured under non-uniform 1319 nm laser illumination with five spot sizes. Two 5.4 mm2 devices were studied. The first used lattice-matched InGaAsP on an InP substrate while the second used lattice-mismatched InGaAs grown on GaAs with a step-graded metamorphic buffer. The maximum measured efficiencies were 52.9% at a laser power of 353 mW and 48.8% at 413 mW for the lattice-matched and -mismatched designs respectively. Both maximal efficiencies were measured with a spot size of 2.3 mm, the largest and most uniform laser-spot applied in this study. The devices were insensitive to the illumination uniformity for input powers < 100 mW, exhibiting a logarithmic relationship between open-circuit voltage and short-circuit current density consistent with the non-ideal diode equation. At higher powers, deviations were observed from this trend and both devices exhibited better performance for larger spot sizes. Distributed circuit modeling (DCM), which uses a two-diode model and accounts for lateral current flow and resistive losses, was used to explore the mechanisms responsible for the measured beam-size dependence. Agreement was achieved between the DCM and experimental data measured under broadband uniform illumination. Under a Gaussian laser-illumination profile, comparison between the DCM and experimental data suggested that both resistive losses and localized heating likely contributed to the performance reductions under non-uniform illumination. Better performance at higher illumination powers could be achieved by engineering a more uniform illumination profile, optimizing the front metallization, or adopting multi-junction device architectures.

978-1-7281-6118-1/22 $31.00 © 2022 IEEE

Post-annealing Treatment on Hydrothermally Grown Sb2(S,Se)3 Thin Films for Efficient Solar Cells

Suman Rijal, Zhaoning Song, Deng-Bing Li, Jaehoon Chung, Sandip S Bista, Dipendra Pokhrel, Sabin Neupane, Randy Ellingson, Yanfa Yan

The University of Toledo, Toledo, OH, United States

In this work, we fabricate antimony selenosulfide (Sb2(S, Se)3) thin film solar cells by a hydrothermal method followed by a post-deposition annealing process at different temperatures. The effects of the annealing temperature on the morphological and structural properties of the Sb2(S, Se)3 films are systematically investigated by scanning electron microscopy and X-ray diffraction analyses. We find that a proper annealing temperature leads to a high-quality Sb2(S, Se)3 film with large crystal grains, proper stoichiometry, and high crystallinity. After optimizing the process, we obtained Sb2(S,Se)3 solar cells with an improved power conversion efficiency from 2.04 to 8.48%.

Wafer-Scale Pulsed Laser Deposition of ITO for Silicon Heterojunction Solar Cells: Reduced Damage vs Interfacial Resistance

Yury Smirnov, Pierre-Alexis Repecaud, Leonard Tutsch, Ileana Florea, Pere Roca i Cabarrocas, Martin Bivour, Monica Morales-Masis

University of Twente, Enschede, Netherlands

Fraunhofer Institute for Solar Energy Systems, Freiburg, Germany

Ecole Polytechnique, CNRS, Palaiseau, France

Pulsed laser deposition (PLD) has been recently proposed as an alternative low-damage physical vapor deposition technique to deposit Transparent Conducting Oxides (TCOs) onto sensitive functional layers such as required in various solar cells. Here we studied the role of deposition pressure during PLD of TCO with identical sheet resistance (60 $\Omega/\square$) on final silicon heterojunction (SHJ) solar cell performance. Solar cells with PLD Sn-doped In2O3 (ITO) at all conditions maintained high passivation quality but increased pressures lead to high series resistance. Transmission electron microscopy spectrometry revealed the formation of a parasitic SiOx at the ITO/a-Si:H interface of the SHJ cell causing a transport barrier. The optimized ITO films with highest carrier density allows obtaining SHJ efficiency >21% with 75-nm-thick PLD ITO. We furthermore demonstrate high efficiency SHJ cells (>22%) with reduced indium consumption by combining an optimized thin ITO film by PLD and a TiOx capping layer.

Potential Capacity and Targeted Costs for Floating Photovoltaics in North America

Leonardo Micheli

University of Jaén, Jaén, Spain

Floating Photovoltaics (FPV) is getting significant attention worldwide, pushed by the lower rent fees of water surfaces compared to land and by the lack of competition with agriculture. This work makes use of previously presented methodologies to analyze the FPV potential in the northern American countries, estimating the yields of systems installed on the inland water basins. The FPV cost of electricity is also discussed and compared with that of traditional land-based photovoltaic (LPV) systems. This way, it is possible to estimate the capital expenditure that FPV should target to be cost-competitive with LPV.

Techno-economic Analysis of Novel PV Plant Designs for Extreme Cost Reductions

Nicholas Pilot, Robin Bedilion, Daniel Fregosi, Sean Hackett, Michael Bolen, and Joseph Stekli

Electric Power Research Institute (EPRI), Palo Alto, CA, 94304, USA

Abstract— A techno-economic analysis is performed examining the cost and performance of future large-scale photovoltaic (PV) plant components, including bifacial modules, tandem modules, increased plant voltage architectures, and module-level power electronics. PV plant designs incorporating these components are compared with current PV technologies based on levelized cost of electricity (LCOE). Baseline models are developed and validated with plant performance data. Expected performance and costs of PV technologies are incorporated into the baseline models. An evolutionary algorithm is utilized to optimize PV plant configurations and LCOE. This follow-on analysis focuses on tandem modules, increased plant voltage architectures, and module-level power electronics.

Keywords—bifacial modules, tandem modules, voltage architectures, module-level power electronics, photovoltaic, technoeconomic analysis, levelized cost of electricity, utility-scale, optimization, System Advisor Model.

I. EVALUATION ABSTRACT

A. Introduction

Solar PV has experienced a precipitous decline in costs over the past decade, the bulk of which can be attributed to cost reductions and efficiency improvements of PV modules. More efficient modules also reduce the needed amount of land, racking, and mounting equipment that reduce overall plant cost per Watt-dc. However, with the cost of PV modules falling below \$0.30/Wdc globally and efficiencies of crystalline silicon (c-Si) cells approaching theoretical limits, cost reductions and efficiency improvements in traditional c-Si PV modules are anticipated to asymptote.

In an effort to continue PV's decreasing cost trends, research has focused on other individual aspects of plants, including new PV cell and module technologies to further increase efficiency and power output, and increased voltages and module-level power electronics for reduced energy loss. However, more research is needed on how these individual innovations can best come together to provide the lowest cost PV electricity. As with other increasingly constrained and optimized systems, trade-offs need to be made. For PV plants, technology selection decisions balance cost, power output, and reliability to achieve the lowest levelized cost of electricity (LCOE). It is not readily apparent how low of a LCOE can be achieved by any given combination of technologies.

There are two broad areas for opportunity for innovation in PV plant design. The first is improvements in the individual components that make up a PV plant. Table I outlines various pieces of a PV plant and innovations that will impact the cost and performance of a PV plant. The second opportunity for reducing cost through PV plant design is in optimizing the integration of the plant components as a whole. Currently, most equipment in a PV plant is optimized at the component level with relatively little consideration of how it may interact with other components or functions of the plant.

The project seeks to gain a deeper understanding of the cost and performance of the future PV plant components outlined in Table I and how they may be integrated into new PV plant designs that significantly reduce the LCOE of PV. Baseline models for three irradiance profiles – Southwest, Southeast, and Midwest – are developed within the National Renewable Energy Laboratory (NREL) System Advisor Model (SAM) and validated against actual PV plant performance data. Cost and performance data of future PV technologies, based upon a comprehensive literature review and responses gathered from industry experts during informational interviews, are incorporated into these baseline models. An evolutionary algorithm is then implemented to determine an optimal PV plant configuration and technology combination for each location based on expected performance and cost of these new technologies. Utility-scale PV plant LCOE results are presented, identifying plant-level configurations that may have the greatest impact on future plant cost and performance and discussing development opportunities for future technology and cost improvements.

TABLE I. PV PLANT TECHNOLOGIES FOR EXPLORATION AND ASSOCIATED DESIGN CONSIDERATIONS

PV Plant Technology	Example Plant Design Considerations
Bifacial modules	Added energy from increasing module height vs. increased racking and wiring cost; added energy from increasing ground albedo vs. cost of solution
Tandem modules	Added energy from increasing efficiency vs. increased wiring and balance-of-plant costs
Increased plant voltages above 1500 Vdc	Reduced energy losses vs. increased component costs
Module-level power electronics for large-scale plants	Reduced energy losses and potential for lower cost per inverter vs. increased upfront and maintenance costs

B. Methodology and Analysis

Baseline models were developed based on three PV plants with varying plant configurations and weather profiles within NREL's SAM. To accurately calibrate each baseline model, plant data were gathered from three existing utility-scale PV plants in the Southeast, Midwest, and Southwest ranging from 1-MWac to 50-MWac and compared to baseline model performance. The economic performance of these baseline PV plants was then analyzed using SAM and compared to power purchase agreements for similar plants. After calibrating the baseline SAM models, the performance and cost metrics for modules, inverters, and system losses were updated to match current manufacturer specifications and plant configurations to better compare modern PV technologies with the novel PV technologies investigated. These updated models serve as the baseline for the remainder of the study and are used for comparison with the LCOEs of the novel PV technologies.

A comprehensive literature review and informational interviews provided insights on the characteristics and current state of the technologies outlined in Table I, including the trade-off between increased performance/heat dissipation with increased bifacial module array height versus increased racking costs and performance saturation at array heights above 2 meters [1][2][3], the potential benefits of microinverters over string inverters [4], and challenges associated with increasing plant voltages to 2000 Vdc were collected [5].

Parameters for each of the technologies listed in Table I were incorporated into SAM and compared with current PV technologies based on LCOE analysis. A custom four-terminal (4T) perovskite-silicon tandem module was developed within SAM based on industry literature, existing module specifications, and informational interview responses. 4T tandem modules allow for a mechanical stacking approach to developing a utility-scale tandem module within SAM, while two-terminal (2T) tandem modules require voltage or current matching between each tandem layer, which is difficult to model given SAM's custom module input limitations. For this reason, a mechanically stacked 4T tandem module was developed for investigation. Because this 4T configuration would result in two different stringing voltages associated with the top layer and bottom layer, which SAM's system design, inverter specifications, and cost parameters are not designed to capture, this analysis assumed module-level power electronics that would be able to combine these two voltages, yielding 2T output for a 4T tandem module that could be strung back to the inverter. These module-level power electronics are typically a fraction of a cent for tandem modules and are assumed negligible in cost relative to the cost of tandem modules.

Specifications from the baseline c-Si module were utilized to establish the framework for the custom monofacial 4T tandem module, using the monofacial 4T tandem module efficiency of 26.5% reported by Coletti et al. (2020) [6] to establish a scale factor between the baseline module and a tandem module. This scale factor was applied to each of the baseline module's specifications, with the temperature coefficient of V_{oc} assumed to receive the full benefit of the tandem structure and temperature coefficient of I_{sc} remaining constant. All other module specifications were scaled equally.

This custom tandem module was then incorporated into each plant model within SAM.

Plant architectures of 2000V, 2500V, and 3000V were investigated to understand the impact of increased plant voltage over a range of increasing voltages. This analysis focused primarily on the inverters required to operate a PV plant at these higher voltages. Because there are currently no inverters in the market designed to operate at these increased voltages, custom inverters were developed to conduct this analysis.

To estimate the new characteristics for these inverters, over 90 inverter datasheets from five different manufacturers were compiled to inform the inverter specifications of the custom 2000V, 2500V, and 3000V inverters. Inverters were grouped by maximum DC voltage and plotted as their maximum AC power output versus maximum DC current (see Fig. 1) and linear regressions were calculated for each existing inverter voltage at 600V, 1000V, and 1500V. Utilizing a scale factor between the slopes of these regression lines, new regression lines for 2000V, 2500V, and 3000V inverters were calculated. From there, the maximum power outputs of these conceptual inverters were calculated assuming the same DC current as the baseline inverters.

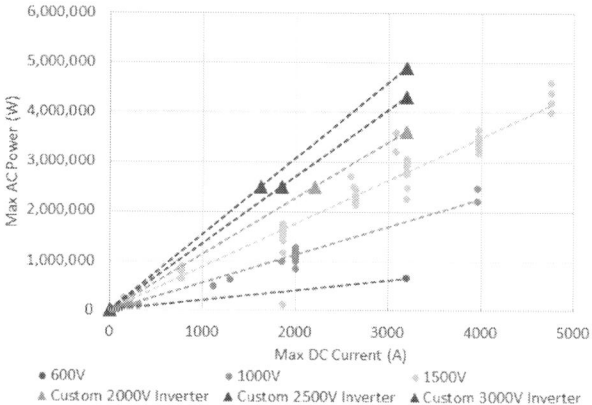

Fig.1 Relationship Between the Max DC Current and Max AC Output Power for Existing 600V, 1000V, and 1500V Inverters and Custom 2000V, 2500V, and 3000V Inverters

Finally, two different module level power electronics technologies were considered for analysis: microinverters and DC-DC optimizers. A custom inverter was created within SAM based on existing microinverter specifications and applied to the baseline models for comparison, along with system changes to module mismatch losses. As a separate analysis, DC-DC optimizers were incorporated into the baseline SAM models through adjustments of module mismatch and system losses.

Following the integration of innovative PV components, an evolutionary algorithm was utilized, where possible, to determine an optimal PV plant configuration and technology combination for each location based on expected performance and cost of the new technology configurations and the technology design tradeoffs listed in Table I and LCOE results for these technologies were calculated. These results identify plant-level configurations that may have the greatest impact on future plant cost and performance. Development opportunities

978-1-7281-6118-1/22 $31.00 © 2022 IEEE

for future technology and cost improvements are also identified. [*Further details will be included in the full manuscript.*]

C. Project Significance

The current PV project development and operator market is highly competitive, which has reduced or eliminated profit margins. In turn, there is a lack of financial and human capital available to invest in and pilot innovative technologies. By identifying those potential technologies and designs and their potential impact on cost reduction, this project serves to highlight opportunities in new PV plant design and focus developers' resources on those technologies and designs that are likely to have the greatest impact in reducing cost.

Furthermore, this project may serve as an important starting point for the development of a PV plant roadmap. For example, the International Technology Roadmap for PV Modules has been useful for identifying known opportunities for reducing costs and/or increasing efficiency, signaling what research is needed, and forecasting commercialization timelines [7]. A similar roadmap for plants is anticipated to aid decisions made by financiers, performance modeling software vendors, engineering, procurement, and construction (EPC) firms, and owner/operators.

D. Summary

The future PV technologies listed in Table I have been modeled and, where possible, optimized and evaluated on a LCOE basis compared to current state-of-the-art PV plants, identifying plant-level configurations that may have the greatest impact on future plant cost and performance. Using the selected optimization algorithm allows for assessment of the effects of coordinated modifications to plant configuration such as ground clearance height, GCR, and albedo, based on expected performance and estimated costs associated with these plant changes, showing the value of strategic coordination between the technologies listed in Table I and plant design aspects. As these future PV technologies continue to develop, this work may be utilized as a roadmap to optimize future PV plant LCOE based on component-level performance alongside system costs and locational characteristics.

REFERENCES

[1] M. T. Patel, R. A. Vijayan, R. Asadpour, M. Varadharajaperumal, M. Ryyan Khan, and M. A. Alam. "Temperature Dependent Energy Gain of Bifacial PV Farms: A Global Perspective." Applied Energy. vol. 276. October 2020.

[2] J. Lopez-Garcia, D. Pavenello, and T. Sample. "Analysis of Temperature Coefficients of Bifacial Crystalline Silicon PV Modules". IEEE Journal of Photovoltaics, pp. 1-9. May 2018.

[3] U. A. Yusufoglu, T. M. Pletzer, L. J. Koduvelikulathu, C. Comparotto, R. Kopecek, and H. Kurz. "Analysis of the Annual Performance of Bifacial Modules and Optimization Methods". IEEE Journal of Photovoltaics. vol.5. November 2014.

[4] C. Deline, J. Meydbray, M. Donovan, and J. Forrest. "Partial shade evaluation of distributed power electronics for photovoltaic systems". National Renewable Energy Laboratory (NREL). June 2012.

[5] L. Scarpa, G. Chicco, F. Spertino, P. M. Tumino, and M. Nunnari. "Technical Solutions and Standards Upgrade for Photovoltaic Systems Operated Over 1500 Vdc". 2018 IEEE 4th International Forum on Research and Technology for Society and Industry (RTSI). September 2018.

[6] G. Coletti, S.L. Luxembourg, L.J. Geerligs, V. Rosca, A.R. Burgers, Y. Wu, L. Okel, M. Kloos, F.J.K. Danzl, M. Najafi, D. Zhang, I. Dogan, V. Zardetto, F. Di Giacomo, J. Kroon, T. Aernouts, J. Hüpkes, C.H. Burgess, M. Creatore, R. Andriessen, and S. Veenstra. "Bifacial Four-Terminal Perovskite/Silicon Tandem Solar Cells and Modules." ACS Energy Letters 2020 5 (5), 1676-1680. April 2020.

[7] M. Fischer, M. Woodhouse, S. Herritsch, and J. Trube. "International Technology Roadmap for Photovoltaic (ITRPV) 11. Edition." April 2020.

Acknowledgment: This material is based upon work supported by the U.S. Department of Energy's Office of Energy Efficiency and Renewable Energy (EERE) under the Solar Energy Technologies Office (SETO) Award Number DE-EE0008981.

Disclaimer: This report was prepared as an account of work sponsored by an agency of the United States Government. Neither the United States Government nor any agency thereof, nor any of their employees, makes any warranty, express or implied, or assumes any legal liability or responsibility for the accuracy, completeness, or usefulness of any information, apparatus, product, or process disclosed, or represents that its use would not infringe privately owned rights. Reference herein to any specific commercial product, process, or service by trade name, trademark, manufacturer, or otherwise does not necessarily constitute or imply its endorsement, recommendation, or favoring by the United States Government or any agency thereof. The views and opinions of authors expressed herein do not necessarily state or reflect those of the United States Government or any agency thereof.

Potentiostatic Photoluminescence Imaging of Charge Extraction in Perovskite Solar Cells

Lukas Wagner, Patrick Schygulla, Jan Philipp Herterich, Mohamed Elshamy, Dmitry Bogachuk, Salma Zouhair, Simone Mastroianni, Uli Würfel, Yuhang Liu, Shaik M. Zakeeruddin, Michael Grätzel, Andreas Hinsch, Stefan W. Glunz

Fraunhofer Institute for Solar Energy Systems ISE, Freiburg, Germany

Laboratory for Photovoltaic Energy Conversion, University of Freiburg, Freiburg, Germany

Freiburg Materials Research Center FMF, University of Freiburg, Freiburg, Germany

Abdelmalek Essaadi University, Tangier, Morocco

Laboratory of Photonics and Interfaces (LPI), École Polytechnique Fédérale de Lausanne, Lausanne, Switzerland

We propose a novel approach for microscopically resolved photocurrent imaging. The method is based on the notion that electrical bias-dependent photoluminescence images reveal fundamental information on charge extraction of photovoltaic devices. The approach is derived from basic physical principles and verified by means of a near-to-ideal III-V solar cell. It is demonstrated that the approach is of special relevance for liquid-processed perovskite solar cells. We outline the potential to investigate the local charge extraction efficiency, which can be related to fundamental charge extraction properties associated with the quality of interfaces and morphological defects from device processing. The method demonstrates that photoluminescence imaging can be a powerful tool for device optimization as well as fundamental studies when carried out at different bias voltages.

Flexible GaAs Solar Cell Using Water-Soluble Sacrificial Layer for Epitaxial Lift-Off Process

SAHIL SHARMA, CARLOS A FAVELA, BO YU, EDUARD GALSTYAN, VENKAT SELVAMANICKAM

Material Science Engineering, University of Houston. , Houston, TX, United States

Mechanical Engineering, University of Houston, Houston, TX, United States

Advanced Manufacturing Institute, University of Houston, Houston, TX, United States

Texas Center of Superconductivity, University of Houston, Houston, TX, United States

Cost reduction of III-V semiconductor solar cells can be achieved by the epitaxial liftoff method which involves selective etching of a sacrificial layer and reusing the substrate. For GaAs epitaxial lift-off, lattice-matched AlAs is used as a sacrificial layer to separate the grown GaAs device from the GaAs wafer. Since selective etching of AlAs requires the use of concentrated hydrofluoric (HF) or hydrochloric acid (HCl) for an extended period of time, it could damage the GaAs wafer underneath. As a result, the wafer has to undergo expensive and intensive chemical mechanical polishing processes. In our work, we addressed this issue using a water-soluble lattice-matched sacrificial layer. Water-soluble alkaline earth compounds that possesses structural symmetry for epitaxial growth of zinc blende GaAs film were used. The best results were obtained with a 3-layer structure consisting of buffer layers of alkaline earth compounds tuned for lattice match with GaAs, sandwiching another water-soluble alkaline earth compound sacrificial layer. 2D X-ray Diffraction (XRD) measurement showed epitaxial growth of GaAs device film grown by metal-organic chemical vapor deposition on lattice-matched buffer film. Single junction solar cell devices have been fabricated with this architecture. Measurements at one sun showed minimal degradation in device performance before and after liftoff.

An Evaluation of Empirical Models for use in Normalizing PV Plant Performance Data

Daniel Fregosi and Michael Bolen

Electric Power Research Institute, Charlotte, North Carolina, 28262, USA

Abstract— **Monitoring PV plant health informs operations and maintenance activities. Models of the plant are used to calculate expected power, given the meteorological conditions. The expected power is used to normalize plant performance. This work evaluates the accuracy of various empirical, or data-driven, models and compares them to physics-based models by measuring how closely the calculated expected power matches actual power. It was found that empirical models offer greater accuracy and ease of setup than physics-based models. Methods to improve model performance are proposed and evaluated.**

Keywords—Photovoltaics (PV), performance monitoring, normalization, empirical, data-driven, regression

I. INTRODUCTION

As PV plants gradually lose capacity over time due to various factors including the accumulation of small faults and failures from normal wear-and-tear, both reversible and irreversible soiling covering the modules' surface, and various degradation mechanisms occurring inside the module. It is important to monitor the plant health to keep track of the plant capacity. Knowledge of the state of a plant's health informs performance predictions, guides maintenance activities, and can trigger further investigation if excessive degradation is found.

To evaluate plant health, one compares how much power the plant produces with the amount one would expect given the weather conditions, as measured by a meteorological station on-site. The expected power is defined here as the amount of power the plant should produce given full health and given the specific weather conditions (measured by the plane-off-array irradiance, ambient temperature, and wind speed). When comparing actual power with expected power, relatively small changes in plant performance over time are highlighted. Reductions in output are visible when a PV plant experiences soiling or degradation. Improvements in performance, while rarer, can be seen when maintenance activities are performed or when modules are cleaned either by washing or rainfall.

The focus of this work is the evaluation of models used to calculate expected power. The models are evaluated based on their accuracy and ease of setup, such that they can effectively be deployed to monitor a PV fleet. Specifically, the use of data-driven, empirical models is tested as an alternative to physics-based models. Empirical models offer ease of implementation, as setup can be automated, while offering potentially greater accuracy, as each of the model parameters are tuned to fit each individual sub-array. A range of models including statistical regression, machine learning, and physics-based ones are

evaluated. In addition to evaluating various models, variations on training length and data filtering are considered.

II. MODELLING OVERVIEW

The parameters of an empirical model are entirely determined by finding the values that most closely match the model to the data. Since each model is tailored to the specific plant or inverter-based sub-array, empirical models can often achieve higher accuracy than physics models based on plant specifications. In addition, empirical model setup can be automated, so it can be scaled to an entire fleet and periodic updates can be made without having to manually re-train or re-configure the models. One downside of empirical models is that care must be taken in selecting the training data. If a model is presented with a condition outside the window of its training, the results is unpredictable. Physics-based models are more generalizable to all conditions since they are based on a physical understanding of the system.

The inputs used in this study are only the ones measured most commonly at large-scale plants: power, plane-of-array irradiance, ambient temperature, and wind speed. Power is taken at the inverter-level rather than the plant-level because the irregular performance filters work much better at the inverter-level granularity. Additionally, the differences in performance from one inverter-based array to another are informative. The methods could be applied just as well to combiner-level measurements, for an even more granular analysis if desired. Models work for both tracking and fixed tilt systems since plane-of-array irradiance is used.

If additional measurements are available, such as from soiling sensors, back-of-module temperature sensors, rain gauges, and so on, they can be added as additional inputs to further improve model performance. The angle of incidence of the sun to the module surface is another useful input that can be found with the help of code libraries such as pvlib and knowledge of the location, module orientation, and timestamp [1].

A few additional key concepts regarding the system model are useful to consider here. The model should calculate expected performance by reproducing the initial, as-built system performance. The model should capture the dynamics of the plant as it is constructed, including any idiosyncrasies whether they are in the design or not. This is because we are interested in the most accurate performance predictions based on how the plant actually performs. This is distinct from comparing the as-built performance with the as-designed system, in which case physics-based models like SAM and PVsyst are very useful. Physics-based models can be tuned by adjusting the loss factors until the expected power matches actual power for the initial period of performance. However, it is difficult to tune the many parameters in the non-linear physics models to further increase accuracy. In linear regression models, many parameters are easily tuned simultaneously. In this work, models are applied, and tuned, at the inverter-level rather than the plant-level to increase accuracy.

This convention of tuning the model to the as-built system is useful for evaluating how plant performance changes over time, but not necessarily useful for evaluating the construction quality as is done in commissioning. While not the focus of this work, the as-built model can be used to calculate an implied nameplate by feeding in standard test conditions (or typical weather profiles) in place of the actual weather into the model and evaluating the output. The result shows the true capacity of the plant including losses such as inverter and wiring loss.

When selecting data to serve as the initial baseline performance to train the model, it is important to check that start-up challenges have been worked through and that the plant is operating at full capacity. In this work, the model remains fixed after the initial baselining or training is done. Any subsequent changes in PV plant performance result in a deviation in actual versus expected power. Depending on the application, it may be desirable to periodically update or retrain the model, such as when the system model is used to forecast production on a yearly or monthly basis. Model training is covered in more detail later.

Lastly, the setup explored in this work assumes that models are time-insensitive, such that they do not anticipate or predict deviations in performance like degradation or soiling. The date of a particular timestamp has no impact on the model prediction. The goal of the model is to reproduce the regular, deterministic, behavior of the plant given the external (exogenous) weather inputs. Long-term trends and patterns are then analyzed in subsequent analysis steps. Time-series modelling techniques, like autoregressive moving average (ARMA) models, can be used to recognize and decompose time-series patterns into long-term trends, seasonal components, anomalies, and noise. These will be applied in future work.

III. MODEL EVALUATION

A. Data Peparation and Training

The accuracy of various models and training methods was tested by applying each model to a dataset consisting of 79 inverters from 4 PV plants. The models were judged by how well the expected power calculation matched the actual power. A test set of 30% of the data is withheld from training and used for evaluating model fit. When selecting data to serve as the initial baseline performance to train the model, it is important to check for data quality and that start-up challenges in the plant have been worked through, such that the plant is at full capacity.

Training is used to calibrate the model parameters. An initial period is selected as the performance baseline. For this study, the first year of plant operation is taken as the training period. A full year is selected so that the training set contains a range of seasonal effects. The impacts of shorter training periods are tested in the model variations section later.

While the first year of performance approximates pristine plant health, short-term performance excursions likely exist. These irregularities have the potential to alter the model if their prevalence is too high. Fortunately, prominent irregularities can be detected with simple filters and isolated during model training. In addition to, it is important to remove bad data due to sensor error and stuck or interpolated values. Irradiance sensor error is common. In systems with multiple irradiance sensors, error can be minimized by comparing quantities as in [2].

Another important preprocessing step for model training is to remove clipped points. There are two reasons to exclude clipped points from the analysis. First, we are interested in evaluating the health of a plant, but during clipped periods there is no information about the plant performance since the power is artificially limited by the inverter. Second, clipping is a non-linear behavior that can be difficult for linear models to replicate. Aside from clipping, the physics of PV plants can be reasonably approximated by linear equations. Since clipping is relatively easy to detect, clipped times can simply be removed from the dataset when evaluating plant health.

In this study, model fit is used to compare models against each other. The root mean squared error (RMSE) is used here as the metric for success. A lower RMSE corresponds to a more accurate model. In the RMSE calculation, errors are squared, thus large outliers are penalized disproportionally high compared to small errors. The method used for training is ordinary least squares (OLS) which finds the model parameters that minimize the square or errors. In practice, evaluating model fit may not be necessary, but it could be useful information to know the margin of error to expect when evaluating PV plant health.

B. Model Variations

Model Name	Equivalent Industry Model	Unique inputs	Combination order, count
1st Order	PPI	Plane-of-array Irradiance (POA), Ambient Temperature, Wind Speed	1, 4
2nd Order Basic	ASTM 2848	POA, Ambient Temperature, Wind Speed	2, 10
2nd Order Additional Inputs		POA, log(POA), Ambient Temperature, Wind Speed, Angle of Incidence (aoi), cos(aoi), Minute of Day	2, 36
3rd Order		Same as above	3, 120
4th Order		Same as above	4, 330

TABLE 1 LIST OF REGRESSION MODELS AND INPUT VARIATIONS

978-1-7281-6118-1/22 $31.00 © 2022 IEEE

Three categories of models were evaluated: linear regression, physics, and neural networks. Neural networks can capture non-linear behavior when a non-linear activation function is selected. Table 1 describes the regression models evaluated here. The first order model is a simple combination of the three inputs (plane of array irradiance, ambient temperature, and wind speed) along with an additional offset term to eliminate constant bias. This model is equivalent to the Power Performance Index (PPI). The second order basic model, which is equivalent to the ASTM 2848 model, takes the same inputs and combines them up to second order. By multiplying inputs to create new ones, some non-linear behavior can be captured while still using a linear regression model. The new inputs are calculated at each timestamp before the model is trained. This process is computationally efficient. An example of second order combination of two inputs is shown in Eq. 1.

$$[a,b] \rightarrow [1, a, b, ab, a^2, b^2] \qquad (1)$$

The second order additional inputs model adds the angle of incidence and minute of day as inputs to compensate for physical phenomena that are dependent on the sun's angle in the sky versus the module surface. This model also uses two inputs derived by transforming existing inputs. The logarithm of the irradiance and the cosine of the angle of incidence are computed before the model is trained.

C. Evaluation Results

Models were first tested using synthetic data generated by simulating PV plants with physics-based models. By testing models with synthetic data, they are evaluated under ideal conditions and the best-case theoretical performance of the model can be measured. This test shows how well the empirical models can possibly capture the underlying physics of the plant. Actual datasets contain sensor error and stochastic physical phenomena like soiling and small faults, whereas the synthetic data is entirely deterministic. These results set the floor for model error.

128 synthetic datasets were generated by simulating plants at 8 locations. Irradiance, temperature, and wind speed patterns were taken from historical data and scrambled by selecting 5 randomized years of historical weather data for each synthetic dataset. In addition to the changing location, datasets varied by the following plant design parameters: albedo, mounting (fixed/tracking), dc/ac ratio, module technology (thin film/crystalline), tilt and azimuth angle, temperature coefficient, and ground coverage ratio. Models were each evaluated on the same 128 datasets, and the results are pictured in Fig. 1. The box and whisker plots show the error for each model across the 128 datasets. RMSE is computed hourly and as a percentage of the DC nameplate of the array. The results show increasing model performance as complexity increases from left to right.

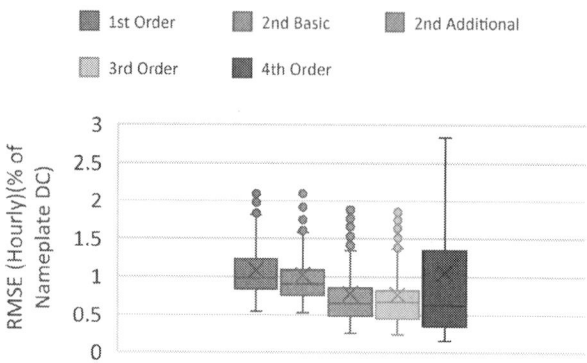

Fig. 1. Comparison of regression models, by increasing complexity (left-to-right) when evaluated on synthetic performance data

Overall, model error is low, with a median error below 1% hourly RMSE in all models. A slight improvement is made by increasing from first order to second order. A more dramatic improvement is realized as additional inputs are added to the second order model, reducing median error from 0.9% to 0.65%. Minimal accuracy improvement is realized in moving to a third order model. Finally, a fourth order model shows worsening performance, perhaps due to instability from too many inputs.

Next the models are tested on 79 individual inverter-based arrays from 4 PV plants. In addition to the regression models which were tested on the synthetic data, physics-based and neural network models were evaluated. Hourly RMSE grouped by model is given in Fig. 2. The overall model error raises from below 1% in the synthetic data to 3-5% here. As in the synthetic data, model accuracy increases for the four regression models as complexity increases up to third order. The physics-based model performs about as well as the first order regression model and worse than higher order regression models. Lastly, the neural network, which is essentially a non-linear regression model, performs about as well as the third order regression model.

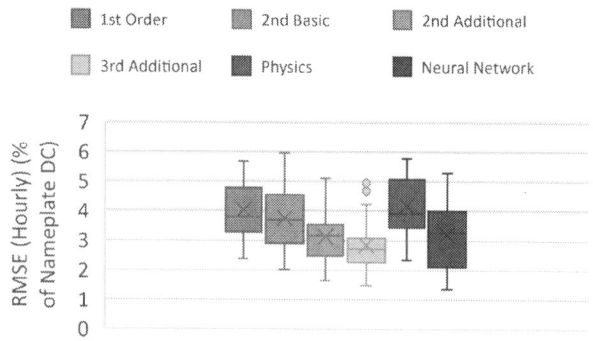

Fig. 2. Comparison of model performance (from left-to-right: four regressions, physics, neural network) evaluated on data from 79 arrays

The neural network used is the standard multi-layer perceptron regressor from the scikit-learn python package, with a hidden layer size of 100. The additional, derived inputs are used for the neural network regressor, and the inputs are scaled using the scikit-learn StandardScalar function. Scaling inputs is not necessary for the regression models since ordinary least squares optimization is insensitive to input scaling. For the physics model, the pvlib python package is used simulate the system performance with the given plant specs. For each inverter-based array, a scale factor was applied to tune the model to the data and remove any proportional bias.

The accuracy of expected performance calculations can be increased by narrowing the range of data to include times where the model is expected perform better. One example considered here is times of clear sky and times of high irradiance. The trade-off for narrowing the datasets to increase accuracy is that there is reduced visibility when evaluating plant health.

When taking a closer look at expected performance versus actual performance, it is clear that the model is most accurate during clear skies. One major reason for this is that when variable clouds are passing, they do not uniformly and simultaneously cover the full array and the irradiance sensor. An example is shown in Fig. 3. where the normalized power is overlaid on the raw power. Normalized power is centered near unity during clear skies and is highly scattered during variable clouds.

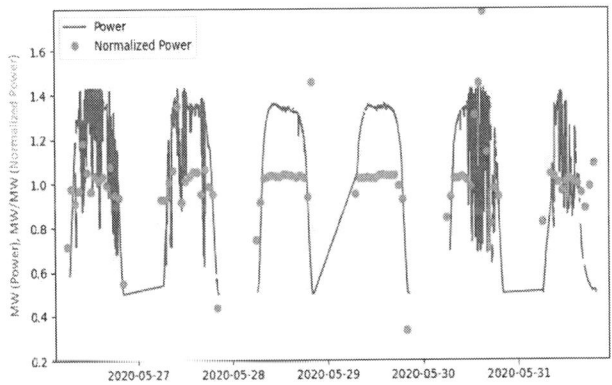

Fig. 3. Illustration of model accuracy in normalizing power during clear skies and variable cloudiness (normalized power should remain steady near 1)

The model error for three scenarios regarding clear sky filtering are shown in Fig. 4. The first set is the baseline hourly RMSE considering all points. Next, if only the clear sky points are evaluated, the RMSE reduces by just over a percentage point. Finally, the third option considers a sensor-less approach that is enabled when only clear sky times are considered. Here, a lookup-table of clear sky irradiance, using pvlib, is utilized for the irradiance input rather than the on-site sensor. This option is useful if sensor data is unavailable, if there are concerns about sensor accuracy, or if the application is very sensitive to long-term sensor drift such as in degradation analysis. These results show that the sensor-less option is only marginally less-accurate than the sensor option when considering clear sky times. For the clear sky options, not only should evaluation be limited to the clear sky times, but training

data should be limited to clear sky times as well rather than all points. Methods to detect clear sky periods are demonstrated and published [3].

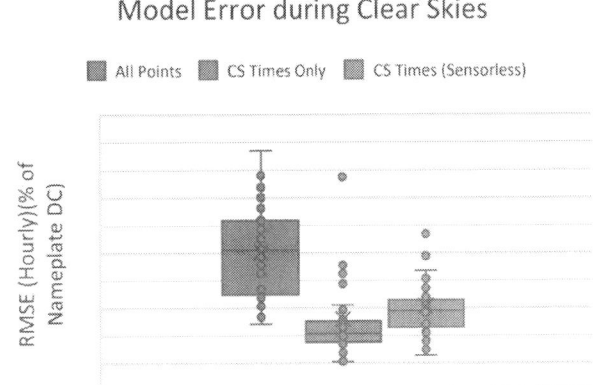

Fig. 4. Comparison of model accuracy during clear skies with (middle) and without (right) irradiance sensing

The next variation considered is training data length. This tested the impact of using less than a full year of data, in the case that a plant has not yet been fully operational for a year. The model error for training lengths ranging from one week to one year are shown in Fig. 5. As expected, model error decreases with longer training intervals. There are dramatic improvements up to 1 month and modest improvements up to six months. Interestingly, for physics-based models the training length beyond one week does not impact model accuracy when there is only one scalar parameter to tune. Seasonal effects are already built in to the physics-based model. The RMSE is shown side by side for regression (left) and physics-based models (right) in Figure 5. For training intervals 1 month or greater, the regression models outperform the physics-based models.

Lastly, the effects adding a minimum irradiance threshold were studied. The idea is that PV performance is more nonlinear at low irradiance than at higher irradiance due to phenomena like row-to-row self-shading and reflection loss for low angles of incidence. It was found that the empirical model accuracy improves marginally when a minimum irradiance threshold is used. A small but steady improvement was observed as the minimum irradiance threshold was increased from 100 to 700 w/m^2, as seen in Fig. 6. The downside of using minimum irradiance threshold is the reduction of data points for evaluating the plant health. The improvement in model accuracy is likely not worth the reduction in datapoints for most applications. The error here is calculated as normalization error rather than residual error since residual error would be biased by the higher power levels present in the data with high irradiance thresholds.

Error by Training Length (Empirical)

Error by Training Length (Physics)

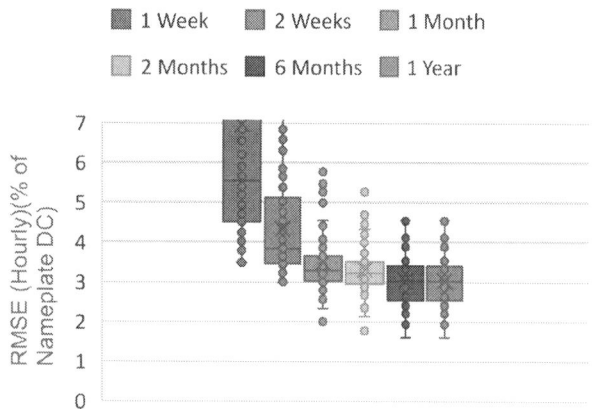
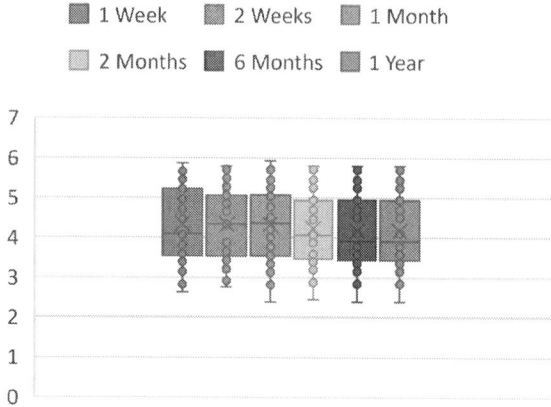

Fig. 5. Comparison of model accuracy by training length for empirical (left) and physics-based (right) models

IV. ANALYSIS AND CONCLUSIONS

In summary, it was found that linear regression models calculated calculate expected power with more accuracy than physics-based models, on the 79 datasets tested here. Regression model accuracy is improved by adding additional inputs such as angle of incidence and time of day, transforming inputs, and combining inputs up to second order. Neural network models offer similar performance to linear regression models.

By limiting the window of times in which the health of the plant is evaluated, the model can achieve higher accuracy. For clear sky times, error is reduced from 3.5% to 2.3%, however, clear sky times make up only about 15% of the data.

The significance of the results is that plant operators can replace physics-based models with empirical models for monitoring applications to increase accuracy. Since regression model setup can be automated, fleet-wide implementation can

be achieved down to the inverter or even combiner-level. For further reading, the white paper in [4] gives an overview of tools and methods for PV plant health evaluation and performance loss rate analysis.

REFERENCES

[1] William F. Holmgren, Clifford W. Hansen, and Mark A. Mikofski. "pvlib python: a python package for modeling solar energy systems." Journal of Open Source Software, 3(29), 884, (2018). https://doi.org/10.21105/joss.00884

[2] D. Fregosi, M. Bolen and B. Paudyal, "An Assessment of In-Field Irradiance Sensor Accuracy and Error Mitigation Techniques," 2021 IEEE 48th Photovoltaic Specialists Conference (PVSC), 2021, pp. 1430-1436, doi: 10.1109/PVSC43889.2021.9518813.

[3] Reno, M.J. and C.W. Hansen, "Identification of periods of clear sky irradiance in time series of GHI measurements" Renewable Energy, v90, p. 520-531, 2016.

[4] D. Fregosi, M. Bolen, "Analyzing Performance Loss Rates in PV Plants Using Operational (SCADA) Data", EPRI. Palo Alto, CA: 2022. 3002021060

Model Error with Minimum Irradiance Thresholds

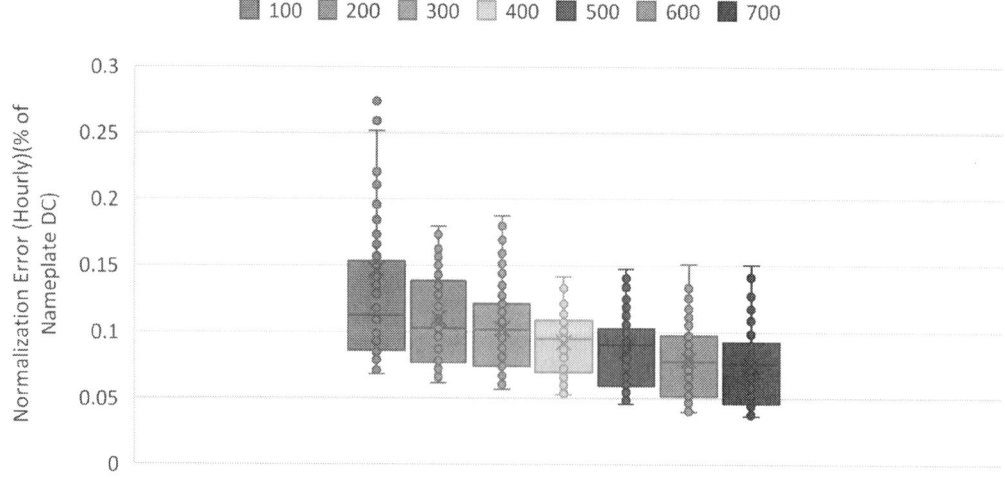

Fig. 6. Comparison of model accuracy by training length for empirical (left) and physics-based (right) models

Performance investigation of batteries supporting solar power in U.S.

Farzan ZareAfifi, Daniel Baerwaldt, Socheata Hour, Yi Hao Xie, Sarah Kurtz

University of California Merced, Merced, CA, 95343, USA

Abstract—The use of batteries in energy storage power plants in the United States has increased significantly. A prime objective of the plants is to provide power in times of peak demand after being charged with renewable sources, mainly solar power. Modeling of solar adoption is largely dependent on understanding how solar and batteries work together, including the need to quantify the battery efficiency. In this study, the efficiency of the energy storage plants in U.S. was calculated based on U.S. Energy Information Administration (EIA) data. A mathematical model is proposed relating the efficiency and the number of cycles per month, grouping plants by installation year. We conclude that newer plants show higher efficiencies, and in the case of experiencing an average of one cycle per day, the newer plants show efficiencies of around 90%. Also, we see a lower efficiency for plants cycled less than five times per month. The efficiency is observed to be between 80% and 90% for batteries experiencing more than five full cycles each month.

I. INTRODUCTION

In support of the extensive solar generation, the United States has been experiencing growth in the use of batteries. California is the leading state in deploying batteries. Fig.1 shows how batteries were mainly used for ancillary services (*e.g.* load following and frequency regulation) in California in 2019 and 2020, but starting in 2021, batteries have been used to store electricity from renewable resources according to California Independent System Operator (CAISO) data [1]. Comparing Figs. 1 and 2, we see that the batteries start charging around 8 am as solar output increases and discharge starting around 6 pm when the solar output is disappearing.

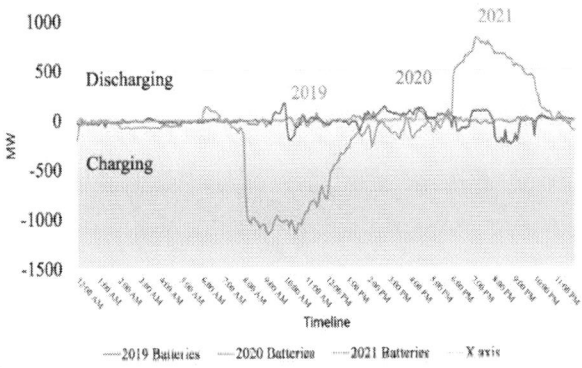

Figure 1. Battery utilization on August 18th in 2019, 2020, 2021 shows a dramatic increase in 2021 according to CAISO data [1,2]

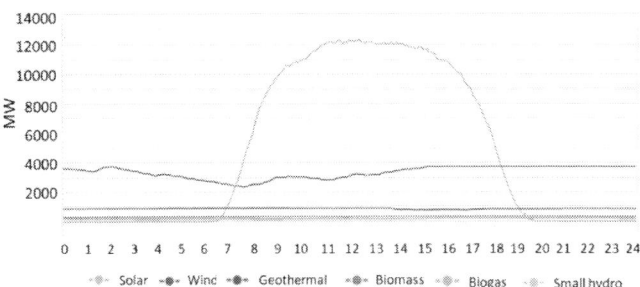

Figure 2. Renewable trend in California on August 18th, 2021, shows that significant portion of renewables is solar power [1]

As storage is increasingly coupled with solar, understanding battery efficiency becomes important. While it is well established that the efficiency may be highly variable, the use of batteries has grown to the point where we can use the publicly reported (either EIA or CAISO) data to quantify the average efficiency that is observed. As the batteries are mainly used to store solar power and will be discharged on the same day after sunset, analyzing the batteries' efficiency when experiencing one cycle per day is of prime importance.

Another performance metric in batteries is energy capacity. Atalay et al. [3] showed that the greater the number of cycles a battery experiences (more age), the more significant the capacity reduction will be. Due to the lack of data, the capacity is not studied here.

In this study, the data reported in EIA923 [4] and EIA860 [5] datasets have been used to calculate the number of monthly cycles and the efficiency of energy storage plants in U.S. In section II, a mathematical model for predicting the efficiencies of the batteries has been introduced. In section III, we have applied the model to the real-world data from the datasets.

II. THE PROPOSED MODEL

In this study, only plants reported in both EIA923 [4] and EIA860 [5] are considered since the information provided in both are needed for the analysis. We investigated plants with a prime mover identified as batteries. Some plants reported significant outages, which are not considered in this analysis (see appendix). Almost all the plants use lithium batteries

978-1-7281-6118-1/22 $31.00 © 2022 IEEE

(LIBs); thus, battery type analysis is not in the scope of this study. Also, factors of charging and discharging rate, depth of discharge, and temperature conditions are not analyzed. Therefore, the proposed equation here predicts the efficiencies of the batteries only in terms of the monthly number of cycles they are experiencing.

For this study, we proposed equation (1) for correlating efficiency with the plants' monthly cycles.

$$\eta = C_1 - C_2 \times \frac{30}{Number\ of\ Monthly\ Cycles}$$

$$= C_1 - \frac{C_2}{Number\ of\ Average\ Daily\ Cyles} \quad (1)$$

The efficiency obtained from this equation (η) is for one month. C_1 shows the maximum efficiency that can be achieved using the LIBs. C_2 shows the average efficiency reduction if a plant experiences 30 cycles in one month. More cycles lead to higher efficiency, according to the equation. As a result, C_2 may be a loss related to the batteries' idle loss. It is also possible that C_2 reflects another loss like inclusion of the energy needed to run an air conditioner.

According to equation (1), the efficiency of a plant experiencing an average of one cycle per day will be C_1-C_2.

III. RESULTS AND DISCUSSION

Efficiency versus the number of monthly cycles for different months and plants is shown in figure 3. In this figure, each point represents one month. According to the figure, as the monthly cycles increases, the efficiency also increases, consistent with the proposed model.

According to eq. 1, as C_2 increases, the loss will be more significant, and the efficiency will decrease. According to the EIA data, the efficiency of each plant stayed the same from year to year. Consequently, although the age of a battery adversely affects its capacity, we observe that it does not significantly affect its efficiency (data not shown). Also, we observe no trend for C_2 in different seasons; thus, apparently, the operating conditions of the batteries are independent of the outdoor temperature and cannot be analyzed with the available data.

Figure 4 shows the box plot of the data for the 5-cycles-per-month intervals. The graph is for the number of monthly cycles smaller than one hundred since most of our data are in this region. This graph shows a sudden increase in efficiencies from zero to five monthly cycles. For plants experiencing > 5 monthly cycles the efficiencies are between 80% to 90%.

Figures 5 to 9 show curve fits for plants with initial operating years of 2016 to 2020, respectively. In fig (7), the excluded points belong to the 2020 months showing efficiencies above 90%, while data for the plant in the previous months show efficiencies around 70%. Thus, most probably, the plant has been updated, and the plant's 2020 data are considered in fig (9).

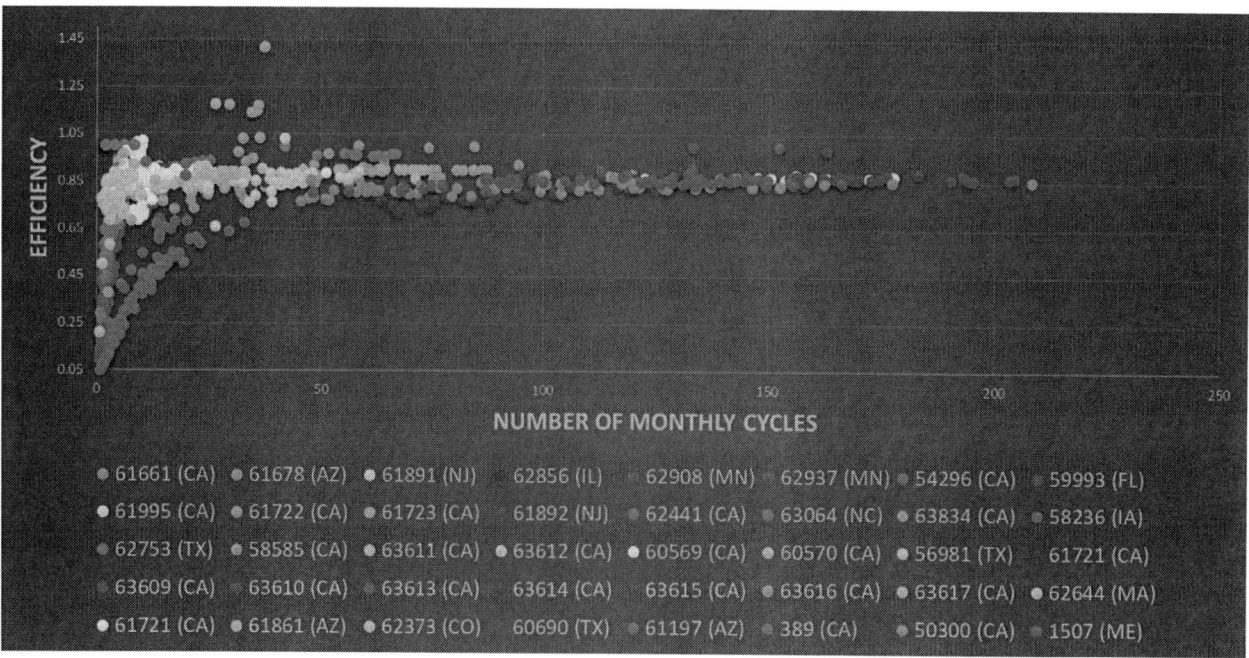

Figure 3. Efficiency versus number of monthly cycles for forty plants based on data from EIA datasets; The location (state) of each plant has been indicated in the parentheses next to the plant IDs below the graph.

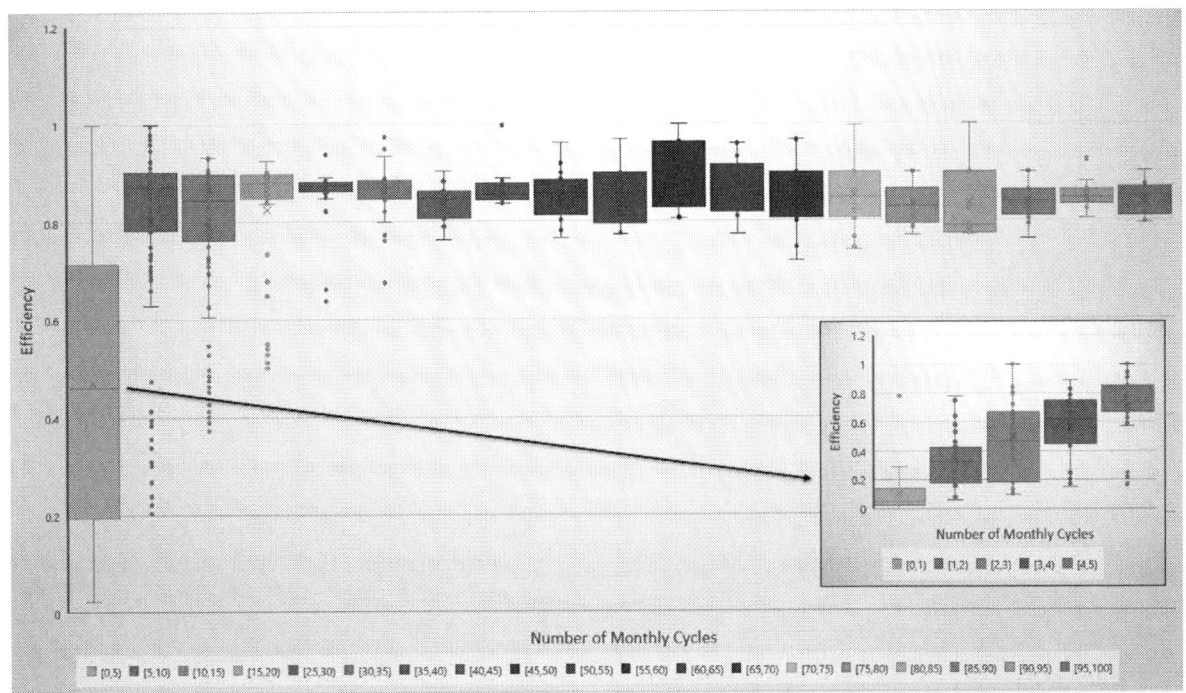

Figure 4. Box and Whisker plot for all the plants for 5-cycles-per-month intervals

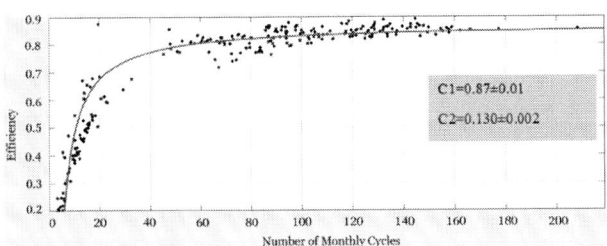

Figure 5. The best curve fit for plants with the installation year of 2016

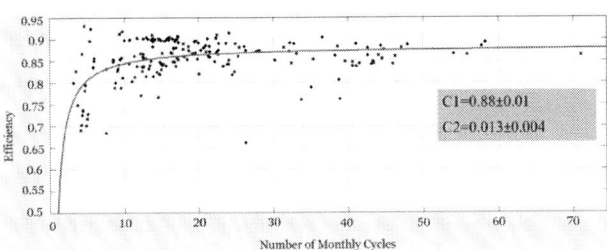

Figure 6. The best curve fit for plants with the installation year of 2017

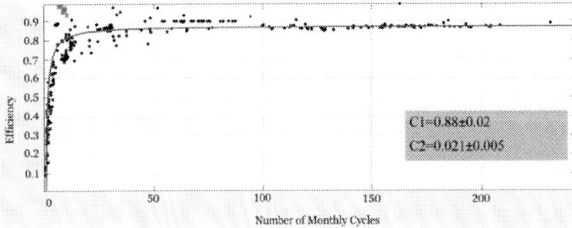

Figure 7. The best curve fit for plants with the installation year of 2018

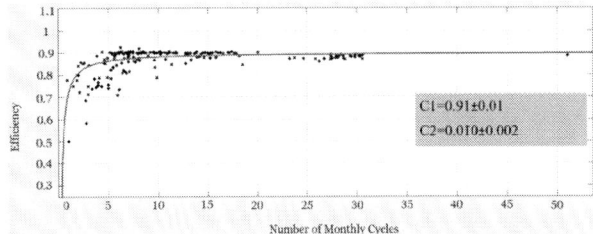

Figure 8. The best curve fit for plants with the installation year of 2019

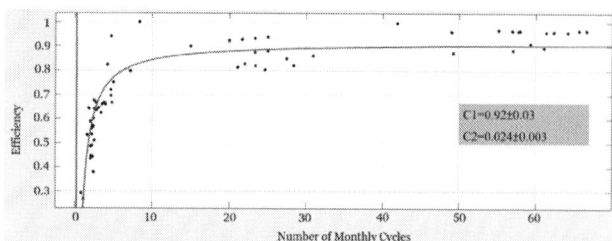

Figure 9. The best curve fit for plants with the installation year of 2020

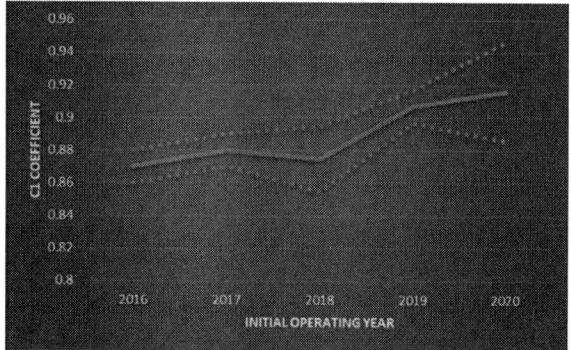

Figure 10. The calculated C_1 coefficients versus initial operating year. The dotted lines indicate the uncertainty of the fit.

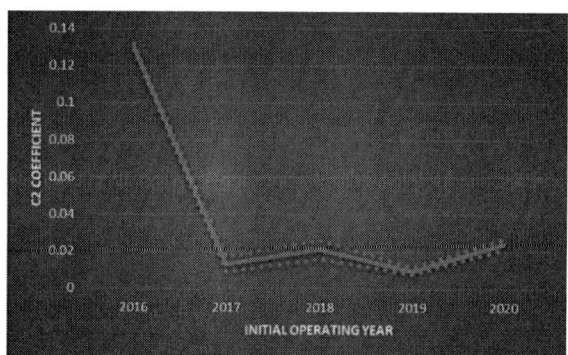

Figure 11. The calculated C_2 coefficients versus initial operating year. The dotted lines indicate the uncertainty of the fit.

The fit coefficients are summarized in Figs (10) and (11). The newer plants show higher efficiencies and experience low level of losses. In other words, plants with older installation years generally require more monthly cycles to reach high efficiencies of 80% or 90%. We do not know the cause of the improved performance for newer systems, but we surmise that the improved performance is from technological advances.

As mentioned before, evaluating one cycle per day is important (the predicted efficiency, in this case, is C1-C2). Figure 12 demonstrates that the newer installed plants will experience efficiencies of around 90% if they experience one cycle per day (an average of 30 cycles per month).

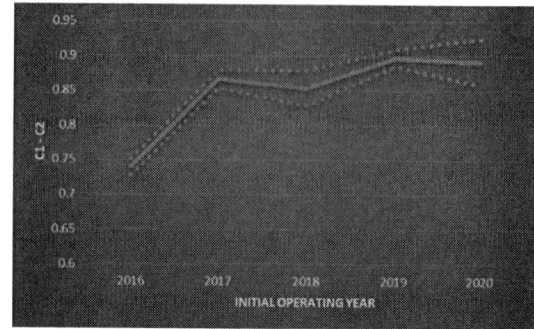

Figure 12. The predicted efficiencies for the average of one cycle per day case. The dotted lines indicate the uncertainty of the fits.

IV. CONCLUSIONS

This study analyzed the efficiency of U.S. energy storage power plants, providing real-world data for quantifying the efficiency of today's batteries. A mathematical model for the correlation between efficiency and the number of monthly cycles for the plants was proposed to include two coefficients representing the loss and the maximum efficiency. Since a greater number of cycles is correlated with higher efficiencies, the loss predicted by the model might be directly related to the idle loss of the batteries, or it may be that it represents other losses like air conditioning. The plants were categorized into five groups based on their initial operating years: from 2016 to 2020. The two coefficients (for the best curve fit based on the model) were found for each group. By comparing the fit coefficients, we concluded that the maximum efficiency has an increasing trend from the older to the newer plants. This trend may be attributed to the technological progress in making batteries. The newer plants exhibit efficiencies of around 90% if they experience an average of one cycle per day. By analyzing the data, we saw that the efficiencies are increasing rapidly from zero to five monthly cycles. For more than five monthly cycles, most plants show efficiencies between 80% and 90%. The number of energy storage plants using batteries has dramatically increased in 2021. Therefore, for future work, it will be beneficial to analyze the plants established in 2021. The EIA data for plants with the operating year of 2021 will be released in summer 2022.

REFERENCES

[1] http://www.caiso.com/TodaysOutlook/Pages/index.html

[2] Daniel Baerwaldt, Socheata Hour, Yi Hao Xie, Pedro Sanchez, Sarah Kurtz (2021). A Quantitative Analysis of Batteries in California, Poster presented at University of California Merced

[3] Atalay, S., Sheikh, M., Mariani, A., Merla, Y., Bower, E., & Widanage, W. D. (2020). Theory of battery ageing in a lithium-ion battery: Capacity fade, nonlinear ageing and lifetime prediction. *Journal of Power Sources*, *478*, 229026. https://doi.org/10.1016/j.jpowsour.2020.229026

[4] https://www.eia.gov/electricity/data/eia923/

[5] https://www.eia.gov/electricity/data/eia860/

APPENDIX

Some data were excluded from the analysis, or they were considered with a different initial operating year. These plants are as follows:

Plant Code in EIA Datasets	Decision	Explanation
62381	Not Considered	Efficiencies much higher than one
62382	Not Considered	Efficiencies much higher than one
62682	Not Considered	Significant outages in the datasets
62683	Not Considered	Significant outages in the datasets
61892	Considered in the 2018 group	In the dataset, the operating year is 2019, but there are some reported data for the plant in 2018
60690	Considered in the 2016 group	In the dataset, the operating year is 2017, but the incorporation date of the plant is 13 April 2016 (Reference)
56981	Considered in the 2017 group	A new generator with similar description was reported in 2017 (Reference)

Numerical modeling of capacitance signatures of perovskite solar cells

Rasha Awni, Zhaoning Song, Chongwen Li, Lei Chen, Suman Rijal, Sandip Bista, Tao Zhu, Xiaoming Wang, Yanfa Yan

Department of Physics and Astronomy, and The Wright Center for Photovoltaics Innovation and Commercialization (PVIC), University of Toledo, Toledo, OH, United States

Perovskite solar cells (PSCs) have drawn attention owing to their high-power conversion efficiency (PCE) and low manufacturing cost. Notable efforts have been made to further improve solar cells performance. However, an accurate assessment and resolving the issues responsible for the power conversion efficiency loss is required for future progress. Capacitance-based techniques such as capacitance-voltage (C-V) measurements are widely used to characterize the depletion region properties of PSCs, such as doping density and their distribution, built-in potential, and depletion region width can be extracted. Here, we present numerical simulation using the one-dimensional (1D) solar cell capacitance simulator (SCAPS) software to show that several factors must be taken into consideration in order to appropriately estimate the depletion region properties using C-V measurements. We find that the capacitance due to charge injection from the charge transport layers (CTLs) and charge accumulation at interfaces may significantly contribute to the measured capacitance of PSCs. We show that these contributions are very pronounced when the carrier density is low and the perovskite layer is thin. The low mobility and poor conductivity of CTLs may lead to voltage drop and the formation of built-in potential inside the CTLs. Accordingly, a measured built-in potential may represent the summation of voltage drop over the serially connected CTLs and perovskite layers. Our results provide important guidance for the suitable estimation of depletion region properties of PCSs using C-V measurements.

978-1-7281-6118-1/22 $31.00 © 2022 IEEE

Reverse Energy Injustice On Molokai Island To The Underserved Communities With 100% Energy From The Sun (Light & Heat) For Energy Cost Savings Equity

John O. Borland

J.O.B. Technologies, Aiea, Hawaii, 96701, USA

Abstract— To reverse energy injustice on Molokai island to the underserved families, I proposed three solutions: 1) Grid-Tie homes: Solar + Storage Island Nano-Grid for 100% energy from the Sun (Light & Heat) and NEM Equivalent Savings™ for $25/month bill with energy cost savings equity of 81-93%. 2) Grid-Tie rental Section-8 homes or apartments/condos without rooftop solar access: Battery only or Battery + Solar-Balcony/Awning/Patio using Bi-facial panels and batteries sized 10, 20 or 30 kWh for 100% of the overnight energy usage to reduce bill to $31/month with energy cost savings equity of 77-92%. 3) Off-Grid homes using gas generators: replace "dirty" gas generators with Solar + Storage Island Nano-Grid and Nano-Grid Clusters for 100% energy from the Sun (Light & Heat) to improve Quality of Life and save up to 95% in electricity costs due to high cost of gasoline.

Keywords— *energy injustice, energy cost savings equity, NEM equivalent savings, solar + storage, battery + balcony-solar, island nano-grid and clusters*

I. INTRODUCTION

The global carbon emission numbers continue to climb even with COVID-19 lockdowns in 2020 and 2021. In May 2022 CO_2 level measured on Mauna Loa observatory was up another 1.1PPM (+0.3%) from 2021 to 420.2PPM. China continues to lead with highest carbon emissions accounting for 28.9% of the world carbon emissions in 2020 up from 27.7% in 2018, #2 US share grew to 15.0% after dropping to 14.5% in 2018 and #3 India at 7.4% was up from 7.0% in 2018. To reverse this trend requires adopting the "Triple-D", Decarbonize, Decentralize and Digitize. The State of Hawaii leads the nation at Decarbonization with 38% renewable clean energy for 2021 and a goal of 100% by 2045. The residential solar program in Hawaii has been very successful with 1/3 of homes (~75K) having rooftop solar accounting for 41% of Hawaii's 38% renewable clean energy. But this has resulted in: 1) the grid having a severe Duck Curve due to excess daytime PV grid-export followed by high evening peak demand and 2) created a severe social economic "Energy Divide" with affluent families benefiting from the 80-90% energy cost savings reducing their energy burden from ~6.8% of income to <0.3%. With the State and Federal solar energy incentives, a payback of 3-4 years is achievable. The underserved communities are left behind with increase energy bill from Hawaiian Electric (HECO). Hardest hit with energy injustice are low-to-moderate income (LMI) families with an energy burden >19% of income. With Energy Justice and 80-90% savings, their energy burden would be reduced to <2% of income, a significant relief when living pay-check to pay-check and "Life Changing". In Hawaii, the greatest energy insecurity and social inequity is on Molokai Island where a Civil Beat article on 9/22/2021 reported 62% are Native Hawaiian and many residents live without a connection to the grid because of the high electricity rates and connection fees [1]. The article quotes "When the sun sets on Molokai, hundreds of homes go dark. …forcing many families to choose between using power at night or feeding their children." The HECO solar-PV hosting capacity for Molokai Island is shown in Fig.1 with the majority of locations in red prohibiting grid-connection with Grid-Export.

Fig.1: Hawaiian Electric solar-PV hosting capacity for Molokai.

II. 100% ENERGY FROM THE SUN 6 YEARS CASE STUDY

The 6 years case study residential solar + storage (battery and thermal) on Oahu, installed in June 2016, achieved 100% energy from the Sun since April 2017 on sunny and partially sunny days >90% of the time or >330 days a year. The monthly pre-solar HECO bill is shown in Fig. 2. Between Jan 2012 and May 2016 averaged $396/month (1250kWh/month) while post-solar + storage self-consumption achieved NEM Equivalent Savings™ of <$25/month 7 months a year since July 2017, averaging $30.31/month (<99kWh/month) for a reduction of 92.3%. The yearly energy cost savings is shown in Fig.3 at $4,390/year for a payback/ROI (return on investment) of 3.1 years and 25 year savings of ~$109K. The actual payback after 6 years compared to the estimated payback are shown in Fig.4 with a 6 month shift in the data due to only half year data in 2016. These energy savings were realized by home energy digitization and implementing the energy conservation behavioral lifestyle changes listed in Fig.5 including demand response load shifting with time-of-use (TOU) rates and home energy efficiency improvements, improving Quality of Life [2].

This allowed 100% energy from the Sun with Sun heat for hot thermal storage (hot water) and Sun light for PV-generation for daytime energy demand, overnight battery storage and cold thermal storage (room/space cooling) as shown in Fig.6.

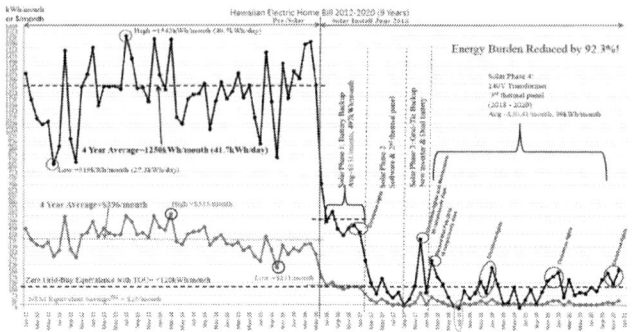

Fig.2: Case study HECO monthly electricity bill from Jan 2012 to Dec 2020.

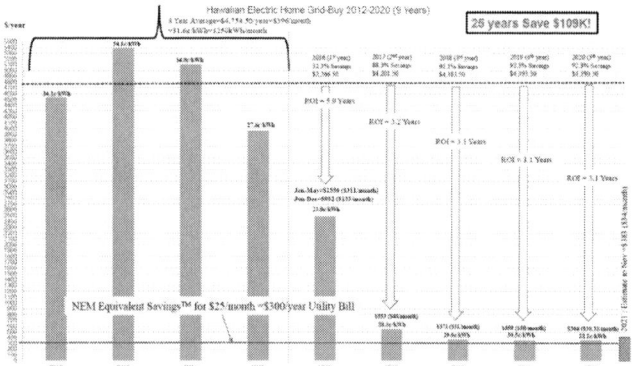

Fig.3: Case study HECO yearly electricity bill from 2012 to 2020.

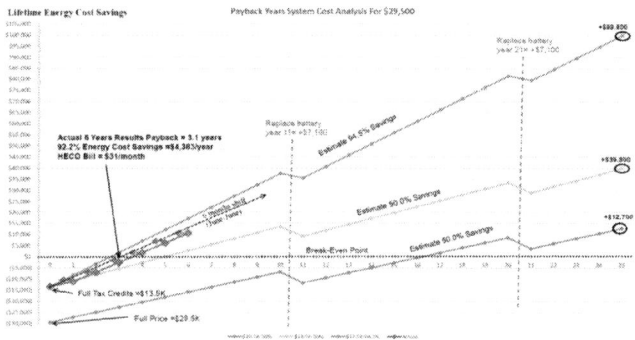

Fig.4: Payback years versus lifetime energy cost savings estimate versus actual 6 year results.

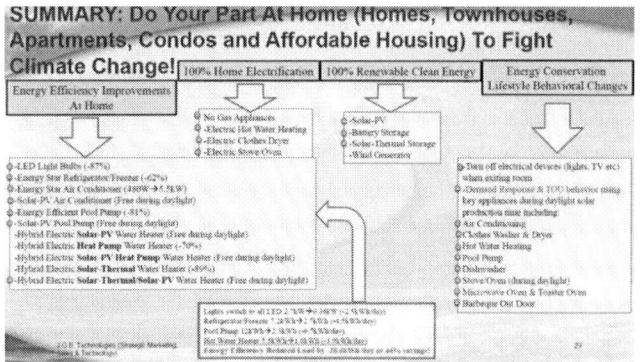

Fig.5: Energy conservation lifestyle behavioral changes include demand response load shifting, TOU rates and energy efficient appliances.

Getting 100% Energy From The Sun (Heat & Light)

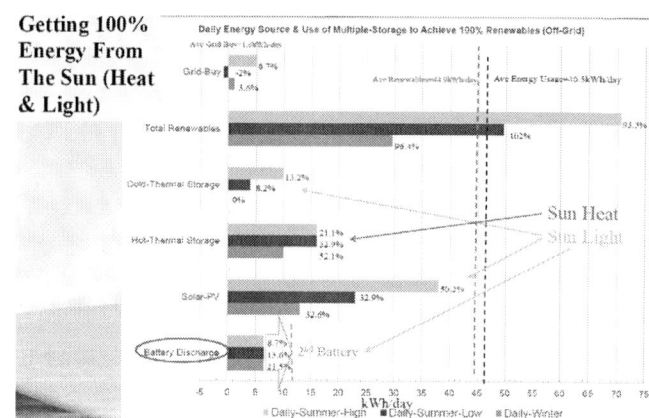

Fig.6: Breakdown of 100% energy from the Sun into light for PV used for daytime usage, battery storage and cold thermal storage while heat for hot thermal storage.

This required home energy digitization which started in Jan 2018, adding smart IoT devices for level-1&2 autonomous home energy management (real-time energy monitoring with connected devices). Level-3 (insight assisted change) was developed Jan to Aug 2020 using various Raspberry Pi devices with temperature sensors and on/off relay and circuit breakers to automate demand response load shifting for the hybrid solar/electric hot water thermal storage shown in Fig. 7 [3]. Sunny day 12/1/21 used solar thermal heating to 162°F while rainy day 12/6/21 solar thermal heating only reached 84°F requiring electric hot water heating from solar-PV, battery discharge or Grid-Buy to reach ~135°F. Space/room cooling used demand response load shifting for solar AC with cold thermal storage. Today, level-4 (full home optimization) adds weather sensors, Sky-cam and local weather info for weather and PV-generation forecasting based on seasonal Digital Twin (Fig.8). Using the seasonal Digital Twin as reference we can determine if the reduced PV-gen is due to weather of PV-curtailment. From 2017 to 2020 we reduced PV-curtailment by 43% and thereby boosted PV-gen by 176% from 15.6kWh/day to 27.6kWh/day (+12.0kWh/day). 40+ IoT devices now monitor, control and balance home energy usage. Data analytics and Pareto analysis of the top 7 weekly household appliance showed appliances used for heating or cooling purposes accounted for 75% winter-time and 90% summer-time energy usage (Fig.9). This allowed customized and flexible behavioral demand response load shifting to maximize daytime energy usage at lowest time-of-use (TOU) rates, a 92.3% energy cost savings since 2018. This also improved Quality of Life when automated. Energy monitoring devices show during the summer, daytime energy usage 6AM-6PM was 20.0kWh (69% daily usage) and overnight 6PM-6AM was 9.0kWh (31% daily usage) while winter, daytime energy usage 6AM-6PM was 17.4kWh (50% daily usage) and overnight 6PM-6AM was 17.1kWh (50% daily usage). A home energy audit showed energy efficiency improvements including switching to all LED light bulbs and Energy Star appliances reduced usage by ~44% (Fig.5).

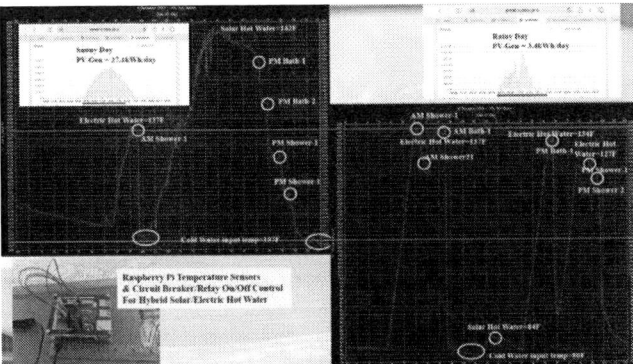

Fig. 7: Digitized hybrid solar/electric hot water storage system using Raspberry Pi for monitor and control. Comparison for sunny day 12/1/21 to rainy day 12/6/21.

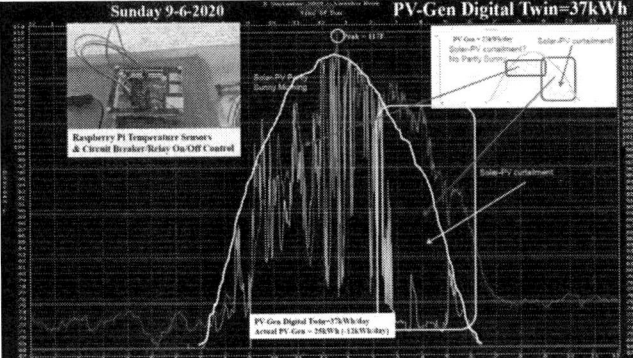

Fig. 8: Digitized PV-generation for Digital Twin analysis of weather or PV-curtailment effects.

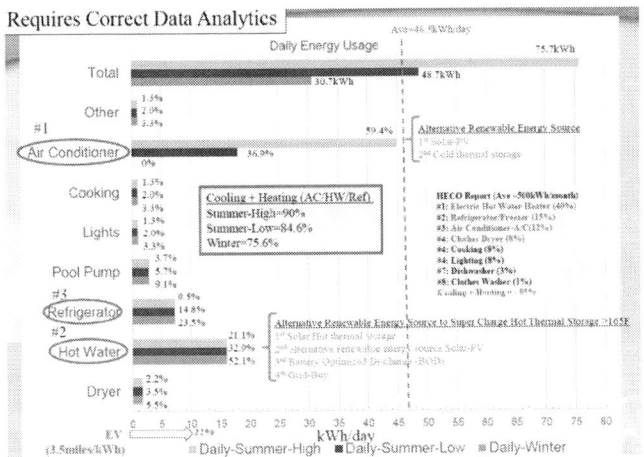

Fig.9: Seasonal Pareto analysis of the top 7 daily home appliance energy usage.

III. RESULTS

By applying these best practices and lessons learned from making the Oahu case study residential solar + storage (battery & thermal) self-consumption with no Grid-Export, energy independent and resilient from the grid, we performed an energy cost savings analysis to estimate the potential cost savings equity by reversing energy injustice on Molokai. Using the 6 years residential solar + storage case study results the data analysis uses the actual 4.0 hours per day for solar PV generation to estimate PV sizing and $3.00/W for PV system costs. For Li-ion battery storage costs, 75¢/W to size the hybrid battery capacity system cost including inverter.

A. Oahu Energy Cost Savings Analysis

HECO electricity rate on Oahu used in the calculations is 31.7¢/kWh and the low daytime TOU rate is 13¢/kWh. The monthly electricity usage in the analysis is based on the case study results in Fig.2 of 1250kWh/month (41.6kWh/day) and $396/month. Affluent family annual income is defined as >$100K/year for an energy burden of <4.8% of income while LMI family is defined as <$24K/year with an energy burden of >19.8% of income as shown in Table 1. The rooftop solar-PV sizing required to meet 100% energy from the Sun would be 10.5kW and the battery sizing of 27.5kWh is based on 66% overnight energy usage and 34% daytime usage for self-consumption, no Grid-Export. An 80 gallon electric hot water system is standard. NEM Equivalent Savings™ to reduce the monthly HECO bill to $25/month for a 93.7% savings. The energy burden for LMI families is reduced to 1.3% with a payback of 5.3 years. Switching to a 50 gallon electric heat pump hot water system would lower the battery overnight capacity needs to 19.5kWh and therefore the solar-PV sizing to 8.3kW reducing payback to 4.3 years due to lower total system costs. A 120 gallon solar thermal hot water system would reduce battery capacity sizing to 11.5kWh and solar-PV sizing to 6.3kW and the payback would be 3.2 years. For verification and validation, the actual case study results are also included in Table 1 showing with solar thermal hot water the solar-PV sizing was 7.1kW with a 14.8kWh battery for a 92.5% energy cost savings and 3.1 years payback as shown earlier in Fig.4.

Families living in apartment/condos and Section-8 homes without access to rooftop solar can also benefit with energy cost savings by copying the battery only programs offered in MA and CA even without the State or Federal incentives. HECO offers TOU rate to all residential customers in Hawaii with slightly different low daytime rates that vary between 13-18¢/kWh but in general >59% lower than standard rates. As shown in Table 1 "Battery Only" with a capacity of 27.5kWh and no incentives except TOU rates would reduce LMI families energy burden to 8.1% of income with a 7.4 years payback and a new monthly HECO bill of $162.24/month, a savings of 59%. Adding "Solar-Balcony/Awning/Patio" of >1000W would allow the State and Federal energy incentives, reducing payback to 3.6 years with slight reduction in energy burden to 7.8% and 60.6% savings. Switching to a heat pump hot water system has the greatest impact by reducing battery capacity size to 19.5kWh for a payback of 2.9 years and energy burden of 6.9% of income with a 65% energy cost savings.

B. Molokai Energy Cost Savings Analysis

The analysis for Molokai is shown in Table 2. The average family uses only about 325kWh/month (10.8kWh/day) for a HECO monthly bill of $135/month. Rooftop solar-PV sizing would be 2.7kW to meet 100% energy from the Sun with overnight battery capacity sizing of 7.2kWh to meet overnight energy needs of 66%. NEM Equivalent Savings™ HECO bill of $25/month would reduce LMI family energy burden from 6.8% to 1.3% for a savings of 81.5% and payback of 4.6 years. Adding solar thermal hot water would reduce payback to 4.1 years. Apartment/condo and affordable housing (section-8) LMI families with "Battery Only" would save 61.9% reducing HECO monthly bill to $51.48/month, reduce energy burden to 2.6% with payback of 4.0 years. Battery +Solar-

Balcony/Awning/Patio would save 66.2% reducing HECO bill to $45.57/month and energy burden to 2.3% of income with 3.6 years payback. Adding a heat pump hot water system would reduce battery capacity to 4.0kWh for a payback of 2.6 years and savings of 77.30% for a HECO bill of $30.69/month and energy burden of 1.5%. Off-Grid homes, solar + storage allows replacement of "dirty" gas generators, achieving 100% energy from the Sun with improved Quality of Life as reported in Puerto Rico by Borland and Tanaka [4]. Today's high gasoline prices of $6/gallon on Molokai makes gas generator electricity cost prohibitive at $500-728/month ($1.66-2.42/kWh) compared to Solar + Storage at $33/month (11¢/kWh) for an energy cost savings 92-95%. Clustering of Island Nano-Grids to form Nano-Grid Clusters with energy sharing offers additional neighborhood community resilience as shown in Fig. 10.

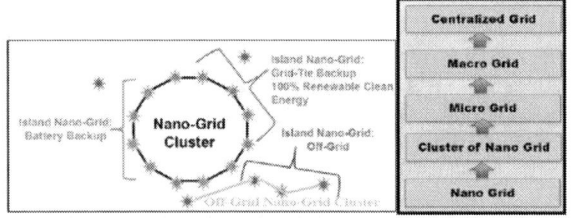

Fig.10: Off-grid Island Nano-Grid and Nano-Grid clusters for energy sharing and community resilience.

IV. SUMMARY

In summary, these best practices developed over the past 6-years provide the tools to reverse energy injustice to the underserved community on Molokai. This requires the Triple-D: 1) Decarbonization with 100% home electrification and no gas appliances, 2) Decentralization using Island Nano-Grid technology with switchable On/Off Grid mode of operation for resilience and energy independence, and 3) Digitization for home energy automation, control and balance using energy efficient household appliances. Energy sharing and Digital Twin analysis maximizes daily and monthly energy cost savings for NEM Equivalent Savings™ (<$25/month).

REFERENCES

[1] L. Teruya, Civil Beat article, https://www.civilbeat.org2021/09/molokai-has-an-electricity-problem-this-co-op-wants-to-change-that/.

[2] J. Borland, book chapter, "Residential Island Nano-Grid for 100% Renewable Clean Energy", in the book *Accelerating the Transition to a 100% Renewable Energy Era*, editor Tanay Sidki Uyar, Springer International Publishing, p. 507-528, June 2020.

[3] J. Borland, "Smart IoT Devices to Control & Balance Home Energy Ecosystem for Grid Backup", Solar Power and Energy Storage International 2020, virtual September 2020.

[4] J. Borland and T. Tanaka, "Solar Plus Multi-Storage Restores Power to Families in Puerto Rico", Renewable Energy World, May 1, 2018.

TABLE 1: Oahu energy cost savings analysis and family energy burden based on 1250kWh/month and $396/month baseline.

Oahu Large Family (1250kWh/month)	HECO Bill (31.7¢/kWh)	Energy Burden	PV-size ($3/W)	Battery Size	Hot Water Heating	Energy Equity New HECO Bill	Energy Savings	Energy Burden	Payback
Singel Family Home (Solar + Storage)									
Affluent Family (>$100K/year)	$396	4.80%	10.5kW	27.5kWh	Electric	$25	93.70%	0.30%	4.6 years
Low-Moderate-Income Family (<$24K/year)	$396	19.80%	10.5kW	27.5kWh	Electric	$25	93.70%	1.30%	4.6 years
Add Heat Pump Hot Water									
Affluent Family (>$100K/year)	$396	4.80%	8.3kW	19.5kWh	Heat Pump	$25	93.70%	0.30%	4.1 years
Low-Moderate-Income Family (<$24K/year)	$396	19.80%	8.3kW	19.5kWh	Heat Pump	$25	93.70%	1.30%	4.1 years
Add Solar Thermal Hot Water									
Affluent Family (>$100K/year)	$396	4.80%	6.3kW	11.5kWh	Solar-Thermal	$25	93.70%	0.30%	4.1 years
Low-Moderate-Income Family (<$24K/year)	$396	19.80%	6.3kW	11.5kWh	Solar-Thermal	$25	93.70%	1.30%	4.1 years
Actual Aiea Case Study Results	$396		7.1kW	14.8kWh	Electric	$136	65.70%		3.8 years
Actual Aiea Case Study Results	$396		7.1kW	14.8kWh	Solar-Thermal	$25	92.50%		3.1 years
Apartment/Condo/Affordable Housing									
Affluent Family (>$100K/year)	$396	4.80%	No PV	27.5kWh	Electric	$162.24	59.00%	1.90%	7.4 years
Low-Moderate-Income Family (<$24K/year)	$396	19.80%	No PV	27.5kWh	Electric	$162.24	59.00%	8.10%	7.4 years
Add Balcony-Solar (+Balcony-PV)									
Affluent Family (>$100K/year)	$396	4.80%	800W	27.5kWh	Electric	$156.00	60.60%	1.87%	3.6 years
Low-Moderate-Income Family (<$24K/year)	$396	19.80%	800W	27.5kWh	Electric	$156.00	60.60%	7.80%	3.6 years
Add Heat Pump Hot Water									
Affluent Family (>$100K/year)	$396	4.80%	800W	27.5kWh	Heat Pump	$138.45	65.00%	1.66%	2.9 years
Low-Moderate-Income Family (<$24K/year)	$396	19.80%	800W	27.5kWh	Heat Pump	$138.45	65.00%	6.92%	2.9 years

TABLE 2: Molokai energy cost savings analysis based on 325kWh/month.

Molokai Family (325kWh/month)	HECO Bill (38¢/kWh)	Energy Burden	PV-size	Battery size	Hot Water Heating	Energy Equity New HECO Bill	Energy Savings	Energy Burden	Payback
Singel Family Home (Solar + Storage)									
Affluent Family (>$100K/year)	$135	1.60%	2.7kW	7.2kWh	Electric	$25	81.50%	0.30%	4.6 years
Low-Middle-Income Family (<$24K/year)	$135	6.80%	2.7kW	7.2kWh	Electric	$25	81.50%	1.30%	4.6 years
Add Solar Thermal Hot Water									
Affluent Family (>$100K/year)	$135	1.60%	1.7kW	7.2kWh	Solar-Thermal	$25	81.50%	0.30%	4.1 years
Low-Middle-Income Family (<$24K/year)	$135	6.80%	1.7kW	7.2kWh	Solar-Thermal	$25	81.50%	1.30%	4.1 years
Apartment/Condo/Affordable Housing									
Affluent Family (>$100K/year)	$135	1.60%	No PV	7.2kWh	Electric	$51.48	61.90%	0.60%	4.0 years
Low-Middle-Income Family (<$24K/year)	$135	6.80%	No PV	7.2kWh	Electric	$51.48	61.90%	2.60%	4.0 years
Add Balcony-Solar (+Balcony-PV)									
Affluent Family (>$100K/year)	$135	1.60%	800W	7.2kWh	Electric	$45.57	66.20%	0.50%	3.6 years
Low-Middle-Income Family (<$24K/year)	$135	6.80%	800W	7.2kWh	Electric	$45.57	66.20%	2.30%	3.6 years
Add Heat Pump Hot Water									
Affluent Family (>$100K/year)	$135	1.60%	800W	7.2kWh	Heat Pump	$30.69	77.30%	0.40%	2.6 years
Low-Middle-Income Family (<$24K/year)	$135	6.80%	800W	7.2kWh	Heat Pump	$30.69	77.30%	1.50%	2.6 years

Performance loss rate estimation for systems affected by potential induced degradation

Panagiotis Goumenos, Andreas Livera, Michalis Florides and George E. Georghiou

Photovoltaic Technology Laboratory, FOSS Research Centre for Sustainable Energy,
Department of Electrical and Computer Engineering, University of Cyprus
Panepistimiou 1 Avenue, P.O. Box 20537, Nicosia, 1678, Cyprus

Abstract—The impact of potential induced degradation (PID) on the performance degradation is an area which has not been fully explored yet. The scope of this work is to investigate the impact of PID on the performance loss rate (PLR) estimation of fielded photovoltaic (PV) systems. To extract the PLR, three common statistical methods were used; linear regression (LR) with ordinary least squares, locally estimated scatterplot smoothing (LOESS) and robust principal component analysis (RPCA). The results demonstrated PLR values higher than -1 %/year for PID affected systems, while values lower than -0.5 %/year were obtained for non-affected PID systems. Additionally, PID affected PV systems exhibited higher PLR in the early years of operation in contrast to the obtained PLR value in the subsequent years.

Keywords—*monitoring, performance loss rate, performance ratio, photovoltaic systems, potential induced degradation*

I. INTRODUCTION

The accurate evaluation of the performance loss rate (PLR) of photovoltaic (PV) systems is crucial for reducing investment risks and further increasing the bankability of the technology [1]. In particular, the PLR estimation of PV systems is influenced by different failures and degradation mechanisms occurring during the lifetime operation of the system. A major degradation mechanism in PV modules is potential induced degradation (PID), which results from the use of increased voltage of up to 1000 V. The factors that can cause PID are related to the potential difference of the module with respect to earth. The high potential difference combined with elevated temperature and humidity leads to the motion of sodium ions from the front glass to the PV cells of the module [2]. The accumulation of sodium in the cell junction leads to electrical shunting and, hence, power loss [3]. Recently, particular attention has been given to PID since it is a degradation factor responsible for significant power production losses (higher than 25% in some cases) [4]–[7].

In this work the impact of PID in PLR estimation of fielded PV systems was examined. The PLR was extracted using common statistical and comparative methods such as linear regression (LR) with ordinary least squares, locally estimated scatterplot smoothing (LOESS) and robust principal component analysis (RPCA). Additionally, the performance of these techniques was assessed to derive the most robust technique for PLR estimation for PV systems affected by PID.

II. METHODOLOGY

A. Experimental setup

The analysis was performed on four test PV arrays (A-D) installed at the University of Cyprus (UCY) in Nicosia, Cyprus. Two of the examined PV systems (arrays C and D) were affected by PID as verified by indoor electroluminescence (EL) tests. Each test PV array comprised of 2 poly-crystalline silicon (poly c-Si) PV modules of 240 W_p, that are connected in series. The PV modules were installed in an open-field mounting arrangement at an inclination angle of 27.5° due South.

Meteorological and electrical measurements for the investigated PV arrays were acquired according to the requirements set by the IEC 61724-1 [8]. The meteorological measurements include the in-plane irradiance (G_I), ambient (T_{amb}) and module (T_{mod}) temperature, while the electrical measurements include the array DC current (I_A), voltage (V_A) and power (P_A). Measurements were acquired since April 2015 and throughout the evaluation period (April 2015 – April 2020), all the PV arrays and sensors were kept clean to minimise the effect of soiling.

B. Data filtering and performance time series construction

The meteorological and electrical data were measured at a resolution of one second and stored as sixty-minute averages. Initially, the recorded data were pre-processed to identify missing data by searching for missing and erroneous values (application of boxplot outlier rule) [9]. The measurements were restricted to daylight hours by filtering out irradiance values lower than 200 W/m².

The G_I and P_A measurements were then used to construct the PV array monthly performance ratio (PR) time series over the evaluation period (from April 2015 to April 2020), as depicted in Fig. 1. The investigated PV systems exhibit a seasonal behaviour, with higher PR in the winter and lower in the summer. Furthermore, PID affected PV systems (C and D), exhibit lower PR values compare to the non-affected PV systems (A and B).

978-1-7281-6118-1/22 $31.00 © 2022 IEEE

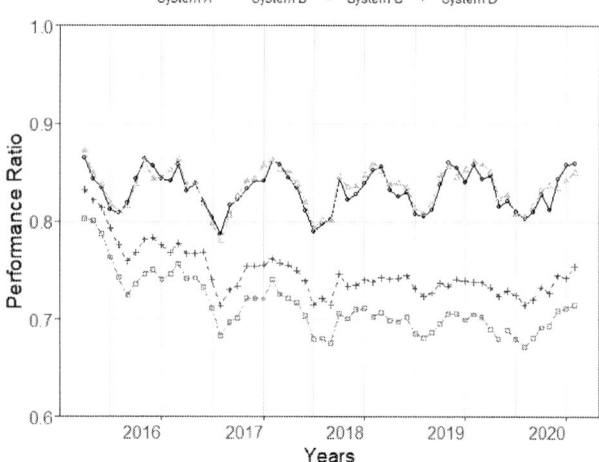

Fig. 1. Monthly performance ratio (PR) time series of the c-Si PV systems under study over the evaluation period from April 2015 to April 2020.

C. Time series analysis

The annual PLR of the PV systems was calculated by applying different statistical and comparative techniques on the monthly constructed PR time series [10]. The goal of the statistical analysis was to extract the trend of the PV performance time series and translate the slope of the trend to the PLR, in units of %/year [10]. Conversely, comparative techniques perform linear comparisons on subsequent datasets (e.g., monthly PR in this case) to extract the annual PLR.

For all the techniques, the following issues were considered:

- Length of available data: Minimum 5-year time series to perform a reliable evaluation [11].
- Erroneous values and outliers: The accuracy of the several methodologies is strongly affected by invalid field measurements. For this reason, special care should be taken in adopting appropriate filtering techniques to eliminate outliers and remove/correct erroneous measurements at the data pre-processing step.
- Missing data: Missing measurements (or outages) can greatly affect the value of the annual PLR. Appropriate dealing with missing data, by applying data deletion and imputation techniques, should be considered before calculating the PLR of PV arrays.
- Construction of the performance time series: The performance metric used to assess the PLR can influence the performed analysis. In this work, the PR (at a monthly resolution) was selected as the performance metric in this investigation because it is a normalized parameter and a key performance indicator (KPI), typically used to characterize PV plant performance for acceptance and operations testing.

D. Performance loss rate estimation methods

The preliminary analysis focused on linear PLR extraction using three common statistical methods. The LR method was employed due to its simplicity, while the LOESS method was selected because it provides robust estimates of the trend and seasonal components that are not distorted by outliers and missing values [12]. The RPCA method was used because it does not produce any uncertainty during the calculation and extracts the outliers from the time series.

LR can be expressed as (1):

$$y = A + Bx \qquad (1)$$

where, x is the explanatory variable and y is the dependent variable. The annual PLR was then calculated by multiplying the slope estimator (B) by twelve (12) due to the monthly aggregated time series [13]. The expanded uncertainty was used for expressing the uncertainty of the measurement results.

LOESS is a nonparametric method that is used to smooth a series of data, for which no assumptions regarding the data's underlying structure are made. In this work, the LR was used to fit a smooth curve through a scatterplot of data of the created trend from the LOESS method.

The RPCA was solved with the use of augmented Lagrange multiplier method as (2):

$$A = D + P \qquad (2)$$

where D is the low rank matrix, which in the context of data analysis designates that the number of characteristic features dominating the original data in A must be smaller than its size, and P is the unknown sparse perturbation matrix, which is the matrix supposed to cause the outliers.

III. RESULTS

The benchmarking results for linear PLR are presented in Fig. 2. By applying the LR method on the PR trend, the linear PLR ranged from -0.013 to -1.91 %/year (system A: -0.081 ± 0.053 %/year, system B: -0.04 ± 0.027 %/year, system C: -1.54 ± 0.37 %/year and system D: -1.28 ± 0.33 %/year). Likewise, the LOESS method yielded PLR values in the range of -0.004 to -1.79 %/year (system A: -0.069 ± 0.065 %/year, system B: -0.045 ± 0.0312 %/year and system C: -1.55 ± 0.24 %/year and system D: -1.30 ± 0.23 %/year). Lastly, the RPCA application resulted in PLR values ranging from -0.20 to -1.46 %/year (system A: -0.34 %/year, system B: -0.20 %/year, system C: -1.46 %/year and system D: -1.24 %/year). PV systems affected by PID (C and D) exhibited higher PLR values than the non-affected PV systems.

The LR method demonstrated the highest uncertainty between the investigated methods; though the obtained median value for the LR method was similar with the one obtained by the LOESS method. The results from the RPCA application were in good agreement to the results obtained by the other two methods (i.e., LR and LOESS) when computing the PLR for PV systems C and D. However, for PV systems A and B, the RPCA method yielded higher PLR values compared to the other two methods.

In general, the three statistical methods demonstrated PLR values higher than -1 %/year for PID affected systems, while values lower than -0.5 %/year were obtained for non-

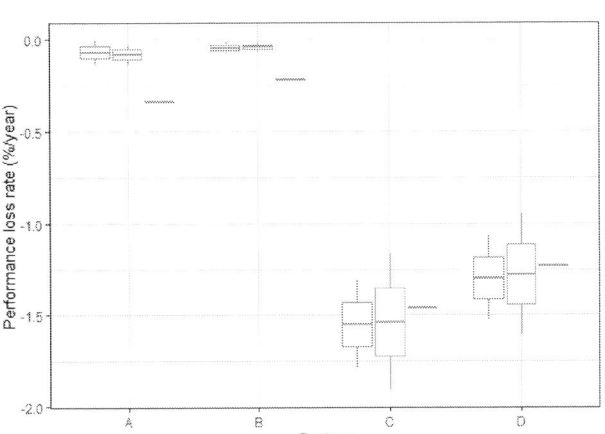

Fig. 2. Comparison of different statistical methods for extracting the linear performance loss rate (PLR) of fielded PV systems. Measurement uncertainty is given by the boxplots. For the RPCA method, no uncertainty was computed and thus it is represented with a straight line.

affected PID systems, that are align with PLR values reported in the literature for c-Si PV systems.

The PLR was then computed for each PV system for every year of operation as shown in Fig. 3. It can be observed that PV systems non-affected by PID (A and B) exhibit PLR values between *0 %/year* to *-0.5 %/year* for every year. For PID affected PV systems (systems C and D), higher PLR values were obtained at the initial years of operation, exhibiting PLR values higher than -3 %/year in year 2, followed by a decreasing trend in the following years.

IV. CONCLUSION

The estimates for the investigated c-Si PV systems clearly demonstrated that PID can greatly affect the value of the annual PLR. Linear PLR, extracted by LR, LOESS and RPCA methods, higher than -1 %/year was obtained for PID affected PV systems, while PV systems without PID showed

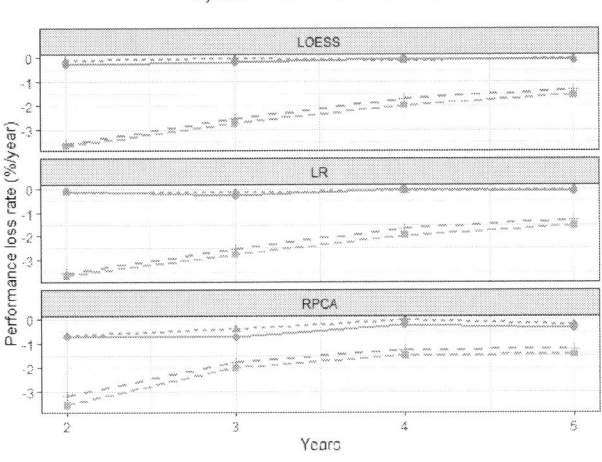

Fig. 3. Performance loss rate (PLR) per year by using the LOESS, LR and RPCA for the investigated c-Si PV systems.

PLR values lower than -0.5 %/year.

The performed analysis also revealed that the performance degradation of PID affected PV modules is higher in the early years of operation; PLR values higher than -3 %/year were obtained in the second year followed by a decreasing trend in the following years.

ACKNOWLEDGMENT

This work was funded through the ROM-PV project, which is supported under the umbrella of SOLAR-ERA.NET cofunded by the General Secretariat for Research and Technology, the Ministry of Economy, Industry and Competitiveness-State Research Agency (MINECO-AEI) and the Research and Innovation Foundation (RIF) of Cyprus.

REFERENCES

[1] D. C. Jordan and S. R. Kurtz, "Photovoltaic degradation rates - An analytical review," *Progress in Photovoltaics: Research and Applications*, Volume 21, Number 1, pp. 12–29, Jan. 2013.

[2] J. Carolus, W. De Ceuninck, and M. Daenen, "Irreversible damage at high levels of potential-induced degradation on photovoltaic modules: A test campaign," *IEEE International Reliability Physics Symposium Proceedings*, pp. 2F5.1-2F5.6, May 2017.

[3] D. Lausch *et al.*, "Potential-induced degradation (PID): Introduction of a novel test approach and explanation of increased depletion region recombination," *IEEE Journal of Photovoltaics*, Volume 4, Number 3, pp. 834–840, 2014.

[4] P. Hacke, M. Kempe, J. Wohlgemuth, J. Li, and Y.-C. Shen, "Potential-Induced Degradation-Delamination Mode in Crystalline Silicon Modules: Preprint."

[5] P. Hacke *et al.*, "Characterization of Multicrystalline Silicon Modules with System Bias Voltage Applied in Damp Heat," 2010.

[6] M. Florides, G. Makrides, and G. E. Georghiou, "Early Detection of Potential Induced Degradation by Measurement of the Forward DC Resistance in Crystalline PV Cells," *IEEE Journal of Photovoltaics*, Volume 9, Number 4, pp. 942–950, Jul. 2019.

[7] M. Florides, G. Makrides, and G. E. Georghiou, "Electrical and Temperature Behavior of the Forward DC Resistance With Potential Induced Degradation of the Shunting Type in Crystalline Silicon Photovoltaic Cells and Modules," *IEEE Journal of Photovoltaics*, pp. 1–10, 2020.

[8] IEC 61724-1:2017, "Photovoltaic system performance - Part 1: Monitoring," 2017.

[9] M. Theristis, A. Livera, C. B. Jones, G. Makrides, G. E. Georghiou, and J. S. Stein, "Nonlinear Photovoltaic Degradation Rates: Modeling and Comparison Against Conventional Methods," *IEEE Journal of Photovoltaics*, Volume 10, Number 4, pp. 1112–1118, 2020.

[10] A. Phinikarides, N. Kindyni, G. Makrides, and G. E. Georghiou, "Review of photovoltaic degradation rate methodologies," *Renewable and Sustainable Energy Reviews*, Volume 40, pp. 143–152, Dec. 2014.

[11] G. Makrides, M. Theristis, J. Bratcher, J. Pratt, and G. E. Georghiou, "Five-year performance and reliability analysis of monocrystalline photovoltaic modules with different backsheet materials," *Solar Energy*, Volume 171, Number June, pp. 491–499, 2018.

[12] R. Cleveland, W. Cleveland, J. McRae, and I. Terpenning, "STL: A seasonal-trend decomposition procedure based on Loess," *Journal of Official Statistics*, Volume 6, Number 1, pp. 3–73, 1990.

[13] A. Phinikarides, N. Philippou, G. Makrides, and G. E. Georghiou, "Performance loss rates of different photovoltaic technologies after eight years of operation under warm climate conditions," *29th European Photovoltaic Solar Energy Conference and Exhibition (EU PVSEC)*, Number June, pp. 2664–2668, 2014.

Assessing and Optimizing Free Space Luminescent Solar Concentrators for Urban Façade Installation

Shweta Pal[1] and Rebecca Saive[1]

[1]University of Twente, Overijssel, Enschede, 7522NB, the Netherlands.

Abstract— **Net-zero energy buildings (NZEBs) for urban settings require novel building-integrated photovoltaic systems, to enable optimal use of available land. Free-space luminescent solar concentrators (FSLSCs) can concentrate incoming irradiance into a smaller cone, such that, when optimally positioned around a bifacial module, can enhance photovoltaic output. In this work, we assess and optimize an FSLSC façade for enhancing the yield of a bifacial module-based fence. Using a reverse ray-tracing based algorithm, we calculated the short-circuit current density, the total power per unit area and the module current mismatch induced by the FSLSC façade. Our calculations show a relative power per unit area enhancement of ~36% and ~111% by an optimized FSLSC façade, as compared to a white-painted diffuse and an ideal black façade, respectively.**

Keywords—Free-space luminescent solar concentrator, Luminescent solar concentrator, Exotic reflectors, Albedo, Yield modeling, Bifacial solar cells, Building integrated photovoltaics, BIPV, Reverse ray-tracing, Mismatch, Net-zero energy buildings.

I. INTRODUCTION

Net-zero energy buildings (NZEB) sustainably generate energy equal to their annual consumption. The booming market demand for NZEBs [1] require innovative building-integrated photovoltaic (BIPV) designs. Novel BIPV configurations are particularly important for urban settings where the energy demand is high, but free land is scarce and instead, building façades constitute the most abundant type of surface. However, irradiance on façades is often not ideal. A lower cost alternative to directly placing solar panels could be to redirect and concentrate light from façades onto standard (bifacial) solar panels placed in the vicinity of the building (see Fig. 1d). The bifacial module can capture light from both faces, *i.e.*, front (from the sun/sky) and rear (from the façade). The incoming irradiance consists of direct (unscattered) and diffuse (scattered) sunlight. Direct irradiance concentration using lenses and mirrors is a well-understood and mature technology [2-5]. But concentration of diffuse light poses an interesting challenge and is particularly important for cloudy countries with high fraction of diffuse irradiance or for façades applications which do not allow for solar tracking. Focusing such diffuse or randomized light requires energy input to decrease the entropy, *i.e.*, bringing order in a disordered system.

A free-space luminescent solar concentrator (FSLSC) [6] provides a clever way to focus direct and diffuse light onto a solar cell. It consists of a nanophotonic-coating and luminophore embedded waveguide surrounded by Lambertian walls (see Fig. 1a.). The luminophores down-convert the high energy photons

(i.e., expends energy) to achieve control. The absorbance and emission peak of a luminophore dye is shown in Fig. 1b. The nanophotonic coating is designed such that it is transparent to the photons within the dye's absorbance peak (λ_{ab}) coming from every direction, and to photons within the dye's emission peak (λ_{em}) only in a small escape cone. The high energy photons (λ_{ab}) enter the structure from every angle, then are red-shifted (λ_{em}) by the dye, finally, leaving through the escape cone. Fig. 1c. shows the spectral and angular reflectance of an ideal optimized nanophotonic coating [7] with a $\pm20°$ wide escape cone for the wavelength range indicated by λ_{es}. Incoming photons lying between 600 nm and 700 nm (λ_{es}) that are incident within the cone are accepted and emitted again in the same cone (shown by blue region). Whereas if they are incident from outside the cone, they are specularly reflected (shown by the yellow region). All other wavelengths are specularly reflected for all the angles (shown by the yellow area). The Lambertian coating reflects all the photons back into the system diffusively, and prevents leakage or trapping of light due to total internal reflection. Strategic orientation of the FSLSC's escape cone towards a bifacial module can improve the photovoltaic yield. For best results, a) the absorbance range (λ_{ab}) and the quantum yield (QY) of the dye must be maximized, b) λ_{es}, λ_{em} and the wavelengths for which the module's spectral response is the highest must overlap. FSLSCs can also help in yield

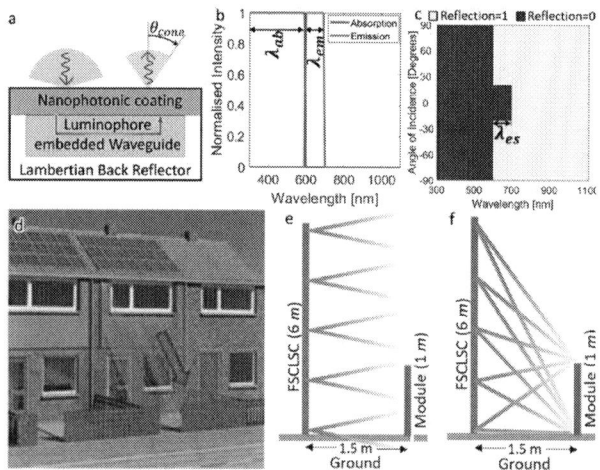

Fig. 1. a) Schematic of a free-space luminescent solar concentrator (FSLSC) with an upright escape cone. b) Spectral absorption and emission profile of an ideal luminophore, c) Spectral and angular reflectance of an ideal nanophotonic coating with a $\pm20°$ escape cone about 0° with respect to the surface normal. d) Façade integrated FSLSC behind a bifacial module fence. e-f) Schematic of a façade-integrated configuration using unoptimized and optimized façade, respectively.

978-1-7281-6118-1/22 $31.00 ©2022 IEEE

improvement while keeping the temperature relatively low, as the incoming photons are of lower energy and cannot contribute to heat due to thermalization losses.

FSLSCs can potentially lead to innovative BIPV designs for urban NZEBs. One interesting design involves using an FSLSC as a façade material and positioning a vertical bifacial module in the front as a fence (see Fig. 1d). But a façade fully covered with an FSLSC having fixed, upright cone (see Fig. 1e) will lead to the following issues: 1) the FSLSC points far away from the module will not contribute to power generation as the light from the escape cone will not reach the module, 2) Some FSLSC points will focus light on a passerby or undesirable places like the street or neighboring buildings, 3) FSLSC points near the ground will focus some light on the ground instead of the module. Such a situation leads to wastage of materials and loss of potential yield enhancement. Thus, the cone size and cone tilt of the FSLSC (or nanophotonic coating) must be optimized such that every point on the façade has its cone oriented towards the module only (see Fig. 1f).

For this report, we assess and optimize the performance of an FSLSC façade in front of bifacial module-based fence using a reverse-ray tracing approach. We found a relative power per unit area enhancement of ~36% and ~111% due to an optimized FSLSC facade, as compared to a white-painted diffuse (Reflectance = 0.85[11]) and an ideal black façade, respectively.

II. OUTPUT ENHANCEMENT BY AN OPTIMISED FSLSC FAÇADE

We computationally analyze the output of a bifacial module fence due to an optimized FSLSC façade (as shown in Fig. 1f). We assume the dye properties shown in 1b and a quantum yield of 100%. The absorbed wavelengths were evenly distributed over the emission range (λ_{em}) displayed in Fig. 1b. The short-circuit current density (J_{sc}), total power per unit area (P_{total}) and current mismatch of the module was calculated using a 2D reverse ray-tracing methodology discussed in [8]. The module height was set to be 1 m, the house height was set to be 6 m (a two-story building), and the separation between them was set to be 1.5 m (see Fig. 3a). The module was divided into 100 pixels,

where module pixel = 1 is the closest to the ground. Similarly, the FSLSC was divided into 600 pixels. AM1.5G [9] was used as the input irradiance and the angle of incidence (AoI) was varied from 0° (vertical relative to the ground) to 90° (horizontal relative to the ground), i.e., solar noon to sunrise or sunset. The incoming irradiance is assumed to consist of parallel rays. For the module, we used a silicon heterojunction bifacial module[10] with bifaciality = 98.8%.

We begin the discussion by evaluating the short-circuit current density (J_{sc}) due to an optimized FSLSC façade. Fig. 2a shows the J_{sc} of the bifacial module fence due to the front illumination (the sun/sky) and the rear illumination (the façade), as a function of the changing angle of the sun. The rear J_{sc} due to an ideal diffuse façade (Reflectance= 1), white-painted façade (Reflectance = 0.85), ideal specular and an unoptimized FSLSC (material properties shown in Fig. 1b and 1c) are also shown for comparison.

At $AoI = 0°$, the irradiance is vertically incident on the ground. Hence, the rays pass tangentially, without interacting with the module and the façade, which leads to zero output. Now, as the sun goes away from the solar noon, the front J_{sc} increases, reaching a maxima at $AoI = 90°$. This is because, at $AoI = 90°$, the sun rays are normally incident on the front face of the cell, thus leading to a maximum output. For a specular reflector, the output becomes non-zero for $AoI \geq 15°$, because specularly reflected rays reach the module only for $AoI > 14.3°$ (see Fig 1c- dashed line). For an unoptimized FSLSC façade, the down-converted photons from 20° cone reach the module which leads to non-zero output. This is only contributed by façade pixels below 1.55 m. An optimized FSLSC façade also contributes down-converted photons but from every pixel due to cone size and tilt optimization. For $AoI > 14.3°$, specular reflection due to unoptimized and optimized FSLSC façade start to contribute as well, thus leading to an increase in output. The rear J_{sc} due to a specular, unoptimized, and optimized FSLSC façade reach a maximum at $AoI = 70°$, since beyond that, shading due to module sets in (see Fig. 2c- solid line). The shading increases with the increasing AoI, which leads to a decline in output. For diffuse reflectors, ideal and white-paint, the output increases as the AoI increases reaching a maximum at $AoI = 55°$. This because shading sets in at $AoI \geq 56.3°$ (see Fig. 2d).

To completely evaluate the feasibility of a configuration, one must calculate the total power output. For this, the front and rear J_{sc} are added to obtain total J_{sc}, which is used to calculate the total power per unit area of the system. The resulting power per unit area for various façade-integrated configurations are plotted in Fig. 2b, and the relative enhancement, i.e., the ratio of area under the curve, was calculated. We observe a relative yield improvement of ~36% due to an optimized FSLSC facade as opposed to a white-painted diffuse façade. On comparison with an ideal black façade (or only front illumination), an optimized FSLSC results in a relative enhancement of ~111%.

III. MISMATCH DUE TO AN OPTIMIZED FSLSC FAÇADE

To rigorously discuss the merit of a concentrator, one must also calculate the current mismatch introduced on the module by it. Current mismatch is the spatially varying current generation

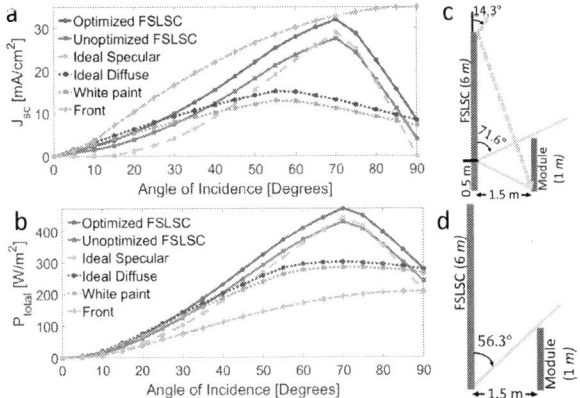

Fig. 2. a) The short-circuit current density (J_{sc}) and b) the total power per unit (P_{total}) of a bifacial module fence due to an optimised FSLSC façade as a function of changing angle of incidence (AoI). c) Schematic showing the minimum AoI for which the specular reflection reaches the module (dashed line) and the minimum AoI which causes shading for a specular façade (solid line). d) Schematic showing the minimum AoI which which causes shading for a diffuse façade.

in the module due to spatially varying illumination incident on it. A module, generally, consists of many solar cells connected in series. Therefore, the cells must obey the current matching condition, *i.e.*, every cell must produce the same current. Thus, all the cells drop their current values to match the minimum value. The excess is lost as heat and if this heat is excessive, it can damage the cells. As an FSLSC has a spectrum and angle-dependent reflectance, it can redirect light inhomogeneously onto the module, causing mismatch. The spatially varying short-circuit current density generation along the length of the module due to an optimized FSLSC façade is shown in Fig. 3d.

Before we begin the discussion, it is to be noted that there are two cosine terms associated with every reflected photon (see Fig. 3a and 3b). Namely, 1) $\cos(\theta_R)$ where θ_R is the angle of reflection relative to the façade's normal, 2) $\cos(\theta_M)$ where θ_M is the angle at which the module receives the photon relative to its own surface normal. Here, since the façade and the module are parallel to each other, θ_R and θ_M are equal. Also, along the surface normal, the module has the highest external quantum efficiency (EQE), which is the efficiency of absorbed photon to electron conversion at a given wavelength. The EQE eventually reaches zero as the incoming photon angle approaches the surface tangent.

In Fig. 3d, for $AoI = 0°$, the light passes tangentially and does not get redirected to the module. Hence, every pixel on the module generates zero output, causing no mismatch. For $AoI > 0°$, the incoming irradiance starts getting redirected to the module by the façade. Module pixel=100 always generates the highest output at a given AoI. This is because: 1) the illuminated façade pixels emit and reflect at angles for which the $\cos(\theta_R)$ is higher (see Fig. 3a), 2) the module receives the redirected irradiance at a better angle, i.e., higher $\cos(\theta_M)$ (see Fig. 3a), 3) The EQE is better for those θ_M values. The contribution of these three factors reduces as one moves down along the module (see Fig. 3b), thus leading to a spatially varying short-circuit current density. For $AoI > 70°$, shading of the façade by the module sets in. Because of this, there are no specularly reflected rays reaching near the bottom of the module, thus causing 'self-shading'. The self-shading increases as the AoI increases. For $AoI = 90°$, the mismatch is quite low (range~$0.09 \, mA/cm^2$). This is because the specular reflection contribution is zero as either the rays are reflected back in the same direction or there is shading (see Fig. 3c). The mismatch is only due to down-converted photons.

Quantifying the increase in power output and the mismatch can help in assessing the performance and economic feasibility of an FSLSC façade, and further optimize the configuration for desired output. Optimization algorithms can be implemented to improve the final total output or the mismatch introduced by the FSLSC, depending upon the application.

IV. CONCLUSION

In this report, we computationally assessed and optimized the performance of an FSLSC as a façade for a bifacial module-based fence. We showed that, when optimized, such a façade can lead to a relative output enhancement of ~36% as compared to a white-painted façade (Reflectance=0.85) and ~111% as compared to an ideal black façade. Optimizing the cone size and cone tilt can boost the implementation of FSLSCs as façade materials. Knowledge of the cone size and tilt can also serve as motivation to innovate better nanophotonic coatings, or FSLSC configurations (flexible or tilted FSLSCs). Another improvement can involve maximizing the range of λ_{ab} and quantum yield of the dye, i.e., down-converting more photons for higher output. This can be achieved by using a better dye or using multiple dyes. Our approach enables rigorous assessment and optimizations of FSLSC façade for urban NZEBs, thus bridging the gap between the lab-level devices and their real world applications.

REFERENCES

[1] Global Industry Analysts, "Global Net-Zero Energy Buildings (NZEBs) Industry Market Report," Global Industry Analysts, July 2021.

[2] M. Abdelhamid, B. K. Widyolar, L. Jiang, R. Winston, E. Yablonovitch, G. Scranton, D. Cygan, H. Abbasi and A. Kozlov, "Novel double-stage high-concentrated solar hybrid photovoltaic/thermal (PV/T) collector with nonimaging optics and GaAs solar cells reflector," *Applied Energy*, vol. 182, pp. 68-79, November 2016.

[3] K.K. Chong, S.L. Lau, T.K. Yew and P. C.L. Tan, "Design and development in optics of concentrator photovoltaic system," *Renewable and Sustainable Energy Reviews*, vol. 19, pp. 598-612, March 2013.

[4] Chung-Yu Tsai, "Improved irradiance distribution on high concentration solar cell using free-form concentrator," *Solar Energy*, vol. 115, pp. 694-707, May 2015.

[5] Z. Zhuang and F. Yu, "Optimization design of hybrid Fresnel-based concentrator for generating uniformity irradiance with the broad solar spectrum," *Optics & Laser Technology*, vol. 60, pp. 27-33, August 2014.

[6] L. Einhaus and R. Saive, "Free-Space Concentration of Diffused Light for Photovoltaics," in *47th IEEE PVSC*, 2020.

[7] G. Heres, L. Einhaus and R. Saive, "Analytical Model for the Performance of a Free-Space Luminescent Solar Concentrator," in *IEEE Photovoltaic Specialists Conference (PVSC)*, online, 2021.

[8] S. Pal and R. Saive, "Investigating Bifacial Photovoltaic Output Under Diffuse, Specular and Glossy Albedo," in *OSA PVLED - Optical Devices and Materials for Solar Energy and Solid-state Lighting*, Online, July 2021.

[9] "E891-87 American Society for Testing and Materials, Annual Book of ASTM Standards," Standard Tables for Terrestrial Direct Normal Solar Spectral Irradiance for Air Mass 1.5.

[10] S. Pal, A. Reinders and R. Saive, "Simulation of Bifacial and Monofacial Silicon Solar Cell Short-Circuit Current Density Under Measured Spectro-Angular Solar Irradiance," *IEEE Journal of Photovoltaics*, vol. 10, no. 6, pp. 1803-1815, 2020.

[11] S. Sharples and S. Mahambrey, "Reflectance distributions and atrium daylight levels: a model study," *International Journal of Lighting Research and Technology*, vol. 31, no. 4, pp. 165-170, 1999.

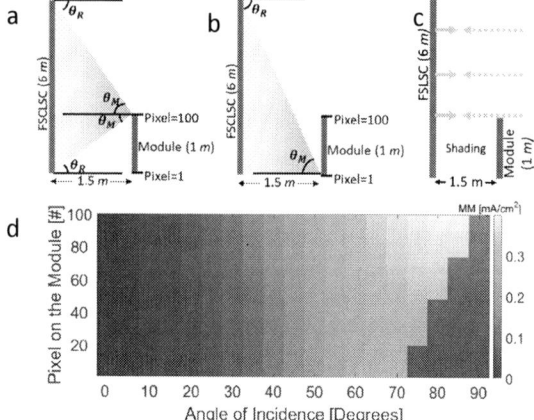

Fig. 3. a-c) Schematic showing the configuration and the angles. d) The spatially varying short-circuit current denisty along the length of a bifacial module fence, as a function of changing angle of incidence, due to an optimised FSLSC façade.

Curvilinear Prismatic Window Which Eliminates Glare and Reduces Front-Surface Reflections for PV Modules and Other Surfaces

Mark O'Neill[1] and Chris Youtsey[2]

[1]Mark O'Neill, LLC, Fort Worth, TX, 76244, USA and [2]MicroLink Devices, Inc., Niles, IL 60714, USA

Abstract—A novel transparent window for PV modules and other surfaces has recently been developed to both eliminate glare and minimize front-surface reflections. The new window technology can be implemented with a thin, ultra-light curvilinear prismatic film which can be attached directly to PV cells or to another planar window layer such as glass over the PV cells. The new curvilinear prismatic window technology can be applied to both space and ground PV modules using appropriate materials and coatings. Analysis and testing have demonstrated the performance gains offered by the new technology as well as the elimination of glare.

Keywords—PV module window, anti-reflection, anti-glare, performance enhancement

I. INTRODUCTION AND BACKGROUND

Conventional PV module windows typically comprise planar glass or polymer layers providing protection for the underlying PV cells from the environment. In space, the environment can include charge particles, solar ultraviolet radiation, atomic oxygen, etc. On the ground, the environment can include rain, snow, sleet, and hail, etc. Front-surface reflection losses from such conventional windows reduce the current and power output of the PV cells, especially for high solar ray incidence angles. The flat surfaces of conventional windows can also lead to glare due to the specular component of the reflected sunlight. Glare problems have led to cancellations of some terrestrial PV systems near airports, highways, and occupied buildings. Glare problems from constellations of spacecraft in low earth orbit (LEO) have also caused problems for ground-based telescopes. A new curvilinear prismatic window has recently been developed which overcomes the glare problem and minimizes the front-surface reflection power losses for PV modules both in space and on the ground [1].

The optical and thermal benefits of texturing window layers for PV modules have been recognized by many previous researchers [2 and 3]. Various textures have been analyzed and tested for transmittance improvements due to reduced front-surface reflections. Various textures have also been analyzed and tested for enhanced front-surface waste heat rejection due to the greater front-surface area of textured windows compared to flat windows. But these previous textured surface geometries have not incorporated curved surfaces which eliminate glare.

II. DESCRIPTION OF THE CURVILINEAR PRISMATIC WINDOW

The curvilinear prismatic window is a variation of the linear prismatic window achieved by changing the prismatic path from a straight line to a curved line. Fig. 1 shows the basic configuration with greatly exaggerated prism size. The curvilinear prisms follow a curvilinear path with the orientation such that a liquid can run from top to bottom without encountering a barrier. For terrestrial applications, this liquid could be rain or cleaning water. For space applications, this liquid could be a cleaning fluid such as isopropyl alcohol used before launch. Other textured surface geometries such as inverted pyramids can trap liquids and dirt in regions where barriers to flow exist.

Fig. 1 also shows the triangular cross-sectional geometry of each prism. The sun direction relative to the prismatic pattern is defined by the azimuth (az) and elevation (el) angles shown in Fig. 1. Compared to a linear prismatic pattern, the new curvilinear prismatic pattern offers major performance improvements for very small values of both az and el angles.

The curvilinear prismatic window can be mass-produced by thermally embossing a thermoplastic polymer film or by casting a different polymer against an embossed thermoplastic polymer film. For terrestrial applications, two preferred materials for the embossed polymer film are acrylic or aliphatic thermoplastic polyurethane (TPU). For space applications, the curvilinear prisms can be formed in space-grade materials including glass, silicones, or fluoropolymers. The preferred manufacturing

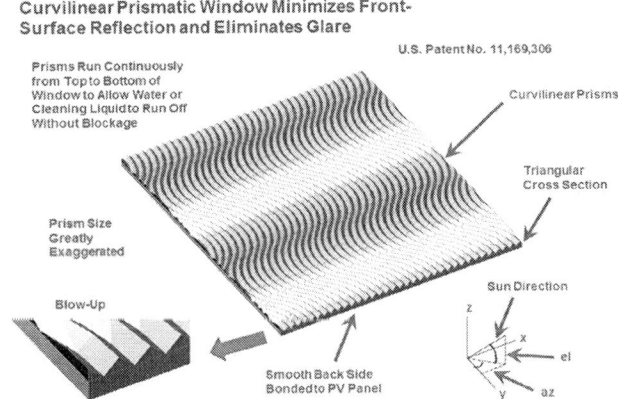

Fig. 1.   Curvilinear Prismatic Window.

approach for a space PV module is still under evaluation. The preferred manufacturing approach for a terrestrial PV module is to laminate an acrylic pressure sensitive adhesive (PSA) to the embossed acrylic curvilinear prismatic film to enable easy attachment to the PV module on top of the existing flat glass window. For terrestrial applications, the curvilinear prismatic film could be attached in the factory or in the field for already deployed arrays.

III. How the New Curvilinear Prismatic Film Works

The curvilinear prismatic film works in two ways:

- The triangular prisms minimize front-surface reflections for both large and small sun elevation angles for all sun azimuth angles

- The curvilinear prismatic path eliminates glare by spreading the reflected light in all directions due to the curvature

Fig. 2 shows how the prismatic film works for large sun elevation angles. The prisms shown have 45° faces which work very well to minimize reflection losses for solar rays arriving at large lateral incidence angles. This minimization occurs because the tilted faces of each prism intercept the incoming rays at a more normal incidence angle than a flat window. The rays which enter each prism make their way to the PV cell below the window either directly or after total internal reflection (TIR) from the opposing prism face.

High Lateral Incidence Angle Light Is Efficiently Captured by 45° Prisms

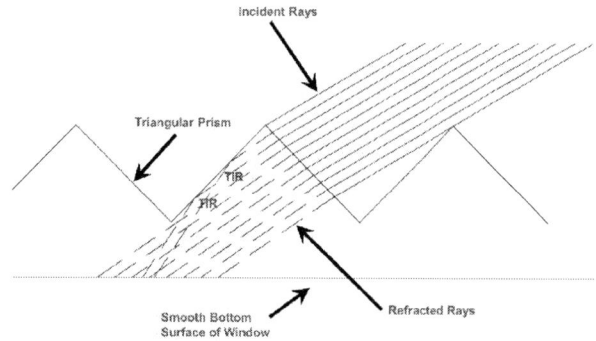

Fig. 2.   Performance Gain for High Lateral Incidence Angle.

Fig. 3 shows how the prismatic film works for small sun elevation angles. The prisms shown have 45° faces which work very well to minimize reflection losses for solar rays arriving at near normal incidence angles. This minimization occurs because the rays which are reflected by the outer surface of the tilted faces of each prism intercept the neighboring prisms which recover most of the reflected light and deliver it to the PV cell.

The curvilinear path of the prisms eliminates glare by spreading the reflected light into a wide range of departing angles, as shown in Fig. 4. Since there are no flat surfaces on the exposed face of the curvilinear prismatic window, there can be no glare from rays reflected by the exposed face of the window. Furthermore, rays which are reflected from the PV

Small Lateral Incidence Angle Light Is Efficiently Captured by 45° Prisms Including First Surface Reflections

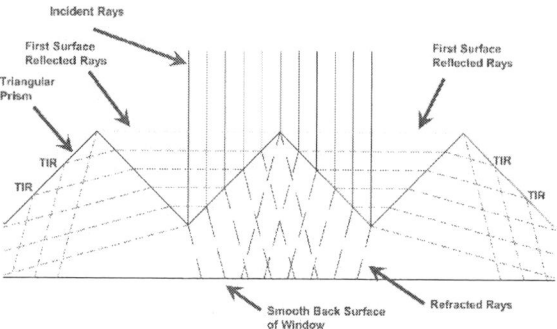

Fig. 3.   Performance Gain for Smal Lateral Incidence Angle.

Glare Elimination Is Accomplished by Curved Surfaces Which Spread Light

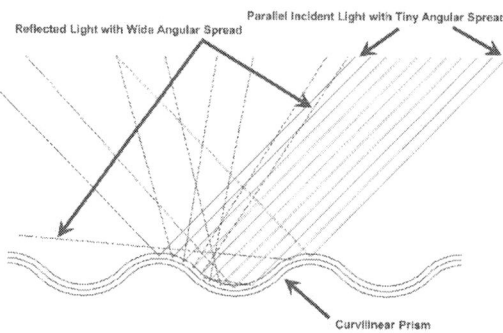

Fig. 4. Glare Elimination with Curvilinear Prisms.

cells below the window are refracted by the curved surfaces of the new prismatic window and spread into a wide range of departing angles, eliminating glare from these reflections too.

The new curvilinear prismatic window also eliminates the "weak spot" for first-surface reflections from a linear prismatic window. The "weak spot" occurs at sun azimuth angles near zero, i.e., with the rays approaching parallel to the prisms. When the sun azimuth angle is also near zero, the linear prisms allow a path for the rays to escape by proceeding along the prisms and between the prisms from end to end of the window. The curvilinear prisms eliminate this "weak spot" and provide much better optical performance for small combined sun azimuth and elevation angles.

IV. Performance

The new curvilinear prismatic window provides excellent performance over the full range of possible sun azimuth and elevation angles of incidence. Fig. 5 shows results of a recently refined parametric optical analysis of the preferred geometry of the new window made of silicone (1.4 refractive index). Note that the net transmittance into the window is much higher than for a flat ceria-doped microsheet glass (CMG) window over the full range of sun azimuth and elevation angles. The performance gain is greatest for small sun elevation angles, but still appreciable for high sun elevation angles including normal incidence.

Parametric Results for Curvilinear Prismatic Window Using Cosine Path with ±45° Max Slope

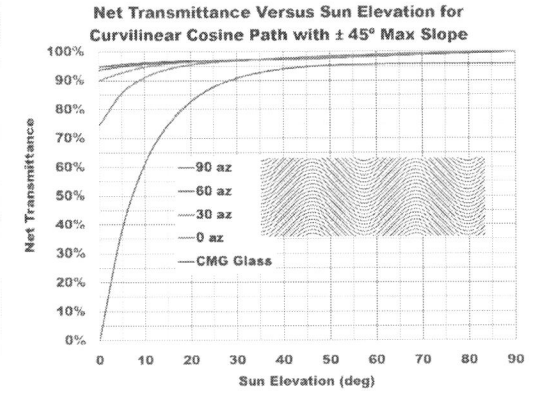

Fig. 5.   Net Transmittance into Curvilinear Prismatic Window.

For comparison, Fig. 6 shows comparable results for a linear prismatic window. The "weak spot" is pointed out by the arrow. Indeed, at zero sun azimuth and zero sun elevation angle, the linear prismatic window has 0% net transmittance, like the flat CMG window. The curvilinear prismatic window eliminates this weak spot as shown in Fig. 5.

Parametric Results for Linear Prismatic Window Using Linear Path with ±0° Max Slope

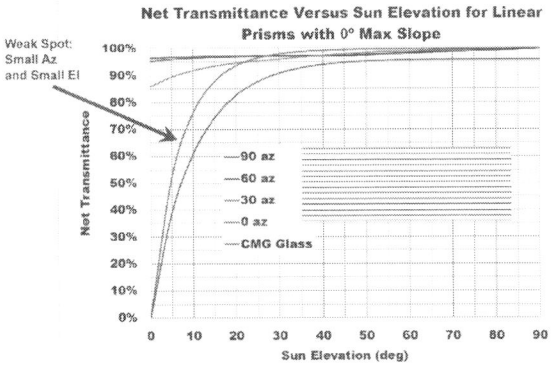

Fig. 6.   Net Transmittance into Linear Prismatic Window.

Before we developed the tooling to make the curvilinear prismatic window pattern, we instead ran experiments with linear prismatic windows over triple-junction PV cells to validate the predicted results shown in Fig. 6. The measured results have shown outstanding agreement with predictions. The same ray trace models are used for both linear and curvilinear prismatic windows, so we are very confident in the results predicted in Fig. 5.

We have recently received prototype acrylic molding tools to make the first prototype silicone curvilinear prismatic windows. We have performed free-standing prismatic film transmittance measurements using the simple test approach shown in Fig. 7. For comparison, we also measured a commercially available linear prismatic film from 3M known as OLF-2405 and plano-plano film made from the same silicone. All three films perform close to their theoretical maximum transmittance values (within ± 1% experimental error margins).

Free-Standing Prismatic Film Transmittance Test (Theoretical Max Value 95.6%)

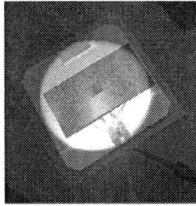

- LED Lamp Illuminating 1 cm Square Aperture over 2 cm Square Multi-Junction Cell

- Near-ISC Measurements Made with and without Sample Over Aperture (Total and Diffuse, Lamp On and Off)

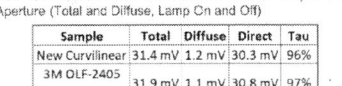
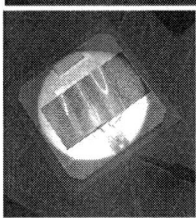

Sample	Total	Diffuse	Direct	Tau
New Curvilinear	31.4 mV	1.2 mV	30.3 mV	96%
3M OLF-2405 Linear Prisms	31.9 mV	1.1 mV	30.8 mV	97%
Plano-Plano	31.5 mV	1.5 mV	30.0 mV	95%
Bare	33.2 mV	1.6 mV	31.6 mV	100%

Fig. 7.   Free-Standing Prismatic Film Test and Initial Results.

The free-standing film has a significantly lower transmittance than the same film bonded to a solar cell or other surface because of losses at the smooth bottom surface for the former, as shown in Fig. 8.

Silicone Prismatic Window Theoretical Transmittance at Normal Incidence

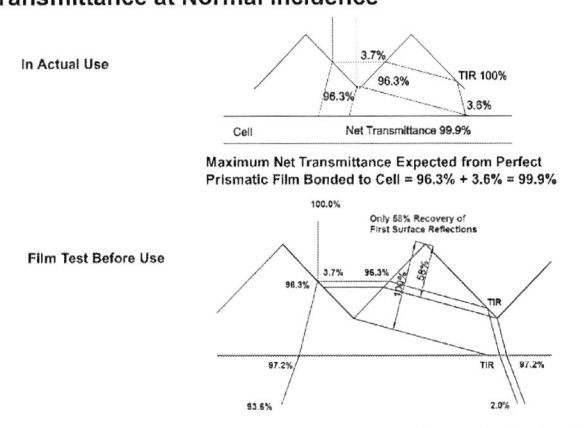

Fig. 8.   Transmittance into Bonded and Free-Standing Prismatic Film.

We have also just assembled and evaluated the first prototype of a multi-junction cell employing a curvilinear prismatic window, with results shown in Fig. 9. For comparison Fig. 9 also includes results for a cell employing a flat FEP window. Since the prismatic window was thicker than the FEP window, we painted the edges of the prismatic window black to minimize any stray light entry at high incidence angles. The azimuth angle of the incident light was approximately 0° as defined in Fig. 1. Since the two cells were not perfectly current-matched in their bare state and the collimated halogen lamp did not provide a full AM0 spectrum, we have normalized the results shown in Fig. 9 to show the ratio of short-circuit current to irradiance for various incidence angles relative to a normal incidence angle. The measured results for both cells are in relatively good agreement with our predicted results from ray trace models. Note the substantial performance gain provided by the curvilinear prismatic window compared to the flat FEP window at large incidence angles. FEP has a much lower refractive index than ceria-doped microsheet glass (CMG), resulting in significantly lower reflection losses and higher transmittance than previously shown in Fig. 5.

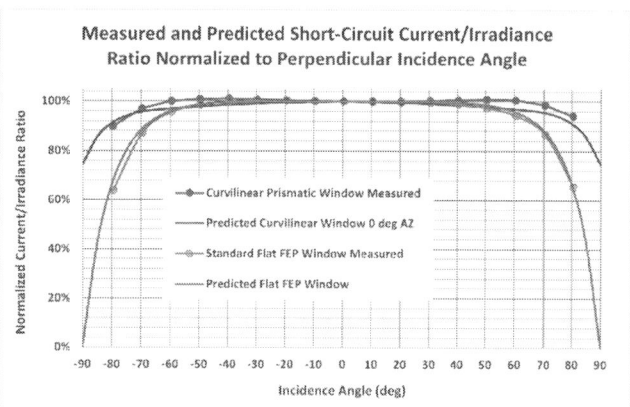

Fig. 9. Measured and Predicted Results for First Complete Prototype.

V. DEMONSTRATION OF GLARE ELIMINATION

We have used a piece of the first silicone prototype prismatic parts to demonstrate the glare elimination attribute of the curvilinear prismatic window. As shown in Fig. 10, we bonded this prismatic film with transparent adhesive to a piece of glass with the backside of the glass painted flat black to simulate a PV cell or module. A circular LED lamp illuminated the model at about 60° incidence, and we rotated the sample to simulate different sun azimuth angles. The round white lamp reflection glare from the top surface of the glass is readily seen, but the quasi-curvilinear prismatic window sample eliminated the glare as expected for all simulated sun azimuth angles.

We also demonstrated the anti-glare properties of the new curvilinear prismatic window on non-solar-panel surfaces by bonding a piece to an optical mirror, as shown in Fig. 11. The reflected light still emerges from the curvilinear prismatic window, but it has changed from specular to diffuse reflection,

Anti-Glare Demonstration

- 10 cm Square Glass Plate Painted Flat Black on Bottom Side to Simulate Solar Panel

- LED Lamp Light Shows Reflected Glare Off Top Surface of Glass

- 5 cm Square Silicone Curvilinear Prismatic Window Bonded to Top Surface of Glass Eliminates Glare for All Incidence Angles

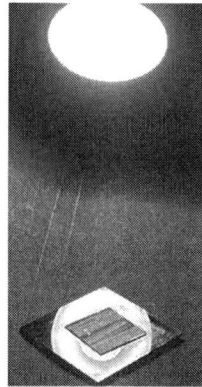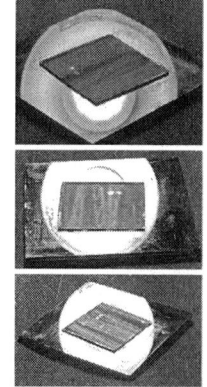

Fig. 10. Demonstration of Glare Elimination.

Anti-Glare Demonstration on Mirror

- 10 cm x 15 cm Mirror Simulates Reflective Spacecraft Surface

- LED Lamp Light Shows Reflected Glare Off Mirror

- 5 cm Square Silicone Curvilinear Prismatic Window Bonded to Top Surface of Mirror Eliminates Glare for All Incidence Angles

- Curvilinear Prismatic Window Still Allows Diffuse Reflection from Mirror, Thus Not Changing Solar Absorptance but Eliminating Glare

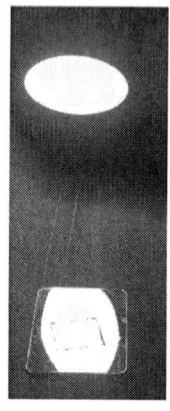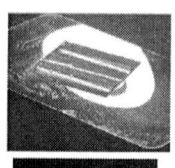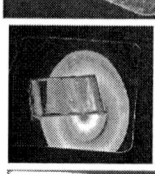

Fig. 11. Anti-Glare Demonstration on Mirror.

thereby maintaining the same optical performance but eliminating glare. This shows that a reflective spacecraft surface could still maintain a high solar reflectance (and corresponding lower surface temperature) on orbit while eliminating glare with the new curvilinear prismatic window.

In addition to optical performance benefits, the new curvilinear prismatic window provides some unexpected benefits in thermal performance. For space applications, the prismatic surface provides an increase in effective emittance for the window compared to a conventional flat window surface. It is easily shown using radiosity analysis that the effective emittance is defined by the following equation:

$$\epsilon_{effective} = \frac{1}{\left(1 + \frac{A_{aperture}}{\varepsilon \, A_{surface}} - \frac{A_{aperture}}{A_{surface}}\right)}$$

The flat aperture area of the window ($A_{aperture}$) is smaller than the prismatic surface area ($A_{surface}$) of the window which causes the effective emittance ($\varepsilon_{effective}$) to be larger than the basic material emittance (ε) of the window. For example, if the basic window material emittance is 80%, the effective emittance for the curvilinear prismatic window will be 85%. This small increase in effective emittance equates to about 3°C reduction in cell operating temperature at the hottest point (subsolar) of a LEO solar array at 400 km orbital altitude, resulting in about 0.6% higher power output for a 30% multi-junction cell.

For ground-based applications, the greater surface area of the curvilinear prismatic window offers both better convective heat transfer and radiative heat transfer, leading to slightly lower cell operating temperatures. Others have previously demonstrated this thermal performance advantage for ground-based linear prismatic windows on photovoltaic panels [2].

One major advantage of the new curvilinear prismatic window for space applications relates to solar photovoltaic array specific power. The optimal thickness of the window to provide radiation protection to the underlying photovoltaic cell depends on the mission-specific orbit and lifetime. A parametric analysis of the effect of window thickness on solar cell equivalent 1 MeV electron fluence for four different missions is summarized in Fig. 12. These results were obtained using the online tool known

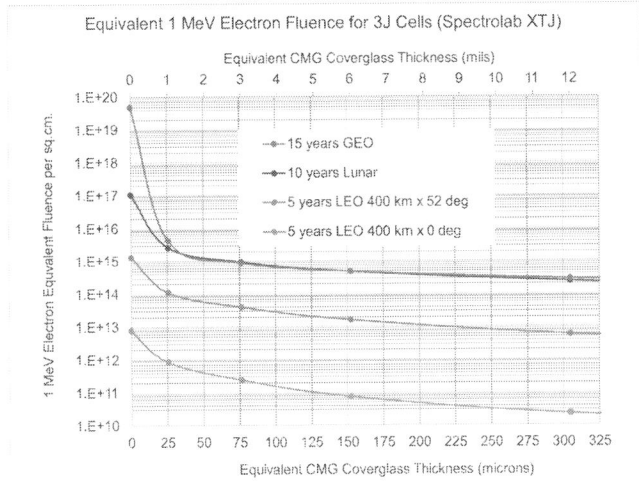

Fig. 12. Equivalent 1 MeV Electron Dose Versus Window Thickness.

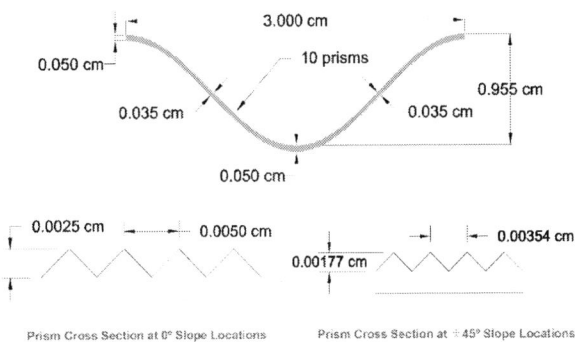

Fig. 13. Prismatic Geometry Selected for Prototypes.

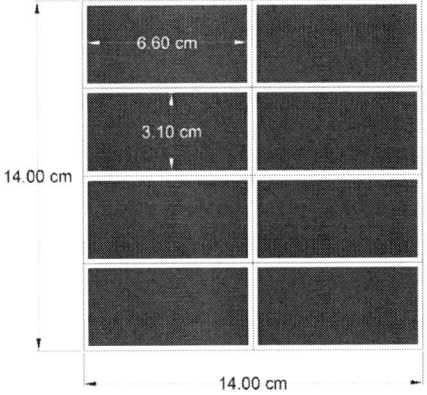

Fig. 14. Possible Prototype Module Size.

as SPENVIS (Space Environment Information System) managed by the European Space Agency (ESA) at https://www.spenvis.oma.be/regulation.php. Note the rapid falloff in dose with equivalent ceria-doped microsheet glass (CMG) window thickness for all the missions considered.

For the curvilinear prismatic window, the thickness can be selected at any value since the window is cast or molded from polymers, such as space-qualified silicone material. For conventional CMG windows, only specific values of the window thickness can be selected, and these are limited to values that provide reasonable tolerance for handling without breakage, typically above 75 microns. In addition, CMG windows require a minimum thickness of silicone adhesive, typically about 50 microns, to bond the window to the cell. Since silicone has a density of only 40% of CMG, its effective thickness is only 40% of CMG. So, a 75-micron CMG window with 50 microns of silicone adhesive has an equivalent thickness of about 95 microns. For many missions, especially LEO missions, a thinner window would offer substantial benefits in terms of mass and specific power. The new curvilinear prismatic window offers these benefits.

VI. PRESENT DEVELOPMENT STATUS

A prototype design has been selected with prisms shown in Fig. 13. Prototype tooling is under development to produce prototype silicone windows about 15 cm x 15 cm in area. This area will enable multi-cell module encapsulation with a single curvilinear prismatic window as shown in Fig. 14.

We have just received the first plastic prototype tooling and we expect additional prototype cells and modules to be equipped with curvilinear prismatic windows and evaluated in the next few months. We have made and tested the first prototype single-cell articles with performance improvements matching expectations as presented above.

VII. GLARE MITIGATION APPLICATIONS FOR THE NEW WINDOW

Glare is a big problem for both ground and space applications, as shown for example in Fig. 15. Ground-based solar arrays near buildings, roadways, and airports can cause glare which is unpleasant and sometimes hazardous.

Terrestrial Solar Collectors Cause Glare Problems (Especially Near Roadways and Airports)

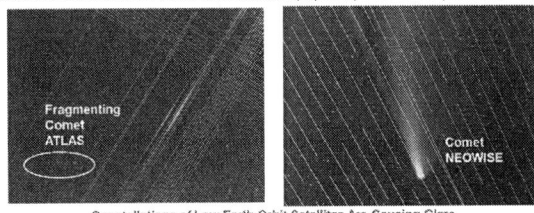

Constellations of Low Earth Orbit Satellites Are Causing Glare Problems for Astronomy (Starlink Streaks Shown Above)

Fig. 15. Glare Problem Examples.

Constellations of many thousands of LEO spacecraft can disrupt ground-based astronomy due to glare especially near sunrise and sunset. Such spacecraft glare is not always due to solar arrays but instead to other specularly reflecting surfaces. Fortunately, the new curvilinear prismatic window can be applied not only to solar arrays but also to other reflecting surfaces, as shown in Fig. 16. The preferred materials for the

This Same Design Could Be Used for Space and Ground Applications

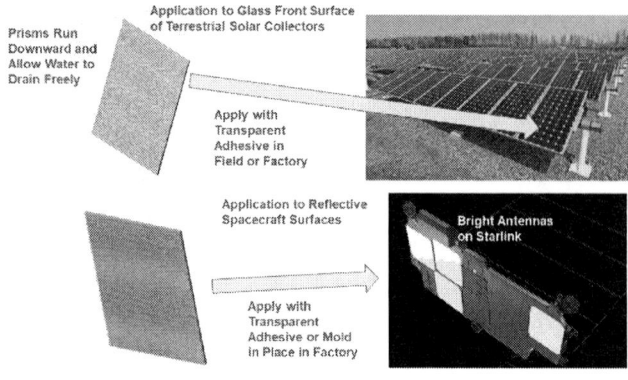

Fig. 16. Curvilinear Prismatic Windows for Glare Elimination.

curvilinear prismatic window will be different for space and ground applications, but the functionality will be the same.

VIII. SUMMARY AND SIGNIFICANCE

The new curvilinear prismatic window offers power output advantages over conventional PV module windows for both space and ground applications. It also eliminates glare, which is a significant and growing problems both on orbit [4] and on the ground [5]. The new window also provides advantages in heat rejection, lowering cell operating temperature a few degrees compared to flat windows both in space and on the ground. The new window also offers advantages in optimizing radiation shielding thickness to maximize specific power. Very thin silicone windows can be made using processes previously developed for ultra-thin space Fresnel lenses [6]. The new window also offers a number of other advantages which will be presented in future publications on the new technology.

Mass-production processes have been identified and development efforts are ongoing to advance the technology readiness level of the new window. One key advantage of the 45-degree prism faces is that the prismatic pattern is the same for both positive (molding tool) and negative (molded copy) versions of the pattern. Therefore, copies of masters can be used as second-generation masters, simplifying tooling replication.

IX. ACKNOWLEDGEMENT

Development of the new curvilinear prismatic window is being supported by NASA under a Phase II SBIR contract with MicroLink Devices.

X. REFERENCES

[1] M. O'Neill, "Curvilinear prismatic film which eliminates glare and reduces front-surface reflections for solar panels and other surfaces," U.S. Patent No. 11,169,306, November 9, 2021.

[2] M. Duell, M. Ebert, M. Muller, B. Li, M. Koch, T. Christian, R.F. Perdichizzi, B. Marion, S. Kurtz, D.M.J. Doble, "Impact of structured glass on light transmission, temperature and power of PV modules," 25th European Photovoltaic Solar Energy Conference and Exhibition / 5th World Conference on Photovoltaic Energy Conversion, 6-10 September 2010, Valencia, Spain.

[3] Z. Zhou, Y. Jiang, N. Ekins-Daukes, M. Keevers and M. A. Green, "Optical and thermal emission benefits of differently textured glass for photovoltaic modules," in IEEE Journal of Photovoltaics, vol. 11, no. 1, pp. 131-137, Jan. 2021.

[4] K. Whitt, "Satellites versus stars: which will dominate the sky?," https://earthsky.org/space/satellites-versus-stars-night-sky-kessler-syndrome/, September 28, 2021.

[5] K. Ulenhuth, "Glare study prompts utility Evergy to cancel solar array at Kansas City airport," https://energynews.us/2020/12/10/glare-study-prompts-utility-evergy-to-cancel-solar-array-at-kansas-city-airport/, December 10, 2020.

[6] M. O'Neill et al., "Space PV concentrators for outer planet and near-sun missions, using ultra-light Fresnel lenses made with vanishing tools," 49th IEEE Photovoltaic Specialists Conference, June 19, 2019, Chicago.

Monochromatic Light Trapping in Photonic Power Converters

Nicholas P. Irvin, Neda Nouri, Chaomin Zhang, Christopher E. Valdivia, Karin Hinzer, Richard R. King, Christiana B. Honsberg

Arizona State University, TEMPE, AZ, United States

University of Ottawa, Ottawa, ON, Canada

Photonic power converters (PPC) are photovoltaic devices that convert monochromatic light into electricity. Recent records with III-V PPC have achieved 68.9% but demand high carrier lifetimes. This report shows that the introduction of light trapping can yield substantially higher efficiencies, as all the incident light is weakly absorbed and thus amenable to absorption enhancement. For PPCs with low carrier lifetimes (i.e., 1-100 ns for moderately-doped GaAs), light trapping yields significant gains by enabling thinning of the material and reduction of recombination. In this report, multiple light trapping designs are compared, including nanostructures, rear diffuse reflectors, and angular selective filters. While some light-trapping designs achieve higher ideal PPC performance, other designs are seen to be more tolerant to dimensional variation in fabricated structures and incident angles.

978-1-7281-6118-1/22 $31.00 © 2022 IEEE

Implications of Agriculturally Co-Located Solar PV Installations on the FEW Nexus in the Central Valley

Jacob T Stid, Siddharth Shukla, Annick Anctil, Anthony D Kendall, David W Hyndman

Michigan State University, East Lansing, MI, United States

The University of Texas at Dallas, Richardson, TX, United States

Understanding agriculturally co-located solar PV installation practices and preferences is imperative to foster a future where solar power and agriculture co-exist with limited impact on agricultural production. We investigate the impacts of adjacently co-locating solar PV and agriculture on agricultural fields in California' Central Valley. We recently developed a comprehensive remotely-sensed dataset of 694 arrays (2,052 MW) which are agriculturally adjacent co-located. We calculated the food production, electricity generation, and change in water consumption relative to the prior agricultural land use for the expected 25 year lifespan of each array. We calculated that by 2042, these arrays which converted 34 km2 of cropland would remove 1.7 trillion kcal of crop from production. Assuming cropland irrigation was forgone rather than redistributed, the total forgone irrigation water use exceeded operation and maintenance water use by a factor of 7. We also estimated the expected value of generated electricity and show that these installations are profitable, typically exceeding lost revenue from agricultural production by a factor of 15. With its profitability, agricultural co-location will likely continue to expand. Unregulated conversion of high value land could have impacts on future crop prices and availability. Thus, our research suggests the need to account for location-specific food and water resources when co-locating solar PV to reduce impacts on U.S. agricultural production and water as solar becomes more prevalent. Our results also indicate a potential use of renewable energy as a method for agricultural risk management in regions of high water stress and years of drought.

Photovoltaic Investigation on the Lunar Surface (PILS): Design Considerations and Ground Testing

Jeremiah S McNatt, Timothy J Peshek, Norman F Prokop, Greeta J Thaikattil, Michael J Krasowski, Amy R Stalker, Brian J Tomko, Mathew R Deminico

NASA Glenn Research Center, Cleveland, OH, United States

The PILS (Photovoltaic Investigation on the Lunar Surface) platform consists of flight demonstrations of multiple solar cell technologies from multiple companies that could be used for future lunar missions. It also includes a solar charging experiment to shape design considerations of high voltage solar arrays on the Moon that could power in-situ resource utilization systems and other lunar surface assets. This paper describes the design considerations of the PILS platform and ground testing performed prior to spacecraft integration. The platform is scheduled to operate on the lunar surface at Lacus Mortis in late 2022.

Reference Cell Performance and Modeling on a One-Axis Tracking Surface

Frank Vignola[1], Josh Peterson[1], Rich Kessler[1], Sean Snider[2], Peter Gotseff[3], Manajit Sengupta[3], Aron Habte[3], Afshin Andreas[3], and Fotis Mavromatakis[4]

[1]Material Science Institute/University of Oregon, Eugene, Oregon, 97403 (USA)

[2]St. Mary's High School, Medford, Oregon, 97504, (USA)

[3]National Renewable Energy Laboratory, Golden, Colorado, 80401 (USA)

[4]Department of Electrical and Computer Engineering/Hellenic Mediterranean University, Heraklion 71004 Crete Greece

Abstract—Performance of five silicon-based reference cells is examined on a single-axis tracking surface. The reference cells' output are modeled using one-minute sampled spectral irradiance and reference cell temperature measurements. The model also incorporates spectral responsivity data for the reference cell. The transmission of light through the glazing multiplies the sum over all appropriate wavelengths of temperature adjusted reference cell responsivity times the measured spectral irradiance. Modeled reference cell output is compared with measured reference cell output under clear sky and totally cloudy sky conditions. For each reference cell the ratio of modeled to measured output varies by less than 2% over the year.

Keywords—Solar Reference Cells, Spectral Irradiance, Modeling

I. INTRODUCTION

Reference cells can be used to monitor and evaluate the performance of photovoltaic systems in the field because they have spectral and transmission characteristics similar to photovoltaic modules. Thermopile-based pyranometers are excellent instruments used to measure incident radiation in the field and are minimally affected by changes in the distribution of incident spectral irradiance and by the angle-of-incidence (*AOI*) of incoming irradiance. To estimate photovoltaic system performance using data from pyranometers, the spectral and angle-of-incidence effects seen by photovoltaic modules have to be modeled. Many models and programs have been developed and tested to estimate photovoltaic system performance using irradiance data. To estimate incident radiation with reference cells for comparison to irradiance estimates from satellites or for use at different tilts and orientations, models are needed to translate the reference cell measurements into irradiance values.

The overarching goal of this project is to understand, characterize, and evaluate the measurements from reference cells so that they can be used to produce useful estimates of the incident irradiance and build confidence in reference cell measurements for the analysis of photovoltaic (PV) system performance. To avoid introducing systematic biases inherent in pyranometer measurements into irradiance estimates from reference cells, a model was developed and is being tested to emulate the measurements of reference cells. Once confidence in the model is established, the model components can then be used to estimate the incident irradiance and the uncertainties in the irradiance estimates from reference cells.

The three main factors affecting the modeling of reference cells are:

1. Dependence on the incident spectral distribution,
2. Effect of transmission of light through the glazing (angle-of-incidence effects), and
3. Influence of reference cell temperature on the output values.

In previous studies, [1, 2, 3] the output of reference cells on a two-axis tracking surface was modeled using measured spectral irradiance and the temperature of the reference cells. The use of the two-axis tracking surface minimized the angle-of-incident effects on the measurements. Since the temperature effects are small, the study on the two-axis tracking surface enabled testing the importance of the spectral irradiance distribution on the output of reference cells. The current study expands the original analysis by studying the output of five different reference cells and evaluating the performance on a one-axis tracking surface. The angle-of-incident effects become important because, on a one-axis tracking surface, the angle-of-incidence varies considerably over the day and over the year.

The instruments used in the experiment are discussed first. Then the model used to emulate the performance of reference cells is described. Next the relationship between the reference cell and model output is evaluated. After examining the magnitude of different components of the model, a summary of results is presented along with a discussion of next steps.

II. EXPERIMENTAL SETUP

The equipment used in this experiment are located at the Solar Radiation Research Laboratory (SRRL) in Golden, Colorado. Fig. 1 is a photograph of the EKO one-axis tracker with the pyranometers, reference cells, and spectroradiometers.

Mounted on the platform are the EKO Weiser spectroradiometer, a Kipp & Zonen CMP 22 pyranometer along with an SP-Lite 2, a Li-Cor pyranometer, and five reference cells. The reference cells are:

1. Atonometrics (ATO)
2. EETS (EET)
3. IKS Photovoltaik (IKS)
4. IMT (IMT)
5. NES (NES)

978-1-7281-6118-1/22 $31.00 © 2022 IEEE

Fig. 1. Experimental setup at SRRL in Golden, Colorado. Photo from NREL.

These mono-crystalline reference cells were calibrated at the NREL Cell Lab under a standard lamp perpendicular to the reference cell. The factory calibrations were used in the experiment. The lab calibration have a 0.9% uncertainty at the 95% level of confidence [4]. Most of the lab calibrations were 0.8% lower than the factory calibrations, but the lab calibration for the EET calibration was 1.0% higher than the factory calibration value. These values fall within the range of uncertainties of 1.4% - 3.0% quoted by the manufacturers.

The spectral responsivity of the reference cells were derived from the quantum efficiency values measured at the NREL Cell Lab for each reference cell. Spectral responsivities were then calculated and normalized to 1 at the peak response wavelength. Because the spectral responsivities were relative values, a scale factor has to be determined for each reference cell.

III. MODEL DESCRIPTION

The model proposes that the measured output of the reference cell (RC) is proportional to the average transmission of light through the glazing times the sum over all wavelengths of the reference cell spectral responsivity $R(T)$ adjusted for temperature times the incident radiation I_λ (1).

Because the spectral responsivity is a relative value a calibration constant, K, is need to produce an output value. This is similar to the responsivity used to determine the output of a pyranometer. The reference cell (RC) output is modeled by the right side of (1).

$$RC_{model} = F(AOI)/K \cdot \sum_{\lambda=280nm}^{4000nm} R_\lambda(T) \cdot I_\lambda \qquad (1)$$

where $F(AOI)$, the average transmission of light through the reference cell glazing, is determined using the Marion model [5] and the broadband beam and diffuse components of the incident radiation. In this model, $F(AOI)$ is assumed to be independent of wavelength. The scaling factor, K, relates the model estimates to the measured values. The scaling factor is needed because the spectral responsivities are relative values and not absolute values.

The model uses the reference cell spectral response, $R_\lambda(T)$, determined from the reference cell normalized quantum efficiency adjusted for the measured reference cell temperature, T. The adjustment to the spectral responsivity $R_\lambda(T)$ was determined using the Hishikawa model [6]. For wavelengths below the peak wavelength, the spectral responsivity was not adjusted. For wavelength above the peak wavelength, the spectral responsivity was assumed to be the spectral responsivity of the wavelength of λ plus a shift. The shift was calculated using (2).

$$Shift = 0.45 \cdot (T - 25) \qquad (2)$$

Shift values are rounded to the nearest one nm. A factor of 0.45 times the difference between the reference cell temperature and 25C was used to emulate the shift suggested by Hishikawa [6] for single crystalline silicon. For example, if the reference cell temperature was 45C, then the *Shift* would be 9 nm and the spectral responsivity at 1000 nm would be give the value of the reference cell responsivity at 991 nm.

The spectral irradiance, I_λ, is measured at one-nm wavelengths from 350 nm to 1650 nm using an EKO Weiser spectroradiometer mounted on the one-axis tracker. The EKO Weiser spectroradiometer used on the one-axis tracker consists of two spectroradiometers one of which measures wavelengths from 350 nm to 1100 nm and another that measures wavelengths from 900 nm to 1650 nm. The wavelengths values from 900 nm to 1100 nm are determined from algorithms using measurements from both instruments. Spectral intensity from 300 nm to 350 nm was estimated using a linear fit between 0 Wm^{-2} at 300 nm to the measured irradiance at 350 nm.

To obtain the angle-of-incidence function, the incident radiation is separated into the beam irradiance and diffuse irradiance components on the one-axis tracking surface. The various diffuse components are obtained from the Perez model [8]. The diffuse irradiance is separated into circumsolar, dome, horizon, and ground reflected components. The transmission of light through the glazing $F(AOI)$ of the various irradiance components were obtain from Marion model [5].

Five reference cells were compared, the ATO, IMT, IKS, EET, and NES. These are all monocrystalline reference cells and factory calibration values were used to obtain measured output.

For a one-axis tracker, the tilt of the tracker is given by (3) where *SZA* and *AZM* are the solar zenith angle and the azimuthal angle of the sun respectively [7]

$$Tilt = \tan^{-1}[\tan(SZA) \cdot \sin(AZM - 180)] \qquad (3)$$

The angle-of-incidence (*AOI*) is given in (4)

$$AOI = \cos^{-1}[-\sin(SZA)\sin(AZM) \qquad (4)$$
$$\cdot \sin(Tilt) + \cos(SZA) \cdot \cos(Tilt)]$$

IV. RESULTS

An initial evaluation of the modeled and measured output of the EET reference cell is shown in Fig. 2, a plot of reference cell output verses angle-of-incidence (*AOI*). The plot is shown for the clear day of September 12, 2020. The left vertical axis is the measure reference cell output and the right vertical axis is the modeled reference cell output. The reference cell obtained its calibration from the factory. Since the model components do not have an established calibration methodology, the output from the reference cell has to serve to determine the calibration factor (*K*). The scale on the right hand axis is 0.56 time the scale on the left hand axis and this gives a first estimate of the model's calibration factor.

To evaluate the relationship between the measured and modeled values, a more detailed examination is done by looking at the ratio of the modeled and measured output of the reference cell, see Fig. 3. This ratio yields the calibration factor *K* from (1). Because there was no well-defined method to obtain field calibration values that relate reference cell output to broadband irradiance, the *K* are specific to the reference cell and the factory calibration value used. If a standard calibration methodology is obtained for reference cells, then a more defined specification for *K* values can be obtained.

One-minute data was used and sample results from all five reference cells are plotted in Fig. 3 for September, 12, 2020. Although the *K* ratios differ by about 4%, they have similar patterns over the day.

To give a common reference point, the *K* factors for the reference cells were normalized to the EET reference cells at a solar zenith angle of 45° on September 12, 2020. For each type of reference cell, the normalization factor (NF) was calculated from the ratio of the K factors at a SZA of 45° on September 12, 2020 (5).

$$NF = K(RC, 45°)/K(EET, 45°) \qquad (5)$$

where the normalization factor (*NF*) multiplies all the *K* factors for reference cell type *RC*.

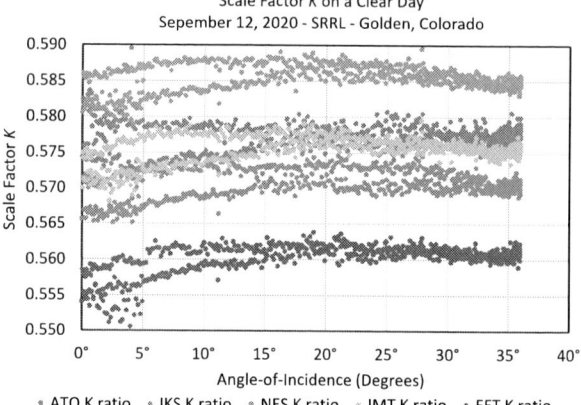

Fig. 3. Scaling factor, *K*, needed to equate modeled reference cell output to measured reference cell output on a one-axis tracking surface. 9/12/2020

In doing this, plot of *K'* factors, K factor multiplied by the ratio, for all reference cells show them almost falling on top of each other and exhibit 1% spread over the day. This 1% spread over the day is typical of the individual *K* values in Fig. 3. For most of the day, the values are within 1% of the noontime value. Each modeled reference cell shows a small split in the *K* values determined in the morning and afternoon as a function of *AOI*.

To evaluate if the results observed on September 12, 2020 are consistent throughout the year, Fig. 4 and 5 plot the *K'* factor for the clear day on July 1, 2020 and December 25, 2020. This is an attempt to discuss universal characteristics of *K* values, the ratios were normalized to the EET ratio at a solar zenith angle of 45° on September 12, 2020. The goal of the normalization is to help provide a visual basis for the discussion of the uncertainties and biases of the model and minimize the differences caused by different calibration methodologies.

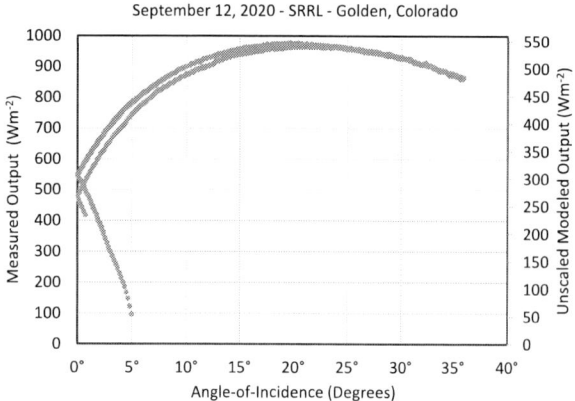

Fig. 2. Unscaled modeled output and measured output of an EET reference cell on a sunny day, Sept. 12, 2020. Vertical scale on left is for the measure RC output and the vertical scale on the right is for the unscaled modeled RC output.

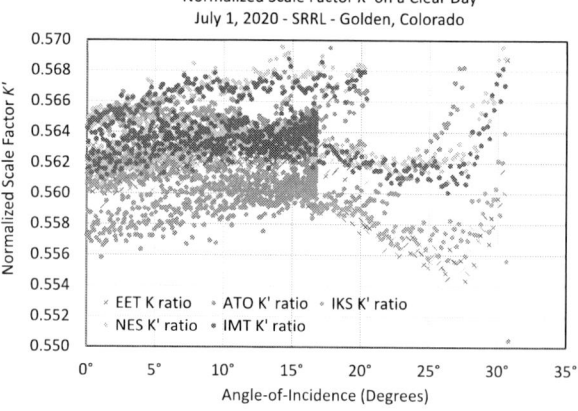

Fig. 4. Normalized scale factor *K'* relating modeled output to the measured output on a one-axis tracking surface. *K* factors were normalized using the value that made all reference cells agree with the EET output on 9/12/2020 at a SZA of 45°.

In Fig. 4, the difference between the various reference cells is about 1.7% over the day. Some of this difference is brought about by the lower values of the normalized ATO K values. AOI greater than 20° correspond to early morning values. The patterns are different for groups of reference cells and the source of this difference has not been identified.

In Fig, 5, a similar clear sky plot is obtained for December 25, 2020. The spread in K values is about 2% at the largest AOIs. The largest AOI values occur at this time of year when the one-axis tracking surface is near the horizontal position. Near solar noon, there is a peak in the K values. As opposed to the data in Fig. 4, the ATO K values in Fig. 5 are now larger than the other K values.

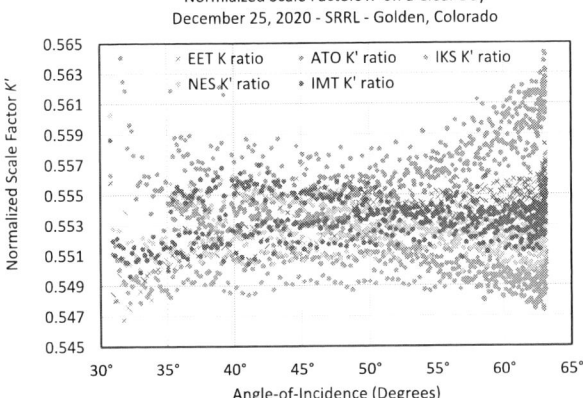

Fig. 5. Normalized scale factor relating modeled output to the measured output on a one-axis tracking surface. K factors were normalized using the value that made all reference cells agree with the EET output on 9/12/2020 at a SZA of 45°.

To examine the changes in behavior over the year, Fig. 6 plots the clear sky values of the reference cells over the 2020 through 2021 time period. The clear sky values were determined by comparing the clear sky models for global horizontal irradiance (GHI) against measured GHI values and using percent cloudiness values of 15% or less. In addition direct normal irradiance (DNI) values were used to help insure the skies were not significantly affected by thin clouds.

The NES reference cells shown in Fig. 6 exhibit a 2% difference in the ratio K over the year. The ATO reference cells exhibit smaller change over the year, but have about a 2% variation over the day. These examples are typical of other reference cells studied. The sinusoidal variation over the year is under investigation.

In Fig. 7, the performance of the NES reference cells over totally cloudy periods is studied. For the model to be most useful, it has to work under all sky conditions. It is difficult to make comparisons over partially cloudy skies because the irradiance can vary significantly over short intervals and spectral measurements are made over a short time span (one to five seconds). To counter this, sample reference cell measurements were used instead of minute averages. This is fine for clear sky comparisons, but is difficult for partially cloudy skies because timing issues of when measurements were made and when clouds affect a sensor become important.

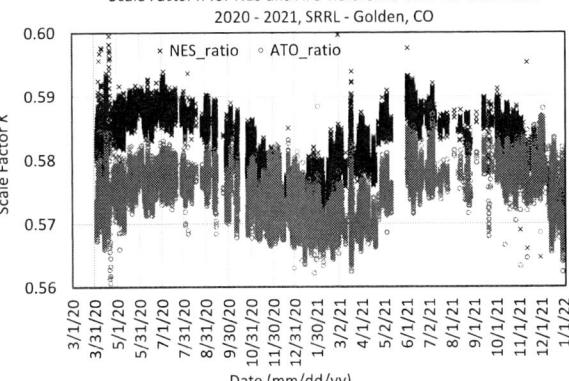

Fig. 6. Scale factor K relating modeled to measured output for NES (x) and ATO (o) reference cells for 2020 and 2021 on a one-axis tracking surface. K has a peak in summer and a minimum in winter. This is about a 2% difference over the year.

Therefore total cloudy skies are studied. For the data from SRRL, periods where cloud cover were greater than 75% and DNI measurements of less than 100 Wm^{-2} were selected.

K varies about 4% on cloudy days and doesn't show a marked trend over the year. Some of the variation shown in Fig.7 results from other factors such as maintenance or problems with other instruments. The ratios for the NES instrument are typical of plots for other instruments over the year.

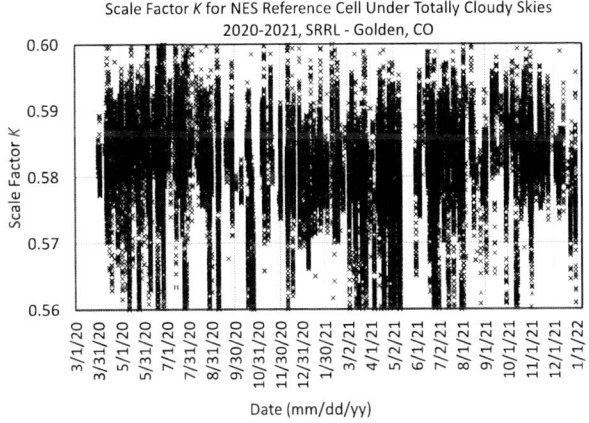

Fig. 7. Scale factor K relating modeled to measured output for a NES reference cell on a one-axis tracking surface for 2020 under totally cloudy skies.

A. Discussion of Results

The average values of the K factors over year has been calculated using the average $F(AOI)$s and the sum of the reference cell spectral responsivity times incident spectral irradiance and dividing by the reference cell measurement. This is based on the assumption that the modeled reference cell output is equal to the measured reference cell output once the scale factor, K has been determined. The comparison between the model estimates and the measured reference measurements agree to better than 2% as illustrated in Figs. 2 to 7. Is this difference a limitation of the model or is the variation associated with the measurements? The next subsections look at some of the possible sources of biases associated with the various model components.

1) Uncertainties associated with temperature measurements

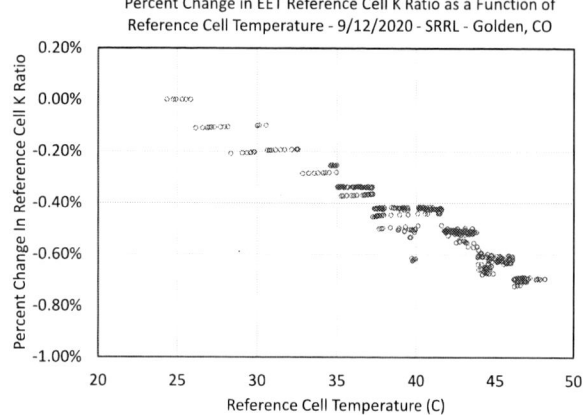

Fig. 8. Study of the effect of the temperature adjustment on the ratio of modeled to measured reference cell output on September 12, 2020. The plot is $(K_{WithoutTemperature} - K_{WithTemperature})/K_{WithTemperature}$.

To start, the influence temperature measurements in the temperature adjustment is examined. With PV systems, temperature is important but with reference cells, the temperature dependence is minimal because the measurements are made using short circuit current which has minimal dependence on temperature. A quick estimate of the importance of temperature can be obtained by examining the difference between the modeled K values with and without an adjustment for temperature. Fig. 8 plots the difference between K values with and without the temperature adjustment. This plot is showing that as an extreme, the overall change in the K value as a function of temperature is about 0.03% per degree C or about 0.7% over the 20C range.

With the estimate of the dependence of K on temperature, one can estimate the uncertainty introduced to the results by the uncertainty in the temperature measurement. There is a significant difference in temperature measurements between reference cells. On a sunny day, the temperature difference between the hottest and coldest reference cell can be 10C or more with the standard deviation between all reference cells about 5C. If one assumes a 5C uncertainty in the temperature measurement, this translate into 0.15% uncertainty in K.

In addition to the difference between measured temperature of individual reference cells, reference cell temperatures sometimes vary from one minute to the next. This can either be related incident radiation, wind speed, or stability of the temperature measurement. A plot of IMT reference cell temperature is shown in Fig. 9 alongside a plot of IMT output. The IMT output is fairly consistent and the temperature of the reference cells can vary by as much as 10C in a manner of minutes. From Fig. 8, a 10C difference would amount to about a 0.3% change in the K ratio. This temperature variation adds to the variation of the K values obtained.

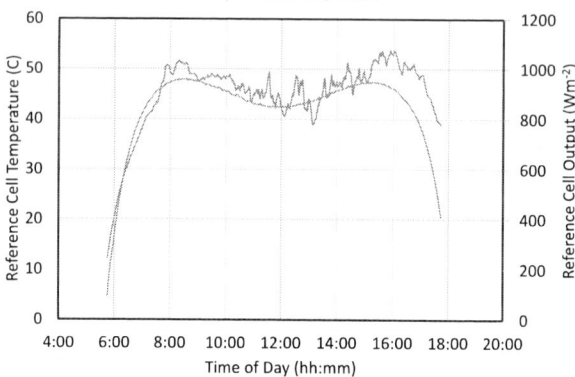

Fig. 9. Variation in the IMT reference cell temperature over the day on September 12, 2020. Also plotted in the IMT reference cell output. Wind speed varies from 2 m/s to 3 m/s over the day.

2) Uncertainties associated with F(AOI)

The uncertainty in the amount of light transmitted through the glazing is dependent on the angle-of-incidence and the irradiance, the $F(AOI)$ for DNI is 1. The $F(AOI)s$ for the various irradiance components are plotted against AOI on September 12, 2020 in Fig.10.

As shown in Fig. 10, diffuse radiation from the dome is fairly constant over the day and is about 0.96. The circumsolar transmission factor is close to 1 and decreases slightly as the AOI increases. When the tracker is horizontal, the angle-of-incidence for the horizon irradiance and the ground reflected irradiance is ~90 degrees. At these AOI values, 37° on 9/12/2020, the modeled horizon and ground reflected $F(AOI)$ goes to zero. This is an approximation for the horizon brightening as the horizon brightening is actually slightly above the horizon. Both the horizon brightening transmission and the ground reflected transmission increase significantly as the surface tilts and the average AOI of the surface decreases relative magnitudes of the direct, diffuse, and ground reflected irradiance. For a surface perpendicular to the incident beam, the

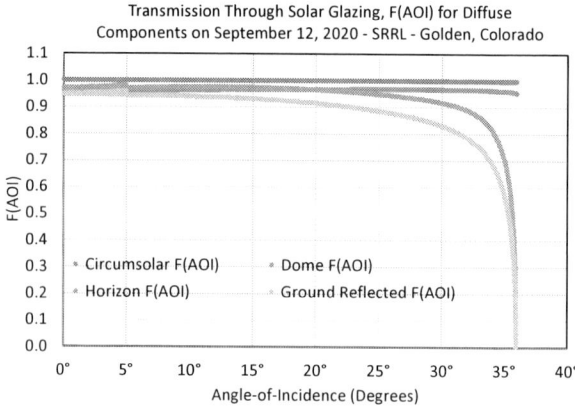

Fig. 10. Transmission through a solar glazing for various diffuse components [8] as a function of AOI on 9/12/2020.

average *F(AOI)* is obtained by weighing the *F(AOI)*s times the magnitude of the irradiance components. On a clear day in September, the average *F(AOI)* varies from close to 1 to 0.9945 under clear skies, see Fig. 11. The average *F(AOI)* is dominated by the beam and circumsolar irradiance that have *F(AOI)*s close to 1. The sharp decrease in average *F(AOI)* at an *AOI* around 37° is the result of the rapid rise in the ground and horizon brightening *F(AOI)*s combined with the sudden addition of these irradiance components to the total diffuse irradiance. The modeled average *F(AOI)* drops from above 0.997 to slightly below 0.995. For an *AOI* about 35° and the rise in the horizon and ground reflected starts to moderate and the relationship between average *F(AOI)* and *AOI* becomes more linear. The morning and afternoon average *F(AOI)* differ because the relative magnitude of DNI and the diffuse irradiance is different.

Also shown in Fig. 11 is the average *F(AOI)* for December 25, 2020. At larger *AOI*s, near solar noon, less light is transmitted through the glazing and in December, the *AOI* values are much larger than September and the transmission of light is less. For the clear day on December 25, 2020, *F(AOI)* varies from 0.998 near sunrise to 0.947 at noon. Therefore, any biases associated with *F(AOI)* are most important in the winter months when the *AOI* become greater than 40° to 45°.

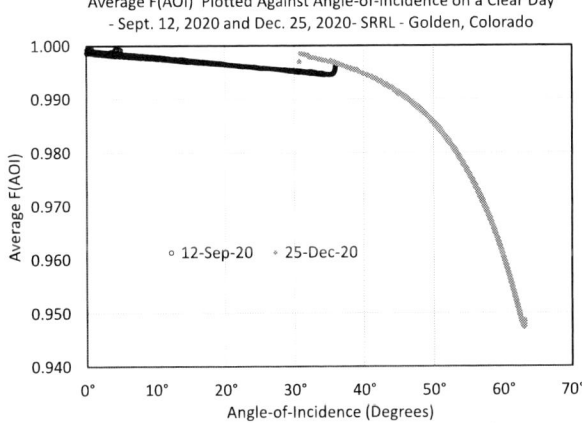

Fig. 11. Average transmission of light through a solar glazing, *F(AOI)*, plotted against angle-of-incidence – September 12, 2020 and December 25, 2020. Different proportions of DNI and DHI values account for the differences in *F(AOI)* between 30° and 37°

3) *Biases in estimating the diffuse components*

There are also biases associated with separating the diffuse radiation into the diffuse components. Separating the diffuse components and projecting them onto a one-axis tracking surface offers a unique test of the Perez model [8] that separates the irradiance into the diffuse and ground reflected components. The orientation of the surface varies over the day resulting in many orientations not usually experienced.

The modeled diffuse components on a one-axis tracking surface under clear skies on September 12, 2020 are shown in Fig. 12. The components were calculated using the Perez model [8] with the tilt and orientation calculated each minute. The calculated total diffuse on a one-axis tracking surface is about

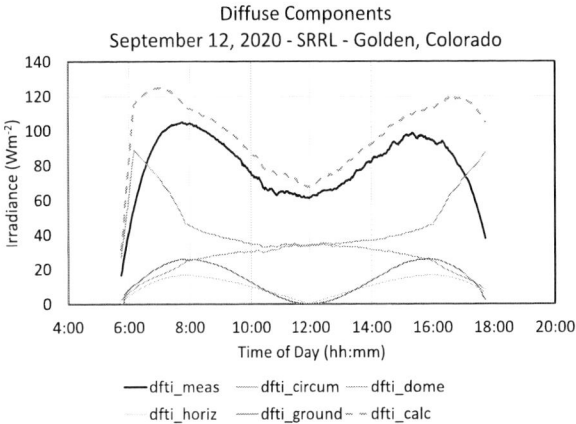

Fig. 12. Diffuse and ground reflected components on September 12, 2020. Solid black line is the diffuse irradiance obtained by subtracting DNI projected onto a one-axis tracking surface from the measured GTI. The dashed green line is the sum of the diffuse and ground reflected components on a one-axis tracking surface.

10 Wm⁻² higher than the "measured" diffuse for much of the day. However, the modeled diffuse overestimates the diffuse from the measured data in the morning and afternoon hours.

The diffuse components for December 25, 2020 are plotted in Fig. 13. The 25ᵗʰ was chosen because it was a clear day with minimal snow cover on the ground. At solar noon, the measured and calculated diffuse irradiance are within 10 to 15 Wm⁻² of each other. In the morning, during the clearest part of the day, the difference between the measured and calculated total diffuse is the largest. At solar noon when the one-axis tracking surface is horizontal, the GTI measurements is about 10 Wm⁻² below other diffuse measurements at SRRL. This is within the specifications of the pyranometer. However, this doesn't explain the larger differences in the morning. The probable cause for this difference is the overestimate of one or more of the diffuse components. One possibility is that the circumsolar diffuse irradiance is assumed to be evenly distributed around

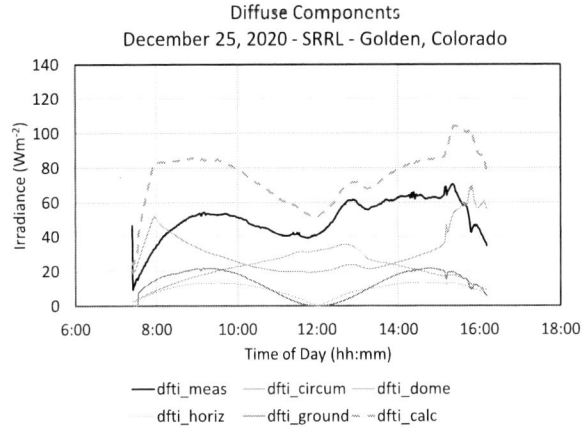

Fig. 13. Diffuse and ground reflected components on December 25, 2020. Solid black line is the diffuse irradiance obtained by subtracting DNI projected onto a one-axis tracking surface from the measured GTI. The dashed green line is the sum of the diffuse and ground reflected components on a one-axis tracking surface.

the sun. When the sun is low on the horizon as during the winter or near sunrise or sunset, some of this circumsolar may be obscured by the horizon. Modeling of the reference cell output might be improved with a better understanding of how to separate the diffuse irradiance into it various components.

4) Influence of biases in the diffuse components

The average $F(AOI)$ is dependent upon the relative magnitudes of the beam and diffuse components on the tilted surface. The average diffuse $F(AOI)$ was obtained by averaging the product of the modeled diffuse components times the $F(AOI)$ for each component (6). To minimize the bias that results from the difference between the sum of the diffuse tilted components and the measured tilted diffuse, the sum of the modeled $DfTI$ component, $DfTI_{calculated}$, was used to obtain the $F(AOI)_{avgdiff}$.

$$F(AOI)_{avgdiff} = \quad [DfTI_{circumsolar} \cdot F(AOI)_{circum} + \quad (6)$$
$$DfTI_{dome} \cdot F(AOI)_{dome} +$$
$$DfTI_{horizon} \cdot F(AOI)_{horizon} +$$
$$GRI_{ground} \cdot F(AOI)_{ground}]/DfTI_{calculated}$$

where $F(AOI)_{avgdiff}$ is the average of the tilted diffuse components. The diffuse components on the tilted surface are $DfTI_{circumsolar}$ the diffuse circumsolar component, $DfTI_{dome}$ the diffuse dome component, $DfTI_{horizon}$ the diffuse horizon brightening component, and GRI_{ground} the ground reflected component. A similar naming convention applies to the $F(AOI)$ components.

The average $F(AOI)_{avg}$ for all irradiance is given by combining the beam with the diffuse components (7).

$$F(AOI)_{avg} = [DfTI_{meas} \cdot F(AOI)_{avgdiff} + \quad (7)$$
$$DNI \cdot cos(AOI) \cdot F(AOI)_{circum}]/GTI$$

where GTI is the total measured irradiance on a one-axis tracking surface. When $F(AOI)_{avg}$ is calculated, $F(AOI)_{avgdiff}$ is multiplied by the $DfTI_{meas}$, the diffuse value obtained by subtracting DNI projected onto the tilted surface from GTI (8).

$$DfTI_{meas} = GTI - DNI \cdot cos(AOI) \quad (8)$$

Obtaining $DfTI_{meas}$ using (8) is another source of uncertainty because the measured GTI and DNI also have well characterized uncertainties and biases that increase the uncertainty in DHI_{meas} obtained this way. Uncertainties in the $DfTI_{meas}$ can be ±10 to 20 Wm^{-2}.

Evaluating (7), the uncertainty in DHI_{meas} only becomes important if $F(AOI)_{avgdiff}$ is much less than one. Since, under clear skies, the average $F(AOI)$s on a one-axis tracking surface change very little except during the winter months. The relative difference between the modeled diffuse components and DHI_{meas} aren't significant. The same cannot be said for the winter months, when the AOI is large and the individual diffuse components have greater influence on $F(AOI)_{avg}$.

V. SIGNIFICANCE OF FINDINGS

The proposed reference cell model uses spectral and temperature measurements, spectral responsivity data for the reference cell, and broadband irradiance measurements. The model estimates of the reference cell output to within 1% to 2% of measure reference cell output. This information should be

sufficient to estimate the incident irradiance being measured by the reference cell. Two critical pieces of information are missing before practical use of this model can be made. First, a definition of standard conditions is needed along with a quantified methodology for calibrating the reference cell.

To illustrate the need for standard conditions, a comparison between the measured reference cell output and measured GTI on a one-axis tracking surface using a Kipp & Zonen CMP 22 pyranometer is shown in Fig. 14. The range of differences between reference cell measurements and the high quality pyranometer readings range from +6% to -4% with some reference cells experiencing a smaller range. By taking into account the transmission of irradiance through the glazing, the temperature of the reference cell, and the changing spectral irradiance over the day, it is possible to significantly reduce this difference. First the reference cells needs to be calibrated using a standard methodology. Under these conditions, a direct link between reference cell output and incident radiation would be obtained. Using the data and model under these conditions, the model's K value can be determined. Currently there is an ambiguity here and similar reference cells produce different values under identical conditions, see Fig. 3.

This study has shown how the reference cell measurements are dependent on the incident spectral irradiance, the transmission of light through the glazing, and the reference cell temperature. As the conditions experienced by the reference cell in the field move away from the standard conditions, the relationship between irradiance and reference cell output changes. The changing relationship between "measured" GTI and reference cell output is shown in Fig. 14. The GTI measurements are from a CMP 22 pyranometer that also has some dependence on conditions under which the measurement were made. However, these are small compared to the dependences (systematic biases) of the reference cells. Knowing the dependencies of the reference cell to changing conditions can be taken into account and it is possible to make adjustments to the reference cell output to obtain better estimates of incident radiation. This is similar to adjustments

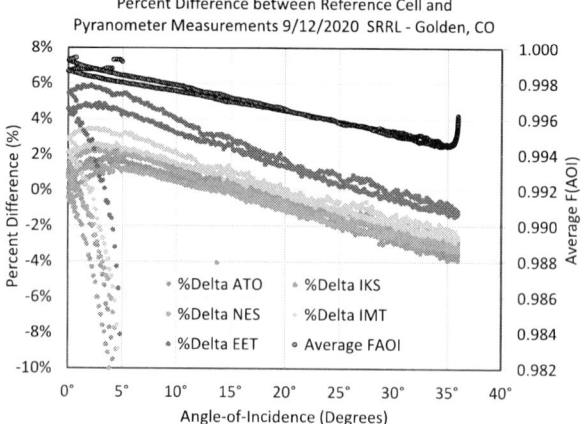

Fig. 14. Percent difference between reference cell and pyranometer measurements (CMP 22) on a one-axis tracking surface on September 12, 2020. The scale for the average F(AOI) is the right hand axis. Factory calibrations were used for the reference cell measurements.

made to rotating shadowband radiometers to obtain better estimates of beam and diffuse irradiance.

The methodology used in this study should enable reliable estimates of irradiance. In practice, the spectral irradiance would have to be modeled instead of measured because spectral irradiance measurements are rarely available. This spectral modeling needs testing. More work is necessary to validate the model and determine how well irradiance values can be obtained from reference cell measurements. The spread of reference cell outputs is an example of this issue. Once standard conditions and calibrations are obtained, it is possible to compare pyranometer readings and reference cell measurements by reducing the systematic biases of the reference cell biases from the measurements.

Figs. 3 through 6, illustrate that it is possible to model the conditions that result in the reference cell measurements to within 1% to 2% of the reference cell output. By normalizing the reference cell output to standard conditions it should be possible to specify the reference cell output under those conditions. Under the standard conditions, the reference cell is calibration to match the incident broadband irradiance. The normalized reference cell output would give a much better estimate to the incident radiation.

It is important to evaluate results from different locations and on a horizontal or fixed tilted surfaces. So far the model has only been tested with *AOI* up to 60°. The *F(AOI)* for *AOI* between 60° and 90° are expected to change significantly and these angles have not been tested in this data. Different locations are also expected to expand the sky conditions being tested and are needed to help insure the universal applicability of the model.

Pyranometers measurements have their own set of uncertainties and systematic biases as do reference cell measurements. The uncertainties and biases associated with pyranometer measurements are well defined and procedures to obtain the uncertainty values are well established. If reference cells measurements are to be compared with pyranometer measurements, calibration procedures and methodology for reference cells have to be established. The model used to emulate the performance of reference cells provides some guidance on information that needs to be taken into account. The uncertainties incorporated in the model need to be better defined and explored. The uncertainties introduced by the various components of the model need to be determined. This is not easy even with well-defined calibration procedures.

The small uncertainties obtained during this study are dependent on the spectral irradiance measurements. If the spectral irradiance is modeled, then the uncertainties will increase because there are larger uncertainties in models spectral values. This is an aspect for future work.

Overall, the goal has been to produce useful results that will improve the understanding and performance estimates of photovoltaic systems. The model replicates the performance of reference cells and it should be possible to estimate incident radiation. The opposite is also true, that given the incident radiation, the model with adjustments to match the max power point, should be able to estimate the performance of PV systems. With better understanding and more reliable information, the risks of deploying photovoltaic systems is reduced and the profits from PV systems will be determined with more reliability.

VI. ACKNOWLEDGEMENTS

The University of Oregon Solar Radiation Monitoring Laboratory would like to thank the National Renewable Energy Laboratory as well as the Murdoch Family Trust for funding the project. We also thank the other sponsors of the University of Oregon Solar Radiation Monitoring Laboratory, the Bonneville Power Administration, the Energy Trust of Oregon, and Portland General Electric.

This work was authored in part by Alliance for Sustainable Energy, LLC, the manager and operator of the National Renewable Energy Laboratory for the U.S. Department of Energy (DOE) under Contract No. DE-AC36-08GO28308. Funding provided by U.S. Department of Energy Office of Energy Efficiency and Renewable Energy Solar Energy Technologies Office. The views expressed in the article do not necessarily represent the views of the DOE or the U.S. Government. The U.S. Government retains and the publisher, by accepting the article for publication, acknowledges that the U.S. Government retains a nonexclusive, paid-up, irrevocable, worldwide license to publish or reproduce the published form of this work, or allow others to do so, for U.S. Government purposes.

REFERENCES

[1] F. Vignola, J. Peterson, R. Kessler, S. Snider, P. Gotseff, M. Sengupta, A. Habte, A. Andreas, and F. Mavromatakis, "Influence of Diffuse and Ground-Reflected Irradiance on the Spectral Modeling of Solar Reference Cells," Proceedings of the American Solar Energy Society, Boulder, CO., 2021

[2] F. Vignola, J. Peterson, R. Kessler, S. Snider, A. Andreas, A. Habte, P. Gotseff, M. Sengupta, and F. Mavromatakis, "Evaluation of Reference Solar Cells on a Two-Axis Tracking Using Spectral Measurements," SolarPACES, September 27–October 1, 2020

[3] F. Vignola, J. Peterson, R. Kessler, V. Sandhu, S. Snider, A. Habte, P. Gotseff, A. Andreas, M. Sengupta, and F. Mavromatakis, "Improved Field Evaluation of Reference Cell Using Spectral Measurements," Solar Energy 215, (2021) pp. 482-491

[4] M. Sengupta, A. Habte, Y. Xie, P. Gotseff, M. Kutchenreiter, A. Afshin, I. Reda, F. Vignola, A. Driesse, C. Gueymard, S. Bandyopadhyay, and A. Denhard, "Solar Radiation Research Laboratory (SRRL) Final Report: Fiscal Years 2019–2021," 2022

[5] B. Marion, 2017. "Numerical method for angle-of-incidence correction factors for diffuse radiation incident photovoltaic modules," Solar Energy 147 344–348, 2017

[6] Y. Hishikawa, M. Yoshita, H. Ohshima, K. Yamagoe, H. Shimura, A. Sasaki, and T. Ueda, "Temperature dependence of the short circuit current and spectral responsivity of various kinds of crystalline silicon photovoltaic devices," Japanese Journal of Applied Physics 57, 08RG17 (2018)

[7] W. Marion and A. P. Dobos, "Rotation Angle for the Optimum Tracking of One-Axis Trackers," NREL/TP-6A20-58891 July 2013 https://www.nrel.gov/docs/fy13osti/58891.pdf].

[8] R. Perez, P. Ineichen, R. Seals, J. Michalsky, "Modeling daylight availability and irradiance components from direct and global irradiance," Solar Energy 44, 271–289, 1990

Optimization of CdTe Solar Cells using Co-sputtered CdSeTe

Deng-Bing Li, Sandip Singh Bista, Neupane Sabin, Xiaomeng Duan, Manoj K Jamarkattel, Abdul Quader, Adam Philips, Michael Heben, Randall J Ellingson, Yanfa Yan

Department of Physics and Astronomy, and Wright Center for Photovoltaics Innovation and Commercialization (PVIC), University of Toledo, Toledo, OH, United States

Co-deposited CdSeTe has been demonstrated to be an efficient way to improve the short-circuited current density (JSC) in CdSeTe solar cells. In this work, co-sputtered CdSeTe film with different compositions is investigated. The effect of CdSeTe layer thickness and the CdCl2 treatment temperatures are systematically investigated to improve the JSC and the power conversion efficiency. We found that insufficient diffusion of Se will result in poor carrier collection efficiencies at blue wavelength and interior front junction, yielding low VOC and JSC. When Se is over diffused at the front interface, the carrier collection efficiencies at a long wavelength will be reduced. Finally, an impressive JSC of 28.3 mA/cm2 was demonstrated. The optimization of copper activation is currently undertaken, and higher device performance is promised.

Parametric study of building-integrated photovoltaic windows

Yuan Gao, Jacob Jonsson, Charlie Curcija

Lawrence Berkeley National Laboratory, Berkeley, CA, United States

Building integrated photovoltaic (BIPV), as a distributed energy resource, can cover a part of the building energy demands and even help achieve the idea of net-zero energy buildings. By connecting with energy storage and grid, the entire BIPV systems have a high demand flexibility potential and can improve building resilience against power outages. Roof BIPVs, though considered as the mainstream, have limited area in high-rise buildings compared with windows, where semi-transparent PVs can play a significant role of energy resources given the considerable vertical window areas in modern urban environment. Material scientists have developed various semi-transparent solar cells with a wide range of power conversion efficiencies (PCEs), and solar and visible transmittance. However, it is not clear about the optimal configurations of semi-transparent solar cells for different types of buildings and climates. To tackle this problem, we conducted a parametric study on PV windows in a reference commercial building considering variables including PCE, solar transmittance, solar absorptance, U factor, daylighting control, window orientations, and climate types. Our model considers the thermal effects of PV windows, i.e., a load or grid connected PV window turns partial solar absorption into electricity instead of heat. The first finding, which differs from roof PVs, is that the vertical solar radiation on east and west facing windows is comparable to that on the south facing windows because the special 90° title angle results in more uniform POA irradiance in different orientations. It means the combination of PV windows in different orientations provide more stable power generation for the building. Results show that the PCE of PV windows dominates the energy saving despite other variables. The balance between solar transmittance and absorptance is also important for energy saving. Slightly higher visible transmittance (0.1) benefits the building energy saving when daylighting control is applied.

Impact of Daily Irradiance Profiles on Intra-Day Solar Forecasting

Javier Lopez-Lorente, Spyros Theocharides, George Makrides, and George E. Georghiou

PV Technology Laboratory, FOSS Research Centre for Sustainable Energy, Department of Electrical and Computer Engineering, University of Cyprus, Nicosia, 1678, Cyprus

Abstract—The analysis of solar energy integration requires capturing the temporal variability of solar irradiance, which can highly increase the errors of solar forecasting models. In this paper, a classification of day types for solar energy applications is investigated and the impact on intra-day solar generation forecasting is assessed. The proposed classification approach uses unsupervised learning based on a combination of self-organized maps and mean-shift clustering with six location-independent metrics related to irradiance variability and energy yield. Two types of forecasts (a deterministic forecast and a probabilistic one) are used as a basis to illustrate the impact of the resulting daily irradiance profiles. The forecasts are emulated by adding white noise to historic hourly irradiance observations and their performance is compared to state-of-the-art benchmarking models. The results illustrate the magnitude of dispersion between intra-hour observations and hourly forecasts, where deterministic forecasts observed 16% to 64% of deviation and probabilistic forecasts observed prediction intervals from 11% to 36%, depending on the daily typology. The findings of this study provide useful information on the impact of intra-day irradiance variability on the performance of forecasting models and how these can be adapted based on each day type to support the integration of solar generation in electricity systems.

Keywords—irradiation profiles, irradiance variability, self-organized map classification, solar forecasting, unsupervised machine learning.

I. INTRODUCTION

Solar photovoltaic (PV) generation has exhibited the highest growth rate among renewable energy technologies since 2016 [1]. However, the variable nature of solar irradiance renders the accurate forecasting of solar generation challenging. Prior research efforts have focused on solar irradiance and PV generation forecasting to reduce forecasting uncertainties at different time-scales of interest [2].

Solar forecasting has developed intensively during the last decade and notorious work and literature reviews (e.g., [3]) have been produced in this sub-domain of energy forecasting. Within solar irradiance forecasting, there are two main approaches: deterministic (or point) forecasts and probabilistic forecasts, where the latter is currently at an early stage of maturity [4]. Despite the extensive body of literature, frameworks for the standardization and benchmarking of solar forecasts continue to be proposed. In deterministic forecasts, the 24-hour clear-sky persistence model is a well-accepted and extended baseline model in the forecasting community [5]. However, new universal benchmarking models for probabilistic forecasts have been proposed in recent years. For example, the Complete-History Persistence Ensemble (CH-PeEn) [6] or the clear-sky dependent climatology (CSD-CLIM) model [7].

At an intra-day level, the variability of solar resources is affected by the diurnal cycle. This effect can be closely estimated under cloudless conditions using clear-sky irradiance models. However, cloud cover is responsible for the greatest intra-day variability in the available solar irradiation at a given location and its forecast is more challenging. Local cloud cover in days with variable skies can lead to irradiance differences in the order of 600 W/m^2 in nearby areas (3-5 km) [8], which can be a challenge to account for when using gridded satellite data of higher spatial resolutions. The relationship between solar irradiance variability and cloud properties or weather patterns is a subtopic of solar forecasting currently under investigation (e.g., [9]) with the aim to develop irradiance variability forecasting models to decrease the uncertainty of day-ahead and intra-day forecasts.

In this context, the assessment of daily solar irradiance profiling classifications, applicable for solar PV integration in power networks, is of interest to understand the evolution of solar irradiance and PV power variability over time. Several daily irradiance classifications were proposed in the literature. Stein *et al.* [10] studied daily irradiance patterns and proposed 4 categories of irradiance conditions attending to the variability index (VI) and the daily clearness index (K_t), which resembled an arrowhead. The intra-hourly variability of PV production was assessed in [11] using 3 categories of irradiance profiles based on the average daily cloud cover. Similarly, 3 categories (i.e., high yield, low yield and highly variable) were used in [12] to assess high-frequency PV power fluctuations based on daily PV energy yield and a normalized VI to the day with the highest observed variability (VI_n). Theocharides *et al.* [13] developed a day-ahead PV forecasting model based on artificial neural networks (ANNs), which included k-means clustering of forecasted global horizontal irradiance (GHI) and resulted in 5 clusters. Hartmann [14] recently reviewed 6 solar irradiance categorization methods, where it was illustrated that most classifications include 3 to 5 categories and overpopulated groups and misclassification issues were identified. However, the author concluded that the number of categories is highly influenced by the variables assessed. To that end, there are multiple studies whose findings are above that range. For example, 10 clusters were identified in [15] when analysing GHI and ramp-rate observations. [16] focused on direct normal

---

This work was co-funded by the European Regional Development Fund and the Republic of Cyprus through the Research and Innovation Foundation in the framework of the project "ELECTRA" (protocol number: INTEGRATED/0918).

irradiance (DNI) observations and found 8 categories when considering the temporal variability in hourly intervals and the intensity of irradiance variability. In [17], up to 20 daily categories were identified from GHI and DNI observations using probabilistic approaches based on the clearness index, diffuse fraction and the probabilistic variability score. Table I provides a summary of the main characteristics of studies addressing the classification of daily irradiance profiles in terms of solar components studied, the daily metrics used as input, classification technique and the resulting number of groups or clusters identified.

TABLE I. TAXONOMY TABLE OF STUDIES ADDRESSING THE CLASSIFICATION OF DAILY IRRADIANCE PROFILES.

Solar Component	Input Daily Metrics	Classification Method	Number of groups	Ref.
GHI	VI; clearness index	Deterministic	4	[10]
GHI	Mean cloud cover	Deterministic (manual classification[a])	3	[11]
GHI	PV yield; normalized VI	Deterministic	3	[12]
GHI	Hourly GHI forecast	k-means	5	[13]
GHI	Hourly GHI; hourly ramp-rate	Combination k-POP/k-means	10	[15]
DNI	1-min variablity of beam clear-sky index	Deterministic (manual classification[a])	8	[16]
GHI & DNI	Clearness index; diffuse fraction; variability score	Combination Deterministic/ Probabilistic	Up to 20	[17]

[a.] Classification is determined from observations in the data.

In this paper, a categorization of the daily irradiance profiles is proposed considering six location-independent metrics related to energy yield and variability of solar radiation and using unsupervised self-organized maps (SOM) combined with mean-shift clustering. These metrics are: (1) the daily *VI* as proposed by [10] for GHI and DNI measurements; (2) daily global horizontal and beam irradiation, (3) daily mean clearness index (K_t); and (4) the maximum daily ramp rate of 1-minute clear-sky index observations as defined by [18]. This study differentiates from previous work by using a different methodology and an extended set of metrics related to both GHI and DNI observations. The use of SOM permits a low-dimensional representation that preserves the relationships of the multi-dimensional input dataset. While SOM has already been used in atmospheric and Earth observation research as a technique for cloud type classification [19] and cloud masking identification [20] from satellite observations, it is a novel technique for daily irradiance profiling. The combination of SOM and mean-shift facilitates the determination of cluster numbers through a technique whose input is not a predetermined number of clusters as in other techniques (e.g., k-means clustering).

Two types of forecasts are used as a basis to illustrate the impact of the resulting daily irradiance profiles. The first one consists of an emulated deterministic forecast produced by adding white noise to the historic mean hourly irradiance observations of GHI and DNI. The second one is a probabilistic

forecast produced in a similar way, where the white noise is instead added to the historic median (P50) hourly irradiance data. Furthermore, the response of state-of-the-art benchmarking forecasting models as a function of daily irradiance profile types is also evaluated, i.e. the 24-hour clear-sky persistence forecast for the deterministic approach and the CH-PeEn for the probabilistic forecast. The study evaluates a selected site with a temperate climate using high-quality data from a BSRN station. The results illustrate how the different solar irradiance clusters affect the performance of irradiance forecasting models in terms of deviation and dispersion within hourly intervals. The findings of this paper help develop an understanding of how daily irradiance variability impacts the performance of forecasting models, which can benefit the effective integration of PV generation in electricity systems.

The remainder of this paper is structured as follows. Section II introduces the methodological approach and describes the data sources used. Section III presents the results of the daily irradiance clustering and the effect thereof in emulated deterministic and probabilistic forecasts. Section IV discusses the findings of this work. Finally, Section V concludes this paper.

II. METHODOLOGY AND DATA SOURCES

This section describes the methodology followed to classify the daily GHI and DNI irradiance profiles, the emulation of solar deterministic and probabilistic forecasts and the evaluation thereof. An overview of the methodology as a flowchart is illustrated in Fig. 1. The methodology can be considered a fivefold process: (1) quality control of irradiance data; (2) estimation of the location-independent variables related to daily solar irradiance; (3) classification of daily irradiance profiles; (4) emulation of forecasting models with state-of-the-art accuracies; and (5) evaluation of the performance of the forecasts as a function of the identified daily categories of irradiance profiles.

A. Quality Control of Irradiance

Ensuring high quality irradiance observations requires a preliminary quality control check of the initial timeseries of GHI and DNI. The quality control routine for irradiance data checked

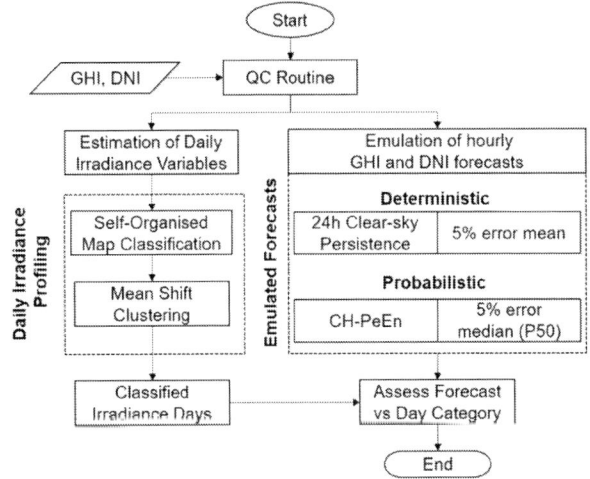

Fig. 1. Flowchart of the methodology for the classification of daily irradiance profiles and emulation of hourly solar forecasts for evaluation.

for physical limits on each irradiance component (global horizontal G and direct normal G_b) using the criteria of the QCRad methodology [21] given by (1) and (2):

$$-4 \text{ W/m}^2 < G < S_a \cdot 1.5 \cdot \mu_0^{1.2} + 100 \text{ W/m}^2 \qquad (1)$$

$$-4 \text{ W/m}^2 < G_b < S_a \qquad (2)$$

where S_a denotes the solar constant at the mean Earth-Sun distance (i.e., 1,368 W/m²) and μ_0 denotes the cosine of the solar zenith angle.

B. Daily Irradiance-related Variables

The classification of daily irradiance profiles is based on six location-independent metrics related to solar energy yield, irradiance variability and ramps at a daily scale, including both global horizontal and direct normal data. The six proposed location-independent metrics used are: (1) the variability index of DNI observations; (2) the variability index for GHI observations; (3) daily global horizontal energy yield; (4) daily direct energy yield; (5) the mean daily clearness index; and (6) the maximum daily ramp of the clear-sky index of the 1-minute observations.

The variability index (*VI*) [10] is a dimensionless metric, which compares the length of the GHI or DNI, to quantify intra-day irradiance variability that can be applied to GHI observations as given by (3) and to DNI observation using (4):

$$VI_{GHI} = \frac{\sum_{k=2}^{n} \sqrt{(GHI_k - GHI_{k-1})^2 + \Delta t^2}}{\sum_{k=2}^{n} \sqrt{(GHI_{0k} - GHI_{0k-1})^2 + \Delta t^2}} \qquad (3)$$

$$VI_{DNI} = \frac{\sum_{k=2}^{n} \sqrt{(DNI_k - DNI_{k-1})^2 + \Delta t^2}}{\sum_{k=2}^{n} \sqrt{(DNI_{0k} - DNI_{k-1})^2 + \Delta t^2}} \qquad (4)$$

where GHI, GHI_0, DNI, and DNI_0 are vectors for the global horizontal, clear-sky global horizontal, direct normal, and clear-sky direct normal irradiance, respectively, of length n, averaged at a time interval (Δt). Other metrics were the daily global horizontal energy yield (H_g) and direct normal energy yield (H_d), which were estimated using the integral surface integral of the observations. Another metric was the mean daily clearness index (K_t), which is a dimensionless number between 0 and 1 representing the solar radiation transmitted through the atmosphere until ground level as given by (5):

$$K_t = H_g / H_0 \qquad (5)$$

It is defined as the surface radiation divided by the extra-terrestrial radiation (H_0). The final metric was the maximum daily ramp rate observed in the clear-sky index (max. $\Delta K_t{}^*_{\Delta t}$) [18], [22]. The clear-sky index denoted as $K_t{}^*$, is the ratio of GHI to clear-sky GHI. For the estimation of the variability indices and clear-sky index, the clear-sky radiation model by Ineichen and Perez was used in this case [23].

C. Daily Irradiance Profiling and Classification

The classification of the daily irradiance profiles is addressed using the unsupervised machine learning technique

Kohonen Map or Self-Organized Map (SOM) in combination to mean-shift clustering. SOM are a type of artificial neural network that apply dimensionality reduction and enable conversion of a dataset with multiple dimensions into a two-dimensional map [24]. The application of SOM permits estimating a lower-dimensional representation that preserves the topology relations of the input dataset [19]. In this case, the six-dimensional dataset of daily irradiance-related variables including metrics related to both global and direct normal observations resulted in a two-dimensional dataset. The SOM was applied with a learning rate of 0.01, random initialization of the weights and a size of 20 x 20 nodes. Mean-shift clustering was applied to the obtained SOM in order to determine the number of characteristic categories or profiles. Mean-shift is a non-parametric clustering technique using Kernel density estimation and gradient thereof to generate a mean-shift vector pointing toward the direction of the maximum increase in the density, which after successive computation returns the number of centroids (clusters) observed in the dataset [25], [26]. This method was selected over other techniques (e.g., k-means clustering) as the number of clusters is data-driven, rather than a pre-determined input into the algorithm. A quantile value of 20% as a hyperparameter was used for the mean-shift model.

D. Emulation of Forecasting Models

To evaluate the effect of the different daily irradiance profiles in the solar forecasting models, two high accuracy forecasting models, a deterministic and a probabilistic forecast, were emulated making use of the actual irradiance values observed. In addition, baseline benchmarking models were implemented to be used as a reference in each of the forecasting approaches.

1) Deterministic forecast

a) Persistent forecast – 24h clear-sky persistence

The baseline model to evaluate the deterministic forecast model was a 24-hour clear-sky persistence model, where the irradiance observations correspond to the distribution of the clear-sky index observed the previous day [5].

$$G_{t+24} = G_t \cdot K_t{}^* \qquad (6)$$

where G_{t+24} is the day-ahead irradiance of the persistence model and G_t is the irradiance at time t.

b) Emulation of high accuracy forecast

The forecasting models for GHI and DNI were emulated by adding white noise following a normal distribution ($\mu=3.5$, $\sigma=1$) to the hourly mean GHI and DNI values. The values for the normal distribution were those returning a normalized root mean square error of 5%, which was selected with the intention to represent a forecast with high accuracy (HA) for current state-of-the-art acceptable performance levels. A second normal distribution ($\mu=1$, $\sigma=0.1$) was used to provide the sign of the emulated error in the forecasts. The white noise ensures a forecast without autocorrelation; thus, removing the influence of any seasonality factors [27].

2) Probabilistic forecast

a) Baseline model – CH-PeEn

New benchmarking models for probabilistic forecasts have been proposed in the recent years. Among them, the Complete-History Persistence Ensemble (CH-PeEn) [6] is one of the benchmarking methods more extended in the solar forecasting community. The CH-PeEn baseline model utilizes past measurements to form predictive distributions conditioned by each hour or period of the day. There are however other methods such as the clear-sky dependent climatology model (CSD-CLIM) [7], which form the predictive distributions between the relationship of the historic irradiance observations and their estimated clear-sky irradiance.

In this paper, the CH-PeEn model was reproduced for GHI and DNI observations. The CH-PeEn was constructed using 1-year of data at 1-hour resolution. For the complete details of this model, the reader is referred to [6], where a sample code for the method is also provided.

b) Emulation of high accuracy forecast

The HA probabilistic forecast for GHI and DNI was emulated using as reference the historic hourly median (P50) values. In order to have a P50 forecast different from the median, white noise with normal distributions (μ=3.5, σ =1) was included in a similar manner to the deterministic approach.

E. Evaluation Metrics

The purpose of the study is to evaluate the deviation or dispersion between the baseline and emulated hourly forecasts and the absolute range of intra-hourly (1-minute) observations as a function of the categorized days. For the deterministic forecasts, the maximum relative percentage deviation between the most dispersed value ($x_{max,t}$) from the 1-minute observations within an hour t and the forecast for that hour ($x_{forecast,t}$) was computed as given by (7), which was then averaged daily.

$$max.\,deviation = 100 \cdot \left| \frac{x_{max,t} - x_{forecast,t}}{x_{forecast,t}} \right| \qquad (7)$$

For the probabilistic forecasts, the cumulative distribution function of the emulated forecast was used to estimate the prediction interval (PI) centered on the median with a coverage (1-α) of 100% as given by (8) [4]. The minimum and maximum values of the 1-minute observations for each hour were used as lower ($\hat{q}_{\tau=\alpha/2}$) and upper ($\hat{q}_{\tau=1-\alpha/2}$) bounds for the quantiles ($\hat{q}_\tau$), respectively. The hourly PI was then averaged daily.

$$\widehat{PI}_{(1-\alpha)100\%} = \left[\hat{q}_{\tau=\alpha/2}, \hat{q}_{\tau=1-\alpha/2} \right] \qquad (8)$$

F. Data Sources

The data used in this study correspond to a well-maintained meteorological station belonging to the Baseline Surface Radiation Network (BSRN) of the Work Radiation Monitoring Center. In particular, the station used was Carpentras (CAR) in France [28], which is the classified hot-summer Mediterranean climate (*CSa*) according to the Köppen climate classification [29]. The data used are those available for the year 2018.

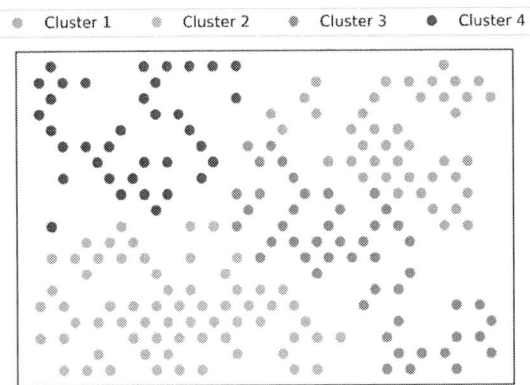

Fig. 2. Resulting two-dimensional SOM with the daily irradiance categories estimated using mean-shift clustering. Each data point represents a day type, where days with the similar characteristics are associated the same location in the SOM technique (total of 193 day types). Data for BSRN station Carpentras, 2018.

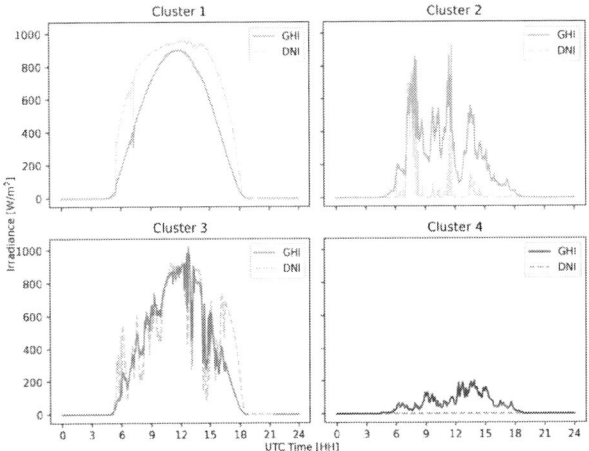

Fig. 3. Sample days within each of the four categorized clusters. Cluster 1 represents clear-sky days; cluster 2 includes days with moderate variability and high ramp rates; cluster 3 comprises days with high variability; and cluster 4 includes cloudy days. Data for 24th April (cluster 1), 1st May (cluster 2), 16th April (cluster 3), and 13th May (cluster 4).

III. RESULTS

The results are introduced below in three separate sections covering: (i) the results of the daily irradiance clustering or profiling; (ii) the effect of each daily category in deterministic forecasts; and (iii) the effect of each daily category in probabilistic forecasts.

A. Daily Irradiance Clustering

The classification of daily irradiance profiles with the proposed SOM and mean-shift technique resulted in four different clusters when assessing the six location-independent metrics related to energy yield and variability for global horizontal and direct normal irradiance observations. The two-dimensional representation of SOM is illustrated in Fig. 2, where each data point corresponds to a day type as result of the SOM dimensionality reduction. A total of 193 different day types were the result of the SOM technique, which were then categorized with the mean-shift method into the four identified clusters. The days were classified as follows: 97 days were classified as cluster 1; 105 days as cluster 2; 79 days as cluster 3; and 83 days as cluster 4. An example of a day in each of the categories is illustrated in Fig. 3. The selected days were the closest to the median within each group, and help understand the differences identified by the algorithm. It can be observed how days classified in Cluster 1 are clear-sky days or days with primarily sunny conditions. Therefore, characterized by high energy yield (both global horizontal and beam irradiation), low variability and ramp rates, and a high daily clearness index. In Cluster 2, days with moderate-to-high variability and high ramp rates are found. Cluster 3 includes days with high variability, high ramp rates and mid-high energy yields. Finally, Cluster 4 includes cloudy days with primarily overcast sky conditions and reduced direct irradiation.

Quantitatively, the mean and standard deviation of the six location-independent metrics for each daily category are presented in Table II, which helps obtain additional insights about each group. It illustrates the differences between Cluster 2 and 3, both highly variable; however, days in Cluster 3 have the highest variability with a mean VI_{GHI} of 7.5 and VI_{DNI} of 10. While both categories observe similar maximum clear-sky index ramps, an average of 0.54 and 0.52 for Cluster 2 and 3, respectively, days in Cluster 3 have far higher energy yield and relatively daily clearness index compared to Cluster 2. Table II also verifies some of the characteristics visually observed in Fig. 3. The lowest variability in both GHI and DNI is found in Cluster 1 (clear-sky days), even lower than the typical overcast days of Cluster 4. As expected for clear-sky days, Cluster 1 has the highest clearness index. The relationships among the six location-independent metrics used to assess daily irradiance variability and energy yield are shown in Fig. 4, which presents a matrix of scatter plots with the identified clusters and a histogram for each of the metrics. This graph permits the visualization of the different clusters, whose distributions can be compared to the categorization of other studies (see Section IV).

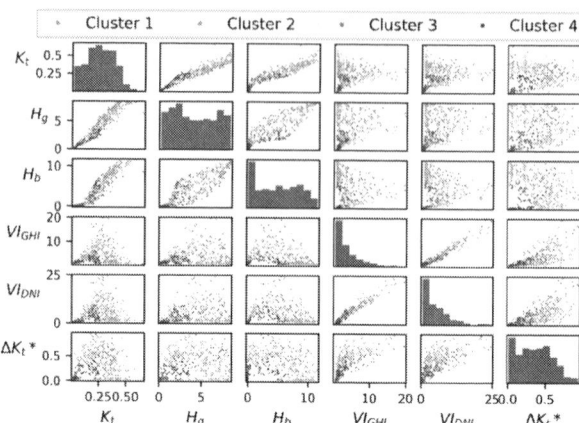

Fig. 4. Scatter matrix with identified clusters of the six location-independent metrics used as input in the SOM technique.

of each solar component with respect to the range of dispersion observed in the 1-minute observations. The identification of the daily mean deviation per cluster permits obtaining insights about the impact of each day category on the forecasts. For the HA

TABLE II. CHARACTERISTICS IDENTIFIED (MEAN AND STANDARD DEVIATION) IN THE DAILY IRRADIANCE PROFILES RESULTED FROM THE CLASSIFICATION OF SOM COMBINED WITH MEAN-SHIFT TECHNIQUE.

Daily Location-Independent Variable	Daily category/cluster (μ, σ)			
	1	2	3	4
Clearness Index (K_t)	0.39 (0.07)	0.17 (0.06)	0.32 (0.06)	0.14 (0.08)
Global Energy Yield (H_g)	6.55 (1.47)	2.55 (1.28)	5.78 (1.46)	1.69 (1.12)
Beam Energy Yield (H_b)	8.89 (1.43)	2.56 (1.97)	6.27 (2.14)	1.96 (2.16)
VI in GHI (VI_{GHI})	1.55 (0.83)	4.92 (2.44)	7.51 (4.72)	2.15 (1.29)
VI in DNI (VI_{DNI})	2.27 (1.42)	6.66 (3.51)	10.03 (5.89)	2.67 (2.30)
Max. rate-of-change Clear-sky Index ($\Delta K_t{*}_{1min}$)	0.21 (0.17)	0.54 (0.13)	0.52 (0.20)	0.18 (0.12)

B. Effect of Daily Profile in Deterministic Forecasts

The deviation of the deterministic forecasts for GHI and DNI are evaluated in this section. Fig. 5 shows the timeseries of daily mean percentage deviation for the high accuracy (HA) forecasts

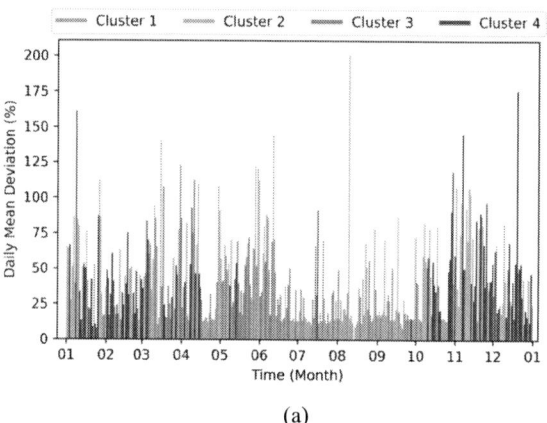

(a)

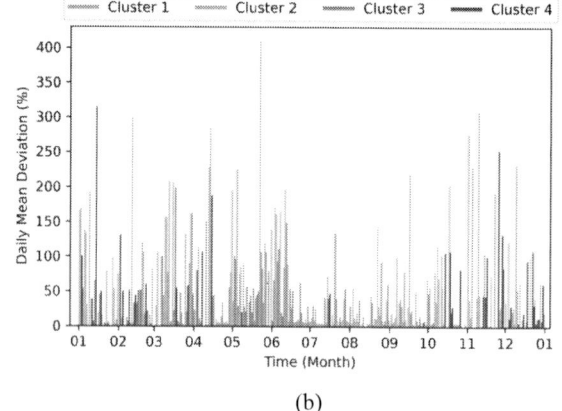

(b)

Fig. 5. Daily mean percentage deviation between the HA hourly deterministic forecasts and the 1-minute observations: (a) GHI forecasts and (b) DNI forecasts. Data identified by daily irradiance category. Cluster 1 (clear-sky days) reports consistently lower percentage deviation (below 50%). Cluster 2 to 4 register higher values throughout the year, whose mean hourly deviation exceed 100% in multiple occasions.

forecast of GHI, Cluster 1 representing clear-sky days reports the lowest percentage of deviation. This is expected as the intra-hour irradiance variability on sunny days is reduced to that of the diurnal cycle. The daily mean deviation is over 50% in most of the days in Clusters 2 to 4. For DNI forecasts, Clusters 1 and 4 have the lowest deviation. This is also an expected behaviour as DNI variability on sunny days is that of the diurnal cycle and there is limited direct energy yield during overcast days. It is worth noting the maximum values of daily average deviations, which expand further in DNI forecasts as observed on the y-axis of each subplot. Table III presents the mean and standard deviation of the daily mean deviation per cluster for the HA forecasts and the 24h clear-sky persistence baseline model. The results of the HA forecasts are those for the timeseries of Fig. 5, where it can be observed how Cluster 2 presents the highest deviations on average (63.7% for GHI and 104% for DNI), followed by Cluster 4 (49.3% for GHI and 63.2% for DNI). Regarding the performance of the 24-hour persistence baseline model, it can be observed that the average values of the daily mean deviation of the hourly forecasts are far higher than the emulated HA forecasts. However, there is an exception in the persistence DNI forecasts of days in Cluster 4, which have an average deviation of 1.5%, despite the large standard deviation. This can be understood when looking at the location within the year of days in Cluster 4 of Fig. 5, which are mostly grouped in periods of several cloudy/overcast days, rather than single days.

TABLE III. DAILY MEAN PERCENTAGE DEVIATION (MEAN AND STANDARD DEVIATION) BETWEEN DETERMINISTIC HOURLY FORECASTS AND 1-MINUTE OBSERVATIONS.

Deterministic Forecast	Daily category/cluster (μ, σ)			
	1	2	3	4
HA Forecast GHI	16.6 (17.4)	63.7 (40.8)	40.3 (32.2)	49.3 (28.5)
HA Forecast DNI	9.3 (14.4)	104.0 (77.6)	48.9 (47.0)	63.2 (43.0)
Clear-sky Persistence GHI	77.8 (110.3)	282.2 (323.7)	121.7 (117.9)	66.7 (99.6)
Clear-sky Persistence DNI	63.7 (99.6)	148.9 (260.2)	73.1 (100.3)	1.5 (70.9)

C. Effect of Daily Profile in Probabilistic Forecasts

The results for probabilistic forecasts are evaluated below using the prediction interval (PI) of the hourly probabilistic forecasts, which includes the maximum variations observed in the intra-hourly (1-minute) data. Fig. 6 introduces the timeseries of daily mean PI for the emulated hourly HA probabilistic forecasts with the clusters identified. As found with the deterministic forecasts, days in Cluster 1 (clear-sky days) report lower deviations. Similarly, the deviation as shown in the PI values centered on the median is higher in the DNI forecasts than in the GHI forecasts. Table IV complements Fig. 6 by presenting the average and standard deviation of the daily mean PIs. As anticipated by the figure, the HA forecasts of days in Cluster 1 report the lowest PIs (11.6% for GHI and 13.2% for DNI), followed by Cluster 4, which presents almost the same results for both solar components (15.4% for GHI and 15.5% for DNI). The PIs observed in Cluster 2 and 3, which are the daily profiles with high irradiance variability, present the highest PIs for GHI (25.7% for Cluster 2 and 26.1% for Cluster 3) and DNI forecasts (33.2% for Cluster 2 and 35.5% for Cluster 3).

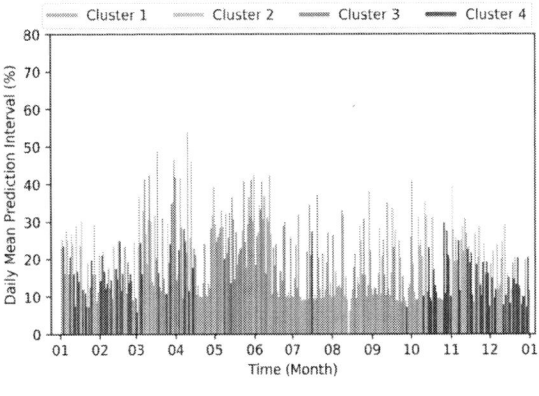

(a)

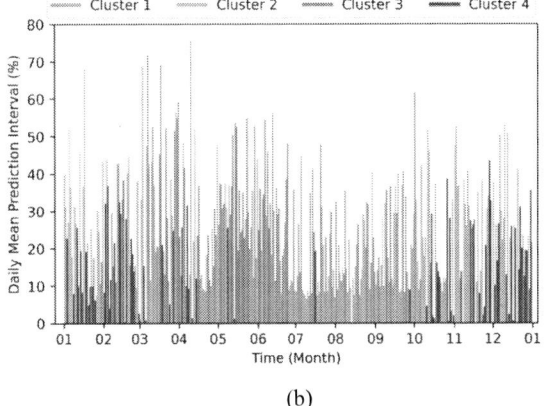

(b)

Fig. 6. Daily mean prediction intervals observed between the HA hourly probabilistic forecasts and the 1-minute observations: (a) GHI forecasts and (b) DNI forecasts. Data identified by daily irradiance category. Cluster 1 (clear-sky days) report ranges circa 10-15% on average. Cluster 2 to 3 observe higher probabilistic ranges around 25% for GHI to 35% for DNI from P50 estimations.

With regards to the probabilistic baseline models, the CH-PeEn model observed slightly higher PIs centered in the median (P50) for GHI forecasts than the emulated forecasts. However, for DNI forecasts, the CH-PeEn model reported lower average and standard deviation PIs for day types other than sunny/clear-sky days (Cluster 1) as presented in Table IV, where the maximum difference is found in days of Cluster 2 (i.e., variable days towards cloudy) with a PI of 25.7% for the CH-PeEn and 33.2% for the emulated forecast.

TABLE IV. DAILY MEAN PREDICTION INTERVALS EXPRESSED IN PERCENTAGE (MEAN AND STANDARD DEVIATION) BETWEEN PROBABILISTIC HOURLY FORECASTS AND 1-MINUTE OBSERVATIONS.

Probabilistic Forecast	Daily category/cluster (μ, σ)			
	1	2	3	4
HA Forecast GHI	11.6 (8.3)	25.7 (13.2)	26.1 (12.9)	15.4 (6.9)
HA Forecast DNI	13.2 (10.2)	33.2 (22.1)	35.5 (20.3)	15.5 (10.2)
Baseline CH-PeEn GHI	17.9 (9.6)	29.0 (15.9)	35.2 (17.4)	15.7 (8.7)
Baseline CH-PeEn DNI	14.6 (9.3)	25.7 (15.3)	30.0 (16.4)	13.3 (8.9)

IV. DISCUSSION

The results are discussed below to analyze the findings and relevance thereof. The classification of the daily irradiance profiles was implemented by using SOM in combination with mean-shift clustering, which represented a novel methodology when compared to previous literature addressing this topic as illustrated by the taxonomy table (Table I). Previous work had proposed classifications of GHI and/or DNI profiles based on up to 3 metrics. In this work, the input metrics for the classification problem were 6 location-independent related to irradiance variability and energy yield, some of which are common to those used in the literature. However, the main difference of the proposed approach is accounting for metrics integrating energy yield, irradiance ramp-rates and variability for GHI and DNI daily profiles. The classification with SOM plus mean-shift resulted in 4 groups, which is aligned with the literature with 3-5 clusters [14]. The groups identified are: (Cluster 1) clear-sky or primarily clear-sky days; (Cluster 2) highly variable days towards cloudy conditions, which are characterized by moderate variability and high rates of change and generally have low energy yield; (Cluster 3) highly variable days towards sunny conditions, characterized by high variability and high ramp rates; and (Cluster 4) cloudy or overcast days.

By using common metrics with other literature, the daily irradiance profiles identified can be compared to the literature. In [10], VI_{GHI} and K_t were used to classify daily irradiance profiles for GHI resulting in 4 categories (i.e., overcast, clear-sky, high-variable all day, and days with variable skies). Similarly, in [12] 3 categories (i.e., high yield, low yield, and highly variable) were identified by using a normalized VI_{GHI} and H_g. In this work, the proposed Cluster 1 (sunny or primarily sunny conditions) would be equivalent to the 'clear-sky' and 'high yield' categories in [10] and [12], respectively. Cluster 4 (cloudy/overcast conditions) would also have equivalence with the 'overcast' and 'low yield' categories in [10] and [12], respectively. Considering that the type 'high-variable all day' of [10] and 'highly variable' of [12] are very similar, the differences with the proposed Clusters 2 and 3 can be commented upon. The distribution of VI_{GHI} v. K_t used in [10], which resembles an arrowhead, is illustrated in the scatter matrix (Fig. 4 - top row, 4th column) and it can be observed that the classification by the SOM plus mean-shift algorithm returns as Cluster 2 (in blue) those values with K_t below the midpoint of the y-axis approximately, and as Cluster 3 (in green) those values over the midpoint of the y-axis. Thus, characterising differently the daily categories 'high-variable all day', and 'days with variable skies' proposed by [10]. As described in Section III, the differences between Clusters 2 and 3 are also with regards to maximum ramp rate of change as observed in the distributions of K_t v. ΔK_t^*; H_g v. ΔK_t^*; and H_b v. ΔK_t^* (see Fig. 4: bottom row, columns 1 to 3).

The effect of these daily irradiance categories in deterministic and probabilistic forecasting models was then evaluated. For deterministic models, it was shown how the daily average irradiance observed intra-hourly can exceed over 100% of high accuracy forecasts in non-sunny days (Clusters 2 to 4). The dispersion between intra-hourly observations and hourly forecasts was higher in DNI forecasts, which is expected as intermittent cloud cover can completely change the DNI levels.

Regarding the 24-hour persistence model, it was observed that the model is not able to deal with the intra-hourly variability of solar resources with the exception of overcast/cloudy days grouped under Cluster 4, which reported a daily mean deviation of hourly forecasts of 1.5%; it is, however, worth noting the standard deviation of 70.9%. This lower dispersion observed by the deterministic baseline model can be understood as full-overcast conditions are not isolated events and occurred during a few consecutive days (e.g., produced by an anticyclone or a storm) as observed by the timeseries of day types (see Fig. 5). For probabilistic forecasts, the daily mean value of hourly PI required to capture the intra-hourly observations was used a metric for the dispersion of the forecasts. The results showed that days in Cluster 1 (clear-sky) have the lowest PIs for the emulated HA forecasts of any solar component, followed by days in Cluster 4, which had very similar results for GHI and DNI. Clusters 2 and 3, which cover high variable days, reported daily mean PIs below 36% for any of the cases, including the emulated HA and baseline CH-PeEn forecasts. It is worth mentioning that the CH-PeEn model reported lower daily mean PIs for DNI than those of the emulated forecasts for any day type different from those in Cluster 1.

Since probabilistic forecasts are usually defined in prediction intervals from the median or with probabilities of exceedance (e.g., P10, P25 or P90), probabilistic forecasts are intrinsically more suitable to deal with deviations from the median or any other reference quantile. The results showed that PIs of 1-minute observations with respect to HA forecasts can be below 80%, both for GHI and DNI. Therefore, a probabilistic forecast expressed with probabilities of exceedance (e.g., P10, P50 and P90) would cover a wider range than the mean PIs observed in the data. This also highlights the importance of the findings, where the probabilities of exceedance used to give the forecasts could be adapted to each of the daily categories for higher confidence intervals. For example, for days in Cluster 1, an hourly GHI probabilistic forecast with probabilities of exceedance (P40, P50 and P60) would be sufficient to cover the daily mean PI with 1 standard deviation observed in the 1-minute observations (i.e., μ=11.6, σ=8.3; PI≈20 with lower bound in quantile q_{40} and upper bound quantile q_{60}).

V. SUMMARY OF THE WORK

A categorization of the daily irradiance profiles attending to six location-independent variability and energy yield metrics related to GHI and DNI was proposed in this work. The classification of daily irradiance profiles used SOM and the mean-shift technique as a differentiating methodology compared to previous related work. Four daily irradiance categories were identified: clear-sky days; variable days towards sunny conditions; variable days towards cloudy conditions and cloudy or overcast days. In addition, the impact of each daily category on hourly deterministic and probabilistic forecasts in terms of intra-hour variability was evaluated.

High accuracy deterministic (or point) and probabilistic forecasts were emulated by adding white noise, which ensured a forecast without autocorrelation or seasonality factors. These forecasts were used as a basis to illustrate the impact of the resulting daily irradiance profiles. Moreover, state-of-the-art benchmarking models for forecasts were also evaluated: the 24-

hour persistence model for deterministic forecasts and the CH-PeEn for the probabilistic forecasts. The results showed that the mean daily dispersion of intra-hourly observations can reach from 16.6% to 63.7% on average depending on the daily GHI profile, exceeding over 100% at times, for high accuracy GHI deterministic forecasts; and range from 9.3% to 104%, exceeding 200-300% at times, for DNI deterministic forecasts. Thus, illustrating the potential intra-hourly irradiance variability that deterministic forecasts cannot capture. The results showed that hourly probabilistic forecasts required daily average prediction intervals ranged from 11% to 35.5% to capture all the irradiance variability in 1-minute GHI and DNI observations occurring within the hourly intervals regardless of the daily categorization. Thus, suggesting that probabilistic forecasts are better suited to capture the variability of solar irradiance occurring at increased temporal resolutions.

Overall, this study provides useful insights on the deviation that hourly deterministic and probabilistic forecasts can observe in an intra-day and intra-hourly resolution. These findings illustrate how the production of solar forecasts could be adapted considering each daily category to support the integration of solar energy in power systems.

ACKNOWLEDGMENT

This work was co-funded by the European Regional Development Fund and the Republic of Cyprus through the Research and Innovation Foundation in the framework of the project ELECTRA with protocol number INTEGRATED/0918/0071.

REFERENCES

[1] IRENA, "Renewable capacity statistics 2021," Abu Dhabi, 2021. [Online]. Available: https://www.irena.org/publications/2021/March/Renewable-Capacity-Statistics-2021.

[2] IRENA, "Innovation landscape brief: Advanced forecasting of variable renewable power generation," Abu Dhabi, 2020.

[3] Y. Chu, M. Li, C. F. M. Coimbra, D. Feng, and H. Wang, "Intra-hour irradiance forecasting techniques for solar power integration: a review," *iScience*, vol. 24, no. 10, p. 103136, 2021, doi: 10.1016/j.isci.2021.103136.

[4] P. Lauret, M. David, and P. Pinson, "Verification of solar irradiance probabilistic forecasts," *Sol. Energy*, vol. 194, pp. 254–271, 2019, doi: 10.1016/j.solener.2019.10.041.

[5] D. Yang et al., "Verification of deterministic solar forecasts," *Sol. Energy*, vol. 210, pp. 20–37, 2020, doi: 10.1016/j.solener.2020.04.019.

[6] D. Yang, "A universal benchmarking method for probabilistic solar irradiance forecasting," *Sol. Energy*, vol. 184, pp. 410–416, 2019, doi: 10.1016/j.solener.2019.04.018.

[7] J. Le Gal La Salle, M. David, and P. Lauret, "A new climatology reference model to benchmark probabilistic solar forecasts," *Sol. Energy*, vol. 223, pp. 398–414, 2021, doi: 10.1016/j.solener.2021.05.037.

[8] J. Lopez Lorente, X. A. Liu, and D. J. Morrow, "Worldwide evaluation and correction of irradiance measurements from personal weather stations under all-sky conditions," *Sol. Energy*, vol. 207, pp. 925–936, 2020, doi: 10.1016/j.solener.2020.06.073.

[9] L. D. Riihimaki, X. Li, Z. Hou, and L. K. Berg, "Improving prediction of surface solar irradiance variability by integrating observed cloud characteristics and machine learning," *Sol. Energy*, vol. 225, pp. 275–285, 2021, doi: 10.1016/j.solener.2021.07.047.

[10] J. S. Stein, C. W. Hansen, and M. J. Reno, "The variability index: A new and novel metric for quantifying irradiance and PV output variability," in *World Renewable Energy Congress*, 2012, pp. 1–7,

doi: 10.1111/j.1472-765x.2009.02724.x.

[11] J. Lopez Lorente, X. Liu, D. J. Morrow, and P. V. Brogan, "Potential for crowdsourced weather stations to assess intra-hourly variability of photovoltaic systems," in *36th European Photovoltaic Solar Energy Conference and Exhibition*, 2019, pp. 1410–1416, doi: 10.4229/EUPVSEC20192019-5DO.2.2.

[12] F. P. M. Kreuwel, W. H. Knap, L. R. Visser, W. G. J. H. M. van Sark, J. Vilà-Guerau de Arellano, and C. C. van Heerwaarden, "Analysis of high frequency photovoltaic solar energy fluctuations," *Sol. Energy*, vol. 206, pp. 381–389, 2020, doi: 10.1016/j.solener.2020.05.093.

[13] S. Theocharides, G. Makrides, A. Livera, M. Theristis, P. Kaimakis, and G. E. Georghiou, "Day-ahead photovoltaic power production forecasting methodology based on machine learning and statistical post-processing," *Appl. Energy*, vol. 268, p. 115203, 2020, doi: 10.1016/j.apenergy.2020.115023.

[14] B. Hartmann, "Comparing various solar irradiance categorization methods – A critique on robustness," *Renew. Energy*, vol. 154, pp. 661–671, 2020, doi: 10.1016/j.renene.2020.03.055.

[15] S. Theocharides, V. Venizelou, G. Makrides, and G. E. Georghiou, "Daily solar irradiance profile characterization and ramp rate analysis at different time resolutions," in *IEEE 44th Photovoltaic Specialist Conference (PVSC)*, 2017, pp. 1163–1168, doi: 10.1109/PVSC.2017.8366517.

[16] M. Schroedter-Homscheidt, M. Kosmale, S. Jung, and J. Kleissl, "Classifying ground-measured 1 minute temporal variability within hourly intervals for direct normal irradiances," *Energy Meteorol.*, vol. 27, no. 2, pp. 161–179, 2018, doi: 10.1127/metz/2018/0875.

[17] A. Castillejo-Cuberos and R. Escobar, "Understanding solar resource variability: An in-depth analysis, using Chile as a case of study," *Renew. Sustain. Energy Rev.*, vol. 120, p. 109664, 2020, doi: 10.1016/j.rser.2019.109664.

[18] T. E. Hoff and R. Perez, "Quantifying PV power output variability," *Sol. Energy*, vol. 84, no. 10, pp. 1782–1793, 2010, doi: 10.1016/j.solener.2010.07.003.

[19] W. Zhang, J. Wang, D. Jin, L. Oreopoulos, and Z. Z., "A deterministic Self-Organizing Map approach and its application on satellite data based cloud type classification," in *IEEE International Conference on Big Data*, 2018, pp. 2027–2034, doi: 10.1109/BigData.2018.8622558.

[20] V. Kristollari and V. Karathanassi, "Fine-tuning Self-Organizing Maps for Sentinel-2 imagery: Separating clouds from bright surfaces," *Remote Sens.*, vol. 12, no. 12, p. 1923, 2020, doi: 10.3390/rs12121923.

[21] C. N. Long and Y. Shi, "An automated quality assessment and control algorithm for surface radiation measurements," *Open Atmos. Sci. J.*, vol. 2, pp. 23–37, 2008, doi: 10.2174/1874282300802010023.

[22] R. Perez et al., "Spatial and temporal variability of solar energy," *Found. Trends Renew. Energy*, vol. 1, no. 1, pp. 1–44, 2016, doi: 10.1561/2700000006.

[23] P. Ineichen and R. Perez, "A new airmass independent formulation for the Linke turbidity coefficient," *Sol. Energy*, vol. 73, pp. 151–157, 2002, doi: 10.1016/S0038-092X(02)00045-2.

[24] T. Kohonen, "The self-organizing map," *Proc. IEEE*, vol. 78, no. 9, pp. 1464–1480, 1990, doi: 10.1109/5.58325.

[25] Y. Cheng, "Mean shift, mode seeking, and Clustering," *IEEE Trans. Pattern Anal. Mach. Intell.*, vol. 17, no. 8, pp. 790–799, 1995, doi: 10.1109/34.400568.

[26] D. Comaniciu and P. Meer, "Mean shift: a robust approach toward feature space analysis," *IEEE Trans. Pattern Anal. Mach. Intell.*, vol. 24, no. 5, pp. 603–619, 2002, doi: 10.1109/34.1000236.

[27] I. U. Moffat and E. A. Akpan, "White noise analysis: a measure of time series model adequacy," *Appl. Math.*, vol. 10, no. 11, pp. 989–1003, 2019, doi: 10.4236/am.2019.1011069.

[28] L. Brunier, "Basic measurements of radiation at station Carpentras (2018-12)," *Centre Radiometrique, PANGAEA*, 2019. https://doi.org/10.1594/PANGAEA.898286 (accessed Apr. 01, 2022).

[29] M. C. Peel, B. L. Finlayson, and T. A. McMahon, "Updated world map of the Köppen-Geiger climate classification," *Hydrol. Earth Syst. Sci.*, vol. 11, pp. 1633–1644, 2007, doi: 10.5194/hess-11-1633-2007.

Hyperspectral Imaging of Localized, Optically-Active Defects in GaAs Solar Cells

Behrang H. Hamadani[1], Margaret A. Stevens[2], Brianna Conrad[1], Matthew P. Lumb[3], Eric Armour[4], and Kenneth J. Schmieder[5]

[1]National Institute of Standards and Technology, Gaithersburg, MD 20899 USA
[2]NRC Postdoc Residing at NRL, Washington, DC 20375 USA
[3]Formerly with George Washington University, Washington, DC 20052 USA
[4]Veeco Instruments, Somerset, NJ 08873 USA
[5]U.S. Naval Research Laboratory, Washington, DC 20375 USA

Abstract—A novel hyperspectral imaging instrument with dual electroluminescence and photoluminescence capability was used to image and analyze localized radiative defects in rear-junction GaAs solar cells grown at 1 μm/min. The absolute photon emission rates were measured over a wide spectral range encompassing both the sub-band gap region and the higher photon energies, resolving spatial features on the order of 1 μm with complete spectral information. We find that some radiative defects are pinholes in the GaAs active layer, transmitting luminescence signal from other buried device layers. More notably, we observe the formation of a high-intensity, halo-like defect band in the vicinity of some of these processing-related defects. Such defects show a characteristic double-peak emission with maxima at 870 nm and 894 nm at room temperature corresponding to the band-to-band and the band-to-impurity optical transitions respectively. These sub-band gap radiative regions are likely gallium antisite defects and are thought to form shallow impurity bands within the band gap.

Keywords—*hyperspectral imaging, defects, GaAs solar cells, radiative defects, photoluminescence.*

I. Introduction

Hyperspectral (HS) imaging is a convenient wide-field technique that combines spectroscopy and imaging to obtain a spectrum for each pixel of an image [1], [2]. Given the appropriate spectral and spatial resolution, HS imaging in either electroluminescence (EL) or photoluminescence (PL) modes can be used to obtain the full spectral emission information of visualized local defects or regions of heterogeneity across the active layer of a solar cell [3]. Traditional mapping techniques such as micro-PL can provide high-resolution images under a scanning optical microscope; however, such techniques are time-consuming and injection levels are locally very high [4]. Given appropriate illumination and collection optics and camera pixel size and sensitivity, HS wide-field imaging can be performed at intensities well under 10 mW/cm^2, spatial resolution of < 1 μm, and a spectral resolution of ≈ 1 nm. If the HS imager is calibrated by use of radiometric standards to measure the absolute photon emission rates, i.e., photons/m^2·s·eV for each photon energy E, then the generalized Planck law can be used to calculate local Fermi level splitting energy, local radiative efficiency (if incident photon flux is known) and other important device-related parameters [5], [6].

Although luminescence-based imaging techniques will only reveal the location and the intensity of radiative transitions in semiconductor materials, such observations could also indicate nonradiative centers or local shunts across a device and will reveal how the voltage or current flow is affected around a particular local defect or region. In this paper, we report a novel radiative defect halo that is (mostly) formed at locations of processing-related defects within the active GaAs emitter layer of rear-junction solar cells. The high growth rate of the active layer makes these cells particularly susceptible to local defect formations. We believe these defects are gallium antisite or vacancy defects that are formed in large concentrations at the locations shown. These types of defects are usually more prominent at lower temperatures due to the temperature-activated nature of the transition; however, with high enough concentration, they can be spectrally resolved even at room temperature. The temperature evolution of one such defect will also be briefly discussed.

II. Methods

A. Device fabrication and structure

The devices were grown using a K475i Veeco MOCVD tool [7]. The precursors were trimethylgallium (TMGa), trimethylindium (TMIn), trimethylaluminum (TMAl), arsine (AsH$_3$), phosphine (PH$_3$), disilane (Si$_2$H$_6$), diethyltellurium (DETe), carbon tetrabromide (CBr$_4$), and dimethylzinc (DMZn). GaAs emitter and base growth rate was approximately 1 μm/min. The GaAs V/III ratio was 9. Growth temperature varied between 640 °C to 680 °C, and chamber pressure was held at 42 Torr. GaAs substrates were (100) 5° offcut toward <011> p-type. Solar cells were grown above an AlGaAs distributed Bragg reflector (DBR) with a stop-band nominally centered at 850 nm. Devices were fabricated with standard lithographic and wet etch processes into mesas of 5 mm×5 mm. Device structure is shown in the inset of Fig. 2. Further structure details and process information can be found in reference [6].

B. Device measurements and characterization

The solar cells were wire-bonded and characterized using current vs voltage (*I-V*), external quantum efficiency (EQE), and external radiative efficiency (ERE) measurements. The typical

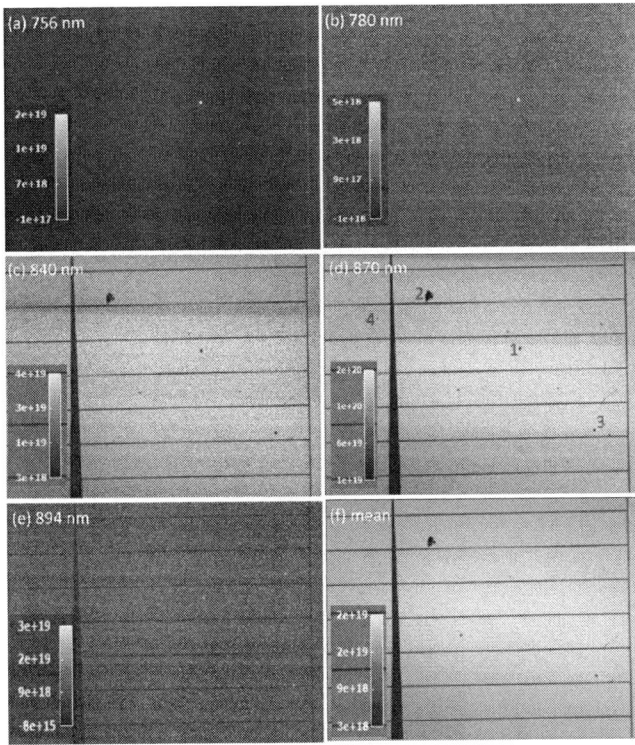

Fig. 1. Absolute PL images of a rear-junction GaAs solar cell at several wavelengths (a-e) and the mean PL image (f) obtained under the 5× objective. The FOV is ≈ 3.5 mm. Emission rates are in photons/(m²·s·eV).

open circuit voltage V_{oc} and the short circuit current I_{sc} under air mass 1.5 G illumination for these cells were 1.032 V and 26.8 mA/cm², respectively, with the power conversion efficiency (PCE) at ≈ 22.55 % but higher performance has been reported previously [8], [9]. The EQE at 840 nm is at 93.7 %, and the average device ERE at a laser illumination intensity of ≈ 700 W/m² is about 1 %, which is comparable with other GaAs devices of similar PCE that have been tested in our lab. Our expanded uncertainty in PCE measurements is about 1 % and the ERE measurements are accurate to about ± 15 %.

C. Hyperspectral characterization

HS imaging in both PL and EL modes was performed using a wide-field imaging system with capability to scan the spectral region of 400 nm-1600 nm using two camera systems. In this paper, only PL images are shown and discussed. The various objective lenses allow for a field of view (FOV) ranging from 160 mm to less than 500 μm. In the 400 nm to 1000 nm region, the spectral resolution is better than 2 nm. For PL excitation, a 532 nm laser is used to uniformly illuminate the entire FOV, with laser intensities depending on the objective, neutral density filter, and the laser power setting. The reported data here are all at laser intensities in the range of 100 mW/cm² to 800 mW/cm². The absolute calibration of the imager is discussed elsewhere [5]. Temperature dependent imaging was performed with a liquid nitrogen-flow optical cryostat under vacuum.

III. RESULTS

Figure 1(a-e) shows a series of absolute PL images, taken with a 5 × objective at several wavelengths, with the colorbar

showing the absolute photon emission rates. The dark horizontal gridlines and the thicker vertical line are the top electrode contact, with 350 μm spacing between the horizontal lines. Since the band-to-band (BB) transition in GaAs is at ≈ 870 nm (1.425 eV), no emission should be visible in the 756 nm (Fig. 1(a)) image. A closer examination however reveals several bright spots where a significant signal is recorded. These spots mostly fade out at 780 nm. The 840 nm and the 870 nm images show large bright areas across the device corresponding to the BB signal with emission rates strongest at 870 nm. The 756-nm bright spots are now mostly dark, indicating little BB signal at those sites. There are also additional defects appearing dark in these images.

Examination of Fig. 1(e) at 894 nm reveals a faint, doughnut-shaped signal at the location of some of the dark spots, particularly those present at 756 nm, which is discussed next. One final observation here is that the mean PL signal (averaged over all the wavelengths from 700 nm to 950 nm), as shown in Fig. 1(f) is not sufficient to show all the nuances that the hyperspectral imaging has revealed.

The complete emission spectra from inside the 4 localized defects as labelled in Fig. 1(d) and also the whole-cell PL signal are shown in Fig. 2. The most prominent of these defects is *defect 1*, which shows a substantial peak at 756 nm, much more intense than its peak at 870 nm, which is the reason why it appears very bright in Fig. 1(a) but dark in Fig. 1(d). Others such as *defect 4* also show the 756 nm peak, though at a smaller rate. We have concluded that these defects are pinhole defects that transmit the PL signal associated with the buried DBR layer up to the top, as verified by independent DBR-only PL measurements. It is likely that the pinholes originate from imperfections in the photoresist mask during the fabrication process, but they could also be related to defect formation during the fast growth conditions of these devices. Some other defects such as *defect 2* and *defect 3* show little to none of the 756 nm signal, and hence are not pinholes but likely areas of low-quality GaAs growth. The whole-cell signal does not seem to be affected much by the deep pinholes since there are only a few of

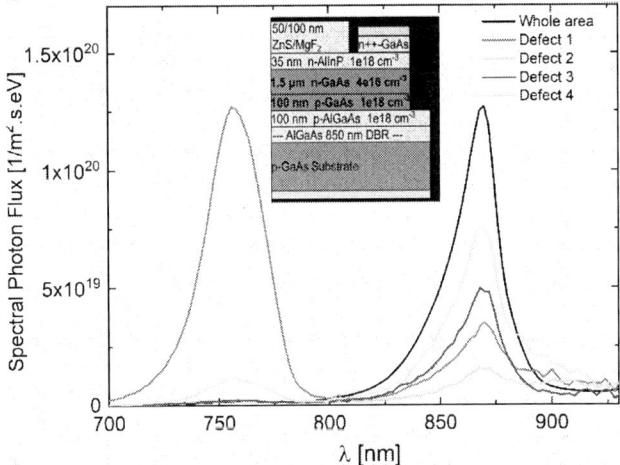

Fig. 2. (Inset) device structure. (main) The PL emission spectra from several defect sites labelled in Fig. 1(d) and also the whole-area PL signal.

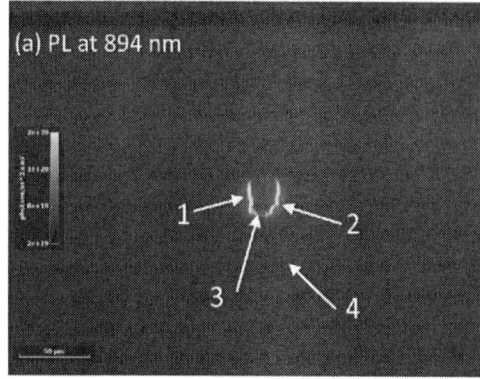

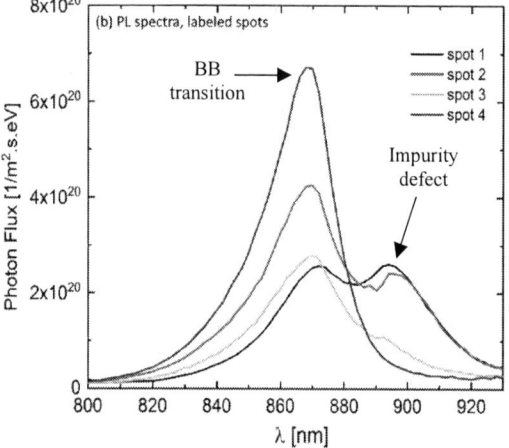

Fig. 3. (a) The PL image of a localized defect under the 20× objective at 894 nm and at 297 K, showing a strong radiative signal from a halo band of what is believed to be the site of Ga_{As} antisite defects. (b) Spectra from several labelled locations (tip of the arrow).

them, hence showing only the BB signal peaked at 870 nm at 297 K.

The PL curve below the band gap energy (>870 nm) sampled at the defect sites does not drop off to zero but rather displays a shallow tail distribution. In Fig. 3(a), we have magnified *defect 1* with a 20× objective for a better visualization. This image clearly reveals the previously-mentioned bright halo around the center of the defect at 894 nm. Spectrum sampling

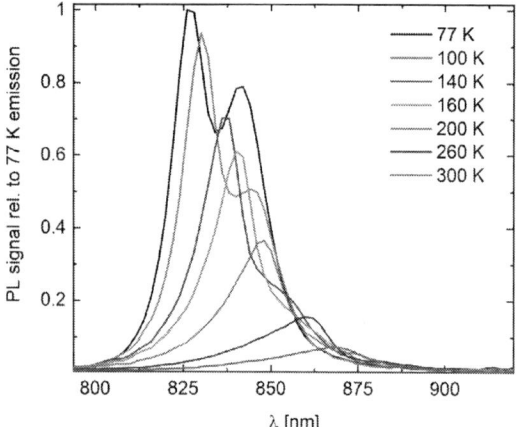

Fig. 4. Temperature dependence of a single defect spectra, showing the prominence of a sub-band gap defect peak at 77 K.

(Fig. 3(b)) of a few spots directly on and off the halo shows a double-peak distribution on the bright regions and reduced intensities at other areas. Therefore, the 870 nm peak is the BB transition, and the 894 nm peak indicates the formation of a defect sub-band gap transition at these particular locations.

Room temperature measurements alone are not sufficient to identify the nature of these sub-band gap defects and GaAs radiative peaks are usually studied at low temperatures, particularly at 77 K. We have performed preliminary temperature dependent measurements to better understand the evolution of the defect peak with temperature. These measurements, which have been performed with the 5× objective, are shown in Fig. 4 where we plot the normalized PL signal (relative to the 77 K PL) at one such defect site for several temperatures. Here, the PL signal is an average signal from the defect region and hence does not show a distinct double peak at 300 K. However, a double peak is clearly visible at 77 K, with the higher-wavelength peak emanating from the defect emission. From this plot, it is clear that the defect peak has a substantial temperature dependence and shows an activated behavior. The PL intensity of the BB peak usually increases substantially at lower T, as shown here, and the peak distribution becomes sharper due to the freezing of nonradiative recombination. The peak position also shifts to higher energy (1.50 eV, or 827 nm) due to the temperature dependence of the band gap energy [10].

Furthermore, we observe that the defect peak increases in intensity as T is lowered, at a rate faster than the BB peak, and reaches an almost comparable emission intensity as the BB peak at 77 K (dark curve). The defect peak position at 77 K is 1.473 eV (842 nm). This peak, particularly in Si-doped GaAs samples has been associated with gallium antisite defects (Ga_{As} or a Ga atom at an As site) or gallium vacancies that in large concentrations can act as a band of compensating acceptors causing the optical transition to shift from BB to band-to-impurity (i.e., electrons in the conduction band recombine with holes trapped on these acceptors) [11], [12]. This type of transition is often referred to as free-to-bound (FB) transition[13], [14]. Detailed temperature-dependent PL intensity measurements can reveal significant information regarding these defect transitions and will be discussed in a future work. Our preliminary results indicate an activation energy of about 41 meV for this transition. The peak position at 77 K can be slightly dependent on the Si dopant concentration with values reported between 1.45 eV and 1.47 eV from lightly doped to heavily doped materials. Since the defect peak in our samples is at 1.473 eV, we speculate a higher concentration of Si doping in the vicinity of these defects.

IV. CONCLUSIONS

Hyperspectral imaging in PL mode has been successfully used to study local radiative defects in GaAs solar cells. In addition to pinhole defect sites, a halo-like, radiative transition with peak emission at 894 nm at room T has been observed at several sites including around the pinhole defects. Low temperature imaging indicates that these defects are likely composed of Ga antisite point defects, energetically located about 41 meV above the valance band edge.

REFERENCES

[1] A. Delamarre, L. Lombez, and J. F. Guillemoles, "Characterization of solar cells using electroluminescence and photoluminescence hyperspectral images," in *Proc. of SPIE*, Feb. 2012, vol. 8256, p. 825614, doi: 10.1117/12.906859.

[2] E. Gaufrès *et al.*, "Hyperspectral Raman imaging using Bragg tunable filters of graphene and other low-dimensional materials," *J. Raman Spectrosc.*, vol. 49, no. 1, pp. 174–182, Jan. 2018, doi: 10.1002/jrs.5298.

[3] I. Burud, T. Mehl, A. Flo, D. Lausch, and E. Olsen, "Hyperspectral photoluminescence imaging of defects in solar cells," *J. Spectr. Imaging*, vol. 5, no. August, pp. 1–5, 2016, doi: 10.1255/jsi.2016.a8.

[4] Q. Chen *et al.*, "Impact of Individual Structural Defects in GaAs Solar Cells: A Correlative and In Operando Investigation of Signatures, Structures, and Effects," *Adv. Opt. Mater.*, vol. 9, no. 2, p. 2001487, Jan. 2021, doi: 10.1002/adom.202001487.

[5] S. M. Chavali, J. Roller, M. Dagenais, and B. H. Hamadani, "A comparative study of subcell optoelectronic properties and energy losses in multijunction solar cells," *Sol. Energy Mater. Sol. Cells*, vol. 236, no. August 2021, p. 111543, Mar. 2022, doi: 10.1016/j.solmat.2021.111543.

[6] B. H. Hamadani, "Understanding photovoltaic energy losses under indoor lighting conditions," *Appl. Phys. Lett.*, vol. 117, no. 4, p. 043904, Jul. 2020, doi: 10.1063/5.0017890.

[7] NIST Disclaimer: "Certain commercial equipment, instruments, software, or materials are identified in this paper to specify the experimental procedure adequately. Such identification is not intended to imply recommendation or endorsement by the National Institute of Standards and Technology, nor is it intended to imply that the materials or equipment identified are necessarily the best available for the purpose".

[8] M. A. Stevens *et al.*, "High Growth Rate Rear-Junction GaAs Solar Cell with a Distributed Bragg Reflector," in *2021 IEEE 48th Photovoltaic Specialists Conference (PVSC)*, Jun. 2021, pp. 0342–0345, doi: 10.1109/PVSC43889.2021.9518409.

[9] K. J. Schmieder *et al.*, "Effect of Growth Temperature on GaAs Solar Cells at High MOCVD Growth Rates," *IEEE J. Photovoltaics*, vol. 7, no. 1, pp. 340–346, Jan. 2017, doi: 10.1109/JPHOTOV.2016.2614346.

[10] I. Vurgaftman, M. P. Lumb, and J. R. Meyer, *Bands and Photons in III-V Semiconductor Quantum Structures*. Oxford University Press, 2020.

[11] L. Pavesi, M. Henini, and D. Johnston, "Influence of the As overpressure during the molecular beam epitaxy growth of Si-doped (211)A and (311)A GaAs," *Appl. Phys. Lett.*, vol. 2846, no. November 1994, p. 2846, 1995, doi: 10.1063/1.113449.

[12] N. H. Ky and F. K. Reinhart, "Amphoteric native defect reactions in Si-doped GaAs," *J. Appl. Phys.*, vol. 83, no. 2, pp. 718–724, Jan. 1998, doi: 10.1063/1.366743.

[13] T. Unold and L. Gütay, "Photoluminescence Analysis of Thin-Film Solar Cells," in *Advanced Characterization Techniques for Thin Film Solar Cells*, vol. 1–2, Weinheim, Germany: Wiley-VCH Verlag GmbH & Co. KGaA, 2016, pp. 275–297.

[14] S. Levcenko, V. E. Tezlevan, E. Arushanov, S. Schorr, and T. Unold, "Free-to-bound recombination in near stoichiometric Cu2ZnSnS4 single crystals," *Phys. Rev. B*, vol. 86, no. 4, p. 045206, Jul. 2012, doi: 10.1103/PhysRevB.86.045206.

Achieving Global Decarbonization by Photovoltaic Electrification: Impact of Disruptive Technologies

Billy J. Stanbery, Jao van de Lagemaat

HelioSourceTech, Tucson, AZ, United States

National Renewable Energy Laboratory, Golden, CO, United States

In this contribution, we investigate the role of disruptive technology on the complete decarbonization of the world' energy system by looking at plausible scenarios to provide the majority of the world' energy by photovoltaic technologies between 2050 and 2060. To provide enough energy, the world would need to install more than 60 TW of nameplate capacity in just a few decades and would need to rapidly ramp production capacity to achieve this followed by a fast ramp down when the target is reached to a level needed to sustain the then existing capacity and modest subsequent growth. We show how such a trajectory could be achieved with current technologies while largely avoiding stranded manufacturing assets. We then attempt to answer the questions of whether a technology that provides low capital intensity including thin-film technologies such as perovskites or which provides higher efficiency such as tandems can have an influence on the deployment trajectory and the total cost of both installed production capacity and of the total cost of goods sold. We show that disruptive technologies can lower the total cost of goods by 100' of Billions of dollars and that there exists a over 1 T$ market opportunity for such disruptive technologies.

Harsh sequential stress tests for improved PV durability

Jean Patrice Rakotoniaina; Romain Couderc; Eszter Voroshazi; Jérémie Aimé

University of Grenoble Alpes, CEA, LITEN, DTS, INES, F-38000, France

Abstract—**Accelerated aging is required to verify within a few months whether photovoltaic modules are capable of withstanding outdoor conditions for a minimum of 25 years. The IEC 61215 standard is designed to identify the least robust modules through a series of tests. However, these tests are now commonly accepted as insufficient to test the reliability of modules. CEA developed a harsh sequential stress test for PV called STROKE. The goal of the STROKE is to reveal the degradation mechanisms of photovoltaic modules by limiting the duration and number of tests. This approach differs from the approaches of the independent laboratory PVEL and the recent IEC 63209 standard, since the sequential aspect of the tests are not taken into account in both. A series of tests on commercial references of modules covering the current market allowed to validate the STROKE approach. The results show a significant time saving on humidity, UV and thermal cycling tests. The total time required for STROKE is 90 days.**

Keywords—*PV modules, sequential accelerated ageing, PV module degradations, PV durability.*

I. INTRODUCTION

The IEC 61215 [1] reference standard is designed to exclude modules subject to early problems through a series of tests. Many methods had been proposed so far [2-8]. One of the most successful now is to keep the standard conditions and to increase the duration of each test: 2000 hours of damp heat (DH) instead of 1000h, 600 thermal cycles (TC) instead of 200 cycles. The recent IEC 63209 standard is largely inspired by this method and reinforce the sequential branch UV/CT/HF in two approaching sequences. One is dedicated to mechanical stress without UV and the second to backsheet degradation with UV on the backside of the module in order to reveal backsheets cracks. However, the damage caused by UV from the front side are not evaluated and the synergy between DML and DH is not tested nor the one between DH and UV on the front side. The CEA has developed a sequence in order to test it (Figure 1). It has already been claimed that sequential testing is very effective for acceleration and for reproducing degradation phenomena [9-11].

II. THE STROKE SEQUENCE

STROKE is a sequence of five tests more restrictive than those of the standard (Figure 1. In the standard IEC61215, the only sequential test performed is the UV/TC/HF (humidity freeze) sequence. DML (dynamic mechanical load) and reinforced DH (XDH) tests are added in STROKE before a reinforced UV/TC/HF sequence (XUV/XTC/XHF). They stress the module on aspects that are expected to interfere with the XUV/XTC/XHF sequence. DML stresses the module interfaces, interconnections and forms cracks in the cells. This preconditioning is could be very demanding when combined with XTC. Thanks to XDH, humidity comes in the module. It could be harmful for interfaces and change the photochemistry occurring in the module under illumination.

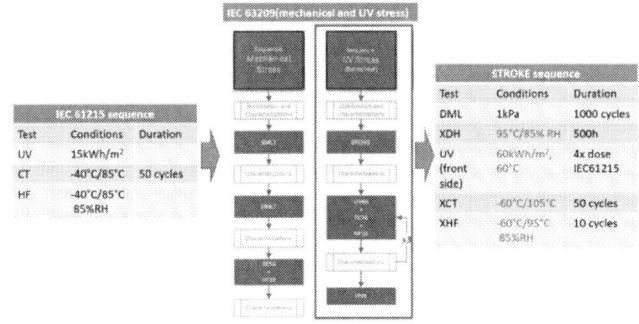

Figure 1: Comparison of the IEC 61215 UV/CT/HF, IEC63209 branch mechanical and UV stress sequence and STROKE

DH test is performed at 85°C and 85% relative humidity while XDH test is at 95°C and 85% relative humidity. The diffusion of water in the modules is accelerated without modifying too much the properties of the polymers,.
- The UV ageing allows us to highlight the sensitivity of cells, encapsulants and module interconnections to UV radiation on the front side.

The reinforced XUV test means to multiply by 4 the total energy received by the module from 15kWh/m² to 60kWh/m² [13].

- The TC (temperature cycle) test simulates the thermal cycles that modules undergo in real conditions due to the day/night cycles. The most common degradation modes observed following these tests are the degradation of interconnections and the delamination of the module.
The STROKE reinforced XTC thermal cycle test means a temperature cycle of -60° to 105°C instead of the standard -40/85°C. The increase in the temperature difference between the maximum and minimum temperature makes it possible to increase the stress and obtain information in fewer cycles.

The HF (humidity freeze) test shows the corrosion of the cells and interconnections as well as the delamination of the module. The reinforced HF humidity-freeze test means a cycle at -60/105°C instead of the standard -40/85°C at 85% relative humidity.

III. RESULTS AND DISCUSSIONS

Before to use XDH and XTC in the STROKE sequence, tests were carried out in order to verify that no unexpected phenomena occur because of the higher temperatures involved in the reinforced tests. DH, XDH, TC and XTC tests were carried out on three different commercial module configurations (>1.5m²). Two modules of each configuration were used in each test.

Each reinforced test has been validated separately before putting it in the STROKE sequence.

Validation of XDH

Fig. 2 shows EL pictures of the module with DH3000 (a) and XDH1500 (b) respectively. The EL pictures together with the analysis of the electrical parameter, XDH and DH show similar degradation phenomena, namely the humidity penetration impacting mostly the FF.

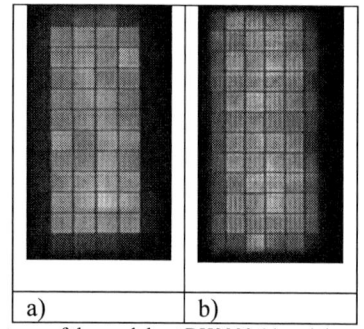

Figure 2: EL pictures of the module at DH3000 (a) and the same type at XDH 1500 (b)

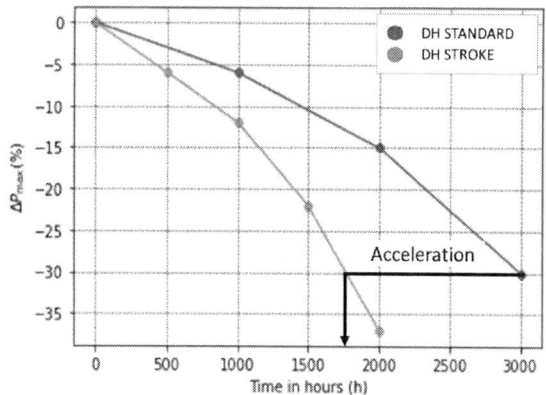

Figure 3: Normalized Power loss of DH(85°C/85%RH) vs. WDH(95°C,85%RH)

Fig. 3 shows the normalized power loss in damp heat for DH and XDH for one module type. The normalized power loss of a DH of 3000h corresponds to a XDH of 1750h. Only one configuration is presented here for sake of brevity but results leading to the same conclusion are obtained for the three configuration with different

The explanation of the acceleration factor via the activation energy can be found in [12].

Validation of XCT

The test occur in the same chamber as for CT. An example is shown in Figure. 4 using the comparison of the normalized Pmax. The Pmax loss of 600 CT is similar to XCT 300 for this module type. The degradation is driven by the FF loss due to increase of the serial resistance. EL pictures shown in Figure 5 confirms the interconnection degradation.

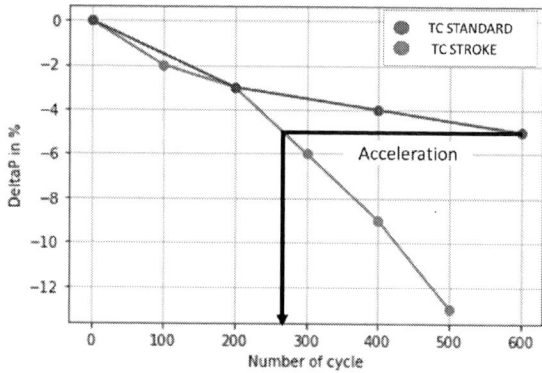

Figure 4: normalized Pmax losses for CT vs STROKE XCT

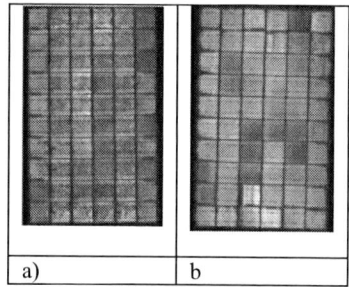

Figure 5: EL pictures of the module at CT600 and the same type at XCT300

A. STROKE RESULTS

The STROKE sequence according to Figure1 has been applied to four types of commercial modules (Table(1)). Three modules per type endured the STROKE sequence. The obtained results are always similar for each module of one type.

Table 1: Types of modules used in STROKE

Reference	Cells technology	Bifaciality	Architecture
1	HET	YES	Full cells
2	HET	YES	Full cells
3	PERC	NO	Half cells
4	PERC	NO	Shingling

Results are presented in Fig. 6 showing the losses of = P_{max}, I_{sc}, V_{oc} and FF during the STROKE sequence.
DML does not significantly affect P_{max} for any module. After XDH, a P_{max} loss is visible (<4%). Mostly from I_{sc} and FF losses coming from the ingress of humidity. After XUV, an important P_{max} loss occurs for 3 types of modules. Only reference 2 resisted well to UV. The encapsulant of this reference is the only one incorporating UV absorbers. After XTC, the parameters remain stable. After XHF, a slight increase of V_{oc}

978-1-7281-6118-1/22 $31.00 © 2022 IEEE

for the most degraded references (1, 3 and 4) indicates that the V_{oc} degradation caused by UV is partially reversible thanks to XHF (Figure 7). Further investigations are in progress to better understand the phenomena.

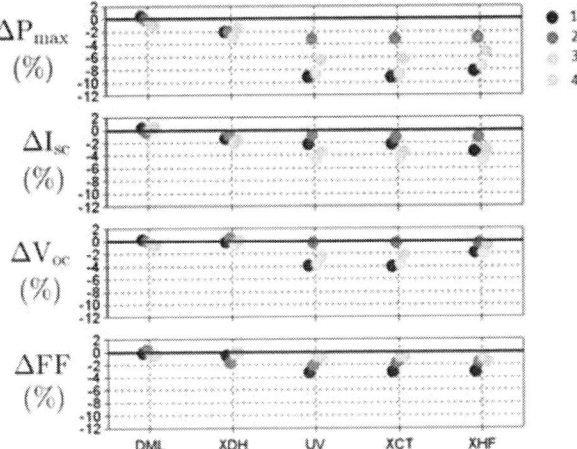

Figure 6: Normalized Pmax, Isc, Voc and FF losses for whole STROKE sequence on four type of modules references.

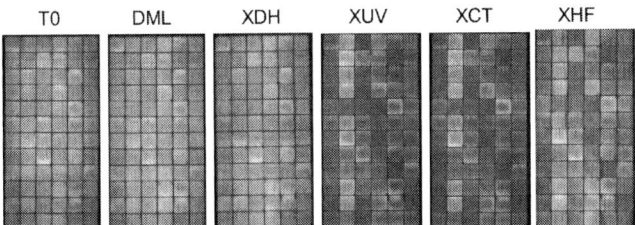

Figure 7: EL pictures Reference 1 in the STROKE sequence

IV. CONCLUSIONS AND OUTLOOK

The enhanced XDH (95°C, 85%RH) proposed in STROKE showed an acceleration factor of 1.7. XDH 1750 shows normalized power loss similar to DH 3000 while keeping the same degradation phenomena. This gain in acceleration factor could be explained by the activation energy changing the chamber temperature from 85°C to 95°C.

Similar to the XDH, for the enhanced XCT (-60°C to 105°C°), XTC with 300 cycle provided similar results to TC with 600 cycles while keeping same the degradation type, in this case the interconnection degradation.

For the STROKE sequence, it seems that the XUV with 60 kWh/m2 (four times the standard dosis) was the test which give more stress to the modules.

Those results demonstrate the acceleration of the degradation mechanisms by the application of the STROKE approach. The time saving is significant with a reduction in the number of cycles to highlight the mechanisms and a causal link with the activation energy to determine the most impacting factors. The STROKE sequences require 90 days in total, enabling an access more quickly to the degradation mechanisms of photovoltaic modules

REFERENCES

[1] IEC 61215: Crystalline silicon terrestrial photovoltaic(PV) modules–Design qualification and type approval(2005, 2nd edition);

[2] C. R. Osterwald and T. J. McMahon, Prog. Photovoltaics 17, 11 (2009).

[3] D. DeGraaff, S. Caldwell, R. Lacerda, G. Bunea, A. Terao, and D. Rose, Proc. 25th European Photovoltaic Solar Energy Conf. Exhib./5th World Conf. Photovoltaic Energy Conversion, 2010, p. 3722.

[4] J. H. Wohlgemuth and S. Kurtz, Proc. 37th IEEE Photovoltaic Specialists Conf., 2011, p. 3601.

[5] Y. Aoki, M. Okamoto, A. Masuda, T. Doi, and T. Tanahashi, Jpn. J. Appl. Phys. 51, 10NF13 (2012).

[6] A. Takano, H. Yanase, T. Sakai, H. Nishihara, T. Nakamura, and S. Fujikake, Proc. 28th European Photovoltaic Solar Energy Conf. Exhib., 2013, p. 2159.

[7] J. Wohlgemuth and S. Kurtz, Proc. 40th IEEE Photovoltaic Specialists Conf., 2014, p. 3589.

[8] S. Suzuki, T. Tanahashi, T. Doi, and A. Masuda, Jpn. J. Appl. Phys. 55, 022302 (2016).

[9] T. Ngo, Y. Heta, T. Doi, and A. Masuda, to be published in Jpn. J. Appl. Phys.

[10] T. Doi, H. Morita, T. Amioka, T. Shioda, S. Suzuki, and A. Masuda, Tech. Dig. 6th World Conf. Photovoltaic Energy Conversion, 2014, p. 1015.

[11] William Gambogi, Thomas Felder; Steven MacMaster; Kaushik Roy-Choudhury; Bao-Ling Yu; Katherine Stika; Hongjie Hu; Nancy Phillips; T. John Trout, Sequential Stress Testing to Predict Photovoltaic Module Durability, 45th IEEE 2018, pp1593-1596

[12] Koehl et al. Prog. Photovolt: Res. Appl. 2017; 25:175–183

[13] J.F.Lelièvre,R.Couderc,N.Pinochet,L.Sicot,D.Munoz,P.Ferrada,A.Marzo ,F.Valencia,E.Urrejola, Solar Energy Materials and Solar Cells,Volume 236, March 2022, 111508

PV Module Operating Temperature Model Equivalence and Parameter Translation

Anton Driesse[1], Marios Theristis[2], Joshua S. Stein[2]

[1]PV Performance Labs, Freiburg, Germany,
[2]Sandia National Laboratories, Albuquerque, NM, United States

Abstract—**PV module operating temperature is the second most important factor influencing system yield, after irradiance. A variety of temperature models are used within yield simulation software to predict module operating temperature, which then determines operating efficiency. Four temperature models are frequently used: PVsyst, Faiman, SAPM and SAM NOCT. Although these models are similar, their parameter values are not directly interchangeable. In this work we demonstrate the equivalence or near-equivalence of these four temperature models, and from there we develop equations to convert their parameter values back and forth. This is more than a convenience for users of simulation software. We use this capability, for example, to compare and analyze the typical and default values preset for different model/software combinations. The functions to perform the parameter conversions are made available as open-source software in pvlib-python.**

Keywords—photovoltaic system performance, module temperature, thermal model, PV system simulation

I. INTRODUCTION

It is well known that temperature is the second most influential factor in photovoltaic (PV) performance. Temperature effects on PV module efficiency are well-documented [1] and many models exist for calculating module temperature [2]. Module temperature models vary in complexity, from three-dimensional finite element analysis [3] to simpler and more practical empirical approaches in one dimension [4]–[7]. Although three-dimensions and finite element analysis may be more accurate with respect to modeling the heat transfer behavior of a PV module, it is computationally prohibitive when it comes to long-term, fine resolution energy yield analyses and/or real-time monitoring. Furthermore, defining realistic simulation boundary conditions for even a small system is a complex task.

Therefore, empirical approaches such as those available in pvlib-python [8] are commonly used by the industry. These empirical models use irradiance and meteorological variables as inputs (mainly plane-of-array irradiance, ambient temperature, wind speed) and assume or fit parameters to represent the heat transfer mechanisms between a module and its surroundings.

In this work we demonstrate that four of the most common models, Faiman, Sandia Array Performance Model (SAPM), PVsyst and System Advisor Model Nominal Operation Cell Temperature (SAM NOCT), have identical or very similar characteristics and differ primarily in their parameterization. As a result, it is possible to transform the parameters for one model for use with another model and obtain substantially the same predictions of module temperature. To evaluate the overall impact of the different module temperature

estimates on annual yields we use the six climate profiles of IEC-61853-4 [9]. These are hourly average data files that contain a representative year of weather data for six diverse climates around the globe.

II. BACKGROUND

Although "thermal balance" or "heat transfer" labels are often used to tag more advanced modeling efforts that go back to first principles, all the simple lumped parameter models in use actually describe a thermal balance between a PV module and its environment. A high-level thermal balance (steady-state, disregarding thermal capacitance) is given by the following equation, which expresses simply that the heat gains must balance the losses:

$$q_{sun} - q_{elec} - q_{rad} - q_{conv} - q_{cond} = 0 \qquad (1)$$

where q_{sun} is the energy flux from the sun, q_{elec} the energy flux removed as electricity, and the remaining three terms are the heat loss fluxes by radiation, convection and conduction (all in W/m^2). Conduction is usually negligible since the area of contact between the frame and mounting structure is very small, and the remaining radiative and convective losses are lumped together as q_{cr} to produce the much simpler thermal balance equation:

$$q_{sun} - q_{elec} - q_{cr} = 0 \qquad (2)$$

Although q_{cr} actually depends on a large number of parameters and properties, its main tendency is to increase as the temperature of the module (T_m) rises above the temperature of the environment (T_a). Using a one-dimensional, linear approximation of this tendency produces the thermal balance equation:

$$q_{sun} - q_{elec} - U_{cr} \cdot (T_m - T_a) = 0 \qquad (3)$$

where U_{cr} is the combined or lumped heat transfer coefficient. This can easily be rearranged to calculate module temperature:

$$T_m = T_a + \frac{q_{sun} - q_{elec}}{U_{cr}} \qquad (4)$$

Since q_{elec} is approximately proportional to q_{sun} we can further simplify by scaling U_{cr} to U:

$$T_m = T_a + \frac{q_{sun}}{U} \qquad (5)$$

978-1-7281-6118-1/22 $31.00 © 2022 IEEE

The four models discussed in this paper are all variations of this base model equation.

One final question to consider is whether q_{sun} should be equal to plane-of-array (POA) irradiance (G_{poa}). For this work we will adjust the models where necessary so that they all use G_{poa} as input.

The final base model equation is then:

$$T_m = T_a + \frac{G_{poa}}{U} \qquad (6)$$

III. MODEL EQUATIONS

In this section we briefly present the four model equations followed by the expressions for the heat transfer coefficient U for each model. These are obtained by simply rearranging a portion of each model equation. All models use G_{poa} and wind speed (WS) as inputs.

The Faiman model has two empirical parameters u_0 and u_1:

$$T_m = T_a + \frac{G_{poa}}{u_0 + u_1 \cdot WS} \qquad (7)$$

$$U = u_0 + u_1 \cdot WS \qquad (8)$$

The SAPM module temperature model also has two empirical parameters, a and b:

$$T_m = T_a + G_{poa} \cdot e^a \cdot e^{b \cdot WS} \qquad (9)$$

$$U = e^{-a} \cdot e^{-b \cdot WS} \qquad (10)$$

The two empirical parameters in the PVsyst model, u_c and u_v, are accompanied by two additional model inputs, module efficiency η and optical absorptance α:

$$T_m = T_a + \frac{G_{poa} \cdot \alpha \cdot (1 - \eta)}{u_c + u_v \cdot WS}, \quad \alpha = 0.9 \qquad (11)$$

$$U = \frac{u_c + u_v \cdot WS}{\alpha \cdot (1 - \eta)}, \quad \alpha = 0.9 \qquad (12)$$

When the model is used for simulation within PVsyst the absorptance remains constant at a nominal value whereas the efficiency is recalculated at every time step to represent actual operating conditions.

The fourth model, SAM NOCT, has one empirical parameter T_{noct} and two additional inputs, module efficiency η and the transmittance-absorptance product $(\tau\alpha)$ [1]. Frequently, models based on NOCT are shown as a single equation, but we prefer to present this in two parts for greater clarity. The first equation converts the empirical T_{noct} into an empirical lumped heat transfer coefficient U_{noct}:

$$U_{noct} = \frac{800 \cdot ((\tau\alpha)_t - \eta_t)}{T_{noct} - 20}, \quad \eta_t = 0 \qquad (13)$$

---

[1] The parentheses are part of the tau-alpha notation.

The subscript t is used to represent conditions during NOCT testing. When we cast U_{noct} as the primary model parameter, the SAM NOCT model actually looks quite similar to the PVsyst model. The key difference is that SAM uses a fixed ratio to set the wind speed coefficient. Interestingly, this empirical ratio was derived from measurements made exactly 100 years ago, using a square copper plate in a wind tunnel [10]. In the SAM documentation the ratio is presented as a fraction as shown here:

$$T_m = T_a + \frac{G_{poa} \cdot ((\tau\alpha) - \eta)}{U_{noct} \dfrac{5.7 + 3.8 \cdot WS}{9.5}}, \quad \eta = \eta_{STC} \qquad (14)$$

The ratio and its purpose are easier to recognize when presented in decimal form, which we use for the final model equation and heat transfer coefficient equation:

$$T_m = T_a + \frac{G_{poa}((\tau\alpha) - \eta)}{U_{noct} \cdot (0.6 + 0.4 \cdot WS)}, \quad \eta = \eta_{STC} \qquad (15)$$

$$U = \frac{U_{noct} \cdot (0.6 + 0.4 \cdot WS)}{(\tau\alpha) - \eta}, \quad \eta = \eta_{STC} \qquad (16)$$

The single-equation representation of the SAM NOCT model is a combinations of (13) and (14) where $(\tau\alpha)$ and $(\tau\alpha)_t$ are equated. When the model is used for simulation within SAM the efficiency input remains constant at the STC value whereas $(\tau\alpha)$ is recalculated at every time step to represent actual operating conditions.

IV. MODEL OBSERVATIONS

A. Absorptance, transmittance and reflectance

It is evident that the portion of incident radiation that is reflected at the front cover cannot contribute to either heating or electricity generation. However, module electrical efficiency is evaluated and specified based on the total incident irradiance at normal incidence measured *prior* to reflections. The PVsyst model equation therefore contains a small logical error because it calculates the electrical energy based on absorbed irradiance $G_{poa}\alpha$ rather than total incident irradiance G_{poa}.

Similar to PVsyst, SAM uses a transmittance-absorptance product $(\tau\alpha)$ to represent the absorptance of the module, and correctly calculates the electrical efficiency based on the total incident irradiance. The quantity $(\tau\alpha)$ was developed to represent the effective absorptance of a thermal absorber under a cover system where multiple reflections occur [11]. While multiple reflections may occur within PV module material layers as well, the situation is different because heat absorbed in all layers contributes to the module heating; in other words, the cover is part of the absorber. Thus, it would seem more appropriate to use a simple absorptance nomenclature, as is done in the PVsyst model.

A difficult aspect of the optical properties is their dependence on the incidence angle of the incoming solar radiation. Since the temperature models are empirical their parameters are determined from a series of outdoor measurements where the incidence angle changes over time. However, a subset of medium to high irradiance points is usually chosen to extract the final parameter values, and such conditions occur primarily at low incidence angles where reflectance losses are low and absorptance is high. Thus, using the temperature model parameters for high-incidence angles is a form of extrapolation.

During a PV system simulation the value of G_{poa} is typically reduced by the incidence angle modifier (*IAM*) at each time step so that module electrical output and thermal input are both reduced simultaneously for reflection losses. SAM takes a different approach and reduces the transmittance-absorptance product ($\tau\alpha$) in the temperature model at higher incidence angles. This causes two problems: first, the electrical output in the thermal balance is not reduced; and second, since SAM equates ($\tau\alpha$) and ($\tau\alpha$)$_{noct}$, the implicit empirical parameter U_o is also changed at every time step. It seems likely that this is unintended behavior of the SAM software, and if so, it can be corrected easily by subtracting reflection losses from G_{poa} prior to running the temperature model.

B. Wind speed

Each model has a coefficient that multiplies wind speed. In practice, wind speed increases with height above the ground in a non-linear manner, therefore the model coefficients are valid for a specific combination of module installation height and wind speed measurement height. The Faiman model coefficients and NOCT values are typically determined using wind speed measured near module height, but the modules are not necessarily near ground level. SAPM and PVsyst coefficients are specified for use with wind speed data at the standard 10 m height, but that information is only useful if the module height corresponding to those coefficients is also clearly specified.

To accommodate a different installation or measurement height, either the wind speed data or the wind speed coefficient can be adjusted. The basis for such an adjustment is often either the *log law* or the *power law* which describe the wind speed profile as a function of height using an empirical parameter related to the surface roughness; however, these profiles are not applicable close to the ground or close to the level of the objects that contribute to the roughness. [12] For the adjustment from 10 m to 2 m, a nominal height for a ground-mounted array, some reduction ratios found in the literature are: 0.51 [7], 0.56 [13], 0.67 [14] and 0.725 [15]. This range gives an indication of the level of uncertainty associated with such adjustments.

The SAM NOCT model uses a predetermined factor of 0.51 to reduce 10 m wind speed data to ground-mounted, module-level wind speed, but unfortunately the origin of this value is undocumented. The need for adjustments with other models to or from different heights should be evaluated by the user of the models on a case by case basis taking into consideration the both the source of the empirical parameters and the source of the wind speed data.

C. Cell vs. back-of-module temperature

Heat is conducted away from the cells to the front and back surfaces of the module where it is dissipated by convection and radiation. The thermal resistances of the PV module materials lead to a temperature gradient in the direction of the heat flow, hence the daytime back-of-module temperature is generally a bit lower than the internal cell temperature. King [5] proposes an adjustment term for the SAPM model to account for this:

$$T_c = T_m + \frac{G_{poa}}{1000} \cdot \Delta T \qquad (17)$$

where ΔT is the temperature difference between cell and module observed under 1000 W/m^2 illumination. King reports a difference of 3 °C for both glass-glass and glass-polymer modules in open-rack configuration.

TABLE I. PARAMETER VALUES FOR GLASS-POLYMER MODULES IN TILTED OPEN RACK CONFIGURATION FOR FOUR MODELS

	Extra inputs	Commonly used parameters	Calculated equivalents
Faiman		$u_0 = 25.0$ $u_1 = 6.84$	$u_0 = 34.27$ $u_1 = 3.42$
SAPM		$a = -3.56$ $b = -0.075$	$a = -3.56$ $b = -0.075$
PVsyst	$\eta = 12\%$ $\alpha = 0.9$	$u_c = 29.0$ $u_v = 0.0$	$u_c = 27.14$ $u_v = 2.71$
NOCT SAM	$\eta = 12\%$ $(\tau\alpha) = 0.9$	$T_{noct} = 45.0$ $(U_o = 25.0)$	$T_{noct} = 42.53$ $(U_o = 27.70)$

Using the measured thickness and thermal conductivity of the EVA and polymer back sheet of a typical module listed in [16] we calculate the thermal conductance of the rear layers to be approximately 370 W/m^2/K. Furthermore, under 1000 W/m^2 illumination we estimate that typically 10% would be reflected, 15% converted to electricity and of the 75% that is absorbed as thermal energy at most half would be dissipated through the back (375 W/m^2). The expected value of ΔT should be therefore at most 375/370 or 1.0 °C. Since the ΔT observed by King is three times the expected value, it seems likely that much of the observed difference was attributable to the temperature sensor design or to the attachment method.

Differences on the order of 1 °C are well below the uncertainty of all the module temperature models; therefore, from this point on we will refer only to module temperature and consider it to be approximately equal to cell temperature.

V. TRANSLATION OF PARAMETER VALUES

In section III we have demonstrated that all four temperature models have the same basic structure with the differences appearing in the formulation of the heat transfer coefficient U, which is a function of wind speed. Since U is independent of irradiance and ambient temperature, it can be plotted quite simply against wind speed in order to compare the models and their empirical parameters visually. (See Fig. 1.)

A variety of published parameter values are available for use with the four temperature models. The first comparison uses default or recommended values that are frequently seen in examples and are listed in Table I. All values are for poly/mono-crystalline silicon modules with glass fronts and polymer back sheets that are installed at fixed tilt on open ground-mounted racks. We assumed fairly low module efficiency of 12% for this comparison because these parameter values were all determined for older modules operating at temperatures necessarily well above 25°C.

Fig. 1 shows that at moderate wind speed of around 2 m/s, the four models all predict similar heat transfer coefficients, but the variation with wind speed differs dramatically. The gray histogram in the background represents the distribution of available solar energy by wind speed[2] in the six IEC-61853 climate profiles combined. This makes clear that the large differences at very high wind speed are not

---

[2] The wind speed shown has been reconverted to 10 m measurement height by undoing the conversion from 10 m to 2 m reported in [15].

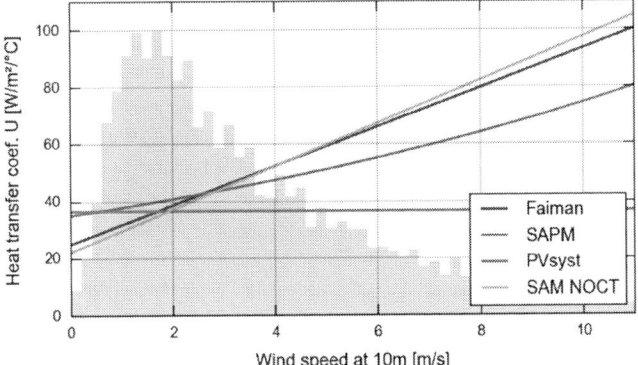

Fig. 1. Heat transfer coefficients calculated by four models using commonly available parameter values. The histogram in the background represents the amount of solar energy available by wind speed based the six climate profiles of IEC 61853-4. Differences among heat transfer coefficients are substantial, but they are smallest at modereate windspeeds where most of the solar energy is available.

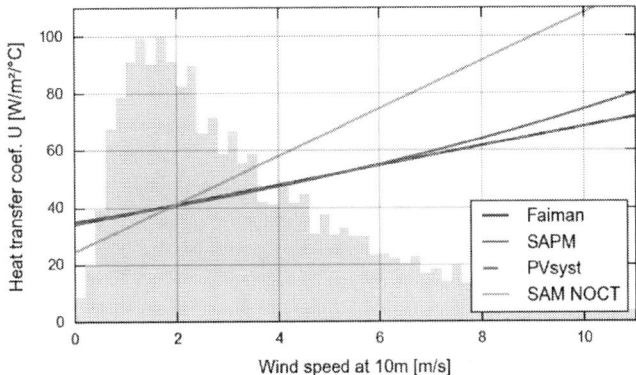

Fig. 2. After translation, the Faiman and PVsyst heat transfer coefficients closely match the SAPM over the windspeed range where most solar energy is available. All three models are equal at 1.4 and 5.4 m/s. The NOCT SAM model has only one degree of freedom and cannot fully match the others.

necessarily going to have a large impact on energy yield predictions, but on the other hand a small difference at moderate wind speed could be significant.

We now propose that two models can be considered equivalent if parameter values can be found such that both models predict the same heat transfer coefficients at all wind speeds. It can be observed that the Faiman and PVsyst heat transfer coefficients graphs are straight lines and their two parameters set the slope and y-intercept of those two lines, thus, one model can always be brought into perfect alignment with the other. SAM NOCT also produces a straight line but has only a single parameter that controls both the slope and intercept. With one degree of freedom it cannot generate the same heat transfer coefficients as the Faiman and PVsyst models for all their possible parameter combinations. But, the reverse *is* possible. Finally, the two SAPM model parameters also control the slope and y-intercept of its heat transfer coefficient graph. However, the exponential function imposes a slight curvature on the line so it can at best provide a near approximation of the other three (and vice versa).

Now that we know what is possible, we can consider how parameters could be translated from a source model to a target model to have them predict the same—or nearly the same—temperature coefficients. If the target model has two parameters, the source model is evaluated at two wind speeds to create a system of two equations. These are then solved for the target model parameters. The example in (18) is the system of equations for finding Faiman model parameters where U_i and U_j are calculated using any of the other models.

$$U_i = u_0 + u_1 \cdot WS_i \qquad (18)$$
$$U_j = u_0 + u_1 \cdot WS_j$$

The solution becomes trivial if we choose wind speed values of 0 and 1 m/s, but this is not appropriate for the SAPM where we must calculate an *approximate* equivalence. Based on the available energy by wind speed in the six climatic profiles (in the background of Fig. 1) we choose the 20% and 80% points of the cumulative distribution function, which occur at 1.4 m/s and 5.4 m/s. This ensures that the SAPM curve crosses its straight-line counterparts at those two wind speed values and remains a close approximation over the range where most solar energy is available.

Since the SAM NOCT model has a single parameter, T_{noct}, only a single equation is needed. We found that using the NOCT standard

wind speed of 1 m/s at module height, which SAM equates to 1.96 m/s at 10 m produces a good compromise on the predicted module temperatures as well as energy gains and losses (this is seen in the following section).

The right-most column in Table I lists parameter values that were calculated in the manner described above, and that allow all four models to predict equal heat transfer coefficients or as close as the model allow, using the SAPM values as a starting point. The results are plotted in Fig. 2, illustrating both how minor the difference between the two-parameter models is, and how much of a handicap it is for SAM NOCT to have just a single parameter. In the next section we will also look at differences in temperature and yield predictions that result from this.

In this paper we show a single translation example, but translations are possible between any pair of models, and all combinations have been tested.

VI. COMPARISON OF PREDICTED TEMPERATURES AND ENERGY YIELDS

In this section we compare the four models to see how they would differ when used for system simulation. The weather data selected for this test consists of the six climatic profiles of IEC 61853-4, which are designed to represent a wide variety of global climate conditions.

A full simulation is not needed for this relative comparison; only the following steps are needed:

1. Calculate module temperatures using the four different models and parameter values from Table I.

2. Calculate POA effective irradiance, G_{eff}, by applying an incidence angle modifier. We use the Martin-Ruiz model [17] with default parameters for simplicity.

3. Calculate electrical gains/losses due to temperature as:

$$E_{gain} = (T_m - 25) \cdot \gamma \cdot G_{eff}, \quad \gamma = -0.5\%/°C \qquad (19)$$

4. Calculate percentage change in annual yield:

$$\Delta_{yield} = \frac{\sum E_{gain}}{\sum G_{eff}} \cdot 100\% \qquad (20)$$

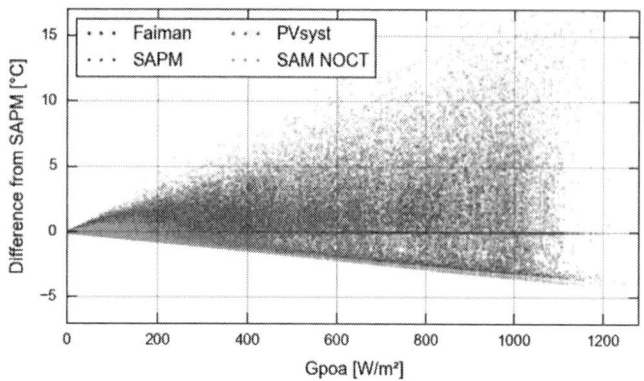

Fig. 3. Differences in predicted module temperatures between SAPM and the other three models using commonly available parameter values (not implying that SAPM is better). Faiman and PVsyst frequently have higher temperatures than SAPM while SAM NOCT predicts consistently lower at high irradiance..

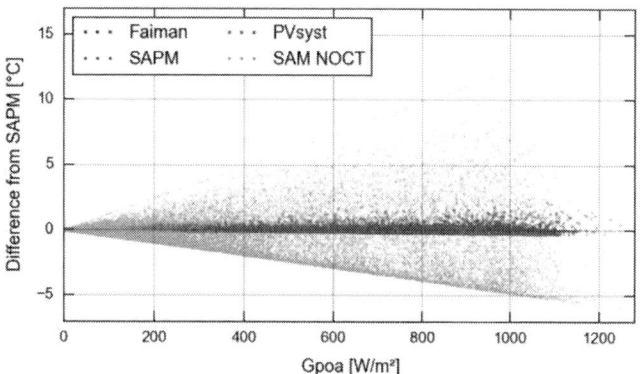

Fig. 4. After parameter translation the temperature differences between the first three models are minimal whereas the SAM NOCT model—as already seen in Fig. 2—cannot match the others.

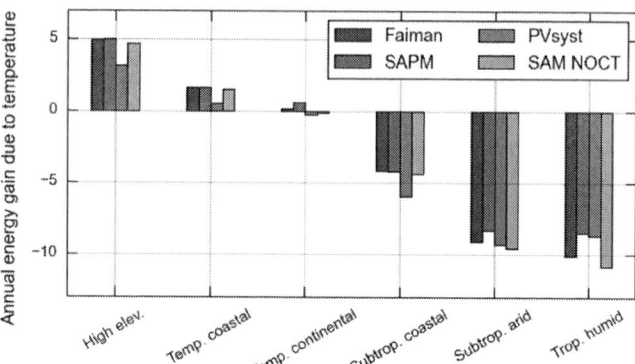

Fig. 5. Impact on annual yield due to temperature as estimated by four models using commonly available parameter values. Differences between models are as high as 2% and the ranking of the models changes with climate.

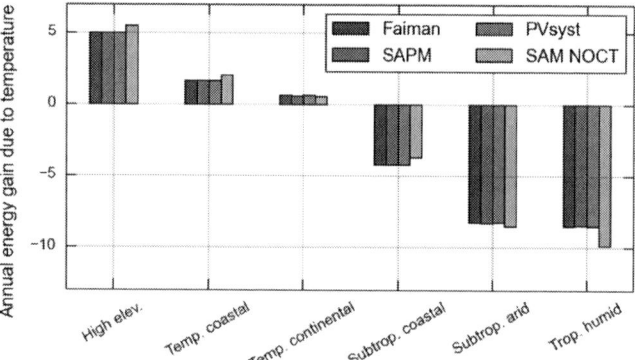

Fig. 6. After parameter translation the first three models converge on the virtually the same annual yield impact for each climate. The translated parameter for SAM NOCT do predict the same overal gain for all six climates combined (within 0.1%), but in individual climates differences remain.

Fig. 3 shows that there is a wide spread in module temperature predictions using the first set of parameter values. For Fig. 4 the translated parameters are used, with the result that the SAPM, PVsyst and Faiman models all predict very nearly the same temperatures. SAM NOCT shows large deviations because its wind speed coefficient cannot be separately adjusted.

The temperature differences look substantial, but what often matters more is the total annual yield. Using (19) and (20) we can easily convert the temperature differences into yield differences, which are plotted in Fig. 5 and Fig. 6. Using the first set of parameter values the four models differ from each other by as much as 2.4%, which is substantial. When the translated parameters are used the yield predictions of SAPM, PVsyst and Faiman converge, as expected, to within ±0.1%. The SAM NOCT model is still an outlier, deviating as much as 1.4% from the other three.

VII. SUMMARY AND CONCUSIONS

We have shown that four popular module temperature models can be analyzed and compared in three ways: by effective heat transfer coefficients, by module temperature differences and by impact on annual yield predictions. Substantial differences in annual yield of more than 2% can result when using commonly available published or default parameter values, providing strong motivation for further investigation.

Uniformly presented equations for the heat transfer coefficients in this work make the translation of parameter values from one model to another possible. When using such translated parameters the temperature and yield predictions from the two-parameter models align very well with each other. The SAM NOCT model remains an outlier due to the fixed ratio for its internal wind speed coefficient.

All four module temperature models discussed in this paper are already available in the open source pvlib-python package hosted on github [8]. The methods developed in this work to translate parameters are open and available in pvlib-python as well, and are linked to feature request #1442 [18]. The translated parameters obtained using these functions can be used with any open source or commercial PV system simulation software.

The task of PV module temperature modeling for real PV installations remains difficult due to the many variables that interact; therefore, much more guidance is needed for the appropriate empirical parameter values in different mounting configurations and climatic conditions. Parameter values reported in the literature often lack the level of context and detail needed to assess their applicability in other situations. We hope that reports on model parameters will increase in both quantity and quality as this topic regains attention. The methods developed in this work will amplify the reach of new guidance by making parameter recommendations usable not just with one model and one software program, but with many.

VIII. ACKNOWLEDGEMENT

This work was supported by the U.S. Department of Energy's Office of Energy Efficiency and Renewable Energy (EERE) under the Solar Energy Technologies Office Award Number 38267. Sandia National Laboratories is a multimission laboratory managed and operated by National Technology & Engineering Solutions of Sandia, LLC, a wholly owned subsidiary of Honeywell International Inc., for the U.S. Department of Energy's National Nuclear Security Administration under contract DE-NA0003525. This paper describes objective technical results and analysis. Any subjective views or opinions that might be expressed in the paper do not necessarily represent the views of the U.S. Department of Energy or the United States Government.

IX. REFERENCES

[1] E. Skoplaki and J. A. Palyvos, "On the temperature dependence of photovoltaic module electrical performance: A review of efficiency/power correlations," *Solar Energy*, vol. 83, no. 5, pp. 614–624, May 2009, doi: 10.1016/j.solener.2008.10.008.

[2] E. Skoplaki and J. A. Palyvos, "Operating temperature of photovoltaic modules: A survey of pertinent correlations," *Renewable Energy*, vol. 34, no. 1, pp. 23–29, Jan. 2009, doi: 10.1016/j.renene.2008.04.009.

[3] J. Zhou, Q. Yi, Y. Wang, and Z. Ye, "Temperature distribution of photovoltaic module based on finite element simulation," *Solar Energy*, vol. 111, pp. 97–103, Jan. 2015, doi: 10.1016/j.solener.2014.10.040.

[4] D. Faiman, "Assessing the outdoor operating temperature of photovoltaic modules," *Prog. Photovolt: Res. Appl.*, vol. 16, no. 4, pp. 307–315, Jun. 2008, doi: 10.1002/pip.813.

[5] D. L. King, W. E. Boyson, and J. A. Kratochvil, "Photovoltaic array performance model.," SAND2004-3535, 919131, Aug. 2004. doi: 10.2172/919131.

[6] PVsyst SA, "Project design > Array and system losses > Array Thermal losses," *PVsyst 7 Help*, Jan. 25, 2022. https://www.pvsyst.com/help/thermal_loss.htm (accessed Jan. 25, 2022).

[7] P. Gilman, N. A. DiOrio, J. M. Freeman, S. Janzou, A. Dobos, and D. Ryberg, "SAM Photovoltaic Model Technical Reference Update," NREL/TP--6A20-67399, 1429291, Mar. 2018. doi: 10.2172/1429291.

[8] W. F. Holmgren, C. W. Hansen, and M. A. Mikofski, "pvlib python: a python package for modeling solar energy systems," *Journal of Open Source Software*, vol. 3, no. 29, p. 884, Sep. 2018, doi: 10.21105/joss.00884.

[9] IEC, "IEC61853-4 Ed. 1.0: Photovoltaic (PV) module performance testing and energy rating - Part IV: Standard reference climatic profiles," IEC, Geneva, First edition, 2019.

[10] J. A. Palyvos, "A survey of wind convection coefficient correlations for building envelope energy systems' modeling," *Applied Thermal Engineering*, vol. 28, no. 8–9, pp. 801–808, Jun. 2008, doi: 10.1016/j.applthermaleng.2007.12.005.

[11] J. A. Duffie and W. A. Beckman, *Solar engineering of thermal processes*, 3rd ed. Hoboken, N.J: Wiley, 2006.

[12] J. F. Manwell, J. G. McGowan, and A. L. Rogers, *Wind energy explained: theory, design and application*, 2nd ed. Chichester, U.K: Wiley, 2009.

[13] M. Prilliman, J. S. Stein, D. Riley, and G. Tamizhmani, "Transient Weighted Moving-Average Model of Photovoltaic Module Back-Surface Temperature," *IEEE J. Photovoltaics*, vol. 10, no. 4, pp. 1053–1060, Jul. 2020, doi: 10.1109/JPHOTOV.2020.2992351.

[14] E. Skoplaki, A. G. Boudouvis, and J. A. Palyvos, "A simple correlation for the operating temperature of photovoltaic modules of arbitrary mounting," *Solar Energy Materials and Solar Cells*, vol. 92, no. 11, pp. 1393–1402, Nov. 2008, doi: 10.1016/j.solmat.2008.05.016.

[15] T. Huld and A. Amillo, "Estimating PV Module Performance over Large Geographical Regions: The Role of Irradiance, Air Temperature, Wind Speed and Solar Spectrum," *Energies*, vol. 8, no. 6, pp. 5159–5181, Jun. 2015, doi: 10.3390/en8065159.

[16] T. J. Silverman *et al.*, "Reducing Operating Temperature in Photovoltaic Modules," *IEEE J. Photovoltaics*, vol. 8, no. 2, pp. 532–540, Mar. 2018, doi: 10.1109/JPHOTOV.2017.2779842.

[17] N. Martin and J. M. Ruiz, "Calculation of the PV modules angular losses under field conditions by means of an analytical model," *Solar Energy Materials*, p. 14, 2001.

[18] Anton Driesse, "Temperature model parameter translation · Issue #1442 · pvlib/pvlib-python," *GitHub*. https://github.com/pvlib/pvlib-python/issues/1442 (accessed May 30, 2022).

Accelerating Simulation for High-Fidelity PV Inverter System Reliability Assessment with High-Performance Computing

Liwei Wang[1], Ramanathan Thiagarajan[2], Shuangshuang Jin[1], and Zheyu Zhang[1]

[1]Clemson University, North Charleston, SC, 29405, USA

[2]National Renewable Energy Laboratory, Golden, CO, 80401, USA

Abstract—The overall cost of photovoltaic (PV) systems has shown a downward trend during the last decade; however, PV inverter failures account for the highest cost of operation and maintenance. To address this, reliability tools with powerful computation and better accuracy are required for the lifetime prediction and degradation evaluation of PV inverters. This paper proposes an event-driven parallel computing-based simulator. The proposed simulator applies high-performance computing techniques and other accessory optimization techniques—including cluster merging, adaptive model updates, and steady-state identification—to make reliability assessments for PV inverters under given input mission profiles and operating conditions with high efficiency and high fidelity. The main idea of the simulator and its workflow are introduced. Then, a demo PV inverter system simulator is implemented, and the speedup of the total simulations of the switching model reaches 123.03 times.

Keywords—PV System, Reliability Assessment, Power Electronics, High-performance Computing

I. INTRODUCTION

The U.S. Department of Energy estimates the photovoltaic (PV) penetration in the U.S. power market to be upward of 18% by 2050, with a $1/watt price point. As the price of PV modules decreases, the price of inverters becomes more important. Inverters now constitute 8%–12% of the total PV lifetime cost [1]. One key price driver of the inverter components is inverter reliability. PV modules are offered up to 30-year warranties. In contrast, typical warranties on inverters last only 5–10 years. Even with the most optimistic view of inverter lifetime, it will be essential to replace or repair an inverter multiple times during the lifetime of a PV module [2]. Further, field data from PV power plant operators demonstrate that power electronics converters contribute the most to operation-and-maintenance events, which are responsible for between 43% and 70% of the service calls [3]. This leads to increased maintenance costs, lost energy production, and, finally, higher levelized cost of energy. Consequently, it is critical to have a generic tool from a third party for PV inverter reliability assessment.

Various tools have been developed to provide reliability assessments for power electronics systems. References [4], [5], [6], [7] introduced physics-to-failure-based analysis tools that the designer could use to predict the corresponding reliability

interactively with the design change. But even if these tools can provide decent assessment results, they have some limitations, such as lacking a comprehensive investigation of system-level and/or component-level degradation and not considering long-term actual mission profiles. The Design for Reliability and Robustness (DfR2) tool [8] was proposed as a complete assessment tool for power electronics components and systems. DfR2 integrates advanced reliability models, such as the electrical model and thermal model that consider statistical distribution for lifetime assessment, which enable the thorough investigation of significant factors associated with component degradation.

The current state of the art presents a trend: The accuracy requirements for reliability assessment tools are increasing because more factors are considered in the analytical model and more data are collected in the long-term mission profiles for the data analysis. It would be time-consuming to perform massive computations with high precision in complex evaluation models in a conventional way. High-performance computing (HPC) employing distributed and/or parallel computing mechanisms holds the promise to achieving the expected high computational efficiency. HPC systems are designed for aggregating large-scale computational resources to solve massive scientific problems. By leveraging the techniques spanning hardware, software, network, and security, HPC systems could provide well-optimized performance for scientific computations by efficiently using high-performance processors. With the rapid development of multiprocessor architectures and the emergence of new ideas such as heterogeneous architectures and cloud computing, HPC has incredible potential to provide great quantities of computational power. Even though HPC has been widely used in power systems simulations, power electronics simulations find very few applications in the current state-of-art.

This work proposes an event-driven parallel computing-based simulator to distribute the computational tasks to multiple machines to accelerate the computational speed. The cases with the same operating conditions are regarded as one event. All diverse events are then simulated in parallel. Adaptive model updates are also performed to guarantee the accuracy of the model. As a result, the computational efficiency of the simulations is significantly improved without sacrificing fidelity. The design and main workflow are introduced in

978-1-7281-6118-1/22 $31.00 © 2022 IEEE

Section II; the performance and corresponding analysis are demonstrated in Section III; and Section IV gives a summary and conclusion.

II. DESIGN FOR HIGH-SPEED, HIGH-FIDELITY SIMULATOR

The Tool for Reliability Assessment of Critical Electronics in PV (TRACE-PV) is proposed for predicting the lifetime of PV inverters based on the physics-to-failure mechanisms. Compared to the state of the art [9], the TRACE-PV tool builds the degradation model considering all components with high failure rates and evaluates the reliability from different aspects, such as system configuration and grid disturbance. The summary diagram for TRACE-PV is shown in Figure 1.

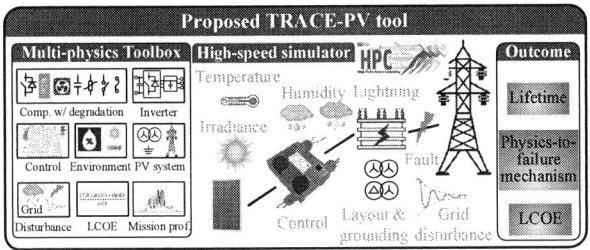

Fig. 1. Summary diagram of TRACE-PV

Figure 2 shows the overall workflow of the HPC simulator. The mission profile and the circuit design are inputs for the simulator. The first step of the simulator is to perform the cluster merging, which is used to preprocess the mission profile data to reduce the number of computations that need to occur. Then the simulator runs the simulation model for each set of data from the mission profile and determines stressors over time. The stressors are used as inputs for the degradation model to predict the lifetime and degradation status. A condition check for detecting whether a critical degradation happens follows. The model parameters are adaptively updated if a check is triggered. The simulator needs to undertake a complex computational workload and adaptively update the model according to the current operating conditions to reach high efficiency and high fidelity. To achieve these goals, several strategies are applied.

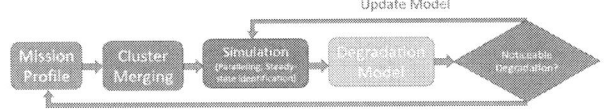

Fig. 2. Overall workflow of the simulator

A. Cluster merging

Mission profiles include environmental information, such as temperature, solar irradiance, and relative humidity; and operational condition information, including power (kW and kVA), voltage, and power factor. Each record in the mission profile indicates the operation condition and the external environmental factors that may affect the runnability of PV inverters during each time segment. This time-series information demonstrates the variance of PV inverters' operation conditions as the time change. Traditionally, sequentially performing simulations for the accumulated history record causes a long simulation time; however, it is observed that extensive records in the mission profile have high similarity. The weather information regularly changes following climate change and other natural factors over some time. The operation information barely changes in the adjacent time period. And the obvious change in the operating condition is usually caused by unexpected incidents or the occurrence of long accumulated degradation.

Therefore, merging similar records could eliminate redundant records and reduce the computational load. The similarity between each record in the mission profile needs to be evaluated. For each data field, only the two records whose difference between them is below a predefined threshold are regarded as similar records. Considering the efficiency of locating the data and the economy of storage, a lookup table is created to store the unique record, and each record will be transferred to a hash value calculated by the hash function. For each record, if it does not have a similar record existing in the lookup table, this record will be stored in the lookup table and will be treated as a new cluster. Otherwise, this record is merged into the existing cluster. The index of this record will be logged in the lookup table for the convenience of further locating it. The reduced number of records and the reduction rate corresponding to the difference threshold in the mission profile after the cluster merge is shown in Table 1. Considering the trade-off between simulation accuracy and reduction rate, the difference threshold is selected as 1%.

TABLE I. REDUCTION RATE FOR DIFFERENT THRESHOLDS

Difference Threshold (%)	Number of categories in 2016–2019 (348,000 records in total)	Reduction Rate (%)
50%	43212	12.73%
30%	60242	17.63%
10%	105412	30.61%
5%	129236	37.45%
1%	142501	41.26%
0.5%	142651	41.31%
0.1%	142765	41.34%

B. Parallel computing

For each mission profile record, one simulation gives out with corresponding stressors. A long-term mission profile will be used to calculate stressors over time. To further reduce the computational time of the simulation, parallel computing techniques are used. Task-level parallelism is applied to partition the large-scale computational workload. As a result of the heavy data dependency within the PV system model simulation, each model simulation is treated as a task. A task takes one mission profile record as the input and gives the over-time stressor as the output. Multiple computer processors are aggregated to execute the computational tasks in parallel. As a benefit from the least data dependency among tasks, each task is independently assigned to a processor. Figure 3 depicts the working pattern of this approach compared to sequential computing. Before the simulation for each case starts, the simulator checks whether the stressors of the corresponding record have been calculated. The simulation for the current case can be skipped if the results from similar inputs already exist in the lookup table without triggering the critical degradation condition. If the simulation model parameter update is triggered by critical degradation, the stressors need to be recalculated, and the corresponding information needs to be updated in the lookup table. Then multiple cases simultaneously simulate round by round, showing a big savings in the overall computational time compared to sequential computing.

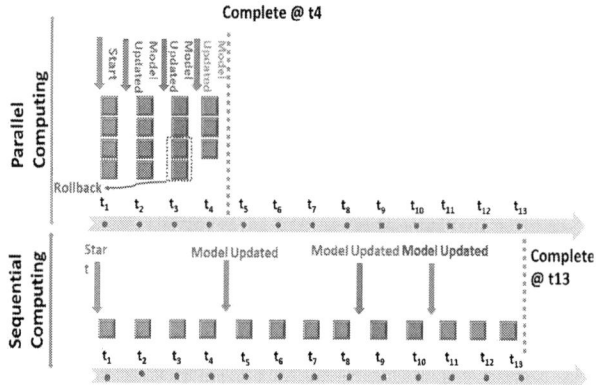

Fig. 3. Timeline comparison of parallel computing with four processors and sequential computing

C. Adaptive model update

As time goes by, the durability of the components degrades. Even if under the same environmental and operating conditions, the components might endure heavier pressure. With the accumulated stressors, the components will experience critical degradation at some point. This change in component parameters is reflected in the voltage and power. To accommodate this change, the simulator sets a condition check to detect whether noticeable degradation happens. Once the change in the operating condition reaches a predefined threshold, the parameters of the simulator will be updated so that it can properly evaluate the current degradation condition of the components. To address the issue that some processors are likely running the outdated simulation model while a certain processor might have triggered the noticeable degradation check, the model implements a rollback function to keep itself updated.

As shown in the parallel computing part of Figure 3, four processors are fully used to simultaneously run four simulation cases at each moment. At each round, only one simulation runs on each processor. When these four processors finish one round of simulations, the simulation results will be checked to see whether noticeable degradation happens. Each simulation result triggering the check will be abandoned, and the corresponding cases will be rolled back to execute the simulation again with the updated model. To minimize the number of simulation results that need to be discarded, a log for the time-series simulation result is recorded in the lookup table. When a check is triggered, only the first case requiring an update needs to be discarded, instead of all the simulation results being abandoned in a row. When the predefined threshold is triggered, the value threshold will be updated according to the degradation condition variance compared to the previous level.

D. Steady-state identification

In addition to the mentioned HPC techniques, another optimization method is applied. For each simulation case, which is limited by the data gap between two consecutive time steps, the simulation will keep running to feed the interval before the next mission profile point arrives; however, the stressors remain the same once the simulation steady state is reached. The simulator is thus designed to perform a steady-state detection by tracking the change in the stressors with the longest time constant (e.g., temperature information originating from the thermal model of the inverters) to save simulation time without affecting accuracy. Figure 4 shows the temperature variance curve of the thermal model of the inverters. The temperature is used as an indicator to demonstrate the operation condition of the inverter. It assumes that the system reaches steady state when the consecutive three temperature differences at the neighboring two time stamps are the same (1% error tolerance). Based on the calculation of the temperature difference, the data point at 110 seconds is detected as a steady-state point, and the simulation will be terminated at this data point. According to the shown curve, we believe the calculated steady-state point can indicate the steady-state status of the system.

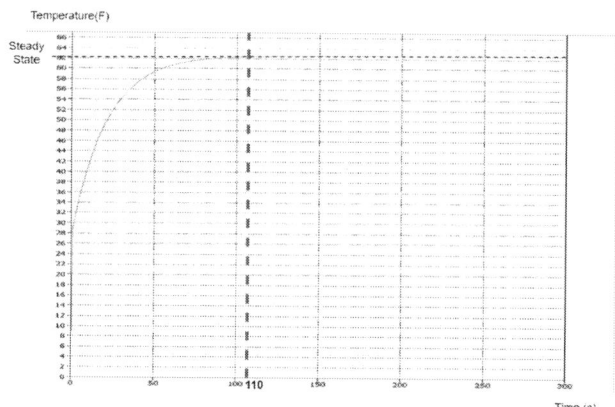

Fig. 4. Steady-state detection of the temperature of the thermal model

III. CASE STUDY AND PERFORMANCE ANALYSIS

To prove the effectiveness of the simulator, a demo simulator was designed for the PV inverters considering the PV system (e.g., PV panel, step-up transformer, and grid configuration). The demo is implemented based on MATLAB Simulink and Open MPI. The mission profiles contain historical environmental parameters that include temperature, relative humidity, solar irradiance from the National Renewable Energy Laboratory's National Solar Radiation Database (NSRDB) [10], and operational field conditions that include AC power, three-phase line-to-line voltages with a 5-minute resolution for an actual PV plant [11]. The demo model simulation is based on MATLAB Simulink, and each case of simulation is treated as an independent task, which runs on a single processor. With the help of the C++ MEX function [12], the model implemented by the MATLAB Simulink model can be called in the C++ program without consuming too much efficiency. MATLAB can accept inputs from the C++ code and return the simulation outputs to the C++ code. The parallel computing was implemented using the Open MPI on C++ [13]. Because running simulations of the

978-1-7281-6118-1/22 $31.00 © 2022 IEEE

data several years; long is time consuming, as a demo, 2016 cases are used as a preventative situation for 1 week. After performing the cluster merging, the number of cases was reduced to 963, 48% compared to the number before the cluster merging. In this case, steady-state detection was not used because the thermal model was not implemented in the demo model.

Figure 4 shows the speedup of the simulation after each optimization method. By applying the cluster merging, the number of cases that needs to be calculated has been reduced to 963, only 48% of the original number, and it achieves 2.09 times the speedup compared to the simulation time without any optimization. Then, with the use of parallel computing, the simulation time proportionally decreases when multiple processors are assigned. As the trend in the orange arrow shows, when 8 processors are used to do the simulation, the simulation performance is improved by 7.97 times. When 64 cores are assigned, the computational efficiency increases by 42.33 times compared to the sequential version. The simulation speed increases up to 58.77 times when 100 processors are used. The total simulation time will be reduced by 123 times with all the optimizations applied. Caused by` the native parallelism of the computation, the computational performance of the simulation is highly scalable with an increasing number of processors. The simulation performance can be further improved if more computing resources can be used. And it is noticeable that the increased rate of speed up after using more than 64 cores starts to decrease. Even if the data communications between each task is the least, the overhead consumption on the processors' schedule has more impact on the performance with more processors applied.

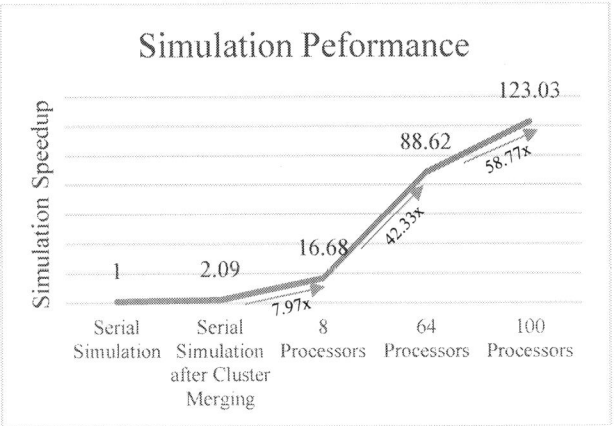

Fig. 5. Comparison of the simulation speedup contributed by each optimization method

IV. CONCLUSION AND FUTURE WORK

With the high demand for accuracy and real-time capability of PV system degradation evaluation [8], future simulation tools are required to consider more parameters and more historical environmental and operation data. This paper proposes an event-driven HPC-based simulator to accelerate the PV system degradation assessment simulations of both the average model and the switch model. In addition to the parallel computing techniques, other optimization methods—such as cluster merging, model updates, and steady-state identification—are

applied to reduce the computational workload and increase the computational efficiency as well as maintain the accuracy of the simulation results. **Our main contribution is designing a complete prototyping PV system degradation assessment simulator with high fidelity and high efficiency.** To better present the framework and validate the effectiveness, one PV system demo case study is implemented based on MATLAB Simulink and Open MPI on C++ according to our design. With all optimization methods applied, the computational efficiency of the simulation of the switch model reaches almost 123 times under the guaranteed fidelity compared to the traditional sequential simulator. Performance boosts are also promising with increased computing resources employed.

Though the decent performance gain has benefited from these optimization techniques, there is space for further improvements. Because the demo model is implemented in MATLAB Simulink and the simulator is implemented in C++, the initialization and transfer overhead caused by communication between C++ and MATLAB have a noticeable impact on the performance. To remove this overhead, a C++-based simulation model will be implemented in the future. More finely tuned partition strategies will be applied corresponding to the new implementation instead of treating the simulation model as a black box. Two-level parallelism can be applied: one for data-level parallelism within each model simulation and the other for simultaneously executing multiple simulation cases.

ACKNOWLEDGMENTS

This work was authored in part by the National Renewable Energy Laboratory, operated by Alliance for Sustainable Energy, LLC, for the U.S. Department of Energy (DOE) under Contract No. DE-AC36-08GO28308. Funding provided by U.S. Department of Energy Office of Energy Efficiency and Renewable Energy Solar Energy Technologies Office. The views expressed in the article do not necessarily represent the views of the DOE or the U.S Government. The U.S. Government retains and the publisher, by accepting the article for publication, acknowledges that the U.S. Government retains a nonexclusive, paid-up, irrevocable, worldwide license to publish or reproduce the published form of this work, or allow others to do so, for U.S. Government purposes.

REFERENCES

[1] U. D. o. Energy, "$1/Watt Photovoltaic Systems White Paper," 2010.

[2] R. Kaplar, Jack Flicker, Reliability of power conversion systems in photovoltaic applications, 2015.

[3] P. Hacke, et al. "A status review of photovoltaic power conversion equipment reliability, safety, and quality assurance protocols," Renewable and Sustainable Energy Reviews, vol. 82, pp. 1097-1112, 2018.

[4] "Center for Advanced Life Cycle Engineering," [Online]. Available: http://www.calce.umd.edu/software. [Accessed 5 August 2021].

[5] "DfR Solutions," [Online]. Available: http://www.dfrsolutions.com. [Accessed 5 Aug 2021].

[6] I. Vernica, K. Ma, F. Blaabjerg, "Optimal derating strategy of power electronics converter for maximum wind energy production with lifetime information of power devices," J. Emerg. Sel. Topics Power Electron, vol. 6, no. 1, pp. 1-10, 2018.

[7] I. Vernica, K. Ma, F. Blaabjerg, "Reliability assessment platform for the power semiconductor devices – Study case on 3-phase grid-connected inverter application," Microelectron. Rel, Vols. 76-77, pp. 31-37, 2017.

[8] I. Vernica, H. Wang, F. Blaabjerg, "Design for reliability and robustness tool platform for power electronic systems—Study case on motor drive applications," in 2018 IEEE Applied Power Electronics Conference and Exposition (APEC), 2018.

[9] A. Nagarajan, R. Thiagarajan, I. Repins, P. Hacke, "Photovoltaic inverter reliability assessment," National Renewable Energy Lab. (NREL), Golden, CO (United States)., 2019.

[10] "National Renewable Energy Laboratory's developer network," National Renewable Energy Laboratory, [Online]. Available: https://developer.nrel.gov/.

[11] Yaskawa selectria solar, [Online]: https://www.solrenview.com/srvp/FSC/SRV/main.php?siteId=3886.

[12] *C++ MEX Functions - MATLAB & Simulink*. (2022). C++ MEX Functions. https://www.mathworks.com/help/matlab/matlab_external/c-mex-functions.html

[13] Gabriel, Edgar, et al. "Open MPI: Goals, concept, and design of a next generation MPI implementation." European Parallel Virtual Machine/Message Passing Interface Users' Group Meeting. Springer, Berlin, Heidelberg, 2004.

Inverter Reliability Estimation for Advanced Inverter Functionality

Jack Flicker[1], Jay Johnson[1], Matthew J. Reno[1], Joseph A. Azzolini[1], Peter Hacke[2] and Ramanathan Thiagarajan [2]

[1]Sandia National Laboratories, Albuquerque, NM, 87185, United States

[2]National Renewable Energy Laboratory, Golden, CO, 80401, United States

Abstract—**In the near future, grid operators are expected to regularly use advanced distributed energy resource (DER) functions, defined in IEEE 1547-2018, to perform a range of grid-support operations. Many of these functions adjust the active and reactive power of the device through commanded or autonomous modes, which will produce new stresses on the grid-interfacing power electronics components, such as DC/AC inverters. In previous work, multiple DER devices were instrumented to evaluate additional component stress under multiple reactive power setpoints. We utilize quasi-static time-series simulations to determine voltage-reactive power mode (volt-var) mission profile of inverters in an active power system. Mission profiles and loss estimates are then combined to estimate the reduction of the useful life of inverters from different reactive power profiles. It was found that the average lifetime reduction was approximately 0.15% for an inverter between standard unity power factor operation and the IEEE 1547 default volt-var curve based on thermal damage due to switching in the power transistors. For an inverter with an expected 20-year lifetime, the 1547 volt-var curve would reduce the expected life of the device by 12 days. This framework for determining an inverter's useful life from experimental and modeling data can be applied to any failure mechanism and advanced inverter operation.**

Keywords—*inverter reliability, grid-support functions, stress, lifetime, component degradation*

I. Introduction

In response to increasing photovoltaic (PV) penetration levels on the grid, distributed energy resource (DER) devices—e.g., photovoltaic inverters and energy storage systems (ESS)—are now required to include control modes to improve power quality on the grid, in accordance with the U.S. interconnection standard, IEEE 1547-2018 [1]. The set of control modes in IEEE 1547 includes fixed power factor, voltage-reactive power mode, limit active power, etc. which provide grid operators with new ways to support distribution feeder voltages [2], participate in load-generation balancing [3], and stabilize the system during power system faults [4].

While the impact of advanced inverter functionality on grid stability and power quality has been well studied [5], the impact of this behavior on inverter lifetime and reliability has not. Typically, power electronics devices operating with grid-support modes (e.g., non-unity power factor) typically, without design modifications to mitigate, experience increased switching losses or other component stresses, thereby reducing the overall system lifetime of the inverter [6]. Significant barriers exist in quantifying this reduction in unit lifetime using component-level stresses. The first barrier is instrumenting and monitoring switch stress across the inverter's entire active and reactive power operating envelope. The other is determining the appropriate lifetime profiles to estimate lifetime reductions.

In previous work [7], we addressed the first of those challenges and describe a method of instrumenting and autonomously measuring inverter component stress for a variety of different advanced inverter operating conditions using the System Validation Platform (SVP). Originally, the SVP was developed as a flexible framework to autonomously measure system-level inverter operations for certification processes [8]. We altered the SVP framework to measure, process, and store component-level loss maps to calculate inverter lifetimes when operating with advanced inverter functions. We used this open framework to measure changes in switching loss in the inverter as a function of the PV system power factor and incident irradiance (Figure 1).

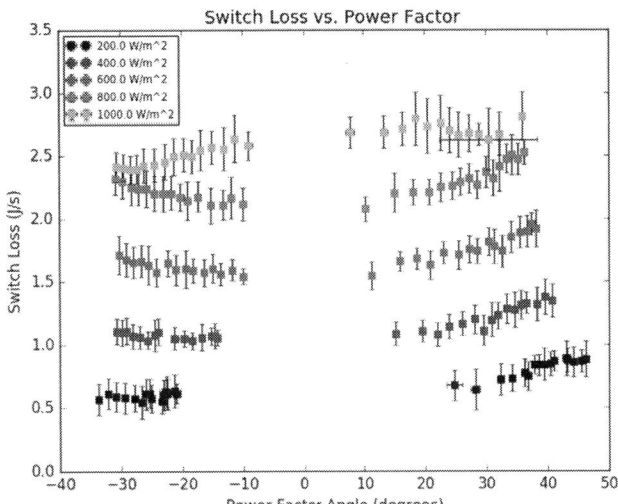

Figure 1: PV inverter switch losses for a range of PV irradiance levels and power factors.

The collected data shows an increasing switching loss with irradiance until the curtailment regime is reached (e.g., with the

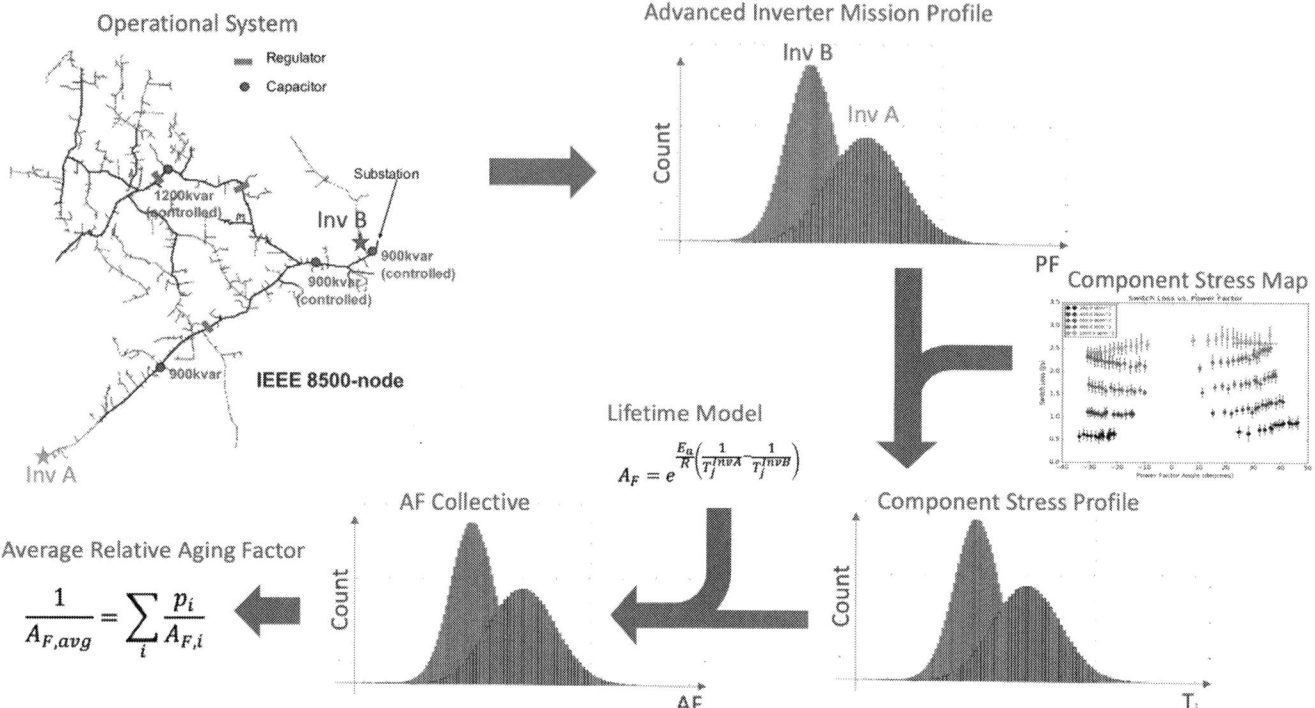

Figure 2: Framework for determination of average inverter acceleration factor. An inverter in a system will have a given profile of advanced inverter functionality. This profile can be paired with a component loss map determined experimentally to derive an ensemble of component stress over the profile period. This stress can be translated into a collective of acceleration factors and an average acceleration factor can be derived.

1000 W/m² turquoise points). Also, the switch loss is shown to be a function of the power factor angle with a minimum loss at zero power factor angle and a maximum at larger positive and negative power factor angles. It is clear from experimental data that switch loss is at a minimum for unity power factor operation and increases for non-unity power factor operation. It would be expected that increased loss internal to the inverter will yield some reduced lifetime (due to increased operational temperature, in this case). However, the extent of this reduction in useful life is not only dependent on the increase in component stress as a function of advanced operating mode, but on the amount of time an inverter will utilize these advanced operating functions in fielded use as well as the specific acceleration factor for the applicable failure mechanisms.

In this work, we utilize data collected previously [7] to calculate a possible reduction in useful life of inverters due to thermal stress (driven by increased loss) in a high-side switch in the H-bridge. . Although this paper focuses on the additional thermal stress from volt-var, the process introduced here is generally applicable to any failure mechanisms through the substitution of appropriate experimental data and lifetime models. Section II introduces the lifetime estimation framework with application to static advanced inverter operation. Section III introduces simulations to estimate dynamic inverter advanced inverter mission profile—a simplified representation of the power or thermal profile over the life of the deployed product—in a fielded system. Section IV evaluates expected inverter reduction in useful life based on these dynamic advanced inverter mission profiles

II. LIFETIME ESTIMATION

Lifetime modeling of any electronic component is typically a pairwise comparison between states. It can compare two similar devices under test that are subject to different operational conditions or the theoretical comparison between the same device subject to different operational conditions. We have developed a framework for evaluating the relative lifetime of two inverters (these can be two distinct inverters or subject to different operational conditions) or nominally the same inverter. This framework, described in Figure 2 utilizes experimental component level stress, paired with the expected operation of the unit in the field, and an applicable lifetime model, to compare the relative degradation between two different inverters.

We consider two inverters: Inverter A, denoted by a purple star, which operates on one bus in a simulated system (the IEEE 8500-node test feeder [9] in this case), as shown in Figure 2, and Inverter B, denoted by a blue star operating at some other location on the system. Due to their different locations in the system, these two inverters have different operating points over a given period. This is because many autonomous grid-support functions, such as volt-var, act on local measurements, such as system voltage, and not global throughout the system, such as frequency. This operational mission profile can be a time-domain measurement or histogram of operational states (as shown in Figure 2). These operational mission profiles can be paired with the experimental component stress maps described previously [7] to transform the advanced inverter mission profile into a component stress profile. This component stress can then be paired with an appropriate lifetime model to transform the component stress profile into a collection of acceleration factors. A simple weighted average can then be used to calculate an average acceleration factor comparison between the inverter units.

This process is generic to any failure mechanism that impacts inverter lifetime, although, for the failure mechanisms to impact overall system lifetime, the example mechanism must be the relevant limiting failure mechanisms for inverter lifetime. We use this framework to derive the expected reduction in inverter useful life via the experimental measurement of switch loss as a function of the power factor coupled with a thermally driven failure mechanism determined by switch junction temperature. Although switch failure is a significant failure mechanism in inverters [10], there are many other competing failure mechanisms (e.g. capacitor wear out, contactor failure). Whether this specific failure mechanism is the predominant failure mechanism depends on many factors including actual environmental stress (thermal and electrical), system design, and device usage. However, the framework for evaluation shown in Figure 2 is broadly applicable to different operating modes and failure mechanisms when incorporating the appropriate measurements, fielded operational states, and reliability models.

In this paper, we assume inverter failures are due to switch loss from thermal damage. For a thermally driven process in the switch, the temperature rise in the junction at a steady-state condition due to loss can be calculated by (1).

$$T_j(t) = T_{amb}(t) + P_{loss}(t) \cdot R_{th} \tag{1}$$

where T_j is the junction temperature in Kelvin, T_{amb} is the ambient temperature in Kelvin, P_{loss} is the switch loss in Watts, and R^{th} is the junction to case thermal resistance in K/W.

If T_j is calculated for two systems (T_j^1 and T_j^2), then a relative acceleration factor (A_F) can be found, assuming an Arrhenius model (2).

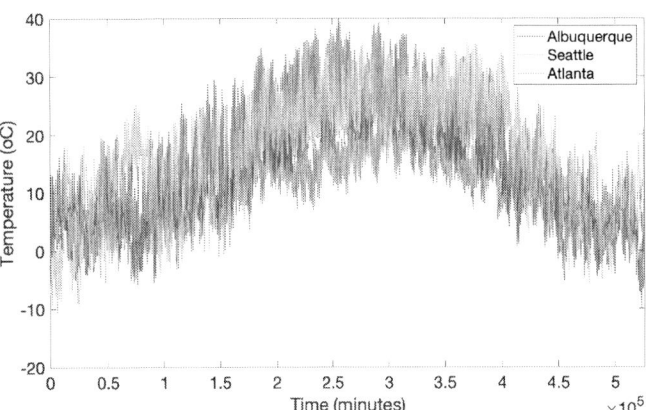

Figure 3: Ambient temperature data in 15-minute intervals over a year for Albuquerque, Atlanta, and Seattle.

$$A_F(t) = e^{\frac{-E_a}{k}\left(\frac{1}{T_j^1} - \frac{1}{T_j^2}\right)} \tag{2}$$

where k is Boltzmann's constant (8.617×10^{-5} eV/k) and E_a is the activation energy for the failure mechanism (in eV).

For a varying junction temperature, (2) will yield a varying acceleration factor. This distribution of acceleration factors is known as a stress collective [11]. A weighted average is used to determine an average acceleration factor (3).

$$A_F^{avg} = \sum_{t_{initial}}^{t_{final}} \frac{1}{N \cdot A_F(t)} \tag{3}$$

Even without detailed knowledge of inverter fielded advanced inverter operation, we can use this framework to calculate the relative change in acceleration factor between two inverters in a simple case, i.e., two inverters operating at a static power factor. In this case, we have one inverter operating in a PF=1 condition and another equivalent inverter operating at a PF=+0.85 condition. We consider the acceleration factor between the two inverters for three locations (Albuquerque, NM, Atlanta, GA, and Seattle, WA) using 15-minute ambient temperature and

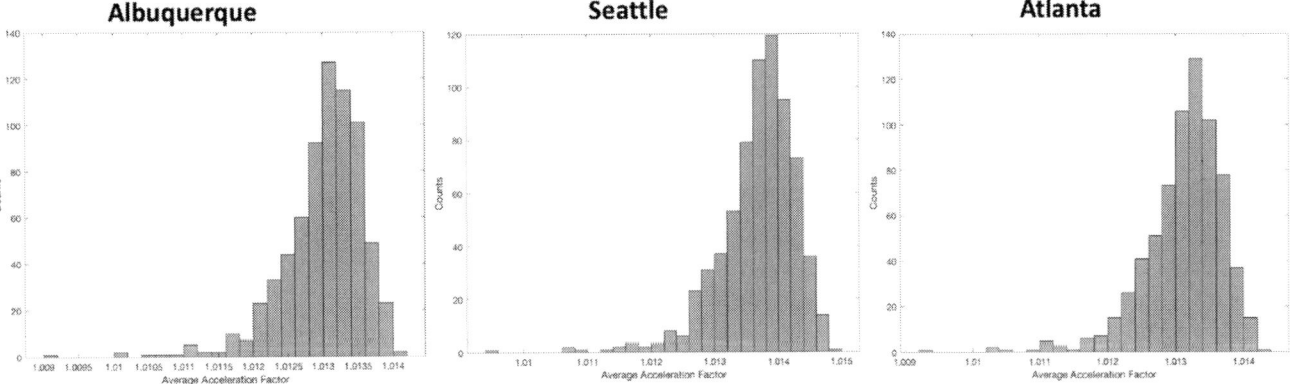

Figure 4: Histogram of acceleration factors for Albuquerque (left), Seattle (center), and Atlanta (right) for comparison between unity PF operation and constant +0.85 PF operation.

978-1-7281-6118-1/22 $31.00 © 2022 IEEE

irradiance data from the TMY2019 data from the National Solar Radiation Database [12]. A thermal resistance value of 0.5 K/W is assumed (which represents an excellent thermal dissipation) and P_{loss} at a given irradiance is calculated through interpolation of the data in Figure 3. Although activation energy is process specific, an activation energy of 0.8 eV is assumed, which is consistent with failure analysis for semiconductors (e.g. MIL-HDBK-217F [13]).

Using these values, a mean acceleration factor (based on a curve fit using a Generalized Extreme Value Distribution) between the two devices is calculated to be ~1.0014 (Figure 4), which indicates a 1.4% faster degradation for a unit operating at PF=+0.85 compared to unity. If a standard inverter is assumed to have a 20-year lifetime, this would mean that the inverter operating at PF=+0.85 would fail 3-months earlier on average. The distribution and mean do not vary significantly between locations.

III. QSTS SIMULATIONS

Of course, devices which operate at fixed power factors (other than unity) are unusual in the field and the analysis for inverters operating at set power factors is simplistic. Autonomous advanced inverter operations measure the local grid conditions and autonomously respond. Given a set of grid conditions, the inverter will produce a known response, which alters system conditions based on inverter location, grid voltage stiffness, and X/R ratio in a feedback loop. Therefore, it is not possible *a priori* to determine the inverter mission profile in such a non-linear system. To address this second challenge and determine a *realistic* inverter advanced operational profile in a system, we select a representative use case and study year-long quasi-static time-series simulations (QSTS) of a 701-inverter distribution system to determine inverter mission profiles. These time-based mission profiles for 701 inverters are then used with the experimental loss maps to characterize the reduction in useful life for a given volt-var curve.

QSTS simulations solve a series of sequential steady-state power-flow equations in which the converged state of each iteration is used as the beginning state of the next [14]. QSTS is used to study equipment control operation, voltage regulation, and reactive power management for DER like solar PV systems [15].

We implemented QSTS simulations [5] on a modified version of the EPRI Ckt5 test feeder (an actual 12.47 kV distribution circuit) with 1,379 residential customers, a maximum bus distance of 3.24 miles from the substation, and a total of 701 residential PV systems to develop year-long, 15 minute active and reactive time series profiles for the inverters at different locations in a power distribution system [16] (Figure 5). Multiple simulations can be carried out utilizing different programmed volt-var curves in each scenario [17]. This determines inverter mission profiles subject to different programmed volt-var curves. These simulations result in 15-minute time series data for each inverter in the system based on the given volt-var curves.

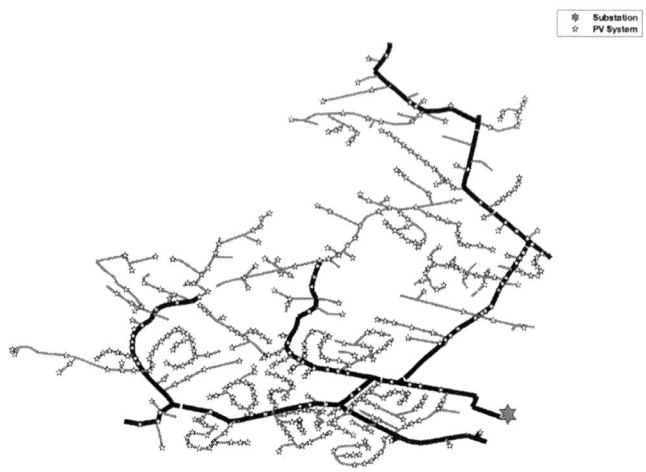

Figure 5: EPRI Ckt5 circuit with 701 operational inverters utilized for QSTS simulations to determine inverter volt-var mission profile.

IV. RESULTS

We have used this framework to calculate the relative change in average acceleration factor for the 701 inverters operating in the EPRI Ckt5 test feeder using the general framework outlined in Figure 2. The average acceleration factor compares the 701 inverters operating at a static power factor (PF=1) vs. the IEEE 1547 Category B [1] default volt-var curve (Figure 6).

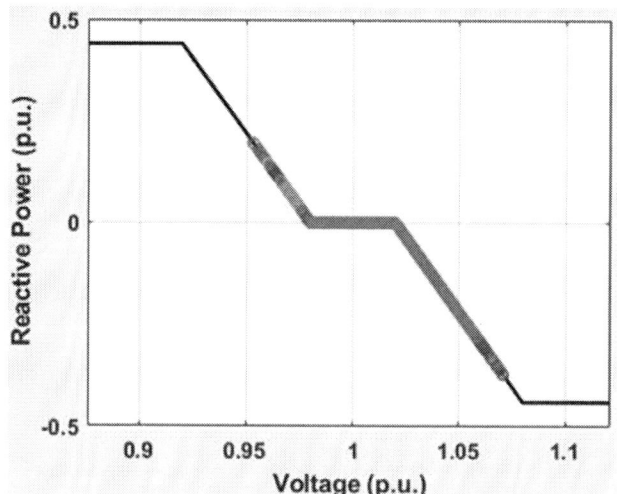

Figure 6: Reactive Power vs. Voltage for a single time point of the system overlaid on the IEEE 1547 Category B [1] default volt-var curve. Each point indicates one of the 701 inverters in the system.

We consider this system in three different geographic locations (Albuquerque, NM, Atlanta, GA, and Seattle, WA) using 15-minute ambient temperature data from the TMY2019 data from the National Solar Radiation Database [3]. As before, a thermal resistance value of 0.5 K/W with an activation energy of 0.8 eV is assumed.

Using these values, an average acceleration factor can be calculated for each inverter in the population by comparing the two cases (unity power factor and 1547 Category B). Although the specific acceleration factor for an inverter is a complicated convolution of factors (location, phase, line impedance, coupling with other inverters, etc.), the population of all inverters will have a distribution of acceleration factors (Figure 7). We have calculated the average acceleration factors for all 701 inverters in the system and found the mean to be 1.0015 ± 0.00033 in the Albuquerque thermal environment, 1.0016 ± 0.00035 for the system in the Seattle thermal environment, and 1.0015 ± 0.00033 for the system in the Atlanta thermal environment. The similarity of these results indicates that ambient temperature is a weak driver for switch loss compared to power loss from the switching itself. It also indicates that incorporation of the IEEE 1547 Category B volt-var curve only results in a 0.15% faster degradation rate for the median inverter. In this case, if a standard inverter is assumed to have a 20-year lifetime, this would mean that an average inverter operating with the IEEE 1547 Category B curve would fail only 12 days earlier on average. This indicates that for this failure mechanism under the usage conditions studied here, inverter volt-var capability does not significantly decrease inverter lifetime.

The results here assume a semiconductor device with excellent heat dissipation (R_{th} = 0.5 K/W) as well as activation energy consistent with MIL-HDBK-217F. In reality, thermal dissipation can vary significantly, from 0.5 to upwards of 4 K/W [18] while activation energy for thermally activated failure mechanisms can vary from 0.5 to 2 eV [19]. To evaluate the sensitivity of the acceleration factor to these two parameters, the framework in Figure 2 was carried out for different values of R_{th} and E_a. The distribution of acceleration factors for the 701 inverters was fit using a generalized extreme value (GEV) distribution. Figure 8 shows the median of the histogram distribution for the different simulations as a function of changes in R_{th} and Ea. The inset in the figures shows the change in standard deviation. In both cases, the acceleration factor is linear with both E_a and R_{th}. For extreme values of E_a and R_{th} (1.2 eV and 4 K/W, respectively), the acceleration factor would be 1.8%, which represents an absolute worst-case scenario for this failure mechanism.

In addition to the mean of the distribution, a sensitivity study of the standard deviation of the distribution was also carried out. The standard deviation of the distribution is also linear with respect to the parameters studied. However, it increases at approximately the same rate as the distribution mean, so that the ratio between the mean and the standard deviation stays relatively constant. This indicates that the shape of the distribution is constant as R_{th} and E_a are varied, which can be seen from the distribution insets. As the parameters increase, there may be a slight increase in distribution asymmetry due to a longer tail at the high A_F end, which would explain the slight increase in standard deviation. However, this difference is extremely slight, if present at all over the range of parameter variation studied.

Since we have average acceleration factor values for 701 different inverters, it is possible to try and determine if there are correlations between the area on the feeder and the acceleration factor. By plotting the distance from the substation vs acceleration factor for all the inverters (A, B, and C-phases are considered separately), we can clearly see a dependence on the aging rate as a function of distance from the substation (Figure 9). The linear fit to this distribution indicates an additional aging of approximately -0.25 %/km from the substation, so that inverters closer to the substation age at a quicker rate than those downstream of the substation.

In the system studied, inverters closer to the substation age faster than those further away. This is counter-intuitive to many systems, where PV inverters far from the substation experience larger voltage deviations than those close to the feeder. However, in this system, inverters close to the substation age faster most likely because the feeder operates at high voltage, typically 1.05 pu (at the high end of ANSI range A to offset line losses and low voltages at the ends of the feeder), and is very stiff. This means that inverters near the substation work more often to lower the voltage *and* are trying to do this against a stiff source (the substation interconnect) so that they are operating in this manner for longer lengths of time. In short, in this system, inverters near the substation operate at non-unity more often and at higher PF than the inverters at the ends of the

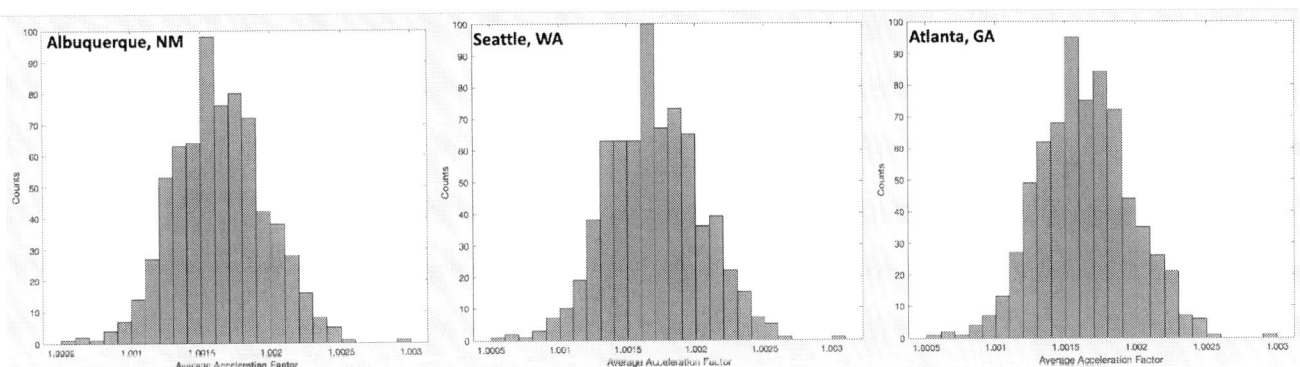

Figure 7: Average acceleration factor for 701 inverters in the Ckt5 system based on ambient temperature data in three locations (left) Albuquerque, NM (center) Seattle, WA (right) Atlanta, GA.

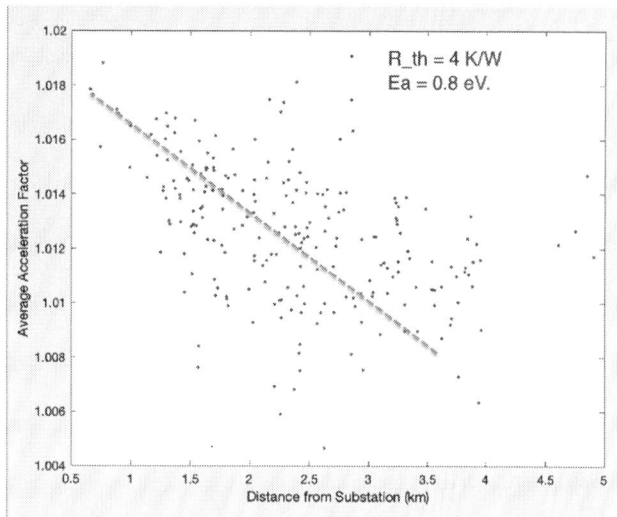

Figure 9: Average acceleration factor for inverters in Ckt5 system vs distance from the substation on the A-phase. Inverter closer to the substation have a higher acceleration

distribution feeder. This offers another impetus for running substation voltages at a lower level (in addition to operational considerations, such as Conservation Voltage Reduction, CVR [20])

This relationship is not necessarily generalizable to all systems, because it is complicated by a variety of factors: feeder voltage, length of feeder, presence of voltage devices on the system, inverter-inverter interactions, number/distribution of DER, power injected by DER, local voltage condition, etc.

V. CONCLUSION

In this work, several advancements were made regarding the reliability impacts of advanced inverter operation. A framework was developed to couple experimental component-level stress measurement data with system-level operational data and derive a reduction in useful life due to inverter fielded operational modes. QSTS simulations were used to model expected inverter power factor operation and a reduction in useful life was calculated based on experimental SVP data. It was found that with standard assumptions, the mean additional aging rate due to power factor operation for inverters was only ~0.15%. Sensitivity analysis on assumptions in the reliability model gave a maximum (corner case) mean aging due to advanced inverter functions of 1.8%, which would represent a worst-case reduction in useful life of 4.5 months for an inverter expected to last 20 years. For reasonable assumptions, the mean reduction in useful life is significantly less, indicating that for the failure mechanism studied in detail (thermal aging in the switch), advanced inverter operation is not a significant contributor to early failure based on added stress from vol-var operation.

ACKNOWLEDGMENT

Sandia National Laboratories is a multimission laboratory managed and operated by National Technology & Engineering Solutions of Sandia, LLC, a wholly owned subsidiary of Honeywell International Inc., for the U.S. Department of Energy's National Nuclear Security Administration under contract DE-NA0003525.

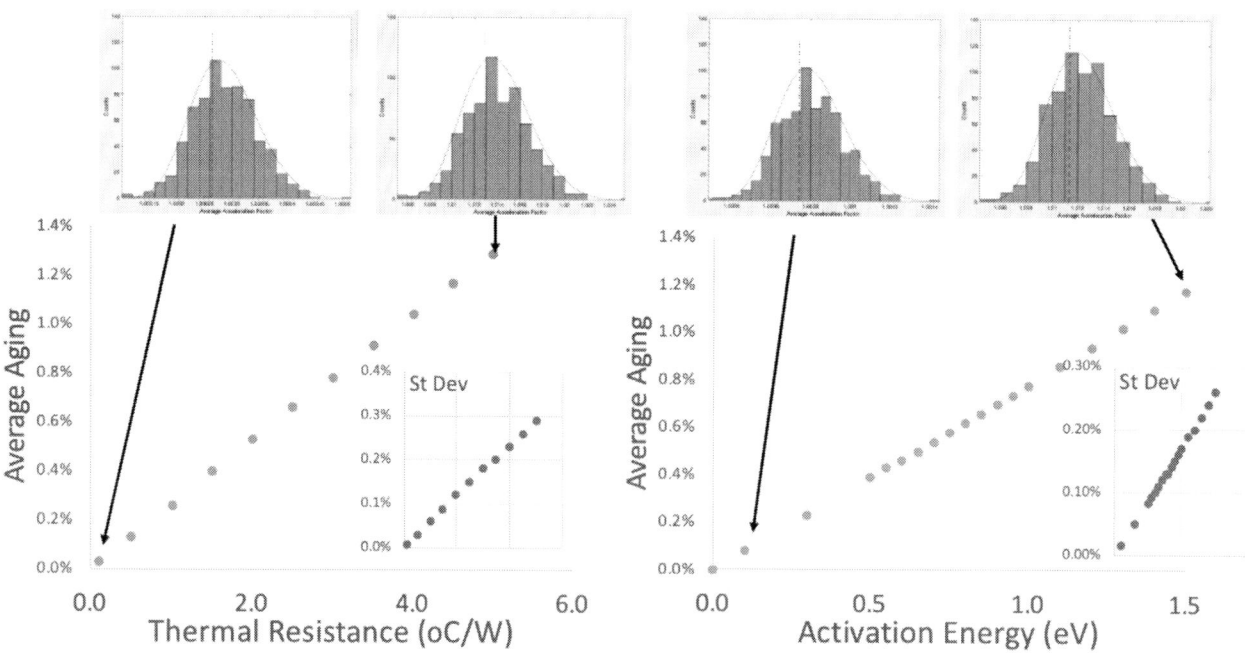

Figure 8: Acceleration factor as a function of thermal resistance (left) and activation energy (right). The acceleration factor is linear with both parameters. The inset in each figure shows the standard deviation of the distribution, which is also linear. The distribution and fit are denoted for the low and high values of the parameter.

REFERENCES

[1] *IEEE Standard for Interconnection and Interoperability of Distributed Energy Resources with Associated Electric Power Systems Interfaces*, I. o. E. a. E. E. (IEEE), 2018.

[2] A. Summers, J. Johnson, R. Darbali-Zamora, C. Hansen, J. Anandan, and C. Showalter, "A comparison of DER voltage regulation technologies using real-time simulations," *Energies*, vol. 13, no. 14, p. 3562, 2020.

[3] R. W. Kenyon *et al.*, "Stability and control of power systems with high penetrations of inverter-based resources: An accessible review of current knowledge and open questions," *Solar Energy*, vol. 210, pp. 149-168, 2020.

[4] M. Talha, S. Raihan, and N. Abd Rahim, "PV inverter with decoupled active and reactive power control to mitigate grid faults," *Renewable Energy*, vol. 162, pp. 877-892, 2020.

[5] J. Smith, W. Sunderman, R. Dugan, and B. Seal, "Smart inverter volt/var control functions for high penetration of PV on distribution systems," in *2011 IEEE/PES Power Systems Conference and Exposition*, 2011: IEEE, pp. 1-6.

[6] J. Flicker and S. Gonzalez, "Performance and reliability of PV inverter component and systems due to advanced inverter functionality," 2015: IEEE, pp. 1-5.

[7] J. Flicker, J. Johnson, P. Hacke, and R. Thiagarajan, "Automating Component-Level Stress Measurements for Inverter Reliability Estimation," in *IEEE Photovoltaic Specialists Conference (PVSC)*, Virtual, 2021.

[8] J. Johnson and B. Fox, "Automating the Sandia advanced interoperability test protocols," *40th IEEE PVSC, Denver, CO*, pp. 8-13, 2014.

[9] R. Dugan and R. Arritt, "The IEEE 8500-node test feeder," *Electric Power Research Institute, Palo Alto, CA, USA*, 2010.

[10] A. Golnas, "PV System Reliability: An Operator's Perspective," presented at the 38th Photovoltaic Specialists Conference, Austin, TX, Aug 15, 2012.

[11] R. Stadler and A. Maurer, "Methods for Durability Testing and Lifetime Estimation of Thermal Interface Materials in Batteries," *Batteries*, vol. 5, no. 1, p. 34, 2019. [Online]. Available: https://www.mdpi.com/2313-0105/5/1/34.

[12] M. Sengupta, Y. Xie, A. Lopez, A. Habte, G. Maclaurin, and J. Shelby, "The national solar radiation data base (NSRDB)," *Renewable and Sustainable Energy Reviews*, vol. 89, pp. 51-60, 2018.

[13] *Military Handbook: Reliability prediction of electronic equipment*, 1991.

[14] M. J. Reno, J. Deboever, and B. Mather, "Motivation and requirements for quasi-static time series (QSTS) for distribution system analysis," in *2017 IEEE Power & Energy Society General Meeting*, 2017: IEEE, pp. 1-5.

[15] D. G. Photovoltaics and E. Storage, "IEEE guide for conducting distribution impact studies for distributed resource interconnection," 2014.

[16] "Summary of EPRI test circuits," EPRI, [online] Available: http://svn.code.sf.net/p/electricdsslcode/trunkiDistrib/EPRITestCircuits/.

[17] J. Seuss, M. J. Reno, R. J. Broderick, and S. Grijalva, "Analysis of PV advanced inverter functions and setpoints under time series simulation," *Sandia National Laboratories SAND2016-4856*, 2016.

[18] Y. Liu, *Power electronic packaging: design, assembly process, reliability and modeling*. Springer Science & Business Media, 2012.

[19] K. Puschkarsky, H. Reisinger, C. Schluender, W. Gustin, and T. Grasser, "Voltage-dependent activation energy maps for analytic lifetime modeling of NBTI without time extrapolation," *IEEE Trans. Electr. Dev.*, vol. 65, no. 11, pp. 4764-4771, 2018.

[20] C. Miller, P. Carroll, and A. Bell, "Smart Grid Demonstration Project," National Rural Electric Cooperative Association, Arlington, VA (United States), 2015.

978-1-7281-6118-1/22 $31.00 © 2022 IEEE

Grid-Forming and Grid-Following Inverter Comparison of Droop Response

Nicholas S. Gurule[1], Javier Hernandez Alvidrez[1], Matthew J. Reno[1], Wei Du[2], and Kevin Schneider[2]
[1]Sandia National Laboratories, Albuquerque, New Mexico, 87185, USA
[2]Pacific Northwest National Laboratory, Richland, Washington, 99354, USA

Abstract—**With the increase in penetration of inverter-based resources (IBRs) in the electrical power system, the ability of these devices to provide grid support to the system has become a necessity. With standards previously developed for the interconnection requirements of grid-following inverters (GFLI) (most commonly photovoltaic inverters), it has been well documented how these inverters "should" respond to changes in voltage and frequency. However, with other IBRs such as grid-forming inverters (GFMIs) (used for energy storage systems, standalone systems, and as uninterruptable power supplies) these requirements are either: not yet documented, or require a more in deep analysis. With the increased interest in microgrids, GFMIs that can be paralleled onto a distribution system have become desired. With the proper control schemes, a GFMI can help maintain grid stability through fast response compared to rotating machines. This paper will present an experimental comparison of commercially available GFMI and GFLI ' responses to voltage and frequency deviation, as well as the GFMI operating as a standalone system and subjected to various changes in loads.**

Keywords— **Grid Support, Inverter, Droop Control, Volt-Var, Freq-Watt, Distributed Energy Resource, Microgrid**

I. INTRODUCTION

With the growing amount of renewable energy being added to the distribution system, the overall penetration of inverter-based resources (IBRs) has increased significantly. This leads to lower overall inertia in the system and can lead to instability in the power system [1]. With modern interconnection standards requiring grid support functionalities such as volt-var (VV) and frequency-watt (FW), the stability is improved, but grid-following inverters (GFLIs) used for these resources still do not provide any inertia and require the utility to be present to operate. Grid-forming inverters (GFMIs) on the other hand can operate in parallel with the grid like a GFLI, however more interestingly they can operate autonomously or as the main voltage reference of a microgrid (and provide an emulated inertia). GFMIs can provide a fast response to change in voltage and frequency reference, or deviation of load when compared to synchronous machines, and can achieve load sharing and grid support functionality through droop characteristics if their control schemes allow it. This enables the capability to maintain grid stability when paralleled with other generating units or ensure proper load sharing when operating in a microgrid.

This study analyzes the grid support capabilities of a GFLI and two GFMIs. The inverters are subjected to testing to determine exactly how fast each inverter can respond to an event, as well as to determine how tunable these parameters are. While IBRs can respond very quickly, a rapid response is not always what is desired.

II. BACKGROUND

In the early stages of the adoption of photovoltaic (PV) inverters, utilities called for the standardization of distributed energy resources (DER) requirements when interconnecting to the grid. This led to the development of IEEE Std 1547-2003, which required all PV inverters to trip following any "major" voltage or frequency disturbance. Additionally, these IBRs were either: not capable, or not permitted to provide any support functionality such as VV or FW.

With the lack of grid support functionalities and the low level of fault current capability [2], many utility operators have been hesitant to accept IBRs as an acceptable DER. Synchronous generators (SG) have been widely accepted as a viable source for local loads and available fault current over the years, but as the growing demand for clean, and lower cost energy increases, utilities have been driven to accept these devices. However, to allow for this acceptance, more was to be required from IBRs.

With the rapid adoption of GFLIs, particularly in California, it was found that there was a need for a change to their interconnection requirements. In 2014, California amended their interconnection requirements, Rule 21 of the California Public Utilities Commission, due to the ever-growing amount of renewable DER being added to their system and the distant revision of the IEEE 1547 standard. VV functionality has been shown to be a great way for GFLIs to assist in stabilizing the grid voltage level [1], and thus the amendment to Rule 21 called for all new inverters to be interconnected into the California distribution system to have VV functionality (with a reactive power priority) and allowed for the option of other grid support functionalities, such as FW. Additionally, inverters were not required to trip during grid disturbances as rapidly as seen previously in IEEE Std 1547-2003, thus allowing the inverter to ride through different levels of low or high voltage disturbances. This change in California's requirements lead to an addendum to IEEE Std 1547-2003, resulting in IEEE Std 1547a-2014, allowing for these newly desired support functionalities, and adding the option for voltage and frequency ride-through capability.

More recently, IEEE Std 1547-2018 [3] was approved for publication, which requires all inverters to have primary grid support functionalities (VV and FW), as well as additional ones such as volt-watt and watt-var. This new standard reflects how over the past two decades many utilities have transitioned from not wanting support from DER to accepting the increase in rapidly growing technology and utilizing its capability to promote a stable and grid reliability.

While IEEE Std 1547-2018 covers the interconnection requirements for GFLIs, as well as synchronous machines, there is no clear area where grid-forming IBRs fall. GFMIs can utilize several control schemes that replicate the droop characteristics

TABLE I: GFLI FREQUENCY-WATT RESPONSE TIME

Open Loop Response Time:	ROCOF (Hz/s):	Settled Power (p.u.):	Power Response, 1% (s):	Response Time Error (%):
10.0s	3	0.1857	7.332	26.9
5.0s	3	0.1852	4.901	2.0
1.0s	3	0.1851	0.880	12.0
0.2s	3	0.1851	0.687	245.0
	3	0.1861	0.512	N/A
0.0s	10	0.1857	0.501	N/A
	100	0.1859	0.570	N/A

of synchronous machines, and act as an analog replacement for electromechanical generation [4]. However, due to their fast electronic devices, GFMIs can respond quicker than conventional forms of generation, such as synchronous machines, and should be able to respond as fast as GFLIs when desired.

Additionally, with the look into microgrids as a solution option for resilient power, GFMIs are an attractive source for islanded systems. These devices can create a voltage and frequency reference of the system and dampen the system through fast injection of real and reactive power to the microgrid [5]. Higher penetration of GFLIs can lead to fluctuation in system frequency and voltage due to their rapid response. The characteristics of GFMIs can help to stabilize the system from tripping during variation of load and generation and allow for a more desirable response in a low-inertia system than synchronous machines [6].

III. EXPERIMENTAL SETUP

For this study two different commercially available three-phase inverters were used, a 24 kW GFLI, and a 100 kW GFMI. The GFLI is a conventional PV inverter with VV and FW functionality. The GFMI is an energy storage inverter that utilizes a virtual synchronous machine (VSM) control scheme for its grid-forming functionality and droop characteristics. For the grid-tied responses, the inverters were connected directly to an Ametek RS-90 regenerative power grid-emulator. This allowed for precise control of RMS voltage and frequency. Additionally, the GFMI were tested for their islanded response, i.e., if the inverter was acting as the main source within an islanded microgrid. The GFMI was operated in its droop control and connected to a balanced resistive load to determine its frequency response, and a balanced inductive load to determine its voltage response. Fig. 1 shows the islanded and grid-tied test setups used for this study.

For DC sources, the GFLI was connected to eight Ametek TerraSAS PV simulators, set to operate at a total of 1.15 p.u. of the rated power of the inverter. This is to ensure that the GFLI's output is not limited by the DC supply. GFLIs do not output much current above their output rating, and do not utilize additional headroom for overload capability. The GFMI was connected to an NH Research 9300 battery test system as the DC source. Note that the PV system is floating when the GFLI is not operating, but is grounded when it is. The battery system is always ground referenced.

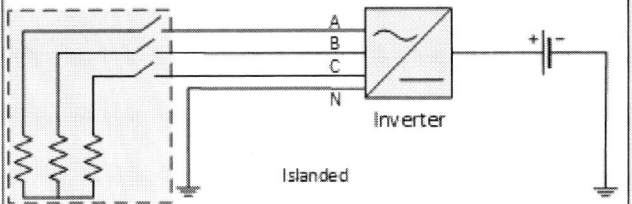
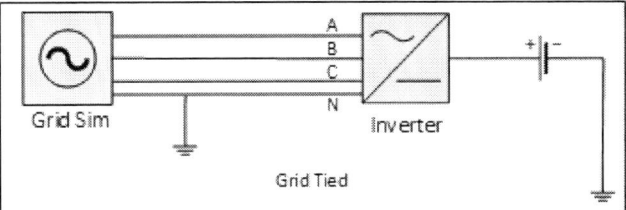

Fig. 1: Setup for Islanded and Grid-Tied Testing

The GFLI was evaluated for both voltage and frequency support functionality response time. This was achieved by programming the grid simulator to change the voltage or frequency level to a predetermined amount without varying the second parameter. For the frequency response, the source was programmed to ramp to 61 Hz at various rates. For the voltage response, the voltage was stepped to either 0.98 p.u. or 1.02 p.u., of nominal voltage, to go to the extreme ends of the VV slope but not go into the constant var portions. Testing was performed with both: a positive, and a negative shift from nominal voltage to determine if there is a response difference for production and absorption of reactive power. Additionally, the variable open-loop response time(s) were adjusted to determine the accuracy of this parameter. The FW and VV profiles used were the most aggressive allowed by IEEE Std 1547-2018.

The GFMI was initially tested for its frequency response. The GFMI was first interconnected to the grid simulator to see its response to different rate of changes of frequency (ROCOF) at a fixed frequency droop of 2% and a change in ±0.6 Hz (1% change in frequency). Following this, the inverter was tested in an islanded state to verify the response time of different sizes of block loads. For all these tests, the inertia time constant was adjusted, as well as the frequency droop value, to determine the correlation between these two parameters and the time response. Following the initial testing, the inertia time constant (ITC) of the droop control, as well as the droop setpoint were changed to determine how these affect the response time of the device. For

TABLE II: GFLI VOLT-VAR RESPONSE TIME

Open Loop Response Time:	Voltage Setpoint (p.u.):	Settled Power (p.u.):	Power Response, 1% (s):	Response Time Error (%):
10.0s	0.98	0.4373	12.347	23.5
	1.02	-0.4388	10.357	3.6
5.0s	0.98	0.4404	6.624	32.5
	1.02	-0.4388	5.233	4.7
2.0s	0.98	0.4387	3.106	55.3
	1.02	-0.4388	2.163	8.2

TABLE III: GFMI Grid-Tied Frequency Response, Default Inertia Parameters, 2% Droop

ROCOF:	Frequency Setpoint (Hz):	Settled Power (p.u.):	Power Response, 1% (s):
3Hz/s	59.4	0.5087	0.818
	60.6	-0.5436	0.777
10Hz/s	59.4	0.5084	0.687
	60.6	-0.5437	0.713
100Hz/s	59.4	0.5083	0.717
	60.6	-0.5431	0.670

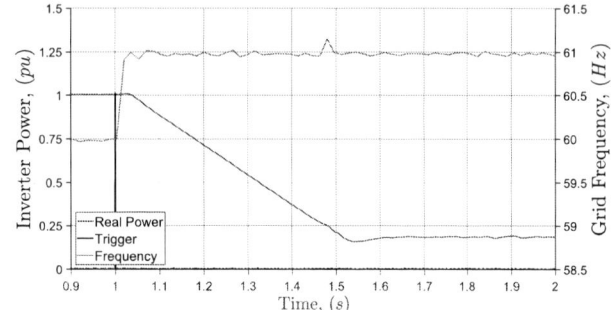

Fig. 2: GFLI Frequency Response, 100 Hz/s Ramp to 61 Hz

these variable parameter tests, a load or frequency setpoint was utilized such that the inverter supplied roughly 0.50p.u. of real power.

The GFMI was also tested for its voltage response for both grid-tied and islanded conditions. For the grid-tied evaluation, two droop setpoints were used, as well as three separate ITCs. The voltage setpoints were selected to demand around 0.50p.u. of reactive power from the GFMI. Additionally, a 0.50p.u. inductive load was used for the islanded scenario.

To determine the response time of each test, the output power of the inverter was calculated, and the average settling power of the inverter was determined. The settling time was then determined when the power reached and stayed within 1% of the settling power.

The data was captured for this study using a custom LabView data acquisition system capable of collecting data at rates up to 60k samples per second per channel. All waveform data was then post-processed with a Matlab script and frequency was verified with a dq-frame-based Phase-Locked Loop (PLL) model in Simulink.

IV. EXPERIMENTAL RESULTS

A. GFLI, Frequency Response

With the grid simulator set to 60 Hz and 277 V L-N/480 V L-L, the inverter was allowed to operate at rated power. For the

initial tests, the frequency was ramped to 61 Hz at a rate of change of 3 Hz/s, which is the max ROCOF allowable by IEEE Std 1547-2018. This frequency was held for 30 s to ensure that the inverter response settles to a stable power level. The open-loop tests were set to 10 s, 5 s, 1s, and 200 ms. As can be seen in Table I, when setting the open-loop response time to 1 s or greater, the response time of the inverter was faster than expected from the response time setpoint. However, when the inverter was set to a 200 ms response time, the actual response time was nearly two and a half times greater than programmed. This is expected since the frequency ramp takes one-third of a second to achieve (1 Hz at 3 Hz/s). Something to note is that the response of the GFLI is very linear, with the power ramping at a constant rate following the change in frequency. This can be seen in Fig. 2, where the ROCOF was set to 100 Hz/s, yet the inverter evenly ramps power over the span of around 500 ms, with a slight undershoot before hitting steady state.

To look into the control limitations of the inverter response, the open-loop response time was set to 0 s, and the inverter was subjected to increasing ROCOFs. As a baseline, the inverter was tested at the 3 Hz/s as previously and did result in a quicker response than when set to 200 ms response (about 180 ms faster) but still took over a 510 ms to reach steady state. The ROCOF was then increased to 10 and 100 Hz/s. The 10 Hz/s test resulted in slightly faster settling times (just over 500ms), but at the higher rate of 100 Hz/s took over 570 ms to reach stability. This may be happening due to interference from the rapid change in frequency and the PLL controller. Overall, it can be said that we are hitting the limit of the FW control at about 500 ms.

B. GFLI, Voltage Response

Due to limitations of how the VV parameters are set on the inverter tested, an open-loop response time of 2s was the minimum value that could be set, and the maximum reactive power magnitude that could be set was 0.44p.u. (IEEE Std 1547-2018 allows for 1s open-loop response time and ±1.00p.u. of reactive power, where CA Rule 21 only required VV functionality with 0.30 p.u. of reactive power and a 5 s response

TABLE IV: GFMI Grid-Tied Frequency Response, Variable Droop, and Inertia Time Constant

Time Constant:	Droop (%):	Frequency Setpoint (Hz):	Settled Power (p.u.):	Power Response, 5% (s):	Power Response, 1% (s):
1000ms	2	59.4	0.4972	0.1771	0.6414
	5	58.5	0.4962	0.1100	0.5975
	10	57.0	0.4970	0.1046	0.5219
2000ms	2	59.4	0.4968	0.1738	0.6159
	5	58.5	0.4962	0.1364	0.5643
	10	57.0	0.4970	0.1415	0.5939
5000ms	2	59.4	0.4972	0.2375	0.8446
	5	58.5	0.4963	0.1806	0.6198
	10	57.0	0.4972	0.2350	0.7173
10000ms	2	59.4	0.4971	0.4223	1.2786
	5	58.5	0.4966	0.2152	0.8352
	10	57.0	0.4972	0.3933	0.8806

TABLE V: GFMI ISLANDED FREQUENCY RESPONSE, DEFAULT INERTIA PARAMETERS, 2% DROOP

Block Load:	Settled Frequency (Hz):	Power Response, 1% (s):	Settled Power (p.u.):
0.20p.u.	59.68	0.020	0.2040
0.40p.u.	58.45	0.152	0.3946
0.60p.u.	59.20	0.211	0.5847

time). The inverter's open-loop response times were tested at 10 s, 5 s, and 2 s on both sides of the VV profile.

From Table II it can be seen that when the voltage rises and subsequently causes the production of inductive reactive power, the actual response time is relatively close to the set open-loop response time, with less than 10% error with the worst case being when set to 2 s. however, with a change in the opposite direction, the error was far greater, between 20 and 60%. It should also be noted that for each test case, the settling time was greater than that of the set response time, unlike the frequency response times. Unfortunately, we could not set the VV parameters in such a way that we could find the control limitations of the response time.

C. GFMI, Grid-Tied Frequency Response

With the GFMI in a grid-forming droop mode, and synchronized to the grid simulator, similar tests can be performed on the device as with the GFLI. For this study, the inverter was subjected to various ROCOFs without changing any inverter parameters. The ROCOFs used were the same as the GFLI tests.

As depicted in Table III, the response time of the inverter does decrease with an increase in ROCOF, however the overall response time is not very quick; greater than that of the fastest response of the GFLI. The slow response may be due to the control scheme of the inverter, using a virtual synchronous machine control. This mimics the operation of conventional SG, which are known to have a slow response time. A curious note is that in the positive changes in frequency, there is a larger power magnitude. This is due to the transformer integrated with the inverter, and the measurement being taken between the transformer and grid simulator, not the transformer and inverter.

With the control scheme used for the GFMI, there are a few parameters that can be adjusted to affect the response time of the IBR. In discussing with the manufacturer of the GFMI tested, the theoretical response time of the IBR can be determined by assuming that the ITC is the time required for the IBR to respond to a 100% change in frequency. Therefore, increasing this parameter, or increasing the amount the frequency changes at a specific load (droop) will affect the response time.

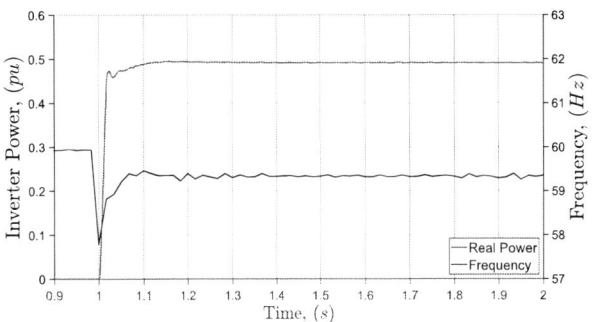

Fig. 3: GFMI, Islanded Frequency Response, 2% Droop, 1000 ms Inertia Time Constant

For these tests, three droop setpoints were used (2%, 5%, and 10%), and the ITC was adjusted from 1000 to 10000 ms. For each droop setpoint, the frequency of the grid simulator was set appropriately to request 0.50 p.u. from the GFMI. In all cases, a ROCOF of 100 Hz/s was used since this provided the quickest response previously. The results for these tests are shown in Table IV. An interesting observation is that the 2% cases all respond slower than the 5% or 10% cases. Additionally, the response time was not able to be greatly increased, even with the inertia time constant increased ten-fold. Lastly, while the 10% droop cases did respond faster than the 2% droop cases, before these cases reached steady state, the power output had large overshoots up to 0.95 p.u. when the power should settle at 0.50 p.u. Additionally in Table IV, the time that the GFMI takes to reach 5% of the settled power level is shown. To reach this threshold, the inverter takes significantly less time, between 18-44% of the time it takes to reach the 1% threshold. This is unlike the GFLI's frequency response that is linear and does not reach near the settle power until nearly reaching steady state.

D. GFMI, Islanded Frequency Response

The initial test was to verify how increasing block loads affected the settling time of the GFMI. With a frequency droop of 2% set, the inverter was subjected to block loads of 0.20, 0.40,

TABLE VI: GFMI ISLANDED FREQUENCY RESPONSE, VARIABLE DROOP AND INERTIA TIME CONSTANT

Time Constant:	Droop (%):	Settled Frequency (Hz):	Frequency Response, 0.2% (s):	Settled Power (p.u.):	Power Response, 5% (s):	Power Response, 1% (s):
1000 ms	2%	59.3360	0.0642	0.4855	0.0200	0.0626
	5%	58.4207	0.0993	0.4855	0.0200	0.1934
	10%	56.9018	0.1259	0.4854	0.0200	0.2001
2000 ms	2%	59.3355	0.2199	0.4853	0.0200	0.2983
	5%	58.4180	0.1583	0.4863	0.1585	0.2724
	10%	56.9030	0.2103	0.4854	0.0200	0.2992
5000 ms	2%	59.3353	0.5142	0.4852	0.6075	0.6863
	5%	58.4187	0.3040	0.4862	0.3134	0.4005
	10%	56.9039	0.4470	0.4852	0.0200	0.5232
10000 ms	2%	59.3359	1.7483	0.4850	4.7369	4.8197
	5%	58.4185	0.3569	0.4861	0.4331	0.5234
	10%	56.9019	0.8152	0.4853	0.7530	0.8757
30000 ms	2%	59.1302	9.6622	0.5622	9.9138	9.3298
	5%	58.4190	0.7288	0.4862	0.7389	0.8307
	10%	56.9044	2.1639	0.4849	1.9134	2.3167

TABLE VII: GFMI GRID-TIED VOLTAGE RESPONSE, VARIABLE DROOP AND INERTIA TIME CONSTANT

Time Constant:	Droop (%):	Voltage Set Point (p.u.):	Settled Power (p.u.):	Power Response (5%):	Power Response (1%):
1000 ms	1%	0.975	0.4622	1.3257	1.9909
	1%	1.025	-0.5109	1.7232	2.7462
	5%	0.95	0.4611	0.6379	0.8343
	5%	1.05	-0.5328	0.7887	0.9406
2000 ms	1%	0.975	0.4619	1.3543	2.0348
	1%	1.025	-0.5138	1.6914	2.6375
	5%	0.95	0.4579	0.6271	0.8166
	5%	1.05	-0.5304	0.7743	0.9246
5000 ms	1%	0.975	0.4664	1.3903	2.1522
	1%	1.025	-0.5112	1.6745	2.7632
	5%	0.95	0.4604	0.6583	0.8671
	5%	1.05	-0.5309	0.7421	0.9728

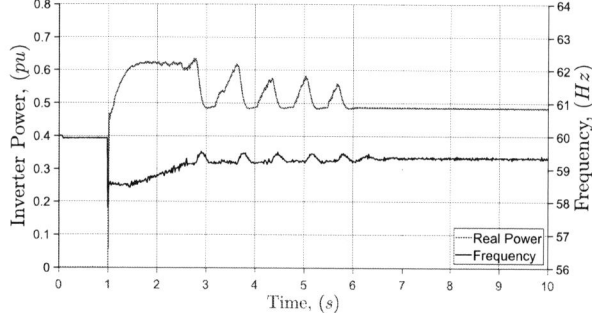

Fig. 4: GFMI Islanded Frequency Response, Initial Sign of Instability

and 0.60 p.u. As can be seen in Table V the inverter responds very quickly to block loads, especially at low powers.

While the GFMI is fast enough in an islanded state to match the response time of the GFLI, not all DER can respond as fast as IBRs. Due to the rotating mass of a SG, the maximum required ROCOF is 1 Hz/s for a SG without tripping, according to IEEE Std 1547-2018. SGs have protection devices that determine if the change in frequency is too great and isolate the SG from the system so as to not cause undue harm to the rotating equipment. Because of this, it is important to find a balance of rapid load pickup and slow ROCOF to ensure that all DER stay online, and the loads are fully supported [7]. This is where the adjustability of IBRs can greatly help an islanded system.

Similar to the grid-tied tests, both the droop setpoint and the ITC were varied to see how the response time of voltage and frequency were affected. For these tests, a 0.50 p.u. load was used for evaluation. Table VI shows the results from this series of tests. Additionally, Fig. 3 shows the response of the inverter with the GFMI set to 2% droop and 1000 ms ITC. A couple of observations were made from the results obtained. The first item of interest is the power output of the IBR before reaching steady state. In most cases, the power output of the inverter overshot the settled power by 5%+ (this issue is the result of a voltage rise of the IBRs output following the onset of the block load resulting in an over-power of the fixed resistance load). At lower ITCs, this over-power was minimal, but was still on the order of 250

ms, with larger magnitudes at higher ITCs and lengths exceeding 750 ms. Additionally, at an ITC of 10000 ms, the response of the inverter started to exhibit oscillations for the 2% droop case, with the power oscillating between 0.45-0.65 p.u. several times before finally settling after 4 s. Such oscillations are depicted in Fig. 4. Furthermore, at an ITC of 30000 ms, stability was never reached following the oscillatory dynamics for the 2% droop case.

The aforementioned oscillations and instability can be explained as follows. As the onset of the load block demands a fast reaction from the control scheme of the inverter, the large ITC values limit the speed of the converter, which is based on the dynamics of a SG. The initial oscillations are the first indication that the converter is not providing enough frequency damping due to the large ITC values. Such oscillations are also present in the voltage control loop. For even larger ITC values (30000 ms), the converter just cannot provide sufficient damping to withstand the oscillations, and thus instability is reached. Furthermore, the relatively small frequency droop seems that is also constraining the dynamics of the emulated rotor angle and directly influencing the corresponding swing equation [8], which in turn can drive the virtual synchronous machine out of step.

While the goal was to slow down the response time of the IBR, the observations described above now pose new considerations. While these tests are an idealized system (a load directly connected to the IBR) and do not perfectly represent what will occur in a microgrid, these results provide insight into what might occur. The overvoltage event seen following the onset of the block load could cause damage to loads if the voltage was to exceed a critical threshold. Additionally, to drastically slow down the response time, higher droop setpoints are required, and thus larger changes in frequency are observed. This can be detrimental to motor loads if low frequencies are sustained. Lastly, it is crucial to know when the IBR will go unstable.

E. GFMI, Grid-Tied Voltage Response

Similar to the GFLI, the GFMI was subjected to a step-change in voltage in both directions from nominal to observe the

TABLE VIII: GFMI ISLANDED VOLTAGE RESPONSE, VARIABLE DROOP AND INERTIA TIME CONSTANT

Time Constant:	Droop (%):	Settled Voltage (p.u.):	Voltage Response, 0.5% (s):	Settled Power (p.u.):	Power Response, 5% (s):	Power Response, 1% (s):
1000 ms	1%	0.9708	0.3021	0.4923	0.1457	0.2338
	5%	0.9506	0.8622	0.4721	0.1356	0.2328
2000 ms	1%	0.9661	5.0077	0.4879	0.4158	3.1035
	5%	0.9506	0.6502	0.4724	0.1830	0.5588
5000 ms	1%	0.9665	3.8611	0.4882	0.9100	3.3318
	5%	0.9512	2.0604	0.4722	0.4806	1.5721

response of both positive and negative reactive power. Droops of 1% and 5% were selected for this series of testing. While these droops suggest a voltage change of 0.005 p.u. and 0.025 p.u. would result in a reactive power level of 0.50 p.u., this is not the case. A virtual impedance is programmed into the VSM control of the IBR, an increase in voltage loss is observed as a greater amount of current is pushed by the GFMI. Because of this characteristic, a ±2.5% change in voltage was used for the 1% droop cases, and a ±5% change in voltage was used for the 5% case to achieve a 0.50 p.u. reactive power draw from the GFMI, thus at these setpoints, the IBR is emulating an equivalent droop of around 5% and 10%. Because the droop setting does not match the actual droop, this could be very misleading to utility engineers and operators and lead to confusion.

Table VII shows the results from this series of tests. Similar to its grid-tied frequency response, the GFMI responded slower at a more aggressive droop (contrary to that suggested by the manufacturer). The 1% droop cases all were nearly or greater than 2 s to respond to the change in voltage while the 5% droop cases all responded in under 1 s. No instability or overshooting of reactive power output was observed, and therefore this can only be contributed to the control scheme. Additionally, although the ITC was adjusted from 1000 ms to 5000 ms, no notable change in response time was seen between similar test cases. While the 5% cases all responded quicker than the 1% cases, the quickest response to within 1% of settled reactive power was still greater than 800 ms, and responses to within 5% of settled power were all over 600 ms. The voltage response of this IBR is significantly slower than its frequency response at the default 1000 ms ITC.

F. GFMI, Islanded Voltage Response

Voltage control in an islanded system is as crucial as frequency control. While frequency deviation is solely dependent on the resistive load to generation balance, voltage is additionally dependent on reactive load. While there is limited capacitive load on any system, a steady inductive load can come in the form of transformer magnetization current and line impedance, with dynamic inductance such as motor starts and transformer inrush current. So, while testing the GFMI under a fixed inductive load is not a perfect representation of an actual system, it will provide some insight into the response of the inverter.

For these tests, the same droop and ITC setpoints were used as for the grid-tied tests. To draw reactive power from the GFMI a 0.50 p.u. inductive load was used. Because a discrete reactive load was used, there was a resistive element to the inductance. The peak of the real power provided to the inductive load was less than 0.05p.u. with a settled power of less than 0.02 p.u. Therefore, the real power component of the load should have little effect on the response time of the GFMI. The results from this series of tests can be seen in Table VIII. With the voltage droop set to 1%, a voltage drop of around 0.03 p.u., and a drop of 0.05 p.u. when set to a 5% droop, similar to the values used for the grid-tied testing. At the default ITC of 1000ms the GFMI responds quite fast to the inductive block load, reaching within 1% of the settled power in under 250 ms for both droop cases. However, at higher ITCs, the 1% droop case starts to drastically increase in response time, with over a 3 s response for both the 2000ms and 5000ms ITC, yet no sign of instability is seen. The

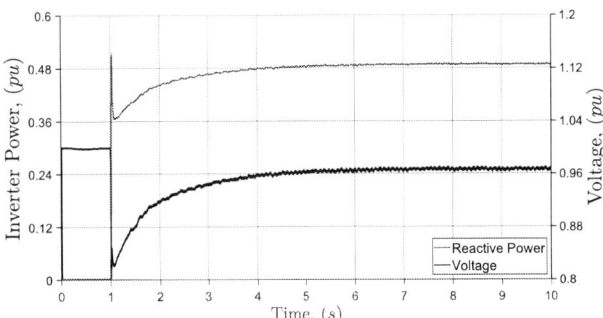

Fig. 5: GFMI, Islanded Voltage Response, 1% droop, 5000 ms Inertia Time Constant, 50kVAr Load

5% droop case responded rather well to the increase to around 500 ms and 1.5 s as the ITC was increased.

When looking at the data, a few things can be seen. First, the reactive power of the inverter quickly reaches near the settled power before dropping around 0.10 p.u. and resettling. Additionally, the voltage drop at the onset of the load goes to nearly 0.80 p.u. and is below the settled voltage for upwards of 5 s. This change in voltage is very large in comparison to the stabilized voltage, and a low voltage is sustained for a significant amount of time due to the slow response time. Fig. 5 shows the 1% droop, 5000 ms ITC test case. Lastly, when looking at the waveform, at the onset of the load, the initial current output of the GFMI has a DC offset.

V. CONCLUSION

While the switching components of IBRs have the capability to respond to grid variability, a major limitation to these devices is the control scheme that drives the device. In the case of the GFLI tested, this limitation only allowed for a response of 500 ms to be achieved. With the GFMI, the results were more dependent on whether the IBR was grid-tied or islanded. Under an islanded scenario the GFMI was able to pick up most of the load within 20 ms with a full load response time of around 200 ms; but, while grid-tied, this response was increased to 200 ms for 95% of the load pickup with and around a 600 ms full load support. In a purely IBR system, fast response is a benefit, however, the utility and many islanded systems still utilize rotating equipment. Because of this, it is also crucial to ensure that inverters can have a slower response, that is in the order of the inertia time constants of the rotary machines, to ensure stability on a system with legacy equipment.

With modern standards, GFLIs have very strict requirements for response time, having to be able to be set as fast as 200ms for frequency response, or as slow as 90 s for voltage response. An additional way that these devices can be set to promote grid stability is to set a deadband in the voltage and frequency response so that the device only responds when the voltage and frequency reference exceeds a threshold.

GFMIs are a crucial technology, either as the main source and voltage reference in a purely IBR system, or as an emulated inertia source in a system with high IBR penetration. Depending on the control scheme, these devices can respond to a load start very quickly but have dampened response after picking up most of the load. With the GFMI used for this series of tests, it was found that understanding the control of the IBR was important, as too much of a change from the default setting would lead to instability when islanded.

With IBRs being utilized for renewable sources such as PV and wind resources, as well as energy storage systems, it is vital to understand how these devices can be programmed to ensure a stable grid. As rotating generation is removed from the utility, other grid-forming devices will need to take their place and balance the inertia of the system to respond to the load demand, and not cause a crash whenever load is gained or lost. This study gives us a look into the dynamics of these devices so that we can plan for a future with 100% penetration of IBRs.

VI. ACKOWLEDGEMENT

Sandia National Laboratories is a multi-mission laboratory managed and operated by National Technology and Engineering Solutions of Sandia, LLC., a wholly owned subsidiary of Honeywell International, Inc., for the U.S. Department of Energy's National Nuclear Security Administration under contract DE-NA-0003525.

VII. REFERENCES

[1] J. W. Smith, W. Sunderman, R. Dugan and B. Seal, "Smart inverter volt/var control functions for high penetration of PV on distribution systems," 2011 IEEE/PES Power Systems Conference and Exposition, 2011, pp. 1-6, doi: 10.1109/PSCE.2011.5772598.

[2] S. Gonzalez, N. Gurule, M. J. Reno, and J. Johnson "Fault Current Experimental Results of Photovoltaic Inverters Operating with Grid-Support Functionality," IEEE Photovoltaic Specialists Conference (PVSC), 2018.

[3] IEEE Standard for Interconnection and Interoperability of Distributed Energy Resources with Associated Electric Power Systems Interfaces," in IEEE Std 1547-2018 (Revision of IEEE Std 1547-2003), pp.1-138, 6 April 2018.

[4] Lin, Yashen, Joseph H. Eto, Brian B. Johnson, Jack D. Flicker, Robert H. Lasseter, Hugo N. Villegas Pico, Gab-Su Seo, Brian J. Pierre, and Abraham Ellis. 2020. Research Roadmap on Grid-Forming Inverters. Golden, CO: National Renewable Energy Laboratory. NREL/TP-5D00-73476. https://www.nrel.gov/docs/fy21osti/73476.pdf.

[5] R. H. Lasseter, Z. Chen and D. Pattabiraman, "Grid-Forming Inverters: A Critical Asset for the Power Grid," in IEEE Journal of Emerging and Selected Topics in Power Electronics, vol. 8, no. 2, pp. 925-935, June 2020, doi: 10.1109/JESTPE.2019.2959271.

[6] B. J. Pierre et al., "Bulk Power System Dynamics with Varying Levels of Synchronous Generators and Grid-Forming Power Inverters," 2019 IEEE 46th Photovoltaic Specialists Conference (PVSC), 2019, pp. 0880-0886, doi: 10.1109/PVSC40753.2019.8980733.

[7] J. Matevosyan et al., "Grid-Forming Inverters: Are They the Key for High Renewable Penetration?," in IEEE Power and Energy Magazine, vol. 17, no. 6, pp. 89-98, Nov.-Dec. 2019, doi: 10.1109/MPE.2019.2933072.

[8] Kundur, P., Balu, N. J., & Lauby, M. G. (1994). *Power system stability and control*. New York: McGraw-Hill.

[9] A. D. Paquette, M. J. Reno, R. G. Harley, and D. M. Divan, "Transient Load Sharing Between Inverters and Synchronous Generators in Islanded Microgrids," in Energy Conversion Congress and Exposition (ECCE), 2012.

[10] J. Hernández-Alvídrez, N. S. Gurule, R. Darbali-Zamora, M. J. Reno, and J. D. Flicker, "Modeling Grid-Forming Inverters Dynamics Under Ground Fault Scenarios Using Experimental Data From Commercially Available Equipment", IEEE Photovoltaic Specialists Conference (PVSC), 2021.

[11] J. Hernandez-Alvidrez, A. Summers, M. J. Reno, J. Flicker, N. Pragallapati "Simulation of Grid-Forming Inverters Dynamic Models using a Power Hardware-in-the-Loop Testbed," IEEE Photovoltaic Specialists Conference (PVSC), 2019.

[12] J. Alvidrez, A. Summers, N. Pragallapati, M. J. Reno, S. Ranade, J. Johnson, S. Brahma, and J. Quiroz "PV-Inverter Dynamic Model Validation and Comparison Under Fault Scenarios Using a Power Hardware-in-the-Loop Testbed," IEEE Photovoltaic Specialists Conference (PVSC), 2018.

Inferring PV System Specifications from Net Load

Upama Nakarmi, Thomas E. Hoff, Marc Perez, Philip Gruenhagen

Clean Power Research, Napa, CA, United States

The increasing penetration of grid-connected solar photovoltaics (PV) has challenged the existing method of planning and operating the distribution grid. Electric utilities increasingly require visibility on the Behind-The-Meter (BTM) PV installations within their territories. They are interested in knowing the physical characteristics of the installed PV systems as well as their energy production profiles. This paper provides a novel method to infer the azimuth, tilt, and size of a PV system using only net load data and PV system location. Net load (total load minus PV production) represents the combined effect of consumption and PV system production. Our method utilizes SolarAnywhere® to produce plane of array irradiance for various azimuth and tilt combinations which serves as the ground truth data. The optimal PV system specification for each individual PV system is inferred by matching characteristics derived from net load data to the ones derived from the ground truth data. The proposed method is demonstrated on a test site in California and then implemented for all residential PV systems in a utility' service territory. All locations have both net load and PV production data. We use the net load data to infer the system specifications and then the PV production data to validate the method' accuracy.

978-1-7281-6118-1/22 $31.00 © 2022 IEEE

Planarizing HVPE Growth on GaAs Substrates Produced by Controlled Spalling

Anna K Braun, William E McMahon, Allison N Perna, Kevin L Schulte, Corinne E Packard, Aaron J Ptak

Colorado School of Mines, Golden, CO, United States

National Renewable Energy Laboratory, Golden, CO, United States

In this work, we show hydride vapor phase epitaxy (HVPE) overgrowth behavior of two growth conditions on three different facet morphologies produced by controlled spalling of (100) GaAs. In situ planarization of the surface through overgrowth has potential to overcome the significant challenge facets present to direct regrowth of photovoltaic devices and enabling low-cost substrate reuse. Substrate offcut and spall depth affect the surface morphology of the facet face, and this morphology plays an important role in facilitating planarizing overgrowth. We also show that growth conditions can be tuned to improve planarization efficiency on different surfaces. These results are critical for understanding the kinetics that allow planarizing growth to enable direct reuse of spalled (100) GaAs substrates.

Lamination Process Induced Residual Stress in Glass-Glass vs. Glass-Backsheet Modules

Farhan Rahman, Ian M. Slauch, Rico Meier, Jared Tracy, Elizabeth C. Palmiotti, Mariana I. Bertoni, James Y. Hartley

Sandia National Laboratories, Albuquerque, NM, United States

Fulton School of Engineering, Arizona State University, Tempe, AZ, United States

DuPont Photovoltaic and Advanced Materials, Wilmington, DE, United States

Manufacturing process induces residual stress which can cause immediate or delayed cell breakage. Therefore, to reduce the levelized cost of energy for photovoltaic (PV) modules, it is important to analyze the effect of manufacturing processes on residual cell stress development. In present study, Finite element analysis (FEA) was performed to investigate the effects of photovoltaic module architecture: glass-glass (GG) or glass-backsheet (GB) on residual cell stress. During PV module manufacturing, the soldering of interconnects and lamination impart residual stresses onto the cell. However, for both module configurations the stress states at the beginning of the lamination process are same. FEA was performed using a high-fidelity mesh, to simulate the cooling of the heated encapsulant during lamination. Plasticity of solder and ribbons as well as viscoelasticity of encapsulant material were considered in the simulation. Higher stress in the GG configuration in comparison to the GB configuration was found. In GG, the maximum cell stress was at the front edge of the top ribbon while for GB it was along the ribbon footprint (side-edge of interconnect) on the wafer' back-side (backsheet side). The performance of the different module configurations and the locations of peak stresses from current study agreed qualitatively with cell stress values obtained earlier using X-ray Topography (XRT) experiments. Separate soldering simulation indicated lamination step to be dominant in imparting residual stress during manufacturing. To perform a quantitative comparison between FEA and XRT results, the post-soldering stress values will be incorporated as initial condition for lamination simulation in future research.

978-1-7281-6118-1/22 $31.00 © 2022 IEEE

Near-Busbar Degradation of Screen-Printed Metallization in Silicon Photovoltaic Modules

Dana B. Sulas-Kern,* Helio Moutinho,* Tristan Erion-Lorico,# and Steve Johnston*

*National Renewable Energy Laboratory, Golden, CO, 80401, USA
#PV Evolution Labs, Napa, CA 94558 USA

Abstract— We study photovoltaic (PV) module degradation after extended accelerated stress testing including 2000 hours of damp heat followed by a current-injection procedure meant to stabilize defects linked to light-induced degradation. In addition to de-stabilization/recovery of light-induced defects, we observe severe series resistance due to loss of contact between the Si cell and near-busbar screen-printed metallization (i.e. grid finger delamination). Using scanning electron microscopy and energy dispersive x-ray spectroscopy on cell fragments cored from the module, we show poor contact is caused by a gap between the screen-printed Ag metallization and Si due to missing glass frit.

Keywords—silicon, photovoltaics, metallization, degradation

I. INTRODUCTION

With a goal of producing reliable silicon photovoltaic (PV) modules having low degradation rates (~0.5%/year) and long operating lifetimes (up to 50 years), module manufacturers are employing extended accelerated stress testing procedures above and beyond the typical IEC 61215 certification. In this study, we discuss a module degradation process resulting from extended damp heat followed by forward-bias current injection at elevated temperature. We discuss two processes, including de-stabilization and subsequent recovery of defects related to possible light-induced degradations (e.g. LID, LeTID) along with degradation of the contact between the silicon cell and the near-busbar screen-printed metallization. We core cell fragments out of the PV module for microscopic study of the metal contact degradation, and we show that much of the glass frit is missing at the interface between the silicon and Ag fingers. We show frit compositions including oxides of Pb, Te, Zn, and Mg. Our continued studies will investigate the possible leaching and electrochemical reactions of these frit materials into acidic solution that may exist in the module upon water ingress and encapsulant decomposition.

II. EXPERIMENTAL DETAILS

A. Accelerated Stress Testing

Damp heat (DH) was carried out at 85C and 85%RH for 1000 hrs, repeated twice. Next, the modules were contacted at 85C with current injection of I_{SC} for 48 hours, designed to stabilize defects linked to light-induced degradation (LID)[1], though other defects may also be affected such as light and elevated temperature induced degradation (LeTID)[2]. A second stabilization step was initiated due to lack of full power recovery that would be expected if LID was the only degradation process affecting the modules. At this time, the climate chamber experienced a brief spike to 115C. After shutdown and inspection of the modules, the stabilization step was repeated at 85C and I_{SC} injection for 48 hrs.

B. Imaging

Electroluminescence (EL) was imaged at I_{SC} injection and photoluminescence (PL) used 808 nm laser diodes at 0.25 Sun. EL and PL were detected with a Princeton Instruments PIXIS silicon CCD camera with an InP long-pass filter. UVF was excited by scanning a 360 nm laser with galvanometer mirrors at 0.1 and 79 Hz and recorded with the PIXIS and a 400 nm long-pass filter. For modules, the camera view was one cell, and images were assembled from automated module motion. Dark lock-in thermography (DLIT) was excited with I_{SC} injection at 0.5Hz and imaged through the backsheet using a Cedip-FLIR InSb camera with lock-in detection.

C. Scanning Electron Microscopy (SEM) and Energy-Dispersive X-ray Spectroscopy (EDS)

SEM and EDS used a ThermoFisher Nova NanoSEM 630 field-emission SEM and an AZtecLive microanalysis system with Ultim Max 170 analytical silicon drift detector from Oxford Instruments.

III. RESULTS AND DISCUSSION

A. Accelerated Stress Testing

Fig. 1 shows the changes in maximum power (P_{MAX}), open-circuit voltage (V_{OC}), short-circuit current (I_{SC}), and fill factor (FF) during accelerated stress testing. DH2000 caused a large drop in P_{MAX} and decrease in all performance metrics. EL imaging shows emergence of a patch-work pattern in which there is greater relative brightness contrast among the cells of the module. This may indicate de-stabilization of previously-stabilized light-induced defects [2]. The possible de-stabilization would affect P_{MAX} via decrease in V_{OC} and I_{SC}. In addition, the DH2000 image also shows cell darkening near the cell edges, likely caused by metallization corrosion upon moisture ingress which would further affect P_{MAX} via FF loss.

The subsequent 85C/I_{SC}-injection stabilization procedure then caused improvements in both I_{SC} and V_{OC}, suggesting recovery of light-induced defects. Consistent with LID or LeTID recovery, the EL images show more uniform cell-to-cell EL intensities and overall relative EL brightening [1,2]. However, FF continued to decrease, and EL imaging showed grid finger disconnections along the busbars that emerged at

This work was authored in part by Alliance for Sustainable Energy, LLC, the manager and operator of the National Renewable Energy Laboratory for the U.S. Department of Energy (DOE) under contract No. DE-AC36-08GO28308. Funding was provided by U.S. Department of Energy Office of Energy Efficiency and Renewable Energy Solar Energy Technologies Office and by DuraMAT. The views expressed in this article do not necessarily represent the view of the DOE or the U.S. Government. The publisher, by accepting the article for publication, acknowledges that the U.S. Government retains a nonexclusive, paid-up, irrevocable, worldwide license to publish or reproduce the published form of this work, or allow others to do so, for U.S. Government purposes.

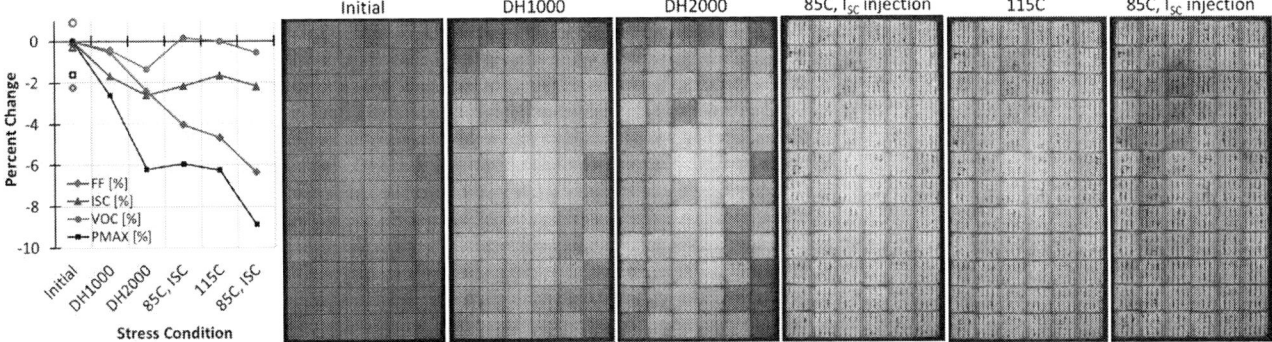

Fig. 1. Changes in module performance during accelerated stress testing relative to initial performance. Open markers indicate differences in initial performance compared to nameplate. Electroluminescence images show module uniformity or defects after completion of each accelerated stress condition.

the 85C/I_{SC}-injection stabilization step. While V_{OC} and I_{SC} stabilized during subsequent stress, FF continued to decline. In this paper, we investigate the underlying cause of further FF loss by increasing series resistance.

B. Post-Stress Imaging

Fig. 2 shows a comparison of DLIT, EL, PL, and UVF images after completion of the accelerated stress process, along with a zoomed-in example of a representative cell. The dark areas along the busbars in EL imaging indicate disconnected metallization where current is not injected into the module. In PL, these same regions appear bright as the poor metal contact hinders charge extraction from these areas via lateral balancing currents. In DLIT, the areas with disconnected metallization are cold. These areas are not visible using UVF, unlike previous reports in *fielded* modules with similar degradation [3]. Rather, UVF shows some fluorescent residual material (possibly solder flux) along the busbars with some pooling at the end of the busbars.

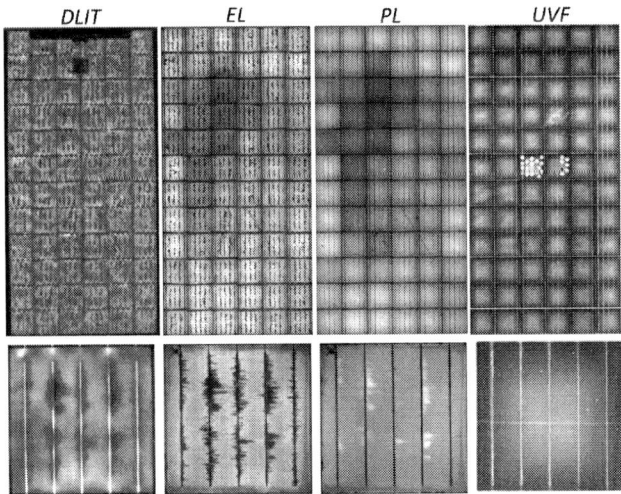

Fig. 2. Dark lock-in thermography (DLIT), electroluminescence (EL), photoluminescenec (PL) and UV-fluroescence (UVF). Note – UVF images collected after coring near the center of the module.

C. Coring

Guided by the images, we cored cell fragments from the modules as described in [4]. Fig. 3 shows an area of the

module after a cell fragment was extracted, viewed from the rear side of the module with the backsheet removed. We found that the ribbons, busbars, and some screen-printed fingers remained attached to the module's front-side encapsulant. This indicates a weak connection between the PV cell and the screen printing in these areas. For all pieces cored, the busbars always remained embedded in the front-side encapsulant. However, we optimized the ability to keep most of the degraded fingers attached to the cell fragments by gentle heating of the metal coring stub using a soldering iron.

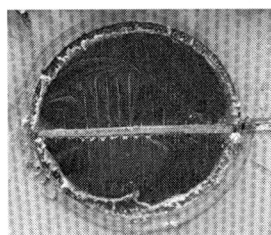

Fig. 3. Area cored from module, showing front-side busbar and metallization remaining embedded in encapsulant.

Fig. 4a shows PL of a cored cell fragment extracted from the module, where bright areas along some fingers indicate degraded electrical contact of the screen-printed metallization, similar to the cell and module PL images before coring (see Fig. 2). Figure 4b gives a zoomed-in view of the near busbar region showing that the finger areas nearest to the busbars were peeled off the sample during coring.

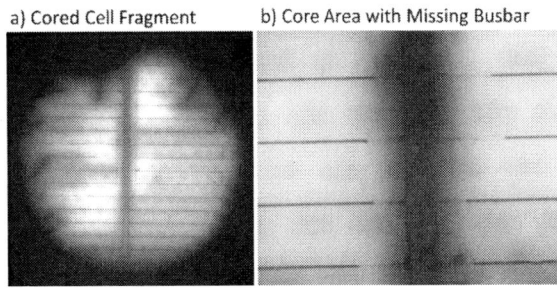

Fig. 4. PL of cored cell fragment, showing (a) pattern of brighter PL around degraded fingers and (b) missing busbar and missing near-busbar fingers.

Despite the missing busbar, we contact individual grid fingers on the cored cell pieces and use EL to identify either non-degraded fingers with good electrical contact (Fig. 5a) or degraded fingers having poor electrical contact (Fig. 5b).

978-1-7281-6118-1/22 $31.00 © 2022 IEEE

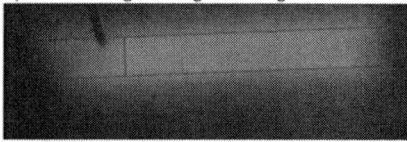

a) EL contacting non-degraded finger

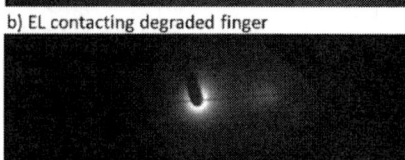

b) EL contacting degraded finger

Fig. 5. Examples of EL when contacting (a) non-degraded fingres and (b) degraded fingers of a cored cell fragment.

D. Microscopy

Fig. 6 shows glass frit remaining on the Si surface in the areas where the busbar and fingers were peeled off. The frit includes oxides of Te, Zn, Mg, and Pb. We also observe Sn along the busbar in a pattern different from the residual frit. The Sn originates from the solder that attaches the Cu ribbon to the Ag screen printing. The solder/Ag interface appears stronger than the Ag/frit/cell interface, causing the Ag metallization to remain attached to the ribbon and peel off the cell. Therefore, the degradation mechanism in these cells is not a solder-related degradation mechanism. Rather, it is frit degradation causing disconnection of the Ag paste from the cell, despite a remaining strong contact between solder/Ag. Other studies similarly show concerns about degradation of lead-rich frit especially in modules having EVA [3].

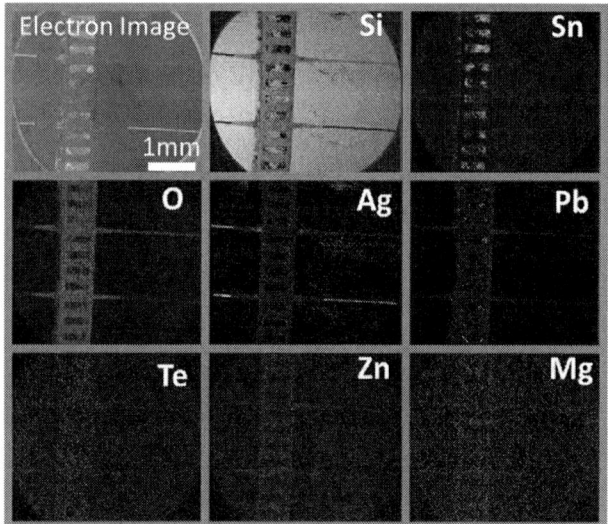

Fig. 6. EDS maps of a cored cell fragment with missing busbar/grid fingers.

We prepared cross sections to investigate the Ag/frit/cell interfaces. Consistent with Fig. 6, the cross sections (Fig. 7-8) show glass frit at the interface between the Ag fingers and Si cell primarily composed of lead oxide and a smaller amount of Te, Zn, and Mg oxides, similar to some previously observed frit compositions [3]. Fingers with good contact were identified by EL (see Fig. 5), and show a continuous frit layer at the Ag/Si interface (Fig. 7). However, the fingers with high resistance according to EL images show a break at the Ag/frit/Si interface (Fig. 8). In these degraded fingers, any

remaining frit material appears to be attached to the Si surface. The break in contact occurs between the Ag and oxides.

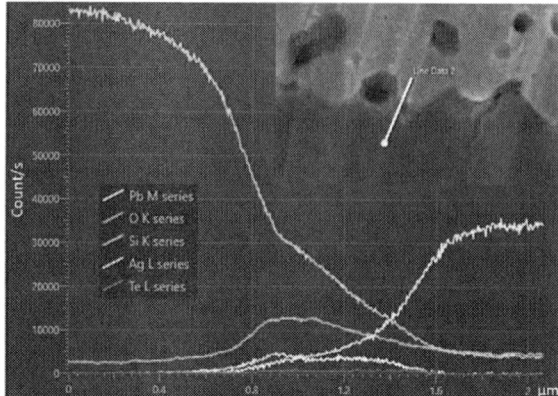

Fig. 7. EDS linescan of non-degraded finger.

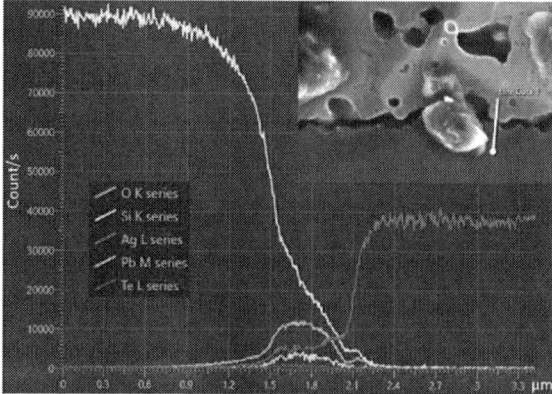

Fig. 8. EDS linescan of degraded finger.

IV. CONCLUSIONS

Here, we showed degradation of near-busbar metallization in Si PV modules, emerging after DH2000 followed by light-induced defect stabilization with I_{SC} injection at 85C. We identified that the weakest interface in these cells was between the Ag paste and oxides that make up the glass frit, primarily lead oxide. The fit is missing in degraded fingers, leaving an open gap between the Ag paste and silicon that causes high resistance. Further studies will continue to investigate the interaction of this frit material with acidic solutions.

REFERENCES

[1] I.L. Repins, F. Kersten, B. Hallam, K. VanSant, M.B. Koentopp "Stabilization of light-induced effects in Si modules for IEC 61215 design qualification" *Solar Energy* vol. 208, pp. 894-904, **2020**.

[2] M.G. Deceglie, T.J. Silverman, S.W. Johnston, J.A. Rand, M.J. Reed, R. Flottemesch, and I.L. Repins "Light and elevated temperature induced degradation (LeTID) in a utility-scale photovoltaic system" *IEEE J. Photovolt.* vol. 10, pp. 1084-1092, **2020**.

[3] N. Iqbal, D.J. Colvin, E.J. Schneller, T.S. Sakthivel, R. Ristau, B.D. Huey, B.X.J. Yu, J-N. Jaubert, A.J. Curran, M.Wang, S. Seal, R.H. French, K.O. Davis "Characterization of front contact degradation in monocrystalline and multicrystalline silicon photovoltaic modules following damp heat exposure" *Solar Energy Mater. And Solar Cells* vol. 235, pp. 111468, **2022**.

[4] H. Moutinho, B. To, D. Sulas-Kern, C.-S. Jiang, M. Al-Jassim, S. Johnston "Advances in coring procedures of silicon photovoltaic modules" *IEEE 47th Photovoltaic Specialists Conference (PVSC)*, pp. 1449-1453, **2020**.

This page intentionally left blank.

Improving Behind-the-Meter PV Impact Studies with Data-Driven Modeling and Analysis

Joseph A. Azzolini, Samuel Talkington, Matthew J. Reno, Santiago Grijalva, Logan Blakely, David Pinney, Stanley McHann

Sandia National Laboratories, Albuquerque, NM, United States

Georgia Institute of Technology, Atlanta, GA, United States

National Rural Electric Cooperative Association, Arlington, VA, United States

Frequent changes in penetration levels of distributed energy resources (DERs) and grid control objectives have caused the maintenance of accurate and reliable grid models for behind-the-meter (BTM) photovoltaic (PV) system impact studies to become an increasingly challenging task. At the same time, high adoption rates of advanced metering infrastructure (AMI) devices have improved load modeling techniques and have enabled the application of machine learning algorithms to a wide variety of model calibration tasks. Therefore, we propose that these algorithms can be applied to improve the quality of the input data and grid models used for PV impact studies. In this paper, these potential improvements were assessed for their ability to improve the accuracy of locational BTM PV hosting capacity analysis (HCA). Specifically, the voltage- and thermal-constrained hosting capacities of every customer location on a distribution feeder (1,379 in total) were calculated every 15 minutes for an entire year before and after each calibration algorithm or load modeling technique was applied. Overall, the HCA results were found to be highly sensitive to the various modeling deficiencies under investigation, illustrating the opportunity for more data-centric/model-free approaches to PV impact studies.

A Model to Predict Daily Snow Albedo Change Over Time

Christopher Pike[1], Daniel Riley[2], and Laurie Burnham[2]

[1]University of Alaska Fairbanks, Fairbanks, Alaska, 99775, [2]Sandia National Lab, Albuquerque, New Mexico, 87123

Abstract— As solar photovoltaic technology is deployed in snowier and more northern locations and bifacial module use continues to increase, ground albedo from snow plays a significant role in the energy output of these systems. Accurately knowing the ground albedo allows for more accurate production models. Here we use multiple years of measured snow and albedo data from 6 locations to develop a snow albedo change model that uses time and temperature as the major inputs.

Keywords—albedo, snow, modeling

I. INTRODUCTION

Solar photovoltaic (PV) technology is continuing to evolve both in the locations where it is being deployed as well as with module construction. Increasing amounts of solar are being deployed in snowy and high latitude locations, using steeper tilt angles than arrays found at middle latitudes [1][2] . In addition, newly constructed arrays are increasingly utilizing bifacial module technology to maximize production and lower the levelized cost of energy (LCOE)[3][4]. As the price of bifacial modules drops in relation to monofacial PV technology this trend is expected to continue and the International Technology Roadmap for Photovoltaic notes that by 2030 bifacial PV cells are expected to account for 70% of the total world PV cell market[5].

To accurately model the energy output of bifacial PV, rear side plane of array irradiance is needed and ground albedo is an important input needed to accurately model rear side irradiance[6][7]. In addition, ground albedo is also an important input to improve models of high tilt angle monofacial solar PV that is typically found at high latitudes [8]. Ground albedo can change rapidly depending on the presence of snow as well as the age and characteristics of the snow [9]. In the absence of measured albedo data, ground albedo is usually represented in PV models by .2 when no snow is present and .8 when snow is present. This is an obvious over simplification. The ability to accurately model how the albedo of snow changes over time and temperature will improve the accuracy of the models in cold places where snow can be present early in the fall and extend well into the spring season. This long duration snow coverage can result in significant ground reflected irradiance and bifacial gains [10]. This paper will present a method to model the albedo of snow based on location, temperature, and the number of days since the last snowfall.

II. EVALUATION ABSTRACT

A. Summary of the work

In order to better understand the snow albedo change over time six albedo data sets were used for model development with the goal of utilizing long duration data sets whenever possible as well as using a geographically diverse range of data sets. Three albedo data sets located in Fairbanks, Alaska, Willow, Alaska, and Fairlee, Vermont were measured specifically for the development and validation of this model. Three other datasets from Brookings, South Dakota, Flagstaff, Arizona, and Rosemount, Minnesota were downloaded from the NREL Duramat website and are part of the AmeriFlux data network. Details of the datasets are shown below in **Error! Reference source not found.**.

Data Processing

The data sets used in this project were either collected at 1 minute resolution, or 1 hour resolution as specified in Table 1. For this analysis all data sets were converted to 1 hour resolution when necessary. Data was scrubbed when any of the following criteria were met:

- Global horizontal irradiance (GHI) > 1200 W/m2
- Reflected horizontal irradiance (RHI) > 1100 W/m2
- Albedo values > 0.95 or <.1
- Solar elevation < 5 degrees
- Other known shading or non-standard events

Snow coverage observations included daily snowfall and snow depth and were obtained either from onsite image acquisition or from the NOAA national Climate Data Center data set closest to the albedo data collection site. The exact location of the snow observations is shown in Table 1.

Analysis

The average daily albedo and the daily snowfall data sets were combined to observe how snow albedo changes over time and specifically how albedo changes as time passes after a snow event. Data was divided by snow events where an event was defined as a time period with three or more days since measurable snowfall as seen in Figure 2 and Figure 3.

This work supported by the U.S. Department of Energy
Solar Energy Technologies Office Award Number 34363 and 38527

Table 1. Location and time span of albedo and snowfall measurements.

Albedometer locations	Albedometer Coordinates	Time Span	Data time resolution	Snow Observation Locations*	Distance Apart (km)
Fairbanks, Alaska	64.853, -147.8603	1/2019-5/2022	one min	64.8039, -147.876	6
Willow, Alaska	61.688, -149.968	3/2020-5/2022	one min	Camera at albedometer	Same Location
Fairlee, Vermont	43.86, -72.22	2/2021-4/2021	one min	43.7917, -72.2578	8
**Brookings, South Dakota	44.3453, -96.8362	4/2004-3/2010	one hour	44.32503, -96.7686	6
**Flagstaff, Arizona	35.4454, -111.772	12/2005-1/2011	one hour	35.19037, -111.674	30
**Rosemount, Minnesota	44.6781, -93.0723	11/2015-12/2018	one hour	44.8831, -93.2289	26
*Weather collected from NOAA National Climate Data Center ** All three datasets downloaded from the duramat website and are from the AmeriFlux data network https://ameriflux.lbl.gov/data/data-policy/					

Each snowfall event is then examined through the lens of what we refer to in this paper as melt degree hours (MDH). The (MDH) approach incorporates time and temperature by modifying the degree day metric used in building science and in some snow melt models [9]. Since days with cold nights and warm days could still have an average daily temperature near, or even below 0°C it was necessary to average hourly temperatures rather than use daily averages, this way the melting effect of the warm daytime temperatures is part of the model. MDH were calculated as the difference between average hourly temperatures and 0°C. To simplify the model, this is then converted to a binary value so that average hourly temps greater than 0°C are assigned 1 and classified as one melt hour, and average hourly temperatures less than 0°C are assigned 0 and are not included in the melt hour total. For each snow event average daily albedo can then be graphed based on the number of melt hours since the last snowfall as show in figure 1. Events with MH equal to zero (where the temp stayed below 0°C for the entire event) were separated from events with MH greater than 0.

Next, an additional filtering process was applied so that events with max albedo values less than 0.5 were filtered out. The theory here was that fresh snow generally has an albedo value of about .8 so events with maximum albedos of 0.5 or less were either the result of faulty data or incomplete snow coverage. During the analysis each event was shifted so that the max event albedo was equal to 1 to allow for the straightforward comparison of events and focus on the albedo change over time. An example of the shifted albedo versus melt hours since snowfall graph is shown in figure 2.

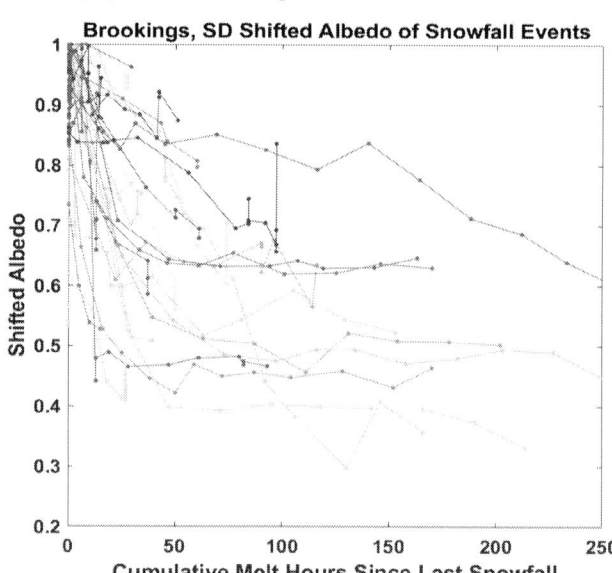

Figure 2. Albedo change as a function of melt hours since the most recent snowfall is shown for snow events in Brookings, SD. Albedo data has been cleaned compared to the data shown in Figure 1 so that snow events with maximum albedos less than .5 have been scrubbed. The remaining events have been shifted so that the maximum albedo during each event is represented as 1.

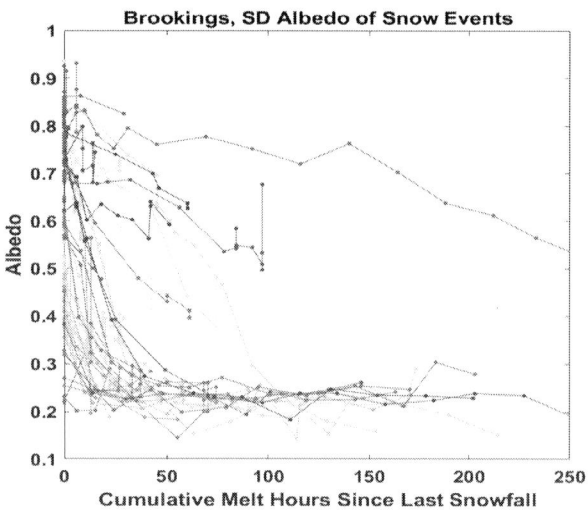

Figure 1. Albedo change as a function of cumulative melt hours since the last snowfall are shown for snow events in Brookings, SD.

B. Analysis of the results

Model development resulting from the data collection and analysis presented above is still in development. In the full paper a mathematical model that describes the albedo change of snow based on time and temperature since snowfall will be presented. Analysis to date shows two distinct relationships between albedo and melt hours since snowfall as seen in figure 3. These relationships are currently being investigated and integrated into the model.

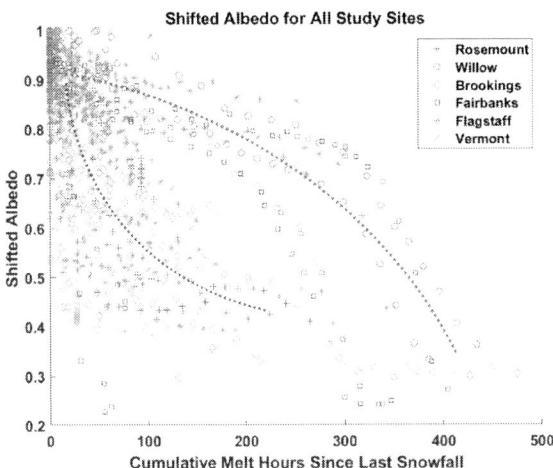

Figure 3. Snow events from all study sites are shown with 2 sample best fit lines to demonstrate the relationships between albedo and melt hours since snowfall.

C. Significance of the work for the field

In the absence of albedo data most PV models use a binary value for snow albedo (.8 if snow is present and .2 if snow is not present). In the real world, snow albedo is often between these two values. The snow albedo change model presented in the full paper and presentation can be integrated in existing PV energy performance models to improve the accuracy of the models for PV arrays being developed in locations that receive snowfall.

III. REFERENCES

[1] Burnham, Laurie. "PV Performance and Reliability in Snowy Climates: Opportunities and Challenges." (2019).

[2] Burnham, Laurie, Daniel Riley, Bevan Walker, and Joshua M. Pearce. "Performance of bifacial photovoltaic modules on a dual-axis tracker in a high-latitude, high-albedo environment." In *2019 IEEE 46th Photovoltaic Specialists Conference (PVSC)*, pp. 1320-1327. IEEE, 2019.

[3] Rodríguez-Gallegos, Carlos D., Haohui Liu, Oktoviano Gandhi, Jai Prakash Singh, Vijay Krishnamurthy, Abhishek Kumar, Joshua S. Stein et al. "Global techno-economic performance of bifacial and tracking photovoltaic systems." *Joule* 4, no. 7 (2020): 1514-1541.

[4] Deline, Chris, Silvana Ayala Pelaez, Sara MacAlpine, and Carlos Olalla. "Estimating and parameterizing mismatch power loss in bifacial photovoltaic systems." *Progress in Photovoltaics: Research and Applications* 28, no. 7 (2020): 691-703.

[5] International Technology Roadmap for Photovoltaic (ITRPV)—Results 2019. Available online: http://itrpv.vdma.org/ (accessed on 10 February 2021).

[6] Marion, Bill. "Measured and satellite-derived albedo data for estimating bifacial photovoltaic system performance." *Solar Energy* 215 (2021): 321-327.

[7] Chiodetti, Matthieu, Jinsuk Kang, Christian Reise, and Amy Lindsay. "Predicting yields of bifacial PV power plants–What accuracy is possible?." *System* 2 (2018): 1.

[8] Kotak, Y., M. S. Gul, T. Muneer, and S. M. Ivanova. "Investigating the impact of ground albedo on the performance of pv systems." In *Proceedings of the CIBSE Technical Symposium, London, UK*, pp. 16-17. 2015.

[9] Kane, Douglas L., Robert E. Gieck, and Larry D. Hinzman. "Snowmelt modeling at small Alaskan Arctic watershed." *Journal of Hydrologic Engineering* 2, no. 4 (1997): 204-210.

[10] Pike, Christopher, Erin Whitney, Michelle Wilber, and Joshua S. Stein. "Field Performance of South-Facing and East-West Facing Bifacial Modules in the Arctic." *Energies* 14, no. 4 (2021): 1210.

High-specific-power Schottky-junction photovoltaics from CVD-grown MoS2

Timothy Ismael, Kazi M. Islam, Muhammad A. Abbas, George B. Ingrish, Claire E. Luthy, Orhan Kizilkaya, Carlos M. Gutierrez, Meghan E. Bush, Jeremiah S. McNatt, Anthony J. Hoffman, Matthew D. Escarra

Tulane University, New Orleans, LA, United States

Louisiana State University, Baton Rouge, LA, United States

University of Notre Dame, Notre Dame, IN, United States

NASA Glenn Research Center, Cleveland, OH, United States

High-specific-power ultralight photovoltaics (PV) can provide space missions with solar power generation at minimal weight and volume. Two-dimensional (2D) transition metal dichalcogenides exhibit high absorption at sub-nanometer thickness, enabling one to three orders of magnitude higher power density than top existing ultrathin PV cells. However, quality and scalability hinder the production of ultrathin and ultralightweight 2D PV. To obtain high quality monolayer films, we optimized the chemical vapor deposition process with resulting films of >cm2 showing 85% uniformity when analyzed via Raman spatial mapping. To show promise towards the fabrication of >cm2 scale PV devices, we present 25 mm2 scale Schottky PV, fabricated by transferring monolayer MoS2 films onto asymmetric Ti and Pt interlocking finger contacts. Under 1 sun AM1.5D illumination, 2.2 kW/kg specific power was achieved with 0.65 nm thick MoS2. Modeling of the MoS2 Schottky junction solar cell showed that 69.9 kW/kg specific power is attainable with a single monolayer device. Furthermore, we studied the absorption of single and stacked MoS2 monolayers computationally, using the transfer matrix method, and experimentally, by incrementally stacking and characterizing the monolayers. This stacking maintains an enhanced absorption per layer, not observed in directly synthesized multilayer films, pointing to an expected proportional increase in the efficiency of the MoS2 Schottky PV. Lastly, we propose a full-scale, 2D material-based 60 W solar array for a 6U CubeSat and investigate feasibility with preliminary sub-scale prototypes.

Highly Stretchable, Durable and Lightweight Lego®-style 3-Dimensional Photovoltaic

Min Ju Yun[1], Yeon Hyang Sim[1,2], Dong Yoon Lee[1], and Seung I. Cha[*1,2]

1. Korea Electrotechnology Research Institute, Changwon-si, Gyeongsangnam-do, 51543, Korea
2. University of Science and Technology, Changwon-si, Gyeongsangnam-do, 51543, Korea

Abstract— Solar PV can cover applications that consume high power to applications that consume low power. And with diversification of applications under urban environment, it is changing the point of view of one-size-one-fits based on watt-per-cost concepts to customization-fits with energy-yield-per-watt. With this view point, to integrated PV to application like BIPV, VIPV and device-integrated PV a flexible and foldable solar cell has been on the rise as one way for a system. Accord with flexible solar cell, additional functions of light weight, stretchability and low cost are attracting attention, and durability as well. For these demands, we have proposed Lego®-style assembly module that shows high photovoltaic performance according to modulization without any degradation with average 20% energy conversion efficiency. And it has great durability under compression stress of 5000N, and maintain the photovoltaic performance under high temperature and humidity environment for 500 hours. In addition, application to BIPV, VIPV or device-integrated PV requires installation on arbitrary surface, deformation on 3-D structure, having freedom of design and aesthetic effect, Lego®-style assembly module is suitable on these demands.

Keywords—stretchable, 3-D module, silicon solar cell, VIPV, Lego®-style

I. INTRODUCTION

Among the renewable energy sources, research of providing electricity through solar energy have been studied for a long time and it can be installed and used worldwide from utility-scale system including solar photovoltaic (PV) agriculture to integrating system like building-integrated PV (BIPV), vehicle-integrated PV (VIPV) and device-integrated PV. That is solar PV can cover applications that consume high power to low power. To integrate into these application, a flexible and foldable solar cell has been on the rise as one way for a system. In flexible solar cell, many concepts based on organic solar cells or pervoskite solar cells are proposed based on flexible electrodes, but these have limitations on high energy conversion efficiency, durability and the modulization [1-3]. In order to compensate for these limitation, a lot of research is doing the works to make flexible solar cells using silicon solar cells. Based on silicon solar cell, singled solar cell and curved solar cell are presented for securing light weight and durability by replacing glass and for flexibility. But the curvature angle is too small and there is a limit to be applied to an actual curved surface such as an arbitrary surface.

To overcome these shortcoming, we have proposed Lego®-style assembly module that have actual concept of foldable and stretchable, high energy conversion efficiency based on silicon solar cell, mechanical strength, durability, and stable modulization. This module based on rigid-island concept and tessellation structure. A stretchable interconnector is placed between the solar cells for a stretchable device while retaining the high photovoltaic performance, stability, and rigidity of the original characteristics of silicon solar cell. Lego®-style assembly module shows high durability under high temperature and humidity environment also it shows stable modulization and electrical connection with average conversion efficiency 21% without any degradation.

II. EXPERIMENTAL DETAILS

A. Fabrication of Lego®-style assembled modules

For the encapsulated solar-cell unit, passivated emitter rear cells (PERC; LWM5BB, Lightway) were laser-cut into triangular shapes with a side length of 30 mm. In the case of hexagon-shaped solar-cell building blocks used for a flower-like-shaped Lego®-style assembled module, the center hexagons' side length was 20 mm and the small hexagons' side length was 15 mm. Circular solar units (Fig. S1) were 35 mm in diameter. Each was cut into a shaped bare solar cell, placed into a dipping machine (KD Keukdong, 60Φ), and took it out after 1 s. On the pre-coated busbar location of the solar cell, ribbons were connected by a soldering method (FX-951, Hakko). The Lego®-style assembled modules were constructed using multiple solar-cell building blocks assembled with an encapsulated solar-cell unit and a stretchable interconnector via a simple soldering method. The Lego®-style modules were assembled using 2 to 16 solar-cell building blocks connected in series.

B. Characterization

978-1-7281-6118-1/22 $31.00 © 2022 IEEE

For measurement of the adhesion strength between pre-coated solder on a solar cell and ribbon, a 90° ribbon peel-off test (TEST RIG 5kN-90, Hegewald & Peschke) was conducted. Mechanical compressive tests were performed using a universal testing machine (AGS-X, 5kN, SHIMADZU) to confirm the reliability of the solar-cell unit. The same machine was used to measure the tensile strength of the wavy structure, stretchable polymer resin, and the stretchable interconnector to characterize their stretching behaviors. To characterize the stretchable interconnector's electrical stability, its electrical resistance was measured during deformation and repeated stretching to 40% using a Keithley 2636B source meter connected to a custom-built automatic stretching tester (SnM).

The PV performance of the Lego®-style assembled modules was confirmed by measuring their energy conversion efficiency with a solar simulator (Sun 2000, 1000 W Xe source; Abet Technologies; 2400 Keithley source meter with a KG-3 filter). Before the measurement, we calibrated the solar simulator using an NREL-certified reference cell and set the simulator to 1 Sun and 1.5 AM. To characterize the durability of the Lego®-style assembled module, an endurance test was conducted at 85% relative humidity and 85 °C for 500 h in a constant-temperature, constant-humidity chamber (AL HMUV, ALISTA). The ribbon peel-off test and the durability tests of modules were conducted by an external institute, Chungbuk Technopark, to ensure the reliability of the test data. Damp Heat test in 500 h of the module in accordance with KS C IEC 61215, 10.13.

III. RESULTS AND DISCUSSION

Lego®-style assembly module was proposed in this study to become a base step for technological and commercial development for integrated-PV applications on arbitrary surface. For the Lego®-style assembly module, one solar cell is completely encapsulated with front cover and back frame to have durability characteristic, and then the encapsulated solar cell units are connected to the circuit so it can be a flexible module like rigid-island structure. Furthermore, the interconnector used for circuit connection has a stretchable structure so a stretchable and flexible module with excellent durability can be proposed and fabricated.

The encapsulated solar cell unit can be diversified without restriction in shape, such as hexagon or circle. The back frame is designed so that there is a space for the interconnector to fit in, and encapsulated solar cell unit and the encapsulated stretchable interconnector can be easily assembled like Lego® blocks to form one solar cell building block as shown in Fig. 1 (a). Building blocks can be assembled like Lego®-style according to a desired building block's number, design and connection circuit with various way like only series or parallel or mixed series and parallel on a substrate of any shape and material to make a module as shown in Fig. 1 (b).

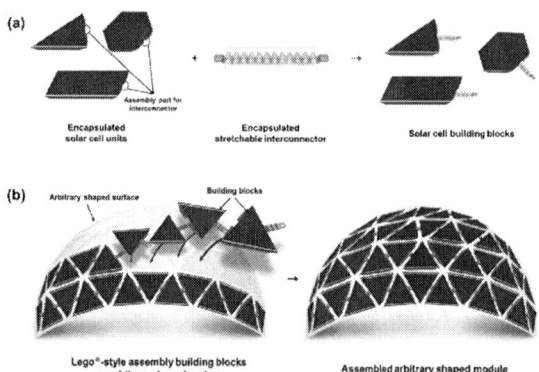

Fig. 1. Schematic illustration of (a) fabrication solar cell building blocks combining encapsulated solar cell unit and stretchable interconnector like Lego® block and (b) assembly solar cell building blocks on arbitrary shaped surface and assemble module.

Encapsulation was performed by coating the front side with silicone material and for the back frame, epoxy was utilized and it has a great advantage in durability, and another advantage is that it can implement various colors using dyes, which can even create an aesthetic impact. For the stretchable interconnector, extensible wavy structure was formed by imprinting a copper ribbon with a 25 μm thickness and 2 mm width using a cogwheel rack gear and pinion manufactured by using a 3-D printer. Wavy structured copper ribbon could be recovered to its original state after extension for the stretchable characteristic, so it was cured and demolded in a mold after being embedded in stretchable polymer resin.

The encapsulated solar cell unit and stretchable interconnector unit that is fabricated through each fabrication process are assembled like Lego® blocks in the interconnector space of the encapsulated solar cell's back frame to form a single solar cell building block. The photographs transformed by stretching, folding, and twisting of a 2-cell sub-module connected one solar cell building block and one encapsulated solar cell unit is shown in Fig. 2 (a). Various deformations are possible freely by the encapsulated stretchable interconnector, and as a result of comparing the 2-cell sub-module performance before and after stretching deformation, it confirmed that the performance was maintained (Fig. 2 (b)). In case of 1-cell, the average energy conversion value is 21.7%, for 2-cell module the average value is 20.7%, for 4-cell module the average value is 20.5%, for 8-cell module the average value is 21%,

978-1-7281-6118-1/22 $31.00 © 2022 IEEE

and for 16-cell module the average value is 20.5%, so there is almost no decrease in energy conversion efficiency value, which means the electrical connection is very stable without contact loss between the solar cell building blocks. Additional, aesthetic impact can be given by adding dyes to the epoxy resin for the back frame coloring. Any color that desired is possible, and in this study, dark gray, dark brown and dark pink colors were used for color effect (Fig. 2 (c)).

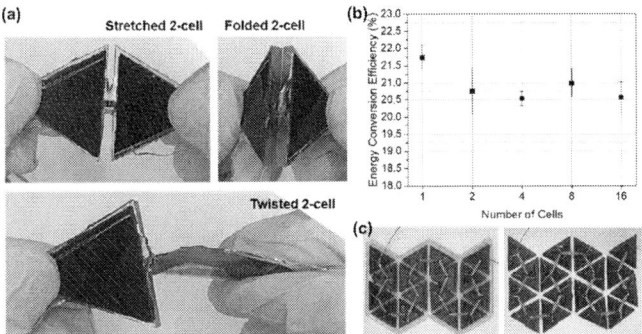

Fig. 2.   (a) Photographs of deformed as stretching, folding and twisting of 2-cell Lego®-style assembly module. (b) Energy conversion efficiency Lego®-style assembly module according to number of solar cell units. The average value of each three modules is shown. (c) Photographs of 16-cell Lego®-style assembly module with dark gray, dark brown and dark pink colored back frame.

Fig. 3 (a) shows the transformation into various 3-D arbitrary shaped modules with 16-cell Lego®-style assembled module. Through this the module shows free for folding or covering on curved surface. In order to figure out the effect of diversifying the shape of encapsulated solar cell unit and module design, a flower-like shaped Lego®-style assembled module was fabricated using hexagon-shaped solar cell building blocks as shown in Fig. 3 (b).

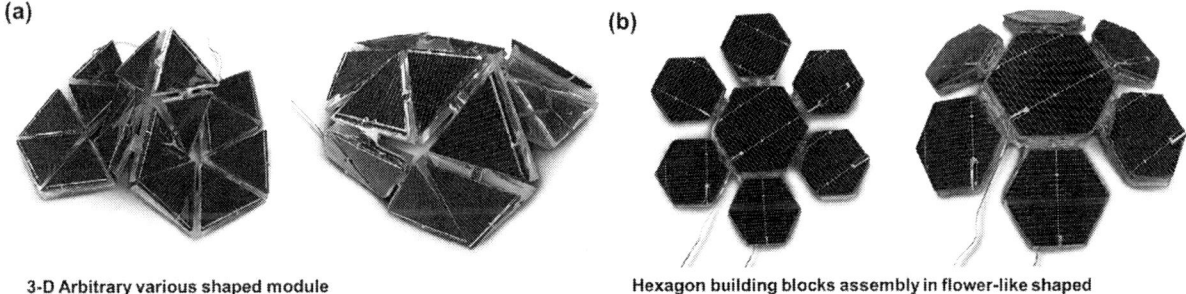

Fig. 3.   Example of readable plot using different colors and line styles for clarity. (*figure caption*)

This proposed Lego®-style assembly module has high performance on stretchability and durability both of them. We have expected that it could be a base step for integrated-PV application development that needs installable on arbitrary surface, deformation on 3-D structure, having freedom of design and aesthetic effect.

ACKNOWLEDGMENT

This research was supported by the Korea Electrotechnology Research Institute (KERI) Primary Research Program through the National Research Council of Science & Technology (NST) funded by the Ministry of Science and ICT (MSIT) (No. 22A01005).

REFERENCES

[1]   M. Jo, S. Bae, I. Oh, J. Jeong, B. Kang, S. J. Hwang, S. S. Lee, H. J. Son, B. Moon, M. J. Ko and L. Phillip, "3d printer-based encapsulated origami electronics for extreme system stretchability and high areal coverage," ACS Nano, vol. 13, pp. 12500–12510, 2019.

[2]   16.   F. Kassaei, R. Rafiei and F. Torabi, "Inflexible silicon solar cell encapsulation process on curved surfaces: Experimental investigation," Environ. Prog. Sustain. Energy, vol. 40, e13513, 2021.

[3]   17.   H. Li, X. Li, W. Wang, J. Huang, J. Li, Y. Lu, J. Chang, J. Fang and W. Song, "Highly foldable and efficient paper‐based perovskite solar cells," Solar RRL, vol. 3, 1800317, 2019.

Optimal Strategy for Using Biomass to enable California High Penetration Solar

Mahmoud Y. Abido, Sarah R. Kurtz

University of California Merced, Merced, CA, United States

Cairo University, Giza, Egypt

Slowing climate change requires reducing emissions of carbon dioxide. So, moving towards a 100% renewable energy grid is paramount. Solar energy has great potential but identifying how to provide power when the sun sets and during the winter is a challenge. Here we focus on how to use biomethane to reduce the size of that challenge. We analyze strategies for using biomethane and identify the novel Allam cycle technology instead of the current Combined Cycle Gas Turbines (CCGT). We use a simple approach to identify the best combination between solar and the Allam cycle to enable high penetration of solar.

Automatic Crack Segmentation in Electroluminescence Images of Solar Modules and Maximum Inactive Area Prediction

Xin Chen, Todd Karin, Anubhav Jain

Lawrence Berkeley National Laboratory, Berkeley, CA, United States

Cracks on solar cells can cause degradation of photovoltaic (PV) modules, and electroluminescence (EL) images are a common technique for identifying cracks. However, to process a large number of such images it is necessary to develop automated routines for analysis. This article introduces a fast semantic segmentation method (~0.18s/cell) to segment crack lines, cross cracks, busbars, and dark areas on EL images of PV modules. We trained a UNet neural network model on a training set of 1,272 images, and we evaluated its performance on a validation set of 206 images and a testing set of 359 images. We report the performance on the testing set with an average F1 score of 0.875 and an IoU score of 0.782. We introduce our algorithm of predicting the worst-case degradation area with cracks detected. We also demonstrate our automatic preprocessing tool of cropping individual cell images from EL images of PV modules (~0.72s/module). Our methods are published as open-source software and might be used to segment other kinds of defects or similar types of images by transfer learning in the future.

Experimental assessment of temperature estimation models of bifacial photovoltaic modules

Gaetano Mannino[1], Giuseppe Marco Tina[1], Mario Cacciato[1], Lorenzo Todaro[2], Fabrizio Bizzarri[2], Andrea Canino[2]

[1]DIEEI, University of Catania, 95124 Catania, Italy
[2]Enel Green Power SpA, Viale Regina Margherita, 125, 00198, Rome, Italy

Abstract— **The operating temperature significantly affects the efficiency and degradation of the photovoltaic modules. Higher temperatures contribute to various degradation mechanisms, thermal cycling also have a further impact on degradation. For the estimation of temperatures for the monofacial photovoltaic modules various models have been proposed, however, there are still not many studies on temperature estimation for the bifacial modules, which are going through a period of expansion in the market. In the present study, some steady-state models initially elaborated for the monofacial modules have been applied to the bifacial ones, evaluating the consistency of the estimation results with the measured temperature values. Some adjustments to monofacial temperature models are also proposed to take into account the irradiance on the back of the module. Finally, a temperature estimation model for bifacial modules, based on experimental data, is proposed.**

Keywords: bifacial, monofacial photovoltaic module, thermal model.

I. INTRODUCTION

Considering the environmental conditions and the technical characteristics of the photovoltaic modules, in the literature there are various studies on models that can estimate the temperature of the photovoltaic modules, however, recent studies have dealt with the evaluation of the temperature of bifacial photovoltaic modules [1] [2] [3]. In this paper, some models proposed for the estimation of the temperatures in the case of the monofacial modules have been used in the case of the bifacial modules to verify which ones can best adapt.
Then, some modifications have been made to the models taken into consideration. Finally, an accurate model for the estimation of the temperatures for bifacial modules has been proposed starting from the real data of a plant equipped with modules with monocrystalline silicon cells. Have been conducted the tests on data collected with 1 minute of resolution and considering both day and night, for this reason, the steady-state models taken into consideration will still be subject to a greater error than when they are used exclusively during the day and within certain irradiation ranges.

II. THERMAL MODEL OF MONO AND BIFACIAL MODULES

In the following, temperature models for bifacial PV modules are described.

Monofacial module temperature

Sandia model [4]: $\quad T_{mod} = T_a + (G_f) \cdot (e^{a+bWs})$ $\quad$ (1)

Where T_{mod} is the module temperature [°C], T_a is the ambient temperature [°C], G_f is the front plane of array irradiance [W/m^2], a and b are two empirical coefficients, Ws is the wind speed [m/s]

Faiman model [5]: $\quad T_{mod} = T_a + \frac{G_f}{U_0+U_1W_s}$ $\quad$ (2)

Where $U_0[W/(m^2K)]$ a $U_1[W/(m^3sK)]$ are empirical coefficients.
Once the module temperature is calculated, it is possible to estimate the cell temperature using [4]:

$$T_{cell} = T_{mod} + \Delta T \cdot \frac{G_f}{G_{stc}} \quad (3)$$

ΔT is the temperature difference parameter, it values typically from 2 to 3 °C for flat-plate modules in an open-rack mount [4]. To directly estimate cell temperature, it is possible to use:

Servant model [6]: $\quad T_{cell} = T_a + 0.0138 \cdot G_f \cdot (1 + 0.031 \cdot T_a) \cdot (1 - 0.042Ws)$ $\quad$ (4)

PVsyst model [7]: $\quad T_{cell} = T_a + G_f \frac{\alpha(1-\eta_{STC})}{U_c+U_vW_s}$ $\quad$ (5)

Where α is the absorption coefficient (0.9 default in PVsyst), η_{STC} is the efficiency at standard test conditions, U_c and U_v two empirical constants as in eq.(2).

Bifacial module temperature

Sandia bPV model [1]: $T_{mod} = T_a+(G_f + G_r) \cdot (e^{a+bWs})$ $\quad$ (6)

Where G_r is the rear irradiance [W/m^2].
Starting with the Sandia model, a coefficient c of G_r is introduced

$$T_{mod} = T_a + (G_f + cG_r) \cdot (e^{a+bWs}) \quad (7)$$

Further, the new term (1-nstc) is added to take into account the variation of T$_{mod}$ due to the transformation of solar energy into electricity:

$$T_{mod} = T_a + (1 - \eta_{STC})(G_f + cG_r) \cdot (e^{a+bWs}) \quad (8)$$

Multi-variable polynomial model - MVPM

To provide models with sufficient accuracy, three Multi-Variable Polynomial Models (named MVPM-xx , where xx is the number of parameters) are proposed, whose coefficients are found using the experimental data. The input variables in all three cases are: the front irradiance G_f, back-module irradiance Gr, ambient temperature T_a and wind speed w_s. In the first model, MVPM-4, only first-degree terms are considered, in the model MVPM-8, first and second degree terms are considered, and in the third model, MVPM-11, the mixed terms are also added.

$$T_{mod} = a_1 G_f + a_2 G_r + a_3 T_a + a_4 w_s \quad (9)$$

$$T_{mod} = a_1 G_f + a_2 G_r + a_3 T_a + a_4 w_s + a_5 G_f{}^2 + a_6 G_r{}^2 + a_7 T_a{}^2 + a_8 w_s{}^2 \quad (10)$$

$$T_{mod} = a_1 G_f + a_2 G_r + a_3 T_a + a_4 w_s + a_5 G_f{}^2 + a_6 G_r{}^2 + a_7 T_a{}^2 + a_8 w_s{}^2 + a_9 G_f \cdot T_a + a_{10} G_f \cdot w_s + a_{11} T_a \cdot w_s \quad (11)$$

III. EXPERIMENTAL SET-UP

The 2.84 kW photovoltaic system taken into consideration consists of seven bifacial PERC monocrystalline silicon modules with a nominal power of 405W each. It faces south with an inclination of 35.7 °.

Considered PV system

Figure 1 Bifacial PV plant, Catania (37.4°,15°)

For the analysis of the models and the elaboration of the multi-variable polynomial model, data from the following sensors was used: a pyranometer inclined with the same angle of the modules 37.5°, a reference cell located on the back of the modules in the central area of the photovoltaic system, a thermohygrometer located in the same area of the system, a PT 100 is attached by conductive adhesive tape to the central area of one module of the system. An anemometer placed 10 m above the ground and about 300 meters away from PV system is used to acquire the wind speed. The time sampling adopted for the acquired data is 1 minute for all quantities, except the wind speed which was sampled every 10 minutes. Through linear interpolation, the value of the wind speed was obtained every minute.

Figure 2 Rear module view and anemometric tower. Reference cell in red circle, PT100 in blue circle.

IV. RESULTS AND DISCUSSION

To test the temperature estimation models, the data acquired during the week from 19/05/2021 to 26/05/2021 were used; the modules considered had been installed few days before.
Except for the experimental model, other models (which are studied to evaluate the temperature only during the day) will have error values that will be greater than those they would have by considering only the daytime or with filtering the irradiation within a certain range of values.

TABLE I. Coefficients and statistical indices of physical thermal models to calculate T_{mod}

MODEL	Coef. 1	Coef. 2	Coef. 3	RMSE	R2
Sandia (1)	a= -3.47	b= -0.0594	—	2.56	0.97
Sandia (1) optimal a, b	a= 3.71	b = 0.0356	—	2.08	0.97
Sandia bPV(6)	a= -3.47	b= -0.0594	—	3.74	0.92
Sandia bPV(6) optimal a, b	a= -3.8589	b= -0.0297	—	2.21	0.97
Sandia bPV (7) Optimal, a,b,c	a=-3.2662	b= -0.0546	c = -1.0324	2.01	0.98
Sandia bPV (8) Optimal c	a=-3.47	b=-0.0594	c= -1.0219	2.01	0.98
Faiman (2)	U_0 =25	U_1=6.84	—	2.58	0.96
Faiman (2) Optimal U0, U1	U_0=33.33	U_1=3.42	—	2.12	0.97
Servant* (3)	—	—	—	2.62	0.96
PVsyst* (5)	U_c=25	U_v=1.2	—	2.09	0.97

*To estimate the module temperature with Servant and pvsyst model, eq. 3 resolved with respect to Tmod was used and inserting the result respectively of eq 4 and eq 5 instead of Tcell

As indicated in [1], Sandia model coefficients a and b can be "fine-tuned" according to measurements after that the system has been installed in the way to reduce error due to site differences and different anemometer height from standard meteorological practice. To select the optimal values of a and b, the MATLAB function fmincon was used, with the aim of reducing R2. The same procedure has been adopted to find U0 and U1 optimal parameters, coefficient c in Sandia model for bifacial, coefficients in the experimental model.

TABLE II. Coefficients and statistical indices of MVPM thermal models to calculate T_{mod}

coefficients	variables	MVPM-4	MVPM-8	MVPM-11
a1	G_f	0.0231	0.0344	0.0356
a2	G_r	0.0145	-0.004	-0.0076
a3	T_a	0.9564	0.5878	0.5645
a4	w_s	-0.451	0.8534	0.9016
a5	G_f^2	—	-9.9136E-06	-9.10E-06
a6	G_r^2	—	-2.71E-04	-2.36E-04
a7	T_a^2	—	0.014	0.0157
a8	w_s^2	—	-0.1655	-0.1152
a9	$G_f \cdot T_a$	—	—	-3.09E-05
a10	$G_f \cdot w_s$	—	—	-3.26E-04
a11	$T_a \cdot w_s$	—	—	-1.27E-02
RMSE		1.69	1.40	1.40
R2		0.98	0.98	0.98

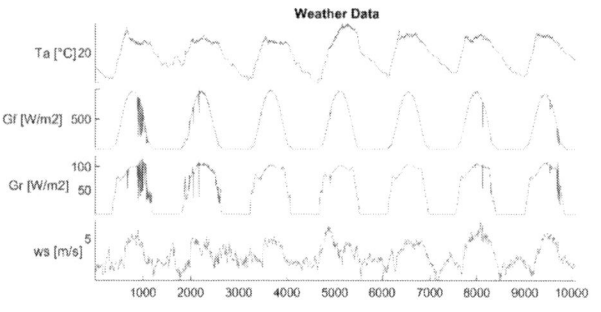

Figure 3 Weather data, from top to down: Ambient temperature, Front irradiance, Rear irradiance, wind speed

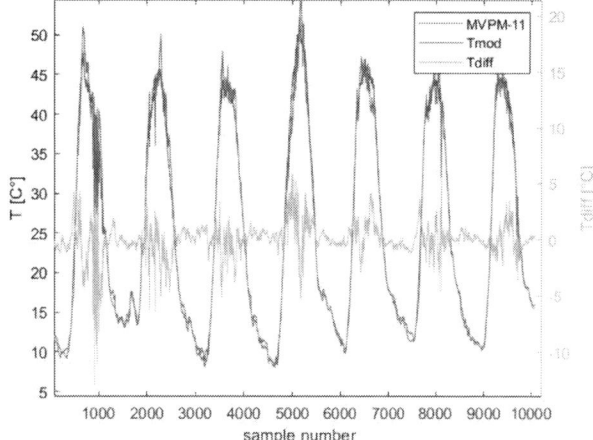

Figure 4 Temperature curves of a bifacial module and error: temperature calculated by MVPV-11 model (blue line), measured temperature (red line), error (yellow line)

In the temperature models analysis, it was highlighted that:

- even in the case of monofacial modules the albedo can be different depending on the location, therefore also in thermal models processed for monofacial modules, a percentage of irradiation on the back of the module is implicitly considered, but which is not quantified.
- In bifacial modules, differently by monofacial modules equipped with tedlar back-module surface, there is a dark surface that heats up more than a white surface such as that of tedlar, however part of the radiation hitting the rear module is transformed into electrical energy.

CONCLUSIONS

Some monofacial temperature estimation model applied to bifacial modules have been investigated; then the improvements in terms of RMSE and R2 provided by searching for the optimal parameters a and b for the Sandia model, U0 and U1 for the Faiman model have been quantified.

Improved models for bifacial modules, based on Sandia bifacial temperature model have been proposed, improvements have been obtained by inserting a third coefficient c which multiplies the rear irradiance.

The MVPMs provide the best values of RMSE and R2, being elaborated from the experimental data. Further studies will aim to find a better way to express a new equation that takes into account the specific construction aspects of the modules, meteorological characteristics of the site and the seasonality of the environmental parameters.

ACKNOWLEDGMENT

The authors would like to thank Enel Green Power technician Giuseppe Walter Gangemi for technical support. The authors would like to thank Enel Green Power for making the facilities available for research purposes.

REFERENCES

[1] C. Hansen, D. Riley, C. Deline, F. Toor, & J. Stein, "A Detailed Performance Model for Bifacial PV Modules "(No. SAND2017-11013C). Sandia National Lab.(SNL-NM), Albuquerque, NM (United States),2017.

[2] G. M. Tina, F. B. Scavo, & A. Gagliano. "Multilayer thermal model for evaluating the performances of monofacial and bifacial photovoltaic modules", *IEEE Journal of Photovoltaics*, 10(4), 1035-1043, 2020.

[3] M. W. P. E. Lamers, E. Özkalay, R. S. R. Gali, G. J. M. Janssen, A. W. Weeber, I. G. Romijn, & B. B. Van Aken, "Temperature effects of bifacial modules: Hotter or cooler?", *Solar Energy Materials and Solar Cells*, 185, 192-197, 2018.

[4] D. L. King, J. A. Kratochvil, and W. E. Boyson, "Photovoltaic array performance model", United States, Department of Energy, 2004.

[5] D. Faiman, "Assessing the outdoor operating temperature of photovoltaic modules", *Progress in Photovoltaics: Research and Applications*, 16(4), 307-315, 2008.

[6] J. M. Servant, "Calculation of the cell temperature for photovoltaic modules from climatic data." Intersol Eighty Five. Pergamon, 1640-16, 1986.

[7] https://www.pvsyst.com/help/index.html, accessed on 12/01/2022

[8] F. K. Nyarko, G. Takyi, E. H. Amalu, & M. S. Adaramola, "Generating temperature cycle profile from in-situ climatic condition for accurate prediction of thermo-mechanical degradation of c-Si photovoltaic module", *Engineering Science and Technology, an International Journal*, 22(2), 502-514, 2019.

[9] Sharma, & S. S.Chandel, "Performance and degradation analysis for long term reliability of solar photovoltaic systems: A review" *Renewable and sustainable energy reviews*, 27, 753-767, 2013.

Seasonal dependence of diurnal efficiency degradation and recovery in perovskite mini-modules during outdoor testing

Vasiliki Paraskeva[1], Maria Hadjipanayi[1], Matthew Norton[1], Aranzazu Aguirre[2], Afshin Hadipour[2], Rita Ebner[3] and George E. Georghiou[1]

[1]FOSS Research Centre for Sustainable Energy, Department of Electrical and Computer Engineering, University of Cyprus, 75 Kallipoleos St., Nicosia, 1678, Cyprus

[2]imec, Kapeldreef 75, 3001 Leuven, Belgium

[3]AIT Austrian Institute of Technology, Center for Energy, Giefinggasse 2, 1210 Vienna

Abstract—The diurnal efficiency degradation and subsequent recovery of several identical perovskite mini-modules has been investigated during outdoor testing over different seasons. Seasonal dependence of recovery and diurnal efficiency degradation in perovskite devices has been demonstrated. The higher irradiation and ambient temperatures during summer months were found to enhance the diurnal efficiency degradation-to-recovery ratio over the first days of testing, leading to significant accelerated performance degradation in the perovskite modules tested in those conditions. Diurnal efficiency degradation and performance recovery were found to be higher at early degradation stages, where higher absolute efficiency values are present.

Keywords—perovskites, mini-modules, performance measurements, recovery, outdoor testing

I. INTRODUCTION

In less than a decade, perovskite technology has emerged with immense promise as potentially the cheapest alternative to the present commercially available photovoltaic technologies. However, despite an expeditious rise in their efficiency, perovskite-based solar cells are still far from commercialization because of inadequate stability [1]. Reversible/temporal changes in perovskite performances have also been observed creating doubts about the amount of real degradation of perovskites under light [2]. Several papers report that during recovery in the dark, the loss in perovskite device performance under previous light exposure can be fully or partially reversed. Perovskite recovery was found to depend on cell aging with higher recovery to be obtained in perovskite cells at early stages of degradation [3]. Further studies of performance recovery demonstrated that this is dependent from the bias conditions of the device under test [4]. Because at least in part, the perovskite degradation is due to the migration of mobile ionic species, the recovery time of the perovskite is manipulated by an electric bias as well. Since limited experience of recovery of perovskites is available so far,

this study aims to give insight into the seasonal dependence of diurnal efficiency degradation and recovery in identical perovskite modules exposed outdoors for several months. Furthermore, performance degradation, voltage and current losses were investigated in an attempt to establish the impact of environmental conditions during the long-term operation of perovskite mini-modules. Diurnal efficiency degradation-to-recovery ratio that determines the overall performance degradation of the perovskite device was also examined for modules tested outdoors at different seasons.

II. EXPERIMENTAL APPROACH

Five (5) identical perovskite modules have been mounted outdoors in a fixed plane array at different seasons and current-voltage (I-V) measurements have been collected at regular intervals over several months of testing. The first batch of modules (Batch 1) consisted of 3 identical modules and was located outdoors from the middle of January until the beginning of May. The second batch of modules (Batch 2) consisted of another two modules and was located outdoors from the end of August until December. The active layer of the perovskite modules under test is a two-cation perovskite ($Cs_{0.18}Fa_{0.82}PbI_{2.82}Br_{0.18}$). An identical testing procedure was applied to the two batches of modules for the collection of their electrical parameters. Open-circuit loading was applied between the I-V scans. Forward and reverse voltage sweeps have been applied during each I-V curve. Alongside the I-V traces from the devices, environmental sensors have been used to collect solar irradiance in the plane of array, ambient and device temperature, wind velocity and humidity/precipitation levels. The electrical measurements have been acquired by a single current-voltage source-meter multiplexed to take sequential measurements from the devices under test. LabVIEW software was designed to record the I-V-traces every 5 minutes at high Global Normal Irradiance (GNI) conditions (GNI>400 W/m^2). Both forward

978-1-7281-6118-1/22 $31.00 © 2022 IEEE

(<0V to > open-circuit voltage (Voc)) and reverse (> Voc to < 0V) voltage sweeps have been applied to the devices. Forward-first voltage sweeps have been used at all instances. The voltage sweep rate was chosen to be 1 V/sec. Alongside with perovskite modules studied outdoors, control modules of identical structure were kept indoors under dark, controlled conditions to help distinguish the degradation that occurs outdoors from the aging degradation processes occurring inherently within the device. The current-voltage characteristic of the control modules was measured with the TRI-SOL Solar Simulator, Class AAA, Xe Arc Lamp with 100 mW/cm^2 power output. The current at the contacts of the cell was collected by using a Keithley 2651A source meter and by applying the ivRider software. The temperature of the control samples was kept at 25°C.

III. RESULTS AND DISCUSSION

A. Power Conversion Efficiency (PCE) Degradation

The Batch 2 modules exposed outdoors during the summer period presented a rapid drop in performance. For a better understanding of the performance degradation of the modules over the first days of outdoor exposure, the power conversion efficiency loss was calculated for each module over the first week of testing. The graph in Fig.1. demonstrates that modules located outdoors during the summer period presented significant reduction in their efficiency from the first week of outdoor exposure. During the first week of outdoor exposure for Batch 1 modules the mean module temperature was around 28°C and the total weekly solar irradiation applied on the modules was 20.22 kWh/m^2. On the other hand, during the first week of exposure for Batch 2 modules the mean module temperature was found to be 49.6°C and the total weekly solar irradiation applied on the modules 41.25 kWh/m^2. This module temperature was collected only during the collection of current-voltage characteristics of the modules at irradiance values higher than 400W/m^2.

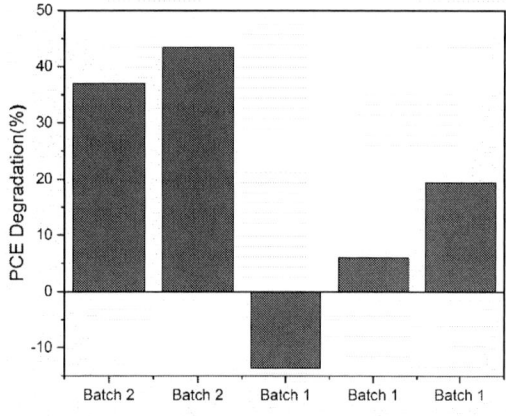

Fig. 1. Efficiency degradation of modules in batches 1 and 2 after one week of outdoor testing. Batch 1 was located outdoors during winter months while Batch 2 modules located outdoors during summer months.

The results indicate that higher levels of irradiance and temperature present during summer months are likely to be the major causes of accelerated performance degradation in the perovskite mini-modules. The normalized PCE degradation of the two batches of modules over time demonstrates that perovskite devices tested in winter present very small performance degradation and sometimes efficiency gain the first days of exposure with large deviations in efficiency (Fig.2). On the other, perovskites tested during summer time present a continuous and significant drop of their efficiency until it stabiizes after the 60th day of exposure. The mean value of efficiency the first day of exposure was set as the starting point for the normalization.

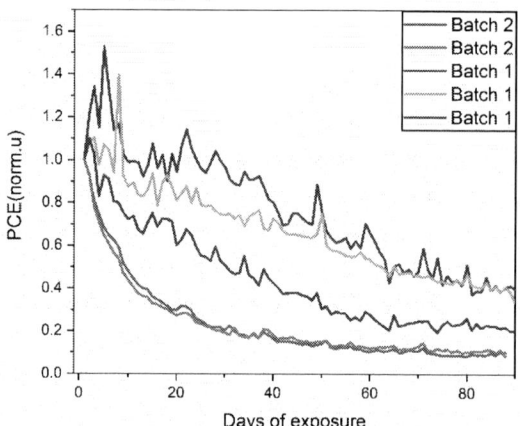

Fig. 2. Efficiency degradation of modules in batches 1 and 2 against time of exposure. Batch 1 was located outdoors during winter months while Batch 2 modules were located outdoors during summer months.

To identify the root cause of efficiency degradation of the modules, the electrical parameters of the modules have been studied over the testing period. The decrease in power conversion efficiency of the modules is linked to significant reduction in current output. The current losses presented in the Batch 2 modules were significantly higher than those obtained in Batch 1 modules (see Fig.3a). The mean value of current in the first day of exposure was set as the starting point for the normalization. Fig. 3(a) demonstrates that the modules of Batch 1 present current output enhancement intiallly and reach their higher current output after 4-7 days of operation in the field. After that period current starts to reduce slightly and with large fluctuations. Devices of Batch 2 tested in summer present a current exponential reduction from their first day of operation outdoors with insignificant fluctuations. Data of Fig. 3(a) demonstrates that current reduction of 50% occurs in Batch 1 devices after the 80th day of exposure while in Batch 2 modules occurs only in 14 days of outdoor testing. Voltage losses were also studied in perovskite devices (Fig.3b). Voltage losses were much lower than current losses in all modules under test. Small differences in voltage losses are apparent between the two batches of modules. It is worth to note that no visible degradation due to humidity ingress inside the encapsulant material was observed in all modules under test.

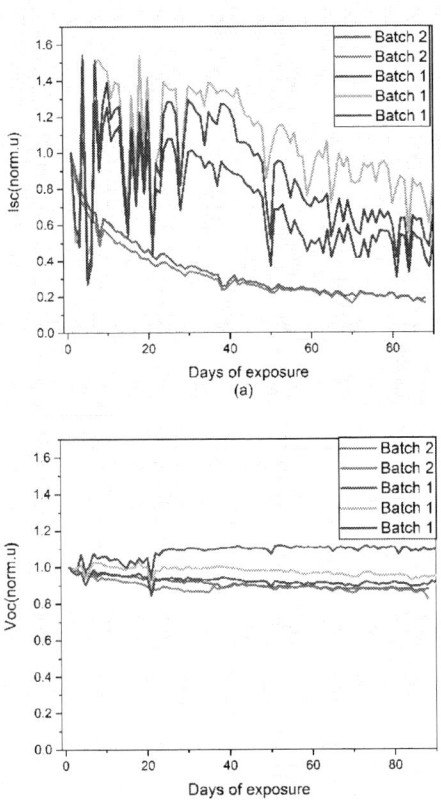

Fig. 3. (a) Short-circuit current and (b) open-circuit voltage evolution over time for the two batches of modules tested. Significant current losses were detected in all modules under test.

B. Diurnal efficiency degradation

The efficiency of the perovskite modules changes over the day. The diurnal efficiency degradation of all modules under test was calculated for each day in the field to investigate any possible trend between the two batches of modules tested at different seasons. The diurnal efficiency degradation was calculated based on the efficiency values at the beginning and the end of each day. Absolute values of diurnal efficiency degradation over time for the two batches of modules tested are depicted in Fig.4. Higher values of the diurnal efficiency degradation are apparent the first days of outdoor exposure in both batches of modules and this is irrelevant from the environmental conditions present at the time of outdoor testing. A fast drop of the diurnal efficiency degradation values towards zero occurs in Batch 2 while in Batch 1 the reduction of the diurnal efficiency degradation values over time is quite low. Almost in all instances the diurnal efficiency degradation values of the modules tested in winter are higher. Only the first week of exposure more instances of higher diurnal efficiency degradation are obtained in Batch 2 which is exposed to higher irradiance and temperature levels. The accelerated performance degradation occurs in Batch 2 modules from the first days of exposure leaving the modules with low overall efficiency which show lower diurnal changes in efficiency and recovery. The diurnal efficiency degradation was then studied over the

efficiency of the modules to find a correlation between both parameters. The results are depicted in Fig.5.

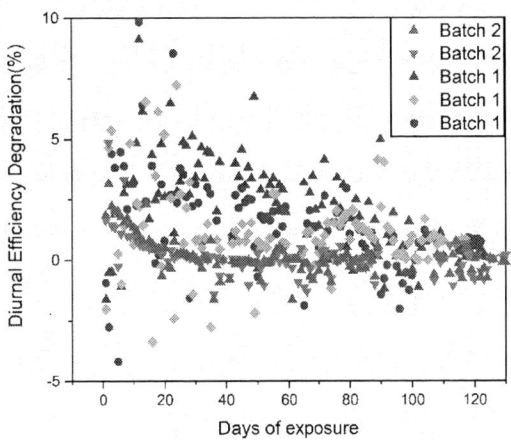

Fig. 4. Absolute values of diurnal efficiency degradation of the two batches of modules for 120 days of outdoor exposure.

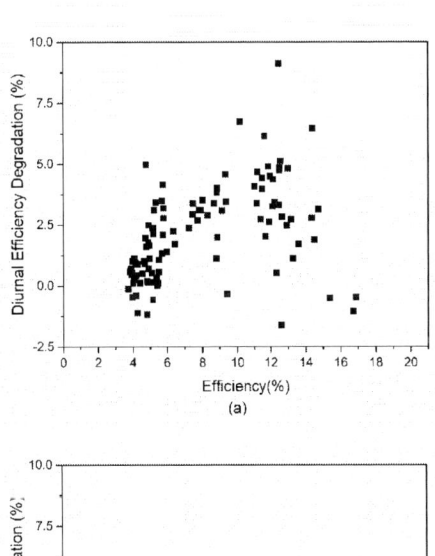

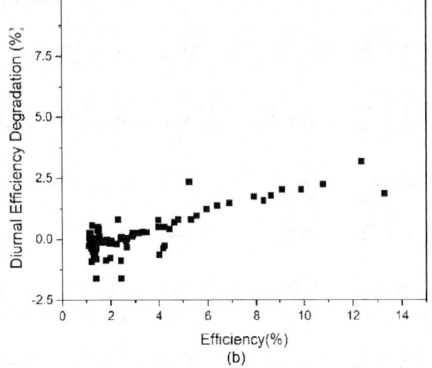

Fig. 5. Diurnal efficeincy degradation against absolute efficiency for modules of (a) Batch 1 and (b) Batch 2.

Modules tested at different seasons present similar trend in their diurnal performance degradation: higher diurnal degradation values present at higher efficiencies. An almost linear

978-1-7281-6118-1/22 $31.00 © 2022 IEEE

relationship occurs between diurnal efficiency degradation and absolute efficiency in modules tested during summer months (Batch 2). Non-linear relationship of the diurnal performance degradation with efficiency presents in Batch 1 modules which might be attributed to the larger values of the absolute efficiency present in these modules for a longer time interval arising from slower overall performance degradation of the modules.

C. Performance recovery

Performance recovery was analyzed next for the two different batches of modules for the whole period of exposure (see Fig.6). The recovery was calculated by considering the final efficiency value of the previous day and the initial efficiency value of the next day. A clear classification of the two batches of modules can be obtained based on their recovery values over the testing period. In agreement with diurnal efficiency degradation values the performance recovery in Batch 1 is significantly higher for all days of testing apart from the first week. Furthermore, reduction of performance recovery is obtained over time of exposure outdoors for both batches under test. That fact indicates higher performance recovery at the early stages of degradation in agreement with previous studies. The rate of performance recovery reduction depends on the outdoor testing conditions with higher irradiance and temperature levels to cause quicker reduction in performance recovery. Faster reduction in diurnal efficiency degradation and performance recovery in modules tested in summer give evidence of the accelerated overall performance degradation that occurs in Batch 2 modules.

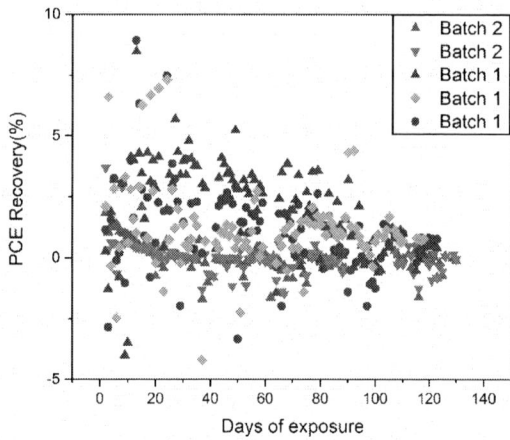

Fig. 6. Absolute values of performance recovery of the two batches of modules for 120 days of outdoor exposure.

Performance recovery studies present in Fig.6 indicate that modules tested in summer, after they lost significant amount of performance, they exhibit smaller values of recovery in agreement with their diurnal efficiency degradation data. It is worth to note that diurnal efficiency degradation and efficiency recovery are driven mainly by the diurnal current degradation and current recovery. Diurnal voltage degradation and subsequent recovery are insignificant.

Investigation of the performance recovery against absolute efficiency was then implemented (see Fig.7). Similar to diurnal

efficiency degradation results the performance recovery presents higher values at higher absolute efficiencies. A more linear relationship between recovery and efficiency is obtained in the Batch 2 modules tested in summer months in agreement with the diurnal efficiency results of Fig.5. The linear relationship between those parameters is attributed to the quick loss of the overall performance of the modules which leads to quick loss of the performance recovery of the modules as well. More pronounced recovery occurs at early stages of degradation where efficiency is higher in both batches under study.

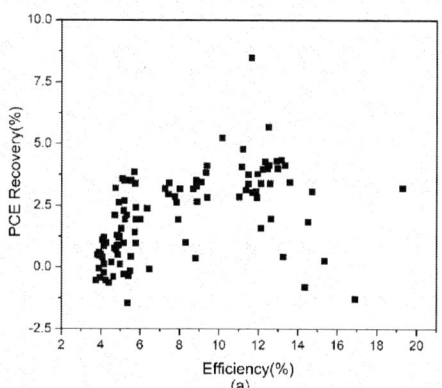

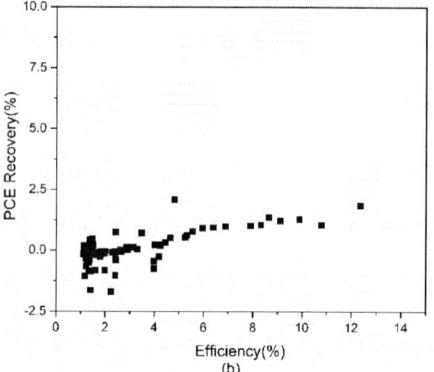

Fig. 7. PCE recovery against absolute efficiency for a modules of (a) Batch 1 and (b) Batch 2.

Non-linear relationship between recovery and efficiency is present in modules tested in winter. In Batch 1 high values of efficiency remain for a prolonged time interval and this leads to higher performance recovery values for extended time periods. Furthermore, the presence of cloudy conditions adds larger dispersion of the data since the uncertainty regarding the initial and final efficiency of the modules is higher.

D. Diurnal Performance Degradation-to-Recovery Ratio

Diurnal performance degradation-to-performance recovery ratio was investigated next for the two different batches of modules. This ratio determines the overall performance degradation of the module and merits investigation. The ratio for a module of Batch 1 and a module of Batch 2 for the whole period of outdoor testing is demonstrated in Fig.8.

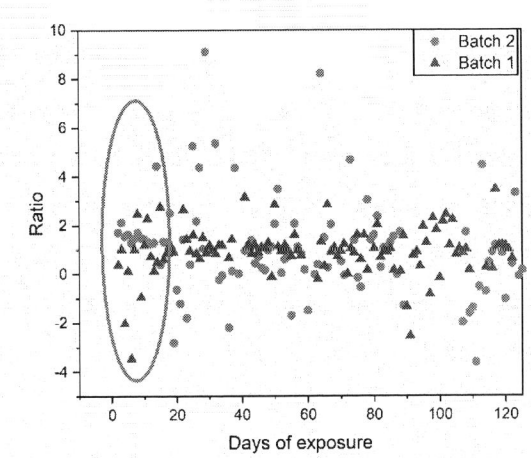

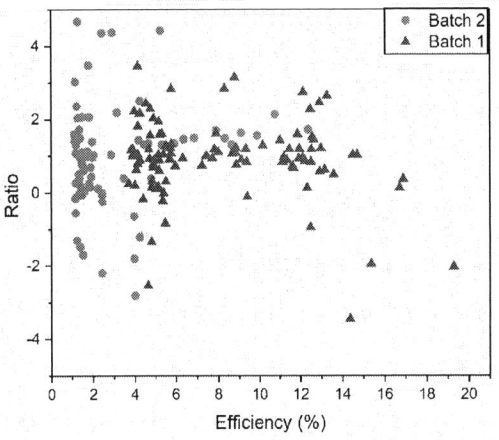

Fig. 8. Diurnal performance degradation-to-performance recovery ratio for a module of Batch 1 and Batch 2 for 120 days of outdoor exposure.

Fig. 9. Diurnal performance degradation-to-performance recovery ratio for a module of Batch 1 and Batch 2 against absolute efficiency.

The ratios in Fig.8 demonstrate a clear classification for the 2 Batches of modules only the first 15 days of outdoor exposure (indicated by the red circle in Fig.8). During this time frame the diurnal efficiency degradation-to- recovery ratio is higher for Batch 2 modules with a ratio value to be higher than unity. This gives evidence that diurnal efficiency degradation is higher than recovery which leads to overall performance reduction. At the same time period the ratio for Batch 1 modules is lower than unity indicating the dominance of recovery compared to diurnal efficiency degradation and the absence of significant performance degradation of the modules. For the rest of the testing period no classification of the two batches can be achieved according to their diurnal efficiency degradation-to- performance recovery values.

For a better understanding of the diurnal efficiency degradation-to-recovery ratio at different seasons, the ratio was plotted over efficiency (see Fig.9). At higher efficiency values the ratio of Batch 2 lies at higher values. This gives evidence that at early degradation stages where efficiency is high the diurnal degradation-to-recovery ratio is higher for Batch 2 modules. The ratio values of Batch 1 modules in more instances are lower than unity. Furthermore, a lot of ratio instances are present at efficiency values lower than 2% in Batch 2. The accelerated performance degradation the first days of exposure leaves the modules with low values of efficiency for the rest of the testing period.

In an attempt to compare the degradation of the modules tested outdoors with the concurrent degradation of the control modules kept indoors, the control modules were tested regularly under the solar simulator over the timespan of the outdoor studies. Great stability over time was obtained for all samples under test. No performance degradation was observed between the initial and final values of efficiency. The same holds for both the short-circuit current and open-circuit voltage characteristics of the reference samples.

The environmental conditions present over the testing period of outdoor exposure were collected at regular time intervals. Particularly, data such Global Normal Irradiance (GNI), ambient temperature, relative humidity were acquired every 5 minutes. The total irradiation received by the devices over time and the average ambient temperature present during the outdoor exposure of the two batches are plotted in Fig.10. Fig.10 demonstrates the higher total irradiation applied to Batch 2 and also the higher temperature levels present over the first 80 days of outdoor testing of Batch 2 modules. By comparing the degradation levels of the modules after the application of the same amount of irradiance it can be observed that the performance degradation of Batch 2 exposed at higher temperatures is higher. This fact supports the conclusion that the temperature and irradiance values at early degradation stages determines the daily efficiency degradation and recovery processes in perovskite mini-modules outdoors.

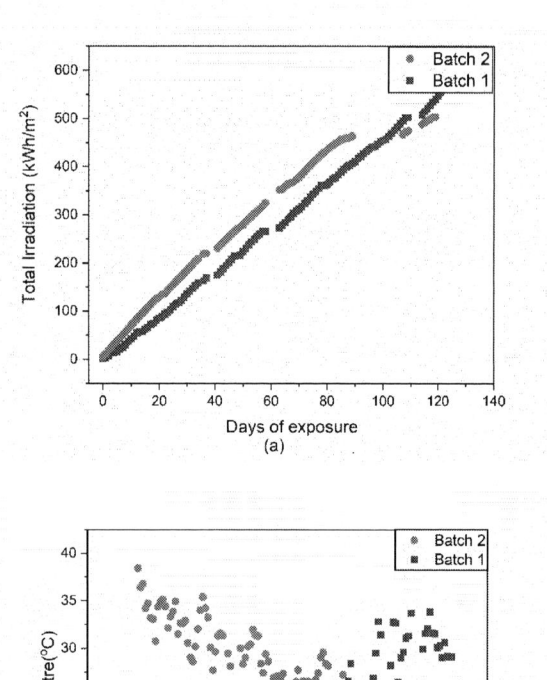

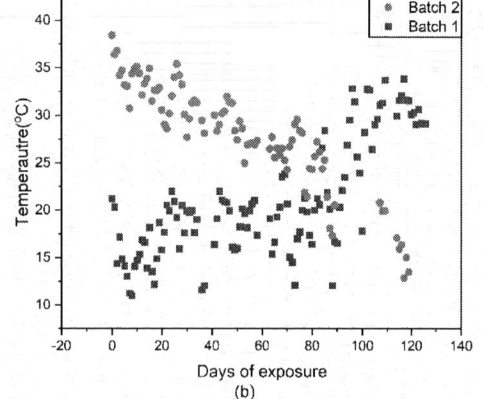

Fig. 10. (a) Total irradiance applied on the two batches of modules and (b) mean temperature levels present during the field testing of the batches.

IV. CONCLUSIONS

The long-term outdoor performance of identical perovskite mini-modules installed side-by-side at different seasons was studied. The higher irradiation and ambient temperature present during summer months were found to cause higher diurnal efficiency degradation-to-recovery ratio over the first days of testing, resulting in accelerated efficiency degradation in modules. This fact give evidence that module lifetime maybe determined from the temperature and irradiance levels the first days of testing and their values are critical for the perovskite modules operation. In all modules under test, diurnal efficiency degradation and performance recovery are higher at early degradation stages where absolute efficiency values from the modules are higher. In modules tested during summer months an almost linear relationship was found between diurnal efficiency degradation and performance recovery against absolute efficiency values. In modules tested during summer months, after the first weeks of outdoor exposure when the modules lost significant performance, low diurnal efficiency changes and recovery are obtained for the rest of the period.

Voltage losses were not significant for all samples under test. No significant differences were found between the voltage values of modules tested at different periods. Performance degradation is correlated mainly with current losses in the samples. More pronounced current losses were obtained in samples tested during summer months.

ACKNOWLEDGMENT

This work was funded through the European Regional Development Fund and the Republic of Cyprus in the framework of the project "DegradationLab" with grant number INFRASTRUCTURES/1216/0043.

REFERENCES

[1] S. Mazumdar, Y. Zhao, and X. Zhang, "Stability of Perovskite Solar Cells : Degradation Mechanisms and Remedies," *Front. Electron.*, vol. 2, no. August, pp. 1–34, 2021, doi: 10.3389/felec.2021.712785.

[2] S.-W. Lee *et al.*, "UV Degradation and Recovery of Perovskite Solar Cells.," *Sci. Rep.*, vol. 6, p. 38150, Dec. 2016, doi: 10.1038/srep38150.

[3] M. V. Khenkin *et al.*, "Dynamics of Photoinduced Degradation of Perovskite Photovoltaics: From Reversible to Irreversible Processes," *ACS Appl. Energy Mater.*, vol. 1, no. 2, pp. 799–806, 2018, doi: 10.1021/acsaem.7b00256.

[4] M. Prete *et al.*, "Bias-Dependent Dynamics of Degradation and Recovery in Perovskite Solar Cells," *Appl. Energy Mater.*, vol. 4, pp. 6562–6573, 2021, doi: 10.1021/acsaem.1c00588.

Applying unsupervised machine learning for the detection of shading on a portfolio of commercial roof-top power plants in Germany

Nicolas Holland*, Klaus Kiefer*, Christian Reise*, E.A. Sarquis Filho**, Bernd Kollosch***, Björn Müller****

* Fraunhofer Institute for Solar Energy Systems ISE, Heidenhofstrasse 2, 79110 Freiburg, Germany
** IDMEC, Instituto Superior Técnico, Universidade de Lisboa, Av. Rovisco Pais, 1049-001 Lisboa, Portugal
*** Pohlen Solar GmbH, Am Pannhaus 2-10, 52511 Geilenkirchen, Germany
**** Enmova GmbH, Basler Str. 115, 79115 Freiburg, Germany

Abstract—Obstacles that cast shading on commercially operated PV power plants can lead to a variety of issues, besides causing less energy , like false alerts in failure detection systems or skewed performace ratios. The detection and monitoring of shading effects using on-site inspections can be challenging, especially when one handles a large portfolio of power plants over a period of many years, since shading behaviours can also change over time. We apply an unsupervised method for detecting shading directly from power measurements to create so called shading masks, which make binary statements over whether or not a power plant or subplant is subject to shading at a given time. The shading masks are compared with the results of on-site inspections and they are used to create loss estimates.

Index Terms—shading, performance ratio, monitoring

I. INTRODUCTION

Shading on commercially operated PV power plants causes fewer produced energy and can cause a variety of issues. It can be the cause for fault monitoring systems to produce false alerts, as they may interpret power loss due to shading as inverter failures or in cases where operators are contractually obligated to guarantee certain performance ratios it can skew their evaluations.

We have been working with operators of a large portfolio of PV power plants in Germany, whose shading estimation often includes on-site visits and manual inspection of obstacles and their potential shading effect. Manual inspection however requires a lot of effort, especially when considering growing portfolios and the need to monitor changes in shading behaviour over time, e.g. due to growing trees. That is what creates the need for a method rooted in data analysis, to identify both shaded moments in time and shading caused energy loss. We revise the shading detection method developed in [1], which was applied on irradiance measurement sensors and transfer it to power plants. Since ground truth knowledge of shading events is not easily available, we rely on unsupervised learning to cluster measurement data into shading affected and unaffected classes. The method does

This work was partly financed by the Federal Ministry for Economic Affairs and Energy of Germany (BMWi) within the framework of the research program "Innovations for the energy transition" funded under the contract number 03EE1058.

Fig. 1. Model of a commercial roof-top PV power plant with partial shading. Shading is caused largely by trees and affects the lower right of the plant. Image by Google earth Version 9.152.0.1. In the result section we refer to this power plant as A.

take into account the reality of partial shading where only some inverters see losses due to shading while the power plant as a whole remains largely unaffected. This is especially useful for failure detection methods that compare individual inverters with each other, like examined in [2] and [3]. The result of our approach are shading masks, which correspond to solar angles, meaning moments in time, with shading labels, meaning statements of whether there is shading for a given solar position. For our evaluation we compare those data generated shading masks with on-site observations and look into shading loss estimations derived from the shading masks. Figure 1 shows a model of one of the portfolios power plants, with visible partial shading. This power plant is also included in this publications analysis and evaluation.

II. DATASET

In this publication we focus on the analysis done on commercially operated power plants with installed capacities ranging from 60 kWp to 110 kWp. All power plants are built with multiple inverters that experience different amounts shading and they were all reported to have energy loss issues due to shading. We refer to the power plants described here as A, B and C. The measurement data of the plants include 5 years of AC power measurements for the inverters as well as on-site irradiance measurements both having 5 minute values.

In addition to the on-site measurements we include modelled irradiance data, derived from EUMETSAT satellite images using the heliosat method [4]. The satellite derived irradiance is given as global horizontal irradiance (GHI) and is converted to plane-of-array irradiance (POA) using the direct/diffuse separation model by Ineichen et. al. [5] and the transposition model by Perez et. al. [6] so that it can be compared with both on-site irradiance and power measurements. For both models we use the implementations available in the open-source PVLIB software [7]. While satellite derived data is less correlated with the power measurements than the on-site irradiance sensor, there can be issues eg when the sensors are also affected by shading or otherwise malfunctioned, which can be detected by comparisons with satellite data.

III. METHOD

A. Shading Detection

The core idea of the method is to relate power measurements with solar positions and detecting shading by observing its negative impact on power production. Our method for detecting measurements affected by shading and deriving shading masks from it, is first conceived in [1], where it is used to detect shading in pyranometer measurements. We like to give a brief summary of how the method is used there before showing how we revised it to our use case:

1) Irradiance measurements are converted to clearsky index values.
2) These are then filtered for clearsky conditions.
3) And then they are sorted in a grid by solar azimuth and elevation.
4) On this grid the data is divided into shaded and unshaded clusters using k-means clustering [8].

One major revision when transfering this method to power measurements is to use performance ratio (PR) instead of clearsky indices. In that regard the method is quite similar to Bognar et. al. [9], who train support vector machines (SVM) on performance ratios between simulated and measured power to predicted solar angles for which power measurements are affected by shading. In our case the performance ratio is defined as

$$PR = \frac{\frac{P_{AC}}{P_{nom}}}{\frac{G_{poa}}{1000}} \qquad (1)$$

with P_{AC} being the measured AC power, either for the entire power plant or each inverter individually, P_{nom} the nominal power and G_{poa} the irradiance in plane-of-array. Note that for G_{poa} one can use either the measured irradiance or the modelled irradiance. When using modelled irradiance we refer to the performance ratio as PR_{sat}. Since the measurement data was available in 5 minute values, the PR can also be computed for 5 minute intervals. The performance ratio is sorted into bins along solar azimuth and elevation, where each bin has a size of $0.5°$ and all datapoints within one bin are averaged

to a single value. Outliers and small artefacts are removed by smoothing the PR data along the solar elevation axis

$$PR_{a,e}^{smo} = \frac{1}{2r+1} \sum_{\sigma=e-r}^{e+r} PR_{a,\sigma} \qquad (2)$$

with a, e being the position of PR in the azimuth, elevation grid and r the range over which the box smoothing is done. We set r to 5 bins or $2.5°$ elevation.

The classification of datapoints into shaded or unshaded is done by applying K-means on the smoothed PR grid. Since k-means is an unsupervised machine learning method, it is able to make predictions over whether an azimuth, elevation pair is likely affected by shading using just the PR without any prior assumptions. The choice of k=2 represents the idea that there are only two classes, shaded and unshaded. Additional classes like "likely shaded" and "unlikely shaded" can be achieved by setting k>2. Further robustness against outliers and artefacts is achieved by smoothing the resulting grid of classified bins using a Gaussian filter. It is this smoothed grid of classified bins that we refer to as a shading mask. Since azimuth, elevations bins can easily be converted back into timestamps / intervals, any given shading mask contains all information necessary to describe the shading situation for any point in time for a given power plant / inverter. Note that since for low solar elevation angles PR often times is less reliable, the resulting shading masks are also more likely to have artifacts.

B. Application of Shading Detection

We observe several applications for our shading detection. For tracking changes in shading behaviour over time one can compute shading masks for different years and compare changes in shading clusters. A growing tree for example will cause the respective shading cluster to grow.

The detection of shading and inverter failures can be mutually beneficial and has a "Chicken-Egg" aspect to it, in the sense that failure detection is improved by having shading masks and shading detection is improved by detecting failures first. We describe the effect of failure detection on shading masks in our quality control section. Here we want to mention, that in cases where failure detection is implemented by comparing individual inverters against ech other or a reference model, a sudden drop in power production in one inverter can easily be mistaken for a failure in that inverter, when in fact it may be caused by partial shading instead. Removing or flagging measurements at which shading is known to happen can therefore reduce false positive alerts in failure detection systems.

And the estimation of energy loss due to shading can be achieved by comparing inverters with detected shading and their unshaded counterparts. In cases where reliable simulations of a power plant are available, the difference between simulated and measured power during shaded periods is also a viable option for assessing energy loss. Without a reliable simulation the loss can be estimated by eg computing a factor

f similar to a performance ratio through dividing the measured AC energy $E_{P,u}$ with the irradiance energy density $E_{G,u}$ for time periods that are unshaded. By using this factor on the energy density on the shaded time periods $E_{G,s}$, we get an estimate of what the plant should have produced without shading, and the difference to the actual measurements $E_{P,s}$ yields the loss estimate

$$f := \frac{E_{P,u}}{E_{G,u}} \qquad (3)$$

$$Loss \approx f \cdot E_{G,s} - E_{P,s} \qquad (4)$$

C. Quality Control

The quality of the shading analysis is directly influenced by the quality of the measurements. One can easily imagine how it is impossible to detect shading in the performance ratio when the sensor used for irradiance measurements is effected by shading itself or if there are other failures which cause the performance ratio to be abnormal. We adopted these quality control measures for computing shading masks.

1) To assure accuracy of sensor measurements, they are compared with satellite derived irradiance in a scatter plot. Unusual biases, root-mean-square errors or visual artifacts are treated as an indication that the irradiance measurement might have issues. Figure 2 shows the comparison of irradiance values for power plant A. The layout of the power plant in figure 3 shows the location of the measurment sensor, so we can already assume that the irradiance measurment should not be affected by shading. The RMSE is 23.6% and the bias is 7.8%, both with respect to mean measurement of $281 W/m^2$, and both metrics were computed for hourly mean values. In cases where the sensor measurements appear to be distorted, satellite derived irradiance can be used as an alternative to sensor measurments for computing performance ratio. However, since the intrinsic deviations between ground and satellite based measurements, the resulting performance ratio as well as the shading mask will be of reduced quality and should also be subject to manual inspection before using in a real-time service.

2) We use the fault detection methods described in [3], which was another result of the overall project to identify and remove measurements that appear to be affected by inverter failures, because inverter failures which reduce PR can accidentally be identified as shading. Especially when they appear over a long period of time there is a good chance that the smoothing filters will not be able to remove them before applying the clustering. It is therefore always recommended to use failure filtered power measurements when computing PR time series.

3) And lastly, the resulting shading masks are subject to quality control themselves, which is usually done by manual inspection of the shading mask in form of a visualization. With a dataset containing measurements for all possible solar angles, one can start computing

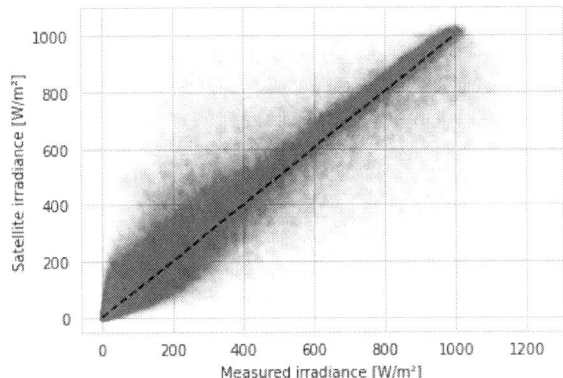

Fig. 2. Scatter plot of satellite derived irradiance and irradiance measurements for power plant A. Data from January 2015 to December 2020.

shading masks, which in extreme cases can be as little as 6 months of data. Using averages of multiple years of data has the advantage that the shading masks are more robust against outliers and artifacts, but one can miss changes over time. We therefore compute both multiple year averages for real-time services as well as shading masks from individual years and compare them.

IV. RESULTS

In the result section we go through results for different aspects of our analysis, namely the comparison with on-site assessments, energy loss estimations and observed shading mask changes over time for a few power plants from the portfolio.

1) Results for power plant A

For the power plant in figure 1 there is a layout of the plant including estimates for which areas are affected by shading made manually on site. Figure 3 shows the layout including two lines indicating the affected areas for given dates. The power plant consists of 5 inverters relating to 5 sub-areas ie. module strings. It is reported that two inverters are not affected by shading while the other three are affected to different degrees. The inverter connected to modules in the lower right of the plant is heavily affected through out the year. Our data derived shading masks are in agreement with this manual observation. Figure 4 shows the PR and shading masks for the most and least affected inverters, which are highlighted in figure 3. The heavily shaded area in the lower right only has PR values > 0.6 during the summer months with sufficiently large solar elevation angles. In figure 5 we compare power measurements of those two inverters on April 21. 2019, which happened to be a day with mostly clearsky conditions. The potential shading loss in the affected inverter is apparent and only during the peak of the day when the solar angles are slightly above the shading mask, does the power measurement appear unaffected.

2) Results for power plant B

978-1-7281-6118-1/22 $31.00 © 2022 IEEE

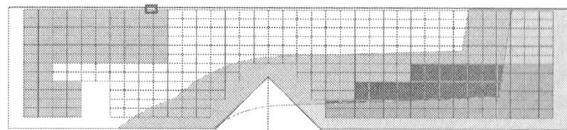

Fig. 3. Layout of a power plant that is affected by shading due to trees. An on-site inspection resulted in shading estimates in the form of lines / areas that are strongly affected given certain solar positions / times of year. The pink line is the estimate for 21. April and August, the grey line for 21 March and September. Highlighted are the modules connected to the two inverters referenced in Figure 4. The red square indicates the position of the irradiance measurement sensor.

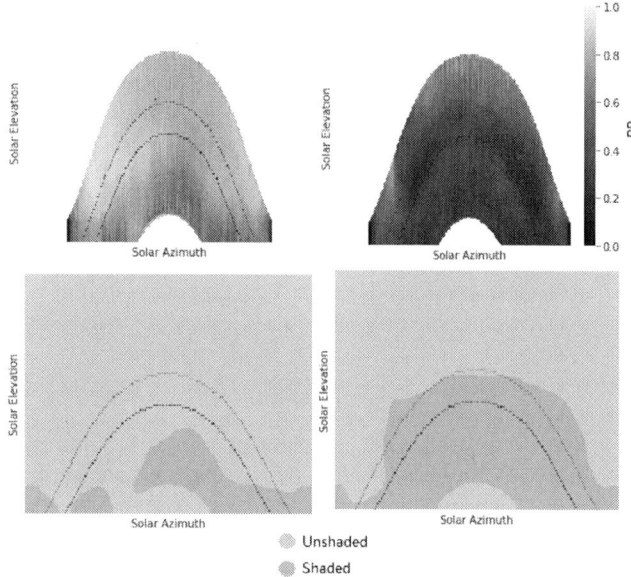

Fig. 4. PR grids and binary shading masks for inverters 1 (left) and 5 (right) of the power plant in figure 3. The solar positions for the dates referenced in the layout are highlighted.

Another interesting power plant that was affected by shading is seen in figure 6. Parts of the power plant in the lower left side are affected by shading in the morning due to a building, while parts in right side are affected by trees. In the afternoon, those left side modules remain unaffected, while the right side modules are seeing shading from that building. The shading masks for the inverters connecting those modules are shown in figure 7, and they capture the behaviour, that was observed by inspections on site.

How the shading affects the power production is shown in figure 8, which presents the AC output of the two inverters referenced in figure 7 for a day in July 2019 that had mostly clearsky condition. The highlighted parts show the timerange for which the shading masks predict shading and are derived from the shading mask. We use the formula (4) to estimate the shading loss of the two inverters for that day. The inverter referenced in the upper plot has an estimated loss of 6.8kWh which is about 8.7% of that days production and the lower

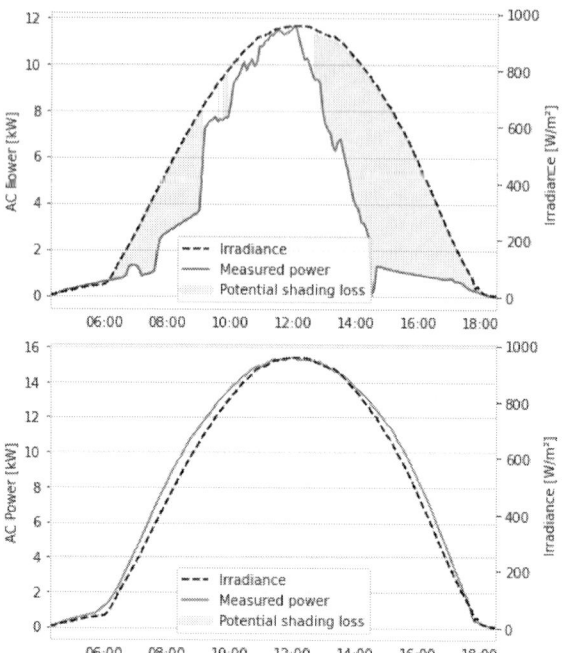

Fig. 5. Power and irradiance measurements for power plant A on April 21, 2019. The measurments appear to agree with our shading masks from figure 4.

Fig. 6. Model of a power plant that has part of its modules being affected by shading in the morning due to a building and another part being affected in the morning by trees and in the afternoon by that building. Image by Google earth Version 9.152.0.1. We refer to this power plant as B.

inverter has an estimated loss of 3.3kWh which is about 4.4%.

3) Results for power plant C

Lastly we want to look at a 60kWp roof-top plant that had an interesting behaviour over time. The power plant is depicted in figure 9 and on left hand side there is a few trees that can cast a significant amount of shading onto the power plant. When computing shading masks for the plant over time, we found that the shading masks appeard to have grown over time, which does agree with on-site observations, stating that the amount of shading appears to have become more of a problem over time. Figure 10 shows two shading masks for the years 2018 and 2019. The mask for 2019 is about 11.5 degree taller then what was found for 2018.

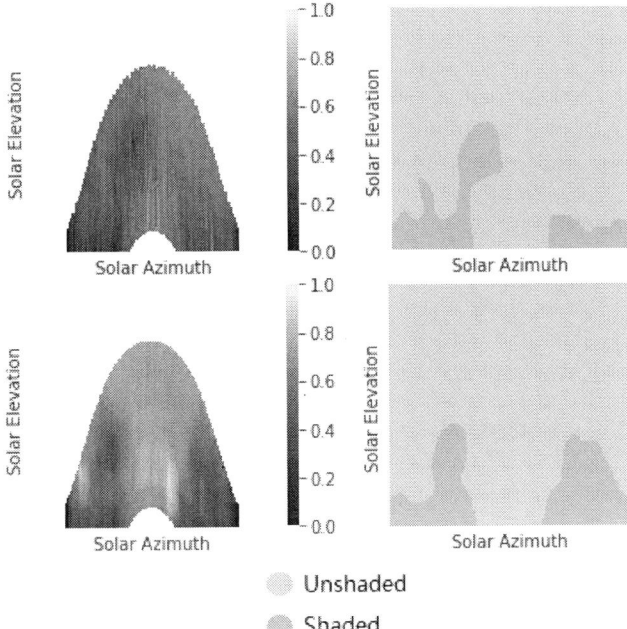

Fig. 9. Model of a power plant with trees on the left hand side. While on the 3D model the trees do not appear like a big problem, on-site visits have confirmed that they can cast a significant amount of shading onto the power plant. Image by Google earth Version 9.152.0.1. We refer to this power plant as C.

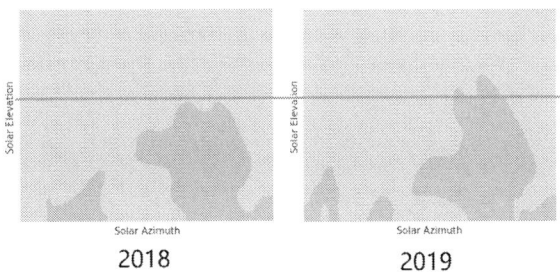

2018 2019

Fig. 7. PR grids and binary shading masks for two inverters of plant B in figure 6. The upper grids show the inverter connecting the modules in the lower left of the plant, the lower grids show the inverter connecting the modules on the right side of the plant.

Fig. 10. Two shading masks of power plant C, computed with data from 2018 and 2019. The orange line indicates how the shading mask has grown from one year to the next, by about 11.5 degree elevation. Note that the shading detected in the right hand side corresponds to shading occuring when the sun is in the west, which matches what we observe from the tree locations. Note that the masks were computed with satellite irradiance instead of on-site measurements.

REFERENCES

[1] Lorenz, Elke, et al. "High resolution measurement network of global horizontal and tilted solar irradiance in southern Germany with a new quality control scheme." Solar Energy 231 (2022): 593-606.

[2] Sarquis Filho, Eduardo A., et al. "Analysis of automatic fault detection methods for commercially operated pv power plants." Ratio 7: 18.

[3] Sarquis Filho, Abdon, et al. "Practical Recommendations for the Design of Automatic Fault Detection Algorithms Based on Experiments with Field Monitoring Data." arXiv e-prints (2022): arXiv-2203.

[4] Hammer A., Heinemann D., Hoyer C., Kuhlemann R., Lorenz E., M R. Solar energy assessment using remote sensing technologies Remote Sens. Environ., 86 (3) (2003), pp. 423-432, 10.1016/S0034-4257(03)00083-X

[5] Ineichen P., Perez R.R., Seal R.D., Maxwell E.L., Zalenka A. Dynamic global-to-direct irradiance conversion models ASHRAE Trans., 98 (1) (1992), pp. 354-369

[6] Perez R., Ineichen P., Seals R., Michalsky J., Stewart R. Modeling daylight availability and irradiance components from direct and global irradiance Sol. Energy, 44 (5) (1990), pp. 271-289, 10.1016/0038-092X(90)90055-H

[7] Holmgren W.F., Hansen C.W., Mikofski M.A. Pvlib python: a python package for modeling solar energy systems J. Open Source Softw., 3 (29) (2018), p. 884, 10.21105/joss.00884

[8] Lloyd S. Least squares quantization in PCM IEEE Trans. Inform. Theory, 28 (2) (1982), pp. 129-137

[9] Bognár A., Loonen R., Valckenborg R., Hensen J. An unsupervised method for identifying local PV shading based on AC power and regional irradiance data Sol. Energy, 174 (2018), pp. 1068-1077, 10.1016/j.solener.2018.10.007

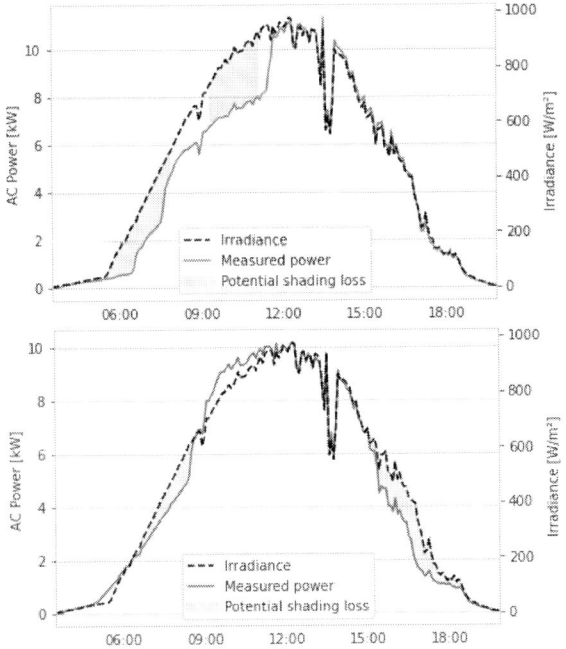

Fig. 8. Power production and irradiance measurements along with predictions by the shading mask for a clearsky day in July 2019 for power plant B. The two time series shown refer to the two inverters from figure 7 in the same order from up to down.

Electron selective TiO$_x$ contact for ultrathin amorphous germanium solar cells

Norbert Osterthun, Hosni Meddeb, Nils Neugebohrn, Kai Gehrke, Martin Vehse

German Aerospace Centre (DLR) Institute of Networked Energy Systems, Oldenburg, 26129, Carl-von-Ossietzky-Str. 15, Germany

Abstract — **Ultrathin amorphous germanium (a-Ge:H) absorbers enable innovative concepts like fully amorphous multi-quantum well solar cells, switchable PV windows or spectrally selective solar cells. Like in other solar cell technologies, TiO$_x$ is a suitable electron-selective and high-bandgap contact, enabling low parasitic absorption. However, due to quantum confinement effect the thickness of the ultrathin a-Ge:H absorber layer becomes an important parameter for the proper band alignment. We show that an absorber thickness of 5 nm leads to an s-shape in the J-V behavior, which can be eliminated by reducing the absorber size to 2 nm.**

Keywords—Titanium oxide, charge carrier selective material, amorphous germanium, quantum confinement, ultrathin solar cell numerical modeling

I. INTRODUCTION

Solar cells based on an ultrathin amorphous germanium nanoabsorber with a thickness below 20 nm enable various applications such as an optically switchable concept for solar windows [1, 2] or spectrally selective solar cells for greenhouses [2-4]. Due to the large Bohr radius of ~24 nm for germanium, quantum size effects can be exploited for layer thickness below this range [5-7]. The bandgap widening in the nanoabsorber from 0.98 eV for 20 nm to 1.56 eV for 2 nm allows a major gain in the open circuit-voltage and fill factor [6]. This enables a drastic reduction of the nanoabsorber thickness, while maintaining comparable power conversion efficiency level [6].

However, since such technology employs optical resonant nanocavity as light trapping scheme, further reduction of the parasitic absorption losses in the carrier transport layers is desired. This could enable not only an improvement of the photocurrent but also an enhancement of the light transmission in semitransparent solar cell configuration. One way to overcome the absorption losses is to integrate metal oxides as charge carrier selective contacts [8, 9]. These materials exhibit low absorption due to their high optical band gap. One of the most studied metal oxides in this context is titanium oxide (TiO$_x$). It is successfully applied in organic and dopant-free asymmetric heterocontact (DASH) silicon solar cells [10-13]. TiO$_x$ exhibits a wide range of work functions and doping levels [13-15]. Therefore it can be utilized as electron or hole selective contact depending on several criteria such as work function, deposition conditions, post-deposition treatments, negative fixed charge and capping material [10, 11].

In this study, we show that sputtered TiO$_x$ is a suitable electron selective contact for a-Ge:H solar cells and we examine the influence of the quantum confinement in the a-Ge:H on the extraction of charge carriers, when TiO$_x$ is used as electron carrier selective contact.

II. METHODS

A. Solar cell fabrication

A 10 x 10 cm² glass with a 1000 nm thick aluminum doped zinc oxide front contact (AZO) was used as substrate. The TiOx was deposited in a drive-by process via reactive magnetron sputtering using titanium tube targets using the process parameters shown in table 1. During the process, argon and oxygen where used as process gases. The argon flow was kept constant, while the oxygen flow was continuously adjusted with a closed-loop partial pressure control to achieve a constant oxygen partial pressure. The deposition rate of TiO$_x$ was determined by a single layer deposition on glass. The subsequent silicon and germanium layers were deposited by plasma enhanced vapor deposition and the silver back contact was added by electron beam evaporation. The detail process parameters for the other layers can be found in a previous publication [2].

Table 1 Deposition parameters for the TiOx contact.

	Power (W)	Ar flow (sccm)	Oxygen partial pressure (10-5 mbar)	Pressure (10-3 mbar)	Passes	Speed (m/min)
TiOx	1800	300	4	3.3	4	0.55

B. Electrical modeling

Electrical modeling was performed with the numerical software tool Afors-Het. The basic model for the a-Ge:H solar cell was introduced by Meddeb et al. and the modeling parameters for the silicon and germanium can be found there as well [6]. The n-doped silicon in the model was replaced by a TiOx layer, for which work functions of 3.8 and 4 eV were considered and an electron doping of 1E18 was assumed, which is consistent with literature [15]. The parameters can be found in Table 2.

Table 2 Modeling parameters for TiOx layer of a-Ge:H QW solar cell

Variable	TiOx
Thickness (nm)	7
Dielectric constant	11.9
Electron affinity (eV)	4
Band gap (eV)	3.4
Effective (Ec / Ev) density (cm^{-3})	1×10^{22}
Electron mobility (cm2/V·s)	5
Hole mobility (cm2/V·s)	1
Hole concentration (cm^{-3})	100
Electron concentration (cm^{-3})	1×10^{18}
Dangling bonds acceptor/donor (cm^3)	1×10^{17}
Urbach tails width Vb/Cb (eV)	0.03
Interface defects density (cm^{-2})	-
Interface model	Drift/ diffusion

III. RESULTS AND DISCUSSION

The layers structure of an a-Ge:H solar cell with TiO$_x$ as electron selective contact is shown in figure 1. The TiO$_x$ layer replaces an n-doped a-Si:H layer and is deposited directly on top of the AZO front contact. It is followed by intrinsic silicon layers sandwiching the a-Ge:H absorber. The solar cell is completed by a p-doped μc-Si:H and a silver back contact. The absorber thickness of the a-Ge:H absorber quantum well (QW) was chosen to be 2 nm and 5 nm in order to achieve different band gaps and an upward shifting of the conduction band in the a-Ge:H at reduced layer thickness [6]. For comparison with traditional thin film solar cells, a third solar cell was fabricated in which the a-Ge:H absorber was replaced with a 70 nm-thick a-Si:H absorber.

Glass	
AZO	1000 nm
TiOx	7 nm
i a-Si:H	5 nm
i a-Ge:H	2 or 5 nm
i μc-Si:H	5 nm
p μc-Si:H	7 nm
Ag	300 nm

Fig. 1 Layer stack of the a-Ge:H QW solar cell with TiOx as electron selective contact.

Figure 2 shows the experimental J-V-curves and the corresponding solar cell parameters of the three solar cells. The solar cell with the silicon absorber is s-shape free. Therefore, electron extraction with TiO$_x$ as charge carrier selective contact was successful and no barrier is formed at the TiO$_x$/a-Si:H interface. In contrast, the J-V-curve of the solar cell with 5 nm a-Ge:H absorber shows an s-shape, which does not occur in a-Ge:H solar cells with a conventional n-doped a-

Si:H as window layer [6]. This indicates a large offset in the band alignment, most likely at the interface between the amorphous silicon (a-Si:H) buffer and the germanium absorber. Another reason could be due to the requirement imposed by the charge transport in QW nanostructures where a sufficient internal electric field is crucial for the extraction and the sweeping of the charge carriers out, and hence prevent the capture and recombination of the escaped carriers from QW [16]. Interestingly, the s-shape can be removed when the a-Ge:H thickness is reduced to 2 nm. This suggests that the upward shift of the conduction band in the a-Ge:H absorber due to quantum confinement effects leads to an improved carrier collection by tunneling and thermal emission processes due to boost of the extraction and the sweeping of the charge carriers out of the a-Ge:H QW.

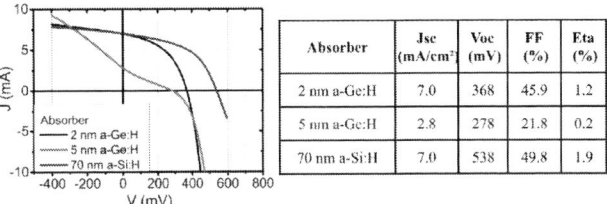

Absorber	Jsc (mA/cm²)	Voc (mV)	FF (%)	Eta (%)
2 nm a-Ge:H	7.0	368	45.9	1.2
5 nm a-Ge:H	2.8	278	21.8	0.2
70 nm a-Si:H	7.0	538	49.8	1.9

Fig. 2 Measured JV-curves and corresponding solar cell parameters

In order to further understand the difference between the 2 nm and 5 nm a-Ge:H absorber, we performed electrical modeling for the two a-Ge:H solar cells. The calculated band diagrams under equilibrium conditions are shown in figure 3. For the modeling, an electron affinity of 4 eV and an electron doping concentration of 1×10^{18} cm^{-3} is used for the TiO$_x$. For both absorber-thicknesses, a continuous conduction band can be observed at the interface between TiO$_x$ and the a-Si:H buffer. However, a conduction band offset ΔE_C occurs at the interface between the a-Si:H buffer and the a-Ge:H absorber. This effect is also known for a-Ge:H QW with n-doped silicon as contact [6]. Thus, the charge carriers are extracted not only by thermionic emission but also via tunneling. A non-optimal work function of TiO$_x$ as electron contact could influence the charge transport and hamper the selectivity [11]. The widening of the space-charge region width increases the tunneling distance compared to an n-doped silicon contact. This can lead to limited charge carrier extraction, resulting in an s-shape [17].

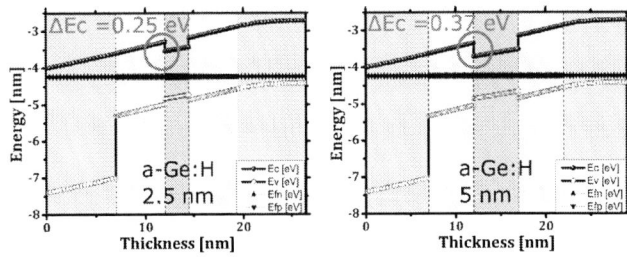

Fig. 3 Band-diagrams for the solar cells with 2.5 nm (a) and 5 nm (b) a-Ge:H and TiOx as charge carrier selective contact.

The calculated IV-curves are shown in figure 3. For an electron affinity with 4 eV an s-shape for the absorber thickness of 5 nm can be observed, while no s-shape for the 2 nm absorber thickness occurs. These results are in good agreement with measured results. One way to improve the solar cell performances is to reduce the electron affinity of the TiO_x to 3.6 eV. This would lead to an improved FF for the solar cell with the 2 nm-absorber and the s-shape issue of the solar cell with the 5 nm-absorber would be solved.

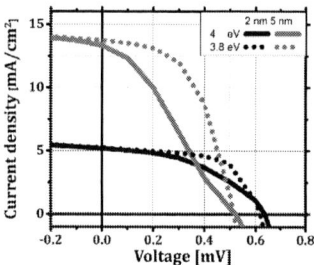

Fig. 4 Simulated jV-curves for solar cells with 2 and 5 nm absorber and for two different work functions (3.8 and 4 eV)

IV. CONCLUSION

In this work, we demonstrate, that TiO_x is a suitable electron selective contact for the ultrathin a-Ge:H solar cell. The replacement of the conventional n-doped silicon with TiO_x does not lead to a barrier at the interface between the TiO_x and the a-Si:H buffer. We found, that an s-shape in J-V curves occurs for a 5 nm-thick a-Ge:H absorber. This detrimental aspect can be prevented by reducing the absorber thickness down to 2 nm due to the size-dependent bandgap. Furthermore, we show, that an improvement of the TiO_x work function toward lower values down to 3.8 eV can lead to enhanced solar cells characteristics.

ACKNOWLEDGMENT

The authors thank C. Lattyak, M. Götz-Köhler, D. Berends, and U. Banik for helpful discussion. This work was funded by the Energy branch of German Aerospace Centre (DLR).

REFERENCES

[1] M. Götz, M. Lengert, N. Osterthun, K. Gehrke, M. Vehse, and C. Agert, "Switchable Photocurrent Generation in an Ultrathin Resonant Cavity Solar Cell," *ACS Photonics,* vol. 7, no. 4, pp. 1022-1029, 2020, doi: 10.1021/acsphotonics.9b01734.

[2] M. Götz, N. Osterthun, K. Gehrke, M. Vehse, and C. Agert, "Ultrathin Nano-Absorbers in Photovoltaics: Prospects and Innovative Applications," *Coatings,* vol. 10, no. 3, p. 218, 2020, doi: 10.3390/coatings10030218.

[3] N. Osterthun *et al.*, "Influence of spectrally selective solar cells on microalgae growth in photo-bioreactors," in *AIP Conference Proceedings*, 2021, vol. 2361, no. 1: AIP Publishing LLC, p. 070001.

[4] N. Osterthun, N. Neugebohrn, K. Gehrke, M. Vehse, and C. Agert, "Spectral engineering of ultrathin germanium solar cells for combined photovoltaic and photosynthesis," *Optics Express,* vol. 29, no. 2, pp. 938-950, 2021.

[5] E. G. Barbagiovanni, D. J. Lockwood, P. J. Simpson, and L. V. Goncharova, "Quantum confinement in Si and Ge nanostructures," *Journal of Applied Physics,* vol. 111, no. 3, 2012, doi: 10.1063/1.3680884.

[6] H. Meddeb *et al.*, "Quantum confinement-tunable solar cell based on ultrathin amorphous germanium," *Nano Energy,* vol. 76, p. 105048, 2020, doi: 10.1016/j.nanoen.2020.105048.

[7] H. Meddeb *et al.*, "Quantum Well Solar Cell Using Ultrathin Germanium Nanoabsorber," in *2020 47th IEEE Photovoltaic Specialists Conference (PVSC)*, 2020: IEEE, pp. 1149-1152.

[8] T. G. Allen, J. Bullock, X. Yang, A. Javey, and S. De Wolf, "Passivating contacts for crystalline silicon solar cells," *Nature Energy,* pp. 1-15, 2019.

[9] P. Gao *et al.*, "Dopant-Free and Carrier-Selective Heterocontacts for Silicon Solar Cells: Recent Advances and Perspectives," *Advanced science,* vol. 5, no. 3, p. 1700547, Mar 2018, doi: 10.1002/advs.201700547.

[10] T. Matsui, M. Bivour, P. Ndione, P. Hettich, and M. Hermle, "Investigation of atomic-layer-deposited TiOx as selective electron and hole contacts to crystalline silicon," *Energy Procedia,* vol. 124, pp. 628-634, 2017.

[11] T. Matsui, M. Bivour, P. F. Ndione, R. S. Bonilla, and M. Hermle, "Origin of the tunable carrier selectivity of atomic-layer-deposited TiOx nanolayers in crystalline silicon solar cells," *Solar Energy Materials and Solar Cells,* vol. 209, p. 110461, 2020.

[12] M. Mirsafaei *et al.*, "Sputter-Deposited Titanium Oxide Layers as Efficient Electron Selective Contacts in Organic Photovoltaic Devices," *ACS Applied Energy Materials,* vol. 3, no. 1, pp. 253-259, 2019, doi: 10.1021/acsaem.9b01454.

[13] J. Bullock *et al.*, "Dopant‐Free Partial Rear Contacts Enabling 23% Silicon Solar Cells," *Advanced Energy Materials,* vol. 9, no. 9, 2019, doi: 10.1002/aenm.201803367.

[14] X. Yin *et al.*, "19.2% Efficient InP Heterojunction Solar Cell with Electron-Selective TiO2 Contact," *ACS Photonics,* vol. 1, no. 12, pp. 1245-1250, Dec 17 2014, doi: 10.1021/ph500153c.

[15] M. C. K. Sellers and E. G. Seebauer, "Measurement method for carrier concentration in TiO2 via the Mott–Schottky approach," *Thin Solid Films,* vol. 519, no. 7, pp. 2103-2110, 2011, doi: 10.1016/j.tsf.2010.10.071.

[16] J. Nelson, M. Paxman, K. Barnham, J. Roberts, and C. Button, "Steady-state carrier escape from single quantum wells," *IEEE Journal of Quantum Electronics,* vol. 29, no. 6, pp. 1460-1468, 1993.

[17] R. Saive, "S-Shaped Current–Voltage Characteristics in Solar Cells: A Review," *IEEE Journal of Photovoltaics,* vol. 9, no. 6, pp. 1477-1484, 2019, doi: 10.1109/jphotov.2019.2930409.

Magnetic Field Imaging (MFI) of Shingle Solar Modules

Julian Weber, Stephan Hoffmann, Kai Kaufmann, Angela De Rose

Fraunhofer Institute for Solar Energy Systems, Freiburg, Germany

DENKweit GmbH, Halle (Saale), Germany

Within the last years, a new characterization method for solar modules called Magnetic Field Imaging (MFI) has been introduced. MFI reveals the strength as well as the direction of currents flowing within solar modules by analyzing the magnetic field distribution and thereby allows to trace back electrical defects. Within our work, we demonstrate how MFI can be exploited to characterize solar modules that contain stripe-like solar cells interconnected in a roof tile manner - so called shingle solar modules. In comparison to conventional solar modules, shingle modules yield the potential for higher power densities, exhibit a more homogeneous appearance and are less prone to power losses when being partially shaded. Due to these advantages, shingle solar modules are expected to gain a growing share of the solar module market within the next years, reaching 11% by 2027. As we demonstrate, MFI allows to detect different defects that are typical as well as unique for shingle solar modules, namely a local shunt by smudged electrically conductive adhesive (ECA), a systematic tilt between individual shingles, and locally poor shingle interconnection. Each defect type shows a fingerprint-like characteristic within MFI measurements and thus can be identified without ambiguity. Compared to electroluminescence (EL), MFI tends to display the defects more prominent and clearer. Furthermore, we demonstrate that MFI can be used to study the response of shingle modules to partial shading and observe that the module current bypasses shaded regions, as predicted by theory. We conclude that MFI would allow to investigate any kind of module under partial shading revealing the current flow through the solar cells, the connectors, and the bypass diodes.

978-1-7281-6118-1/22 $31.00 © 2022 IEEE

Evaluating Electroluminescence Imaging and Image Processing as a Quantitative Solar Cell Characterization Method

Meghan E Bush, Timothy J Peshek, Erica N Montbach

NASA Glenn Research Center, Cleveland, OH, United States

Mitigating dust accumulation on the surface of solar arrays is crucial for maintaining maximum power output. We propose investigating electroluminescence imaging paired with image processing as a means of evaluating various dust mitigation techniques. Image processing was able to clearly differentiate between pristine and dusted solar cells. Paired with traditional analysis techniques, this method proves to be a quick and powerful characterization tool.

CdTe-based photovoltaics using a CdTe/CdSe/CdTe absorber layer structure.

Jacob F Leaver, Ken Durose, Jonathan D Major

University of Liverpool, Liverpool, United Kingdom

One of the recent improvements in CdTe photovoltaics has been the inclusion of a $CdSe_xTe_{1-x}$ (CST) layer at the front interface of the absorber. The CST layer has a lower bandgap than CdTe, leading to increased current, whilst there is evidence that Se aids defect passivation which improves voltage. In this work we make use of a CdTe(A)/CdSe/CdTe(B) "andwich" structure with the aim of producing CdTe-based devices with a U-shaped bandgap profile. Highest device performances were achieved for CdTe(A) thicknesses in the range 0 nm to 100 nm, above which performance declined due to delamination from the substrate. Whilst the EQE curves did not indicate a U-shaped profile, the sandwich structure did enable higher CdSe incorporation into the CST layer.

Effects of Growth Temperature on Electrical Conductivity in Low-Dimensional, Ruddlesden-Popper Perovskite Thin Films Deposited by RIR-MAPLE

Niara E. Wright, Adrienne D. Stiff-Roberts

Duke University, Durham, NC, United States

Low-dimensional hybrid organic-inorganic perovskite (HOIPs) are being developed as wide bandgap, passivating layers on top of three-dimensional (3D) perovskite absorber layers to improve long-term device stability due to their greater resistance to moisture compared to the 3D HOIPs [1]. However, the same ligands that resist moisture, are also a barrier to charge transport, which can overshadow the significant advantages of using these materials as passivation layers in perovskite solar cells. Current work to overcome the charge transport challenge in low-dimensional perovskites focuses on: 1) achieving vertical orientation of quantum wells [2] and 2) increasing conductivity across the R-ligands and van der Waals gap via molecular modification [3]. Most studies focus on the first strategy and demonstrate vertical orientation, but do not show increased conductivity through the material, which is relevant for solar cell integration, thereby leaving a gap in the knowledge base. This work seeks to establish process-structure-property relationships to correlate a unique process (resonant infrared, matrix-assisted pulsed laser evaporation (RIR-MAPLE)) to the HOIP structure (grain size and orientation), and subsequently, a relevant solar cell property (conductivity) in the specific case of n = 1, 2D perovskites. The deposition technique used here - RIR-MAPLE - is a modified pulsed laser deposition technique that has been previously demonstrated for deposition of 3D HOIPs [4], 2D (n = 1) Ruddlesden-Popper HOIPs [5], 2D (n = 1) Dion-Jacobson HOIPs [6]. In this study, $(PEA)_2PbI_4$ thin films are deposited by RIR-MAPLE at four growth temperatures (~10oC, 25oC, 50oC, and 75oC). Williamson-Hall analysis is used forthe average grain size of each film and scanning electron microscopy is used for changes in planar and cross-sectional morphology as a function of growth temperature. J-V measurements in the vertical and lateral directions of the film are used to investigate conductivity. References [1]   G. Grancini and M. K. Nazeeruddin, Nat. Rev. Mater., vol. 4, no. 1, pp. 4-22, Jan. 2019. [2]   F. Zhang, H. Lu, J. Tong, J. J. Berry, M. C. Beard, and K. Zhu, Energy Environ. Sci., vol. 13, no. 4, pp. 1154-1186, Apr. 2020. [3]   J. V. Passarelli et al., J. Am. Chem. Soc., vol. 140, no. 23, pp. 7313-7323, Jun. 2018. [4]   E. T. Barraza, W. A. Dunlap-Shohl, D. B. Mitzi, and A. D. Stiff-Roberts, J. Electron. Mater., vol. 47, no. 2, pp. 917-926, 2018. [5]   A. D. Stiff-Roberts and N. E. Wright, in Low-Dimensional Materials and Devices 2019, 2019, vol. 11085. [6]   W. A. Dunlap-Shohl et al., Mater. Horizons, vol. 6, no. 8, pp. 1707-1716, Oct. 2019.

GeCl₄-based High Quality Ge epitaxy on Engineered Ge Substrates for Thin Multi-junction Solar Cells

Jinyoun Cho [1], Clément Porret [2], Valérie Depauw [2], Guillaume Courtois [1], Daniel McDermott [2], Roger Loo [2], Kristof Dessein [1], and Rufi Kurstjens [1]

[1] Umicore, Olen, Belgium, [2] imec (partner in EnergyVille), Leuven, Belgium

Abstract— **Germanium, a critical raw material, is used as a template for III-V epitaxial growth and as a bottom cell in multi-junction solar cells. To reduce the amount of germanium used, a detachable substrate is very interesting, especially if the Ge foil thickness can be adjusted as needed. In this study, the potential of GeCl₄-based epitaxy was demonstrated. A growth rate up to 190 nm/min and a thickness up to 15 μm were achieved. Detachable foils were then formed by porosification, annealing and epitaxial growth. The effective minority-carrier lifetime in the surface-passivated foil could be measured and proved quite high: over 25 μs. Results presented in this contribution confirm that the so-prepared foils constitute a suitable platform for the fabrication of high-performance multi-junction solar cells.**

Keywords—Germanium, GeCl₄, Ge epitaxy, Ge-on-Nothing, lift-off

INTRODUCTION

Today, sustainability in technology development is getting more attention to limit the environmental impact. In this respect, the efficient use and recycling of materials are becoming more important. Ge is a critical raw material [1] and its consumption for solar cells needs to be reduced because the Ge substrate is a hotspot for mineral depletion [2].

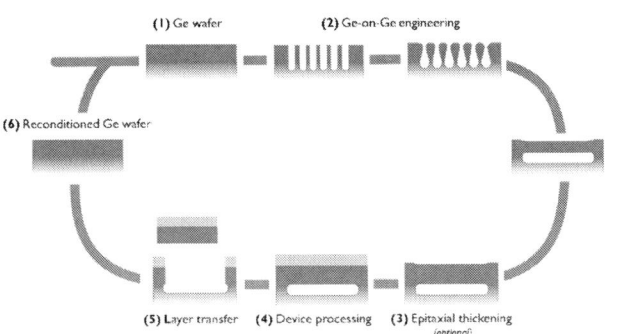

Figure 1. A schematic process flow for the Ge substrate engineering and re-use.

To meet these goals, engineered Ge substrates which enable substrate re-use have been developed [3], [4]. These substrates are suitable for making lightweight solar cells, and there is a potential for cost reduction by using less Ge.

Our approach, as shown in Figure 1, presented at the preceding two PVSC conferences, consists in i) forming a porous layer, and ii) annealing the sample to make about 1 μm thick foils [3], [4]. However, additional thickness is required to absorb enough IR light in the Ge bottom cell of a multijunction solar cell [5]. As a consequence, a relatively thick epi-layer must be grown fast and be of high quality at the same time.

Ge epitaxy on engineered substrates can be an alternative for thick Ge epi layer. Many studies about Ge epitaxy were carried out using GeH₄ as a precursor. However, the relatively low growth rate obtained with GeH₄ and Ge deposition on the cold reactor tube is a barrier for depositing μm-scale thick Ge films. GeCl4 is considered as an alternative precursor as it can be used for the deposition of thick epi layers with a limited risk for unwanted deposition on the cold reactor sidewalls. Compared to a growth rate of 36 nm/min for GeH₄, a higher growth rate of about 48 nm/min for GeCl₄ was achieved [6]. This increase, however, was not considered high enough.

In this study, a much higher Ge growth rate approaching 0.2 μm/min was demonstrated based on a GeCl₄ precursor, and a 15μm thick Ge layer was deposited on the engineered substrate. Then, this foil was passivated and the minority carrier lifetime therein was measured to evaluate the quality of epi layer.

GECL₄-BASED GE EPITAXY

A. Growth rate

As test vehicles for checking growth rate, epitaxial Ge layers were grown on p-type Si (001) substrates, with a thickness of 700 μm and a diameter of 200 mm. Prior to the epi processing, the wafers were patterned with oxide stripes, which allows to estimate growth rates and process uniformities. An ASM-Epsilon® 2000 system was used for the epitaxy process. H₂ was used as a carrier gas for GeCl₄. The flow rates mentioned in Figure 2 are nominal numbers which is expected as a mg/min however, those values are not verified yet.

978-1-7281-6118-1/22 $31.00 © 2022 IEEE

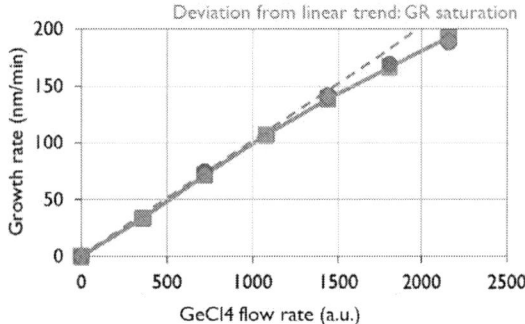

Figure 2. Ge epitaxial growth rate as a function GeCl₄ flow rate used during epitaxial growth. Data points from different runs are plotted on the graph and indicate a good process reproducibility. The blue dashed line corresponds to a linear trendline based on data points obtained for GeCl₄ flow rates ≤ 1000 a.u..

A deposition temperature of 550°C was in a mass-transport limited regime [7], thus, the growth rate scales (nearly) linearly with precursor flow (Figure 2). The epitaxial growth was performed by changing the precursor flow from 360 a.u. to 2160 a.u. while the pressure and temperature were fixed at 760 Torr and 550°C, respectively.

Although a small deviation from the dashed linear line (guide to the eye) was shown, a fairly linear relationship was obtained and reproduced. A maximum growth rate of about 190 nm/min is obtained for a GeCl₄ flow rate of 2160 a.u., limit set by our current hardware. This rate is about four times larger than the value reported in [6]. Furthermore, it can be expected that an even higher growth rate can be achieved with higher GeCl₄ flow rates because full saturation (plateau) in the curve was not yet reached.

B. Surface smoothness

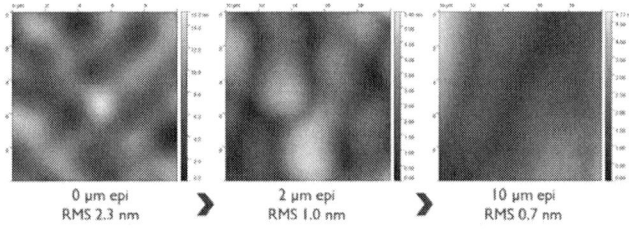

Figure 3. Root mean square (RMS) roughness measured by AFM (a) after surface restructuring shown in Figure 1, (b) after epitaxy of 2 μm, and (c) 10μm thick Ge layers.

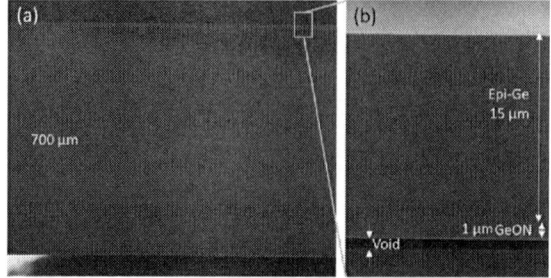

Figure 4. Cross section SEM images of (a) an engineered substrate including Ge-foil and (b) zoomed-in inspection focusing on the thickened Ge foil (1μm from the restructured layer + 15μm from Ge epitaxy).

For fabricating engineered Ge substrates, single side polished, p-type, 700 μm-thick, 8-inch (200 mm) round Ge wafers ((100), 6° offcut, (ρ) 8-30 mΩ·cm) were used.

The surface quality is an essential characteristic for the subsequent epitaxy of Ge and III-V layer. Especially, surface roughness after Ge epitaxy needs to be sufficiently low to enable epitaxial growth of high-quality III-V layers. Figure 3 describes the evolution of surface roughness with increasing the epitaxial thickness. Currently, the 8-inch Ge wafers have a roughness specification of less than 1 nm in RMS (root mean square) for the polished side. 2.3 nm in RMS roughness was measured after restructuring the porous structures (shortly, GeON: Ge-on-Nothing). However, the roughness reduces with increasing thickness of the Ge epitaxial layer. After 10 μm thick layer growth, the surface roughness returned to the level of the original bulk substrate.

General sample cross-sections are illustrated in Figure 4. About 16 μm thick foil (1 μm of GeON + 15 μm of Ge-epi) was well observed thanks to the void between the bulk and the foil (Figure 4 (a-b)).

FOIL QUALITY CHARACTERIZATION

Thick Ge epitaxial foil exhibited low surface roughness and good thickness uniformity [7]. The bulk part of the foil should be a high quality to be used as a bottom cell of a multijunction solar cell. In this respect, measuring an effective minority carrier lifetime (τ_{eff}) serves as a proxy to get access to the foil bulk quality (τ_{bulk}). τ_{eff} renders the contribution of the surface recombination velocity (S), the wafer thickness (W), and the bulk lifetime (τ_{bulk}). In the case of a wafer featuring asymmetrical surfaces, τ_{eff} can be expressed as in Equation (1).

$$1/\tau_{eff} = S_{Front}/W + S_{Rear}/W + 1/\tau_{bulk} \qquad (1)$$

In order to use τ_{eff} as an indication of τ_{bulk}, τ_{eff} should not be limited by high S. Thus, surface passivation of the foil is required. However, detaching and handling a free-standing 16-μm-thick foil is not straightforward. For the front side passivation process, the mother substrate provides the necessary mechanical support. For the rear-side passivation, bonding with a rigid carrier substrate and bonding adhesives were used.

European Space Agency (ESA) as part of the project with Contract No. 4000129924/20/NL/FE.

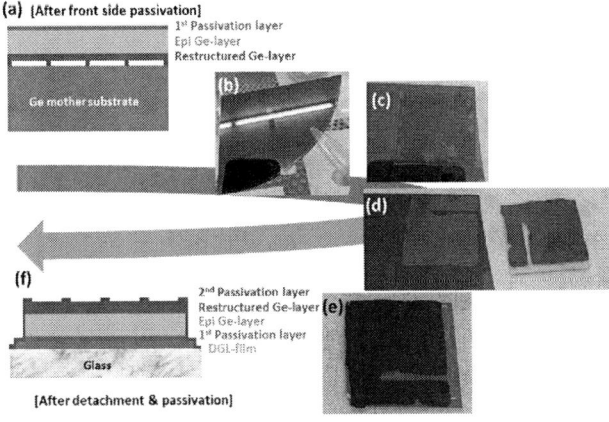

Figure 5. Sample images as per process progress. Schematic cross-section of the sample (a) after front side passivation and (f) after detachment and the rear side passivation. Sample pictures (b) after the front side passivation (≈ cross-section is shown in (a)). (c) after applying DGL-film on engineered area, (d) after loading glass substrate on DGL-film and detaching foil, and (e) after the rear side passivation (cross-section is shown in (f)).

Hydrogenated amorphous silicon (a-Si:H) is a good passivation material for Ge and low surface recombination velocity was reported [8]. An intrinsic/p-type a-Si:H stack was deposited by PECVD (Plasma-enhanced chemical vapor deposition) as a passivation stack for the experiment. To provide mechanical support to the foil, bonding adhesive film (DGL-film from Gel-Pak) and a 700 μm thick carrier glass substrate were used.

As shown in Figure 5 (a), a sample with an undoped, 15 μm thick epitaxial Ge layer on the restructured Ge layer of 1 μm was used for the test (hence a total thickness of 16 μm). We assume the epilayer as an intrinsic since no doping gas was used during growth. However, background doping might exist. Figure 5 (b-e) shows the progress of the bonding process using DGL-film. Although there was a crack in the foil, the fragment was sufficiently large for the lifetime characterization. Moreover, the rear side passivation was successfully achieved. In general, the GeON layer (restructured Ge layer) has a doping concentration similar to that of the bulk substrate, and the doped layer has a shorter effective lifetime compared to the undoped layer. Thus, the front and rear sides of the foil were not identical. More specifically, it is different in terms of the doping type (p-type (N_A 10^{17}-10^{18}cm^{-3} for reorganized layer) vs. intrinsic (N_i 2×10^{13} cm^{-3} for the epitaxial layer) and roughness (mirror-like surface vs. rougher surface with many broken support structures, i.e., pillars) (see Figure 5 (f)).

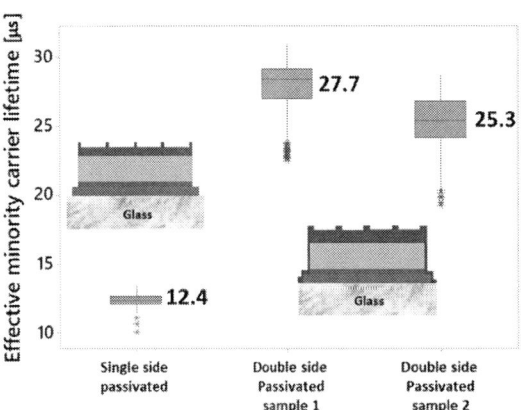

Figure 6. Effective lifetime comparison between single- and double-side passivated detached foils (the average value on the graph).

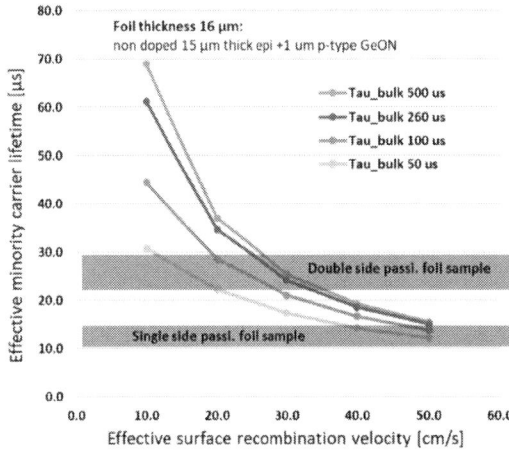

Figure 7. A simulated effective lifetime of the single- and double-side passivated samples, based on the foil thickness, surface recombination velocities, and bulk lifetimes.

Two types of samples were prepared in order to see the differences depending on the sample structure. The τ_{eff} of the foils was measured by Microwave detected Photo Conductance Decay (μ-PCD). Figure 6 shows that a single-side passivated sample exhibits about half of the lifetime of the double-side passivated samples. Based on this, it was found that the lifetime measurement tool is sensitive to the sample structures and showed a reasonable trend. However, the measured lifetime values of the foil was high compared to the effective 3-10 μs for the passivated bulk Ge reference wafer (p-type, Ga doped, 170 μm, ρ 10-20 mΩ·cm, lifetime data is not included here). Although the lower doping concentration results in the higher effective lifetime, it was necessary to check whether the measured high value is a theoretically meaningful number.

In order to compute the attainable effective minority carrier lifetime, reference values for a bulk lifetime and surface recombination velocity (S) were sought for. The literature [9] reported that a τ_{bulk} of Ge is to be larger than 500 μs for FZ (Floating-zone, N_A 7×10^{13} cm^{-3}) material and larger than 260 μs for Cz (Czochralski, N_A 1-1.6×10^{14} cm^{-3}) materials, respectively. According to the literature [8], a S of 17 cm/s was achieved on the lowly doped surface of N_A 2×10^{14} cm^{-3} based on the same

PECVD a-Si:H passivation. The current test foil consisted of a stack of a 15 μm thick undoped Ge epi-layer and a 1 μm thick GeON. Accordingly, most of the foil bulk assumed as an intrinsic. Considering possible τ_{bulk}, S and foil thickness, the measured effective lifetime values of the detached foil were in a reasonable range. As future work, understanding a bulk lifetime and surface recombination velocity in the foil will be characterized to understand the foil characteristics in details.

CONCLUSION

Ge-On-Nothing (GeON) approach is interesting approach for the substrate reuse. However, foil thickness after GeON process is not thick enough to absorb all illuminated light in the Ge bottom cell in multijunction solar cells. Accordingly, Ge epitaxy for thickening the Ge layer is necessary. A high growth rate of 190 nm/ min using $GeCl_4$ precursor was achieved and a further increase is likely possible. In addition, a high average effective minority carrier lifetime above 25 μs was demonstrated for a 15 μm thick Ge epi-layer grown on top of a GeON template. It implies that the restructured Ge is suitable for the subsequent epitaxy of high-quality Ge.

ACKNOWLEDGEMENT

This project was financially supported by European Space Agency (ESA) with a contract no. 4000129924/20/NL/FE.

REFERENCES

[1] E. Commission, "Critical Raw Materials Resilience: Charting a Path towards greater Security and Sustainability," Brussels: Europeancommission, 2020, pp. 69–82. [Online]. Available: https://ec.europa.eu/docsroom/documents/42849

[2] GreenSat_Consortium, "GREENSAT: EXECUTIVE SUMMARY REPORT [ESA contract no. TAS-GREENSAT-ESR]," 2019. [Online]. Available: https://nebula.esa.int/sites/default/files/neb_study/2506/C4000121203ExS.pdf

[3] G. Courtois, R. Kurstjens, J. Cho, K. Dessein, I. Garcia, I. Rey-Stolle, C. Algora, V. Depauw, C. Porret, and R. Loo, "Development of germanium-on-germanium engineered substrates for III-V multijunction solar cells," in *2020 47th IEEE Photovoltaic Specialists Conference (PVSC)*, Jun. 2020, vol. 2020-June, pp. 1053–1055. doi: 10.1109/PVSC45281.2020.9300462.

[4] R. Kurstjens, G. Courtois, J. Cho, K. Dessein, I. Garcia, I. Rey-Stolle, C. Algora, V. Depauw, C. Porret, and R. Loo, "GaInP solar cells grown on Ge-on-Ge engineered substrates," in *2021 IEEE 48th Photovoltaic Specialists Conference (PVSC)*, Jun. 2021, pp. 0175–0177. doi: 10.1109/PVSC43889.2021.9518407.

[5] I. Lombardero, M. Ochoa, N. Miyashita, Y. Okada, and C. Algora, "Theoretical and experimental assessment of thinned germanium substrates for III–V multijunction solar cells," *Prog. Photovoltaics Res. Appl.*, vol. 28, no. 11, pp. 1097–1106, 2020, doi: 10.1002/pip.3281.

[6] J. S. Park, M. Curtin, C. Major, S. Bengtson, M. Carroll, and A. Lochtefeld, "Reduced-pressure chemical vapor deposition of epitaxial Ge films on Si(001) substrates using GeCl4," *Electrochem. Solid-State Lett.*, vol. 10, no. 11, pp. 313–316, 2007, doi: 10.1149/1.2771069.

[7] D. McDermott, C. Porret, V. Depauw, G. Courtois, R. Kurstjens, A. Mac Raighne, C. Grogan, D. O'Brien, R. Langer, and R. Loo, "Germanium epitaxy using GeCl4: growth kinetics and material properties," 2021.

[8] N. E. Posthuma, G. Flamand, W. Geens, and J. Poortmans, "Surface passivation for germanium photovoltaic cells," *Sol. Energy Mater. Sol. Cells*, vol. 88, no. 1, pp. 37–45, Jun. 2005, doi: 10.1016/j.solmat.2004.10.005.

[9] E. Gaubas and J. Vanhellemont, "Dependence of carrier lifetime in germanium on resisitivity and carrier injection level," *Appl. Phys. Lett.*, vol. 89, no. 14, pp. 14–17, 2006, doi: 10.1063/1.2358967.

Investigation of Degradation Mechanisms in Carbon-Based Perovskite Solar Cells Exposed to Damp-Heat Conditions

Nikoleta Kyranaki, Cynthia Farha, Lara Perrin, Lionel Flandin, Emilie Planès, Lukas Wagner, Karima Saddedine, David Martineau, Stéphane Cros

Commissariat à l'Energie Atomique, CEA Grenoble, DRT/LITEN/DTS/SCPV/LCT, Le Bourget-du-Lac, France

Univ. Grenoble Alpes, Univ. Savoie Mont Blanc, CNRS, Grenoble INP, LEPMI, Grenoble, France

Fraunhofer Institute for Solar Energy Systems ISE, Freiburg, Germany

Solaronix S.A., Aubonne, Switzerland

The efficiency of perovskite solar cells is continuously increasing in the last decade, reaching up to 25.5%. However, further investigation is required to improve the stability of such devices, starting with the identification of the degradation mechanisms upon exposure to harsh environments. For an increased durability of the devices at high levels of moisture and temperature, a mesoporous structure was adopted, utilising carbon electrode and omitting the hole transport layer, which is susceptible to heat. Furthermore, an adequate encapsulation is required for additional protection against moisture ingress. This scientific work aims at identifying the degradation paths followed by perovskite cells when they are subjected to high temperature and humidity levels. To achieve this goal, full carbon-based perovskite (c-PSC) cells and modified devices were encapsulated with various concepts, and aged either at damp-heat conditions (85% RH / 85 °C) or at high temperature (85 °C) in nitrogen environment. Moreover, different characterisation techniques were compared for the identification of the degradation mechanisms. The results indicate that perovskite reacts with moisture after 1000 hours of damp-heat exposure, while heat degradation occurs much earlier. The latter is detrimental and requires further improvement of the durability of the perovskite absorber to heat.

Selective Etching of 6.1 Å Materials for Transfer-Printed Devices

Margaret A. Stevens[1], Jill A. Nolde[2], Shawn Mack[2], Kenneth J. Schmieder[2]

[1]NRC Postdoctoral Fellow residing at Naval Research Laboratory, Washington DC, 20375, USA
[2]U.S. Naval Research Laboratory, Washington, DC 20375 USA

Abstract— **To enable the micro-transfer printing of GaSb-based multijunction solar cells, we explored methods to maximize etch selectivity between InAsSb and AlGa(As)Sb layers lattice matched to GaSb. We report the etch rates and morphology of InAsSb and AlGa(As)Sb in dilute HF and citric acid solutions. Etch solution temperature and composition were varied to maximize the etch selectivity between the etch stop layer and sacrificial layer. We found a mixture of citric acid and hydrogen peroxide yielded the highest etch selectivity ratio, 852.3, between InAsSb and AlGa(As)Sb. We additionally found that unlike higher fractions of Al, $Al_{0.34}GaAs_{0.03}Sb$ did not appear to oxidize in dilute concentrations of HF and citric acid solutions, making it suitable as an etch stop layer.**

Keywords—GaSb, molecular beam epitaxy, epitaxial liftoff, etch selectivity, infrared materials.

I. Introduction

III-V semiconductors with lattice constants near 6.1 Å, lattice matched to GaSb, are important infrared absorbers and emitters. When employed in multijunction solar cells, they enable full spectrum solar energy harvesting, out to 2500 nm. For example, a mechanically stacked, 5J photovoltaic cell employing GaSb and InGaAsSb subcells achieved 44.5% efficiency with 744 suns solar intensity [1]. To construct this cell, a commercial 3J GaAs-based solar cell was removed from its host substrate and transfer-printed onto a 2J GaSb-based cell. To enable further module development of these high efficiency stacked cells, it would be advantageous to separate from the GaSb host substrate and transfer the entire stacked 5J to a low-cost, semi-insulating, and/or flexible handle.

Epitaxial liftoff (ELO), where an etch layer is selectively removed to separate a device from its growth substrate [2], can be used for bottom-up assembly of subcells with ideal band gap combinations [1], transform rigid solar cells into flexible form factors [3], or integrate cells with a thermally conductive backplane [4]. For GaSb-based devices, heterogeneous integration has been achieved by inverting the device, bonding the epitaxial surface to a new handle, and etching through the substrate [5]. However, this prevents the substrate from being reused and eliminates the additive manufacturing advantage of micro-transfer printing techniques. While complete substrate removal is undesirable in some aspects, it is typically the most successful method of separating a sample from its substrate due to the low etch selectivity between 6.1 Å semiconductors.

To enable successful ELO of GaAs-based devices, the near-infinite etch selectivity of GaAs and AlAs or InAlP allows the etch layer to be removed while preserving the device layers. In the antimonide family, etch selectivity ratios are typically on the order of ~100 and are typically paired with slow etch rates, ranging from ~0.1-10 nm/min [6]. This creates difficulties during device separation, as the etching solution will etch the device appreciably alongside the sacrificial layer. Despite low etch selectivity, there has been previous successful demonstrations of ELO for solar cells and detectors by exploiting the etch selectivity between lattice-mismatched InAs and AlGaSb with dilute hydrofluoric (HF) acid solutions [7], [8]. In addition to dilute HF, other potentially promising etch solutions include dilute ammonium hydroxide [9], [10] and citric acid based etchants [6], [11]. Citric acid-based etchants have demonstrated high etch selectivity, on the order of ~1000s, between InAs and $AlGa_{0.5}Sb$ [6]. However, high Al-containing layers are typically not desirable as etch-stop layers due to oxidation. Besides chemistry, etch solution temperature is another variable that can be adjusted to impact etch selectivity [12].

In this work, we explored the design space between etch selectivity, etch rate, and surface morphology to identify the ideal materials and chemistries to support lattice-matched GaSb-epitaxial lift off and transfer printing for solar cells. We report the etch rates and post-etch morphology of InAsSb and AlGaAsSb in solutions of dilute HF and of citric acid. The vertical etch rates, as an approximation for lateral etch rates, were measured as a function of temperature and etch solution composition to extract the etch selectivity ratio. Morphology of the epitaxial surface after etching was also documented, monitoring for increased surface roughness and for signs of oxidation on Al-containing layers.

II. Methods

Vertical etch test structures were grown by solid source molecular beam epitaxy (MBE) on a Riber 21 Compact DZ. In situ wafer temperature was monitored using a pyrometer calibrated to the (2×5) to (1×3) transition while heating GaSb under a Sb_2 overpressure equivalent to 1 ML/s (414 °C). The oxide was removed from epi-ready (100) GaSb substrate by ramping to 545-550 °C under Sb flux. The sample was cooled to 490 °C and a 500 nm GaSb buffer was grown at 1 ML/s with a V/III flux ratio of 1.14. $InAs_xSb_{1-x}$ layers were grown at 435 °C using an In growth rate of 1 ML/s, and a V/III flux ratio of

1.29, with the ratio of As to Sb tuned to achieve lattice matching to GaSb (x=0.09). $Al_xGa_{1-x}As_{1-y}Sb_y$ layers were grown at 495 °C using a total group III growth rate of 1 ML/s and a V/III flux ratio of 1.14. Both lattice mismatched (y=1) and lattice matched (y<1) were explored, with the flux ratio of Sb to As tuned to achieve lattice matching to GaSb. Samples for vertical etch tests were 200-1000 nm thick. After growth, samples were characterized using x-ray diffraction to measure lattice constant and strain state. Differential interference contrast (Nomarski) microscopy was used to monitor the surface quality and atomic force microscopy (AFM) was used to characterize surface roughness.

Samples were prepared for vertical etch tests using photolithography. Square openings with edge lengths of 200 µm aligned with the [110] and [1$\bar{1}$0] crystallographic directions were patterned across the material to measure etch depths. Etching chemicals included combinations of 49% hydrofluoric acid (HF), citric acid comprising of equal parts $C_6H_8O_7$ mixed with deionized water, and 30% hydrogen peroxide (H_2O_2). Etch solutions were mixed in a beaker surrounded by a water bath inside of an ultrasonic cleaning unit with temperature control. The temperature of the water bath was monitored with a thermometer and agreed with the temperature of the etch solution within +/− 5°C. To achieve the desired etch temperatures, the water bath was either heated using the temperature controls on the ultrasonic unit, or cooled with the addition of ice to the bath. The etch solution was mixed and placed in the water bath for 1 hour prior to etching in order for the temperature to stabilize. Samples were etched for a set period of time and then rinsed with DI water and dried with N_2. Etch depths were measured using profilometry and Nomarski images were recorded after the maximum etch depth.

III. RESULTS

To test the efficacy of dilute HF etches, $InAs_{0.91}Sb_{0.09}$ (hereafter known as InAsSb) and $Al_{0.34}Ga_{0.66}As_{0.03}Sb_{0.97}$ (hereafter known as AlGaAsSb) were etched in solutions of $HF:H_2O_2:H_2O$. Figure 1 shows the results of an etch solution of ratio 1:1:100 at temperatures of 25 °C and 10 °C for both InAsSb and AlGaAsSb. At 25 °C, the etch rate of AlGaAsSb was 11,484 nm/min while the etch rate of InAsSb was 88 nm/min. When the solution was cooled to 10 °C, the etch rates of both AlGaAsSb and InAsSb were slower than at 25 °C. The AlGaAsSb etch rate was found to be 8230 nm/min and the InAsSb etch rate was found to be 48 nm/min. This was a 28% difference from the etch rate at 25 °C for AlGaAsSb, and a 45% difference for InAsSb. Therefore, cooling the solution is an effective method of increasing the etch selectivity ratio for dilute-HF solutions. Ultimately, the etch selectivity ratio for AlGaAsSb/InAsSb in 1:1:100 $HF:H_2O_2:H_2O$ at 10 °C was found to be 171.5.

Figure 1 shows AFM images for InAsSb at both 10 °C and 25 °C. All samples were imaged with Nomarksi pre-etching and were found to be smooth and without high densities of surface defects. While cooling the etch temperature to 10 °C slowed the etch rate of InAsSb, it did not impact the surface morphology. The surface roughness of the InAsSb sample immersed in 10 °C solution was 1.04 nm RMS, while the roughness of the sample immersed in 25 °C solution was 0.996 nm. While either solution temperature would yield surfaces appropriate for transfer

printing, the higher etch selectivity ratios enabled by etching at 10 °C make it a better choice for ELO.

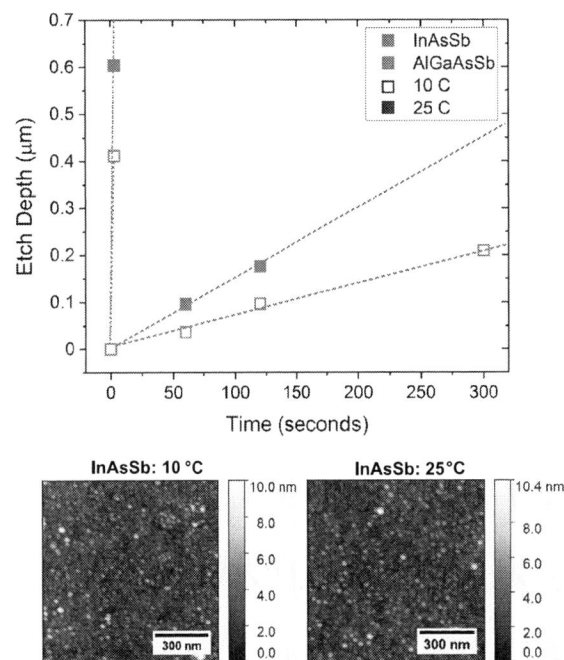

Fig. 1. Etch depth as a funciton of time for AlGaAsSb (teal symbols) and InAsSb (orange symbols) for solution temperatures of 25 °C (closed symbols) and 10 °C (open symbols). The dashed lines are guide to the eye. AFM images of InAsSb etched 209 nm at 10 °C (left), and InAsSb etched 176 nm at 25 °C (right).

Next, the concentration of the etch solution was varied while the temperature was held constant at 10 °C. The concentrations tested were 1:1:100, 1:10:100, and 10:1:100. Increasing the H_2O_2 fraction, 1:10:100, resulted in a faster etch rate of both InAsSb and AlGaAsSb, leading to an etch selectivity of only 62. Increasing the HF fraction, 10:1:100, resulted in slower etch rates for both InAsSb and AlGaAsSb, leading to an etch selectivity of 135. Thus, the highest etch selectivity, 171.5, was achieved with the 1:1:100 ratio solution.

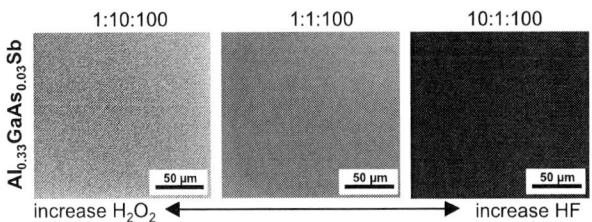

Fig. 2. Nomarski images of AlGaAsSb etched in different concentration of $HF:H_2O_2:H_2O$ at 10 °C.

Figure 2 shows Nomarski images of the AlGaAsSb layer when etched at 10 °C with the various etch solution concentrations. Increasing the hydrogen peroxide fraction, 1:10:100, resulted in increased surface roughness that was visible in the Nomarski images. Increasing the HF fraction, 10:1:100, resulted in a black AlGaAsSb layer when viewed with

Nomarski imaging. This could be due to dramatically increased surface roughness, or due to oxidation of the Al-containing layer. AFM measurements of the black surface showed a RMS roughness of 8.32 nm. Therefore, the black surface is likely due to oxidation. The morphology of the InAsSb surface did not vary significantly with solution concentration based on Nomarski imaging.

Etch rates were also tested as a function of Al composition. $Al_xGa_{1-x}Sb$ were grown with x=0.33, 0.63, and 0.85 and were etched in 1:1:100 $HF:H_2O_2:H_2O$ at 10 °C. Nomarski images of the surface are shown in Figure 3. The etch rate decreased as a function of Al composition. The sample with x=0.63 appeared very dark and rough in the Nomarski image after 3 seconds of etching in 1:1:100. The un-etched area did not undergo any surface changes. On the other hand, the sample with x=0.85 appeared equally dark both in areas that had been etched for 3 seconds and un-etched areas. This likely indicates that oxidation began to occur for x=0.85 before it was placed in solution, whereas x=0.63 seemed to oxidize dramatically during the etch. Oxidation of the surface appears to have impacted the etch rate, similar to what was seen for the 10:1:100 solution concentration. Based on the results shown in Figure 3, AlGa(As)Sb with x=0.34 is the best option to achieve the highest etch selectivity while maintaining the smoothest surface.

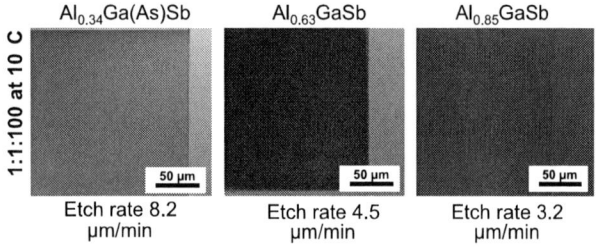

Fig. 3. Nomarski images of AlGa(As)Sb etched in 1:1:100 $HF:H_2O_2:H_2O$ at 10 °C. From left to right, the Al content of the alloy increases.

Since the Al-containing samples with x ≤ 0.34 etched with low-HF concentrations did not appear black under Nomarski imaging, they could be suitable as an etch stop layer with InAsSb acting as the sacrificial layer. Previous work has shown that the etch selectivity in citric acid between InAs and $AlGa_{0.5}Sb$ was 2949 with 1:1 $C_6H_8O_7:H_2O_2$ and 3892 with 5:1 [6]; however, high Al fractions are prone to rapid oxidation in air and solution. In this work, we investigated how lower Al fractions of AlGaAsSb, x=0.34, would react when etched with citric acid.

Figure 4(a) shows the etch rate of AlGaAsSb and InAsSb in 1:1 $C_6H_8O_7:H_2O_2$ at 25 °C. The etch rates in citric acid are considerably slower than that in dilute HF solutions. InAsSb etched at a rate of 137.8 nm/min and AlGaAsSb etched at a rate of 0.57 nm/min. This results in a selectivity ratio of 242. The inset AFM image shows the surface of AlGaAsSb after etching for 75 minutes. The surface roughness was 4.95 nm RMS. After 75 minutes of etching, the AlGaAsSb surface did not appear to be oxidized, but surface texture was present that was not there before etching. The surface roughness of InAsSb, the sacrificial layer in this chemistry, was 1.14 nm (AFM not shown). In all of the etch chemistries tested, the etched surface of InAsSb had a lower RMS than AlGaAsSb.

Figure 4(a) also shows the impact that decreasing temperature has on the etch rates of InAsSb and AlGaAsSb in 1:1 $C_6H_8O_7:H_2O_2$. Temperature was reduced from 25 °C to 5 °C by adding ice to the bath. In the 5 °C solution, the AlGaAsSb etch rate was 0.18 nm/min, while the InAsSb etch rate was 39 nm/min, resulting in an etch selectivity ratio of 215. Compared to the previously measured selectivity ratio of 242 at 25 °C, this shows it is not advantageous to decrease the solution temperature for etches involving citric acid and peroxide.

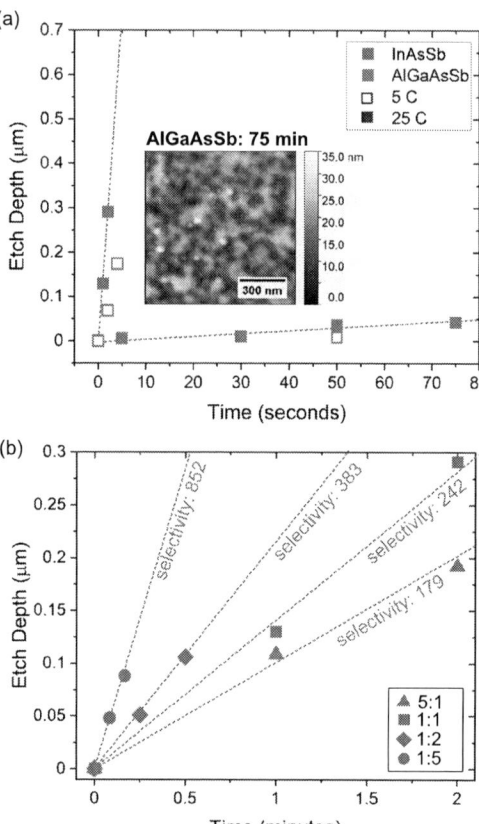

Fig. 4. (a) Etch depth as a function of time for 1:1 $C_6H_8O_7:H_2O_2$ at 25 °C and 5 °C for InAsSb (orange symbols) and AlGaAsSb (teal symbols). The inset shows an AFM image of AlGaAsSb after the maximume etch time. (b) Etch depth as a function of etching time for $C_6H_8O_7:H_2O_2$ concentrations of 5:1, 1:1, 1:2, and 1:5.

To maximize etch selectivity in solutions of citric acid and peroxide, we varied the solution concentration. Figure 4(b) shows the etch depth as a function of etch time for concentrations of 5:1, 1:1, 1:2, and 1:5 $C_6H_8O_7:H_2O_2$. Previous studies have shown that the etch selectivity between InAsSb and AlGaSb increases with increasing citric acid fraction [6]; however, we find the opposite to be true. Increasing the citric acid fraction from 1:1 to 5:1 decreased the selectivity from 242 to 179. Alternatively, changing the $C_6H_8O_7:H_2O_2$ fraction to 1:5 increased the selectivity to 850. This change in selectivity ratio mainly arises from an increase in the etch rate of InAsSb. For 1:1 the etch rate of InAsSb is 137.8 nm/min. At 5:1 concentration the etch rate decreases to 102.3 nm/min, while at 1:5 the etch rate increases to 553.9 nm/min. Therefore, we find that higher fractions of H_2O_2 in the solution results in faster etching of InAsSb, and as a result higher etch selectivity when etched against AlGaAsSb.

Table 1 summarizes the measured etch rates and selectivity ratios for promising etch chemistries found through the optimizations explored in this work. If achieving the highest etch selectivity is paramount, then the 1:5 $C_6H_8O_7$:H_2O_2 at 25 °C, with an etch selectivity of 852.3 between InAsSb/AlGaAsSb is the best choice. If low surface roughness of the etch-stop layer is the priority, then either 1:1:100 HF:H_2O_2:H_2O at 25 °C or 10 °C, with InAsSb surface roughness of ~ 1 nm RMS, is the best choice. Finally, if high etch rate of the sacrificial layer is desired, then 1:1:100 HF:H_2O_2:H_2O at 25 °C, with an AlGaAsSb etch rate of 11,484 nm/min, is the best choice. While each etch solution chemistry has its merits, 1:5 $C_6H_8O_7$:H_2O_2 at 25 °C can provide high etch selectivity, moderate sacrificial layer etch rate, and ~6 nm RMS etch-stop surface roughness, and therefore would be useful for ELO and micro-transfer printing of GaSb-based solar cells.

TABLE I. MEASURED ETCH RATES AND SELECTIVITY RATIOS

	InAs$_{0.91}$Sb nm/min	Al$_{0.34}$Ga(As)Sb nm/min	Selectivity
1:5 $C_6H_8O_7$:H_2O_2 25 °C	554	0.65	852.3
1:1:100 HF:H_2O_2:H_2O 10 °C	48	8230	171.5
1:1:100 HF:H_2O_2:H_2O 25 °C	88	11,484	130.5

IV. CONCLUSIONS

To support epitaxial liftoff of GaSb-based multijunction solar cells, we explored methods to maximize etch selectivity ratios between lattice matched layers of InAsSb and AlGaAsSb. We found for HF-based etch solutions, low HF concentrations lead to higher etch selectivity ratios and decreased surface roughness of the sacrificial layer. Al-containing sacrificial layers with x ≤ 0.34 were found to be stable with low concentrations of HF, but appeared to oxidize with higher HF concentrations. Increasing the Al content of the AlGaAsSb layer resulted in a decreased selectivity ratio, likely due to oxidation of the surface before or during the etching process. Decreasing the etch solution temperature also increased the selectivity ratio, and through these optimizations a selectivity ratio of 171.5 was achieved for AlGaAsSb/InAsSb in 1:1:100 HF:H_2O_2:H_2O at 10 °C. Overall, the highest etch selectivity

ratio achieved in this work utilized a citric-acid based etch, achieving a selectivity ratio of 850 for InAsSb/AlGaAsSb etched in 1:5 $C_6H_8O_7$:H_2O_2 at 25 °C. We find that increasing the H_2O_2 fraction results in faster etching of InAsSb, leading to a higher selectivity ratio between InAsSb/AlGaAsSb. Future work will apply these etches to perform lateral undercutting and transfer printing of GaSb membranes to validate results from the vertical etch tests.

ACKNOWLEDGMENT

This work was supported by the Office of Naval Research

REFERENCES

[1] M. P. Lumb et al., "GaSb-Based Solar Cells for Full Solar Spectrum Energy Harvesting," Adv. Energy Mater., vol. 7, no. 20, pp. 1–9, 2017, doi: 10.1002/aenm.201700345.

[2] E. Yablonovitch, T. Gmitter, J. P. Harbison, and R. Bhat, "Extreme selectivity in the lift-off of epitaxial GaAs films," Appl. Phys. Lett., vol. 51, no. 26, pp. 2222–2224, Dec. 1987, doi: 10.1063/1.98946.

[3] J. Yoon et al., "GaAs photovoltaics and optoelectronics using releasable multilayer epitaxial assemblies," Nature, vol. 465, no. 7296, pp. 329–333, May 2010, doi: 10.1038/nature09054.

[4] D. Kang, S. M. Lee, A. Kwong, and J. Yoon, "Dramatically Enhanced Performance of Flexible Micro-VCSELs via Thermally Engineered Heterogeneous Composite Assemblies," Adv. Opt. Mater., vol. 3, no. 8, pp. 1072–1078, 2015, doi: 10.1002/adom.201400521.

[5] Y. Zhang, A. Haddadi, R. Chevallier, A. Dehzangi, and M. Razeghi, "Thin-Film Antimonide-Based Photodetectors Integrated on Si," IEEE J. Quantum Electron., vol. 54, no. 2, pp. 1–7, 2018, doi: 10.1109/JQE.2018.2808405.

[6] G. C. DeSalvo, R. Kaspi, and C. A. Bozada, "Citric Acid Etching of GaAs1 − x Sb x , Al0.5Ga0.5Sb , and InAs for Heterostructure Device Fabrication," J. Electrochem. Soc., vol. 141, no. 12, pp. 3526–3531, 1994, doi: 10.1149/1.2059365.

[7] M. Zamiri et al., "Indium-bump-free antimonide superlattice membrane detectors on silicon substrates," Appl. Phys. Lett., vol. 108, no. 9, 2016, doi: 10.1063/1.4943248.

[8] V. S. Mangu et al., "Pixelated GaSb solar cells on silicon by membrane bonding," Appl. Phys. Lett., vol. 113, no. 12, p. 123502, Sep. 2018, doi: 10.1063/1.5037800.

[9] K. Yoh, K. Kiyomi, A. Nishida, and M. Inoue, "Indium arsenide quantum wires fabricated by electron beam lithography and wet-chemical etching," Jpn. J. Appl. Phys., vol. 31, no. 12 S, pp. 4515–4519, 1992, doi: 10.1143/JJAP.31.4515.

[10] K. Takei et al., "Nanoscale InGaSb heterostructure membranes on Si substrates for high hole mobility transistors," Nano Lett., vol. 12, no. 4, pp. 2060–2066, 2012, doi: 10.1021/nl300228b.

[11] O. Dier, L. Lin Chun, M. Grau, and M. C. Amann, "Selective and non-selective wet-chemical etchants for GaSb-based materials," Semicond. Sci. Technol., vol. 19, no. 11, pp. 1250–1253, 2004, doi: 10.1088/0268-1242/19/11/006.

[12] J. O'Callaghan et al., "Comparison of InGaAs and InAlAs sacrificial layers for release of InP-based devices," Opt. Mater. Express, vol. 7, no. 12, p. 4408, 2017, doi: 10.1364/ome.7.004408.

Inorganic perovskite solar cells with very high voltage and excellent stability against thermal and environmental degradation

Saba Sharikadze, Junhao Zhu, Ranjith Kottokkaran, Arkadi Akopian and Vikram Dalal

Iowa State University, Ames, Iowa, 50011, USA

Abstract

We report on inorganic CsPbBr₃ solar cells with very high open circuit voltages and excellent environmental stability. The cells were fabricated using vapor deposition. We show that by using an interfacial n-doped CdS (CdS:In) layer between the cell TiO2, we can obtain voltages of ~1.68 V, the highest ever reported in vapor deposited CsPbBr₃ material. A surprising phenomenon was that the crystal structure of the material, and the apparent bandgap, changed when a thicker CdS:In layer was used as the n layer. We also show that there is little environmental degradation in performance for a cell kept for 600 hours in room air, and even for a cell kept at 200 °C for 24 hours in air. The cells were deposited using sequential deposition in vacuum followed by anneals at 450 °C. We study both organic p layers (P3HT) and inorganic p layers (paste coated C).

Keywords—Perovskite solar cells, thermal and environmental stability, fundamental properties, high open circuit voltage

I. INTRODUCTION

Large bandgap perovskite solar cells are of significant importance for making tandem junction cells with Si acting as the bottom cell [1]. A major problem with this material has been a low open circuit voltage. In spite of the bandgap of the material being 2.3 eV, the best open-circuit voltage has been limited to ~1.62V [2-4]. This low voltage has been attributed to various causes, such as interfacial band misalignment at the heterojunction between the perovskite and the doped contact layers, excessive grain boundary recombination etc [5]. In this paper, we show that by changing the heterostructure between the normal n-conducting layer (TiO₂) and the perovskite layer, we can significantly change the open-circuit voltage. In particular, the use of doped CdS layers between the perovskite and TiO₂ leads to significant changes (by 60 meV) in open circuit voltage, with the voltage increasing as the thickness of CdS layer increases. We analyze this strange behavior, and show, by detailed device measurements, that the crystal structure of the perovskite itself is function of the interfacial layer, and that the bandgap of the perovskite changes as the thickness of the CdS layer increases.

We also study the thermal and environmental stability of the inorganic perovskite solar cells. It is well known that inorganic perovskite cells containing iodine decompose in moisture. In contrast, we show that Br containing perovskite cells are no affected by moisture, and we show that our cells show no degradation when exposed to air at room temperature for 500+ hours. Even ahigh temperature anneal at 200 °C in air leads to no perceived degradation in performance. We also show that one can make an all-inorganic cell by using C paste as the p layer, eliminating the organic P3HT p-heterojunction layer.

II. MATERIAL FABRICATION

The cells were fabricated on compact TiO₂ layer deposited on FTO glass substrates. Following the TiO₂ layer, an intermediate n-doped CdS layer was evaporated [6] prior to the deposition of the perovskite layer. The perovskite layer was deposited using sequential deposition [7] from CsBr and PbBr₂ sources, with the PbBr₂ layer being sandwiched between two CsBr layers. After deposition, the layers were annealed at 450 °C for 20 minutes followed by 40 mins at 350 °C. The p layer was either solution grown P3HT, or a C layer coated from a C paste. A final gold layer completed the cell. Fig. 1 shows the device structure.

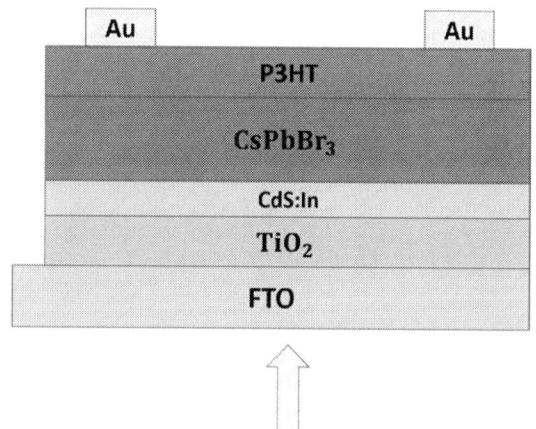

Fig.1. Schematic diagram of the device.

III. SOLAR CELL PROPERTIES

The cells were measured for their electrical performance using an ABET full-spectrum AM1.5 solar simulator. The cells were measured both inside and outside the glove box. Fig. 2 shows the I-V curves for our cells with varying thicknesses of the CdS:In layer. Note how increasing the thickness of the CdS:In layer increases the open circuit voltage - a very surprising development. We also study the quantum efficiency of the device, in particular the fall-off in QE at around 540 nm. The data is shown in Fig. 3, where we show that as the thickness of the CdS:In layer is increased, the fall-off becomes more pronounced, and shifts to 520 nm for larger thickness. This is a very remarkable and surprising result, which suggests that the effective bandgap of the material is changing, with the bandgap becoming slightly larger as the thickness of the CdS:In layer increases.

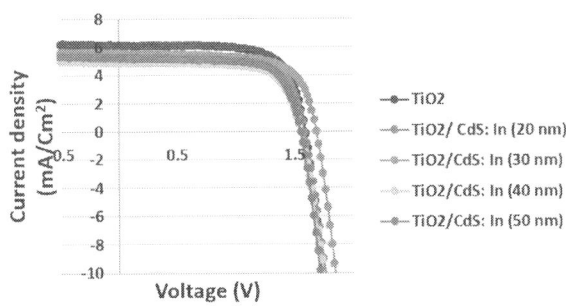

Fig.2. Light IV curves of the devices with various thickness of CdS:In

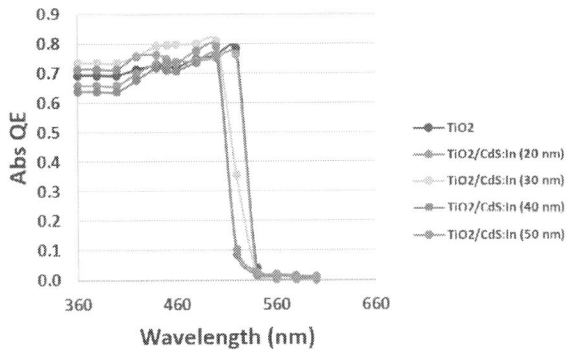

Fig.3. QE of the devices with various thickness of CdS:In

To understand this behavior, we study the crystal structure of the perovskite material, using x-ray diffraction. The data is shown in Fig.4. Note how additional planes appear in the spectrum as the thickness of the CdS:In layer is increased.

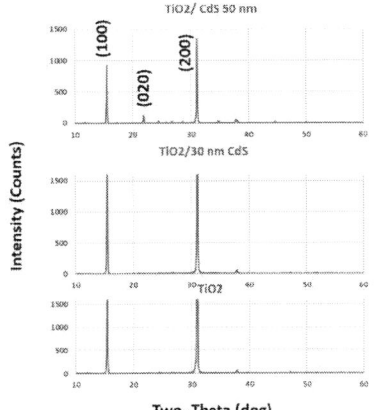

Fig.4. XRD spectrum of the perovskites

IV. STABILITY OF THE DEVICE AGAINST MOISTURE

The device was exposed to air for 1000 hours, and measured periodically for tis properties. In Fig. 5 we show the I-V curves for initial and 500 hours of exposure. From the figure, it is clear that there is virtually no degradation of the device when exposed to room air. We also studied the influence of higher temperatures on stability against moisture by subjecting a cell to room air at a cell temperature of 200 °C for 24 hours. The data is shown in Fig. 6. From Fig. 6, we can see that there is no degradation in cell performance even at 200 °C, thus confirming that both the material and the cell are stable at high temperatures, even when exposed to room air.

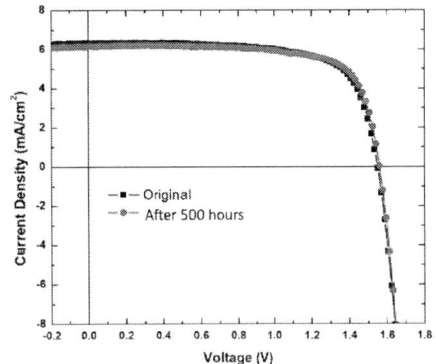

Fig.5. Light IV curves before and after 500 hours of air exposure.

978-1-7281-6118-1/22 $31.00 © 2022 IEEE

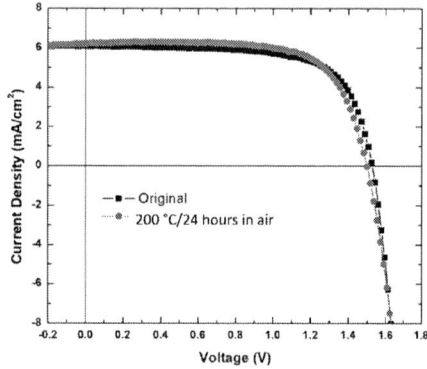

Fig.6. Light Iv curves before and after 200 °C for 24 hrs in air.

V. Devices With C Paste As P-layer

We studied the use of C paste as the p-heterojunction layer. the C paste was diluted with 1-Ethoxy-2- propanol and then coted using a MTI Carbon paste coater. After coating, the C was annealed in air at 130 °C for 30 minutes. In Fig. 7, we show the results on one of our first C paste device, showing a good voltage and a good low resistance contact. The results of further optimization and stability of such, all inorganic, devices are under progress and will be reported at the conference.

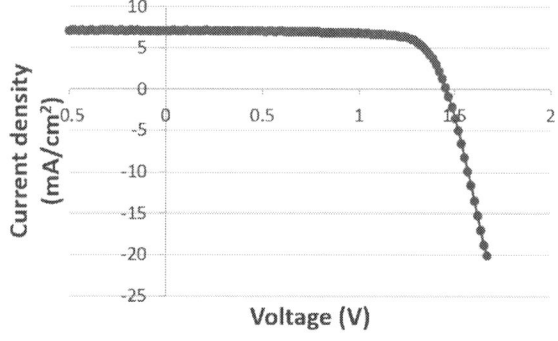

Fig.7. Light Iv curve of the device with carbon as P layer

VI. Conclusions

In summary, we report on an inorganic perovskite material and a device with record high open-circuit voltages, and excellent environmental stability, even at high temperatures. We show that interfacial layers change the crystal structure, the apparent bandgap, and the open circuit voltage of the device. we also show an all-inorganic device, with inorganic heterojunctions, with good device properties. The material was deposited using vacuum deposition techniques.

VII. Acknowledgments

We acknowledge funding from NSF for this project. we also thank Max Noack for his technical help.

VIII. References

[1] Saad Ullah et al. "All-inorganic CsPbBr3 perovskite: a promising choice for photovoltaics" *Mater. Adv.*, 2, 646-683, 2021.

[2] X. Li et al., "All-Inorganic CsPbBr 3 Perovskite Solar Cells with 10.45% Efficiency by Evaporation-Assisted Deposition and Setting Intermediate Energy Levels" *ACS Appl. Mater. Interfaces,* 11 , 29746 —29752, 2019.

[3] J. Zhu, R. Kottokkaran, S. Sharikadze, L. -P. Poly and V. Dalal, "Inorganic perovskite solar cells with high voltage and excellent stability against thermal and environmental degradation," *2021 IEEE 48th Photovoltaic Specialists Conference (PVSC)*, 2021, pp. 0425-0428

[4] Y. Zhao et al., "Using SnO2 QDs and CsMBr3 (M = Sn, Bi, Cu) QDs as Charge-Transporting Materials for 10.6%-Efficiency All-Inorganic CsPbBr3 Perovskite Solar Cells with an Ultrahigh Open-Circuit Voltage of 1.610 V" *Sol. RRL,* 3 , 507 1800284, 2019.

[5] H. Gaonkar, J. Zhu, R. Kottokkaran, M. Noack and V. Dalal, "Thermally Stable Inorganic Perovskite Solar Cells," *2020 47th IEEE Photovoltaic Specialists Conference (PVSC), 2020*, pp. 0167-0169.

[6] H. Gaonkar, J. Zhu, R. Kottokkaran, B. Bhageri, M. Noack and V. Dalal, "Thermally Stable, Efficient, Vapor Deposited Inorganic Perovskite Solar Cells" *ACS Applied Energy Materials* 3 (4), 3497-3503, 2020

[7] R. Kottokkaran, H. Gaonkar, B. Bagheri, and Vikram L. Dalal, "Efficient p-i-n inorganic CsPbI3 perovskite solar cell deposited using layer-by-layer vacuum deposition" *Journal of Vacuum Science & Technology A* 36, 041201, 2018

23.5% Efficiency GaAs Solar Cells Fabricated with Low-cost, Non-vacuum Processing

Phillip R Jahelka, Harry A Atwater, Aaron Ptak, Christiane Frank-Rotsch, Frank Kiessling, Cora Went, Michael Kelzenberg

California Institute of Technology, Pasadena, CA, United States

National Renewable Energy Laboratory, Golden, CO, United States

Leibniz-Institut für Kristallzüchtung, Berlin, Germany

We report advances in low-cost GaAs photovoltaic device processing. First, we developed an open-tube zinc diffusion technique for forming p-type layers with sheet resistance less than 100 ohms per square on GaAs and InP. Second, we use this technique along with chemical surface passivation to fabricate GaAs solar cells that, uncertified, achieve V_{oc} = 964 mV V_{oc}, FF = 82% FF, J_{sc}, = 29.8 mAcm-2 J_{sc}, and 23.5% efficiency. Third, we discovered GaAs cells with long hole diffusion length can be made byfabricated from wafers growing obtained from GaAs ingots grown in a part of the phase diagram thate minimizes the EL2 defect. Fourth, we show an in-air, precious metal-free process for making ohmic contacts to n-type GaAs.

Metallic Lead Recovery Via Electrowinning from Lead Acetate for Silicon Solar Module Recycling

Natalie Click and Meng Tao

Arizona State University, Tempe, AZ 85287, USA
Nmann6@asu.edu   Meng.Tao@asu.edu

Abstract— A new chemistry for lead recovery from silicon module waste is proposed. It involves leaching lead in dilute acetic acid followed by electrowinning to directly recover metallic lead. Cyclic voltammetry reveals that the lead reduction potential is −0.39 V vs silver/silver chloride in a 0.01 M aqueous lead acetate solution. Structural and chemical analyses confirm the presence of lead on the working electrode after one hour of chronoamperometry at −0.8 V. A small amount of a deposit, most likely lead oxide, is found on the counter electrode.

Keywords—silicon module recycling; lead electrowinning; lead recovery; acetic acid; lead leaching

I. INTRODUCTION

In 2016, a report published by the International Renewable Energy Agency (IRENA) projected that the world would see between 60–78 million tons of solar module waste cumulative by 2050 [1]. Not only is it impractical to landfill this amount of solar waste but landfilling modules would counteract their 'green' intentions. The creation of a recycling technology for solar modules would not only engender new jobs but could also close the loop for solar modules making them circular. IRENA estimated that the raw materials recovered from solar module waste could generate $15 billion in revenue [1]. With a more advanced recycling technology, we estimate the revenue to be around $50 billion [2].

Out of all the metals present in silicon modules, lead (Pb) poses the greatest risks to humans. Lead is a known carcinogen and teratogen, meaning improper disposal of silicon modules in landfills could be a liability. Lead recovery from silicon modules has been realized by first leaching it in nitric acid (HNO_3) and then recovering it by precipitation. The recovery rate by this method is low at 93% [3]. The formation of lead precipitates requires further processing to recover metallic lead, which adds to the cost for recycling. A possible alternative to this method is the use of electrowinning, which may result in the direct recovery of metallic lead. Our study shows that lead can be electrowon from a nitric leachate simultaneously with copper at a 99% recovery rate [4]. However, the recovered lead is lead oxide (PbO) requiring further processing for metallic lead. Additionally, nitric acid is highly dangerous and releases harmful nitric and nitrous oxides. Therefore, electrolytic recovery of metallic lead from a 'safer' chemical is preferred.

In this paper, dilute acetic acid (CH_3COOH) is proposed as an alternative medium for the recovery of lead from solar module waste. Cyclic voltammetry is used to identify the reduction potential for lead in a lead acetate system, and SEM and EDX confirm metallic lead formation on the working electrode.

II. EXPERIMENTAL

A cyclic voltammogram was generated for a 0.01 M $Pb(CH_3COO)_2$ solution. This was done in order to determine the redox potential for lead in the acetate leachate. 0.1626 g of lead(II) acetate trihydrate (Alfa-Aesar, 99.0–103.0%, granular) was dissolved in 50 mL DI water in a glass beaker. The working electrode was a thin copper rod (~0.6 mm diameter), and the counter electrode was a thicker copper rod (~2 mm diameter). The reference electrode was silver/silver chloride (Ag/AgCl). A Gamry Reference 3000 potentiostat was used, and the scan rate for the cyclic voltammogram was 100 mV/s.

A Hitachi S-4700 field emission scanning electron microscope (SEM) equipped with an energy-dispersive x-ray spectrometer (EDX) was used to perform morphological and compositional analyses for working and counter electrodes before and after lead electrowinning. 0.1628 g of lead(II) acetate trihydrate was dissolved in 50 mL DI water in a glass beaker. A piece of copper wire measuring 50.8 mm long and 0.9 mm in diameter was used for the working electrode. A copper rod measuring 50.8 mm long and 1.9 mm in diameter was used as the counter electrode. The lead deposit for analysis was obtained by applying −0.8 V vs. Ag/AgCl on the copper working electrode for 1 hour.

A third experiment was run to increase the lead recovery rate. 0.1624 g of lead(II) acetate trihydrate was dissolved in 50 mL DI water in a glass beaker. A copper wire (0.9 mm diameter) was used as the working electrode, and a copper rod (2 mm diameter) was used as the counter electrode. A thick coating of lead was deposited on the working electrode by applying −0.8 V vs. Ag/AgCl for 5 hours.

III. RESULTS AND DISCUSSION

A. Cyclic Voltammetry

A cyclic voltammogram was created to identify the redox potential window for the lead(II) acetic ($Pb(CH_3COO)_2$) system (Fig. 1). The standard potential for the reaction

$$Pb^{2+} + 2e^- \rightarrow Pb \qquad (1)$$

is −0.1262 V vs. standard hydrogen electrode (SHE) [5]. Kraft et al [6] reported that the reduction potential of lead dissolved in 1% and 10% acetic acid is at –0.3 V vs. normal hydrogen electrode. As shown in Fig. 1, a distinct peak is observed at –0.39 V vs. Ag/AgCl on the forward scan (from –0.1 V to –0.6 V). The second peak starts around –0.48 V and could indicate a consecutive reduction reaction, though further electrochemical analyses are needed [7]. A strong reverse sweep peak is observed at –0.22 V.

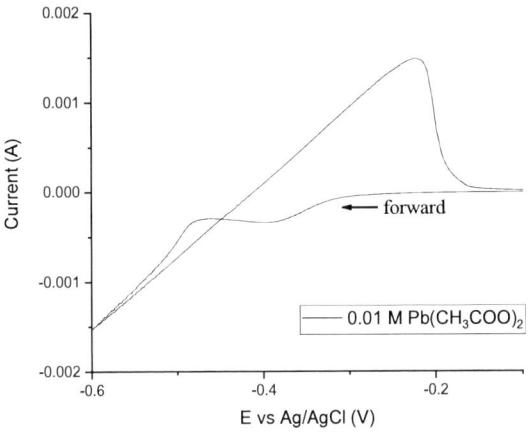

Figure 1: Cyclic voltammogram for 0.01 M Pb(CH₃COO)₂ system

The peak at –0.39 V is likely the lead reduction peak. To support this hypothesis, the Nernst Equation was used to estimate the reduction potential of lead with the following assumptions: the activity of metallic lead $a_R = 1$, the activity of dissolved lead ions $a_O = 0.01$ M, temperature T = 298.15 K, the valence of dissolved lead ions n = 2 (Pb²⁺), and the standard reduction potential of Pb²⁺ $E° = -0.1262$ V vs. SHE [5]. It was found that such a lead acetate solution has a reduction potential of –0.185 V vs. SHE. The difference in standard reduction potential between Ag/AgCl and SHE is 0.20 V [8], so the experimental lead reduction peak at –0.39 V vs. Ag/AgCl agrees with the theory.

B. SEM and EDX Analyses

SEM was used to study the morphology of the deposit formed on the working and counter electrodes. Fig. 2 shows the copper working electrode before lead electrowinning and Fig. 3 shows it after lead deposition. The grooves seen in Fig. 2 are from manufacturing of the copper wire. The crystallites seen in Fig. 3 were identified as lead using EDX. The surface composition of the working electrode after lead electrowinning reported from EDX is shown in Table I. Lead accounts for 30.81% of the surface composition. The recovery rate of lead was estimated by weighing the working electrode before and after electrowinning. The net weight of lead recovered was 0.0023 g, or a 2.22% recovery rate, after 1 hour of electrowinning. For comparison, the composition of the copper working electrode before lead electrowinning is given in Table II. No lead is found before electrowinning. The counter electrode is also copper and the compositions before and after lead deposition on the working electrode are shown in Tables

III and IV. Figs. 4 and 5 show the surface morphology of the counter electrode before and after lead deposition.

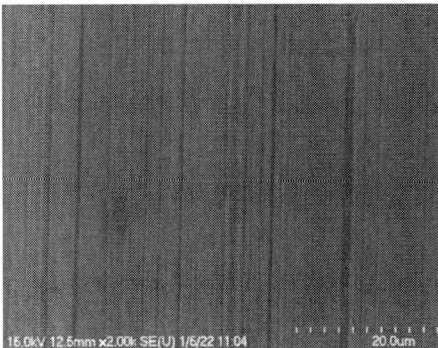

Figure 2: SEM image of the copper working electrode before lead deposition.

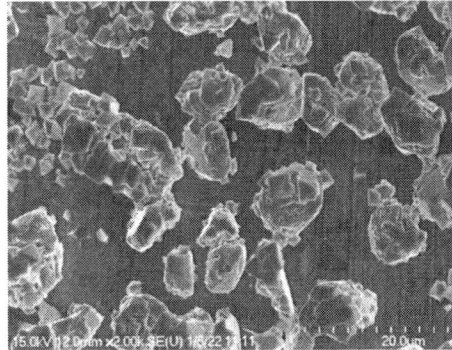

Figure 3: SEM image of the copper working electrode after lead deposition with lead crystallites.

TABLE I: SURFACE COMPOSITION OF THE COPPER WORKING ELECTRODE AFTER LEAD ELECTROWINNING BY EDX.

Element	Weight %
Cu	69.19
Pb	30.81

TABLE III: SURFACE COMPOSITION OF THE COPPER WORKING ELECTRODE BEFORE LEAD ELECTROWINNING BY EDX

Element	Weight %
Cu	100.00

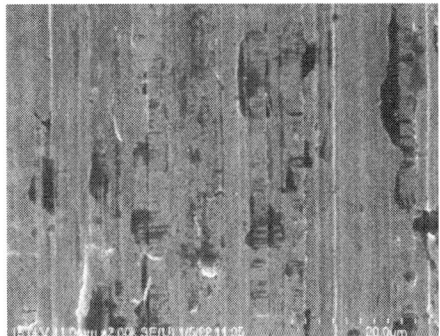

Figure 4: SEM image of the copper counter electrode before lead electrowinning.

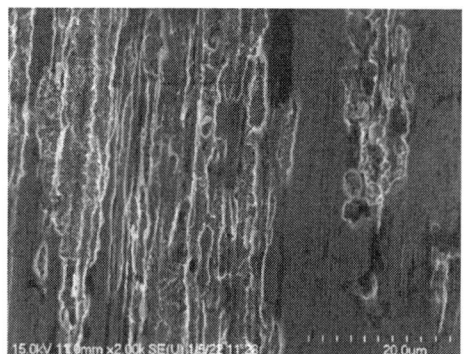

Figure 4: SEM image of the counter electrode after lead electrowinning.

TABLE III: SURFACE COMPOSITION OF THE COOPER COUNTER ELECTRODE AFTER LEAD DEPOSITION ON THE WORKING ELECTRODE

Element	Weight %
Cu	100.00

TABLE IV: SURFACE COMPOSITION OF THE COPPER COUNTER ELECTRODE BEFORE LEAD DEPOSITION ON THE WORKING ELECTRODE

Element	Weight %
Cu	100.00

C. Deposition on the Counter Electrode

While electrowinning at –0.8 V for 1 hour showed lead can successfully be deposited on a copper working electrode, the recovery rate is not high enough. Further experimentation was performed in order to increase recovery rates. After electrowinning for 5 hours, the lead recovery rate increased to 32.48%. However when this was done, a small amount of a deposit also formed on the copper counter electrode and the solution turned blue. Additionally, a black deposit formed on the working electrode along with lead dendrites (Fig. 5). The thin deposit on the counter electrode corresponds to the discoloration seen in Fig. 6. It is hypothesized that the deposit on the counter electrode is a lead oxide species, which was observed on the counter electrode during copper electrowinning in a nitric leachate [4]. Further analysis is needed to confirm it. More importantly, it is desirable to have lead deposit only on the working electrode as metallic lead.

Figure 5: Working electrode (left) with lead deposit and counter electrode (right) after electrowinning at –0.8 V vs. Ag/AgCl for 5 hours.

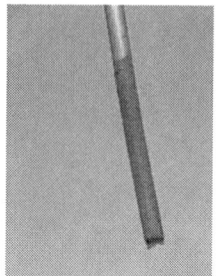

Figure 6: Counter electrode after lead deposition on the working electrode at –0.8 v vs. Ag/AgCl for 5 hours.

IV. CONCLUSION

Acetic acid is proposed as an alternative medium for lead recovery from silicon module waste. The purpose is to avoid more dangerous chemicals and simplify the recovery process for metallic lead. The reduction potential for 0.01 M Pb^{2+} in an acetate leachate was found to be around –0.39 V vs. Ag/AgCl, which agrees with the Nernst equation. Metallic lead was successfully deposited on a copper working electrode from the acetate leachate. This supports the notion of using acetic acid for lead recovery from silicon module waste. A small amount of lead, most likely in the form of lead oxide, was found to deposit on the counter electrode. More studies are needed to eliminate the deposit on the counter electrode and achieve 100% recovery of metallic lead from the system.

ACKNOWLEDGMENT

This work is supported by the U.S. National Science Foundation (grant number 1904544).

REFERENCES

[1] S. Weckend, A. Wade and G. Heath, "End-Of-Life Management Solar Photovoltaic Panels", IRENA and IEA-PVPS, June 2016, pp. 23-35.

[2] M. Tao, et. al. "Major Challenges and Opportunities in Silicon Solar Module Recycling," Progress in Photovoltaics, vol. 28, no. 10, October 2020, pp. 1077-1088.

[3] B. Jung, J. Park, D. Seo and N. Park, "Sustainable System for Raw-Metal Recovery from Crystalline Silicon Solar Panels: From Noble-Metal Extraction to Lead Removal," ACS Sustainable Chem. Eng., vol. 4, 2016, pp. 4079-4083.

[4] W-H Huang, W. J. Shin, L. Wang, W-C Sun, and M. Tao, "Strategy and Technology to Recycle Wafer-Silicon Solar Modules," Solar Energy, vol. 144, 2017, pp. 22-31.

[5] W. M. Haynes, Handbook of Chemistry and Physics, 95th ed. CRC Press: Boca Raton, 2014, pp. 5-83.

[6] A. Kraft, et. al. "Investigation of Acetic Acid Corrosion Impact on Printed Solar Cell Contacts," IEEE J. of Photovoltaics, vol. 5, no. 3, May 2015, pp. 736-743.

[7] A. J. Bard and L. R. Faulkner, Electrochemical Methods, 2nd ed. John Wiley & Sons: USA, 2001, pp. 31, 239-255.

[8] R. G. Bates and J. B. Macaskill, "Standard Potential of the Silver-Silver Chloride Electrode", IUPAC Pure & Appl. Chem., vol. 50. Pergamon Press: Great Britain, 1978, pp. 1701-1706.

Rear Junction Bifacial Screen-Printed Double Side Passivated Contact Si Solar Cells

Young-Woo Ok[1], Vijaykumar D Upadhyaya[1], Brian Rounsaville[1], Ajay D Upadhyaya[1], Wook-Jin Choi[1], Ajeet Rohatgi[1], Gabby De Luna[2], John Derek Arcebal[2], Pradeep Padhamnath[2], Shubham Duttagupta[2]

[1]Georgia Institute of Technology, Atlanta, Georgia, 30332, USA

[2]Solar Energy Research Institute of Singapore, 7 Engineering Drive 1, 117574, Singapore

Abstract—We report on the fabrication of fully screen-printed bifacial large area (244 cm^2) ~20.6% efficient rear junction cells with full area poly-Si/SiO$_x$ passivating contacts on both sides. A full area thin *n*-TOPCon (~35 nm) was deposited on the textured front side in conjunction with thick *p*-TOPCon (~250 nm) on the planar rear side using LPCVD grown intrinsic poly-Si and ex-situ doping technology. Excellent iV$_{oc}$ of ~730 mV was achieved on the cell precursor after SiN$_x$ capping layers on both sides, prior to metallization. However, iVoc dropped from 730 to 715 mV after a simulated high-temperature firing process (>700 °C) without metallization. This was mainly due to the degradation of textured *n*-TOPCon passivation quality after the firing cycle without the metal paste. After fire through screen-printed contact formation, a preliminary efficiency of ~20.6% was achieved with full area double side TOPCon (DS-TOPCon) cell structure with a high V$_{oc}$ of ~700 mV and excellent contact resistance. Device modelling shows that further optimization of paste and firing can lead to much higher efficiency.

Keywords—DS-TOPCon, passivating contact, n-TOPCon, p-TOPCon, screen printed contact, bifacial large area cells

I. INTRODUCTION

Poly-Si/SiO$_x$ carrier selective passivating contacts are considered an ideal candidate for next-generation commercial Si- solar cells because they can give excellent passivation quality as well contact resistance after a high-temperature firing of screen-printed fire through contact. Several groups have reported efficiencies exceeding >22.5% on a large area, screen-printed n-type Si solar cells by employing n-TOPCon (Tunnel Oxide Passivated Contact) or POLO (Polycrystalline silicon on oxide) structures [1-4]. However, TOPCon has been confined primarily to the rear of the cell because of high parasitic absorption losses in the short wavelength range when the poly-Si is placed on front side. Therefore, for the double-side TOPCon cells (DS-TOPCon), poly-Si thickness on the front should be minimized (20-40nm) without compromising contact quality and metal-induced recombination in TOPCon. To bypass this challenge, several groups are trying to make either selective thick poly-Si/SiO$_x$ passivating contacts only under the front metal grid or a full area thin poly-Si on the front in combination with transparent conductive oxide (TCO) layer and low-temperature thick screen printed contacts. Recently, an efficiency of >22% was reported with selective thick front TOPCon layers or with the TCO/thin-poly-Si combination [5-6]. However, both these approaches require additional and expensive processing steps. Therefore, in this paper, we report

on our first attempt to make DS-TOPCon with ~35nm full area front thin poly-Si with traditional fire-through contacts without any TCO. Our simulation results showed that ~25 % efficient rear junction DS-TOPCon cells can be achieved if screen-printed contacts can be directly made on 20 nm poly-Si without any TCO [7]. In this work, we report on our initial attempt that resulted in efficiencies approaching ~21% for fully screen printed large area bifacial DS-TOPCon rear junction cells with the full area ~35nm *n*-TOPCon on front and ~250 nm *p*-TOPCon on the rear. Work is in progress to raise this efficiency by reducing poly-Si thickness to ~20 nm and optimizing the paste and firing conditions.

II. CELL FABRICATION PROCESS

Figure 1 shows the cell structure and process sequence for DS-TOPCon Si solar cells with a full area thin *n*-TOPCon on the front and a thick *p*-TOPCon on the rear. After saw damage etching followed by RCA cleaning and HF treatment, tunnel oxide (iOx: ~1.5nm) was grown followed by intrinsic poly-Si (i-poly-Si:~250nm) deposition by LPCVD process (two stages-single step). Boron dibromide (BBr$_3$) diffusion was done for ex-situ *p*+ doping (formation of *p*-TOPCon). After removing the glass, the front side was textured with a masking layer on the *p*-TOPCon at the rear side. After that, iOx and the i-poly-Si (~35nm) layers were again grown and deposited respectively, using LPCVD tool, followed by a phosphorus diffusion process using a conventional POCl$_3$ tube furnace (formation of *n*-TOPCon). Finally, all glass and mask layers were removed by HF, followed by full cleaning process. PECVD SiO$_x$/SiN$_x$ anti-reflection layers were deposited on the front *n*-TOPCon, and SiN$_x$ was deposited on the rear *p*-TOPCon layer. Ag metal paste was screen-printed on front *n*-TOPCon using 106 lines/35 μm screen openings in combination with five floating busbars. Ag/Al paste was screen-printed using 200 grid lines without busbars on the rear. All samples were fired in a conventional belt furnace with peak temperatures in the range of 700-800°C. In addition, symmetric *p*-TOPCon and *n*-TOPCon test structures, without any metallization, were fabricated to monitor recombination current density (J$_0$) and implied V$_{oc}$ (iV$_{oc}$) after each processing step using QSSPC measurements.

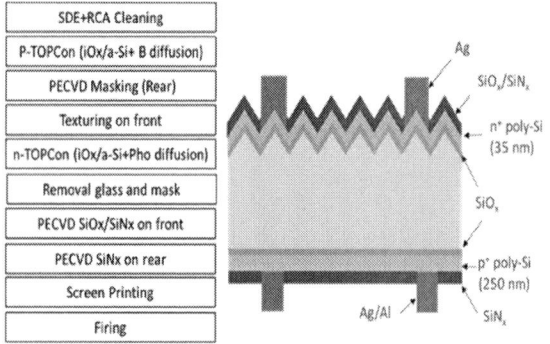

Figure 1 Schematic cell structure and process sequences for DS-TOPCon

III. RESULTS AND DISCUSSION

Figure 2 shows the measured iV_{oc} (mV)and iFF (%) of the unmetallized cell precursors (*n*-TOPCon/*n*-Si substrate/*p*-TOPCon) after each processing step. To monitor the impact of firing temperature on passivation quality, the unmetallized cell precursors were subjected to simulated firing process in a belt furnace with peak temperatures in the range of 700-800°C. High iV_{oc} (~730 mV) and iFF (85.6%) were achieved after PECVD SiN_x deposition on both sides, prior to firing process, due to hydrogen passivation of the iOx/poly-Si interface defects. However, iV_{oc} decreased to $\leq$715 mV after the simulated high-temperature firing process (>700°C).

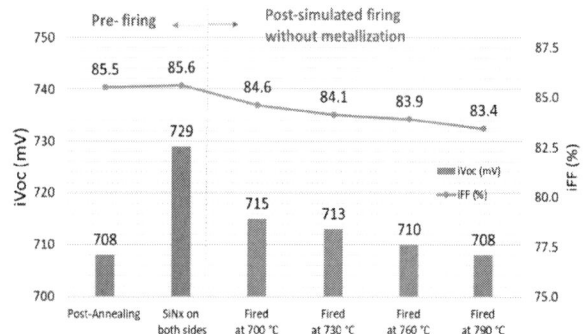

Figure 2 Measured iVoc and iFF in each step and effect of firing temperature

Detailed analysis using symmetric *n*- and *p*-TOPCon structures revealed that the observed 10~15mV drop in iVoc resulted from degradation in passivation quality of the textured front *n*-TOPCon and not from the rear planar *p*-TOPCon. As shown in table 1, J_0 of symmetric n-TOPCon sample increased from 8 to 20 fA/cm^2, but J_0 of symmetric p-TOPCon sample decreased from ~14 to ~12 fA/cm^2 after high temperature firing process.

TABLE I. MEASURED J0 FROM SYMMETRIC N- AND P-TOPCON STRUCTURES BEFORE AND AFTER FIRING PROCESS

Measured Jo (fA/cm^2)	2Jo before firing	2Jo after firing
n-TOPCon symmetric	~8	~20
p-TOPCon symmetric	~14	~12

It was also found that the iV_{oc} and iFF became progressively worse with the increase in peak firing temperature in the range of 700 to 790 °C. Thus, lower firing temperature is desirable from the viewpoint of iV_{oc} and iFF for the high cell efficiency. Next, we investigated the cell performance and the contact quality as a function of firing temperature.

Table 2 shows the measured cell performance after screen-printed fire-though contact as functions of firing temperature. To understand and quantify ohmic contact properties of the front and rear TOPCon, we measured front and back contact resistance using TLM patterns on the cells (Table 3). The cell fired at ~710 °C showed the highest cell V_{oc} (~710 mV) and J_{sc} (38.8 mA/cm^2), but the FF was very low due to high total series resistance (R_s) of 1.46 ohm-cm^2. High Rs was due to higher contact resistance on both sides, as shown in table 3. However, when the firing temperature increased to 730°C, total Rs decreased from 1.46 to 0.65 ohm-cm^2 in conjunction with improved FF (75.2% to 78.6%) and slight decrease in cell V_{oc} (710 to 700 mV). The FF increase is attributed to improved contact resistances on both sides, as shown in table 3. Increased firing temperature in the range of 760-790°C further improved the contact resistance on both sides (Rs~0.5-0.6 ohm-cm^2), but the cell V_{oc} dropped significantly to ~680 mV due to increased metallized J_{0e}. Thus, for our current DS-TOPCon cell with 35 nm front n-poly, we need to stay at lower firing temperatures. We are investigating pastes that can give lower contact resistivity and metal-induced recombination at lower firing temperatures. We also observed a drop in J_{sc} from 38.8 to 37.5 mA/cm^2 when the sample was fired at the peak temperature of ~730°C. The reason for this is currently not understood. Finally, we were able to achieve~20.6% DS-TOPCon cells with V_{oc} of ~700 mV and J_{sc} of 37.5%. Modelling shows that J_{sc} can be improved appreciably by decreasing the front *n*-TOPCon thickness to ~20 nm and eliminating the firing induced loss in Jsc.

TABLE II. MEASURED LIV FROM DS-TOPCON SI SOLAR CELLS FIRED AT DIFFERENT FIRING TEMPERATURES

Firing Temp. (°C)	Voc (mV)	Jsc (mA/cm²)	FF (%)	Eff. (%)	n factor	Rs_ (ohm-cm²)	Rsh_ (ohm-cm²)	pFF (%)
~710	710	38.8	75.2	20.7	1.08	1.46	18175	82.8
~730	699	37.5	78.6	20.6	1.18	0.65	7767	82.0
~760	683	37.3	79.2	20.2	1.17	0.58	10668	82.2
~790	680	37.2	79.2	20.0	1.17	0.56	9522	82.1

TABLE III. MEASURED CONTACT RESISTANCES FROM *N*-TOPCON ON THE FRONT AND *P*-TOPCON ON THE REAR IN DS-TOPCON CELLS

Contact Resistance (mΩ-cm²) from TLM pattern		
Firing Temp. (°C)	Front Tx-*n*-TOPCon	Rear PL *p*-TOPCon
~710	15-20	20-40
~730	3-7	10-15
~760	1-3	5-10

Double-Side Passivated Contacts Solar Cells', IEEE Journal of Photovoltaics, 2021

IV. CONCLUSIONS

In this work, we report on the successful fabrication of ~20.6% efficient screen-printed large area high-efficiency bifacial DS-TOPCon cell with 35 nm thick full area textured n-TOPCon on front and a thick (~250 nm) p-TOPCon on the rear side. Unmetallized SiN_x capped DS-TOPCon rear junction cell precursors gave excellent iV_{oc}~ 730 mV and iFF (85.6%), but iVoc decreased to ~715 mV after the simulated firing process without metallization. This loss was attributed to some degradation in the passivation quality of front textured n-TOPCon. Increased firing temperature in the range of 710 to 790 °C decreased cell V_{oc} and J_{sc} but improved FF. Our initial attempt resulted in ~20.6% efficiency DS-TOPCon cells with ~700 mV. More work is in progress to raise the cell efficiency by contact optimization and reduced poly-Si thickness on front. Modelling shows that close to 25% efficiency can be achieved with this structure with ~20 nm poly on front.

ACKNOWLEDGMENT

This material is based upon work supported by the U.S. Department of Energy's Office of Energy Efficiency and Renewable Energy (EERE) under Solar Energy Technologies Office (SETO) Agreement Number DE-EE0009350, DE-EE0008562 and DE-EE0008975

REFERENCES

[1] Chen, D., Chen, Y., Wang, Z., Gong, J., Liu, C., Zou, Y., He, Y., Wang, Y., Yuan, L., and Lin, W.: '24.58% total area efficiency of screen-printed, large area industrial silicon solar cells with the tunnel oxide passivated contacts (i-TOPCon) design', Solar Energy Materials and Solar Cells, 2020, 206, pp. 110258.

[2] Padhamnath, P., Khanna, A., Balaji, N., Shanmugam, V., Nandakumar, N., Wang, D., Sun, Q., Huang, M., Huang, S., and Fan, B.: 'Progress in screen-printed metallization of industrial solar cells with SiOx/poly-Si passivating contacts', Solar Energy Materials and Solar Cells, 2020, 218, pp. 110751

[3] Min, B., Wehmeier, N., Brendemuehl, T., Haase, F., Larionova, Y., Nasebandt, L., Schulte-Huxel, H., Peibst, R., and Brendel, R.: '716 mV Open - Circuit Voltage with Fully Screen - Printed p - Type Back Junction Solar Cells Featuring an Aluminum Front Grid and a Passivating Polysilicon on Oxide Contact at the Rear Side', Solar RRL, 2021, 5, (1), pp. 2000703

[4] Huang, Y.-Y., Ok, Y.-W., Madani, K., Choi, W., Upadhyaya, A., Upadhyaya, V., Rounsaville, B., Chandrasekaran, V., and Rohatgi, A.: '~23% rear side poly-Si/SiO2 passivated silicon solar cell with optimized ion-implanted boron emitter and screen-printed contacts', Solar Energy Materials and Solar Cells, 2021, 230, pp. 111183

[5] Larionova, Y., Schulte-Huxel, H., Min, B., Schäfer, S., Kluge, T., Mehlich, H., Brendel, R., and Peibst, R.: 'Ultra - Thin Poly - Si Layers: Passivation Quality, Utilization of Charge Carriers Generated in the Poly - Si and Application on Screen - Printed Double - Side Contacted Polycrystalline Si on Oxide Cells', Solar RRL, 2020, 4, (10), pp. 2000177

[6] Yu, B., Shi, J., Li, F., Wang, H., Pang, L., Liu, K., Zhang, D., Wu, C., Liu, Y., and Chen, J.: 'Selective tunnel oxide passivated contact on the emitter of large-size n-type TOPCon bifacial solar cells', Journal of Alloys and Compounds, 2021, 870, pp. 159679

[7] Jain, A., Choi, W.-J., Huang, Y.-Y., Klein, B., and Rohatgi, A.: 'Design, Optimization, and In-Depth Understanding of Front and Rear Junction

Micro-Scale III-V/Ge Multijunction Solar Cell with Through Cell Via Contacts

Mathieu de Lafontaine, Guillaume Gay, Erwine Pargon, Camille Petit-Etienne, Romain Stricher, Serge Ecoffey, Artur Turala, Maïté Volatier, Abdelatif Jaouad, Simon Fafard, Vincent Aimez, Maxime Darnon

Laboratoire Nanotechnologies Nanosystèmes (LN2) CNRS IRL-3463 Institut Interdisciplinaire d'Innovation Technologique (3IT), Sherbrooke, QC, Canada

Université Grenoble Alpes, CNRS, CEA/LETI-Minatec, Grenoble INP, LTM, Grenoble, France

There has been a growing interest for micro-scale concentrator photovoltaics (micro-CPV) over the past few years. The main goal is to reduce the cell size to a sub-millimetric range to gain several benefits such as better handling, better thermal management and reduced series resistance losses. In this work, we present the microfabrication of micro-scale III-V/Ge triple junction solar cells with through cell via contacts (TCVC) for micro-CPV applications. This contact architecture transfers the front side contact to the back side by using isolated and metallized vias to reduce the shading and the resistive losses. A process was developed to fabricate 180x180 µm2 solar cells with a single through cell via contact. The first prototypes have been successfully fabricated and the electrical characterization shows good performance. A 1-sun open-circuit voltage of 2.28 V was obtained, which is high considering the small device size and the high perimeter-to-area ratio (232cm-1). This new architecture represent a key improvement for micro-scale solar cells. With standard busbar and grid line contacts, the metallized area to device area ratio keeps increasing as the device size is reduced. This reduces the power yield per wafer and hinders the benefits of micro-CPV. With TCVC, the metallization ratio remains the same (2.8%), regardless of the device size. As a result, this new architecture could increase the wafer area yield by up to 267% for 180x180 µm2 solar cells.

Spatiotemporal Modeling of Real World Backsheets Field Survey Data: Hierarchical (Multilevel) Generalized Additive Models

Raymond J. Wieser*, Zelin(Zack) Li*, Stephanie L. Moffitt†, Ruben Zabalza‡, Evan Boucher‡, Silvana Ayala§,
Matthew Brown§, Xiaohong Gu†, Liang Ji‡, Colleen O'Brien‡, Adam W. Hauser‖,
Greg S. O'Brien‖, Xuanji Yu*, Roger H. French*, Micheal D. Kempe§, Jared Tracy¶,
Kausik R. Choudhury¶, William J. Gambogi¶, Laura S. Bruckman*, Kenneth P. Boyce‡
* Case Western Reserve University, Cleveland, Ohio, 44106, USA
† National Insitute of Standards and Technology, Gaithersburg, Maryland, 20878, USA
‡ Underwriter Laboratories, Northbrook, Illinois, 60062, USA
§ National Renewable Energy Laboratory, Golden, Colorado, 80401, USA
¶ DuPont Photovoltaic Solutions, Wilmington DE, 19803, USA
‖ Arkema, King of Prussia, Pennsylvania, 19406, USA

Abstract—Assessing photovoltaic module backsheet durability is critical to increasing module lifetime. Laboratory-based accelerating testing has recently failed to predict large scale failures of widely adopted polymeric materials. Additionally, there is a growing concern on characterizing the non-uniformity of field exposure. Therefore, data from field surveys are critical to assess the performance of component lifetimes. Using a documented field survey protocol, 19 field surveys were conducted. The focus of this survey strategy is to investigate spatial continuity in degradation modes. By combining field survey data with real-time satellite weather data, stressor / response models have been trained. Generalized additive Models (GAM) model was created to predict the value of degradation based on measured predictors. Two different GAM constructions were testing using different implementations of basis splines. The model includes variables on the environmental stressors of the system and the location of each measurement in the PV mounting structure. The incorporation of hierarchical structure into the models allowed for material specific degradation rates, while maintaining the assumption of a global trend. The model performed well with an adjusted R^2 of 0.975 for yellowness index prediction.

Index Terms—Backsheet, Degradation, Spatio-temporal, Modeling, Field Survey,

I. INTRODUCTION

To be cost effective, utility-scale power providers are reliant on Photovoltaic (PV) module service lifetimes exceeding 20 years of operation in a variety of exposure environments. PV backsheets are affected by multiple environmental stressors

This material is based upon work supported by the U.S. Department of Energy's Office of Energy Efficiency and Renewable Energy (EERE) under Solar Energy Technologies Office (SETO) Agreement Number DE-EE-0008748 The views expressed herein do not necessarily represent the views of the U.S. Department of Energy or the United States Government. This work made use of the Rider High Performance Computing Resource in the Core Facility for Advanced Research Computing at Case Western Reserve University.

and degrade due to synergistic effects in the field. The study of outdoor backsheets degradation is necessary to understand real-world PV failure and improve accelerated testing protocols. However, current research on outdoor backsheet degradation is much scarcer than that of PV cells or encapsulants [1], [2].

Backsheets in installed PV modules experience various synergistic stressors, including irradiance, temperature, humidity, abrasion, and other factors. Ultraviolet (UV) light reflected from the ground can cause chain scission and loss of mechanical strength [3], [4]. The diurnal and seasonal thermal cycles create thermal-mechanical stress, contributing to backsheet failures [5]. The presence of moisture can also have dramatic effects on the degradation of backsheet materials [6]–[11]. These stressors also vary spatially according to array geometry, changes in ground albedo, and shading due to external objects. It was found that the edge modules experienced higher degradation rates, attributed to an increase of rear-side irradiance [12], [13].

Field survey data has recently been used to illuminate effects observed in long-term outdoor exposures. Measurements of the color and gloss of PV modules after years of outdoor exposure showed differential degradation patterns based on the PV mounting structure [1]. Additionally, this observed trend was modeled using generalized additive modeling (GAM), resulting in an equation that predicted degradation patterns across the surface of the PV backsheet based on its location in the mounting structure [13]. It was found that the edges of PV mounting systems experience faster rates of apparent degradation than the modules located at the center. Moreover, it was found that the distance from the ground was another factor in the degradation rate. An improved GAM was developed using data from 14 individual field surveys. However,

978-1-7281-6118-1/22 $31.00 © 2022 IEEE

the model struggled to differentiate the synergistic effects of climate stressors [14].

This study improves on past GAM modeling efforts with a new hierarchical modeling framework, which allows for pooled model coefficients. Hierarchical (Multilevel / Mixed Effect) modeling has widely been used to model spatio-temporal data in the fields of ecology, finanace, and real-estate [15] [16] [17] [18] [19]. To better parse out the synergistic effects of multiple climatic stressors, a field survey was conducted where multiple backsheet brands and materials were exposed under the same environment. This survey will allow a comparison of the material-specific degradation modes that are excited by the same levels of environmental stressors.

II. METHODOLOGY

A. Field Survey Protocol

A field survey protocol was defined to standardize the measurement locations and the measurement types recorded to produce a uniform data set. This protocol has been applied to 19 PV systems. The current protocol dictates that individual rows of PV modules will be measured systematically, with uniform separation of measurements along the length of the row. It is recommended that a minimum of 12 modules should be recorded for each row measured, but for long rows, the number of modules should increase. Additionally, to observe the 'edge effect', additional modules are surveyed on the row ends. Each module is measured in 6 locations across the surface of its backsheet. Data is collected on the Yellowness Index (ATSM E313), Gloss (ASTM D523), and FTIR spectra of every module surveyed without cleaning the module. The exposure conditions of the field are also noted.

This study presents measurements of four different types of air-side backsheet materials. Polyvinylidene diflouride (PVDF), and Polyvinyl Fluoride (PVF) are fluoro-polymer based films. Fluoroethylene vinyl ether (FEVE) is a spray type fluoro-polymer coating applied to backsheet films. Polyethylene naphthalate (PEN) is a non-fluoropolymer film.

B. Data

Weather data for PV sites was gathered using SolarGIS. The data consisted of temperature, relative humidity, global horizontal irradiance, diffuse horizontal irradiance, and wind speed. Measurements were recorded at five-minute intervals for up to four years. The climate data was obtained from the closest available weather station and ingested into a high performance computer. When available, weather data collected at the specific PV installation is used.

After conducting a field survey, the instrument data is processed through cleaning scripts that label and supplement each observation with meta-data. After the data is cleaned, it is ingested into Case Western's High-Performance Computing Cluster and stored.

C. Modeling

1) Inference By Eye: Inference by Eye is a statistical technique developed to quickly visualize and test for the statistical significance at the 0.05 level, of both null hypothesis testing and two sample t-tests [20]–[22]. With a 95% confidence interval (CI), it can be used to test if a module has undergone a change over time, such that the null hypothesis test fails and the sample has undergone significant change at a 0.05 significance level. And it can be used to compare among many samples, using a two sample t-test and 83.5% CIs, that two samples significantly different from each other at the 0.05 significance level. The basic principle of this technique relies on a simple 2-sample independent students t-test of the difference between two means ($p = 0.05$, $t = 1.96$). A more in-depth discussion of inference by eye can be found in the following paper by Wieser et al. [14].

2) Generalized Additive Modeling: To predict the values of degradation, GAMs have been created to relate the observed responses to real world stressors on the field survey data. Smoothing splines have been integrated as the basis functions for the spatial variability of the observed data. More information on the principles of applying GAM to field survey data can be found in Wieser et al. [14].

The hierarchical extension of a GAM allows for group-specific variations in the overall trend of the model. Hierarchical models allow for flexibility in the fit parameters based on the structure of the data. In particular, the hierarchical model allows for different coefficients for each of the material types in the study while maintaining the assumption that all the observed responses have a shared global trend.

Two approaches to GAM modeling will be discussed. The first approach uses the natural spline model as a regressor in the GAM model (Equation 1), whereas the other approach incorporates and penalizes the spline fitting into the model (Equation 2). In other words, the first approach takes an existing predetermined spline function based off of the co-variate as a variable in the model. The second approach models and minimizes the variance in spline terms while determining the coefficients for each term. The advantage of the second model structure is its ability to capture the hierarchical differences in the structure of the field data. The individual smoothing splines are created as an ensemble of functions that can be penalized through different constraints. For the purposes of this analysis, each material will be considered its own group, and the model will allow for individual smoothing parameters for each material. This results in a model that acknowledges a global shared trend for the spatial dependence of material degradation, but allows material specific flexibility to account for differential rates of degradation.

Along with the information gathered from the field surveys, weather data is also being incorporated into the GAM model. Years' worth of time-series weather data are aggregated, generating general statistics on the climate of each location.

The spatio model of degradation contains variables for the position of the module in the row structure, Length ($\mathbf{L}$) and Depth ($\mathbf{D}$) and interaction between material (M_i) and exposure condition (E_j). The model change point locations a_1 and a_2 determine the position of the knots in the regression form of the GAM. The temporal model includes the following parameters: time of contact wetness (**CW**), cumulative yearly irradiance (**IRR**), time at elevated temperature (**TAET**), and time (**t**).

The general form of the spatio- equation then can be expressed as:

$$Y(L, D, M) = \beta_0 + \beta_1 L + \beta_2 L^2 + \beta_3 L^3 + \beta_4 (L - a_1)$$
$$+ \beta_5 (L - a_2) + \beta_6 D + \beta_7 D^2 + \beta_{M,E} \quad (1)$$

or

$$Y(L, D, M) = \beta_0 + f(L) + f(D) + \beta_{M,E} \quad (2)$$

where $f(L)$ and $f(D)$ are spline terms defined as in Equation 3.

$$f(X) = \sum_{q=0}^{Q} \beta_q \cdot b_q(X) =$$
$$\beta_0 + \beta_1 b_1(X) + \beta_2 b_2(X) + \ldots + \beta_Q b_Q(X) \quad (3)$$

With the general form of the temporal equation:

$$Y(CW, IRR, TAET, t) = \beta_1(CW \times t) + \beta_2(IRR \times t)$$
$$+ \beta_3(TAET \times t) + \beta_4(M_i \times t) \quad (4)$$

The GAM creates all the components separately using a linear combination of all the unique equations for each variable.

Model validation was conducted using the Akaike Information Criterion (AIC), which allows the comparison of GAMs with different combinations of effects. The models were built on subsetted data-sets with 80% of the data allocated to the training of the model. Root Mean Squared Estimates (RMSE) and Adj. R^2 values were used to evaluate the model fit. Two baseline models (linear and piece-wise) were developed to evaluate the performance of the GAMs.

III. RESULTS / DISCUSSION

A. Spatial Modeling

GAMs were created using both approaches. The overall fit of both models provided additional insight into the spatial dependence of degradation. However, the natural spline regression GAM struggled to capture the different mechanisms of degradation. Fluoropolymer based materials are shown to degrade to values of negative YI, whereas other backsheet materials tend to increase their YI. The natural spline model does not differentiate between these mechanisms, and fits the edges of the spatial function to be concave down (becoming more negative), due to the presence of fluoropolymers. However, the penalized smoothing GAM intrinsically captures group specific effects and allows for different functional responses. This allows the penalized smoothing GAM to differentiate between materials that have different degradation responses.

TABLE I: Approximate Significance of the Penalized Splines

Term	Material	Est. Deg. of Freedom	Ref. Deg. of Freedom	Statistic	P-Value
$f(L)$	PVF	0.0002484	36	0.0000016	0.7527733
$f(L)$	FEVE	0.8294896	39	0.0665848	0.0472572
$f(L)$	PEN	22.9198381	26	7.9638536	<2 E -16
$f(L)$	PVDF	1.0228291	39	0.0493353	0.1271306
$f(D)$	PVF	2.2568605	8	0.5338434	0.0999844
$f(D)$	FEVE	0.0114537	7	0.0030046	<2 E -16
$f(D)$	PEN	3.0013728	6	5.6328328	<2 E -16
$f(D)$	PVDF	0.9892293	7	0.2419582	0.0076261

Therefore, the penalized smoothing GAM was able to better model the spatial component of the degradation response (Figure 2). However, there are certain cases where the penalization of the smoothing splines resulted in the loss of information of high frequency spatial dependence. While the penalized smoothing splines fit the overall shape of the data, the natural spline's knots allowed for increased flexibility to capture more of the spatial variance of the observed signal (Figure 1). The resultant smoothing splines from the penalized GAM model can be used to investigate material specific spatial trends. It was shown that the different materials exhibit varying degrees of spatial dependence. The polymers PEN and FEVE where shown to have a strong spatial dependence, where as the fluoro-polymer films did not have a significant dependence on the spatial location of measurement (Table I). A lack of spatial dependence indicates that the material is less sensitive to micro-climatic effects.

The location of the row in the overall field was also analyzed to determine if the row location had a significant effect on the level of degradation observed. The models include separate intercept terms for the different combinations of material type and row exposure conditions. The row exposure of "Fully Shaded" was used as the base case for the model. Interaction terms without data where excluded from the table. Both GAM models found significant effects in the interaction between row exposure and material type. PEN and PVF based backsheets showed varying levels of degradation for different row locations. Frontside unshaded rows exhibited less severe levels of degradation compared to the fully shaded rows.

B. Temporal Modeling

The spatial component of the GAMs captures the anisotropic degradation of backsheet materials. This allows the temporal components to be modeled by removing the variance due to the differences in local exposure conditions. The temporal coefficients can be divided into climatic stressors and material aging effects.

Both GAMs found significant contributions between the interaction terms of climatic stressor and length of exposure (Table III). The annual total dose of irradiance is the strongest contributor to the observed degradation. The other climatic stressor terms had smaller contributions to the overall level of degradation observed. The presence of dew formation and exposure to elevated temperature (T <35°C) were found to

TABLE II: Significance of interaction between row exposure environment and material type, as compared to the base case of "Fully Shaded"

Model	Term	Material	Row Environment	Estimate	Std. Error	t value	P-Value
Natural Spline GAM	$\beta_{M,E}$	PVF	Frontside Unshaded	1.314e+00	1.181e-01	11.124	<2e-16
	$\beta_{M,E}$	FEVE	Frontside Unshaded	-2.959e-02	1.717e-01	-0.250	0.803
	$\beta_{M,E}$	PEN	Frontside Unshaded	-5.272e-01	1.238e-01	-4.285	1.87e-05
	$\beta_{M,E}$	PVDF	Frontside Unshaded	-4.897e-02	8.992e-02	0.247	0.805
	$\beta_{M,E}$	FEVE	Rearside Unshaded	9.443e-02	1.716e-01	0.507	0.612
	$\beta_{M,E}$	PEN	Rearside Unshaded	-2.124e+00	1.568e-01	-13.551	<2e-16
	$\beta_{M,E}$	PVDF	Rearside Unshaded	-1.367e-01	8.933e-02	-0.712	0.476
	$\beta_{M,E}$	PVF	Tracker	-1.364e+00	1.177e-01	-11.581	<2e-16
Penalized Smooth GAM	$\beta_{M,E}$	PVF	Frontside Unshaded	1.088e+00	6.354e-02	17.123	<2e-16
	$\beta_{M,E}$	FEVE	Frontside Unshaded	-2.862e-02	1.397e-01	-0.231	0.817439
	$\beta_{M,E}$	PEN	Frontside Unshaded	-4.043e-01	1.057e-01	-3.823	0.000134
	$\beta_{M,E}$	PVDF	Frontside Unshaded	1.486e-02	7.316e-02	0.308	0.757991
	$\beta_{M,E}$	FEVE	Rearside Unshaded	8.903e-02	1.397e-01	0.644	0.519766
	$\beta_{M,E}$	PEN	Rearside Unshaded	-5.405e-01	1.578e-01	-3.430	0.000610
	$\beta_{M,E}$	PVDF	Rearside Unshaded	-4.385e-02	7.259e-02	-0.476	0.633842
	$\beta_{M,E}$	PVF	Tracker	-1.136e+00	6.305e-02	-18.055	<2e-16

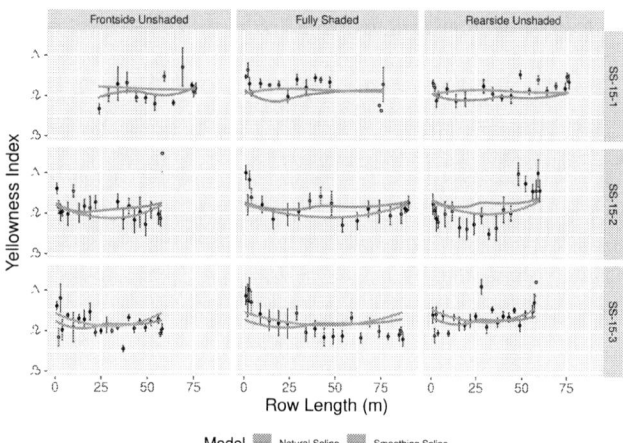

Fig. 1: Results of predicting the trained model on the SS-15 Sites. This model was fitted using an 80% training set.

TABLE III: Interaction terms between length of exposure and climatic stressor.

Model	Term	Stressor	Estimate	Std. Error	T-Value	P-Value
NS GAM	β_1	CW	-3.798e-03	6.938e-04	-5.475	4.65e-08
	β_2	IRR	1.714e-06	9.941e-08	17.241	<2e-16
	β_3	TAET	-4.787e-04	7.794e-05	-6.143	8.92e-10
PS GAM	β_1	CW	-8.479e-03	4.882e-04	-17.367	<2e-16
	β_2	IRR	2.309e-06	7.099e-08	32.525	<2e-16
	β_3	TAET	-8.563e-04	5.565e-05	-15.388	<2e-16

TABLE IV: Interaction terms between length of exposure and material type. These values can be used to determine the types of degradation products formed and susceptibility of the material to exposure.

Model	Term	Material	Estimate	Std. Error	t-Value	P-Value
NS GAM	β_4	PVF	-4.939e-01	1.867e-02	-26.454	<2e-16
	β_4	FEVE	-8.597e-01	4.738e-02	-18.147	<2e-16
	β_4	PEN	2.023e+00	2.338e-02	86.523	<2e-16
	β_4	PVDF	-5.599e-01	4.119e-02	-13.592	<2e-16
PS GAM	β_4	PVF	-4.738e-01	2.683e-02	-17.656	<2e-16
	β_4	FEVE	-6.791e-01	6.710e-02	-10.120	<2e-16
	β_4	PEN	2.216e+00	3.199e-02	69.266	<2e-16
	β_4	PVDF	-7.864e-01	3.428e-02	-22.938	<2e-16

the temporal equation. Over the length of the exposure the fluoro-polymer based materials tend to decrease in the value of YI (turning blue). This observed response to exposure is well captured by both GAMs. The magnitude of these coefficients also provides information on the susceptibility of a material to the formation of degradation products. The fluoro-polymer based backsheets have low coefficients which is related to their overall stability under outdoor exposure. PEN based backsheets exhibited the highest rates of the formation of degradation products IV.

C. Model Fit

Overall the models fit the highly variable data well. The spline terms were able to capture some of the spatial variability in the data. The penalized smoothing spline exhibited lower RMSE and higher adj.R^2. The increase of flexibility of the penalized smoothing model allowed for a better model of the spatial variance of the data. There was little difference between the RMSE values of the testing and training data-sets for both models, indicating that they did not over-fit the training data-set. Although not discussed, a simple linear model and a piece-wise model were used as baseline models. The model accuracy parameters can be found in Table V.

decrease the observed values of degradation. This could relate to the formation of differently colored degradation products. However, the observed effects could be the result of the GAM overfitting some of the variance of the material degradation behaviors.

Material specific degradation rates were also modeled using

978-1-7281-6118-1/22 $31.00 © 2022 IEEE

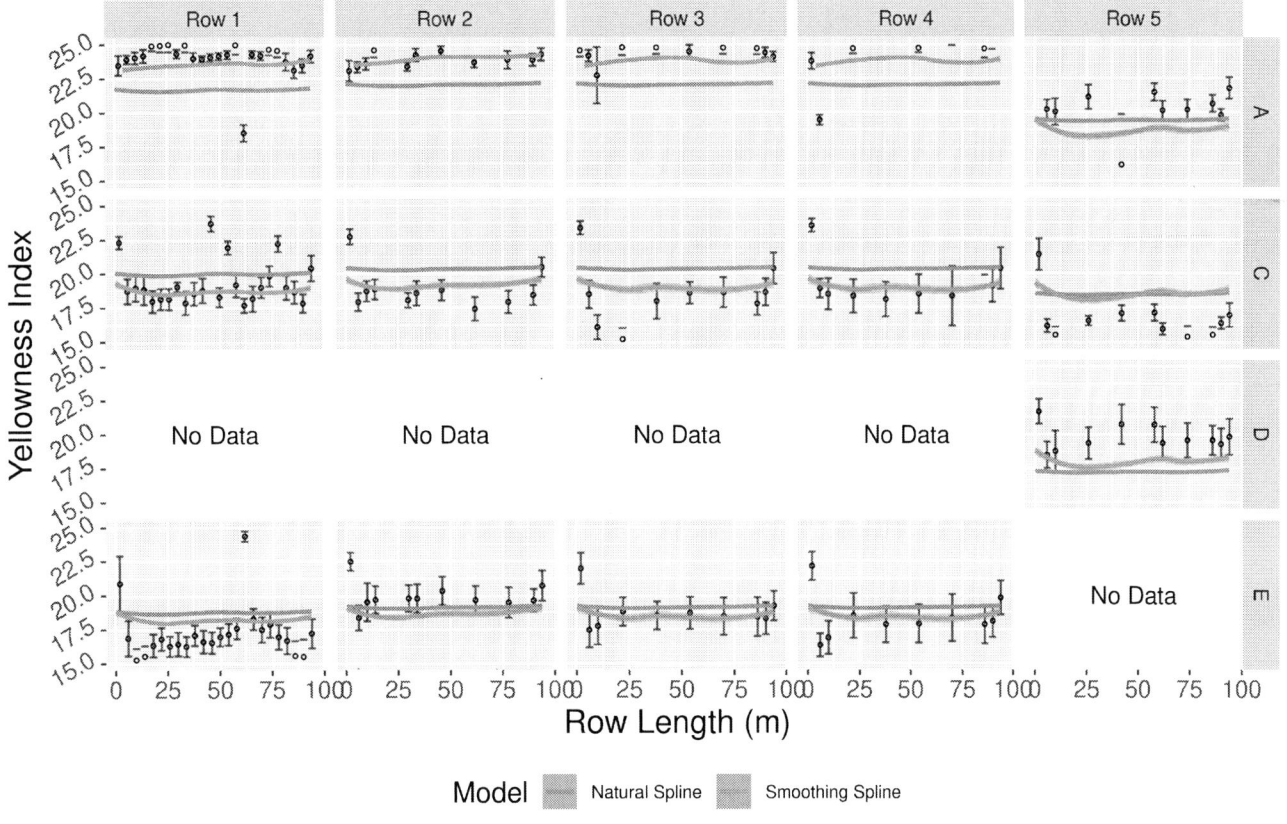

Fig. 2: Results of predicting the trained model on the SS-16-1 Site. This model was fitted using a 80% training set.

IV. CONCLUSIONS

GAMs present a highly flexible framework to analyze hierarchically structure problems. Through careful survey design, a multitude of different exposure related degradation modes can be quantified. Interaction terms between length of exposure and climatic stressor allow for a deeper understanding of the pertinent climatic variables that affect material longevity. In addition, the stability of materials during outdoor exposure can be directly compared. The sensitivity to changes in microclimatic exposure can be understood through the contribution of smoothing splines. It was found that the fluoro-polymer based backsheets were the least susceptible to degradation and the least sensitive to changes in environment.

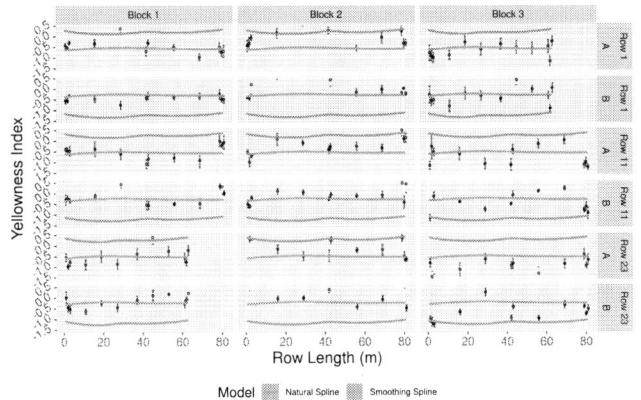

Fig. 3: Results of predicting the trained model on the SS-16-2 Site. This model was fitted using a 80% training set.

TABLE V: Model Accuracy Parameters

Model	$adj R^2$	RMSE Training	RMSE Testing
Linear Model	0.834	2.35	3.35
Piecewise Model	0.883	2.12	2.34
Natural Spline	0.962	1.675	1.626
Penalized Smoothing	0.975	1.339	1.355

REFERENCES

[1] A. Fairbrother, M. Boyd, Y. Lyu, J. Avenet, P. Illich, Y. Wang, M. Kempe, B. Dougherty, L. Bruckman, and X. Gu, "Differential degradation patterns of photovoltaic backsheets at the array level," *Solar Energy*, vol. 163, pp. 62–69, Mar. 2018.

[2] W. Gambogi, Y. Heta, K. Hashimoto, J. Kopchick, T. Felder, S. MacMaster, A. Bradley, B. Hamzavytehrany, L. Garreau-Iles, T. Aoki, K. Stika, T. J. Trout, and T. Sample, "A Comparison of Key PV Backsheet and Module Performance from Fielded Module Exposures and Accelerated Tests," *IEEE Journal of Photovoltaics*, vol. 4, no. 3, pp. 935–941, May 2014.

[3] Y. Lyu, J. H. Kim, A. Fairbrother, and X. Gu, "Degradation and Cracking Behavior of Polyamide-Based Backsheet Subjected to Sequential Fragmentation Test," *IEEE Journal of Photovoltaics*, pp. 1–6, 2018.

[4] C.-C. Lin, Y. Lyu, L.-C. Yu, and X. Gu, "Correlation between mechanical and chemical degradation after outdoor and accelerated laboratory aging for multilayer photovoltaic backsheets," in *Reliability of Photovoltaic Cells, Modules, Components, and Systems IX*, vol. 9938. International Society for Optics and Photonics, Sep. 2016, p. 99380H. [Online]. Available: https://doi.org/10.1117/12.2238216

[5] A. Omazic, G. Oreski, M. Halwachs, G. C. Eder, C. Hirschl, L. Neumaier, G. Pinter, and M. Erceg, "Relation between degradation of polymeric components in crystalline silicon PV module and climatic conditions: A literature review," *Solar Energy Materials and Solar Cells*, vol. 192, pp. 123–133, Apr. 2019, citation Key Alias: omazicRelationDegradationPolymeric2019a. [Online]. Available: http://www.sciencedirect.com/science/article/pii/S0927024818305956

[6] International Electrotechnical Commission, "IEC 61215 Terrestrial photovoltaic (PV) modules - Design qualification and type approval," International Electrotechnical Commission, International Standard, 2016. [Online]. Available: https://webstore.iec.ch/publication/24312

[7] M. D. Kempe and J. H. Wohlgemuth, "Evaluation of temperature and humidity on PV module component degradation," in *2013 IEEE 39th Photovoltaic Specialists Conference (PVSC)*, Jun. 2013, pp. 0120–0125.

[8] E. Wang, H. E. Yang, J. Yen, S. Chi, and C. Wang, "Failure Modes Evaluation of PV Module via Materials Degradation Approach," *Energy Procedia*, vol. 33, pp. 256–264, Jan. 2013. [Online]. Available: http://www.sciencedirect.com/science/article/pii/S1876610213000763

[9] N. Kim, H. Kang, K.-J. Hwang, C. Han, W. S. Hong, D. Kim, E. Lyu, and H. Kim, "Study on the degradation of different types of backsheets used in PV module under accelerated conditions," *Solar Energy Materials and Solar Cells*, vol. 120, pp. 543–548, Jan. 2014. [Online]. Available: https://doi.org/10.1016/j.solmat.2013.09.036

[10] W. Gambogi, Y. Heta, K. Hashimoto, J. Kopchick, T. Felder, S. MacMaster, A. Bradley, B. Hamzavytehraney, V. Felix, T. Aoki, K. Stika, L. Garreau-Illes, and T. J. Trout, "Weathering and durability of PV backsheets and impact on PV module performance," in *Reliability of Photovoltaic Cells, Modules, Components, and Systems VI*, vol. 8825. International Society for Optics and Photonics, Sep. 2013, p. 88250B. [Online]. Available: https://www.spiedigitallibrary.org/conference-proceedings-of-spie/8825/88250B/Weathering-and-durability-of-PV-backsheets-and-impact-on-PV/10.1117/12.2024491.short

[11] F. Liu, L. Jiang, and S. Yang, "Ultra-violet degradation behavior of polymeric backsheets for photovoltaic modules," *Solar Energy*, vol. 108, pp. 88–100, Oct. 2014. [Online]. Available: http://www.sciencedirect.com/science/article/pii/S0038092X14003260

[12] A. Fairbrother, M. Boyd, Y. Lyu, J. Avenet, P. Illich, Y. Wang, M. Kempe, B. Dougherty, L. Bruckman, and X. Gu, "Differential degradation patterns of photovoltaic backsheets at the array level," *Solar Energy*, vol. 163, Feb. 2018. [Online]. Available: https://doi.org/10.1016/j.solener.2018.01.072

[13] Y. Wang, W.-H. Huang, A. Fairbrother, L. S. Fridman, A. J. Curran, N. R. Wheeler, S. Napoli, A. W. Hauser, S. Julien, X. Gu, G. S. O'Brien, K.-T. Wan, L. Ji, M. D. Kempe, K. P. Boyce, R. H. French, and L. S. Bruckman, "Generalized Spatio-Temporal Model of Backsheet Degradation From Field Surveys of Photovoltaic Modules," *IEEE Journal of Photovoltaics*, vol. 9, no. 5, pp. 1374–1381, Sep. 2019.

[14] R. J. Wieser, K. Rath, S. L. Moffitt, R. Zabalza, E. Boucher, S. Ayala, M. Brown, X. Gu, L. Ji, C. O'Brien, A. W. Hauser, G. S. O'Brien, R. H. French, M. D. Kempe, J. Tracy, K. R. Choudhury, W. J. Gambogi, L. S. Bruckman, and K. P. Boyce, "Spatio-Temporal Modeling of Field Surveyed Backsheet Degradation," in *2021 IEEE 48th Photovoltaic Specialists Conference (PVSC)*, Jun. 2021, pp. 1383–1388, iSSN: 0160-8371.

[15] J. M. Diez and H. R. Pulliam, "Hierarchical Analysis of Species Distributions and Abundance Across Environmental Gradients," *Ecology*, vol. 88, no. 12, pp. 3144–3152, 2007, _eprint: https://onlinelibrary.wiley.com/doi/pdf/10.1890/07-0047.1. [Online]. Available: https://onlinelibrary.wiley.com/doi/abs/10.1890/07-0047.1

[16] S. M. McMahon and J. M. Diez, "Scales of association: hierarchical linear models and the measurement of ecological systems," *Ecology Letters*, vol. 10, no. 6, pp. 437–452, 2007, _eprint: https://onlinelibrary.wiley.com/doi/pdf/10.1111/j.1461-0248.2007.01036.x. [Online]. Available: https://onlinelibrary.wiley.com/doi/abs/10.1111/j.1461-0248.2007.01036.x

[17] R. A. Chisholm, H. C. Muller-Landau, K. Abdul Rahman, D. P. Bebber, Y. Bin, S. A. Bohlman, N. A. Bourg, J. Brinks, S. Bunyavejchewin, N. Butt, H. Cao, M. Cao, D. Cárdenas, L.-W. Chang, J.-M. Chiang, G. Chuyong, R. Condit, H. S. Dattaraja, S. Davies, A. Duque, C. Fletcher, N. Gunatilleke, S. Gunatilleke, Z. Hao, R. D. Harrison, R. Howe, C.-F. Hsieh, S. P. Hubbell, A. Itoh, D. Kenfack, S. Kiratiprayoon, A. J. Larson, J. Lian, D. Lin, H. Liu, J. A. Lutz, K. Ma, Y. Malhi, S. McMahon, W. McShea, M. Meegaskumbura, S. Mohd. Razman, M. D. Morecroft, C. J. Nytch, A. Oliveira, G. G. Parker, S. Pulla, R. Punchi-Manage, H. Romero-Saltos, W. Sang, J. Schurman, S.-H. Su, R. Sukumar, I.-F. Sun, H. S. Suresh, S. Tan, D. Thomas, S. Thomas, J. Thompson, R. Valencia, A. Wolf, S. Yap, W. Ye, Z. Yuan, and J. K. Zimmerman, "Scale-dependent relationships between tree species richness and ecosystem function in forests," *Journal of Ecology*, vol. 101, no. 5, pp. 1214–1224, 2013, _eprint: https://onlinelibrary.wiley.com/doi/pdf/10.1111/1365-2745.12132. [Online]. Available: https://onlinelibrary.wiley.com/doi/abs/10.1111/1365-2745.12132

[18] J. D. McCabe, J. D. Clare, T. A. Miller, T. E. Katzner, J. Cooper, S. Somershoe, D. Hanni, C. A. Kelly, R. Sargent, E. C. Soehren, C. Threadgill, M. Maddox, J. Stober, M. Martell, T. Salo, A. Berry, M. J. Lanzone, M. A. Braham, and C. J. W. McClure, "Resource selection functions based on hierarchical generalized additive models provide new insights into individual animal variation and species distributions," *Ecography*, vol. n/a, no. n/a, _eprint: https://onlinelibrary.wiley.com/doi/pdf/10.1111/ecog.06058. [Online]. Available: https://onlinelibrary.wiley.com/doi/abs/10.1111/ecog.06058

[19] E. J. Pedersen, D. L. Miller, G. L. Simpson, and N. Ross, "Hierarchical generalized additive models in ecology: an introduction with mgcv," *PeerJ*, vol. 7, p. e6876, 2019.

[20] G. Cumming and S. Finch, "Inference by Eye: Confidence Intervals and How to Read Pictures of Data," *American Psychologist*, vol. 60, no. 2, pp. 170–180, 2005, place: US Publisher: American Psychological Association.

[21] A. Hazra, "Using the confidence interval confidently," *Journal of Thoracic Disease*, vol. 9, no. 10, pp. 4124–4129, Oct. 2017, tex.ids= hazraUsingConfidenceInterval2017 publisher: AME Publishing Company. [Online]. Available: http://jtd.amegroups.com/article/view/16406/13455

[22] J. Leppink, "A Pragmatic Approach to Statistical Testing and Estimation (PASTE)," *Health Professions Education*, vol. 4, no. 4, pp. 329–339, Dec. 2018. [Online]. Available: https://www.sciencedirect.com/science/article/pii/S2452301117301487

Mapping of Local Defects and Voltages in Solar Cells using Non-Contact Electrostatic Voltmeter Method

Hamza Ahmad Raza, Govindasamy TamizhMani

Photovoltaic Reliability Laboratory, Arizona State University (ASU-PRL), Mesa, AZ, USA

Abstract— Underperforming cells in a photovoltaic (PV) module or the modules in a PV string are typically detected and mapped using electroluminescence (EL) infrared (IR) imaging, and current voltage (IV) curve techniques. In the current work, a non-contact electrostatic voltmeter (ESV) technique is presented to detect and map the underperforming spots in a cell and the cells in a module. The ESV technique relies on the voltage mapping of the charged surface of the superstrate glass. The voltage values obtained using ESV at various good and poor performing spots of the cells have been validated using the voltage values obtained in EL analysis. The difference between EL-derived voltage and ESV-measured voltage is determined to be less than 2%. In this work, we combine the strengths of two complementary techniques of ESV (strength: quantitative) and EL (strength: spatial mapping) to obtain a quantitative spatial mapping of defects. This work is further extendable to detect poor performing modules in PV power plants.

*Index Terms—*Crack detection, electrostatic voltmeter, electroluminescence, solar cell characterization.

I. INTRODUCTION

The current voltage (IV) curve analysis and the Electroluminescence (EL) imaging are the most used characterization techniques by photovoltaic (PV) industry for detecting the defects in the solar cells. The IV curve analysis provides the overall performance of the PV module but the cracked cells identification, types of cracks and the power lost caused by the cracked cell isn't possible to determine through this curve analysis technique. EL imaging techniques for crack detection in solar cells useful to detect defect and show three modes (A,B and C) of cracks. Mode A cracks are just cosmetic, Mode B cracks partially reduce power, whereas mode C cracks appearing very dark and show cell dead areas. But there is no direct quantitative analysis to conclude the total breakage and separation of that dark area from the cell [1]. EL imaging requires module disconnection from the string and dark environment. Also, the EL approach provides a spatial analysis of the cell, and there are some assumptions involved to get a quantitative analysis of the cell using EL images. Therefore, a non-intrusive solar cell characterization tool is needed to mitigate the deficiencies of existing processes and techniques. In this paper, we present a non-intrusive/non-contact quantitative technique using an electrostatic voltmeter (ESV) to determine the locations and effects of defects in a cell or module. Hishikawa tested ESV as a non-contact measurement of surface electric potential for solar cell application [2]. He introduced the artificial defect in the cell and placed a surface potential sensor at the center of the cell to measure the defective cell output voltage. [3]. This study extends the use of the non-contact electric potential method to map surface voltage to find

the location of the low voltage areas which also gave information about the size, location, and the position of crack by placing the probe sensor at the different positions on the cell. Furthermore, the voltage mapping through EL imaging is performed, and ESV results are compared with it to conclude the effectiveness of this quantitative method for solar cell characterization.

II. EXPERIMENTAL SETUP

In first experiment, high biased and low biased EL images of single isolated clean crystalline silicon (C-Si) solar cell were taken and converted into voltage map by using the technique presented by M. Kontges et al. [4]. Twenty-six spots on cell were selected and voltage mapping was performed using ESV (model 320C TREK manufactured) and EL and ESV voltages were compared. In 2nd experiment, a cracked cell was identified from an EL image of 96 cells (C-Si) module to identify the different cracked areas using TREK ESV model 344. ESV measures voltage without any charge transfer and operates with an external probe sensor and internal data acquisition system, which amplifies, modulates, and demodulates the incoming signal and displays the output as shown in Fig.1 [5],[6],[7]. The current sensed by the ESV probe is amplified and converted into a voltage proportional to the current. As a graphical representation shown in Fig.2, the electric field move out from the cell surface and strike the ESV probe sensor for voltage detection, as shown by black arrows.

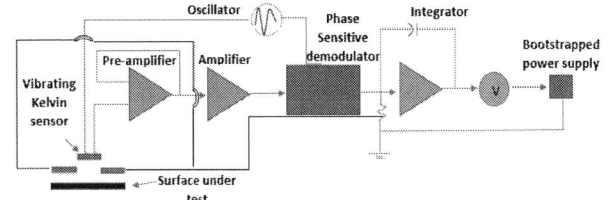

Fig. 1: ESV circuitry.

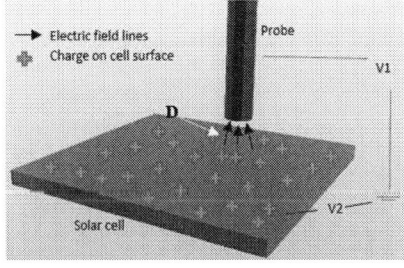

Fig. 2: Probe (shown in black) sensing the cell surface voltage where V1 and V2 is the probe and cell voltage to ground.

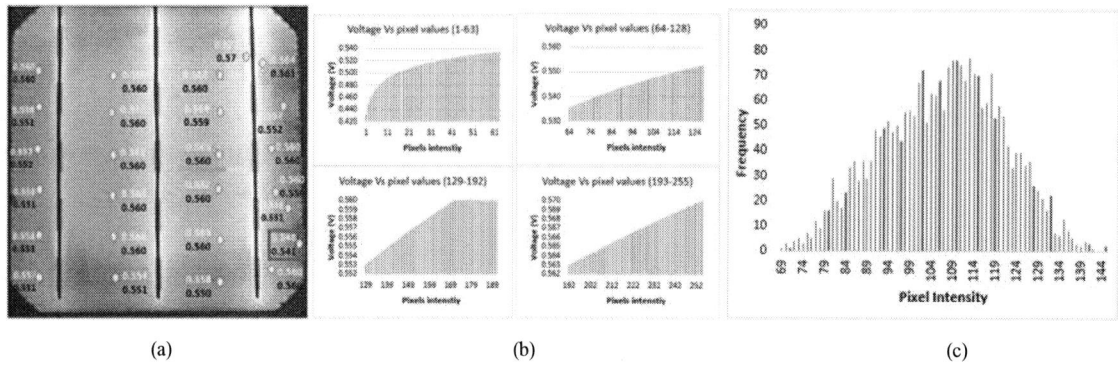

Fig. 3 (a) EL and ESV voltages on high biased EL image. (b) voltage values against every pixel intensity. (c) frequency of the pixels

III. RESULTS AND DISCUSSION

EL images at full and low biased condition were taken and ESV probe was placed directly on the selected positions on cell and voltage sensed is shown by black fonts in Fig.3(a). Furthermore, the EL images were converted into voltage map using the (1) and the EL image converted voltage for the selected points is shown by yellow fonts in Fig. 2(a).

$$\Phi(x) = C(x)e^{\frac{U(x)}{V(t)}} \tag{1}$$

$\Phi(x)$ represents greyscale intensity, $C(x)$ is the calibration constant containing material and optical properties of the material, $U(x)$ is the applied voltage, and $V(t)$ is the thermal voltage of the cell. EL image into voltage conversion equation converts each pixel intensity into voltage but each selected position was having 3165 number of pixels. It is because of the probe sensor larger area than the area covered by one pixel on cell EL image. First, intensity of all 3165 pixels was used and converted into voltage map. Pixel intensities on the EL image ranges from 0 to 255 for 8-bit image where 0 means the darkest area and 255 means brightest area and it is shown in Fig.3(b). Secondly, a point (out of 26 points) on the cell was chosen where ESV showed lowest voltage. It is shown on bottom right corner and named as point 6 on cell. At this point the minimum pixel intensity was 69 and maximum went to 145 and total number of pixels were still 3165. The EL voltages at pixel intensities 69 and 145 were 0.533 and 0.556 volts. The ESV voltage of 0.541V lies within this upper and lower limit of the EL converted voltages. It shows that neither the use of highest pixel is useful in converting the EL image into voltage map nor the lowest one. It is because highest pixel intensity of 145 appeared just two times and lowest pixel intensity 69 appeared one time. Frequency graph of all pixels on point 6 is shown in Fig.3(c). The lowest difference of EL image converted voltage with the ESV voltage appeared at the pixel intensity of 110 which is very close to the average of the 3165-pixel intensities. EL voltage appeared 0.548 and it has a 7mV difference with the ESV voltage. Conversion for EL image to voltage for all 26 points was performed using the average intensity of the pixels and it was observed that the maximum difference between measured voltage values through ESV and EL technology differs by less than 2%.

Fig.4 is showing the EL image of the module. The string of twelve cells was isolated from the module because it had the required cracked cell with three different regions: bright, grey, and dark regions. The ESV probe placed at different positions mapped the high, low, and medium voltage areas on the cell and Fig.5 (a) showing these areas by white, black, and yellow spots. The white circle shows the highest voltage of 3.79 volts. Black circles on the top and right side of the cell are representing minimum voltage, which ranges between 3.69 to 3.79 volts. Yellow circles are showing the medium voltage level of 3.76 to 3.79 volts. This difference of 60-100 mV between the minimum and maximum voltage shows that ESV can detect the low and high voltage producing areas on the cell. The Fig.5(b) represents the EL image of the cell chosen from the 12 cells string, and it shows the different pixel intensity areas on the cell. The EL image shows the dark, white, and grey areas exactly where ESV is reading maximum, minimum, and average voltage values. The bright area on the EL image is falling nearly on the region marked with the white spots on the ESV voltage image. Comparison of Fig.5 (a) and (b) shows the effectiveness of ESV as characterization tool for solar cells. The dark areas on the EL images do not necessarily correspond to completely dead areas. The cell's EL image had visible dark and grey areas, which reflect the power loss at these positions. Usually, in the literature, the dark areas (produced due to the crack) on the EL image are considered dead areas with zero power output [1], but direct measurement of the local cell voltage through ESV provided in-depth information about the severity of the crack. The results indicated the drop in voltage on the dark area, which points to the defected part of the cell. But due to the cell's fingers in contact, a complete loss of voltage did not appear at this point. A cell area with zero volts (100% voltage drop) will be the one with broken fingers and zero electrical contact with the rest of the cell. The visual analysis of the EL images shows that the dark cell area may produce no power, but ESV results proved the presence of an electrically connected path in the cell determining the ESV efficacy in detecting the cracked areas. The EL image analysis showed that pixel intensities at the dark area on the cell range from 5 to 26, and the average value was 20 due to which a voltage appeared there. ESV reads a maximum of 3.73 volts and a minimum of 3.69 volts in the dark area. The output voltage of the last cell was 3.145 volts, and the dead-looking area of the observed cell showed a voltage range between 3.69 to 3.73 volts. It explains that the appearance of 0.545 to 0.585 volts at the dead area of an observed cell is witnessing some electrical connection between the good and dark (dead-looking) area. In the third step, the EL image conversion into the voltage map produced another voltage set. Results were compared with ESV values for analysis and the quantitative analysis through EL image produced a voltage of 3.735 at the pixel intensity of 20, where the ESV voltage reading was 3.730.

978-1-7281-6118-1/22 $31.00 © 2022 IEEE

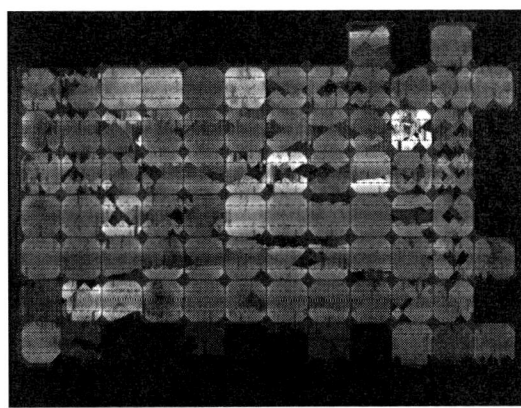

Fig. 4: EL image of module

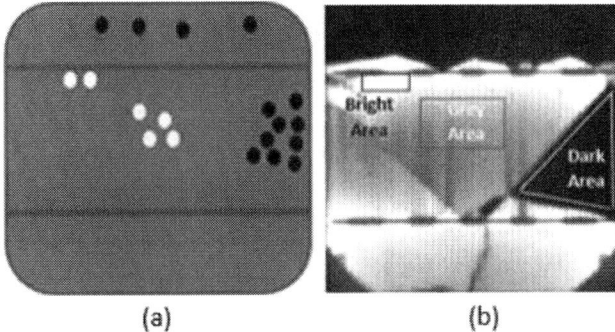

(a)   (b)

Fig. 5: Comparison of the defective areas detected by ESV and EL (a) ESV mapping on the cell (b) defective areas on EL image.

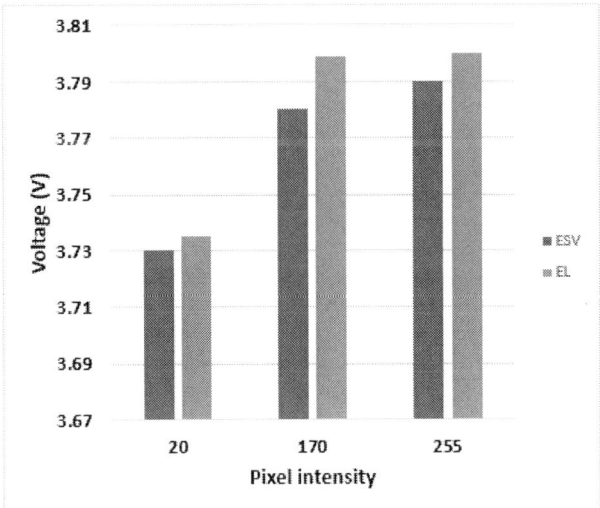

Fig. 6: The ESV and EL voltage values at different pixel intensity values

In Fig. 6, the orange and blue bar is showing EL and ESV voltage, respectively. The comparison of two voltages shows that the ESV sensed voltage is within the 1% range for bright and dark areas of the cell.

IV. CONCLUSION AND FUTURE WORK

Electrostatic voltmeter has been explored as a direct and quantitative characterization/diagnostic tool to map the defect locations and performance impacts of the detected locations on cells and modules.

First, the ESV technique was able to quantitatively differentiate the good areas from the bad areas of the cell. The ESV technique was applied on an individual solar cell and the voltage difference of as high as 100 mV was observed between good and bad areas of the cell.

Second, the directly measured ESV voltages closely matched with the indirectly calculated EL voltages. The ESV-measured voltage was found to be less than 11mV in magnitude compared to the EL-calculated voltage.

Third, the ESV-measured voltages clearly differentiated the performance impacts of bright, gray, and dark areas identified by the corresponding EL images. EL images showed complete dark, apparently isolated, areas on some parts of a cell while ESV measured some voltages in those apparently isolated dark areas indicating the presence of some electrical connection of those dark areas with the rest of the cell. It can be concluded that the ESV technique can potentially be used as a tool to directly differentiate the impacts of different types (A, B or C) of defects identified the EL technique. .

Finally, it may be worth mentioning that the ESV technique can be extended to directly, quantitatively, and non-intrusively detect the poor performing cells in a module and modules in a PV power plant. In this work, the ESV probe was moved manually on the cell surface, that is a time-consuming process. For better repeatability mapping of the whole cell surface, a robotic arm with precise probe movement is recommended.

REFERENCES

[1] I. E. Comission, "Photovoltaic devices - Part 13: Electroluminescence of photovoltaic modules," 2018. [Online]. Available: https://webstore.iec.ch/publication/26703. [Accessed: 01-Feb-2018].

[2] Y. Hishikawa, K. Yamagoe, and T. Onuma, "Non-contact measurement of electric potential of photovoltaic cells in a module and novel characterization technologies," Jpn. J. Appl. Phys., vol. 54, no. 8S1, p. 08KG05, 2015.

[3] S. Miyajima, K. Nishioka, and Y. Hishikawa, "Non-contact Voltage Measurement of Solar Cell with Electrostatic Voltmeter," in *2017 IEEE 44th Photovoltaic Specialist Conference (PVSC)*, 2017, pp. 481–483.

[4] M. Köntges, I. Kunze, S. Kajari-Schröder, X. Breitenmoser, and B. Bjørneklett, "The risk of power loss in crystalline silicon based photovoltaic modules due to micro-cracks," *Sol. Energy Mater. Sol. Cells*, vol. 95, no. 4, pp. 1131–1137, 2011.

[5] B. Zhang, W. Gao, Z. Qi, Q. Wang, and G. Zhang, "Inversion algorithm to calculate charge density on solid dielectric surface based on surface potential measurement," *IEEE Trans. Instrum. Meas.*, vol. 66, no. 12, pp. 3316–3326, 2017.

[6] M. A. Noras, "Non-contact surface charge/voltage measurements," in *Trek application note No. 3001*, 2002.

[7] M. A. Noras, "Non-contact surface charge/voltage measurements Capacitive probe-principle of operation, Trek Application Note No. 3001." 2002.

978-1-7281-6118-1/22 $31.00 © 2022 IEEE

Tellurium Oxide as a Back-Contact Buffer layer for CdTe Solar Cells

Camden Kasik, Ramesh Pandey, Akash Shah, and James Sites
Colorado State University, Fort Collins, Colorado, 80523, United States

Abstract—**Tellurium oxide (TeO$_x$) was deposited at the back of CdSeTe/CdTe substrates to create a buffer layer to reduce recombination and create a high performing dopant free cell. In these experiments we study the effects TeO$_x$ thickness, CdCl$_2$ treatment, and copper doping have on device performance. Current voltage measurements show efficiencies and open circuit voltages up to 17.5% and 829 mV without the use of copper doping. Furthermore, the TRPL lifetime showed a very significant increase in the absence of copper.**

Index Terms—**CdTe, Thin films, Tellurium Oxide**

I. INTRODUCTION

Cadmium telluride (CdTe) solar devices have made significant advancements in recent years, with efficiencies reaching 22.1% [1]. These recent advancements can be attributed to work done at the front contact of devices, incorporating an MgZnO buffer layer and graded CdSeTe/CdTe absorber at the front of the cell. While the short circuit current density (J_{sc}) and fill factor are close the Schokley-Queisser limits, device performance still falls well short of the theoretical maximum of 33%. The limitation on CdTe device performance is attributed to low open circuit voltages (V$_{oc}$) compared to the Schokley-Queisser limit [2].

The back interface of CdTe devices remains the main issue limiting the V$_{oc}$. A lack of passivation of the back interface in CdTe devices causes recombination, limiting the device performance [3]. The deep valance band of CdTe also presents problems. The CdTe valance band being lower than most contacting metals causes downward band bending at the back of the device, limiting the voltage carriers can be extracted at.

Copper doping has been used to increase the V$_{oc}$, but copper is highly mobile in CdTe which causes long-term degradation of modules, decreasing the efficiency and V$_{oc}$ over time [2],[4]. By creating high performing cells without the need for copper doping the degradation in these cells could be sharply reduced. Work with silicon devices has shown the use of metal-oxides as a hole selective transport layer for dopant free devices [5]. A metal-oxide buffer layer may help with surface passivation, band bending, and doping requirements. In this work we explore the use of tellurium oxide as a hole selective back contact layer for cadmium telluried solar devices.

II. EXPERIMENT

A. Device Structures

The TeO$_x$ thickness, CdCl$_2$ treatment, and copper doping were all varied to explore the functionality of a TeO$_x$ buffer layer. All devices fabricated in this experiment had a RF

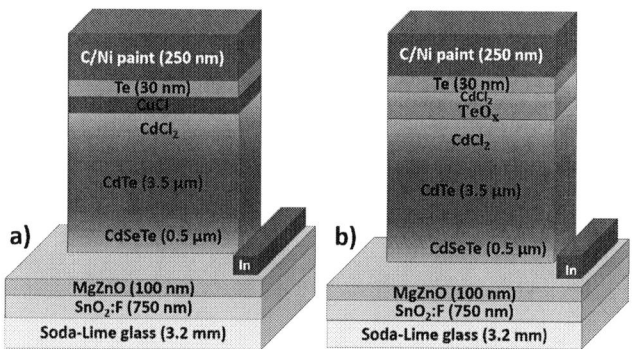

Fig. 1. a) Baseline device used for comparison to experimental devices. b) Example structure of a TeO$_x$ device used in these experiments.

sputtered MgZnO buffer layer at the front of the cell described here [6]. An absorber consisting of $0.5\mu m$ cadmium selenium telluride (CST) and $3.5\mu m$ CdTe were deposited via closed space sublimation (CSS) without any intentional grading. The deposition temperatures were 575°C and 555°C for CST and CdTe respectively. An aggressive CdCl$_2$ passivation treatment followed without a vacuum break.

After the absorber deposition, either a CuCl treatment or a TeO$_x$ layer followed as shown in figure 1. Copper doping was done via CSS and the TeO$_x$ layer was deposited by RF sputtering onto the absorber layer. TeO$_x$ thicknesses of 5 nm, 10 nm, and 20 nm were used to explore the effect of thickness on performance. Devices with and without copper were fabricated to see if copper doping was essential to provide respectable V$_{oc}$ and efficiencies for TeO$_x$ devices. The copper doping in these cases was done after the second CdCl$_2$ treatment to avoid Cu migration due to the 450°C CdCl$_2$ deposition temperature.

To study the effects of CdCl$_2$ on TeO$_x$, devices were made with two different passivation treatments. The first being an aggressive 900 s CdCl$_2$ treatment being done before the TeO$_x$ layer, to passivate the bulk of the absorber, followed by another 900 s treatment done after the layer of TeO$_x$ was deposited. The second CdCl$_2$ sequence tested involved no CdCl$_2$ treatment prior to the TeO$_x$ layer and instead only received the 900 s treatment after the TeO$_x$ deposition. This was done to see if performing a passivation treatment only after the buffer layer would passivate both the CdTe/TeO$_x$ interface and the bulk of the absorber. All devices were finished with an evaporated 30 nm Te layer, carbon paint

978-1-7281-6118-1/22 $31.00 © 2022 IEEE

polymer binding, and Ni paint. Twenty five devices were then delineated from the manufactured plate for measurements.

B. Characterization

Performance of the devices was measured with current voltage (J-V), under standard conditions, and admittance measurements. Temperature dependent J-V measurements (JVT) were also done to inspect the back contact barrier of devices with TeO_x and different dopants. Photoluminescence (PL), with an excitation laser of 520 nm, and time resolved photoluminescence (TRPL) measurements were used to investigate the carrier lifetime.

III. RESULTS AND DISCUSSION

A. Thickness and copper doping

Undoped and doped devices were made with different thicknesses of TeO_x to compare both the impact thickness has on performance, and if thickness matters for the impact of copper doping. The J-V parameters were measured and the results of the undoped devices are shown in figure 2. On average the efficiencies for devices without copper doping outperformed devices with copper doping. Devices with TeO_x demonstrate a significant improvement when there is no copper doping. For a 5 nm TeO_x layer the average efficiencies go from 14.8% to 17.1% while V_{oc} goes from 779 mV to 808 mV when copper is excluded. The same favorable trend for TeO_x devices without copper was shown for the 10 nm devices. TeO_x devices without copper not only have better overall performance than doped devices, but the uniformity across the twenty five device plate is significantly better.

Based on the results from these devices, it is clear the 5 nm devices outperformed the thicker 10 nm devices in every parameter, with the best 5 nm device reaching 17.5% efficiency with a V_{oc} of 829 mV. While the J_{sc} between the 5 nm and 10 nm devices seems unaffected, the efficiency improvement can be attributed to higher fill factor and V_{oc}. Increasing the thickness of TeO_x to 20 nm further reduced device performance.

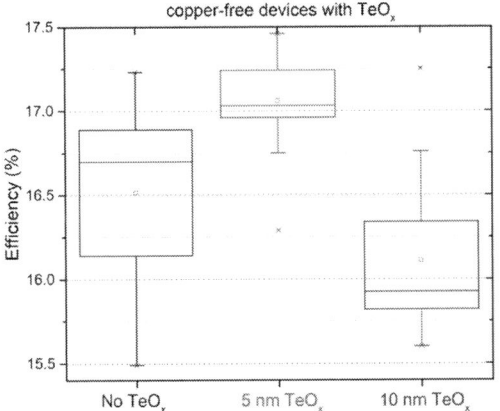

Fig. 2. Efficiency plots comparing devices of different TeO_x thicknesses without intentional Cu doping

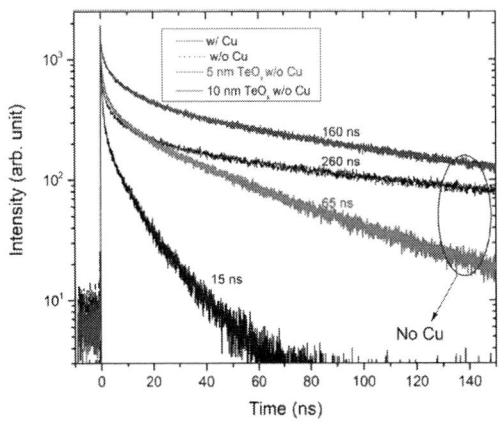

Fig. 3. TRPL plots for devices with 5 and 10nm TeO_x layers, as well as a doped and undoped baseline device for comparison.

The carrier lifetimes of undoped TeO_x devices are shown in figure 3. It is well known copper doping limits carrier lifetime [7]. Figure 3 shows this is true with TeO_x devices as well. The lifetimes are lower than the undoped cell without TeO_x, but show serious improvement compared to the copper doped device.

Temperature dependent J-V measurements, plots shown in figure 4 were made to inspect the back barrier of TeO_x devices. As the temperature is decreased the device with copper doping exhibits a rollover effect, indicative of the formation of a back contact barrier inhibiting current flow. The device without intentional doping does not demonstrate this rollover effect. This suggests TeO_x does not create a large hole barrier by itself, but the intentional copper doping has some current limitation in forward bias.

B. $CdCl_2$ Treatment

The $CdCl_2$ variations were made on devices without intentional copper doping after finding the results of the previous section. The single $CdCl_2$ was performed after the TeO_x with the goal to passivate both the $CdTe/TeO_x$ interface, and the bulk CdTe. Devices with only a single $CdCl_2$ treatment performed far worse than those with a double passivation treatment. Average efficiencies went from 8.5% and 7.6%, for 5 nm and 10 nm respectively, to 17.1% and 16% when passivation was done before and after the TeO_x layer. A major improvement to performance uniformity was also noticed when including two $CdCl_2$ treatments. These improvements were present in all J-V parameters measured.

Admittance measurements were used to create carrier density plots shown in figure 5. Devices with two $CdCl_2$ treatments had an order of magnitude increase, from 1E14 to 1E15, in carrier density compared to single treatment devices. The zero bias point for the double $CdCl_2$ devices is at the bottom of the curve, while it is on the right branch of the single treated devices. This indicates the absorbers become fully depleted earlier in voltage bias when only a single $CdCl_2$ treatment is used.

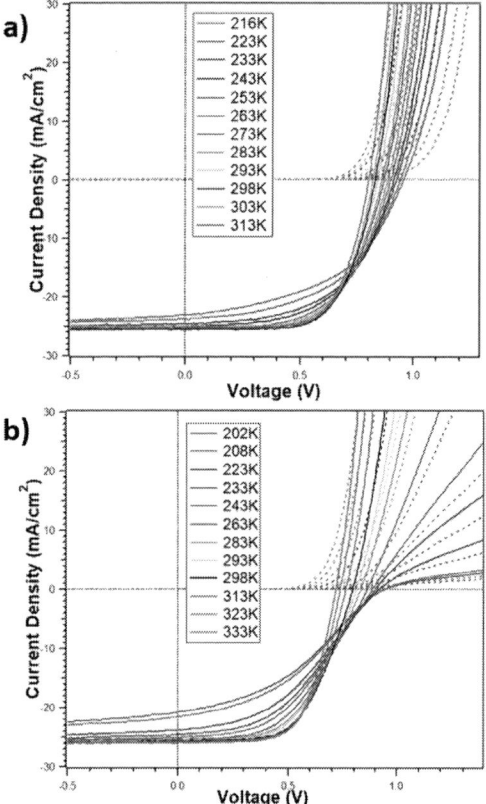

Fig. 4. J-V curves measured at different temperature to inspect the affect of doping in TeO_x on the back barrier. a) No intentional Cu doping b) Device doped with Cu after TeO_x layer.

These results show the importance of passivating before the TeO_x layer to ensure bulk passivation, and treating after the TeO_x for interface passivation. A single $CdCl_2$ passivation treatment done after the TeO_x layer is not adequate to passivate both the interface and the bulk absorber.

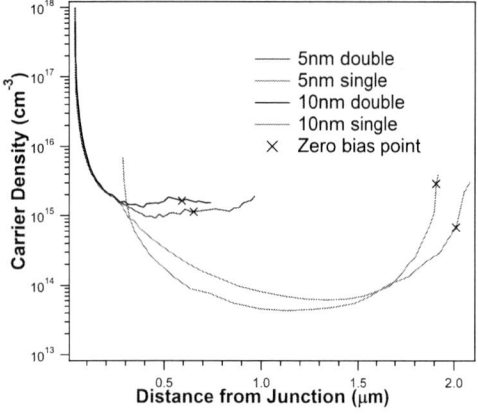

Fig. 5. Carrier density plots for devices with different $CdCl_2$ treatments. This is used to show the effect a second passivation treatment has on TeO_x devices.

IV. CONCLUSION

Copper doping has been widely used in CdTe devices to produce high efficiency devices, but this comes with the cost of lower carrier lifetimes and faster module degradation. Using TeO_x as a back contact buffer layer we have demonstrated devices up to 17.5% without intentional doping. Creating high performing devices without intentional doping could increase module lifetime, and cost effectiveness, of CdTe solar cells. A 5 nm TeO_x layer, with a $CdCl_2$ passivation treatment before and after the layer, produces the highest performing devices in our experiments. JVT measurements suggest the inclusion of copper doping in TeO_x devices contributes to the formation of a hole barrier in forward bias. We believe this TeO_x layer favorably bends the energy bands at the back of the cell, helping hole collection and reducing recombination at the back interface. Photoelectron spectroscopy will be used to determine the band structure around the TeO_x layer. The large bandgap of TeO_x makes it a viable candidate for bifacial and tandem solar cells, which is an area we hope to explore in the future.

ACKNOWLEDGMENT

The authors would like to thank Dr. W.S. Sampath for use of his deposition systems and numerous helpful discussions, Pascal Jundt for TRPL measurements, and 5N Plus for deposition materials.

REFERENCES

[1] "Best research-cell efficiency chart," *NREL.gov,* 17-Nov-2021. [Online]. Available: https://www.nrel.gov/pv/cell-efficiency.html.

[2] J. M. Burst et al., "CdTe solar cells with open-circuit voltage breaking the 1v barrier," *Nature Energy, vol. 1, no. 3,* 2016.

[3] Alexandra M. Huss, Jennifer A. Drayton, and James R. Sites. "Front and Back Interface Recombination of MZO/CdTe/TeSolar Cells," *2018 IEEE 7thWorld Conference on Photovoltaic Energy Conversion(WCPEC) (A Joint Conference of 45th IEEE PVSC, 28th PVSEC 34th EU PVSEC),* 2018.

[4] M. Nardone and D. S. Albin, "Degradation of CdTe solar cells: Simulation and Experiment," *IEEE Journal of Photovoltaics, vol. 5, no. 3, pp. 962-967,* 2015.

[5] C. Battaglia, S. M. de Nicolas, S. de Wolf, X. Yin, M. Zheng, C. Ballif, and A. Javey, "Hole selective moox contact for silicon heterojunction solar cells," *2014 IEEE 40th Photovoltaic Specialist Conference (PVSC),* 2014.

[6] A. H. Munshi et al., "Polycrystalline cdsete/cdte absorber cells with 28 ma/CM2 short-circuit current," *IEEE Journal of Photovoltaics, vol. 8, no. 1, pp. 310–314,* 2018.

[7] D. Kuciauskas et al., "The impact of CU on recombination in high voltage CdTe solar cells," *Applied Physics Letters, vol. 107, no. 24, p. 243906,* 2015.

Intelligent Cloud-Based Monitoring and Control Digital Twin for Photovoltaic Power Plants

Andreas Livera [1], George Paphitis [1], Loucas Pikolos [1], Ioannis Papadopoulos [1], Jesús Montes-Romero [2], Javier Lopez-Lorente [1], George Makrides [1], Juergen Sutterlueti [3] and George E. Georghiou [1]

[1] PV Technology Laboratory, FOSS Research Centre for Sustainable Energy, Department of Electrical and Computer Engineering, University of Cyprus (UCY), 1678 Nicosia, Cyprus

[2] Advances in Photovoltaic Technology, CEACTEMA, University of Jaén (UJA), 23071 Jaén, Spain

[3] Gantner Instruments GmbH, Montafonerstraße 4, 6780 Schruns, Austria

Abstract—A main challenge in the scope of integrating higher shares of photovoltaic (PV) systems is to ensure optimal operations. This can be achieved through next-generation monitoring with automatic data-driven functionalities. This work aims to address this fundamental challenge by presenting the stage of implementation of an advanced cloud-based monitoring platform and a control digital twin for PV power plants (MW scale). The platform is fully equipped with a multitude of artificial intelligent (AI) algorithms for health-state diagnostics and analytics. The performance of the digital twin to act as a health-state monitor was validated against field and synthetic data from PV systems at different locations and demonstrated high accuracies for PV performance modelling and fault diagnosis.

Keywords—data quality, digital twin, machine learning, monitoring, performance, photovoltaic

I. INTRODUCTION

Entering an era towards a decarbonised future requires a radical energy and digital transition supported by intelligent and fully flexible solutions, enabling higher shares of solar and distributed energy resource (DER) technologies. Along this context, the increase of lifetime production and sustainability of photovoltaic (PV) systems renders imperative the alignment with new communicative, automated and interactive developments such as Solar 3.0, Internet of Things (IoT) and Industry 4.0 concepts [1]. Consequently, enhanced PV power plant monitoring and operational control through automated functionalities and intelligent grid-edge solutions present a very important feature for plant operators and utilities in the direction of optimal operation management.

Recent advances in the field of artificial intelligence (AI), cloud computing and IoT [2], enable PV monitoring systems to evolve and integrate data-driven diagnostic and predictive analytics [3]. AI-driven capabilities are utilized to optimize PV plant performance by quickly identifying and accurately quantifying the factors behind various failure/loss mechanisms and timely mitigating systemic issues [4]–[6]. Within this framework, an important element of an intelligent monitoring system is the concept of digital twin (i.e., a virtual system that accurately replicates the physical asset, such as a PV system, and

its performance) for health-state monitoring and diagnostics [7]. Digital twins can use real-time IoT data and apply data quality and AI-driven analytics to schedule predictive maintenance, implement performance optimizations and trigger alarms in case of systemic issues and faults [7]. These capabilities add up to the fact that the digital twin model can change in real-time as the state of the physical system changes during the operation. Thus, the utilization of the digital twin technology can help PV operators to identify maintenance needs more quickly and as a result lower costs and minimize downtimes, while delivering "extra-returns" (e.g., increasing revenues for PV plant owners and/or maximizing the return on investment).

At the core of the implemented PV digital twins for fault diagnostics is the comparative analysis of the performance of the digital twin (by leveraging a multitude of AI techniques, including ensemble machine learning models, neural networks, support vector machine, etc.) with real-time assets to check for performance deviations (indicating faults and triggering alarms - see Fig. 1) and provide new insights for better decisions and timely maintenance [8].

Even though, a lot of parts of PV systems are already computerized and acquire massive amounts of data, the

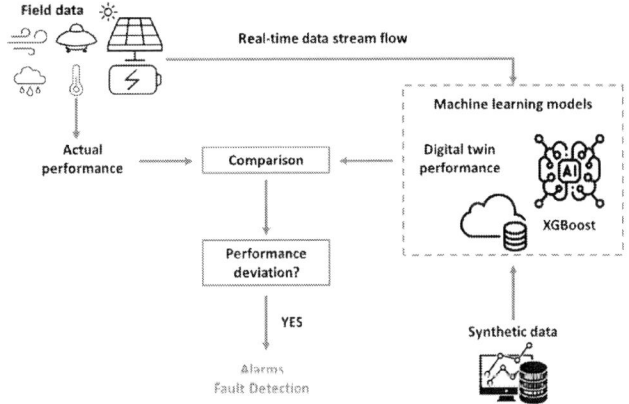

Fig. 1. Schematic of digital twin approach for PV fault diagnostics and health-state monitoring.

utilization of those data to build accurate predictive capabilities (via digital twins) is an area of continuous improvement [9]. To this end, the lack of a unified health-state approach that does not rely on the availability of on-site data, is the main reason why there is still not a generalized and standardized method to accurately construct digital twin models of PV systems [7].

The purpose of this paper is to address the fundamental challenges of developing robust, scalable, and accurate monitoring system and control solutions for large-scale PV systems (MW scale). The proposed digital twin enables the digital-enhancement of PV power plants for real-time asset observability and control (using grid-edge Linux devices and communication interfaces and industrial protocols/standards like Modbus TCP, EtherCat, OPC-UA, etc., to stream high-resolution data to an advanced cloud database), which acts as a high-level health-state monitor to timely prognose/diagnose failures. Ultimately, useful information is further provided on the effectiveness of the digital twin when trained using field and/or synthetic generated data of varying granularity and applied to the latest machine learning (ML) regressor (i.e., the eXtreme Gradient Boosting).

II. EXPERIMENTAL SETUP

Historical field measurements obtained from two sites with different climates were used to demonstrate the functionalities of the platform and to validate the performance of the digital twin: (1) the DER-Grid smart infrastructure of the University of Cyprus (UCY)-FOSS nanogrid (nG) in Cyprus (Köppen-Geiger-Photovoltaic [KGPV] climate classification CH; steppe climate with high irradiation) [10]; and (2) a utility-scale PV power plant administered by Gantner Instruments (GI) in the Netherlands (KGPV climate classification DM; temperate climate with medium irradiation) [10].

The DER-Grid smart infrastructure of UCY-FOSS is a flexible and scalable testing and demonstration platform for smart grid technologies. The infrastructure includes solar-plus storage and test-bed PV systems. In this work, a test-bench PV system of 1.025 kW$_p$ nominal capacity (that is connected to the grid using a smart inverter) was used for the construction and validation of the digital twin. The test-bench PV array comprises of five crystalline-Silicon (c-Si) modules connected in series. The system is installed in an open-field mounting arrangement at a tilt angle of 27° due South.

Similarly, the utility-scale PV power plant administered by GI is a 54.36 MW$_p$ c-Si ground mounted decentral system that comprises of 651 multi-string inverters. The field measurements (obtained at the inverter and system level) from the PV power plant were used to verify the site-independence and transferability of the proposed digital twin.

The main PV operational parameters along with the prevailing meteorological conditions of the two PV sites are recorded using a high-performance edge controller (i.e., the Gantner Instruments Q.station-XT) for high-speed data acquisition and real-time control. Data recording and storage (at a resolution of 1 second and 15-, 30- and 60-minute granularity) is according to the requirements set by the IEC 61724-1 [11].

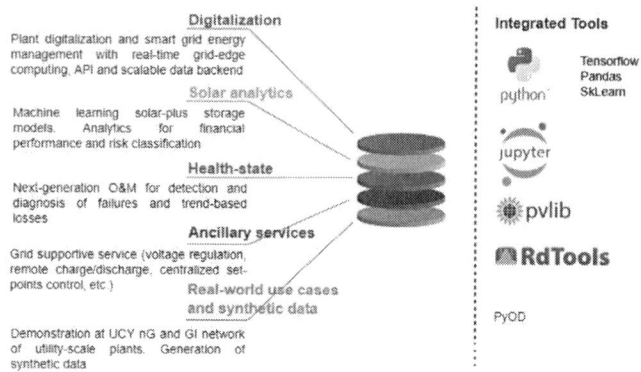

Fig. 2. Integrated functionalities of the advanced cloud-based monitoring platform for PV systems.

III. METHODOLOGY

An intelligent cloud-based platform was developed in this work for real-time monitoring and control in PV power plants of MW scale. The platform incorporates a digital twin for PV performance assessment and fault diagnosis. Solar analytic functionalities were also integrated to the platform (see Fig. 2) by leveraging statistical and ML techniques. Such functionalities entail the diagnostic power of AI and mechanistic approaches to overcome data mining issues and capture the actual behaviour of PV systems, to provide insights and effectively diagnose performance issues proactively, thus enhancing the efficiency and profitability of PV power plants.

A. Digitalization

The advanced monitoring platform utilizes adaptable and scalable Linux edge computing devices with cloud connectivity that act as power plant controllers for real-time data-acquisition (sampling frequency lower than 1 second and 1 kHz cycles internally for control), high-speed triggered data logging (sampling frequency 1 Hz) and analytics. The devices are configured with developed interoperable interfaces and tools utilizing popular open-access communication protocols (i.e., Modbus TCP/IP and SunSpec) [4]. This facilitates the implementation of a unified digitalized information flow architecture for future PV power plants enabling remote and real-time observability, management and reliable process control (see Fig. 3).

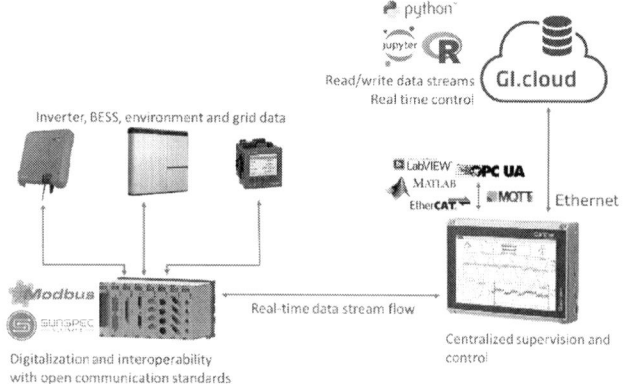

Fig. 3. Next-generation multi-service monitoring system architecture for grid-edge control and AI-driven smart grid services.

The demanding streaming of big data is performed through secure websockets. The data are stored by the "GI.cloud" advanced platform, that integrates high-resolution sampling, high-performance big data analytics and easy data accessibility.

The platform combines Gantner's edge-type monitoring and control units, an adaptive and scalable cloud backend and a comprehensive user interface and applications with state-of-the-art application programming interfaces (API).

B. Real-world use cases and synthetic data

To demonstrate the effectiveness of the monitoring platform and control digital twin, actual PV datasets containing field meteorological and electrical measurements (obtained from the two sites) were used. The datasets were constructed from the recorded field measurements of in-plane irradiance (G_I), ambient temperature (T_{amb}) and DC array/ inverter power (P_A). The solar azimuth (φ_s) and elevation (α) angles were also estimated using solar position algorithms [12]. Finally, additional yields and performance parameters such as the PV array energy yield (Y_A), the final PV system yield (Y_f), the reference yield (Y_r) and the monthly DC performance ratio (PR) were calculated [3]. Lastly, weather data that were unavailable at the power plants (e.g., rainfall measurements) were sourced from Modern-Era Retrospective analysis for Research and Applications, Version 2 (MERRA-2) [13].

Data quality routines (DQRs) were then applied to the constructed PV datasets to ensure data fidelity [14]. The DQRs methodology incorporates statistical and comparative algorithms to detect data issues, such as sensor malfunctions and/or invalid (erroneous and missing) values, as well as a sequence of filtering (e.g., $G_I > 50$ W/m^2 and $P_A > 1\%$ of rated system power in this study) and aggregation (i.e., daily, weekly and monthly) stages for time series preparation for further performance analysis [14]. Data deletion, correction and inference techniques were not considered in this work to fully capture the exhibited patterns during fault conditions.

In parallel, synthetic data were generated for the two test PV installations by employing the Sandia PV Array Performance Model (SAPM) [15]. The SAPM was used to simulate the power output of the investigated PV system by utilizing the historical measurements of G_I, T_{amb} and system-specific metadata. One of the biggest drawbacks of the model is that SAPM coefficients lack of physical meaning [16]. The mechanistic performance model (MPM) could also be used as a robust alternative solution for synthesising PV performance data [17].

C. Solar analytics (PV System Predictive Models)

The digital twin was developed by applying the MPM and a ML model to both field and synthetic data (over a 1-year period). The MPM is an improved version of the parametric model [17] applied on normalised PV performance data [9]. MPM's advantages include its simplicity and high predictive accuracy, while also yielding meaningful model coefficients [9], that can be used for detecting PV degradation modes and failures [17].

In parallel, the eXtreme Gradient Boosting (XGBoost) regressor was selected to predict the PV performance since it is one of the most popular ML algorithms these days [18]. The XGBoost has reported the best performance in both prediction and classification problems [18]. It is an ensemble algorithm that combines decision trees (or weak learners) using a gradient boosting architecture to construct an enhanced model and optimize the output prediction (i.e., the power at the DC side in this work) [7].

For the XGBoost model development, the available input features of G_I, T_{amb} were used along with the calculated parameters of φ_s and α. The model hyper-parameters were subsequently derived as reported in [7]. To construct an optimally performing XGBoost digital twin, a supervised learning procedure was followed. Specifically, the 1-year field and synthetic time series were split into train and test subsets. Different train set partitions (10%, 30%, 50% and 70% shares of the entire time series) and two data split approaches were then examined. The first approach (i.e., the sequential) utilized successive samples, while the second (i.e., the random) included random samples taken from the entire time series.

The optimal derived digital twin can be then used for PV performance prediction, timely and proactive detection of faults (by triggering alarms in case of performance deviations) [7].

D. Health-state (Diagnosis of failures and losses)

PV system health-state tools are used to monitor the state condition of PV assets in real-time and detect underperformance conditions (e.g., failures and trend-based losses) in PV systems. Failures (causing power reductions from 20% to 100% compared to the predicted power [19]) can be detected by applying either supervised or unsupervised ML algorithms on the available time series data. Supervised algorithms are preferred in cases of historic measurements and maintenance log availability and when fault emulation and/or data labelling is possible. The procedure for developing an accurate XGBoost digital twin for failure diagnosis based on a supervised process (using both field and synthetic data) was described in detail by Livera *et al.* [7]. Fault detection and classification results from the application of the XGBoost classifier on data from the test-bench PV system were also presented in [7]. The obtained results proved that the proposed ML classifier was capable of detecting field-emulated failures and synthetic power-loss events (that may even cause small power loss – relative power reduction of 5%), achieving high sensitivity and specificity indicators (> 82%) [7].

For practical PV diagnostic applications (e.g., in cases of maintenance logs and labelling information unavailability), unsupervised procedures are particularly suited. In this context, fault conditions can be detected by applying outlier detection algorithms (e.g., boxplots, Isolation Forest, etc.) on the available time series data. The utilization of unsupervised algorithms enables the universal application (i.e., algorithms are applicable to any PV system, and they are data-, location- and installation-independent) of health-state tools.

In this work, the *PyOD* python library was used [20], that includes more than 30 detection algorithms. Therefore, an initial investigation using synthetic generated data and imputed fault conditions (i.e., reduced power measurements) was performed to select the best performing algorithm for detecting outliers in PV performance time series data. Fault patterns were introduced to the synthetic generated data by declining the array DC power at a relative magnitude of 10% or 20% during high irradiance conditions ($G_I > 600$ W/m^2). The introduction of fault conditions (that were imputed at a 10% share per year) is necessary for the training procedure and performance assessment phase of the fault detection and classification algorithms.

In parallel, trend-based performance losses (e.g., soiling, snow loss and degradation) were detected by leveraging open-source tools and statistical techniques. Trend-based losses refer to linear and nonlinear drops in performance time series and profiles that may reduce the produced power of a PV system by up to 20% [3]. Such phenomena can result in reversible (e.g., soiling) and irreversible (e.g., module degradation) performance loss based on the caused damage [3].

Soiling losses were extracted by utilizing the *RdTools* open-source python library [21] and the stochastic rate and recovery (SRR) method [22], that can quantify the soiling loss directly from energy yield/performance data. Statistical techniques were then used for estimating the linear or nonlinear (by taking the weighted average of the segments) performance loss rate (PLR). Common techniques for linear PLR estimation include the linear regression with ordinary least squares (OLS), classical seasonal decomposition (CSD), seasonal and trend decomposition using locally weighted scatterplot smoothing, LOWESS, (STL), and Holt-Winters (HW) triple-exponential smoothing. In addition, the year-on-year (YOY) method, that is available through the python library *RdTools* [21], was also used. For nonlinear PLR estimation, the Facebook prophet (FBP) [23] was selected to model the monthly PR trends and subsequent change-points (CPs). The FBP detects the number and location(s) of CPs by capturing statistical changes in the slopes of defined segments. It distributes "potential" CPs uniformly along the selected range of the time series' trend and then compares the slopes to extract the significant CPs. The FBP can also be used for optimal scheduling of operation and maintenance (O&M) activities, since it has the capability of differentiating faults from reversible and irreversible loss mechanisms, by adjusting its flexibility parameter [19].

The application of health-state tools provides insights into plant performance bottlenecks. Addressing the detected performance issues in a timely manner, will increase plant's productivity (and revenue) and will minimize downtimes.

E. Performance Metrics

The normalised root mean square error (nRMSE), that describes the standard deviation of the prediction errors relative to the nominal capacity of the PV system ($P_{nominal}$), was used to evaluate the predictive accuracy of the digital twin model [7]. The nRMSE was calculated as follows:

$$nRMSE = \frac{100}{P_{nominal}} \sqrt{\frac{\sum_1^n (y_{measured,i} - y_{predicted,i})^2}{n}} \quad (1)$$

where n is the amount of data points, $y_{measured,i}$, $y_{predicted,i}$ is the measured and predicted power, respectively.

To evaluate the performance of the fault classification algorithms, labels were inserted to the PV datasets that contain emulated power-loss conditions. A confusion matrix of a binary classification problem (Class 0 represents normal PV operation, while Class 1 indicates fault incidents) was used along with the sensitivity and precision metrics [7].

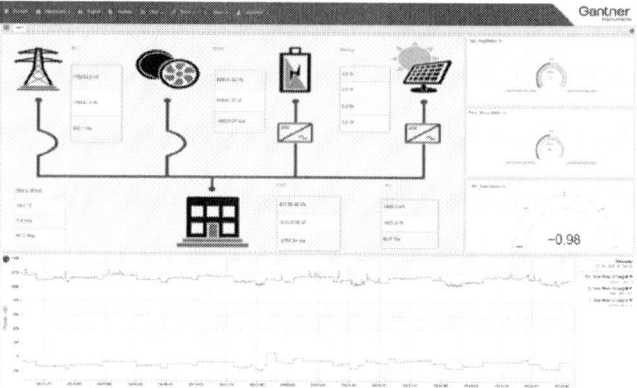

Fig. 4. Supervisory dashboard visualizing real-time data streams from a solar-plus storage system installed at the UCY-FOSS nG in Cyprus.

IV. RESULTS

In this paper, focus is shed on the results obtained from the digital twin implementation for PV performance modelling and fault diagnostics. The operations of the digital twin were validated at the test-bench PV system in Cyprus and verified at a utility-scale PV system of GI's network, in the Netherlands.

A. Cloud-based Monitoring Platform

An advanced cloud-based platform for real-time PV performance monitoring and reliable control was developed. The monitoring platform provides real-time supervision and observability of Distributed Energy Resource (DER) assets through the features of data quality, digital twin replica, health-state monitor, event-triggered fault detection and power quality, smart grid energy services and control alerts.

As a snapshot of the commenced work, Fig. 4 shows the GI cloud dashboard designed to visualize 1-second data streams acquired from the solar-plus storage system in Cyprus. The monitoring data (an example is provided in Fig. 5, showing the UCY campus monitoring) provide the necessary real-time observability of all assets and allow operators to assess the health-state of DER components. Automatic alerts and O&M recommendations are provided in case of systemic issues and failures.

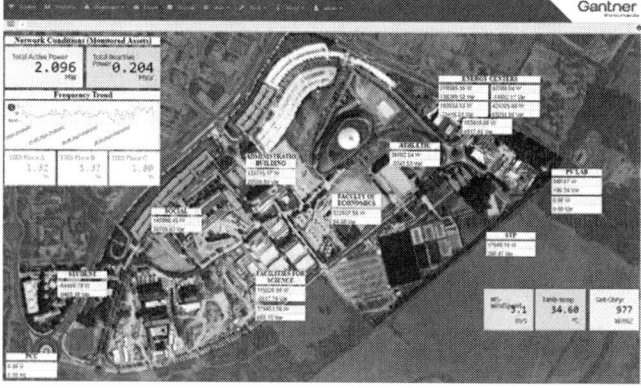

Fig. 5. Supervisory dashboard visualizing UCY campus monitoring for smart grid services and control.

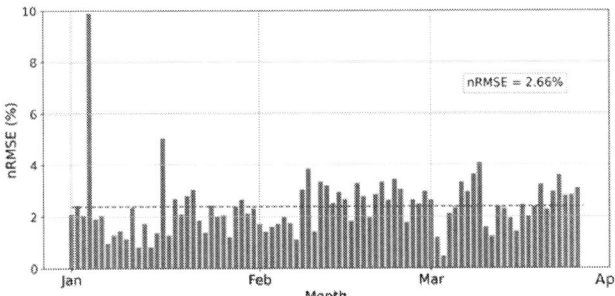

Fig. 6. XGBoost digital twin predictive accuracy (given by daily nRMSE) for the test-bench PV system over the test set period (January to April). The model was trained with 60-minute field data and a sequential 70:30% train and test set approach. The train set included days from June to December.

B. PV System Predictive Models

The work commenced to construct the optimal digital twin included an extensive evaluation of applying the XGBoost regressor to both field and synthetic data (over a 1-year period) of the test-bench PV system at different supervised learning regimes. The results demonstrated that the digital twin, trained with field data (70:30% train and test set approach using successive samples), exhibited high predictive accuracies (nRMSE < 3.17%) at granularities of 30- and 60-minute. The daily *nRMSE* when applying the digital twin model (trained using successive samples from June to December and 60-minute field data) to the test set data is shown in Fig. 6, demonstrating an average *nRMSE* of 2.66%. When using random samples for the training process (70:30% train and test approach), the model's accuracy was improved, achieving a nRMSE of 2.07% and 1.92% at granularities of 30- and 60-minute, respectively.

In case of field data unavailability, training the model with synthetic generated data proved to be a robust alternative. For 60-minute synthetic data, an average nRMSE of 3.83% was obtained for the model when using successive samples (70:30% train and test set approach). For the respective case of random sampling, the nRMSE was reduced to 3.22%.

To prove the robustness of the model on low fractions of field data, the model was then developed by using a learning procedure applied to four different train sets. These were generated from the actual PV dataset (containing 60-minute field measurements) by partitioning it sequentially into train subsets of 10%, 30%, 50%, and 70%. The results of the different data train partitions using both successive and random samples are summarised in Table I. When trained with different field data partitions from 10-70% shares, the model yielded nRMSE in the range of 1.92-2.91% and 2.66-4.65% for random and successive samples, respectively.

TABLE I. PREDICTION ACCURACY OF THE XGBOOST MODEL AT DIFFERENT SUPERVISED LEARNING TRAIN PARTITIONING REGIMES FOR THE TEST-BENCH PV SYSTEM IN CYPRUS

Supervised learning regime partition	nRMSE (%)	
	Random samples	*Successive samples*
10% train set and 30% test set	2.91	4.65
30% train set and 30% test set	2.50	4.34
50% train set and 30% test set	2.16	2.92
70% train set and 30% test set	1.92	2.66

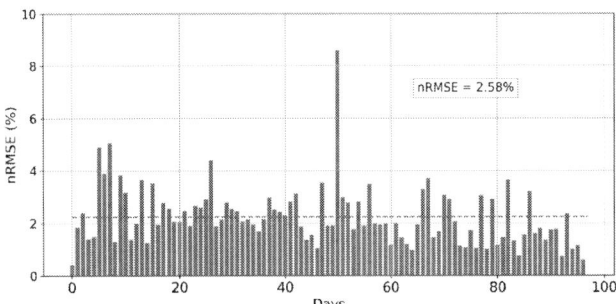

Fig. 7. XGBoost digital twin predictive accuracy (given by daily nRMSE) for PV power plant (inverter level) data over the test set period. The model was trained with 15-minute field data and a random 70:30% train and test set approach (i.e., the days in the subsets were randomly selected from the yearly dataset, thus including days from all four seasons).

Historical field measurements (15-minute granularity) from the PV power plant were used for the benchmarking procedure in respect to data granularity-, location- and system-independence. The XGBoost digital twin model was trained randomly using a 70% portion of field measurements and yielded an average nRMSE of 2.58% (see Fig. 7).

Likewise, the MPM digital twin trained by a random 70:30% train and test set approach, exhibited an average nRMSE of 2.95% (see Fig. 8). By fitting the data using the MPM, physically significant coefficients were obtained (C1=71.83%, C2=-0.34%/K, C3=14.42%, C4=0.46%, and C5=1%/ms^{-1}); C1 represents the maximum power - overall quality of the PV system (tolerance dependency), C2 indicates the maximum power temperature coefficient "gamma", which aligns with the stated gamma value (i.e., -0,0037) in the manufacturer's datasheet, C3 gives the low light dependency (due to open-circuit voltage and the shunt resistance of the module), C4 represents the high light dependency (due to module series resistance) and C5 indicates the wind speed dependency. These coefficients are also important for fault diagnostics (e.g., coefficients that glitch or are not expected values may indicate fault conditions); C1 + C4 gives the expected performance (~72%) at Standard Test Conditions (STC), C2 is important for degradation studies (e.g., quantify changes with MPM coefficients over time), C3 and C4 are loss characterisation and identification coefficients.

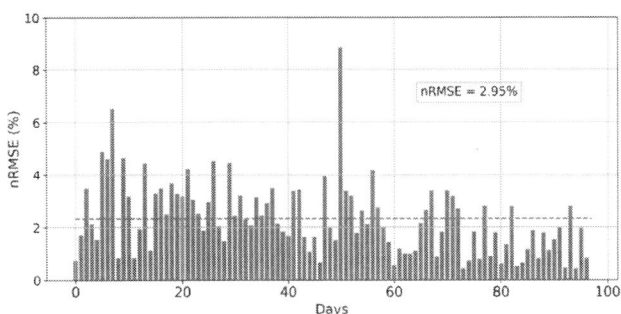

Fig. 8. MPM digital twin predictive accuracy (given by daily nRMSE) for PV power plant (inverter level) data over the test set period. The MPM was trained with 15-minute field data and a random 70:30% train and test set approach (i.e., the days in the subsets were randomly selected from the yearly dataset, including days from all four seasons).

The obtained results demonstrated that both models can provide high predictive accuracy using 15-, 30- and/or 60-minute granularity measurements. In addition, the XGBoost yielded lower nRMSE values than the MPM. Though, the MPM is a simplistic model, robust at low field data training partitions [9] and it provides meaningful coefficients for PV performance and fault diagnostics [17]. Therefore, the selection of the digital twin model depends on the choice of the final user (i.e., PV plant owner/operator) and it is dependent on several parameters (e.g., the availability of measured parameters and monitoring period, model complexity and accuracy, extracted coefficients, etc.).

In the scope of providing guidelines for optimally training the predictive models, the obtained results provided adequate information that models should be trained using random data samples. Furthermore, a greater amount of field measurements for the train set is preferable to achieve more accurate results. Even in cases of field data unavailability, robust digital twin models can be built using synthetic generated data. Finally, the results from both sites provide evidence of the effectiveness of digital twin models for PV performance modelling at varying data granularities, different system installations and locations, verifying its transferability and suitability for any PV system.

C. Diagnosis of fault conditions (failures and trend losses)

A comparative assessment between different unsupervised outlier detection algorithms, implemented in the PyOD python toolbox [20], was initially performed using daily aggregated data. As such, yearly synthetic data from the test-bench PV system with emulated fault patterns (i.e., reduced power measurements by 10% magnitude) at a 10% share in the time series (i.e., 30 data points in total) were used for assessing the performance of the outlier detection algorithms. The

benchmarking results demonstrated that the Angle-Based Outlier Detection (ABOD) was the best performing algorithm for detecting outliers in the daily power time series. The ABOD correctly detected 26 outlying data points out of 30 (see Fig. 9), achieving a detection accuracy of 86.67%. The ABOD was then applied on the historic field measurements of the test PV systems, that were aggregated daily. The algorithm detected 11 outliers in the yearly PV dataset of the test-bench PV system, while 17 outliers (indicated by black circles in Fig. 10) were detected for the PV power plant. The accuracy of the ABOD algorithm could not be assessed though due to unavailability of maintenance logs.

Hence, the actual PV dataset of the test-bench PV system was processed, and emulated fault conditions were introduced to the electrical measurements. Labels were then inserted to the dataset, indicating normal or fault conditions (e.g., reduced voltage or power), to assist the fault classification phase. Voltage- and power-related faults were emulated to the PV dataset by randomly selecting data points (10% from the entire dataset) and reducing the voltage or power measurements at a relative magnitude of 10% or 20% during high irradiance conditions ($G_I > 600$ W/m^2).

A supervised learning procedure based on a random 70:30% train and test set approach was thus followed for diagnosing normal and fault conditions using the XGBoost classifier. The XGBoost algorithm achieved high sensitivity and precision indicators of 98.6% and 100% (100%, 91.9%, 8.1% and 0% were classified as TP, TN, FN and FP data points) for reduced power measurements of 10%. Likewise, the XGBoost classifier achieved a sensitivity of 99.3% and a precision of 99.5% (99.5%, 95.9%, 4.1% and 0.5% of the given data points as TP,

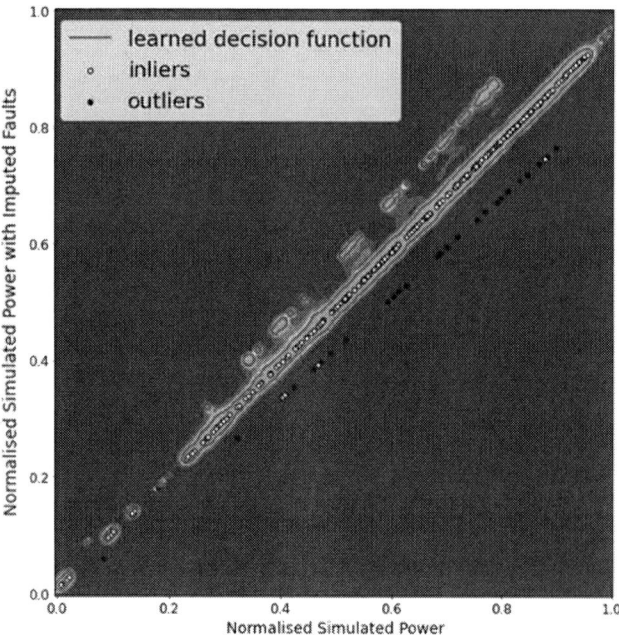

Fig. 9. Contour plot demonstrating the outliers detected by the ABOD algorithm in the synthetic generated power time series with emulated fault patterns. Outliers and inliers are depicted in black and white colour, respectively, while the red solid lines (orange shaded area) indicate the learned decision function.

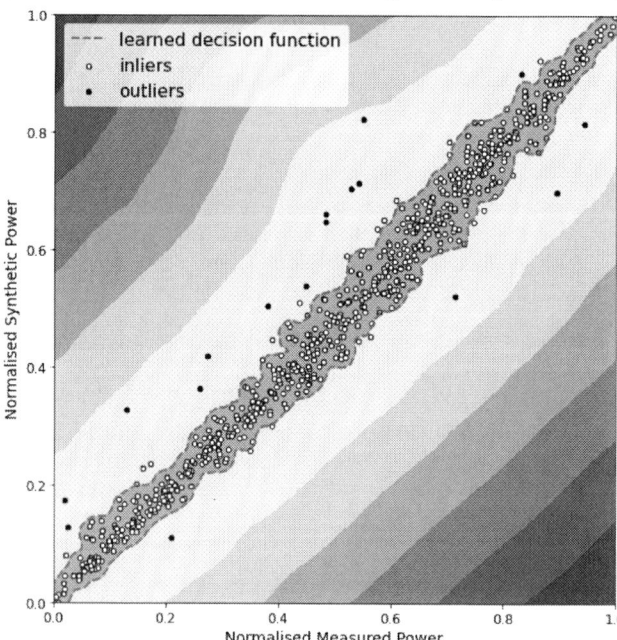

Fig. 10. Contour plot demonstrating the outliers detected by the ABOD algorithm in the power time series of the test PV power plant (MW scale) during the period from January 2019 to December 2021. Outliers and inliers are depicted in black and white colour, respectively, while the learned decision function is indicated by a red dashed line (orange shaded area).

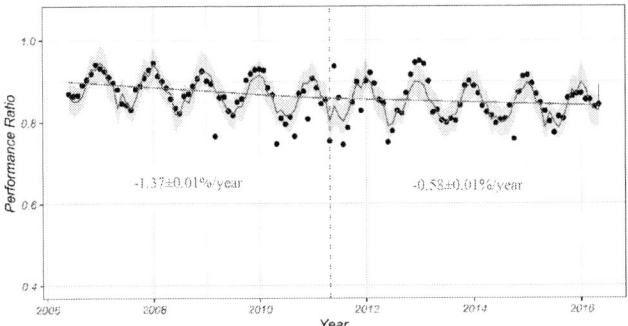

Fig. 11. PR time series (black dots) of the test-bench PV system along with the extracted trend colored in red. The blue solid line is the FBP fit, while the blue shaded area indicates the uncertainty. The detected CP is indicated with a red dashed vertical line.

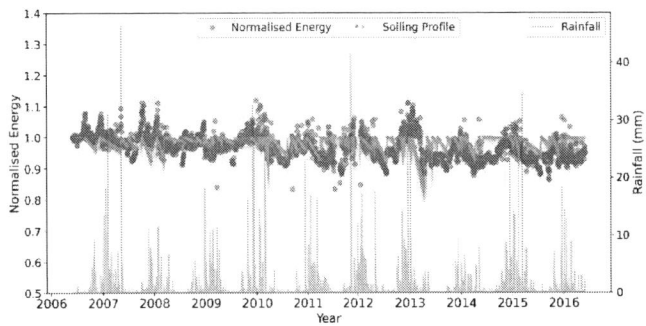

Fig. 12. Soiling losses experienced by the test-bench PV system over the evaluation period (June 2006 to May 2016). Rainfall measurements were downloaded from MERRA-2 [13].

TN, FN and FP) for the incidents causing a 20% relative power reduction. Similar results were obtained for the emulated fault conditions of reduced voltage. More specifically, the XGBoost classified 88.2% and 100% of the given data points as TN (correct fault conditions predictions), for the case of 10% and 20% relative reduction of voltage, respectively.

The *RdTools* library and statistical techniques were then used for evaluating the reliability of the test-bench PV system, over the period from June 2006 to May 2016. The PLR of the PV system was estimated by applying different statistical techniques on the monthly PR time series. Over the evaluation period, an annual linear PLR ranging from -0.54 to -0.72%/year was obtained by the different statistical techniques (i.e., OLS, CSD, STL, HW). In addition, by applying the YOY method (available in *RdTools* [21]), that is being accepted and used by the industry, a PLR of -0.74%/year (with a confidence interval from -0.78%/year to -0.64%/year) was obtained. The FBP was subsequently used for modelling the PR trend, determining nonlinear drops in PV performance and estimating either the linear or nonlinear PLR. A CP in the trend was detected (see Fig. 11), indicating a nonlinear power decline. By applying the OLS method on the FBP extracted trend and subsequent two segments, a PLR of -1.37%/year and -0.58%/year was obtained. Finally, the flexibility parameter of FBP was re-adjusted (and it was set to 0.04) [24] to capture only degradation rate (R_D) changes and a R_D of -0.47%/year was obtained.

In parallel, the SRR model was used for soiling loss extraction and detection of cleaning events [22]. The SRR model was fed with the available meteorological and P_A measurements and the normalised energy (i.e., ratio of modelled and measured power production at the DC side) was exported (see blue markers in Fig. 12) [22]. The cleaning events were identified from the positive shifts in the DC performance profile generated by the Monte Carlo simulation (orange lines in Fig. 12). For the test-bench PV system, 57 soiling periods were detected. Further, the obtained results demonstrated soiling rates ranging from 0 to -0.57%/day (see Fig. 12) over the evaluation period (June 2006 - May 2016). The test-bench PV system experienced medium soiling losses, ranging from 2.42% to 3.53%.

V. SUMMARY OF THE WORK

The implementation of a cloud-based monitoring platform,

that incorporates an accurate control digital twin for MW scale PV power plants was presented in this work. The developed platform acquires real-time data (sampling frequency of 1 Hz), stores high resolution data (frequency < 1 second) and hosts novel AI-driven algorithms for analysing the data and deciding on event actions. It operates entirely on the available field measurements and acts as an enabling technology towards digitalization, smart control and PV performance optimization.

The control digital twin utilizes an XGBoost regressor for PV performance modelling. The model is optimally trained on low amounts of field data using minimal features and coupled to DQRs for data sanity. In case of field data unavailability, a high-performing digital twin model can be constructed using synthetic data. The effectiveness of the digital twin was experimentally validated using field data from a test-bench PV system in Cyprus. The results demonstrated high accuracies for the XGBoost model, exhibiting errors < 2.91%, even when trained with low field data shares. Moreover, the application of synthetic data proved to be a robust alternative for training the digital twin model since it exhibited nRMSE < 3.83%.

The verification results of applying the digital twin to field data of a utility-scale PV power plant (MW scale), showed high predictive accuracies < 3% for both XGBoost and MPM digital twins. The results demonstrate the effectiveness of models for PV performance modelling at different locations and installations (up to MW scale), verifying the transferability and scalability of the constructed digital twins.

The robustness of the digital twin to detect faults was then evaluated. The results provide proof that robust health-state and control architectures can be built using synthetic data and emulated fault patterns. The results also showed that both supervised and unsupervised procedures can be used for detecting fault conditions in PV systems. For the supervised procedure, high sensitivity and precision indicators were obtained from the XGBoost's classifier application, verifying its applicability for PV fault diagnostics, even for MW scale.

Overall, the benchmarking results provide useful information for developing accurate, transferable and location- and installation-independent digital twins, that can be integrated into a cloud-based platform for real-time monitoring and reliable control in PV systems. The proposed work is expected to have significant impact on the value chain of the PV technology given the reduction of electricity costs by increasing the lifetime output and enabling higher shares.

ACKNOWLEDGMENT

This work was funded through the PV-ANALYTIC project, which is supported under the umbrella of SOLAR-ERA.NET Cofund 2 by the Austrian Research Promotion Agency (FFG, 873782) and the Cyprus Research & Innovation Foundation (RIF, P2P/SOLAR/0818/0012). SOLAR-ERA.NET is supported by the European Commission within the EU Framework Programme for Research and Innovation HORIZON 2020 (Cofund ERA-NET Action, N° 786483).

REFERENCES

[1] SolarPower Europe [SPE], "Operation & Maintenance Best Practice Guidelines / Version 4.0," 2019.

[2] S. Samara and E. Natsheh, "Intelligent Real-Time Photovoltaic Panel Monitoring System Using Artificial Neural Networks," *IEEE Access*, vol. 7, pp. 50287–50299, 2019, doi: 10.1109/ACCESS.2019.2911250.

[3] A. Livera, M. Theristis, L. Micheli, E. F. Fernández, J. S. Stein, and G. E. Georghiou, "Operation and maintenance decision support system for photovoltaic systems," *IEEE Access*, vol. 10, pp. 42481–42496, 2022, doi: 10.1109/ACCESS.2022.3168140.

[4] A. Livera, M. Theristis, G. Makrides, and G. E. Georghiou, "Recent advances in failure diagnosis techniques based on performance data analysis for grid-connected photovoltaic systems," *Renew. Energy*, vol. 133, pp. 126–143, 2019, doi: 10.1016/j.renene.2018.09.101.

[5] K. M. Sundaram, A. Hussain, P. Sanjeevikumar, J. B. Holm-Nielsen, V. K. Kaliappan, and B. K. Santhoshi, "Deep Learning for Fault Diagnostics in Bearings, Insulators, PV Panels, Power Lines, and Electric Vehicle Applications - The State-of-the-Art Approaches," *IEEE Access*, vol. 9, pp. 41246–41260, 2021, doi: 10.1109/ACCESS.2021.3064360.

[6] K. Dhibi *et al.*, "Reduced Kernel Random Forest Technique for Fault Detection and Classification in Grid-Tied PV Systems," *IEEE J. Photovoltaics*, vol. 10, no. 6, pp. 1864–1871, 2020, doi: 10.1109/JPHOTOV.2020.3011068.

[7] A. Livera, G. Paphitis, M. Theristis, J. Lopez-Lorente, G. Makrides, and E. George, "Photovoltaic system health-state architecture for data-driven failure detection," *Solar*, vol. 2, no. 1, pp. 81–89, 2022, doi: https://doi.org/10.3390/solar2010006.

[8] K. Arafet and R. Berlanga, "Digital twins in solar farms: An approach through time series and deep learning," *Algorithms*, vol. 14, no. 5, 2021, doi: 10.3390/a14050156.

[9] A. Livera, M. Theristis, G. Makrides, J. Sutterlueti, S. Ransome, and G. E. Georghiou, "Performance analysis of mechanistic and machine learning models for photovoltaic energy yield prediction," in *36th European Photovoltaic Solar Energy Conference (EU PVSEC)*, 2019, pp. 1272–1277, doi: 10.4229/EUPVSEC20192019-5BO.5.2.

[10] J. Ascencio-Vásquez, K. Brecl, and M. Topič, "Methodology of Köppen-Geiger-Photovoltaic climate classification and implications to worldwide mapping of PV system performance," *Sol. Energy*, vol. 191, no. August,

pp. 672–685, 2019, doi: 10.1016/j.solener.2019.08.072.

[11] IEC, "IEC 61724-1:2021: Photovoltaic system performance - Part 1: Monitoring," 2021.

[12] I. Reda and A. Afshin, "Solar Position Algorithm for Solar Radiation Applications, National Renewable Energy Laboratory Technical Report, NREL/Tp-560-34302, 2008," doi: 10.2172/15003974.

[13] R. Gelaro *et al.*, "The modern-era retrospective analysis for research and applications, version 2 (MERRA-2)," *J. Clim.*, vol. 30, no. 14, pp. 5419–5454, 2017, doi: 10.1175/JCLI-D-16-0758.1.

[14] A. Livera *et al.*, "Data processing and quality verification for improved photovoltaic performance and reliability analytics," *Prog. Photovoltaics Res. Appl.*, vol. 29, pp. 143– 158, 2021, doi: 10.1002/pip.3349.

[15] D. L. King, W. E. Boyson, and J. A. Kratochvil, "Photovoltaic array performance model," 2004.

[16] J. S. Stein, J. Sutterlueti, S. Ransome, C. W. Hansen, and B. H. King, "Outdoor PV Performance Evaluation of Three Different Models: Single-Diode, SAPM and Loss Factor Model," *28th Eur. Photovolt. Sol. Energy Conf. Exhib.*, pp. 2865–2871, 2013.

[17] S. Ransome and J. Sutterlueti, "Quantifying Long Term PV Performance and Degradation under Real Outdoor and IEC 61853 Test Conditions Using High Quality Module IV Measurements," in *37th European Photovoltaic Solar Energy Conference (EU PVSEC)*, 2019, pp. 1640–1645.

[18] T. Chen and C. Guestrin, "XGBoost: A Scalable Tree Boosting System," in *KDD '16: Proceedings of the 22nd ACM SIGKDD International Conference on Knowledge Discovery and Data Mining*, 2016.

[19] A. Livera, M. Theristis, J. S. Stein, and G. E. Georghiou, "Failure Diagnosis and Trend-Based Performance Losses Routines for the Detection and Classification of Incidents in Large-Scale Photovoltaic Systems," in *38h European Photovoltaic Solar Energy Conference (EU PVSEC)*, 2021, pp. 973–978, doi: 10.4229/EUPVSEC20212021-5CO.9.3.

[20] Y. Zhao, Z. Nasrullah, and Z. Li, "PyOD: A python toolbox for scalable outlier detection," *J. Mach. Learn. Res.*, vol. 20, pp. 1–7, 2019.

[21] M. G. Deceglie, D. Jordan, A. Shinn, and C. Deline, "RdTools : An Open Source Python Library for PV Degradation Analysis degradation rate," pp. 1–15, 2018.

[22] M. G. Deceglie, L. Micheli, and M. Muller, "Quantifying Soiling Loss Directly from PV Yield," *IEEE J. Photovoltaics*, vol. 8, no. 2, pp. 547–551, 2018, doi: 10.1109/JPHOTOV.2017.2784682.

[23] M. Theristis, A. Livera, C. B. Jones, G. Makrides, G. E. Georghiou, and J. S. Stein, "Nonlinear Photovoltaic Degradation Rates: Modeling and Comparison Against Conventional Methods," *IEEE J. Photovoltaics*, vol. 10, no. 4, pp. 1112–1118, 2020, doi: 10.1109/JPHOTOV.2020.2992432.

[24] A. Livera, M. Theristis, L. Micheli, J. S. Stein, and G. E. Georghiou, "Failure diagnosis and trend-based performance losses routines for the detection and classification of incidents in large-scale photovoltaic systems," *Prog. Photovoltaics Res. Appl.*, no. Special Issue, pp. 1–17, 2022, doi: 10.1002/pip.3578.

Predicting solar cell recombination from C-V-f fingerprints using machine learning

Isaac K. Lam, Austin G. Kuba, Nathan J. Rollins, William N. Shafarman

Materials Science & Engineering, University of Delaware, Newark, DE, United States

Institute of Energy Conversion, University of Delaware, Newark, DE, United States

Independent Researcher, Boston, MA, United States

Capacitance measurement techniques are powerful methods for characterizing semiconductor devices. Voltage dependent admittance spectroscopy (C-V-f) has recently been used to characterize electronic loss mechanisms in CIGS solar cells. In this work, drift-diffusion simulations of devices are used to create a large dataset of C-V-f loss map images that provide a fingerprint for the electronic loss mechanisms of a solar cell. Analytic extraction of electronic properties from these loss maps is difficult, so a machine learning method for characterizing measured C-V-f profiles of real devices is developed to identify dominant loss mechanisms. The method is demonstrated with a perovskite solar cell. Various properties are simulated including contact work functions, doping concentrations, series resistance, bulk defect concentrations, and interface defect concentrations. To reduce computational complexity, the simulations focus primarily on MAPI bulk defects and C60/MAPI/CuPC interface defects. Principal component analysis is used to verify that different features observed in the loss maps can be represented independently of each other. Although the simulated data appears to be a good candidate for modelling, there could be issues reconciling simulated and experimental data due to factors such as experimental noise, variation in measurement intensity, and contributions not accounted for in the simulation such as perovskite ion migration.

Current & Future Photovoltaic System Impacts on City-Wide Grid Performance & Neighborhood Microgrids

C. Birk Jones, William F. Vining, & Thad Haines

Sandia National Laboratory, Albuquerque, NM, 87123, U.S.A

Abstract—An accurate understanding of the electric grid's performance subject to existing and future photovoltaic (PV) arrays requires a city-wide simulation of distribution and sub-transmission systems. This work used existing PV array locations and adoption predictions from past work as inputs to model PV integration in an entire city. Distribution and sub-transmission systems were combined into a single simulation environment that allowed for each to be emulated simultaneously. Simulations of PV adoptions at 2020, 2030, 2040, and 2050 levels were performed and the net power profiles and voltages were output. The 2020 year simulations used actual permit data to identify PV locations, and future years used predictions that assumed linear trends at different spatial regions. Assuming that adoptions follow the same linear adoption rate, the city's distribution grids won't see significant voltage violations until 2040 and 2050, where some areas of the grid had simulated maximum voltages above 1.05 PU. The sub-transmission simulation results showed very little change in system voltage over the next 30 years. When considering the distributed PVs' contribution to microgrid operations, most of the city will likely need more generation capabilities to power microgrids whose extents were defined by the maximum modularity community detection algorithm.

Index Terms—photovoltaic, hosting capacity, microgrids

I. INTRODUCTION

A city-wide hosting capacity evaluation provides a realistic impact assessment of distributed photovoltaic (PV) systems on both sub-transmission and distribution electric grid networks now and in the future. This assessment, presented here, considers existing PV locations based on city and county permit records [1]. Future integration levels are understood by implementing a linear least-squares adoption forecast for roof-top PV systems. This forecast considers income and building use diversity throughout the region of study to predict future penetration levels and locations [2]. This paper explores PV adoptions levels now and 10, 20, and 30 years into the future. The PV locations are then associated with loads in the city-wide model and included in the simulations that output system voltage throughout the entire system consisting of sub-transmission and distribution networks.

This region-based simulation effort is important for coordinating the transition from fossil fuel based electrical power generation to renewable energy sources. Mandates, such as the "Energy Transition Act" in New Mexico (NM) [3], require the elimination of carbon emissions in the electrical power sector, and it demands that 80% of generated power come from renewable energy and 50% of the energy be produced by solar PV. This crucial transition depends on multiple new power generation sources to make up for the decommissioning of fossil fuel dependent systems, such as a large coal power plant in Northern NM. However, it is unclear what percentage roof-top solar will contribute and if it will create grid performance issues within or throughout city-wide grid networks.

Distributed PV (dPV) power generation provides a unique contribution to the overall electric grid. Utilities may not appreciate the thousands of PV systems located at commercial and residential buildings since they don't have direct visibility or understanding of their operations. However, distributing small-scale PV throughout the grid requires less oversight and risk when compared to large PV plants at a central location [4]. Also, dPV, embedded inside communities, offers the opportunity to power isolated systems during a contingency event, assuming a grid-forming (GFM) inverter is available to set the voltage and frequency. This means that not only are dPV systems useful for reducing carbon emissions they can also be leveraged ruing a contingency event to power microgrids within neighborhoods.

Existing literature documents various hosting capacity studies that consider the impact of new dPV systems [5]. The research studies include single-point simulations to understand location impacts [6] and infrastructure upgrade needs required to support a new PV system [7]. Typically, these past approaches use an iterative process that loops through various PV sizes until a violation occurs and then moves to a new bus where the process repeats [8]. Other studies employ a stochastic hosting capacity approach that iterates through different penetration levels and randomly places the PV systems [9] or Electric Vehicles (EV) [10] throughout the grid.

Instead of relying on iterations that randomly place PV at certain locations, this simulation effort used actual PV system locations to understand current operations. Then, future adoption predictions at different points in the future were added to provide realistic hosting capacity results for dPV throughout an entire city.

A secondary, and final, assessment considered the potential contribution of the dPV on operations of embedded neighborhood microgrids. This analysis compared the energy consumption and generation for a single day within electrical groups defined by the maximum modularity community detection algorithm [11]. Each of these assessments used a synthetic

978-1-7281-6118-1/22 $31.00 © 2022 IEEE

electric grid that represented the power system in Santa Fe, New Mexico.

II. ELECTRIC POWER SYSTEM SIMULATION

A. Test System

The Electric Power System (EPS) model, shown in Fig. 1, was used to test the city-wide assessment of dPV. The EPS represented the City of Santa Fe [12] and was downloaded from [13]. Fig. 1 depicts the medium voltage power lines that connect the substations to the transmission system using black solid lines. The distribution systems connected to the eight substations are depicted using different colors in Fig. 1. The low-voltage distribution lines support over 84,000 customers. The lines extend over 1,921 km through two transmission substations, eight sub-transmission substations and finally to 28 feeders. These feeders power 38,590 buildings located throughout the surrounding area and in more densely populated areas within the city.

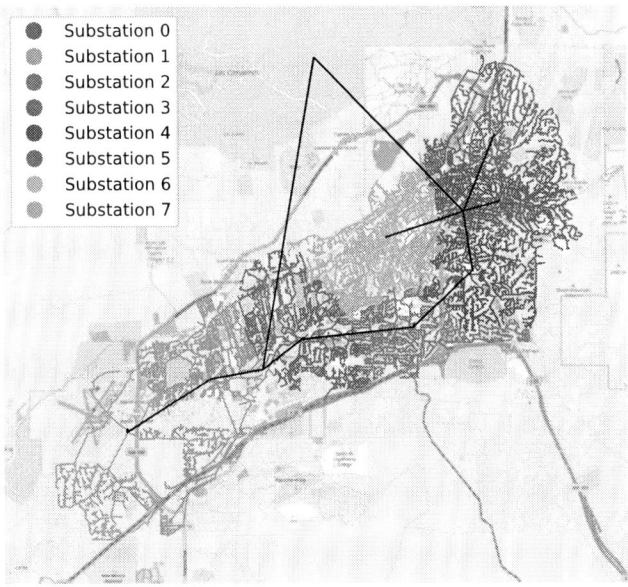

Fig. 1. This map shows the layout of the sub-transmission, in black, and distribution systems connected to the eight substations throughout the City of Santa Fe, New Mexico.

B. Connected Loads

A reasonable representation of the loads was achieved by assigning demand profiles that represent different building use types to loads inside the associated zoning districts. Maps that designate the allowable building use types were provided by the municipalities zoning office. Load profiles that represent residential, commercial, shopping, mixed-use, and others were assigned loads within each of their respective districts. This data, plotted for a single week in Fig. 2, was collected from a Typical Meteorological Year 3 (TMY3) data set released by NREL in 2014 [14].

Note that the electric power usage information corresponding to Santa Fe buildings were normalized to the week's

maximum, as shown in Fig. 2, to be applied as a scaling factor to the default model load values. Half hour load data was generated from the hourly data using a quadratic interpolation method.

Also, the simulations for 10, 20, and 30 years into the future did not include changes to the load. This assessment only focused on changes in dPV adoptions and held everything else constant. But, more realistic simulations will include changes in load demand due to energy efficiency and increases in consumption from more electric vehicles charging on the grid.

C. Current & Future Photovoltaic Adoptions

The existing and future PV systems were modeled using the OpenDSS features. Each system was set to be 135% of the rated load defined by the OpenDSS model. This sizing of the PV system was described in past work to size PV systems for a cybersecurity impact study [4] because it was

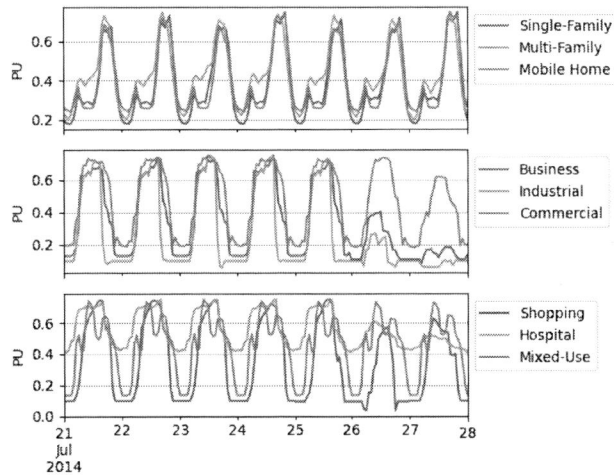

Fig. 2. The three plots show the demand profiles for residential, commercial, business, industrial, shopping, hospital, and mixed-use building types. The profiles were assigned to all of the loads in the grid model and used to emulate electrical operations throughout the entire city.

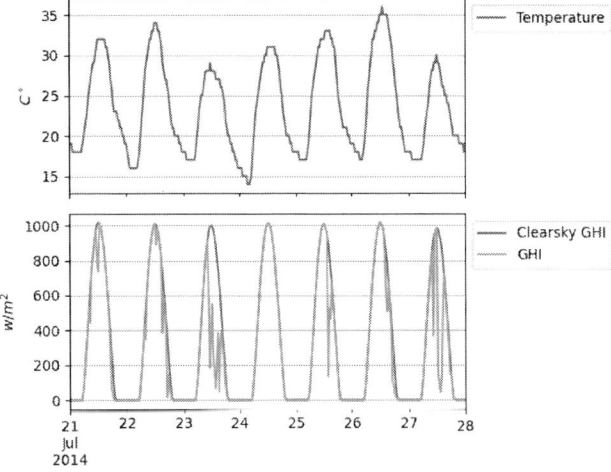

Fig. 3. Irradiance and temperature data for the week July 14, 2014 from NSRDB.

found, on average, to offset 100% of the building loads annual energy consumption. Data from the National Solar Radiation Database (NSRDB) [15], shown in Fig. 3, provided necessary irradiance and temperature values to simulate the PV arrays. The Global Horizontal Irradiance (GHI) values, provided by NSRDB, were normalized to 1000 W/m² = 1 PU and provided to the OpenDSS model.

Figs. 4(a) and 4(b) depict the changes in PV adoptions from 2010 to 2020 using density plots. Fig. 4(c) describes the future estimate of PV systems in 2050. The 2030, 2040, and 2050 predictions are based on least-squares linear models for various sections of the city as explained in [2]. The prediction models consider 2009-2020 PV permit data (collected from the City of Santa Fe), census median income data, and building zone area designations as inputs. The models then output the number of loads in each region with a co-located PV array.

Historical records show that a majority of the existing PV arrays were installed in areas where the income was above $50,000. If the city follows similar trends observed between 2009 and 2020, low-income areas will have limited PV array installations. This is evident in Fig. 4(c) where the regions colored in blue, for median incomes less than $40,000, have low densities of PV systems on building roof-tops.

III. PHOTOVOLTAIC SYSTEM IMPACTS ON THE GRID

The EPS simulations evaluated the integration of realistic PV adoption levels dispersed throughout the city of Santa Fe. The simulation effort began by producing and assigning load profile data to different building types, as described in Sec. II-B. The load profiles were mapped to loads in the EPS model based on the zoning area map. Week-long irradiance profiles, provided by National Solar Radiation Database (NSRDB) [15], were used as input to the OpenDSS PV models.

The PV and load data were used by the distribution feeder models and run alongside the sub-transmission system in a co-simulation environment. The week-long simulation results output the distribution and sub-transmission power demand and voltages throughout the EPSs over the single week. To understand the existing and future PV integration impacts, a spatial and temporal review was performed to understand impacts throughout the system and the results are described in V.

IV. NEIGHBORHOOD MICROGRID POTENTIAL

During a contingency event, individual electrical neighborhoods may be islanded and become resilience zones that support local loads during a grid outage that does not impact local distribution lines. These zones will need extra resources like electric battery storage and a GFM inverter. But, they could leverage existing PV resources already on the grid that were originally designed and constructed to offset fossil fuel based energy production.

To estimate the potential contribution of dPV now and in the future, this work aggregated the time-series simulation results by resilience zones. The aggregation involved a summation

of the generation and consumption energy within pre-defined clusters. In this case, the clusters were defined using the modularity maximization [11]. Modularity maximization only considered the connectivity of the network to define zones.

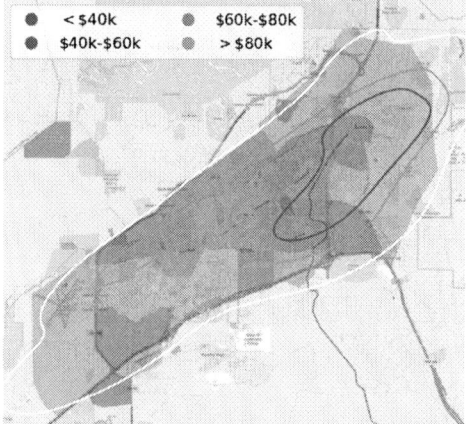

(a) 2010 PV Adoption Density Plot

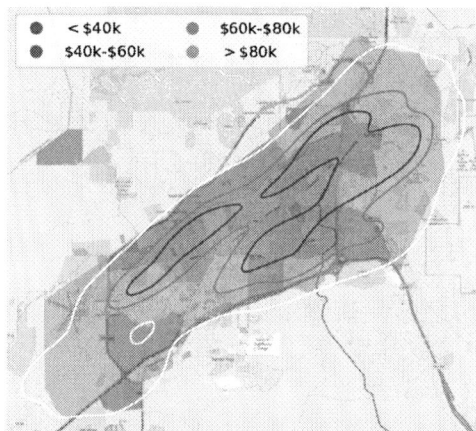

(b) 2020 PV Adoption Density Plot

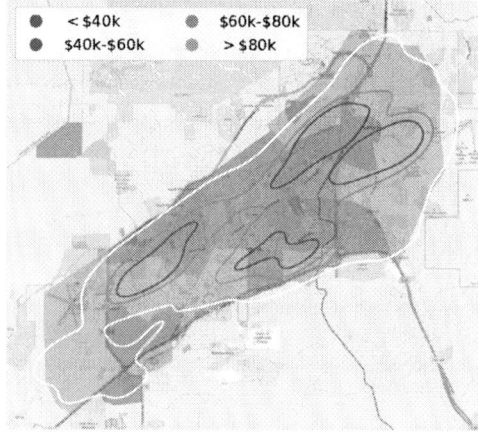

(c) 2050 PV Adoption Density Plot

Fig. 4. This figures show the actual and predicted PV adoption densities in 2010, 2020, and 2050. The background of each density plot includes census median income data.

978-1-7281-6118-1/22 $31.00 © 2022 IEEE

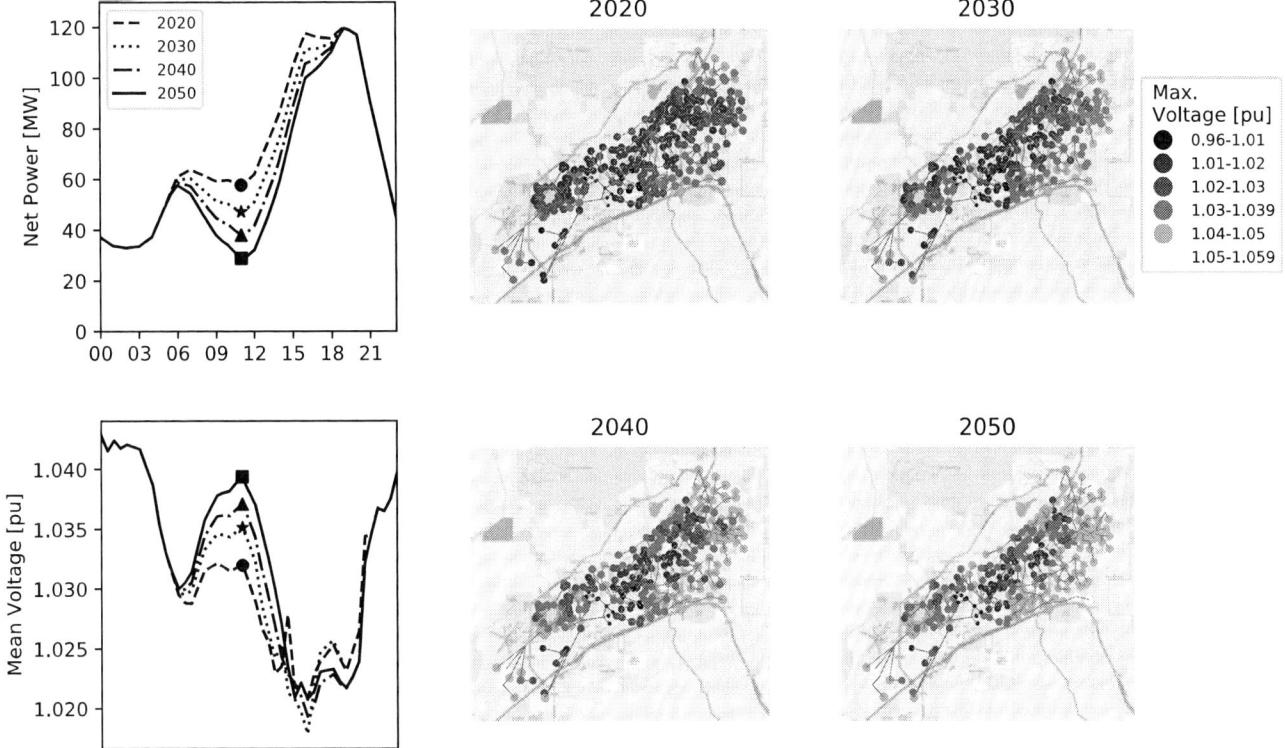

Fig. 5. This figure shows how the net power can change over the next 30 years as more PV systems are installed throughout the city. Because PV adoptions are not consistent spatially, the two maps of the City of Santa Fe show that some areas of the grid will experience different changes in the maximum voltage.

It did not factor in the typically energy consumption or PV generation to find an optimal sections or groups that could act as a microgrid without significant upfront costs. Future simulation studies could identify resilient communities using spatial-temporal net energy [16] or other methods at different PV penetration levels now and into the future.

V. RESULTS

A. Distribution Performance

The simulation emulated operations over a seven day period and Fig. 5 describes results for a single day within the simulation week. The plots on the left side of Fig. 5 show the net power and mean voltage across all of the distribution systems. Only PV adoptions in 2050 are predicted to decrease the net power below the current, 2020 year, night time demand.

The decrease in net power caused an increase in the mean voltage of about 0.0025 Per Unit (PU) for each 10 year period (between 2020 and 2050). As a result, the mean voltage will likely increase by about 0.1 PU from 2020 to 2050. This increase was found not to cause significant violations throughout the network, but it may present other challenges associated with management of load and generation during large ramp rates when the sunsets.

The four maps to the right of the time-series plot in Fig. 5 depict the spatial variation in maximum voltage throughout all of the distribution networks for 2020, 2030, 2040, and 2050 penetration levels. It is evident that the maximum voltage increases in areas that experience the high PV adoption rates and are at the edges of the grid. For instances, in 2020 the maximum voltages measured where in the northeast, southeast, and southwest edges of the grid. This trend remained about the same but expanded spatially and the voltage increased at PV penetration levels predicted to occur in 2030, 2040, and 2050. There are some maximum voltages that exceed 1.05 ANSI standard thresholds in 2040 and 2050 years.

Some areas within the center of the city did not experience significant changes in voltage on the distribution networks. This can be attributed to the location of the nodes in relation to the substation; many of these nodes may be close to the substation and therefore not experience a major swing in voltage with the integration of new PV systems. Some areas, like low-income housing districts, may not have nor will they have significant penetrations of PV on residential and commercial roof-tops. In these areas the voltage will likely not change too much.

B. Sub-Transmission Performance

Fig. 6 shows sub-transmission net power demand, PV power, and the maximum voltage at 2020, 2030, 2040, and 2050 PV adoption levels. The simulation utilized the global horizontal irradiance shape, which included variable generation on most days and one clear sky day. Under clear sky conditions, the peak PV power generation exceeded 40 MW in 2050, which in some instances was about half of the power

978-1-7281-6118-1/22 $31.00 © 2022 IEEE

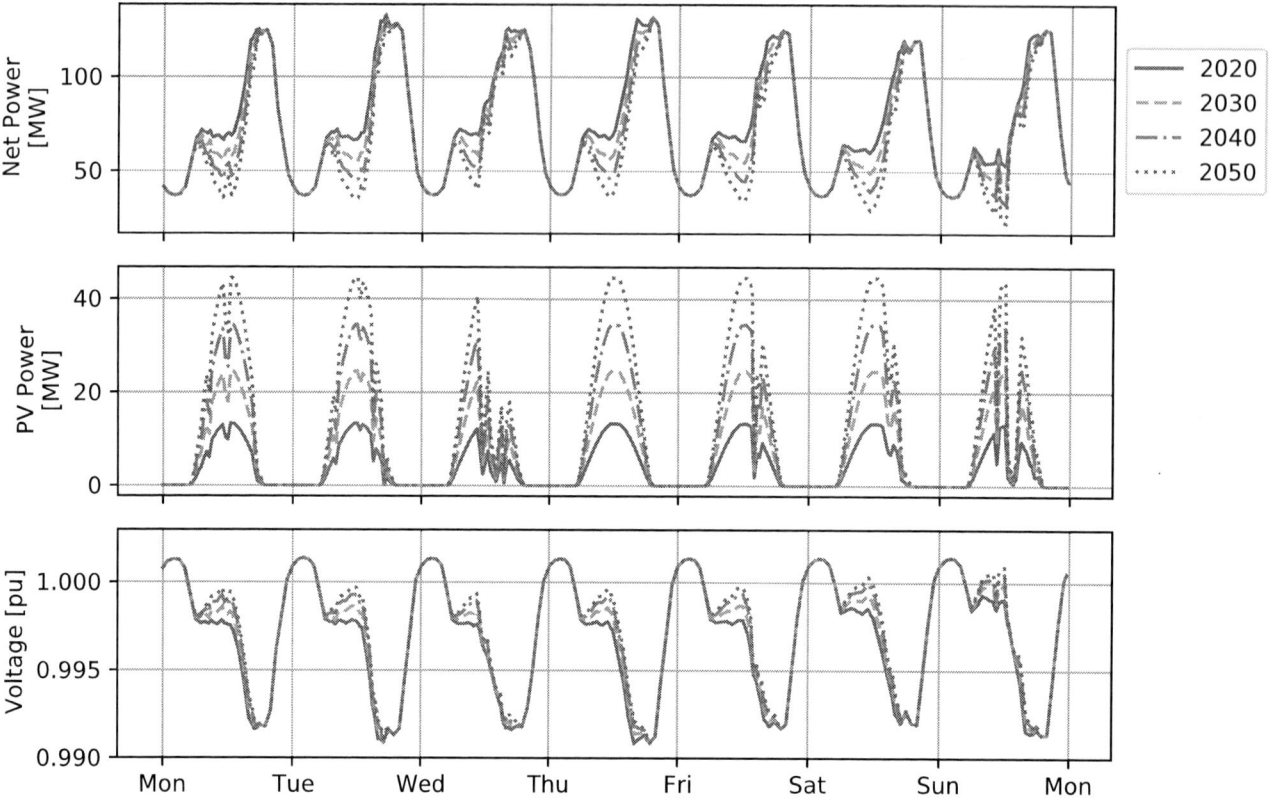

Fig. 6. This figure plots the net power and maximum voltage measured on the sub-transmission system. During the day, the PV power made a significant portion of the demand. Because of this, the voltage had a noticeable change for the different adoption rates. However, the change in maximum voltage was not significant enough to create an issues.

demand. When the PV generation decreased at the end of each day, the system net power ramp rate (i.e., change in power over time) increased with higher PV penetration.

The dPV generation had little effect on the sub-transmission maximum voltages as shown in the bottom plot in Fig. 6. The sub-transmission was able to maintain a voltage between 0.99 and 1.005 pu during the entire week-long simulation and for each of the PV adoption scenarios. These voltage plots, for each simulation year, had very smooth behavior with very small deviations away from 1.0 PU. Voltage regulators included in the distribution systems likely allowed the voltage to not deviate too far from the desired values.

These results indicate that significant penetrations of PV scattered throughout the city on distribution grids will not cause sub-transmission operations issues by themselves. Even at penetration levels where nearly half of the load was served by dPV, the system voltage changed very little and did not approach any violation thresholds.

C. Neighborhood Microgrids

Electrically connected communities may include dPV systems that can help offset the energy consumption while in microgrid mode during a contingency event. These microgrids, shown in Fig. 7, are represented by the different colors. In this case, the community detection method (maximum modularity

algorithm) identified 312 different microgrid systems with densely connected nodes.

In order for each system to work they would need a device, like a GFM inverter, to provide a voltage source. It would also

Fig. 7. This map of the electric grid indicates the assumed embedded microgrid locations and sizes with different colors. This groups were defined using the maximum modularity algorithm that considers the connectivity of the electrical connections to define communities.

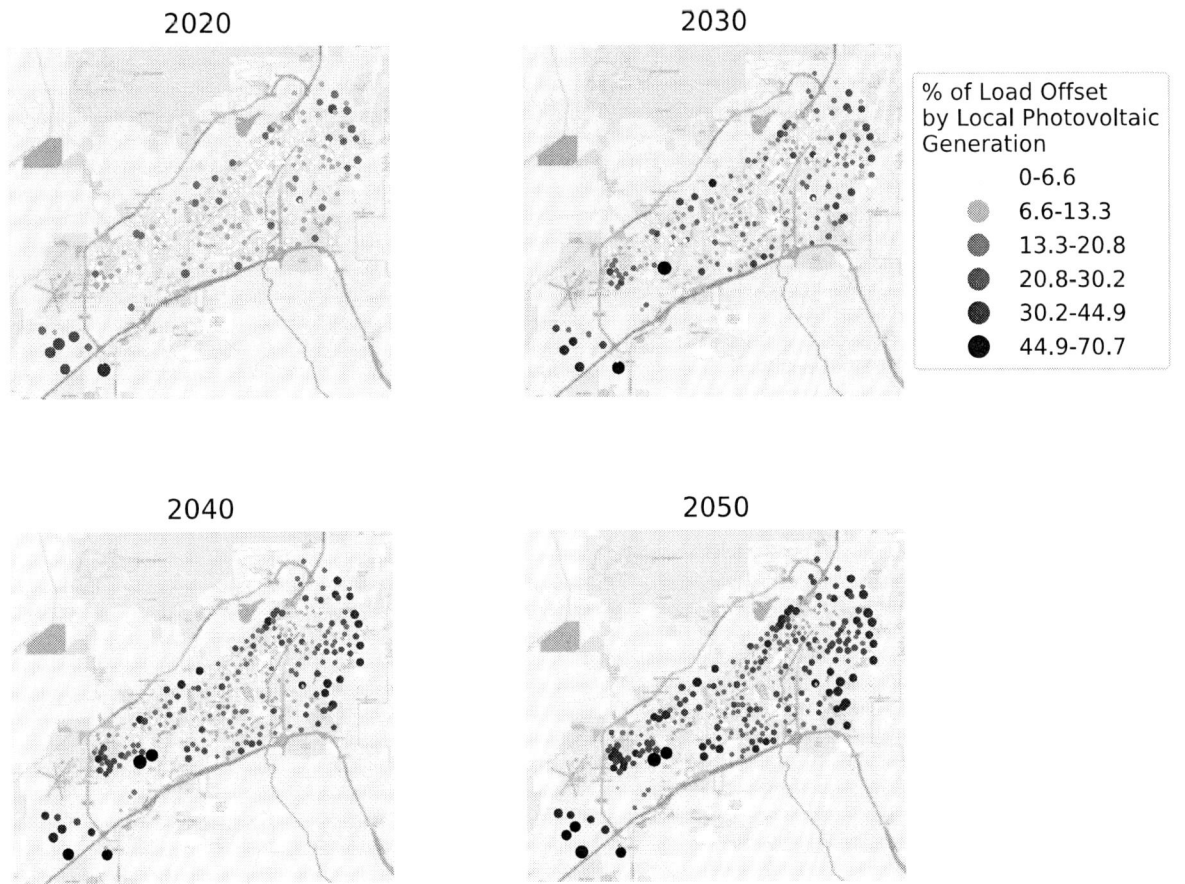

Fig. 8. These four maps depict what the percentage of the load's energy was offset by local photovoltaic (PV) generation over a single day of operations. There are noticeable differences in the percentage of load offset by PV from 2020 to 2050. Most notably are areas in the northeast, and west sectors of the city. Note that the size of the circles represent the demand in each microgrid and, as described by the legend, the color represents the percentage of load offset by local PV.

need a battery storage system to support the GFM inverter operations. But to charge the GFM inverter's battery, the existing PV systems could provide some or all of the necessary energy.

This assessment of daily energy provides a simple review for how each microgrid can support typical (non-resilience event) demand behaviors for a single day of operations. If it cannot support the loads, the assessment identifies the approximate addition generation required to support the load throughout a single day.

Fig. 8 shows how the amount of dPV generation over a single day at 2020, 2030, 2040, and 2050 levels compares to the energy consumption in the same microgrid. The Fig. 8 map represents each of the microgrids shown in Fig. 7 with a circle at the centroid of each system. The circle's color, in Fig. 8, described the percentage of load that is offset by the local dPV generation and the size describes the amount of energy consumption.

For the day examined here, the percentage of load offset by dPV varied between 0 and 70.7% for 2020, 2030, 2040, and 2050 PV adoption scenarios, as shown in Fig. 8. At 2020 dPV integration levels, the largest offset of the local load occurred in the southwest portion of the city. Other areas of the city showed an increase in the overall offset of load in 2030, 2040, and 2050 that was mostly in the northeast where the median income was high. In 2050, there were some high offset so energy consumption in the center of the city, but none where in low-income areas. To better serve these low-income areas, equity-based incentives and community-based PV areas are likely needed to increase dPV adoptions.

VI. CONCLUSION

The city-wide simulation effort with realistic PV locations identified the potential changes in performance at the distribution and sub-transmission system levels. Although the changes in demand were noticeable, the impact on voltage and line loading were minimal at 10, 20, and 30 years into the future. It wasn't until 2050 that a noticeable number of locations had maximum voltages near or above 1.05 PU. Yet, these high voltages did not occur for a significant amount of time and likely can be mitigated with reactive power support at the point of common coupling.

A review of dPV's contribution to neighborhood resilience found that the predicted penetration levels will likely not be sufficient to support a large number of microgrids. In addition to the inclusion of new battery storage and GFM inverters, neighborhoods will need to invest in more solar generation to offset daily energy consumption.

Future work is required to provide a deeper review and understanding of embedded microgrids that utilize existing resources. Optimization and graph theory techniques will be used to identify proper grouping of loads and generation to efficiently segregate the system into microgrids.

ACKNOWLEDGMENT

The paper describes objective technical results and analysis. Any subjective views or opinions that might be expressed in the paper do not necessarily represent the views of the U.S. Department of Energy or the United States Government.

Sandia National Laboratories is a multimission laboratory managed and operated by National and Engineering Solutions of Sandia, LLC., a wholly owned subsidiary of Honeywell International, Inc., for the U.S. Department of Energy's National Nuclear Security Administration under contract DE-NA0003525.

This work was supported by the 'Distributed Energy Systems (DER) Modeling' Laboratory Directed Research and Development (LDRD) project.

REFERENCES

[1] "City of Santa Fe, New Mexico," https://www.santafenm.gov/building_permits, 2021.

[2] C. Jones, W. Vining, and J. Haines, "City-Wide Distributed Roof-Top Photovoltaic System Adoption Forecast, Grid Impacts, & Neighborhood Microgrid Contributions." Sandia National Lab. (SNL-NM), Albuquerque, NM (United States), Tech. Rep. SAND2021-12108, Sep. 2021.

[3] J. Candelaria, N. P. Small, M. Stewart, P. R. Caballero, and B. Egolf, "ENERGY TRANSITION ACT," 2019.

[4] C. B. Jones, M. Lave, M. J. Reno, R. Darbali-Zamora, A. Summers, and S. Hossain-McKenzie, "Volt-Var Curve Reactive Power Control Requirements and Risks for Feeders with Distributed Roof-Top Photovoltaic Systems," *Energies*, vol. 13, no. 17, p. 4303, Jan. 2020.

[5] S. M. Ismael, S. H. E. Abdel Aleem, A. Y. Abdelaziz, and A. F. Zobaa, "State-of-the-art of hosting capacity in modern power systems with distributed generation," *Renewable Energy*, vol. 130, pp. 1002–1020, Jan. 2019.

[6] K. Coogan, M. J. Reno, S. Grijalva, and R. J. Broderick, "Locational dependence of PV hosting capacity correlated with feeder load," in *2014 IEEE PES T D Conference and Exposition*, Apr. 2014, pp. 1–5.

[7] S. Stanfield and S. Stephanie, "Optimizing the Grid: Regulator's Guide to Hosting Capacity Analyses for Distributed Energy Resources," Interstate Renewable Energy Council, Tech. Rep., Dec. 2017.

[8] M. J. Reno, K. Coogan, J. Seuss, and R. J. Broderick, "Novel Methods to Determine Feeder Locational PV Hosting Capacity and PV Impact Signatures," Sandia National Lab. (SNL-NM), Albuquerque, NM (United States), Tech. Rep. SAND2017-4954, May 2017.

[9] J. Smith, "Stochastic Analysis to Determine Feeder Hosting Capacity for Distributed Solar PV," Electric Power Research Institute, Palo Alto, CA, Tech. Rep. 1026640, Dec. 2012.

[10] C. B. Jones, M. Lave, and R. Darbali-Zamora, "Overall Capacity Assessment of Distribution Feeders with Different Electric Vehicle Adoptions," in *2020 IEEE Power Energy Society General Meeting (PESGM)*, Aug. 2020, pp. 1–5.

[11] G. Lin, S. Liu, A. Zhou, J. Dai, B. Chai, B. Zhang, H. Qiu, K. Gao, Y. Song, and R. Chen, "Community detection in power grids based on Louvain heuristic algorithm," in *2017 IEEE Conference on Energy Internet and Energy System Integration (EI2)*, Nov. 2017, pp. 1–4.

[12] V. Krishnan, B. Bugbee, T. Elgindy, C. Mateo, P. Duenas, F. Postigo, J.-S. Lacroix, T. G. S. Roman, and B. Palmintier, "Validation of Synthetic U.S. Electric Power Distribution System Data Sets," *IEEE Transactions on Smart Grid*, vol. 11, no. 5, pp. 4477–4489, Sep. 2020.

[13] T. Elgindy, C. Mateo Garcia, P. Duenas Martinez, B. Palmintier, F. Postigo Marcos, T. Gomez San Roman, F. de Cuadra García, V. Krishnan, and N. Gensollen, "Santa_Fe_Synthetic_Network," Jun. 2020.

[14] E. Wilson, "Open Energy Data Initiative (OEDI)," 2021.

[15] NREL, "National Solar Radiation Database," 2021.

[16] S. Xie, H. Wang, S. Wang, H. Lu, Y. Hong, D. Jin, and Q. Liu, "Discovering communities for microgrids with spatial-temporal net energy," *Journal of Modern Power Systems and Clean Energy*, vol. 7, no. 6, pp. 1536–1546, Nov. 2019.

Passivating Surface Iodide Defects Slows the CsPbI3 Phase Transformation

Jeffrey A Christians, Jonathan Outen, Rory M Campagna, Zachery R Wylie, Peter Ruffolo

Hope College, Holland, MI, United States

Efforts to improve halide perovskite solar cell stability require detailed understanding of all of the various degradation modes that can occur in a device. One degradation mode of note is the perovskite to nonperovskite phase transition. In this work, we demonstrate a simple method to track the kinetics of this in CsPbI3 thin films. We demonstrate that this phase transition is first order with respect to atmospheric water. Moreover, we show the ability to decrease the phase transition rate from 7.5×10^{-3} s-1 to 1.4×10^{-3} s-1 by surface iodide treatment of the films. This insight will help the design of more robust perovskite films and continued understanding of this phase transformation will likely be important for efforts aimed at extrapolating accelerated lifetime tests.

Which potential for Kesterite absorbers in tandem solar cells: a quantitative modelling approach

Alex Jimenez, Alejandro Navarro, Sergio Girlado, Kunal Jogendra Tiwari, Marcel Placidi, Lorenzo Calvo-Barrio, Joaquim Puigdollers, Edgardo Saucedo, Zacharie Jehl Li-Kao

Polytechnic University of Catalonia (UPC), Barcelona, Spain

Catalonia Institute for Energy Research (IREC), Barcelona, Spain

CCiTUB, Universitat de Barcelona, Barcelona, Spain

The potential of Kesterite absorbers used both as top or bottom cell, in combination with crystalline silicon bottom cell and a Perovskite top cell respectively, is investigated using a combination of optical and electrical modelling. Using a transfer matrix approach to determine the transmission of a given top cell, the electrical behavior of the bottom cell in tandem condition is evaluated. Unlike past studies on a related topic, the results reported here are deemed close to quantitative, relying on a consistent set of experimental data for both the optical and electrical model. After demonstrating the closeness of a simulated CZTSe baseline solar cell with its experimental counterpart, an incremental set of experimentally realistic optimizations are investigated to further enhance the PV performance. A combination of a 21%-Perovskite subcell with a 17%-CZTSe subcell is found sufficient to overcome the single junction detailed balance limit and approach the 30% efficiency threshold. Following a similar approach, a wide bandgap CZG(S,Se) top cell is evaluated in combination with a state-of-the-art c-Si bottom cell. Such design is found markedly more challenging for the Kesterite top cell with the necessary use of innovative selective contacts and a reduction of the bulk defect density by two orders of magnitude to approach the 30% efficiency threshold. Each specific optimization will be discussed in the context of current experimental trends in Kesterite solar cells, and this work will conclude by offering perspectives for full Kesterite tandem solar cells as well as multijunction devices with 3 subcells or more. This work offers, for the first time, a reliably quantified overview of the potential of Kesterite absorbers in multijunction devices, and will help experimentalists identifying and focusing their efforts toward the current bottlenecks of this technology.

Measuring Global, Direct, Diffuse, and Ground-Reflected Irradiance Using a Reference Cell Array

Michael Gostein[1], Adam Hoffman[2], Bruce H. King[3], Audrey Marquis[1]

[1]Atonometrics, Austin, USA; [2]Maxeon, San Jose, USA; [3]Sandia National Laboratories, Albuquerque, USA

Abstract **We report on an updated field trial of a novel method for measuring global horizontal irradiance (GHI), direct normal irradiance (DNI), diffuse horizontal irradiance (DHI), and reflected horizontal irradiance (RHI) using an array of reference cells with no moving parts. The technique, similar to previously explored methods using multi-pyranometer array concepts, uses multiple reference cells facing different directions together with an analysis model that fits the data from the sensors to determine GHI, DNI, DHI, and RHI. In previous work we reported on a first test of this approach at a single site. In this work we report further progress with this method, including an updated configuration of the reference cell array that offers improved sensitivity, an updated analysis that does not require site-specific training, and early field test results from prototypes installed at two different locations with different latitude.**

Index Terms — **photovoltaic systems, resource assessment, pyranometer, reference cell**

I. INTRODUCTION

Key steps in financing, commissioning, and monitoring commercial solar photovoltaic (PV) power plants require measurements of solar irradiance to predict and assess plant performance. Solar irradiance has multiple components, including direct, diffuse, and ground-reflected light. As PV systems become more sophisticated, especially for bifacial systems, there is an increasing need for more sophisticated irradiance measurements that resolve all irradiance components.

The most complete assessments of solar irradiance include determination of beam, diffuse, and ground-reflected components – e.g., direct normal irradiance (DNI), diffuse horizontal irradiance (DHI), and reflected horizontal irradiance (RHI) – which allows determination of irradiance on any plane via transposition. However, due to the cost and maintenance requirements of specialized equipment for DNI and DHI measurement, which usually involves moving parts for sun tracking, these measurements are typically omitted.

Pre-construction resource assessment campaigns typically only measure global horizontal irradiance (GHI), using a horizontally oriented thermopile pyranometer, and (with the increasing importance of bifacial PV systems) reflected horizontal irradiance (RHI), using a downfacing pyranometer or reference cell. Performance modeling software then employs decomposition models [1], such as DIRINT [2], to estimate DNI and DHI from GHI, followed by transposition models [1] to determine plane-of-array (POA) irradiance. However, the uncertainty of resource assessments and performance predictions may be increased due to reliance on decomposition models to determine diffuse irradiance.

Meteorological stations at built PV power plants typically include direct measurements of POA in addition to GHI [3], with GHI used for connection to plant design predictions and POA for monitoring of resource on the module plane. However, explicit measurement of direct and diffuse irradiance could reduce uncertainties in performance assessment, by allowing more accurate treatment of angle of incidence losses and, for bifacial systems, enabling better predictions of rear-side irradiance.

Here we discuss a novel method to provide a more complete irradiance assessment at lower cost. The method uses an array of multiple reference cells to determine GHI, DNI, DHI, and RHI, with no moving parts. The method is similar to that used in previous work on multi-pyranometer arrays [4][5][6]. However, in this work we use PV reference cells instead of pyranometers, we introduce a correction for incidence angle dependence of the sensors, and we explicitly measure and correct for ground-reflected irradiance instead of blocking it with a horizon blocking ring. In addition, we have implemented a fast calculation using a combination of simple models. Since our calculation determines the beam component of irradiance, it allows the effect of the sensor's incidence angle response

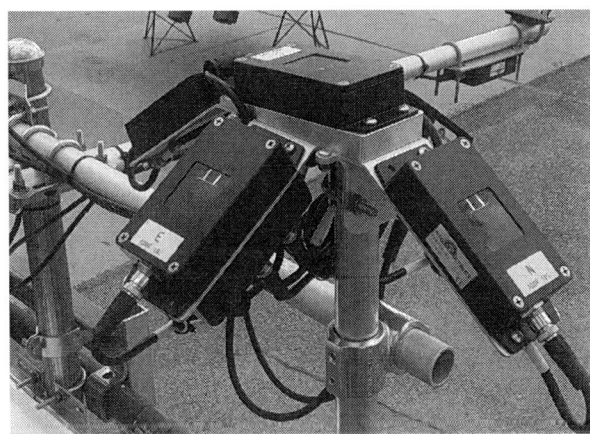

Fig. 1. Updated reference cell array prototype installed at Sandia National Laboratories, with five sky-facing reference cells (center, north, east, south, west) and one downfacing cell (on extension arm).

[7][8][9] to be automatically compensated, enabling accurate measurements of GHI with reference cells.

Previous work [10] reported on a first field test of a prototype reference cell array with measurements for direct and diffuse irradiance compared to research-grade systems. The study relied on a neural network model that was tuned for the specific site under study. We have since updated the reference cell array configuration to offer better sensitivity at all latitudes and have replaced the neural network model with an explicit calculation that is site independent.

Here we report early field test results of the new concept, with prototypes deployed at two different locations. Results are evaluated with several metrics, including comparison to research-grade measurements of direct and diffuse light and assessment of accuracy of PV system plane-of-array irradiance prediction.

II. EXPERIMENT

The new reference cell array concept includes 5 sky-facing reference cells, with four cells facing nominally north, east, south, and west on a 35-degree tilt and a central cell at a 5-degree south-facing tilt. Using a sensitivity analysis that considered the minimum difference in solar angle of incidence between any pair of reference cells, we determined that this configuration would have good sensitivity at all latitudes corresponding to typical installation locations. The central cell is intended to be nominally horizontal, with a 5-degree tilt for water roll off.

A prototype design is shown in Fig. 1. The prototype is assembled from existing PV reference cells (Atonometrics, RC18) and is intended to provide proof-of-concept. Following successful proof-of-concept a more compact productized version will be produced. The prototype also includes an additional downfacing reference cell mounted on an extension arm for measuring reflected horizontal irradiance (RHI) and albedo (RHI / GHI).

In March 2022, two similar prototypes were installed for field testing, one at the SunPower research and development facility near Davis, California, and one at Sandia National Laboratories in Albuquerque, New Mexico. Both sites also include a research-grade two-axis sun-tracking system (Kipp & Zonen, SOLYS) with a pyrheliometer and a shaded pyranometer for DNI and DHI measurement, as well as separate pyranometers for GHI measurement. The output of the reference cell arrays is evaluated against data from these reference instruments.

All the RC18 reference cells were calibrated at Atonometrics against an NREL-traceable reference cell on a class A solar simulator prior to the start of data collection.

In both installations the reference cell array is mounted at the edge of an elevated meteorological station platform approximately 4 meters off the ground. In the SunPower installation the platform edge is aligned east-west, the north reference cell faces the open side of the platform, and the downfacing reference cell is mounted directly underneath the five sky-facing reference cells. In the Sandia installation, the platform edge is aligned north-south, the west reference cell faces the open edge of the platform, and the down-facing reference cell is mounted on an arm extending off the platform over a paved area, as shown in Fig. 1.

The reference cell tilt angles were measured immediately following installation and confirmed to be within 1 degree of the design values. Reference cell azimuth angles were measured at installation using a compass; however, adjustments to azimuthal orientation values were made in post-processing the data by performing a best-fit of simulated to measured data on a single clear day at each site.

Data presented here were collected between mid-March and mid-April 2022 over about five weeks at each site.

Data are collected from all instruments on a one-minute or five-minute scan rate. Raw data from the reference cell array are recorded and then analyzed offline to determine results for DNI, DHI, GHI, and RHI. Offline calculations run on a personal computer and are implemented in MATLAB using the Sandia PVLIB toolbox [11].

III. ANALYSIS METHOD

A. Nomenclature

Refer to Table 1 for a summary of nomenclature.

B. Basic model

Our goal is to determine DNI, DHI, and RHI at each time point by analyzing the readings of the six reference cells in the reference cell array.

At each time point we collect the irradiance readings of the six reference cells into a vector B, where B_i is the reading of the ith reference cell. Now we relate the unknowns DNI, DHI, and RHI to the B_i values by

$$\begin{bmatrix} B_1 \\ B_2 \\ B_3 \\ B_4 \\ B_5 \\ B_6 \end{bmatrix} = \begin{bmatrix} A_{11} & A_{12} & A_{13} \\ A_{21} & A_{22} & A_{23} \\ A_{31} & A_{32} & A_{33} \\ A_{41} & A_{42} & A_{43} \\ A_{51} & A_{52} & A_{53} \\ A_{61} & A_{62} & A_{63} \end{bmatrix} \cdot \begin{bmatrix} DNI \\ DHI \\ RHI \end{bmatrix} \quad (1)$$

where A is a matrix of sensitivity factors. Represented more simply,

$$B = A \cdot x \quad (2)$$

where x is the vector (DNI, RHI, RHI).

The columns of A quantify, respectively, sensitivity of each reference cell to DNI, DHI, and RHI, and each row of A corresponds to reference cell i. Values of the sensitivity factors in A depend on the reference cell orientation and the solar position at each moment in time.

We calculate the left-most (DNI) column of A as

$$A_{i1} = IAM(\theta_i) \cdot \cos \theta_i \qquad (3)$$

where θ_i is the solar angle of incidence on reference cell i, and $IAM(\theta_i)$ is the incidence angle modifier function of the reference cell for beam radiation. IAM expresses the deviation of an irradiance sensor from pure cosine dependence [7][8] and it should be determined experimentally, as discussed below.

For the middle (DHI) column of A, we begin by assuming diffuse light is isotropically distributed in the sky dome, such that

$$A_{i2} = IAM_{diff} \cdot \left(\frac{1 + \cos \beta_i}{2}\right) \qquad (4)$$

where the term in parentheses quantifies the reduction in diffuse light received from the sky as the reference cell's tilt angle β_i increases, and IAM_{diff} quantifies the reduction in a reference cell's reading due to the distribution of incidence angles from the isotropically distributed diffuse light. IAM_{diff} is calculated by averaging $IAM(\theta_i)$ over the sky dome such as in [12].

We calculate the final (RHI) column of A by assuming ground-reflected light is diffuse and isotropically distributed (Lambertian reflectance), such that

$$A_{i3} = IAM_{diff} \cdot \left(\frac{1 - \cos \beta_i}{2}\right) \qquad (5)$$

where the term in parentheses quantifies the fraction of ground-reflected diffuse light that is captured by the reference cell as a function of its tilt angle.

At each time point we calculate the solar position and corresponding angle of incidences, calculate the matrix A per above, and calculate

$$x = B/A \qquad (6)$$

using a least-squares solution method [13], yielding results for DNI, DHI, and RHI.

C. Anisotropic Diffuse Model

To improve accuracy for DHI determination, we consider the anisotropic distribution of diffuse light in the sky dome, replacing eq. (4) with

$$A_{i2} = IAM_{diff} \cdot f_P(DNI, DHI, H, Z, \gamma, AM, \beta_i, \alpha_i) \qquad (7)$$

where f_P is calculated using the empirical Perez sky diffuse model [14] in the PVLIB toolbox [15]. To calculate f_P, we use the Perez sky diffuse model to calculate the diffuse irradiance falling on the plane with tilt β_i and azimuth α_i (i.e. the reference cell plane) for given DNI, DHI, and other input

Table 1. Summary of nomenclature

Global horizontal irradiance	GHI
Direct normal irradiance	DNI
Diffuse horizontal irradiance	DHI
Reflected horizontal irradiance	RHI
Albedo	a
Solar zenith	Z
Solar azimuth	γ
Reference cell i tilt angle	β_i
Reference cell i azimuth angle	α_i
Solar angle of incidence to reference cell i	θ_i
Incidence angle modifier, beam radiation	$IAM(\theta_i)$
Incidence angle modifier, diffuse radiation	IAM_{diff}
Vector of irradiance readings	B
Matrix of sensitivities	A
Vector of unknowns (DNI, DHI, RHI)	x
Extraterrestrial normal irradiance	H
Air mass	AM
Sensitivity to DHI in anisotropic model	f_P

parameters listed and then normalize by the DHI value. This provides the slope of A_{i3} with respect to small changes in DHI.

However, note that the calculation of f_P requires DNI and DHI, which are two of our three unknowns. Therefore, we perform the complete analysis in two stages. In the first stage, we construct A using the isotropic diffuse model in eq. (4) and then solve eq. (6), yielding an initial estimate for x. In the second stage, we use the estimates for the DNI and DHI components of x to construct A again using eq. (7) and then repeat the solution per eq. (6).

D. GHI

Once DNI and DHI are determined, we calculate GHI per

$$GHI = (\cos Z) \cdot DNI + DHI \qquad (8)$$

E. Using Assumed Albedo

In place of directly solving for RHI, we can also modify the equations to use an assumed albedo a, which is the ratio RHI/GHI.

To do this, we construct x as the two-element vector (DNI, DHI) and eliminate the third column of A (see eq. (1)). Then we add

$$IAM_{diff} \cdot a \cdot (\cos Z) \cdot \left(\frac{1 - \cos \beta_i}{2}\right) \qquad (9)$$

to eq. (3) and add

$$IAM_{diff} \cdot a \cdot \left(\frac{1 - \cos \beta_i}{2}\right) \qquad (10)$$

to eq. (4) and eq. (7).

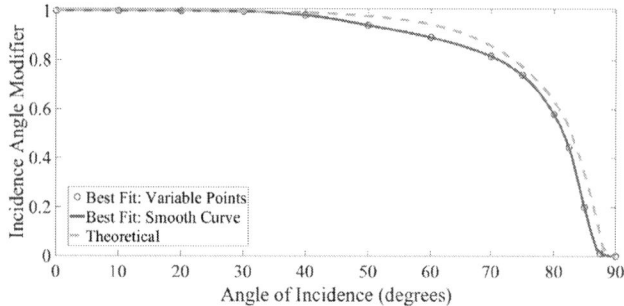

Fig. 2. Best fit vs. theoretical curves for reference cell incidence angle modifier $IAM(\theta_i)$.

IV. IAM FUNCTION

Accurate results require precise determination of the reference cell *IAM* functions. The primary factor determining $IAM(\theta_i)$ is reflection losses from the reference cell surface which can be calculated theoretically using the known indices of refraction of the reference cell glass, encapsulation, and silicon. However, geometric and other factors also modify the function. Our previous work [10], [16], [17] used a theoretically calculated curve for $IAM(\theta_i)$ which attempted to take into account both the optical and geometric properties of the RC18 reference cells. However, in this work, we empirically determined a new curve for $IAM(\theta_i)$ which produces a best fit to the tracking diffusometer data for DHI on a single clear day for the reference cell array installed at the SunPower location. Results are presented in Fig. 2. Differences between the theoretical and best-fit $IAM(\theta_i)$ curve likely explain some anomalies observed in previous work.

V. RESULTS

A. Time Series

Fig. 3 displays an example set of results for GHI, DNI, DHI, and RHI determined from the reference cell array at the SunPower location over a four-day period. These data illustrate sensitivity to a range of diffuse irradiance levels on clear and cloudy days. The figure shows general agreement between the reference cell array results and those of the reference instruments. (Note there was no reference instrumentation for RHI.)

Results from the Sandia test site were similar, albeit with an anomaly that causes an asymmetry in the DHI profile on very clear days, resulting in DHI readings about 10-30 W/m² higher than expected during late afternoon hours. We suspect that the asymmetry is caused by the specific configuration of the reference cell array at the Sandia site as described in section II. The downfacing reference cell is located on an extension arm overhanging the meteorology platform to the west; therefore it sees shade in the morning and reflected light from the platform

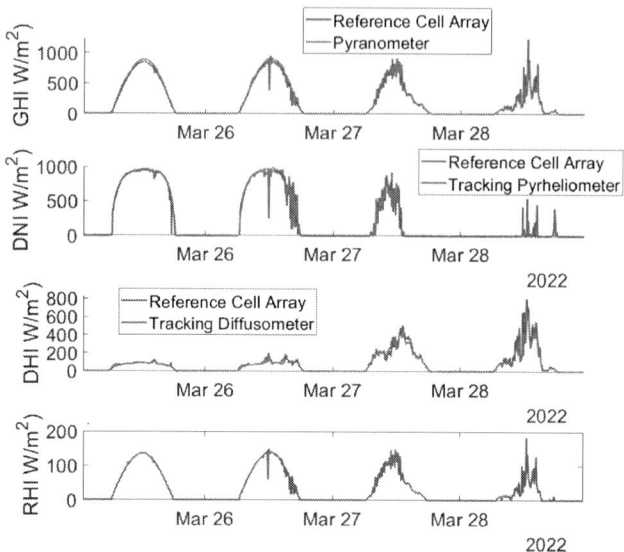

Fig. 3. Example reference cell array results from the SunPower site over a four-day period and comparison to reference instruments.

building wall in the afternoon. Furthermore, the west-facing cell of the array looks out over an open area west of the platform which may have higher albedo than the area to the east on top of the platform itself. These effects are under further investigation. However, as shown in the tables below, results from the Sandia station still generally agree well with the reference instrumentation, especially at higher DHI values corresponding to cloudy days.

B. Correlation to Reference Instruments

Table 3 summarizes the comparison of the reference cell array results to those from the reference instruments over the approximately five-week period at each site. The table shows mean-bias deviation (MBD) and root-mean-square deviation (RMSD) for hourly-averaged data, where the reference cell array results for GHI, DNI, and DHI are correlated against, respectively, data from the thermopile pyranometer, tracking pyrheliometer, and tracking diffusometer. For comparison, we also include statistics for correlation of DNI and DHI derived from the pyranometer GHI measurements via DIRINT decomposition [1] versus the DNI and DHI from the reference instruments.

The table generally shows good performance for the reference cell array, with low MBD in the range 0-2% for GHI, DNI, and DHI. One exception is the elevated MBD of 4.6% (5.6 W/m²) for the DHI results at the Sandia site. This is likely due to the elevated reference cell array DHI results at this site on clear days during afternoon hours, as discussed above, which is under investigation. The reference cell array results generally correlate better to the reference instruments than those from the DIRINT decomposition. We expect that the relative benefit of the reference cell array versus decomposition alone would

978-1-7281-6118-1/22 $31.00 © 2022 IEEE   288

Table 2: Correlation of hourly-averaged reference cell array and decomposition results with reference instrumentation over five-week trial.

Parameter	Reference Instrument	Reference Cell Array		DIRINT Decomposition	
		MBD W/m² (%)	RMSD W/m² (%)	MBD W/m² (%)	RMSD W/m² (%)
SunPower					
GHI	Thermopile Pyranometer	-2.3 (-0.5)	17.7 (3.8)	n/a	n/a
DNI	Tracking Pyrheliometer	1.2 (0.2)	47.6 (7.7)	43.5 (7.2)	64.6 (10.7)
DHI	Tracking Diffusometer	-0.9 (-0.6)	29.6 (21.8)	-24.0 (-17.7)	33.3 (24.5)
Sandia					
GHI	Thermopile Pyranometer	-8.6 (-1.7)	33.6 (6.5)	n/a	n/a
DNI	Tracking Pyrheliometer	-12.6 (-1.9)	42.1 (6.3)	12.4 (1.9)	65.0 (9.8)
DHI	Tracking Diffusometer	5.6 (4.6)	36.3 (29.5)	1.0 (0.8)	36.3 (29.4)

increase when evaluating data over longer time periods with more seasonal variation. Note, for example, that the decomposition-derived DHI in Table 3 shows a relatively poor MBD at the SunPower site (-17.7%) but an excellent value at the Sandia site (0.8%), while the reference cell array results for DHI are much more consistent at both sites, with MBD of -0.6% and 4.6%, respectively.

C. POA Irradiance Prediction

Since the goal of measuring DNI and DHI is to predict PV performance by calculating irradiance in a specific plane of array (POA), at each site we have evaluated the results by using the DNI and DHI to predict the POA irradiance measured by a reference cell on a tilted plane elsewhere on the site. The three DNI and DHI sources are the reference cell array, the reference instruments (tracking pyrheliometer and diffusometer), and DIRINT decomposition. The POA reference cell was south-facing at 25-degree tilt at the Sandia site and south-facing with 20-degree tilt at SunPower.

Table 3: Correlation of hourly-averaged predicted POA irradiance, using three different DNI and DHI sources, with measured POA reference cell irradiance measurement.

DNI & DHI Source for POA Prediction	Correlation to POA Reference Cell	
	MBD W/m² (%)	RMSD W/m² (%)
SunPower		
Reference Cell Array	-2.2 (-0.4)	15.4 (2.7)
Tracking Instruments	5.5 (1.0)	11.5 (2.1)
DIRINT Decomposition	5.1 (0.9)	14.0 (2.5)
Sandia		
Reference Cell Array	0.2 (0.0)	17.2 (3.3)
Tracking Instruments	6.7 (1.3)	25.5 (4.8)
DIRINT Decomposition	8.0 (1.5)	25.5 (4.8)

Table 2 presents the results of these tests. Predictions made using all three DNI and DHI sources have similar quality of correlation against the measured POA data. A close inspection of the data showed that small differences in RMSD were dominated by discrepancies between the instruments on cloudy days with high irradiance fluctuations, when differences in sensor position and acquisition timing result in non-correlation. To draw significant conclusions about the relative benefit of the reference cell array results versus decomposition analysis alone will require collection of data over a longer time.

In this work we assessed only the prediction of front-side POA. As we continue this investigation, we intend to assess the utility of the reference cell array data for predicting also rear-side POA irradiance that would be relevant to bifacial modules. Here consistency of DHI determination will be much more important.

VI. ONGOING WORK

Data presented here were acquired over a limited period of approximately 5 weeks at each site. Data collection and analysis are continuing.

For this work, we determined a best-fit IAM function for the reference cells by optimizing the agreement between the reference cell and tracking diffusometer DHI results on one clear day at one site. However, we plan to directly measure the reference cell IAM outdoors as in [9] or with similar protocols. This will provide an independent basis for evaluating the reference cell array analysis and comparing results between sites. In addition, it could elucidate any anisotropic aspects of the reference cell angle of incidence response, which could then permit more accurate analysis.

Soiling could have a significant impact on the results by altering the IAM at high incidence angles. The potential uncertainty from this effect could be studied using a modeling approach, by considering angle-dependent shading due to dust particles at different soiling levels as in [18].

As discussed above, reference cell array results for DHI at Sandia showed anomalously high readings for late afternoon hours on very clear days. We are investigating potential reconfiguration of the array to eliminate this effect.

VI. CONCLUSIONS

We have demonstrated the potential to use a reference cell array, with no moving parts, to determine global, direct, diffuse, and ground-reflected radiation components. Updating on previous work with this concept [10], the analysis is now based on explicit calculation rather than training to reference data. The only site-specific data used for calibrating the technique were the data used to establish reference cell incidence angle response, and in upcoming work we plan to measure this response independently. Results at the two sites studied show good correlation between the reference cell array and high-quality reference instrumentation for GHI, DNI, and DHI.

Further work is ongoing. A successful outcome of this project would enable using a reference cell array to add capability to resource assessment campaigns and solar power plant meteorological stations at low cost.

ACKNOWLEDGEMENT

This material is based upon work supported by the U.S. Department of Energy's Solar Energy Technologies Office under Award Number DE-SC0020831.

Sandia National Laboratories is a multimission laboratory managed and operated by National Technology and Engineering Solutions of Sandia, LLC, a wholly owned subsidiary of Honeywell International Inc., for the U.S. Department of Energy's National Nuclear Security Administration under contract DE-NA0003525.

REFERENCES

[1] M. Lave, W. Hayes, A. Pohl, and C. W. Hansen, "Evaluation of global horizontal irradiance to plane-of-array irradiance models at locations across the United States," *IEEE J. Photovoltaics*, vol. 5, no. 2, pp. 597–606, Mar. 2015, doi: 10.1109/JPHOTOV.2015.2392938.

[2] P. Ineichen, R. Perez, R. Seal, E. Maxwell, and A. Zalenka, "Dynamic global-to-direct irradiance conversion models," *ASHRAE Trans.*, vol. 98, no. 1, pp. 354–369, 1992.

[3] "IEC 61724-1 Ed. 1.0 en:2017 - Photovoltaic system performance - Part 1: Monitoring."

[4] D. Faiman, D. Feuermann, and A. Zemel, "Site-independent algorithm for obtaining the direct beam insolation from a multipyranometer instrument," *Sol. Energy*, vol. 50, no. 1, pp. 53–57, Jan. 1993, doi: 10.1016/0038-092X(93)90007-B.

[5] B. Marion, "Multi-Pyranometer Array Design and Performance Summary," *Proc. 1998 Am. Sol. Energy Annu. Conf.*, 1998.

[6] J. C. Baltazar, Y. Sun, and J. Haberl, "Improved methodology to evaluate clear-sky direct normal irradiance with a multi-pyranometer array," *Sol. Energy*, vol. 121, pp. 123–130, Nov. 2015, doi: 10.1016/j.solener.2015.07.015.

[7] N. Martin and J. M. Ruiz, "Calculation of the PV modules angular losses under field conditions by means of an analytical model," *Sol. Energy Mater. Sol. Cells*, vol. 70, no. 1, pp. 25–38, Dec. 2001, doi: 10.1016/S0927-0248(00)00408-6.

[8] B. H. King and C. D. Robinson, "Differential Analysis of the Angle of Incidence Response of Utility-Grade PV Modules,"

in *2019 IEEE 46th Photovoltaic Specialists Conference (PVSC)*, Jun. 2019, pp. 77–81, doi: 10.1109/PVSC40753.2019.8981355.

[9] B. H. King, D. Riley, C. D. Robinson, and L. Pratt, "Recent advancements in outdoor measurement techniques for angle of incidence effects," *2015 IEEE 42nd Photovolt. Spec. Conf. PVSC 2015*, Dec. 2015, doi: 10.1109/PVSC.2015.7355849.

[10] M. Gostein, A. Hoffman, F. Farina, and B. Stueve, "Measuring Global, Direct, and Diffuse Irradiance Using a Reference Cell Array," in *Conference Record of the IEEE Photovoltaic Specialists Conference*, Jun. 2021, pp. 923–927, doi: 10.1109/PVSC43889.2021.9518735.

[11] "PV Performance Modeling Collaborative | PV_LIB Toolbox." https://pvpmc.sandia.gov/applications/pv_lib-toolbox/ (accessed Mar. 15, 2021).

[12] B. Marion, "Numerical method for angle-of-incidence correction factors for diffuse radiation incident photovoltaic modules," *Sol. Energy*, vol. 147, pp. 344–348, May 2017, doi: 10.1016/j.solener.2017.03.027.

[13] "Solve systems of linear equations Ax = B for x - MATLAB mldivide \." https://www.mathworks.com/help/matlab/ref/mldivide.html?s_tid=srchtitle (accessed May 19, 2022).

[14] R. Perez, P. Ineichen, R. Seals, J. Michalsky, and R. Stewart, "Modeling daylight availability and irradiance components from direct and global irradiance," *Sol. Energy*, vol. 44, no. 5, pp. 271–289, Jan. 1990, doi: 10.1016/0038-092X(90)90055-H.

[15] "PV Performance Modeling Collaborative | Perez Sky Diffuse Model." https://pvpmc.sandia.gov/modeling-steps/1-weather-design-inputs/plane-of-array-poa-irradiance/calculating-poa-irradiance/poa-sky-diffuse/perez-sky-diffuse-model/ (accessed May 19, 2022).

[16] M. Gostein, B. Stueve, R. Clark, P. Keelin, M. Grammatico, and M. Reusser, "Field Trial of Meteorological Station Using PV Reference Cells," in *Conference Record of the IEEE Photovoltaic Specialists Conference*, Jun. 2020, vol. 2020-June, pp. 0520–0523, doi: 10.1109/PVSC45281.2020.9300799.

[17] M. Gostein, R. Clark, M. Grammatico, P. Keelin, M. Reusser, and B. Stueve, "Updated Trial of Meteorological Station Using PV Reference Cell Array," *Conf. Rec. IEEE Photovolt. Spec. Conf.*, pp. 2357–2359, Jun. 2021, doi: 10.1109/PVSC43889.2021.9518743.

[18] J. Zorrilla-Casanova *et al.*, "Losses produced by soiling in the incoming radiation to photovoltaic modules," 2012, doi: 10.1002/pip.1258.

Validation of In-Situ I-V Measurement Unit for PV System Monitoring Applications

Audrey Marquis[1], Michael Gostein[1], and Bruce H. King[2]

[1]Atonometrics, Inc., Austin, TX, USA; [2]Sandia National Laboratories, Albuqurque, NM, USA

Abstract—**Electrical output measurement of PV reference modules in an array is used for a variety of applications in PV system monitoring, including measurements of irradiance and soiling as well as detection of degradation and faults. Using currently available measurement equipment, reference modules are typically standalone modules installed alongside the array, which results in extra costs for system engineering and racking. New products will allow measuring I-V curves on power-producing PV modules in-situ within a PV array, simplifying implementation. However, for in-situ use it is critical that the I-V measurement unit does not disturb inverter operation or reduce energy output. In this work we developed a test protocol to quantify impacts of the I-V unit on inverter energy output and evaluated a new in-situ I-V measurement product in multiple arrays. In initial results, we found no adverse impact of the I-V unit on inverter operation, with minimal power loss approximately as expected from the series resistance and duty rate of the I-V measurement.**

Keywords—photovoltaic systems, monitoring, I-V curves

I. INTRODUCTION

Various applications in PV system monitoring use current-voltage (I-V) measurement on PV reference modules – modules identical to those of the PV array but designated for measurement and characterization purposes – to extract system performance data. Examples include determining matched irradiance by measuring short-circuit current of a calibrated reference module [1]; determining soiling loss by comparing short-circuit current or maximum-power output of clean and dirty reference module pairs [2]; identifying faults by classification of reference module I-V curves [3]; measuring module degradation rates; and determining effective total irradiance for bifacial systems from a reference module within an array [4].

However, due to the capabilities of currently used I-V curve tracer products, PV reference modules must typically be installed standalone – at the array but not within it. Reference modules are therefore typically installed on extension arms above or to the side of a row of modules or on separate racking. Alternatively, standalone reference modules sometimes occupy a standard module position with neighboring modules wired around them. These standalone configurations result in significant extra costs for engineering and racking.

It would be advantageous to have I-V tracers that could be deployed in-situ within an array on power-producing modules, simplifying deployment and lowering overall costs. This has been explored by a number of groups [5][6]. Here we report on a new product for in-situ I-V measurement developed by Atonometrics, the RDE300i, and on our study to validate the RDE300i product for in-situ application.

For in-situ application it is critical to know that the I-V unit will not disturb inverter function and that any energy loss will be negligible. We have begun a test program at Sandia National Laboratories to evaluate the RDE300i and quantify any impact on inverter function and energy harvest.

II. EXPERIMENT

A. I-V System

The RDE300i, shown in the top panel of Fig. 1, is designed to be installed between a selected module and the rest of the array, for in-situ operation. The module is connected to the

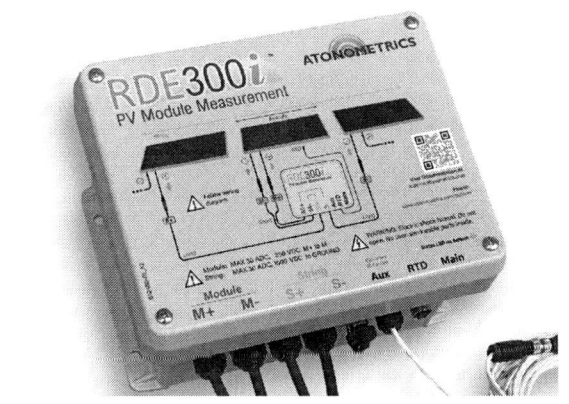

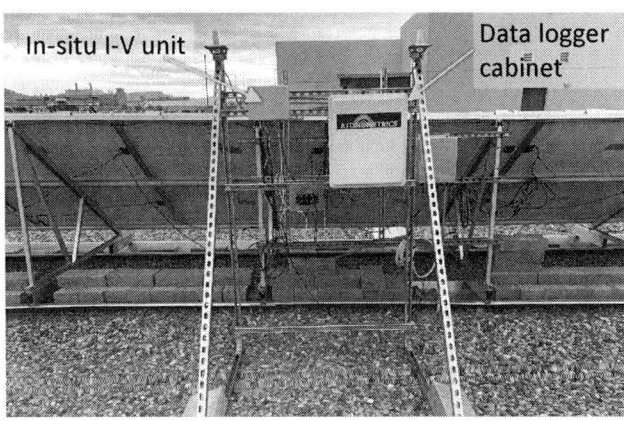

Fig. 1. Top: RDE300i PV module in-situ I-V tracer. Bottom: RDE300i I-V tracer on a test stand at Sandia National Laboratories wired into a PV string.

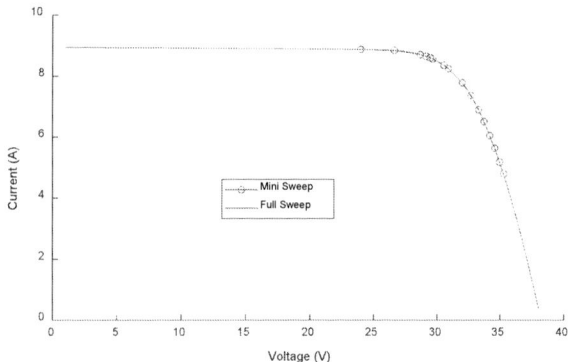

Fig. 2. Example of I-V curves showing two operation modes.

RDE300i which is then connected to the neighboring modules on the PV string. I-V sweeps are performed at periodic intervals. The RDE300i supports two I-V sweep modes, illustrated by Fig. 2. In one mode the RDE300i momentarily disconnects the module from the string, performs a full I-V sweep, and reconnects the module. While the module is disconnected, string current goes around the module via an integrated bypass circuit. In another mode, the RDE300i does not disconnect the module but performs a mini-sweep, as illustrated in Fig. 2, to measure the module's maximum power point without disconnection. The mini-sweep mode permits more frequent measurements while minimizing power loss. In between I-V sweeps the RDE300i directly connects the module to the string for pass-through operation. RDE300i can measure I-V sweeps up to 30 A, 250 V, and 1500 W.

The bottom panel of Fig. 1 shows a prototype RDE300i I-V unit wired in-situ a string of crystalline silicon PV modules in a test array at Sandia National Laboratories. For this study the I-V unit is mounted on a test stand that may be easily moved to different arrays that use different modules and inverters. However, in typical installations the RDE300i unit would likely be mounted underneath the module racking.

For these experiments we are using a data logger, mounted in the cabinet shown to the right of the I-V unit in Fig. 1, to capture the I-V curves and the fit results. We measure a full I-V sweep every 1 minute and a mini-sweep every 10 seconds. From the full I-V sweeps the I-V unit automatically determines short-circuit current (Isc), open-circuit voltage (Voc), maximum power (Pmax), and voltage and current of the maximum power point (Vmp and Imp). From the mini sweeps it determines Imp, Vmp, and Pmax. In addition, in between the I-V sweeps the unit records the output voltage and current (Vout and Iout), which are the voltage and current operating point of the module governed by the inverter's maximum power point tracking of the entire string. Fig. 3 depicts an example of data outputs of the RDE300i over a ten-day period.

B. PV Arrays

We have begun a test campaign at the Sandia National Laboratories Photovoltaic Systems Evaluation Laboratory (PSEL) in Albuquerque, New Mexico, which has numerous test arrays and inverter systems. For the testing performed so far, we selected three arrays representing different module and inverter types. Two of the arrays use crystalline silicon Suniva 270 W modules and one uses crystalline silicon LG 400 W modules. Inverters included models from Power One and SMA. Strings selected for testing have output power up to 4.8 kW. As the study proceeds, we plan to test with additional module types, including monofacial and bifacial silicon as well as CdTe thin film modules, and additional inverters from multiple manufacturers.

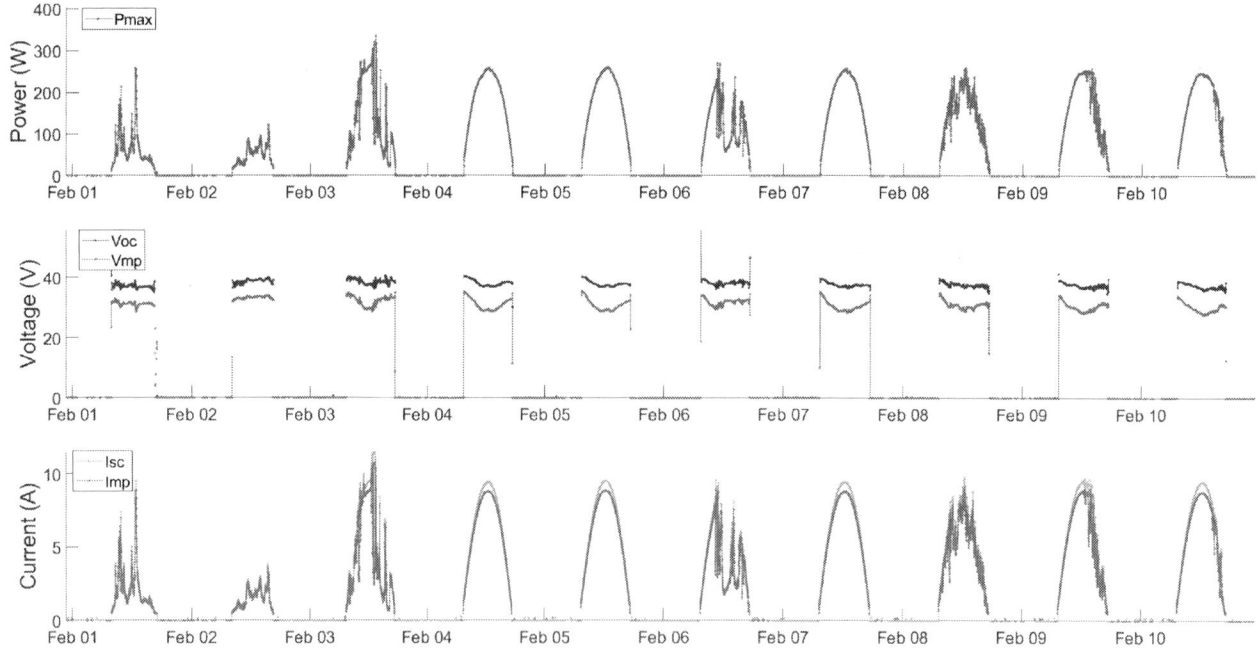

Fig. 3. Example data outputs from RDE300i in-situ I-V tracer, showing measurements of module power, voltage, and current parameters.

Table 1: Summary of functionality and energy harvest test results

System	Hours of Collected Data	Minutes of Inverter Fault	Performance Index	Expected Power Losses	Observed Power Losses
System 1	916	0	0.9984 ± 0.0017	1.8 W	1.02 W
System 2	648	0	0.9983 ± 0.0017	1.5 W	1.23 W
System 3	570	0	0.9963 ± 0.0009	3.4 W	4.99 W

C. Functionality Test

The first part of our test protocol is a functionality test to identify and quantify fault events and determine if the I-V sweeps performed by the RDE300i unit adversely affect inverter operation, for example causing the inverter to register a fault or temporarily disengage from maximum power point tracking.

To perform this test, we insert the RDE300i into one of the PV array test systems and record the module's maximum power, measured by the RDE300i unit, together with the array DC power output, measured by a separate array-level monitoring system, once per minute for 2-4 weeks. We then correlate the two data sets and look for outliers. Any points in which the array is not producing power, as determined by the array monitoring system, but the module itself has power-producing capability, as determined by the P_{max} measured by the RDE300i, would indicate that the inverter is not operating, even though it should be. These points could indicate a momentary inverter fault condition caused by an adverse inverter response to the I-V sweep.

D. Energy Harvest Test

The second part of our test protocol, the energy harvest test, is designed to quantify the power loss from the array introduced by the insertion of the RDE300i in-situ I-V measurement unit. A small amount of power loss will occur due to added series resistance of the measurement device and cables as well as from the lost operation time of the module during I-V sweeps. However, these losses should be negligible. Of greater concern is whether the periodic I-V operation subtly disrupts optimal maximum power point tracking and therefore results in lower power output. The energy harvest test is designed to put an upper limit on any such energy loss.

To perform this test, we use two identical strings, one with the RDE300i and one without, each with its own inverter channel. The DC output of each string is measured by an array-level monitoring system at the inverter inputs. We then compare the DC energy input to the two inverters over 2-4 weeks. Since the two strings and inverters may have different efficiencies, we compare the strings both before and after insertion of the RDE300i. From these data we calculate a performance index

$$ PI = \frac{(\sum P_2 \,/\, \sum P_1)_{after}}{(\sum P_2 \,/\, \sum P_1)_{before}} \qquad (1) $$

where subscripts 1 and 2 refer to the two strings used in the test, P_1 and P_2 are the 1-minute average power outputs of each string,

summations are taken over all minutes in the test, and "before" and "after" refer to data acquired before and after putting the RDE300i into string 2. The objective is for the performance index to be close to 1.00; to the extent that it is less than 1.00, the difference equals an energy loss factor.

III. RESULTS

Table 1 summarizes results of both the functionality and energy harvest tests on the three systems tested to date.

Using the functionality test protocol described above, we found zero instances of inverter fault on any of the three systems while in operation with RDE300i. During this testing the RDE300i unit was performing mini-sweeps once every 10 seconds and full sweeps once every 60 seconds. These sweeps did not cause any of the tested inverters to detect a fault condition and shut down even temporarily.

Performance indices evaluated for all three systems using eq. (1) are shown in Table 1. For this analysis, data were filtered to exclude days with low insolation when performance of the strings could change due to non-linearities. To determine an uncertainty for each performance index resulting from random fluctuations in irradiance, temperature, and other conditions, we evaluated the performance index with eq. (1) using subsets of the full data for each system, where each subset comprises a single day from the "before" period paired with a single day from the "after" period, and then calculated the standard deviation σ of the results for all subsets of day-pairs. This allowed us to estimate the uncertainty of the result using the complete data set as $\sigma/\sqrt{N}$ where N is the number of day pairs in the full data set.

As shown in Table 1, performance indices for all three systems were very close to 1.00, indicating minimal energy loss.

As discussed above, some minimal level of power loss is expected even if inverter max power point tracking is unaffected. The series resistance of the RDE300i unit and its associated PV cables is approximately 60 milliohms. In addition, due to full I-V sweeps conducted once per minute, the PV module connected to RDE300i is off-line up to 0.8% of the time. Using these two factors we estimated the expected average power loss in each system test due to series resistance and measurement time alone. We can then compare this with observed power losses calculated using the performance index results.

Table 1 shows the expected and actual observed average power loss due to RDE300i for each of the three systems tested. The observed losses are very close to the expected ones. This indicates no detectible adverse effect of RDE300i on the inverter

978-1-7281-6118-1/22 $31.00 © 2022 IEEE

maximum power point tracking, and certainly no fault conditions that resulted in gross power loss.

IV. Conclusions

We have developed a protocol for validating I-V tracing equipment for in-situ use within PV arrays to enable efficient use of PV reference modules for system monitoring applications. Our protocol was designed to identify any fault conditions caused by I-V sweeps and to quantify any power losses due to adverse impact on inverter maximum power point tracking. In our evaluation of the RDE300i product on three systems, comprising different inverter types, we observed no adverse impact on the I-V unit on inverter operation, with only minimal power losses as expected from the insertion of the I-V unit. This suggests that the I-V unit is well suited for PV power plant deployment. Our study is ongoing; we continue to test additional module and inverter types.

Acknowledgment

This material is based upon work supported by the U.S. Department of Energy's Solar Energy Technologies Office under Award Number DE-SC0020012.

Sandia National Laboratories is a multimission laboratory managed and operated by National Technology and Engineering Solutions of Sandia, LLC, a wholly owned subsidiary of Honeywell International Inc., for the U.S. Department of Energy's National Nuclear Security Administration under contract DE-NA0003525.

References

[1] J. Polo, W. G. Fernandez-Neira, and M. C. Alonso-García, "On the use of reference modules as irradiance sensor for monitoring and modelling rooftop PV systems," *Renew. Energy*, vol. 106, pp. 186–191, Jun. 2017, doi: 10.1016/J.RENENE.2017.01.026.

[2] M. Gostein, J. R. Caron, and B. Littmann, "Measuring soiling losses at utility-scale PV power plants," in *2014 IEEE 40th Photovoltaic Specialist Conference, PVSC 2014*, 2014, pp. 885–890, doi: 10.1109/PVSC.2014.6925056.

[3] C. B. Jones *et al.*, "Automatic fault classification of photovoltaic strings based on an in situ IV characterization system and a Gaussian process algorithm," in *2017 IEEE 44th Photovoltaic Specialist Conference, PVSC 2017*, 2017, pp. 1264–1267, doi: 10.1109/PVSC.2017.8366372.

[4] M. Gostein *et al.*, "Measuring Irradiance for Bifacial PV Systems," *Conf. Rec. IEEE Photovolt. Spec. Conf.*, pp. 896–903, Jun. 2021, doi: 10.1109/PVSC43889.2021.9518601.

[5] J. E. Quiroz, J. S. Stein, C. K. Carmignani, and K. Gillispie, "In-situ module-level I–V tracers for novel PV monitoring," *2015 IEEE 42nd Photovolt. Spec. Conf. PVSC 2015*, Dec. 2015, doi: 10.1109/PVSC.2015.7355608.

[6] J. I. Morales-Aragonés *et al.*, "Low-Cost Electronics for Online I-V Tracing at Photovoltaic Module Level: Development of Two Strategies and Comparison between Them," *Electron. 2021, Vol. 10, Page 671*, vol. 10, no. 6, p. 671, Mar. 2021, doi: 10.3390/ELECTRONICS10060671.

Effect of near-interface compensation of CdSeTe absorber layers on solar cell performance

Brian Good, Eric Colegrove, and Matthew O. Reese

National Renewable Energy Laboratory, Golden, CO 80401, USA

University of Illinois at Chicago, Chicago, IL 60607, USA

Abstract—**Arsenic has been shown to be an effective p-type dopant of CdTe. However, challenges remain in the fabrication of high efficiency CdTe solar cells using As. As-doped CdTe is prone to self-compensation and observed accumulation of dopant atoms in CdTe near the interface with MgZnO (MZO) suggests that this could be occurring. In this study, we use SCAPS 1-D modeling software to investigate the effect of near-interface compensation. We consider three cases: shallow donors, deep recombination centers, and a thin layer of excess positive charge accumulation. All of these cases are shown to have significant effects on the current-voltage characteristics, while the thin charge layer also affects capacitance-voltage measurements.**

Keywords—*cadmium telluride, CdSeTe, arsenic, compensation, doping, modeling, simulation*

I. INTRODUCTION

CdTe solar cells have undergone significant improvements in recent years, reaching record efficiencies of ~22%. However, open-circuit voltages (V_{oc}) remain well below the detailed balance limit of ~1.2 V [1]. One method of increasing V_{oc} is to improve the doping in the absorber. Specifically, the goal of achieving higher open-circuit voltages through Group V doping has been of significant interest. Arsenic has been found to be an effective dopant of CdTe, with greater stability and dopant activation ratio than copper [2]. However, As also can self-compensate in CdTe, especially once it exceeds the mid-10^{17} cm^{-3} range [2-5].

Secondary ion mass spectroscopy (SIMS) has shown a significant accumulation of arsenic at the front interface of CdSeTe absorbers (Fig. 1). Within 500 nm of the junction with

MgZnO (MZO), corresponding to the Se-rich region, As content increases from ~3x10^{18} to ~2x10^{20} cm^{-3}. Similar results have been previously reported in literature [2]. Since net acceptor levels of As in px-CdTe have been shown to saturate at ~1-2x10^{17} cm^{-3} with an activation ratio of about 1% [2,6], we expect that a significant excess of As atoms will result in compensation of the acceptor density. In fact, trap densities in CdSeTe:As have been estimated at 10^{17}–10^{18} cm^{-3} by varying injection levels in time-resolved photoluminescence measurements [7]. Current-voltage (JV) and capacitance-voltage (CV) parameters for devices with different processing conditions are given in Table I. Devices A and B were fabricated at NREL, while C is from Metzger et. al [2]. This shows one can obtain a wide range of V_{oc} while maintaining a similar carrier concentration, leading us to speculate about the effect of near-interface compensation.

Additionally, inconsistencies in CV measurements have been observed in As-doped devices. Depletion widths (W) in some devices are lower than would be expected from the corresponding carrier concentration as calculated from first principles [6]. This could also be the result of dopant accumulation near the interface, particularly in a layer that is too thin to raise the effective carrier concentration.

SCAPS 1-D simulation software [8] was used to assess the impact of interfacial As accumulation. We investigate the cases of shallow donors compensating the doping level and deep levels acting as recombination centers. We also introduce a thin "charge layer" at the interface. The effects of all of these cases on device performance, particularly V_{oc}, are considered.

II. SUMMARY OF MODEL

SCAPS 1-D software was used for all simulations. The device stack used was SnO$_2$:F/SnO$_2$/MZO/CdSeTe/CdTe/Au (Fig. 2). Shallow donor density (N_D) was set to 10^{20} cm^{-3} for SnO$_2$:F, 10^{18} cm^{-3} for SnO$_2$, and 10^{14} cm^{-3} for MZO. Low n-type doping is assumed in MZO, as our MZO films are too

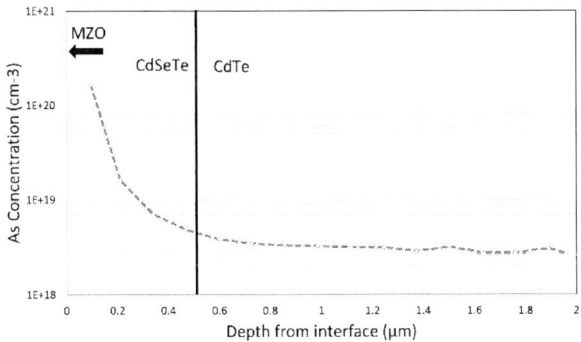

Fig. 1. Secondary ion mass spectroscopy (SIMS) data for arsenic in CdSeTe/CdTe. Distance is measured from front interface (MZO/CdSeTe).

TABLE I. EXPERIMENTAL DEVICE RESULTS

Device	V_{oc} (mV)	J_{sc} (mA/cm^2)	FF (%)	Eff (%)	W (μm)	CC (cm^{-3})
A	639	25.5	70.8	11.5	0.28	1.2E+16
B	729	28.6	73.3	15.3	0.34	1.2E+16
C [2]	856	30.5	79.8	20.8	0.20	1.0E+16

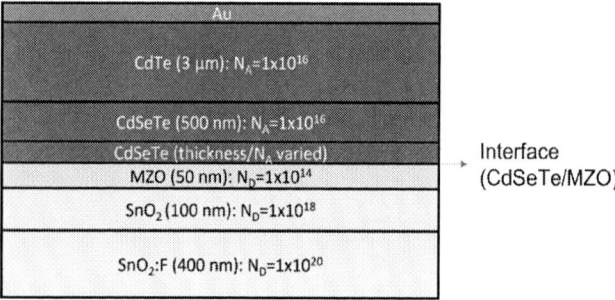

Fig. 2. Device stack used for simulations.

resistive to be reliably measured by the Hall method. Shallow acceptor density (N_A) in CdSeTe and CdTe was set to 10^{16} cm^{-3}, corresponding to the doping density achieved with As. An interface layer was introduced between MZO and CdSeTe, with interface recombination velocity (S) assumed to be 10^4 cm/s [9]. The conduction band offset (CBO) at the front interface is assumed to be 0.2 eV, which is near optimum alignment for junctions with CdTe [1,10].

In order to simulate As accumulation near the interface, the CdSeTe layer is split into a region of constant doping and an interfacial region in which doping is varied. Parameters for the varied layer are shown in Table II. Arsenic is known to form both shallow and deep compensating donors in CdTe; notable examples are $(As_iAs_i)^{2+}$ and $(As_{Te}As_i)$ complexes, respectively [4]. In this work, we consider shallow and deep levels separately, although both may be present. For the case of shallow compensating donors, N_D is varied from $10^{15} - 10^{17}$ cm^{-3}, to cover cases ranging from low compensation ($N_D \ll N_A$) to type conversion ($N_D > N_A$). For deep recombination centers, a mid-gap state is introduced in this region with densities of $10^{14} - 10^{18}$ cm^{-3}. In both cases, thickness of the compensated layer (d) was varied from 10-100 nm. The "charge layer" is then introduced as a highly p-type doped CdSeTe layer with d = 1 nm, while the main CdSeTe layer is kept at 500 nm thickness and p = 10^{16} cm^{-3}.

III. RESULTS AND DISCUSSION

A. Shallow donors

Figure 3 shows simulated V_{oc}, short circuit current (J_{sc}), fill factor, and efficiency for the specified range of compensated layer thickness and compensating donor density. The V_{oc} improves with increasing N_D, corresponding to reduced doping near the interface. For a given donor density, increasing thickness tends to slightly increase the effect, but V_{oc} is more strongly dependent on the defect density. J_{sc} improves very slightly as thickness increases. Fill factor follows the same trend as V_{oc}, and therefore efficiency also improves with increased thickness and N_D. The best results

TABLE II. PROPERTIES OF VARIED CdSeTe LAYER

Defect	Thickness (nm)	N_A (cm^{-3})	N_D (cm^{-3})
Shallow donor	10-100	1×10^{16}	1×10^{15}-1×10^{17}
Deep donor	10-100	1×10^{16}	1×10^{14}-1×10^{18}
Positive charge layer	1	1×10^{16}-1×10^{20}	0

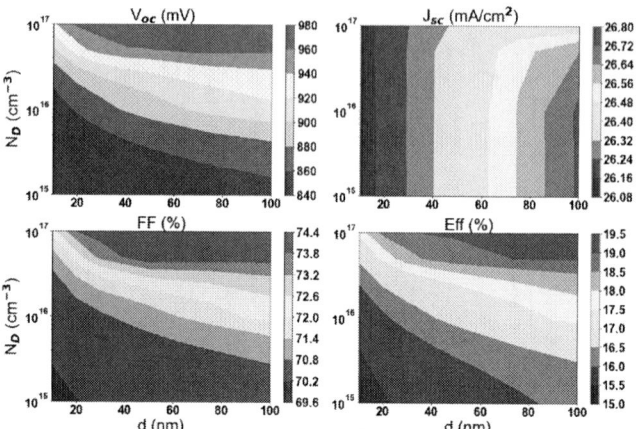

Fig. 3. Simulated open-circuit voltage, short-circuit current, fill factor, and efficiency for devices with shallow compensating donors near interface.

occur when $N_D > N_A$ corresponding to n-type doping near the interface. It should be noted that peak efficiencies are lower (~19.5%) than record device performance; this is largely due to the low emitter doping.

Another consideration that will need to be investigated further is the difference in lifetime between the CdTe and CdSeTe layers. In the simplified model used in this work, CdTe and CdSeTe are only distinguished by their bandgaps (1.5 eV and 1.4 eV, respectively). However, the alloying of Se in CdTe can additionally increase carrier lifetime [11]. Future work will involve adding more recombination centers in the CdTe bulk in order to reduce its lifetime relative to CdSeTe.

B. Deep defects

Figure 4 shows the same performance parameters when a deep donor is introduced in this region. This corresponds to an effective recombination center, which results in decreased carrier lifetime. The trend in V_{oc} is reversed compared to shallow compensating donors: V_{oc} decreases with increased thickness and defect density. A similar trend occurs in the J_{sc} and fill factor, so the lowest efficiencies correspond to the thickest and most defective layers. It is clear from these results that deep defects can be quite detrimental to device performance, even if concentrated within 100 nm of the front interface. If the concentration is kept low ($N_D < 10^{16}$ cm^{-3} in these simulations), the effect is minimal and less dependent on

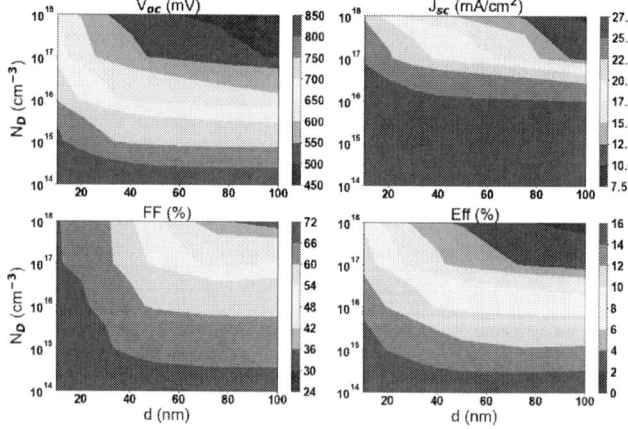

Fig. 4. Simulated open-circuit voltage, short-circuit current, fill factor, and efficiency for devices with mid-gap recombination centers near interface.

the thickness of the defective layer. It should also be noted that the capture cross section (σ) for these deep defects was kept at 10^{-12} cm^{-2}; varying σ will inversely change the defect concentration required to see these effects.

C. Charge accumulation

Next, we consider the case for a very thin, highly doped CdSeTe layer at the interface. We treat this as an accumulation of positive charge that could be affect device performance without significantly changing the measured carrier concentration. We vary the p-type doping of the layer and the conduction band offset with MZO, while keeping the thickness constant at 1 nm (Fig. 5).

V_{oc} has a maximum closer to flat bands for a high interface charge concentration, compared to the 0.2 eV optimum previously shown. All performance parameters are drastically reduced with high charge concentration and positive CBO.

Another notable effect of this charge layer occurs in the CV profile. Table III shows the V_{oc}, depletion width (W), and carrier concentration (CC) for simulations with and without the charge layer. The hole density in the charge layer for these simulations was set at 10^{19} cm^{-3}. While the carrier concentration is unaffected by the addition of the charge layer, there is a substantial increase in V_{oc} and decrease in W. This replicates the previously described effect in our As-doped devices, showing that it is consistent with significant interfacial charge accumulation. Treating the junction as an idealized parallel plate capacitor with C=Q/V=Aϵ/d and d=W, increasing charge proportionally more than voltage should result in a greater capacitance and therefore smaller distance, corresponding to a smaller depletion width.

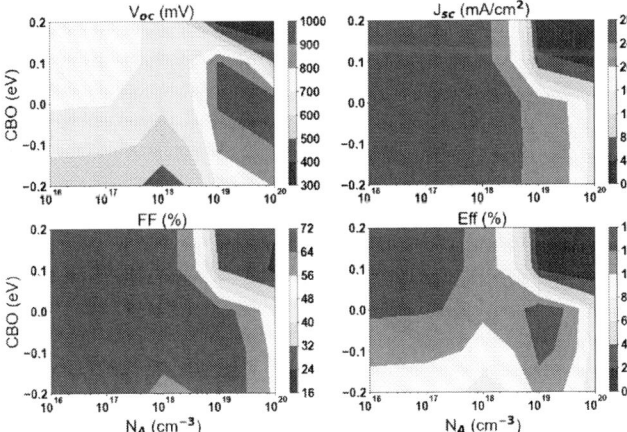

Fig. 5. Simulated open-circuit voltage, short-circuit current, fill factor, and efficiency for devices with thin charge layer at interface.

TABLE III. EFFECT OF CHARGE LAYER ON V_{oc} AND CV CHARACTERISTICS

Charge layer	Voc (mV)	W (µm)	CC (cm^{-3})
N	703	0.40	1×10^{16}
Y	914	0.22	1×10^{16}

IV. CONCLUSIONS

Depending on the nature of compensating defects, we illustrate how they can bolster or hinder device performance. In the case of shallow compensating donors, V_{oc} is increased with increasing compensation. This effect could be similarly achieved by simply reducing the As concentration near the interface, resulting in lower net acceptor density. Conversely, deep defects act as effective recombination centers and can severely decrease device performance if concentrated near the interface. A thin layer of charge at the interface has also been shown to significantly impact both JV and CV characteristics.

We have shown that dopant atom accumulation near the CdSeTe/MZO interface can have a significant impact on device performance, notably V_{oc}. Keeping As atoms away from the front interface can create a benefit by lowering net acceptor density near the junction as well as a reduction of recombination centers. This suggests a mechanism for some of the observations of reduced performance in highly p-doped devices and a potential path for increasing V_{oc}.

ACKNOWLEDGMENT

This work was authored by the National Renewable Energy Laboratory, operated by Alliance for Sustainable Energy, LLC, for the U.S. Department of Energy (DOE) under Contract No. DE-AC36-08GO28308. Funding was provided by the United States Department of Energy Office of Energy and Renewable Energy's Solar Energy Technologies Office under agreement 38257.

REFERENCES

[1] T. Ablekim, E. Colegrove, and W.K. Metzger, "Interface Engineering for 25% CdTe Solar Cells," ACS Appl. En. Mater., vol. 1, no. 10, pp 5135-5139, September 2018.

[2] W.K. Metzger et al., "Exceeding 20% efficiency with in situ group V doping in polycrystalline CdTe solar cells," Nature Energy, vol. 4 pp. 837-845, August 2019.

[3] T. Ablekim et al., "Self-compensation in arsenic doping of CdTe," Sci. Reports, vol. 7, no. 4563, July 2017.

[4] D. Krasikov and I. Sankin, "Beyond thermodynamic defect models: A kinetic simulation of arsenic activation in CdTe," Phys. Rev. Materials, vol. 2, no. 103803, October 2018.

[5] A. Nagaoka, D. Kuciauskas, J. McCoy, and M. Scarpulla, "High p-type doping, mobility, and photocarrier lifetime in arsenic-doped CdTe single crystals," Appl. Phys. Lett. vol. 112 no. 192101, May 2018.

[6] E. Colegrove et al., "Investigating the role of Copper in Arsenic doped CdSeTe Photovoltaics," unpublished.

[7] D.L. McGott, B. Good, B. Fluegel, J. Duenow, C. Wolden, and M.O. Reese, "Dual-Wavelength Time-Resolved Photoluminescence Study of CdSexTe1-x Surface Passivation via MgyZn1-yO and Al2O3," JPV vol. 12, issue 1, pp 309-315, November 2021.

[8] M. Burgelman, P. Nollett, S. Degrave, "Modelling polycrystalline semiconductor solar cells," Thin Solid Films, 361-362, pp. 527-532, 2000.

[9] R.Pandey, T. Shimpey, A. Munshi, and J. Sites, "Impact of Carrier Concentration and Carrier Lifetime on MgZnO/CdSeTe/CdTe Solar Cells," IEEE J. Photovoltaics, vol. 10, no. 6, November 2020.

[10] T. Song, A. Kanevce, and J.R. Sites, "Emitter/absorber interface of CdTe solar cells," J. Appl. Phys., vol. 119, no. 233104, June 2016.

[11] X. Zheng, E. Colegrove, J.N. Duenow, T. Moseley, and W.K. Metzger, "Roles of bandgrading, lifetime, band alignment, and carrier concentration in high-efficiency CdSeTe solar cells," J. Appl. Phys. vol. 128, no. 053102, August 2010.

Millions of Small Pressure Cycles Drive Damage in Cracked Solar Cells

Timothy J Silverman, Nick Bosco, Michael Owen-Bellini, Cara Libby, Michael G Deceglie

National Renewable Energy Laboratory, Golden, CO, United States

Electric Power Research Institute, Washington, DC, United States

We applied time-varying air pressure to a PV module containing newly cracked cells. The test used a new dynamic mechanical acceleration (DMX) apparatus. We applied pressure cycles similar to natural, wind-driven cycles. Compared to standard dynamic mechanical load (DML) tests, we applied much lower pressure (10 Pa to 300 Pa RMS) and many more cycles (one million at each of four pressure levels). We present a case study on a single cell in a commercial module. We monitored electrical continuity loss across cracks using electroluminescence (EL) imaging. 30 Pa pressure cycles caused permanent damage that continued worsening even after tens of thousands of cycles. After one million 30 Pa cycles, a series of 100 Pa cycles still caused new, permanent damage to existing cracks. 300 Pa cycles caused further worsening and introduced new cracks.

Quantifying Energy flows in PV Circularity Processes

Heather M. Mirletz[1,2], Silvana Ovaitt[1], Ashley Gaulding[1], Seetharaman Sridhar[3], Teresa Barnes[1]

[1]National Renewable Energy Laboratory, Golden CO 80401 US,
[2]Advanced Energy Systems Graduate Program, Colorado School of Mines, Golden CO 80401,
[3]Arizona State University, Research and Innovation Fulton Schools of Engineering, Tempe, AZ 85281 USA

Abstract—As sustainable deployment and end-of-life management become a hot topic to timely address in the PV community, a dynamic comparative evaluation of the benefits of circular pathways such as reuse, and remanufacturing, recycling has not been performed holistically beyond material flows or LCA analysis. Energy flows are critical for evaluating energy generation technologies. Previously they have been used to compare renewables to fossil generation and then between PV technologies. This paper quantifies energy flows to evaluate circular pathways for PV. The energy flows tracking manufacturing, generation, and losses complementary to the mass flows of silicon are quantified leveraging the PV ICE framework.

Keywords—circular economy, photovoltaics, energy return on investment (EROI), remanufacture, recycle, energy flows

I. INTRODUCTION

To avoid catastrophic climate change, we need to decarbonize our energy sectors by deploying renewables, such as PV. But this requires unprecedented scale-up for manufacturing and deployment. For example, the US currently has 100 GW on-grid and 5.5 GW of PV manufacturing capacity, and the recent Solar Futures study from NREL [1] projects that 1 TW is required by 2035 and 1.5 TW by 2050. This entails 30 GW annual deployment through 2025 and 60 GW annually

through 2030. But this scale-up must occur globally, which will further strain global supply chains, as seen for glass and polysilicon. Alternate and transparent supply chains, such as those enabled by the circular economy (CE), are an opportunity to move away from price shocks, shortages, and negative environmental and social impacts. CE is an economic and industrial model for materials circulation, minimizing wastes and decoupling economic growth from the use of virgin resources, and underpinned by decarbonization [2]–[5].

Manufacturing is notoriously the most environmentally impactful step in a PV system lifecycle. PERC module (~80% of market share) manufacturing contributes to approximately 24 gCO_2 eq/kWh, or 92% of the technology's lifecycle global warming potential [6]. Manufacturing steps are an excellent opportunity to reduce the footprint of PV modules through steps like improved efficiencies, reducing embodied material mass, and alternate supply chains enabled by CE. For example, glass cullet presents an opportunity to reduce waste and lower energy requirements. Silicon kerf loss is also an opportunity for reduced costs, energy and manufacturing scrap, through use of diamond wire sawing and kerf slurry recycling.

	R-Strategies	Generalized Description	Proposed PV Specific
Smarter Product Design	R0: Refuse	Make redundant/eliminate need	Refuse Fossil-Fuels and Carbon intensive materials **Refuse Virgin and Conflict Materials**
	R1: Rethink	Multifunctional, more use intensive	Future proofing/backward compatible Design for Repair Integrated PV High energy yield PV systems
	R2: Reduce	Efficient product manufacture, Design for longevity	**Reduce Material usage/W_p** Material substitution Manufacturing Yields **Design for longevity**
Extend lifespan of product and its parts	R3: Reuse	Re-use if good condition	Merchant Tail, Resell in secondary market
	R4: Repair	Repair and maintenance for extended life	Onsite fix to power problem
	R5: Refurbish	Restore older to updated functionality	Demount for more extensive repair Repower site with new modules
	R6: Remanufacture	Use parts in new product for same function	Disassemble and Intact component recovery
	R7: Repurpose	Use parts in new product with different function	Repower on disturbed sites
Useful application of materials	R8: Recycle	Process materials, high or low quality	**Material recovery at varying quality levels**
	R9: Remine	Landfill mining	Bulldoze PV system in place, mine landfilled modules
	R10: Recover	Energy recovery through incineration	Burn component materials for energy generation

Fig. 1. Circular economy provides various value retention options or pathways. These pathways can reduce the environmental intensity of decarbonization-scale PV deployment, but an evlauation not only of the virgin material needs and waste reductions but of the energy flows is necessary.

Traditional Si module   **Perovskite ABC Module**

Assuming 100% of glass   Assuming 100% of glass
Is recycled into close-loop   Is remanufactured

Fig. 2   Diagram of the two evaluated scenarios: 100% recycled modules, vs 100% remanufactured modules. Glass remanufacture is potentially enabled by technology designs such as the perovskite all-back-contact architectures. Scenarios are evaluated on material and energy flows of glass.

Most energy transition studies focus on critical or near-critical material demands, such as Ag, Cu, In, Te [7]. Suggested mitigation strategies for material supply shortages include recycling and decreased embodied material mass, but impacts on energy demands and generation are not quantified. Circularity is not correlated to reduced environmental impacts [7], [8]. The energy demands of recycling have been studied, but there is a lack of studies comparing the energy demands of other circular pathways, such as those shown in Fig. 1.

Glass is an ideal target for increasing PV material circularity. Glass represents ~80% of a module's weight and accounts for significant energy demands in manufacturing (second to silicon) [9]. PV glass is high-quality, low-iron, and not currently closed-loop recycled. Closing the loop would reduce sand mining requirements—but quality and lifetime must be maintained to maintain capacity and avoid increased deployment [10]. Currently, closed-loop recycling is cost and energetically expensive. Alternatively, PV glass remanufacturing could be enabled by reliable and innovative module designs.

This paper establishes a methodology for capturing PV technology energy requirements, generation, and losses considering manufacturing evolutions and emerging technology trends. Two module technologies are compared: a traditional G/G Si module with closed-loop glass recycling, and a hypothetical perovskite module with all-back-contact (ABC) architecture enabling front glass remanufacturing (Fig. 2). The energy analysis is run parallel to material flows for each step in a circular photovoltaic lifecycle. Finally, a discussion of the tradeoffs of different pathways is presented.

II.   METHODOLOGY

A. PV ICE Tool Overview

PV in Circular Economy (PV ICE) [11] is an open-source tool providing decision-makers with a data-backed, energy and mass-flow-based evaluation of CE pathway decisions for PV. Given known (historic) and expected (future) average bill of materials and processes for deployed PV, it can estimate material demands, wastes, installed capacities, and energy expenditure, accounting for changes in PV designs, performance, and market shares. Here, we use PV ICE's ability to track energy and mass flows to estimate the impact of various

circular pathways available for PV. We compare recycling versus remanufacture of glass.

Energy flows are quantified for each lifecycle stage; virgin material extraction, manufacturing, use phase, EoL, and circular pathways. Where possible, technology-based energy flows are used. For example, the energy demand to grow monocrystalline silicon ingots has decreased with time. Similarly, the energy requirements to cast mc-Si are not the same as mono-Si ingot growth—the energy demands will be weighted by technology market share. Technology evolution is reflected in the dynamic annual energy flows.

The recycling process considers how many modules are recycled, if each module material(s) is a recycling target, and the yield and the energy needs associated with recycling of that material. Quality and disposition (i.e., closed-loop) of the recycled material are also considered. Recycling energies for low- and high-qualities and closed-loops are calculated. This is compared to module collection and material(s) remanufacturing. The yield of the remanufacturing process accounts for losses due to reverse logistics, contamination, and broken components for remanufacturing and the energetic needs.

For this paper, lifetime assumptions (including reliability and economic project lifetime) are the same for both technologies. Collection efficiencies are also assumed to be the same at 100%. Reuse and repair are set to 0% to explore only recycling and remanufacture pathways. In the first case, 100% of the modules are recycled, with 100% of the glass in those modules recycled at variable yield between 50-80%. For the second case, 50-80% of the glass+TCO is recovered intact for remanufacture.

B. PV Energy Framework

Renewable energy technologies generate power over their useful life, offsetting energy required for manufacturing in a quantity known as Energy Return on Investment (EROI). For this tool, EROI is defined as:

$$EROI = \frac{\sum E_{out}}{\sum E_{in}}$$

In addition to nameplate installed capacity, the PV ICE tool calculates effective capacity in the use phase by considering both newly installed modules and module degradation by years in service. This consideration of degradation along with the bottom-up material and energy accounting enables evaluation of circular decisions such as field repair, off-site refurbishment, reuse, or recycling in both energy and material dimensions.

DISCUSSION

As modules with higher efficiency and power become the new norm and PV deployment increases, understanding the impacts of virgin material extraction, waste, and energies can enable stakeholders to plan for improved circularity or energy efficiency. Fig. 3 shows diagrams material and energetic needs of a Si module with glass recycling, where glass is crushed and remelted, vs. a hypothetical ABC perovskite module where the glass is directly offset at the manufacturing stage, avoiding recycling or extraction energy needs.

978-1-7281-6118-1/22 $31.00 © 2022 IEEE

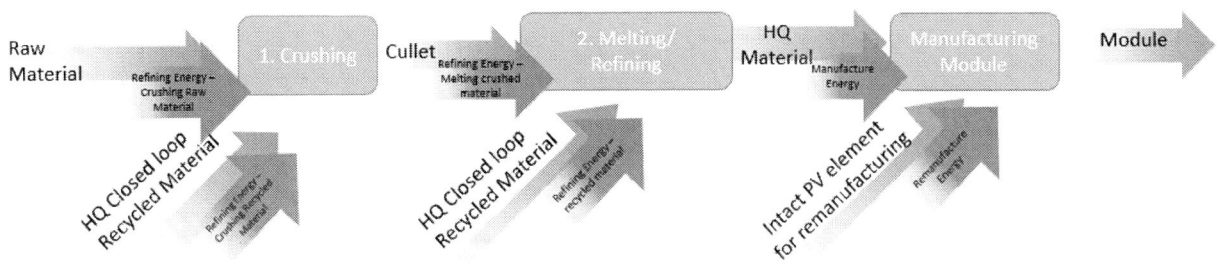

Fig. 3 Detail of the main energy pathways for recycled and remanufactured glass. PV ICE evaluates closed-loop energy offsets accounting for the manufacturing process; for example, energies for crushing and melting are captured separately to account for different circular feedstocks. Remanufactured PV glass is removed intact and feeds directly into the manufacturing of the module.

Technology advances can improve efficiencies throughout the PV module lifecycle. Policies can shape research focus and economically viable pathways through incentives. Research and policy priorities should be data-driven, and priorities can be clarified in terms of EROI and material benefits. We use the PV ICE tool to highlight key areas for improvements and tradeoffs:

- Design with alternative materials.

- O&M for increased system lifetime.

- In-field and out-of-field repairs vs repowering.

- Repair reliability, re-certification and warranty.

- Improved end-of-life module collection

- Industry targets for recycling yields and energies.

- Reducing material per module while maintaining or increasing production and operational efficiency.

- Recycling complexity and design for remanufacture.

- Identify complementary/symbiotic sectors for recycled PV materials of varying quality

- Understand current repowering practices.

III. SUMMARY

The PV ICE tool is flexible and scalable to accommodate temporal and geographical information, with dynamic flows that can consider multiple materials and energy flows. Future work includes implementing different energy technologies within this framework. For the conference, targeted efficiency improvements and potential CE pathways in the PV lifecycle will be evaluated on best material efficiency and EROI.

ACKNOWLEDGMENTS

This work was authored [in part] by the National Renewable Energy Laboratory, operated by Alliance for Sustainable Energy, LLC, for the U.S. Department of Energy (DOE) under Contract No. DE-AC36-08GO28308. The views expressed in the article do not necessarily represent the views of the DOE or the U.S. Government.

REFERENCES

[1] K. Ardani et al., "Solar Futures Study," EERE DOE, Sep. 2021. [Online]. Available: https://www.energy.gov/eere/solar/solar-futures-study

[2] E. A. Olivetti and J. M. Cullen, "Toward a sustainable materials system," Science, vol. 360, no. 6396, pp. 1396–1398, Jun. 2018, doi: 10.1126/science.aat6821.

[3] D. Smith and S. Jones, "Circularity Indicators: An Approach to Measuring Circularity: Methodology," Ellen MacArthur Foundation, 2019.

[4] J. Kirchherr, D. Reike, and M. P. Hekkert, "Conceptualizing the Circular Economy: An Analysis of 114 Definitions," SSRN Electron. J., vol. 127, Jan. 2017, doi: 10.2139/ssrn.3037579.

[5] P. van Loon, D. Diener, and S. Harris, "Circular products and business models and environmental impact reductions: Current knowledge and knowledge gaps," J. Clean. Prod., vol. 288, p. 125627, Mar. 2021, doi: 10.1016/j.jclepro.2020.125627.

[6] M. M. Lunardi, J. P. Alvarez-Gaitan, J. I. Bilbao, and R. Corkish, "A Review of Recycling Processes for Photovoltaic Modules," in Solar Panels and Photovoltaic Materials, B. Zaidi, Ed. InTech, 2018. doi: 10.5772/intechopen.74390.

[7] E. Gervais, S. Shammugam, L. Friedrich, and T. Schlegl, "Raw material needs for the large-scale deployment of photovoltaics – Effects of innovation-driven roadmaps on material constraints until 2050," Renew. Sustain. Energy Rev., vol. 137, p. 110589, Mar. 2021, doi: 10.1016/j.rser.2020.110589.

[8] S. Harris, M. Martin, and D. Diener, "Circularity for circularity's sake? Scoping review of assessment methods for environmental performance in the circular economy.," Sustain. Prod. Consum., vol. 26, pp. 172–186, Apr. 2021, doi: 10.1016/j.spc.2020.09.018.

[9] D. D. Furszyfer Del Rio et al., "Decarbonizing the glass industry: A critical and systematic review of developments, sociotechnical systems and policy options," Renew. Sustain. Energy Rev., p. 111885, Dec. 2021, doi: 10.1016/j.rser.2021.111885.

[10] H. Mirletz, S. Ovaitt, S. Sridhar, and T. Barnes, "PV Lifetime and Recycling to Mitigate Material Impacts of Decarbonization," (submitted).

[11] S. Ovaitt & H. Mirletz, S. Seetharaman, and T. Barnes, "PV in the Circular Economy, A Dynamic Framework Analyzing Technology Evolution and Reliability Impacts," ISCIENCE, Jan. 2022, doi: https://doi.org/10.1016/j.isci.2021.103488.

Uncertainty Quantification of Bifacial Performance Modeling

Matthew J. Prilliman, Janine M. Freeman Keith

National Renewable Energy Laboratory, Golden, Colorado, 80401-3305, USA

Abstract—Analysis on uncertainty in the annual energy of PV systems that can be attributed to parameters of particular importance to bifacial PV modules is presented. Monte Carlo simulations are used to evaluate the effect of uncertain module bifaciality factors, module transmission fractions, albedo values, and ground clearance. The analyses cover a wide spectrum of potential PV array archetypes through variation of installation parameters. The results of the Monte Carlo analysis reveal that the uncertainty is largely dependent on albedo uncertainty, but more simulations are needed to identify trends across system archetypes. The simulations are aimed at attributing an annual energy uncertainty factor for bifacial considerations that can be applied in post-processing of project probability of exceedance analysis.

Keywords—photovoltaic, energy modeling, uncertainty.

I. INTRODUCTION

Uncertainty in project PV project performance is a key consideration in project financing that must be accounted for to avoid overestimation of annual energy and insufficient project payback. This uncertainty can come from any step in the performance modeling pipeline, such as uncertainty in solar resource measurement or uncertainty in performance models themselves. One area of potential modeling uncertainty that has not been fully investigated is uncertainty in the performance of bifacial PV modules, where irradiance is absorbed and converted to electricity on both the front and rear surfaces of the module. As bifacial modules continue to gain market share, the uncertainty in the rear-side contribution to the annual energy output must be quantified to give investors a better understanding of the expected project performance.

The uncertainty of annual energy is often expressed through the application of annual energy factors unique to the individual sources of uncertainty [1]. Factors for sources of uncertainty such as solar resource measurement uncertainty are regularly calculated as part of an independent engineer's uncertainty calculation for prospective PV projects. For bifacial systems, however, there is not much consensus on how to calculate said uncertainty factor for the bifacial energy added to the system output. This work presents a unique approach to quantifying the uncertainty of bifacial energy output through analysis of input parameter distributions of particular importance to bifacial module energy production. These values are the bifaciality factor, the fraction of light transmitted through the module glass, the clearance height of the module from the ground, and the measured ground albedo.

II. BACKGROUND

Bifacial PV systems differ from monofacial PV systems in their ability to convert incident irradiance from both the front and back surfaces of the PV module into electricity. The additional rear-side output of the module is similar to the front-side output in that it is primarily dependent on the magnitude of incident irradiance, but there are different considerations for input variable sensitivity and uncertainty in these bifacial systems. Variables such as the ground albedo and module ground clearance height can have an outsized effect on rear-side insolation as compared to front-side insolation. Previous efforts at quantifying rear-side insolation model sensitivity have revealed that the insolation is mainly dependent on albedo when the albedo range is high, and on ground coverage ratio, or the ratio of panel length to row-to-row distance, when dealing with low albedo ranges [2].

Modeling bifacial PV modules involves the calculation of plane-of-array (POA) irradiance hitting the rear-side of the module at a given time step. The rear-side incident irradiance can vary depending on the location of the sun relative to the module, panel installation parameters such as tilt and azimuth, and the ground albedo. While many of these parameters are known before the system is installed, or easily measurable, others are unknown or difficult to measure and can thus lead to unaccounted for uncertainty in bifacial system performance. Such variables include the ground albedo, or ratio of light that is reflected from the ground. Often the seasonal changes in albedo are not considered in system performance despite the outsized effect these changes can have on bifacial system annual output due to increased ground reflection on the back surface. The ground clearance height of the module also determines the amount of direct and diffuse light received on the rear surface of the module due to changing angles between the module and the sun's ray along with changing shading conditions due to module position. The ground clearance height that a system is installed at can vary across a wide range of heights based on project-specific ground reflection and shading considerations. Higher ground clearance heights often increase bifacial energy output by allowing for increased ground reflected irradiance on the backside of the module. Other module-specific variables such as the transmission fraction and bifaciality factor are often not reported on module specification sheets. The transmission fraction describes the amount of light that passes through the top glass of the module's rear surface (primarily between gaps in PV cells) [3]. The bifaciality factor is the ratio of the bifacial module's rear surface output to the front surface output for the

---

This work was authored by the National Renewable Energy Laboratory, operated by Alliance for Sustainable Energy, LLC, for the U.S. Department of Energy (DOE) under Contract No. DE-AC36-08GO28308. Funding provided by U.S. Department of Energy Office of Energy Efficiency and Renewable Energy Solar Energy Technologies Office under the Solar Energy Technologies Office Award Number DE-EE00036530.

978-1-7281-6118-1/22 $31.00 © 2022 IEEE

same STC conditions [3]. This bifaciality factor is usually determined through standardized indoor testing procedures but is often not included in module specification sheets. Bifaciality factors have been reported to be anywhere from 0.65 to 0.95 depending on the cell type and manufacturer [4]. Each of these variables can introduce uncertainty into the bifacial gain, or additional energy provided by the module's rear-side, and ultimately the annual energy that is of primary concern to the PV project investment. The uncertainty in these models stemming from these variables must be analyzed as these are variables that are often overlooked or not directly measurable when installing new systems.

III. METHODOLOGY

Annual energy simulations of bifacial PV systems were performed using the PySAM Python wrapper of the performance models provided by NREL's System Advisor Model (SAM) [5]. A system with a capacity of 50 MW and system specifications provided in Table I is simulated with assumed bifacial PV module behavior. The annual energy production in kWh is evaluated for a variety of PV system archetypes to gain an understanding of the bifacial uncertainty as it relates to systems of different module tilts, azimuth angles, ground coverage ratios, and other system installation parameters. The bifacial gain of the system is also evaluated for each system archetype by subtracting the monofacial system annual energy output from the bifacial system annual energy output for each simulation to isolate the energy produced from the rear surface of the module. These archetypes are defined based on the following system parameters and specified ranges shown in Table II. For each system archetype, a reference or "true" annual energy value is used to compare against a distribution of 100 annual energy and bifacial gain values derived using values for ground clearance height, albedo, bifaciality factor, and transmission factor that have been stochastically sampled from a normal distribution around the mean value given for the reference case. The distribution on each of these four variables is defined with one standard deviation being ± 10% from the reference value for each variable. While not every system archetype from the variable ranges in Table II was simulated, enough simulations were performed to investigate trends in annual energy and bifacial gain changes for both fixed-tilt and single-axis tracking systems. The distribution space of the four variables is canvased through Monte Carlo sampling in which random samples of each variable are taken from each probability range before re-calculating the annual energy to determine each variable's effect on the annual energy uncertainty in each system archetype. The relative difference between the annual energy reference value and Monte Carlo sampled annual energy results at each of the 100 samples taken was found using the following relation:

$$D_{r,E} = \frac{E_{MC} - E_R}{E_R} \quad (1)$$

$$D_{r,B} = \frac{B_{MC} - B_R}{B_R} \quad (2)$$

Where E_{MC} is the annual energy output from a given Monte Carlo simulation, B_R is the bifacial gain of the reference simulation, $D_{r,E}$ is the relative difference between reference and sampled annual energy, and is the $D_{r,B}$ is the relative difference

in bifacial gain between reference and sample simulations. The $D_{r,E}$ and $D_{r,B}$ values from each sample were then averaged to find the mean bias error (MBE) of the simulation and the standard deviation of the samples from said the reference energy value was also evaluated.

TABLE I. BIFACIAL SYSTEM SPECIFICATIONS

System Spec	Value
Weather location	Phoenix, AZ
Module efficiency	19.02%
DC:AC Ratio	1.33
Inverter efficiency	96.8%
DC Losses	4.4%
AC Wiring Losses	1.0%
Annual Degradation	0.5%/year

Recognizable trends in the MBE and standard deviation across the archetype grid motivated the parameters used in subsequent archetype simulations in order to examine as many variable boundaries as possible.

TABLE II. VARIABLE RANGES FOR BIFACIAL SYSTEM ARCHYTPES

Variable	Units	Lower bound	Increment	Upper bound
Tilt angle	deg	0	10	40
Azimuth	deg	90	90	270
Ground clearance height	meters	1	1	4
Ground coverage ratio	none	0.3	0.1	0.6
Albedo	none	0.2	0.2	0.8
Bifaciality	none	0.65	0.1	0.95
Transmission fraction	none	0	0.013	0.026
Tracking type	0/1	0 (fixed tilt)	1	1 (single-axis)

IV. RESULTS

Annual energy mean bias errors for the Monte Carlo simulations of the different fixed-tilt system archetypes are plotted against the annual energy standard deviations of said simulations in Figure 1. This Figure and all figures presented in this manuscript were generated using the Plotly Python package [6]. Both the mean bias error and standard deviation are presented as percentages of the reference annual energy for the system archetype. The color scale in the plot represents the different annual albedo values used in the simulation, while the symbols represent different ground clearance heights of the modules. Analysis of this Figure reveals that deviation from the reference annual energy values increase for increasing albedo. This can be attributed to changes in annual energy production from both the front and rear surfaces of the modules, as changing albedo impacts the ground reflected irradiance that reaches both modules. Further trends of the bifacial specific sensitivities can be seen in Figure 2, which shows the bifacial annual energy gain calculated from the difference between monofacial and bifacial systems for each system archetype. The results in this Figure, normalized to a percentage of the reference bifacial annual energy gain in a manner similar to that of Figure 1, effectively isolate the bifacial portion of the annual energy production and the sensitivities of the bifacial gain to the key variables being investigated in this work. Analysis of this Figure reveals that the standard deviation in bifacial gain increases for decreasing albedo, likely due to the decreased backside ground reflected irradiance. The standard deviation is also generally lower at 3-meter ground clearance height (squares) The bounds of bias error for the bifacial gain is within ± 0.04% for all archetypes.

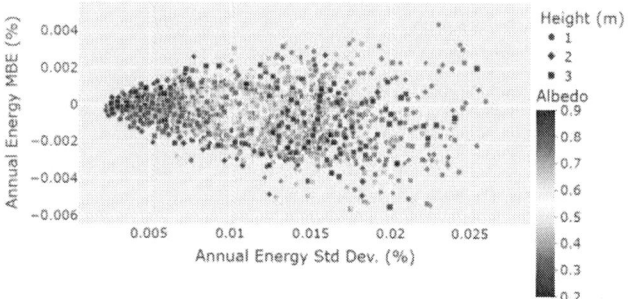

FIGURE 1. ANNUAL ENERGY MBE AND STANDARD DEVIATION FOR FIXED-TILT BIFACIAL SYSTEM SIMULATIONS

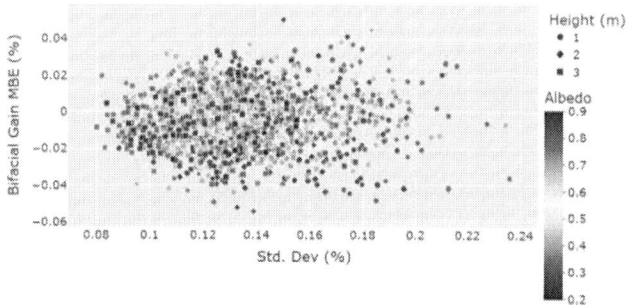

FIGURE 2. BIFACIAL GAIN MBE AND STANDARD DEVIATION FOR FIXED-TILT BIFACIAL SYSTEM SIMULATIONS

The results for non-tracking archetypes were further categorized by system archetype parameters such as tilt angle, azimuth angle, ground coverage ratio, and ground clearance height to further investigate trends in bifacial system behavior. These results are shown in Figures 3-6. Figure 3 shows increased standard deviation for systems with a fixed tilt angle of 0° that would have poor bifacial energy gain due to minimal view factors between the sun and the rear surface of the module that is parallel to the horizontal ground. Figure 4 shows the bifacial gain dependence based on azimuth angle, with higher standard deviation being seen for systems oriented facing East or West as compared to the typical South facing configuration. Figure 5 shows increased bifacial gain deviation from the reference value for higher GCR values such as 0.6 or 0.8 likely because of

increased row-to-row shading. Figure 6 divides the bifacial gain values based on ground clearance height and reveals more variance in results for the lowest height of 1 meter due to lower view factors for irradiance to hit the rear surface of the module.

Bifacial system archetypes based on single-axis tracking systems were simulated as well, with the annual energy MBE and standard deviation results being shown in Figure 7. The trends in annual energy bias and standard deviations are similar to those for the bifacial systems due to the dependence of both front and rear module surface energy production on system parameters such as albedo, ground coverage ratio, and ground clearance height. When evaluating the bias error and standard deviation in bifacial gain for single-axis tracking systems as shown in Figure 8, no clear trends in parameter sensitivity can be identified. The bounds of the mean bias error for these single axis-tracking simulations fall within ± 0.03%.

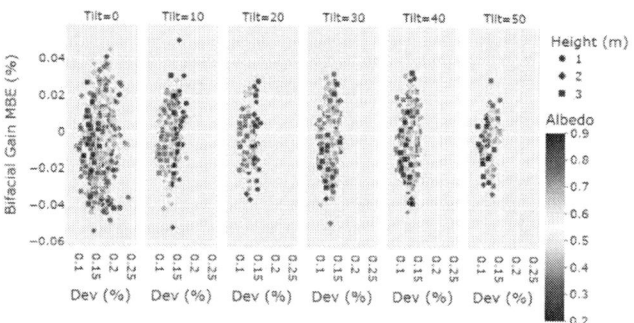

FIGURE 3. BIFACIAL GAIN MBE AND STANDARD DEVIATION FOR FIXED-TILT SYSTEMS SEPARATED BY TILT ANGLE

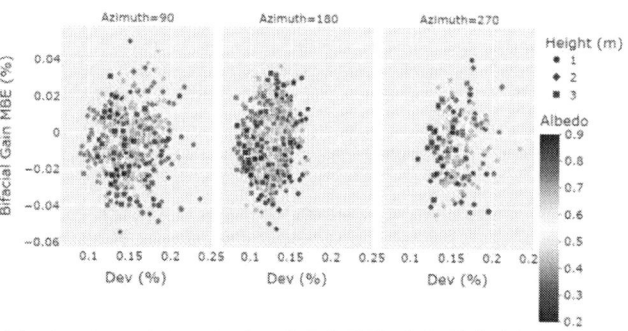

FIGURE 4. BIFACIAL GAIN MBE AND STANDARD DEVIATION FOR FIXED-TILT SYSTEMS SEPARATED BY AZIMUTH ANGLE

The single-axis tracking simulations were also evaluated based on the system archetype parameters of GCR and ground clearance height. Different tracker tilt angles and azimuth angles were not evaluated in the simulations as east to west tracking systems on horizontal trackers are the predominant system archetype of interest. In Figure 7, we see similar annual energy bias errors and deviations as for fixed tilt systems; however, in Figure 8, the increase in standard deviation of the bifacial gain with decreased albedo that was present for fixed tilt systems has disappeared. Future work should investigate whether this is a result of the weather file used (Phoenix, AZ), backtracking in the single-axis tracking system, a result of the bifacial model algorithm, or something else. The GCR dependence shown in

Figure 9 is similar to that of the fixed-tilt systems, with higher GCR values leading to more row shading and more deviation from the reference bifacial gain value. The ground clearance height dependence shown in Figure 10 does not reveal any clear trends in bifacial gain bias error or deviation from the reference value.

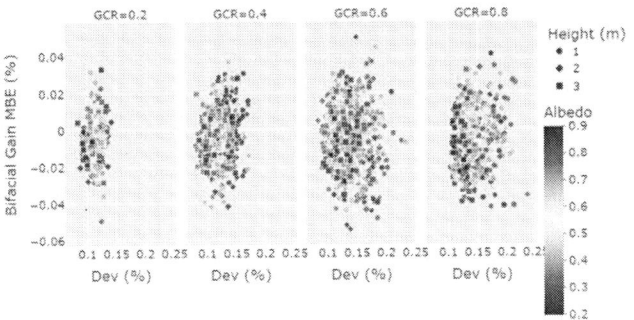

FIGURE 5. BIFACIAL GAIN MBE AND STANDARD DEVIATION FOR FIXED-TILT SYSTEMS SEPARATED BY GROUND COVERAGE RATIO (GCR)

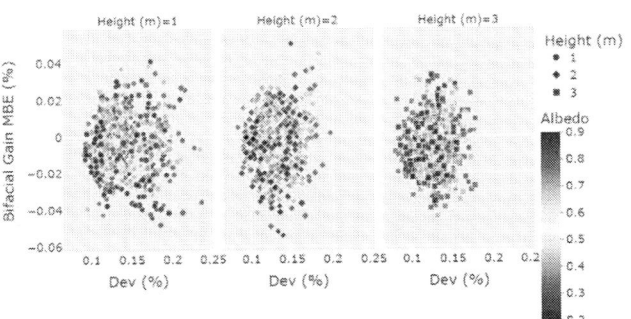

FIGURE 6. BIFACIAL GAIN MBE AND STANDARD DEVIATION FOR FIXED-TILT SYSTEMS SEPARATED BY GROUND CLEARANCE HEIGHT

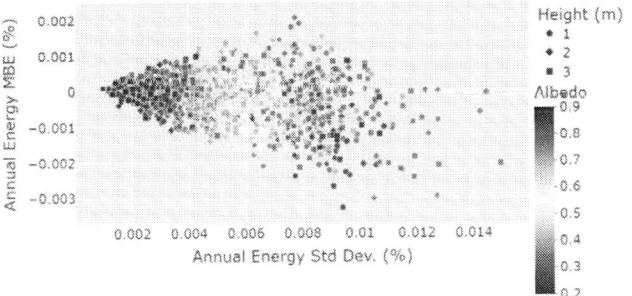

FIGURE 7. ANNUAL ENERGY MBE AND STANDARD DEVIATION FOR SINGLE-AXIS TRACKING BIFACIAL SYSTEM SIMULATIONS

V. SUMMARY AND CONCLUSIONS

This manuscript contains bifacial system simulations that quantify the sensitivity of bifacial performance gain to various PV system parameters that are of particular importance to bifacial systems. Monte Carlo simulations of numerous bifacial system archetypes in which normal distributions of albedo, ground clearance height, bifaciality, and transmission fraction are used to determine sets of MBE and standard deviation from a reference annual energy value. Results of these simulations show clear sensitivity in albedo for annual energy production and bifacial gain. The ranges of bias error for bifacial gain have been found to be $\pm 0.02\%$ for fixed-tilt systems and $\pm 0.03\%$ for single-axis tracking systems. These results can be used to inform uncertainty quantifications for bifacial models used in bifacial system performance estimates based on the approach described in [1]. Scripts for the PySAM implementation of this sensitivity analysis will be made publicly available in the near future. Future work in this area could include simulations of more bifacial system archetypes such as vertical fixed-tilt bifacial systems.

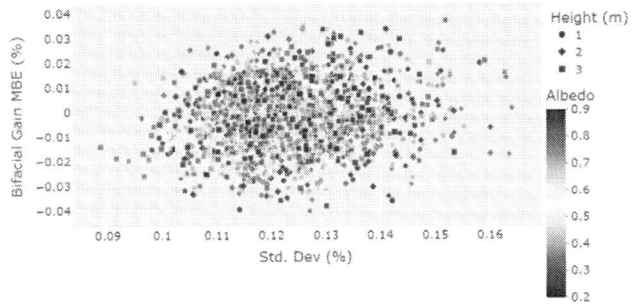

FIGURE 8. BIFACIAL GAIN MBE AND STANDARD DEVIATION FOR SINGLE-AXIS TRACKING BIFACIAL SYSTEM SIMULATIONS

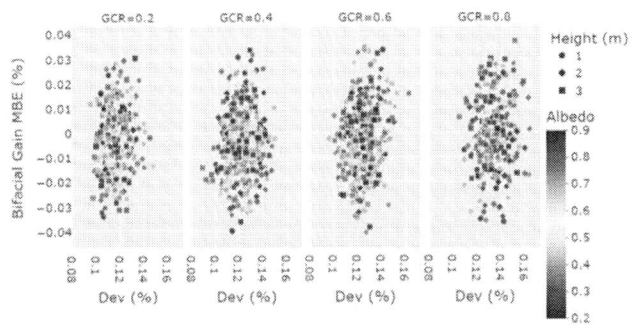

FIGURE 9. BIFACIAL GAIN MBE AND STANDARD DEVIATION FOR SINGLE-AXIS TRACKING SYSTEMS SEPARATED BY GROUND COVERAGE RATIO (GCR)

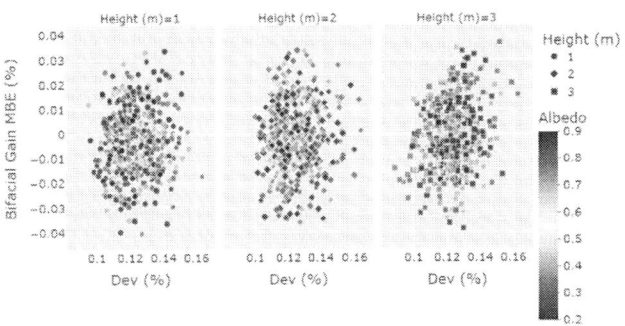

FIGURE 9. BIFACIAL GAIN MBE AND STANDARD DEVIATION FOR SINGLE-AXIS TRACKING SYSTEMS SEPARATED BY GROUND CLEARANCE HEIGHT

VI. ACKNOWLEDGMENT

This work was authored by the National Renewable Energy Laboratory, operated by Alliance for Sustainable Energy, LLC, for the U.S. Department of Energy (DOE) under Contract No. DE-AC36-08GO28308. Funding provided by U.S. Department of Energy Office of Energy Efficiency and Renewable Energy Solar Energy Technologies Office under the Solar Energy Technologies Office Award Number DE-EE00036530.

VII. REFERENCES

[1] C. Reise, B. Müller. "Uncertainties in PV System Yield Predictions and Assessments," Intl. Enegy Agency Report IEA-PVPS T13-12:2018, 2018. ISBN 978-3-906042-51-0.

[2] J.S. Stein, M. Prilliman, C. Stark, J. Nagyvary, S. Ayala Pelaez, C. Deline, Bifacial Performance Optimization Studies using Bifacial Radiance and High Performance Computing, Presented in: 36th EU PVSEC, Marseille France, Poster.

[3] P. Gilman, A. Dobos, N. DiOrio, J. Freeman, S. Janzou, D. Ryberg, "SAM Photovoltaic Model Technical Reference Update". 93 pp.; NREL/TP-6A20-67399, 2018.

[4] C. Deline, S. Pelaez, B. Marion, B. Sekulic, M. Woodhouse, J. Stein, "Bifacial PV System Performance: Separating Fact from Fiction," Presented in: 46th PVSC, Chicaco, IL, Presentation.

[5] PySAM . National Renewable Energy Laboratory. Golden, CO. Accessed. https://github.com/nrel/pysam

[6] Plotly Technologies Inc. Collaborative data science. Montréal, QC, 2015. https://plot.ly.

Field Experience Detecting PV Underperformance in Real Time Using Existing Instrumentation

Scott Sheppard[1], Tim Cook[1], Daniel Fregosi[2], Christopher Perullo[1], Michael Bolen[2]

[1]Turbine Logic, Atlanta, Georgia, 30308, USA
[2]Electric Power Research Institute, Charlotte, North Carolina, 28262, USA

Abstract—Maintenance at large-scale photovoltaic plants employs a mix of preventative and corrective maintenance practices. Large outages, such as an inverter tripping offline, are often easy to detect. More subtle sub-inverter faults and failures can accumulate and go unnoticed for months or years. A software-based fault detection method has been developed to analyze commonly measured data from large-scale PV plants for more timely detection of subtle underperformance. The method has been demonstrated on eight datasets from large-scale plants with high accuracy of detection. Results are validated using aerial infrared scanning. String outages are detected with a true positive rate of 73 percent and tracker issues are detected with a true positive rate of 88 percent. The developed method can be uniformly applied to photovoltaic plants across a range of scales and configurations to assess performance, quickly detect underperformance, and determine the source and location of failures. The results inform and improve operations and maintenance at PV plants, ultimately aiding in improved affordability, reliability, availability, and resiliency of solar electricity.

Keywords—Solar, Photovoltaics (PV), Data Analysis, Operations, Maintenance, Monitoring, Detection, Failures, Faults, Underperformance

I. Introduction

Monitoring and Diagnostics (M&D) is of increasing importance for all forms of power generating assets with respect to improved operations, maintenance, reliability, and affordability. Advances in data analytics and algorithms, affordable data storage and computational power, ease of information transfer and communication, and sensor and instrumentation capabilities and affordability all offer new capabilities and insights to plant owners and operators [1], [2]. Utilizing and integrating each of these technological advancements to their fullest potential is an active research area.

Maintenance at many photovoltaic (PV) plants utilizes a mix of preventative and corrective maintenance practices. Corrective (a.k.a., reactive, break-fix) maintenance resolves failures after they occur. Preventative maintenance involves performing tasks before equipment fails and/or becomes unsafe. When preventative maintenance occurs, it can be based on multiple factors, generally categorized as time-based (e.g., annual) or condition-based (e.g., predictive analytics). Developing the basis for condition-based maintenance often requires analyzing data from the plant. Detecting when major pieces of equipment fail, such as inverters, is straightforward based on data collected from existing, intrinsic sensors embedded by the equipment manufacturer [3].

Detecting more subtle failures that occur within the DC collector field is more difficult. Failures that impact a single module, or even an entire string, are often undetectable through monitoring production yield alone. These failures can still be identified through various field inspection techniques which are not conducive to continuous online monitoring [4]. Some manual electrical testing methods, such as field I-V curve tracing, may take days or weeks to complete. Varying meteorological conditions over this time period can often mask any issues that may be present [5].

Time-based aerial infrared (IR) imaging is one common detection method used to detect failures in the collector field. This is usually performed on an annual cadence. Failures detected include electrical and mechanical issues such as module underperformance, module hot spots, broken modules, string outages, and broken or misaligned trackers (if applicable). While it depends on the configuration of the specific plant, these types of failures each individually represent a loss of less than 1% of the power output of a single inverter. The ability to supplement physical inspection with continuous monitoring based on available data would greatly improve plant reliability and affordability. The goal of the present work is to leverage existing sensor suites and data at large-scale PV plants to aid in operation and maintenance by improving the ability to detect and diagnose subtle failures in relative real-time. The focus is on the DC collector side of the plant where there is an abundance of data available for analysis and the faults and failures are typically more subtle and harder to detect than those impacting central inverters.

II. Monitoring and Diagnostic Challenges

Due to the variability in site architecture at PV plants, developing a standardized M&D approach can be very difficult. The numbers of modules, strings, combiner boxes, and inverters vary; panel and inverter hardware vary; solar irradiation and other weather varies; racking (e.g., fixed, single-axis tracking) and orientation varies. Determining how a plant should perform from one day to the next becomes an onerous M&D activity that is often outside the expertise or time constraints of most utility M&D centers. Considering current, voltage, and tracker channels (if applicable) at a large-scale PV plant, monitoring centers are typically evaluating signals from tens of thousands of sensors or more.

978-1-7281-6118-1/22 $31.00 © 2022 IEEE

Many subtle issues go unnoticed either indefinitely or until they become severe enough that they must be fixed. Due to the physical size and remote location of many PV plants, routine inspections are not performed frequently as they require a large amount of employee hours to perform [5]. As a result, even when degradation is expected, locating the source of the power loss to a specific section of the plant can be a nontrivial task. Periodic IR and visual overscans help in this effort, but as they are often performed on an annual basis many failures go undetected for weeks to months at a time.

The developed fault detection method addresses these challenges by comparing measured data from the plant to performance calculated by a physical model of the plant coupled with feature extraction. This modeling approach provides the opportunity to leverage all the data currently being measured to find periods and locations of underperformance. While the proposed method is still reactive, it moves maintenance towards an analytics-driven, condition-based approach.

III. Fault Detection with PV Plant Data

A. Performance Model Framework

Fig. 1 shows the general workflow of the developed fault detection model. There are three key components to the model: the physical plant model, cleanup of measured data, and comparative analysis between the model and measured performance and among similar components at the plant. The physical models have been developed using pvlib-python, an open-sourced software administered by Sandia National Lab [6], [7]. This software package was chosen due to its extensive and robust hardware library that allows for accurate replication of a commercially operating plant. Additionally, the pvlib-python package allows the user to specify plant configuration information, such as the number of modules per string and number of strings per array, which allows for performance calculations specific to the user defined system. When site specific weather data is available, as is the case at most large-scale PV plants, the models can be used to calculate expected current, voltage, and power for each modeled subarray. This functionality allows for site specific modeled performance that accounts for weather conditions experienced at the plant rather than relying on clear sky models to calculate expected performance.

B. Data Cleanup

The next key step of the analysis process is cleanup of the measured data. The following simple rules were arrived at through iteration with the models, are easy to apply, and reduce false alarms:

- Remove points with plane of array irradiance below 500W/m²

- Remove points with solar elevation below 30°

- Remove known bad data points

These filters have the greatest impact when applied in tandem with each other. The first two filters primarily remove points early and late in the day (and overnight). At low solar elevation, adjacent rows tend to self-shade which adds a large degree of noise to the data for these time periods. Removing

This work is funded in part by the U.S. Department of Energy Solar Energy Technologies Office, under award number DE-EE-0008976.

Fig. 1. Physical model framework for detecting faults at PV plants.

known bad values from the dataset also significantly reduces noise in the data. These values are typically values that are outside the bounds of the sensors measuring capabilities and any null readings that come from any of the sensing hardware (for example, many data historians have a "time out" value that indicates that no data is being logged). Other data quality filters than can be applied are detailed in [8].

Typically, inverter max power point trackers will shift the DC operating point of an array to match the power rating of the inverter hardware, resulting in DC clipping. As a result, the appearance of a fault is present in the DC power output of an array since the DC power output will be low relative to its expected output. If analysis is being done on the DC power output of an array, it is recommended that any points where the DC power is 99% or greater than the inverter rating be removed from analysis.

C. Cloud Detection

Another data cleanup step that is used is a cloud detection filter on the data. Transient cloud cover adds a significant amount of noise to PV data, and removing this data provides a much cleaner dataset for further analysis. This functionality is provided within pvlib-python and is described in [9] and [10]. The application of the pvlib-python cloud detection thresholds was modified to increase performance across differing site architectures.

Through further usage of the modified cloud detection algorithm from pvlib-python, the method proved to be too aggressive in its filtering, often filtering seemingly clear sky irradiance data as "cloudy". The main driver for this overclassification of cloudy data was determined to be caused by a significant difference in the clear sky model of irradiance used by pvlib-python to perform cloud detection and the on-site measurements. An example of this difference is shown in Fig. 2. Due to this difference, a new method for filtering data based on cloud cover through comparison to historical, locally measured irradiance data was developed.

The new filter uses statistical methods and predictions based on a polynomial regression of historical plane of array (POA) irradiance measurements to filter out cloudy data. The

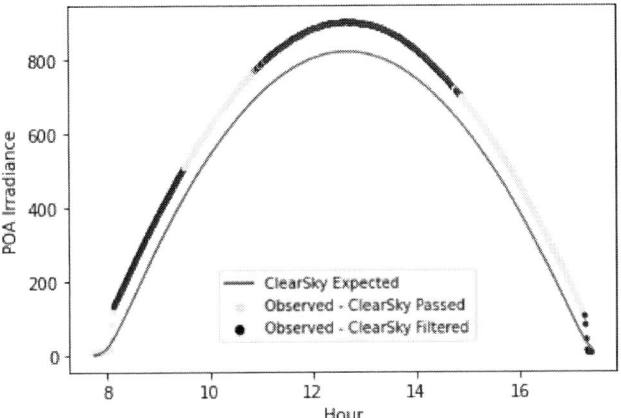

Fig. 2. Discrepancy between measured irradiance and the clear sky model.

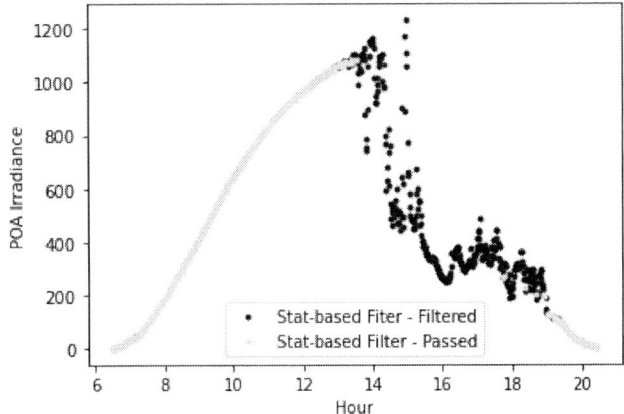

Fig. 4. POA irradiance for an example day highlighting the impacts of the initial statistics-based filter.

architecture for the new clear sky filter is shown in Fig. 3. For a given day, the POA timeseries is first statistically filtered to eliminate points with erratic or rapidly changing POA signal. The underlying assumption is that POA measurements should vary smoothly with time in the absence of cloud cover. The remaining data is then regressed with a 4th-order polynomial to calculate a prediction of the clear sky irradiance. Depending on the amount of data removed from the first round of filtering, data from previous days is overlaid on the current day and included in the regression. Then, a second filtering step eliminates data points which deviate from both the predicted clear sky irradiance value and slope by a significant margin at each point in time, adding datapoints to the initial filter. Optionally, this prediction-filter step can run in a loop, adding points to the filter with each iteration.

To create the cloud detection filter for a given day, first, the derivative of the POA irradiance values is calculated for each datapoint over the course of a day. Then, the standard deviation of the derivative is calculated for a five-minute moving window centered on each time step. This standard deviation is normalized by the maximum POA value in the entire timeseries, and all points with a standard deviation greater than 1% of the maximum POA irradiance value are filtered out. This initial statistics-based filter is applied to each day in the dataset. The effect of this statistics-based filter on an example day is shown in Fig. 4.

For each day, this filtered irradiance data is used to create a prediction of clear sky POA irradiance based on a polynomial regression of data. The 4th-order polynomial is suitable because it can closely fit nearly any daily solar profile, regardless of location, tracking system, or orientation. The previous six days of filtered POA data is also included in the regression,

dramatically improving the accuracy and reliability of the prediction on cloudy days. Due to changes in sunrise and sunset from day to day, six historical days of data was chosen as the upper limit for the time window for data inclusion to keep the endpoints in the measured daily POA curves from drifting too far apart from each other. An exception is made for individual days in which more than 75% of data remains after the initial statistics-based filter. On these days, the regression uses only data from the current day to regress the 4th-order polynomial.

An example of the improvements to the regression provided by increasing the number of days in the regression is shown in Fig. 5. This figure shows three regressions overlaid on the filtered data shown in Fig. 4. Each line represents a regression model including differing amounts of historical data in the irradiance regression. As Fig. 5 shows, a larger amount of data leads to a more accurate prediction of clear sky irradiance.

The trained 4th-order regression model is used to create a prediction of the clear sky POA for the current day. To further improve the predicted clear sky value, the filtered data is subject to an additional round of filtering, incorporating the results from the clear sky regression. First, the slopes of expected clear sky irradiance and measured irradiance are calculated over a nine-

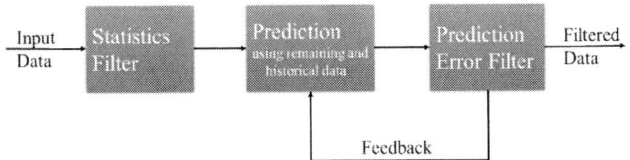

Fig. 3. Process for filtering measured POA data to create a data-based expected clear sky irradiance prediction.

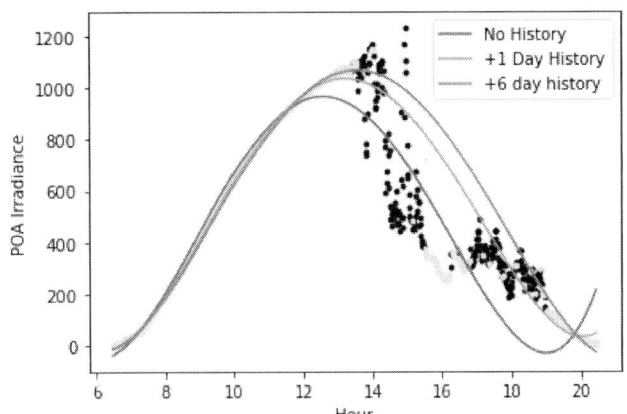

Fig. 5. Results of the 4th-order regression of POA irradiance as the number of days included in the regression increases.

minute interval, centered at each time step. The final filter assesses the difference in both values and slopes of the measured POA and the predicted clear sky POA at each time step. Values which fall more than 20% below the predicted value (normalized to the maximum value in the dataset) are removed. Time steps with a slope deviating by more than 50° from the predicted slope are also removed.

The remaining values in the timeseries are again regressed in a 4th-order regression model to update the prediction and iterate the second stage of filtering. Through testing of this clear sky modeling approach, it was found that only one iteration of the regression and 2nd stage of filtering was necessary for the clear sky filter to reach a stable state.

Fig. 6 shows the results of the prediction-based filter, with the red line approximating the 20% difference threshold for the difference in expected and measured POA values. The points in red represent those that passed the original statistics-based filter (shown in Fig. 4) but fall outside the thresholds for either predicted value or slope differences and were removed by the prediction filter. Near the peak of the curve, a small number of datapoints are eliminated based on the deviation in slope though they are still close to the predicted value.

This new filtering approach shows an improvement in performance over original detection method from pvlib-python. Comparison on a full year of POA data from a single site, counting only daytime data, shows that the two filters yield the same result (pass or remove) for 77.5% of the entire series. The filter results differ on the remaining data points, with 16.2% of data points filtered only by the pvlib-python approach, and only 6.3% of data points remove by the new filtering approach. These results are listed in Table I. A significant portion of the data filtered only by the pvlib-python method represent inappropriate filtering of clear sky irradiance values. This is observable in Fig. 7a for a relatively clear day, in which the green datapoints represent datapoints which were removed by pvlib-python filter but passed the new filter. Similarly, many of datapoints that were not filtered by the pvlib-python method should have been removed. This is most often the case for short duration cloud events, exemplified as the blue points in Fig. 7b.

TABLE I.   SUMMARY AND COMPARISON OF CLEAR SKY FILTERS APPLIED TO DAYTIME DATA.

		New Filter Classification		
		Cloudy	*Clear*	
Pvlib-python Filter Classification	*Cloudy*	28%	16.2%	**44.2%**
	Clear	6.3%	49.5%	**55.8%**
		34.3%	**65.7%**	

The 4th order polynomial regression provides a significantly better expected clear sky POA signal based on historical data. This result greatly reduces false positive filtering of clear datapoints and provides more available data, post-filtering. The regression-based filter also slightly improves filtering of short duration cloud events.

D. Feature Extraction

Finally, feature extraction analysis is performed between cleaned, measured data and the expected, model prediction. The pvlib-python model provides estimated values of expected operating voltage, current, and power as well estimates of each subsystem's short-circuit current (Isc) and open-circuit voltage (Voc). This information can be used to construct a rough

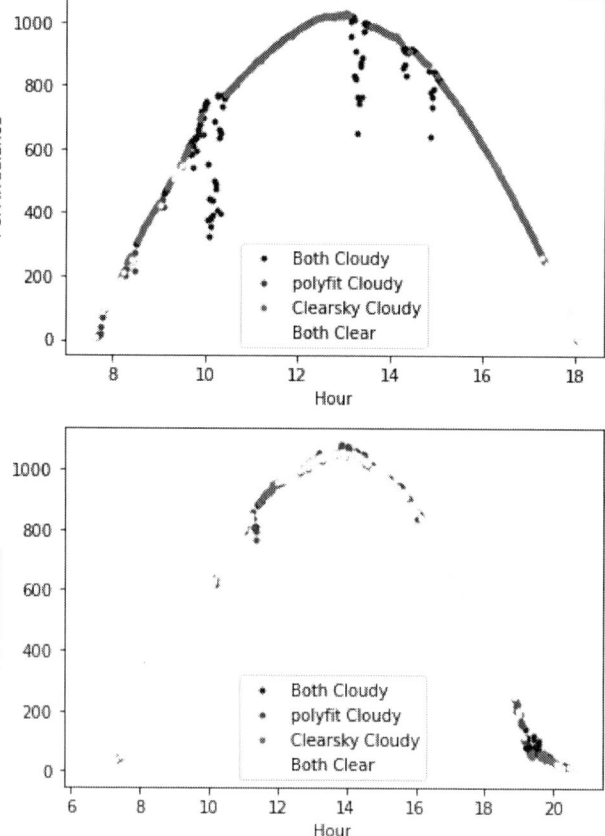

Fig. 7. Sample results comparing the original pvlib-python cloud filter to the new filter for a) a day where the original filter misclassified a majority of the data as cloudy and b) a day where the original filter missed some periods of cloud cover.

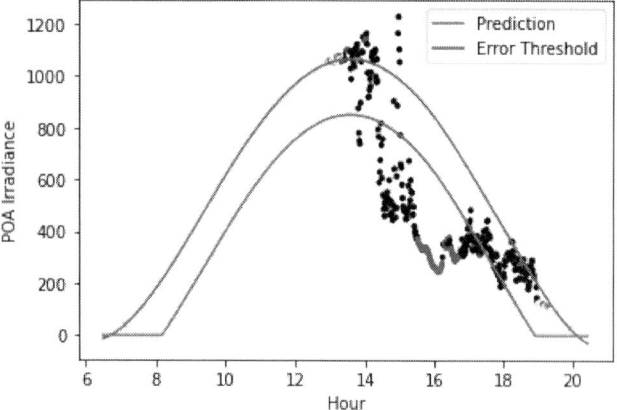

Fig. 6.   The results of the additional filtering on expected cloudy points.

approximation of the expected I-V curve of the modeled subsystem. As different hardware failures have real impacts to the subarray's measured current or voltage (or both), it follows that the shape of the subarray's I-V curve would be impacted as a result of the failure [11]. The two sources of information are used to calculate derived metrics that are used in further anomaly detection. These derived metrics are compared among similar hardware components, where any detected anomalies are assumed to be indicative of a fault present at the PV plant. This analysis is performed on a day-by-day basis to ensure that adequate data is available to perform the necessary fault detection. This fault detection method has been applied to several datasets from many plants of varying sizes and geographic locations with success.

IV. RESULTS

A. Historical Datasets

Historical data was provided for six different large-scall PV plants across the southern US, representing a wide range of plant architectures. These plants covered a broad range of several defining characteristics, including:

- Location – southeastern US to southwestern US

- Power generating capacities – 10+ MW to 200+ MW

- Number of inverters on site – 20+ to 120+

- Number of combiner boxes per inverter – <10 per inverter to 20+ per inverter

- Number of strings per combiner box – <10 to 90+ per combiner box

- Specific inverter hardware

- Specific module hardware

Each of these sites has single-axis trackers across the DC collector field.

The data provided for each site consisted of an extremely large number of data channels, spanning several months to more than a year of data per site. The following measurements were extracted from site historians and were available for fault detection analysis:

- Inverter DC power, current, and voltage for all inverters on site

- Combiner box DC current for all combiner boxes on site

- Meteorological measurements (POA irradiance, ambient temperature, and wind speed) for all meteorological stations at each site

- Tracker positions for all tracker controllers at each site

For each plant, a physical model was built to match the site architecture to calculate expected performance using the meteorological data provided at each plant. The modeled performance data was then compared to measured data at each combiner box to perform the fault detection analysis.

B. Validation

To validate the fault detection method, historical data was extracted from each plant for time periods concurrent with previously performed aerial IR and visual scans. These aerial scans were performed by a third party to notify the plant of failures impacting their hardware. These scans identified a wide range of failures, including tracker-related failures, sectional, string, and partial string outages, and a number of failures that singularly impact individual modules. These failures were aggregated to the combiner box level to match the granularity of sensor data that was available at each site. Each combiner box was then labeled by the failure that was anticipated to have the most impact to the sensor measurements. In general, tracker-related failures have the greatest impact to power input, followed by sectional and string outages, while module and sub-module failures will have the smallest impact to the combiner box measurements.

For validation purposes, it was assumed that the failures found by the aerial scans were completely correct, both in terms of failures being present and absent. Each dataset was processed through its respective model to calculate expected current and voltage measurements for each combiner box at the plant. The potential faults identified by the physical model were then validated against the failures identified by the aerial IR scans to determine true and false positive rates. Two of the sites had scans available for two consecutive years. In all, this allowed for eight different validation datasets.

C. Confusion Matrix Development

For each plant, the respective fault detection model was run on site data concurrent to the aerial scan to determine if a fault may be detectable at the combiner box level using historian data alone. The results from the fault detection model were then joined to the results from the aerial scans. Using the fault detection results and the aerial scans, a confusion matrix was constructed for each dataset to assess the accuracy of the fault detection model. For the purposes of this paper, the confusion matrix results are defined as follows:

- True positive – both the aerial scans and the fault detection model noted a fault, within a given combiner box

- False positive – the fault detection model noted a fault while the aerial scans did not, within a given combiner box

- True negative – neither the aerial scans nor the fault detection model noted a fault, within a given combiner box

- False negative – the aerial scans noted a fault while the fault detection model did not, within a given combiner box

The fault detection model detects the presence of faults on a day-to-day basis. Sample results for a single inverter are shown in Fig. 8. This figure shows how the confusion matrix classification changes for each combiner box over the course of a week. One observation is the generally steady state of the fault detection results; the model does not repeatedly bounce between faulted and unfaulted states. This time component has not been

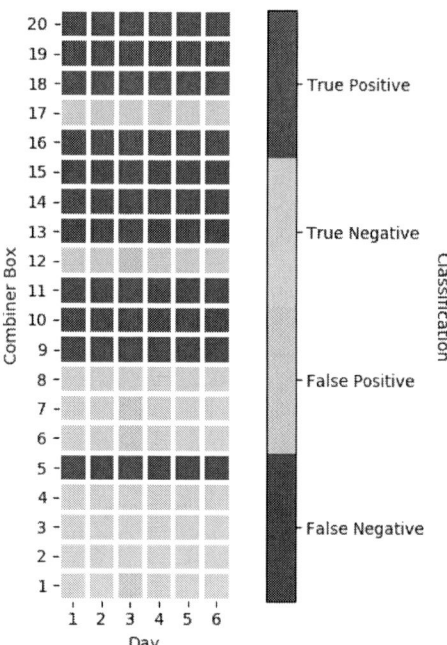

Fig. 8. True positive, true negative, false positive, and false negative classifications for fault detection for a single inverter. The aerial scan was performed on Day 5.

heavily analyzed to-date but could be used as part of the detection model to determine the presence (or lack) of faults. Additionally, since the presence of faults are evaluated on a day-by-day basis, this analysis method allows for the detection of the onset of faults. If implemented in a near real-time environment, this method could notify operators of faults as they happen, an improvement over the typically annual aerial inspections.

Since the aerial scans only reflect the state of the hardware on a single day, the remaining analysis in this paper is focused only on the fault detection results on the day of the scan. A potential next step is to consider how the time-based change in fault detection could be included to improve the accuracy of the fault detection model.

D. Fleet Summary

For each dataset, the results of the fault detection model were joined with the results of the aerial scans. To characterize the types of failures that were accurately detected by the model, the aerial scan results were categorized into three types: minor failures, string outages, and tracker failures. A listing of the failures in each of these categories is shown in Table II. In short, minor failures impact one to a few modules, string outages impact one to a few strings, and tracker failures are failure of individual tracker controllers and impact multiple strings.

The results from all datasets are summarized in Table III. Note that several sites in the summary are lacking tracker failures. Depending on the site, this is due to either a lack of labeling of tracker failures or a lack of tracker failures.

As shown in Table III, this fault detection method has consistently shown true positive rates at 75% or higher for string outages when validated against aerial IR scans performed the

TABLE II. SUMMARY OF LABELS APPLIED TO EACH FAILURE.

Summary Category	Failures
Minor	Cell Faults
	Hot Spots
	Module Anomalies
	Sub-module Anomalies
String Outage	String Outage
	Sectional Outage
Tracker	Tracker at stow
	Tracker anomaly

TABLE III. FAULT DETECTION TRUE AND FALSE POSITIVE RATES.

Site	True and False Positive Rate Summary		
	Fault Type	TPR	FPR
1	Normal	-	36%
	Minor	34%	-
	String	100%	-
	Tracker	77%	-
2a	Normal	-	39%
	Minor	44%	-
	String	88%	-
	Tracker	74%	-
2b	Normal	-	44%
	Minor	44%	-
	String	90%	-
3	Normal	-	67%
	Minor	86%	-
	String	67%	-
4	Normal	-	32%
	Minor	41%	-
	String	54%	-
	Tracker	56%	-
5a	Normal	-	7%
	Minor	92%	-
	String	88%	-
5b	Normal	-	30%
	Minor	33%	-
	String	91%	-
6	Normal	-	50%
	Minor	82%	-
	String	92%	-

same day. Depending on the site architecture, these faults individually represent approximately a 1% loss of power output at the inverter level and are the types of faults that are small enough to go unnoticed when monitoring inverter and plant power outputs and, when considered in aggregate, are one of the largest sources of power loss detected by aerial overscans. The fault detection method shows good detection rates for these faults, but, unlike aerial scans, is able to perform fault detection analysis on an ongoing basis. An overall summary of the fault detection method is shown in Table IV. The overall true positive rate across all faults and all eight datasets was 66%. The true positive rates for the faults with the largest impact to plant power production are significantly higher: 88% for string outages and 73% for tracker faults. Even the minor faults, which only impact a single module, where detected with approximately a 50% true positive rate. With additional tuning of internal detection thresholds, these results are expected to improve.

V. Conclusions

The results discussed above show the strength of using a physical model in conjunction with real PV data to perform fault detection on DC collector field data. Currently, many sites only perform aerial scans annually, which limits their insight into the presence of subtle, small-scale faults. The developed fault detection method operates on data that is typically available at M&D centers on a continuous basis and is capable of providing reliable fault detection year-round. Additionally, the detection locates faults to specific combiner boxes, reducing the field inspection effort of locating the fault when maintenance needs to be performed. With real time results, M&D centers could leverage this fault detection model to schedule maintenance as issues arise, rather than wait for the next annual scan to notify

them of hardware failure. Production, and therefore revenue, is reduced when these failures are undetected and uncorrected.

To ensure that operators do not become burdened by attempting to diagnose false alarms, future work is aimed at reducing the false positive rate while keeping the true positive rate high. One approach will be to consider how day-to-day variations in the fault detection results can be leveraged to determine if a fault is truly present. Additionally, more tuning of the internal fault thresholds could lead to potential improvements in true and false positive rates.

References

[1] "Chapter 2: Energy sectors and systems," Quadrennial Technology Review 2015, US Department of Energy, 2015.

[2] "Chapter 4: Advancing Clean Electric Power Technologies," Quadrennial Technology Review 2015, US Department of Energy, 2015.

[3] W. Brooks, T. Basso, and M. Coddington, "Field guide for testing existing photovoltaic systems for ground faults and installing equipment to mitigate fire hazards," United States, https://doi.org/10.2172/1225963.

[4] R. R. Hill, G. T. Klise. And J. R. Balfour, "Precursor report of data needs and recommended practices for PV plant availability, operations, and maintenance reporting". https://osti.gov/search/identifier:1169447.

[5] Walker, H. A. "Best practices for operation and maintenance of photovoltaic and energy storage systems; 3rd Edition," United States. https://doi.org/10.2172/1489002.

[6] W. F. Holmgren, C. W. Hansen, and M. A. Mikofski. "Pvlib-python: a python package for modeling solar energy systems." Journal of Open Source Software, 3(29), 884, (2018). https://doi.org/10.21105/joss.00884.

[7] W. F. Holmgren et al, pvlib/pvlib-python: v0.7.2 (v0.7.2). Zenodo. https://doi.org/10.5281/zenodo3762635.

[8] "Photovoltaic systems performance – Part 3: Energy evaluation method," IEC Technical Specification, IEC TS 61724-3, ISBN 978-2-8322-3531-7.

[9] M. Reno and C. Hansen, "Identification of periods of clear sky irradiance in time series of GHI measurements," Renewable Energy, vol. 90, 2016, pp. 520-531.

[10] B. Ellis, M. Deceglie and A. Jain, "Automatic detection of clear-sky periods using ground and satellite based solar resource Data," 2018 IEEE 7th World Conference on Photovoltaic Energy Conversion (WCPEC) (A Joint Conference of 45th IEEE PVSC, 28th PVSEC & 34th EU PVSEC), 2018, pp. 2293-2298.

[11] M. Kontges, et al. "Performance and reliability of photovoltaic systems, subtask 3.2: Review on failures of photovoltaic modules," IEA Photovoltaic Power Systems Programme, Task 13, 2013, ISBN 978-3-906042-16-9.

TABLE IV.   OVERALL SUMMARY OF THE FAULT DETECTION MODEL.

Fault Type	TPR	FPR
Overall	66%	28%
Normal	-	28%
Minor	45%	-
String	88%	-
Tracker	73%	-

Enhancing temporal variability of 5-minute satellite-derived solar irradiance data

[1]Jing Huang, [2]Richard Perez, [2]James Schlemmer, [1]Alex Kubiniec, [1]Marc Perez, [1]Akanksha Bhat and [1]Patrick Keelin

[1]Clean Power Research, Napa, CA, USA
[2]Atmospheric Sciences Research Center, SUNY, Albany, NY, USA

Abstract—**Satellite-derived solar irradiance data are known to underestimate temporal variability compared to point measurements because of their pixel-averaging nature. In this study, we apply an algorithm imposing random noise to enhance the temporal variability of 5-minute satellite-derived solar irradiance data. We show that the resulting product, termed as True Dynamics, has clear-sky exceedance events and the frequency of large ramp events closer to observation. In addition, the increase of temporal resolution of irradiance data significantly reduces the underestimation error of power inverter clipping under high DC:AC capacity ratios conditions.**

Keywords—solar irradiance, SolarAnywhere, variability, satellite, photovoltaics, inverter clipping

I. INTRODUCTION

Spatial and temporal variability caused by events such as cloud and aerosol passage is one of the inherent characteristics of solar irradiance data. This has cascading effects on estimating and forecasting solar power generation at various aggregation levels. Ideally, the solar industry desires high-quality solar resource data at fine spatial (e.g. sub-km) and temporal (e.g. sub-minute) scales to capture variability patterns for accurate bottom-up solar power modeling. Although this goal has been gradually approached mainly by development in remote sensing and image processing techniques, complementary empirical approaches attempting to reconstruct spatiotemporal solar variabilities are still useful.

Perez et al. [1] parameterized intra-hourly solar variabilities using four key metrics distilled from measurement data at 24 sites across the United States, which includes the standard deviation of the global irradiance clear sky index, and the mean index change from one time interval to the next, as well as the maximum and standard deviation of the latter. The clear-sky index, or *kt*, which is defined as Global Horizontal Irradiance (GHI) divided by GHI at a presumptive clear-sky condition. In this study, we apply this approach to the latest SolarAnywhere® (SA hereafter) [6] V3.5 data product, which will be operational at 5-min interval covering the entire United States. We show that this approach can artificially enhance the temporal variability and create clear-sky exceedance events that match better with observation at 10 ground stations in the eastern US. In addition, we quantify the effects of both irradiance temporal resolution and empirical variability enhancement on the estimation of solar photovoltaic (PV) power as a function of PV systems' DC:AC capacity ratios.

A. Ground stations for validation

Figure 1 and Table 1 provide information on the reference ground stations used in this study. The measurement data span the entire year of 2020 with a temporal resolution of 1 minute, which can then be down-sampled to a resolution of 5 minutes to match the resolution of the SA data.

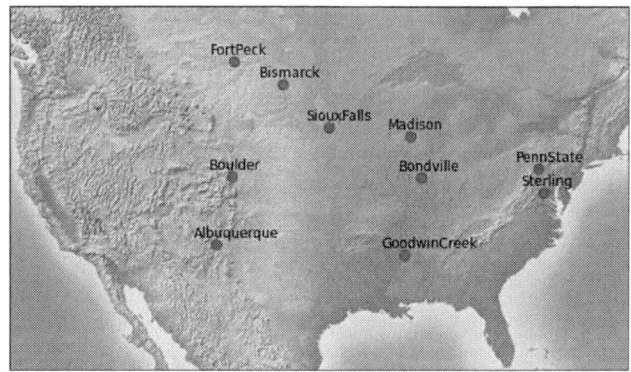

Figure 1. 10 reference ground stations in the eastern United States are used in this study.

Table 1. Metadata of the 10 reference ground stations

Station	Network	Latitude	Longitude	Elevation
Boulder	SURFRAD	40.13	-105.23	1661 m
Penn State	SURFRAD	40.72	-77.93	400 m
Fort Peck	SURFRAD	48.31	-105.10	686 m
Goodwin Creek	SURFRAD	34.25	-89.87	104 m
Sioux Falls	SURFRAD	43.73	-96.62	475 m
Bondiville	SURFRAD	40.05	-88.37	212 m
Madison	SOLRAD	43.07	-89.41	285 m
Bismarck	SOLRAD	46.77	-100.76	508 m
Sterling	SOLRAD	38.98	-77.48	93 m
Albuquerque	SOLRAD	35.04	-106.62	1546 m

B. SolarAnywhere® V3.5

SA provides bankable solar irradiance data to support the growth of solar industry globally. The latest version V3.5 which was introduced May 2021, uses 3-hourly aerosol optical depth

(AOD) data from MERRA-2 to account for aerosol variability in modeling clear-sky irradiance. This represents a key operational improvement compared to the previous version V3.4 that used only monthly climatological aerosol data. The resulting difference in clear-sky Direct Normal Irradiance (DNI) is significant particularly during onset and passage of intense aerosol events such as wildfires as shown for the Boulder site from 08/2020 to 10/2020. In addition, the temporal resolution of SA V3.5 will increase from 30 minutes to 5 minutes operationally soon.

Thus, we evaluate the 5-min V3.5 product in this study. Because of the improved clear-sky modeling in V3.5, the resulting *kt* has a better representation of cloud opacity. Figure 3 shows a good match between SolarAnywhere and ground measurements. The deviations from the 1:1 line are genesymmetric, except when GHI is high (i.e. > 900 W m^{-2}). This is because the SA irradiance model, like other semi-empirical satellite models, assume that irradiance quantities are capped at their clear-sky values [4,5] whilst the measurments do contain clear-sky exceedence events.

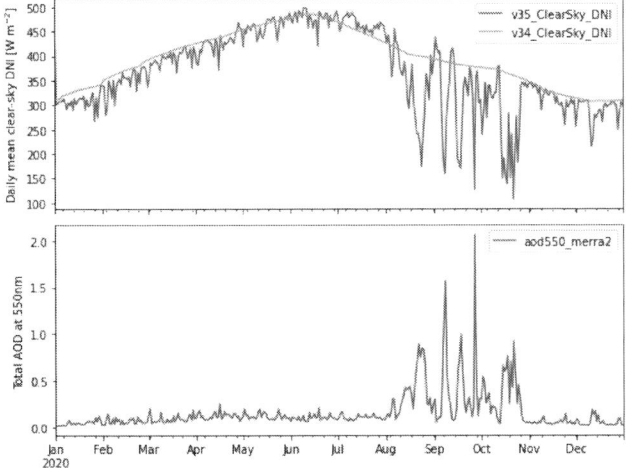

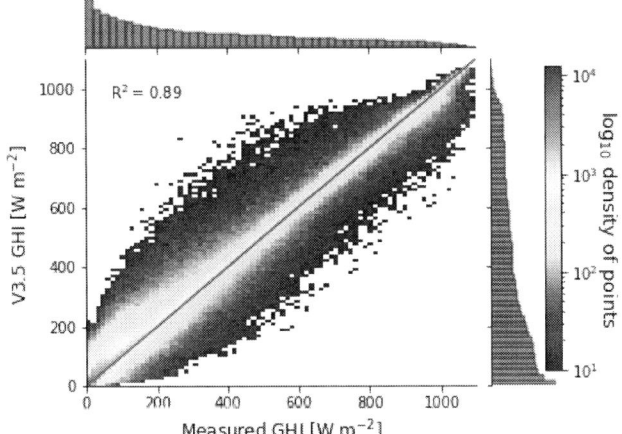

Figure 2. (top) Annual variations of daily mean clear-sky DNI for the Boulder site in 2020. The smoother line is V3.4 and the more variable line is V3.5; (bottom) The corresponding daily mean AOD at 550nm retrieved from MERRA-2.

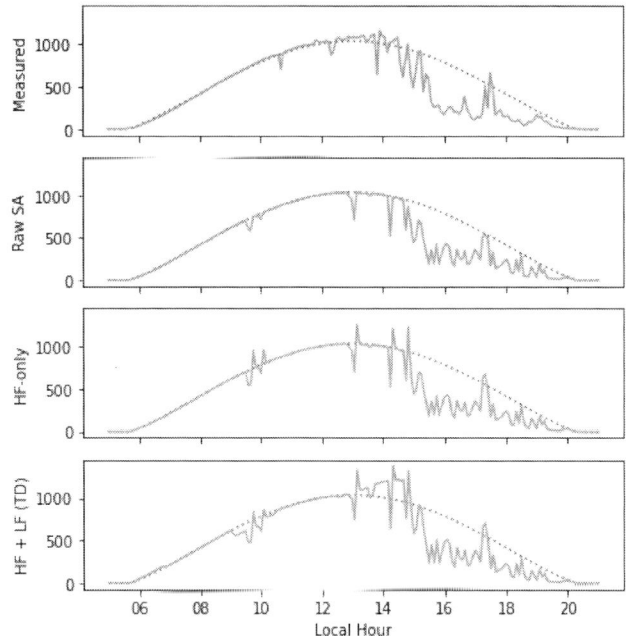

Figure 3. 2D histogram between 5-min SolarAnywhere® V3.5 Global Horizontal Irradiance (GHI) and 5-min pyranometer-measured GHI for the aggregation of 10 stations.

C. Enhancing variability of global irradiance at 5-min

To reconcile the extent of temporal variability of the SA V3.5 with measurements which includes clear-sky exceedance events, we impose random noises onto SA V3.5 irradiance data based on parameterizations in Perez et al. [1]. These clear sky exceedance events are modeled by randomly enhancing the satellite-derived clearness index changes from one five-minute interval to the next (Δkt) under conditions that exhibit both high variabilities as determined from the satellite-derived Δkt and a relatively high irradiance level, as determined by the value of *kt*. Action thresholds are defined for both Δkt and *kt*, above which high-frequency (HF) random enhancement is applied. The amplitude of the enhancement is fitted to observations so that amplified Δkt distributions approach observations.

In addition to the HF variability enhancement, low-frequency (LF) background amplification is also conducted. This step is designed to enhance the effectiveness of accurately capturing losses in all inverter-limited (DC:AC) cases, not by only accounting for short-term spikes but also accounting for prolonged conditions when irradiance can remain higher than clear sky (e.g., from stable cloud enhancement) and vice-versa without affecting overall bias.

To distinguish from the raw 5-min SA data, here the variability enhanced 5-min product is termed as True Dynamics (TD). A preview of TD is provided in Figure 4. By visual inspection this product adds clear-sky exceedance events whilst enhancing its temporal variability noticeably. In particular, note the sustained higher irradiance levels near 2pm at bottom compared to the HF-only enhanced model at bottom middle.

Figure 4. An example day (Jun 29th, 2020) of 5-min GHI time series at Boulder comparing GHI time series between (top) measurement, (top middle) raw SA, (bottom middle) high-frequency only amplified SA and (bottom) fully amplified SA with both high-frequency and low-frequency components. The unit of GHI is W m⁻².

II. THE EFFECTS OF VARIABILITY ENHANCEMENT

After the variability-enhanced time series data pass visual inspection, they need to be examined in more details from the perspectives of distribution and accuracy.

A. Distributions

Figure 5 illustrates the effects of variability enhancement in terms of distribution. It can be discerned that SA generates kt more frequently in the range of (0.75, 1] whilst the variability enhancement procedure is able to alleviate this over-population. In addition, the variability-enhanced kt has occurrences in the clear-sky exceedance bin (1, 1.25] that do not exist in the raw SA data.

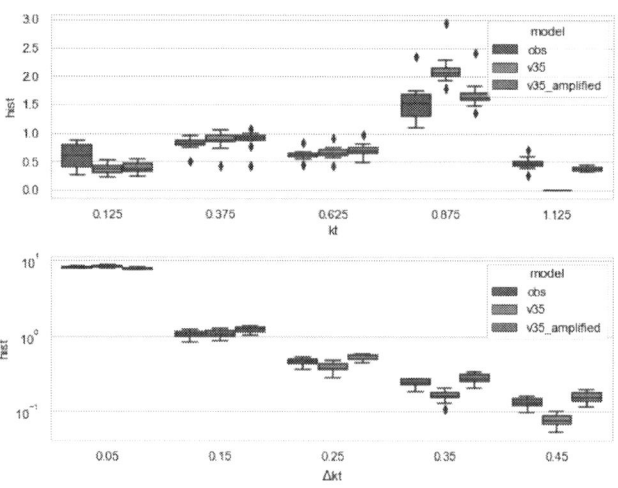

Figure 5. Histogram box plots of 10 sites for (top) kt and (bottom) Δkt. The bins for kt are (0, 0.25], (0.25, 0.5], (0.5, 0.75], (0.75, 1], (1, 1.25]. The bins for Δkt are (0, 0.1], (0.1, 0.2], (0.2, 0.3], (0.3, 0.4], (0.4, 0.5]. Note that the histogram of Δkt is plotted in log scale to highlight the comparison of large ramp events.

An important characteristic of temporal variability is the frequency of ramp events. It is clear from the Δkt plot in Figure 5 that whilst the raw SA kt has less frequent large ramp events than measurements, the variability-enhancing algorithm is able to increase the frequency of large ramp events to the observed extent. This feature would be useful for scenarios where ramp events play a key role such as hybrid PV-battery systems providing pre-scheduled power generation or stability and adequacy testing for ISO system planning.

B. Error metrics

It is promising that our artificial enhancement of variability has created clear-sky exceedance events that do not exist in raw satellite data and increased the frequency of large ramp events. However, due to the random nature of modifying the satellite-derived kt in the algorithm, the amplified kt becomes inherently less accurate than the raw kt. To quantify the tradeoff between variability and accuracy, we employ Taylor Diagram on the raw kt and the modified kt signal.

Taylor Diagram was invented by Taylor (2001) originally for geophysical studies. It displays three important and connected metrics within just one graph. Assuming we have a general reference field r, and a test field t, the three metrics are connected by the following equation:

$$\sigma_\Delta^2 = \sigma_r^2 + \sigma_t^2 - 2\sigma_r\sigma_t\rho, \qquad (1)$$

where σ_Δ, σ_r, σ_t are the standard deviation of the error, the reference field, and the test field, respectively, and ρ is the correlation coefficient between the reference and the test field.

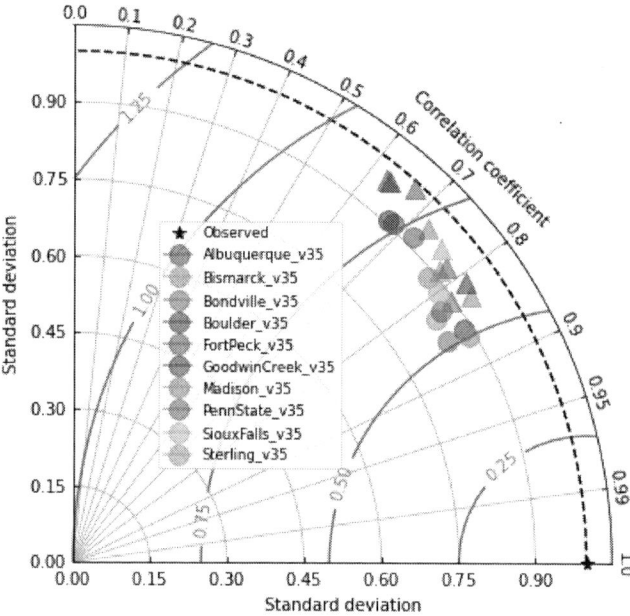

Figure 6. Taylor diagram illustrates three metrics (i.e. correlation coefficients, centered RMSE and standard deviation) of the raw SA kt (circles) and the amplified SA kt (triangles).

In our case, the reference field is kt derived from measurements. Note that the measured kt is normalized to have a unity standard deviation such that the star position always represents observation. The test field is the raw SA kt and the amplified kt. It is clear from Figure 6 that the visual effect of manually enhancing the variability of kt is to move the position of a site generally upward in the Taylor diagram. This move corresponds to an increased σ_t and σ_Δ, and a decreased ρ. Since an increasing σ_t implies higher temporal variability and an increasing σ_Δ and a decreasing ρ imply reduced accuracy, it can be drawn that for all 10 sites, this step has served its purpose of design to enhance temporal variability but at a price of reduced accuracy.

III. POWER INVERTER CLIPPING

The industry-standard data requirement for solar energy resource assessment has traditionally been hourly resolution. However, this has been shown to be insufficient to inform critical financial decisions particularly under scenarios where the capacity of AC inverters is smaller than that of DC power generation [3]. Indeed, the lack of both intra-hour variability and clear sky exceedance events, implies an underestimation of clipping losses: actual data spikes near and above clear sky are curtailed while modeled data remaining below clear sky are not. As such, we examine what the availability of 5-min satellite-derived irradiance data (raw and variability enhanced) imply to addressing this issue.

To quantify the effect of time resolution on power estimation, we build a power model using PVLIB for a hypothetical horizontal single-axis tracking PV system at all 10 locations. This is a prevailing choice for new installation of utility-scale solar PV farms due to its relative low cost and excellent mechanical durability. The maximum rotation angle of the tracking system is specified at 70 degrees with backtracking capability. The PVWatts model is used for DC and AC with default parameter setting [6]. To model the temperature derating effect, we use temperature site measurement and a coefficient of -0.37% $^\circ$C^{-1} which is representative of common multi-crystalline PV cells. The separation of GHI into direct and diffuse components are performed using the DIRINT model [7]. The overestimation of power by the use of hourly irradiance data during variable irradiance and inverter clipping period is illustrated in Figure 7.

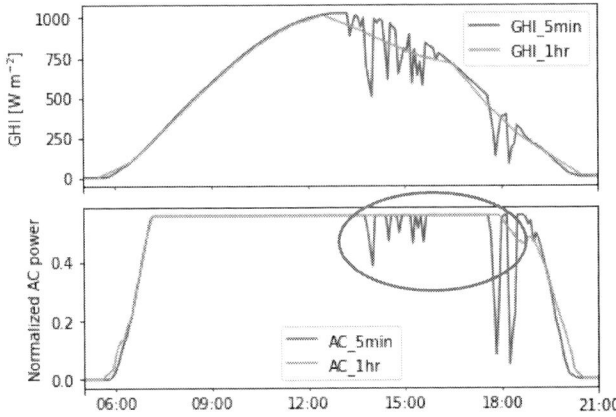

Figure 7. Comparison of hourly averaged and 5-minute (top) GHI time series data; (bottom) AC power yield normalized by the capacity of the inverter. A DC:AC ratio of 1.8 is used for irradiance-to-power conversion. The red circle indicates the period when the sub-hourly variability of AC power is not accounted for by using hourly irradiance data.

Using the proposed configuration for the power model, we are able to estimate AC power yield at various time resolutions. To overcome inaccuracies due to numerical issues we first linearly interpolate and align all data sources to 1-min resolution. And then we run the power model for various DC:AC scenarios. The curtailed power is calculated as P(DC:AC=1.0) − P(DC:AC), where P is the output of the PVLIB power model. It should be noted that the bulk bias in irradiance plays an increasingly important role in the estimation of the curtailed power as DC:AC increases. However, site-specific debiasing is beyond the scope of this study and its effectiveness is not always warranted. As such, we do not perform any debiasing operations and rather assess the performance of the raw 5-min SA data and the variability-enhanced 5-min SA data (i.e. TD). We regard the power estimation using the 1-min measurement data as truth and

also use its total production to normalize the error in power curtailment estimation.

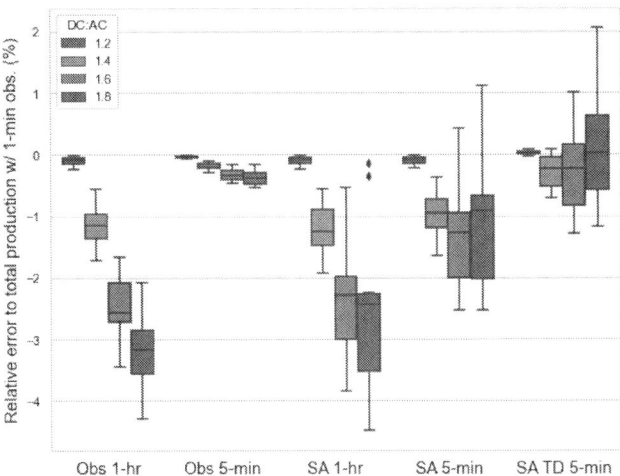

Figure 8. Box plots of relative error of curtailed AC power estimation using 1-hour observation, 5-min observation, 1-hour SA, 5-min SA, 5-min TD.

Figure 8 illustrates the effects of data sources and time resolutions on the accuracy of estimating curtailed power due to the limited capacity of AC power inverter. Compared to using hourly averaged irradiance data, the use of 5-minute data is able to generally reduce the underestimation of power inverter clipping significantly. For example, under DC:AC=1.8, the 10-site average of the clipped power error decreases from -3% when using hourly SA data to -1% when using 5-min SA data. Furthermore, with our variability treatment, the 5-min TD data present an error of only 0.3% for 10-site average and DC:AC=1.8 with an associated standard deviation of around 1%. Although it is inevitable that the bias in SA data plays a role in this comparison, in particular under high DC:AC scenarios, it is clear that our new 5-min SA data and our variability enhancement technology does help reduce the error in estimating power inverter clipping and hence total power production.

IV. DISCUSSION

1. We have succeeded in implementing a variability enhancing algorithm on 5-minute satellite-derived solar irradiance data. As shown quantitatively, the algorithm has created clear-sky exceedance events more commensurate with measurement and increased the frequency of large ramp events. However, it also inevitably reduces short-term accuracy as measured by correlation and RMSE. The amplified irradiance data are suitable for scenarios where variability characteristics are important. For example, the extent and frequency of power ramp events play a key role in the battery life and stability of a hybrid PV-battery system.

2. As shown, the amplified irradiance data (i.e. TD) are also suitable for long-term PV yield estimation when the DC:AC ratio is high. As an indication, the error of

estimating power clipping loss is only 0.3% for a 10-site average under DC:AC=1.8.

3. We note that no debiasing schemes have been applied for individual sites. With efforts dedicated to addressing this, the 5-minute SA data and variability enhancements have the potential to further improve the performance of power clipping loss and the total PV yield estimation and other applications where the information of intra-hourly variabilities are required.

REFERENCES

[1] R. Perez, S. Kivalov, J. Schlemmer, K. Hemker, and T. Hoff, "Parameterization of site-specific short-term irradiance variability," Sol. Energy, 85 (7), 1343-1353, 2011

[2] K.E. Taylor, (2001). "Summarizing multiple aspects of model performance in a single diagram". J. Geophys. Res. 106, 7183–7192, 2001

[3] K. Bradford, R. Walker, D. Moon and M. Ibanez, "A regression model to correct for intra-hourly irradiance variability bias in solar energy models", 2020 47th IEEE Photovoltaic Specialists Conference (PVSC), 2020, pp. 2679-2682.

[4] Perez R., P. Ineichen, K. Moore, M. Kmiecik, C. Chain, R. George and F. Vignola, (2002): A New Operational Satellite-to-Irradiance Model. Solar Energy 73, 5, pp. 307-317.

[5] Perez R., P. Ineichen, M. Kmiecik, K. Moore, R. George and D. Renne, (2004): Producing satellite-derived irradiances in complex arid terrain. Solar Energy 77, 4, 363-370

[6] Dobos A.P., (2014) "PVWatts Version 5 Manual", NREL. https://pvwatts.nrel.gov/downloads/pvwattsv5.pdf

[7] Perez, R., P. Ineichen, E. Maxwell, R. Seals and A. Zelenka, (1992). "Dynamic Global-to-Direct Irradiance Conversion Models". ASHRAE Transactions-Research Series, pp. 354-369

The Materials Degradation in Encapsulants for Application in Glass/Glass PV Modules After Accelerated Aging

Sona Ulicna, Archana Sinha, David C. Miller, Laura T. Schelhas, Michael Owen-Bellini

SLAC National Accelerator Laboratory, Menlo Park, CA, United States

National Renewable Energy Laboratory (NREL), Golden, CO, United States

This work compares material properties of various encapsulant resins and formulations for application in glass/glass PV modules. UV weathering resulted in encapsulant discoloration and loss of optical transmittance through the coupons. This was accompanied with chemical degradation of the encapsulant, in particular at the center of the coupon. Damp heat weathering led to temporary moisture trapping within the glass/glass coupons and changes in polymer crystallinity resulting from annealing at elevated chamber temperatures, but negligible degradation of the chemistry of the encapsulants. Sequential weathering resulted in similar chemical degradation as in the UV weathering, but coupons experienced greater loss in transmittance. Out of the studied set of encapsulant types and formulations, POE encapsulants showed greater durability compared to EVAs and TPOs. A change in crystallinity was observed for the TPOs that may explain the evolution of optical haze observed for those encapsulants.

Perovskite PV Design for Stable Space Operation

Kaitlyn T. VanSant, Ahmad R. Kirmani, Jay B. Patel, Laura E. Mundt, David P. Ostrowski, Brian M. Wieliczka, Gabriella D. Lahti, Michael D. McGehee, Laura T. Schelhas, Joseph M. Luther, Timothy J. Peshek, Lyndsey B. McMillon-Brown

NASA Glenn Research Center, Cleveland, OH, United States

National Renewable Energy Laboratory, Golden, CO, United States

University of Colorado, Boulder, Boulder, CO, United States

SLAC National Accelerator Laboratory, Menlo Park, CA, United States

Metal halide perovskites are an emerging technology area for photovoltaic (PV) space applications. The goal of our research is to design a perovskite solar cell (PSC) that can exhibit stable performance, when exposed to space-relevant stress conditions. This presentation will focus on the down-selection of both the contact layers and the encapsulation scheme for potentially space-compatible PSCs.

Thermoradiative Cell Technology: Analysis and Loss Mechanisms

Geoffrey A. Landis

NASA John Glenn Research Center, 21000 Brookpark Road, Cleveland, OH, 44017, U.S.A.

Abstract— **The thermoradiative cell is a solid-state device for conversion of heat energy to electrical power. The maximum power point bias for a thermoradiative cell is derived from detailed balance considerations, and the effects of parasitic thermal emissivity on the conversion efficiency is considered.**

Keywords— *thermoradiative cell*

I. INTRODUCTION

The thermoradiative cell is a concept for conversion of heat energy to electrical power. Conceptually, a thermoradiative cell is identical to a photovoltaic cell operating with the thermodynamic input and output reversed: a semiconductor diode that takes thermal energy as input, and produces electrical power and emitting waste heat in the form of (infrared) light.

The concept of a thermoradiative cell was introduced by Standberg [1] and by Santhanam and Fan [2] in 2015 and 2016, based on the concept that an ideal photovoltaic cell is a heat engine, operating on a temperature difference between photons (*e.g.*, from the sun) as a high temperature source, and the external environment as a low temperature sink. Since an ideal heat engine operates when the high and low temperature sides of the engine were reversed, they showed that a photovoltaic cell (or a device identical in structure to a photovoltaic cell) would also operate with sources reversed. Thus, the thermoradiative cell has heat as the energy input and photons as the waste heat output (figure 1). The cells radiate heat to a lower temperature, assumed to be the low temperature of deep space.

Analyses of efficiency limits for thermoradiative conversion using various ideal assumptions has subsequently been done by a number of researchers [1-7]. Applications proposed include terrestrial power [1], conversion of waste heat to energy [8], and use as for converting heat from isotope or nuclear power sources to electrical power for spacecraft [9-10].

Thermoradiative cells are typically designed to operate at a heat source temperature that may be as high as 1000 to 1500 K (*e.g.*, for nuclear heat sources), or as low as 300 K (*e.g.*, for recovering energy from waste heat). Since these temperatures are low compared to the equivalent photon temperature of solar energy (~6000 K), the optimum bandgap for thermoradiative cells is correspondingly lower than that for solar cells. Thus, thermoradiative cells are necessarily low-bandgap devices. However, they differ from thermophotovoltaic (TPV) cells in that thermoradiative cells operate at high temperature, while TPV cells operate at lower temperatures.

As will all heat engines (including solar cells), a thermoradiative cell is limited to a maximum possible efficiency by the Carnot efficiency limit,

$$\eta_{carnot} = (1 - T_c/T_h) \tag{1}$$

where T_h is the (absolute) operating temperature of the cell and T_c is the (absolute) temperature of the heat sink, *i.e.*, the temperature of deep space (for a cell radiating to deep space), or the spectrum-averaged effective temperature of the sky (for a cell operating terrestrially). However, in the real-world cells are not likely to approach the Carnot efficiency limit.

Operating in the reverse direction from photovoltaic cells, thermoradiative cells utilize the thermal dark current, and reject the radiation from electron-hole recombination as waste heat in the form of infrared radiation. The waste heat rejection in the form of infrared radiation means that the thermoradiative cell must have an unimpeded view to the sky (or to some other low-temperature heat sink).

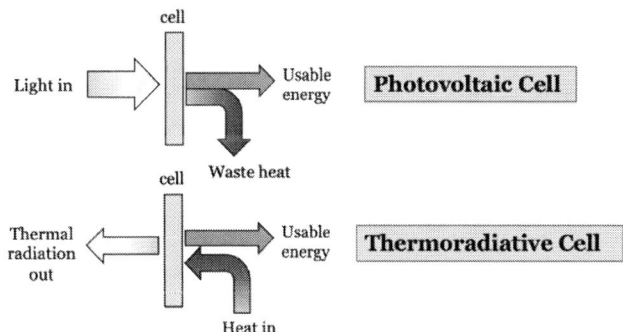

Fig. 1. Photovoltaic and thermoradiative cells compared

II. OPERATING VOLTAGE

The current through a diode is the sum of two components, forward and reverse current. Reverse current consists of majority carriers which are thermally excited "up" across the junction (*i.e.*, the electrons injected to the p side from the n side, and holes injected to the n side from the p) and subsequently recombine. Forward current is carried by minority carriers that are generated by absorption of photons from the environment (the reverse of recombination) and move "down" across the junction. At zero voltage, in thermal equilibrium and in the dark, the principle of detailed balance requires these two currents to be exactly equal, and thus the rate of recombination of carriers thermally injected across the junction exactly equals the rate of generation of carriers by absorption of photons from the environment.

978-1-7281-6118-1/22 $31.00 © 2022 IEEE

Outside of thermal equilibrium, if the external temperature is higher than the diode temperature, more carriers are generated from absorption of thermal photons than are injected across the junction, and hence the forward current exceeds the reverse current. This results in thermophotovoltaic operation. On the other hand, if the external environment is lower in temperature than the diode, the reverse current is greater than the forward current, resulting in thermoradiative operation.

The schematic band diagram of a thermoradiative cell is shown in figure 2.

Thermoradiative cells thus produce power in reverse bias (*i.e.,* the 2nd quadrant of the IV curve), rather than in forward bias (the 4th quadrant of the IV curve), as photovoltaic cells do. This has the result that the bias voltage at the maximum efficiency point is not the same as the bias voltage for maximum power output.

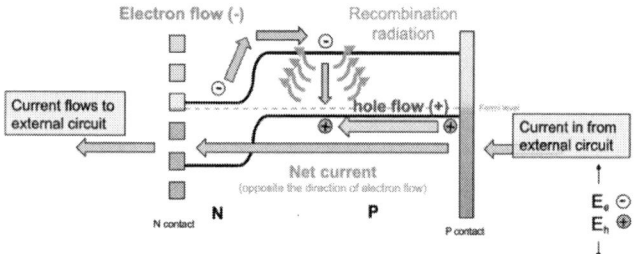

Fig. 2. A simplified energy band-diagram schematic of a diode operating as a thermoradiative cell.

A. Operating Point in the Detailed Balance Limit

The maximum power operating point can be calculated in the detailed-balance (Shockley-Queisser) limit, in which only the losses intrinsic to the process are considered. In this limit, all of the thermal emissivity from the cell is due to band-to-band recombination of carriers.

In this analysis, we consider an ideal diode. At a bias voltage V, the thermal dark current of the ideal diode is:

$$I(V) = I_o e^{-(qV/kT)} \qquad (2)$$

where the applied (bias) voltage is V, I_o the dark saturation current, q the electron charge, k is the Boltzmann constant, and T the diode temperature in K.

The total current through the device is the thermal dark current plus the photovoltaic current. In the dark, the photovoltaic current is equal to the current generated by the IR photons emitted from the surrounding environment. In thermal equilibrium, detailed balance tells us that at V=0 the photovoltaic current must exactly equal the thermal dark current, and hence, as expected, the total current is zero at zero bias.

For the ideal thermoradiative cell, we remove the assumption of thermal equilibrium of the diode with its surroundings, and consider a diode at a higher temperature than the background. In the limit that the background temperature is negligible compared to the diode temperature, then the photovoltaic current can be neglected. The power output is then simply the current times the voltage:

$$P(V) = VI_o e^{-(qV/kT)} \qquad (3)$$

The output power as a function of voltage is shown in figure 3.

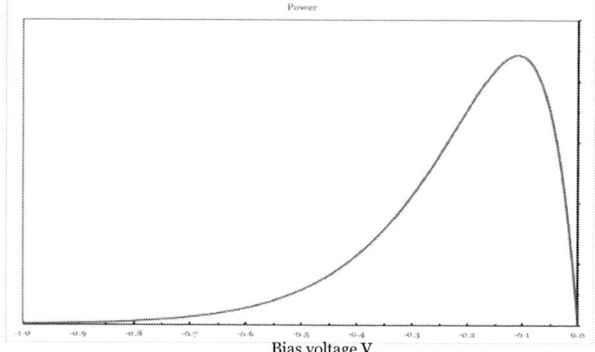

Fig. 3. Output power as a function of bias voltage for ideal thermoradiative diode at 1000°C, assuming that the heat-sink temperature T_c is negligible.

The maximum power voltage is found by taking the derivative with respect to voltage and setting this to zero. The result gives the bias voltage V_{mp} for the maximum output:

$$V_{mp} = kT/q \qquad (4)$$

In this ideal case, then, the operating voltage for maximum power will be ~ 25 mV for a thermoradiative cell operating at room temperature, rising to ~ 100 mV for a cell operating at 900°C.

For the case where the external temperature is not negligible, the thermophotovoltaic current must be subtracted from the dark current in equation 2. In general, for cells operating in space, the thermal sink temperature will be low enough that this term will not be important for operation near the maximum power point. Including the radiation from the environment shifts the maximum power point to a slightly lower voltage.

B. Theoretical Current

The dark current I_o is a function of the semiconductor parameters. However, the maximum generated current can be seen from noting that for the generated current, each carrier releases one photon of recombination radiation. But the recombination radiation is limited by thermodynamics, and cannot be greater than the blackbody radiation. This is equivalent to an assumption that outgoing radiation is equal to the Planck spectrum, with an emissivity of 1 for photons of energy greater than the bandgap, and emissivity of 0 for photons of energy less than the bandgap. The maximum current I is then qN, where N is the number of photons emitted, found by integrating the number of photons from the Planck blackbody spectrum, from energy of the bandgap to infinity. Thus, this current I_o is the same as the current of an (ideal) thermophotovoltaic cell of the same bandgap operating at the same temperature.

Since the number of photons in the blackbody spectrum is proportional to T^3, and the voltage at the maximum power point proportional to T, the total power output is proportional to T^4. Thus, although the efficiency has only a minor dependence on temperature, the power output depends strongly on temperature.

C. Theoretical Efficiency

In the detailed balance model, the only efficiency loss is the energy in the recombination photons radiated to space. This allows a quick calculation of the maximum possible efficiency.

A minority carrier electron will have, from the Boltzmann[*] distribution, an average thermal energy $^3/_2kT$ above the conduction band edge. This will recombine with a hole that has an average energy of $^3/_2kT$ below the valence band.

Thus, the average radiated photon will have an energy of E_g+3kT for each carrier pair recombined.[†] From this, we see that the maximum possible efficiency depends on bias voltage V:

$$\eta_{ideal}= qV/(qV+E_g+3kT) \qquad (5)$$

In the previous section, it was shown that at the maximum power point, the bias voltage is kT/q, so the ideal efficiency at maximum power point bias is:

$$\eta_{(max\ power)} = kT/(E_g+4kT) \qquad (6)$$

Since E_g cannot be less than zero, this implies a maximum efficiency of 25% at the maximum power point. More typically, E_g is ~kT, so for this bandgap the maximum possible efficiency at maximum power would be 20%.

It should again be emphasized that the maximum power point is not the maximum efficiency point, and higher efficiency could be achieved at a higher negative bias, although with a lower power output. This is because the energy generated per injected carrier is proportional to the voltage, while the power per carrier lost to radiation is independent of the voltage. Thus, for example, a bias voltage of 2kT would produce an ideal efficiency of $2kT/(E_g+5kT)$, with a maximum ideal efficiency of 40%, while only reducing the power by a factor of 2/e, 74%. Beyond this, though, the exponential decrease in current at high negative bias decreases power rapidly.

However, it is unlikely that real-world thermoradiative cells would operate to the left of the ideal maximum power point, because other losses that are independent of bias voltage will drive the actual maximum efficiency point to lower (negative) voltages, as discussed in the next section.

This efficiency limit assumes that all the recombination radiation escapes the cell. It would be possible to put selective reflectors on the surface to limit the outgoing radiation spectrum, in the limiting case, only allowing outgoing radiation at exactly the bandgap energy. In this case, the energy of the recombination radiation at energies greater than E_g is recaptured by the cell, and does not contribute to the efficiency loss, at the cost of lowering the current produced (and hence the power) proportionately to the number of photons reflected.

III. EFFICIENCY LOSSES

The power calculated from the detailed balance limit of an ideal diode does not incorporate losses due to mechanisms that are not intrinsic to the operation. Several of these mechanisms have been addressed by previous analyses, for example, the effect of non-radiative recombination [6], such as Shockley-Read-Hall (SRH) recombination or Auger recombination, reduce efficiency. This is because every pathway for minority carriers to recombine releasing heat must also have the reverse pathway for heat to thermally excite carriers from the majority-carrier into the minority-carrier band. Since the density of majority carriers is by definition greater than the density of minority carriers, this thermal generation current will necessarily be greater than the recombination current. Since this thermal current is in the opposite direction from the power generation, it will reduce efficiency.

Efficiency is calculated as the amount of electrical power produced divided by the total power input, where the total power includes both the generated power as well as the power radiated to space. The power radiated to space includes the band-to-band recombination radiation, accounted for in the ideal model, but also includes all other sources of thermal emissivity, which will be referred to as "parasitic emissivity".

A. Parasitic Emissivity

Although we have assumed so far that the external environment is at a temperature low enough that radiation from the environment has a negligible effect on the performance, external sources will not be the only source of thermal radiation. Outside of semiconductor itself, all materials in contact with the cell will be potential sources of radiation. This includes the front surface encapsulation, contact metallization, and any window layers, as well as the rear surfaces and rear contact metallization.

While metals typically have high reflectivity and hence low emissivity in the thermal infrared band, the emissivity is not zero, and even relatively low thermal emissivity can have a significant effect on efficiency.

Emission is shown in figure 4, where red arrows indicate long wavelength emission, and blue arrows shorter radiation.

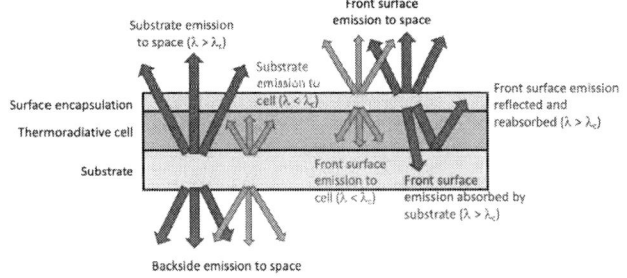

Fig. 4. Thermal emissivity from front and rear surfaces

A critical parameter is whether the parasitic emission is at wavelength greater than, or less than, the bandgap cut-off wavelength λ_c, the wavelength below which the semiconductor material absorbs light. This is related to the bandgap E_g by the

---

[*] Boltzmann distribution is used for convenience. Far from the Fermi level, the Fermi-Dirac distribution is indistinguishable from the Boltzmann distribution to the level of approximation here.
[†] The minority carriers are not necessarily in thermal equilibrium with the lattice, since they are radiatively coupled to the external heat

sink, and hence may have thermally relaxed to a temperature less than T. In the analysis of efficiency, this is not important, since the energy loss to space is the same regardless of whether it is done in a single recombination, or in thermal relaxation followed by recombination.

Einstein relation:

$$\lambda_c = hc/E_g \qquad (7)$$

where h is the Planck constant and c the speed of light. For a simplified analysis, a reasonable assumption is that all of the radiation of $\lambda < \lambda_c$ that passes into the semiconductor is absorbed, while the semiconductor is transparent to radiation of $\lambda > \lambda_c$ passes through the cell. (A detailed analysis would incorporate a wavelength-dependent absorption constant.)

The emissivity at $\lambda < \lambda_c$ that is absorbed by the active region of the cell creates a thermophotovoltaic current which subtracts from the thermoradiative current and hence reduces power. This results in a power loss that is directly proportional to the bias voltage. The effect of this on output power is shown in figure 5.

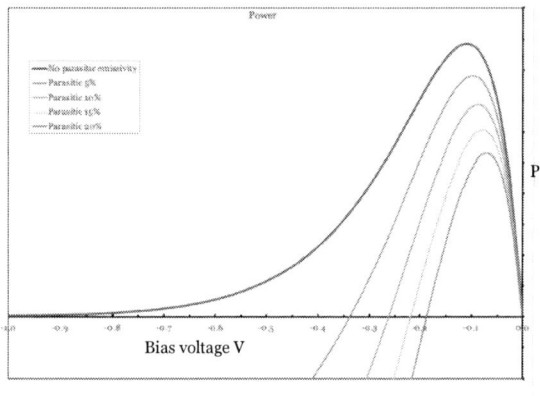

Fig. 5. Power output as a function of bias voltage for a thermoradiative cell as 1000°C, showing the effect of parasitic emissivity at wavelengths shorter than λ_c, ranging from 5% to 20% emissivity. The power output decreases and the maximum power point moves to lower (negative) voltage.

On the other hand, emissivity at wavelengths longer than λ_c, which does not create light-generated pairs in the diode, and emissivity which is directed outward, rather than toward the diode, will results in a loss that is independent of the operating point. This will reduce the efficiency, since it represents heat energy that is not converted into electrical power. This does not change power output if the temperature is constant. For the case of constant thermal power input, the increased radiative cooling will decrease the operating temperature, resulting in reduced power.

Therefore, not merely the total thermal emissivity, but the thermal emissivity as a function of wavelength is of interest.

B. Thermal Loss to Space

Parasitic emissivity is proportional to the blackbody emission times the thermal emissivity of the material, ε. In general, ε will depend on wavelength.

Parasitic emissivity sources include emissivity of the substrate, emissivity due to defects, free-carrier emissivity, and emissivity of the front surface, including contact metallization, surface coatings of the semiconductor, and encapsulation (if used).

C. Parasitic Emissivity: rear and front surfaces

For thermal losses, the emission from the rear and the front surfaces of the thermoradiative cell is of particular concern.

The thermoradiative cell will be on a substrate, which provides the thermal contact to the heat source, the back-surface electrical contact, and serves as mechanical support. Since the substrate will be at the same or higher temperature as the diode, the blackbody peak for the thermal emission will be close to the bandgap wavelength λ_c of the thermoradiative diode. The semiconductor is assumed transparent to portion of the substrate thermal emission at wavelengths $\lambda > \lambda_c$ and so at these wavelengths, radiation from the substrate will transmit through the semiconductor to space, and must be accounted for as thermal loss. As noted above, the portion of the thermal emission at shorter wavelengths is absorbed by the semiconductor, and hence is not a thermal loss, but subtracts from the thermoradiative current, reducing the power. Emission from the rear side of the substrate escapes to space regardless of wavelength, as shown in figure 4.

Since the back surface metallization coverage is typically much larger than the front surface metallization, and the cell will be on a substrate, a low emissivity back contact or use of a reflective interface layer between the substrate and the cell is critical, using materials with low absorption (and emission) near the cut-off wavelength. Thus, like thermophotovoltaic cells, a back surface reflector (BSR) is desirable. The reflectivity of the back-surface reflector on a thermoradiative cell must be very high. For example, a back-surface reflector of 90% reflectivity near the band edge would result in 10% absorption, and hence thermophotovoltaic carrier generation, that could swamp the thermoradiative current.

On the side not facing the semiconductor, to minimize radiative losses, the rear surface of the substrate must be a low emissivity surface, or else insulation added to prevent it from losing heat, for example, by multi-layer insulation (MLI).

The infrared transparency of front surface coatings of the cell and the encapsulation to protect it from the environment is also of concern. Encapsulation can be a significant issue: conventional solar cell encapsulation techniques using glass, for example, will block the outgoing recombination radiation, since glass is opaque to the infrared wavelengths radiated by the cell. The materials used to protect thermoradiative cells from environmental degradation should be infrared transparent.

Thermal emissivity from the top surface (i.e., the encapsulation) is also shown in figure 4. Emission from the front surface of the encapsulation escapes directly to space, while the rear side of the front encapsulation will emit back to the thermoradiative cell, where the portion with wavelength $\lambda < \lambda_c$ is absorbed

D. Losses by conduction

In addition to losses by radiation, an array will require structural attachments as well as conductors to transfer the generated current to the user. All of these will represent conductive losses of thermal energy. Applications

Thermoradiative conversion could, in principle, be used for energy conversion for any thermal source. The original proposals [1,2] looked at terrestrial applications. However, the Earth's atmosphere is not IR transparent, and so effective temperature of the heat sink is at best 200K (for a clear night sky with low humidity), and in most applications much higher. This results in a downward infrared flux which cancels out much of the upward radiation driving the cell current. Earth also has a surface environment in which convective cooling competes with radiative cooling for heat transfer, making it a non-ideal environment for thermoradiative energy conversion.

Space applications are well suited for this technology [8-10], since with no atmosphere the cell radiates to the heat sink of deep space. The heat source would be a radioisotope (*e.g.*, Pu-238) or nuclear reactor.

Space applications to bodies with atmospheres (*i.e.*, Venus, Mars, Titan) are, like Earth, not ideal. Again, atmospheric convection will be significant source of heat loss, and the infrared opacity of the atmospheres means that the cold side temperature will be the atmospheric temperature, not the temperature of deep space.

For bodies with no atmosphere, such as the lunar surface, the radiating surface must be in the dark, as well as shielded from other sources of infrared, such as the Earth, and the hot lunar surface.

In this application, the technology competes with other energy conversion technologies, such as heat engines, thermoelectric, and thermoradiative conversion. An interesting point is that for all other forms of thermal generation, the waste heat radiator is on the cold side of the system, and high radiator temperature decreases efficiency and power output. For thermoradiative cells, the radiator is the hot side of the system, and high radiator temperature increases power output without decreasing efficiency. Since thermoradiative cells operate best at high temperature, radiators run hotter, and hence smaller radiators can be used.[12]

IV. CONCLUSIONS

Thermoradiative cells represent a new concept for converting heat energy into electrical power using a semiconductor diode which radiates waste heat into space. A simplified model is presented in which the maximum power point and the ideal efficiency is calculated from the detailed balance limit.

The detailed balance efficiency neglects a number of real-world effects. In particular, it assumes that the emissivity of infrared to space comprises only band-to-band recombination of thermally-generated minority carriers. In real devices, a number of other sources of emissivity will reduce the heat, and will produce thermophotovoltaic carriers that reduce the output efficiency.

ACKNOWLEDGMENT

This work was supported in part by the NASA Glenn Center Innovation Fund. I would also like to acknowledge discussions with Prof. Jamie Phillips (now at University of Delaware), and work with Advanced Cooling, Inc., funded by the NASA Small Business Innovative Research (SBIR) program.

REFERENCES

[1] R. Strandberg (2015) "Theoretical efficiency limits for thermoradiative energy conversion," *Journal of Applied Physics, 117*, 055105; https://doi.org/10.1063/1.4907392

[2] P. Santhanam and S. Fan (2016) "Thermal-to-electrical energy conversion by diodes under negative illumination," *Physical Review B 93*, 161410(R) (2016)

[3] W-C. Hsu, et al. (2016) "Entropic and Near-Field Improvements of Thermoradiative Cells," *Scientific Reports, 6*, 34837; https://doi.org/10.1038/srep34837

[4] X. Zhang, *et al.* (2017) "Parametric design criteria of an updated thermoradiative cell operating at optimal states," *J. Applied Physics, 122*, 174505; https://doi.org/10.1063/1.4998002

[5] J. J. Fernández (2017) "Thermoradiative Energy Conversion With Quasi-Fermi Level Variations," *IEEE Trans. Electron Dev., 64*, No. 1, 250-255; doi: 10.1109/TED.2016.2627605.

[6] N. J. Ekins-Daukes, *et al.* (2020) "Generating Power at Night Using a Thermoradiative Diode, How is this Possible?" *47th IEEE Photovoltaic Specialists Conference*, 15 June - 21 August 2020. https://doi.org/10.1109/PVSC45281.2020.9300980

[7] S. Fan (2020) "Thermodynamics of light and its implication for harvesting solar energy and the coldness of the universe," *47th IEEE Photovoltaic Specialists Conference*, 15 June-21 Aug. 2020.

[8] J. Wang, C-H. Chen, R. Bonner, and W. G. Anderson (2019) "Thermo-Radiative Cell- A New Waste Heat Recovery Technology for Space Power Applications", *AIAA Propulsion and Energy Forum*, August 19-22, 2019, Indianapolis, IN.

[9] G. Landis, J. Wang and J. Phillips (2021) "Thermoradiative Conversion: Introduction and Mission Applications," *AIAA Propulsion and Energy Forum 2021*, August 9-11, 2021.

[10] G. Landis (2021) "Thermoradiative Cells: Photovoltaics in the Dark," *Space Photovoltaic Research and Technology 26*, NASA Glenn Research Center, Oct. 19-21 2021.

[11] M. Green (1982). "Dark Characteristics", section 4.6, *Solar Cells: Operating Principles, Technology and System Applications*, pp. 72-76, Prentice Hall.

[12] G. Landis, "Thermoradiative Arrays: a New Technology for Conversion of Heat into Electrical Power," *Nuclear & Emerging Technology for Space Conference*, Cleveland OH, May 8-12, 2022.

Estimation of Shade Losses in Unlabeled PV Data

Bennet E. Meyers, David J. F. Rodriguez

SLAC National Accelerator Laboratory, Menlo Park, CA, United States

Stanford University, Stanford, CA, United States

We provide a methodology for estimating the losses due to shading for photovoltaic (PV) systems. We focus this work on estimating the losses from historical power production data that are unlabeled, i.e. power measurements with time stamps, but no other information such as site configuration or meteorological data. This approach is appropriate for analyzing field production data from fleets of distributed rooftop systems and is highly automatic, allowing for scaling to large fleets of heterogeneous PV systems.

Contribution of Na+ from Glass to PID-s in Solar Modules: Na Migration in EVA

Jacob A. Clenney, Erick Martinez Loran, Guillaume von Gastrow, Tanguy Terlier, David P. Fenning, Rico Meier, Mariana I. Bertoni

Arizona State Univeristy, Phoenix, AZ, United States

University of California San Diego, La Jolla, CA, United States

SunPower Corp, San Jose, CA, United States

Rice University, Houston, TX, United States

Sodium induced shunting continues to be a challenging issue in crystalline Si solar modules. Potential-Induced Degradation of the Shunting type (PID-s) has been linked to Na, but the source is unclear. In this paper we evaluate the ion migration kinetics in encapsulant material under operational conditions. Analysis of Na migration profiles reveal the diffusivity constant and activation energy of Na in EVA. Implementing these results in breakthrough time simulations indicates that Na migration from the glass through EVA is too slow to account for experimentally observed PID-s degradation indicating contamination during production as the most likely source.

Importance of ideality factors in perovskite/Si tandem solar cell design

Benjamin Williams, Benjamin Daiber, Chris Case

Oxford Photovoltaics, YARNTON, United Kingdom

For perovskite-on-silicon tandem technology to achieve even higher efficiencies than have been demonstrated in recent work[1], it is paramount to identify fundamental electrical bottlenecks that are present in the state-of-the-art devices. One such bottleneck, an overly high diode ideality factor (IF) of the perovskite cell, limits the achievable Fill Factor (FF), and it is driven by non-favorable charge recombination characteristics within the perovskite bulk and interfaces. As a result, there have been recent efforts to quantify perovskite-on-silicon IFs using optical techniques[2], and earlier work evaluating different methods to accurately determine perovskite IF in single junction solar cells[3]. In this work we extend the importance of the IF to being a key variable for the design of tandem perovskite-on-silicon solar cells, due to its strong influence over whether tandem sub-cells should be current-matched or intentional current mismatched. Calculations are presented to quantify these links, and its implications on cell design discussed, in view of achieving record R&D cells, and how this may differ for real world applications. We ultimately conclude that minimizing the perovskite cell IF not only increases the achievable efficiency by >3% absolute, but also reduces the requirement to intentionally design mismatched tandems, thereby allowing for improved performance across locations and seasonally dependent spectra. We then evaluate different electrical methods to rapidly quantify sub-cell IFs so that it can be used as a tool to screen for cell improvements with statistically significant results and we discuss its use in terms of being able to fingerprint device regions that have excessive non-radiative charge carrier recombination. A number of case studies from Oxford PV devices are used throughout the work.

A deep learning approach to increase luminescence image resolution of solar cells

Priya Dwivedi, Robert Lee Chin, Thorsten Trupke, Ziv Hameiri

University of New South Wales (UNSW), Sydney, Australia

Luminescence imaging is an indispensable method to inspect solar cells and determine their critical electrical parameters. Low spatial resolution of images may limit the identification of small features and defects. On the other hand, high spatial resolution images often require expensive imaging systems. To that end, we present a deep learning approach to increase the spatial resolution of luminescence images with no additional hardware requirements. This study provides a simple and cost-effective method to enhance the capability of existing luminescence imaging systems using deep learning.

Agrivoltaics Using Bi-Facial PVs for Permaculture in Utility-Scale Projects

P.M. Jansson[1,2] (*Senior Member*), M.G. Newberry III[2], and S.M. Myers[2]

[1]Electrical and Computer Engineering Department, Bucknell University, Lewisburg, PA, USA
[2]Bucknell Center for Sustainability & the Environment (BCSE), Bucknell University, Lewisburg, PA, USA

Abstract— **As photovoltaic (PV) technology is rapidly becoming the most affordable, reliable, predictable and easily deployable electric generation technology of the modern world, we see continued growth in the scale and number of systems being deployed each year. The growth of market share in utility scale systems has rapidly eclipsed roof mounted over the past decade and is poised to become the most economic and environmentally friendly electric power source in the next few years. As this demand for utility scale system sites grows, developers turn increasingly to cleared lands and rich, agricultural lands to site their ever-larger solar farms. Recent press suggests that there is a growing resistance to PV systems, especially at these large, utility-scale project sizes when they take over farmland, creating a potential conflict with domestic food production. Our research attempts to demonstrate that there are multiple permaculture applications suitable under these large, utility-scale arrays which could potentially defuse this resistance by proving affordable permaculture production in these agrivoltaic settings. Based upon our spectroscopic analysis it is clear that certain crops will flourish better under bi-facial modules in utility-scale agrivoltaic applications when compared to standard PV modules. It is our hope that through more research with our student teams, we will demonstrate the success of plant growth and increased yields under these bi-facial arrays and identify a host of plant species that will thrive in agrivoltaic settings. Species that take advantage of the altered spectrum, will be favored, and while labor increases in the planting and harvesting beneath these agrivoltaic arrays, that the increases in organic product output may justify financially the use of rich farmland for both solar and plant farming.**

Keywords— *agrivoltaics, bi-facial modules, utility-scale PV arrays, permaculture, organic farming, spectral analysis*

I. OVERVIEW AND MOTIVATION

Just last year the U.S. Lawrence Berkeley National Lab [1] reported that the photovoltaic "utility-scale sector added 14 GW (or 73%) of all new solar capacity of 2020. It was the year with the greatest utility-scale solar capacity expansion in the United States so far, representing a year-over-year growth of 65%." Further, their report stated that "Utility-scale solar accounts for 61% of cumulative solar capacity." This trend continues to dominate further into the future as shown clearly in Figure 1. Over the past decade we have seen large, utility-scale PV systems levelized cost of energy plummet. According to the same LBNL report "utility-scale PV's average LCOE (levelized cost of electricity) has fallen by about 85% since 2010, to $34/MWh in 2020. This is so very significant given that PJM's "Load-weighted average LMP (locational marginal price) for 2021 is $39.86/MWh." [2] PJM is the largest electrical interconnection in the U.S. and

we know from temporal analysis that higher loads during the

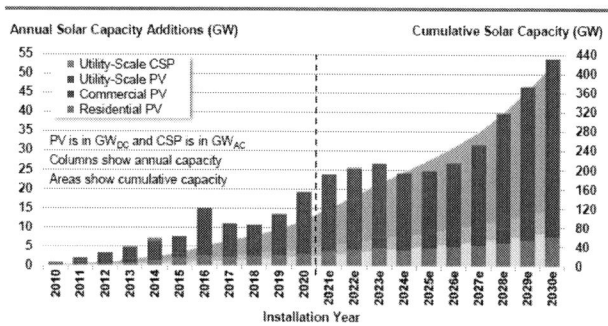

Fig. 1 – U.S. Photovoltaic Capacity Additions

daylight hours on the electric grid means that average prices are significantly higher when photovoltaic systems are generating than during the night when they are not. Figure 2 demonstrates that not only have large scale systems become the least costly over the past two decades, but they continue to decline in real terms assuring that they will become (if they are not already) the lowest cost and lowest carbon source of electricity for the U.S. grid. [3] And since "Installed prices in the United States are more than double those in most other countries, as reported by IRENA" [3] it is likely that these downward trends will continue for large scale systems.

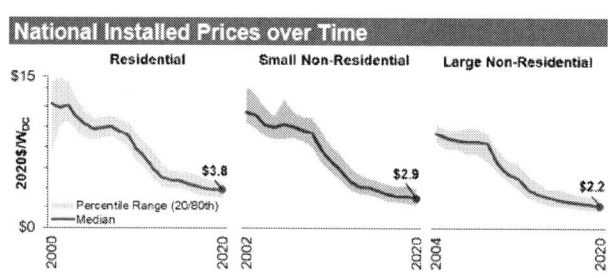

Fig. 2 – U.S. Installed Photovoltaic Costs [3]

However, "Large-scale development of solar-generated electricity is hindered in some regions of the U.S. by land use competition and localized social resistance." [4,5] Much of that resistance comes from the fact that areas often ideal for farming are also excellent for solar-farming. [6] A host of researchers before our team have successfully grappled with the challenges and competition between generating food and clean electricity [7-13] and have agreed that agrivoltaics provide a solution: "The potential for dual-use, agrivoltaic systems may alleviate land competition or other spatial constraints for solar power development, creating a

978-1-7281-6118-1/22 $31.00 © 2022 IEEE

significant opportunity for future energy sustainability. Global energy demand would be offset by solar production if even less than 1% of cropland were converted to an agrivoltaic system." [6] The many benefits of agrivoltaics include: increased land productivity by up to 70% [7], reduction in water needs for crops [8], protection from heat stress and drought [9], an over 30% increase in economic value from farms deploying agrivoltaic systems instead of conventional agriculture [10], potential for year round cultivation in cold climates [11], optimal spectral sharing between plants and bi-facial modules [12], and the creation of smart greenhouse technology using agrivoltaics to optimize plant microclimates [13]. Our work at Bucknell University began with the engagement of student engineers in efforts to maximize some of these benefits in the context of very large-scale arrays and using our sustainable gardens and university farm as potential demonstration sites.

II. INITIAL SPECTRAL ANALYSIS (2020)

In 2019 students working with the Center for Sustainability and the Environment and the University Farm at Bucknell began their investigations into the compatibility of bi-facial photovoltaic modules with permaculture. [14] Using the UPRTek PG100N Handheld PAR Meter [15], a spectrum analysis was done under differing environments: under bifacial PV modules, under traditional PV modules, and under full sun. What was observed was that the spectrum requirements of many agricultural plants were compatible with that available under many bi-facial modules (see Figures 3 and 4). The bi-facial modules tested (near edge) provides plants with a light distribution more optimal than direct sun.

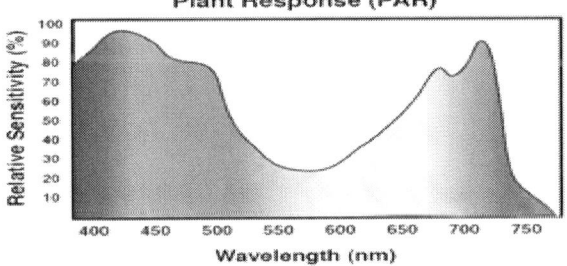

Fig. 3 – Plant Optimal Response

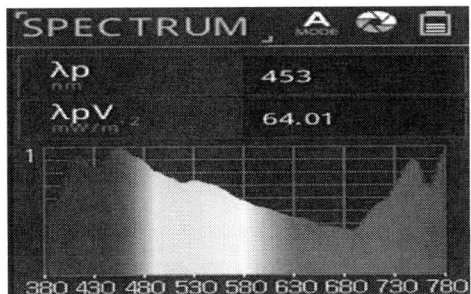

Fig. 4 – Under Bi-facial PVs (near edge)

There is a significant (though not complete) attenuation of light intensity in the blue spectrum when readings are taken directly beneath the bi-facial array – away from all module edges (see Figure 5).

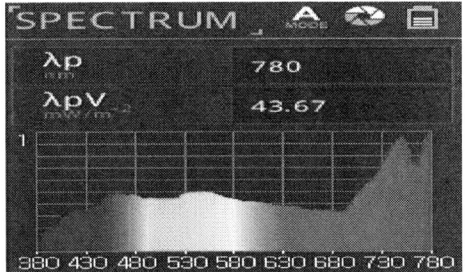

Fig. 5 – Under Full Shade of Bi-facial PVs

III. CURRENT SPECTRAL OBSERVATIONS (2021-22)

Recently, our team continued our investigations of permaculture methods that are compatible with bi-facial photovoltaic modules with the collection of empirical evidence for comparison with the 2019 data. Using the UPRTek Handheld PAR Meter [15], a spectrum analysis was performed within the BCSE's Living Greenhouse as well as under a traditional utility-scale (1.6 MW-DCp) PV module array (e.g., the 7-acre site of our university's new solar farm on campus). Both sites have potential for plant propagation pending the plant genera. We observed that the Living Greenhouse met spectrum needs of many agricultural and agronomical plants used in small-scale and large agricultural operations (see Figure 6).

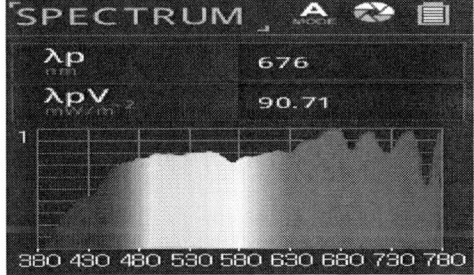

Fig. 6 – Inside BCSE Living Greenhouse

The observations made under the traditional PV modules at the large-scale (1.6 MW-DCp) solar farm recorded spectrum results (see Figure 7) less optimal for plant growth. With a significant loss of red-light spectrum (i.e., freq. > 630nm).

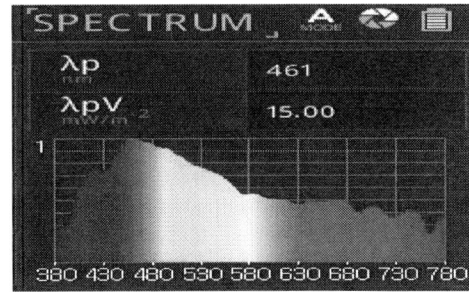

Fig. 7 – Under Tradition Utility-Scale PV Array

Although, this environment provided a light distribution that would affect specific physiological responses in plants [16] it can serve as beneficial to some plant genera [16,17]. When comparing Figures 4 and 7 of our preliminary spectral results we see near-edge bifacial light is of a range and quality similar to optimal plant response (Figure 3) and not lacking in red-light frequencies demonstrated under standard utility-scale PV arrays. While blue-light attenuation occurs under full shade with the bifacial module we tested, we believe that pure glass on glass bi-facials (without selective rear coatings may limit this attenuation. Crop plants exhibit higher relative action from the frequencies above 550 nm. (Figure 8) [18].

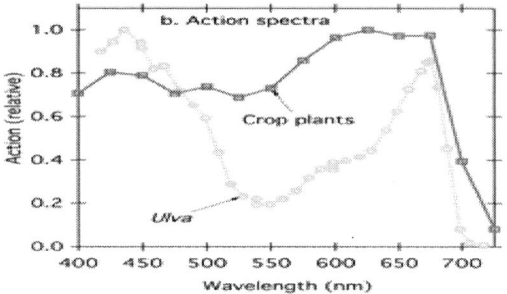

Fig. 8 – Crop Plants' Action Spectra

IV. PLANTS FAVORING RED/BLUE LIGHT

Our review of the literature shows that other research teams have now demonstrated a host of crops (plants, fruit bushes, etc.) compatible for permaculture and have growing heights that are also suitable for under-array farming in utility-scale agrivoltaic settings (See Table 1).

Table 1 – Spectral Frequencies Optimal by Plant Type

Light Color (Frequency Range)	Plants Favoring the Frequencies
Far red (>700 nm)	N/A
Red (600 – 700 nm)	Corn, cucumbers, kale, lettuce, radish, red lettuce. spinach, sweet peppers, tomatoes, wheat
Blue-Green (420 – 540 nm)	Alfalfa, bilberries, blueberries, Chinese cabbage, cucumbers, raspberries, rice, strawberries
Shade (predominantly blue-green light with short periods of red light)	Allegheny serviceberry, beets, carrots, cauliflower, chokeberry, garlic, onions wild ginger

V. PLANNED STUDENT RESEARCH

The next steps with our student research teams are to design/build experimental greenhouses in the backyard of the Sustainable Experiential Learning Laboratory Site (i.e., the "SELL Site") with bi-facial PV modules mounted as the roof structures. Students will grow specific crops within these greenhouses and determine the effect of the various bi-facial modules on plant growth, vitality and yield. A farm on campus will serve as the control for the experimental bi-facial greenhouse plots. Since the university has access to our large-scale array, we also intend to plant experimental plots of crops under these panels to serve as a comparison with the other experimental plots. We plan to have key partners come together to facilitate the use of the farm as a control site and the solar farm as one experimental area with the small bi-facial greenhouse kits as the second experiment. We will instruct our students how to grow specific crops under the traditional PV modules and within the greenhouses and monitor environmental conditions over time to determine growth rates (e.g., stem and leaf size, yield, if applicable).

VI. ACKNOWLEDGMENTS

The authors wish to thank our Center for Sustainability and the Environment (BCSE) and the William Corrington Renewable Energy Fund without whose support and funding this work would not be possible.

VII. REFERENCES

[1] M. Bolinger, et al., "Utility-Scale Solar, (2021 Edition) Empirical Trends in Deployment, Technology, Cost, Performance, PPA Pricing, and Value in the United States", Lawrence Berkeley National Laboratory, October 2021

[2] PJM Market Report – December 2021 [online] Available at: https://www.pjm.com/-/media/committees-groups/committees/mc/2021/20211213-webinar/20211213-item-06a-markets-report.ashx

[3] G. Barbose, et al., "Tracking the Sun: Pricing and Design Trends for Distributed Photovoltaic Systems in the United States (2021 Edition)", Lawrence Berkeley National Laboratory, September 2021

[4] A.S. Pascaris, et al., "Integrating solar energy with agriculture: Industry perspectives on the market, community, and socio-political dimensions of agrivoltaics", Energy Research & Social Science 75 (2021) 102023

[5] J. Brooker, 'The country's biggest solar farm is coming to one of the coal friendliest states", Grist, Jan 5, 2022, online [Available] https://grist.org/energy/the-countrys-biggest-solar-farm-is-coming-to-one-of-the-coal-friendliest-states/

[6] E.H. Adehl, et al., "Solar PV Power Potential is Greatest Over Croplands", Scientific Reports, Nature Research, 07 Aug 2019.

[7] A. Weseleki, et al. "Agrophotovoltaic systems: applications, challenges, and opportunities. A review", Agronomy for Sustainable Development (2019) 39: 35 https://doi.org/10.1007/s13593-019-0581-3

[8] S. Parkinson and J. Hunt, "Economic Potential for Rainfed Agrivoltaics in Groundwater-Stressed Regions" Environ. Sci. Technol. Lett. 2020, 7, 525−531

[9] B. Willockx, et al., "Theoretical potential of agrovoltaic systems in Europe: a preliminary study with winter wheat", 2020, 47th IEEE Photovoltaic Specialists Conference (PVSC)

[10] H. Dinesh and J.M. Pearce, "The potential of agrivoltaic systems", Renewable and Sustainable Energy Reviews 54 (2016) pp. 299–308

[11] M. Hernandez Velasco, "Enabling Year-round Cultivation in the Nordics-Agrivoltaics and Adaptive LED Lighting Control of Daily Light Integral", Agriculture 2021,11, 1255. https://doi.org/10.3390/agriculture11121255

[12] R.A. Vijayan, et. al, "Optimizing the spectral sharing in a vertical bifacial agrivoltaics farm", J. Phys. D: Appl. Phys. 54 (2021) 304004 (7pp)

[13] M.A. Minandal, et al., "Design and Simulation of Smart Greenhouse for Agrivoltaics Microclimates Optimization", 2021 International Symposium on Electronics and Smart Devices

[14] K. Fox, P.M. Jansson, et al., "Light Quality Analysis of Bifacial Photovoltaic Module Shading and a Potential Integration with Agrivoltaic Systems", [Poster], 7th Sustainability Symposium, Bucknell University, 2019

[15] UPRTek PG100N Handheld Spectral PAR Meter, online [Available] https://www.uprtek.com/en/product/spectrometers/pg200n

[16] L. Zoratti, et al., "Modification of Sunlight Radiation through Colored Photo-Selective Nets Affect Anthocyanin Profile in *Vaccinium spp.* Berries", PLOS ONE 2015, 10, 8, 1-17

[17] I. G. Tarakanov, et al., "Effects of Light Spectral Quality on the Micropropagated Raspberry Plants during Ex Vitro Adaptation", Plants, 2021, 10, 2071, 1-24

[18] H.L. Gorton, "Biological Action Spectra", Online [Available], http://photobiology.info/Gorton.html

Dedicated cold-climate field laboratory for photovoltaic system and component studies: the Michigan Regional Test Center as a case study

Laurie Burnham[1], Daniel Riley[1], Bruce H. King[1]. Jennifer Braid[1], Paul Dice[2], Ana Dyreson[2], William Snyder[1], Christopher Pike[3]

[1]Sandia National Laboratories, Albuquerque, NM 87185, [2]Michigan Technological University, Houghton, MI 49931, [3]University of Alaska Fairbanks, Fairbanks, AK 99775

Abstract—Snow and ice accumulation on photovoltaic (PV) panels is a recognized—but poorly quantified—contributor to PV performance, not only in geographic areas that see persistent snow in winter but also at lower latitudes, where frozen precipitation and "snowmageddon" events can wreak havoc with the solar infrastructure. In addition, research on the impact of snow and cold on PV systems has not kept pace with the proliferation of new technologies, the rapid deployment of PV in northern latitudes, and experiences with long-term field performance. This paper describes the value of a dedicated outdoor research facility for longitudinal performance and reliability studies of emerging technologies in cold climates.

Keywords—component, formatting, style, styling, insert (key words)

I. INTRODUCTION

PV installations are proliferating rapidly at northern latitudes, a trend driven by the rapid drop in cost per installed watt, solar-friendly policies and generally cooler operating temperatures that increase conversion efficiencies. Even in geographic areas that were once considered marginal for solar, solar now accounts for the majority of newly installed generating capacity. But longitudinal field data for modules and systems exposed repeatedly to extreme combinations of snow, wind and temperature loading are lacking. In addition, the quantification and prediction of energy losses attributable to snow shading of PV systems remains rudimentary, despite the impact of such losses on levelized-cost-of-energy (LCOE) calculations and insurance rates.

The increase in frequency and severity of extreme weather events, including record-breaking snowfalls, underscores the importance of field studies. The fact that snow can adhere to PV arrays for days at a time introduces significant uncertainty regarding both resource availability and average energy yields [1]. Another, less-well understood, issue is the impact of extreme winter weather on component reliability. Few data exist, for example, to identify the boundary conditions under which heavy snow and ice loads deform or destroy the integrity of module frames and racks; even less is known

about the invisible and long-term damage to solar-cell integrity. These unknown concerns are magnified in the context of technological evolution and the rapid proliferation of technologies that range from thinner and larger solar cells to new architectures and materials to jumbo-sized modules. What is known, however, is that the risk of extreme weather, including heavy seasonal snowstorms, which can drop three feet of more of snow in a single event [2] [3], is increasingly factored into the risk calculations of insurers and investors: insurance rates for renewable energy projects are seeing steep price increases (as much as 400% in the last two years) and some insurers are refusing to cover cell cracking, at any price [4]. Field sites with high-fidelity monitoring instrumentation can help by providing real-world data needed to inform financial models and attract long-term equity capital, reduce insurance risk, and make better technological and design choices.

II. IMPORTANCE OF FIELD DATA FOR NORTHERN LATITUDES

Simulations and accelerated testing protocols, while excellent at identifying early failure modes, do not adequately capture the complexities and realities of fielded PV systems, regardless of the operating climate. They especially fall short for geographic regions that see persistent snow in winter, where the multiple variables that influence snow shading and snow shedding, including snow load that varies in depth, weight and spatial distribution, are hard to accurately measure. In addition, the uneven distribution and partial shading of both modules and rows results in energy losses that are hard to measure. From a reliability perspective, the non-uniform shedding of snow from a tilted module can impose uneven stress on a module not captured by any model. What is more, simulations cannot fully capture encapsulant behavior, including its increased brittleness, at low temperatures [5], or how the combination of extreme cold and snow load can differentially stress a module's components and transfer that stress to the solar cells and interconnect circuitry [6]. Yet accurate predictions of the performance of photovoltaic systems based on models that assume typical meteorological year data are the norm and basis for return-on-investment calculations.

III. WINTER-SPECIFIC PERFORMANCE AND RELIABILITY CHALLENGES

In addition to the above challenges, research on the impact of snow and cold on PV systems has not kept pace with the proliferation of new technologies. Studies comparing shedding rates for modules at different tilt angles and orientations have been conducted but are of limited duration and impact [7]. And while some data exists on bifacial energy gain in winter [8], performance indices for emerging bifacial technologies (e.g., large-format modules, M12 cell size, transparent backsheets) are unknown, and existing snow-shedding models do not take into account key variables such as albedo, module orientation or the non-uniformity of snow. Investigations of cell cracking under thermomechanical load have established a general principle, but little is known about cell or polymer behavior in response to actual and uneven snow load or the prolonged low-temperature-exposure typical of a fielded installation. In addition, virtually nothing is known about how different module technologies or tracking systems handle snow and cold loading.

A strong case can therefore be made for a dedicated northern field laboratory, equipped with high-fidelity analytical and measurement tools to generate performance and reliability data that can inform:

- Module choice and the relative advantages/disadvantages from both a performance and reliability perspective of different architectural features, such as half-cut vs shingled cells; multiwire vs traditional three-busbar interconnects; 166mm vs 210mm cells.
- Validated performance models to reduce uncertainty in LCOE calculations.
- O&M protocols and best practices aimed at managing and mitigating snow losses.
- The commercial potential and bankability of new technologies specifically designed to improve system efficiencies in winter (e.g. snowphobic coatings [9].

IV. RATIONALE FOR A DEDICATED NORTHERN RESEARCH CENTER

Data are vitally important in this era of rapid solar innovation, with its outpouring of new materials and higher-efficiency cell and module designs, especially in northern regions, where little or no baseline reliability data exist and the prevailing climatic stressors, including combined snow, wind and cold loading, are not adequately captured by accelerated laboratory testing. Accelerated testing has value in identifying early failure modes, manufacturing defects and design flaws that might be climate-dependent. Multi-year field reliability testing under realistic climate conditions should be routinely integrated into the product-development cycle, especially for new cell types and module architectures [10]. Such testing should include high-quality monitoring of components and systems to enable comparative performance

assessments and to capture early failures as well as power output over time.

V. COLD CLIMATE SITE IN NORTHERN MICHIGAN

In 2021, the US Department of Energy (DOE) added a fifth site to its portfolio of Regional Test Centers [10], which enable the cross-climate field validation of emerging solar technologies and are managed by Sandia National Laboratories but operated by their site owners. The newly-commissioned Michigan site, located on Michigan Technological University property, supports demonstration studies and research specific to the cold-weather performance of PV components and systems. This site is a model of what an outdoor research site for the evaluation of PV technologies should be and is described here in some detail to raise awareness for its capabilities and enable its replication in other regions of the US and the world.

A. Site Description

The Michigan Regional Test Center (MI RTC) is located in Calumet, Michigan, 47°N/88°W, on the Keweenaw Peninsula with Lake Superior less than ten miles to the east and west. The site, which allows for cross-technological comparisons under identical and relatively predictable conditions, provides ideal field conditions for evaluating solar technologies in winter: snow and low temperatures here are both predictable and persistent: the average annual snowfall is 202 inches and can be as much 300 inches or more. In addition to heavy lake-effect snowfall, the temperature in this region of the Upper Peninsula is influenced by the site's proximity to Lake Superior (the lake's stored heat tends to keep the ambient air temperature highs around 20 F in the winter and 70 F in the summer; if the lake freezes over, temperatures drop routinely).

B. Infrastructure and Instrumentation

Like the other Regional Test Centers, the MI RTC has high-fidelity instrumentation to measure and track data (see Table 1) and also pre-installed grid-tied racking to support the efficient installation of experimental systems. Onsite measurements include module- and string-level DC and AC power, ground-based irradiance and other meteorological data, snow depth and snow load sensors and cameras to capture time-series image of snow coverage and snow-shedding rates. In addition, the MI RTC has a flexible design that allows for the installation of different module types and sizes in different orientations on fixed-tilt racks but can also support single- and dual-axis tracking systems. The site also has mock rooftops and other structures to support building-integrated PV studies.

C. Other Contributors to the Site's Successful Rollout

While the MI RTC is still evolving, several features--apart from its infrastructure and technical capabilities—help distinguish the site and are likely to be important to its long-term success:

- Partnership with Sandia National Laboratories that 000access to the expertise and technical capabilities of the largest national lab in the US, and

---

This work is funded in part or whole by the U.S. Department of Energy Solar Energy Technologies Office, under Award Numbers 34363 and 38527.

978-1-7281-6118-1/22 $31.00 © 2022 IEEE

opportunities for collaboration above and beyond the RTC.

- **Commitment by site owner to the mission.** The MI RTC is located on property provided by Michigan Technological University, is staffed by university employees and supports the research interests of faculty and graduate students. The engagement of members of the university with a vested interest in the site's impact and effectiveness has been an essential driver of the site's success.
- **Project Flexibility.** The site has an a*pplied focus* but supports other research activities and projects, with Sandia, and independent of Sandia, including industry- and government-funded investigations as well as student research.
- Positioned as a focal point for cold-climate data- and information-sharing
- Supports the standardization of monitoring systems and O&M protocols
- Expandable and adaptable to novel and future technologies

D. Current research activities at the MI RTC

The MI RTC supports numerous research projects aimed at reducing risk in the US solar sector, including:

- Thermo-mechanical loading as a factor in long-term reliability (longitudinal studies of cells exposed to repeated cold snow loading)
- Reliability and performance analyses of emerging technologies
- Design optimization and performance of bifacial systems
- Single-axis tracker optimization
- Multi-variate performance modeling
- Development and validation of an energy-loss estimation tool for partial snow coverage

VI. SUMMARY

Accurate predictions of the long-term performance and reliability of PV systems in regions that see persistent snow in winter are largely lacking. The MI RTC, which represents a collaboration between Sandia National Laboratorie0s and Michigan Technological University, provides 1) a technical platform, including high-fidelity instrumentation to enable the accurate valuation of prospective PV plants at northern latitudes; 2) data to support optimized designs and engineered solutions that increase PV performance and reliability in snowy regions, thereby reducing the levelized cost of energy (LCOE) and giving stakeholders greater confidence performance projections; and 3) longitudinal data on module and cell damage to support more robust module designs, inform standards, and lower insurance premiums.

Table 1. Meteorological Instrumentation at the MI RTC

Description	Manufacturer	Model
Met Tower		
• 10-meter tower	Campbell Scientific	UT30
• Wind speed and direction	RM Young	03002-L
• Humidity and temperature	Vaisala	HMP60-L
Irradiance Tracker		
Solar tracker	Kipp & Zonen	SOLYS2
Pyranometer, global	Kipp & Zonen	CMP 11
Ventilator	Kipp & Zonen	CVF 3
Pyranometer, diffuse	Eppley	8-48
Ventilator	Epley	VEN
Pyrheliometer	Kipp & Zonen	CHP 1
Rain Gauge, heated	Met One	375
Data Logger		
Logger enclosure	Campbell Scientific	ENC12/14-TM-ES
Data logger	Campbell Scientific	CR1000-XT-SW-NC
Ethernet interface	Campbell Scientific	NL120-XT-SW
Power supply	Campbell Scientific	PS200-SW
Barometric pressure	Vaisala	PTB110

ACKNOWLEDGMENT

This work is funded in part or whole by the U.S. Department of Energy Solar Energy Technologies Office, under Award Numbers 34363 and 38527.

REFERENCES

[1] Solar Generation Index, 2021

[2] Climate change and extreme snow in the US, NOAA National Centers for Environmental Information: https://www.ncdc.noaa.gov/news/climate-change-and-extreme-snow-us

[3] J. Lawrimire, T.R. Karl, M. Squires, D.A> Robinson, et al, Trends and variability in severe snowstorms east of the Rockies, J. Hydrometeorology, 2015, 15 (5): 1762-1777; https://doi.org/10.1175/JHM-D-13-068.1K.

[4] T. Sylvia, Insurance for renewable energy projects adapts to more frequent and destructive disasters; PV Magazine, 3/9/2021. https://pv-magazine-usa.com/2021/03/09/insurance-for-renewable-energy-projects-adapts-to-more-frequent-and-destructive-disasters/

[5] N.Bosco, M. Springer, and X. He, "Viscoelastic Material Characterization and Modeling of Photovoltaic Module Packaging Materials for Direct Finite-Element Method Input," IEEE Journal of Photovoltaics, vol. 10, no. 5, pp. 1424-1440, 2020. doi:10.1109/JPHOTOV.2020.3005086.

[6] H. Seigneur, E. Schneller, J. Lincoln, H. Ebrahimi, R. Ghosh, A. Gabor, M. Rowell and V. Huayamave, 2019, June. Microcrack formation in silicon solar cells during cold temperatures. In 2019 IEEE 46th Photovoltaic Specialists Conference (PVSC) (Vol. 2, pp. 1-6). IEEE. doi: 10.1109/PVSC40753.2019.9198968

[7] T. Townsend and L. Powers, "Photovoltaics and snow: An update from two winters of measurements in the SIERRA," in 2011 37th IEEE Photovoltaic Specialists Conference, 2011, pp. 003231–003236

[8] J. Stein, D. Riley, M. Lave, C. Deline, F. Toor and C. Hansen, "Outdoor field performance from bifacial photovoltaic modules and systems," Sandia report. SAND2017-10254C

[9] A. Dhyani, C. Pike, J.L. Braid, E. Whitney, L. Burnham, A.Tuteja, Facilitating Large-Scale Snow Shedding from In-Field Solar Arrays using Icephobic Surfaces with Low-Interfacial Toughness; Advanced Materials Technologies; DOI 10.1002/admt.202101032.

[10] H. Wirth, D. Moser, M. Topic, M. van Iseghem, U. Jahn, G. Adinolfi, et al. PVSC Research Challenges in PV Reliability, Secretariat of the European Technology and Innovation Platform for Photovoltaics, Oct. 28 2020; 7pp.

[11] https://energy.sandia.gov/programs/renewable-energy/photovoltaics/regional-test-cente

Efficient self-protected thin film c-Si solar cell against reverse-biasing condition: A simulation study

Omar M. Saif [1,2], Abdelhalim Zekry[1], Ahmed Shaker[3], M. Abouelatta[1], and Ahmed Saeed[4]

[1]Department of Electronics and Communications, Faculty of Engineering, Ain Shams University, Cairo, Egypt
[2]Department of Electronics and Communications, Engineering School, Canadian International College, Zayed Campus, Giza, Egypt.
[3]Engineering Physics and Mathematics Department, Faculty of Engineering, Ain Shams University, Cairo, Egypt
[4]Electrical Engineering Department, Future University in Egypt, Cairo, Egypt

Abstract—**In order to protect the solar panel against reverse biasing, bypass Schottky diodes are usually connected in parallel with a string of cells which makes the circuit more complex and costly. This study presents a novel thin-film crystalline silicon device design using physics-based TCAD simulations. In our proposed solar cell, a P$^+$ reverse conducting layer is introduced to form an N^{++}/P$^+$/P structure. The novel device structure provides two different functions depending on the type of applied bias. It operates as an efficient solar cell when operating at forward biasing while it acts as a backward diode when reverse biased. Therefore, such a proposed device is self-protected against the reverse current. Also, it can reduce circuit complexity and manufacturing cost.**

Keywords— self-protected, c-Si solar cell, reverse-conducting-solar cell, backward diode, conversion efficiency,

I. INTRODUCTION

Silicon solar cells have represented the dominant Photovoltaic (PV) technology in the dynamic photovoltaics industries for several decades [1]. That is because silicon is non-toxic in all its natural forms and has a bandgap within the optimal range for PV conversion [2]. Today, more than 60% of the PV module employs strings of crystalline silicon (c-Si) solar cells [3] connected in series to generate the desired output power. A large number of series-connected c-Si cells makes the PV systems sensitive to the mismatch of cells. Mismatch conditions may appear due to malfunction of c-Si cells or partial shading conditions caused by nearby buildings, trees, or cloud movements. Thus, the faulty or shadowed cell gets reverse-biased and acts as a load to the illuminated strings and hence electric power is dissipated. As a result, the temperature of these cells increases and causes localized overheating or hot spots [4]. Hence, a mismatch condition not only impacts the solar cell conversion efficiency and reduces its performance but also adversely affects the reliability of the PV module [5].

One possible solution to the mismatch condition was to connect a bypass diode in parallel with each cell [6]. This approach was not favorable for the c-Si solar cells as it required a considerable high cost and could be even detrimental in terms of power production when many bypass diodes are activated because of their power consumption [7]. Recently, the main protection method against reverse biasing conditions is a passive bypass diode that is placed in parallel with a string of PV cells [8, 9]. In addition, recent studies focus on mitigating the impact of hot spots and increasing the reliability of the PV module by modifying the bypass (protection) circuit [10-13]. All these studies increase the number of components in the protection circuit that may consume more power and increase the complexity and the overall PV module cost. So, finding an innovative way, not only to optimize the conversion efficiency of c-Si solar cells but also to overcome the reverse biasing condition, is extremely important for future PV technology.

This work presents a state-of-the-art design for a self-protected thin-film c-Si solar cell by introducing a P$^+$ reverse conducting layer (RCL). At forward biasing, the device operates as an efficient solar cell while it operates as a backward diode when applying reverse biasing. Thus, no need for bypass diodes or the modified protection circuit. A physics-based TCAD simulation of our proposed cell is carried out. The performance of such a novel device is compared with the performance of a normal design without RCL.

II. DEVICE STRUCTURE

The construction of the state-of-the-art single-junction self-protected thin-film c-Si solar cell is shown in Fig. 1. It is an N^{++}/P$^+$/P structure with full back metallization and partially front metalized contact to allow the passage of light to the cell while collecting the cell current. The typical construction parameters of the single-junction thin-film c-Si solar cell are: p-layer substrate thickness T_P = 20 µm, thickness of the Reverse-Conducting p-layer T_{PRCL} = 12 nm, Emitter layer (n^{++}-layer) thickness T_N = 100 nm. The doping concentration of p-layer substrate is selected to be 1e17 /cm^3 [14], and the initial doping concentration of the P$^+$ RCL is 1e19 cm^{-3} while the doping concentration of the emitter is selected in such a way to allow reaching the solid solubility limit [15].

The RCL plays an important role in determining the functionality of the device when applying forward bias and small reverse bias as will be shown thereafter. The back surface field (BSF) layer with a basic high doping concentration of 5e18 cm^{-3} and thickness T_{BSF} of 1000 nm is introduced to passivate the rear surface. The interface between the base and the BSF layers behaves like a p-n junction. An electric field forms at the interface, which introduces a barrier to the electrons flow to the rear surface. So, the electron concentration is maintained at a higher level and the BSF has a net effect of passivating the rear surface [16]. That in turn has a direct effect on increasing the short-circuit-current, J_{SC}, and the open-circuit voltage, V_{OC}, for the thin-film c-Si solar cell [17] and hence, increasing its conversion efficiency. Finally, the cathode electrode is designed to minimize the shadowing effect.

978-1-7281-6118-1/22 $31.00 © 2022 IEEE

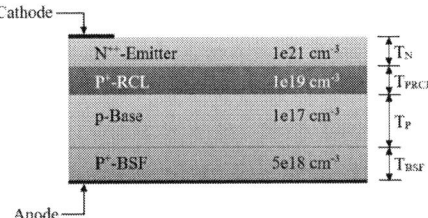

Fig. 1. Schematic diagram of a novel self-protected thin-film c-Si solar cell with a reverse conducting layer

III. TCAD SIMULATION MODEL

In this work, all numerical simulations are carried out using Silvaco TCAD Tools. All electrical simulations are performed using drift-diffusion transport models where electrostatic Poisson's equation and continuity equations are solved for both electrons and holes. For all simulations, the Shockley-Read-Hall recombination model with doping dependence (CONSRH) is enabled. In addition, for silicon, Auger recombination is notable around 1e19 cm^{-3} or higher [18]. Also, in the presence of these heavy doping regions (>1e18 cm^{-3}) [19]; so, the bandgap narrowing effects are enabled in ATLAS using BGN model. Regarding the mobility models, FLDMOB model turns on the electric field dependent mobility while CONMOB model specifies the concentration-dependent mobility which is based on experimental results and valid for a temperature of 300 K. In order to model the forward and reverse tunneling currents of degenerately doped p-n junctions, the non-local band to band (BTB) tunneling model (BBT.NONLOCAL) is used [20]. For optical simulation, we adopted the transfer matrix method (TMM) to calculate the intensity distribution and photogeneration profiles in the solar cell under illumination AM1.5G solar spectrum. The reflectance and transmittance are computed by TMM in terms of the incident light properties such as intensity, wavelength, and the semiconductor properties such as complex refractive index. The refractive (n) and extinction coefficient (k) for c-Si that depends on wavelength are utilized [21].

IV. EFFECT OF REVERSE CONDUCTING LAYER

As mentioned earlier, the self-protected thin-film c-Si solar cell utilizes a P$^+$ RCL between the emitter and the base that has a significant impact on the device performance. In order to demonstrate this impact, we present the electrical and optical performance of two devices, the novel basic self-protected thin-film c-Si cell (novel design) shown in Fig. 1 and the same cell without RCL (normal structure) utilizing two biasing conditions, forward biasing and low reverse biasing.

Optically, the response of both novel design and normal design to the incident radiation is approximately the same as shown in Fig. 2a which illustrates the EQE for both novel and normal designs. It is clear that the area under EQE curve leads to approximate equal current density. So, it would be expected that both cells have the same short circuit current density.

Electrically, at forward biasing, V_{OC} and J_{SC} generated using a novel design are almost the same as those introduced by the normal structure (see Fig. 2b). This is because the shift of quasi fermi level toward the band edges is extremely the same for both novel and normal structures. That leads to equal contact

differences in potential. Hence, the conversion efficiency and maximum power transferred for novel and normal structures are almost the same. Thereby, the RCL has no adverse effect on the electrical performance of the solar cell at forward biasing. Table 1 summarizes the performance simulated results for both structures at the illuminated condition.

However, when applying a small reverse biasing, the novel design functions differently from the normal structure. Fig. 3a shows the performance of the device at both forward and reverse bias conditions. The I-V curve of the novel design at the reverse portion shows that there is heavy conduction that takes place at a few millivolts. The most important factor that determines the performance of the device when applying reverse biasing is the reverse voltage that corresponds to the short-circuit current (the ON-voltage of the backward diode). It is found from the I-V characteristics that the reverse ON-voltage of the novel device is extremely smaller than the ON-voltage of the best bypass diode (Schottky diode) which is around 0.4 V integrated on the PV panels. This results in a diode operating as a highly efficient rectifier over a limited voltage range but in the backward direction "backward diode". Thus, the novel self-protected thin-film c-Si solar cell shows the ability to mitigate the effect of the reverse bias voltage due to malfunctioning cells or partial shading and thereby the ability to prevent the overheating or hot spots in the solar cell.

Further, the novel design is a self-protected solar cell, i.e., each cell has its own protection to reverse current by the maximum value of short-circuit-current. This is another advantage of the novel design over the bypass diode as it protects the cell by itself, and the bypass diode protects the string of connected cells. Therefore, at reverse biasing and by using the novel design, the power loss will be due to malfunctioning cells only. On the other hand, by using the bypass diodes, the power loss will be due to the string of cells. Therefore, the power dissipated through the novel design is lower than the power dissipated when utilizing bypass diodes.

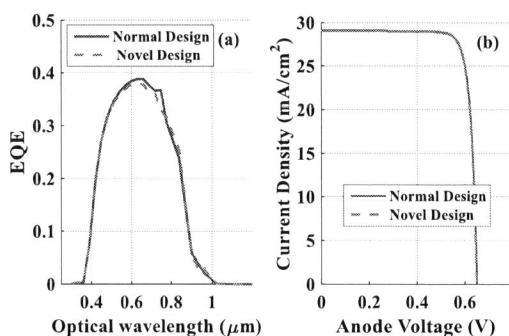

Fig. 2 Optical and Electrical performance of both Novel and Normal Designs at illumination (a) I-V characteristics and (b) external quantum efficiency.

TABLE I. COMPARISON BETWEEN THE PERFORMANCES OF THE NOVEL BASIC DESIGN AND THE NORMAL STRUCTURE

	V_{OC} [V]	J_{SC} [mA/cm^2]	FF [%]	PCE [%]	P_{max} [mW/cm^2]
Novel Design	0.651	29.07	83.26	15.77	15.78
Normal Design	0.652	29.06	83.5	15.8	15.8

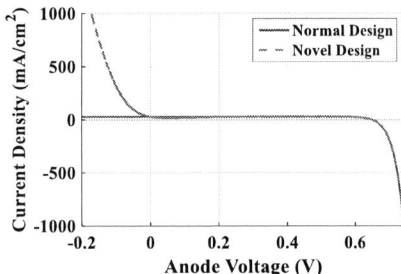

Fig. 3. Comparison between electrical performance for novel design and normal structure.

Next, we investigate the impact of the doping concentration and the thickness of the RCL on the performance of the novel solar cell. The most appropriate doping concentration and thickness are those at which the device operates with the highest efficiency in the forward condition while operating as a perfect backward diode in the reverse biasing condition. Fig. 4 shows that the best thickness between the selected range (10 nm to 100 nm) is 10 nm at which the highest efficiency could be obtained. However, at the reverse biasing and with the same thickness, the device does not operate as a backward diode. So, we select a thickness of 12 nm which is the minimum thickness of the reverse conducting layer that guarantees a backward diode when applying a reverse bias. In addition, the doping concentration is varied between 1e18 /cm^3 to 2e19 /cm^3. It is found that the novel device operates at reverse biasing as a backward diode when the doping concentration of RCL is greater than 7e18 /cm^3 to 1e19 /cm^3 while increasing the doping concentration over 1e19 /cm^3 makes the device act as a tunnel diode (Fig. 4b). As well, operating the device as a thin-film solar cell, the efficiency will increase by increasing the doping concentration of RCL.

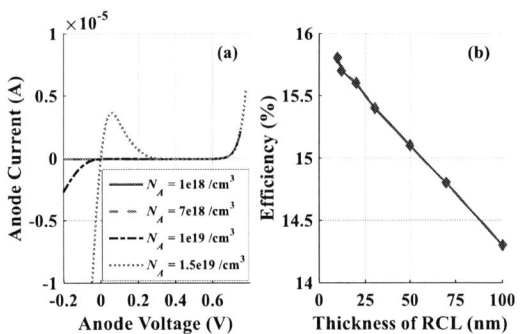

Fig. 4. Effect of doping concentration and thickness of RCL on the performance of the novel device.

CONCLUSION

TCAD modeling and simulation of a state-of-the-art self-protected thin-film c-Si device with a reverse-conducting P$^+$ layer is carried out. The introduction of the reverse-conducing layer has shown that the device operates as an efficient solar cell by the application of forward biasing. Meanwhile, by the application of the reverse bias, it generates a considerable high reverse current within extremely low ON-voltage. That is, the cell can protect itself against the reverse biasing without adding much more cost comparing with the available protection circuit in the market.

REFERENCES

[1] S. Bhattacharya, S. John, "Beyond 30% Conversion Efficiency in Silicon Solar Cells: A Numerical Demonstration". *Sci. Rep.* vol. 9, 2019.

[2] P. M. Ushasree, B. Bora, "CHAPTER 1 Silicon Solar Cells", Solar Energy Capture Materials, *The Royal Society of Chemistry*, pp.1-55, 2019

[3] Mechanical Engineering Industry Association, "International technology roadmap for photovoltaic", 11th ed, VDMA, Frankfurt, Germany, 2019

[4] A. Zekry, A. Shaker, M. Salem, "Solar Cells and Arrays", in "Advances in Renewable Energies and Power Technologies", 1st ed, 2018, pp.3–56.

[5] G.Oreskia, G.C. Eder, Y. Voronko, A. Omazic, L. Neumaier, W. Mühleisen, G. Ujvari, R. Ebner, M. Edler, "Reduction of PV module temperature using thermally conductive backsheets," *IEEE J. Photovolt.*, vol. 8, no. 5, Sep. 2018, pp. 1160–1167

[6] K. Chen, D. Chen, Y. Zhu, H. Shen, "Study of crystalline silicon solar cells with integrated bypass diodes", *Sci. China Technol. Sci.* vol. 55, no. 3, 2012, pp. 594–599.

[7] S. Daliento, L. Mele, E. Bobeico, L. Lancellotti, P. Morvillo, "Analytical modelling and minority current measurements for the determination of the emitter surface recombination velocity in silicon solar cells," *Sol. Energy Mater. Sol. Cells*, vol. 91, no. 8, 2007, pp. 707–713

[8] Z. Pezeshki, A. Zekry, (2021). "State-of-the-Art and Prospective of Solar Cells", in "Fundamentals of Solar Cell Design (eds Inamuddin, M.I. Ahamed, R. Boddula and M. Rezakazemi).

[9] S. Silvestre, A. Boronat, A. Chouder, "Study of bypass diodes configuration on PV modules", *Appl. Energy*, vol. 86, no. 9, 2009, pp. 1632–1640

[10] R. Witteck, M. Siebert, S. Blankemeyer, H. Schulte-Huxel, M. Köntges, "Three Bypass Diodes Architecture at the Limit," in *IEEE Journal of Photovoltaics*, vol. 10, no. 6, pp. 1828-1838, Nov. 2020

[11] S. Ghosh, V. K. Yadav and V. Mukherjee, "A Novel Hot Spot Mitigation Circuit for Improved Reliability of PV Module," in *IEEE Transactions on Device and Materials Reliability*, vol. 20, no. 1, pp. 191-198, March 2020

[12] M. Dhimish, V. Holmes, P. Mather and M. Sibley, "Novel hot spot mitigation technique to enhance photovoltaic solar panels output power performance," *Sol. Energy Mater Sol. Cells*, vol. 179, pp. 72–79, 2018

[13] B. K. Karmakar and A. K. Pradhan, "Detection and Classification of Faults in Solar PV Array Using Thevenin Equivalent Resistance," in *IEEE Journal of Photovoltaics*, vol. 10, no. 2, pp. 644-654, March 2020

[14] A. Zekry and W. Gerlach, "Reduction of the current gain of the n-p-n transistor component of a thyristor due to the doping concentration of the p-base," *IEEE Transactions on Electron Devices*, vol. 35, no. 3, pp. 365–372, 1988.

[15] A. Zekry, A. Shaker, M. Ossaimee, M. S. Salem and M. abouelatta, "A comprehensive semi-analytical model of the polysilicon emitter contact in bipolar transistors," *J Comput Electron*, vol. 17, pp. 246–255, 2018.

[16] J. G. Fossum, "Physical operation of back-surface-field silicon solar cells", *IEEE Transactions on Electron Devices*, vol. 24, pp. 322-325, 1977

[17] O. V. Roos, "A simple theory of back surface field (BSF) solar cells", *J. Appl. Phys*, vol. 49, pp. 3503-3511, 1978

[18] J. Dziewior and W. Schmid, "Auger coefficients for highly doped and highly excited silicon," *Appl. Phys. Lett*, vol. 31, no. 5, pp. 346–348, 1977

[19] J. W. Slotboom, "The PN Product in Silicon", *Solid State Electronics*, vol. 20, pp. 279-283, 1977

[20] Silvaco International, "ATLAS user's manual, Device Simulation Software," 2016

[21] SOPRA N&K Database [available online]. http://refractiveindex.info/. Accessed 03 Nov 2021

Dynamic Simulation of a Load-Matching Photovoltaic System for Green Hydrogen Production

Christian R. Polo
Arizona State University
Tempe, USA
CRPolo@asu.edu

William J. Parquette
Arizona State University
Tempe, USA
WParquet@asu.edu

Kelvin Tan
Arizona State University
Tempe, USA
KTan13@asu.edu

Meng Tao
Arizona State University
Tempe, USA
Meng.Tao@asu.edu

Abstract—**This paper presents the electrolytic application of a load-matching PV system to produce green hydrogen. The system has proven its viability with purely resistive loads, and a static analysis has shown the performance potential of the system for electrolytic applications. This paper focuses on dynamic simulation of the load-matching PV system for green hydrogen production in SIMULINK. It is shown that an over 99% energy transfer efficiency from the PV array's available energy to the electrolytic loads can be achieved under dynamic conditions for the system. The design parameters to optimize include the number of hydrogen cells per stack, the stack resistance, and the number of available stacks in the system. This system provides a simple but efficient approach for large-scale photovoltaic hydrogen production.**

Keywords—*maximum power point tracking, load management, load matching, photovoltaic hydrogen production, green hydrogen*

I. INTRODUCTION

The current, popular method of connecting a photovoltaic (PV) array to its load is through a maximum power point tracker (MPPT) that maximizes the system power output over the course of the day. The MPPT is often incorporated into a power converter that introduces additional costs and power losses to the system. Our group's previous research has shown success in an alternative, load-matching PV system. The key to our system is that the multiple loads are directly coupled to the PV array, thereby removing the power converter, and the number of loads connected to the PV array varies throughout the day to match the maximum available PV power. The load-matching PV system has shown above 99% energy transfer efficiency for resistive loads [1,2], and significant work has gone into optimizing the control algorithm of the system [3,4]. Within the scope of this paper, the term "energy transfer efficiency" is used to describe how much energy is consumed by the loads compared to the maximum available energy output of the PV array.

With the significant push for green hydrogen [5], this paper reports our research on electrolytic applications of the load-matching PV system where PV power is directly applied to the electrolytic production of hydrogen. Our previous research [6] revealed the performance potential of the load-matching PV system for electrolytic applications, but no dynamic performance of the PV-powered hydrogen production system was presented. The goal of this research is to understand this system under various operating conditions using SIMULINK simulations.

This paper will begin with a brief overview of the load-matching PV system, explain the PV and electrolyzer models

used in simulation, describe the approach used to optimize the system, discuss the results, and finish with a summary of the results and their importance. It will be shown that with the intermittent PV power and the dynamic operation of electrolyzers, the load-matching PV system is able to efficiently and economically produce hydrogen.

II. THE LOAD-MATCHING PV SYSTEM

The load-matching PV system is a load-varying system, where a varying number of loads are connected in parallel to the PV array, as shown in Fig. 1. For electrolytic hydrogen production, all the loads are identical electrolytic stacks. The loads connect to the PV array through switches. Varying the number of connected loads changes the total load resistance, which creates multiple different load lines for the system to operate on [1]. The controller manages the switches so that the active load line transfers the maximum power at a given irradiance. A visual of the load-line analysis will be presented later in this paper.

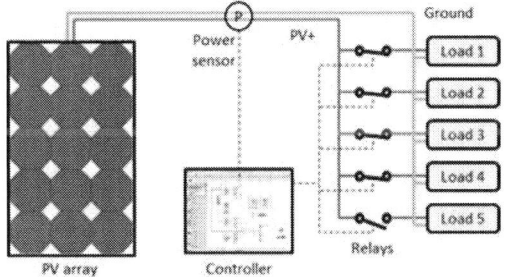

Fig. 1. The basic load-matching PV system with five loads [3,4].

The controller uses a feedback-based algorithm for managing the loads. It switches a load on or off and then compares the powers before and after the switch to determine whether to keep the load switched or undo the switch [3,4]. Fig. 2 shows a simplified block diagram of the algorithm used in the controller, and previous research has optimized the algorithm [4]. In this research, the optimized parameters for the algorithm such as the 350 Hz sampling frequency and 30 second delay time between power measurements are used.

978-1-7281-6118-1/22 $31.00 © 2022 IEEE

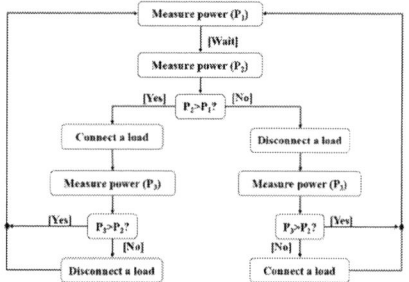

Fig. 2. The basic control algorithm for the load-managing PV system [3,4].

III. SYSTEM MODELS

A. PV Array and Component Models

The system in Fig. 1 with electrolytic loads is implemented and simulated in SIMULINK. For the PV array, the system involves a single module with a maximum power point voltage of 31.7 V and a maximum power point current of 9.13 A under an irradiance of 1 kW/m^2, but it can be easily expanded to a large array with many strings of PV modules. Two irradiance profiles from a sunny day and a cloudy day are taken from Sandia National Laboratory's PV Lib toolbox [7] to simulate the system under different weather conditions. The irradiance profiles used are given in Fig. 3. With the selected PV module and assuming two-axis tracking, the sunny day irradiance profile produces a maximum PV energy of 2.1912 kWh/day and the cloudy day profile 2.0433 kWh/day. The rest of the simulation reflects an ideal system by using ideal switches, non-intrusive power measurements, and ideal electrolytic loads.

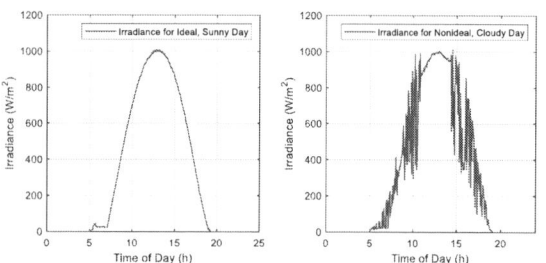

Fig. 3. Irradiance profiles of a sunny day (left) and a cloudy day (right) used in this research.

B. Electrolytic Load Model

To mimic an electrolytic load, a model was developed for a polymer electrolyte membrane (PEM) cell from the polarization curve provided by Nel Hydrogen. As shown in Fig. 4, the electrical behavior of the PEM cell shows a threshold voltage of 1.56 V followed by a linear, active region with a specific cell resistance of 0.214 Ω-cm^2. Therefore, an electrolytic load is represented with an ideal diode in series with a resistor in SIMULINK simulations.

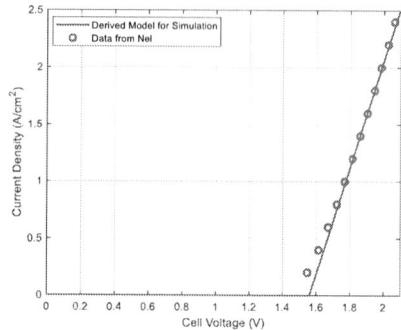

Fig. 4. Voltage and current-density characteristic of a polymer electrolyte membrane cell from Nel Hydrogen and the derived model for simulation.

In the load-matching system, each load is a stack of multiple electrolytic cells connected in series. In the simulation, the ideal diode sets the threshold voltage, which is a multiple of 1.56 V and determined by the number of cells per stack. The resistance of a stack in the active region relates to the cell cathode size and the number of cells per stack. In this research, the cell cathode size is determined by the desired resistance of the stack. For real systems, it is likely that the PV array will be adjusted to match the electrolytic stacks available on the market.

C. Coupling PV Array and Electrolytic Loads

Fig. 5 shows the output characteristics of the PV array and the load lines of a load-matching system with four electrolytic loads. The top figure shows the current-voltage (I-V) curves of the PV array at different irradiances and the available load lines of the system. The points where the array I-V curves and load lines intersect are the possible operating points of the system at each irradiance. The number of possible operating points is given by the number of available loads, and in this case, there are four. The bottom figure shows the output power of the PV array, where the maximum point for each irradiance is indicated with a circle. Each load line approximately intersects the maximum power point of the PV array for one of the plotted irradiances. As the irradiance changes throughout the day, the system connects the number of loads that results in the closest load line to the maximum power point.

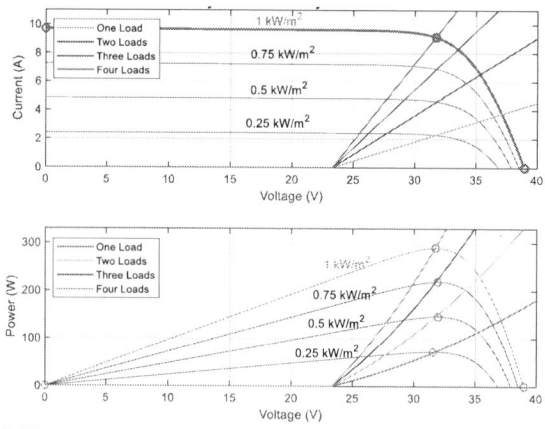

Fig. 5. PV output curves and electrolytic load lines for the one-module array used for simulation. Top: the current-voltage characteristics; Bottom: the power-voltage characteristics.

IV. System Optimization

To optimize the system and achieve an energy transfer efficiency of 99%, the number of cells per stack and the number of stacks in the system are varied. Since the PV array is fixed, the PEM cell cathode size must be modified to reach the desired energy transfer efficiency.

A. Electrolytic Stack Model

As mentioned above, the parameters to set for the stack model are the number of cells per stack and the stack resistance. The number of cells should be chosen for the cells to operate within their specified voltage range. According to Nel Hydrogen, the maximum operating voltage per cell is 2.6 V, and the minimum voltage is the threshold voltage of 1.56 V. The specified range of a PEM cell is between 2 V to 2.2 V.

As the number of cells in a stack increases, the threshold voltage increases. This pushes the load lines towards the right on the voltage axis in Fig. 5, so the load lines and their operating points get squeezed together.

After determining the number of cells in a stack, the combined resistance of all the stacks in the system can be determined by the slope of the straight line between the threshold voltage of a stack and the maximum power point voltage under an irradiance of 1 kW/m² in the top figure of Fig. 5. For the remainder of this paper, a "matched" system will refer to a system where the combined resistance of all the loads has been designed this way. However, the resistance for each stack is not known until the number of loads is determined.

When the resistance of a stack and the number of cells per stack are known, the PEM cell cathode size can be calculated using Fig. 4. It is also important to analyze how variations in stack resistance affect the system performance because of variability and degradation in the PEM cells. It will be shown that slight variations in cell or stack resistance do not impact the overall energy transfer efficiency.

B. Number of Stacks in the System

After the stack is developed, the number of stacks in the system is adjusted to achieve an energy transfer efficiency of 99% for both sunny and cloudy days. For both the number of cells per stack and the number of stacks in the system, there is a cost-to-performance consideration. This is because more cells per stack and more stacks in a system incur higher system costs, especially if the performance gains are marginal. This is particularly prevalent to the number of stacks in the system.

V. Simulation Results

A. Number of Cells per Stack

The number of cells in a stack was changed between 14 and 17 cells, and the operating voltage of the stack was calculated throughout the day (Fig. 6). Because the cells are serially connected, the specified voltage range and voltage limits for the stack are given by the individual cell range and limits multiplied by the number of cells in the stack. These plots were developed using six loads and the sunny irradiance profile. The 15-cell stack keeps the operating voltage within the specified range for most of the day, while the 14-cell stack overshoots the specified

range and higher numbers, like the 16 and 17-cell stacks, undershoot.

The resistance for the 15-cell stack is 5.5 Ω. With six stacks in the system, the combined resistance of all six loads is 0.92 Ω which provides a matched system.

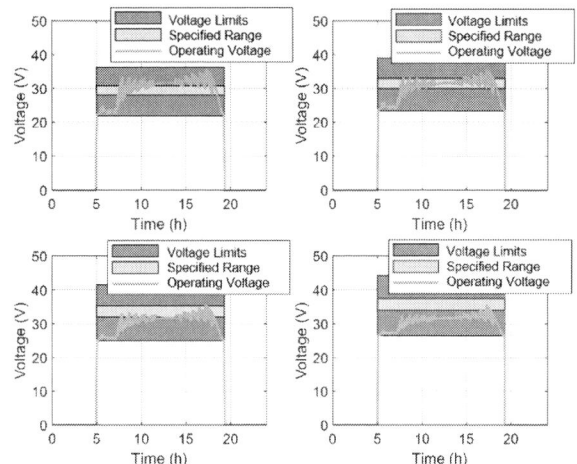

Fig. 6. Operating voltage throughout the day for different numbers of cells in a stack.

The 15-cell stack also possesses stability in its performance for slight variations in its resistance. Fig. 7 shows the energy transfer efficiency for the 15-cell stack with the stack resistance varying between 2 Ω and 8 Ω, while the matched resistance is 5.5 Ω. For a ±10% variation in stack resistance (5–6 Ω), there is nearly no impact on system performance. Lower resistances are less impactful, but once the stack resistance is above 10% larger, the performance degrades significantly.

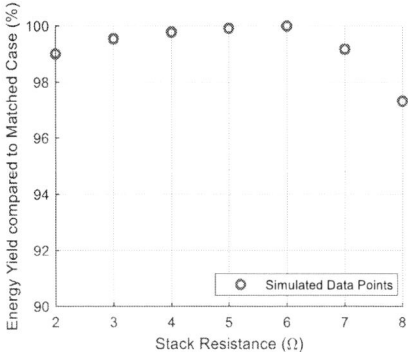

Fig. 7. Energy transfer efficiency as a function of stack resistance in a six-load system. The matched stack resistance is 5.5 Ω.

B. A Four-Stack System

To reduce the system cost, a four-stack system was investigated while each stack still contained 15 cells. The resistance of each stack is 3.68 Ω for a matched system. Our previous research suggests that a load-matching PV system with four electrolytic loads can achieve an energy transfer efficiency of 99% [6]. In this research, it is found that on the sunny day (Fig. 8), the system has a 99.2% energy transfer efficiency with an energy output of 2.176 kWh, and on the cloudy day (Fig. 9), there is a 99.0% efficiency with an output of 2.033 kWh.

In Figs. 8 and 9, the voltage, current, power, and number of loads connected to the PV array are displayed to get a full picture of how the system performs throughout the day. In the sunny day, all the plots clearly show the four different connections of loads throughout the day. When the system changes the number of connected loads, the voltage and current jump up and down. On the other hand, the cloudy day shows highly-fluctuating voltage, current, and power following the irradiance pattern in Fig. 3. Despite the noisy pattern, the system performs well given its high efficiency during the cloudy day.

As designed, all four loads are connected during the hours of peak irradiance. Note that the current and power traces in Fig. 8 appear asymmetrical, but as shown in the plot showing the number of connected loads, the system behavior is symmetrical. The switching in the system is bound to accrue some energy losses, but they are minimal.

To calculate the amount of hydrogen produced by such a system, the total charge passed through all the PEM cells in the sunny day is 3.74×10^6 C. If the quantum efficiency of the PEM cells is 100%, the total charge would produce about 38.9 g of H_2. This is not a realistic expectation of the system performance but provides a figure for analysis.

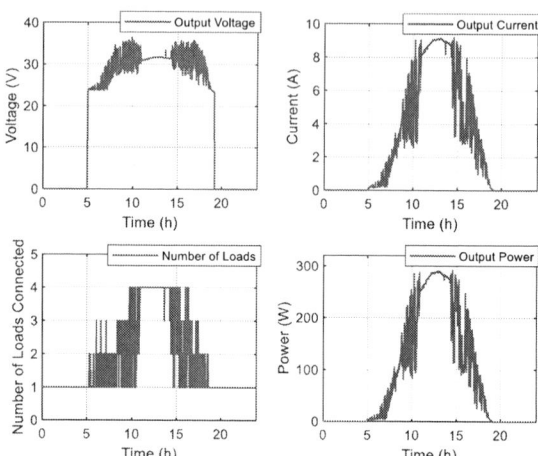

Fig. 9. Voltage, current, power, and number of connected loads of the four-load system throughout the cloudy day resulting in a 99.0% efficiency.

To examine the performance of each stack in the four-stack system, Fig. 10 displays the current through each stack throughout the day. Because all the stacks are identical, the current for each stack exhibits a similar trend with the only difference being the active time of the stack. As expected, the first stack experiences the most active time, while the fourth stack sees a fraction of the total active time. In real systems, the algorithm should rotate the order of the stacks so that, on average, all the stacks are powered equally throughout the lifetime of the system for equal degradation.

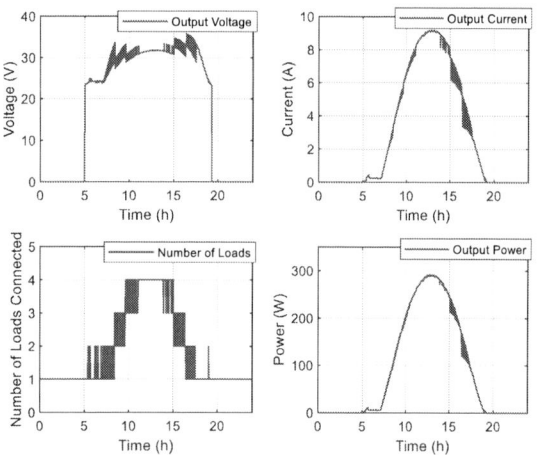

Fig. 8. Voltage, current, power, and number of connected loads of the four-load system throughout the sunny day resulting in a 99.2% efficiency.

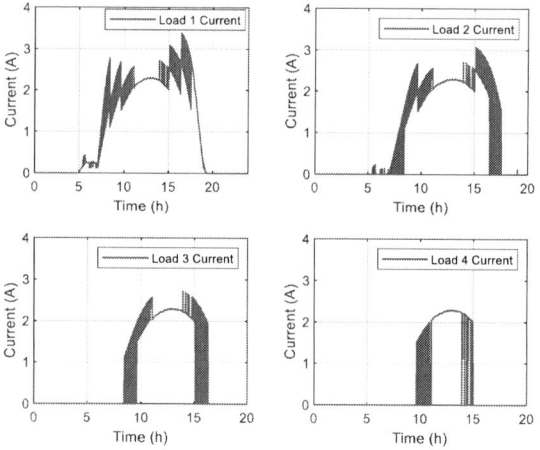

Fig. 10. Current through each of the four stacks in the system throughout the sunny day.

C. Four Loads vs Six Loads

In a sunny day, the irradiance profile exhibits a near-sinusoidal pattern, and an efficient PV system should have the power consumed by the loads follow that pattern as closely as possible. The moments of difference between the sinusoidal irradiance and the power curves are when the system changes the number of connected loads. These moments result in power losses that lower the system's energy transfer efficiency.

Most of the attempts by the system to change the number of connected loads are unsuccessful (Figs. 8 and 9). The successful attempts can be noticed for the four-load system in Fig. 11 at 9 hours, 175 W and again at 11.5 hours, 250 W. The moments correspond to the switch from two connected loads to three connected loads, and then to four connected loads, respectively. In comparison, the successful attempts for the six-load system, also shown in Fig. 11, are much more subtle because it has more possible operating points. As a result, the six-load system has an efficiency of 99.6% in the sunny day and 99.2% in the cloudy day. As discussed earlier, having more available loads in the system allows for a more efficient system.

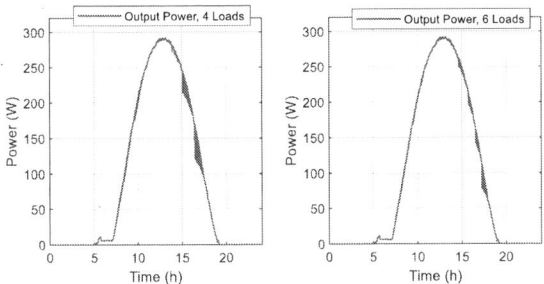

Fig. 11. Power consumed by loads in the four-load system (left) and the six-load system (right) throughout the sunny day.

Although there are differences in the energy transfer efficiency between four loads and six loads, they are small given that there is only a 0.4% difference in the sunny day and 0.2% in the cloudy day, and the four-load system is still able to deliver 99% of the maximum available PV energy to the loads. The results reveal that going from four loads to six loads, the electrolyzer cost increases by 50% but the system energy efficiency gain is only 0.2–0.4%. For a cost-performance perspective, the four-load system is far more cost effective.

VI. CONCLUSION

This paper reports dynamic simulations of a load-matching PV system for green hydrogen production. It is revealed that a system with four electrolytic loads delivers above 99% energy transfer efficiency from the maximum available PV energy. For a single-module PV array with a maximum power point voltage of 31.7 V, each electrolytic stack should contain 15 PEM cells. It is also found that the cell and stack resistance can vary ±10% without affecting the system efficiency. The simulation outlined in this paper can be easily expanded to much larger PV-powered hydrogen production systems.

REFERENCES

[1] J. A. Azzolini and M. Tao, "A control strategy for improved efficiency in direct-coupled photovoltaic systems through load management," Applied Energy, vol. 231, pp. 926-936 (2018).

[2] M. Tao, "Digital load management for variable output energy systems," US patent no. 10,399,441 (2019).

[3] J. A. Azzolini and M. Tao, "Maximum power point tracking through load management," US provisional patent application no. 63,126,053 (2020).

[4] K. Tan, J. Azzolini, W. Parquette, C. Polo and M. Tao, "Algorithms for Maximum Power Point Tracking through Load Management," IEEE 48th Photovoltaic Specialists Conference (2021).

[5] U.S. Department of Energy, "U.S. Department of Energy Hydrogen Program Plan," (2020).

[6] J. A. Azzolini, M. Tao, K. Ayers and J. Vacek, "A Load-Managing Photovoltaic System for Driving Hydrogen Production," *2020 47th IEEE Photovoltaic Specialists Conference (PVSC)* pp. 1927-1932 (2020).

[7] J. S. Stein, D. Riley, and C. W. Hensen, "PV LIB Toolbox (Version 1.1)," Sandia National Laboratories (2014).

Embodied Energy and CO2 from the Manufacture of Cadmium Telluride and Silicon Photovoltaics

Hope Wikoff, Samantha B Reese, Matthew O Reese

National Renewable Energy Laboratory, Golden, CO, United States

As the US DOE pushes for an installation of 1 TW of solar PV by 2035, the need to compare utility scale technologies - Si and CdTe PV - becomes clear. Here, we employ a cradle to gate Life Cycle Assessment modeling tool, the Material Flows through Industry tool, to quantify the overall Energy and Carbon Equivalents involved in the manufacture of state-of-the-art solar modules from leading manufacturers. We also perform an analysis to identify which components within the solar module are contributing the most to these impacts. Our component analysis suggests that CdTe module impacts, which are overall less than Si, come from glass and frame materials, while it is the silicon semiconductor that contributes most to the impacts of a silicon module. Finally, we show that when scaled up to DOE goals the implementation of CdTe over Si could save 804 billion kgCO2eq, roughly 4X the total 2019 US electricity generation carbon emissions.

Collection of Heat Loss in Photovoltaic System by Parallely Connected Thermoelectric Network

Joel Erickson, and Jing Bai

University of Minnesota Duluth Electrical Engineering Department, Duluth, Minnesota, 55811, United States (Joel Erickson: Master's Student, Jing Bai: Professor and Department Head)

Abstract—**The goal of this work is to increase solar cell efficiency by efficiently combining the electric power of a solar cell and a thermoelectric generator into a single two terminal hybrid device. This work presents a method of achieving this by dividing the thermoelectric generator into smaller thermoelectric generators, forming a parallel connected network with them, and connecting this network in series with the solar cell. An equivalent circuit model was developed for this device scheme, and compared with experimental data. The data show some support of the model, but fine evaluation of the model's accuracy was hindered by limitations in the experimental setup. If thermoelectric generator efficiency increases in the future, it may become practical to combine thermoelectric generators with solar cells. Providing a method for combining the two power sources at the cellular level may be important for simplifying and improving systems that use these photovoltaic/thermoelectric hybrids.**

I. INTRODUCTION

In photovoltaics (PV), a significant portion of energy is lost as waste heat due to the mismatch between the energy of incident photons and the band structure of the solar cell. One possible method of using this waste heat is make a temperature differential and extract power from it using a thermoelectric generator (TEG). TEGs use the thermoelectric effect and a junction of two different materials to create a voltage differential from a temperature differential.

Currently, TEGs are far too inefficient to make this a practical application, but much research is going into improving them. This includes optimizing device structure, creating new meta materials, and employing nano-technology [3, 6]. High efficiency thermoelectric generators may be an available technology in the near future.

An additional problem with combining TEGs and PV is that the system will now have two power sources and require additional power electronics to process the power. The work in this paper seeks to address this problem by investigating a method for creating a two terminal device that will output an efficient combination of the TEG and PV power. This would eliminate the need for additional power electronics and simplify any PV/TEG hybrid system.

A significant amount of work has been done on the PV/TEG hybrid system. In most of this work, the PV and TEG power are kept separated [4, 5]. A handful of papers have been published that involve combining the PV and TEG power, but none seem to present a generally efficient way of doing it [1, 8, 9, 10]. In some of this prior work, combining the PV and TEG sources in series and in parallel are examined and an equivalent circuit model of the series connection was created. This was done by combining the single diode PV model with shunt and series resistances with the equivalent circuit model of a TEG (a voltage source with series resistance) [1].

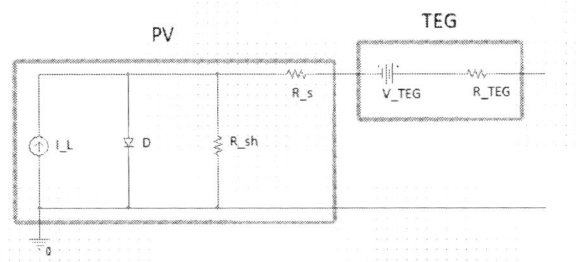

Fig. 1. Circuit model of TEG in series with PV cell. I_L is the light generated current, I_o is the diode's reverse saturation current, R_sh is the PV shunt resistance, R_s is the PV series resistance, V_TEG is the TEG voltage, and R_TEG is the TEG series resistance.

The circuit model in figure 1 can be described by the following equation. V_T is a term known as the thermal voltage, and is a combination of Boltzmann's constant, the charge of an electron, and the PV cell temperature.

$$I = I_L - I_o \left[\exp\left(\frac{V + V_{TEG} + I(R_s + R_{TEG})}{V_T} \right) - 1 \right] - \frac{V + V_{TEG} + I(R_s + R_{TEG})}{R_{sh}} \quad (1)$$

The TEG in series increases the open circuit voltage of the device, but also presents a problem. The additional series resistance from the TEG adds to the series resistance from the PV single diode model. This decreases the fill factor of the hybrid device's IV curve, which often results in a lower max power than just the PV alone [7].

II. THEORETICAL WORK

The solution that this paper presents is to split the TEG into a parallel network of smaller TEGs. In a TEG, the thermoelectric effect creates a voltage difference between a junction of two dissimilar materials called legs which are also the source of the TEG's resistance [2]. These junctions are added in series to sum their voltages and a single TEG module will consist of many of these junctions [2]. If the pairs of legs are approximated as being identical, then dividing the number of junctions in series will divide the voltage and resistance by the same amount. The divisions of the TEG can be placed in parallel, which further decreases the net contribution of series resistance by the TEG portion of the hybrid device.

978-1-7281-6118-1/22 $31.00 © 2022 IEEE

In this work, a modification of the prior PV/TEG hybrid series connection model is made to create a model of a TEG parallel network in series with the PV. The new model is also generalized for n divisions of the TEG component. The input parameters are the single diode model parameters (light generated current, diode reverse saturation current, temperature, shunt resistance, and series resistance) of the PV and the equivalent circuit model parameters (voltage, and series resistance) of the unsplit TEG.

This modification uses linear methods (Thevenin and Superposition) to reduce the TEG parallel network to an equivalent voltage source and resistance. Since the rest of the model contains a diode, it is not linear and the use of these methods is an approximation. The model using this approximation was compared with a PSPICE model not using the approximation. The two models were nearly identical, which validates the use of the approximation.

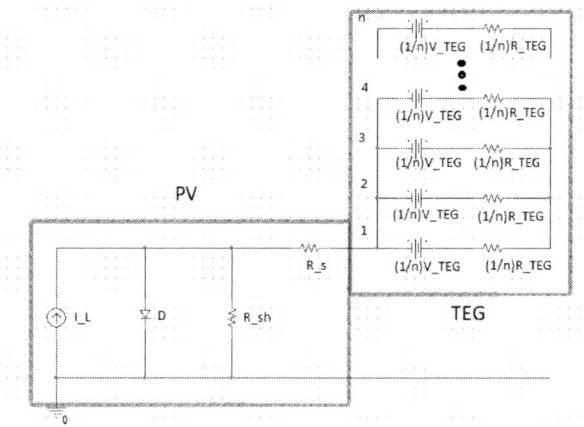

Fig. 2. Circuit model of PV cell in series with parallel network of TEG split into n pieces.

The circuit model in figure 2 can be described by the following equation by using the linear approximation.

$$I = I_L - I_o \left[\exp\left(\frac{V + \frac{V_{TEG}}{n} + I\left(R_s + \frac{R_{TEG}}{n^2}\right)}{V_T} \right) - 1 \right] - \frac{V + \frac{V_{TEG}}{n} + I\left(R_s + \frac{R_{TEG}}{n^2}\right)}{R_{sh}} \quad (2)$$

The model was run for different PV and TEG circuit parameters, and the power from each source was always able to be efficiently combined given the proper number of divisions in the TEG. This provides empirical evidence for the success of this method in combining the PV and TEG power into a single device with two terminals and its own IV curve.

III. EXPERIMENTAL WORK

Experimental data was collected to test the model. In the experimental setup, commercially available TEG modules were attached to commercially available PV cells and connected in series to make hybrid PV/TEG devices. 6 TEG modules were connected in series to form an initial unsplit TEG. The modules were then divided and connected in parallel to split the initial TEG by divisions of 2, 3, and 6. A full spectrum light bulb was used to provide a consistent light source.

The IV curves were plotted by manually collecting points using multimeters and a potentiometer as a variable resistor. IV

data of the PV and TEG by themselves, along with temperature measurements, were used to gather the parameters of the equivalent circuit models for each. These parameters were used to construct the model of the hybrid devices which were then compared with the collected data of the hybrid devices. Chi squared values were used to assess the goodness of fit between the data and the model.

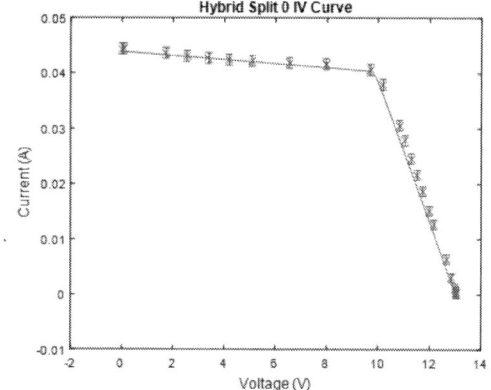

Fig. 3. Model data comparison for hybrid device with no splitting of TEG. The blue crosses are the data points with error bars and the orange line is the model prediction. This is the poorest fitting data with a chi squared value of 74.9398. There are 25 data points which correspond to an average deviation of nearly 3 standard deviations.

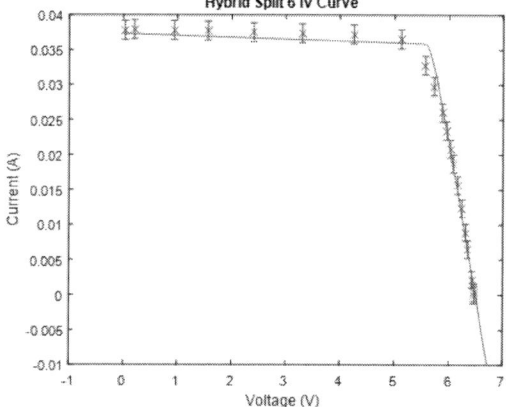

Fig. 4. Model data comparison for hybrid device with TEG split into 6 pieces. The blue crosses are the data points with error bars and the orange line is the model prediction. This is the best fitting data with a chi squared value of 18.7713. There are 21 data points which correspond to an average deviation of nearly 0.9 standard deviations.

The collected data show some agreeance with the model, but the chi squared values of some runs showed an average deviation from 1 to 3 standard deviations which indicates a poor fit. The poorness of the fit is reasonable, considering the unevenness of the splitting of the TEGs. The model assumes that the TEGs will by split evenly, but the experimental setup was too crude to do this well. Efforts were made to keep the light conditions on each PV/TEG module even, but this was too difficult and the TEG parallel network always had unequal light conditions on each piece. This resulted in significantly unequal voltages and resistances on each piece of the TEG parallel network.

978-1-7281-6118-1/22 $31.00 © 2022 IEEE

IV. SUMMARY

In summary, a parallel network of TEGs connected in series appears to be a promising method for combining PV and TEGs into a single two terminal device. A model for this system is built off of well supported constituent models, and shows the effectiveness of the method. Experimental data offers some support of the model, but a more sophisticated setup is needed to verify fine accuracy of the model in predicting device IV characteristics. If TEGs become more efficient and PV/TEG hybrids become practical, this work will be useful in constructing single cell hybrid devices and understanding their IV characteristics.

References

[1] H. Chen, N. Wang, H. He, "Equivalent circuit analysis of photovoltaic-thermoelectric hybrid device with different TE module structure", *Advances in condensed Matter Physics*, vol. 2014, no. 824038, 2014.

[2] H.J. Goldsmid, "Introduction to Thermoelectricity", Second Edition, Springer, New York (NY), 2016.

[3] D. Li, Y. Gong, Y. Chen, J. Lin, Q. Kahn, Y. Zhang, Y. Li, H. Zhang, H. Xie, "Recent progress of two-dimensional thermoelectric materials", *Nano-Micro Letters*, vol. 12, no. 36, 2020.

[4] P.D. Raut, V.V. Shukla, S.S. Joshi, "Recent developments in photovoltaic-thermoelectric combined system", *International Journal of Engineering & Technology*, vol. 7, no. 4, pp. 2619-2627, 2018.

[5] A.Z. Sahin, K.G. Ismaila, B.S. Yilbas, A. Al-Sharafi, "A review on the performance of photovoltaic/thermoelectric hybrid generators", *International Journal of Energy*, vol. 44, pp. 3365-3394, 2020.

[6] S. Shittu, G. Li, X. Zhao, X. Ma, "Review of thermoelectric geometry and structure optimization for performance enhancement", *Applied Energy*, vol. 268, no. 115075, 2020.

[7] A. Smets, K. Jager, O. Isabella, R.V. Swaaij, M. Zeman, "Solar Energy". UIT Cambridge Ltd, Cambridge England, 2016.

[8] N. Wang, L. Han, H. He, N. Park, K. Koumoto, "A novel high-performance photovoltaic-thermoelectric hybrid device", *Energy and Environmental Science*, vol. 4, pp. 3676-3679, 2011.

[9] L. Xu, Y. Xiong, A. Mei, Y. Hu, Y. Rong, Y. Zhou, B. Hu, H. Han, "Efficient perovskite photovoltaic-thermoelectric hybrid device". *Adv. Energy Mater*, vol. 8, no. 1702937, 2018.

[10] J. Zhang, Y. Xuan, "The electric feature synergy in the photovoltaic – thermoelectric hybrid system", *Energy*, vol. 181, pp. 387-394, 2019.

Ultra-Thin and Lightweight CdS/CdTe Solar Cell Fabricated on Ceramic Substrate for Space Applications

Manoj K. Jamarkattel[1], Adam B. Phillips[1], Geethika K. Liyanage[1], Fadhil K. Alfadhili[1], Ebin Bastola[1], Victor V. Plotnikov[2], Alvin D. Compaan[2], Randy J. Ellingson[1], and Michael J. Heben[1].

[1]Wright Center for Photovoltaic Innovation and Commercialization, Department of Physics and Astronomy, University of Toledo, Toledo, OH, 43606, USA

[2]Lucintech, Inc., 1775 Progress Dr., Perrysburg, OH, 43551, USA

Abstract— **Ultra-thin and light weight cadmium telluride (CdTe) solar cells were fabricated on 20-micron thick yttria-stabilized zirconia (3YSZ) substrate in superstrate configuration. Optimization of $CdCl_2$ treatment and copper diffusion were done to enhance the preformation of the device. Due to high reflectance off the substrate surface, anti-reflecting layer was deposited on the front of the device to reduce the reflectance which increase current density. Here, we present ultra-thin and light weight CdS/CdTe solar cells with conversion efficiency of 11.2 % and specific power > 6 kW/kg. This could make CdTe based solar cells applicable for space applications.**

Keywords— Ultra-thin, Flexible, CdTe, Space application

I. INTRODUCTION

Solar energy is one of the major power sources for space vehicles and satellites, and is being developed for unmanned aerial vehicles. Both single junction and multi junction GaAs solar cells have been commonly used for space applications [1]. CdTe based solar cells offer another promising approach due to the robust nature of the materials and the high optical absorption coefficient with a suitable band gap for optimal photon power conversion. Current manufacturing techniques allow for mass production of high efficiency CdTe photovoltaic (PV) modules at low cost per Watt. Today's CdTe solar cells hold a record efficiency of 22.1%[2]. However, most of these devices are made on thick glass substrates and include a second glass sheet on the back, making them heavy and rigid. This limits their application to terrestrial usage.

For space applications, some of the important requirements are radiation hardness and high specific power (W/kg). CdTe has been demonstrated to be stable against proton and electron irradiation [3]. To increase the specific power, light weight substrates are needed. Device fabrication on thin glass and polyimide substrates have been reported. AM1.5 efficiencies ranging from 14 to 16.4 % [4-6] have been reported for devices fabricated on 100 μm thick Corning® Willow® Glass substrates, and 8.6 % has been reported for polyimide [7].

Most high efficiency CdTe devices have been fabricated using higher temperature deposition methods such as closed-space sublimation and vapor transport deposition at 500-650 °C. Such temperatures are too high for polymer substrates. The lightest commercially available substrate reported to date for high temperature growth of CdTe is Willow® Glass. 3 mole % yttria-stabilized tetragonal zirconia substrate (3YSZ), commercialized by ENrG Inc., could be a potential new substrate option. 3YSZ has a bending radius of 2.2 cm [8] and is compatible with high temperature processing [9]. While 3YSZ has been used for IR nano-optics [8], very little work has been carried out using it for fabricating solar cells. Forabe *et al.* demonstrated greater than 17% efficient copper-indium-gallium diselenide (CIGS) solar cells fabricated on 3YSZ in the substrate configuration [10]. The CdTe manufacturing process, on the other hand, is typically done in the so-called "superstrate" configuration, where the sunlight is admitted to the cell through the substrate on which the depositions are done.

Here, we present results on ultra-thin and light weight CdTe solar cells made on commercially available 20 μm thick 3YSZ ceramic substrate. The measured specific power was 6 kW/kg and the champion cell of 11.2 % efficiency. This result demonstrated possibilities of CdTe solar cells towards space application.

II. EXPERIMENTAL DETAILS

3YSZ substrates were cut into 3 " x 3 " pieces using a 532 nm laser and cleaned using Micro-90 detergent and deionized water with sonication. Aluminum doped zinc oxide (AZO) was RF sputtered at $250°C$ as the transparent conducting oxide (TCO) and then an intrinsic zinc oxide (i-ZO) layer was deposited to form a high resistive transparent layer (HRT) layer. An 100 nm thick cadmium sulfide (CdS) layer and 2 microns of CdTe were deposited by RF-magnetron sputtering to form the window and absorber layers, respectively. The devices were further cut into 1.5 " x 1.5 " and activated after applying saturated $CdCl_2$ solution made in anhydrous methanol and then heating in dry air for 25 min at $387°C$. The excess $CdCl_2$ was removed by rinsing in methanol. Cu doping

978-1-7281-6118-1/22 $31.00 © 2022 IEEE

was performed using $CuCl_2$ solution made in methanol and annealed at 150 °C. The devices were completed by evaporating 40 nm of Au through a mask with an area of 0.08 cm^2. For the device with an anti-reflective coating, 100 nm of magnesium fluoride (MgF_2) was deposited on the front surface of the substrate.

The reflection of the front of the device was measured using a PerkinElmer Lambda 1050 UV/Vis/NIR spectrophotometer. Current density-voltage (JV) characteristics were measured using a Keithley 2440 digital source meter and a solar simulator with AM1.5G illumination. External quantum efficiencies (EQE) were measured using PV Instruments system (model IVQE8-C). Scanning electron microscopy (SEM) was carried out using a Hitachi S-4800 UHR scanning electron microscope.

III. RESULTS AND DISCUSSION

A cross-sectional SEM image of a completed device with the structure of 3YSZ/AZO-iZO/CdS/CdTe is shown in Fig. 1. XRD measurements showed that the CdTe films were polycrystalline with the cubic zincblende structure with

Fig. 1. Cross section SEM image of device stacks

preferred orientation in the (111) direction. Fig. 2. shows side and front views of the completed devices samples. Circular

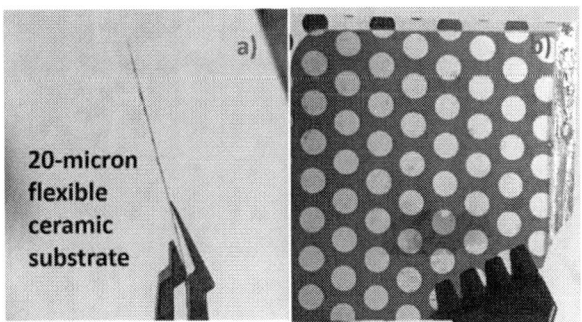

Fig. 2. Images showing a) side view and b) front view of completed devices.

yellow dots as shown in Fig. 2b) are individual solar cells with gold as a back electrode.

Fig. 3. shows the statistics for the current-voltage (JV) properties and a typical JV curves is presented in Fig. 4(b). The average short circuit current density (J_{sc}, 18.5 mAcm^{-2}) was 2-3 mAcm^{-2} lower than the reported value for sputtered CdS/CdTe devices [11]. As shown in Fig. 4., in addition to the

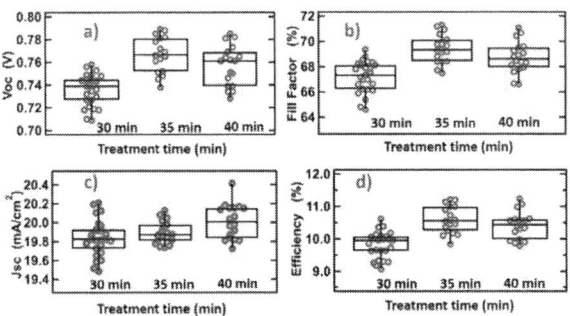

Fig. 3. PV parameters for different Cu annealing duration at 150 °C

typical notch in the External Quantum Efficiency (EQE) data observed for absorption and poor collection in CdS for the wavelength range of 400 to 600 nm, the majority of the J_{sc} loss can be attributed to low EQE across the entire wavelength range with of ~70% even with application of a MgF_2 anti-reflecting (AR) layer. Reflection measurements (shown as 1-R in graph) showed that substrate itself has high reflection off the surface which reduced the current collection. With MgF_2 applied the champion device showed a V_{oc} of 785 mV, a J_{sc} of 20.2 mAcm^{-2}, a FF 70.9%, and a efficiency of 11.2 %. Even

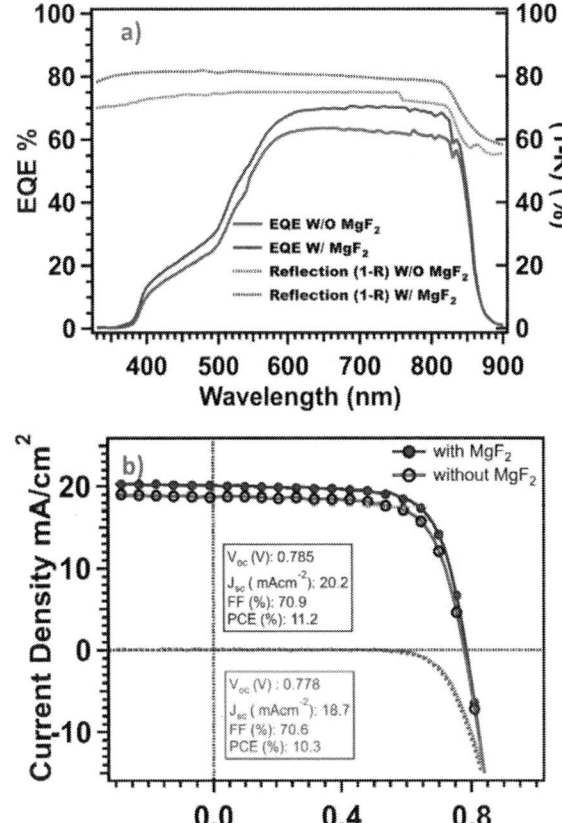

Fig. 4. a) EQE and b) JV measurements of Champion cells.

with an anti-refelection layer the J_{oc} value is still ~10% lower than values reported for sputtered CdS/CdTe devices fabricated on 3.2 mm glass substrate [11]. This clearly indicates that YSZ has limitation for higher current collections as compared to glass substrate, mostly due to high reflection.

Since specific power is key factor for space applications, here, we determined the specific power of the solar cells fabricated using 20-micron thick 3YSZ, 100-micron thick Willow® Glass, and 3.2 mm thick F-doped SnO2/glass [12](Tec 12D; Pilkington North America) in our laboratory by weighing each sample. From Table 1, 3YSZ solar cells have the highest specific power of 6.3 kW/kg. This can be further increased to ~8 kW/kg by increasing the device efficiency to 14 % by incorporating Se incorporation and improving light coupling.

TABLE I.   SPECIFIC POWER CALCULATION

Substrate	3YSZ	Willow® Glass	3.2 mm TEC
Device Efficiency	11.2 %	14.4 %[6]	17.46% [13]
Specific Power	6.3 kW/kg	5.8 kW/kg	0.28 kW/kg

IV. CONCLUSION

In this work we showed that, ultra-thin and light weight CdTe solar cell can be fabricated on 20 microns thick 3YSZ ceramic substrates. An efficiency of 11.2 % was measured with a specific power of 6.3 kW/kg. This value could be further increased to 8 kW/kg through Se incorporation and better light coupling.

ACKNOWLEDGEMENT

This report is based on research sponsored by the U.S. DOE's Office of Energy Efficiency and Renewable Energy (EERE) under Solar Energy Technologies Office (SETO) Agreement DE-EE0008974 and Air Force Research Laboratory under agreement numbers FA9453-18-2-0037, FA9453-21-C-0056 and FA9453-19-C-1002. The U.S. Government is authorized to reproduce and distribute reprints for Governmental purposes not withstanding any copyright notation thereon. The views and conclusions contained herein are those of the authors and should not be interpreted as necessarily representing the official policies or endorsements, either expressed or implied, of Air Force Research Laboratory or the U.S. Government. The Authors like to thank Pilkington North America for providing TEC12D glasses.

REFERENCES

[1]   N. Papez, R. Dallaev, S. Talu, and J. Kastyl, "Overview of the Current State of Gallium Arsenide-Based Solar Cells," *Materials (Basel),* vol. 14, no. 11, Jun 4 2021.

[2]   M. A. Green, Y. Hishikawa, E. D. Dunlop, D. H. Levi, J. Hohl-Ebinger, and A. W. Y. Ho-Baillie, "Solar cell efficiency tables (version 52)," (in English), *Prog Photovoltaics,* vol. 26, no. 7, pp. 427-436, Jul 2018, doi: 10.1002/pip.3040.

[3]   A.Romeo, D. L. Bätzner, H. Zogg, and A. N. Tiwari, "Potential of CdTe thin film solar cells for space applications," *Proceedings of the 17th European Photovoltaic Conference and Exhibition,* pp. pp 2183-2186, 2001.

[4]   H. P. Mahabaduge *et al.*, "High-efficiency, flexible CdTe solar cells on ultra-thin glass substrates," *Applied Physics Letters,* vol. 106, no. 13, 2015, doi: 10.1063/1.4916634.

[5]   W. L. Rance *et al.*, "14%-efficient flexible CdTe solar cells on ultra-thin glass substrates," *Applied Physics Letters,* vol. 104, no. 14, 2014.

[6]   G. K. Liyanage *et al.*, "RF-Sputtered Cd2SnO4 for Flexible Glass CdTe Solar Cells," in *2016 Ieee 43rd Photovoltaic Specialists Conference,* (IEEE Photovoltaic Specialists Conference, 2016, pp. 450-453.

[7]   A. N. Tiwari, A. Romeo, D. Baetzner, and H. Zogg, "Flexible CdTe solar cells on polymer films," *Progress in Photovoltaics: Research and Applications,* vol. 9, no. 3, pp. 211-215, 2001, doi: 10.1002/pip.374.

[8]   K. K. Gopalan, D. Rodrigo, B. Paulillo, K. K. Soni, and V. Pruneri, "Ultrathin Yttria-Stabilized Zirconia as a Flexible and Stable Substrate for Infrared Nano-Optics," *Advanced Optical Materials,* vol. 7, no. 2, 2019.

[9]   S. R. Casolco, J. Xu, and J. E. Garay, "Transparent/translucent polycrystalline nanostructured yttria stabilized zirconia with varying colors," *Scripta Materialia,* vol. 58, no. 6, pp. 516-519, 2008/03/01/ 2008.

[10]   D. Fobare *et al.*, "Novel application of Yttria Stabilized Zirconia as a substrate for thin film CIGS solar cells," in *2014 IEEE 40th Photovoltaic Specialist Conference (PVSC),* 8-13 June 2014 2014, pp. 0341-0344.

[11]   E. Bastola *et al.*, "Open-circuit Voltage Exceeding 840 mV for All-Sputtered CdS/CdTe Devices," in *2020 47th IEEE Photovoltaic Specialists Conference (PVSC),* 15 June-21 Aug. 2020 2020, pp. 2513-2518.

[12]   Pilkington, "SolarEnergy brochure." [Online]. Available: https://www.pilkington.com/en/us/products/product-categories/solar-energy/nsg-tec-for-solar-applications#brochures.

[13]   M. K. Jamarkattel *et al.*, "Improving CdSeTe Devices With a Back Buffer Layer of Cu_xAlO_y," *IEEE Journal of Photovoltaics,* pp. 1-6, 2021.

Continuous Flash Sublimation of Inorganic Halide Perovskites: Enabling Industrially Compatible Deposition Rates

Tobias Abzieher, Christopher P. Muzzillo, Mirzo Mirzokarimov, Ahmad R. Kirmani, Gabriella Lahti, Wylie Kau, Daniel M. Kroupa, Joseph M. Luther, David T. Moore

National Renewable Energy Laboratory (NREL), Golden, CO, United States

BlueDot Photonics, Inc., Seattle, WA, United States

Despite the outstanding progress in performance of halide perovskite solar cell absorbers fabricated via vapor-based approaches, increasing deposition rate in these approaches has been neglected nearly completely. In fact, today' deposition times for the fabrication of high performing absorbers are typically in the range of up to several hours, being orders of magnitude away from industrially reasonable deposition times. In this work, continuous flash sublimation of inorganic halide perovskite absorbers is introduced as a concept to overcome this fundamental limitation of state-of-the-art approaches. Solar cells employing flash-sublimed CsPbI2Br absorbers reach power conversion efficiencies as high as 12.3%, making them one of the most efficient and, with deposition times below five minutes, at the same time the fastest vapor-processed inorganic perovskite absorbers reported so far. Besides the promising performance and significantly higher deposition rates, the approach enables continuous deposition, thus, circumventing another major bottleneck of established vapor deposition approaches toward commercialization.

Temperature- and Illumination-Dependent Characterization of Wide Bandgap Sulfide CIGS and CZTS Solar Cells

Simon M.F. Zhang, Guojun He, Chang Yan, Kaiwen Sun, Xiaojing Hao, Ivan Perez-Wurfl, Ziv Hameiri

University of New South Wales, Sydney, Australia

Photovoltaic devices are exposed to a wide range of temperatures and illumination conditions in the field. Using temperature-dependent current-voltage and Suns-VOC measurements, we investigate the performance of two types of wide-bandgap pure-sulfide inorganic thin film solar cells: CuInGaS2 (CIGS) and Cu2ZnSnS4 (CZTS). We find that both technologies behave linearly with temperature, with one-sun sensitivities close to that of crystalline silicon cells. In addition, we find that the open-circuit voltage of the investigated cells becomes significantly more sensitive to temperature at low light intensities as compared to one-sun. To our knowledge, these results are being presented for the first time, and we expect them to be instructional to the field and indoor yield prediction as well as tandem cell design.

Fill Factor Prediction of Modern Industrial Cells: Potential Gaps and Improvements

Gaia Maria N Javier, Priya Dwivedi, Yoann Buratti, Thorsten Trupke, Ziv Hameiri

University of New South Wales, Sydney, Australia

Extracting solar cell electrical parameters directly from luminescence images, instead of the common current-voltage (I-V) measurements, can significantly increase the throughput and reduce the operation cost of photovoltaic production lines. This study investigates the capability of obtaining the fill factor (FF) from luminescence images by assessing the accuracy of published empirical expressions for the FF. The fitting approach for empirical coefficients was first modified. The resulting coefficients marginally improved the fit for the electrical range suggested in the literature as well as of current state-of-the-art solar cells. Nevertheless, through a dataset of 15,000 I-V measurements of industrial cells, a gap between the predicted and measured FF was observed. The impact of the effective ideality factor, edge recombination, and non-uniform recombination on the estimated FF were therefore investigated. Results show that adding information on the ideality factor or edge recombination increases the prediction accuracy. Moreover, the expressions tend to overestimate the FF for non-uniform cells. This study provides insights on the accurate estimation of FF through metrics that can be captured from luminescence images. This paves the way to improving the analysis of luminescence images for end-of-line characterization in industrial manufacturing lines.

Real-time Prediction Algorithms to Detect Clouds and Forecast Photovoltaic System Performance

Maqsood Ali Mughal, Habeebullah Adua, Muhammad Hammad Uddin, Evan Sauter, Stephen Natale, Timothy Lewis, and Jonathan G. Ferreira

Electrical & Computer Engineering Department, Worcester Polytechnic Institute, Worcester, Massachusetts, 01609, USA

Abstract— When clouds roll over photovoltaics (PV) arrays, PV systems' output power fluctuates. As a result, the PV penetration into the power grid causes energy imbalance and technical problems such as voltage instability in distribution networks. We designed an intelligent Cloud Motion Vector Sensor (CMVS) system model that utilizes gradient matrix and peak detection algorithms to compute cloud motion parameters such as cloud speed and direction. A predictive PV System model is also designed in Simulink that uses a perturb and observes maximum power point tracking (MPPT) algorithm to predict the performance of the PV system. Power management companies may benefit from the two models to better manage grid integration of renewables and non-renewables in the energy mix.

Keywords—Cloud detection, solar forecasting, prediction algorithms, photovoltaics, voltage fluctuation

I. INTRODUCTION

The performance of the PV system is susceptible to weather conditions such as cloud motion and the associated shadows moving over PV arrays, causing irradiance transitions. PV losses can be up to 80% during such transitions caused by an overpassing cloud shadow [1]. As a result, the energy supply can fluctuate below the actual demand. With PV penetration increasing in distribution networks, this fluctuation can cause significant inefficiencies in power supply. According to the 2021 U.S. Solar Market insight report by Solar Energy Industries Association (SEIA), the utility sector registered its largest third-quarter growth with 3.8 GWdc installed, despite the turmoil around supply chain constraints and rising equipment costs due to COVID-19 [2]. This increase in distributed PV power generation systems is causing energy imbalance and technical problems such as reverse power flows and voltage fluctuation in distribution feeders. Using the line voltage regulators, switched capacitor banks, and backup generators to mitigate the power loss is inefficient because switching these devices to operate more frequently than usual could damage the device, cause voltage flicker, and equipment to malfunction [3]. For example, utility companies report an increase in tap changers replacement in distribution transformers since the integration of PV systems into the power grid [4]. Hence, there is an essential need for power management companies to predict variability in PV systems performance and balance energy generation to minimize disruptions and the overall cost of operation[5], [6].

Thus, it becomes critical to detect clouds in advance and forecast the PV system output power to prevent voltage fluctuations and better integrate renewables and non-renewables in the energy mix. This is essential to make informed decisions about the amount of available PV energy for smart management of utilities and grid operations.

Data from National Weather Service helps plan future PV projects, but the information satellites provide about clouds varies seasonally, regionally, with the time of day, and with retrieval techniques [7]. In the past, various classical threshold-based algorithms [8] and machine learning techniques [9]–[11] have been employed, such as the linear combination model [11], multiple neural networks [12], and pixel-based classification [13] to detect clouds. However, each method has drawbacks regarding robustness to work accurately in different weather conditions. For example, threshold-based algorithms look for a series of proper thresholds of brightness temperatures via specific channels for different sensors and achieve cloud detection. But the threshold values vary by season, surfaces, and sun elevation, which causes variation in the threshold, ultimately influencing cloud detection results of the satellite images due to poor universality [14], [15]. Similarly, machine learning techniques are not consistent because model training depends upon the existing data, and the result may be misleading [12], [13], [16]. Also, predicting local and real-short-time information about clouds is challenging [17] while using satellite data.

In this paper, we introduced two models: (1) Cloud Motion Vector Sensor (CMVS) system model that utilizes gradient matrix and peak detection algorithms to detect clouds and compute cloud motion parameters such as cloud speed, direction, depth as well as the time of arrival of cloud at the PV site, and (2) a predictive model using perturb and observe (P&O) maximum power point tracking (MPPT) algorithm in Simulink/MATLAB that takes in real-time temperature and irradiance measurements to track the maximum power and forecast PV system performance.

The model can also trigger a signal to an alternate energy source such as a battery for convenient switching and smart energy management in case of PV output power going below a certain level. The performance of the two models was validated by an experimental setup at the WPI's 1 kW system PV site located at 37 Dean Street (East-Hall Parking Garage), WPI, Worcester, MA, U.S.

II. METHODOLOGY

A. CMVS System Model

A.1. CMVS System Components

The hardware design of the CMVS system consists of nine TSL2591 light sensors by Adafruit, two TCA9548A 1-to-8 I2C multiplexers by Adafruit, an ESP32 MCU by Espressif Systems, and a power circuit to power all the electronics.

We designed 3D-printed enclosures to protect light sensors from the environment. The enclosures are designed with a sliding jacket to fit a UV-VIS grade fused quartz glass plate on top of the sensor, as shown in Fig 1. The design also includes holes to maintain a fixed position. Each sensor was separated by an equal distance and angle and the wiring cables were protected with heat shrinking and weatherproof tubing.

A.2. CMVS System Design

The CMVS design is a cluster of nine-light sensors designed in a circular arrangement. The pictorial illustration of the CMVS design is shown in Fig 2. Eight sensors are positioned 45 degrees apart and 12 ft away from the base light sensor, which is placed at the origin (or center) and mounted on top of the CMVS PCB development board. The board houses two multiplexers and an ESP32 microcontroller (MCU) as shown in Fig 3. ESP32 is a low-cost development platform for embedded applications with integrated Wi-Fi and Bluetooth technology and was used collect irradiance measurements from all nine light sensors. For this study, the CMVS system design dimensions were chosen given the amount of space available on the roof, however, these dimensions are easily scalable. Similarly, the distance between the CMVS system and the PV site can be scaled to best support the short-term forecasting needs of the power management companies. Multiple of these light sensor clusters can form a big network of sensor clusters and if spread throughout a PV Site (or city) will forecast the PV output power of multiple PV sites within a short distance. This will bring communities a step closer to a concept of a smart city with more efficient and reliable distributed energy systems.

A.3. Working of CMVS System Model

The CMVS system model block diagram is shown in Fig 4. CMVS model will gather ambient light data transmitted using multiplexers to ESP32 MCU via I2C protocol. The data is then transferred onto ThingSpeak via WiFi connection for

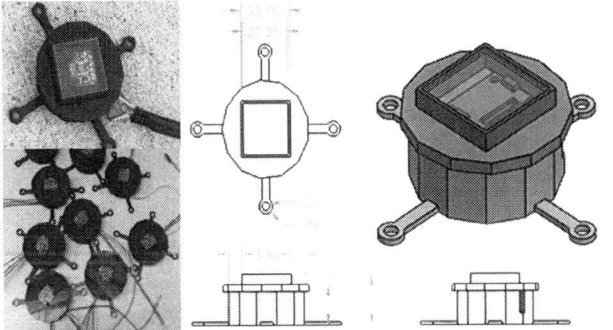

Fig. 1. 3D-printed protective enclosure for light sensors

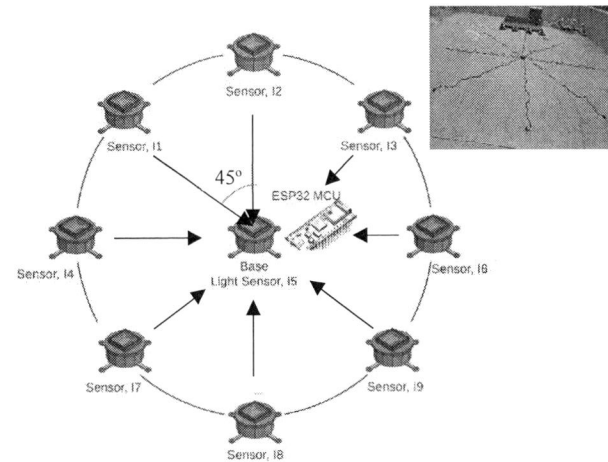

Fig. 2. CMVS design – a base light sensor at the origin with eight light sensors positioned in circular arrangement, 45 degrees apart. The distance between a light sensor and base light sensor is 12 ft.

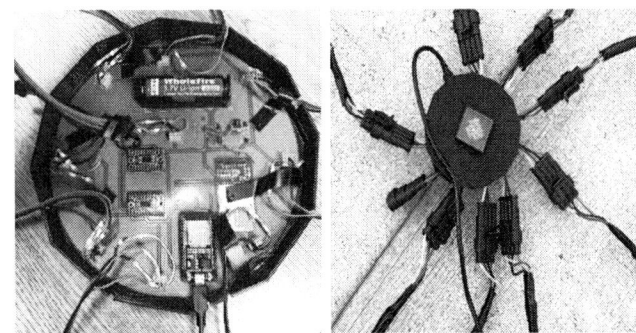

Fig. 3. CMVS PCB Development Board houses multiplexers and ESP32 MCU that collects irradiance measurements from TSL2591 light sensors and transfers it to ThingSpeak, an API-based IoT platform, for algorithmic processing and visualization.

algorithmic processing. The cloud detection algorithms are designed in a Matlab environment, and ThingSpeak enables to schedule and run Matlab scripts, which then computes cloud motion parameters.

A.4. Light Sensor Data Collection

CMVS design utilizes TSL2591, a luminosity sensor with an I2C interface and a built-in ADC that transforms light intensity into a digital signal output. This output is then fed into the ESP32 MCU via I2C protocol for data transfer and processing. TSL2591 draws 0.5 mA of current when actively sensing and less than 15 µA when in low-power mode, making the sensor great for low-power data-logging systems.

To convert lux data from the light sensor into irradiance (W/m^2), an Arduino sketch was modified using a conversion guide and Adafruit libraries. The accuracy of TSL2591 was tested against the LI-200R Pyranometer, the device widely used in the solar industry to measure global solar irradiation. The two devices performed remarkably well when tested indoors but varied widely outdoors in bright light, especially during transient light intensity changes. Using the curve fitting

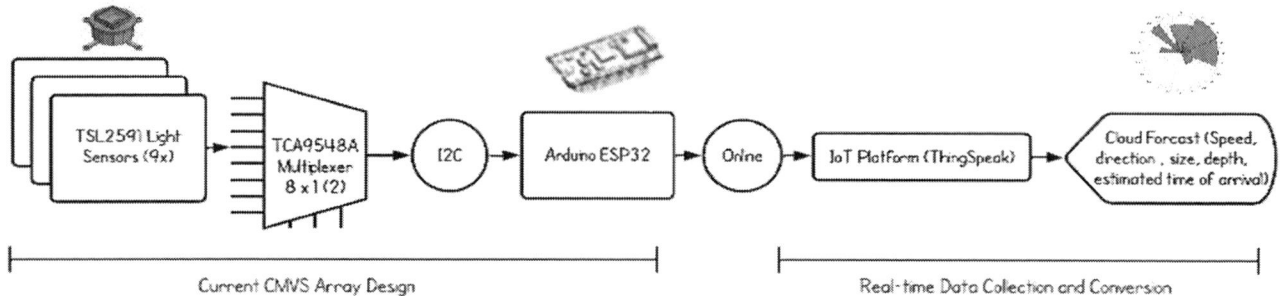

Fig. 4. CMVS system model block diagram

process, the TSL2591 sensor values were calibrated to assure the correct translation of the response of the TSL2591 into solar irradiance values. The lux and irradiance values were recorded for both TSL2591 and LI-200R in different scenarios and weather conditions. These values were plotted against the calibrated device to obtain the trend line and multiplier for the relation, as shown in Fig. 5.

Furthermore, to evaluate the accuracy of the TSL2591 light sensors, we performed a comparative analysis of the irradiance measurements of TSL2591 and LI-200R pyranometer by logging data at the same rate for 7 hours. The analysis revealed an average accuracy of > 97 % between the two devices as demonstrated in Fig.6. We further obtained some synthetic

datasets using the Monte Carlo simulation, which we used to train and test the CMVS algorithm.

Using two TCA9548A 1-to-8 multiplexers, irradiance measurements from TSL2591 light sensors were transmitted via the I2C channel address by connecting all sensors through data lines (SDA to SDA) and clock lines (SCL to SCL) using green and blue wires as shown in Fig. 7. The ESP32 MCU chip comes with 48 pins with pin GPIO 21 and GPIO 22 for SDA and SCL connected to the I2C connection of the multiplexers to enable data transfer. The chip also supports Wi-Fi, making it possible to log our sensor data directly to ThingSpeak, and we set the data rate to 1s to enable real-time data transfer. Lastly, Matlab scripts iterate through the TSL2591 light sensors connected to the multiplexers and record the measurements for algorithmic operations to compute cloud motion parameters.

A.5. Real-time Prediction Algorithm

Following the research work by [18], where they experimented with using three solar irradiance sensors within a single system, cloud shadow reaches two sensors when a sensor pair is parallel to the cloud edge. If the cloud motion vector is parallel to the cloud edge, the shadow will remain at the origin and never reach the other two sensors. In this work, we further developed the approach by using as many as nine-light sensors, where eight of the sensors can detect both point cloud and cloud edges. However, the eight sensors determine the cloud edges relative to the sensor at the origin. The method and approach used in these algorithms are described below:

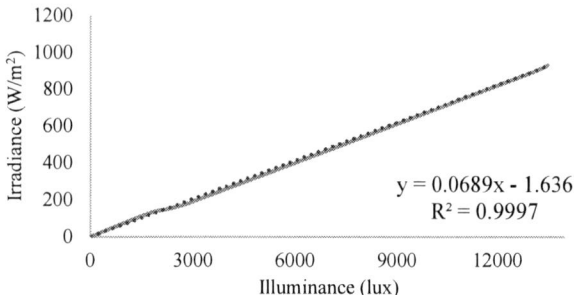

Fig. 5. The trend-line fitting using the curve fitting process for conversion of lux to irradiance

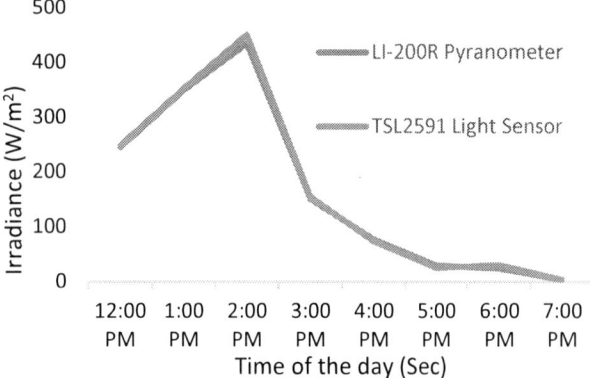

Fig. 6. LI-200R vs. TSL2591 sensor values

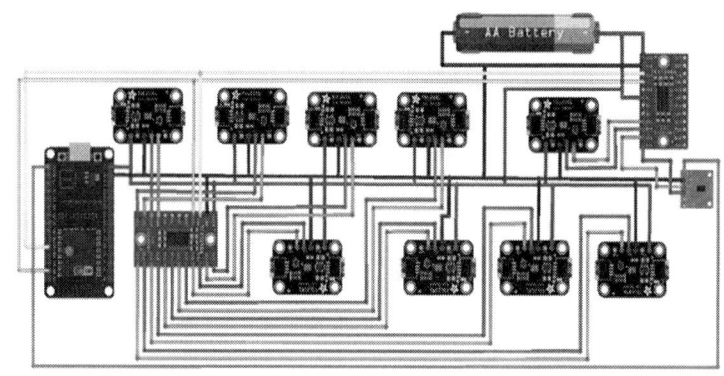

Fig. 7. CMVS System Data Collection Circuit Diagram

TABLE I. NOMENCLATURE IN THE RESEARCH WORK

Symbol	Parameter	Unit
υ_{xy}	Speed of the cloud	ms^{-1}
θ	Direction of the cloud shadow vector	deg.
t_{xy}	Time lag between the cloud coverage of light	s
d	Distance between two light sensors	m
L_c	Length of the cloud shadow	m
Δt_c	Coverage time	s
D_{cs}	Cloud shadow depth	%
ir_n	Average value of the irradiance	W/m^2
Φ	Angle of separation b/w PV site & sensor cluster	deg.
d_θ	Distance between PV Site and sensor cluster	m
$t_{arrival}$	Time-of-arrival of cloud at the PV Site	s

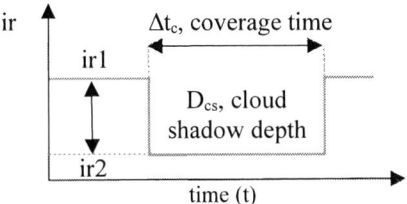

Fig.8. Change in irradiance to determine the length and depth of cloud shadow length and depth of cloud shadow.

A.5.3. Cloud's Time-of-Arrival at the PV Site

With the angle and speed of the cloud shadow known, along with the distance of the PV site and its angle of separation (Φ) from the CMVS design, we calculated the time-of-arrival ($t_{arrival}$) forecast for the cloud shadow to reach the PV site using the (5) below:

$$t_{arrival} = \frac{d_\theta}{\upsilon \cos(\Phi - \theta)} \qquad (5)$$

B. PV System Model

The predictive PV system model is designed in Simulink/MATLAB®. The model utilizes an intelligent perturb and observe (P&O) maximum power point tracking (MPPT) algorithm to track the maximum power and predict the performance of the PV system as a function of solar irradiance and temperature measurements. The two measurements are acquired using a LI-200R pyranometer by LI-COR Inc. and a TMP102 digital temperature sensor by Sparkfun operating at 1 Hz frequency.

The block diagram of the PV system model is shown in Fig. 9 and will be discussed in the following sections. For this study, the irradiance measurements are recorded using the LI-200R pyranometer, however, an equivalent mean average of the nine TSL2591 light sensors can also be used given the high accuracy of sensor values as validated and discussed previously.

B.1. PV System Design and Specifications

The nominal capacity of the actual PV system at standard temperature and conditions (STC) installed at the WPI campus is 960 Watt. The system comprises three PV arrays each with a nominal capacity of 330 W. The Trina Solar 330 W (TSM DD06M.05(II)) modules were installed and their configuration is given in Table II. The system has a charge controller (FM60-150VDC) and an Inverter module Advanced PLC AGM battery as an energy storage device connected to an LED signboard (a load). The system is configured through the MATE3S system display and controller through which the daily performance of the actual PV system can be monitored on-site as well as remotely using an online energy management tool, OPTICS RE, which can be found through the following link https://opticsre.com/Home/#/system/dashboard. The performance reports can be downloaded through an SD card installed inside the MATE3S controller or by signing up online to receive alerts with a downloadable link to the performance report.

A.5.1. Cloud Speed and Direction

We used the Gradient Matrix Method to calculate cloud shadow speed and direction. This method is based-upon cloud edge detection. In this method, the irradiance measurements from nine-light sensors are transformed into a 3x3 matrix and are processed much the same as image analysis. Once the 3x3 matrix is created, it is convoluted with the Sobel Filter. The Sobel filter is a matrix used in image processing and digital vision, specifically with edge detection algorithms. Once these two matrices are convolved, the result will be two magnitudes labeled A_x and A_y. Now taking these two magnitudes and using the (1), the angle to determine cloud shadow direction (θ) was obtained.

$$\theta = \tan^{-1} \frac{A_y}{A_x} \qquad (1)$$

We further used the Peak Matching Method, which also integrates the angle calculated from the Gradient Matrix Method to calculate the speed of the cloud υ_{xy}. This method utilizes the data from a pair of sensors and uses the correlation between speed, distance, and time as shown in (2).

$$v_{xy} = \frac{d}{t_{xy}} \qquad (2)$$

A.5.2. Cloud Shadow Size

To calculate the length of the cloud shadow L_c moving in a direction, we used (2) and irradiance data from the light sensors. Knowing the cloud coverage time (see Fig. 8) from the reduction in irradiance values, (3) can be used to calculate the length of the cloud shadow.

$$L_C = v \cdot \Delta t_c \qquad (3)$$

Similarly, we calculated the cloud shadow depth D_{cs} using the reduction rate in irradiance values from the cloud cover from (4).

$$D_{CS} (\%) = \left(1 - \frac{ir_2}{ir_1}\right) X\ 100 \qquad (4)$$

978-1-7281-6118-1/22 $31.00 © 2022 IEEE

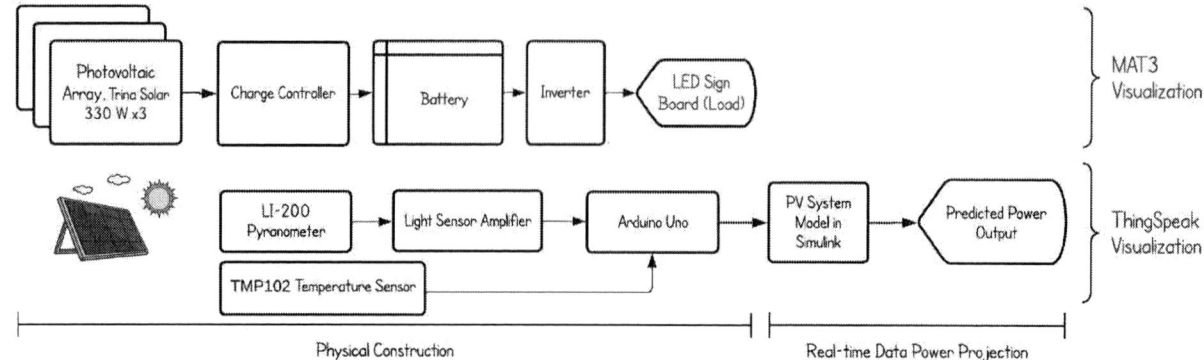

Fig.9. PV System Model block diagram

TABLE II CONFIGURATION OF TRINA SOLAR 330 W MODULE AT STC

Parameters	Value
Solar Cells	Monocrystalline
Cell Orientation	120 Cells
Peak Power Watt	330 W
Maximum Power Voltage	33.8 V
Maximum Power Current	9.76 A
Open Circuit Voltage	40.6 V
Short Circuit Current	10.40 A
Maximum Efficiency	19.0 %

B.2. Data Collection for PV System Model

Both LI-200R pyranometer and TMP102 temperature sensors were installed and positioned next to the base light sensor at the WPI PV site. The pyranometer measurements are imported into the PV system model using a 2420 Light Sensor Amplifier from LI-COR. The amplifier converts the current (µA) signal from the irradiation sensor into a voltage that can be read by Arduino. The gain settings of the amplifier were optimized by calculating the ideal amplifier gain to accommodate a variety of full-scale light intensities and full-scale voltage ranges by setting switches on the amplifier. The amplifier has a gain accuracy of ±0.5% overall ranges at operating temperatures of −40 to 50 °C. The amplifier's gain was set to 0.05 V µA−1 to match the 5 V range of the Arduino's analog-to-digital converter. Using this gain factor, the voltage multiplier was used to convert the voltage measured by the pyranometer into units of irradiance (W/m^2). Finally, using the ThingSpeak Support Toolbox, the real-time irradiance measurements were imported into the PV system model. The same toolbox also enabled importing temperature sensor readings from TMP102.

B.3. Operation of PV System Model

The design, configuration, and performance of the PV system were studied from the actual PV system as given in Table II to design the PV system model that resembles the actual PV system. A single-phase inverter topology with the MPPT technique was used to build and configure the PV system model in Simulink®. This topology has been addressed in [19]. The model is composed of a PV array block from the Simulink library configured using information from Table II. The PV array block has two inputs, solar irradiance (W/m^2) and

TABLE III NOMENCLATURE USED IN (6)

Symbol	Parameter	Unit
I_{pv}	PV current of the array in Ampere	A
I_0	saturation current of the array in Ampere	A.
V_t	Thermal Voltage of the array=$N_s kT/q$	V
N_s	Number of solar cells connected in series	
k	Boltzmann constant	J/K
q	The elementary charge on an electron	C
R_s	Equivalent Series resistance of PV cells	Ω
R_p	Equivalent Parallel resistance of PV cells	Ω
T	Temperature of the solar cell	K
V	Output voltage from PV cell	V
P	Power	W

TABLE IV PARAMETERS FOR THE SIMULATION OF PV SYSTEM MODEL

Parameters	Value
Input Filter Capacitor	1000 µF
Boost Inductor	3.5 µH
Output Capacitor	184.4 µ F
DC Bus Capacitor	1640 C F
LCL Filter Capacitor	100 µF, 3 µF and 100 µF
Full Bridge Converter	Average Model (U$_{ref}$ Controlled)
PWM Frequency	10,000 Hz
MPPT Controller	Perturb & Observe Algorithm

temperature (°C), and the output is the open-circuit voltage (V$_{oc}$) terminals. The PV array block uses a single diode model to generate the PV power as shown in Fig. 10. As can be seen in the model a current source is fed by irradiation and temperature. The generated current of the solar cell can be determined by (6). This model current when flows through internal resistance as shown in the single diode model produces a voltage that appears as the open-circuit voltage at the output terminal. The variables

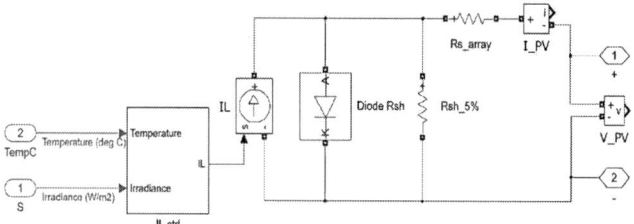

Fig. 10 The Single Diode Model for PV cell

used in (6) are listed above in Table III. The power can be obtained by the product of voltage and current values from the output using voltage and current sensor [5], [20].

$$I = I_{pv} - I_0 \left[exp\left(\frac{V+R_s I}{V_t \alpha}\right) - 1 \right] - \frac{V+R_s I}{R_p} \quad (6)$$

This predictive PV system model utilizes the standard P&O MPPT algorithm, which has been used in the past to track the maximum power of the PV system model by [21], [22]. This P&O algorithm takes the value of the voltage and current from the PV model and perturbs the voltage to get the maximum power at given irradiance and temperature. The model uses a two-stage inverter design to generate power output from the PV array configured. In the first stage, a boost converter is used to increase voltage, and then a two-level inverter topology is used to convert direct current power to alternating current power. A Boost converter with proportional plus integral (PI) control is used to implement the two-stage inverter followed by sine wave pulse width modulation (SPWM) control for the two-level inverter configuration. The Boost converter is controlled through the P&O MPPT controller. The resistive load is used at the inverter output terminal. The components of the PV system model and corresponding values are mentioned in Table IV. Irradiance and temperature sensor measurements were supplied to the PV system model in real-time. The predictive PV system model built in Simulink is shown in Fig 11.

III. RESULTS AND DISCUSSION

A. PV System Model Performance

To evaluate the performance of the PV system model, the PV output power values collected from 11:00 to 17:00 EST for two random days were used as shown in Fig. 12 and 13. The

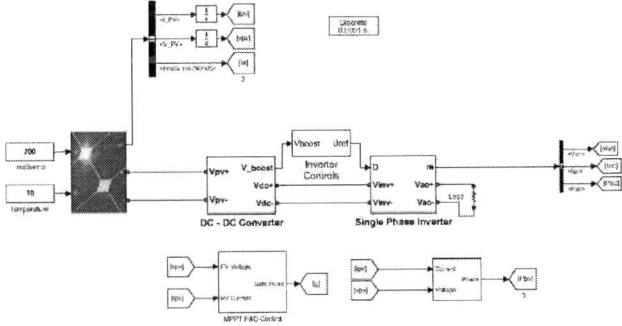

Fig. 11 The Simulation Setup for PV System Model

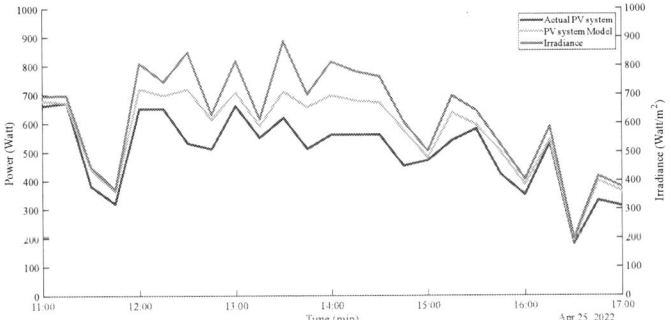

Fig. 12 The PV system performance comparison for Day 1

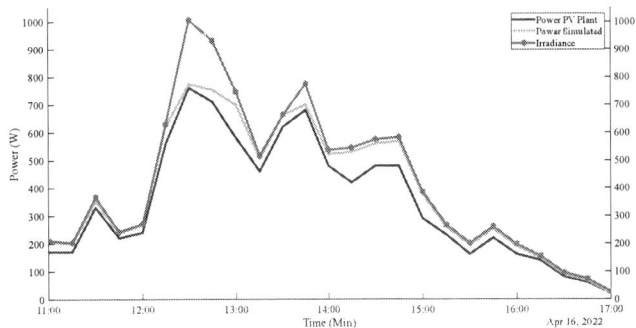

Fig. 13 The PV system performance comparison for Day 2

predicted power values from the PV system model were compared against the power values (available through performance reports from OPTICS RE) produced by the actual PV system. In this experiment, the weather in Worcester, MA played a vital role, as we were experiencing days with cloudy weather conditions during the week. The overall performance of the PV system model was satisfactory with an accuracy of 84%.

B. CMVS System Model Performance

The synthetic dataset was used to train and test the performance and accuracy of the CMVS model. Every angle responded correctly, as depicted in Fig. 15a. We further test the accuracy using real-world conditions by placing objects near each sensor in other to cast shade on them. Fig. 15b shows the equivalent histogram obtained from a real-world dataset. We realized that both the high-level and low-level cloud directions deflect toward the respective sensor angles. The cloud depth, length, and arrival time were not computable in Fig. 15a because of the discrete nature of the test.

Also, cloud parameters with negative values indicate either the system is under a static shade or PVSite_phi and PVSite_distance were set to 0 from the Matlab function. To test the performance of the PV model, the mean average of the CMVS irradiance value was input to the PV model using Thingspeak input. The model power significantly decreases when shade is cast on any sensor. This explains PV power reduction when a panel is induced to partial object shading. Generally, shading a cell in a PV module will cause the output power of the whole module to fall to half or more no matter how many cells there are in the string.

We further installed the CMVS model 100ft away from the site PV panel and modified the PVSite_phi to 90' and PVSite_distance to 30m. The 90' angle is the direction of the first sensor towards the installed PV panel. Several cloud events were considered for testing the performance of both the CMVS model and PV system model power. The high and low-level clouds' directions disperse widely and generate small bin indications for clear skies other than their primary directional angles as shown in Fig. 15 c and d. Even though the brightness is not constant, the illumination around each sensor is almost equivalent. This justifies short-term cloud instability as the reason for the directional disperse of the algorithm. Also, the cloud shadow parameters are often different for every page refresh or cloud event. The speed and length of the cloud were computable in this scenario. However, the cloud depth sometimes produces NAN or infinitesimal values because the algorithm calculates a series of zero or infinite, making it

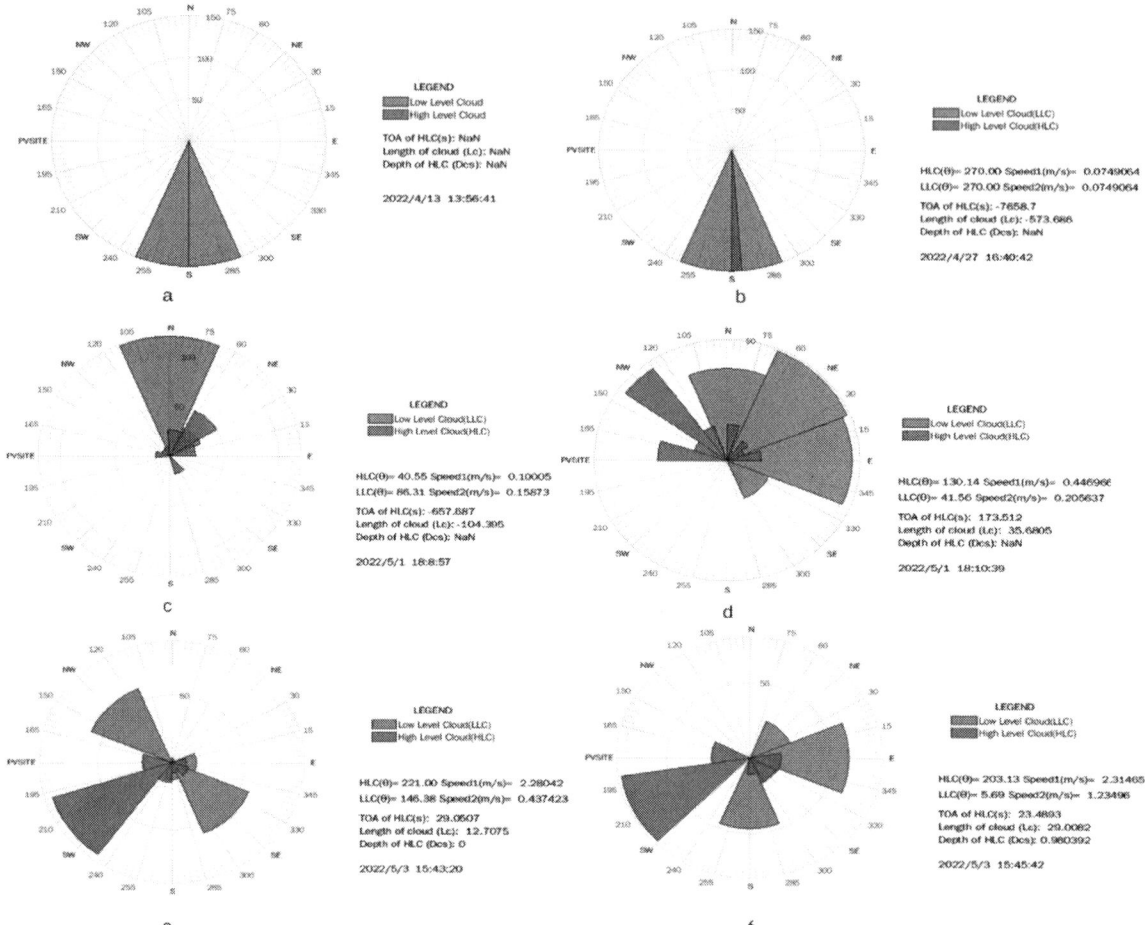

Fig. 14 The CMVS cloud parameters

difficult for the algorithm to compute. Instances of the negative cloud parameters indicate that the cloud shadow is moving faster than the system can detect due to the high resolution of the algorithm.

We also monitor the system on regular and cloudy days, noticing obtainable differences in the cloud directions and other cloud parameters. The high-level cloud and other cloud parameters were easily determined, and eye inspection can quickly tell the cloud direction is performing efficiently. The TOA, length of the cloud, and depth were obtained from the high-level clouds. They had no relation to the installed PV system in instances when the directions point away from the PV site. However, for cloud events with high-level cloud pointing toward the installed PV panel, the cloud speed is the speed the algorithm determines for the point cloud will take to reach the PV panel from our 100 ft installation point. We record each value of the cloud parameters relative to the PV system model output power for further analysis. Fig. 14 e&f shows the simulation results on a regular day.

The PV System Model power in this experimental series was assumed to be the predicted power of the system. It indicates the power the PV System Model will produce if an actual cloud event occurs at the panel heads. We obtained the predicted power from Thingpeak data export, while collecting data from

the outback was tedious, especially in obtaining the required resolution. We recorded the cloud parameters and predicted power from our visualization platform while we compared the predicted power to the Outback PV system performance. There were no significant differences in the performances because of the small nature of the project and installation proximity. However, the power differences indicate the variable cloudiness between the two installation positions. We are only compiling a dataset for cloud events when the cloud moves towards the PV site and the respective predicted power.

C. Visualization and Energy Management

Thingspeak, an API-based IoT platform service, was used for data collection from sensors, algorithmic processing, visualization and analyzing of real-time data in the cloud. Matlab scripts iterate through the TSL2591 light sensors to record the irradiance measurements for algorithmic operations and compute cloud motion parameters. ThingSpeak enables to schedule and run Matlab scripts at regular interval of time.

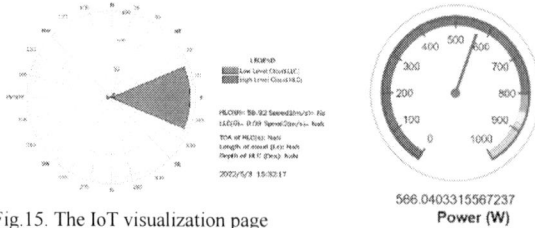

Fig.15. The IoT visualization page

Similarly, the predictive PV output power values from the Simulink model were sent to ThingSpeak for visualization. The pages autorefresh itself every 10 seconds using the Metarefresh HTML function to monitor cloud motion and output power of the PV system, as shown in Fig.15.

IV. CONCLUSION

The paper discusses a design and development of a unique CMVS system model to detect cloud and compute cloud motion parameters. The model can predict local and real-short-time information about clouds that grid operators can use to manage better grid integration of renewables and non-renewables in the energy mix. One significant advantage of the CMVS system design is the scalable deployment of the low-cost TSL2591 light sensor clusters forming a big network of sensor clusters, which, if spread throughout the PV Site (or city) can help forecast PV output power of multiple PV systems within a short distance. This brings society a step closer to a concept of a smart city with more efficient and reliable microgrid systems. A predictive PV System model is also designed in Simulink that uses perturb and observes maximum power point tracking (MPPT) algorithm to predict the performance of the PV system. Although there are a few limitations to the model in terms of accuracy falling short when there is a significant change in irradiance in a short time from cloud cover, the overall performance of the model was satisfactorily exhibiting an accuracy of 84% when compared to the performance of the actual PV system. Future work will repeat this same process by using four CMVS models placed at the four cardinal positions of the installed PV system. We plan to deploy an onsite computer vision at the farm center to capture real-time cloud conditions to validate the CMVS forecast and set a threshold for when to mount the backups.

V. ACKNOWLEDGMENT

This work was supported by the MassCEC Catalyst program. We also appreciate the help of Umair Zia from Eversource Energy for his support and feedback.

REFERENCES

[1] K. Rahimi, S. Omran, M. Dilek, and R. Broadwater, "Computation of voltage flicker with cloud motion simulator," IEEE Transactions on Industry Applications, vol. 54, no. 3, pp. 2628–2636, May 2018, doi: 10.1109/TIA.2017.2787621.

[2] "Solar Market Insight Report 2021 Q4, Solar Energy Industries Association(SEIA). 2021" https://www.seia.org/research-resources/solar-market-insight-report-2021-q4 (accessed May 06, 2022).

[3] M. A. Mughal et al., "Cloud motion vector system to detect clouds and forecast real-time photovoltaic system performance; cloud motion vector system to detect clouds and forecast real-time photovoltaic system performance," 2021, doi: 10.1109/PVSC43889.2021.9518910.

[4] L. Hülsmann, "Evaluation of two distribution grids in terms of PV penetration limits and effectiveness of reactive power controls," 2016.

[5] T. Ma, H. Yang, and L. Lu, "Solar photovoltaic system modeling and performance prediction," Renewable and Sustainable Energy Reviews, vol. 36, pp. 304–315, Aug. 2014, doi: 10.1016/J.RSER.2014.04.057.

[6] M. A. Khallat and S. Rahman, "A probabilistic approach to photovoltaic generator performance prediction," IEEE Transactions on Energy Conversion, vol. EC-1, no. 3, pp. 34–40, 1986, doi: 10.1109/TEC.1986.4765731.

[7] S. Mahajan and B. Fataniya, "Cloud detection methodologies: variants and development—a review," Complex & Intelligent Systems, vol. 6, no. 2, pp. 251–261, Jul. 2020, doi: 10.1007/s40747-019-00128-0.

[8] H. Fu et al., "Cloud detection for FV meteorology satellite based on ensemble thresholds and random forests approach," Remote Sensing, vol. 11, no. 1, p. 44, Dec. 2018, doi: 10.3390/rs11010044.

[9] K. Tan, Y. Zhang, and X. Tong, "Cloud extraction from Chinese high-resolution satellite imagery by probabilistic latent semantic analysis and object-based machine learning," Remote Sensing, vol. 8, no. 11, p. 963, Nov. 2016, doi: 10.3390/rs8110963.

[10] L. Gomez-Chova, G. Mateo-Garcia, J. Munoz-Mari, and G. Camps-Valls, "Cloud detection machine learning algorithms for PROBA-V," in 2017 IEEE International Geoscience and Remote Sensing Symposium (IGARSS), Jul. 2017, pp. 2251–2254. doi: 10.1109/IGARSS.2017.8127437.

[11] L. Liu, J. Li, Y. Wang, Y. Xiao, W. Zhang, and S. Zhang, "Thin cloud detection method by linear combination model of cloud image," The International Archives of the Photogrammetry, Remote Sensing and Spatial Information Sciences, vol. XLII–3, pp. 1079–1083, Apr. 2018, doi: 10.5194/isprs-archives-XLII-3-1079-2018.

[12] Y. Chen, R. Fan, M. Bilal, X. Yang, J. Wang, and W. Li, "Multilevel cloud detection for high-resolution remote sensing imagery using multiple convolutional neural networks," ISPRS International Journal of Geo-Information, vol. 7, no. 5, p. 181, May 2018, doi: 10.3390/ijgi7050181.

[13] W. Wu, J. Luo, X. Hu, H. Yang, and Y. Yang, "A thin-cloud mask method for remote sensing images based on sparse dark pixel region detection," Remote Sensing, vol. 10, no. 4, p. 617, Apr. 2018, doi: 10.3390/rs10040617.

[14] C. Li, J. Ma, P. Yang, and Z. Li, "Detection of cloud cover using dynamic thresholds and radiative transfer models from the polarization satellite image," Journal of Quantitative Spectroscopy and Radiative Transfer, vol. 222–223, pp. 196–214, Jan. 2019, doi: 10.1016/j.jqsrt.2018.10.026.

[15] A. Dybbroe, K.-G. Karlsson, and A. Thoss, "NWCSAF AVHRR cloud detection and analysis using dynamic thresholds and radiative t modeling. Part I: algorithm description," Journal of Applied Meteorology, vol. 44, no. 1, pp. 39–54, Jan. 2005, doi: 10.1175/JAM-2188.1.

[16] Y. Changhui, Y. Yuan, M. Minjing, and Z. Menglu, "Cloud detection method based on feature extraction in remote sensing images," The International Archives of the Photogrammetry, Remote Sensing and Spatial Information Sciences, vol. XL-2/W1, pp. 173–177, May 2013, doi: 10.5194/isprsarchives-XL-2-W1-173-2013.

[17] P. Shah, "Development and hardware implementation of an efficient algorithm for cloud detection from satellite images," Signal & Image Processing : An International Journal, vol. 7, no. 2, pp. 73–80, Apr. 2016, doi: 10.5121/sipij.2016.7205.

[18] J. L. Bosch and J. Kleissl, "Cloud motion vectors from a network of ground sensors in a solar power plant," Solar Energy, vol. 95, pp. 13–20, Sep. 2013, doi: 10.1016/j.solener.2013.05.027.

[19] F. Gao, D. Li, P. C. Loh, Y. Tang, and P. Wang, "Indirect DC-link voltage control of two-stage single-phase PV inverter," 2009 IEEE Energy Conversion Congress and Exposition, ECCE 2009, pp. 1166–1172, 2009, doi: 10.1109/ECCE.2009.5316399.

[20] M. G. Villalva, J. R. Gazoli, and E. R. Filho, "Comprehensive approach to modeling and simulation of photovoltaic arrays," IEEE Transactions on Power Electronics, vol. 24, no. 5, pp. 1198–1208, 2009, doi: 10.1109/TPEL.2009.2013862.

[21] N. Femia, G. Petrone, G. Spagnuolo, and M. Vitelli, "A technique for improving P&O MPPT performances of double-stage grid-connected photovoltaic systems," IEEE Transactions on Industrial Electronics, vol. 56, no. 11, pp. 4473–4482, 2009, doi: 10.1109/TIE.2009.2029589.

[22] M. H. Uddin, M. A. Baig, and M. Ali, "Comparision of 'perturb & observe' and 'incremental conductance', maximum power point tracking algorithms on real environmental conditions," 2016 International Conference on Computing, Electronic and Electrical Engineering, ICE Cube 2016 - Proceedings, pp. 313–317, Jun. 2016, doi: 10.1109/ICECUBE.2016.7495244.

Surrogate Modeling for Rapid Prediction of Energy Yield from Vehicle-Integrated Photovoltaics

Timofey Golubev

ThermoAnalytics, Inc., Calumet, MI, 49913, USA

Abstract—This work demonstrates the development of a machine learning-based surrogate model for rapid prediction of energy yield from vehicle-integrated photovoltaic (VIPV) systems. The surrogate model was trained using results from thermal-electrical simulations that couple a commercial heat transfer solver with temperature-dependent electrical models. The trained model was tested on unseen data also generated from the thermal-electrical simulations. The surrogate model was used to perform a sensitivity analysis of the impact of location and meteorological conditions on VIPV energy production.

Index Terms—vehicle-integrated photovoltaics, surrogate modeling, energy yield, electric vehicles, thermal-electrical

I. INTRODUCTION

Vehicle-integrated photovoltaics (VIPVs) may benefit electric vehicle (EV) performance by extending the driving range, reducing the frequency of charging, and powering secondary electronic systems [1]–[3]. Evaluating the electricity production of VIPVs through simulation is necessary to efficiently design these systems and estimate their energy yield. In previous work, we developed a VIPV modeling approach that considers thermal effects by coupling a commercial heat transfer software with electrical photovoltaic (PV) models [4], [5]. While this approach allows for careful consideration of the vehicle's geometry and the system's thermal and electrical properties, it takes more than one hour to run a year-long hourly energy production simulation for a single location. For comprehensive studies of energy yield under varying geographical and meteorological boundary conditions, faster simulations are desirable. In this work, we develop an approach for rapid prediction of VIPV energy yield through surrogate modeling.

II. METHODOLOGY

A. Thermal-Electrical Model

The surrogate model was trained using results from a coupled thermal-electrical model of a Jeep Grand Cherokee with on-board PVs, which we developed in previous work [4], [5]. The model was created in the commercial heat transfer software TAITherm [6], which uses a numerical, finite volume method based on first principles physics to solve for heat transfer due to conduction, convection, and radiation (Fig. 1a). A PV equivalent circuit electrical model was implemented in Python and coupled with the thermal model. The irradiance and temperature dependent parameters of the electrical model are recomputed at each time-step of the thermal simulation, using the newly-computed module temperatures and incident

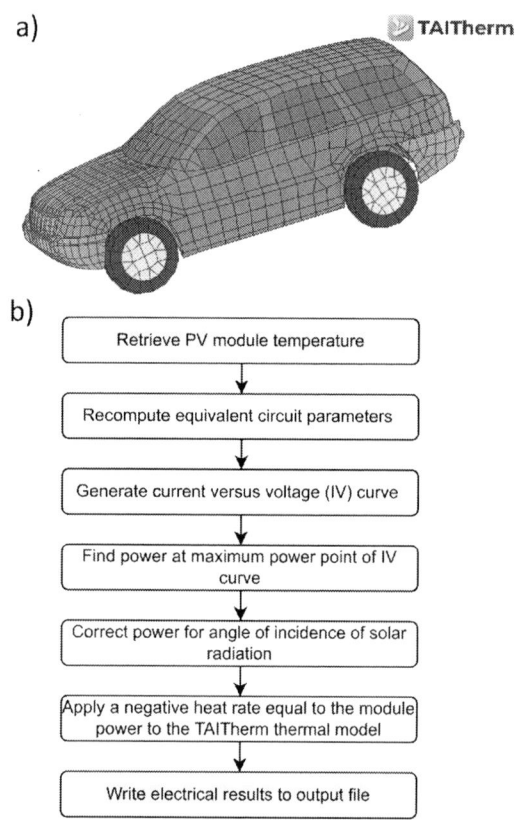

Fig. 1. a) TAITherm vehicle model geometry. b) Flow chart of the calculations at each time-step of the coupled thermal-electric PV model.

irradiance from TAITherm. In contrast with our previous work, which used the five-parameter De Soto model [7], here we use the six-parameter California Energy Commission (CEC) model [8]. The CEC model enables more robust and accurate derivation of the equivalent circuit parameters from PV module datasheets.

Fig. 1b shows the PV electrical calculations that are performed at the end of each time-step of the thermal simulation. After recomputing the equivalent circuit parameters, a non-linear equation is solved for the current versus voltage (IV) behavior of the solar modules. The modules are assumed to operate at the maximum power point of the IV curve; therefore, the PV power production is simply the product of current and voltage along the IV curve that gives the maximum value. Then, the power is corrected by an incidence angle

978-1-7281-6118-1/22 $31.00 © 2022 IEEE

TABLE I
EQUIVALENT CIRCUIT PARAMETERS USED TO MODEL THE VIPVS

Parameter	Description	Value
a_{ref}	Modified nonideality factor	2.5654 (dimensionless)
$I_{L,ref}$	Light current	6.5395 A
$I_{0,ref}$	Diode reverse saturation current	1.2992×10^{-12} A
$R_{s,ref}$	Series resistance	0.69760 Ω
$R_{sh,ref}$	Shunt (parallel) resistance	232.66 Ω
Adjust	Adjustment to temp.-coeff. of I_{sc}	14.930%

TABLE III
SUMMARY STATISTICS OF TUNED MODEL ERRORS IN VIPV POWER (W)
PREDICTION OVER FIVE CROSS-VALIDATION FOLDS

	Elastic Net	Support Vector Machine	Random Forest	Gradient Boosting	Artificial Neural Network
MAE	3.03	0.67	1.71	1.35	0.80
Stdev.	0.15	0.03	0.08	0.06	0.05

modifier to account for the effect of the angle of incidence of incoming solar radiation on power output. Finally, a negative heat rate equal to the module power output is imposed on the PV modules in the TAITherm vehicle model to account for the portion of solar radiation that is converted to electricity instead of heat.

For this study, we model the integration of SunPower Maxeon Gen III solar panels into the 2.8 m^2 area of the roof of the Jeep. We derive the equivalent circuit parameters from the module datasheet [9] using the method described in [8] (Table I). The computed power for the 1.77 m^2 module is scaled to represent the power that would be produced by a VIPV system with a 2.8 m^2 area. We assume that the vehicle is kept outside on asphalt (infrared emissivity of 0.94 and solar absorptivity of 0.85) in an unshaded area. To simulate the climatic conditions of the different locations, we apply typical meteorological year (TMY) weather data from the National Solar Radiation Database (NSRDB) [10] as boundary conditions.

B. Surrogate Model

To generate the training dataset for the surrogate model creation, we ran a year-long TAITherm simulation of the VIPV system in Seattle, Washington using one hour time-steps. We randomly selected 80% of the resulting data for training and 20% for a test set. We use the following model inputs (i.e., features): global horizontal irradiance (GHI), direct normal irradiance (DNI), direct horizontal irradiance (DHI), solar azimuth, solar zenith, and ambient air temperature. The

target variable is the power output of the VIPV in each hour of the year. We generated the surrogate model using an in-house automated model selection and hyperparameter tuning code, which compared the performance of elastic net, support vector regression (SVR), random forest, gradient boosting, and artificial neural network multi-layer perceptron (ANN) machine learning algorithms from the Scikit-Learn Python package (version 0.24.2) [11]. In all cases, the features were scaled prior to model training. Standardization was used for all models except ANN, for which we used min-max scaling. We tuned the hyperparameters for each of the algorithms by conducting a grid search over the hyperparameter space with five-fold cross-validation (CV) on the training dataset. Table II shows the hyperparameter search space and the final hyperparameter values that were chosen by the optimization for each model. The model selection process completed in one hour using 20 CPU threads on an Intel i9-10900K.

Table III shows the summary statistics of the mean absolute error (MAE) over five CV folds of the machine learning models after hyperparameter tuning. The best two performing models were found to be SVR and ANN. The best performing SVR used a radial basis function (rbf) kernel with a regularization parameter, C, of 1000. The best performing ANN used a logistic activation function, 3 hidden layers with 100 neurons in each, an L2 penalty parameter of 10^{-5}, and the `lbfgs` solver with 5000 maximum iterations. Scikit-Learn default values were used for all other parameters.

TABLE II
HYPERPARAMETER OPTIMIZATION OF THE SURROGATE MODELS

Model	Parameter	Search Space	Final Values
Elastic Net	L1 ratio	[0.1, 0.5, 0.7, 0.9, 0.95, 0.99, 1]	1
Support Vector	Kernel	[linear, polynomial (degree [2, 3, 4, 5]), rbf]	rbf
	Kernel coefficient (gamma)	[scale, auto]	auto
	Regularization (C)	[0.001, 0.01, 0.1, 10, 100, 1000]	1000
	Epsilon	[0.1, 0.5, 1, 1.5]	0.1
Random Forest	Number of Estimators	[10, 50, 100, 500, 1000]	1000
	Max Features	[2, 3, 4, 5]	5
Gradient Boost	Number of Estimators	[10, 50, 100, 500, 1000]	1000
	Learning Rate	[0.0001, 0.001, 0.01, 0.1, 1.0]	0.1
Artificial Neural Net	Activation	[identity, logistic, tanh, relu]	logistic
	L2 Regularization (alpha)	[10^{-6}, 10^{-5}, 10^{-4}, 10^{-3}]	10^{-5}
	Hidden Layers Configuration	[(10, 10, 10), (30, 30, 30), (50, 50, 50), (50, 100, 50), (100, 100, 100)]	(100, 100, 100)

III. RESULTS AND DISCUSSION

Fig. 2 shows comparisons of the PV power output prediction of the surrogate models and the TAITherm thermal-electrical model on the Seattle test dataset (the 20% of the TAITherm results that we excluded from the training data) and full-year hourly datasets from Phoenix, Arizona, and Philadelphia, Pennsylvania. Note that the surrogate models were trained only with data from Seattle. The mean absolute error (MAE), root mean square error (RMSE), mean absolute percent error (MAPE), and coefficient of determination (R^2) were used to evaluate model performance. While the SVR has slightly better performance scores than the ANN on test data from the same location as the models were trained on (Seattle), the ANN generalizes better to new locations, with less than 3% error and $R^2 > 0.99$. Therefore, we used the ANN for the remainder of this study.

The TAITherm model took 1.7 hours to run an hour-by-hour annual energy yield simulation using 8 threads on an Intel Xeon W-2145. Predicting energy yield with the ANN model was 5 orders of magnitude faster, taking only 0.014 seconds. Training the ANN on the test dataset consisting of 7008 data points (80% of the 8760 hours in a year) took 107 seconds. This speedup with only a small loss in accuracy enables comprehensive studies of the performance of a particular VIPV system in different regions, weather conditions, and EV usage patterns (e.g., time parked outside versus in a garage). For example, the simulation time for a study of 100 locations is reduced from 170 hours with the thermal-electrical approach to less than 2 hours, most of which is time for running the TAITherm model to generate training data.

As a demonstration of an analysis that can be performed with the surrogate model, we apply the ANN to predict the energy production of the VIPV system in four cities: Seattle, Washington; Phoenix, Arizona; Philadelphia, Pennsylvania; and Miami, Florida. Fig. 3 shows the sum of the hourly energy yield predictions for each month. We find that the ANN slightly under-predicts (by < 5%) the energy yields for Phoenix, Philadelphia, and Miami, which are locations that were not included in the training data. Since the ANN's prediction error is much smaller than the seasonal variability in energy yield, it is still suitable for comparative analysis. The ANN's generalizability to new locations could possibly be improved by including geographically and meteorologically diverse locations in the training dataset.

The surrogate model can also be used to quantify the uncertainty in VIPV energy production in a given location due to year-to-year variations in weather patterns. As an example, we used the ANN and 20 years of historical weather data from the NSRDB to predict energy production with hourly resolution in four cities (Fig. 4). The standard deviation of annual yields is expected to be 3.3%, 1.1%, 4.3%, and 3% in Seattle, Phoenix, Philadelphia, and Miami, respectively. This small year-to-year variability suggests that using TMY weather data is sufficient for realistic VIPV yield predictions.

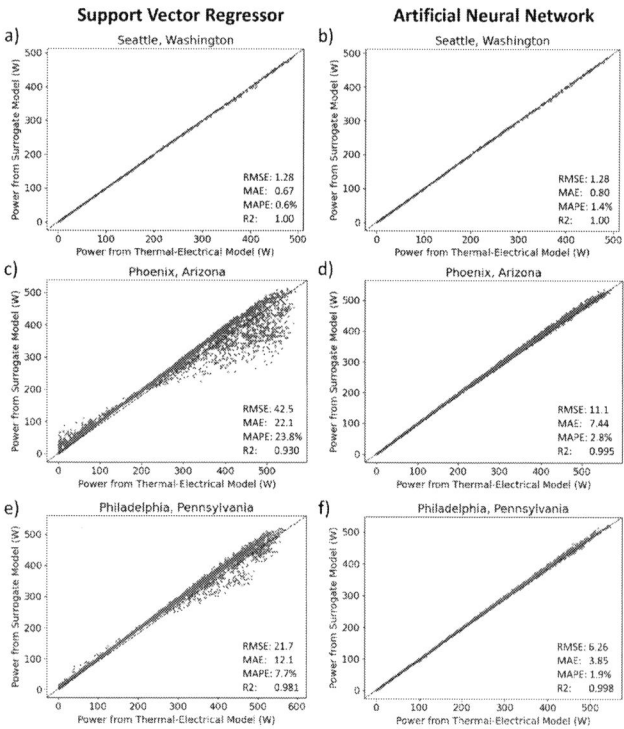

Fig. 2. Comparison of the VIPV power output prediction of the SVR (a,c,e) and ANN (b,d,f) surrogate models on unseen data versus the thermal-electrical model for three locations. The diagonal dotted line is included as a visual aid and represents an exact match between the models.

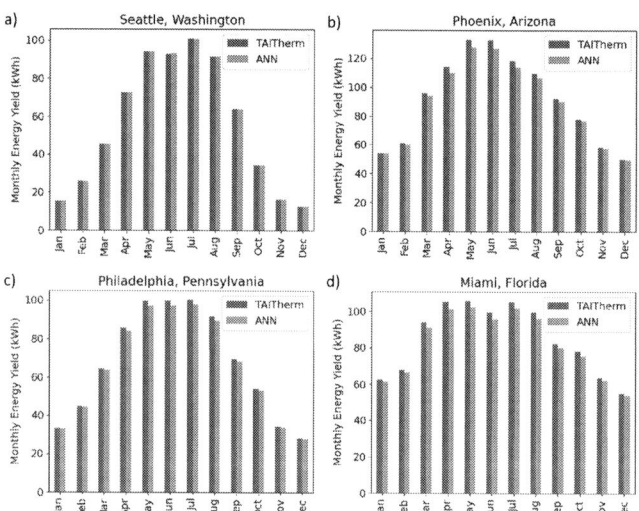

Fig. 3. Monthly energy yield of the VIPV system in four locations as predicted by the optimized ANN and the TAITherm thermal-electrical model.

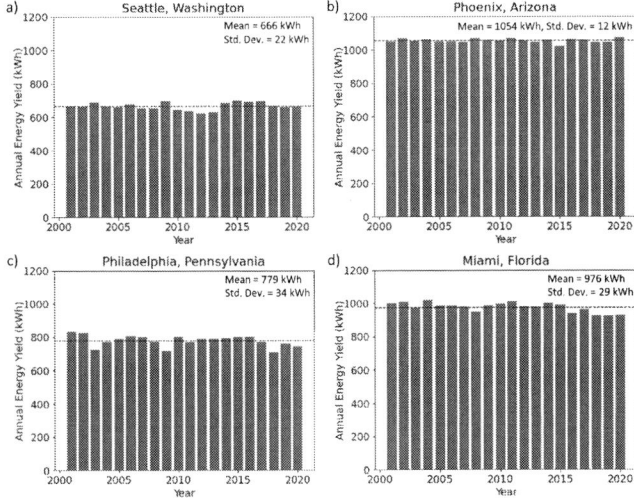

Fig. 4. Surrogate model predictions of the annual energy yield of the VIPV system in four cities based on historical weather data from 2001-2020. The dashed horizontal line represents the mean annual yield.

IV. CONCLUSION

This work demonstrates an approach for accelerating VIPV energy yield studies by training a surrogate model on a physics-based thermal-electrical simulation of a particular VIPV system. The surrogate model can then be used to predict the impact of geographic location and weather on that system's energy production with 5 orders of magnitude speedup and less than 3% error when compared to the physics-based simulation. In this study, the surrogate model was applied for energy predictions in four cities. The results show that while the year-to-year energy yield variability at a single location is small ($< 5\%$), the variability between seasons and locations is significant. Therefore, comprehensive simulation studies over many regions should be performed to provide realistic estimates of a VIPV system's energy production. In future work, the surrogate model's prediction accuracy could be improved by using data from several geographic locations for training. Additionally, training the model to predict the impact of changing the VIPV system's thermal and electrical properties on energy production should be explored. Finally, the model could be combined with EV energy consumption simulations to study the potential impact of VIPVs on driving range in different regions of the world.

ACKNOWLEDGMENT

The author thanks David M. Less for useful discussions.

REFERENCES

[1] M. Heinrich, C. Kutter, F. Basler, M. Mittag, L.E. Alanis, D. Eberlein, A. Schmid, C. Reise, T. Kroyer, D.H. Neuhaus, H. Wirth, "Potential and challenges of vehicle integrated photovoltaics for passenger cars," in 37th European PV Solar Energy Conference and Exhibition (EU PVSEC), 2020.

[2] New Energy and Industrial Technology Development Organization (NEDO), "PV-powered vehicle strategy committee interim report," 2018. [Online] Available: https://www.nedo.go.jp/content/100885778.pdf. [Accessed: 12-May-2021].

[3] B. Commault, T. Duigou, V. Maneval, J. Gaume, F. Chabuel, E. Voroshazi, "Overview and Perspectives for Vehicle-Integrated Photovoltaics," Appl. Sci. vol. 11, 2021.

[4] T. Golubev, "Multi-physics modeling and simulation of photovoltaic devices and systems," Ph.D thesis, Michigan State University, 2020.

[5] T. Golubev, R.R. Lunt, "Evaluating the Electricity Production of Electric Vehicle-Integrated Photovoltaics via a Coupled Modeling Approach" in 2021 48th Int. Photovoltaics Specialists Conference, IEEE, 2021.

[6] *TAITherm* (2021.2.0.). ThermoAnalytics, Inc.

[7] W. De Soto, S.A. Klein, and W.A. Beckman, "Improvement and validation of a model for photovoltaic array performance," Sol. Energy, vol. 80, no. 1, pp. 78–88, 2006.

[8] A. Dobos, "An Improved Coefficient Calculator for the California Energy Commission 6 Parameter Photovoltaic Module Model," J. Sol. Energy Eng., vol. 134, 2012.

[9] Sunpower. "Maxeon 3 BLK". [Online]. Available: https://sunpower.maxeon.com/int/sites/default/files/2020-09/sp_mst_MAX3-375BLK_355BLK_ds_en_a4_mc4_532497.pdf [Accessed: 21-Nov-2021].

[10] M. Sengupta, Y. Xie, A. Lopez, A. Habte, G. Maclaurin, and J. Shelby, "The national solar radiation database (NSRDB)," Renew. Sustain. Energy Rev., vol. 89, pp. 51–60, 2018.

[11] F. Pedregosa et al., "Scikit-learn: Machine Learning in Python," Journal of Machine Learning Research, vol. 12, pp. 2825–2830, 2011.

Novel Laser Oxidation for Screen-Printed Selective Area Front Poly-Silicon Contacts for TOPCon Cells

Sagnik Dasgupta, Young-Woo Ok, Vijaykumar D. Upadhyaya, Wook-Jin Choi, Ying-Yuan Huang, Shubham Duttagupta, Ajeet Rohatgi

Georgia Institute of Technology, Atlanta, GA, United States

Solar Energy Research Institute of Singapore, Singapore, Singapore

National Yang Ming Chiao Tung University, Tainan City, Taiwan

The efficiency potential of double-side tunnel oxide passivated contact (DS-TOPCon) solar cells is limited by parasitic absorption in the front poly-Si layer, despite excellent passivation and high VOC. The use of patterned poly-Si only under the front metal grid lines can significantly reduce the parasitic absorption loss without sacrificing voltage. In this work, we demonstrate a simple, manufacturing-friendly method of patterning the front poly-Si using a nanosecond UV (355 nm) laser. We found that with laser powers ³3 W at a 400 mm/s scan speed, a 1-4 nm thick stoichiometric SiO2 layer was grown on TOPCon. This served as a mask for KOH-etching of 200 nm poly-Si, allowing for patterning of poly-Si fingers required for selective TOPCon. While laser powers above 3 W caused substantial deterioration in passivation quality, the resulting damage in J0 was largely recovered by subsequent PECVD SiNx deposition. At 3 W, the full area J0 was found to be 36.8 fA.cm-2. This translates to 1.68 fA.cm-2 for 4.48% coverage from the wing area of the poly-finger lines (100 lines-100 wide, 30 metal) contributing to a total front J0 of ~10 fA.cm-2, well suited for 25% efficient solar cells.

Contactless Determination of Emitter Sheet Resistance for Diffused Silicon Wafers

Yan Zhu, Thorsten Trupke, Ziv Hameiri

University of New South Wales, Sydney, Australia

The emitter sheet resistance is one of the essential parameters for silicon solar cells with diffused layers. Conventional measurement methods of emitter sheet resistance either require electrical contacts or are impacted by the bulk resistivity. In this paper, a novel method based on the combination of eddy-current conductance and photoluminescence imaging is developed for a contactless determination of the emitter sheet resistance as well as bulk resistivity. The accuracy of the method is demonstrated by both numerical simulation and experimental validation. The contactless nature of this method makes it an attractive proposition for inline inspection of diffused layers in solar cell manufacturing.

Photodoping causes inconsistencies in the injection-dependent lifetimes of perovskite thin films

Robert A Lee Chin, Arman Soufiani, Jianghui Zheng, Paul Fassl, Anita Ho-Baillie, Ulrich Paetzold, Thorsten Trupke, Ziv Hameiri

SPREE, UNSW, Sydney, Australia

School of Physics, UNSW, Sydney, Australia

Sydney Nano, University of Sydney, Sydney, Australia

Light Technology Institute, Karlsruhe Institute of Technology, Karlsruhe, Germany

Institute of Microstructure Technology, Karlsruhe Institute of Technology, Eggenstein-Leopoldshafen, Germany

The carrier lifetime (τ) is an essential parameter for quantifying the recombination in perovskite thin films (PTFs). τ is often measured from photoluminescence (PL) techniques such as the transient photoluminescence (TRPL) decay. However, there is no standard analysis for the PL-based lifetime, leading to inconsistencies regarding the bulk doping and charge carrier trapping. This study aims to elucidate these inconsistencies and thus help to determine the most appropriate lifetime model for PTFs. The common models for the TRPL are presented. Next, the injection-dependent apparent lifetime methodology is discussed with respect to the photo-doping effect. Finally, we apply these models to the lifetime measured using the transient and steady-state (SS) excitation modes to resolve inconsistencies in the lifetime analysis caused by the photo-doping.

Investigating the impurity gettering rate in polycrystalline-silicon based passivating contacts

Zhongshu Yang, Jan Krügener, Frank Feldmann, Jana-Isabelle Polzin, Bernd Steinhauser, Tien T. Le, Daniel MacDonald, AnYao Liu

School of Engineering, The Australian National University, Canberra, Australia

Institute of Electronic Materials and Device, Leibniz University Hannover, Hannover, Germany

Fraunhofer Institute for Solar EnergySystems ISE, Freiburg, Germany

Polycrystalline-silicon/oxide (poly-Si/SiOx) passivating contacts for high efficiency solar cells exhibit excellent surface passivation, carrier selectivity, and impurity gettering effects. However, the ultrathin SiOx interlayer can act as a diffusion barrier for metal impurities and this potentially slows down the overall gettering rate of the poly-Si/SiOx structures. Herein, the factors that determine the blocking effects of the SiOx interlayers are identified and investigated by examining two general types of the SiOx interlayers: 1.3-nm ultrathin tunneling SiOx with negligible pinholes and 2.5-nm SiOx with thermally created pinholes. Iron is used as a tracer impurity in silicon to quantify the gettering rate. By fitting the experimental gettering kinetics by a diffusion-limited segregation gettering model, the blocking effects of the SiOx interlayers are quantified by a transport parameter. Both the oxide stoichiometry and pinhole density affect the effective transport of iron through SiOx interlayers. The oxide stoichiometry depends strongly on the oxidation method, while the pinhole density is affected by the activation temperature, doping concentration, doping technique, and possibly the dopant type as well. To enable a fast gettering process during typical high-temperature formation of the poly-Si/SiOx structures, a SiOx interlayer that is less stoichiometric or with a higher pinhole density is preferred.

978-1-7281-6118-1/22 $31.00 © 2022 IEEE

Differences of CIGS cell performance with Zn(O,S)/(Zn,Mg)O or CdS/i-ZnO buffers system explored by numerical simulations

Giovanna Sozzi[1], Dimitrios Hariskos[2], Wolfram Witte[2]

[1]Department of Engineering and Architecture, University of Parma, Parco Area delle Scienze 181A, 43124 Parma, Italy

[2]Zentrum für Sonnenenergie- und Wasserstoff-Forschung Baden-Württemberg (ZSW), Meitnerstraße 1, 70563 Stuttgart, Germany

Abstract—Starting from the standard layer sequence of Mo/Cu(In,Ga)Se$_2$/CdS/i-ZnO/ZnO:Al the cell n-side has been modified by replacing the CdS/i-ZnO with Zn(O,S) buffer in combination with (Zn,Mg)O as high-resistive layer, without changing the CIGS bulk. Measurements show a reduction of the cell performance compared to CdS/i-ZnO structure. In order to investigate the observed behavior, numerical simulations are used to examine the effect of the CIGS/Zn(O,S) interface properties on the cell performance. In particular, since the two sets of cells share the same CIGS, the effects on the solar cell's figures of merit of variations of the conduction band offset and defects properties at the buffer/absorber interface are analyzed.

Keywords— *CIGS, conduction band-offset, Zn$_{1-x}$Mg$_x$O high-resistive layer, Zn(O,S) buffer layer.*

I. INTRODUCTION

Thin-film solar cells based on Cu(In,Ga)Se$_2$ (CIGS) or Cu(In,Ga)(S,Se)$_2$ (CIGSSe) with efficiencies over 22% have been obtained using a CdS buffer layer [1], but also alternative materials with larger bandgap energy have been extensively examined such as ZnS, Zn(O,S), (Cd,Zn)S, (Zn,Mg)O (ZMO), and it has been reported that the world record 23.35%-efficient Cd-free CIGSSe solar cell is attained with a ZnS(O,OH) buffer [2].

Replacing the CdS buffer layer with a material of wider bandgap, e. g. Zn(O,S), will result in a higher short-circuit current density J$_{SC}$, due to the reduced light absorption in the short wavelength range in the n-side layers of the cell.

However, the band alignment at the Zn(O,S)/CIGS interface that depends on the oxygen and sulfur content [3], as well as the interface quality can affect the cell performance [4].

This paper compares experimental data of CIGS cells with CdS/i-ZnO and Zn(O,S)/ZMO buffer systems with numerical simulations, mainly focusing on the effect of CIGS/Zn(O,S) interface properties on the cell behavior.

The final aim is to help interpreting the experimental findings and give useful information to guide the improvement of the cell performance.

II. MATERIALS AND METHODS

A. Samples and Measurements

Solar cells with same CIGS absorber and different buffer and high-resistive (HR) layers were fabricated at ZSW, namely, the standard n-side stack CdS/i-ZnO/ZnO:Al (AZO) and the alternative stack Zn(O,S)/(Zn,Mg)O/ZnO:Al, both without anti-reflective coating. Cells with a sputtered Zn$_{1-x}$Mg$_x$O HR layer with composition x = 0.15 and solution-grown Zn(O,S) buffer with a fixed S/(S+O) ratio around 0.7 are considered in the present work.

The CIGS did not undergo any post-deposition treatment. The completed cells were measured as-fabricated without any additional post-anneal or heat-light soak treatment, which mostly improves the performance of this kind of devices.

Table I reports the median (best) values of the measured parameters of the two sets of cells: despite the best Zn(O,S)-

TABLE I.   MEDIANS (BEST) OF MEASURED CELL PARAMETERS

Buffer/HR	Voc (mV)	Jsc (mA/cm²)	FF (%)	η (%)
CdS/ZnO	703 (711)	32.35 (32.45)	74.05 (75.8)	16.72 (17.49)
Zn(O,S)/Zn$_{0.85}$Mg$_{0.15}$O	617 (622)	29.95 (30.1)	69.1 (75.3)	13.09 (14.11)

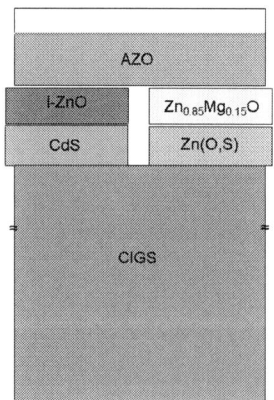

Fig. 1. Schematic cross-section of the CIGS solar cell (not to scale).

buffered cells have efficiencies around 14%, the median efficiency value is lower than CdS buffered cells, mainly due to a lower V_{OC} (about 80 mV less) and FF (about 5% absolute less).

In this particular case the measured J_{SC} is also lower in Zn(O,S)-buffered cells compared to the CdS-buffered cells, despite the larger energy gap and lower thickness of Zn(O,S)/ZMO layers. In order to investigate this behavior, simulations of cells with same CIGS bulk and different n-side and CIGS/Zn(O,S) interface properties have been performed, as detailed in the following.

B. Simulations

We simulated the solar cell both in dark and under AM1.5G light using the Synopsys Sentaurus-Tcad suite. Details about simulations can be found in [5,6]. The cell geometry is depicted in Fig. 1.

The main parameters of the layers used in simulations are listed in Table II, together with the Conduction Band Offset (CBO) at the different hetero-interfaces, inferred from data in [7-9].

TABLE II.   MATERIAL PARAMETERS USED IN SIMULATION

Material	E_g (eV)	Thickness (nm)	Doping (cm^{-3})	Material Interface	CBO (eV)
AZO	3.3	280	$4 \cdot 10^{19}$	AZO/ $Zn_{0.85}Mg_{0.15}O$	-0.17
i-ZnO	3.3	80	10^{17}	$Zn_{0.85}Mg_{0.15}O$/ Zn(O,S)	-0.16
$Zn_{0.85}Mg_{0.15}O$	3.6	40	10^{17}	Zn(O,S)/CIGS	variable
Zn(O,S)	2.85	25	10^{17}	i-ZnO/CdS	-0.2
CdS	2.4	50	10^{17}	CdS/CIGS	0.1
CIGS	graded	2000	$4 \cdot 10^{16}$		

The dependence of CIGS bandgap on the [Ga]/([Ga]+[In]), GGI ratio, is accounted for in simulation by loading the measured GGI profile into the cell model and calculating the depth-dependent energy-gap, E_g, with the formula in [10]; the corresponding depth-dependent complex refractive indexes in

the absorber are varied consistently, as explained in [11]. No grain boundaries are considered in the cell model.

The standard CIGS/CdS/i-ZnO/ZnO:Al has been simulated and optimized at first by comparing simulated current-voltage (J-V) curves and parameters with measurements.

Starting from this cell model, the n-side has been replaced with Zn(O,S)/ZMO/ZnO:Al without changing the CIGS bulk properties, thus focusing the analysis on the effect of CIGS/Zn(O,S) interface properties on the cell behavior.

In particular, in a first set of simulations, the CBO (spike-like, i.e. conduction band in the absorber lower than in the buffer) at the CIGS/Zn(O,S) interface (CBO$_{CIGS/Zn(O,S)}$) has been varied between 0.1-0.6 eV.

Then, since no CBO allowed to obtain a good match between measured and simulated parameters, we explored the effect of interfacial recombination by considering an acceptor defect located at midgap [12] with density, N_{TA}, covering the range 10^{10}-10^{14} cm^{-2}.

III. RESULTS AND DISCUSSION

A. CIGS/CdS/i-ZnO/ZnO:Al cell

At first, the model of the standard CdS-buffered cell has been realized and optimized by comparing the simulated cell parameters and current in dark and under light with measurements.

A surface recombination velocity $S_{rec}=1.5 \cdot 10^2$ cm/s at the CIGS/CdS interface allows to obtain simulated J-V curves and

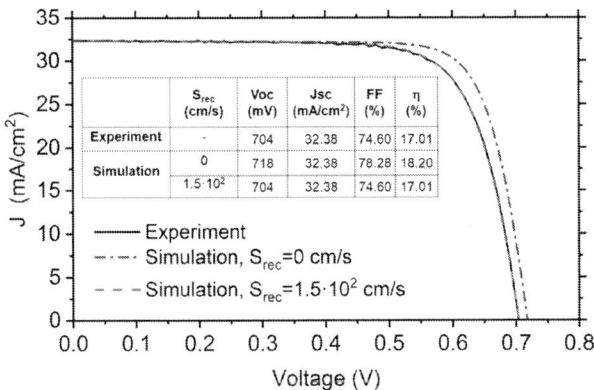

Fig. 2. Measured and simulated J-V curve under AM1.5G light and corresponding figures of merit for a cell with CdS buffer.

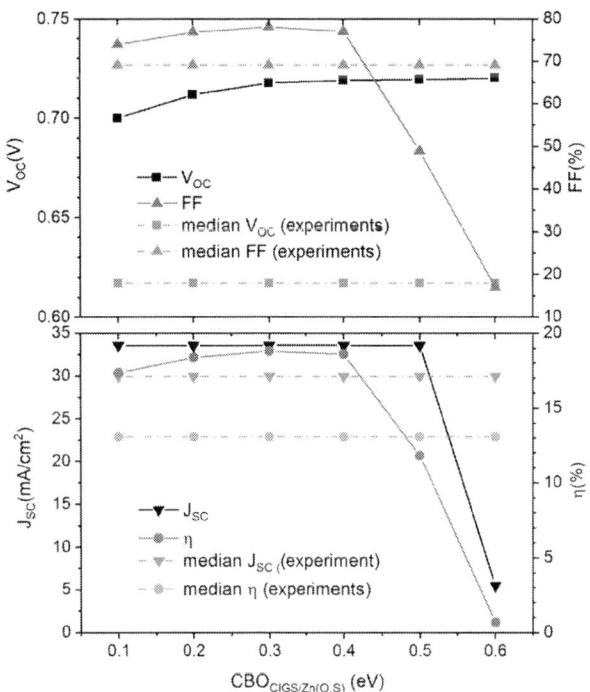

Fig. 3. Figures of merit for the CIGS cells with Zn(O,S)/ZMO as a function of the CBO at CIGS/Zn(O,S) interface.

parameters in excellent agreement with the experimental data, as shown in Fig. 2.

B. CIGS/Zn(O,S)/Zn$_{0.85}$Mg$_{0.15}$O/ZnO:Al cell: variable CBO at Zn(O,S)/CIGS

With the same CIGS parameters used in the standard CdS cell described in section III A, the n-side of the cell has been replaced with Zn(O,S)/Zn$_{0.85}$Mg$_{0.15}$O/ZnO:Al, and the CBO$_{Zn(O,S)/CIGS}$ has been varied between 0.1 and 0.6 eV. The simulated figures of merit are plotted in Fig. 3, together with the medians of measured parameters as reference.

The simulated V$_{OC}$ is always higher than the measured one, while the FF and J$_{SC}$ fall below the experimental values for a CBO$_{Zn(O,S)/CIGS}$ larger than 0.4 and 0.5 eV, respectively.

The larger simulated V$_{OC}$ and FF can be explained in terms of a better energy band alignment of the different materials.

The increase of simulated J$_{SC}$ = 33.56 mA/cm^2, about 1.18 mA/cm^2 more than in the cell with CdS, instead, is due to reduced light absorption in the n-side of the cell: the Zn(O,S)/ZMO layers have larger energy gap and reduced thickness than the CdS/i-ZnO buffer system.

However, this current behavior is not observed in this particular experiment, where J$_{SC}$ reduces compared to the CdS-buffered reference cells.

Since the experimental cells were fabricated on the same CIGS, we concentrate on the absorber/buffer interface. Despite different factors can affect the interface properties (i.e., the Cd diffusion coefficient higher than for Zn could affect the CIGS surface), we focus our analysis on the effect of variation of the recombination velocity at CIGS/Zn(O,S) interface, as described in the following section.

C. CIGS/Zn(O,S)/Zn$_{0.85}$Mg$_{0.15}$O/ZnO:Al cell: variable buffer/absorber interface recombination velocity

The recombination velocity at CIGS/Zn(O,S) interface is S$_{rec}$=N$_{TA}$*σ*v$_{th}$, where N$_{TA}$ is the defect density at interface (cm^{-2}), σ is the electron and hole capture cross section (cm^2) and v$_{th}$ the thermal velocity (10^7 cm/s).

In the following, same hole and electron capture cross section is considered. Figures of merit shown in Fig. 4 refer to a σ=8·10^{-17} cm^2 and a CBO$_{CIGS/ Zn(O,S)}$=0.2 eV.

In order to have simulated cell parameters similar to the experimental ones (experimental reference curves in Fig. 4), a large density of defects N$_{TA}$, between 10^{12} and 10^{13} cm^{-2} (light blue area in Fig. 4) has to be considered at buffer/absorber interface.

These values correspond to S$_{rec}$ in the range 8·10^2 - 8·10^3 cm/s, i.e., larger than in the CdS case. As shown in Table III, a N$_{TA}$ = 4·10^{12}cm^{-2} (i.e., S$_{rec}$=3.2·10^3 cm/s) allows to obtain a better match between simulations and the median values of experimental parameters.

TABLE III. SIMULATED PARAMETERS AND MEDIANS OF MEASURED PARAMETERS OF ZN(O,S)/ZN$_{0.85}$MG$_{0.15}$O CELL. SIMULATION WITH N$_{TA}$=4·10^{12}CM^{-2}, σ=8·10^{-17} CM^2.

	Voc (mV)	Jsc (mA/cm^2)	FF (%)	η (%)
Meas. (median)	617	29.95	69.1	13.09
Sim.	617	30.65	69.06	13.05

However, it is worth observing that the same S$_{rec}$ can be obtained with a lower defect density with larger capture cross section.

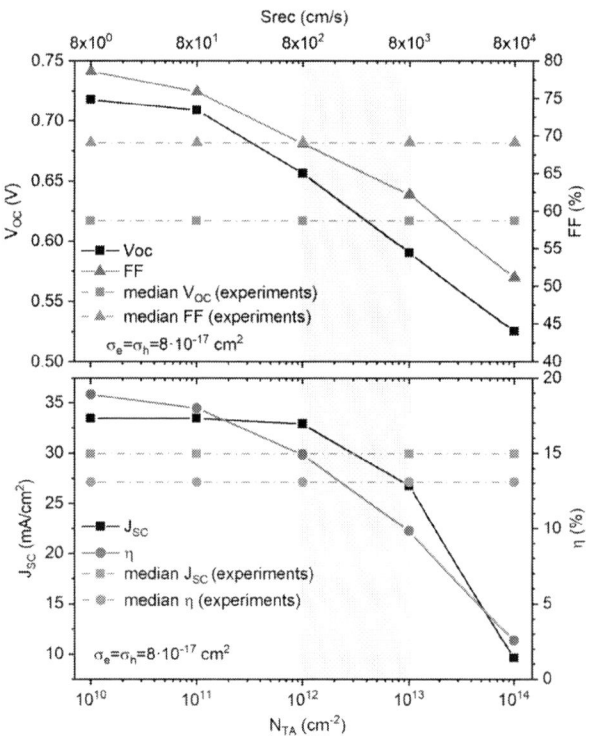

Fig. 4. Figures of merit for the CIGS cells with Zn(O,S)/ZMO as a function of the acceptor trap density at CIGS/Zn(O,S) interface.

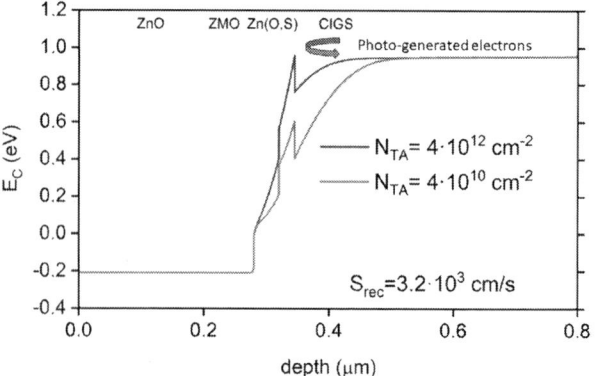

Fig. 5. Conduction band profiles along a vertical line in the middle of the cell at short-circuit condition, for two different density of acceptor defect, N$_{TA}$, at Zn(O,S)/CIGS interface.

The conduction band profiles along a vertical line in the middle of the cell are shown in Fig. 5 for the same $S_{rec}=3.2 \cdot 10^3$ cm/s at the Zn(O,S)/CIGS interface, but for two different acceptor defect densities and carrier capture cross sections, namely $N_{TA}= 4 \cdot 10^{12}$ cm^{-2} ($\sigma= 8 \cdot 10^{-17}$cm^2) and $4 \cdot 10^{10}$ cm^{-2} ($\sigma= 8 \cdot 10^{-15}$cm^2): the ionized negative acceptor charge alters the energy bands at the buffer/absorber interface, pushing the energy bands upwards.

As a consequence, the energy barrier seen at the Zn(O,S)/CIGS heterojunction by the photo-generated electrons leaving the absorber towards the window increases, thus affecting the J_{SC} and the FF [13], the more, the larger the N_{TA} (see Fig. 5).

As shown by the simulated PV parameters reported in Table IV for a constant $S_{rec}=3.2 \cdot 10^3$ cm/s and variable N_{TA} and s, a high density of acceptor trap of low capture cross section is thus more harmful for the cell performance than having less defects with larger capture cross-section.

Considering a density of acceptor defects at the Zn(O,S)/CIGS interface of $N_{TA}=4 \cdot 10^{12}$ cm^{-2} the simulation PV parameters are in agreement with experimental data.

TABLE IV.   SIMULATED PARAMETERS OF ZN(O,S)/ZN$_{0.85}$MG$_{0.15}$O CELL, FOR CONSTANT S$_{REC}$=3.2·10^3 CM/S AND VARIABLE N$_{TA}$ AND σ.

N_{TA} (cm^{-2})	σ (cm^2)	V_{oc} (mV)	J_{sc} (mA/cm^2)	FF (%)	η (%)
$4 \cdot 10^{12}$	$8 \cdot 10^{-17}$	617	30.65	69.06	13.05
$4 \cdot 10^{10}$	$8 \cdot 10^{-15}$	613	33.47	74.01	15.19
$4 \cdot 10^{8}$	$8 \cdot 10^{-13}$	637	33.47	75.51	16.11

IV. CONCLUSIONS

We have analyzed the behavior of two sets of cells with the same Cu(In,Ga)Se$_2$ absorber but different buffer and high-resistive layer, namely the standard Mo/CIGS/CdS/i-ZnO/ZnO:Al sequence and the Mo/CIGS/Zn(O,S)/Zn$_{0.85}$Mg$_{0.15}$O/ZnO;Al alternative.

Measurements show a reduction of the cell performance of as-fabricated Zn(O,S)/ZMO cells compared to the CdS/i-ZnO ones. Since the two sets of cells have the same CIGS, the analysis has been initially focused on the effect of CIGS/Zn(O,S) on the cell performance.

Numerical simulations show that the variation of CBO at CIGS/Zn(O,S) interface cannot alone explain the experimental data. However, if a large amount of deep acceptor traps is also considered at the interface, the enhanced recombination, and the variation of energy barriers due to the ionized charge allow to obtain simulated solar cell parameters in good agreement with the experimental measurements.

REFERENCES

[1] P. Jackson, R. Wuerz, D. Hariskos, E. Lotter, W. Witte, M. Powalla, Effects of heavy alkali elements in Cu(In,Ga)Se 2 solar cells with efficiencies up to 22.6%, Phys. Status Solidi - Rapid Res. Lett. 586 (2016) 583–586.

[2] Nakamura, M.; Yamaguchi, K.; Kimoto, Y.; Yasaki, Y.; Kato, T.; Sugimoto, H. Cd-free Cu(In, Ga)(Se, S)2 Thin-Film Solar Cell with Record Efficiency of 23.35%. IEEE J. Photovolt. 2019, 9, 1863−1867

[3] C. Persson, C. Platzer-Björkman, J. Malmström, T. Törndahl, M. Edoff, Strong valence-band offset bowing of ZnO1-xSx enhances p-type nitrogen doping of ZnO-like alloys, Phys. Rev. Lett. 97 (2006) 1–4.

[4] W.J. Lee, H.J. Yu, J.H. Wi, D.H. Cho, W.S. Han, J. Yoo, Y. Yi, J.H. Song, Y.D. Chung, Behavior of Photocarriers in the Light-Induced Metastable State in the p-n Heterojunction of a Cu(In,Ga)Se2 Solar Cell with CBD-ZnS Buffer Layer, ACS Appl. Mater. Interfaces. 8 (2016) 22151–22158.

[5] G. Sozzi, D. Pignoloni, R. Menozzi, F. Pianezzi, P. Reinhard, B. Bissig, S. Buecheler, A.N. Tiwari, Designing CIGS solar cells with front-side point contacts, 2015 IEEE 42nd Photovolt. Spec. Conf., IEEE, 2015, pp. 1–5.

[6] G. Sozzi, S. Di Napoli, R. Menozzi, B. Bissig, S. Buecheler, A.N. Tiwari, Impact of front-side point contact/passivation geometry on thin-film solar cell performance, Sol. Energy Mater. Sol. Cells. 165 (2017) 94–102.

[7] T. Minemoto, Y. Hashimoto, T. Satoh, T. Negami, H. Takakura, Y. Hamakawa, Cu(In,Ga)Se2 solar cells with controlled conduction band offset of window/Cu(In,Ga)Se2 layers, J. Appl. Phys. 89 (2001) 8327–8330.

[8] B.K. Meyer, A. Polity, B. Farangis, Y. He, D. Hasselkamp, T. Krämer, C. Wang, Structural properties and bandgap bowing of ZnO1-xSx thin films deposited by reactive sputtering, Appl. Phys. Lett. 85 (2004) 4929–4931.

[9] S.C. Su, Y.M. Lu, Z.Z. Zhang, C.X. Shan, B.H. Li, D.Z. Shen, B. Yao, J.Y. Zhang, D.X. Zhao, X.W. Fan, Valence band offset of ZnO/Zn0.85Mg0.15 O heterojunction measured by x-ray photoelectron spectroscopy, Appl. Phys. Lett. 93 (2008) 1–4.

[10] S.-H. Wei, S.B. Zhang, A. Zunger, Effects of Ga addition to CuInSe2 on its electronic, structural, and defect properties, Appl. Phys. Lett. 72 (1998) 3199–3201.

[11] G. Sozzi, S. Di Napoli, R. Menozzi, R. Carron, E. Avancini, B. Bissig, S. Buecheler, A.N. Tiwari, Analysis of Ga grading in CIGS absorbers with different Cu content, 2016 IEEE 43rd Photovolt. Spec. Conf., IEEE, 2016, pp. 2279–2282.

[12] S. Lany, A. Zunger, Light- and bias-induced metastabilities in Cu(In,Ga)Se2 based solar cells caused by the (VSe-VCu) vacancy complex, J. Appl. Phys. 100 (2006).

[13] G. Sozzi, S. Di Napoli, R. Menozzi, F. Werner, S. Siebentritt, P. Jackson, W. Witte, Influence of Conduction Band Offsets at Window/Buffer and Buffer/Absorber Interfaces on the Roll-Over of J-V Curves of CIGS Solar Cells, 2017 IEEE 44th Photovolt. Spec. Conf., IEEE, 2017, pp. 2205–2208.

Fabricating high aspect ratio front contacts for solar cells by string-printing

Mathis Van de Voorde, Rebecca Saive

University of Twente, Enschede, Netherlands

European School of Chemistry, Polymers and Materials Science, Strasbourg, France

We are developing a method for the fabrication of high aspect ratio and high throughput solar cell front contacts, called string-printing. For this, a thread coated with silver paste approaches a silicon substrate until contact is made and then is pulled away to form high aspect ratio, ideally triangular-shaped silver contacts. Here, we describe the fabrication method and show first results. So far, we have been able to fabricate structures with an aspect ratio of 1 whereas we noticed a strong dependence on the thread diameter and the paste viscosity. Furthermore, we also suspect a dependence on the withdrawal speed of the thread. Our approach is a highly scalable, low temperature process that can boost the performance of solar cell metallization for the terawatt future.

Benchmarking PV performance models with high quality IEC 61853 Matrix measurements (Bilinear interpolation, SAPM, PVGIS, MLFM and 1-diode)

Steve Ransome

SRCL (Steve Ransome Consulting Ltd.) , #99 KT2 6AF, UK

Abstract—High quality indoor IEC 61853 Matrix measurements for 9 c-Si and HIT modules have been used to benchmark fits by several PV performance models including the Bilinear interpolation, SAPM, PVGIS (6-parameter) and Mechanistic Loss Factors Model (MLFM). The de Soto 1-diode has not yet been fitted. Residuals have been analysed to show limitations from each of the models and improvements that could be made fitting p_mp, i_sc, i_mp, v_mp and v_oc. Residuals from smoothed models also help quantify indoor matrix measurement scatter (e.g. '±0.5% rmse') or if there are measurement inaccuracies. Added random scatter errors were also fitted by the models to analyse their robustness. Recommendations will be made on optimizing matrix data fitting, coefficient extraction e.g. LIC/STC or gamma (= -1/p_mp * dp_mp / dtemp_module) and degradation analysis (for subsequent measurements after tests).

Keywords—energy, modeling, photovoltaic systems, power, simulation, degradation

I. INTRODUCTION

IEC 61853 indoor matrices measure values of p_mp (or pr_dc) and often some or all of v_mp, i_mp, i_sc and v_oc at up to 28 points usually at 7 different irradiances and 4 module temperatures as shown in fig 1.

Definitions used in this work are listed in Appendix A.

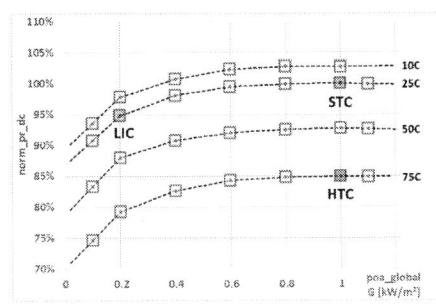

Fig. 1. Typical high quality IEC 61853 performance measurements (squares) and some marked conditions STC = grey, LIC = blue, HTC = red.

II. FITTING PR_DC FOR DIFFERENT MODULES

PV performance models are fitted to matrix measurements by a solver or curve fitter to minimise the RMSE. Fig 2 shows a close-up of the model fits calculated every 0.02kW/m² (smooth coloured lines) and linear interpolation (dashed) vs. the measured data (squares). This is done for each performance model for every parameter and module.

The discrepancies have been analysed to determine how good are the fits and whether they are random (e.g. noise), vary per module or parameter or they are systematic (for example always overestimating at high irradiance).

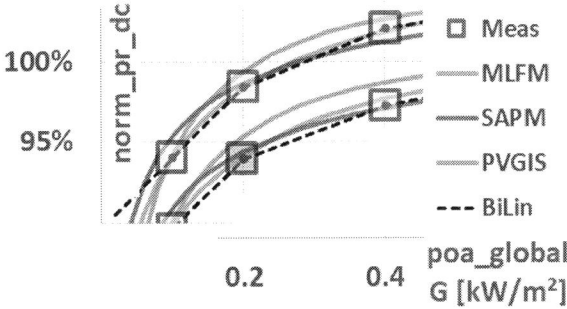

Fig. 2. Model fits (lines) showing discrepancies as vertical offsets vs. measured data (squares).

Fig 3 plots pr_dc measured vs. fits for four models for four different c-Si modules #1, #4, #5 and #8. Systematic residuals can be seen by model which are highlighted with the model's name and arrows showing direction of discrepancy and listed in table I.

TABLE I. SUMMARY OF MODEL DISCREPANCIES IN FIG 3.

Model	Model behaviour
SAPM	Overestimates low and high irradiances, underestimates mid ranges e.g. "↑↓↑".
PVGIS	Discrepancies are mostly at mid light levels. They vary each module; it fits module #5 well, underestimates #1 and 4, overestimates #8.
BILIN	Overestimates interpolating to lowest light levels, underestimates at mid-range. The discrepancy extrapolating at high light level varies due to random noise on the last two points.
MLFM4	Seems to fit all the data well with no obvious discrepancies.

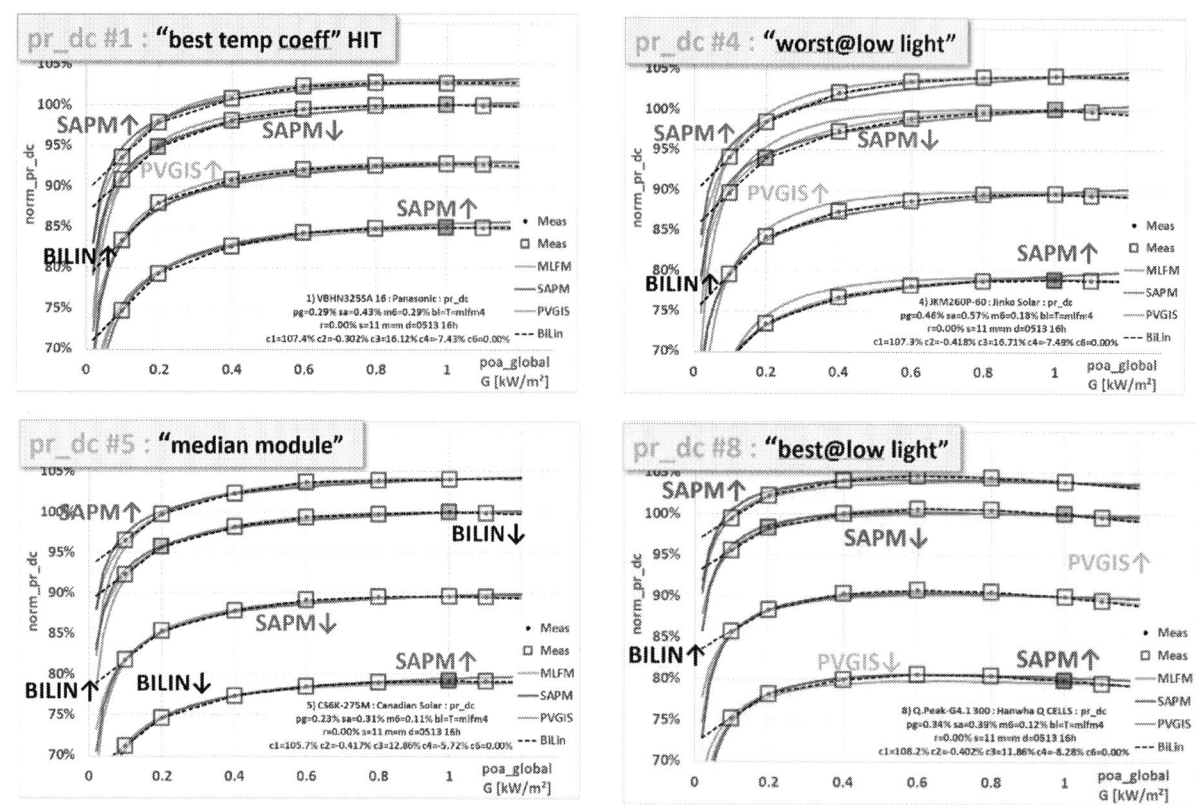

Fig. 3.   pr_dc for modules #1, #4, #5, #8 Measured vs. fitted by four models to good indoor IEC 61853 matrix measurements

III.   FITTING OTHER MODULE PARAMETERS

SAPM, BILIN and MLFM4 can fit more than just pr_dc. PVGIS only claims to fit pr_dc but here this equation has been used to test its fit against other parameters from the iv curve as shown in fig 4. All the parameters have been normalised against their stc values.

Fig 5 shows fits against measured data (as in fig 3) for four normalised parameters norm_i_mp, norm_v_mp, norm_v_oc and pr_dc for a typical module #5 (definitions of parameters are given in the header of each graph and in appendix A).

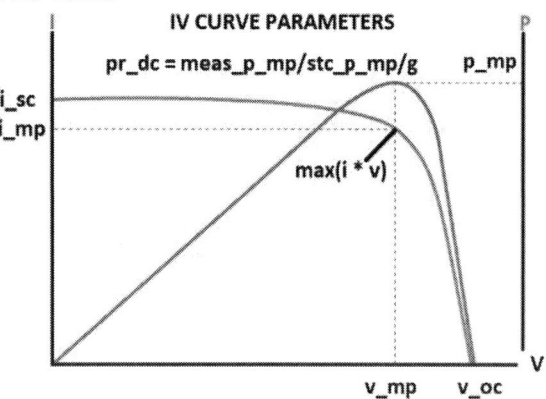

Fig 4   Definitions of fit parameters from IV and PV curves

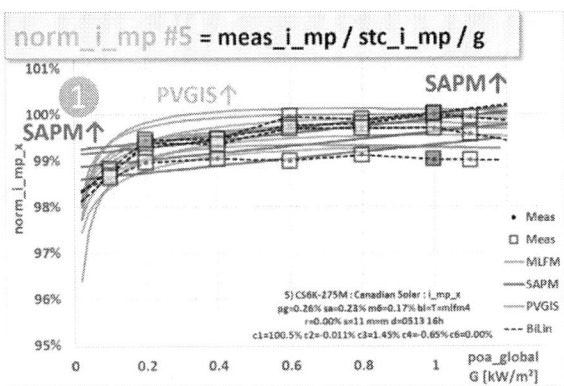
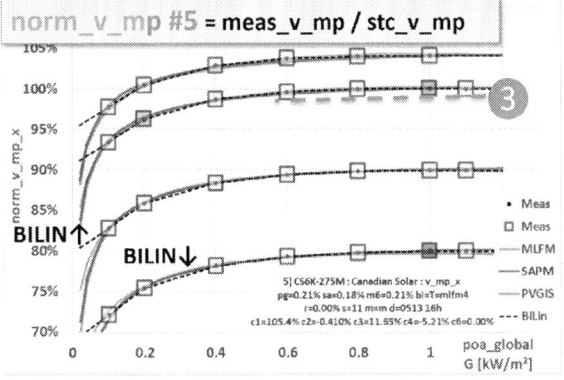
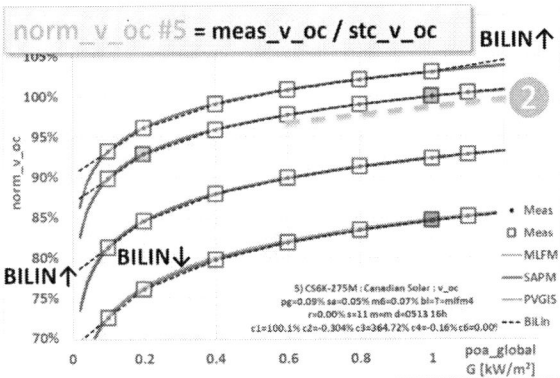
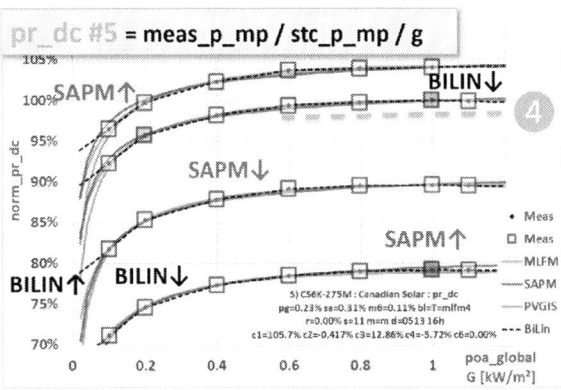

Fig. 4.   Performance fits for four parameters for module #5 by mlfm4, sapm, pvgis and bilin with high quality indoor IEC 61853 matrix measurements

The "Slope at high irradiance" (= d/dg |g>0.6kW/m²) differentiates the parameters due to i².r_series loss which limits v_mp and pr_dc (as i=i_mp) and thus flattens the slope but not v_oc (as i=0) where the value continually increases with irradiance.

Table II summarises parameter fits as plotted in fig 4.

TABLE II.   SUMMARY OF MODEL DISCREPANCIES IN FIG 4.

Param-eter	Comments
❶ i_mp	Only varies by ~1% so only has a small effect on pr_dc (=norm_i_mp*norm_v_mp).
❷ v_oc	At v_oc, i=0 so "i².r_series loss" = 0, curves are still rising at high irradiance. SAPM and PVGIS fit v_oc closer than pr_dc.
❸ v_mp	Has an "i_mp².r_series loss" which flattens curves at high g
❹ pr_dc	pr_dc curve shape is dominated by shape of v_mp.

BiLin always underestimates curved parts of the curve (0.1<g<0.6) and usually overestimates extrapolations (depending on random noise).

As SAPM and PVGIS can both fit v_oc better than they do v_mp or pr_dc, it suggests that the discrepancy might be dure to the i^2.r_series loss which affects v_mp and pr_dc but not v_oc. This would need a loss term dependent on g, neither

have this term but the MLFM4 has a "c_4" term giving it a better fit than the other models.

IV.   BENCHMARKING MODELS WITH RMSE FITS

Table III gives average rmse, table IV lists stdev rmse of the model fits to the parameters of 9 modules.

SAPM has separate equations to calculate i_mp and v_mp, these have been fitted separately (see appendix A).

MLFM4 had up to now used the same equation for all parameters (1) but as its v_oc fit was worse than the SAPM and the PVGIS models an updated equation has been used for v_oc (2) to improve its fit.

MLFM4:
$$\text{norm\_pr\_dc} =$$
$$c\_1 + c\_2*dt + c\_3*log10(g) + c\_4*g \qquad (1)$$
$$\text{norm\_v\_oc} =$$
$$c\_1 + c\_2*dt + c\_3*log10(g)*t\_K/t\_stc\_K + c\_4*g \qquad (2)$$

TABLE III.   AVE RMSE FIT ERRORS PER MODEL AND PARAMETER

Parameters	A) MLFM4	B) SAPM	C) PVGIS
norm_i_sc	0.16%	0.42%	0.36%
norm_v_oc *	0.08%	0.07%	0.12%
norm_i_mp	0.23%	0.29%	0.32%
norm_v_mp	0.20%	0.25%	0.27%
norm_ff	0.31%	0.56%	0.23%
pr_dc	0.17%	0.40%	0.32%
Avg all params	0.19%	0.33%	0.27%

978-1-7281-6118-1/22 $31.00 © 2022 IEEE

TABLE IV. StDev RMSE Fit Errors per Model and Parameter

Parameters	A) MLFM4	B) SAPM	C) PVGIS
norm_i_sc	0.05%	0.08%	0.11%
norm_v_oc *	0.01%	0.03%	0.06%
norm_i_mp	0.05%	0.08%	0.14%
norm_v_mp	0.05%	0.06%	0.07%
norm_ff	0.07%	0.11%	0.05%
pr_dc	0.05%	0.09%	0.09%
Avg all params	0.04%	0.06%	0.07%

MLFM4 (with modified its v_oc equations) is seen to have much lower average rmse than the SAPM or the PVGIS. The stdev of the fits is also better than the SAPM or PVGIS (which tends to fit modules sometimes with unphysical behaviour and/or non-linearities).

V. DERIVING TEMPERATURE COEFFICIENTS DIRECTLY FROM MODEL FITS

IEC 61853 takes a separate set of measurements at a fixed irradiance 1000W/m^2 varying the module temperature to derive temperature coefficients as trends (for alpha_isc and _imp, beta_voc and _vmp, gamma_pmp etc.) as in fig 4.

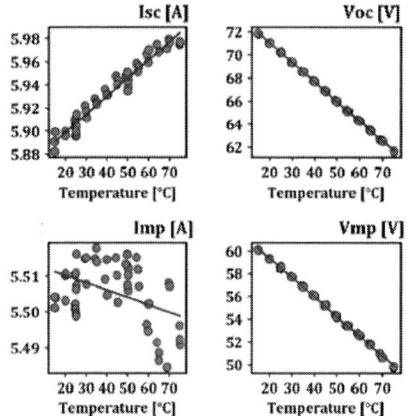

Fig. 5.   Typical IEC measurements trend fits

However, fitting good models to all the points of a well measured matrix simultaneously means that average temperature coefficients are derived without needing extra measurements. Table III shows that modelled temperature coefficients vs. IEC trend fit with extra points.

TABLE V   Temp Coeffs (%/K) Trend fit vs. Models

IEC trend temp. coeffs.	Parameters	A) MLFM4	B) SAPM	C) PVGIS	Common
0.03%	norm_i_sc	0.00%	0.00%	0.03%	alpha
-0.24%	norm_v_oc *	-0.01%	0.00%	0.02%	beta
0.00%	norm_i_mp	0.00%	0.00%	0.05%	
-0.29%	norm_v_mp	-0.01%	0.00%	0.04%	
-0.08%	norm_ff	-0.01%	0.00%	0.05%	
-0.30%	pr_dc	-0.01%	0.00%	0.05%	gamma
1/K	Residual error	(1) < +/- 0.01%	(1) < +/- 0.01%	(2) ~ +0.02-5%	

MLFM4 and SAPM both fit measured IEC better than <0.01%/K), PVGIS underestimates by about 0.03%/K.

$$\text{Temperature coefficient} = \text{fit}(g{=}1, t{=}26C) / \text{fit}(g{=}1, t{=}25C) - 1 \quad /K \qquad (3)$$

PVGIS fits often give non-linear temperature coefficients whereas measured ones are usually linear.

A suggested procedure for optimized temperature coefficient finding is to fit the matrix with MLFM4 then use its value of c_2. The SAPM could also be used and report its aimp or bvmp0 terms (note aimp is relative 1/K but bvmp0 is absolute V/K).

VI. BILINEAR FITTING IEC 61853 MATRIX PR_DC

IEC 61853 suggests bilinearly interpolating and extrapolating between measured matrix values. However measured performance is non-linear with irradiance as in fig 1 and interpolated points will have underperformance errors between measurement points.

Fig 6 illustrates this with a "bilinear fit - smoothed pr_dc" (y axis) vs. irradiance and temperature surface plot for a typical c-Si module. This is synthesised smooth data (no measurement scatter) in this plot to understand the linear errors better.

a) errors are worst at extrapolated points (below 0.1kW/m^2).

b) zero error at measured points e.g. (=0.2kW/m^2, T=25C)

c) worst errors midway between measured values (e.g. 0.35% at 0.15kW/m^2 but are lower at higher irradiance (d).

e) Extrapolations >1.1 kW/m^2 show a linear increase in fit error although it is still small (and happens rarely outdoors)

f) Extrapolation for both irradiance and temperature simultaneously have quite a high error.

g) There is no measurement point for (g=1.1, t=15C) so extrapolations need to be done for both with higher errors.

Most of the time in field data (g, t) values will be within the measurement grid so calculated energy yield errors will mostly depend on interpolations.

Interpolated errors here are <0.35% but they are all the same sign and are under predicting.

Fig 7 shows a similar plot but now with 1% RMSE random noise added to each iv point to show how bi-linear interpolations and extrapolations can't fit noisy data well.

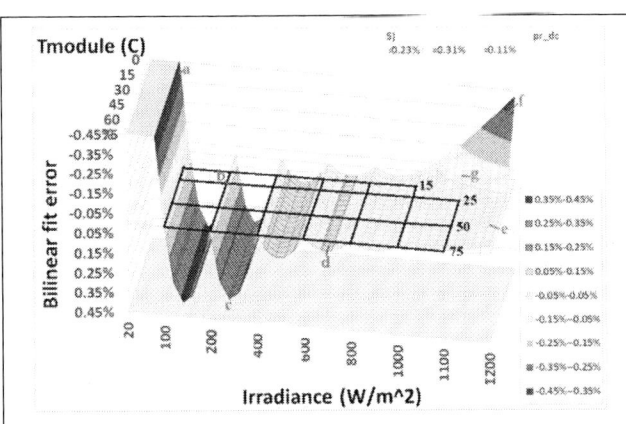

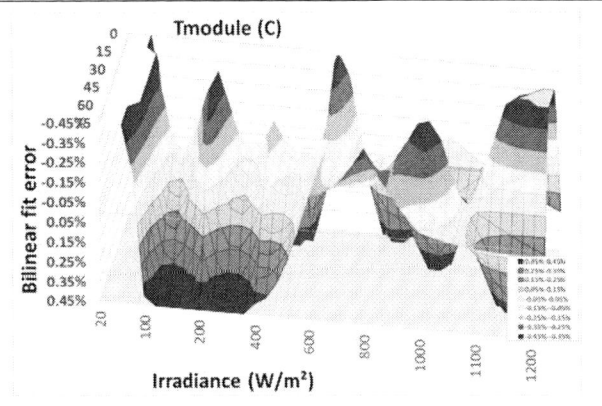

Fig. 6. Bilinear fit error vs. noiseless synthesized matrix measurements ($\updownarrow$) vs. irradiance ($\leftrightarrow$) and module temperature ($\swarrow$) for a typical c-Si module.

Fig. 7. Bilinear fit error vs. data as fig. 5 but with added 1% rmse noise.

VII. FITTING NOISY DATA

The measured data in figs 3 and 4 so far have been very smooth and enabled the optimum model parameters to be found for best fitting. However some matrices may be more noisy, outdoor data will also suffer noise due to angle of incidence, direct fraction, soiling, spectrum and ageing effects and will not usually be so smooth.

Fig 8 show the fits to data from figs 3 and 4 with added random noise, 1% rmse to i and to v points separately. The smooth raw measured data are yellow squares and the random noisy data are the black points (two different seed values).

MLFM4 and SAPM fit physically well as they are constricted to 'sensible behaviour' by their mechanistic coefficients, e.g. the fitted temperature coefficients and low light coefficients will be linear.

PVGIS and BILIN are not restricted by mechanistic effects and in the case of PVGIS can give unrealistic non-linear behaviour, BILIN will just fit the random noise by interpolation or extrapolation giving unrealistic behaviour.

With Noisy data is is useful to have some mechanistic approachs to the model which restrict the fit to sensible behaviour.

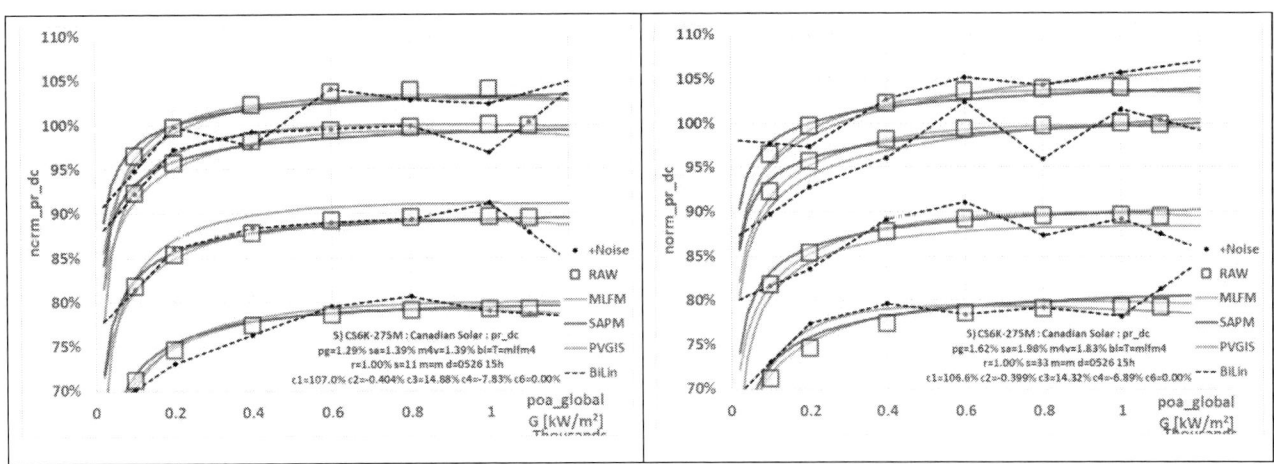

Fig. 8. Fitting Noisy data 1% rms added to i and v separately, Two different random seeds 11 and 33. Compare fits with #5 pr_dc without noise

VIII. CONCLUSIONS

- Optimised models need an $i^2.r$_series loss term for best fits (which is in MLFM4 but not SAPM or PVGIS)
- Accurate temperature coefficients can be better found by just fitting the matrix data with a good

model rather than taking extra points and trend fitting.
- Mechanistic models are better behaved fitting noisy data as they are constrained to realistic behaviour rather than unrealistic non linearities.
- The best model (MLFM4) has about 50% of RMSE of other models tested.

Summary of data findings

MODEL RESIDUALS vs. IRRADIANCE AND FIT PARAMETER (=fit, ↑overestimate, ↓underestimate)

Model name	A) MLFM4+			B) SAPM			C) PVGIS			D) Bi-Linear			E)
Irradiance range g (kW/m2)	<0.2	0.2 - 0.6	>0.6	<0.2	0.2 - 0.6	>0.6	<0.2	0.2 - 0.6	>0.6	<0.2	0.2 - 0.6	>0.6	
pr_dc (r_series loss)	=	=	↑	↑	↓	↑	=	↑	↕	↑	=	↓	
v_mp (r_series loss)	=	=	=	=↑	=	=↑	=	↑	=↓	↑	↓	=	
imp (smaller so noisier)	=	=	=	↑	↓	↑	=	↑	=	fits noise			
voc (no r_series loss)	=	=	=	=	=	=	=	=	=	↑	↓	↑	

AVERAGE RESIDUAL FIT ERROR (pr_dc FOR 9 MODULES)

Avg nRMSE pr_dc	0.17%	0.40%	0.32%	n/a
Std nRMSE pr_dc	0.05%	0.09%	0.09%	n/a

SUMMARY OF MODEL PERFORMANCE

Voc fit now improved	Residuals depend on g	Residuals depend on g	No coefficients
Useful orthogonal coeffs	Overestimates high g	Residuals vary by module	Poor fit to noisy data
Good temp coeffs	Good temp coeffs	Unphysical fits	Poor with extrapolated
Best model overall	Physical fits	Non-linear temp coeffs	Can't fit very dense matrix

IX. REFERENCES

[1] IEC IEC 61853 1-4

[2] SAPM https://pvpmc.sandia.gov/modeling-steps/2-dc-module-iv/point-value-models/sandia-pv-array-performance-model/

[3] PVGIS

[4] Bi Linear fit : https://webstore.iec.ch/preview/info_iec61853-1%7Bed1.0%7Db.pdf

[5] Accurate module performance characterisation using novel outdoor matrix methods VIRTUAL PVSC-48 June 2021

[6] S. Ransome and J. Sutterlueti "Checking the new IEC 61853.1-4 with high quality 3rd party data to benchmark its practical relevance in energy yield prediction" PVSC-46 Chicago, 2019"

[7] W. De Soto et al., "Improvement and validation of a model for photovoltaic array performance", Solar Energy, vol 80, pp. 78-88, (2006)

[8] PVPMC Holmgren, W. C. Hansen and M. Mikofski (2018). "pvlib Python: A python package for modeling solar energy systems." Journal of Open Source Software 3(29): 884.

X. APPENDIX A.

```
# CODE : nomenclature and definitions in this work - python
# g = measured poa irradiance 0.1 - 1.1 [kW/m^2]
# t = measured module temperature 15, 25, 50 and 75 [C]
# percentages are values ~1 e.g. 80% = 0.80 not 80!
#                                             unit (eqt'n)
g_stc = 1                                     # [kW/m^2]
g                                             # [kW/m^2] poa irradiance
t = t_mod -25                                 # [C]
# module efficiency
stc_eta_mod = (stc_p_mp / mod_area_m2 / g_stc / 1000) # [%]


# calculated from measured data
meas_p_mp = meas_i_mp * meas_v_mp             # [W]
meas_ff = meas_p_mp / meas_i_sc / meas_v_oc   # [%]
meas_eta_mod = (meas_p_mp / mod_area_m2 / g / 1000) # [%]


# normalised data for easier fitting and understanding
# use if possible rather than A, V or W
norm_i_sc = meas_i_sc / stc_i_sc / g          # [%]
norm_i_mp = meas_i_mp / stc_i_mp / g          # [%]
norm_v_oc = meas_v_oc / stc_v_oc              # [%]
norm_v_mp = meas_v_mp / stc_v_mp              # [%]
norm_pr_dc = meas_p_mp / stc_p_mp / g         # [%]
```

Equations used by the PV Performance models

```
  MLFM4: mechanistic performance model, 4 meaningful, normalised coefficients
      param  = c_1 +c_2*t +c_3*log10(g) +c_4*g # not for v_oc # c_4*g to fit r_series loss
    * v_oc   = c_1 +c_2*t +c_3*log10(g)*t_K/t_stc_K +c_4*g # v_oc only


  SAPM: "partly mechanistic" dimensioned
      v_mp   = vmpo +c2*s*d*ln(g) +c3*s*(d*ln(g))^2 + bvmpo*t # no term by g for r_series
      i_mp   = impo *(c0*g+c1*(g)^2)*(1+aimp*t)
      pr_dc  = v_mp * i_mp / p_mp_stc / g ; d=N *kb *(T+273.15)/q
      v_oc   = voco +c8*s*d*ln(g) +bvoco*t


  PVGIS: 6-7 mostly empirical coefficients, no g term for r_series
      param  = k_0 +k_1*ln(g) +k_2*ln(g)^2 +k_3*t +k_4*t*ln(g) +k_5*t*ln(g)^2 +k_6*t^2


  Bi-lin: just linear interpolation and extrapolation to any matrix
```

Determining the decomposition voltage of $Cu(In_{1-x}Ga_x)Se_2$

Klaas Bakker[1,4], Joaquin Coll Matas[1], Johan Bosman[1], Nicolas Barreau[2], Arthur Weeber[3,4] and Mirjam Theelen[1]

[1]TNO - Solliance, High Tech Campus 21, 5656 AE Eindhoven, The Netherlands
[2]Université de Nantes, Institut des Matériaux Jean Rouxel, 2 rue de la Houssinière, BP 32229, 44322 Nantes Cedex 3 France.
[3]TNO Energy Transition - Solar Energy, Westerduinweg 3, 1755 LE, Petten, The Netherlands
[4]Delft university of Technology, PVMD, Mekelweg 4, 2628 CD Delft. The Netherlands

Abstract—Partial shading of CIGS modules can lead to permanent damage of the module in the shaded area. This is caused by harmful reverse bias voltages in the shaded area which lead to reverse bias induced defects, also known as wormlike defects. A lot is already known about the origin and propagation of wormlike defects. However, the fundamental question; why is CIGS so sensitive to reverse bias damage? has not yet been answered. In this study we show that CIGS semiconductor material in the presence of an electric field will spontaneously decompose.

Keywords—*CIGS, reverse bias, decomposition, partial shading, reliability*

I. INTRODUCTION

Partial shading of CIGS modules can lead to permanent damage in the shaded area. A recent literature review [1] showed that in all studies that measured electroluminescence (EL) measurements after performing shading tests on CIGS modules permanent damage was found. In EL these defects appear as shunts. Closer visual inspection learns that the defects have a very distinct appearance and are therefore often called wormlike defects. Wormlike defects are caused by a large negative voltage (reverse bias) and are trails of damaged material left behind by a propagating hotspot. They act as local shunts that permanently decrease module performance.

Research to the origin of these defects [2], [3] showed that a chemical reaction is responsible for a change in composition, and that the reaction propagates to a new spot when the material is consumed. In a previous study [3] it was observed that the TCO conductivity influences the propagation patterns and it was concluded that the electric field has an important contribution in the formation and propagation of wormlike defects. Further evidence for the influence of the electric field was found in a study between the relation of absorber thickness and reverse bias voltage required to start wormlike defects [4]. In this study it was shown that thinner cells required a much smaller voltage to form wormlike defects.

The observed dependency on the electric field made us hypothesize that the chalcogenide CIGS structure decomposes under influence of a large electric field to the copper poor ordered vacancy compound (OVC) and copper selenide. The reaction would be $Cu(In_{1-x}Ga_x)Se_2$ decomposes in $Cu(In_{1-x}Ga_x)_3Se_5$ + Cu_2Se or $Cu(In_{1-x}Ga_x)_5Se_8$ + $2Cu_2Se$. This hypothesis is further supported by the fact that several studies on the compositional changes in wormlike defects observed Cu rich islands in the damaged material [5], [6].

The conditions of the initial reaction that forms wormlike defects in CIGS solar cells is complicated to detect. In this study the decomposition of CIGS semiconductor material used in solar cells under the influence of an electric field is determined, using a dedicated sample configuration. With this sample configuration the conductivity of the CIGS material can be measured using a simple one dimensional approach. By varying the electric field changes in conductivity can be detected and linked to changes in the CIGS material. To the best of our knowledge this is the first time that it is shown that CIGS decomposes when a large electric field is applied. This is an important step towards understanding the complex mechanism behind the formation of wormlike defects.

II. EXPERIMENTAL

A dedicated sample configuration was developed to measure the conductivity of CIGS semiconductor material with different compositions. Fig. 1 shows a schematic representation of the sample layout as well as a microscope picture of the molybdenum electrodes. The process steps shown in Fig. 1 (a) are:

1. Deposition and laser structuring of the 400 nm Mo layer.

2. Coevaporation of 2 µm 3-stage CIGS and sputtering of a 150 nm i-ZnO capping layer.

3. Isolating of CIGS and opening the contacts by manually removing CIGS with a scalpel. Contacting with silver paint.

The i-Zno caping layer is added to protect the CIGS and does not play a role in the electrical characterization. The laser pattern is designed to leave a gap between rounded electrodes, that has a defined distance. Two different laser patterns where used with a gap between electrodes of 32 or 86 µm. Electrical characterization was performed by an *IV* sweep from 0 to 50 V

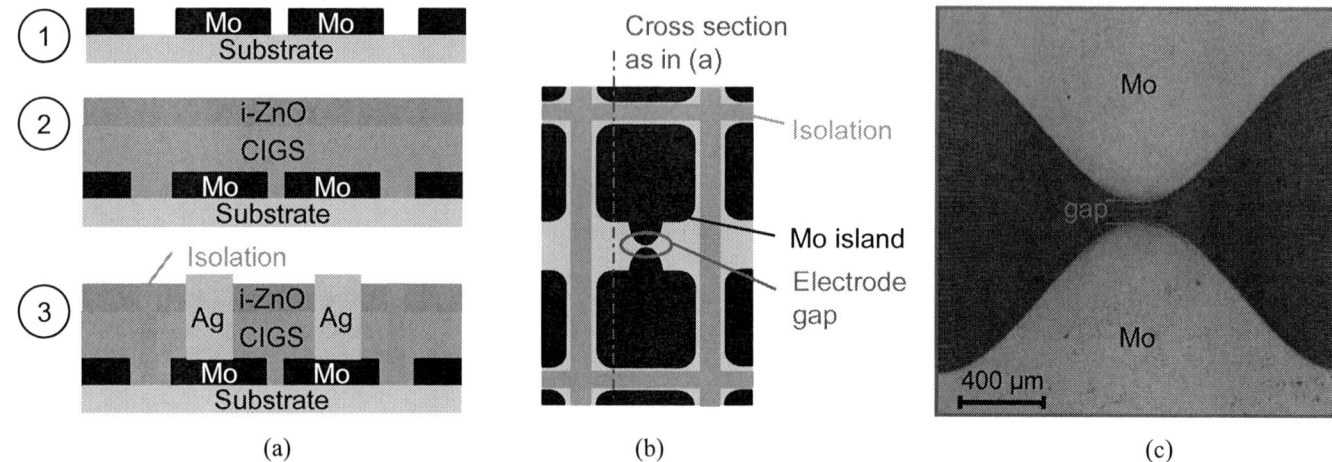

Fig. 1. Schematic and microscope photo of sample layout. In (a) a schematic cross section is given with from top to bottom the process 1 to 3 steps. A schematic top view is drawn in (b) to indicate the approximate shape of the Mo islands after step 1 and the position of the isolation scribes after step 3. (c) Shows a microscope image of the electrode gap after laser scribing (step 1).

in the dark with a Keithley 2400 source measure unit controlled by Rera tracer III software. After contacting, the sample was kept in the dark for a minimum of 60 seconds before performing the *IV* sweep.

III. RESULTS AND DISCUSSION

Fig. 2 shows a typical *IV* curve of an electrode pair with a gap of 86 μm. At low voltages the *IV* curve follows a straight line, showing Ohmic behavior. At higher voltages the behavior deviates from the straight line, followed by a sharp change in current. This change in current could be both positive or negative and indicates a change in material properties. Therefore, we propose the term *decomposition voltage* for this point on the *IV* curve, because after this point visual compositional changes appear similar to wormlike defects.

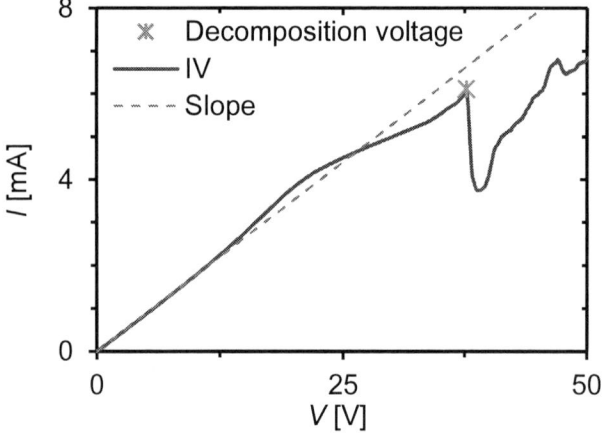

Fig. 2. *IV* curve of sample with a 86 μm electrode gap. The measured *IV* curve is plotted as a blue solid line. The red dotted line is a superpolation of the average slope netween 0 and 10 V to indicate ohmic behavior. The decomposition voltage is indicated with a green asterix

The sharp change in current observed during the *IV* sweeps is very similar to the increase in current during the formation of wormlike defects [4]. The visual appearance after decomposition is also very similar to the visual appearance of

wormlike defects. A microscope picture of a sample after an *IV* sweep can be found in Fig. 3. The composition of the defects has not been studied in detail yet. However, initial microscopy and Raman measurements showed similarities with wormlike defects, created by reverse bias in CIGS solar cells. Just like in studies of wormlike defects no traces of OVC or Cu_2Se have been found. The reaction is so aggressive that it is likely that the initial reaction products (OVC + Cu_2Se) are consumed in a sequential reaction that gives the wormlike defects its distinct appearance.

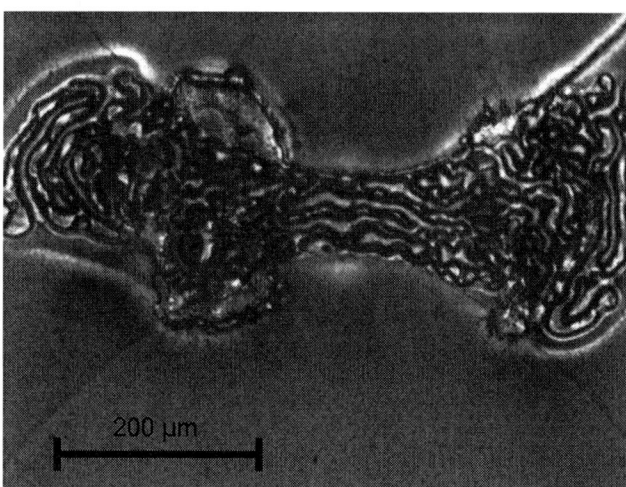

Fig. 3. Microscope picture of decomposed CIGS between two electrodes with a gap of 86 μm.

Because of the 1D approach the measurements are very reproducible. A large number of samples with CGI (copper/(gallium + indium)) varying from 0.795 to 0.894 was examined. All measurements showed the same behavior, as can be seen in Fig. 4 where all measurements with a 32 μm electrode gap are plotted up to the decomposition voltage. The different shades of blue represent the copper content.

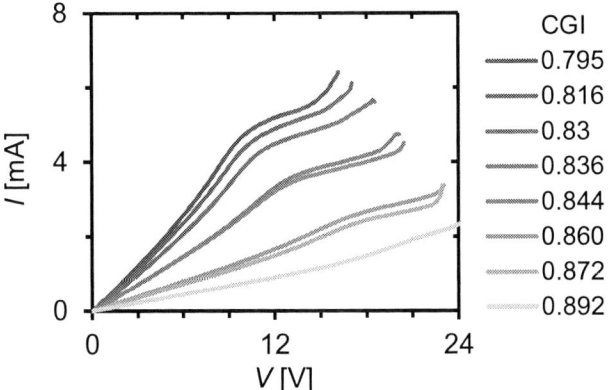

Fig. 4. IV curves of all samples with a 32 μm electrode gap up to the decomposition voltage. The CGI of each sample is plotted in different shades of blue, with the darkest color being the lowest CGI.

From Fig. 4 it can be seen that both slope and decomposition voltage depend on the CGI. Therefore, the dependency of the normalized resistance and normalized decomposition voltage on the CGI is given in Fig. 5. The normalized resistance has a logarithmic dependency on CGI. The normalized decomposition voltage, or electric field, is depending linearly on the CGI, with the larger gap being less sensitive to the difference in copper concentration. This might be a geometric effect caused by the shape of the electrodes.

The exact relationships are still unclear. However, the conductivity is depending on carrier density and mobility so it is not unlikely that the semiconductor parameters have a large influence on the decomposition properties.

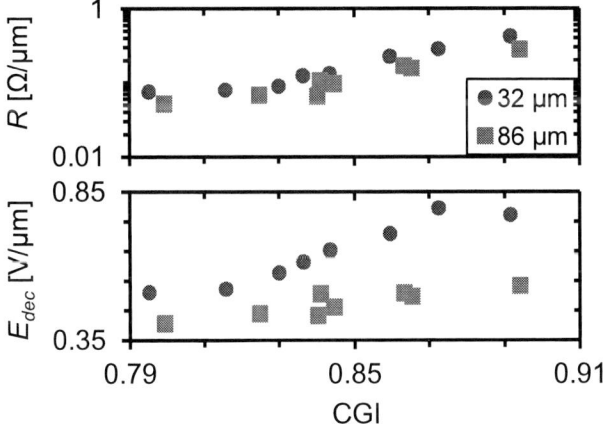

Fig. 5. Resistance and decomposition voltage normalized to electrode distance plotted against CGI for two different electrode gaps. In all graphs the 32 and 86 μm gaps are represented by dark blue circles and red squares, respectively.

IV. Summary

The formation of wormlike defects during reverse bias is a major reliability concern for CIGS solar cells and modules. One of the unknown parameters required to predict the formation and propagation of wormlike defects is the sensitivity of CIGS towards an electric field. This is difficult to determine in the traditional rather complicated CIGS solar cell stack. Therefore, a dedicated device structure was developed to measure the influence of an electric field on the CIGS absorber material used in solar cell. *IV* measurements on these structures showed that CIGS material is spontaneously decomposing. We proposed the term decomposition voltage for the electrical voltage required for decomposition. The decomposition voltage was found to be depending on CGI. The approach used in this study changed the puzzle of decomposition of CIGS from the complicated solar cell stack to a simple one-dimensional approach and proved that CIGS is unstable in the presence of a large electric field. Therefore, CIGS solar cells need to be protected against the harmful effects of reverse bias.

References

[1] K. Bakker, A. Weeber, and M. Theelen, "Reliability implications of partial shading on CIGS photovoltaic devices: A literature review," *J. Mater. Res.*, vol. 34, no. 24, pp. 3977–3987, Dec. 2019.

[2] K. Bakker, H. Nilsson Ahman, K. Aantjes, N. Barreau, A. Weeber, and M. Theelen, "Material Property Changes in Defects Caused by Reverse Bias Exposure of CIGS Solar Cells," *IEEE J. Photovoltaics*, vol. 9, no. 6, pp. 1868–1872, Nov. 2019.

[3] K. Bakker, H. N. Åhman, T. Burgers, N. Barreau, A. Weeber, and M. Theelen, "Propagation mechanism of reverse bias induced defects in Cu(In,Ga)Se$_2$ solar cells," *Sol. Energy Mater. Sol. Cells*, vol. 205, p. 110249, Feb. 2020.

[4] K. Bakker, A. Rasia, S. Assen, B. Ben Said Aflouat, A. Weeber, and M. Theelen, "How the absorber thickness influences the formation of reverse bias induced defects in CIGS solar cells," *EPJ Photovoltaics*, vol. 11, no. 9, Nov. 2020.

[5] P. O. Westin, U. Zimmermann, L. Stolt, and M. Edoff, "Reverse Bias Damage in CIGS Modules," in *24th European Photovoltaic Solar Energy Conference*, 2009, pp. 2967–2970.

[6] H. Guthrey *et al.*, "Characterization and modeling of reverse-bias breakdown in Cu(In,Ga)Se 2 photovoltaic devices," *Prog. Photovoltaics Res. Appl.*, vol. 27, no. 9, pp. 812–823, Sep. 2019.

A Combined Shading and Radiation Simulation Tool for Defining Agrivoltaic Systems

Haomiao Wang, Henry J. Williams, Xiaotong Bu, and K. Max Zhang

Cornell University, Ithaca, NY, 14850, United States

Abstract—**Agrivoltaic systems have the potential to resolve rapidly rising global food and energy challenges by co-locating agriculture and solar photovoltaics (PV). In the United States, Massachusetts created the Solar Massachusetts Renewable Target (SMART) Program to incentivize agrivoltaic development. The program relies on a shading-only simulation tool to differentiate agrivoltaic sites from traditional solar farms. In this paper, we demonstrate that radiation must be considered along with shading to identify land suitable for agricultural activity in agrivoltaic systems. To this end, we present a combined shading and radiation simulation tool and show that percent shade does not singularly determine land available for crop growth. Thus, we recommend the SMART Program update their current method for defining agrivoltaic systems to include radiation modeling.**

Keywords—Agrivoltaic, shading tool, radiation model

I. Introduction

Global energy demand is rapidly expanding as the world population is estimated to reach 9.8 billion by 2050, and the added challenge of climate change is driving new strategies for energy production [1]. Solar photovoltaics (PV) is the fastest growing clean energy source, but its rapid development is limited by land requirements. By co-locating PV with agriculture, agrivoltaics creates a dual-use system that alleviates land-use competition and increases renewable energy production. The state of Massachusetts (MA) would rely on up to 100% of agricultural land to meet statewide energy demands in a 100% solar-powered economy scenario [2]. The challenge of agrivoltaic farming is optimizing shade management so that crops receive sufficient light to reach maximum growing potential, while also leaving enough light for solar PV conversion. Shade and radiation models are therefore necessary to understand the relationship between panel management and crop yield for a given system.

For Massachusetts, a dual-use shading tool has been created to enable solar developers and farmers to co-design qualifying projects under The Solar Massachusetts Renewable Target (SMART) Program, an incentive program established to support solar development [3]. The shading tool determines percent shade for a given solar farm design in a general location, and it guides the SMART Program in creating standards for agrivoltaic projects. The tool does not, however, predict ground-level Photosynthetically Active Radiation (PAR) values, and it only considers three general locations: Western MA, Central MA, and Eastern MA. Thus, we maintain that the model is insufficient in determining land availability for crops within an agrivoltaic system.

New York State Energy Research and Development Authority (NYSERDA)

In this paper, we present an open-source tool that predicts both percent shade and PAR values for an agrivoltaic system, given a site layout and global coordinates. The principal aim of this tool is to accurately guide agrivoltaic regulatory standards, as well as serve as a platform for solar developers and farmers to co-design agrivoltaic systems. This tool has been evaluated against field data, and it is accessible on a public domain website.

II. Methodology

The proposed model determines shading conditions and solar irradiance levels available at ground level for a given solar farm layout. Input parameters include the latitude of the solar farm, height of solar panels, tilt angle, width, length, cardinal orientation, the gap between rows of solar panels, and local weather data.

A. Calculation of Shading Condition

The shadow area cast by the solar panels is defined by angles describing the sun's position in the sky (zenith angle (θ_z), solar altitude angle (α_s), solar azimuth angle (γ_s), and declination (δ)), and angles which describe the geometric relationship between the sunbeam and the solar panels (surface azimuth angle (γ), hour angle (ω), and angle of incidence (θ)) [4]. Given a solar site's exact location, solar time (T_{solar}) is calculated with respect to the local standard time (T_{loc}) to obtain the accurate hour angle for a solar site. The calculation of solar time considers the angle (B) of which the earth has moved from the solstice by day (n) of the year and the equation of time (E), which represents the difference between solar time and clock time. The shadow corner coordinates are then calculated using projection matrices. Table I shows the equations for shading calculations.

B. Calculation of Photosynthetically Active Radiation

Photosynthetically Active Radiation (PAR) is calculated given shadow coverage. Since PAR is active from 400-700 nm, the solar constant ($G_{sc.400-700}$) in the photosynthetically active radiation range is 2450 $\mu mol/m^2$-s [5]. Hourly extraterrestrial solar radiation (G_o) is derived from solar radiation on a plane normal to the sunbeam [5, 6]. With the input of local weather data, an accurate hourly sky clearness index (K_t) is calculated. This determines the fraction of solar radiation available given cloud coverage in a particular region [6, 7]. The ratio of diffuse radiation (I_d) to global radiation (I) is calculated according to the sky clearness index (K_t) [7]. This provides a measure of diffuse radiation in shaded regions which do not experience global solar radiation. Table I shows the equations for PAR calculations.

TABLE I. EQUATIONS USED TO CALCULATE SHADING AND PHOTOSYNTHETICALLY ACTIVE RADIATION

$\theta_z = \cos^{-1}(\cos(\phi)\cos(\delta)\cos(\omega) + \sin(\phi)\sin(\delta))$	$B = (n\text{-}1)\dfrac{2\pi}{365}$
$\alpha_s = 90 - \theta_z$	$G_{sc,\,400\text{-}700} = (f_{0\text{-}700} - f_{0\text{-}400})G_{sc}$
$\gamma_s = \sin(\omega)\left\|\cos^{-1}\left(\dfrac{\cos(\theta_z)\sin(\phi) - \sin(\delta)}{\sin(\theta_z)\cos(\phi)}\right)\right\|$	$\begin{bmatrix}1 & 0 & \tan(\theta_z)\sin(\gamma_s) \\ 0 & 1 & \tan(\theta_z)\cos(\gamma_s)\end{bmatrix}\begin{bmatrix}x \\ y \\ z\end{bmatrix} = \begin{bmatrix}x^{'} \\ y^{'} \\ z^{'}\end{bmatrix}$
$\omega = \dfrac{2\pi}{24}(T_{solar}\text{-}12)$	$G_o = G_{sc}(1+0.033\cos(\dfrac{360n}{365}))(\cos(\phi)\cos(\delta)\cos(\omega) + \sin(\phi)\sin(\delta))$
$T_{solar} = \dfrac{4(L_{st} - L_{loc})+E}{60} + T_{st}$	$K_t = \dfrac{I}{3600G_o}$
$E = 229.2(0.000075+0.001868\cos(B) \text{-}0.032077\sin(B) - 0.014615\cos(2B) - 0.04089\sin(2B))$	
$\dfrac{I_d}{I} = \begin{cases} 1.0 - 0.09K_t & \text{for } K_t \leq 0.22 \\ 0.9511 - 0.1604K_t + 4.388K_t^2 - 16.638K_t^3 + 12.336K_t^4 & \text{for } 0.22 < K_t \leq 0.80 \\ 0.165 & \text{for } K_t > 0.80 \end{cases}$	

III. RESULT AND DISCUSSION

A. Validation

As shown in Fig. 1, we validated our model with measured data from a solar farm in Tucson, Arizona, where an experimental agrivoltaic site provides daily average global and midday diffuse radiation measurements [8]. Local weather data from May 2017 from the National Solar Radiation Database (NSRDB) is used in our simulation tool [9]. The monthly average measured diffuse radiation is 8.32% greater than simulated data, and the monthly average measured global radiation is 7.23% less than simulated data. Differences are attributed to the 5% experimental error in the sensor. Thus, we maintain that our simulation tool performs well for the purpose of defining an agrivoltaic system based on suitability for crops.

B. Proposed Solar Design in Western MA

We used our simulation tool to analyze a proposed south-facing fixed tilt solar farm design in Northampton, MA (42.33 N, 72.62 W). This location is chosen to represent the Western

MA region as defined by the SMART Program tool. Table II shows the parameters of the proposed solar farm. The model uses local weather data from 2019 from the NSRDB [9]. The aim is to determine whether this solar design can be counted as an agrivoltaic system according to the criteria from the SMART Program by analyzing shading conditions and PAR level in the growing season (March to October).

C. Shading Analysis

Fig. 2(a) shows the shading percentage map of the proposed solar farm design in the growing season. On the map, the outlined boxes represent locations of solar arrays. Throughout the growing season, heavily shaded regions extend past the North edges of the solar panels, whereas non-shaded regions adjoin the South edges. The average shading percentage of this design is 56.39%. This value is taken in a region within the solar array grid to prevent edge effects.

Using the SMART Program shading tool, the same design results in 59% shade, which is just 2.61% higher than our model. Therefore, our shading model is comparable to the SMART Program tool for the purpose of shading analysis. Our model is beneficial because it can be used to predict shading and solar radiation conditions in any location by inputting the local longitude and latitude, while the SMART Program tool can only be used to analyze a solar farm design in Western MA, Central MA, or Eastern MA.

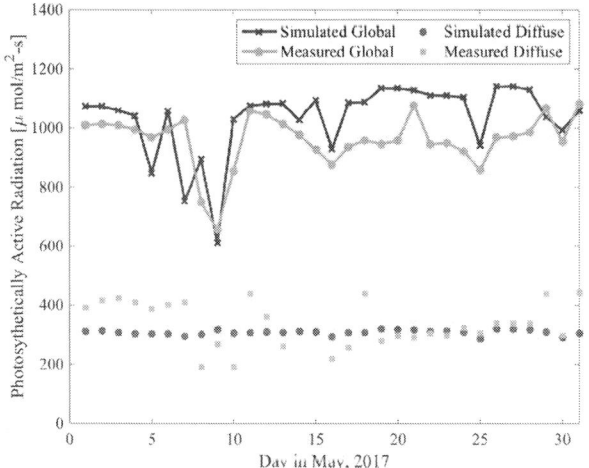

Fig. 1. Model validation using midday measured global and diffuse PAR compared to simulated data for Tucson, Arizona in May, 2017.

TABLE II. PARAMETERS OF THE STUDIED SOLAR FARM DESIGN

Length of Panels [m]	4
Tilt Angle [°]	21
Length of Rows [m]	20
Ground Clearance [m]	1
Gap Distance between Rows [m]	4
Number of Rows	4
Length of Panels [m]	4

978-1-7281-6118-1/22 $31.00 © 2022 IEEE

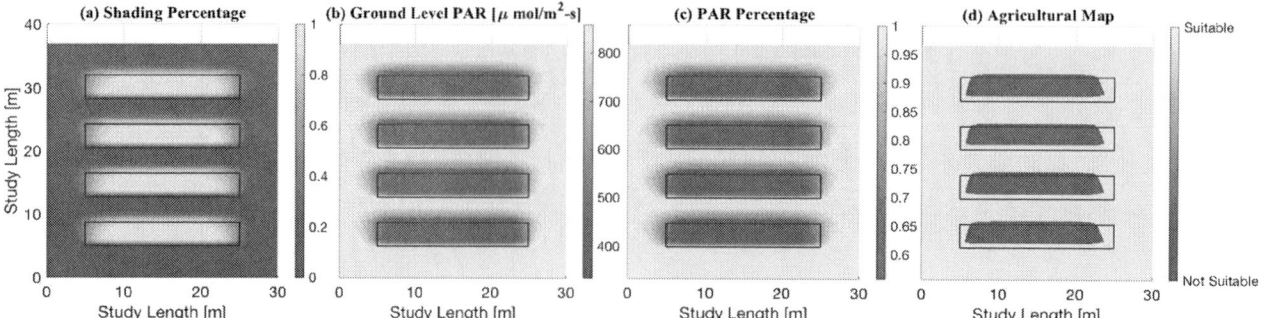

Fig. 2. (a) Percentage of shading on the ground, (b) ground level PAR value, (c) percentage of PAR received on the ground, (d) areas suitable for growing lettuce throughout the growing season for the proposed solar farm design, with outlined boxes representing locations of solar panels

The SMART Program requires that a solar farm has less than 50% shade to be considered an agrivoltaic site, as determined by their shading tool [3]. This standard aims to require sufficient light for plants below the solar panels, but their shading tool considers only the effect of direct sunlight. Thus, the solar farm design proposed in this study would not be considered an agrivoltaic site under this rule.

D. Photosynthetically Active Radiation Analysis

Fig.2 (b) shows the ground level PAR map, and Fig. 2(c) shows the PAR percentage map of the proposed solar farm design. Because diffuse radiation has horizontal direction, there is ground level PAR in fully shaded areas. This tool estimates that this solar farm design only blocks 25.39% of PAR and the average ground level PAR in this solar is 556.53 µmol/m^2-s.

Fig. 2(d) shows a map of areas suitable for agricultural activity. This crop map example uses lettuce, which requires a PAR value of 420 µmol/m^2-s to grow [10]. Areas that receive a PAR value suitable for growing lettuce are marked in yellow. The model estimates that 55.17% of the land in this solar farm is suitable for growing lettuce.

The current method for defining an agrivoltaic system in Massachusetts SMART program is insufficient, as it neglects to consider percent PAR for a solar farm design and therefore does not illustrate how much area in this solar farm is suitable for agricultural activity. As shown in Fig. 2, area suitable for agricultural activity can be higher than 50%, even when the solar farm blocks more than 50% of direct sunlight. This is due to diffuse PAR that reaches the crops beneath the solar panels.

IV. SUMMARY

We developed a combined radiation model and shading tool to explore a holistic method for differentiating an agrivoltaic system from a traditional solar farm. The tool is evaluated against measured radiation data in Tucson, Arizona. We compared our tool to the SMART Program shading analysis tool using a proposed solar farm design in Northampton, Massachusetts. Our simulation tool considers global radiation as well as diffuse radiation, which provides an accurate estimation for the radiation conditions under solar panels, and thus the area suitable for agricultural activity. Our tool supports the advancement of agrivoltaic systems by providing shading and

radiation estimations and optimizing PAR conditions for various crops through site design.

With our newly developed tool, incentive programs and regulatory organizations can create more accurate standards for agrivoltaic systems. This tool can help solar developers and farmers to co-design agrivoltaic farms for mutual benefit as global food and energy needs rise in conjunction with rapidly increasing land constraints.

ACKNOWLEDGMENT

We would like to acknowledge the support from the New York State Energy Research and Development Authority (NYSERDA).

REFERENCES

[1] United Nations, Department of Economic and Social Affairs, Population Division, "United population prospect: the 2017 revision, key findings and advance tables," 2017.

[2] C. K. Miskin, Y. Li, A. Perna, R. G. Ellis, E. K. Grubbs, P. Bermel, and R. Agrawal, "Sustainable co-production of food and solar power to relax land-use constraints," *Nature Sustainability*, vol.2, pp.972-980, 2019.

[3] Commonwealth of Massachusetts, Executive Offics of Energy and Environmental Affairs, Department of Energy Resources, and Department of Agricultural Resources, "Solar Massachusetts Renewable Target Program: Guildeline regarding the definition of agricultural solar tariff gerenation units," 2018.

[4] J.W. Spencer, "Fourier Series Representation of the Position of the Sun," *Search*, vol. 2, pp.162-172, 1971.

[5] M. Iqbal, An Introduction to Solar Radiation. Toronto: Academic, 1983.

[6] D. G. Erbs, S. A. Klein, and W. A. Beckman, "Estimation of Degree-Days and Ambient Temperature Bin Data from Monthly Average Temperatures," *ASHRAE J.*, vol. 25:6, pp.60-66, 1983.

[7] J. A. Duffie and W. A. Beckman, Solar Engineering of Thermal Processes, 4th ed., Wiley, 2013.

[8] G. A. Barron-Gaffor, M. A. Pavao-Zuckerman, R. L. Minor, L. F. Sutter, I. Barnett-Moreno, D. T. Blackett, et al., "Agrivoltaics Provide Mutual Benefit across the Food-Energy-Water Nexus in Drylands," *Nature Sustainability*, vol. 2, pp.848-855, 2019.

[9] M. Sengupta, Y. Xie, A. Lopez, A. Habte, G. Maclaurin, and J. Shelby, "The National Solar Radiation Data Base (NSRDB)." *Renewable and Sustainable Energy Reviews*, vol.89, pp.51-60, 2018.

[10] S. Tazawa, "Effects of various radiant sources on plant growth (part 1)," *Japan Agricultural Research Quarterly*, vol. 33, pp. 163-176, 1999.

Measurement of Band Alignment between ZnO Based Front Emitters and CdCl$_2$ Treated CdSeTe/CdTe Absorbers

Xiaolei Liu, Luke Jones, Luksa Kujovic, Nicholas Hunwick, Luis Infante-Ortega, Michael Walls
CREST, Loughborough University, Loughborough, Leicestershire, LE11 3TU, UK.

Tushar Shimpi, Walajabad Sampath, Kurt Barth*
NSF I/UCRC for Next Generation Photovoltaics Colorado State University, Fort Collins, Co 80523, US

Stephen Jones, Ochai Oklobia, Stuart Irvine
CSER, Swansea University, Swansea, SA2 8PP, UK

Abstract—Thin film CdSeTe/CdTe solar cells have achieved > 22% record efficiency and generated solar electricity at a cost as low as 3 US cents per kW.hr in large scale utilities. In this work, the use of ZnO based n-type emitters is considered with the aim of improving the thin film CdSeTe/CdTe photovoltaic device efficiency still further. The measured conduction band offsets (CBOs) between ZnO and CdSeTe are determined to be in the "cliff" conformation. These CBOs will be optimized to achieve a "spike" conformation by incorporating suitable dopants to achieve high device efficiency. This work identifies new pathways to highly efficient ZnO based n-type emitters for arsenic doped CdTe solar cells. In particular, we have identified Sn, Ce and Cs as suitable dopants to generate the preferred 'spike' in the band alignment between the emitter layer and the CdSeTe absorber. Suitably doped emitter layers will be used in device research to reduce the deficit in open circuit voltage.

Keywords—CdTe solar cell, ZnO based front emitter, conduction band offset, interface

* Professor Kurt Barth's present affiliation is CREST, Loughborough University, UK.

I. INTRODUCTION

Thin film CdTe photovoltaic (PV) modules accounted for about 80% of thin film PV manufacturing in 2020 [1]. Its production capacity is forecast to continue to increase. The cost of solar electricity generated from CdTe PV technology is already as low as 3 US cents per kW.hr in large scale utilities. As the most successful second-generation PV technology, CdTe solar cells have achieved > 22% record device efficiency [2]. Further increasing the module efficiency will lower the cost of solar electricity to consumers and lead to faster substitution of fossil fuels. The power sector generates a quarter of global greenhouse gas emissions. Therefore, the development of thin film CdTe PV technology will help accelerate the power generation transition from fossil fuels to renewable energy and contribute to the net zero carbon emission target by 2050 to limit the global temperature increase to 1.5 degrees [3].

The key hurdle to further improve the CdTe device efficiency is the open circuit voltage deficit. One reason for the limitation is the low carrier concentration to 10^{14} cm^{-3} achieved by copper doping in the CdTe absorber. Recently, researchers have used Group V elements, such as arsenic (As), to increase the carrier concentration of CdTe absorbers to 10^{16} cm^{-3} and reported a conversion efficiency of 22% [4]. Given the increased carrier concentration of the CdTe absorber, the front n-type emitter between the transparent conducting oxide and absorber becomes more critical to device efficiency. It is vital that the carrier concentration in the emitter layer matches that in the CdTe absorber. The conduction band offset (CBO) at the emitter/absorber interface also plays a critical role in achieving high device efficiency [5]. However, disparate CBOs have been reported [6-8]. These inconsistent values may be caused by the different deposition methods and surface preparation methods used since these may modify the interface band structure. We report on new measurements from materials deposited by two deposition methods with the aim of providing more consistent values. This then enables the most promising dopants to be identified with more confidence. In this paper, X-ray Photoelectron Spectroscopy (XPS) and the Kraut Method are used to measure and calculate the band alignment between ZnO and CdSeTe. Here we report on results from a comprehensive investigation, which will shed light on the optimization of the CBO (i.e., +0.3~+0.4 "spike" conformation) between ZnO and CdSeTe. This work will identify new pathways to highly efficient ZnO based n-type emitters for arsenic doped CdTe solar cells with the device structure of glass/FTO/ZnO based n-type emitter/ (arsenic doped CdSeTe/CdTe absorber)/Te/carbon and Ni back contact.

II. EXPERIMENTAL DETAILS

The Kraut method requires the use of three samples: a 150 nm thick ZnO film; an interfacial ZnO/CdSeTe sample (i.e. 3 nm thin ZnO overlayer on top of a 300 nm thick CdSeTe substrate layer); a clean 300 nm thick CdSeTe film. The 150 nm thick ZnO film was deposited onto soda lime glass in an Orion 8 HV magnetron sputtering system (AJA International Inc., USA). 1% O$_2$ in O$_2$+Ar sputtering gases was applied in the deposition process. The 300 nm thick CdSeTe film was provided by CSU and deposited onto NSG TEC 10 glass using Close-Spaced Sublimation (CSS) and CdCl₂ treated at 400 °C. The interfacial ZnO/CdSeTe sample enables the XPS core-levels of CdSeTe to be made visible alongside the core-levels of ZnO.

All spectra were recorded using a Thermo Scientific K-Alpha XPS with a monochromated Aluminium Kα source. The sample surface was characterized using survey scans and high-resolution scans with the following parameters: Surveys were measured using 200eV pass energy, 1eV step size, 10mS dwell time averaged over 10 Scans. High-resolution scans were measured using 50eV pass energy, 0.1eV step size, 50mS dwell time averaged over 5 scans for each element of interest. All results were charge corrected to carbon C1s at 284.8eV. Additionally, to perform the Kraut Method, valence band scans were conducted at 50eV pass energy, 0.2eV step size, 50ms dwell time averaged over 20 scans between a binding energy range of 0 to 25eV.

III. RESULTS AND DISCUSSION

From the Kraut Method, the valence band offset (VBO) is calculated by using the separations of the core-levels and the valence band maximum of the thick ZnO and CdSeTe films, and the relative difference between the same core-levels in the interfacial ZnO/CdSeTe sample (equation 1) [9, 10]. As given in equation 2, the CBO is determined by incorporating the band gaps of the thick ZnO and CdSeTe films.

$$\Delta E_v = (E_B - E_v)_{over} - (E_B - E_v)_{sub} - (E_B^{over} - E_B^{sub}) \quad (1)$$

$$\Delta E_c = \Delta E_v + E_g^{over} - E_g^{sub} \quad (2)$$

Where E_B is the binding energy of core-levels, E_v is the valence band maximum, and E_g is the band gap.

Fig. 1 shows the Zn 2p, O 1s core-levels and valence band maximum spectra of the 150 nm thick ZnO film. The Zn 2p core-levels are fitted with a single doublet, representing the spin-orbit splitting of the level [10]. The O 1s core-levels are fitted with two peaks. The larger peak represents Zn-O bonds. The shoulder is fitted with a second peak, which is assigned to organic oxide (e.g., C-O, C=O) from adventitious carbon contamination on the surface. The valence band maximum of ZnO is determined by linear extrapolation of the leading edge to the baseline, resulting in E_v of 1.71 eV.

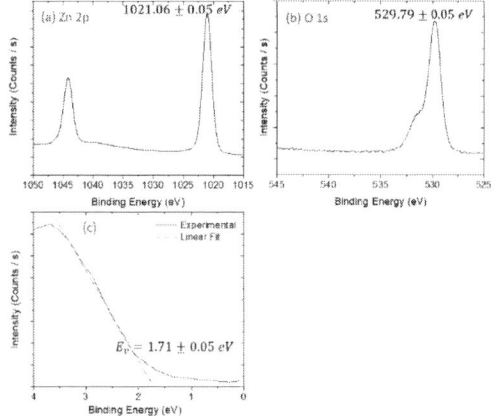

Fig. 1. XPS core levels of the 150 nm thick ZnO film (a) Zn 2p (b) O 1s (c) the valence band maximum.

The XPS spectra of the interfacial ZnO/CdSeTe sample is shown in Fig. 2. It is observed that the XPS peaks of Zn 2p and O 1s are shifted by 0.6 and 1.56 eV, respectively. This might be due to small size effect of the ZnO islands formed on CdSeTe films [11]. The core-levels of Cd 3d, Se 3d and Te 3d are fitted with single doublets due to spin-orbit splitting, which are consistent to the reported binding energies. The two shifted higher energy peaks in Fig. 2(e) are assigned to Te⁴⁺, representing the formation of a thin layer of TeO₂ on the CdSeTe surface [12].

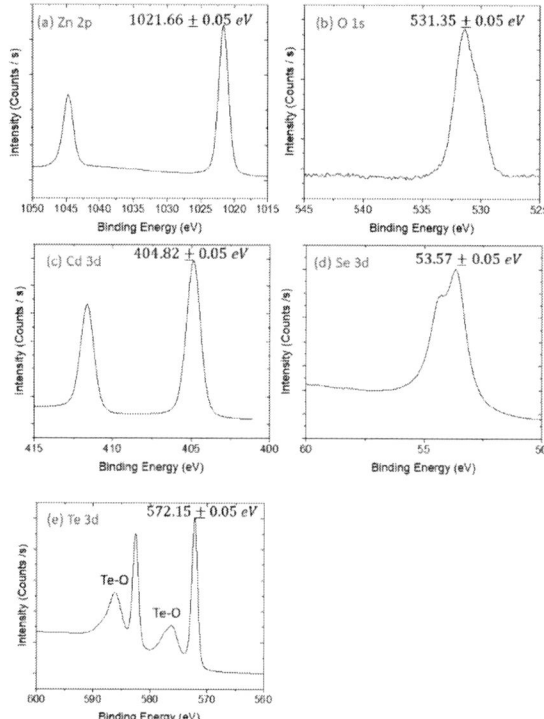

Fig. 2. XPS core levels of the interfacial ZnO/CdSeTe sample (a) Zn 2p (b) O 1s (c) Cd 3d (d) Se 3d (e) Te 3d.

Fig. 3 shows the Cd 3d, Se 3d, Te 3d core-levels and valence band maximum spectra of the 300 nm thick CdSeTe film. The core-levels of Cd 3d, Se 3d and Te 3d are fitted with single doublets. It is observed that the ratio of Se 3d₃/₂ to 3d₅/₂ is reduced from 1.86 to 1.45, in contrast to the interfacial ZnO/CdSeTe sample. The valence band spectrum of CdSeTe is linearly fitted, resulting in E_v of 0.55 eV.

The Kraut method is used to calculate the band alignment at the ZnO/CdSeTe interface from the binding energies and valence band maximum in Fig. 1, 2 and 3. The band gap energies of ZnO and CdSeTe are extracted from Tauc plots, which are 3.54 and 1.39 eV, respectively. The calculation parameters and resulting band offsets are tabulated in Table I. The final valence band offset (VBO) and conduction band offset (CBO) are obtained from the arithmetic mean of the VBOs and CBOs of each core-level. Therefore, the final VBO and CBO are -2.47 and -0.32 eV "cliff" conformation, respectively.

The XPS and Kraut Method have also been used to measure and calculate the band alignment between ZnO and CdSeTe provided by Swansea University [13]. The Swansea's CdSeTe samples were deposited using metal-organic chemical vapor deposition (MOCVD) and CdCl$_2$ treated. The VBO and CBO for Swansea's samples is -2.34 and -0.22 eV "cliff" conformation, respectively. The small disparity between the CdSeTe samples of CSU and Swansea may be due to the different processing methods, which modifies the chemical composition of the CdSeTe surface and alters the interface band structure.

IV. CONCLUSION

In this work, the CBOs between ZnO and CdSeTe thin films using separate and different deposition systems have been determined to be -0.32 and -0.22 eV in the "cliff" conformation, These CBOs will be optimized to +0.3~+0.4 eV "spike" conformation by adding suitable dopants, such as Sn, Ce, and Cs, which are required to achieve high device efficiency. The Mg doped ZnO (MZO) has been used to adjust the CBO to the optimal range. However, it has been observed that MZO is not stable in humid air due to hydrolysis [14]. Therefore, the effect of alternative dopants (e.g., Sn, Ce, Cs) on CBO and carrier concentration of ZnO based n-type emitters will be investigated in future research.

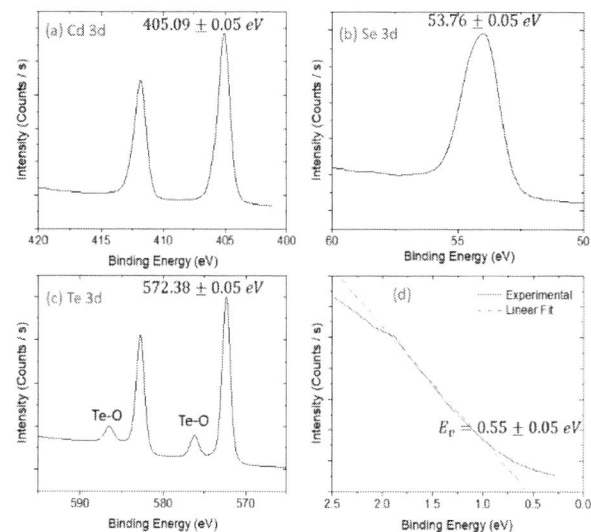

Fig. 3. XPS core levels of the 300 nm thick CdSeTe film (a) Cd 3d (b) Se 3d (c) Te 3d (d) the valence band maximum.

TABLE I CALCULATION PARAMETERS AND RESULTING VBOS AND CBOS FOR THE ZNO/CDSETE INTERFACE.

Sample	Core-level	E_B (eV)	E_v (eV)	E_g (eV)	E_B-E_v (eV)	ΔE_v Zn (eV)	ΔE_v O (eV)	ΔE_c Zn (eV)	ΔE_c O (eV)
ZnO	Zn 2p$_{3/2}$	1021.06	1.71	3.54	1019.35	–	–	–	–
	O 1s	529.79	1.71	3.54	528.08	–	–	–	–
CdSeTe	Cd 3d$_{5/2}$	405.09	0.55	1.39	404.54	-2.03	-2.99	0.11	-0.85
	Se 3d$_{5/2}$	53.76	0.55	1.39	53.21	-1.95	-2.91	0.20	-0.76
	Te 3d$_{5/2}$	572.38	0.55	1.39	571.83	-1.99	-2.95	0.16	-0.80

ACKNOWLEDGMENT

The UK authors are grateful to EPSRC for funding the project through EPW00092X/1.

REFERENCES

1. REN21. "Renewables 2021 Global Status Report", 2021; Available from: http://www.ren21.net/.
2. Martin Green, Ewan Dunlop, Jochen Hohl-Ebinger, Masahiro Yoshita, Nikos Kopidakis, and Xiaojing Hao, "Solar cell efficiency tables (version 57)", Progress in Photovoltaics: Research and Applications, 29(1): p. 3-15, 2021.
3. UN. "UN Climate Change Conference UK 2021", 2021; Available from: https://ukcop26.org/.
4. First Solar. "First Solar Update". in CdTe workshop, 2021. US Manufacturing of Advanced Cadmium Telluride (US-MAC).
5. Tao Song, Ana Kanevce, and James R. Sites, "Emitter/absorber interface of CdTe solar cells", Journal of Applied Physics, 119(23): p. 233104, 2016.
6. A. Kanevce, M. O. Reese, T. M. Barnes, S. A. Jensen, and W. K. Metzger, "The roles of carrier concentration and interface, bulk, and grain-boundary recombination for 25% efficient CdTe solar cells", Journal of Applied Physics, 121(21), 2017.
7. J. M. Kephart, J. W. McCamy, Z. Ma, A. Ganjoo, F. M. Alamgir, and W. S. Sampath, "Band alignment of front contact layers for high-efficiency CdTe solar cells", Solar Energy Materials and Solar Cells, 157: p. 266 275, 2016.

8. Christos Potamialis, Process sensitivities and interface optimisation of CdTe solar cells deposited by close-space sublimation, in CREST, Loughborough University. 2019.
9. E. A. Kraut, R. W. Grant, J. R. Waldrop, and S. P. Kowalczyk, "Precise Determination of the Valence-Band Edge in X-Ray Photoemission Spectra: Application to Measurement of Semiconductor Interface Potentials", Physical Review Letters, 44(24): p. 1620-1623, 1980.
10. James Thomas Gibbon, Band Alignments and Interfaces in Kesterite Photovoltaics. 2018, University of Liverpool.
11. Steffen Oswald, Franziska Thoss, Martin Zier, Martin Hoffmann, Tony Jaumann, Markus Herklotz, Kristian Nikolowski, Frieder Scheiba, Michael Kohl, Lars Giebeler, Daria Mikhailova, and Helmut Ehrenberg, "Binding Energy Referencing for XPS in Alkali Metal-Based Battery Materials Research (II): Application to Complex Composite Electrodes", Batteries, 4(3): p. 36, 2018.
12. D. N. Bose, M. S. Hedge, S. Basu, and K. C. Mandal, "XPS investigation of CdTe surfaces: effect of Ru modification", Semiconductor Science and Technology, 4(10): p. 866-870, 1989.
13. O. Oklobia, G. Kartopu, S. Jones, P. Siderfin, B. Grew, H. K. H. Lee, W. C. Tsoi, Ali Abbas, J. M. Walls, D. L. McGott, M. O. Reese, and S. J. C. Irvine, "Development of arsenic doped Cd(Se,Te) absorbers by MOCVD for thin film solar cells", Solar Energy Materials and Solar Cells, 231: p. 111325, 2021.
14. Francesco Bittau, Shridhar Jagdale, Christos Potamialis, Jake W. Bowers, John M. Walls, Amit H. Munshi, Kurt L. Barth, and Walajabad S. Sampath, "Degradation of Mg-doped zinc oxide buffer layers in thin film CdTe solar cells", Thin Solid Films, 691, 2019.

Performance Investigation and Analysis of Anti-Soiling Coatings in Hot Desert Climate

Hebatalla Alhamadani*, Shaikha Hassan, Gerhard Mathiak, Omar Albadwawi, and Vivian Alberts

Research & Development Center, Dubai Electricity & Water Authority (DEWA), Dubai, United Arab Emirates

*hebatalla.alhamadani@dewa.gov.ae

Abstract— In this paper, the performance of hydrophobic coating on glass coupons and photocatalytic hydrophilic coating on solar modules installed in hot desert climate is analyzed using different test methods. Results reveal that after one month of outdoor exposure, and before cleaning the samples, the transmittance of uncleaned coated coupons reduced by 12.5% in comparison to 17.6% for the uncoated coupon. This result confirms the anti-soiling effect of the coating. After cleaning, the transmittance of the coated glass coupons under light exposure was slightly higher than coated coupons placed under shade, as well as higher in comparison to the uncoated coupons. This phenomenon could be caused by the degradation of the coating layer by light and high temperature. It is also observed that the wetting angle reduces with light exposure. In addition, the external quantum efficiency peak of the uncoated solar module was found to be higher than the coated module by approximately 1%. The I-V curves show higher power losses over the exposure time for coated module due to light soaking. These results provide an insight into the actual performance of different commercial anti-soiling coatings under hot desert conditions.

Keywords— PV reliability, desert climate, anti-soiling coating, transmittance, QE analysis

I. INTRODUCTION

Desert regions are characterized by a high dust concentration [1]. Despite substantial progress in the production of advanced solar cells in recent years, the problem of decreased efficiency due to dust accumulation remains a challenge, particularly in semi-arid or desert areas, where dust deposition on the surface of solar modules has a significant negative impact on their performance over a short period of time [2][3]. For instance, approximately 40% of reduction in efficiency and output power of uncleaned commercial solar module has been observed over a period of 6 months in Saudi Arabia, which is mainly attributed to the accumulation of dust particles over the module surface [2][3][4]. In addition, a study conducted in Dubai revealed soiling rate of 0.7% power loss per day [5].

To optimize solar modules, functional coatings on cover glasses of solar modules are developed. Since a protective glass coating is used as both a cover and a light transmission material, the choice and processing of this glass has an impact on its overall efficiency [6][7]. Optical reflection at the air/glass interface reduces the performance of the devices. Solar modules are often exposed to changing environmental conditions.

Although anti-reflective coatings on the glass can reduce reflection, dust accumulation on the module surfaces causes additional performance degradation [8]. As a result, light incident on a solar module can be lost by reflection by the cover glass as well as scattering or absorption by particles on the solar module. Maintaining a tolerable level of contamination, and thus performance, over time is crucial [9]. Removing contaminants from solar modules in order to retain optimal solar harvesting capabilities is time-consuming and costly.

Self-cleaning solar modules present a potentially cost-effective solution [10]. This is achieved by altering the morphology of solar module surface by introducing the super hydrophobic surface or super hydrophilic surface to create an anti-soiling effect. Surfaces are classified based on how they interact with water droplets. These transparent surfaces are used to minimize the reflection of light and dust collection on solar modules. While most publications present the effect of dust accumulation in reducing the efficiency of solar modules or the transmission of glass modules [11-24], few have studied and evaluated the performance of self-cleaning coatings considering the harsh desert conditions.

Wonwook et al. [25] compared the transmittance of a coated glass PV module to uncoated module in Jeddah, Saudi Arabia. Results showed a 3% increase in transmittance for coated module, from 88.5% to 91.5%, in two months. Similarly, Kivambe et al. [26] assessed the effectiveness of anti-soiling coatings on PV glass in Doha's desert climate. It was observed the average periodic energy transmission of coated glass was 1.5% higher than uncoated glass, for a duration of 12 weeks, suggesting that these coatings had only marginal advantages over uncoated glass samples. Gholami et al. [27] conducted a 70 day test period of nanocoatings on glass samples located in Isfahan, Iran. Results showed transmission coefficient loss reduced from 24% for uncoated glass to 14% for coated glass, resulting in a 10% gain due to anti-soiling effect.

In this context, this paper investigates the performance of anti-soiling coatings in desert climatic conditions. The various samples with or without coatings were studied by transmittance and water contact angle (WCA) measurement, quantum efficiency analysis for non-destructive detection along with electrical and spectral analysis [24]. The results provide an insight into the actual performance of commercial anti-soiling coatings in hot desert regions and assists the solar industry in

978-1-7281-6118-1/22 $31.00 © 2022 IEEE

improving the durability of these coatings and develop coatings tailored to specific site conditions.

II. METHODOLOGY

A. Glass Coupons

In this study, four coated glass coupons (50x50mm) with an organic hydrophobic layer and four uncoated coupons were tested at Dubai electricity and water authority (DEWA)'s R&D Center outdoor testing facility located within Mohammed bin Rashid (MBR) Solar Park, Dubai, UAE. Prior to outdoor installation, all coupons were cleaned with distilled spray water and dried with soft fiber cloth to conduct indoor (reference) glass transmittance test and WCA measurement. For each coupon, the transmittance was measured across three points, and an average was calculated.

All coupons were then placed at the outdoor test facility for one month at a tilt angle of 25° facing south under identical environmental conditions, except for light exposure, as per the following: 4 coupons placed under direct sunlight (2 uncoated, 2 hydrophobic), 4 coupons placed under shade (2 uncoated, 2 hydrophobic), to compare the effect of light exposure on performance of coating.

After one month, the coupons were brought indoors and the glass transmittance test were carried out for the coupons with dust and after cleaning by distilled spray water and soft dried cloth, as per the following: 2 coupons placed under direct sunlight (1 uncoated, 1 hydrophobic), 2 coupons placed under shade (1 uncoated, 1 hydrophobic). Simultaneously, the remaining 4 cleaned coupons were used for the WCA measurement test.

B. Solar Modules

One solar module was coated with an inorganic hydrophilic layer. Initial indoor I-V measurement and External Quantum Efficiency (EQE) were carried out for the coated solar module and one uncoated solar module. Afterwards, the modules were placed outdoors in open circuit condition for 1-2 days to activate the solar cells. The panels were then brought indoors and the same measurement techniques were repeated for both panels. This process was repeated for four times. The purpose was to electrically stabilize the module (e.g. light induced degradation effect) to establish I-V and QE reference parameters. The EQE measurements were performed on a 4cmx1cm area of solar modules, using a spectral response measurement system that allows for non-destructive measurement [17].

III. RESULTS AND DISCUSSIONS

A. Testing of glass coupons

a) Transmittance Analysis

Table I. shows transmittance measurements after one month outdoor exposure. Under direct light conditions, uncoated coupon showed the lowest transmittance (70%), followed by hydrophobic (73.64%). In addition, both coupons showed a lower transmittance in comparison to its reference coupons by 18.95%, and 16.29%, respectively. Similarly, under shaded conditions, the uncoated coupon exhibited the lowest transmittance (72.37%), followed by hydrophobic (77.44%), as shown in Fig.1. In comparison to its reference coupons, the

transmittance of uncoated coupon reduced by 17.57%, and hydrophobic by 12.5%. It was observed that the transmittance for coupons placed under the shade was better than those placed under direct sunlight. The uncoated coupon's performance was higher by 1.38%, whereas the hydrophobic by 3.79%.

After the coupons were cleaned, under direct light conditions, the coupons had similar transmittance with hydrophobic exhibiting the highest (91.62%), followed by uncoated (91.31%), as shown in Fig 2. Moreover, the uncoated and hydrophobic coupons showed a higher transmittance in comparison to its reference coupon by 1.36% and 1.68% respectively. Under shaded conditions, the hydrophobic coupon exhibited the highest transmittance (90.83%), followed by uncoated (90.01%). No significant change in transmittance was observed between the uncoated coupon and the reference coupon. However, the hydrophobic coupon showed a higher transmittance in comparison to its reference coupon by 0.89%. In contradiction to the uncleaned coupons, it was observed that coupons placed under direct sunlight showed a higher transmittance than the coupons placed under shade. The transmittance of the uncoated coupon was higher by 1.29% and the hydrophobic by 0.79%. Additionally, the transmittance of hydrophobic coupons in light and shade conditions was found to be higher than the uncoated coupons by 0.32% and 0.82%.

TABLE I.   TRANSMITTANCE (T) AT 1200 NM MEASUREMENTS AFTER 1 MONTHS OUTDOOR EXPOSURE

Coupon Type	Outdoors					
	Max T(%) Uncleaned Coupons		ΔMax T(%)	Max T(%) Cleaned Coupons		ΔMax T(%)
	Light	Shade	NA	Light	Shade	NA
Uncoated	70.98	72.37	1.39	91.30	90.01	1.29
Hydrophobic	73.64	77.44	3.80	91.62	90.83	0.79
ΔMax T (%)	2.66	5.07	NA	0.32	0.82	NA

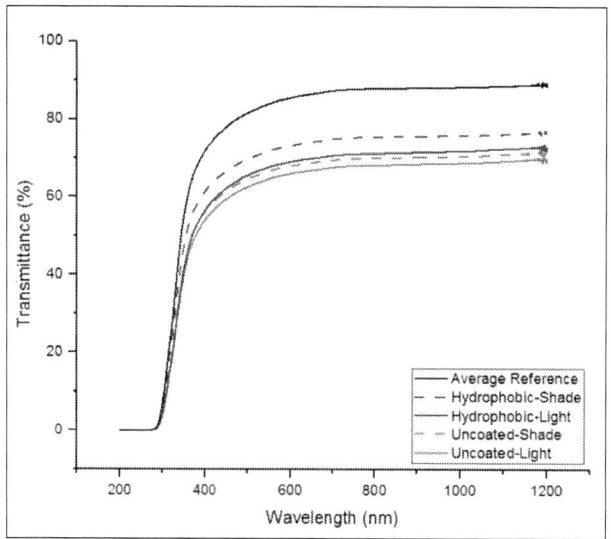

Fig. 1. Transmittance analysis of uncleaned coated and uncoated coupons placed under outdoor shade and light conditions in comparison with unexposed reference

b) Water Contact Angle Measurement Analysis

Table II. displays the water contact angle measurements after one month outdoor exposure, under different light conditions. It was observed that the contact angle for uncoated coupon increased by approximately 44 degrees after being placed outdoor for one month in the shade, and by approximately 25 degrees after being placed outdoor, as shown in Fig 3. However, the contact angle for hydrophobic coupon decreased by approximately 5 degrees after being placed outdoor for one month in the shade, as represented in Fig. 4, and by approximately 41 degrees after being placed outdoor for one month in the light indicating loss of hydrophobic property.

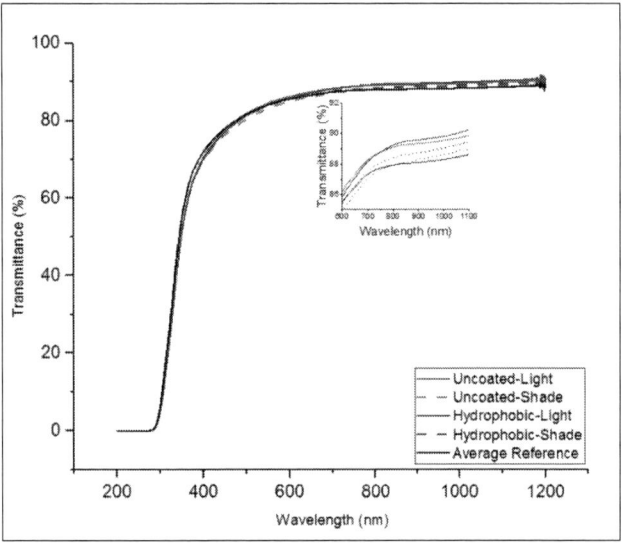

Fig. 2. Transmittance analysis of cleaned coated & uncoated coupons placed under outdoor shade and light conditions

TABLE II. WCA MEASUREMENTS AFTER ONE MONTH OUTDOOR EXPOSURE MEASURED IN THE MIDDLE OF THE COUPON WITH DROPLET SIZE 5μL

	Exposure	Coating	Avg. Angle(°)
Reference	Indoor/no exposure	Uncoated	41.8
	Indoor/no exposure	Hydrophobic	109
1 month	Outdoor-shade	Uncoated	85.4
		Hydrophobic	104.5
	Outdoor-light	Uncoated	67.1
		Hydrophobic	68.4

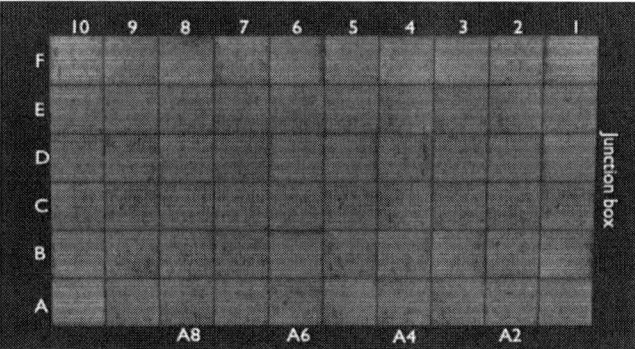

Fig. 3. Transmittance analysis of uncleaned coated and uncoated coupons placed under outdoor shade and light conditions in comparison with unexposed reference

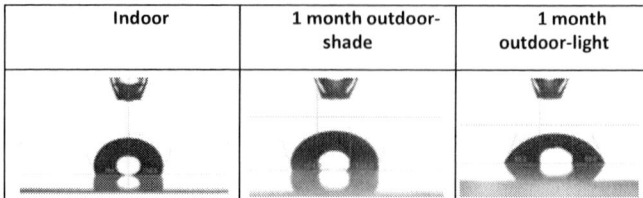

Fig. 4. WCA measurement for hydrophobic coupon, droplet size of 5uL, location: middle of coupon

B. Testing of Coated Solar Modules

a) External Quantum Efficiency (EQE) Analysis

There was no significant variation in the EQE measurements in all four sets of outdoor measurements (see Fig.7). This includes the EQE between the solar cells on the same module, as well as between the coated and uncoated solar module. The EQE increased by approximately 2% in comparison to initial indoor measurement, when both modules were placed outdoors.

Fig. 6 shows the four cell positions in which the EQE was measured for coated and uncoated module. Fig.7 shows EQE curve for all four cell positions taken at the center of the cell for both coated and uncoated modules. For both modules, the EQE peak occurred at a wavelength of 640 nm across all cell positions. Fig. 8 closely examines the EQE peak for both modules where it is observed the EQE peak of the uncoated module was slightly higher (~88%) than the coated module (~87%) at almost all cell positions. No discrepancies in the shape of the curve between the uncoated and coated solar modules.

Fig. 6. EQE cell positioning of the coated module

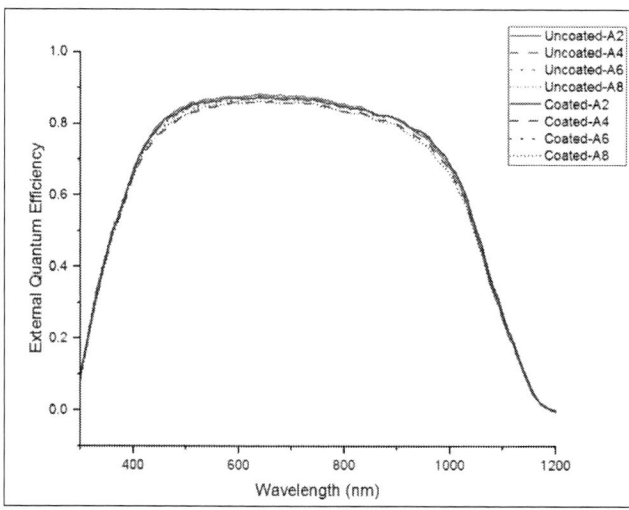

Fig. 7. EQE curves after third set of outdoor measurement

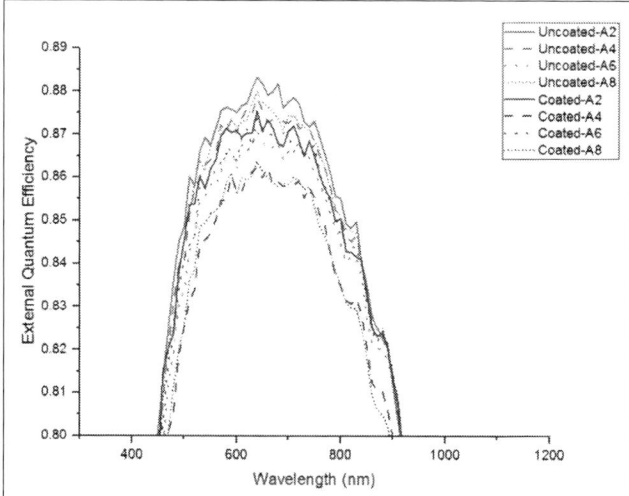

Fig. 8. EQE curves after third set of outdoor measurement- detailed 80- 89%

b) Light Exposure Effect Analysis

The I-V readings for the coated module showed fluctuation after four sets of outdoor measurements, whereas it was stable for the uncoated panel from the second set of outdoor measurement, as shown Fig. 9 below. To establish stable I-V reference parameter for the coated module, a fifth set of measurement was carried out after placing the coated module outdoors for additional four days. The fluctuation effect of the coated solar module has to be investigated further.

IV. CONCLUSION

This paper investigates and analyzes the performance of two different anti-soiling coatings under hot desert conditions. Results reveal the transmittance of uncleaned coated glass coupons was reduced by 12.5% when placed outdoors for 1 months, due to dust accumulation. Nevertheless, the loss of transmittance was lower in comparison to the uncoated coupon (17.57%), a 5% gain due to the anti-soiling effect of the coating. The results are in good agreement with the literature review of commercial coating performance in desert

conditions where a transmittance gain of 3% was observed in Saudi Arabia and

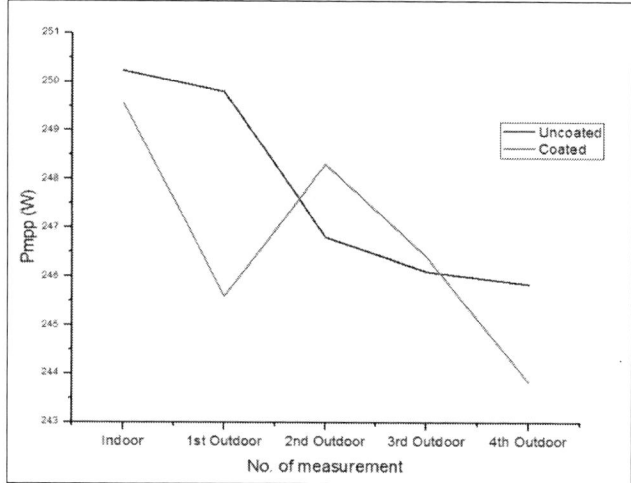

Fig. 9. Power curve after four sets of outdoor exposure. The power loss of the uncoated module is steady, the coated sample is unsteady

10% in Iran due to anti-soiling coating effect. After cleaning, the transmittance of the coated glass coupons under light exposure was slightly higher than without light and higher than the uncoated sample.This can be attributed to the degradation of the coating layer during light exposure. The transmittance of hydrophobic coupons in light and shade conditions was found to be higher than the uncoated coupons by 0.32% and 0.82%. Additionally,the wetting angle reduced with light exposure.

Results also show the EQE peak of the uncoated solar module was higher than the photocatalytic hydrophilic module by approximately 1%, across all cell positions. The I-V curves show power loss over exposure time due to light soaking, in which the effect was higher for the coated sample.

This work provides an insight into the actual performance of commercial anti-soiling coatings in desert climatic conditions and aids the solar industry in developing more comprehensive measurements for evaluation and development of anti-soiling coatings for solar PV applications in desert environment.

ACKNOWLEDGMENT

The authors would like to acknowledge University of Sharjah for granting us access to their advanced materials research lab to conduct water contact angle measurement tests. Authors would also like to acknowledge Shashank Suvarn from DEWA R&D Center for his support in this work.

REFERENCES

[1] Kumar, S. et al., "Comparative Investigation and Analysis of Encapsulant Degradation and Glass Abrasion in Desert Exposed Photovoltaic Modules", 2021 48th IEEE Photovoltaics Specialists Conference, 2021, pp. 793-798.

[2] T. Sarver, A. Al-Qaraghuli, and L. L. Kazmerski, "A comprehensivereview of the impact of dust on the use of solar energy: History, investigations, results, literature, and mitigation approaches," Renew.Sustain. Energy Rev., vol. 22, pp. 698–733, 2013.

[3] G. G. Jang et al., "The anti-soiling performance of highly reflective

978-1-7281-6118-1/22 $31.00 © 2022 IEEE

superhydrophobic nanoparticle-textured mirrors," Nanoscale, vol. 10, no. 30, pp. 14600–14612, 2018.

[4] M. GREEN et al., "Fabrication of highly transparent self-cleaning protection films for photovoltaic systems," Prog. Photovoltaics Res.Appl., vol. 20, no. 6, pp. 1114–1129, 2012.

[5] Elnosh, A. et al., "Estimation of Soiling Rates from PV Modules in the Desert Climate of Dubai", EUPVSC, 2017.

[6] J. Deubener, G. Helsch, A. Moiseev, and H. Bornhöft, "Glasses for solar energy conversion systems," J. Eur. Ceram. Soc., vol. 29, no. 7,pp. 1203–1210, 2009.

[7] X. Li, J. He, and W. Liu, "Broadband anti-reflective and water-repellent coatings on glass substrates for self-cleaning photovoltaic cells," Mater. Res. Bull., vol. 48, no. 7, pp. 2522–2528, 2013.

[8] L. K. Verma et al., "Self-cleaning and antireflective packaging glassfor solar modules," Renew. Energy, vol. 36, no. 9, pp. 2489–2493, 2011.

[9] A. A. Hegazy, "Effect of dust accumulation on solar transmittance through glass covers of plate-type collectors," Renew. energy, vol. 22, no. 4, pp. 525–540, 2001.

[10] F. Li, Q. Li, and H. Kim, "Spray deposition of electrospun TiO 2 nanoparticles with self-cleaning and transparent properties onto glass," Appl. Surf. Sci., vol. 276, pp. 390–396, 2013.

[11] H. A. Kazem, T. Khatib, K. Sopian, F. Buttinger, W. Elmenreich, and A. S. Albusaidi, "Effect of dust deposition on the performance of multi-crystalline photovoltaic modules based on experimental measurements," Int. J. Renew. Energy Res., vol 3, no. 4, pp. 850–853, 2013.

[12] S. A. Sulaiman, H. H. Hussain, N. Siti, H. N. Leh, and M. S. I. Razali,"Effects of Dust on the Performance of PV Panels," Int. J. Mech. Aerospace, Ind. Mechatron. Manuf. Eng., vol. 5, no. 10, pp. 2028– 2033, 2011.

[13] Aïssa, B., Isaifan, R., Madhavan, V. et al. "Structural and physical properties of the dust particles in Qatar and their influence on the PV panel performance". Sci Rep., vol. (6), 31467, 2016.

[14] M.R. Maghami, H. Hizam, C. Gomes, M.A. Radzi, M.I. Rezadad, S. Hajighorbani, Power loss due to soiling on solar panel: a review, Renew. Sustain. Energy Rev. 59, pp. 1307–1316, 2016.

[15] A. Rao, R. Pillai, M. Mani, P. Ramamurthy, Influence of dust deposition on pho- tovoltaic panel performance, Energy Procedia, vol. 54, pp. 690–700, 2014.

[16] S. Mekhilef, R. Saidur, M. Kamalisarvestani, Effect of dust, humidity and air velocity on efficiency of photovoltaic cells, Renew. Sustain. Energy Rev. 16, pp. 2920–2925, 2012.

[17] F. Mejia, J. Kleissl, J.L. Bosch, The effect of dust on solar photovoltaic systems, Energy Procedia, vol. 49, pp. 2370–2376, 2014.

[18] V. Sharma, S.S. Chandel, Performance and degradation analysis for long term re- liability of solar photovoltaic systems: a review, Renew. Sustain. Energy Rev. 27, pp. 753–767, 2013.

[19] S.A. Sulaiman, A.K. Singh, M.M.M. Mokhtar, M.A. Bou-Rabee, Influence of dirt accumulation on performance of PV panels, Energy Procedia, vol. 50, pp. 50–56, 2014.

[20] T. Sarver, A. Al-Qaraghuli, L.L. Kazmerski, A comprehensive review of the impact of dust on the use of solar energy: history, investigations, results, literature, and mitigation approaches, Renew. Sustain. Energy Rev. 22, pp. 698–733, 2013.

[21] S.C.S. Costa, A.S.A.C. Diniz, L.L. Kazmerski, Dust and soiling issues and impacts relating to solar energy systems: literature review update for 2012–2015, Renew. Sustain. Energy Rev. 63, pp. 33–61, 2016.

[22] A. Alheloo et al., "Indoor characterisation of thin-film PV modules installed for 4.6 years in desert conditions," 2020 47th IEEE Photovoltaic Specialists Conference, Calgary, 2020, pp. 1489-1493.

[23] Benjamin Figgis. "Investigation of PV soiling and condensation in desert environments via outdoor microscopy". Mechanics of materials [physics.class-ph]. Université de Strasbourg, 2018.

[24] O. Albadwawi et al., "Investigation of Bifacial PV modules degradation under desert climatic conditions," 2020 47th IEEE Photovoltaic Specialists Conference, Calgary, 2020, pp. 1505-1509.

[25] W. Oh et al., "Evaluation of anti-soiling and anti-reflection coating for photovoltaic modules," J. Nanosci. Nanotechnol., vol. 16, no. 10,pp. 10689–10692, 2016.

[26] Kivambe,M et al., "Performance of antisoiling coatings for enhancement of PV panel performance in Doha", Qatar Foundation Annual Research Conference Proceedings, vol. 2018, no. 1, 2018.

[27] Gholami, Aslan & Alemrajabi, Ali Akbar & Saboonchi, Ahmad, "Experimental study of self-cleaning property of titanium dioxide and nanospray coatings in solar applications", Solar Energy, vol. 15, 2016.

Light Distribution and Uniformity Evaluation of Cross Compound Parabolic Concentrators

Mazin Al-Shidhani[1,2], Mohammad Alnajideen[1,*] and Gao Min[1]

[1] *School of Engineering, Cardiff University, Cardiff, CF24 3AA, Wales, UK*
[2] *Petroleum Development Oman, P.O Box:81, Postal Code 100, Muscat, Sultanate of Oman*

[*] Corresponding author: AlnajideenMI@cardiff.ac.uk

ABSTRACT

Concentrator photovoltaic systems are capable of generating electric power from sunlight more than that produced from unconcentrated systems of the same photovoltaic material and size, thus beneficial to reduce photovoltaic areas and costs. In this work, we propose a Compound Parabolic Concentrator (CPC) for low concentration photovoltaic systems. A symmetrical CPC, which has a square entry and exit apertures known as a Cross CPC (CCPC), has been optically simulated and tested under a standard test condition for different geometrical concentration ratios of 2.9x, 4.0x, 6.0x, 8.0x and 9.0x. These concentrators were coupled to the same monocrystalline laser grooved buried contact silicon solar cell to acquire the characteristic current-voltage (I-V) curves. The design, simulation, fabrication and experiment testing of concentrators are presented. The non-uniformity light distribution across the receiver surface area and its effect on the power output of solar cell is demonstrated and validated. A new batch of five concentrators referred as an improved design has been fabricated and tested to validate the raytracing simulation. A comparative study between the old and new CPC designs is conducted based on the I-V performance curves. It has been proven that high uniformity of over 90% can be achieved by adjusting the solar cell at an optimal position, which varies based on the geometrical aspects of a CPC.

A Comparative Study of 3D Printed Non-Imaging Solar V-Trough and Compound Parabolic Concentrators for Low-Cost, High-Performance CPV Applications

Mohammad Alnajideen[1,2, a], Mazin Al-Shidhani[1,3], and Gao Min[1]

[1] School of Engineering, Cardiff University, Cardiff, CF24 3AA, Wales, UK

[2] Faculty of Engineering, Mutah University, P.O Box: 7, Al-Karak 61710, Jordan

[3] Petroleum Development Oman, P.O Box:81, Postal Code 100, Muscat, Sultanate of Oman

[a] Corresponding author: [a] AlnajideenMI@cardiff.ac.uk

1. Introduction (CPV-18)

V-trough Solar Concentrator (VSC) and Compound Parabolic Concentrator (CPC) are non-imaging concentrators and are employed as primary or secondary optical elements for low concentration photovoltaic (PV) cells. The key advantages of the VSC and CPC, include i) the design compactness, which is preferable for rooftop building applications, ii) high-optical performance and uniform illumination of receivers that is desirable for PV applications, iii) high acceptance angle to the incident light that can eliminate the use of sun-tracking systems, and iv) low-cost due to the ease of fabrication. The basic structure of VSC and CPC concentrators is the same, which consists of two symmetrical reflectors aligned with a photovoltaic (PV) receiver. They can also consist of four reflectors known as Crossed VSC and CPC (CVSC and CCPC), aiming to increase the light concentration onto the PV cell. The architecture of concentrators' reflectors is a flat shape in VSCs while it is a parabolic shape in CPCs. The reflectors are working to increase the incoming solar radiation on the PV area, thus decreasing the total area of the PV, increasing the output power and thereby decreasing the total cost. In the literature [1-11], there is a limited experimental research work conducted a comparative study on the optical and electrical performance of non-imaging concentrators, therefore, this paper aims to compare the performance of the VSC with the CPC in terms of the angular response, light uniformity of receivers, light concentration ratio, geometrical aspects, annual energy yields and the fabrication cost.

2. Abstract

This study presents a comparative simulation and experimental study on the optical and electrical performances of non-imaging concentrators; V-trough Solar Concentrator (VSC) and Compound Parabolic Concentrator (CPC). A number of VSC and CPC of various concentration ratios and acceptance angles were designed, simulated, fabricated and tested for this study. Advanced SolidWorks and TracePro simulations were used to design the geometrical boundaries and simulate the optical behaviour of concentrators. 3D printing technology, high-reflective materials, a mono-crystalline Silicon solar cell were employed for the construction of concentrators. An identical experimental concentrator PV set-up, a standard PV cell characterisation system and a spectroradiometer were used to investigate the performance of the concentrator PV system in an indoor environment at standard test condition of 1kW/m^2, AM1.5G and 25 °C. The results of the evaluation study showed that each type of concentrators has unique characteristics. For instance, the CPC offers a concentration ratio higher than the VSC while the light irradiance on the PV cell of VSC is more uniform than the CPC. The comparison between concentrators provides useful guidelines for achieving higher optical performance and improved electrical yields from those concentrators at affordable prices. The agreement between the simulation and experimental results is found to be less than 1%.

978-1-7281-6118-1/22 $31.00 © 2022 IEEE

References

1. Hollands, K.G.T., *A concentrator for thin-film solar cells.* Solar Energy, 1971. **13**(2): p. 149-163.
2. Winston, R., *Principles of solar concentrators of a novel design.* Solar Energy, 1974. **16**(2): p. 89-95.
3. Al-Najideen, M., M. Al-Shidhani, and G. Min. *Optimum design of V-trough solar concentrator for photovoltaic applications.* in *15th International Conference on Concentrator Photovoltaic Systems, CPV 2019.* 2019. American Institute of Physics Inc.
4. Al-Shidhani, M., et al. *Design and testing of 3D printed cross compound parabolic concentrators for LCPV system.* in *14th International Conference on Concentrator Photovoltaic Systems, CPV 2018.* 2018. American Institute of Physics Inc.
5. Paul, D.I., *Theoretical and Experimental Optical Evaluation and Comparison of Symmetric 2D CPC and V-Trough Collector for Photovoltaic Applications.* International Journal of Photoenergy, 2015. **2015**.
6. Proell, M., et al., *Experimental efficiency of a low concentrating CPC PVT flat plate collector.* Solar Energy, 2017. **147**: p. 463-469.
7. Singh, H., M. Sabry, and D.A.G. Redpath, *Experimental investigations into low concentrating line axis solar concentrators for CPV applications.* Solar Energy, 2016. **136**: p. 421-427.
8. Parupudi, R.V., et al., *Long term performance analysis of low concentrating photovoltaic (LCPV) systems for building retrofit.* Applied Energy, 2021. **300**: p. 117412.
9. Michael, J.J., et al., *Enhanced electrical performance in a solar photovoltaic module using V-trough concentrators.* Energy, 2018. **148**: p. 605-613.
10. Hadavinia, H. and H. Singh, *Modelling and experimental analysis of low concentrating solar panels for use in building integrated and applied photovoltaic (BIPV/BAPV) systems.* Renewable Energy, 2019. **139**: p. 815-829.
11. Ustaoglu, A., U. Ozbey, and H. Torlaklı, *Numerical investigation of concentrating photovoltaic/thermal (CPV/T) system using compound hyperbolic –trumpet, V-trough and compound parabolic concentrators.* Renewable Energy, 2020. **152**: p. 1192-1208.

Development of Photovoltaic Inverter Model with Islanding Detection Using the Sandia Frequency Shift Method

Nelson E. Saavedra-Peña[1], Rachid Darbali-Zamora[2], Edgardo Desarden-Carrero[1], and Erick Aponte-Bezares[1]

[1]University of Puerto Rico-Mayagüez, Mayagüez, Puerto Rico 00682, USA
[2]Sandia National Laboratories, Albuquerque, New Mexico, 87185, USA

Abstract —With increasing interest in renewable energy, more distributed energy resources (DERs) are being connected into the grid. Islanding conditions occur when a DER disconnects from the grid but remains energized. The Sandia Frequency Shift (SFS) is an islanding detection technique that introduces a small perturbation into the current phase angle of the photovoltaic (PV) inverter, with the use of positive feedback, triggering under frequency or over frequency protection when islanding occurs. Herein, this work presents PV inverter model that utilizes the SFS method for islanding detection. *MATLAB/Simulink* is used to test the performance of the developed PV inverter model with the SFS islanding detection capabilities. A comparison is made between PV inverters with different SFS control parameters. Simulation results are obtained for the dynamic response of each PV inverter when subjected to different islanding conditions.

Index Terms — islanding detection, distributed generation, renewable energy, Sandia Frequency Shift

I. INTRODUCTION

As traditional sources of generation are being replaced with inverter-based sources (IBR), more distributed energy resources (DERs) are being connected into the grid [1], [2]. This makes islanding detection a relevant and important topic of discussion. Islanding occurs when a DER disconnects from the grid but remains energized [3], [4]. Novel islanding detection methods are being developed [5]. A non-detection zone (NDZ) is an area of operation where islanding algorithms are not able to identify islanded conditions. The NDZ defines the effectiveness of an islanding detection method [6]. Ideally, these methods would have a NDZ close to zero. Various techniques can reduce significantly the NDZ for a specific method. Non-linear programming optimization has been used in the past to minimize the NDZ of the method by selecting appropriate load parameters and feedback gain [7].

The Sandia frequency shift (SFS) is an active islanding detection method developed by Sandia National Laboratories. The SFS method introduces a small perturbation into the output current phase angle of the photovoltaic (PV) inverter. Positive feedback triggers under-frequency protection (UFP) or over-frequency protection (OFP) when islanding occurs. Time response of the islanding detection will depend on the parameter selection. Improvements to the parameters of the SFS method have been achieved by adaptively tunning to eliminate NDZ using fuzzy load parameter [8]. Other approaches utilize mapping methods that analyze the equilibrium boundaries of the NDZ [9]. There is also interest in understand the interactions of multiple PV inverters [10], [11]. Other research includes islanding detection interactions with grid-support and ride-through capabilities [12], [13], [14]. Hybrid approaches incorporating the SFS, and other methods have also been developed. The islanding detection method demonstrated in [15], uses a combination of the SFS with the rate of change of frequency (ROCOF) to reduce NDZ and improve time response. Previous work demonstrates the effectiveness of the SFS method compared to other methods that use the phase angle criteria [16]. Regardless, there is still a need for developing simulation models of PV inverters with the capability to perform islanding detection methods such as the SFS.

Herein, this work presents a simulation model of a PV inverter with the SFS method for islanding detection. Simulation results obtained using *MATLAB/Simulink* are presented for the PV inverter model subjected to islanding conditions. A comparison was made between PV inverters models with different SFS control parameters. These results provide a glimpse into the interactions between PV inverters and their SFS settings when subjected to islanding conditions.

II. PV INVERTER MODELING DESIGN APPROACH

To demonstrate the capabilities of the SFS method, a mathematical model of a PV inverter with active and reactive power setpoints was developed. Fig. 1 illustrates a circuit diagram describing the islanding detection system used for testing the developed SFS PV inverter. In this diagram, the switch S_2 disconnects the grid and creates the islanding condition for the PV inverter. When the PV inverter detects the islanded condition, the switch S_1 opens, and the PV inverter disconnects. One of the biggest concerns for distribution connected generation is islanding, which led to rigorous operating criteria in the IEEE 1547 Standard [17]. The IEEE 1547 Standard specifies that the DER must detect and disconnect within 2 s (120 cycles) of islanding [18], [19], [20].

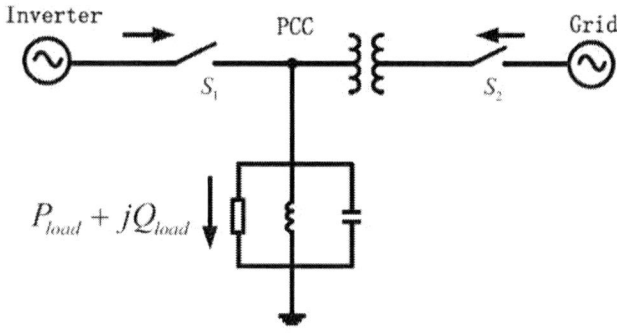

Fig. 1. Circuit diagram describing the islanding detection.

978-1-7281-6118-1/22 $31.00 © 2022 IEEE

A. Photovoltaic Inverter Model

The representation of the PV inverter consists of an averaged power converter model, system transformer, and a current control loop that generates the pulse width modulation (PWM) signal for the averaged power converter model [21]. The PV inverter model utilizes a phase-locked loop (PLL) as a mechanics in order to synchronize with the grid. The PV inverter model with an *LC* output filter was derived using the equivalent per-phase circuit as shown in Fig. 2. Using two axis reference frame theory from [22], the direct (*d*) and quadrature (*q*) axis equivalent inverter circuits are shown Fig. 3 (a) and Fig. 3 (b), respectively. The *dq-frame* circuits are modeled in the same way with exception of a change in the polarities of the dependent current and voltage sources.

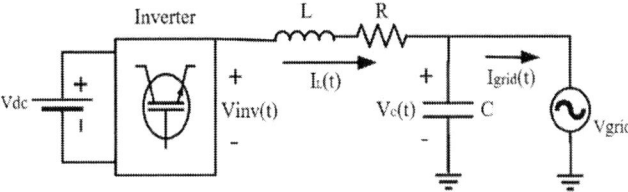

Fig 2. One line diagram of a grid connected PV inverter.

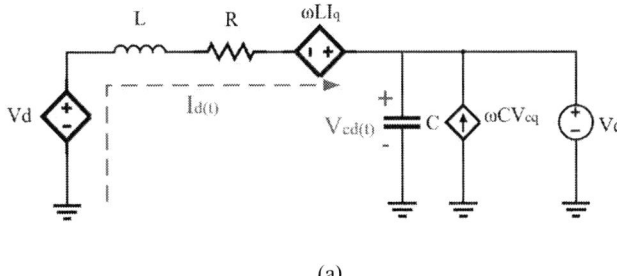

(a)

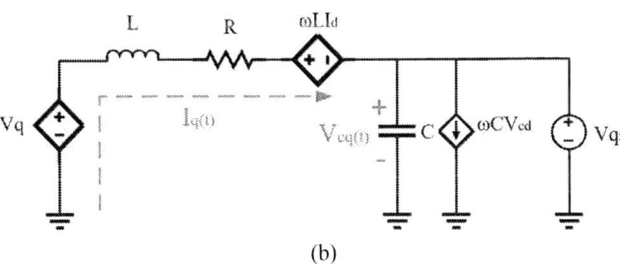

(b)

Fig 3. Equivalent PV inverter circuit schematic. (a) Direct axis PV inverter circuit. (b) Quadrature axis PV inverter circuit.

The state-space model of the inverter using *ABC*-coordinates is shown in matrix form in equation (1).

$$
\begin{bmatrix} \dot{I}_L \\ \dot{V}_C \end{bmatrix} = \begin{bmatrix} -\dfrac{R}{L} & -\dfrac{1}{L} \\ \dfrac{1}{C} & 0 \end{bmatrix} \begin{bmatrix} I_L \\ V_C \end{bmatrix} + \begin{bmatrix} \dfrac{1}{L} \\ 0 \end{bmatrix} \cdot [V_{inv}] \tag{1}
$$

In this equation variables *R*, *L* and *C* represent the PV inverter equivalent series resistance of the filter, filter inductance, and the filter capacitance, respectively. The variables I_L and V_C are the filter inductor current and the filter capacitor voltage. The variable V_{inv} is the inverter voltage measured before the PV

inverter filter. The state-space variables $\dot{I}_L$ and $\dot{V}_C$ are the derivatives of the inductor current and capacitor voltage, respectively. Applying the *dq0* transform to the equivalent per-phase circuit of the inverter, a *dq0* reference frame state-space model is shown in matrix form in equation (2).

$$
\begin{bmatrix} \dot{I}_d \\ \dot{I}_q \\ \dot{V}_{Cd} \\ \dot{V}_{Cq} \end{bmatrix} = \begin{bmatrix} -\dfrac{R}{L} & \omega & -\dfrac{1}{L} & 0 \\ -\omega & -\dfrac{R}{L} & 0 & -\dfrac{1}{L} \\ \dfrac{1}{C} & 0 & 0 & \omega \\ 0 & \dfrac{1}{C} & -\omega & 0 \end{bmatrix} \cdot \begin{bmatrix} I_d \\ I_q \\ V_{Cd} \\ V_{Cq} \end{bmatrix} + \begin{bmatrix} \dfrac{1}{L} & 0 \\ 0 & \dfrac{1}{L} \\ 0 & 0 \\ 0 & 0 \end{bmatrix} \cdot \begin{bmatrix} V_d \\ V_q \end{bmatrix} \tag{2}
$$

In this equation, the variables I_d and I_q are the inverter direct (*d*) and quadrature (*q*) axis currents in the *dq-frame*, respectively. The variables V_{Cd} and V_{Cq} are the *d* and *q* axis capacitor voltages in the *dq-frame*, respectively. The variables V_d and V_q are the inverter *d* and *q* axis voltages in the *dq-frame*, respectively. The variable ω is the angular frequency of the inverter. The state-space variables $\dot{I}_d$ and $\dot{I}_q$ are the derivatives of the *d* and *q* axis inverter currents in the *dq-frame*, respectively. The state-space variables $\dot{V}_{Cd}$ and $\dot{V}_{Cq}$ are the derivatives of the *d* and *q* axis capacitor voltage in the *dq-frame*, respectively. Equation (3) illustrates the PV inverter mathematical model using the modulating voltages from control.

$$
\begin{bmatrix} \dot{I}_d \\ \dot{I}_q \end{bmatrix} = \begin{bmatrix} \dfrac{-R}{L} & 0 \\ 0 & \dfrac{-R}{L} \end{bmatrix} \cdot \begin{bmatrix} I_d \\ I_q \end{bmatrix} + \begin{bmatrix} \dfrac{1}{L} & 0 \\ 0 & \dfrac{1}{L} \end{bmatrix} \cdot \begin{bmatrix} M_d \\ M_q \end{bmatrix} \tag{3}
$$

For the control equations, the dynamics of the capacitor voltage (grid voltage) are modeled as a disturbance and included into the inverter modulated voltages M_d and M_q. The modulated voltages M_d and M_q are shown in equation (4) and equation (5), respectively.

$$
M_d = V_d - \omega \cdot L \cdot I_q + V_{Cd} \tag{4}
$$

$$
M_q = V_q + \omega \cdot L \cdot I_d - V_{Cq} \tag{5}
$$

$$
V_{Cd} = V_{ds} \tag{6}
$$

$$
V_{Cq} = V_{qs} \tag{7}
$$

The active and reactive power of the inverter was controlled by regulating the *d* and *q* axis currents. This was achieved by using a Proportional-Integral (PI) current controller. The *d* axis controller was chosen to be 3.5 times larger than *q* axis gains so that the system reacts faster to changes in active power than to changes in reactive power. This is done since the inverter delivers power at unity power factor (*Q=0*). The equations for total active and reactive power injected to the grid using the Park's transformation are shown in equation (8) and equation (9), respectively.

978-1-7281-6118-1/22 $31.00 © 2022 IEEE

$$P_{ref} = \frac{3}{2} \cdot (V_d \cdot I_d + V_q \cdot I_q) \qquad (8)$$

$$Q_{ref} = \frac{3}{2} \cdot (V_q \cdot I_d - V_d \cdot I_q) \qquad (9)$$

B. Sandia Frequency Shift Islanding Detection Method

The SFS method introduces a small perturbation into the current phase angle of the inverter with the use of positive feedback. When compared with the voltage signal at the point of common coupling (PCC), it produces a shift in the phase of the inverter current. The SFS will have no impact on grid-connected inverter frequency apart from manageable power quality issues that arise because of the injection of current harmonics. When grid-connected, the grid system stiffness eliminates any considerable frequency deviation introduced by the method. On the other hand, the SFS will increase inverter current frequency rapidly when islanded. The ultimate goal of the detection method is to trigger UFP or OFP when islanding occurs. Time response of the islanding detection will depend greatly in the value of the positive feedback gain selected. The performance of the method will also depend on the type of PV inverter controller [23]. Load quality factor and design parameters will also affect the non-detection zone (NDZ) implemented by the SFS [24], [25]. The current-controller block diagram with the implemented SFS active islanding detection method is shown in Fig. 4.

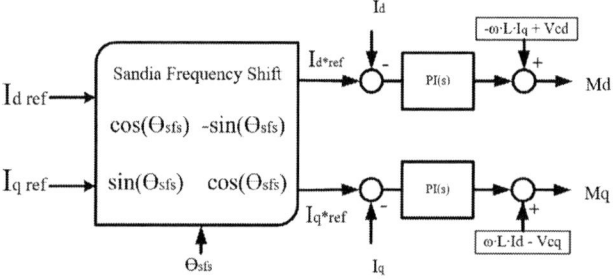

Fig 4. Block diagram of inverter PI current-controller equipped with Sandia frequency shift active islanding detection method.

Equation (10) illustrates the phase angle transformation for the SFS method. In this equation, variables $I^*_{d_{ref}}$ and $I^*_{q_{ref}}$ represent the inverter reference current after the SFS was implemented.

$$\begin{bmatrix} I^*_{d_{ref}} \\ I^*_{q_{ref}} \end{bmatrix} = \begin{bmatrix} \cos(\theta_{sfs}) & -\sin(\theta_{sfs}) \\ \sin(\theta_{sfs}) & \cos(\theta_{sfs}) \end{bmatrix} \cdot \begin{bmatrix} I_{d_{ref}} \\ I_{q_{ref}} \end{bmatrix} \qquad (10)$$

The variables $I_{d_{ref}}$ and $I_{q_{ref}}$ are the *dq* reference currents obtained from the active and reactive power references, respectively. Finally, the variable θ_{sfs} represents the SFS phase angle. The expression shown in equation (11) used to calculate the SFS phase angle between PV inverter current and voltage at the PCC. The SFS current phase angle will have a value close to zero when grid-connected but will deviate considerably if islanding occurs.

$$\theta_{sfs} = \frac{\pi}{2} \cdot [cf_0 + K \cdot (f_{PCC} - f_{nom})] \qquad (11)$$

In this equation, the variable cf_0, represents the chopping factor. The value for chopping factor, cf_0, influences both power quality degradation and detection time for the SFS method. The variable f_{PCC}, is the frequency at the PCC, while f_{nom}, is nominal system frequency (*60 Hz*). In this equation the parameter K is the SFS feedback gain. For this paper, parameter cf_0 was chosen as 0, this was done in order to isolate the influence of SFS feedback gain, K, in the islanding detection time of the PV inverters. Q_f is defined by the relationship shown in equation (12).

$$Q_f = R \cdot \sqrt{\frac{C}{L}} \qquad (12)$$

In this equation, the R, L and C values are the resistive, inductive, and capacitive elements of the load bank. The load was selected such that the PV inverter supplies the power required at unity power factor (p.f.). This scenario is the most difficult for the PV inverter protection to detect because, when the grid is disconnected from the system under study, no change in power will result from the formation of the island. Equation (13) and (14) were used to calculate load active power (P_{Load}) and reactive power (Q_{Load}), respectively.

$$P_{Load} = V^2 \cdot R_{Load} \qquad (13)$$

$$Q_{Load} = \frac{V^2}{\omega \cdot L_{Load}} - V^2 \cdot \omega \cdot C_{Load} \qquad (14)$$

Table I summarizes the load parameters used for testing the developed SFS islanding detection method for the PV inverter models.

TABLE I:
LOAD PARAMETERS DESCRIPTION

Parameter	Description	Value	Units
R_{Load}	Load Resistor	9.6	Ω
C_{Load}	Load Capacitor	14.1	μF
L_{Load}	Load Inductor	500	mH
f_0	Resonant Frequency	59.9	Hz

III. PV INVERTER MODEL DEVELOPMENT

A simplified circuit, like the one shown in Fig. 1, consisting of a programmable load and ideal source was used for testing the developed PV inverter models. Table II summarizes the control and filter parameters for the developed PV inverter models with SFS unintentional islanding detection capabilities. One PV inverter without SFS capability was chosen as the experimental control group. Three SFS equipped PV inverter were compared against the PV inverter without SFS capability. Three different SFS feedback gains were also chosen for each of the SFS equipped PV inverters.

TABLE II:
PHOTOVOLTAIC INVERTER PARAMETERS

Parameter	Description	Value	Units
L_{DG}	Filter Inductor	10	mH
R_{DG}	Filter Resistance	0.01	Ω
C_{DG}	Filter Capacitor	20	μF
fc	Cut-Off Frequency	355	Hz
Fsw	Switching Frequency	5	kHz
K_i (PI)	PI Integral Gain	200	V/A·s
K_p (PI)	PI Proportional Gain	15	V/A
V_{LL}	Line-Line Voltage	480	V_{rms}
S_{nom}	Rated Power	24	kVA
f_{nom}	Nominal Frequency	60	Hz
P_{set}	Active Power Setpoint	24	kW
Q_{set}	Reactive Power Setpoint	0	kVARS

IV. SIMULATION RESULTS

Simulation results were obtained for the PV inverter models subjected to islanding conditions. All PV inverters were subjected to the same islanding scenario to compare the performance of the SFS equipped PV inverters vs. the NO SFS capability PV inverter. In the simulations, the islanding condition was initiated at 0 s.

Fig. 5 illustrates the simulation results obtained by all four PV inverters frequency when subjected to islanding conditions. Notice from the simulation results that PV inverter without SFS did not cease operations even after the island condition was introduced. Simulation demonstrates that deviation of frequency caused by formation of the island is not enough to drive PV inverter frequency with NO SFS capability out of the limits (UFL). This PV inverter without SFS capability continued operating normally resulting in hazardous conditions for load and PV inverter since no frequency or voltage regulation is present. Notice that SFS PV inverter 1 ($K = 0.015$) was able to detect the islanded condition and disconnect after 0.56 s (33.3 cycles) of islanding. SFS PV inverter 2 ($K = 0.03$) detected the islanding condition at 0.33 s (20 cycles), while SFS PV inverter 3 ($K = 0.06$) detected the islanding condition after 0.22 s (13.4 cycles). The SFS equipped PV inverters all deviated from nominal system frequency when disconnected from the grid as expected. Notice that the SFS was able to accomplish this by forcing island frequency onto UFL where protection triggers (UFP) and ultimately trips the PV inverter.

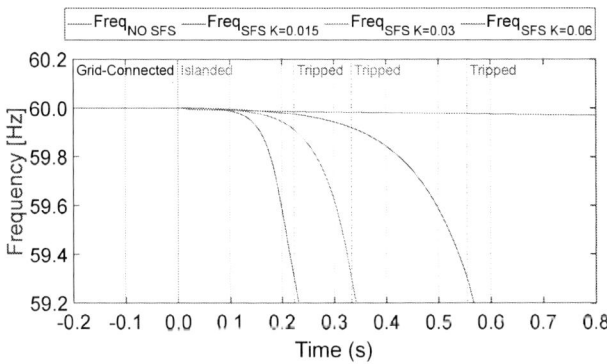

Fig. 5. Simulation results for the PV inverters frequency at varying SFS gain values.

Fig. 6 illustrates simulation results obtained from the PV inverters instantaneous voltage. Notice from these simulation results that as the feedback gain was adjusted, the PV inverter instantaneous voltage response was affected.

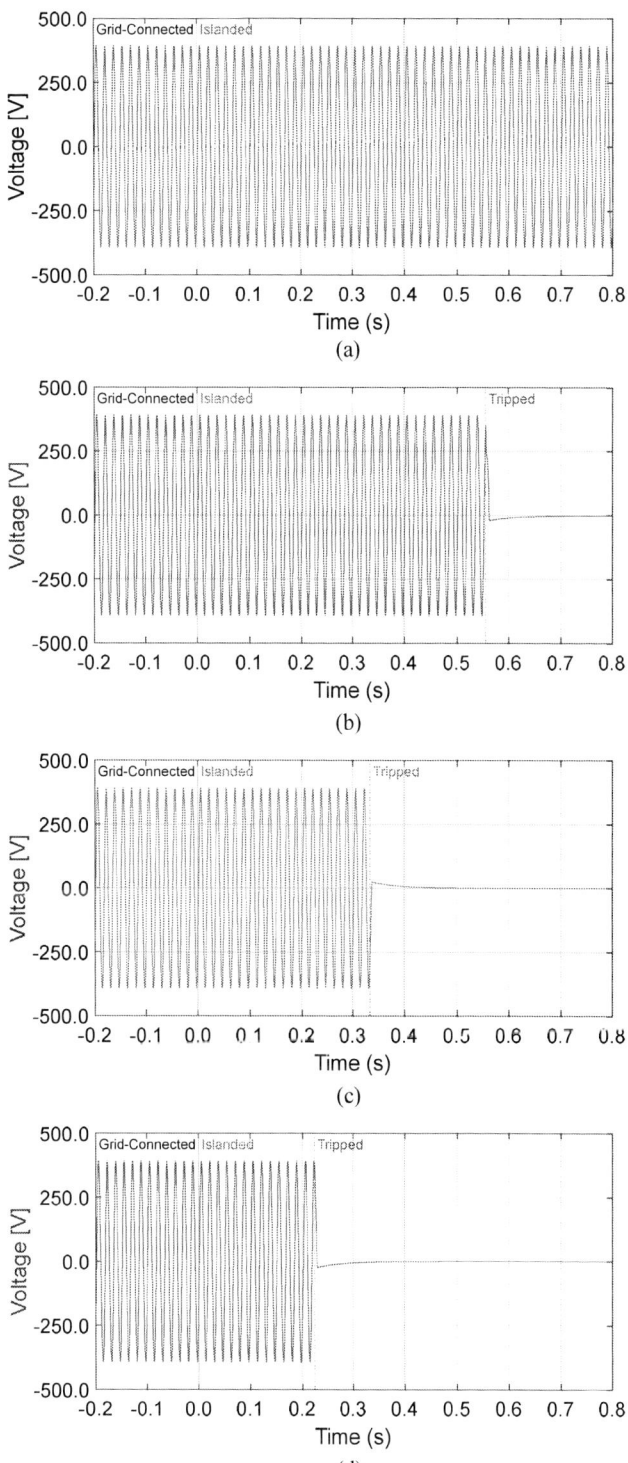

Fig. 6 Simulation results for the four PV inverters instantaneous voltages. (a) PV inverter with no SFS capability. (b) SFS PV inverter with K = 0.015. (c) SFS PV inverter with K = 0.03. (d) SFS PV inverter with K = 0.06.

Fig. 7 illustrates simulation results obtained from the PV inverters instantaneous current. The SFS gain affects deviating the frequency of the system faster as the gain value increases. Therefore, simulation results suggest that as SFS gain increases, time detection by the SFS reduces.

Fig. 8 illustrates the simulation results for the PV inverter RMS voltage. Fig. 9 illustrates the simulation results for the PV inverter RMS current. Simulation results shows that for all PV inverters, RMS voltage and current are both reduced to zero when the PV inverter disconnects. The values oscillated before the PV inverter was ultimately tripped because of the deviation from nominal frequency caused by the SFS islanding detection method. Fig. 10 illustrates the simulation results obtained for the PV inverter active power. Fig. 11 illustrates the simulation results obtained for the PV inverter reactive power. Although the PV inverters are set to provide zero reactive power, when the island condition is introduced, there is a small spike in reactive power. This could be due to a phase shift created when the PV inverter is subjected to an islanding condition.

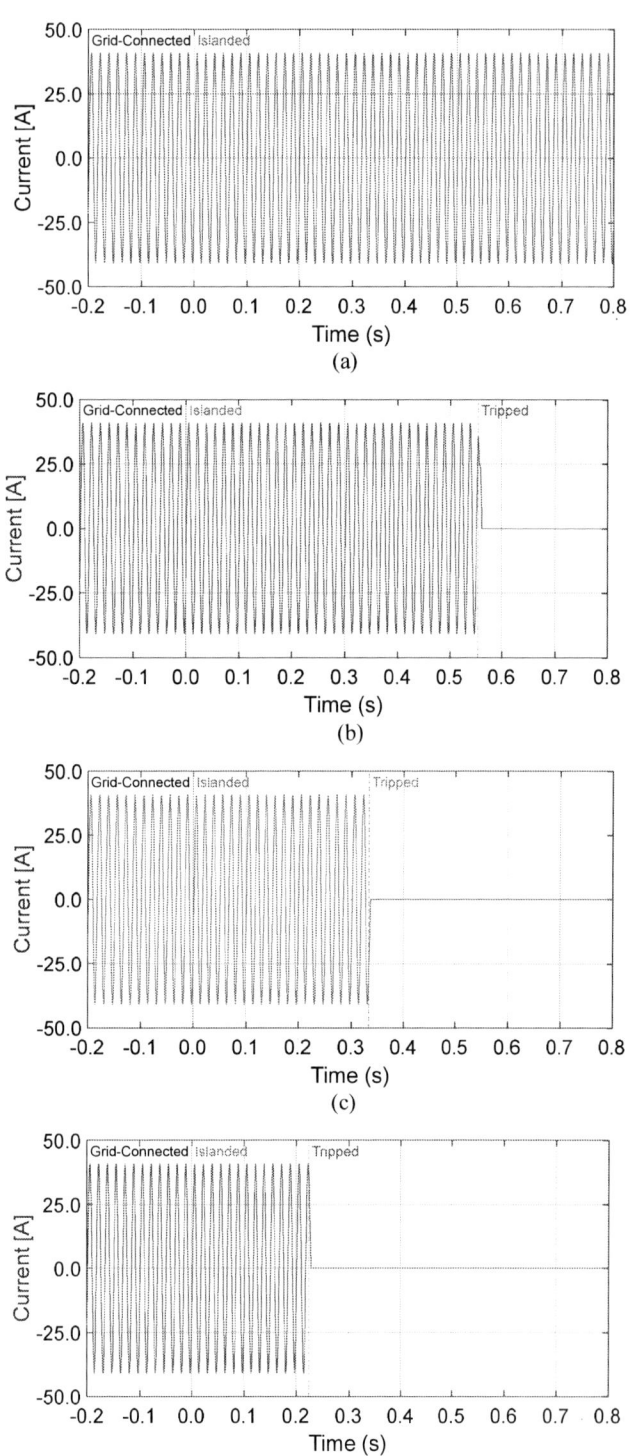

Fig. 7 Simulation results for the four PV inverters instantaneous currents. (a) PV Inverter with no SFS capability. (b) SFS PV inverter with K = 0.015. (c) SFS PV inverter with K = 0.03. (d) SFS PV inverter with K = 0.06.

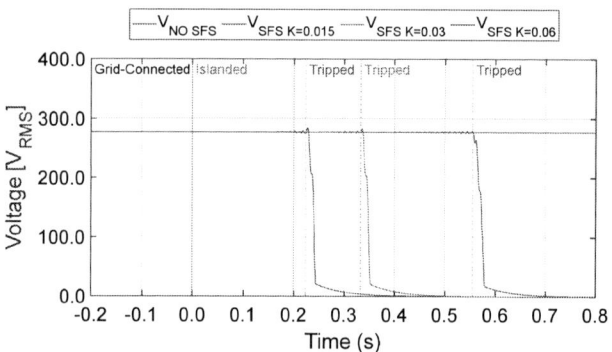

Fig. 8. Simulation results for the PV inverters RMS voltages at varying SFS feedback gain.

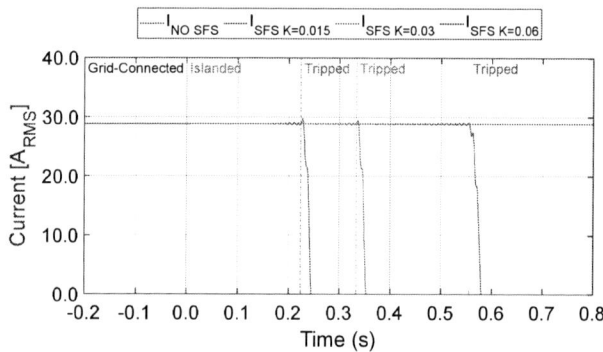

Fig. 9. Simulation results for the PV inverters RMS currents at varying SFS feedback gain.

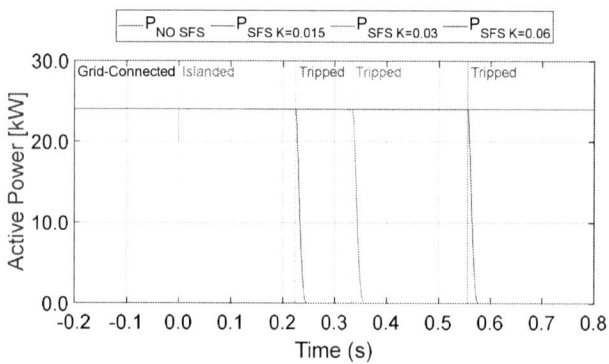

Fig 10. Simulation results for the PV Inverters Active Power at varying SFS feedback gain.

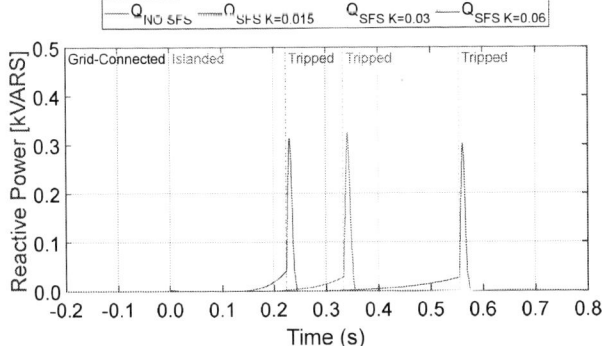

Fig 11. Simulation results for the PV inverters reactive power at varying SFS feedback gain.

Notice from these simulation results that as the PV inverter detected the formation of the island, it ceased to operate and deliver power in compliance with the applicable regulations and standards. The gains selected for the SFS simulations guaranteed an island time detection faster than the 2 s (120 cycles). This demonstrates that the SFS is a functional unintentional islanding detection method and can be designed to meet unintentional islanding detection guidelines depending on load parameters.

Lastly, a simulation scenario was developed where load quality factor, Q_f, was manipulated to derive a relationship between load Q_f value and SFS time detection. An SFS feedback gain sweep was evaluated for different quality factors (Q_f). Load Q_f was calculated using equation (12). Table III illustrates the combination of passive elements used to create a parametric relationship between the SFS feedback gain and the response time. Results demonstrate that as the SFS parameter K is increased, the response time of the PV inverter to detect an islanding condition decreases. Similar simulations are performed to obtain a parametric relation between varying SFS feedback gain K at four different Q_f. These simulations are performed with Q_f values of 0.05 (R=9.6 Ω, L=500 mH and C=14.1 μF), 1.27 (R=9.6 Ω, L=20 mH and C=350 μF), 2.55 (R=9.6 Ω, L=10 mH and C=703 μF) and 3.45 (R=13 Ω, L=10 mH and C=703 μF).

TABLE III:
DESCRIPTION OF THE QUALITY FACTOR ELEMENTS

Q_f	R (Ω)	C (μF)	L (mH)
0.05	9.6	14.1	500
1.27	9.6	350.0	20
2.55	9.6	703.0	10
3.45	13.0	703.0	10

Table IV summarizes the PV inverter response time at different SFS feedback gains and four different Q_f values. In the simulation results, SFS feedback gain values that yield response times longer than 2 s are not presented and are defined as no detection (ND) scenario. Fig. 12 illustrates the PV inverter response time at varying SFS gains and four different Q_f values.

TABLE IV:
PHOTOVOLTAIC INVERTER PARAMETRIC RESULT SUMMARY

Feedback Gain (K)	Response Time (s)			
	$Q_f = 0.05$	$Q_f = 1.27$	$Q_f = 2.55$	$Q_f = 3.45$
0.000	ND	ND	ND	ND
0.010	0.765	ND	ND	ND
0.020	0.449	ND	ND	ND
0.030	0.336	ND	ND	ND
0.040	0.280	ND	ND	ND
0.050	0.248	1.36	ND	ND
0.060	0.229	0.635	0.521	ND
0.070	0.218	0.379	0.335	1.64
0.080	0.214	0.236	0.258	0.512
0.090	0.224	0.135	0.215	0.359
0.100	0.191	0.068	0.187	0.288
0.200	0.106	0.066	0.103	0.132
0.300	0.089	0.062	0.087	0.107

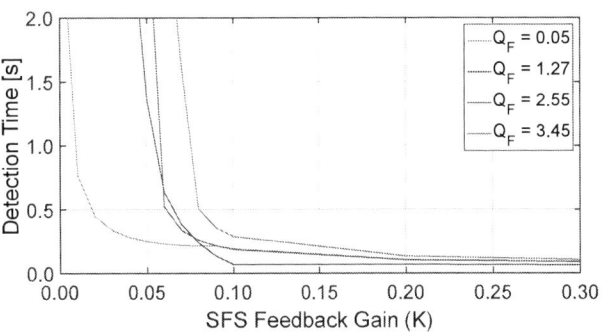

Fig. 12. Parametric Analysis for the PV Inverters Response Time at Varying SFS Feedback Gain and Four Q_f Values.

Notice that as quality factor Q_f is increased, the PV inverters SFS feedback gain K must also be increased to be able detect the formation of an island condition.

V. CONCLUSION

In this paper, a *dq0* current controlled PV inverter simulation model with the SFS islanding detection method was developed. The simulation model successfully disconnected using the SFS method when subjected to an islanding condition. Simulation results illustrate that as the SFS feedback gain became larger, the detection time of the island became smaller. The SFS islanding detection method amplifies the deviation in island frequency from nominal operation value (60 Hz). Frequency deviation will depend on the resonant frequency of the island when no SFS is present. When SFS is equipped into PV inverters controllers, the frequency deviation will be influenced greatly by load Q_f and SFS gain selection. The SFS is an effective method to detect unintentional islanding and can be implemented by using simple trigonometric functions (i.e., cosine, sine) and common controller schemes and techniques.

ACKNOWLEDGEMENT

Sandia National Laboratories is a multi-mission laboratory managed and operated by National Technology and Engineering Solutions of Sandia, LLC., a wholly owned subsidiary of Honeywell International, Inc., for the U.S. Department of Energy's National Nuclear Security Administration under contract DE-NA-0003525.

This work was sponsored in part by the Consortium for Hybrid Resilient Energy Systems (CHRES) under grant number DE-NA0003982 from the National Nuclear Security Administration part of the U.S. Department of Energy.

REFERENCES

[1] J. Johnson *et al.*, "Distribution Voltage Regulation Using Extremum Seeking Control with Power Hardware-in-the-Loop," *IEEE Journal of Photovoltaics*, vol. 8, no. 6, pp. 1824-1832, Nov. 2018.

[2] R. Darbali-Zamora, *et al.*, "Distribution Feeder Fault Comparison Utilizing a Real-Time Power Hardware-in-the-Loop Approach for Photovoltaic System Applications," *IEEE 46th Photovoltaic Specialists Conference (PVSC)*, 2019, pp. 2916-2922.

[3] M. R. Oshiro, R. Barros Godoy, M. A. G. de Brito, and L. Galotto, "Performance Analysis of Active Anti-islanding Techniques for Photovoltaic Application," *IEEE 15th Brazilian Power Electronics Conference and 5th IEEE Southern Power Electronics Conference (COBEP/SPEC)*, 2019, pp. 1-6.

[4] M. V. G. Reis, *et al.*, "Analysis of the Sandia Frequency Shift (SFS) islanding detection method with a single-phase photovoltaic distributed generation system," *IEEE PES Innovative Smart Grid Technologies Latin America (ISGT LATAM)*, 2015, pp. 125-129.

[5] N. E. Saavedra-Pena, R. Darbali-Zamora, E. Desarden-Carrero, E. E. Aponte-Bezares; "Towards an Islanding Detection Method Using a Digital Twin Concept", *48th IEEE Photovoltaic Specialists Conference (PVSC)*, June 20-25, 2021.

[6] E. Desardén-Carrero, R. Darbali-Zamora and E. E. Aponte-Bezares, "Analysis of Commonly Used Local Anti-Islanding Protection Methods in Photovoltaic Systems in Light of the New IEEE 1547-2018 Standard Requirements," *IEEE 46th Photovoltaic Specialists Conference (PVSC)*, 2019, pp. 2962-2969.

[7] H. H. Zeineldin and S. Kennedy, "Sandia Frequency-Shift Parameter Selection to Eliminate Nondetection Zones," *IEEE Transactions on Power Delivery*, vol. 24, no. 1, pp. 486-487, Jan. 2009.

[8] H. Vahedi and M. Karrari, "Adaptive Fuzzy Sandia Frequency-Shift Method for Islanding Protection of Inverter-Based Distributed Generation," *IEEE Transactions on Power Delivery*, vol. 28, no. 1, pp. 84-92, Jan. 2013.

[9] F. Yu, Y. Fan, M. Cheng and G. Hu, "Parameter design optimization for Sandia frequency shift islanding detection method," *3rd IEEE International Symposium on Power Electronics for Distributed Generation Systems (PEDG)*, 2012, pp. 182-186.

[10] L. A. C. Lopes and Y. Zhang, "Islanding Detection Assessment of Multi-Inverter Systems with Active Frequency Drifting Methods," *IEEE Transactions on Power Delivery*, vol. 23, no. 1, pp. 480-486, Jan. 2008.

[11] W. Han, T. Zheng, and X. Wang, "Investigation of multi-inverter distributed generation resident Sandia frequency shift anti-islanding method," *International Conference on Advanced Power System Automation and Protection*, 2011, pp. 935-939.

[12] R. Darbali-Zamora, *et al.*, "Evaluation of Photovoltaic Inverters Under Balanced and Unbalanced Voltage Phase Angle Jump Conditions," *47th IEEE Photovoltaic Specialists Conference (PVSC)*, 2020, pp. 1562-1569.

[13] E. Desarden-Carrero, R. Darbali-Zamora, E. E. Aponte-Bezares; "Analysis of Grid Support Functionality Dynamics Under Ride-Through Requirements Using Power-Hardware-in-the-Loop Implementation", *48th IEEE Photovoltaic Specialists Conference (PVSC)*, June 20-25, 2021.

[14] N. Ninad *et al.*, "PV Inverter Grid Support Function Assessment using Open-Source IEEE P1547.1 Test Package," *47th IEEE Photovoltaic Specialists Conference (PVSC)*, 2020, pp. 1138-1144.

[15] M. Khodaparastan, H. Vahedi, F. Khazaeli and H. Oraee, "A Novel Hybrid Islanding Detection Method for Inverter-Based DGs Using SFS and ROCOF," *IEEE Transactions on Power Delivery*, vol. 32, no. 5, pp. 2162-2170, Oct. 2017.

[16] M. E. Ropp, *et al.*, "Determining the relative effectiveness of islanding detection methods using phase criteria and nondetection zones," *IEEE Transactions on Energy Conversion*, vol. 15, no. 3, pp. 290-296, Sept. 2000.

[17] North American Electric Reliability Corporation (NERC), "1,200 MW Fault Induced Solar Photovoltaic Resource Interruption Disturbance Report: Southern California 8/16/2016 Event", Version 1.0 June 8, 2017.

[18] IEEE Standard for Interconnection and Interoperability of Distributed Energy Resources with Associated Electric Power Systems Interfaces," in IEEE Std 1547-2018 (Revision of IEEE Std 1547-2003), pp.1-138, 6 April 2018.

[19] IEEE Standard Conformance Test Procedures for Equipment Interconnecting Distributed Energy Resources with Electric Power Systems and Associated Interfaces," in IEEE Std 1547.1-2020, pp.1-282, 21 Jan. 2020.

[20] E. Desarden-Carrero, R. Darbali-Zamora, N. S. Gurule, E. Aponte-Bezares and S. Gonzalez, "Evaluation of the IEEE Std 1547.1-2020 Unintentional Islanding Test Using Power Hardware-in-the-Loop," *47th IEEE Photovoltaic Specialists Conference (PVSC)*, 2020, pp. 2262-2269.

[21] V. Purba, B. B. Johnson and S. V. Dhople, "Reduced-order Aggregate Model for Parallel-connected Grid-tied Three-phase Photovoltaic Inverters," *2019 IEEE 46th Photovoltaic Specialists Conference (PVSC)*, 2019, pp. 0724-0729

[22] A. Yazdani and R. Iravani, "Voltage-Sourced Converters in Power Systems: Modeling, Control, and Applications", *Wiley-IEEE Press*, 2010, pp.1-541.

[23] X. Wang, *et al.*, "Impact of DG Interface Controls on the Sandia Frequency Shift Antiislanding Method," *IEEE Transactions on Energy Conversion*, vol. 22, no. 3, pp. 792-794, Sept. 2007.

[24] H. H. Zeineldin and S. Kennedy, "Instability criterion to eliminate the Non-Detection Zone of the Sandia Frequency Shift method," *IEEE/PES Power Systems Conference and Exposition*, 2009, pp. 1-5.

[25] M. Al Hosani, Z. Qu and H. H. Zeineldin, "Scheduled Perturbation to Reduce Nondetection Zone for Low Gain Sandia Frequency Shift Method," *IEEE Transactions on Smart Grid*, vol. 6, no. 6, pp. 3095-3103, Nov. 2015.

Optical Properties of Thin Film Sb2Se3 and Identification of its Electronic Losses in Photovoltaic Devices

Niva K. Jayswal, Suman Rijal, Biwas Subedi, Indra Subedi, Zhaoning Song, Robert W. Collins, Yanfa Yan, Nikolas J. Podraza

University of Toledo, Toledo, OH, United States

Antimony selenide (Sb2Se3) is a highly promising solar cell absorber material with excellent optoelectronic properties including high absorption coefficient in the visible energy range. Here, we investigate the optical and electronic properties of thin film polycrystalline Sb2Se3 deposited on glass. The indirect bandgap of 1.117 ± 0.001 eV, direct optical gap of 1.175 ± 0.002 eV, and Urbach energy of 21.1 ± 0.6 meV are determined from the absorption coefficient spectra obtained from photothermal deflection spectroscopy. Complex dielectric function ($\varepsilon = \varepsilon 1 + i\varepsilon 2$) spectra are determined using through-the-glass spectroscopic ellipsometry measurements in $0.75 - 4$ eV spectral range due to the roughness of the Sb2Se3 film. These spectra in ε along with the solar cell component layer thicknesses are used as input parameters for external quantum efficiency (EQE) simulation to investigate electronic losses in substrate type Sb2Se3 based solar cells. The difference between simulated EQE for Sb2Se3 based PV assuming complete carrier collection in absorber layer Sb2Se3 and measured EQE are evaluated to obtain $97.0 \pm 0.2\%$ carrier collection probability in Sb2Se3 at the heterojunction interface and a 400 nm carrier collection length throughout the Sb2Se3 absorber layer. The difference between measured short circuit current density and that simulated assuming no electronic losses shows that 5.4 mA/cm2 is lost due to incomplete carrier collection.

Racking Reflection and Shading Effects on Single Axis Tracked Bifacial Photovoltaic Modules

Mandy R Lewis, Trevor J Coathup, Annie C J Russell, Javier Guerrero-Perez, Christopher E Valdivia, Karin Hinzer

University of Ottawa, Ottawa, ON, Canada

Soltec Innovations, Murcia, Spain

Bifacial photovoltaics (PV) is predicted to comprise 80% of the silicon PV share within the next ten years. However, bifacial energy yield models are still undergoing validation, and their uncertainty may slow adoption. One of the challenges of single-axis-tracked (SAT) bifacial PV performance modelling is accurately accounting for the effects of racking elements, such as the frame, module supports, and torque tube, on the rear irradiance. In this work, we calculated front and rear irradiances for the center modules of a 2-in-portrait SAT bifacial photovoltaic system from hourly typical meteorological year data for the Bifacial Test Evaluation Center (BiTEC) site in Livermore, California using bifacial_radiance ray tracing software. For every hourly timestamp, we calculated 2D front and rear irradiance maps in three cases: with no racking, absorptive racking, and reflective racking. From these, we calculated three racking effects: shading, reflection, and shading and reflection combined. We also calculated shading and reflection factors as well as rear irradiance non-uniformity for each case. For the PV system modelled, racking reflection is focused in the same areas of the module as racking shading, partially counteracting shading-induced irradiance reduction and irradiance non-uniformity. For example, for a winter day at noon, racking reflection reduces the rear shading factor from -18.4% to -10.8% and the irradiance non-uniformity from 14.8% to 10.8%. The effects of racking, including both shading and reflection, vary by time of day and year. Accounting for these variations, rather than using annual average correction factors, will improve energy yield prediction accuracy for bifacial PV, especially over short time periods.

Spectroscopic Ellipsometry Analysis and Quantum Efficiency Simulation of CuInSe₂ Solar Cells

Dhurba R. Sapkota, Ambalanath Shan, Balaji Ramanujam, Puja Pradhan, Richard Irving, Adam B. Phillips, Michael J. Heben, Randy J. Ellingson, Sylvain Marsillac*, Nikolas J. Podraza, and Robert W. Collins

Wright Center for Photovoltaics Innovation & Commercialization, Univ. Toledo, Toledo, Ohio, 43606, USA; * Virginia Institute of Photovoltaics, Old Dominion Univ., Norfolk, Virginia, 23529, USA

Abstract — CuInSe₂ (CIS) thin-film solar cells in the substrate configuration were fabricated and studied using spectroscopic ellipsometry (SE). Starting with the parameterized complex dielectric functions of the individual component layers, SE analysis was performed using a step-wise procedure that ranks the fitting parameters according to their ability to reduce the mean square error of the fit. The resulting layer thicknesses and dielectric functions were used to simulate the external quantum efficiency (EQE) of the device assuming complete active layer collection. Electronic losses were identified by comparison with the measured EQE. Simulation results show that complete collection would yield a short circuit current density of 42.62 mA/cm². The goal of this study is to develop and optimize 1.0 eV bandgap CIS for use as a bottom cell absorber of multijunction solar cells.

Keywords—Spectroscopic Ellipsometry, CuInSe₂, EQE Simulation.

I. INTRODUCTION

Application of CuInSe₂ (CIS) as a photovoltaic absorber has attracted interest since the first report by Wagner *et al.* in 1974. This report described the fabrication of a hetero-junction CIS/CdS PV detector using a bulk p-type crystalline CIS absorber that yielded a solar conversion efficiency of 5% [1,2]. The first fully thin film CuInSe₂/CdS solar cells were fabricated from absorbers co-evaporated from CuInSe₂ and Se in a single stage [3]. Continuous efficiency improvements have been made since then, resulting in a 14.5% efficient thin-film CIS device by multi-stage thermal co-evaporation [4].

The advances in CIS devices reaching efficiencies of ~ 15% [5,6], in conjunction with the narrow bandgap of CIS (1.02 eV), motivate applications of the material as the bottom cell absorber in multijunction thin film PV devices. The electronic properties of the p-type CIS are controlled by the Cu stoichiometry, indicated as y in Cu_yInSe_2, such that for optimum solar cell performance, $0.85 < y < 0.95$ [7]. The efficiencies of the best CIS devices, however, are still far from those of Cu(In,Ga)Se₂ alloys which have exceeded 22% through reduction of recombination in the depletion and heterojunction regions [8]. In spite of the lower efficiency of CIS single junction devices, progress has been made on tandem devices with a CIS bottom cell [9]. The combination of the CIS or CIGS bottom cell and an ~ 1.6 eV bandgap perovskite top cell has the potential to realize high efficiency

all-thin-film tandems [10-12]. Because of its narrow bandgap, unalloyed CIS is uniquely suited to serve as the bottom solar cell of multijunction thin film solar cells. Perovskite compounds for the top cell were first identified as efficient light harvesters for photoelectrochemical cells and later work progressed rapidly from liquid to solid state junctions [13].

The research reported here focuses on procedures for understanding carrier collection for a CIS device via spectroscopic ellipsometry (SE) analysis and external quantum efficiency (EQE) measurements and simulations. The optical properties, bandgaps, and Urbach tail energies of one-stage CIS thin films for the different compositional ratios y = [Cu]/[In], were studied previously [14]. Applying such results, the optical design of perovskite tandems with CIS bottom cells was undertaken [15]. Photoluminescence studies of the CIS thin films of various ratios y from Ref. [14] have been reported, in addition to the hole concentration and mobility for a sample with y = 0.9 [16,17]. Finally, the results of real time studies of the structural evolution and optical properties of CIS thin films, including two-stage CIS materials, were also presented previously [18,19].

II. EXPERIMENTAL DETAILS

A. CIS Solar Cell Fabrication

The substrates used for the CIS solar cells are soda-lime glass (SLG) 2.5 cm × 7.5 cm × 0.15 cm in size. These substrates are cleaned using Micro-90 solution first, followed by rinsing in deionized water, both steps in an ultrasonic cleaner. After cleaning, the substrates are dried with nitrogen gas and immediately introduced into the vacuum chamber for Mo deposition. Mo is deposited in a two-step sputtering process at a substrate temperature of 250°C. The CIS absorber layer studied here is deposited by thermal co-evaporation in a high vacuum chamber to an ~ 1.7 μm thickness on the Mo surface held at 570°C. A one-stage co-evaporation process was performed from Cu, In, and Se fluxes with an intended CIS composition of y = [Cu]/[In] = 0.90 and an optimum deposition rate of R ~ 3.5 Å/s [19]. Cu and In flux calibration based on real time SE analysis during step-wise deposition of Cu and In₂Se₃ films enable identification of the Cu and In source temperatures needed to obtain the desired y and R

978-1-7281-6118-1/22 $31.00 © 2022 IEEE

values. The Se source temperature was fixed during the deposition, yielding a room temperature Se deposition rate of 20 Å/s at the substrate location. CdS is deposited on the SLG/Mo/CIS structures by chemical bath deposition (CBD) to a 50 nm intended thickness. In this process, an aqueous solution of cadmium acetate and thiourea with ammonium hydroxide catalyst is used, the latter diluted at 1.1 M in the solution. The molar ratio of Cd to S in the CBD solution is [Cd]/[S] = 0.0292, and a 25 → 65°C temperature ramp is used during deposition. A ZnO/ZnO:Al bilayer serving as the transparent conducting top contact is deposited by sputtering at room temperature with intended thicknesses of 50/200 nm. Finally, a Ni/Al/Ni trilayer is deposited by electron beam evaporation through masks to serve as the grids. The sequence of deposition of each layer is presented in Fig. 1 starting from the Mo back contact and ending with the metal grids.

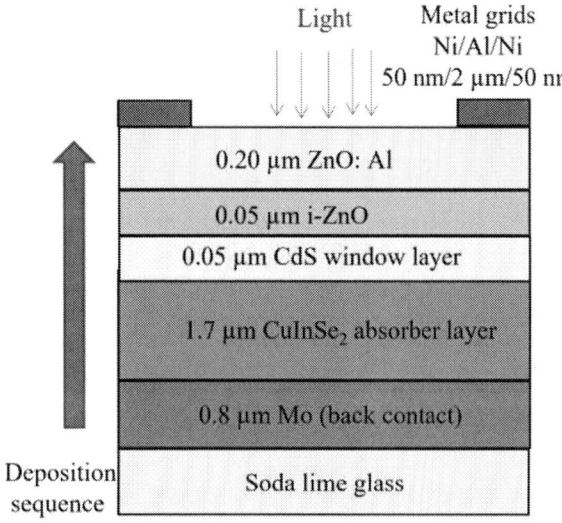

Fig. 1. Schematic of the CIS solar cell structure in the substrate configuration used in this study.

B. CIS Absorber Characterization

The CIS absorber layer was characterized by X-ray diffraction (XRD), energy dispersive X-ray spectroscopy (EDS), and scanning electron microscopy (SEM) for crystal structure and grain size, composition, and layer thickness, respectively. The CIS deposition rate was calculated from the film thickness and deposition time. XRD patterns for the CIS thin film, as measured by a Rigaku XRD instrument with SAXS, were found to be consistent with the tetragonal chalcopyrite, copper indium diselenide [20]. The EDS results show a composition y within the range of 0.89 ± 0.01. The SEM measurement provides the surface morphology as well, presented in the inset of Fig. 2. The average size of coherently scattering domains within the absorber was calculated as ~ 185 Å by the Scherrer equation from the (112), (204), and (312) diffractions of the XRD pattern also presented in Fig. 2.

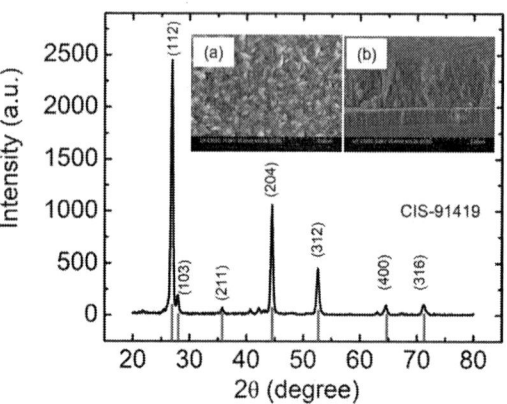

Fig. 2. X-ray diffraction pattern and (a, b) scanning electron microscopy images of the CIS absorber layer.

C. Solar cell results

Solar cells 0.5 cm x 1.0 cm in area were completed from the CIS absorber layer. Figure 3 (a) shows the light and dark current density versus voltage (J-V) characteristics and (b) shows the EQE spectrum for the maximum efficiency device of this study. Table I presents the device parameters of the maximum efficiency solar cell fabricated from the absorber, along with the average of the eight best devices. The table includes the open-circuit voltage, short circuit current density, fill factor, efficiency, shunt resistance, and series resistance.

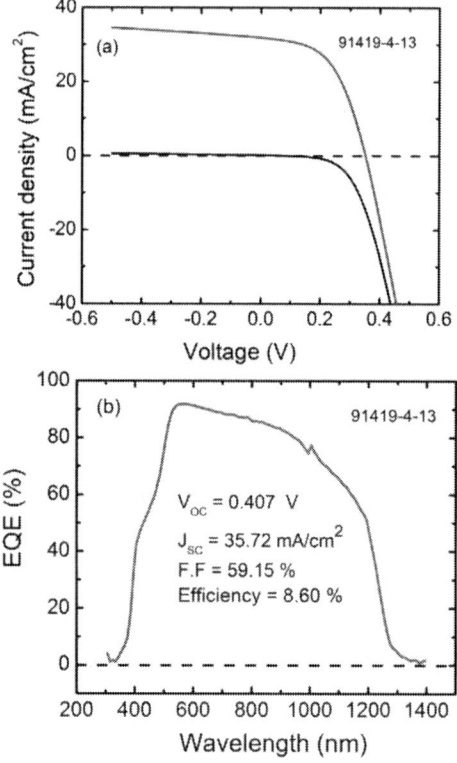

Fig. 3. (a) Light and dark J-V curves and (b) EQE of the CIS device.

TABLE I. DEVICE PARAMETERS OF THE MAXIMUM EFFICIENCY SOLAR CELL AND THE AVERAGE PARAMETERS OF THE EIGHT BEST CELLS FABRICATED FROM THE CIS ABSORBER (91419-4). VALUES IN PARENTHESES ARE BASED ON A J_{sc} MEASUREMENT BY INTEGRATION OF THE EQE SPECTRUM OF FIG. 3(b).

CIS device	V_{oc} (V)	J_{sc} (mA/cm²)	FF (%)	η (%)	R_s (Ω cm²)	R_{sh} (Ω cm²)
Max. η (EQE)	0.407	30.73 **(35.72)**	59.15	7.40 **(8.60)**	2.23	650
Average	0.394	31.82	56.06	6.99	2.70	553

III. SE ANALYSIS AND EQE SIMULATION

Although multiple stage CIS solar cells having absorbers without Ga and efficiencies up to 15% have been reported [5], such processes are more complex and challenging to scale up compared to simple single stage processes. Additional optimization and process steps such as post-deposition treatments are expected to improve the single stage CIS solar cell efficiency above the best cell result shown in Fig. 3 and Table I. To further explore the limitations on the single stage CIS solar cell efficiency, the EQE spectrum of Fig. 3(b) has been modeled in depth. This modeling applies the index of refraction and extinction coefficient spectra (n, k), or correspondingly, the real and imaginary parts of the complex dielectric function spectra $(\varepsilon_1, \varepsilon_2)$ of the solar cell component layers in Fig. 1, including Mo, CIS, CdS, ZnO, and ZnO:Al.

An ex-situ SE measurement was performed on the CIS solar cell of Fig. 3 at a 70° angle of incidence over the photon energy range from 0.75 to 4.0 eV. The SE data are fitted using a multilayer model, and the structural and optical property parameters of the model, the latter including layer volume fractions which control the $(\varepsilon_1, \varepsilon_2)$ spectra, are determined in a step-wise mean square error (MSE) reduction procedure. Each fitting parameter of the model is ordered in two sets according to its ability to reduce the MSE, the first set being the structural parameters and the second set being the optical property parameters [21]. The results of the procedure from which nine structural and six optical property parameters could be deduced are presented in Table II and Fig. 4. From the final best fit spectra in the ellipsometry angles (ψ, Δ), the Mueller matrix spectra N = cos2ψ, C = sin2ψ cosΔ, and S = sin2ψ sinΔ spectra (lines) can be determined as shown along with the data (points) in Fig. 5. The results of this analysis for the layer stack of the CIS solar cell is presented in Fig. 6.

The structural and optical parameters of the layer stack for the CIS solar cell of Fig. 6 are used to simulate and model its EQE. The CIS bulk layer and CIS/CdS interface layer are assigned as the active layers contributing to photogenerated charge carrier collection. The contribution of each of these two active layers to the EQE is presented in Fig. 7, as determined based on the assumption of 100% carrier collection. Figure 8 shows the measured normal incidence EQE spectrum, along with the summed spectrum from Fig. 7 (dashed line) simulated assuming 100% contribution from the two active layers. In this solar cell, electronic losses are evident, resulting in partial collection from the CIS active layer. The inset in Fig. 8 shows the collection model and Fig. 9 shows a step-wise profile of collection from the active layers that gives good agreement between the simulated and measured EQE spectra (solid line). The reflectance of the complete multilayer structure and the optical absorption of each layer in the structure of the CIS cell are shown in Fig. 10. Presented in the Fig. 11 is a comparison of the measured EQE spectra of the CIS device obtained before and after anti-reflection coating together with the simulation of Fig. 7 assuming 100% collection from the two active layers of CIS device. Measured and predicted solar cell short circuit current densities and efficiencies are presented in Table III.

TABLE II. STEP-WISE MSE REDUCTION WITH STRUCTURAL AND OPTICAL PARAMETERS IN SE ANALYSIS OF THE CIS SOLAR CELL.

Types of Parameters	No of Parameters	Parameters yielding greatest reduction in MSE	MSE
Structural	1	CIS bulk layer: thickness	331.43
	2	AZO/Ambient interface layer: thickness	293.80
	3	AZO bulk layer: thickness	197.05
	4	CIS/CdS interface layer: thickness	179.78
	5	Mo/CIS interface layer: thickness	163.42
	6	CdS/ZnO interface layer: thickness	157.46
	7	CdS bulk layer: thickness	143.69
	8	ZnO/AZO interface layer: thickness	143.17
	9	ZnO bulk layer: thickness	143.09
Optical	10	AZO/Ambient interface layer: void volume fraction	131.34
	11	AZO bulk layer: ε_1 constant offset, $\varepsilon_{1,\infty}$	81.71
	12	CIS bulk layer: lowest CP energy, E_1	65.94
	13	ZnO bulk layer: ε_1 constant offset, $\varepsilon_{1,\infty}$	63.70
	14	CIS bulk layer: lowest CP energy amplitude	57.67
	15	ZnO/AZO interface layer: AZO volume fraction	48.44

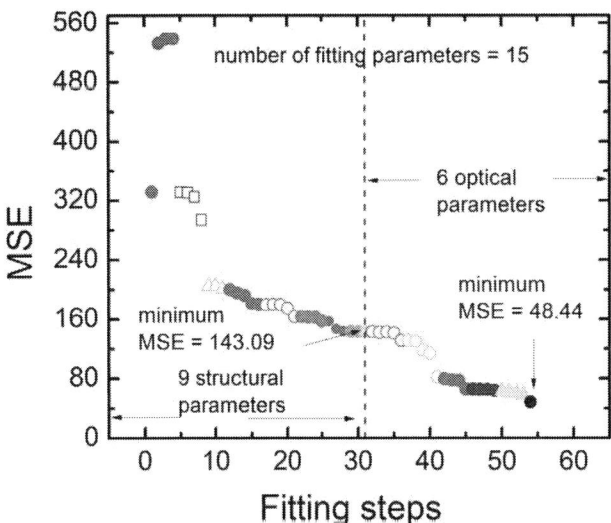

Fig. 4. Step-wise mean square error reduction in the analysis of SE data collected from the CIS solar cell at an angle of incidence of 70°.

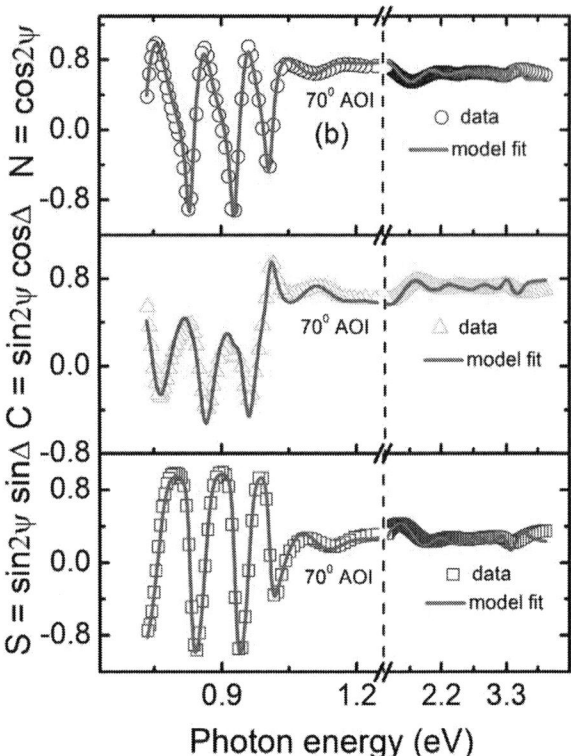

Fig. 5. Experimental Mueller matrix spectra (N, C, S) from a spectroscopic ellipsometry measurement of the CIS solar cell along with the corresponding best fit results using the model of Fig. 6.

layer (composition, properties)	thickness	Effective thickness
surface roughness (f_v = 46.10 ± 0.52%)	58.80 ± 0.74 nm	
ZnO:Al (f_v = 0.00%)	71.16 ± 0.98 nm	ZnO:Al: 153.06 nm
i-ZnO / ZnO:Al (f_{ZnO} = 3.20 ± 2.89%)	51.60 ± 1.73 nm	
i-ZnO (f_v = 0.00%)	37.51 ± 1.74 nm	ZnO: 62.27 nm
CdS / i-ZnO (f_{CdS} = 50.00%)	46.22 ± 0.98 nm	
CdS (f_v = 0.00%)	31.21 ± 1.09 nm	CdS: 82.02 nm
CIS / CdS (f_{CIS} = 50%)	55.40 ± 0.64 nm	
CIS (f_v = 0.00%)	1637.07 ± 2.74 nm	CIS: 1673.45 nm
Mo / CIS (f_{Mo} = 50.00%)	17.36 ± 1.45 nm	Mo
Mo (opaque)		

Fig. 6. Layer stack of the CIS solar cell including bulk, interface, and surface layers from the SE analysis of Figs. 4, 5, and Table II.

IV. CONCLUSIONS AND FUTURE WORK

Based on a multilayer structural analysis of an optimized single stage CIS solar cell by spectroscopic ellipsometry (SE), the EQE spectrum of the cell can be simulated and compared with measured results. Observed deviations help to refine the optical model for the cell, as well as to identify electronic

losses. Complete collection by the active layers of the device, the CIS/CdS interface and CIS absorber layers, is expected to yield a potential short circuit current density of 42.62 mA/cm². The lower experimental value of J_{sc} = 35.72 mA/cm² is an indication of recombination in the absorber layer along with possible effects of a large conduction band offset at the CdS/CIS interface and formation of barrier against collection of the photogenerated electrons in the absorber [22]. Future steps will involve optimization of the CIS EQE performance by reducing the deep absorber recombination and top contact optical losses through variations in the CIS thickness and other layer thicknesses of the device, followed by post-deposition treatments.

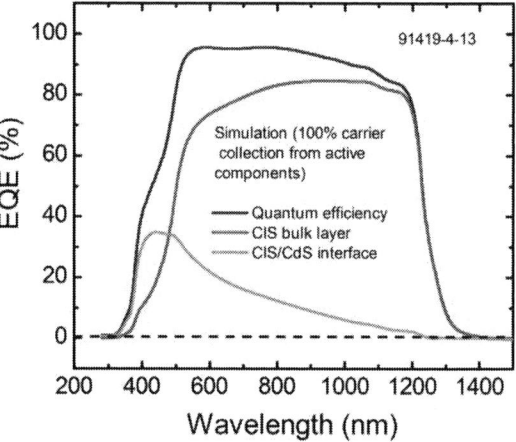

Fig. 7. Components of the EQE spectrum from the two active layers of the CIS solar cell structure of Fig. 6 simulated optically assuming 100% charge carrier collection.

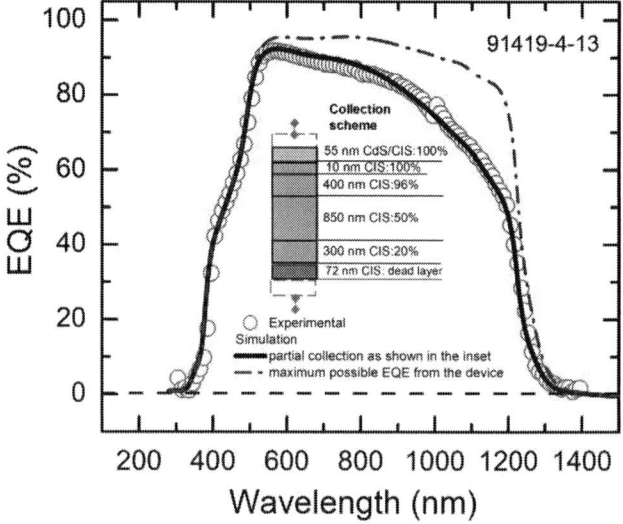

Fig. 8. Measured normal incidence EQE and the ideal simulated EQE (broken line) obtained using the Fig. 6 structural model. Also shown is the simulated EQE (solid line) from the inset collection profile.

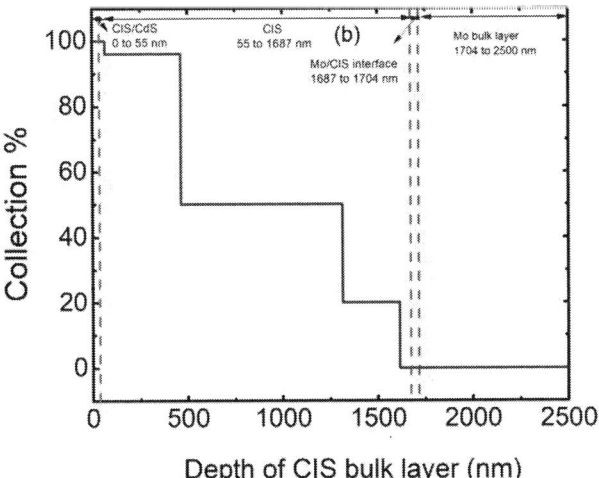

Fig. 9. A depth profile in carrier collection that enables close simulation of the measured EQE spectrum of the CIS solar cell in Fig. 8.

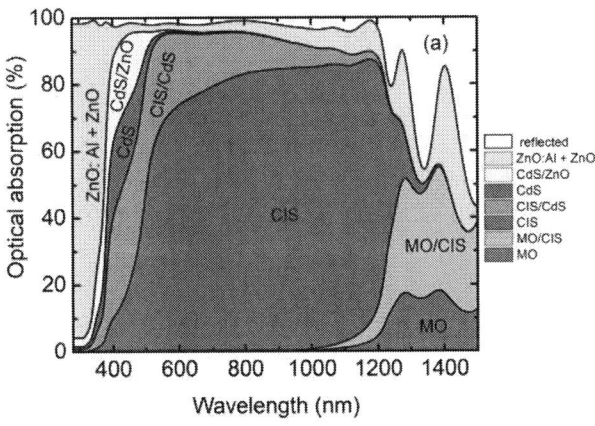

Fig. 10. Simulated reflectance from the complete stack and optical absorption in the individual layers of the CIS solar cell structure.

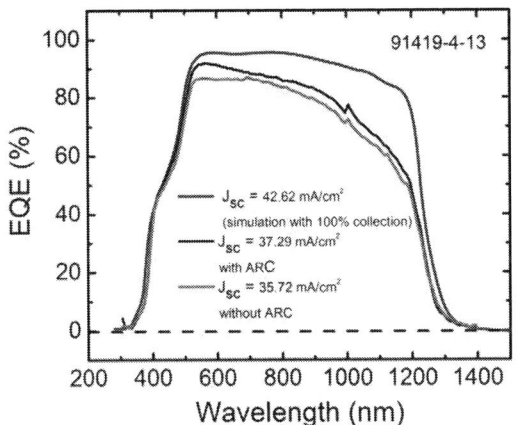

Fig. 11. Measured EQE spectra along with integrated current densities before and after ARC, for comparison with the simulation assuming 100% collection.

TABLE III. COMPARISON OF THE SHORT-CIRCUIT CURRENT DENSITIES FOR A CuInSe$_2$ SOLAR CELL FABRICATED WITH AND WITHOUT AN ARC AND FOR A SIMULATION ASSUMING 100% COLLECTION FROM THE ACTIVE LAYERS.

	J$_{sc}$ from EQE (mA/cm^2)	Predicted η (%)
Before ARC	35.72	8.60
After ARC	37.29	8.98
Simulation with 100% collection from active layers	42.62	9.80

ACKNOWLEDGEMENTS

This material is based on research sponsored by Air Force Research Laboratory under agreement number FA9453-21-C-0056. The U.S. Government is authorized to reproduce and distribute reprints for Governmental purposes notwithstanding any copyright notation thereon. The views expressed are those of the authors and do not reflect the official guidance or position of the United States Government, the Department of Defense or of the United States Air Force. The appearance of external hyperlinks does not constitute endorsement by the United States Department of Defense (DoD) of the linked websites, or the information, products, or services contained therein. The DoD does not exercise any editorial, security, or other control over the information you may find at these locations. Approved for public release; distribution is unlimited. Public Affairs release approval #AFRL-2022-2408.

REFERENCES

[1] S. Wagner, J. Shay, P. Migliorato, and H. Kasper, "CuInSe$_2$/CdS heterojunction photovoltaic detectors," *Applied Physics Letters,* vol. 25, no. 8, pp. 434-435, 1974.

[2] J. Shay, S. Wagner, and H. Kasper, "Efficient CuInSe$_2$/CdS solar cells," *Applied Physics Letters,* vol. 27, no. 2, pp. 89-90, 1975.

[3] L. Kazmerski, F. White, and G. Morgan, "Thin-film CuInSe$_2$/CdS heterojunction solar cells," *Applied Physics Letters,* vol. 29, no. 4, pp. 268-270, 1976.

[4] J. A. AbuShama, S. Johnston, T. Moriarty, G. Teeter, K. Ramanathan, and R. Noufi, "Properties of ZnO/CdS/CuInSe$_2$ solar cells with improved performance," *Progress in Photovoltaics: Research and Applications,* vol. 12, no. 1, pp. 39-45, 2004.

[5] J. AbuShama, R. Noufi, S. Johnston, S. Ward, and X. Wu, "Improved performance in CuInSe$_2$ and surface-modified CuGaSe$_2$ solar cells," in *Conference Record of the Thirty-first IEEE Photovoltaic Specialists Conference, 2005,* 2005: IEEE, pp. 299-302.

[6] Y. Zhang, R. E. Bartolo, S. J. Kwon, and M. Dagenais, "High short-circuit current density in CIS solar cells by a simple two-step selenization process with a KF postdeposition treatment," *IEEE Journal of Photovoltaics,* vol. 7, no. 2, pp. 676-683, 2016.

[7] A. Rockett and R. Birkmire, "CuInSe$_2$ for photovoltaic applications," *Journal of Applied Physics,* vol. 70, no. 7, pp. R81-R97, 1991.

[8] K. F. Tai, R. Kamada, T. Yagioka, T. Kato, and H. Sugimoto, "From 20.9 to 22.3% Cu(In,Ga)(S,Se)$_2$ solar cell: Reduced recombination rate at the heterojunction and the depletion region due to K-treatment," *Japanese Journal of Applied Physics,* vol. 56, no. 8S2, p. 08MC03, 2017.

[9] Y. H. Jang, J. M. Lee, J. W. Seo, I. Kim, and D.-K. Lee, "Monolithic tandem solar cells comprising electrodeposited CuInSe$_2$ and perovskite solar cells with a nanoparticulate ZnO buffer layer," *Journal of Materials Chemistry A,* vol. 5, no. 36, pp. 19439-19446, 2017.

[10] T. Feurer *et al.*, "Single-graded CIGS with narrow bandgap for tandem solar cells," *Science and Technology of Advanced Materials,* vol. 19, no. 1, pp. 263-270, 2018.

[11] Q. Han *et al.*, "High-performance perovskite/Cu(In,Ga)Se$_2$ monolithic tandem solar cells," *Science,* vol. 361, no. 6405, pp. 904-908, 2018.

[12] R. H. Ahangharnejhad *et al.*, "Irradiance and temperature considerations in the design and deployment of high annual energy yield perovskite/CIGS tandems," *Sustainable Energy & Fuels,* vol. 3, no. 7, pp. 1841-1851, 2019.

[13] M. M. Lee, J. Teuscher, T. Miyasaka, T. N. Murakami, and H. J. Snaith, "Efficient hybrid solar cells based on meso-superstructured organometal halide perovskites," *Science,* vol. 338, no. 6107, pp. 643-647, 2012.

[14] D. R. Sapkota *et al.*, "Spectroscopic ellipsometry investigation of CuInSe$_2$ as a narrow bandgap component of thin film tandem solar cells," in *2018 IEEE 7th World Conference on Photovoltaic Energy Conversion (WCPEC) (A Joint Conference of 45th IEEE PVSC, 28th PVSEC & 34th EU PVSEC)*, 2018: IEEE, pp. 1943-1948.

[15] R. H. Ahangharnejhad *et al.*, "Optical design of perovskite solar cells for applications in monolithic tandem configuration with CuInSe$_2$ bottom cells," *MRS Advances,* vol. 3, no. 52, pp. 3111-3119, 2018.

[16] N. Shrestha *et al.*, "Identification of defect levels in copper indium diselenide (CuInSe$_2$) thin films via photoluminescence studies," *MRS advances,* vol. 3, no. 52, pp. 3135-3141, 2018.

[17] P. Uprety *et al.*, "Optical Hall effect of PV device materials," *IEEE Journal of Photovoltaics,* vol. 8, no. 6, pp. 1793-1799, 2018.

[18] D. R. Sapkota *et al.*, "Structural and optical properties of two-stage CuInSe$_2$ thin films studied by real time spectroscopic ellipsometry," in *2019 IEEE 46th Photovoltaic Specialists Conference (PVSC)*, 16-21 June 2019, pp. 0943-0948.

[19] D. R. Sapkota *et al.*, "Evaluation of CuInSe$_2$ materials and solar cells co-evaporated at different rates based on real time spectroscopic ellipsometry calibrations," in *2021 IEEE 48th Photovoltaic Specialists Conference (PVSC)*, 2021: IEEE, pp. 0451-0458.

[20] H. Hahn, G. Frank, W. Klingler, A.-D. Meyer, and G. Störger, "Untersuchungen über ternäre chalkogenide. V. Über einige ternäre Chalkogenide mit Chalko-pyritstruktur," *Zeitschrift für Anorganische und Allgemeine Chemie,* vol. 271, no. 3-4, pp. 153-170, 1953.

[21] J. Chen *et al.*, "Through-the-glass spectroscopic ellipsometry of CdTe solar cells," in *34th IEEE Photovoltaic Specialists Conference (PVSC)*, 2009: IEEE, pp. 001748-001753.

[22] T. Minemoto *et al.*, "Theoretical analysis of the effect of conduction band offset of window/CIS layers on performance of CIS solar cells using device simulation," *Solar Energy Materials and Solar Cells,* vol. 67, no. 1-4, pp. 83-88, 2001.

Novel Interconnection Method for Micro-CPV: 132 Solar Cell Prototype

Norman Jost, Steve Askins, Richard Dixon, Mathieu Ackermann, Cesar Dominguez, Ignacio Anton

Insituto de Energía Solar Universidad Politécnica de Madrid, Madrid, Spain

Dycotec Materials Ltd, Swindon, United Kingdom

Insolight SA, Lausanne, Switzerland

Micro-concentrator photovoltaics (micro-CPV) consists of the reduction in size of the components of the conventional concentrator photovoltaic (CPV) technology, attaining equally high efficiencies and reducing material costs and manufacturing costs. In this publication we focus on the implementation of high throughput manufacturing methods for the interconnection of the solar cells. The goal is to enable large area interconnection of thousands of micro-solar cells with a low cost of $3euro/m2$, considering low cost silver inks and screen printing as used for the research in this publication. The method used is a result of crosslinkage with another industry, the assembly and interconnection of light emitting diodes (LED) for illumination or TV screens. The method consists in the following steps: 1. Screen printing of the interconnection tracks on the glass board and curing at high temperatures of over 100°C. 2. Pick-and-place of the solar cells and reflow soldering for the electrical contact and adhesion. 3. Dielectric syringe printing around the solar cells for electrical insulation of the solar cell lateral and ultraviolet light curing of the dielectric. This step would be disposable if dies are electrically insulated. 4. Screen printing of the final interconnection layer with epoxy silver ink and temperature curing. Step 2 can be reemplaced by other paralell assembly methods such as fluid self assembly or transfer pritning of dies. Step 3 can be avoided by the use of perimiter insulted solar cells which will be available in the future. The current prototype is a 145x190mm2 board with a successful interconnection of 12 rows containing each 11 x 1mm2 inverted metamorphic multi-junction solar cells (132 cells in total). The fill factor (FF) of the full board is 76% at 214 suns. The tecnology is further developed for the purpose of using even smaller solar cells and interconnecting on a larger area.

Understanding the Behavior of Fixed Composition CdSe$_x$Te$_{1-x}$(CST) Solar Cells

Ebin Bastola[1], Adam B. Phillips[1], Abasi Abudulium[1], Vlad Kornienko[2], Manoj K. Jamarkattel[1], Zulkifl H. Rabbani[1], Jared D. Friedl[1], Prabodika N. Kalurachchi[1], Ali Abbas[2], Abdul Quader[1], Xavier Mathew[1,3], Michael Walls[2], Randy J. Ellingson[1], Michael J. Heben[1]

[1]*Wright Center for Photovoltaics Innovation and Commercialization (PVIC), Department of Physics and Astronomy, University of Toledo, Toledo, OH, United States*
[2]*Centre for Renewable Energy Systems Technology (CREST), Loughborough University, Loughborough, United Kingdom*
[3]*Instituto de Energías Renovables, Universidad Nacional Autónoma de México, Temixco, Mexico*

Abstract

Cadmium selenide (CdSe) plays a vital role to achieving the high short-circuit current density (J_{SC}) and passivating the defects in the absorber layer for CdTe photovoltaics necessary to reach high efficiency. Incorporation of CdSe into devices can be done either by fabricating a CdSe/CdTe bilayer or directly depositing the CdSe$_x$Te$_{1-x}$ (CST). While the bilayer results in better device performance, the intrinsic properties of the CST suggest it should be the better absorber material. Here, we fabricated and investigated the structural and opto-electronic properties of fixed composition CST films for varying Se concentrations and report device parameters. The films were produced by leveraging our multisource evaporation chamber, allowing a wide range of Se compositions to be investigated without modification to the system. For fixed compositions CST absorber layers, the minority carrier lifetime is improved with higher Se content though the grain sizes are slightly smaller for higher Se content. Note that all these samples (pure CdTe and CST) have undergone same CdCl$_2$ treatment. The device efficiency for fixed composition CST absorber layer observed is as high as 12.2% while for pure CdTe device (no Se) is 7%. The short circuit current density is high (28 mAcm^{-2}), but CST devices suffer from low open circuit voltage (V_{oc}) and fill factor (FF). For comparison, CdSe/CdTe bilayer devices also fabricated using this system were able to reach efficiency up to 17.7% (V_{OC} 839 mV, J_{SC} 29.0 mAcm^{-2}, FF 72.6%), indicating the system produces good material. We will discuss the material properties of CST and correlate these values to the device performance.

Effects of Novel In+RbF Post-Deposition Treatment on $Cu(In_xGa_{1-x})Se_2$ Solar Cells

Jake Wands, Polyxeni Tsoulka, Thomas Lepetit, Nicolas Barreau, and Angus Rockett
Colorado School of Mines, Golden, CO, 80401, USA
Université de Nantes, Nantes, France

Abstract—Alkali halide post-deposition treatment (PDT) has been shown to improve $Cu(In_xGa_{1-x})Se_2$ (CIGS) photovoltaic devices. In this study temperature-dependent current-voltage (JVT) experiments were used to evaluate a novel PDT performed by co-evaporating indium and RbF under a sulphur atmosphere. Three devices, one as-deposited, one with RbF PDT, and one with In+RbF PDT, were deposited by a three stage co-evaporation process. JVT results suggest the In+RbF PDT increased V_{OC} and efficiency over the other treatments. It was also observed that the recombination rate was decreased and the ideality factor improved which may explain the improvements in V_{OC}.

Index Terms—CIGS, characterization, recombination, thin film PV

I. INTRODUCTION

Having achieved a record cell efficiency of 23.4%, $Cu(In_xGa_{1-x})Se_2$ (CIGS) is a leading candidate material for thin-film photovoltaics [1]. A key component to the success of CIGS is the alkali halide post-deposition treatment (PDT). Sodium and potassium based treatments have been the most common techniques historically, but RbF has gained interest in recent years [2]–[5]. Recently a novel PDT has been developed by substituting the traditional selenium atmosphere with sulfur and co-evaporating indium along with RbF. This technique has been shown to be effective at improving open-circuit voltage (V_{OC}) and fill factor (FF). In this study temperature-dependent current-voltage (JVT) measurements were used to characterize the recombination mechanisms within devices using this PDT technique.

II. EXPERIMENTAL PROCEDURES

The devices in this study were deposited using a three stage co-evaporation process resulting in a CIGS film with a Cu/III ratio of 0.95 and a Ga/III ratio of 0.30. The PDT was performed by co-evaporating the alkali components with sulfur to create a 10nm thick layer on top of the CIGS. The PDT involved a 10 minute anneal at 350°C. More details on the process can be found in [6].

The device performance parameters can be found in Table I. After the RbF PDT the V_{OC} and FF decrease significantly resulting in an overall reduced efficiency. The In+RbF PDT sample exhibited the highest V_{OC} and overall efficiency despite a slight decrease in FF compared with the as deposited sample.

This research was supported by the U.S. Department of Energy's Office of Energy Efficiency and Renewable Energy (EERE) under Solar Energy Technologies Office (SETO) Award Number DE-EE0008755.

TABLE I
DEVICE PERFORMANCE PARAMETERS

	V_{OC} (mV)	Jsc (mA/cm^2)	FF (%)	Efficiency (%)
As-Deposited	643	31.9	65.5	13.5
RbF PDT	593	31.8	56.1	10.6
In+RbF PDT	698	31.8	62.8	14.0

JVT analysis was performed using a solar simulator at an irradiance of 100mW/cm^2 and decreasing the temperature (T) from 350K to 150K in 10K increments. Extracting the reverse saturation current, J_0, and diode ideality factor (n) was performed using the procedures outlined in [7]. Dark JV curves were used to minimize the potential effects of voltage-dependent carrier collection and compared to similar analyses under light to determine how the carrier transport had changed.

III. RESULTS

To better understand the source of recombination within the devices V_{OC} vs temperature curves were used to calculate the diode activation energy (E_A). In diodes with a thermally activated transport mechanism the V_{OC} is related to temperature by ([8]):

$$V_{OC} = E_A/q - (n(T)kT)/q * ln(J_{00}/J_L) \qquad (1)$$

where n(T) is the ideality factor, J_{00} is the reverse saturation current density prefactor, and J_L is the light-induced current density, k is the Boltzmann constant, and q is the electron charge. By applying a linear fit to each dataset, E_A can be identified by the 0 K intercept. If $E_A \approx E_g$ the dominant recombination mechanism is likely in the bulk of the absorber. If $E_A < E_g$ then interface recombination may be the most important factor [8].

As seen in Fig. 1 all three devices provide a good linear fit above 200 K to extract E_A. The as-deposited sample extrapolates to 1.18eV while the RbF and In+RbF devices were 1.19eV and 1.22eV, respectively. The bandgap on all three devices was found to be 1.15eV using photoluminescence. This result suggests that all three devices are dominated by recombination mechanisms within the bulk of the absorber rather than at interfaces.

The sample's behaviors begin to diverge when J_0 is plotted vs temperature (Fig. 2). J_0 is higher across the entire temperature range in the RbF PDT sample compared with

978-1-7281-6118-1/22 $31.00 © 2022 IEEE

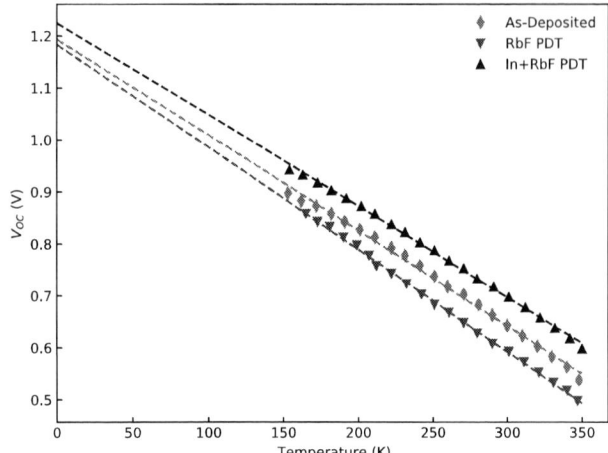

Fig. 1. Open circuit voltage as a function of temperature. The reference sample data is shown in blue, the orange data points are for the In+RbF PDT, and the grey data points for the RbF PDT.

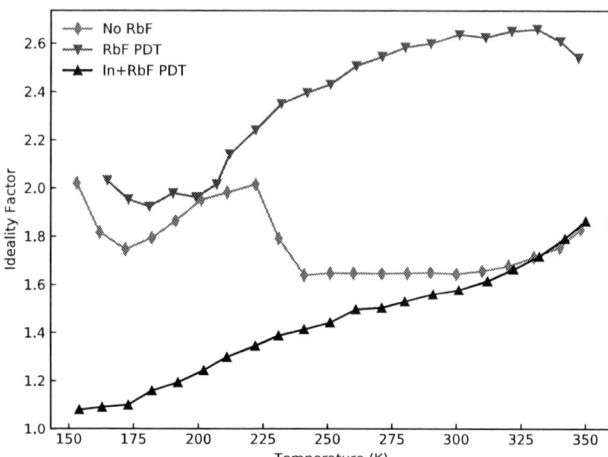

Fig. 3. Diode ideality factor as a function of temperature.

the as-deposited device. This indicates that recombination has increased in the device and results in a lower V_{OC}. The lowest J_0 across all temperatures is found in the In+RbF device, particularly at low temperatures where the performance is significantly improved. The addition of In to the PDT process presumably alters the near surface chemistry of the CIGS, apparently reducing defects responsible for recombination (possibly grain boundaries where the In incorporation would be most significant).

In Fig. 3 the diode ideality factor, n, is plotted against temperature. Above 230K the as-deposited sample is relatively stable with n 1.65. This suggests Shockley-Read-Hall recombination through sub-bandgap defects is the dominant recombination mechanism in this device. The RbF PDT sample shows much different behavior with the ideality factor rising from

2.0 at low temperature to 2.65 at higher temperatures. Ideality factors above 2 are not uncommon in CIGS devices and often suggest a tunneling mechanism is enhancing the recombination rate. This would explain the increased J_0 in the device. In the In+RbF device the ideality factor increases from a value near 1.0 at low temperature to 1.86 at high temperature, consistent with SRH recombination being dominant. It is uncommon for the ideality factor to increase as the temperature rises. Further investigation is needed to identify the defect characteristics in this device.

IV. CONCLUSIONS

In this study a unique CIGS post-deposition treatment using an elemental sulfur atmosphere and In co-evaporated with RbF was characterized to better understand recombination within the device and the potential of the PDT process. Temperature-dependent current-voltage measurements were used to extract V_{OC}, J_0, and ideality factor as temperature changed. Using V_{OC} vs temperature analysis it was found that all three devices are dominated by bulk recombination. J_0 values showed recombination rates are increased after RbF PDT compared with an as-deposited device. However, the In+RbF PDT significantly decreased J_0 showing that the addition of indium reduced recombination within the device. Extracting the ideality factor for each device revealed the as-deposited sample shows classic SRH recombination through sub-bandgap defects. The RbF PDT sample had an ideality factor above 2 which may indicate tunneling is enhancing the recombination rate. Finally, the In+RbF PDT device had an ideality factor near 1 at low temperatures which rose to 1.86 at high temperature. While this may indicate SRH recombination, it is unusual and warrants further investigation into the defect character of the device.

ACKNOWLEDGMENT

The authors thank Dr. Steve Johnston for providing support with the JVT measurement technique.

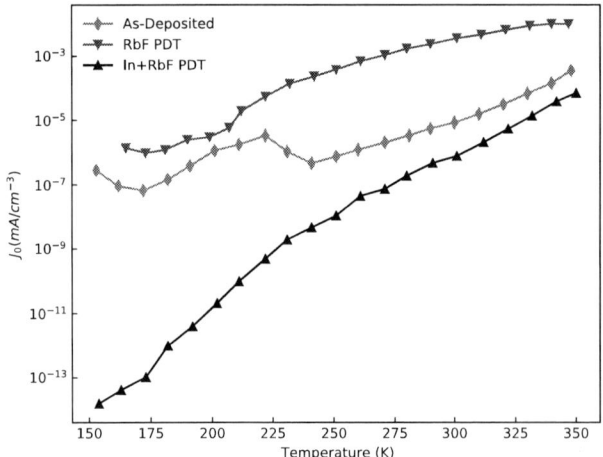

Fig. 2. J_0 data as a function of temperature for the three samples.

REFERENCES

[1] M. Nakamura, K. Yamaguchi, Y. Kimoto, Y. Yasaki, T. Kato, and H. Sugimoto, "Cd-Free Cu(In,Ga)(Se,S)2 Thin-Film Solar Cell With Record Efficiency of 23.35%," *IEEE Journal of Photovoltaics*, vol. 9, no. 6, pp. 1863–1867, Nov. 2019, conference Name: IEEE Journal of Photovoltaics.

[2] P. Jackson, R. Wuerz, D. Hariskos, E. Lotter, W. Witte, and M. Powalla, "Effects of heavy alkali elements in Cu(In,Ga)Se2 solar cells with efficiencies up to 22.6%," *physica status solidi (RRL) – Rapid Research Letters*, vol. 10, no. 8, pp. 583–586, 2016, _eprint: https://onlinelibrary.wiley.com/doi/pdf/10.1002/pssr.201600199. [Online]. Available: https://onlinelibrary.wiley.com/doi/abs/10.1002/pssr.201600199

[3] A. Kanevce, S. Paetel, D. Hariskos, and T. Magorian Friedlmeier, "Impact of RbF-PDT on Cu(In,Ga)Se$_2$ solar cells with CdS and Zn(O,S) buffer layers," *EPJ Photovolt.*, vol. 11, p. 8, 2020. [Online]. Available: https://www.epj-pv.org/10.1051/epjpv/2020005

[4] T. Kodalle, M. D. Heinemann, D. Greiner, H. A. Yetkin, M. Klupsch, C. Li, P. A. van Aken, I. Lauermann, R. Schlatmann, and C. A. Kaufmann, "Elucidating the Mechanism of an RbF Post Deposition Treatment in CIGS Thin Film Solar Cells," *Solar RRL*, vol. 2, no. 9, p. 1800156, 2018, _eprint: https://onlinelibrary.wiley.com/doi/pdf/10.1002/solr.201800156. [Online]. Available: https://onlinelibrary.wiley.com/doi/abs/10.1002/solr.201800156

[5] F. Pianezzi, P. Reinhard, A. Chirilă, B. Bissig, S. Nishiwaki, S. Buecheler, and A. N. Tiwari, "Unveiling the effects of post-deposition treatment with different alkaline elements on the electronic properties of CIGS thin film solar cells," *Phys. Chem. Chem. Phys.*, vol. 16, no. 19, pp. 8843–8851, Apr. 2014, publisher: The Royal Society of Chemistry. [Online]. Available: https://pubs.rsc.org/en/content/articlelanding/2014/cp/c4cp00614c

[6] P. Tsoulka, A. Crossay, L. Arzel, S. Harel, and N. Barreau, "Alternative alkali fluoride post-deposition treatment under elemental sulfur atmosphere for high-efficiency Cu(In,Ga)Se2-based solar cells," *Progress in Photovoltaics: Research and Applications*, vol. n/a, no. n/a, _eprint: https://onlinelibrary.wiley.com/doi/pdf/10.1002/pip.3508. [Online]. Available: http://onlinelibrary.wiley.com/doi/abs/10.1002/pip.3508

[7] S. S. Hegedus and W. N. Shafarman, "Thin-film solar cells: device measurements and analysis," *Progress in Photovoltaics: Research and Applications*, vol. 12, no. 2-3, pp. 155–176, 2004, _eprint: https://onlinelibrary.wiley.com/doi/pdf/10.1002/pip.518. [Online]. Available: http://onlinelibrary.wiley.com/doi/abs/10.1002/pip.518

[8] C. J. Hages, N. J. Carter, R. Agrawal, and T. Unold, "Generalized current-voltage analysis and efficiency limitations in non-ideal solar cells: Case of Cu$_2$ZnSn(S$_x$Se$_{1x}$)$_4$ and Cu$_2$Zn(Sn$_y$Ge$_{1y}$)(S$_x$Se$_{1x}$)$_4$," *Journal of Applied Physics*, vol. 115, no. 23, p. 234504, Jun. 2014. [Online]. Available: http://aip.scitation.org/doi/10.1063/1.4882119

Seasonal Dependence of Bifacial Photovoltaic Array Gain due to Inverter Clipping

Thunchanok Kaewnukultorn, Steven Hegedus

This paper studies the effect of inverter clipping in bi-facial photovoltaic (PV) array under different ground cover reflectivity. The 5 kW bi-facial array was installed at the University of Delaware, USA and was connected to 3.8 kW grid tied inverter. Two parameters, expected PV power and DC energy efficiency, were developed in this research to compare the inverter clipping loss with actual data collected from on-site sensors. The results show that the clipping potentially occurs in summer from 10:30 to 14:30 due to high solar irradiance. The daily generation with the white ground cover experienced higher clipping loss than the gravel and consequently resulted in negligible bi-facial gain. In winter, the bi-facial gain is higher as the clipping loss is absent.

Accuracy of Potential High Limit Estimation for Solar Plants in the Southeast US

William B. Hobbs[1], and David J. Ault[1], Vahan Gevorgian[2], and Govind Saraswat[2]

[1]Southern Company, Birmingham, AL, 35203, USA
[2]NREL, Golden, CO, 80401, USA

Abstract—**Flexible solar operation, where solar photovoltaic (PV) plants follow up- and down-regulation signals, has significant potential to improve integration of solar into power grids. To optimize operation, it is important to accurately estimate the potential maximum power output, or potential high limit (PHL), of a plant in real time during periods where output has been reduced. As the PHL cannot be directly measured while a plant is curtailed, and it is driven by highly variable weather and plant conditions, model-based estimation methods are subject to errors. An estimation method using a subset of a plant as a reference has been developed by NREL. Here, we evaluate a version of that method using data from several utility-scale plants in the Southeast US spanning up to a full year.**

Index Terms—**flexible solar power, grid integration, automatic generation control, AGC**

I. INTRODUCTION

Flexible solar operation, where solar photovoltaic (PV) plants follow up- and down-regulation signals, has significant potential to improve integration of solar into power grids, as has been shown through single-plant demonstrations and several modeling studies [1]–[5]. To optimize operation, it is important to accurately estimate the potential maximum power output, or potential high limit (PHL), of a plant in real time during periods where output has been reduced. As the PHL cannot be directly measured while a plant is curtailed, and it is driven by highly variable weather and plant conditions, model-based estimation methods are subject to errors. Measurements of weather, solar plant performance models (e.g., models implemented in pvlib), and past plant performance could be used to estimate PHL. However, a simpler and potentially more accurate method has been developed by NREL that use an uncurtailed subset of a plant's inverters as a reference. Versions of this method have been demonstrated using plants in the Western US [3], [6], but only for one to four days, and performance has not been characterized in climates in other regions with different climates.

Here, we use data from several utility scale plants in the Southeast US, spanning a full year, to evaluate the accuracy of a version of NREL's method.

II. METHODS

A. Data

Real AC power output from each inverter was typically sampled by the plant SCADA system at 10 second intervals. Data compression was then used by the data historian (e.g., OSISoft PI) to drop values that did not have meaningful changes from the measurement(s) in previous timesteps. For this analysis, 10 second interpolated values were queried from the historian.

B. Filtering

To simplify the analysis, periods of time with inverter data issues, whether due to data collection problems or inverter outages, where excluded.

For each timestep, the number of total inverters and number of references inverters with valid measurements are counted, where inverter power measurements must be valid numbers greater than zero. Any timesteps where the number of remaining inverter measurements or reference inverter measurements are at or below a minimum threshold of necessary inverters are dropped from the analysis. The minimum thresholds of all inverters and reference inverters were set to 90% of the count of all inverters and count of reference inverters, respectively. For example, if a plant has 30 total inverters and 10 are used as reference inverters, a timestep must have valid power measurements for more than 27 (i.e., 28 or more) of all inverters and more than 9 (i.e., all 10) of the reference inverters.

Setting a requirement of power values greater than zero automatically filters analysis results to daytime periods.

After filtering data for individual plants, PHL estimates for the aggregation of plants was performed by filtering to include time intervals for the plant with the highest availability of data. Because the number of intervals included for each plant varies slightly, particularly periods of low power near sunrise and sunset that also have relatively low absolute PHL error, comparisons of error statistics between individual and aggregated plants are only indicative.

C. Potential High Limit Power estimation

Plant-level power was estimated using a subset of inverters, referred to as "reference inverters", by scaling their output in each timestep linearly, similar to methods presented by Gevorgian at NREL [6].

An additional step was added to adjust for the fact that not all inverters will perform the same. Part of that is due to the permanent difference of some inverters having PV arrays with different DC nameplate ratings, but it can also be due to things than change over time like PV string and combiner outages, uneven soiling losses, etc. For a given set of reference inverters, a linear scalar "bias adjustment factor" was

978-1-7281-6118-1/22 $31.00 © 2022 IEEE

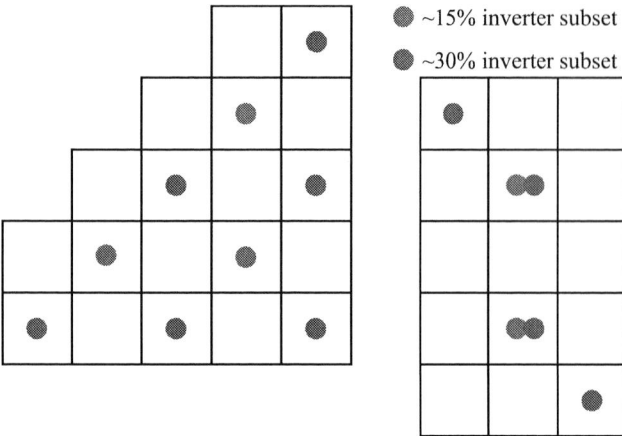

Fig. 1. Hypothetical plant layout with 34 inverters (grid squares), where subsets representing approximately 15% (red, 5 inverters) and 30% (blue, 10 inverters) of inverters have been selected as reference inverters.

calculated using a subset of the time range (e.g., one month) such that the integrated output of the reference inverters, when normalized to their nameplate rating and multiplied by the bias adjustment factor, was within approximately 0.5% of the normalized output of the whole plant. Typical bias adjustment factors range from 0.97 to 1.03.

After applying a bias adjustment factor, estimated plant level power output was limited to the plant nameplate (lesser of sum of inverter AC nameplates or the interconnection limit).

D. Error Calculation

Error was calculated in each timestep and normalized to plant nameplate ac power using:

$$Error(\%) = \frac{Estimated\,Power - Actual\,Power}{Plant\,Nameplate\,Power} \times 100 \quad (1)$$

E. Reference Inverter Selection

Reference inverters were selected manually with a goal of uniform geographic representation. Different numbers of inverters were selected, targeting 15%, 20%, and 30% of plant nameplate capacity. An example using a hypothetical plant layout is shown in Fig. 1.

III. Results: Single-Plant PHL

Single-plant analysis was performed on three plants in the Southeast US, Plant A, Plant B, and Plant C. Approximate specifications for each plant are shown in Table I. Plants A and B were located close enough together that they could be treated as a single plant, with a mix of fixed-tilt and tracking, for some analyses.

A. Plant A – Target 15% Reference Inverters

As an example, here are detailed error statistics for Plant A, using 15% of inverters as reference inverters. Fig. 2 shows the cumulative error distribution for all 10 second intervals used

TABLE I
PLANT SPECIFICATIONS

Plant	Configuration	Approx. nameplate power	Approx. number of inverters	Months of data
A	Fixed-tilt	100 MW	30	4
B	Tracking	25 MW	6	4
C	Fixed-tilt	50 MW	30	12

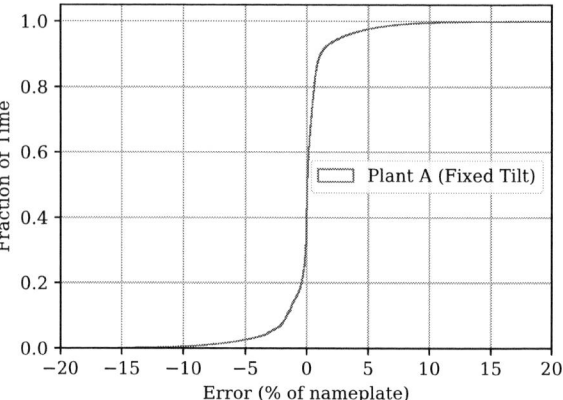

Fig. 2. Cumulative distribution of estimation error for Plant A, 15% reference inverters for a full year of data.

in the analysis. The errors are tightly centered around zero for much of the time, but long "tails" exceeding ±10% error are present for relatively small fractions of time.

The standard deviation of error was 2.57%, and individual percentiles are shown in Table II.

A sample day in January, with a mix of clear sky and variable irradiance, is shown in Fig. 3.

The error range for 95% of daytime intervals was selected as a key metric. This is bounded by the 2.5th and 97.5th

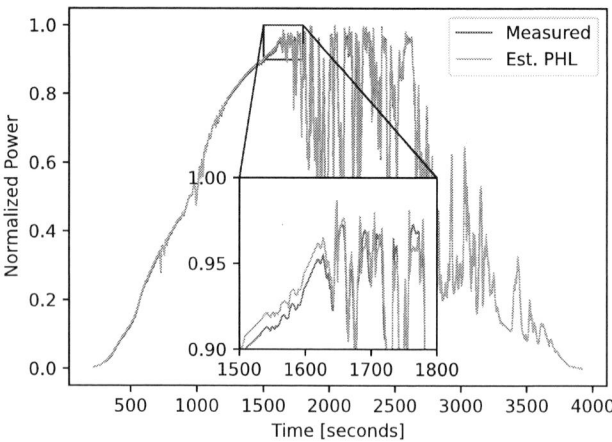

Fig. 3. Normalized measured power (blue) and estimated PHL (orange) for Plant A, 15% reference inverters, on a sample day in January.

TABLE II
ANNUAL ESTIMATION ERROR STATISTICS FOR PLANT A, 15% REFERENCE INVERTERS.

Percentile	Error (%)
00.5	-10.28
02.5	**-5.07**
05.0	-3.02
10.0	-1.59
12.5	-1.22
87.5	0.90
90.0	1.14
95.0	2.78
97.5	**4.85**
99.5	9.46
Max:	27.24
Max:	-35.76
StDev:	2.21

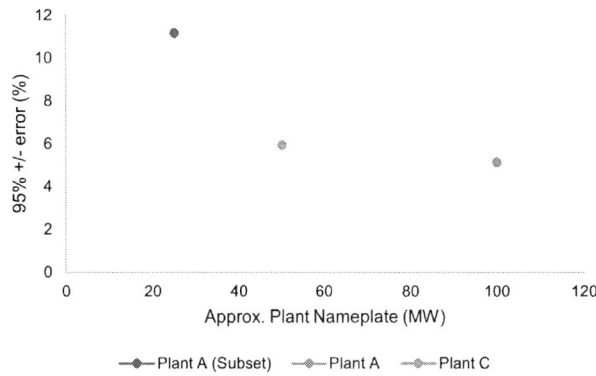

Fig. 5. Annual estimation errors for different effective plant sizes using approximately 15% reference inverters.

95% of error for Plant B (tracking) was within ±13.61% and for a small subset of plant Plant A (fixed) it was ±11.16%. The fixed-tilt system had lower errors by about 20% (relative), which is intuitive as tracking plants produce more energy and therefore have more potential for more frequent, larger errors.

IV. RESULTS: AGGREGATED PHL AND TIME OFFSETS

In addition to individual plant PHL accuracy, grid operators care about the accuracy of aggregated PHL estimates across the fleet of solar facilities . Similarly, accuracy implications of latency caused by AGC communication cycles are relevant to grid operators. To explore these considerations we brought in additional plants to add geographic diversity, with the expectation that it would reduce relative errors, and we introduced time offsets, which have the potential to increase errors.

Data from two additional plants were added, resulting in an aggregation of five plants, totaling approximately 250 MW, with about 150 MW fixed-tilt and 100 MW tracking. The five plants fit into a rectangular region of approximately 140 km by 200 km (87 miles by 124 miles, with the closest two plants being about 80 km (50 mi) from each other. For reference, Southern Company's balancing area in the Southeast US fits in a rectangle of approximately 800 by 500 km.

Preceding results all calculate PHL in the same timestep as reference inverter and plant power, representing a "real-time" process. In practice, PHL calculations in plant controllers could happen in fractions of a second, but responses to AGC systems could take three AGC cycles, which are typically four to seconds seconds (e.g., the plant reports its PHL in one AGC communication cycle, the grid operator calculates the desired response and sends it back to the plant in the next cycle, and the plant's response is measured in the third cycle). To represent this 12-18 second lag, analysis of the aggregated PHL was performed with 10 and 20 second offsets in estimated PHL relative to measured power.

Results for the aggregation of plants with a range of offsets are shown in Table III and Fig. 6.

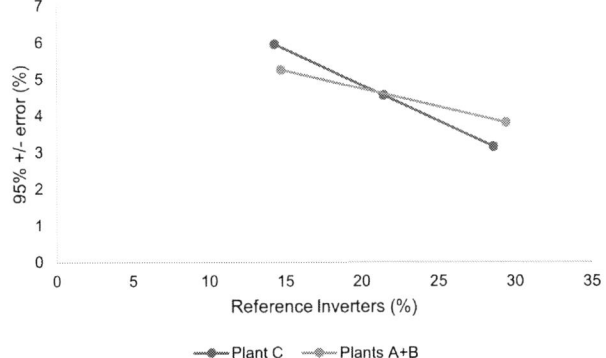

Fig. 4. Annual estimation errors for different amounts of reference inverters at an approximation of two plants.

percentiles of error, highlighted in Table II. The 95th percentile "±" error in this case is 5.13%, the maximum of the absolute value of the 2.5th and 97.5th percentile errors, and a 99th percentile "±" error of 10.34%.

B. Number of reference inverters

By varying the number of reference inverters used, we can see the benefits of higher percentages of inverters being used as reference inverters (which has the trade-off of reducing AGC range). These results, shown in Fig. 4, are based on subsets of inverters at Plant C and the combined Plants A and B.

C. Plant size

A comparison of different subsets of inverters from the fixed-tilt plants, Plants A and C, shows the impact of effective plant size. All scenarios had a target of 15% reference inverters. As shown in Fig. 5, there is a clear reduction in error as plant size increases, although there appear to be diminishing returns after approximately 50 MW.

D. Fixed tilt vs tracking

To compare fixed and tracking configurations, a subset of inverters from Plant A, similar in size to all of Plant B, were compared with Plant B. For a full year of data at both plants,

TABLE III
ANNUAL ESTIMATION ERROR STATISTICS FOR AN AGGREGATION OF FIVE
PLANTS, EACH USING 15% REFERENCE INVERTERS, WITH 0, 10, AND 20
SECONDS OF TIME OFFSET.

Percentile	Error (%)		
	0 s Offset	10 s Offset	20 s Offset
00.5	-5.51	-5.89	-6.60
02.5	**-2.98**	**-3.24**	**-3.63**
05.0	-1.95	-2.14	-2.40
10.0	-1.26	-1.33	-1.44
12.5	-1.02	-1.11	-1.22
87.5	0.98	1.09	1.23
90.0	1.23	1.35	1.51
95.0	2.01	2.17	2.39
97.5	**2.77**	**3.03**	**3.43**
99.5	4.89	5.35	6.12
Max:	13.35	49.62	50.06
Max:	-14.67	-51.12	-52.79
StDev:	1.32	1.48	1.70

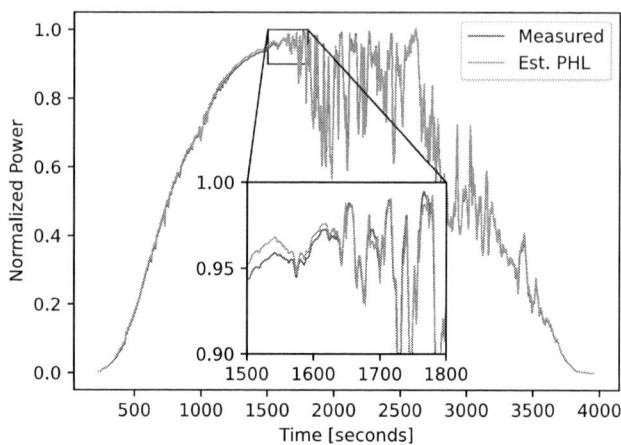

Fig. 7. Normalized measured power (blue) and estimated PHL (orange) for Plant A, 15% reference inverters, on a sample day in January.

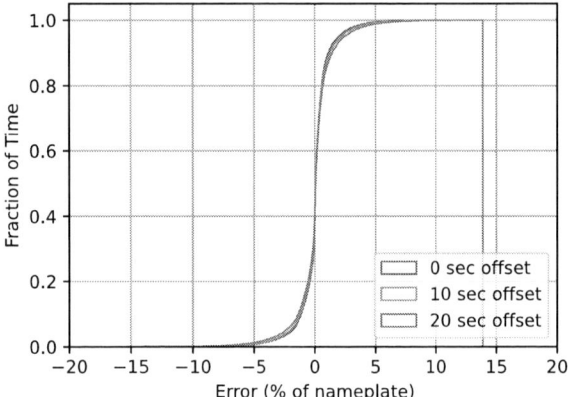

Fig. 6. Cumulative distribution of error for an aggregation of five plants, each using 15% reference inverters, with 0 (blue), 10 (orange), and 20 (green) seconds of time offset..

Aggregated results with no offset for a sample day in January, with a mix of clear sky and variable irradiance, are shown in Fig. 7. This is the same day used in Fig. 3.

V. DISCUSSION

We demonstrated that potential high limit estimation errors can be about ±5% for 95% of daytime in a year for an approximately 50 MW plant using 15% of inverters as reference inverters. We also showed that larger plants are expected to have lower error, and using high fractions of inverters as reference inverters would also reduce error. Tracking plants could increase error by about 20%, relative. A modest aggregation of five plants, spanning only about 7% of Southern Company's balancing area, had an over 50% relative reduction in 95th and 99th percentile "±" error compared to a single 50 MW plant. Adding a 20 second time offset to the aggregated analysis, representing delays from several AGC communication cycles, only increased these errors by about 25%, relative.

Additional considerations related to this analysis include:

- Selection of 15-30% of inverters as reference inverters represents something close to a "worst case" scenario, where the plant may be curtailed by up to 70-85%. Higher percentages of reference inverters would improve performance.
- Similarly, "dynamic" reference inverter selection, where the number of inverters used as reference inverters is increased in real time dependent on the power level that is requested from the plant, is expected to significantly improve performance. For example, if the plant is called for a 10% downward regulation, then 80%+ of inverters could be used as reference inverters.
- Smaller plant sizes are known to have worse estimation accuracy, particularly on a percentage basis, although can have similar errors to larger plants on an absolute (MW) basis, depending on the exact configuration of the plants being compared.
- Some plants have an AC overbuild, meaning that the sum of inverter AC nameplate capacity is greater than plant interconnection capacity. This means that some inverters must be curtailed if weather conditions are such that full power can be produced by all inverters. This complicates post-analysis, i.e. the present work, but if PHL estimation is performed within the plant control system, all relevant control schemes and real-time status would be available to adjust the PHL estimate as needed.
- The linear bias adjustment we applied is a simplified way to address relative differences between reference inverters and whole plants, and is a linear correction to potentially non-linear differences (e.g., small differences in array orientation causing output to be higher and lower depending on time of day and year). More complex methods to adjust for these non-linear differences could show improvements.

REFERENCES

[1] Q. Wang, W. B. Hobbs, A. Tuohy, M. Bello, and D. J. Ault, "Evaluating potential benefits of flexible solar power generation in the southern company system," *IEEE Journal of Photovoltaics*, vol. 12, no. 1, pp. 152–160, 2022. [Online]. Available: https://doi.org/10.1109/JPHOTOV.2021.3126118

[2] Q. Wang and B.-M. Hodge, "Enhancing power system operational flexibility with flexible ramping products: A review," *IEEE Transactions on Industrial Informatics*, vol. 13, no. 4, pp. 1652–1664, 2016. [Online]. Available: https://doi.org/10.1109/TII.2016.2637879

[3] C. Loutan, P. Klauer, S. Chowdhury, S. Hall, M. Morjaria, V. Chadliev, N. Milam, C. Milan, and V. Gevorgian, "Demonstration of essential reliability services by a 300-mw solar photovoltaic power plant," National Renewable Energy Lab (NREL), Golden, CO (United States), Tech. Rep., 2017. [Online]. Available: https://www.nrel.gov/docs/fy17osti/67799.pdf

[4] J. Nelson, S. Kasina, J. Stevens, J. Moore, A. Olson, M. Morjaria, J. Smolenski, and J. Aponte, "Investigating the economic value of flexible solar power plant operation," *Energy and Environmental Economics, Inc*, 2018. [Online]. Available: https://www.ethree.com/wp-content/uploads/2018/10/Investigating-the-Economic-Value-of-Flexible-Solar-Power-Plant-Operation.pdf

[5] S. Dahlke, M. Morjaria, V. Gevorgian, and B. Mather, "The economics of flexible solar for electricity markets in transition," First Solar, Tech. Rep., May 2020, accessed January 2022. [Online]. Available: https://www.firstsolar.com/-/media/First-Solar/Documents/Grid-Evolution/The_Economics_of_Flexible_Solar_for_Electricity_Markets_in_Transition.ashx?la=en

[6] V. Gevorgian, "Highly accurate method for real-time active power reserve estimation for utility-scale photovoltaic power plants," National Renewable Energy Lab.(NREL), Golden, CO (United States), Tech. Rep., 2019. [Online]. Available: https://www.osti.gov/biblio/1505550

Improved Efficiency of non-toxic Cu3BiS3 Thin Film Solar Cell Employing PCBM Electron Transport Layer

Sandip Das

Department of Electrical and Computer Engineering, Kennesaw State University, Marietta, GA, United States

In this work, design and numerical simulation results on a new thin-film solar cell structure using an emerging photovoltaic absorber material - Cu3BiS3, is reported. A new structure with PCBM as the electron transport layer (ETL) exhibited significantly improved power conversion efficiency (PCE) of 11.13% in comparison to a standard reference cell with n-type CdS layer showing only 6.94% efficiency. The improved cell with the configuration of SLG/Mo/Cu3BiS3/PCBM/Al showed a high open-circuit voltage (Voc) of 0.627 V, a short-circuit current density (Jsc) of 23.08 mA/cm2, and a fill factor (FF) of 77%. Our results show high potential of Cu3BiS3 to realize efficient, non-toxic, low-cost solar cells.

978-1-7281-6118-1/22 $31.00 © 2022 IEEE

Evaluating Intrinsic Defects Across CIGS Absorber via X-ray Absorption Near Edge Structures

Srisuda Rojsatien, Tara Nietzold, Niranjana Kumar, Barry Lai, Jeff Bailey, Arun Mannodi-Kanakkithodi, Maria K. Y. Chan, Mariana Bertoni

Arizona State University, Tempe, AZ, United States

Argonne National Laboratory, Argonne, IL, United States

MiaSolé Hi-Tech Corp., Santa Clara, CA, United States

Purdue University, West Lafayette, IN, United States

X-ray Absorption Near Edge Structures (XANES) is a powerful tool to unravel chemical environment as it is sensitive to oxidation state and small structural variations. In this work, we measured XANES spectra and simulated structures to fit the Se local structures, including intrinsic defects around Se atoms, inside the absorber layer of $Cu(In,Ga)Se2$ (CIGS) solar cells. This work reveals for the first time the distributions of point defects in the absorber, across the Mo and CdS interfaces, validating the presence of selenium vacancies (VSe) and copper vacancies (VCu) proposed by multiple authors but also suggesting that there are more clusters of VSe and VCu on the CdS side and more VCu clusters on the Mo side.

Accelerated Durability Evaluation of Emerging Cell Interconnect Technologies

Fang Li[1], Dylan J. Colvin[2], Kristopher O. Davis[2], Andrew Gabor[3], GovindaSamy TamizhMani[1]

[1]Photovoltaic Reliability Laboratory, Arizona State University (ASU-PRL) Mesa, Arizona, 85212, USA
[2]Department of Materials Science and Engineering, University of Central Florida, Orlando, Florida, 32816, USA
[3]BrightSpot Automation LLC, Westford, Massachusetts, 01886, USA

Abstract—In this study, we investigate the durability of four different cell interconnect technologies using three long-term accelerated stress tests. The interconnect technologies investigated in this study are: conventional five busbar, tabbed interconnect ribbons (ribbon-tabbed); 12 busbar, soldered wire interconnects (soldered wire); shingled cell interconnects attached with an electrically conductive adhesive (shingled); and laminated wire interconnects with no busbar (laminated wire). The accelerated stress tests implemented in this study are: dynamic mechanical load followed by thermal cycling and humidity freeze (sequence 1); thermal cycling (500 cycles; sequence 2); damp heat (2000 hours; sequence 3). Overall, the average degradation of both modules in all the three sequences is found to be the lowest for the solder wire technology among the four investigated technologies assuming the influence of encapsulant and backsheet is identical in all the constructions.

Keywords— accelerated stress test, cell interconnection, durability, reliability, ribbon, wire interconnection

I. INTRODUCTION

A reliable and economical cell interconnection technique can improve photovoltaic (PV) module efficiency, reduce the assembly costs, and guarantee a lifetime field operation. Ribbon tabbing is the most common technique and has the longer history of use in the industry, while the weaknesses are widely identified as a result of high resistive losses and the incompatibility with very thin wafers [1]. Consequently, to optimize connection conductivity and durability while minimize shading loss, many novel techniques that have been commercialized include electrically conductive adhesive (ECA) gluing or shingled [2], soldered wire or multiwire [3], laminated wire or SmartWire [4], interdigitated back contact (IBC) [5], metallization wrap through (MWT) [6], and emitter wrap through (EWT) [7].

Few studies focus on the extended accelerated stress testing (EAST) on the cell interconnections. Unlike the standard thermal cycling (TC) test, a rapid TC of ribbon, multiwire and shingle interconnection adopted in 2-cell modules are compared for 200 cycles of TC in around 9 days and found the lowest degradation for shingle modules versus 5 busbar and multiwire modules [8]. Solder joints of soldered wire interconnection in the fabricated two 3-cell-string module are more stable than ribbon interconnection post 1250 cycles of TC [9]. SmartWire-based 60-cell heterojunction (HJT) modules with polyolefin encapsulants can survive (less than 5% power loss) in harsh environments of around 3000 hours of damp heat (DH) and 800 cycles of TC [10].

Therefore, in this paper, we aim to study the interconnect-related failure modes of high efficiency commercial modules made with four different cell interconnection techniques including conventional tabbing ribbon method, and other emerging methods such as laminated wire, shingled, and soldered wire. Multi-characterization approaches were applied to modules subjected to three indoor EAST sequences. Current-voltage (I-V) performance and Electroluminescence (EL) images are briefly discussed.

II. METHODOLOGY

A. Module Characterization

A total of 28 commercial modules, having 4 different cell interconnection techniques with 7 modules each, were tested in three sequences (1 control module and 2 test modules per sequence). A close view of each of the interconnection techniques investigated in this study is shown in Fig.1.

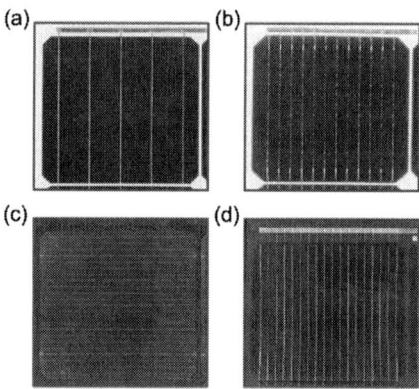

Fig. 1. Representative visual images of cells within modules representing the following interconnection technologies: (a) ribbon-tabbed, (b) soldered wire, (c) shingled, and (d) laminated wire.

A summary of the measurement techniques and equipment used is provided in Table I. A cooled Si charge-coupled device camera with 850nm long pass filter was used for EL imaging. Infrared (IR) imaging was performed after ten minutes stabilization with short-circuit current (I_{SC}) injection. UV fluorescence (UVF) images were acquired using filtered UV flash with dominant wavelength of 395nm. UV-Vis-IR reflectance spectra were obtained at the center of two edge cells

---

This work was funded by the Department of Energy's Solar Energy Technologies Office under grand number DE-EE0008155.

978-1-7281-6118-1/22 $31.00 © 2022 IEEE

using a handheld spectrophotometer. A colorimeter was used to obtain the yellowness index change.

TABLE I. CHARACTERIZATION TECHNIQUES AND EQUIPMENT

Measurement	Hardware
Flash IV @1sun	Spire 5600
Dark IV	Keithley multimeter 2700
Routine EL imaging	Sensovation HR-830 coolSamBa
IR thermography	Fluke Ti55
UVF imaging	Filtered Nikon D3400
UV-Vis-IR reflectance	ASD Fieldspec 4
Colorimetry	Xrite Ci6X

B. Accelerated Stress Tests

After initial 15 kWh/m² light soaking of all 28 modules, the indoor long-term sequential and extended accelerated tests were initiated as shown in Fig. 2. The accelerated stress tests were implemented in three sequences: Sequence 1 - dynamic mechanical load (1000 cycles at 1000 pascals) followed by thermal cycling (50 cycles) and humidity freeze (10 cycles); Sequence 2 - thermal cycling (500 cycles; 200 cycles followed 300 cycles); Sequence 3 - damp heat (2000 hours; 1000 hours followed by 1000 hours). Sequence 1 followed the Qualification Plus protocol of NREL [11]. Sequence 2 and Sequence 3 followed the IEC 63209 standard.

Two modules were tested in sequence 2 and sequence 3 for each of the four cell interconnection technologies. In sequence 1, two modules were tested for each of the tabbed-ribbon and laminated-wire technologies whereas only one module was tested for each of the soldered-wire and shingled technologies due to handling breakages of the two modules. As shown in Fig. 2, the modules were characterized initially before the accelerated tests, intermittently and finally after the accelerated stress tests. The characterization techniques included: light IV (LIV), dark IV (DIV), EL, IR, UVF, UV-Vis-NIR reflectance and colorimetry. Additional details on the accelerated tests and the characterization tools will be provided in the full paper.

III. RESULTS

Due to space limitation of this extended abstract, only power degradation results and EL data are presented here. The results obtained using all other characterization tools will be presented in the full paper or a future JPV paper.

A. Power Degradation

Fig. 3 shows the degradation of maximum power (P_{MP}) for the test modules after all three accelerated stress test sequences. In sequences 1 and 2, all the cell interconnect technologies performed well with a maximum degradation of about 4.5% in the laminated wire technology. In sequence 3, the tabbed ribbon technology experienced the highest degradation of about 12% while the soldered wire technology experienced the lowest degradation of about 4.0%.

A comparative assessment of the four cell interconnection technologies was performed, and a summary of the comparison is presented in Fig. 4 using a four color-coded rank ordering approach with rank 4 (red; ≥ 7.5% loss) being the highest degradation and rank 1 (green; <3.5% loss) being the lowest degradation. Overall, the average degradation of both modules

in all the three sequences is found to be the lowest for the solder wire technology among the four investigated technologies assuming the influence of encapsulant and backsheet is identical in all the constructions.

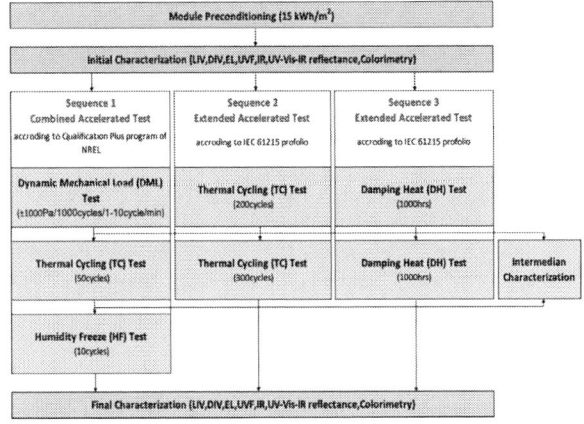

Fig. 2. Three accelerated stress test sequences used in the study. The modules were characterized using flash IV, dark IV, EL, UVF, IR, UV-Vis-IR reflectance, and colorimetry.

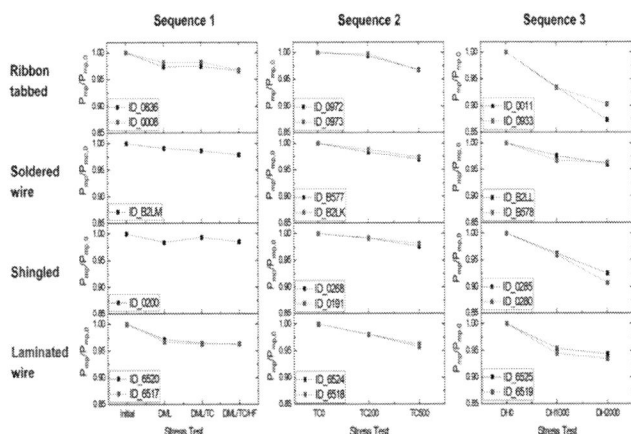

Fig. 3. Time-series P_{MP} losses of 22 test modules with four different cell interconnect techniques (ribbon-tabbed, soldered wire, shingled and laminated wire) post each stage of three test sequences.

Cell Interconn ection	Maximum Power Loss (%)			
	Sequence 1 DML/TC/HF	Sequence 2 TC500	Sequence 3 DH2000	Average
Tabbed ribbon	3.55	3.22	12.66	5.98
	3.39	3.34	9.71	
Soldered wire	3.02	2.98	3.98	3.24
	Handling breakage	2.59	3.63	
Shingled	2.88	2.4	7.5	4.79
	Handling breakage	1.85	9.3	
Laminated wire	4.13	4.28	5.69	4.77
	4.16	3.81	6.56	

Color code:

Rank 1	< 3.5%
Rank 2	3.5-5.0%
Rank 3	5.0-7.5%
Rank 4	≥7.5%

Fig. 4. Ranking table for P_{MP} losses of 22 test modules after each stress test sequence and average power loss in all the sequences for each cell interconnection technology. Four different colors indicate the severity of power degradation.

B. EL Characteristics

The differences in the extent of degradations of the four interconnect technologies subjected to the three stress test sequences can be visualized in the EL images shown in Fig. 5. The ribbon-tabbed and shingled technologies showed the usual increased level of darker pixels in the EL images after the stress tests in all the three sequences. However, the soldered-wire and laminated-wire technologies showed some unusual pattern of darkness. The soldered-wire module showed square dark spots at the center of each cell after the damp heat test sequence (sequence 3) though such pattern is not observed in the other two sequences. The laminated-wire module showed two unusual patterns: two sweep-like dark lobes near each cell center in all three sequences; darkened ring close to each cell perimeter. Potential reasons for these unusual dark patterns in the EL images will be discussed in the full paper.

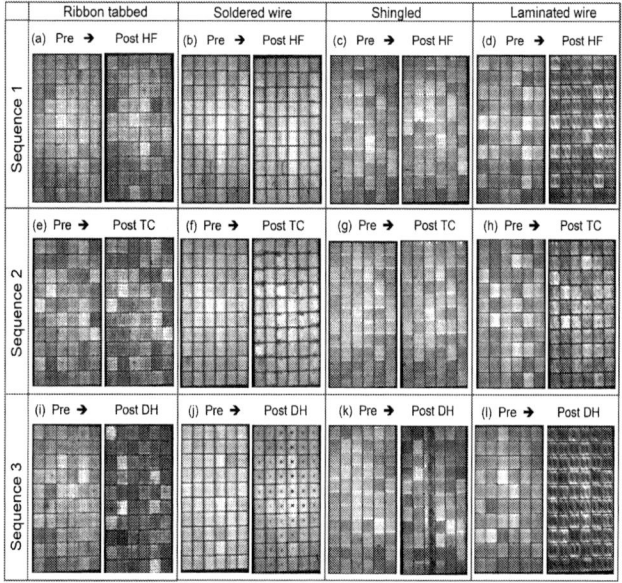

Fig. 5. Pre- and post-sequence EL images of the representative modules employed the four investigated cell interconnections.

IV. SUMMARY AND CONCLUSION

Reliability and durability evaluations of four different cell interconnect technologies have been evaluated using three long-term sequential and extended accelerated stress tests. Overall, the average degradation of all the modules in all three sequences is found to be the lowest for the solder wire technology among the four investigated technologies assuming the influence of encapsulant and backsheet is identical in all the constructions. In the full paper, the bill of material information for encapsulant and backsheet will be included and discussed.

The average degradation in all the sequences follows the following order: soldered wire technology < laminated wire technology ≈ shingled technology < ribbon tabbed technology. The pre-stress and post-stress characterizations revealed a few key durability issues, especially unusual interconnect degradation patterns in EL images of laminated wire technology. Detailed analysis on the performance data in conjunction with the characterization data will be presented in the full paper.

ACKNOWLEDGMENT

The authors would like to thank SolarPTL, LLC for flash I-V measurements of all modules investigated in this study.

REFERENCES

[1] L. Theunissen, B. Willems, J. Burke, D. Tonini, M. Galiazzo, and A. Henckens, "Electrically conductive adhesives as cell interconnection material in shingled module technology," AIP Conference Proceedings, vol. 1999, no. August 2018, 2018.

[2] M. Springer, J. Hartley, and N. Bosco, "Multiscale Modeling of Shingled Cell Photovoltaic Modules for Reliability Assessment of Electrically Conductive Adhesive Cell Interconnects," IEEE Journal of Photo-voltaics, vol. 11, no. 4, pp. 1040–1047, 2021.

[3] J. Walter, M. Tranitz, M. Volk, C. Ebert, and U. Eitner, "Multi-wire interconnection of busbar-free solar cells," Energy Procedia, vol. 55, pp. 380–388, 2014.

[4] A. Faes, A. Lachowicz, A. Bettinelli, P.-j. Ribeyron, D. Mun˜oz, J.-f. Lerat, J. Geissbu¨hler, H.-Y. Li, C. Ballif, and M. Despeisse, "Metalliza-tion and interconnection for high-efficiency bifacial silicon heterojunc-tion solar cells and modules," Photovoltaics International, vol. 41, no. September, p. 65, 2018.

[5] M. K. Mat Desa, S. Sapeai, A. W. Azhari, K. Sopian, M. Y. Sulaiman, N. Amin, and S. H. Zaidi, "Silicon back contact solar cell configuration: A pathway towards higher efficiency," Renewable and Sustainable Energy Reviews, vol. 60, pp. 1516–1532, 2016.

[6] E. Van Kerschaver and G. Beaucarne, "Back-contact solar cells: A review," Progress in Photovoltaics: Research and Applications, vol. 14, no. 2, pp. 107–123, 2006.

[7] M. T. Zarmai, N. N. Ekere, C. F. Oduoza, and E. H.Amalu, "A review of interconnection technologies for improved crystalline silicon solar cell photovoltaic module assembly," Applied Energy, vol. 154, pp. 173–182, 2015.

[8] C. Schiller, "Accelerated TC Test in Comparison with Standard TC Test for PV Modules with Ribbon, Wire and Shingle Interconnection," 36th European PV Solar Energy Conference and Exhibition, no. September, pp. 9–13, 2019.

[9] J. Walter, L. C. Rendler, C. Ebert, A. Kraft, and U. Eitner, "Solder joint stability study of wire-based interconnection compared to ribbon interconnection," Energy Procedia, vol. 124, pp. 515–525, 2017.

[10] P. Papet, S. Ha¨nni, L. Andreetta, etc., "Overlap modules: A unique cell layup using smart wire connection technology," AIP Conference Proceedings, vol. 2147, no. August, pp. 1–6, 2019.

[11] J. Wohlgemuth and S. Kurtz, "Photovoltaic Module Qualification Plus Testing," 2014 IEEE 40th Photovoltaic Specialist Conference, PVSC 2014, no. December, pp. 3589–3594, 2014.

Tracking Se Local Structures Across CdSeTe Absorber with X-ray Microscopy

Srisuda Rojsatien, Niranjana Kumar, Trumann Walker, Barry Lai, Dan Mao, Arun Mannodi-Kanakkithodi, Maria K. Y. Chan, Mariana Bertoni

Arizona State University, Tempe, AZ, United States

Argonne National Laboratory, Argonne, IL, United States

First Solar, Perrysburg, OH, United States

Purdue University, West Lafayette, IN, United States

X-ray microscopy is a powerful tool to study defects in solar cells as it allows to correlate pixel-by-pixel the local environment of selected atoms and the nanoscale electrical performance. In this work, we used X-ray absorption near edge structures (XANES) and X-ray induced current (XBIC) to track Se local structures, particularly changes in Se-Cd bond lengths, across the $CdSe_xTe_{(1-x)}$ absorber layer, and contrasts areas with high and low electrical performance. Even though all the experimental Se K-edge XANES clearly show signature of $CdSe_xTe_{(1-x)}$, there are spectral changes both across the absorber, and at different performing areas, revealing different atomic surrounding of Se atoms which together with XANES at the As K-edge may provide a full picture of the role of defects in the bulk's electrical performance.

Epitaxial growth of detachable GaAs/Ge heterostructure on mesoporous Ge substrate for layer separation and substrate reuse

Nicolas Paupy, Bouraoui Ilahi, Zakaria Oulad Elhmaidi, Valentin Daniel, Tadeáš Hanuš, Roxana Arvinte, Alexandre Heintz, Alex Brice Poungoué Mbeunmi, Thierno Mamoudou Diallo, Richard Arès, Abderraouf Boucherif

Institut Interdisciplinaire d'Innovation Technologique (3IT), Université de Sherbrooke, Sherbrooke, QC, Canada

Laboratoire Nanotechnologies Nano systèmes (LN2) – CNRS UMI-3463 Institut Interdisciplinaire d'Innovation Technologique (3IT), Université de Sherbrooke, Sherbrooke, QC, Canada

Currently, III-V multijunction solar cells holds the highest efficiency. However, they are expensive to produce and this prevents their large-scale use. For single GaAs solar cells, most of the total cost of the cell comes from the substrate. For III-V cells on Ge substrate, the thickness of the substrate is usually between 150 µm and 180 µm, whereas only 1 µm would be sufficient to maintain the performance of the cell. Moreover, 95% of the weight of the cell comes from the substrate which is a problem for the space domain. It is therefore necessary to find an economical and reliable approach for layer separation and reuse of Ge substrate. The approach proposed in this work shows the epitaxial growth of high-quality monocrystalline Germanium template on a 4-inch porous Ge substrate. The porous layer reconstruction leaves voided interface with nanometer scale pillars allowing defect free detachment of the epitaxial layer. Furthermore, the Ge epilayer demonstrates very low surface roughness, comparable to that of the epi-ready bulk Ge substrate with a preserved miscut angle suitable for III-V materials and solar cells heteroepitaxy. Accordingly, high crystal quality GaAs epilayer has been successfully demonstrated on the designed Ge template. Our finding paves the way to a reliable and cost-effective approach towards III-V/Ge multijunction solar cells detachment and substrate reuse that may constitute a technological breakthrough for both space and terrestrial PV applications.

978-1-7281-6118-1/22 $31.00 © 2022 IEEE

Analyzing Hosting Capacity Protection Constraints Under Time-Varying PV Inverter Fault Response

Joseph A. Azzolini, Nicholas S. Gurule, Rachid Darbali-Zamora, and Matthew J. Reno

Sandia National Laboratories, Albuquerque, NM, 87123, USA

Abstract—The proper coordination of power system protective devices is essential for maintaining grid safety and reliability but requires precise knowledge of fault current contributions from generators like solar photovoltaic (PV) systems. PV inverter fault response is known to change with atmospheric conditions, grid conditions, and inverter control settings, but this time-varying behavior may not be fully captured by conventional static fault studies that are used to evaluate protection constraints in PV hosting capacity analyses. To address this knowledge gap, hosting capacity protection constraints were evaluated on a simplified test circuit using both a time-series fault analysis and a conventional static fault study approach. A PV fault contribution model was developed and utilized in the test circuit after being validated by hardware experiments under various irradiances, fault voltages, and advanced inverter control settings. While the results were comparable for certain protection constraints, the time-series fault study identified additional impacts that would not have been captured with the conventional static approach. Overall, while conducting full time-series fault studies may become prohibitively burdensome, these findings indicate that existing fault study practices may be improved by including additional test scenarios to better capture the time-varying impacts of PV on hosting capacity protection constraints.

Keywords—advanced inverters, fault analysis, hosting capacity analysis, power system protection, time-series analysis

I. INTRODUCTION

Before a new solar photovoltaic (PV) system installation is allowed to connect to the power grid, a number of screens and analysis tools are often applied to ensure the safety and reliability of the integrated system. PV hosting capacity analysis (HCA) is one tool that can be used to determine the maximum allowable amount of PV that can be installed on a feeder or at a given location on the grid before certain operating constraints are violated [1], such as exceeding thermal loading limits on power lines or transformers [2], inducing extreme voltages beyond acceptable ranges [3], or causing the miscoordination of protective devices [4]. Like many other types of grid-impact studies, there has been growing interest in evaluating PV hosting capacity as a time-series to better capture the time-varying and time-dependent aspects of both the grid and the PV systems [5]. However, when it comes to evaluating protection constraints for

HCA, conventional static fault studies are used that typically only consider when the PV system is operating at unity power factor (PF) and full output capacity. Since the fault contribution of a PV system depends on several time-varying parameters [6], static fault studies may not be sufficient to capture the full range of PV impacts on protection.

Over time, PV inverter capabilities and design topologies have evolved in response to technological advances and updated interconnection standards [7], which has influenced their behavior during blue-sky conditions as well as their response to fault conditions. PV inverters must be able to ride through a variety of disturbances to voltage, frequency, and phase angle [8], and their response to those disturbances depends on whether they are grid-following or grid-forming inverters [9]. Advanced capabilities also enable PV inverters to provide grid support through autonomous control objectives like Volt-VAR or Volt-WATT, as well as to participate in feeder wide control schemes like conservation voltage reduction (CVR), Volt/VAR optimization (VVO), or other specialized objectives of distributed energy resource management systems (DERMS) [10]. Through these advanced control objectives, PV systems may be required to operate over a wide range of non-unity PFs and/or curtail real power based on time-varying atmospheric or grid conditions. While these control objectives and operating conditions do impact PV inverter fault response, HCA tools rely on a conventional static fault analysis method that typically only considers a small set of PV operating conditions like constant unity PF and pre-fault PV output at zero or full capacity.

Many studies have investigated the impacts of distributed PV on protection systems [11], but the degree to which existing HCA fault study assumptions and practices are suitable to evaluate modern PV inverter impacts remains unclear. This paper explores this knowledge gap by first applying hardware tests to validate a PV inverter fault model, then evaluating protection constraints on a test circuit using both conventional static fault analysis and a time-series fault analysis approach. Specifically, this paper combines yearlong time-series power flow simulations with corresponding fault studies at each time point to evaluate the PV impact on two common HCA protection constraints. The main contributions of this paper include:

- A hardware-validated PV fault contribution model that is applicable for different irradiances, fault voltages, and inverter Volt-VAR functions
- A time-series fault analysis methodology
- A static vs. time-series fault analysis comparison and recommendations to improve existing practices

---

This material is based upon work supported by the U.S. Department of Energy's Office of Energy Efficiency and Renewable Energy (EERE) under the Solar Energy Technologies Office (SETO). Sandia National Laboratories is a multimission laboratory managed and operated by National Technology and Engineering Solutions of Sandia, LLC., a wholly owned subsidiary of Honeywell International, Inc., for the U.S. Department of Energy's National Nuclear Security Administration under contract DE-NA0003525.

978-1-7281-6118-1/22 $31.00 © 2022 IEEE

Overall, this paper highlights the importance of inverter modeling and time-series analysis for achieving accurate PV HCA results and ensuring a safe and reliable electric grid.

II. BACKGROUND

The protection system is designed to rapidly isolate and remove faults on the grid as they occur while minimizing the disconnection of customers. Before PV systems and other distributed energy resources (DERs) were present on the grid, protection systems were designed to recognize and respond to large fault currents flowing from the substation to the fault location. When DERs are installed, they can lead to reverse power flows and fault currents from multiple injection points that either increase or decrease fault currents "seen" by protective devices, meaning the "legacy" protection systems may no longer be able to detect certain faults or may inadvertently disconnect more customers than necessary [12, 13]. Specifically, the scenarios of interest in this work are when faults occur downstream of the PV system where the voltage remains high enough at the PV location that the PV system continues to inject current, or "ride through" the low voltage event. If the fault occurs between the PV system and the substation, the PV will stop injecting current (i.e., it will have no impact on the protection system).

To avoid these types of issues, PV planning and interconnection studies often include static fault studies to evaluate existing protection systems for potential impacts [14], like relay desensitization, loss of coordination, nuisance tripping, sympathetic tripping, etc. The fault studies analyze a variety of fault types (e.g., single-line-to-ground, line-to-line, etc.) at different grid locations and with different resistances. PV systems are typically represented by simplified short-circuit models in which the PV system is modeled as a fixed current source outputting 1.2x its per unit (p.u.) rated current in phase with the voltage (i.e., a grid-following PV inverter operating at unity PF). Protection studies will then coordinate settings for protective devices considering the fault study results for this full-rated PV output contribution and the case of zero PV output (i.e., when a fault occurs during the nighttime when PV is off).

These conventional assumptions used in static fault studies for PV inverter fault response may not always be applicable. First, PV inverters only operate at full output capacity for a fraction of the year. PV output changes throughout the day and seasonally as irradiance and temperature change. Second, the PV inverter fault current is not always in phase with the voltage. Today, PV inverters must be capable of responding to centralized control signals from grid operators and be able to operate in a variety of autonomous grid-support modes [7], meaning an inverter may operate at a wide range of PFs and maintain those PFs during faults [6]. Lastly, PV inverter current limits are not standardized, meaning the 1.2x assumption could be an over- or under-estimation of an inverter's actual output current limit; furthermore, some PV inverters apply their current limit based on pre-fault output current (e.g., about 1.1x in [15]), and unlike synchronous generators, inverter fault response can vary from one manufacturer to another [16]. Novel protection schemes are being developed to accommodate these behaviors and characteristics, but their success is dependent upon the accuracy of the underlying inverter models [17].

III. STEADY-STATE PV INVERTER FAULT MODEL

This work is focused on the impacts of steady-state fault current injections from PV inverters. In general, when a fault occurs on a circuit, the voltage will sag at the output terminals of the PV inverter, or its point of common coupling (PCC), based on the characteristics and proximity of the fault. Initially, a transient current spike is observed that lasts for about 0.1 ms, meaning the energy contained in the spike is small enough to be ignored by protection schemes [18]. The inverter fault current will then settle to its steady-state value, which is proportional to the fault voltage and subject to the internal current limiting characteristics of the inverter. As noted earlier, some inverters do limit their fault current based on pre-fault output current [15]. However, with all else equal, applying a current limit based on the output current rating of an inverter would result in higher fault current magnitudes and would therefore be of more interest from the perspective of protection system impacts.

A. Model Definition

From here on, the phrase "fault current injection" will refer to the magnitude and angle of the steady-state PV inverter positive-sequence current output during a fault, I_F, defined in (1) as:

$$I_F = |I_F| \angle \Theta_I \tag{1}$$

where $|I_F|$ represents the fault positive-sequence current magnitude and Θ_I represents the angle difference between positive-sequence current and positive-sequence voltage. For this paper, it is assumed that PV inverters are grid-following type inverters that do not inject any negative sequence current [19], so all quantities and variables are positive-sequence values from here on. The fault current magnitude is subsequently defined in (2):

$$|I_F| = \left| \left(\frac{P_{Pre}/PF}{V_F} \right) \right| \tag{2}$$

where P_{Pre} is the pre-fault real power output of the inverter, PF is the pre-fault PF, V_F is the PCC fault voltage, and that (2) is subject to the following constraints:

$$|I_F| \le I_{Limit} \tag{3}$$
$$\max(S_{Pre}) \le S_{Rated} \tag{4}$$
$$S_F \le S_{Pre} \tag{5}$$
$$PF_F = PF_{Pre} = PF \tag{6}$$
$$|I_F| = 0, when\ V_F < 0.5\ p.u. \tag{7}$$

Constraint (3) states that the output current magnitude is bound by the current limit of the inverter, I_{Limit}, which is expressed as a p.u. factor of its output current rating. Constraints (4) and (5) describe that apparent power output cannot exceed the apparent power rating of the inverter (expressed in kVA or p.u.), and that the pre-fault output, S_{Pre}, will be maintained during the fault, S_F, unless the current magnitude was limited. Regardless of whether the fault current is limited, the pre-fault PF, PF_{Pre}, will be maintained during the fault [6], PF_F, and can simply be referred to as PF as in (6). One exception to this constraint is if the inverter is set to provide dynamic voltage support; in this case, the inverter would output reactive power to support the voltage during a fault regardless of the pre-fault PF [7] but analysis of this capability is beyond the scope of this work. Lastly, constraint (7) represents the momentary cessation requirement

978-1-7281-6118-1/22 $31.00 © 2022 IEEE

from [7], in which the PV inverter shall cease to inject current if the fault voltage, V_F, at its PCC falls below 0.5 p.u. (meaning $|I_F|$ is zero when V_F is less than 0.5 p.u.).

The injection angle of the PV inverter fault current is defined in (8). Since Constraint (6) also applies to (8), Θ_I will be maintained during the fault as well.

$$\Theta_I = \cos^{-1}(PF) \qquad (8)$$

Each of the dependent variables of I_F may vary through time; P_{Pre} depends on the PV array output which changes with irradiance and temperature, V_F changes with fault characteristics, and PF changes based on the control objective for the PV system. For example, when the inverter is set to operate in autonomous Volt-VAR mode per [7], PF will change based on P_{Pre} and pre-fault PCC voltage.

B. Model Validation and Current Limit Calibration

The steady-state PV inverter fault model defined in Section III. A. was validated through hardware testing of an actual 3-phase 33 kVA PV inverter using the test setup depicted in Fig. 1. The input terminals of the inverter were connected to a PV array simulator and the output terminals were connected to a grid simulator. For all hardware tests, the simulated PV array size was set to 33 kW (i.e., a DC/AC ratio of 1) such that the value of irradiance (in kW/m²) would be equivalent to P_{Pre} in p.u., and the simulated temperature was held constant at 25°C. Overall, this setup enabled each of the time-varying parameters of I_F to be controlled independently.

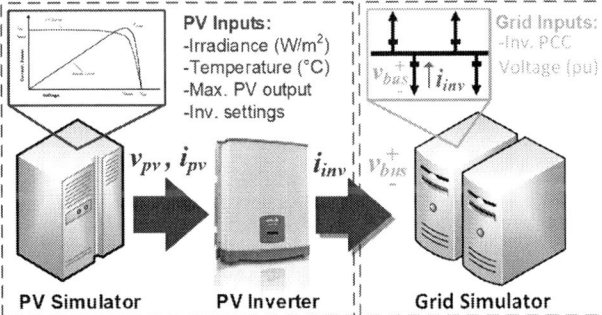

Fig. 1. Experimental setup for PV inverter hardware testing.

The first set of tests was designed to validate the relationship between PV inverter fault current magnitude and fault voltage. For these tests, the inverter was set to operate at unity PF ($PF=1$), and irradiance was held constant (at 0.9 and 0.2 kW/m²) while different fault voltages were applied. During the first voltage sweep with irradiance at 0.9 kW/m², it was observed that the inverter when into momentary cessation when the fault voltage was below 0.5 p.u. as required by [7] and that the output current was limited at 1.625x the nominal current rating of the inverter (i.e., 1.625 p.u.), much higher than conventional assumptions that are around 1.1-1.2 p.u., as observed in [18].

Based on these results, the default inverter model in OpenDSS [20] was modified in several ways to accommodate the 1.625 p.u. current limit and to ensure it was applied according to equations (1) through (8). By default, the current limit in OpenDSS is determined by the reciprocal of the *Vminpu* parameter. This default model does agree with electromagnetic

transient (EMT) inverter models under full PV output and relatively high fault voltages [17] but since the current limit is applied as a function of pre-fault output current, it does not hold under low irradiance and low fault voltage conditions for inverters that limit current as a function of their rated current. Therefore, the proposed model was implemented by dynamically adjusting *Vminpu* as PV and grid conditions changed; specifically, *Vminpu* was kept near zero unless the positive sequence fault current was above 1.625 p.u.

After the inverter model was calibrated, the hardware test setup was replicated in simulation (i.e., connecting a voltage source directly to the PCC terminals of the inverter model and manually adjusting the PV array inputs). The hardware and simulation results are presented in Fig. 2, which shows that, aside from a slight difference in losses, the inverter model sufficiently captures the relationship between fault current magnitude and fault voltage. Since the PV system had a DC/AC ratio of 1 and operated at unity PF, irradiance was equal to the product of voltage and current for any data points that were not current-limited (i.e., apparent power balance was maintained). For illustrative purposes, the momentary cessation feature was not included in the simulation results, and the voltage sweep was repeated after setting the PV array irradiance to 0.5 kW/m².

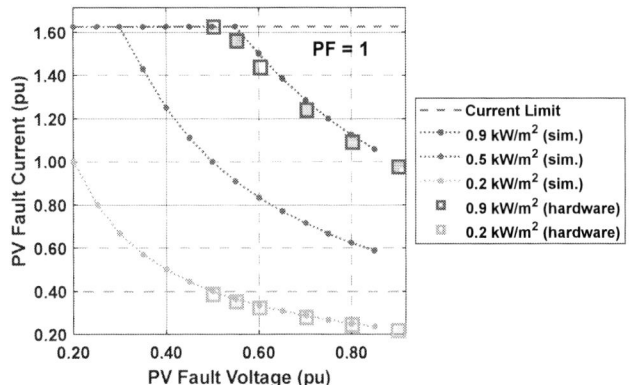

Fig. 2. Inverter fault current magnitude vs. fault voltage (V_F) at unity PF.

The next set of tests was conducted to validate the relationship between fault current magnitude and pre-fault real power output of the inverter, which is directly proportional to irradiance in this case since the DC/AC ratio was set to 1 and temperature was held constant. The inverter was set to unity PF and a fault voltage of 0.55 p.u. was applied after adjusting the input irradiance to various levels. This test was then replicated in simulation, and repeated for fault voltages of 0.45 p.u. and 0.65 p.u. The results are presented in Fig. 3, which show that the inverter model matched the hardware results and the current limit of 1.625 p.u. was once again visible. When voltage and PF were held constant, the linear relationship between fault current and irradiance was observed, as predicted by (2).

The last set of tests investigated the relationship between fault current magnitude and PF. The results are presented in Fig. 4, where the tick marks on the x-axis are spaced according to Θ_I (in degrees). The fault conditions, in this case, included a voltage sag to 0.80 p.u. and a -20° change in positive sequence voltage phase angle, which is within the ride-through requirements [7] and matches values used in prior work [6] that were based on fault simulations on an actual 21.7 km-long distribution feeder model. According to Equation (2) and Constraint (4), the maximum possible fault current magnitude for a fault voltage of 0.80 p.u. is 1.25 p.u., as confirmed by the simulation and hardware results in Fig. 4. This figure shows that when real power is held constant, the fault current magnitude increases with the reactive power associated with the non-unity PFs until the current limit is reached. At this point, the current magnitude remains constant, but the current angle continues to change. The inverter was also able to quickly synchronize to the step-change in phase angle and operate at a very wide range of PFs, even under low irradiance conditions (0.05 kW/m²), meaning it is able to operate at practically any PF required by autonomous or centralized controls.

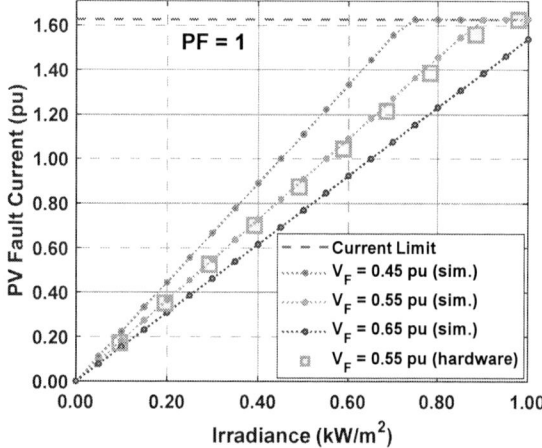

Fig. 3. Inverter fault current magnitude vs. irradiance (i.e., P_{Pre}) at unity PF.

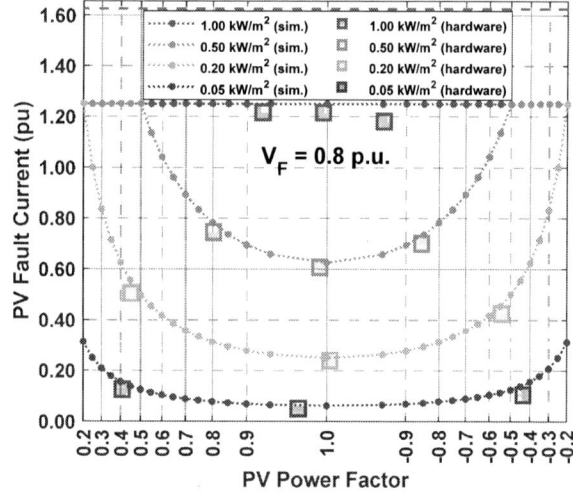

Fig. 4. Inverter fault current magnitude vs. PF ($V_F = 0.8$ p.u.).

While other inverters may have stricter limitations that are more aligned with the minimum requirements in IEEE 1547 [7], the capabilities depicted in Fig. 4 would still represent the more interesting case in terms of protection system impacts since it essentially has the widest possible operating range of P_{Pre} and PF; also, IEEE Std P2800-2022 recommends that large-scale PV inverters be able to meet minimum reactive power capabilities at all active power output levels (including at zero)

[21], so it is conceivable that future iterations of distribution interconnection standards might follow similar trends.

IV. METHODS

The remainder of the paper analyzes the differences between conventional static fault studies and time-series fault studies for evaluating protection system impacts associated with PV HCA. The following subsections describe the test circuit, introduce the protection system impact metrics evaluated, and present the methodologies for the static and time-series fault studies.

A. Test Circuit

The test circuit depicted in Fig. 5 was created to showcase some of the potential impacts of PV on protection devices and how those impacts vary throughout time as the PV output and grid conditions change. Note that the PV system was placed near the substation so that the voltage it experiences during a fault would be high enough for the PV inverter to ride through (i.e., >0.5 p.u.). Also, the closer the PV is to the substation, the greater the potential there is for desensitization of substation relays to occur.

The circuit was modeled in OpenDSS [20] and represents a 5 km long distribution feeder that contains two protective devices (a substation relay and a recloser), an aggregated load at the end of the feeder, and a large PV system with a grid-following inverter connected through a step-up transformer located 1 km downstream of the substation. The feeder is supplied by a voltage source that represents the sub-transmission system and the voltage on the feeder is regulated by a load-tap changer (LTC) on the substation transformer. The fault location (between the recloser and load, as shown in Fig. 5) and characteristics (a 3-line-to-ground, or 3LG, fault with 1-ohm resistance) were held constant for all static and time-series fault studies. This work focuses on 3LG faults since single-line-to-ground faults are generally detected by the ground overcurrent elements. The 3LG faults are the most extreme fault currents seen, and are the most susceptible to any desensitization or impacts due to the balanced PV current injections.

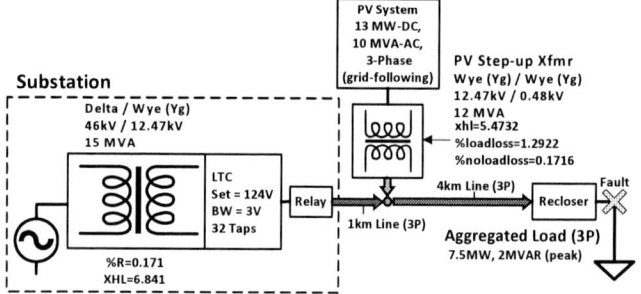

Fig. 5. Diagram of test circuit used for static and time-series fault studies.

To accommodate the time-series simulations, each of the time-varying elements in the circuit was assigned yearlong profiles with 15-minute resolutions that dictate the parameters of each element at every time point of the year. The time-varying elements in the circuit included the substation voltage source, the PV array output, and the aggregated load. For the time-series analyses, the PV inverter was set to operate in autonomous Volt-VAR mode following a slightly modified version of the

978-1-7281-6118-1/22 $31.00 © 2022 IEEE

Category B default settings [7] that allowed for maximum VAR injection and absorption at 0.95 and 1.05 p.u. voltage, respectively, as opposed to 0.92 and 1.08 p.u. This mode was selected to enable the inverter PF to vary over time (as a function of grid voltage and PV array power).

B. Protection System Impact Metrics

Two protection system impact metrics were selected to compare the results from the static and time-series fault studies: *desensitization* and *interrupt capability change*, both of which are measured in amps and are commonly evaluated in PV HCA. It was assumed that the protection devices were perfectly coordinated to detect and isolate faults before the PV system was installed. Therefore, the "PV Off" case represented the baseline fault current magnitude, $|I_{F_{Baseline}}|$. Both metrics were tracked for each of the two protective devices in Fig. 5.

Desensitization refers to the degree to which a protective device is less able to detect a fault. This metric is associated with the "minimum pick-up" current setting of a protective device, which specifies the current magnitude threshold that triggers the device. In other words, the device will not act if it senses currents below this threshold. Generally, distribution protection devices and set so that they will be able to detect and trip for a fault at the end of the feeder. When PV systems inject current during a fault, the fault current seen by a protective device may be reduced, meaning the device has become at least partially *desensitized* to the fault, or completely *desensitized* if the fault current is reduced below its minimum pick-up setting. For this work, $I_{Desensitization}$ is defined in (9) as:

$$I_{Desensitization} = min(|I_F|) - |I_{F_{Baseline}}| \qquad (9)$$

On the other hand, current injections from PV systems can also increase the fault current seen by protective devices. The *interrupt capability* metric is associated with the physical device parameters that determine the largest current magnitude it can safely and reliably *interrupt*. If the fault current exceeds this threshold, a different protective device may respond instead and disconnect more customers than required, or the device will attempt to operate, risking device damage and/or failure. For this work, the *interrupt capability change* ($\Delta I_{Interrupt}$) is defined in (10) as:

$$\Delta I_{Interrupt} = max(|I_F|) - |I_{F_{Baseline}}| \qquad (10)$$

Since the PV system in Fig. 5 has a grid-following inverter, it injects balanced currents across all three phases [19]. Therefore, to simplify the analyses, the protection metrics are reported for one of the phase elements (Phase A).

V. RESULTS

A. Static Fault Analysis

First, the static fault studies were conducted on the test circuit from Fig. 5 to determine the baseline fault current, $|I_{F_{Baseline}}|$, in the circuit (i.e., without any PV current injections). For this case, the fault current in the circuit was 1,837 A; since the relay and recloser are in series with one another and the PV system was off, the fault current was approximately the same for both devices aside from some losses. This result is represented as the horizontal dotted line "PV off" in Fig. 6.

The next test case represented the typical PV parameters used for evaluating protection impacts in HCA, in which the PV system is operating at full rated capacity (Irradiance = 1.0 kW/m^2) and unity PF. The results of each device for this case are represented by the intersection to the vertical dashed line "PV on (HCA)" in Fig. 6. In this case, the PV system injected 751.8 A of fault current in phase with the voltage at the node between the relay and recloser. For the plots on the right, note that the PV system had a DC/AC ratio of 1.3, so the inverter was operating at full capacity for any irradiance above 0.77 kW/m^2. Compared to the baseline scenario, less fault current was supplied from the substation, so the relay fault current was reduced to 1,619 A, or *desensitized* by 218 A (13.47%). The PV fault contribution also led to an increase in fault voltage compared to the baseline scenario, which increased the fault current through the recloser to 1,917 A, resulting in a $\Delta I_{Interrupt}$ of 80 A (4.17%). For context, the EPRI DRIVE tool for streamlined PV HCA [22] uses a default value of 10% to flag deviations in fault currents caused by PV injections.

In addition to the "PV off" and "PV on (HCA)" cases, static fault studies were conducted for a range of PV inverter PF conditions (left column plots in Fig. 6.) and irradiance levels (right column plots in Fig. 6.). Unlike the results in Fig. 2-Fig. 4, the inverter PCC voltage (bottom plots in Fig. 6) was dependent on the fault resistance and location—not directly controlled. These additional fault studies were intended to highlight just a portion of the potential impacts to relay and recloser fault currents, since the PV inverter can operate at essentially any combination of irradiance and PF.

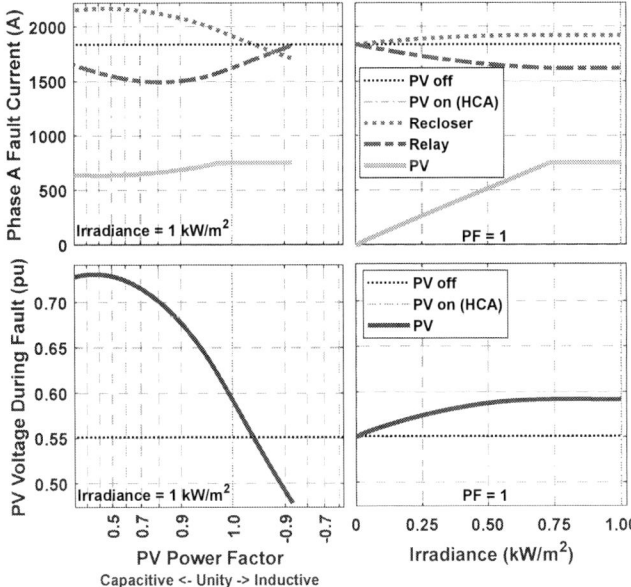

Fig. 6. Static fault study results depicting the impacts of time-varying PV parameters. The "PV on (HCA)" line highlights the conditions used for HCA, where PF=1 and Irradiance=1 kW/m^2 and the intersections with this line represent the results for each device.

As seen in the top left plot of Fig. 6, there were some PF values that caused more relay desensitization than the "PV on (HCA)" case, some that caused less, and others that resulted in larger fault currents through the relay than through the recloser. Some test conditions resulted in fault voltages that would have required momentary cessation (i.e., <0.50 p.u.) [7], but these data points were included to show what would have happened if the inverter kept injecting fault current. Large-scale PV inverters connected to the transmission system should now be able to ride through voltages as low as 0.10 p.u. [21], so if similar requirements are adopted for distribution system interconnections, the range of potential protection systems impacts would expand. This same plot also highlights that the PV fault current angle alone can impact protection devices, which is in line with previous discussions [6]; from unity PF to -0.87, the PV fault current magnitude was at its limit, yet the relay and recloser fault currents continued to change.

B. Time-Series Fault Analysis

The static fault results in Fig. 6 are useful in analyzing the complex relationships between PV parameters and fault currents in a given test circuit, but they do not provide practical bounds for the combinations of irradiance and PF conditions that a PV inverter may operate at. Since the PV inverter PF may be dependent on atmospheric conditions, grid conditions, or centralized control objectives, time-series analysis is required. In this work, the PV inverter was set to operate in autonomous Volt-VAR mode with VAR priority, meaning the PF would change with the grid voltage at the PV inverter terminals as well as with the input power to the PV array.

Overall, the time-series fault analysis was accomplished with a two-step approach. First, a quasi-static time-series (QSTS) simulation was conducted on the circuit with time steps of 15 minutes for every time point of the year (35,040 total time points), but without any faults applied. In QSTS simulations, the solution of one time point serves as the initial state for the next, so the entire year was simulated consecutively while PV parameters and grid state variables were recorded. Thus, the results from this simulation provided the pre-fault conditions for every time point of the year. Next, fault studies were conducted for each of these time points after manually adjusting the circuit corresponding to the pre-fault conditions at that time, while fault currents and voltages throughout the circuit were recorded.

Results from the initial QSTS simulation are presented in Fig. 7, which shows the cumulative distribution function (CDF) plots for PV PF and real power output after filtering out nighttime data points when the PV is off. As anticipated, the PV inverter only operated at its full power output for less than 25% of the daylight hours throughout the year. This figure also shows that the PV inverter operated with PFs as low as 0.2 (inductive) but did not operate at any capacitive PFs. While IEEE 1547 says that the minimum reactive power requirements of an inverter do not require it to be able to handle 0.2 PF, the hardware tests showed that the inverter had no problem operating even very close to a power factor of zero.

An example of the time-series fault study results is presented in Fig. 8, which shows two consecutive days from the analysis selected at random. The first day highlights the variability associated with distributed PV, where the possible fault current rises and falls throughout the day as clouds pass over the PV array. In contrast, the second day depicts clear sky conditions for which the PV is able to inject its maximum current for a significant portion of the daylight hours. However, the most consequential time points on both days were actually in the early

mornings and late afternoons when PV fault current magnitude was relatively low (see the zoomed-in section of Fig. 8); during these time periods, the recloser was briefly desensitized and the relay fault current is increased—the opposite of the behavior over the remaining time periods of the day. In these cases, even though the PV fault current magnitude was low, the PF required by the Volt-VAR control at those times was fairly extreme and ended up having an impact on the protection devices.

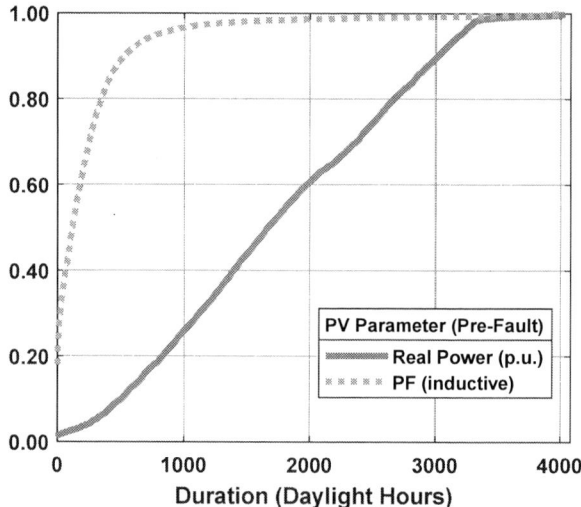

Fig. 7. Cumulative distribution plot of time-varying PV parameters from the pre-fault QSTS simulation.

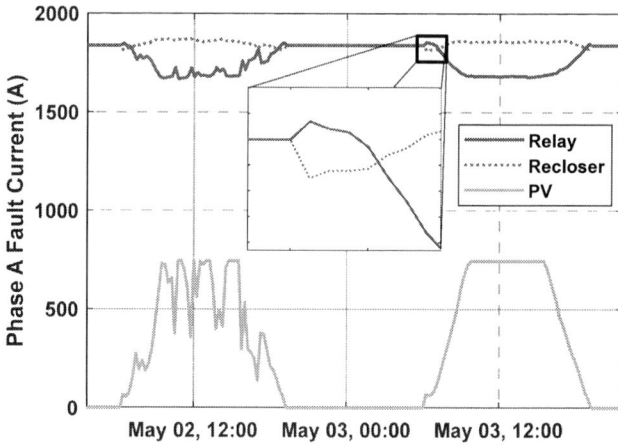

Fig. 8. Time-series fault study results for two random consecutive days.

The time-series fault behavior of the PV inverter for the rest of the year is summarized in Fig. 9. This figure shows all the combinations of fault current magnitudes and angles measured over the year, while the color of the pixel expresses the total duration of that combination (pixels in black represent combinations that did not occur). There were a wide variety of magnitude and angle combinations throughout the year but just a few current-limited combinations made up a significant portion of the total duration. Interestingly, *none* of the current-limited data points occurred at unity PF (i.e., with a fault current angle of 0°).

978-1-7281-6118-1/22 $31.00 © 2022 IEEE

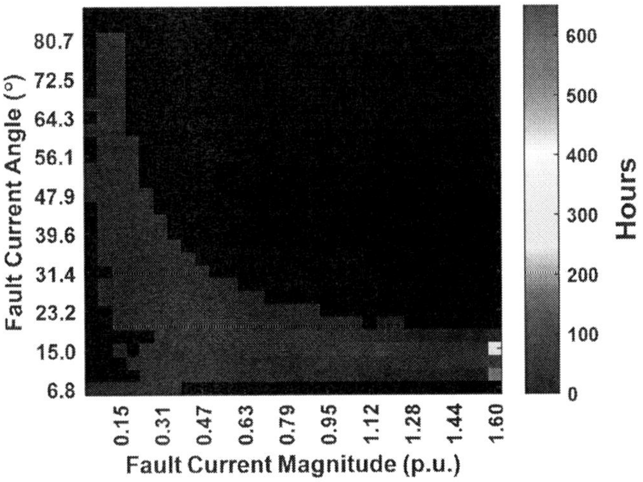

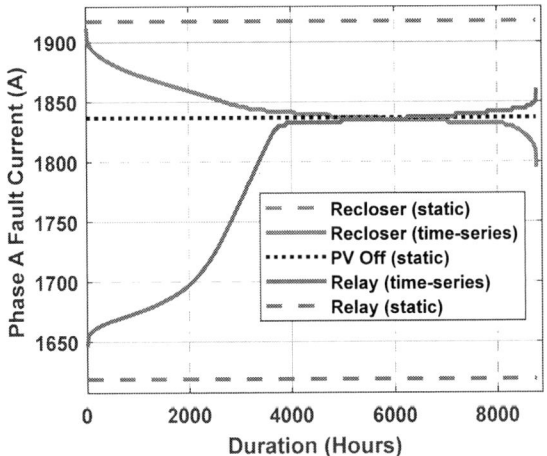

Fig. 9. Heatmap of the amount of time during the year that the PV inverter fault contributions would have been at that magnitude and angle.

C. Static vs. Time-series Analysis

The static and time-series fault study results for the relay and recloser fault currents are presented in Fig. 10, where the static results are represented by the horizontal lines and the time-series results are represented by CDF plots. Since the PV fault contribution generally had the opposite impact on the recloser compared to the relay, the recloser CDF is plotted in descending order to clarify the analyses. The dashed lines for the relay and recloser and black dotted "PV Off (Static)" line correspond to the "PV on (HCA)" and "PV off" results from Fig. 6, respectively. The results in Fig. 10 were used to evaluate the protection impact metrics from Section IV. B., which are summarized in Table I. For the relay desensitization impacts, the static fault study overestimated the time-series results by 27 A. However, this overestimation is likely acceptable since it is in the ballpark of the headroom that is typically applied when coordinating settings for protection devices. For impacts to interrupt capabilities, the static results detected no change, but the time-series study found an increase of 24 A. Although the magnitude of this discrepancy was not large in this case, it does indicate that the conventional methods may not sufficiently capture certain impacts.

Fig. 10. Comparison of PV impacts on protection devices evaluated with static and time-series fault studies.

TABLE I. PROTECTION IMPACT METRIC COMPARISON

Device	$I_{Desensitization}$ (static)	$I_{Desensitization}$ (time-series)	$\Delta I_{Interrupt}$ (static)	$\Delta I_{Interrupt}$ (time-series)
Relay	218 A	191 A	0 A	24 A
Recloser	0 A	42 A	80 A	75 A

The takeaways for the recloser impacts were similar but for the opposite metrics. The static fault study slightly overestimated the change in interrupt capability (by just 5 A), but did not capture any of the desensitizing impacts of the PV fault contribution (42 A). Again, while the differences in magnitudes of the metrics may not be alarming in this case, it may be prudent to include additional scenarios to ensure static fault studies sufficiently capture all relevant protection impacts of distributed PV, particularly when grid-support functions or other advanced control objectives will be implemented.

VI. DISCUSSION

Although this work was not intended to cover all potential PV impacts on protection systems, there were some observations that could be applicable to the broader field. For instance, there was a significant difference between the actual current limit of the inverter tested in Section III. B. (1.625 p.u.) compared to typical values used to represent the current (~1.2 p.u.). So, if the information regarding a current limit of an inverter is missing from an inverter's technical datasheet, conventional assumptions may cause issues. Similarly, certain inverters limit their output current based on pre-fault operating conditions, yet the way in which the current limit is applied is also not required to be explained on datasheets.

The results in this work do indicate that time-series fault studies could provide additional value; the increase in the maximum fault current seen by the relay and the desensitization experienced by the recloser would not have been captured with the conventional approach. While the magnitude of those impacts was not particularly large, it is worth noting that this work was limited to a single test circuit focused on one of many possible inverter control functions (autonomous Volt-VAR). Although the settings for the Volt-VAR curve enabled the inverter to inject or absorb reactive power, the voltages at the inverter PCC over the year only ever required unity PF operation or reactive power absorption (see Fig. 7). According to Fig. 6, the maximum recloser fault current could have increased significantly above the unity PF case if the inverter had injected reactive power, while the minimum relay fault current could have been significantly reduced. For instance, if the distribution system operator had initiated a conservation voltage reduction (CVR) event, the same autonomous Volt-VAR settings would have attempted to boost the voltage back up by injecting reactive power. It is possible that if a fault occurred during a CVR event, the relay would have been desensitized more than the static case due to the angle of the PV fault current injection due to the reactive power injection before the fault. Alternatively, enabling the inverter to provide dynamic voltage support would have also resulted in PV reactive power injections during any fault. In other words, the fact that there was not a large difference between the static and time-series fault studies in this case is partly attributed to the simulation setup and selected controls.

978-1-7281-6118-1/22 $31.00 © 2022 IEEE

As penetration levels of PV and other DERs increase, there will likely be more incentive to leverage advanced inverter capabilities through centralized/feeder-wide controls to improve grid conditions. If existing trends continue, inverters may be expected to ride through more extreme disturbances or even attempt to mitigate them. Thus, it is certainly conceivable that any given inverter could end up injecting its maximum fault current at much more extreme PFs than those shown in Fig. 9. So, while practical challenges of implementing full yearlong time-series fault studies may remain due to their relative complexity and computational burden, additional cases should be included in static fault studies to represent the most extreme expected combinations of fault current magnitudes and angles, with current limits of all simulated inverters set according to their actual hardware or software constraints.

VII. CONCLUSION

Assessing the impact of distributed PV systems and other DER installations on power system protection is becoming more crucial as penetration levels rise and as the utilization of advanced inverter capabilities increases. In this work, a time-series fault study framework was proposed and implemented. The results were compared to the conventional static fault study approach for evaluating common protection constraints of PV HCA. Both studies utilized a validated PV fault model that was developed by testing a commercial off-the-shelf inverter under a variety of irradiances, fault voltages, and control settings. While the conventional static approach captured certain metrics well, not all PV impacts from the time-series results were captured. Overall, conducting time-series fault studies may be prohibitively burdensome in many cases, but the results in this work indicate that the static approach can be improved by including additional scenarios to analyze, ensuring protection systems remain reliable in the presence of distributed PV with advanced inverter functions enabled.

ACKNOWLEDGMENT

Thank you to Michael Ropp and Shibani Ghosh for their review and suggestions on this work.

REFERENCES

[1] S. Stanfield, Y. Zakai, and M. McKerley, "Key Decisions for Hosting Capacity Analysis," IREC, 2021.

[2] "IEEE Guide for Loading Mineral-Oil-Immersed Transformers and Step-Voltage Regulators," *IEEE Std C57.91-2011 (Revision of IEEE Std C57.91-1995)*, pp. 1-123, 2012.

[3] *ANSI C84.1-2020 American National Standard For Electric Power Systems and Equipment - Voltage Ratings (60 Hz)*, ANSI, 2020.

[4] "IEEE Standard for Interconnection and Interoperability of Distributed Energy Resources with Associated Electric Power Systems Interfaces--Amendment 1: To Provide More Flexibility for Adoption of Abnormal Operating Performance Category III," *IEEE Std 1547a-2020 (Amendment to IEEE Std 1547-2018)*, pp. 1-16, 2020, doi: 10.1109/IEEESTD.2020.9069495.

[5] A. K. Jain, K. Horowitz, F. Ding, K. S. Sedzro *et al.*, "Dynamic hosting capacity analysis for distributed photovoltaic resources—Framework and case study," *Applied Energy*, 2020.

[6] N. S. Gurule, J. A. Azzolini, R. Darbali-Zamora, and M. J. Reno, "Impact of Grid Support Functionality on PV Inverter Response

to Faults," in *2021 IEEE 48th Photovoltaic Specialists Conference (PVSC)*, 20-25 June 2021 2021, pp. 1440-1447.

[7] *IEEE 1547 Standard for Interconnection and Interoperability of Distributed Energy Resources with Associated Electric Power Systems Interfaces*, IEEE, 2018.

[8] R. Darbali-Zamora, J. Johnson, N. S. Gurule, M. J. Reno, N. Ninad, and E. Apablaza-Arancibia, "Evaluation of Photovoltaic Inverters Under Balanced and Unbalanced Voltage Phase Angle Jump Conditions," in *2020 47th IEEE Photovoltaic Specialists Conference (PVSC)*, 15 June-21 Aug. 2020 2020, pp. 1562-1569.

[9] N. S. Gurule, J. Hernandez Alvidrez, M. J. Reno, and J. Flicker, "Multiple Inverter Microgrid Experimental Fault Testing," *49th IEEE Photovoltaic Specialists Conference (PVSC 49)*, 2022.

[10] K. Ardani, E. O'Shaughnessy, and P. Schwabe, "Coordinating Distributed Energy Resources for Grid Services: A Case Study of Pacific Gas and Electric," National Renewable Energy Laboratory, Golden, CO, 2018.

[11] H. Yazdanpanahi, Y. W. Li, and W. Xu, "A New Control Strategy to Mitigate the Impact of Inverter-Based DGs on Protection System," *IEEE Transactions on Smart Grid*, vol. 3, no. 3, pp. 1427-1436, 2012.

[12] R. A. Walling, R. Saint, R. C. Dugan, J. Burke, and L. A. Kojovic, "Summary of Distributed Resources Impact on Power Delivery Systems," *IEEE Transactions on Power Delivery*, vol. 23, no. 3, pp. 1636-1644, 2008.

[13] J. Seuss, M. J. Reno, R. J. Broderick, and S. Grijalva, "Determining the Impact of Steady-State PV Fault Current Injections on Distribution Protection," Sandia National Laboratories, SAND2016, 2016.

[14] Y. Tang, T. McDermott, and J. Homer, "Summary of electric distribution system analyses with a focus on DERs," in *2018 IEEE Power & Energy Society Innovative Smart Grid Technologies Conference (ISGT)*, pp. 1-5, 2018.

[15] G. Kou, L. Chen, P. VanSant, F. Velez-Cedeno, and Y. Liu, "Fault Characteristics of Distributed Solar Generation," *IEEE Trans. Power Deliv.*, vol. 35, no. 2, pp. 1062-1064, 2020.

[16] J. Hernandez-Alvidrez, A. Summers, N. Pragallapati, M. J. Reno *et al.*, "PV-Inverter Dynamic Model Validation and Comparison Under Fault Scenarios Using a Power Hardware-in-the-Loop Testbed," in *IEEE 7th World Conference on Photovoltaic Energy Conversion (WCPEC)*, 2018.

[17] Y. N. Velaga, R. Jain, and J. Sawant, "Modeling Distributed Energy Resources for Analyzing Distribution Systems with High Renewable Penetration," presented at the IEEE Rural Electric Power Conference, United States, 2022.

[18] S. Gonzalez, N. Gurule, M. J. Reno, and J. Johnson, "Fault Current Experimental Results of Photovoltaic Inverters Operating with Grid-Support Functionality," in *IEEE 7th World Conference on Photovoltaic Energy Conversion (WCPEC)*, 2018.

[19] M. J. Reno, S. Brahma, A. Bidram, and M. E. Ropp, "Influence of Inverter-Based Resources on Microgrid Protection: Part 1: Microgrids in Radial Distribution Systems," *IEEE Power and Energy Magazine*, vol. 19, no. 3, pp. 36-46, 2021.

[20] EPRI. "Open Distribution System Simulator (OpenDSS)." http://sourceforge.net/projects/electricdss/.

[21] *IEEE Standard for Interconnection and Interoperability of Inverter-Based Resources (IBRs) Interconnecting with Associated Transmission Electric Power Systems*, IEEE Std 2800™-2022, IEEE, 2022.

[22] M. Rylander, J. Smith, and W. Sunderman, "Streamlined Method for Determining Distribution System Hosting Capacity," *IEEE Trans. Ind. Appl.*, vol. 52, no. 1, pp. 105-111, 2016.

Tuning Thermal Induced Porous-Ge Reconstruction for Layer Transfer and Substrate Re-use

Ahmed Ayari, Bouraoui Ilahi, Roxana Arvinte, Tadeas Hanus, Laurie Mouchel, Denis Machon, Abderraouf Boucherif

Institut Interdisciplinaire d'Innovation Technologique (3IT), Université de Sherbrooke, Sherbrooke, QC, Canada

Laboratoire Nanotechnologies Nano systèmes (LN2) – CNRS UMI-3463 Institut Interdisciplinaire d'Innovation Technologique (3IT), Université de Sherbrooke, Sherbrooke, QC, Canada

Owing to their high efficiency, and heat and radiation resistance, III-V semiconductor multi-junction solar cells are dominating the space PV market. However, a lot of work has still to be done in terms of mass and cost reduction. Accordingly, reliable reduction of the substrate thickness can be obtained by solar cell detachment and substrate reuse allowing reducing both solar cells' weight. and cost. The use of porous germanium as weak layer for solar cell detachment is one of the most promising approaches ensuring scalability and cost-effectiveness. In seek of Ge substrate design providing both epitaxial seed layer and voided weak layer underneath suitable for III-V materials growth and subsequent detachment, we provide systematic investigation of thermal induced reorganization of porous germanium with various porosity levels and thicknesses. Indeed, high porosity structure shows fast reconstruction rate with increasing the thermal budget testifying its aptitude to form controllable voided separation layer. Meanwhile, low porosity structure' reconstruction is found to be mediated by pores transformation to faceted small voids, giving rise to monocrystalline material with stable thickness potentially useful as a template for epitaxial growth. Epitaxial template on weak layer design with tunable morphological and mechanical properties has been fabricated by considering a structure with gradual low porosity on top to high porosity in depth. Almost non-porous, suspended thin Ge layer connected to the bulk substrate trough pillars of few tenths of nm in diameter with micrometer scale spacing, has been successfully demonstrated. Our results show that Ge layer with gradual porosity constitute a viable approach for solar cell detachment offering tunable properties depending on the porous layers thicknesses and porosity.

Simulation-Based Determination of Shockley-Read-Hall Recombination Lifetimes in Group-V Doped P-N Junction CdTe Devices

Alexandra M. Bothwell, Darius Kuciauskas

National Renewable Energy Laboratory, Golden, CO, United States

Carrier recombination is a significant limiting mechanism in efficiency development in CdTe-based photovoltaic devices. Therefore, it is necessary to quantify bulk lifetimes for targeted improvement in devices. Time-resolved photoluminescence (TRPL) measurements on double heterostructures with no space-charge field are often used for bulk lifetime determination, while such determination on completed devices has not been consistently demonstrated. In group-V doped CdTe solar cells with hole density >1014 cm-3 and concomitantly stronger space charge fields, determination of bulk lifetime is both necessary and uncertain. In this work simulated TRPL decays and recombination rate components are analyzed for CdTe-based device structures when hole density is ~1015 cm-3, typical for high-efficiency bilayer absorber (CdSeTe/CdTe:As) solar cells fabricated at Colorado State University. We demonstrate that an effective Shockley-Read Hall recombination lifetime (front, bulk, and back) can be determined in devices with a p-n junction. Results support earlier reported >3 μs lifetime attribution to excellent bulk properties and interface passivation and should enable higher open circuit voltages.

Hardware-in-the-Loop Lab for Testing Grid Supporting Functions of Smart Inverters

Thunchanok Kaewnukultorn, Sergio Sepúlveda-Mora, Steven Hegedus

Institute of Energy Conversion & Department of Electrical and Computer Engineering, University of Delaware, Newark, DE, United States

Departamento de Electricidad & Electrónica, Universidad Francisco de Paula Santander, Cúcuta, Columbia

Integration of a high percentage of renewable energy requires smart inverters to contribute to voltage and frequency stabilization of the electric grid. In this work, we describe the implementation of a hardware-in-the-loop lab to test grid supporting functions of smart inverters under different grid and PV conditions. We performed efficiency tests and implemented volt-var controls on two commercially available PV inverters. To adjust the volt-var curve remotely, we created a customized algorithm for one of the inverters because this advanced function was not available via Modbus communication. When inverters operate at full power and the inverters need to inject or absorb large amounts of reactive power, the real power is reduced to keep the apparent power constant; this condition can negatively affect customer revenue and might require incentives to the system owner to compensate for this loss.

A Deep Learning Approach to Denoise Electroluminescence Images of Solar Cells

Grace Liu, Priya Dwivedi, Thorsten Trupke, Ziv Hameiri

University of New South Wales, Sydney, Australia

Luminescence imaging is essential for solar cell performance and reliability analysis. It is used to identify spatial defects and extract key electrical parameters. To reliably identify defects, high quality images are desirable; however, acquiring such images implies a higher cost and lower throughput as they require better imaging systems and longer exposure times. Reducing the exposure time or using cheaper cameras increases the amount of noise. Therefore, this study proposes a deep learning model to significantly reduce the noise in electroluminescence images, hence, improving the quality of their analysis without the need for additional hardware expenses. The proposed deep learning approach improved noisy images by 30.4% and 39.3% in terms of the peak signal-to-noise ratio and structural similarity index, respectively.

Temperature Dependence of Silicon-Dielectric Interface Recombination

Anh Huy Tuan Le, Eduardo Prieto Ochoa, Ruy Sebastian Bonilla, Nino Borojevic, Ziv Hameiri

School of Photovoltaic and Renewable Energy Engineering, University of New South Wales, Sydney, Australia

Department of Materials, University of Oxford, Oxford, United Kingdom

Investigations into the temperature dependence of the surface recombination at the interface between silicon and various dielectrics in modern solar cells are of significant interest as they (a) provide fundamental information regarding the interfaces, and (b) allow to improve predictions regarding the performance of solar cells under actual operating conditions. In this study, we use a novel technique based on external bias voltages to control the carrier population at the silicon-oxide/silicon, silicon-nitride/silicon, and aluminum-oxide/silicon interfaces from heavy accumulation to heavy inversion in the temperature range 25-90 °C. We find that the effective lifetime slightly increases at elevated temperatures when the imbalance of the carrier populations is amplified. In the studied temperature range, it seems that the electron and hole capture cross-sections at all the interfaces are temperature-dependent. The technique offers a simple and versatile manner to separate the chemical passivation from the charge-assisted population control at the silicon/dielectric interface, as a function of temperature. It can be very useful for the optimization of the dielectric layers and the investigation of the fundamental properties of the passivation at this interface.

Optical Simulations of all-Inorganic CsPbBr₃ Perovskite Quantum Dot Intermediate Band Solar Cells (QDIBSCs)

Ola Rashwan and Chase Sasala

Pennsylvania State University - Harrisburg, Middletown, Pennsylvania, 17036, USA

Abstract—This study investigates the optical behavior of the intermediate band solar cell consisting of all-inorganic $CsPbBr_3$ halide perovskite matrix and either lead sulfide quantum dots or germanium quantum dots. The Finite elements method approach is used to solve Maxwell's equations. It is found out that the lead sulfide quantum dots(PbS-QDs) outperforms the Ge-QDs. Additionally, the bigger the physical quantum dot size and the higher the volume concentration of the quantum dots, the higher the absorbance in the visible and infrared ranges. More studies need to be performed on Ge-based QD intermediate band solar cells.

Keywords—Intermediate band solar cells, all-inorganic halide perovskites, lead sulfide quantum dots, germanium quantum dots, optical modeling

I. Introduction (*Heading 1*)

Intermediate band solar cells (IBSCs) offer a great approach to exceed Shockley- Queissar efficiency limit for a single-junction solar cell. The optimized efficiency of the ideal intermediate band solar cell is 63.2%. The concept of the intermediate band solar cells depends on increasing the photogenerated current while preserving the high voltage of the host material. There are two common methods to realize the intermediate band in the bandgap of the host material: impurity doping and quantum dot [1-2]; however, the quantum dots are more feasible. The semiconducting quantum dots are relatively low cost, are tunable in terms of electronic and optical properties, and have versatile processing capabilities.

In the intermediate band solar cells with quantum dots, there is an intermediate band (IB) created between the valance band (VB) and the conduction band (CB)of a wide- bandgap halide perovskite matrix by dispersing quantum dots (QDs) into the perovskite matrix. The confinement of electrons or holes of the quantum dots leads to the creation of discrete energy levels, compared to the continuum of states in bulk material. This allows an intermediate band (IB) of energy between the valance band and conduction band of the bulk absorber material to be created. In the IBSCs, there is the photon absorption which results in the electron transition between the valence band to the conduction band of the perovskite matrix. Additionally, another two photons absorption of lower energies that causes the electrons to move from the valence band to the intermediate band, the from the intermediate band into the conduction band. This results in a substantial increase in the photogenerated current [3]. The halide perovskite matrix must have a wide bandgap to preserve a high open-circuit voltage. The overall efficiency increases significantly.

Hosokawa et al. [4] recently introduced the concept of IB to the perovskite solar cells and discussed the solution-based processing of intermediate band solar cells with methylammonium lead bromide ($CH_3NH_3PbBr_3$) MAPbBr₃ with PbS quantum dots and the efficiency limits of IB solar cells perovskite solar cells. They assumed that to realize the maximum efficiency of the IBSC of 63%, the bandgap of the bulk material must be around 2.3 and the bandgap of the IB material should split the bandgap into 1.5eV and 0.9eV using a quantum dot material that has a bandgap of 1 eV. In this study, $CsPbBr_3$ which is all inorganic halide perovskite with a wide bandgap of 2.4 eV is used as a matrix material to improve the stability of the cell. The effect of the quantum dot material (Ge vs. PBs), their volumetric concentrations, and their sizes on the optical absorbance are investigated.

II. Methods/ Materials

A. Materials

The stack is composed of FTO of 50 nm and TiO2 of 40nm as an electron transport layer, the all inorganic $CsPbBr_3$ of 450 nm with dispersed QDs, SpiroOMeTAD of a 250 nm as a hole transport layer, and a gold electrode of 80 nm. The QDs were modeled as spheres. To model the varying volumetric concentrations, the number of the spheres is changed such that they occupied the required volume. Two volumetric concentrations were calculated and tested using eq.1(1) :5% and 10%. Additionally, the plain CsPbBr3 was tested as a control cell.

A schematic illustration of the modeled solar cell is shown in Fig.1.

$$\% = \frac{\left(\frac{\%V_{PbS/Ge}}{100\%}\right)(V_{Absorber\ Layer})}{\left(1+\left(\frac{\%V_{PbS/Ge}}{100\%}\right)\right)} \quad (1)$$

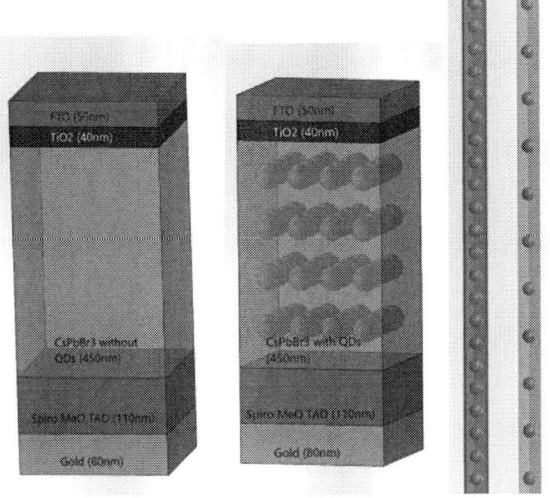

Fig. 1. From left to right: all inorganic CsPbBr3 without QD, all inorganic CsPbBr3 without QD, and modeled perovskite absorber with different QDs concentrations.

B. Optical Modeling

ANSYS High-Frequency Structure Simulator (HFSS) was used to numerically solve Maxwell's Equations in the frequency domain to find the components of the electromagnetic field.

$$A(\lambda o) = \iiint \pi \frac{c}{\lambda_0} \varepsilon''(\lambda) |\boldsymbol{E}|^2 dV \qquad (2)$$

Where $A(\lambda)$ is the absorption, c is the speed of light (m/s), $|\boldsymbol{E}|$ is the magnitude of the Electric Field (N/C), and $\varepsilon''(\lambda)$ is the imaginary part of the experimentally measured frequency-dependent dielectric permittivity. Absorption was calculated as the volume loss density integrated over the perovskite layer with dispersed quantum dots. Only a small part of the solar cell (unit cell) with proper boundary conditions was modeled and analyzed to reduce the computational cost as shown in Fig.1 Special boundary conditions, master, and slave boundary conditions, were applied to model the periodicity of the unit cell in 3D space. The Floquet ports excitations were assigned to the top and bottom faces of the unit cell to simulate the solar light propagation into a unit cell. Air Mass 1.5G incident light was simulated from 250nm to 1090 nm.

III. RESULTS AND DISCUSSION

The aim of the research is to investigate the absorbance of all inorganic halide perovskite with the dispersed Ge and PbS colloidal quantum dots (QDs) which are used to form an intermediate band. The effect of the QDs material Ge vs. PbS, the quantum dot volume concentration (5% and 10 %) and the quantum dots' sizes (8nm and 10 nm) on the absorbance are investigated.

A. The effect of QD mateial: Ge vs. PbS

Lead sulfide (PbS) quantum dots outperforms germanium (Ge) quantum dots. Fig.2 shows that the normalized absorbance of the dispersed PbS quantum dots within CsPbBr3

is higher than the absorbance of the Ge quantum dots across all solar spectra. This behavior is expected due to their large Bohr exciton radius and narrow bulk bandgap. Besides, the excitons in lead sulfide and lead selenide can be easily separated into electrons and holes because of their high dielectric constant and extinction coefficient. The main concern of the lead-based quantum dot is the lead toxicity which arise environmental concern that hinders any technology involved lead constituent from commercialization.

Therefore, utilizing germanium as an alternative for lead is a viable option since it is environmentally benign, abundant, and relatively cheap. Nevertheless, Ge quantum dots have been synthesized recently and the processibility is more complicated than lead chalcogenide quantum dots. More research needs to be done to further optimize the electronic and optical properties of the Ge quantum dots.

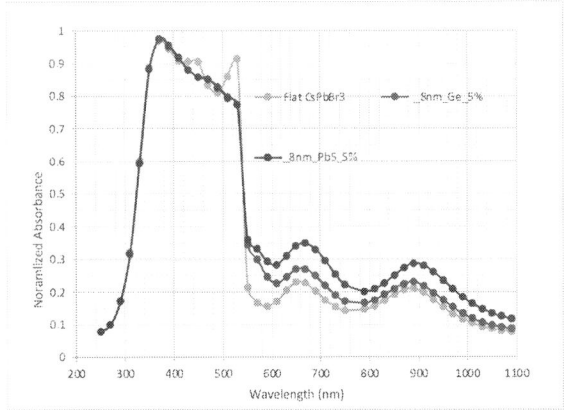

Fig. 2. The differece between GeQDs and PbS QDs compared to the wide bandgap CsPbBr₃ without QDs.

B. The effect of the Volume Concentration

The effect of the volume concentrations on the normalized absorbance is shown in Fig. 3. For both quantum dots materials, as the volume concentration increases from 5% to 10 %, the absorbance increased.

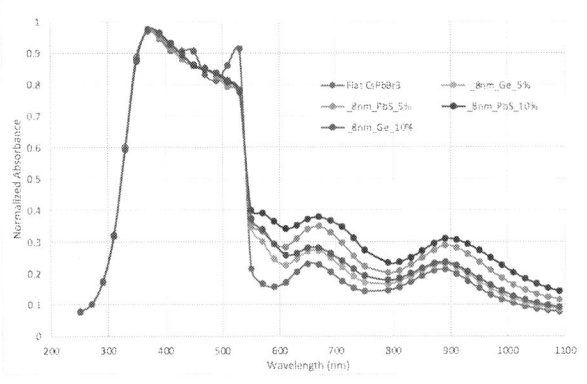

Fig. 3. The effect of the QDs volume concentrations on nomalized absorbance

978-1-7281-6118-1/22 $31.00 © 2022 IEEE

However, there is a slight increase in the Ge quantum dots especially in the Infrared range whereas, there is a substantial increase in the PbS quantum dots especially in the visible regime.

C. The effect of the QD size

The size of the quantum dots is a significant factor in tuning their optical and electronic properties. The bandgap of the quantum dots changes with the physical size of the quantum dots which enabling bandgap engineering. For example, the bandgap of the bulk lead sulfide (PbS) is 0.4 eV which can not be utilized as a photovoltaic material; however, the bandgap of the PbS quantum dots could be tuned between 0.9eV to 1.7eV [].For germanium , the bulk bandgap is 0.67 eV while the Ge quantum dots bandgaps could range from 0.87 eV to 1.15 eV [5]. These tunable bandgap quantum dots are excellent for the photovoltaic applications especially in the visible and infrared ranges.

When the size of a bulk semiconductor is reduced, discrete energy levels exist in the energy band due to the quantum confinement effect. When the quantum dots absorb light, they generate excitons which if they are separated will contribute to the photogenerated current.

In Fig.4, as the quantum dot physical size increased from 8 nm to 10 nm, there is a slight increase in the absorbance for 5% quantum dot volume concentrations. However, when the volume concentrations increases to 10 % , there is a noticeable increase in the absorbance. This is might be explained by increasing the utilization of the solar spectrum. This agreed with the results in [5]. Previous studies showed that the increase in the absorbance and the photocurrent generation is due to the increased mobilities with the increase of the quantum dot size. The carriers might need to make smaller number of hops to move between the generation sites and the extraction layer [6].

Another recent analytical study [8] agreed with our finding showed that the quantum dot size does not affect the open circuit voltage of the intermediate band solar cells; however, the short circuit current density increases with the increase in the physical quantum dot size.

Finally, increasing the quantum dot size would provide a control over the crystallinity of the quantum dots and would improve the cell performance. An experimental study [8] performed an optical characterization of the Ge QDs suggested that the smaller the quantum dot size, the Ge has more tendency to become amorphous which in turns deteriorates the cell efficiency.

IV. Conclusion

In this study, the intermediate band solar cell of all - inorganic halide perovskite CsPbBr$_3$ with dispersed quantum dots of either lead sulfide or germanium was simulated. Two different volume concentrations of the quantum dots semiconductor materials were investigated (5% and 10%), in addition to two sizes of quantum dots (8nm and 10 nm). The effect on the overall absorbance was noted.

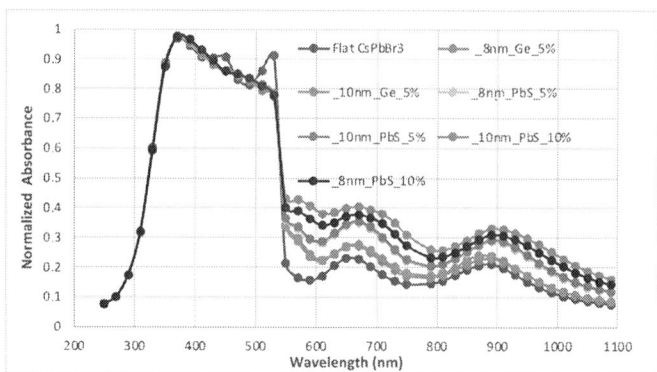

Fig. 4. the Effect of the quantum dot size on the Normalized Absorbance

A substantial increase in absorbance was noticed in the visible and NIR spectra due to the two-step photon absorption process compared with the plain cell without any quantum dots. The cells with the highest volume concentration of 10% of and bigger quantum dot size recorded the best performance. Besides, the PbS QDs-based intermediate band solar cell outperformed the Ge QDs-based intermediate band solar cell. However, the challenge of the PbS QDs IBSC is the lead toxicity. Therefore, more optimization is needed for the Ge QDs- based IBSCs.

Acknowledgment

This work was supported by Penn State Materials Research Institute (MRI) and Penn State Institutes of Energy and Environment (IEE) seed grant.

References

[1] A. Luque and A.Marti, "On certain integrals of Lipschitz-Hankel type involving products of Bessel function Increasing the Efficiency of Ideal Solar Cells by Photon Induced Transitions at Intermediate Levels," Phys.Rev. Lett., vol.78, pp.5014-, 1997.

[2] A. Luque and A.Marti, "The Intermediate Band Solar Cell: Progress Toward the Realization of an Attractive Concept," Adv. Mater., vol.22, pp.160-174, 2010.

[3] A. Luque, A.Marti and C.Stanley, "Understanding intermediate-band solar cells,"Nat.Photonics, vol.6, pp.146-152, 2012.

[4] Hosokawa et al., "Solution processed intermediate-band solar cells with lead sulfide quantum dots and lead halide perovskites, Nature Communication, 10(1) 43, 2019.

[5] C.Church, E. Muthuswamy, G.Zhai, S. Kauzlarich, and S.Carter, "Quantum dot Ge/TiO2 heterojunction photoconductor fabrication and performance," Appl. Phys. Lett. ,vol.103, 223506, 2013.

[6] Y. Liu, M. Gibbs, J. Puthussery, S. Gaik, R. Ihly, H. W. Hillhouse, and M. Law, "Dependence of Carrier Mobility on Nanocrystal Size and Ligand Length in PbSe Nanocrystal Solids," Nano Lett., vol.10, pp.1960-1969, 2010.

[7] A. El Aouami et.al," Numerical modeling of the size effect in CdSe/ZnS and InP/ZnS based Intermediate Band Solar Cells," Phys. Scr., vol. 96, 035502, 2021.

[8] E. Muthuswamy, A. S. Iskandar, M. M. Amador, and S. M. Kauzlarich, "Facile Synthesis of Germanium Nanoparticles with Size Control: Microwave versus Conventional Heating," Chem. Mater., vol. 25, pp.1416-1424, 2013

Life-Cycle Analysis of crystalline-Si "Direct Wafer" and Tandem Perovskite PV Modules and Systems

Enrica Leccisi[1], Adam Lorenz[2], Vasilis Fthenakis[1]

[1]Center for Life Cycle Analysis, Columbia University, 618 Mudd Bldg., 500 W 120th street, NY 10027
[2]CubicPV Inc., 6 Preston Court, Bedford, MA 01730, USA

Abstract—Significant reduction in energy and environmental impact is achievable through the use of kerfless wafers in PV module manufacturing. This paper summarizes a life cycle analysis (LCA) comparing "Direct Wafer" technology produced by CubicPV to conventional Czochralski wafers as well as the use of both wafer types in tandem with perovskite PV. Direct Wafer PV production avoids ingot production, associated electricity consumption and kerf loss and enables reduced wafer thickness. The impact is quantified in the LCA results in terms of life-cycle cumulative energy demand (CED), global warming potential, and Energy Pay Back Time. With Direct Wafer tandem modules, reduction in CED of the installed system is greater than 35%. Keywords— LCA; CED; GWP; kerfless; tandem PV; perovskite; EPBT.

I. INTRODUCTION

Solar photovoltaic (PV) market has been growing rapidly to address the energy transition challenge and to meet the increasing demand for affordable green power world-wide [1]. Photovoltaics entail several technologies, the most mature being crystalline-silicon, accounting for approximately 95% of the total global production [2]. Over recent years, there has been significant improvement in the material and energy utilization in the production of c-Si PV wafers, cells and panels [3-4], but it still relies on energy-intensive ingot crystallization and wasteful sawing into wafers.

Eliminating the current energy-intensive wafering process offers a compelling pathway to lower the environmental impact and broad cost of PV systems. It is now possible to produce a solar wafer directly from molten silicon in a single-step machine, removing the need for ingot growth and subsequent sawing. The key enabler of this ingot-free process is a method for freezing a thin sheet on the surface of a silicon melt and extracting the free-standing sheet (wafer) from the molten bath. A laser trims the sheet to the desired wafer geometry and adds the trimmings back to the melt. This ability to morph molten silicon directly into a wafer, avoiding the ingot and subsequent kerf-loss, is called "Direct Wafer®" manufacturing.

Additional manufacturing and material advancements specific to perovskite solar cells (PSC) constitute one of the most promising developments in photovoltaic technologies [5-10]. Solar cells formed by Direct Wafer product combined with perovskite cells in tandem architectures have the potential to dramatically reduce the levelized cost of solar electricity (LCOE). This cost reduction can be attributed to the economic value of having a very low-cost bottom cell, enabled by ingot-free production, in a tandem device. The bottom layer in tandem PV may account for only one-third of the power generation but still carries the full cost of production, adding weight to the necessity of a low-cost silicon semiconductor produced through Direct Wafer manufacturing. This paper presents the results of life-cycle analyses (LCA) of c-Si-photovoltaic module (PV) production based on the Direct Wafer technology developed by CubicPV in both single-junction and tandem Si/perovskite. The energy use and environmental impacts of these modules are compared with those of conventional sc-Si PV technologies.

II. METHODOLOGY

A. Life-Cycle Analysis

Life Cycle Analysis (LCA) is an analytical methodology adopted to quantify the sustainability of a product or a system through all stages of its life cycle. It is a fundamental framework for examining potential environmental and energy impacts and mitigation strategies, and for guiding sustainability improvements in their development and production. LCA has been developed to investigate the environmental footprint of a product or a system, from the extraction of primary resources and production of raw materials to manufacturing, distribution, use and re-use, maintenance, and end-of-life management.

The discipline was standardized initially by the Society of Environmental Toxicology and Chemistry (SETAC) and subsequently by the International Organization for Standardization (ISO) Standards 14040 and 14044 (ISO, 2006a, 2006b) and the European Commission (European Commission, 2010a, 2010b) [11-14]. In addition, the International Energy Agency (IEA) has issued PV specific LCA guidelines for assuring consistency, balance, and transparency of PV LCA [15]. Both the ISO methodology frameworks and the IEA PV LCA guidelines were followed in the current study.

Besides addressing a number of environmental impact categories, such as global warming potential (GWP), LCA also allows the calculation of the total primary energy harvested from the environment – cumulative energy demand (CED) – normalized per unit of rated electric power output.

In the context of LCA, Energy Payback Time (EPBT) is an estimation of how many years it takes for the PV system to generate an amount of electricity which is equivalent to the primary energy invested through its entire life-cycle. EPBT strongly depends on geographical deployment, in terms of irradiation levels and associated grid mix efficiencies [16-18].

978-1-7281-6118-1/22 $31.00 © 2022 IEEE

The analysis was performed using the LCA software SimaPro 9.

B. System Boundaries

We analyzed the Direct Wafer PV technology alone and in a tandem configuration with perovskite layers and compared it with conventional single-crystalline (sc-Si) PV systems, for which the production is based on the Czochralski (Cz) crystallization process. This LCA details the contributions of each stage of the system life-cycle. The manufacturing steps that are described by the life-cycle inventory (LCI) data of the Direct Wafer and conventional single crystalline-Si PV production are shown in Figure 1. The PV system is completed by the balance of system (BOS), which includes mounting and supporting structures, power electronics and cables.

The tandem PV modules entail a mechanically stacked configuration, where perovskite layers are deposited on the front glass and silicon bottom cells are optically coupled though encapsulation to collect sub-bandgap light passing through the top layer.

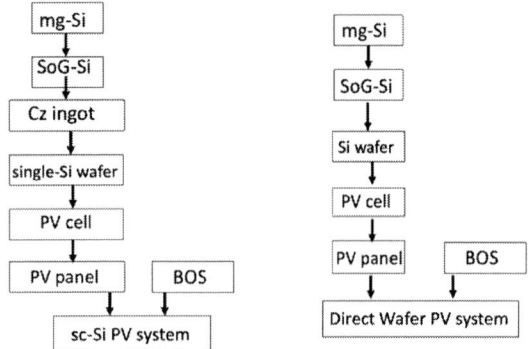

Fig. 1. Flow diagram for manufacturing single-crystalline Si (sc-Si) with Czochralski and Direct Wafer PV systems. MG-Si: metallurgic grade-Silicon; SoG-Si: solar grade-Silicon; BOS: balance of system.

C. Life-cycle Inventory

The Ecoinvent V3 Database (Ecoinvent, Zurich, Switzerland) was used as the source of background data [19].

Each considered PV system is classified by country of production for providing energy and environmental comparisons between technologies and local electric mixtures. US, China and India are the three main producing regions for which the impacts per m² of PV cells are presented.

We updated the Ecoinvent electricity mix data for all PV manufacturing and for the Si supplying countries since they influence the amount of primary energy ultimately required for each production process as well as the associated environmental impacts. We used data available from the U.S. Energy Information Administration [20-22], and used the following foreground inventory data:

- Direct Wafer PV production data for material usage, electricity consumption and emissions provided by

CubicPV Inc. based on measurements from existing equipment, projected for operation of 2 GW per year.

- Cz sc-Si PV LCI data from the 2020 IEA-photovoltaic power systems (PVPS) Task 12 Report [23] for conventional crystalline-silicon PV production, and for the MG-Si, SoG-Si, PV cell and PV panel stages of Direct Wafer production. These data were extracted by the most recent LCA of sc-Si [4] and the SoG silicon is based on the market-dominant Siemens process. The aluminum quantity needed in the PV panel stage for the frame has been updated (from 2.13 kg/m² listed in the IEA PVPS 2020 to 1.578 kg/m2 representing the current state of the art).

- LCI data from LCA of emerging perovskite PV [8-9].

- Typical balance of system (BOS) data representing ground-mounted fixed, latitude-tilt installations [3].

Table 1 shows the key life-cycle inventory parameters of Direct Wafer technologies compared to conventional Cz sc-Si.

In terms of perovskite PV, the main life-cycle inventory data are extracted from the latest estimates for scalable production by Leccisi and Fthenakis [6-9].

TABLE I. KEY LIFE-CYCLE INVENTORY PARAMETERS OF DIRECT WAFER AND CONVENTIONAL CZ SC-SI

	Unit	Direct Wafer	Cz sc-Si
Thickness	μm	140	170
Kerf loss	μm	-	65
Silicon demand for wafering	g/m²	353	595
Electricity for MG-silicon	kWh/kg	11	11
Electricity for SOG-Si	kWh/kg	49	49
Electricity for Cz ingot	kWh/kg	-	32
Electricity for wafering	kWh/m²	7.95	4.76
Electricity for cell production	kWh/m²	17.7	17.7
Electricity for panel production	kWh/m²	14	14

III. RESULTS

Figure 2 shows a comparison of the Direct Wafer and conventional Cz sc-Si technologies in terms of Cumulative Energy Demand (CED), MJ/m², for projected production of PV cells in India, China and the United States. The sc-Si values are based on 170 μm thick Czochralski wafers with PERC cell process, compared to 140 μm thick Direct Wafer 3D product with thicker and stronger edges [24]. As shown by the CED breakdown in this figure, the Direct Wafer technology results in substantial energy demand savings (about 45%) as it eliminates the Ingot stages (contributing from 250 to 340 MJ/m² depending on country) and reduces the burden of the wafer conversion by eliminating kerf loss and reduced consumables (no wire). Specifically, the Si usage in Direct Wafer technology is 0.353 kg/m², while in conventional Cz sc-Si is 0.595 kg/m² [23].

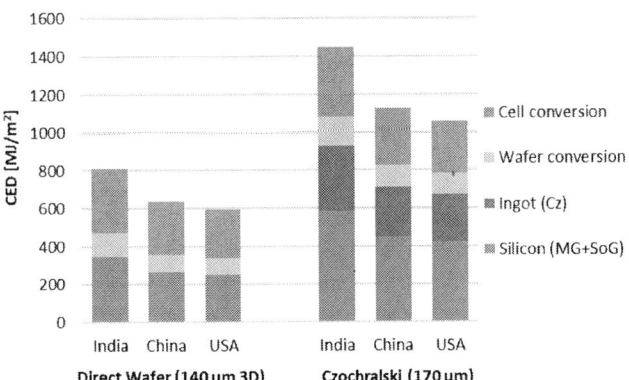

Fig. 2. Cumulative Energy Demand (CED) expressed as MJ/m² comparing single-junction Direct Wafer PV cells produced in India, China and US with conventional Cz sc-Si PV cells produced in the same countries with breakdown of manufacturing stages. PV panel contributions are not included.

Comparing the three countries, results show that India's PV cell production is more energy intensive due to its current electricity grid mix composition. As shown in Table 2, fossil fuel based power generation in India is 75% of the total, while in China and the US is respectively 67% and 61%. Also, nuclear power represents only 3% of the total in India, versus 5% in China and 20% in the US. Data in Table II are extracted from the U.S. Energy Information Administration [20-22], referring to 2019 for US and China and 2020 for India.

The contribution analysis of the Indian's CED grid mix shows that the non renewable fossil technologies requires 12.8 MJ per kWh in India (on a total of 14.2MJ/kWh), while in China it is 8.3 MJ/kWh (on a total of 10.2MJ/kWh).

TABLE II . ELECTRICITY GRID MIXES IN INDIA, CHINA AND USA

Power Plant Types	ELECTRICITY GRID MIX		
	COUNTRIES		
Country Total Demand	INDIA 1,452 10⁹ kWh	CHINA 7,136 10⁹ kWh	USA 4,128 10⁹ kWh
GAS	20%	2%	38%
COAL	55%	65%	23%
OIL	-	-	2%
NUCLEAR	3%	5%	20%
HYDRO	11%	17%	7%
SOLAR	4%	3%	2%
WIND	5%	6%	7%
BIOMASS	2%	2%	1%

It is noted that the lower energy burden atributed to China is due to the higher efficiency of the Chinese coal power plants.

The ecoinvent database descibes coal power plants as 31% efficient in China, 19% efficient in India, and 32% efficient in the US (specifically for the US Northeast Power Coordinating Council grid). Ecoinvent also models the US natural gas plants as single-cycle turbines with life-cycle efficiency of 27%, whereas most of the NG power plants are combined-cycle with average 47% efficiency. This correction was implemented in the results presented herein.

The corresponding Global Warming Potential (GWP) shown in Figure 3 expressed as kgCO2eq/m² with a breakdown of life-cycle stages. This illustrates the importance of raw materials sourcing and carbon intensity of the electricity grid.

Comparing the same technology produced in the three countries, results mainly show the environmental benefit – in terms of avoided CO_2 – for producing PV in the US when it is compared to Indian and Chinese production. Specifically, this reduction is about 50%.

Comparing a DW based production in the US with the Chinese production of Cz sc-Si, which represents the dominant share of the current market, shows an even greater advantage for the former. More specifically, a Direct Wafer product produced with US silicon has 77% lower GWP (18 kgCO2eq/m²) than a Czochralski wafer produced in China (78 kgCO2eq/m²). This is the result of the avoided ingot stage, and the reductions in the Silicon supply chain (MG+SoG), thus the avoided kerf-loss and thinner wafer which reduce the Si use from 0.595 kg/m² to 0.353 kg/m².

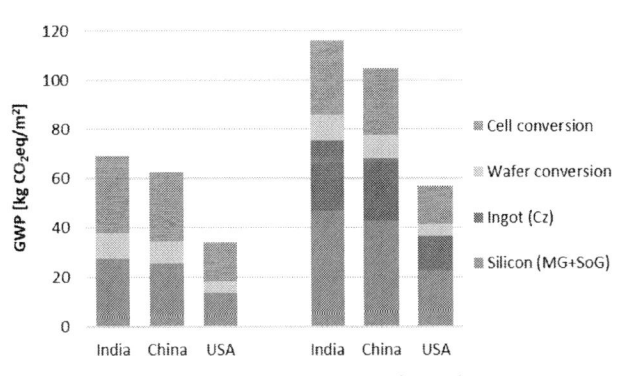

Fig. 3. Global Warming Potential (GWP) expressed as kgCO2eq/m² comparison between single-junction Direct Wafer PV cells produced in India, China, and US with conventional Cz sc-Si-PV cells produced in the same countries with breakdown of life-cycle stages. PV panel contributions are not included and addressed at system level in Fig.4.

At the system level, the CED of cells shown in Figure 2 are incorporated as the building blocks of the module efficiency-dependent CED analysis shown in Figure 4.

An efficiency of 21% was assumed for conventional Cz sc-Si PV based on a 2020 Fraunhofer report [2], and a 20% is assumed for Direct Wafer based PV.

For the Si-perovskite tandems efficiencies have been assumed to be 25%, 25.5%, and 28%, in line with values reported by NREL best research efficiency chart [25] and industry projections [26].

The indicated 1% efficiency delta between single crystal Cz and Direct Wafer product becomes smaller in a tandem module because less light reaches the bottom cell, but the full embedded energy burden still impacts the system.

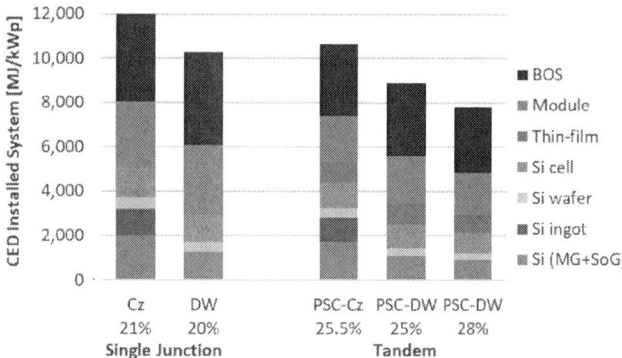

Fig. 4. Cumulative Energy Demand (CED) expressed as MJ/kWp comparison between Direct Wafer-perovkite tandem system with single-junction Direct Wafer PV, conventional Cz sc-Si-PV, and Cz sc-Si-perovskite tandem systems. Module and BOS contributions are included.

The CED [MJ/kWp] from the panel/module conversion is one of the main contributor in all the five PVs, especially in PSC-DW in which it is about 60% of the total. For the materials used in module conversion and BOS, country-specific production data was not available for all the main components (Al, glass, plastic encapsulants and backsheets, steel racking and cabling), so global averages were used that do not fully reflect the differences between manufacturing locations.

Investigating contribution analysis results for the module conversion show the aluminum needed for the frame is the main inputs as it requires 149 MJ/m^2, assuming a global average production, as shown in Figure 5.

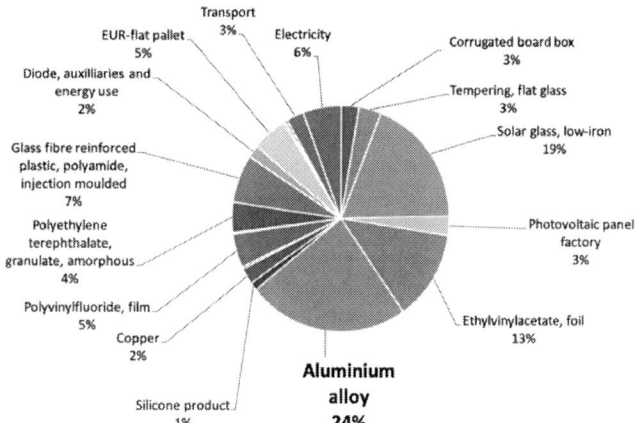

Fig. 5. Contribution analysis of the cumulative Energy Demand (CED) expressed as MJ/m2 for the PV module conversion. Global geographical representation is represented as average.

A further sensitivity analysis on aluminum production shows that producing it in the US would provide a CED savings of approximately 23% compared to the global average one. This improvement is not captured in the analysis conducted for this paper.

The Energy Pay Back Times (EPBT) for operation of systems employing each of these five types of modules in regions of medium and high solar irradiation, are shown in Figure 6. Results show that Direct Wafer EPBTs range from 6 to 5 months – respectively at 1800 and 2300 kWh/(m^2·yr) – compared to conventional Cz sc-Si that range from 8 to 6 months. PSC-Si tandem systems – installed at high irradiation locations - shows an EPBT of 4 months.

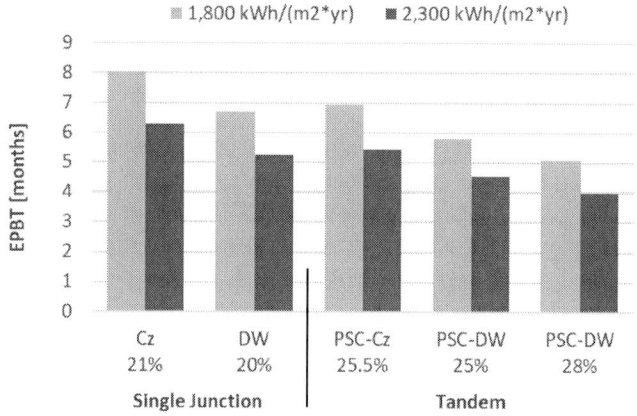

Fig. 6. Energy Pay Back Time [years] comparison between a Direct Wafer-perovkite tandem system with single-junction Direct Wafer PV, conventional Cz sc-Si-PV, and Cz sc-Si-perovskite tandem systems. Module and BOS contributions are included. Irradiation levels: 1800 and 2300 kWh/(m^2·yr).

IV. CONCLUSION

This paper showed the environmental and energy life-cycle impacts of silicon Direct Wafer (DW) and mechanically-stacked Direct Wafer/perovskite- tandem (PSC-DW) devices under development by CubicPV, in comparison with conventional sc-Si wafer technologies, both single-junction and Si-perovskite tandems. Results show that the tandem PSC-DW is 35% less energy demanding than conventional sc-Si systems. When the DW production is entirely based in the US it will generate 77% less greenhouse gases than a conventional sc-Si wafer production in China.

REFERENCES

[1] Haegel NM, Atwater H, Barnes T, Breyer C, Burrell A, Chiang YM, De Wolf S, Dimmler B, Feldman D, Glunz S and Goldschmidt JC. Terawatt-scale photovoltaics: Transform global energy. Science. 2019; 364(6443):836-838. https://doi.org/10.1126/science.aaw1845

[2] Fraunhofer Institute for Solar Energy Systems, Photovoltaics Report, 2020, https://www.ise.fraunhofer.de/content/dam/ise/de/documents/publications/studies/Photovoltaics-Report.pdf, accessed January 2022.

[3] Leccisi E, Raugei M, Fthenakis V. The energy and environmental performance of ground-mounted photovoltaic systems – a timely update. Energies 2016. 9(8):622. https://doi.org/10.3390/en9080622

[4] Fthenakis, V. and Leccisi, E.. Updated sustainability status of crystalline silicon ‐ based photovoltaic systems: Life ‐ cycle energy and environmental impact reduction trends. Progress in Photovoltaics: Research and Applications 2021;29:1068–1077.. https://doi.org/10.1002/pip.3441

[5] Billen P, Leccisi E, Dastidar S, Li S, Lobaton L, Spatari S, Fafarman AT, Fthenakis V and Baxter J. Comparative evaluation of lead emissions and toxicity potential in the life cycle of lead halide perovskite photovoltaics. Energy 2019. 166:1089-1096. https://doi.org/10.1016/j.energy.2018.10.141

[6] Leccisi E and Fthenakis V. Life-cycle environmental impacts of single-junction and tandem perovskite PVs: A critical review and future perspectives. Progress in Energy. 2020. 2:1-24. https://doi.org/10.1088/2516-1083/ab7e84

[7] Leccisi E and Fthenakis V. Critical review of perovskite photovoltaic life cycle environmental impact studies. IEEE 46th Photovoltaic Specialists Conference (PVSC), Chicago, IL, USA 2019. 2:1-6 https://ieeexplore.ieee.org/document/9198977

[8] Leccisi E and Fthenakis V. Life-cycle energy demand and carbon emissions of scalable single-junction and tandem perovskite PV. Progress in Photovoltaics, 2021, 29:1078–1092. https://doi.org/10.1002/pip.3442

[9] Fthenakis, V. and Leccisi, E., "Life-Cycle Analysis of Tandem PV Perovskite-Modules and Systems," 2021 IEEE 48th Photovoltaic Specialists Conference (PVSC), 2021, pp. 1478-1485, http://doi.org/10.1109/PVSC43889.2021.9518752

[10] Z. Li, T. R. Klein, D. H. Kim, M. Yang, J. J. Berry, M. Van Hest, K. Zhu. Scalable fabrication of perovskite solar cells. Nature Reviews Materials, 2018, 3, 1-20.

[11] Guinée JB and Lindeijer E. Handbook on life cycle assessment: operational guide to the ISO standards Vol. 7. Springer Science & Business Media 2002).

[12] Consoli F, Allen D, Boustead I, de Oude N, Fava J, Franklin R, Jensen AA, Parrish R, Perriman R, Postlethwaite D, Quay B, Séguin J, Vigon B. Guidelines for Life-Cycle Assessment: A "Code of Practice". Proceeding of the SETAC Workshop, Sesimbra, Portugal, 1993.

[13] International Organization for Standardization, Standard ISO, 14040 Environmental Management – Life Cycle Assessment – Principles and Framework, 2006, https://www.iso.org/standard/37456.html, January 2022.

[14] International Organization for Standardization, Standard ISO, 14044 Environmental Management – Life Cycle Assessment – Requirements and guidelines, https://www.iso.org/standard/38498.html, accessed January 2022.

[15] Fthenakis V., Frischknecht R, Raugei M, Kim HC, Alsema E, Held M and de Wild-Scholten M. Methodology Guidelines on Life Cycle Assessment of Photovoltaic Electricity, 2nd edition, Int Energy Agency PVPS Task 12, 2011, Report IEA-PVPS T12-03:2011).

[16] Raugei M, Peluso A, Leccisi E, and Fthenakis V. Life-Cycle Carbon Emissions and Energy Return on Investment for 80% Domestic Renewable Electricity with Battery Storage in California (USA). Energies 2020;13(15):3934. https://doi.org/10.3390/en13153934

[17] Leccisi E, Raugei M, and Fthenakis V. The energy performance of potential scenarios with large-scale PV deployment in Chile – a dynamic analysis. 2018 IEEE 7th WCPEC, a joint conference of 45th IEEE PVSC, 28th PVSEC & 34th EU PVSEC, Waikoloa, HI, USA, 2018 2018; 2441-2446), doi: 10.1109/PVSC.2018.8547293. https://ieeexplore.ieee.org/document/8547293

[18] Raugei M, Leccisi E, and Fthenakis V. What Are the Energy and Environmental Impacts of Adding Battery Storage to Photovoltaics? A Generalized Life Cycle Assessment. Energy Technology 2020;8(11)1901146:1-9 https://doi.org/10.1002/ente.201901146

[19] Ecoinvent database, 2022, https://www.ecoinvent.org , accessed January 2022.

[20] US Energy Information Administration (EIA) 2020. China, electricity. Available at: Available at: https://www.eia.gov/international/data/country/CHN/electricity/electricity-generation?pd=2&p=00000000000000000000000000000fvu&u=0&f=A&v=mapbubble&a=-&i=none&vo=value&&t=C&g=none&l=249-38&s=315532800000&e=1546300800000

[21] US Energy Information Administration (EIA) 2020. Electricity data. Available at: https://www.eia.gov/electricity/data/browser/#/topic/0?agg=2,0,1&fuel=vtvv&geo=g&sec=g&linechart=ELEC.GEN.ALL-US-99.A~ELEC.GEN.COW-US-99.A~ELEC.GEN.NG-US-99.A~ELEC.GEN.NUC-US-99.A~ELEC.GEN.HYC-US-99.A~ELEC.GEN.WND-US-99.A~ELEC.GEN.TSN-US-99.A&columnchart=ELEC.GEN.ALL-US-99.A~ELEC.GEN.COW-US-99.A~ELEC.GEN.NG-US-99.A~ELEC.GEN.NUC-US-99.A~ELEC.GEN.HYC-US-99.A~ELEC.GEN.WND-US-99.A&map=ELEC.GEN.ALL-US-99.A&freq=A&ctype=linechart&ltype=pin&rtype=s&maptype=0&rse=0&pin=

[22] US Energy Information Administration (EIA) 2020. Available at: https://www.eia.gov/international/data/country/IND/electricity/electricity-generation?pd=2&p=00000000000000000000000000000fvu&u=0&f=A&v=mapbubble&a=-&i=none&vo=value&&t=C&g=none&l=249--106&s=315532800000&e=1577836800000

[23] Frischknecht R., Stolz P., Krebs L, de Wild-Scholten M, Sinha P, Fthenakis V, Kim HC, Raugei M, Stucki M, 2020 Life Cycle Inventories and Life Cycle Assessment of Photovoltaic Systems, International Energy Agency (IEA) PVPS Task 12, Report T12-19:2020

[24] Lorenz, A., Hofstetter, J., Malkasian, H., Sanderson, L., & van Mierlo, F. 3 Dimensional Direct Wafer product with locally-controlled thickness. In *Proc. 32nd European Photovoltaic Solar Energy Conference and Exhibition* (pp. 310-312). http://doi.org/10.4229/EUPVSEC20162016-2BO.2.5

[25] NREL best research cell efficiency chart. Available at: https://www.nrel.gov/pv/cell-efficiency.html. Last accessed on May 2022.

[26] International Technology Roadmap for Photovoltaic (ITRPV). Available at: https://www.vdma.org/international-technology-roadmap-photovoltaic

Bandgap model using symbolic regression for environmentally compatible lead-free inorganic double perovskites

Ahmer A.B. Baloch, Omar Albadwawi, Badreyya AlShehhi, Vivian Alberts

Research and Development Center, Dubai Electricity and Water Authority, Dubai, United Arab Emirates

Abstract — **Data-driven models have become an essential practice of scientific research in the perovskite field, along with theory and experiments. Material informatics has emerged as a viable alternative method of exploring and formulating novel perovskite compounds using a descriptor-based approach. Herein, we develop a method that includes feature augmentation with symbolic regression to rapidly estimate and screen non-toxic lead-free inorganic double perovskites ($A_2BB'X_6$) using machine learning. Predictive models were created by identifying a physico-chemical relevant descriptor from an extensive pool of augmented features. Using primary atomic and molecular features, a high dimensional space of descriptors (containing $\approx 3\times10^5$ features) was reconstructed using mathematical operators. By increasing the complexity from 1-D to 5-D descriptor, the correlation coefficient was increased from 81.6% to 92.4%. These accurate and interpretable models can then be employed for screening lead-free perovskites with appropriate bandgaps and stability.**

Keywords— Perovskite, Machine Learning, Lead-free, Solar Cell, Descriptor.

I. INTRODUCTION

Nowadays, computational materials discovery (CMD) has become a viable alternative method to discover and design new materials and devices through data-driven approaches [1], [2]. Perovskite Solar Cells (PSC) is one such technology that has greatly benefitted from data-driven materials science, i.e. from the successful prediction of formation energy to effective screening of stable structures [3]. PSCs have received a great deal of interest due to their high efficiency, ease of fabrication, and favorable optoelectronic properties [4]. As a result, the efficiency of PSC has grown dramatically, rising from 3.8% to more than 25% [5].

Substantial effort has been directed toward creating novel perovskite materials that have excellent stability, suitable bandgap, and are non-toxic [6], [7]. Inorganic perovskite materials exhibit more stability when compared to hybrid organic-inorganic materials, and as a consequence, they are the leading candidates for commercialization. In light of these considerations, two primary paths for developing potential perovskite materials have emerged: a) double perovskites and b) mixed perovskites [8]. When it comes to double perovskites, the halide double perovskite is the most common form, with a formula of $A_2BB'X_6$. Double perovskites and mixed perovskites can also be combined to make higher dimensional mixed perovskites, creating new, viable perovskites without lead cations. However, the research of mixed double perovskites remains relatively new and unexplored.

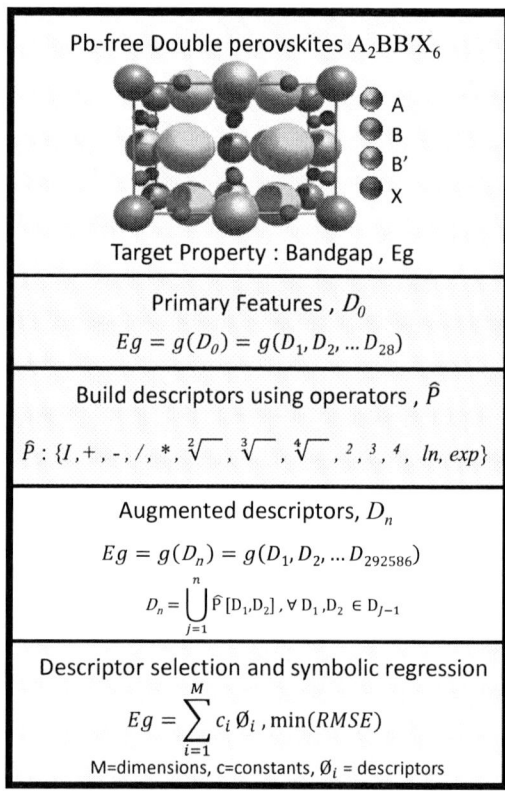

Fig. 1. The method adopted for developing the bandgap model using symbolic regression for lead-free inorganic double perovskites

Recently, material informatics has become a valuable way to find novel or improved materials by using short formulae or so-called "descriptors" containing fundamental physico-chemical properties. These descriptors are essential for high throughput material screening and the identification of novel materials from a large pool of compounds in data repositories. In CMD for double perovskites, regression models for bandgaps have been developed lately using machine learning [9]. However, these models cannot be employed for quick screening and lack interpretability to describe the known experimental data. By developing a descriptor through symbolic regression, we can provide a framework to scan and fabricate lead-free perovskite materials with appropriate bandgaps and better stability to advance the successful implementation of perovskite materials in photovoltaic solar cells. For this purpose, we developed a method that incorporated feature augmentation and regression to identify physico-chemical relevant features. Predictive models were generated from a large pool of possible descriptors using the SISSO algorithm for regression [10].

II. METHODOLOGY

Descriptors are formulae summarizing the essential features to describe the governing physics of material property. The central methodology for the project is from primary feature selection, augmenting features to develop descriptors using mathematical operators, and performing linear regression for the best fit model by minimizing RMSE. Symbolic regression was carried out using a data-driven method called "SISSO" (Sure Independence Screening and Sparsifying Operator) for lead-free inorganic perovskite materials ($A_2BB'X_6$). Data for models and features were extracted from Gao et al [11] and modeling was conducted using MATLAB. Figure 1 highlights the procedure adopted for developing the bandgap model.

The target property of bandgap, E_g, was considered using 28 atomic and molecular features, shown in Table I, such that these Primary Features (D_0) are related: $Eg = g(D_0) = g(D_1, D_2, ... D_{28})$. They include Proton number, Ionic radius, Electronegativity, Valence electrons, Coefficient of polarization, Average bond enthalpy, Enthalpy of formation, Ionization potential, and Electron affinity. As shown in Table 1.

Table I: Atomic and molecular features used for developing descriptor for A2BB'X6 - lead-free inorganic double perovskites.

$A_2BB'X_6$ Features, Symbol	A	B	B'	X	AX	BX	B'X
Ionic radius, R	✓	✓	✓	✓	-	-	-
Proton Number, Z	✓	✓	✓	✓	-	-	-
Electronegativity, X	-	✓	✓	✓	-	-	-
Valence electrons, V	✓	✓	✓	✓	-	-	-
Polarization, P	✓	✓	✓	✓	-	-	-
Enthalpy of formation, E	✓	✓	✓	✓	-	-	-
Electron affinity, A	-	-	-	✓	-	-	-
Ionization Potential, I	✓	-	-	-	-	-	-
Average bond enthalpy, BE	-	-	-	-	✓	✓	✓

From D_0, high dimensional feature space (D_n) for augmented descriptors were constructed (i.e. $\approx 3 \times 10^5$ descriptors) using meaningful combinations via mathematical operators, $\hat{P}$: $\{I, +, -, /, *, \sqrt[2]{\ }, \sqrt[3]{\ }, \sqrt[4]{\ }, {}^2, {}^3, {}^4, ln, exp\}$. As a result, $Eg = g(D_n) = g(D_1, D_2, ... D_{292586})$ such that $D_n = \bigcup_{j=1}^{n} \hat{P}[D_1, D_2], \forall D_1, D_2 \in D_{J-1}$. Linear regression is then

carried out for the best fit to identify a descriptor among a huge set of candidates by minimizing root mean square error (RMSE) for $Eg = \sum_{i=1}^{M} c_i \emptyset_i$ where M=dimensions, c=constants, $\emptyset_i$ = descriptors. Metrics for assessing the performance of the model include:

- Root Mean Square Error, $RMSE = \sqrt{\frac{1}{N} \sum_{i=1}^{N} (Eg_i - \widehat{Eg_i})^2}$

- Correlation Coefficient, $\rho = \sqrt{1 - \frac{\sum(Eg_i - \widehat{Eg_i})^2}{\sum(Eg_i - \overline{Eg_i})^2}}$

Here, N is the sample size, Eg_i is the true bandgap, $(\widehat{Eg_i})$ is the predicted bandgap, and $\overline{Eg_i}$ is the mean of the bandgap.

III. RESULTS AND DISCUSSION

In the model development, the total number of samples employed was N=745 for double perovskites ($A_2BB'X_6$) with a mean bandgap (μ) =1.86 eV and standard deviation (σ) = 1.55 eV as shown in Fig.2. The data contains both narrow and wide bandgaps ranging from 0 eV to 6.49 eV, hence making it more general purpose and to be used in various optoelectronic devices. Table II shows the formulae for lead-free inorganic double perovskites using SISSO from augmented physicochemical descriptors (D_n) (i.e. $\approx 3 \times 10^5$ descriptors). Models were analyzed from one-dimension (1-D) to five-dimension (5-D) where the term dimension refers to a particular combination of primary features as described by $\sum_{i=1}^{M} c_i \emptyset_i$.

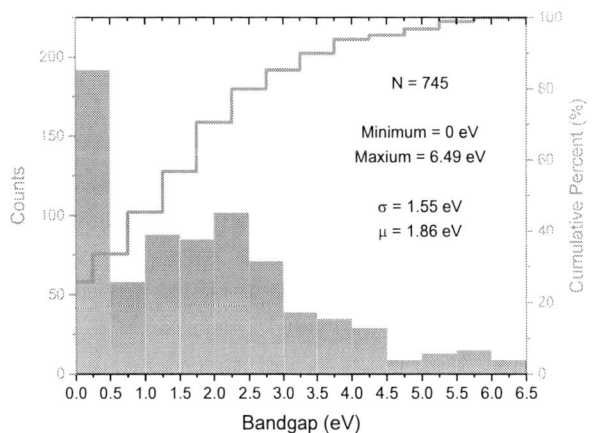

Fig. 2. Statistical distribution of the data set used for developing the bandgap symbolic regression model. Data from Reference [11]

By increasing the complexity from 1-D to 5-D as shown in Fig. 3, the model accuracy (ρ) was increased from 81.6% to 92.4%, showing a better correlation. Here, the Correlation Coefficient, ρ, is an indicator of the linear relation between two predicted ($\widehat{Eg_i}$) and actual bandgap values (Eg_i). In addition, RMSE for bandgap estimates was reduced from 0.89 eV to 0.59 eV from $1D - 5D$, respectively. This is primarily due to higher-order non-linear descriptors, as defined in Table II. Also, more features from each site (A, B, B', X) are included in higher dimensions for better interpretability.

978-1-7281-6118-1/22 $31.00 © 2022 IEEE

In all descriptors, the valence electrons feature, V, was found to be a leading parameter. The cause for this is that the orientation of valence electrons and bond length is integral to the fundamentals of the band structure. For 1D and 2D, we found that $Eg_{1D} = f\left(\frac{1}{R_X^3 V_B^4}\right)$ and $Eg_{2D} = f\left(\frac{1}{R_X^3 V_B^4}, \frac{1}{X_{B'}^2 \sqrt{V_A}}\right)$.

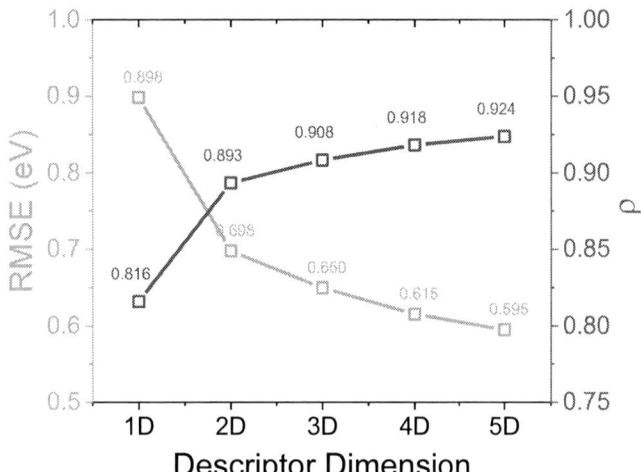

Fig. 3. RMSE and Correlation coefficient (ρ) for descriptor/formula dimensions.

Ionic radii (R), valence electrons (V), and electronegativity (X) were found to be good descriptive variables. Bandgap was found to be primarily affected by the size of X site anion and valence electron/electronegativity for cations A, B. The bandgap was found to narrow as the number of valence electrons increases, V, due to the decrease in bond strength and weak orbital overlapping. Bandgap was also found to decrease with an increase in anion radius, R_X, which agrees well with the results from Deng et al [12]. A study by Liu et al showed that by changing the wave function of valence electrons, the bandgap of perovskites can be reduced from 1.48 eV to 1.33 eV [13]. In the 3D descriptor, anion electronegativity, X_X, is directly proportional to the bandgap value as similarly reported by Castelli et al [14]. In contrast, when electronegativity of cation B', $X_{B'}$, is increased, the bandgap decreases which is in accordance with the study on perovskite metal oxides [15].

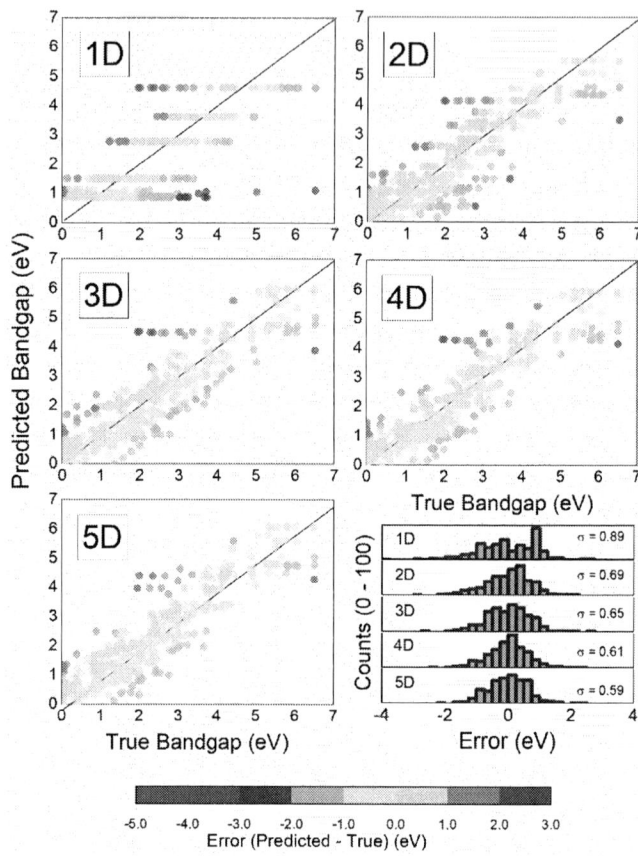

Fig. 4. Predicted bandgap vs true bandgap values for 1D − 5D symbolic regression models with error histograms and standard deviation (σ).

In the higher dimensional models such as 3-D,4-D, and 5-D, the proton number Z is also incorporated as it determines chemical bonding behavior through orbital hybridization. Average bond enthalpy between A and X, BE_{AX}, is another parameter that is introduced for a 5-D model which is similarly related to covalent bond. Figure 4 further shows the results from regression for all cases considered with the error bar. It can be seen that the 5-D descriptor performs best in terms of prediction with the minimum RMSE of 0.59 eV. On the other hand, the 1-D model has a poor fit and does not accurately capture the data. To understand the error distribution, Fig. 4 also presents the error histograms. The standard deviation, σ, for 1-D was higher at 0.89 eV with a random distribution. After increasing the

Table II: Descriptor (Formula) selected from symbolic regression of bandgap for lead-free inorganic double perovskites

Dimension	Bandgap Descriptor, Eg	C_0	C_1	C_2	C_3	C_4	C_5
1D	$C_0 + C_1\left(\frac{1}{R_X^3 V_B^4}\right)$	0.885	1.309				
2D	$C_0 + C_1\left(\frac{1}{R_X^3 V_B^4}\right) + C_2\left(\frac{1}{X_{B'}^2 \sqrt{V_A}}\right)$	-0.218	1.267	3.163			
3D	$C_0 + C_1\left(\frac{X_X^3}{V_B^4}\right) + C_2\left(\frac{1}{X_{B'}^2 \sqrt{V_A}}\right) + C_3\left(\frac{\sqrt{Z_X}}{R_B}\right)$	0.573	0.053	3.509	-0.154		
4D	$C_0 + C_1\left(\frac{1}{R_X^4 V_B^4}\right) + C_2\left(\frac{1}{X_{B'}^2 \sqrt{V_A}}\right) + C_3\left(\frac{\sqrt{Z_X}}{R_B}\right) + C_4\left(\frac{Z_{B'}^4}{X_X^4}\right)$	0.445	0.784	3.877	-0.155	1.1×10^{-6}	
5D	$C_0 + C_1\left(\frac{1}{Z_X V_B^4}\right) + C_2\left(\frac{1}{X_{B'}^2 V_A}\right) + C_3\left(\frac{Z_X^2}{R_B^2}\right) + C_4\left(\frac{Z_{B'}^4}{X_X^3}\right) + C_5(V_{B'} BE_{AX}^3)$	0.359	32.327	3.534	-2.5×10^{-4}	3.5×10^{-7}	-0.001

dimensions, the model showed normal distribution with σ approaching 0.59 eV for the 5-D case indicating fewer outliers.

IV. CONCLUSION

In this work, we present an accurate and interpretable symbolic regression model for environmentally compatible lead-free inorganic double perovskites ($A_2BB'X_6$). By using basic atomic and molecular parameters, an extensive descriptor space was employed for regression analysis. These physicochemical based formulae can then be used for quick screening for suitable bandgaps and better stability.

ACKNOWLEDGMENT

The authors would like to acknowledge the support from DEWA R&D.

REFERENCES

[1] R. Batra, L. Song, and R. Ramprasad, "Emerging materials intelligence ecosystems propelled by machine learning," *Nat. Rev. Mater.*, vol. 6, no. 8, pp. 655–678, Aug. 2021, doi: 10.1038/s41578-020-00255-y.

[2] A. A. B. Baloch, S. M. Alqahtani, F. Mumtaz, A. H. Muqaibel, S. N. Rashkeev, and F. H. Alharbi, "Extending Shannon's ionic radii database using machine learning," *Phys. Rev. Mater.*, vol. 5, no. 4, p. 043804, Apr. 2021, doi: 10.1103/PhysRevMaterials.5.043804.

[3] B. Yılmaz and R. Yıldırım, "Critical review of machine learning applications in perovskite solar research," *Nano Energy*, vol. 80, p. 105546, Feb. 2021, doi: 10.1016/j.nanoen.2020.105546.

[4] A. A. B. Baloch, F. H. Alharbi, G. Grancini, M. I. Hossain, M. K. Nazeeruddin, and N. Tabet, "Analysis of Photocarrier Dynamics at Interfaces in Perovskite Solar Cells by Time-Resolved Photoluminescence," *J. Phys. Chem. C*, vol. 122, no. 47, pp. 26805–26815, 2018, doi: 10.1021/acs.jpcc.8b07069.

[5] M. Green, E. Dunlop, J. Hohl-Ebinger, M. Yoshita, N. Kopidakis, and X. Hao, "Solar cell efficiency tables (version 57)," *Prog. Photovoltaics Res. Appl.*, vol. 29, no. 1, pp. 3–15, Jan. 2021, doi: 10.1002/pip.3371.

[6] A. A. B. Baloch, S. P. Aly, M. I. Hossain, F. El-Mellouhi, N. Tabet,

and F. H. Alharbi, "Full space device optimization for solar cells," *Sci. Rep.*, vol. 7, no. 1, p. 11984, 2017, doi: 10.1038/s41598-017-12158-0.

[7] A. A. B. Baloch, M. I. Hossain, N. Tabet, and F. H. Alharbi, "Practical Efficiency Limit of Methylammonium Lead Iodide Perovskite ($CH_3NH_3PbI_3$) Solar Cells," *J. Phys. Chem. Lett.*, vol. 9, no. 2, pp. 426–434, Jan. 2018, doi: 10.1021/acs.jpclett.7b03343.

[8] L. Meng, J. You, and Y. Yang, "Addressing the stability issue of perovskite solar cells for commercial applications," *Nat. Commun.*, vol. 9, no. 1, pp. 1–4, 2018, doi: 10.1038/s41467-018-07255-1.

[9] Z. Guo and B. Lin, "Machine learning stability and band gap of lead-free halide double perovskite materials for perovskite solar cells," *Sol. Energy*, vol. 228, pp. 689–699, Nov. 2021, doi: 10.1016/j.solener.2021.09.030.

[10] R. Ouyang, S. Curtarolo, E. Ahmetcik, M. Scheffler, and L. M. Ghiringhelli, "SISSO: A compressed-sensing method for identifying the best low-dimensional descriptor in an immensity of offered candidates," *Phys. Rev. Mater.*, vol. 2, no. 8, p. 083802, Aug. 2018, doi: 10.1103/PhysRevMaterials.2.083802.

[11] Z. Gao et al., "Screening for lead-free inorganic double perovskites with suitable band gaps and high stability using combined machine learning and DFT calculation," *Appl. Surf. Sci.*, vol. 568, no. July, p. 150916, 2021, doi: 10.1016/j.apsusc.2021.150916.

[12] Z. Deng, F. Wei, S. Sun, G. Kieslich, A. K. Cheetham, and P. D. Bristowe, "Exploring the properties of lead-free hybrid double perovskites using a combined computational-experimental approach," *J. Mater. Chem. A*, vol. 4, no. 31, pp. 12025–12029, 2016, doi: 10.1039/C6TA05817E.

[13] G. Liu et al., "Pressure-Induced Bandgap Optimization in Lead-Based Perovskites with Prolonged Carrier Lifetime and Ambient Retainability," *Adv. Funct. Mater.*, vol. 27, no. 3, p. 1604208, Jan. 2017, doi: 10.1002/adfm.201604208.

[14] I. E. Castelli, J. M. García-Lastra, K. S. Thygesen, and K. W. Jacobsen, "Bandgap calculations and trends of organometal halide perovskites," *APL Mater.*, vol. 2, no. 8, p. 081514, Aug. 2014, doi: 10.1063/1.4893495.

[15] I. E. Castelli et al., "Computational screening of perovskite metal oxides for optimal solar light capture," *Energy Environ. Sci.*, vol. 5, no. 2, pp. 5814–5819, 2012, doi: 10.1039/C1EE02717D.

Enhancement in the efficiency of rear emitter SHJ solar cells by using a CaF2/ITO double-layer anti-reflective coating.

Muhammad Aleem Zahid, Muhammad Quddamah Khokhar, Youngkuk Kim, Junsin Yi

sungkyunkwan university, suwon, Korea

A method to minimize optical losses in front side of rear emitter silicon heterojunction (SHJ) solar cells by applying double-layer antireflection coating (DLARC) has been examined. The primary aim is to increase the short circuit current density (Jsc) using DLARC that is associated with the cell' absorbance. The optical and electrical properties of single-layer antireflection coating (SLARC) were studied and compared with CaF2/ITO DLARC. OPAL 2 was utilized as a simulation tool to increase the optical properties of CaF2/ITO DLARC and the experimental result verifies its validity. ITO was deposited using a sputtering system while a thermal evaporator was utilized to deposit CaF2 to make DLARC. The SHJ solar cell having DLARC demonstrated an average reflection of 6.55% which was 31.7% less than SLARC that was 9.59% for the wavelength range of 300 to 1100 nm. The fabricated textured solar cell with CaF2/ITO DLARC showed enhancement in external quantum efficiency from 76.89% to 81.06%. It also indicated an improvement in Jsc from 39.91 mA/cm2 to 40.83 mA/cm2 and enhancement of cell efficiency from 20.45% to 21.05%.

A thermal model for bifacial PV panels

Shahzada Pamir Aly, Jim Joseph John, Gerhard Mathiak, Omar Albadwawi, Luis Pomares, Vivian Alberts

Research and Development Center, Dubai Electricity and Water Authority, Dubai, United Arab Emirates

Abstract—A one-dimensional transient thermal model has been developed to predict the through thickness temperature of a bifacial PV panel under varying field operating conditions. The developed model can be used for any bifacial PV technology and can be applied to any location in the world. The model requires inputs of irradiance, ambient temperature and wind speed field data, along with the PV site information, to estimate the PV cells temperature. The model has been validated using experimental data from two different locations, in Italy and Spain, with RMSE less than 2.28 °C for both case studies.

Keywords—PV panel, bifacial, thermal model, one-dimensional.

I. INTRODUCTION

Bifacial PV technology is gaining momentum due to its ability to generate power from both sides, thus helping to realize lower LCOE. Compared to monofacial they have higher energy yields, better performance in low light conditions and a lower temperature coefficient. The bifacial gain can be even up to 30%, depending on the cell technology, ground albedo and the structure height and positioning. The bifacial cells technology share in the world market is expected to grow from present ~30% to ~80% in 2030 [1].

Modeling and simulating the performance of any PV system is of utmost importance to calculate its bankability. The PV performance models for monofacial are well established, but their adaptation for bifacial is still an ongoing process. One of the key inputs for any PV performance model is the PV cell temperature. Thus, accurate prediction of PV cell temperature is an absolute necessity to better predict the energy yield [2].

The variation of temperature within a PV panel in operation is very dynamic. Due to the thermal mass of various layers of a PV panel, steady-state approaches are not the true depiction of their thermal behavior, especially under rapidly varying field conditions [3]. Likewise, it is important to include all the layers which have significant thermal mass to affect the temperature of the PV panel [4]. Tina et al. [5] developed a three-layer, front plus back glass and PV cells layer, thermal model for bifacial PV panels. They also considered each of these layers to be isothermal, where in reality there is temperature profile.

To improve on their model, we have included five-layers in this work (front and back EVA as extra layers). Moreover, using 1-D finite difference (FD) numerical approach, our proposed model can capture more realistic through-thickness temperature gradient within the PV panel.

The developed model is fully dynamic and only needs inputs of incident irradiance, ambient temperature and wind speed data, along with the PV site location information, to simulate the temperature variations within a bifacial PV panel.

II. THERMAL MODEL

A. 1-D FD Approach

Basically, the heat transfer equation is solved to find the temperature distribution within the solid domain of the PV panel. The 1-D heat transfer equation is:

$$\rho C \frac{\partial T(x,t)}{\partial t} = \nabla \cdot k \nabla T(x,t) + Q \tag{1}$$

Where, ρ is the density, k is the thermal conductivity, Q is the internal heat generation, T is the spatial temperature and ∇ is the del operator. As shown in Fig. 1, in this work we considered five layers of a typical bifacial panel. Other thinner layers, such as anti-reflective coatings, were ignored as they do not contribute much to heat gain or heat loss from the panel. The values of the parameters in (1) are taken from [6].

In this work a 1-D FD approach has been used to numerically solve the heat equation. In the FD approach, we basically divide the entire solid domain of the PV panel into a through-thickness 1-D grid (Fig. 1) and then algebraically solve the heat equation for every node (m) of the grid to find the overall through thickness temperature distribution within the PV panel.

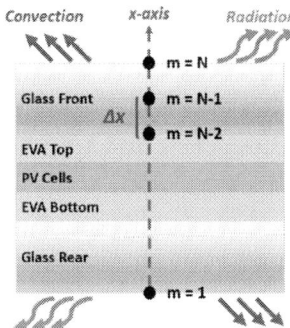

Fig. 1. 1-D FD grid of a typical bifacial PV panel.

B. Governing Equations

It has been assumed that only PV cells and the glass layers absorb the incident irradiance. To estimate the heat generation within the PV cells (Q_{PV}), we first find the absorbed irradiance in the cells (S_{PV}) by the following relation:

$$S_{PV} = \alpha_{PV} \times \tau \times \text{POA} \tag{2}$$

Where, α_{PV} (=0.93 for silicon) is the average absorptivity of the cells over the entire spectrum, τ is the transmittivity through the glass layers based on the incident angle modifier [7] and POA is the incident plane of array irradiance (front or rear). The absorbed irradiance in the glass layers (S_{GLASS}) is found by:

978-1-7281-6118-1/22 $31.00 © 2022 IEEE

$$S_{GLASS} = \alpha \times \text{POA} \qquad (3)$$

Here, α (=0.05 in this study) is the average absorptivity of the glass over the entire spectrum. Moreover, the efficiency of the PV cells (η_{PV}) is a function of temperature, given by:

$$\eta_{PV} = \eta_{PV,ref}[1 - \beta_{ref}(T_{PV} - T_{ref})] \qquad (4)$$

$\eta_{PV,ref}$ is the efficiency and T_{ref} (=25 °C) is the reference temperature. Whereas, β_{ref} is the power temperature coefficient of the PV cells.

The heat generated within the PV cells (Q_{PV}) and the glass (Q_{GLASS}) layers is then found as:

$$Q_{PV} = \frac{S_{PV} \times \text{Area}_{PV,Cells} \times (1 - \eta_{PV})}{\text{Volume}_{PV,Cells}} \qquad (5)$$

$$Q_{GLASS} = \frac{S_{GLASS} \times \text{Area}_{PV,Panel}}{\text{Volume}_{PV,Panel}} \qquad (6)$$

These heat generation terms, along with other physical parameters in (1) are solved for each node, for every time step, using 1-D FD method to find the temperature distribution within the PV panel. From this temperature distribution we can easily identify the PV cell temperature.

As shown in Fig. 1, the heat generated within the PV panel is dissipated to the surroundings via convection and radiation. These convective and radiative losses, mainly due to wind speed and the surface minus ambient temperature difference, can be estimated based on various case specific empirical relations [3].

III. VALIDATION

The developed thermal model was used to simulate the rear glass temperature of the bifacial PV panel and was compared to the experimentally measured values.

A. Experimental data

The experimental data for the model validation was obtained from the literature. One of the chosen locations is Catania (Italy) [5] and other is Barcelona (Spain) [8]. For both these locations, measured POA irradiance (front and rear), ambient temperature and the wind speed are shown in Fig. 2 and Fig. 3. However, since the wind speed data for the location at Barcelona was not provided in [8], it was obtained from 'weather spark' historical meteorological data repository [9]. Details of the site location and characteristics of the used bifacial panels in each case are given in TABLE I. The information which was not directly provided in [5] and [8] was either assumed or inferred, and these values are shown with a steric (*) symbol in TABLE I.

B. Results

The modeled and the experimentally measured back glass temperatures for both locations are shown in Fig. 4. As summarized in TABLE II. the model's accuracy was evaluated based on statistical errors like root mean square error (RMSE), mean bias error (MBE) and the correlation coefficient (r). The lower values of RMSE show that the model has a low forecasting error and the higher values of 'r' indicate that the model follows the experimental trend very well. The positive MBE value shows that the model generally overestimates the measured value and vice versa.

TABLE I. BIFACIAL PV PANELS CHARACTERISTICS

Characteristics	Italy [5]	Spain [8]
Latitude / Longitue (°)	37.50 N / 15.08 E	41.38 N / 2.16 E
Date of Field Data	Aug 03, 2019	Mar 07, 2020
Panel Inclination - β (°)	45	34
Surface Azimuth (°)	0	−17
Efficiency_STC (%)	17	19.5
Temp. coefficient − β_{ref} (%)	−0.44	−0.50*
No.of cells	72*	60
Max Power at STC (W)	340	320
Bifaciality Factor	85	70
Length × Width (m × m)	1.91 × 1.04*	1.66 × 0.99*

*Assumed or inferred values from the literature

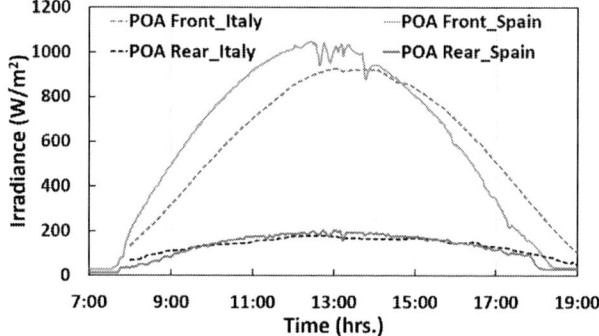

Fig. 2. POA irradiance (front and rear) for Italy and Spain.

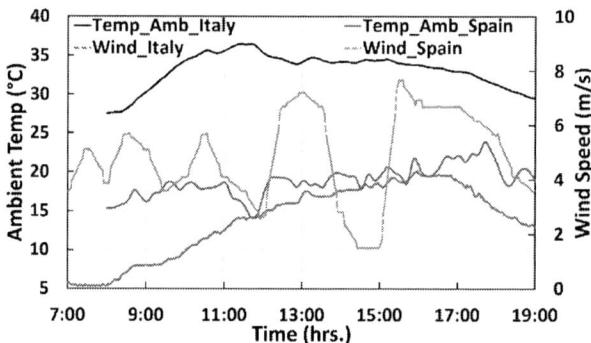

Fig. 3. Ambient temperature and wind speed for Italy and Spain.

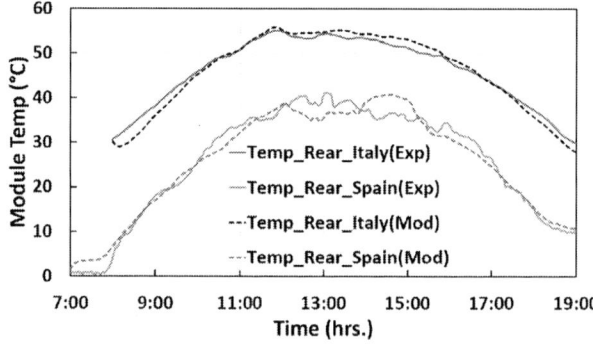

Fig. 4. Experimental vs. modeled back glass temperature for Italy and Spain.

TABLE II. STATISTICAL ERRORS IN PREDICTING PV TEMPERATURE

	Italy	Spain
RMSE (°C)	1.432	2.275
MBE (°C)	−0.141	0.002
r	0.996	0.984

C. Assumptions

Following assumptions have been made for the simulations performed in this work:

- In each bifacial panel, all the PV cells are exactly identical in every aspect.

- The front and rear irradiance are perfectly homogeneous.

- There is no dust deposition or shading, both at front and rear glass.

- The panels loose heat to surroundings only from front and the rear sides.

- Temperature of the PV panel varies only through the thickness and not laterally.

- The wind always flows along the greater length of the PV panel [10].

D. Discussion

The statistical errors in TABLE II show that model's prediction accuracy is already in a well acceptable range; still, the slight deviations in these results could be explained due to the following factors:

- Due to the unavailability of the wind speed data in [8] for Barcelona, the data shown in Fig. 3 (for Spain only) was taken from another nearby location in Barcelona, from an open field area site, measured at a height of 10 meters.

- Error in the experimental data during digitization from the figures in [5] and [8].

- The temperature sensors attached at the back glass might not be in the exact center, while the 1-D modeling approach assumes it to be in the exact center.

- The POA distribution at the rear side might not be exactly homogenous (while the front POA generally is).

- Some of the bifacial PV panels' characteristics were assumed or inferred, as they were not directly given in [5] and [8].

All these factors can affect the accuracy of the model. To verify this, our future work consists of deploying our own bifacial setup, as to minimize any errors in the model due to aforementioned reasons. Moreover, the model should be (and is already under development) extended to at least 2-D, as to cater for the normally inhomogeneous POA distribution at the rear side of the bifacial panel.

IV. CONCLUSIONS

A dynamic thermal model was developed to predict the PV cells temperature variation under field operation. The developed model is validated using experimental data available in the literature from two different locations, one in Barcelona (Spain) and the other in Catania (Italy). These validations show that the developed model can be used for any location in the world to predict the working temperature of fielded bifacial PV panels. The statistical errors show that the maximum RMSE is 2.28 °C for Spain (possibility due to using wind speed data from another nearby site location), which still is a pretty good estimate from modeling point of view. Similarly, the correlation coefficient of more than 0.98 in both cases show that the model is highly efficient in following the experimental trends.

REFERENCES

[1] T. Jutta, "International Technology Roadmap for Photovoltaic (ITRPV) - 2020," Leipzig, Apr. 2021. Accessed: Dec. 25, 2021. [Online]. Available: itrpv.vdma.org

[2] M. T. Patel, R. A. Vijayan, R. Asadpour, M. Varadharajaperumal, M. R. Khan, and M. A. Alam, "Temperature-dependent energy gain of bifacial PV farms: A global perspective," Applied Energy, vol. 276, p. 115405, Oct. 2020, doi: 10.1016/j.apenergy.2020.115405.

[3] S. Armstrong and W. G. Hurley, "A thermal model for photovoltaic panels under varying atmospheric conditions," Applied Thermal Engineering, vol. 30, no. 11–12, pp. 1488–1495, 2010, doi: 10.1016/j.applthermaleng.2010.03.012.

[4] M. Mattei, G. Notton, C. Cristofari, M. Muselli, and P. Poggi, "Calculation of the polycrystalline PV module temperature using a simple method of energy balance," Renewable energy, vol. 31, no. 4, pp. 553–567, 2006, doi: 10.1016/j.renene.2005.03.010.

[5] G. M. Tina, F. B. Scavo, and A. Gagliano, "Multilayer Thermal Model for Evaluating the Performances of Monofacial and Bifacial Photovoltaic Modules," IEEE Journal of Photovoltaics, vol. 10, no. 4, pp. 1035–1043, Jul. 2020, doi: 10.1109/JPHOTOV.2020.2982117.

[6] J. C. Sánchez Barroso, N. Barth, J. P. M. Correia, S. Ahzi, and M. A. Khaleel, "A computational analysis of coupled thermal and electrical behavior of PV panels," Solar Energy Materials and Solar Cells, vol. 148, pp. 73–86, 2016, doi: 10.1016/j.solmat.2015.09.004.

[7] M. J. Carvalho, P. Horta, J. F. Mendes, M. C. Pereira, and W. M. Carbajal, "Incidence Angle Modifiers: A General Approach for Energy Calculations," in Proceedings of ISES World Congress 2007 (Vol. I -- Vol. V): Solar Energy and Human Settlement, D. Y. Goswami and Y. Zhao, Eds. Berlin, Heidelberg: Springer Berlin Heidelberg, 2009, pp. 608–612. doi: 10.1007/978-3-540-75997-3_112.

[8] S. Bouchakour, D. Valencia-Caballero, A. Luna, E. Roman, E. A. K. Boudjelthia, and P. Rodríguez, "Modelling and Simulation of Bifacial PV Production Using Monofacial Electrical Models," Energies, vol. 14, no. 14, pp. 1–16, Jul. 2021, doi: 10.3390/en14144224.

[9] Cedar Lake Ventures, "Weather Spark," 2021. https://weatherspark.com/h/d/47213/2020/3/7/Historical-Weather-on-Saturday-March-7-2020-in-Barcelona-Spain#Figures-WindSpeed (accessed Dec. 27, 2021).

[10] S. P. Aly et al., "Mitigating the effect of heat and dust to enhance solar panels efficiency," in Proceedings International Renewable and Sustainable Energy Conference (IRSEC), 2016, pp. 835–841. doi: 10.1109/IRSEC.2016.7983870.

Planar Transparent Conductive Oxide/Ag Rear Contacts for High Efficiency III-V Photovoltaics

Christopher T. Gregory, Sean J. Babcock, Richard R. King

Arizona State University, Tempe, AZ, United States

Photon recycling in photovoltaic devices can be attained by using highly reflective back surfaces. Some of the highest reflectance surfaces are composed of a plane of dielectric material deposited on a highly reflective metal such as Ag or Au. Although optically effective, the use of a planar dielectric layer complicates electrical contact to the device, leading to approaches such as point contacts. A simple solution-that may not result in significant optical or resistive losses-is to use a planar transparent conductive oxide (TCO) layer instead of a dielectric layer. This work investigates the viability of a such a contact. It is observed that contact resistivities of the TCO/Ag stack on a highly doped AlGaAs or GaAs contact layer are below 0.1 Ω cm2 for TCO doping concentrations on the order of 1019 cm-3. The contact resistivity can be reduced further by increasing the doping in the semiconductor layer. Internal hemispheric reflectances of the proposed contacts are expected to reach up to 98% at the wavelength of interest, facilitating photon recycling. The performance of this contact structure suggests use in technologies such as photonic power converters and thermophotovoltaics.

Quantitative Measurement of Active Dopant Density Distribution in Black Silicon Solar Cell Using Scanning Nonlinear Dielectric Microscopy

Yasuo Cho[1], Beniamino Iandolo[2] and Ole Hansen[2]

[1]Research Institute of Electrical Communication, Tohoku University, Sendai, 980-8577, Japan
[2]DTU Nanolab, Technical University of Denmark, Lyngby, 2800, Denmark

Abstract— We investigated quantitatively the carrier distribution in a phosphorous (P) diffused black Silicon (Si) solar cell using scanning nonlinear dielectric microscopy (SNDM). As a reference, we measured the carrier distribution on a flat Si sample fabricated under the same P diffusion conditions. The precise carrier distributions in the emitter were visualized, which revealed the feature of carrier distribution in the emitter of black Si solar cell. Super-higher-order-SNDM was also employed to perform a quantitative analysis of the depletion layer distribution. It was found that the carrier density profile and the depletion layer thickness is less regular in the black Si than in the flat emitter, suggesting that this fluctuation may affect the power conversion efficiency of black Si solar cell.

Keywords—black silicon solar cell, carrier profiling, depletion layer, scanning nonlinear dielectric microscopy, super-higher-order scanning nonlinear dielectric microscopy

I. INTRODUCTION

Black silicon (Si) is a nanostructured type of Si capable of near-total absorption of light. Therefore, the use of black Si can lead to highly efficient solar cells [1]. The performance of Si solar cells is dependent on the active dopant carrier distribution in emitters. The carrier distribution in black Si solar cells has been difficult to measure, and thus to optimize, due to the irregular, nano-scaled structure of the emitter.

On the other hand, one of the authors previously succeeded in quantitatively analyzing such distributions in monocrystalline Si solar cells using scanning nonlinear dielectric microscopy (SNDM) [2]-[5].

In this paper, we investigated quantitatively the carrier distribution in a phosphorous (P) diffused black Si solar cell using SNDM. As a reference, we also measured the carrier distribution on a flat Si sample fabricated under the same P diffusion conditions. The precise carrier distributions in the emitter were visualized, which revealed the feature of carrier distribution in the emitter of black Si solar cell.

In addition, an enhanced version of this technique, referred to as the super-higher-order (SHO)-SNDM [6][7], was employed to visualize depletion layer distributions in both black and flat Si solar cells.

II. EXPERIMENTAL SETUP AND SAMPLES

Figure1 shows schematic diagram of SNDM. In SNDM, the sensor (SNDM probe) comprises a high-frequency inductance/capacitance (LC) self-oscillator that responds to changes in the capacitance $\Delta C_{ts}(t)$ between a sample and a microscopy tip that is itself conductive, relative to a static capacitance (C_{ts}) [8]. The capacitance between the sample and the tip represents a parallel capacitance relative to the LC tank circuit. Consequently, the oscillation frequency $\Delta f(t)$ varies from the initial value (f_0) as a consequence of $\Delta C_{ts}(t)$, and frequency modulation (FM) can be achieved via the application of a sinusoidal voltage to the sample. An FM demodulator subsequently demodulates the $\Delta f(t)$ signal, and data that are proportional to the change in capacitance with respect to voltage (that is, a dC/dV signal) are obtained via a lock-in amplifier.

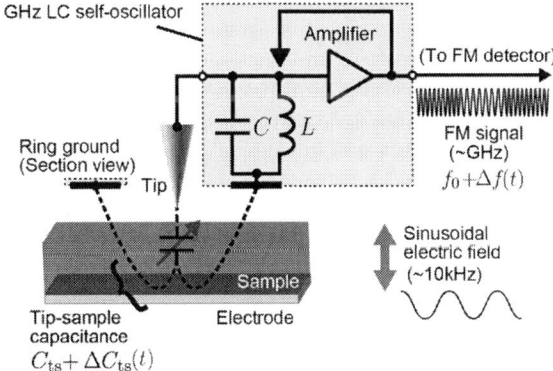

Fig. 1. Schematic diagram of SNDM.

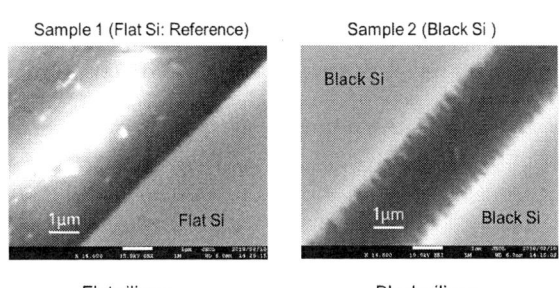

Fig. 2. SEM micrograph of the samples.

Figure 2 shows SEM micrographs of samples which were fabricated under the same P diffusion conditions. Each cell was cross-sectioned, after which the cross section was polished chemomechanically and assessed using SNDM.

III. EXPERIMENTAL RESULTS

The resulting images are provided in Fig. 3. To permit a quantitative comparison of the carrier concentrations in samples, both p-type and n-type standard Si staircase samples were prepared for the calibration of the carrier densities [2]. These samples had seven and six layers, respectively, each with a different carrier density. The carrier density in each layer was determined using secondary ion mass spectrometry (SIMS), and these values ranged from 10^{16} to 10^{20} cm^{-3}. These standard

samples were polished at the same time as the Si solar cell test samples and so it was assumed that the P (phosphorous)-diffused Si solar cells and the standard samples had the same surface characteristics. The standard samples were assessed after the SNDM analyses of the P-diffused Si solar cells.

Using the data obtained from the standard samples, we calibrated the SNDM signal to obtain the quantitative values for the carrier densities.

Figure 3 shows the two-dimensional SNDM images (that is, the raw data), their topographies and one-dimensional carrier density profiles along the white lines in the SNDM images for both flat Si and black Si solar cells. The magnitude of the SNDM signal changes with the carrier density and the electron density

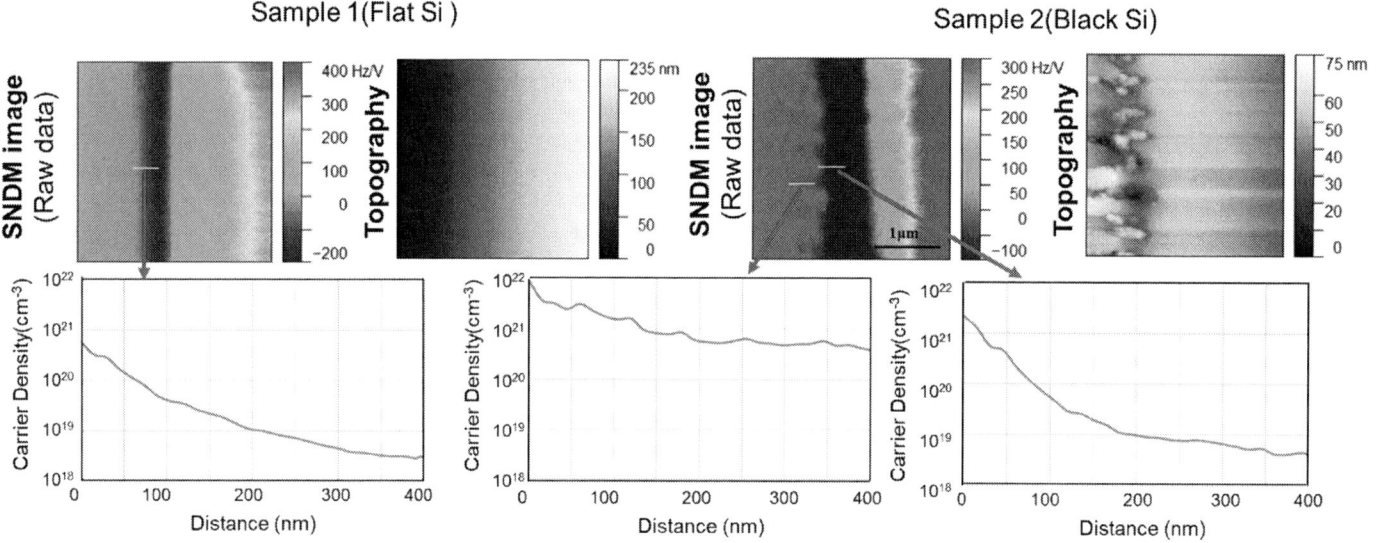

Fig. 3. SNDM Signals and Line Profiles of Carrier Density.

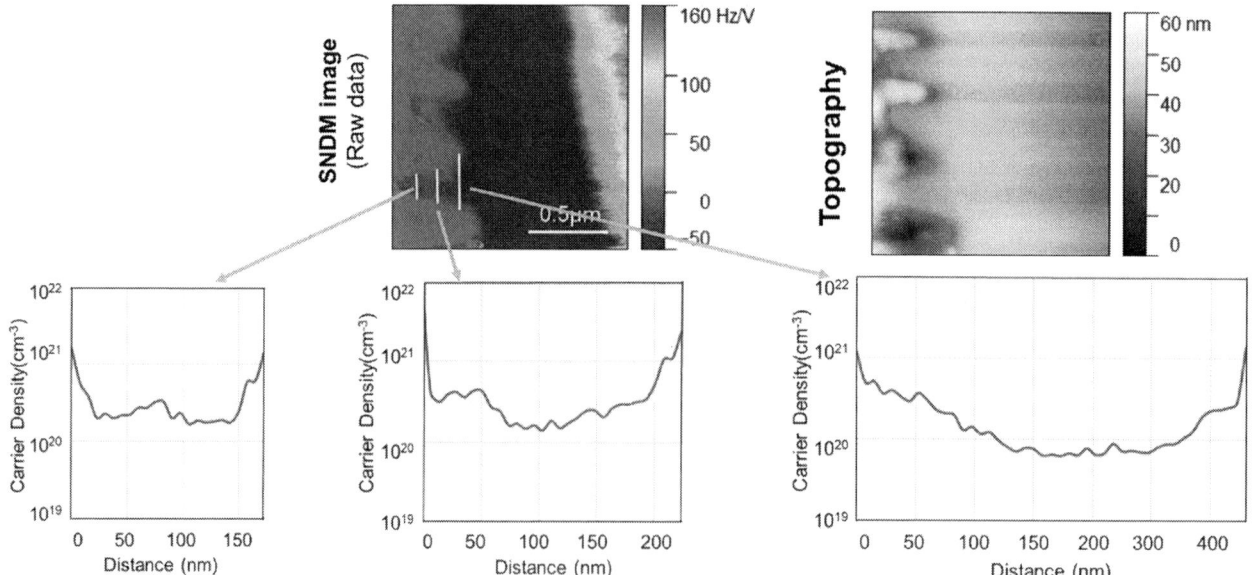

Fig. 4. Cross-sectional carrier distribution profiles in black Si cell.

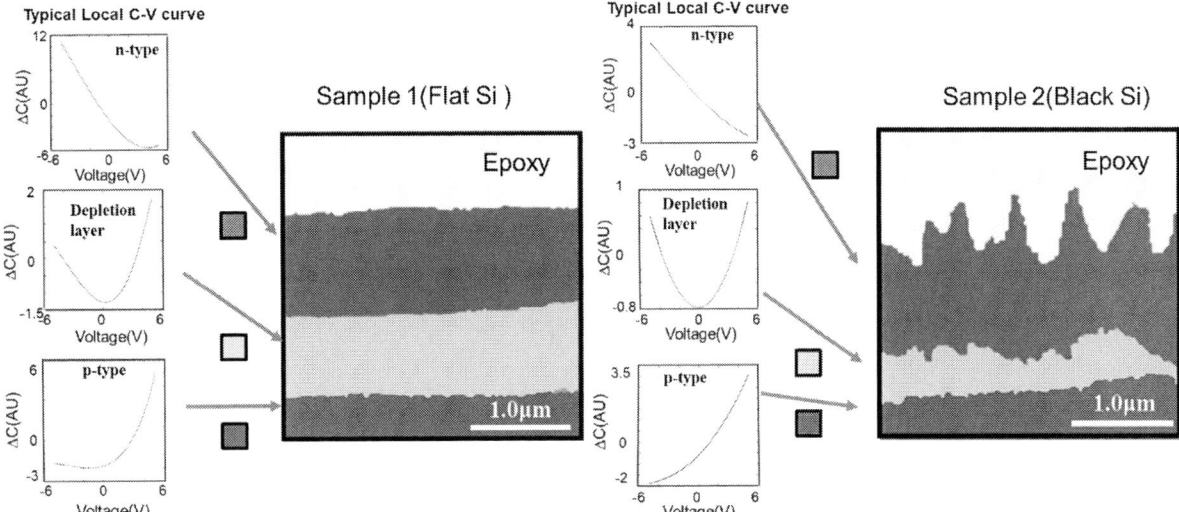

Fig. 5. Depletion layer visualizations using super-higher-order SNDM.

decreases exponentially on going from the surface to the bulk. The results clearly demonstrate that carrier distribution in concave area in black Si solar cell is similar to that in flat Si solar cell and the carrier density in convex area (in the needle) is higher than that in concave area and flat Si.

Figure 4 shows SNDM image (RAW data), topography and its cross-sectional quantitative carrier distribution profiles in black Si solar cell. Cross-sectional quantitative carrier distribution profiles have been successfully obtained. This results indicate that these lateral inhomogeneous carrier distributions may influence the power conversion efficiency of black Si solar cell.

A new method related to SNDM has recently been developed that returns both a dC/dV signal and higher-order differential values. This technique provides data regarding various materials that are significantly more precise and is referred to as SHO-SNDM [6]. ΔC-V plots at individual pixels can be readily reconstructed using SHO-SNDM, such that the analytical power of the method is greatly increased. SHO-SNDM employs an equivalent experimental apparatus to SNDM other than the incorporation of a multi-channel lock-in amplifier. During analyses using this technique, higher harmonic components are detected. This is in contrast to the standard SNDM, which employs solely the first-order dC/dV harmonic component. Using this method, we also examined the distribution of the depletion layer thickness in the cross-section of a flat Si solar cell and black Si solar cell In this SHO-SNDM experiment, harmonic components of capacitance variation up to the 6th were detected and ΔC-V curves were reconstructed from the SHO-SNDM data. Then, we categorized each pixel as either n-type, depletion layer, or p-type, judging from the shape of ΔC-V curves (i.e. monotonically decrease curve: n-type, V-shape curve: depletion layer, monotonically increase curve: p-type) [7]. The results are shown in Fig. 5 with typical ΔC-V curves. The green areas in the images indicate the regions of depletion layer. It was found that fluctuation of depletion layer thickness of black

Si solar cell is larger than that of flat Si solar cell. This fluctuation also may influence the power conversion efficiency of black Si solar cell.

IV. CONLUSION

The precise carrier distributions in the emitter were visualized, which revealed the feature of carrier distribution in the emitter of black Si solar cell. Super-higher-order-SNDM was also employed to perform an analysis of the depletion layer distribution. It was found that the carrier density profile and the depletion layer thickness is less regular in the black Si than in the flat Si, suggesting that this fluctuation may affect the power conversion efficiency of black Si solar cell.

REFERENCES

[1] H. Savin, P. Repo, G. v. Gastrow, P. Ortega, E. Calle, M. Garín, and R. Alcubilla, "Black silicon solar cells with interdigitated back-contacts achieve 22.1% efficiency," Nature Nanotechnology, vol.10, 2015, pp. 624–629.

[2] K. Hirose, K. Tanahashi, H. Takato, and Y. Cho, "Quantitative measurement of active dopant density distribution in phosphorus-implanted monocrystalline silicon solar cell using scanning nonlinear dielectric microscopy," Appl. Phys. Lett., vol. 111, 2017, 032101.

[3] Y. Cho, S. Jonai, and A. Masuda, "A scanning nonlinear dielectric microscopic investigation of potential-induced degradation in monocrystalline silicon solar cell," Appl. Phys. Lett., vol.116, 2020, 182107.

[4] Y. Cho, A. Kirihara, and T. Saeki, "Scanning nonlinear dielectric microscope," Rev. Sci. Instrum., vol.67, 1996, pp. 2297-2303.

[5] Y. Cho, "Scanning Nonlinear Dielectric Microscopy: Investigation of Ferroelectric, Dielectric, and Semiconductor Materials and Devices," Elsevier, 2020; ISBN 978-0-08-102803-2.

[6] N. Chinone, T. Nakamura, and Y. Cho, "Cross-sectional dopant profiling and depletion layer visualization of SiC power double diffused metal-oxide-semiconductor field effect transistor using super-higher-order nonlinear dielectric microscopy," J. Appl. Phys., vol. 116, 2014, 084509.

[7] H. Edwards, R. McGlothlin, R. San Martin, U. Elisa, M. Gribelyuk, R. Mahaffy, C. Ken Shih, R. Scott List, and V. A. Ukraintsev, " Scanning capacitance spectroscopy: An analytical technique for pn-junction delineation in Si devices," Appl. Phys. Lett., vo;.72, 1998, pp.698-700.

[8] Y. Cho, S. Atsumi, and K. Nakamura, "Scanning Nonlinear Dielectric Microscope Using a Lumped Constant Resonator Probe and Its Application to Investigation of Ferroelectric Polarization Distributions", Jpn. J. Appl. Phys., vol. 36, 1997, pp.3152-315.

Flexible and Lightweight CdS/CdTe Solar Cells via a Water-Assisted Lift-Off Process

Sandip S Bista, Deng-Bing Li, Suman Rijal, Sabin Neupane, Rasha A Awni, Randy J. Ellingson, Zhaoning Song, Adam Phillips, Michael Heben, and Yanfa Yan

Department of Physics and Astronomy, and the Wright Center for photovoltaics Innovation and Commercialization, The University of Toledo, Toledo, OH, 43606 USA

Abstract— In this work, CdS/CdTe thin-film solar cells were grown on a mica sheet and delaminated through a water assisted lift-off process and transfer to the substrate of interest. An CdCl₂ vapor treatment was applied for the CdS buffer layer and CdTe absorber layer, individually. We observed that two-time CdCl₂ treated films (one for CdS and second for CdTe) can help enhance the recrystallization, yielding larger grain sizes for both CdS buffer layer and CdTe absorber layer compared to one-time CdCl₂ treatment after CdTe deposition. Finally, we succeeded to fabricate ultrathin, lightweight, and flexible polycrystalline CdTe solar cells with a power conversion efficiency of 12.6 % with a high specific power (W/Kg).

Keywords— flexible solar cell, cadmium telluride (CdTe), cadmium chloride (CdCl₂).

I. INTRODUCTION

Efficient CdTe solar cells are generally fabricated on rigid glass substrates, where the actual functional layers occupy less than 1% of the total weight. Therefore, reducing the glass thickness or using lightweight, flexible substrates can make devices more cost-efficient with wide potential applications in curve surface integration, mobile electronics, and space applications. However, the state-of-the-art flexible solar cell is far below its rigid counterparts. Delaminating devices from the rigid substrate become a promising method to fabricate flexible, lightweight, and efficient polycrystalline CdTe solar cells.

Currently, various techniques have been developed to delaminate the CdTe film, e.g., adding a sacrificial layer[1, 2], mechanical lift-off[3, 4], thermomechanical stress[5-7], and water-assisted lift-off[8]. Recently, Wen and coworkers reported the water-assisted lift-off technique to separate epitaxial CdTe film from a mica ($K_2O.Al_2O_3.SiO_2$) substrate, demonstrating flexible CdTe solar cells with a power conversion efficiency (PCE) of 9.59 %[8]. Mica is a layered flexible material and can be easily cleaved from a bulk material, and it is stable up to 700 ºC.

In this work, the CdS and CdTe film are deposited sequentially on mica. Due to the moisture sensitivity characteristic of mica, a CdCl₂ vapor treatment is introduced for the CdTe film. We find that an additional CdCl₂ vapor treatment for the CdS film before CdTe deposition is helpful for improving the morphology of CdS and the afterward deposited CdTe absorber, resulting in a PCE improvement from 11.11% to 12.6%.

II. EXPERIMENTAL DETAILS

For CdTe device fabrication, a 100 nm CdS film was first thermally evaporated on freshly cleaved mica sheets at room temperature under 1×10^{-7} Torr with a deposition rate of 2.0 nm/s. The deposition rate and the thickness were monitored by a quartz crystal. Prior to the CdTe deposition, a CdCl₂ vapor treatment was carried out for the mica/CdS stack in closed space sublimation system at 400 ºC for 5 mins under 400 Torr in a nitrogen environment. The CdTe film was then deposited by the closed space sublimation (CSS) method with the source and substrate temperatures of 580 ºC and 520 ºC, respectively, under 1 Torr pressure in a 1% oxygen and 99% argon environment. Afterward, CdCl₂ activation was carried out in a closed space sublimation system at a 380 ºC source and a 390 ºC substrate temperature for 30 mins under 400 Torr in dry air. Subsequently, a 3.5 nm copper (Cu) and 50 nm gold (Au) bilayer was thermally evaporated and annealed at 200 ºC for 20 minutes in a nitrogen ambient to facilitate the Cu diffusion.

After Cu diffusion, a flexible substrate aluminum foil was attached on the film stack using epoxy (Loctite EA 0151) as a binder. After drying naturally, the film stack was immersed into DI water to facilitate the delamination process. Within a few minutes, the mica sheet and CdS/CdTe/Cu/Au/epoxy/aluminum foil stack separated. We also transformed the film stack onto SU-8 (3050) photoresist layer. It was observed that the film stack transformed onto SU-8 photoresist layer delaminated much faster, which is less than a minute. The separation is mainly attributed to the weak Van der Waal force between the CdS film and mica, which is weaker than the surface tension exerted by water. Finally, a transparent conduction oxide (IZO) layer was deposited as a front contact to complete the device.

The morphology of film was characterized by high resolution field emission scanning electron microscopy (FE-SEM, Hitachi S-4800). The XRD patterns were measured using a Rigaku Ultima III X-ray diffractometer with a Cu source with a rated tube voltage of 20-60 KV operating at 40 KV and 44 mA. The device PCE was characterized by measuring current density-voltage(J-V) curves under an AM1.5G illumination using a solar simulator and a source meter (keithley-2400). The external quantum efficiency (EQE) spectra were measured using a EQE system from PV Instrument. Both the J-V and EQE systems are calibrated with a standard Si solar cell.

III. RESULTS AND DISCUSSIONS

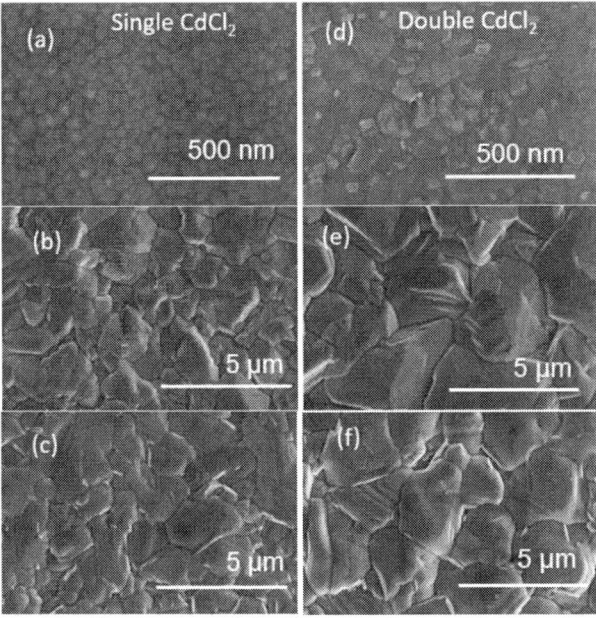

Figure 1 (a), (b), and (c), SEM images of air annealed CdS, as deposited CdTe, and CdCl₂ treated CdTe film, respectively. (d), (e), and (f) are SEM images of CdCl₂ treated CdS, as deposited CdTe, and CdCl₂ treated CdTe film, respectively.

To investigate the effect of the CdCl₂ treatment for the CdS film, a device with CdS annealed at 400 °C for 5 min in air without CdCl₂ treatment is used as a reference. As shown in Figure 1(a-f), the CdS film annealed without CdCl₂ has grains between 30-45 nm (Figure 1a). A significant grain growth to 75-150 nm with polyhedron shape was observed in the CdS film annealed with CdCl₂ (Figure 1d). Furthermore, the CdTe film deposited on CdCl₂ treated CdS film has significantly larger grains (Figure 1e) compared to the CdTe films deposited on air annealed CdS (Figure 1b). After the CdCl₂ treatment for the CdTe film, the surface of both films appears smooth (Figure 1c and 1f). This result implies that the CdCl₂ treatment for the CdS film can promote CdS recrystallization and grain growth of CdS, which further improves the quality of afterward deposited CdTe films and result in larger CdTe grain sizes.

Figures 2a and 2b are top-view SEM images of the surface of delaminated CdS/CdTe films deposited on CdS films without and with the CdCl₂ treatment, respectively. A lower density of pinholes and significantly improved CdS grain size can be clearly observed in the film with CdCl₂ treated CdS. The front (CdS and CdTe interface) and back surfaces CdTe were further investigated through XRD measurement for the

CdS/CdTe films with air annealed and CdCl₂ treated CdS. The front interface of the air annealed film shows strong (002), and (101) crystal orientation for CdS with hexagonal structure, while under CdCl₂ treatment of both CdS and CdTe, the CdS has intense (002), (101), (004), and (202) peaks. The XRD pattern for the back surface for both single and double CdCl₂ film remains the same. At the front surface, the film with CdCl₂ treated CdS and CdTe is more randomly oriented when compared with the film with air annealed CdS. This is unlike the literature report that vacuum evaporated CdS on mica substrate are epitaxial[9].

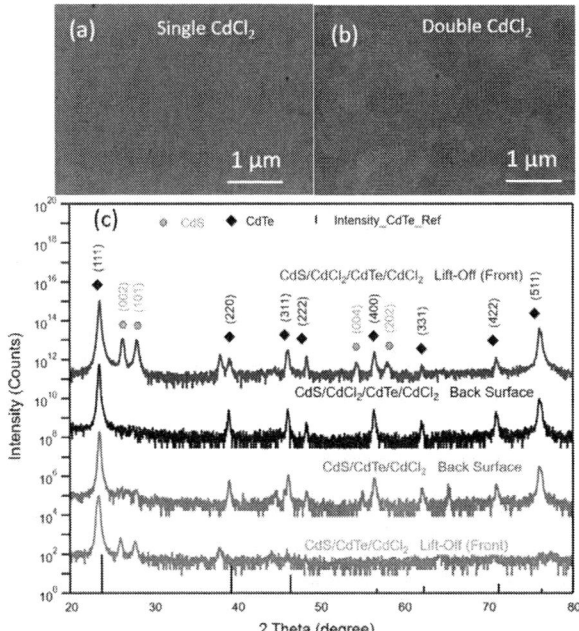

Figure 2 (a) and (b) top-view SEM images of the front surface of delaminated CdS/CdTe films with single and double CdCl₂ treatment, respectively, (c) XRD patterns of back surface and delaminated front interface under single and double CdCl₂ treatment.

The schematic diagram of the film stacks before delamination is shown in Figure 3a. After delamination, the device is finished by depositing a IZO film at the delaminated surface (figure 3b and 3c). Figure 3(d) shows the measured J-V curves of the best performing cells for the devices with and without the first CdCl₂ treatment (for the CdS film). As seen, the flexible solar cell with CdCl₂ treated CdS delivers a maximum PCE of 12.60% with an open-circuited voltage (V_{OC}) of 829 mV, a short-circuited current density (J_{SC}) of 23.64 mA/cm², and a fill factor (FF) of 64.30%. However, without the CdCl₂ treatment for the CdS film, the device has a lower PCE of 11.11 % with a V_{OC} of 788 mV, a J_{SC} of 23.36 mA/cm², and an FF of 60.34%. The improvements in V_{OC} and FF by double CdCl₂ treatments can be attributed to the improved crystallinity, larger grain size, and passivated interface.

Figure 3(e) shows the measured external quantum efficiency (EQE). At the shorter wavelength region, the double CdCl₂ treated device has slightly better spectral response. This might be due to better front interface or thickness variation of mica substrates. Overall, both devices show a similar spectral response at the longer wavelength region. The EQE shows significant spectral losses at the shorter wavelength region, mainly due to the parasitic absorption lost caused by thick CdS window layer.

Further, we studied the specific power of our flexible solar cells. The specific power was calculated using the maximum power generation of the solar cell and its areal density. The device transformed onto SU-8 photoresist layer (50 μm) recorded significantly high specific power (1510 W/Kg). Details of the calculation will be included in the final manuscript.

However, the performance of flexible device is slightly lower than that of the device fabricated on rigid glass substrate. This is mainly due to chloride and oxide residue at the back surface of CdTe. These residues cannot be removed through methanol rinsing or surface etching, a typical process generally done for the devices fabricated on the rigid substrate. The presence of these residues at the back surface is detrimental to the formation of quality back contact between CdTe and Au electrode and also affect Cu diffusion, which is detrimental and leading to high R_S and low R_{SH}. Further work is currently undertaken to improve the front and back interfaces.

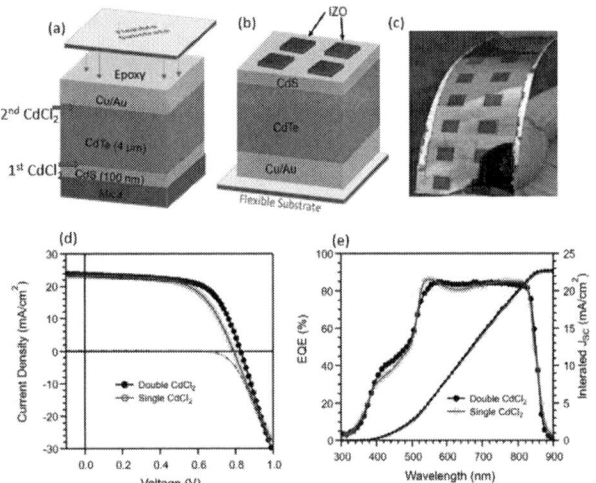

Figure 3 (a) schematic diagram of device stacks before delamination, (b) after delamination, (c) photo image of flexible CdTe solar cell, (e) J-V curve, and (d) EQE curve.

IV. CONCLUSION

We used a water-assisted lift-off technique to delaminate CdS-CdTe film from mica sheet and fabricated ultrathin, lightweight, flexible CdTe solar cells. The lift-off films are smooth and mirror like. We introduced an CdCl₂ vapor treatment for the CdS film, which has been demonstrated beneficial for improving grain sizes of both the CdS and CdTe films. Finally, an ultrathin, lightweight, and flexible polycrystalline CdTe solar cells with a PCE of 12.6% is fabricated.

ACKNOWLEDGMENT

This work is based on research sponsored by Air Force Research Laboratory under agreement number FA9453-19-1002 and FA9453-21-C0056. The U.S. Government is authorized to reproduce and distribute reprints for Governmental purposes notwithstanding any copyright notation thereon.

REFERENCES

1. Seredyński, B., et al., *(Cd,Zn,Mg)Te-based microcavity on MgTe sacrificial buffer: Growth, lift-off, and transmission studies of polaritons.* Physical Review Materials, 2018. **2**(4): p. 043406.
2. Ding, J., et al., *Epitaxial lift-off CdTe/MgCdTe double heterostructures for thin-film and flexible solar cells applications.* Applied Physics Letters, 2021. **118**(18): p. 181101.
3. Pookpanratana, S., et al. *Chemical structure of buried interfaces in CdTe thin film solar cells.* in *2010 35th IEEE Photovoltaic Specialists Conference.* 2010.
4. Albin, D., Y. Yan, and M. Al‐Jassim, *The effect of oxygen on interface microstructure evolution in CdS/CdTe solar cells.* Progress in Photovoltaics: Research and Applications, 2002. **10**(5): p. 309-322.
5. Perkins, C.L., et al., *SnO2-Catalyzed Oxidation in High-Efficiency CdTe Solar Cells.* ACS Applied Materials & Interfaces, 2019. **11**(13): p. 13003-13010.
6. Perkins, C.L., et al., *Two-Dimensional Cadmium Chloride Nanosheets in Cadmium Telluride Solar Cells.* ACS Applied Materials & Interfaces, 2017. **9**(24): p. 20561-20565.
7. McGott, D.L., et al., *Thermomechanical Lift-Off and Recontacting of CdTe Solar Cells.* ACS Appl Mater Interfaces, 2018. **10**(51): p. 44854-44861.
8. Wen, X., et al., *Epitaxial CdTe Thin Films on Mica by Vapor Transport Deposition for Flexible Solar Cells.* ACS Applied Energy Materials, 2020. **3**(5): p. 4589-4599.
9. Yang, Y.-B., et al., *Surface and interface of epitaxial CdTe film on CdS buffered van der Waals mica substrate.* Applied Surface Science, 2017. **413**: p. 219-232.

Public road tests of Toyota Prius equipped with high efficiency PV module with output power of 860 W

Taizo Masuda, Takashi Nakado, Masafumi Yamaguchi, Tatsuya Takamoto, kensuke Nishioka, kazumi Yamada

Toyota Motor Corporation, Susono, Japan

The University of Electro-Communications, Chofu, Japan

Toyota Motor Corporation, Toyota, Japan

Toyota Technological Institute, Nagoya, Japan

Sharp corporation, Yamato-koriyama, Japan

University of Miyazaki, Miyazaki, Japan

We created a photovoltaic (PV) powered plug-in hybrid electric vehicle (PHEV) which is equipped with PV modules that have output power of 860 W. The integrated PV comprised of triple-junction III-V-based solar cells which conversion efficiency is approximately 34%. The public road tests were conducted to experimentally confirm the benefits of the vehicle integrated PV modules for one year. The measured results showed that the number of plug-in charging can be reduced greatly which increase the convenience of the battery-based vehicles by the integrated PV module. Even energy self-sufficiency was achieved for the neighborhood driving patter. In addition, the results suggest that the annual PV-powered driving range reaches 6211 km by the PV module. Since the average annual driving range of the passenger car in Japan is 10000 km, it was experimentally confirmed that 62% of CO_2 emission from passenger car can be reduced by installing the PV module.

Revealing Sub-Cell Degradation of Multi-Junction Solar Cells by Absolute Electroluminescence Imaging

Youyang Wang[1], Liying Li[2], Xiaobo Hu[1], Yun Jia[1], Guoen Weng[1], Xianjia Luo[1], Shaoqiang Chen[1,4], and Hidefumi Akiyama[3]

1) Department of Electronic Engineering, East China Normal University, Shanghai, China.
2) Department of Computer Science and Technology, East China Normal University, Shanghai, China.
3) Institute for Solid State Physics, The University of Tokyo, 5-1-5 Kashiwanoha, Kashiwa, Chiba, Japan.
4) Ministry of Education Nanophotonics & Advanced Instrument Engineering Research Center, East China Normal University, Shanghai, China.

Abstract—**Photovoltaic solar cells degrade or even fail during long-term storage, affecting their conversion efficiency and lifetime. In this work, an industry-standard InGaP/GaAs/InGaAs multi-junction solar cell (MJSC) was monitored in a non-working idle state for more than two years. The sub-cell degradation caused by prolonged storage and its origin have been revealed by using the absolute electroluminescence (EL) imaging technique. Several potential defects were found to gradually exacerbate over time, introducing more non-radiative recombination centers and reducing the sub-cell external radiative efficiency under low-injection conditions. Quantitative evaluation results show that long-term idle resulted in an over 0.2% efficiency degradation and an over 1% fill factor degradation for each sub-cell, which is mainly originated from the non-radiative recombination loss.**

Keywords—*absolute EL imaging, sub-cell degradation, multi-junction solar cells, III-V compound semiconductor*

I. INTRODUCTION

III-V compound multi-junction solar cells (MJSCs) that aid the optimal utilization of the broad solar spectrum have achieved the highest conversion efficiency among the photovoltaic material family [1], [2]. However, each sub-cell needs to be connected in series through epitaxy in a certain structural sequence, which causes a significant increase in manufacturing complexity [3], [4]. As a result, defects are more likely to be introduced during fabrication, bringing serious challenges to the efficiency and lifetime of MJSCs.

Generally, an industrial standard MJSC is expected to have a service life of over 25 years [5]. However, many factors can lead to premature degradation or even failure of MJSCs during the long-term deployment or storage. On the one hand, harsh environmental factors such as solar irradiation, high temperature, high humidity, wind, and dust may introduce more defects and reduce the conversion efficiency of MJSCs [5]. On the other hand, the sub-cell materials also undergo slow changes over time, such as impurity diffusion [6]. This may lead to a gradual deterioration of defects introduced during manufacturing and/or expose some potential defects that seem normal, affecting the overall performance of MJSCs [7]. Therefore, it is necessary to study the long-term degradation behavior of MJSCs. Current research efforts mainly focus on the effect of environmental factors on the MJSC performance degradation [8], [9], whereas few works report the long-term monitoring of sub-cell defects.

Absolute electroluminescence (EL) imaging is a practical technique to visualize defect distributions and evaluate individual sub-cell performance of MJSCs [10]–[12]. By calibrating the sensitivity of the charge-coupled device (CCD), sub-cell performance such as current density-voltage (J-V) relations, series resistance distributions, open-circuit voltage (V_{oc}) mappings, and energy losses can be quantitatively assessed without extra parametric adjustment or fitting [13], [14]. Therefore, absolute EL imaging is not only beneficial for monitoring the change of defect distribution over time, but also for theoretically predicting the sub-cell performance degradation due to the long period of storage time.

In this work, we tracked the sub-cell degradation of an industry-standard multi-junction solar cell (MJSC) with over two years of idle time. The MJSC sample was stored in a drying cabinet without exposure to the outdoors or in working conditions. By using the absolute EL imaging technique, we first compared the defect distributions of sub-cells at different time periods. Based on the measured EL images, we then evaluated the time-induced sub-cell degradation of the MJSC sample, including the EL efficiency, J-V characteristics, and energy loss mechanisms. We clearly revealed that several potential defects were exacerbated and introduced more non-radiative (NR) recombination centers during prolonged storage, leading to an over 0.2% efficiency degradation for each sub-cell.

II. EXPERIMENTS AND EVALUATIONS

A. Basic information of the MJSC sample

An industry-standard flexible InGaP/GaAs/InGaAs MJSC sample with a total size of 2 cm × 4 cm was used in this work, the schematic structure is illustrated in Fig. 1 (a). The MJSC sample was produced in 2019 AD without packaging. After conducting necessary measurements such as J-V, EQE, and absolute EL imaging in 2019, the sample has been stored separately in a drying cabinet for more than two years by

---

This work was supported by the National Natural Science Foundation of China under Grant 61874044; the National Key Research and Development Project of China under Grant 2019YFB1503402; the Fundamental Research Funds for the Central Universities; and ECNU Academic Innovation Promotion Program for Excellent Doctoral Students under Grant YBNLTS2020-043.

December 2021. The internal environment of the drying cabinet is kept at room temperature (295 K) with a humidity of 30%.

B. Absolute EL imaging

The absolute EL images of the MJSC sample were measured by using the absolute EL imaging system, which is presented in Fig. 1 (b). Various forward currents were injected into the cell in a dark box at room temperature. A Si CCD camera with a sensitivity up to 1100 nm was placed vertically above the cell. A 750 nm short-pass or 750 nm long-pass optical filter was inserted in the front of the camera to collect the EL images of InGaP top-cell or GaAs middle-cell separately. To obtain the sub-cell absolute EL emissions, the method of radiant-flux LED standards for calibration [11] was adopted to calibrate the captured EL images. Note that in this work, we only focus on the EL evaluation of InGaP top cell and GaAs middle cell, since the EL emissions of InGaAs bottom cell (~1270 nm) cannot be captured by the Si CCD camera.

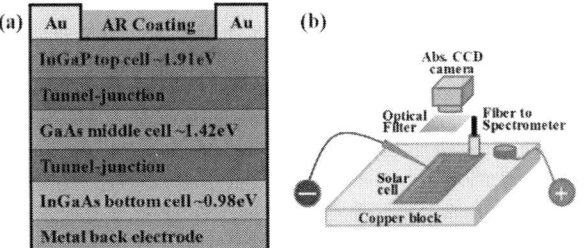

Fig. 1. 1(a) Schematic structure of the InGaP/GaAs/InGaAs MJSC sample. (b) Schematic of the absolute EL imaging system.

C. Solar cell evaluations

The carrier-balance equation can theoretically describe the sub-cell optoelectronic generation and loss processes under arbitrary current-injection conditions [10], which is given by:

$$\left(J_{sun\_i} + J\right)/q + R_{ext\_i-1\to i} = R_{ext\_i} + R_{NR\_i} + R_{ext\_i\to i+1} = \left(1/y_{ext\_i}\right)R_{ext\_i}, \quad (1)$$

where the subscript i = 1, 2, or 3 represents the top, middle, or bottom sub-cell, J_{sun_i} denotes the sub-cell photocurrent derived by EQE measurements, R_{ext_i} is the sub-cell absolute EL intensity, R_{NR_i} is the sub-cell NR recombination, and $R_{ext\_i\to i+1}$ describes the luminescent coupling between sub-cells.

By further introducing the reciprocity relation [15], we can establish the relationship between EL imaging and J-V characteristics:

$$V_i\left(J\right) = \frac{kT}{q}\ln\frac{\left(J_{sun\_i} + J + q\cdot R_{ext\_i-1\to i}\right)\cdot y_{ext\_i}\left(J\right)}{J_0} \quad (2)$$

where J_0 is the radiation rate per unit area in terms of the current density, k, T, and q denote the Boltzmann constant, Kelvin temperature, and the electron charge, respectively. Thus, essential performances such as V_{oc}, filling factor (FF), efficiency, as well as various energy loss mechanisms can be quantitatively predicted.

III. RESULTS AND DISCUSSTIONS

Fig. 2 shows the measured absolute EL images of the top and middle cells under the injection current of 1 mA/cm², in which the upper two images were measured in 2019 AD whereas the lower two were captured in 2021 AD. It is obvious that the overall EL intensity of the MJSC sample decreased after two years of storage. For the 2021-measured EL images, an additional finger interruption can be clearly observed, which is located at the right side of the middle worm-like defect and extended to the edge of the MJSC. This finger interruption blocks the out-diffusion of the injection current, causing more NR recombination in the upper right part of the MJSC. In addition, a point-shape dark area became gradually apparent during the idle process and was found in both sub-cells of the 2021-measured EL images. We warn that this defect may introduce a current sink in both sub-cells, resulting in a significant EL intensity degradation in the surrounding area, which is extremely detrimental to the cell performance.

Fig. 3 illustrates the sub-cell absolute EL images under the injection current of 12 mA/cm². On the one hand, more defects appear in the 2021-measured images, e.g., the worm-like defect at the upper left of the cell. On the other hand, we found that the overall EL intensity measured in different years shows no obvious difference, which can be attributed to the compensation of defect effects under high-injection current conditions. Therefore, the EL behaviors of the MJSC under high-injection conditions are independent of the idle time.

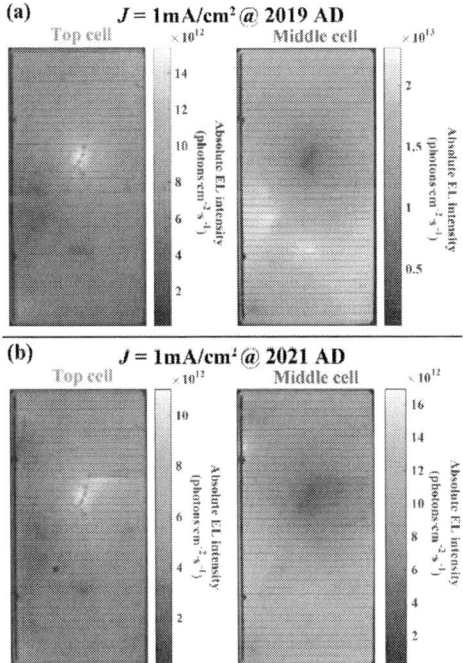

Fig. 2. Absolute EL images of the top and middle cells measured in (a) 2019 AD and (b) 2021 AD under the injection current of 1 mA/cm².

978-1-7281-6118-1/22 $31.00 © 2022 IEEE

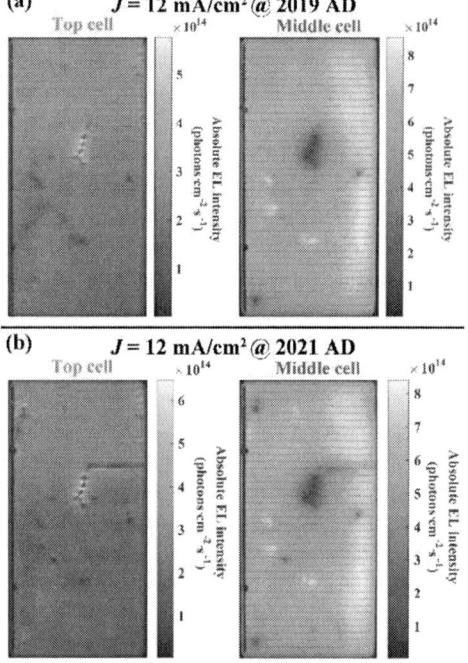

Fig. 3. Absolute EL images of the top and middle cells measured in (a) 2019 AD and (b) 2021 AD under the injection current of 12 mA/cm².

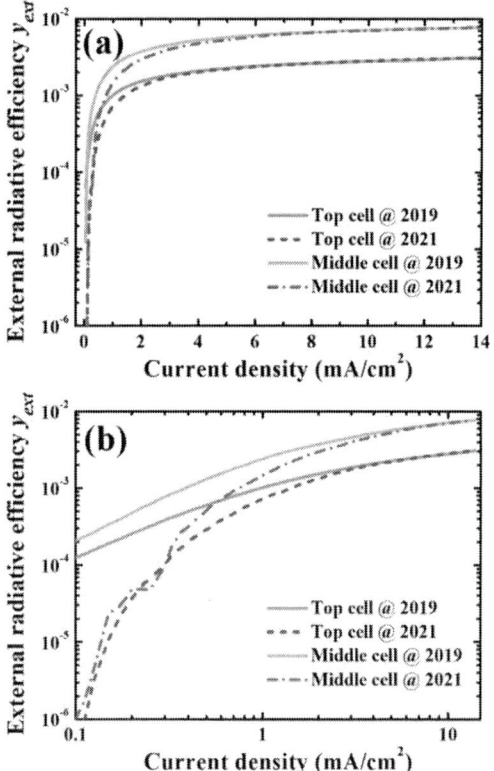

Fig. 4. Injection current density-dependent sub-cell external radiative efficiency (y_{ext}) measured in 2019 AD (solid curves) and 2021 AD (dashed/dash-dotted curves) in (a) linear-log scale and (b) log-log scale.

To evaluate the effect of idle time on the sub-cell EL efficiency of the MJSC sample, we derived the injection current-dependent sub-cell y_{ext} measured in different years, as plotted in Fig. 4. It can be seen that under high-injection conditions ($J > 5$ mA/cm²), the sub-cell y_{ext} measured in 2021 is almost identical to the 2019-measured one, which is consistent with the results obtained from the EL image information. However, under low-injection conditions, the 2021-measured y_{ext} dropped more severely for both sub-cells compared with the 2019-measured one. It can be seen from Fig. 4 (b) that after two years of storage, the sub-cell y_{ext} decreased from 10^{-4} to 10^{-6} under $J = 0.1$ mA/cm², indicating that the newly revealed or deepened defects introduced more NR recombination centers in the MJSC, and further affected the overall solar cell performance.

The solid curves in Fig. 5 show the 2019-measured sub-cell J-V curves evaluated from the absolute EL imaging, and the dotted curves plot the 2021-measured ones. It can be seen that the bending of both sub-cell J-V curves near the maximum power point becomes softer after two years of storage, which lowers the FF of the MJSC sample. The softer bending can be attributed to the significant degradation of sub-cell y_{ext} under low-injection current conditions. Namely, over storage time, more carriers tend to be trapped by defect-induced NR recombination centers, instead of being transported to electrodes for power generation under a fixed working condition. Table 1 summarizes the essential photovoltaic parameters derived from sub-cell J-V curves measured in different years. It is obvious that both sub-cells show an over 1% FF degradation and a 0.2% efficiency degradation, which is mainly caused by the deepening defects during the long-time storage. We also note that the sub-cell V_{oc} does not exhibit degradation behavior with the idle time, indicating that the effect of defects can be neglected under the open-circuit working condition.

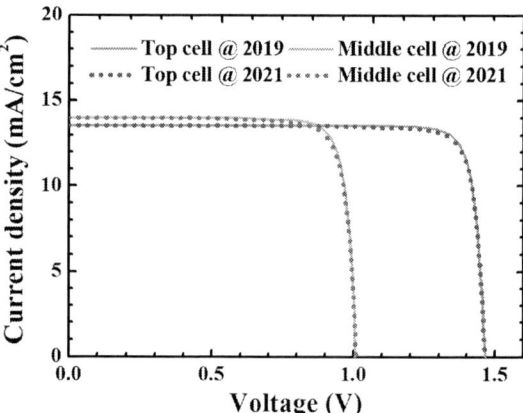

Fig. 5. EL-evaluated sub-cell J-V curves measured in 2019 AD (solid curves) and 2021 AD (dotted curves) under the illumination of AM1.5G 1-sun.

TABLE I. Essential photovoltaic parameters derived from sub-cell *J-V* curves measured in 2019 AD and 2021 AD under the condition of AM1.5G 1-sun.

Sub-cell	Year	V_{oc} (V)	J_{sc} (mA/cm^2)	FF (%)	Eff. (%)
Top	2019 AD	1.466	13.54	87.91	17.45
	2021 AD	1.465	13.54	86.84	17.23
	Deviation	0.001	-	1.07	0.22
Middle	2019 AD	1.008	13.98	86.88	11.85
	2021 AD	1.008	13.98	85.31	11.65
	Deviation	0	-	1.57	0.20

According to the carrier-balance equation, we further evaluate the contribution of sub-cell energy losses under the working condition of maximum-output-point with AM1.5G 1-sun illumination. It can be seen from Fig. 6 that NR loss increases significantly with longer idle time, which can be attributed to the defect-induced recombination centers. However, the other energy loss mechanisms do not show an obvious correlation with idle time. Therefore, NR loss is considered as the main reason for the MJSC efficiency degradation with idle time. The experimental results reveal that even if the MJSC is kept in a non-illuminated idle state, some potential defects may be exacerbated during prolonged storage, leading to significant degradation of the overall performance for MJSCs.

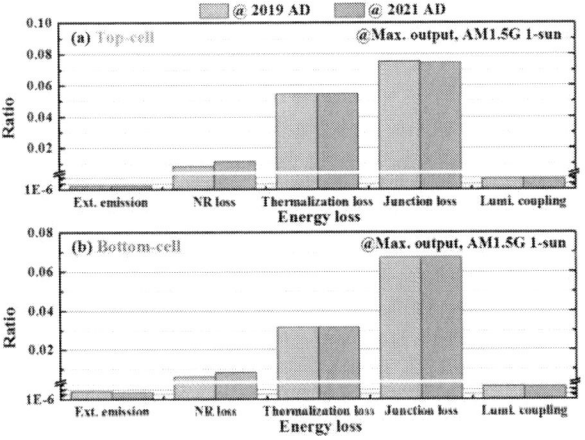

Fig. 6. Comparison of energy losses measured in 2019 AD (yellow columns) and 2021 AD (blue columns) for (a) top cell and (b) middle cell. The working condition is AM1.5G 1-sun at the maximum-output-point.

IV. CONCLUSION

In this work, we monitored an industry-standard InGaP/GaAs/InGaAs MJSC sample in a non-working idle state for more than two years. Absolute EL imaging measurements in 2019 AD and 2021 AD revealed the sub-cell degradation of the MJSC sample with idle time. We found that several potential defects were gradually exacerbated over time, reducing the sub-cell external radiative efficiency under low-injection conditions. Moreover, quantitative evaluations of photovoltaic properties

and energy loss mechanisms showed that long-term idle resulted in an over 0.2% efficiency degradation and an over 1% fill factor degradation for each sub-cell, which is found to be originated from the NR loss.

REFERENCES

[1] M. A. Green, E. D. Dunlop, J. Hohl-Ebinger, M. Yoshita, N. Kopidakis, and X. Hao, "Solar cell efficiency tables (version 59)," *Prog. Photovoltaics Res. Appl.*, vol. 30, no. 1, pp. 3–12, 2022, doi: 10.1002/pip.3506.

[2] J. F. Geisz *et al.*, "Building a Six-Junction Inverted Metamorphic Concentrator Solar Cell," *IEEE J. Photovoltaics*, vol. 8, no. 2, pp. 626–632, 2018, doi: 10.1109/JPHOTOV.2017.2778567.

[3] S. J. Polly, Z. S. Bittner, M. F. Bennett, R. P. Raffaelle, and S. M. Hubbard, "Development of a multi-source solar simulator for spatial uniformity and close spectral matching to AM0 and AM1.5," in *Conference Record of the IEEE Photovoltaic Specialists Conference*, 2011, pp. 001739–001743, doi: 10.1109/PVSC.2011.6186290.

[4] I. Massiot, A. Cattoni, and S. Collin, "Progress and prospects for ultrathin solar cells," *Nat. Energy*, vol. 5, no. 12, pp. 959–972, 2020, doi: 10.1038/s41560-020-00714-4.

[5] V. Sharma and S. S. Chandel, "Performance and degradation analysis for long term reliability of solar photovoltaic systems: A review," *Renew. Sustain. Energy Rev.*, vol. 27, pp. 753–767, 2013, doi: 10.1016/j.rser.2013.07.046.

[6] R. H. Van Leest *et al.*, "Effects of copper diffusion in gallium arsenide solar cells for space applications," *Sol. Energy Mater. Sol. Cells*, vol. 140, pp. 45–53, 2015, doi: 10.1016/j.solmat.2015.03.020.

[7] J. Hong *et al.*, "Absolute electroluminescence imaging with distributed circuit modeling: Excellent for solar-cell defect diagnosis," *Prog. Photovoltaics Res. Appl.*, vol. 28, no. 4, pp. 295–306, 2020, doi: 10.1002/pip.3236.

[8] E. L. Meyer and E. E. Van Dyk, "Assessing the reliability and degradation of photovoltaic module performance parameters," *IEEE Trans. Reliab.*, vol. 53, no. 1, pp. 83–92, 2004, doi: 10.1109/TR.2004.824831.

[9] L. Zhu *et al.*, "Characterizations of radiation damage in multijunction solar cells focused on subcell internal luminescence quantum yields via absolute electroluminescence measurements," *IEEE J. Photovoltaics*, vol. 6, no. 3, pp. 777–782, 2016, doi: 10.1109/JPHOTOV.2016.2540247.

[10] S. Chen *et al.*, "Thorough subcells diagnosis in a multi-junction solar cell via absolute electroluminescence-efficiency measurements," *Sci. Rep.*, vol. 5, pp. 1–6, 2015, doi: 10.1038/srep07836.

[11] M. Yoshita *et al.*, "Accuracy evaluations for standardization of multi-junction solar-cell characterizations via absolute electroluminescence," in *2017 IEEE 44th Photovoltaic Specialist Conferenc (PVSC)*, 2017, pp. 1–4, doi: 10.1109/PVSC.2017.8366847.

[12] Y. Wang *et al.*, "Adaptive automatic solar cell defect detection and classification based on absolute electroluminescence imaging," *Energy*, vol. 229, p. 120606, 2021, doi: 10.1016/j.energy.2021.120606.

[13] L. Zhu *et al.*, "Current leakage and fill factor in multi-junction solar cells linked via absolute electroluminescence characterization," in *2016 IEEE 43rd Photovoltaic Specialists Conference (PVSC)*, 2016, pp. 1239–1243, doi: 10.1109/PVSC.2016.7749812.

[14] T. Mochizuki *et al.*, "Solar-cell radiance standard for absolute electroluminescence measurements and open-circuit voltage mapping of silicon solar modules," *J. Appl. Phys.*, vol. 119, no. 3, p. 034501, 2016, doi: 10.1063/1.4940159.

[15] U. Rau, "Reciprocity relation between photovoltaic quantum efficiency and electroluminescent emission of solar cells," *Phys. Rev. B - Condens. Matter Mater. Phys.*, vol. 76, no. 8, pp. 1–8, 2007, doi: 10.1103/PhysRevB.76.085303.

Using machine learning to predict the complete degradation of accelerated damp heat testing in just 10% of testing time

Zubair Abdullah-Vetter, Priya Dwivedi, Robert Lee Chin, Brendan Wright, Thorsten Trupke, and Ziv Hameiri

University of New South Wales (UNSW), Sydney NSW 2052, Australia

Abstract—The ability to accurately predict the long-term performance of photovoltaic modules would have substantial benefits for the photovoltaic market. If we can precisely determine how new modules will perform after 25-30 years in the field, the reliability and bankability of photovoltaic systems will significantly increase. Keeping this target in mind, this study presents the first step towards achieving more cost-effective degradation monitoring. We develop machine learning models to predict the performance of photovoltaic modules at the end of 1,000 hours of damp heat tests after modules have only spent less than 10% of that time in damp heat conditions. Hence, we investigate the ability of unsupervised neural ordinary differential networks to model the entire dynamics of the degradation during a damp heat test using only the data that is collected in the first 10% of the process. The developed algorithms can significantly reduce the required time for damp heat tests and pave the way to transform the photovoltaic market.

Keywords—unsupervised machine learning, IEC 61215, damp heat

I. INTRODUCTION

The long-term performance, reliability, and therefore bankability of a photovoltaic (PV) module is a key aspect of their expected 25-to-30 years of field operation. It is challenging to determine the performance of PV modules during this long period as real-time outdoor tests take years to conduct [1]. The capability to precisely predict how new modules will perform after 25-30 years in the field will greatly improve the reliability of PV module performance.

When investigating elevated temperature and humidity degradation of PV modules, previous studies have used sparse sequential measurements throughout damp heat tests [2], [3]. As required by the International Electrotechnical Commission (IEC), a damp heat test involves placing a PV module in an environmental chamber set to 85°C and 85% relative humidity (RH) for 1,000 hours [4]. Based on current-voltage (I-V) measurements, analytical expressions have been developed to estimate the 'time to failure', usually defined as a loss of 20% of the initial efficiency [2]. Common degradation modes found in modern PV modules during a damp heat test include delamination and discoloration of the encapsulant and corrosion or breakages of the cell interconnections, resulting in a decrease in power output [2]. To reach these failure points, the damp heat tests are run for multiples of the required 1,000 hours, especially for sturdier glass-glass modules [5]. While 1,000 hours already equates to a staggering 42 days experiment, some damp heat tests have exceeded six months [2], [3] and future tests are expected to take even longer [2], [5] with higher associated costs.

In this study, we propose the use of an unsupervised machine learning (ML) architecture to predict the dynamics of degradation during a damp heat test from only near initial state measurements. As such, the trained model will have the capability to predict the long-term performance of a PV module in less than 10% of the testing time. The ML model is trained on an array of in-situ and periodic measurements. We propose this method as an initial step towards achieving accurate predictions of modules' performance after decades of outdoor operation.

II. DATASET

The samples used in this study are four-cell mini-modules (referred to as 'modules'). The cells are bifacial passivated emitter and rear contact (PERC) with nine busbars taken from a range of efficiency bins (22.6% - 23%) of a state-of-art industrial production line. Additional modules were fabricated using cells rejected from these bins. The modules were fabricated using a 3 mm thick soda lime glass, an ethylene vinyl acetate (EVA; Lushan) encapsulant, and a polyethylene terephthalate (PET) based backsheet (Jolywood Solar).

Damp heat tests are conducted in an ASLI TH-150C environmental chamber with internal dimensions of 500×500×600 mm (width×depth×height, respectively). To maximize the throughput, 11 modules are placed in the chamber for each 1,000 hours test. The ASLI chamber runs the damp heat tests with a fixed 85℃ and 85% RH (±0.5°C and ±2% RH, respectively). An extra Pt100 platinum resistance temperature probe was installed to monitor the internal temperature of the chamber. The results of at least three batches (11 modules each) will be reported in the final manuscript.

Throughout the test, both in-situ and periodic measurements are collected to form the dataset used for the ML training.

This work was supported by the Australian Government through the Australian Renewable Energy Agency (ARENA, Grant 2020/RND016). The views expressed herein are not necessarily the views of the Australian Government, and the Australian Government does not accept responsibility for any information or advice contained.

A. In-situ measurement system

A unique in-situ data collection system was designed to increase the temporal resolution of the dataset, enabling the use of sequential-based ML algorithms that require consistent timestep data [6]. Using this system, dark I-V measurements and electroluminescence (EL) images are collected every five minutes.

For dark I-V measurements, all 11 modules undergoing the test are connected to a Keithley 2561A source measurement unit (SMU). A four-wire measurement configuration is applied with each module having separate relay switches, allowing the SMU to measure the dark I-V of each module in a sequential order. To obtain a high dynamic range of the current from 100 nA to 10 A, different current output ranges of the SMU are applied. Using the two-diode model [7], with fixed ideality factors n_1 and n_2, the measurements are fitted to extract the two diode saturation currents (J_{01} and J_{02}), series resistance (R_s), and shunt (R_{sh}). As an example, an in-situ dark J-V measurement of a representative module is presented in Fig. 1. The voltage was divided by four to represent the average cell voltage (V_{cell}) and the current was divided by the area of a cell to give the current density (J_{cell}). The low voltage range ($V_{cell} < 0.3$ V) is dominated by R_{sh} and J_{02} while the low to mid voltage range (V_{cell} from 0.3 to 0.5 V) is impacted by J_{01}. Both J_{01} and R_s have strong effect at higher voltages ($V_{cell} > 0.5$ V).

Due to space constraints in the environmental chamber, the EL imaging system was installed outside the environmental chamber. An in-house EL imaging system was designed and built to capture in-situ EL images through the chamber's window. This setup also reduces the need to extensively cool the camera within the damp heat environment. The system is comprised of an ASI 2600MM pro-CMOS (complementary metal-oxide semiconductor) sensor placed inside a dark box attached to the chamber's door. Due to the small window aperture, only one module faces the imaging system with a single cell in the camera field-of-view. During EL imaging, the same electrical system used to measure dark I-V is used to drive a forward current of 10 A through the module. To improve the signal to noise ratio, an exposure time of 60 seconds and gain of 50 is used. Background image (measured at zero current, but using the same sensor conditions) subtraction is applied to all

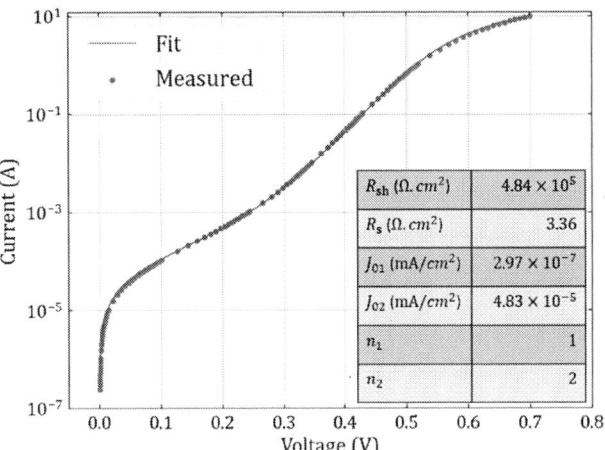

Fig. 1. Averaged dark JV curve of one of the modules collected during the damp heat test (85°C, 85% RH) and the two-diode model fit obtained via manual fitting. The obtained fitting parameters are in the table.

R_{sh} ($\Omega.cm^2$)	4.84×10^5
R_s ($\Omega.cm^2$)	3.36
J_{01} (mA/cm^2)	2.97×10^{-7}
J_{02} (mA/cm^2)	4.83×10^{-5}
n_1	1
n_2	2

images. This background image is renewed before each EL image. Flat field image correction is applied prior to the ML training. The in-situ EL image obtained by the in-situ system and EL image obtained by an industrial line scan module imaging system (M1; BT Imaging) is compared in Fig. 2. The image from the in-situ system is comparable to the industrial system. The same cracks and dark patches are visible in both EL images. The line scan photoluminescence (PL) from the M1 is also included. The dark patches in the EL image appear bright in the PL image, indicating that there is partial or full electronic isolation at those regions.

B. Periodic measurements

In addition to the in-situ measurements, every 48 hours, the chamber is ramped down to 25°C for one hour, such that the modules and chamber reach thermal equilibrium. The samples are then removed to obtain periodic characterization measurements. EL and PL images are collected using a BT Imaging M1 line scanning system at a line scan speed of 40 mm/s with a forward bias of 10 A for EL and an illumination intensity of approximately four suns equivalent photon flux for

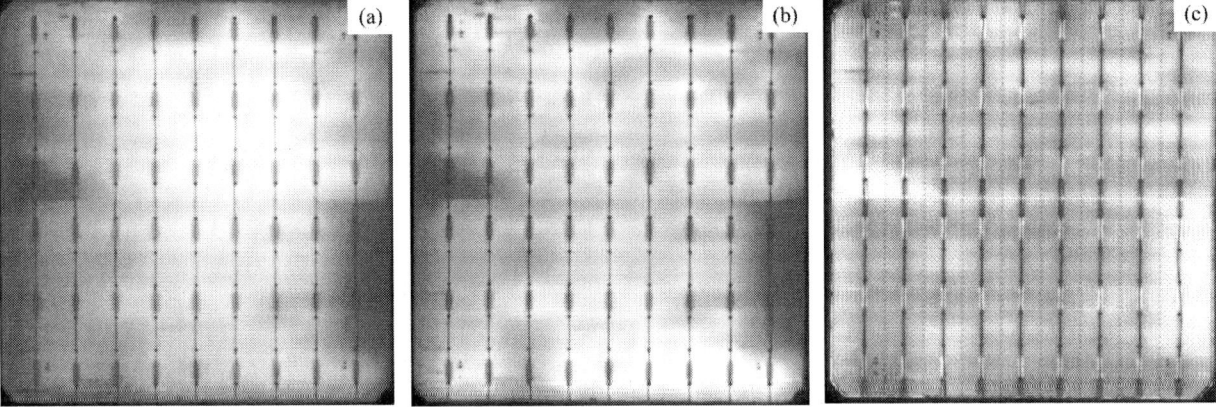

Fig. 2. Different luminescence images of the same cell taken at 25°C after 432 hours of damp heat degradation by (a) the in-situ EL system, (b) line scan EL, and (c) line scan PL. Both line scan images are taken by the BT Imaging M1 system.

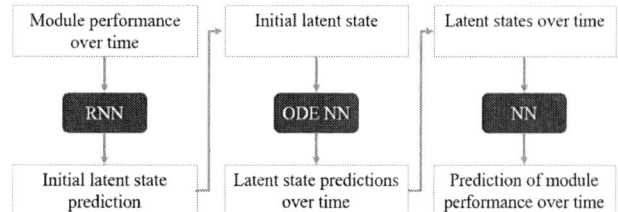

Fig. 3. Diagram of the ML pipeline proposed in this study. Consisting of an input RNN that predicts the initial latent states, an ODE NN to predict the dynamics of degradation in the latent space, and an output NN that predicts the modules' performance during the damp heat test [10].

PL imaging. Representative EL and PL images collected after 432 hours are presented in Fig. 2. Light I-V measurements are performed using a flash module I-V tester (Eternal Sun; Spire) with an AM1.5G spectrum simulated by the tool's xenon lamp solar simulator. While the flash tester does not have temperature control for a sample being measured, it does have an array of temperature probes that are attached to the sample. During the I-V measurement, the temperature of the modules was 25±1.5°C. All I-V measurements were temperature corrected using the temperature coefficients of the cells.

III. MACHINE LEARNING

A popular deep learning architecture that utilizes sequential data is a recurrent neural network (RNN) [6], which has seen success in temporal-based problems such as speech recognition [6]. However, basic RNNs struggle to bridge large time steps in sequential data, such as predicting the 100th word given the first word of a paragraph [6]. As we aim to determine the degradation of modules after 1,000 hours given only their near-initial performance, we develop an unsupervised ML methodology that can overcome this limitation. Unsupervised ML has been increasingly used to model complex non-linear systems [9]. The high dimensionality of the feature space in the collected dataset can also be simplified since unsupervised ML models are often used to transform data into a lower dimensional 'latent' space [9]. The framework of our approach in presented in Fig. 3. A

combination of an RNN and an ordinary differential equation (ODE) neural network (NN) is trained on the dataset of in-situ and periodic measurements during the damp heat test. The trained model is capable of learning a 'smoothed average behavior' of the modules during the test.

We recently successfully modelled the dynamics of light induced degradation (LID) in silicon heterojunction cells (SHJ) using unsupervised ML [10]. This capability is demonstrated in Fig. 4 as it compares the model-based predictions and the measurements of the open-circuit voltage (V_{oc}) and R_s during the degradation process [10]. A similar approach is used to train the architecture using the unique dataset collected over the damp heat tests. As a result, given only the near initial state measurements, the model will have the capability to predict the dynamics of elevated temperature and humidity degradation seen in the PV modules. The results of this training approach will be included in the final manuscript.

IV. CONCLUSION

This study proposes an unsupervised ML architecture to predict the dynamics of elevated temperature and humidity degradation of PV modules. As a result, in less than 10% of the testing time, the model will have the capability to accurately predict the complete degradation expected at the end of a 1,000 hour damp heat test. As the test is representative of a module's expected 25-30 year lifespan, these predictions are the first steps to accurately predicting the performance of new modules after decades of operation. This capability will greatly improve the reliability and bankability of PV modules, revolutionizing the PV industry.

REFERENCES

[1] I. Kaaya, J. Ascencio-Vásquez, K.-A. Weiss, and M. Topič, "Assessment of uncertainties and variations in PV modules degradation rates and lifetime predictions using physical models," *Solar Energy*, vol. 218, pp. 354–367, 2021.

[2] M. Koehl, S. Hoffmann, and S. Wiesmeier, "Evaluation of damp-heat testing of photovoltaic modules," *Progress in Photovoltaics: Research and Applications*, vol. 25, no. 2, pp. 175–183, 2017.

[3] N. Iqbal *et al.*, "Characterization of front contact degradation in monocrystalline and multicrystalline silicon photovoltaic modules following damp heat exposure," *Solar Energy Materials and Solar Cells*, vol. 235, p. 111468, 2022.

[4] International Electrotechnical Commission, "IEC 61215-1:2021." https://webstore.iec.ch/publication/61345.

[5] J. Tang *et al.*, "The performance of double glass photovoltaic modules under composite test conditions," *Energy Procedia*, vol. 130, pp. 87–93, 2017.

[6] R. C. Staudemeyer and E. R. Morris, "Understanding LSTM -- a tutorial into long short-term memory recurrent neural networks," *arXiv:1909.09586*, 2019.

[7] S. Suckow, T. M. Pletzer, and H. Kurz, "Fast and reliable calculation of the two-diode model without simplifications," *Progress in Photovoltaics: Research and Applications*, vol. 22, no. 4, pp. 494–501, 2014.

[8] International Electrotechnical Commission, "IEC TS 60904-1-2:2019." https://webstore.iec.ch/publication/34357.

[9] S. Kim, W. Ji, S. Deng, Y. Ma, and C. Rackauckas, "Stiff neural ordinary differential equations," *Chaos*, vol. 31, no. 9, p. 093122, 2021.

[10] Brendan Wright and Brett Hallam, "Applied machine learning to model LID dynamics in SHJ solar cells," presented at the 11th International Conference on Crystalline Silicon Photovoltaics, 2021.

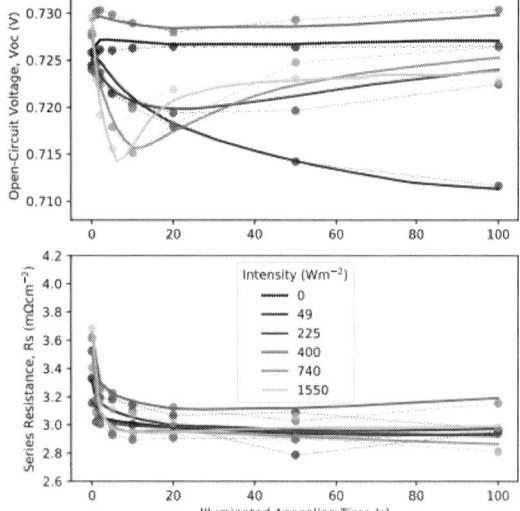

Fig. 4. The RNN ODE model inference of R_s and V_{oc} behavior during LID caused by illuminated annealing. The circles with dotted lines are the measurements whereas the unbroken lines are the predicted dynamics [10].

CuCl Doping Variations in High Efficiency Polycrystalline CdSeTe/CdTe Thin Film Solar Cells

Zachary F. Lustig, Tushar M. Shimpi, Akash Shah, Walajabad S. Sampath

Department of Mechanical Engineering, Colorado State University, Fort Collins, CO, United States

In this paper, the influence of CuCl doping process parameters on the performance of solar cells was studied. The devices were fabricated with graded CdSeTe/CdTe absorber. The critical parameters for CuCl doping process were identified and varied in a 2-level 3-factor statistically designed experiment. The analysis of response data generated from the performance of the devices was analyzed using JMP software. The performances of devices with 19%+ revealed two different optimal processing conditions for CuCl doping.

Automated analysis of internal quantum efficiency using chain order regression

Zubair Abdullah-Vetter, Priya Dwivedi, Yoann Buratti, Alfred Krzywicki, Arcot Sowmya, Thorsten Trupke, and Ziv Hameiri

University of New South Wales (UNSW), Sydney NSW 2052, Australia

Abstract— Spectral analysis of internal quantum efficiency (IQE) measurements of solar cells is a powerful method to identify performance-limiting mechanisms in photovoltaic devices. This analysis is usually performed using complex curve-fitting methods to extract various electrical and optical performance parameters. As these traditional fitting methods are not easy to use and are often sensitive to measurement noise, many users do not utilize the full potential of the IQE measurements to provide the key properties of their solar cells. In this study, we propose a simplified approach to analyze IQE curves of silicon solar cells using machine learning models that are trained to extract valuable information regarding the cell's performance and decoupling the parasitic absorption of the anti-reflection coating. The proposed approach is demonstrated to be a powerful characterization tool for solar cells as machine learning unlocks the full potential of IQE measurements.

Keywords—spectral response, random forest, chain order regression, open-source

I. INTRODUCTION

The electrical response of a photovoltaic (PV) device to the excitation of light of a single wavelength is known as its spectral response and is a measure of the current output at each wavelength of light [1]. This response can be converted to the internal quantum efficiency (IQE), which indicates the ratio between the *collected carriers* and the number of *absorbed photons* by the device [1], [2]. It involves measuring the short circuit current, reflectance, and incident photon flux under variable monochromatic illumination [1]. Comparing the measurement results to a perfect quantum efficiency of 100% across the spectrum provides key information regarding the performance-limiting mechanisms of the device [1], [2].

Traditional analysis methods of silicon (Si) solar cells require fitting specific sections of the measured IQE to complex formulas [2]. A common method is based on plotting the inverse IQE versus the wavelength-dependent absorption depth in Si. The inversed slope of this plot yields the effective diffusion length (L_{eff}) of the minority carriers in the bulk of the device [2]. This particular approach is usually limited to the near-infrared wavelength range and the resulting L_{eff} is strongly impacted by the selected wavelength range and measurement noise [2], [3]. Other quantitative analysis methods involve mathematical models that are used to fit the measured IQE curve [4]–[6]. They

often require manipulation of many parameters within the mathematical models, which can be time-consuming and complex depending on the number of unknown device parameters [4]–[6].

In this study, we propose the use of machine learning (ML) regression models to automatically extract the key electrical and optical parameters of solar cells directly from IQE measurements, across the entire measured spectrum. Furthermore, the trained models also decouple the effect of the anti-reflection coating (ARC) parasitic absorption from IQE measurements, which normally requires referencing a generalized database [7]. Following detailed testing and sensitivity analysis, we plan to share the trained models for the benefit of the PV community, providing an easy to use method to extract key performance parameters of their PV devices.

II. DATASET

To efficiently train multiple regression models with thousands of IQE curves, a dataset of IQE curves was simulated using formulas adapted from Fischer [7]:

$$IQE = X_{EMI} \times \int \eta(z)G(z) \cdot dz, \qquad (1)$$

where X_{EMI} is the carrier collector term, while the bulk term (IQE_{bulk}) is the integral of the bulk collection efficiency, $\eta(z)$, multiplied by the normalised generation profile, $G(z)$, over the cell depth, z. The key independent parameters of X_{EMI} are the width of the emitter (w_e) and the emitter collection efficiency (IQE_0), while $\eta(z)$ is determined by the rear surface recombination velocity (SRV_b) and L_{eff}. The internal reflectance of the rear (RB) and how diffused the light is after rear internal reflection (Λ) impact $G(z)$.

Different combinations of the six parameters were used to generate a dataset of 200,000 labelled IQE curves, each simulated in the wavelength range from 250 to 1,200 nm in 10 nm steps. The IQE difference curves (discrete derivative of each IQE curve) are also computed and added to the IQE curve, forming the feature vector used for training the ML models. Once trained, the ML models are able to extract the parameters from any given IQE measurement.

III. MACHINE LEARNING MODELS

A separate random forest (RF) regression model is used to predict each parameter [8]. Therefore, the 'trained model' is a combination of regression models, each outputting a single

---

This work was supported by the Australian Government through the Australian Renewable Energy Agency (ARENA, Grant 2020/RND016). The views expressed herein are not necessarily the views of the Australian Government, and the Australian Government does not accept responsibility for any information or advice contained.

978-1-7281-6118-1/22 $31.00 © 2022 IEEE   476

parameter. Hence, the proposed approach automatically decouples the front, bulk, and rear regions of an IQE measurement and provides the user with a compact list of the key electrical and optical parameters of the investigated device [9].

The dataset was split into training and test subsets with a 70:30 ratio [10]. The test set results were evaluated using a root mean square relative error (RMSRE) [11]:

$$\text{RMSRE}(y, \hat{y}) = \sqrt{\frac{1}{N} \sum_{i=1}^{N} \left(\frac{y_i - \hat{y}_i}{y_{max}}\right)^2}, \qquad (2)$$

Where $\hat{y}_i$ is the predicted value of the ith sample, and y_i is the corresponding true value out of a total N samples. The relative error is calculated by dividing the absolute error by the maximum value in the parameter dataset, y_{max}. Throughout this paper, the RMSRE is expressed as a percentage such that a perfect prediction results in 0% error. This metric allows easy comparison between the predicted parameters, as each parameter have different units and scale.

To further improve the training of the RF models, chain regression [11] was used. This methodology involves choosing the order of the different RF models to be trained. After the first model is trained, the prediction results are used to train the next

model. The subsequent models then use all the previous prediction results in their training. Therefore, chain regression results in simplifying the training by exploiting any correlations between the solar cell parameters. The chain order used in this study starts with IQE_0, followed by w_e, RB, L_{eff}, Λ, and then SRV_b. This order was chosen by first training the RF models without chain regression, which showed the first four parameters had much lower RMSRE than the last two. Therefore, the selected order leverages the more accurately predicted parameters to reduce the error in the latter parameter predictions. Other regression models and different chain regression orders will be trained with this methodology, and the results will be included in the final manuscript.

IV. RESULTS

The results of the RF chain regression are displayed in Fig. 1. The graphs represent the predicted vs actual cell parameters and are presented in the same order as the chain used to train the model. It can be seen that the proposed method is capable of achieving extremely accurate results for IQE_0, w_e, RB, and L_{eff} with slightly higher errors for Λ and SRV_b.

While the results shown in Fig. 1 are remarkable for predicting the six parameters from a simulated IQE curve, solar cells also have an ARC layer that causes parasitic absorption in the short wavelength region [12]. Typical decoupling of this parasitic effect involves referencing a database of known absorption profiles of silicon nitride (SiN_x) [7].

In this study, we propose the use of the same chain regression methodology to predict the six parameters from IQE curves with SiN_x absorption. A new dataset of 200,000 IQE curves that includes parasitic absorption profiles, which are affected by both the thickness of the ARC layer and the wavelength dependent extinction coefficient (k), was created. An example of applying chain regression to this dataset is shown

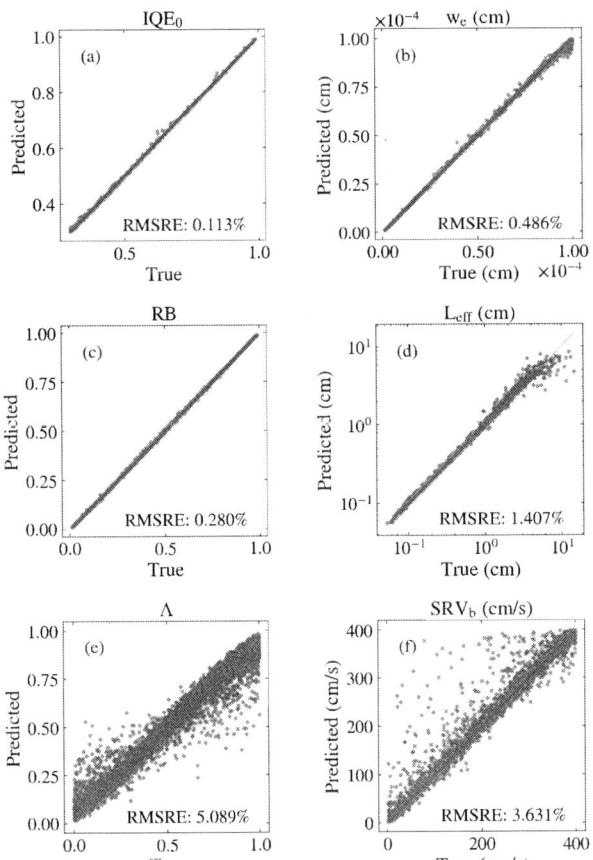

Fig. 1. The RF chain regression results displayed as predicted value vs true value plots and in their chain order (a) IQE_0, (b) w_e, (c) RB, (d) L_{eff}, (e) Λ, and (f) SRV_b. The RMSRE value is also shown on each plot.

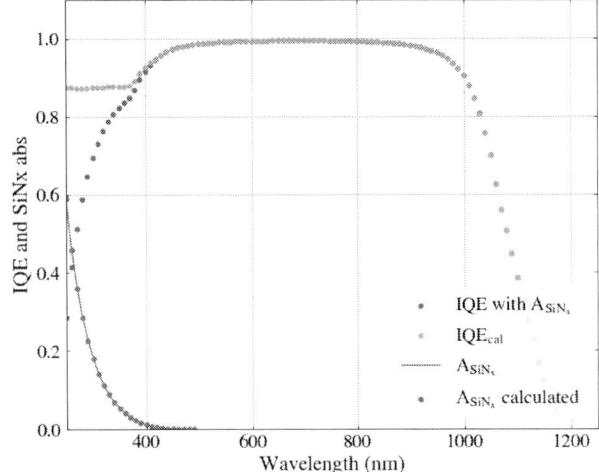

Fig. 2. A representative example of how the absorption in the silicon nitride (A_{SiN_n}) is decoupled. Given the simulated IQE curve with A_{SiN_x} (blue), the RF chain model predicted the six performance parameters, which were input into Equation (1) to calculate IQE_{cal} (orange). Subtracting the simulated from the calculated IQE gives the calculated A_{SiN_x} (red dots), which is compared to the simulated A_{SiN_x} (red line).

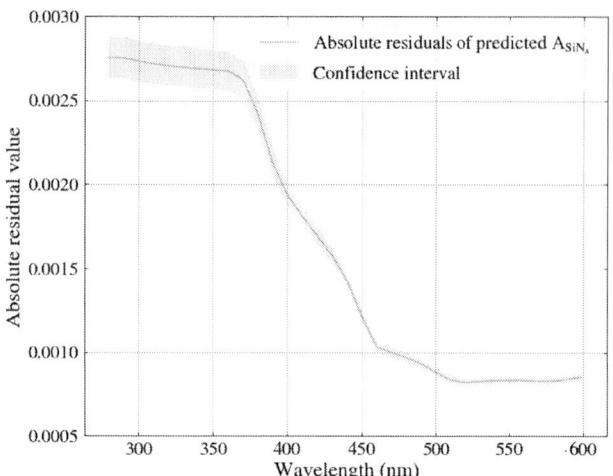

Fig. 3. The absolute residuals of the simulated vs predicted A_{SiN_x} at each wavelength with 95% confidence interval.

in Fig. 2, comparing the predicted and simulated ARC absorption. The six parameters were accurately predicted from the simulated IQE curve with parasitic absorption of the SiN_x. Using Equation (1), the IQE without SiN_x is calculated and the parasitic absorption is accurately decoupled by simple subtraction. This decoupling approach is applied to all the simulated curves in the dataset and can be applied to any IQE measurement.

The results of decoupling the SiN_x from all the IQE curves in the test set are displayed in Fig. 3. After applying the prediction and subtraction as shown in Fig. 2 to every curve, the absolute residuals at each wavelength was calculated with a 95% confidence interval. At most there is an inconsequential difference of < 0.003 between the simulated and predicted SiN_x parasitic absorption, which occurs at the shorter wavelength region. With even lower residuals at longer wavelengths, the proposed method is capable of accurately decoupling and predicting the parasitic absorption in the ARC.

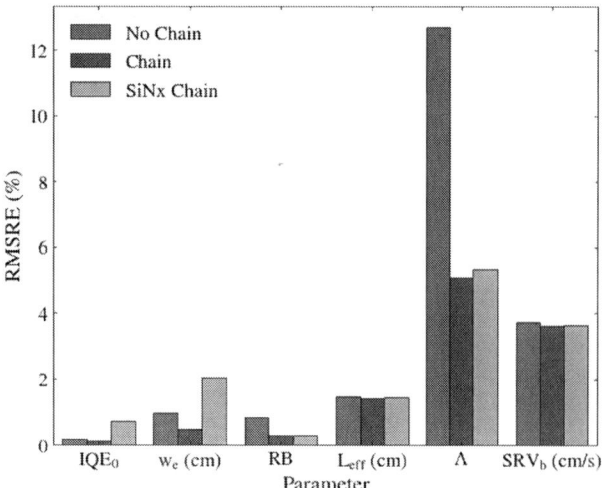

Fig. 4. A comparison of the RMSRE scores for each method used in this study.

The different RMSRE results are summarized in Fig. 4. Comparing with and without chain regression, the RF chain model greatly improves the prediction results for Λ, while also improving the results for the other parameters. Applied on the IQE curves with SiN_x, the RMSRE increases, however, still maintains a relatively low error. Therefore, it is clear that the proposed methodology is capable of predicting the six parameters with relatively high accuracy. As a result, the proposed method is able to predict the key performance parameters from IQE curves, simplifying the fitting of IQE measurements. The application of this methodology to IQE measurements obtained from different Si solar cells will be included in the final manuscript.

V. CONCLUSION

Automated analysis of IQE measurements of Si solar cells using ML was presented. By training ML regression models on a large dataset of simulated IQE curves in a chain order, the capability to extract key performance parameters of PV cells has been developed. Furthermore, when applying this methodology to IQE curves with parasitic absorption in the ARC, the chain regression method still maintains a relatively low error when predicting those key parameters. This allows a simple and accurate decoupling of the parasitic absorption in the ARC layer without referencing a generalized database of different profiles. This study provides an accurate tool that is easier to use than traditional spectral response analysis fitting methods. This method will be tested on experimental data and undergo sensitivity analysis and the results will be reported at the conference. We also plan to share the trained models for the benefit of the PV community.

REFERENCES

[1] J. S. Hartman and M. A. Lind, "Spectral response measurements for solar cells," *Solar Cells*, vol. 7, no. 1, pp. 147–157, 1982.

[2] P. A. Basore, "Extended spectral analysis of internal quantum efficiency," in *23rd IEEE PVSRC*, 1993, pp. 147–152.

[3] B. Fischer, M. Keil, P. Fath, and E. Bucher, "Scanning IQE-measurement for accurate current determination on very large area solar cells," in *29th IEEE PVSRC*, 2002, pp. 454–457.

[4] D. Lan and M. A. Green, "Extended spectral response analysis of conventional and front surface field solar cells," *Solar Energy Materials and Solar Cells*, vol. 134, pp. 346–350, 2015.

[5] W. J. Yang, Z. Q. Ma, X. Tang, C. B. Feng, W. G. Zhao, and P. P. Shi, "Internal quantum efficiency for solar cells," *Solar Energy*, vol. 82, no. 2, pp. 106–110, 2008.

[6] D. A. Clugston and P. A. Basore, "PC1D version 5: 32-bit solar cell modeling on personal computers," in *26th IEEE PVSRC*, 1997.

[7] B. Fischer, "Loss Analysis of Crystalline Silicon Solar Cells Using Photoconductance and Quantum Efficiency Measurements," PhD Thesis, Konstanz University, 2003.

[8] L. Breiman, "Random Forests," *Machine Learning*, vol. 45, no. 1, pp. 5–32, 2001.

[9] P. P. Altermatt, J. Müller, and B. Fischer, "A simple emitter model for quantum efficiency curves and extracting the emitter saturation current," *28th European Photovoltaic Solar Energy Conference and Exhibition*, pp. 840–845, 2013.

[10] S. Raschka and V. Mirjalili, *Python Machine Learning*, 3rd ed, Birmingham - UK: Packt Publishing, 2019.

[11] F. Pedregosa *et al.*, "Scikit-learn: machine learning in python," *Journal of Machine Learning Research*, vol. 12, pp. 2825–2830, 2011.

[12] M. R. Vogt, "Development of physical models for the simulation of optical properties of solar cell modules," PhD Thesis, Leibniz Universität Hannover, 2015.

Photon recycling and luminescent coupling in all-perovskite tandem solar cells quantified by full opto-electronic device simulation

Urs Aeberhard, Simon J. Zeder, Beat Ruhstaller

Fluxim AG, Winterthur, Switzerland

Integrated Systems Lab, ETHZ, Zürich, Switzerland

PV-Lab, EPFL, Neuchatel, Switzerland

Zurich Unversity of Applied Sciences, Winterthur, Switzerland

All-perovskite tandem solar cells are analyzed by means of full opto-electronic device simulation. The optical model based on a novel detailed-balance compliant Green's function dipole emission approach provides a consistent consideration in the wave optics regime of photon-recycling in the individual sub-cells and of luminescent coupling between top- and bottom cells. The electrical model based on drift-diffusion charge transport enables the consideration of non-radiative losses due to transport and recombination and includes the interconnection of the subcells via the recombination junction. The tandem device characteristics resulting from the fully coupled simulation are compared to the optical limit of the detailed balance approach that assumes a uniform quasi-Fermi level splitting, which enables a detailed opto-electronic loss analysis.

The National Solar Radiation Database (NSRDB): Current Status

Aron Habte, Manajit Sengupta, Yu Xie, Grant Buster, Michael Rossol, Paul Edwards, Galen Maclaurin, Evan Rosenlieb, Jaemo Yang, Haiku Sky, Mike Bannister, Billy Roberts

National Renewable Energy Lab., Golden, CO, United States

The National Solar Radiation Database (NSRDB) has evolved significantly since the first release of the point source database in 1992. The NSRDB has become the industry standard for public long-term time-series solar resource data for performance and economic analysis of solar energy projects. NREL continuously updates the scientific methods and spatiotemporal scale of the NSRDB data based on new research and improvements in satellite technology. Currently the NSRDB covers most of the globe and is a publicly available database of high-resolution solar resource data in the U.S. with over 150,000 users annually. NSRDB represents the state of the art in satellite-based estimation of solar resource information and uses a unique physics-based modeling approach that allows improvements in accuracy with the deployment of the next-generation geostationary satellites. This paper will cover the evolution of the NSRDB, current status of the database including estimated uncertainty and spatiotemporal scale and variability, and the advancement in modeling approach.

FTO Delamination for Photovoltaic Module Separation

Jongwon Ko, Soohyun Bae, Yoonmook Kang, Hae-Seok Lee, Donghwan Kim

Korea University, Seoul, South Korea

Korea University, Seoul, South Korea

As the amount of photovoltaic (PV) modules installed around the world increases, the amount of end-of-life PV modules is also expected to increase. In view of this prospect, studies on a method of processing end-of-life PV modules are in progress. The most popular existing method of disposing of end-of-life PV modules is incineration. Glass and solar cells are often broken during this process. Methods such as chemical and heated cutter are being studied to recover the PV module components in an unbroken state. In this study, the transparent conducting oxide (TCO) delamination phenomenon existing in thin-film solar cells was applied to a silicon PV module to try to separate the encapsulant and glass. The fluorine-doped tin oxide (FTO) thin film was used as a sacrificial layer between the glass and the encapsulant. The sacrificial layer was delaminated by applying voltage and heating to the module, and as a result, it was confirmed that the glass and the encapsulant were separated.

Investigation of CsF-Treatment Effects on Cu(In,Ga)(S,Se)2 Solar Cells using Photothermal Atomic Force Microscopy under Various Photoexcitation Conditions

Ayaka Yamada, Takuji Takahashi

Institute of Industrial Science, the University of Tokyo, Tokyo, Japan

NanoQuine, The University of Tokyo, Tokyo, Japan

Performance of a thin film Cu(In,Ga)(S,Se)2 [CIGSSe] solar cell has been well improved by CsF-treatment. In this study, we have investigated the effect of the CsF-treatment by photothermal atomic force microscopy [PT-AFM] through examination of non-radiative recombination properties. The used samples were CdS/CIGSSe materials with or without the CsF-treatment, and PT-AFM were performed under three excitation conditions: (1) standard condition, (2) high-photon-energy condition, where a light with high photon energy was used for shallow excitation near the surface by shortening a penetration depth of light, and (3) high-modulation-frequency condition, where an incident light was modulated at a high frequency to detect the fast thermal expansion caused by non-radiative recombination near the surface. Under conditions (2) and (3), the PT signals are considered to be dominated by the heat generation near the surface. Topographic and PT signal images were taken by PT-AFM on two samples under three excitation conditions, and we have compared the intensities and distributions of the PT signals among them to discuss the effects of the CsF-treatment. First, we found that the PT signal intensity on the CsF-treated sample was weaker than that on the as-grown sample under any excitation condition, indicating that the non-radiative recombination was suppressed by the CsF-treatment. Especially on the CsF-treated sample, the PT signals were weakened under conditions (2) and (3), compared with those under condition (1). Therefore, we can consider that the CsF-treatment effectively passivates the recombination centers near the surface. Moreover, the reduction of the PT signal was very apparent around GBs. From those results, we conclude that the non-radiative recombination centers distributed at the CdS/CIGSSe interface and along GB were effectively passivated by the CsF-treatment.

Polysilicon passivating contact layer for crystalline silicon solar cells: a dopant-grading approach

DUY PHONG PHAM, JUNSIN YI

Sungkyunkwan University, Suwon, Korea

In polycrystalline silicon/crystalline silicon passivating contact (hereafter poly-Si/c-Si), the heavy doping requirement for a poly-Si layer to provide high-quality contact passivation may promote dopant diffusion into c-Si. A dopant-grading approach is proposed to constrain the dopant diffusion from poly-Si into c-Si during the high thermal annealing. The approach is implemented by adjusting the dopant concentration appropriately throughout the thickness of the poly-Si layer. Accordingly, a dopant-grading with a high dopant concentration distributed far from the c-Si interface enhances passivation quality, enabling a 730 mV implied open-circuit voltage and a 5.1 fA/cm2 recombination current density. The grading improves the open-circuit voltage, fill factor, and efficiency of the device by 10 mV, 0.5%, and 0.6%, respectively. A heterojunction cell device including graded-poly-Si at the back side achieves an efficiency of 22%.

Effects of (i)a-Si:H deposition temperature on passivation quality and performance of high-efficiency silicon heterojunction solar cells

Yifeng Zhao, Paul Procel, Arno H. M. Smets, Luana Mazzarella, Can Han, Liqi Cao, Guangtao Yang, Zhirong Yao, Arthur Weeber, Miro Zeman, Olindo Isabella

Photovoltaic Materials and Devices Group, Delft University of Technology, Delft, Netherlands

Excellent surface passivation induced by (i)a-Si:H is critical to achieve high-efficiency silicon heterojunction (SHJ) solar cells. In this study, we investigated the effects of (i)a-Si:H deposition temperature on passivation quality and solar cell performance. Among the deposition temperatures we investigated (140 - 200 °C), lower temperatures seem to result in less dense (i)a-Si:H films, which hinder their surface passivation capabilities. However, with additional hydrogen plasma treatments (HPTs), those (i)a-Si:H layers exhibited significant improvements and better passivation qualities than their higher temperature counterparts. On the other hand, even though we observed the highest VOCs for cells with (i)a-Si:H deposited at the lowest temperature (140 °C), the related FFs are poorer as compared to their higher temperature counterparts. The optimum trade-off between VOC and FF was found with temperatures ranging from 160 °C to 180 °C, which delivered independently certified efficiencies of 23.71%. Thus our study reveals two critical requirements for optimizing the (i)a-Si:H layers in high-efficiency SHJ solar cells: (i) excellent surface passivation quality to reduce losses induced by interface recombinations and (ii) less-defective (i)a-Si:H bulk to improve the charge carrier collections.

Investigation of the crack propensity of co-extruded polypropylene based backsheets

Gernot Oreski[1], Chiara Barretta[1], Astrid Macher[1], Gabriele Eder[2], Lukas Neumaier[3], Markus Feichtner[4] and Minna Aarnio-Winterhof[5]

[1] Polymer Competence Center Leoben, Leoben, 8700, Austria
[2] Austrian Research Institute for Chemistry and Technology, Vienna, 1030, Austria
[3] Silicon Austria Labs, Villach/St.Magdalen, 9524, Austria
[4] Kioto Solar, St. Veit/Glan, 9300, Austria
[5] Borealis Polyolefine, Linz, 4021, Austria

Abstract—Co-extruded backsheets based on polypropylene (PP) are an interesting alternative to laminated backsheets containing polyester films (PET). Backsheet cracking has become a frequent failure mode in the last years, causing not only safety issues but on the long term also reducing the lifetime of PV modules. In this work the crack susceptibility of three different backsheets was investigated using solder bump coupon specimens: two co-extruded backsheets based on polyamide (PA) and PP, together with one laminated backsheet containing a PET core layer and polyvinyl fluoride (PVF) outer layers. The solder bump coupons were exposed to test sequence of exposure to UV light followed by thermal cycles. Overall, the PP as well as the PVF-PET backsheet showed excellent stability and no susceptibility to material embrittlement or cracking. By comparison, the PA based backsheets showed next to significant discoloration also strong cracking after a few test cycles. Overall, the study confirms that co-extruded PP backsheets show great potential to be a valid replacement of standard PET based backsheets in PV modules.

Keywords—backsheet, cracking, polypropylene, solder bump

I. INTRODUCTION

In general, the two main requirements for backsheets of a PV module are (i) protection of the inner layers from external stressors and (ii) electrical insulation. Additionally, the backsheet needs to act as a barrier against water vapor ingress [1]. Several backsheet architectures are in the market, mainly laminated multi-layer or co-extruded films [2–4]. Next to delamination and discoloration, cracking has been reported to be among the most frequent backsheet failures [5–10].

Within the last years, an enhanced occurrence of backsheet cracking was observed in the field (after only 4 to 7 years operational time) and reported at several conferences and publications. Highest failure rate was observed for polyamide (PA) based backsheets, but also for backsheets using poly vinylidene fluoride (PVDF) and polyethylene terephthalate (PET) as outer layers cracking has been reported [7–12].

Most recently, co-extruded backsheets based on polypropylene (PP) were developed and successfully introduced into the market [3,4,11]. The property profile of this material combines the low water vapor transmission rate of PET based backsheets but provides high permeability of acetic acid but also oxygen. The main advantages of co-extruded backsheets are as follows:

i. The backsheet is produced in one step, which also means reduced processing induced material degradation [12]

ii. Likelihood of delamination, which is a major backsheet failure mode, is significantly reduced.

iii. Full back integration allows for easy material modifications regarding additive formulation, fillers or layer geometry

The process of material innovations for PV, such as co-extruded PP backsheets, is complicated due to the complex interactions within a PV module. The advantage of one material may be outweighed by its interaction with another component, leading to unexpected failure modes like backsheet cracking [11].

Several studies reported excellent stability of PP based backsheets against damp heat as well as extensive irradiation exposure [3,4,11]. None of the studies showed any significant deterioration of mechanical properties or sensitivity for embrittlement and cracking. Only slight yellowing was observed, mostly depending on the additive-mix used.

However, in all of these studies only single or dual stress tests have been performed. But the newest results show that backsheet cracking (independent of the materials used) results from material degradation caused by UV, humidity and/or temperature combined with thermo-mechanical loads inducing crack initiation and crack propagation [9,13,14]. Therefore, the main aim of this study is to investigate the crack propensity of co-extruded polypropylene backsheets. For this purpose, solder bump coupons as proposed by Kempe at al. [14] have been produced by using standard ethylene vinyl acetate (EVA) as encapsulant and two different types of backsheets (see Tables 1 and 2). The specimens were exposed to a sequential test consisting of (1) 500 h of xenon exposure and (2) 100 thermal cycles from -40 °C to 85 °C, which was repeated five times. After each cycle the specimen were investigated using infrared spectroscopy and UV/VIS/NIR spectroscopy.

II. EXPERIMENTAL

A. Samples

Two different co-extruded backsheets were investigated: One backsheet based on PP (CPO), and one based on PA (AAA) as known bad material. Additionally, also one standard laminated backsheet using PET as core and polyvinyl fluoride (PVF) as outer layers has been investigated. The exact composition of each backsheet is displayed in Table 1. Solder bump coupons (see Table 2) with a size of 150 x 76 mm were prepared using standard flat glass, a standard EVA encapsulant (F406PS, Hangzhou First Applied Material Co., Ltd) and the backsheets listed in Table 1. The lamination was done according the recommendations given in the data sheet of the EVA.

TABLE I.   INVESTIGATED BACKSHEET TPYES

Name	Composition		
	Cell side	Core	Out
CPO	PP+TiO$_2$	PP	PP+TiO$_2$
AAA	PA12+TiO$_2$	PA12+PP+SiO$_2$	PA12
TPT	PVF	PET	PVF

TABLE II.   SOLDER BUMP SPECIMENS

Specimen	Front	Encapsulant	Backsheet
Coupon 1	Glass	EVA	AAA
Coupon 2	Glass	EVA	CPO
Coupon 3	Glass	EVA	TPT

B. Exposure

After lamination, the specimens were exposed to a total of 5 cycles consisting of 500 h of xenon exposure and 100 thermal cycling. Xenon exposure was done in a Xenontest 440 (Atlas Material Testing Technology GmbH, Germany) to condition A3 of IEC62788-7-2 (0.8 W/m²/nm @ 340 nm, Black Panel temperature = 90 °C, relative humidity = 20%). Thermal cycling was done according IEC61730-2(2016) in a climate chamber (Weiss WLK 64-40, Weiss Technik GmbH, Germany) from -40 °C to 85 °C.

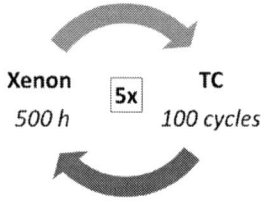

Fig. 1.   Test cycle for solder bump specimens

C. Characterization methods

Changes in the chemical structure of the backsheet surface were identified via Fourier Transform Infrared spectroscopy in attenuated total reflectance mode (FTIR-ATR). The measurements were conducted on a Spectrum GX FTIR spectrometer (Perkin Elmer) equipped with an ATR unit Pike GladiATR (Pike Technologies) using a ZnSe crystal with diamond on top as reflective element. The spectra were recorded between 4000 to 650 cm^{-1}, an average of 16 scans and at a resolution of 4 cm^{-1} for all samples. At least three independent measurements were taken for each sample/coupon. The optical

This project has received funding from the Austrian Research Promotion Agency (FFG)

changes of the backsheets were measured using a Lambda 950 UV/VIS/NIR spectrometer (Perkin Elmer). Spectra were recorded from 250 to 2500 nm in 5 nm steps with an integrating sphere from Labsphere in order to measure hemispherical reflectance. Then the Yellowness Index (YI, 10° observer) was calculated according to ASTM E313.

III. RESULTS AND DISCUSSION

Figure 2 shows the images of the solder test coupons with the co-extruded PA and PP backsheets after 3 and 4 cycles of Xenon and TC exposure. Small cracks in the AAA backsheet appear first after Xenon 1500 h and TC300x, but only when the backsheet is directly exposed to the xenon lamps (not, when irradiation through glass takes place). An additional 500 h of Xenon exposure does not lead to any further crack growth, which confirms that thermal expansion during TC is the main driving factor for crack growth. Interestingly, no cracking was observed above the wire, the cracks are oriented in machine direction of the film. Also, some chalking was observed for the AAA backsheet. By comparison, no cracks or any other sign of material embrittlement of the PP based backsheet was observed (see Fig.2 right). Also, the TPT backsheet did not show any cracks. Figure 3 shows the evolution of the yellowness index through the 5 aging steps.

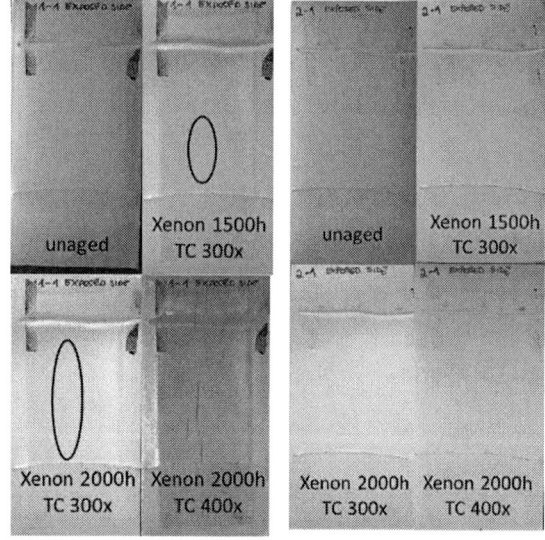

Fig. 2.   Solder bump coupons with AAA backsheet (left) and CPO backsheet (right)

All backsheets the samples show an increase in yellowing during TC and photobleaching during Xenon-weathering, with the AAA backsheet showing the highest yellowing, with an increasing YI after each thermal cycling step. After 2000 h of Xenon exposure and 400 thermal cycles a YI of nearly 19 was calculated. Interestingly, no backsheet yellowing was observable when the glass side was exposed to the Xenon lamps. Moreover, the photobleaching effect does not reverse the discoloration entirely, resulting in a YI higher than 4 after step 9 (2500 h Xenon, 400 thermal cycles). Infrared spectra of the exposed PA surface showed typical signs of photo-oxidation, e.g. an increase in the peak at 1711 cm-1 was observed [9].

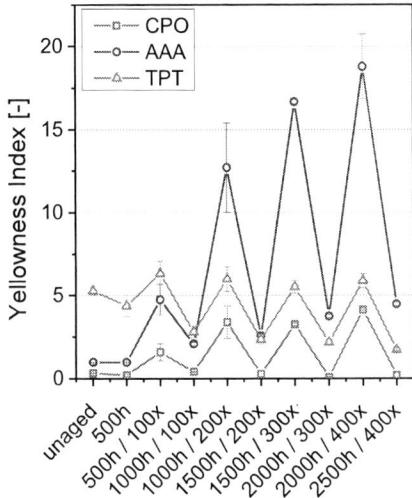

Fig. 3. Yellowness index of test coupons

By comparison, the maximum yellowness index of the PP backsheet was around 4 after 2000 h of Xenon exposure and 400 thermal cycles. Also, photobleaching leads to a full recovery of the samples, resulting in YI values around 0 after each Xenon exposure step. FTIR-ATR spectra, however, indicate surface oxidation of the directly exposed PP layer, a phenomenon that has already been reported in previous studies [3,4]. A similar behavior was observed for the TPT backsheet. Interestingly there the photobleaching leads to a YI value lower than the initial unaged one.

IV. Summary and Conclusion

In this work the crack susceptibility of co-extruded backsheets based on PA and PP was investigated using solder bump coupon specimens. The laminated samples were exposed to a sequential test consisting of 500 h of xenon exposure and 100 thermal cycles from -40 °C to 85 °C, which was repeated five times. Similar to TPT, also PP based backsheets exhibit excellent stability, and so far, there is no indication that the backsheets are susceptible to cracking after 4 test cycles. By comparison, the PA backsheet exhibited first cracking after 3 test cycles (1500 h of Xenon 300 thermal cycles). Overall, the study confirms that co-extruded PP backsheets show great potential to be a valid replacement of standard PET based backsheets in PV modules.

Acknowledgment *(Heading 5)*

This work was conducted as part of the Austrian "Energy Research Program" project "PV Re² - Sustainable Photovoltaics" (FFG No. 867267) funded by the Austrian Climate and Energy Fund and the Austrian Research Promotion Agency (FFG).

V. References

1. Oreski G, Wallner GM. Aging mechanisms of polymeric films for PV encapsulation. *Solar Energy* 2005; **79(6)**: 612–7, DOI: 10.1016/j.solener.2005.02.008.
2. Geretschläger KJ, Wallner GM, Fischer J. Structure and basic properties of photovoltaic module backsheet films. *Solar Energy Materials and Solar Cells* 2016; **144**: 451–6, DOI: 10.1016/j.solmat.2015.09.060.
3. Omazic A, Oreski G, Edler M, Eder GC, Hirschl C, Pinter G, Erceg M. Increased reliability of modified polyolefin backsheet over commonly used polyester backsheets for crystalline PV modules. *J. Appl. Polym. Sci;* **2020(137)**: 48899, DOI: 10.1002/app.48899.
4. Oreski G, Eder GC, Voronko Y, Omazic A, Neumaier L, Mühleisen W, Ujvari G, Ebner R, Edler M. Performance of PV modules using co-extruded backsheets based on polypropylene. *Solar Energy Materials and Solar Cells* 2021; **223**: 110976, DOI: 10.1016/j.solmat.2021.110976.
5. Omazic A, Oreski G, Halwachs M, Eder GC, Hirschl C, Neumaier L, Pinter G, Erceg M. Relation between degradation of polymeric components in crystalline silicon PV module and climatic conditions: A literature review. *Solar Energy Materials and Solar Cells* 2019; **192**: 123–33, DOI: 10.1016/j.solmat.2018.12.027.
6. Köntges M, Oreski G, Jahn U, Hacke P, Weiss K, Razzongles G, Paggi M, Parlevliet D, Tanahashi T, French R, Richter M, Morlier A, Tjengdrawira C, Li H, Berger KA, Makrides G, Herrmann W. *Assessment of*

Photovoltaic Module Failures in the Field: Report IEA-PVPS T13-09:2017; 2017.

7. Halwachs M, Neumaier L, Vollert N, Maul L, Dimitriadis S, Voronko Y, Eder GC, Omazic A, Mühleisen W, Hirschl C, Schwark M, Berger KA, Ebner R. Statistical evaluation of PV system performance and failure data among different climate zones. *Renewable Energy* 2019; **139**: 1040–60, DOI: 10.1016/j.renene.2019.02.135.
8. Gambogi W, Heta Y, Hashimoto K, Kopchick J, Felder T, MacMaster S, Bradley A, Hamzavytehrany B, Garreau-Iles L, Aoki T, Stika K, Trout TJ, Sample T. A comparison of key PV backsheet and module performance from fielded module exposures and accelerated tests. *IEEE J. Photovoltaics* 2014; **4(3)**: 935–41, DOI: 10.1109/JPHOTOV.2014.2305472.
9. Eder GC, Voronko Y, Oreski G, Mühleisen W, Knausz M, Omazic A, Rainer A, Hirschl C, Sonnleitner H. Error analysis of aged modules with cracked polyamide backsheets. *Solar Energy Materials and Solar Cells* 2019; **203**: 110194, DOI: 10.1016/j.solmat.2019.110194.
10. Aghaei M, Fairbrother A, Gok A, Ahmad S, Kazim S, Lobato K, Oreski G, Reinders A, Schmitz J, Theelen M, Yilmaz P, Kettle J. Review of degradation and failure phenomena in photovoltaic modules. *Renewable and Sustainable Energy Reviews* 2022; **159**: 112160, DOI: 10.1016/j.rscr.2022.112160.
11. Oreski G, Stein JS, Eder GC, Berger K, Bruckman L, French R, Vedde J, Weiß KA. Motivation, benefits, and challenges for new photovoltaic material & module

developments. *Prog. Energy* 2022; **4(3)**: 32003, DOI: 10.1088/2516-1083/ac6f3f.

12. Ehrenstein GW, Pongratz S. *Resistance and stability of polymers*. Munich: Hanser Publishers; 2013.

13. Owen-Bellini M, Moffitt SL, Sinha A, Maes AM, Meert JJ, Karin T, Takacs C, Jenket DR, Hartley JY, Miller DC, Hacke P, Schelhas LT. Towards validation of combined-accelerated stress testing through failure analysis of polyamide-based photovoltaic backsheets. *Scientific Reports* 2021; **11(1)**: 2019, DOI: 10.1038/s41598-021-81381-7.

14. M. D. Kempe, T. Lockman, J. Morse. Development of Testing Methods to Predict Cracking in Photovoltaic Backsheets. In: *2019 IEEE 46th Photovoltaic Specialists Conference (PVSC)*; 2019, pp. 2411–2416.

Comparative Life Cycle Assessment of Crystalline Silicon Glass-Sheet based PV Modules and Plastic PV Modules

Sakthi Guhan Somasundaram[a], Xitong Zhu[b], Angele Reinders[a,b]

a) University of Twente, Enschede, Overjissel, 7500 AE, The Netherlands, b) Eindhoven University of Technology, Eindhoven, North Brabant, 5612 AZ, The Netherlands.

Abstract— **A comparative life cycle assessment between plastic PV modules and conventional crystalline silicon glass-sheet based PV modules, shows environmental benefits for plastic PV modules. Plastic PV modules are produced by encapsulating crystalline silicon cells in polymeric materials. Using GaBi software, several LCA models were developed. When considering PV module's production location, GHG emissions for China were 2 times higher than Europe. With China as the production location, the GHG emissions for conventional and plastic PV modules during the production phases were 46 and 45 gCO₂eq/kWh respectively. During the disposal phase, carbon dioxide emissions were 57% lower for plastic PV modules.**

Keywords—*life cycle assessment, photovoltaic modules, plastic encapsulant, environmental impact, greenhouse gas emissions.*

I. INTRODUCTION

In this study, the environmental impact of different designs of PV modules is analyzed by means of life cycle assessment (LCA). The study is focused on the comparison of a glass-sheet based PV module (see Fig. 1) and a plastic PV module (see Fig. 2). Plastic PV modules comprise a new design that is currently in the developmental phase at University of Twente, the Netherlands. The proposed plastic PV module involves plastic manufacturing processes to encapsulate mono-crystalline silicon PV cells with a chosen transparent polymer, in this case Poly(methyl methacrylate) (PMMA) or Polycarbonate (PC). This approach aims at improving the recyclability of PV modules and shortening the production time and its associated labor. The LCA results are intended to help understand the environmental burdens and trade-offs corresponding to the production and end-of-life management options of these plastic PV modules as compared to conventional PV modules which are glass sheet based laminates containing silicon solar cells in between several other sheets like Tedlar foil, Ethylene Vinyl Acetate (EVA) foil, etc.

In previous LCA studies, the green house gas (GHG) emissions were mostly observed to be in the ranges of 25-40 and 70-85 gCO₂eq/kWh for the glass sheet based crystalline-silicon PV modules that were produced in Europe and China respectively [1]–[3]. The differences in the GHG emission ranges depend on the parameters such as the order level of inventory, scope and depth of the LCA study and the databases used in modelling. When it comes to the geography of European Union, LCA studies were mostly focused on Western and Southern countries, that includes Germany, Greece, France Spain, Italy, etc [4]–[6]. Previous LCA studies have shown that the carbon dioxide (CO_2) emissions were greatly reduced by 30% and 40% in Germany and European Union when compared to the PV modules produced in China [7]. This paper emphasizes on the effect of several environmental impact categories as most of the recent studies prioritize only on GHG emissions for a comparison based on the production and installation location of PV modules.

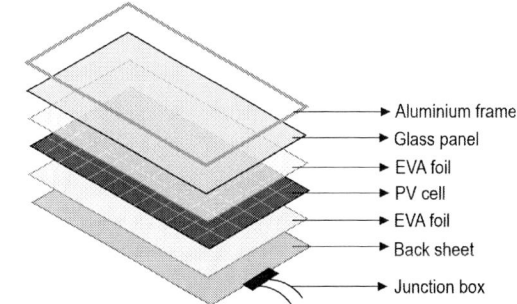

Fig. 1. Structure of a conventional glass-sheet based PV module

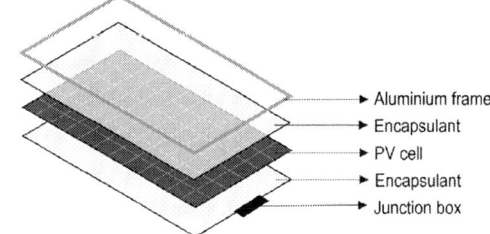

Fig. 2. Structure of a plastic PV module

So far, little attention is given for new variants of PV modules like glass and glass (G-G), glass and plastic (G-P) due to the unavailability of life cycle inventory (LCI) data [7][8]. This LCA study compares the single-crystalline plastic PV module (P-P type) and a conventional G type PV module. Therefore, this LCA study will be useful for the practitioners to understand the effect of a new PV module type on several environmental impact categories. In most LCA studies, the end-of-life management of PV modules is not frequently discussed

978-1-7281-6118-1/22 $31.00 © 2022 IEEE

as their processes and the inventory data is still being developed [8]. But in this study, the disposal phases for both the type of PV modules are discussed by constructing various scenarios based on the material recovery.

II. RESEARCH METHODOLOGY

This research paper is focused on comparing the aforementioned two types of PV modules and finding a better fit in terms of impact on environment and human lives. The research methodology used in this study is life cycle assessment which is schematically presented in Fig. 3.

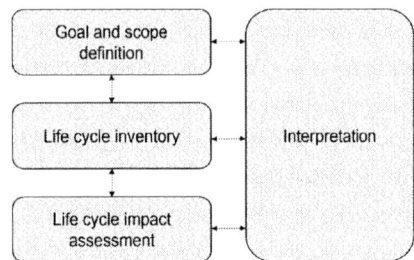

Fig. 3. Schematic representation of the LCA method

Life cycle assessment is a cradle-to-grave method that focuses on studying the comprehensive environmental aspects involved in the life cycle of a product, starting from raw material extraction to using and recycling the product. These studies help in decision making on the basis of informed choices, strategizing businesses and implementing governmental & environmental regulations in practice. ISO 14040 and 14044 provide the framework for LCAs . Usually an LCA consists of 4 phases:

1. Goal and scope definition, see Section A.

2. Life Cycle Inventory (LCI), see Section B.

3. Life Cycle Impact Assessment (LCIA), see Section C.

4. Interpretation, meant to interpret the results and discuss their accuracy, validity and consequences originating from the results.

The establishment of Photovoltaic Power System (PVPS) Programme Task 12 by International Energy Agency (IEA) ensures consistency, balance, reliability, and credibility of LCA results for various PV technologies [8]. For this reason, data for this LCA study has been obtained from PVPS Task 12 report. Generally, the ecoinvent database is used as it contains LCI information on photovoltaics. However, the IEA PVPS Task 12 does not recommend a particular database for an LCA study on photovoltaics [8]. Therefore, it was decided to use the default educational database available in the GaBi software for this study, as the PV technologies used in both the PV module types were the same, resulting in the same effects due to PV cells. The focus of this study is also on the effect of encapsulants on the environmental parameters, in particular their difference in different scenarios, which will be introduced in the following.

A. Goal and scope definition

The objective of an LCA study is defined by the goal, whereas system boundaries and other limitations are assessed in the scope. The comparisons between the two types of PV modules involve the influence of a) polymer type for plastic PV module and b) production location.

a) Polymer type for plastic PV module – PMMA and PC.

b) Production location – Europe (RER) and Asia Pacific (APAC), focusing on China (CN).

The functional unit framed for this study is *'To be able to produce power using a module area of $1m^2$ for 5 hours per day for 25 years'*.

It is assumed that both types of PV modules contain the same type of mono-crystalline silicon PV cells. A lifetime of 25 years was chosen for the PV modules while framing the functional unit. The system boundaries and assumptions that are crucial to this LCA study are listed below:

Production phase:

- The production phase includes raw materials and their extraction from ores. This also includes the usage of energy, water, material and processes to produce the raw materials.
- The transportation of raw materials from the mining site to the factories were considered; modes of transportation in this study are lorries and trains.
- The product packaging was not considered because of its irrelevance in the context of the objectives.

Use phase:

- The use phase includes the power generated by the solar modules. It was assumed that the energy produced by the two types of PV modules is same, because of the usage of same PV technology (mono c-Si).
- A lifetime of 25 years and a PV module efficiency of 20% based on the present standards and practices in the PV module manufacturing industry were assumed in accordance to IEA guidelines.
- A junction box was included but inverters, batteries, cables and other electronics were not considered.
- The support, repair and maintenance of the products were excluded from the LCA modelling.
- The transportation of products from the factories/construction site to the consumer markets were included, whereas the transportation of the used products to the disposal sites were excluded.

Disposal phase:

- Three techniques (recycling, landfill, incineration) were considered and the fractions for disposal criteria were assumed based on literature and relevant sources.

B. Life Cycle Inventory (LCI)

A Life Cycle Inventory (LCI) quantifies all elementary flows associated with individual processes in an LCA, i.e. inputs and outputs. Input flows are mass (materials and resources), energy flows and land use, while output flows are emissions to air, water, soil, and other products of the processes. In this study, the first order level of inventory was considered as it is the common approach practiced by the researchers in LCA. This means that the inflows and outflows that have direct interactions with the product-system were considered, including the raw materials and their production from ores. However, energy and

978-1-7281-6118-1/22 $31.00 © 2022 IEEE

material flows for the machines, water network, land conversion, techno-sphere and resources associated with the production sites were not considered. In this study, 7 LCA scenarios, also called GaBi models, were developed based on the encapsulant material and production location used, see Table I. Two production locations were chosen for comparison, to understand the primary reason for their differences in carbon footprints and implicate the importance of the energy resource type (renewable/non-renewable) used in the production processes.

TABLE I. VARIANT SETS FOR GaBi MODELLING

Phase	Type of PV panel	Location of production
Production	Conventional	China
	Plastic (PMMA)	China
	Plastic (PC)	China
	Conventional	Europe
	Plastic (PMMA)	Europe
Disposal	Conventional	China, Europe
	Plastic (PMMA)	China, Europe

C. Life Cycle Impact Assessment (LCIA)

Life Cycle Impact Assessment (LCIA) is used to understand the relevance of all the inputs and outputs in an environment framework. In an LCIA, the elementary flows that are obtained from the LCI are evaluated using a framework that convert these flows into potential impact indicators. The ReCiPe method is used for evaluating the LCIA results by focusing on the impact indicators, also called as environmental impact categories. There are 16 impact indicators namely, climate change, terrestrial acidification, freshwater eutrophication, ozone depletion, fossil depletion, freshwater ecotoxicity, human toxicity, ionizing radiation, marine ecotoxicity, marine eutrophication, metal depletion, natural land transformation, particulate matter formation, photochemical oxidant formation, terrestrial ecotoxicity and water depletion.

III. CONSTRUCTION OF GABI MODELS

The main framework of LCA includes the production, use and disposal phases of a product and each phase is modelled involving the creation of GaBi parameters - plans, processes and flows. In Fig. 4, the various life cycle stages of a product are represented. It starts with the production phase that includes raw material acquisition, followed by material processing and product manufacturing. The use phase represent the product utilization/application. The last stage of the product is the end-of-life management or disposal phase. In this phase, the products are decommissioned and are sent to waste treatment where they are either recycled, land-filled or incinerated. This generic scheme has been applied to the comparison of PV modules in our study.

A. Production phase:

The production processes of a conventional PV and plastic PV modules consists of 7 and 5 subassemblies respectively. They include PV cells, encapsulant, tabbing wires, aluminum frame, junction box, Ethylene Vinyl Acetate (EVA) films, back sheets and glass panels.

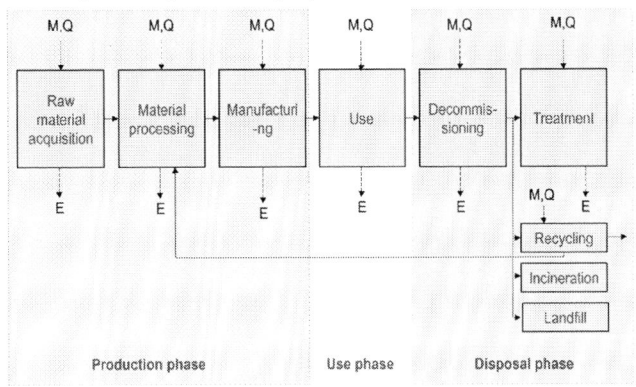

Fig. 4. Generic representation of various stages in a life cycle of a product

The PV cell production process is similar for both the plastic and conventional PV modules. It is based on the Siemens process, see Fig. 5, and modelled according to data available in the GaBi database. Further the materials, processes, energy flows differ only during the production phase of the different PV modules.

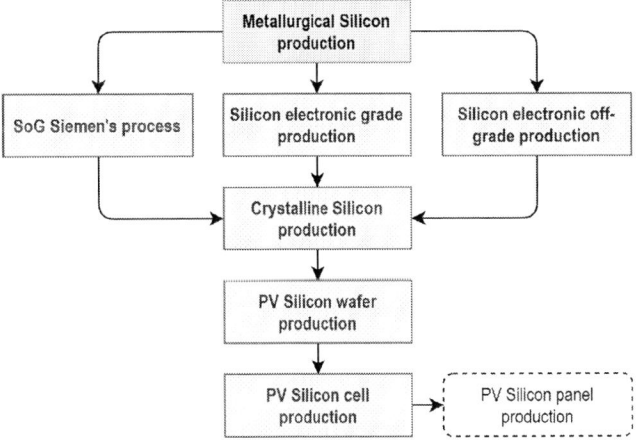

Fig. 5. Stages involved in the production of PV cells before PV module production

TABLE II. PART DESCRIPTION OF A CONVENTIONAL PV MODULE OF 1 M²

S.No.	Part	Sub-part	Material	Quantity
1.	PV cell	PV wafer Anti-reflection coating	Mono C-Si TiO$_2$	1.03 m² 30g
2.	EVA film	-	EVA	438g
3.	Bus/tabbing wire	-	Cu	150g
4.	Back sheet	-	PET	346g

S.No.	Part	Sub-part	Material	Quantity
5.	Glass panel	-	Solar glass	8810g
6.	Aluminum frame	-	Al alloy	2130g
7.	Junction box	Enclosure	PPO	41.2g
		Conducting strip	Cu	10g
		Bypass diode	Si/Se/Ge	12g
		Cable	Olefin, Cu	170g
		MC4 connector	PPO, Cu	21g

Seven sub-assemblies constitute the production of a conventional PV module and the part description is presented in Table II. The energy, water and transport requirements for all the processes were chosen from the PVPS Task 12 report [9]. The production process of the plastic PV module was split up into 5 subassemblies namely PV cell, encapsulant, tabbing wires, junction box and aluminum frame, see Table III. The PV solar cells are tabbed with copper strips. These copper strips were produced using extrusion and rolling processes. The PV cells were then aligned and encapsulated with the polymer, using a plastic manufacturing process. Finally, the product was connected to a junction box and fixed onto an aluminum frame for support and structural integrity. The electricity, water and transport requirements for extrusion and rolling of copper strips, and extrusion of aluminum frame were chosen based on the IEA's PVPS Task12 datasheet [9]. These resources were already mapped to the processes in the inventory. The energy and water requirements for the plastic manufacturing process was manually entered with the data similar to extrusion process due to its unavailability.

TABLE III.   PART DESCRIPTION OF A PLASTIC PV MODULE OF 1M^2

S.No.	Part	Sub-part	Material	Quantity
1.	PV cell	PV wafer	Mono C-Si	1.03 m^2
		Anti-reflection coating	TiO$_2$	30 g
2.	Encapsulant		PMMA	3070 g
3.	Bus/tabbing wire	-	Cu	150 g
4.	Aluminum frame	-	Al alloy	2130 g
5.	Junction box	Enclosure	PPO	41.2 g
		Conducting strips	Cu	10 g
		Bypass diode	Si/Se/Ge	12 g
		Cable	Olefin, Cu	170 g
		MC4 connector	PPO, Cu	21 g

B. Use phase:

The use phase of this LCA study focuses on the electricity production by the PV modules. A PV module efficiency of 20%, module area of 1m^2, performance ratio of 0.8 and a lifespan of 25 years were considered for the solar energy calculation. Energy, water and transportation used for service and maintenance were excluded as they seem to be insignificant for the study. As already mentioned, auxiliary equipment like inverters and batteries, along with distribution and transmission losses in the electricity grid were excluded. It was assumed that both the plastic and conventional PV modules produced the

same amount of electricity due to the same PV technology used. The schematic representation of the use phase is presented in Fig. 6.

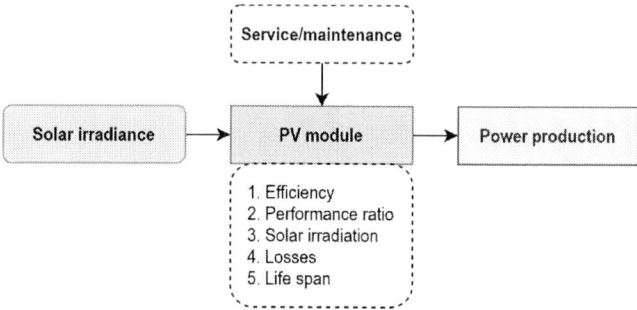

Fig. 6. GaBi modelling of the use phase of glass-sheet based and plastic PV modules

C. Disposal phase:

End-of-Life or waste management of PV modules is of paramount importance during the life cycle of PV modules. IEA has stated that recycling or repurposing the PV modules can unlock 78 million tonnes of raw materials by 2050 and their value could be as high as USD 15 billion [10]. Generally, PV modules can be recycled, incinerated and land-filled. The steps involved in the recycling process of PV panels are presented in Fig. 7.

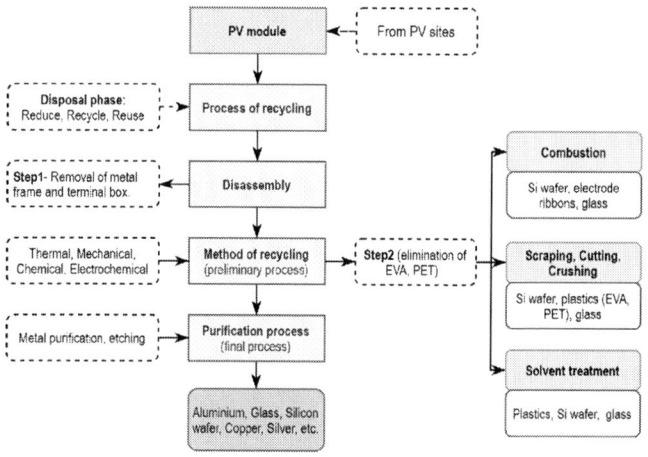

Fig. 7. Stages involved in the end-of-life management of PV module

The waste PV modules from different PV sites are brought together and disassembled. During the first step of this process, the metal frame and the terminal box are removed from the PV panel. The remaining part of the module is then subjected to either one of the following recycling methods. They are i) Thermal, ii) Mechanical, iii) Chemical and iv) Electro-Chemical. During this step, the EVA and PET sheets are eliminated and the silicon wafer, glass, electrode ribbons and plastic materials are separated in the next stage. Finally, these materials are subjected to metal purification and etching to recover aluminum, glass cullet, silicon, copper, silver, etc. In this LCA study, the disposal allocation percentages of recycling, incineration and landfill processes for each part of the PV modules were allocated based on the lifecycle inventories of IEA's PVPS Task 12 report which was collected from four recyclers in Central Europe that process c-Si panels [9].

TABLE IV. DISPOSAL ALLOCATION OF A CONVENTIONAL PV MODULE OF 1M²

Part	Sub-part	Recycling (%)	Incinera -tion (%)	Landfill (%)
PV wafer	-	65	-	35
Bus/tabbing wire	-	4.4	-	95.6
EVA film	-	-	100	-
Back sheet		20	80	-
Glass panel	-	66.2	-	33.8
Aluminum frame	-	12.1	-	87.9
Junction box	Copper strip Bypass diode Plastic (PPO)	4.4 - 82	- - 18	95.6 100 -
Soldering wire	-	-	100	-

TABLE V. DISPOSAL ALLOCATION OF A PLASTIC PV MODULE OF 1M²

Part	Sub-part	Recycling (%)	Incinera -tion (%)	Landfill (%)
PV wafer	-	65	-	35
Bus/tabbing wire	-	4.4	-	95.6
Encapsulant	-	92	-	8
Aluminum frame	-	12.1	-	87.9
Junction box	Copper strip Bypass diode Plastic (PPO)	4.4 - 82	- - 18	95.6 100 -
Soldering wire	-	-	100	-

The allocation percentages are presented in Table IV & Table V. In the process of disposal, PV wafers, copper parts and

aluminum frames can be recycled by 65%, 4.4% and 12.1% respectively. The waste slurry from metal parts - copper and aluminum is diverted to a landfill. The polyphenylene oxide parts (enclosure, casing of MC4 connectors) from junction box has the recyclability factor of 82%. The average electricity used by the recycling plant is assumed to be 0.111 kWh/kg [11]. The disposal allocation factor for the encapsulant was considered as 92% based on the recyclability of PMMA and PC from literature [12].

IV. RESULTS

Initial assessment of the LCA models showed that the plastic PV modules having the encapsulant material PMMA imparted lower impact on maximum number of midpoint categories compared to PC. Therefore, in the remainder of this LCA study, PMMA was selected for further comparisons. The midpoint categories considered here are climate change, ozone depletion, fossil depletion, human toxicity, marine ecotoxicity, particular matter formation and terrestrial ecotoxicity.

A. Impact on climate change:

The energy utilization and the CO_2 emission flows of various processes during the plastic PV module production in China and Europe are highlighted in the Sankey diagrams, see Fig. 8 & Fig. 9. The Czochralski process (CZ-Si) accounts for the highest contribution of emissions due to their highest energy utilization, followed by the SoG-Siemens process or PV panel production process. For the midpoint indicator climate change, the calculated CO_2 emissions for PV module production in China (180 $kgCO_2$eq) were 2 times higher as compared to Europe (92 $kgCO_2$eq), due to a higher share of coal fired plants in the Chinese electricity grid. The same pattern was seen for glass-sheet based conventional PV modules. The composition of the electricity grid of EU28 which was used in GaBi modelling was not available. So, the electricity grid mix data of Netherlands and Germany were used to present an idea of the resource allocation in Europe, see Table VI. The share of renewable energy resources for the electricity grid of Netherlands (13.5%) and Germany (38.4%) are approximately 50 and 144 times higher than that of China.

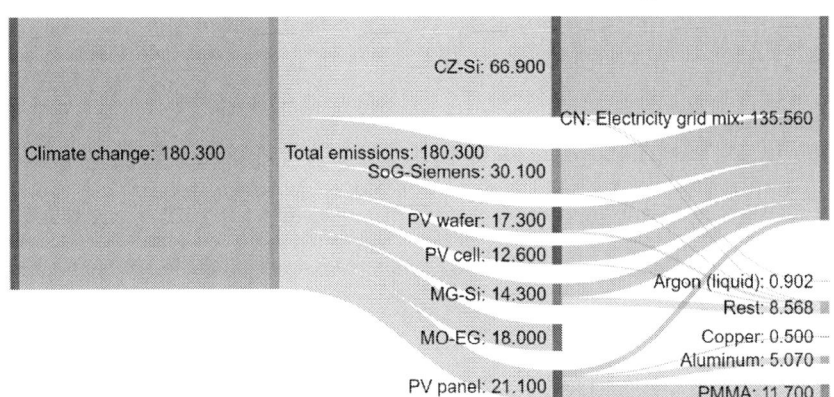

Fig. 8. Sankey diagram of carbon dioxide gas emissions in kgCO₂ for a plastic PV module during the production phase in China (electricity grid – China)

978-1-7281-6118-1/22 $31.00 © 2022 IEEE

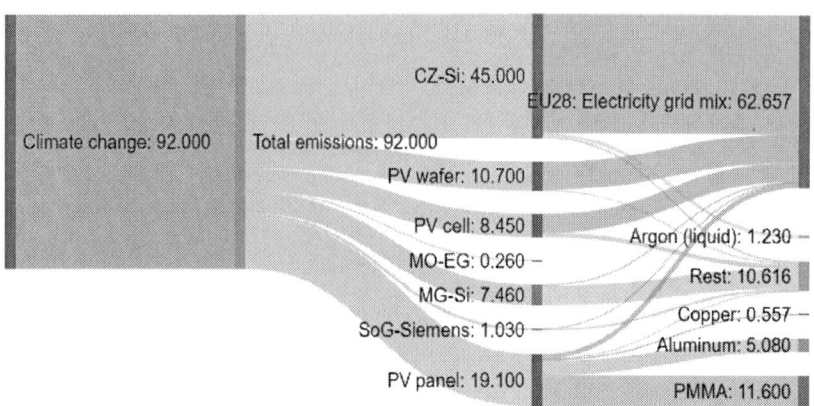

Fig. 9. Sankey diagram of carbon dioxide gas emissions in kgCO$_2$ for a plastic module during the production phase in Europe (electricity grid - European Union)

The highest potential of emission reduction can be achieved with renewable sources [13]. The energy utilization in the various stages of PV module production has changed in the recent years and the growth of renewables has contributed to a positive impact on environment and human lives [14]–[16].

TABLE VI. RESOURCE DISTRIBUTION FOR ELECTRICITY GRID MIX IN GABI (YEAR - 2018)

Type of resource	Energy source	China (%)	Netherlands (%)	Germany (%)
Non-renewable	Coal gases	2.31	2.84	1.77
	Hard coal	95.5	28.45	19.97
	Heavy fuel	0.21	1.85	0.95
	Natural gas	1.54	49.98	10.47
	Waste	0.18	3.43	2.27
	Lignite	-	-	26.20
Renewable	Hydropower	0.17	0.11	4.29
	Nuclear	0.01	3.96	16.35
	Photovoltaics	0.17	0.76	6.07
	Wind	0.01	5.61	9.66
	Biomass	-	3	2

B. Impact on other midpoint indicators:

In addition to climate change indicator, 6 other impact categories are discussed here. Figures 10, 11 and 12 represent the normalized values of 7 midpoint indicators for the complete lifecycles, production and disposal phases for the LCA scenarios - conventional PV (China), plastic PV (China), conventional PV (Europe) and plastic PV (Europe). Ozone depletion is caused by the chemical compounds such as chlorine, bromine, CFCs, etc. It is primarily the thinning of ozone layer that is present in the stratosphere above the Earth's surface. The ozone depletion levels for Plastic PV (PPV) and Conventional PV (ConvPV) modules during the production phases are similar. The fossil depletion impact category is based on the usage of fossil fuels. Similar to ozone depletion, the difference in values is contributed by PMMA that accounts for 6.72 kg oil eq. During the disposal phase, the PPV module accounts for lesser depletion of fossil (16.5% decrease) when compared to ConvPV as the number of individual components that end up in landfill/incineration are comparatively lesser and most of them are recycled.

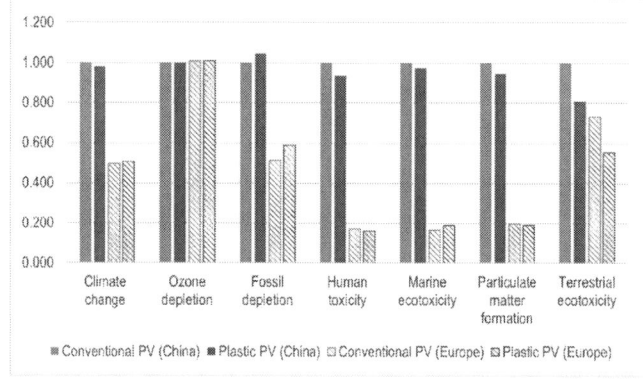

Fig. 10. Normalised values of the environmental impact categories for complete lifecycles of the PV modules evaluated by an LCA in this study

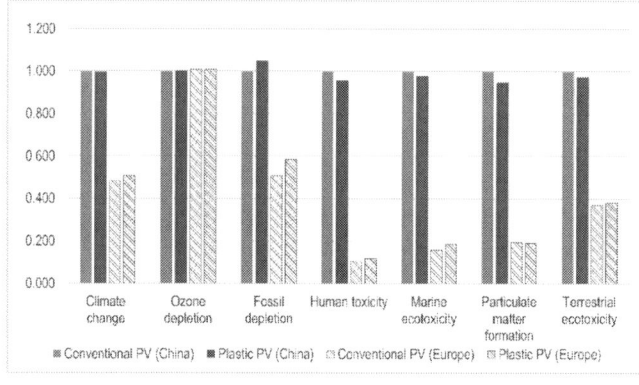

Fig. 11. Normalized values of the environmental impact categories in the production phase of PV modules evaluated by an LCA in this study

Human toxicity refers to the damage caused by toxic substances on the health of human beings. The variables like exposure, potency and severity are used in determining the characterization factors for human health. The human toxicity caused due to ConvPV modules are slightly higher (4.3%) than PPV modules in the production phase. During the disposal phase, there is a reduction of 32.1% for PPV modules when compared to ConvPV modules.

978-1-7281-6118-1/22 $31.00 © 2022 IEEE

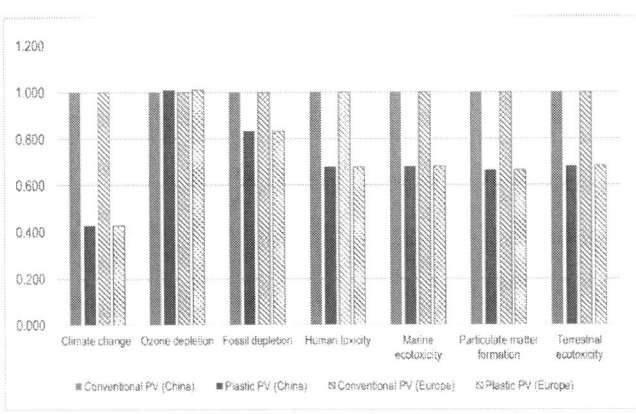

Fig. 12. Normalized values of the environmental impact categories in the disposal phase of PV modules evaluated by an LCA in this study

Moving to the next midpoint indicator, *marine ecotoxicity* is caused due to the discharge of harmful chemicals into marine habitats. It is lower for PPV (China) modules when compared to ConvPV (China) modules during both production and disposal phases. The PPV modules have lower impact in comparison to the ConvPV modules for particulate matter formation. The emissions of nitrogen oxides (NOx), ammonia (NH_3), sulphur dioxide (SO_2) and PM2.5 to the atmosphere contribute to *particulate matter formation*. The stages namely, production of crystalline silicon and SoG Siemen's process during the production phase account for the highest emissions in both the PV modules for this indicator. The study of harmful effects of chemicals on terrestrial organisms is termed as *terrestrial ecotoxicity*. Similar to previous trends, the emissions caused due to ConvPV modules are slightly higher than PPV modules. There is an increase of 2.6% and 31.6% in the emissions from ConvPV modules during the production and disposal phases respectively for this indicator. When the two types of modules are compared based on the production locations, the PPV modules imparted lesser environmental impacts for China and higher for Europe but in general, the enormous environmental improvements caused in Europe instead of China, outperforms this slightly higher impact. The PPV modules fared better in the disposal phase for both the locations.

C. Carbon footprints due to the effect of solar irradiation and PV module's installation locations:

In Europe, higher values of solar irradiation were witnessed in southern countries like Italy and Greece, with the values reaching up to 2000 kWh/m²/y. Lower values of 800 kWh/m²/y can be seen for northern countries [17]. Three countries were chosen as the installation locations for PV modules, based on the latitude to analyze the effect of solar irradiation on the lifecycle of PV systems - Greece (Athens - 37.9838 °N), France (Paris - 48.8566 °N) and Netherlands (Amsterdam - 52.3676 °N). The PV module's efficiency, module area and life span were considered as 20%, 1m² and 25 years respectively. The recent literature studies suggests that PR can vary having values lower than 65% and also higher than 90% [18]. Therefore, a standard value of 80% was assumed for this study. The average

solar irradiation presumed for Greece, France and Netherlands were 1700, 1310 and 1000 (kWh/m²/y) [19] and the calculated values of energy production were 6800, 5240 and 4000 (kWh/m²/y) respectively. From GaBi modelling, the emissions of the plastic PV module were observed as 180 kgCO₂eq in the production phase. With this value, the emission values (in gCO₂eq/kWh) were calculated for the use phase using the below Equation 1.

$$Emissions = \frac{GWP}{G * PR * \eta * r} \qquad (1)$$

GWP is the global warming potential (gCO₂/m²/y), G is the solar irradiation (in kWh/m²/y), PR is the performance ratio (in %), η is the efficiency of PV module (in %) and r is the life span of PV module. Fig. 13 represents the values for the three PV module installation locations. The values were 45.2, 34.6 and 26.6 gCO₂eq/kWh for Netherlands, France and Greece respectively.

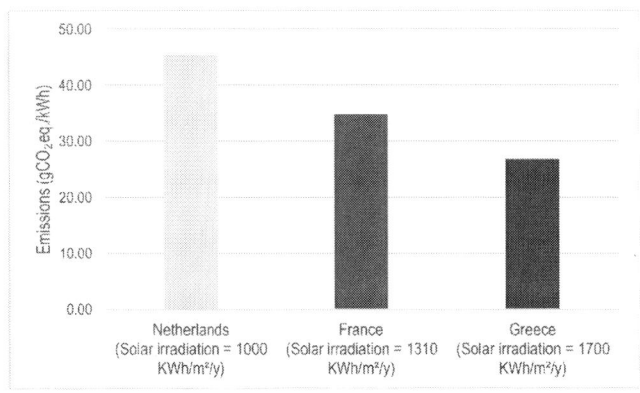

Fig. 13. Effect of CO₂ gas emissions due to change in locations based on latitudinal differences for a plastic PV module

The CO₂ eq. emissions per kWh for Netherlands were higher compared to France and Greece for both the PV module types. Greece emerged with the lowest values on comparison due to higher solar irradiation experienced. This implies that the kgCO₂eq emitted for every kWh of energy produced decreased due to higher solar irradiation.

D. Carbon footprints due to change in disposal allocation factors:

The main objective of the newly proposed plastic PV modules is to reduce the carbon footprint of PV technologies. This comes into play with minimizing the wastage of materials during the production and disposal phases. Based on the varied combination of recycling+landfill allocation factors for PV wafer, 4 scenarios (A, B, C, D) were developed to understand the extent of the emission reduction potential. The different allocation percentages considered for recycling+landfill were 70%+30% (A), 75%+25% (B), 80%+20% (C) and 85%+15% (D). These formulated scenarios were analyzed using GaBi software with a focus on climate change midpoint indicator. Increase of recycling percentages reduced the CO₂ emissions to a greater extent. The highest reduction of 5.3% was obtained for scenario D, having the highest recycling percentage.

V. DISCUSSION AND CONCLUSIONS

A comparative life cycle analysis was carried out between glass-sheet based conventional PV modules and plastic PV modules on the basis of polymer type, production location, use location and end-of-life management. The CO_2 gas emissions during the use phase were compared for Greece, France and the Netherlands. Greece resulted in the lowest life cycle emission of 26.6 gCO_2eq/kWh for a lifetime of 25 years. The CO_2 gas emissions for the Netherlands are 41% and 24% higher than Greece and France respectively. These values tend to be in the same range as previous studies under similar conditions. With the latest PV modules aiming for a lifetime of 30 or even 40 years, the extent of emission reduction is said to be significantly higher – 16% and 37% respectively. The emission reduction potential for the plastic PV module rely greatly on the encapsulant's lifetime. It is to be noted that the encapsulant's lifespan of 25 years was based on the advanced developments available in the current scenario. For more conservative values of 10 or 15 years, the plastic modules may fall behind the glass-sheet based PV modules in the midpoint point categories.

In a plastic PV module, several factors including the choice of polymer, additives, manufacturing method determine the total PV module's lifespan. New developments to increase the lifetime of encapsulants could potentially further increase the emission reduction potential. Since higher recyclability is one of the proposed advantages of plastic PV modules, different recycling percentages were considered in the LCA analysis. The recycling percentages for PV wafers in the disposal phase were increased by 5%, 10% and 15%. As a result, their corresponding normalized CO_2 emission values decreased by 1.6%, 3.8%, 5.3%. The recycling percentages of encapsulant, junction box were based on material manufacturer's data due to unavailability and inconsistency of data in the literature. The LCI inventory used in this study is based on the Task12 report of the year 2020. In recent times, the production of silicon wafer has been significantly improved every year, with a reduction in the cumulative energy demand [6]. Therefore, these uncertainties shall affect the accuracy of the end result of this study. To conclude this LCA study, the findings show that the adverse impact caused by the PV modules were significantly lower when the manufacturing location would not be China but Europe instead. In the near future, the adaptation of plastic PV module manufacturing methods for different technologies like perovskites or copper indium gallium selenide (CIGS) could definitely pave way for the realization of a significant reduction of carbon footprints [20].

REFERENCES

[1] M. J. De Wild-Scholten, "Energy payback time and carbon footprint of commercial photovoltaic systems," Sol. Energy Mater. Sol. Cells, vol. 119, pp. 296–305, 2013, doi: 10.1016/j.solmat.2013.08.037.

[2] D. Yue, F. You, and S. B. Darling, "Corrigendum to 'Domestic and overseas manufacturing scenarios of silicon-based photovoltaics: Life cycle energy and environmental comparative analysis' [Solar Energy 105 (2014) 669-678]," Sol. Energy, vol. 107, pp. 380–380, 2014, doi: 10.1016/j.solener.2014.06.001.

[3] E. Leccisi, M. Raugei, and V. Fthenakis, "The energy and environmental performance of ground-mounted photovoltaic systems - A timely update," Energies, vol. 9, no. 8, 2016, doi: 10.3390/en9080622.

[4] A. Beylot et al., "Environmental impacts of large-scale grid-connected ground-mounted PV installations," Renew. Energy, vol. 61, pp. 2–6, 2014, doi: 10.1016/j.renene.2012.04.051.

[5] A. Dominguez-Ramos, M. Held, R. Aldaco, M. Fischer, and A. Irabien, "Prospective CO2 emissions from energy supplying systems: Photovoltaic systems and conventional grid within Spanish frame conditions," Int. J. Life Cycle Assess., vol. 15, no. 6, pp. 557–566, 2010, doi: 10.1007/s11367-010-0192-3.

[6] V. Fthenakis and E. Leccisi, "Updated sustainability status of crystalline silicon-based photovoltaic systems: Life-cycle energy and environmental impact reduction trends," Prog. Photovoltaics Res. Appl., vol. 29, no. 10, pp. 1068–1077, 2021, doi: 10.1002/pip.3441.

[7] A. Müller, L. Friedrich, C. Reichel, S. Herceg, M. Mittag, and D. H. Neuhaus, "A comparative life cycle assessment of silicon PV modules: Impact of module design, manufacturing location and inventory," Sol. Energy Mater. Sol. Cells, vol. 230, no. April, p. 111277, 2021, doi: 10.1016/j.solmat.2021.111277.

[8] R. Frischknecht, G. Heath, M. Raugei, P. Sinha, M. de Wild - Scholten, V. Fthenakis, H. C. Kim, E. Alsema and M. Held, 2016, Methodology Guidelines on Life Cycle Assessment of Photovoltaic Electricity.

[9] P. Stolz, R. Frischknecht, K. Wambach, P. Sinha, and G. L. B.-S.-F. Heath, Life cycle assessment of current photovoltaic module recycling, vol. 13. 2017.

[10] C. E. L. Latunussa, F. Ardente, G. A. Blengini, and L. Mancini, "Life Cycle Assessment of an innovative recycling process for crystalline silicon photovoltaic panels," Sol. Energy Mater. Sol. Cells, vol. 156, pp. 101–111, 2016, doi: 10.1016/j.solmat.2016.03.020.

[11] IRENA, "Renewables Information 2019 – Analysis - IEA," 2016, [Online]. Available: https://www.irena.org/publications/2016/Jun/End-of-life-management-Solar-Photovoltaic-Panels.

[12] M. Huijbregts et al., "ReCiPe 2016 - A harmonized life cycle impact assessment method at midpoint and endpoint level. Report I: Characterization," Natl. Inst. Public Heal. Environ., p. 194, 2016, [Online]. Available: https://www.rivm.nl/bibliotheek/rapporten/2016-0104.pdf.

[13] O. Kanz, A. Reinders, J. May, and K. Ding, "Environmental impacts of integrated photovoltaic modules in light utility electric vehicles," Energies, vol. 13, no. 19, pp. 1–13, 2020, doi: 10.3390/en13195120.

[14] E. Alsemaa, "PERGAMON Renewable and Sustainable Energy Reviews 1 "0887# 276Ð304 Energy requirements of thin solar cell modules review."

[15] V. M. Fthenakis, Life cycle impact analysis of cadmium in CdTe PV production, vol. 8, no. 4. 2004.

[16] D. A. R. Barkhouse, O. Gunawan, T. Gokmen, T. K. Todorov, and D. B. Mitzi, "Yield predictions for photovoltaic power plants:empirical validation,recent advances and remaining uncertainties," Prog. Photovoltaics Res. Appl., vol. 20, no. 1, pp. 6–11, 2015, doi: 10.1002/pip.

[17] A. Louwen, W. G. J. H. M. van Sark, W. C. Turkenburg, R. E. I. Schropp, and A. C. Faaij, "R&D integrated life cycle assessment: A case study on the R&D of silicon heterojunction (SHJ) solar cell based pv systems," 27th Eur. Photovolt. Sol. Energy Conf. Exhib., no. September, pp. 4673–4678, 2012, doi: 10.4229/27thEUPVSEC2012-6CV.4.15.

[18] E. Bellini, "Nrel's performance ratio method applied to thousands of pv systems," 2020. https://www.pv-magazine.com/2020/10/30/nrels-performance-ratio-method-applied-to-thousands-of-pv-systems/.

[19] A. Müller, L. Friedrich, C. Reichel, S. Herceg, M. Mittag, and D. H. Neuhaus, "A comparative life cycle assessment of silicon PV modules: Impact of module design, manufacturing location and inventory," Sol. Energy Mater. Sol. Cells, vol. 230, no. June, p. 111277, 2021, doi: 10.1016/j.solmat.2021.111277.

[20] E. Leccisi and V. Fthenakis, "Life cycle energy demand and carbon emissions of scalable single-junction and tandem perovskite PV," Prog. Photovoltaics Res. Appl., vol. 29, no. 10, pp. 1078–1092, 2021, doi: 10.1002/pip.3442.

Study of ALD-grown Tin Oxide as an Electron Selective Layer for NIP Perovskite-Based Solar Cells

Félix Gayot[1], Elise Bruhat[1], Matthieu Manceau[1], Eric De Vito[2], Denis Mariolle[3], Stéphane Cros[1]

[1] Univ. Grenoble Alpes, INES, F-73375 Le Bourget du Lac, France - CEA Liten, Department of Solar Technologies, F-73375 Le Bourget du Lac

[2] Univ. Grenoble Alpes, CEA Liten, Department of Nanomaterial technologies, 38054 Grenoble Cedex, France

[3] Univ. Grenoble Alpes, CEA, Leti, F-38000 Grenoble, France

Abstract—This work presents a comparative study between tin(IV) oxide (SnO₂) thin films deposited either by solution process or by Atomic Layer Deposition (ALD) for an application as an electron selective layer in perovskite/silicon tandem solar cells. This study is motivated by the usually lower performances of solar cells using electron selective layer (ESL) made of ALD-grown SnO₂ compared to ones using a solution-processed ESL. Chemical, electrical, optical and topographical properties of each type of film were investigated. In an attempt to link thin film properties to device characteristics single-junction perovskite solar cells and perovskite/silicon tandem solar cells were fabricated. Despite the high-quality conductivity and optical properties of ALD-grown SnO₂, perovskite-based solar cells employing such film showed limited performances. Characterization of perovskite films properties grown on both type of SnO₂ did not rise significant differences and tend to indicate some hindering factors at the ALD-grown SnO₂ interface with perovskite. Specifically, a larger workfunction for ALD-grown SnO₂ may create a potential barrier for electron extraction at perovskite interface

Keywords—perovskite, solar cells, ALD, electron selective layer, tin oxide

I. Introduction

The reduction of thermalization losses enabled by stacking a wide bandgap perovskite (PK) top-cell (~1.65 eV) on a crystalline silicon (c-Si) bottom-cell (1.12 eV) allows perovskite/silicon (PK/c-Si) tandem solar cells to show outstanding theoretical efficiency limits above 35% [1], [2], thereby overcoming the standard c-Si solar cells efficiency limit of 29.8 % [3]. Today, PK/c-Si tandem cells have already reached 29.8 % of power conversion efficiency at lab scale[4]. If c-Si solar cells are the industrial standard that dominates the global market [5], PK single-junction (SJ) solar cells on the other hand are less mature, despite an exceptionally quick increase in efficiencies above 25 % at the lab scale over the last decade [6].

This maturity level difference and the fact that c-Si and PK-based devices rely on completely different fabrication processes rise several technical issues to develop functional PK/c-Si tandem cells. In the case of 2-terminal tandem cells, where the PK and c-Si sub-cells form a single monolithic stack, strict process compatibility is required and subcells combination needs to fulfil current matching to reach high efficiencies [7].

Besides, a lack of industrial technique to deposit PK cell layers on top of large and/or textured areas makes the fabrication of commercially viable PK SJ and PK/c-Si tandem cells yet to be achieved. That is why a lot of research work focuses on developing new fabrication processes for conformal deposition on large and textured areas for each PK cell constitutive layers. While some state-of-the art SJ PK solar cells present a so-called NIP architecture [8]; best results have been achieved so far in the PIN configuration for tandem cells [9], [10]. Nevertheless, NIP tandem architecture shows promising results as well [11], [12].

In NIP-structured PK-based devices, the Electron Selective Layer (ESL) lies below the photoactive PK film. The latter must be able to nucleate and grow as a homogeneous film on top of it. Besides, the ESL must allow electrons extraction from the PK conduction band while preventing such transfer for photogenerated holes. Furthermore, it is essential for the ESL to show high transparency in the 300-750 nm range for SJ PK cells or in the 750-1100 nm for PK/c-Si tandem cells.

Tin(IV) oxide (SnO₂) is currently one of the main candidates to make highly efficient ESL for PK and PK/c-Si tandem solar cells thanks to its large bandgap, high electronic mobility and its n-type nature [13]. Solution-based processes are mainly reported for SnO₂ ESL deposition in NIP devices [8], [14]. However, regarding future scale-up developments, solvent-based processes seem not the most appropriate for PK cell layers deposition on top of large and textured substrates [15]–[17]. Atomic Layer Deposition (ALD) seems particularly attractive for ESL deposition [18]. Its cyclic nature of segregated injections of precursors allows conformal, dense and non-aggressive thin film growth, with a fine control over thicknesses [19].

Although effective ESL were fabricated using ALD-grown SnO₂ (SnO₂^{ALD}) for PK and PK/c-Si tandem solar cells [20]–[24], these studies outline interfacial defects between SnO₂^{ALD} and the PK absorber [24]. Adding another thin film such as C₆₀ to SnO₂^{ALD} layer to form bi-layered ESL has proven to passivate such defects [24], [25]. Therefore, state-of-the-art tandem cells employing SnO₂^{ALD} use bi-layered ESL [4], [9], [10] although it would be preferable to use a monolayer to simplify the fabrication process and limit the cost of the device.

However, in such PIN Pvk-based devices, this Pvk-C_{60} interface is reported as performance-limiting [26]. Furthermore, regarding PK single junction cells, reported SnO_2^{ALD}-based performances remain lower than what can be achieved when solution-processed SnO_2 (SnO_2^{SP}) is employed as ESL [14], [27].

This work aims at better understanding SnO_2^{ALD} limitations when employed as single-layer ESL. A systematic comparison between a SnO_2^{SP} and SnO_2^{ALD} has been conducted. Chemical, electrical, optical and topographical properties of each type of film were investigated. Similarly, some interfacial properties of the ESL/PK junction were compared for each kind of SnO_2. Whenever possible, investigated SnO_2 thin films were integrated in PK single-junction and PK/c-Si tandem solar cells in an attempt to link material properties and device performances.

II. EXPERIMENTAL DETAILS

A. Device fabrication

1) PK SJ solar cell

Single junction PK solar cells were fabricated according to the NIP-type architecture depicted in Fig. 1a. 30 nm-thick SnO_2^{ALD} layers were grown on top of commercial glass/ITO substrates at 150°C from Tetrakisdimethylamino-tin(IV) (TDMASn) and hydrogen peroxide (H_2O_2) as tin precursor and oxidizing reactant, respectively. 30 to 40 nm-thick SnO_2^{SP} layers were prepared by spin-coating of a water-based SnO_2 nanoparticles dispersion on UV-O_3 treated substrates. Spin-coated samples underwent annealing at 80°C for 1 minute. Bi-layered ESL were fabricated by evaporating 10nm of C_{60} on top of SnO_2. The PK precursor solution was subsequently spin-coated under N_2 atmosphere on top of SnO_2 layer to form $Cs_{0.05}FA_{0.95}Pb(I_{0.83}Br0_{.17})_3$ double-cation PK. In the case of SnO_2^{ALD}, substrates require a UV-O_3 treatment during 30 minutes before PK coating to improve surface wettability. 150μL of chlorobenzene was dropped 5s before the spinning of the PK precursors solution ends. Afterwards, films were annealed at 100°C for one hour. Then, doped Poly(triarylamine) (PTAA) was spin-coated on top of PK to form a 200 nm-thick hole selective layer. Finally, a gold top contact was evaporated on PTAA.

1) PK/c-Si tandem solar cell

PK/c-Si tandem solar cells were fabricated according to the NIP-type architecture depicted in Fig. 1.b. From a 6-inch Float-zoned double side polished n-type c-Si wafer, passivating intrinsic and doped hydrogenated amorphous silicon layers (a-Si:H) were grown on each wafer side by Plasma-Enhanced Chemical Vapour Deposition (PECVD). A 12 nm-thick Indium Tin Oxide (ITO) layer was subsequently deposited on a-Si:H (p) to form the recombination junction. The same deposition technique was used to deposit a thicker ITO layer on the a-Si:H (n) to form the back contact. PK top-cell layers were then deposited according to the same process as depicted in section II.A.1) A 100 nm-thick ITO layer was sputtered on top of PTAA to form a transparent top contact. Finally, thermal evaporation of gold metal lines was performed to collect current out of the device. In addition, SnO_2^{ALD} ESL-based cell had their edges laser cut to avoid bottom cell shunting by the ALD deposition.

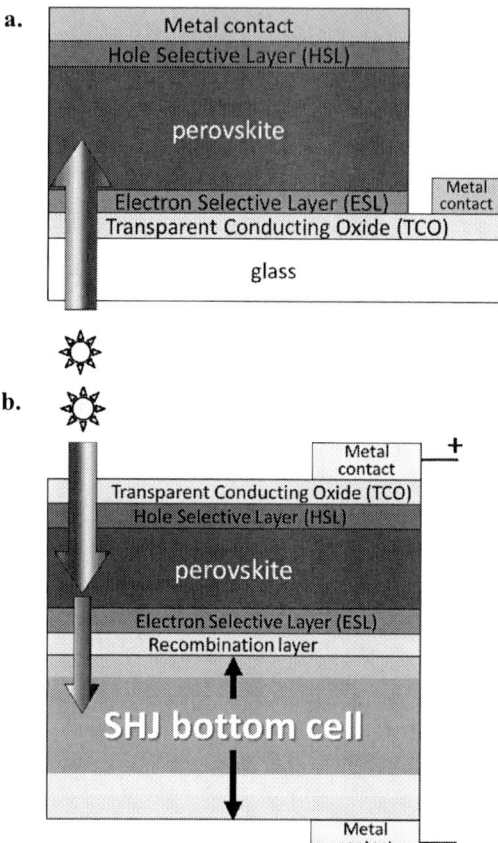

Figure 1 - Schematic of a PK single junction cell (a.) and of a PK/c-Si tandem cell (b.) both with the NIP architecture

B. Characterisation method

SnO_2 thin film chemical environment was examined through Hard X-ray Photoelectron Spectroscopy (HAXPES) with a Cr-Kα X-ray source on a PHI Quantes XPS/HAXPES tool, on glass/SnO_2 and glass/ITO/SnO_2 test samples. Atomic Force Microscopy (AFM) and Kelvin Probe Force Microscopy (KPFM) characterisations were performed with a Bruker Dimension Icon instrument in Tapping Mode with amplitude modulation on glass/ITO/SnO_2 test samples. Scanning Electron Microscope (SEM) images of textured SnO_2 depositied on silicon heterojunction (SHJ) bottom cells samples were recorded in a FEI Nova Nanosem instrument. Optical absorption was measured with a Perkin-Elmer UV-visible-nIR spectrophotometer on glass/ITO/SnO_2 samples. Electrical sheet resistance of glass/SnO_2 samples was measured by 4-probe method on a Napson tool. Hall effect measurement done with an Ecopia instrument from Bridge Technology confirmed the SnO_2^{ALD} resistivity found by 4-probe measurements but could not be performed on SnO_2^{SP} layers which are too resistive. Steady-state photoluminescence spectra were recorded with F-4500 FL spectrophotometer (Hitachi), with a Xenon lamp. Each glass/ITO/SnO_2/PK sample was illuminated during 1 minute prior to measurement to stabilize photoluminescence yield. Device cross-section images were recorded by Transmission Electron Microscopy (TEM) with a TEM Tecnai Osiris

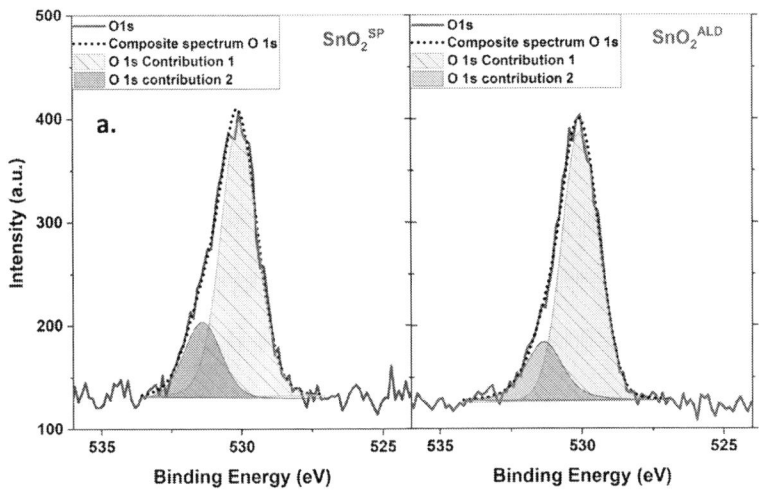
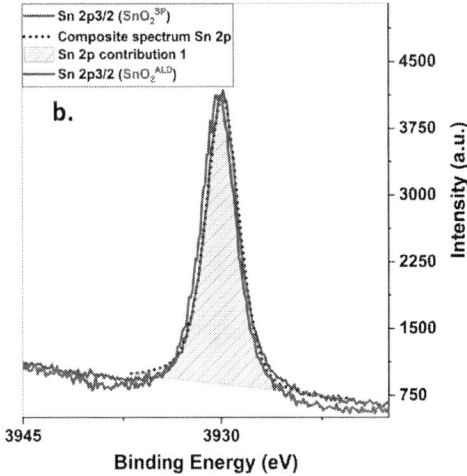

Fig. 2 - Fitted HAXPES spectra centred on the O 1s binding energies region for SnO_2^{SP} and SnO_2^{ALD} (a.) and HAXPES spectra centred on $Sn2p_{3/2}$ binding energies region of SnO_2^{SP} and SnO_2^{ALD} (b.), where only the fitting curve for SnO_2^{SP} sample is shown.

instrument at an acceleration voltage of 200 keV. Elemental detection was performed by Energy Dispersive X-ray (EDX).

III. RESULTS

ALD-grown films were first chemically characterized by HAXPES to ensure the SnO_2 stoichiometry of these films. Their electric, optical properties and morphological properties were then analyzed and compared to the ones of reference SnO_2^{SP} thin films. SnO_2^{ALD} ESL were then integrated in PK single junction and PK/c-Si tandem cells. Based on these cells' properties, further characterizations were conducted to examine the SnO_2^{ALD}-PK interface.

A. SnO₂ thin film properties

HAXPES measurements were performed on 30nm-thick ALD-grown and solution-processed films to check their stoichiometry. The large X-rays energy provided by the Cr lamp can probe the films over a depth of about 20nm. Such in-depth X-ray penetration enables to probe bulk chemical environment. Still, in Fig. 2.a, a clear shoulder at high binding energies is seen for the O 1s peak of both SnO_2^{SP} and SnO_2^{ALD}, which is attributed to -OH species. In Fig. 2.b, Sn $2p_{3/2}$ peak of each SnO_2 shows a single contribution (for the sake of clarity,

only SnO_2^{SP}-related fitting curve is drawn). In addition, the absence of signal in the N 1s and C 1s binding energies regions (Fig. 2c.) testified that no precursor residuals remain in ALD-grown films. Quantitative analysis after peaks fitting allowed us to check SnO_2^{ALD} stoichiometry with an O to Sn ratio equal to two. One can further note in Fig. 2.b that peaks rise at same binding energies for both SnO_2 films, which ensures a similar chemical environment for both types of SnO_2.

As one can see in Table I, SnO_2^{ALD} layers show a close optical absorption with respect to SnO_2^{SP} films. On the contrary, SnO_2^{ALD} layers conductivity is about 40,000 times larger than for SnO_2^{SP} thin films. These measured properties are thus not limiting for a use of $SnO2^{ALD}$ in PK-based solar cells, knowing that reference SnO_2^{SP} has proven to be an efficient ESL in such devices. Moreover, SnO_2^{ALD} much larger conductivity may allow the investigation of a wider range of layer thicknesses without affecting solar cells series resistance.

SEM image in Fig.3.c demonstrates that SnO_2^{SP} films cannot cover c-Si bottom cell textured surface. On the contrary, Fig. 3.a and 3.b highlight ALD capability to grow SnO_2 ESL on top of large and textured substrates such as a c-Si bottom cell. It can be seen that the SnO_2^{ALD} layer deposited on top of the c-

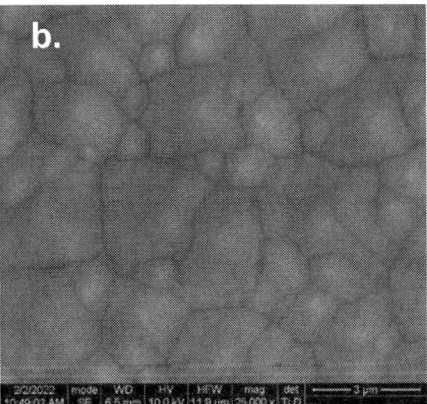
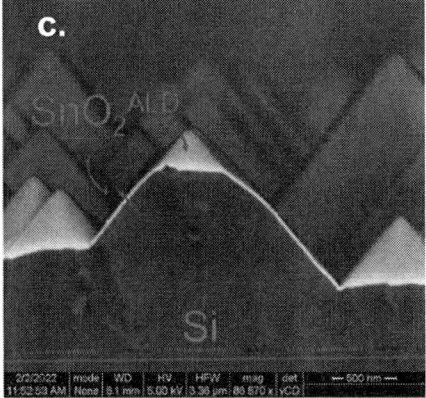

Fig. 3 - Cross-section (a.) and top-view SEM images of a SnO_2^{ALD} (b.) and thin of a SnO_2^{SP} (c.) film grown on top of a textured SHJ bottom cell.

Si bottom cell's pyramidal surface is conformal with a homogeneous thickness of 33 ± 0.5 nm.

B. Integration of SnO_2^{ALD} as an ESL in PK-based solar cells

Based on the characterized SnO_2^{ALD} thin films properties, PK/c-Si tandem cells that integrate a SnO_2^{ALD} ESL were fabricated. Current density - voltage (J(V)) curves in Fig. 4.a show that contrary to the reference with SnO_2^{SP}, the SnO_2^{ALD} ESL-based tandem cell does not feature a proper tandem characteristic. The large open-circuit voltage (V_{OC}) difference from V_{OC} =1.7 V (reference) to V_{OC} = 0.85 V (SnO_2^{ALD} -based tandem) clearly indicates that one of the subcells does not fully contribute to voltage generation. The cell edges cutting (see section 1.1.2) makes unlikely a sub-cell shorting due to the ALD process. Since no reverse breakdown is observed in Fig. 4.a, the SnO_2^{ALD}-based tandem J(V) curve more probably attests of a particularly large shunt of the PK top-cell. Therefore, the SnO_2^{ALD} ESL-based device almost acts as a SHJ

a.

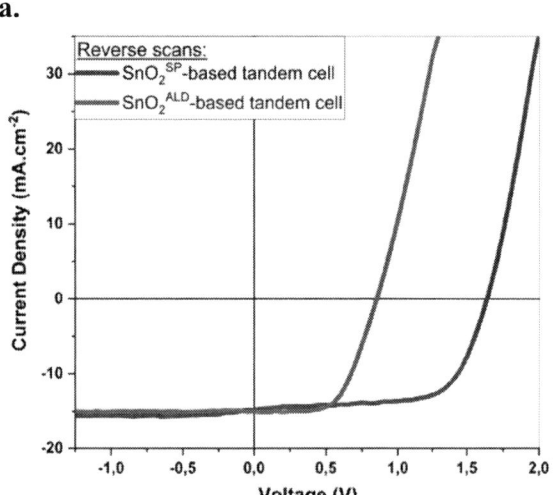

b.

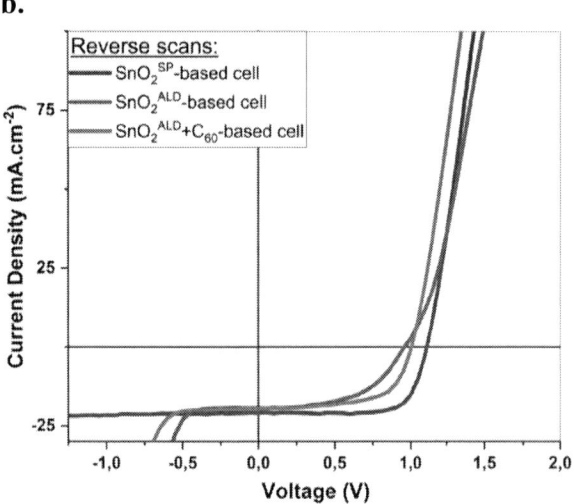

Fig. 4 - Current density-voltage characteristics of a PK/c-Si tandem cells (a.) and of single junction PK cells that use reference SnO_2^{SP} or SnO_2^{ALD} as an ESL or a SnO_2^{ALD}+C_{60} bilayer.

TABLE I. Effective optical absorption (A_{eff}) in the 300-1200nm range and electrical resistivity of both types of SnO_2 on glass.

	A_{eff} (300-800nm) (%)	A_{eff} (800-1200nm) (%)	ρ ($\Omega.m$)
SnO_2^{SP}	2.0	4.3	2.26
SnO_2^{ALD}	2.3	4.0	5.9×10^{-5}

cell, with a slightly enhanced open-circuit voltage coming from the strongly shunted PK top-cell. Such large shunts may occur through a high rate of non-radiative recombinations within the top-cell.

In order to dispose of a simpler test structure to study the impact of using SnO_2^{ALD} ESL, PK single-junction cell were also fabricated. J(V) curves for PK-based solar cells that integrate an ESL made of SnO_2^{ALD} or SnO_2^{SP} can be compared in Fig. 4.b. In agreement with previous reports [22], [24], SnO_2^{ALD}-based cells perform less than reference ones, and all photovoltaics parameters are affected. SnO_2^{ALD} ESL-based cells show average short current density (J_{SC}), V_{OC} and Fill Factor (FF) values of respectively 18.5 mA.cm^{-2}, 0.88 V and 46% whereas reference cells feature on average 21.1 mA.cm^{-2}, 1.08 V and 69% for J_{SC}, V_{OC} and FF respectively. A so-called "s-shape" that strongly degrades the maximum power output of SnO_2^{ALD}-based single junction cells can be observed. It may be the signature of an inadequate energy bands alignment which could involve an electron barrier in the system [28]. These cells also show a very weak reverse breakdown voltage, which can be attributed to poor hole blocking ability of the ALD-grown ESL. These two distinctive features seem to illustrate a lack of charge selectivity. As it is known to be an effective strategy to enhance performances, cells with a bi-layered ESL made of SnO_2^{ALD} and C_{60} were also fabricated to illustrate the origin of performance limitations. It can be seen from Fig. 4.b that adding a C_{60} thin film in between SnO_2^{ALD} and PK improves the device characteristics, agreeing with literature [25], [27]. Although V_{OC}, J_{SC} and reverse breakdown voltage remain quite degraded mainly due to a weak shunt resistance (not explained so far), the elimination of the s-shape improves the FF to an averaged value of 59%. Series resistance seems also reduced with the adjunction of C_{60}, especially around the V_{OC} with the suppression of the "s-shape", and owing for a better charge transport throughout the structure.

All these J(V) characteristics seem to point out a lack of electron selectivity and/or a limited electron extraction from PK by SnO_2^{ALD}. Besides, in a previous work [29], a comparison of PK films grown either on SnO_2^{SP} or on SnO_2^{ALD} did not give rise to any significant differences between analysed properties, thus making unlikely that performance limitations originate from the PK itself.

C. SnO_2-Perovskite interface analysis

In contrast, interfaces between charge selective layers and PK are known to be a critical issues in PK-based photovoltaics [30]. It is of interest to focus on SnO_2^{ALD} and SnO_2^{SP} interface with PK in view of a better understanding concerning the observed device limitations.

TABLE II. Averaged roughness for glass/ITO, glass/ITO/SnO$_2$^{ALD} and glass/ITO/SnO$_2$^{SP} top surface

	Glass/ITO	Glass/ITO/ SnO$_2$^{ALD}	Glass/ITO/ SnO$_2$^{SP}
Roughness (RMS) [nm]	4	4	2

Steady-state photoluminescence measurements were performed on glass/ITO/SnO$_2$/PK samples. Glass/ ITO/ SnO$_2$^{ALD} /PK yielded about 4 times higher photoluminescence intensity than glass/ITO/ SnO$_2$^{SP} /PK (Fig. 5). Since SnO$_2$^{ALD}-based solar cells give lower V$_{OC}$ than reference ones, this increase of photoluminescence intensity supports the assumption of a poor electron extraction at the SnO$_2$^{ALD}/PK interface [31] rather than being caused by a reduced non-radiative recombination rate at this interface. A limited electron extraction at the ESL/PK interface in the SnO$_2$^{ALD} -based device may be related to a different effective contact area between the ESL and the PK film whether the latter is grown on SnO$_2$^{ALD} or SnO$_2$^{SP} [32]. AFM characterization was performed on glass/ITO, glass/ITO/SnO$_2$^{ALD} and glass/ITO/SnO$_2$^{SP} samples showed an averaged surface roughness of 4 nm for the two former and of 2 nm for the latter (see Table 2). In addition of outlining once more that ALD-grown films are totally conformal to their ITO substrate, it shows on the contrary that SnO2 nanoparticles of the solution-processed film tend to fill ITO roughness. Such result unveiled that the ITO/SnO$_2$^{ALD} stack presents a larger surface roughness than ITO/SnO$_2$^{SP} does, at the AFM resolution level, which is about 0.5nm height to about 100nm laterally. Additionally, TEM was performed on c-Si/ITO/ESL/PK stacks for both types of SnO$_2$ layers. As one can see from Fig. 6, SnO$_2$ layers are well covering the ITO substrate and do not show any visible pinhole. However, EDX profile scans revealed the presence of I, Br and Pb in the SnO$_2$^{SP} film, whereas the SnO$_2$^{ALD} dense nature prevent any penetration of these PK elements. Thus, TEM-EDX profiles highlighted the nanoporous nature of SnO2^{SP} thin films. From the cell characteristics (Fig. 2), it is clear that such PK-related elements

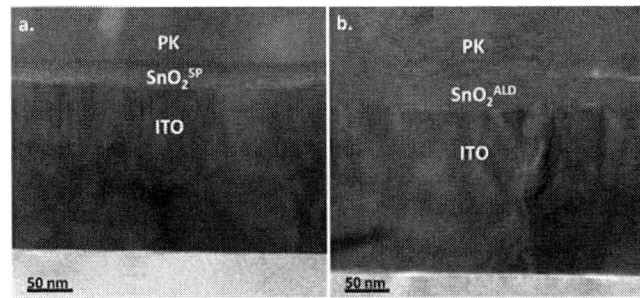

Fig. 6 - TEM cross-section images of a Si/ITO/ SnO$_2$^{SP}/PK (a.) and Si/ITO/SnO$_2$^{ALD}/PK (b.) stacks.

intermixing in the SnO$_2$^{SP} is not performance limiting. It is however difficult to conclude about a possible enhanced electron extraction due to the much larger contact area between PK and SnO$_2$^{SP} nanoparticles, since the phase of these intermixed elements is unknown. For future prospect, one could ask if such phase is crystalline PK or a sum of different chemical complexes such as PbI$_2$, FAI, CsI etc. Answering this question could eventually state if the absence of such interpenetration is detrimental for SnO$_2$^{ALD}-based devices.

On another hand, an energy barrier for electron transport at SnO$_2$^{ALD}/PK interface could be involved. On this matter, KPFM was performed to probe SnO$_2$ workfunction. Table 3 presents workfunction mean values probed for SnO$_2$^{ALD}, SnO$_2$^{SP} and PK by KPFM. SnO$_2$^{ALD} exhibits a 120 meV larger workfunction than SnO$_2$^{SP}. In principle, a low workfunction is preferable for efficient electron extraction [17], [33], to form a negative potential step when going from PK to the ESL. Here both interfaces show a positive workfunction step going from PK to SnO$_2$. However, one can see from Table 3 that PK workfunction appears to be dependent on its substrate when measured by KPFM and that PK to SnO$_2$^{ALD} workfunction step is about 100 meV larger than PK to SnO$_2$^{SP} step. Hence, although such workfunction differences may create a potential barrier hindering electron extraction from PK in both cases, this barrier would be 100 meV larger for the SnO$_2$^{ALD} -based interface. Furthermore, it should be kept in mind that in order to conclude properly, this analysis must be done in view of respective valence band maxima and conduction band minima for both ESL and PK layers [34]. Further characterization by Ultra-Violet Photoelectron Spectroscopy (UPS) on both SnO$_2$ and PK should give a clearer picture of the energetic levels alignment at ESL/PK interface.

IV. CONCLUSION

This work is part of our study that seeks to clarify the causes of PK-based solar cells performance limitations when SnO$_2$^{ALD} is employed as the ESL instead of SnO$_2$^{SP}. SnO$_2$^{ALD} and SnO$_2$^{SP} thin film properties and their interface with PK were comparatively characterized. The systematic comparison of

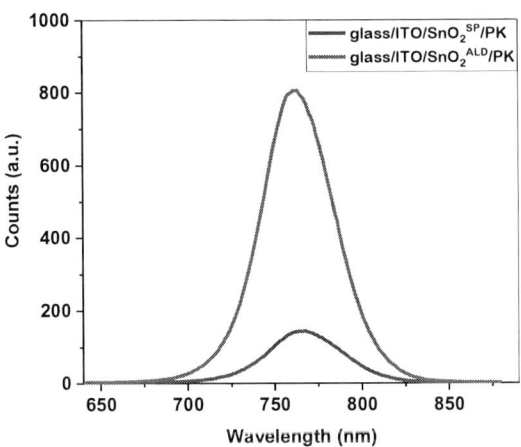

Figure 5 - Absolute intensity recorded from steady-state photoluminescence measurement on glass/ITO/SnO$_2$^{SP}/PK and glass/ITO/SnO$_2$^{ALD}/PK samples.

TABLE III. Averaged workfunction values measured by KPFM on SnO$_2$^{SP}, SnO$_2$^{ALD} and PK surface

	SnO$_2$^{SP}	SnO$_2$^{ALD}	SnO$_2$^{SP}/PK	SnO$_2$^{ALD}/PK
Φ [eV]	4.5±0.1	4.9±0.1	4.3±0.1	4.6±0.1

SnO_2^{ALD} thin films properties to reference SnO_2^{SP} ones ensured the relatively pure SnO_2 growth by ALD, highlighted the substantially higher conductivity of ALD-grown layers and their slightly rougher surface (when deposited onto ITO). Steady-state photoluminescence showed a more intense emission for PK films deposited on top of SnO_2^{ALD} than reference PK films, owing for a reduced SnO_2^{ALD} electron extraction ability. The larger workfunction exhibited by SnO_2^{ALD} may cause a potential barrier that limits electron extraction. Further characterization by UPS shall give a more refined picture of the bands alignment at ESL/PK interface. Besides, TEM-EDX profiling showed the presence of PK-related elements within SnO_2^{SP} films only. This intermixing could hypothetically benefits (passivation effect, enhanced electron extraction...) the SnO_2^{SP} device performances. Nevertheless, investigating this assumption would require a better insight in the chemical structure of this PK-based phase.

ACKNOWLEDGMENT

Part of this work, carried out on the Platform for Nanocharacterisation (PFNC), was supported by the "Recherches Technologiques de Base" program of the French National Research Agency (ANR).

The authors would like to acknowledge ENEL Green Power (EGP) for funding and for the extensive common R&D developments on the HJT technology.

The authors would like also Dr Virginie Brizé (CEA Liten - INES) for helping with SEM imaging.

REFERENCES

[1] P. Löper et al., "Organic–Inorganic Halide Perovskites: Perspectives for Silicon-Based Tandem Solar Cells," *IEEE Journal of Photovoltaics*, vol. 4, no. 6, pp. 1545–1551, Nov. 2014, doi: 10.1109/JPHOTOV.2014.2355421.

[2] Q. Wali, N. K. Elumalai, Y. Iqbal, A. Uddin, and R. Jose, "Tandem perovskite solar cells," *Renewable and Sustainable Energy Reviews*, vol. 84, pp. 89–110, Mar. 2018, doi: 10.1016/j.rser.2018.01.005.

[3] T. Tiedje, E. Yablonovitch, G. D. Cody, and B. G. Brooks, "Limiting efficiency of silicon solar cells," *IEEE Transactions on Electron Devices*, vol. 31, no. 5, pp. 711–716, May 1984, doi: 10.1109/T-ED.1984.21594.

[4] P. Tockhorn et al., "Nano-optical designs enhance monolithic perovskite/silicon tandem solar cells toward 29.8% efficiency." Mar. 28, 2022. doi: 10.21203/rs.3.rs-1439562/v1.

[5] D. D. Smith, P. Cousins, S. Westerberg, R. D. Jesus-Tabajonda, G. Aniero, and Y.-C. Shen, "Toward the Practical Limits of Silicon Solar Cells," *IEEE Journal of Photovoltaics*, vol. 4, no. 6, pp. 1465–1469, Nov. 2014, doi: 10.1109/JPHOTOV.2014.2350695.

[6] M. Green, E. Dunlop, J. Hohl-Ebinger, M. Yoshita, N. Kopidakis, and X. Hao, "Solar cell efficiency tables (version 57)," *Progress in Photovoltaics: Research and Applications*, vol. 29, no. 1, pp. 3–15, 2021, doi: 10.1002/pip.3371.

[7] J. Werner et al., "Efficient Near-Infrared-Transparent Perovskite Solar Cells Enabling Direct Comparison of 4-Terminal and Monolithic Perovskite/Silicon Tandem Cells," *ACS Energy Lett.*, vol. 1, no. 2, pp. 474–480, Aug. 2016, doi: 10.1021/acsenergylett.6b00254.

[8] H. Min et al., "Perovskite solar cells with atomically coherent interlayers on SnO2 electrodes," *Nature*, vol. 598, no. 7881, pp. 444–450, Oct. 2021, doi: 10.1038/s41586-021-03964-8.

[9] A. Al-Ashouri et al., "Monolithic perovskite/silicon tandem solar cell with >29% efficiency by enhanced hole extraction," *Science*, vol. 370, no. 6522, pp. 1300–1309, Dec. 2020, doi: 10.1126/science.abd4016.

[10] F. Sahli et al., "Fully textured monolithic perovskite/silicon tandem solar cells with 25.2% power conversion efficiency," *Nature Mater*, vol. 17, no. 9, pp. 820–826, Sep. 2018, doi: 10.1038/s41563-018-0115-4.

[11] J. Zheng et al., "Large-Area 23%-Efficient Monolithic Perovskite/Homojunction-Silicon Tandem Solar Cell with Enhanced UV Stability Using Down-Shifting Material," *ACS Energy Lett.*, vol. 4, no. 11, pp. 2623–2631, Nov. 2019, doi: 10.1021/acsenergylett.9b01783.

[12] J. Liu et al., "28.2%-efficient, outdoor-stable perovskite/silicon tandem solar cell," *Joule*, vol. 5, no. 12, pp. 3169–3186, 2021, doi: 10.1016/j.joule.2021.11.003.

[13] C. Altinkaya et al., "Tin Oxide Electron-Selective Layers for Efficient, Stable, and Scalable Perovskite Solar Cells," *Advanced Materials*, vol. 33, no. 15, p. 2005504, 2021, doi: https://doi.org/10.1002/adma.202005504.

[14] Q. Jiang et al., "Surface passivation of perovskite film for efficient solar cells," *Nature Photonics*, vol. 13, no. 7, Art. no. 7, Jul. 2019, doi: 10.1038/s41566-019-0398-2.

[15] W. Ke et al., "Effects of annealing temperature of tin oxide electron selective layers on the performance of perovskite solar cells," *J. Mater. Chem. A*, vol. 3, no. 47, pp. 24163–24168, Nov. 2015, doi: 10.1039/C5TA06574G.

[16] Q. Dong et al., "Insight into Perovskite Solar Cells Based on SnO 2 Compact Electron-Selective Layer," *J. Phys. Chem. C*, vol. 119, no. 19, pp. 10212–10217, May 2015, doi: 10.1021/acs.jpcc.5b00541.

[17] T. Hu et al., "Indium-Free Perovskite Solar Cells Enabled by Impermeable Tin-Oxide Electron Extraction Layers," *Advanced Materials*, vol. 29, no. 27, p. 1606656, 2017, doi: 10.1002/adma.201606656.

[18] V. Zardetto et al., "Atomic layer deposition for perovskite solar cells: research status, opportunities and challenges," *Sustainable Energy Fuels*, vol. 1, no. 1, pp. 30–55, 2017, doi: 10.1039/C6SE00076B.

[19] S. M. George, "Atomic Layer Deposition: An Overview," *Chem. Rev.*, vol. 110, no. 1, pp. 111–131, Jan. 2010, doi: 10.1021/cr900056b.

[20] J. P. Correa Baena et al., "Highly efficient planar perovskite solar cells through band alignment engineering," *Energy Environ. Sci.*, vol. 8, no. 10, pp. 2928–2934, 2015, doi: 10.1039/C5EE02608C.

[21] S. Albrecht et al., "Monolithic perovskite/silicon-heterojunction tandem solar cells processed at low temperature," *Energy Environ. Sci.*, vol. 9, no. 1, pp. 81–88, 2016, doi: 10.1039/C5EE02965A.

[22] Y. Lee et al., "Efficient Planar Perovskite Solar Cells Using Passivated Tin Oxide as an Electron Transport Layer," *Adv. Sci.*, vol. 5, no. 6, p. 1800130, Jun. 2018, doi: 10.1002/advs.201800130.

[23] Y. Kuang et al., "Low-Temperature Plasma-Assisted Atomic-Layer-Deposited SnO2 as an Electron Transport Layer in Planar Perovskite Solar Cells," *ACS Appl. Mater. Interfaces*, vol. 10, no. 36, pp. 30367–30378, Sep. 2018, doi: 10.1021/acsami.8b09515.

[24] A. F. Palmstrom et al., "Interfacial Effects of Tin Oxide Atomic Layer Deposition in Metal Halide Perovskite Photovoltaics," *Advanced Energy Materials*, vol. 8, no. 23, p. 1800591, 2018, doi: https://doi.org/10.1002/aenm.201800591.

[25] C. Wang et al., "Low-temperature plasma-enhanced atomic layer deposition of tin oxide electron selective layers for highly efficient planar perovskite solar cells," *Journal of Materials Chemistry A*, vol. 4, no. 31, pp. 12080–12087, 2016, doi: 10.1039/C6TA04503K.

[26] J. Warby et al., "Understanding Performance Limiting Interfacial Recombination in pin Perovskite Solar Cells," *Advanced Energy Materials*, vol. 12, no. 12, p. 2103567, 2022, doi: 10.1002/aenm.202103567.

[27] J. J. Yoo et al., "Efficient perovskite solar cells via improved carrier management," *Nature*, vol. 590, no. 7847, Art. no. 7847, Feb. 2021, doi: 10.1038/s41586-021-03285-w.

[28] R. Scheer and H.-W. Schok, "Appendix A: Frequently Observed Anomalies," in *Chalcogenide Photovoltaics*, John Wiley & Sons, Ltd, 2011, pp. 305–314. doi: 10.1002/9783527633708.ch7.

[29] F. Gayot, E. Bruhat, M. Manceau, E. D. Vito, and S. Cros, "Study of ALD-Grown SnO2 as an Electron Selective Layer for NIP Perovskite-Based Solar Cells," p. 5. doi: 10.4229/EUPVSEC20212021-3BV.1.5.

[30] P. Schulz, D. Cahen, and A. Kahn, "Halide Perovskites: Is It All about the Interfaces?," *Chem. Rev.*, vol. 119, no. 5, pp. 3349–3417, Mar. 2019, doi: 10.1021/acs.chemrev.8b00558.

[31] T. Kirchartz, J. A. Márquez, M. Stolterfoht, and T. Unold, "Photoluminescence-Based Characterization of Halide Perovskites for Photovoltaics," *Advanced Energy Materials*, vol. n/a, no. n/a, p. 1904134, doi: 10.1002/aenm.201904134.

978-1-7281-6118-1/22 $31.00 © 2022 IEEE

[32] T. Leijtens, B. Lauber, G. E. Eperon, S. D. Stranks, and H. J. Snaith, "The Importance of Perovskite Pore Filling in Organometal Mixed Halide Sensitized TiO2-Based Solar Cells," *J. Phys. Chem. Lett.*, vol. 5, no. 7, pp. 1096–1102, Apr. 2014, doi: 10.1021/jz500209g.

[33] J. Zhang *et al.*, "3,5-Difluorophenylboronic acid-modified SnO2 as ETLs for perovskite solar cells: PCE > 22.3%, T82 > 3000 h," *Chemical Engineering Journal*, p. 133744, Nov. 2021, doi: 10.1016/j.cej.2021.133744.

[34] M. A. Martínez-Puente *et al.*, "Unintentional Hydrogen Incorporation into the SnO2Electron Transport Layer by ALD and Its Effect on the Electronic Band Structure," *ACS Applied Energy Materials*, 2021, doi: 10.1021/acsaem.1c01836.

Investigation of lead-free 2D/3D mixed-dimensional tin perovskite solar cell embedded with plasmonic metal nanoparticles

Atanu Purkayastha[*], Manoranjan Minz, Ramesh Kumar Sonkar, and Arun Tej Mallajosyula

Department of Electronics and Electrical, Indian Institute of Technology Guwahati, Guwahati, Assam, 781039, India

Abstract—**In this work, various metal nanoparticles have been used to improve the efficiency of a single layer 2D/3D mixed-dimensional hybrid tin perovskite solar cell, exploiting the surface plasmon effect. Silvaco TCAD and Lumerical FDTD software have been used for this study. As a first step, the thickness of perovskite layer, without the presence of any nanoparticles, has been optimized using experimentally determined optical constants. It has been found that the best possible efficiency of 14.13% is obtained at 180 nm. At this thickness, the size and position of metal nanoparticles in the perovskite layer have been varied. With spherical silver nanoparticles of 15 nm radius placed at the electron transport layer interface, the light absorption has been significantly enhanced for wavelengths between 450-600 nm and above 800 nm. Owing to this, the maximum possible photocurrent density improved by 7% to 29.64 mA.cm^{-2}.**

Index Terms—**Transfer-matrix method, drift-diffusion, finite-difference-time-domain, complex refractive index, total internal reflection, surface plasmons**

I. INTRODUCTION

Perovskite solar cells (PSCs) have shown an immense improvement of power conversion efficiency (PCE) over the last ten years [1]. However, inherent instability and the presence of lead in these devices are major concerns for their commercialization. Tin-based hybrid perovskites can be suitable alternatives to these materials. Among these, FASnI$_3$ has been well studied. It has, relatively, good carrier mobility (67 cm^{-2}V^{-1}s^{-1}), low optical bandgap (1.4 eV), and high absorption coefficient (10^{5}cm^{-1}) [2]. It has been reported that the introduction of a small concentration of 2D phase into FASnI$_3$ reduces the oxidation of Sn^{4+} to Sn^{2+}, thereby enhancing the PCE and stability of the device. Recently, a maximum PCE of 14% has been reported for the single-junction 2D/3D FASnI$_3$ solar cells [3], and the theoretical efficiency limit for the single junction PSCs is predicted to be 30.88% [2]. Clearly, there is still opportunity for a improvement in performance parameters of PSCs.

One of the most promising approaches for improving light absorption in thin-film solar cells is the plasmonic effect. Surface plasmons are a collection of conduction electrons of metallic nanoparticles (NPs) that vibrate in response to light at the metal dielectric contact. At resonance, surface plasmons have a high near-field scattering and a strong far-field scattering. A strong far-field improves active layer absorption by total internal reflection, whereas a strong local field enhances surrounding medium absorption using Poynting's theorem for

power dissipation [4]. The size, location, and material of the NPs must be tuned to match the spectrum features of the solar cell with the plasmon resonance. Otherwise, the parasitic absorption of metal NPs may result in heat generation, which is undesirable for solar cells.Many research groups have reported that using metal NPs increases the PCE of thinner lead perovskite solar cells by more than 10% [5], [6]. However, to our knowledge, there have been no reports on the effect of plasmonic metal nanoparticles in lead-free tin PSCs yet. The optical performance of lead-free tin PSCs has been studied in this work by incorporating metal NPs of varying size and materials at different positions of the active layer.

A schematic of the device structure, with its active layer embedded with Ag nanoparticles at different locations, is shown in Fig. 1(a). The energy band diagram of the illuminated structure under short circuit condition is shown in Fig. 1(b). The parameters used for device simulation are listed in Table I. The background doping density, lifetime, mobility, electron affinity, bandgap, hole and electron density of states are all obtained from the literature [7]–[10].

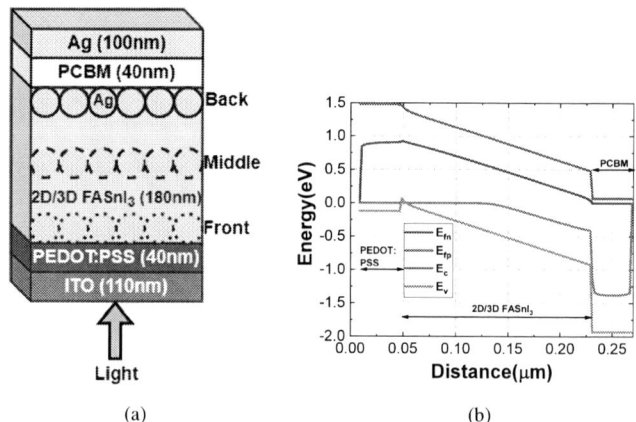

(a) (b)

Fig. 1. (a) Simulated architecture of the 2D/3D FASnI$_3$ solar cell embedded with the Ag NPs, (b) Band diagram of the illuminated structure under short-circuit conditions. Here, E$_{fn}$, E$_{fp}$, E$_v$, and E$_c$, electron quasi-Fermi level, hole quasi-Fermi level, valence band, and conduction band respectively.

The wavelength dependent complex refractive index data of 0.15M PEAI in FASnI$_3$, determined using variable angle spectroscopic ellipsometry, is shown in Fig. 2. In contrast, the wavelength dependent complex refractive indices of PCBM and PEDOT:PSS are extracted from the literature [8]. The

978-1-7281-6118-1/22 $31.00 © 2022 IEEE

TABLE I
SIMULATION PARAMETERS OF LEAD-FREE 2D/3D FASNI$_3$ SOLAR CELL

Parameters	PEDOT:PSS	FASnI$_3$	PCBM
Thickness (μm)	0.04	0.18	0.04
N$_A$ (cm^{-3})	1e19	1e13	-
N$_D$ (cm^{-3})	-	-	1e19
E$_G$ (eV)	1.6	1.4	2.1
Electron affinity (eV)	3.6	3.5	3.9
Relative Dielectric Permitivity	2.2	32	3
N$_C$ (cm^{-3})	2e21	1e18	1e21
N$_V$ (cm^{-3})	2e21	1e18	1e22
Mobility of electron (cm^2/V-s)	1	22	0.01
Mobility of hole (cm^2/V-s)	40	22	0.01
Lifetime of electron (s)	4e-6	1.7e-9	4e-6
Lifetime of hole (s)	4e-6	1.7e-9	4e-6

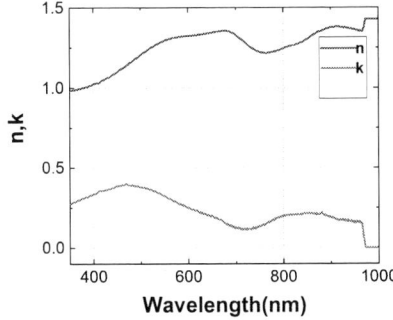

Fig. 2. Experimentally measured optical constants of 2D/3D FASnI$_3$ for a variable wavelength range.

active layer thickness is initially optimized using the SIL-VACO 2D TCAD tool. The simulation is carried out using the transfer-matrix method along with drift-diffusion and SRH recombination models. Using the commercially available Lumerical software package, the absorption spectra and, thus, the maximum possible photocurrent density ($J_{PH|MAX}$) of plasmon assisted PSCs have been determined using the finite-difference-time-domain method (FDTD). $J_{PH|MAX}$ of the device with and without metal NPs has been calculated to quantify the optical absorption of the active layer, which is given as

$$J_{PH|MAX} = q \int_{300}^{900} A_p(\lambda)\phi_{AM1.5G}(\lambda)d\lambda \qquad (1)$$

where $\phi_{AM1.5G}$ is the photon flux density of the AM1.5G sunlight spectrum, q is the charge of electron, and A_p is the absorption coefficient of the active layer. A wavelength (λ) range of 300 to 900 nm is used for the integration. By mapping the $J_{PH|MAX}$ values obtained from Lumerical FDTD to the J_{sc} from Silvaco TCAD for the above mentioned device structure, we obtained a conversion factor of 1.12 i.e, $J_{PH|MAX} = 1.12*J_{sc}$. Using this conversion factor, we estimated the J_{sc} of devices with all NPs.

II. RESULTS AND DISCUSSION

A. Active layer thickness optimization

The thickness of 2D/3D FASnI$_3$ has been varied from 100 to 500 nm, and electrical parameters such as fill factor (FF), short-circuit current density (J_{sc}), open circuit voltage (V_{oc}),

and PCE are analyzed. As shown in Fig. 3(a), the increase in perovskite layer thickness decreased V_{oc} from 0.83 V to 0.73 V.

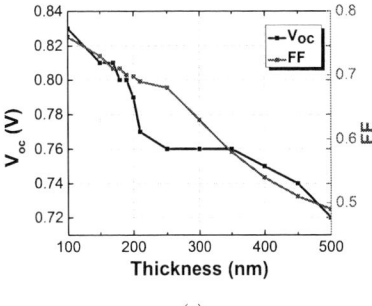

(a)

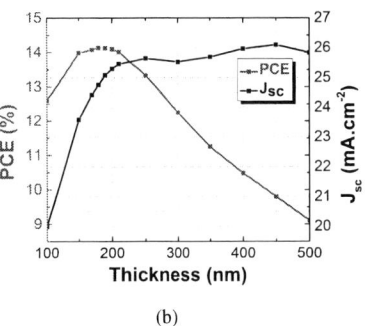

(b)

Fig. 3. Electrical parameters (a) V_{oc}, FF, (b) PCE, J_{sc} as a function of active layer thickness.

Owing to the low carrier lifetime (1.7 ns) of the 2D/3D FASnI$_3$ layer, the carrier recombination rate increased significantly with thickness shown previously [8]. J_{sc} increased from 20 to 26 mA.cm^{-2} due to the increased generation rate. It can be observed from Fig. 3(b) that the PCE of the device has reached the maximum value of 14.13% at an optimal active layer thickness of 180 nm. At this thickness, the $J_{PH|MAX}$ of the device without NPs is calculated to be 27.73 mA.cm^{-2}.

B. Optimization of size and position of Ag nanoparticles

Fig. 4 shows a contour plot of $J_{PH|MAX}$ of the device after spherical Ag NPs with radius ranging from 5 to 30 nm are inserted in various regions of the active layer. When the Ag NPs

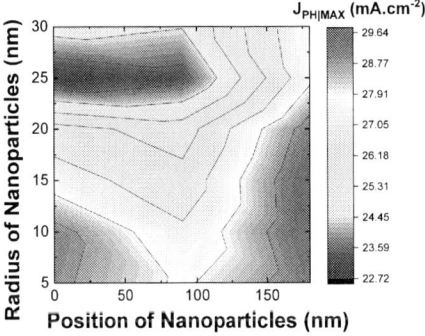

Fig. 4. $J_{PH|MAX}$ of the device with the variation in the sizes and position of Ag NPs.

978-1-7281-6118-1/22 $31.00 © 2022 IEEE   505

have been positioned in the front (near PEDOT:PSS/FASnI$_3$ interface), centre, and back (near PCBM/FASnI$_3$ interface) of the active layer, the optimal radii have been calculated to be 10, 5, and 15 nm respectively. Among all, the maximum increase in $J_{PH|MAX}$ (29.64 mA.cm^{-2}) has been obtained when Ag nanoparticles of 15 nm size are embedded at the rear end. As can be seen from Fig. 5, the fraction of light absorbed is significantly high for wavelengths between 450-620 nm where the solar photon flux is significant. The increase in absorption could be attributed to an increase in electric field intensity caused by the excitation of the localized plasmon resonance and the scattering effect of the metal NPs. The

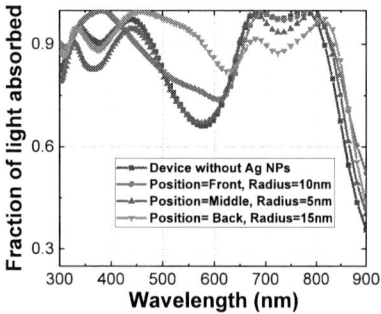

Fig. 5. Fraction of the light absorbed in the device with the variable radius and position of Ag NPs.

use of metallic NPs does not appear to influence the V_{oc} and FF appreciably [6]. As a result, it can be assumed that these electrical parameter values are identical to those of the device without metal NPs. Hence, the PCE of the device has increased to 15.05%.

C. Other metal nanoparticles

According to the previous finding, the 15 nm radius Ag NPs in the rear position of the active layer resulted in the highest absorption. Other NPs materials, such as Au, Al, and Cu have also been investigated for the same radius and position. The device with Al NPs has the worst performance of all, owing to the low absorption of active layer shown in Fig. 6. On the

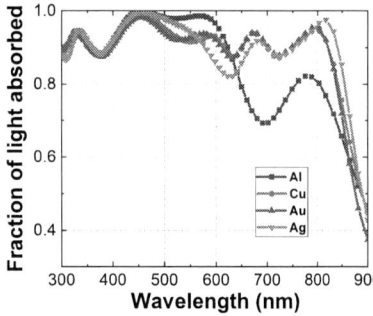

Fig. 6. Fraction of the light absorbed in the device with different metal NPs.

other hand, the highest absorption and enhancement in $J_{PH|MAX}$

is obtained for Ag NPs. When compared to Al NPs, Ag NPs showed absorption enhancement in the spectral range greater than 600 nm. The $J_{PH|MAX}$ value of the device with the Ag, Al, Cu, and Au NPs are 29.64, 27.70, 29.46, and 29.23 mA.cm^{-2}, respectively.

III. SUMMARY

The electrical parameters of the 2D/3D tin PSCs have been found to be maximum for the active layer thickness of 180 nm, according to the simulation analysis. At this thickness, the values of V_{oc}, J_{sc}, FF, and PCE were 0.80 V, 24.76 mA.cm^{-2}, 0.71, and 14.13% respectively. While the FF and V_{oc} both decrease monotonically with thickness, J_{sc} increases with the active layer thickness. Furthermore, embedding the size optimized Ag NPs at the different positions of the active layer improved its light absorption efficiency. The optimum performance parameter values are obtained by positioning the 15 nm radius Ag NPs at the back end of the 180 nm active layer. Under the assumption of V_{oc} and FF not being negatively effected, the increase in $J_{PH|MAX}$ translates to an efficiency of 15.05%.

REFERENCES

[1] Green, Martin, Ewan Dunlop, Jochen Hohl-Ebinger, Masahiro Yoshita, Nikos Kopidakis, and Xiaojing Hao. "Solar cell efficiency tables (version 57)." Progress in photovoltaics: research and applications 29, no. 1 (2021): 3-15.

[2] Purkayastha, Atanu, and Arun Tej Mallajosyula. "Optical modelling of tandem solar cells using hybrid organic-inorganic tin perovskite bottom sub-cell." Solar Energy 218 (2021): 251-261.

[3] Yu, Bin-Bin, Zhenhua Chen, Yudong Zhu, Yiyu Wang, Bing Han, Guocong Chen, Xusheng Zhang, Zheng Du, and Zhubing He. "Heterogeneous 2D/3D Tin-Halides Perovskite Solar Cells with Certified Conversion Efficiency Breaking 14%." Advanced Materials 33, no. 36 (2021): 2102055.

[4] Perrakis, George, George Kakavelakis, George Kenanakis, Constantinos Petridis, Emmanuel Stratakis, Maria Kafesaki, and Emmanuel Kymakis. "Efficient and environmental-friendly perovskite solar cells via embedding plasmonic nanoparticles: an optical simulation study on realistic device architectures." Optics express 27, no. 22 (2019): 31144-31163.

[5] hosein Mohammadi, Mohammad, Mehdi Eskandari, and Davood Fathi. "Effects of the location and size of plasmonic nanoparticles (Ag and Au) in improving the optical absorption and efficiency of perovskite solar cells." Journal of Alloys and Compounds 877 (2021): 160177.

[6] Siavash Moakhar, Roozbeh, Somayeh Gholipour, Saeid Masudy-Panah, Ashkan Seza, Ali Mehdikhani, Nastaran Riahi-Noori, Saeede Tafazoli, Nazanin Timasi, Yee-Fun Lim, and Michael Saliba. "Recent advances in plasmonic perovskite solar cells." Advanced Science 7, no. 13 (2020): 1902448.

[7] Abdelaziz, Saied, A. Zekry, Ahmed Shaker, and M. Abouelatta. "Investigating the performance of formamidinium tin-based perovskite solar cell by SCAPS device simulation." Optical Materials 101 (2020): 109738.

[8] Purkayastha, Atanu, and Arun Tej Mallajosyula. "Effect of active layer thickness and angle of incidence on the efficiency of planar heterojunction lead-free tin perovskite solar cell." In 2021 IEEE 48th Photovoltaic Specialists Conference (PVSC), pp. 1236-1239. IEEE, 2021.

[9] Wang, Fei, Xianyuan Jiang, Hao Chen, Yuequn Shang, Hefei Liu, Jingle Wei, Wenjia Zhou, Hailong He, Weimin Liu, and Zhijun Ning. "2D-quasi-2D-3D hierarchy structure for tin perovskite solar cells with enhanced efficiency and stability." Joule 2, no. 12 (2018): 2732-2743.

[10] Jiang, Xianyuan, Fei Wang, Qi Wei, Hansheng Li, Yuequn Shang, Wenjia Zhou, Cheng Wang et al. "Ultra-high open-circuit voltage of tin perovskite solar cells via an electron transporting layer design." Nature communications 11, no. 1 (2020): 1-7.

High-efficiency solar cell by combining high and low thermal budget for Si passviting contacts

Muhammad Quddamah Khokhar, Shahzada Qamar Hussin, Muhammad Aleem Zahid, Duy Phong Pham, Eun-Chel Cho, Junsin Yi

Department of Electrical and Computer Engineering, Sungkyunkwan University, Suwon Gyeonggi-Do, 16419, Republic of Korea , Suwon, Korea

Department of Physics, COMSATS University Islamabad, Lahore Campus, Lahore 54000, Pakistan, Lahore, Pakistan

Various parameters including optimal surface passivation, carrier selectivity, and low recombination losses are prerequisite for high efficiency in silicon heterojunction solar cells. Herein, the surface passivation quality of crystalline silicon solar cells is improved by a hybrid passivation structure including a silicon heterojunction contact at the front side and a stack of tunneling oxide with n-type nano-crystalline silicon oxide (nc-SiOx(n)) passivating contact at the rear side. In our study, Initially, a low-temperature, energy-efficient, electron-selective hetero-contact was fabricated by using a hydrogenated intrinsic amorphous silicon (a-Si:H(i)) layer as the front surface. Secondly, the poly-silicon layer swapped with nc-SiOx(n) layer to improve the effective surface passivation, electrical properties, and carrier selectivity. The passivation properties of symmetric structures were optimized, and the Boron doped amorphous silicon (a-Si:H(p))/a-Si:H(i) showed a lifetime (τeff) of 1.6 ms and implied open-circuit voltage (i-Voc) of 719 mV, respectively. On the other hand, the nc-SiOx(n)/SiO2 stack shows a τeff of 2.1 ms and i-Voc of 725 mV respectively. The optimized passivated layers were used for the fabrication of hybrid solar cell and showed the performance as; Voc = 724 mV, Jsc = 38.95 mA/cm2, FF of 75.9%, η= 21.4%.

The Role of the European Green Deal for the Photovoltaic Market Growth in the European Union

Arnulf Jäger-Waldau
European Commission
Joint Research Centre (JRC),
Ispra, Italy
arnulf.jaeger-waldau@ec.europa.eu

Georgia Kakoulaki
European Commission
Joint Research Centre (JRC),
Ispra, Italy
Georgia.Kakoulaki@ec.europa.eu

Nigel Taylor
European Commission
Joint Research Centre (JRC),
Ispra, Italy
Nigel.Taylor@ec.europa.eu

Sandor Szábo
European Commission
Joint Research Centre (JRC),
Ispra, Italy
sandor.szabo@ec.europa.eu

Abstract—Since the introduction of the first European Renewable Energy Directive in 2009, PV installations have significantly increased to reach over 165 GWp in the European Union at the end of 2021. The new Green Deal, endorsed by the European Council in December 2020 set new targets for the GHG reductions until 2030, which already now led to a significant growth of the photovoltaic market in the European Union. What are the opportunities and challenges for the further deployment of solar photovoltaics? The lessons learned during the last decade highlight the importance of legal and regulatory continuity and reliability to ensure investor confidence. Will the Green Deal not only lead to a growth in PV capacity but to a revival of an European solar cell and module manufacturing industry?

Keywords— European Renewable Energy Directive, Green Deal, greenhouse gas emission, solar photovoltaics, PV deployment

I. INTRODUCTION

The political goal of the European Union is to transform to a prosperous, modern, competitive and climate neutral economy by 2050. In order to accelerate this transition, a European Green Deal was with a greenhouse gas emission reduction goal of at least 55% by 2030 was agreed at the European Council in December 2020. The COVID-19 Recovery and Resilience Facility together with the European Union's next long term budget represents EUR 2.02 trillion (USD 2.46 trillion) of spending between 2020 and 2027. Each EU Member State had to prepare a national recovery and resilience plan, which outlines their individual reform and investment agendas for the years 2021-2023 in order to be eligible for the Recovery and Resilience Facility. A minimum of 37% of expenditure has to be earmarked for actions to combat climate change.

Since the end of 2010, the capacity of grid-connected solar photovoltaic (PV) systems in the European Union (EU27) has increased from 34.2 GWp to over 165 GWp at the end of 2021 [1].

II. CURRENT SITUATION

The Climate Scenario of the New Energy Outlook (NEO) 2020 by Bloomberg New Energy Finance forecasted an increase of roughly 25% of the electricity demand in Europe by 2030 and a doubling by 2050 compared to 2019 [2]. The main drivers of this increase are the electrification of transport, industrial processes and buildings as well as the use of green hydrogen.

The Impact Assessment for the Green Deal forecasts a electricity generation in the European Union (EU27) between 3100 and 3200 TWh in 2030 [3]. The share of solar photovoltaic electricity in these models is about 14% compared to 5% in 2020. The energy system model simulation used for the Impact Assessment report projections of solar PV capacity as alternating current (ac) capacity to be comparable to the other power generation plants. The required PV capacity to reach the 2030 target was modelled with 360 to $370 GW_{ac}$ capacity.

A requirement of the Renewable Energy Directive II was that the European Member States submitted National Energy and Climate Plans (NECPs) in 2019 outlining the plans of each Member State and how they planned to reach the old 2030 target respectively [4]. However, the pledged PV capacity of 260 GWp in the low case and 340 GWp in the high case are by far insufficient to reach the new targets of the Green Deal.

The recovery funds complemented with the ongoing revision of the renewable energy directive as part of the "Fit for 55" package of EU legislative measures to implement the new 55% GHG reduction target for 2030 will further help to accelerate the deployment of PV. The publication of the 6th IPCC Assessment Report in April 2022 [5] and the geopolitical developments in the first half of 2022 have highlighted the urgency of the clean energy transition. The European Commission had reacted with the REPowerEU Communication and the Solar Strategy Communication in March and May 2022 respectively [6, 7].

In 2021, the photovoltaic market in the European Union grew by more than 25% to about 26 GWp and reached a cumulative installed capacity of over 165 GWp [1] (Fig. 1). The Solar Strategy communication calls for about 450 GWac of new photovoltaic system capacity between 2021 and 2030. Given the current trend of installing 1.25 to 1.3 times the AC capacity in DC, this would bring the total nominal photovoltaic (PV) capacity in the European Union (EU) to approximately 720 GWp. Compared to 2021, this would require an annual market

978-1-7281-6118-1/22 $31.00 © 2022 IEEE

volume increase to over 100 GWp annually by 2030, which is achievable if the current market trend can be maintained.

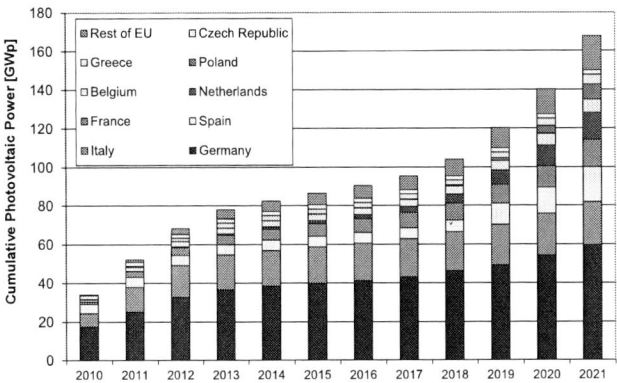

Fig. 1.   Grid-connected PV capacity in EU [1]

With such a market increase, the issue of the security of supply becomes more important. In order to hedge against supply chain disruptions, which could be seen in the last two years during the COVID pandemic, the European Union has to rebuild its local manufacturing capabilities along the full value chain.

The main building blocks to achieve this are a number of initiatives, which should deliver the expected outcome:

1) European Solar Rooftops Initiative
2) Utility scale deployment
3) Solar value for buildings, districts and cities
4) Preparing the energy network for the efficient distribution of solar energy
5) Establishment of a resilient supply chain
6) Supporting investments regarding EU PV manufacturing (de-risking, funding)

Even before the recent publication of the REPowerEU Communication, an increasing number of countries have increased their PV targets for the energy transition or are in the process of a re-evaluation. At the beginning of May 2022, the Energy Ministers from Austria, Belgium, Lithuania, Luxembourg, and Spain signed a letter to the European Commission asking for a more ambition of the EU with regard to solar power – including at least 1 TW of solar capacity in the EU by 2030.

Due to this very recent developments, the currently known announcements listed in Table I do not yet reflect the new targets of the Solar Strategy. However, compared to last year, the 2030 numbers which have been announced increased by about 130 GW and are closing in to reach 600 GW by 2030. However, this still falls short of the targeted capacity above 700 GW. In fact the results of a 2019 EU research [8] showed that 24% of the existing electricity consumption could potentially be produced by solar PV on the EU rooftops (680 TWh), two thirds of which at a lower cost than the current residential tariffs.

TABLE I.   INSTALLED CAPACITIES IN EU FOR 2021, HIGH CASE NECP TARGETS FOR 2030 AND NEW TRENDS

Country	Capacity 2021 [GW]	Capacity 2030 (NECP High) [GW]	Capacity 2030 (New Trends) [GW]
Austria	2.50	12.00	**13.00**
Belgium	6.90	11.00	**22.00**
Bulgaria	1.28	2.90	**3.20**
Croatia	0.15	0.77	0.77
Cyprus	0.38	0.80	0.80
Czechia	2.35	3.98	3.98
Denmark	2.90	7.84	7.84
Estonia	0.20	0.42	0.42
Finland	0.46	1.20	1.20
France	14.00	25.00	**36.00**
Germany	59.50	70.51	**250.00**
Greece	5.00	8.00	**13.00**
Hungary	3.00	6.00	6.50
Ireland	0.10	1.50	1.50
Italy	22.20	51.12	**52.00**
Latvia	0.01	0.50	0.50
Lithuania	0.14	1.53	1.53
Luxembourg	0.19	1.11	1.11
Malta	0.20	0.26	0.26
Netherlands	14.00	36.00	36.00
Poland	7.69	7.30	**20.00**
Portugal	1.77	9.00	10.00
Romania	1.60	5.89	5.89
Slovakia	0.70	1.20	1.20
Slovenia	0.36	1.65	1.65
Spain	18.60	44.00	**92.00**
Sweden	1.80	2.50	3.50
Total	168.00	313.98	585.85

The European Rooftops Initiative mentioned as one of the pillars of the Solar Energy Strategy aims to accelerate the deployment of PV systems on rooftops. Such systems can be deployed very rapidly, as they utilise existing structures and avoid conflicts with other public goods like the environment. To achieve this swiftly, immediate action is necessary by end 2022. The following actions are mentioned:

- Increase the EU 2030 target for renewables share to 45%.
- Limit the length of permitting for rooftop solar installations, including large ones, to a maximum of 3 months.
- Adopt provisions to ensure that all new buildings are "solar ready".
- Ensure that EU legislation is fully implemented in all Member States allowing consumers in multi-

apartment buildings to effectively exercise their right to collective self-consumption, without undue costs[1].

The EU and Member States will work together to support building-integrated PV systems for both new buildings and renovations.In addition the Member States are urged to

- Establish robust support frameworks for rooftop systems, including in combination with energy storage and heat-pumps, based on predictable payback times that are shorter than 10 years.
- Have a massive deployment of rooftop solar energy in high energy consumption buildings (Energy Performance Certificate class D or above).
- Combine solar deployment with roof renovations and energy storage; this should be implemented through a one-stop shop integrating all aspects.
- Install solar energy in all public buildings fit for this purpose by 2025.

The Member States should implement the measures under this initiative as a priority, using available EU funding, in particular the new REPowerEU chapters of their Recovery and Resilience Plans. The Commission will monitor progress in the implementation of this initiative on an annual basis, through the relevant fora, with the sector's stakeholders and the Member States.

If fully implemented, the Solar Strategy Initiative, as part of the REPowerEU plan, can accelerate rooftop installations and add 19 TWh of electricity after the first year of its implementation (36% more than expected in the Fit for 55 projections). By 2025, it will result in 58 TWh of additional electricity generated (more than double the Fit for 55 projections).

III. OUTLOOK

Despite the COVID19 pandemic, the annual market continued with more than 25% growth in the European Union to over 26 GWp in 2021 [1]. Seven countries countries installed more than 1GWp, namely Germany (5.4 to 5.5 GWp), Spain (4.6 to 4.8 GWp), Poland (3.7 to 3.8 GWp), the Netherlands (3.2 to 3.4 GWp), France (2.7 to 2.9 GWp), Greece (1.1to 1.2 GWp) and Denmark (1.1 to 1.3 GWp). The EU is a leading installer of PV per capita with 400 Wp/capita on average, having ten EU members in the first 15 countries in this ranking (with Australia, Japan, S Korea, Taiwan and US from outside EU) [9].

The introduction of green hydrogen and the European Commission's hydrogen strategy will play an important role for the future need of solar PV capacity as current hydrogen production is still 97% based on fossil fuels [10]. The Communication related to the hydrogen strategy for a climate-neutral Europe mentions a first phase of at least 6 GW of green hydrogen electrolysers by 2024 and at least 40 GW of RES-powered electrolysers by 2030 [11]. The full-time operation of this level of electrolyser capacity would require the equivalent electricity output of 256 TWh. Under the assumption that PV generated electricity would need to supply about half of this electricity and this capacity is fairly distributed close to the hydrogen demand this would require approximately 115 GW of PV capacity.

Additional demand for green electricity and rooftop PV systems will come from the foreseen Renovation Wave and the transformation of the Energy Performance of Building Directive with the concept of Nearly Zero-Energy Buildings (NZEBs) in national legislation [12] At municipal and regional level in the EU there is an ongoing discussion to make the installation of renewable energy systems in new buildings mandatory. According to the European Central Bank, approximately 1.6 million new residential buildings were constructed in 2016 in the EU [13]. If on average a 4 kW PV system would be needed on each building this would add more than 6 GWp per year or another 60 GWp until 2030. In case the annual number of housing completions rebounds to that of the 2000-2009 (2.2-2.8 million annually), new buildings could host up 11 GWp of solar PV, annually.

The cumulative PV power capacity additions under the Solar Strategy requires a compound annual growth rates (CAGR) of 17.5%, which is slightly lower than current market trend with 18%. To reach 1 TW by 2030 CAGR would have to increase to 22%, which is feasibible according to industry analysts. Figure 3 illustrates the projection until 2030.

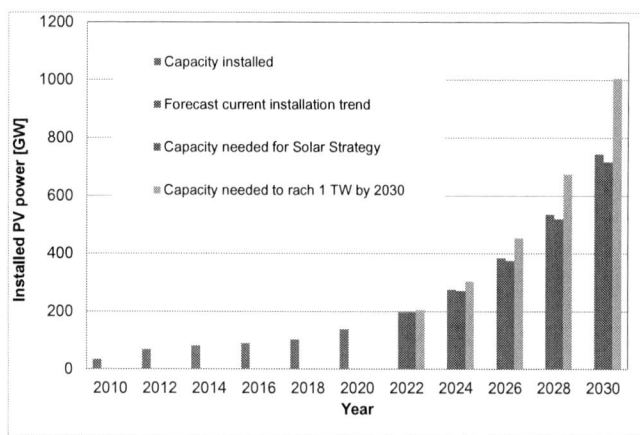

Fig. 2. Actual and projected photovoltaic installations from 2010 to 2030 in the EU

An invigorated EU PV market may need a development strategy for the full PV value chain, supported by research and innovation. This should include new cell and module

---

[1] Both the Directive (EU) 2018/2001 of the European Parliament and of the Council of 11 December 2018 on the promotion of the use of energy from renewable sources and the Directive (EU) 2019/944 of the European Parliament and of the Council of 5 June 2019 on common rules for the internal market for electricity contain provisions on collective self-consumption.

manufacturing in the EU, with significant job creation potential. Here the Green Deal's circular economy action plan is highly relevant, with its aim of promoting sustainability across the whole value chain and encouraging businesses to offer reusable, durable and repairable products. In addition the proposed 'renovation wave' of public and private buildings can also be an important stimulus for using PV products to achieve near-zero energy buildings.

Last but not least, the development of the PV markets need a dimension for a just transition by ensuring a significant share of decentralised PV to provide local jobs and ensure citizen participation.

Disclaimer: The scientific output expressed is based on the current information available to the authors, and does not imply a policy position of the European Commission.

IV. REFERENCES

[1] Jäger-Waldau A., Snapshot of Photovoltaics – February 2022, EPJ Photovoltaics 13, 9 (2022), doi 10.1051/epjpv/2022010

[2] Bloomberg New Energy Finance, New Energy Outlook 2020, October 2020

[3] European Commission. Impact Assessment on Stepping up Europe's 2030 climate ambition Investing in a climate-neutral future for the benefit of our people. Brussels, Belgium: 2020

[4] European Commission. Renewable Energy Progress Report. Brussels, Belgium: 2020

[5] IPCC Sixth Assessment Report, WG III, 2022, https://www.ipcc.ch/report/ar6/wg3/

[6] European Commission, REPowerEU Communication, 08.03.2022, COM(2022) 108 final

[7] European Commission, Solar Strategy Communication, 18.05.2022, COM(2022) 221 final

[8] Bódis K, Kougias I, Jäger-Waldau A, Taylor N, Szabó S. A high-resolution geospatial assessment of the rooftop solar photovoltaic potential in the European Union. *Renewable and Sustainable Energy Reviews,* Volume 114, October 2019, 109309, doi:10.1016/j.rser.2019.109309

[9] https://www.statista.com/statistics/612412/installed-solar-photovoltaics-capacity-eu/#:~:text=Solar%20photovoltaics%20capacity%20installed%20per,EU%2D27%202021%2C%20by%20country&text=At%20815%20watts%20per%20inhabitant,545%20watts%20per%20inhabitant%2C%20respectively

[10] Kakoulaki G, Kougias I, Taylor N, Dolci F, Moya J, Jäger-Waldau A. Green Hydrogen in Europe - a regional assessment: Substituting Existing Production with Electrolysis Powered by Renewables. Energy Convers Manag 2020;113649

[11] European Commission. A hydrogen strategy for a climate-neutral Europe. vol. 53. Brussels, Belgium: 2020. https://doi.org/10.1017/CBO9781107415324.004

[12] Official Journal of the European Union. Directive (EU) 2018/844 amending Directive 2010/31/EU on the energy performance of buildings and Directive 2012/27/EU on energy efficiency. Off J Eur Union 2018;L 156/75. [4] Official Journal of the European Union. Directive (EU) 2018/844 amending Directive 2010/31/EU on the energy performance of buildings and Directive 2012/27/EU on energy efficiency. Off J Eur Union 2018;L 156/75.

[13] European Central Bank. Statistical Data Warehouse. SHI Struct Hous Indic Stat 2020. https://sdw.ecb.europa.eu/browse.do?node=70499

978-1-7281-6118-1/22 $31.00 © 2022 IEEE

Assessment of mechanical robustness of conventional and CFRP-based lightweight PV module architectures under static loads

Umang Desai [a,b], Aparna Singh[a,*]

a: Department of metallurgical engineering and materials science, Indian Institute of Technology Bombay, Mumbai-400076, India.

b: National centre for photovoltaic research and education, Indian Institute of Technology Bombay, Mumbai-400076, India.

*: corresponding author (A. Singh, Aparna_s@iitb.ac.in)

Abstract—This work compares the deformation and stresses generated in the solar cells in the conventional and lightweight PV modules due to application of combined static load of 2400 Pa and self-weight of the structures. The maximum displacement in the conventional design and lightweight design is found to be 19.5 mm and 35.4 mm, respectively. Maximum principal stress in the solar cells for the lightweight design is 121 MPa, which is less than the value of maximum principal stress of 133 MPa for the conventional design. The results presented here indicate that the lightweight design has better performance than the conventional design when both the self-weight and static load of 2400 Pa are applied.

Keywords—finite element analysis, lightweight PV, CFRP, stresses, PV module reliability

I. INTRODUCTION

The conventional terrestrial photovoltaic (PV) modules have an architecture of glass/ethylene vinyl acetate (EVA)/Si-cell/EVA/backsheet. While glass is an essential component for the mechanical stiffness of a PV module, it makes the structure bulky owing to its high density (~2500 kg/m2). Therefore, the bulky conventional design is not ideal for building-integrated PV (BIPV) applications requiring rugged mechanical mounting. Installation of multiple modules on a fragile and old structure may also become challenging. Furthermore, the conventional architecture can add to the transportation cost due to the increased freight weight. To overcome these limitations of the PV modules' conventional architecture, the lightweight design has recently gained the focus of the PV community. The essential requirement of the lightweight modules is the replacement of the bulky glass with optically transparent polymers (such as ethylene tetrafluoroethylene (ETFE)[1], polyethene terephthalate (PET)[2], polycarbonate[3], Poly (methyl methacrylate) (PMMA)[4], acrylic[5], etc.). However, since these polymeric materials do not provide enough mechanical stiffness, a layer of composite material (i.e., fibre-

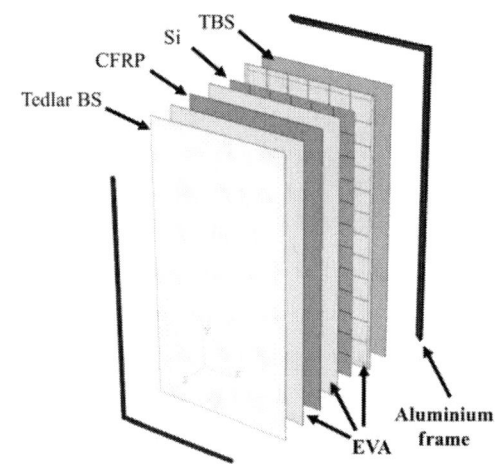

Fig. 1. FE model of lightweight PV module in this study.

reinforced polymers) is used as a substrate to safeguard against external mechanical stressors. Previous studies have reported the successful use of carbon fibre reinforced polymers (CFRP) [6] and glass fibre reinforced polymers (GFRP)[7] as composite substrates in lightweight designs. The materials mentioned above can reduce the area density of the conventional PV architecture from 12-16 kg/m2 to about 6.5 kg/m2 [1]. The other potential applications of the lightweight PV architecture include integration on vehicles and portable electronics as well as other niche applications[8]. Previous studies have demonstrated the durability of lightweight architecture against environmental stressors[9][10]. However, a comparison between the conventional and the lightweight structures for durability against mechanical stressors is not yet reported.

Therefore, in this study, we have performed finite element simulations to understand the generation of stress due to the combined load of 2400 Pa and self-weight of the structure for both conventional and lightweight designs. The lightweight structure presented in this work has the architecture of transparent backsheet (TBS)/EVA/Si-cell/EVA/CFRP/EVA/polyvinyl fluoride (PVF) backsheet (BS) which is mounted in the aluminium frame as shown in Fig. 1. The model that represents the conventional design has layers of glass/EVA/Si-cell/EVA/BS. The details on the FE modelling approach have been discussed in the subsequent sections of the manuscript.

TABLE I. PROPERTIES OF MATERIALS [13] AND THICKNESS OF EACH LAYER

Materials	Young's modulus (GPa)	Poisson's ratio	Mass density (Kg/m³)	Thickness (mm)
Ethylene vinyl acetate (EVA)	0.067	0.41	1030	0.3
Silicon cell	112	0.28	2329	0.2
Transparent backsheet	3.5	0.27	1100	0.5
PVF backsheet	3.5	0.27	1100	0.5
CFRP	-	-	1750	1.6

II. FINITE ELEMENT (FE) ANALYSIS

This section discusses the geometry, material properties, FE mesh and boundary conditions for the FE model of the conventional and lightweight structures. The FE simulations have been performed using Abaqus 6.14 package.

A. Geometry of the module

The lightweight and conventional PV modules in the FE model are 1980 mm in length and 995 mm in width and contains solar cells with dimensions of 156 mm X 156 mm. The thickness of each layer in PV modules is as shown in Table 1. The solar cells have been embedded between two layers of EVA just like an actual module. However, the PV ribbons that connect the two adjacent solar cells in the modules are not modelled to reduce the complexity. Furthermore, the modules have aluminium frames along their periphery for mechanical handling and fastening with the mounting structure. Therefore, in this work, modelling is done such that the modules rest in the cavity of the aluminium frames (Fig. 1).

B. Material properties

The properties of materials (TBS, EVA, silicon and BS) have been listed in Table 1 and are modelled as isotropic materials. On the other hand, the composite substrate (carbon fiber reinforced polymer) has been modelled as an orthogonal material for which definition of engineering constants has been described below[11]:

$$
\begin{Bmatrix} \sigma_{11} \\ \sigma_{22} \\ \sigma_{33} \\ \sigma_{12} \\ \sigma_{13} \\ \sigma_{23} \end{Bmatrix} = \begin{pmatrix} D_{1111} & D_{1122} & D_{1133} & 0 & 0 & 0 \\ D_{1122} & D_{2222} & D_{2233} & 0 & 0 & 0 \\ D_{1133} & D_{2233} & D_{3333} & 0 & 0 & 0 \\ 0 & 0 & 0 & D_{4444} & 0 & 0 \\ 0 & 0 & 0 & 0 & D_{5555} & 0 \\ 0 & 0 & 0 & 0 & 0 & D_{6666} \end{pmatrix} \begin{Bmatrix} \epsilon_{11} \\ \epsilon_{22} \\ \epsilon_{33} \\ \epsilon_{12} \\ \epsilon_{13} \\ \epsilon_{23} \end{Bmatrix}
$$

Where,
$$D_{1111} = E_1(1 - v_{23}v_{32})\alpha, \quad D_{2222} = E_2(1 - v_{13}v_{31})\alpha,$$

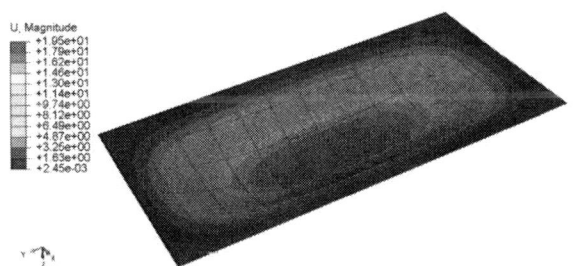

(a): Contours of max. displacement for conventional design

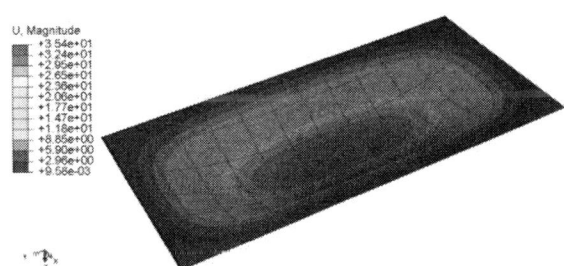

(b): Contours of max. displacement for lightweight design

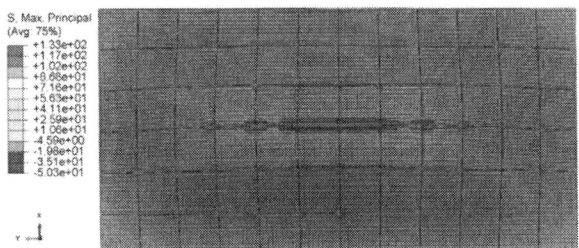

(c): Contours of max. principal stress for conventional design

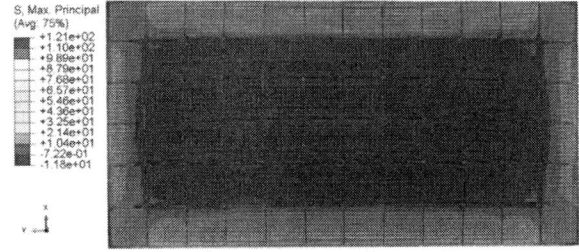

(d): Contours of max. principal stress for lightweight design

Fig. 2. Displacement profiles for cells in (a) the conventional and (b) the lightweight modules; and the contours of maximum principal stress for cells in (c) the conventional and (d) the lightweight modules.

978-1-7281-6118-1/22 $31.00 © 2022 IEEE

$D_{3333} = E_3(1 - \nu_{12}\nu_{21})\alpha, D_{1122} = E_1(\nu_{21} - \nu_{31}\nu_{23})\alpha,$
$D_{1133} = E_1(\nu_{31} - \nu_{21}\nu_{32})\alpha, D_{2233} = E_2(\nu_{32} - \nu_{12}\nu_{31})\alpha,$
$D_{1212} = G_{12}, D_{1313} = G_{13}, D_{2323} = G_{23},$
$\alpha = \dfrac{1}{1 - \nu_{12}\nu_{21} - \nu_{23}\nu_{32} - \nu_{31}\nu_{13} - 2\nu_{21}\nu_{32}\nu_{13}}.$

The values for various elastic constants are described as follows[12]: $E_1 = 103.9\ GPa; E_2 = 6.62\ GPa; E_3 = 6.62\ GPa;\ \nu_{12} = \nu_{21} = 0.25;\ \nu_{13} = \nu_{31} = 0.33; \nu_{23} = \nu_{32} = 0.25; G_{12} = G_{12} = 4.88\ GPa;\ G_{23} = 2.48\ GPa.$

The composite substrate has been modelled comprising eight laminae, each layer of 200 μm thickness, with alternate ply orientation of 0° and 90°. The orientation of 90° is along the longer edge of the module.

TABLE II.   COMPARISON OF MAXIMUM DISPLACEMENT AND STRESSES IN SOLAR CELLS FOR THE TWO MODULE DESIGNS

Design	Max. displacement (mm) due to:			Max. Principal stress (MPa) due to:		
	Self-weight	2400 Pa	Self-weight and 2400 Pa	Self-weight	2400 Pa	Self-weight and 2400 Pa
Conventional	0.61	18.9	19.5	4.19	128	133
Lightweight	0.4	35	35.4	1.37	120	121

C. FE mesh and boundary conditions

The FE mesh used in this work has 8-node brick (hexahedral) elements with incompatible modes to avoid the shear locking during the bending of the elements[11]. The simulations are done such that the frame of the module is held fixed and a surface traction of 2400 Pa has been applied on the transparent backsheet (or glass) of the module to represent the wind load on the module as per the IEC standard[14]. Furthermore, the effect of the self-weight of the structure has also been considered for both the designs. A PV module has various interfaces such as TBS/ EVA, EVA/cell, EVA/CFRP and EVA/BS. The relative motion among these layers is constricted in a module and to implement this, the tie boundary condition is applied at these interfaces in the FE model. The tie boundary condition is also applied between edges of the layers and the cavity of the frame to restrain the relative motion.

III. RESULTS AND DISCUSSION

A. Displacement and stress profile in the layer of solar cells due to static loading of 2400 Pa and self-weight

The simulation results for displacement and generation of maximum principal stresses due to application of combination of the static load of 2400 Pa and self-weight as well as individual loads in the array of solar cells has been presented in Table II. The representative contours of principal stresses and displacement for the array of the solar cells are presented in Fig. 2. The maximum displacement in the conventional and lightweight design is found to be 19.5 mm (Fig. 2 a) and 35.4 mm (Fig. 2 b), respectively. Furthermore, the maximum stress in the solar cells encapsulated in the conventional design is

found to be 133 MPa (Fig. 2 c), which is close to the value of maximum principal stress of 121 MPa in lightweight design (Fig. 2 d). The maximum stresses are generated in the cells which are close to the longer edge of the module for the lightweight design while for the conventional design, the most heavily stressed cells are found at the center of the module. The maximum stresses generated in both the designs described in this work are well below the fracture strength of silicon cell which is about 200 MPa[15]. Therefore, it can be argued that the lightweight design will be able to sustain the self-weight and static load of 2400 Pa, which simulates wind load in the field.

IV. CONCLUSIONS AND FUTURE WORK

This work compares the effect of self-weight and static load of 2400 Pa on the lightweight and conventional design of PV modules through FE simulations. The value of max. principal stress generated in the solar cells for the lightweight design is less than the conventional design, which indicates the ability of a CFRP based composite substrate to safeguard the PV module against a combined load of 2400 Pa and self-weight. However, the maximum displacement in the lightweight structure was found to be considerably higher (almost double) than the conventional design.

ACKNOWLEDGMENT

The authors acknowledge the financial support provided by the National Centre for Photovoltaic Research and Education (NCPRE) Phase-II (Grant No. 16MNRE002) by the Ministry of New and Renewable Energy (MNRE), Government of India towards this work

REFERENCES

[1] A. C. Martins, V. Chapuis, A. Virtuani, L.-E. Perret-Aebi, and C. Ballif, "Hail Resistance of Composite-Based Glass-Free Lightweight Modules for Building Integrated Photovoltaics Applications," in *33rd European Photovoltaic Solar Energy Conference and Exhibition (PVSEC 2017)*, 2017, pp. 2604–2608.

[2] S. Smith *et al.*, "Transparent backsheets for bifacial photovoltaic (PV) modules: Material characterization and accelerated laboratory testing," *Prog. Photovoltaics Res. Appl.*, pp. 1901–1906, 2021.

[3] A. S. Budiman *et al.*, "Enabling curvable silicon photovoltaics technology using polycarbonate-sandwiched laminate design," *Sol. Energy*, vol. 220, no. November 2020, pp. 462–472, 2021.

[4] D. C. Miller *et al.*, "An investigation of the changes in poly(methyl methacrylate) specimens after exposure to ultra-violet light, heat, and humidity," *Sol. Energy Mater. Sol. Cells*, vol. 111, pp. 165–180, 2013.

[5] T. Kajisa *et al.*, "Novel lighter weight crystalline silicon photovoltaic module using acrylic-film as a cover sheet," *Jpn. J. Appl. Phys.*, vol. 53, no. 9, 2014.

[6] K. K. Hung and I. Chasiotis, "Control of substrate strain transfer to thin film photovoltaics via interface design," *Sol. Energy*, vol. 225, no. June, pp. 643–655, 2021.

[7] C. Kutter, F. Basler, L. E. Alanis, J. Markert, M. Heinrich, and D. H. Neuhaus, "Integrated Lightweight, Glass-Free PV Module Technology For Box Bodies Of Commercial Trucks," *37th Eur. Photovolt. Sol. Energy Conf.*, no. November, pp. 1711–1718, 2020.

[8] Y. Liu, A. Mukherjee, H. Wu, P. Yu, Y. Li, and C. Hsieh, "Accelerated Test for Light-Weight Photovoltaic Module Encapsulants," *Conf. Rec. IEEE Photovolt. Spec. Conf.*, vol. 2020-June, pp. 0707–0709, 2020.

[9] P. Grygiel *et al.*, "Prototype design and development of low-load-roof photovoltaic modules for applications in on-grid systems," *Sol. Energy Mater. Sol. Cells*, vol. 233, 2021.

[10] G. Imbuluzqueta, N. Yurrita, J. Aizpurua, F. J. Cano, and O. Zubillaga, "Composite material with enhanced ultraviolet performance stability for photovoltaic modules," *Sol. Energy Mater.*

Sol. Cells, vol. 200, no. May, p. 109947, 2019.

[11] D. S. Simulia, "ABAQUS/Standard User's Manual, Version 6.14.," Providence, RI., 2014.

[12] A. P. S. and T. P. S. Prashob P. S, "Determination of orthotropic properties of carbon fiber reinforced polymer by tensile tests and matrix digestion," in *International Conference on Composite Materials and Structures- ICCMS 2017*, 2017.

[13] U. Desai, D. P. Vasudevan, A. Kottantharayil, and A. Singh, "Prediction of vibration induced damage in photovoltaic modules during transportation: finite element model and field study," *Eng. Res. Express*, vol. 3, no. 4, p. 045045, 2021.

[14] IEC-61215, "Terrestrial Photovoltaic (PV) Modules—Design Qualification and Type Approval."

[15] F. Kaule, W. Wang, and S. Schoenfelder, "Modeling and testing the mechanical strength of solar cells," *Sol. Energy Mater. Sol. Cells*, vol. 120, no. PART A, pp. 441–447, 2014.

Exploring the Role of Temperature and Hole Transport Layer on the Ribbon Orientation and Efficiency of Sb2Se3 cells Deposited via Thermal Evaporation

Ryan Voyce, Stephen Campbell, Oliver S. Hutter, Guillaume Zoppi, Neil S. Beattie, Elizabeth A. Gibson, Vincent Barrioz

Northumbria University, Newcastle upon Tyne, United Kingdom

Newcastle University, Newcastle upon Tyne, United Kingdom

Antimony selenide (Sb2Se3) has emerged as a promising candidate for next generation solar cell devices due to its non-toxicity, low cost, and earth abundance. Coupling these factors with its promising optoelectrical properties of its high absorption coefficient and almost ideal band gap for single-junction cells yields an incredibly attractive absorber material. Issues in the material come from poor carrier management, particularly in the mobility of photogenerated carriers within the absorber layer and through the immediate interfaces. The orientation of the $(Sb_4Se_6)_n$ ribbons grown via thermal evaporation was investigated by varying the deposition temperature and the post-annealing treatment. 300 °C as the deposition temperature was most conducive to promoting ribbon orientations which were perpendicular to the substrate. Annealing effects were shown to be able to induce crystallinity in films at a lower temperature than the deposition temperature as well as being able to influence orientation away from (hk0) orientations. Using these parameters, the effect of 15 nm thick NiOX and MoOX as Hole Transport Layer (HTL) materials deposited via electron-beam evaporation in antimony selenide solar cells is investigated in superstrate and substrate configurations. Notable improvements were found to the efficiency of the devices when NiOX was considered as the HTL, but a degradation occurred when fabricated with MoOX in superstrate configuration; substrate configuration was only viable with a NiOX HTL.

Decentralized BESS Control on a Real Low Voltage System with a Large Number of Prosumers

Bruno Cortes*[a], Ricardo Torquato[a], Tiago R. Ricciardi[a], Fernanda C. L. Trindade[a], Walmir Freitas[a], Victor B. Riboldi[b], and Kunlin Wu[b]

[a]Department of Systems and Energy, University of Campinas, Campinas, São Paulo, 13083-852, Brazil
[b]CPFL Energia, Campinas, São Paulo, 13087-397, Brazil
*brcortes@unicamp.br

Abstract—Recent advances in technology and financial costs reduction are allowing an anticipated insertion of Battery Energy Storage Systems (BESSs) in the low-voltage (LV) distribution systems, especially in customers owning a photovoltaic (PV) generation system. On the other hand, utilities are also installing medium-size BESSs on LV systems to assist in system operation. This massive integration of medium and small-sized BESS into PV-rich LV systems can create adverse impacts on the circuit operation as each BESS can operate with a different control mode, with different control settings. Therefore, since there are distinct control modes of operation, it is necessary to analyze them and how their parametrization and combined operation affect the grid indices. The parametrization and impacts of multiple BESS operating in an LV system are investigated in this paper through quasi-static time series simulations on a real LV gated community network with high PV penetration. The results reveal the need for properly coordinating the operation of the multiple BESS, otherwise, operating indices of the grid may deteriorate significantly.

Keywords—Distribution system, energy storage systems, low voltage network, PV generation.

I. INTRODUCTION

The world has been seeking more sustainable sources, which is emphasized by the Organization for Economic Co-operation and Development (OECD) support for the development of better environmental, social, and governmental practices [1]. Specifically, in the electric power sector, the recent uptake of decentralized prosumers on the distribution grid may deeply impact the overall operation of the system, which is requiring utilities and national agencies to revise standards and practices that were once proven to be true [1]–[3].

This is explained as the PV generation brings uncertainty to the generation side, besides allowing the possible change in the direction of the power flow, which may lead to unexpected operation of the system [2], [3]. Additionally, the increasing number of PV generation has already been particularly associated with the deterioration of the power quality, e.g., the rise of overvoltage, unbalance, and overloading of both distribution transformers and conductors [4], [5]. Meanwhile, despite the consumers having been known to present some characteristic load profiles along the days, their consumption is also related to uncertainties in distribution systems [6]–[8].

This work was supported in part by São Paulo Research Foundation (FAPESP), grants 2020/10523-4, 2018/24018-0, and 2016/08645-9, in part by National Council for Scientific and Technological Development (CNPq), grant 304373/2020-6, and in part by CPFL Energia, grant PD-02937-3018/2016.

Under this outcome, one of the solutions is the energy management at the service line of each consumer by installing BESS. BESS has presented a gradual cost reduction in the latter years, and there is a rational expectation of 28 to 58% of capital cost reduction from now until 2030 [9]. The latest advances have made BESS a feasible solution to reduce the transformer load factor and enhance the power quality indices near its connection. It can also work as a backup power source during a blackout and postpone reinforcements investments, avoiding a considerable financial expenditure.

Regarding control strategies, five different voltage-dependent ones were analyzed in [10] aiming to achieve a high self-consumption for a household with PV and BESS under minimum voltage violations. However, the authors have pointed out that it would be tricky to define the critical voltage threshold as the inclusion of a voltage regulation might reduce the economic benefit for the BESS owner, while the voltage profile would be improved at the bus around its location without any financial reward. Distinctly, conventional features usually available in PV inverters were evaluated in [11] to show their influence on the overall PV hosting capacity, while associated with BESS integration. The results have shown that the inclusion of BESS has a positive impact on the overall hosting capacity, but it may range severely depending on the selected inverter features.

Particularly, an adaptive control strategy that changes BESS charge and discharge rate along the day was proposed in [12] to reduce the voltage and thermal issues in a distribution network. It was also highlighted that high PV penetration in LV networks might substantially lead to congested lines on the medium-voltage side. Finally, business models considering the Brazilian regulatory context were evaluated in [13] aiming to turn the integration of prosumers with BESS economically viable. The inclusion of a battery degradation model in the study has presented non-negligible impacts on the results, highlighting the importance of establishing a control strategy that limits the charge and discharge cycles of the BESS throughout the day.

In this way, this paper first presents the impacts of an uncoordinated operation of small and medium-size BESS in an LV system, confirming the importance of developing suitable control modes for operating in a real LV network with high PV+BESS penetration. Then, a methodology is developed to establish adequate setpoints for the medium-size BESS controllers, and two new advanced strategies are proposed to enhance the overall operation. The studies were run through the DSS Python module of the DSS Extensions implementation of the OpenDSS [14], which allows faster simulations, among other notable improvements. Lastly, the quasi-static time-series simulations were run with a 15-minute time-step.

II. System Description

A. Electrical Network

The LV circuit used in the simulations refers to a gated community and is composed of a step-down transformer, 11.4kV/220V, Δ-Y$_g$, 150 kVA; 47 households, of which 11 are two-phase connected and 36 are three-phase connected; 27 households have PV generation installed with 2.5 kWp, two-phase connected, presenting a total installed PV capacity of 67.5 kWp, corresponding to 45% of the transformer capacity.

Additionally, it is considered the installation of a two-phase, 5 kW/12.5 kWh, small-size BESS at each of the 27 households with PV generation and a unique three-phase, 100 kW/225 kWh, medium-size BESS connected at the secondary side of the step-down transformer. Each BESS presents distinct modes of operation, as detailed below. Despite using 2.5 kWp PV generators, the installation of 5 kW small-size BESS was considered to allow the prosumer to achieve a 100% surplus of power in relation to the maximum PV generation, aiming to obtain a higher financial gain that would help out the prosumer on the economic viability of the equipment installation. The basic idea for the small-size BESS is to charge it during the day along the off-peak hours (lower tariffs) and discharge at the intermediary and peak hours, taking advantage of the energy cost distinction during these periods.

In this work, a real Brazilian utility and its respective tariff defined by the regulatory agency were considered, consisting of (*i*) *peak tariff tier* (from 18:00 to 21:00) of 1.026 R\$/kWh; (*ii*) *intermediary tariff tier* (from 16:00 to 18:00, and 21:00 to 22:00) of 0.672 R\$/kWh; and (*iii*) *off-peak tariff tier* of 0.49 R\$/kWh for the remaining hours of the day.

Regarding the load consumption, the load profile of each household was determined by considering the total amount of energy billed on a month and using typical load profiles with 15-minute time resolution, following the proposal of [6]. On the other hand, the PV generation profile was measured with a 1-minute time resolution and the missing data was filled by the following: (*i*) if the lack of measurement occurs during less than 2 hours on the analyzed day, the considered period is filled with the interpolation of the existing data; (*ii*) otherwise, i.e., if the lack of measurement occurs during longer than 2 hours, the whole day is replaced by the average day of the analyzed month. After that, the PV generation data is transformed into a 15-minute time interval curve by calculating the average generation in each 15-minute time window, which is also the period adopted for the load consumption profiles.

B. Typical BESS Modes of Operation

Conventionally, the Peak-Shaving and Time Modes are typical operation modes for dispatching a BESS and can be implemented during the procedure of both charging or discharging, as configured by the user [15]. However, when referring to the gated community electrical network and the location of each BESS, two goals need to be achieved separately by these modes. On the prosumer side, by considering the adoption of the time-of-use tariff, the energy price changes at specific hourly intervals during the day [16]. So, by establishing that the same energy price is applied for the consumption and the generation, the highest revenue will always be achieved by charging (consuming power) the BESS at each prosumer during the day and discharging (injecting power) it during the time interval of highest price, but note that some variations on these intervals will be analyzed at some point in Section IV. So, the Time Mode of operation is the one employed for this purpose, for the small-size customer BESS (at the prosumers).

Under this circumstance, the charge rate to obtain SOC=100% in the small-size BESS at 16:00 (the starting time of the intermediary tariff tier) is estimated through (1), where $C_{estimated}$ is the estimated charge rate, ΔT_{Charge} is the time available for charging (in this case, it is 8 hours since the small-size BESS charges from 8:00 to 16:00), $P_{nominal}$ is the rated active power of the BESS (in this case, 5 kW), $E_{nominal}$ is the rated energy capacity of the BESS (in this case, 12.5 kWh), η_{Charge} is the efficiency of the BESS during the charging process (90%), and $E_{initial}$ is the energy stored in the BESS at the beginning of the charge (i.e., at 8:00). The necessary discharge rate to ensure that the discharge process will occur during only the interval when the tariffs are most expensive (i.e., between 16:00 and 22:00) is calculated through (2), where $D_{estimated}$ is the estimated discharge rate, $\Delta T_{Discharge}$ is the duration time of the discharge process (in this case, it is 6 hours because the BESS discharges at intermediary and peak tariff tier periods: from 16:00 to 22:00), $\eta_{Discharge}$ is the efficiency of the BESS during the discharging process (90%), and E_{final} is the energy stored in the BESS at the end of the discharge (in this study, the objective is that the stored energy at the end of the discharge is the minimum recommended by the manufacturer, avoiding the deterioration of the life cycle of the BESS – in general, about 20% of the rated capacity).

$$C_{estimated} = \frac{(E_{nominal} - E_{initial})}{\Delta T_{Charge} \cdot P_{nominal} \cdot \eta_{Charge}} \tag{1}$$

$$D_{estimated} = \frac{\left(E_{nominal} - E_{final}\right)}{\Delta T_{Discharge} \cdot P_{nominal} \cdot \left(\eta_{Discharge}\right)^{-1}} \tag{2}$$

On the other hand, the medium-size BESS is particularly located at the transformer secondary terminals to precisely limit the direct and reverse active power flowing through it. Herein, it is considered that the medium-size BESS is owned by the utility, which would like to limit the impact of the gated community on its grid. Therefore, the Peak-Shaving Mode of operation is essential to achieving this goal. The ideal scenario might be to adjust both lower and upper setpoints as small as they can be (ideally close to zero), virtually islanding the LV circuit. Despite this utopia, it must be pointed out that it would not be possible as both targets affect the start time, end time, and also the duration of both charge and discharge processes of the BESS, i.e., it limits the capability of energy management that this equipment provides. So, the setpoints definition remains necessary and it is addressed next.

III. PROPOSED STRATEGIES FOR THE MEDIUM-SIZE BESS OPERATION

Following, different strategies are proposed and implemented to properly coordinate the operation of the medium-size BESS with the operation of the small-size BESSs, which are assumed to operate according to the time mode.

Because the consumption measured in the transformer is the total aggregate value of the LV network and does not change considerably between days, the use of active power measurements from the previous days is useful to estimate targets of charging and discharging for the medium-size BESS [17]. Therefore, a simple alternative is to obtain the charge and discharge targets, T_{low} and T_{high}, respectively, based on the observation of the historical measurements of active power flow through the distribution transformer. T_{Low} is the lower active power target (BESS charges if the active power becomes lower than T_{Low}) and T_{High} is the higher active power target (BESS discharges if the active power flow surpasses T_{High}). In this work, the adopted values of T_{High} and T_{Low} for the medium-size BESS are 8 kW and 0 kW, respectively. So, these settings establish what is called Method 0 in the studies that will be presented in Section IV. Additionally, these settings are used as the basis for the conceptualization of the strategies described as Method 1 and Method 2 that are shown next, and it will serve for comparison.

A. Method 1 (requires communication)

A typical scheme implemented in Germany to avoid an excessive injection of energy into the grid by the photovoltaic generation is to use only the measurement of the power generated in each prosumer as a basis for the control of the associated BESS. According to [18] and [19], in Germany, the control of the small-size BESS is implemented so that the BESS starts charging when the power generated by the prosumer exceeds 70% of the rated capacity of the photovoltaic generation, limiting the power injection. In this situation, the discharging process occurs when there is no more power injected by photovoltaics.

Based on that, Method 1 proposes keeping T_{High} and changing T_{Low} from 0 to 1/3 of the medium-size BESS rated power (-33.3 kW) when the total energy injected by the n_PV photovoltaic generators at the LV network (E_{total}, from (3)) in a specific day surpasses 70% of the BESS energy capacity (157,5 kWh) (as in (4)). This change allows a greater reverse power-flow during periods with high photovoltaic generation and prevents the medium-size equipment from being fully charged before the period with the intermediate and peak tariffs, which starts at 16:00.

With this purpose, it is also necessary to define the instant in which T_{Low} changes from -33.3 kW back to 0. At the time interval when the relationship defined in (4) is satisfied ($t_{initial}$), the change on the reverse power-flow adjustment (T_{Low}) of the medium-size BESS occurs. Such T_{Low} adjustment must hold for 80% of the period between the start time ($t_{initial}$) and the beginning of the intermediate tariff period (16:00), i.e., the duration ΔT of the modified T_{Low} is shown in (5). As the

prosumers will use the small-size BESS to discharge during the high tariff tier period, high power injection is expected from the prosumers between 16:00 and around 18:00, as there is power being injected by both the small-size BESS and by the photovoltaic generators.

Although this period from 16:00 to 18:00 is used to charge the medium-size BESS (which is expected to be fully charged around 17:00-18:00 to meet the subsequent peak), it may not be long enough to fully charge the equipment. Therefore, the threshold of 80% of the period ($16 - t_{initial}$) is set to guarantee that the medium-size BESS will have more time to fully charge, i.e., after 80% of the period ($16 - t_{initial}$) the reverse power-flow setpoint returns to 0 kW to force the equipment to charge, if necessary. From these definitions, the final instant (t_{final}) of the period in which T_{Low} remains in -33.3 kW is calculated by (6).

$$E^{total} = \sum_{h=1}^{current\_hour} \sum_{pros=1}^{n\_PV} P_{pros,h}^{PV} \qquad (3)$$

$$E^{total} > 70\% \times E_{nom}^{BESS_{transformer}} \qquad (4)$$

$$\Delta T = 0.8 \times (16 - t_{initial}) \qquad (5)$$

$$t_{final} = t_{initial} + \Delta T \qquad (6)$$

Although the proposed strategy helps the medium-size BESS not to have a large amount of stored energy at the end of the day, it may be insufficient to prevent the accumulation of energy throughout several days, which eventually inhibits its correct operation. To ensure that the equipment is fully discharged at the end of the day, its operating mode is changed to Time Mode at 22:00, after the end of the intermediate tariff period, allowing the equipment to discharge so that the desired state of charge value (i.e., SOC=40%) is always obtained by the end of the day (24:00).

B. Method 2 (does not require communication)

Like the previous method, the second proposal, i.e., Method 2, also assumes that, over some time, the BESS reverse power-flow target defined in Method 0 is changed to one-third of the medium-size BESS rated power, allowing power injection from the gated community to the utility grid. The difference between this method and Method 1 is in which period of the day the reverse power-flow is allowed. In Method 1, this period starts when the energy injected by the photovoltaic generators exceeds 70% of the capacity of the medium-size BESS. In Method 2, this period is between 16:00 and 22:00 (a total of 6 hours), when intermediate and peak tariff tiers occur. It can be pointed out, therefore, that Method 2 does not require any communication between prosumers and the medium-size BESS (at the MV:LV transformer). Instead, it takes advantage of the tariff distinction during the day.

Again, to ensure that the medium-size BESS is fully discharged at the end of the day, its operation mode is changed to Time Mode at 22:00, after the end of the intermediate tariff period, causing the equipment to discharge until SOC=40%.

IV. RESULTS AND DISCUSSION

This session starts by presenting results of the test system neglecting the methods 0, 1, and 2 to show how the uncoordinated operation of the BESS can affect the performance of the complete energy storage system (Section IV.A). Section IV.B shows how the operation of the energy storage system can be improved using the proposed approaches.

A. Assessement of the BESS Performance under Conventional Operation

This subsection presents the studies of the BESS operation, highlights the parametrization importance, and shows the main impacts on the gated community system. For this initial study, the Base Case refers to the case without any BESS in the LV network. For the remaining cases, the setpoints of the Peak-Shaving Mode at the medium-size BESS were fixed to 10 kW for direct active power and -2 kW for the reverse one. Studies were conducted considering variations in the start time of both charge (Case 1) and discharge (Case 2) periods and on the rate of discharge (Case 3) of all small-size BESS at the households. The base case values for these four parameters are 8:00, 15:00, and 20%, respectively.

By varying the charging start time from 8:00 to 12:00 with 1-hour steps, the results in Fig. 1 show an unusual reverse peak around 13:00 and 15:00. This occurred because the medium-size BESS was fully charged early in the day and could not absorb all PV energy as the small-size BESSs have contributed for a shorter time.

In contrast, by varying the starting of discharge from 15:00 to 19:00 with 1-hour steps, the results in Fig. 2 show an uncommon reverse peak after 16:00. It occurred because the small-size BESS started to discharge at 15:00, acting as a huge generation that led the medium-size BESS to be fully charged for some time until the peak demand raises around 18:00.

Finally, Fig. 3 shows that, by changing the rate of discharge from 20% to 40% with steps of 5%, a reverse peak may occur around 16h, as the small-size BESS will work as a huge source of generation during a period when the load is not high. These results highlight the importance of proper coordination among the BESSs in a scenario (such as an LV system) with a high BESS penetration (in this case, there are 27 small BESSs and 1 medium-size BESS in the circuit). If they are uncoordinated, they might have a significant negative impact on the circuit.

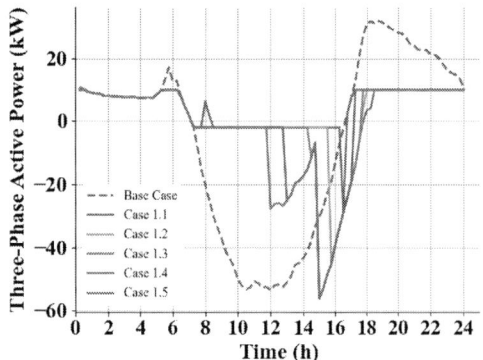

Fig. 1. Transformer loading curve – Case 1 (varying the charging start time from 8:00 to 12:00 with 1-hour steps)

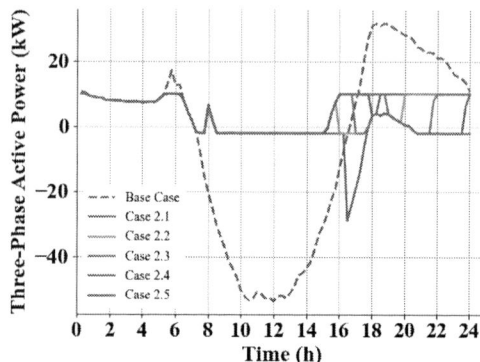

Fig. 2. Transformer loading curve – Case 2 (varying the starting of discharge from 15:00 to 19:00 with 1-hour steps)

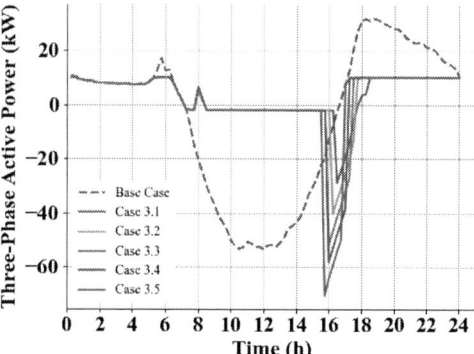

Fig. 3. Transformer loading curve – Case 3 (varying the rate of discharge from 20% to 40% with steps of 5%)

TABLE I shows that the variations done in Cases 1 and 3 have considerably affected the transformer demand factor, leading to higher peak demand as the variations between the subcases were done. However, the transformer load factor (ratio between average and maximum demand) presented a more stable value, with lower variations for all three cases. Additionally, the medium-size BESS was not found fully charged during the simulation period only when the discharge of the small-size BESSs occurred later than 15:00. This was expected as the closer to the peak demand period (between 16-22h), the lower the net energy being injected by the prosumers, avoiding the medium-size BESS to charge. Note that the Base Case values refer to the situation where no BESS is installed in the circuit. In contrast, some subcases have shown values for both load and demand factors higher than ones of the base case, which highlights how severe can be the impact of the parameters (that were changed in each subcase) in the results and unexpected adverse impacts may occur if the combined operation of multiple BESS is not properly considered.

In addition to these adverse impacts on the system loading, the state of charge of the BESS must also be properly monitored and managed. For instance, taking Case 2.2 as an example, Fig. 4 shows that the medium-size BESS on Peak-Shaving Mode stocks accumulative energy at the end of sequential days, leading to a fully charged battery on undesired intervals. This hinders the BESS capability of energy management after a few days.

TABLE I RESULTS FOR CASES 1, 2 AND 3

Case	Demand Factor [%]	Load Factor [%]	Maximum SOC [%]
Base Case	71.36	44.85	100.00
1.1	38.36	23.04	100.00
1.2	60.29	17.52	100.00
1.3	74.88	17.29	100.00
1.4	74.89	19.27	100.00
1.5	74.89	20.48	100.00
2.1	38.36	23.04	100.00
2.2	13.35	53.77	93.96
2.3	13.35	50.65	78.67
2.4	13.35	47.31	78.67
2.5	13.35	48.6	78.67
3.1	38.36	23.04	100.00
3.2	53.75	18.85	100.00
3.3	68.86	16.3	100.00
3.4	77.36	14.78	100.00
3.5	94.22	13.07	100.00

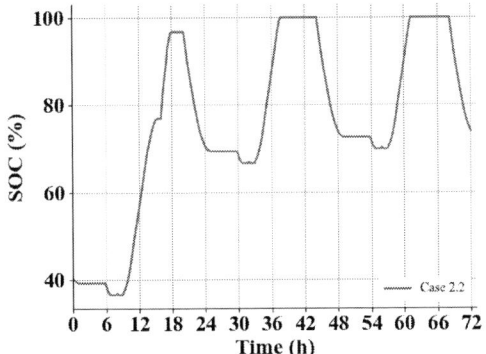

Fig. 4. State of Charge for the Medium-Size BESS – Case 2.2

Regarding the voltage variation, it must be mentioned that the lack of coordination among BESSs also caused voltage deterioration. For instance, the maximum voltage increased by 0.55% in Case 3.5 and the minimum voltage decreased by 2.15% in Case 2.5, which is a significant impact. This further confirms that, if there are multiple BESSs in the system, there should be some level of coordination among them.

B. Enhanced Operation under the Proposed Strategies

After implementing the proposals presented in Section III, Fig. 5 shows the SOC behavior during three days of operation. In contrast to Method 0, it can be noted that the BESS state of charge for Method 1 and Method 2 is stable across multiple days, i.e., it does not increase indefinitely. In addition, during the operation, as shown in Fig. 6, the equipment was able to limit the direct and reverse active power that is injected into the distribution grid most of the time when using both proposed strategies. Note that, in Method 2, the BESS has also supported the grid by injecting energy during the intermediate and peak tariff tiers (between 16:00 and 22:00) when the rest of the network may have to deal with higher loading conditions. Thus, under these assumptions, the inverse peak of generation is delayed to the period with peak loading of the grid, but it also

presents a lower value, impacting positively both the transformer load and demand factors.

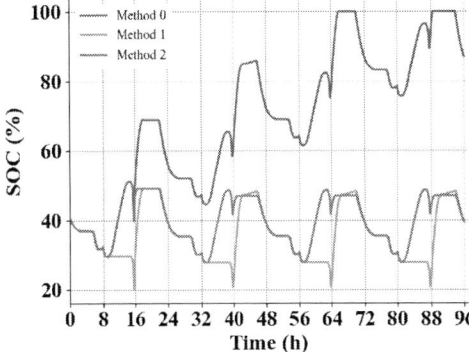

Fig. 5. State of Charge for the Medium-size BESS – Proposed Strategies

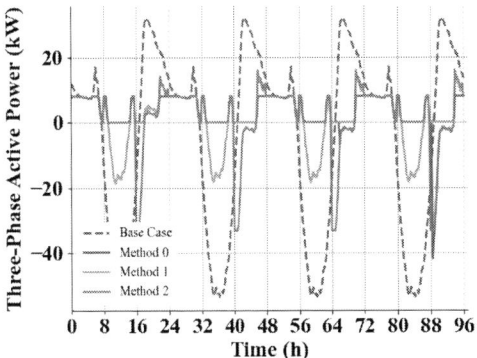

Fig. 6. Transformer Loading Curve – Proposed Strategies

TABLE II RESULTS FOR METHODS 0, 1 AND 2

Method	Demand Factor [%]	Load Factor [%]	Maximum SOC [%]
0	55.85	10.60	100.00
1	25.02	36.38	49.22
2	44.44	20.09	51.12

It can be seen from Fig. 6 and TABLE II that the peak demand for Method 2 is higher than the one for Method 1. As the managed energy was not so different for each method, this higher demand factor has led to a lower load factor for Method 2 but note that this load factor value was still higher than the one found for Method 0. Regarding the maximum SOC, the conventional Peak Shaving Mode led to a fully charged medium-size BESS, while the proposed methods could present lower values, around 50% for the SOC. This way, the implementation of both strategies would require a BESS with lower capacity and, thus, cheaper than the one installed, requiring lower installation costs that will positively impact the payback time when considering the investment.

Finally, it can be seen in TABLE III that the gated community would pay a much lower energy cost if Method 2 is considered. Since this method aims on taking advantage of the different costs at each tariff tier, a better financial performance was expected. Note that, despite the base case showing that an energy credit of R$ 26.89 would be obtained, no energy management was considered, which would affect the power quality indices of the circuit and the overall operation of the gated community. Thus, the base case is a situation that must be avoided due to the huge penetration level of photovoltaics.

TABLE III ENERGY TARIFF OF THE GATED COMMUNITY

Method	Consumed Energy [R$]	Produced Energy [R$]	Total Net Cost [R$]
Base Case	660.76	687.65	-26.89
0	187.46	46.47	140.99
1	202.20	129.13	73.07
2	199.75	169.49	30.27

1) Impact of the number of prosumers with BESS

Considering the high cost of acquiring a BESS, this subsection aims to present the main impacts that different penetration level of the small-size BESS has in the circuit. It was considered that the small-size BESS is deactivated from 0 (all prosumers have the equipment) to 27 (there is no BESS in the circuit at all) in each simulation. In general, the lower the penetration of small-size BESS, the greater the participation of the medium-size BESS in the management of network consumption. Therefore, for low penetrations of small-size BESSs, the medium-size BESS is not capable of managing the network consumption and, as expected, the benefits of the equipment are smaller. Also, it was noted that the operation of the medium-size BESS in Method 0 was more impacted by the penetration level of small-size BESSs. This occurred because the fewer the prosumers with small BESSs, the lower will be the energy absorbed (injected by the photovoltaic generators) locally by them throughout the day, requiring the medium-size BESS to charge, even more, leading it to be fully charged faster.

For all 28 subcases that were simulated, Fig. 7 shows how much the amount of BESS in the prosumers impacts the demand factor and consequently the maximum demand of the LV circuit. It is notable that Method 1 presented the lowest demand factor. In an intermediate range, from 13 to 22 prosumers with BESS, the performance of Method 2 was very close to the one found for Method 1. In general, Method 2 presented a higher maximum demand than Method 1 because the discharge of energy stored in the medium-size BESS starts at 16:00 and its demand factor is limited to the lower target (-33.3 kW), regardless of the discrepancy between the stored and consumed energy, resulting in a demand factor close to 44.4% since the rated power of the transformer is 75 kVA. Therefore, through these results, it is concluded that the use of all methods was able to reduce the maximum demand of the LV system transformer and Method 1 was more effective in this aspect for most of the evaluated range.

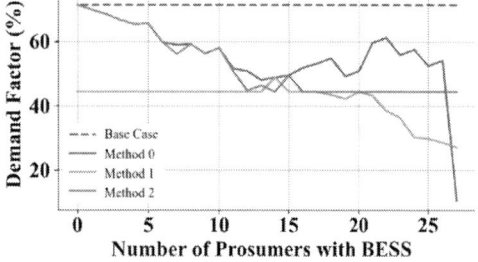

Fig. 7. Maximum Demand Factor

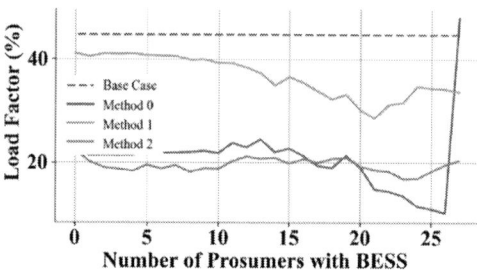

Fig. 8. Maximum Load Factor

With the reduction of the maximum demand and the energy management that Peak-Shaving Mode offers to the circuit, it was expected that the average demand would also be reduced. However, the number of prosumers with small-size BESSs has a direct impact on the charging and discharging process of the medium-size BESS throughout the day. This occurs because the lack of BESSs in the prosumers changes the power flow that reaches the transformer and, consequently, the operation of the medium-size BESS. From Fig. 7 and Fig. 8, it can be seen that the number of prosumers with BESS does not affect so much the average demand when Method 2 is applied with at least 13 prosumers with BESS. As the load factor of Method 2 is always around 20%, below 13 prosumers with BESS, the average demand grew proportionally to the increase of the maximum demand (associated with the maximum demand factor of Fig. 7). As for Method 1, its use has always led to a higher load factor than Method 2, with values close to 40% when up to 13 prosumers with BESS were considered.

Remarkably, as the load profile was impacted in the three methods, the other variables of interest will also be. To clearly present these impacts, below, only the maximum values found over the simulation period for each of the variables are considered and they are compared to the result obtained in the Base Case (without any BESS in the network) when possible.

The values found for the maximum medium-size BESS state of charge are shown in Fig. 9 for each method. As it is known, the medium-size BESS is fully charged at the end of the third day for Method 0. Thus, this method is not suitable for operating the medium-size BESS. Notably, Method 1 proved capable of ensuring that the medium-size BESS is not fully charged even when no prosumer has a BESS, while with Method 2 at least 20 small-size BESS are needed to avoid the loss of its operational flexibility. Thus, as observed in Fig. 9, it can be concluded that the operation in Method 1 allows the required nominal capacity of the medium-size BESS to be reasonably lower than the capacity when using Method 2. However, the energy capacity should not be the only matter of importance to be considered when choosing one method over the other.

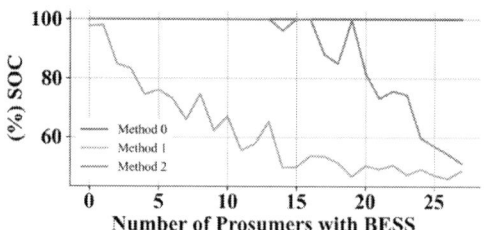

Fig. 9. Maximum SOC

978-1-7281-6118-1/22 $31.00 © 2022 IEEE

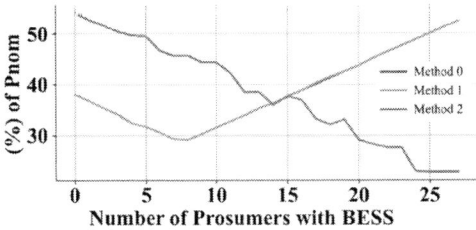

Fig. 10. Maximum Charge Rate

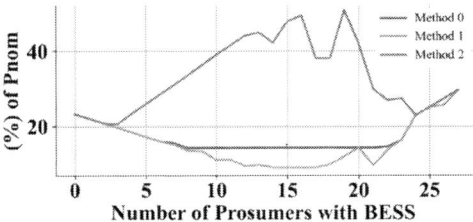

Fig. 11. Maximum Discharge Rate

Other extremely important parameters for the BESS are the charge and discharge rates as low values are desirable to avoid equipment degradation. Fig. 10 and Fig. 11 show the maximum values found, where Method 2 was able to guarantee lower charge rates than Method 1 from at least 15 prosumers with small-size BESS, while Method 1 proved to be the best option below 14 prosumers with small-size BESS. On the other hand, the number of prosumers with small-size BESS showed to severely impact the maximum discharge rate of the medium-size BESS with Method 2, while a much smaller variation was observed for Method 1. Note that, when all prosumers have small-size BESS, Method 2 has presented the lowest rates of charge and discharge.

Regarding the total network losses, the three methods were able to reduce losses as shown in Fig. 12. As the small-size BESSs help reduce the current flowing in the cables, the more amount of BESS, the lower the losses. Finally, Fig. 13 shows the results for the electricity bill paid by the gated community. In summary, Method 1 was more helpful when there are up to 23 prosumers with small-size BESS, while Method 2 presented similar results to Method 0. However, Method 2 was the one with the lowest cost for the circuit when all prosumers have a BESS. Note that the differences between the results found for Methods 1 and 2 are small and both methods performed better than Method 0, especially for at least half of the prosumers having a small-size BESS, i.e., above 13.

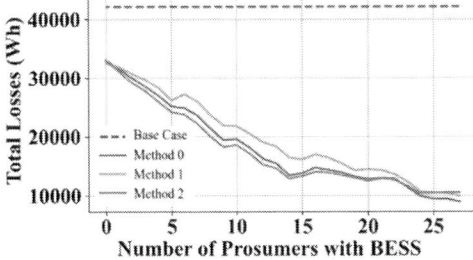

Fig. 12. Total Losses

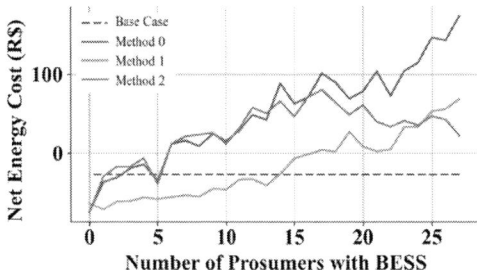

Fig. 13. Energy Tariff of the Gated Community

V. CONCLUSIONS

This paper showed that, in a scenario with high penetration of BESS, there must exist coordination among their operation. It was shown that if no coordination is considered, the peak loading level of the circuit and its voltage levels can deteriorate, and the BESS can become fully charged and lose its energy management capability after a few days. Besides, some potential impacts of the uncoordinated operation of multiple BESSs were detailed. Two composite BESS controls were also proposed and tested to avoid the highlighted problems. To exemplify the proposed coordination methods, it was shown that, by using one of the proposed strategies, the medium-size BESS was able to successfully reduce the reverse peak power injection around noon due to the PV generators and support the grid during its peak loading hours. In addition, the SOC and both discharge and charge rates of the BESS also remained within acceptable limits across multiple days.

REFERENCES

[1] R. Boffo and R. Patalano, "Esg investing: Practices, progress and challenges," OECD, Paris, 2020.

[2] K. Schneider et al., "Analytic Considerations and Design Basis for the IEEE Distribution Test Feeders," *IEEE Trans. Power Syst.*, vol. 33, no. 3, pp. 3181–3188, May 2018, doi: 10.1109/TPWRS.2017.2760011.

[3] H. Sun et al., "Review of Challenges and Research Opportunities for Voltage Control in Smart Grids," *IEEE Trans. Power Syst.*, vol. 34, no. 4, pp. 2790–2801, Jul. 2019, doi: 10.1109/TPWRS.2019.2897948.

[4] F. C. L. Trindade, T. S. Ferreira, M. G. Lopes, and W. Freitas, "Mitigation of fast voltage variations during cloud transients in distribution systems with PV solar farms," *IEEE Trans. Power Deliv.*, vol. 32, no. 2, pp. 921–932, 2016.

[5] R. Torquato, D. Salles, C. Oriente Pereira, P. C. M. Meira, and W. Freitas, "A Comprehensive Assessment of PV Hosting Capacity on Low-Voltage Distribution Systems," *IEEE Trans. Power Deliv.*, vol. 33, no. 2, pp. 1002–1012, Apr. 2018, doi: 10.1109/TPWRD.2018.2798707.

[6] J. A. Jardini, C. M. V. Tahan, M. R. Gouvea, S. U. Ahn, and F. M. Figueiredo, "Daily load profiles for residential, commercial and industrial low voltage consumers," *IEEE Trans. Power Deliv.*, vol. 15, no. 1, pp. 375–380, Jan. 2000, doi: 10.1109/61.847276.

[7] Y. Wang, Q. Chen, C. Kang, and Q. Xia, "Clustering of Electricity Consumption Behavior Dynamics Toward Big Data Applications," *IEEE Trans. Smart Grid*, vol. 7, no. 5, pp. 2437–2447, Sep. 2016.

[8] B. R. Lopes, F. C. L. Trindade, and T. R. Ricciardi, "Metodologia para Modelagem Probabilística de Curvas de Carga de Consumidores Comerciais de Baixa Tensão," *Simpósio Bras. Sist. Elétricos - SBSE*, vol. 1, no. 1, Art. no. 1, 2020, doi: 10.48011/sbse.v1i1.2403.

[9] W. Cole, A. W. Frazier, and C. Augustine, "Cost Projections for Utility-Scale Battery Storage: 2021 Update," National Renewable Energy Lab.(NREL), Golden, CO (United States), 2021.

[10] J. von Appen, T. Stetz, M. Braun, and A. Schmiegel, "Local Voltage Control Strategies for PV Storage Systems in Distribution Grids," *IEEE*

Trans. Smart Grid, vol. 5, no. 2, pp. 1002–1009, Mar. 2014, doi: 10.1109/TSG.2013.2291116.

[11] J. F. Sousa, C. L. Borges, and J. Mitra, "PV hosting capacity of LV distribution networks using smart inverters and storage systems: a practical margin," *IET Renew. Power Gener.*, vol. 14, no. 8, pp. 1332–1339, 2020.

[12] A. T. Procopiou, K. Petrou, L. F. Ochoa, T. Langstaff, and J. Theunissen, "Adaptive Decentralized Control of Residential Storage in PV-Rich MV–LV Networks," *IEEE Trans. Power Syst.*, vol. 34, no. 3, pp. 2378–2389, May 2019, doi: 10.1109/TPWRS.2018.2889843.

[13] L. Deotti, W. Guedes, B. Dias, and T. Soares, "Technical and Economic Analysis of Battery Storage for Residential Solar Photovoltaic Systems in the Brazilian Regulatory Context," *Energies*, vol. 13, no. 24, p. 6517, 2020, doi: 10.3390/en13246517.

[14] P. Meira and D. Krishnamurthy, "DSS-Extensions: Multi-platform OpenDSS extensions," 2021, Accessed: Oct. 15, 2021. [Online]. Available: https://dss-extensions.org/

[15] R. Dugan, J. A. Taylor, and D. Montenegro, "Energy Storage Modeling for Distribution Planning," *IEEE Trans. Ind. Appl.*, vol. PP, no. 99, pp. 1–1, 2016, doi: 10.1109/REPC.2016.11.

[16] R. S. Ferreira, L. A. Barroso, P. R. Lino, P. Valenzuela, and M. M. Carvalho, "Time-of-use tariffs in Brazil: Design and implementation issues," in *2013 IEEE PES Conference on Innovative Smart Grid Technologies (ISGT Latin America)*, Apr. 2013, pp. 1–8. doi: 10.1109/ISGT-LA.2013.6554486.

[17] C. O. Pereira *et al.*, "Pre-Installation Studies of a BESS in a Real LV Network with High PV Penetration," in *2019 IEEE PES Innovative Smart Grid Technologies Conference - Latin America (ISGT Latin America)*, Sep. 2019, pp. 1–6. doi: 10.1109/ISGT-LA.2019.8895326.

[18] T. Stetz, F. Marten, and M. Braun, "Improved low voltage grid-integration of photovoltaic systems in Germany," *IEEE Trans. Sustain. Energy*, vol. 4, no. 2, pp. 534–542, 2012.

[19] K. D. Pippi, T. A. Papadopoulos, and G. C. Kryonidis, "Impact assessment framework of PV-BES systems to active distribution networks," *IET Renew. Power Gener.*, vol. 16, no. 1, pp. 33–47, 2022, doi: 10.1049/rpg2.12313.

Hydrogen complexes present after different firing profiles and their influence on LeTID degradation

Benjamin Hammann, Nicole Assmann, Philip M. Weiser, Wolfram Kwapil, Tim Niewelt, Florian Schindler, Rune Søndenå, Eduard V. Monakhov, Martin C. Schubert

Fraunhofer Institute for Solar Energy ISE, Freiburg, Germany

Centre for Materials Science and Nanotechnology, Department of Physics, University of Oslo, Oslo, Norway

Laboratory for Photovoltaic Energy Conversion, Department of Sustainable Systems Engineering (INATECH), University of Freiburg, Freiburg, Germany

School of Engineering, University of Warwick, Coventry, United Kingdom

Institute for Energy Technology, Kjeller, Norway

The influence of the cooling rate during the fast-firing process and of the sample thickness on the initial hydrogen (complex) distribution in p- and n-type silicon wafers is investigated using low-temperature Fourier Transform-Infrared (FT-IR) spectroscopy. The impact of the introduced hydrogen on the formation of defects during dark annealing and light soaking is then studied by resistivity and charge carrier lifetime measurements. We observe a lower overall hydrogen concentration for thinner wafers or slower cooling rates. This is especially pronounced for the concentration of the hydrogen molecule H2A. We observe a weak signature of light- and elevated-temperature-induced degradation (LeTID) during dark annealing accompanied by a significant increase in BH pair concentration. Interestingly, the extent of degradation does not correlate with the chosen process variations. Regeneration of the carrier lifetime occurs earlier in thinner wafers and in fast-fired samples. During light soaking, the LeTID extent clearly correlates with the initial hydrogen (H2A) content, while the BH pair formation appears to be suppressed. In addition to H2A and BH-pairs, the dark annealing experiments indicate that at least one more source of hydrogen is present in the initial wafers.

Superior Performance of Two-Phase Triple Halide Inorganic Perovskites

Deniz N. Cakan, Rishi E. Kumar, Connor Dolan, Moses Kodur, Yanqi Luo, Tao Zhou, Zhonghou Cai, Barry Lai, Martin Holt, David P. Fenning

University of California, San Diego, La Jolla, CA, United States

Argonne National Laboratory, Lemont, IL, United States

Inorganic halide perovskites are attractive for achieving the wide bandgap optimal for a high-efficiency perovskite-perovskite tandem photovoltaic based on today' Pb-Sn low bandgap compositions. However, they have suffered from lower photoluminescent quantum yield relative to hybrid compositions and phase instability. To improve upon metastable CsPbI3, we explore triple-halide alloying of minor amounts of Br and Cl with I. In agreement with previous reports for hybrid analogues, we observe a chlorine solubility limit in the majority iodine-bromine all-inorganic perovskite lattice. Past this solubility limit we observe the perovskite forming a split phase of iodine-bromine-rich and bromine-chlorine-rich clusters. Interestingly, these dual-phase thin films show superior and long lasting PL-intensity under 40-sun equivalent 633 nm laser intensity, which hints at possible synergistic effects of this chemical heterogeneity. We leverage multi-modal synchrotron microscopy and correlative spectroscopic micro-photoluminescence (µPL) on all-inorganic triple halide perovskites CsPbX3 (X-site: I/Br/Cl) films to elucidate mechanisms for superior performance in the face of phase segregation. The results suggest that a greater focus on harnessing the flexibility of the inorganic perovskite material system holds promise to retrace the outstanding performance and stability gains made in hybrid analogues.

Thermally evaporated titanium dioxide film as an electron-selective contact for silicon solar cells

Changhyun Lee, Soohyun Bae, Hyunju Lee, Yoonmook Kang, Hae-Seok Lee, Donghwan Kim

Materials Science and Engineering, Korea University , Seoul, Korea

Semiconductor lab, Toyota Technological Institute, Seoul, Japan

KU-KIST Green School, Graduate School of Energy and Environment, Korea University, Nagoya, Korea

We conducted research on TiOx to create an electron selective contact structure that does not require a high-temperature heat-treatment process. Titanium metal was deposited by thermal evaporator and an additional oxidation process was conducted to form titanium oxide. The chemical composition and phase of the titanium dioxide layers were analyzed using X-ray diffraction and X-ray photoelectron spectroscopy. Passivation effects of each titanium oxide layer were measured by quasi-steady-state photoconductance. Electron selectivity of titanium oxide layers and band alignment of TiOx/Si was demonstrated by UV photoelectron spectroscopy (UPS) and UV-vis spectroscopy analyses. With this oxidized titanium oxide on the silicon surface, band offsets of the conduction and valence bands were analyzed to confirm the selectivity of the layers.

978-1-7281-6118-1/22 $31.00 © 2022 IEEE

3 MeV Proton Radiation Tolerance Study of Ultra-thin Gallium Arsenide Solar Cells for Space Applications

Larkin Sayre, Armin Barthel, Andrew Johnson, Louise C Hirst

University of Cambridge, Cambridge, United Kingdom

IQE plc., Cardiff, United Kingdom

Ultra-thin single-junction GaAs solar cells with an 80 nm absorber layer were fabricated and exposed to 3 MeV proton radiation at a range of fluences. Off-wafer devices with an integrated Ag back surface mirror were irradiated as well as on-wafer devices with no mirror. Both 80 nm device designs exhibited high tolerance to extremely high proton fluences with no decrease in short circuit current. They also showed higher absolute maximum power values once moderate to high fluences were achieved when compared to devices with absorber layers of 800 and 3500 nm.

Monolithic Perovskite/Silicon Tandem Solar Cells on p-type POLO/PERC Silicon Bottom Cells

Silvia Mariotti, Klaus Jäger, Marvin Diederich, Marlene S. Härtel, Bor Li, Kári Sveinbjörnsson, Eike Köhnen, Rolf Brendel, Sarah Kajari-Schröder, Robby Peibst, Steve Albrecht, Lars Korte, Tobias Wietler

Helmholtz-Zentrum Berlin für Materialien und Energie GmbH (HZB), Berlin, Germany

Institute for Solar Energy Research (ISFH) GmbH, Emmerthal, Germany

Technische Universität Berlin - Fakultät Elektrotechnik und Informatik, Berlin, Germany

Leibniz University Hannover, Hannover, Germany

We report on proof-of-concept perovskite/silicon tandem solar cells on bottom cells featuring a polycrystalline silicon on oxide (POLO) front junction and a PERC-type passivated rear side with local aluminium-p+ contacts. We implement a process flow which is compatible with industrial, mainstream PERC technology. The top and bottom cells are connected via an indium tin oxide (ITO) layer, and the perovskite top cell is then monolithically integrated in a p-i-n architecture. The tunnel recombination junction between the two sub-cells, as well as the perovskite top cell are adapted from high efficiency perovskite/silicon heterojunction-based tandems. For the perovskite absorber layer, we use a mixed cation, mixed halide perovskite with a band gap of 1.68 eV. The proof-of-concept tandem cells demonstrate a power conversion efficiency (PCE) of 21.3%. We identify a potential for major performance enhancements by process and layer optimizations. Supported by optical simulations, we estimate a PCE potential of 29.5% for this tandem stack based on POLO/PERC bottom cells. Thus, we demonstrate that the large technology base of p-type PERC production has significant potential for an upgrade to highly efficient perovskite/POLO/PERC tandem solar cells.

Multiple substrate reuse: a straightforward reconditioning of Ge wafers after porous separation

Alexandre Chapotot, Javier Arias-Zapata, Tadeáš Hanuš, Bouraoui Ilahi, Nicolas Paupy, Valentin Daniel, Zakaria Oulad El Hmaidi, Jérémie Chrétien, Gwenaëlle Hamon, Maxime Darnon, Abderraouf Boucherif

Institut Interdisciplinaire d'Innovation Technologique (3IT), Université de Sherbrooke, Sherbrooke, QC, Canada

Laboratoire Nanotechnologies Nanosystèmes (LN2)—CNRS UMI-3463, Institut Interdisciplinaire d'Innovation Technologique (3IT), Université de Sherbrooke, Sherbrooke, QC, Canada

Epitaxial thin film detachment and substrate reuse is one of the promising approaches to reduce the weight and the cost of triple junction (3J) solar cells on Ge substrate for both terrestrial and space PV. This approach is based on epitaxial growth of high-quality solar cell materials on porosified Ge substrate. The mesoporous layer created by electrochemical etching undergoes thermal induced reconstruction leading to the formation of voided weak layer suitable for epilayers detachment. This approach is low-cost, scalable to large surfaces and allows the substrate reuse for several epitaxial cycles upon appropriate reconditioning. Accordingly, the success of the reconditioning step is conditional to both reliability and cost-effectiveness. In this context, we report the first successful proof-of-concept of Ge substrate reuse for epitaxy after the epilayer detachment. We demonstrate that chemical etching with HF-based mixture allows to recondition the detached substrate providing a low surface roughness of 1.3 nm without any CMP step. The reconditioned substrate was then porosified giving rise to homogenous porous layer suitable for epitaxial regrowth. A second growth cycle has been successively performed on the reconditioned and reporosified substrate. The epitaxial Ge layer from the second cycle is found to have high crystalline quality and low surface roughness as revealed by X-ray diffraction and atomic force microscopy investigations. Our results demonstrate a CMP-free reliable Ge substrate reconditioning process for epitaxy, which paves the way to the substrate multi-reuse for triple junction solar cell cost-reduction.

Short Drying Processes for Silicon Solar Cells

Daniel Ourinson, Michael Linse, Markus Klawitter, Andreas Lorenz

Fraunhofer ISE, Freiburg, Germany

This work successfully demonstrates very short drying processes for the screen-printed electrodes of passivated emitter and rear cells. The short drying processes are conducted in an inline heating system with vertical-cavity surface-emitting lasers (VCSEL) as its heat source, being a compact alternative to the conventional heat chamber. While the heating time of the conventional industrial drying process is believed to be about 20 s (to the best of the authors' knowledge), the VCSEL system allows a significant reduction down to 0.15 s while maintaining a similar power conversion efficiency and contact adhesion quality.

Sn4+-free, stable tin perovskite films for lead-free perovskite solar cells

Ajay Singh, Jeremy Hieulle, Himanshu Phirke, Joana A. F. Machado, Sevan Gharabeiki, Rukhsar Ahmad, Susanne Siebentritt, Alex Redinger

Department of Physics and Materials Science, University of Luxembourg, Luxembourg, Luxembourg

Organic-inorganic hybrid perovskite solar cells (PSCs) achieved already more than 25% power conversion efficiency making them one of the fastest-growing solar cell technologies. However, most state-of-art PSCs employ lead (Pb) as a "B" cation in the ABX3 type crystalline absorbers. Sn has been considered to be a potential replacement for Pb. However, chemically processed Sn-perovskite films often exhibit a high number of Sn vacancies and a high density of Sn4+ species, leading to poor stability and low power conversion efficiency of the tin-based PSCs. Herein we report co-evaporation of CH3NH3I and SnI2 to obtain Sn4+-free CH3NH3SnI3 perovskite films with excellent optoelectronic properties and improved stability. XRD measurements confirm the polycrystalline films, mainly oriented in the (100) plane of a cubic crystal. XPS analysis confirms that the films consist of Sn only in the 2+ oxidation state. AFM and KPFM analysis reveal smooth topology and uniform surface workfunction. The films show very good stability under heating and photodegradation confirmed by XRD and photoluminescence (PL) degradation measurements. The co-evaporated CH3NH3SnI3 films exhibit PL quantum yields up to 9x10-4 translating in a quasi Fermi-level splitting of 844 meV under one sun equivalent conditions. Extracted low values of Urbach energies (16 meV) suggest that the films exhibit a low number of defects near the band edges showing high promise in developing lead-free, high efficiency, and stable PSCs.

Fill factor losses in Cu(In,Ga)Se2 based solar cells due to metastabel defects – the effect of Ag addition

Thomas P. Weiss, Omar Ramirez, Taowen Wang, Valentina Serrano-Escalante, Stefan Paetel, Wolfram Witte, Jiro Nishinaga, Thomas Feurer, Ayodhya N. Tiwari, Susanne Siebentritt

University of Luxembourg, Department of Physics and Materials Science, 4263 Esch-sur-Alzette, Luxembourg

Zentrum für Sonnenenergie- und Wasserstoff-Forschung Baden-Württemberg (ZSW), 70563 Stuttgart, Germany

Research Institute for Energy Conservation, National Institute of Advanced Industrial Science and Technology (AIST), Tsukuba, Japan

Laboratory for Thin Films and Photovoltaics, Empa - Swiss Federal Laboratories for Materials Science and Technology, 8600 Dübendorf, Switzerland

The fill factor in state-of-the-art Cu(In,Ga)Se2 based solar cells is still relatively low as a consequence of diode factors greater than the ideal value of 1. We show that the increased diode factor results from metastable defects, also responsible for the persistent photoconductivity, which increase the net doping upon electron injection. We present measurements from photoluminescence, capacitance-voltage and current-voltage characteristics in corroboration of simulations including a metastable defect, which all consistently describe the observed diode factor greater than 1. It is demonstrated that the addition of Ag to Cu(In,Ga)Se2 decreases metastable defects and that fill factors as high as 81.0% are achieved.

Analysis of the Soiling Effects on Commissioning of Photovoltaic Systems: Short-Circuit Current Correction

Dênio Alves Cassini[1], Suellen C. Silva Costa[1], Antonia Sonia A.C. Diniz[1], and Lawrence L. Kazmerski[1,2]

[1]Pontifícia Universidade Católica de Minas Gerais (PUC Minas), Belo Horizonte, Brasil
[2]Renewable and Sustainable Energy Institute (RASEI), University of Colorado Boulder, Colorado USA

Abstract—*This study presents a methodology to correct short-circuit current data measured during cold commissioning of two Brazil PV power plants in the cities of Januária (semi-arid climate) and Paracatu (Equatorial climate), specifically for the soiling conditions of the PV modules. For this, the soiling ratio (SRatio) was determined for each PV string in the installation, and this factor was used in the proposed model for correction of the measured short-circuit current in order to derive the clean operating condition of the modules. The results reinforce the importance of evaluating the cold commissioning of the PV plant correcting for the conditions observed in visual inspection. Our studies have also shown differences between the soiling among the strings. The analysis shows the importance of examining all strings independently to correct more accurately for the clean performance metrics.*

Keywords—Commissioning, Photovoltaic Plants, Soiling.

I. INTRODUCTION

One of the last stages of a PV power plant approval consists of cold commissioning to ensure that it is safe, meets the project objectives, and operates and produces energy in accordance with the expectations defined in the contract. During this pre-grid connection commissioning phase, visual inspections and specified electrical and thermal tests are performed in order to identify any faults in the PV plant components. The visual inspection identifies if the panels are not in their ideal or clean condition indicating soiling ratios (SRatio) less than unity. Thus, a methodology must be applied to correct the measured electrical parameters to account for the state of cleanliness of the modules. This avoids errors to the measured I_{sc} and V_{oc} defined in the standard [1,2]. The methodologies used to determine the soiling ratio are: (i) the comparison between electrical and thermal parameters measured for clean and naturally soiled PV modules allowing to quantify SRatio and soiling rate (SRate) [3-6], and (ii) in the determination of the slope of the performance metric from PV plant during dry periods, obtaining SRate values [4,7,8]. However, both methodologies require long-term monitoring. The soiling is the result of the interaction of a series of factors related to the environment and the local climate, the system configuration, and the design of the modules [9-12]. For these reasons, soiling effects can change from place to place and over time (e.g., seasonally or even hourly). In some cases, different soiling losses can be experienced among different strings and modules even within the same PV plant. Furthermore, the soiling effects can be different for different PV module technologies [3,13].

The objective of this research is to establish a method to account for the soiling in the cold commissioning of the PV power plant, taking into account the potential differences among the strings without long-term monitoring. The clean surface condition of a PV module is one of the factors for the cold conditioning procedure to obtain the required performance of this device. However, there are challenges in maintaining this condition in practice. The specificity of this loss factor requires a methodology to be applied to correct the electrical parameters measured during cold commissioning tests to ensure proper collected data. We present a model to correct the electrical parameters measured during cold commissioning for the soiling conditions encountered. We apply the model to the case of two PV power plants in the cities of Januária (4.39-MWp) and Paracatu (3.41-MWp) in Minas Gerais (MG) State of Brazil. These represent two different climate zones (semi-arid and Equitorial, respectively) to compare the results of these commissioning case studies.

II. METHODOLOGY

A. Measuring electrical and thermal parameters from PV Modules – Commissioning test

During the cold commissioning, a complete visual inspection of all strings of each PV plant was completed in order to evaluate the condition of metallic support structures, photovoltaic modules conditions, combiner boxes, cabling, connections, conduits, and among other system components.

Data regarding open circuit voltage (V_{oc}), short-circuit current (I_{sc}), polarity (+/-), and continuity condition–including connection continuity to the main earth terminal–were collected using Seaward *Solar Utility Pro* equipment, having an uncertainty of ±5%. The I_{sc} test is performed on each photovoltaic string in order to check for serious faults in the PV array wiring, being the measured data are not considered as a measure of PV performance. The V_{oc} test is performed in order to verify that the series of modules are correctly connected and, specifically, if the expected and required number of modules is connected in series.

The measured panel V_{oc} and I_{sc} must be compared with the expected datasheet values. The strings pass these tests if the measured values vary within 5% of the expected value [1,2]. On the other hand, the insulation resistance (R_{iso}) of the positive

(+) and negative (-) poles and the Earth contact resistance (R_{PE}) in all strings that make up the PV generator were measured using a Megabras Digital Megohmeter (Model MD-5060x, uncertainty ±5% of reading between 1-MΩ and 1-TΩ). A calibrated Seaward Solar Survey SS 200R Si reference cell (uncertainty of ±5%) positioned with the same tilt and orientation as the tested photovoltaic modules provided the solar irradiance data. The operating temperatures of the modules and ambient were monitored through the use of thermocouple sensors. The module temperature probes were fixed above a cell at the center-rear of the panel. Another ambient temperature probe was positioned near that region but without any physical contact with the modules or structure. Data were collected only with irradiance conditions >700 W/m² and ambient T ≤ 40 ºC, and clear sky [1,14].

B. Correction model of the short-circuit current for soiling PV modules conditions

The PV module performance losses due to the soiling layer deposition can be quantified by determining the soiling ratio (SRatio), the ratio between the short-circuit current (I_{sc}) extracted from the soiled PV module and the I_{sc} measured from the clean module under the same operating conditions. The SRatio is a dimensionless parameter that varies from 1 to 0, with the clean module "1" and severe soiling situations approaching "0". Because the soiling layer reflects the incoming light, it decreased the current generated by the module. This is a source of error in the cold commissioning tests. Short-circuit current data were considered in developing and evaluating the methodology for the correction in these current studies. Considering the data collected during the commissioning, the SRatio was determined using Eq. (1) [3,6]:

$$ \text{SRatio}_{I_{sc}} = \frac{I_{sc_{sujo}}}{I_{sc_0} * \left(1 + \alpha * \left(T_{c_{sujo}} - T_0\right)\right) * \left(\frac{POA}{G_0}\right)} \quad (1) $$

where $I_{sc_{sujo}}$ is the short-circuit current measured in the module with natural soiling deposition, I_{sc_0} is the short-circuit current of the module in the reference condition (STC), α is the coefficient of temperature for short-circuit current, $T_{c_{sujo}}$ is the temperature from soiled module, T_0 and G_0 are the temperature and solar irradiance at the reference condition (1000 W/m² and 25 °C), and POA is the plane-of-array irradiance measured by the reference cell.

After determining the SRatio, the measured current must be corrected considering this loss factor according to Eq. (2). In this way, the corrected measure current will indicate the value for this parameter for clean condition of the PV module when the SRatio is equal to "1".

$$ I_{sc_{corrigida}} = I_{sc_{med}} * \left(1 + \left(1 - \text{SRatio}_{I_{sc}}\right)\right) \quad (2) $$

where $I_{sc_{med}}$ is short-circuit current measured.

III. RESULTS

The methodology was applied and evaluated for data measured during the commissioning of the 2-PV plants installed in the cities of Januária and Paracatu, MG, Brazil.

A. PV plants installed in Januária (semi-arid climate)

The PV plant in Januária has three generation units (GU), with an installed generating capacity of 4.39-MWp. The GU₁ and GU₃ have the same configuration, each containing a generator composed of 3,360 monocrystalline silicon modules with a nominal power of 435-Wp each, totaling an installed power of 1.4616-MWp. GU₂ has 3,360 modules, however, 1,932 modules are of 435-Wp and 1,428 are of 440-Wp, Both use monocrystalline silicon technology, with GU₂ totaling 1.4687-MWp. In all GUs, the PV modules were installed with azimuth equal to 0º and with tilt of the 15º. Each GU has eight inverters with a nominal power of 125 kW, with 15 strings of 28 modules connected to each inverter through a combiner box. Thus, GU₁ and GU₃ have an overload, due to the ratio between PV generator power versus inverter power, of 46.16%, while GU₂ has an overload of 46.87%.

Figure 1 shows the variation in the SRatio for the strings connected to the eight inverters of each of the generation units (GUs) from PV plants installed in Januária. It can be observed that the SRatio for GU₂ is lower compared to other GUs, which indicates *greater* soiling deposition on its PV modules strings. This is likely due to this GU being parallel to an unpaved road used for local transit. The results obtained for GU₁ showed greater soiling deposition in the strings connected to inverter 1 and 8, in relation to the SRatio data obtained for GU₂.

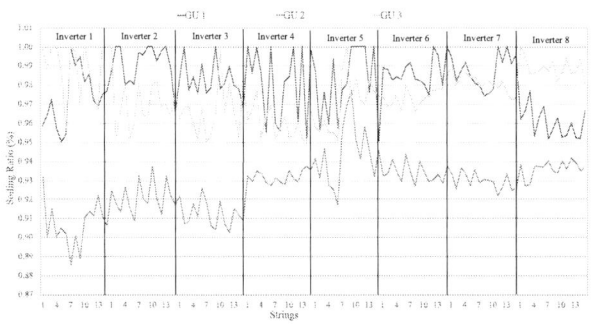

Fig. 1. SRatio calculated for each strings connected to the eight inverters of the three generation units (GU) from PV plant installed in Januária (semi-arid climate).

Figure 2 shows the percentage difference obtained by ratio the measured and expected I_{sc} (yellow line), and the difference between the corrected measured and expected I_{sc} (blue line) for GUs from PV plant in Januária. The red dashed-lines are the limits (minimum and maximum) relative to acceptable variation according to technical standards [1,2]. The current measured in the strings with soiling was corrected using the SRatio factor (Eq. 2) that indicates the level of soiling in each series of modules analyzed. Thus, the corrected value refers to the I_{sc} measured for the needed clean module condition.

Figure 2a shows: (1) Before the correction, the percentage difference of the current for GU₁ was between the acceptable variation according to standards, reaching values below 100% in almost all strings. This indicates that the measured current was smaller than expected due to the negative impact of the soiling. And (2) After correcting the I_{sc} in relation to the soiling condition found in each PV string (SRatio), the values of this electrical parameter showed an increase showing that the measured current was close to expected, around 100% in almost all strings, when the soiling effects are treated). The same occurred for GU₃, as indicated in Fig. 2c.

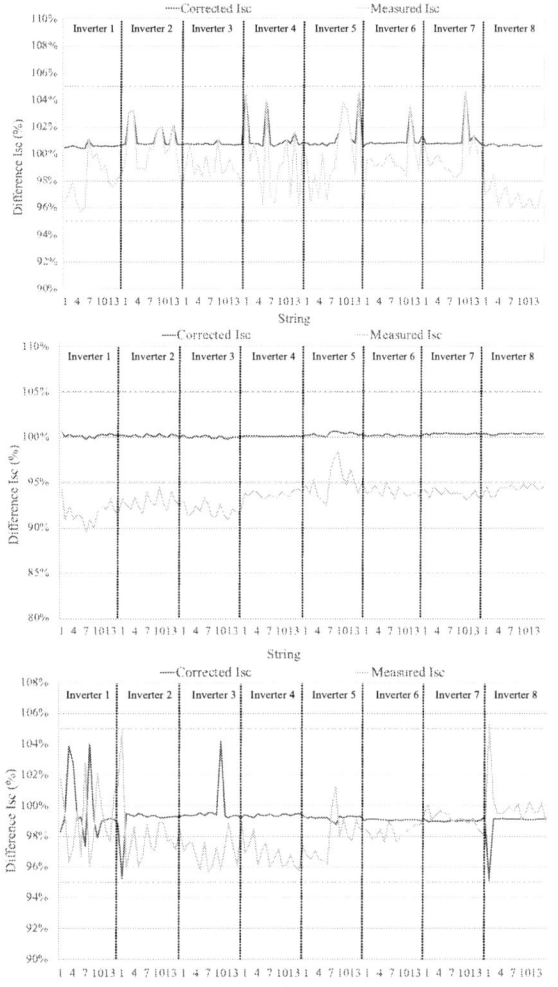

Fig. 2. Percentage differences between measured and expectd I_{sc} from GU₁ (a), GU₂ (b) and GU₃ (c) from PV plants in Januária: (1) yellow line – ratio the measured versus expected short-circuit current and (2) blue line – ratio the corrected measure versus short-circuit current.

In the case of GU₂, Fig. 2b, the difference between the measure and the expected (yellow line) yields values below 95%, not indicating compliance to the technical standards. However, after correcting the current (blue line), these values remained close to 100% in all strings. This shows that the measured current for clean condition was close to expected, indicating that there is no fault related to wiring and connections. This is an example of how the soiling deposition on the PV modules can lead to misinterpretations regarding possible flaws in the series evaluated.

The strings in all GUs have identical configurations. Thus, it is expected that the behavior is similar for all strings for stable irradiance conditions. However, it was noted that the difference for the measured and expected I_{sc} fluctuated significantly in relation to the different evaluated strings, as well as between GUs. This demonstrates that the modules of the strings have different soiling densities, as well as possible deposition non-uniformities, as indicated in Fig. 1.

B. PV plants installed in Paracatu (equatorial climate)

The PV plant installed in Paracatu is composed of 3-GUs with a total capacity of 3.41-MWp. GU₁ and GU₂ have different configurations. GU₃, having four inverters with nominal power of 125-Wp, being connected to each of these 14 strings of 28 monocrystalline silicon modules with nominal power of 435-Wp, resulting in an overload of 36.42%. Each or the other two GUs is composed of eight arrangements of 14 strings of 28 modules, with the same series and manufacturer used in UG₁. These arrangements are connected to an inverter (125-Wp), presenting an overload of 36.42%. All modules were installed with azimuth equal to 0º and tilt of the 17º. Figure 3 shows the SRatio identified for each string connected to the eight inverters belonging to the three GUs of the PV in Paracatu. It can be noted that the deposited soiling layer caused the reduction of the SRatio to below 0.95 for almost all strings, with the exception some strings from GU₁. The SRatio values for GU₃ were smaller, characterizing greater soiling, followed by GU₂.

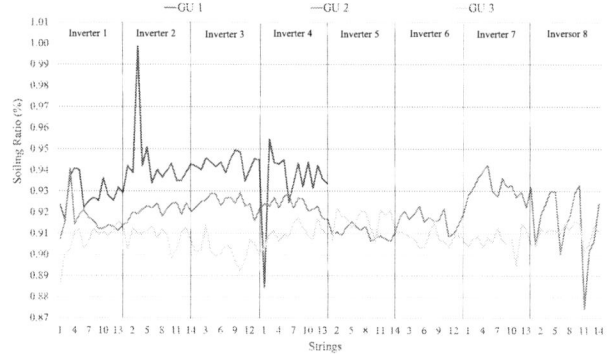

Fig. 3. SRatio calculated for each strings connected to the eight inverters of the three generation units (GU) from PV plant installed in Paracatu (equatorial climate).

Figure 4 represents the variation of the percentage difference obtained by the ratio between measured and expected I_{sc} (yellow line) and corrected measure and expected measured I_{sc} (blue line). It can be noted the differences of the I_{sc} for all GUs presented values below the minimum limit established in technical standards, that is below 95%. This means that the measured current was below the acceptable expected value due to the greater density of soiling accumulated in the PV modules of this UFV–a result confirmed by the SRatio data identified and indicated in Fig. 3.

Although all strings have the same configuration, it is possible to observe that the variation in the difference between measured and expected I_{sc} was greater for GU₂ and GU₃, in addition to greater variation in the differences between strings of the same GU. According to standard [1], for systems with multiple identical PV series and stable irradiance conditions, the current measurements of each series must be compared. This comparison is recommended in order to verify that the connection conditions are adequate, as expected and designed. However, it was seen that the soiling deposition and the non-uniformity between the deposited layer densities in the different

978-1-7281-6118-1/22 $31.00 © 2022 IEEE

strings of the GUs can cause noise capable of causing misinterpretation of the monitored results. In this case, it is always recommended to evaluate the cleaning conditions of each string of the respective GUs during the visual inspection.

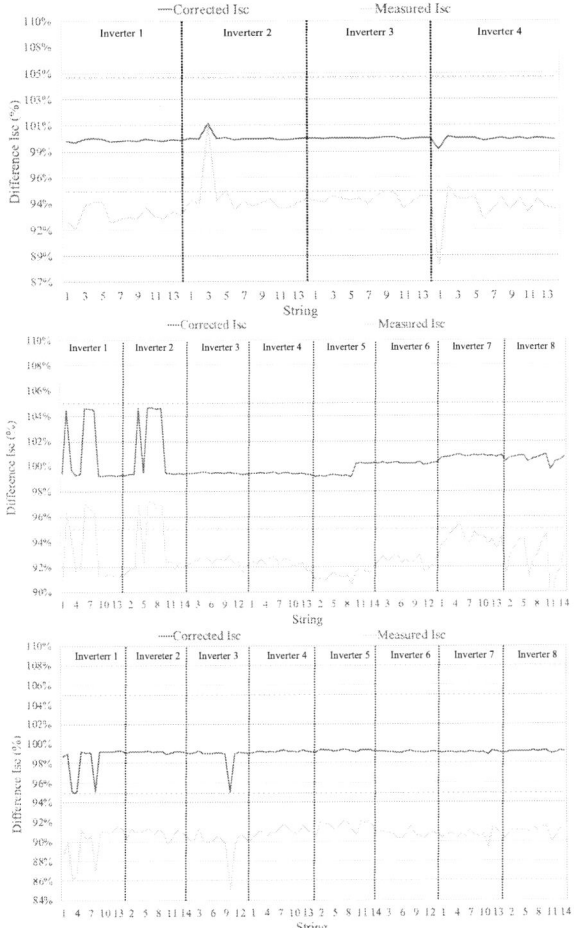

Fig. 4. Percentage differences between measured and expectd I_{sc} from GU$_1$ (a), GU$_2$ (b) and GU$_3$ (c) from PV plants in Paracatu: (1) yellow line – ratio the measured versus expected short-circuit current and (2) blue line – ratio the corrected measure versus short-circuit current.

IV. CONCLUSIONS

The major objective of this case study was to present a methodology to correct short-circuit current data measured in modules with different soiling deposition conditions during the cold commissioning. For this, the SRatio for each string from two PV plants installed in Januária (Semi-Arid climate) and Paracatu (Equatorial climate) in Brazil was identified, and this factor was used for correcting the electrical parameters measured for clean panel condition. Cold commissioning is done with the objective of proving the correct functioning of all system components, including the PV strings, checking and validating the PV plant, in order to identify problems that could compromise the safety and efficiency of power generation through the

realization of tests described in standards [1,2]. Thus, soiling can cause misinterpretation as to possible failures in the PV plant, demonstrating the importance of associating information obtained from visual inspection together with data from electrical and thermal tests. The method applied in this study proved to be effective, allowing the evaluation of the operation of the PV plants for the cleaning condition of the modules.

ACKNOWLEDGMENT

The authors gratefully acknowledge the support and technical assistance of CAPES (*Coordenação de Aperfeiçoamento de Pessoal de Nível Superior*), CEMIG (*Companhia Energética de Minas Gerais*), and the Renewable and Sustainable Energy Insititute - RASEI (a joint institute between the University of Colorado Boulder and the National Renewable Energy Laboratory - NREL).

REFERENCES

[1] Associação Brasileira de Normas Técnicas (ABNT). NBR 16274 – Sistemas fotovoltaicos conectados à rede – Requisitos mínimos para documentação, ensaios de comissionamento, inspeção e avaliação de desempenho, 2014.

[2] International Electrotechnical Commission (IEC). IEC 62446-1: Photovoltaic (PV) systems – Requirements for testing, documentation and maintenance – Part 1: Grid connected systems – Minimum requirements for system documentation, commissioning tests and inspection, 2014.

[3] Costa, S. C. S.; Kazmerski, L. L.; Diniz, A. S. A. C. Impact of soiling on Si and CdTe PV modules: Case study in different Brazil climate zones. Energy Conversion and Management:X, v. 10, 2021.

[4] Costa, S. C. S. Avaliação do potencial dos sistemas fotovoltaicos conectados à rede elétrica com geradores de diversas tecnologias. 2011. 82 f. Tese (Doutorado). Pontifícia Universidade Católica de Minas Gerais – Pós-Graduação Engenharia Mecânica, Belo Horizonte, Minas Gerais.

[5] Micheli, L.; Muller, M. An investigation of the key parameters for predicting PV soiling losses. Progress in Photovoltaics: Research and Applications, v. 25, p. 291-307, 2017.

[6] Gostein, M.; Duster, T.; Thuman, C. Accurately measuring PV soiling losses with soiling station employing module power measurements. 2015 IEEE 42nd Photovoltaic Specialist Conference (PVSC), 2015.

[7] Costa, S. C. S.; Kazmerski, L. L.; Diniz, A. S. A. C. Estimate of soiling rates based on soiling monitoring station and PV system data: case study for Equatorial-Climate Brazil. IEEE Journal of Photovoltaics, v. 11, p. 461 – 468, 2021.

[8] Deceglie, M. G.; Muller, M.; Defreitas, Z.; Kurtz, S. A scalable method for extracting soiling rates from PV production data. 2016 IEEE 43 rd Photovoltaic Specialists Conference (PVSC), p. 2061-2065, 2016.

[9] Bessa, J. G.; Micheli, L.; Almonacid, F.; Fernández, E. F. Monitoring photovoltaic soiling: assessment, challenges, ans perspectives of current and potential strategies. iScience, v. 24, 2021.

[10] Mani, M.; Pillai, R. Impact of dust on solar photovoltaic (PV) performance: Research status, challenges and recommendations. Renewable and Sustainable Energy Reviews, v. 14, p. 3124 – 3131, 2010.

[11] Appels, R.; Lefevre, B.; Herteleer, B.; Goverde, H.; Beerten, A.; Paesen, R.; Medts, K.; Driesen, J.; Poortmans, J. Effect of soiling on photovoltaic modules. Solar Energy, v. 96, p. 283 – 291, 2013.

[12] Øvrum, Ø.; Marchetti, J.M.; Kelesoglu, S.; Marstein, E.S. Comparative analysis of site-specific soiling losses on PV power production. IEEE Journal of Photovoltaics, v. 11, p. 158 – 163, 2021.

[13] Qasem, H.; Betts, T. R.; Müllejans, H.; Albusairi, H.; Gottschalg, R. Dust-induced shading on photovoltaic modules. Progress in Photovoltaics: Research and Applications, v. 22, p. 218 – 226, 2014.

[14] International Electrotechnical Commission (IEC). IEC 60904-1: Photovoltaic devices – Part 1: Measurement of photovoltaic current-voltage characteristics, 2020.

Optical absorption of MoS2 based ultrathin solar cells

Carlos Bueno-Blanco, Simon Aurel Svatek, Elisa Antolin

Instituto de Energía Solar. Universidad Politécnica de Madrid., Madrid, Spain

Transition metal dichalcogenides (TMDCs) are promising candidates for ultrathin solar cells because of their strong light absorption and chemical stability. Among the different TMDC structures proposed for solar energy harvesting applications, MoS2 homojunctions exhibit the largest open-circuit voltages so far. However, these devices typically lose approximately half of the incident light due to reflection. Here, we calculate the absorbance of MoS2-based structures including h-BN, SiO2 or MgF2 as antireflection layers and an Ag back mirror, using a model that considers multiple reflections and interference. We find a considerable reduction of reflectance losses with the addition of an h-BN layer. h-BN is a layered material like MoS2, and therefore, it can be transferred onto the solar cell with the same techniques as for constructing the MoS2 junction. We find that combining an h-BN antireflection layer and an Ag back mirror, a MoS2 slab of 97 nm increases its absorbance to reach 89% in the photon energy range between 300 and 700 nm. This result can be further improved using a cavity structure consisting of h-BN/MoS2/h-BN/Ag. Inside the cavity, a MoS2 absorber that is 57 nm thick absorbs 91% of the photons. Finally, we find that the absorber thickness inside the cavity structure can be reduced drastically with only a small loss to the absorbance, namely 87% absorbance at a thickness of only 7 nm. This comprehensive study of absorbance contributes to the design of a new generation of high-efficiency ultrathin solar cells based on TMDC materials.

Microinverter testing update using high power modules: Efficiency, yield, and conformity to a new "estimation formula" for variation of PV panel size

Stefan Krauter, Jörg Bendfeld, Marius Möller

Paderborn University, EET-NEK, Pohlweg 55, D-33098 Paderborn, Germany, E-mail: Stefan.Krauter@upb.de

Abstract—The market for microinverters is growing, especially in Europe. Driven by the strongly rising prices for electricity, many small photovoltaic energy systems are being installed. Since monitoring for these plants is often quite costly, their yields are often not logged. Since 2014, microinverters have been studied at Paderborn University. The investigations are divided into indoor and outdoor tests. In the indoor area conversion efficiencies as a function of load have been measured with high accuracy and ranked according to Euro- and CEC weightings. In the outdoor laboratory, the behavior in the real world is tested: Energy yields have been measured outdoors using identical and calibrated crystalline silicon PV modules. Investigations were carried out with modules of the power of 215 W$_p$ until the year 2020. Because of the increasing module power nowadays, modules with an output of 360 W$_p$ are now being used. To assess the influence of PV module size, two extremes have been investigated: A rather small module with 215 W$_p$ - as it has been used 10 years ago, and a new module (2021) offering 360 W$_p$. Both types of modules contain 60 solar cells in series connection. Appling the low-power modules, the challenge for the different micro-inverters has been during weak-light conditions, using the high-power modules, some inverters temporarily reach their power limits and yield is reduced. A method using a reference configuration of inverter & module and a linear equation resulting in the actual yield, any module & inverter configuration can be characterized by just the two coefficients.

Keywords—Microinverter, Conversion Efficiency, Energy rating, Weak-Light Performance, Overload

I. INTRODUCTION

Microinverters are inverters that are connected primarily to a single PV module (occasionally to two modules, as indicated in the tables; very few are available for four modules, but these are not considered here), so each module–inverter combination acts as an independent power plant. The microinverter consists of a maximum power point tracker (MPPT), the DC-AC inverter, and an islanding protection unit (see e.g., [1]). For higher power requirements, several module-inverter combinations are interconnected in parallel on the AC output side. This configuration offers various advantages: Easier planning and easier installation, easy up- and downscaling of a power plant, including extensions or repair that could be carried out even during power plant operation. Logistics is simplified.

Effect of shadowing is very limited, and due to low system voltages, potential induced degradation (PID) does not occur. An excellent overview of the development and the advantages of microinverters has been compiled by H. Oldenkamp [2]. However, costs of power plants based on micro inverters are about 10–20% higher. Some of the inverters cannot be operated by themselves and require a control unit (often combined with a remote shutdown option and a monitoring system), or a protective device for grid interfacing (depending on national regulations), thus adding extra costs. Also, conversion efficiency may not be as high as for central inverters. Due to smart master–slave concepts centralized solutions with multiple but relatively large inverters may offer higher yields under weak light conditions. [3] is giving a performance comparison of a microinverter, a power-optimizer, and a central inverter.

II. MEASUREMENT CAMPAIGN

A. Indoor Test

Due to the reproducible test conditions in the indoor lab, the inverters have been examined individually with predefined and controlled input data. Input has been a PV module simulator with data being set corresponding to the modules used in the outdoor test. The main output data being recorded is the delivered AC power of the inverters. Besides input power, output is also a function of input voltage. If input voltage is getting too low, the inverters even stop operating. The following examinations are based on the possible range of input data (including voltage) given the specific PV module also used for the outdoor investigation

Peak efficiency is often reached close to the maximum load of the inverter. Peak efficiency (often promoted in data sheets) is not a helpful value since most of the time the inverters operate in the range of 20% to 40% of their rated power – at least under non-arid conditions. Consequently, an adequately weighted efficiency is a more adequate value to rate conversion devices. One type of weighted efficiency is the so-called "European Efficiency" η_{Euro} which it is calculated according to:

$$\eta_{Euro} = 0.03 \cdot \eta_{5\%} + 0.06 \cdot \eta_{10\%} + 0.13 \cdot \eta_{20\%} \qquad (1)$$
$$+ 0.1 \cdot \eta_{30\%} + 0.48 \cdot \eta_{50\%} + 0.2 \cdot \eta_{100\%}$$

The other is the "CEC efficiency" by the California Energy Commission (CEC). CEC efficiency is computed as an average value of DC–AC conversion efficiencies at six pre-defined relative output values between 10% and 100% of its rated power (with an emphasis on higher irradiance levels) is determined by:

$$\eta_{CEC} = 0.04 \cdot \eta_{10\%} + 0.05 \cdot \eta_{20\%} + 0.12 \cdot \eta_{30\%} \quad (2)$$
$$+ 0.21 \cdot \eta_{50\%} + 0.53 \cdot \eta_{75\%} + 0.05 \cdot \eta_{100\%}$$

For the "European Efficiency", weighting factors for high relative power values are lower.

The output power values used for the inverters (adjusted by controlling the DC input current) are continuously increased in 1024 steps from 0 to maximum. Each step takes eight seconds while the measurement duration is 500 ms. Figure 1 shows an example for the measuring procedure.

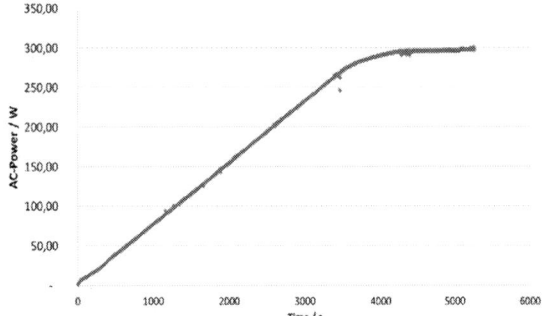

Figure 1: Measured AC power output (in Watt) as a function of measurement duration (in seconds) for linear increasing DC input current.

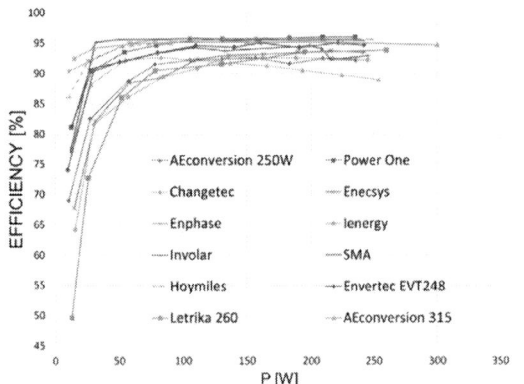

Figure 2: Measured DC-AC conversion efficiencies as a function of power output, for microinverters with single PV module inputs

The measured DC-AC conversion efficiencies of all inverters are shown in Fig. 2 and Fig. 3. Based on those measurements, the European (EU) efficiencies and the CEC efficiencies for the micro inverters have been calculated according to (1) and (2), eleven micro-inverters are designed for single modules, five inverters have inputs for 2 modules: Involar MAC 500, the APS Y 500, the Envertech EVT 560, the WVC 600 and the WVC

700. The WVC 600* inverter stopped operating at a measured power of 250 W. After a test run on higher temperatures the inverter failed constantly. Since the documentation of the WVC 600 and WVC 700 inverter has been extremely poor, its rated power has been assumed. For this reason, the WVC 700 is shown twice. First, with the assumed rated power of 600 W, then with the assumed rated power of 700 W. The maximum measured power of the WVC 700 inverter was 600 W only.

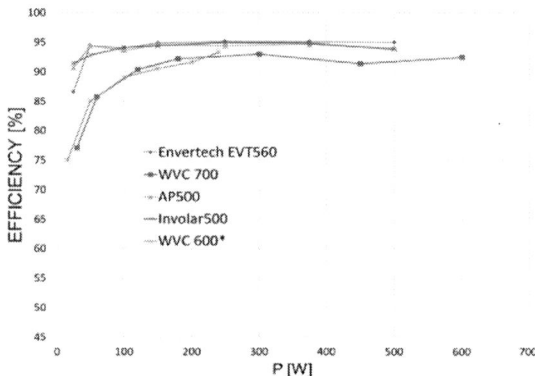

Figure 3: Measured DC-AC conversion efficiencies as a function of power output, for five inverters that have inputs for two PV modules

The ranking considering the European conversion efficiency is shown in Table I: AEconversion/Aptronic 250 W and Enecsys have the same EU-efficiency, thus sharing rank number 13.

TABLE I. RANKING OF ALL TESTED MICROINVERTERS BY "EUROPEAN CONVERSION EFFICIENCY", ACCORDING TO (1), SAME TYPES AS LISTED IN TABLE III

Rank	Manufacturer	European-Efficiency	Relative eff. vs #1
1	SMA	95.4 %	100.0 %
2	Enphase	95.2 %	99.8 %
3	Hoymiles	95.0 %	99.5 %
4	AEconversion315	94.9 %	99.5 %
5	Envertech560	94.61 %	99.2 %
6	Power One	94.6 %	99.2 %
7	Involar 500	94.3 %	98.8 %
8	APSystems 500	94.1 %	98.6 %
9	Envertech	93.2 %	97.7 %
10	Involar	92.7 %	97.2 %
11	WVC 700 (600W)	91.6 %	96.0 %
12	Changetech	90.9 %	95.3 %
13	Aptronic	90.4 %	94.7 %
13	Enecsys	90.4 %	94.7 %
14	Ienergy	89.9 %	94.3 %

15	Letrika 260	88.7 %	93.0 %
16	WVC 700 (700W)	73.3 %	76.8 %
17	WVC 600*		0.0 %

*failed.

Table II shown the same with the CEC-efficiency formula (2) applied.

TABLE II. RANKING OF ALL TESTED MICROINVERTERS BY "CEC EFFICIENCY", ACCORDING TO (2), SAME TYPES AS LISTED IN TABLE III

Rank	Manufacturer	CEC-Efficiency	Relative eff. vs #1
1	Enphase	95.55 %	100.0%
2	Power one	95.46 %	99.9%
3	Hoymiles	95.40 %	99.8 %
4	SMA	95.06 %	99.5 %
5	AEconversion 315	94.98 %	99.4 %
6	Envertech 560	94.64 %	99.0 %
7	Involar 500	94.60 %	99.0 %
8	APSystems 500	94.50 %	98.9 %
9	Envertech	94.05 %	98.4 %
10	Involar	93.86 %	98.2 %
11	Enecsys	92.00 %	96.3 %
12	WVC 700 (600W)	91.60 %	95.9 %
13	Letrika 260	91.50 %	95.8 %
14	Ienergy	91.45 %	95.7 %
15	AEconversion 250	91.21 %	95.5 %
16	Changetech	90.90 %	95.1 %
17	WVC 700 (700W)	87.50 %	91.6 %
18	WVC 600*	*	0.0 %

* failed

B. Outdoor Measurements

The new configuration for the tests, using ten 360 W_p modules (lower row, from left), is shown in Figure 4. Modules have been manufactured by Solarwatt®, the power output at STC of each module has been measured in the factory in Dresden (Germany). Additionally, one module has been sent for a precision measurement to the testing laboratory ISFH in Hameln (Germany). It turned out that the factory measurements have been very accurate (362 W_p vs. 359.34 W_p ±3% at ISFH in July 2021).

Besides the effects already observed with the 215 W_p modules, such as distinct conversion efficiencies at different irradiance levels, speed of MPPT algorithms, minimum thresholds for initiating operation; additionally, temporal saturation effects are observed at some inverters with the new 360 W_p modules applied.

The resulting electrical energy yields during the course of a day for the different microinverters and module configurations are shown in Figure 5 for a daily course and in Table IV over a longer period of time for the 215 W_p modules. To some extent, the above-mentioned effects can be observed.

Figure 4: Configuration of PV outdoor laboratory for electrical energy yield comparison of microinverters using eight equal, calibrated PV modules (of 360 W_p each) as inputs.

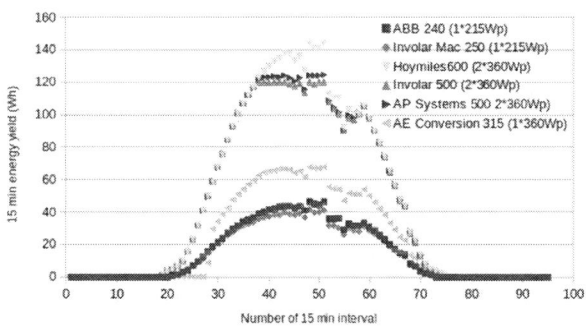

Figure 5: Electrical energy yields (during an interval of 15 minutes) of different inverters and 2 different PV module sizes during a mostly clear day (some clouds in the afternoon).

TABLE III. RANKING OF MICROINVERTERS BY ENERGY YIELD, USING THE 215 W_P MODULES [4]

Rank	Manufacturer	Type	Relative yield vs. ABB (#1)
1	Involar	MAC 500	100.7 %
2	Power One/Aurora/ABB	Micro-0.25-i	100.0 %
3	APSystems	YC 500	99.3 %
4*	Hoymiles	MI 600	97.4%*
5	SMA	Sunny Boy 240	95.2 %

6	Enphase	M 215	95.2 %
7	Involar	MAC 250	94.2 %
8*	Envertech	EVT 300	94.0%*
9*	WVC 700	700 (600 W)	91.7%*
10	AEconversion/Aptronic	INV 250-45	92.5 %
11	Envertech	EVT 248	92.1 %
12	IEnergy	GT 260	91.5 %
13	Enecsys	SMI-S-240W	88.7 %
14	Hoymiles	MI 250	78.4 %
15	Changetec	ELV 300-25	75.6 %

Measurements indicated by * are preliminary only.
WVC 600 failed, insufficient data has been recorded.
AEconversion 315 is still under investigation.

While the different types of effects make it quite cumbersome to predict an energy yield for a certain configuration at a certain location, a more consumer-friendly yield-predicting method has been elaborated by performing some yield data analysis.

Each microinverter has been directly connected to a calibrated electrical energy meter with a S_0-interface. To secure an accurate yield measurement, the calibrated electrical energy meters are replaced on a regular base with new freshly calibrated ones. All S_0-interfaces have been connected to a server-based data acquisition system.

C. Method to describe yield characteristics

To ease the characterization of a specific combination of PV module & microinverter, a linear equation has been applied to a well investigated reference characteristics of a very good inverter without issues for low irradiance, MPPT, and saturation. The inverter chosen as a reference has been the Enphase M 215, which ranked #1 at the CEC-efficiency rankings, see [4].

Plotting a function of the actual yield (y) over the reference yield (x), that function would be $y = a x + b$ with the trivial coefficients $a = 1$ and $b = 0$ for the reference configuration (Enphase M 215 with the Q-cells 215 W_p module). Figure 6 shows the original configuration with the inverters for single modules and the 215 W_p modules attached. The coefficients of the different inverters for the relative yield equation $y = ax + b$ have been elaborated in Table IV: It can be observed that for low daily yields Involar MAC 250 is performing a little bit better than the reference, so b is above 0, for high yields its performance is decreasing (relative to the reference), so a is above 1. For the Envertech EVT 300 the characteristics is vice versa: Performance at low yields is worse than the reference, so b is negative; relative performance is increasing towards high reference yields, so steepness of curve is higher, resulting in an $a > 1$.

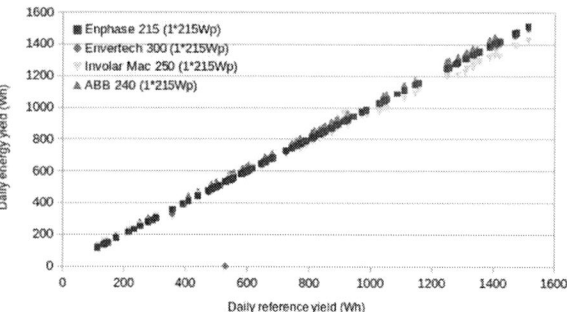

Figure 6: Electrical energy yields of different inverters for single modules with a 215 W_p module attached. Daily reference yield (x-axis) is the electrical energy yield (AC) achieved by the Enphase 215 inverter with a single 215 W_p module applied.

TABLE IV. COEFFICIENTS FOR RELATIVE DAILY YIELD $Y = A\,X + B$ (REFERENCED TO ENPHASE 215, ALL WITH A SINGLE 215 W_P MODULE)

Manu-facturer	Type	a	b (Wh)
Involar	MAC 500	0.923	+43.35
Power One / Aurora / ABB	Micro-0.25-i	1.011	+25.90
Envertech	EVT 300	1.020	-33.45
Enphase	M 215	1.000	±0.00

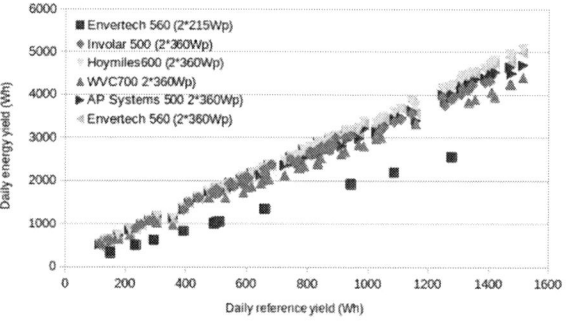

Figure 7: Electrical energy yields of different inverters for two modules with two 215 W_p or 360 W_p modules attached. Daily reference yield (x-axis) is the yield achieved by the Enphase 215 with a single 215 W_p module applied.

Figure 7 shows the characteristics of different microinverters that can serve two modules, either with two 215 W_p (older measurements) or two 360 W_p modules (latest measurements). Table V shows the corresponding coefficients a (for "steepness") and b (for "offset") of the relative daily yield curve.

The coefficients of determination R^2 for all regressions of the measurement values to determine the coefficients a and b have been in the vicinity of 0.99 or above.

TABLE V.   COEFFICIENTS FOR REL. DAILY YIELD $Y = AX + B$ FOR MICROINVERTERS SERVING TWO MODULES, EITHER 215 W_P OR 360 W_P TYPES (REFERRED TO ENPHASE 215 WITH A 215 W_P MODULE)

Manu-facturer	Type (module power)	a	b (Wh)
Envertech	EVT 560 (2 x 215 W_p)	1.983	+37.80
Envertech	EVT 560 (2 x 360 W_p)	3.227	+109.97
Hoymiles	HM 600 (2 x 360 W_p)	3.217	+152.23
Involar	MAC 500 (2 x 360 W_p)	2.889	+180.70
AP Systems	YC 500 (2 x 360 W_p)	2.953	+254.77
WVC	WVC 700 (2 x 360 W_p)	2.750	+172.39

III. CONCLUSION AND OUTLOOK

The use of a reference configuration together with the two coefficients of a linear equation enables a simple method to describe quite accurately the daily yield performance of any microinverter in combination with any PV module, even with under- or oversized ones. While prices of PV modules are decreasing at a higher pace than prices for microinverters, we will see more configurations with oversized modules and more saturated microinverters more often in the future. This underlines the necessity of a method (e.g., as described) to extrapolate energy yield.

REFERENCES

[1] W-F. Lai, S-M. Chen, T-J. Liang, K-W. Lee, A. Ioinovici: "Design and Implementation of Grid Connection Photovoltaic Inverter", Proceedings IEEE Energy Conversion Congress and Exposition (2012) 2426.

[2] H. Oldenkamp, I.J. de Jong: "The return of the AC-module inverter", Proceedings 24th European Photovoltaic Solar Energy Conference (2009) 3101

[3] D. Stellbogen, P. Lechner, M. Senger: "Field and laboratory performance characterisation of microinverter and Power optimizer systems", Proceedings 32nd European Photovoltaic Solar Energy Conference (2016) 1654.

[4] S. Krauter, J. Bendfeld: "Micro-Inverters: An Update and Comparison of Conversion Efficiencies and Energy Yields". Proceedings of 37th European Photovoltaic Solar Energy Conference (2020) 935.

Glare potential evaluation of structured PV glass based on gonioreflectometry

Markus Babin, Sune Thorsteinsson, Adrian A. Santamaria Lancia, Michael L. Jakobsen, Sergiu V. Spataru

Department of Photonics Engineering, Technical University of Denmark, Roskilde, Denmark

Estimating glare is becoming an increasingly important step in the planning stage of PV installations. This is especially true for building-integrated photovoltaics (BIPV), where atypical orientations and tilt angles can cause reflections in unusual locations. Currently, simulation tools perform simplified ray-tracing calculations to determine specular reflections from PV installations in their environment, often based on reflectance fit functions. In this work, gonioreflectometric measurements are used to determine the bi-directional reflectance distribution function (BRDF) of PV mini-modules with different surface glasses. They show significant differences between smooth and structured glass surfaces, with satinated glass showing overall lowest reflectance. At medium to high incidence angles, however, reflections are not only higher than at normal incidence, but may also peak at different view angles than the specular reflection angles. This indicates, that the common methodology of simulating only specular reflections may underestimate glare. It furthermore suggests, that a low number of reflectance fit functions may be insufficient to describe possible glare from structured surfaces, as they can exhibit vastly different BRDFs, despite possibly similar specular reflectance fit functions. In addition, calculations based on retinal irradiance threshold values for eye damage show that even for PV modules with flat glass, temporary retinal damage in form of flash blindness is highly unlikely. In any case, retinal irradiance values are far below values required for retinal burn damage, limiting reflections from PV installations to disability or discomfort glare levels. Even for the lowest reflecting surfaces, i.e. satinated glass, reflectances at incidence angles beyond 50° are high enough to possibly cause uncomfortable reflections, requiring ray-tracing simulations for accurate estimations of glare.

Upstream-Downstream Optimization of Volt-Var Control in Smart Grids

Laura R. Fardin, Christiano Lyra, and Fernanda C. L. Trindade

School of Electrical and Computer Engineering, University of Campinas, Campinas, São Paulo, 13083-852, Brazil

Abstract—With the invention of smart inverters, customer-owned PV systems have the potential to support voltage and reactive power control. These inverters can inject reactive power into the network to reduce losses and ensure an adequate voltage profile for all the buses in the system. This work proposes an upstream-downstream optimization approach for volt-var control (VVC) integrating the action of traditional grid equipment and smart inverters. The proposed algorithm is based on dynamic programming concepts and is validated in a 136-bus test system. The results show that the proposed VVC can successfully reduce such voltage levels to suitable operating levels.

Index Terms—Dynamic Programming; PV Systems; Smart Grids; Volt-Var Control

NOMENCLATURE

χ_k	Line reactance
$\mathcal{R}$	Set of bars
$\overline{V}$	Upper limit of voltage magnitude
$\underline{V}$	Lower limit of voltage magnitude
A_k	Set of kl lines connected to bus k
CAP	Set of capacitor banks
cap	Index representing the elements of CAP
DG	Set of distributed generations
dg	Index representing the elements of DG
k	Bus index
LD	Set of loads
ld	Index representing the elements of LD
$Losses_k$	Technical losses on a k-line of a radial distribution feeder
P_k	Active power entering bus k
P_k^{dg}	Active power injected by distributed generation
P_k^{ld}	Active power consumed by the load
Q_k	Reactive power entering bus k
Q_k^{cap}	Reactive power injected by capacitor bank
Q_k^{dg}	Reactive power injected by distributed generation
Q_k^{ld}	Reactive power consumed by the load
r_k	Line resistance
U_k	Set of feasible decision variables
u_k	Control variables
V_k	Voltage at the end point k of the line
X_k	Set of feasible state variables
x_k	State variables
x_{kl}	State variables in "buses-lines" pairs kl
N_{pv}	Number of PV systems present in the network
NB_{el}	Number of buses eligible for the installation of a PV system
NP_{pv}	Photovoltaic penetration level

This work was supported by Coordenação de Aperfeiçoamento de Pessoal de Nível Superior - Brazil (CAPES), by Conselho Nacional de Desenvolvimento Científico e Tecnológico (CNPq), grants 88887.510356/2020-00 and 304373/2020-6, and by São Paulo Research Foundation (FAPESP), grant 2021/04726-2 and 2020/10523-4.

I. INTRODUCTION

Due to current policies aimed at sustainability and decreasing dependence on fossil fuels, the penetration of wind and solar generation sources in electric power distribution networks is increasing exponentially. The Brazilian iNDC (*Intended Nationally Determined Contributions*) proposes, by 2030, an expansion in the use of renewable energies in the Brazilian energy matrix by approximately 33% [1]. In recent years, there has been significant growth in households with distributed PV sources. According to a recent report by the Brazilian Electricity Regulatory Agency (ANEEL), the generation distributed in Brazil reached in August 2021 the mark of 6,919 MW installed in a total of 582,433 units, resulting in a growth of 92.4% of installed capacity compared to the previous year [2].

Several other countries are also establishing legislation and incentive policies to reduce dependence on fossil fuels gradually. In the UK, in 2020, most of the energy consumed, around 55%, came from low-carbon sources (including renewable sources). The majority of PV installations, around 88%, are made up of small roof generating structures, with an average capacity of 250 kW. Regarded as a world reference in adopting incentive policies for renewable energy sources, Germany started its programs in 1990 and periodically updated them [3]. One of these programs envisages reducing CO2 emissions by 40% by 2020. After the Fukushima nuclear accident in 2011, Japan has committed to increasing energy from renewable sources to 24% by 2030 [3].

The growing installation of distributed PV generators has imposed new challenges for the control of voltage and reactive power (also called Volt-Var Control (VVC)) and providing new possibilities for carrying out such control. Alternatives to obtain optimal power flows in distribution networks using VVC strategies have been explored since the 1990s [4], [5].

However, despite the combinatorial aspect arising from the discrete values for taps of transformers and capacitors, before the introduction of distributed generation in distribution networks, the optimization of VVC was a relatively simple problem, with a tenuous coupling between voltage management and the reduction of reactive power flows. As the voltages fell from the substations to the farthest buses from the feeders, the adjustment of taps of substation transformers and voltage regulators was usually sufficient to keep the voltages within allowable values. As such, the optimization of the control of capacitors occurred to reduce the losses resulting from the reactive powe flows without the need for voltage adjustments, as it only helped reduce the voltage drop. Few cases led to the need for readjustment transformer taps and voltage regulators, with little effect on reactive optimization. As in real situations, the number of switchable capacitors in the feeders is usually minimal, the reactive optimization problems by capacitor tap adjustments were of little difficulty. The enumeration of alternatives was the strategy used in most cases (except that the number of tap changes added some complexity [4], [6]). VVC strategies by simple rules, as proposed by Baran and Hsu [7], could be adequate for real problems.

Over the past few years, the introduction of distributed generation into grids has significantly increased the coupling and complexity of VVC. Generators installed across distribution systems lead to changes in power flows and can cause over-voltages [8], [9]; consequently, the reactive control alternatives no longer just help maintain a good voltage profile across the feeders. On the other hand, intelligent inverters associated with generation sources can act in the VVC, expanding the possibilities of action to optimize merit criteria expressed in well-defined "objective functions". This work proposes a strategy for VVC in distribution networks using dynamic programming concepts.

II. UPSTREAM-DOWNSTREAM DYNAMIC PROGRAMMING

The methodology for VVC optimization has as its main point the development of an optimization strategy for non-convex control problems with discrete variables for optimization of energy flows in smart grids. The optimization strategy, represented by the acronym UD-DP (upstream-downstream dynamic programming), is based on extensions of dynamic programming ideas [10]–[12] associated with the optimization of power flows [13]. These ideas mix dynamic programming and network optimization concepts in a methodology to optimize flows with discrete control variables. They bring the perspective of computational efficiency and quality of solutions in optimizing VVC in distribution networks. Strategies where a central agent frequently acts (at intervals of a few minutes), modifying the states of controllable equipment, are attractive benefits for distribution networks permeated by PV generations.

Optimization processes based on dynamic programming aim to find the solution, for all stages and all feasible states, of the Hamilton-Jacobi-Bellman (HJB) Recursive Equation [10]. UD-DP generalizes the HJB recursive equation by including internal stream optimizations at points where there are branches. Focusing on solving the HJB equation at these points (where there are branches in the feeders), we have:

$$F_k(x_k) = Min_{u_k \in \mathcal{U}_k}[e_k(x_k, u_k) + F_{k+1}(x_{k+1})], \quad \forall k \in \mathcal{R} \tag{1}$$

where $k \in \mathcal{R}$ is the associated index associated with the "slash-line" pair (line connecting the slash to components in the direction of substation), $F_k(X_k)$ is the function of *accumulated optimal cost* from *state* x_k to all terminal *sheets* (buses) of the feeder, u_k are the actions of *control* on bus k and $e_k(x_k, u_k)$ is the *elementary cost* (associated with the "rowslash" pair k). States and controls belong to the sets $\mathcal{X}_k$ and $\mathcal{U}_k$ ($x_k \in \mathcal{X}_k$ and $u_k \in \mathcal{U}_k$).

Through the following equations, which perform the optimization of flows, we obtain the value of the function $F_{k+1}(x_{k+1})$:

$$x_{k+1} = f_k(x_k, u_k) \tag{2}$$

$$x_{k+1} = \sum_{kl \in A_k} x_{kl} \tag{3}$$

$$F_{k+1}(x_{k+1}) = Min_{\forall x_{kl}, kl \in A_k}[\sum_{kl \in A_k} F_{kl}(x_{kl})], \quad x_{kl} \in \mathcal{X}_{kl} \tag{4}$$

The solution of HJB recursive equations in dynamic programming provides the functions *optimal policies* $\pi^* = \{\mu_k^*, \forall k\}$, which define the *optimal controls* u^* associated with each of the *feasible states* ($u_k^* = \mu_k^*(x_k)$). The UD-DP's *optimal policies* functions are augmented with the *optimal flow distributions* $\mathcal{O}_k(x_k, u_k^*)$, which define the optimal flows x_{kl}^* for all branches kl that branch from the bus k ($kl \in A_k$):

$$\mu_k^*(x_k) = (u_k^*, \mathcal{O}_k(x_k, u_k^*)) \tag{5}$$

$$\mathcal{O}_k(x_k, u_k^*) = (x_{k1}^*, x_{k2}^*, \cdots, x_{kp}^*) \tag{6}$$

The identification process of "row-buses" sets with branches ($k \in \mathcal{R}$) and the optimization of flows uses concepts and data structures developed in the area of optimization of flows in grids. This knowledge allows the implicit ordering of feeder buses in *pre-order* [13]. The *pre-order* ensures that all values of the functions $F_{kl}(x_{kl})$ are well defined for the calculation of $F_{k+1}(x_{k+1})$, in the internal flow optimization process.

Analogous to dynamic programming, UD-DP optimization strategies correspond to a set of concepts that adapt to specific applications. The most important aspect is the design of each of the components of *states* (x_k) and *controls*, considering the trade-off between the benefits of information for VVC and the computational complexity of the optimal policy acquisition process.

978-1-7281-6118-1/22 $31.00 © 2022 IEEE

III. PROBLEM FORMULATION

A. MATHEMATICAL MODEL

Using the previously presented theory, it is possible to determine a mathematical optimization model to represent the VVC problem in smart grids with PV generation. Given a radial network, with power offers and demands at each node, together with the technical losses present in each branch, the problem to be addressed here is finding the best option for total network operation to reduce such losses and ensure that voltages and power factor are within specified limits. For this, the control of available devices is used, such as capacitor banks, voltage regulators, and inverters of PV generators.

Technical losses on any k line of a radial distribution feeder are calculated as follows:

$$Losses_k = R_k \cdot \left[\frac{(P_k)^2 + (Q_k)^2}{(V_k)^2} \right] \quad (7)$$

Equation (7) represents the mathematical equation to be minimized by dynamic programming, subject to:

$$\underline{V} \leq V_k \leq \overline{V} \quad (8)$$

It is possible to present such a problem in the following way:

$$Min_{u_0, u_1, \ldots, u_{n-1}} \left[\sum_{k=0}^{n-1} e_k(x_k, u_k) + \psi(x_n) \right]$$
$$x_{k+1} = f_k(x_k, u_k) \quad (9)$$
$$x_k \in X_k$$
$$u_k \in U_k$$

In the approach proposed by dynamic programming, the following components stand out:

- The stages k represent each node of the analyzed grid;
- The states x_k, represented by a one-dimensional vector, represent the state variable available for each node k:

$$x_k = \left[x_k^1 \right] = \left[Q_k \right] \quad (10)$$

- The decisions u_k, represented by a two-dimensional vector, represent the control variables of the optimization problem:

$$u_k = \begin{bmatrix} u_k^1 \\ u_k^2 \end{bmatrix} = \begin{bmatrix} Q_k^{dg} \\ Q_k^{cap} \end{bmatrix} \quad (11)$$

- The elementary costs $e_k(x_k, u_k)$, associated with the states x_k and decisions u_k, indicate the technical losses in a node k to be minimized:

$$e_k(x_k, u_k) = R_k \cdot \left[\frac{(P_k)^2 + (x_k)^2}{(V_k)^2} \right] \quad (12)$$

Subject to:

$$P_k - P_k^{ld} + P_k^{dg} = \sum_{j \in A_k} [P_{k+1,j} + $$
$$+ R_{k+1,j} \cdot \left(\frac{(P_{k+1,j})^2 + (x_{k+1,j})^2}{(V_{k+1,j})^2} \right) \quad (13)$$

$$x_k - Q_k^{ld} - u_k^1 - u_k^2 = \sum_{j \in A_k} [x_{k+1,j} + $$
$$+ \chi_{k+1,j} \cdot \left(\frac{(P_{k+1,j})^2 + (x_{k+1,j})^2}{(V_{k+1,j})^2} \right) \quad (14)$$

$$(V_{k+1,j})^2 = (V_k)^2 - 2 \cdot (R_{k+1,j} \cdot P_{k+1,j} + $$
$$+ \chi_{k+1,j} \cdot x_{k+1,j}) \quad (15)$$

Equations (13) and (14) represent the power balance equations of the active and reactive power flow for each node k. Where we have that the active and reactive power flows that leave node k has the subscript $k+1$. In order to maintain the balance between the powers, every inflow is equal to the outflow plus losses. Each node k receives a flow P_k and Q_k from the parent node and P_k^{dg} and Q_k^{dg} if it has the presence of distributed generation or Q_k^{cap} for the case of the presence of a capacitor bank. The flow that leaves bus k comes from the loads (P_k^{ld} and Q_k^{ld}), if there is any, and from the child nodes, if there is any. If there is generation distributed in the hypothetical k bus, depending on the time of day and the current generation and consumption, it can behave as a source or customer of active power. Equation (15) represents voltage Kirchhoff's law for bus k.

B. OPTIMIZATION PROCEDURE

The proposed backward process determines for each system bus the set of optimized values for the state (see (10)) and control (see (11)) variables. For this, it uses the Hamilton-Jacobi-Belman recursive equations (see (9)), starting from the furthest node in the network. In summary, the optimization model proposed in this work is:

minimize: (12)
subject to: (8) and (13) to (15)

For this, we will use the iterative process proposed by [11], which consists of four basic steps:

1) Solves the optimization problem, minimizing the (12), assuming that the voltages are equal to 1 pu;
2) Calculates the voltages for each member of the system, using (15);
3) Solves the optimization problem again, using now the values of voltages obtained in the previous item;
4) Compares the two results obtained. If the solutions are different, go back to step 2. Otherwise, we have the optimal solution to the problem.

978-1-7281-6118-1/22 $31.00 © 2022 IEEE

IV. SIMULATIONS

This work adopts a real distribution network with 136 buses and 152 lines located in the city of Sete Lagoas (Brazil). The system operates with a rated voltage of 13.8 kV, with a maximum active and reactive power demand of 18,313 MW and 9,384 MVar, respectively, and a short circuit level of 100 MVA. The network also has three 800 kVar switchable capacitor banks at buses 47, 92, and 124. For our case study, we use the single-phase equivalent of this system. Details of this distribution network can be found in [14] and are thus omitted in this paper due to space limitations.

Fig. 1 illustrates typical consumer load curves and their generation curves over a typical day considered for the simulation. In this case, we assume that all customers in this system have the same load profile (shown in Fig. 1). In addition, we use the characteristic values of solar radiation for the city of Sete Lagoas for the photovoltaic generation curve.

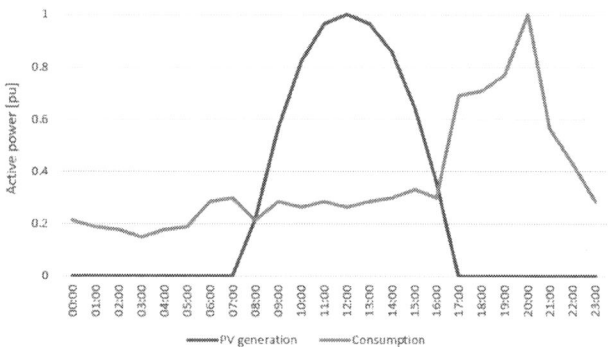

Fig. 1. Normalized PV generation and power consumption data profiles.

We performed the analysis for photovoltaic penetration values ranging from 5 to 50%, is defined according to the equation below:

$$NP_{pv} = \frac{N_{pv}}{NB_{el}} \qquad (16)$$

The allocation of photovoltaic systems was random, while the dimensioning considered a compensation of 100% of the consumption. For each penetration level, we simulated the value of losses without VVC and with VVC. Therefore, we consider two different cases for the sizing of inverters: sized with the same rated capacity of the PV system and oversized by 15%. We compare with the system value without distributed generation for photovoltaic energy. The results are depicted in Fig. 2.

Fig. 2 shows that initially when inserting distributed generation into the grid, there is a reduction in total active losses, as expected. But as we further increase the number of photovoltaic systems, there is a gradual increase in losses. This happens after a certain penetration level, which depends on the characteristics of each network. The increase in losses occurred with 35% of photovoltaic penetration for the case analyzed.

The increase in total active losses is due to the bidirectional energy flow. When we insert distributed generators into the system, the losses naturally decrease due to the presence of generators closer to the loads. Therefore, the active power that would flow from the substation to the load, covering in some cases long paths, is now readily provided by the photovoltaic system. However, as we increase the number of these generators, there is an increase in the current flowing in the network and, consequently, an increase in the total active losses.

Fig. 2 also demonstrates that the losses with VVC are lower when compared to the losses of the system without VVC, as expected. However, this difference increases as the penetration of photovoltaic energy increases. Although, when we are dealing with a slightly oversized inverter, the results are better. For example, for the system with 40% penetration, when we deal with the oversized inverter, we reduce losses to levels close to the ones of the system without distributed generation. For the system with 50% of PV penetration level, the losses reduction is 13.2% for the system with oversized PV inverters against 4.3% in case the system has inverters sized with the same rated capacity of the PV system.

A second analysis of this real 136-bus network sought to assess the impact of different oversizing levels in active power losses. As indicated in Fig. 3, greater oversizing does not influence the losses, as systems with 45% oversizing had total active losses very close to the system with 15% oversizing. As the photovoltaic inverter is the most expensive part of the entire PV system, it is not feasible for the customer to purchase an inverter much larger than the sized one. Thus, a slight oversizing can be enough for a more effective reduction of losses.

Fig. 4 shows us the voltage profile for bus 115 of the 136-bus real power system for 50% PV penetration. This specific bus has a photovoltaic generator. When there is only the photovoltaic system inserted, there is a significant voltage increase in the moments of maximum generation. During certain times of the day, the voltage upper the limit established by ANEEL as adequate (1.05 pu). However, with VVC, this increase is controlled and does not exceed levels established by the Brazilian regulation.

V. CONCLUSION

This paper proposed an algorithm based on dynamic programming to apply VVC concepts in distribution networks with high penetration of photovoltaic distributed generation. The proposed approach successfully reduced total active losses. We also demonstrated that as the number of photovoltaic generators increases, there is a more significant reduction in losses since more smart inverters participate of the VVC.

The simulations also proved that there is a more significant reduction in total losses when there is a slight oversizing of the inverter. An oversized inverter, even at peak generation, can inject more reactive power into the system than a correctly sized inverter. The results also show that slight oversize is

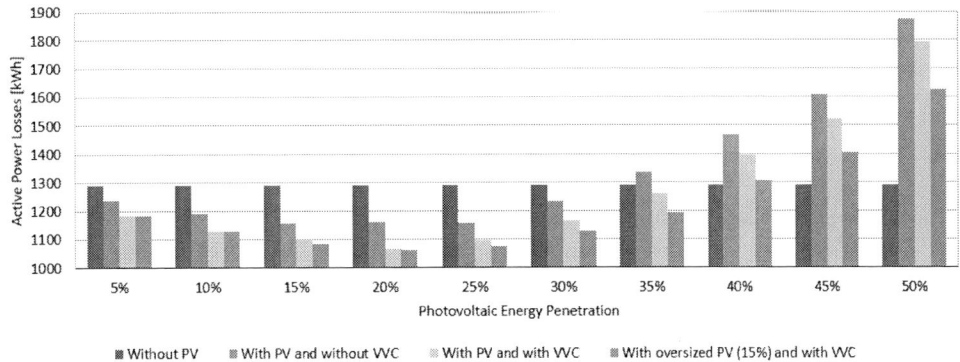

Fig. 2. PV charging and power consumption data profiles.

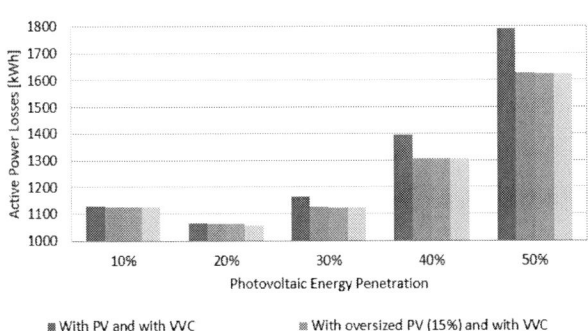

Fig. 3. Comparison of different levels of inverter oversizing.

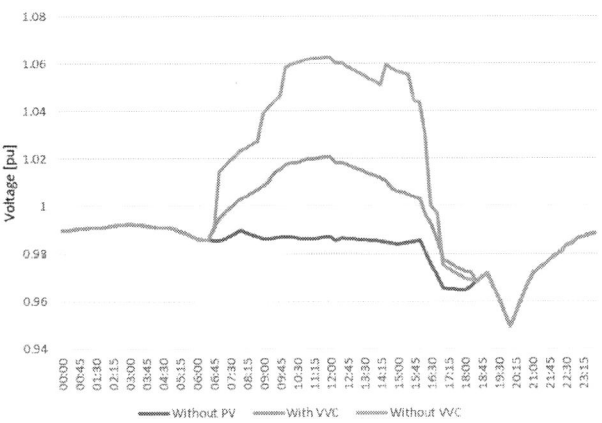

Fig. 4. Voltage level at 115 bus for different system configurations.

enough to improve the VVC, as higher levels of oversizing did not caused significant changes in energy losses.

Lastly, VVC also proved to maintain the proper voltage profile. In systems with high photovoltaic penetration, there are many cases of overvoltage associated with reverse power flow. The VVC can reduce such voltage levels to suitable operating levels through reactive power control.

REFERENCES

[1] Federative Republic of Brazil, *Intended nationally determined contribution towards achieving the objective of the united nations framework convention on climate change* [Online]. Available: https://www4.unfccc.int/sites/ndcstaging/PublishedDocuments/Brazil%20First/BRAZIL%20iNDC%20english%20FINAL.pdf

[2] Ministério de Minas e Energia, *Boletim Mensal de Monitoramento do Sistema Elétrico Brasileiro Agosto/2021* [Online]. Available: https://www.gov.br/mme/pt-br/assuntos/secretarias/energia-eletrica/publicacoes/boletim-de-monitoramento-do-sistema-eletrico/2021/boletim-de-monitoramento-do-sistema-eletrico-agosto-2021.pdf

[3] N. de Castro and G. Dantas, Experiências Internacionais em Geração Distribuída: Motivações, Impactos e Ajustes, 1st ed. Publit Soluções Editoriais, 2018, Available: http://gesel.ie.ufrj.br/app/webroot/files/IFES/BV/livro_experiencias_internacionais_em_gd.pdf

[4] Y.-Y. Hsu, "Dispatch of capacitors on distribution system using dynamic programming," IEE Proceedings C (Generation, Transmission and Distribution), vol. 140, pp. 433–438, November 1993.

[5] I. Roytelman, B. K. Wee, and R. L. Lugtu, "Volt/var control algorithm for modern distribution management system," IEEE Transactions on Power Systems, vol. 10, no. 3, pp. 1454–1460, 1995.

[6] M. B. Liu, C. A. Canizares, and W. Huang, "Reactive power and voltage control in distribution systems with limited switching operations," IEEE Transactions on Power Systems, vol. 24, no. 2, pp. 889–899, 2009.

[7] M. E. Baran and Ming-Yung Hsu, "Volt/var control at distribution substations," IEEE Transactions on Power Systems, vol. 14, no. 1, pp. 312–318, 1999.

[8] P. M. S. Carvalho, P. F. Correia, and L. A. F. M. Ferreira, "Distributed reactive power generation control for voltage rise mitigation in distribution networks," IEEE Transactions on Power Systems, vol. 23, no. 2, pp. 766–772, 2008.

[9] M. E. Baran, H. Hooshyar, Z. Shen, and A. Huang, "Accommodating high pv penetration on distribution feeders," IEEE Transactions on Smart Grid, vol. 3, no. 2, pp. 1039–1046, 2012.

[10] D. P. Bertsekas, Dynamic Programming and Optimal Control – Volume 1, 3rd ed. Athena Scientific, 2005.

[11] J. F. V. Gonzalez, C. Lyra, and F. L. Usberti, "A pseudo-polynomial algorithm for optimal capacitor placement on electric power distribution networks," European Journal of Operational Research, vol. 222, no. 1, pp. 149–156, 2012.

[12] J. C. López, P. P. Vergara, C. Lyra, M. J. Rider, and L. C. P. da Silva, "Optimal operation of radial distribution systems using extended dynamic programming," IEEE Transactions on Power Systems, vol. 33, no. 2, pp. 1352–1363, 2018.

[13] R. K. Ahuja, J. B. Orlin, and T. L. Magnanti, Network Flows: Theory, Algorithms, and Applications. Prentice Hall, 1993.

[14] M. A. O. Leite, "Reconfiguração de Redes de Distribuição Primária de Energia Elétrica para Redução de Perdas Técnicas," Federal University of Minas Gerais, Belo Horizonte - MG - Brazil, 2014.

Growth of GaAs on Ge/Si (001) nanovoided virtual substrate

Jonathan Henriques, Alexandre Heintz, Bouraoui Ilahi, Richard Arès, Abderraouf Boucherif

Institut Interdisciplinaire d'Innovation Technologique (3IT), Université de Sherbrooke, Sherbrooke, QC, Canada

Laboratoire Nanotechnologies Nanosystèmes (LN2) — CNRS UMI-3463, Institut Interdisciplinaire d'Innovation Technologique (3IT), Université de Sherbrooke, Sherbrooke, QC, Canada

The integration of III-V compounds on Si substrate is very promising for photovoltaic applications. This would be an alternative to obtain low cost and high efficiency solar cells. Currently, III-V solar cells are produced on Ge substrate, which engages high production costs. However, the heteroepitaxy of these materials on silicon implies the appearance of defects and dislocations related to the difference in lattice parameter and thermal expansion coefficient. Ge is commonly employed as an intermediate buffer layer to integrate such materials. This process involves high Ge thickness and several postgrowth annealing steps to reduce the dislocation density down to 106 cm-2 which is still too high. Recently, an innovative approach using dislocation-selective electrochemical deep etching, to create nanovoid inside the germanium epilayer on silicon has shown efficiency to trap and annihilate the dislocations reducing their density down to 104 cm-2. In this work, we report on the growth of GaAs on virtual Ge/Si (001) substrate following a new approach based on direct growth of Ge buffer layer on porous Ge/Si substrate. The thermally induced reorganization of the porous Ge (PGe) leaves high density of nanoscale voids within the Ge buffer layer leading to the enhancement of the optical and structural properties compared to that directly grown on Ge/Si. Our results show that the nano-voided Ge/Si virtual substrate is potentially interesting for direct growth of III-V solar cells on Si (001) substrate.

Monte Carlo evaluation of multijunction solar systems in tandem and 4-terminal configurations

Roberto Corso[1,2], Marco Leonardi[1,2], Andrea Scuto[1] and Salvatore A. Lombardo[1]

[1] Institute for Microelectronics and Microsystems (IMM), National Research Council (CNR), Catania, 95121, Italy
[2] Department of Physics and Astronomy, University of Catania, Catania, 95123, Italy

Abstract—**Multijunction systems are a promising alternative to overcome the efficiency limit of single-junction cells. A comparison between the two connection schemes, tandem and 4-terminal, is necessary for the development of the applications of this technology. To this purpose, we employed a Monte Carlo simulation algorithm, validated on experimental data, to evaluate the power conversion efficiency of a multijunction system in both configurations by obtaining the absorption spectrum of each layer and calculating the current-voltage characteristics for different thicknesses of each layer in the top cell.**

Index Terms—**Monte Carlo, solar cell modeling, multijunction solar system**

I. INTRODUCTION

The multijunction system design is one of the diverse technologies that are being developed to overcome the efficiency limit of single-junction silicon photovoltaics, and it is the one currently resulting in the highest efficiencies recorded [1]. Multijunction systems are made up by two or more single-junction cells usually stacked on top of one another; the semiconductors are chosen with decreasing bandgap energy (from top to bottom) so that each cell absorbs the portion of solar spectrum it is most suited to and allows transmission of the remaining light to the cell below. Therefore, a careful choice of semiconductors, conductive layers and thicknesses of the various layers is crucial and this topic is being the center of an ever-increasing research, with relevant efforts dedicated towards the development of accurate simulation tools.

Throughout the years, a variety of models have been proposed to analyze the performance of photovoltaic (PV) cells. Most of these models focus on the prediction of the behavior of carriers within the cell [2], [3] and adopt the Transfer Matrix (TM) approach to determine the light absorption profile; however, this approach can be correctly applied only to the case of one-dimensional systems and it is not suited to the analysis of three-dimensional geometric structures, such as the surface texturing of silicon solar cells. To overcome this limitation, simplified mixed approaches have been adopted [4] that apply the TM method to a limited fraction of the possible paths of light in the system; however, a single textured surface can lead to many different paths whose probabilities depend on the texturing structure and on the angle of incidence of light [5], and the number of possible paths increases as more layers are added to the system

The present research was, in part, funded by PON Ricerca e Innovazione 2014–2020, under Decreto Direttoriale di concessione dell'agevolazione del 21 May 2019 Prot. N. 991, Contract Code ARS01_00519, BEST-4U Project.

For this reason, rather than analytical models, fully numerical algorithms are more suited to study these systems, and in this work we present a model based on geometric optics which performs Monte Carlo ray tracing of light traveling through the layers of a PV cell. The model has been validated by comparison with experimental data, and has been applied to the case of a multijunction system. Generally, the system can be arranged by either connecting sub-cells in series (tandem configuration) or connecting modules in parallel (4-terminal or 4T configuration). The model has been employed to evaluate the power conversion efficiency (PCE) for different thicknesses of each layer in the top cell in both configurations.

II. MODEL DESCRIPTION

The optical behaviour of a PV cell is completely determined by its layers, and in particular, for each layer j, by its dispersive complex refractive index $n_j(\lambda) + ik_j(\lambda)$, its thickness t_j, and by its layer texturization geometry, if present. The cell is assumed infinite along the x- and y-axes. Flat surfaces are represented by (001) planes, while the silicon bottom cell is textured, with pyramids whose faces are $\{111\}$ planes. The unit-cell is defined by the $\{110\}$ planes which include the edges of the base of the pyramid, as depicted in Fig. 1. The unit-cell approach has been employed because it allows to extend the textured surface along the x- and y-axes

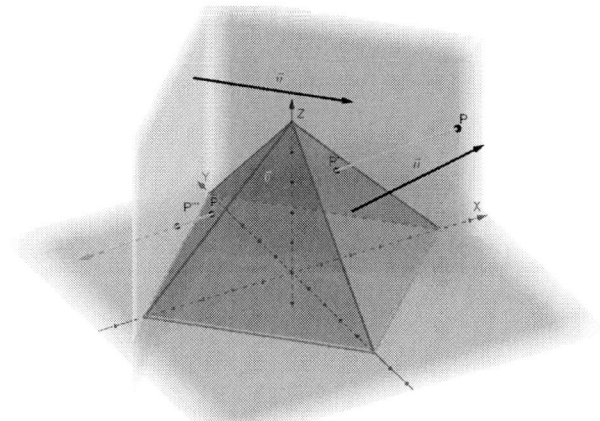

Fig. 1. View of a single pyramid and the surrounding unit-cell. For better visualization, only two planes are shown. Vectors $\vec{n}$ and $\vec{u}$ are normal to their respective planes. Of the three intersection points $\vec{p'}$, $\vec{p''}$ and $\vec{p'''}$, only $\vec{p'}$ will be used in the following steps as it is the closest to the starting position $\vec{p}$.

without being computationally costly. Furthermore, the model can also simulate over-etched pyramids, which are obtained by replacing the tip of the pyramid with a smooth curve.

Each photon is initialized in the first layer with a direction $\vec{v} = (0, 0, -1)$ and a random position $\vec{p} = (p_x, p_y, p_z)$. The general steps of the algorithm are the following:

- When a photon approaches a pyramid, for each face $\vec{n}$ of the pyramid and each face $\vec{u}$ of the unit-cell the intersection point $\vec{p'}$ with $\vec{v}$ is calculated, and the closest one is selected.
- Then, the absorption probability $\alpha(\lambda, d)$, where $d = \|\vec{p'} - \vec{p}\|$, is calculated from the Lambert-Beer law. A random number is generated, and if it is lower than α the photon is considered absorbed in the current layer and a new photon is initialized; otherwise, the position of the photon is updated with the coordinates of $\vec{p'}$.
- If the photon reaches an interface, reflection and transmission probability are determined by the Fresnel equations, the direction of reflected and transmitted light are determined by Snell's law and another random number is generated to decide between reflection and transmission. Then $\vec{v}$ and the current layer are updated accordingly. In the case where the photon reaches a face of the unit-cell, periodic boundary conditions are applied by making the photon re-enter the unit-cell from the opposite face.
- If the photon reaches the opposite surface of a layer and the surface is textured, new p_x and p_y are randomly generated. This step reflects the fact that the chemical etching process produces randomly positioned pyramids, and the positions of the pyramids on one side of the layer are uncorrelated to those on the other side.
- Finally, if the photon is in the first or last layer, traveling away from the cell and there is no possible intersection with the layer, the photon is considered reflected or transmitted, respectively.

With this algorithm each photon is tracked until it is either reflected, transmitted or absorbed. An user-defined number of photons are simulated for each wavelength, obtaining the total reflectance $R(\lambda)$ and total transmittance $T(\lambda)$ of the system and the absorbance $A_j(\lambda)$ of each layer. Then, the ideal current-voltage (I-V) characteristics of the cell can be calculated as in [6], with the short circuit current being:

$$I_{sc} = qS \int A_s(\lambda) \cdot \phi(\lambda) d\lambda \qquad (1)$$

where q is the electron charge, S the cell area, $A_s(\lambda)$ the semiconductor's absorbance and $\phi(\lambda)$ the photon flux (the ASTM G-173-03 solar spectrum has been used throughout this work). The dark saturation current I_0, as specified in [6], is calculated from the blackbody emission of the cell, and the I-V curve is calculated according to:

$$I(V) = I_{sc} - I_0 \cdot \left(e^{\frac{qV}{n k_b T_c}} - 1 \right) \qquad (2)$$

where n is the cell ideality factor, k_b is the Boltzmann constant and T_c is the temperature of the cell.

The model has been validated by comparing the reflection, transmission and absorption spectra of a textured Si cell covered by indium tin oxide (ITO) with experimental data, as reported in Fig. 2.

III. MULTIJUNCTION SYSTEM SIMULATION

The proposed algorithm has been used to predict the ideal PCE of a multijunction solar system. Silicon has been chosen for the bottom cell; the optimal bandgap energy to be paired with silicon is about 1.7 eV [7]. Among the many materials that can reach this energy (III-V semiconductors, perovskites, chalcogenides, kesterites, organic semiconductors, etc.), the refractive index of copper gallium selenide has been assumed for the top cell semiconductor (TCS). The silicon bottom cell presents pyramidal texturing, whereas the top cell is flat. Each sub-cell is positioned between two layers of transparent conductive oxide (TCO), either aluminum zinc oxide (AZO) or ITO for the top cell and ITO for the bottom cell. A dieletric interlayer has been positioned between the sub-cells. A diagram of the layers has been reported in Fig. 3.

Fig. 4 reports the absorption spectra of the sub-cells, showing how the two semiconductor work on different portions of the solar spectrum (separated at the bandgap energy of TCS) as well as a parasitic absorption of blue and infrared light from the TCO layers.

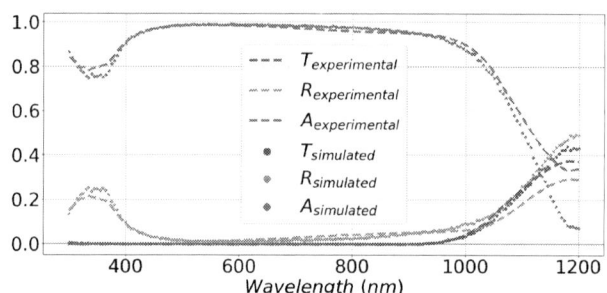

Fig. 2. Comparison between experimental and simulated data of reflectance, transimttance and absorption of a textured Si cell with ITO coating.

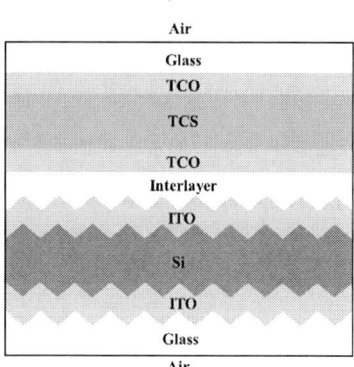

Fig. 3. Diagram of the multijunction system. Air is assumed to be semi-infinite on both sides. Thicknesses not in scale.

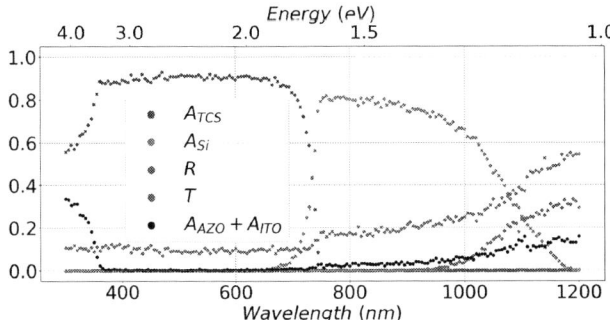

Fig. 4. Reflection and transmission spectra of the multijunction solar system and absorption spectra of the sub-cells.

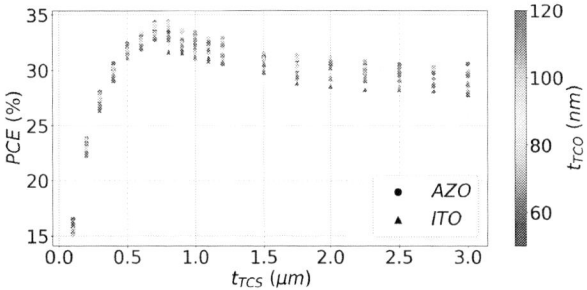

Fig. 6. Dependance of the ideal PCE of the multijunction system in the tandem configuration on the thicknesses of the layers of the top cell.

In general, sub-cells in a multijunction system can be connected either in a tandem configuration (with sub-cells connected in series) or in a 4T configuration (with sub-cells connected in parallel). The main difference between these two schemes is that the series connection is heavily dependent on the current matching of the sub-cells, which in turn depends on the thicknesses of the semiconductor layers and on the illumination conditions under which the cell operates. On the other hand, the parallel connection depends on the voltage matching of the sub-cells, but it does not impose constraints on the layer thicknesses and it is more impervious to spectral variations, as the V_{oc} depends on the solar radiation intensity logarithmically.

The I-V characteristics of the multijunction system have been simulated as described in [6] under standard test conditions while changing the thicknesses of the layers of the top cell. From the results reported in Fig. 5, it can be seen that PCE in the 4T configuration generally increases with the TCS thickness, confirming that the parallel connection is not impaired by the thicknesses of the sub-cells. Moreover, thicker TCO layers result in lower PCE values. On the contrary, Fig. 6 shows how, as a consequence of the current matching required by the series connection, the PCE in the tandem configuration has an absolute maximum. Figs 5 and 6 show a superiority in terms of PCE of the parallel connection, and it has to be noted that the values of Fig. 6 relative to

the tandem configuration have been reached assuming the ASTM G-173-03 solar spectrum, while real spectra can be significantly different during the day, influencing the current matching between the two sub-cells.

IV. CONCLUSIONS

In this work we have carried out a comparison of a multijunction system in the tandem and 4-terminal configurations. To this purpose, we developed an algorithm to perform Monte Carlo simulations of photons reflected, transmitted and absorbed by the system. The thicknesses of the layers in the top cell have been varied to optimize the power conversion efficiency of the system. The results of the simulations show that, in contrast to the 4-terminal approach, the tandem connection imposes an optimal thickness to the semiconductor of the top cell. Moreover, the 4-terminal configuration reaches higher values than the tandem configuration.

REFERENCES

[1] https://www.nrel.gov/pv/cell-efficiency.html last accessed in 09/01/2022
[2] Burgelman, M., Nollet, P., Degrave, S., 2000. Modelling polycrystalline semiconductor solar cells. Thin Solid Films 361, 527–532. https://doi.org/10.1016/S0040-6090(99)00825-1
[3] R. Varache, C. Leendertz, M. E. Gueunier-Farret, J. Haschke, D. Muñoz, and L. Korte. "Investigation of Selective Junctions Using a Newly Developed Tunnel Current Model for Solar Cell Applications." Solar Energy Materials and Solar Cells 141 (2015) 14–23. doi:10.1016/j.solmat.2015.05.014.
[4] Nishigaki, Y., Nagai, T., Nishiwaki, M., Aizawa, T., Kozawa, M., Hanzawa, K., Kato, Y., Sai, H., Hiramatsu, H., Hosono, H., Fujiwara, H., 2020. Extraordinary Strong Band-Edge Absorption in Distorted Chalcogenide Perovskites. Sol. RRL 4, 1–8. https://doi.org/10.1002/solr.201900555
[5] Chen, Q., Liu, Y., Wang, Y., Chen, W., Wu, J., Zhao, Y., Du, X., 2019. Optical properties of a random inverted pyramid textured silicon surface studied by the ray tracing method. Sol. Energy 186, 392–397. https://doi.org/10.1016/j.solener.2019.05.031
[6] Shockley, W., Queisser, H.J., 1961. Detailed balance limit of efficiency of p-n junction solar cells. J. Appl. Phys. 32, 510–519. https://doi.org/10.1063/1.1736034
[7] Yu, Z., Leilaeioun, M., Holman, Z., 2016. Selecting tandem partners for silicon solar cells. Nat. Energy 1. https://doi.org/10.1038/nenergy.2016.137

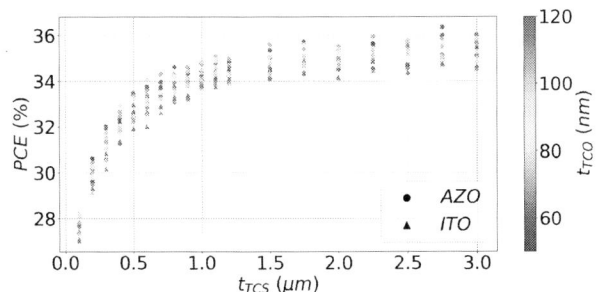

Fig. 5. Dependance of the ideal PCE of the multijunction system in the 4-terminal configuration on the thicknesses of the layers of the top cell.

978-1-7281-6118-1/22 $31.00 © 2022 IEEE

Outdoor energy performances for standard and bifacial modules as well on the failure modes observed in outdoor conditions

Ottanà A.(1, 3); Rametta F. (1); Gangemi W. (2); Colletti C. (1); Di Stefano A. (2); Canino A. (2); Foti M. (1); Gerardi C.(1); Bizzarri F. (3);

1) EGP - Enel Green Power, Contrada Blocco Torrazze Zona Industriale 95121, Catania, Italy, 2) EGP - Enel Green Power, Contrada Passo Martino Zona Industriale 95121, Catania, Italy, 3) EGP - Enel Green Power Viale Regina Margherita 125, 00198 Roma, Italy

Abstract— **In this article is reported the outdoor performance evaluation of 3SUN bifacial HJT PV modules carried out in the Enel Green Power Innovation Lab of Catania – Passo Martino. In the analysis in terms of Performace Ratio, a monofacial PV technology is used as a reference, and in order to ensure a fair comparison different commercial module technologies have been selected and monitored. The present paper aims at demonstrating the superior production of the heterojunction modules due to their higher passivation, better thermal coefficients and highlight the bifacial gain.**

Keywords— *Heterojunction, Bifacial, Outdoor Performance, Performance Ratio*

I. Introduction

In this article is presented the outdoor performance analysis of 3SUN "BEST4U Initial Generation" heterojunction PV modules [1-3] conducted at the Enel Green Power Innovation Lab of Catania – Passo Martino.

The present work aims to demonstrate an increased energy yield up to 5% in sunny regions compared to Al-BSF solar cells, and to demonstrate that bifaciality can bring energy yield in line with expectation of albedo and module spacing (i.e. typically 15-20% more energy than an equivalent monofacial module) [4-6]. For this purpose, the performance evaluation is carried out in comparison with a monofacial PV module technology, considered in this document as a "world widespread" reference. Moreover, in order to ensure a fair comparison with the 3SUN modules, different commercial bifacial module technologies have been selected and tested.

II. Testing facility and tested PV modules

The test facility dedicated to double-sided modules was designed and built in the Innovation Lab of Catania – Passo Martino in 2015.

All the exposed modules have carried out the same process of characterization imposed by the 61215 IEC standard: visual inspection (MQT 01) and electroluminescence, Performance at STC (MQT 6.1), initial stabilization (MQT 19.1), electrical insulation test in a dry and wet environment (MQT 03, MQT15), which will be repeated at the end of the bench test period to evaluate the structural degradation of the modules of each technology.

For each module technology, independent systems of approximately 3 kW have been created, using a single-phase inverter with 1 MPP tracker for each string and dedicated transducers for detecting the current and voltage in DC Systems. The systems began to acquire data from the middle of 2019. The monitoring equipment is composed of:

- 8 module temperature sensors
- 1 front side pyranometer
- 1 back side pyranometer
- DC voltage and current sensors for each system

Monitoring is entrusted to precision instrumentation that sends data 24 hours a day to a server, directly accessible via web portal.

Table I reports the list of the PV modules installed and monitored at Passo Martino. Monofacial Poly p-Si modules are used as a "world widespread" reference and, in addition to 3SUN dual-glass bifacial modules based on ECA-busbar interconnections, the outdoor performance has been evaluated also for different commercial module technologies in order to ensure a fair comparison. Other bifacial dual-glass modules were tested: two based on HJT technology (one of which with Smart Wire Connection Technology) and one based on PERT solar cells technology. As depicted in Fig. 1, Bifi HJT A / Bifi PERT and Bifi HJT 3SUN / Bifi HJT SWCT technologies have been installed adjacent to each other for direct comparison. The Monofacial Poly p-Si string is positioned in the top row of the installation and, for reasons of space, it is divided into 2 rows belonging to the same inverter.

III. Analysis of outdoor performance at Enel innovation lab Passo Martino

In this section is reported the outdoor performance evaluation of the different PV module technologies, carried out at the Enel Green Power Innovation Lab of Catania – Passo Martino.

The monitoring period goes from May 2019 to April 2020, with the exception of Bifi HJT A modules, which were installed in October 2019.

TABLE I.　Pmpp Gain After Modules Stabilization (MQT 19.1 IEC 61215-2) And Installation Date

Module type	N. of cells	N. of modules	Avg Pmpp @ t0 [W]	Avg Pmpp @ tf [W]	Gain after stab. [%]
Bifi HJT 3SUN	72	8	358.5	360.4	0.55
Bifi HJT SWCT	72	8	355.8	358.1	0.63
Bifi HJT A	60	10	306.8	307.9	0.36
Bifi PERT	60	10	301.1	301.4	0.09
Monofacial Poly p-Si	60	11	275.9	273.7	-0.78

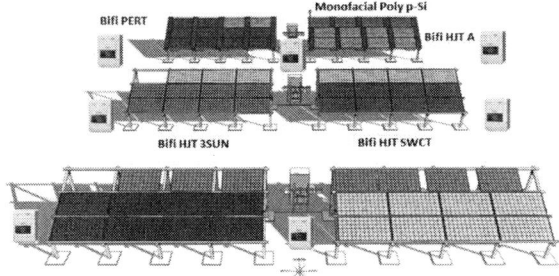

Fig. 1.　Installation final configuration

A. Performance indicators

The performance evaluation of each single technology can be carried out by following two paths: the first involves calculating the Performance Ratio (PR) [7], given by the ratio between the energy produced and the expected energy. Specifically, the PR is calculated with the total energy produced by both the front and rear sides of the module, divided only by the radiation captured by the front of the module in order to compare the performance of the single and double-sided modules. The second performance calculation system evaluates the Equivalent Operating Hours, i.e. the actual hours of production of the strings.

B. Analysis from May-19 to April-20

Fig. 2 and Fig. 3 show the average PR and EOH that go from the installation period of the various technologies involved in the test bench until the beginning of April. Fig. 4 and Fig. 5 show a comparative graph of the average gain in PR over the Monofacial Poly p-Si reference, both of Bifi HJT 3SUN and Bifi HJT SWCT technologies, respectively. The PR comparison and gain between Monofacial Poly p-Si modules and the Bifi PERT and Bifi HJT A technologies are shown in Fig. 6 and Fig. 7, respectively.

C. Module failure modes

For the arrangement of the modules and the period of their installation, at the moment there is a fairly linear performance trend, but the most reliable result can be obtained at the end of the annual cycle, where the weather conditions that influence the data so much, with good approximation will align to the initial ones and real degradation will be seen.

Installation differences affect how much the modules, in the back and in the front, perceive sunlight. In fact, when the solar angle is zero with respect to the normal of the module surface, the acquired low irradiance data become relevant. The

installation of the following modules occurred quite in the same period, May 2019:
- monofacial polycristalline p-Si
- bifacial HJT with Smart Wire
- bifacial HJT 3SUN
- bifacial PERT

while the following modules were added in October:
- bifacial HJT A

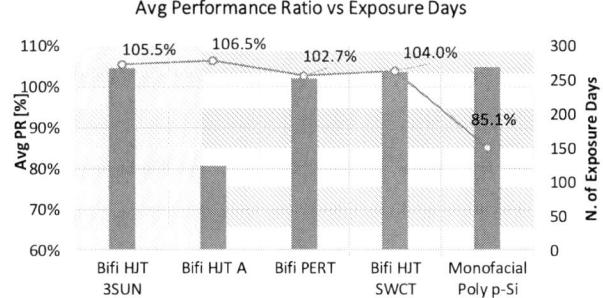

Fig. 2.　Average PR and exposures days from May 2019 to April 2020

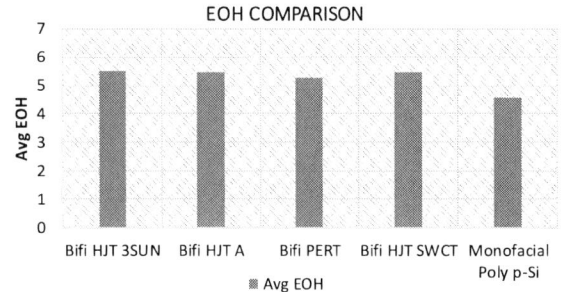

Fig. 3.　Average EOH from May 2019 to April 2020

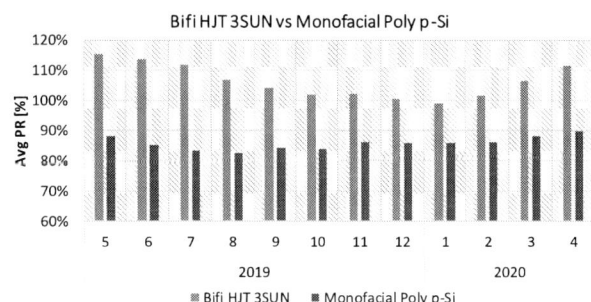

Fig. 4.　PR comparison between Bifi HJT SWCT and Monofacial Poly p-Si from May 2019 to April 2020

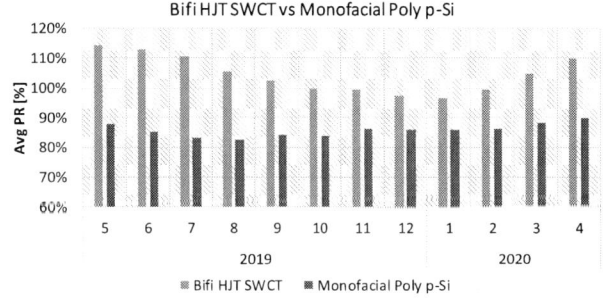

Fig. 5. PR comparison between Bifi HJT SWCT and Monofacial Poly p-Si from May 2019 to April 2020

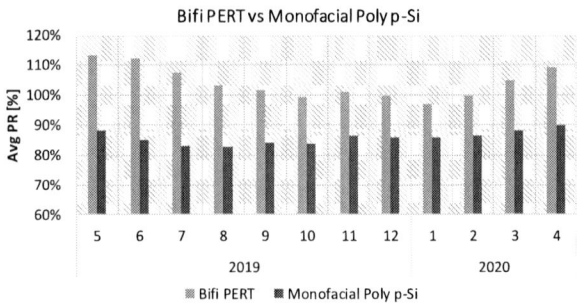

Fig. 6. PR comparison and Gain between Bifi PERT and Monofacial Poly p-Si from May 2019 to April 2020

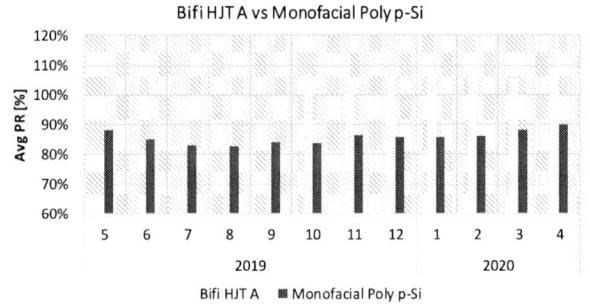

Fig. 7. PR comparison and Gain between Bifi HJT A and Monofacial Poly p-Si from May 2019 to April 2020

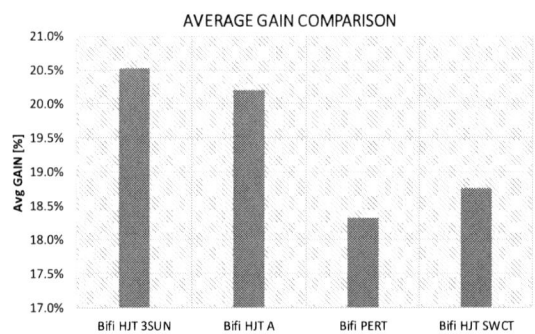

Fig. 8. Average gain of each module set vs Monofacial Poly p-Si

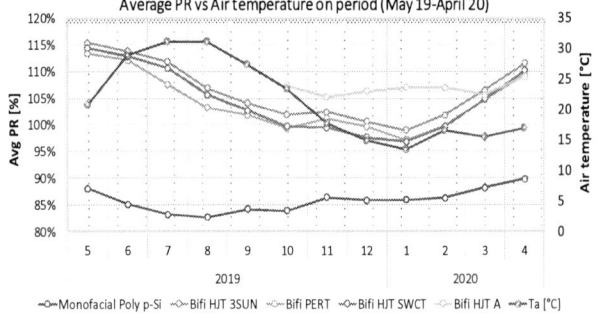

Fig. 9. Average PR in relation to the ambient temperature in the period May'19-April'20

Fig. 9 reports the average PR of all the technologies in relation to the ambient temperature. From this graph, Bifi HJT A technology appears to be the one with the highest performance. However, the lowest degradation related to its installation must be considered. Therefore, taking into account the installation timeline, it can be said that the most performing technology is HJT 3SUN.

IV. SUMMARY

Different modules technologies have been installed and monitored over almost a year, and the outdoor performance evaluation is conducted in terms of comparison the Performace Ratio parameter.

The initial goal of demonstrating an increased energy yield up to 5% in sunny regions compared to Al-BSF solar cells, and to demonstrate that bifaciality can bring energy yield in line with expectation of albedo, and module spacing (i.e. typically more 15-20% more energy than an equivalent mono-facial module) has been achieved. Indeed, the average gain of bifacial 3SUN "BEST4U Initial Generation" modules with respect to the "world widespread" monofacial reference technology is 20.5%. In conclusion, it has been successfully demonstrated the superior energy production of the 3SUN "BEST4U Initial Generation" bifacial heterojunction modules.

ACKNOWLEDGMENTS

This work has been supported by the PON Ricerca e Innovazione 2014-2020 project BEST4U, funded under Decreto Direttoriale di concessione dell'agevolazione del 21-05-2019 prot. n. 991, contract code ARS01_00519.

REFERENCES

[1] W. Favre et al, "25% efficient large area silicon solar cell: paving the way for premium PV manufacturing in Europe" in EU PVSEC, 11 September 2020

[2] G. Condorelli et al., "High Efficiency Hetero-Junction: From Pilot Line To Industrial Production," 2018 IEEE 7th World Conference on Photovoltaic Energy Conversion (WCPEC) (A Joint Conference of 45th IEEE PVSC, 28th PVSEC & 34th EU PVSEC), Waikoloa Village, HI, 2018, pp. 1970-1973, doi: 10.1109/PVSC.2018.8548197.

[3] G. Condorelli et al., "Initial Results of Enel Green Power Silicon Heterojunction Factory and Strategies for Improvements," 2020 47th IEEE Photovoltaic Specialists Conference (PVSC), Calgary, OR, 2020, pp. 1702-1705, doi: 10.1109/PVSC45281.2020.9300806.

[4] G. Razongles, L. Sicot, M. Joanny, E. Gerritsen, P. Lefillastre, S. Schroder, andP. Lay, "Bifacial photovoltaic modules: Measurement challenges",Energy Procedia, vol. 92, pp. 188–198, 2016.

[5] J. S. Stein, D. Riley, M. Lave, C. Hansen, C. Deline, F. Toor, "Outdoor Field Performance from Bifacial Photovoltaic Modules and Systems," 2017 IEEE 44th Photovoltaic Specialist Conference (PVSC), 2017, pp. 3184-3189, doi: 10.1109/PVSC.2017.8366042.

[6] H. Park, S. Chang, S. Park, W. K. Kim, "Outdoor performance test of bifacial n-type silicon photovoltaic modules," Sustainability,vol. 11, no. 22, 2019, Art. no. 6234, doi: 10.3390/su11226234.

[7] Janssen, G.J.; Van Aken, B.B.; Carr, A.J.; Mewe, A.A. "Outdoor Performance of Bifacial Modules by Measurements and Modelling," Energy Procedia2015, 77, 364–373, doi:10.1016/j.egypro.2015.07.051.

978-1-7281-6118-1/22 $31.00 © 2022 IEEE

Nanoabsorbers for Semitransparent Photovoltaics

Maximilian Götz-Köhler, Hosni Meddeb, Norbert Osterthun, Nils Neugebohrn, Kai Gehrke, Martin Vehse

German Aerospace Center (DLR) Institute of Networked Energy Systems, 26122 Oldenburg

Abstract— **Semitransparent solar cells based on ultrathin nanoabsorbers that allow partial transmission of visible light are attracting interest due to their promising applications in photovoltaic windows. To simultaneously achieve considerable visible transparency and high power conversion efficiency, proper adjustment and spectral engineering of the photoactive materials and functional layers are required. Mainly, the tuning of the thickness and the optoelectronic properties in different nanoabsorber material systems can drastically alter the solar cell device outputs. In this work, we analyze the influence of the nanoabsorber thickness from a selection of conventional semiconductor materials with different absorption coefficients on semitransparent solar cells. Using optical modeling, the design rules related to several characteristics are addressed including the average visible transmission (AVT), photocurrent generation, light utilization efficiency (LUE), color rendering index (CRI).**

Keywords— *Nanoabsorbers, transparent PV, PV windows, optical modelling*

I. INTRODUCTION

Transparent photovoltaics are increasingly attracting attention as they can open up new areas of application for solar cells such as PV windows, greenhouses and indoor PV [1, 2]. Especially for photoactive materials which cannot compete against commercially available crystalline silicon wafer technology in terms of efficiency and cost effectiveness, visibly transparent applications often become an advantageous area of study and a potential niche market [3]. Hereby, an appropriate trade-off between power conversion efficiency (PCE), average visible transparency (AVT) and color rendering index (CRI) is required to meet the specifications of a target application [2, 4]. Furthermore, the necessary reduction of nanoabsorber thickness to achieve transparency can be beneficial in terms of material consumption manufacturing cost and production throughput. In the special case of absorbers with thickness below the materials Bohr radius it is also possible to tune the bandgap of the absorber due to quantum confinement effects [5, 6]. Especially, when shaping the spectrum of transmitted light to achieve color neutrality, the capability of tuning the optical response of single layers becomes increasingly important. Therefore, the absorption coefficient α of the photoactive material in solar cell is a dominating factor for the characteristics of semitransparent PV. While α governs the optical response of the solar cell, the electron mobility and the carrier lifetime influence the electrical characteristics of the solar cell [7].

The difference in absorption coefficients of various semiconductor's absorbers allows for a wide controllability of the visible transparency and color perception of PV devices.

This work was funded by the Energy branch of the German Aerospace Center

Different studies already assess the maximum achievable PCE of transparent and semitransparent PV depending on the bandgap of the absorber by means of detailed balance analysis [2, 8], but omit realistic absorption profiles due to refractive index changes and parasitic optical losses in the surrounding functional solar cell layers. In this contribution, we study the influence of the refractive index and the thickness of ultrathin nanoabsorber layers ($t<200$ nm) in a simple semitransparent solar cell configuration on the visible transmission and light utilization efficiency (LUE). The results show that the suitable selection of the absorber can significantly increase the transparency and the local absorption of light. Furthermore, the design rules for an optimal configuration of refractive index in the nanoabsorber material are determined based on the light utilization efficiency .

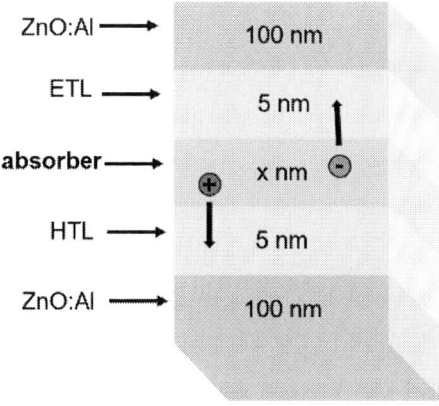

Fig. 1. Solar cell layer stack studied in this work: Front contact, electron transport layer (ETL), absorber, hole transport layer (HTL) and back electrode.

II. RESULTS & DISCUSSION

A. Absorber Materials

The application of an ultrathin deeply subwavelength ($d<\lambda/10$) absorbers allows partial transmission of light due to the reduced Lambert-Beer absorption, while providing the chance to create optical nanocavities for enhanced absorption at selected wavelength regions [9]. This work addresses the impact of the selection of different photoactive materials and their thickness, on both the absorption and the transmission of light. To simplify the analysis in this abstract, we consider a layer stack shown in Fig. 1, with a 100 nm aluminum-doped zinc-oxide (AZO) front contact. For the electron transport layer (ETL) and hole transport layer (HTL) we adopt the refractive index data of

978-1-7281-6118-1/22 $31.00 © 2022 IEEE 557

titanium dioxide (TiO₂) and molybdenum oxide (MoOₓ), respectively. Both materials provide high transparency and carrier selective conductivity to realize a working solar cell. The layer stack allows for multi-resonant absorption of light due to high changes of refractive index within the solar cell. We do not consider other absorption enhancement techniques such as Lambertian scattering in this work, but recommend other publications on this topic [6].

The photoactive materials studied in this work are amorphous germanium (a-Ge:H), gallium arsenide (GaAs), Cu(In,Ga)(S,Se)₂, the perovskite structure CH₃NH₃PbI₃ (MAPI), the polymer PTB7:PC₇₁BM and amorphous silicon (a-Si:H). It is clear that these semiconductors materials cannot be introduced into the same device structures due to the electronic, structural and morphological restrictions imposed by the functional layers and the substrates. For instance, GaAs-based solar cell devices are typically epitaxially grown, and to the best of our knowledge, no semitransparent PV have been built based on III-V absorber due to economic and technological hindrances. However, it was already demonstrated that the photoactive materials can be transferred to a glass substrate and non-epitaxial carrier selective contacts can be applied [10–12]. Nevertheless, the comparison of the optical parameters still gives us important information on how the interplay between absorption coefficient and thickness can be exploited for transparent PV. The thickness of the studied absorbers is chosen to be between 0 nm and 200 nm to analyze different subwavelength regimes.

The absorption coefficient data for each material is retrieved from literature: PTB7:PC₇₁BM from [13], GaAs from [14], CIGS from [15], CH3NH3PbI3 from [16] and a-Si:H and a-Ge:H from [17]. The absorption coefficients are shown in Fig 2. The optical simulations were performed with 1D transfer matrix method in the software package CODE/Scout and with the python package tmm v0.1.8.

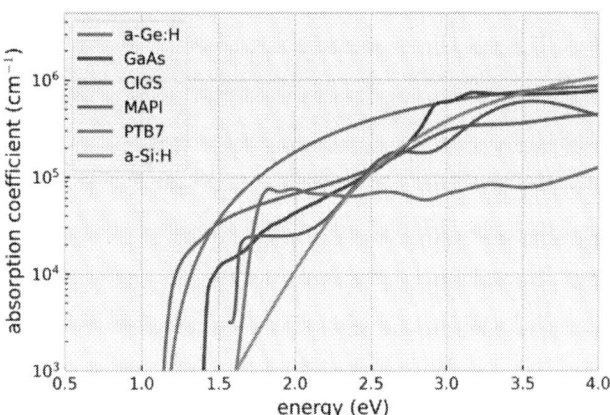

Fig. 2. Absorption coefficients of the studied materials

B. Average Visible Transmittance

The first interesting aspect to consider is the average visible transmittance (AVT) of the solar cell stack depending on the thickness of the absorber layer. The AVT is calculated as the transmitted light of the AM1.5G spectrum convoluted with the photopic response of the human eye.

$$AVT = \frac{\int_{300}^{900} T(\lambda) G(\lambda) P(\lambda) d\lambda}{\int_{300}^{900} G(\lambda) P(\lambda) d\lambda}$$

Here, $G(\lambda)$ is the radiation of the AM1.5G spectrum and $P(\lambda)$ the photopic response of the human eye. Compared to opaque solar cells, the AVT is an additional parameter which has to be evaluated in order to characterize semitransparent PV.

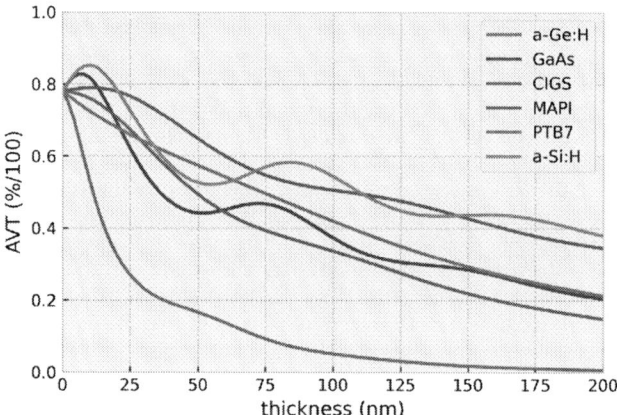

Fig. 3. Average visible transmittance (AVT) of the solar cell stacks with different absorbers. The AVT is analysed for different thicknesses of the absorber layer.

Fig. 3 shows the AVT of the solar cell stack depending on the thickness of the different absorbers. It immediately becomes clear, that less light is transmitted through the a-Ge:H layer stack than all the other configurations for all considered layer thicknesses. This is due to the high absorption coefficient and low bandgap of a-Ge:H. Overall, the AVT is decreasing with increasing absorber thickness, as expected. Materials like MAPI and a-Si:H with smaller α still reach an AVT of over 30% for absorber thicknesses of 200 nm. Thickness dependent oscillations of the AVT for a-Si:H and GaAs originate from internal resonant interference in the solar cell stack.

An AVT of at least 50% is needed for PV window applications [2]. Therefore, we take a closer look at the spectral transmittance of the solar cell stacks at an AVT of 50%. Here, we can see that most solar cell configurations have a similar transmittance spectrum, independent of their absorbance and bandgap. Only the organic absorber PTB7:PC₇₁BM stands out. Fig.4 shows the spectral transmittance together with the photopic response of the human eye and the AM1.5G spectrum. It is interesting to notice the thickness of the absorbers required to achieve an AVT of 50%: For a-Ge:H the AVT is reduced to 50% at a thickness of only 12 nm, while the thickness of MAPI and a-Si:H absorbers can be increased to above 100 nm in order to reach AVT=50% (see Table II-1).

Table II-1: Thickness of absorbers at AVT=50% and at max. LUE. Open circuit voltage (Voc) and fill factor (FF) used to calculate LUE

Material	Thickness at AVT=50% (nm)	Thickness at max. LUE (nm)	AVT at max. LUE (%)	Voc (V)	FF (%)
a-Ge:H	12	8	59	0.65	84
GaAs	36	82	45	1.12	89
CIGS	50	100	34	0.7	84
MAPI	106	128	47	0.8	86
PTB7	74	96	43	0.75	85
a-Si:H	110	88	58	0.87	87

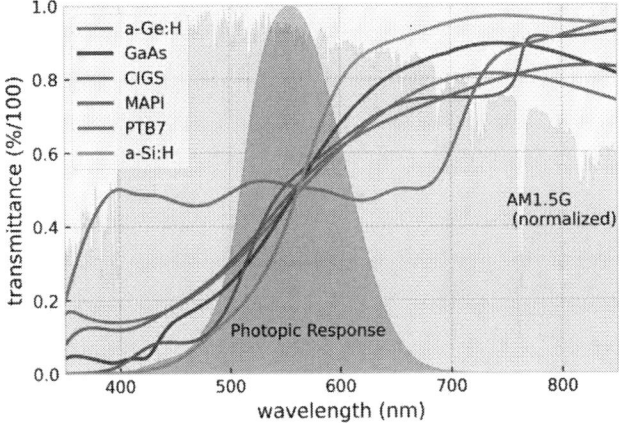

Fig. 4. Spectral transmittance of solar cells with different absorbers at an AVT of 50%

This shows that due to the differences in the refractive indices of the absorbers, the achievable transparency depends strongly on the layer's thickness. Overall, only the steep decrease in the transmittance for most absorbers within the range of the photopic response leads to a color impression of the transmitted light, which will be discussed in a later section.

Most light is transmitted in the spectral region of the color red, while blue light is only transmitted for the organic absorber.

C. Short Circuit Current

Besides the transmission of light, semi-transparent solar cells also need to absorb solar irradiance in order to generate electricity. Hereby, the absorption must take place predominantly in the semiconductor layer to achieve a high photocurrent generation. Parasitic absorption in electrode and carrier selective layers should be avoided. For the analysis of the photocurrent generation, the local absorption (a) of light in the absorber materials for varying thicknesses is calculated. From the local absorption the short circuit current can be studied using this formula:

$$J_{sc} = e \cdot \int_0^\infty \eta(E) a(E) G(E) dE$$

Here, e is the elementary charge, $a(E)$ the local absorption of light depending on the photon energy E and $G(E)$ the photon

flux of the sun. The parameter $\eta(E)$ is the photoconversion efficiency. We assume here $\eta(E) = 1$.

Fig. 5 presents the evolution of J_{sc} depending on the thickness of the different absorber materials. The two materials with the lowest bandgap, a-Ge:H and CIGS reach the highest Jsc throughout the complete parameter variation. This is due to their high absorption coefficient allowing them to absorb a higher percentage of the incoming light than the other semiconductors. All materials show a steep increase in photocurrent for the first 25 nm. Afterwards, the increase of photocurrent becomes lesser with increasing thickness. No material reaches a point of saturation within 200 nm layer thickness. This is mainly due to the remaining transparency in the NIR region close to the respective bandgaps. It is also worth noting, that the evolution of Jsc is very similar for GaAs, MAPI and PTB7, even though the absorption coefficients of these materials differ significantly from each other.

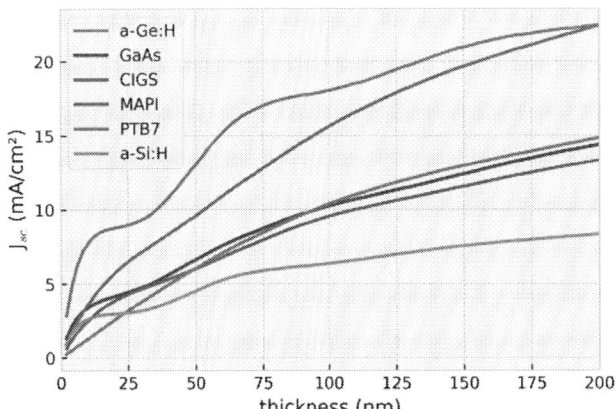

Fig. 5. Short Circuit current density of different solar cells stacks depending on absorber thickness

D. Light Utilization Efficiency

After studying the transmittance and absorbance of light for different absorbers in the solar cell stack, we want to assess the efficiency with which these materials utilize the solar irradiation. Therefore, the power conversion efficiency of the solar cell has to be estimated. From the previous results we know the maximum Jsc of the solar cell. With the Voc and FF from Table II-1 we can calculate the efficiency as product of these factors. The Voc values include radiative and non radiative losses calculated in [18]. Multiplying the PCE with the AVT caluclated before, results in the LUE:

$$LUE = AVT \cdot \frac{J_{sc} \cdot V_{oc} \cdot FF}{P_{in}}$$

Fig.6 presents the LUE over absorber thickness for the six different absorber materials in the layerstack from Fig. 1. The Figure can be seperated into three different sections. In the ultra-thin regime below 25 nm thickness, only GaAs and a-Ge:H show a comparable high LUE. While semitransparent solar cells of using GaAs with this thickness are not known yet, a-Ge:H solar cells in this thickness regime exist [1]. In the second region between 25 and 120 nm thickness, the LUE of all materials increases except for a-Ge:H. In this region, GaAs reaches the total maximum with an LUE of above 4. In the region of layer thicknesses above 120 nm, the LUE of all materials decreases, only a-Si:H, MAPI and GaAs remain at a high level, with MAPI reaching the highest LUE at 200 nm thickness.

Each material reaches its max. LUE at a different thickness, also shown in Table II-1. While a-Ge:H is already at the highest value for a thickness of only 12 nm, all other materials reach their peak LUE at thicknesses above 80 nm. The highest achievable LUE for each material lies within the range of values of LUE=2.5 to 3.5, with a-Ge:H being the exception with a max LUE of 2.3. The interference patterns from GaAs and a-Si:H, which were also present in the AVT plot (Fig.3) can be seen again in Fig. 6, where they determine the max. LUE position.

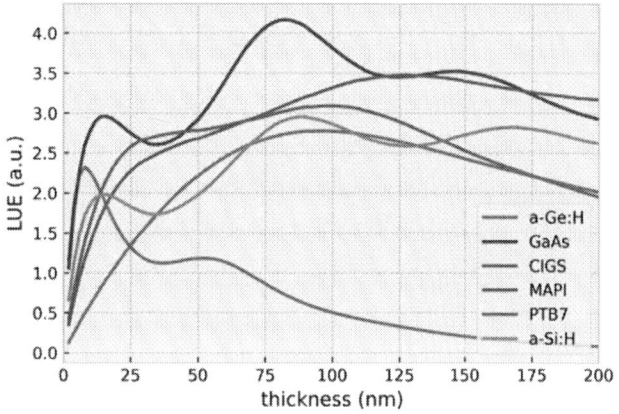

Fig. 6. Light utilization efficiency (LUE) deoending on layer thickness for different solar cell stacks

E. Color Rendering Index

To apply the studied semitransparent solar cells as PV windows, they have to achieve high color neutrality. This can be quantified by the color rendering index and with the help of color patches as shown in Fig. 7. The CRI values are calculated from the AVT spectra of the solar cells. The color patch is generated in the software package CODE/Scout and uses the D65 illumination spectrum.

Fig. 7 shows the CRI of the six different solar cells at their highest LUE. We can see that PTB7 achieves by far the highest CRI of 96, while the other materials only reach values slightly above 50. Germanium with a CRI of 74 reaches the second highest value.

III. CONCLUSION

It is striking to see, that each absorber material shows strengths and weaknesses in different categories for the application in transparent PV. Amorphous silicon has a comparable high bandgap and low absorption coefficient. This makes it a good

material to show high AVT for a broad range of different thicknesses. It is especially interesting also for its application as doped layer in ultrathin solar cells, as has been shown in different works [17]. The application of a-Si:H as nanoabsorber for ultrathin semitransparent solar cells seems to be feasible for thicknesses above 150 nm. Hereby, additional functional layers such as structural color filters could be applied to improve the color neutrality. The other amorphous material studied in this work is a-Ge:H. It has the highest absorption coefficient and a comparable low bandgap energy. This allows the material to reach a relatively high LUE at a thickness of only 12 nm, which is strikingly less than required for all other materials.

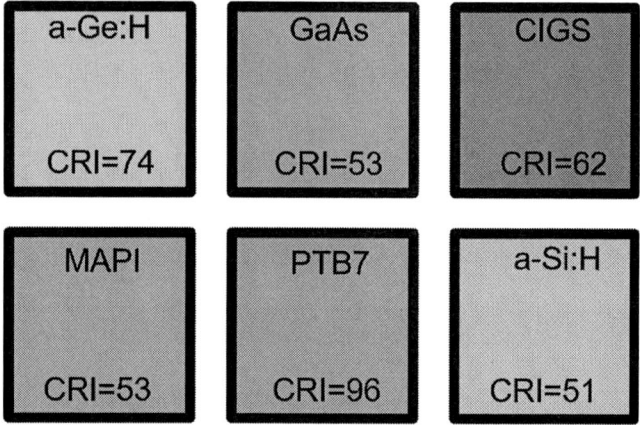

Fig. 7. Color appearance and CRI of transmitted light for all six absorber materials with the thickness corresponding to maximum LUE.

Additionally, a-Ge:H has a better color neutrality than all the other materials, except for the organic absorber. The downside of a-Ge:H is its overall low LUE, determined by the high internal losses (low V_{oc}, low FF). CIGS and GaAs are two materials, which are unlikely to be realized in the studied layer stack, but nevertheless give interesting results in how their layer thickness influences their solar cell performance. CIGS have the smallest bandgap of all studied materials. This, in combination with the high absorption coefficient leads to a very high Jsc throughout the entire studied thickness range. Between 25 nm and 50 nm layer thickness, CIGS even achieve the highest LUE of all materials due to its good combination of high AVT and PCE. GaAs on the other hand, achieves overall the highest LUE due to its high material quality (low losses, high Voc, high FF). Hereby, only the rather bad color neutrality is a downside. The organic absorber PTB7:PC$_{71}$BM reaches the highest color neutrality due to its absorption coefficient, but only has a reduced LUE. The perovskite structure studied in this work shows next to a-Si the highest AVT values and achieves the best LUE for thicknesses above 175 nm. Therefore, the material consumption will be higher compared to a-Ge:H, but also the achievable LUE is better.

Overall, the results presented in this work show, that the choice of absorber plays a crucial role for semitransparent PV. Different absorbers might be appropriate for different applications, depending on the required AVT, color neutrality or efficiency. This work gives an analysis on how these parameters can be studied and shows an overview of the advantages and disadvantages of selected absorber materials.

REFERENCES

[1] H. Meddeb, M. Götz‑Köhler, N. Neugebohrn, U. Banik, N. Osterthun, O. Sergeev, D. Berends, C. Lattyak, K. Gehrke, and M. Vehse, "Tunable Photovoltaics: Adapting Solar Cell Technologies to Versatile Applications," *Advanced Energy Materials*, p. 2200713, 2022.

[2] C. J. Traverse, R. Pandey, M. C. Barr, and R. R. Lunt, "Emergence of highly transparent photovoltaics for distributed applications," *Nat Energy*, vol. 2, no. 11, pp. 849–860, 2017.

[3] O. Almora, D. Baran, G. C. Bazan, C. Berger, C. I. Cabrera, K. R. Catchpole, S. Erten‑Ela, F. Guo, J. Hauch, A. W. Y. Ho‑Baillie, T. J. Jacobsson, R. A. J. Janssen, T. Kirchartz, N. Kopidakis, Y. Li, M. A. Loi, R. R. Lunt, X. Mathew, M. D. McGehee, J. Min, D. B. Mitzi, M. K. Nazeeruddin, J. Nelson, A. F. Nogueira, U. W. Paetzold, N.-G. Park, B. P. Rand, U. Rau, H. J. Snaith, E. Unger, L. Vaillant‑Roca, H.-L. Yip, and C. J. Brabec, "Device Performance of Emerging Photovoltaic Materials (Version 1)," *Adv. Energy Mater.*, vol. 11, no. 11, p. 2002774, 2021.

[4] C. Yang, D. Liu, M. Bates, M. C. Barr, and R. R. Lunt, "How to Accurately Report Transparent Solar Cells," *Joule*, vol. 3, no. 8, pp. 1803–1809, 2019.

[5] H. Meddeb, N. Osterthun, M. Götz, O. Sergeev, K. Gehrke, M. Vehse, and C. Agert, "Quantum confinement-tunable solar cell based on ultrathin amorphous germanium," *Nano Energy*, vol. 76, p. 105048, 2020.

[6] I. Massiot, A. Cattoni, and S. Collin, "Progress and prospects for ultrathin solar cells," *Nat Energy*, vol. 5, no. 12, pp. 959–972, 2020.

[7] P. Kaienburg, L. Krückemeier, D. Lübke, J. Nelson, U. Rau, and T. Kirchartz, "How solar cell efficiency is governed by the αμτ product," *Phys. Rev. Research*, vol. 2, no. 2, 2020.

[8] L. M. Wheeler and V. M. Wheeler, "Detailed Balance Analysis of Photovoltaic Windows," *ACS Energy Lett.*, vol. 4, no. 9, pp. 2130–2136, 2019.

[9] M. A. Kats and F. Capasso, "Optical absorbers based on strong interference in ultra‑thin films," *Laser & Photonics Reviews*, vol. 10, no. 5, pp. 735–749, 2016.

[10] N. Vandamme, H.-L. Chen, A. Gaucher, B. Behaghel, A. Lemaitre, A. Cattoni, C. Dupuis, N. Bardou, J.-F. Guillemoles, and S. Collin, "Ultrathin GaAs Solar Cells With a Silver Back Mirror," *IEEE J. Photovoltaics*, vol. 5, no. 2, pp. 565–570, 2015.

[11] V. Raj, H. H. Tan, and C. Jagadish, "Non-epitaxial carrier selective contacts for III-V solar cells: A review," *Applied Materials Today*, vol. 18, p. 100503, 2020.

[12] V. Raj, T. Haggren, J. Tournet, H. H. Tan, and C. Jagadish, "Electron-Selective Contact for GaAs Solar Cells," *ACS Appl. Energy Mater.*, vol. 4, no. 2, pp. 1356–1364, 2021.

[13] C. Stelling, C. R. Singh, M. Karg, T. A. F. König, M. Thelakkat, and M. Retsch, "Plasmonic nanomeshes: their ambivalent role as transparent electrodes in organic solar cells," (eng), *Scientific reports*, vol. 7, p. 42530, 2017.

[14] M. D. Sturge, "Optical Absorption of Gallium Arsenide between 0.6 and 2.75 eV," *Phys. Rev.*, vol. 127, no. 3, pp. 768–773, 1962.

[15] A. Loubat, C. Eypert, F. Mollica, M. Bouttemy, N. Naghavi, D. Lincot, and A. Etcheberry, "Optical properties of ultrathin CIGS films studied by spectroscopic ellipsometry assisted by chemical engineering," *Applied Surface Science*, vol. 421, pp. 643–650, 2017.

[16] A. M. A. Leguy, Y. Hu, M. Campoy-Quiles, M. I. Alonso, O. J. Weber, P. Azarhoosh, M. van Schilfgaarde, M. T. Weller, T. Bein, J. Nelson, P. Docampo, and P. R. F. Barnes, "Reversible Hydration of CH 3 NH 3 PbI 3 in Films, Single Crystals, and Solar Cells," *Chem. Mater.*, vol. 27, no. 9, pp. 3397–3407, 2015.

[17] M. Götz, N. Osterthun, K. Gehrke, M. Vehse, and C. Agert, "Ultrathin Nano-Absorbers in Photovoltaics: Prospects and Innovative Applications," *Coatings*, vol. 10, no. 3, p. 218, 2020.

[18] T. Kirchartz and U. Rau, "What Makes a Good Solar Cell?," *Adv. Energy Mater.*, vol. 8, no. 28, p. 1703385, 2018.

State of the Art of Modelling Soiling and Snow Losses in PV Systems

Sébastien ARBARETAZ[1], Murielle STEPEC[2], Eszter VOROSHAZI[2]

[1]Uni. Savoie Mont Blanc, CEA, Liten, INES, 73375 Le Bourget du Lac, France, sebastien.arbaretaz@cea.fr
[2]CEA, Liten, INES, 73375, Le Bourget du Lac, France, murielle.stepec@cea.fr, eszter.voroshazi@cea.fr

Abstract—This paper summarizes and compares the work on modelling soiling and snow losses in photovoltaic (PV) systems in the last few years. Various modelling methods and their parameters are detailed such as Markov chain model and Bill Marion's model. A comparison of the different models is made. For this purpose, accuracy and the influence of parameters is studied. Most of the models were trained and tested with data from the same location or country, thus they described the environment in which they were constructed. These models need to be trained and tested in various environment in order to evaluate their flexibility.

Keywords—soiling, dust, snow, modelling, photovoltaics

I. INTRODUCTION

The cost of photovoltaic energy has considerably decreased in recent years and is becoming more and more cost-effective. As a result, the installed PV capacity has increased significantly and countries with low irradiance and high snowfall are now investing in PV. In this context, many studies are focusing on the optimization of production and the modelling of losses due to soiling and snow.

This paper provides a state of the art of soiling and snow modelling methods. It highlights the useful information and influencing factors of the various models presented. This state of the art also provides new trends in the techniques used for the realization of even more reliable models such as artificial intelligence, Multiple Linear Regression (MLR) method and data inference taking into account the available data of the neighboring PV plants. Thus, we have tried to give an accurate representation of the main principles of soiling and snow models and we have described some of the most important factors that influence the performance of PV plants. We have also presented the new technological trends.

II. SOILING

During the last decade, a large variety of models has been developed to estimate soiling losses due to dust. Many methods are used in the publications, which were found during our research for this state of the art. The distribution of these methods is represented in Fig. 1. As we can see, the most common methods are Artificial Neural Network (ANN), MLR, physical and numerical simulation. It is noteworthy that some publications compare different methods for the same data. Each method has its advantages and its limits.

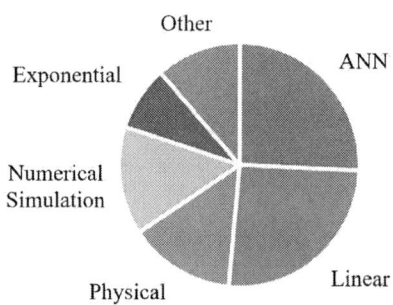

Fig. 1. Type's distribution of soiling losses models in literature during the last years.

A. Major Model to Estimate Dust Soiling Losses

Laarabi et al. [1] developed a model with an ANN methodology. The authors used irradiance, wind speed, wind direction, ambient temperature, relative humidity and rainfall as inputs. The sensitivity analysis showed that the most influential parameters were relative humidity follow by wind direction, wind speed and irradiance. Ambient temperature and rainfall had a lower influence.

Several authors compared ANN methodology with MLR methodology. Javed et al. [2] used in inputs wind speed, wind direction, relative humidity, ambient temperature, PM_{10}, frequency of wind gustiness and cumulative exposure time. They also added in inputs wind speed, PM_{10} and relative humidity of previous day. As output, they calculated a Cleanness Index with the same methodology as Guo et al. [3] described further. Hammad et al. [4] compared an ANN based model and a MLR based model to estimate the daily system conversion efficiency from exposure days and ambient temperature.

Some models have been developed with a physical approach, Guo et al. [3] developed a MLR based model using PM_{10}, wind speed and relative humidity in inputs and they calculated a Cleanness Index based on the temperature-corrected performance ratio for output. They compared it with a physical model constructed as proportional to the daily average dust deposition flux minus the daily average dust resuspension flux. It took as variables PM_{10}, wind speed, relative humidity, gravitational settling velocity of particle, threshold friction velocity of dust resuspension and kinematic viscosity of air.

978-1-7281-6118-1/22 $31.00 © 2022 IEEE

Coello et al. [5] developed a physical model and compared three deposition velocity equations. They used them to calculate the mass accumulation of dust with PM_{10}, $PM_{2.5}$ and tilt angle as inputs. Mass accumulation was related to transmittance with the equation of Hegazy [6]. They took into account rainfall by defining a cleaning threshold.

Wu et al. [7] elaborated a numerical model by computational fluid dynamics simulation. They modelled dust accumulation in terms of wind direction and wind speed for particles in the range from 10 to 100 µm. Three PV panels lined up behind each other were simulated to study dust repartition.

Xu et al. [8] defined a linear equation based on experimental data for dust accumulation depending on tilt angle and dust accumulation on a horizontal surface. Mass accumulation was then related to transmittance with the equation of Hegazy [6].

Cheema et al. [9] built a Markov chain model which take into consideration ambient temperature, solar irradiance, dust accumulation and rate of dust accumulation to model dust accumulation and power losses. The outcome of the model was virtually generated scenarios.

B. Comparison of Models' Results

The most common parameters used as inputs are ambient temperature, irradiance, wind speed, wind direction, relative humidity, PM, rainfall and exposure time. The tilt angle is often fixed so it is not taken as a parameter. However, some publications investigated soiling level for different tilt angle. Some publication also presented a sensitivity analysis to determine the degree of influence of each factor.

Accuracy of models cited above is summarize in Table I. Most of them have a good accuracy. However, they were validated with data from the same location than training. It results that current models describe the environment in which they were developed and need to be tested in various environment.

TABLE I. COMPARISON OF MODELS' ACCURACY

Publication	Model	Output	Location	Accuracy
Laarabi et al. [1]	ANN	Transmitted irradiance	Rabat, Morocco	$R^2 = 0.9286$
Javed et al. [2]	ANN and MLR	Daily Cleanness Index	Doha, Qatar	ANN: $R^2 = 0.537$ MLR: $R^2 = 0.167$
Hammad et al. [4]	ANN and MLR	Daily system conversion efficiency	Zarqa, Jordan	ANN: $R^2 = 0.892$ MLR: $R^2 = 0.864$
Guo et al. [3]	MLR and Semi-Physical	Daily Cleanness Index	Doha, Qatar	MLR: $R^2 = 0.0949$ Semi-Physical: $R^2 = 0.1774$
Wu et al. [7]	Numerical Simulation	Dust accumulation	Liverpool, England	Average error between 0.2 and 0.5%
Xu et al. [8]	Linear	Dust accumulation	Minia, Egypt	Deviation between 0.33 and 26.67%
Cheema et al. [9]	Markov chain	Power	Arizona City, USA	Average error < 5%

III. SNOW

A. Dynamic Snow Loss Model in PVSIM: Modelling Impact of Snow on PV Production

This paper presents the method implemented in SunPower's PVSim software for modelling production losses due to snow [10], annual losses due to snow are estimated at 5% in average.

The method was based on 60 PV systems among 22 sites with various technologies: tracker, rooftop, grounds mounted. Various criteria was taken into account as tilt angle, 30-years history of annual losses, meteorological data of snow on the ground from météonorm7 and free space around module. In this method, irradiance, temperature and energy produced by the converter was measured every 15 minutes for each site. The performance ratio (PR) is given by

$$PR = (P / Ir) (Ir_{max} / P_{max}) \qquad (1)$$

where P, P_{max}, Ir and Ir_{max} are the power, the maximum power, the irradiance and the maximum irradiance, respectively.

The PV module is considered free of snow if PR is higher than 50% or if there are no snow on the ground. The model of sliding snow used a threshold, critical tilt and free space around PV module. This model was evaluated by comparison of measured and modelled annual losses (RMSE = 0.9).

B. Measured and Modeled Photovoltaic System Energy Losses from Snow for Colorado and Wisconsin Locations

This paper presents Bill Marion's Snow energy loss model [11], which is often considered as a reference.

This model used historical weather data and measurements from Colorado and Wisconsin during 2011 and 2012 winters. It considered various PV systems (residential with remote support, commercial rack mounted PV power plant on flat roof and ground mounted PV power plant). It took into account, irradiance, tilt angle, daily snow depth and 30-year historical and simulated weather data in Wisconsin. This model was evaluated by two comparisons. First one was energy produced without snow versus measured energy produced and second one was measured losses versus modelled losses.

The result considered two years of measurements. For residential sector, the modelled losses were 1.5% smaller than measured ones. For nonresidential, the modelled losses had an absolute deviation of 0.5% against measurements. The standard deviation between measured and modelled losses was 10.5%.

C. Snow-Induced PV Loss Modeling Using Production-Data Inferred PV System Models (Sweden Case)

In some snow-rich regions with cold winters, PV systems suffer annual energy production losses up to 34% for some years, and monthly losses up to 100%. In this context, the modelling of snow losses is an important issue.

Current snow loss models are based on a limited number of sites, winter seasons and climatic diversity. This paper presents a method that overcomes this obstacle by using PV system performance data and freely available meteorological data [12]. This paper presents a model that uses data from power plants located around the considered plant, 263 neighbored sites in northern Sweden were considered (Fig.2). This method increased the input data and allowed increasing the accuracy.

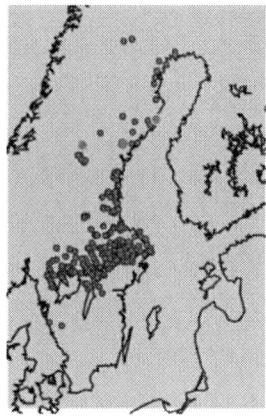

Fig. 2. Map of Sweden showing consideration sites (green) and neighbored sites (red). Reference [12]

This method took into account various data: peak power, tilt angle, azimuth, orientation, number of modules, cleaning, times series data of AC converter and snow covering pictures. The historical data were increased by using meteorological and snow cover data coming from European dataset from 1961 to 2019.

The accuracy of model was increased by using snow loss simulation (using PVlib software) and systems characteristics (tilt, azimuth, shading factors). This increasing data allowed the model to be even more accurate than Bill Marion's model.

D. Synergistic Optimization of Renewable Energy Installations through Evolution Strategy

This paper presents a model for managing the storage of renewable energy and reducing the conventional electricity generation [13]. The model is called OREES (Optimized Renewable Energy by Evolution Strategy). It finds the optimal implementation of PV panels and wind turbines.

This model used the spatiotemporal variability of wind, solar and hydropower to better align electricity production with electricity consumption and to reduce the need for storage. It took into account site criteria (slope, elevation, glacier, persistent snow cover) and historical consumption during the year.

Regarding Switzerland, PV plants play an important part, especially if they are located in mountain that offer high insolation in winter. The production of PV systems in snow environment can be increased to three times and so reduces the electricity storage need of 80%.

IV. CONCLUSION

This paper provides a review of current soiling and snow losses modelling methods for PV systems. Several models for dust and snow losses along with their parameters in input and output were described. The most common parameters were ambient temperature, irradiance, wind, relative humidity, PM, rainfall and snowfall. Tilt angle was not taken as a parameter in most of the dust soiling models but it was a major parameter for snow losses models. Methods of modelling used in publications

and main conclusions were presented. We compared the accuracy of models and the data used in input with each other. The degree of influence of factors was also studied. The main conclusion is that models were trained and validated with data from a specific location or country. A possible improvement could be to train and test models in various environment in order to evaluate their ability to work in other environment.

ACKNOWLEDGMENT

The authors would like to thank Jérémie Aimé, Ioannis Tsanakas and Eric Pilat from CEA/INES for their support in the literature research.

Part of this work has been carried out in the framework of the European H2020 project SERENDI-PV. This project has received funding from the European Union's Horizon 2020 research and innovation program under grant agreement 953016.

REFERENCES

[1] B. Laarabi, O. May Tzuc, D. Dahlioui, A. Bassam, M. Flota-Bañuelos, and A. Barhdadi, "Artificial neural network modeling and sensitivity analysis for soiling effects on photovoltaic panels in Morocco," *Superlattices Microstruct.*, vol. 127, no. 2019, pp. 139–150, 2019.

[2] W. Javed, B. Guo, and B. Figgis, "Modeling of photovoltaic soiling loss as a function of environmental variables," *Sol. Energy*, vol. 157, no. May, pp. 397–407, 2017.

[3] B. Guo, W. Javed, S. Khan, B. Figgis, and T. Mirza, "Models for prediction of soiling-caused photovoltaic power output degradation based on environmental variables in Doha, Qatar," *ASME 2016 10th Int. Conf. Energy Sustain. ES 2016, collocated with ASME 2016 Power Conf. ASME 2016 14th Int. Conf. Fuel Cell Sci. Eng. Technol.*, vol. 1, no. October 2020, 2016.

[4] B. Hammad, M. Al-Abed, A. Al-Ghandoor, A. Al-Sardeah, and A. Al-Bashir, "Modeling and analysis of dust and temperature effects on photovoltaic systems' performance and optimal cleaning frequency: Jordan case study," *Renew. Sustain. Energy Rev.*, vol. 82, no. August 2017, pp. 2218–2234, 2018.

[5] M. Coello and L. Boyle, "Simple model for predicting time series soiling of photovoltaic panels," *IEEE J. Photovoltaics*, vol. 9, no. 5, pp. 1382–1387, 2019.

[6] A. A. Hegazy, "Effect of dust accumulation on solar transmittance through glass covers of plate-type collectors," *Renew. energy*, vol. 22, no. 4, pp. 525–540, 2001.

[7] Z. Wu, Z. Zhou, and M. Alkahtani, "Time-effective dust deposition analysis of PV modules based on finite element simulation for candidate site determination," *IEEE Access*, vol. 8, pp. 65137–65147, 2020.

[8] R. Xu, K. Ni, Y. Hu, J. Si, H. Wen, and D. Yu, "Analysis of the optimum tilt angle for a soiled PV panel," *Energy Convers. Manag.*, vol. 148, pp. 100–109, 2017.

[9] A. Cheema, M. F. Shaaban, and M. H. Ismail, "A novel stochastic dynamic modeling for photovoltaic systems considering dust and cleaning," *Appl. Energy*, vol. 300, no. June, p. 117399, 2021.

[10] D. Gun, M. Anderson, G. Kimball, and B. Bourne, "Dynamic snow loss model in PVSim: modeling impact of snow on PV production," *Conference Record of the World Conference on Photovoltaic Energy Conversion*, 2018.

[11] B. Marion, R. Schaefer, H. Caine, and G. Sanchez, "Measured and modeled photovoltaic system energy losses from snow for Colorado and Wisconsin locations," *Sol. Energy*, vol. 97, pp. 112–121, 2013.

[12] M. Van Noord, T. Landelius, and S. Andersson, "Snow-induced PV loss modeling using production-data inferred PV system models," *Energies*, vol. 14, no. 6, pp. 1–19, 2021.

[13] J. Dujardin, A. Kahl, and M. Lehning, "Synergistic optimization of renewable energy installations through evolution strategy," *Environ. Res. Lett.*, vol. 16, no. 6, 2021.

Paving the Way to Building-Integrated Translucent Tandem Photovoltaics: Process Optimization and Transfer to Perovskite-Perovskite 2-Terminal Tandem Cells

David Benedikt Ritzer, Marco Alejandro Ruiz-Preciado, Bahram Abdollahi Nejand, Tobias Abzieher, Ulrich Wilhelm Paetzold

Institute of Microstructure Technology, Eggenstein-Leopoldsh, Germany

Light Technology Institute, Karlsruhe, Germany

While conventional opaque PV is hardly applicable to more than rooftops, transparent PV (TPV) promises energy harvesting at optimized average visible transmittance (AVT) and power conversion efficiency (PCE) without adversely affecting underlying facades, windows and buildings' residents. However, for successful market penetration, TPVs require an improvement in PCEs at optimized application-dependent AVTs to strengthen economic incentives. In particular, translucent PV, which is based on the segmentation of conventional opaque solar cells on transparent substrates and is thus distinguished by its technological flexibility, essential color neutrality, and ease of transparency variations, shows significantly over-proportional efficiency losses at AVTs above 20%. Furthermore, enhancement of neutral color rendering, sharpness of view and design flexibility is pivotal to increase its public acceptance. First, a high-throughput laser scribing setup is employed to micro-pattern translucent perovskite solar cells and submodules of up to 51 cm2 aperture area, enabling versatile transparent area formats and transparency variations. An in-depth analysis of electrical and optical performance using current-density-voltage-characteristics, laser-beam-induced current mapping and photoluminescence measurements result in optimal scribing parameters and transparent area formats, mitigating over proportional losses even at AVT levels above 20%. The resulting optimized translucent PV devices exhibit PCEs of up to 16.2% and 8.0% at 8.1% and 38.6% AVT, respectively. Varying transparent area formats and their spatial distribution, ideal designs regarding perception are validated by characterization of view through images regarding color rendering via UV/VIS-photospectroscopy and optical distortion via blind/referenceless image spatial quality evaluator algorithm (BRISQUE). Color rendering indices of up to 94 proof color neutrality at high perceptual quality of the corresponding view through images, ensuring a colorfast and crisp view for inhabitants when later applied in facades. Finally, the optimized micro-patterning process is transferred for the first time to 2-terminal perovskite-perovskite tandem cells, demonstrating feasibility as well as the future potential of high-efficiency translucent tandem PV by exploitation of their elevated Shockley-Queisser limit.

Terawatt-Scale Photovoltaics Enabled by Technological Learning

Lukas Wagner, Robert Pietzcker, Lorenz Friedrich, Jan Christoph Goldschmidt

1Fraunhofer Institute for Solar Energy Systems, Freiburg, Germany

Potsdam Institute for Climate Impact Research, Potsdam, Germany

Philipps-University Marburg, Marburg, Germany

Cost efficient climate change mitigation requires installing up to 170 TWp photovoltaic (PV) electricity production capacity until the year 2100. The question is, whether and how such growth is possible from a resource perspective. We have assessed the resource demand of such multi-TW-scale PV. Given the long time scale of our analysis, we did not focus on any specific technology. Instead, we looked at the fundamental resources energy, float-glass, and capital investments that will be necessary independently from which PV technology will dominate in the future. In our analysis, we considered via a learning rate approach that PV technology is continuously improving. Conversion efficiency is increasing, while cost, and energy consumption during production are continuously decreasing. We found that without further technological learning, serious resource constraints will limit the growth of PV industry. On the other hand, continued technological learning at current rates would enable rapid growth within reasonable boundaries of resource demand. With such technological learning, energy demand for production will correspond to 2-5% of global energy consumption leading to cumulative greenhouse gas emissions of 4-11% of the 1.5°C emission budget. Glass demand will exceed current float-glass production, requiring rapid capacity expansion. Installations costs would be in the range of 300-600 billion $US2020 per year. Technological solutions enabling such learning are foreseeable. Especially perovskite-based tandem solar cells promise to reach efficiencies, energy, and costs targets that allow for staying on the development paths obtained from extrapolating current learning rates. The specific material demands, however, of such technologies need to be analyzed carefully and the development steered towards using abundant and non-toxic materials to reach real sustainability.

978-1-7281-6118-1/22 $31.00 © 2022 IEEE

Effects of solar spectrum and albedo on the performance of bifacial Si heterojunction mini-modules

Marco Leonardi[1,2], Roberto Corso[1,2], Andrea Scuto[1], Gabriella Milazzo[1], Carmelo Connelli[3], Marina Foti[3], Cosimo Gerardi[3], Fabrizio Bizzarri[4], Stefania M. S. Privitera[1] and Salvatore A. Lombardo[1]

1 National Research Council, Institute for Microelectronics and Microsystems (CNR-IMM), Catania, 95121, Italy
2 Department of Physics and Astronomy, University of Catania, Catania, 95123, Italy
3 Enel Green Power, Catania, 95121, Italy
4 Enel Green Power, Rome, 00198, Italy

Abstract—We compare the dependencies of performance of bifacial and monofacial mini-modules on the impinging solar radiation intensity, spectrum and the ground albedo. Monofacial and bifacial mini-modules realized with n-type Si heterojunction solar cells were tested in outdoor conditions. Two different types of ground, with a low and high albedo, have been used. By studying the correlation between short circuit current and power at the maximum power point of the modules with the solar radiation intensity, spectrum and with the ground albedo, we provide understanding on the measured peculiar time evolution of the bifacial module performance along the daytime.

Index Terms—Bifacial solar cells, Heterojunction, Solar spectrum

I. INTRODUCTION

Bifacial photovoltaics (BPV) represents a compelling technology for the future development of photovoltaics (PV). Bifacial solar cells simultaneously collect incident solar radiation from the sun on the front side, like conventional monofacial PV (MPV), and the albedo radiation from the ground on the back side. Therefore, compared to the standard MPV, BPV can effectively increase the energy yield of PV modules and strings at reduced costs. Several BPV technologies are available on the market. These include the passivated emitter rear contact (PERC) cells, the passivated emitter rear totally diffused (PERT) cells and the silicon heterojunction technology (SHJ) cells. N-type SHJ cells are a promising technology for BPV due to their high short-circuit current (I_{sc}), open-circuit voltage (V_{oc}) and bifaciality factor (BF) compared to the most common p-type cells [1]. A significant parameter to be taken into account to optimize the bifacial module (BM) installation is the ground albedo. Reference [2] studied how to improve the power output of BMs compared to monofacial module (MM) installations by choosing different terrain materials. Additionally, the tuning of rear passivation can enhance the responsivity to infrared light [3], which is efficiently reflected by most terrains, such as common asphalt

[4], and this effect is further amplified in low irradiance conditions [5]. Reference [6] reported that BMs produce more than MMs in the early and late hours of the day, when the sun is approximately parallel to the plane of the modules, whilst around midday the difference is negligible, as in the early morning and late afternoon little direct light can reach the modules and all energy is generated by indirect light, which is most favorable for BMs.

We have recently proposed and experimentally validated a numerical 3D model which calculates the performance of bifacial PV modules and strings, able to evaluate temperature, I-V characteristics, and power at the MPPT of the PV modules, based on the evaluation of the impinging radiation collected by the front, as in the conventional monofacial cells, and by the back of the cells, coming from the ground albedo [7]–[9]. This model has been validated on experimental data and applied for evaluating the energy yield of BPV systems. We have shown that the increase of energy yield in the case of bifacial operation is mainly due to the higher short-circuit current. Moreover, the bifaciality advantage increases in PV technologies with lower (in absolute value) temperature coefficients, as it is in the case of Si HJT modules [9].

With the expected expansion of bifacial solar farms in the following years, it is crucial to understand the complexity of the parameter combination on the output of BMs, which arises from the combined effects of light irradiation on the front and back side of the module, spectral response and efficiency losses from thermal effects. These aspects have to be taken into account for a realistic assessment of BPV regarding the location and technology specifications. In this work, we present an experimental study focusing on solar spectrum dependence and direct comparison of monofacial and bifacial modules. Given their low temperature coefficient, we study the case of SHJ solar cells. We experimentally evaluated and compared under the same outdoor conditions two mini modules of three solar cells operating in monofacial and bifacial mode, determining the main dependence on background radiation and solar spectrum.

The present research was, in part, funded by PON Ricerca e Innovazione 2014–2020, under Decreto Direttoriale di concessione dell'agevolazione del 21 May 2019 Prot. N. 991, Contract Code ARS01_00519, BEST-4U Project.

II. Experimental

Two identical mini-modules of 3 bifacial n-type SHJ cells in series were prepared. These SHJ cells, developed by Enel Green Power at its 3SUN facility, show a V_{oc} of 730 mV, an I_{sc} of 9.3 A, a BF of 90%, and a power conversion cell efficiency of 22.7% under standard test conditions. One of the modules was covered with white cardboard on the back to completely prevent the albedo light collection.

The mini-modules were tested in outdoor conditions in Catania, Italy (37°26' N, 15°4' E). The data were collected on sunny days with no clouds in February 2021. An image of the mini-modules during the experiments is shown in Fig. 1. The module height (55 cm) and orientation (35° inclination from the ground, modules facing south) were chosen to optimize solar radiation and ground albedo collection, according to our previously developed model [4], [7]–[9].

The solar irradiation was measured by using an MS-40 EKO pyranometer. The pyranometer was tilted at 35° like the mini-modules. The solar spectrum was monitored with a spectrophotometer. The current-voltage (I-V) characteristics of both mini-modules have been measured in a Kelvin configuration simultaneously and independently. The current is evaluated by the voltage drop across a known resistor in series to the MM and the BM, respectively. The voltage across the mini-module is swept by a variable load, obtained by a powerMOS driven from OFF to ON condition. All the sensors' data were acquired through a National Instrument data logger. Two conditions of ground albedo were studied: asphalt and a white plastic sheet (WPS), as shown in Fig. 1.

III. Results and discussion

A. Irradiance effect on I_{sc} and power output

Fig. 2a and Fig. 2b show cumulative data of I_{sc} collected over the measurement period as a function of irradiance and daily time for MM and BM on WPS, respectively. I_{sc} increases with the solar irradiance, but the MM has lower values under the same solar irradiance, with a difference of about 2 A. Besides, while the MM shows no difference in I_{sc} between morning and afternoon, the BM is characterized by higher I_{sc} values in the afternoon; this effect is even more visible with the WPS owing to its high albedo.

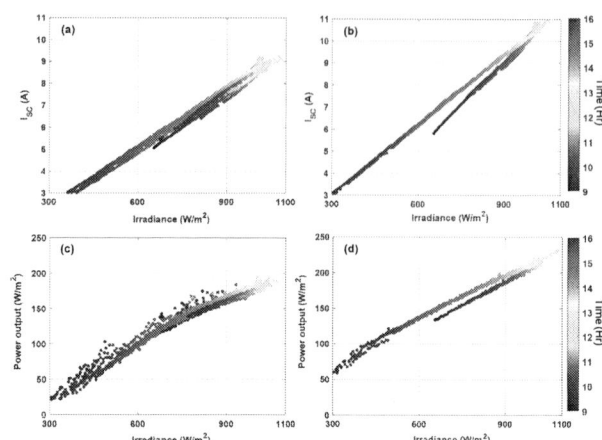

Fig. 2. Isc as a function of irradiance for MM (a) and BM on WPS (b); Power output as a function of time for MM (c) and BM on WPS (d).

Fig. 2c and Fig. 2d show the power output as a function of irradiance for MM, BM on WPS. As expected, the power output of the MM is lower than the BM output. Similarly to what has been observed for the I_{sc}, the BM power versus irradiance graphs shows the same difference in values between morning and afternoon. A noticeable drop in power output for the MM in the afternoon is observed. This is due to the sunlight's low angle of incidence. On the contrary, the BM harvests the albedo irradiance on its rear face, therefore producing more power and for a longer time.

We propose that the I_{sc} difference results from the cooperation of two effects: the ability of bifacial modules to collect diffused light from the ground and the behavior of the solar spectrum during the day.

B. Albedo influence on module performance

To investigate the first contribution, we have measured the albedo, for this particular case defined as the ratio of the photocurrents detected by two calibrated silicon solar cells, one facing up, towards the sky, oriented like the PV modules, and the other exactly parallel, but facing down, seeing the ground. Data collected during one day are reported in Fig. 3, showing that the ratio is higher in the afternoon. This effect

Fig. 1. Picture of the experimental setup.

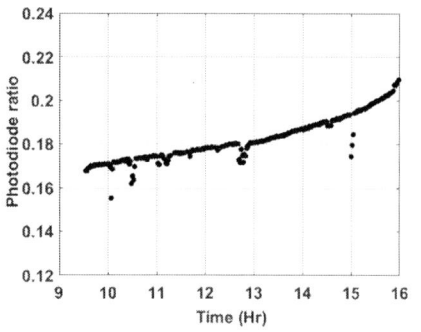

Fig. 3. Ratio of the intensities registered by the photodiodes over one day.

arises from the consideration that the reflectivity of asphalt is prominently high in infrared region [4], and the solar spectrum shifts to a higher wavelength in the afternoon. Therefore, the light impinging on the back side of the BM in the afternoon is higher than at midday and this causes the BM to produce higher currents in the afternoon given the same frontal irradiance, as observed.

C. Solar spectrum evolution

As previously anticipated, we have measured the time evolution of the solar spectrum during the daytime. Fig. 4 shows how the central spectrum wavelength (CSW), calculated as the average wavelength weighted by the solar spectrum intensity, increases in the afternoon. There is a clear variation of the solar spectrum shape during the day. The trend is almost symmetrical with the center at noon but shows a rapid increase after 3 pm. The CSW reaches its maximum of 728 nm at 4 pm, with a difference of over 40 nm compared to noon.

The redshift is expected to increase efficiency based on the responsivity of the SHJ cells. Fig. 5 shows the responsivity of the front and back side of a single SHJ cell. The produced current per impinging watt rises with the radiation wavelength and it reaches its maximum at 1030 nm. These data explain the increase of module efficiency in the afternoon, when more infrared photons contribute to the module output, as demonstrated in Fig. 4. The effect is even more robust in the BM than the MM because the former absorbs light from both sides of the cells.

IV. CONCLUSIONS

In this work, we have shown and discussed the results of outdoor tests on two PV mini-modules based on Si hetero-junction technology.

We have continuously monitored the behaviour of the modules from the open- to the short-circuit condition. This has allowed us to monitor the V_{oc} and I_{sc} values and the maximum power point region for each operation cycle. Using this approach, we have directly compared monofacial and bifacial operations, evaluating the correlation between the PV electrical performance and the solar irradiance, albedo and solar spectrum.

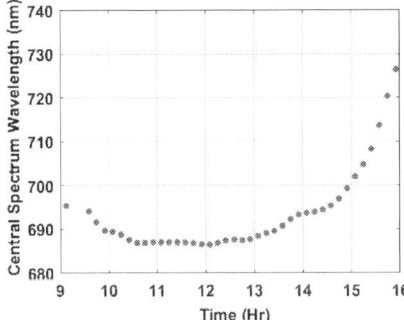

Fig. 4. Mean central wavelength of the solar spectrum during the day.

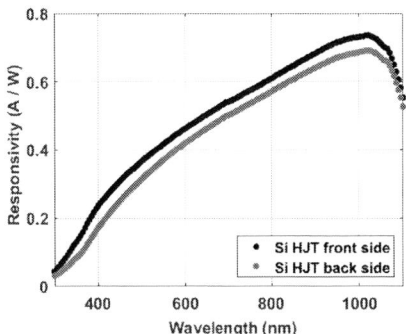

Fig. 5. Spectral responsivity of a single Si HJT solar cell.

The results reveal that the observed increase in power output in the case of bifacial systems is mainly due to the higher short-circuit current primarily influenced by higher albedo in the afternoon and higher responsivity of the cells to infrared light. The results confirm the effectiveness of bifaciality in improving the power output, proving that this technology plays a crucial role throughout the whole day, both in low and high irradiance conditions.

REFERENCES

[1] B. Yu, D. Song, Z. Sun, K. Liu, Y. Zhang, D. Rong, L. Liu, A study on electrical performance of N-type bifacial PV modules, Sol. Energy. 137 (2016) 129–133. https://doi.org/10.1016/j.solener.2016.08.011.

[2] G.J.M. Janssen, B.B. Van Aken, A.J. Carr, A.A. Mewe, Outdoor Performance of Bifacial Modules by Measurements and Modelling, Energy Procedia. 77 (2015) 364–373. https://doi.org/10.1016/j.egypro.2015.07.051.

[3] R. Hezel, Novel Applications of Bifacial Solar Cells, Prog. Photovoltaics Res. Appl. 11 (2003) 549–556. https://doi.org/10.1002/pip.510.

[4] S.M.S. Privitera, M. Muller, W. Zwaygardt, M. Carmo, R.G. Milazzo, P. Zani, M. Leonardi, F. Maita, A. Canino, M. Foti, F. Bizzarri, C. Gerardi, S.A. Lombardo, Highly efficient solar hydrogen production through the use of bifacial photovoltaics and membrane electrolysis, J. Power Sources. 473 (2020) 228619. https://doi.org/10.1016/j.jpowsour.2020.228619.

[5] B. Robles-Ocampo, E. Ruíz-Vasquez, H. Canseco-Sánchez, R.C. Cornejo-Meza, G. Trápaga-Martínez, F.J. García-Rodriguez, J. González-Hernández, Y. V. Vorobiev, Photovoltaic/thermal solar hybrid system with bifacial PV module and transparent plane collector, Sol. Energy Mater. Sol. Cells. 91 (2007) 1966–1971. https://doi.org/10.1016/j.solmat.2007.08.005.

[6] B.B. Van Aken, M. Jansen, A.J. Carr, G.J.M. Janssen, A.A. Mewe, Relation Between Indoor Flash Testing and Outdoor Performance of Bifacial Modules, 29th Eur. Photovolt. Sol. Energy Conf. Exhib. (2014) 2399–2402. https://doi.org/10.4229/EUPVSEC20142014-5CO.16.1.

[7] 10. Galluzzo, F.R.; Canino, A.; Gerardi, C.; Lombardo, S.A. A new model for predicting bifacial PV modules performance: First validation results. In Proceedings of the 2019 IEEE 46th Photovoltaic Specialists Conference (PVSC), Chicago, IL, USA, 16–21 June 2019; pp. 1293–1297. https://doi.org/10.1109/pvsc40753.2019.8980925

[8] 11. Galluzzo, F.R.; Zani, P.E.; Foti, M.; Canino, A.; Gerardi, C.; Lombardo, S. Numerical Modeling of Bifacial PV String Performance: Perimeter Effect and Influence of Uniaxial Solar Trackers. Energies 2020, 13, 869. https://doi.org/10.3390/en13040869.

[9] Leonardi, M.; Corso, R.; Milazzo, R.G.; Connelli, C.; Foti, M.; Gerardi, C.; Bizzarri, F.; Privitera, S.M.S.; Lombardo, S.A. The Effects of Module Temperature on the Energy Yield of Bifacial Photovoltaics: Data and Model. Energies 2022, 15, 22. https://doi.org/10.3390/en15010022

Role of back-side indium tin oxide on the degradation mechanism of silicon heterojunction solar cells

[1,2]Gbenga D. Obikoya, [1,2]Anishkumar Soman, [1,2]Ujjwal K. Das, [1,2]Steven S. Hegedus

[1]Department of Electrical and Computer Engineering, University of Delaware, Newark, DE 19716, USA

[2]Institute of Energy Conversion, University of Delaware, Newark, DE 19716, USA

Abstract— **Silicon heterojunction (SHJ) solar cells with only front-side indium tin oxide (ITO) and those with both sides ITO were subjected to accelerated degradation in argon ambient, at temperature of 90°C and illumination intensity of 100mW/cm². Between 0 and 500 hours of exposure, the cells were removed iteratively for current density-voltage (JV), quantum efficiency (QE) and Suns-V_{OC} measurements. While cells with both sides ITO demonstrated stability in performance after 500 hours of exposure, the cells without the rear-side ITO degraded in performance, with an average open-circuit voltage (V_{OC}) loss of about 12% and an average efficiency (η) loss of about 16%.**

Keywords—silicon heterojunction (SHJ), accelerated thermal stress (ATS), both-side ITO (BSI), front-side ITO (FSI), degradation

I. INTRODUCTION

Silicon heterojunction (SHJ) solar cell has attracted a significant attention in recent years due to excellent surface passivation quality which offers a remarkably high open-circuit voltage (V_{OC}) up to about 750 mV [1] and a record efficiency over 26% [2], Furthermore, the low temperature coefficient associated with the V_{OC} in SHJ solar cells compared to the diffused homojunction solar cells [3] makes them a better choice for installation in high temperature climates. In such hot climates, the SHJ solar cells can provide 10% power increase relative to the c-Si homojunction solar cells [4].

However, any device structure with an a-Si layer could raise a concern of long-term stability and reliability. Indium tin oxide (ITO) layer deposition condition such as varying thicknesses has been reported to affect the passivation quality of a-Si/c-Si heterojunction solar cell [5], [6], indicating the role that the ITO plays on the recombination mechanism at the a-Si/c-Si interfaces and consequently, on the stability of SHJ cells over time.

The market share of SHJ solar cell in the solar energy market is expected to rise to 10% by 2024 and to 17% by 2030 and more so, the module for special environment such as tropical climates and deserts may account for about 25% around the same period [7]. Therefore, research into the long-term stability of SHJ solar cells is of utmost importance in one part to provide insight into how to mitigate the potential loss mechanism and in other part to make them competitive with Si homojunction counterparts which have a lifetime of about 25 years under field operation.

In this paper, SHJ solar cells with no back-side ITO but only front-side ITO (FSI) and those with front- and back-sides ITO i.e. with both sides ITO (BSI), as shown in Fig. 1(a) and (b) respectively, were investigated for possible degradation under stress temperature of 90°C, at 1-Sun illumination intensity and in argon ambient for 500 hours. The BSI solar cells were almost stable while the FSI cells degraded in electrical performance, indicating that the cause of degradation can be ascribed to lack of rear-side ITO in SHJ solar cells.

II. EXPERIMENTAL SET UP

A. Solar cells fabrication

Front junction SHJ solar cells (both FSI and BSI) were fabricated with a-Si deposited on textured 150μm thick n-type Czochralski wafer using a plasma-enhanced chemical vapor deposition (PECVD) reactor. The complete deposition process and device fabrication steps are described elsewhere [8].

The ITO with thickness of 80nm was sputtered unto the front sides of the FSI cell and on both sides of the BSI cell. Four cells of area of 0.56 cm² were defined by ITO shadow mask on the front of 1 sq inch silicon pieces. The rear ITO covered the back surface and was larger than the cell area. The front and the back metal contacts were formed on both the FSI and the BSI cells using electron-beam evaporation. Cells were annealed for 15 minutes at 150°C after metallization. The complete solar cell structures are shown in Fig. 1(a) and (b) for FSI cell and BSI cell respectively.

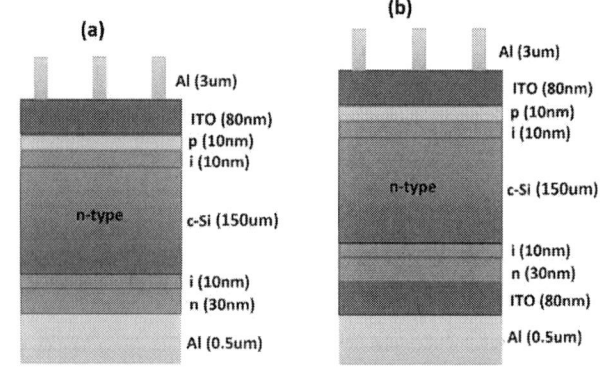

Fig. 1. SHJ cell structures (a) FSI – with no back-ITO (b) BSI – with back-ITO

B. Accelerated thermal stress experiment

Total of eight unencapsulated SHJ cells (four FSI and four BSI) were subjected to same stress conditions, with temperature of about 90°C at 100 mW/cm² illumination intensity in argon ambient for 500 hours, inside a test chamber designed and constructed at the Institute of Energy Conversion (IEC). The chamber underwent several pump-purge cycles with argon backfill to reduce oxygen and moisture before stressing. The initial performances of the stressed solar cells are shown in Table I.

TABLE I. INITIAL AVERAGE JV PERFORMANCE OF SHJ SOLAR CELLS

# of Cell	Cell Area (cm²)	ITO Condition	Initial J-V Performance			
			V_{OC} (V)	J_{SC} (mA/cm²)	FF (%)	η (%)
4	0.56	FSI (no back ITO)	661 ± 3	34.5 ± 0.3	75.4 ± 0.5	17.2 ± 0.1
4	0.56	BSI (both front and back ITO)	671 ± 4	35.8 ± 0.4	72.8 ± 0.6	17.5 ± 0.3

C. Characterizations

Current density-voltage (J-V), quantum efficiency (QE) and Suns-V_{OC} measurements were conducted on all cells at 0 hour, before they were loaded into the stress chamber. Between 0 and 500 hours, the cells were removed 4 times and ex-situ characterizations were repeated.

The dark and light J-V measurements were carried out using a class AAA solar simulator from Optical Association Inc. (OAI). The JV scans were collected from reverse-bias to forward-bias voltage direction, both in the dark and in the light at ambient temperature using a 4-point probe system.

The QE measurements were conducted in the dark with no voltage bias using a QE set up in conjunction with 4-point probe system. The QE set up consists of a 200W quartz tungsten halogen projector lamp, a monochromator, a light chopper and collimating lenses.

Suns-V_{OC} characterizations were conducted using Sinton Instruments equipped with xenon flashlamp, fine-point voltage probe, and photodiode. Suns-V_{OC} measurements were obtained both under a blue light (using 472 nm band-pass filter) and a red light (using long-pass filter with a cutoff wavelength of about 723 nm).

III. RESULTS AND DISCUSSION

The light J-V parameters versus the stress time are shown in Fig. 2(a) – (d). BSI cells were observed to be nearly stable while the FSI cells degraded in performance, dominated by V_{OC} degradation, during the accelerated thermal stress (ATS) experiment. Each data point is the average of 4 cells.

For BSI cells in Fig. 2, the average V_{OC} was observed to improve by about 0.8% for the first 65 hours and average fill factor (FF) improved by about 1.2% for the beginning 25 hours of stress duration. Some authors ascribed the initial increase in performance to decrease in the density of recombination-active

interface states, which has been observed in SHJ modules some hours after they were installed to operate in the field. [9], [10].

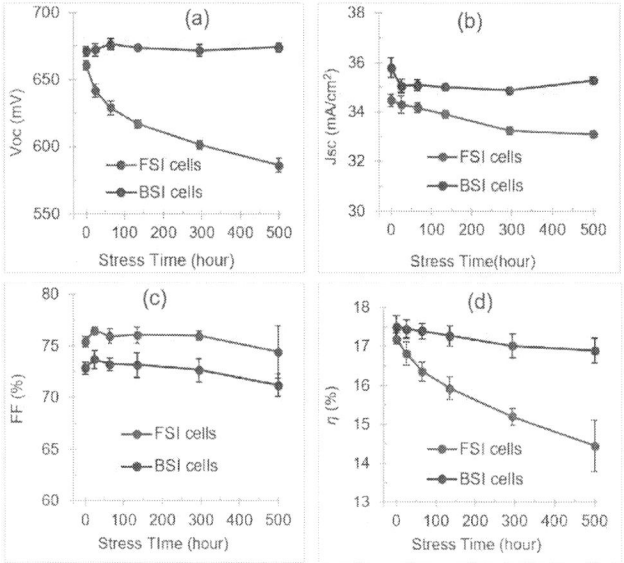

Fig. 2. Performance of FSI cells (blue curves) and BSI cells (red curves) exposed to ATS in argon ambient environment at 90°C under illumination intensity of 100mW/cm² (a) V_{OC} vs. stress time (b) J_{SC} vs. stress time (c) FF vs. stress time (d) η vs. stress time

For the FSI solar cells, Fig. 2(a) shows that the V_{OC} degradation kinetics was higher at the beginning of the stress experiment and later slowed down over time. After 500 hours under ATS at 90°C, as summarized in Table II, significant degradation was observed in FSI cells while the BSI cells displayed almost stable performance.

TABLE II. RELATIVE CHANGE IN PERFORMANCE AFTER 500 HOURS OF LIGHT SOAKING

Cell Type	Relative ΔV_{OC}	Relative ΔJ_{SC}	Relative ΔFF	Relative $\Delta\eta$
BSI	+0.5	-1.5%	-2.2%	-3.4%
FSI	-12%	-4.0%	-1.3%	-16%

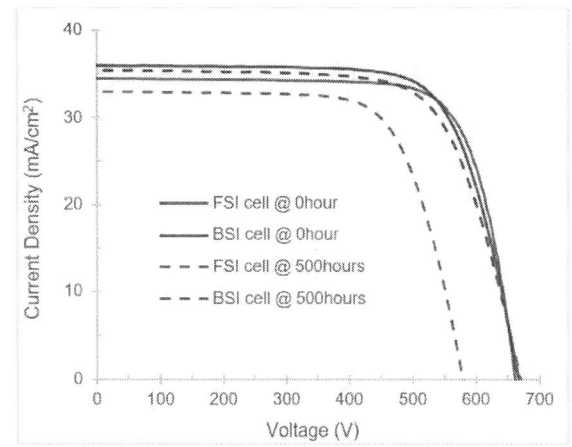

Fig. 3. JV curves taken at 0 and 500 hours of ATS experiment in argon ambient at 90°C under 1-Sun illumination intensity for FSI cells (blue curves) and BSI cells (red curves)

The relative change in J_{SC} and V_{OC} for FSI and BSI cells can also be seen in Fig. 3, for current density versus voltage curves at 0 and 500 hours. Slight drop was observed in J_{SC} for both FSI and BSI cells, while significant drop in V_{OC} was observed only in FSI cells. However, the FF appeared to be less sensitive to ATS. Therefore, the negligible gain in V_{OC} with slight drop in J_{SC} resulted in almost stable efficiency for BSI cells, whereas significant drop in V_{OC} with drop in J_{SC} led to significant drop in efficiency for FSI cells.

Suns-V_{OC} measurements under blue and red lights, as seen in Fig. 4(a) and (b), were obtained to gain an insight into region where passivation loss is more dominant in SHJ solar cells. Before the ATS experiment at 0 hour for both cell cases, Suns-V_{OC} curves showed no strong dependency on wavelength i.e. the blue curve/blue light and red curve/red light as seen in Figure 4(a) and (b) are overlapping at 0 hour. This serves as an indication of uniform passivation both at the front and the back side of the SHJ cells.

After 500 hours of ATS experiment, Fig. 4(a) shows the stable passivation quality, which is inferred from the shape of the curve, for the BSI cells. In this cell, the same illumination intensity from the xenon flash lamp produced the same V_{OC} at either 0 hour or at 500 hours, with both blue and red lights. This is an indication of stable passivation quality and stable recombination mechanism.

However in FSI cells, there was a change in shape and shift in V_{OC} which denote loss of passivation over time, as seen in Fig. 4(b). Here, the same illumination intensity from the xenon flash lamp produced higher V_{OC} at 0 hour than at 500 hours, and this is an indication of increased in trap density in the SHJ cells resulting from loss of passivation quality over time. The V_{OC} decreased with red light more than with blue light, indicating passivation loss is more prominent at the rear side of the cell.

The drop in J_{SC} can further be understood by considering the QE profiles taken under zero voltage bias in the dark, as shown in Fig. 5. Higher QE response was observed in BSI cell than in FSI cell. As previously detailed by Ahmed *et. al.* [11], the rear-side ITO/Al in BSI cells served as a better back reflector than the Al alone in FSI cells. This enhances optical absorption of long wavelength (>950 nm) photons, as evident by higher QE at 0 hour in BSI cell compared to FSI cell. Since FSI cell lack rear-side ITO layer between the amorphous layers and the aluminum back contact, parasitic absorption occurs whereby long wavelength photons reaching the rear side were not reflected internally back into the solar cells but rather got absorbed into the back metal contact. This causes decrease in QE at the longer wavelength, and consequently led in J_{SC} loss for FSI cells at 0 hour.

After 500 hours of stress experiment, further reduction in QE response at longer wavelength for FSI cell was observed in Fig. 5. The QE reduction is likely due to recombination loss caused by the reduced passivation quality over time. Further study is needed to confirm the exact cause of J_{SC} loss. However, the QE response, which remained almost constant after 500 hours of ATS experiment, can be linked to the presence of rear-side ITO between the metal back contact and the amorphous layer in BSI cell.

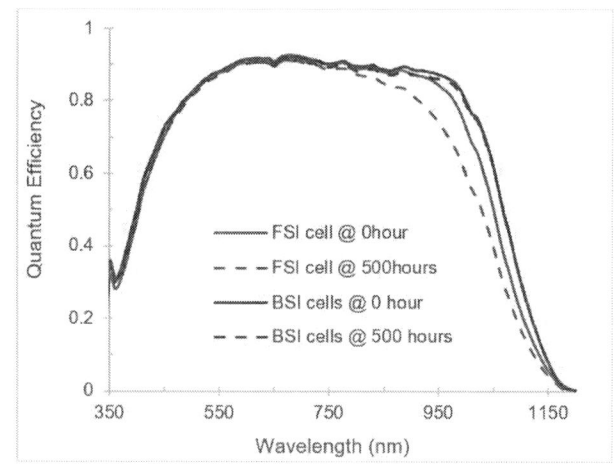

Fig. 5. QE measurements taken at 0 and 500 hours of ATS experiment in argon ambient at 90°C under 1-Sun illumination for FSI cells (blue curves) and BSI cells (red curves).

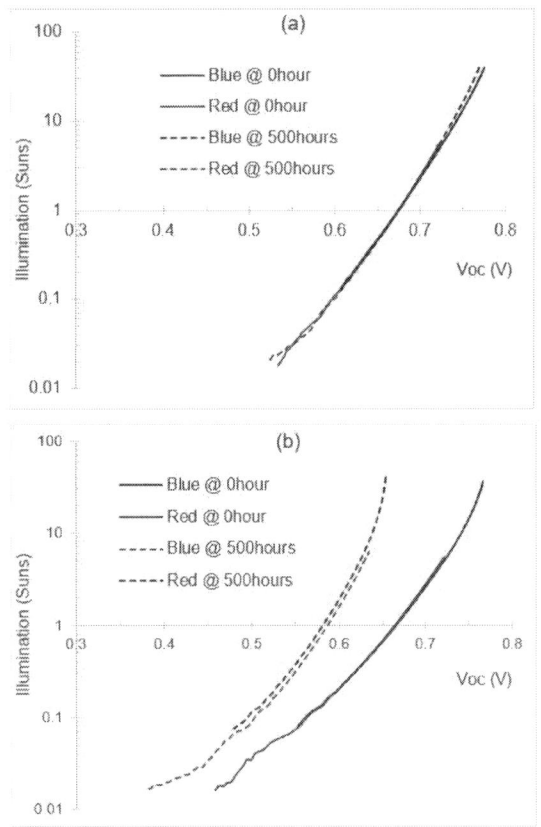

Fig. 4. Suns-V_{OC} measurement taken with red and blue lights (red and blue curves, respectively) at 0 and 500 hours of ATS experiment in argon ambient at 90°C under 1-Sun illumination intensity (a) BSI cell – no loss of passivation quality and stable V_{OC} (b) FSI cell – loss of passivation quality and V_{OC} loss

IV. SIGNIFICANCE

Understanding the cause of instability in SHJ solar cells will facilitate the possible solution needed to mitigate or eradicate the potential loss mechanism. It has been observed in this study that the performance loss was caused by lack of rear side ITO; and the presence of ITO on both the front and the rear sides of the SHJ solar cells can help to ensure improved stability. Therefore, long-term stability can increase the market share for SHJ solar cells and bring about their competition with silicon homojunction solar cells in terms of reliability, especially their installation in hot climate, where they can provide 10% power increase relative to the silicon homojunction solar cells [4].

V. Summary

The role of rear ITO contact in the stability of two categories of SHJ cell structures (FSI and BSI cells) was investigated under ATS experiment in argon ambient, at exposure temperature of 90°C and illumination intensity of 100mW/cm^2 for 500 hours. While the BSI cells displayed nearly stable performance, the FSI cells degraded in performance. The presence of a rear ITO layer between the aluminum and the doped a-Si layer appears to have greatly minimized degradation in BSI cells. V_{OC} loss in FSI cells during the ATS experiment likely arose from the possibility of aluminum diffusing into the n-type doped a-Si and perhaps into the intrinsic a-Si thereby, resulting in loss of passivation quality and consequently led to increased recombination of photo-generated carriers. Such aluminum diffusion scenario has been reported to cause loss mechanisms in thin-film amorphous silicon solar cells [12], [13]. The loss of passivation quality is observed to be higher at the rear side of the cells than the front side. The lower J_{SC} before the stress experiment for FSI cells stemmed from higher optical loss due to parasitic absorption. After 500hours, the loss in J_{SC} under the ATS experiment for FSI cells was likely caused by increased recombination that arose from the possibility of aluminum diffusion into the amorphous layers. Further study is needed to understand the exact cause of J_{SC} loss.

Acknowledgment

The authors are thankful to Chris Thompson and Nuha Ahmed for their assistance in characterizations at the beginning of this research. The author would also like to thank Shannon Fields for helping with ITO sputtering and the maintenance of the test chamber. This work was partially supported by the U.S. Department of Energy's Office of Energy Efficiency and Renewable Energy (EERE) under Solar Energy Technologies Office (SETO) agreement Number DE-EE0007534.

References

[1] M. Taguchi et al., "24.7% record efficiency HIT solar cell on thin silicon wafer," IEEE Journal of photovoltaics vol. 4, no. 1, pp. 96-99, 2013.

[2] K. Yoshikawa et al., "Silicon heterojunction solar cell with interdigitated back contacts for a photoconversion efficiency over 26%." Nature energy vol. 2, no. 5, pp. 1-8, 2017.

[3] D. L. Bätzner et al., "Properties of high efficiency silicon heterojunction cells." Energy Procedia vol. 8, pp. 153-159, 2018.

[4] A. Abdallah, D. Martinez, B. Figgis and O. El Daif. "Performance of Silicon Heterojunction Photovoltaic modules in Qatar climatic conditions." Renewable Energy 97 (2016): 860-865.

[5] A. H. T. Le et al., "Damage to passivation contact in silicon heterojunction solar cells by ITO sputtering under various plasma excitation modes." Solar Energy Materials and Solar Cells vol.192, pp. 36-43, 2019.

[6] M. Semma et al., "Impact of deposition of indium tin oxide double layers on hydrogenated amorphous silicon/crystalline silicon heterojunction." AIP Advances vol. 10, no. 6: 065008, pp. 1-10, 2020.

[7] ITRPV, International Technology Roadmap for Photovoltaic (ITRPV), eleventh ed., pp. 35-54, 2020.

[8] Z. Shu, U. Das, J. Allen, R. Birkmire and S. Hegedus. "Experimental and simulated analysis of front versus all‐back‐contact silicon heterojunction solar cells: effect of interface and doped a‐Si: H layer defects." Progress in photovoltaics: Research and Applications vol. 23, no. 1, pp. 78-93, 2015.

[9] J. Cattin et al., "Influence of light soaking on silicon heterojunction solar cells with various architectures." IEEE Journal of Photovoltaics vol. 11, no. 3, 575-583, 2021.

[10] B. Wright, C. Madumelu, A. Soeriyadi, M. Wright and B. Hallam. "Evidence for a Light‐Induced Degradation Mechanism at Elevated Temperatures in Commercial N‐Type Silicon Heterojunction Solar Cells." Solar RRL vol. 4, no. 11: 2000214, pp. 1-5, 2020.

[11] N. Ahmed, L. Zhang, G. Sriramagiri, U. Das and S. Hegedus. "Electroluminescence analysis for spatial characterization of parasitic optical losses in silicon heterojunction solar cells." *Journal of Applied Physics* 123, no. 14 (2018): 143103.

[12] M. S. Haque, H. A. Naseem and W. D. Brown. "Aluminum-induced degradation and failure mechanisms of a-Si: H solar cells." Solar Energy Materials and Solar Cells vol. 41, pp. 543-555, 1996.

[13] M. S. Haque, H. A. Naseem and W. D. Brown. "Degradation and failure mechanisms of a-Si: H solar cells with aluminum contacts." In Proceedings of 1994 IEEE 1st World Conference on Photovoltaic Energy Conversion-WCPEC (A Joint Conference of PVSC, PVSEC and PSEC), vol. 1, pp. 642-645. IEEE, 1994.

Vinyl acetate content tailoring in ethylene vinyl acetate improves the resilience against environmental stressors

Umang Desai, Bhuwanesh Kumar Sharma, Aparna Singh

Department of metallurgical engineering and materials science, Indian Institute of Technology Bombay, Mumbai, India

National centre for photovoltaic research and education, Indian Institute of Technology Bomby, Mumbai, India

Faculty of Science, Department of Chemistry, MUIS, Ganpat University, Mehsana, India

The most common encapsulant used in the photovoltaic (PV) modules is ethylene vinyl acetate (EVA). The properties of EVA can be tailored by modulating its vinyl acetate (VA) content. As a rule of thumb, EVA containing 28 to 33% VA content is used as an encapsulant in PV industry. It is well known that the EVA films degrade due to the environmental stressors like humidity, temperature and solar radiation during the field operation of the modules. The immediate consequence of the EVA' degradation is the loss of its optical properties which ultimately reduce the electrical performance of the PV modules. In this work, we summarize the effect of damp-heat and ultraviolet ageing on the free-standing films of EVA that contain the VA content outside the most commonly used and accepted range. The VA contents discussed here include 18, 24, 33, and 40 wt. % with the necessary additives. These films have been cured at 150 C under vacuum and subsequently subjected to accelerated damp-heat (85 C and 85% RH) aging and UV aging at a wavelength of 340 nm for 1000 and 2000 h. Tensile strength, thermal stability and degree of crosslinking have been found to be greater for EVA containing lower VA content (18 and 24%) and diminished for EVA with higher VA content (33 and 40%) due to DH aging. However, inadequate % transmittance and relatively high stiffness after 2000 h of DH aging discourages EVA18 as an encapsulant. Moreover, EVAs with higher VA content (33 and 40%) have been degraded significantly after 2000 h of DH aging resulting in significant loss in transmittance and mechanical integrity. The effect of UV aging up to 2000 h suggests that the optimum range of VA content in EVA should be between 18 and 33% by weight. VA content beyond 40% degrades almost all properties needed for an encapsulate material after aging of only 2000 h. VA content of around 18% is the most stable under UV aging conditions but has a slightly lower value of transmittance for the unaged sample although the difference in transmittance between different specimens decreases with UV aging. Therefore, from the findings of the above-mentioned studies, it is recommended to the PV community to decrease the VA content in EVA films up to 24 % by weight for better performance against the environmental stressors.

Poisson drift diffusion modeling of valley photovoltaic devices

Daixi Xia, Hassan Allami, Jacob J. Krich

Department of Physics, University of Ottawa, Ottawa, ON, Canada

School of Electrical Engineering and Computer Science, University of Ottawa, Ottawa, ON, Canada

We present Poisson/drift-diffusion (PDD) modeling of valley photovoltaics (VPV). VPV devices have the potential to enable long-lasting hot carrier populations in satellite valleys of the conduction band, exploiting intervalley scattering effects at high electric field, similar to the Gunn effect. Hot carrier effects are hard to include in quasi-equilibrium PDD models. We present a mapping from the electric-field dependence of valley scattering rates calculated using the ensemble Monte Carlo (EMC) method to an effective electric field. This mapping gives valley scattering rates that agree both with the EMC simulations and also with equilibrium detailed balance. This effective electric field is the key to using the computationally inexpensive PDD modeling for strongly nonequilibrium devices like VPV.

2T Mechanically Stacked Perovskite/Si tandem Cells Beyond 28%: the Role of 2D Materials in Perovskite Top Cells Coupled with a Commercially Available Bifacial c-Si Heterojunction Cell.

Antonio Agresti, Sara Pescetelli, Fabio Matteocci, Erica Magliano, Elisa Nonni, Giuseppe Bengasi, Carmelo Connelli, Cosimo Gerardi, Hanna Pazniak, Sebastiano Bellani, Francesco Bonaccorso, Fabrizio Bizzarri, Marina Foti, Aldo Di Carlo

C.H.O.S.E. (Center for Hybrid and Organic Solar Energy), Electronic Engineerng Department, University of Rome Tor Vergata, Roma, Italy

Enel Green Power (EGP) SpA, Catania, Italy

Université Grenoble Alpes, CNRS, Grenoble INP, LMGP, Grenoble, France

BeDimensional Spa., Genova, Italy

Istituto di Struttura della Materia (CNR-ISM) National Research Council, Roma, Italy

Perovskite/Silicon tandem technology represents a promising route to achieve 30% power conversion efficiency, by ensuring low levelized costs energy while being competitive with the already commercialized photovoltaic (PV) technologies. Despite the impressive results demonstrated employing a two-terminal (2T) monolithic architecture, the use of record efficiency amorphous/crystalline silicon heterojunction (Si-HJT) cells with micrometer-sized textured front surface, strongly limits the possibility to perform high-temperature and solution processing of the top perovskite cell. To overcome this limitation, we develop a tandem device structure consisting in a mechanically stacked 2T perovskite/silicon tandem solar cell, with the sub-cells independently fabricated, optimized, and subsequently coupled by contacting the back electrode of the mesoscopic perovskite top cell with the texturized and metalized front contact of the silicon bottom cell. The possibility to separately optimize the two sub-cells allows to carefully choose the most promising device structure for both top and bottom cells. Indeed, semi-transparent perovskite top cell performance is boosted through a rational use of bi-dimensional materials (graphene, MXenes and functionalized MoS2) to tune the device interfaces. In addition, a protective buffer layer (PBL) based on MoO3 thin film is used to prevent damages induced by the transparent electrode sputtering deposition over the hole transporting layer. At the same time, a textured amorphous/crystalline silicon heterojunction (c-Si HTJ) cell fabricated with an in-line production process is used as state of art bottom cell. The tandem perovskite/Si tandem device demonstrates remarkable power conversion efficiency of 28%.

CdTe:In - Post-Growth Doping and Proposals for Photovoltaic Devices

Luke Thomas, Theo DC Hobaon, Laurie J Phillips, Kieran J Cheetham, Neil Tarbuck, Mark Isaacs, Huw Sheil, Vin Dhanak, Tim D Veal, Stephen Campbell, Vincent Barrioz, Jon D Major, Ken Durose

Unviersity of Liverpool, Liverpool, United Kingdom

STFC, Daresbury, Daresbury, United Kingdom

Harwell XPS, Didcot, United Kingdom

University of Northumbria, Newcastle - upon - Tyne, United Kingdom

Imperial College, London, United Kingdom

PV devices having n- rather than p-type CdTe absorbers offer the potential advantages of lower contact barriers and higher doping levels. In this work we present the results of post-growth doping trials using indium metal and indium chloride. The chemical incorporation of indium was measured by quantitative SIMS and the surface contamination resulting from indium chloride doping was evaluated by XRD and XPS. Carrier lifetimes were estimated by TRPL and the surface Fermi levels were measured by XPS. Trial junction structures were fabricated based on CdS and CdTe:In. These gave weak junctions as expected, but chlorine treatment gave PCE' of up to 10.3%. Such structures are not expected to out-perform n-CdS/p-CdTe junctions and we therefore propose a number of other junction designs for future studies which we assess here with SCAPS models.

Multiple Inverter Microgrid Experimental Fault Testing

Nicholas S. Gurule, Javier Hernandez Alvidrez, Matthew J. Reno, and Jack Flicker
Sandia National Laboratories, Albuquerque, New Mexico, 87185, USA

Abstract—**For the resiliency of both small and large distribution systems, the concept of microgrids is arising. The ability for sections of the distribution system to be "self-sufficient" and operate under their own energy generation is a desirable concept. This would allow for only small sections of the system to be without power after being affected by abnormal events such as a fault or a natural disaster, and allow for a greater number of consumers to go through their lives as normal. Research is needed to determine how different forms of generation will perform in a microgrid, as well as how to properly protect an islanded system. While synchronous generators are well understood and generally accepted amongst utility operators, inverter-based resources (IBRs) are less common. An IBR's fault characteristic varies between manufacturers and is heavily based on the internal control scheme. Additionally, with the internal protections of these devices to not damage the switching components, IBRs are usually limited to only 1.1-2.5p.u. of the rated current, depending on the technology. This results in traditional protection methods such as overcurrent devices being unable to "trip" in a microgrid with high IBR penetration. Moreover, grid-following inverters (commonly used for photovoltaic systems) require a voltage source to synchronize with before operating. Also, these inverters do not provide any inertia to a system. On the other hand, grid-forming inverters can operate as a primary voltage source, and provide an "emulated inertia" to the system. This study will look at a small islanded system with a grid-forming inverter, and a grid-following inverter subjected to a line-to-ground fault.**

Keywords— **Microgrid, Inverter, Fault, Grid-Forming, Distributed Energy Resource, Volt-Var, Frequency-Watt**

I. INTRODUCTION

It has been demonstrated that inverter-based resources (IBRs) are usually not capable of producing the necessary amounts of fault current, removing overcurrent protection as a viable option for heavily inverter-based microgrids. Other options such as negative-sequence current detection are becoming a more popular option as a protection scheme. However, dependent on the control scheme of an IBR, negative- or zero-sequence current may not be produced by a device during a fault. Specifically, it has been shown that a grid-following inverter (GFLI), commonly used for photovoltaic (PV) installations, that operates as a near-ideal, balanced current source, does not output a significant amount of negative- or zero-sequence current due to the controls. Grid-forming inverters (GFMIs) can be capable of outputting unbalanced currents depending on the control scheme, a needed feature if the intent for one of these devices is to operate as the primary source of a system.

While some work has been done to investigate the fault characteristics of a single device, or with multiple simulated/prototype devices [1]-[6], this study intends to look into the interaction of a commercially available GFMI and GFLI in a small islanded system after a single line to ground fault is applied to the system. While the GFMI will operate as the primary source and be set to operate with a droop characteristic (similar to that of a synchronous generator), the GFLI will be operating as an auxiliary source; either set to a fixed power or grid support functions (GSF) in the form of volt-var (VV) and frequency-watt (FW) control.

II. BACKGROUND

GFMIs have been around for decades, originally being used for remote, standalone systems. However, these devices never gained popularity as they were too large for localized generation and could not operate with the grid, only working after the power went out. However, with an increase in electricity costs and tax incentives for the installation of PV, GFLIs have been placed throughout the distribution system. As the name implies, GFLIs follow the grid voltage, and therefore require a voltage and frequency reference to be present to operate. When the power goes out the inverter will not provide power to the system. Now with "modern" control schemes for GFMIs, grid synchronization is possible, and the device can act as either a grid support unit (providing real and reactive power when there is a deviation in the voltage and current) or operate as a primary source for a local system. Depending on the size of the GFMI, this system could be a single home, a neighborhood, or small town or city.

Both of these technologies have their own advantages and disadvantages. The growing penetration of PV inverters on the distribution system has resulted in the increased requirements for grid interconnection of GFLIs. This includes the addition of GSFs to the capability of the inverters (as well as voltage and frequency ride-through capability). GSFs help to maintain a stable grid under normal operating conditions, while event ride-through allows for GFLIs to continue operating (up to a certain point) during abnormal operating conditions (such as faults) to ensure that a potentially large amount of generation does not trip prematurely. GFMIs usually use some form of droop characteristic to maintain synchronization with the utility or other grid-forming devices [7],[8] and through this droop can also provide a form of grid support similar to that of the GFLIs. Both these systems are also capable of responding to events very quickly, much faster than traditional rotating generation. This quick response capability is very beneficial in a high inertia system, like the utility.

While the advantages of IBRs are attractive, there are a few shortcomings that should not be overlooked. First, GFLIs are commonly used with intermittent resources. Solar and wind resources are not always present, and thus are not a reliable form of generation. A GFMI is usually used for an energy storage system (ESS), and therefore as long as the ESS is charged (charged when excess generation is available and only

---

978-1-7281-6118-1/22 $31.00 © 2022 IEEE

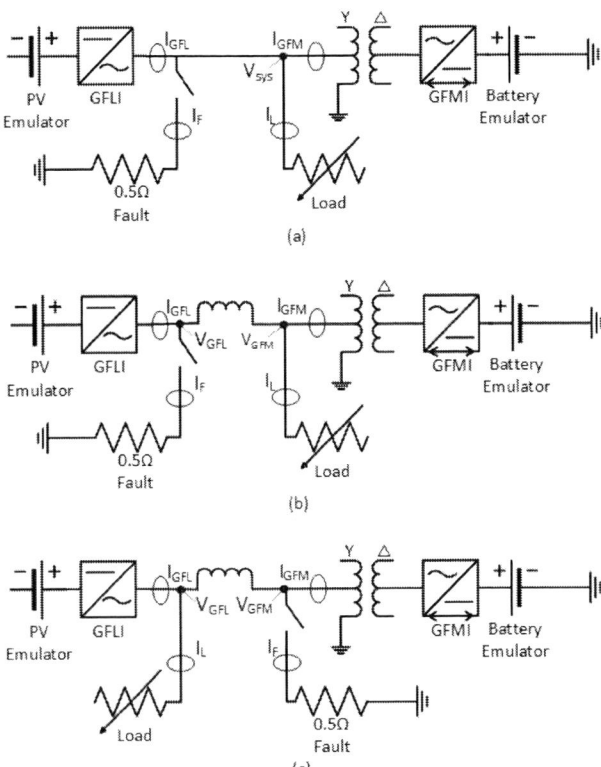

Fig. 1: Experimental Lab Test Configurations, Single Bus Fault (a), GFLI Side Fault (b), and GFMI Side Fault (c)

discharged when needed), the resource is reliable. Additionally, a disadvantage to an IBR is the sensitive components that they are comprised of. The circuitry of these devices has a maximum current limit, and thus has internal controls to protect these components. This results in devices that can only output slightly above rated current to usually no more than 3 p.u. during faults, and only if the headroom is available.

While the slow response of synchronous generators is not preferred, this response is well known and has been modeled and observed on both grid-tied and islanded systems. The effects of increased penetration of IBRs have been studied, as well as their effectiveness as a grid support device in both grid-tied [9] and microgrid applications. However, little work has been performed to determine how commercially available GFLIs and GFMIs interact in a high-IBR penetration islanded system (or in a fully IBR system).

III. EXPERIMENTAL SETUP

For this study, a commercially available 24 kW GFLI and 100 kW GFMI were used as the IBRs. A 0-150 kW programmable resistive delta configured load was used as an aggregate load for the system. A 150 kW single-phase load was used as a fault (~0.5 ohms to ground).

Three test configurations were utilized for this study as shown in Fig. 1. The first was a single bus system, where all sources and loads were connected directly together, with no additional line impedance added to the system and all lines being less than 50 m. The other two configurations had a 1 mH inductance added to each of the three-phase lines between the

two IBRs to emulate an extended line length, roughly 500 m of the largest conductor used in the system. For each of these configurations, the locations of the load and fault was reversed, i.e., the first configuration had the fault on the GFLI side with the load on the GFMI side, and the second the opposite.

For DC sources, the GFLI was connected to an Ametek TerraSAS 100 kW photovoltaic emulators, set to operate 1.25 p.u. of the rated power of the inverter. The GFMI was connected to an NH Research 9300 100 kW battery emulator as the DC source.

The GFMI requires an external delta/wye transformer for both isolation as well as achieving common system voltage. This IBR operated as the voltage source of the system was set to have a droop of 1% for both voltage and current, with a priority for real power output. No additional bias was set to the voltage and frequency, so at nominal system voltage (277/480 V) and frequency (60 Hz), the inverter should not output any power.

For each test configuration, the GFLI was first tested with all GSF off, followed by having both VV and FW functions enabled. The device is factory set to have reactive power priority, and can not be set for real power priority. With GSF enabled, the slopes were set to 1% with no deadband to match the characteristics of the GFMI. Additionally, the FW profile was evaluated at 0.5 p.u. and 1.0 p.u. initial output powers when operating at nominal frequency. These are seen in Fig. 2.

The load of the system was also varied for each test configuration. For each case, at least 24 kW of load was present on the system (to match the power rating of the GFLI). Further, tests were performed with an additional 25 kW load to the system to represent a quarter of the power rating of the GFMI rating. Additional load cases would have been evaluated,

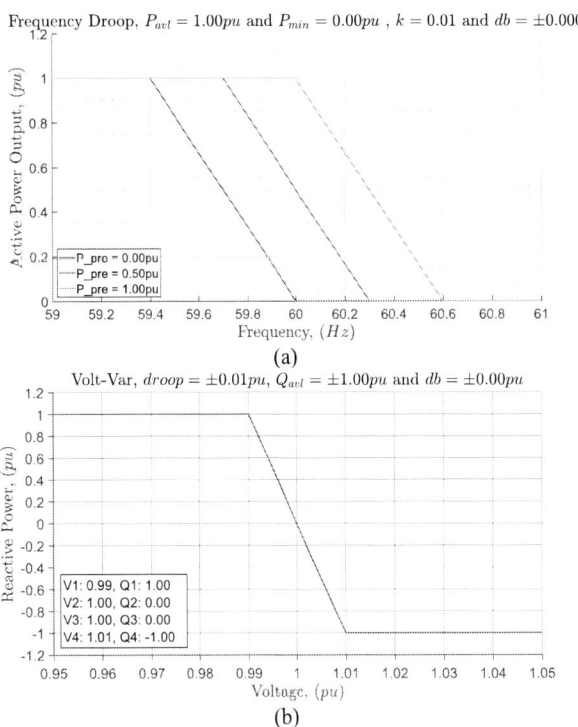

Fig. 2: GFLI Grid Support Function Profiles for a) frequency-watt, and b) volt-var

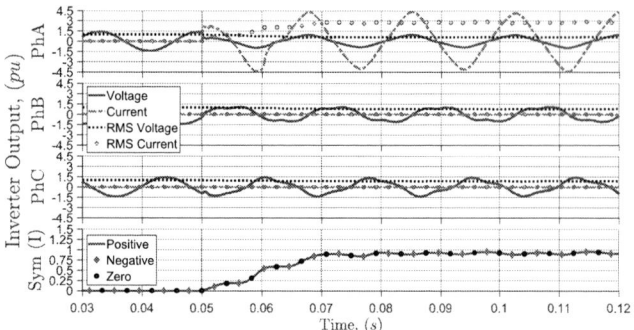

Fig. 3: GFMI Subjected to 150kW Single Line Fault

however the battery emulator tripped and stopped exporting power due to overloading during the fault event.

For each test case, a single-line-to-ground Phase A fault was applied to the system for at least 0.5s, but no longer than 2s. A single line-to-ground fault was chosen for this testing as it is the most prevalent fault to occur within any system. This was to ensure that the system could reach steady-state during the fault but not inadvertently trip the GFLI due to a low voltage or frequency event.

IV. EXPERIMENTAL RESULTS

A. Configuration A

To first see if GFLIs can help an islanded system during a fault, the GFMI was first subjected to the fault without the GFLI present. For this initial test, no additional load was applied to the GFMI. Fig. 3 shows the response of the GFMI. An initial observation is that althrough the fault was only applied to phase A on the GFMI, all three phases were affected by the event. This is a trend seen throughout the tests to follow. Once the inverter response reached stability, the Phase A voltage of the GFMI was measured at an average of 0.536 p.u. with an average fault current contribution of 2.462 p.u. and a peak of 2.960 p.u. immediately after the onset of the fault. This is on par with what is known for the overload capability of GFMIs. Sensitive components internal to the IBR would need to be greatly oversized for higher levels of fault current to be achieved, so inverter controls are designed to protect the components by limiting the output current.

The GFLI was then added to the system, and a load of 24kW was initially placed onto the system to offset the power output of the GFLI. This allowed the GFMI to remain unloaded for a fair comparison to the previous test. For this first test with the GFLI, no GSFs were enabled. However, at the onset of the fault, the GFLI went into momentary cessation due to its low voltage ride-through settings. Because of this, the Phase A voltage drops to 0.511 p.u. and the GFMI must support both the load and fault, requiring a current output of 2.473 p.u. Following the removal of the fault, the inverter recovered to prefault conditions within 500ms. While this no longer posed as a good comparison with the previous test, it does show that knowing how an inverter responds to faults is crucial. For the tests to follow, the low voltage ride-through settings were adjusted to ensure that the GFLI would continue to operate during the fault.

With new settings, the test was rerun to get the desired comparison. Now that the GFLI does not go into momentary cessation, as seen in Fig. 4, an increase in system voltage, up to

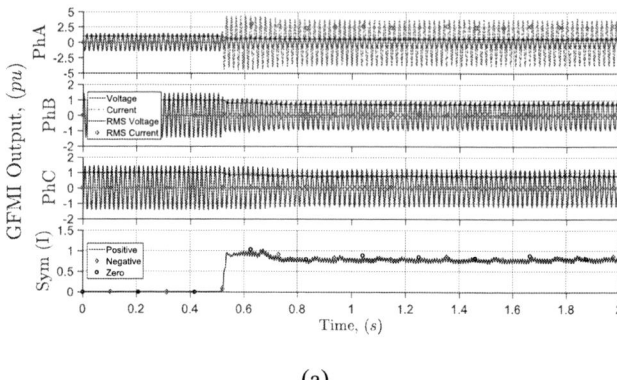

(a)

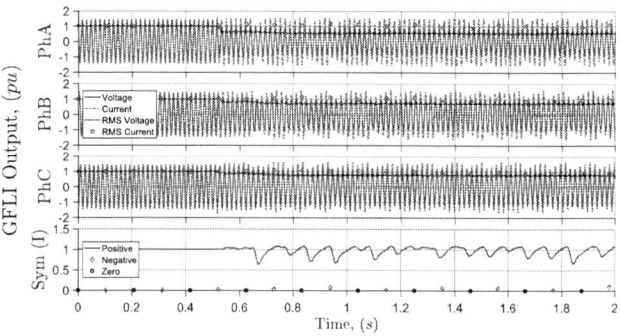

(b)

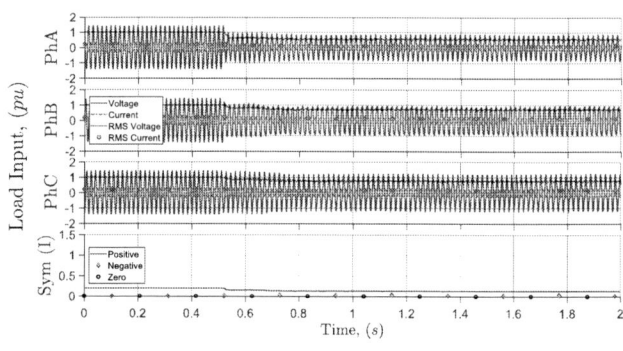

(c)

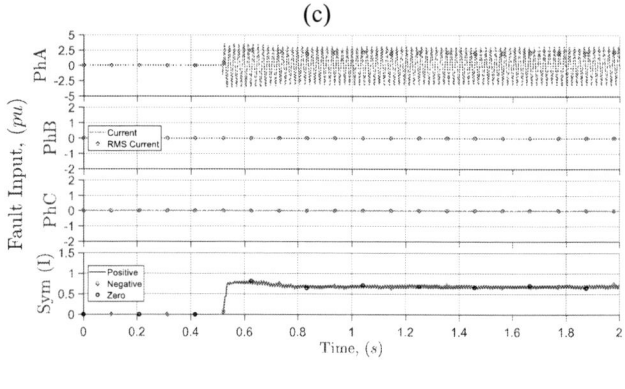

(d)

Fig. 4: Configuration A, 24 kW total load, GSF Disabled, 0.5-ohm single-line-to-ground fault on phase A applied at 0.5 seconds with the currents measured at the: GFMI (a), GFLI (b), Load (c), Fault (d)

978-1-7281-6118-1/22 $31.00 © 2022 IEEE

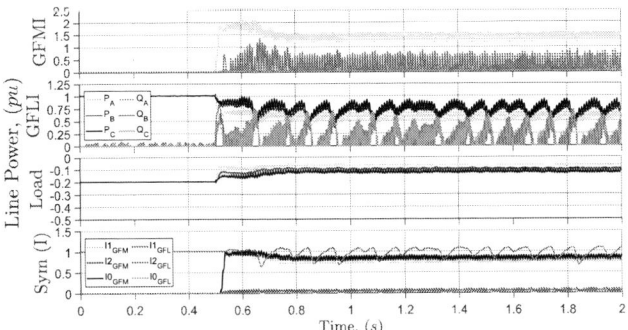

Fig. 5: Configuration A, GSF Disabled, 24kW Load, IBR, Load Power and Sequence Current Injections

0.563 p.u. on Phase A was seen and an overall fault current contribution of 2.588 p.u. (with relation to the GFMI rated line current), and a GFMI contribution of 2.501 p.u. Therefore, the GFLI was able to contribute 0.36 p.u. of its rated line current to the fault with the remainder supporting the full load. While the GFLI support was minimal, the system was able to provide more current to the fault than without the GFLI, and sustain a slightly higher system voltage. Additionally, it can be seen that the GFMI provided practically all of the negative- and zero-sequence current with the GFLI only outputting a minimal amount (less than 0.05 p.u.). In a system with low short-circuit current capability, such as a purely IBR system, non-positive-sequence currents can play a vital role in protection schemes as they are a good indicator for unbalanced faults [10],[11]. Note that the current from the GFMI is measured after the delta-wye transformer, so the zero-sequence current is actually being provided by the transformer and not the GFMI. A curious note is that the GFLI's output current is modulated throughout the fault. This has been seen in previous tests [9], such as low voltage ride-through testing and phase jumps when this IBR is near the point of tripping. This is due to the fact that the phase-locked loop used to synchronize the GFLI is not able to sustain synchronization. This can be seen in Fig. 5, where the GFLI is intermittently producing reactive power while the VV function is disabled.

For the remainder of the tests, 49 kW of total resistive load was used. At this point, all GSFs were still disabled within the GFLI. Fig. 6 shows the results from this test case. Due to the increase in load, the system voltage dropped to 0.481 p.u., and had a total fault current of 2.329 p.u. of the GFMI line current rating. Additionally, the GFMI contributed 2.307 p.u. to the

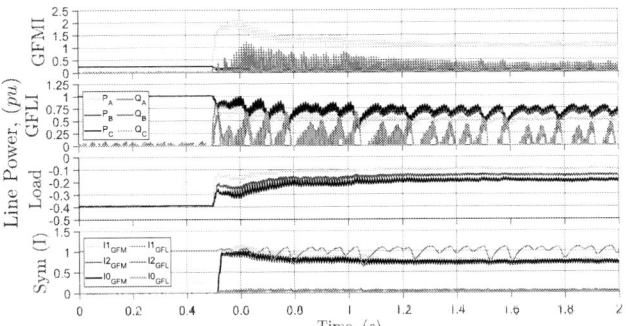

Fig. 6: Configuration A, GSF Disabled, 49kW Load, IBR, Load Power and Sequence Current Injections

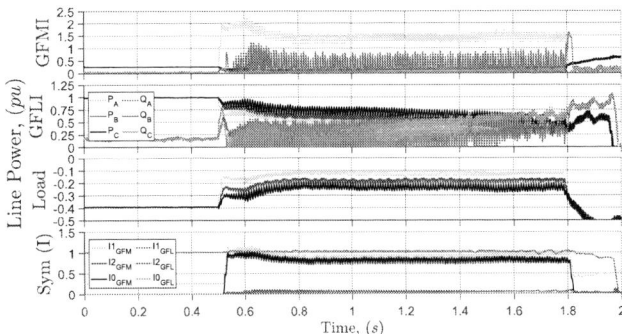

Fig. 7: Configuration A, GSF Enabled, 49kW Load, IBR, Load Power and Sequence Current Injections

fault. Therefore the GFLI contributed little to the fault, however, most of the load was supported by the GFLI, allowing the GFMI to provide more current to the fault.

Next, VV and FW were enabled on the GFLI. With the FW curve operating such that at nominal frequency the GFLI operates at rated power, the initial response of the inverter was similar to the tests with GSF disabled. However, as the fault progressed the VV function began to kick in and produce reactive power to try to boost the system voltage, reducing the real power generation of the GFLI. With a FW curve that sets the GFLI to operate at 0.5 p.u. at nominal frequency, at the initialization of the fault, the inverter quickly jumps up in real power until the VV catches up and starts driving the inverter to inject reactive power into the system. Because the two FW cases were very comparable due to the reactive power priority of the device, future tests with GSFs enabled only refer to the rated power operation case. When comparing what the voltage settled to during the fault, it was noticed that there was around a 6.5% increase in bus voltage when GSF was enabled. The total fault current was 2.545 p.u. of the GFMI line current rating, and was fully supported by the GFMI. The current magnitude of the GFMI was greater than the fault current, due to the absorption of reactive power produced by the GFLI, covering a small portion of the load. While this increase is minimal, it does show that even with a low GFLI to GFMI ratio the GSF could assist the main generation during a fault. Interestingly following the removal of the fault, the GFLI tripped due to an overvoltage event, which did not occur in the tests without GSF. The VV function of the GFLI still produced reactive power following the fault, as seen in Fig. 7, and drove the GFMI voltage up until the point that the GFLI tripped. This is a good demonstration of how too slow of a response or too aggressive of a grid support function could be detrimental in a low inertia system.

B. Configuration B

Following the single bus tests, the IBRs were separated into two buses using an inductor bank to emulate a 500 m line length between the devices. To start, the fault was implemented on the GFLI bus, with the load on the GFMI bus. Like the single bus tests, the GFLI was initially operated without GSF enabled, and set to operate at rated power. The first observation is that the GFMI supplied less current to the fault than in the Configuration A tests. The total fault current was 1.958 p.u. of the GFMI rated line current, and was fully supplied by the GFMI, with the GFMI providing very little to the load. This is seen in Fig. 8 where the GFMI Phase A real power peaked at 1.5 p.u. of the line rating,

978-1-7281-6118-1/22 $31.00 © 2022 IEEE

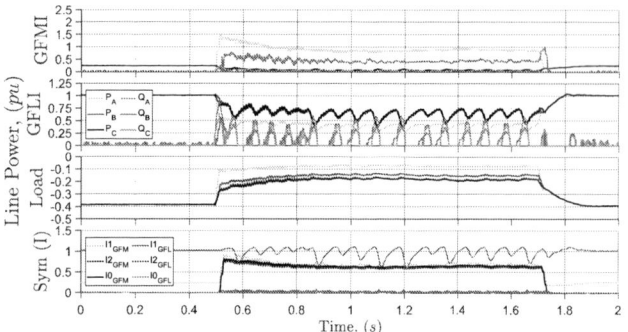

Fig. 8: Configuration B, GSF Disabled, 49kW Load, IBR, Load Power and Sequence Current Injections

compared to the 2 p.u. output of the single bus scenario. Additionally, the settled power level of the GFMI was less than the single bus cases. Looking at the results in more detail, it was found that while the GFMI bus voltage was identical to the single bus test, however the GFLI bus voltage was over 6% lower due to the additional line length. Due to the fixed resistance of the fault and the drop in bus voltage, the observed fault current is reduced when compared to tests performed in Configuration A. Beyond these observations, the dynamics of the IBRs were very consistent with those seen in the Configuration A tests and continued on this trend throughout the remainder of the testing.

The same system was then evaluated with GSF enabled. The results for this test are shown in Fig. 9. Like the Configuration A test, the same VV profile was used, however only the rated power FW profile was used. The bus voltages were once again analyzed to determine how the GSF assisted during the fault. In this test case, the bus voltages were greater than the case without GSF enabled, with the voltage increased by around 6% on both busses. With the GSF enabled, the GFLI bus voltage was now near the Configuration A disabled GSF fault voltage with the GFMI bus being sustained at a greater voltage. Additionally, the total fault current was 2.238 p.u. of the GFMI rated line current, which is lower than either of the Configuration A tests. So, while having GSFs enabled helps the system voltage, if the fault is furthest away from the primary source, little benefit is seen.

C. Configuration C

For the final series of tests, the locations of the load and fault were swapped. Once again the system was set with the GFLI to operate at rated power with GSFs disabled. Fig. 10 shows the results for the GFMI side fault without GSFs enabled. With the

fault applied, it was observed that the voltage of both buses was between 3.5-4.5% greater than the Configuration A GSF-disabled case, and both bus voltages were quite close to each other. The total fault current observed was 2.404 p.u. of the GFMI rated line current being fully supported by the GFMI. Like the Configuration B disabled GSF test, little support was given to the load from the GFMI.

After enabling the GSFs, the voltage level of the busses saw an additional 1.5-4.5% increase during the fault. Moreover, the voltages of both buses were very balanced with little deviation between the two. With a fault current of 2.497 p.u. of the GFMI rated line current, the system produced nearly as much fault current as the Configuration A case with less load. Interestingly, because the VV function was enabled, the GFLI could not fully support the load as it was supplying reactive power. So, although there was current being provided to the load by the GFMI and subsequently through the inductor bank, the reactive power compensation from the GFLI was able to help boost both bus voltages. The results for this final test case are shown in Fig. 11.

V. CONCLUSION

While we are still a long way away from a purely IBR utility, it is important to know how these devices act under abnormal conditions. By understanding how these devices interact during faults, we can figure out how to tune different technologies to assist during a fault and determine what needs to be looked for to protect the utility. Because IBRs are current-limited, traditional protection schemes cannot be used.

From this study, a few things were observed. First, if a GFLI does not have GSFs and the fault is far away from the main source, the bus voltage is significantly reduced, even when the

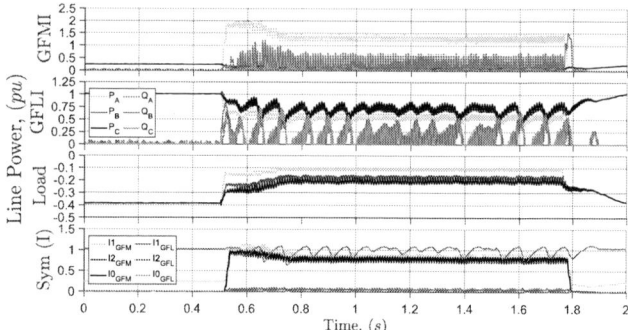

Fig. 10: Configuration C, GSF Disabled, 49kW Load, IBR, Load Power and Sequence Current Injections

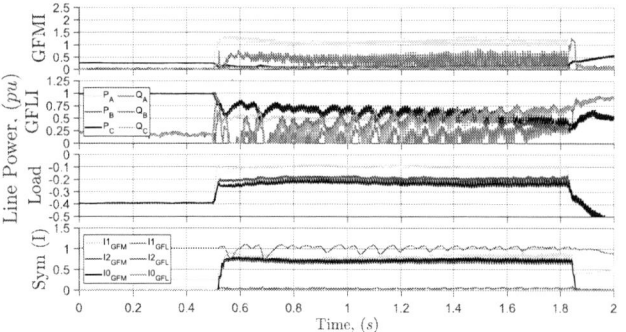

Fig. 9: Configuration B, GSF Enabled, 49kW Load, IBR, Load Power and Sequence Current Injections

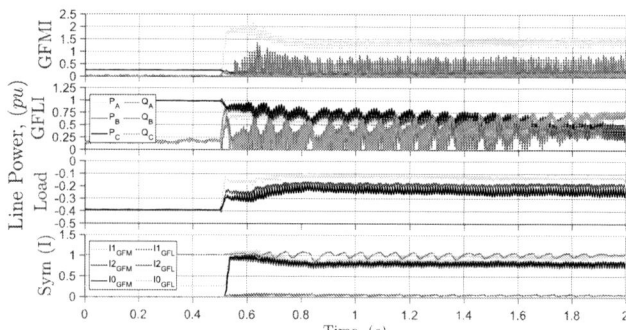

Fig. 11: Configuration C, GSF Enabled, 49kW Load, IBR, Load Power and Sequence Current Injections

GFLI is on the same bus and producing power. Following this, if GSFs can be enabled, not only can the GFLI help support the bus voltage, but also help support the system overall. Furthermore, if the fault is closer to the main source, the current sourcing nature of the GFLI can help support the load and allow for the main source to provide fault current.

While GSFs can potentially provide support during a fault on a low inertia system, it should be noted that the recovery from the fault is as important as the fault itself. When a fault is cleared or removed, it is crucial that the voltage and frequency reference return to pre-fault conditions as smoothly as possible. When GSFs are disabled, only the main source is trying to regulate the reference. However, when GFLIs are using GSFs such as FW and VV, they are also trying to regulate the reference as well. If The GFLI is set to respond very slowly, or the GSFs are too aggressive when the fault is cleared, the GFLI will continue to output reactive current that was meant to boost the voltage during the fault and subsequently continue to drive the voltage of a low inertia system until the VV profile catches back up or significant GFLIs trip and the system can reach stability. Since GFLIs may be the majority for generation on an IBR system, it is crucial that these devices do not trip following the clearing of a fault, as the system could collapse if the load is too great. Therefore having an understanding of how GSFs should be set so that GFLIs can both support the system during a fault and recover properly afterward is very important.

VI. Acknowledgment

Sandia National Laboratories is a multi-mission laboratory managed and operated by National Technology and Engineering Solutions of Sandia, LLC., a wholly owned subsidiary of Honeywell International, Inc., for the U.S. Department of Energy's National Nuclear Security Administration under contract DE-NA-0003525.

VII. References

[1] James Keller and B. Kroposki. Understanding fault characteristics of inverter-based distributed energy resources. No. NREL/TP-550-46698. National Renewable Energy Lab. (NREL), Golden, CO (United States), 2010.

[2] S. Gonzalez, N. Gurule, M. J. Reno, and J. Johnson, "Fault Current Experimental Results of Photovoltaic Inverters Operating with Grid-Support Functionality," 2018 IEEE 7th World Conference on Photovoltaic Energy Conversion (WCPEC) (A Joint Conference of 45th IEEE PVSC, 28th PVSEC & 34th EU PVSEC), 2018, pp. 1406-1411, doi: 10.1109/PVSC.2018.8547449.

[3] N. S. Gurule, J. Hernandez-Alvidrez, M. J. Reno, A. Summers, S. Gonzalez, and J. Flicker, "Grid-forming Inverter Experimental Testing of Fault Current Contributions," 2019 IEEE 46th Photovoltaic Specialists Conference (PVSC), 2019, pp. 3150-3155, doi: 10.1109/PVSC40753.2019.8980892.

[4] N. S. Gurule, J. Hernandez-Alvidrez, R. Darbali-Zamora, M. J. Reno, and J. D. Flicker, "Experimental Evaluation of Grid-Forming Inverters Under Unbalanced and Fault Conditions," IECON 2020 The 46th Annual Conference of the IEEE Industrial Electronics Society, 2020, pp. 4057-4062, doi: 10.1109/IECON43393.2020.9254562.

[5] R. Darbali-Zamora, J. Johnson, N. S. Gurule, M. J. Reno, N. Ninad, and E. Apablaza-Arancibia, "Evaluation of Photovoltaic Inverters Under Balanced and Unbalanced Voltage Phase Angle Jump Conditions," 2020 47th IEEE Photovoltaic Specialists Conference (PVSC), 2020, pp. 1562-1569, doi: 10.1109/PVSC45281.2020.9300604.

[6] N. S. Gurule, J. A. Azzolini, R. Darbali-Zamora, and M. J. Reno, "Impact of Grid Support Functionality on PV Inverter Response to Faults," 2021 IEEE 48th Photovoltaic Specialists Conference (PVSC), 2021, pp. 1440-1447, doi: 10.1109/PVSC43889.2021.9518953.

[7] Lin, Yashen, et al. Research roadmap on grid-forming inverters. No. NREL/TP-5D00-73476. National Renewable Energy Lab.(NREL), Golden, CO (United States), 2020.

[8] W. Du et al., "Modeling of Grid-Forming and Grid-Following Inverters for Dynamic Simulation of Large-Scale Distribution Systems," in IEEE Transactions on Power Delivery, vol. 36, no. 4, pp. 2035-2045, Aug. 2021, doi: 10.1109/TPWRD.2020.3018647.

[9] J. Johnson et al., "Distribution Voltage Regulation Using Extremum Seeking Control With Power Hardware-in-the-Loop," in IEEE Journal of Photovoltaics, vol. 8, no. 6, pp. 1824-1832, Nov. 2018, doi: 10.1109/JPHOTOV.2018.2869758.

[10] Buigues, G., et al. "Microgrid Protection: Technical challenges and existing techniques." International Conference on Renewable Energies and Power Quality. Vol. 1. 2013.

[11] Hassan Nikkhajoei and Robert H. Lasseter. "Microgrid fault protection based on symmetrical and differential current components." Power system engineering research center (2006): 71-74.

[12] M. J. Reno, S. Brahma, A. Bidram, and M. E. Ropp, "Influence of Inverter-Based Resources on Microgrid Protection: Part 1: Microgrids in Radial Distribution Systems," IEEE Power and Energy Magazine, 2021.

[13] S. S. Venkata, M. J. Reno, W. Bower, S. Manson, J. Reilly and G. W. Sey Jr. "Microgrid Protection: Advancing the State of the Art," Sandia National Laboratories, SAND2019-3167, 2019.

[14] B. Reimer, T. Khalili, A Bidram, M. J. Reno, and R. C. Matthews, "Optimal Protection Relay Placement in Microgrids," IEEE Kansas Power and Energy Conference (KPEC), 2020.

[15] J. A. Azzolini and M. J. Reno, "Analyzing PV Hosting Capacity under Time-Varying Inverter Fault Response", IEEE Photovoltaic Specialists Conference (PVSC), 2022.

Reactive Anisotropic Conductive Adhesive Wafer Bonding for Solar Cells

Eric M. Rehder
Spectrolab
Los Angeles,, USA
eric.m.rehder@boeing.com

Shoghig Mesropian
Spectrolab
Los Angeles, USA
shoghig.mesropian@boeing.com

Xing-Quan Liu
Spectrolab
Los Angeles, USA
xingquan.liu2@boeing.com

Integration of dissimilar materials is an important technology for future space solar cells. Adhesive bonding is used to integrate dissimilar materials while widely spaced, reactive conducting particles are used to achieve electrical conduction. Indium is shown to be highly reactive with Au leading to alloyed junctions. This reactive anisotropic conducting adhesive structure reliably achieves low resistance III-V and Si bonded structures with low bow and without cracking.

Keywords—multijunction solar cell, wafer bonding, anisotropic conducting adhesive

I. INTRODUCTION

Multijunction (MJ) III-V solar cells for space applications have been in production for 25 years. The efficiency of these devices is now exceeding 32% in AM0 for solar cells >70 cm^2 in area [1]. Advancements in solar cell and solar panel manufacturing continue to increase the W/area power production. Increases in solar cell efficiency and capability are sought through the integration of MJ solar cells with other materials. Increased solar cell efficiency can be achieved by having more solar cells that better match the solar spectrum. Alternatively, integration of a MJ solar cell with power management circuitry in a Si wafer can increase power production in the complex, space environment.

The space environment has widely varying conditions of temperature and radiation. Thus the voltage output of the solar cells will change greatly during a mission while the voltage input need of the battery system is largely fixed. The integration of silicon electronics to manage the voltage and current output of the solar cell can improve system level efficiencies. We will demonstrate wafer bonding that is able to integrate III-V and silicon materials. This is an adhesive bond process utilizing a low density of conducting particles, where particles are selected to match to the adhesive bond thickness, which is an anisotropic conducting adhesive (ACA). Figure 1(a) shows a cross-sectional diagram of this structure. Electrical conduction between semiconductors is achieved through single particles. The adhesive process minimizes stresses from the coefficient of thermal expansion (CTE) mismatch between Si and III-V materials and eliminates cracks. This ACA assembly utilizes metal coated polymer particles. The metals coating the particle, III-V, and Si wafers are selected to achieve a reactive metal bond. This will be shown to establish continuous metal joints. Electrical conductivity is achieved through each individual particle which can be widely dispersed in the adhesive. This maintains the ease of manufacturing and mechanical benefits of an adhesive bond. Each particle is electrically isolated thus a

broad array of independent connections can be engineered across a wafer enabling far more complex structures than demonstrated in this example.

This structure can be extended to optical integration. Direct wafer bonding can achieve this, Figure 1 (b), but it is difficult to achieve. Mechanically stacking solar cells with insulating adhesive is achievable, Figure 1(c), but the electrical connections become very complex. By exchanging the metal bonding interface layers with transparent conducting coatings (TCC) we achieve optical integration, Figure 1 (d). The particle density can be < 1% allowing high optical transmission for stacked optical structures. This assembly provides for the mechanical, electrical, and optical integration of dissimilar materials that is highly manufacturable at the wafer scale [2].

II. EXPERIMENTAL

A first set of samples were prepared consisting of IMM structures bonded to a Si wafer. IMM solar cells were epitaxially deposited on 100 mm GaAs substrates as previously described [3]. These had 3 junctions with bandgaps of 1.9, 1.4, 1.1 eV or 4 junctions with bandgaps of 1.95, 1.45,

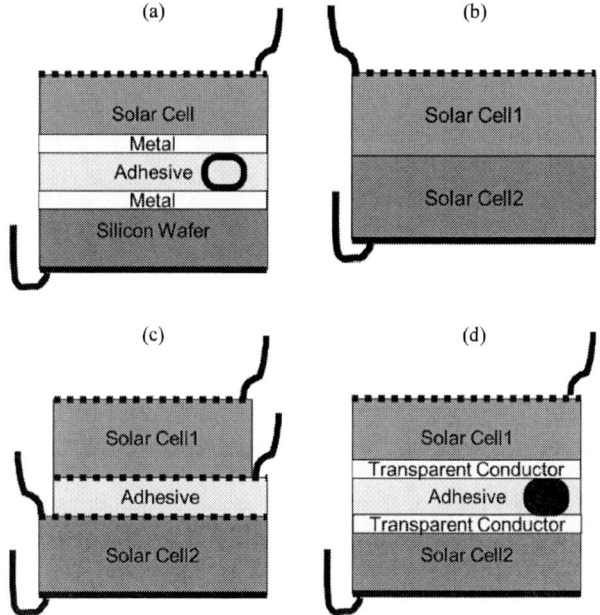

Figure 1. Schematic of diverse approaches to stacking multijunction solar cell structures including the metal particles imbedded in the adhesive layer.

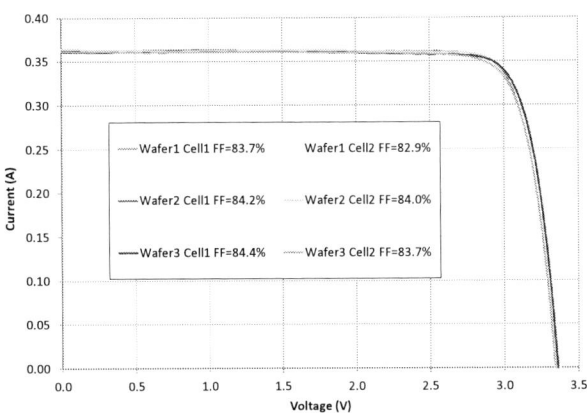

Figure 2. LIV data of IMM cells bonded to Si wafer with uniform and high fill factors.

1.05, 0.7 eV. The surface of these wafers had metal evaporated to form the back contact of the solar cell containing 50 Å Ti, 5000 Å Au, and 100 Å Ti. Two types of bonding were performed with different Si metallizations. A 100 mm Si wafer was prepared with either non-reactive metal layers with 50 Å Ti, 5000 Å Au, and 100 Å Ti or reactive metal layers 800 Å Ti, 1000 Å Pd, 21,700 Å In, 1000 Å Au and 50 Å Ti. Cyclotene 3022-63 was mixed with 15 µm particles. These particles have a 15 µm diameter polymer core with Ni and Au coatings. The AP3000 adhesion promoter was spin coated and baked onto both wafers. Then the Cyclotene mixture was spin coated onto the Si wafer. A spin speed of 3000 RPM was used to target an adhesive thickness of 13 µm. Wafers were bonded by hand in air and loaded into a wafer bonder. Vacuum was applied with pressure and temperature. Bonded wafer pairs were loaded into an oven to achieve a handling cure. The silicon wafer backside was coated with photoresist for protection and the GaAs wafer was etched chemically. This etching removed the GaAs wafer and stopped at an InGaP etch stop, which was then removed with HCl. The photoresist coating on the backside of the Si wafer was removed.

Complete curing of the bonding adhesive was carried out according to the DOW application note [4]. The solar cell assembly then has an n-type semiconductor surface able to proceed through device processing similar to a conventional MJ solar cell wafer with thin photopatterned metal grid lines and anti-reflection coating. Several 100 mm wafers with two 24 cm² solar cells on each wafer were fabricated.

A focused ion beam was used to prepare cross section samples that were imaged in a secondary electron microscope (FIB-SEM) in order to investigate the metal particles. Solar cell testing included light current-voltage (LIV) measurement under a solar simulator that is calibrated using the relevant IMM3J or IMM4J isotypes to the AM0 spectrum. Measurements were taken at 25°C. Cells were forward biased at one third of Isc to excite the electroluminescence to inspect for cracks and photographs taken.

Optically integrated structures were prepared similarly. The top wafer was a 6 junction, III-V IMM structure grown on a 100 mm GaAs wafer. This was bonded to a 100 mm Ge wafer with a diffused Ge PN junction and an n-GaAs contact layer. Instead of the evaporated metal contact layers a ZnO:Al TCC was sputtered onto both wafers. ACA was achieved with solid indium particles having a 10-20 µm diameter mixed into the Cyclotene adhesive. Wafer processing followed the steps previously described. Internal quantum efficiency measurements were taken to demonstrate the optical, electrical, and mechanical attachment.

III. RESULTS

The wafer bonding of III-V and Si wafers was mechanically successful, routinely producing wafers and 24 cm² cells without cracks. Wafer bow was less than 1 mm and wafers were processed through wafer handling and vacuum hold down systems without special accommodations. The III-V/Si devices with non-reactive metal produced inconsistent electrical results. Poor electrical contact resulted in a series resistance and a low device fill factor in the LIV testing. The fill factors of cells with non-reactive metal varied from 84% to 75% indicating a series resistance. In some cases cell testing at the wafer level would achieve 84% fill factor. Retesting the cell after dicing it out of the wafer would result in the fill factor dropping to 75%.

Figure 2 shows test results from a 3 wafer batch with 2 solar cells from each wafer all with reactive metal. Each of the 6, 24 cm² solar cells achieved a fill factor of 83-84%. The III-V/Si integration can cause high stresses and crack formation which was inspected with forward bias electroluminescence. Figure 3 shows a typical example of a reactive bonded solar cell that is free of cracks and uniformly illuminated.

FIB-SEM was used to prepare and observe cross section images of the conducting particles. Figure 4 is a micrograph of a solar cell with non-reactive metallization. The polymer particle is solid but voids develop during the FIB process and should be ignored. The particle is in the Cyclotene adhesive layer with multijunction solar cell above and Si wafer below. Smooth metal layers are observed on both wafers as well as the metal coating surrounding the particle. The particle is highly compressed yet the metal layers are not touching. Multiple images at different FIB cross section depth repeated this observation.

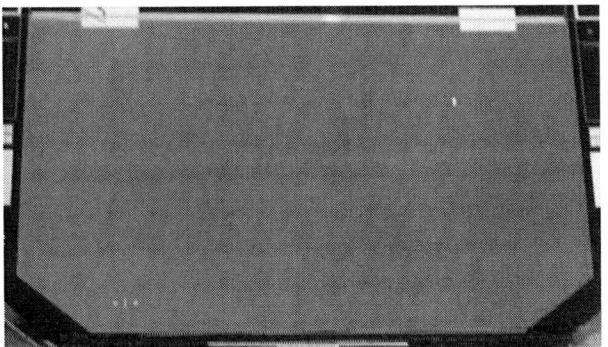

Figure 3. Photograph of electroluminescence of IMM cell bonded to Si wafer without any cracks.

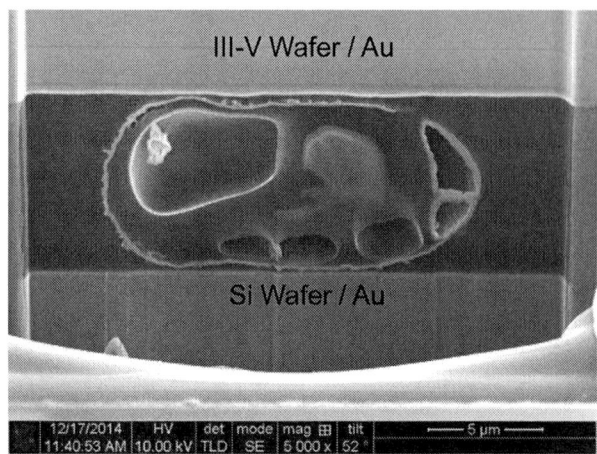

Figure 4. FIB-SEM cross section image of non-reactive particle with inconsistent metal connections.

In Figure 5 the structure using the reactive metallization is shown with a dramatically different appearance The top III-V wafer and particle are unchanged from the nonreactive structure. The difference is the indium film on the lower Si wafer, which has made contact with the metal skin of the particle. The metal skin is now much thicker than in the nonreactive version. Also, the solar cell metallization is identical in the two versions but after bonding to reactive metal the solar cell metallization is thicker and rougher. The wafer and particle metallizations are no longer separate but have merged without clear interfaces.

Optical integration across the bonded interface is examined in Figure 6. The internal quantum efficiency is plotted across the spectral range of the 7 junction cell. A response from each junction is observed. The 6 lowest wavelength cells were based on the IMM wafer with the longest wavelength cell being the Ge cell. The integrated quantum efficiency is also plotted.

IV. DISCUSSION

The successful mechanical integration of the III-V and Si materials is demonstrated by the LIV data and the forward biased image. This is particularly noteworthy to be achieved

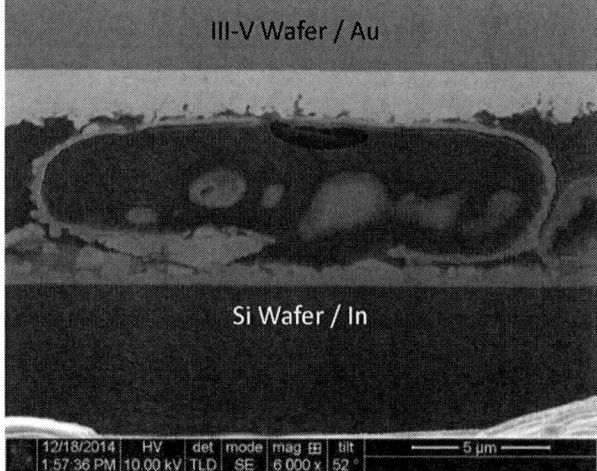

Figure 5. LIV data of IMM cells bonded to Si wafer with uniform and high fill factors.

with low cost wafer scale processing leading to 24 cm² solar cells.

The electrical integration is best judged by the fill factor from the LIV data. Poor conductivity will lead to a series resistance, which shifts the knee region of the LIV curve to lower voltages. The two sample types both achieve fill factors of 84%. Cells with reactive metallization achieved this on each solar cell. The non-reactive metallization had wider variation in fill factor and series resistance. The drop in fill factor when the cell is diced from the wafer indicates a variation across the wafer. The metal at the bond interface has sufficient lateral conductivity to carry the current to any low resistance connection points across the wafer. Dicing can separate the cell from these low resistance points increasing the series resistance. The reactive metal-based cells do not have this behavior indicating a more uniform, more conducting series of particles in the adhesive layer.

The non-reactive ACA structure relies upon the metal skin of the particles making contact to the wafer metallization. Inspection of multiple particles and multiple FIB-SEM planes has shown that the particle can be highly compressed but the Cyclotene adhesive tends to remain between the particles and both metalized wafers. This is consistent with the electrical measurements indicating variable and high series resistance. The devices are operational but a high and variable series resistance degrades their output power.

This metal connection is transformed by the reactive ACA. The FIB-SEM shows that the rough indium layer is able to make contact to the conducting particle. With the adhesive barrier layer penetrated the indium diffuses across the metal skin of the particle and continues into the metal layer of the III-V wafer causing both to grow in thickness. The migration of indium across the structure forms a continuous metal junction with substantial cross section.

The reactive ACA process results in a metal junction that is more mechanically and electrically stable. Solar cells made with the reactive metallization have high fill factors not degraded by a series resistance from the bonding process. This process is able to integrate III-V and Si materials mechanically, over large areas. In this structure the Si wafer is simply a conducting mechanical support. But the Si wafer could have various power control functions built in to control the solar cell power output.

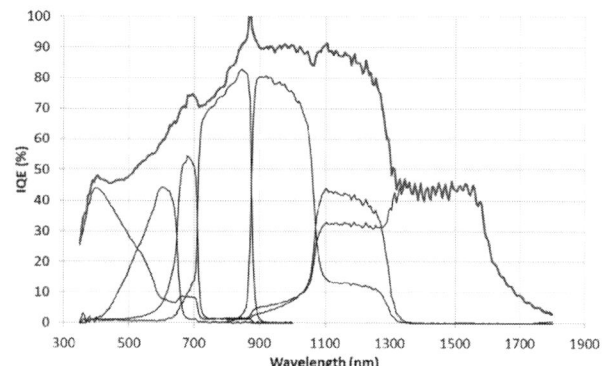

Figure 6. Internal quantum efficiency data of 6 junction IMM cell bonded to Ge cell.

This process functions at the wafer scale thus in a single bonding operation a large number of small solar cells could be independently connected to the Si wafer which could have any variety of circuitry built in to control the delivery of power to the panel. This applies equally to non-solar cell devices as the III-V could be an array of optical detectors, emitters, or any type of electronic device such as monolithic microwave integrated circuits (MMIC). This process allows large areas of III-V and Si electronics to be integrated with independent electrical connections taking place at the wafer scale without the parasitic losses of chip level integration.

Optical integration in addition to mechanical and electrical was carried out with TCC. The conduction through single particles allows a low density of these particles to be used (<1%). This allows light to pass through the bonding adhesive while maintaining electrical conductivity. The IQE measurement in Fig 7 shows the conductivity is achieved and a response from all the solar cells is measured. This affirms the combination of mechanical, electrical, and optical integration in this device.

V. CONCLUSION

A conducting adhesive bond mechanically and electrically attaches III-V and Si wafers to create high performance solar cells. This reactive metal bond is achieved by using indium's high solid-state mobility to fuse the metallization of each wafer to metal coated polymer particles. This fused assembly creates low resistance pathways between the wafers. Additionally, the process successfully integrates III-V and Si materials without cracking and with low bow. This reactive ACA process can be used for a wide variety of device integration needs.

REFERENCES

[1] D. C. Law, P. T. Chiu, C. M. Fetzer, M. Haddad, S. Mesropin, R. Cravens, P. Hebert, J. H. Ermer, and J. P. Krogan, , "Development of XTJ Targeted Environment (XTE) Solar Cells for Specific Space Applications," Proc. Of 7th World Conference on Photovoltaic Energy Conversion (WCPEC), (2018).

[2] E. M. Rehder, "Bonding Using Conductive Particles in Conducting Adhesives," U.S. Patent 10,177,265, Jan. 8, 2019.

[3] E. M. Rehder, B. Jun, P. Chiu, S. Wierman, K. Edmondson, X.-Q. Liu, S. Mesropian, P. Pien, J. Boisvert, and N. H. Karam, "Environmental testing of inverted metamorphic solar cells for space", Proc. Of 40th Photovoltaic Specialist Conference (PVSC), p 3608-3611, (2014).

[4] "Processing Procedures for CYCLOTENE 3000 Series Resins," Dow, 2005.

The Effect of Dust Hygroscopicity on Soiling and Self-Cleaning Processes in a Condensing Environment

Jordan Eidlisz[1], Nadera Sultana[2], Illya Nayshevsky[1,3], QianFeng Xu[1,4], and Alan M. Lyons[1,3,4]

1. College of Staten Island, City University of New York, Staten Island, 10314
2. Graduate Center, City University of New York, New York, 10016
3. PhD Program in Chemistry at the Graduate Center, City University of New York, New York, 10016
4. ARL Designs, 125th Street, New York, NY 10027

Abstract— **Dew promotes chemical interactions between soilants and glass that lead to increased soiling rates and cleaning costs. Anti-soiling coatings have been developed to address these issues, and prior experiments have quantified the soiling impact of several categories of particle chemistries. In this paper, the impact that the hygroscopicity of a soilant has on soiling and cleaning values was measured on hydrophobic coated glass and compared to bare glass samples. Results will be presented from UV-visible direct transmittance and optical image processing measurements to characterize soiling and self-cleaning of surfaces as a function of particle hygroscopicity in a condensing environment, mimicking natural dew conditions.**

Keywords—soiling, hygroscopic, hydrophilic, hydrophobic coating, self-cleaning, anti-soiling, solar cover glass

I. INTRODUCTION

Dew can promote chemical interactions between dust and the solar cover glass of photovoltaic panels which can lead to increased cleaning difficulty and expenses [1]. Accordingly, the severity of soiling is reported to be greatest when dew is present [2]. Even in dry and arid environments, a significant amount of dew can form, as the lower nightly temperatures facilitate dew formation [3]. For example, dew formation has been reported to occur around 200 nights per year in the northern Negev desert and on 73% of measured nights in the Taklimakan Desert in China [4,5].

To combat these phenomenon, various anti-soiling coatings have been developed. Our group has shown that Hydrophobic surfaces reduce soiling compared to hydrophilic Bare Glass surfaces [6]. This lower soiling rate is caused by a dust herding mechanism, where the highly mobile drops on the Hydrophobic surface can sweep dust particles into multiple concentrated piles as the drop evaporates. In addition, we have also shown that water collection from dew can be increased by over 36% by creating a hybrid surface, containing an array of hydrophilic features at the top, with the remainder of the surface being hydrophobic [7]. These optimized hybrid surfaces were shown to increase cleaning efficacy for several soilants tested compared to hydrophilic or uniformly coated Hydrophobic surfaces [8]. However, an experiment has not yet been performed to determine whether the hygroscopic nature of soilants effects the soiling and cleaning rates of a surface.

A useful metric to evaluate the hygroscopicity of a mineral is deliquescence relative humidity (DRH). The DRH of a substance is the relative humidity value at which the mineral will absorb sufficient water to spontaneously become an aqueous solution. Hygroscopic minerals are abundant in nature and usage. NaCl deposition occurs frequently in areas near oceans, and $CaCl_2$ deposition is common in agricultural areas. Our hypothesized salt soiling mechanism is as follows. Salts are typically deposited onto solar glass surfaces via wind from the ocean or agricultural fields. After dew condenses on the soilant, with more dew condensing on more hygroscopic soilants, the soilant dissolves. The water then evaporates as the sun comes up, causing the salts to precipitate into larger particles, while also forming a thin film haze on the surface, both of which obstruct sunlight and reduce solar panel efficacy.

II. EXPERIMENTAL PROCEDURE

A. Materials

Low iron glass substrates from Pilkington, 3 mm thick, were cut to 50 mm x 57 mm and thoroughly washed. Hydrophobic coatings on glass were prepared by thermally laminating a fluorinated ethylene-propylene (FEP) layer onto the glass, followed by peeling as described previously [6,7,8]. An array of hydrophilic channels was formed in Hybrid samples by selectively abrading away the coating as shown schematically in Fig. 1. Contact angle measurements were taken of every sample using a model 250-f1 contact angle goniometer (ramé-hart Instrument Co.) Ten measurements per sample were automatically performed using 5 μL droplets of DI water to determine the contact angle. The soilants used were Arizona Test Dust, $CaCO_3$ NaCl, and $CaCl_2$ as described in Table 1. Prior studies have shown that particles soiled on photovoltaic panels usually have an average diameter of 16 μm [9]. All soilants used were milled to ensure large outlier particles were not present. The soilant particle size distribution was determined by taking several images using a Nikon SMZ 1500 with an INFINITY2-1C camera after the soilant was applied to a Bare Glass surface.

TABLE I.   SOILANT TYPE, SOLUBILITY, DRH, REACTIVITY, AND SOURCE

Soilant	Solubility in Water	DRH	Source	Median Particle Diameter
Arizona Test Dust	Negligible	Negligible	PTI Inc.	5 μm
$CaCO_3$	0.047 g/L	Negligible	Sigma Aldrich	5 μm
NaCl	360 g/L	76%	Table Salt	6 μm
$CaCl_2$	811 g/L	31%	Cabisco Chem	6 μm

B. Experimental Methods

Glass surfaces were coated with one of the selected soilants in the accelerated soiling chamber [6]. This apparatus was designed to replicate soiling conditions in the Arizona desert [10]. The dust holder was surrounded by a 50°C heater to prevent the hygroscopic soilants from absorbing moisture and dissolving within the dust holder. For each trail, 40. mg of soilant was weighed and placed in the holder. Two glass samples were placed adjacent to one another on a Pelitier plate at the bottom of the soiling chamber, and four soiling cycles were run as detailed in previous articles [6,7,8]. The environment was kept at 70% RH, and water was allowed to condense on the glass surface at 10°C for two minutes prior to dust ejection. For each self-cleaning measurement, a single soiled sample was placed in the artificial dew condensation chamber for two hours as detailed in previous articles [6,7,8].

C. Analysis

Transmittance measurements from 350 - 850nm were taken with a Lambda 650 UV-vis spectrophotometer (PerkinElmer) at three locations (4 mm x 15 mm) on each sample (Fig. 2) at three different time points: before any soiling cycles; after 4 soiling cycles had been completed; and after the cleaning process had been completed. The representative solar weighted transmittance (RSWT) spectrum was calculated from the transmittance measurements. [11]

Microscopy images were taken of each sample at five locations (Fig. 2) using a Celestron 5 MP Digital Microscope concurrent with the UV-vis measurements. The size of each microscopy image was 80mm² (10mm x 8mm). A python program was used to analyze these images and quantify the surface area covered (SAC) by dust.

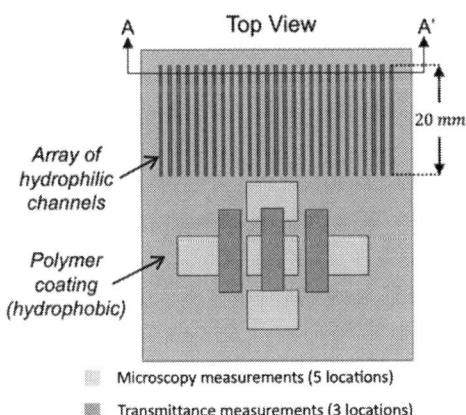

Figure 1. Hybrid surface with microscopy measurement locations (orange) and transmittance measurement locations (green).

III. RESULTS

A. Surface Charecterization

The average contact angle for the Hydrophobic surfaces was 126.2° ± 1.8°, demonstrating the non-wetting nature of the surfaces. Previous measurements taken on hybrid surfaces revealed that the average distance between adjacent hybrid features was 1.6 mm, and the average width of hydrophilic features was 0.4 mm.

B. Artificial Soiling

The percent direct transmittance (%T) loss from soiling depends on both surface coating and soilant type. The overall decrease in %T after four applications of each soilant type is shown in Fig. 2. On Hydrophobic surfaces, soiling levels were significantly lower than on Bare Glass surfaces. The average %T loss for Hydrophobic surfaces was 3.4% ± 0.7%, while the average %T loss for Bare Glass surfaces was 6.6% ± 2.1%.

On Bare Glass surfaces the %T loss increases with increasing hygroscopicity of the soilants; Arizona Test Dust resulted in the lowest soiling (-4.8% ± 0.2%) whereas CaCl₂, the most hygroscopic soilant, resulted in the largest %T decrease (-10.2% ± 0.6%). On Hydrophobic surfaces, all types of particles soiled similarly, however %T reduction on CaCO₃ (-4.5% ± 0.2%) was larger than the other soilants (average of -3.0%).

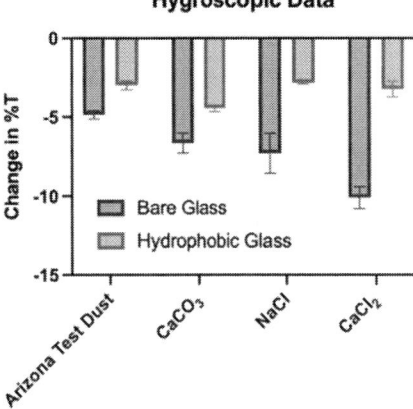

Figure 2. Change in RSWT %T as a function of surface coating and soilant type after four soiling cycles, averaged over n = 4 trials.

Soiling was also evaluated using microscope images as shown for both types of surfaces and the four types of soilants in Fig. 3. Confirming the %T measurements, Bare Glass surfaces qualitatively appear more highly soiled than Hydrophobic surfaces. In addition, the trend between soiling level and the hygroscopic nature of the soilant can be seen on Bare Glass surfaces. On these Bare Glass surfaces, the average size of the highly scattering areas was 16 μm², 22 μm², 37 μm², and 121 μm² for Arizona Test Dust, CaCO₃, NaCl, and CaCl₂ respectively. On Hydrophobic surfaces the same trend is observed, but on a smaller scale, with the average size being 15 μm², 19 μm², 20 μm², and 25 μm² for Arizona Test Dust, CaCO₃, NaCl, and CaCl₂ respectively.

C. Soiling Reproducability and Uniformity

Both transmittance and SAC data demonstrated soiling uniformity within samples, and reproducibility between samples. The average difference in %T within samples was 1.9% ± 1.2%, and the average difference in %T between samples soiled using the same soilant was 1.0% ± 0.8%, inferring both high uniformity and high reproducibility. The average difference in SAC between samples soiled using the same soilant was 4.6% ± 1.5%. As the hygroscopic nature of the soilant increased the reproducibility and uniformity decreased.

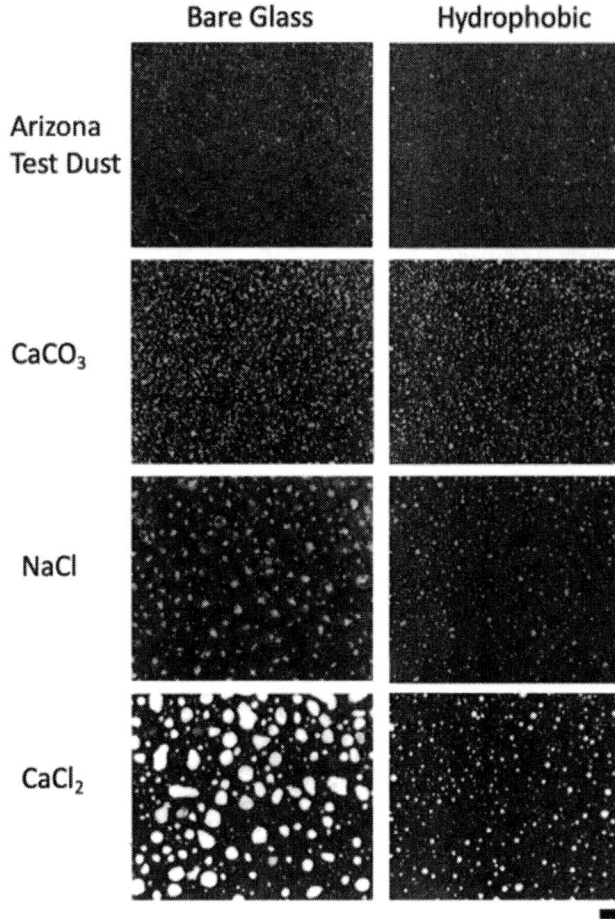

Figure 3. Microscopy images of each soilant type after four dust applications on Bare Glass and Hydrophobic surfaces.

D. Artificial Dew Cleaning

Artificial dew cleaning trials are underway and will be presented at the conference.

IV. CONCLUSION

A correlation between wetting properties of a glass surface and soiling levels was observed. Surfaces with a Hydrophobic FEP coating soiled significantly less than Bare Glass hydrophilic surfaces by both %T and SAC metrics. Using Hydrophobic surfaces as opposed to Bare Glass reduced the %T loss by as much as 68%, from 10.1% to 3.2%, when using the most hygroscopic soilant, $CaCl_2$. A correlation between soilant hygroscopicity and soiling level was also observed. On all surfaces a decrease in %T, and increase in SAC, was observed as the DRH of soilants increased. However, this effect was more apparent on Bare Glass than Hydrophobic surfaces.

We hypothesize that the superior performance of hydrophobic coatings results from the condensation mechanism and droplet mobility. Water condenses in a dropwise manner on Hydrophobic surfaces, as opposed to a filmwise manner on

hydrophilic Bare Glass surfaces. On a Hydrophobic surface, the droplets shrink laterally as the liquid evaporates, further consolidating the soilant by the previously reported dust herding mechanism [7]. Thus the fraction of surface obscured by the soilant is relatively low and the %T decrease is minimized. On Bare Glass, the more soluble salts fully dissolve in the larger area liquid films. The salts precipitate to form larger crystals and/or films as the surface is dried. Unlike liquid drops on hydrophobic surfaces that shrink laterally, the area covered by liquid does not shrink during evaporation on Bare Glass because the liquid-solid contact line is pinned to the hydrophilic surface. The increase in soilant coverage as the DRH of a soilant increases is readily apparent in the optical micrographs, especially on Bare Glass substrates.

ACKNOWLEDGMENT

The authors acknowledge National Science Foundation grant no. CBET- 1805179 for supporting this work. Jordan Eidlisz would also like to thank the generous support of the Macaulay Honors College at the College of Staten Island.

REFERENCES

[1] M. Melcher, R. Wiesinger, and M. Schreiner, "Degradation of glass artifacts: Application of modern surface analytical techniques," *Accounts Chem. Res.*, vol. 43, no. 6, pp. 916–926, 2010.

[2] K. Ilse, B. Figgis, M.Z. Khan, V. Naumann, C. Hagendorf, Dew as a Detrimental Influencing Factor for Soiling of PV Modules, IEEE J. Photovoltaics. (2018) 1–8. doi:10.1109/JPHOTOV .2018.2882649.

[3] Ilse, K. K., Figgis, B. W., Naumann, V., Hagendorf, C. & Bagdahn, J. Fundamentals of soiling processes on photovoltaic modules. *Renewable and Sustainable Energy Reviews* **98**, 239–254 (2018).

[4] D. Beysens *et al.*,"Application of passive radiative cooling for dew condensation," *Energy*, vol. 31, no. 13, pp. 2303–2315, 2006.

[5] Hao, X.-M.; Li, C.; Guo, B.; Ma, J.-X.; Ayup, M.; Chen, Z.-S. Dew Formation and Its Long-Term Trend in a Desert Riparian Forest Ecosystem on the Eastern Edge of the Taklimakan Desert in China. *Journal of Hydrology* **2012**, *472-473*, 90–98.

[6] Nayshevsky, I.; Xu, Q.F.; G. Barahman, Lyons, A.M., Fluoropolymer coatings for solar cover glass: Anti-soiling mechanisms in the presence of dew, *Solar Energy Materials and Solar Cells*, **2020**, 206, 110281

[7] Nayshevsky, Illya, *et al.* "Hydrophobic–Hydrophilic Surfaces Exhibiting Dropwise Condensation for Anti-Soiling Applications." *IEEE Journal of Photovoltaics*, vol. 9, no. 1, 2019, pp. 302–307., doi:10.1109/jphotov.2018.2882636.

[8] Nayshevsky, I.; Xu, Q.F.; Newkirk, J.M.; Furhang, D.; Miller, D.C.; Lyons, A.M., Self-cleaning hybrid hydrophobic-hydrophilic surfaces: durability and effect of artificial soilant particle type, *IEEE J. Photovoltaics*, **2020**, 10, p.577-584

[9] I. Nayshevsky, Q. Xu, and A. Lyons, "Literature Survey of Dust Particle Dimensions on Soiled Solar Panel Modules," in *2018 International PV Soiling Workshop*, 2018. http://www.condensationexperiments.com/WorldBubbleMap/citations.html

[10] W. Herrmann, M. Schweiger, G. Tamizhmani, B. Shisler, and C. Kamalaksha, "Soiling and self-cleaning of PV modules under the weather conditions of two locations in Arizona and South-East India," *2015 IEEE 42nd Photovolt. Spec. Conf. PVSC 2015*, pp. 1–5, 2015.

[11] Miller, David C., et al. "Examination of an Optical Transmittance Test for Photovoltaic Encapsulation Materials." *Reliability of Photovoltaic Cells, Modules, Components, and Systems VI*, 2013, doi:10.1117/12.2024372.

[12] M.-A. Simard and C. Jolicoeur, "Chemical admixture- cement interactions: Phenomenology and physico-chemical concepts," *Cem. Concr. Compos.*, vol. 20, no. 2–3, pp. 87– 101, 1998.

On the Stability of Indium Tin Oxide with Functional Layers Back Contact Applications in Semitransparent Cu(In,Ga)Se2 Solar Cells

Robert Fonoll-Rubio, Marcel Placidi, Torsten Hoelscher, Angelica Thomere, Zacharie Jehl Li-Kao, Maxim Guc, Victor Izquierdo-Roca, Roland Scheer, Alejandro Pérez-Rodríguez

IREC, Sant Adrià de Besòs, Spain

Electronic Engineering Department, Polytechnic University of Catalonia (UPC), Barcelona, Spain

Martin-Luther-University (MLU), Halle, Germany

Departament d'Enginyeria Electronica i Biomedica, IN2UB, Universitat de Barcelona, Barcelona, Spain

Cu(In,Ga)Se2 (CIGSe) solar cells grown on semitransparent back contacts offer different applications such as bifacial devices, semitransparent building-integrated photovoltaics (BIPV), and tandem solar cells. Focusing on BIPV, solar modules can be an integral part of windows, skylights, glass-based facades, blinds and other related structures. Additionally, a fast increase of interest in integration of semitransparent PV modules is observed in agrovoltaics. Currently, semitransparent BIPV products based on amorphous silicon (a-Si) are being commercialized. However, the energy conversion efficiency of a-Si solar cells has been stuck around 10%. In this regard CIGSe absorbers offer an alternative for the BIPV market that promises higher efficiencies than a-Si without reducing or even with higher flexibility in application strategies. One of the key elements for the high efficiency device development is the back contact, which should maintain its properties (transparency and electrical conductivity) under the conditions of the absorber layer deposition. In this work, the stability of different ITO based back contact configurations under the co-evaporation processes used for the synthesis of CIGSe is studied. Bare ITO contacts and contacts functionalized with nanometric Mo and MoSe2 layers were subjected to the analysis, which allowed to achieve a relative solar cell efficiency higher than 80% of the efficiency of reference CIGSe devices fabricated with a standard Mo back contact (which had energy conversion efficiencies up to 16.6%). The stability of the semitransparent back contact was analyzed by Raman scattering spectroscopy allowing to define the maximum critical processing temperature of CIGSe which has no harmful effect on the ITO. Variation of the thickness of the functionalizing nanometric layers allowed to improve the device efficiency through improved electrical quality of the CIGSe/ITO interface, without significant loss of the transparency of the back contact.

978-1-7281-6118-1/22 $31.00 © 2022 IEEE

Bulk lifetime study of p-type Czochralski silicon with different processing history using quinhydrone-methanol surface passivation

Tasnim K. Mouri[1,2], Ajay Upadhyaya[3], Ajeet Rohatgi[3], William N Shafarman[1,2], Ujjwal K. Das[1,2]

[1]Institute of Energy Conversion, University of Delaware, Newark, United States of America
[2]Materials Science & Engineering, University of Delaware, Newark, United States of America
[3]School of Electrical and Computer Engineering, Georgia Institute of Technology, Atlanta, United States of America

Abstract— The bulk lifetimes of p-type textured Cz-Si wafers with different thermal processing history were assessed using quinhydrone-methanol (QH-MeOH) surface passivation. The wafer bulk quality quantified by effective carrier lifetime (τ_{eff}) and implied voltage (iV_{oc}) depends on the thermal processing of different surface treatment and passivation materials. It was observed that wafers which had H$_2$S reaction at 550ºC demonstrated higher τ_{eff} = 235 µs and iV_{oc} = 693 mV implying bulk quality improvement after the H$_2$S reaction process.

Keywords—p-type silicon, surface passivation, hydrogen sulfide reaction, bulk lifetime, quinhydrone methanol passivation

I. INTRODUCTION

One of the critical requirements for a high efficiency silicon (Si) solar cell is to obtain an efficient surface passivation, while maintaining its good bulk quality after going through all device fabricating steps at different thermal and chemical processes. The record-breaking efficiencies of 23.56% for p-type passivated emitter rear contact Si (PERC) [1] and 26% for p-type tunnel oxide passivating contact (TOPCon) were achieved by minimizing surface recombination losses, resulting in high open circuit voltages ($\approx$ 690 mV in PERC and $\geq$ 730 mV in TOPCon) [2]. Conventionally used aluminum oxide (Al$_2$O$_3$) and silicon nitride (SiN$_x$) have demonstrated good passivation quality on heavily doped p$^+$ and n$^+$ diffused surfaces at relatively low temperatures of ~400°C and ~450°C, respectively. Thermally grown Si oxide (SiO$_2$) at > 850°C is also widely used for n-type Si surface passivation. In our previous work, we have demonstrated that industrial Czochralski-grown (Cz) planar and textured Si wafer surfaces can also be effectively passivated by hydrogen sulfide (H$_2$S) gas reaction at 550°C to attain surface recombination velocities (SRV) < 2 cm/s and < 10 cm/s on 2 Ω.cm n-type and p-type Si, respectively [3,4]. Such different thermo-chemical surface treatments and passivation can alter the bulk electronic quality of the Si wafers, as observed by phosphorus diffusion-induced impurity gettering, hydrogen diffusion in Si, and additional bulk defect formation on p-type wafer during thermal oxidation processing, to name a few.

In this work, we studied the effect of the different thermo-chemical surface passivation processes on the bulk qualities of p-type Cz Si wafers. The relative changes of effective minority carrier lifetime (τ_{eff}) are measured by the quasi-steady-state photoconductance (QSSPC) method, after wet-chemical removal of all surface layers followed by the bare Si surface passivation with quinhydrone-methanol (QH-MeOH) solution. QH-MeOH is one of the less hazardous processes that provides a relatively stable chemical passivation at room temperature [5].

II. EXPERIMENTAL

For these experiments, 1" x 1" boron doped p-type <100> Cz Si wafers with resistivity of 2.0 ± 0.1 Ω.cm and thickness 160 ± 10 µm were used. The wafers were cut from 6" pseudo-square wafers with one side planar and the other side random pyramid texture. The wafers were divided into 3 sets for different surface passivation treatments. Sample set (a) received no surface processes as reference samples. Set (b) had phosphoryl chloride (POCl$_3$) diffusion (n$^+$ diffused layer) at 860°C in the textured side with sheet resistance of 74 Ω/sq and the planar side passivated by Al$_2$O$_3$ using atmospheric pressure chemical vapor deposition (APCVD) at 530°C followed by SiN$_x$ deposition at 450 °C using low frequency (~ 40KHz) PECVD. Set (c) had n$^+$ diffusion same as in (b) with textured surface passivation by thermal oxide (SiO$_2$) grown at 855°C and planar surface passivation by thermal oxide/SiN$_x$ stack. The sets (b) and (c) can be considered as the PERC cell precursor structures with Al$_2$O$_3$ and SiO$_2$ passivation, respectively. These structures were then subjected to chemical etching in 10% HF solution for 45 min to remove all surface passivation layers.

The samples were then further divided into 2 groups for each set. The group 1 wafers were set aside, while the group 2 samples were subjected to H$_2$S gas reaction at 550°C for 60 mins in an atmospheric pressure thermal chemical vapor deposition (APTCVD) reactor in a H$_2$S-Ar gas mixture with 3.5% H$_2$S gas-phase concentration. All wafers were then etched in 10% HF for 1 min followed by in an acid mixture of 100:1 nitric acid to HF (HNA) solution for 5 min. This etching step completely removes all surface impurities, n$^+$ diffused layer, and 3 µm of Si, exposing the Si bulk that had undergone different processing history for different sets. Complete removal of the n$^+$ layer was ensured by sheet resistance measurements that matched the sheet resistance of reference samples with no n^+ layer. Since all wafers have undergone the same wet-chemical etching steps, the samples were assumed to have same surface structures differing only in their processing history.

Effective minority carrier lifetime (τ_{eff}) were measured with surface passivated by QH-MeOH assumed to elucidate the changes in bulk quality with different processing history. The QH-MeOH solution was prepared by dissolving 0.436 gm quinhydrone powder in 100 ml methanol, resulting in molarity of 0.02M, and put into a zip-lock plastic bag. The Si wafers were cleaned in 10% HF solution for 1 min prior to putting them into a QH-MeOH containing bag. A Sinton WCT-100 tool was used to measure the injection-level dependent τ_{eff} using the QSSPC technique [6,7]. In this work, the τ_{eff} and implied V_{OC} (iV_{OC}) values are reported at excess carrier density (Δn) of 1×10^{15} cm^{-3} and at 1-sun intensity, respectively [5]. Table 1 lists the different processing history for each sample set and the group with schematic structures.

TABLE I. THERMAL AND CHEMICAL HISTORY FOR EACH TYPE OF STRUCTURES ON P-TYPE TEXTURED Si WAFERS

Wafer type		
Bare wafer (Ref)	**PERC precursors with Al$_2$O$_3$ (n$^+$/Al$_2$O$_3$)**	**PERC precursors with SiO$_2$ (n$^+$/SiO$_2$)**
1. Starting structure		
p-type Cz-Si	p-type Cz-Si, n$^+$ layer, Al$_2$O$_x$, SiN$_x$	p-type Cz-Si, n$^+$ layer, Th. SiO$_2$, SiN$_x$
2. 45 min HF etching to remove both side passivation layers		
p-type Cz-Si	p-type Cz-Si, n$^+$ layer	p-type Cz-Si, n$^+$ layer
3. Half of the samples of each type were reacted in H$_2$S, while the other half skipped to step 4 (schematics below are for H$_2$S reacted wafers only)		
p-type Cz-Si, S, S	p-type Cz-Si, n$^+$ layer, S-layer	p-type Cz-Si, n$^+$ layer, S-layer
4. 1 min HF + 5 min HNA etching to remove S-reacted surface and n$^+$ diffused layer		
	p-type Cz-Si	
5. QH-MeOH surface Passivation and lifetime test (effective lifetime values reported in this work are only in this step)		
	Sealable bag, QH-MeOH solution, Si wafers	

III. RESULTS AND DISCUSSION

Measurement of τ_{eff} and iV_{oc} of all the p-type textured Si wafers immersed in the QH-MeOH solution in a zip lock bag,

gave an insight about process-induced changes in wafer bulk quality. The quinhydrone (mixture of p-benzoquinone and hydroquinone) forms strong bonds with hydrogen and with -OCHy molecules from the methanol solution at the Si surface [8]. As shown in Fig. 1, lifetime increased for the first 20 min and stayed stable thereafter for at least another 30 min. Therefore, the τ_{eff} and iV_{oc} values reported here were measured after at least 20 min in QH-MeOH solution in subsequent experiments.

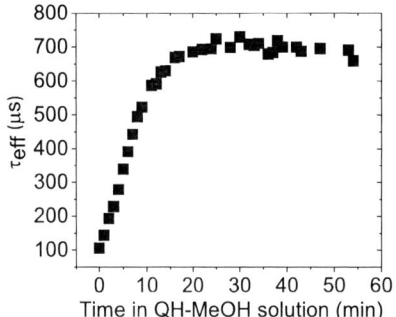

Fig. 1: τ_{eff} as a function of time in QH-MeOH solution measured at $\Delta n = 10^{15}$ cm^{-3} for n-type textured Si.

Figs. 2(a) and 2(b), show the values of τ_{eff} and iV_{oc} at 1-sun intensity for each wafer set and group with different thermo-chemical surface processing steps, after removal of all surface passivation and diffused layer, and surface passivated by QH-MeOH solution (step 5 in Table I).

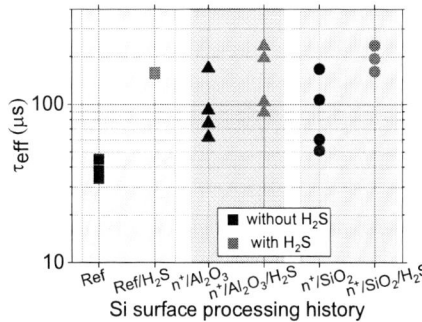

Fig. 2(a): Effective minority carrier lifetime at $\Delta n = 10^{15}$ cm^{-3} in QH-MeOH solution for each wafer group with different processing history. Black and red symbols represent wafers processed without and with intermediate H$_2$S reaction, respectively.

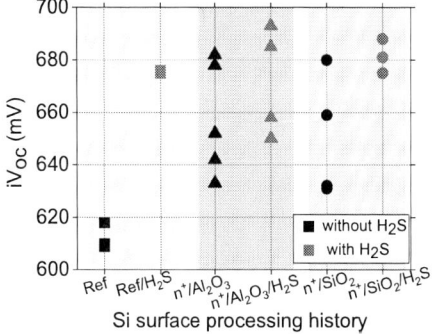

Fig. 2(b): 1-sun implied V_{OC} (iV_{oc}) for each wafer group with different processing history. Black and red symbols represent wafers processed without and with intermediate H$_2$S reaction, respectively.

978-1-7281-6118-1/22 $31.00 © 2022 IEEE 593

Figure 2(a) shows that the n^+ diffusion process (n^+/Al_2O_3 and n^+/SiO_2, black triangles and circles) leads to higher τ_{eff} (>150 µs) than the Ref samples (< 50 µs, black squares). This is likely due to the known bulk quality improvement by the impurity gettering process typically observed in n^+ diffusion and glass removal steps. It is notable that H_2S passivation reaction on the Ref samples without the history of n^+ diffusion (red squares) also improved the bulk quality with τ_{eff} >150 µs. The samples with processing history of both n^+ diffusion and H_2S reaction (red triangles and circles) achieved τ_{eff} ~250 µs. The iV_{OC} values estimated form the injection-level dependent lifetime curves, plotted in Fig.2(b), show a similar trend as τ_{eff} for the samples with different processing history. The iV_{OC} increased for the samples with H_2S reaction history; from 618 to 676 mV for the Ref, 682 to 693 mV for the "n^+/Al_2O_3" and 678 to 688 mV for the "n^+/SiO_2" processing history.

Fig. 3 shows the injection-level-dependent τ_{eff} curves for the Ref, n^+/Al_2O_3 and n^+/SiO_2 samples with and without the H_2S reaction history. The figure shows improvement of lifetime over the Δn range of 10^{14} to 10^{16} cm^{-3} for all sample structures with H_2S reaction history. This suggests improvement of bulk quality of the wafers during H_2S reaction. Experimentally, H_2S gas exposure to Si(100) surfaces in an ultra-high vacuum (base pressure ~4 x 10^{-11} (Torr) chamber is shown to result in dissociative adsorption ($H_2S \rightarrow H+HS$) at temperatures ranging from -145 to $425^{O}C$ [9]. Temperature-programmed desorption (TPD) and Auger electron spectroscopy (AES) measurements showed a simultaneous desorption of hydrogen, accompanied by S diffusion into the Si crystal over the temperature range 525–625°C and the formation of Si–S–Si bonds by breaking the

of atomic H generated by the dissociation of H_2S into the Si bulk that passivate bulk defects, and 2) a temperature-induced improvement of bulk properties.

To test these hypotheses, we repeated the experiments as illustrated in Table I with the modification of step 3, where half of the samples were subjected to an Ar annealing process at 550°C for 60 mins – same as the H_2S reaction process, but without the H_2S gas flow. The wafers were then characterized by the same approaches described above. Ar annealing achieved $\tau_{eff} \approx 245$ µs with an enhancement of iVoc from 615 mV to 705 mV for the Ref, 618 mV to 698 mV for "n^+/Al_2O_3" and 623 mV to 700 mV for the "n^+/SiO_2" processing history. Fig. 4 shows the injection-level dependent τ_{eff} curves for the Ref, n^+/Al_2O_3 and n^+/SiO_2 samples with and without the Ar annealing. τ_{eff} improved over the Δn range of 10^{14} to 10^{16} cm^{-3} for all sample structures with Ar annealing. These results indicate that just the thermal annealing in an inert Ar atmosphere can improve the bulk quality even without the presence of S and/or hydrogen. So, based on these results, we can suggest that H_2S reaction provides the surface passivation [4] but the improvement of bulk quality occurs due to the temperature annealing effect during H_2S reaction.

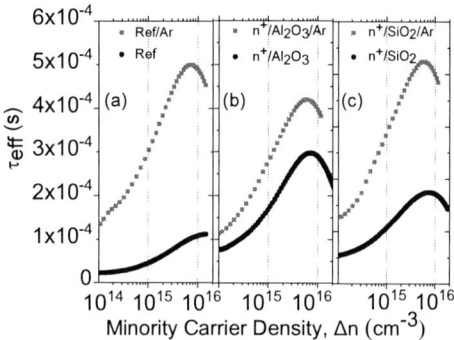

Fig. 4: Comparison of injection level (Dn) dependent effective lifetime curves of p-type Cz wafers with QH-MeOH surface passivation with and without Ar annealing for 3 different history of surface thermo-chemical processing (a) Ref and Ref/Ar; (b) n^+/Al_2O_3 and n^+/Al_2O_3/Ar; (c) n^+/SiO_2 and n^+/SiO_2/Ar.

IV. CONCLUSION

In this work, we investigated the bulk quality of p-type Cz-Si wafers with different thermal and chemical processing histories. The bulk quality was elucidated by the measured τ_{eff} values with the surface passivated by quinhydrone-methanol solution at room temperature. The results indicate that the n^+ diffusion, Al_2O_3, and/or thermal oxide growth processes all improve bulk quality by varying degrees. The H_2S reaction process additionally improves the bulk quality for all the wafers studied here. The increase in bulk lifetime during H_2S reaction process is not due to S and/or H diffusion in the bulk but is likely due to temperature-induced changes of the bulk quality, since similar bulk quality improvement is observed even for wafers annealed in inert Ar ambient without any H_2S gas flow.

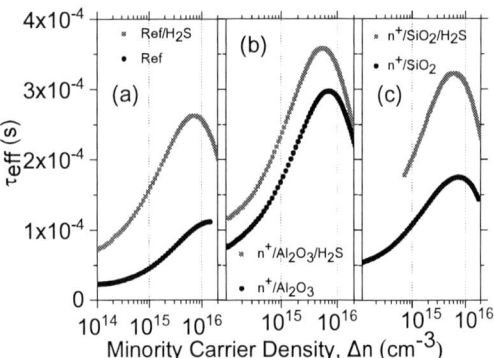

Fig. 3: Injection level (Δn) dependent τ_{eff} curves of p-type Cz wafers with QH-MeOH surface passivation with history of different surface thermo-chemical processing (a) Ref and Ref/H_2S; (b) n^+/Al_2O_3 and n^+/Al_2O_3/H_2S; (c) n^+/SiO_2 and n^+/SiO_2/H_2S for their respective highest lifetime and iV_{oc} values.

Si dimer bonds. However, S-diffusion into the Si bulk at the similar time-temperature range was found to be limited to < 5 nm [8] from the surface. Han et. al. used the mean free path of the sulfur Auger electron (LMM at 154 eV) to estimate the depth of S diffusion to be ~2.5 monolayers, [8]. Therefore, significant bulk defect passivation by S is unlikely. Two other possible reasons are: 1) the temperature-dependent diffusion

ACKNOWLEDGMENT

This work is supported by the US Department of Energy's Office of Energy Efficiency and Renewable Energy (EERE) under Solar Energy Technologies Office (SETO) Agreement Number DE-EE0008554.

REFERENCES

[1] http://taiyangnews.info/technology/trina-solars-23-56-efficiency-for-p-type-perc-cell/

[2] S. Mack, D. Herrmann, M. Lenes, M. Renes, A. Wolf, "Progress in p‑type Tunnel Oxide‑Passivated Contact Solar Cells with Screen‑Printed Contacts," Solar RRL 5, 2100152 2021.

[3] U. K. Das, T. K. Mouri, R. Theisen, Y. W. Ok, A. Upadhyaya and A. Rohatgi, "Dopant diffused Si surface passivation by H2S gas reaction and quinhydrone-methanol treatment," 2021 IEEE 48th Photovoltaic Specialists Conference (PVSC), pp. 1355-1358 2021.

[4] U. Das, R. Theisen, G. Hanket, A. Upadhyaya, A. Rohatgi, A. Hua, L. Weinhardt, D. Hauschild, and C. Heske, "Sulfurization as a promising surface passivation approach for both n- and p-type Si," Proc. 47th IEEE PVSC Conf., pp. 1167 – 1170, 2020.

[5] N. A. Kotulak, M. Chen, N. Schreiber, K. Jones, R. L. Opila, "Examining the free radical bonding mechanism of benzoquinone– and hydroquinone–methanol passivation of silicon surfaces," Applied Surface Science 354 469–474 2015.

[6] R A Sinton and A Cuevas, "Contactless determination of current–voltage characteristics and minority carrier lifetimes in semiconductors from photoconductance data," Appl. Phys. Lett. 69 2510 1996.

[7] A. Cuevas and D. Macdonald, "Measuring and interpreting the lifetime of silicon wafers," Sol. Energy 76 255 2004.

[8] R. Har-Lavan, R. Schreiber, O. Yaffe, and D. Cahen, "Molecular field effect passivation: Quinhydrone/methanol treatment of n-Si(100)," Journal of Applied Physics 113, 084909 2013.

[9] M. Han, Y. Luo, N. Camillone, III, and R. M. Osgood, Jr., "Reaction of H2S with Si(100)," J. Phys. Chem. B, Vol. 104, No. 28, 2000.

PV Module Degradation Due to Frequent and Prolonged Inverter Clipping: A Preliminary Study

Manjunath Matam*, Ryan M. Smith†, and Hubert Seigneur*
*Florida Solar Energy Center, University of Central Florida, USA
†Pordis LLC, Austin, Texas, USA

Abstract—Photovoltaic (PV) plant owners usually install PV arrays that are larger than the inverter's rated capacity. With a DC to AC ratio greater than unity, the PV array occasionally operates at a sub-optimal point imposed by the inverter. This is common practice in the USA, and is known as 'clipping.' Clipping may be implemented by operating the array at voltages below the maximum power point (MPP), or more commonly, above the MPP. Advantages of clipping operation are the utilization of an inverter's full capacity and improved financial break-even time. Regulatory requirements may also mandate reduced power output from an array in a forced clipping situation known as curtailment. However, the long-term impact on the PV array's life, array degradation, aging, hot spots, and module warranty has not been adequately investigated for prolonged off-MPP operation. This paper presents preliminary studies of the potential long-term impact of clipping on the PV array beyond energy production. For this study, the module and string level data, primarily through I-V curves and IR imaging, is investigated along with analysis of inverter electrical data. This preliminary work shows that the PV array exhibits significantly different temperature signatures if operated below the maximum power point as opposed to above, conditions which could be experienced when clipping. In addition, the initial data revealed unique IR imaging patterns: checkerboard patterns at voltages below the MPP and uniform elevated temperature patterns at voltages above the MPP.

Index Terms—photovoltaic, PV, module, clipping, DC/AC ratio, curtailment

I. INTRODUCTION

Prolonged inverter clipping and frequent curtailments could be a severe threat to USA's PV 50-useful years goal if not investigated. Inverter clipping depends on the PV plant DC/AC ratio [1]. Curtailment, forcing the PV plant to reduce its output power, is dependent on grid conditions and financial decisions [2]. DC/AC ratios from 4591 US utility-scale plants commissioned over the last decade reveal a concerning trend (Figure 1) [3]: it has been increasing without a complete understanding the potential impact to plant health.

Investors and owners desire a high DC/AC ratio plant to achieve increased net income, a high internal rate of return (IRR), and an optimal levelized cost of energy (LCOE) [4], [5]. Apart from the cost-benefit economics, the DC/AC ratio is also influenced by the plant latitude, module technology, array orientation, and inverter efficiency [6]. As a result, the prevalence of high DC/AC ratios is increasing. For instance, countries have stipulated a minimum or standard DC/AC ratio

This material is based upon work funded by the U.S. Department of Energy's Solar Energy Technologies Office Award Number DE-EE0008157. The opinions, findings, and conclusions stated herein are those of the authors and do not necessarily reflect those of the U.S. Department of Energy.

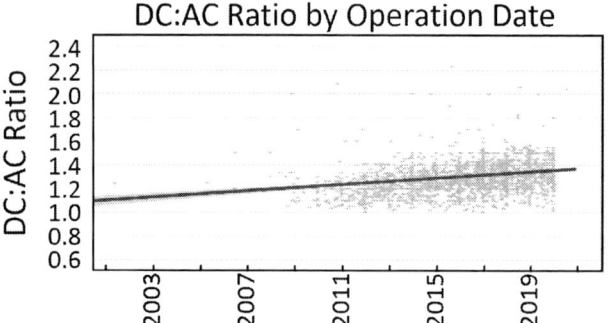

Fig. 1. The DC/AC ratio of 4,591 utility-scale PV plants installed in USA show an increasing trend, where one sample denotes one PV plant.

for new plants [7]; for example, Germany prescribes 1.42 [6], and Australia defines 1.31 for crystalline-silicon (c-Si) and 1.12 for thin-film (TF) systems [8]. However, no research institutions have investigated the impact on module degradation or plant life; an exception is a few simulation studies on inverter reliability and performance masking [9]. Power curtailment, an equally common occurrence, is a direct consequence of increased penetration of variable renewable energy (VRE) on the grid [10]. Though clipping and curtailment differ in terminology, the effects are the same: operation of the array at a non-ideal operating point.

With the highest domestic utility-scale PV plant penetration, California curtails some plant's output daily, up to ten times in a few cases, to prevent the excess generation, over-voltage, avoid congestion, and are paid for this through bidding [11]; curtailment may therefore be economically beneficial to plant owners but not necessarily to plant health [11]. For example, using a 1618 GW PV installation and 346 GW storage scenario across the USA by 2050, a 6.6% annual curtailment leads to a 7% increase in LCOE [10] based on a $0.02/kWh LCOE and standard module degradation rates, but may be higher if off-peak operation conditions are considered. Incidentally, the USA aims for a $ 0.02/kWh LCOE by 2030, which may cause curtailment consequences to be realized earlier and more frequently.

A. Key Concerns Associated with Increasing DC/AC Ratio

Prolonged clipping via off-MPP high-voltage operation are a direct consequence of a high DC/AC ratio and curtailment. A failure-ticketing analysis of 350 utility-scale PV systems with DC/AC ratios ranging from 1 to 1.3 shows that the inverter triggered the ticket in 43% of cases [12]. However, module failure tickets were raised only when significant power loss was detected at the inverter level; unfortunately, detecting module failure becomes more difficult with a high DC/AC ratio [13]. In addition, field studies show a direct and inverse relationship between the DC/AC ratio and inverter reliability [9]; a high ratio leads to increased inverter thermal loading, mean time between failures (MTBF), and replacement costs, negatively impacting LCOE. As a result, inverter companies recently began stipulating a maximum DC/AC ratio to minimize thermal loading, increase MTBF, maintain performance, efficiency, and warranties [14]. However, to our knowledge, no PV module manufacturers have raised concerns about DC/AC ratio, and hence I-V operating point, affecting degradation rates and warranties.

Curtailment level and frequency depend on storage reserves, particularly traditional energy generators (coal, gas, nuclear), present on the grid [10]. Not all countries consider curtailment an economically viable option; for instance, Germany, with relatively similar PV penetration as the US, penalizes curtailment. Moreover, most PV installations in Germany are in the distribution sector, unlike in the US, where more than half of installations are in the utility sector. Researchers worldwide continue to view curtailment as wasted energy [15] but not a threat to plant health. In particular, the curtailment slew rate which is the rate at which output power is reduced, may impact the inverter components' thermal loading [14] and, ultimately, plant life.

This paper examines the data collected from a 6.2 kW grid-tied PV system installed in Cocoa, FL at the Florida Solar Energy Center, a research center of the University of Central Florida. It analyzes various measurements including infrared (IR) imaging, module temperatures, and inverter-collected DC electrical data at various simulated DC to AC ratios. Throughout this paper, we use the general term clipping to describe any off-MPP operation of the array, such as through traditional inverter power limiting or curtailment. When appropriate, clipping may be described as high voltage or low voltage and describes the voltage relative to the maximum power point voltage - high meaning above V_{MPP}, and low meaning below V_{MPP}. A description of the experimental setup, strategy adopted for implementing the high and low voltage clipping on a commercial inverter, and clipping levels studied are presented in Section II. Data filters and assumptions in the analysis, experimental results are presented in the Section III.

II. LOW AND HIGH VOLTAGE CLIPPING EXPERIMENTS

The objective of the experimental tests is to simulate clipping conditions while collecting inverter electrical data, module temperatures, and infrared images of the selected modules. Although the impact of clipping on inverters should not neglected, this paper is limited to the impact of clipping on PV module long-term performance and reliability. The authors have noted that inverters usually implement clipping by moving the operating point from the MPP to a point at a higher voltage. However, it is possible to achieve the same level of clipping by transitioning from the MPP to a lower voltage operating point. Therefore, we investigate both low- and high-voltage clipping operation.

A. Experimental setup, Array configuration, Equipment

A 6.2 kW grid-tied PV plant was chosen to perform a set of clipping experiments. The setup was installed and has been operating since November of 2018. The PV array is composed of two strings (named String-1 and String-2 throughout) consisting of 12 series connected 270W modules per string. Specifications of the PV modules and complete schematic details are available in [16]. Each string feeds one of the two independent maximum power point trackers (MPPT) of a 12kW SMA Sunny Boy 12000 TL-US-10 grid-tied inverter. Inverter electrical data, plane-of-array (POA) irradiance (Licor pyranometer), and module temperatures (type-T thermocouple) are collected at 5-second intervals through a Campbell Scientific data logger. Between the PV array and inverter is an in-situ I-V curve tracer configured for capturing string and module I-V curves every 30 min when the incident irradiance is above 200 W/m². Each string I-V curve has around 1000 points and there is about a 4-minute time difference between string traces.

To understand the thermal response of modules with common faults when the array operates under clipping conditions, two laboratory-prepared modules with induced cell cracks and two with interconnect failures were installed into String-2. Indoor electroluminescence (EL) images of the cracked and interconnect-failed modules are shown in Figure 2.

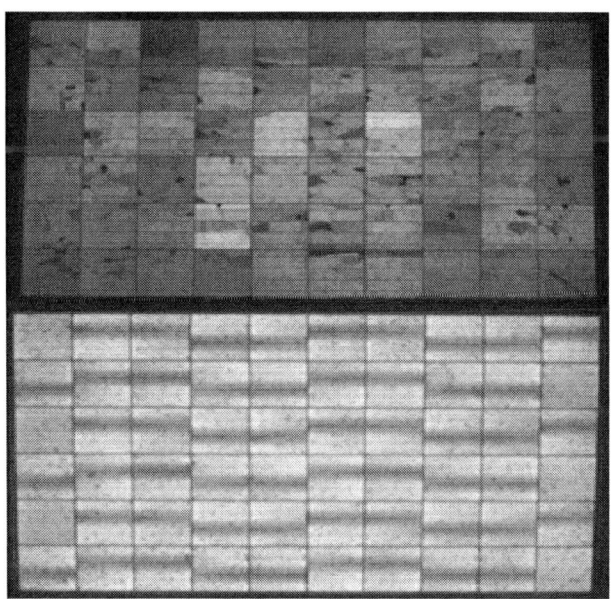

Fig. 2. EL image of the cracked (top) and interconnect failed (bottom) PV module.

Two types of measurements were collected for analyzing, quantifying, and assessing the thermal impact of clipping on the PV modules. First, type-T thermocouples were attached to the backsheets of three PV modules within each string. Instantaneous temperature data was recorded with no averaging. With

978-1-7281-6118-1/22 $31.00 © 2022 IEEE

this, it was possible to compare the temperature changes as a result of each clipping condition on the individual PV modules and on each string's aggregate temperature. To understand the impact of clipping on modules with different induced internal failures, IR imaging of a normal, cracked, and interconnect-failed module was performed. Analyzing these exemplars is critical to understanding the behavior and presentation of module faults when the utility-scale PV arrays are subjected to clipping. An IR camera supported by a tripod was placed normal to the module surface and the images were collected at a 1-second interval. IR measurements were conducted by following accepted practices: irradiance $>700W/m^2$, camera positioned within a 60-90° angle to the module plane (90° is perpendicular to the module), with a low ambient temperature and wind speed [17]. IR imaging shows the temperature distribution across the module.

B. Approach to low and high voltage clipping using constant voltage mode

Power versus voltage (P-V) curves of the experimental PV array measured at various irradiance on a clear-sky day are shown in Figure 3. Note that the array P-V curve is obtained by combining the individual P-V curves of both experimental strings. The irradiance difference between the two string P-V curves is $11.47(\pm 6.35)$ W/m^2; therefore no mathematical corrections were performed. Under normal conditions, the inverter maintains the PV array at the maximum power point (solid dot markers in the figure). As the experimental PV array is of a fixed 6480W nameplate rating, we computed the voltage limits for different DC/AC ratios to limit the AC output power. Computed low- and high-voltage clipping limits for various DC/AC ratios are graphically shown in the Figure 3 (dotted horizontal lines) and tabulated in Table I.

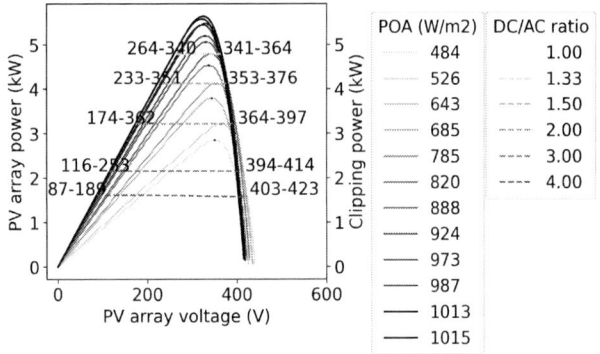

Fig. 3. Power-Voltage curve showing high- and low-voltage range of the inverter DC input for various DC/AC ratios.

C. Experimental conditions and challenges

To our knowledge, commercial inverters universally force the PV array to operate above V_{MPP} to implement clipping and curtailment operations, which we term high-voltage clipping. Alternatively, the inverter could operate the array below V_{MPP} to achieve the same limitation on AC output power, termed low-voltage clipping. Unfortunately, commercial inverters do

not usually have a provision to enable low-voltage clipping operation. However, to facilitate a comparison between the two clipping conditions, high- and low-voltage, we adopted a constant voltage (CV) approach which is allowed by the inverter associated with this experiment. Importantly, the inverter allows for independent operation of each input, and thus one string is operated at a high constant voltage while the other string is operated at a low constant voltage, thus facilitating simulated high- and low-voltage clipping for the study. Note that the lower limit on the inverter in constant voltage mode was 153V.

At the time of experimentation, sky conditions at the test site were mostly cloudy. Florida receives an annual average 77% moderate to severely cloudy days. However, clear sky days are critical to understanding the impact of clipping by capturing the impact on module temperatures. Therefore, experiments were conducted over a series of days, and in some cases for just a few hours a day when the sky was mostly clear, in order to build a more complete data set for analysis. A list of the experiments, experimental conditions, type, and location of the measurements collected are tabulated in Table II.

Apart from the irradiance conditions, a power difference between both strings was anticipated although efforts were made to achieve a ¡5% power difference between strings throughout the experiments. This was due to the limitations imposed by the manually set constant voltage mode for the low- and high-voltage simulated clipping.

III. EXPERIMENTAL RESULTS AND POTENTIAL IMPACT ON PV MODULE RELIABILITY

After analyzing the results of various experimental tests conducted over a range of days, we down-selected to four days, the details of which are detailed in Table II. In this section, we present the data filters, assumptions, and analysis of the results.

A. Data filters and assumptions in the analysis

Time-synchronized data was processed for detecting, filtering, and removing unreliable and outlier measurements. The data was filtered to eliminate high rate-of-change irradiance conditions, where the POA irradiance changed greater than $15W/m^2/minute$, and is referred to as the one minute sky-stability filter. The data was then filtered by the power difference between the String-1 and String-2 with data removed if the string-to-string power difference was more than 5%. These filters are effective in dropping unreliable data and illuminating the differences between the high- and low- voltage clipping conditions.

B. Time-series temperature results and analysis

The time-series data indicates that modules experience a lower overall temperature when operated at low-voltage clipping compared to the high-voltage clipping, as shown by the box plots of Figure 4 and Figure5. Figure 4 shows the difference in temperature between both strings for each time-stamp. Negligible string-to-string difference was observed when both strings were operating under the same condition

TABLE I. HIGH- AND LOW- VOLTAGE CLIPPING LIMITS OF VARIOUS DC/AC RATIOS FOR THE EXPERIMENTAL PV ARRAY

	PV array power (W)	DC/AC ratio	Inv. AC power limit (W)	Clipping power (W)	Voltage range for clipping			
					Low voltage		High voltage	
1	6480	1.00	6480.00	6480.00	-	-	-	-
2	6480	1.33	4872.18	4872.18	264.0	340.0	341.0	364.0
3	6480	1.50	4320.00	4320.00	233.0	351.0	353.0	376.0
4	6480	2.00	3240.00	3240.00	174.0	362.0	364.0	397.0
5	6480	3.00	2160.00	2160.00	116.0	253.0	394.0	414.0
6	6480	4.00	1620.00	1620.00	87.0	189.0	403.0	423.0

TABLE II. EXPERIMENTAL CONDITIONS AND MEASUREMENTS COLLECTED ACROSS MULTIPLE TEST DAYS

Exp.	Day	Experimental condition	S1 Temp (mean, °C) M1, M6, M10	S2 Temp (mean, °C) M15, M20, M23	IR Imaging		
					Nominal (M16)	C. Crack (M18)	Interconnect (M17)
1	April 4, 2022	S2-MPPT; S2-390V; S2-153V	-	-	-	✓	-
2	April 12, 2022	S2-MPPT; S2-390V; S2-153V	-	-	✓	-	-
3	April 19, 2022	S1,S2 MPPT; then S1-153 V, S2-390V	✓	✓	-	-	-
4	May 17, 2022	S1, S2-MPPT; S1-375 V, S2-153 V; vice versa	✓	✓	-	-	✓

(MPP or VOC). However, when one of the strings operated above MPP (375V) and the other below MPP (153 V), the lower voltage string was on average 2.5°C cooler even though the both strings were producing same power. This supports one potential advantage of clipping to a voltage below MPP.

Figure 5 shows the measured temperature of the String-1 and String-2 modules under the same conditions. Although the temperature is effectively the same when both strings were operating under the same condition (MPP or VOC), there is a difference in their absolute value which may be explained by the experimental conditions. Both the MPPT and V_{OC} data was recorded at an average 700 W/m^2 irradiance. The additional V_{OC} temperature increase is attributed to thermalization and recombination within the cells [18]. The simulated clipping experiment, labeled '$S1_{high}S2_{low}$' in the figure, was conducted at an average irradiance of 800 W/m^2. The higher irradiance results in a higher overall temperature for the modules compared to the MPPT and V_{OC} conditions. Although we believe the experiments indicate a temperature difference between the strings under simulated low- and high-voltage clipping conditions, longer duration tests are needed to understand and quantify the factors responsible for the temperature differences observed.

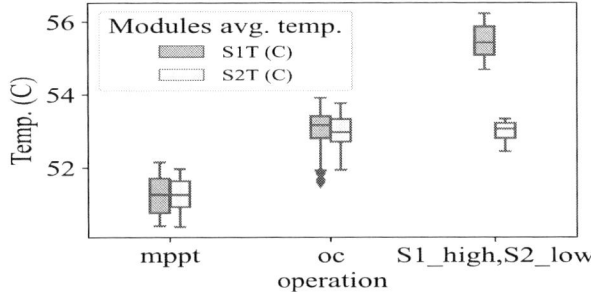

Fig. 5. Box plot of measured String-1 and String-2 module temperatures by operating condition.

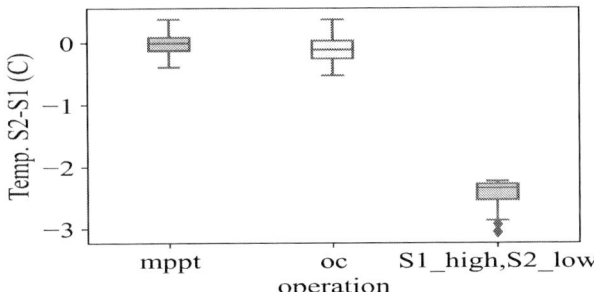

Fig. 4. Box plot of the temperature difference between String-1 and String-2 modules by operating condition.

C. IR imaging results and analysis of select PV modules

The infrared camera provided TIFF formatted raw images. A Python program was used to process the raw images and produce 1200 DPI JPG images with consistent temperature scales. The temperature scale chosen varied based on the day and time of imaging. However, the same temperature scale was used for comparing images of each specific module. Throughout the experiments, the IR images considered for analysis were captured after five to ten minutes of stabilization at the target operating condition. Analysis of the IR images of normal, cracked, and interconnect-failed modules collected under different operating conditions are presented below and are tabulated in Table II.

1) IR imaging of a normal PV module: Normal module IR images are shown in Figure 6. Simulated low-voltage clipping exhibited clear checkerboard patterns. The high temperature cells were operating approximately 15°C higher than neighboring cells which themselves were at the lowest temperature compared to the MPP and high-voltage clipping condition. At MPP, the modules were operating at a lower temperature than when operating under the high-voltage clipping condition. The PV module showed a fairly uniform distribution and an average 5°C higher temperature under the high-voltage clipping condition.

2) IR imaging of a cracked cell PV module: Infrared images of a module with induced cracked cells, shown in Figure 7, showed earlier stage checker patterns, wherein, a few cells showed an average 5°C higher temperature compared to the neighbor cells. Again, the module showed a less apparent infrared pattern at MPP condition and an average 5°C higher and

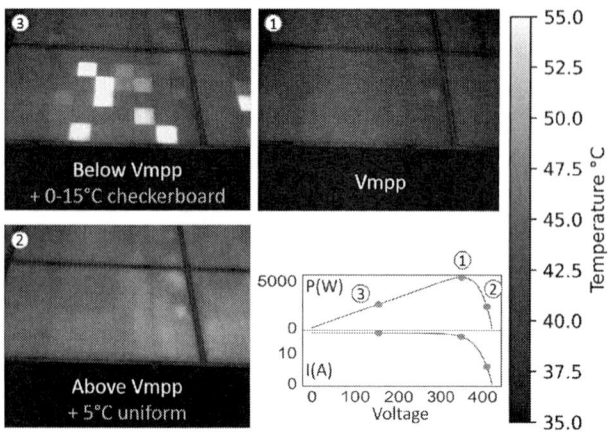

Fig. 6. IR image of a nominal PV module when the string is operated at the three indicated conditions.

uniformly distributed temperature under high-voltage clipping conditions. It is noted that the IR image at MPP condition exhibits similar patterns seen in the EL image of the module, Figure 2. The corresponding areas of interest in both IR and EL are highlighted in Figure 7.

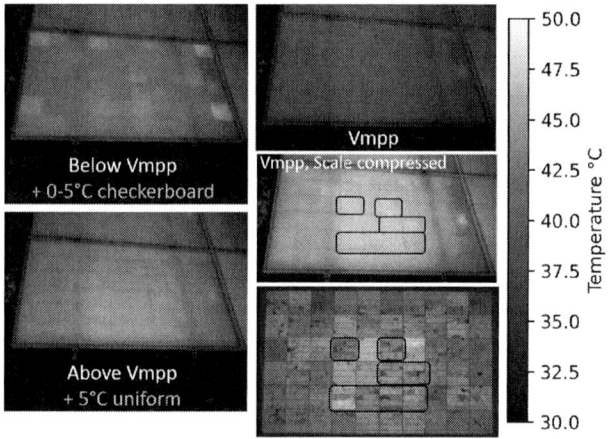

Fig. 7. IR image of a induced cracked cell PV module when the string is operated at the three indicated conditions.

3) IR imaging of a interconnect-failed PV module: Images of the interconnect-failed module, shown in Figure 8, shows higher and more uniform temperature distribution under the high-voltage clipping condition. Interestingly, the same cells showed low temperature and high temperature bands at the high-voltage clipping condition. As the module used in these tests was modified in the laboratory, the interconnect failures are known. The darker region had the interconnect cut at both ends of the cells and brighter region had no cuts at the ends. The low-voltage and MPP conditions do not show the brighter or darker regions. Instead, the former shows early checker patterns and the latter shows a more uniform temperature pattern. Portions of the IR image collected at high-voltage show similar patterns as of EL image, but these are not seen at MPP as was in the crack-module. The corresponding areas of interest in both IR and EL are highlighted in Figure 8

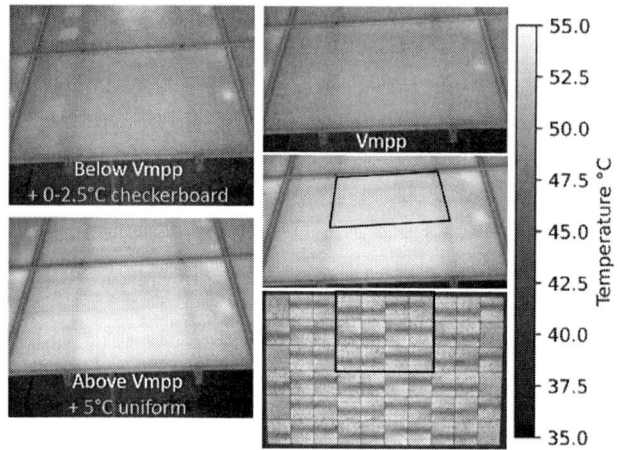

Fig. 8. IR image of a module with induced interconnect failures when the string is operated at the three indicated conditions.

D. IR imaging results and analysis of the PV array

IR imaging of the complete PV array under three operating conditions was performed and is shown in Figure 9. Several of the modules without induced failures showed similar patterns and trends. At low-voltage clipping, a few modules of both String-1 and String-2 showed the checkered patterns, some of which spread uniformly across the module might imply manufacturing-related defects within the PV module. However, a majority of the modules showed no signs of elevated temperatures. At MPP, all modules of the both strings consistently showed expected uniform temperatures. It is to be noted that full power is extracted at MPP from all modules reducing carrier recombination or cell mismatch. Under simulated high-voltage clipping, a uniformly distributed elevated temperature was noted across all the modules of the PV array.

E. Degradation analysis of the PV modules

The PV array's IR imaging showed checkered patterns across a few modules. To determine if the checker pattern was a result of degradation, an analysis of the maximum power point data from in-situ I-V curves was performed. As stated previously, I-V curve tracing equipment has been collecting module-level I-V curves every 30 min for the past 3 years [19]. I-V curve data was used to compute each modules P_{mpp} at each time-stamp with POA irradiance and back-of-module temperature data included in the analysis. A one minute sky-stability filter was applied to the irradiance data. The *Python* language based *rdtools* software was used to calculate each module's degradation in P_{MPP}. Four modules of String-1 were considered for the analysis (M1, M4, M5, and M12). Of these, M1 and M12 showed the normal temperature profile while M4 and M5 showed a checkered temperature profile. The M1 temperature data was applied to the other modules due to a lack of back-of-module temperature sensors on the three modules. The degradation results are shown in Figure 10. The two normal modules showed an average 0.725% annual degradation rate, where as the two checkered pattern modules showed an average 0.795% annual degradation rate. The results

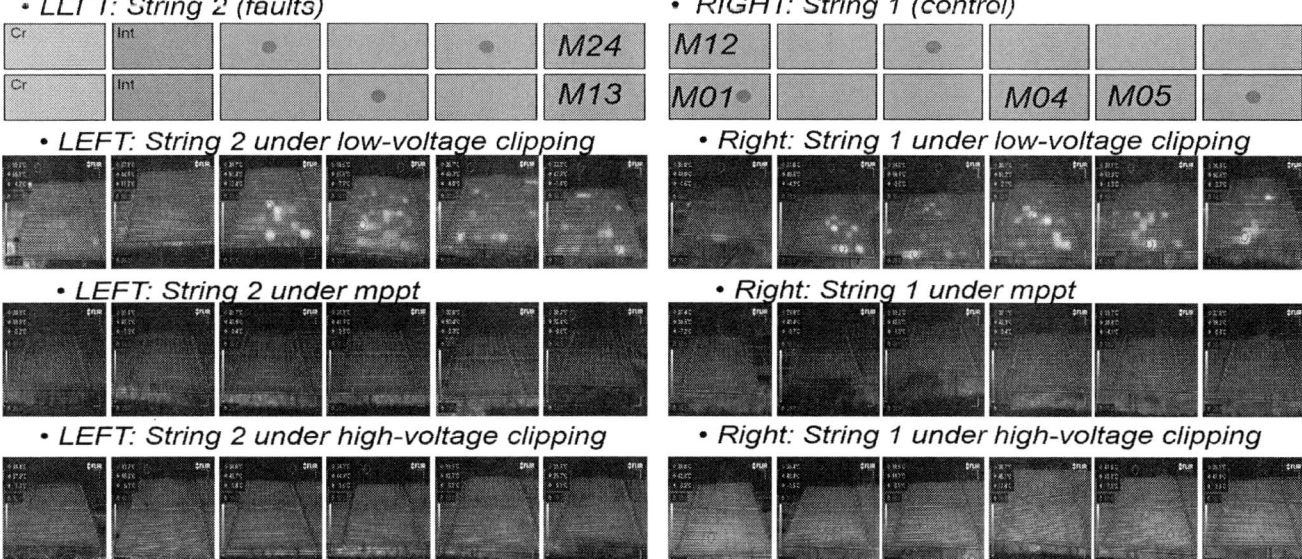

Fig. 9. IR images of the PV array under different operating conditions. First row: field position and module numbering in String-1 and String-2; Second row: modules at simulated low-voltage clipping; Third row: modules at maximum power point (MPP) operation; Fourth row: modules at simulated high-voltage clipping.

indicate no statistically significant difference between the module degradation rates. However, further long term studies are needed to understand if the checkered patterns are a result of degradation or may become more apparent after further field exposure.

IV. CONCLUSION

The 4591 utility-scale PV plants installed in the USA over the past decade show a trend: PV plants' DC/AC ratios have been increasing over time. DC/AC ratios above unity and more frequent curtailment translate to an increased occurance of clipping, wherein the PV array is operated at a voltage above the maximum power point for an extended duration, and raises unanswered questions: how will clipping affect module degradation rates, PV plant aging, PV modules' 50-year life, and financial realities. Our short-duration simulated clipping experiments, performed on a 6.2 kW grid PV array, showed that modules exhibit significantly different temperature signatures if operated at low-voltage clipping instead of high-voltage clipping. Time series analysis of the data collected at the long-term and short-term duration tests reveal an interesting observation: high-voltage clipping leads to higher module temperature compared to the low-voltage clipping. The initial data also indicates an effect on the severity of hot spots depending on the clipping voltage. Infrared imaging of the modules confirmed checkerboard patterns under low-voltage clipping conditions and uniformly distributed higher temperature patterns at the high-voltage clipping conditions. An analysis of module-level in-situ I-V curves showed no statistically significant difference in degradation rates between nominal and checkerboard pattern modules in spite of the observed IR pattern. Further work is needed to thoroughly understand any impact on module degradation rates, or degradation mechanisms, and to answer whether it is beneficial to operate the PV arrays through low-voltage instead of high-voltage clipping.

REFERENCES

[1] A. Allik, H. Lill, and A. Annuk, "Effects of price developments on photovoltaic panel to inverter power ratios," in *8th ICRERA*, 2019, pp. 371–376.

[2] M. Bilenko, I. Buratynskyi, I. Leshchenko, T. Nechaieva, and S. Shulzhenko, *Nonlinear mathematical model of optimal solar photovoltaic station design*, ser. Systems, Decision and Control in Energy II. Springer International Publishing, 2021.

[3] U. E. I. Administration. (2021, June) Eia - 860 solar generator data. [Online]. Available: https://www.eia.gov/electricity/data/eia860/

[4] S. Chen, P. Li, D. Brady, and B. Lehman, "Determining the optimum grid-connected photovoltaic inverter size," *Solar Energy*, vol. 87, pp. 96–116, 2013.

[5] H. Wang, M. A. Muñoz-García, G. P. Moreda, and M. Alonso-García, "Optimum inverter sizing of grid-connected photovoltaic systems based on energetic and economic considerations," *Renewable Energy*, 2018.

[6] J. Väisänen, A. Kosonen, J. Ahola, T. Sallinen, and T. Hannula, "Optimal sizing ratio of a solar pv inverter for minimizing the levelized cost of electricity in finnish irradiation conditions," *Solar Energy*, vol. 185, pp. 350–362, 2019.

[7] M. A. Ramli, A. Hiendro, K. Sedraoui, and S. Twaha, "Optimal sizing of grid-connected photovoltaic energy system in saudi arabia," *Renewable Energy*, 2015.

[8] M. Z. Hussin, A. M. Omar, S. Shaari, and N. D. M. Sin, "Review of state-of-the-art: Inverter-to-array power ratio for thin – film sizing technique," *Renewable & Sustainable Energy Reviews*, vol. 74, pp. 265–277, 2017.

[9] A. Sangwongwanich, Y. Yang, D. Sera, F. Blaabjerg, and D. Zhou, "On the impacts of pv array sizing on the inverter reliability and lifetime," *IEEE Transactions on Industry Applications*, vol. 54, no. 4, pp. 3656–3667, 2018.

[10] B. Frew, W. J. Cole, P. Denholm, A. W. Frazier, N. M. Vincent, and R. Margolis, "Sunny with a chance of curtailment: Operating the us grid with very high levels of solar photovoltaics." *iScience*, 2019.

[11] C. ISO. (2021, June) Curtailment: Managing oversupply. [Online]. Available: http://www.caiso.com/informed/Pages/ManagingOversupply.aspx

[12] A. Golnas, "Pv system reliability: An operator's perspective," *IEEE Journal of Photovoltaics*, vol. 3, no. 1, pp. 416–421, 2013.

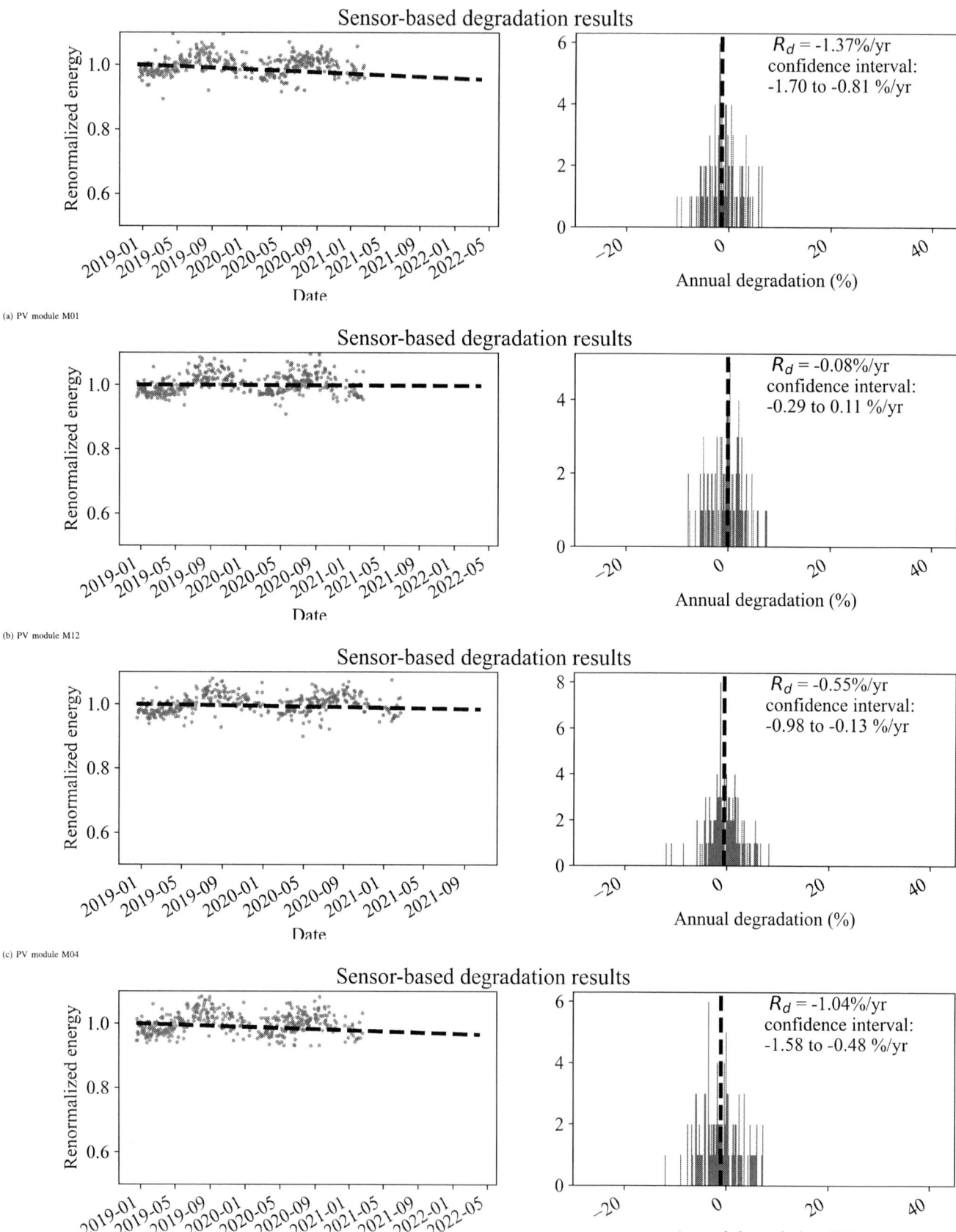

(a) PV module M01

(b) PV module M12

(c) PV module M04

(d) PV module M05

Fig. 10. Degradation results of the indicated modules. Degradation is computed from each module's P_{mpp} extracted from periodic in-situ I-V curves.

[13] M. Emmanuel, J. Giraldez, P. Gotseff, and A. Hoke, "Estimation of solar photovoltaic energy curtailment due to volt–watt control," *Iet Renewable Power Generation*, 2020.

[14] S. Inverter. (2020, Aug.). [Online]. Available: https://www.sma-america.com/shadefix/whitepaper-download.html

[15] N. Stringer, N. Haghdadi, A. Bruce, and I. MacGill, "Fair consumer outcomes in the balance: Data driven analysis of distributed PV curtailment," *Renewable Energy*, vol. 173, no. C, pp. 972–986, 2021.

[16] R. M. Smith, M. Matam, and H. Seigneur, "Mismatch losses in a pv system due to shortened strings," *Energy Conversion and Management*, vol. 250, p. 114891, 2021.

[17] I. E. A.-P. programme. (2014, Mar.) Review of failures of photovoltaic modules. [Online]. Available: https://iea-pvps.org/key-topics/review-of-failures-of-photovoltaic-modules-final/

[18] R. Couderc, M. Amara, and M. Lemiti, "In-depth analysis of heat generation in silicon solar cells," *IEEE Journal of Photovoltaics*, vol. 6, no. 5, pp. 1123–1131, 2016.

[19] M. Matam and H. Seigneur, "An algorithm for filtering the time-series i-v curves of a pv plant," in *IEEE 48th PVSC*, 2021, pp. 0147–0149.

ETFE and its role in the fabrication of lightweight c-Si solar modules

Fabiana Lisco, Farwa Bukhari, Luke Jones, Adam Law, John Michael Walls, Christophe Ballif

École Polytechnique Fédérale de Lausanne (EPFL), Neuchatel, Switzerland

Centre for Renewable Energy Systems Technology (CREST), Loughborough, United Kingdom

Glass-free, lightweight (LW), photovoltaic modules have the potential to enable new uses of solar in building integrated and vehicle integrated applications. Glass-free modules have the advantages of reduced weight, lower-cost mounting solutions and reduced transportation costs. ETFE is a promising candidate to replace glass as the front cover sheet material in module fabrication. In this paper, we report on its optical, chemical, self-cleaning, and morphological properties. Durability is a key issue especially for building-integrated applications since harsh conditions are prevalent outdoors. We report on the effect of sequences of environmental stress tests (UV exposure, and Damp Heat). The module cover material is also subject to regular cleaning, which can cause surface damage. We report on the use of linear abrasion testing to assess the abrasion resistance of ETFE to a variety of abrasive materials. The results provide useful data on the advantages and disadvantages of replacing glass with a durable polymer for front sheets in solar modules.

Thermal Stability of 2D/3D Halide Perovskites

Jeffrey A Christians, Josephine L Surel, Elizabeth V Cutlip

Hope College, Holland, MI, United States

Halide perovskite solar cell stability is a major topic of research for the field and has explored intrinsic material stability, device architecture, and packaging. Efforts at improving halide perovskite material stability have recently seen success with mixed dimensionality, or 2D/3D, halide perovskites where a long-chain ammonium cation can be added to the material and improve thermal stability and moisture resistance. In this report, we explore a series of long-chain ammonium cations and investigate their effects on the thermal stability of methylammonium lead iodide films. We see that even small changes in the long-chain cation structure can lead to significant thermal stability differences, even when only small amounts of this cation are added to the perovskite thin films. This work helps to uncover thermal degradation mechanisms in mixed 2D/3D halide perovskites, and should lead to more thermally stable perovskite solar cells.

Life Cycle Assessment analysis of thin-film, flexible solar panels produced in the Netherlands

Gianluca Limodio, Seba Makhlouf, Edward Hamers, Arno Smets

Delft University of Technology, Delft, Netherlands

HyET Solar Netherlands B.V, Arnhem, Netherlands

This work focuses mainly on LCA of thin-film solar panels produced in the Netherlands. Using all the input real-data from the factory, we are able to estimate all the environmental impact of the manufacturing steps. Moreover, we evaluate the impact of the technological advancement (Higher efficiency, different deposition rate of the bottom cell in tandem configuration). Finally, large scale utilities and location analyses have been carried out.

Front SiON/TCO Stacks Development for Double Side Poly-Si/SiOX Passivated Contacts Solar Cells

Charles Seron, Thibaut Desrues, Christine Denis, Raphaël Cabal, Frédéric Jay, Adeline Lanterne, Quentin Rafhay, Anne Kaminski, Sébastien Dubois

Univ. Grenoble Alpes, CEA, LITEN, Campus INES, 73375 Le Bourget du Lac, France, LE BOURGET DU LAC, France

Univ. Grenoble Alpes, Univ. Savoie Mt-Blanc, CNRS, Grenoble INP, IMEP-LAHC, 38000 Grenoble, France, GRENOBLE, France

This work reports on the use of a front side SiON/TCO bilayer in double side poly-Si/SiOX-based passivated contacts solar cells. This approach presents the advantage of a low indium consumption either by reducing the indium-based transparent conductive oxide (TCO) thickness or by enhancing its substitution with a Zinc-based TCO, such as AZO (Aluminum-doped Zinc Oxide). An electrical study with a TCO thickness reduced to 20 nm on textured surfaces has shown excellent responses for SiON/AZO stacks, especially regarding the contact resistivity. The developed SiON/TCO bilayers were finally integrated in complete solar cells. Interestingly, the substitution of the standard 70 nm-thick ITO:H layer by a 20 nm-thick ITO:H film covered by SiON led to an efficiency gain of +0.5% abs. Regarding AZO, the replacement of the standard 70 nm-thick AZO layer by a 20 nm-thick AZO film covered by SiON conducted to a JSC gain of +0.8 mA/cm2. These gains in performances could be raised with further post-treatments still under investigation. However, the current results already confirm the possibility to optimize thin-poly-Si based passivated contacts solar cells towards In-free fabrication processes.

978-1-7281-6118-1/22 $31.00 © 2022 IEEE

Comparing the Accuracy of Horizon Shade Modelling Based on Digital Surface Models Versus Fisheye Sky Imaging

Daniel Alvarez Mira, Martin Bartholomäus, Sebastian Poessl, Peter B. Poulsen, Sergiu V. Spataru
Technical University of Denmark, Department of Photonics Engineering, Roskilde, Sjælland, 4000, Denmark

Abstract—**Fisheye sky imaging is a well established and accurate method for estimating shading in photovoltaic (PV) applications, however, it requires physical presence and measurements at the studied location. On the other hand, Digital Surface Models (DSM) can be used to estimate horizon shading remotely, and presents a more scalable method for evaluating potential PV installation site efficiently. Nowadays, high resolution DSM data is readily available in many countries, especially for urban environments. This has opened up new potential applications for more efficient shading estimation for aiding the deployment of PV in urban environments where horizon and nearby shading is significant. In this work, the accuracy of DSM-based methods to characterize horizon shading is assessed and compared to that obtained with fisheye imaging based methods. DSMs with different resolutions are studied and results show that high resolution DSMs are comparable. However, the accuracy of lower resolution DSMs will be highly influenced by the characteristics of the studied location.**

Index Terms—**digital surface model, shading analysis, horizon shading characterization, site assessment, fisheye imaging**

I. Introduction

Onsite assessment of the available solar resource is a crucial step in the feasibility analysis and design phase of any PV system, from large scale PV plants to small solar PV powered products. Local topography, surrounding buildings or nearby vegetation and trees can block part of the incoming sunlight for certain periods of the year, causing shading and irradiance losses. Accurately assessing the shading losses for a potential installation site is a fine balance between the size of the project, available resources and the shading potential of the landscape, fx. a field versus an urban environment.

Historically, for utility scale PV plants or rooftop systems with little to no shading elements in the surrounding, the estimation of horizon shadings is done by 3D modelling of the obstruction elements in the surroundings. Most PV design software, such as PVsyst or System Advisor Modelling (SAM), include this possibility.

On the other side, for PV systems installed in urban environments, where the amount of potential shading elements in the surroundings is considerable, 3D modelling becomes a tedious and laborious task, proving itself an unfeasible methodology. Instead, hemispherical sky image-based methods have been used to measure the shading impact of the surroundings on the solar irradiance reaching a specific surface. These are implemented using fisheye camera lenses [1] or sky dome

projection imaging [2], and are well established site-specific shading analysis methods.

However, depending on the relative distance of the obstruction elements (e.g. buildings or trees), the characterization of the surroundings, and therefore the estimated shaded irradiance, may differ drastically at two points separated a few meters apart. This implies that a fisheye image needs to be taken at any possible installation location. This poses a challenge for applications such as solar street lighting or building integrated PV deployed in urban environments, where nearby shading can have a big impact on the available solar irradiance, and hemispherical sky imaging of the possible installation locations might be tedious manual labour or not possible.

On the contrary, digital surface model (DSM) data used in urban planning and visualization is often made publicly available for large cities or entire countries [3, 4]. The geospatial information contained in these models can be used to estimate available irradiance on urban surfaces as shown in [5, 6]. These do not require on-site measurements and allow assessment of a large number of possible installation locations with minimal effort. However, the accuracy of DSM-based methodologies in the characterization of horizon shading elements for shading and irradiance loss modelling has not been assessed yet.

To estimate the accuracy of DSM-based models, we use publicly available DSM data [7] with different resolutions to characterize horizon shading at several locations and compare it to on-site measured horizon shading with a hemispherical sky (fisheye) image shading analysis tool.

II. Digital surface models

Digital Surface Models (DSM) are geospatial digitized maps representing the objects' height. Opposite to Digital Elevation Models (DEM) and Digital Terrain Models (DTM), they include not only the Earth's surface elevation but natural and man-made objects (such as buildings and vegetation) [8, 9]. The most common applications of DSMs are urban/architectural planning and navigation systems where their used is preferred to DEMs or DTMs as they include building and other objects information. However, in recent years DSMs are also used to estimate solar irradiance on building surfaces for indoor climate and photovoltaic potential purposes.

DSMs are usually created using Light Detection and Ranging (LiDAR) or stereo photogrammetry [14]. The former

978-1-7281-6118-1/22 $31.00 © 2022 IEEE

TABLE I: Commercial and open access geospatial digitized maps

Data set	Type	Coverage	Resolution	Acquisition method	Acquisition period
DK-DEM [7]	DSM	Denmark	0.4 m	airborne LiDAR	2018 (1/5 renewed yearly)
Lantmäteriet [4]	DSM	Sweden	0.5 - 1 m	Aerial photos	2019 (1/3 renewed yearly)
ALOS PRISM L1B [10]	DSM	Worldwide	2.5 m at nadir	Satellite images	2006 - 2011
CartoSat-1 (IRS-P5) [11]	DSM	Worldwide	2.5 m	Satellite images	2005 - 2019
TanDEM-X [12]	DEM	Worldwide	12 m	Radar	2010 - 2015
SRTM [13]	DEM	Almost worldwide	30 / 90 m	Radar	2000

requires the use of LiDAR sensors mounted on airplanes or automobiles while the latter derives DSMs from aerial or satellite stereo photographs and stereo matching algorithms such as the Semi-Global Matching algorithm [15].

The coverage, resolution and other parameters of these DSMs are greatly influenced by the acquisition/creation method used and therefore, DSMs can be found with a wide range of different coverage and resolutions. Table I lists a few of these models and DEMs available including their main characteristics without claim of completeness.

For this work, the Danish Elevation Model (DK-DEM) - available to download at the Danish Agency for Data Supply and Efficiency (SDFE) website ([16]) is used. To create this model, and to avoid corrupted measurements, airborne LiDAR on four planes were used on cloud-free days between 2014 and 2015. However, all elevation models produced by the SDFE are updated in a five-year cycle covering 1/5 of Denmark's land area since 2018 [17].

The DK-DEM contains around 415 billion data points over the entire area of Denmark, stored in a raster map with a 0.4 m by 0.4 m cell size (resolution) and, horizontal and vertical accuracy of 0.15 m and 0.05 m respectively. To reduce the file sizes for analyzing, the full DK-DEM is split into 1 km x 1 km tiles which can individually be downloaded for further processing [16, 17, 7, 3]. Figure 1 shows the DSM raster file clipped to the area studied in this work.

III. HORIZON SHADING CHARACTERIZATION

Modelling and analyzing the irradiance losses caused by shading on a given surface, first requires the characterization of the shading potential at that location.

The shading potential of a site is typically represented in the form of a horizon line (HL) diagram, that represents sunlight

Fig. 1: DSM raster of the studied area.

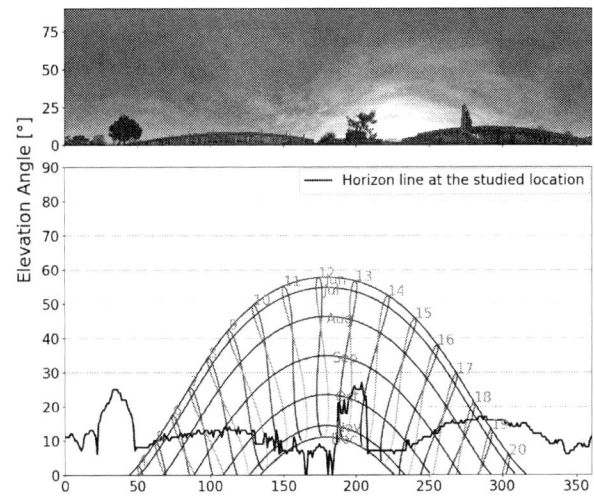

Azimuth Angle [°] (East = 90 ; South = 180 ; West = 270)

Fig. 2: Panoramic image of the surroundings and the Sun's path combined with the horizon line at that location. The panoramic representation is an imperfect transformation from a fisheye image and its use is only for illustrative purposes. The horizon line was obtained following the fisheye methodology (SunEye).

obstruction, both near and far. For each azimuth direction it shows the maximum elevation angle of such elements. This information combined with the Sun's path at a studied location (Figure 2) will determine the instances where the direct irradiance is blocked; and the percentage of the sky dome seen at the location is correlated with the amount of diffuse irradiance reaching the site. In this work, we have used two methods to extract the horizon lines at a specific location, yet there are other methodologies to compute them. Both methodologies are presented below.

A. Characterization Based on Hemispherical Sky Imaging

Shading analysis based on hemispherical sky imaging is often implemented using cameras equipped with 180° fisheye lenses, such as the Solmetric SunEye shading analysis tool [1], which are widely used in the industry. Determining the horizon shading line involves specific steps of camera position calibration and image processing that have been described previously [18, 19, 20]. Figure 4 illustrates fisheye sky images taken at two locations that will be discussed further.

978-1-7281-6118-1/22 $31.00 © 2022 IEEE

B. Characterization Based on DSM and GIS Modelling

The open source software QGIS is used in conjunction with GRASS GIS [21], to process the DK-DEM data and for computing the horizon shading as follows. The tile containing the corresponding part of the DK-DEM is imported to QGIS and clipped to fit the study area by using the internal raster extraction tool. In the next step, the clipped DSM is imported to GRASS and used as an input for a raster tool called *r.horizon* alongside the coordinates of the investigated location, which calculates the maximum elevation of all obstructions from 0° to 360° azimuth angle relative to the point of analysis. This function returns a horizon line for each of the studied locations.

IV. DSM RESOLUTION ANALYSIS

As shown in Table I, the DK-DEM is extraordinarily accurate compared to other available DSMs which do not have such a high resolution. Access to other DSMs in this paper's study area are often only available upon request or otherwise restricted. Therefore, to categorize the effects of using DSMs to model irradiance, we down-sampled the DK-DEM data set to lower resolutions, closer to those of other DSMs with worldwide coverage, and quantified its influence in the horizon shading characterization.

The DK-DEM is provided as a raster-layer with a 0.4 m horizontal resolution. It can be understood as a grey-scale picture with a 0.4 m pixel-length, where the color of each pixel represents a height value. Several down-sampling methods are implemented in GRASS GIS [21], and for the purpose of this study we use the average aggregation [22]. When averaging, the aggregate is computed over all of the input cells whose centers lie within the output cell. The resolution is halved (pixel length doubled) between each step, creating maps with a resolution of 0.4 m (original), 0.8 m, 1.6 m, 3.2 m, 6.4 m and 12.8 m. Figure 3 shows a visual representation of some of the output DSMs further used in this work.

The process described in Section III-B is repeated for all the down-sampled data sets to obtain horizon lines at the studied location similar to those that would be obtained when using DSMs with lower resolutions.

V. HEMISPHERICAL SKY IMAGE VERSUS DSM BASED HORIZON SHADING ESTIMATIONS

To model shaded irradiance in urban environments it is necessary to properly characterize the obstruction elements in the surroundings of a potential PV installation site. Afterwards, correction factors can be applied to the direct and diffuse components of solar irradiance in accordance with the horizon line of the location. On one hand, from the horizon line, a Sky View Factor (SVF) indicating the fraction of the sky visible from the location can be calculated as explained in a previous study [20]. This SVF will determine the amount of diffuse irradiance reaching the location in the irradiance modelling.

On the other hand, for each time stamp during the year, the solar elevation angle has to be compared to the elevation of the obstruction elements at the solar azimuth direction to

(a) 0.4 meters

(b) 3.2 meters

(c) 12.8 meters

Fig. 3: 3D visualization of the studied area DSM with different resolutions after down-sampling the original dataset.

determine if the direct solar irradiance will be blocked. Thus, the accuracy of the horizon line at each azimuth direction, meaning how precisely it estimates the elevation angle of the obstruction elements in all directions, will determine to which accuracy the direct solar irradiance can be modelled.

Therefore, to quantify the HL estimation accuracy of the DSM-based methods the absolute SVF difference and the obstruction elevation error (represented by the root mean square error - RMSE) compared to the fisheye-based method are assessed at two different locations.

A. Study locations

We compare horizon lines obtained with the DK-DEM model to those obtained with the commercially available Solmetric SunEye 210 at two different built-up locations at DTU Risø campus, Roskilde, Denmark. Location 1 presents a low shading profile (Figure 4a) while Location 2 shows more severe shadings (Figure 4b). Afterwards, the horizon lines obtained onsite with the hemispherical sky imaging method, are compared to those derived from the DK-DEM down-

(a) Location 1.

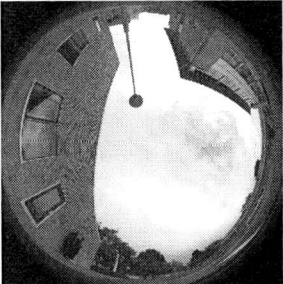

(b) Location 2.

Fig. 4: Fisheye images taken at the studied locations.

sampled versions as well, to evaluate the suitability of lower resolution DSMs for horizon shading characterization.

B. Hemispherical Sky Image versus DSM Based Results

On a first look (Figure 5), the horizon lines obtained with the SunEye and DK-DEM at both locations show good agreement. However, some elements show differences in their representation such as the the tree at Location 1 and the light pole at Location 2, both in the south direction (see the top center region in Figures 4a & 4b). In the case of the low shading profile location (Location 1), the differences seems negligible whilst in the heavily shaded one (Location 2), these differences are more observable. Nonetheless, both methods show similar capabilities recognizing obstructions elements, including their relative direction and magnitude. Differences in the SVF between both HLs are less than 1% for Location 1 and around 4% for Location 2. However these differences will only influence the modelling of the diffuse irradiance, while the effects on the modelling of the direct irradiance requires further investigation.

C. Impact of DSM resolution

As seen in the previous section, the characterization of the surroundings by the DK-DSM seems to be comparable to that one performed by the fisheye-based commercial tool. However, the DK-DEM shows an extraordinarily accurate representation of the Danish landscape. In order to understand the implications of using lower resolution DSMs to model horizon shadings, the down-sampled DSMs presented in Section IV are compared to the fisheye-based tool characterization of the studied locations. The horizon lines extracted from the down-sampled DSMs are represented in Figure 6.

One should take into consideration that the panoramic images displayed in Figure 6 do not represent the studied locations accurately. They are derived from fisheye images where the exact projection type is unknown and could only be estimated, which leads to small deviations from the real shading profile at the studied locations. Its representation is intended as a visualization tool that helps spotting the sources of errors and understanding the intrinsic differences shown by lower resolution DSM when representing certain elements such as buildings, trees or street lights.

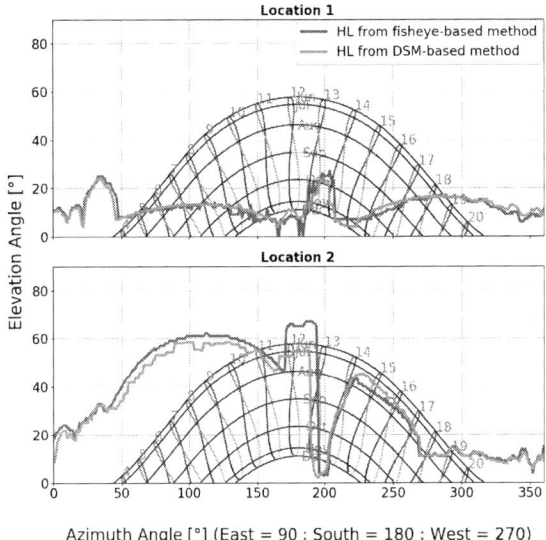

Azimuth Angle [°] (East = 90 ; South = 180 ; West = 270)

Fig. 5: Solar path and obtained horizon lines at the studied locations.

At the location with low shading profile it can be seen that all horizon lines from the different DSM data sets show good comparability in the regions with buildings, while in the regions will small trees there is a gradual reduction in the HL height. This behaviour is due to the grid size of the down-sampled DSMs. As the grid size increases, more points are contained within each of the cells in this model. For example at Location 1, the two trees are not close to any other element taller than them and therefore, as the resolution of the DSM decreases, their elevation is reduced by the natural effects of the average. A similar effect can be seen in the street light located in the south direction at Location 2.

The implications of this effect is that the SVF is reduced as shown in Figure 7. However, for a location with low shading profiles the impact on the SVF is almost negligible. The absolute difference in SVF compared to the one obtained with the fisheye-based method never exceeds 0.35% as shown in Figure 7. The average difference to the reference horizon line (fisheye-based) in each azimuth direction is represented by the Root Mean Square Error (RMSE), also shown in Figure 7. This indicator shows that the average error per azimuth direction increases from 2° with the DK-DEM to 6° in the worst scenario (DSM with 12.8 m resolution).

If the characterization of the horizon shading is analyzed at a location with greater obstruction elements, such as Location 2, the results vary significantly. Not only are the magnitudes of the SVF difference and the RMSE noticeably higher but, they both show an observable increase for DSM resolutions lower than 3.2 m. It is believed that these errors are caused by the use of the averaging method to down-sample the DK-DEM. When averaging for DSM grid sizes lower than the distance from the analyzed location to considerable obstruction elements, the elevation at that site and the horizon shading elements may be

978-1-7281-6118-1/22 $31.00 © 2022 IEEE

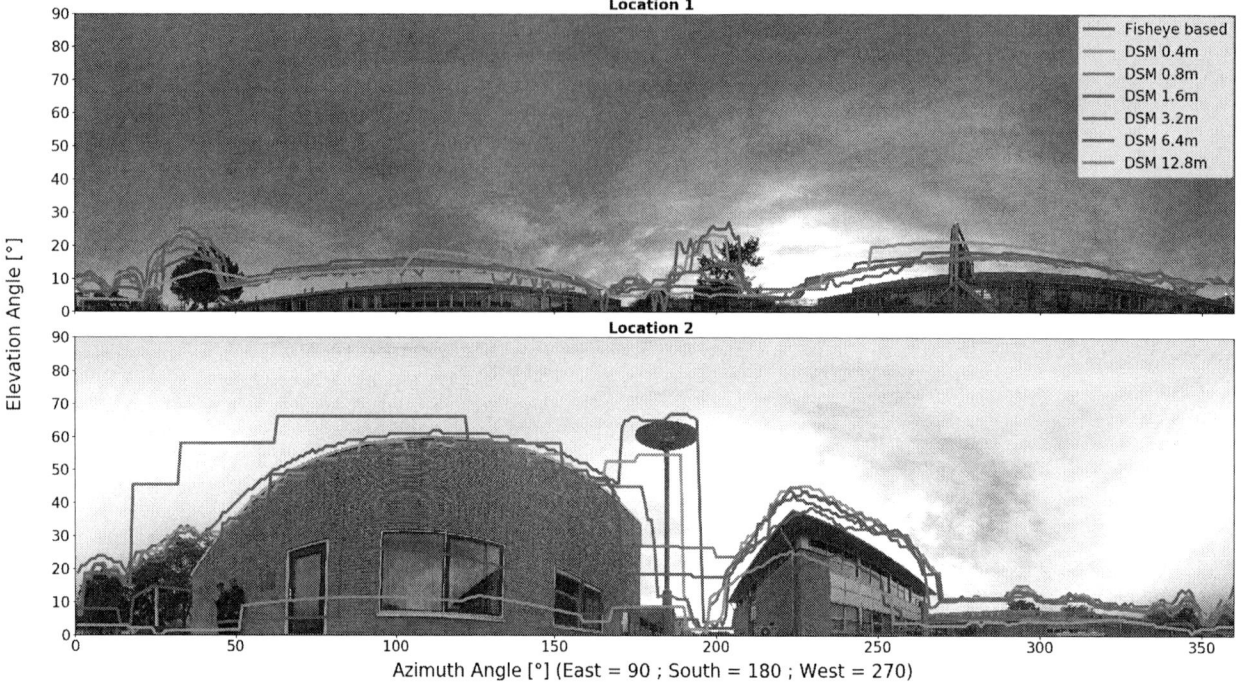

Fig. 6: Visual comparison of the horizon lines obtained at the studied locations. The panoramic representation of the studied locations is a non-perfect transformation from fisheye images only intended as a visualization tool.

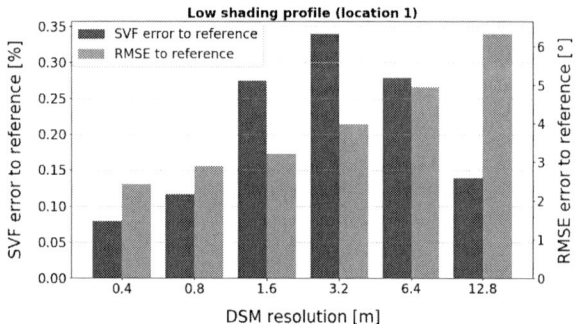

Fig. 7: Horizon line errors to fisheye-based reference in a low shading profile location.

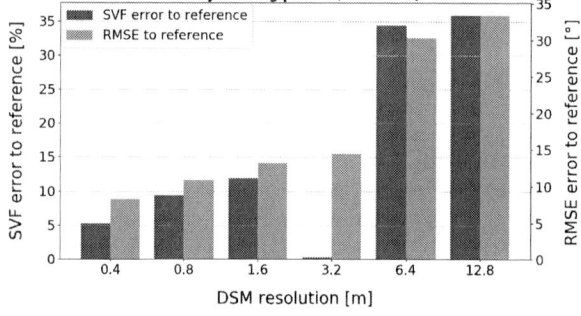

Fig. 8: Horizon line errors to fisheye-based reference in a heavily shaded profile location.

accurately represented. On the contrary, for grid sizes greater than the distance to taller objects, the elevation of the studied location increases due to the inclusion of those obstruction elements in the averaging process at its DSM cell. This will produce an inaccurate representation, where the elevation of the obstruction elements in comparison to that of the studied location is reduced.

It is worth mentioning as well, that even though the 3.2 m resolution DSM seem to show almost no SVF difference, this is just a random effect of overestimations in the horizon line at certain sky regions being balanced out by underestimations at other regions, as visible in Figure 3.

VI. SUMMARY & CONCLUSIONS

Currently, the availability, resolution and thereby applicability of Digital Surface Models is increasing. Initially used for urban planning and visualization, DSM data is now being employed in the solar resource assessment or shading modelling as it can accurately represent the shape and configuration of elements in the surrounding that could potentially block solar irradiance. However, the accuracy of DSMs to characterize the obstruction elements in the vicinity of potential PV installation sites had not been evaluated before.

In this work we evaluated the Danish Elevation Model (DK-DEM), as well as lower reosultion DSMs derived from it, compared to traditionally used hemispheric sky imaging. The

978-1-7281-6118-1/22 $31.00 © 2022 IEEE

differences between both methodologies and multiple DSM data sets has been evaluated over two different locations with different horizon shading profiles.

The Danish Elevation Model has proved to be comparable to the traditional hemispheric sky imaging based methodologies in terms of horizon shading characterization for both slightly and heavily shaded locations.

When comparing DSMs with lower resolutions than the DK-DEM, the results differ depending on the nature of the location analyzed. Locations with low horizon shading profiles show small to no difference to the fisheye-based methodology up to 12.6 m resolution (the lowest analyzed resolution in this paper).

On the other side, locations with greater obstruction elements in the vicinity show noticeable differences when DSMs with lower than 3.2 m of resolution are employed. Our hypothesis is that the use of the averaging method to downsample the DK-DEM lead to a noticeable increase in the elevation of the investigated location for grid sizes greater than the distance from the location to the nearest considerable obstruction elements.

To properly evaluate the effects of using lower resolution DSMs in the horizon shading characterization, more locations should be analyzed. Furthermore, to estimate the error that these inaccuracies may cause in the irradiance modelling in urban environments, an irradiance analysis must be performed that includes a comparison to fisheye-based methods.

REFERENCES

[1] Inc. Solmetric. *Solmetric SunEye 210 user's guide.* 2011. URL: https://www.solmetric.com/downloads-suneye.html (visited on 01/10/2022).

[2] Solar Pathfinder. "Instruction manual for the Solar Pathfinder™". In: *Linden, TN. Available online at: http://www.solarpathfinder.com/pdf/pathfinder-manual.pdf* (2008).

[3] SDFE. *Danmarks Højdemodel - Overflade.* Styrelsen for dataforsyning og effektivisering. URL: https://dataforsyningen.dk/data/928#description. accessed: 23.05.2022.

[4] Lantmäteriet. *Surface Model Download, from aerial photographs.* Swedish Land Survey. URL: https://www.lantmateriet.se/en/maps-and-geographic-information/geodata-products/product-list/surface-model-download-from-aerial-photographs/#steg=4. accessed: 23.05.2022.

[5] Miguel Centeno Brito et al. "3D solar potential in the urban environment: A case study in lisbon". In: *Energies* 12.18 (2019), p. 3457.

[6] Paula Redweik, Cristina Catita, and Miguel Brito. "Solar energy potential on roofs and facades in an urban landscape". In: *Solar Energy* 97 (2013), pp. 332–341.

[7] SDFE. *Danmarks Højdemodel.* Styrelsen for dataforsyning og effektivisering. URL: https://sdfe.dk/hent-data/danmarks-hoejdemodel/. accessed: 23.05.2022.

[8] Geospatial Agency Innoter. *Digital Surface Model.* URL: https://innoter.com/en/products/spatial-data/dsm-generation/ (visited on 01/11/2022).

[9] Inc. Geodetics. *DEM, DSM & DTM: Digital Elevation Model – Why It's Important.* URL: https://geodetics.com/dem-dsm-dtm-digital-elevation-models/ (visited on 12/26/2021).

[10] *ALOS PRISM L1B.* URL: https://earth.esa.int/eogateway/catalog/alos-prism-l1b?text=alos. accessed: 23.05.2022.

[11] *CartoSat-1 archive and Euro-Maps 3D Digital Surface Model.* URL: https://earth.esa.int/eogateway/catalog/cartosat-1-archive-and-euro-maps-3d-digital-surface-model. accessed: 23.05.2022.

[12] *TanDEM-X - Digital Elevation Model (DEM) - Global, 12m.* URL: https://data.europa.eu/data/datasets/5eecdf4c-de57-4624-99e9-60086b032aea?locale=es. accessed: 23.05.2022.

[13] NASA. *U.S. Releases Enhanced Shuttle Land Elevation Data.* URL: https://www2.jpl.nasa.gov/srtm/. accessed: 23.05.2022.

[14] UP42 GmbH. *Everything you need to know about Digital Elevation Models (DEMs), Digital Surface Models (DSMs), and Digital Terrain Models (DTMs).* 2021. URL: https://up42.com/blog/tech/everything-you-need-to-know-about-digital-elevation-models-dem-digital (visited on 01/16/2021).

[15] Heiko Hirschmüller. "Semi-global matching-motivation, developments and applications". In: *Photogrammetric Week 11* (2011), pp. 173–184.

[16] SDFE. *Kortforsyningen download.* Styrelsen for dataforsyning og effektivisering. URL: https://download.kortforsyningen.dk/. accessed: 23.05.2022.

[17] *DHM product specification v1.0.0.* Styrelsen for Dataforsyning og Effektivisering. Oct. 2020. URL: https://dataforsyningen.dk/asset/PDF/produkt_dokumentation/dhm-prodspec-v1.0.0.pdf.

[18] Roberto Carrasco-Hernandez. *Calculation of patterns of solar radiation within urban geometries.* The University of Manchester (United Kingdom), 2015.

[19] Ziyu Liu et al. "Towards feasibility of photovoltaic road for urban traffic-solar energy estimation using street view image". In: *Journal of Cleaner Production* 228 (2019), pp. 303–318.

[20] Daniel Alvarez Mira et al. "Accuracy Evaluation of Horizon Shading Estimation Based on Fisheye Sky Imaging". In: *2021 IEEE 48th Photovoltaic Specialists Conference (PVSC).* IEEE. 2021, pp. 2052–2059.

[21] *GRASS GIS.* OSGeo - The open source geospatial foundation. URL: https://www.osgeo.org/projects/grass-gis/. accessed: 09.01.2022.

[22] Glynn Clements. *r.resamp.stats.* GRASS Development Team. 2022. URL: https://grass.osgeo.org/grass80/manuals/r.resamp.stats.html. accessed 18.05.2022.

Ongoing performance assessment strategies & operational challenges when managing hundreds of distributed photovoltaic assets across Asia

André M. Nobre, Anusha Agarwal and Sai Pranav

Cleantech Solar, 25 Church Street #03-04, Capital Square Three, 049482, Singapore

Abstract — **When managing a few solar farms of tens of MWp, asset owners are occupied with a handful of these systems within their delimited area. However, picture managing hundreds of small systems (below 1 MWp/site) across a vast area encompassing several developing countries. Challenges range from assessing the performance of the photovoltaic systems to maintaining these in a cost-effective manner. In this paper, we further attack two major points which were previously presented at two past IEEE PVSC events – (1) determining how best to gauge performance ratios of PV systems in a highly distributed setting given the availability and boundary conditions of irradiation sources, and (2) analyzing how assets have concurrently performed in a diversified portfolio. Asia is a hot spot for PV system deployment, yet little research exists when compared to other matured markets in Europe and America. This work aims at further contributing knowledge on PV systems behavior in this part of the world.**

I. INTRODUCTION

Photovoltaic (PV) systems deployment continues to pick up worldwide. In 2020, cumulative capacity reached 759 GWp [1], with potential 2021 additions of ~190 GWp [2], bringing global PV volumes closer to the 1 TWp mark. When observing markets (as per 2020 figures), more than half of systems are situated in the Asian continent. Yet, most scientific work continues to be focused on Europe and USA. Even when faced with major COVID pandemic impacts, India likely deployed ~7.4 GWp in 2021 [3], a year-on-year growth rate of +70%.

Solar farms continue to be deployed worldwide, breaking levelized cost of electricity (LCOE) records [4]. However, the rooftop distributed segment continues to gain momentum [5]. Nevertheless, these smaller and scattered systems present operational challenges, as previously published in [6, 7].

This work is aimed at expanding Asian PV knowledge, tackling both the main challenges of those publications with regards to determining baselines for performance ratio (PR) calculations and how systems are performing in the field (plus challenges observed from running a vast number of systems from a developer's perspective). An example is extreme soiling, see Fig. 1, making it difficult, to gauge performance of systems with ground sensors at certain locations. Nearly 500 PV systems (~400 MWp) shown here (see Fig. 2) had operated for a full year (May-21 to Apr-22 time frame) and were considered for the analysis. This study allows for a larger pool of systems to undergo zoomed-in checks beyond what is normally found in literature, which usually considers single-system evaluations or benchmarking of a handful of systems at different climatic zones. It is hoped that the sharing of such results prompt other large portfolio system owners to publish similar works.

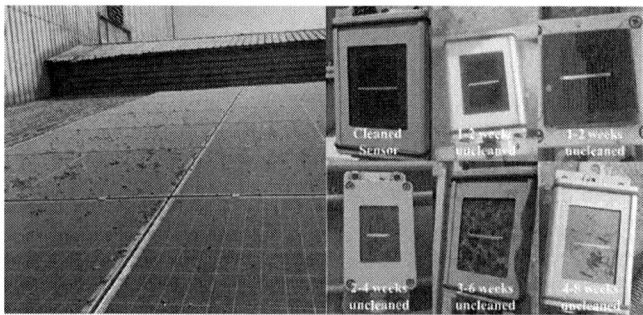

Fig. 1. Left – Soiled PV system in Thailand after two months without cleaning activities due to COVID restrictions. The type of industry where system is located may accelerate soiling accumulation (here animal ration factory, which attracts birds). Right – Various irradiance sensors in India after weeks without cleaning.

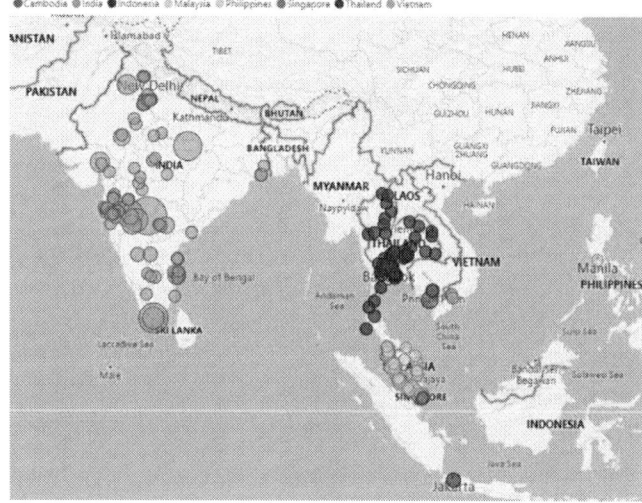

Fig. 2. Map of India and Southeast Asia showing a portfolio with ~400 MWp at nearly 500 PV sites in eight countries (as of Apr-22).

II. METHOD

The work proposes two areas of investigation:

A. Performance assessment strategies

In [6], it was shown that the use of ground sensors is severely compromised for performance ratio calculation purposes (due to rapid soiling accumulation – Fig. 1 (right) – and distributed PV management challenges). It was further demonstrated that the use of satellite data enhances the ability to gauge how systems are accurately performing. However, what this

978-1-7281-6118-1/22 $31.00 © 2022 IEEE

previous work did not tackle was the fact that the cost of satellite data point becomes prohibitive if every site requires a data set. When operating a multitude of smaller PV systems (several systems below ~100 kWp and even under ~50 kWp), purchasing exclusive satellite data is not a cost-effective solution. As data points become limited for smooth operations, a compromise must be reached. Fig. 3 illustrates how with increased distance between the source of irradiation and the PV asset, the values of PR drift and become non-representative of the true technical quality of the system. This is especially the case in a tropical climate setting (here Malaysia), with frequent cloud cover conditions.

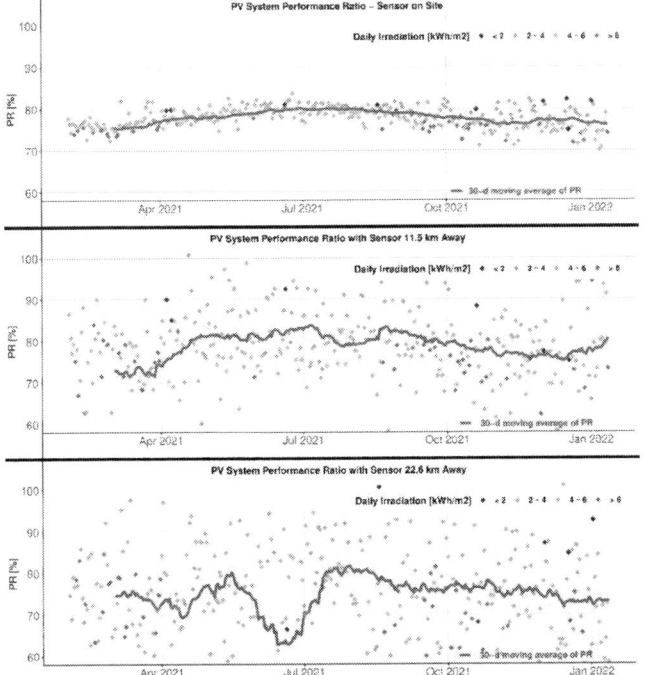

Fig. 3. Top – Performance ratio (PR) evolution of a PV system with a sensor located in situ. Middle – A second PV system PR is calculated but with a sensor available only at ~12 km of distance. Bottom – Similar analysis but with a sensor located ~23 km away from system.

We propose extrapolation exercises to determine acceptable range of separation between irradiation sources and sites. We use Malaysia as a test country due to the sheer volume of small assets present (200+ of circa 50 kWp of size). The number of irradiance sensors utilized was 31 (installed at the time the figure was generated). The root mean square error (RMSE) is calculated by comparing measured values at reference stations (31 of them) against extrapolated station-pairs (30 of them), thus a total of 930 "station-pairs" were available for the extrapolation exercise. The formula for the RMSE follows:

$$RMSE\% = 100 \cdot \frac{\sqrt{\frac{1}{N}\sum_{j=1}^{N}(Ei-Mi)^2}}{\frac{1}{N}\sum_{i=1}^{N}(Mi)} \qquad (1)$$

where Mi is the measured daily irradiation at a site and Ei is the extrapolated value from the "station-pair". The value of N is equivalent to the number of days of data used, which was 30 days for this study.

Additionally, four sites with dedicated meteorological stations were used and contrasted with the satellite data from the same latitude/longitude of these locations. The goal here is to gauge the RMSEs of a ground vs. satellite source in relation, as well, to the other hundreds of pairs of extrapolated stations.

The objective of such study is to find an acceptable distance whereby a new PV system can confidently draw its performance from, without the need of a dedicated sensor, nor the purchase of a satellite data point.

B. Performance assessment execution and ongoing identified challenges

In this section, we gauge the performance ratio of the systems contained in the analysis using methods described in [7], for example, using global horizontal irradiation from satellite sources as baseline for PV systems installed at shallow tilt angles (< 5 degrees) instead of potentially compromised soiled sensors. Observations are made with regards to PRs across the portfolio and challenges faced during the operational year are further studied.

Based on the experience of managing these assets over the years of their lifetime, we propose to tackle the three most negatively-contributing challenges observed causing poor field performance, namely:

a) Advanced soiling impact: the distributed systems of the portfolio present in this study encompass a variety of industrial sites, from different backgrounds, some notably with tremendous soiling conditions found on location, such as cement factories and animal ration production. Fig. 4 illustrates one of the examples covered in this work, with a cement factory with two PV systems – one further away from the dust zone (which happens to be a floating PV system, left) and one in the heart of the factory, near its dusty activities (right).

Fig. 4. Left – Floating PV system located at a pond in a cement factory premises (process water pond with ~3 m depth). This system is ~900 m from the core of the factory, where dusty industrial activities take place. Right – PV modules heavily affected by dusty activities.

978-1-7281-6118-1/22 $31.00 © 2022 IEEE

b) <u>Extraordinary shading impact</u>: as the portfolio contains systems located in busy city centers, such as Singapore, Kuala Lumpur, Penang, Bangkok, to name a few, it is expected that the heavily-built surrounding environment of commercial & industrial buildings may pose shading obstacles to an optimal generation. We tackle examples where nearby conditions seriously hinder optimum system performance, primarily across a portfolio of systems deployed at petrol station roofs, which are notably lower in height vs. nearby taller structures, even trees. Fig. 5 shows the proposed methodology to be adopted for gauging of these extreme losses across affected systems. It draws from the method developed in [8], whereby it is necessary to filter out instances of cloudy conditions not to mix those with under-generation. A ±15% filter versus a clear-sky day (for the PV system typical performance) is proposed, which allows for the removal of cloudy conditions. The exercise is performed at four different portions of the year, with 30 days of data each (executed at the solstices and equinoxes to observe varying sun position conditions). Fig. 5 exemplifies the algorithm created, where the system clear-sky AC power output average is shown, together with the ±15% threshold band. Points that land inside the interval "survive" and are averaged to determine the generation potential for the system for that time of the year. The resulting curve is compared against a "clear-sky" optimum (if the system had no shading impact) so that losses can be ultimate gauged.

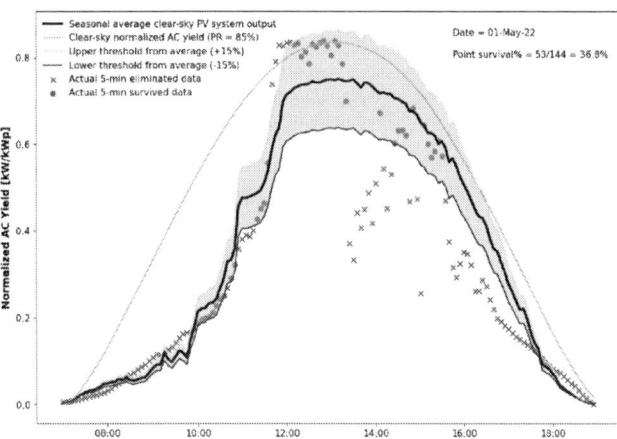

Fig. 5. Proposed method of eliminating cloudy conditions of a given day's PV system AC output. Here, this Singapore system at a petrol station roof is strongly shaded in the morning by a tree located at the east side of the array. Late afternoon shading is also observed.

c) <u>High curtailment levels</u>: in certain markets, most notably in Thailand, PV systems are not allowed to export the excess energy (greater than the loads of the PV circuit) into the grid. Active curtailment devices must be present at such sites. With COVID lockdowns registered in

various parts of Asia, sites suffered particularly in 2020 and parts of 2021. Since the analysis of this work covered the most recent 12 months of system operations (May-21 to Apr-22), we have observed a diminished impact from COVID restrictions across the portfolio. Nonetheless, for a particular country portfolio of systems located at supermarket stores in Thailand, we observed varying degrees of curtailment which brought asset performance down to concerning levels, thus presenting relevant data for our work.

III. RESULTS & DISCUSSIONS

For the first portion of the investigation, the extrapolation exercises for Malaysia are shown in Fig. 6 for 31 stations plus satellite vs. ground checks. As expected, with increased distance, root mean square errors climb when using stations further apart from each other. For the RMSEs between satellite source vs. ground, values ranged from 10-13%. For Malaysia, we determined a desired separation of ~10-15 km between a PV system and its reference irradiance source (ground or satellite). Due to its proximity and similar weather, we applied the same philosophy when deploying PV systems at petrol stations in Singapore, keeping a slighter maximum separation of ~10 km for the city-state, given its much smaller country area.

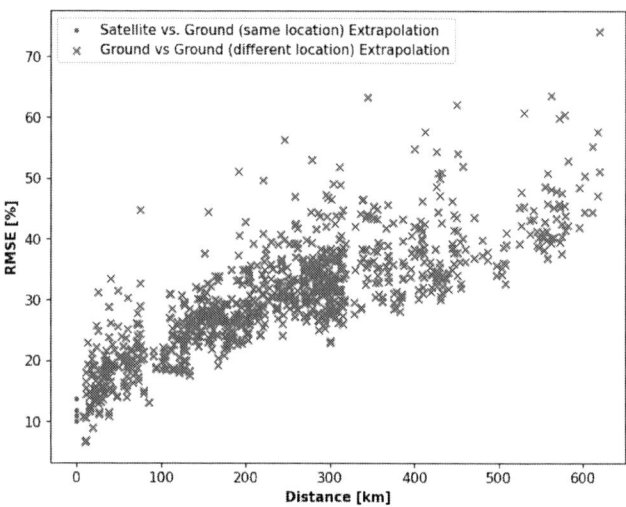

Fig. 6. RMSEs observed when extrapolating values between stations where irradiance sensors are found in Malaysia. Satellite vs. ground checks against meteorological stations are also plotted.

It is noticed that for a tropical location such as found in Malaysia that using sensors apart by greater than 20 km seems to lose fidelity on the performance ratio calculations, as seen in Fig. 3 (middle). It is likely that certain parts of the world with noticeably clearer sky conditions, that sensors beyond 20 km and even as far as 40 km could be used in between sites. However, for this portfolio, we opted to restrict new sensor purchases when stations were greater than ~10 km away

(Singapore) and ~10-15 km (Malaysia) from its nearest neighbor with desirable success in maintaining a good agreement on system performance calculations with such arrangements. For a universe of circa 215 petrol stations deployed (for the Malaysia portfolio), only about 40 sensors have been installed to date (~18% of the sites). The ~175 sensors saved plus other installation costs avoided (cabling, labor) are estimated to have saved ~100 kUSD to the developer.

For the Singapore market, a country of circa 610 km² of area, results matched with our observations from the capital of Malaysia (Kuala Lumpur, area = 2,200 km²). However, due to the smaller area of Singapore, we decided to ultimately deploy sensors at shorter separation (~10 km) to cater our fleet of operational PV systems (see Fig. 7). With the techniques developed, the number of irradiance devices deployed was kept to a reasonable minimum, which represented considerable savings in the thousands of dollars of instrumentation by avoiding instruments at several of the sites (sensors which are not even possible to be regularly maintained due to logistics/cost pressures).

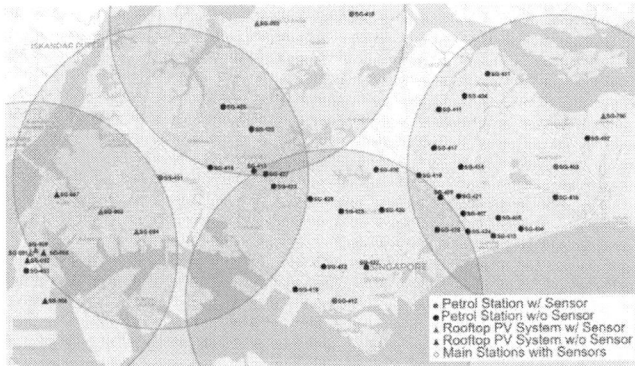

Fig. 7.   Map of Singapore showing existing PV systems of the entire country portfolio. Only a handful of sensors were needed for an optimum country coverage.

For the second part of this work, we assessed the performance as per [7], but now with a considerable higher number of systems. Fig. 8 showcases our fleet performance ratios for the operational year ranging from May-21 to Apr-22. We note that several systems, especially in India (IN) and Thailand (TH), suffered from COVID lockdown restrictions, as they could not be cleaned due to various government controls. Some sites, for example, could not be cleaned for three months during peak dry season, with observed performance ratios falling to sub 50% levels (not even visualized in Fig. 8 as it would make the graph more difficult to read with a wider PR variance).

Malaysia (MY) and Singapore (SG) fleets operated closer to business model expectations (dotted green PR line at about 75%). This is made possible given the fact that constant rain showers are present at both countries, without system cleaning needs. Any underperformance could be linked to equipment failure and inability to go to a site (e.g. client COVID

restrictions prevented swap of damaged inverters or breakers at certain parts of the operational year).

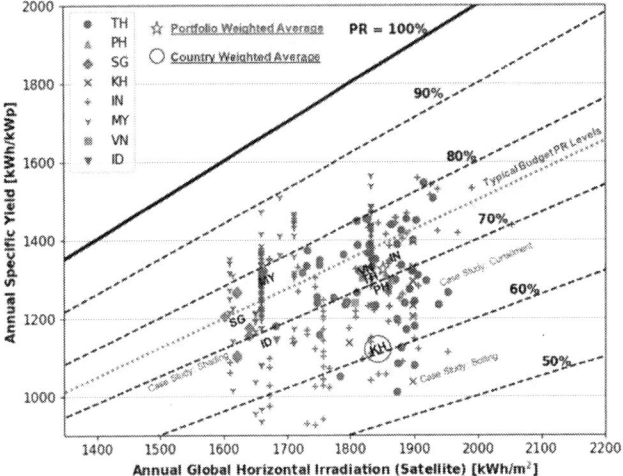

Fig. 8.   Scatter plot of annual specific yield (kWh/kWp) vs. annual global horizontal irradiation (satellite- or ground-based, in kWh/m²) for nearly 500 PV systems. Some systems had PR below 50% and are outside of the field of view of the graph. Case studies are marked and discussed in this work.

Three case studies are marked in Fig. 8 and which are addressed next, namely a) a shading case study of a PV system at a petrol station, b) a heavy soiling case study at a cement factory and c) a curtailment case study.

For case study a), Fig. 9 illustrates a PV system at a petrol station deeply surrounded by taller residential buildings. The views from the two system images are shown and the direction of the photo indicated at the bird-eye view of the figure (bottom right), with the north indicated. The residential buildings to the west of the system are at least 20 stories in height, with the petrol station at nearly street level.

Fig. 9.   Example of PV system atop a petrol station at nearly street level in Singapore (1°N). Taller residential buildings surround the PV system, especially posing shading obstacles in the afternoon.

978-1-7281-6118-1/22 $31.00 © 2022 IEEE

When using the proposed algorithm for removal of cloudy conditions (see Fig. 10), it is found that the system loses ~11% of its annual generation potential due to the shading obstacles found in the afternoon. The losses vary throughout the year as seen from the curves but are similar in magnitude especially given the system is near the Equator, thus having comparable operational conditions across the year.

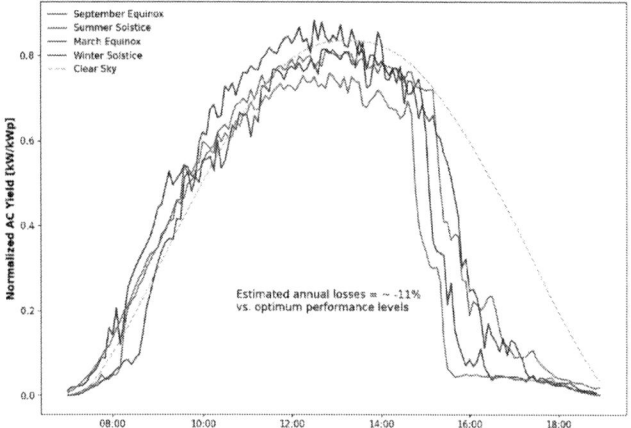

Fig. 10. Expected clear-sky PV system output for a ~82% performance ratio day (orange dotted bell curve). The four resulting actual PV system AC outputs for the shaded site are shown for four different portions of the year (multi colors).

The PV systems which are part of the petrol station portfolio program across Malaysia and Singapore were studied for similar shading conditions found elsewhere. Table I summarizes the average behavior of these systems. The Singapore portfolio is exposed to a more densely-built environment, with losses found to be higher (-7.9% vs. 2.9% for Malaysia). A larger number of systems in Singapore (nearly a third) also under-performed beyond -10% away from expected P50 performance levels than their counterparts in Malaysia.

Tab. I. Two petrol station portfolios in Malaysia and Singapore with their respective metrics for their operation.

Performance Benchmark	Malaysia	Singapore
Number of Petrol Stations	189	21
Total Size [MWp]	8.79	1.44
Average Size [kWp/site]	47	66
Avg. Yield [% against P50]	-2.5%	-7.9%
% Sites Yield Var. [> -10%]	14%	29%

For case study b), we observe advanced soiling conditions found at a cement factory in Cambodia. Fig. 11 illustrates the performance ratio evolution across various years at a floating PV system located ~900 m away from the "clinker zone" of the factory, its heart and where dustiest conditions are found. The cement adhesion to PV panels adds not only a blockage to the light, but also its removal is a labor-intensive process which is needed daily, even during rainy season. Three distinct episodes

are highlighted in the figure: "teething of the system", when cleaning was not performed as site construction was ongoing; a period of cleaning contractor change, which caused delays during the onboarding of a new company, and thus further cement accumulation on panels; and a third episode when the rooftop section of the systems went through a safety retrofit program and cleaning was halted (thus a new episode of cement soiling accumulation took place). At the bottom of Fig. 11, one can observe that the PV section near the heart of the factory underperforms regularly at about 10% below the pond system. One could argue that self-cooling from floating systems could justify that, but this assumption is dispelled in [9].

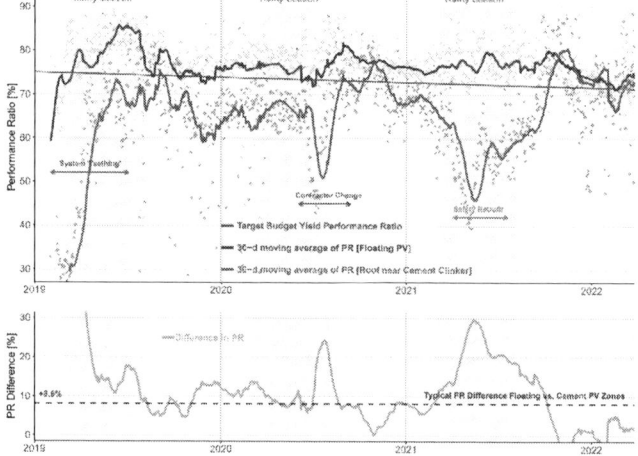

Fig. 11. Observed performance ratio (PR) at two distinct system sections – one a floating PV system ~900 m from the cement factory core vs. another at a roof located in the middle of the dusty factory activities. The difference in PR is shown at the bottom of the figure as well as other periods of interest.

Finally, in case study c), we discuss the behavior of a fleet of PV systems at supermarket stores across Thailand. The country suffered various COVID restrictions from 2020 and well into 2022. That caused, for example, less flow of people into the supermarkets and thus, fewer cooling loads are needed to keep the store's temperature and of its produce. Additionally, we observed variances in ambient temperature (longer winter periods at certain parts of North of Thailand), which caused, once again, savings in air-conditioning or cooling machines (freezers) needs. With less loads at the supermarkets, more solar electricity was curtailed as grid exports are not permitted in the country. Beyond those two examples, it was further investigated that portfolio #2 of these stores faced extra levels of curtailment. The management of the supermarket chain further confirmed that many tenants within the stores were changed during the operational year of study, thus less loads were drawn. Additionally, tourists from Laos (to the North) diminished during the restrictive period, reducing footfall at the supermarkets of that portfolio. Fig. 12 captures the lifetime of these three supermarket portfolios and their varying curtailment degrees. Portfolio #2 is responsible for several underperformance points found in Fig. 8 ("TH" blue circles below 75% threshold of performance ratios).

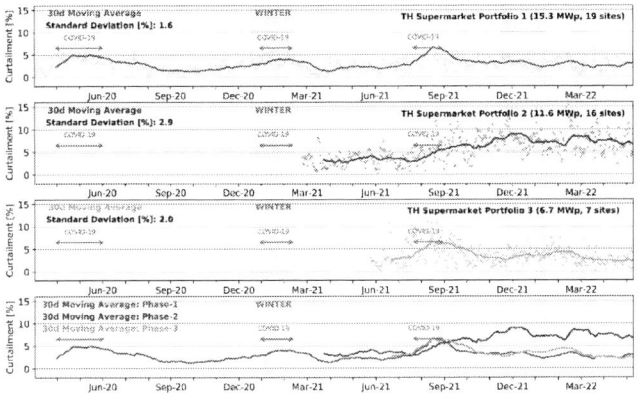

Fig. 12. Curtailment levels observed across a portfolio of PV systems located at supermarket stores across Thailand. Curtailment levels (due to non-exports allowed) vary across seasons but are also a function of COVID restrictions as well as other special events (see text).

In general for the portfolio, COVID challenges ranged from site access restriction (inability to clean or swap damaged parts), but also lack of loads in markets where exports are not permitted. A weighted portfolio performance ratio of ~71% was found to be satisfactory against target budget expectations, given worldwide pandemic impacts. Such a result represented an improvement from 2020 levels, highlighting a strong recovery of PV against COVID-related challenges.

IV. CONCLUSIONS

When attempting to find records of performance levels for PV systems in the Asian region, it becomes clear publications are scarce. A few investigations are mostly available as single site analysis (such as [10]) or containing a handful of systems. Information on fleet performance of assets is usually kept confidential by developers. Therefore, it is aimed that such work published here can bring light to the status quo of Asian systems' performance and barriers faced when operating assets which are bound to dominate future renewable energy grids of the world.

Challenges observed at not exclusive to Asian regions. The fact that a challenging portfolio of an approaching 0.5 GWp in size can nearly reach intended performance levels, with sites located at various industrial conditions, heavily densely-built surroundings is further proof that solar photovoltaics are a trusted source of electricity for now and the future. Improvements in operation & maintenance practices as well as proper modelling of business expectations (returns on investment aimed by stakeholders) go hand in hand in the distributed PV segment.

ACKNOWLEDGEMENT

The authors would like to acknowledge the support of other Cleantech Solar team members across geographies: Rohit Jaswal (India), Thornthanut Pakdeepinyo & Jason Ward (Thailand), Adam Loh & Thinesh Rao (Malaysia), Meas Chansovathdy (Cambodia) and Darren Ong, Bristo Paulose, Lim Zhee Hong & Tan Jun Kiat (Singapore). Further thanks are extended to Raju Shukla and Holger Eick for the continuous support to the advances of knowledge in photovoltaic system performance & long-term behavior in Asian locations.

REFERENCES

[1] *Snapshot of Global PV Markets 2021*. International Energy Agency – Photovoltaic Power Systems Programme. 21 pages.

[2] *Solar deployment to reach 191GW in 2021 but fall far short of 2030 ambitions, BloombergNEF.* https://www.pv-tech.org/solar-deployment-to-reach-191gw-in-2021-but-fall-far-short-of-2030-ambitions-bloombergnef-says/. Accessed on 27-Dec-2021.

[3] *India set to record best year ever for new PV additions in 2021.* https://www.pv-magazine.com/2021/12/10/india-set-to-record-best-year-ever-for-new-pv-additions-in-2021/. Accessed on 27-Dec-2021.

[4] *Utility scale solar reaches LCOE of $0.028-$0.041/kWh in the US, Lazard finds.* https://www.pv-magazine.com/2021/11/05/utility-scale-solar-reaches-lcoe-of-0-028-0-041-kwh-in-the-us-lazard-finds/. Accessed on 27-Dec-2021.

[5] *India added 1.3 GW of rooftop solar in first nine months of 2021.* https://www.pv-magazine.com/2021/12/22/india-added-1-3-gw-of-rooftop-solar-in-first-nine-months-of-2021/. Accessed on 27-Dec-2021.

[6] A.M. Nobre, S. Karthik, C. Liu, R. Jaswal, R. Baker, R. Malhotra, A. Khor. *Irradiance Measurement Considerations for System Performance Assessment when managing fleets of photovoltaic assets across Asia.* 44th IEEE Photovoltaic Specialists Conference, Washington DC, USA, 2017.

[7] A.M. Nobre, S. Karthik, Wan Y.L., R. Baker, R. Malhotra, A. Khor. *Performance Evaluation of a Fleet of Photovoltaic Systems Across India and Southeast Asia.* 46th IEEE Photovoltaic Specialists Conference, Chicago, USA, 2019.

[8] A.M. Nobre, S. Karthik, H. Liu, D. Yang, F.R. Martins, E.B. Pereira, R. Rüther, T. Reindl, I.M. Peters. *On the impact of haze on the yield of photovoltaic systems in Singapore.* Renewable Energy 89 (2016).

[9] I.M. Peters, A.M. Nobre. *Deciphering the thermal behavior of floating photovoltaic installations.* Solar Energy Advances 2 (2022).

[10] N. Anang, et al. *Performance analysis of a grid-connected rooftop solar PV system in Kuala Terengganu, Malaysia.* Energy and Buildings, Volume 248.

Development of Highly Uniform and Reproducible DI-O₃ layers for Photovoltaic Applications and Beyond

Munan Gao[*1], Vibhor Kumar[1], and Ngwe Zin[*12]

[1]Rutgers, the State University of New Jersey, New Brunswick
[2]Department of Electrical and Computer Engineering, Rutgers

Abstract— We demonstrate the development and characterization of ozonated oxide (DI-O₃) for applications in photovoltaics and beyond. The growth of DI-O₃ is investigated by varying 1) the sample orientation in a wafer cassette, 2) growth time and 3) ozone concentration in parts-per-million (ppm); while the thickness and uniformity of DI-O₃ layer on the entire 100 mm diameter are characterized by the multi-wavelength ellipsometer equipped with a wafer mapping capability. DI-O₃ samples grown horizontally in the wafer cassette provide better uniformity than vertically grown counterparts (i.e., ±1.62% compared to ±8.02% across 100 mm diameter). A modest increase in DI-O₃ thickness is observed with increasing growth time (i.e., from 1 min to 3 min and 10 min). An average of 0.06 nm and 0.10 nm of DI-O₃ thickness, respectively, is increased when the growth time is increased by 2 min and 7 min. Varying ozone concentration from 3.5 ppm to 15 ppm, on the other hand, is observed with a minimal effect on DI-O₃ growth. Thanks to the growth technique (horizontal orientation and continuous dissolved ozone flow) and parameters (15 ppm, 420 ml/min flow and 30°C of dissolved ozone) used in this contribution, we demonstrate the reproducibility of growing DI-O₃ with 1.53±0.01 nm and < 3% uniformity on multiple 100 diameter wafers.

Keywords—Dissolved Ozone, ozonated oxide, DI-O₃, ppm, ellipsometry, TMAH, passivated contact.

I. INTRODUCTION

Ozone generated oxide is a simple but versatile technique used in many applications [1-5]. The use of ozone oxide for cleaning, passivation and tunneling layer have been widely presented. The ozone method has proved to provide a comparable cleaning efficiency as the standard RCA cleaning, but with the reduced cost and simpler process [6-8]. A stack of DI-O₃/AlO$_x$ as a passivation layer was also shown promising as a superior passivation dielectric stack [9-12]. The role of a thin oxide as a tunneling layer, additionally, has attracted increased interests in recent years due to its potential of building high-efficiency solar cells [13-15]. But the detailed examination of the oxide layer itself is limited with focusing on electronic properties [16, 17]. Here we present the investigation of developing an ultrathin oxide by dissolved ozone in deionized water with varied experimental conditions to achieve a desired thickness with high uniformity and reproducibility. The state-of-the-art uniformity and thickness control may find usefulness in processes that are sensitive to the properties such as passivating and tunneling.

II. MATERIALS AND METHODS

Single side polished (SSP) p-type <100> 1-10 Ω-cm 100 mm diameter (dia) wafers and double side polished (DSP) n-type <100> 1-10 Ω-cm 100 mm dia wafers are used in this contribution. 25% Tetramethylammonium hydroxide (TMAH) solution at 80°C is used to etch off silicon. 5% HF concentration is used, unless otherwise stated. A dissolved ozone (DO) system capable of generating up to 5 gallons per minute of dissolved ozone solution is used. The ozonated oxide (DI-O₃) growth process took place in a tank with samples placed in a wafer cassette for vertical and horizontal orientations (see Figure 1). Continuous DO flow of 420 mL/min at 30°C is maintained during DI-O₃ growth. Thickness and uniformity are characterized by a multi-wavelength ellipsometer equipped with the wafer mapping capability featuring up to 49-points measurement pattern for each measurement with 3 mm of edge exclusion.

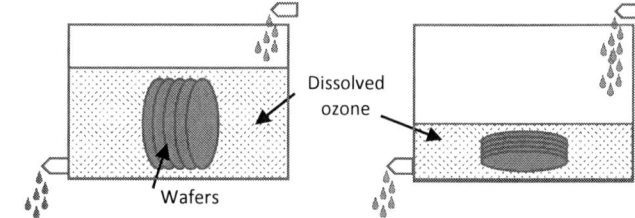

Fig. 1. The setup of the dissolved ozone tank, featuring wafers placed in vertical and horizontal orientations.

III. EXPERIMENTS AND DISCUSSIONS

P-type samples were first etched in TMAH at 80°C for 5 mins, then rinsed in deionized water (DIW) and dipped in HF to make the surface hydrophobic. Samples placed horizontally and vertically, respectively, in the carrier cassette were then lowered into the DO solution tank (shown in Figure 1) for 5 mins to grow DI-O₃ layers. Samples were then removed out of the tank, rinsed in DIW, dried and measured for thickness and uniformity by the ellipsometer. Fig. 2. shows that although the thickness of DI-O₃ for both samples grown vertically and horizontally have relatively the same thickness (i.e., ~1.3 nm), vertically placed samples were grown with the notably thicker DI-O₃ film at the bottom side of wafers, thus resulting in poorer uniformity than horizontally placed counterparts (i.e., ±1.62% compared to ±8.02%). Vertically grown samples resulting with the notably thicker DI-O₃ layer could be due to the retention of substantial

*Corresponding Author

978-1-7281-6118-1/22 $31.00 © 2022 IEEE

dissolved ozone solution on the bottom side of the wafer sample, and also possibly specific to the set up used in this experiment. Horizontally grown DI-O₃ samples were instead used in the following sections.

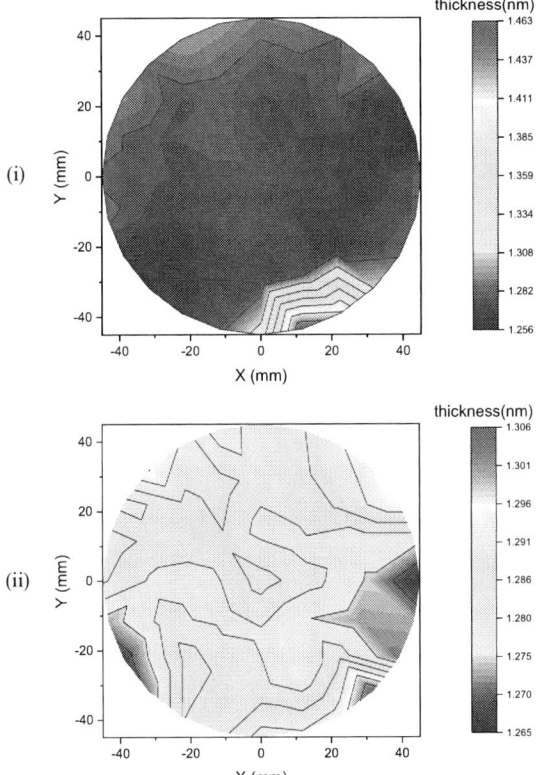

Fig. 2. Surface profile of DI-O₃ grown samples grown in vertical (i) and horizontal (ii) orientations. Vertically grown samples have an average thickness of 1.281 nm with 8.02% of uniformity, horizontally grown samples have 1.282 nm with 1.62% uniformity.

A set of DSP n-type wafers were again etched in TMAH for 10 mins, rinsed in DIW and dipped in HF to make hydrophobic surface. Wafers were then split into three groups: the first group was processed for 1 min of DI-O₃ growth, the second for 3 min and the last for 10 min. All wafers were later processed in 3.5 ppm, 7 ppm and 15 ppm of ozone concentration. Thickness of each individual wafer grown with DI-O₃ were then measured by the ellipsometer. As shown in Fig.3, a modest increase in DI-O₃ thickness is observed with increasing growth time (i.e., from 1 min to 3 min and 10 min). An average of 0.06 nm and 0.10 nm of DI-O₃ thickness, respectively, is increased when the growth time is increased by 2 min and 7 min. 10 min of growth time results in ~ 1.5 nm of DI-O₃, but further DI-O₃ growth in thickness appears to saturate beyond 1.5 nm. Varying ozone concentration from 3.5 ppm to 7 ppm and 15 ppm, on the other hand, is observed with a minimal effect on DI-O₃ growth. Based on these results, ozone concentration as low as 3.5 ppm is sufficient to grow DI-O₃ with the thickness > 1.3 nm, although the length of growth duration can further increase DI-O₃ thickness.

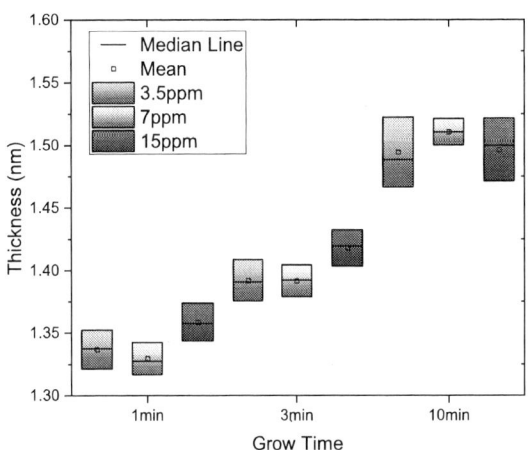

Fig. 3. Thickness of DI_O₃ grown in relation to the length of growth time and ozone concentration in ppm.

Based on the growth technique and parameters such as the horizontal placement of wafers in the cassette, continuous dissolved ozone flow at 420 ml/min, ozone concentration at 15 ppm, and 30°C of dissolved ozone operating temperature; multiple DSP n-type wafers were then grown in the dissolved ozone solution for 10 min on separate batches and the ellipsometer was then used to measure the thickness and uniformity. As shown in Fig. 4, wafers were grown with an average of 1.53±0.01 nm of DI-O₃ layers with < 3% uniformity, and in some cases with < 2% uniformity. We can also DI-O₃ films with different thicknesses by using specific growth time. Growth techniques and parameters used in this investigation demonstrate that we are able to develop ultra-thin and highly reproducible DI-O₃ layers on multiple 100 mm dia wafers.

IV. CONCLUSION

We presented the development of DI-O₃ layers by wafer orientation (i.e., horizontal and vertical), growth time and ozone concentration. Horizontally grown wafers were resulted with the better uniformity (i.e., ±1.62% compared to ±8.02% across 100 mm diameter) than vertically grown counterparts. However, with further investigations vertically grown wafers could achieve the uniformity comparable to horizontally grown wafers. Although increasing the growth time increases DI-O₃ thickness, varying ozone concentration has minimal or no effect on increasing DI-O₃ thickness. With the specific set of growth techniques and parameters, including horizontal wafer orientation, constant ozone flow of 420 ml/min, 15 ppm ozone, 30°C of DO; we demonstrated the development of ultra-thin, highly uniform and reproducible DI-O₃ layers with desired thicknesses.

ACKNOWLEDGEMENT

This material is based upon work supported by the U.S. Department of Energy's Office of Energy Efficiency and Renewable Energy (EERE) under the Solar Energy Technologies Office Agreement Number DEEE0009367. Weeks Hall Nanofabrication Core Facility, School of Engineering, Rutgers is central to the works presented in this contribution.

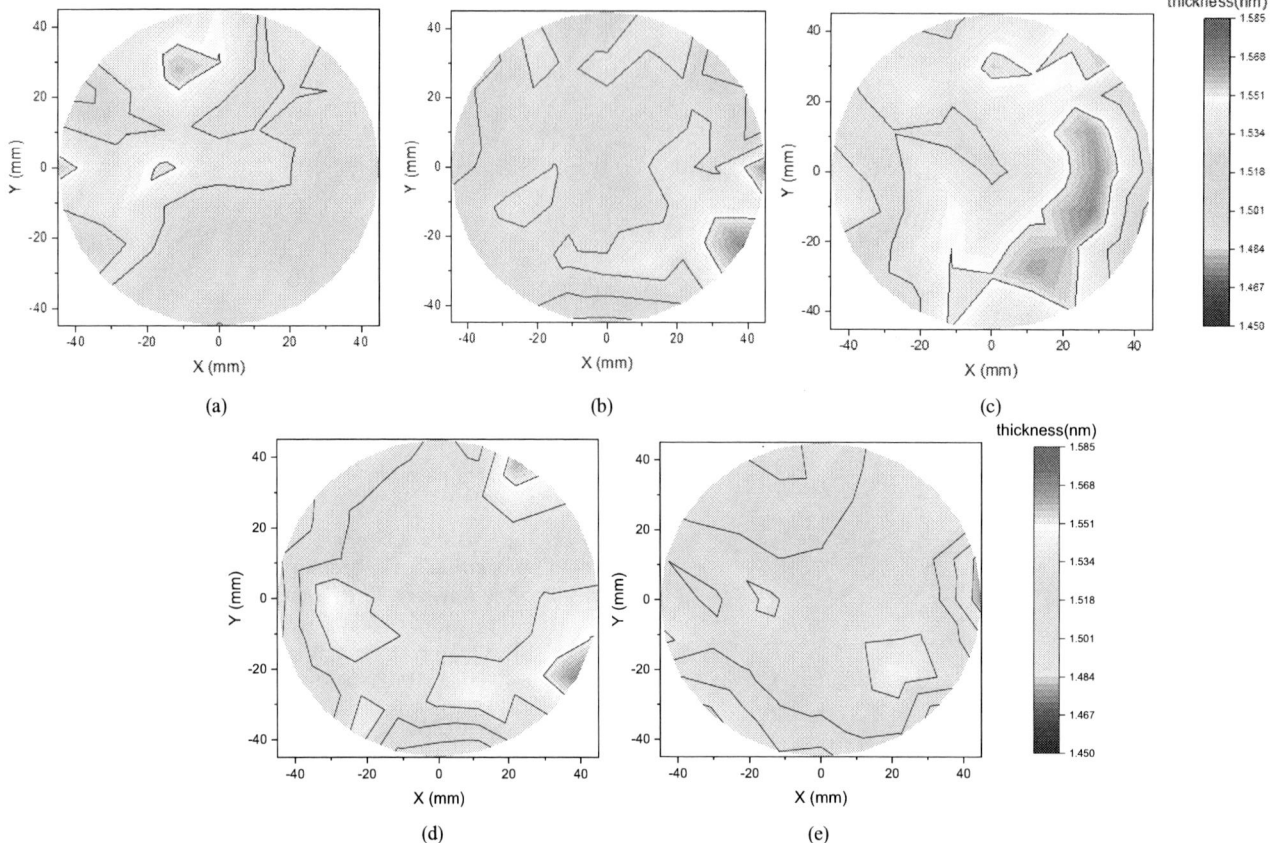

Fig. 4.Surface Profiles of DI-O3 grown samples measured by the multiwavelength ellipsometer featuring 49 points mapping capability across 100 mm dia meter. (a) 1.53 nm with ±1.59% uniformity, (b) 1.53 nm with ±2.38% uniformity, (c) 1.54 nm with ±1.73% uniformity, (d) 1.53 nm with ±2.63% uniformity, and (e) 1.52 nm with ±2.39% uniformity.

REFERENCES

[1] Bakhshi, S., et al., *Simple and versatile UV-ozone oxide for silicon solar cell applications.* Solar Energy Materials and Solar Cells, 2018. **185**: p. 505-510.

[2] Siemens, W., *Ueber die elektrostatische Induction und die Verzögerung des Stroms in Flaschendrähten.* Annalen der Physik, 1857. **178**(9): p. 66-122.

[3] Zhang, H. and J. Ouyang, *High-efficiency inverted bulk heterojunction polymer solar cells with UV ozone-treated ultrathin aluminum interlayer.* Applied Physics Letters, 2010. **97**(6): p. 063509.

[4] Xia, Y., et al., *Graphene Oxide by UV-Ozone Treatment as an Efficient Hole Extraction Layer for Highly Efficient and Stable Polymer Solar Cells.* ACS Applied Materials & Interfaces, 2017. **9**(31): p. 26252-26256.

[5] Turren-Cruz, S.-H., et al., *Enhanced charge carrier mobility and lifetime suppress hysteresis and improve efficiency in planar perovskite solar cells.* Energy & Environmental Science, 2018. **11**(1): p. 78-86.

[6] Chen, G. *The applications of DI-O3 water on wafer surface preparation.* in *International Conference on Wafer Rinsing, Water Reclamation and Environmental Technology for Semiconductor Manufacturing, Industrial Technology Research Institute, Hsinchu, Taiwan.* 1999.

[7] Ali, H., et al., *Transmission electron microscopy based interface analysis of the origin of the variation in surface recombination of silicon for different surface preparation methods and passivation materials.* physica status solidi (a), 2017. **214**(10): p. 1700286.

[8] Moldovan, A., et al., *Simple cleaning and conditioning of silicon surfaces with UV/ozone sources.* Energy Procedia, 2014. **55**: p. 834-844.

[9] Hezel, R. and K. Jaeger, *Low-temperature surface passivation of silicon for solar cells.* Journal of the Electrochemical Society, 1989. **136**(2): p. 518-523.

[10] König, D., et al., *Evidence of a high density of fixed negative charges in an insulation layer compound on silicon.* Thin Solid Films, 2001. **385**(1-2): p. 126-131.

[11] Saint-Cast, P., et al., *High-efficiency c-Si solar cells passivated with ALD and PECVD aluminum oxide.* IEEE Electron Device Letters, 2010. **31**(7): p. 695-697.

[12] Bakhshi, S., et al., *Improving silicon surface passivation with a silicon oxide layer grown via ozonated deionized water*, in *Proceedings of the 44th IEEE Photovoltaic Specialists Conference.* 2017: Washington DC.

[13] Moldovan, A., et al. *Tunnel oxide passivated carrier-selective contacts based on ultra-thin SiO2 layers grown by photo-oxidation or wet-chemical oxidation in ozonized water.* in *2015 IEEE 42nd Photovoltaic Specialist Conference (PVSC).* 2015. IEEE.

[14] Zhang, Z., et al., *Carrier transport through the ultrathin silicon-oxide layer in tunnel oxide passivated contact (TOPCon) c-Si solar cells.* Solar Energy Materials and Solar Cells, 2018. **187**: p. 113-122.

[15] Mousumi, J.F., et al., *Phosphorus-doped polysilicon passivating contacts deposited by atmospheric pressure chemical vapor deposition.* Journal of Physics D: Applied Physics, 2021. **54**(38): p. 384003.

[16] Angermann, H., et al., *Electronic interface properties of silicon substrates after ozone based wet-chemical oxidation studied by SPV measurements.* Applied surface science, 2012. **258**(21): p. 8387-8396.

[17] Angermann, H., et al. *Surface charge and interface state density on silicon substrates after Ozone based wet-chemical oxidation and Hydrogen-termination.* in *Solid State Phenomena.* 2013. Trans Tech Publ.

Characterizing the capacitance of different c-Si PV cell technologies using impedance spectroscopy

David A. van Nijen, Patrizio Manganiello, Mirco Muttillo, Miro Zeman, Olindo Isabella

TU Delft, Delft, Netherlands

In conditions of partial shading, the deployment of sub-module maximum power point tracking is known to increase the energy yield of crystalline silicon (c-Si) based photovoltaic (PV) modules. To facilitate PV module designs endowed with an increasing granularity of sub-module power converters, it could be advantageous to exploit the impedance of the c-Si PV cells. For example, it is well known that c-Si PV cells exhibit capacitive effects. However, for such applications, it is critical that the self-capacitance of the PV cells is large enough. As of yet, there are no reports that give a clear overview of the self-capacitance that can be expected for modern industrial c-Si cell technologies. In this work, we report the capacitance of four different industrial c-Si PV cell technologies, namely Al-BSF, PERC, IBC, and SHJ. These capacitance values are obtained by fitting a dynamic model of the solar cell to measurements obtained using impedance spectroscopy in dark conditions. It was found that depending on the cell technology, capacitances in the range of 0.5-6 mF/cm^2 can be expected for operation at maximum power point. Our results show that there is a high potential to remove the input capacitors from sub-module power converters in PV applications, due to the fact that the ripple voltage at the input of the converter is naturally suppressed by the PV self-capacitance.

978-1-7281-6118-1/22 $31.00 © 2022 IEEE

Demonstration of a Monolithically Integrated Hybrid Electroabsorptive Modulator/Photovoltaic Device for Bidirectional Free Space Optical Communication at 1.55 µm

Emily Kessler-Lewis, Stephen J. Polly, Elijah Sacchitella, Seth M. Hubbard, Raymond Hoheisel

Rochester Institute of Technology, Rochester, NY, United States

Blacksky Aerospace LLC, Arlington, VA, United States

A three-terminal, monolithically integrated device is presented combining both power generation and optical communication at 1.55 µm. A discrete InP cell without an anti-reflection coating with an AM0 efficiency of 12.8 % and an InGaAs/InAlAs electroabsorption modulator with a peak ON/OFF ratio of 3.0 have been developed. A monolithic device has been grown and fabricated, and is undergoing simultaneous PV collection/free space optical communication testing.

Glued III-V on Si tandem solar cells using hybrid transparent conductive layers

Phuong-Linh Nguyen, Jeronimo Buencuerpo, Philippe Baranek, Oliver Hoehn, David Lackner, Frank Dimroth, Marco Faustini, Stephane Collin, Andrea Cattoni

EDF R&D, EFESE, Technologie du Solaire, Palaiseau, France

Institut Photovoltaique d'Ile-de-France (IPVF), Palaiseau, France

Centre de Nanosciences and Nanotechnologies (C2N), CNRS, Paris-Saclay University, Palaiseau, France

Fraunhofer Institute for Solar Energy Systems (ISE), Freiburg, Germany

Laboratoire Chimie de la Matiere Condensee de Paris, Sorbonne Universite, CNRS, Paris, France

The photovoltaic market is dominated by c-Si solar cells, with a record efficiency of 26.7% which approaches the detailed balance limit. Tandem solar cells combining a III-V semiconductor top cell with a Si bottom cell are one of the most studied routes to exceed 30% efficiency. Direct growth of III-V on Si has achieved 25.9% efficiency but remains a challenge. Direct wafer bonding has demonstrated state of the art efficiencies of up to 35.9%, but this technique is not suitable for industrial production as it requires extremely low roughness, low particle contamination, surface activation and bonding under vacuum. The integration of III-V on Si using conductive transparent glues may circumvent these constraints. In this contribution, we design and fabricate a 2-terminal AlGaAs/ARC/Glue/ARC/TOPCon tandem cell using sol-gel derived TiO_2 ARCs and PEDOT:PSS-based glue. The ARCs ensure Ohmic contact with the sub-cells (replacing the tunnelling recombination junction) as well as transmission of red photons from the III-V to the Si cell. The gluing layer ensures the electrical interconnection between sub-cells and, thanks to its lower refractive index, an enhanced photon recycling in the top cell (up to 0.9% improvement of the tandem' efficiency). We use electromagnetic simulations to optimize the ARC/Glue/ARC stack and achieve current matching; current losses from reflection (1.1 $mA/cm2$) remaining essentially equal to those of the direct bonding architecture (1.0 $mA/cm2$). TiO_2 ARCs and PEDOT:PSS-based glue are synthesized and deposited by spin-coating. The lamination process is performed in air, using a simple hydraulic hot press at low curing temperature (120 °C). The 1st-generation cells show a promising VOC (1.8 V), while the efficiency is currently limited by high series resistance. The 2nd generation is currently under fabrication using an improved lamination process to solve this problem.

978-1-7281-6118-1/22 $31.00 © 2022 IEEE

Flexible All-Perovskite Tandem Solar Cells with High Specific Power

Zhaoning Song, Cong Chen, Chongwen Li, Lei Chen, Yanfa Yan

Wright Center for Photovoltaics Innovation and Commercialization, Department of Physics and Astronomy, The University of Toledo, Toledo, OH, United States

High specific power cost-effective thin-film tandem solar cells are desired for space applications. Metal halide perovskites have recently emerged as an ideal candidate for constructing multi-junction thin-film tandem solar cells on lightweight and flexible substrates that promise high specific powers, owing to their tunable bandgap, outstanding optical properties, and low-temperature solution processibility. Here, we report on the in-house preparation of radio-frequency magnetron sputtered indium tin oxide for both the transparent conductive electrode and the interconnecting layer of monolithically integrated all-perovskite tandem solar cells. Our proof-of-concept lightweight and flexible all-perovskite tandem solar cells deliver a stabilized power conversion efficiency of more than 20% and a specific power density of more than 870 W/kg.

Impact of Thermal Annealing on the Mechanical Properties of Ge epilayer on Mesoporous Germanium for Layer Separation and Substrate Re-use

Firas Zouaghi, Ahmed Ayari, Bouraoui Ilahi, Jeremie Chretien, Tadeas Hanus, Nicolas Paupy, Nicolas Quaegebeur, Abderraouf Boucherif

Institut Interdisciplinaire d'Innovation Technologique (3IT), Université de Sherbrooke, Sherbrooke, QC, Canada

Laboratoire Nanotechnologies Nano systèmes (LN2) – CNRS UMI-3463 Institut Interdisciplinaire d'Innovation Technologique (3IT), Université de Sherbrooke, Sherbrooke, QC, Canada

GAUS, Department of Mechanical Engineering, Université de Sherbrooke, Sherbrooke, QC, Canada

III-V multi-junction solar cells on Ge substrate are widely used for space PV applications. However, challenges related to weight and cost reduction are still encountered. Porous Ge (PGe) lift-off is one of the reliable approaches allowing solar cell detachment (mass reduction) while offering substrate multiple reuses (cost reduction). Accordingly, the adhesion of the epilayer to the Ge substrate through PGe reconstructed weak layer (nm scale pillars) needs to be investigated and optimized to offer scalable and controllable detachment. In this context, detachable Ge epilayer on porosified 4" Ge substrate has been grown. The weak layer mechanical properties dependence on the post-growth thermal annealing at various temperatures (from 600ºC to 750ºC) is experimentally and numerically investigated. Pull test experiments were performed to measure the adhesion of the weak layer for each annealing temperature. Pillar densities and diameters were determined precisely for each temperature using plan view SEM observations of the detached substrate. The pillars density is found to decrease in favor of an overall diameter increase as the annealing temperature increases. The values are used as input for a 3D pull test model to simulate the detachment process and to study the influence of the pillar characteristics on the epilayer adhesion. Our results show that the epitaxial Ge template is very stable even at high annealing temperature while keeping low adhesion force allowing mechanical detachment which is very suitable to grow detachable multi-junction solar cell.

Development of a Novel Soiling Chamber for Testing Antisoiling Coatings

Matthew T Muller

National Renewable Energy Laboratory, golden, CO, United States

This study presents the development and validation of a novel soiling chamber. The chamber is novel in that it includes wind induced soiling, feedback from a low-cost particulate monitor, and in-situ Isc measurements. Validation with side-by-side identical modules within the chamber produced soiling losses of 7% over 19 hours while the soiling ratio was always within 0.5% between the two modules. Initial side-by-side testing of an anti-soiling coated module versus and uncoated module demonstrated significant wind induced cleaning of the coated module. Specifically, the coated module showed only 0.8% soiling loss while the uncoated module reached as much as 10.5% soiling loss.

Progress and Demonstration of Micro-CPV Module with Integrated Planar Tracking and Diffuse Light Collection

Steve A Askins, Guido Vallerotto, César Dominguez, Mathilde Duchemin, Gaël Nardin, Mathieu Ackermann, Delphine Petri, Matthieu Despeisse, Jacques Levrat, Xavier Niquille, Christophe Ballif, Juan F Martinez, Marc Steiner, Gerald Siefer, Ignacio Antón

Instituto de Energia Solar - UPM , Madrid, Spain

Insolight SA, Renens, Switzerland

CSEM PV-Center, Neuchâtel, Switzerland

EPFL PVLab, Lausanne, Switzerland

Fraunhofer ISE, Freiburg, Germany

Several technologies have been identified that could produce a new type of high-performance solar product optimized for space-constrained applications: micro-CPV, planar microtracking, and diffuse capture. The Swiss start-up Insolight and the Hiperion consortium are bringing such a device to the industrial level. In this work we share the latest results for full-scale modules, discuss improvements to the design and resulting performance gains, and will report the results from pilot installations in Madrid and Freiburg.

A Deep Learning Approach for PV Failure Mode Detection in Infrared Images: First Insights

Daniel Rocha[1,2,3], Miguel Lopes[7], Jennifer P. Teixeira[1], Paulo A. Fernandes[1,4,5], Modesto Morais[7] and Pedro M. P. Salomé[1,6]

[1]INL - International Iberian Nanotechnology Laboratory, Braga, 4715-330, Portugal

[2]Algoritmi R&D, University of Minho, Guimarães, Portugal

[3]2Ai, School of Technology, Polytechnic Institute of Cávado and Ave, Barcelos, Portugal

[4]I3N, Universidade de Aveiro, Campus Universitário de Santiago, 3810-193, Aveiro, Portugal

[5]CIETI, Departamento de Física, Instituto Superior de Engenharia do Porto, Instituto Politécnico do Porto, Porto 4200-072, Portugal

[6]Departamento de Física, Universidade de Aveiro, Campus Universitário Santiago, 3810-193, Aveiro, Portugal

[7]IEP - Instituto Electrotécnico Português, Custóias 4460-817 , Portugal

Abstract— **Large-scale solar power plants require cheap and quick inspections, for this unmanned aerial vehicle (UAV's) for high resolution optical and infrared imaging were introduced in the past years. While using UAV's is fast for image acquisition, image is a time-consuming process where the best of practice today is still for an expert to individually analyze each image. As such, in this work we use computer vision to accelerate this process. We performed an instance segmentation assessment using a pre-trained mask R-CNN for the segmentation of defective modules, and cells, as well as for segmentation and classification of failures. This method was chosen due its good past performance. In this work we created a database from a solar power plant consisting of 42048 modules and an expert analyzed the images. Later on, our computer algorithm results were benchmarked against the expert. Our algorithm achieved a mean average precision (mAP) in defective module segmentation mask of 72.1 % and 47.9 % in segmentation mask of failure type with an intersection over union threshold (IoU) of 0.50, without human interference. The presented preliminary results allow to assess the methodology advantages and drawbacks to increase performance and pave the way to a large-scale study.**

Keywords—deep learning, instance segmentation, fault detection, solar module detection, photovoltaic system, thermographic inspection

I. Introduction

In recent years, there has been an increased demand for photovoltaic (PV) solar energy and due to scale benefits, solar plants have increased in size from a few MW to GW. The sheer size of these installations makes operation and maintenance processes (O&M) complex, expensive and can compromise the plant's return of investment.

Currently, the monitoring systems are excessively human dependent, and the continuous increase of the solar power plants, requires the development of fast and cheap systems, that simultaneously allow for an efficient inspection. Thus, this topic has increased the interest of the scientific community in the recognition of defects in IR images from inspections carried out by Unmanned Aerial Vehicles (UAV). These resources are the cheapest alternative for thermographic inspection in large PV plants, optimizing labor and time costs. The recent use of deep learning techniques in the identification and classification of defects in PV modules allowed to overcome limitations of classical image processing methods [1]–[4]. In this sense, deep learning techniques, namely instance segmentation, have been used for module detection and/or fault localization [4]–[6].

In this work we compare two improved approaches based in these previous works. In the first experiment, we use instance segmentation as in [6] for fault localization and improve it with the fault classification. In our second experiment, we use instance segmentation as in [4] for module localization but only detect defective PV modules, allowing in the future for the classification. Both approaches are preliminary steps that need further developments.

II. Thermographic Inspection

Defective solar cells have a quite different IR radiation emission depending on its source defect, and as such, IR imaging is used to analyze fail modes of PV systems. The IEC TS 62446-3: 2017 standard [5] defines the external conditions for the thermographic inspection of PV modules [7].

A. Environment Conditions

To carry out the local IR measurements, it is necessary to consider the environmental conditions. In this way, the measurements should only be performed under the following parameters:

- Irradiation minimum: 600 W/m^2.
- Wind Speed maximum: 28 km/h.

- Cloud Coverage Maximum: 2 okta (unit of measurement to describe cloud cover).
- Soiling: no visible soiling or dust on the modules.

B. Technical Requirements

Image acquisition is performed using a UAV holding two cameras, one for IR images and one for visible (VIS) images. The image acquisition is performed at a constant speed of 3 m/s, with a resolution up to 3 cm/px. To minimize the effects of the optical reflection, the angle between the modules and the camera should be higher than 30 °.

III. METHODOLOGY

A. Dataset

Based on the environmental conditions and technical requirements, we performed a thermographic inspection using a UAV with a Workswell WIRIS 2nd gen camera which contains a VIS and thermal lens. The plant consists of 42048 modules (CSUN240-60P and GOOSUN GS230) with an inclination of 30 ° located in Beja, Portugal, installed between 2010 and 2012 with an installed peak power value close to 10 MWp. Table 1 shows the used environment and technical conditions of the thermographic and visual inspection that led to the creation of the database.

TABLE 1 – THERMOGRAPHIC AND VISUAL INSPECTION DETAILS

Modules Emissivity	0.95
Air Temperature (ºC)	35
Wind Velocity (km/h)	10~15
Distance (m)	23 ~25
Camera	Thermal & Visual
Irradiation (W/m^2)	700 ~ 1000
Thermal Camera Resolution (pixels)	640 x 512
Visible Camera Resolution (pixels)	1600 x 1200
UAV Speed (m/s)	3

The 1368 images from thermographic inspection were selected and labeled by an expert. The dataset has 204 images with defects consisting of 414 identified and classified anomalies. The categories of anomalies as well as the number of defects present in the dataset are presented in Table 2 and shown in Fig. 1.

TABLE 2 – TYPE AND NUMBER OF DEFECTS

Type of defect	Number of Defects
1 x Substring in open circuit	135
2 x Substring in open circuit	13
Single cell with difference in temperature	208
Modules in open circuit	35
Heat module junction	16
Module in short circuit	7

As each image consists of several modules, each image could had more than one anomaly. A major drawback from the dataset is the presence of reflections, as shown in Fig. 2, shadows and bird drops where there is a possibility of assessment error or misinterpretation during labeling. In order to minimize the error propagation during the image analysis, several images have been discarded from the original dataset during training and evaluation of the algorithm. In total, 46 images were removed with reflection issues. Thus, only 158 images were considered.

Fig. 1 – Examples of defective PV modules from dataset. a) Modules in open circuit; b) 1 x Substring in open circuit; c) 2 x Substring in open circuit; d) Module in short circuit; e) Heat module junction; f) Single cell with difference in temperature;

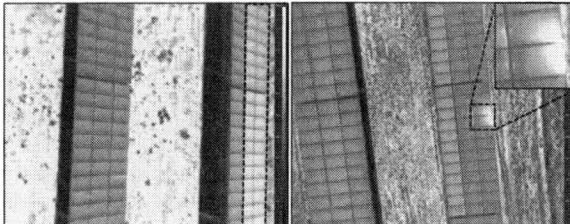

Fig. 2 – Images with reflection effect

B. Deep Learning Model

State-of-the-art studies show the effectiveness of using instance segmentation, namely, mask R-CNN [8], for PV module segmentation or failure detection. We joined the relevant results from [4]–[6] and evaluate the effectiveness of mask R-CNN in the detection of defective modules, as well as in characterization and localization of the failures. Developed on top of faster R-CNN [9], the Mask R-CNN provides an additional branch for mask prediction to the existing branch for bounding box prediction. Using Detectron 2, a pytorch library for detection and segmentation algorithms, we used the weighs file mask_rcnn_R_50_FPN_3x with a ResNet50 as backbone for our implementation through the transfer learning.

We performed two experiences, one for segmentation of PV module defective and other for segmentation of failure type. The dataset was split in test and validation with 80 % and 20 % for each one, respectively, to evaluate how well our model performs.

C. Evaluation Procedure

We evaluated the performance of the proposed approach in terms of mean average precision (mAP) measured in reference an Intersection over Union (IoU) threshold of 0.50.

$$IoU = \frac{\text{Area of Overlap}}{\text{Area of Union}} \qquad (1)$$

For the IoU was calculated the precision (2) and recall (3).

$$Precison\ (P) = \frac{TP}{TP + FP} \qquad (2)$$

$$Recall\ (R) = \frac{TP}{TP + FN} \qquad (3)$$

Where TP, FP and FN represent, the number of True Positives, False Positives, and False Negatives, respectively. The AP is the area under the resultant precison-recal curve where the mAP score is calculated by taking the mean AP over all classes.

IV. RESULTS AND DISCUSSIONS

In the first experiment we fine tune the pre-trained mask R-CNN only for detecting the PV defective modules. In the second experiment we segment and classify the failure inside the module. The main advantage comparatively to the previous approach is that for PV modules with more than one defect can be identified separately. As discussed in the previous section we evaluated our model in terms of average precision for an IoU threshold of 0.50.

TABLE 3 – EVALUATION OF DIFFERENT TYPES OF SEGMENTATION

Segmentantion mode	Module Segmentation	Segmentation of failure	Identification of failure	mAP at IoU = .50	
				Bounding Box	Segmentation Mask
Experiment 1: Segmentantion of PV Module defective	X			0.721	0.721
Experiment 2: Segmentantion of Failure Type		X	X	0.466	0.479

The Mask R-CNN loss is defined as the sum of the different losses from the two branches in the network where it is included the classification loss, bounding box loss and mask loss. In total our model achieved 0.31 of loss for the first experiment and 0.54 for the second experiment.

The main differences between the experiments reside in the fact that in Experiment 2 different segmentation of failure modes and classification are performed. In addition, there is an unbalanced number of images between the types of defects that contribute to a major model loss.

The unbalanced dataset between the types of defects for Experiment 2 is shown in the results when comparing an individual AP of the two most representative defects, substring in open circuit and module in open circuit, with 0.50 and 0.60, respectively. Thus, a larger and more balanced dataset with defects known will allow for better results. Moreover, the annotation process revealed its importance in the last experiment when corrections to the boxing annotations around the failure were refined. Additionally, images that contain minimal noise can be interpreted as hot spots or modules in open circuit.

V. CONCLUSIONS

In this work, we propose a novel approach for PV fault detection using instance segmentation for the detection and classification of the fail modes. The preliminary results show

that despite trained drone operators following a standardized but complex image acquisition procedure, the collected images still have limitations that are difficult be evaluated such as reflections and shadows which inevitably impair in a small extent the human expert but into a greater extent the automatic fault detection process. With the current used algorithms, higher standards and more high-quality training of drone operators will be necessary to mitigate these failures. On the other hand, intelligent flights with reflection angle correction capability will lead to better image quality. Consequently, it will contribute to a better performance when applying fault detection using neural networks due to noise minimization. A dataset was built to demonstrate that instance segmentation can effectively detect and locate defects in infrared images. This approach has the major advantage to overcome the necessity of segment the image between foreground and background and also avoiding the segmentation of all PV modules.

The preliminary results also provide good insights for a future work considering the small size of our dataset and a larger dataset will improve the results.

ACKNOWLEDGMENT

We acknowledge the European Union (FEDER funds through COMPETE 2020 - POCI-01-0247-FEDER-068919).

REFERENCES

[1] R. Pierdicca, E. S. Malinverni, F. Piccinini, M. Paolanti, A. Felicetti, and P. Zingaretti, in *International Archives of the Photogrammetry, Remote Sensing and Spatial Information Sciences - ISPRS Archives*, 2018, vol. 42, no. 2, pp. 893–900, doi: 10.5194/isprs-archives-XLII-2-893-2018.

[2] A. Oliveira, M. Aghaei, and R. Rüther, "Automatic Fault Detection of Photovoltaic Arrays by Convolutional Neural Networks During Aerial Infrared Thermography," 2019.

[3] C. Dunderdale, W. Brettenny, C. Clohessy, and E. E. van Dyk, *Prog. Photovoltaics Res. Appl.*, vol. 28, no. 3, pp. 177–188, 2020, doi: 10.1002/pip.3191.

[4] L. Bommes, T. Pickel, C. Buerhop-Lutz, J. Hauch, C. Brabec, and I. M. Peters, *Prog. Photovoltaics Res. Appl.*, 2021, doi: 10.1002/pip.3448.

[5] J. J. Vega Díaz, M. Vlaminck, D. Lefkaditis, S. A. Orjuela Vargas, and H. Luong, *Sensors*, vol. 20, no. 21, 2020, doi: 10.3390/s20216219.

[6] R. Pierdicca, M. Paolanti, A. Felicetti, F. Piccinini, and P. Zingaretti, *Energies*, vol. 13, no. 24, 2020, doi: 10.3390/en13246496.

[7] "IEC TS 62446-3:2017 | IEC Webstore." https://webstore.iec.ch/publication/28628 (accessed Oct. 04, 2021).

[8] K. He, G. Gkioxari, P. Dollár, R. Girshick, Mar. 2017, Accessed: Dec. 02, 2021. Available: https://arxiv.org/abs/1703.06870.

[9] S. Ren, K. He, R. Girshick, and J. Sun, "Faster R-CNN: Towards Real-Time Object Detection with Region Proposal Networks," Jun. 2015, Accessed: Dec. 02, 2021. [Online]. Available: http://arxiv.org/abs/1506.01497.

What Are PVDF-Based Backsheets Made Of?

Chiara Barretta[1], Eric Helfer[1], Astrid E. Macher[1], Gernot Oreski[1]

[1]) Polymer Competence Center Leoben GmbH (PCCL), Leoben, 8700, Austria

Abstract—**It is important to have a deep knowledge of a material's formulation to understand the degradation mechanisms taking place when the materials are exposed to external stresses. In the last years, cracking of backsheets with polyvinylidene fluoride (PVDF) has become an issue. PVDF based backsheets are widely used, but there is still quite some uncertainty regarding their exact formulation and the mechanisms behind the degradation. In this study seven PVDF films were investigated to get a deeper knowledge regarding their formulations and their principal properties.**

Keywords—backsheet, cracks, degradation, PVDF.

I. INTRODUCTION

Backsheets are a fundamental part in photovoltaic modules because they protect the internal components from environmental stress factors and provide electrical insulation to the structure. Photovoltaic (PV) backsheets might be subject to degradation modes such as discoloration, delamination, cracks, bubble formation and chalking, just to name a few. Especially backsheet cracks represent a critical issue for a PV module because they can reduce the electrical insulation properties of the backsheets and become a safety issue. Modules with cracked backsheets in most cases need to be replaced and this can lead to warranty claims and disputes. It is very well known nowadays the problematic cracking behavior of AAA backsheets, polyamide (PA) based, that significantly developed cracks in the first years of operation. Substantial effort has been carried out in the PV community to understand the degradation mechanisms leading to the cracks development [1-3] and even ageing tests able to reproduce artificially the appearance of the cracks were developed [4]. The observation of crack formation in polyvinylidene fluoride (PVDF) based backsheets significantly increased [5] in the last two to three years. Among the PV modules investigated (corresponding to about 3 GW), almost 20% of modules with PVDF-based backsheets developed cracks in the first 6-10 years of operation, apparently regardless of the location were the modules were installed. The figure becomes particularly relevant when thinking that in 2019 PVDF accounted for about 64% of the whole PV backsheet market share [6].

Even though PVDF-based backsheets are widely diffused on the market and in the field, there are still many questions open, especially regarding the exact composition of these materials, the features that are responsible for the formation of cracks and the mechanisms leading to these phenomena. It was reported in [7] that embrittlement of PVDF films after damp heat (DH) test could be attributed to physical ageing due to the occurrence of post-crystallization phenomena that lead to an increase of α-crystalline phases. Recent studies [8] showed that even mild storage conditions, namely at room temperature in the dark, led to significant decrease of mechanical properties of PVDF materials after 17 years due to physical ageing. Moreover, it was reported that backsheets with PVDF layer and acrylates/PVDF blends experienced a degradation of the acrylate moieties under combined exposure to UV, high temperature and humidity (75 °C/50% relative humidity) [9] and decrease of mechanical properties.

This study aims to get a better understanding of the composition of PVDF films used for photovoltaic backsheets and to investigate the variability in terms of composition.

II. EXPERIMENTAL

Seven PVDF films used to produce commercially available backsheets were used in this study. Two transparent films (PVDF_T1 and PVDF_T2) as well as five white films (PVDF_W1 to PVDF_W5) were object of the investigation and their thicknesses spanned from 22 μm to 31 μm. The materials were characterized using the following methods: (I) Fourier Transform Infrared Spectroscopy in Total Attenuated Reflectance (FTIR ATR) mode for the identification of the material components and crystalline phases, (II) Thermogravimetric Analysis (TGA) to identify materials' composition and fillers, (III) Differential Scanning Calorimetry (DSC) for the analysis of thermal behavior and morphology, (IV) tensile test to investigate anisotropic materials behavior and mechanical properties.

III. RESULTS AND DISCUSSION

The materials were analyzed by means of FTIR ATR spectroscopy to investigate their chemical composition at the surface. The results, Fig. 1, show differences between the spectra, meaning that the chemical composition is not the same for all the PVDFs investigated. The displayed spectra are obtained averaging 6 measurements for each material in different positions of the specimen, a baseline correction is not applied. Typical peaks of associated to PVDF can be detected and additional peaks could be identified for some materials. In particular, a peak at around $1730\ cm^{-1}$ can be seen for PVDF_T1, PVDF_W1, PVDF_W3, PVDF_W4 and PVDF_W5. This band can be associated to stretching vibration of C=O bonds possibly connected to acrylate-based units. In case of PVDF_T2, there is a peak of smaller intensity at the same wavenumber and it might be due to a lower ratio of acrylate/PVDF of the formulation. However, clear conclusion based only on FTIR ATR results cannot be made. Additionally, at about $650\ cm^{-1}$ it is possible to notice that some materials have a much higher absorbance than others. The higher absorbance might be due to a presence of TiO_2 as filler that typically presents a strong absorption band at $690\ cm^{-1}$. The samples PVDF_T1, PVDF_T2 and PVDF_W2 showed low

---

This work was conducted as part of the Austrian "e!MISSION.at – Energy Mission Austria" project "PV40+" (FFG No. 881868) funded by the Austrian Climate and Energy Fund and the Austrian Research Promotion Agency (FFG).

absorbance values at 650 cm^{-1}. The first two samples mentioned are transparent and the result would well correlate with the absence of fillers. The sample PVDF_W2, instead, might have a different filler than TiO$_2$ in its formulation. Additional FTIR measurements were carried out in transmittance mode to analyze the materials through their thickness. The results showed that the samples PVDF_W2 presented the peak at around 1730 cm^{-1} correlated to the acrylate-based moieties and high absorbance at 650 cm^{-1} correlated possibly TiO$_2$ filler. The results suggested that the sample could be characterized by a multilayer structure with pure PVDF on the outer sides and a blend of acrylate/PVDF/TiO$_2$ on the inner side. Microscopy images confirmed the multilayer structure hypothesis. FTIR ATR spectra can be also used to differentiate between the crystalline phases characteristic of PVDF materials. The materials analyzed showed presence of both crystalline phases: (I) α-type visible at 974 cm^{-1}, 854 cm^{-1}, 794 cm^{-1} and 764 cm^{-1}, (II) β-type visible at 1422 cm^{-1}, 1279 cm^{-1} and 842 cm^{-1}. As described earlier, storage conditions might alter the type of crystalline phases and the ratio between α- and β-type will be a parameter to monitor during exposure of the materials to artificial ageing tests in planned future work.

The results of TGA, Fig. 2, showed that the samples PVDF_T1 and PVDF_T2 have a two-steps decomposition thermogram characteristic of acrylate/PVDF blends [10], characterized by a maximum weight loss rate at about 380 °C and about 460 °C, respectively. The sample PVDF_T2 has a lower weight loss with respect to PVDF_T1 in correspondence of the acrylate decomposition (~385 °C), thus potentially confirming the results of FTIR ATR measurements of a possible lower ratio of acrylate/PVDF for this material. Additionally, the sample PVDF_T1 has a weight loss with maximum rate at about 252 °C, indicating that there are additional substances decomposing at this temperature. Further analysis, such as evolved gas analysis (EGA) coupled to FTIR or mass spectrometry (MS) would be necessary to draw more clear conclusions regarding the exact composition of these two materials. The final residue for both materials is lower compared to the others and this is a confirmation of the absence of additional fillers. The presence of the residue is due to a

formation of decomposition products that cannot be further thermally decomposed (char). The samples PVDF_W1 and PVDF_W3 show a similar behavior, with a two-step decomposition thermogram. The presence of a maximum weight loss rate at about 388 °C is due to the fact that TiO$_2$ acts as a catalyzer for the PVDF decomposition. Additionally, these materials show a second decomposition step at about 555 °C. The identification of the substance decomposing in this range requires further investigation and it might be due to the presence of an additional substance in the material formulation. The samples PVDF_W4 and PVDF_W5 present a typical behavior of a blend including TiO$_2$ as filler and do not present an additional weight loss step at about 555 °C. The sample PVDF_W2, instead, is characterized by a one-step decomposition curve, with maximum rate at about 360 °C. This might be an indication of a filler different than TiO$_2$ in the formulation and/or a different ratio between PVDF and acrylate. Furthermore, it is not possible to make a clear calculation of the filler content for all the materials, because the final residue of the filled samples depends not only on the filler content, but also on the char produced during the decomposition. The amount of char itself might be also influenced by the type of filler and its content [10].

Based on the presented results, it seems that there are at least three classes, possibly four, that can describe the PVDF materials investigated: (I) acrylate/PVDF blends without filler, (II) acrylate/PVDF blends filled with TiO$_2$ and possibly an additional substance, (III) multilayer structure with pure PVDF and acrylate/PVDF/TiO$_2$ blend.

The results of DSC analysis showed that the materials have a crystallinity between about 30% (PVDF_W3) and 38% (PVDF_T2). Additionally, the melting temperature is between 166 °C (PVDF_W2) and 169.6 °C (PVDF_W4).

The evaluation of the mechanical properties was carried out by means of tensile tests, see Table 1. The materials that showed the highest elastic modulus were PVDF_W1 and PVDF_W2 and they showed also the highest values for stress at break. All the materials investigated showed anisotropic behavior with

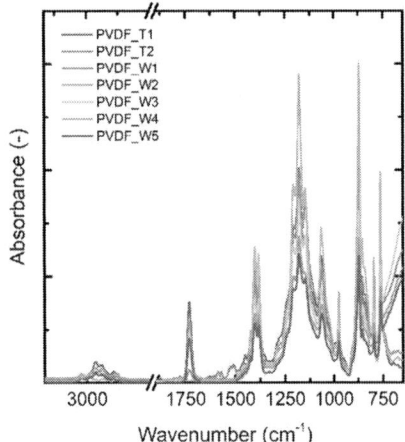

Fig. 1. FTIR ATR measured spectra of PVDF samples.

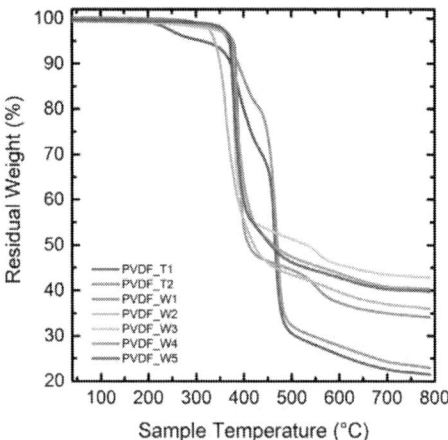

Fig. 2. Thermograms of PVDF samples measured by means of TGA.

significant differences for the strain at break, where the values for transverse direction (TD) are much lower than in machine direction (MD), except for PVDF_T1, where also a high standard deviation of the results is observable. Additionally, a strain at break ratio (Table 2), calculated as $\varepsilon_{B\text{-}TD}/\varepsilon_{B\text{-}MD}$, was considered as an indicator for the anisotropic behavior of the samples investigated. Almost all the PVDFs showed a value lower than 0.5, while the sample PVDF_T1 showed a value greater than 1.

TABLE 1. RESULTS OF TENSILE TESTS ON PVDF SAMPLES.

Sample		Elastic modulus, E_t, MPa	Stress at break, σ_B, MPa	Strain at break, ε_B, %
PVDF_T1	MD	792.1±19	27.5±1.0	98.6±10.3
	TD	996.6±33	18.4±0.7	153±52.1
PVDF_T2	MD	1280±70	40.9±2.7	85.7±10.9
	TD	1610±68	21.9±10.3	25.2±11.8
PVDF_W1	MD	1710±98	42.5±1.5	102±11.1
	TD	1766±72	24.8±1.8	27.8±12.5
PVDF_W2	MD	1980±23	36.3±0.9	116±32.7
	TD	2153±52	24.0±11.9	14.9±4.9
PVDF_W3	MD	1001±40	32.9±0.7	66.8±10.0
	TD	1148±35	20.3±0.8	30.2±30.5
PVDF_W4	MD	756.9±24	23.5±1.0	91.7±29.5
	TD	864.7±24	14.1±6.4	38.4±16.6
PVDF_W5	MD	994.9±18	24.2±1.0	187±25.8
	TD	1120±26	18.7±1.1	22.2±7.3

MD: machine direction, TD: transverse direction

The information about the PVDFs characteristics obtained so far are summarized in Table 2.

TABLE 2. SUMMARY OF CHARACTERISTICS OF INVESTIGATED PVDFs

Sample	Blend	Decomposition behavior	Filler	Strain at break ratio (TD/MD)
PVDF_T1	Yes, acrylate-based	A, (+1 step at 252 °C)	No	1.55±0.69
PVDF_T2	Yes, acrylate-based, possibly lower amount	A	No	0.29±0.18
PVDF_W1	Yes	B	TiO$_2$, possible additional	0.28±0.15
PVDF_W2	Multilayer	C	TiO$_2$, possible additional	0.13±0.08
PVDF_W3	Yes, acrylate-based	B	TiO$_2$, possible additional	0.45±0.52
PVDF_W4	Yes, acrylate-based	D	TiO$_2$	0.42±0.32
PVDF_W5	Yes, acrylate-based	D	TiO$_2$	0.12±0.06

IV. CONCLUSIONS

To summarize, seven PVDF films were investigated to better understand their chemical composition and their main characteristics in terms of thermal and mechanical properties. The results showed that at least three groups could be identified based on presence of acrylate-based polymer/PVDF blends, presence of filler(s) and type of filler(s). Further work will be focused on subjecting the materials to artificial ageing tests to better understand how the different material characteristics might be related to changes in mechanical properties to link material features to tendency of cracks formation.

REFERENCES

[1] Y. Lyu et al., "Drivers for the cracking of multilayer polyamide-based backsheets in field photovoltaic modules: In-depth degradation mapping analysis," Prog Photovolt Res Appl, no. 28, pp. 704–716, 2020, doi: 10.1002/pip.3260.

[2] Y. Lyu, J. H. Kim, A. Fairbrother, and X. Gu, "Degradation and cracking behavior of polyamide-based backsheet subjected to sequential fragmentation test," IEEE J. Photovoltaics, vol. 8, no. 6, pp. 1748–1753, 2018, doi: 10.1109/JPHOTOV.2018.2863789.

[3] G. C. Eder et al., "Error analysis of aged modules with cracked polyamide backsheets," Solar Energy Materials and Solar Cells, vol. 203, p. 110194, 2019, doi: 10.1016/j.solmat.2019.110194.

[4] M. Owen-Bellini et al., "Advancing reliability assessments of photovoltaic modules and materials using combined-accelerated stress testing," Prog Photovolt Res Appl, vol. 29, no. 1, pp. 64–82, 2021, doi: 10.1002/pip.3342.

[5] DuPont Photovoltaic Solutions, "DuPont global PV reliability: 2020 Field analysis," 2020. [Online]. Available: https://www.dupont.com/news/20200512-2020-global-pv-reliability-report.html

[6] M. Schmela and S. K. Chunduri, "Market Survey Backsheets & Encapsulation 2020," 2020. [Online]. Available: http://taiyangnews.info/reports/market-survey-backsheets-encapsulation-2020

[7] G. Oreski and G. M. Wallner, "Aging mechanisms of polymeric films for PV encapsulation," *Solar Energy*, vol. 79, no. 6, pp. 612–617, 2005, doi: 10.1016/j.solener.2005.02.008

[8] G. Oreski, A. Macher and K. Resch-Fauster, "Low temperature induced physical aging effects of backsheet materials," 2021 IEEE 48th Photovoltaic Specialists Conference (PVSC), 2021, pp. 0818-0821

[9] S. Smith *et al.*, "Transparent backsheets for bifacial photovoltaic (PV) modules: Material characterization and accelerated laboratory testing". *Prog Photovolt Res Appl.* 2021; 1- 11.

[10] W. Li., H. Li, and YM. Zhang., "Preparation and investigation of PVDF/PMMA/TiO2 composite film", *J Mater Sci,* 44, 2009,. pp. 2977–29

Excess current due to embedded superlattices in Graphene/Ox/n-GaAs solar cells, at 50 Suns and above

AC Varonides, Loyola Science Center 151, Physics & Engineering Department, University of Scranton, 204 Monroe Ave, Scranton, PA 18510, USA

Abstract - **We demonstrate the advantage of current density increase in graphene/n-GaAs Schottky barrier solar cells with concentrated-light trapping in a twenty-period AlAs-GaAs superlattice (SL) embedded in the bulk of an n-GaAs layer. Such a layer traps photo-excited carrier leading to maximum absorption through optical-gap increase from 1.42 to 2.129 eV. We focus on carrier escape from quantum wells, due to thermionic emission and conclude that excess carriers, thermally escaping from quantum wells of the superlattice region, contribute additional current depending on temperature and solar concentration. Retaining ideal photodiode conditions and ignoring contributions from the bulk and the depletion regions, we show the feasibility of high current generation under solar concentration, due to electrons from the superlattice region of the graphene-GaAs/Alloy Schottky cell.**

Keywords - GaAs solar cells, superlattices, graphene, Schottky contacts.

I. INTRODUCTION

Graphene-Oxide-Semiconductor (GOS) cells [1,2,3] with n-GaAs the semiconductor and with graphene replacing the metal in a metal/insulator/semiconductor Schottky configuration for a solar cell [4], leads to better device performance due to (a) GaAs wider optical gap and (b) graphene's superior optical transmittance properties. Additionally, more than one bandgap in the semiconductor layer could offer wider optical gap options for solar photon absorption, if two or more media could be involved in the cell design.

In this communication, we propose a short superlattice (SL) layer (small number of repeat distances or periods), embedded in the bulk of a Schottky barrier graphene/n-GaAs solar cell. In such a structure, illumination generates carriers in all regions of the device. Specifically, graphene is a transparent window for the cell in mind, and a source of excited electrons that thermionically escape to the semiconductor side. Photo-excitation occurs in the n-GaAs layer as well in the form of (i) minority photo-holes in the bulk of the GaAs layer (ii) minority holes in the depletion region between the oxide and the semiconductor and (iii)

majority photo-excited electrons from the quantum wells of the superlattice layer. Photo-electrons from the quantum wells of the superlattice, overcome potential barriers by thermionic emission and produce thermionic current (J_{SL}). Thermionic emission generates electrons from the graphene layer as well. In this work, we use collective thermionic current out of quantum wells (GaAs) through excess electron populations occupying eigen-energies. We treat the device as a Schottky barrier solar cell [4, 9, 10], where a (mono) layer graphene replaces the metal of the Schottky diode and the contact between the graphene window and the semiconductor can be achieved either directly or via a thin oxide layer. Energy bands of the n-doped GaAs layer bend upwards forming a depletion region W. Under illumination Fermi level splitting leads to open-circuit voltage.

The G-Ox-S/n-GaAs solar cell

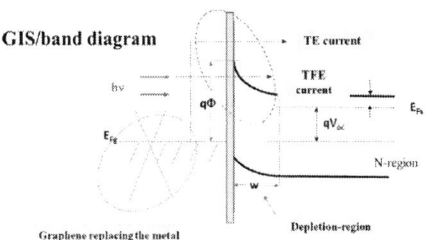

Figure 1: Graphene-based Schottky Barrier cell

The superlattice (SL) contributes excess current component J_{SL} added to minority carrier current components from the bulk region of the semiconductor. Such a hybrid design allowing a synergy between quantum size effects in the quantum well traps and the bulk of the semiconductor, offers a reduction of recombination losses in both superlattice and depletion regions due to effective mass separation of carriers (in the SL region) and the electrostatic field at the depletion region. We therefore consider carrier contribution to current J_L as the sum of three terms $J_L = J_{SL} + J_{PL} + J_{PW}$, related to superlattice-generated photocarriers, minority photoholes from the bulk semiconductor region and

978-1-7281-6118-1/22 $31.00 © 2022 IEEE

photoholes from the depletion region respectively. The Schottky cell J-V characteristic, expressed as J = J$_{oo}$ (exp (qV/kT)-1) − J$_L$, includes a prefactor related to the media involved (here graphene with its own density of states [5]) and photocurrent J$_L$ as the sum of three photocurrent terms as mentioned above.

II. EXCESS CURRENT FROM THE SUPERLATTICE

A proposed, graphene-based, Schottky contact solar cell, in Figure 1, contains six different layers, with graphene as the window to solar photons [4, 9, 10]. The bulk of the device is the n-GaAs layer with an embedded AlAs/GaAs SL region. The latter is "grown" at some distance from the edge of the depletion region of the n-GaAs layer. Such region (SL) in the bulk generates excess current from the hybrid G-Ox-n/GaAs/Alloy cell due to (thermionic) electron escape from quantum wells. The proposed short superlattice region (Figure 2) is a SL with 20 periods along the growth direction including 2nm-GaAs quantum wells and AlAs quantum barriers (> 100nm) for negligible tunneling. Location of the SL region in the bulk n-GaAs is selected anywhere, past the depletion width region.

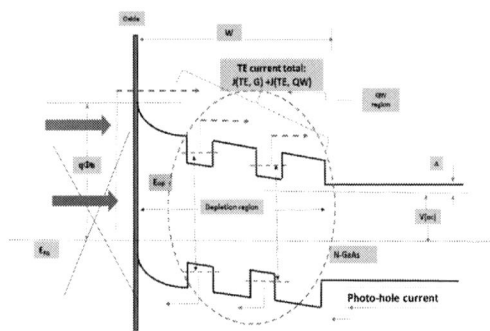

Figure 2: G/Ox/n-GaAs/Alloy hybrid solar cell

Under illumination, photo-excited electrons and holes occupy eigen-states in quantum wells (qw) from which thermionic emission generates excess current. Inserted lattice-matched (AlAs/GaAs/AlAs) multilayer includes two different energy gaps (2.16eV[4]/1.42eV/2.16eV) and causes (a) cumulative excess current density J (SL) and (b) optical gap increase. The latter guarantees (optical gap) photon absorption at photon energies above 1.42eV (GaAs). Recombination at the SL region is expected to be reduced through effective mass separation between photo-excited electrons and holes in the continuum of the conduction and valence bands respectively (see Figure 3). Figure 3 describes superlattice carrier excitation in the band continuum (e.g. electrons) by suitable quantum well width choice.

QW geometry

$\bullet$ $E_j = \left(\frac{\hbar^2 \pi^2}{2 m_n L_w^2}\right) j^2 ; j = 1, 2$

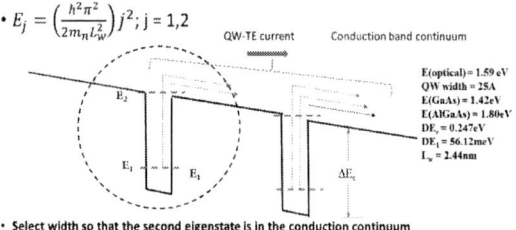

$\bullet$ Select width so that the second eigenstate is in the conduction continuum

Figure 3: G/Ox/n-GaAs/Alloy hybrid solar cell

The oxide layer (sufficiently thin for carrier tunneling) bridges the graphene and semiconductor layers, allowing electron transport through tunneling and reducing recombination losses at the GaAs junction. Current density J$_{SL}$ due to photo-excited electrons thermionically escaping from quantum wells to the conduction band in the bulk of the device is [5]:

$$J(SL) = \int_{Ec2}^{\infty} dE g(E) f(E) v(E)$$

$$= q \int_{Ec2}^{\infty} dE \left[\frac{m^*}{\pi \hbar^2 L_w}\right] e^{-\frac{E-E_F}{kT}} \sqrt{\frac{2(E-E_{c2})}{m^*}} \quad (1)$$

Where the bracketed term represents SL-region's two-dimensional density of states (DOS), f(E) is the Fermi-Dirac distribution function turned to Maxwell-Boltzmann for all energy levels above 3kT, v(E) is electron velocity of kinetically active electrons in quantum wells, E$_{c2}$ is the conduction band edge of the wide-gap layer (AlAs). The same expression includes the electronic charge q and effective mass m* of photoelectrons in GaAs layers, while L$_w$ is the width of an individual quantum well (in nm). The radical term in (1) represents velocity of kinetically active and thermionically excited electrons *overcoming* the potential barriers of quantum wells. Standard calculation steps lead (1) to an explicit form:

$$J_{SL} = \left(\frac{q\sqrt{2\pi m^* k^3}}{\pi \hbar^2 L_w}\right) T^{\frac{3}{2}} \exp\left(-\frac{\Delta E_c}{kT}\right) \exp\left(-\frac{E_{c1}-E_{Fn}}{kT}\right) \quad (2)$$

Where E$_{c2}$ (E$_{c1}$) stand for wide-gap (low-gap) edges of AlAs (GaAs) layers respectively and E$_{Fn}$ is the n-region quasi-Fermi level. Noting that E$_{c2}$ − E$_{Fn}$ = (E$_{c2}$ − E$_{c1}$) + (E$_{c1}$ − E$_{Fn}$) = ΔE$_c$ + (E$_{c1}$ − E$_{Fn}$), where E$_{c2}$ − E$_{c1}$ = ΔE$_C$ (the conduction band discontinuity). Note in (2) the effect of the SL region on thermionic current via temperature as a T$^{3/2}$–dependence instead of the usual T^2- dependence in the bulk [5, 6]. Solar photons generate additional carrier concentration in quantum wells δn$_{ph}$ (λ) in addition to dark electron concentration n$_{qwd}$. Total electron concentration in each quantum well is the sum of dark n$_{qwd}$, photo-generated δn$_{ph}$ (λ) carriers and diffusion n$_{ph, w}$ (λ) carriers n$_{diff}$ + n$_{qwd}$ + δn$_{ph}$ (λ). Of these three terms, only the illumination-related term

δn_{ph} is numerically significant at room temperature (i.e. n_{diff}, $n_{qwd} \ll \delta n_{ph}$, mainly due to the X solar factor) [5, 6, 7]. Therefore, total photo-electron population in individual quantum wells can be approximated as

$$n_{ph,qw} \cong \sqrt{\frac{X\alpha\Phi(\lambda)}{b}}e^{-\frac{\alpha x}{2}}(cm^{-3}) \qquad (3)$$

Here, $\Phi(\lambda)$ represents incident photon flux ($cm^{-2}s^{-1}$), $\alpha(\lambda)$ (cm^{-1}) is the absorption coefficient of the low gap medium (GaAs), and b is the radiation recombination coefficient ($cm^3 s^{-1}$) [11]. Expression (2) via (3) describes thermionic emission current off individual quantum wells ($J_{SL, qw}$) as a strong function of conduction band discontinuity, absorption coefficient (GaAs) and solar photon flux Φ:

$$J_{SL,qw} = \frac{\sqrt{\pi}qh}{m^*L_w}\sqrt{\frac{X\alpha\Phi(\lambda)}{b}}e^{-\frac{\Delta E_c}{kT}}e^{-\frac{\alpha x}{2}} \qquad (4)$$

Cumulative contribution from the superlattice layer derives by summing over all quantum wells, relative to the first one, a distance x_1 away from the edge of the depletion region in the n-GaAs bulk. For a number N of quantum wells, current contribution J_{SL} is equivalent to summing over all terms above:

$$J_{SL} = \frac{\sqrt{\pi}qh}{m^*L_w}\sqrt{\frac{X\alpha F\Phi(\lambda)}{b}}e^{-\frac{\Delta E_c}{kT}}e^{-\frac{\alpha x_1}{2}}\left(\frac{1-e^{-N\alpha d}}{1-e^{-\alpha d}}\right) \quad (5)$$

the last term in (5) is a geometric series of N terms (i.e. superlattice periods of thickness d), finalizing the total thermionic emission current out of a superlattice region. SL-values selected are L_w=2nm, barrier thickness above 50nm leading to d=52nm repeat distance (SL thickness near a micron for 20 periods). Total SL-photocurrent depends on the number N of repeat distances in the SL layer. Note that (a) positioning of the SL region in the bulk includes the exponential factor containing the absorption coefficient and (b) the conduction band-discontinuity ΔE_c plays a pivotal role as seen from the exponential term in (7). Conduction/valence band offsets depend on characteristic heterojunction factors. In this study we adopt the general ΔE_c = (2/3) of ΔE_g band discontinuity rule [13, 14, 15]. The total current depends on the position of the SL factor: the closer to the depletion edge, the higher the value of the exponential factor. In the next, we take such considerations and compute total SL current from the superlattice region. Conveniently, III-V type heterostructures form type–I superlattices in solar cells with band offsets determined by the ΔE_c factor. Based on effective mass approximation, we adopt band offset connection according to the rule $\Delta E_c/\Delta E_g$ = 0.65, obtaining band discontinuity $\Delta E_{c, AlAs/GaAs}$ = 0.481eV, $\Delta E_{v, AlAs/GaAs}$ = 0.259eV. Electron-to-hole ground states occur at E_{1e} = 0.470eV and E_{1h} = 0.239eV respectively (both measured from the bottom of the quantum wells to the ground states E_{1e} and E_{1h}) and for quantum wells as narrow as 2nm. The effective optical gap is now increased from 1.42eV (GaAs) to $E_{opt} = |E_{1e}-E_{1h}|$ = 2.129eV, corresponding to 582nm of solar wavelength, indicating maximum absorption at solar irradiance I_{rr} = 1.3729W/m^2 or solar photon flux Φ_λ = I_{rr}/hν = (1.3729)/ (2.129) (1.6) (10^{-19}) (10^4) = (4.03) (10^{14}) $cm^{-2}s^{-1}$.

Figure 4 depicts photocurrent generated from the SL photo-carrier reservoir (SL):

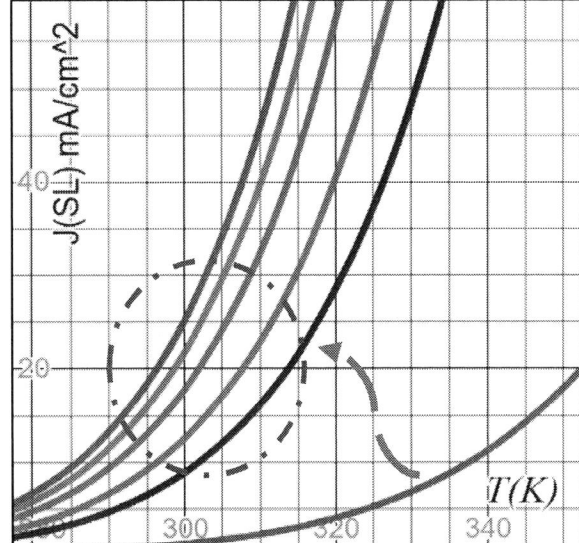

Figure 4: SL-Photocurrent vs T(K) at different solar concentrations (X= 1 => **50, 100, 200, 300, and 400**). Short-circuit currents are J(SL, mA/cm^2) = 9.704, 13.723, 19.408, 23.769 and 27.446.

As seen from the figure above, J_{SL} ranges from 1.372mA/cm^2 at X=1 to 27.44 mA/cm^2 at X=400, at 300K. Table 1 below shows open-circuit voltage and photocurrent from the illuminated superlattice region.

III. OPEN-CIRCUIT VOLTAGE

Photo-current J_L is the total short-circuit current as the sum of three components (i) photocurrent J_{SL} due to superlattice excitation (ii) photocurrent J_{pw} due to minority holes in the depletion region and (iii) photocurrent J_{PL} from the bulk of the n-GaAs layer due to minority excess photo-excited holes under solar exposure Φ_λ. Open-circuit voltage (V_{oc}) calculation is possible via standard solar cell modeling. Open-circuit voltage depends on photocurrent and temperature according to the relation ($J_{SL} \gg J_P, J_{PL}$):

$$V_{oc} = V_t \ln[1+(J_{SL}/J_o)] \qquad (6)$$

The current prefactor J_o is the saturation current of the hybrid G/Ox/n-GaAs Schottky diode and is given explicitly below [5]:

$$J_o = \left(\frac{A^{**}}{d}\right)|t|T^{3/2}e^{-q\Phi_B/kT} \qquad (7)$$

Thermionic carrier transport from the graphene side to the semiconductor is feasible as long as electrons can surmount the Schottky barrier. Such modeling has been presented elsewhere, based on the Landauer formula [4,5,10]. A^{**} is a *new* Richardson's [4, 5, 10] constant for the graphene layer, d is graphene layer's thickness (here taken to be one nanometer), |t| is an average probability tunneling probability of electrons through the thin oxide layer [10], $q\Phi_B$ is the Schottky barrier height at the G/Oxide interface. Note also the $T^{3/2}$ instead of T^2 dependence of the current prefactor due to replacing the metal with a graphene layer in Schottky diodes. Based on expressions (8) and (9), open-circuit voltage becomes: In order to emphasize the superlattice effect, we neglect both photohole currents (preliminary calculations show J_{pw} and $J_{PL} \sim 4 mA/cm^2$) in the bulk and the depletion region of the junction. Both hole-current components are heavily affected by the transmission probability [12] through the oxide layer ($\sim 10^{-3}$) much lower compared to J(SL) indicating SL photocurrent domination at least in the numerical sense. Such values dominate at one-sun exposure ($J_{SL} = 1.372 mA/cm^2 < J_{pw}$ and J_{PL}) and at dark conditions where $J_{SL}^{dark} < J_{pw}$ and J_{PL}, but are much lower at 50 suns exposure and beyond. Without loss of the main part of the argument made here (i.e. *superlattice regions under solar concentration cause sharp photocurrent increase*), expression (6) via (7) takes the form of a G-Ox-S open-circuit voltage [4], but with its own new parameters:

$$V_{oc} = V_t \ln\left(1 + \frac{J_{SL}}{J_{oo}}\right) = V_t \ln\left(\frac{J_{SL}}{(A^{**}/d)|t|T^{\frac{3}{2}}}\right) + \Phi_B \quad (8)$$

Sampled computed open-circuit voltage values are depicted on Table 1 below:

TABLE 1 **OC-voltage and SL photocurrent**

X(suns)	V(oc) (V)	J(SL) mA/cm²
1	0.805	1.372
50	0.838	9.704
100	0.846	13.723
200	0.853	19.408

As seen from (8), open-circuit voltage depends on superlattice excess photocurrent, graphene properties through the new [5] Richardson's constant A** and thickness d (nm), on temperature T and on Schottky barrier height through tunneling probability. The numerical value for the (A^{**}/d) [5,6,7] ratio is 3.35×10^{-4} mA/cm²/°K$^{3/2}$ (with d=1nm). Barrier height value for undoped graphene/GaAs is at 0.79eV [17,18]. Open-circuit voltage depends strongly on temperature (see (8) above) through thermal voltage V_t, and the Schottky current prefactor including the $T^{3/2}$ factor [5]. Note also that SL currents depend on temperature as seen from (7).

IV. CONCLUSION

We have demonstrated Schottky solar cell performance improvement by proposing (a) replacing the metal window material with a monolayer graphene layer and (b) "growing" a III-V lattice-matched superlattice embedded in the bulk region of the semiconductor. The diode's specific geometry is Graphene/Oxide/n-GaAs/AlAs-GaAs (SL) Schottky cell. The role of the embedded superlattice is to absorb, trap and thermionically emit photoelectrons in the conduction band continuum as excess carriers for current formation in the cell. To demonstrate the importance of the design, we neglected excess minority holes from the depletion region and the n-GaAs region. Excess minority carrier populations

are present of course but they are numerically and computationally inferior to the photoelectron concentration in the quantum wells of the embedded superlattice. We find that photocurrent due to electrons becomes significant at high solar concentrations. Qualitatively, the device generates current (a) due to minority photo-generated holes from the semiconductor (under one sun) and (b) due to thermionically escaping electrons from the superlattice embedded near the depletion region at the Oxide-semiconductor interface. The superlattice serves as a reservoir of majority electrons in the bulk of the semiconductor layer. On the other hand, proposing graphene instead of a metal as the window of the cell provides minimal absorption of solar photons at the window layer and causes thermionic current from the graphene with electrons surmounting the Schottky barrier (under assumed minimum reflection). Photo-excitation sets the device in a forward bias mode: graphene's and the semiconductor quasi-Fermi levels split producing open circuit voltage under different solar illumination levels. In the process, it becomes clear that solar concentration sharply improves short circuit current and additionally overall device performance in terms of both open circuit voltage power collection efficiency. Under varying solar concentration levels, the superlattice acts as photocarrier reservoir where electrons accumulate under direct light absorption and excitation to eigen-energy levels in quantum wells. By means of thermionic theory arguments, there is non-zero probability for photo-electrons in eigen-energy levels, to surmount the barriers and escape to the continuum of the conduction band. The advantage of a superlattice is that a cloud of electrons diffuses to the junction as majority carrier current. Simultaneously, electrons from the illuminated graphene side cross the junction forming net electron current. The proposed diode current-voltage characteristic follows typical Schottky diode patterns but with two changes in the current pre-factor (i) a new Richardson's constant and (ii) $T^{3/2}$ – temperature dependence instead of the T^2 one. Keeping photocurrent values due to thermally escaping electrons from the quantum well of the superlattice regions only, our modeling depicts (ideal conditions, 300K) high current and improved open-circuit voltage. For instance, at 50-sun concentration levels, short circuit current density, open circuit voltage and power collection efficiency (PCE) values are 9.704 mA/cm², 0.838V and 7.29% respectively. Ongoing projection work at-100 suns-indicates 13.723 mA/cm², 0.846V and 10.367% performance and at 200suns overall PCE exceeds 19%. The proposed design exploits quantum size effects in graphene-based Schottky solar cells. By introducing a superlattice region in the n-GaAs bulk semiconductor layer and by replacing the metal with a graphene layer, high absorption is feasible and and improved efficiency. Ultimately, by (a) replacing metal with graphene layer(s) (b) inserting a SL layer in the bulk of the semiconductor and (c) solar concentration above 50 suns, one expects

(i) High mobility photo-carriers
(ii) High transparency
(iii) Sharp increase in current
(iv) Improved open-circuit voltage and
(v) Improved PCE

978-1-7281-6118-1/22 $31.00 © 2022 IEEE

V. REFERENCES

[1] AK Geim, KS Novoselov, The rise of graphene, Nature Mater.2007, 6, 183-191.

[2] H Raza, Graphene Nanoelectronics: Metrology, Properties and Applications, Springer Verlag, Berlin, 2012.

[3] C Berger, Z Song, X Li et al., Electronic Confinement and Coherence in Patterned Epitaxial Graphene, Science 2006, 312 1191-1196.

[4] SM Sze, Physics of Semiconductor Devices, Second Edition, A Wiley-Interscience publication, 1981.

[5] AC Varonides, Combined thermionic and field emission reverse current for ideal graphene/n-Si Schottky contacts in a modified Landauer formalism, Phys. Status Solidi C 13, No. 10–12, 1040–1044 (2016) /DOI 10.1002/pssc.201600096.

[6] AC Varonides, Fermi level splitting and thermionic current improvement in low-dimensional multiquantum-well (MQW) p-i-n structures, Thin Solid Films 511-512 (2006), 89-92.

[7] AC Varonides, High Efficiency Multijunction Solar Cells with Finely-tuned Quantum Wells, S Logothetidis (ed.) Nanostructured Materials and Their Applications, Springer-Verlag, Berlin Heidelberg, pp 85-103 (2012).

[8] Yi Song, X Li, C Mackin et al, Role of Interfacial Oxide in High-Efficiency Graphene-Silicon Schottky Barrier Solar Cells, Nano Lett 2015, 15, 2104-2110.

[9] Ashcroft and Mermin, Solid State Physics, Saunders College Publishers, 1976.

[10] KK Ng and HC Card, Asymmetry in the SiO_2 tunneling barriers to electrons and holes, J. Appl. Phys. 51(4), April 1980.

[11] Dieter Schroeder, Semiconductor material and device characterization, IEEE Press, Wiley-Interscience, 3d edition (2006).

[12] A Modinos, Field, Thermionic and secondary electron emission spectroscopy, Springer-Verlag, NY, 1984.

[13] L Ross, SP Svensson, P Luigi (Eds.), Pseudomorphic HEMT Technology and Applications, NATO ASI series, p. 26, Kluwer Publishers, 1996.

[14] Y Fu, "Physical Models of Semiconductor Quantum Devices, Springer, 2nd edition, 2014.

[15] R Magnanini et al, Investigation of GaAs/InGaP superlattices for quantum well solar cells, Thin Solid Films, Vol 516, Issue 20, 2008, pp 6734-6738.

[16] AC Varonides, work in progress.

[17] Di Bartolomeo, Graphene Schottky Diodes: an experimental review of the rectifying graphene/semiconductor heterojunction, Physics Reports, Vol 606, 2016, pp 1-58.

[18] S Tongay, M Lemaitre et al, Rectification at Graphene-Semiconductor Interfaces: Zero-Gap Semiconductor-based Diodes, Phys Rev X, 2 (2012).

[19] D Tomer et al, Inhomogeneity in barrier height at Graphene/Si (GaAs) Schottky junctions, 2015 Nanotechnology, 26, 215702.

[20] Dikai Xu, Xuegong Yu, Lifei Yang, Deren Yang, Interface engineering of Graphene Silicon heterojunction solar cells, Superlattices and Microstructures, 2016 Elsevier Ltd.

[21] W Jie, F Zheng, and J Hao, Graphene/gallium arsenide-based Schottky junction solar cells, Appl Phys Lett 103 (23), 233111 (2013).

[22] X Li et al, Graphene/h-BN sandwich diode as solar cell and photodetector, Optics express 24(1), 2016

[23] X Li, et al, 18.5% efficient graphene/GaAs van der Waals heterostructure solar cell, Elsevier, Nano Energy, Vol 16, September 2015, 310-319 (2015).

[24] Dasaradha Rao Lambada et al, Investigation of Illumination Effects on the Electrical Properties of Au/GO/p-InP Heterojunction with a Graphene Oxide layer, Nanomanufacturing and Metrology (2020) 3:269-281.

[25] Subash Adhikari et al, Minimizing Trap Charge Density towards an Ideal Diode in Graphene-Silicon Schottky Solar Cell, ACS Appl Mater Interfaces (2019) Jan9; (1) 880-888.

[26] Kumar A Kashid et al, Enhanced Thermionic Emission and Low 1/f Noise in Exfoliated Graphene/GaN Schottky Barrier Diode, ACS Appl Mater Interfaces, 2016 Mar;8(12):8213-23.

[27] Dub M Sai, et al, Graphene as a Schottky Barrier contact to Al/GaN Heterostructures, Materials (Basel). 2020 Sep 17;13(18):4140.

[28] Lee M, Vu TKO, Lee KS, Kim EK, Park S., Electronic Transport Mechanism for Schottky Diodes Formed by Au/HVPE a-Plane GaN Templates Grown via In Situ GaN Nanodot Formation, Nanomaterials (Basel). 2018 Jun 2;8(6):397.

[29] GaN.Lee M, Ahn CW, Vu TKO, Lee HU, Jeong Y, Hahm MG, Kim EK, Park S., Current Transport Mechanism in Palladium Schottky Contact on Si-Based Freestanding, Nanomaterials (Basel). 2020 Feb 10;10(2):297.

978-1-7281-6118-1/22 $31.00 © 2022 IEEE

Stability of Silicon Heterojunction solar cells having hydrogen plasma treated intrinsic layer

Anishkumar Soman[1,2], Gbenga Obikoya[1,2], Steve Johnston[3], Steven Harvey[3], Ujjwal Das[2], Steven Hegedus[1,2]

[1] Department of Electrical and Computer Engineering, University of Delaware, Newark, DE, USA 19716
[2] Institute of Energy Conversion, University of Delaware, Newark, DE, USA 19716
[3] National Renewable Energy Laboratory (NREL), Golden, CO, USA 80401

Abstract—The open-circuit voltage of Silicon Heterojunction (HJ) cells can be improved by using extrinsic hydrogen plasma treatment (HPT) to reduce the trap states (N_T) which we have verified using Deep Level Transient Spectroscopy (DLTS). However, hydrogen has been associated with long-term degradation. We have investigated the stability of HJ cells with and without HPT using accelerated degradation (1-sun illumination, 90°C temperature, argon ambient for 1000 hours). The stability has been studied using DLTS, current-voltage, and Suns-V_{OC}. Our results show HPT cells are stable and maintain their higher V_{OC} and lower N_T after accelerated degradation.

Keywords— degradation, hydrogen passivation, heterojunction, silicon solar cells, reliability, light soaking.

I. INTRODUCTION

As the industry transitions to n-type silicon substrates, there are different architectures of silicon solar cells like Tunnel Oxide Passivated Contacts (TOPCon) [1], Heterojunction technology (HJT) [2] and Inter-digitated Back Contacts (IBC) [3] concepts that have been gaining momentum for increased production. Silicon Heterojunction solar cells has been one of the top contenders for commercial manufacturing due to the multifarious advantages which includes record single-junction efficiency [4], high open circuit voltage (V_{OC}) and very low temperature co-efficient. The high V_{OC} stems from the superior passivation of the dangling bonds due to the ultra-thin intrinsic amorphous silicon layer (a-Si:H) over crystalline silicon (c-Si). To further enhance the interface passivation different methodologies like extrinsic hydrogen plasma treatment (HPT) where hydrogen bonds to interface silicon dangling bonds has been proposed by different groups [5],[6]. However, hydrogen migration has been associated with long-term degradation in silicon solar cells [7],[8]. In our previous work, we had demonstrated a direct-current (d.c.) plasma enhanced chemical vapor deposition (PECVD) based extrinsic post-HPT method which can result in high implied V_{OC} of 755mV in front heterojunction device architecture [9]. To understand if such methods can be viable for industrial application and to have increased market share for HJT, it's imperative to know of any possible degradation mechanisms due to this additional hydrogen introduction. To the best of our knowledge there are no studies which looks into the long-term stability of silicon heterojunction solar cells having intrinsic layer with HPT. In this work, we study the stability of Si HJ cells with and without HPT

when they are subjected to elevated temperature (90°C) and 1-sun illumination for 1000 hours. The cells are then characterized for V_{OC} loss or change in series resistance (R_S) after the light soaking experiments. We have demonstrated that the previously obtained higher V_{OC} due to HPT comes from reduced trap states and the HPT method doesn't introduce additional degradation in the long-term performance of the cells.

II. EXPERIMENTAL DETAIL

The Front HJ cells in this study are fabricated on 140μm thick, <100> oriented n-type CZ wafer having resistivity of 1-10Ω.cm. The substrates have undergone standard cleaning and random pyramid texturization [10]. The details of the deposition condition of the intrinsic layer with HPT and doped layers by d.c.PECVD are given elsewhere [11],[12]. The front as well as back ITO was deposited by rf-sputtering and Ag/Al by e-beam evaporation for contacts. The cells have a configuration of Ag front grid/indium tin oxide (ITO)/p-a-Si:H/i-a-Si:H/n-c-Si/i-a-Si:H/n-a-Si:H/ITO/Al back contact with a cell area of 2.5cm².

Two batches of unencapsulated cells with and without HPT intrinsic a-Si:H layer were subjected to accelerated degradation in an in-house built stress chamber at elevated temperature of 90°C and 1-sun (100mW/cm²) illumination intensity in argon ambient (to mitigate any degradation due to moisture or oxygen) for 1000 hours. The current density-voltage (J-V) characterization of the cells were performed using a class AAA AM1.5 simulator from Optical Association Inc. (OAI). The illumination intensity dependent V_{OC} of the cells were measured using Sinton Suns V_{OC} tool to negate the effect of any resistive losses related to series Resistance (R_S). The J-V parameters were extracted using a single-diode model by Hegedus et al. [13]. A commercial DLTS system with Fourier transform was used to characterize defects. All cells for DLTS were laser scribed to smaller area, subjected to same reverse bias of -1V, with filling pulse set to 0.8V (forward) for 50μs filling pulse width. A time window of 5ms was used to capture the peaks.

978-1-7281-6118-1/22 $31.00 © 2022 IEEE

III. RESULTS AND DISCUSSION

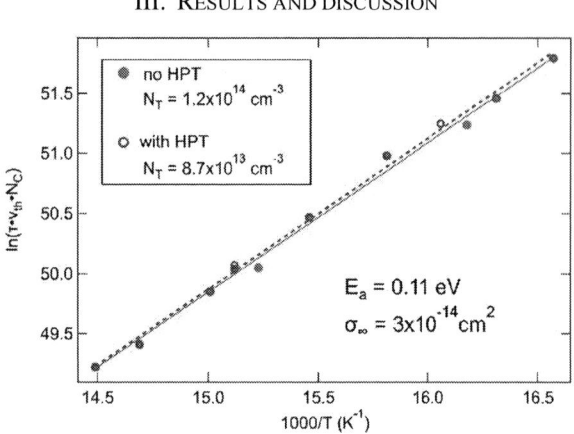

Fig. 1. DLTS Arrhenius plot of the peaks for cells with and without HPT. The vertical axis shows the log of carrier lifetime (τ), thermal velocity (v_{th}) and density of states (N_C)

Post HPT of intrinsic (i-layer) has been previously reported to have higher minority carrier lifetime and implied V_{OC} on cells before metallization [12]. We have first verified that HPT improves the interface passivation by reducing defect density with the help of DLTS. The trap density (N_T) and activation energy were calculated using standard method [14]. The Arrhenius plot of the peak in Fig. 1 exhibits the decrease in number of trap states (N_T) from 1.2×10^{14} cm^{-3} for no HPT to 8.7×10^{13} cm^{-3} for i-layer with HPT. Additionally, the analysis shows a shallow electron trap with an activation energy (E_a) of 0.11 eV from the slope of the curve whereas the intercept of T tending to infinity gives a capture cross section (σ_∞) of 3×10^{-14} cm^{-2}. Thus, it clearly shows that HPT can help reduce the active electronic trap states which can lead to reduced recombination thereby translating to higher carrier lifetime and V_{OC}.

Fig. 2 shows the trend for different J-V parameters for a light soaking duration of 1000 hours. From Fig. 2a we can observe that the V_{OC} of the cells show a slight jump at ~60 hours which gradually decreases over time. This jump appears irrespective of intrinsic layer having HPT or not. The high standard deviation occurs since we have taken the average of 6 cells for each condition having different efficiencies to remove any sample-to-sample process-induced variation. Though the absolute value changes, the trend remains the same. The average drop in V_{OC} of 2mV for HPT samples is comparable to the average drop in V_{OC} of 3mV for non-HPT samples. The short-circuit current density (J_{SC}) shows negligible variation whereas the Fill Factor (FF) shows a similar trend as V_{OC} with a slight initial bump followed by a gradual decrease. The overall drop in efficiency of ~0.4% for HPT samples is comparable to ~0.5% of non-HPT samples which suggests the stable performance of HJ cells with HPT. The marginal loss in efficiency due to decrease in V_{OC} and FF irrespective of any HPT is similar to observations made in previous studies [15]. Diode analysis of the J-V curve of cells before and after the light soaking experiment can help reveal if the loss in FF can be attributed to any resistive losses. Table 1 shows the ideality factor (n), series resistance (R_S) and total recombination current density (J_O) of HPT and non HPT HJ cells before and after 1000 hours of stress experiment.

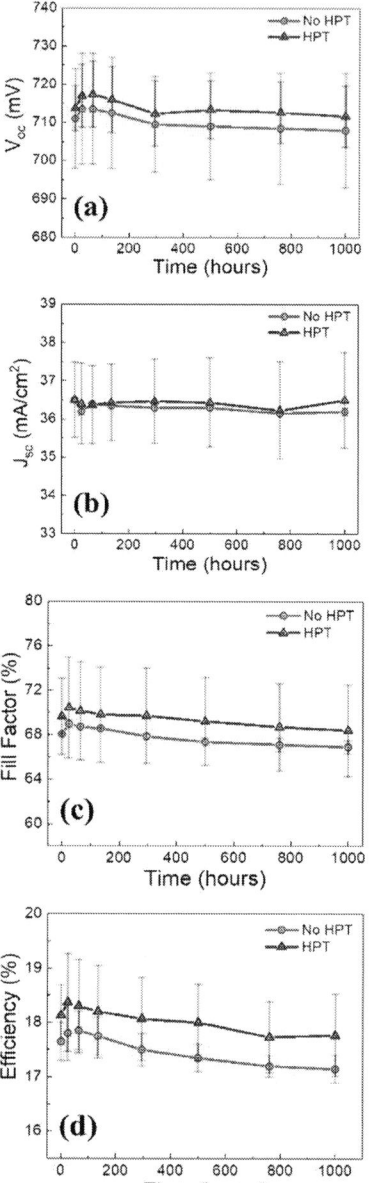

Fig. 2. J-V characteristic (a) V_{OC} (b) J_{SC} (c) Fill Factor and (d) Efficiency of Front HJ cells with and without HPT for a light soaking duration of 1000 hours at 1-sun illumination and 90°C

TABLE I. DIODE ANALYSIS PARAMETERS FROM J-V CURVES OF HJ CELLS WITH AND WITHOUT HPT BEFORE AND AFTER LIGHT SOAKING

Sample	n	J_0 (A/cm²)	R_S (Ω.cm²)
No HPT – 0 hours	1.82	1.03×10^{-8}	1.92
HPT – 0 hours	1.82	1.03×10^{-8}	2.28
No HPT – 1000 hours	1.75	7.87×10^{-9}	2.43
HPT – 1000 hours	1.92	2.43×10^{-9}	2.31

From Table 1 we can observe that for cells with no HPT, there is an increase in R_S which could result in drop in FF whereas for cells with HPT the increase in R_S is negligible. For cell with HPT the increase in ideality factor could contribute to reduction in FF. The Suns-V_{OC} measurement which doesn't account for the resistive losses due to R_S in Fig. 3 is used to understand the V_{OC} loss in cells with and without HPT before and after 1000 hours of light soaking and elevated temperature exposure.

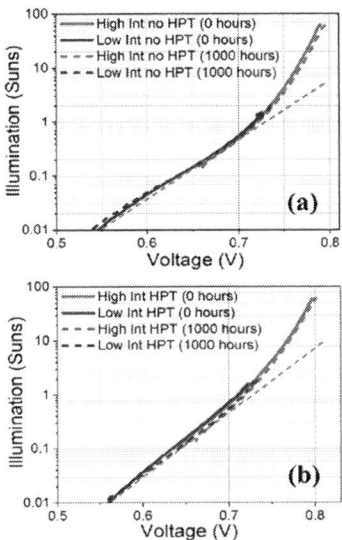

Fig. 3. Suns-V_{OC} results of HJ cells (a) without HPT (b) with HPT under low and high intensity white light illumination before and after 1000 hours of light soaking experiment

The results from Fig. 3a & b which shows V_{OC} as a function of illumination exhibits a linear trend for low intensity while V_{OC} reduces at a higher intensity regardless of HPT of the intrinsic layer. The light soaking doesn't introduce any additional back bending of the logarithmic curve which corroborates with the stable performance of the cells.

Additional experiments are underway using Time-of-Flight Secondary Ion Mass Spectrometry (ToF-SIMS) and Deep Level Transient Spectroscopy (DLTS) to look into the hydrogen depth profile and electronically active defect states in HJ cells with HPT after undergoing accelerated degradation.

IV. SUMMARY

The V_{OC} of Silicon HJ cells can be enhanced using hydrogen plasma treatment for the intrinsic layer which passivates the interface dangling bonds. This study has shown that HPT assists in reducing density of trap states which can help achieve higher V_{OC}. However, recently hydrogen has been associated with long-term degradation in silicon solar cells. In this study we have examined the stability of Si HJ cells which has undergone HPT and subjected to accelerated degradation by light soaking in 1-sun and elevated temperature of 90°C for 1000 hours and compared it's results with cells which hasn't undergone HPT. Our initial results show that such extrinsic hydrogenation doesn't introduce any additional loss mechanism under the subjected stress conditions. No change in the Suns-V_{OC} curve after light soaking affirms the stable nature of the HPT HJ cells. We have confirmed there are no additional defect states caused due to hydrogen after light soaking which will be reported later.

Thus, HPT can be a viable technique for integration into commercially manufacturing long term stable Si HJ solar cells.

ACKNOWLEDGMENT

The authors would like to acknowledge support from the Graduate College at the University of Delaware for the Dissertation Fellowship and the US Department of Energy Sunshot Program PVRD Award DE-0007534. The authors would also like to thank Dirk Jordan, Andrew Norman, Harvey Guthrey and Dana Kern from National Renewable Energy Laboratory (NREL) for valuable discussions.

REFERENCES

[1] F. Feldmann, M. Bivour, C. Reichel, M. Hermle, and S. W. Glunz, "Passivated rear contacts for high-efficiency n-type Si solar cells providing high interface passivation quality and excellent transport characteristics," *Sol. Energy Mater. Sol. Cells*, vol. 120, no. PART A, pp. 270–274, Jan. 2014.

[2] H. Sakata *et al.*, "20.7% highest efficiency large area (100.5 cm2) HIT™ cell," *Conf. Rec. IEEE Photovolt. Spec. Conf.*, vol. 2000-January, pp. 7–12, 2000.

[3] M. Lu, S. Bowden, U. Das, and R. Birkmire, "Interdigitated back contact silicon heterojunction solar cell and the effect of front surface passivation," *Appl. Phys. Lett.*, vol. 91, no. 6, p. 063507, Aug. 2007.

[4] K. Yoshikawa *et al.*, "Exceeding conversion efficiency of 26% by heterojunction interdigitated back contact solar cell with thin film Si technology," *Sol. Energy Mater. Sol. Cells*, vol. 173, pp. 37–42, Dec. 2017.

[5] A. Descoeudres *et al.*, "Improved amorphous/crystalline silicon interface passivation by hydrogen plasma treatment," *Appl. Phys. Lett.*, vol. 99, no. 12, p. 123506, 2011.

[6] A. Antony and A. Soman, "Hydrogen Plasma Treatment to Enhance a-Si/c-Si Interface Passivation," in *32nd European Photovoltaic Solar Energy Conference and Exhibition*, 2016, pp. 738–741.

[7] P. Mahtani, R. Varache, B. Jovet, C. Longeaud, J. P. Kleider, and N. P. Kherani, "Light induced changes in the amorphous—crystalline silicon heterointerface," *J. Appl. Phys.*, vol. 114, no. 12, p. 124503, Sep. 2013.

[8] A. C. N. Wenham *et al.*, "Hydrogen-Induced Degradation," *2018 IEEE 7th World Conf. Photovolt. Energy Conversion, WCPEC 2018 - A Jt. Conf. 45th IEEE PVSC, 28th PVSEC 34th EU PVSEC*, pp. 1–8, Nov. 2018.

[9] A. Soman, U. Nsofor, U. Das, T. Gu, and S. Hegedus, "A-Si:H/c-Si interface hydrogenation for implied Voc = 755 mV in Silicon heterojunction solar cell," in *Conference Record of the IEEE Photovoltaic Specialists Conference*, 2019, pp. 1927–1930.

[10] U. J. Nsotor *et al.*, "Analysis of silicon wafer surface preparation for heterojunction solar cells using X-ray photoelectron spectroscopy and effective minority carrier lifetime," *Sol. Energy Mater. Sol. Cells*, vol. 183, pp. 205–210, Aug. 2018.

[11] Z. Shu, U. Das, J. Allen, R. Birkmire, and S. Hegedus, "Experimental and simulated analysis of front versus all-back-contact silicon heterojunction solar cells: effect of interface and doped a-Si:H layer defects," *Prog. Photovoltaics Res. Appl.*, vol. 23, no. 1, pp. 78–93, Jan. 2015.

[12] A. Soman, U. Nsofor, U. Das, T. Gu, and S. Hegedus, "Correlation between in Situ Diagnostics of the Hydrogen Plasma and the Interface Passivation Quality of Hydrogen Plasma Post-Treated a-Si:H in Silicon Heterojunction Solar Cells," *ACS Appl. Mater. Interfaces*, p. acsami.9b01686, Apr. 2019.

[13] S. S. Hegedus and W. N. Shafarman, "Thin-film solar cells: device measurements and analysis," *Prog. Photovoltaics Res. Appl.*, vol. 12, no. 2–3, pp. 155–176, Mar. 2004.

[14] D. V. Lang, "Deep-level transient spectroscopy: A new method to characterize traps in semiconductors," *J. Appl. Phys.*, vol. 45, no. 7, p. 3023, Oct. 2003.

[15] D. C. Jordan *et al.*, "Silicon heterojunction system field performance," *IEEE J. Photovoltaics*, vol. 8, no. 1, pp. 177–182, Jan. 2018.

Snow Shedding properties of Bifacial PV Panels

Ajay Singh, Derek Jones
Campbell Scientific Inc., Logan, UT, 84321, USA

Abstract—Solar photovoltaic (PV) systems are frequently installed in climates with significant snowfall. Loss of energy production due to snow on pv panels is an important issue. It has been recognized for some time that bifacial PV panels have better snow shedding capabilities. In this paper we present a study of comparison between the snow shedding characteristics of bifacial and monofacial pv panels. As snow accumulates on, around, and underneath PV panels the site albedo can increase to as much as 75% of the incident solar radiation. This additional light reflected off the ground is absorbed by the bifacial panel, resulting in both heat and electrical generation. The additional heat increases the rate of snowmelt on the front side of the panel. In contrast monofacial panels do not absorb solar energy from the rear side and no significant heat generation takes place. Results of this study illustrate how bifacial panels are advantageous over monofacial panels in climates where snow is expected due to their ability to produce energy from the additional reflected irradiance and a quicker recovery time (snowmelt) than monofacial panels.

Index Terms—snow shedding, bifacial pv panels, monofacial pv panels

I. INTRODUCTION

Photovoltaic (PV) panels are rapidly emerging as most viable form of renewable energy in the fight against climate change. In 2020 the deployed capacity of PV plants was about 700 GW. PV panels are also getting deployed in high latitude areas. In these regions during winter, snowfall greatly diminishes the electricity generation of installed PV panels in these areas. In these regions wintercan be cold, long, and accompanied by abundant solar resources and large snow falls. The PV panels are installed in open areas leaving them vulnerable to being covered with snow which will have a significant impact onelectricity production until the snow is removed by some means. Andrews et al. and Andenaes et al. [1], [3] studied the effect of snow on solar PV panels. It is believed that upto 15% of electricity generation is lost due to snow. Sunny days can accompany extended periods of subfreezing air temperatures following a snowfall event. Snow is a highly reflective material and even a thin layer can result in substantial reduction of incident light intensity reaching the solar cells. Snow and ice accumulation on PV panels also creates problems related to loading, drainage of snowmelt and safe removal of snow from the pv panels. To mitigate this, snow is often removed mechanically [5]. Research is being done on special coatings for PV panel surfaces that inhibit the formation or acuumulation of snow and ice, melting accumulated snow by passing a current through pv panels thus utilizing the panel as an electrical load [2]. Riley et al. [4] studied differences in snow shedding from framed vs frameless pv panels.

Bifacial photovoltaic panels offer the promise of generating 10–15% more electrical energy by absorbing light reflected off of ground from the rear side of these panels. These panels use glass or another transparent back sheet to capture reflected light. More and more PV manufacturers are offering these panels and they are increasingly accepted in the marketplace. There has been some evidence that these panels shed snow faster than monofacial panels, but we have yet to find apublished a systematic study on this topic. Pike et al. [6] reported on a study of energy production from vertically mounted bifacial panels in Alaska compared to production from PV panels mounted at fixed tilt facing south. The annual energy production from vertically mounted PV panels was unaffected due to snow and overall similar annual energy was produced by all panels. In this paper we present a study of comparison of snow shedding characteristics of a bifacial and monofacial panels and present some evidence of the mechanism.

II. EXPERIMENTAL ARRANGEMENTS AND METHODOLOGY

The experiments were conducted at the headquarters of Campbell Scientific Inc. in Logan, Utah, USA. One 380 W half cut bifacial solar panel and one 215 W monofacial panel mounted horizontally was the setup used in these experiments. Two silicon (Si) pyranometers (EKO, model ML02), were also installed, one facing up and another facing down. This specific silicon pyranometer was chosen for its compact size and ease of installation and its closer spectral response to the PV material of the panels. The downward facing pyranometer was mounted directly at the center of the rearside of the bifacial panel to measure the reflected light reaching the back side of the panel. The small size of the pyranometer and the construction of half cut pv panel facilitated the measurement of reflected light intensity without blocking anyactive portion of the PV panel. The upward facing pyranometer measured incident radiation on the front surface of the pv panel, which is also the Global Horizontal Irradiance (GHI) for our installation geometry. Short-circuit current from each panel was measured every minute and the pyranometers were measured every second, saving the one-minute averages for further processing. The Campbell Scientific soiling solution CR–PVS2 was used to measure the short circuit current. The CR–PVS2 incoporates back of module temperature sensors, the CS241, also manufactured by Campbell Scientific to measure the back of module temperature of both panels. Air temperature, wind speed and wind direction at the site were measured using an all–in–one weather sensor, the MetSense500 made by Gill instruments. Snow accumulation on ground was measured using Campbell Scientific SR50A.

978-1-7281-6118-1/22 $31.00 © 2022 IEEE

III. RESULTS AND DISCUSSIONS

Fig. 1 shows measurements from December 2, before any snow accumulation at the study site to provide a baseline under no snow conditions. The top panel in Fig. 1 shows the effective irradiance received by the monofacial and bifacial PV panels, the incident solar radiation as well as the reflected light intensity. The lower shows the Ambient air temperature, and the temperature of the monofacial and the bifacial PV panels. The second panel also shows the snow depth as measured by Campbell Scientific snow depth sensor model SR 50A. There was no snow on the ground on December 2. The peak GHI incident irradiance is observed to be $\sim$ 430 W/m^2 with peak reflected light intensity of about $\sim$ 88 W/m^2. The the albedo is about 20% of the incident light intensity without the snow on the ground. After the first snow activity on Dec 15-16 the top of the PV panels got covered with snow.There was about 15 cms of snow on the ground, Fig. 2 shows the same quantities for Dec 16. Dec 16–18 was overcast sky with not much solar irradiance. But the albedo increased drastically due to the high reflectivity from snow on the ground. This is evident from the reflected light intensity of up to almost 75% of the incident intensity. The increased albedo can be seen during December 20 as the sky cleared up as shown in Fig. 3. We also observe that the panel temperature of the bifacial panels is substantially higher than the amibient air temperature, where as the monofacial panel temperature is close to the ambient air temperature.

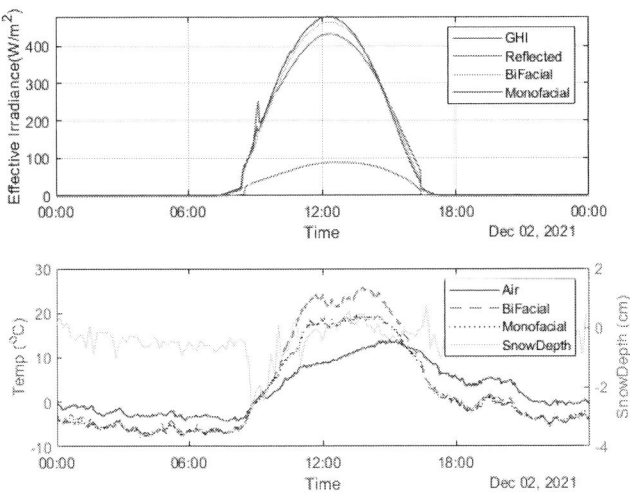

Fig. 1. Measureements from december 2, 2021, before any major snowfall. Top panel shows horizontal irradiance, reflected light intensity receved by the bifacial panels, effective irradiance recived byt he Bi facial as well as monofacial panels. Bottom panel shows back of module temperatures as measured by a Pt 1000 temperature sensor, ambient air temperature and snow depth at the experiment site.

During the period of Dec 16–Dec 20, back of module temperature for the bifacial panel was observed to be about 10°C higher than the air temperature. The temperature difference increases to almost 20°C in subsequent days. The

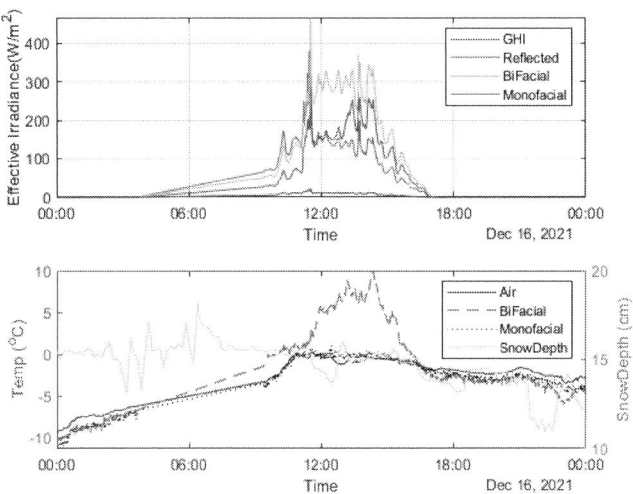

Fig. 2. Measurements from December 16,2021. There was a snowfall between Dec 15-16. Top panel shows horizontal irradiance, reflected light intensity receved by the bifacial panels, effective irradiance recived byt he Bi facial as well as monofacial panels. Bottom panel shows back of module temperatures as measured by a Pt 1000 temperature sensor, ambient air temperature and snow depth at the experiment site.

monofacial panel remained at the air temperature, which was close to freezing temperature. Thus snow on monofacial panels remained for another two-three days, while the snow on Bifacial panel melted as seen visually in the picture shown in Fig. 4. This picture was taken on Dec 20, 2021. The heat generated by absorption of the reflected light from the rear side of the bifacial panel seems to be the reason for improved snow shedding property of the bifacial panels. This effect will be even more effective if the panels are mounted at a fixed tilt or on a single axis tracker. This offers another advantage for the use of bifacial panels.

IV. CONCLUSIONS

We have presented a systematic study of snow shedding properties of bifacial and monofacial PV panel at our site in Logan, UT. The presented data clearly demonstrate the PV panel temperatures in the excess of 20°C compared to the ambient air temperature, during the periods of high intensity of reflected light due to snow on the ground. The monofacial panels remained close to the air temperature during this period. Thus, Bifacial pv panels have a clear advantage in regions with high snow fall.

REFERENCES

[1] Rob W. Andrews, Andrew Pollard, Joshua M. Pearce, Solar Energy, 92, (2013), 8497.
[2] Chenyue Yana, Minglu Qua, Yuan Chenb, and Min Fenga, Solar Energy, 206. 2020, 374–380.
[3] Erlend Andenæs, Bjørn PetterJellea, Kristin Ramlo, Tore Kolas, Josefine Selj, and Sean Erik Foss, Solar Energy, 159, (2018) 318–328.
[4] D. Riley, L. Burnham, B. Walker and J. M. Pearce, "Differences in Snow Shedding in Photovoltaic Systems with Framed and Frameless panels," 2019 IEEE 46th Photovoltaic Specialists Conference (PVSC), 2019, pp. 0558-0561, doi: 10.1109/PVSC40753.2019.8981389.

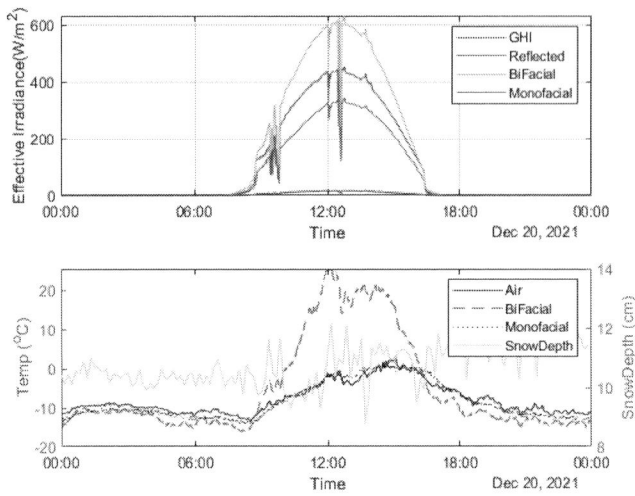

Fig. 3. Same quantities as Figure 1 and 2 above, but from measurements on December 20, 2021.

Fig. 4. A picture of the solar panels taken on morning of December 20, 2021. We can clearly see bifacial panels completely free of snow, while the monofacial panel is still fully covered fully with about 3 cm of snow, even after 4 days of snowfall.

[5] Gao, Y., 2013. Talking about the cleaning method of photovoltaic panels. Sol. Energy 9, 63–64.

[6] Christopher Pike, Erin Whitney, Michelle Wilber and Joshua S. Stein, Energies, 14, (2021), 1210.

Glass-glass PV modules: Characterization of chemical and mechanical degradation

Laura Spinella, Sona Ulicna, Archana Sinha, Dana B. Sulas-Kern, Michael Owen-Bellini, Steve Johnston, Laura T. Schelhas

National Renewable Energy Laboratory (NREL), Golden, CO, United States

SLAC National Accelerator Laboratory, Menlo Park, CA, United States

Glass-glass (G/G) photovoltaic modules are quickly rising in popularity, but the durability of modern G/G packaging has not yet been established. In this work, we examine the interfacial degradation modes in G/G modules under damp heat (DH) with and without bias voltage, comparing emerging polyolefin elastomers (POE) and industry-standard poly(ethylene-co-vinyl acetate) (EVA) encapsulants. The transport of ionic species at cell/encapsulant interface is investigated, demonstrating that the POE used in this study limits both sodium and silver ion migration compared with the EVA. Changes to the chemical structures of the encapsulants at the cell/encapsulant interfaces demonstrate both POE and EVA are more susceptible to degrade in modules with transparent backsheet than in the G/G configuration. While the POE structure was shown to degrade, this degradation did not appear to hamper its ability to limit ion migration. Adhesion testing revealed that the POE and EVA had comparable critical debond energies after the DH exposures regardless of applied bias polarities. The results of this study indicate that the interfacial degradation mechanisms of G/G appear to be similar to that of conventional glass/backsheet modules. For emerging materials, our results demonstrate that POE offers advantages over EVA, but that transparent backsheets may accelerate encapsulant degradation due to increased moisture ingress. This indicates that G/G may have some advantage over G/TB.

Model of an Automous PV Home using a Hybrid Storage System based on Li-ion batteries and Hydrogen Storage with Waste Heat Utilization

Marius Möller, Stefan Krauter

University of Paderborn EET–NEK, 33098 Paderborn, Germany

Abstract— This paper provides a hybrid energy system model created in Matlab/Simulink which is based on photovoltaics as its main energy source. The model includes a hybrid energy storage which consists of a short-term lithium-ion battery and hydrogen as long-term storage to ensure energy autonomy even during periods of low PV production (e.g., in winter). The sectors heat and electricity are coupled by using the waste heat generated by production and reconversion of hydrogen through an electrolyser respectively a fuel cell. A heat pump has been considered to cover the residual heat demand for well insulated homes. Within this paper a model of the space heating system as well as the hot water heating system is presented. The model is designed for the simulation and analysis of a whole year energy flow by using a time series of loads, weather and heat profiles as input. Moreover, results of the energy balance within the energy system by simulation of a complete year by varying the orientation (elevation and azimuth) of the PV system and the component sizing, such as the lithium-ion battery capacity, are presented. It turned out that a high amount of heating energy can be saved by using the waste heat generated by the electrolyser and the fuel cell. The model is well suited for the analysis of the effects of different component dimensionings in a hydrogen-based energy system on the overall energy balance within the residential sector.

Keywords— Hybrid energy system, hydrogen storage, lithium-ion battery, electrical and heating sector, power management, renewable energies (RE), Matlab/Simulink.

I. INTRODUCTION

Due to the rapidly increasing global warming issues caused by the emission of greenhouse gases into the atmosphere from a largely fossil fuel-based energy supply, there is a need for alternative technologies to cover the world's energy demand. Besides the shift towards environmentally friendly renewable energies, sector coupling by using the synergetic aspects between the electricity, residential, and mobility sector is getting increasingly important. The residential sector in Germany is responsible for 26.5% of the total energy demand and is still predominantly covered by fossil energy sources [1], [2]. Moreover, the highest energy demand occurs predominantly during winter which coincides with low PV generation. This intermittent nature of renewable energy sources (RES) necessitates efficient and easy scalable long-term storages such as hydrogen for use during periods of low energy production from renewable energies [3]. Hydrogen could be one promising approach for the success of the energy change, especially when it is produced by RES.

Within this paper, a model of a hybrid energy system based on hydrogen and a lithium-ion battery whose energy demand is mainly covered via photovoltaics is presented (see Fig. 1). The lithium-ion battery acts as short-term energy storage with the aim to ensure the coverage of the energy demand even during rapid load changes. If the lithium-ion battery is fully charged and the generated PV energy still exceeds the energy demand, an electrolyser is switched on for the production of hydrogen. The produced hydrogen is subsequently stored inside a hydrogen tank via a compressor. If the energy demand exceeds the energy provided by the PV system, a fuel cell is used for the reconversion of hydrogen. As the electrical efficiency of an electrolyser, respectively a fuel cell, is relatively low, the overall efficiency can be increased by the usage of the generated waste heat in the heating sector. Therefore, heat exchangers have been introduced into the energy system model, which extract the heat generated inside the electrolyser respectively the fuel cell and feed it into the heating system.

As the produced PV energy in combination with a lithium-ion battery achieves a high self-sufficiency during midyear, the fuel cell is predominantly used in winter which coincidences with the heating period. Therefore, an integration of the waste heat into the space heating system is preferable. In contrast, the electrolyser is commonly operating during the summer period where typically only hot water heating demand occurs. Therefore, the waste heat generated by the electrolyser is feed into the hot water heating system.

Within this energy system, a Proton Exchange Membrane (PEM) electrolyser is used due to its advantages such as the intrinsic ability to cope with fast transient electrical power variations which is especially important when fed by fluctuating renewable energies like PV in the designated use case [4]. Moreover, they have, compared to other electrolyser types, a smaller dimension and mass, lower operating temperatures and lower power consumption and are therefore better suited for small-scale applications [5]. Low-temperature PEM electrolysers operate at 50°C to 80°C and have to be cooled in order to protect them against degradation caused by a temperature increase above 80°C [6]. The heat extracted from the electrolyser can be used as additional heat for the hot water heating system. The residual heat demand is covered by a hot water heat pump. Such a heat pump typically achieves water temperatures of around 65°C maximum. Additionally, the temperature inside a hot water storage shall always exceed 50°C.

978-1-7281-6118-1/22 $31.00 © 2022 IEEE

If the temperature of the coolant water extracted during operation of the electrolyser exceeds the actual hot water storage temperature, the heat can be used for heating up the water inside the hot water storage.

Wetterdienst (DWD) which were provided with a time resolution of one hour [12]. The data used were recorded in 2015 in Wuerzburg (Germany). The resolution was converted to a time resolution of 15 minutes using linear interpolation. As the

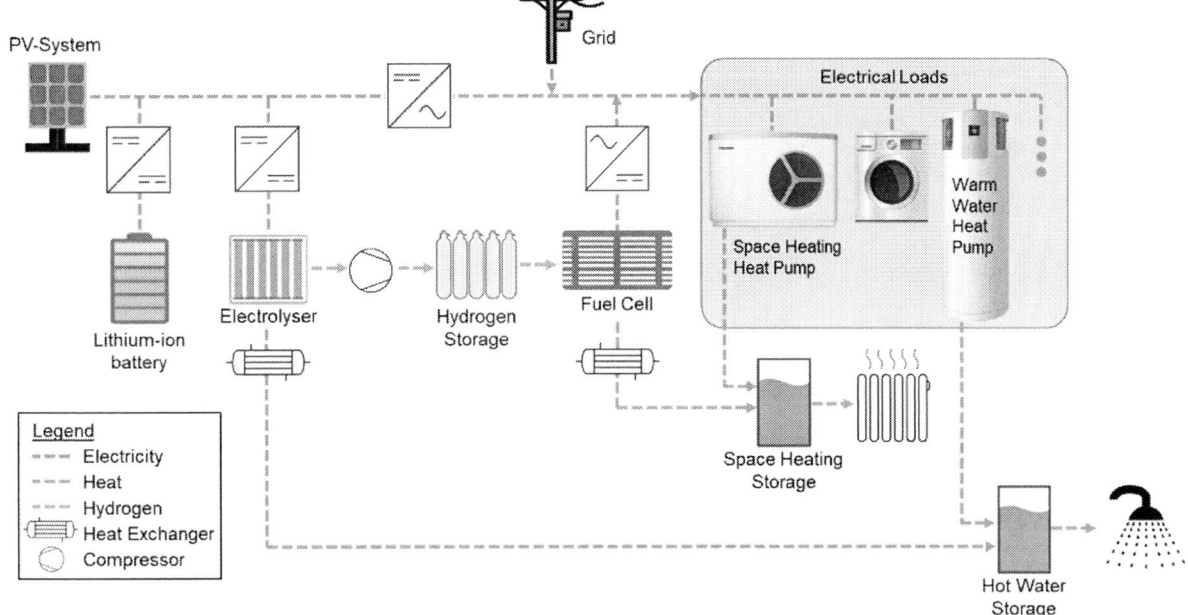

Fig. 1. System architecture of the household energy system (Image sources: [7], [8], [9], [10])

The model as shown in Fig. 2 was created in Matlab/Simulink and was designed for the analysis of real data series over an entire year with a time resolution of 15 minutes. A representative dataset recorded in Switzerland in 2012 was used as household load profile [11]. The load profile originally had a time resolution of 1 second which was converted to a time resolution of 15 minutes by averaging the values. The irradiation and temperature profile are received from the Deutscher

heating system is also considered, a synthetic heat demand profile was generated using the reference load profiles from the VDI 4655 standard [13]. Using the temperature profile for the year 2015 and an excel tool created by Hessen [14], which uses the reference load profiles from the VDI 6455 standard, a space heating demand profile and a hot water heating demand profile with a time resolution of 15 minutes was generated.

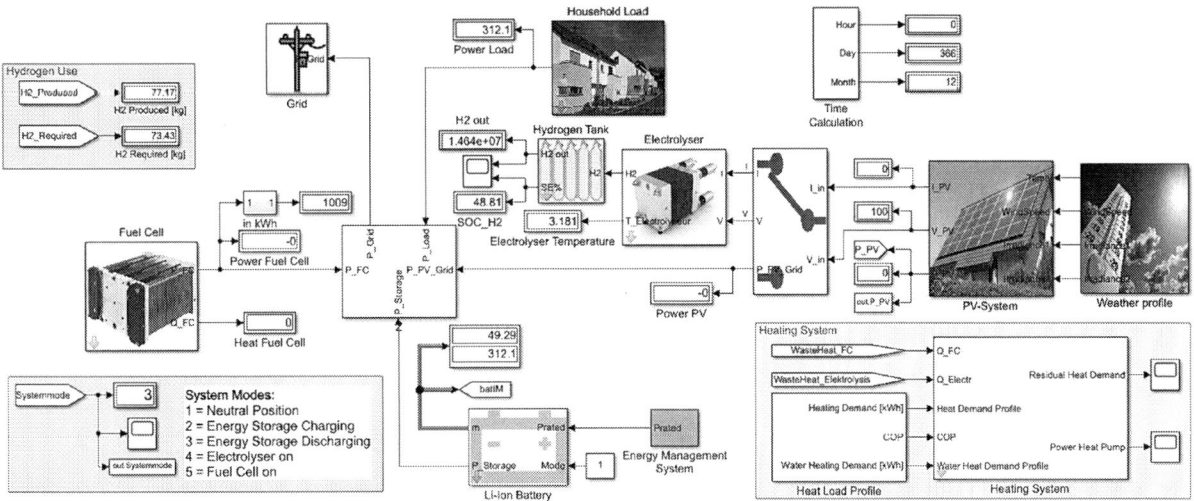

Fig. 2. Simulink Model of the overall energy system (Image sources: [9], [10], [15], [16], [17], [18], [19]])

In [20], some parts of the developed model have already been validated by comparing the simulation results with well-established simulation results from the HOMER energy simulation software package.

II. RESULTS

The simulation has been executed by varying the PV system orientation. An orientation towards South using 22 modules with an elevation angle of 45° and a peak power of 310 W_p has been compared to a split orientation towards east and west with 11 modules each. The simulation by usage of an orientation towards south calculated that 3851.94 kWh energy were used inside the electrolyser for production of hydrogen, which leads to a total of 73.43 kg produced hydrogen within one year. In return, the fuel cell required 57.65 kg hydrogen for production of 654.55 kWh electricity. By changing the PV orientation towards east and west, the electrolyser receives 3147.03 kWh/a of electricity for the production of 59.91 kg hydrogen whereas the fuel cell requires 72.97 kg of hydrogen for the production of 1002.30 kWh of electrical energy.

Using an orientation towards south leads to a space heating demand reduction of 579.16 kWh by using the waste heat produced within the fuel cell. The space heating demand was reduced by 650.94 kWh by usage of an East-West orientation. Additionally, an East-West orientation leads to a much less workload of the electrolyser during operation compared to an orientation towards South. This results in lower temperature increase inside the electrolyser whereby a waste heat utilization in the hot water heating system can be used less often.

Changing the elevation angle of the PV systems also has a significant impact on the energy balance. An elevation angle of 30° leads to a higher energy production during summer compared to an angle of 45° while an elevation angle of 60° leads to a higher energy production during winter. Table 1 shows the results of the simulation by using different elevation angles. The required hydrogen doesn't change significantly by using an elevation angle of 60° or 45°. However, 10 kg more hydrogen have been produced by usage of an elevation angle of 45°. In contrast the required hydrogen inside the fuel cell has been increased to 50.22 kg whereas more hydrogen has been produced inside the electrolyser by usage of an elevation angle of 30°. The results show that an elevation angle of 30° leads to an increased storage effort compared to the other elevation angles.

TABLE 1. EFFECTS OF DIFFERENT ELEVATION ANGLES ON HYDROGEN PRODUCTION AND REQUIREMENT

Orientation	Elevation Angle	PV Production [kWh]	Required Hydrogen [kg]	Produced Hydrogen [kg]
South	60°	6756,20	47,95	63,54
South	45°	7314,60	47,65	73,43
South	30°	7522,44	50,22	78,12
East/West	45°	6092,83	72,97	59,91

REFERENCES

[1] Umweltbundesamt, "Energieverbrauch privater Haushalte," Available online: https://www.umweltbundesamt.de/daten/private-haushalte-konsum/wohnen/energieverbrauch-privater-haushalte#hochster-anteil-am-energieverbrauch-zum-heizen (accessed on Jan 06, 2022).

[2] Umweltbundesamt, "Energieverbrauch nach Energieträgern und Sektoren," Available online: https://www.umweltbundesamt.de/daten/energie/energieverbrauch-nach-energietraegern-sektoren#allgemeine-entwicklung-und-einflussfaktoren (accessed on Jan 06, 2022).

[3] S. Boulmrharj et al, "Performance Assessment of a Hybrid System with Hydrogen Storage and Fuel Cell for Cogeneration in Buildings," MDPI, 2020.

[4] A. Smith and M. Newborough, "Low-cost polymer electrolysers and electrolyser implementation scenarios for carbon abatement," Report to the carbon trust and ITM-power plc. 2004.

[5] C. Wang, "Modeling and Control of Hybrid Wind/Photovoltaic/Fuel Cell Distributed Generation Systems," Ph.D. thesis, the Montana State University, Montana, 2006.

[6] P. Martinez, M. Serra and R. Costa-Castelló, "Modeling and control of HTPEMFC based Combined Heat and Power for confort control", Institut de Robòtica i Informàtica Industrial, CSIC-UPC, 2017.

[7] "Die Top 10 der Stromverbraucher". Available online: https://energieeffizienz-gefaellt-mir.info/die-top-10-der-stromverbraucher/, (accessed on Jan 06, 2022).

[8] STIEBEL ELTRON, "Luft/Wasser-Wärmepumpe WPL 25 AC". Available online: https://heizung-billiger.de/69540-stiebel-eltron-luft-wasser-warmepumpe-wpl-25-ac-stiebel-236645-4017212366455.html (accessed on Jan 06, 2022).

[9] Flaticon, "Strommast kostenlos Icon". Available online: https://www.flaticon.com/de/kostenloses-icon/strommast_1965065?related_id=1965065&origin=search, (accessed on Jan 06, 2022).

[10] Ina Matthes, "Brandenburger erzeugt eigenen Strom mit Wasserstoff". Available online: https://www.moz.de/nachrichten/wirtschaft/energie-brandenburger-erzeugt-eigenen-strom-mit-wasserstoff-50383121.html, (accessed on Jan 06, 2022).

[11] W. Kleiminger and C. Beckel, "ECO data set (Electricity Consumption & Occupancy) A Research Project of the Distributed Systems Group," ETH Zürich, 2016. Available online: https://rossa-prod-ap21.ethz.ch:8443/delivery/DeliveryManagerServlet?dps_pid=IE594964 (accessed on Jan 06, 2022).

[12] Deutscher Wetterdienst (DWD), "Climate Data Center," Available online: https://opendata.dwd.de/climate_environment/CDC/ (accessed on Jan 06, 2022).

[13] Verein Deutscher Ingenieure (VDI), "VDI 4655 – Reference load profiles of single-family and multi-family houses for the use of CHP systems," 2008.

[14] H. Heesen, "Synthese von Strom- und Wärmeprofilen nach VDI 4655 (2020)," Hochschule Trier. Available online: umwelt-campus.de/energietools (accessed on Jan 06, 2022).

[15] "ElringKlinger kommt mit neuen Produkten zur IAA," Available online: https://www.electrive.net/2018/09/14/elringklinger-kommt-mit-neuen-produkten-zur-iaa/, (accessed on Jan 06, 2022).

[16] H-Tec Systems, "Elektrolyse-Stacks SERIES S30," Available online: https://www.h-tec.com/produkte/, (accessed on Jan 06, 2022).

[17] The North Africa Post, "Ghana: 12 MW Rooftop Solar PV to be constructed," Available online: https://northafricapost.com/28211-ghana-12-mw-rooftop-solar-pv-to-be-constructed.html, (accessed on Jan 06, 2022).

[18] Frielinghaus Schüren Architekten, "Neubau: Wohnsiedlung in Wetter mit 40 Einfamilienhäusern," Available online: https://frielinghaus-schueren.de/neubau-wohnsiedlung-40-efh-wetter/, (accessed on Jan 06, 2022).

[19] "KLIMA und WETTER: Chile," Available online: https://franks-travelbox.com/suedamerika/chile/klima-und-wetter-chile/, (accessed on Jan 06, 2022).

[20] M. Möller; S. Krauter, "Hybrid Energy System Model in Matlab/Simulink based on Solar Energy, Lithium-Ion Battery and Hydrogen," MDPI *Energies*, 2022.

Towards High Efficiency All-Perovskite Tandem Solar Cell by Preventing Performance Loss Arising from Physically Mixed Interfacial Layers

Biwas Subedi, Alex Bordovalos, Lei Chen, Zhaoning Song, Cong Chen, Yanfa Yan, Nikolas J Podraza

University of Toledo, Toledo, OH, United States

Physically mixed layers are major sources of optical and electronic losses in inorganic-organic metal halide perovskite-based photovoltaics (PV). Physical mixing of adjacent component layers at interfaces within a two-terminal all-perovskite tandem PV is identified by through-the-glass spectroscopic ellipsometry. Using measured device structures and optical properties of component layers including these physical mixtures, external quantum efficiency (EQE) spectra are simulated and compared to experimental EQE. Deviations between experimental and simulated EQE help to identify physically mixed layers as sources of optical and electronic losses and to quantify these losses. A physically mixed layer formed between the silver (Ag) electrical contact and electron transport fullerene (C60) / hole blocking bathocuproine (BCP) layers shows parasitic absorption losses of 2.3 mA/cm2 in the terms of short circuit current density (JSC). Similarly, electronic loss of 1.5 mA/cm2 is observed in the bottom subcell due to only 66% collection of photogenerated carriers stemming from poor passivation in the narrow-Eg perovskite nucleation layer formed near the hole transport layer. Optimized combinations of wide-Eg (1.74 eV) and narrow-Eg (1.25 eV) perovskite absorber thicknesses to maximize the tandem JSC are determined. Preventing physical mixing of component materials at interfaces and using optimized absorber layer thicknesses will help in fabricating higher efficiency all-perovskite tandem PV.

Nonparametric Temporal Downscaling of GHI Clear-sky Indices using Gaussian Copula

Jing Huang[1], Marc Perez[1], Richard Perez[2], Dazhi Yang[3], Patrick Keelin[1] and Tom Hoff[1]

[1]Clean Power Research, Napa, CA, USA
[2]Atmospheric Sciences Research Center, SUNY, Albany, NY, USA
[3]School of Electrical Engineering and Automation, Harbin Institute of Technology, Harbin, Heilongjiang, China

Abstract—**Small-scale variabilities of solar irradiance are important for many applications. Downscaling approaches need to be developed where only the averaged state of solar irradiance is known. In this study, we investigate the use of copula for temporally downscaling GHI clear-sky indices. With the correlation structure and distribution information derived from measurements at 10 stations across the United States, the copula approach is capable of downscaling clear-sky indices from hourly averages to any arbitrary fine scale whilst preserving its original power spectra.**

Keywords—solar irradiance, clear-sky index, variability, photovoltaics, downscaling, copula, power spectra

I. METHODOLOGY

It generally holds that modeling accuracy increases with the granularity of available data feed. For solar applications, the requirement for granular solar datasets is often not met by either sparse ground measurement stations or geostationary satellites. For example, the solar industry discovers that minute-level solar irradiance data are preferred to inform critical investment decision-making of utility-scale solar photovoltaics (PV) farms particularly when the DC capacity of solar power inverters is much greater than the AC capacity of PV modules [1]. The distributed nature of solar power generators and the increasing need for customized and accurate control of hybrid systems as envisioned by [2] also call for high-resolution solar data. As such, downscaling methods are potentially useful to produce realistic small-scale variabilities.

There are many methods which have been proposed for synthetic solar data generation ranging from Fourier time series and Markov chain models to computation-intensive machine learning (see [3] for a review). In this study, we adopt a statistical tool called copula, which has been recently applied to model the clear-sky index (CSI) of solar irradiance recently. Widén and Munkhammar [4] downscaled hourly CSI by modeling its distribution as a two-state Gaussian mixture model and assuming an exponential decay of correlation. In contrast, we employ a nonparametric downscaling approach with the key information of distribution and correlation being derived from measurements at 10 reference ground stations.

A. Ground stations for reference and validation

Figure 1 provides information on the reference ground stations used in this study. The measurement global horizontal irradiance (GHI) data span the entire year of 2020 with a temporal resolution of 1 minute. In addition, we obtain the corresponding clear-sky GHI at all the reference stations from SolarAnywhere® V3.5 [5]. The CSI, or *kt*, is then calculated as the ratio of GHI to the clear-sky GHI. We cap *kt* values at 1.3.

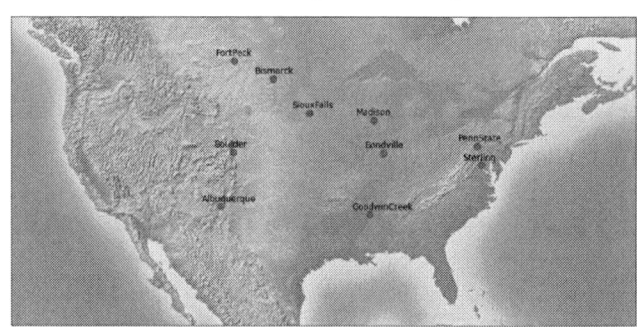

Figure 1. 10 reference ground stations in the eastern United States are used in this study.

B. Gaussian Copula

The concept of copula was originally proposed by Sklar [6]. A copula is a joint multivariate cumulative distribution function (CDF) where all marginal probability distributions are uniform within [0, 1] and Gaussian copula simply implies that the joint CDF is assumed to be Gaussian.

Consider M related random variables of CSI (*kt*), KT_t, $KT_{t+\Delta t}$, ..., $KT_{t+(M-1)\Delta t}$ where each random variable represents a relative temporal point in a time period of $M\Delta t$ for an arbitrary site. For example, if we split 1-min *kt* values at Boulder into hourly blocks, KT_t represents a random variable denoting the *kt* values at the first minute of an hourly block and $KT_{t+(M-1)\Delta t}$ represents that at the last minute, and $M=60$. Denoting the marginal CDF of KT as

$$F_t(kt_\alpha) = \alpha, \tag{1}$$

the M-dimensional joint CDF $\mathbf{F}$ can be linked to their marginal distribution via a Gaussian copula,

$$\mathbf{F}\left(KT_t, KT_{t+\Delta t}, \dots, KT_{t+(M-1)\Delta t}\right) = \mathbf{C}(F_t, F_{t+\Delta t}, \dots, F_{t+(M-1)\Delta t}). \tag{2}$$

Since copula requires all marginal distributions to be uniform within [0,1], it can be further expressed as

978-1-7281-6118-1/22 $31.00 © 2022 IEEE

$$C\left(U_t, U_{t+\Delta t}, \ldots, U_{t+(M-1)\Delta t}\right) =$$
$$\mathbf{F}_\Sigma\left(N^{-1}(U_t), N^{-1}(U_{t+\Delta t}), \ldots, N^{-1}(U_{t+(M-1)\Delta t})\right). \quad (3)$$

where U_t are uniformly distributed random variables transformed from F_t and N^{-1} is the inverse CDF of a standard univariate normal distribution and $\mathbf{F}_\Sigma$ is the joint CDF of multivariate normal distribution constrained by the M×M correlation matrix Σ. Since random and correlated samples can be drawn from a multivariate gaussian generator (e.g. using `random.multivariate_normal` in Python `numpy` package), they can then be converted to scenarios of kt via inverse marginal CDF F_t^{-1}.

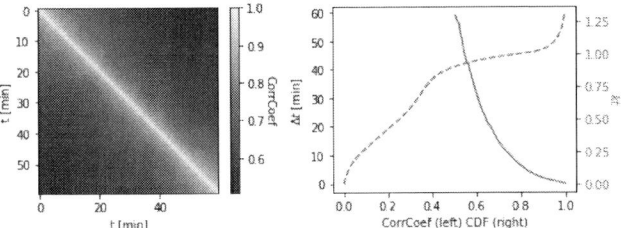

Figure 2. Boulder site: (left) Color plot of correlation matrix of kt; (right) The decay of correlation coefficient with Δt on the left y-axis and the CDF of kt on the right y-axis.

As shown in Figure 2, the correlation matrix Σ and the CDF of kt can be obtained from the measured GHI time series and the SolarAnywhere clear-sky GHI. The functional decay of correlation $\rho(\Delta t)$ is further determined empirically by aggregating and taking medians of the correlation coefficients with the same Δt from Σ.

C. Nonparametric Downscaling

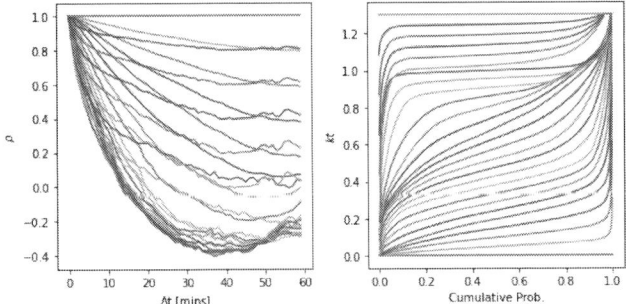

Figure 3. Boulder site: (left) $\rho(\Delta t)$ and (right) CDF plots for $\overline{kt} = 0.05, 0.1, \ldots, 1.3$.

As mentioned earlier, the main difference of this study from [3] in terms of downscaling is that we take a nonparametric approach. Specifically for one station: (1) we calculate the hourly-averaged CSI values, denoted as $\overline{kt}$; (2) we divide $\overline{kt}$ into 10 quantile groups of equal size (i.e. with quantile ranges of [0, 0.1], …, [0.9, 1]); (3) for each group, we calculate the corresponding $\rho(\Delta t)$ and CDF individually; (4) we linearly interpolate $\rho(\Delta t)$ and CDF for 26 $\overline{kt}$ values, i.e. 0.05, 0.1, …, 1.3. Note that we set the boundary values $\rho\left(\Delta t, \overline{kt} = 0\right) =$

$\rho\left(\Delta t, \overline{kt} = 1.3\right) = 1$, and $\text{CDF}\left(\overline{kt} = 0\right) = 0$ and $\text{CDF}\left(\overline{kt} = 1.3\right) = 1.3$.

We repeat the above procedures for all 10 reference stations. Figure 4 shows the convergence of 10 sites for $\rho(\Delta t)$ and CDF at $\overline{kt} = 0.5$. The convergence is generally good except when $\overline{kt}$ is marginal and thus data density is low. It is possible to further model $\rho(\Delta t)$ and CDF based on location (e.g., coordinates) but we are not pursuing that direction for model simplicity. Instead, we obtain the location-agnostic autocorrelation and CDF information by taking the median of all sites as shown in Figure 5.

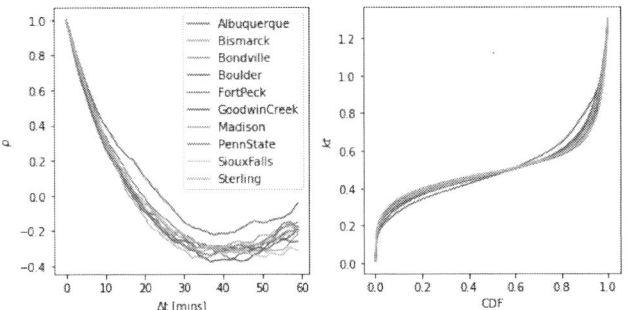

Figure 4. Autocorrelation functions $\rho(\Delta t)$ and CDF at $\overline{kt} = 0.5$ for 10 individual sites.

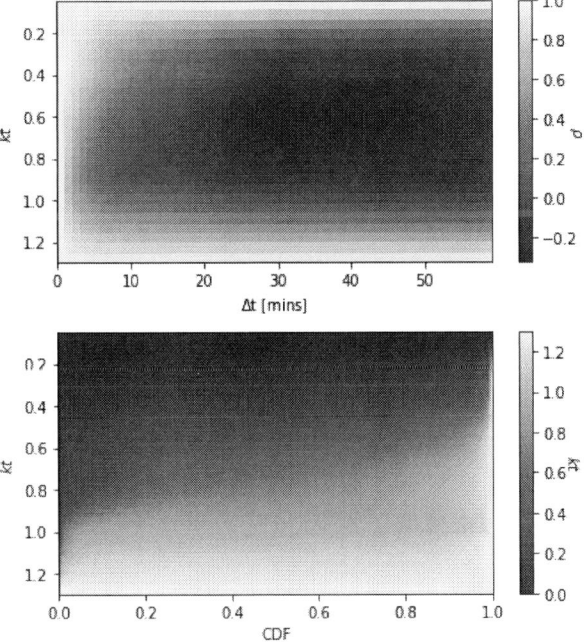

Figure 5. Median of all sites: (top) $\rho(\Delta t)$ and (bottom) CDF color plots for $\overline{kt} = 0.05, 0.1, \ldots, 1.3$.

Then, downscaling the hourly average $\overline{kt}$ simply involves interpolating $\overline{kt}$ from Figure 5 to get the corresponding $\rho(\Delta t)$ and CDF and then generating the downscaled time series using Equation (1) and (3). In practice, we generate 100 samples for each $\overline{kt}$ and then select one sample based on two criteria: (1) the transition from the previous hourly segment to the generated

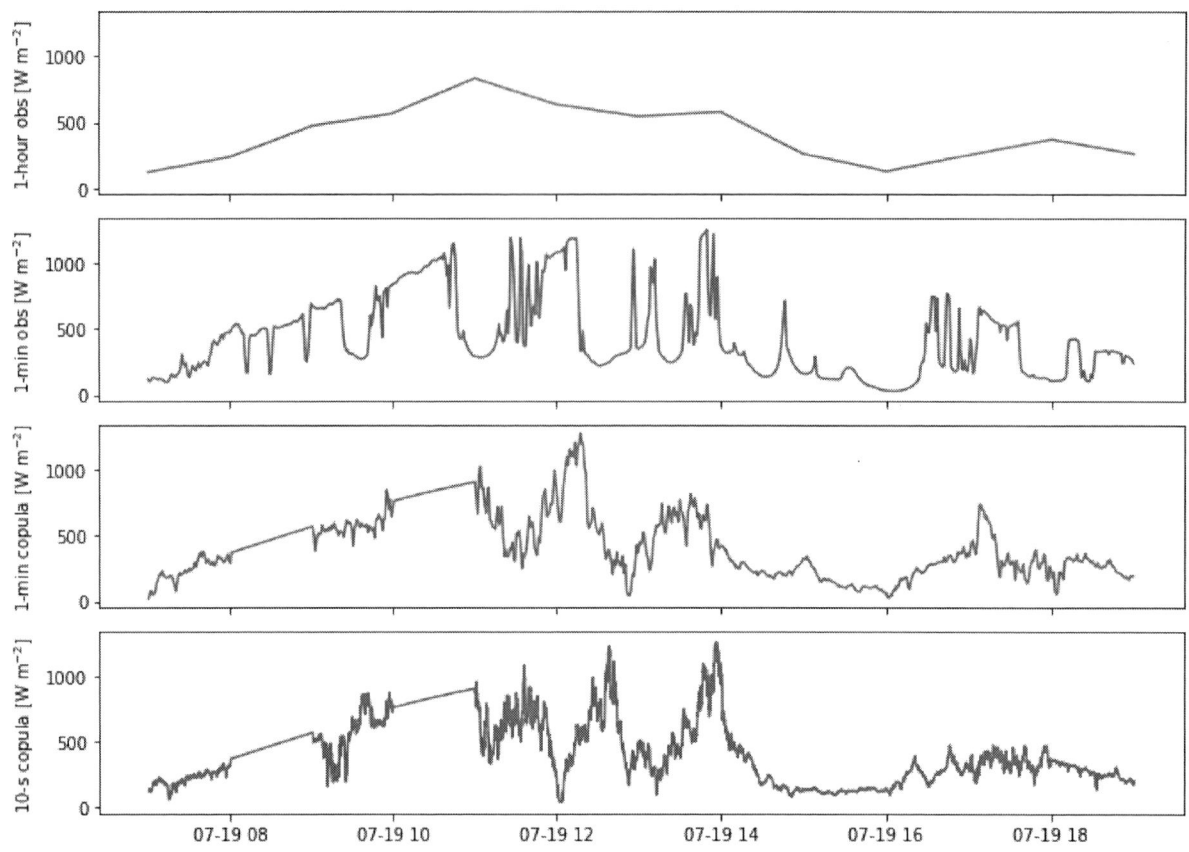

Figure 6. One example day at Boulder: (top) 1-hour GHI measurement (middle top) 1-min GHI measurement; (middle bottom) 1-min downscaled GHI from hourly averages; and (bottom) 10-s downscaled GHI from hourly averages.

hourly segment should be smooth; (2) the average of the generated hourly segment should be as close to $\overline{kt}$ as possible.

II. VALIDATION

A downscaled GHI time series is shown in Figure 6. By interpolating ρ and CDF to the given $\overline{kt}$, this nonparametric procedure is capable of downscaling $\overline{kt}$ to an arbitrarily fine time scale. The simulated time series in 1-min and 10-s resolution are similar to the 1-min measurement in the sense that (1) they share the same large-scale trends (i.e. hourly means) and (2) the extent of variability changes with $\overline{kt}$ with a peak when $\overline{kt}$ is intermediate. In addition, we quantify the cross-scale variabilities by calculating the power spectra of kt and show that the power spectra of the 1-min measurement and 1-min simulation match each other closely across the frequency range down to the Nyquist frequency (i.e. 0.5 min⁻¹).

In addition, we also model the power generation resulted from observed and copula-downscaled GHI to validate its performance. We use PVLIB to model a hypothetical horizontal single-axis tracking PV system, and then calculate the power loss due to inverter clipping under various DC:AC ratios. We follow the same technical procedures as described in our companion paper [7]. Figure 8 demonstrates the superiority of the downscaled 1-min time series over the hourly averaged

observation data for those scenarios with high DC:AC ratios. When the clipping loss error of hourly observation data

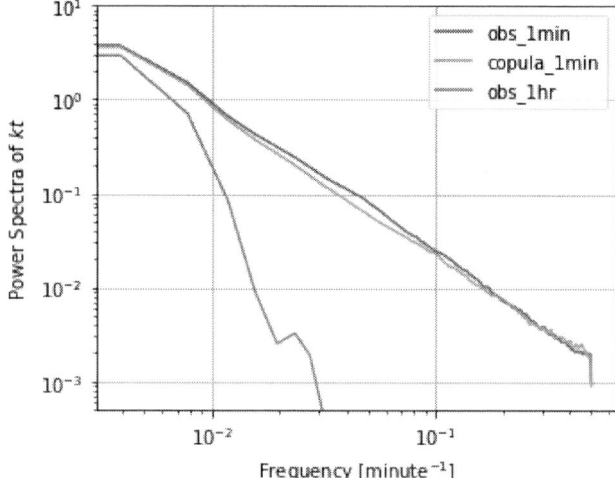

Figure 7. Power spectra of kt for 1-min and 1-hour measurement and 1-min simulation.

monotonically increases with DC:AC ratios, it plateaus around 1% of underestimation for DC:AC > 1.6 for copula-downscaled data, which verifies the capability of our copula approach to faithfully reproduce intra-hour variabilities.

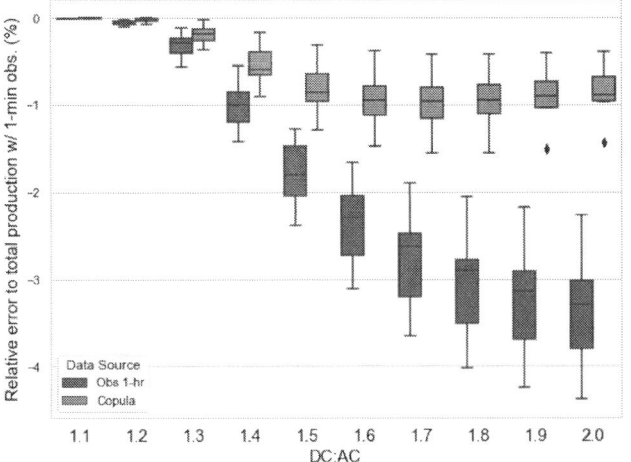

Figure 8. Box plots of relative error of curtailed AC power estimation using interpolated hourly-averaged observation and 1-min copula-downscaled time series for DC:AC ranging from 1.1 to 2.0.

III. CONCLUSION

In this study, we have applied Gaussian copula techniques for temporal downscaling of solar irradiance, which have already been proven very useful for wind energy applications. Built on previous studies, we have proposed a pure nonparametric approach with the autocorrelation and distribution functions heuristically derived from 10 measurement stations across the continental United States. We have demonstrated that the simulated 1-min time series is able to accurately reproduce the power spectra of the original 1-min observation data. In addition, the copula-downscaled data generally perform well in estimating AC power when the DC:AC ratio is high. This is in clear contrast to using hourly-averaged observations, which leads to a monotonic increase in error with increasing DC:AC ratio.

It is possible to further improve this method. For example, the copula parameters as shown in Figure 5 can be further regressed from location information or be tuned for individual applications such as power estimation. In general, these results demonstrate the effectiveness of the proposed methodology to meet customer needs for high-frequency and high-quality irradiance data.

REFERENCES

[1] K. Bradford, R. Walker, D. Moon and M. Ibanez, "A regression model to correct for intra-hourly irradiance variability bias in solar energy models", 2020 47th IEEE Photovoltaic Specialists Conference (PVSC), 2020, pp. 2679-2682.

[2] M. Ahlstrom, J. Mays, E. Gimon, A. Gelston, C. Murphy, P. Denholm, and G. Nemet, "Hybrid Resources: Challenges, Implications, Opportunities, and Innovation," in IEEE Power and Energy Magazine, vol. 19, no. 6, pp. 37-44, Nov.-Dec. 2021

[3] Munkhammar, J. and Widén, J. (2021), "Established mathematical approaches for synthetic solar irradiance data generation," in Bright, J. M. (ed.), Synthetic Solar Irradiance: Modeling Solar Data, Melville, New York: AIP Publishing, pp. 3-1–3-36

[4] Widén, J. and Munkhammar, J. (2019), "Spatio-temporal downscaling of hourly solar irradiance data using Gaussian copulas," Proceedings of the 46th IEEE Photovoltaic Specialists Conference (PVSC), Chicago, USA, 16–21 June 2019.

[5] P. Keelin, A. Kubiniec, A. Bhat, M. Perez, J. Dise, R. Perez and J. Schlemmer, (2021) "Quantifying the solar impacts of wildfire smoke in western North America," 2021 IEEE 48th Photovoltaic Specialists Conference (PVSC), 2021, pp. 1401-1404, doi: 10.1109/PVSC43889.2021.9518440.

[6] Sklar, A. (1959) 'Fonctions de répartition à n dimensions et leurs marges', Publ. Inst. Statist. Univ. Paris, 8, pp. 229–231.

[7] J. Huang, R. Perez, J. Schlemmer, A. Kubiniec, M. Perez, A. Bhat and P. Keelin, (2022) "Enhancing temporal variability of 5-minute satellite-derived solar irradiance data", 2022 IEEE 49th Photovoltaic Specialists Conference (PVSC), Philadelphia, PA, USA.

Ga-doping of MZO in CdSeTe/CdTe Thin Film Solar Cells

Mustafa Togay[1], Tushar Shimpi[2], Sampath S. Walajabad[2], Kurt L. Barth[2*], Eric Don[3], Gabor Parada[4], J. Michael Walls[1] and Jake W. Bowers[1]

[1]CREST, Wolfson School of Mechanical, Electrical and Manufacturing Engineering, Loughborough University, Loughborough, Leicestershire, LE11 3TU, United Kingdom

[2]NSF I/UCRC for Next Generation Photovoltaics, Colorado State University, Fort Collins, CO 80526 United States

[3]SemiMetrics Ltd., PO Box 36, Kings Langley, WD4 9WB, United Kingdom

[4]Semilab Co. Ltd., Prielle Kornélia u. 4/A. H-1117 Budapest, Hungary

*Note: Present affiliation is CREST, Loughborough University

Abstract— **Metastable effects in high efficiency MZO/CdSeTe/CdTe solar cells have been studied in an attempt to recover the device performance. Devices with the MZO buffer layer have shown an `S` shaped behaviour in the J-V characteristics before any preconditioning. This is removed after light soaking under 1000 Wm^{-2} at 25 °C. However, this recovery remained only for a short period of time while the devices were stored under vacuum in the dark. Recent studies with Ga doping of the buffer MZO has shown the removal of this metastability in the J-V characteristics of CdTe devices can be achieved without light soaking. A significant improvement in conductivity and Hall signal has been measured with Ga doped MZO layers compared to previously measured MZO films. However, a gradual decrease in the Hall signal has been observed over time after films were light soaked and removed from the desiccator.**

Keywords—metastabilities, light soaking, preconditioning, degradation, Hall effect measurements, MZO, GMZO, CdSeTe, CdTe

I. INTRODUCTION

Currently, polycrystalline cadmium telluride (CdTe) thin film solar cells have achieved champion efficiencies ~22% (First Solar) [1]. The optimum band gap (E_g = 1.5 eV) and a high absorption coefficient (10^5 cm^{-1}) of CdTe material makes it an attractive absorber for low-cost and high-efficiency thin film solar cells [2].

The conventional cadmium telluride (CdTe) solar cells consist of a p-n junction containing a p-doped CdTe layer with an n-doped cadmium sulphide (CdS) layer [3]. The absorption of light in the CdS layer of photons with energy above the CdS bandgap (2.45 eV) prevents some of the usable photons reaching the absorber layer. This is a major drawback as it causes photogenerated carriers not to be collected and the light absorbed here is wasted. This parasitic absorption limits the

short-circuit current density (J_{sc}). Recent research has focused on introducing new materials for the buffer layer and overcoming this limitation using a high bandgap material. Magnesium-doped zinc oxide (MZO) has a bandgap of $E_g > 3.3$ eV which can transmit a larger fraction of the solar spectrum compared to CdS [4][5]. One of the advantages of using MZO is the tuneability of the conduction band alignment with CdTe [6]. This can be achieved by changing the concentration of Mg in the film and the ratio of MgO/ZnO, varying the bandgap from MgO (7.8 eV) to ZnO (3.1 eV) depending on the metal ratios [7]. Devices with MZO as a buffer layer along with graded CdSeTe/CdTe absorbers have achieved power conversion efficiencies (PCEs) around 19% [8]. However, modifying the device architecture has introduced more complexity and difficulties in understanding the factors that limit performance, stability issues and transient metastable behavior of PV devices. Current research has found that the devices with MZO buffer layers suffer from metastability effects in J-V characteristics showing an `S` shaped behaviour (often called `kink` type) before any preconditioning, such as light soaking [9][10][11][12]. Other work has shown that degradation of the MZO layer occurs if the layer is exposed to humidity in the atmosphere [13].

Alternatively, doping of the MZO layer can enhance the film conductivity. As-deposited MZO thin films are electrically insulating, hence MZO as an n-type buffer layer should not work in the device. It is therefore clear that another mechanism is present which enables the MZO to function as a window/buffer/emitter layer. It is believed that oxygen vacancies created within the MZO film during CdCl$_2$ activation treatment result in improved carrier concentration in the film, which makes the MZO function. Ga doping of MZO has been reported to eliminate the `S` shaped behaviour in J-V characteristics without requiring any exposure of light onto the devices at 1000 Wm^{-2} [12]. A key advantage of Ga (from other dopants such as B, Al, In, etc…) is that it is more moisture resistant and stable in an oxidising environment [14][15]. Hence, a small concentration of Ga could be a suitable dopant for the MZO. The addition of Ga should enhance the carrier concentration of the film sufficiently for the Ga doped MZO to act as a good transparent buffer/emitter layer.

This paper reports on the conductivity and the metastable behaviour, before and after, any preconditioning of Ga doped MZO films (GMZO) measured using a high sensitivity Parallel Dipole Line (PDL) Hall effect measurement system from SemiLab. It also reports on the preconditioning approaches that have been used on the GMZO films to investigate the conductivity and stability of the material. These approaches include annealing and light soaking under 1000 Wm^{-2} at 25 °C for 60 and 120 minutes.

A. Sample preparation and light soaking

The GMZO films used for this study were fabricated at Colorado State University (CSU) using an RF magnetron sputtering system [4] on NSG-Pilkington TEC^{TM} SB glass (includes a sodium barrier layer with a thin SiO$_2$ overcoat) with film thickness of 520nm. Two sets of samples, as deposited, and annealed were provided. The GMZO films were annealed in Rapid Temperature Cycling (RTC) for 30 minutes [16] and the substrate temperature measured under the pyrometer at the end of the process was 480°C.

The films were light soaked using an ABET solar simulator under 1000 Wm^{-2} at 25°C with Class A AM1.5G spectral match using a xenon arc lamp. The simulator has an in-house built temperature stage using an Adaptive® JUNIOR PID controller to maintain the sample temperature constant at 25°C. Light soaking was performed for 60 and 120 minutes.

B. PDL Hall effect measurements

The Hall effect measurements were performed using a High sensitivity Parallel Dipole Line (PDL) [17][18] Hall effect measurement system from SemiLab. This system is capable of working in both AC and DC Hall measurement modes, where the AC field can be used for materials with mobility of below 0.1 cm^2/Vs. It uses a master and slave magnet with a large magnetic field approximately 2.5 T, peak to peak. Hall measurement yields resistivity, sheet resistance, mobility, carrier concentration and carrier type of the material. The measurement follows 3 important steps; the contact check, the sheet resistance measurement, and the Hall resistance measurement in order to extract reliable and accurate carrier concentration and mobility. The details of the measurements steps have been provided elsewhere [10].

II. RESULTS AND DISCUSSION

As deposited and annealed GMZO films were still too resistive for further Hall effect measurements consistent with the MZO films measured previously [9] , thus no useful Hall signal was detected. Although both films were resistive, annealed GMZO films showed improved I-V response from the Hall measurements. Light soaking experiments have shown no improvements in conductivity for as deposited GMZO films. However, annealed GMZO films showed a large improvement in conductivity after light soaking. Table 1 shows the effect of the light soaking on conductivity measurements for the annealed GMZO films.

TABLE I. THE EFFECT OF LIGHT SOAKING ON CONDUCTIVITY MEASUREMENTS IN ANNEALED GMZO FILMS.

Light soaking conditions	2-point Resistance [Ω]	Sheet Resistance [Ω/□]	Sample Resistivity [Ω.cm]
Initial (no light soaking)	3.96×10^8	2.92×10^9	1.64×10^5
1 hour of light soaking	4.85×10^7	3.28×10^7	4.08×10^3
2 hours of light soaking	3.57×10^6	1.83×10^6	3.26×10^2

After 1 hour of light soaking, measurements showed an improvement in the conductivity and linearity of I-V response. Moreover, further improvements have been observed after 2 hours of light soaking, while using a longer light soaking duration did not show any further improvements either in conductivity or the linearity of the I-V response. Nonetheless, as deposited GMZO films showed no improvements in conductivity after 1 or 2 hours of light soaking. Figure 1 shows the variation in resistivity (after 1 hour of light soaking) for annealed GMZO films which were exposed to the atmosphere for 3 days. Light soaking causes a very rapid increase in conductivity. Following the initial light soak, the film was kept in the dark and the resistivity increases when the film was measured the following day. This trend in increase in resistivity continued for the next 3 days.

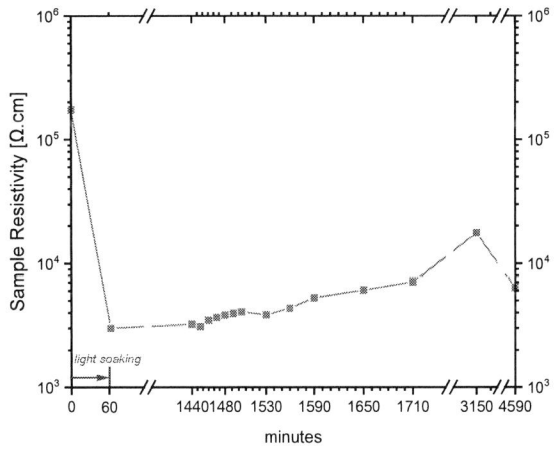

Fig. 1. The analysis of the resistivity of annealed GMZO films exposed to atmosphere after 1 hour of light soaking.

Measuring the Hall effect on films which are highly resistive and exhibit metastability (when exposed to the atmosphere) can be quite challenging. The duration of the measurements with these metastable films is crucial as the films degrade over time, as shown in Figure 1. On an attempt of measuring films after 1 hour of light soaking, a gradual decrease in Hall signal was observed, and the resistivity returned to its initial value after the measurement was completed. However, after 2 hours of light soaking, the Hall signal was more stable throughout the measurements. Many attempts have been made to extract the carrier concentration and mobility by varying the duration of the measurement time, drive current, and the speed of the rotating magnets. Although the Hall signal was fully resolved after 2 hours of light soaking, the results from the Lock-in and Fourier Transform (FT) analysis in step 3 were not in close agreement, and no clear FT peak has been observed. Hence, the measurements were not successful. The decrease in resistivity does indicate some increase in carrier concentration. Since the resistivity has improved by 3 orders of magnitude, the carrier concentration was expected to increase as well.

III. CONCLUSIONS AND FUTURE WORK

With an addition of Ga in to the MZO films there was an improvement in conductivity and Hall signal compared to undoped MZO films. 2 hours of light soaking resulted in a significant improvement in conductivity. However, a gradual decrease in the signal was observed over time after the films were light soaked and removed from the desiccator. Thus, no reliable and accurate results could be extracted for the carrier concentration and mobility. The full paper will include further analysis of light soaking on full CdSeTe/CdTe devices incorporating a GMZO buffer layer. We will also report on the effect of the CdCl$_2$ treatment on GMZO films on the Hall effect measurements.

Acknowledgements

The authors are grateful to EPSRC grant for funding through EP/P02484X/1 and EP/N510014/1 to support this work.

References

[1] F. Hussin, G. Issabayeva, and M. K. Aroua, "Solar photovoltaic applications: Opportunities and challenges," Rev. Chem. Eng., vol. 34, no. 4, pp. 503–528, 2018.

[2] T. Makino, Y. Segawe, M. Kawasaki, A. Ohtomo, R. Shiroki, K. Tamura, T. Yasuda, and H. Koinuma., "Band gap engineering based on Mg$_x$Zn$_{1-x}$O and Cd$_y$Zn$_{1-y}$O ternary alloy films," Appl. Phys. Lett., vol. 78, no. 9, pp. 1237–1239, 2001.

[3] I. M. Dharmadasa, P. A. Bingham, O. K. Echendu, H. I. Salim, T. Druffel, R. Dharmadasa, G. U. Sumanasekera, R. R. Dharmasena, M. B. Dergacheva, K. A. Mit, K. A. Urazov, L. Bowen, M. Walls and A. Abbas, "Fabrication of CdS/CdTe-Based Thin Film Solar Cells Using an Electrochemical Technique," Coatings, vol. 4, no. 3, pp. 380–415, 2014.

[4] T. Makino et al., "Band gap engineering based on MgxZn1-xO and CdyZn1-yO ternary alloy films," Appl. Phys. Lett., vol. 78, no. 9, pp. 1237–1239, 2001.

[5] J. M. Kephart, J. W. McCamy, Z. Ma, A. Ganjoo, F. M. Alamgir, and W. S. Sampath, "Band alignment of front contact layers for high-efficiency CdTe solar cells," Sol. Energy Mater. Sol. Cells, vol. 157, pp. 266–275, 2016.

[6] S. Paul, C. Swartz, S. Sohal, C. Grice, S. S. Bista, D. B. Li, Y. Yan, M. Holtz, and J. V. Lic., "Buffer/absorber interface recombination reduction and improvement of back-contact barrier height in CdTe solar cells," Thin Solid Films, vol. 685, no. July, pp. 385–392, 2019.

[7] T. Song, A. Kanevce, and J. R. Sites, "Emitter/absorber interface of CdTe solar cells," J. Appl. Phys., vol. 119, no. 23, 2016.

[8] F. Bittau, C. Potamialis, M. Togay, A. Abbas, P. Isherwood, J. Bowers, and J. Walls, "Analysis and optimisation of the glass/TCO/MZO stack for thin film CdTe solar cells," Sol. Energy Mater. Sol. Cells, vol. 187, pp. 15–22, 2018.

[9] A. H. Munshi, J. M.Kephart, A. Abbas, A. Danielson, G. Gelinas J. N. Beaudry, K. L. Barth, J. M. Walls, and W. S. Sampath, "Effect of CdCl$_2$ passivation treatment on microstructure and performance of CdSeTe/CdTe thin-film photovoltaic devices," Sol. Energy Mater. Sol. Cells, vol. 186, pp. 259–265, 2018.

[10] M. Togay, R. Greenhalgh T. Shimpi, S. S. Walajabad, K. L. Barth, J. M. Walls and J. W. Bowers, "Transient Metastable Behaviour in Highly Efficient MZO/CdSeTe /CdTe Thin Film Solar Cells, 2021 IEEE 48th Photovoltaic Specialists Conference (PVSC), pp. 0637-0642, 2021."

[11] T. Ablekim, C. Perkins, X. Zheng, C. Reich,D. Swanson, E. Colegrove,N. Joel, D. Albin, S. Nanayakkara, M. Reese,T. Shimpi,W. Sampath, andW. Metzger, "Tailoring MgZnO/Cd(Se) Te Interfaces for Photovoltaics," IEEE J. Photovoltaics, 2018.

[12] D. B. Li, R. Awni, S. Bista, N. Shrestha, C. Grince, L. Chen, G. Liyagane, M. Razooqi, A. Phillips, M. Heben, R. Ellingson, and Y. Yan, "Eliminating S-Kink to Maximize the Performance of MgZnO/CdTe Solar Cells," ACS Appl. Energy Mater., vol. 2, no. 4, pp. 2896–2903, 2019.

[13] R. Pandey, A. Shah, A. Munshi, T. Shimpi, P. Jundt, J. Guo, R. F. Klie, W. Sampath, and J. R. Sites, "Mitigation of J–V distortion in CdTe solar cells by Ga-doping of MgZnO emitter," Sol. Energy Mater. Sol. Cells, vol. 232, p. 111324, 2021.

[14] F. Bittau, S. Jagdale, C. Potamialis, J. W. Bowers, J. M. Walls, A. H. Munshi, K. L. Barth, W. S. Sampath, "Degradation of Mg-doped zinc oxide buffer layers in thin film CdTe solar cells," Thin Solid Films, vol. 691, no. September, p. 137556, 2019.

[15] W. Lee, S. Shin, D. Jung, J. Kim, C. Nahm, T. Moon,and B. Park, "Investigation of electronic and optical properties in Al-Ga codoped ZnO thin films," Curr. Appl. Phys., vol. 12, no. 3, pp. 628–631, 2012.

[16] H. Liu, V. Avrutin, N. Izyumskaya, Ü. Özgr, and H. Morkoç, "Transparent conducting oxides for electrode applications in light emitting and absorbing devices," Superlattices Microstruct., vol. 48, no. 5, pp. 458–484, 2010.

[17] D. E. Swanson, J. M. Kephart, P. S. Kobyakov, K. Walters, K. C. Cameron, J. Drayton, J. R. Sites, K. L. Barth, and W. S. Sampath, "Single vacuum chamber with multiple close space sublimation sources to fabricate CdTe solar cells," J. Vac. Sci. Technol. A Vacuum, Surfaces, Film., vol. 34, no. 2, p. 021202, 2016.

[18] O. Gunawan, Y. Virgus, and K. F. Tai, "A parallel dipole line system," Appl. Phys. Lett., vol. 106, no. 6, 2015. O. Gunawan et al., "Carrier-resolved photo-Hall effect," Nature, vol. 575, no. 7781, pp. 151–155, 2019.

Firm PV Power Generation in Switzerland

Jan Remund[1], Marc Perez[2] and Richard Perez[3]

[1]Meteotest AG, Bern, Switzerland
[2]Clean Power Research, Napa, CA, USA
[3]Atmospheric Sciences Research Center, SUNY, Albany, NY, USA

Abstract— **We investigate whether PV can effectively and economically contribute to a massively renewable energy (RE) power generation future for Switzerland. Taking advantage of the country's flexible hydropower resources, we determine the optimum PV/battery configurations that can meet the country's growing electrical demand firmly 24x365 at the least possible cost while entirely phasing out nuclear power generation. We examine several ultra-high RE scenarios where PV and hydro would meet the bulk of the country's demand. Depending on future cost predictions for PV and batteries, and a small contribution from domestic or imported dispatchable resources, we show that power production costs on the Swiss grid would range from 6 to 10 US cents per kWh. This is well in line with current market prices on the regional TSOs.**

Keywords—storage, implicit storage, firm power generation, photovoltaics, grid integration, high penetration renewables

I. METHODOLOGY: FIRM SOLAR POWER GENERATION

24/365 firm power availability is a prerequisite for intermittent solar and wind resources if they are to evolve from their current position at the margin of a core of dispatchable generation to a grid-dominant position.

It is now well understood that the least expensive way to transform intermittent renewables into firm power generators entails: (1) applying implicit storage – i.e., overbuilding and dynamically curtailing the resources [1-3] to keep real energy storage requirements at economically reasonable levels – and (2) optimally combining renewable resources that may have different daily and seasonal availabilities [4].

The Clean Power Research CPT model [5] we apply in the present investigation was designed to derive the least cost combination of intermittent renewables (PV, wind) and storage – real and implicit – for any location/region. The model also accounts for region or policy-specific operational contexts, such as any allowance for dispatchable supply-side generation (e.g., thermal generation from natural gas or e-fuels), the availability of other renewables (e.g., hydro), or the application of demand-side load management strategies.

Inputs to the model include the Capital Expenses (CapEx) and Operating Expenses (OpEx) of all considered generation, storage, and load management resources as well as multi-year hourly site/time-specific time series of renewable electrical production and demand. The main output of the model is the levelized production Cost Of firm Electricity (LCOE) of the optimized resources' blend. The model also produces the optimal amounts of real and implicit storage required to achieve this optimum LCOE.

Results of previous investigations in the continental US (CONUS) and tropical island power grids indicate that a 95% optimized wind/solar blend and an allowance for 5% supply-side flexibility via natural gas could yield firm 24/365 LCOEs below 4 cents per kWh by 2040, with PV/wind overbuild of the order of 50% [4, 6, 7].

A. The case of Switzerland

The situation in Switzerland is markedly different from our previous CONUS and tropical island case studies. It is characterized by both unique assets and unique challenges.

- Assets: Switzerland possesses a large existing hydro/hydro storage resource, including 'classical' run-of-river hydropower and two types of storage systems: conventional pumped hydro, and one-way seasonal buffers (lakes holding large quantities of water released on demand). The storage systems are currently applied to maximize market economics (e.g., arbitrage in neighboring European markets). The specs of the hydropower assets are reported in Table 1 along with the other energy generating resources currently available in the country.

- Challenges: (1) it is environmentally difficult to deploy new wind, so large-scale natural wind-solar complementarities cannot be fully tapped. (2) The solar resource is highly seasonal with very low wintertime (Nov – mid Feb) solar production when electrical demand peaks.

Table 1. Current (2018-2020 average) power generation resources in Switzerland

	Installed capacity	Annual energy yield
Nuclear	3 GW	24.2 TWh
Run-of-river hydro	4.2 GW	17.6 TWh
One-way hydro buffer	8.2 GW	18 TWh*
Pumped hydro	2.9 GW	4.2Twh^
PV	2.4 GW	2.2 TWh
Wind	0.1 GW	0.1 TWh
Non-renewable thermal generation	0.42 GW	1.17 TWh
Renewable thermal generation	0.55 GW	1.84 TWh
Imports	7 GW	10.77 TWh
Exports	10 GW	-10 TWh

*the full-to-empty buffer system has a capacity of 10Wh.. Total output includes river-flow
^ pumped hydro output -- net production in zero.

The annual (2018) dispatching of these resources is illustrated in Figure 1. 30-day running means have been plotted to remove short-term fluctuations and improve visualization. The top edge of the graph represents demand on the Swiss grid. Note that the Swiss production is insufficient in winter and early spring, requiring imports from the rest of Europe. However, production exceeds demand in summer and is exported, mainly to the summer-peaking Italian grid.

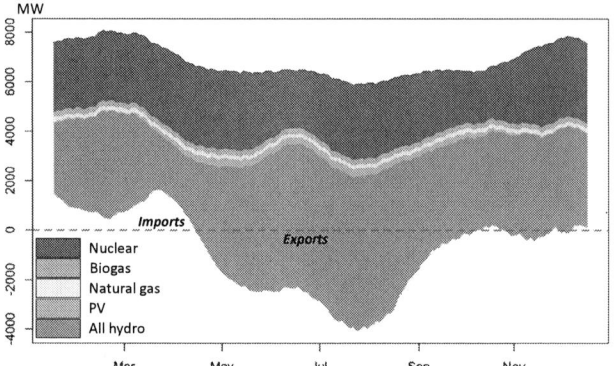

Figure 1. Annual dispatch of Swiss-based of supply-side resources for the year 2020. The top line of the stacked graph represents the Swiss grid load [8]

II. CASE STUDY

We explore six scenarios at the 2050 horizon where the PV resource will be the central part of a high renewable energy (RE) firm power delivery system for the Swiss power grid. All scenarios include a phasing out nuclear power generation.

- Scenario #1 – *Energy perspectives 2050+* [9]. This scenario retains the current small contributions from thermal energy production and adds 2.12 GW of wind power generation amounting to a total wind energy production of 4.3 TWh/year. Pumped hydro capacity is increased from 2.9 to 5.7 GW with a commensurate increase in energy storage reserve. One-way buffer hydro storage capacity is increased from 8.2 to 9 GW, with a 2 TWh increase in full-to-empty long term energy reserve.

- Scenario #2 – *10% imports, no restrictions*. This scenario also retains the small current contributions from thermal production but adds only 1.02 GW of wind (2.15 TWh/yr). It allows for net imports from the European grid to total 8.25 TWh/yr. Pumped hydro capacity is only increased to 4.42 GW, while the buffer hydro storage capacity is increased to 8.5 GW with a full-to-empty energy reserve increase of 1 TWh.

- Scenario #3 – *No imports, natural gas*. This scenario is identical to scenario #2 but replaces the 8.25 TWh of imports with new thermal generation from natural gas (2.8 GW new capacity; prices including CO_2 certificates). Net-zero import/export are limited to 3 GW with 10 TWh exchanged annually each way.

- Scenario #4 – *No imports, e-fuels*. This scenario is identical to scenario # 3, but replaces natural gas by e-fuels, produced either domestically or abroad. Note that this scenario is 100% renewable.

- Scenario #5 – *Imports and e-fuels*. This scenario retains the level of net imports from scenario # 2 and includes roughly half of the e-fuel thermal generation from scenario #4 (1.7 GW, 4.97 TWh/yr).

- Scenario #6 – *Imports, e-fuels, and agri-PV*. This scenario is identical to scenario #5, but with a slightly lower capital cost for new PV resulting from extensive agri-PV deployments (see below).

For all scenarios, the considered 2050 Swiss electrical demand is assumed to be 30% higher than current (from transportation/building electrification) and nuclear generation is eliminated. The new load demand profile is extrapolated upward from current load (+30% for all hours). The energy demand balance not met by hydro, or the wind/thermal/import resources identified above is met by new firm PV generation. Figure2 summarizes the contribution all supply-side energy sources in each scenario compared to current. It clearly illustrates the central role to be played by new firm PV generation, ranging from 35% of total generation in scenarios #4 and 5 to 46% in scenario #1.

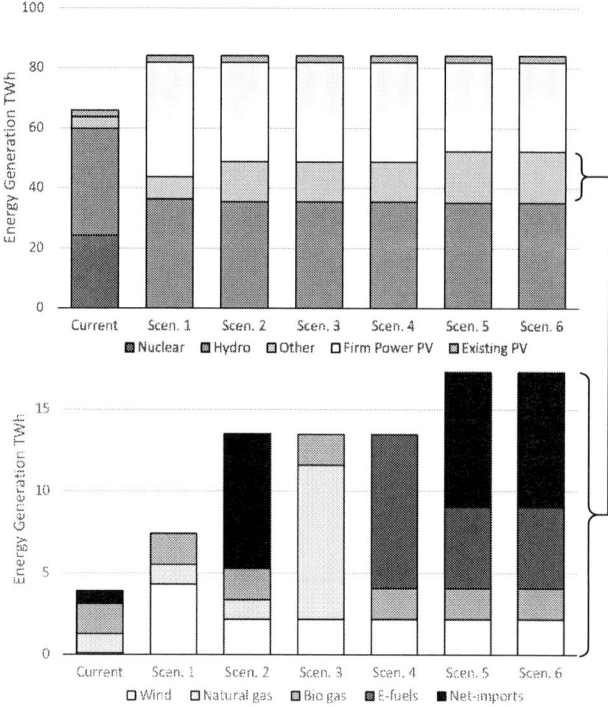

Figure 2. Supply-side electrical energy resources for all scenarios compared to current. The bottom part of the figure provides details for the source labeled as 'other' in the top part. Note that scenario #4 is the only scenario that does not include non-renewable (nat. gas) or possibly non-renewable (imports) resources.

For each scenario, we investigate two sets of assumptions regarding (1) Switzerland interconnectivity with the larger European grid, and (2) the cost of new PV and electrochemical storage (battery) technologies.

For interconnectivity, we look at two configurations:

(1) net-zero imports where the Swiss grid would continue to operate interconnected with the larger European grid and,

(2) an extreme limit case where the grid would operate in full autonomy.

In the net-zero case, supply-side flexibility is provided by allowing 10 TWh/yr to be both exported from and imported onto the Swiss grid with a maximum capacity of 3 GW (note that this is above and beyond the supply-side-only imports identified in scenarios #2, 5, & 6) In the autonomous case, Switzerland would operate independently from the larger European grid to the exception of one-way imports considered in scenarios #2, 5 and 6. We note that this autonomous configuration is unlikely given the country's natural interconnectedness, but this limit case is nevertheless informative in quantifying extreme resiliency conditions.

Regarding technology, we consider two assumptions for PV and electrochemical storage CapEx at the 2050 horizon based conservative or optimistic predictions from the NREL technology roadmap.

(1) The conservative assumption sets turnkey PV at $860/kW and electrochemical storage at $330/kWh.

(2) The optimistic assumption prices these technologies at $390/kW and $45/kWh, respectively.

The first assumption reflects small scale systems (e.g., user-sited) likely to be prevalent in the Swiss PV/storage build-up, while the second reflects utility-size systems that may also be [partially] considered. In both cases we use $4/kW/yr for PV OpEx, and 0.25%/yr for battery OpEx. Note that for scenario #6, we apply a smaller CapEx for the Swiss assumption – $786/kW – that is reflective of the larger proportion of agri-PV deployment assumed in this scenario.

All supply-side resources to the exception of new PV are considered as either dispatchable or must-run. Their financial impact is captured through their electrical generation costs identified in Table 2, i.e., we assume that these market-based prices embody both their CapEx and Opex.

Table 2. Assumed 2050 power generation cost of supply-side and storage resources on the Swiss grid

		Generation cost (¢/kWh)	Notes
Dispatchable Resources	Hydro storage	7/6.5	7 ¢/kWh for Scenario #1 only
	Pumped Hydro	6/5.8	6 ¢/kWh for Scenarios #1-4 --discharge cost
	Thermal Natural gas	11.1/14.1	14.1 ¢/kWh for Sceario #3 (E-certification fee)
	Thermal Bio gas	11.1	
	Thermal e-fuels	19.7/17.9	17.9 ¢/kWh for Scenario #4 only
	Imports	6	
	Exports	-5	
Must-run Resources	Run-of-River hydro	5	
	Existing Wind	15	
	New Wind	12/11	12 ¢/kWh for Scenario #1 only
	Existing PV	6.9	

We apply the Clean Power Transformation model to determine the optimum PV and battery resources needed to meet demand firmly at the least possible cost while dispatchable resources are optimally deployed toward this minimum cost/firm power generation objective. The results of this optimization include the required quantities of new battery storage, new PV, curtailed PV output (implicit storage), the electricity generation cost of the optimum supply-side/storage blend that will supply Swiss demand 24x365.

We apply three years' worth (2018-2020) of experimental data consisting of hourly electrical demand, and measured hourly production of nuclear, PV, wind, and hydropower resources. Future new hourly PV generation is extrapolated from current measured production, prorating to new capacity.

III. RESULTS

In figure 3, we report the new PV capacity, curtailed PV output (implicit storage), and battery storage required in each scenario to firmly meet demand on the Swiss power grid.

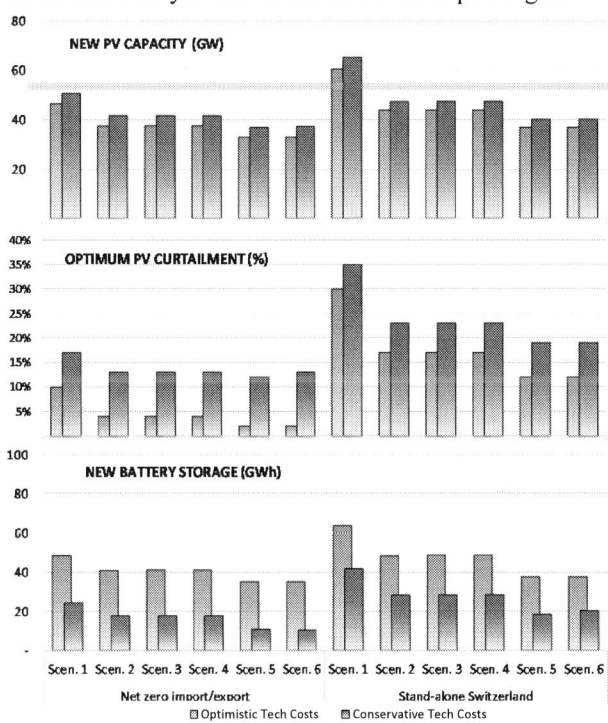

Figure 3: New PV capacity, proactive curtailment, and battery storage required to meet the new Swiss demand 24x365 in each of the six scenarios, and each technology cost and grid interconnectivity assumption.

New PV capacities (figure 3, top) range from 32.9 GW (scenario #5 & #6 with net-zero interconnectivity and optimistic technology costs) to 65.2 GW (scenario #1 stand-alone grid and conservative costs). Applying optimistic cost assumptions reduces new PV requirements by about 9% overall compared to conservative costs. Operating the Swiss grid stand-alone would require 17% more PV to be built than allowing net-zero interconnectivity. We plotted a "max acceptable" line indicating the maximum amount of new PV that could be reasonably deployed in the country. This amount is the result of a comprehensive analysis from Remund et al. [10] that considered all deployable in-country options (including roof space, exclusion zones, farmland, etc.) given current PV efficiencies. Importantly, all but one scenario (#1 autonomous grid) fall under this upper limit.

PV output curtailment (figure 3, middle) ranges from 2 % (scenario #5 & #6 with net-zero interconnectivity and optimistic tech costs) to 35% (scenario #1 autonomous grid and conservative costs). Technology cost assumptions have a strong influence on required curtailment. Applying optimistic cost reduces the need for curtailment by an average of 41%. Stand-alone grid operation, without net-zero flexibility would increase operational curtailment by 130%.

New battery storage requirements (figure 3, bottom) range from 10.5 GWh (scenario #6 conservative cost assumptions) to 64 GWh (scenario #1 with stand-alone grid and optimistic tech costs). Applying optimistic cost assumptions leads to two times more battery storage overall. This significant difference is because future utility-scale NREL battery cost predictions are very low compared to the conservative small-scale estimates (8 times less) while the difference for PV between the two estimates amounts only to a factor of two. Interestingly, autonomous operation of the Swiss grid would only require 32% more battery storage than net-zero interconnected operation. In all cases, required battery storage is low, amounting to 0.3 hours of full PV capacity in the case of conservative cost assumptions, and ~1.2 hours in the case of optimistic cost assumptions. The bottom line is that no new long-term storage beyond the small addition to the existing buffer hydro system (+10% for scenarios #2-6, +20% for scenario #1), as is often assumed when envisaging ultra-high PV or wind penetration. This observation corroborates results obtained in the US [4].

Figure 4 reports the blended all-resources power generation LCOEs on the Swiss power grid.

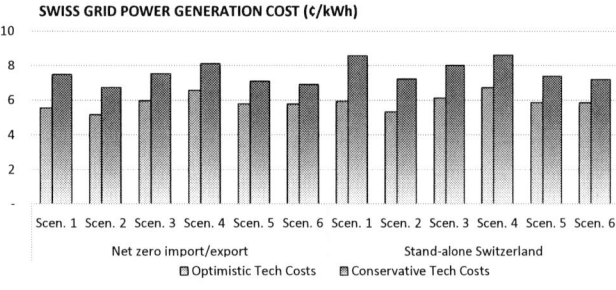

Figure 4: Electric power generation cost on the Swiss grid in each of the six scenarios, and each technology cost and grid autonomy case.

Electricity production costs range from 5.2 ¢/kWh (scenario #2, net-zero interconnectivity, optimistic technology costs) to 8.9 ¢/kWh (scenario #1 autonomous grid operation and conservative, small-scale tech costs). Applying optimistic

utility scale storage/PV cost assumptions reduces generation costs by an average of 22%. Importantly, as unlikely as this configuration may be, stand-alone grid operation would only increase these costs by an average of 5.5% i.e., not constituting a showstopper.

The new annual dispatch of all resources is illustrated in Figure 6 for the 100% RE (r-fuel) scenario #4. The top graph illustrates the net-zero import/export grid configuration, while the bottom graph illustrates the autonomous grid configuration. As in Figure 1, 30-day running mean have been plotted to remove short-term fluctuations and improve visualization.

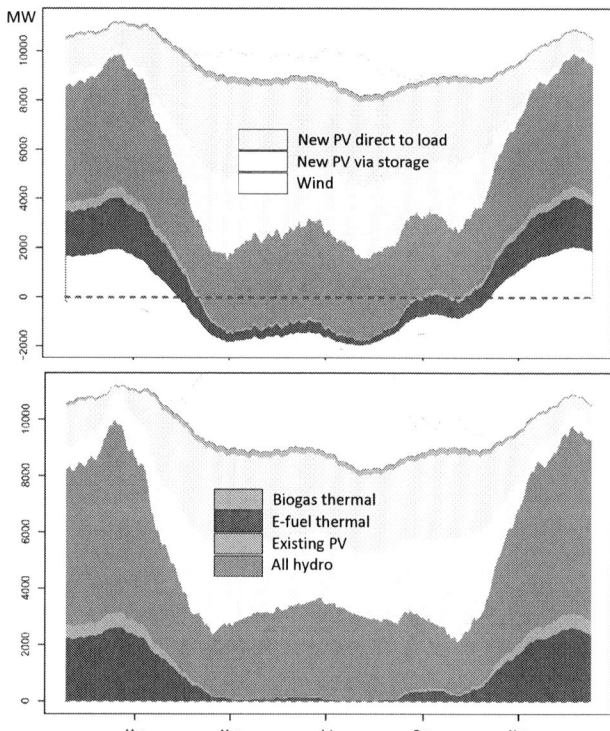

Figure 5: Annual dispatch of supply-side resources for the year 2020 illustrated for the 100% renewable scenario with e-fuels (#4). The top graph represents the net-zero interconnected configuration where winter imports are energetically matched to summer export amounting to net-zero. The bottom graph corresponds to the extreme stand-alone grid configuration.

Implicit storage impact: Figure 6 illustrates the importance of overbuilding and operationally curtailing the PV resource on the bottom line: production costs would be an average of 63% higher across all scenarios for the net-zero interconnected configuration, and 450% higher in the autonomous grid configuration. The main factor for this cost difference is the amount of new battery storage required that would respectively be 1300% and 7500% higher without PV oversize/curtailment.

Sensitivity analysis: The three years, analyzed independently, lead to very comparable firm power production cost results overall as seen in in Figure 7 for the 100% renewable scenario #4.

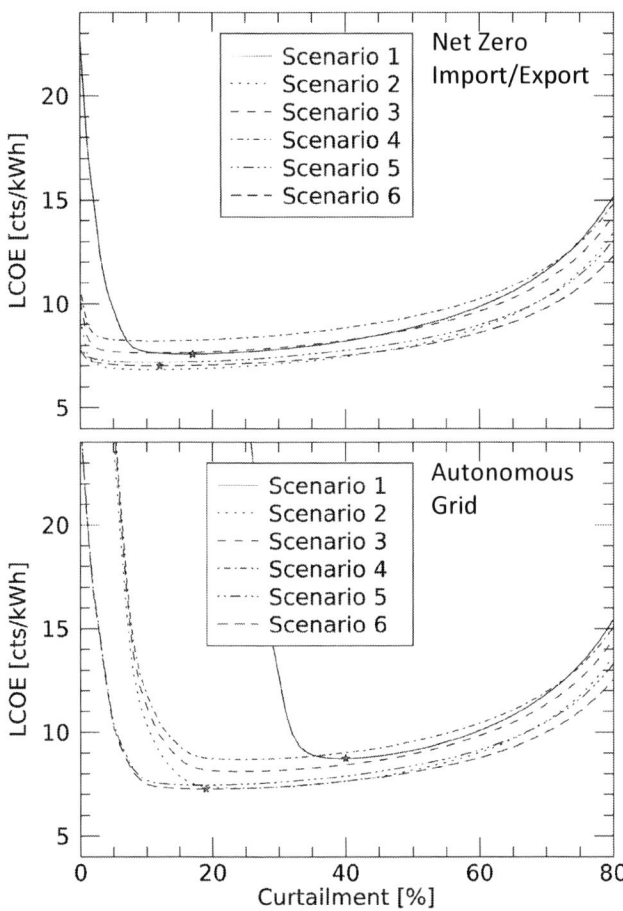

Figure 6: Electricity production cost on the Swiss power grid as a function of PV output curtailment for all scenarios. The top graph corresponds to the interconnected grid configuration with net-zero import/exports with the larger European grid. The bottom graph represents autonomous grid configuration.

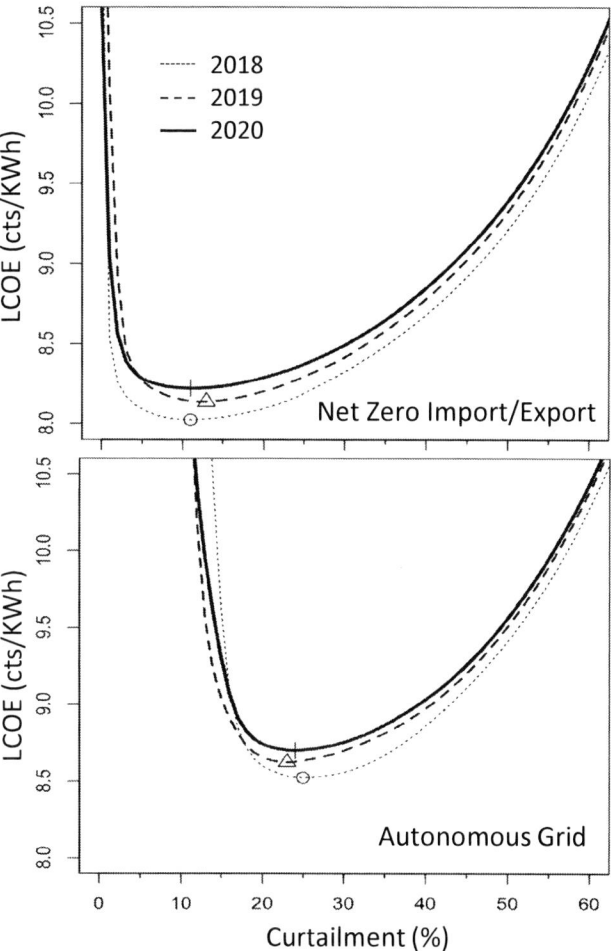

Figure 7: Comparing 2018, 2019 and 2020 electricity production cost on the Swiss power grid as a function of PV output curtailment for scenario #4. The top graph corresponds to the interconnected grid configuration with net-zero import/exports with the larger European grid. The bottom graph represents autonomous grid configuration.

IV. DISCUSSION

Our investigation shows that high-RE solutions for Switzerland, with PV playing a central role as a complementary resource to the Country's hydropower system, are both physically and economically reasonable, despite the minor role wind power can play, and the mediocre PV resource in winter months.

It is important to state that operational costs in all considered scenarios are reasonable compared to current wholesale market prices in Switzerland (these have been well above 20 ¢/kWh the last five months [12]) The present ultra-high RE costs are even reasonable when compared to earlier pre-crisis TSO wholesale prices (4-6 ¢/kWh) noting that these earlier TSO prices do not fully factor-in environmental or strategic externalities which, as we see today with international tensions, can be consequential.

Another particularly important observation is the result obtained for the 100% RE scenario (#4). Not only are operational generation costs reasonable (6½-8½ ¢/kWh depending on technology and autonomy assumptions), but they show the supply-side flexibility catalyst role that e-fuels can play, even as expensive as they are expected to be at 18-20 ¢/kWh.

Finally, we stress the importance of implicit storage (i.e., optimally overbuilding the PV resources). Not implementing this deployment strategy would result in higher prices on the network. It is therefore important to operationalize optimal overbuilding and curtailment early-on, by e.g., implementing appropriate regulations that would lead to firm power monetization, instead of current run-of-the-whether PV production.

978-1-7281-6118-1/22 $31.00 © 2022 IEEE

V. Acknowledgement

This work has been funded by the Swiss Federal Office of Energy (contract number SI/502264-01).

References

[1] Perez, M. R. Perez & T. Hoff, (2021): IMPLICIT STORAGE – Optimally Achieving Lowest-Cost 100% Renewable Power generation. Solar World Congress. International Energy Society.

[2] O'Shaughnessy, Eric, Jesse Cruce, and Kaifeng Xu. 2021. Solar PV Curtailment in Changing Grid and Technological Contexts: Preprint. Golden, CO: National Renewable Energy Laboratory. NREL/CP-6A20-74176. https://www.nrel.gov/docs/fy21osti/74176.pdf.

[3] Tong, D., Farnham, D.J., Duan, L. et al. Geophysical constraints on the reliability of solar and wind power worldwide. Nat Commun 12, 6146 (2021). https://doi.org/10.1038/s41467-021-26355-z

[4] Perez M., (2020): Pathways to 100% Renewables across the MISO Region. http://mnsolarpathways

[5] Perez, M., R. Perez, K. Rabago & M. Putnam, (2019): Overbuilding & curtailment: The cost-effective enablers of firm PV generation. Solar Energy 180, 412-422

[6] Perez R., M.Perez, J. Schlemmer, J. Dise, T. E. Hoff, A. Swierc, P. Keelin, M. Pierro & C. Cornaro, (2020): From Firm Solar Power Forecasts to Firm Solar Power Generation an Effective Path to Ultra-High Renewable Penetration a New York Case Study. Energies 2020, 13, 4489.

[7] Tapaches, E., M. Perez, R. Perez, T. Chamarande, P. Lauret & M. David, (2020) : Synthese du Rapport Technique du Projet PEPS Reunion. ADEME, France. https://librairie.ademe.fr/energies-renouvelables-reseaux-et-stockage/53-projet-peps-reunion.html

[8] European Network Transmission System Operator, ENTSO-E (2018-2020): https://transparency.entsoe.eu/

[9] Swiss Federal Office of Energy, 2021: Swiss Energy Perspectives 2050+. https://www.bfe.admin.ch/bfe/en/home/policy/energy-perspectives-2050-plus.html

[10] Remund, J., Albrecht, S. & Stickelberger, D. Das Schweizer PV-Potenzial basierend auf jedem Gebäude. in *Photovoltaik Symposium Bad Staffelstein* (2019).

[11] Swiss Federal Office of Energy, 2021: Swiss Energy Perspectives 2050+. https://www.bfe.admin.ch/bfe/en/home/policy/energy-perspectives-2050-plus.html

[12] TNO & Fraunhofer ISE, (2022): Swiss Energy Charts. https://energy-charts.info/charts/price_spot_market/chart.htm?l=en&c=CH&interval=year&year=2022&legendItems=0000100000

Demonstration of Point Contact Geometry for Solar Cells Using Single Walled Carbon Nanotube

Fadhil K. Alfadhili, Adam B. Phillips, Manoj K. Jamarkattel, Bhuiyan M. Anwar, Prabodika N. Kaluarachchi, Zahrah S. Almutawah, Abdul Quader, Deng-Bing Li, Yanfa Yan, Randy J. Ellingson, Michael J. Heben

Wright Center for Photovoltaics Innovation and Commercialization, Department of Physics and Astronomy, University of Toledo, Toledo, OH, United States

Passivation of back interfaces of CdTe-based devices will be critical to achieve the highest efficiency device performance. While AlOx has been shown to passivate the interface of CdTe films, it can also lead to a blocking layer when used as a back contact. Here we provide a pathway for hole transport through the passivating AlOX by incorporating single walled carbon nanotubes (SWCNTs) to form a point contact geometry. By varying the concentration of the AlOx precursor, SWCNTs, and surfactant, we are able to change the shape of the current density-voltage curve from having a strong s-kink behavior, indicative of blocking behavior, with low SWCNT ratios to no s-kink behavior with high SWCNT, low surfactant concentrations. We use back illuminated device measurements to show that both the AlOx and SWCNTs are necessary to achieve the best current collection.

Accelerate Cycles of Learning: Unencapsulated Silicon Photovoltaic Cells to Environmental Stressors

Nafis Iqbal[1], Nitin K. Chockalingam[2], Kehley A. Coleman[2], Jeffrie Fina[2], Kristopher O. Davis[1], Laura S. Bruckman[2], and Ina T. Martin[2]

1. University of Central Florida, Orlando, FL, 32816, USA
2. Case Western Reserve University, Cleveland, OH, 44106, USA

Abstract—We present materials and device performance data from distinct commercial technologies. Passivated emitter and rear contact (PERC) and silicon heterojunction (SHJ) solar cells are subjected to accelerated aging methods: damp heat and acetic acid exposures of unencapsulated devices. The unencapsulated devices are exposed to these accelerated aging conditions to provide information on materials and device changes, separate from the encapsulants, and to rapidly screen for degradation susceptibility from heat, moisture, and low pH. Ultimately, this approach can help accelerate cycles of learning in PV reliability and durability for novel cell types.

Keywords—Silicon heterojunction, PERC, accelerated aging, acetic acid, ITO, XPS

I. INTRODUCTION

Advanced crystalline silicon photovoltaic (PV) cell architectures are designed to mitigate energy conversion losses present in traditional architectures. In recent years, passivated emitter and rear contact (PERC) cells have displaced full area aluminum back surface field (Al-BSF) cells as the dominant PV cell technology due to their ability to lower rear surface recombination and improve light trapping. These cells feature a relatively heavily phosphorus doped front surface to form the n^+p homojunction with the boron or gallium doped p-type wafer. Silver (Ag) and aluminum (Al) pastes are printed on the n^+ front and p rear, putting metal in direct contact with the c-Si absorber. This leads to high contact recombination, limiting the overall open-circuit voltage (V_{OC}) of PERC cells. In contrast, silicon heterojunction (SHJ) cells can achieve world record open-circuit voltages by eliminating contact recombination. However, the use of new materials and processes introduce the potential for new failure modes [1],[2].

As the name implies, SHJ cells feature a heterojunction. This cell architecture decouples the metal contact from the absorber to form a passivating, carrier selective contact, thus limiting contact recombination. Typically, this is formed by depositing a multilayer stack of intrinsic, hydrogenated amorphous silicon (a-Si:H) to passivated dangling bonds at the n-type c-Si absorber surface, followed by doped a-Si:H to provide carrier selectivity, and an appropriate transparent conductive oxide (TCO) to provide lateral transport and ensure good contact with the metal. Indium tin oxide (ITO) is the typical TCO of choice, and the ITO film can also serve as an

antireflection coating (ARC) and assist in light trapping at the rear. In the past, SHJ cells have maintained a small market share, there is growing interest in shifting more production to SHJ due to the market push for higher efficiencies [1].

Traditionally, comparison of performance data from different climate zones and accelerated aging exposures provides insight into degradation pathways [3], and the materials and design features that constrain silicon device lifetime. Whereas most accelerated aging testing comprises encapsulated devices, PV manufacturers have a variety of significantly faster methods to test unencapsulated devices, including boiling them in water, and exposure to acetic acid. Similarly, our accelerated aging of unencapsulated devices will allow rapid feedback on performance loss.

SHJ cells are sensitive to stressors such as damp heat and irradiance, and degradation varies across manufacturers. Jordan et al. compared the degradation of fielded SHJ and PERC modules, and while their initial results found that the degradation rates of the modules were on par with conventional cell technologies, the mechanisms by which degradation occurred were different [4]. Traditional cells primarily saw performance losses due to a reduction of I_{SC}; SHJ devices also experienced reductions in V_{OC} and FF. Sinha et al. published an initial comparison of the effect of UV exposure on the performance of different unencapsulated Si architectures; overall, they found that degradation was actually more pronounced in emerging cell technologies, including SHJ devices, compared to classic Al-BSF devices [5]. Changes in surface hydrogen composition after degradation were also observed. The authors hypothesized that the performance change was due to an increase in ITO resistance or deterioration of the passivating interfaces. However, Jordan et al. found that increases in ITO resistance were not observed in fielded SHJ cells as compared to control cells. Further, Bertoni's work [6] with surface recombination velocity measurements of c-Si/a-Si:H stacks supports the degradation of the surface passivation layer as a root cause of the observed loss performance over time. Thus, the exact cause of the performance changes in SHJ cells is not well understood - likely because of the expression of multiple degradation modes that are affected by variations in materials and/or stressors. Understanding these variations is critical to de-risking SHJs for high-volume manufacturing. These previous works demonstrate the effectiveness of

978-1-7281-6118-1/22 $31.00 © 2022 IEEE

incorporating accelerated aging exposures of unencapsulated devices into durability studies.

Here, we exposed unencapsulated PERC and SHJ cells to accelerated aging conditions that encompass heat, moisture, and acidic conditions. Device performance was characterized before and after accelerated aging. Further, the ITO and Ag gridlines of the SHJ cells were characterized via X-ray photoelectron spectroscopy (XPS) and optical profilometry. SHJ devices demonstrated stable I_{SC}, V_{OC}, and FF values over 1750 h of DH exposure. PERC devices were similarly stable through 1500 h of DH exposure, but the I_{SC} and FF decreased at longer exposure times (>1500 h for I_{SC}, and >2000 h for FF). In contrast, acetic acid exposure results in decreased V_{OC} from recombination losses for both PERC and SHJ. Further, acetic acid exposure resulted in increased front contact resistivity for both cell types; the effect was more pronounced for PERC devices. XPS of the DH exposed SHJ front contact shows changes to the ITO and Ag gridlines consistent with moisture induced degradation, but there was no detectable macroscopic roughening of the Ag gridlines up to 1000 h of DH exposure (as measured by optical profilometry).

II. EXPERIMENTAL

A. Cell descriptions and exposures

Fig. 1 shows the two solar cell architectures used in this study: (a) monocrystalline *p*-type silicon PERC cell featuring a n^+p homojunction, Ag front contacts, and Al rear contacts featuring a local p^+p hole contact, and (b) monocrystalline *n*-type silicon SHJ cell featuring intrinsic and doped a-Si:H layers, ITO, and Ag contacts on the front and rear. Two accelerated aging exposure protocols were used in this study: damp heat (DH, 85 °C, 85 % relative humidity), and soaking in 5 vol.% (0.84 M) acetic acid solution. Mini-cells and transmission line measurement (TLM) strips were prepared from full cells using laser scribe and cleave method [7]. Full cells were lightly scribed from the backside using a 1064 nm fiber laser followed by cleaving. This process helps minimize laser damages and/or shunting. However, due to damage at the sample edges, the performance of the mini-cells can slightly vary/decrease compared to the full cells.

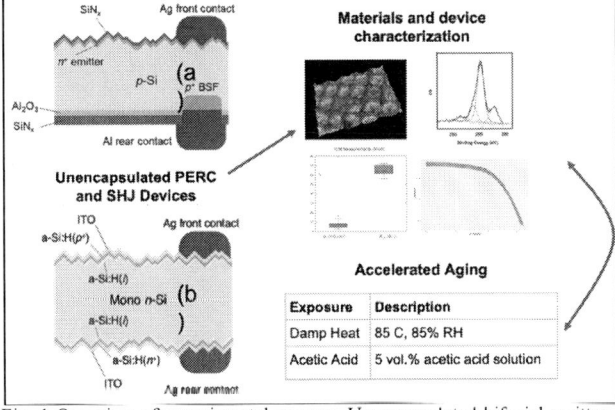

Fig. 1 Overview of experimental process. Unencapsulated bifacial emitter and rear contact (PERC, above) and silicon heterojunction (SHJ, below) solar cells are characterized using a combination of materials and device characterization methods. Samples are then exposed to accelerated aging conditions for set time periods and characterized between steps.

B. Material and device characterization

Current-voltage (I-V) characteristics of the devices were collected using an All Real Apollo Solar Simulator to illuminate the samples. A subset of SHJ samples were diced into 1 cm wide rectangles, with 30 gridlines running across the ~ 6 cm length. TLM measurements of 22 samples resulted in a sheet resistance of 65 ± 5 Ω/sq and a contact resistivity of 7.3 ± 3.5 mΩ cm^2. In addition to the TLM strips, ~ 7.5 cm x 4 cm mini-cells were prepared from full cells for accelerated aging exposure. For acetic acid treatments, samples were exposed to acetic acid solution at room temperature for 60, 120, and 180 min. Changes in contact resistivity and sheet resistance before and after exposure were tracked using the TLM method. In addition, Suns-V_{OC} and photoluminescence (PL) imaging at 1 Sun condition were performed at each step of exposure to track the change in V_{OC}, saturation current density (J_0) and PL counts. Damp heat (DH) exposures were performed using the damp heat (DH) protocol of the International Electrotechnical Commission (IEC 61646) for PV modules (85% humidity and 85 °C for 1000 h), then extended to longer times.

A Zygo NewView 7300 optical profilometer was used to measure surface topography of the front Ag gridlines. XPS measurements were made on a PHI Versaprobe 5000 scanning X-ray photoelectron spectrometer. Spectra were collected using a monochromatic Al K X-ray source (1486.6 eV focused source operated at 50 W, 15 kV and rastered at 200 um x 200 um), hemispherical analyzer, and multichannel detector. A low-energy (~1 eV) electron flood gun and 10 eV argon ion flood gun were used for dual charge neutralization. Survey spectra were collected using an analyzer pass energy and step size of 93.9 eV and 0.5 eV/step, respectively. High-resolution spectra were collected using a pass energy of 23.5 eV and a step size of 0.1 eV/step. High-resolution spectra were charge referenced by setting the C 1s hydrocarbon peak to 284.8 eV.

III. RESULTS AND ANALYSIS

Fig. 1 shows a schematic of the experimental process. Unencapsulated PERC and SHJ solar cells were characterized using both materials and device characterization methods. Samples were exposed to accelerated aging conditions for set time periods, and re-characterized between steps.

A. PERC cell results

Fig. 2 shows the 83.4 % confidence intervals (CI) of the V_{OC} (A), I_{SC} (B), and Fill Factor (C) values of the devices during periodic intervals of DH exposure, beyond 2000 h. V_{OC} values do not change significantly in this time period. I_{SC} and FF values decrease from the control to the last step because the confidence intervals do not overlap. The 83.4 % CI represents a visualization of a two sample t-test; therefore, if the CI do not overlap, then the samples can be considered different [8], [9].

Fig. 3 shows the effects of acetic acid exposure on the PERC devices. The PERC cells show no significant degradation in V_{OC}, J_0, and PL counts after the initial acetic acid exposure; this suggests that recombination losses are minor due to acid treatment. In contrast, there is a significant change in front contact resistivity due to acid treatment. The front contact resistivity mean increases ~5 times after the final step of acid exposure. This is likely due to the degradation of the interfacial

978-1-7281-6118-1/22 $31.00 © 2022 IEEE

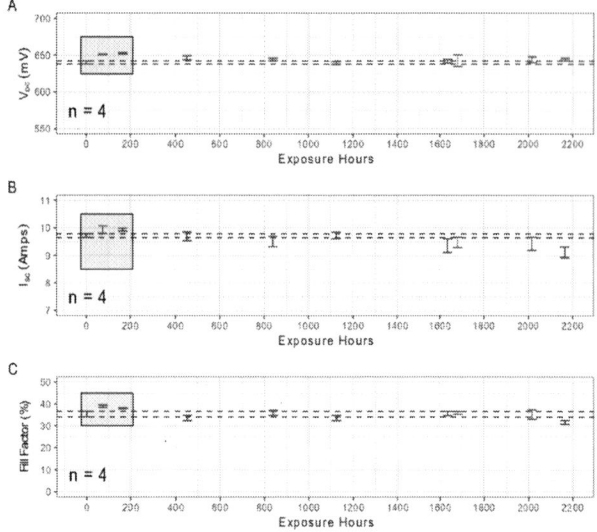

Fig. 2. IV characteristics of PERC cells under DH exposure including (A) V_{OC}, (B) I_{SC}, and (C) Fill Factor. There are 4 samples in each step and the 83.4 % confidence interval (CI) is displayed for each step. The dashed blue line shows the control CI.

glass layer by acetic acid, causing contact/adhesion losses [10], [11], [12], [13]. The rear contacts of the PERC devices also show some minor degradation after 180 min of acid treatment (shown by widening of the CI). As the rear contacts do not have a glass frit interface, they are more robust to acetic acid exposure. Previous studies have shown that control PERC devices soaked in water for the same amount of time did not show these changes, indicating that they are related to the acid exposure, not the moisture [14].

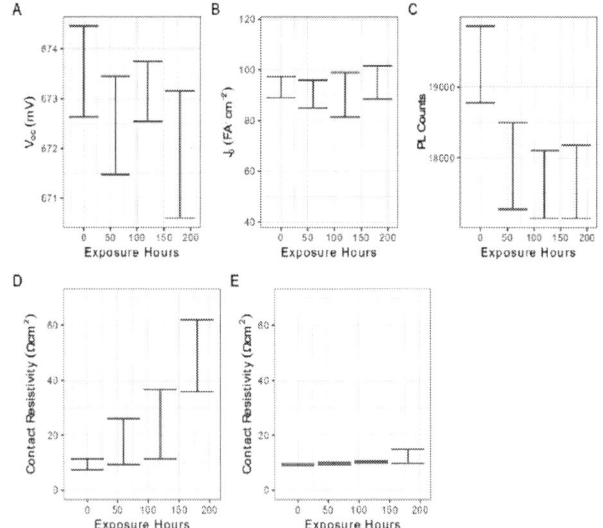

Fig. 3. PERC cell characterization after acetic acid exposure including (A) V_{OC}, (B) J_0, (C) PL counts, (D) front contact resistivity, and (E) rear contact resistivity; error bars show 83.4 % CI, with 8 samples per exposure step.

B. SHJ cell results

Fig. 4 shows the normalized (A) V_{OC}, (B) I_{SC}, and (C) FF values of SHJ devices during periodic intervals of DH exposure, beyond 1600 h. Values are normalized to time 0 baseline measurements, and CI are not shown because of the limited sample numbers per exposure step. V_{OC}, I_{SC}, and FF values do not change significantly in this time period, i.e., they do not deviate significantly from the control step represented by the black dashed line.

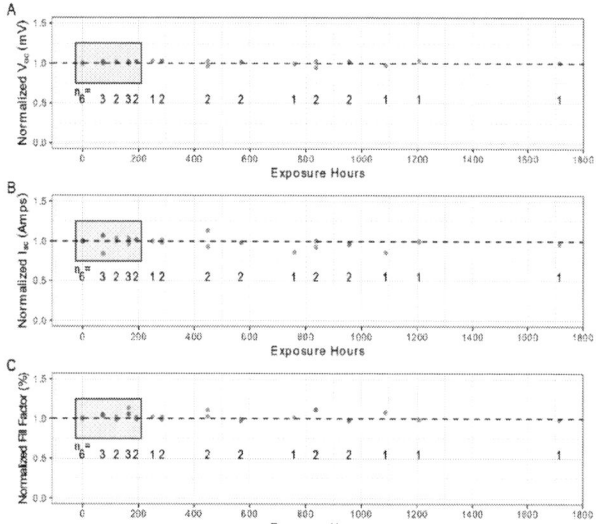

Fig. 4. Normalized IV characteristics of SHJ cells under DH exposure: (A) V_{OC}, (B) I_{SC} and (C) Fill Factor. The number of samples per step are listed below the data. The dashed line shows the normalized control value of 1.

In contrast, Fig. 5 shows that SHJ devices exposed to acetic acid solution experience a decrease in mean V_{OC} (~2%), and a corresponding increase in J_0 (B) after the final exposure step. Additionally, there was an ~11% drop in mean PL counts (C), with the 83.4 % CI not overlapping between the initial and final exposure, suggesting that recombination degradation is taking place due to acid treatment. As with PERC, control devices soaked in water for the same time did not show these changes.

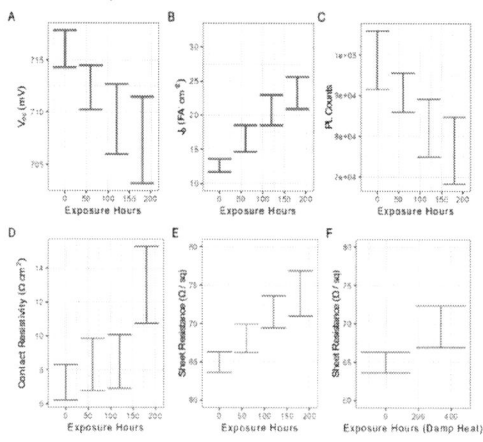

Figure 5. SHJ cell characterization after acetic acid exposure: (A) V_{OC}, (B) J_0, (C) PL counts, (D) front contact resistivity, (E), sheet resistance, and (F) sheet resistance after DH exposure with 83.4 % CI and 8 samples per exposure step.

TLM and four-point probe measurements of the SHJ devices provide additional information on the contacts (Fig. 5). Acetic acid exposure results in an increase (~8% of the mean) in contact resistivity (D) for the SHJ samples, which could be related to the oxidation/degradation of the Ag metal contacts. The increase occurred at the last exposure step. Additionally, there was an increase (~11% of the mean) in ITO sheet resistance (E) for the SHJ samples which is indicative of the ITO being etched by acetic acid or some other form of degradation that leads to an increase in the ITO resistivity [15]. Similarly, four-point probe measurements show an increase in sheet resistance of the ITO front contact with DH exposure (F). The 336 h of aging results in an average increase of 5% in the mean in DH with no overlap in the CI.

XPS was used to monitor the changes in the front contact of the SHJ cells during DH exposure. Fig. 6 shows the survey XPS spectra of the SHJ ITO front surface with (a) and without (b) the Ag gridline. As expected, the ITO surface shows the presence of In, Sn, O, and C. The measurement that includes the Ag gridline has these elements, and the addition of Ag. Analysis of the high-resolution XPS spectra of the individual elements yields information on their binding environments. In the unaged samples, the In 3d5/2 and Sn 3d5/2 peaks were each fit with a single peak at 444.4 eV and 486.7 eV, respectively. These values correspond to In^{3+} and Sn^{4+}, as expected for In_2O_3 and SnO_2 in ITO [14]. The Ag 3d5/2 peak was fitted with a single component at 368 eV, consistent with metallic Ag (Ag0) [16].

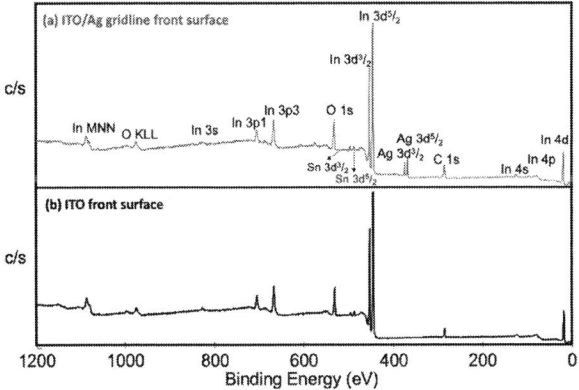

Fig. 6. Survey XPS spectra of the SHJ ITO front surface with (a) and without (b) the Ag gridline

In addition to the metallic elements, the O 1s and C 1s spectra are also analyzed. Figure 7 shows the deconvolution of high-resolution O 1s (a and c) and C 1s (b and d) XPS spectra of the ITO and ITO/Ag of the SHJ cell before exposure. The fits are consistent with literature reports of similar materials. The O 1s envelope was fit using three components, following Donley et al. [17]. Peak I (529.5 eV) is consistent with O^{2-} anions in the TCO lattice; a decrease in this peak is indicative of lattice degradation. Peak II (530.6 eV) is related to the oxygen-vacancy concentration; degradation of this feature is correlated to decreased carrier concentration and conductivity. Peak III (531.5 eV) is related to hydroxyl groups and/or O-containing adsorbates on the ITO surface; an increase in area is related to degradation products. The fits are similar between

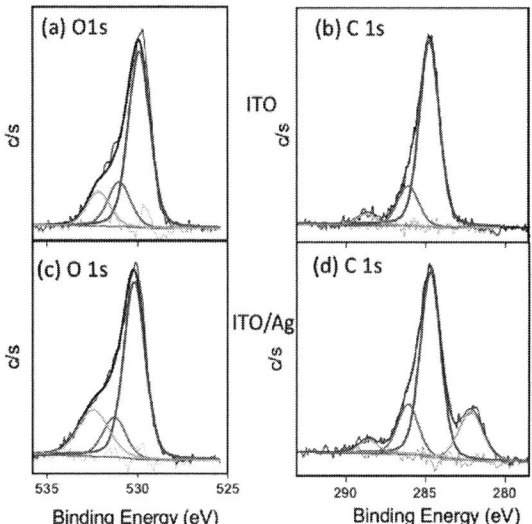

Fig. 7. Deconvolution of the high-resolution XPS spectra of the O 1s (left) and C 1s (right) spectra measured for the baseline SHJ cells front surface on the ITO (a and b), and ITO/Ag (c and d).

points on ITO vs. ITO/Ag. The C 1s spectra show C-C/C-H, C-O, C=O moieties for both surfaces, likely due to C in the film and adventitious C. The measurement on the ITO/Ag also has a lower bonding energy peak, attributed to the presence of carbide bonding.

Fig. 8 shows the In 3d5/2, O 1s, Sn 3d5/2, and C 1s spectra of the ITO, at different steps of DH exposure. The elemental spectra shift and/or broaden as the samples are exposed to DH. The Sn 3d5/2 spectra are noisier than other elements, due to the lower concentration of Sn in the ITO. Both In 3d5/2 and O 1s gain signal at higher binding energies. Fitting In in ITO is non-trivial due to the presence of multiple overlapping species; however, the shift to a higher binding energy is consistent with degradation of the ITO [12].

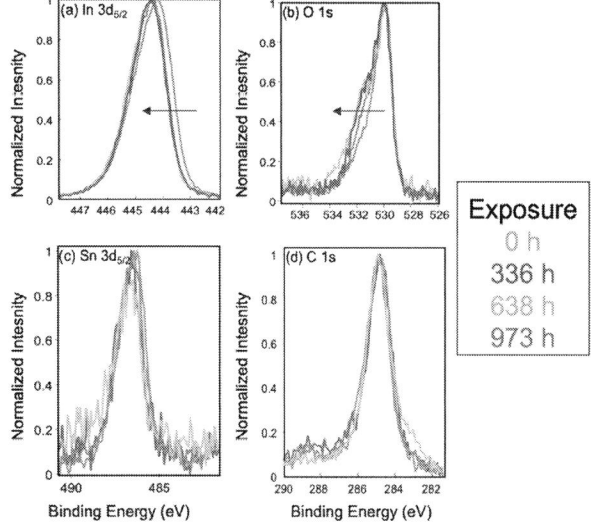

Fig. 8. XPS high-resolution spectra of the ITO surface of SHJ cells exposed to DH: (a) In 3d5/2, (b) O 1s, (c) Sn 3d5/2, and (d) C 1s.

Table 1 shows the change in component fits for the O 1s peak at 0 and 973 h of exposure. Fits of the three components of the O1s peak show an increase in relative % of the O 1s component Peak 3, and a decrease in O 1s component peak I, both of which indicate surface degradation of the ITO.

TABLE I. Deconvolution of high-resolution XPS spectra for select elements measured on the front contact of SHJ cells after 0 h and 973 h of DH exposure. The reported values are the average and standard deviation of three measurements on a sample.

DH exposure (h)	O 1s			Ag 3d5/2	
	%O_I	%O_{II}	%O_{III}	%Ag^0	%AgO_x
0	65.0 ± 1.2	21.3 ± 1.2	13.8 ± 0.1	100	0
970	54.3 ± 6.3	22.3 ± 1.0	23.5 ± 6.4	91.1 ± 2.6	8.9 ± 2.6

XPS measurements of the Ag gridline/ITO show similar changes to the In 3d5/2, O 1s, and Sn 3d5/2 with DH exposure. Further there are changes to the Ag 3d5/2 and C 1s spectra, with DH exposure, as shown in Fig. 9. The Ag 3d5/2 spectrum has a new peak at 367 eV with DH exposure, consistent with silver oxide formation. After 973 h of DH exposure, the silver oxide peak makes up 9 ± 3% of the Ag composition. Lastly, the carbide component of the C 1s envelope decreases with DH exposure, indicating a change in the Ag paste.

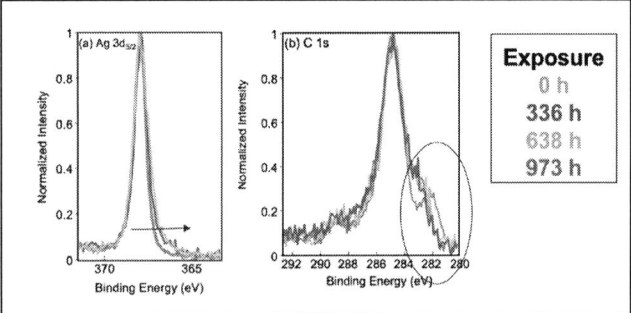

Fig. 9. XPS high-resolution spectra of the Ag gridline/ITO surface of SHJ cells exposed to DH: (a) Ag 3d5/2 and (b) C 1s.

IV. DISCUSSION

Modules installed in the field are exposed to a combination of stressors that result in loss of performance over time. These stressors can include, but are not limited to, moisture, heat, and acetic acid. In this experiment, the effects of different stressors are assessed via cell level DH and acetic acid exposures. Both PERC and SHJ devices are remarkably stable to DH, with only a slight I_{SC} and FF degradation in PERC devices after >1500 h DH exposure. In contrast, acetic acid exposure resulted in decreased V_{OC} values for SHJ cells, and degradation of the front contacts in both PERC and SHJs, but with greater degradation in the PERC. Acetic acid seems to be the main driving force behind recombination losses in these studies.

When comparing between the cell technologies, the SHJ cells show more degradation of the V_{OC}, likely due to acetic acid attacking the thin film layers (ITO, a-Si;H). In contrast, PERC cells appeared more stable, with minor recombination degradation. All the stressors affect some aspect of the contact properties; DH results in decreased I_{SC} in the PERC cells, in very long exposures (> 1500 h). TLM measurements show increased front contact resistance of the PERC and SHJ with acetic acid exposure, with a greater effect on the PERC cells.

For the PERC cells, the major degradation in front contact resistivity is due to acetic acid treatment. This could be due to the well-known acetic acid degradation of the interfacial glass layer/metal-Si interface of the front Ag contacts. Acetic acid strongly attacks the glass layer resulting in adhesion loss/delamination [10], [11], [12], [14].

Signatures of surface degradation of the ITO and the Ag gridlines are present in the XPS data and suggest the surface of the front contact is affected by DH exposure. XPS characterization is highly surface sensitive, sampling the top ~5 nm of the front contact under the conditions used here. Thus, although degradation signatures of the ITO and Ag are present after ~1000 h of DH exposure, the degradation is not sufficient to decrease device performance (i.e., there is no corresponding change in I_{SC}). There is, however, an increase in the ITO sheet resistance (Figure 5F), after 336 h of DH exposure, potentially signaling the beginning of a change in electrical performance.

Changes to the surface are not necessarily changes to the device performance, as the ITO remains sufficiently conductive to move charges to the gridlines, the ITO/Ag interface is likely not affected at this degree of DH exposure, and oxidation of the gridlines is minor enough to not affect the device measurements. Additional samples are necessary to identify trends in the performance of these cell types. Further work using time-of-flight secondary ion mass spectrometry (ToF SIMS) would be useful to distinguish between changes in the ITO surface, vs. changes at the c-Si/a-Si:H and a-Si:H/ITO interfaces.

SUMMARY AND FUTURE WORK

Current methods of determining reliability and durability center on accelerated aging of encapsulated modules, and retrieval of fielded modules. Multiscale characterization comprises module level measurements, and moves into microscopy of cored out regions. Interaction between cell optimization and module durability efforts is critical to prevent research silos, which can lead to failure prone cells being integrated into modules for extensive environmental testing. This leads to wasted time and resources and the perception of new technologies as unreliable, both of which create barriers to these new technologies that clearly offer efficiency advantages.

The introduction of new materials and processes introduce new failure modes. Ideally, new materials degradation pathways can be vetted during device optimization; parallel efforts are needed to rapidly establish degradation mechanisms. This work is the first step towards creating a structure that combines device and test structures, accelerated and field aging, and a comprehensive metrology plan to bridge device and module studies, with the aim of creating a reliable screening process needed for development of robust products. Ultimately, we will leverage fundamental materials science to understand, predict, and mitigate device failure.

ACKNOWLEDGMENT

Part of this work is supported by the U.S. Department of Energy's Office of Energy Efficiency and Renewable Energy (EERE) under the Solar Energy Technologies Office Agreement Number DE-EE0008155. Authors also acknowledge Austin Stoezter and Nikko Whatley from the University of Central Florida for helping with the acetic acid experiment.

REFERENCES

[1] G. Oreski, J. S. Stein, G. C. Eder, K. Berger, L. Bruckman, R. French, J. Vedde, and K. A. Weiß, "Motivation, benefits, and challenges for new photovoltaic material & module developments," *Prog. Energy*, vol. 4, no. 3, p. 032003, May 2022, doi: 10.1088/2516-1083/ac6f3f.

[2] T. J. Peshek, J. S. Fada, and I. T. Martin, "4 - Degradation processes in photovoltaic cells," in *Durability and Reliability of Polymers and Other Materials in Photovoltaic Modules*, H. E. Yang, R. H. French, and L. S. Bruckman, Eds. William Andrew Publishing, 2019, pp. 97–118. doi: 10.1016/B978-0-12-811545-9.00004-5.

[3] S. N. Venkat, J. Liu, J. Wegmueller, B. Yu, B. Gould, X. Li, J.-N. Jaubert, J. L. Braid, L. S. Bruckman, and R. H. French, "Degradation pathway modeling of pv minimodule variants with different packaging materials under indoor accelerated exposures," in *2021 IEEE 48th Photovoltaic Specialists Conference (PVSC)*, Jun. 2021, pp. 1725–1731. doi: 10.1109/PVSC43889.2021.9518926.

[4] "International technology roadmap for photovoltaic (itrpv) (12th edition)." 2021. [Online]. Available: https://itrpv.vdma.org/ viewer/-/v2article/render/73707020

[5] C. Sainsbury, Z. Hameiri, O. Kunz, R. Bhoopathy, I. Repins, C. Deline, A. G. Norman, M. Young, C. S. Jiang, C. Xiao, H. R. Moutinho, S. Johnston, D. B. Sulas-Kern, and D. C. Jordan, "High efficiency module degradation – from atoms to systems," *37th Eur. Photovolt. Sol. Energy Conf. Exhib.*, pp. 828–833, Oct. 2020, doi: 10.4229/EUPVSEC20202020-4BO.14.2.

[6] S. Bernardini and M. I. Bertoni, "Insights into the degradation of amorphous silicon passivation layer for heterojunction solar cells," *Phys. Status Solidi A*, vol. 216, no. 4, p. 1800705, 2019, doi: 10.1002/pssa.201800705.

[7] S. Guo, G. Gregory, A. M. Gabor, W. V. Schoenfeld, and K. O. Davis, "Detailed investigation of TLM contact resistance measurements on crystalline silicon solar cells," *Sol. Energy*, vol. 151, pp. 163–172, Jul. 2017, doi: 10.1016/j.solener.2017.05.015.

[8] G. Cumming and S. Finch, "Inference by eye: confidence intervals and how to read pictures of data," *Am. Psychol.*, vol. 60, no. 2, pp. 170–180, Mar. 2005, doi: 10.1037/0003-066X.60.2.170.

[9] R. J. Wieser, K. Rath, S. L. Moffitt, R. Zabalza, E. Boucher, S. Ayala, M. Brown, X. Gu, L. Ji, C. O'Brien, A. W. Hauser, G. S. O'Brien, R. H. French, M. D. Kempe, J. Tracy, K. R. Choudhury, W. J. Gambogi, L. S. Bruckman, and K. P. Boyce, "Spatio-temporal modeling of field surveyed backsheet degradation," in *2021 IEEE 48th Photovoltaic Specialists Conference (PVSC)*, Jun. 2021, pp. 1383–1388. doi: 10.1109/PVSC43889.2021.9519128.

[10] A. Sinha, J. Qian, K. Hurst, S. L. Moffitt, Laura. T. Schelhas, D. C. Miller, and P. Hacke, "UV-induced degradation of high-efficiency solar cells with different architectures," in *2020 47th IEEE Photovoltaic Specialists Conference (PVSC)*, Jun. 2020, pp. 1990–1991. doi: 10.1109/PVSC45281.2020.9300993.

[11] C. Peike, S. Hoffmann, P. Hülsmann, B. Thaidigsmann, K. A. Weiß, M. Koehl, and P. Bentz, "Origin of damp-heat induced cell degradation," *Sol. Energy Mater. Sol. Cells*, vol. 116, pp. 49–54, Sep. 2013, doi: 10.1016/j.solmat.2013.03.022.

[12] A. Kraft, L. Labusch, T. Ensslen, I. Dürr, J. Bartsch, M. Glatthaar, S. Glunz, and H. Reinecke, "Investigation of acetic acid corrosion impact on printed solar cell contacts," *IEEE J. Photovolt.*, vol. 5, no. 3, pp. 736–743, May 2015, doi: 10.1109/JPHOTOV.2015.2395146.

[13] E. A. Gaulding, J. S. Mangum, S. W. Johnston, C.-S. Jiang, H. Moutinho, M. J. Reed, J. A. Rand, R. Flottemesch, T. J. Silverman, and M. G. Deceglie, "Differences in printed contacts lead to susceptibility of silicon cells to series resistance degradation," *IEEE J. Photovolt.*, vol. 12, no. 3, pp. 690–695, May 2022, doi: 10.1109/JPHOTOV.2022.3150727.

[14] N. Iqbal, D. J. Colvin, E. J. Schneller, T. S. Sakthivel, R. Ristau, B. D. Huey, B. X. J. Yu, J.-N. Jaubert, A. J. Curran, M. Wang, S. Seal, R. H. French, and K. O. Davis, "Characterization of front contact degradation in monocrystalline and multicrystalline silicon photovoltaic modules following damp heat exposure," *Sol. Energy Mater. Sol. Cells*, vol. 235, p. 111468, Jan. 2022, doi: 10.1016/j.solmat.2021.111468.

[15] A. Rahmawati, K. A. Kuncoro, S. Ismadji, J.-C. Liu, A. Rahmawati, K. A. Kuncoro, S. Ismadji, and J.-C. Liu, "Subcritical water extraction of indium from indium tin oxide scrap using organic acid solutions," *Environ. Chem.*, vol. 17, no. 2, pp. 158–162, Oct. 2019, doi: 10.1071/EN19233.

[16] A. W. Hains, J. Liu, A. B. F. Martinson, M. D. Irwin, and T. J. Marks, "Anode interfacial tuning via electron-blocking/hole-transport layers and indium tin oxide surface treatment in bulk-heterojunction organic photovoltaic cells," *Adv. Funct. Mater.*, vol. 20, no. 4, pp. 595–606, 2010, doi: 10.1002/adfm.200901045.

[17] C. Donley, D. Dunphy, D. Paine, C. Carter, K. Nebesny, P. Lee, D. Alloway, and N. R. Armstrong, "Characterization of indium−tin oxide interfaces using x-ray photoelectron spectroscopy and redox processes of a chemisorbed probe molecule: effect of surface

pretreatment conditions," *Langmuir*, vol. 18, no. 2, pp. 450–457, Jan. 2002, doi: 10.1021/la011101t.

End of Use, Circularity, and Sustainability Considerations in Solar Photovoltaic Module Design and Product Development and Support

Chris Powicki,[1] Wayne Li,[2] and Cara Libby[2]

[1] Water Energy Ecology Information & Design Services, Inc., Brewster, Massachusetts, 02631 USA

[2] Electric Power Research Institute (EPRI), Palo Alto, California, 94304 USA

Abstract—Solar photovoltaic (PV) cells and modules have advanced for decades driven by end-user needs and societal goals, often environmental. This study explores module design trends and environmental considerations based on published materials and industry surveys and interviews. Augmenting the traditional emphasis on improving performance and cost, technology innovation has increasingly focused on realizing efficiencies in materials and manufacturing to gain competitive edge. With PV industry scale-up well under way and increasing recognition that the volume of defective, damaged, and spent modules will explode in the decades ahead, an additional design objective—addressing end of use, circularity, and sustainability—is emerging. Technology learning and market pull in these areas can help in reducing PV's levelized cost of energy and improving the value proposition offered by module manufacturers.

Keywords—photovoltaic modules, end of life, recycling, circularity

I. Introduction

Today's commercial solar photovoltaic (PV) modules represent complex and highly engineered assemblages of glass, semiconductor, and other materials designed to supply electricity on a reliable basis for periods of 25 years and longer. During normal operations, PV technologies offer the environmental advantages of requiring no resource inputs and producing no air pollution, greenhouse gas emissions, solid waste, or wastewater. In 2022, cumulative global PV deployment will exceed 1 terawatt (TW) of capacity, and PV will represent the least-cost option for generating on-site electricity and supplying bulk power to the grid in many world regions. Large-scale energy-economy modeling studies forecast that PV and wind will represent the leading U.S. and global sources of capacity additions out to 2050, depending on scenario. [1-2]

Between now and then, most PV projects deployed to date will reach end of commercial lifetime, and even currently fielded modules that find a second life will be retired. Just as annual global capacity additions expanded rapidly early in this century and especially in the last decade, so too will the volume of spent modules—but lagging by about 25 years as shown in

Figure 1. [3] Globally, billions of end-of-use modules will require disposition before 2050.

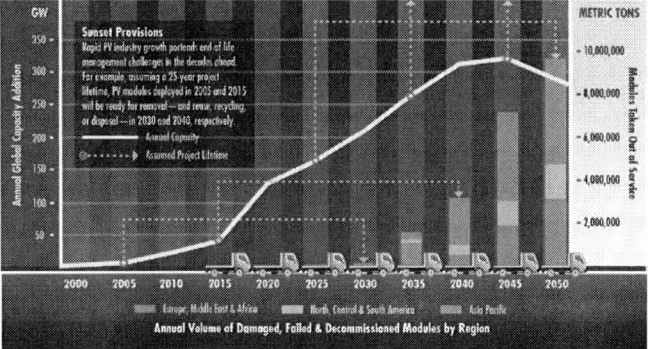

Fig. 1. Global PV deployment in gigawatts (GW) and spent module tonnage by region, 2000-50 [3]

PV industry, government, and other stakeholders recognize the societal imperative for energy solutions that do not create legacy problems. Recycling is global best practice for spent modules and required in many nation states, while a broader class of sustainability considerations is focusing attention on circular economy principles. However, modules are not designed with final disposition in mind, and collection and processing infrastructure is only beginning to emerge due to low and unpredictable volumes of spent and damaged modules. Recycling costs substantially exceed salvage values as a result of challenges involved both in deconstructing modules designed and engineered for long-term operation and in meeting quality or purity requirements for recovered materials. Current recycling technologies also require resource inputs and generate residual waste.

As shown in Figure 2, glass is the predominant constituent for both crystalline silicon (c-Si) and cadmium-telluride (CdTe) modules, with polymers, semiconductors, metals, and trace constituents making up the remainder. [3-4] In the United States under present federal regulations, end-of-use modules can be recycled or disposed of in conventional landfills, unless they fail required toxicity testing and must be managed as hazardous waste. Studies performed using the toxicity characteristic leaching procedure specified by the U.S. Environmental

Protection Agency identify lead as the primary concern within market-leading c-Si modules, but they also can contain metals and organic compounds with known or potential health risks. For thin-film CdTe modules, cadmium is the main concern, but other hazardous materials also may be present.

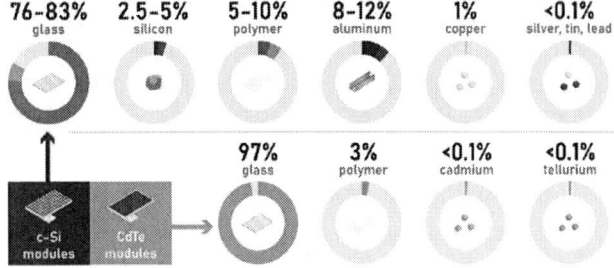

Fig. 2. Approximate composition of leading module technologies [4]

While cost and performance improvement and materials and manufacturing efficiency will remain key drivers for PV module technology innovation, a third dimension is emerging. This paper summarizes findings from an EPRI project exploring the extent to which end of use, circularity, and sustainability considerations are being addressed during module design and product development and support.

II. APPROACH

EPRI's study included a review of content and information resources available from major PV module manufacturers through their websites, as well as published literature. In addition, an online survey was created, and interviews were completed with manufacturers, other stakeholders, and utilities that own and operate PV fleets and are active both in managing small volumes of end-of-use modules and in planning for the future.

Web-based information focusing on module technologies and corporate messaging were reviewed for about 40 suppliers variously considered "Tier 1" according to bankability, product quality, or other metrics, including the following: AE Solar, Boviet, Canadian Solar, Chint/Astronergy, DMEGC, Eging, Enel/3Sun, ET Solar, Exiom, First Solar, GCL Systems, Haitai Solar, Hansol Technics, Hanwha Q Cells, Heliene, HT-SAAE, JA Solar, Jetion, Jinko, Jinergy, Jolywood, Leapton Energy, LONGi, Neo Solar Power, Renesola Yixing, Risen Energy, S-Energy, Seraphim/SEG, Sharp, Shinsung, Sumec/Phono Solar, SunPower/Maxeon, Swelect, Talesun, Trina Solar, Ulica Solar, Vikram Solar, VSUN Solar, Waaree, and ZNShine.

In addition, an online survey was sent to these suppliers in November 2021, seeking information on modules, technologies, products, and services with potential for reducing end-of-use risks and costs and for adding life-cycle value beyond capital cost and field performance. The survey included questions addressing whether and how individual suppliers are

- Meeting RoHS2 requirements for lead and reducing or eliminating use of other toxic metals, conflict minerals, fluoropolymers, and plastics

- Developing, testing, and applying novel and substitute module materials

- Integrating recycled content and recovered cell/module materials, including manufacturing waste

- Facilitating module repair/reuse, disassembly, and materials recovery and recycling

- Providing materials composition, toxicity testing, and waste characterization data

- Offering reusable or biodegradable packaging

- Achieving product-level and corporate standards, certifications, and endorsements; and

- Offering module take-back, reuse/donation, and recycling support services

Response to the online survey was limited due to myriad challenges associated with reaching appropriate technical experts within individual suppliers located around the world—and then getting them to share potentially proprietary information. Recent EPRI publications and interviews conducted with leading c-Si and CdTe manufacturers and large utilities provided additional important perspective on existing and emerging drivers and trends relating to PV module technology development and procurement. [3,5]

Across the industry and the literature, the annually updated International Technology Roadmap for Photovoltaics (ITRPV) represents the most comprehensive publicly available resource addressing worldwide developments in c-Si cell and module materials, manufacturing processes, and products, with proven success in engaging suppliers in assessing the state of the art and facilitating continuous technology improvement. The 12th ITRPV, published in 2021 based on responses to an extensive questionnaire provided by 56 leading industry suppliers, manufacturers, research institutes, and consultancies, quantified many recent and anticipated trends having environmental implications. [6] The 13th edition, published in 2022 reflecting input from 62 organizations, provides updated content, and supplemental ITRPV R&D survey results highlight growing industry interest in end-of-use considerations. [7-8]

III. RESULTS AND DISCUSSION

Since their invention, PV cells and then modules have been designed and engineered to achieve the levels of performance and cost needed to find real-world applications, create markets, and then to continuously improve—initially in space and remote locations where conventional energy supply options are impractical, and eventually in grid-connected rooftop and ground-mount installations that supplement or supplant incumbent power generation technologies. Over the past 15 years, industry scale-up and global competition have placed a premium on design innovations that deliver the cost reductions required to support continued growth in deployment of PV products offering better performance and thus improved economics over longer lifetimes.

A. Design Trends and Environmental Considerations

Generally, two cost-based design and development objectives have driven the PV manufacturing industry's evolution and maturation:

1. **Cost and Performance Improvement (CPI):** Optimizing cell and module materials, components, designs, and processes to increase solar-to-electricity conversion efficiency and long-term reliability.

2. **Materials and Manufacturing Efficiency (MME):** Optimizing cell and module designs and production processes to increase scale and throughput while reducing inputs of raw materials and other resources.

CPI innovations boost peak output per watt (W_P) of module capacity—as well as life-cycle productivity—without increasing manufacturing costs. MME innovations drive down these costs without adversely impacting performance, and they reduce the environmental impacts of module manufacturing. Since the U.S. government initiated PV "bulk buys" in the mid 1970s, progress in CPI has been the primary factor underlying the linear log-log relationship between the increasing annual global volume of module shipments in megawatts (MW_P) and the falling average sale price for modules ($\$/W_P$) at year's end. MME innovations have helped maintain this technology learning trend since the mid 2000s as global shipments have soared from a handful to more than 100 GW annually.

Meanwhile, life-cycle assessments and other studies have identified PV as environmentally advantageous, delivering improved metrics as the technology has evolved over time, and having further potential for progress. Module manufacturing and end of use—together encompassing materials sourcing, processing into components, and final disposition after removal from service—are the life-cycle stages with the greatest impacts, as assessed in carbon dioxide equivalents (CO_{2eq}) per kilowatt-hour (kWh) of electricity generation. Over the long term, CPI innovations have increased the denominator in this formulation by enabling higher-efficiency modules with lower rates of in-service performance degradation to produce greater amounts of non-emitting energy—often while directly displacing fossil-fuel-fired generation. Recent MME innovations have reduced the numerator by minimizing usage of raw materials and consumption of energy, water, and other resources throughout cell and module production.

Concerns relating to resource consumption and climate change continue to motivate many module end users, both directly and by leveraging varied government support mechanisms, but economic advantages alone position PV as a leading energy source for decades ahead: Unsubsidized PV is at or near grid parity due largely to low module prices and complementary but less substantial reductions in balance-of-system and soft costs. In quantifying the long-term relationship between annual module shipments and selling prices, ITRPV differentiates between *efficiency learning*, analogous to designing for CPI, and *per-piece learning*, analogous to designing for MME. The 2022 ITRPV notes that achievement of additional increases in W_P is likely to put upward pressure on manufacturing costs, that per-piece learning has abated, and that market factors, such as changes in polysilicon supply, can have significant impacts on module pricing. [7] While these developments suggest challenges ahead in extending the technology learning curve, modules account for a smaller share of up-front installation costs for PV systems than they did a decade ago.

Focusing on CPI then adding MME as a critical driver have not only mitigated the influence of module pricing on PV economics and industry evolution but also increased the impact of end-of-life management on levelized cost of energy (LCOE), a metric of considerable interest to large-scale solar power producers and purchasers. The global PV manufacturing industry now has multidimensional opportunity to adopt a third major design driver, as described below:

3. **End of Use (EOU):** Optimizing cell and module materials, designs, and production processes to mitigate environmental and human health risks, enhance recyclability and circularity, and promote sustainability.

B. Environmental, Circularity, and Sustainability Considerations

Figure 3 illustrates the mutually reinforcing interplay between CPI- and MME-driven design as a means of reducing module selling prices and achieving scaling, and it positions designing for EOU as an opportunity for optimizing life-cycle cost, performance, and resource efficiency within a circular economy framework.

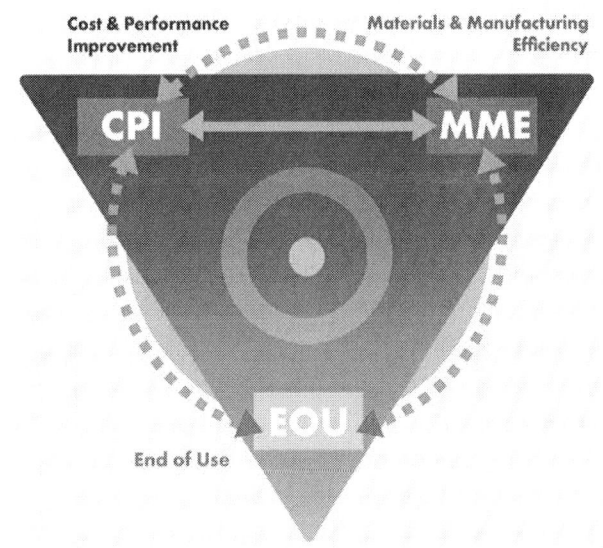

Fig. 3. Established and emerging PV module design trends

Historically, PV end users—from homeowners to utility-scale power producers—have been responsible for final disposition of end-of-use modules consistent with applicable waste management and disposal requirements, which vary by jurisdiction. The Waste Electrical and Electronic Equipment (WEEE) Directive adopted by the European Union (EU) in 2012 designated modules as e-waste not suitable for disposal in conventional landfills and introduced the concept of extended producer responsibility (EPR) to the PV industry, requiring manufacturers and other entities selling modules into the EU market to finance collection and recycling. Similar takeback policies have been implemented elsewhere, including by the U.S. state of Washington.

EPR compels module suppliers to internalize the risks of end-of-life management and the costs of recycling by qualified service providers and processors and thereby creates incentives to reduce them—but only when participating in applicable markets, which are limited beyond the EU. Elsewhere, risks and costs are externalized and borne by end users. Low-cost landfill disposal is allowable across the United States except in Washington and in California, where spent and damaged modules are designated for universal waste management and recycling. The lack of harmonized policy drivers represents a key barrier to global adoption of circular economy practices across the PV industry value chain—and especially to designing modules for EOU.

For CdTe technology, First Solar is in a unique position as the leading supplier *and* recycler of modules having a heavy metal element with well-characterized toxicity as a critical constituent for converting photons into electrons. To mitigate environmental risks and concerns, the company launched a voluntary global takeback program for all customers in 2005, initially by pre-funding collection and recycling as a component in module pricing. Since 2012, "pay-as-you-go" end-of-life service agreements have been available to customers in markets not subject to EPR requirements. These only cover recycling costs, not the reverse logistics involved in delivering damaged, underperforming, or spent modules to recycling facilities co-located with manufacturing plants.

By handling a consistent volume of manufacturing waste over the long term, First Solar has developed substantial experience with recycling process operation and optimization, which combined with innovations in module design and manufacturing has enabled the incorporation of recovered semiconductor materials in new modules. The company's expert recycling team provides input to its change management system for module design and manufacturing, but CPI remains the over-riding R&D priority, guided by an imperative to maintain and improve the high levels of stability and long-term reliability achieved in today's CdTe products.

Across all module technologies, the need to avoid cost increases and ensure field durability represents a second key barrier to designing for EOU: Reducing or eliminating the use of certain toxic substances may be impractical, and modules designed for recyclability—for example, to facilitate backsheet removal, or the separation of glass, encapsulant, semiconductor, and other layers without cross-contamination—may degrade long-term performance under field conditions.

Among suppliers considered in this study, First Solar is the most proactive in promoting module recycling and quantifying materials recovery rates via web-based content and resources. Major c-Si manufacturers typically provide little or no content on recycling. In general, module suppliers communicate from CPI and MME perspectives, touting cost-effectiveness and high levels of performance and reliability—and sometimes providing detail on design, manufacturing, and technology innovations offering differentiation in a competitive market. A number of manufacturers post product specification sheets online (rather than make them available on request), but these documents do not address module composition in detail, if at all.

Typically, the featured environmental benefits of modules involve harnessing solar energy to generate renewable electricity without emitting CO_2 or conventional air pollutants. Many suppliers highlight MME-driven process efficiencies and additional measures, such as operating production facilities using onsite or purchased renewable energy, to burnish the environmental credentials of their modules. Some go a step further, listing eco-labeling standards and certifications achieved for individual products or organizationally. At present, however, no universally accepted environmental quality criteria, standards, or certifications applicable specifically to PV modules exist—another significant barrier to designing for EOU.

Many module suppliers also promote their corporate environmental, sustainability, and governance (ESG) commitments and credentials, highlighting goals, policies, and progress in areas such as worker protection, conflict minerals avoidance, climate mitigation, resource efficiency, circularity, diversity, equity, and community building. Increasing focus on these ESG considerations is evidenced largely by corporate statements and dedicated reports and dashboards. Some module suppliers have gone further, preparing comprehensive life-cycle analyses and environmental disclosure statements for individual products and making them publicly available as a commitment to transparency. These documents can include detailed information on module composition and other EOU-relevant considerations.

The CdTe and c-Si module manufacturers interviewed during the course of this study identify corporate strategy and external factors as vital for establishing EOU as a third major design driver, complementing the historical focus on CPI and subsequent addition of MME. They indicate that forward-thinking suppliers and end users are helping push the industry in this direction, that market pull from a broader customer base is necessary to accelerate the process, and that harmonized EOU policies and broadly accepted standards are critical in creating a level playing field across geographic regions and module technologies.

Interviews with utility end users of PV technology explored the influence of life-cycle costs and risks, as well as ESG commitments, in making procurement decisions and managing assets across expanding project portfolios. One utility noted that the liabilities associated with managing end-of-use modules are taking on increased weight. Another described how circularity and sustainability objectives and criteria are embedded across the enterprise and communicated to suppliers of modules and other components—initially as an incentive and eventually as a requirement. Notably, ESG considerations are now given substantial influence during procurement, in the form of a credit that can favor compliant suppliers over lower-cost options. The future corporate, shareholder, and societal value derived is expected to exceed the additional investment.

C. Anticipated Trends and Next Steps

The 13th ITRPV features the usual c-Si technology trends and developments, many involving resource efficiencies and material substitutions and thus implicating the costs and risks associated with end of life. For example, manufacturers are reducing use of solar-grade silicon, silver, and aluminum, the three most expensive module materials. These trends will decrease the volume and value of recoverable materials at end of use and the risks of environmental exposure due to silver, a contaminant with known health risks. Broadly, trends and changes in wafer format, cell type, module size, bifaciality, glass thickness, coating, conductor, encapsulant, backsheet, frame, junction box, and other design elements can have net positive, net negative, or neutral impacts on end-of-life costs and risks and on the emergence of a circular PV economy.

2022's ITRPV results also include industry respondent data addressing a series of R&D questions, including several that directly implicate designing for EOU. Across diverse possible routes to low-cost or high-value PV, the strongest influence on competitiveness within the next 1 to 2 years is projected to be decreased material use in module frames and mounting systems and within 10 years new semiconductor materials, such as perovskites, for which end-of-life issues are not well understood. The third-leading factor 10 years out is anticipated to be new and recyclable encapsulants. Industry stakeholders selected reliable life extension, low-impact materials and MME innovations, and recyclability as important for promoting sustainability through 2024. Designing for EOU is identified as the clear strategic sustainability priority across the next decade.

EPRI is interested in developing a better understanding of the impacts of observed module technology and design trends on end-of-life management costs and risks, the potential of EOU-driven module designs and circular PV economies in meeting customer needs and creating new sources of value, and the opportunity for *environmental learning* to join efficiency learning and per-piece learning in leading the PV industry's evolution and continued expansion. A 2022 white paper will provide more complete discussion.

REFERENCES

[1] U.S. Energy Information Administration, *Annual Energy Outlook 2021*. Washington, DC, 2021.

[2] International Energy Administration, *World Energy Outlook 2021*. Paris, France, 2021.

[3] EPRI, *Decommissioning Plans for Large-Scale Solar Plants: Issues, Uncertainties, and Opportunities*. Palo Alto, CA: 3002019363, 2020.

[4] EPRI. Solar Photovoltaics End-of-Life Management Infographic. Palo Alto, CA: 3002021132, 2021.

[5] EPRI. Solar Power Fact Book, 12th Edition: Volume 1—Solar Photovoltaics (PV). Palo Alto, CA: 3002021427, 2022.

[6] M. Fischer, M. Woodhouse, S. Herritsch, and J. Trube, *International Technology Roadmap for Photovoltaics, 12th Edition*. Frankfurt, Germany: VDMA e.V., 2021.

[7] M. Fischer, M. Woodhouse, S. Herritsch, and J. Trube, *International Technology Roadmap for Photovoltaics, 2022 Results, 13th Edition*. Frankfurt, Germany: VDMA e.V., 2022.

[8] M. Fischer, M. Woodhouse, S. Herritsch, and J. Trube, *International Technology Roadmap for Photovoltaics, 2022 R&D Results*. Frankfurt, Germany: VDMA e.V., 2022.

Degradation of Crystalline Silicon Photovoltaic Modules Installed in Different Climates

Chiara Barretta[1], Astrid E. Macher[1], Julián Ascencio-Vásquez[2,3], Marc Köntges[4], Marko Topič[3], Gernot Oreski[1]

[1] Polymer Competence Center Leoben GmbH (PCCL), Leoben, 8700, Austria, [2] Envision Digital, Redwood City, CA, 94065, USA, [3] Laboratory of Photovoltaics and Optoelectronics, Faculty of Electrical Engineering, University of Ljubljana, Ljubljana, 1000, Slovenia, [4] Institut für Solarenergieforschung, Emmerthal, 31860, Germany

Abstract—**The study presents a degradation analysis carried out on photovoltaic modules with the same bill of materials exposed in different climates: moderate (Germany) and tropical (the Caribbean). The modules exposed in tropical climate experienced severe power degradation after about 7 years of exposure, mainly due to the occurrence of acetic acid-related degradation modes. Evidence of acetic acid presence could be seen in the electroluminescence images and ion chromatography. Reduction of molar mass could be detected by means of differential scanning calorimetry, but thermogravimetric analysis and infrared spectroscopy were proven to be not suitable methods to detect chain scission phenomena.**

Keywords—*acetic acid, climate, corrosion, degradation, encapsulant, ethylene vinyl acetate (EVA), crystalline silicon photovoltaic, power loss.*

I. INTRODUCTION

In recent years, more and more attention has been given to the importance of testing photovoltaic (PV) modules in different climates as well as to provide guidelines about climate specific PV module designs. Jordan et al. [1] first published a literature review regarding changes in power output of PV systems all over the world, trying to identify and correlate power losses to specific climate-related degradation modes. However, over the last decade many studies dealing with monitoring of PV systems' performances in different climate zones were published and more clear correlations between climatic conditions and PV degradation were found [2].

Omazic et al. [3] reviewed literature regarding field aged PV modules and identified the main degradation modes according to 5 different climate zones. Hot and humid conditions in tropical climates result as the most harmful for PV modules and encapsulant discoloration, extensive delamination at different interfaces and corrosion are the most relevant degradation modes. Production of acetic acid is favored by high humidity and high temperature conditions and all the acetic acid-related degradation modes occur with higher rates compared to other climates. Ascencio-Vásquez et al. [4] developed a worldwide map of degradation mechanisms and degradation rates (%/year) for crystalline silicon (c-Si) PV modules based on three mechanisms: ethylene vinyl acetate (EVA) hydrolysis, thermomechanical and photo-degradation, driven by temperature, temperature differences, relative humidity and ultraviolet (UV) irradiation [5]. In their study, the tropical climates are the most un-favorable for c-Si PV modules because of the high humidity and high temperature conditions, combined with irradiation.

However, most of the published studies include investigations of electrical performances [6-8] of field aged PV modules or characterization via non-destructive methods [9-10]. Only few studies report destructive investigations on PV materials to better understand the degradation mechanisms that took place over the years [11-12].

II. EXPERIMENTAL

Four PV modules with the same bill of materials were object of the study. One module, considered as reference, was kept in the dark while the other three modules were exposed outdoor and operated in two different locations for about 7 years. The analyzed PV modules are named as follows:

- M1, reference module, stored in the dark;

- M2, module exposed in moderate climate (Germany) for about 7 years;

- M3 and M4, modules exposed in tropical climate (the Caribbean) for about 7 years.

The PV modules were flashed to measure I-V curves and extract the parameters of interest. Electroluminescence images were taken to further investigate the modules and the associated degradation modes. For each module, encapsulant samples were withdrawn from five different positions to evaluate the influence of the microclimate on polymer behavior and to assess interactions between the encapsulant and the different PV module components (see Fig. 1):

- Back encapsulant, in contact with the cell and the backsheet, indirectly exposed to outdoor conditions (P1 and P2),

- Back encapsulant between cell and backsheet (P3), below the adhesive layer connecting the junction box,

- Back encapsulant between cell and backsheet (P4) in the area inside the junction box,

---

This project has received funding from the European Union's Horizon 2020 research and innovation programme under the Marie Sklodowska-Curie grant agreement No. 721452.

978-1-7281-6118-1/22 $31.00 © 2022 IEEE

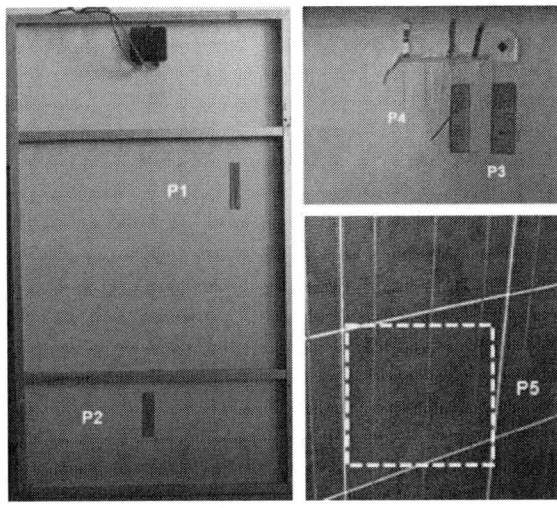

Fig. 1. Areas of withdrawal of encapsulant materials from the PV modules.

• Front encapsulant (P5), between glass and the solar cell approximatively in the center of the module.

The extracted materials were analyzed by means of: Fourier Transform Infrared spectroscopy in Attenuated Total Reflectance mode (FTIR ATR) to evaluate chemical changes, Differential Scanning Calorimetry (DSC) to evaluate changes in morphology, thermal behavior and molar mass, Thermogravimetric Analysis (TGA) to assess changes in thermal stability Thermal Desorption Gas Chromatography coupled to Mass Spectrometry (TD-GCMS) to qualitatively analyze consumption of stabilizers. Additionally, quantitative acetic acid measurements were performed on EVA samples taken from M1 and from modules exposed in tropical climate that showed power degradation between 31% and 53% by means of ion chromatography.

III. RESULTS AND DISCUSSION

The results of electrical measurements showed that M3 and M4 lost 45% and 10% of their power with respect to the nameplate value, respectively. The power loss was mainly associated to a drop in fill factor linked to corrosion of metallization and cell degradation. The electroluminescence images showed evidence of acetic acid corrosion on the cells, characterized by a typical pattern with dark areas on the cells, Fig. 2. The reference module M1 did not show power degradation with respect to the nameplate value and the electroluminescence images did not show significant anomalies.

The qualitative additive analysis results performed on the encapsulant showed that the reference module had a primary, a secondary antioxidant and an UV absorber. The primary antioxidant could not be detected in P5 in all modules. The modules exposed to tropical climate, additionally showed a partial consumption of the primary antioxidant in P3 and P4 (back encapsulant in the junction box area) and the UV absorber was no longer detectable in P5 (front encapsulant). The encapsulants extracted from M2, module exposed in moderate climate, did not show differences in the stabilizers distribution compared to the reference module.

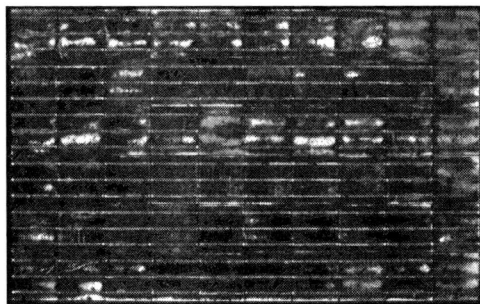

Fig. 2. Electroluminescence image of M3 at short circuit current (I_{SC}) showing typical acetic acid-related degradation pattern.

Quantitative acetic acid measurements were performed on encapsulant extracted from M1 and from modules that showed 31% and 53% power loss associated with drop in fill factor and the same EL features as M3 and M4. The concentration of acetic acid was equal to 82 ± 14 µg/g $_{EVA}$ for the reference module (M1), to 498 ± 87 µg/g $_{EVA}$ for the module with 31% power loss and to 827 ± 144 µg/g $_{EVA}$ for the module with 53% power loss. It is reasonable to assume that the values for M3 and M4 would have been included within the range mentioned above.

Results of the FTIR ATR spectroscopy measurements and TGA did not show evidence of deacetylation even though it was proved that deacetylation was the main mechanism taking place for the modules exposed in the tropical climate. The vibrations of the groups associated with vinyl acetate units in the FTIR ATR spectra for the modules exposed in the tropical climate did not show significant differences with respect to the reference module. At the same time, T5 and T40 values extracted from TGA measurements, namely the temperature at which 5% and 40% weight loss take place, did not allow to draw significant conclusions. The limit of detection for the two methods mentioned above might be too high and not appropriate to describe the changes happening to the materials.

The melting enthalpy of the second heating run in DSC experiments was considered as further indicator of changes in molar mass for the analyzed encapsulant samples, Fig. 3. The

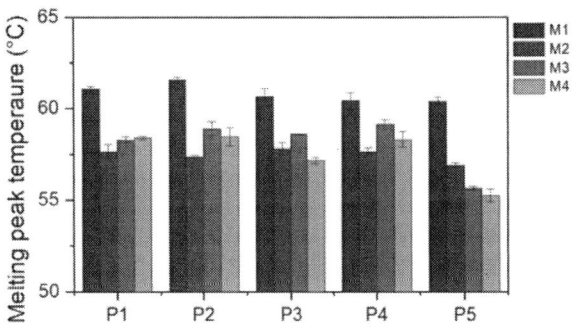

Fig. 3. Melting peak temperatures extracted from the second heating curves of DSC thermograms for the EVA withdrawn from M1 to M4 in the different positions, back encapsulant (P1 and P2), back encapsulant beneath the junction box (P3 and P4), front encapsulant (P5).

978-1-7281-6118-1/22 $31.00 © 2022 IEEE

material taken from the modules exposed in Germany and the Caribbean showed lower values compared to the reference, but did not show a clear trend. The only relevant difference could be observed for the materials extracted in P5 (front encapsulant) for the modules exposed in tropical climate. The samples extracted from the modules exposed in the Caribbean showed a melting peak temperature of about 55 °C, whereas the material taken from M2 showed a value of about 57 °C and the reference module a value of about 60.4 °C. This behavior might be explained from the higher temperature reached by the encapsulant above the solar cell due to the effect of the solar radiation during the exposure. The almost constantly higher module temperature experienced by the modules in the Caribbean associated with high humidity values and UV doses most likely caused a reduction of the EVA molecular mass, resulting in a decrease of the melting temperature.

IV. CONCLUSIONS

The modules exposed in tropical climate showed very strong power loss (about 45% for M3 and about 10% for M4) compared to the reference sample. No significant performance loss was reported from the PV system owner for the module operating in moderate climate (M2). The power loss was associated with a drop of fill factor and with a decrease of short circuit current. The decrease in fill factor was associated with corrosion of metallization and cell degradation. The acetic acid measurements showed that the modules that experience power losses between 31% and 53% were characterized by acetic acid concentration 6-10 times higher than the reference module (M1). Interestingly, even though significant acetic acid amount was detected via ion chromatography and clear evidences could be seen from electroluminescence images, no relevant signs of deacetylation could be determined via FTIR ATR spectroscopy or TGA for the materials extracted from the modules exposed in tropical climate (M3 and M4). The most significant difference that could be detected was related to the thermal properties when observing the melting peak of the second heating run of DSC measurements. The results showed a decrease of the temperature for all the field exposed modules with respect to the reference. In particular, the front encapsulant (P5) extracted from the modules exposed in tropical climate (M3 and M4) showed a further decrease of the melting peak temperature, and about 5 °C of difference could be detected with respect to the reference module. The depletion of the UV absorber in the front encapsulant might have been the cause for EVA degradation and acetic acid production in the modules exposed in the tropical climate due to the simultaneous effect of high temperatures, humidity and irradiation. Based on this evidence, it seems that DSC would be the most sensitive method, compared to FTIR ATR spectroscopy and TGA, to detect changes in molar mass for EVA encapsulant. Ion chromatography remains the main characterization method able to provide quantitative values of acetic acid concentrations. Additive analysis, though qualitative, was very useful to understand the possible root causes behind encapsulant degradation.

REFERENCES

[1] D. C. Jordan and S. R. Kurtz, "Photovoltaic Degradation Rates-an Analytical Review," *Prog. Photovolt: Res. Appl.*, vol. 21, no. 1, pp. 12–29, 2013, doi: 10.1002/pip.1182

[2] M. Halwachs *et al.*, "Statistical evaluation of PV system performance and failure data among different climate zones," *Renewable Energy*, vol. 139, pp. 1040–1060, 2019, doi: 10.1016/j.renene.2019.02.135.

[3] A. Omazic *et al.*, "Relation between degradation of polymeric components in crystalline silicon PV module and climatic conditions: A literature review," *Solar Energy Materials and Solar Cells*, vol. 192, pp. 123–133, 2019, doi: 10.1016/j.solmat.2018.12.027.

[4] J. Ascencio-Vásquez, I. Kaaya, K. Brecl, K.-A. Weiss, and M. Topič, "Global Climate Data Processing and Mapping of Degradation Mechanisms and Degradation Rates of PV Modules," *Energies*, vol. 12, no. 24, p. 4749, 2019, doi: 10.3390/en12244749.

[5] I. Kaaya, M. Koehl, A. P. Mehilli, S. de Cardona Mariano, and K. A. Weiss, "Modeling Outdoor Service Lifetime Prediction of PV Modules: Effects of Combined Climatic Stressors on PV Module Power Degradation," *IEEE J. Photovoltaics*, vol. 9, no. 4, pp. 1105–1112, 2019, doi: 10.1109/JPHOTOV.2019.2916197.

[6] A. Pozza and T. Sample, "Crystalline silicon PV module degradation after 20 years of field exposure studied by electrical tests, electroluminescence, and LBIC," *Progress in Photovoltaics: Research and Application*, pp. 368–378, 2016, doi: 10.1002/pip.2717.

[7] K.Yedidi, S.Tatapudi, J.Mallineni, B.Knisely, K.Kutiche, G.TamizhMani, "Failure and Degradation Modes and Rates of PV Modules in a Hot-Dry Climate: Results after 16 years of field exposure," in *Photovoltaic Specialist Conference (PVSC), 2014 IEEE 40th*, pp. 3245–3247.

[8] S. W. Adler, M. S. Wiig, A. Skomedal, H. Haug, and E. S. Marstein, "Degradation Analysis of Utility-Scale PV Plants in Different Climate Zones," *IEEE J. Photovoltaics*, vol. 11, no. 2, pp. 513–518, 2021, doi: 10.1109/JPHOTOV.2020.3043120.

[9] E. Annigoni, A. Virtuani, M. Caccivio, G. Friesen, D. Chianese, and C. Ballif, "35 years of photovoltaics: Analysis of the TISO-10-kW solar plant, lessons learnt in safety and performance—Part 2," vol. 27, no. 9, pp. 760–778, 2019, doi: 10.1002/PIP.3146.

[10] N. Bansal, P. Pany, and G. Singh, "Visual degradation and performance evaluation of utility scale solar photovoltaic power plant in hot and dry climate in western India," *Case Studies in Thermal Engineering*, vol. 26, p. 101010, 2021, doi: 10.1016/j.csite.2021.101010.

[11] M. C. C. d. Oliveira *et al.*, "Comparison and analysis of performance and degradation differences of crystalline-Si photovoltaic modules after 15-years of field operation," *Solar Energy*, vol. 191, pp. 235–250, 2019, doi: 10.1016/j.solener.2019.08.051.

[12] K. Hara and Y. Chiba, "Spectroscopic investigation of long-term outdoor-exposed crystalline silicon photovoltaic modules," *Journal of Photochemistry and Photobiology A: Chemistry*, vol. 404, p. 112891, 2021, doi: 10.1016/j.jphotochem.2020.112891.

Comparing Fluorinated and Non-Fluorinated Anti-Soiling Coatings for Solar Panel Cover Glass

Luke O. Jones, Adam M. Law, Gary Critchlow, John M. Walls

Centre for Renewable Energy Systems Technology (CREST), Loughborough, United Kingdom

Loughborough University, Loughborough, United Kingdom

Dust, dirt, debris, and biological matter collect on the surface of solar panel cover glass and attenuate the light entering the solar cell, reducing electricity generation and power output. Hydrophobic coatings are a passive anti-soiling method that utilizes low surface energy materials to force liquid droplets to cohere together which create a 'elf-cleaning' effect. As many hydrophobic coatings are fluorinated, this study evaluates the effectiveness of fluorine-free coatings as an alternative due to the environmental and biological risks posed by fluorine containing materials. A fluoroalkylsilane-based coating and a polydimethylsiloxane-based coating was deposited on soda-lime glass slide and exposed to 400 hours of UV and damp heat accelerated ageing tests. The two coatings were shown to be resistant to the accelerated ageing, with the fluorine-free coating marginally outperforming the fluorinated coating. For both coatings, little reduction in optical transmittance was observed, and each coating retained their hydrophobic properties in water contact angle tests. Surface chemical characterisation using X-ray photoelectron spectroscopy showed that the fluorinated coating was in the starting phase of degradation with carbonyl and methyl groups replacing trifluoromethyl groups, reducing fluorine surface content, and reducing the effectiveness as an anti-soiling coating.

AUTHOR INDEX

Aarnio-Winterhof, Minna 485
Abad, Eduardo Camarillo 795
Abbas, A. 705
Abbas, Ali 63, 414, 786, 838, 900
Abbas, Muhammad A. 208
Abbott, Malcolm D. 1033
Abdallah, Amir A. 52, 1151
Abdullah-Vetter, Zubair 472, 476
Abe, Adedoyin 1279
Abido, Mahmoud Y. 212
Ablinger, Ron 725
Abouelatta, M. 336
Abouelatta, Mohamed 1230
Abraham, Sherin Ann 967
Abudulimu, Abasi 414, 792, 972, 1088, 1190
Abzieher, Tobias 351, 565
Ackermann, Mathieu 413, 629
Adawi, Mohamed 1279
Adua, Habeebullah 354
Aeberhard, Urs 479, 1339
Afshari, Hadi 836
Agarwal, Anusha 614
Agarwal, Sumit 996
Agrawal, Rakesh 1333
Agresti, Antonio 576
Aguirre, Aranzazu 217
Ahangharnejhad, Ramez Hosseinian 701
Ahmad, Rukhsar 532
Ahmed, M. Sojib 1182
Ahmed, N. 811
Aiken, Dan 1332
Aimé, Jérémie 169
Aimez, Vincent 254
Aïssa, Brahim 52, 98, 1151
Akhtar, M. Shaheer 1292
Akhtar, Naureen 1237
Akiyama, Hidefumi 468
Akopian, Arkadi 244
Al Hasan, Naila M. 961
Al Katrib, Mirella 1136
Al-Jassim, M. M. 872
Al-Jassim, Mowafak M. 819
Al-Jassim, Mowafak 75, 1321
Al-Modaf, Fhad 1248
Al-Shidhani, Mazin 395, 396
Alam, Muhammad A. 843, 1182
Albadwawi, Omar 390, 452, 457
Alberts, Vivian 390, 452, 457

Albin, David 722, 819
Albrecht, Steve 529
Alfadhili, Fadhil K. 348, 667
Alhamadani, Hebatalla 390
Alhammadi, Aisha 12
Ali, Adnan 98
Ali, Md. Mahbub 1139
Alkhayat, Rabee B. 1252
Allami, Hassan 575
Almache, Estefania 1363
Almeida, Carlos Frederico Meschini 839
Almenabawy, Sara M. 74
Almenabawy, Sara 854
Almutawah, Zahrah S. 667
Alnajideen, Mohammad 395, 396
Alnaqbi, Wafa 12
Alom, Md Zahangir 976, 988, 1101
Alshehhi, Badreyya 452
Alvarez, Genesis 693
Alvi, Md. Shifain Mahathir 1139
Alvidrez, Javier Hernandez 190, 578
Aly, Shahzada Pamir 457
Amer, Fathy Z. 1230
Amin, H. M. Noman 1235
Anadkat, Nisheka 1133
Ananthanarayanan, Divya 80
Anctil, Annick 144, 1028, 1060, 1313
Anderberg, Allan 766
Anderson, Kevin S. 714
Anderson, Kevin 733
Anderson, Nick 1346
Andreas, Afshin 146
Aneja, Saurabh 733, 1033
Anitat, Remi 740
Anjum, Sara 1083
Ankireddy, Krishnamraju 937
Antolin, Elisa 538, 1100, 1340
Antón, Ignacio 413, 629, 739
Anwar, Bhuiyan M. 667
Aponte-Bezares, Erick E. 1091
Aponte-Bezares, Erick 398, 916
Araki, Kenji 58
Arbaretaz, Sébastien 562
Arcebal, John Derek 251
Arehart, Aaron R. 1315
Arès, Richard 430, 550
Arias-Zapata, Javier 530, 1291
Armour, Eric 164

Arnold, Rachael 961
Arredondo-Orozco, Carlos A. 778
Artuk, Kerem 1119
Arvinte, Roxana 430, 439
Ascencio-Vásquez, Julián 680
Askins, Steve A 629
Askins, Steve 413, 739
Assmann, Nicole 525
Atcitty, Stanley 1201
Athanasopoulos, Stavros 1305
Atwater, Harry A. 247, 820, 1083, 1087
Augustine, Sijo 1201
Augusto, André 1045
Ault, David J. 419
Avasthi, Sushobhan 1133, 1266
Awni, Rasha A. 464, 761, 1170
Awni, Rasha 126
Ayala, Silvana 255
Ayari, Ahmed 439, 627
Ayon, Arturo A. 1079
Ayon, Arturo 1069
Azzolini, Joseph A. 62, 183, 204, 431
Ba, Fatimata 1262
Babcock, Sean J. 460
Babin, Markus 544
Bae, Soohyun 481, 527
Baerwaldt, Daniel 121
Bafti, Arijeta 1196
Bagshaw, Heath 1299
Bai, Jing 345
Bailey, Jeff 425
Bakker, Klaas 381, 740, 748
Ballif, Christophe 604, 629, 1119, 1120, 1132
Baloch, Ahmer A. B. 452
Bannister, Mike 480
Bansal, Shubhra 1099
Baranek, Philippe 625
Baribeau, Laurier S. 1031
Barnes, Teresa 299
Barreau, Nicolas 381, 415, 941
Barretta, Chiara 485, 633, 680
Barrioz, Vincent 73, 516, 577, 1311
Barron-Gafford, Greg 1189
Barth, Kurt L. 658
Barth, Kurt 63, 387
Barthel, Armin 528
Bartholomäus, Martin 608
Bashardoust, Sattar 18
Bastola, Ebin 348, 414, 701, 761, 792, 828, 884,
.... 972, 1088, 1169
Bauhuis, Gerard J. 1280
Bauhuis, Gerard 875

Baumgarten, Katrina 1320
Bauser, Haley C. 820
Baxter, Jason B. 43
Bayrakci-Boz, Mesude 847, 948
Beal, Richard 1346
Beattie, Meghan N. 107, 1295
Beattie, Neil S. 516, 1311
Beattie, Neil 73
Bedilion, Robin 111
Belfore, Benjamin 1204, 1209, 1214, 1219, 1224
Bellani, Sebastiano 576
Belledin, Udo 18
Bendfeld, Jörg 539
Bengasi, Giuseppe 576
Berlinguette, Curtis P. 1072
Berry, Joseph J. 1043
Berry, Joseph 806
Berson, Solenn 1072
Bertagnolli, Justin 814
Bertomeu, Joan 1363
Bertoni, Mariana I. 8, 199, 327, 914, 1004, 1019, 1314
Bertoni, Mariana 425, 429, 708, 924
Bessa, João Gabriel 1294
Betak, Juraj 945
Bhat, A. 876
Bhat, Akanksha 314
Bhattacharya, Swastik 1173
Bicer, B. 879
Bista, Sandip S. 108, 464, 953, 1170
Bista, Sandip Singh 154
Bista, Sandip 126
Biswas, Dhrubes 991
Bittner, Zac 1332
Bivour, Martin 109
Bizzarri, F. 554
Bizzarri, Fabrizio 214, 567, 576, 1360
Blaauw, D. 811
Blakely, Logan 204
Bläsi, Benedikt 38
Bliss, Martin 838, 900
Boccard, Mathieu 1119
Bogachuk, Dmitry 114
Bogner, Brandon M. 1167, 1314
Bogner, Brandon 991
Bohémier, Cédric 883
Bojorquez, Jose Raul Montes 1069
Bolen, Michael 111, 116, 307, 802
Bonaccorso, Francesco 576
Bonilla, Ruy Sebastian 443
Booker, Edward P. 1072
Bordovalos, Alex 653, 1032
Borland, John O. 127

Borojevic, Nino...443
Bosco, Nick 106, 298, 783
Bosman, Johan...381
Bothwell, Alexandra M....................440, 1315
Boucher, Evan 255
Boucherif, Abderraouf....... 430, 439, 530, 550, 627,
....770, 1291
Bowden, Stuart G................... 1045, 1178, 1189
Bowden, Stuart 904
Bowers, Jake W. 658, 857
Bowers, Jake 63, 900
Bowersox, David A. 964
Bowman, Mitch 1033
Boyce, Kenneth P. 255
Boyer, Jacob T 867
Brabec, Christoph 31
Bracamonte, Maria Fernanda Villa 1069
Braid, Jennifer L. 805, 980, 1020, 1073
Braid, Jennifer 333
Brantl, Matthew 915
Brau, Tyler R. 828
Brau, Tyler 1055
Braun, Anna K. 198, 975, 1265
Brendel, Rolf 529
Brewster, Charles..741
Brlekovic, Filip 1196
Bromley-Dulfano, Isaac 907
Brown, Lance...1033
Brown, Matthew 255, 992
Brown, Scott 1310
Browne, Jack 1159
Brownell, Brenton.......................................980
Bruckman, Laura S. 255, 668, 796, 980, 1020
Bruhat, Elise 497
Bründlinger, Roland 725
Bu, Xiaotong...384
Buencuerpo, Jeronimo 625
Bueno-Blanco, Carlos.................538, 1100, 1340
Bukhari, F. 705
Bukhari, Farwa 604, 786
Buratti, Yoann 353, 476
Burgos, Rolando 693
Burnham, Laurie................205, 333, 1343
Bush, Meghan E. 208, 232
Buster, Grant...480
Cabal, Raphaël 607
Cabarrocas, Pere Roca I 109
Cacciato, Mario...214
Cai, Zhonghou 526
Cain, Joe 1322
Cakan, Deniz N. 526, 1070, 1071
Cal, Raul Bayoan 1319

Calaf, Marc 1319
Calvo-Barrio, Lorenzo 284
Camilo, Henrique Fernandes........................839
Campagna, Rory M 283
Campbell, Stephen 73, 516, 577, 1311
Campos, Christopher 751
Canino, A. 554
Canino, Andrea 214, 1360
Caño-Prades, Ivan 1304
Cao, Liqi 484
Cappelle, Jan 1162
Carbonaro, Agata 1360
Carr, Anna...1306
Carron, Romain 1145
Case, Chris 328
Caselles, Jaime 739
Cassini, Dênio Alves 534
Castillo, Jasmine Martinez 904
Cattoni, Andrea 625, 722
Cebecauer, Tomas 945
Celik, Ilke 1284
Cha, Seung I. 209
Chacon, Sergio A 836
Chan, Maria K. Y. 425, 429
Chan, Maria...924
Chapotot, Alexandre 530, 1291
Chard, Julie 684
Chatratin, Intuon 808
Chaudhary, Jatin...1173
Chavez, Andre...783
Cheetham, Kieran J 577
Chen, Cong 626, 653
Chen, Jie...975
Chen, Kejun 996
Chen, Lei...................126, 626, 653, 953, 1032
Chen, Shangshang 1338
Chen, Shaoqiang 468
Chen, Tao..1126
Chen, Xin 213, 1236
Chenenko, Jason 975
Cheng, Kai 7
Chestek, C. 811
Chia, Keaton 1310
Chin, Robert A Lee 368
Chin, Robert Lee 329, 472
Chin, Xin Yu 1119
Cho, Ben 1332
Cho, Eun-Chel...507
Cho, Jinyoun 235
Cho, Yasuo..461
Chockalingam, Nitin K. 668
Choi, Wook-Jin 251, 366, 1068

Choudhury, Kaushik Roy 980, 1065
Choudhury, Kausik R. 255
Chowdhury, Subhra 991
Chrétien, Jérémie 530
Chretien, Jeremie 627, 770
Christians, Jeffrey A 283, 605
Chung, Jaehoon 108, 898, 1055
Clenney, Jacob A. 327
Click, Natalie 248
Coathup, Trevor J 406
Coco, Fabrizio 1360
Colegrove, Eric 295
Coleman, Kehley A. 668
Coletti, Gianluca 1306
Coll, Pablo Guimera 1314
Colletti, C. 554
Collin, Stéphane 625, 722
Collins, Robert W. 405, 407, 833
Colombara, Diego 897
Colvin, Dylan J. 426, 1046
Compaan, Alvin D. 348
Connelli, Carmelo 567, 576
Conrad, Brianna 164, 837
Contractor, Ardeshir 1275
Cook, Tim 307
Corbett, Brian 1159
Cordon, Jazmine 904
Corrado, Casey 1110
Corso, Roberto 551, 567
Cortes, Bruno 517
Cortes, Jorge 1310
Costa, Suellen C. Silva 534
Costals, Eloi Ros 1363
Costello, Shannon E. 828
Couderc, Romain 169
Courtois, Guillaume 235
Covino, Maurice 1046
Cowles, Gabriel 1201
Crawford, Rose 836
Crawford, Zachary 1062
Critchlow, Gary 683
Cros, Stéphane 239, 497, 1072
Crowley, Kyle M 1109
Cruz, Florencia A 1294
Csank, Jeffrey T 754
Curcija, Charlie 155
Cutlip, Elizabeth V 605
D'Rozario, Julia 913
Daenen, Michaël 748
Dai, Xuezeng 1338
Daiber, Benjamin 328
Dalal, Vikram 244

Dancza, Viktor 7
Daniel, Valentin 430, 530, 770
Danielson, Adam 773, 819, 887
Darbali-Zamora, Rachid 398, 431, 754, 1091, 1335
Darnon, Maxime 254, 530, 770
Das, Sandip 424
Das, Ujjwal K 570, 592, 998, 999
Das, Ujjwal 76, 643
Dasgupta, Sagnik 366, 1168
Dauskardt, Reinhold H 106
Davis, Benjamin E. 1323
Davis, Kristopher O. 426, 668, 796, 1046
Davis, Melissa A 1293
De Angelis, Valerio 1201
De Callafon, Raymond 1310
De La Cruz, Rosa 1305
De Lafontaine, Mathieu 254, 770
De Luna, Gabby 251
De Rose, Angela 231
De Vito, Eric 497
De Vrijer, Thierry 1300
De, Shoubhik 859
Deans, Gordon 77, 80
Debije, Michael G. 89
Deceglie, M. G. 872
Deceglie, Michael G 298, 806, 855, 997
Deline, Chris 992, 1037
Deminico, Mathew R 145
Denafas, Julius 1245
Deng, Yehao 1338
Denis, Christine 607
Denz, Cornelia 69
Depauw, Valérie 235
Derkacs, Daniel 1332
Desai, Jal 899
Desai, Umang 512, 574
Desarden-Carrero, Edgardo 398, 916
Despeisse, Matthieu 629
Desrues, Thibaut 607
Dessein, Kristof 235
Dhanak, Vin 577
Dhanak, Vinod R 1312
Dharmadasa, Ruvini 937
Di Carlo, Aldo 576
Di Matteo, Alfredo 1306
Di Stefano, A 554
Di Stefano, Agnese 1360
Diallo, Thierno Mamoudou 430
Dias, Pablo R 1047, 1177
Dice, Paul 333, 1343
Diederich, Marvin 529
Diggs, Andrew 1061

Digregorio, Steven 1004
Dimroth, Frank 625
Ding, Jia 1007
Diniz, Antonia Sonia A. C. 534
Dione, Babou 1262
Diop, Pape 1262
Dittmann, Sebastian 1146
Dixon, Richard 413
Dobson, Kevin D 998, 999
Dolan, Connor 526, 1070, 1071
Dominguez, César 413, 629
Don, Christopher H 1312
Don, Eric 658
Dorman, Kyle R 732
Dow, Andrew R. R. 754
Dowd, Misha 1320
Driesse, Anton 172
Druffel, Thad 937
Drury, Storm 1177
Du, Wei 190
Duan, Xiaomeng 154
Dubois, Sébastien 607
Duchamp, Martial 73
Duchemin, Mathilde 629
Duenow, Joel 722
Dulal, Prabin 774
Duran, Ines 1340
Durant, Brandon K. 836
Durose, Ken 233, 577, 1312
Duttagupta, Shubham 251, 366, 1068
Dvorak, Adam 1105
Dwivedi, Priya 329, 353, 442, 472, 476
Dykes, Myla 904
Dyreson, Ana 333
Dziechciarz, Mikolaj 740
Ebner, Rita 217
Ebong, Abasifreke 937
Eckstein, Klaus 1190
Ecoffey, Serge 254
Eder, Gabriele 485
Edwards, Paul 480
Eggink, Wouter 731
Eidlisz, Jordan 588
Ekins-Daukes, Nicholas J 959
El Gemayel, Mirella 1190
El Hmaidi, Zakaria Oulad 530, 770
El-Atab, Nazek 1248
Elahi, Sheikh Tawsif 976, 1101
Elhmaidi, Zakaria Oulad 430
Ellingson, Randall J 154
Ellingson, Randy J. 348, 407, 414, 464, 667, 701,
.... 774, 792, 828, 953, 972, 1055, 1088, 1169, 1170, 1190

Ellingson, Randy 108
Elshamy, Mohamed 114
Elsworth, James 1322
Emshadi, Khalid 1332
Eperon, Giles E 836
Erickson, Joel 345
Erion-Lorico, Tristan 200
Escarra, Matthew D 208
Escobar, D. Martínez 1059
Esmaielpour, Hamidreza 732
Espenlaub, Andrew 1332
Estrada, Joseph 741
Exilhomme, Amina 1350
Fafard, Simon 254
Fafarman, Aaron T. 43
Fai, Calvin 769
Fairbrother, Andrew 1120
Falkenberg, Gerald 1145
Fan, Yangxin 796
Fanego, Vicente Lara 923
Fardin, Laura R. 545
Farha, Cynthia 59, 239
Farrell, John 884
Farshchi, Rouin 1315
Fassl, Paul 368
Faustini, Marco 625
Favela, Carlos A 115
Fedorenko, Anastasiia 991
Fei, Chengbin 1338
Feichtner, Markus 485
Feldmann, Frank 369
Fencl, Frank 1332
Fenning, David P 327, 526, 914, 1003, 1019, 1070, 1071
Ferekides, Chris 976, 988, 1101
Ferguson, Andrew J. 1283
Fernandes, Paulo A. 630
Fernández, Eduardo F 1294
Fernández, Johjan Stiven Zea 893
Ferreira, Jonathan G. 354
Ferry, David K. 732
Feurer, Thomas 533
Fevola, Giovanni 1145
Fiala, Peter 1119
Fievez, Mathilde 1072
Filho, E. A. Sarquis 223
Fina, Jeffrie 668
Fisher, Kathryn C. 1118
Flandin, Lionel 59, 239
Fleming, Robert A. 1279
Flicker, Jack D. 754
Flicker, Jack 183, 578
Flood, Andrew G 1264

Florea, Ileana	109
Florides, Michalis	131
Flottemesch, R.	872
Flottemesch, Robert	855, 997
Fonoll-Rubio, Robert	591
Forcade, Gavin P.	107, 1359
Ford, Bethan	1311
Ford, Ethan	1191
Ford, Jody	1037
Fortmann, Charles M.	1350
Foti, M.	554
Foti, Marina	567, 576, 1306, 1360
France, Ryan M.	1283
Frank-Rotsch, Christiane	247
Frasson, Nicola	1360
Fregosi, Daniel	111, 116, 307, 802
Freitas, Walmir	83, 517
French, Roger H.	255, 796, 980, 1020
Friedl, Jared D.	414, 701, 761, 828
Friedman, Daniel J.	766
Friedrich, Lorenz	566
Friesen, Hal	814
Frouin, Bérengère	722
Fthenakis, Vasilis M.	43
Fthenakis, Vasilis	447
Fuke, Pavan	859
Fusaro, Daniel	753
Gabor, Andrew M.	1046
Gabor, Andrew	426
Galiana, Beatriz	1305
Galiazzo, Marco	1360
Gallon, Josh	766
Galstyan, Eduard	115
Gambogi, William J.	255, 868
Gambogi, William	914
Gangemi, W.	554
Gao, Munan	620
Gao, Yuan	155, 1229
Garcia, Juan Lopez	52, 1151
García-Tabares, Elisa	1305
Gardner, Mathew	693
Garrevoet, Jan	1145
Gauding, E. A.	872
Gaulding, Ashley	299
Gaulding, E. Ashley	855, 997
Gay, Guillaume	254
Gayoso, Natalie	899
Gayot, Félix	497
Geerligs, Bart	748
Gehrke, Kai	228, 557
Geisz, John F.	975
Genç, Ezgi	1132

Georghiou, George E.	131, 156, 217, 267
Gerardi, C.	554
Gerardi, Cosimo	567, 576, 1306, 1360
Gevorgian, Vahan	419
Ghahremani, Amir Hossein	1055
Gharabeiki, Sevan	532
Ghimire, Kiran	752
Ghosh, Shibani	967
Gibbs, Jacob M.	828
Gibson, Elizabeth A.	516
Giraldo, Sergio	284, 1304
Glunz, Stefan W.	114
Gnocchi, Luca	1120
Godfrey, Tim	741
Goga, Adam	1062
Goldschmidt, Jan Christoph	566
Golobostanfard, Mohammadreza	1119
Golubev, Timofey	362
González, Carlos Ernesto Arrieta	893
Gonzalez, Pilar Espinet	1083
Good, Brian	295
Goodnick, Stephen M.	1010, 1178
Goodnick, Stephen	1061, 1159
Gorman, Will	1123
Gostein, Michael	285, 291
Gotseff, Peter	146
Gottschalg, Ralph	1146
Götz-Köhler, Maximilian	557
Goumenos, Panagiotis	131
Grätzel, Michael	114
Green, Martin	1126
Greenhalgh, Rachael C.	857
Greenhalgh, Rachael	900
Gregory, Christopher T.	460
Grijalva, Santiago	204
Grover, Sachit	1355
Gruenhagen, Philip	197
Gruginskie, Natasha	875
Gu, Jianli	967
Gu, Xiaohong	255, 1044
Guc, Maxim	591
Guerrero-Perez, Javier	406
Guesnay, Quentin	1119
Gueymard, Christian A.	72, 945
Guglielmino, Alfredo	1360
Guillevin, Nicolas	1306
Gunda, Thushara	899
Guo, Bing	1334
Gupta, Mool	1168
Gupta, Rajesh	1310
Gurian, Patrick	43
Gurule, Nicholas S.	190, 431, 578

Gurule, Nicholas 1335
Guthrey, Harvey 996, 1321
Gutierrez, Carlos M. 208
Habte, Aron 146, 480
Hacke, Peter 15, 183, 1065
Hackett, Sean 111
Hadipour, Afshin 217
Hadjipanayi, Maria 217
Hages, Charles J. 769
Haines, Thad 276
Haley, T. 876
Hallam, Brett 1047, 1177
Halloran, Liam J. 1341
Hamadani, Behrang H. 164, 837
Hameiri, Ziv 329, 352, 353, 367, 368, 442, 443, 472, 476, 1045
Hamers, Edward 606
Hammann, Benjamin 525
Hammann, Liv 95
Hamon, Gwenaëlle 530, 770
Han, Can 484
Han, Sang M. 783
Hanna, Francis 1313
Hansen, Clifford W. 22
Hansen, Clifford 1335
Hansen, Ole 461
Hanuš, Tadeáš 430, 439, 530, 627, 770, 1291
Hao, Xiaojing 352, 1126
Haque, Anisul 1182
Harding, Alexander J 998, 999
Hariskos, Dimitrios 370
Hart, John T 1332
Härtel, Marlene S. 529
Hartley, James Y. 199
Hartweg, Barry B. 1118
Harvey, Steven 643
Hassan, Shaikha 390
Hauch, Jens 31
Haug, Franz-Josef 1132
Hauschild, Dirk 76
Hauser, Adam W. 255
Hauser, Hubert 38
Hayes, Christopher 858
Haysom, Joan E. 883, 1366
Hazari, J. 876
He, Guojun 352
Heben, Michael J. 348, 407, 414, 667, 701, 761, 774, 792, 806, 828, 972, 1055, 1088, 1169, 1190
Heben, Michael 154, 464, 884, 1170
Hegedus, Steven S. 570
Hegedus, Steven 418, 441, 643
Heikkonen, Jukka 1173

Heintz, Alexandre 430, 550
Helfer, Eric 633
Helmers, Henning 38, 107, 1235
Henriques, Jonathan 550
Hense, Jan 1145
Hernández, M Johann A. 778, 893
Hernandez, Michael 904
Hertel, Tobias 1190
Herterich, Jan Philipp 114
Heske, Clemens 76
Hessler-Wyser, Aïcha 1119
Hieulle, Jeremy 532
Hildreth, Owen 1004
Hinsch, Andreas 114
Hinzer, Karin 8, 107, 143, 406, 883, 1031, 1295, 1346, 1359, 1366
Hirst, Louise C. 528, 795
Ho-Baillie, Anita 368
Höahn, Oliver 107
Hobaon, Theo DC 577
Hobbs, William B. 419, 714
Hobson, Theodore D C 1312
Hodge, Bri-Mathias 1038
Hoehn, Oliver 625
Hoek, Eelko 1306
Hoelscher, Torsten 591
Hoex, Bram 1126, 1191
Hofelmann, Ariana B. 862
Hoff, Thomas E. 197
Hoff, Tom 654
Hoffman, Adam 285
Hoffman, Anthony J. 208
Hoffmann, Stephan 231
Hoheisel, Raymond 624
Höhn, Oliver 38, 1235
Hoke, Andy 1000, 1038
Holland, Nicolas 223
Holman, Zach 773
Holman, Zachary C. 887, 1118
Holman, Zachary 1329
Holmgren, William F. 714, 995
Holt, Emily 1110
Holt, Martin 526
Honsberg, Christiana B. 143, 1059, 1178, 1189
Hour, Socheata 121
Howard, Kassidy H. 819
Hsu, Yu-Lin 1185
Hu, Xiaobo 468
Hua, Amandee 76
Huang, Jialiang 1126
Huang, Jinchao 1201
Huang, Jing 314, 654

Huang, Jinsong 806, 1338
Huang, Liangyi 796
Huang, Ying-Yuan 366
Hubbard, Seth M 624, 991, 1167, 1314
Hubbard, Seth 913
Huddy, Julia E 28
Huneycutt, Sandra 937
Huntamer, Ryo 915
Hunter, Robert F. H. 1031
Hunwick, Nicholas 387, 827
Huque, Aminul 725, 741, 1013
Hussin, Shahzada Qamar 507
Hutter, Oliver S 516
Hyndman, David W 144
Iandolo, Beniamino 461
Ilahi, Bouraoui 430, 439, 530, 550, 627, 770, 1291
Imaizumi, Mitsuru 1257
Imbrock, Jörg 69
Infante-Ortega, Luis 387
Ingrish, George B 208
Iqbal, Nafis 668
Irvin, Nicholas P. 143, 1059
Irvine, Stuart J. C. 838
Irvine, Stuart 63, 387
Irving, Richard 407
Isaacs, Mark 577
Isabella, Olindo 484, 623
Isherwood, Patrick J. M. 827
Ishikawa, Yasuaki 44
Islam, Kazi M. 208
Islam, Mohaimenul 1139
Ismael, Timothy 208
Izquierdo-Roca, Victor 591
Jackson, Nicole D 899
Jacobs, Daniel 1119
Jacobs, Deborah 1044
Jäger, Klaus 529
Jäger-Waldau, Arnulf 508
Jahandardoost, Mohsen 1099
Jahangir, Jabir Bin 843
Jahelka, Phillip R 247
Jain, Anubhav 213, 1236
Jakobsen, Michael L. 544
Jamarkattel, Manoj K 154, 348, 414, 667, 828,
.... 972, 1088, 1169, 1170
Jamarkattel, Manoj 884
Janotti, Anderson 808
Jansson, P. M. 330
Jaouad, Abdelatif 254
Jaramillo, M Adolfo A 778
Jaubert, Jean-Nicolas 980, 1020
Javed, Wasim 1334

Javier, Gaia Maria N 353
Jay, Frédéric 607
Jayswal, Niva K. 405
Jeangros, Quentin 1119
Jeffries, April 783
Jelinek, Alexander 1195
Jhang, Song-Syun 1044
Ji, Liang 255
Jia, Yun 468
Jiang, C.-S. 872
Jiang, Chenhui 1126
Jiang, Chun-Sheng 15, 819, 1321
Jiang, Nan 76
Jimenez, Alex 284
Jimenez, Maykel 1363
Jin, Shuangshuang 178
John, Jim Joseph 457
Johnson, Andrew 528
Johnson, Jay 183, 1335
Johnson, Samuel A 1043
Johnston, S. 872
Johnston, Steve W 855, 997
Johnston, Steve 200, 643, 649, 975, 1321
Jones, C. Birk 276, 1091
Jones, Derek 646
Jones, Kevin 693
Jones, L. 705
Jones, Leanne A H 1312
Jones, Luke O 683, 786
Jones, Luke 387, 604
Jones, Michael 73, 1311
Jones, Russell K 27
Jones, Stephen 387
Jones, Steve 63, 838
Jonsson, Jacob 155
Jordan, Michelle 904
Jost, Norman 413
Joyce, Hannah J 795
Ju, Zheng 1007, 1288
Junda, Maxwell M 752
Jundt, Pascal 47
Kaewnukultorn, Thunchanok 418, 441
Kahane, Ben 1033
Kahrl, Fredrich 1123
Kajari-Schröder, Sarah 529
Kakoulaki, Georgia 508
Kaluarachchi, Prabodika N. 414, 667, 828
Kamino, Brett 1119
Kaminski, Anne 607
Kanakkithodi, Arun K. M. 924
Kanaya, Shusaku 1257
Kanevce, Ana 1315

Kang, Yoonmook..481, 527
Kannakithodi, Arun Kumar Mannodi 1003
Kanth, Rajeev 1173
Kanyinda-Malu, Clement..1305
Karas, Joseph 1342
Karin, Todd..213, 1236
Kartopu, Giray 63, 838
Kasik, Camden L. 702
Kasik, Camden...264
Kau, Wylie..351
Kaufmann, Kai...231
Kazmerski, Lawrence L. 534
Keelin, Patrick 314, 654
Keith, Janine M. F. 1319
Keith, Janine M. Freeman..302
Kellar, Jon J. 1284
Kelzenberg, Michael D 1087
Kelzenberg, Michael...247
Kempe, Michael D. 255, 992
Kendall, Anthony D 144
Kenyon, Rick Wallace 1038
Kern, Dana...1321
Kessler, Rich...146
Kessler-Lewis, Emily..624
Khadka, Dhruba B. 1, 4
Khalifa, Sherif A. 43
Khan, M. Rezwan 1182
Khan, M. Ryyan 1182
Kharait, Rounak A. 714
Kharait, Rounak 858, 995
Kherani, Nazir P. 74, 854, 1264
Khokhar, Muhammad Quddamah..............................456, 507
Khurram, Adil 1310
Kiefer, Klaus...223
Kiener, Daniel 1195
Kiessling, Frank..247
Killam, Alex 904
Kim, Donghwan..481, 527
Kim, Eun-Gyeong..1292
Kim, James Hyungkwan 1123
Kim, Moonyong..1047, 1177
Kim, Tae-Gwan 1292
Kim, Youngkuk..456
King, Bruce H..............................285, 291, 333, 961, 1073
King, Bruce..1343
King, Richard R. 143, 460, 1059, 1178
King, Richard..1159
Kingma, Aldo 740, 1258
Kirmani, Ahmad R..........................320, 351, 1087
Kizilkaya, Orhan..208
Klawitter, Markus...531
Kleissl, Jan..1310

Klie, Robert F..884
Knapp, Steve 1073
Knodle, Phillip J...1046
Ko, Jongwon 481
Kodur, Moses....................................526, 1003, 1070, 1071
Köhnen, Eike...529
Kollosch, Bernd 223
Komoll, Felix 807
Komoto, Keiichi..1179
Köntges, Marc..680, 1245
Kopidakis, Nikos..............................766, 806, 814
Kornienko, Vlad..414
Kornienko, Vladislav 63, 838, 900
Korte, Lars 529
Kossen, Eric 1306
Kothari, Mallika...1060
Kottantharayil, Anil..859
Kottokkaran, Ranjith 244
Kovtun, Michael..1062
Krajne, Andraž 1196
Krasowski, Michael J 145
Krauter, Stefan...539, 650
Krich, Jacob J..............................575, 1295, 1359
Krishnan, Vidya 1329
Kroupa, Daniel M. 351
Krügener, Jan 369
Krzywicki, Alfred 476
Kuba, Austin G. 275, 998, 999
Kubiniec, A. 876
Kubiniec, Alex 314
Kuciauskas, Darius..440, 1315
Kujovic, Luksa 387
Kulicek, Jaroslav..1088
Kumar, Akash 1065
Kumar, Niranjana Mohan...924
Kumar, Niranjana..425, 429
Kumar, Rishi E..............526, 914, 1003, 1019, 1070, 1071
Kumar, Sandeep..1133
Kumar, Vibhor 620
Kumari, Madhuri 888
Kundu, Soumen 1266
Kurstjens, Rufi 235
Kurtz, Sarah R..212, 1059
Kurtz, Sarah 27, 121, 960
Kwapil, Wolfram 525
Kyranaki, Nikoleta 239
Lackner, David 38, 107, 625, 1235
Ladd, Anthony J. C. 769
Lahti, Gabriella D..320
Lahti, Gabriella 351
Lai, Barry 425, 429, 526, 924, 1070, 1071
Lam, Isaac K. 275

Lam, Issac	76
Lancia, Adrian A. Santamaria	544
Landis, Geoffrey A.	321
Lanterne, Adeline	607
Lao, Yao	1320
Lara-Fanego, Vicente	945
Laurikenas, Paulius	1245
Lave, Matthew S.	1091
Lavrova, Olga	1201
Law, A.	705
Law, Adam M.	683, 786
Law, Adam	604
Le, Anh Huy Tuan	443, 1045
Le, Tien T.	369
Leaver, Jacob F	233
Leccisi, Enrica	447
Lee, Changhyun	527
Lee, Dong Yoon	209
Lee, Hae-Seok	481, 527
Lee, Hoon Jeong	862
Lee, Hyunjong	1059
Lee, Hyunju	527
Lee, J.	811
Lee, Kyumin	753
Lee, Ross	721
Lehmann, Mario	1132
Leijtens, Tomas	1059
Lennon, Alison	1047
Leonardi, Marco	551, 567
Lepetit, Thomas	415, 941
Leshin, Jeremy	1332
Letner, J.	811
Levrat, Jacques	629
Lewis, Mandy R.	406
Lewis, Timothy	354
Li, Baojie	1236
Li, Bor.	529
Li, Chongwen	126, 626, 898
Li, Deng-Bing	108, 154, 464, 667, 701, 972, 1170
Li, Dinica	77
Li, F.	879
Li, Fang	426
Li, Jianjun	1126
Li, Liying	468
Li, Mengjie	796
Li, Wayne	675
Li, Xiaoping	838, 1355
Li, Xinjun	980, 1020
Li, Zelin Zack	255
Li-Kao, Zacharie Jehl	284, 591
Libby, C.	879
Libby, Cara	298, 675

Libraro, Sofia	1132
Lightfoote, Stephen	1163, 1347
Limodio, Gianluca	606
Lin, Der-Yuh	1100, 1340
Lin, Renxing	1229
Lin, Yida	862
Linse, Michael	531
Lisco, Fabiana	604
Litrico, Grazia	1360
Liu, Anyao	369
Liu, Grace	442
Liu, Jiqi	796, 980, 1020
Liu, Julie B.	1350
Liu, Simon H.	1320
Liu, Xiaolei	387, 827
Liu, Xing-Quan	584
Liu, Yuhang	114
Livera, Andreas	131, 267
Liyanage, Geethika K.	348
Loh, Joel Y. Y.	1264
Lombardo, Salvatore A.	551, 567
Loo, Roger	235
Lopes, Miguel	630
Lopez, Hector	1121
Lopez, Julià	1363
Lopez, Kevin	1320
Lopez-Lorente, Javier	156, 267
Loran, Erick Martinez	327
Lorenz, Adam.	447
Lorenz, Andreas	531
Lu, Dingyuan	838
Lu, Shengyuan	1359
Lüer, Larry	1190
Lumb, Matthew P.	164
Luna, Monica	1100
Luna-Delrisco, Mario	893
Lunis, Paul	826
Luo, Xianjia	468
Luo, Xin	1229
Luo, Yanqi	526, 1070, 1071
Lustig, Zachary F.	475
Luther, Joseph M.	320, 351, 1087
Luthy, Claire E.	208
Lyons, Alan M.	588
Lyra, Christiano	545
Ma, Yiwei	725, 741
Maasen, Jeroen	1280
Macalpine, Sara M.	964
Macdonald, Daniel	75, 369
Machado, Joana A. F.	532
Macher, Astrid E.	633, 680
Macher, Astrid	485

Machon, Denis 439
Mack, Charles 766, 814
Mack, Sebastian 18
Mack, Shawn 240
Mackenzie, Devin 806
Maclaurin, Galen 480
Macleod, Benjamin P. 1072
Madani, Keeya 1068
Madden, Ryan 828
Magginetti, David J. 1185
Magliano, Erica 576
Mahaffey, Mason 773, 1329
Mahmood, Farrukh Ibne 1198
Mahmud, Rasel 1000
Mainali, Madan K 752
Major, Jon D 577
Major, Jonathan D 233, 1299, 1312
Makhlouf, Seba 606
Makrides, George 156, 267
Mallajosyula, Arun Tej 504
Mallick, Rajni 838
Mallineni, Jaya 1073
Manceau, Matthieu 497, 1072
Mandic, Vilko 1122, 1195, 1196, 1197
Manganiello, Patrizio 623
Mangum, J. 872
Mangum, John S. 855, 975
Mann, Colin J. 1320
Mannino, Gaetano 214
Mannodi-Kanakkithodi, Arun 425, 429
Manshanden, Petra 1306
Mansoori, Ahmad 1332
Manzoor, Salman 708
Mao, Chengliang 1264
Mao, Dan 429, 924
Marchand, Jorge I. T. 1037
Mariam, Tamanna 792, 1055
Mariolle, Denis 497
Mariotti, Silvia 529
Marquis, Audrey 285, 291
Marsillac, Sylvain 407, 1204, 1209, 1214, 1219, 1224
Marti, Antonio 1100, 1340
Martí, David 739
Martin, Ina T. 668
Martineau, David 59, 239
Martinez, Juan F 629
Martinez, Mario 1340
Martinez, Sonia 1310
Martinez-Szewczyk, Michal W 1004
Martinez-Szewczyk, Michael 8
Masmitjà, Gerard 1363
Masters, Adam 1204, 1209, 1214, 1219, 1224

Mastroianni, Simone 114
Masuda, Taizo 58, 467
Matam, Manjunath 596, 692
Matas, Joaquin Coll 381
Mather, Barry 907
Matheron, Muriel 1072
Mathew, Nini Rose 792
Mathew, Xavier 414, 761, 792
Mathiak, Gerhard 390, 457
Matteocci, Fabio 576
Matthew, Xavier 972
Mavromatakis, Fotis 146
Mazzarella, Luana 484
Mbeunmi, Alex Brice Poungoué 430
McDermott, Daniel 235
McGehee, Michael D. 320, 1043
McHann, Stanley 204
McIntosh, Keith R. 1033
McMahon, William E. 198
McMillon-Brown, Lyndsey B. 320, 1109
McMinn, Allison 1288
McNatt, Jeremiah S. 145, 208
Meddeb, Hosni 228, 557
Meidanshahi, Reza Vatan 1010
Meier, Rico 199, 327, 914, 1019
Meier, Sebastian B 1190
Meira, Paulo 83
Mendis, Budhika G 1263
Merkel, Milena 69
Mesropian, Shoghig 584
Meßmer, Marius 18
Metzger, Wyatt K. 1355
Meyer, Abigail. R. 996
Meyers, Bennet E. 326, 1048
Meyers, Bennet 930
Meza, Carlos 1146
Micheli, Leonardo 110, 1294
Mikofski, Mark A. 714, 995
Mikofski, Mark 858
Milazzo, Gabriella 567
Miller, David C. 319, 961
Miller, David 15
Miller, Emily J 898
Miller, Michael F. 1315
Miller, Nate 1332
Miller, Wes 838
Mills, Andrew 1123, 1158
Min, Gao 395, 396
Minz, Manoranjan 504
Mira, Daniel Alvarez 608
Mirletz, Heather M. 299
Mirzokarimov, Mirzo 351

Mishima, Tetsuya D.	732
Mishra, Aditya	1310
Mitra, Anirban	98
Mitterhofer, Stefan	1044
Miyano, Kenjiro	1, 4
Moffitt, Stephanie L.	255
Moffitt, Stephanie	1044
Molesky, Sean	1359
Möller, Marius	539, 650
Monakhov, Eduard V.	525
Moncada, Sebastián Villegas	893
Montañés, Cristina Crespo	1123
Montbach, Erica N.	232
Montes-Bojorquez, Jose Raul	1079
Montes-Romero, Jesús	267
Moore, David T.	351
Morais, Modesto	630
Morales-Masis, Monica	109
Morisset, Audrey	1132
Morris, Kerrie M.	857
Morris, Kerrie	900
Morse, Joshua	15
Moseley, John	722
Mouchel, Laurie	439
Mouri, Tasnim K.	76, 592
Mousa, Mohamed	1230
Moutinho, Helio R.	819
Moutinho, Helio	200
Moutinho, R.	872
Mubarak, Roaa I.	1230
Mughal, Maqsood Ali	354
Mukherjee, Shagorika	808
Mulder, Peter	875
Müller, Björn	223
Muller, Matthew T.	628
Muller, Matthew	709
Mundt, Laura E.	320
Muñoz-Rojas, David	722
Muñoz-Santiuste, Juan Enrique	1305
Munshi, Amit	819, 1086
Muttillo, Mirco	623
Muzzillo, Christopher P.	351
Myers, S. M.	330
Najafi, Mehrdad	748
Nakado, Takashi	58, 467
Nakamura, Tetsuya	1257
Nakarmi, Upama	197
Namal, Imge	1190
Namjil, Enebish	1179
Nanda, Govind	854
Nardin, Gaël	629
Nardone, Marco	819

Nassif, Alexandre B.	1127
Natale, Stephen	354
Navarro, Alejandro	284
Navarro-Güell, Alejandro	1304
Nayfeh, Ammar	12
Naylor, Matthew C.	1311
Nayshevsky, Illya	588
Nazginov, Elizabeth	1350
Ndione, Paul	806
Needell, David R.	820
Nejand, Bahram Abdollahi	565
Nemeth, William	996
Ness, Jon	1322
Neto, João Cardoso Das Neves	839
Neugebohrn, Nils	228, 557
Neumaier, Lukas	485
Neupane, Sabin	108, 464, 1170
Newberry, M. G.	330
Newkirk, Jimmy	961
Newmiller, Jeff	995
Nguyen, Dong C.	44
Nguyen, Hieu	75
Nguyen, Phuong-Linh	625
Nicholas, Norm	1033
Nielsen, Michael P	959
Nieschwitz, Kristof J.	828
Nietzold, Tara	425, 924
Niewelt, Tim	525
Niquille, Xavier	629
Nishinaga, Jiro	533
Nishioka, Kensuke	58, 467
Nobre, André M.	614
Nocerino, John	1320
Nolde, Jill A.	240
Nolde, Kristian	1033
Nonni, Elisa	576
Norman, Andrew	1321
Norton, Matthew	217
Nouri, Neda	143, 1295
Nuzzo, Ralph G.	820
O'Brien, Colleen	255
O'Brien, Greg S.	255
O'Neill, Mark	137
Obikoya, Gbenga D.	570
Obikoya, Gbenga	643
Ochoa, Eduardo Prieto	443
Ok, Young-Woo	251, 366, 1068
Okel, Lars	1306
Oklobia, Ochai	63, 387, 838
Okumura, Kenichi	58
Okumura, Teppei	1257
Olhmann, Jens	1280

Oltjen, William C. 796
Onno, Arthur 773, 887, 1329
Oreski, Gernot 485, 633, 680
Orozco, Carlos A Arredondo 893
Ortega, Pablo 1363
Ortiz, Jonathan 1320
Ortiz-Rivera, Eduardo I. 916
Ossig, Christina 1145
Osterthun, Norbert 228, 557
Ostrowski, David P. 320
Ota, Yasuyuki 58
Othman, Mostafa 1119
Ottanà, A. 554
Ottoson, Larry 766
Ourinson, Daniel 531
Outen, Jonathan 283
Ovaitt, Silvana 299, 992
Owen-Bellini, Michael 298, 319, 649, 806
Packard, Corinne E. 198, 975, 1265
Padhamnath, Pradeep 251
Paetel, Stefan 533, 1315
Paetzold, Ulrich Wilhelm 565
Paetzold, Ulrich 368
Page, Matthew 996
Pal, Shweta 134
Palacios, Felipe 754
Palariya, Anuj Kumar 1133
Palekis, Vasilios 976, 988, 1101
Palmiotti, Elizabeth C. 199
Palmiotti, Elizabeth 941, 1204, 1219, 1224
Palmstrom, Axel F. 1043
Pancari, Mia 1350
Pandey, Ramesh 264
Paniagua, Donny Campos 1271
Panžic, Ivana 1122, 1195, 1196, 1197
Papadopoulos, Ioannis 267
Papaioannou, Estefania 721
Paphitis, George 267
Parada, Gabor 658
Paraskeva, Vasiliki 217
Pargon, Erwine 254
Parikh, Abhishek 858
Park, Alex 904
Parquette, William J. 62, 339
Pasmans, Peter 961
Passow, Kendra 753
Patel, Aesha P. 828, 953
Patel, Aesha 1088
Patel, Jay B. 320
Patel, Muhammed Tahir 843
Patel, Pravin 1332
Patjens, Svenja 1145

Pato, Pedro A. V. 83
Paudyal, Bijaya 1271
Paupy, Nicolas 430, 530, 627, 770
Pavgi, Ashwini 1065
Pavic, Luka 1196
Pawar, Vani 1133
Paynabar, Kamran 802
Pazniak, Hanna 576
Pearce, Phoebe M. 959
Peibst, Robby 529
Pelland, Sophie 72
Perea, Gracia Belén 1305
Pereira, Rui N. 98
Perez, Marc 197, 314, 654, 661, 876
Perez, Richard 314, 654, 661
Pérez-Rodríguez, Alejandro 591
Perez-Wurfl, Ivan 352
Perna, Allison N. 198, 1265
Perrin, Greg 15
Perrin, Lara 59, 239, 1136
Perry, Albert 1332
Perry, Kirsten R. 714
Perry, Kirsten 709, 751, 1037
Perry, Lakesha 1044
Perullo, Christopher 307
Pescetelli, Sara 576
Peshek, Timothy J. 145, 232, 320, 1109
Peters, Ian Marius 31, 1191
Peterson, Josh 146, 684
Petit-Etienne, Camille 254
Petri, Delphine 629
Pettengill, Alexandra 1320
Pham, Duy Phong 483, 507
Phelan, Megan E. 820
Phillips, Adam 154, 464, 1055, 1170
Phillips, Adam B. 348, 407, 414, 667, 701, 761,
.... 774, 792, 828, 972, 1088, 1169, 1190
Phillips, J. 811
Phillips, Jamie D. 9
Phillips, Laurie J 577, 1312
Phirke, Himanshu 532
Pierce, Benjamin G 805
Pietzcker, Robert 566
Pike, Christopher 205, 333
Pikolos, Loucas 267
Pilor, Modou 1262
Pilot, Nicholas 111
Pinney, David 204
Pita, Alondra 904
Placidi, Marcel 284, 591, 1304
Planès, Emilie 59, 239, 1136
Plotnikov, Victor V. 348

Podraza, Nikolas J. ...... 405, 407, 653, 752, 774, 833, ... 898, 972, 1032, 1169
Poessl, Sebastian 608
Pokhrel, Dipendra 108, 792, 1088, 1169
Polizzotti, Alex 1355
Polly, Stephen J. 624, 991, 1167, 1314
Polly, Steve 913
Polo, Christian R. 62, 339
Polzin, Jana-Isabelle 369
Pomares, Luis 457
Porret, Clément 235
Potter, Maggie M. 820
Poudel, Deewakar 1204, 1209, 1214, 1219, 1224
Poulsen, Peter B. 608, 871
Powell, Kaden M. 1185
Powicki, Chris 675
Prabaswara, Aditya 1159
Pradhan, Puja 407
Pranav, Sai 614
Prasanna, Rohit 1358
Prilliman, Matthew J. 302
Prilliman, Matthew 1319
Prinja, Rajiv 74, 854
Privitera, Stefania M. S. 567
Procel, Paul 484
Prokop, Norman F 145
Provost, Philippe-Olivier 1291
Ptak, Aaron J. 198, 867, 975, 1265
Ptak, Aaron 247
Puigdollers, Joaquim 284, 1304, 1363
Purkayastha, Atanu 504
Pusay, Benjamin 1363
Pusch, Andreas 959
Pyrlik, Niklas 1145
Qi, Xin 1007, 1288
Qian, Chen 1126
Qu, Yongtao 73, 1311
Quader, Abdul 154, 414, 667, 1169
Quaegebeur, Nicolas 627
Quispe, David 1329
Rabbani, Zulkifl H. 414, 701
Radovanovic-Peric, Floren 1122, 1195, 1197
Rafhay, Quentin 607
Rago, Emily 961
Ragonesi, Antonino 1306, 1360
Rahman, Farhan 199
Rahman, Mahfujur 1037
Rahman, Md. Mosaddequr 1139
Rahman, Nahian 1350
Rajakaruna, Manoj 953, 1055
Raka, Rawnak Reza 1139
Rakotoniaina, Jean Patrice 169

Ramanujam, Balaji 407
Rametta, F. 554
Rametta, Francesco 1306, 1360
Ramirez, Omar 533
Ranade, Sathishkumar 1201
Ranalli, Joseph 32, 847
Rand, James A. 855, 997
Rane, Karan 1275
Ransome, Steve 375
Rashwan, Ola 444
Rath, Kunal 1020
Rath, Thomas 1122
Rau, Uwe 807, 856
Raupp, Christopher 1073
Raza, Hamza Ahmad 261
Reagan, Jeremiah B. 960
Rebouças, Lara Barros 1280
Redinger, Alex 532
Reece, Peter J 959
Reed, Mason J. 855, 997
Reese, Matthew O. 295, 344
Reese, Samantha B 344
Rehder, Eric M. 584
Reich, Carey L. 887
Reich, Carey 773
Reinders, Angèle H. M. E. 89
Reinders, Angele 489, 731
Reise, Christian 223
Remund, Jan. 661
Ren, Wei 1013
Reno, Matthew J. 183, 190, 204, 431, 578
Repecaud, Pierre-Alexis 109
Repins, I. L. 872
Repins, Ingrid L. 855
Repins, Ingrid 806, 1342
Rezek, Bohuslav 1088
Rezk, Ayman 12
Rhee, Kurt 29
Riboldi, Victor B. 517
Ricciardi, Tiago R. 83, 517
Rice, Anthony D. 975
Richardson, Raphael 998
Riedel-Lyngskær, Nicholas 871
Rigby, Oliver M 1263
Rijal, Suman 108, 126, 405, 464, 792, 1170
Riley, Daniel 205, 333, 805, 1343
Ritzer, David Benedikt 565
Roberts, Billy 480
Robinson, Gerald 1322
Robinson, Justin 684, 1073
Robledo, Maryan 904
Rocha, Daniel 630

Rock, Nathan D 1086
Rockett, Angus 415, 941, 1204, 1209, 1214, 1219, 1224
Rodriguez, Alejandro W 1359
Rodriguez, David J. F. 326
Rodriguez, David Jose Florez 1048
Rodriguez-Garcia, Gonzalo 1284
Rodriguez-Peña, Micaela 1100
Rohatgi, Ajeet 76, 251, 366, 592, 783, 1068, 1168
Rojsatien, Srisuda 425, 429, 924
Rollins, Nathan J 275
Rolston, Nicholas J 1082
Roosen, Dorrit 1258
Rosenblatt, Nathan 1355
Rosenlieb, Evan 480
Rossol, Michael 480
Rounsaville, Brian 251, 783
Rout, Bibhudutta 836
Roy, Etee Kawna 1185
Ruffolo, Peter 283
Ruhstaller, Beat 479
Ruiz-Arias, Jose A. 945
Ruiz-Preciado, Marco Alejandro 565
Rummel, Brian 783
Russell, Annie C. J. 406, 1366
Russell, Annie 883
Ryczek, Catherine N. 820
Saavedra-Peña, Nelson E. 398
Sabin, Neupane 154
Sacchitella, Elijah J. 1314
Sacchitella, Elijah 624
Saddedine, Karima 239
Saeed, Ahmed 336, 1230
Sætre, Tor Oskar 1237
Sahli, Florent 1119
Saif, Omar M. 336
Saive, Rebecca 134, 374
Sala, Jacopo 748
Salah, Mostafa M. 1230
Salas, Eduardo 1305
Salome, Pedro M. P. 630
Samake, Papa Monzon 1262
Sampath, Walajabad S. 475, 887
Sampath, Walajabad 387, 773, 819
Sampson, Robert 1189
Sánchez, Hugo 1146
Santala, Annikki 1059
Santiwipharat, Chaiwarut 999
Santos, Michael B. 732
Sapkota, Dhurba R. 407, 1252
Saraswat, Govind 419
Sasala, Chase 444
Satou, Akinori 58

Saucedo, Edgardo 284, 1304, 1363
Sauter, Evan 354
Sayre, Larkin 528
Sazzad, Muhammad H 959
Scarpulla, Michael A 1086
Schaefer, Stephen 1007, 1288
Schaffer, Kevin G 828
Schall, Jackson W. 1321
Schambach, Lauren 1110
Schauerte, Meike 38, 1235
Scheer, Roland 591
Scheideler, William J 28
Scheiner, Aaron 22
Schelhas, Laura T 319, 320, 649, 806, 961
Schermer, John J. 1280
Schermer, John 875
Schindler, Florian 525
Schlemmer, James 314
Schmieder, Kenneth J 164, 240
Schneider, Kevin 190
Schriemer, Henry 1346
Schroer, Christian G 1145
Schropp, Andreas 1145
Schubert, Martin C. 525
Schulte, Kevin L. 198, 867, 1265
Schygulla, Patrick 38, 114
Sciuto, Marcello 1306, 1360
Scuto, Andrea 551, 567
Sedzro, Kwami Senam 967
Seel, Joachim 1158
Segbefia, Oscar Kwame 1237
Seibert, Samuel 953
Seiboth, Frank 1145
Seigneur, Hubert 596, 692, 796, 826
Sckkat, Abderrahime 722
Sellers, Ian R. 732, 836, 1178
Selvamanickam, Venkat 115
Semichaevsky, Andrey 92
Sengupta, Manajit 146, 480
Sepúlveda-Mora, Sergio 441
Seren, Sven 18
Seron, Charles 607
Serrano-Escalante, Valentina 533
Seum, Abu Niem 1139
Seymour, K. 876
Seyrich, Martin 1145
Shafarman, William N. 275, 592, 998, 999
Shafarman, William 1315
Shah, Akash 264, 475
Shah, Sanket 753
Shaker, Ahmed 336, 1230
Shalvey, Thomas P. 1299

Shan, Ambalanath 407
Sharikadze, Saba 244
Sharma, Bhuwanesh Kumar 574
Sharma, Bosky 1119
Sharma, Geetu 854, 1264
Sharma, Sahil 115
Sharp, Jon 753
Shaw, Daniel Z. 702
Shaw, S. 879
Sheil, Huw 577
Sheppard, Scott 307
Shi, Xiaojie 725
Shih, Nathan 1037
Shimpi, Tushar M. 475
Shimpi, Tushar 387, 658, 819
Shiradkar, Narendra 859, 1275
Shirai, Yasuhiro 1, 4
Shirazi, Eli 731
Shrestha, Bishal 833
Shukla, Siddharth 144
Sidawi, Tala 914, 1019
Siebentritt, Susanne 532, 533
Siebert, Michael 1245
Siefer, Gerald 629
Sillerud, Colin 806
Silva, Brandon 826
Silverman, Tim 1319
Silverman, Timothy J. 298, 806, 855, 997
Sim, Yeon Hyang 209
Simon, John 867, 1265
Simpson, Lin J. 915, 1037
Sindermann, Andrew B. 1314
Singh, Ajay 532, 646
Singh, Aparna 512, 574
Singh, Pritpal 721
Sinha, Archana 319, 649, 961
Sinha, Arpan 1168
Sinha, Parikhit 95
Sinton, Ronald A. 925
Sites, James R. 702
Sites, James 47, 264
Sky, Haiku 480
Slauch, Ian M. 199, 914, 1019
Smets, Arno H. M. 484
Smets, Arno 606, 1300
Smiles, Matthew J. 1312
Smirnov, Yury 109
Smith, Ryan M. 596, 692
Smith, Sarah 1319
Smith, Soshana 1044
Snider, Sean 146
Snyder, William 333, 1343

Solas, Álvaro F. 1294
Soman, Anishkumar 570, 643
Somasundaram, Sakthi Guhan 489
Søndenå, Rune 525
Song, Aaron 1350
Song, Tao 766, 814
Song, Zhaoning 108, 126, 405, 464, 626, 653, 701,
........................752, 792, 953, 972, 1032, 1055, 1170
Sonkar, Ramesh Kumar 504
Soppe, Wim 1258
Sorower, Nur Jahan Beanta 1139
Soufiani, Arman 368
Sovernigo, Enrico 1360
Sowmya, Arcot 476
Sozzi, Giovanna 370
Spampinato, Antonio 1306, 1360
Spatari, Sabrina 43
Spataru, Sergiu V. 544, 608
Spataru, Sergiu 871
Spinella, Laura 649
Sridhar, Seetharaman 299
Srinivasa, Apoorva 1045
St-Gelais, Raphael 1359
Stalker, Amy R. 145
Stanbery, Billy J. 168
Stanislawski, Brooke 1319
Stein, Joshua S. 172, 805, 806, 888, 1073
Steiner, Marc 629
Steiner, Myles A. 1283
Steinhauser, Bernd 369
Stekli, Joseph 111
Stepec, Murielle 562
Stevens, Margaret A. 164, 240
Stid, Jacob T. 144
Stiff-Roberts, Adrienne D. 234
Stradins, Paul 996
Strandwitz, Nicholas C. 1323
Stricher, Romain 254
Stuckelberger, Josua 75
Stuckelberger, Michael E. 1145
Su, Shicheng 1248
Subedi, Biwas 405, 653, 898, 1032
Subedi, Indra 405, 752, 774, 833, 972, 1169
Subedi, Kamala Khanal 774, 1169
Sudbury, Ben A. 1033
Sulas-Kern, D. B. 872
Sulas-Kern, Dana B. 200, 649, 806, 997
Sultana, Nadera 588
Sumita, Taishi 1257
Sun, Kaiwen 352, 1126
Sun, Y. 811
Sung, Li-Piin 1044

Surel, Josephine L..605
Sutterlueti, Juergen 267
Svatek, Simon A. 1100, 1340
Svatek, Simon Aurel.......................................538
Sveinbjörnsson, Kári.......................................529
Swatton, Nicole..92
Sweat, Rebekah... 1293
Szablewski, Marek....................................... 1263
Szábo, Sandor...508
Taherimakhsousi, Nina............................... 1072
Takahashi, Takuji...482
Takamoto, Tatsuya 467
Talkington, Samuel...204
Tamizhmani, G..879
Tamizhmani, Govindasamy 261, 426, 1065, 1198
Tan, Hairen... 1229
Tan, Kelvin...62, 339
Tang, Rongfeng 1126
Tao, Meng... 62, 248, 339
Tarbuck, Neil...577
Tardio, Miguel Modesto 1305
Tatavarti, Rao 913
Tatsiankou, Viktar....................................... 1346
Tawsif, Sheikh Elahi......................................988
Taylor, Nigel...508
Teague, Ian K. 1037
Teeter, Glenn 819
Teixeira, Jennifer P. 630
Terlier, Tanguy 327
Tervo, Eric J. 1283
Terwilliger, Kent.......................................15, 961
Thaikattil, Greeta J..145
Thakur, Pardeep K 1312
Theelen, Mirjam 381, 740
Theingi, San..996
Thellen, Christopher 961
Theocharides, Spyros.....................................156
Theristis, Marios.............................. 172, 826, 888
Thiagarajan, Ramanathan 178, 183
Thom, Daniel...967
Thomas, Luke 577, 1312
Thomere, Angelica 591
Thon, Susanna M. 862
Thornton, Patrick 106
Thorsteinsson, Sune.......................................544
Thuis, Michael 961
Tibbits, Thomas 1235
Till, Micah J..693
Tina, Giuseppe Marco.....................................214
Tiwari, Ayodhya N. 533
Tiwari, Kunal Jogendra...................................284
To, Bobby 15

Todaro, Lorenzo.. 214
Togay, Mustafa 658, 857
Tom, Thomas... 1363
Tomko, Brian J... 145
Tonita, Erin M... 8
Topic, Marko.. 680
Torquato, Ricardo 517
Tracy, Jared.................................199, 255, 914
Traore, Papa Touty...................................... 1262
Treglia, Andrew C. 702
Trindade, Fernanda C. L. 83, 517, 545, 1127
Trujillo, Marena 1038
Truong, Thien.. 75
Trupke, Thorsten................. 329, 353, 367, 368, 442, 472, 476
Tsai, Hsinhan... 1321
Tse, Yau Yau... 63
Tsoulka, Polyxeni.....................................415, 941
Tumusange, Marie S.................................... 1032
Turala, Artur 254
Türkay, Deniz.. 1119
Tutsch, Leonard 109
Tytov, Serhii... 1248
Udaeta, Miguel Edgar Morales 839
Uddin, Muhammad Hammad............................ 354
Ulicná, Sona...319, 649, 961
Underwood, Robert..................................... 1177
Unruh, Davis...1010, 1061
Upadhyaya, Ajay D.. 251
Upadhyaya, Ajay.......................................76, 592
Upadhyaya, Vijaykumar D............................251, 366
Uprety, Prakash 752
Valdes, Nicholas 1315
Valdivia, Christopher E....................8, 107, 143, 406, 883,
....1031, 1295, 1359, 1366
Vallerotto, Guido 629, 739
Van Aken, Bas... 1306
Van De Lagemaat, Jao 168
Van De Voorde, Mathis.................................. 374
Van Den Berg, Joran 740
Van Herpt, Javier 739
Van Nijen, David A...................................... 623
Van Zandt, Devin.. 1013
Vansant, Kaitlyn T. 320, 1087
Vansant, Kaitlyn.. 1109
Varonides, AC 637
Vatan, Reza.. 1061
Veal, Tim D 577, 1312
Vedde, Jan... 871
Vehse, Martin...228, 557
Venkat, Sameera Nalin................................980, 1020
Verma, Navni.. 1275
Victor, David.. 1310

Victoria, Marta	1105
Vignola, Frank	146
Villa, Simona	740
Villa-Bracamonte, Maria Fernanda	1079
Vining, William F.	276
Virtuani, Alessandro	1120
Vlieg, Elias	875, 1280
Volatier, Maïté	254
Von Gastrow, Guillaume	327
Voroshazi, Eszter	169, 562
Voss, Stephen	1347
Voss, Steve	1163
Voyce, Ryan	516
Voz, Cristobal	1363
Wagner, Lukas	114, 239, 566
Walajabad, Sampath S.	658
Walker, Andy	899
Walker, Don	1320
Walker, Janine	1332
Walker, Trumann	429, 924
Walling, Reigh	1013
Walls, J. M.	705
Walls, J. Michael	658
Walls, John M.	683, 827, 857
Walls, John Michael	604, 786
Walls, Michael	63, 387, 414, 838, 900
Walmsley, John	814
Wands, Jake	415
Wang, Biqi	693
Wang, Changlei	752
Wang, Haomiao	384
Wang, Li	1177
Wang, Liwei	178
Wang, Taowen	533
Wang, Wei	976, 988, 1101
Wang, Wenbo	967
Wang, Wenzong	1013
Wang, Xiaoming	126
Wang, Youyang	468
Warner, Cody	1158
Watson, Stephanie	1044
Weber, Julian	231
Weeber, Arthur	381, 484
Wegmueller, Jakob	1020
Weideman, Kyle G	1333
Weinhardt, Lothar	76
Weiser, Philip M.	525
Weiss, Thomas P.	533
Welser, Roger E.	1167
Wen, Bo	693
Wen, Jin	1229
Weng, Guoen	468

Went, Cora	247
Westbrook, Owen W.	964
Wheeler, Aaron	1059
Whiteside, Vincent R.	732
Wieliczka, Brian M.	320
Wieser, Raymond J.	255
Wietler, Tobias	529, 1245
Wikoff, Hope	344
Williams, Benjamin	328
Williams, Henry J.	384
Williams, Rafell	766, 814
Willockx, Brecht	1162
Wilson, Mickey	806
Wilson, Samantha S.	1347
Wilson, Samantha	1163
Wiser, Ryan	1123
Witte, Wolfram	370, 533
Witteck, Robert	1245
Wolf, Andreas	18
Wolff, Christian M.	1119
Wong, Evan	975
Wong, Johnson	77, 80
Wood, Allen	1338
Wright, Brendan	472
Wright, Niara E.	234
Wu, Kunlin	517
Wu, Yinghui	796
Würfel, Uli	114
Wylie, Zachery R	283
Xia, Daixi	575
Xiao, C.	872
Xiao, Chuanxiao	15, 819
Xiao, Ke	1229
Xiao, Xusheng	796
Xie, Yi Hao	121
Xie, Yu	480
Xiong, Gang	838
Xu, Qianfeng	588
Xu, Xinya	1311
Yamada, Ayaka	482
Yamada, Kazumi	58, 467
Yamaguchi, Masafumi	58, 467
Yan, Chang	352
Yan, Yanfa	108, 126, 154, 405, 464, 626, 653, 667, 701, 752, 761, 792, 806, 898, 953, 972, 1032, 1055, 1170
Yanagida, Masatoshi	1, 4
Yang, Dazhi	654
Yang, Guang	1338
Yang, Guangtao	484
Yang, Jaemo	480
Yang, O-Bong	1292
Yang, Wanli	76

Yang, Wei 802
Yang, Zhongshu 369
Yao, Yiyun 967
Yao, Zhirong 484
Yau, Gemini 1038
Ye, Youxiong 28
Yi, Junsin 456, 483, 507
Yilmaz, Pelin 740
Yoon, Heayoung P. 1185
Young, David. L 996
Young, Matthew 75
Youtsey, Chris 137
Yu, Bo 115
Yu, Li 1000, 1038
Yu, Xuanji 255, 796, 980, 1020
Yu, Zhengshan J. 1118
Yu, Zhengshan Jason 1329
Yu, Zhibin 1293
Yuan, Luyao 1028
Yue, Xu 1229
Yun, Min Ju 209
Zabalza, Ruben 255
Zahid, Muhammad Aleem 456, 507
Zakeeruddin, Shaik M. 114
Zamora, Rachid Darbali 916
Zareafifi, Farzan 121
Zeder, Simon J. 479
Zekry, A. 1230
Zekry, Abdelhalim 336
Zeman, Miro 484, 623
Zhang, Chaomin 143
Zhang, K. Max 384
Zhang, Simon M. F. 352
Zhang, Yibo 74, 1264
Zhang, Yong-Hang 1007, 1288
Zhang, Zheyu 178
Zhao, Yifeng 484
Zhao, Zitong 1010, 1061
Zheng, Jianghui 368
Zhou, Tao 526
Zhu, Junhao 244
Zhu, Tao 126, 1055
Zhu, Xiangqi 907
Zhu, Xitong 89, 489, 731
Zhu, Yan 367
Zhu, Zhengtao 1284
Zilouchian, Ali 1121
Zimanyi, Gergely T. 1010, 1062
Zimanyı, Gergely 1061
Zin, Ngwe 620
Ziska, Catharina 1145
Zoppi, Guillaume 73, 516, 1311

Zou, Min 1279
Zou, Yongjie 1159
Zouaghi, Firas 627
Zouhair, Salma 114
Zuiker, Steve 904
Zunft, Heiko 18

IEEE
445 Hoes Lane
Piscataway, NJ 08854-4141

ISBN 978-1-7281-6118-1

2022 IEEE 49th Photovoltaics Specialists Conference (PVSC 2022)

Philadelphia, Pennsylvania, USA
5-10 June 2022

Pages 684-1369

IEEE Catalog Number: CFP22PSC-POD
ISBN: 978-1-7281-6118-1

2022 IEEE 49th Photovoltaics Specialists Conference (PVSC 2022)

Philadelphia, Pennsylvania, USA
5-10 June 2022

Pages 684-1369

IEEE Catalog Number: CFP22PSC-POD
ISBN: 978-1-7281-6118-1

**Copyright © 2022 by the Institute of Electrical and Electronics Engineers, Inc.
All Rights Reserved**

Copyright and Reprint Permissions: Abstracting is permitted with credit to the source. Libraries are permitted to photocopy beyond the limit of U.S. copyright law for private use of patrons those articles in this volume that carry a code at the bottom of the first page, provided the per-copy fee indicated in the code is paid through Copyright Clearance Center, 222 Rosewood Drive, Danvers, MA 01923.

For other copying, reprint or republication permission, write to IEEE Copyrights Manager, IEEE Service Center, 445 Hoes Lane, Piscataway, NJ 08854. All rights reserved.

*** This is a print representation of what appears in the IEEE Digital Library. Some format issues inherent in the e-media version may also appear in this print version.

IEEE Catalog Number: CFP22PSC-POD
ISBN (Print-On-Demand): 978-1-7281-6118-1
ISBN (Online):   978-1-7281-6117-4

Additional Copies of This Publication Are Available From:

Curran Associates, Inc
57 Morehouse Lane
Red Hook, NY 12571 USA
Phone: (845) 758-0400
Fax: (845) 758-2633
E-mail: curran@proceedings.com
Web: www.proceedings.com

TABLE OF CONTENTS

Investigation of Degradation Kinetics of Perovskite Solar Cells by Accelerated Aging......................................1
  Dhruba B. Khadka, Yasuhiro Shirai, Masatoshi Yanagida, Kenjiro Miyano

Effect of Phenethylammonium Thiocyanate Additive in Tin Perovskite for Efficient and Stable Pb-Free Perovskite Solar Cells 4
  Dhruba B. Khadka, Yasuhiro Shirai, Masatoshi Yanagida, Kenjiro Miyano

Investigation of the Circular Economy Approach in Asian Solar PV Manufacturing......................................7
  Viktor Dancza, Kai Cheng

Three General Methods for Predicting Bifacial Photovoltaic Performance Including Spectral Albedo 8
  Erin M. Tonita, Christopher E. Valdivia, Michael Martinez-Szewczyk, Mariana I. Bertoni, Karin Hinzer

Evaluation of Auger Limited Behavior in Thermoradiative Cells......................................9
  Jamie D. Phillips

Fabrication of a Chemically Exfoliated 2D MoS2 Nanoparticle Based Solar Cell 12
  Wafa Alnaqbi, Ayman Rezk, Aisha Alhammadi, Ammar Nayfeh

Evaluating the Durability of Balance of Systems Components Using Combined-Accelerated Stress Testing......................................15
  David Miller, Greg Perrin, Kent Terwilliger, Joshua Morse, Chuanxiao Xiao, Bobby To, Chun-Sheng Jiang, Peter Hacke

High Throughput Boron Emitter Formation from Pre-Deposited APCVD BSG Layers for TOPCon Solar Cells 18
  Marius Meßmer, Sattar Bashardoust, Udo Belledin, Sven Seren, Heiko Zunft, Sebastian Mack, Andreas Wolf

Uncertainty in Annual Energy Resulting from Uncertain Irradiance Measurements 22
  Clifford W. Hansen, Aaron Scheiner

Impact of Photovoltaic Plant Tilt on the Need for Storage......................................27
  Russell K Jones, Sarah Kurtz

Precursor Ink Design for Scalable Fabrication of Perovskite Solar Cells Via High-Speed Flexography......................................28
  Julia E Huddy, Youxiong Ye, William J Scheideler

Terrain Aware Backtracking Via Forward Ray Tracing......................................29
  Kurt Rhee

Sustainability of PV Repowering 31
  Ian Marius Peters, Jens Hauch, Christoph Brabec

Correlations in Spatial Variability When Accounting for Cloud Advection 32
  Joseph Ranalli

Realization of Ultrathin GaAs Photonic Power Converters with Rear-Side Metal Grating on Full 4" Wafers 38

Oliver Höhn, Meike Schauerte, Patrick Schygulla, Hubert Hauser, David Lackner, Benedikt Bläsi, Henning Helmers

Human Health Risk Assessment of Solvents and Lead Toxicity in Emerging Perovskite Solar Cells 43

Sherif A. Khalifa, Sabrina Spatari, Aaron T. Fafarman, Vasilis M. Fthenakis, Patrick Gurian, Jason B. Baxter

Spectral Shape Changes the Optimal Perovskite Thickness of the 2-Terminal Perovskite/Silicon Tandem Solar Cell 44

Dong C. Nguyen, Yasuaki Ishikawa

Leveraging Undoped CdSeTe for >950 mV 47

Pascal Jundt, James Sites

Low-Temperature PECVD Deposition of Highly Conductive N-Type Microcrystalline Silicon Thin Films for Optoelectronic Applications 52

Brahim Aissa, Amir A. Abdallah, Juan Lopez Garcia

Analysis for Solar Coverage and CO2 Emission Reduction of Photovoltaic-Powered Vehicles 58

Masafumi Yamaguchi, Taizo Masuda, Takashi Nakado, Kazumi Yamada, Kenichi Okumura, Akinori Satou, Yasuyuki Ota, Kenji Araki, Kensuke Nishioka

Interrelated Characterizations of 2D/3D Perovskite Solar Cells Aged Under Damp Heat Conditions 59

Cynthia Farha, Emilie Planès, Lara Perrin, David Martineau, Lionel Flandin

An Intelligent Algorithm for Maximum Power Point Tracking in PV Systems Through Load Management 62

Kelvin Tan, Joseph A. Azzolini, William J. Parquette, Christian R. Polo, Meng Tao

Large Area Survey Grain Size and Texture Optimization for Thin Film CdTe Solar Cells Using Xenon-Plasma Focused Ion Beam (PFIB) 63

Vladislav Kornienko, Ochai Oklobia, Stuart Irvine, Steve Jones, Giray Kartopu, Ali Abbas, Yau Yau Tse, Jake Bowers, Kurt Barth, Michael Walls

Diffraction-Optimized Surface Structures for Enhanced Light Harvesting in Organic Solar Cells 69

Milena Merkel, Jörg Imbrock, Cornelia Denz

Validation of Photovoltaic Spectral Effects Derived from Satellite-Based Solar Irradiance Products 72

Sophie Pelland, Christian A. Gueymard

Elimination of the Carbon-Rich Layer in Cu2ZnSn(S, Se)4 Absorbers Prepared from Nanoparticle Inks 73

Stephen Campbell, Martial Duchamp, Neil Beattie, Michael Jones, Guillaume Zoppi, Vincent Barrioz, Yongtao Qu

Light Trapping Characteristics of Photonic Crystal Constructs and Randomly Textured Thin Silicon 74

Sara M. Almenabawy, Yibo Zhang, Rajiv Prinja, Nazir P. Kherani

Gallium-Boron Spin-On Co-Doping for Polycrystalline Silicon Passivating Contacts 75

Thien Truong, Matthew Young, Mowafak Al-Jassim, Daniel Macdonald, Hieu Nguyen, Josua Stuckelberger

Chemical Surface and Interface Structure of Sulfur-Passivated Silicon with a SiNx Capping Layer 76
Amandee Hua, Nan Jiang, Ajay Upadhyaya, Issac Lam, Tasnim K Mouri, Dirk Hauschild, Lothar Weinhardt, Wanli Yang, Ajeet Rohatgi, Ujjwal Das, Clemens Heske

Data Mining of Solar Cells Production Data Using Factorial Analysis 77
Johnson Wong, Dinica Li, Gordon Deans

Statistical and Engineering Process Control of Phosphorus Diffused Solar Wafers Using Contactless Infrared Reflectometry 80
Johnson Wong, Divya Ananthanarayanan, Gordon Deans

Proposal of Connection Assessment Diagrams to Speed Up the Studies of Hosting Capacity of PV Generators in MV Distribution Systems.................. 83
Pedro A. V. Pato, Fernanda C. L. Trindade, Tiago R. Ricciardi, Paulo Meira, Walmir Freitas

Ray Tracing of Bent Applications of Luminescent Solar Concentrator PV Modules 89
Xitong Zhu, Michael G. Debije, Angèle H. M. E. Reinders

Optical Modeling of Light Trapping Using an ITO-Based Electrodynamic Dust Shield Structure 92
Nicole Swatton, Andrey Semichaevsky

Assessing the Alignment of Solar Facilities with Global Climate Goals 95
Parikhit Sinha, Liv Hammann

Preparation of Plasmonic Ag and Au Nanoparticle Interfaces for Photocurrent Enhancement in Si Solar Cells 98
Brahim Aïssa, Adnan Ali, Rui N. Pereira, Anirban Mitra

The Natural and Accelerated Evolution of EVA Adhesion Through Intermediate Exposures 106
Patrick Thornton, Nick Bosco, Reinhold H Dauskardt

High-Performance O- Band Photonic Power Converters Under Non-Uniform Laser Illumination 107
Meghan N. Beattie, Henning Helmers, Gavin P. Forcade, Christopher E. Valdivia, David Lackner, Oliver Höhn, Karin Hinzer

Post-Annealing Treatment on Hydrothermally Grown Sb2(S, Se)3 Thin Films for Efficient Solar Cells.................. 108
Suman Rijal, Zhaoning Song, Deng-Bing Li, Jaehoon Chung, Sandip S Bista, Dipendra Pokhrel, Sabin Neupane, Randy Ellingson, Yanfa Yan

Wafer-Scale Pulsed Laser Deposition of ITO for Silicon Heterojunction Solar Cells: Reduced Damage Vs Interfacial Resistance 109
Yury Smirnov, Pierre-Alexis Repecaud, Leonard Tutsch, Ileana Florea, Pere Roca I Cabarrocas, Martin Bivour, Monica Morales-Masis

Potential Capacity and Targeted Costs for Floating Photovoltaics in North America.................. 110
Leonardo Micheli

Techno-Economic Analysis of Novel PV Plant Designs for Extreme Cost Reductions 111
Nicholas Pilot, Robin Bedilion, Daniel Fregosi, Sean Hackett, Michael Bolen, Joseph Stekli

Potentiostatic Photoluminescence Imaging of Charge Extraction in Perovskite Solar Cells.................. 114
Lukas Wagner, Patrick Schygulla, Jan Philipp Herterich, Mohamed Elshamy, Dmitry Bogachuk, Salma Zouhair, Simone Mastroianni, Uli Wurfel, Yuhang Liu, Shaik M. Zakeeruddin, Michael Grätzel, Andreas Hinsch, Stefan W. Glunz

Flexible GaAs Solar Cell Using Water-Soluble Sacrificial Layer for Epitaxial Lift-Off Process.................... 115
  Sahil Sharma, Carlos A Favela, Bo Yu, Eduard Galstyan, Venkat Selvamanickam

An Evaluation of Empirical Models for Use in Normalizing PV Plant Performance Data............................ 116
  Daniel Fregosi, Michael Bolen

Performance Investigation of Batteries Supporting Solar Power in U.S. 121
  Farzan Zareafifi, Daniel Baerwaldt, Socheata Hour, Yi Hao Xie, Sarah Kurtz

Numerical Modeling of Capacitance Signatures of Perovskite Solar Cells.. 126
  Rasha Awni, Zhaoning Song, Chongwen Li, Lei Chen, Suman Rijal, Sandip Bista, Tao Zhu, Xiaoming Wang, Yanfa Yan

Reverse Energy Injustice on Molokai Island to the Underserved Communities with 100% Energy
from the Sun (Light & Heat) for Energy Cost Savings Equity.. 127
  John O. Borland

Performance Loss Rate Estimation for Systems Affected by Potential Induced Degradation 131
  Panagiotis Goumenos, Andreas Livera, Michalis Florides, George E. Georghiou

Assessing and Optimizing Free Space Luminescent Solar Concentrators for Urban Façade
Installation 134
  Shweta Pal, Rebecca Saive

Curvilinear Prismatic Window Which Eliminates Glare and Reduces Front-Surface Reflections for
PV Modules and Other Surfaces 137
  Mark O'Neill, Chris Youtsey

Monochromatic Light Trapping in Photonic Power Converters.. 143
  Nicholas P. Irvin, Neda Nouri, Chaomin Zhang, Christopher E. Valdivia, Karin Hinzer, Richard R. King, Christiana B. Honsberg

Implications of Agriculturally Co-Located Solar PV Installations on the FEW Nexus in the Central
Valley 144
  Jacob T Stid, Siddharth Shukla, Annick Anctil, Anthony D Kendall, David W Hyndman

Photovoltaic Investigation on the Lunar Surface (PILS): Design Considerations and Ground
Testing 145
  Jeremiah S McNatt, Timothy J Peshek, Norman F Prokop, Greeta J Thaikattil, Michael J Krasowski, Amy R Stalker, Brian J Tomko, Mathew R Deminico

Reference Cell Performance and Modeling on a One-Axis Tracking Surface 146
  Frank Vignola, Josh Peterson, Rich Kessler, Sean Snider, Peter Gotseff, Manajit Sengupta, Aron Habte, Afshin Andreas, Fotis Mavromatakis

Optimization of CdTe Solar Cells Using Co-Sputtered CdSeTe.. 154
  Deng-Bing Li, Sandip Singh Bista, Neupane Sabin, Xiaomeng Duan, Manoj K Jamarkattel, Abdul Quader, Adam Phillips, Michael Heben, Randall J Ellingson, Yanfa Yan

Parametric Study of Building-Integrated Photovoltaic Windows.. 155
  Yuan Gao, Jacob Jonsson, Charlie Curcija

Impact of Daily Irradiance Profiles on Intra-Day Solar Forecasting 156
  Javier Lopez-Lorente, Spyros Theocharides, George Makrides, George E. Georghiou

Hyperspectral Imaging of Localized, Optically-Active Defects in GaAs Solar Cells 164
Behrang H. Hamadani, Margaret A. Stevens, Brianna Conrad, Matthew P. Lumb, Eric Armour, Kenneth J. Schmieder

Achieving Global Decarbonization by Photovoltaic Electrification: Impact of Disruptive Technologies 168
Billy J. Stanbery, Jao Van De Lagemaat

Harsh Sequential Stress Tests for Improved PV Durability 169
Jean Patrice Rakotoniaina, Romain Couderc, Eszter Voroshazi, Jérémie Aimé

PV Module Operating Temperature Model Equivalence and Parameter Translation 172
Anton Driesse, Marios Theristis, Joshua S. Stein

Accelerating Simulation for High-Fidelity PV Inverter System Reliability Assessment with High-Performance Computing 178
Liwei Wang, Ramanathan Thiagarajan, Shuangshuang Jin, Zheyu Zhang

Inverter Reliability Estimation for Advanced Inverter Functionality 183
Jack Flicker, Jay Johnson, Matthew J. Reno, Joseph A. Azzolini, Peter Hacke, Ramanathan Thiagarajan

Grid-Forming and Grid-Following Inverter Comparison of Droop Response 190
Nicholas S. Gurule, Javier Hernandez Alvidrez, Matthew J. Reno, Wei Du, Kevin Schneider

Inferring PV System Specifications from Net Load 197
Upama Nakarmi, Thomas E. Hoff, Marc Perez, Philip Gruenhagen

Planarizing HVPE Growth on GaAs Substrates Produced by Controlled Spalling 198
Anna K Braun, William E McMahon, Allison N Perna, Kevin L Schulte, Corinne E Packard, Aaron J Ptak

Lamination Process Induced Residual Stress in Glass-Glass Vs. Glass-Backsheet Modules 199
Farhan Rahman, Ian M. Slauch, Rico Meier, Jared Tracy, Elizabeth C. Palmiotti, Mariana I. Bertoni, James Y. Hartley

Near-Busbar Degradation of Screen-Printed Metallization in Silicon Photovoltaic Modules 200
Dana B. Sulas-Kern, Helio Moutinho, Tristan Erion-Lorico, Steve Johnston

Improving Behind-The-Meter PV Impact Studies with Data-Driven Modeling and Analysis 204
Joseph A. Azzolini, Samuel Talkington, Matthew J. Reno, Santiago Grijalva, Logan Blakely, David Pinney, Stanley McHann

A Model to Predict Daily Snow Albedo Change Over Time 205
Christopher Pike, Daniel Riley, Laurie Burnham

High-Specific-Power Schottky-Junction Photovoltaics from CVD-Grown MoS2 208
Timothy Ismael, Kazi M. Islam, Muhammad A. Abbas, George B. Ingrish, Claire E. Luthy, Orhan Kizilkaya, Carlos M. Gutierrez, Meghan E. Bush, Jeremiah S. McNatt, Anthony J. Hoffman, Matthew D. Escarra

Highly Stretchable, Durable and Lightweight Lego®-Style 3-Dimensional Photovoltaic 209
Min Ju Yun, Yeon Hyang Sim, Dong Yoon Lee, Seung I. Cha

Optimal Strategy for Using Biomass to Enable California High Penetration Solar 212
Mahmoud Y. Abido, Sarah R. Kurtz

Automatic Crack Segmentation in Electroluminescence Images of Solar Modules and Maximum Inactive Area Prediction 213

Xin Chen, Todd Karin, Anubhav Jain

Experimental Assessment of Temperature Estimation Models of Bifacial Photovoltaic Modules 214

Gaetano Mannino, Giuseppe Marco Tina, Mario Cacciato, Lorenzo Todaro, Fabrizio Bizzarri, Andrea Canino

Seasonal Dependence of Diurnal Efficiency Degradation and Recovery in Perovskite Mini-Modules During Outdoor Testing 217

Vasiliki Paraskeva, Maria Hadjipanayi, Matthew Norton, Aranzazu Aguirre, Afshin Hadipour, Rita Ebner, George E. Georghiou

Applying Unsupervised Machine Learning for the Detection of Shading on a Portfolio of Commercial Roof-Top Power Plants in Germany 223

Nicolas Holland, Klaus Kiefer, Christian Reise, E. A. Sarquis Filho, Bernd Kollosch, Björn Müller

Electron Selective TiO$_x$ Contact for Ultrathin Amorphous Germanium Solar Cells 228

Norbert Osterthun, Hosni Meddeb, Nils Neugebohrn, Kai Gehrke, Martin Vehse

Magnetic Field Imaging (MFI) of Shingle Solar Modules 231

Julian Weber, Stephan Hoffmann, Kai Kaufmann, Angela De Rose

Evaluating Electroluminescence Imaging and Image Processing as a Quantitative Solar Cell Characterization Method 232

Meghan E Bush, Timothy J Peshek, Erica N Montbach

CdTe-Based Photovoltaics Using a CdTe/CdSe/CdTe Absorber Layer Structure 233

Jacob F Leaver, Ken Durose, Jonathan D Major

Effects of Growth Temperature on Electrical Conductivity in Low-Dimensional, Ruddlesden-Popper Perovskite Thin Films Deposited by RIR-MAPLE 234

Niara E. Wright, Adrienne D. Stiff-Roberts

GeCl$_4$-Based High Quality Ge Epitaxy on Engineered Ge Substrates for Thin Multi-Junction Solar Cells 235

Jinyoun Cho, Clément Porret, Valérie Depauw, Guillaume Courtois, Daniel McDermott, Roger Loo, Kristof Dessein, Rufi Kurstjens

Investigation of Degradation Mechanisms in Carbon-Based Perovskite Solar Cells Exposed to Damp-Heat Conditions 239

Nikoleta Kyranaki, Cynthia Farha, Lara Perrin, Lionel Flandin, Emilie Planès, Lukas Wagner, Karima Saddedine, David Martineau, Stéphane Cros

Selective Etching of 6.1 Å Materials for Transfer-Printed Devices 240

Margaret A. Stevens, Jill A. Nolde, Shawn Mack, Kenneth J. Schmieder

Inorganic Perovskite Solar Cells with Very High Voltage and Excellent Stability Against Thermal and Environmental Degradation 244

Saba Sharikadze, Junhao Zhu, Ranjith Kottokkaran, Arkadi Akopian, Vikram Dalal

23.5% Efficiency GaAs Solar Cells Fabricated with Low-Cost, Non-Vacuum Processing 247

Phillip R Jahelka, Harry A Atwater, Aaron Ptak, Christiane Frank-Rotsch, Frank Kiessling, Cora Went, Michael Kelzenberg

Metallic Lead Recovery Via Electrowinning from Lead Acetate for Silicon Solar Module Recycling............248
Natalie Click, Meng Tao

Rear Junction Bifacial Screen-Printed Double Side Passivated Contact Si Solar Cells 251
Young-Woo Ok, Vijaykumar D Upadhyaya, Brian Rounsaville, Ajay D Upadhyaya, Wook-Jin Choi, Ajeet Rohatgi, Gabby De Luna, John Derek Arcebal, Pradeep Padhamnath, Shubham Duttagupta

Micro-Scale III-V/Ge Multijunction Solar Cell with Through Cell Via Contacts............254
Mathieu De Lafontaine, Guillaume Gay, Erwine Pargon, Camille Petit-Etienne, Romain Stricher, Serge Ecoffey, Artur Turala, Maïté Volatier, Abdelatif Jaouad, Simon Fafard, Vincent Aimez, Maxime Darnon

Spatiotemporal Modeling of Real World Backsheets Field Survey Data: Hierarchical (Multilevel) Generalized Additive Models............255
Raymond J. Wieser, Zelin Zack Li, Stephanie L. Moffitt, Ruben Zabalza, Evan Boucher, Silvana Ayala, Matthew Brown, Xiaohong Gu, Liang Ji, Colleen O'Brien, Adam W. Hauser, Greg S. O'Brien, Xuanji Yu, Roger H. French, Michael D. Kempe, Jared Tracy, Kausik R. Choudhury, William J. Gambogi, Laura S. Bruckman, Kenneth P. Boyce

Mapping of Local Defects and Voltages in Solar Cells Using Non-Contact Electrostatic Voltmeter Method 261
Hamza Ahmad Raza, Govindasamy Tamizhmani

Tellurium Oxide as a Back-Contact Buffer Layer for CdTe Solar Cells............264
Camden Kasik, Ramesh Pandey, Akash Shah, James Sites

Intelligent Cloud-Based Monitoring and Control Digital Twin for Photovoltaic Power Plants............267
Andreas Livera, George Paphitis, Loucas Pikolos, Ioannis Papadopoulos, Jesús Montes-Romero, Javier Lopez-Lorente, George Makrides, Juergen Sutterlueti, George E. Georghiou

Predicting Solar Cell Recombination from C-V-F Fingerprints Using Machine Learning 275
Isaac K. Lam, Austin G. Kuba, Nathan J. Rollins, William N. Shafarman

Current & Future Photovoltaic System Impacts on City-Wide Grid Performance & Neighborhood Microgrids 276
C. Birk Jones, William F. Vining, Thad Haines

Passivating Surface Iodide Defects Slows the CsPbI3 Phase Transformation 283
Jeffrey A Christians, Jonathan Outen, Rory M Campagna, Zachery R Wylie, Peter Ruffolo

Which Potential for Kesterite Absorbers in Tandem Solar Cells: A Quantitative Modelling Approach 284
Alex Jimenez, Alejandro Navarro, Sergio Giraldo, Kunal Jogendra Tiwari, Marcel Placidi, Lorenzo Calvo-Barrio, Joaquim Puigdollers, Edgardo Saucedo, Zacharie Jehl Li-Kao

Measuring Global, Direct, Diffuse, and Ground-Reflected Irradiance Using a Reference Cell Array............285
Michael Gostein, Adam Hoffman, Bruce H. King, Audrey Marquis

Validation of In-Situ I-V Measurement Unit for PV System Monitoring Applications............291
Audrey Marquis, Michael Gostein, Bruce H. King

Effect of Near-Interface Compensation of CdSeTe Absorber Layers on Solar Cell Performance............295
Brian Good, Eric Colegrove, Matthew O. Reese

Millions of Small Pressure Cycles Drive Damage in Cracked Solar Cells.. 298
  Timothy J Silverman, Nick Bosco, Michael Owen-Bellini, Cara Libby, Michael G Deceglie

Quantifying Energy Flows in PV Circularity Processes.. 299
  Heather M. Mirletz, Silvana Ovaitt, Ashley Gaulding, Seetharaman Sridhar, Teresa Barnes

Uncertainty Quantification of Bifacial Performance Modeling.. 302
  Matthew J. Prilliman, Janine M. Freeman Keith

Field Experience Detecting PV Underperformance in Real Time Using Existing Instrumentation................ 307
  Scott Sheppard, Tim Cook, Daniel Fregosi, Christopher Perullo, Michael Bolen

Enhancing Temporal Variability of 5-Minute Satellite-Derived Solar Irradiance Data 314
  Jing Huang, Richard Perez, James Schlemmer, Marc Perez, Akanksha Bhat, Patrick Keelin,
  Alex Kubiniec

The Materials Degradation in Encapsulants for Application in Glass/Glass PV Modules After
Accelerated Aging.. 319
  Sona Ulicna, Archana Sinha, David C. Miller, Laura T. Schelhas, Michael Owen-Bellini

Perovskite PV Design for Stable Space Operation... 320
  Kaitlyn T. Vansant, Ahmad R. Kirmani, Jay B. Patel, Laura E. Mundt, David P. Ostrowski,
  Brian M. Wieliczka, Gabriella D. Lahti, Michael D. McGehee, Laura T. Schelhas, Joseph M.
  Luther, Timothy J. Peshek, Lyndsey B. McMillon-Brown

Thermoradiative Cell Technology: Analysis and Loss Mechanisms.. 321
  Geoffrey A. Landis

Estimation of Shade Losses in Unlabeled PV Data.. 326
  Bennet E. Meyers, David J. F. Rodriguez

Contribution of Na+ from Glass to PID-S in Solar Modules: Na Migration in EVA.................................... 327
  Jacob A. Clenney, Erick Martinez Loran, Guillaume Von Gastrow, Tanguy Terlier, David P.
  Fenning, Rico Meier, Mariana I. Bertoni

Importance of Ideality Factors in perovskite/Si Tandem Solar Cell Design 328
  Benjamin Williams, Benjamin Daiber, Chris Case

A Deep Learning Approach to Increase Luminescence Image Resolution of Solar Cells 329
  Priya Dwivedi, Robert Lee Chin, Thorsten Trupke, Ziv Hameiri

Agrivoltaics Using Bi-Facial PVs for Permaculture in Utility-Scale Projects 330
  P. M. Jansson, M. G. Newberry, S. M. Myers

Dedicated Cold-Climate Field Laboratory for Photovoltaic System and Component Studies: The
Michigan Regional Test Center as a Case Study.. 333
  Laurie Burnham, Daniel Riley, Bruce H. King, Jennifer Braid, Paul Dice, Ana Dyreson,
  William Snyder, Christopher Pike

Efficient Self-Protected Thin Film c-Si Solar Cell Against Reverse-Biasing Condition: A
Simulation Study 336
  Omar M. Saif, Abdelhalim Zekry, Ahmed Shaker, M. Abouelatta, Ahmed Saeed

Dynamic Simulation of a Load-Matching Photovoltaic System for Green Hydrogen Production.................. 339
  Christian R. Polo, William J. Parquette, Kelvin Tan, Meng Tao

Embodied Energy and CO2 from the Manufacture of Cadmium Telluride and Silicon Photovoltaics............ 344
  Hope Wikoff, Samantha B Reese, Matthew O Reese

Collection of Heat Loss in Photovoltaic System by Parallely Connected Thermoelectric Network 345
  Joel Erickson, Jing Bai

Ultra-Thin and Lightweight CdS/CdTe Solar Cell Fabricated on Ceramic Substrate for Space
Applications... 348
  *Manoj K. Jamarkattel, Adam B. Phillips, Geethika K. Liyanage, Fadhil K. Alfadhili, Ebin
  Bastola, Victor V. Plotnikov, Alvin D. Compaan, Randy J. Ellingson, Michael J. Heben*

Continuous Flash Sublimation of Inorganic Halide Perovskites: Enabling Industrially Compatible
Deposition Rates.. 351
  *Tobias Abzieher, Christopher P. Muzzillo, Mirzo Mirzokarimov, Ahmad R. Kirmani,
  Gabriella Lahti, Wylie Kau, Daniel M. Kroupa, Joseph M. Luther, David T. Moore*

Temperature- And Illumination-Dependent Characterization of Wide Bandgap Sulfide CIGS and
CZTS Solar Cells.. 352
  *Simon M. F. Zhang, Guojun He, Chang Yan, Kaiwen Sun, Xiaojing Hao, Ivan Perez-Wurfl,
  Ziv Hameiri*

Fill Factor Prediction of Modern Industrial Cells: Potential Gaps and Improvements.................................... 353
  Gaia Maria N Javier, Priya Dwivedi, Yoann Buratti, Thorsten Trupke, Ziv Hameiri

Real-Time Prediction Algorithms to Detect Clouds and Forecast Photovoltaic System Performance............ 354
  *Maqsood Ali Mughal, Habeebullah Adua, Muhammad Hammad Uddin, Evan Sauter, Stephen
  Natale, Timothy Lewis, Jonathan G. Ferreira*

Surrogate Modeling for Rapid Prediction of Energy Yield from Vehicle-Integrated Photovoltaics 362
  Timofey Golubev

Novel Laser Oxidation for Screen-Printed Selective Area Front Poly-Silicon Contacts for TOPCon
Cells.. 366
  *Sagnik Dasgupta, Young-Woo Ok, Vijaykumar D. Upadhyaya, Wook-Jin Choi, Ying-Yuan
  Huang, Shubham Duttagupta, Ajeet Rohatgi*

Contactless Determination of Emitter Sheet Resistance for Diffused Silicon Wafers 367
  Yan Zhu, Thorsten Trupke, Ziv Hameiri

Photodoping Causes Inconsistencies in the Injection-Dependent Lifetimes of Perovskite Thin Films............ 368
  *Robert A Lee Chin, Arman Soufiani, Jianghui Zheng, Paul Fassl, Anita Ho-Baillie, Ulrich
  Paetzold, Thorsten Trupke, Ziv Hameiri*

Investigating the Impurity Gettering Rate in Polycrystalline-Silicon Based Passivating Contacts 369
  *Zhongshu Yang, Jan Krügener, Frank Feldmann, Jana-Isabelle Polzin, Bernd Steinhauser,
  Tien T. Le, Daniel Macdonald, Anyao Liu*

Differences of CIGS Cell Performance with Zn(O, S)/(Zn, Mg)O Or CdS/i-ZnO Buffers System
Explored by Numerical Simulations.. 370
  Giovanna Sozzi, Dimitrios Hariskos, Wolfram Witte

Fabricating High Aspect Ratio Front Contacts for Solar Cells by String-Printing.. 374
  Mathis Van De Voorde, Rebecca Saive

Benchmarking PV Performance Models with High Quality IEC 61853 Matrix Measurements
(Bilinear Interpolation, SAPM, PVGIS, MLFM and 1-Diode) 375
  Steve Ransome

Determining the Decomposition Voltage of $Cu(In_{1-x}Ga_x)Se_2$ 381
  Klaas Bakker, Joaquin Coll Matas, Johan Bosman, Nicolas Barreau, Arthur Weeber, Mirjam Theelen

A Combined Shading and Radiation Simulation Tool for Defining Agrivoltaic Systems 384
  Haomiao Wang, Henry J. Williams, Xiaotong Bu, K. Max Zhang

Measurement of Band Alignment Between ZnO Based Front Emitters and $CdCl_2$ Treated CdSeTe/CdTe Absorbers 387
  Xiaolei Liu, Luke Jones, Luksa Kujovic, Nicholas Hunwick, Luis Infante-Ortega, Michael Walls, Tushar Shimpi, Walajabad Sampath, Kurt Barth, Stephen Jones, Ochai Oklobia, Stuart Irvine

Performance Investigation and Analysis of Anti-Soiling Coatings in Hot Desert Climate 390
  Hebatalla Alhamadani, Shaikha Hassan, Gerhard Mathiak, Omar Albadwawi, Vivian Alberts

Light Distribution and Uniformity Evaluation of Cross Compound Parabolic Concentrators 395
  Mazin Al-Shidhani, Mohammad Alnajideen, Gao Min

A Comparative Study of 3D Printed Non-Imaging Solar V-Trough and Compound Parabolic Concentrators for Low-Cost, High-Performance CPV Applications 396
  Mohammad Alnajideen, Mazin Al-Shidhani, Gao Min

Development of Photovoltaic Inverter Model with Islanding Detection Using the Sandia Frequency Shift Method 398
  Nelson E. Saavedra-Peña, Rachid Darbali-Zamora, Edgardo Desarden-Carrero, Erick Aponte-Bezares

Optical Properties of Thin Film Sb2Se3 and Identification of Its Electronic Losses in Photovoltaic Devices 405
  Niva K. Jayswal, Suman Rijal, Biwas Subedi, Indra Subedi, Zhaoning Song, Robert W. Collins, Yanfa Yan, Nikolas J. Podraza

Racking Reflection and Shading Effects on Single Axis Tracked Bifacial Photovoltaic Modules 406
  Mandy R Lewis, Trevor J Coathup, Annie C J Russell, Javier Guerrero-Perez, Christopher E Valdivia, Karin Hinzer

Spectroscopic Ellipsometry Analysis and Quantum Efficiency Simulation of $CuInSe_2$ Solar Cells 407
  Dhurba R. Sapkota, Ambalanath Shan, Balaji Ramanujam, Puja Pradhan, Richard Irving, Adam B. Phillips, Michael J. Heben, Randy J. Ellingson, Sylvain Marsillac, Nikolas J. Podraza, Robert W. Collins

Novel Interconnection Method for Micro-CPV: 132 Solar Cell Prototype 413
  Norman Jost, Steve Askins, Richard Dixon, Mathieu Ackermann, Cesar Dominguez, Ignacio Anton

Understanding the Behavior of Fixed Composition $CdSe_xTe_{1-x}$(CST) Solar Cells 414
  Ebin Bastola, Adam B. Phillips, Abasi Abudulimu, Vlad Kornienko, Manoj K. Jamarkattel, Zulkifl H. Rabbani, Jared D. Friedl, Prabodika N. Kaluarachchi, Ali Abbas, Abdul Quader, Xavier Mathew, Michael Walls, Randy J. Ellingson, Michael J. Heben

Effects of Novel In+RbF Post-Deposition Treatment on $Cu(In_xGa_{1-x})Se_2$ Solar Cells 415
  Jake Wands, Polyxeni Tsoulka, Thomas Lepetit, Nicolas Barreau, Angus Rockett

Seasonal Dependence of Bifacial Photovoltaic Array Gain Due to Inverter Clipping 418
  Thunchanok Kaewnukultorn, Steven Hegedus

Accuracy of Potential High Limit Estimation for Solar Plants in the Southeast US 419
William B. Hobbs, David J. Ault, Vahan Gevorgian, Govind Saraswat

Improved Efficiency of Non-Toxic Cu3BiS3 Thin Film Solar Cell Employing PCBM Electron
Transport Layer 424
Sandip Das

Evaluating Intrinsic Defects Across CIGS Absorber Via X-Ray Absorption Near Edge Structures 425
Srisuda Rojsatien, Tara Nietzold, Niranjana Kumar, Barry Lai, Jeff Bailey, Arun Mannodi-Kanakkithodi, Maria K. Y. Chan, Mariana Bertoni

Accelerated Durability Evaluation of Emerging Cell Interconnect Technologies........................... 426
Fang Li, Dylan J. Colvin, Kristopher O. Davis, Andrew Gabor, Govindasamy Tamizhmani

Tracking Se Local Structures Across CdSeTe Absorber with X-Ray Microscopy 429
Srisuda Rojsatien, Niranjana Kumar, Trumann Walker, Barry Lai, Dan Mao, Arun Mannodi-Kanakkithodi, Maria K. Y. Chan, Mariana Bertoni

Epitaxial Growth of Detachable GaAs/Ge Heterostructure on Mesoporous Ge Substrate for Layer
Separation and Substrate Reuse.. 430
*Nicolas Paupy, Bouraoui Ilahi, Zakaria Oulad Elhmaidi, Valentin Daniel, Tadeáš Hanuš,
Roxana Arvinte, Alexandre Heintz, Alex Brice Poungoué Mbeunmi, Thierno Mamoudou
Diallo, Richard Arès, Abderraouf Boucherif*

Analyzing Hosting Capacity Protection Constraints Under Time-Varying PV Inverter Fault
Response.. 431
Joseph A. Azzolini, Nicholas S. Gurule, Rachid Darbali-Zamora, Matthew J. Reno

Tuning Thermal Induced Porous-Ge Reconstruction for Layer Transfer and Substrate Re-Use 439
*Ahmed Ayari, Bouraoui Ilahi, Roxana Arvinte, Tadeas Hanus, Laurie Mouchel, Denis
Machon, Abderraouf Boucherif*

Simulation-Based Determination of Shockley-Read-Hall Recombination Lifetimes in Group-V
Doped P-N Junction CdTe Devices.. 440
Alexandra M. Bothwell, Darius Kuciauskas

Hardware-In-The-Loop Lab for Testing Grid Supporting Functions of Smart Inverters 441
Thunchanok Kaewnukultorn, Sergio Sepúlveda-Mora, Steven Hegedus

A Deep Learning Approach to Denoise Electroluminescence Images of Solar Cells 442
Grace Liu, Priya Dwivedi, Thorsten Trupke, Ziv Hameiri

Temperature Dependence of Silicon-Dielectric Interface Recombination 443
*Anh Huy Tuan Le, Eduardo Prieto Ochoa, Ruy Sebastian Bonilla, Nino Borojevic, Ziv
Hameiri*

Optical Simulations of All-Inorganic CsPbBr3 Perovskite Quantum Dot Intermediate Band Solar
Cells (QDIBSCs)... 444
Ola Rashwan, Chase Sasala

Life-Cycle Analysis of crystalline-Si "Direct Wafer" and Tandem Perovskite PV Modules and
Systems.. 447
Enrica Leccisi, Adam Lorenz, Vasilis Fthenakis

Bandgap Model Using Symbolic Regression for Environmentally Compatible Lead-Free Inorganic
Double Perovskites.. 452
Ahmer A. B. Baloch, Omar Albadwawi, Badreyya Alshehhi, Vivian Alberts

Enhancement in the Efficiency of Rear Emitter SHJ Solar Cells by Using a CaF2/ITO Double-Layer Anti-Reflective Coating 456
  Muhammad Aleem Zahid, Muhammad Quddamah Khokhar, Youngkuk Kim, Junsin Yi

A Thermal Model for Bifacial PV Panels.. 457
  Shahzada Pamir Aly, Jim Joseph John, Gerhard Mathiak, Omar Albadwawi, Luis Pomares, Vivian Alberts

Planar Transparent Conductive Oxide/Ag Rear Contacts for High Efficiency III-V Photovoltaics................ 460
  Christopher T. Gregory, Sean J. Babcock, Richard R. King

Quantitative Measurement of Active Dopant Density Distribution in Black Silicon Solar Cell Using Scanning Nonlinear Dielectric Microscopy.. 461
  Yasuo Cho, Beniamino Iandolo, Ole Hansen

Flexible and Lightweight CdS/CdTe Solar Cells Via a Water-Assisted Lift-Off Process.............................. 464
  Sandip S Bista, Deng-Bing Li, Suman Rijal, Sabin Neupane, Rasha A Awni, Randy J. Ellingson, Zhaoning Song, Adam Phillips, Michael Heben, Yanfa Yan

Public Road Tests of Toyota Prius Equipped with High Efficiency PV Module with Output Power of 860W 467
  Taizo Masuda, Takashi Nakado, Masafumi Yamaguchi, Tatsuya Takamoto, Kensuke Nishioka, Kazumi Yamada

Revealing Sub-Cell Degradation of Multi-Junction Solar Cells by Absolute Electroluminescence Imaging.. 468
  Youyang Wang, Liying Li, Xiaobo Hu, Yun Jia, Guoen Weng, Xianjia Luo, Shaoqiang Chen, Hidefumi Akiyama

Using Machine Learning to Predict the Complete Degradation of Accelerated Damp Heat Testing in Just 10% of Testing Time.. 472
  Zubair Abdullah-Vetter, Priya Dwivedi, Robert Lee Chin, Brendan Wright, Thorsten Trupke, Ziv Hameiri

CuCl Doping Variations in High Efficiency Polycrystalline CdSeTe/CdTe Thin Film Solar Cells 475
  Zachary F. Lustig, Tushar M. Shimpi, Akash Shah, Walajabad S. Sampath

Automated Analysis of Internal Quantum Efficiency Using Chain Order Regression............................... 476
  Zubair Abdullah-Vetter, Priya Dwivedi, Yoann Buratti, Alfred Krzywicki, Arcot Sowmya, Thorsten Trupke, Ziv Hameiri

Photon Recycling and Luminescent Coupling in All-Perovskite Tandem Solar Cells Quantified by Full Opto-Electronic Device Simulation 479
  Urs Aeberhard, Simon J. Zeder, Beat Ruhstaller

The National Solar Radiation Database (NSRDB): Current Status... 480
  Aron Habte, Manajit Sengupta, Yu Xie, Grant Buster, Michael Rossol, Paul Edwards, Galen Maclaurin, Evan Rosenlieb, Jaemo Yang, Haiku Sky, Mike Bannister, Billy Roberts

FTO Delamination for Photovoltaic Module Separation 481
  Jongwon Ko, Soohyun Bae, Yoonmook Kang, Hae-Seok Lee, Donghwan Kim

Investigation of CsF - Treatment Effects on Cu(In,Ga)(S,Se)2 Solar Cells Using Photothermal Atomic Force Microscopy Under Various Photoexcitation Conditions 482
  Ayaka Yamada, Takuji Takahashi

Polysilicon Passivating Contact Layer for Crystalline Silicon Solar Cells: A Dopant-Grading Approach 483
Duy Phong Pham, Junsin Yi

Effects of (i)a-Si:H Deposition Temperature on Passivation Quality and Performance of High-Efficiency Silicon Heterojunction Solar Cells 484
Yifeng Zhao, Paul Procel, Arno H. M. Smets, Luana Mazzarella, Can Han, Liqi Cao, Guangtao Yang, Zhirong Yao, Arthur Weeber, Miro Zeman, Olindo Isabella

Investigation of the Crack Propensity of Co-Extruded Polypropylene Based Backsheets 485
Gernot Oreski, Chiara Barretta, Astrid Macher, Gabriele Eder, Lukas Neumaier, Markus Feichtner, Minna Aarnio-Winterhof

Comparative Life Cycle Assessment of Crystalline Silicon Glass-Sheet Based PV Modules and Plastic PV Modules 489
Sakthi Guhan Somasundaram, Xitong Zhu, Angele Reinders

Study of ALD-Grown Tin Oxide as an Electron Selective Layer for NIP Perovskite-Based Solar Cells 497
Félix Gayot, Elise Bruhat, Matthieu Manceau, Eric De Vito, Denis Mariolle, Stéphane Cros

Investigation of Lead-Free 2D/3D Mixed-Dimensional Tin Perovskite Solar Cell Embedded with Plasmonic Metal Nanoparticles 504
Atanu Purkayastha, Manoranjan Minz, Ramesh Kumar Sonkar, Arun Tej Mallajosyula

High-Efficiency Solar Cell by Combining High and Low Thermal Budget for Si Passviting Contacts 507
Muhammad Quddamah Khokhar, Shahzada Qamar Hussin, Muhammad Aleem Zahid, Duy Phong Pham, Eun-Chel Cho, Junsin Yi

The Role of the European Green Deal for the Photovoltaic Market Growth in the European Union 508
Arnulf Jäger-Waldau, Georgia Kakoulaki, Nigel Taylor, Sandor Szábo

Assessment of Mechanical Robustness of Conventional and CFRP-Based Lightweight PV Module Architectures Under Static Loads 512
Umang Desai, Aparna Singh

Exploring the Role of Temperature and Hole Transport Layer on the Ribbon Orientation and Efficiency of Sb2Se3 Cells Deposited Via Thermal Evaporation 516
Ryan Voyce, Stephen Campbell, Oliver S. Hutter, Guillaume Zoppi, Neil S. Beattie, Elizabeth A. Gibson, Vincent Barrioz

Decentralized BESS Control on a Real Low Voltage System with a Large Number of Prosumers 517
Bruno Cortes, Ricardo Torquato, Tiago R. Ricciardi, Fernanda C. L. Trindade, Walmir Freitas, Victor B. Riboldi, Kunlin Wu

Hydrogen Complexes Present After Different Firing Profiles and Their Influence on LeTID Degradation 525
Benjamin Hammann, Nicole Assmann, Philip M. Weiser, Wolfram Kwapil, Tim Niewelt, Florian Schindler, Rune Søndenå, Eduard V. Monakhov, Martin C. Schubert

Superior Performance of Two-Phase Triple Halide Inorganic Perovskites 526
Deniz N. Cakan, Rishi F. Kumar, Connor Dolan, Moses Kodur, Yanqi Luo, Tao Zhou, Zhonghou Cai, Barry Lai, Martin Holt, David P. Fenning

Thermally Evaporated Titanium Dioxide Film as an Electron-Selective Contact for Silicon Solar Cells.. 527
> Changhyun Lee, Soohyun Bae, Hyunju Lee, Yoonmook Kang, Hae-Seok Lee, Donghwan Kim

3 MeV Proton Radiation Tolerance Study of Ultra-Thin Gallium Arsenide Solar Cells for Space Applications... 528
> Larkin Sayre, Armin Barthel, Andrew Johnson, Louise C Hirst

Monolithic Perovskite/Silicon Tandem Solar Cells on P-Type POLO/PERC Silicon Bottom Cells 529
> Silvia Mariotti, Klaus Jäger, Marvin Diederich, Marlene S. Härtel, Bor Li, Kári Sveinbjörnsson, Eike Köhnen, Rolf Brendel, Sarah Kajari-Schröder, Robby Peibst, Steve Albrecht, Lars Korte, Tobias Wietler

Multiple Substrate Reuse: A Straightforward Reconditioning of Ge Wafers After Porous Separation 530
> Alexandre Chapotot, Javier Arias-Zapata, Tadeáš Hanuš, Bouraoui Ilahi, Nicolas Paupy, Valentin Daniel, Zakaria Oulad El Hmaidi, Jérémie Chrétien, Gwenaëlle Hamon, Maxime Darnon, Abderraouf Boucherif

Short Drying Processes for Silicon Solar Cells 531
> Daniel Ourinson, Michael Linse, Markus Klawitter, Andreas Lorenz

Sn4+-Free, Stable Tin Perovskite Films for Lead-Free Perovskite Solar Cells................................. 532
> Ajay Singh, Jeremy Hieulle, Himanshu Phirke, Joana A. F. Machado, Sevan Gharabeiki, Rukhsar Ahmad, Susanne Siebentritt, Alex Redinger

Fill Factor Losses in Cu(In,Ga)Se2 Based Solar Cells Due to Metastabel Defects — the Effect of Ag Addition... 533
> Thomas P. Weiss, Omar Ramirez, Taowen Wang, Valentina Serrano-Escalante, Stefan Paetel, Wolfram Witte, Jiro Nishinaga, Thomas Feurer, Ayodhya N. Tiwari, Susanne Siebentritt

Analysis of the Soiling Effects on Commissioning of Photovoltaic Systems: Short-Circuit Current Correction... 534
> Dênio Alves Cassini, Suellen C. Silva Costa, Antonia Sonia A. C. Diniz, Lawrence L. Kazmerski

Optical Absorption of MoS2 Based Ultrathin Solar Cells.. 538
> Carlos Bueno-Blanco, Simon Aurel Svatek, Elisa Antolin

Microinverter Testing Update Using High Power Modules: Efficiency, Yield, and Conformity to a New "Estimation Formula" for Variation of PV Panel Size... 539
> Stefan Krauter, Jörg Bendfeld, Marius Möller

Glare Potential Evaluation of Structured PV Glass Based on Gonioreflectometry 544
> Markus Babin, Sune Thorsteinsson, Adrian A. Santamaria Lancia, Michael L. Jakobsen, Sergiu V. Spataru

Upstream-Downstream Optimization of Volt-Var Control in Smart Grids..................................... 545
> Laura R. Fardin, Christiano Lyra, Fernanda C. L. Trindade

Growth of GaAs on Ge/Si (001) Nanovoided Virtual Substrate 550
> Jonathan Henriques, Alexandre Heintz, Bouraoui Ilahi, Richard Arès, Abderraouf Boucherif

Monte Carlo Evaluation of Multijunction Solar Systems in Tandem and 4-Terminal Configurations 551
> Roberto Corso, Marco Leonardi, Andrea Scuto, Salvatore A. Lombardo

Outdoor Energy Performances for Standard and Bi-Facial Modules as Well on the Failure Modes Observed in Outdoor Conditions 554
A. Ottanà, F. Rametta, W. Gangemi, C. Colletti, A. Di Stefano, A. Canino, M. Foti, C. Gerardi, F. Bizzarri

Nanoabsorbers for Semitransparent Photovoltaics 557
Maximilian Götz-Köhler, Hosni Meddeb, Norbert Osterthun, Nils Neugebohrn, Kai Gehrke, Martin Vehse

State of the Art of Modelling Soiling and Snow Losses in PV Systems 562
Sébastien Arbaretaz, Murielle Stepec, Eszter Voroshazi

Paving the Way to Building-Integrated Translucent Tandem Photovoltaics: Process Optimization and Transfer to Perovskite-Perovskite 2-Terminal Tandem Cells 565
David Benedikt Ritzer, Marco Alejandro Ruiz-Preciado, Bahram Abdollahi Nejand, Tobias Abzieher, Ulrich Wilhelm Paetzold

Terawatt-Scale Photovoltaics Enabled by Technological Learning 566
Lukas Wagner, Robert Pietzcker, Lorenz Friedrich, Jan Christoph Goldschmidt

Effects of Solar Spectrum and Albedo on the Performance of Bifacial Si Heterojunction Mini-Modules 567
Marco Leonardi, Roberto Corso, Andrea Scuto, Gabriella Milazzo, Carmelo Connelli, Marina Foti, Cosimo Gerardi, Fabrizio Bizzarri, Stefania M. S. Privitera, Salvatore A. Lombardo

Role of Back-Side Indium Tin Oxide on the Degradation Mechanism of Silicon Heterojunction Solar Cells 570
Gbenga D. Obikoya, Anishkumar Soman, Ujjwal K. Das, Steven S. Hegedus

Vinyl Acetate Content Tailoring in Ethylene Vinyl Acetate Improves the Resilience Against Environmental Stressors 574
Umang Desai, Bhuwanesh Kumar Sharma, Aparna Singh

Poisson Drift Diffusion Modeling of Valley Photovoltaic Devices 575
Daixi Xia, Hassan Allami, Jacob J. Krich

2T Mechanically Stacked Perovskite/Si Tandem Cells Beyond 28%: The Role of 2D Materials in Perovskite Top Cells Coupled with a Commercially Available Bifacial c-Si Heterojunction Cell 576
Antonio Agresti, Sara Pescetelli, Fabio Matteocci, Erica Magliano, Elisa Nonni, Giuseppe Bengasi, Carmelo Connelli, Cosimo Gerardi, Hanna Pazniak, Sebastiano Bellani, Francesco Bonaccorso, Fabrizio Bizzarri, Marina Foti, Aldo Di Carlo

CdTe:In - Post-Growth Doping and Proposals for Photovoltaic Devices 577
Luke Thomas, Theo DC Hobaon, Laurie J Phillips, Kieran J Cheetham, Neil Tarbuck, Mark Isaacs, Huw Sheil, Vin Dhanak, Tim D Veal, Stephen Campbell, Vincent Barrioz, Jon D Major, Ken Durose

Multiple Inverter Microgrid Experimental Fault Testing 578
Nicholas S. Gurule, Javier Hernandez Alvidrez, Matthew J. Reno, Jack Flicker

Reactive Anisotropic Conductive Adhesive Wafer Bonding for Solar Cells 584
Eric M. Rehder, Shoghig Mesropian, Xing-Quan Liu

The Effect of Dust Hygroscopicity on Soiling and Self-Cleaning Processes in a Condensing Environment 588
Jordan Eidlisz, Nadera Sultana, Illya Nayshevsky, Qianfeng Xu, Alan M. Lyons

On the Stability of Indium Tin Oxide with Functional Layers Back Contact Applications in Semitransparent Cu(In,Ga)Se2 Solar Cells 591

Robert Fonoll-Rubio, Marcel Placidi, Torsten Hoelscher, Angelica Thomere, Zacharie Jehl Li-Kao, Maxim Guc, Victor Izquierdo-Roca, Roland Scheer, Alejandro Pérez-Rodríguez

Bulk Lifetime Study of P-Type Czochralski Silicon with Different Processing History Using Quinhydrone-Methanol Surface Passivation 592

Tasnim K. Mouri, Ajay Upadhyaya, Ajeet Rohatgi, William N Shafarman, Ujjwal K. Das

PV Module Degradation Due to Frequent and Prolonged Inverter Clipping: A Preliminary Study 596

Manjunath Matam, Ryan M. Smith, Hubert Seigneur

ETFE and Its Role in the Fabrication of Lightweight c-Si Solar Modules 604

Fabiana Lisco, Farwa Bukhari, Luke Jones, Adam Law, John Michael Walls, Christophe Ballif

Thermal Stability of 2D/3D Halide Perovskites 605

Jeffrey A Christians, Josephine L Surel, Elizabeth V Cutlip

Life Cycle Assessment Analysis of Thin-Film, Flexible Solar Panels Produced in the Netherlands 606

Gianluca Limodio, Seba Makhlouf, Edward Hamers, Arno Smets

Front SiON/TCO Stacks Development for Double Side Poly-Si/SiOX Passivated Contacts Solar Cells 607

Charles Seron, Thibaut Desrues, Christine Denis, Raphaël Cabal, Frédéric Jay, Adeline Lanterne, Quentin Rafhay, Anne Kaminski, Sébastien Dubois

Comparing the Accuracy of Horizon Shade Modelling Based on Digital Surface Models Versus Fisheye Sky Imaging 608

Daniel Alvarez Mira, Martin Bartholomäus, Sebastian Poessl, Peter B. Poulsen, Sergiu V. Spataru

Ongoing Performance Assessment Strategies & Operational Challenges When Managing Hundreds of Distributed Photovoltaic Assets Across Asia 614

André M. Nobre, Anusha Agarwal, Sai Pranav

Development of Highly Uniform and Reproducible DI-O_3 Layers for Photovoltaic Applications and Beyond 620

Munan Gao, Vibhor Kumar, Ngwe Zin

Characterizing the Capacitance of Different c-Si PV Cell Technologies Using Impedance Spectroscopy 623

David A. Van Nijen, Patrizio Manganiello, Mirco Muttillo, Miro Zeman, Olindo Isabella

Demonstration of a Monolithically Integrated Hybrid Electroabsorptive Modulator/Photovoltaic Device for Bidirectional Free Space Optical Communication at 1.55 μm 624

Emily Kessler-Lewis, Stephen J. Polly, Elijah Sacchitella, Seth M. Hubbard, Raymond Hoheisel

Glued III-V on Si Tandem Solar Cells Using Hybrid Transparent Conductive Layers 625

Phuong-Linh Nguyen, Jeronimo Buencuerpo, Philippe Baranek, Oliver Hoehn, David Lackner, Frank Dimroth, Marco Faustini, Stephane Collin, Andrea Cattoni

Flexible All-Perovskite Tandem Solar Cells with High Specific Power 626

Zhaoning Song, Cong Chen, Chongwen Li, Lei Chen, Yanfa Yan

Impact of Thermal Annealing on the Mechanical Properties of Ge Epilayer on Mesoporous Germanium for Layer Separation and Substrate Re-Use 627
    Firas Zouaghi, Ahmed Ayari, Bouraoui Ilahi, Jeremie Chretien, Tadeas Hanus, Nicolas Paupy, Nicolas Quaegebeur, Abderraouf Boucherif

Development of a Novel Soiling Chamber for Testing Antisoiling Coatings 628
    Matthew T Muller

Progress and Demonstration of Micro-CPV Module with Integrated Planar Tracking and Diffuse Light Collection 629
    Steve A Askins, Guido Vallerotto, César Dominguez, Mathilde Duchemin, Gaël Nardin, Mathieu Ackermann, Delphine Petri, Matthieu Despeisse, Jacques Levrat, Xavier Niquille, Christophe Ballif, Juan F Martinez, Marc Steiner, Gerald Siefer, Ignacio Antón

A Deep Learning Approach for PV Failure Mode Detection in Infrared Images: First Insights 630
    Daniel Rocha, Miguel Lopes, Jennifer P. Teixeira, Paulo A. Fernandes, Modesto Morais, Pedro M. P. Salome

What Are PVDF-Based Backsheets Made Of? 633
    Chiara Barretta, Eric Helfer, Astrid E. Macher, Gernot Oreski

Excess Current Due to Embedded Superlattices in Graphene/Ox/n-GaAs Solar Cells, at 50 Suns and Above 637
    AC Varonides

Stability of Silicon Heterojunction Solar Cells Having Hydrogen Plasma Treated Intrinsic Layer 643
    Anishkumar Soman, Gbenga Obikoya, Steve Johnston, Steven Harvey, Ujjwal Das, Steven Hegedus

Snow Shedding Properties of Bifacial PV Panels 646
    Ajay Singh, Derek Jones

Glass-Glass PV Modules: Characterization of Chemical and Mechanical Degradation 649
    Laura Spinella, Sona Ulicna, Archana Sinha, Dana B. Sulas-Kern, Michael Owen-Bellini, Steve Johnston, Laura T. Schelhas

Model of an Automous PV Home Using a Hybrid Storage System Based on Li-Ion Batteries and Hydrogen Storage with Waste Heat Utilization 650
    Marius Möller, Stefan Krauter

Towards High Efficiency All-Perovskite Tandem Solar Cell by Preventing Performance Loss Arising from Physically Mixed Interfacial Layers 653
    Biwas Subedi, Alex Bordovalos, Lei Chen, Zhaoning Song, Cong Chen, Yanfa Yan, Nikolas J Podraza

Nonparametric Temporal Downscaling of GHI Clear-Sky Indices Using Gaussian Copula 654
    Jing Huang, Marc Perez, Richard Perez, Dazhi Yang, Patrick Keelin, Tom Hoff

Ga-Doping of MZO in CdSeTe/CdTe Thin Film Solar Cells 658
    Mustafa Togay, Tushar Shimpi, Sampath S. Walajabad, Kurt L. Barth, Eric Don, Gabor Parada, J. Michael Walls, Jake W. Bowers

Firm PV Power Generation in Switzerland 661
    Jan Remund, Marc Perez, Richard Perez

Demonstration of Point Contact Geometry for Solar Cells Using Single Walled Carbon Nanotube 667
  Fadhil K. Alfadhili, Adam B. Phillips, Manoj K. Jamarkattel, Bhuiyan M. Anwar, Prabodika N. Kaluarachchi, Zahrah S. Almutawah, Abdul Quader, Deng-Bing Li, Yanfa Yan, Randy J. Ellingson, Michael J. Heben

Accelerate Cycles of Learning: Unencapsulated Silicon Photovoltaic Cells to Environmental Stressors 668
  Nafis Iqbal, Nitin K. Chockalingam, Kehley A. Coleman, Jeffrie Fina, Kristopher O. Davis, Laura S. Bruckman, Ina T. Martin

End of Use, Circularity, and Sustainability Considerations in Solar Photovoltaic Module Design and Product Development and Support 675
  Chris Powicki, Wayne Li, Cara Libby

Degradation of Crystalline Silicon Photovoltaic Modules Installed in Different Climates 680
  Chiara Barretta, Astrid E. Macher, Julián Ascencio-Vásquez, Marc Köntges, Marko Topic, Gernot Oreski

Comparing Fluorinated and Non-Fluorinated Anti-Soiling Coatings for Solar Panel Cover Glass 683
  Luke O. Jones, Adam M. Law, Gary Critchlow, John M. Walls

Extraction of Prevailing Soiling Rates from Soiling Measurement Data 684
  Josh Peterson, Julie Chard, Justin Robinson

Mismatch Losses in Simulated Commercial and Utility-Scale PV Arrays Due to Shortened Strings 692
  Ryan M. Smith, Manjunath Matam, Hubert Seigneur

Fault Analysis and Relay Assessment on a Substation System with High Penetration of PV Generation 693
  Biqi Wang, Genesis Alvarez, Micah J. Till, Kevin Jones, Mathew Gardner, Rolando Burgos, Bo Wen

Determining Surface Recombination Velocity and Band Bending at the Back Interface of CdTe Devices Using Back Illuminated Quantum Efficiency 701
  Adam B. Phillips, Jared D. Friedl, Zhaoning Song, Ramez Hosseinian Ahangharnejhad, Ebin Bastola, Zulkifl H. Rabbani, Deng-Bing Li, Yanfa Yan, Randy J. Ellingson, Michael J. Heben

Use of a Selenium-Telluride Alloy as a Back Interface for CdTe-Based Cells 702
  Daniel Z. Shaw, Camden L. Kasik, Andrew C. Treglia, James R. Sites

Effect of Dilute Acid Exposure on Sol-Gel Porous Silica Anti-Reflection Coatings 705
  F. Bukhari, L. Jones, A. Law, A. Abbas, J. M. Walls

Insights into the Stability of Amorphous/Crystalline Silicon Interface Under Light and Temperature 708
  Salman Manzoor, Mariana Bertoni

Automated Shift Detection in Sensor-Based PV Power and Irradiance Time Series 709
  Kirsten Perry, Matthew Muller

The Effect of Inverter Loading Ratio on Energy Estimate Bias 714
  Kevin S. Anderson, William B. Hobbs, William F. Holmgren, Kirsten R. Perry, Mark A. Mikofski, Rounak A. Kharait

Life Cycle Assessment of High-Efficiency Si Solar Modules 721
  Estefania Papaioannou, Pritpal Singh, Ross Lee

Alternative Rear Contacts for Ultrathin CdSe$_x$Te$_{1-x}$ Solar Cells........722
Bérengère Frouin, Andrea Cattoni, John Moseley, David Albin, Joel Duenow, Abderrahime Sekkat, David Muñoz-Rojas, Stéphane Collin

Feeder Open-Phase Detection by Smart Inverters........725
Yiwei Ma, Xiaojie Shi, Aminul Huque, Roland Bründlinger, Ron Ablinger

Design with Integrated PV Technologies in Various Products and Environments........731
Eli Shirazi, Wouter Eggink, Xitong Zhu, Angele Reinders

Electric Field and Its Effect on Hot Carriers in InGaAs Valley Photovoltaic Devices........732
Kyle R. Dorman, Vincent R. Whiteside, David K. Ferry, Tetsuya D. Mishima, Hamidreza Esmaielpour, Michael B. Santos, Ian R. Sellers

Single-Axis Tracker Control Optimization Potential for the Contiguous United States........733
Kevin Anderson, Saurabh Aneja

Outdoor Characterization of Hybrid HCPV- T Module Featuring a Passive Tracking System........739
Guido Vallerotto, Steve Askins, Javier Van Herpt, David Martí, Jaime Caselles, Ignacio Antón

CIGS Degradation Due to Water Ingress: Post-Mortem Analysis of a Field-Exposed PV Module........740
Simona Villa, Remi Anitat, Pelin Yilmaz, Aldo Kingma, Mikolaj Dziechciarz, Joran Van Den Berg, Klaas Bakker, Mirjam Theelen

Evaluation of Cellular Based DER Direct Transfer Trip (DTT) Technologies........741
Yiwei Ma, Aminul Huque, Joseph Estrada, Tim Godfrey, Charles Brewster

Towards a Shade Tolerant Monolithically Interconnected Perovskite Module for Use in Four Terminal Tandem Devices........748
Klaas Bakker, Jacopo Sala, Mehrdad Najafi, Michaël Daenen, Bart Geerligs

Panel Segmentation: A Python Package for Automated Solar Array Metadata Extraction Using Satellite Imagery........751
Kirsten Perry, Christopher Campos

Optical Determination of Carrier Concentrations in ITO, PEDOT:PSS, and (FASnI3)0.6(MAPbI3)0.4 Within a PV Device........752
Madan K Mainali, Prakash Uprety, Zhaoning Song, Changlei Wang, Indra Subedi, Kiran Ghimire, Maxwell M Junda, Yanfa Yan, Nikolas J Podraza

Strategies to Optimize and Validate Tracking Performance of Single-Axis Trackers on Diffuse Sites........753
Kendra Passow, Kyumin Lee, Sanket Shah, Daniel Fusaro, Jon Sharp

Development of Hierarchical Control for a Lunar Habitat DC Microgrid Model Using Power Hardware-In-The-Loop........754
Andrew R. R. Dow, Rachid Darbali-Zamora, Jack D. Flicker, Felipe Palacios, Jeffrey T. Csank

Influence of Se Grading on the Free Carrier Profile of CdSeTe/CdTe Solar Cells........761
Jared D. Friedl, Ebin Bastola, Rasha A. Awni, Xavier Mathew, Adam B. Phillips, Yanfa Yan, Michael J. Heben

Statistical Performance Analysis on ≈ 320 Perovskite Single- And Two-Junction Solar Cells and Modules from >30 Global Sources............ 766
Tao Song, Charles Mack, Rafell Williams, Josh Gallon, Allan Anderberg, Larry Ottoson, Daniel J. Friedman, Nikos Kopidakis

GPU-Accelerated Machine Learning for Analysis of Time-Resolved Photoluminescence Data............ 769
Calvin Fai, Anthony J. C. Ladd, Charles J. Hages

Micro-Fabrication and Transfer of a Detachable Ge Epitaxial Layer Grown on Porous Germanium............ 770
Valentin Daniel, Jeremie Chretien, Gwenaelle Hamon, Mathieu De Lafontaine, Nicolas Paupy, Zakaria Oulad El Hmaidi, Bouraoui Ilahi, Tadeàš Hanus, Maxime Darnon, Abderraouf Boucherif

Native Oxide Growth on CdSeTe for Improved Back Surface Passivation............ 773
Adam Danielson, Carey Reich, Mason Mahaffey, Arthur Onno, Zach Holman, Walajabad Sampath

Optical Characterization of Thin Film Cu_xAlO_y in the CdTe Device Configuration............ 774
Indra Subedi, Kamala Khanal Subedi, Prabin Dulal, Adam B. Phillips, Michael J. Heben, Randy J. Ellingson, Nikolas J. Podraza

A Tool for the Simulation, Evaluation and Teaching the Operation of Low Power Microgrids............ 778
Johann A. Hernández M, Adolfo A. Jaramillo M, Carlos A. Arredondo-Orozco

Translating Material-Level Characterization of Carbon-Nanotube-Reinforced Composite Gridlines to Module-Level Degradation............ 783
Andre Chavez, Brian Rummel, April Jeffries, Sang M. Han, Nick Bosco, Brian Rounsaville, Ajeet Rohatgi

Testing the Abrasion Resistance of Porous SiO_2 Anti-Reflection Coatings for Solar Cover Glass............ 786
Adam M Law, Farwa Bukhari, Luke O Jones, Ali Abbas, John Michael Walls

Hydrothermally Deposited Antimony Sulfide Solar Cells with V_{OC} Approaching 800 mV............ 792
Dipendra Pokhrel, Nini Rose Mathew, Suman Rijal, Ebin Bastola, Abasi Abudulimu, Tamanna Mariam, Xavier Mathew, Adam B Phillips, Michael J Heben, Zhaoning Song, Yanfa Yan, Randy J Ellingson

Multiresonant Light Trapping in Ultra-Thin Solar Cells with Transparent Quasi-Random Structures............ 795
Eduardo Camarillo Abad, Hannah J. Joyce, Louise C. Hirst

FAIRification, Quality Assessment, and Missingness Pattern Discovery for Spatiotemporal Photovoltaic Data............ 796
William C. Oltjen, Yangxin Fan, Jiqi Liu, Liangyi Huang, Xuanji Yu, Mengjie Li, Hubert Seigneur, Xusheng Xiao, Kristopher O. Davis, Laura S. Bruckman, Yinghui Wu, Roger H. French

A Sparse and Low Rank Penalized Signal Decomposition Model with Constraints: Anomaly Detection in PV Systems............ 802
Wei Yang, Daniel Fregosi, Michael Bolen, Kamran Paynabar

Cloud Segmentation and Motion Tracking in Sky Images............ 805
Benjamin G Pierce, Joshua S Stein, Jennifer L Braid, Daniel Riley

Towards Standardization of Accelerated Stress Testing Protocols for Metal-Halide Perovskite Photovoltaic Modules 806

Michael Owen-Bellini, Timothy J Silverman, Michael G. Deceglie, Paul Ndione, Nikos Kopidakis, Ingrid Repins, Mickey Wilson, Dana B. Sulas-Kern, Joseph Berry, Laura T. Schelhas, Colin Sillerud, Jinsong Huang, Michael J. Heben, Yanfa Yan, Devin Mackenzie, Joshua S. Stein

The Balance of Thermodynamic Potentials in Solar Cells Investigated by Numerical Device Simulations 807

Felix Komoll, Uwe Rau

Hybrid Functional Calculations for Antimony Doping in CdTe 808

Intuon Chatratin, Shagorika Mukherjee, Anderson Janotti

GaAs-Based Photovoltaic Infrared Energy Harvesting for Microscale Biomedical Implants 811

Y. Sun, J. Letner, J. Lee, N. Ahmed, C. Chestek, D. Blaauw, J. Phillips

Evaluation of an LED Simulator for Single- And Multi-Junction PV Cell Performance Testing 814

Nikos Kopidakis, Tao Song, Charles Mack, Rafell Williams, Hal Friesen, Justin Bertagnolli, John Walmsley

Metastability and Degradation of CdTe Solar Cells Investigated by nm-Scale Electrical Potential Imaging 819

Chun-Sheng Jiang, David Albin, Marco Nardone, Kassidy H. Howard, Adam Danielson, Amit Munshi, Tushar Shimpi, Walajabad Sampath, Chuanxiao Xiao, Helio R. Moutinho, Mowafak M. Al-Jassim, Glenn Teeter

Photovoltaic Thermal Management in Luminescent Solar Concentrators 820

Megan E. Phelan, David R. Needell, Maggie M. Potter, Haley C. Bauser, Catherine N. Ryczek, Ralph G. Nuzzo, Harry A. Atwater

PVRPM in Python: An Overview of New Capabilities 826

Paul Lunis, Brandon Silva, Marios Theristis, Hubert Seigneur

Oxygen and Temperature Effects on NiO Buffer Layers for CdTe Solar Cells 827

Nicholas Hunwick, Xiaolei Liu, Patrick J. M. Isherwood, John. M. Walls

Optimizing $CdCl_2$ Treatment on CdTe Solar Cells Using Spray Deposition Method 828

Prabodika N. Kaluarachchi, Shannon E. Costello, Ryan Madden, Jacob M. Gibbs, Tyler R. Brau, Aesha P. Patel, Manoj K. Jamarkattel, Jared D. Friedl, Kevin G. Schaffer, Kristof J. Nieschwitz, Ebin Bastola, Adam B. Phillips, Randy J. Ellingson, Michael J. Heben

External Quantum Efficiency and Device Reflectance of CIGS PV for Terrestrial and Space Based Applications 833

Bishal Shrestha, Indra Subedi, Robert W. Collins, Nikolas J. Podraza

Radiation Tolerance, High Temperature Stability, and Self-Healing of Triple Halide Perovskite Solar Cells 836

Hadi Afshari, Sergio A Chacon, Brandon K Durant, Rose Crawford, Bibhudutta Rout, Giles E Eperon, Ian R Sellers

Hyperspectral Luminescence Imaging Analysis of Solar Cells with Localized Radiative Defects 837

Brianna Conrad, Behrang H. Hamadani

Impact of In-Situ Cd Saturation MOCVD Grown CdTe Solar Cells on as Doping and VOC 838
Ochai Oklobia, Steve Jones, Giray Kartopu, Dingyuan Lu, Wes Miller, Rajni Mallick,
Xiaoping Li, Gang Xiong, Vladislav Kornienko, Martin Bliss, Ali Abbas, Michael Walls,
Stuart J. C. Irvine

Insertion of Photovoltaic Generation in the Planning of Electricity Distribution Systems Based on
Its Economic Potential.. 839
João Cardoso Das Neves Neto, Miguel Edgar Morales Udaeta, Carlos Frederico Meschini
Almeida, Henrique Fernandes Camilo

Physics-Guided Machine Learning Identifies 5 Optimum Test Locations to Predict Global PV
Energy Yield for Arbitrary Farm Topologies.. 843
Jabir Bin Jahangir, Muhammed Tahir Patel, Muhammad A. Alam

Analyzing Effects of Solar Variability and System Location on LMP Prices..................................... 847
Mesude Bayrakci-Boz, Joseph Ranalli

Demystifying the Effect of Hydrogen Treatment on Silicon Photovoltaics 854
Govind Nanda, Sara Almenabawy, Rajiv Prinja, Geetu Sharma, Nazir P. Kherani

Bill of Materials Variation and Module Degradation in Utility-Scale PV Systems 855
Michael G. Deceglie, E. Ashley Gaulding, John S. Mangum, Timothy J Silverman, Steve W.
Johnston, James A. Rand, Mason J. Reed, Robert Flottemesch, Ingrid L. Repins

The Thermodynamics Behind the Photovoltage Generation and Photocurrent Collection in Solar
Cells... 856
Uwe Rau

Tuning the Band Gap of Magnesium Zinc Oxide to Enhance Band Alignment with CdTe Based
Photovoltaic Devices.. 857
Kerrie M Morris, Mustafa Togay, Rachael C Greenhalgh, Jake W Bowers, John M Walls

Sensitivity of Sub-Hourly Modeling Error to Project Size.. 858
Christopher Hayes, Abhishek Parikh, Mark Mikofski, Rounak Kharait

Energy-Based Soiling Loss Monitoring Approach for Solar PV System 859
Pavan Fuke, Shoubhik De, Narendra Shiradkar, Anil Kottantharayil

Predicting Materials Parameters in Colloidal Quantum Dot Photovoltaic Devices Using Machine
Learning Models Trained on Experimental Data 862
Hoon Jeong Lee, Ariana B. Hofelmann, Yida Lin, Susanna M. Thon

Development of HVPE-Grown III-V Solar Cells Passivated with AlInP.. 867
Jacob T Boyer, Kevin L Schulte, Aaron J Ptak, John Simon

Progress in PV Material Durability Test Methodologies... 868
William J. Gambogi

Measuring Irradiance with Bifacial Reference Panels.. 871
Nicholas Riedel-Lyngskær, Jan Vedde, Peter B. Poulsen, Sergiu Spataru

Local nm-Scale Imaging of Electrical Contact for Series Resistance Degradation of Silicon Solar
Cells... 872
C.-S. Jiang, S. Johnston, E. A. Gauding, M. G. Deceglie, R. Flottemesch, C. Xiao, R.
Moutinho, D. B. Sulas-Kern, J. Mangum, M. M. Al-Jassim, I. L. Repins

Perimeter Recombination in GaAs Solar Cells with Different Geometries 875
Natasha Gruginskie, Gerard Bauhuis, Peter Mulder, Elias Vlieg, John Schermer

Global Ranking of Losses to Photovoltaic Power 876
A. Kubiniec, K. Seymour, A. Bhat, J. Hazari, T. Haley, Marc Perez

PV Module Toxicity Testing Methods and Results: A Literature Review 879
F. Li, S. Shaw, C. Libby, B. Bicer, G. Tamizhmani

Validation of Novel Bifacial Photovoltaic Performance Model with 3D Shading for Fixed-Tilt and
Single-Axis Tracked Systems.. 883
Annie Russell, Christopher E. Valdivia, Cédric Bohémier, Joan E. Haysom, Karin Hinzer

Characterizing the Back-Contact Interface of Bi-Facial Poly-Crystalline CdTe Devices Using
Transmission Electron Microscopy 884
John Farrell, Ebin Bastola, Manoj Jamarkattel, Michael Heben, Robert F. Klie

Photon Management in CdSeTe Absorber Solar Cells: The Case for Increased Attention to Optical
Cell Design 887
Carey L. Reich, Arthur Onno, Adam Danielson, Zachary C. Holman, Walajabad S. Sampath

Geographic Analysis for Determining the Value of Different Photovoltaic Performance Factors.................. 888
Madhuri Kumari, Marios Theristis, Joshua S. Stein

Computerized Tool for Students Training in Solar Geometry.. 893
*Johjan Stiven Zea Fernández, Mario Luna-Delrisco, Sebastián Villegas Moncada, Carlos
Ernesto Arrieta González, Johann A. Hernández M, Carlos A Arredondo Orozco*

Anisotropy-Induced Fluctuations in Cu(In, Ga)Se2 897
Diego Colombara

Development of Spatial Mapping and Degradation Monitoring for Perovskite Films...................... 898
Emily J Miller, Biwas Subedi, Jaehoon Chung, Chongwen Li, Yanfa Yan, Nikolas J Podraza

PV+ Storage Operation and Maintenance 899
Natalie Gayoso, Nicole D Jackson, Thushara Gunda, Jal Desai, Andy Walker

Effect of Microstructure on the Photoactivity of Thin Film CdSe 900
*Rachael Greenhalgh, Kerrie Morris, Vladislav Kornienko, Martin Bliss, Ali Abbas, Jake
Bowers, Michael Walls*

AgriPV Citizen Science Lab: A Collaborative Model for Engineers, Youth Scholars and
Communities 904
*Stuart Bowden, Jazmine Cordon, Myla Dykes, Michael Hernandez, Michelle Jordan, Alex
Killam, Jasmine Martinez Castillo, Alex Park, Alondra Pita, Maryan Robledo, Steve Zuiker*

Behavioral and Population Data-Driven Distribution System Load Modeling 907
Isaac Bromley-Dulfano, Xiangqi Zhu, Barry Mather

Thin-Film Multijunction Inverted Metamorphic Solar Cells with Light Management for Space
Applications.. 913
Julia D'Rozario, Steve Polly, Rao Tatavarti, Seth Hubbard

Probing Dynamic Influence of Moisture Ingress on Cell Deflection in Photovoltaic Modules 914
*Ian M Slauch, Rishi E Kumar, Tala Sidawi, Jared Tracy, William Gambogi, Rico Meier,
David P Fenning, Mariana I Bertoni*

Potovoltaic Module R&D Considerations for Soiling Mitigation 915
  Lin J. Simpson, Matthew Brantl, Ryo Huntamer

Seven-Level Cascaded H-Bridge Multilevel Single-Phase Inverter Implemented with an ATMEGA
Microprocessor 916
  Edgardo Desarden-Carrero, Rachid Darbali Zamora, Erick Aponte-Bezares, Eduardo I. Ortiz-Rivera

Annual Energy Production Uncertainty of Bifacial PV Plants Caused by Inaccuracies in Albedo
Data: Case Studies Using SAM 923
  Vicente Lara Fanego

Arsenic Doped CdSeTe Solar Cells: Charge Collection and Defects.. 924
  Niranjana Mohan Kumar, Srisuda Rojsatien, Trumann Walker, Tara Nietzold, Barry Lai, Arun K. M. Kanakkithodi, Maria Chan, Dan Mao, Mariana Bertoni

Observations on a Colorado Electric-Utility Resource Plan for Increasing Renewables from 55% to
80% by 2030.. 925
  Ronald A. Sinton

Estimation of Soiling Losses in Unlabeled PV Data 930
  Bennet Meyers

Understanding the Solar Cell Contacts with Atmospheric Screen-Printed Copper 937
  Sandra Huneycutt, Abasifreke Ebong, Krishnamraju Ankireddy, Ruvini Dharmadasa, Thad Druffel

Na Diffusion and Device Performance of AgBr Treated CuGaSe$_2$ Thin Films 941
  Elizabeth Palmiotti, Polyxeni Tsoulka, Thomas Lepetit, Nicolas Barreau, Angus Rockett

Extensive Evaluation and Uncertainty Estimation of Albedo Data Sources 945
  Vicente Lara-Fanego, Christian A. Gueymard, Jose A. Ruiz-Arias, Tomas Cebecauer, Juraj Betak

Sampling Solar Irradiance with Copula.. 948
  Mesude Bayrakci-Boz

Analysis of Temperature Dependence of Solar Cell Performance Through Light Soaking...................... 953
  Samuel Seibert, Aesha P. Patel, Manoj Rajakaruna, Sandip S. Bista, Lei Chen, Randy J. Ellingson, Yanfa Yan, Zhaoning Song

Demonstrating the Thermoradiative Diode: Generating Electrical Power Through Radiative
Emission 959
  Nicholas J Ekins-Daukes, Michael P Nielsen, Andreas Pusch, Muhammad H Sazzad, Phoebe M Pearce, Peter J Reece

Vertical Bifacial Solar Panels as a Candidate for Solar Canal Design 960
  Jeremiah B Reagan, Sarah Kurtz

Arrhenius Analysis of the Degradation Modes in Emerging Photovoltaic Backsheets............................ 961
  Naila M. Al Hasan, Rachael Arnold, David C. Miller, Jimmy Newkirk, Emily Rago, Michael Thuis, Bruce H. King, Laura T. Schelhas, Archana Sinha, Kent Terwilliger, Sona Ulicná, Peter Pasmans, Christopher Thellen

Comparison of Measured and Modeled Snow Losses for Photovoltaic Systems in Colorado 964
  Owen W. Westbrook, Sara M. Macalpine, David A. Bowersox

PV Hosting Capacity Estimation: Experiences with Scalable Framework...................................... 967
  Wenbo Wang, Daniel Thom, Kwami Senam Sedzro, Sherin Ann Abraham, Yiyun Yao, Jianli Gu, Shibani Ghosh

Properties of Co-Sputtered $(In_xGa_{(1-x)})_2O_3$ Layers Used in CdTe Solar Cells 972
  Manoj K. Jamarkattel, Adam B. Phillips, Indra Subedi, Abasi Abudulimu, Ebin Bastola, Deng-Bing Li, Zhaoning Song, Xavier Matthew, Yanfa Yan, Randy J. Ellingson, Nikolas J. Podraza, Michael J. Heben

High Efficiency Solar Cells Grown on Spalled Germanium Without Polishing 975
  John S. Mangum, Anthony D. Rice, Jie Chen, Jason Chenenko, Evan Wong, Anna K. Braun, Steve Johnston, John F. Geisz, Aaron J. Ptak, Corinne E. Packard

The Effect of $CdSe_xTe_{1-x}$ Thickness on the $CdSe_xTe_{1-x}$/CdTe Solar Cell Performance........................ 976
  Md Zahangir Alom, Sheikh Tawsif Elahi, Vasilios Palekis, Wei Wang, Chris Ferekides

Overall Performance Losses and Activated Mechanisms in Double Glass and Glass-Backsheet
Photovoltaic Modules with Monofacial and Bifacial PERC Cells, Under Accelerated Exposures................. 980
  Jiqi Liu, Sameera Nalin Venkat, Jennifer L. Braid, Xuanji Yu, Brenton Brownell, Xinjun Li, Jean-Nicolas Jaubert, Kaushik Roy Choudhury, Laura S. Bruckman, Roger H. French

Chlorine Doped n-Type CdTe Solar Cells... 988
  Wei Wang, Vasilios Palekis, Md Zahangir Alom, Sheikh Elahi Tawsif, Chris Ferekides

26.7% AM0, 30.2% AM1.5G Dual Junction Solar Cell with 50x InGaAs Quantum Wells, GaAsP
Strain Compensation, and Distributed Bragg Reflector 991
  Stephen J Polly, Brandon Bogner, Anastasiia Fedorenko, Subhra Chowdhury, Dhrubes Biswas, Seth M Hubbard

Spectral Rear Irradiance Testing and Modeling for Degradation and Performance of Solar Fields................. 992
  Silvana Ovaitt, Matthew Brown, Chris Deline, Michael D. Kempe

Effects of Satellite Sampling on Subhourly Modeling Errors 995
  Mark A. Mikofski, William F. Holmgren, Jeff Newmiller, Rounak Kharait

Room Temperature, Dip Coating Organic Passivation for c-Si Surface 996
  Kejun Chen, Abigail. R Meyer, Harvey Guthrey, William Nemeth, San Theingi, Matthew Page, Sumit Agarwal, David. L Young, Paul Stradins

Investigation of Underperformance in Fielded N-Type Monocrystalline Silicon Photovoltaic
Modules.. 997
  E. Ashley Gaulding, Steve W. Johnston, Dana B. Sulas-Kern, Mason J. Reed, James A. Rand, Robert Flottemesch, Timothy J Silverman, Michael G. Deceglie

Fill Factor Loss in Perovskite Solar Cells Using Fullerene ETLs Caused by Air Exposure 998
  Austin G Kuba, Alexander J Harding, Raphael Richardson, Ujjwal K Das, Kevin D Dobson, William N Shafarman

The Effect of Residual $PbI2$ on 2-Step Vapor-Processed P-I-N and N-I-P MAPbI3 Solar Cells.................... 999
  Austin G Kuba, Alexander J Harding, Chaiwarut Santiwipharat, Ujjwal K Das, Kevin D Dobson, William N Shafarman

Characterization of DER Momentary Cessation and Rate-Of-Change-Of-Frequency Response................. 1000
  Rasel Mahmud, Li Yu, Andy Hoke

No Time to Waste: Quickly Optimizing Perovskite Composition with Off-The-Shelf Active Learning Methods.. 1003
    Rishi E Kumar, Moses Kodur, Arun Kumar Mannodi Kannakithodi, David P Fenning

Reactive Silver Inks as Front Electrodes for TCO Coated Solar Cells.. 1004
    Michael W. Martinez-Szewczyk, Steven Digregorio, Owen Hildreth, Mariana I. Bertoni

Flexible CdTe/MgCdTe Double-Heterostructure Solar Cells Made from Epitaxial Lift-Off Thin Films... 1007
    Xin Qi, Jia Ding, Zheng Ju, Stephen Schaefer, Yong-Hang Zhang

Machine Learning Driven Studies of Performance Degradation in a-Si:H/C-Si Heterojunction Solar Cells... 1010
    Davis Unruh, Reza Vatan Meidanshahi, Zitong Zhao, Stephen M. Goodnick, Gergely T. Zimanyi

Stability Analysis and Volt-Watt Control Setting Guideline for Distributed Energy Resources 1013
    Wenzong Wang, Wei Ren, Aminul Huque, Devin Van Zandt, Reigh Walling

Moisture Ingress and Distribution in Bifacial Silicon Photovoltaics... 1019
    Rishi E Kumar, Tala Sidawi, Ian M Slauch, Rico Meier, Mariana I Bertoni, David P Fenning

Evaluation of PV Module Packaging Strategies of Monofacial and Bifacial PERC Using Degradation Pathway Network Modeling 1020
    Sameera Nalin Venkat, Jiqi Liu, Xuanji Yu, Jakob Wegmueller, Kunal Rath, Xinjun Li, Jean-Nicolas Jaubert, Jennifer L. Braid, Roger H. French, Laura S. Bruckman

Material Use and Life Cycle Impact of Crystalline Silicon PV Modules Over Time 1028
    Luyao Yuan, Annick Anctil

Drift-Diffusion Modelling of Four-Junction InGaP/InGaAs/SiGeSn/Ge Solar Cells 1031
    Laurier S. Baribeau, Robert F. H. Hunter, Christopher E. Valdivia, Karin Hinzer

Impact of Humidity, Temperature, and Oxygen on the Stability of FA0.7MA0.3Sn0.5Pb0.5I3 Perovskites.. 1032
    Alex Bordovalos, Marie S Tumusange, Biwas Subedi, Lei Chen, Zhaoning Song, Yanfa Yan, Nikolas J Podraza

The Influence of Wind and Module Tilt on the Operating Temperature of Single-Axis Trackers................ 1033
    Keith R. McIntosh, Malcolm D. Abbott, Ben A. Sudbury, Saurabh Aneja, Mitch Bowman, Lance Brown, Ben Kahane, Norm Nicholas, Kristian Nolde

Impacts of Nonuniform Soiling on Photovoltiac Production 1037
    Lin J. Simpson, Ian K. Teague, Jody Ford, Nathan Shih, Mahfujur Rahman, Jorge I. T. Marchand, Kirsten Perry, Chris Deline

Operability of a Power System with Synchronous Condensers and Grid-Following Inverters..................... 1038
    Marena Trujillo, Rick Wallace Kenyon, Gemini Yau, Li Yu, Andy Hoke, Bri-Mathias Hodge

Vapor Treatment for Growth of High-Quality Oxide Barriers Within P-I-N Perovskite Solar Cells and Tandems.. 1043
    Samuel A. Johnson, Michael D. McGehee, Joseph J. Berry, Axel F. Palmstrom

Long-Term UV Durability of Laminated Glass/Transparent Backsheet Coupons for Bifacial Photovoltaics: Backsheet Side Exposure... 1044
    Soshana Smith, Stephanie Moffitt, Stefan Mitterhofer, Song-Syun Jhang, Stephanie Watson, Li-Piin Sung, Lakesha Perry, Deborah Jacobs, Xiaohong Gu

Silicon Heterojunction Solar Cells with High Bulk Resistivities Over 1,000 $\Omega \cdot cm$ in Relevant Field Conditions of Illumination and Temperature 1045
Anh Huy Tuan Le, Apoorva Srinivasa, Stuart G. Bowden, Ziv Hameiri, André Augusto

Electroluminescence Analysis and Grading of Hail Damaged Solar Panels 1046
Andrew M. Gabor, Phillip J. Knodle, Maurice Covino, Dylan J. Colvin, Kristopher O. Davis

What is the Role of Recycling in the Solar Terawatt Future? 1047
Pablo R Dias, Moonyong Kim, Alison Lennon, Brett Hallam

Solar Panel Power Simulation for Shade Detection 1048
David Jose Florez Rodriguez, Bennet E. Meyers

Evaluation and Demonstration of Slot-Die Coating for Perovskite Thin Film Mini-Modules for Space Photovoltaics 1055
Manoj Rajakaruna, Amir Hossein Ghahremani, Tao Zhu, Jaehoon Chung, Tamanna Mariam, Tyler Brau, Adam Phillips, Michael J. Heben, Zhaoning Song, Randy J. Ellingson, Yanfa Yan

Deleterious Effect of Light Trapping on the Temperatures of Solar Modules 1059
Nicholas P. Irvin, D. Martínez Escobar, Aaron Wheeler, Tomas Leijtens, Hyunjong Lee, Annikki Santala, Richard R. King, Christiana B. Honsberg, Sarah R. Kurtz

Evaluating the Environmental Benefit of Residential Photovoltaic Modules Early Retirement in California 1060
Mallika Kothari, Annick Anctil

From Femtoseconds to Gigaseconds: The SolDeg Project to Analyze Si Heterojunction Cell Degradation with Machine Learning 1061
Gergely Zimanyi, Davis Unruh, Reza Vatan, Zitong Zhao, Andrew Diggs, Stephen Goodnick

Critical Transport Behavior in Quantum Dot Solids 1062
Michael Kovtun, Zachary Crawford, Adam Goga, Gergely T. Zimanyi

Extended Accelerated Stress Testing (EAST) of Glass/Glass, Glass/Backsheet and Glass/Transparent Backsheet PV Modules: Influence of EVA and POE Encapsulants 1065
Akash Kumar, Ashwini Pavgi, Peter Hacke, Kaushik Roy Choudhury, Govindasamy Tamizhmani

Development of a Co-Anneal Process for Double-Side TOPCon Precursor Fabricated by Ex-Situ POCl3 and APCVD Boron Diffusion 1068
Wook-Jin Choi, Young-Woo Ok, Keeya Madani, Shubham Duttagupta, Ajeet Rohatgi

Complex Refractive Index and Complex Dielectric Function Modeling of Film Stack in Perovskite Solar Cells Using Spectroscopic Ellipsometry 1069
Maria Fernanda Villa Bracamonte, Jose Raul Montes Bojorquez, Arturo Ayon

Spatially-Resolved X-Ray Excited Optical Luminescence of Metal Halide Perovskites 1070
Connor Dolan, Deniz N. Cakan, Rishi E. Kumar, Moses Kodur, Yanqi Luo, Barry Lai, David P. Fenning

Exploring the Composition Space of Wide Band-Gap Absorbers for Silicon-Perovskite Tandems 1071
Moses Kodur, Rishi E. Kumar, Deniz N. Cakan, Connor Dolan, Yanqi Luo, Barry Lai, David P. Fenning

A Machine Vision Tool for Facilitating the Optimization of Large-Area Perovskite Photovoltaics 1072
Mathilde Fievez, Nina Taherimakhsousi, Benjamin P. Macleod, Edward P. Booker, Muriel Matheron, Matthieu Manceau, Stéphane Cros, Solenn Berson, Curtis P. Berlinguette

Effective Irradiance Monitoring Using Reference Modules 1073
Jennifer L. Braid, Joshua S. Stein, Bruce H. King, Christopher Raupp, Jaya Mallineni, Justin Robinson, Steve Knapp

Designing a Multi-Quantum-Dot Array for Efficient Light-Harvesting in Solar Cells 1079
Jose Raul Montes-Bojorquez, Maria Fernanda Villa-Bracamonte, Arturo A. Ayon

Chemomechanics of Halide Perovskites: Linking Mechanical Behavior with Reliability 1082
Nicholas J Rolston

Planar and Nanowire InP Thin Solar Cells for Ultralight Space Power Applications 1083
Sara Anjum, Pilar Espinet Gonzalez, Harry A. Atwater

Effective Passivation of CdTe Rear Interface Via Thin Selenium Interface Layer Indicated by
Surface Photovoltage Spectroscopy 1086
Michael A Scarpulla, Nathan D Rock, Amit Munshi

Study of Perovskite Solar Cells Under High-Fluence, Low-Energy Proton Radiation 1087
Michael D Kelzenberg, Ahmad R. Kirmani, Kaitlyn T. Vansant, Joseph M. Luther, Harry A. Atwater

Photophysical Properties of CdSe/CdTe Bilayer Solar Cells: A Confocal Raman and
Photoluminescence Microscopy Study 1088
Abasi Abudulimu, Jaroslav Kulicek, Ebin Bastola, Adam B Phillips, Aesha Patel, Dipendra Pokhrel, Manoj K. Jamarkattel, Michael J Heben, Bohuslav Rezek, Randy J Ellingson

The Capability of a Grid-Forming Inverter to Support Dynamic Microgrids with High Penetrations
of Photovoltaics Systems 1091
Rachid Darbali-Zamora, C. Birk Jones, Matthew S. Lave, Erick E. Aponte-Bezares

Parametric Analysis of Capacitance-Voltage Data for In-Situ Heat and Light Soaking Behavior of
CIGS Solar Cells 1099
Shubhra Bansal, Mohsen Jahandardoost

MoS2 Solar Cell with 120 Nm-Absorber and 3.8% AM1.5G Efficiency 1100
Elisa Antolin, Simon A. Svatek, Carlos Bueno-Blanco, Antonio Marti, Der-Yuh Lin, Micaela Rodriguez-Peña, Monica Luna

Effects of Arsenic Doping on $CdSe_xTe_{1-x}$/CdTe Solar Cells 1101
Sheikh Tawsif Elahi, Md Zahangir Alom, Wei Wang, Vasilios Palekis, Chris Ferekides

What is the Optimal Electricity Share for Very Inexpensive Solar PV? 1105
Adam Dvorak, Marta Victoria

Transparent Oxides as a Protective Encapsulant for Perovskite Solar Cells in Low Earth Orbit ... 1109
Kyle M Crowley, Kaitlyn Vansant, Timothy J Peshek, Lyndsey B McMillon-Brown

Barriers to Solar Photovoltaic (PV) Adoption on a National Scale in the United States 1110
Casey Corrado, Emily Holt, Lauren Schambach

Laser-Weld Qualification Methods for Al Foil Interconnection of Back-Contacted Cells to Predict
Module Reliability 1118
Barry B. Hartweg, Kathryn C. Fisher, Zhengshan J. Yu, Zachary C. Holman

Highly Efficient Perovskite-On-Silicon Tandem Solar Cells on Planar and Textured Silicon...................... 1119
  Christian M. Wolff, Xin Yu Chin, Deniz Türkay, Kerem Artuk, Mohammadreza Golobostanfard, Florent Sahli, Daniel Jacobs, Quentin Guesnay, Peter Fiala, Mostafa Othman, Bosky Sharma, Brett Kamino, Aïcha Hessler-Wyser, Mathieu Boccard, Quentin Jeangros, Christophe Ballif

Corrosion Testing of Solar Cells: Insights to Wear-Out Mechanisms.. 1120
  Andrew Fairbrother, Luca Gnocchi, Christophe Ballif, Alessandro Virtuani

NRG-X-Change and Cooperative Game Strategies as an Alternative to Net-Metering for Solar
Generation 1121
  Hector Lopez, Ali Zilouchian

Nanostructured ZnO Electron Transporting Materials for Hysteresis-Free Perovskite Solar Cell 1122
  Vilko Mandic, Ivana Panzic, Floren Radovanovic-Peric, Thomas Rath

Variable Renewable Energy Participation in U.S. Ancillary Services Markets: Economic
Evaluation and Key Issues... 1123
  James Hyungkwan Kim, Fredrich Kahrl, Andrew Mills, Ryan Wiser, Cristina Crespo Montañés, Will Gorman

Amorphous Manganese Sulfide Enables Efficient and Stable All-Inorganic Antimony
Selenosulfide Solar Cells... 1126
  Chen Qian, Jianjun Li, Kaiwen Sun, Chenhui Jiang, Jialiang Huang, Rongfeng Tang, Martin Green, Bram Hoex, Tao Chen, Xiaojing Hao

A Data-Driven Feeder Selection Method for Distribution System Planning Studies 1127
  Alexandre B. Nassif, Fernanda C. L. Trindade

Fireable Passivating Tunnel Oxide Contacts for Crystalline Silicon Solar Cell.. 1132
  Franz-Josef Haug, Mario Lehmann, Sofia Libraro, Ezgi Genç, Audrey Morisset, Christophe Ballif

Antisolvent Effect on Acetamidinium Substituted 2D Ruddlesden-Popper Perovskite Solar Cells............... 1133
  Vani Pawar, Anuj Kumar Palariya, Nisheka Anadkat, Sandeep Kumar, Sushobhan Avasthi

Performance and Stability of Electrodeposited Mixed Perovskites $MAPbI_{3-x}Cl_x$ and
$MA_{1-y}FA_yPbI_{3-x}Br_x$ 1136
  Mirella Al Katrib, Lara Perrin, Emilie Planes

Performance Assessment of a Residential Building Integrated Photovoltaic (BIPV) System in
Dhaka City... 1139
  Md. Mahbub Ali, Nur Jahan Beanta Sorower, Abu Niem Seum, Md. Shifain Mahathir Alvi, Rawnak Reza Raka, Mohaimenul Islam, Md. Mosaddequr Rahman

Elucidating Materials Paradigm of CIGS by Structure--Composition-- Performance Correlations............... 1145
  Niklas Pyrlik, Christina Ossig, Giovanni Fevola, Svenja Patjens, Jan Hense, Catharina Ziska, Martin Seyrich, Frank Seiboth, Andreas Schropp, Jan Garrevoet, Gerald Falkenberg, Christian G. Schroer, Romain Carron, Michael E. Stuckelberger

An Experimental Comparison Between View Factor and Ray Tracing Models for Energy
Estimation of Bifacial Modules.. 1146
  Hugo Sánchez, Sebastian Dittmann, Carlos Meza, Ralph Gottschalg

Field Assessment of Transparent Conductive Oxides Stability Under Outdoor Conditions 1151
  Brahim Aïssa, Amir A. Abdallah, Juan Lopez Garcia

Influence of Business Models on PV-Battery Dispatch Decisions and Market Value: A Pilot Study of Operating Plants 1158
Joachim Seel, Cody Warner, Andrew Mills

Wide Bandgap AlGaInP-Based Photovoltaic Cell for Indoor Ambient Energy Harvesting 1159
Aditya Prabaswara, Jack Browne, Yongjie Zou, Richard King, Stephen Goodnick, Brian Corbett

A Comparison Study of the Performance of Vertical Vs Single Axis Tracking Bifacial Agrivoltaic Systems in Belgium 1162
Brecht Willockx, Jan Cappelle

Power Factors 2022 PV System Efficiency Benchmarks 1163
Stephen Lightfoote, Samantha Wilson, Steve Voss

Thin, Radiation-Resilient III-V PV Devices Utilizing Quantum Structures and Epitaxial Light Reflectors 1167
Brandon M. Bogner, Stephen J. Polly, Seth M. Hubbard, Roger E. Welser

Rapid Thermal Annealing (RTA) of Hydrogenated Poly-Si Under Air and Nitrogen and Blister Formation 1168
Arpan Sinha, Sagnik Dasgupta, Ajeet Rohatgi, Mool Gupta

Investigation of High Open Circuit Voltage in CdTe-Based Solar Cells Using Oxide Back Buffer Layers 1169
Abdul Quader, Manoj K. Jamarkattel, Ebin Bastola, Kamala Khanal Subedi, Dipendra Pokhrel, Indra Subedi, Adam B. Phillips, Nikolas J. Podraza, Randy J. Ellingson, Michael J. Heben

Solution-Processed Copper Selenium Oxide (CuSeO$_3$) as Hole Transport Layer for CdS/CdTe Solar Cells 1170
Sandip S Bista, Deng-Bing Li, Suman Rijal, Sabin Neupane, Manoj K Jamarkattel, Rasha A Awni, Zhaoning Song, Adam Phillips, Michael Heben, Randy J. Ellingson, Yanfa Yan

Prediction of Electron Band Gap of A$_2$XY$_6$ Perovskite Compounds Using Machine Learning 1173
Jatin Chaudhary, Swastik Bhattacharya, Jukka Heikkonen, Rajeev Kanth

A Silicon Learning Curve and Polysilicon Requirements for Broad-Electrification with Photovoltaics by 2050 1177
Brett Hallam, Moonyong Kim, Robert Underwood, Storm Drury, Li Wang, Pablo R Dias

Photovoltaic Surfaces to Reverse Global Warming 1178
Christiana B. Honsberg, Stuart G. Bowden, Ian R. Sellers, Richard R. King, Stephen M. Goodnick

Strategies for Implementing of Very Large Scale Solar and Wind Power Plants in the Gobi Desert for the Northeast Asia Regional Energy Market 1179
Enebish Namjil, Keiichi Komoto

Interposed Versus Juxtaposed Solar Array Configurations for Agrivoltaics 1182
M. Sojib Ahmed, M. Rezwan Khan, Anisul Haque, Muhammad A. Alam, M. Ryyan Khan

Fabrication of Microscale Back-Contact Arrays for Local Charge Transport Measurements 1185
Kaden M. Powell, Yu-Lin Hsu, Etee Kawna Roy, David J. Magginetti, Heayoung P. Yoon

Agrivoltaic Modules Optimizing Light for Crops in Dryland Regions 1189
Christiana B. Honsberg, Greg Barron-Gafford, Stuart G. Bowden, Robert Sampson

Understanding Device Performance Limiting Factors by Reproducing the Current-Voltage Characteristics from Transient Optoelectrical Measurements 1190
Abasi Abudulimu, Klaus Eckstein, Mirella El Gemayel, Imge Namal, Adam B Phillips, Michael J Heben, Tobias Hertel, Sebastian B Meier, Larry Lüer, Randy J Ellingson

Impact of the 2019-2020 Australian Black Summer Wildfires on Photovoltaic Energy Production 1191
Ethan Ford, Bram Hoex, Ian Marius Peters

Combining Nanoscale 3D Printing with Spark Ablation to Achieve Novel Nanostructured Surfaces for Photovoltaic Applications 1195
Ivana Panzic, Alexander Jelinek, Floren Radovanovic-Peric, Daniel Kiener, Vilko Mandic

Understanding Configuration of Geopolymer Materials for Application in Solar-Cells 1196
Arijeta Bafti, Filip Brlekovic, Vilko Mandic, Luka Pavic, Ivana Panžic, Andraž Krajne

The Potential Use of Spark Ablation in Development of AgNP Decorated Copper Oxide Thin Films for Photodetection Applications 1197
Floren Radovanovic-Peric, Vilko Mandic, Ivana Panžic

Impact of Anti-Soiling Coating on Potential Induced Degradation of Silicon PV Modules 1198
Farrukh Ibne Mahmood, Govindasamy Tamizhmani

Extreme Solar: Towards 24–7 Renewable Energy 1201
Sijo Augustine, Sathishkumar Ranade, Valerio De Angelis, Gabriel Cowles, Jinchao Huang, Olga Lavrova, Stanley Atcitty

Impact of Indium Chloride Treatment on the Properties of $CuInSe_2$ Thin Films 1204
Deewakar Poudel, Adam Masters, Benjamin Belfore, Elizabeth Palmiotti, Angus Rockett, Sylvain Marsillac

On the Effect of Indium Chloride Dose on the Recrystallization of $Cu(In,Ga)Se_2$ Thin Films and Associated Devices 1209
Deewakar Poudel, Adam Masters, Benjamin Belfore, Angus Rockett, Sylvain Marsillac

Effect of Metal Halides Treatment on High Throughput Low Temperature CIGS Solar Cells 1214
Deewakar Poudel, Benjamin Belfore, Adam Masters, Angus Rockett, Sylvain Marsillac

Grain Enhancement in Polycrystalline $CuGaSe_2$ by AgBr Vapor Treatment 1219
Deewakar Poudel, Benjamin Belfore, Adam Masters, Elizabeth Palmiotti, Angus Rockett, Sylvain Marsillac

Post-Deposition Metal Halide Treatment of $CuGaSe_2$ for Photovoltaic Application 1224
Deewakar Poudel, Benjamin Belfore, Adam Masters, Elizabeth Palmiotti, Angus Rockett, Sylvain Marsillac

Current or Power Matching? A Third Option for Monolithic All-Perovskite Tandem Solar Cells 1229
Yuan Gao, Renxing Lin, Ke Xiao, Xin Luo, Jin Wen, Xu Yue, Hairen Tan

Simulation of High Open-Circuit Voltage Perovskite/CIGS-GeTe Tandem Cell 1230
Mohamed Mousa, Mostafa M. Salah, A. Zekry, Mohamed Abouelatta, Ahmed Shaker, Fathy Z. Amer, Roaa I. Mubarak, Ahmed Saeed

Unlocking 1550 nm Laser Power Conversion by InGaAs Single- And Multi-Junction PV Cells 1235
Henning Helmers, Oliver Höhn, Thomas Tibbits, Meike Schauerte, H. M. Noman Amin, David Lackner

Estimation and Degradation Analysis of Physics-Based Circuit Parameters for PV Systems Using Only DC Operation and Weather Data 1236
Baojie Li, Xin Chen, Todd Karin, Anubhav Jain

The Effect of Moisture Ingress on Titania Antireflection Coatings in Field-Aged Photovoltaic Modules 1237
Oscar Kwame Segbefia, Naureen Akhtar, Tor Oskar Sætre

Improved STC and Energy Yield Performance of Bifacial Modules with White-Grid Rear Reflectors 1245
Robert Witteck, Michael Siebert, Tobias Wietler, Marc Köntges, Paulius Laurikenas, Julius Denafas

4D- Printed Shape Memory Polymer Based Solar Tracker 1248
Serhii Tytov, Fhad Al-Modaf, Shicheng Su, Nazek El-Atab

Study of Degradation of Cu(In,Ga)Se$_2$ Solar Cell Parameters Due to Temperature 1252
Rabee B. Alkhayat, Dhurba R. Sapkota

Flight Demonstration Test of State-Of-The-Art Photovoltaic Devices on JAXA's New ISS Transfer Vehicle HTV-X 1257
Mitsuru Imaizumi, Teppei Okumura, Tetsuya Nakamura, Shusaku Kanaya, Taishi Sumita

Mechanical Degradation Studies on Flexible CIGS Cells and Modules for Floating PV 1258
Wim Soppe, Aldo Kingma, Dorrit Roosen

Influence of Temperature and Magnetic Field on the Transient Voltage Decay of a Silicon Solar Cell with Parallel Vertical Junction in Open Circuit 1262
Pape Diop, Papa Touty Traore, Papa Monzon Samake, Babou Dione, Fatimata Ba, Modou Pilor

Progression in Grain Size of Novel Photoferroic Absorber Bournonite (CuPbSbS3) 1263
Oliver M Rigby, Budhika G Mendis, Marek Szablewski

Direct Observation of an Atomic Thin Inversion Layer at the Native Oxide/ n-Si Interface 1264
Yibo Zhang, Joel Y. Y. Loh, Andrew G. Flood, Chengliang Mao, Geetu Sharma, Nazir P. Kherani

Ultrathin III-V Solar Cells with Light-Trapping Structures Fabricated in Situ Using an HVPE Reactor 1265
Allison N. Perna, Anna K. Braun, Kevin L. Schulte, John Simon, Corinne E. Packard, Aaron J. Ptak

New Substituted Small a cation(Acetamidinium) Based Tin Perovskite Solar Cell 1266
Soumen Kundu, Sushobhan Avasthi

Significance of Power and Energy Ratings of Modules in Large-Scale PV Plants 1271
Bijaya Paudyal, Donny Campos Paniagua

FEM Based Thermal Model of an Agrivoltaic System 1275
Karan Rane, Navni Verma, Ardeshir Contractor, Narendra Shiradkar

Towards Understanding of Cementation of Particulate Soils on PV Cover Glass Materials 1279
Mohamed Adawi, Adedoyin Abe, Min Zou, Robert A. Fleming

Thin-Film Solar Cells with MgF$_2$/Ag Back Mirror Patterning for Improved near-IR Reflectance 1280
Lara Barros Rebouças, Gerard J. Bauhuis, Jens Olhmann, Jeroen Maasen, Elias Vlieg, John J. Schermer

InAs Thermophotovoltaic Cells with Low Reverse Saturation Current 1283
Eric J. Tervo, Andrew J. Ferguson, Myles A. Steiner, Ryan M. France

Toxicity Assessment of Lead and Other Metals Used in Perovskite Solar Panels 1284
Gonzalo Rodriguez-Garcia, Jon J. Kellar, Zhengtao Zhu, Ilke Celik

Molecular Beam Epitaxy Growth of CdSe for Si-Based Tandem Cell Application 1288
Stephen Schaefer, Zheng Ju, Allison McMinn, Xin Qi, Yong-Hang Zhang

Fabrication of Ultrathin Ge Template for Growth of Multijunction Solar Cells Based on Wafer-Scale Porous Ge 1291
Tadeáš Hanuš, Javier Arias-Zapata, Bouraoui Ilahi, Philippe-Olivier Provost, Alexandre Chapotot, Abderraouf Boucherif

Prediction of Novel Phosphors Using Machine Learning for Efficiency Enhancement of Silicon Solar Cells 1292
Tae-Gwan Kim, Eun-Gyeong Kim, M. Shaheer Akhtar, O-Bong Yang

Predictive Modeling of Cracks Within Flexible Perovskite Thin Films.. 1293
Melissa A Davis, Rebekah Sweat, Zhibin Yu

Results of Environmental-Based PV Soiling Models After Extreme Dust Events: The Case of Saharan Dust Intrusions in Southern Spain 1294
João Gabriel Bessa, Álvaro F Solas, Florencia A Cruz, Eduardo F Fernández, Leonardo Micheli

Optical Design Considerations for Thin Photonic Power Converters with Textured Back Reflector 1295
Neda Nouri, Christopher E. Valdivia, Meghan N. Beattie, Jacob J. Krich, Karin Hinzer

Interrelation of CdTe Grain Size, Post-Growth Processing and Window Layer Selection on Solar Cell Performance.. 1299
Thomas P Shalvey, Heath Bagshaw, Jonathan D Major

DIrect Sunlight into CO Conversion 1300
Thierry De Vrijer, Arno Smets

Novel 1D Van Der Waals SbSeI Micro-Columnar Solar Cells by a Self-Catalyzed High Pressure Process... 1304
Ivan Caño-Prades, Alejandro Navarro-Güell, Sergio Giraldo, Joaquim Puigdollers, Marcel Placidi, Edgardo Saucedo

Developement of Phosphors by Magnetron Sputtering for Solar Cells Improvement 1305
Eduardo Salas, Miguel Modesto Tardio, Elisa García-Tabares, Gracia Belén Perea, Rosa De La Cruz, Stavros Athanasopoulos, Clement Kanyinda-Malu, Juan Enrique Muñoz-Santiuste, Beatriz Galiana

High Efficiency Silicon Heterojunction Metal Wrap Through Produced in Industrial Pilot Line................ 1306
Marina Foti, Nicolas Guillevin, Eric Kossen, Lars Okel, Eelko Hoek, Anna Carr, Bas Van Aken, Petra Manshanden, Francesco Rametta, Marcello Sciuto, Antonio Spampinato, Alfredo Di Matteo, Antonino Ragonesi, Gianluca Coletti, Cosimo Gerardi

DERConnect – a Distributed Energy Resources Testbed for Solar Power Integration 1310
   Jan Kleissl, Adil Khurram, Keaton Chia, Scott Brown, Aditya Mishra, Jorge Cortes, Raymond
   De Callafon, Rajesh Gupta, Sonia Martinez, David Victor

Slot-Die Fabrication of Solution-Processed Kesterite Solar Cells for Product Integrated
Photovoltaics 1311
   Xinya Xu, Matthew C Naylor, Michael Jones, Bethan Ford, Stephen Campbell, Yongtao Qu,
   Vincent Barrioz, Guillaume Zoppi, Neil S Beattie

N-Type CdTe Thin Films Via In-Situ Indium Doping 1312
   Theodore D C Hobson, Luke Thomas, Laurie J Phillips, Leanne A H Jones, Matthew J Smiles,
   Christopher H Don, Pardeep K Thakur, Vinod R Dhanak, Tim D Veal, Jonathan D Major,
   Ken Durose

Tellurium Availability for the PV Industry Using a System Dynamics Approach........................ 1313
   Francis Hanna, Annick Anctil

Progress on Substrate Reuse Using Sonic Lift-Off for GaAs- Based Photovoltaics 1314
   Andrew B. Sindermann, Stephen J. Polly, Pablo Guimera Coll, Elijah J. Sacchitella, Brandon
   M. Bogner, Mariana I. Bertoni, Seth M. Hubbard

Bandgap Dependence of Near-Conduction Band State in $(Ag_yCu_{1-y})(In_xGa_{1-x})Se_2$ Solar Cells 1315
   Michael F. Miller, Alexandra M. Bothwell, Nicholas Valdes, Stefan Paetel, Rouin Farshchi,
   Ana Kanevce, William Shafarman, Darius Kuciauskas, Aaron R. Arehart

Improvement in PV Plant Performance for Convection Heat Transfer Changes from Altered Plant
Layout... 1319
   Matthew Prilliman, Sarah Smith, Brooke Stanislawski, Marc Calaf, Raul Bayoan Cal, Tim
   Silverman, Janine M. F. Keith

High Altitude Flight Results Using Selenium, a PV Measurement Ecosystem............................ 1320
   Don Walker, Colin J. Mann, John Nocerino, Kevin Lopez, Alexandra Pettengill, Jonathan
   Ortiz, Katrina Baumgarten, Misha Dowd, Yao Lao, Simon H. Liu

Time-Evolving Electroluminescence Imaging in Perovskite Solar Cells 1321
   Jackson W. Schall, Hsinhan Tsai, Harvey Guthrey, Chun-Sheng Jiang, Steve Johnston, Dana
   Kern, Andrew Norman, Mowafak Al-Jassim

The Nuts and Bolts of PV: Maturing Solar PV Racking and Module Mounting Critical Bolted Joint
Technologies for LCOE Reductions and Increased Reliability 1322
   James Elsworth, Gerald Robinson, Jon Ness, Joe Cain

Atomic Layer Deposited Bilayers and the Influence on Metal-Insulator-Semiconductor Schottky
Barriers 1323
   Benjamin E. Davis, Nicholas C. Strandwitz

Optimizing Perovskite Solar Cells by Understanding the Bulk Properties of Contact Layers 1329
   Mason Mahaffey, Zhengshan Jason Yu, Vidya Krishnan, David Quispe, Arthur Onno,
   Zachary Holman

Development and Qualification of IMMβ and Z4J+, Radiation Hard III-V Solar Cells 1332
   John T Hart, Dan Aiken, Zac Bittner, Ben Cho, Daniel Derkacs, Khalid Emshadi, Andrew
   Espenlaub, Frank Fencl, Jeremy Leshin, Ahmad Mansoori, Nate Miller, Pravin Patel, Albert
   Perry, Janine Walker

The Profound Influence of Substrate Thermal Resistance on the Photovoltaic Properties of Solution-Processed Cu(In,Ga)Se2 1333
Kyle G Weideman, Rakesh Agrawal

Soiling Measurement Based on Checkered Pattern Image Analysis 1334
Bing Guo, Wasim Javed

Modeling Efficiency of Inverters with Multiple Inputs 1335
Clifford Hansen, Jay Johnson, Rachid Darbali-Zamora, Nicholas Gurule

Pathways to High Efficiency Perovskite Monolithic Solar Modules 1338
Xuezeng Dai, Shangshang Chen, Yehao Deng, Allen Wood, Guang Yang, Chengbin Fei, Jinsong Huang

Simulation of Hot-Carrier Filtering in InAs-InP Nanowire Heterostructures 1339
Urs Aeberhard

A Simple Approach to Ohmic Contacts for Transition Metal Dichalcogenide Solar Cells 1340
Mario Martinez, Simon A. Svatek, Carlos Bueno-Blanco, Der-Yuh Lin, Ines Duran, Antonio Marti, Elisa Antolin

Luminescent Solar Concentrators for Building Integrated Photovoltaic Devices 1341
Liam J. Halloran

LETID in Legacy and Modern PV Modules: Accelerated Testing and Field Deployment 1342
Joseph Karas, Ingrid Repins

Measurement of Snow Loading on a Tilted PV Module in Northern Michigan 1343
Daniel Riley, Laurie Burnham, William Snyder, Bruce King, Paul Dice

Probabilistic Assessment of Narrowband Vs Broadband Solar Irradiance Temporal Variability in Ottawa 1346
Nick Anderson, Viktar Tatsiankou, Karin Hinzer, Richard Beal, Henry Schriemer

Evaluation of Solar Capacity Factor of ~2000 Solar Plants Across the United States Using Multilayer Perceptron Regressor Models 1347
Samantha S. Wilson, Stephen Lightfoote, Stephen Voss

Radiant/Non-Radiant Lifetime Switching in Chlorophyll and Application to Energy Storing Photovoltaic Cells 1350
Julie B. Liu, Nahian Rahman, Aaron Song, Elizabeth Nazginov, Mia Pancari, Amina Exilhomme, Charles M. Fortmann

Measuring Carrier Concentration on the Back Side of Thin Film Solar Cells 1355
Nathan Rosenblatt, Alex Polizzotti, Sachit Grover, Xiaoping Li, Wyatt K. Metzger

Reproducibility and Photostability of High-Efficiency Perovskite Solar Cells in Scalable Manufacturing 1358
Rohit Prasanna

Optimized Near-Field Thermophotovoltaic Cell Using InAs and InAsSbP 1359
Gavin P Forcade, Christopher E Valdivia, Sean Molesky, Shengyuan Lu, Alejandro W Rodriguez, Jacob J Krich, Raphael St-Gelais, Karin Hinzer

22% Efficiency Module Combining Silicon Heterojunction Solar and Shingle Interconnection................... 1360
Marina Foti, Marco Galiazzo, Enrico Sovernigo, Nicola Frasson, Cosimo Gerardi, Alfredo Guglielmino, Grazia Litrico, Marcello Sciuto, Antonio Spampinato, Antonino Ragonesi, Francesco Rametta, Andrea Canino, Agata Carbonaro, Fabrizio Coco, Agnese Di Stefano, Fabrizio Bizzarri

Polyethienimine Interface Dipole Tuning for Electron Selective Contacts.. 1363
Eloi Ros Costals, Thomas Tom, Gerard Masmitjà, Benjamin Pusay, Estefania Almache, Maykel Jimenez, Julià Lopez, Edgardo Saucedo, Pablo Ortega, Joan Bertomeu, Joaquim Puigdollers, Cristobal Voz

Impact of Snow Depth on Single-Axis Tracked Bifacial Photovoltaic System Performance 1366
Annie C. J. Russell, Christopher E. Valdivia, Joan E. Haysom, Karin Hinzer

Author Index

Extraction of Prevailing Soiling Rates from Soiling Measurement Data

Josh Peterson, Julie Chard, and Justin Robinson

Groundwork Renewables Inc., Sand City, CA 93955, United States

Abstract—Soiling, the accumulation of airborne particulate matter on PV module surfaces, reduces PV module output and can significantly impact PV project economics. Soiling measurement data are often difficult to interpret due to inherent noise, incomplete or inconsistent cleaning events, and misalignment between paired PV modules. GroundWork has developed a soiling data processing method specific to preconstruction measurement data sets that builds upon prior art. The method employs robust data filters to minimize errors. Monthly and annualized soiling rates are generated, along with their corresponding uncertainty. This paper describes key features of the data processing algorithm and how the GroundWork method overcomes challenges associated with soiling rate calculations.

Keywords—soiling rate, uncertainty, data filtering

I. INTRODUCTION

Photovoltaic (PV) arrays are subject to local weather and atmospheric conditions. The accumulation of airborne particulate matter such as dust, pollen, other naturally occurring substances, or point-source materials on PV module surfaces is referred to as soiling.

There are various ways in which soiling rates can be derived [1]. Some of the most common include: measurement using a pair of matched PV modules, measurement using optical sensors, measurement by image analysis, deriving soiling losses from production data, and estimation via measurement of environmental parameters. The underlying premise of all these techniques is to determine the amount of soiling at a site as it varies with time.

GroundWork Renewables, Inc. (GroundWork) has carried out soiling measurements at hundreds of project sites during preconstruction resource assessments. GroundWork soiling measurement kits employ a pair of matched PV modules. Soiling is measured by comparing the short circuit current of a "clean" control module to that of a "soiled" module. The clean module is washed during every maintenance visit. The soiled module is left to accumulate particulate matter on the module surface. The soiled module is to some extent cleaned naturally by wind, dew and rainfall and is cleaned manually when impacted by noticeable discrete foreign substances such as bird droppings. Soiling data are recorded at one-minute intervals and are post-processed at the end of the measurement campaign.

One well known soiling data post-processing approach is the stochastic rate and recovery (SRR) method [2]. The SRR method estimates PV power production losses due to soiling from PV yield data. GroundWork sought to adopt and adapt the SRR method to post-process soiling measurement data. The GroundWork soiling data processing algorithm (SDPA) implements the SRR approach up to the stochastic step. In preconstruction the target measurement metric is the soiling rate, which can be used as a model input to more accurately predict site-specific soiling.

In adapting the SRR method to GroundWork's needs, several challenges arose.

This paper focuses on the challenges associated with deriving soiling rates from measurements made using a pair of matched PV modules, and the methodology used by GroundWork to overcome these challenges.

Key challenges addressed in this paper include:

- Filtering unwanted data points
- Minimizing errors caused by PV module misalignments
- Dealing with a partially dirty "clean module" between maintenance visits
- Obtaining an uncertainty estimate for soiling rate data

An overview of the GroundWork soiling measurement approach and output data is given and is followed by a description of how this work builds upon previous work. Soiling data filtering steps are described in the context of addressing the challenges listed above. The soiling rate is computed from changes in the measured soiling ratio. From this soiling rate analysis, an uncertainty estimate in the soiling rate is obtained. Examples of data processing results are given, followed by a preliminary study of the soiling rate uncertainty values from across the United States.

II. BASICS OF SOILING MEASUREMENTS

GroundWork soiling measurements are made with a pair of matched PV modules. One module is cleaned on a regular basis and the other module is allowed to naturally soil. The two modules are configured to output a short circuit current (Isc).

A measure of soiling at one moment in time is the soiling ratio, a unitless parameter defined by (1).

978-1-7281-6118-1/22 $31.00 © 2022 IEEE

$$Soiling\ Ratio = \frac{Isc\_Soiled}{Isc\_Clean} \qquad (1)$$

Changing soiling conditions will result in a changing soiling ratio. The rate at which soiling material is accumulating on the modules is determined by measuring the soiling ratio at two points in time. The soiling rate is determined using (2).

$$Soiling\ Rate = \frac{SoilingRatio_{Final} - SoilingRatio_{Initial}}{Time_{Final} - Time_{Initial}} \quad (2)$$

The two modules are matched as well as possible at installation. A mismatch between the modules' Isc outputs under the same environmental conditions will result in a systematic shift in the soiling ratio. However, since the purpose of the experiment is to generate soiling rates according to (2), systematic shifts in the soiling ratio are mathematically canceled out.

GroundWork sought to quantify the uncertainty associated with measured soiling rates. Determining the uncertainty in the soiling rate is difficult when utilizing the SRR method because much of the variation in the more granular soiling ratio data set is lost when the data are binned to daily average values. Due to this limitation, the data used in this publication are analyzed at the one-minute level. In doing this, new challenges were encountered. These new challenges forced GroundWork to apply more stringent filters to the data at the minute level.

Prior to filtering the data, the SDPA hones the dataset in two ways: 1) Because soiling ratios are best computed for fixed-tilt systems from data collected in the middle of the day when incident angles are smallest [4], only data that are within ±2 hours of solar noon are considered by the SDPA. 2) The SDPA is designed to ingest the GroundWork data file structure. Files that are processed have already been subjected to GroundWork QC flagging [3] to identify obviously erroneous data. The SDPA removes flagged data prior to applying additional filters that are specifically designed for the processing of soiling data.

III. APPLYING A SOILING RATIO "FLATTENING" ADJUSTMENT

In some data sets the soiling ratio changes drastically over the course of the day, only to repeat the process again the next day. This pattern is not a measure of soiling but instead is indicative of a measurement bias that can be adjusted for.

Fig. 1A shows a sample of four consecutive days of soiling ratio data vs azimuthal angle of the sun. Each day is represented by a different color. Notice that for each day the soiling ratio begins high and continuously drops throughout the day. This behavior is indicative of an experimental setup that favors one of the modules at certain times of the day, and the other module at other times of the day. In this example, in the morning the soiled module generates a larger Isc relative

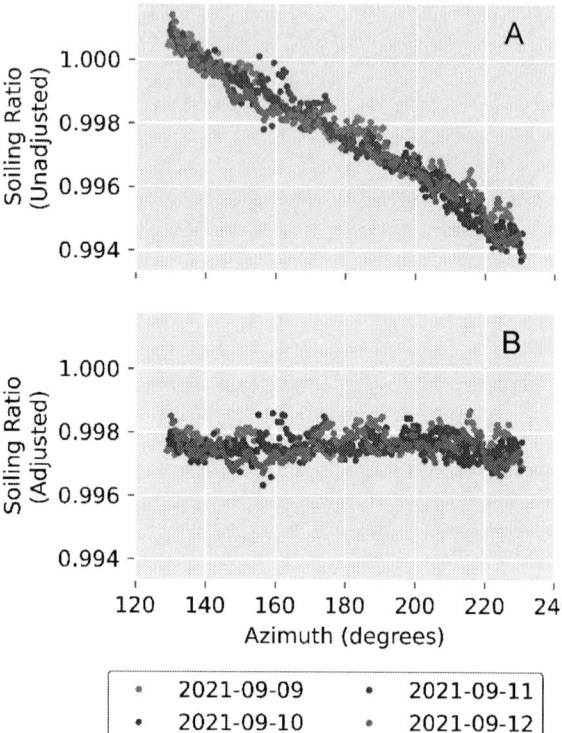

Fig. 1. Soiling ratio vs solar azimuthal angle (AZM). The soiling ratio of four continuous days is shown. In the upper plot the soiling ratio consistently decreases throughout the day. In the lower plot the soiling ratio has been adjusted such that the overall slope of each day is zero. Note the noise in the soiling ratio is preserved. This is most visible on day 2021-09-11 at AZM values ner 160 degrees.

to the clean module. In the afternoon the soiled module generates a smaller Isc relative to the clean module.

This repeating downward trend is likely caused by a misalignment of the modules [4]. In an internal study, GroundWork found that angular displacement between the two modules on the order of 0.8° will cause such trends. GroundWork has improved the soiling kit support structure, and field technicians have improved module alignment during installation, but misalignment is still difficult to avoid over the course of an outdoor measurement campaign.

To account for module misalignment, a "flattening" adjustment was incorporated into the SDPA. In effect the adjustment takes sloping downward curves such as those in Fig. 1A and flattens them, so they are horizontal for the day as seen in Fig. 1B. The noise that is inherent in the data set is preserved, but the overall spread in the soiling ratio data for each day is reduced.

It is clear in Fig. 1 that the daily spread in the soiling ratio has been significantly reduced after the data are adjusted. In this example, the variation in the soiling ratio over the course of a day has been decreased by nearly a factor of 3, from 0.006 to 0.002. Datasets analyzed using methodologies where the entire day of data is wrapped up into a single daily value may appear to be less noisy, but the noise, which is related to the uncertainty, in each daily value is not visible.

The flattening adjustment is performed after preliminary filtering steps are done, and prior to filtering the data using the forward/backward filter (discussed in the next section). The flattening adjustment is done on a daily basis. The flattening adjustment uses the azimuthal angle of each data point as the x-value of the fit. The soiling ratio at each minute is the y-value of the fit. From this list of x and y values, a linear fit is generated.

The flattened data are computed using (3).

$$SR_{flattend} = SR_{measured} - fit(AZM) + fit(180) \quad (3)$$

Where $SR_{flattened}$ is the soiling ratio data after the adjustment. The $SR_{measured}$ value are the measured soiling ratio values. The *fit* is the linear fit value, which is a function of the azimuthal angle of the sun (*AZM*). In (3), term *(SRmeasured – fit(AZM))* preserves the noise in the data. The *fit(180)* maintains the soiling ratio at solar noon.

IV. FILTERING DATA

Soiling measurement data are inherently noisy. In order to remove anomalous data points, five filters are applied to the one-minute data.

Filter 1. Irradiance test

The first filter applied is a global horizontal irradiance (GHI) filter. This filter only passes one-minute soiling data points when a concurrently measured, co-located GHI value is within acceptable limits. The ground-based GHI value must satisfy (4) and (5). Equation 4 requires the GHI data points to pass all GroundWork QC tests [3]. (5) requires GHI values to be within acceptable irradiance limits that are dynamic and a function of the measurement location.

$$GHI\_QC == Good \quad (4)$$

$$Max(200 \ W/m^2, 50\% \ ETR) < GHI < 1500 \ W/m^2 \quad (5)$$

ETR is the modeled extra-terrestrial GHI value. The ETR is almost always greater than 200 W/m² around solar noon under typical circumstances, but under non-typical circumstances (for example extreme northern locations) it may dip below 200 which is why a hard lower limit is also in place. One-minute data points that fail the GHI filter test are not considered in subsequent calculations.

The GHI filter is used to eliminate one-minute data points when the sky conditions are too dark. The measured GHI values are also used to weight the individual one-minute soiling ratios that feed into the daily soiling ratio, which is used to determine the natural reset events as described in Section 5.

Filter 2. Soiling reasonableness test

The second filter applied is a soiling ratio reasonableness test. The one-minute soiling ratio must be greater than or equal to 0.7 and less than or equal to 1.1. This filter eliminates obviously bad soiling data points. Extreme soiling values outside this range can occur if one of the modules is malfunctioning or has localized discrete impacts such as bird droppings.

Filter 3. Daily spread test

Soiling is typically a slow process where the soiling ratio does not change much over the course of a day. The daily spread test finds data points that deviate from the typical range of data for a day. Each day is considered independently. Only data points that passed the first two filters are considered.

Equation 6 defines the acceptable spread in soiling ratio values for each day. In (6), SRatio is the one-minute soiling ratio, P50 is the median value, P5 is the 5% value, and P95 is the 95% value.

$$P50 - 2 * (P50-P5) < SRatio < P50 + 2 * (P95 - P50) \quad (6)$$

First the difference in values from the median to the 5% level value is determined. Then this difference is doubled. By doubling the value, points near the 95% are included, but true outliers are eliminated. The same is done to determine the upper limit.

Filter 4. Forward/backward filter

The soiling ratio can change drastically if either of the modules changes for some reason, for example if the soiled module was cleaned due to rain. However, assuming that such an event does not occur, the soiling ratio from one day to the next should be smoothly varying in time. For example, if the soiling ratio over the previous week was in the range of 0.999 to 1.001, a soiling ratio in that same range would be expected the following day.

The forward/backward filter performs two tests; one looking forward and one looking backward. If the data points on a given day do not match the trend in either direction they are eliminated. A similar approach for removing outliers was employed by Skomedal [5].

The backward filter performs the following analysis: For a day in question, the previous 16 days are selected. 16 days is used here because it is long enough to include weekly trends, typical of a weekly visit by the local technician, but short enough to exclude seasonal variations in the soiling ratio. Of the 16 days, data points are not included if they occur beyond a manual reset event. These are events where both modules are cleaned by the maintenance technician. Up to this point in the SDPA, the natural reset events are not yet determined, so these cannot be used as a boundary.

From the 16 days before the day in question, only points that passed the first three filters are used. A linear fit of the soiling ratio vs time is made. The linear fit is extrapolated forward to the day in question. For the linear fit, the residual value of the fit is determined. The residual is a sum of all absolute differences in the fit from the data points. This is a measure of the noise in the backward fit. From the residual an expected variation is computed according to (7).

$$EV = 3 * \sqrt{\frac{residual}{N}} \quad (7)$$

Where EV is the expected variation, and N is the number of points in the backward fit. A multiplier of 3 is used to increase the allowable range and was determined through experience in looking at the data sets.

From the backward fit and the backward expected variation an upper and lower limit in soiling values is determined according to (8) and (9).

$$Lower\_backward = SR\_fit - EV \qquad (8)$$

$$Upper\_backward = SR\_fit + EV \qquad (9)$$

Where the SR_fit is the expected soiling ratio determined by the linear fit when the fit is extrapolated to the day in question.

The forward filter operates similarly, extrapolating the fit backwards to the day in question. Thus there are two upper limits, and two lower limits. The forward and backward limits are combined using (10) and (11).

$$LL = Min(Lower\_backward, Lower\_forward) \qquad (10)$$

$$UL = Max(Upper\_backward, Upper\_forward) \qquad (11)$$

Where LL is the lower limit, and UL is the upper limit. In this way, data points are considered good if they are within either the forward or backward expected trends. This accounts for jumps in the soiling ratio that occur during the solar noon window.

The forward/backward filter is performed iteratively. During the first iteration, only points that passed the first three filters are selected. During the second iteration and all subsequent iterations, the data points being selected must have passed the first three filters and also have passed the previous forward/backward filter in previous iterations. Once a point has failed a forward/backward test, it is eliminated from all analysis. The loop is repeated until all points pass the forward backward filter.

Filter 5. Immediately after a cleaning.

The final filter applied to the data requires knowledge of times when the clean module was actually clean. GroundWork technicians typically clean on a weekly basis. For a pyranometer this is sufficient. However, for a sensitive study such as soiling, this poses a challenge. The soiling ratio is a measure of how much dust is on the soiled module relative to the clean module. If both the soiled and clean module have some amount of dust, the study is invalid. Therefore, only data collected soon after the clean module was cleaned are selected for further analysis.

Fig. 2 shows the soiling ratio at a site vs time. At this site, there is significant amount soiling. Maintenance events, when the clean module was cleaned, are indicated with vertical

dashed lines. Maintenance events occur at regular weekly intervals. The date and time of each maintenance event is recorded in the maintenance log. Natural reset events, when the soiled and clean modules are both cleaned naturally, cause

an abrupt jump up in the soiling ratio. Natural resets are detected using the SRR method as described in Section V. A natural reset event occurred on 2021-09-26 and is indicated by a solid black vertical line. In Fig. 2, notice the steps in the soiling ratio. The step down always occurs immediately after a maintenance event. This makes sense, as prior to the maintenance visit, the clean and soiled modules accumulated dust at the same rate. When the clean module was cleaned, the previous week's worth of dust was removed and the soiling ratio jumped down to a new level. This process was repeated for multiple weeks. At sites with less soiling, the steps in the soiling ratio are less visible.

To reiterate the five filtering steps, data points that have passed all the filters: 1) have sufficient irradiance, 2) have a reasonable soiling ratio value, 3) are close to the expected ratio for the day, 4) are close to the expected ratio for neighboring days, and 5) are immediately after a cleaning event.

A visualization of data as it progresses through the filtering steps is shown in Fig. 3. Each panel of Fig. 3 represents the data before and after the corresponding filter. The data points that passed a filter are shown as black. Data that did not pass the filter are shown in red. Data that failed a previous filter are shown in pink. The vertical scale of Filter 2 is expanded to show data that failed this filter. The data used in Filters 1, 2, and 3 are unadjusted data. The data used in Filters 4 and 5 have been flattened. The filter limits for Filters 3 and 4 are shown as gold error bars.

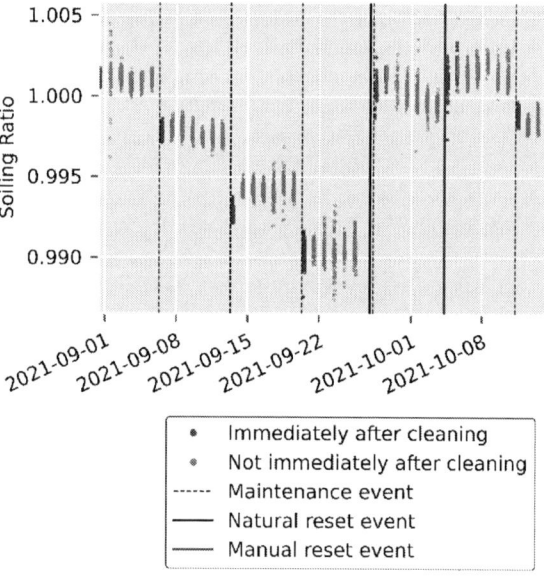

Fig. 2. Steps in the soiling ratio. Steps occur when the clean module is cleaned and are most pronounced at sites with significant soiling.

VI. DETERMINING PAIRS OF POINTS

The ultimate goal of the SDPA is to determine monthly, site-specific soiling rates with their associated uncertainties. The soiling rate between two one-minute soiling ratio data points is the slope of the line connecting those two points, as calculated using (2). The SDPA draws lines between many pairs of one-minute soiling ratios to generate daily and monthly soiling rate values and their associated uncertainties.

When using (2) with one-minute data, the fraction of a day is included in the time difference and is recorded as a decimal. The pairs of points calculation is performed by looping through all of the one-minute data points that have passed the filtering tests and evaluating individual pairs of data points. Each pair must satisfy four criteria:

1. Pairs of points must be in the same soiling period.

2. Pairs of points must be at least two days apart

3. Pairs of points must be fewer than 30 days apart.

4. Pairs of points must be within ±10 minutes of each other in terms of time of day.

It is commonly seen that the soiling ratio systematically varies throughout the day. The purpose of criterion #4 is to further reduce the systematic variation in the soiling ratio throughout the day, similar to the adjustment described in Section 3. Matching pairs of points that occur at the same time of day, for example pairing an 11:30 minute with the minutes between 11:20 and 11:40 on subsequent days, further reduces errors when daily variations are non-linear.

Fig. 4 demonstrates how the minute level data is used to compute the slopes between pairs of points. The black data points are good data that have passed all five filter tests. The red data points have failed a filter test. The gold lines are a sample of the pairs of points lines that join valid pairs. In Fig. 4 only a subset of the lines are drawn for clarity. In the actual data analysis, all pairs are drawn. Notice that the lines do not cross the natural reset or manual reset event boundaries. From Fig. 4, one can visualize the typical slope during a timeframe and also the spread in the slopes present.

VII. COMPUTING THE DAILY SOILING RATES AND UNCERTAINTIES

With the pairs of points array determined, the one-minute soiling rate data can be aggregated to larger time frames. GroundWork generates four soiling rate metrics for each data set: daily, monthly (x2), and yearly. The monthly aggregation has two versions; one where repeating calendar months from different years are combined, and one where each month is treated independently.

From the entire list of pairs of points, the pairs that interact with the timeframe in question are selected. Statistics are performed on this subset. For each timeframe, the median, P5 and P95 values are determined as well as the mean and standard deviation. To be clear, the soiling rate is reported as the change in the soiling ratio that occurs over a day, as dictated by (2). For the monthly and yearly aggregations, the daily soiling rate for the entire month (year) is reported.

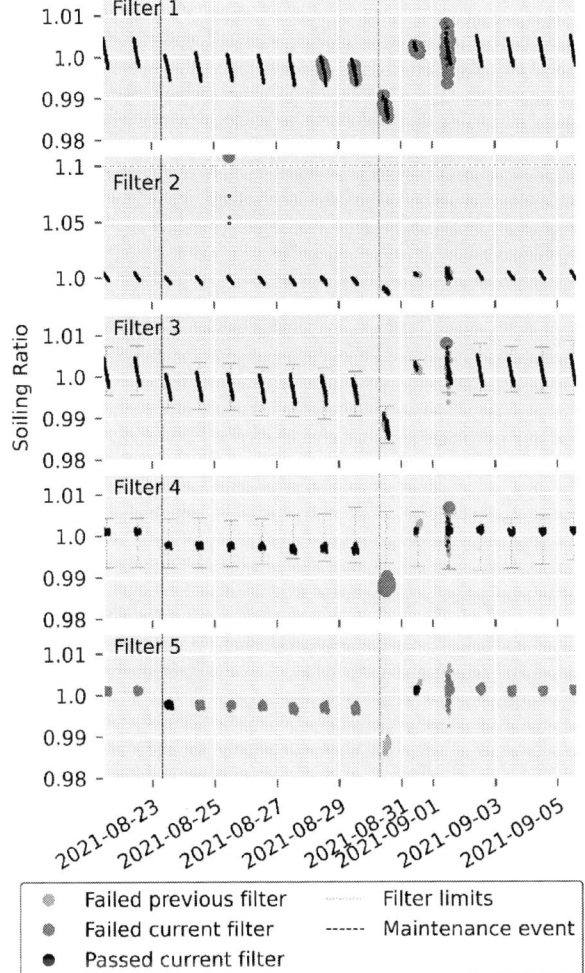

Fig. 3. Progression of data through filtering steps.

Filtering ensures that only valid soiling ratio values are used in the analysis, thereby reducing the overall uncertainty in the calculated soiling rates.

V. COMPUTE NATURAL RESET EVENTS

The one-minute soiling ratio values jump significantly when a natural reset event occurs. Typically, a natural reset event is thought of as cleaning caused by rain, however cleaning from other sources can exist [1, 6, 7]. The natural reset events are detected from the soiling ratio and are not directly linked with rainfall data.

Natural reset events are determined following the SRR method. Three variables must first be determined; a running median of the daily average soiling ratio values, a difference between the running median values on two successive days, and a difference threshold related to the interquartile range of the data (IQR).

The cleaning threshold is determined using a modified version of the IQR method. When the day-to-day difference between two successive running median values is greater than the difference threshold, a cleaning event is triggered.

978-1-7281-6118-1/22 $31.00 © 2022 IEEE

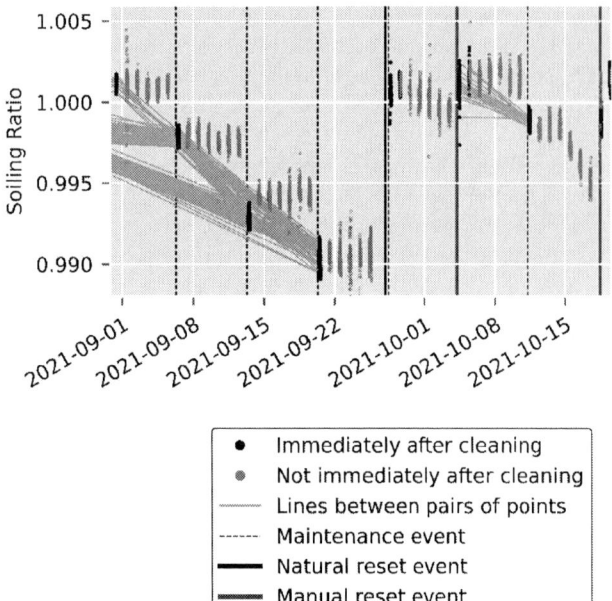

Fig. 4. Demonstration of slopes between pairs of points.

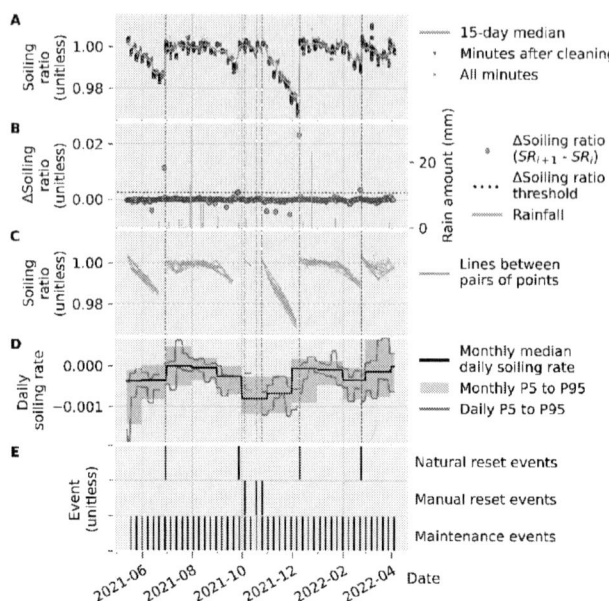

Fig. 5. Time series output file. The example shown here shows significant soiling in the fall of 2021. Most stations do not exhibit this level of soiling.

The P5 and the P95 values are a quantitative indicator of the noise in the soiling rate measurement. Using the collection of lines between pairs of points in Fig. 4 as an example, a median slope and P5 and P95 slopes can easily be calculated. The P95 value can be considered to be the upper limit to the soiling rate while the P5 value can be considered the lower limit. The spread around the median, as reported by the P95 and P5 values, represents the uncertainty in that median value.

It should be mentioned that the combined uncertainties of the individual pieces of measurement hardware are not independently considered. Measurement uncertainty due to hardware is captured in the inherent noise in the soiling ratio data, so is already built into the analysis.

VIII. RESULTS

The results from a sample soiling measurement campaign are shown as time series plots in Fig. 5. There are five subplots: (A) soiling ratio, (B) difference in soiling ratio, (C) pairs of points lines, (D) soiling rate, and (E) events.

The upper plot, labeled (A), is a time series plot showing the soiling ratio vs time. The one-minute data collected immediately after a cleaning event are shown as black circles. These are the data points that passed all the filters discussed in Section 4 and are used in the pairs of points analysis (the black points in Fig. 3, Filter 5). The small black x's indicate one-minute soiling ratio data when the clean module may not actually be clean (not right after a cleaning or reset event). Data points that failed one of the first four filters are not shown.

In (A), the gold line indicates the running daily median. The running median is computed from 7 days of data. The dashed vertical lines indicate the reset events, both natural and manual. These lines are intended to guide the eye. The

locations of the natural reset events should result in a noticeable increase in the soiling ratio.

Plot (B) is the difference between adjacent daily running median values, along the gold line in (A). Differences greater than the cleaning threshold, shown as a horizontal dashed line, trigger a natural reset event. Notice that large upward shifts in the soiling ratio data in (A) represent large positive differences. When the difference is greater than the threshold limit, a natural reset event is identified. Also shown in (B) is rainfall. The right axis in (B) indicates the daily rain total in millimeters.

Plot (C) shows lines between pairs of points. To keep the figure relatively clean, only 2000 randomly selected lines are drawn. Sometimes may not have lines that are shown.

Plot (D) shows the soiling rate vs time. This is the aggregated slopes of the lines in (C). It is seen that the horizontal lines in (C) around 2022-01 result in soiling rate values near zero in (D). The downward sloping lines around 2021-11 result in negative soiling rates in (D).

In Plot (D), the black line is the monthly median daily soiling rate. Notice that the monthly value is constant for entire months. The gold band in (D) represents the P5 and P95 values in the monthly soiling rates. The gray lines represent the P5 to P95 spread in the data when aggregated at the daily level. The daily spread can vary wildly throughout the month and is influenced by temporary deviations more heavily than the monthly spread. Gaps in the daily spread curve indicate that daily soiling rate data were not available for a given day. This is due to timing of the cleaning and reset events.

Plot (E) shows key events in the data set. The dates of natural reset events and manual reset events are shown. Also shown in (E) are the maintenance events.

978-1-7281-6118-1/22 $31.00 © 2022 IEEE   689

IX. AGGREGATING THE SOILING RATE DATA TO MONTHLY VALUES

In an effort to visualize the statistical significance of the soiling rate for each month, the slopes of the lines between pairs of points (Fig. 5C) are binned according to their magnitude. The frequency of occurrence of slopes (soiling rates) of varying magnitudes is plotted as a histogram.

Histograms for each month are shown in Fig. 6, which corresponds to the data set shown in Fig. 5. There are 13 histograms going down the page; one for each of the 12 months in a year and an additional histogram at the bottom for the entire campaign. When a data set spans more than a 12-month period, slopes occurring during the same calendar month but in different years are consolidated.

In both Figs. 5 and 6, it is seen that when the soiling ratio is noisy at the one-minute level, the soiling rate has a larger spread. The flattening step described in Section 3 was implemented in an effort to minimize the variation in the soiling ratio data. The intended purpose was to minimize the uncertainty in the soiling rate data, but because the soiling ratio data are inherently noisy at the one-minute level, the uncertainty can only be reduced so far.

The vertical axes of each subplot in Fig. 6 are a count of the number of slopes included in each bin for that month. The vertical axis is a base-10 log scale. The horizontal axis is the range of daily soilng rates resulting from the analysis.

The median daily soiling rate is represented in each month's histogram as a solid, vertical blue line. The median value is less sensitive to outliers than the mean. The extreme values of P5 and P95 are shown as dashed, vertical blue lines. Similarly, the mean value is drawn as a solid, vertical red line. One standard deviation away from the mean is shown as dashed, vertical red lines. The median and mean soiling rates are limited to negative values, meaning that the largest value they are allowed to obtain is zero. The P5, P95, and standard deviation values are not limited to negative values.

The data that are used to generate the histogram originate from the one-minute data, as opposed to previous soiling rate methods which first aggregate the one-minute data into daily soiling rates. Using the one-minute data preserves the spread in the soiling rate data. The spread in the soiling rate is a measure of the uncertainty in the soiling rate for a particular month.

X. TYPICAL SOILING RATE UNCERTAINTY VALUES

Data from 32 GroundWork soiling stations were analyzed using the SDPA. The stations covered much of the continental United States (Fig. 7).

For each station, the results from the 12-month aggregation were used. From these results, the soiling rate P95 value and the P5 values were obtained for each month. By subtracting these two values a measure of the spread in the daily soiling rate was obtained according to (12).

$$\text{Monthly Spread} = P95 - P5 \qquad (12)$$

The spread in the soiling rate was determined for each month of data for each of the 32 stations analyzed. For several of the stations, a portion of the soiling data was not usable due to maintenance/snow related issues. These months were eliminated from the study (10 months total).

The spread in the data over the course of a year is shown in Fig. 8. The individual station data are shown as tick marks. Data from the western region are color coded as purple. Data from the southeast are color coded as blue. Data from the northeast are color coded as green.

For each month the median spread was computed, shown as the solid gold line. Also for each month, the 95% level of spread was determined. There is not a significant amount of variation in the spread throughout the year.

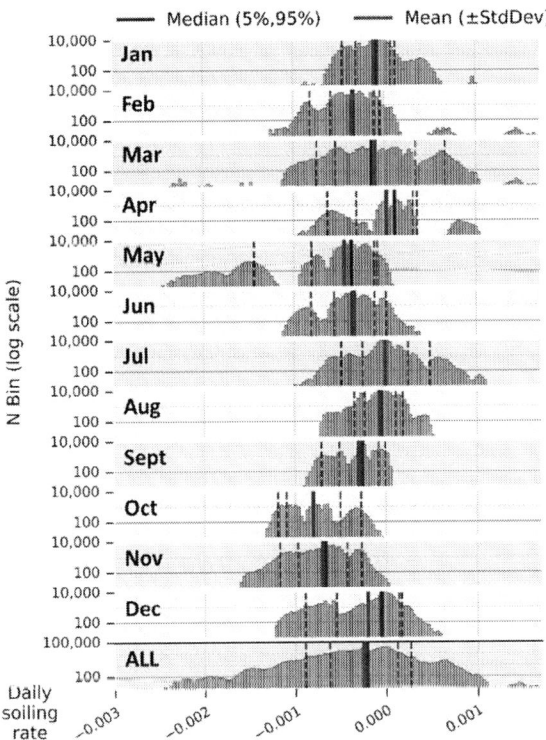

Fig. 6. Histogram of soiling rate values binned according to their magnitude.

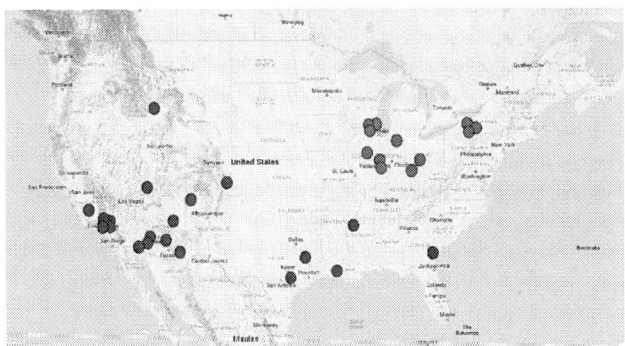

Fig. 7. Map of stations used in study. The stations were divided into different regions of the country, Western, Southeast, Northeast.

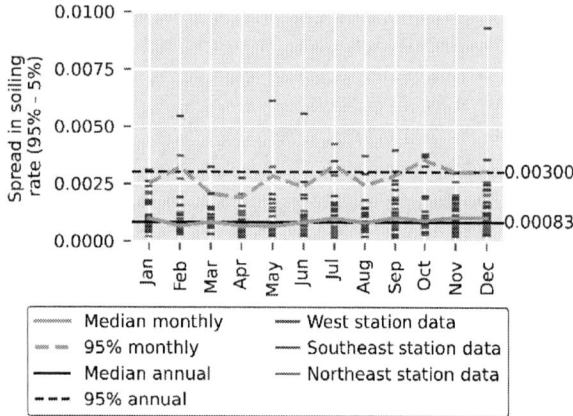

Fig. 8. Spread in the soiling rate for a range of stations across the continental United States

In addition, the median of all spread values for all months was computed and is represented by the solid black line. The median spread in the daily soiling rate for all stations is 0.00083 (8.3x10^{-4}). The 95% upper limit to the spread on an annual basis is 0.00300 (3.0x10^{-3}) and the 5% lower limit is 0.00032 (3.2x10^{-4}).

To state the results another way, for data from a GroundWork-maintained soiling kit that are processed using the SDPA, the typical uncertainty in the daily soiling rate is 8.3x10^{-4} and the uncertainty is expected to fall within the 3.0x10^{-3} to 3.2x10^{-4} range.

Table I gives The P5, median and P95 values when the 32 stations are grouped by region. The total number of months used to compute statistics for each region (N) is also given. The P5 values for all three regions are comparable. The median uncertainty of the western stations is larger than the eastern stations.

XI. CONCLUSION/FUTURE WORK

In conclusion, GroundWork Renewables has developed a soiling data processing algorithm (SDPA). The SPDA processes one-minute data to generate daily, monthly, and yearly soiling rates and their uncertainties.

Data filters eliminate one-minute soiling ratio data points that are outliers within a given day, that do not match trends in prior and future days, and that do not follow a cleaning event. In addition, the soiling ratio is adjusted to reduce the systematic error introduced by module misalignment.

From this set of filtered data, pairs of points are determined. The slope between a given pair of points is a measure of the soiling rate during this time period. By collecting the soiling rate values during a selected time period, the median soiling rate for that time period is determined. In addition, the spread in the data, represented by

TABLE I. SOILING RATE UNCERTAINTY VALUES OF FOR THREE DIFFERENT REGIONS OF THE UNITED STATES.

	N	P5	Median	P95
West	157	3.89 x10^{-4}	1.06 x10^{-3}	3.51 x10^{-3}
Southeast	59	2.27 x10^{-4}	5.86 x10^{-4}	1.58 x10^{-3}
Northeast	136	3.62 x10^{-4}	7.74 x10^{-4}	2.64 x10^{-3}

the distance between the P95 and P5 values, is also determined. The P95 and P5 are a measure of the uncertainty in the soiling rate during this time period.

To get an understanding of how the SDPA operates on a collection of stations, data from 32 stations across the United States were analyzed. From these stations an estimate of typical uncertainty spreads was made. The typical spread in the soiling rate over all 32 stations was 8.3x10^{-4} per day.

GroundWork Renewables continues to develop and refine the SDPA, with a goal to benchmark this approach against other data filters and models. Future plans include:

- Correlate detected reset events with rain data to derive site-specific cleaning thresholds.

- Correlate measured soiling rates to PV production loss

ACKNOWLEDGMENT

This work was funded by GroundWork Renewables, Inc. GroundWork's mission is to accelerate the success of solar in leading utility-scale energy production. GroundWork aims to maximize solar plant value with defensible and actionable reference solar data.

REFERENCES

[1] J. G. Bessa, L. Micheli, J. Montes-Romero, F. Almonacid, and E. F. Fernández, "Estimation of photovoltaic soiling using environmental parameters: A comparative analysis of existing models." *Advanced Sustainable Systems*, 2100335, 2022.

[2] M. G. Deceglie, L. Micheli, and M. Muller, "Quantifying soiling loss directly from PV yield." *IEEE J.Photovolt.*, vol. 8, no. 2, pp. 547-551, Mar. 2018.

[3] GroundWork Renewables, Inc. "Quality control tests for solar meteorological data," unpublished.

[4] M. Gostein, J. R. Caron, and B. Littmann, "Measuring soiling losses at utility-scale PV power plants," in *Proc. IEEE 40th Photovolt. Spec. Conf.*, vol. 6, pp. 0885–0890, 2014.

[5] Å. Skomedal, H. Haug, and E. S. Marstein, "Endogenous soiling rate determination and detection of cleaning events in utility-scale PV plants," *IEEE J. Photovolt.*, vol. 9, no. 3, pp. 858–863, May 2019.

[6] L. Micheli, M. Theristis, A. Livera, J. S. Stein, G. E. Georghiou, M. Muller, et al. "Improved PV soiling extraction through the detection of cleanings and change points." *IEEE J. Photovolt.*, vol. 11, no. 2, pp. 519-526, Mar. 2021.

[7] M. Muller, K. Perry, L. Micheli, F. Almonacid, and E. F. Fernández. "Automated detection of photovoltaic cleaning events: A performance comparison of techniques as applied to a broad set of labeled photovoltaic data sets." *Progress in Photovoltaics: Research and Applications*, vol. 30, no. 5, pp 567-577, Feb. 2022.

978-1-7281-6118-1/22 $31.00 © 2022 IEEE

Mismatch Losses in Simulated Commercial andUtility-scale PV Arrays due to Shortened Strings

Ryan M. Smith, Manjunath Matam, Hubert Seigneur

Pordis LLC, Austin, TX, United States

University of Central Florida, Orlando, FL, United States

The intentional removal of one or more photovoltaic modules from a string, thus shortening the length of the string relative to others within the array, may occur for a variety of reasons. The result is a mismatch in string length which our previous work has shown to impact the operation of the array by 1) shifting the ideal maximum power point of the array, and 2) inducing reverse currents in the shortened strings at VOC, a condition experienced by arrays under normal operation and during some maintenance activities. This work takes the experimentally verified simulation results of our previous small-scale studies and expands the simulations to elucidate behaviors at commercial and utility scales.

Fault Analysis and Relay Assessment on a Substation System with High Penetration of PV Generation

Biqi Wang[1], Genesis Alvarez[2], Micah J. Till[2], Kevin Jones [2], Mathew Gardner[2], Rolando Burgos[1], Bo Wen[1]

1. Center for Power Electronics Systems, Blacksburg,VA, 24061, United States of America
2. Dominion Energy, Richmond, VA, 23220, United States of America

Abstract—With the dramatic transition in distribution power systems from centralized generation to distributed generation (DG), the system fault responses, especially fault current profiles, are going to experience some major changes. These are likely to cause issues in the traditional overcurrent relay such as overlapping coordination and sensitivity deterioration. This paper, based on a Dominion Energy virtual substation system with a high solar energy penetration level, studies both the photovoltaic (PV) inverter fault responses and the distribution system fault-current profile under various fault scenarios. These include analyzing three-phase and single-phase faults under different fault severities and at different locations within the substation. The corresponding performance of an overcurrent protective relay in the substation system was also assessed. A hardware-in-the-loop (HIL) testbed was implemented for validation purposes, with an actual SEL-451 interfacing the real-time simulation in the RTDS platform. The HIL simulation results verify the effectiveness of the presented fault analysis methodology.

Keywords—Hardware-in-the-loop (HIL), real-time digital simulator (RTDS), Power system protection, SEL-451 protective relay, Photovoltaic

I. INTRODUCTION

To ensure the secure operation of a distribution system, it is essential to assess the performance of protective relay devices prior to installing them in the field [1][3][4]. The overcurrent relay is one of the most extensively applied relays [1][2]. Nowadays the distribution power grid is experiencing a dramatically increasing penetration from Distributed Energy Resources (DERs), such as PV farms. Integrating utility-scale DERs into an existing distribution network can cause considerable changes in the fault current signals sensed by the overcurrent relay device. These changes may defeat traditional protection schemes, which were originally designed for radial distribution feeders. Such changes in system fault current profiles will likely create technical issues that could threaten the reliable operation of relays, such as sensitivity deterioration and overlapping coordination [5]. Therefore, it's essential to study system fault characteristics and assess the performance of protective relays in a DER-dominated distribution system.

The real-time hardware-in-the-loop (HIL) simulation incorporates the merits of both software simulation and actual hardware testing, providing good test fidelity as well as wide test coverage [6][7]. In the case of relay performance validation, the HIL testbed places all potentially dangerous power equipment in a real-time simulation, then interfaces the simulation with an actual relay device. This mimics the real system and allows testing under a full range of operation conditions without the risk of damaging any power hardware.

In recent years, some literature has been focused on relay assessment under high DER penetration using HIL testing, while little has studied the system fault current profile based on the fault characteristics of a PV inverter. Ref. [8] evaluated the combined impact of a voltage source converter (VSC) and synchronous condenser on distance protection through HIL testing in a converter-dominated power system. Ref. [9] proposed a control strategy to mitigate the sensitivity deterioration issues of overcurrent protection in a distribution network with high DG penetration. Ref. [10] demonstrated the desensitization phenomenon in overcurrent relay and investigated the SEL-351S relay behavior with high DG penetration, using the OPAL-RT digital simulator. An adaptive overcurrent protection with a high presence of DG was designed in [11] to overcome the reach reduction caused by DG penetration, which required a great amount of recalculation.

This paper analyzes fault characteristics and assesses overcurrent relay schemes in a Dominion Energy virtual substation system with a high PV penetration level. The substation system was modeled in the RTDS platform, and a pre-programmed SEL-451 relay was tested under various fault scenarios in real-time HIL simulations.

Section II of this paper gives a brief introduction to the utilized HIL testbed topology. Section III focuses on the PV inverter control scheme. Section IV presents the SEL-451 main relay schemes. Section V discusses PV inverter fault characteristics, while the fault current responses sensed by the relay and the relay performances under different scenarios are analyzed in Section VI. Finally, in Section VII, conclusions are drawn and future work is outlined.

978-1-7281-6118-1/22 $31.00 © 2022 IEEE

II. HIL TESTBED TOPOLOGY

The HIL relay testbed topology is shown in Fig. 1. The 35 MW substation was implemented in RTDS, where a 34.5 kV bus was supplied by a transmission system equivalent model and two utility-scale PV farms through grid-tracking controlled inverters. The solar facilities have a total power capacity of 21 MW.

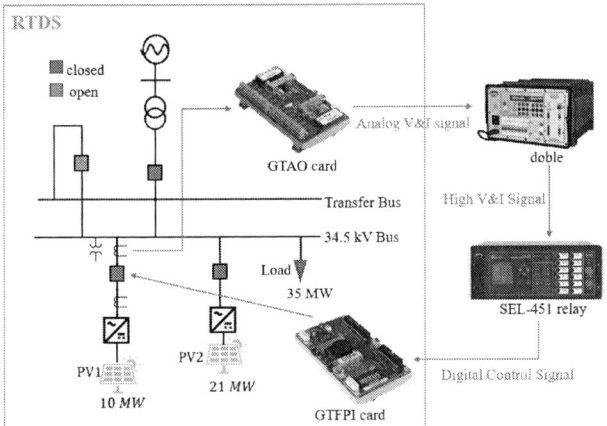

Fig. 1. HIL testing platform of virtual substation model

Under normal operating conditions, PV1 is interconnected to the 34.5 kV bus through a circuit breaker in closed status. The HIL testing discussed in the rest of the paper focuses on the relay logic which controls the circuit breaker that interconnects PV1 to the 34.5 kV bus. Two Current Transformers (CTs) were installed to provide current measurements for the relay device, one at the bus-side of the circuit breaker and another at the line-side. A Potential Transformer (PT) was installed on the 34.5 kV bus to provide bus voltage signal. The tested circuit breaker would be opened under certain fault scenarios to isolate the faulted part from rest of the substation power system.

During the real-time simulation, two sets of three-phase current measurements and one set of three phase voltage measurements are taken at the secondary side of the PT and CTs. The Giga-Transceiver Analogue Output (GTAO) card was utilized to provide analog voltage and current signals to the Doble amplifier, and the amplified analog signals were sent to the digital relay hardware. The SEL-451 relay was able to monitor the substation model and generate the tripping signal under faults. The digital control signal was sent to the RTDS through the Gigabit-Transceiver Front Panel Interface (GTFPI) to trip the circuit breaker between PV1 and the 34.5 kV bus, in order to disconnect the PV farm from the rest of substation system.

III. PV INVERTER TOPOLOGY AND CONTROL

The PV inverter topology is illustrated in Fig. 2. The PV array is interconnected to the grid side through a three-phase half-bridge inverter and a DC-link capacitor. A two-stage LCL filter was applied to filter out the switching ripples and harmonics. The leakage inductor of the step-up transformer

played the role of the grid side inductor L_2. The LCL filter design methodology can be found in [13]. Average models were deployed to model the PV inverter for a faster simulation in the time domain, while still preserving an accurate-enough dynamic behavior. The inverter average model main parameters for PV1 and PV2 are shown in Table I.

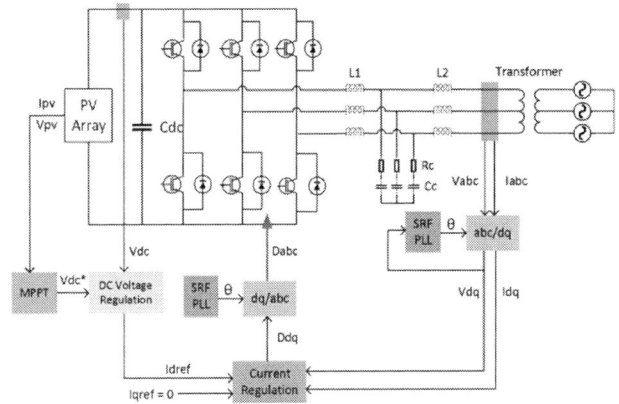

Fig. 2. PV inverter topology and control scheme

TABLE I. PV INVERTER AVERAGE MODEL MAIN PARAMETERS

Parameters	Values
Rated DC Voltage	1.2 kV
Rated Inverter Apparent Power	10 MW/21 MW
Rated AC Voltage (line-to-line)	480 V
Sun irradiance	1000 W/m^2
Cell temperature	25 °C
PV Module Open circuit voltage (V_{oc})	21.7 V
PV Module Short-circuit current (I_{sc})	3.35 A
PV Module Maximum power point voltage (V_{mpp})	17.4 V
PV Module Maximum power point current (I_{mpp})	3.05 A
DC-link Capacitance (C_{dc})	5000 μF
Inverter side inductance (L_1)	19.6 μH / 9.33 μH
LCL filter capacitance (C_c)	2500 μF

The PV inverter grid tracking mode controller is shown in Fig. 3. A synchronous-reference-frame phase-locked-loop (SRF PLL) was used to maintain the inverter controller frame's synchronization with the grid at the point of common coupling (PCC). The Lambert function approximation technique was deployed as the maximum power point tracking (MPPT) algorithm to generate a reference signal for the DC voltage regulator, which ensured a maximum power extraction from the PV array. The PV inverter also operates at unity power-factor mode, which gives no reactive power injection to the grid side and thus forced the I_{qref} to be zero. With I_{dref} provided from the outer DC voltage loop, the dq frame current controller generated duty cycles for the power stage operation. To prevent semiconductor switches in PV inverters from overcurrent damage, a saturation limiter was

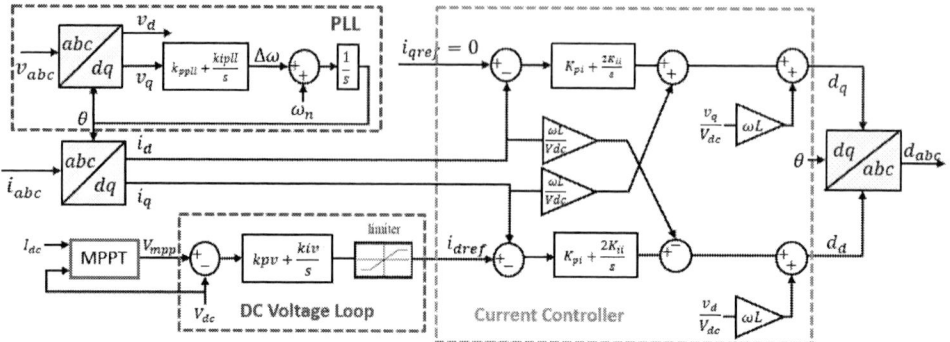

Fig. 3. PV inverter controller diagram

imposed on current reference values, which limited the inverter output current magnitude under 1.5 $p.u.$

IV. SEL-451 RELAY TRIPPING LOGIC

A. Overcurrent Relay

Overcurrent relay is one of the most widely used relay schemes in distribution networks. The main tripping logic corresponding to the circuit breaker (indicated in Fig. 1) was overcurrent protection that was pre-programmed to the relay hardware with SEL-451 relay functions. Both the instantaneous overcurrent protection element (ANSI 50) and the inverse-time overcurrent element (ANSI 51) were included. The tripping of all overcurrent functions is based on the current measurements sent from the bus-side CT. Under normal operation, the rated CT second-side current RMS is 0.418 Amps, with a CT turns ratio of 1:400.

A tripping signal from the instantaneous overcurrent element 67P1T will be generated immediately if the maximum RMS among three-phase analog current signals received on the SEL-451 relay device side exceeded the preset threshold.

The tripping time of the inverse-time overcurrent element is developed using the time-current equations conform to IEEE Standard C37.112-2018[12]. The inverse-time characteristic curve could be categorized into several different types by the curve inverse degree. The U.S. very inverse curve (U3) is applied in the protection schemes tested in this paper. The U3 inverse-time curve can be described as

$$T_p = TD \left(\frac{3.88}{(\frac{I}{I_s})^2 - 1} + 0.0963 \right) \tag{1}$$

where T_p represents the relay function operating time, TD is the time dial setting, I_s is the pickup current, and I is the actual maximum current RMS among three phases. The inverse-time overcurrent element 51S1T operated with the time delay given in (1) as long as the current RMS received from the bus-side CT exceeded the preset pickup current.

Zero-sequence overcurrent elements provided protection against single-phase faults. The residual-ground instantaneous overcurrent element 67G1T operated immediately when the neutral-to-ground current RMS sent from the bus-side CT

exceeded its pickup current. The zero-sequence inverse-time overcurrent element 51S2T operated with a time delay given by the U3 curve shown in (1) when the received zero-sequence current RMS surpassed the pickup value.

Pickup current settings for the four different types of overcurrent relay element are shown in Table II.

TABLE II. OVERCURRENT RELAY ELEMENT PICKUP CURRENTS

Overcurrent Element Types	Pick Up Currents
Instantaneous overcurrent (67P1T)	12.83 A
Inverse-time overcurrent (51S1T)	1.80 A
Residual-ground instantaneous overcurrent (67G1T)	11.75 A
Zero-sequence inverse-time overcurrent (51S2T)	1.70 A

B. Circuit Breaker Internal Fault Detection

The detection of a circuit-breaker fault is based on current measurement results from both the bus-side and PV-side CT. Under system normal operation conditions, the two CTs will sense currents with the same magnitude and different directions. During a circuit-breaker internal fault, the fault currents increase at both sides of the circuit breaker. However, the bus side fault current increase will be much larger than the PV side, due to the 1.5 $p.u.$ saturation limit imposed on the inverter current control loop. The detection methodology used was a logic combination of overcurrent elements from both CTs. The logic diagram is shown in Fig. 4. When only the bus-side CT instantaneous overcurrent element 51S1 operated, the tripping signal was generated from the internal fault detection logic to trip the circuit breaker.

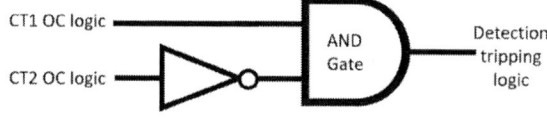

Fig. 4. Circuit breaker internal fault detection logic

V. PV INVERTER FAULT RESPONSE

In order to understand the impact of the integration of large-scale PV farms on protective relay strategies, it is necessary to first study the fault response of the PV inverter. A three-phase fault was tested to occur on the bus to which PV1 is connected, where different fault-to-ground resistances were applied to emulate differing fault severity. All the PV inverter fault characteristics discussed in this section are based on this specific scenario.

A three-phase fault instantly brought down the faulted bus voltage. Under the grid-tracking control mode, after the fault transient, the PV inverter PCC voltage on the q axis remained at zero; the d axis PCC voltage v_d was brought down. The v_d fault responses are shown in Fig. 5. It could be observed that a lower fault resistance, R_f, caused a greater voltage magnitude drop at the PV PCC. The PV is required to cease energizing the substation within 21 seconds when the voltage drops below 0.88 $p.u.$, per IEEE Standard 1547-2018 [14]. For the two most severe fault cases ($R_f = 1,3\ \Omega$), the PV inverter voltage fell into the mandatory operation capability region and was required to trip within 21 seconds by default; For less severe scenarios ($R_f = 5,10,20\ \Omega$), the PV generation was allowed to ride through the system fault.

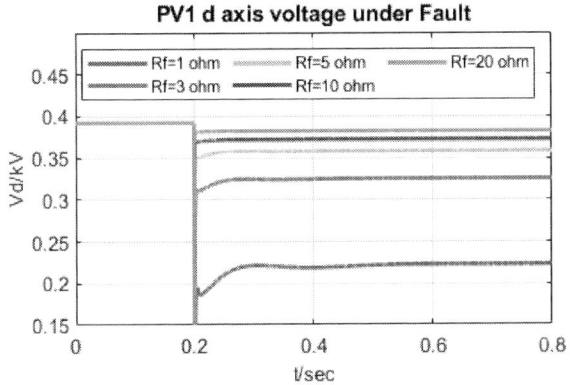

Fig. 5. PV1 d axis voltage fault responses

Fig. 6 illustrates the fault responses of the PV inverter DC link voltage under different fault severities. Under the most severe case, in which $R_f = 1\ \Omega$, the DC regulator was unable to track the reference V_{mpp} anymore. The DC link voltage maintained a higher value after a fault transient stage. The outer loop failed after the fault occurrence. Hence, the solar module's operating point was pushed off from the maximum

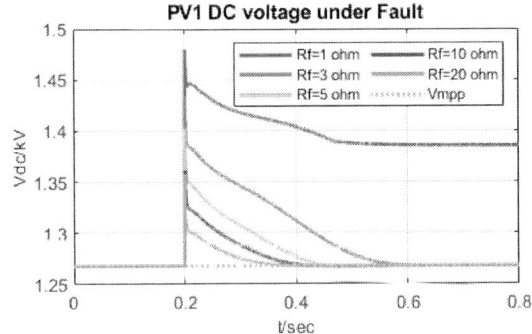

Fig. 6. PV1 DC link voltage fault responses

power point (MPP) and moved to the negative sloping side of the P-V curve, which resulted in a reduced steady-state PV harnessed power, as indicated in Fig. 7. However, in scenarios with higher fault impedance ($R_f = 3,5,10,20\ \Omega$), the outer loop controller was able to regulate the DC link voltage back to V_{mpp} after a transient period, therefore bringing the PV array's operation back to the MPP. The DC link voltage entered the steady state faster under less severe faults (higher controller bandwidth). This tallies with the simulation results in Fig. 7, where the PV1 harnessed power could restore to 10 MW faster after faults' occurrence in less severe scenarios.

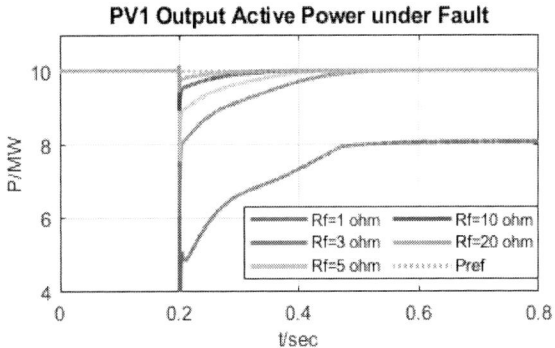

Fig. 7. PV1 active power fault responses

Under the grid-tracking control, after the controller frame synchronized with the grid side, the steady state d axis current could be estimated using

$$i_d = \frac{P_{out}}{v_d} \tag{3}$$

where P_{out} denotes the inverter output active power and v_d denotes the inverter PCC d axis voltage. Due to the PCC voltage drop during a fault, the DC voltage loop output I_{dref} increased and reached steady state after a fault transient. When the steady state P_{out} remained unchanged, under less severe fault conditions v_d was higher and thus caused a lower steady state i_d and a shorter transient period (higher controller bandwidth). This tallies with the i_d fault response results

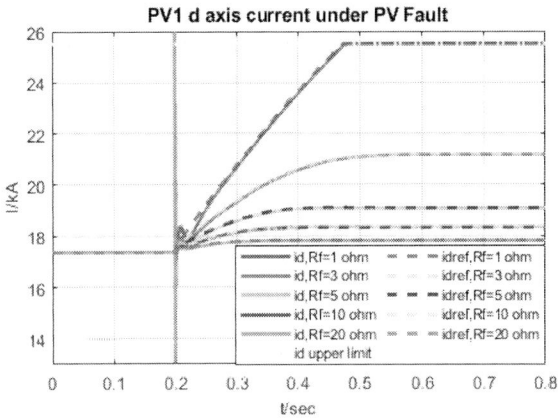

Fig. 8. PV1 d axis current fault response

978-1-7281-6118-1/22 $31.00 © 2022 IEEE

shown in Fig. 8. For the most severe fault scenario ($R_f = 1\ \Omega$), where the steady state P_{out} failed to eventually reach the PV array MPP, I_{dref} saturated at 1.5 $p.u.$ after the fault transient dynamic. Under all scenarios, the d axis instantaneous current was able to track I_{dref}, and the inner AC current loop remained effective.

VI. Relay Fault Current Profile Analysis

A wide range of fault scenarios was performed on the HIL testbed shown in Fig.1 to verify the reliability of the SEL-451 relay tripping logic. It was assumed that the relay scheme programmed into the device would be able to detect the fault occurrence and correctly trip the circuit breaker under different scenarios. Both three-phase and single-phase transient faults were applied at the PV inverter PCC, between CT_1 and CT_2, and also on the 34.5 kV bus, in order to emulate a PV internal fault, a circuit-breaker internal fault, and a bus fault. Different fault-to-ground impedances were also applied to emulate different fault severity. The fault current response at the relay's location is analyzed in this section.

The relay elements operation results recorded by the SEL-451 relay device during faults are shown in Appendix A.

A. Three-Phase Faults

Under a three-phase, short-circuit fault inside the PV facility, the fault current flowed from the bus to the fault location. CT_1 and CT_2 sensed fault currents with the same magnitude and opposite direction. The relay element's response was recorded by the SEL-451 during the fault event, in which a tripping signal was generated with a corresponding time delay to isolate the faulted PV from the substation. The CT-current RMS under a three-phase PV internal fault with different fault resistances is shown in Fig. 9. For the two most severe fault scenarios ($R_f = 1,3\ \Omega$), the current measurement signals from both CTs immediately exceeded the instantaneous overcurrent element threshold, which caused the SEL-451 relay function 67P1T to operate promptly. For the cases with a higher fault impedance ($R_f = 5,10,20\ \Omega$), the current RMS on bus-side CT exceeded the inverse-time overcurrent element threshold, while staying below the instantaneous overcurrent threshold. As a result, the inverse-time overcurrent element 51S1T tripped with the time delay given in (1). The delay time of the tripping signal generation was longer under less severe fault (see Appendix A).

For a circuit-breaker internal fault, the fault current magnitude measured on CT_1 and CT_2 differed tremendously. As shown in Fig. 10, under all levels of tested fault resistance, the current RMS on the bus-side CT had a surge and surpassed the inverse-time overcurrent element 51S1 threshold after the fault occurrence. By contrast, the PV side fault current only increased slightly due to the saturation of inverter current controller. It stayed below the threshold of 51S1. According to the circuit breaker's internal fault detection methodology seen in Fig. 4, overcurrent was only detected on the bus-side CT, and the circuit-breaker internal fault detection element PSV14 tripped. For the case with a

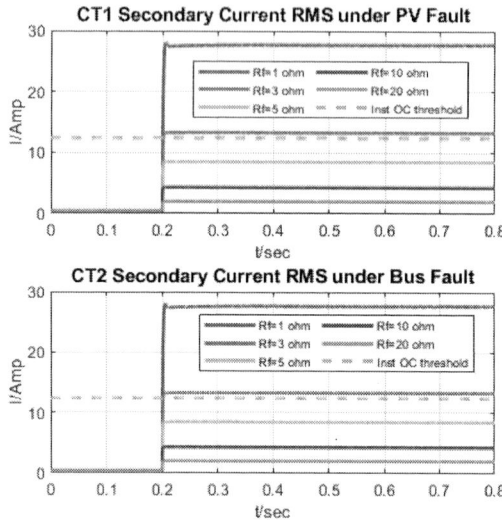

Fig. 9. CT current RMS under three-phase PV fault

higher fault impedance ($R_f = 5,10\ \Omega$), the circuit-breaker internal fault detection element PSV14 operated immediately, while the inverse-time overcurrent element 51S1T operated with a time delay; the digital control signal to trip the circuit breaker was generated at the moment when PSV14 operated, which is prior to the overcurrent element operation (see Appendix A).

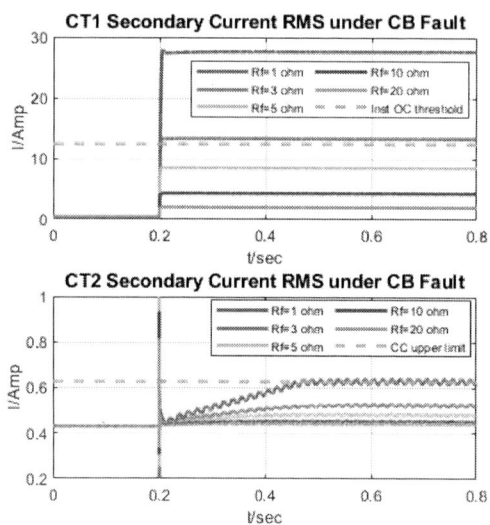

Fig. 10. CT current RMS under three-phase circuit breaker fault

When a three-phase fault occurred on the 34.5 kV bus, the fault current flowed through the two CTs from the PV inverter PCC to the fault location. The CT fault current behaviors under a three-phase bus fault is shown in Fig. 11. The steady-state current RMS on both CTs increased with a 1.5 $p.u.$ upper limit, due to the impact of the PV inverter current controller saturation. The corresponding escalation of fault severity boosted the increase in relay-location-fault current RMS. However, the increase of the steady-state current RMS saturated and stayed constant once the fault resistance was low enough, for example when $R_f = 1\,\Omega$. Also, the transient speed was faster for higher R_f, which supports our analysis in Section V. At all times, the fault current signals received by the SEL-451 hardware remained below the pickup current of the inverse-time overcurrent element 51S1. This meant that the overcurrent protection elements would not operate, and the relay hardware was unable record the fault event on the bus. No trip signal for the circuit breaker was generated, and the PV farm rode through all abnormal conditions and continued to supply the substation power system.

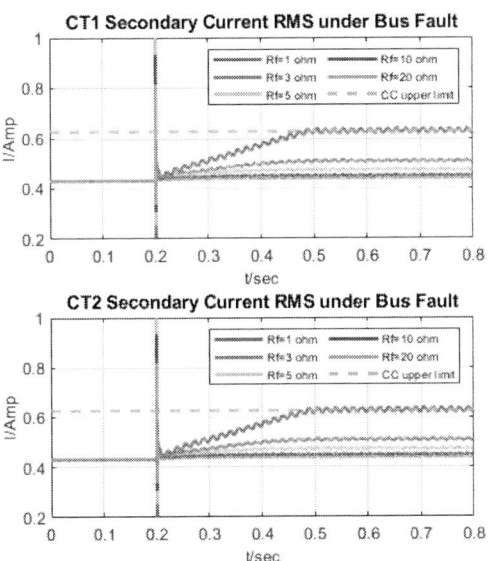

Fig. 11. CT current RMS under three-phase bus fault

However, based on the requirements of DER response under abnormal voltage conditions according to IEEE Standard 1547-2018 [14], PV1 was required to trip with PCC voltage below $0.88\ p.u.$ (corresponding to the two most severe fault cases where $R_f = 1, 3\,\Omega$), and by default the circuit breaker needed to disconnect PV from the distribution system within 21 seconds. To meet the IEEE DER tripping requirements, revision of the protective relay threshold settings is recommended.

B. Single-Phase Faults

When a single-phase fault occurs, the system bus voltages and line currents becomes unbalanced, as well as the PV inverter fault currents. Fig. 12 shows zero-sequence CT current RMS fault behaviors under single-phase PV internal faults. For the two most severe fault scenarios ($R_f = 1, 3\,\Omega$), the steady state zero-sequence current RMS on both CTs exceeded the residual-ground instantaneous OC element 67G1T's threshold. 67G1T operated immediately after the fault occurrence, along with 67P1T, as shown in Appendix A. For less severe fault case such as $R_f = 5\Omega$, the CT secondary-side zero-sequence current RMS on the bus-side CT rose

between the 51S2 threshold and the 67G1 threshold. The 51S2T element was enabled with a time delay given by (1), which triggered the control signal to trip the circuit breaker. 51S2T occurred prior to the 51S1T, which was enabled by the excessive fault phase current (see Appendix A). These show that the zero sequence OC relay was able to improve the protection scheme sensitivity during high impedance single-phase faults.

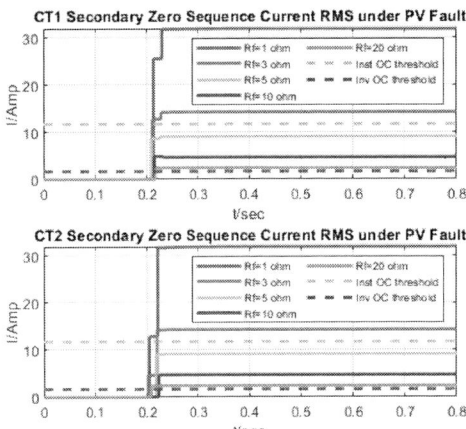

Fig. 12. CT current RMS under a single-phase PV fault

For a single-phase circuit breaker internal fault, the maximum fault current magnitude measured on CT_1 and CT_2 differed significantly. As shown in Fig. 13, the zero-sequence current RMS on the bus-side CT increased sharply and surpassed the pickup value of 51S2 under each tested fault resistance. The increase on PV side fault current was relatively minor and stayed below the preset 51S2 threshold. The circuit breaker's internal fault detection technique was designed to trip element PSV14, which operated immediately during a single-phase circuit breaker internal fault, before the 51S2T element was enabled by current increase on the faulted phase (see Appendix A).

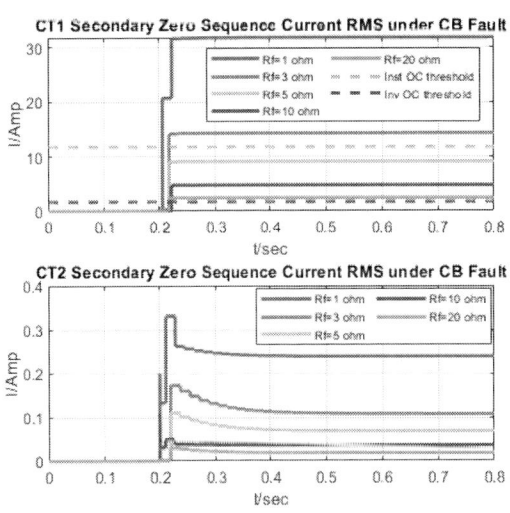

Fig. 13. CT current RMS under a single-phase circuit-breaker fault

When a single-phase fault occurred on the 34.5 kV bus, the zero-sequence fault current through the bus-side and PV-side CTs both had an increase since the inverter current was unbalanced; however, they stayed below the 51S2 element threshold under all tested fault resistances due to the limited fault current contribution from PV1, shown in Fig. 14. The overcurrent element 51S2 did not operate, and the SEL-451 relay device was unable to detect the fault event. No trip signal was generated under such fault scenarios. The PV facility rode through these abnormal conditions and kept energizing the substation system. Similar to the three-phase bus fault, the test results contradicted the requirement that the PV shall not ride through the abnormal conditions under low fault impedance. As the relay performance did not meet the IEEE requirements, a revision of the relay threshold setting is recommended.

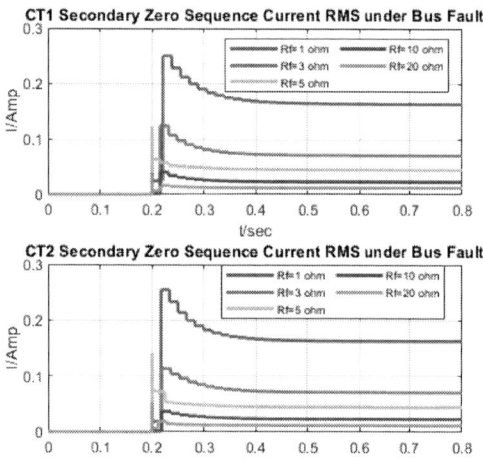

Fig. 14. CT current RMS under single-phase bus fault

VII. CONCLUSION

In summary, this paper focused on evaluating the fault current profile analysis and the system overcurrent relay when a distribution system is interconnected with utility-scale PV generation, based on test results from a real-time hardware-in-the-loop substation system testbed. The PV inverter fault response and the relay installation location fault current behaviors were studied under various fault scenarios, including both three-phase and single-phase faults at different locations within the substation and under different fault severity. Tested locations included inside the PV facility, inside the circuit breaker, and on the 34.5 kV bus.

The grid-tracking controlled PV inverter had a limited fault current contribution to the distribution system because of the saturation limits placed on the current controller; the inverter steady-state fault current magnitude saturated under low impedance faults.

The effectiveness of the tested protective relay was verified under PV internal faults and circuit-breaker internal faults. The circuit breaker will ride through the PV internal fault

when the fault impedance is high enough. The relay sensitivity under high impedance single-phase faults is improved with the zero-sequence overcurrent relay elements. The circuit-breaker fault detection element allows the prompt tripping of the circuit breaker without a time delay, prior to the overcurrent element's operation, even during high impedance faults.

The tested relay performance did not meet the IEEE Standard 1547-2018 DER fault ride-through requirements under low impedance bus faults due to the limited fault current contribution from the PV inverter. A revision of the existing relay settings is suggested.

The fault current contributions from utility-scale PVs are also likely to cause coordination issues, which are not discussed in this paper since only the relay scheme for one circuit breaker was tested. HIL tests involving multiple relay devices in a more complicated system will be conducted for next steps in order to evaluate a high PV penetration's impact on the relays' coordination. Given the increasing DER penetration level, the fault analysis and relay assessment methodology in this paper provides a very practical reference for designing and revising protective relays in DER-dominated distribution systems.

Appendix A

The tripping behaviors of protection functions that were recorded in the SEL-451 relay hardware fault event history under all tested fault scenarios are included in this section.

Relay functions tripping under three-phase PV internal faults and circuit breaker internal faults are shown in Fig. 15 and Fig.16, respectively. Relay functions tripping under single-phase PV internal faults and circuit breaker internal faults are respectively shown in Fig. 17 and Fig.18,

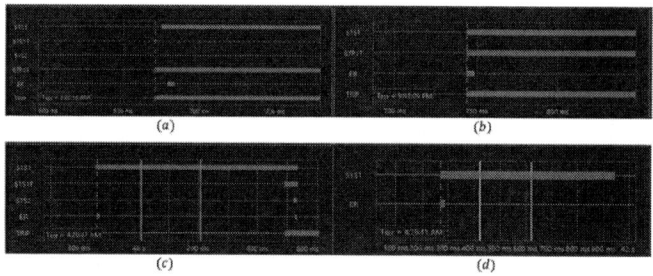

Fig. 15. Relay functions tripping under three-phase PV internal faults (a) $R_f = 1\ \Omega$; (b) $R_f = 3\ \Omega$; (c) $R_f = 5\ \Omega$; (d) $R_f = 10\ \Omega$

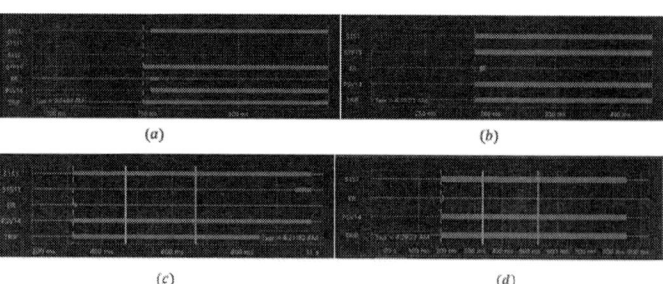

Fig. 16. Relay functions tripping under three-phase CB internal faults (a) $R_f = 1\ \Omega$; (b) $R_f = 3\ \Omega$; (c) $R_f = 5\ \Omega$; (d) $R_f = 10\ \Omega$

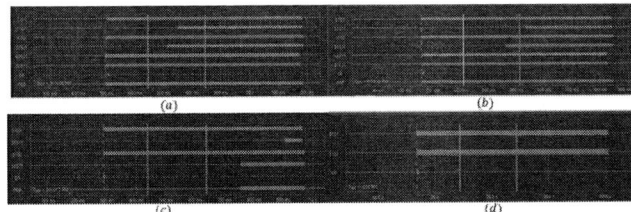

Fig. 17. Relay functions tripping under single-phase PV internal faults (a) $R_f = 1\,\Omega$; (b) $R_f = 3\,\Omega$; (c) $R_f = 5\,\Omega$; (d) $R_f = 10\,\Omega$

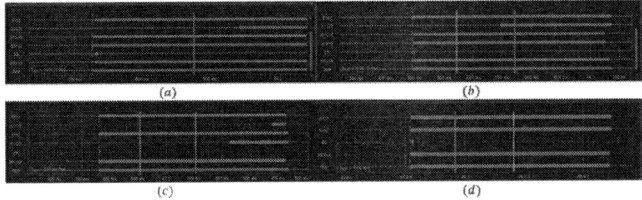

Fig. 18. Relay functions tripping under single-phase CB internal faults (a) $R_f = 1\,\Omega$; (b) $R_f = 3\,\Omega$; (c) $R_f = 5\,\Omega$; (d) $R_f = 10\,\Omega$

respectively, where the red dashed lines represent the detection/pickup moments of fault events. The relay element operation results during bus faults and under the scenarios in which $R_f = 20\,\Omega$ (the most severe fault case) are not included, since no trip signals were generated in these cases. Also, the enablement of "TRIP" is not shown in the fault cases when $R_f = 10\,\Omega$, since the delay imposed on the inverse-time OC element is too long to trigger the digital control signal within the recording range.

REFERENCES

[1] M. S. Almas, R. Leelaruji and L. Vanfretti, "Over-Current Relay Model Implementation For Real Time Simulation & Hardware-In-The-Loop (HIL) Validation," *IECON 2012 - 38th Annual Conference on IEEE Industrial Electronics Society,* 2012, pp. 4789-4796.

[2] A. S. Makhzani, M. Zarghami, B. Falahati and M. Vaziri, "Hardware-In-The-Loop Testing Of Protection Relays In Distribution Feeders With High Penetration of DGs," *2017 North American Power Symposium (NAPS),* 2017, pp. 1-6.

[3] M. S. Almas, R. Leelaruji and L. Vanfretti, "Over-Current Relay Model Implementation For Real Time Simulation & Hardware-In-The-Loop (HIL) Validation," *IECON 2012 - 38th Annual Conference on IEEE Industrial Electronics Society,* 2012, pp. 4789-4796.

[4] A. S. Makhzani, M. Zarghami, B. Falahati and M. Vaziri, "Hardware-In-The-Loop Testing Of Protection Relays In Distribution Feeders With High Penetration of DGs," *2017 North American Power Symposium (NAPS),* 2017, pp. 1-6.

[5] C. S. Edrington, M. Steurer, J. Langston, T. El-Mezyani and K. Schoder, "Role of Power Hardware in the Loop in Modeling and Simulation for Experimentation in Power and Energy Systems," in *Proceedings of the IEEE,* 2015, vol. 103, no. 12, pp. 2401-2409.

[6] G. Li, D. Zhang, Y. Xin, S. Jiang, W. Wang and J. Du, "Design of MMC Hardware-In-The-Loop Platform and Controller Test Scheme," in *CPSS Transactions on Power Electronics and Applications,* 2019, vol. 4, no. 2, pp. 143-151.

[7] B. Wang, R. Burgos, Y. Tang and B. Wen, "Fault Characteristics Analysis on 56-Bus Distribution System with Penetration of Utility-Scale PV Generation," *2021 6th IEEE Workshop on the Electronic Grid (eGRID),* 2021, pp. 01-08.

[8] R. O. Salcedo, J. K. Nowocin, C. L. Smith, R. P. Rekha, E. G. Corbett, E. R. Limpaecher, and J. M. LaPenta, "Development Of A Real-Time Hardware-In-The-Loop Power Systems Simulation Platform To Evaluate Commercial Microgrid Controllers." MIT Lincoln Laboratory, Lexington MA United States, 2016.

[9] R. Stanev, K. Viglov, K. Nakov and T. Asenov, "A Real Time Power Hardware in the Loop Test Bed for Power System Stability Studies," *12th Electrical Engineering Faculty Conference (BulEF),* 2020, pp. 1-5.

[10] J. Jia, G. Yang, A. H. Nielsen and P. Rønne-Hansen, "Impact of VSC Control Strategies and Incorporation of Synchronous Condensers on Distance Protection Under Unbalanced Faults," in *IEEE Transactions on Industrial Electronics,* vol. 66, no. 2, pp. 1108-1118, Feb. 2019.

[11] M. Yousaf, K. M. Muttaqi and D. Sutanto, "A Control Strategy to Mitigate the Sensitivity Deterioration of Overcurrent Protection in Distribution Networks with the Higher Concentration of Synchronous and Inverter Based DG Units," *2020 IEEE Industry Applications Society Annual Meeting,* pp. 1-6.

[12] A. S. Makhzani, M. Zarghami, B. Falahati and M. Vaziri, "Hardware-in-the-loop testing of protection relays in distribution feeders with high penetration of DGs," *2017 North American Power Symposium (NAPS),* 2017, pp. 1-6.

[13] R. Liu, R. Sun and M. Tania, "Hardware-in-the-Loop Relay Testing in Dominion's Blackstart Plan," *2017 IEEE Power & Energy Society General Meeting,* 2017, pp. 1-5.

[14] "IEEE Standard for Inverse-Time Characteristics Equations for Overcurrent Relays," in *IEEE Std C37.112-2018* (Revision of IEEE Std C37.112-1996) , vol., no., pp.1-25, 5 Feb. 2019.

[15] Y. Jiao and F. C. Lee, "LCL Filter Design and Inductor Current Ripple Analysis for a Three-Level NPC Grid Interface Converter," in *IEEE Transactions on Power Electronics,* vol. 30, no. 9, pp. 4659-4668, Sept. 2015

[16] IEEE-The Institute of Electrical and Electronics Engineers. "2018. IEEE Standard for Interconnection and Interoperability of Distributed Energy Resources with Associated Electric Power Systems Interfaces," pp. 1–138, *IEEE Std 1547-2018* (Revision of IEEE Std 1547-2003).

Determining Surface Recombination Velocity and Band Bending at the Back Interface of CdTe Devices Using Back Illuminated Quantum Efficiency

Adam B. Phillips, Jared D. Friedl, Zhaoning Song, Ramez Hosseinian Ahangharnejhad, Ebin Bastola, Zulkifl H. Rabbani, Deng-Bing Li, Yanfa Yan, Randy J. Ellingson, Michael J. Heben

In CdTe-based devices the measured open circuit voltage is significantly below the thermodynamic limit. This suggests that significant recombination occurs at one or more of the front interface, back interface, or bulk and that improving device performance will require reduction of this recombination. It is difficult to determine where the device limiting recombination occurs, so attempts to improve the device typically focus on modifying one interface. Unfortunately, if this interface is not limiting the device, improvements will not be observed. A better method would be to measure the key parameters of an interface then use those parameters to determine if that interface is limiting. With this in mind, we developed a method to measure the back surface recombination velocity (SB) and the materials dependent band bending (BB0) using back illuminated quantum efficiency measurements. To determine both parameters, we must fabricate two devices with varying absorber thickness. For a thin device, the back interface is approximately the depletion width from the front interface, so the amount of band bending at the back interface is near zero. As a result, the internal quantum efficiency (IQE) of this device provides a direct measurement of SB. In the thicker device the band bending equals BB0, so the IQE from the thick device and SB determined from the thin device gives BB0. Acquiring these values is the first step in determining the location of limiting recombination.

Use of a Selenium-Telluride Alloy as a Back Interface for CdTe-Based Cells

Daniel Z. Shaw, Camden L. Kasik, Andrew C. Treglia, and James R. Sites

Colorado State University, Fort Collins, CO 80523, United States

Abstract — A variable tellurium content in a selenium telluride alloy ($Se_{1-x}Te_x$) was applied to thin film CdTe based photovoltaic devices as an electron-reflector layer. Comparing tellurium contents of 10%, 50%, and 70%, and varying the thickness of the layer between 15 and 30nm; led to an initial optimization of the layer. The resulting devices showed that a thicker SeTe layer would result in lower efficiencies, and a lower tellurium content (10%) performed better in terms of V_{oc} and fill factor most notably. These selenium rich devices show great initial promise, reaching efficiencies of 19.1%.

Keywords — CdTe, CdSeTe, SeTe, electron reflector, evaporation, close-space sublimation

I. INTRODUCTION

Cadmium telluride (CdTe) based thin-film photovoltaics have proven to be a viable and cost-effective solution energy solution [1]. While advancements have been made thanks to the high current density of the devices, a limiting factor has proven to be the difficulty in raising the open-circuit voltage. One approach is to use band bending at the back of the device and create an "electron reflector" [2]. The theory behind this layer is that the photo-generated holes will be more easily extracted via the back contact and reduce recombination with the created photoelectrons. This reduces the impact of the Schottky barrier formed by metal back contacts with lower work functions. For this to be effective, the back layer should have a higher band gap and an intermediary valance band maximum (VBM) between CdTe (5.8 eV below the vacuum level) and the work function of the back contact (5.2 eV for Ni) [2].

The alloy selenium telluride ($Se_{1-x}Te_x$) has been shown to have a non-linear tunable band gap [3]. A wide range of tellurium mole fractions have a favorable VBM to perform a band bending to aid the hole extraction. While tellurium's band gap is fairly narrow (0.35 eV), pure selenium has a large direct band gap of ~1.8 eV [3] and has proven to be a useful dopant in CdTe [4]. Selenium's inclusion in the back layer would theoretically provide the VBM step between CdTe and the proven Te back layer [2]. The larger band gap should provide the desired electron reflector behavior that is not present in the Te layer.

This work is a preliminary investigation into the viability of the $Se_{1-x}Te_x$ back layer. The mole fractions of x =0.1, 0.5, and 0.7 were synthesized and evaporated onto the superstrate already optimized here at Colorado State University [4]. The thickness of the $Se_{1-x}Te_x$ was alternated between 15 nm and 30 nm, to ascertain the amount of material necessary to produce the desired effects, without producing a level of resistivity.

II. EXPERIMENTAL

A. Device Fabrication

The device structures were fabricated at Colorado State University in order to study the effect of $Se_{1-x}Te_x$ alloys, where the doping concentration was varied along with the thickness of the alloy. Using commercial soda-lime glass coated with 400 nm of SnO_2:F TCO layer (TEC 10), 100 nm of 11% (Mg,Zn)O (MZO) n-type layer was deposited at room temperature via RF magnetron sputtering. The CdSeTe, CdTe, and $CdCl_2$ treatments were deposited by close-space sublimation (CSS) in a multi-source system [5]. The plates were then removed from vacuum, rinsed with deionized water to remove excess $CdCl_2$. The plates were then placed inside a different CSS system and the CuCl treatment was applied according to the optimization performed in Ref. 6.

The $Se_{1-x}Te_x$ alloy and tellurium layers were evaporated onto the CdSeTe/CdTe film, with half of the plate receiving the Te buffer layer. A carbon/nickel paint back contact was applied before delineation of the cells for each device. The full structure

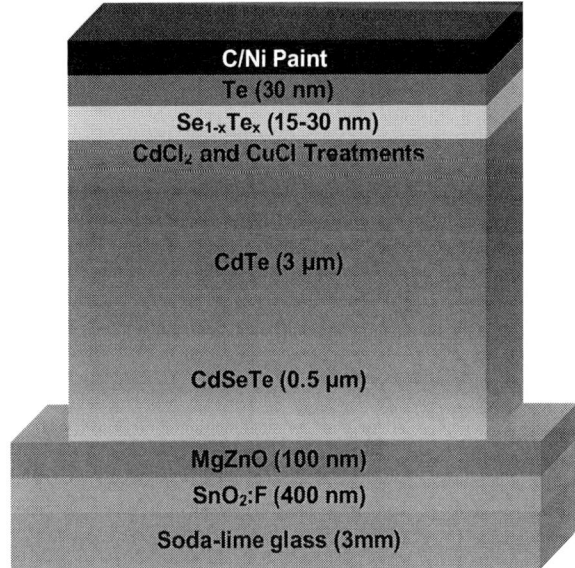

Fig. 1. Device Structure (not to scale). MZO, CdSeTe, CdTe, $CdCl_2$, CuCl and Te layers were consistent for each device. The only variation was the thickness of the $Se_{1-x}Te_x$ layer, and the presence of an additional tellurium layer.

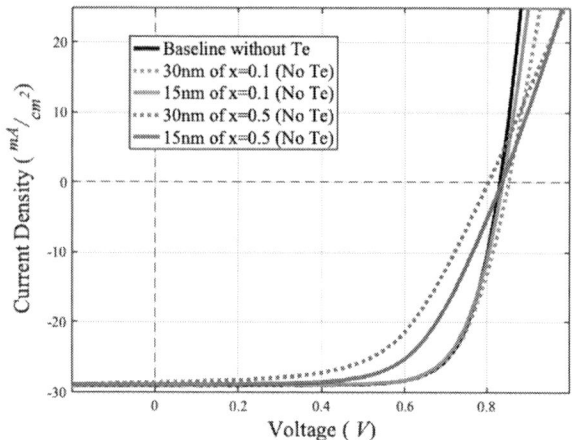

Fig. 2. J-V comparison of 30 nm and 15 nm thicknesses.

is outlined in Fig. 1. The structure chosen as the baseline for this experiment was based upon optimization from previous CdSeTe/CdTe devices [4].

B. Alloy synthesis and Characterization

The alloys used in evaporation were initially synthesized following the recipe in Ref. 3 by Andrew Treglia. Measurements of current-density versus voltage (J-V) were taken under standard test conditions and at room temperature. The photoluminescence (PL) was measured from the glass side of the device using a 520-nm excitation laser, along with 570-nm bypass filter.

III. RESULTS

A. $Se_{1-x}Te_x$ Thickness Variation

For each mole fraction of tellurium, plate of 15 s and 30 nm of $Se_{1-x}Te_x$ were fabricated. As a general trend, it was observed that the 15 nm devices performed better, in terms of J_{sc}, V_{oc}, fill factor, and efficiency, than their 30 nm counterparts. An example of this is shown in Fig. 2, where the thicker $Se_{1-x}Te_x$ layers are indicated by dashed lines. Consistently, the 30 nm devices would show a higher series resistance than the 15 nm.

B. Optimization of the Mole Fraction of Tellurium

The rational for the higher tellurium content of the $Se_{1-x}Te_x$ alloy was the benefit of tellurium's greater conductivity. The concern was with too much selenium at the back, the resistance would be too great to achieve reasonable current. However, across the mole fractions x = 0.1, 0.5, and 0.7 there was no appreciable drop in current density with greater selenium content. With this information, benefits can be seen with a lower mole fraction, likely from the increased band gap.

In Ref. 3, the authors determined the band gaps for multiple mole fractions from absorption spectra data. Their work suggests that for mole fraction x = 0.1, the band gap should be ~1.45 eV with a VBM of ~5.6 eV. Compare this to a band gap of ~0.75 eV and VBM of ~5.4 eV for x = 0.5, and the band gap ~0.6 eV and VBM of ~5.3 eV for x=0.7 [3]. Though it should

be noted their mole fractions were more specific and verified by powdered x-ray diffraction. The valence band maximums of x = 0.1 and 0.5 were below that of CdTe (5.8 eV) and above tellurium (5.40-5.45 eV) [2], which would lead to the desired intermediary behavior. The increased band gap with the higher selenium content suggests that range (below mole fraction x = 0.4) would be a fruitful area to optimize the process.

C. Inclusion of a Tellurium Back Coating

For each device fabricated in the run, there was a coating of tellurium applied to half of the plate after the $Se_{1-x}Te_x$ layer. This was achieved by covering half of the plate during evaporation. There was a clear trend (see Fig. 3) for every mole fraction that the cells with tellurium after the $Se_{1-x}Te_x$ layer performed markedly better than without the tellurium layer. This agrees with the hypothesis that the VBM of the tellurium is needed to bridge the gap with the work function of the back contact.

A side effect that is under review is that of the tellurium layer on the band gap as revealed in photoluminescence measurements. In Fig. 4, it was consistent throughout the devices that the side of the plate which received the tellurium back layer saw a noticeable shift in the peaks to a lower wavelength. Intuitively the layer should not have such a pronounced effect but could be indicative of unforeseen effect brought about by the deposition process.

The incorporation of the tellurium layer also showed minimal variability among cells on the same substrate given device. Drawing attention to Fig. 5 where the spread of efficiencies across each plate is shown. Even including those

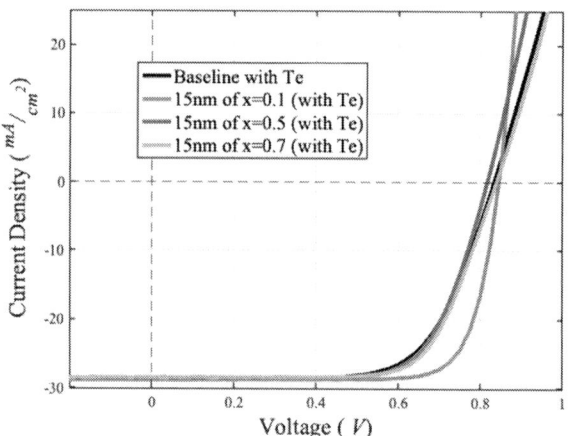

Fig. 3. J-V of 15nm $Se_{1-x}Te_x$ best devices with x = 0.1, 0.5, and 0.7 dopant levels.

TABLE I
J-V PARAMETERS FOR BEST 15NM DEVICES

Device	V_{oc} (V)	Jsc (mA/cm^2)	Fill Factor (%)	Efficiency (%)
Baseline with Te	833	28.7	67	16
15 nm of x=0.7	839	28.5	70	16.7
15 nm of x=0.5	817	28.5	70.3	16.4
15 nm of x=0.1	846	28.9	78.2	19.1

with and without the tellurium layer, for the better performing plates we see a relatively small spread. Where there is a great deal of variability is on the baseline device, which contained no alloy layer, and on the thicker (30 nm) and tellurium rich (x =0.5, 0.7) devices. This is another factor in favoring the 15 nm $Se_{1-x}Te_x$ layers in future devices.

IV. FUTURE WORK

A. Continuation of Mole Fraction Optimization

Now that the concern about conductivity has been at least partially addressed, the focus of this project will be to optimize the mole fraction of the $Se_{1-x}Te_x$. The key will be to maintain the intermediary behavior of the valence band, while maintaining or improving the reflector behavior of the conduction band.

B. Characterization and Stucture

With the promise of this initial work, it is now key to understand how the film is being deposited. This will include ensuring that the correct mole fraction is being deposited during evaporation. This will be compared to using close-space sublimation of two independent sources of selenium and tellurium. From this we can assertain the ideal deposition method and compare difference if any. Once we have achieved an optimal device, it will be important to understand how the polycrystaline structure is deposited and what effect grain boundaries might have. This would will be performed using scanning transmission electron microscopy with our collaborators.

C. Potential Use as an Absorber

In Ref. [3] the $Se_{1-x}Te_x$ alloy was used as a primary absorber in a photovoltaic device. Currently our group is working on "thin absorbers", where instead of a 4 μm-thick absorber, there would be 0.5 μm of CdSeTe and 1.0 μm of CdTe. The layer behind this absorber had been CdMgTe in the work performed by our group. What is currently being explored is CdZnTe for the similar band bending effect in this work. The plan is to also explore different mole fractions of $Se_{1-x}Te_x$ to see if it can provide any benefit.

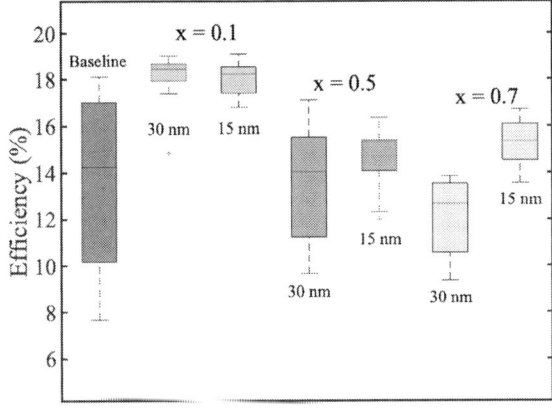

Fig. 5.   Efficiency box plots for all devices (includes Te and Non-Te layers across devices)

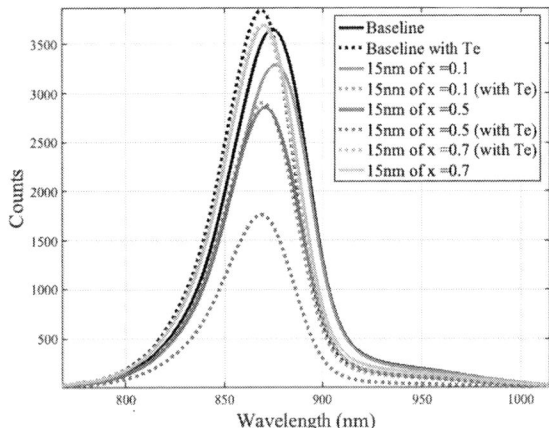

Fig. 4.   Photoluminescence counts for varying dopant levels. There is a consistent shift in the peaks, note the shoulder around 940 nm.

ACKNOWLEDGMENT

The authors would like to thank Prof. W.S. Sampath for use and assistance with the deposition systems as well as Prof. Jamie Neilson for equipment use for the alloy synthesis, and 5N Plus for the deposition materials.

REFERENCES

[1] Miller, C. A., Peters, I. M., & Zaveri, S. (n.d.). (rep.). Thin Film CdTe Photovoltaics and the U.S. Energy Transition in 2020.

[2] Song, T., Moore, A., & Sites, J. R. (2018). Te layer to reduce the CdTe back-contact barrier. *IEEE Journal of Photovoltaics, 8*(1), 293–298. https://doi.org/10.1109/JPHOTOV.2017.2768965

[3] Hadar, I., Hu, X., Luo, Z. Z., Dravid, V. P., & Kanatzidis, M. G. (2019). Nonlinear Band Gap Tunability in Selenium-Tellurium Alloys and Its Utilization in Solar Cells. *ACS Energy Letters, 4*(9), 2137–2143. https://doi.org/10.1021/acsenergylett.9b01619

[4] Munshi, Amit et al. (2017). Polycrystalline CdSeTe/CdTe Absorber Cells With 28 mA/cm2 Short-Circuit Current. IEEE Journal of Photovoltaics. PP. 1-5. 10.1109/JPHOTOV.2017.2775139.

[5] D.E. Swanson, J.M. Kephart, P S Kobyakov, K. Walters, K.C. Cameron, K.L. Barth, W.S. Sampath, J. Drayton, and J.R. Sites, "Single vacuum chamber with multiple close space sublimation sources to fabricate CdTe solar cells," Journal of Vacuum Science & Technology A, vol. 34, issue 2, 021202, 2016.

[6] A. Wojtowicz, A.M. Huss, J.A. Drayton, and J.R. Sites, "Effects of CdCl2 passivation on thin CdTe absorbers fabricated by close-space sublimation," in 44th IEEE Photovoltaic Specialists Conference, 2017

Effect of Dilute Acid Exposure on Sol-gel Porous Silica Anti-Reflection Coatings

F. Bukhari, L. Jones, A. Law, A. Abbas and J.M. Walls

Centre for Renewable Energy Systems Technology (CREST), Loughborough University.

Loughborough Leicestershire, LE11 3TU, UK.

Abstract— The build-up of algae and moss on the outer glass surface of solar modules causes serious attenuation of the incident light. Biocides are effective in removing this surface contamination to restore performance. These biocides are acid based and there is some concern that their use may affect the integrity of the porous silica anti-reflection coatings (ARC). Here we report on the effect of a commercial biocide treatment on sol-gel porous silica AR coatings. The biocide used was SFC Eco which is already deployed on solar utilities. Our initial studies show the effectiveness of nonanoic acid-based biocide as an algae cleaning solution. The treatment using a dilute solution does slightly etch the morphology of the coating. Similar effects could also be caused by acids in rainfall, so we have also conducted tests using sulphuric, nitric and carbonic acids at similar dilutions. These tests showed that these tests also resulted in similar etching but that there no significant change in the transmittance or reflectance of the coating. Our studies are continuing to investigate if long term exposure to acidic environment or cyclic applications of biocides will cause significant degradation.

Keywords— *Anti-reflection, cleaning, nonanoic acid, biofouling, acid rain.*

I. INTRODUCTION

Solar photovoltaic (PV) modules placed outdoors are subject to a wide range of environmental conditions during their lifetime, even in milder climates. A typical manufacturer warranty is 25 or 30 years, so modules are expected to provide good service during this guarantee period. Most solar modules are provided with an anti-reflection (AR) coating applied to the cover glass surface to reduce reflection losses. These AR coatings are effective and reduce reflections for the front surface by about 70% compared to uncoated glass. The environmental durability of these coatings is critical as degradation would lead to power losses. Re-application of the coating in the field is currently not viable.

Soiling of the cover glass is a serious problem that reduces the power output by about 5% depending on location. The soiling attenuates the light transmitted into the module and reduces the current density produced. Power losses of 11% over 18 months have previously been observed [1] as a direct result of biofouling.

Biofouling is the accumulation of organic matter on a surface in a manner that is detrimental. A typical contamination route is deposition of spores on the surface followed by germination fueled by moisture and nutrients from organic matter. [2]

Cementation of the organic matter to the surface can then occur through inorganic contaminants such as salts. [3]

Regular cleaning of PV modules can remove organic matter, although the brushes used by most solar asset managers can cause abrasion damage to the coatings.

Fig. 1. The transmission of light into solar modules is seriously attenuated by the growth of algae and moss.

Algae and moss contaminants can be removed using biocide treatments with dilute solutions substances containing, for example, nonanoic acid . However, there is a concern that the use of dilute acid cleaning solutions may affect the porous silica AR coatings since it is known that the voids in porous silica coatings makes them vulnerable to humidity. [4] Likewise, there is also the issue of acid rain and how this may affect the porous silica coating. Acid rain is caused by pollutants, most notably sulphur, nitrogen and carbon dioxide, present in the atmosphere which react with water molecules to form sulphuric , nitric and carbonic acids. The inclusion of these acids in rainwater lowers the pH in the range ~3.5-5 and can causes damage to plant life, aquatic environments, and corrosion of outdoor metallic structures [5]. Acid rain is present in most parts of the industrialized world, including Europe, North America, and South East Asia [6]. It is composed of all 3 of the acids in varying quantities and produced mainly from human activities. High population density regions with electricity generated by burning coal higher sulphate production so the acid rain has higher sulphuric acid content This occurs in regions in East Asia including China and India. Suburban areas with higher concentrations of industrial and power plants can produce higher amounts of nitrates and therefore can produce acid rain with a higher ratio of nitric acid, such as the east coast of North America. [7]

The impact of acid rain on PV modules has not been thoroughly investigated, although there have been a few studies on the impact on some anti-soiling coatings [8][1] with some fluoropolymer-based coatings vulnerable to acid attack.[1]

II. EXPERIMENTAL DETAILS

A. Acid exposure Tests:

Six coupons with sol-gel porous silica anti-reflective coating were treated with SFC Eco a biocide used for algae and moss removal. Six soda-lime glass slides (uncoated) were used as control samples.

All coupons were cleaned using a 600mL acetone bath for 20mins in ultrasonic bath before tests. A coated and reference glass coupon were set aside and used as a control. 130mL of cleaning solution was prepared by mixing 10mL of the SFC Eco herbicide with 120mL of deionised (DI) water. Coupons were submerged in this solution for 15 seconds, removed from bath and then exposed to outdoor conditions in Loughborough, UK (during British summer conditions) for 5 days. One coated and one reference coupon was removed each day over a 5-day period for analysis. Changes in the coating surface chemistry and surface microstructure/ were then studied.

A separate indoor study was performed to establish how other acids present naturally in acid rain affect the sol-gel porous silica anti-reflection coating. After cleaning, four porous AR coated coupons were immersed in DI water (at pH 7) as a control and immersed in acids HNO_3 (Ph 4.0), H_2SO_4 (Ph 4.0) and H_2CO_3 (Ph 5.1) for 15 seconds. These acid strengths were chosen to simulate the acidic nature of rain in parts of India, South America, and East Asia. After exposure to the acid solutions, the coupons were stored in the laboratory at room temperature.

B. Characterisation Techniques:

Surface chemical analysis was conducted using X-ray photoelectron spectroscopy (XPS) using a Thermo Scientific K-Alpha XPS with monochromated Aluminium Kα radiation X-ray source. Scanning Electron Microscopic (SEM) images were obtained using a JEOL 7800F FEGSEM. Transmittance/reflectance was measured across a wavelength range of 200-1200 nm using a Varian Cary5000 UV-VIS spectrophotometer with an integrating sphere attachment.

III. RESULTS AND DISCUSSION

Visual observation of the coupons treated with the biocide cleaning solution after outdoor exposure suggested minimal to no damage to the coating or to the control coupons (reference glass). XPS revealed the primary ingredient of the SFC Eco biocide, nonanoic acid degrades quickly, oxidizing to form intermediate compounds, which degrade further in a two-step process resulting in the release of carbon dioxide and water.

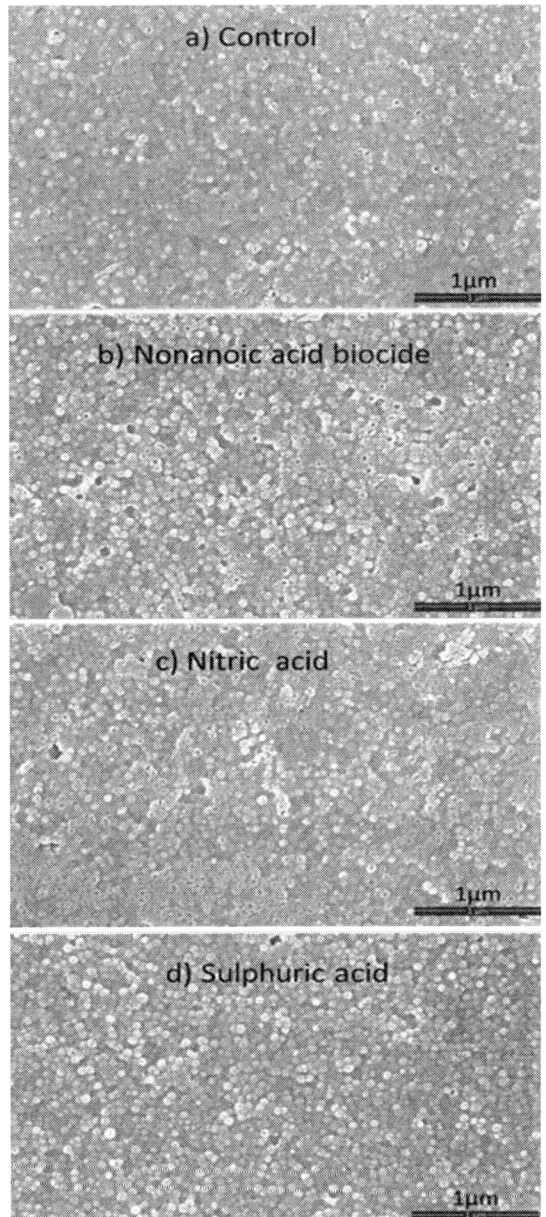

Fig. 2. SEM images of a) ARC -As received coupon, b) Nonanoic acid treated, c) Nitric acid and d) Sulphuric acid treated after 5 days of exposure.

SEM images reveal that changes have occurred to the surface of the porous silica coating. Some etching has taken place as revealed by the increased contrast visible in the surface features. It is possible the etching has modified the external capping layer.

The weighted average reflectance (WAR) for control and treated samples is 8.14% and remained within ±0.04% over a 5-day period during which the biocide disintegrated. Such a change is minor and would not affect the efficiency of the module.

The acid rain simulation tests on these coated coupons also revealed an etching effect very similar to the effects caused by the nonanoic acid in the biocide. Surface assessment showed increased visible porosity of the coating after the 5-day period outdoor exposure. The SEM image of the control coating is smooth compared to the etching caused by sulphuric, nitric and carbonic acid solutions which also may have affected the capping layer.

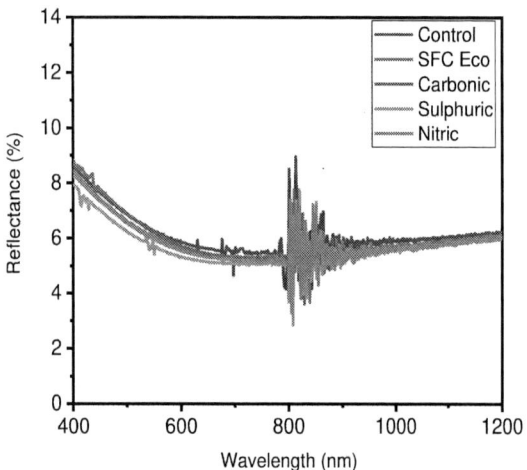

Fig. 3. Reflectance, R (%) measurements of acid exposed and control samples in comparison with a soda-lime glass coupon.

IV. CONCLUSION

Algae and moss soiling can cause serious losses in efficiency from solar modules. This soiling problem can be addressed using a biocide. In this study we have used SFC Eco (Belchim), with an active ingredient of nonanoic acid. We have investigated the effect of the biocide on the porous silica sol gel anti reflection coating. We have also studied the effects of dilute sulphuric, nitric and carbonic acid to simulate the effects of acid rain. XPS measurements reveal no significant changes to the surface chemistry of the coatings. Reflection measurements show a slight reduction in reflection from the coupon treated with sulphuric acid corresponding to a small change in refractive index. Use of the nonanoic acid and the three acid rain simulations showed some degree of surface etching by increasing the contrast of surface features and topography. More details of the porosity could be discerned. The use of nonanoic acid is effective in removing algae and moss from solar cover glass. Its use does not appear to degrade the porous silica AR coating any more than would be expected from exposure to acid rain.

REFERENCES

[1] S. Bhaduri, A. Alath, S. Mallick, N. S. Shiradkar, and A. Kottantharayil, "Identification of Stressors Leading to Degradation of Antisoiling Coating in Warm and Humid Climate Zones," IEEE J. Photovoltaics, vol. 10, no. 1, pp. 166–172, 2020.

[2] S. Toth et al., Soiling and cleaning: Initial observations from 5-year photovoltaic glass coating durability study, vol. 185. 2018.

[3] G. C. Oehler et al., "Testing the durability of anti-soiling coatings for solar cover glass by outdoor exposure in Denmark," Energies, vol. 13, no. 2, 2020.

[4] G. Womack, K. Isbilir, F. Lisco, G. Durand, A. Taylor, and J. M. Walls, "The performance and durability of single-layer sol-gel anti-reflection coatings applied to solar module cover glass," Surf. Coatings Technol., vol. 358, pp. 76–83, 2019.

[5] D. A. Burns, J. Aherne, D. A. Gay, and C. M. B. Lehmann, "Acid rain and its environmental effects: Recent scientific advances," Atmos. Environ., vol. 146, pp. 1–4, Dec. 2016.

[6] H. Mohajan and H. K. Mohajan, "Acid Rain is a Local Environment Pollution but Global Concern," Open Sci. J. Anal. Chem., vol. 3, no. 5, pp. 47–55, 2018.

[7] H. Y. Chen, L. F. Hsu, S. Z. Huang, and L. Zheng, "Assessment of the components and sources of acid deposition in northeast asia: A case study of the coastal and metropolitan cities in Northern Taiwan," Atmosphere (Basel)., vol. 11, no. 9, 2020.

[8] S. Bhaduri, R. Bajhal, S. Mallick, N. Shiradkar, and A. Kottantharayil, "Degradation of anti-soiling coatings: Mechanical impact of rainfall," Conf. Rec. IEEE Photovolt. Spec. Conf., vol. 2020-June, pp. 1098–1101, 2020.

Insights into the stability of amorphous/crystalline silicon interface under light and temperature

Salman Manzoor, Mariana Bertoni

Arizona State University, Tempe, AZ, United States

With silicon heterojunction (SHJ) solar cells inching closer to their practical efficiency limit, long-term reliability, and stability are the key to improve their market adoption. The origins of passivation degradation over time under field operating conditions at the amorphous silicon (a-Si:H)/crystalline silicon (c-Si) interface remains unclear. Herein, we investigate the passivation quality of a-Si:H/c-Si by extracting surface recombination velocity (SRV) from temperature- and injection-dependent lifetime spectroscopy measurements and retrieving interface defect density and charge density at the interface. Our results show a-Si:H/c-Si interface passivation degrades over time due to increase in SRV originating from failing chemical passivation exhibited by an increase in defect density at the interface while field passivation remained the same. Moreover, microstructural analysis of a-Si:H/c-Si with Fourier transform infrared spectroscopy (FTIR) shows increase in defect density at the interface originates due to changes at the interface and not in the bulk of a-Si:H.

Automated Shift Detection in Sensor-Based PV Power and Irradiance Time Series

Kirsten Perry[1],
Matthew Muller[1]
[1]NREL, Golden, CO, 80401, USA

Abstract—PV power and irradiance sensor-based measurements are prone to error, resulting in issues such as abrupt time series data shifts. These shifts, which are usually unintentional, may be caused by software or hardware configuration changes on a PV system, and do not reflect an actual change in overall system performance. Locating these shifts and segmenting the associated time series aids in more accurate future PV analysis. In this research, an offline changepoint detection (CPD) algorithm that automatically detects these abrupt data shifts in sensor-based time series is introduced. Data shift periods in 101 daily PV power and irradiance time series were labeled manually by two solar experts. These data streams represent sensor-based measurements, and display a variety of data shift behaviors. A changepoint detection algorithm was tuned using the 101 labeled data streams, with each model configuration's ability to detect labeled changepoints benchmarked using metrics such as F1-score, recall, and Rand Index. Best performing models on seasonality-corrected data streams include the Pruned Exact Linear (PELT) method, the Binary Segmentation method, and the Bottom-Up method, all scoring an average F1-score of 0.76 or greater at detecting labeled changepoints within a 30-day window for the labeled data sets. To promote further research in this space, we are releasing the labeled data shift sets on U.S. Department of Energy's (DOE) DuraMAT Data Hub, and the associated algorithm in the Python PVAnalytics package.

Index Terms—data shift, changepoint detection, solar, irradiance, power, data quality

I. INTRODUCTION

The use of high-quality PV data is paramount for effective monitoring of photovoltaic (PV) systems, including running advanced analytics routines to estimate system performance and degradation rate [1]. When poor underlying data is fed into PV models, results may be inaccurate and lead to uninformed business decisions that further impact system health and longevity. Consequently, it is paramount to use high-quality and valid data when performing these analytics routines.

Invalid solar time series data, which includes missing data periods and outliers, may occur as result of power outages, equipment failures, and communication issues [1]. One particular issue in power and irradiance data streams is abrupt data shifts. Some example data shifts, taken from sensor-based time series data, are shown in Figure 1. Data shifts such as these are frequently introduced unintentionally, as a result of replacing hardware or by software configuration changes [2]. It is important to note that these shifts do not reflect an actual change in system performance, but are generally a result of data acquisition issues; for example, by changing the scale factor for a particular data output, or by converting an

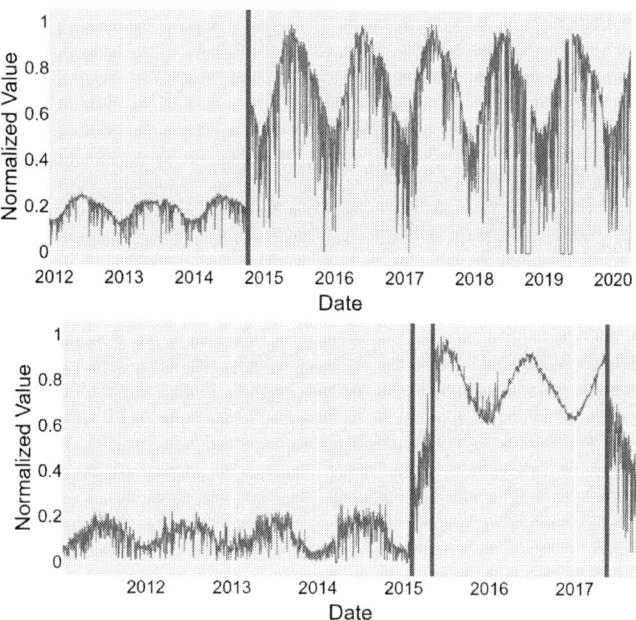

Fig. 1. Two example daily time series with data shifts, taken from data streams in the NREL PV Fleets project. Daily data points are represented in purple, and vertical green lines represent manually labeled changepoints.

AC energy data stream to an AC power data stream and not adequately documenting the change.

Some limited research has been performed exploring the consequences of performing PV analysis using time series with data shifts. In particular, Jordan et al. [2], [3] have examined the effects of abrupt data shifts on degradation estimates in PV time series data. [3] corrects artificially introduced data shifts in a sensor-based temperature time series via a scaling factor, to obtain accurate degradation estimates for a system. Similarly, [2] uses multiple techniques such as standard least squares regression (SLS) and a year-on-year (YoY) approach to correct data shifts to accurately estimate degradation rates.

Lindig et al. [4] also addresses data shifts in the context of data quality filtering for accurate solar performance loss and useful-lifetime calculations. This research specifically recommends filtering out data shift periods in time series where the cause of the data shift is unknown. However, [4] does not provide any process for detection of data shifts in PV time series data, instead relying on manual inspection by a trained analyst to identify shift periods.

The methods previosuly described require manual identification of data shift periods, which may not be feasible if hundreds or even thousands of data streams must be analyzed. This research aims to automate the process of detecting abrupt data shifts in PV data automatically via offline changepoint detection (CPD), without making any data assumptions. The resulting time series segments can then be analyzed separately, or the shortest segment can be removed during later analysis, similar to the process recommended by [4].

CPD is the process of detecting changes in an underlying signal, in particular a time series [5]. The concept of CPD dates back to the 1950's [6], and has various applications in speech processing [7], climatology [8], and financial analysis [9], among others. Offline CPD is a subcategory of CPD, where changepoints are detected in a signal after all data points have been collected [5]. Offline CPD can be formulated as a model selection problem, where we want the best possible segmentation of a signal with a specific quantitative criterion minimized [5]. CPD can be described as a combination of three elements: a cost function, which is a measure of the homogeneity between separate time series segments; a search method, which is the particular procedure employed to solve the optimization problem in question; and a constraint, which is either the number of changepoints in a sequence (if known) or penalty value associated with the goodness-of-fit term [5].

CPD has been applied to PV time series data previously, for different applications. Specifically, Theristis et al. [10], [11] applied the Facebook Prophet CPD algorithm to performance ratio (PR) time series with a non-linear degradation rate, with the intent of detecting degradation rate changes. This research differs from ours in that we are not looking to detect changes in degradation rates across a time series; rather, we are attempting to detect issues with the raw data itself, which occurs as a result of data acquisition problems. To our knowledge, this is the first attempt to automatically identify this particular issue in sensor-based PV data.

II. METHODS

A. Data Sets

To build and validate the shift detection algorithm, 101 data sets representing unique sensor-based irradiance and power data streams were collected and labeled. Time series were collected from multiple PV solar installations, available via the NREL PV Fleets Initiative [12]. The PV Fleets Initiative is a US Department of Energy-funded project, where operational PV plant data is aggregated into a centralized cloud repository for the purpose of large-scale degradation analysis across the US. This database contains sensor-based time series data for over 1700 sites across the United States.

Data streams representing a variety of data shift behaviors, including scaling issues, were selected for labeling. Each time series was summed over a daily basis.

Two experts manually labeled data shifts in each of the 101 daily summed time series. A binary labeling strategy was used.

Each individual data point in the time series was either labeled as a "data shift" point, where a major data shift occurs in the time series, or as a "regular" point where no change occurs. This labeling strategy was used to facilitate finding the specific point in a time series where a shift occurs, so data can be partitioned into individual issue-free segments.

B. Data Pre-Processing

Before applying the shift detection algorithm, each daily summed time series data set was cleaned, with the intent of removing egregious single-point outliers and anomalies. Specifically, the following steps were performed:

- All negative data days were removed.
- Stale data readings, i.e. consecutive repeat daily readings, were identified and removed from the time series. A consecutive repeat window of 6 readings or more was used to identify stale reading periods.
- All values less than the 1st percentile of data and greater than the 99th percentile of data were removed.
- Each daily time series was min-max normalized.

Irradiance and PV power time series can show extreme seasonality year-over-year. Removing seasonality helps to make the time series more stationary, and aids in detecting data shifts more accurately. Seasonality was removed from each time series via the following logic:

- The median value of each day of the year was calculated, resulting in a 365 day-long time series, with a median value for each day in the year. So, for example, in a three-year long time series, the three values occurring at January 1st in the time series are used to calculate the median value of January 1st.
- At each day in the time series, the median day value calculated in the previous step is subtracted from the normalized time series value.

An example time series, pre- and post-seasonality removal, is shown in Figure 2. It is important to note that all time series must be at least two years in length for this strategy to work, or seasonality cannot be calculated and removed.

C. Shift Detection Algorithm

The Python Ruptures changepoint detection package was used to develop and tune the shift detection algorithm [13]. An offline, unsupervised CPD algorithm was tuned using the manually labeled data sets. The following changepoint algorithm parameters were varied during grid search, to find the best-performing algorithm combination on the labeled data:

- Search method: Binary Segmentation, Window-based, Pruned Exact Linear Time (PELT), and Bottom-Up methods
- Cost function: radial basis function (rbf), L1, and L2
- Penalty: value between 10 and 100 inclusive, at intervals of 10. Higher penalty values cause heavier filtering of changepoints, resulting in fewer total detected changepoints. Because we are attempting to automatically detect

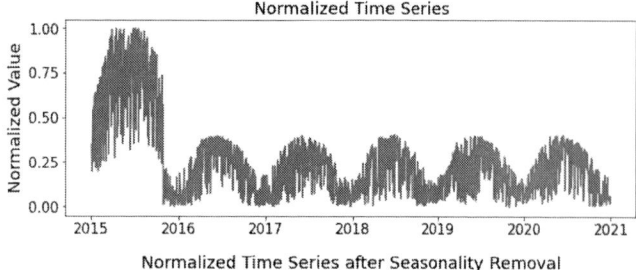

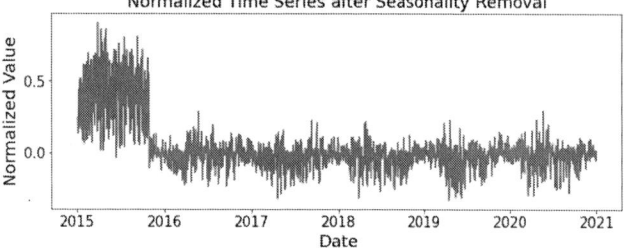

Fig. 2. An example time series before seasonality removal, and after seasonality removal.

data shifts with no prior knowledge of the time series, we do not pass a specific number of changepoints to find, and instead rely on a penalty threshold.

- Width: value between 10 and 110 inclusive, at intervals of 20. Width only applies to the window-based method, and acts as the length of the sliding window.

In addition to running seasonality-corrected data through the CPD algorithm, normalized time series data (with no seasonality correction) was also run. This was to identify the best parameter combination for situations where seasonality cannot be removed, i.e. time series shorter than 2 years in length.

D. Benchmarking Algorithm Performance

The ability of each model to successfully detect change points in the labeled time series was assessed, using CPD-specific F1-score and precision metrics developed by the Alan Turing Institute [14].

F1-score and recall are defined via the following equations, respectively [14]:

$$\text{F-score} = \frac{2 \cdot \text{recall} \cdot \text{precision}}{\text{recall} + \text{precision}} \qquad (1)$$

$$\text{recall} = \frac{TP}{TP + FN} \qquad (2)$$

where TP represents the number of true positives, and FN represents the number of false negatives. A true positive is defined as a detected changepoint within 30 days of a labeled changepoint. This 30-day period acts as a margin-of-error period, to allow for small discrepancies where the detected changepoint may be a few days off from the actual labeled changepoint. Recall is the fraction of relevant changepoints detected out of all correctly-labeled changepoints. A recall of

1 indicates that all labeled changepoints were found. An F1-score of 1 indicates perfect precision and recall, where the only changepoints detected by the algorithm are true positives.

In addition to measuring F1-score and recall, the Rand Index was measured for each test case. The Ruptures Python package implementation for the Rand Index was used [13]. The Rand Index measures the similarity between data clusters; in the case of changepoint detection, it analyzes the similarity between changepoint-separated time series segments [13]. For this research, the Rand Index is a valuable metric because we are most concerned with having consistent time series segments as a final output, not necessarily the changepoints themselves. The Rand Index is defined via the following equation [13]:

$$\text{Rand} = \frac{N_0 + N_1}{T(T+1)/2} \qquad (3)$$

where N_0 represents the number of pairs of samples that belong to the same segment in a sequence T that has been split into segments T_1 and T_2, and N_1 is the number of pairs of samples that belong to different segments according to T_1 and T_2. The Rand Index is normalized between 0 and 1, where 0 indicates complete disagreement and 1 indicates complete agreement.

III. RESULTS

Tables I and II show the five best CPD model configurations based on average F1 score for seasonality-corrected and normalized data streams, respectively. Table III shows Rand Index scores for these particular models. Generally, better overall metric scores were achieved when seasonality-corrected data was used.

Using seasonality-corrected data, the best-performing CPD model, a PELT model, achieved an F1-score of .767 and a Rand index value of 0.871. For normalized-only data, the best performing model, a window-based model, achieved an F1-score of .745 and a Rand Index value of .848. It is noteworthy that the best performing model overall (PELT model with seasonality-removed data) had one of the slowest overall run times, with an average run time of 50.81 seconds per a data stream, with the average data stream length of approximately 2300 data points. Several models achieved slightly lower average F1-scores on the data set, but have far faster average run times. When looking at model performance in terms of both time efficiency and accuracy, we recommend using the Bottom-Up model for seasonality-corrected data (average F1-score of 0.76 and Rand Index of .878, with an average run time of 0.26 seconds). For normalized-only data, we recommend using the Window-based model with the highest average F1-score (F1-score of .745 and Rand Index of .848, with a run time of .2 seconds).

The Rand Index scores for the highest-scoring models are particularly promising (values greater than .8), as they indicate that the final time series outputs are well-segmented. This is particularly important, as we want to perform analysis on data periods that are consistent and free of massive data shifts.

TABLE I
TOP 5 PERFORMING CPD CONFIGURATIONS ON LABELED, SEASONALITY-CORRECTED DATA

Model	Cost	Penalty	Recall	F1	Run Time (s)
PELT	rbf	40	.734	.767	50.81
Binary Seg	rbf	50	.705	.763	2.24
Binary Seg	rbf	40	.726	.762	2.37
PELT	rbf	50	.708	.761	54.99
Bottom-Up	rbf	40	.729	.760	.26

TABLE II
TOP 5 PERFORMING CPD CONFIGURATIONS ON LABELED, NORMALIZED DATA (NO SEASONALITY CORRECTION)

Model	Cost	Penalty	Width	Recall	F1	Run Time (s)
Window	rbf	30	50	.698	.745	.200
Window	rbf	40	50	.671	.741	.199
Window	rbf	20	30	.736	.739	.191
Window	rbf	20	50	.747	.737	.199
Window	rbf	50	50	.654	.736	.206

TABLE III
AVERAGE RAND INDEX FOR THE TOP PERFORMING MODELS

Data Type	Model	Cost	Penalty	Width	Rand
Season-Removed	PELT	rbf	40	NA	.871
Season-Removed	Binary Seg	rbf	50	NA	.864
Season-Removed	Binary Seg	rbf	40	NA	.867
Season-Removed	PELT	rbf	50	NA	.859
Season-Removed	Bottom-Up	rbf	40	NA	.878
Normalized	Window	rbf	30	50	.848
Normalized	Window	rbf	40	50	.840
Normalized	Window	rbf	20	30	.857
Normalized	Window	rbf	20	50	.838
Normalized	Window	rbf	50	50	.828

IV. PVANALYTICS INTEGRATION & FURTHER RESEARCH

The models developed in this research are available for public use via the Python PVAnalytics package [15]. In addition to automated data shift detection, the package includes functionality for identifying the longest continuous time series segment that is free of data shifts and isolating it for further analysis. This detection-and-segmentation process is illustrated in Figure 3.

Our logic for selecting the longest time series segment for further analysis is currently rudimentary, and we plan to develop more advanced processes for analysing and selecting the "best" time series segment for further analysis, based on each segment's overall data quality and availability. We also plan to further investigate whether data shifts caused by scaling issues or similar can be identified and corrected (rather than eliminated), without compromising the overall quality of the time series and biasing future analyses.

ACKNOWLEDGMENT

This work was authored in part by Alliance for Sustainable Energy, LLC, the manager and operator of the National Renewable Energy Laboratory for the U.S. Department of Energy (DOE) under Contract No. DE-AC36-08GO28308. Funding provided by the U.S. Department of Energy's Office of Energy Efficiency and Renewable Energy (EERE) under Solar Energy Technologies Office (SETO) Agreement Numbers 38258.

REFERENCES

[1] A. Livera, M. Theristis, E. Koumpli, S. Theocharides, G. Makrides, J. Sutterlueti, J. S. Stein, and G. E. Georghiou, "Data processing and quality verification for improved photovoltaic performance and reliability analytics," *Progress in Photovoltaics: Research and Applications*, vol. 29, no. 2, pp. 143–158, 2021. [Online]. Available: https://onlinelibrary.wiley.com/doi/abs/10.1002/pip.3349

[2] D. C. Jordan, C. Deline, S. R. Kurtz, G. M. Kimball, and M. Anderson, "Robust pv degradation methodology and application," *IEEE Journal of Photovoltaics*, vol. 8, no. 2, pp. 525–531, 2018.

[3] D. C. Jordan and S. R. Kurtz, "Analytical improvements in pv degradation rate determination," *2010 35th IEEE Photovoltaic Specialists Conference*, pp. 002 688–002 693, 2010.

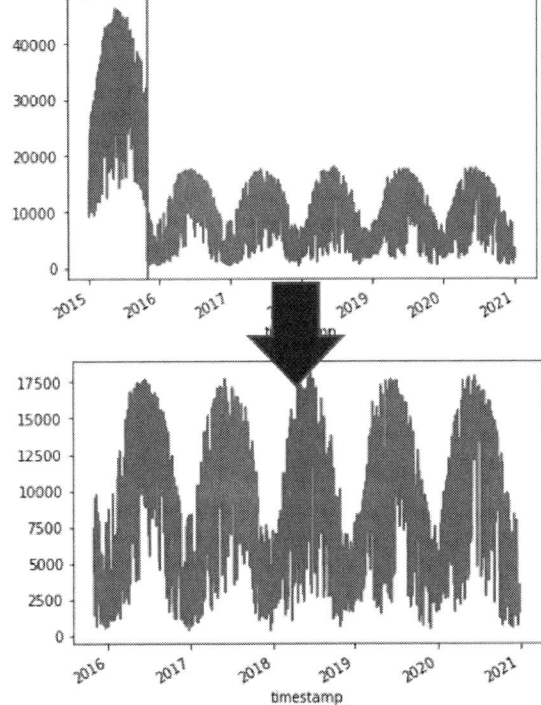

Fig. 3. PVAnalytics pipeline for automated detection of data shifts in time series, and segmentation of the longest sequence free of data shifts. In the upper image, data shifts are automatically detected (see green line) via the PVAnalytics detect_data_shifts() function. In the lower image, the longest time series sequence free of data shifts is segmented, using the PVAnalytics get_longest_shift_segment_dates() function.

[4] S. Lindig, A. Louwen, D. Moser, and M. Topic, "Outdoor pv system monitoring—input data quality, data imputation and filtering approaches," *Energies*, vol. 13, no. 19, 2020. [Online]. Available: https://www.mdpi.com/1996-1073/13/19/5099

[5] C. Truong, L. Oudre, and N. Vayatis, "Selective review of offline change point detection methods," *Signal Processing*, vol. 167, p. 107299, 2020. [Online]. Available: https://www.sciencedirect.com/science/article/pii/S0165168419303494

[6] E. S. Page, "A test for a change in a parameter occurring at an unknown point," *Biometrika*, vol. 42, no. 3/4, pp. 523–527, 1955. [Online]. Available: http://www.jstor.org/stable/2333401

[7] Z. Harchaoui, F. Vallet, A. Lung-Yut-Fong, and O. Cappe, "A regularized kernel-based approach to unsupervised audio segmentation," in *2009 IEEE International Conference on Acoustics, Speech and Signal Processing*, 2009, pp. 1665–1668.

[8] J. Reeves, J. Chen, X. L. Wang, R. Lund, and Q. Q. Lu, "A

review and comparison of changepoint detection techniques for climate data," *Journal of Applied Meteorology and Climatology*, vol. 46, no. 6, pp. 900 – 915, 2007. [Online]. Available: https://journals.ametsoc.org/view/journals/apme/46/6/jam2493.1.xml

[9] M. Lavielle and G. Teyssière, *Adaptive Detection of Multiple Change-Points in Asset Price Volatility*. Berlin, Heidelberg: Springer Berlin Heidelberg, 2007, pp. 129–156. [Online]. Available: https://doi.org/10.1007/978-3-540-34625-8_5

[10] M. Theristis, A. Livera, L. Micheli, C. B. Jones, G. Makrides, G. E. Georghiou, and J. S. Stein, "Modeling nonlinear photovoltaic degradation rates," in *2020 47th IEEE Photovoltaic Specialists Conference (PVSC)*, 2020, pp. 0208–0212.

[11] M. Theristis, A. Livera, C. B. Jones, G. Makrides, G. E. Georghiou, and J. S. Stein, "Nonlinear photovoltaic degradation rates: Modeling and comparison against conventional methods," *IEEE Journal of Photovoltaics*, vol. 10, no. 4, pp. 1112–1118, 2020.

[12] D. C. Jordan, K. Anderson, K. Perry, M. Muller, M. Deceglie, R. White, and C. Deline, "Photovoltaic fleet degradation insights," *Progress in Photovoltaics: Research and Applications*, vol. n/a, no. n/a. [Online]. Available: https://onlinelibrary.wiley.com/doi/abs/10.1002/pip.3566

[13] C. Truong, "ruptures: change point detection in Python," https://github.com/deepcharles/ruptures, 2018.

[14] G. J. J. van den Burg and C. K. I. Williams, "An evaluation of change point detection algorithms," 2020.

[15] PVLib, "PVAnalytics," https://github.com/pvlib/pvanalytics, 2020.

978-1-7281-6118-1/22 $31.00 © 2022 IEEE

The Effect of Inverter Loading Ratio on Energy Estimate Bias

Kevin S. Anderson[1], William B. Hobbs[2], William F. Holmgren[3], Kirsten R. Perry[1],
Mark A. Mikofski[3], and Rounak A. Kharait[3]
[1]NREL, Golden, CO, 80401, USA
[2]Southern Company, Birmingham, AL, 35203, USA
[3]DNV, San Diego, CA, 92123, USA

Abstract—**Subhourly effects, particularly variability in solar irradiance, can lead to underestimation of inverter clipping losses and overestimation of energy in hourly photovoltaic system performance models, particularly for systems with high inverter loading ratios. Direct simulation of this error can be complicated by factors such as the representation of spatial and temporal variability in hourly weather data and transient system conditions. In this work we take an alternative approach using real system power measurements to show that energy predictions from typical industry models suffer from a bias that increases with inverter loading ratio. We also show that this loading ratio-dependent bias is strongly correlated with an empirical subhourly inverter clipping bias derived from real power plant data. Finally, we show that this bias is not necessarily specific to any one model or weather dataset by recreating similar biases with alternatives of each.**

Index Terms—**photovoltaic, inverter, clipping, modeling, high-frequency, subhourly, irradiance, variability**

I. INTRODUCTION

Utility-scale photovoltaic (PV) system design is increasingly trending over time to larger inverter loading ratios (ILR), also referred to as DC:AC ratios [1]. PV inverters with high loading ratios must force their arrays into reduced-efficiency operation in sunny conditions to prevent the total array power output from exceeding the inverter's maximum-rated input power. This power-limiting behavior is called clipping because it disrupts the linear relationship between irradiance and output power, resulting in curtailed performance in high irradiance conditions. An inverter might clip for several hours continuously on a clear-sky day, or only intermittently on days with highly variable irradiance when high-irradiance spikes might last for less than one minute.

The detailed system performance models used by industry developers and financiers to forecast project revenue usually include adjustments for inverter clipping and many other system performance effects. However, the conventional practice of modeling system performance at hourly scale renders these models incapable of directly simulating short-duration effects like subhourly clipping. In effect, these models assume static operating conditions over each hourly simulation interval, causing the effects of subhourly irradiance variability to be overlooked. This causes an otherwise accurate performance model to overestimate production, especially for systems with large loading ratios in climates with high irradiance variability.

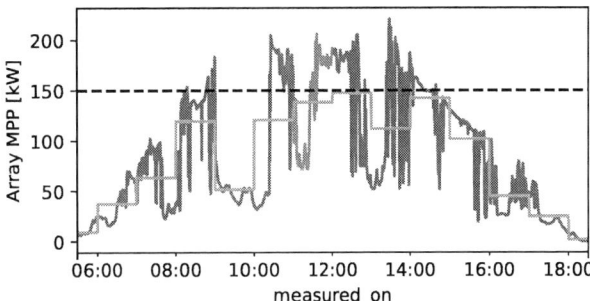

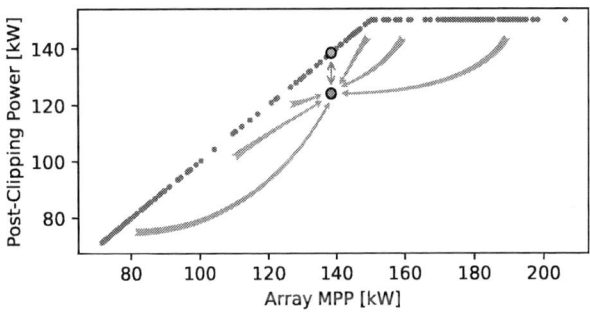

Fig. 1. Conceptual visualization of subhourly clipping. Upper subplot: high-resolution array maximum power point (MPP) data (blue line), and the corresponding average hourly values (orange line). Green line shows a particular hourly interval. Dashed black line shows a hypothetical inverter clipping point. Many MPP values lie above the clipping point at high resolution, but the hourly averages are all below the clipping point. Lower subplot: visualization of the 11:00–12:00 interval shown in green in upper subplot, with high-resolution values (blue dots), the true average of the high-resolution values (green circle), and the naive average that applies clipping at hourly scale (orange circle). Applying clipping at hourly scale overlooks the subhourly clipping loss at higher irradiance, leading to overestimated power output (red arrow).

It is worth emphasizing that this subhourly clipping bias is not a recoverable loss caused by poor system operation, but rather a failure of conventional modeling techniques to fully capture real-world PV system behavior.

A. What is subhourly clipping?

The mathematical foundation of subhourly clipping error is relatively straightforward: inverter clipping is a strongly nonlinear behavior, and the average value a nonlinear function takes across some interval is, in general, not equal to the value the function takes at the average of that interval. Thus,

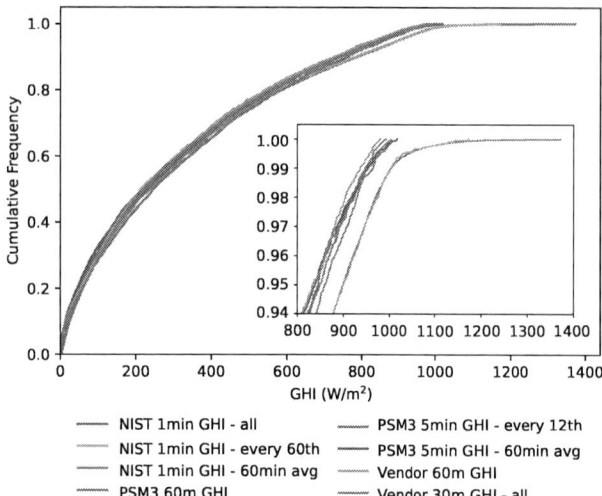

Fig. 2. Cumulative distributions of one year of Global Horizontal Irradiance (GHI) values from minute-interval ground measurements and two satellite datasets for the NIST site, along with a zoomed-in view of the upper end of the CDFs (inset). Regardless of data source (ground measurement and satellite), hourly values do not show the high-irradiance tail observed in 1-minute ground measurements.

to the extent that a particular hourly interval engages with that nonlinearity, the simulated inverter output at the average irradiance will be different from the inverter output averaged across all irradiance values in the interval. Note that similar arguments apply to the other nonlinearities in PV system response, meaning subhourly clipping is not the only contributor to hourly modeling error [2], [3]. Fig. 1 visualizes this effect, showing an example where subhourly irradiance spikes would cause intermittent clipping loss but the average irradiance is low enough for an hourly model to ignore clipping entirely.

B. What are challenges with using satellite data?

Two limitations of satellite-based irradiance datasets are worth mentioning here. The first is that, because these datasets are derived in part from geostationary satellite imagery, their spatial resolution is limited by the the satellite's imaging optics and sensors. In particular, today's datasets are limited to kilometer-scale resolution, which is large compared with individual PV arrays. The second is the limited temporal frequency, which again is constrained by the corresponding imaging limits of the satellite. Together, these two limitations mean that our current satellite-based irradiance datasets are unable to recreate irradiance at the scale of individual PV arrays, at least without a statistical rescaling or similar downscaling step. Some combination of spatial resolution limitations and steps taken to convert imagery to irradiance results in a distribution of irradiance values that do not completely recreate the distribution of ground measurements, even when the satellite data are not time averages.

This is shown in Fig. 2, where maximum Global Horizontal Irradiance (GHI) values are clearly higher in both full and sub-sampled ground measurements than for averaged and "instantaneous" satellite-based data from both PSM3 and a

commercial vendor. This means that, despite previous work suggesting that using "instantaneous" satellite-based irradiance data can partially avoid subhourly clipping bias [4], these satellite-based datasets are still missing a key characteristic needed to fully model the effect of inverter clipping, regardless of temporal resolution.

C. This work

In the absence of an industry-wide shift to higher resolution models and weather data, previous efforts to correct this bias have focused on estimating post-hoc adjustments to forecasts from conventional models. Several approaches for generating these correction factors have been explored, including temperature-corrected insolation ratios [5], machine learning [6]–[8], and direct simulation using ground-measured 1 minute irradiance [9] or NSRDB PSM3 5 minute irradiance [4], [10], [11]. However, it is not obvious how to rigorously validate these correction methods: direct comparison of modeled and measured clipping loss is usually only possible in small research installations with different performance characteristics from utility-scale systems, while comparisons of overall power output are confounded by a myriad of unrelated modeling errors and real-world plant performance issues.

Furthermore, it may be premature to even attempt to correct a subhourly clipping bias, as there remains doubt in the community that real-world revenue projections are meaningfully affected by this error [11]. In this work, we seek to bring clarity to the issue by recreating typical industry modeling procedures, comparing the model output with measured data from real systems to investigate how the overall model bias varies with ILR, and showing that the model bias variation with ILR is consistent with the bias expected from subhourly clipping. This is made possible by two key insights: first, we can generate hourly production for a population of systems spanning a range of ILRs but with otherwise identical performance characteristics by artificially applying inverter clipping to high-resolution power data from a low-ILR system; and second, we can use the same high-resolution data to calculate an empirical ILR-dependent subhourly clipping bias to compare with the overall ILR-dependent model bias. Together, these insights allow us to avoid the confounding effects inherent to a more conventional multi-system analysis.

II. METHODS

A. Weather Datasets

Several satellite-based weather datasets are commonly used for PV performance modeling. Although publicly available satellite-based irradiance datasets exist (for example, the Physical Solar Model v3 (PSM3), part of the National Solar Radiation Data Base (NSRDB) [12]), commercial systems are typically financed using models based on proprietary commercial irradiance data. Here we use hourly interval data from a popular commercial data vendor. We choose single-year weather datasets so that we can directly compare modeled system output with measured system output for the same time periods. This is a divergence from the typical modeling procedure which uses a hypothetical "typical meteorological year" (TMY) dataset. However, because TMY datasets are simply combinations of subsets of single-year datasets, we expect single-year and TMY datasets to share whatever characteristics are relevant to subhourly modeling and therefore do not expect this difference to materially affect our conclusions.

B. PV System Datasets

We examine datasets from six PV systems in the United States representing a range of climates and configurations, summarized in Table I. In all cases, the high-resolution power data are 60 second averages of higher frequency data sampled at 1–15 s intervals depending on the system. Note that, because we perform this analysis at the inverter level (instead of the more typical plant level), the system metadata in Table I reflects the configuration of the chosen inverter rather than the plant as a whole. To add relevant geographical and climate details for the sites, particularly for the anonymous commercial plants, we also included Solar Forecast Arbiter Climate Zone (SFACZ) [13] and Solar Variability Zone (SVZ) [14] for each site.

In each case we analyze data from a single calendar year to facilitate associating the ground-measured data with the corresponding satellite-based weather dataset. The specific calendar year chosen for each systems varies in order to minimize the effect of data gaps and substantial performance issues. However, some data were still excluded because of abnormal operation (three days for Commercial Plant 2 for tracking issues; 41 days for NREL for snow coverage on the array). Additionally, any hourly intervals in the measured power containing zero, negative, or null values were dropped from both the measured and modeled datasets.

Although most of these datasets are proprietary and commercial, the underlying high-frequency power data are publicly available for the National Renewable Energy Laboratory (NREL) [15] and National Institute of Standards and Technology (NIST) [16], [17] systems. The public names for these two systems are "[1283] NREL Research Support Facility II" and "NIST Ground Mount Array", respectively.

C. Empirical Subhourly Clipping Bias

Evaluating the magnitude of subhourly clipping bias from measured data is not straightforward. In principle it could be modeled using a sufficiently accurate performance model and high-resolution weather data, essentially by taking the difference in the model's predictions when running at hourly and (approximately) instantaneous scales. This difference in modeling predictions has been referred to as "clip, then average" versus "average, then clip" to indicate the timescale at which inverter clipping is applied to the "unclipped" array output [18]–[20].

However, it is not clear that current modeling approaches are sufficiently accurate at such a high resolution to generate the "unclipped" power signal, especially for the purpose of

predicting the effects of irradiance variability: spatial nonuniformity of irradiance, thermal transience, and other short-duration effects complicate modeling efforts and few model validation studies are done at the short timescales relevant here. Additionally, satellite-based data often represent instantaneous measurements over some geographic area [21], and hourly instantaneous measurements have been demonstrated to result in less bias than hourly averaged measurements [4], further complicating this issue.

We propose an alternative approach: instead of attempting to recreate an "unclipped" array output via conventional weather-to-power modeling, we instead use real power measurements from an inverter with low loading ratio. Because the inverter has low ILR, it rarely if ever clips, meaning its power output is a perfectly realistic "unclipped" signal we can then use with the "clip, then average" and "average, then clip" approach of estimating subhourly clipping bias at any loading ratio of interest. Crucially, this lets us estimate an "empirical" subhourly clipping bias that varies with ILR without fear of model error and while holding all other system parameters constant. This empirical bias can then be compared with the actual bias of an hourly model, again controlling for all effects except ILR. As mentioned above, the empirical subhourly clipping bias is calculated using the "average, then clip" vs "clip, then average" method:

$$\text{bias} = \frac{E_{\text{AtC}} - E_{\text{CtA}}}{E_{\text{CtA}}} \qquad (1)$$

where E_{AtC} is total energy calculated by averaging measured data to 60 minute intervals and then artificially clipping (analogous to what a conventional hourly simulation model would calculate) and E_{CtA} is energy calculated by first clipping at 1 minute intervals and then averaging to 60 minutes (analogous to what a real system would do). For each system, this bias is evaluated at ILRs of 1.2, 1.3, 1.4, and 1.5 (with the exception of Commercial Plant 1, which does not have low enough nominal ILR for 1.2 to be relevant), corresponding to the typical range of ILRs seen in new systems today [1].

Although this approach of calculating a "true" subhourly clipping bias avoids the majority of possible model bias, the artificial clipping is still a potential source of error. The artificial clipping is applied using a straightforward numerical threshold where power values are clamped to not exceed the desired AC capacity. Although this is a reasonable approximation of the clipping behavior of many real-world inverters, it does ignore secondary effects like thermal throttling and dynamic plant control that cause the inverter's clipping limit to vary with time. We consider this approximation acceptable, as thermal throttling is considered unusual for most inverters, and the systems analyzed here did not have notable overbuilds of inverter capacity relative to interconnection limits, and therefore did not reflect impacts of dynamic plant controllers.

To eliminate the effect of other modeling biases (discussed more in the next section), each system's "unclipped" power signal was rescaled to have zero bias with respect to its

978-1-7281-6118-1/22 $31.00 © 2022 IEEE

TABLE I
SUMMARY OF PV SYSTEM CONFIGURATIONS

System	Size [kW$_{dc}$]	ILR	Rack	Location	Year	SFACZ[1]	SVZ[2]
NIST	271.0	1.04	Fixed Tilt	Gaithersburg, Maryland	2018	7	Moderate (low)
Commercial Plant 1	4609.2	1.15	Single-Axis Tracking	Southeast US	2018	6	Moderate (low)
Commercial Plant 2	594.4	1.19	Single-Axis Tracking	Southeast US	2020	7	Low
NREL	204.1	0.82	Fixed Tilt	Golden, Colorado	2020	4	Moderate (high)
SSRC 1-Axis	2.4	0.80	Single-Axis Tracking	Birmingham, Alabama	2019	6	Moderate (low)
SSRC 30S	2.4	0.80	Fixed Tilt	Birmingham, Alabama	2019	6	Moderate (low)

[1] Solar Forecast Arbiter Climate Zone [13]
[2] Solar Variability Zone [14]

nominal PVsyst model output. By doing this we can view the rescaled "unclipped" power data as the 1 minute analog of the hourly PVsyst model. The rescaling coefficients for each system are as follows, where a coefficient of 1.0 indicates no difference: 1.03 (NIST), 0.96 (Commercial Plant 1), 0.97 (Commercial Plant 2), 1.10 (NREL), 1.02 (SSRC 1-Axis), 1.03 (SSRC 30S). It is unclear why the NREL system data required such a large rescaling to match the output of its nominal PVsyst model.

D. Hourly Performance Models

In the authors' experience, the commercial simulation software PVsyst [22] drives the majority of utility-scale PV system financing. In typical usage a PVsyst model describing the system configuration is applied to an hourly weather dataset and produces a corresponding hourly production time series. We do the same here to mimic typical usage.

We used PVsyst version 7.1.4 to create models of each PV system in this study. We used PAN module and OND inverter files supplied by PVsyst when available and created PAN and OND files from component specifications when needed. We created a "nominal" model for each system that accurately describes the as-built system, including layout for shading. To simulate higher ILRs, we created new variants of the nominal system. For most systems, we increased the ILR by increasing the number of strings in parallel until the ILR was within 0.01 of the desired ILR. For the smaller SSRC systems with only one string of 10 modules, we instead increased the number of modules in series to reach an ILR close to but above the desired ILR (e.g. 1.36 for a desired 1.30), and then adjusted the module quality factor loss parameter to match the desired ILR. All other losses were left at their default settings. Several variants produced a voltage or current higher than the inverter specifications and OND file allow for. In these cases we increased the maximum inverter voltage or current such that PVsyst allowed the model to run and the inverter over-current and over-voltage losses were 0.0%. This represents a minor deviation from the as-built system and does not jeopardize the conclusions of this work. Finally, although these systems likely experienced gradual performance loss in the field for several years prior to the time period used in this analysis, we did not explicitly include the effect of this cumulative performance degradation in the PVsyst models. Instead, we rely on the rescaling procedure described in Section II-C to account for this capacity loss.

Analogous to Eq. 1, the model bias is calculated as:

$$\text{bias} = \frac{E_{\text{PVsyst}} - E_{\text{CtA}}}{E_{\text{CtA}}} \qquad (2)$$

where E_{PVsyst} is PVsyst's modeled output using hourly satellite data and the ILR corresponding to E_{CtA}.

We also used a PVWatts-style [23] model in pvlib [24], [25] with both vendor and PSM3 [12] weather data for a single site, NIST. This relatively naive model, with two sets of weather data, was selected to demonstrate that this bias issue is not unique a specific performance model or hourly satellite-derived weather data source.

III. RESULTS

Fig. 3 compares the PVsyst bias and empirical subhourly clipping bias for each system and ILR. For completeness, the biases are also listed in tabular form in Table II. The two biases have a roughly 1:1 relationship, with root mean squared error of 0.80% and mean bias error of 0.44%. Note that these error statistics are in the original bias units (percent of annual production) and reflect absolute difference of the two biases, not relative difference.

Fig. 4 and Table III show the same results for the NIST site from Fig. 3, with the addition of model bias values from the PVWatts-style pvlib model using both PSM and commercial vendor weather data. The simple pvlib model's bias varies similarly to the PVsyst and empirical biases, although this pvlib model's bias is somewhat higher, even when using the same weather dataset as PVsyst.

IV. DISCUSSION

The strong positive correlation between model bias and ILR in Fig. 3, as well as the rough range of model bias (0–4%) for these ILRs, are consistent with predictions from simulation-based studies [18]–[20]. Fig. 4 uses the NIST site to demonstrate that this model bias is not unique to PVsyst or the particular commercial vendor's weather data.

Contrary to what one might expect, these results indicate that subhourly clipping bias is not restricted to humid climates with high variability like those of the Southeast US; even the semi-arid climate of the NREL system (approximately

978-1-7281-6118-1/22 $31.00 © 2022 IEEE

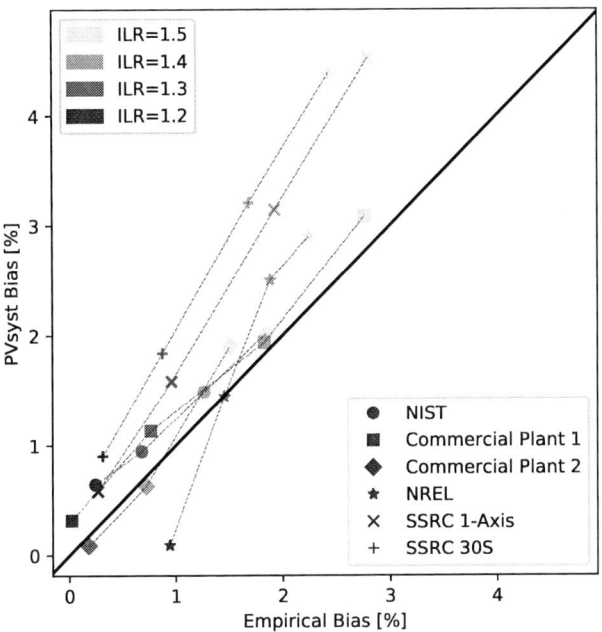

Fig. 3. Comparison of the ILR-dependent PVsyst and empirical biases. Colors indicate the ILR and symbols indicate the system for each pair of biases. The PVsyst ILR-dependent model bias is generally well-predicted by the empirical subhourly clipping bias derived from the measured system power data.

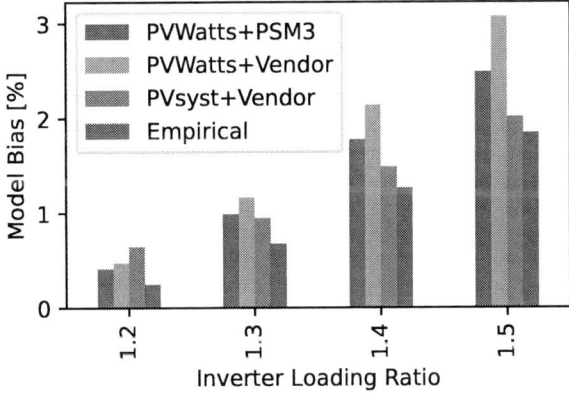

Fig. 4. Comparison of empirical subhourly clipping bias with hourly model bias for three modeling approaches: PVsyst with vendor irradiance, PVWatts with vendor irradiance, and PVWatts with PSM3 irradiance. The two PVWatts models are biased somewhat high compared to the PVsyst model, but follow the same general trend.

1750 meters above sea level) shows 2–3% bias at ILR=1.5. Conceptually, the error only requires that the system operates both below and above the clipping point for a portion of the hour, but the error is independent of the ordering of the points or the number of transitions i.e. the variability. However, we have not characterized the variability at these sites beyond listing their pre-existing zone classifications in Table I. Future work may include a more detailed investigation of the relationship between subhourly irradiance variability and subhourly clipping bias.

TABLE II
RESULTS: ILR-DEPENDENT BIASES (PVSYST+VENDOR)

System	ILR	PVsyst Bias [%]	Empirical Bias [%]
NIST	1.2	0.6	0.2
NIST	1.3	0.9	0.7
NIST	1.4	1.5	1.3
NIST	1.5	2.0	1.8
Commercial Plant 1	1.2	0.3	0.0
Commercial Plant 1	1.3	1.1	0.8
Commercial Plant 1	1.4	1.9	1.8
Commercial Plant 1	1.5	3.1	2.8
Commercial Plant 2	1.3	0.1	0.2
Commercial Plant 2	1.4	0.6	0.7
Commercial Plant 2	1.5	1.9	1.5
NREL	1.2	0.1	0.9
NREL	1.3	1.4	1.5
NREL	1.4	2.5	1.9
NREL	1.5	2.9	2.2
SSRC 1-Axis	1.2	0.6	0.3
SSRC 1-Axis	1.3	1.6	1.0
SSRC 1-Axis	1.4	3.1	1.9
SSRC 1-Axis	1.5	4.5	2.8
SSRC 30S	1.2	0.9	0.3
SSRC 30S	1.3	1.8	0.9
SSRC 30S	1.4	3.2	1.7
SSRC 30S	1.5	4.4	2.4

TABLE III
ILR-DEPENDENT BIASES FOR NIST WITH DIFFERENT MODELS AND WEATHER DATASETS.

ILR	PVWatts+ PSM3 [%]	PVWatts+ Vendor [%]	PVsyst+ Vendor [%]	Empirical [%]
1.2	0.4	0.5	0.6	0.2
1.3	1.0	1.2	0.9	0.7
1.4	1.8	2.1	1.5	1.3
1.5	2.5	3.1	2.0	1.8

It is noteworthy that the largest divergences from a perfect 1:1 relationship between the PVsyst and empirical biases are from the smallest systems, SSRC 1-Axis and SSRC 30S. One possible explanation for this is related to the spatial variability of irradiance: the smaller the array, the less spatial averaging and thus the more output variability it experiences [26]. This suggests that 1 minute averaged data are not able to fully resolve the output variability of these small arrays [26] and the empirical bias calculated here is an underestimate of the true value. It is also possible that the satellite-based weather dataset used for the SSRC systems (they are co-located and covered by the same satellite pixel) has some undiscovered characteristic that disrupts this analysis. In any case, the agreement of the PVsyst and empirical biases is much better for the four larger systems.

Imperfect datasets are another source of uncertainty in the empirical subhourly clipping bias. The two commercial systems have higher loading ratios than the other systems and their power datasets do include some clipping. Similarly to the spatial averaging issue discussed above, this could cause the empirical bias to be somewhat underestimated.

Finally, as mentioned in Section II-C, the simple clipping

model we applied to the measured "unclipped" power is an imperfect approximation of how real inverters behave. In particular, the clipping point of the inverter from one of the commercial systems is known to decrease slightly with increasing temperature.

The PVsyst models have some untracked uncertainty as well. Although we rescaled the "unclipped" power to have zero bias compared with the nominal PVsyst model, varying the model's ILR might introduce some small bias from other model nonlinearities. For example, inverter efficiency is generally not constant over the inverter's power range and increasing the ILR will tend to shift the distribution of operating points towards a different efficiency at the higher end of the efficiency curve.

The similar model biases shown in Fig. 4 and listed in Table III indicate that subhourly clipping model bias is not necessarily specific to PVsyst or the commercial weather dataset used in the primary analysis. This is consistent with expectations; any hourly performance model without some kind of subhourly adjustment might be expected to suffer from this bias, and as satellite-based datasets tend to draw from the same underlying data sources, one might expect little difference there as well.

Solar Variability Zones could be an intuitive reference for subhourly clipping error, but the Commercial Plant 2 site serves as a notable counterexample: its solar variability zone classification [14] is "low" variability (the scale includes one lower classification, "very low") but still exhibits a PVsyst bias of 1.9% at an ILR of 1.5. It is also worth noting that the Solar Variability Zones were developed based on hourly, 10 km gridded NSRDB data and only seven sites in the Western continental US (plus two in Hawaii). Regions with different climates (e.g., higher frequencies of clouds that are smaller than 10 km) may not be as well represented in this dataset.

V. Conclusion

We have shown that energy models using the approach that industry developers and financiers use for financing systems suffer from a material positive bias that grows with increasing inverter loading ratio. Especially for large systems, this bias is a close match to an empirical subhourly clipping bias determined from real-world system data, as well as what previous model-based studies have predicted. We also show that a similar bias is recreated using an alternative hourly simulation tool and weather dataset, indicating that this bias is not unique to a specific model or dataset.

Acknowledgment

The authors wish to acknowledge and thank our colleagues at NREL, NIST, EPRI, and our commercial partner for providing the PV datasets used in this work, as well as Johan Kemnitz for providing the baseline model for the NIST system.

This work was authored in part by Alliance for Sustainable Energy, LLC, the manager and operator of the National Renewable Energy Laboratory for the U.S. Department of Energy (DOE) under Contract No. DE-AC36-08GO28308. Funding

provided by the U.S. Department of Energy's Office of Energy Efficiency and Renewable Energy (EERE) under Solar Energy Technologies Office (SETO) Agreement Numbers 34348. The views expressed in the article do not necessarily represent the views of the DOE or the U.S. Government. The U.S. Government retains and the publisher, by accepting the article for publication, acknowledges that the U.S. Government retains a nonexclusive, paid-up, irrevocable, worldwide license to publish or reproduce the published form of this work, or allow others to do so, for U.S. Government purposes.

References

[1] M. Bolinger, J. Seel, C. Warner, and D. Robson, "Utility-scale solar, 2021 edition: Empirical trends in deployment, technology, cost, performance, PPA pricing, and value in the United States," Lawrence Berkely National Laboratory, Tech. Rep., 10 2021. [Online]. Available: https://www.osti.gov/biblio/1823604

[2] S. Ransome and P. Funtan, "Why hourly averaged measurement data is insufficient to model PV system performance accurately," in *20th European Photovoltaic Solar Energy Conference*, 2005.

[3] C. W. Hansen, J. S. Stein, and D. Riley, "Effect of time scale on analysis of PV system performance," Sandia National Lab, Albuquerque, NM, Tech. Rep. SAND2012-1099, 2012.

[4] D. A. Bowersox and S. M. MacAlpine, "Predicting subhourly clipping losses for utility-scale PV systems," in *2021 IEEE 48th Photovoltaic Specialists Conference (PVSC)*. IEEE, Jun. 2021. [Online]. Available: https://doi.org/10.1109/pvsc43889.2021.9518956

[5] D. Cormode, N. Croft, R. Hamilton, and S. Kottmer, "A method for error compensation of modeled annual energy production estimates introduced by intra-hour irradiance variability at PV power plants with a high DC to AC ratio," in *2019 IEEE 46th Photovoltaic Specialists Conference (PVSC)*. IEEE, Jun. 2019. [Online]. Available: https://doi.org/10.1109/pvsc40753.2019.8981206

[6] K. Bradford, R. Walker, D. Moon, and M. Ibanez, "A regression model to correct for intra-hourly irradiance variability bias in solar energy models," in *2020 47th IEEE Photovoltaic Specialists Conference (PVSC)*. IEEE, Jun. 2020. [Online]. Available: https://doi.org/10.1109/pvsc45281.2020.9300613

[7] K. Anderson and K. Perry, "Estimating subhourly inverter clipping loss from satellite-derived irradiance data," in *2020 47th IEEE Photovoltaic Specialists Conference (PVSC)*. IEEE, Jun. 2020. [Online]. Available: https://doi.org/10.1109/pvsc45281.2020.9300750

[8] A. Parikh, K. Perry, K. Anderson, W. B. Hobbs, R. Kharait, and M. A. Mikofski, "Validation of subhourly clipping loss error corrections," in *2021 IEEE 48th Photovoltaic Specialists Conference (PVSC)*. IEEE, Jun. 2021. [Online]. Available: https://doi.org/10.1109/pvsc43889.2021.9518564

[9] R. Kharait, S. Raju, A. Parikh, M. A. Mikofski, and J. Newmiller, "Energy yield and clipping loss corrections for hourly inputs in climates with solar variability," in *2020 47th IEEE Photovoltaic Specialists Conference (PVSC)*. IEEE, Jun. 2020. [Online]. Available: https://doi.org/10.1109/pvsc45281.2020.9300911

[10] A. Berlinsky, M. Reusser, and S. Vaishnav, "Modeling sub-hourly clipping losses using available irradiance and power data," in *Proceedings of the PV Reliability Workshop, February 22-26, 2021*, no. NREL/CP-5K00-80055, 6 2021. [Online]. Available: https://www.osti.gov/biblio/1797569

[11] C. Bordonaro, "A study of sub-hourly clipping losses for various solar PV plant designs across the United States," in *Proceedings of the PV Reliability Workshop, February 22-26, 2021*, no. NREL/CP-5K00-80055, 6 2021. [Online]. Available: https://www.osti.gov/biblio/1797569

[12] M. Sengupta, Y. Xie, A. Lopez, A. Habte, G. Maclaurin, and J. Shelby, "The national solar radiation data base (NSRDB)," *Renewable and Sustainable Energy Reviews*, vol. 89, pp. 51–60, 2018. [Online]. Available: https://doi.org/10.1016/j.rser.2018.03.003

[13] C. W. Hansen, W. F. Holmgren, A. Tuohy, J. Sharp, A. T. Lorenzo, L. J. Boeman, and A. Golnas, "The solar forecast arbiter: An open source evaluation framework for solar forecasting," in *2019 IEEE 46th Photovoltaic Specialists Conference (PVSC)*, 2019, pp. 2452–2457. [Online]. Available: https://doi.org/10.1109/PVSC40753.2019.8980713

[14] M. Lave, R. J. Broderick, and M. J. Reno, "Solar variability zones: Satellite-derived zones that represent high-frequency ground variability," *Solar Energy*, vol. 151, pp. 119–128, 2017. [Online]. Available: https://doi.org/10.1016/j.solener.2017.05.005

[15] C. Deline, K. Perry, M. Deceglie, M. Muller, W. Sekulic, and D. Jordan, "Photovoltaic data acquisition (PVDAQ) public datasets," Dec 2021. [Online]. Available: https://doi.org/10.25984/1846021

[16] M. Boyd, T. Chen, and B. Dougherty. (2017) NIST campus photovoltaic (PV) arrays and weather station data sets. National Institute of Standards and Technology. U.S. Department of Commerce, Washington, D.C. [Online]. Available: https://doi.org/10.18434/M3S67G

[17] M. Boyd, "Performance data from the NIST photovoltaic arrays and weather station," *Journal of Research of the National Institute of Standards and Technology*, vol. 122, Nov. 2017. [Online]. Available: https://doi.org/10.6028/jres.122.040

[18] J. O. Allen and W. B. Hobbs, "Effect of time-averaging on PV production estimates on systems with high DC to AC ratios," Sandia PV Performance Modeling Collaborative, 2015.

[19] J. O. Allen, W. B. Hobbs, and M. Bolen, "The effect of short-term inverter saturation on PV performance modeling," Sandia PV Performance Modeling Collaborative, 2018. [Online]. Available: https://pvpmc.sandia.gov/download/6707/

[20] ——, "Predicting the effect of short-term inverter saturation on PV performance modeling," Sandia PV Performance Modeling Collaborative, 2019. [Online]. Available: https://pvpmc.sandia.gov/download/7415/

[21] S. M. Wilcox, "National solar radiation database 1991-2010 update: User's manual," National Renewable Energy Lab.(NREL), Golden, CO (United States), Tech. Rep., 2012.

[22] A. Mermoud and B. Wittmer, "PVSYST user's manual," PVsyst SA, 2014.

[23] A. P. Dobos, "PVWatts version 5 manual," National Renewable Energy Laboratory, Golden, CO, Tech. Rep. NREL/TP-6A20-62641, 2014. [Online]. Available: https://doi.org/10.2172/1158421

[24] W. F. Holmgren, C. W. Hansen, and M. A. Mikofski, "pvlib python: a python package for modeling solar energy systems," *Journal of Open Source Software*, vol. 3, no. 29, p. 884, 2018. [Online]. Available: http://doi.org/10.21105/joss.00884

[25] W. Holmgren *et al.*, "pvlib/pvlib-python: v0.9.1," Mar. 2022. [Online]. Available: https://doi.org/10.5281/zenodo.6395177

[26] J. Marcos, L. Marroyo, E. Lorenzo, D. Alvira, and E. Izco, "From irradiance to output power fluctuations: the PV plant as a low pass filter," *Progress in Photovoltaics: Research and Applications*, vol. 19, no. 5, pp. 505–510, 2011. [Online]. Available: https://doi.org/10.1002/pip.1063

Life Cycle Assessment of High-Efficiency Si Solar Modules

Estefania Papaioannou, Pritpal Singh, Ross Lee

Villanova University, Villanova, PA, United States

A Life Cycle Assessment (LCA) is an analytical framework that quantifies the environmental impacts associated with the lifecycle stages of a product. This LCA compares the impacts of 1 kWh of electricity (functional unit) produced by two different hypothetical solar modules: a standard PERC crystalline silicon module and a luminescent solar concentrator (LSC)/PERC tandem module. An LSC-PERC tandem device contains luminescent materials that absorb incoming light with a wide frequency range and re-emits the energy as light in a narrow wavelength range that can be absorbed by silicon, increasing the module power conversion efficiency from 20% to 24%. This 'cradle to grave' LCA includes the following life stages: materials acquisition and module fabrication in China, transport to the hypothetical installation site in New Jersey, installation, use, electricity generated, maintenance, disassembly, and end of life. The results indicate that the tandem module will have less overall environmental impacts per unit of electricity generated than the PERC module. The carbon footprint calculated for the PERC module is 31.3 g CO_2 eq/kWh, compared with 18.0 g CO_2 eq/kWh for the Tandem module. Other impact categories analyzed include ozone depletion, smog, acidification, eutrophication, and human health effects. Several Sustainable Product Innovation (SPI) improvements were proposed and evaluated, including a reduction in the thickness of the silicon cells, replacement of virgin aluminum with recycled aluminum in the frames, removal of antimony in the formulation of the LSC device, and transfer of manufacturing from China to the USA. These SPIs further reduced the environmental impacts in the tandem modules and are reported on in this paper.

978-1-7281-6118-1/22 $31.00 © 2022 IEEE

Alternative Rear Contacts for Ultrathin CdSe$_x$Te$_{1-x}$ Solar Cells

Bérengère Frouin[1], Andrea Cattoni[1], John Moseley[2], David Albin[2], Joel Duenow[2],
Abderrahime Sekkat[3], David Muñoz-Rojas[3] and Stéphane Collin[1]

[1]Centre de Nanosciences et Nanotechnologies (C2N), CNRS, Université Paris-Saclay, Palaiseau, France
[2]National Renewable Energy Laboratory (NREL), Boulder, USA
[3]Université Grenoble Alpes, CNRS, Grenoble INP, Laboratoire des Matériaux et du Génie Physique (LMGP), Grenoble, France

Abstract—The back contact is a key component for further developments of CdSeTe solar cells. In this contribution, we investigate several materials and designs that could help the development of highly-reflective, ohmic back contacts for conventional (3.45 μm) and thin (800 nm) CdSeTe solar cells. We also explore numerically, nanostructured metal back contacts for light-trapping in ultrathin (200 nm) CdTe solar cells.

Index Terms—ultrathin CdSeTe, back contact, nanostructuration

I. INTRODUCTION

CdTe thin-film solar cells have a very low levelized cost of energy and provide an attractive alternative to Si cells. They have reached a remarkable efficiency of 22.1% [1] thanks to several key improvements over the years. First, annealing under CdCl2 atmosphere resulted in a huge efficiency increase as compared to as-grown CdTe through grain interior and grain boundary passivation. Second, the CdS buffer layer was replaced by a more transparent material MgZnO. Finally, the incorporation of Selenium has enabled a significant rise of the short-circuit current and a decrease of the open-circuit voltage deficiency [2].

Despite these device achievements, there are known shortcomings that still limit the efficiency (32% is achievable according to Shockley-Queisser limit) and need to be addressed. The open-circuit voltage is limited below 900 mV, partly due to the low hole density in the active layer and the back contact technology requires optimisation. Indeed, CdTe has a high electron affinity around 4.4 eV and most metals tend to form a Schottky contact with a large hole energy barrier. Only high work function metals such as gold form a quasi-ohmic contact. To overcome this technical difficulty, many CdTe back contacts use small amount of copper which diffuses in the CdTe layer to increase the carrier concentration. However, the stability of such device can be compromised in time. On the industrial side, one of the mid-term challenges is to decrease the absorber layer thickness [3]. Reducing the CdTe thickness, usually 3-4 μm, would decrease even more the fabrication cost and possibly reduce the relatively high bandgap-voltage offset. Nonetheless, it usually comes at the expense of light absorption. Thus, it requires light trapping combined with a highly reflective mirror such as silver to preserve the Jsc. However, because of the CdTe high electron affinity and

low doping, replacing the back contact while keeping high performances is not trivial.

In this contribution, we explore alternative contacts for ultrathin CdSeTe solar cells compatible with highly reflective, nanostructured back mirrors for different absorber thicknesses. We take advantage of the team's knowledge on ultrathin solar cells whose technique was successfully implemented on a 200 nm-thick GaAs layer reaching a 19.9% record efficiency [4]. We present our recent experimental results where several back contacts are investigated and compared. Meanwhile, we show the optical simulation results of CdTe solar cells with various nanostructured back contacts. Finally, we propose optimized architectures and perspectives for ultra-thin CdTe solar cells.

II. EXPERIMENTAL RESULTS

This study is carried out with CdSeTe solar cells supplied by the NREL (US) in superstrate configuration. MgZnO buffer layer was sputtered on TEC12D commercial glass. 150 nm of CdSe and two different CdTe thicknesses: 3.3 μm and 0.65 μm were thermally evaporated sequentially. The device underwent CdCl$_2$ passivation treatment as well as CuCl$_2$ diffusion annealing.

A. Rear contact deposition

We first investigate the electrical properties of solar cells with several metal back contacts that could also act as back mirrors. Gold and silver metallic back contacts are first deposited on thick (3.45 μm) CdSeTe solar cells. The best JV curves for each type of contacts are plotted in figure 1a, and the performance of relevant parameters (short-circuit current (Jsc) extracted from external quantum efficiency (EQE) measurements, open-circuit voltage (Voc), fill factor (FF) and efficiency) are summarized in figure 1b. The reference devices with Au contact reveal promising performances even though the shape of the JV curves shows a rollover. It is attributed to a Schottky energy barrier formed at the CdTe/metal back interface, limiting also the fill factor. The silver contact, which is considered due to its valuable reflective properties, exhibits a good Jsc but a stronger rollover effect and a lower FF.

B. Alumina passivation layer

Al$_2$O$_3$ layers have been added at the CdTe/Au interface to act as passivation layers. Their potential has been demon-

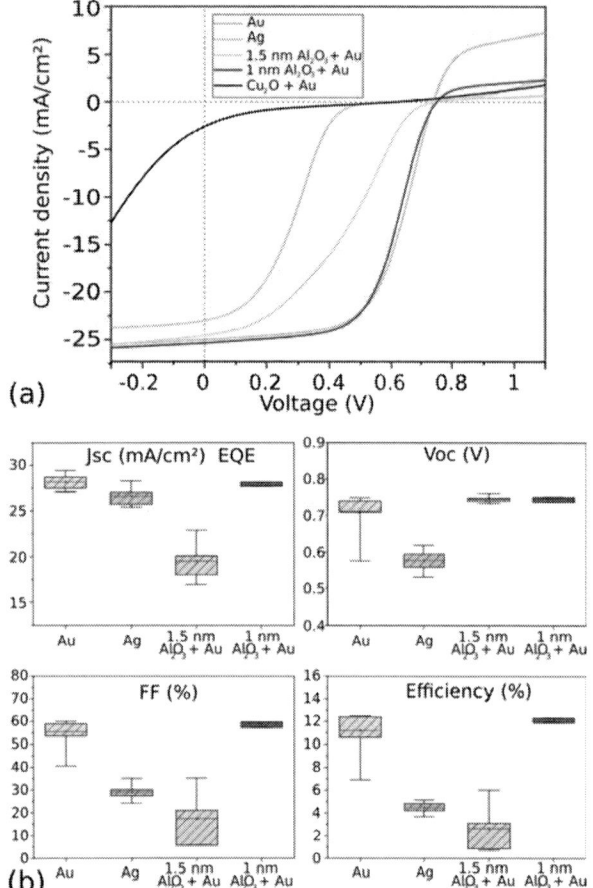

Fig. 1. (a) Current density as a function of the voltage under one sun illumination plotted for the best cells measured with different types of back contact. J–V parameters depict the respective impact of each rear contact. Jsc values were extracted from EQE measurement.

strated through the increase of carrier lifetime [5]. Very thin layers should still enable carrier collection through tunneling. With 1.5 nm of Al_2O_3, we observe a slight improvement of Voc with less dispersed values, but the strong decrease of both FF and Jsc reveals a barrier for hole collection. The Al_2O_3 layer thickness is then reduced to 1 nm. It leads to a slightly more pronounced rollover effect (figure 1a), and overall performances similar to the reference cells made of CdTe/Au back contacts (figure 1b).

C. P-type transparent conductive oxide back contact

We have also investigated Cu_2O as a potential candidate to act as a p-type, transparent, conductive back contact. It was deposited by spatial atomic layer deposition (SALD) under atmospheric conditions. This transparent conductive oxide features a high mobility of 92 cm² V⁻¹s⁻¹ [6], and could provide a low contact resistance on CdTe. Preliminary results of as-deposited Cu_2O with a gold contact are shown in figure 1a. Voc is comparable to the previous results, but a very low FF and Jsc reveal carrier collection issues. Additional analysis and

post-deposition treatments of the Cu_2O layer are in progress and will be presented.

D. Thin CdSeTe solar cells

800 nm-thick CdSeTe solar cells have been fabricated with Au back contacts and are compared to thicker (3.45 μm) devices. The spectral responses (EQE) are plotted in figure 2. For the thinner solar cells, high short-circuit currents (28 mA/cm²) have been obtained despite the strong thickness reduction, with Voc=0.67 V and FF=66%. As expected, the absorption of the thicker device (figure 2b) slightly outperforms the thinner one at high wavelengths, from 700 nm to 900 nm.

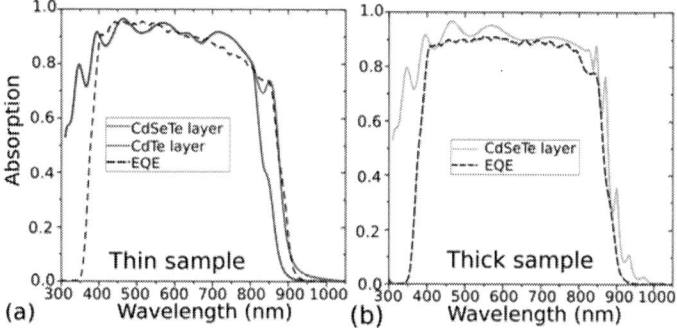

Fig. 2. Experimental EQE performed on CdSeTe samples, with a thickness of active layer equal to (a) 800 nm and (b) 3450 nm respectively (dotted lines). Measurements are compared with optical simulation of CdTe or CdSeTe planar layer with different thicknesses with Au contacts (solid lines).

III. NUMERICAL SIMULATIONS

Our goal is to further reduce the thickness of the active layer without absorption losses. It requires an accurate model for the absorption in thin CdSeTe layers and the design of novel back contacts for light-trapping in very thin CdSeTe layers. In the following, we first compare our simulation results with EQE measurements, and we present numerical results of ultrathin CdTe solar cells with nanostructured back contacts.

A. Absorption calculation

Figure 2 shows a good agreement between numerical simulations of absorption spectra and EQE measured on both CdSeTe samples. It suggests an internal quantum efficiency close to 1. The disagreement at short wavelengths is due to the absorption in the window layer. Above 400 nm wavelength, the numerical results are in very good agreement with the EQE for the thicker cell. The simulated absorption has additional oscillations that are due to resonance effects in the front transparent conductive oxide (TCO) and window layer (400-800 nm) or in the absorber layer close to the bandgap (800-900 nm), depending on the thickness. The latter are smoothed in the spectral response because of the roughness of the CdSeTe polycrystalline layers.

For the thinner device (figure 2a), we investigate the influence of the Se alloys by comparing EQE to different simulation results. First, we consider a 800 nm-thick absorber made of CdTe only. The simulated absorption spectrum drops

978-1-7281-6118-1/22 $31.00 © 2022 IEEE

at wavelengths much shorter than the EQE. Second, we replace the first 320 nm of CdTe by a CdSe$_x$Te$_{1-x}$ layer with x= 0.42, in accordance with the deposition and annealing process. It results in an excellent agreement with EQE measurements, even close to the bandgap. These results also show the impact of the Selenium on the absorption properties at high wavelengths.

B. Light trapping simulation

To gain more insight into the impact of the thickness reduction on the absorption, we have plotted absorption spectra calculated for different CdTe thicknesses with a gold back contact (figure 3a). As expected, significant absorption losses are observed below a thickness of 800 nm, with a drop of the short-circuit current density of ΔJth=7.4 mA/cm² for a thickness of 200 nm.

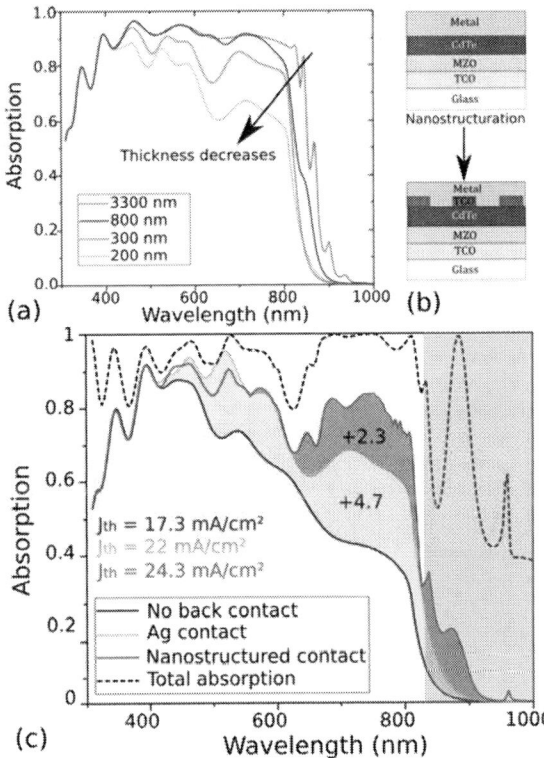

Fig. 3. (a) Calculated absorption plotted as a function of the CdTe thickness. (b) CdTe solar cell scheme in superstrate configuration before and after the nanostructuration with a TCO layer. Optimized calculation of optical absorption in 200 nm-thick CdTe for a CdTe device (blue curve) with no metallic contact, with a flat Ag back mirror (green curve) and with a nanostructured Ag back mirror (red curve). A plot of 1-R (total absorption) for the nanostructured device is also shown with the black dotted curve. The shaded area corresponds to the spectral region of photon energies below the CdTe bandgap.

To compensate for the absorption loss, a new architecture is proposed in figure 3b. The flat back contact is replaced by a nanostructured metallic contact which acts as a diffraction grating. Numerical simulations are performed with the Rigorous Coupled Wave Analysis (RCWA) method, which computes the electromagnetic fields and absorption in each layer of the

solar cell structures. The structure of the rear side features a 2D square periodical pattern similar to the nanostructured mirror introduced in 200 nm-thick GaAs solar cells [4]. The metal is embedded in a transparent conductive oxide layer whose optical index is chosen close to the Cu$_2$O material. The exact geometry was optimized by varying the period, width and height of the nanostructures.

Simulation results are shown in figure 3c. The absorption spectra of CdTe with different back contacts are compared: a planar contact (with and without metal) and the nanostructured mirror made of silver. The beneficial effect of the highly-reflective silver mirror is observed over the whole spectral range. Light-trapping effects are evidenced in the long-wavelength range ($\lambda > 600$ nm), and result in an additional gain of 2.3 mA/cm² for the Jth as compared to the Ag planar mirror. Absorption enhancement is attributed to several overlapping resonances close to the bandgap.

The black dotted line represents the total absorption for a nanostructured cell and reveals nearly perfect absorption in the 650-800 nm wavelength range. It shows that there is still room for improvement. These numerical results are currently limited by parasitic optical losses in the nanostructured silver mirror. Alternative designs are under consideration to circumvent this issue.

IV. CONCLUSION

Several alternative back contacts for CdSeTe solar cells have been investigated, targeting passivated and reflective CdTe/metal back interfaces. We have studied several metals with and without alumina passivation layers, and Cu$_2$O back contacts. Thin (800 nm) CdSeTe solar cells have been fabricated and compared to thicker devices and to simulation results. Nanostructured back contacts have also been designed to further enhance absorption by light-trapping in ultrathin (200 nm) CdTe solar cells. These first results pave the way for novel back contacts that could help increase the efficiency of CdSeTe solar cells and enable even thinner devices.

REFERENCES

[1] M. A. Green et al., "Solar cell efficiency tables (Version 58)," 2021.
[2] J. D. Poplawsky et al., "Structural and compositional dependence of the CdSe$_x$Te$_{1-x}$ alloy layer photoactivity in CdTe-based solar cells," Nat. Commun., vol. 7, pp. 1—10, 2016.
[3] M. Woodhouse et al., "Perspectives on the pathways for cadmium telluride photovoltaic module manufacturers to address expected increases in the price for tellurium," Sol. Energy Mater. Sol. Cells, vol. 115, pp. 199—212, Aug. 2013.
[4] H. Chen et al., "A 19.9%-efficient ultrathin solar cell based on a 205-nm-thick GaAs absorber and a silver nanostructured back mirror," Nat. Energy, vol. 4, pp. 761—767, 2019.
[5] J. M. Kephart et al., "Sputter-Deposited Oxides for Interface Passivation of CdTe Photovoltaics," IEEE J. Photovoltaics, vol. 8, no. 2, pp. 587-593, 2018.
[6] A. Sekkat et al., "Open-air printing of Cu2O thin films with high hole mobility for semitransparent solar harvesters," Commun. Mater., vol. 2, no. 1, pp. 1—10, 2021.

Feeder Open-Phase Detection by Smart Inverters

Yiwei Ma[1], Xiaojie Shi[1], Aminul Huque[1], Roland Bründlinger[2], Ron Ablinger[2]

[1]Electric Power Research Institute, Knoxville, TN, 37932, USA
[2]Austrian Institute of Technology, Vienna, 1210, Austria

Abstract—Inadvertent open phase condition is a concern for distributed energy resources (DER) integration, due to the unregulated voltage on the opened phase. Although IEEE 1547-2018 standard mandates DER to detect and trip for open phase condition at its reference point of applicability (RPA), it may be challenging for DER to detect a feeder (high side of interconnection transformer) open phase condition. This paper presents an improved feeder open phase detection (OPD) method that only utilizes the solar photovoltaic (PV) or energy storage inverter's onboard resources. Controller hardware-in-the-loop (CHIL) results are shown to demonstrate the effectiveness of the proposed OPD algorithm. It is found that the OPD method can successfully detect a feeder open phase condition for Δ/Yg, Yg/Yg, and Δ/Δ transformer, but not with Yg/Δ transformer.

Keywords—Distributed energy resources (DER), Open phase, Smart inverter, Hardware-in-the-loop, Sequence component

I. INTRODUCTION

An open phase event is an abnormal operating condition on three-phase power systems, where one or two out of the three phases are disconnected. These events might be initiated by blown fuses or single-phase switching in distribution circuits that utilize individual phase protections or even broken/downed conductors.

With growing number of distributed energy resources (DERs) like solar PV and energy storage on the distribution circuits, consequences of unintentional open phase conditions become a concern to distribution utilities, since the DER and/or its interconnection transformer may recreate unregulated voltage on the opened phases. Common issues include safety concerns to the public and line workers, as well as equipment damage due to overvoltage or ferroresonance. These open phase events are not rare, especially in systems with individual phase protections. One open phase event captured in field is presented in [1]-[2], which is caused by a blown fuse, and led to 1.22 pu overvoltage on the opened phase. Utilities have also reported plant commissioning delays due to the failure of the open phase detection in the field test.

Fig. 1 shows two example open phase scenarios at an upstream feeder location and DER's reference point of applicability (RPA). IEEE 1547-2018 requires DER to detect and cease to energize and trip for any open phase condition occurring at the RPA. However, it is typically not required and tested in the certification process for DER to detect and trip upon any open phase condition on the upstream feeder and/or with any local loads. For example, IEEE 1547.1-2020 standard conformance tests only included type test procedure for open phase condition with no local loads. Depending on system

generation and load condition and the DER interconnection transformer configuration, the recreated voltage at the DER side may be within the normal operating range, becoming a challenge for the DER to detect the open phase condition.

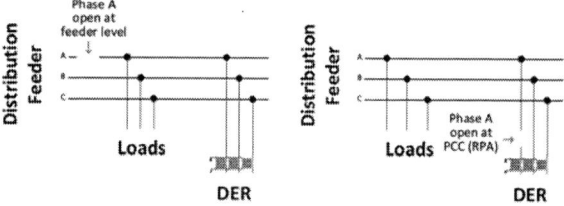

(a) Open phase on upstream feeder   (b) Open phase at RPA

Fig. 1. Example open phase scenarios

Typical solutions to detect the open phase events utilize protective relays on the medium voltage side [3]-[6]. The proposed solutions include zero sequence voltage and current based detections, negative sequence voltage and current based detections, harmonic distortion based detections, and waveform zero-crossing based detections. These solutions are reliable but increase DER interconnection cost due to extra hardware with the medium voltage withstand capability, such as sensors. In addition, most of the works focus on the open phase detection (OPD) at RPA, and may not be applicable if feeder open phase condition needs to be detected.

Reference [1] proposed a feeder open phase detection (OPD) method relying exclusively on the inverter located at the low voltage side. However, it is found that the detection effectiveness drops at low DER power generation levels. This paper improves the OPD method from [1] and reduces the Non-Detection Zone (NDZ). Controller hardware-in-the-loop (CHIL) tests were performed to verify the effectiveness of the proposed OPD algorithm.

II. IMPROVED INVERTER ONBOARD OPEN PHASE DETECTION METHOD

The proposed inverter on-board OPD method is based on the negative sequence voltage, which is usually generated on both the primary and secondary sides of the interconnection transformer during an open phase event. Since the magnitude of negative sequence voltage under open phase conditions varies depending on the configuration of the interconnection transformer and the ratio between generation and load at the opened phase, the OPD scheme is equipped with passive and active methods.

978-1-7281-6118-1/22 $31.00 © 2022 IEEE

A. Passive detection

Passive detection is the foundation of the proposed OPD method. If OPD is enabled, for all configurations, the inverter keeps monitoring the negative sequence voltage at its terminal (V_{neg}) and trips if V_{neg} exceeds the threshold (*Neg_th*) continuously for the delay time (*Neg_td*). The illustrative flowchart is shown in Fig. 2.

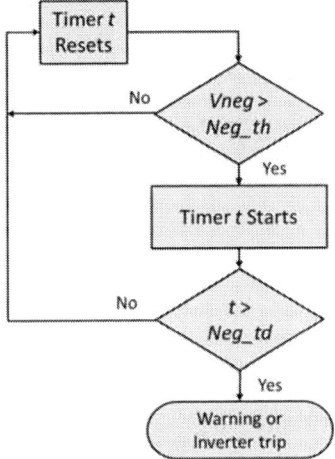

Fig. 2. Passive open phase detecion (OPD) method

With only the passive detection activated, the inverter only measures its V_{neg} without any active perturbation. It may be desirable from the grid perspective, since there will be no perturbation that may negatively impact grid power quality. It also might be easier to implement from the inverter vendors' perspective, since it only requires the V_{neg} measurement and associated trip mechanism and does not require any modification on its control scheme.

On the other hand, there can be a considerable NDZ, especially when in situations where generation and load are similar and create a quasi-equilibrium. If the inverter tripping is not required, the passive detection scheme can also be used to inform the distribution operator of the open-phase condition for further action.

B. Active detection

To reduce the NDZ, an active detection method was developed. If enabled, the inverter will actively inject negative sequence current (I_{negref}) perturbation. When an open phase event occurs, the open phase voltage(s) at the inverter terminals is(are) not regulated by the main utility feeder anymore. Thus, this perturbation can further elevate the negative sequence voltage (V_{neg}), such that it is more likely to reach the negative sequence voltage trip setting *(Neg_th)*.

Two active injection mechanisms were developed: (1) pulse perturbation, and (2) positive feedback. Both mechanisms work together to push the negative sequence voltage to the trip threshold during the open phase condition, while minimizing the power quality degradation in normal operating conditions. The illustration of the active detection method is shown in Fig. 3.

The first mechanism periodically injects pulses of negative sequence currents, which aims to force the negative sequence voltage to exceed a predefined deadband (*db_neg*), which in turn triggers the positive feedback. The magnitude of the pulses is determined by the level of active power output. Since it is identified earlier that the OPD method has a larger NDZ at a low power generation range when there is no load [1], the magnitude is at the highest level (*NegSeqMag*) when the active power is below a settable threshold (P_1), and linearly reduces to 0 at another settable threshold (P_{th}).

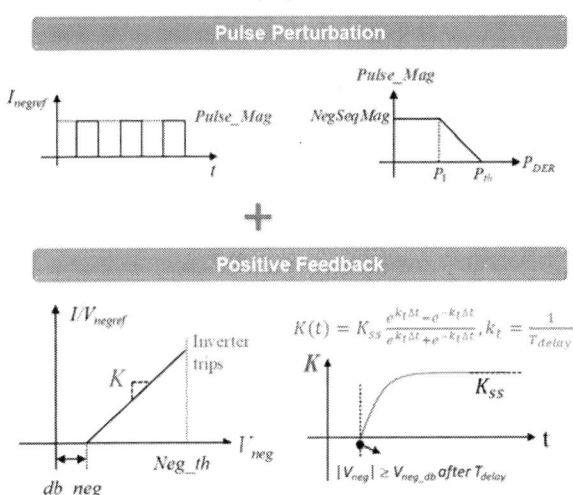

Fig. 3. Active OPD methods (output equals to the superposition of pulse perturbation and perturbation through positive feedback)

The second mechanism relies on the positive feedback of the negative sequence injection, which is superposed to the pulse perturbation. The magnitude of the injection, in addition to the pulse perturbation, is related to the measured negative sequence voltage at the terminal of the inverter (V_{neg}). When it is outside of the deadband (*db_neg*), regardless of being triggered by the pulse perturbation or mismatch between generation and loads in the open phase area, the injection magnitude increases as the V_{neg} increases, following a time-variant factor $K(t)$. The factor $K(t)$ stays 0 for an intentional time delay (T_{delay}), and gradually rises to a steady-state value (K_{ss}), so that it has high immunity to temporary open phase event where no action may be preferred. During an open phase event, the active injection can raise the inverter terminal negative sequence voltage, which in turn increases the magnitude of injection. This positive feedback mechanism ensures a fast and reliable detection of open phase events.

To sum up, the NDZ of the proposed OPD is minimized through two active injection methods described above. The perturbation pulses help to trigger a positive feedback process even when the voltages are within the normal range during an open phase event. Once the negative sequence voltage exceeds the deadband, the positive feedback process further pushes the negative sequence voltage, which eventually leads to a trip of the passive OPD scheme.

On the other hand, if active OPD detection is enabled, the positive feedback mechanism destabilizes the open phase section, and causes the inverter disconnection from grid. If the distribution system operator only needs to be informed of the feeder open phase condition, and inverter is not required to trip, the active OPD injection may not be applicable. While passive detection has a considerable NDZ, further improvements to the OPD algorithm can be made as a follow-up research topic.

III. CONTROLLER HARDWARE IN THE LOOP (CHIL) TESTING

CHIL tests are conducted to verify the proposed OPD algorithm. It is one step closer to the real-world power application than software simulation. All the power system components, including the utility grid, transmission and distribution line, transformer, as well as the inverter power stage, are modeled in a digital real time simulator (DRTS). But the inverter controller board and embedded control algorithms under evaluation is physical. It communicates with the DRTS in real-time through digital and analog input and output (I/O) ports, similar to how an actual inverter controller communicates with its physical power stage circuit.

The CHIL test platform is shown in Fig. 4. The proposed OPD algorithms were implemented in the existing controller of Austrian Institute of Technology (AIT)'s Smart Grid Converter (SGC) [6]. The Typhoon HIL604 simulator serves as DRTS. It simulates the distribution circuit, inverter power stage circuit, as well as PV panel. The inverter circuit model takes the pulse width modulation (PWM) control signal generated from the SGC controller through the digital inputs, and based on it, generates the inverter terminal voltages and currents as analog outputs to the SGC controller.

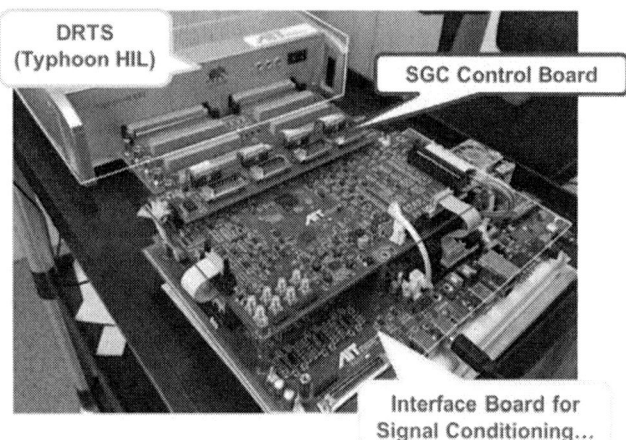

Fig. 4. CHIL test platform to evaluate OPD algorithm implemented in AIT SOC, with Typhoon HIL for circuit simulation.

The simulated distribution circuit diagram is shown in Fig. 5. The utility grid is represented as a voltage source with a series connected R-L impedance. A controllable contactor on phase B is added to initiate the feeder open phase event. As shown in the figure, the inverter connects to the grid through a Yg/Yg interconnection transformer. Other transformer configurations can be tested by modifying the circuit for sensitivity analysis. Load on MV side is modeled as a constant impedance. Load consumption and transformer winding configurations can be

varied for OPD performance evaluation in different system conditions. Detailed circuit parameters are shown in Table I.

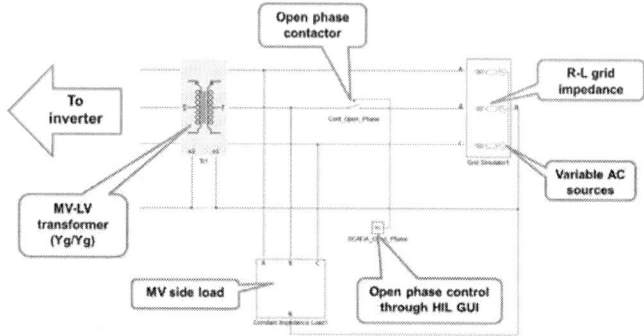

Fig. 5: Power system model for OPD algorithm evaluation

The existing AIT SGC controller already had negative sequence current closed-loop control implemented. Thus, the active OPD injection algorithm was directly implemented as its control reference. The passive detection method was also included in the inverter trip mechanisms. The default OPD trip settings and active control parameters are shown in Table II. The execution frequency of the OPD algorithm is 200 Hz, as a compromise between a reasonable dynamic response and acceptable computing resources taking up.

TABLE I.   CHIL CIRCUIT PARAMETERS FOR OPD EVALUATION

Parameter	Value	Unit
Grid (MV)		
Grid short-circuit ratio	50	-
Short-circuit capacity	1.725	MVA
X/R ratio	3	-
Rated voltage	12.47	kV
Rated frequency	60	Hz
MV/LV Interconnection transformer		
Rating	34.5	kVA
Primary 1 voltage	12.47	kV
Secondary 2 voltage	480	V
Winding configuration	Δ/Yg; Yg/Yg; Yg/Δ; Δ/Δ (MV/LV)	-
Winding resistances	.007	pu
Winding stray inductances	.019	pu
Short-circuit voltage	4.0	%
Short-circuit losses	2350	W
Core model	Magnetizing winding neglected Saturation not modeled	-
Inverter		
Rated apparent power	34.5	kVA
Rated voltage	480	V
Rated frequency	60	Hz
Rated current	41.5	A
Trip settings	IEEE 1547-2018 Cat III	-

TABLE II.   OPD PARAMETERS IN CHIL TESTS

Parameter name	Description	Default value
Neg_td	OPD negative sequence voltage trip time setting	100 (ms)
Neg_th	OPD negative sequence voltage trip voltage setting	0.1 (pu)
$NegSeqMag$	Maximum magnitude of the OPD active injection pulse perturbation	0.05 (pu)
$DutyCycle$	Duty cycle of the OPD active injection pulse perturbation	0.5
$T_{perturb}$	Duration period of OPD active injection pulse perturbation	1 (s)
P_1	Determines the magnitude of OPD active injection pulse perturbation. When the inverter output power is greater than this value, magnitude starts to decrease from the max value.	0.3 (pu)
P_{th}	Determines the magnitude of OPD active injection pulse perturbation. When the inverter output power is greater than this value, magnitude equals 0.	1 (pu)
db_neg	OPD active injection positive feedback deadband	0.03 (pu)
K_{ss}	OPD active injection positive feedback steady-state gain	1.5
T_{delay}	OPD active injection positive feedback time delay	0.4 (s)

CHIL tests are conducted to evaluate the OPD performance under different parameter settings and circuit conditions. Test results are captured through the DRTS. The waveform plots include three-phase rms voltages on the LV (inverter) side, negative sequence and zero sequences of inverter terminal voltage and current, as well as the connection status of the open phase contactor and the DER.

Fig. 6 shows the test results with only passive OPD enabled, with load to generation ratio (LGR) of 0.68. The inverter interconnection transformer configuration is Δ/Yg (MV/LV). In this case, the measured negative sequence voltage is greater than the trip setting of 0.1 pu. The inverter trips in 0.14s after the inception of open phase condition.

If without the negative sequence based passive OPD, the inverter can only rely on its over- and under-voltage protection. For the same circuit condition, the over-voltage trip may take 13 seconds to detect the open phase condition with IEEE 1547-2018 Cat III default trip setting. When LGR is closer to 1, the inverter may even stay connected indefinitely. It is observed through multiple tests that the non-detection zone (NDZ) for the proposed passive OPD only is between LGR of 0.7-1.6.

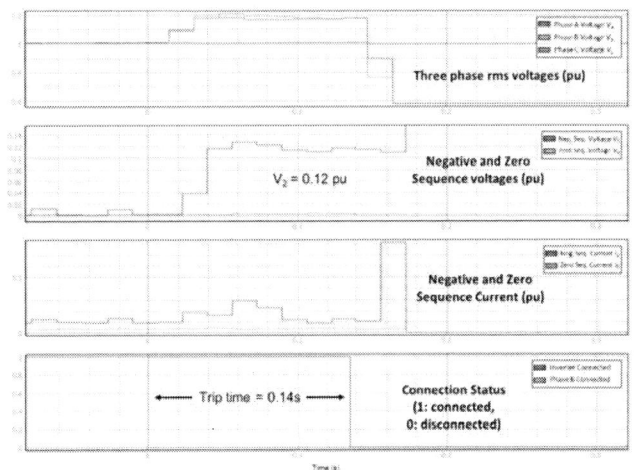

Fig. 6.   CHIL test results with passive OPD enabled, Δ/Yg transformer, P_{load} = 0.5 pu, P_{gen} = 0.74pu

With active injections, the NDZ can be mostly eliminated depending on the system configuration and OPD parameters. Fig. 7 shows an example test result generation and load match condition. The inverter is able to detect the open phase condition in 0.75s.

In grid-connected condition, without an open phase event, the pulse injection on negative sequence current does not change the system voltage. So, there would not be concerns for power quality degradations in normal operation. On the other hand, in open phase condition, the pulse injection creates negative sequence voltage, which helps to triggers the positive feedback mechanism. In the parameter design, the positive feedback time delay T_{delay} is shorter than the duration of the pulse perturbation ($T_{perturb} \times DutyCycle$), such that the positive feedback can be triggered by the perturbation. The system voltage eventually passes the passive OPD trip setting, and the inverter is disconnected. This test result demonstrates that the proposed OPD algorithm can detect the open phase condition even with generation and load match, eliminating the NDZ.

Similar tests are conducted with Yg/Yg, and Δ/Δ transformer configurations. Results have verified the proposed OPD can successfully detect feeder open phase condition in all LGR conditions. However, the proposed OPD is found ineffective with Yg/Δ transformer configuration. Fig. 8 shows an example test result with a LGR of 0.5, when the pulse perturbation is unpractically increased to 0.1pu. The negative sequence voltage is still not high enough to trigger the positive feedback. Further research is planned to be conducted to analyze the effectiveness of the proposed OPD method under Yg/Δ transformer.

978-1-7281-6118-1/22 $31.00 © 2022 IEEE

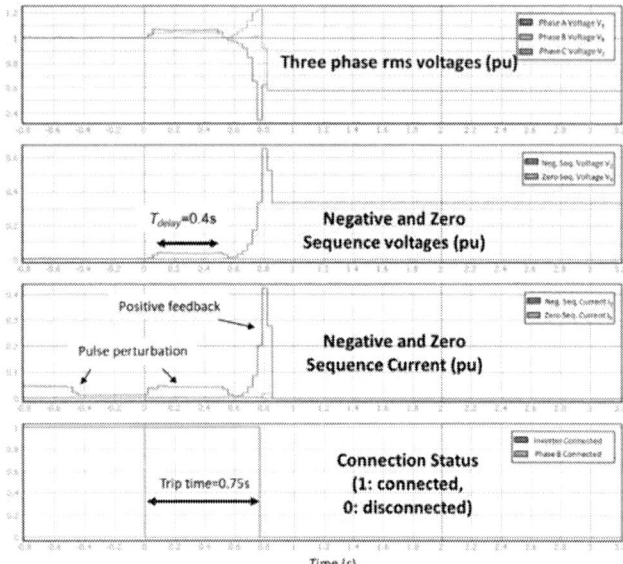

Fig. 7. CHIL test results with both active and passive OPD enabled, Δ/Yg transformer, $P_{load} = P_{gen} = 0.3$pu

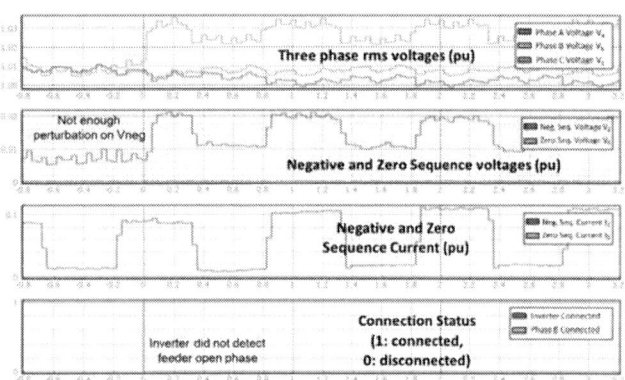

Fig. 8. CHIL test results with both active and passive OPD enabled, increased pulse perturbation magnitude, Yg/Δ transformer, $P_{load} = 0.3$pu, $P_{gen} = 0.6$pu

The OPD algorithm implementation and CHIL tests also identified a few other cases that require further analysis and/or algorithm improvements, specifically:

- Time variant gain $K(t)$ (see Fig. 3) for the positive feedback mechanism is complex to be implemented into the inverter controller firmware and may lead to numerical instability which causes CPU crash. The equation can be simplified in the future.

- The detection effectiveness of OPD depends on the control settings. For example, with a smaller pulse perturbation magnitude *NegSegMag*, the OPD may be ineffective with LGR close to 1. Parameter design for different circuit conditions may be further investigated.

- It has also been identified that in some cases, the pulse perturbation reduces the magnitude of negative sequence voltage, as shown in Fig. 9. It indicates the angle direction of the pulse perturbation may be further improved.

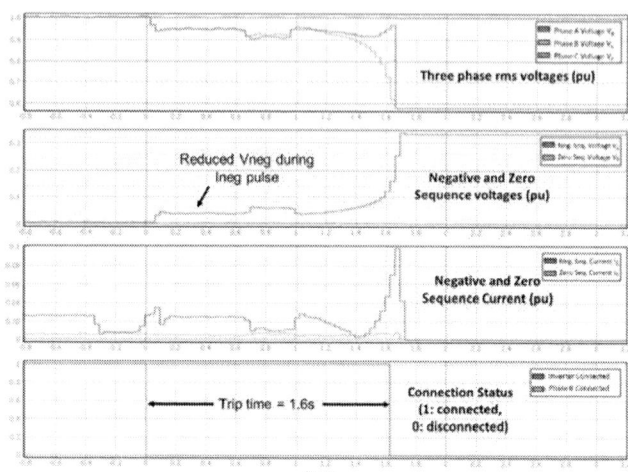

Fig. 9. CHIL test results with both active and passive OPD enabled, *DutyCycle* = 0.7, Δ/Yg transformer, $P_{load} = 0.3pu$, $P_{gen} = 0.24$pu.

To summarize the CHIL test results:

- Even with only the OPD passive detection, the inverter can detect open phase conditions for some cases that are not detectable if only using undervoltage or overvoltage relays;

- OPD active injection has a narrower NDZ than the passive detection. The inverter can detect open phase events in a much wider range of circuit and operating conditions depending on the parameter settings;

- With proper parameter settings, the OPD method can successfully detect a feeder open phase condition for Δ/Yg, Yg/Yg, and Δ/Δ transformer configurations with no NDZ. However, open phase conditions with Yg/Δ transformer cannot be detected, even with extreme settings of OPD parameters.

- Lessons are learned both in the algorithm implementation stage and CHIL testing stage, which will guide the continued research and advancement of the OPD algorithm.

IV. CONCLUSIONS

An improved open phase detection (OPD) method was proposed based on [1]. The method consists of both passive detection of negative sequence voltage, as well as active injection of pulse perturbation and positive feedback. Using only inverter on-board resources at low voltage side, the inverter is able to detect feeder open phase condition at medium voltage.

The algorithms were implemented into an actual inverter controller for Solar PV application and tested on a CHIL platform. The results highlighted the effectiveness of the proposed OPD algorithm. The work also demonstrated the feasibility of the algorithm implementation on a typical converter controller platform.

Future research includes improving the OPD algorithm based on the CHIL test results, verifying the algorithm with different inverter functions enabled, such as volt-var, as well as Power HIL (PHIL) testing with the actual inverter power stage.

REFERENCES

[1] X. Shi, T. Key, W. Wang, A. Huque, "Detection of Feeder Open Phase Events Using Smart Inverters," in Proc. *IEEE PVSC*, 2020.

[2] *Distributed Energy Resources Field Experience: Open Phase*, EPRI, Palo Alto, CA: 2019. 3002015949

[3] *Method for detecting an open-phase condition of a transformer*. Patent, 2014. WO2015126412A1.

[4] A. Norouzi, "Open phase conditions in transformers analysis and protection algorithm," in Proc. *IEEE Annual Conference for Protective Relay Engineers*, 2013.

[5] D. Cox and H. Chaluvadi, "Open-phase detection for station auxiliary transformers," in Proc. *Annual Western Protective Relay Conference*, 2018.

[6] J. Gahan, B. Cockerham, R. Chowdhury, and J. Town, "Field Experience with Open-Phase Testing at Sites with Inverter-Based Resources," in Proc. *IEEE Annual Conference for Protective Relay Engineers*, 2021.

[7] https://www.ait.ac.at/en/solutions/power-system-technologies-development-validation/power-electronics-solutions

Design with Integrated PV Technologies in Various Products and Environments

Eli Shirazi, Wouter Eggink, Xitong Zhu, Angele Reinders

University of Twente, Enschede, Netherlands

Eindhoven University of Technology, Eindhoven, Netherlands

This study focuses on the design of solar powered objects which can be integrated in the landscapes and/or the built environment by optimally using design features of solar technologies, such as color, transparency, surface patterns and form giving. The objects presented in this study are various conceptual designs created by student teams of University of Twente in 2021, covering a wide range of applications connecting to users and reducing environmental impact, that are: (part of) a building, a mobility product, a cityscape, natural landscape, or a newly designed thing which seamlessly fits in its environment.

Electric Field and its Effect on Hot Carriers in InGaAs Valley Photovoltaic Devices

Kyle R. Dorman, Vincent R. Whiteside, David K. Ferry, Tetsuya D. Mishima, Hamidreza Esmaielpour, Michael B. Santos, Ian R. Sellers

Homer L. Dodge Department of Physics & Astronomy, University of Oklahoma, Norman, OK, United States

School of Electrical, Computer, and Energy Engineering, Arizona State University, Tempe, AZ, United States

Walter Schottky Institut, Technische Universität München, Garching, Germany

To maintain a hot carrier population in a robust manner under standard operating conditions, a valley photovoltaics device employs the mechanisms of photoexcited and field aided intervalley scattering to transfer carriers to upper metastable valleys of the band structure. Prior work has demonstrated transfer and storage of enhanced carrier temperatures, a robust effect even at low illumination in comparison to the phonon bottleneck method of hot carrier maintenance. In order to deconvolve the photoexcited intervalley scattering and electric field-aided transfer mechanisms, the previous 20 nm n+-In0.52Al0.48As /250 nm n-In0.53Ga0.47As /1000 nm p+-In0.52Al0.48As /p-InP substrate proof-of-concept device structure was altered to create a set of 25 nm and 100 nm absorber thickness devices, grown by molecular beam epitaxy, then enhanced by a 150 nm ITO top reflective coating. The alteration of the absorber thickness was modeled in NRL Multibands® and shows that as they become thinner, the samples substantially enhance the electric field produced around the interfaces as a result of the doping density. This allows for a series of illumination power dependent photoluminescence measurements with high power 532 nm and 1064 nm laser light, to excite carriers both above and below the upper L valley. This was performed at a range of applied bias voltages to further modulate the electric field strength and investigate its impact, guided by accompanying current density-voltage measurements. From the natural logarithm of the high energy tail of the photoluminescence spectrum, a temperature indicative of the carrier distribution can be determined, revealing both the enhanced carrier temperature even at low illumination seen in prior valley photovoltaic devices and, additionally, a phonon bottleneck-type increase in temperature with increasing power.

Single-Axis Tracker Control Optimization Potential for the Contiguous United States

Kevin Anderson[1] and Saurabh Aneja[2]
[1]National Renewable Energy Laboratory, Golden, CO, 80401, USA
[2]FTC Solar, Austin, TX, 78759, USA

Abstract—**Conventional tracker control algorithms maximize collection of direct irradiance with no regard for collection of diffuse irradiance. Therefore, a tracker control algorithm that optimizes for maximal total irradiance, not just direct, might realize improved insolation collection. Using weather data gridded at 0.25° by 0.25° latitude/longitude spacing covering the contiguous United States, we evaluate the insolation gain of two alternative control algorithms optimized for improved total irradiance collection in monofacial arrays and present annual and monthly geographic heatmaps showing the gains across the contiguous United States. Certain locations show potential annual insolation gains approaching 1.0%, but most locations with recently-built tracker systems show annual gains between 0.1% and 0.4%. We also demonstrate a relationship between a climate's annualized diffuse insolation fraction and its potential tracker optimization insolation gain.**

Index Terms—**photovoltaic, single-axis, tracking, optimization, insolation, diffuse**

I. INTRODUCTION

Minimizing the angle of incidence is the foundation of conventional single-axis tracking strategies. The angle of incidence (AOI), a measure of the degree of alignment between a collector and the sun, is the angle between the collector orientation and solar position vectors. The collector's exposure to direct irradiance from the sun is maximized when AOI is zero, prompting the use of AOI minimization as the basis of tracking. Furthermore, closed-form expressions based on solar position and tracker geometry are available for calculating the tracker rotation that minimizes AOI [1]–[3], making an AOI-based strategy straightforward to implement in practice. Trackers that follow the AOI-minimized rotation are said to "true-track" (often with "backtracking" adjustments to prevent self-shading) and have until recently been the main focus of the single-axis tracker market. This strategy is also called the "astronomical" strategy for its focus on the solar disc's position in the sky.

However, optimizing for direct irradiance alone is suspect in the context of non-concentrating photovoltaic (PV) systems. In contrast to concentrating collectors, typical flat-plate photovoltaic arrays harvest both direct and diffuse irradiance, motivating a tracking strategy that maximizes total in-plane irradiance, not just the direct component. This is made most evident by considering overcast sky conditions when diffuse irradiance dominates: there is nothing to gain by prioritizing alignment with solar position in the absence of direct irradiance, so attention should instead be turned

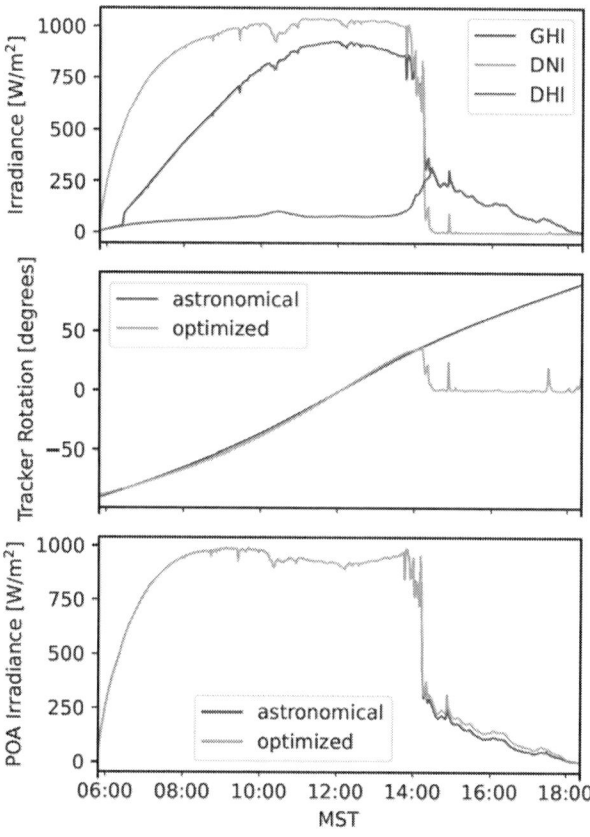

Fig. 1. Example comparison of rotation and simulated irradiance curves for the conventional astronomical tracking strategy and a tracking strategy optimized for total irradiance capture. This alternative strategy can outperform the conventional astronomical strategy, especially in diffuse conditions.

towards the diffuse components originating from the sky dome and reflected from the ground. Even under clear skies these diffuse components still contribute to total incident irradiance, meaning the irradiance-maximized rotation may differ from the true-tracking rotation under all sky conditions.

Fig. 1 compares the simulated in-plane irradiances corresponding to a tracker using the astronomical strategy with one maximizing global incident irradiance. Under the clear skies in the morning, the optimized strategy only very slightly outperforms the astronomical strategy. In this case the difference is so small it is not visible in the figure. However, the astronomical

978-1-7281-6118-1/22 $31.00 © 2022 IEEE

strategy is noticeably outperformed by the optimized strategy under cloudy skies in the later part of the day. The irradiance data in this example came from the Baseline Measurement System at the National Renewable Energy Laboratory's Solar Radiation Research Laboratory [4].

The general concept of considering diffuse irradiance in dual- and single-axis tracker control has been previously investigated for select ground station locations [5]–[9], often corresponding to locations in the Baseline Surface Radiation Network (BSRN) [10], a global network of high-quality irradiance measurement stations. It is worth emphasizing that multiple styles of diffuse-aware tracking have been considered in the literature; most existing studies used horizontal orientation as the alternative to conventional tracking, but some studies [8] instead opted to use the tracker rotation that maximizes incident irradiance according to a particular irradiance transposition model. Some studies have also considered this optimization in the context of bifacial PV arrays [11], [12]; in this work we focus exclusively on front-side irradiance.

In this work we evaluate the potential insolation gain of two single-axis tracker optimization strategies. In contrast to previous studies based on ground station data, the insolation gain is evaluated using gridded weather data, allowing detailed geographic gain comparison across the contiguous United States. Finally, the potential gain is evaluated at locations of existing utility-scale tracking systems.

II. METHODS

The approach is relatively straightforward: for every gridded location in our area of interest, apply each tracker strategy to a weather dataset for that location and model the total annual insolations. The details of each step are as follows:

A. Tracking Strategies

We consider three tracking strategies:

1) *Astronomical*: the conventional AOI-minimized rotation as implemented by common tracker controllers today.
2) *All-times*: the rotation that maximizes incident irradiance, identified via brute force search, considering all times of day.
3) *Between-backtracking*: similar to the "all-times" strategy, except only non-backtracking times (i.e., the portion of the day between the morning and evening backtracking periods) are considered.

The motivation for the between-backtracking strategy may deserve explanation. In certain conditions or situations, perceived or practical limitations may prevent system operators from extending the irradiance optimization methods presented here from being used during backtracking hours. For example, operators may determine that irradiance forecast data needed for rotation determination has less certainty than backtracking processes already employed at existing systems. For this reason, we also calculated insolation gains from times periods outside of backtracking hours only.

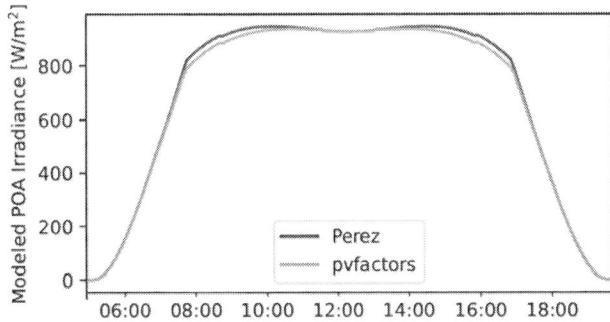

Fig. 2. Example comparison of simulated plane of array (POA) irradiance using the base Perez transposition model and unshaded ground and the same accounting for sky diffuse shading and ground shading using pvfactors.

TABLE I
PVFACTORS MODELING ASSUMPTIONS

Parameter	Value
Row Count	5
Axis Alignment	North-South
Axis Tilt [degrees]	0
Axis Height [m]	3.0
Collector Width [m]	4.0
Ground Coverage Ratio	0.3
Ground Albedo	0.2

Insolation gains for the "all-times" and "between-backtracking" strategies are calculated relative to the astronomical method. In each case the rotation angle is constrained [1], [3] so that rows do not subject each other to direct shading, assuming uniform flat terrain.

B. Incident Irradiance Model

Detailed investigation of tracker rotation optimization warrants a detailed incident irradiance model. In this work we use pvfactors, a detailed transposition and view factor model [13], [14]. By including the effect of row-to-row sky diffuse shading, ground shadows, and module-module reflections as pvfactors does, the modeled irradiance should be more representative of the irradiance incident on an interior row in a PV array, which will generally be somewhat less than the irradiance measured by an isolated radiometer. Fig. 2 shows an example of this difference, comparing the output of the pvfactors model with the output of the Perez irradiance transposition model [15] on which it is based.

In this work we simulated irradiance for the center row of an array with the specifications listed in Table I.

C. Weather Data

Annual weather files for a 0.25° by 0.25° grid covering the contiguous United States were retrieved from the Physical Solar Model v3 (PSM3), part of the National Solar Radiation Data Base (NSRDB) [16]. For this work we select hourly Typical Meteorological Year (TMY) data from the latest "tmy-2020" dataset. TMYs exclude unusual weather conditions, meaning a TMY-based analysis is less likely to be influenced

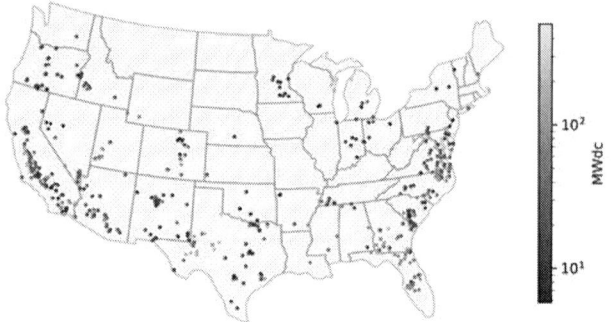

Fig. 3. Locations of utility-scale single-axis tracking PV systems in the 2021 LBNL utility-scale solar database. Color indicates system DC capacity in MW on a logarithmic scale.

by prolonged atypical cloudy or clear periods. In total, this dataset covers over 13000 locations and requires roughly 11 GB to store in uncompressed CSV form.

D. Tracking System Locations

The economic arguments for preferring some regions over others when deploying tracking systems are well established. Thus, it is of interest to examine the modeled insolation gains considering only the subset of locations deemed suitable for tracking systems. We identify suitable locations using the locations of existing tracking systems as a proxy. The locations of existing single-axis tracking PV systems were retrieved from a database of utility-scale systems assembled by Lawrence Berkeley National Laboratory (LBNL) [17]. The 2021 edition of this database includes 660 utility-scale single-axis tracking systems ranging in size from 5 to 500 MWdc. Fig. 3 shows the locations and DC capacities of these systems.

The spatial distribution of these tracking systems is also evolving over time. For example, the database's commercial operation date (COD) records go back to 2007 but the oldest of the 35 systems in Oregon reached COD only in 2016. It seems reasonable to expect this trend of tracking systems expanding into more diffuse climates to continue in the future. Because the spatial distribution of tracking system locations is evolving over time, so too is the distribution of insolation gains at those locations. For this reason we examine not only the modeled insolation gains for all tracker system locations but also the insolation gains partitioned by system COD year.

III. Results

Figs. 4 and 5 show the annualized and monthly potential insolation gain for the all-times strategy. The insolation gains for the between-backtracking times are very similar to those of the all-times strategy (see Fig. 6) and therefore not shown here.

Fig. 7 shows the distribution of insolation gains partitioned by hour of day. The largest gains are concentrated around mid-morning and mid-afternoon, consistent with the intuition that the largest gains come from times of day when the astronomical algorithm results in steep tracker rotations and

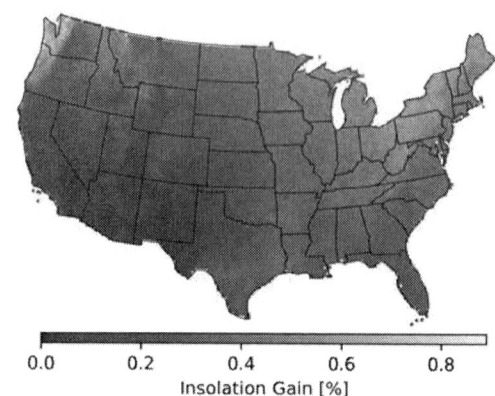

Fig. 4. Annual insolation gain for the all-times strategy. Color shows the increase (in percent) in annual insolation relative to the astronomical tracking strategy.

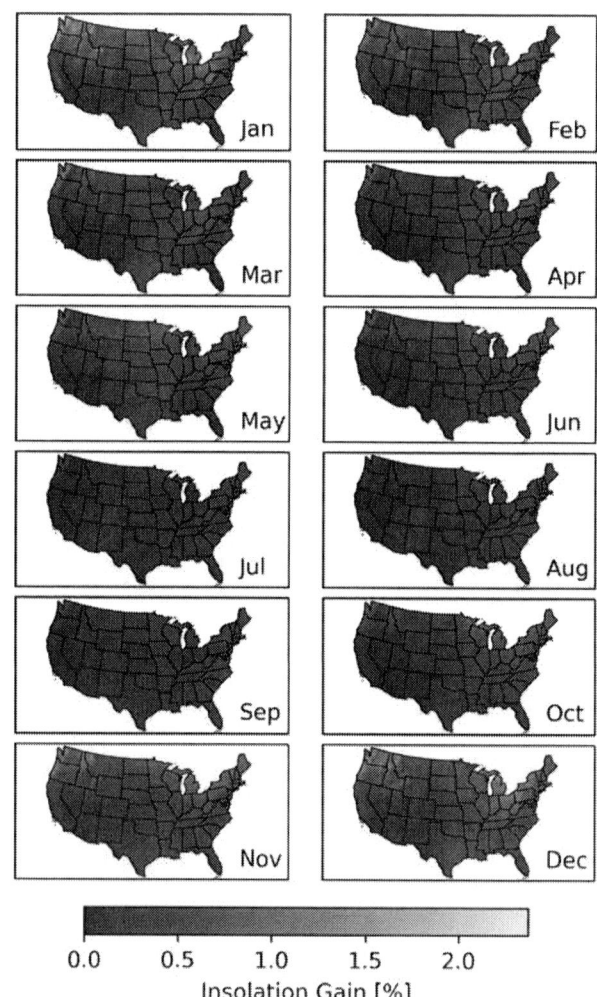

Fig. 5. Monthly insolation gains for the all-times strategy. Color shows the increase in insolation relative to the astronomical tracking strategy, with the largest relative gains being achieved in winter.

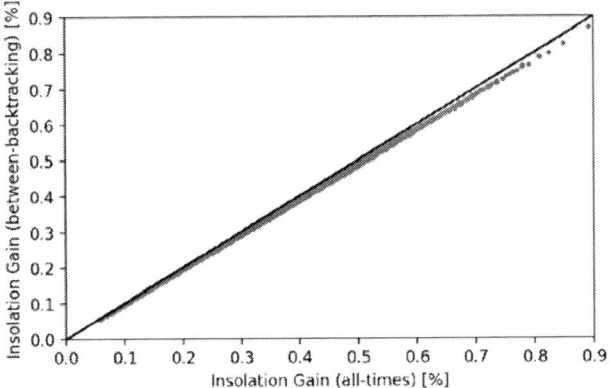

Fig. 6. Annual insolation gain comparison for the all-times and between-backtracking strategies, showing only a marginal decrease (less than 3% relative/.01% absolute) in annual insolation gain by excluding backtracking times. The black line indicates a perfect $y = x$ relationship for reference.

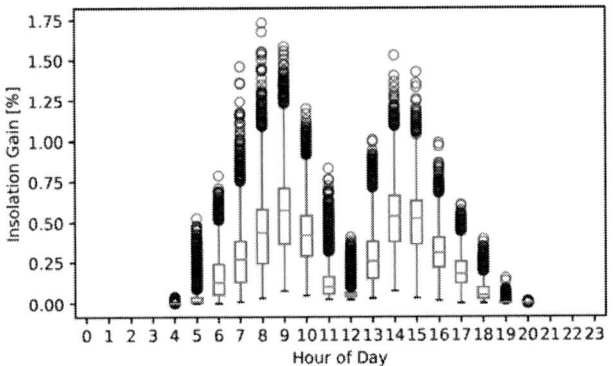

Fig. 7. Distribution of annual insolation gains for all locations, partitioned by hour of day, for the all-times strategy. For example, the hour=9 distribution reflects the annual insolation gain considering only the 365 hour long intervals from 9–10 AM (local standard time) in each year. Each value in the distribution represents the annual partitioned gain for a single location. Gains are concentrated around times of day where the astronomical strategy would orient modules at steep tilts.

is therefore most different from a diffuse-optimized horizontal rotation.

Fig. 8 shows the distribution of potential gains for the all-times strategy at locations of tracking systems in the LBNL utility-scale dataset, partitioned by COD year.

Finally, Fig. 9 shows how, for each location in the 0.25° by 0.25° grid, the potential all-times insolation gain varies with that location's diffuse insolation fraction. The diffuse insolation fraction is the fraction of annual global horizontal insolation (GHI) contributed by diffuse horizontal insolation (DHI). The cloudier the climate, the generally higher the diffuse fraction will be.

IV. DISCUSSION

Consistent with previous studies [7], Fig. 4 shows that insolation gains are strongly dependent on climate; potential

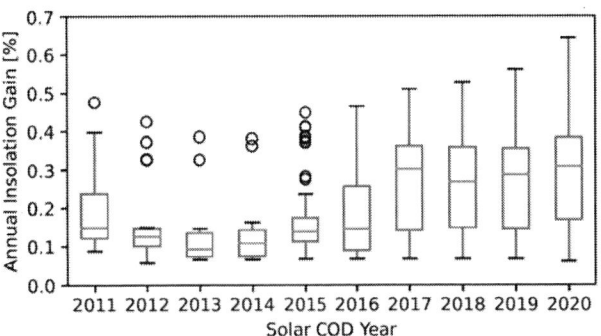

Fig. 8. Distribution of annual insolation gains using the all-times optimization for locations of tracking systems in the LBNL database, partitioned by system commercial operation year. The annual gains show a general upwards trend with time. Years before 2011 are not shown due to their small number of systems.

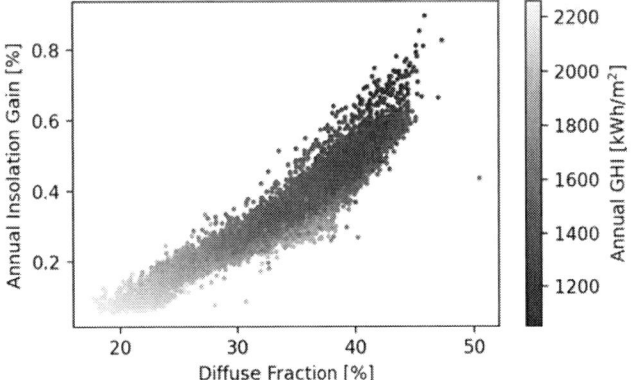

Fig. 9. Relationship between annual all-times insolation gain and annual diffuse fraction, colored by total annual global horizontal insolation (GHI). Locations with more diffuse skies (larger diffuse fraction) tend to have larger potential tracker optimization gain.

gain in the desert Southwest is negligible, but other parts of the country see potential gains from 0.5% to a peak of over 0.8% in the Pacific Northwest for the all-times strategy. The potential gains for the between-backtracking strategy show nearly identical spatial trends and are only marginally reduced compared to the all-times strategy, suggesting that the large majority of potential insolation gains occur outside of backtracking times. This finding is further confirmed in Fig. 7.

The monthly heatmaps in Fig. 5 show that the largest relative insolation gains occur in the North American winter, reaching monthly values approaching 2.5% in some locations. However, it is worth pointing out that large relative winter gains may be smaller in an absolute sense than small relative summer gains due to overall seasonality in solar resource.

Although Fig. 8 does show a trend of increasing potential insolation gain over time, the trend has stagnated since 2017. Together with Fig. 9, showing that diffuse climates tend to have larger potential gains, this suggests that new projects have not been pushing into more diffuse climates since 2017. This finding is consistent with the conclusions of the LBNL report itself [17], which reports stable global horizontal insolation at all utility-scale solar sites (not just tracking PV) since 2017.

Now we discuss several caveats in these results. The ground-reflected component of incident irradiance depends strongly on ground albedo. Ground albedo varies not only with substrate type but also with weather conditions, primarily snowfall. In this work we have assumed a static albedo of 0.2 for all conditions. A more detailed analysis would consider not only geographic variation in ground albedo (e.g. bare soil versus grass) but also the effect of snowfall. A secondary consideration is that this optimization neglects the varying spectral mismatch of each irradiance component with the spectral response of PV absorbers.

An additional concern is that hourly TMYs may lack the resolution needed to fully capture the potential for irradiance-based tracker optimization. A full discussion of the limitations of satellite-based irradiance is out of scope here, but this is a problem both spatially (due to the kilometer-scale size of each satellite pixel) as well as temporally (hourly data cannot resolve subhourly variability, which can be significant in some climates). Future work may include comparing the gains modeled using TMYs with the gains modeled using higher resolution satellite data or ground station measurements.

V. Conclusions

Based on hourly TMY data, we find that the potential insolation gain over the conventional astronomical tracking algorithm depends strongly on local climate, especially the fraction of diffuse annual insolation, with potential annual gains ranging from near-zero in the desert Southwest to over 0.8% in the Pacific Northwest. Typical gains for locations with recently-installed tracking systems range from 0.2–0.4%. The gains also show significant seasonality, with monthly gains as high as 2.4% in winter for some locations. Finally, these results show that omitting backtracking times of day from the irradiance optimization reduces the potential gain only slightly.

Acknowledgment

The authors wish to thank the maintainers of the NSRDB and the LBNL utility-scale solar database. The authors additionally thank the developers of the open-source software packages used in this work (pvfactors, pvlib, and the broader scientific python ecosystem) and highlight the value that open-source software brings to the research community.

This work was authored in part by Alliance for Sustainable Energy, LLC, the manager and operator of the National Renewable Energy Laboratory for the U.S. Department of Energy (DOE) under Contract No. DE-AC36-08GO28308. Funding provided by the U.S. Department of Energy's Office of Energy Efficiency and Renewable Energy (EERE) under Solar Energy Technologies Office (SETO) Agreement Numbers 34348. The views expressed in the article do not necessarily represent the views of the DOE or the U.S. Government. The U.S.

Government retains and the publisher, by accepting the article for publication, acknowledges that the U.S. Government retains a nonexclusive, paid-up, irrevocable, worldwide license to publish or reproduce the published form of this work, or allow others to do so, for U.S. Government purposes.

References

[1] E. Lorenzo, L. Narvarte, and J. Muñoz, "Tracking and back-tracking," *Progress in Photovoltaics: Research and Applications*, vol. 19, no. 6, pp. 747–753, Feb. 2011. [Online]. Available: https://doi.org/10.1002/pip.1085

[2] W. F. Marion and A. P. Dobos, "Rotation angle for the optimum tracking of one-axis trackers," National Renewable Energy Laboratory, Tech. Rep. NREL/TP-6A20-58891, 2013. [Online]. Available: https://doi.org/10.2172/1089596

[3] K. Anderson and M. Mikofski, "Slope-aware backtracking for single-axis trackers," National Renewable Energy Laboratory, Tech. Rep. NREL/TP-5K00-76626, Jul. 2020. [Online]. Available: https://doi.org/10.2172/1660126

[4] A. Andreas and T. Stoffel, "NREL Solar Radiation Research Laboratory (SRRL): Baseline Measurement System (BMS)," National Renewable Energy Laboratory, Tech. Rep. DA-5500-56488, 1981. [Online]. Available: http://dx.doi.org/10.5439/1052221

[5] N. A. Kelly and T. L. Gibson, "Improved photovoltaic energy output for cloudy conditions with a solar tracking system," *Solar Energy*, vol. 83, no. 11, pp. 2092–2102, Nov. 2009. [Online]. Available: https://doi.org/10.1016/j.solener.2009.08.009

[6] ——, "Increasing the solar photovoltaic energy capture on sunny and cloudy days," *Solar Energy*, vol. 85, no. 1, pp. 111–125, Jan. 2011. [Online]. Available: https://doi.org/10.1016/j.solener.2010.10.015

[7] J. Antonanzas, R. Urraca, F. M. de Pison, and F. Antonanzas, "Optimal solar tracking strategy to increase irradiance in the plane of array under cloudy conditions: A study across Europe," *Solar Energy*, vol. 163, pp. 122 – 130, 2018. [Online]. Available: https://doi.org/10.1016/j.solener.2018.01.080

[8] C. D. Rodriguez-Gallegos, O. Gandhi, S. K. Panda, and T. Reindl, "On the PV tracker performance: Tracking the sun versus tracking the best orientation," *IEEE Journal of Photovoltaics*, vol. 10, no. 5, pp. 1474–1480, Sep. 2020. [Online]. Available: https://doi.org/10.1109/jphotov.2020.3006994

[9] G. Quesada, L. Guillon, D. R. Rousse, M. Mehrtash, Y. Dutil, and P.-L. Paradis, "Tracking strategy for photovoltaic solar systems in high latitudes," *Energy Conversion and Management*, vol. 103, pp. 147–156, Oct. 2015. [Online]. Available: https://doi.org/10.1016/j.enconman.2015.06.041

[10] A. Driemel et al., "Baseline surface radiation network (BSRN): structure and data description (1992–2017)," *Earth Syst. Sci. Data*, no. 10, pp. 1491–1501, 2018. [Online]. Available: https://doi.org/10.5194/essd-10-1491-2018

[11] S. A. Pelaez, C. Deline, P. Greenberg, J. S. Stein, and R. K. Kostuk, "Model and validation of single-axis tracking with bifacial PV," *IEEE Journal of Photovoltaics*, vol. 9, no. 3, pp. 715–721, May 2019. [Online]. Available: https://doi.org/10.1109/jphotov.2019.2892872

[12] K. R. McIntosh, M. D. Abbott, and B. A. Sudbury, "The optimal tilt angle of monofacial and bifacial modules on single-axis trackers," *IEEE Journal of Photovoltaics*, vol. 12, no. 1, pp. 397–405, Jan. 2022. [Online]. Available: https://doi.org/10.1109/jphotov.2021.3126115

[13] M. A. Anoma, D. Jacob, B. C. Bourne, J. A. Scholl, D. M. Riley, and C. W. Hansen, "View factor model and validation for bifacial PV and diffuse shade on single-axis trackers," in *2017 IEEE 44th Photovoltaic Specialist Conference (PVSC)*. IEEE, Jun. 2017. [Online]. Available: https://doi.org/10.1109/pvsc.2017.8366704

[14] "pvfactors v1.5.2 (GitHub Repository)," 2021. [Online]. Available: https://github.com/sunpower/pvfactors

[15] R. Perez, P. Ineichen, R. Seals, J. Michalsky, and R. Stewart, "Modeling daylight availability and irradiance components from direct and global irradiance," *Solar Energy*, vol. 44, no. 5, pp. 271–289, 1990. [Online]. Available: https://doi.org/10.1016/0038-092X(90)90055-H

[16] M. Sengupta, Y. Xie, A. Lopez, A. Habte, G. Maclaurin, and J. Shelby, "The national solar radiation data base (NSRDB)," *Renewable and Sustainable Energy Reviews*, vol. 89, pp. 51–60, 2018. [Online]. Available: https://doi.org/10.1016/j.rser.2018.03.003

978-1-7281-6118-1/22 $31.00 © 2022 IEEE

[17] M. Bolinger, J. Seel, C. Warner, and D. Robson, "Utility-scale solar, 2021 edition: Empirical trends in deployment, technology, cost, performance, PPA pricing, and value in the United States," Lawrence Berkely National Laboratory, Tech. Rep., 10 2021. [Online]. Available: https://www.osti.gov/biblio/1823604

Outdoor Characterization of Hybrid HCPV-T Module Featuring a Passive Tracking System

Guido Vallerotto, Steve Askins, Javier van Herpt, David Martí, Jaime Caselles, Ignacio Antón

Instituto de Energía Solar (IES), Madrid, Spain

SolaRays Energy, Pedreguer, Spain

High Concentration PV devices are ideal for adaptation to hybrid production of electricity and thermal energy, but the requirement of a two-axis tracking system is an impediment to use these systems in residential or rooftop applications where Sanitary Hot Water (SHW) has value. The Spanish company SolaRays has developed a low-profile, rooftop compatible system using passive solar tracking, where heat from the sun is used to directly drive tracking motion, offering reliability through simplicity, and allowing for production of both electricity and hot water in a single device at point of use. We present an outdoor characterization of a pilot installation. Results demonstrate that the tracking accuracy achieved is sufficient to ensure losses below 1% in comparison with a conventional electromechanical tracker.

CIGS degradation due to water ingress: post-mortem analysis of a field-exposed PV module

Simona Villa, Remi Anitat, Pelin Yilmaz, Aldo Kingma, Mikolaj Dziechciarz, Joran van den Berg, Klaas Bakker, Mirjam Theelen

TNO partner of Solliance, Eindhoven, Netherlands

University of Twente, Enschede, Netherlands

To bridge the two worlds of the lab and the field, we propose an approach based on coring, unpackaging and post-mortem analysis. We show how in-depth lab-based characterization techniques can be applied on field-deployed PV modules to study the real outdoor degradation mechanisms. In this case, we aim to make a step forward in the understanding of CIGS degradation due to water ingress. A local analysis of both electrical and material properties on a severely degraded commercial module was performed, allowing to trace back the occurring water-induced degradation mechanisms. The degradation mainly affected the TCO layer, through the formation of Zn-based hydroxides, and the Mo back contact layer, which oxidized up to its complete corrosion of the P3 scribes. This resulted in a deterioration of the electrical properties, especially in terms of Rs increase, Voc reduction and Jsc limitation, before ruining the performance entirely at a more advanced stage of degradation.

Evaluation of Cellular Based DER Direct Transfer Trip (DTT) Technologies

Yiwei Ma, Aminul Huque, Joseph Estrada III, Tim Godfrey, Charles Brewster

Electric Power Research Institute, Knoxville, TN, 37932, USA

Abstract—In recent years Direct Transfer Trip (DTT) over commercial cellular communication network is drawing more attention as an alternative to fiber and private copper line due to its lower cost and availability. This paper presents the cellular based DTT topologies, key components, and overall system architectures. In addition, a commercial solution has also been commissioned and evaluated at EPRI laboratory. DTT trip times are measured under various operating conditions. Finally, selected utility pilot projects are listed to provide examples of the technology adoption landscape.

Keywords—Distributed energy resources (DER), Anti-Islanding, Direct transfer trip (DTT), Cellular communication, 4G LTE

I. INTRODUCTION

Island, according to the definition in IEEE 1547-2018 [1], is *"a condition in which a portion of an Area Electric Power system (EPS) is energized solely by one or more Local EPSs through the associated Point-of-Common-Couplings (PCCs) while that portion of the Area EPS is electrically separated from the rest of the Area EPS on all phases to which the distributed energy resources (DER) is connected."*

Consequences of unintentional island may include overvoltage concerns, risk of equipment damage due to out-of-phase reclosing, and safety risks to the line workers and general public. As the penetration of DER increases, the associated risks related with unintentional island pose challenges to the distribution system operators.

To supplement the existing DER inverter's on-board islanding detection capabilities, direct transfer trip (DTT) has been required by many utilities as part of the interconnection requirement, especially for larger scale plants. Its role is to quickly disconnect the DER whenever the feeder protection switch that the DTT is configured to monitor opens. Utilities typically determine the need for DTT based on DER size and/or aggregate DER generation capacity to minimum feeder load ratio. Fig. 1 shows the basic concept of DTT. If any of the substation and/or feeder protection switches open for any reason, a signal will be passed from DTT transmitter to the receiver at the DER site. The DER can be tripped upon the DTT signal, and thus prevent the island to sustain for a longer period.

There are a variety of communication technologies that can be used for transmitting DTT signal. According to a survey conducted in 2019, most of the utilities prefer optical fiber as their DTT communication medium because of its high reliability [2]. However, due to its high cost, especially with remote DERs

that require long distance communication, low-cost alternatives are being sought after.

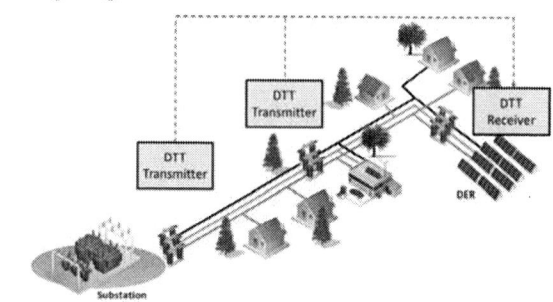

Fig. 1. Concept of direct transfer trip (DTT)

Advances in cellular technology have resulted in improvements in reliability and speed. Long-Term Evolution (LTE) is now a mature 4G technology as the transition to 5G is underway. Commercial cellular networks have increased their capacity which improves speed and availability for all users. Although private LTE [3] is gaining traction, commercial cellular remains a well-established technology for utilities due to its cost-effectiveness and quick deployment. Utilities have been using cellular communication for non-time-critical monitoring and controls, such as SCADA. For DTT, pilot projects are being initiated to integrate cellular communication channels for DER integrations.

This paper first reviews the cellular based DTT technologies, including its key components and system architectures. Then, the performance evaluation results of a commercially available cellular based DTT technology presented. Finally, selected utility pilot projects are discussed to provide examples of the technology adoption landscape

II. REVIEW OF CELLULAR BASED DTT ARCHITECTURE AND KEY COMPONENTS

Several cellular based DTT technologies, developed by utilities internally, or through collaboration with equipment vendors, have been deployed in the field. This section reviews the technology landscape by providing brief descriptions of example cellular based DTT system.

A. Cellular Standards

The first standard defining 4G LTE was completed in 2008. It was the first cellular technology to be based exclusively on Internet Protocol (IP) rather than the voice-oriented circuit switched architecture of previous generations. Within 10 years,

978-1-7281-6118-1/22 $31.00 © 2022 IEEE

LTE was deployed globally. In 2021 GSMA estimates 57% of the 5 billion mobile subscribers are using LTE, with the remainder on 2G, 3G, and 5G. This paper focuses on 4G LTE technologies.

Although 4G LTE is able to support the majority of today's utility use cases, 5G advances cellular standards in three directions: Enhanced Mobile Broadband (eMBB), Massive Machine-Type Communication (mMTC), and Ultra-Reliable Low Latency Communication (URLLC). Since URLLC may offer even lower latency on the order of 1 ms, it has the potential to offer significant value in the DTT use case and will be the subject of future research.

B. Cellular Network Resource Management

Commercial cellular networks implement network management protocols to support large numbers of connected devices. An operator's network has many more devices connected than it could concurrently support if they were all active. Normally, only a percentage of devices are active at the same time. Since an active device consumes limited network resources, the network can serve more customers by putting devices into an idle state when possible. The management of the user devices is controlled by the Radio Resource Control (RRC) layer in LTE. While a device is connected and registered on a network, it will transition between the Connected state and the Idle State. It must be in the connected state to send or receive data. If no data transfer takes place for a period of time, the network will put the device into the Idle state. That time interval is under the control of the cellular network operator.

Transitioning between Idle and Connected states requires RRC signaling and takes time. This increases latency. In the DTT application, the impact of the modem transitioning to Idle state can be seen by an increase in latency when an event occurs after minutes of inactivity. The timeout for the Idle transition varies based on the cellular carrier and the network conditions but is typically in the range of a few minutes. The carrier has to balance the network resource savings from Idle devices with the additional radio overhead for the RRC signaling needed for the transitions.

C. DTT Transmitter and Receiver

A DTT system requires transmitters and receivers to operate. DTT transmitter is installed at the substation or feeder protection switch site, to monitor its status and generate DTT signal. And DTT receiver is installed at the DER site to trip it offline upon receiving the DTT signal.

Fig. 2 shows the diagram of a possible configuration of a DTT transmitter and receiver pair. In this configuration, DTT transmitter and receiver are installed in addition to the existing relay controllers. They both consist of modem(s) to communicate via cellular network, and relay controller that process the DTT logic.

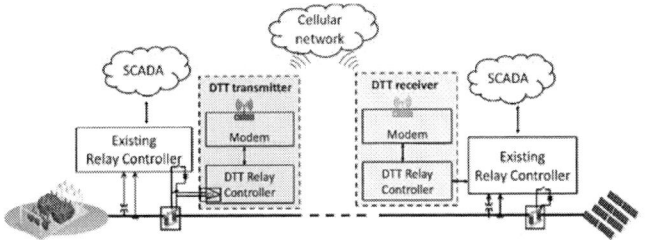

Fig. 2. Possible configuration of a DTT transmitter and receiver pair

The DTT relay controller at the transmitter side directly monitors the control output of the existing relay controller to the feeder protection switch, using its digital input (DI) port. Once the switch opens for any reason, the DTT relay controller can detect the control signal change, generate the DTT signal, and pass it to the modem for further communication. The existing relay controller does not need to change its control logic for the DTT system.

On the DTT receiver side, it requires a cellular modem to receive the DTT signal, and pass to the relay controller. The relay controller can thus generate an open command to the existing relay controller of the DER interconnection switch, using digital input/output ports or other communication protocols. Depending on utility's requirement, other control logics may exist, such as loss-of-communication trip, or direct transfer close (DTC). In this configuration, the existing relay controller at the DER site needs to change its control logic to accept input from the DTT receiver.

There can be other possible configurations for individual DTT controller, such as including the DTT logic inside of the existing relay controller without a dedicated one. These configurations can be similar to the ones used in other communication technologies, such as fiber optics or radio, by replacing the cellular modem to the associated communication interface device.

D. Cellular Network Connection Scheme

The individual DTT transmitters and receivers communicate with each other through cellular network. The simplest connection scheme is through one single cellular network. In this case, all communications are handled by the single cellular network and do not have redundancy. Communication outage would be declared if there is any failure of the cellular system.

To provide backup, the DTT system can utilize double cellular network switch-over technology using dual sim modems. If one cellular network has connection issue, the modem can automatically switch-over to the second network, as shown in Fig.3. This scheme only allows one cellular network connection at any moment.

978-1-7281-6118-1/22 $31.00 © 2022 IEEE

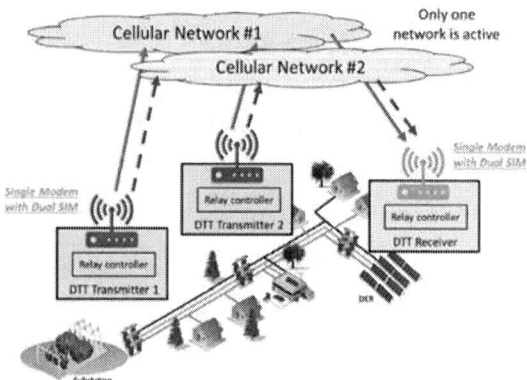

Fig. 3. Double cellular network switch-over connection scheme

Another cellular network connection scheme is the first-to-arrive logic, having two cellular networks active simultaneously. It can be achieved by having two cellular modems connect to the DTT relay controller, as shown in Fig. 4, or alternatively utilize a modem that supports connection to both cellular networks simultaneously.

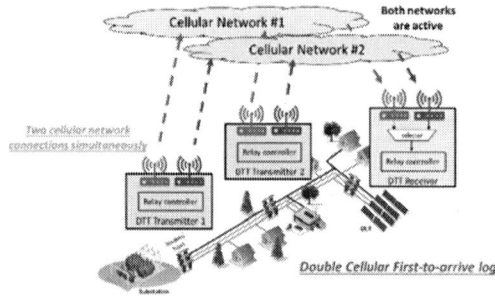

Fig. 4. Double cellular network first-to-arrive connection scheme (logic implemented in relay controller)

E. Communication Protocols

There can be various communication protocols used for cellular based DTT technology. For faster communication, most of the technologies under review utilize IEC 61850 GOOSE (Generic Object Oriented Substation Events) as the DTT communication protocol.

The time sequence of GOOSE message transmission is shown in Fig. 5. The GOOSE messages do not have any handshake mechanism, in order to improve communication speed. If there is no event, the transmitter will broadcast a heartbeat signal continuously with a long cycle time in a steady state to maintain the communication link. But as soon as an event happens, the signal will be sent out immediately without needing to wait until the next data transmission cycle. This signal would be sent multiple times in a short time, and then gradually slow down.

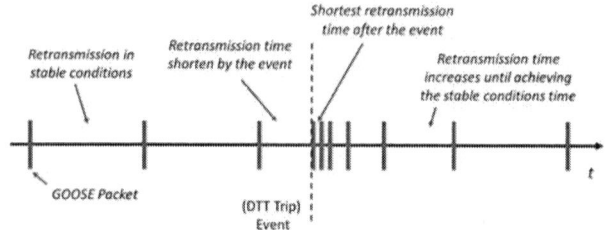

Fig. 5. Time sequence of IEC 61850 GOOSE message transmission

Each GOOSE message in the retransmission sequence carries a time allowed to live (TAL) data that informs the receiver of the maximum time to wait for the next retransmission. The TAL data also defines the maximum time that the GOOSE message remains alive after transmission. On DTT receiver side, loss-of-communication can be declared if no more message is received within the defined TAL interval. And on DTT transmitter side, loss-of-communication can be declared if GOOSE message is still alive after the TAL interval. The detailed implementation of TAL parameter can be different by different vendors. But it is generally multiples of the GOOSE message retransmission time.

GOOSE is a layer 2 (data link) protocol that does not carry IP address information. In contrast, LTE (and 5G) are IP-only networks. They require all transported data to use the IP protocol (IPv4 or IPv6). Therefore, it becomes necessary to encapsulate the GOOSE message packets into IP packets so they can be routed through the cellular network. A Layer 2 Tunneling Protocol (L2TP) is a common approach for encapsulating a GOOSE message to send it through a cellular modem.

Other protocols may also be implemented in the cellular modem used for the DTT use case, such as IPSec, Multi Protocol Label Switching (MPLS), SEL Mirrored Bits protocol, etc.

F. Cellular Based DTT System Architecture

Combining the components described in previous sections, a cellular based DTT system can be established. If there are multiple DTT transmitters and receivers, the control architecture can be separated as distributed control and centralized control, depending on where the DTT control logics reside.

For distributed DTT control architecture, DTT channels need to be established between each DTT transmitter to its downstream DTT receiver. Each transmitter or receiver station needs to have its own control logic developed, such that the DTT receiver can be aware of the status of feeder protection switches. This design has the advantage that true peer-to-peer communication links can be established between all DTT devices, and hence provides the lowest latency for cellular DTT applications.

On the other hand, for centralized DTT control architecture, most of the control logics are within the substation or a central location. The feeder protection switch only needs to report its status to the substation. And the central controller can issue the trip commands according to the system topology information, and trip the downstream DER as needed. From control complexity point of view, centralized control has all its control logic hosted at one location, which may be easier to be maintained.

III. LAB EVALUATION OF CELLULAR BASED DTT TECHNOLOGY

A commercially available cellular based DTT solution has been commissioned and being tested at EPRI laboratory in Knoxville, TN. The technology utilizes a double cellular first-to-arrive logic, with two modems connected to one relay controller for both DTT transmitter and DTT receiver.

The DTT system has a distributed control architecture, where there is a communication channel between each protection switch and the DER that requires DTT control. The system components allow the integration of multiple DTT transmitters to a single DTT receiver, as well as a single DTT transmitter to multiple DTT receivers. Since there is not a concern for system scalability with this approach, EPRI focuses to test the performance with a single DTT transmitter and receiver pair.

The provided solution can produce loss-of-communication warnings at both DTT transmitter and receiver sides for further reactions by system operator. The detection is based on TAL feature of the GOOSE messages. An optional feature also allows to trip the DER under loss-of-communication condition, with a configurable waiting period. The loss-of-communication trip feature was also enabled and tested.

The solution also contains another optional feature called "direct transfer close (DTC)." It allows the DER to automatically reconnect to the grid without human intervention, if the system returns to a normal healthy state. The DTC triggering event includes communication restored, upstream breaker closed, hotline tag removed, or any other events and their combinations. The DTC logic was enabled during the commissioning process at the EPRI lab for the ease of repetitive testing. It is configured such that the DTT receiver closes Contactor 2, when DTT transmitter detects the event when Contactor 1 closes.

The lab test set-up is shown in Fig. 3. Contactor 1 represents the upstream feeder protection switch, and Contactor 2 represents the DER interconnection switch. Data recorder measures the voltage and currents in the test system, as well as the control input of the contactors. The measured channels are indicated as circled numbers.

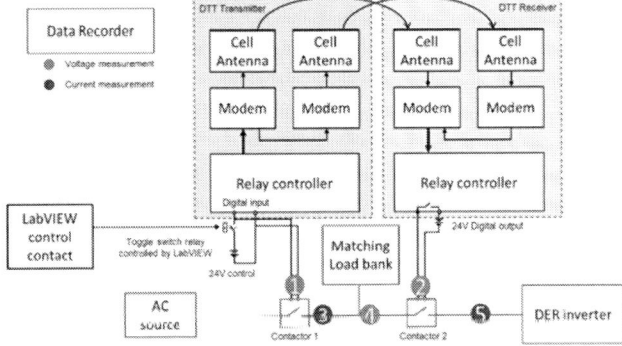

Fig. 6. Lab test set-up to evaluate cellular based DTT performance

Fig. 7 shows an example test waveform captured by the data recorder. Before the islanding event, generation and load match is achieved by adjusting DER inverter output power. The current on Contactor 1, measured by data recorder channel ③, is close to 0. The DER inverter forms the island and maintains the voltage after contactor 1 is opened. After 0.0863s, contactor 2 is opened by DTT control, currents on contactor 2 stops, and voltages at island reduce to 0. It is demonstrated that the DTT control can successfully prevent island formation by tripping the DER interconnection switch in the emulated circuit.

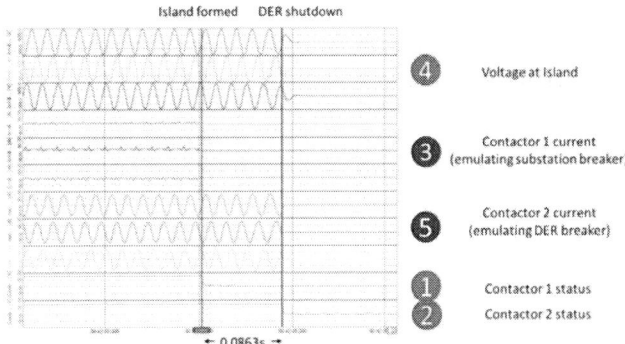

Fig. 7. Islanding test waveform: DTT trip time 0.0863s

Fig. 8 gives another example when the DTT trip time is 1.52s. Before the DTT signal arrives, the DER inverter detects the island, cease to energize, and trip itself in around 700ms, less than the requirements in the IEEE 1547-2018. This result shows the DER inverter's capability to detect the island and shut down by itself without the help of DTT.

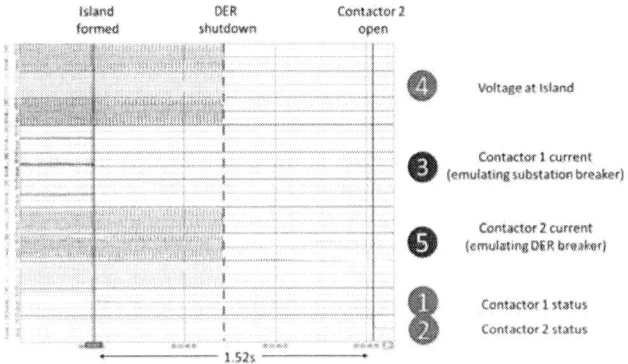

Fig. 8. Islanding test waveform: DTT trip time 1.52s

To obtain statistics on DTT trip time and evaluate its performance, multiple tests have to be performed. Fig. 9 shows the control signals generated by a LabVIEW controlled contact. Contactor 1, which represents the upstream feeder protection switch, is controlled to open for 10 seconds, with a randomized interval of around 1 hour. Since the tests only require the DTT trip time data, only contactors are kept in the test circuit. The rest of the ac circuit components are not included.

978-1-7281-6118-1/22 $31.00 © 2022 IEEE

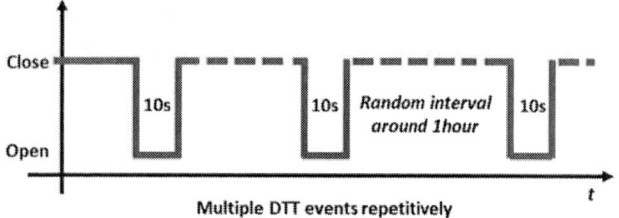

Fig. 9. Control signal for Contactor 1 to trigger DTT tests

In most of the DTT tests, Contactor 2 is opened shortly after Contactor 1 opens. The time delay is measured as DTT trip time. When Contactor 1 is back to the closed condition, Contactor 2 is also back to closed, through the DTC logic configured during the commissioning process. This is noted as Success in the following result analyses. However, in very rare conditions, especially when the cellular signal is intentionally attenuated, there are a few failure modes observed. An illustration of DTT trip time test normal operation and its failure modes are shown in Fig. 10.

The first failure mode is <u>long trip time</u>. It is identified the DTT trip time is longer than 2 seconds, which is the default DER trip time requirement by IEEE 1547-2018. The second one is <u>loss-of-communication trip</u>. This optional function is enabled to trip the DER when the DTT receiver does not receive the GOOSE heartbeat signal for a configurable time. In the test, the intentional delay is set to 1 minute. The third failure mode is <u>DTT fail</u>. This is when the DER does not trip when the feeder breaker opens. It is extremely rare that only one instance was observed out of more than 1000 DTT tests performed. And the instance is when there is only one cellular carrier connected, and its signal was intentionally attenuated. And the last failure mode is <u>DTC fail</u>. This is when the DER does not reconnect when the feeder breaker closes after a DTT test. Since the DER remains disconnected, the following DTT test is not counted. Note that there can be different causes of DTC fail. One possible cause is that the cellular communication is lost at the moment of DTC signal.

Fig. 11 compares the interquartile range (IQR) of the measured DTT trip times, as well as the numbers of DTT success operations and failures. The boxes indicate the 75th, 50th (median), and 25th percentiles of the data. The whiskers indicate

the maximum and minimum values within the range of 1.5×IQR below 25th and above 75th percentiles of the data. The orange line indicates the mean value, and the dots are outliers. The names of cellular network carriers are concealed because the purpose is not to compare the performance of different carriers.

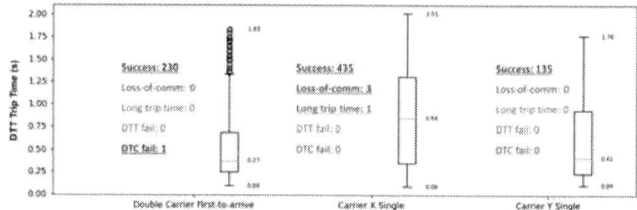

Fig. 11. Interquartile range of measured DTT trip times using default and reduced GOOSE message retransmission time.

By comparing the statistics of measured DTT trip time, the complete solution with double carrier first-to-arrive logic has less DTT trip time. It is expected because the trip action is based on the fastest signal. In addition, the complete solution has a better success rate than carrier X alone. Since the DTT signals are passed through the two carriers simultaneously, the DTT receiver can operate if one trip signal is successfully transmitted. However, it is unknown why DTC has failed once with double carrier first-to arrive logic. The event log does not indicate any loss-of-communication detected for this DTC fail event.

To investigate how are the DTT performance metrics impacted by cellular signal attenuation and GOOSE communication retransmission time settings, tests are conducted with only single cellular network active. Fig. 12 shows the IQR results comparing normal operation and the case when cellular signal is attenuated by 30dB. Table I shows the cellular signal strength metrics.

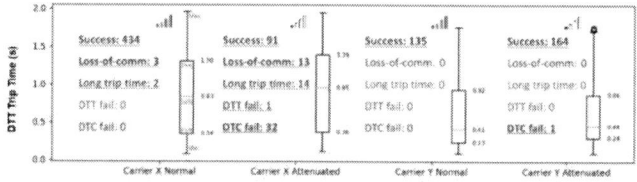

Fig. 12. Interquartile range of measured DTT trip times with normal and attenuated cellular signal.

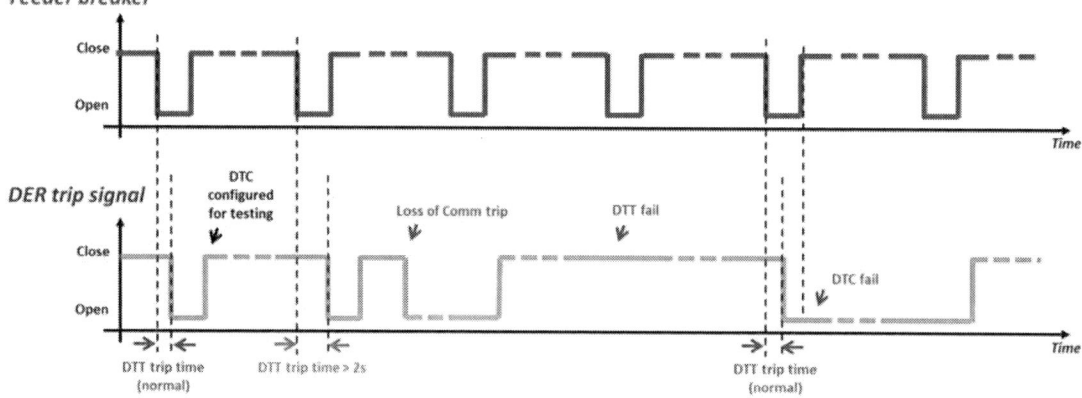

Fig. 10. Illustration of DTT test success operations and failure modes

TABLE I. CELLULAR SIGNAL STRENGTH MEASUREMENTS FOR NORMAL AND ATTENUATED CONDITIONS.

	Carrier X		Carrier Y	
	Normal	Attenuated	Normal	Attenuated
Received Signal Strength Indicator (RSSI)	~-60dB	~-90dB	~-60dB	~-90dB
Signal to Interference & Noise Ratio (SINR)	9dB	9dB	9dB	9dB
Reference Signal Received Quality (RSRQ)	~-12dB	~-16dB	~-8dB	~-13dB
Reference Signal Received Power (RSRP)	~-85dB	~-115dB	~-90dB	~-110dB

It can be seen that the DTT trip time is not significantly impacted by cellular signal attenuation. But the failure rate may increase. It is worth noting that the 32 instances of DTC fail for Carrier X attenuated case were consecutive, and all following one loss-of-communication trip event. This may indicate prolonged cellular communication loss that lasts for more than a day when the cellular signal is low.

On the other hand, the GOOSE message retransmission rate may impact the DTT trip time significantly. The default configuration has the shortest retransmission time after an event is 500ms for both cellular networks, and the retransmission time in stable condition is 38s and 39s. Two sets of reduced GOOSE retransmission time settings are tested. Test configuration #1 has the shortest retransmission time of 500ms and stable retransmission time of 1s. And test configuration #2 has the shortest retransmission time of 100ms and stable retransmission time of 1s. Fig. 13 shows the IQR plot with the comparison of default settings. It is observed that the DTT trip time has significantly reduced.

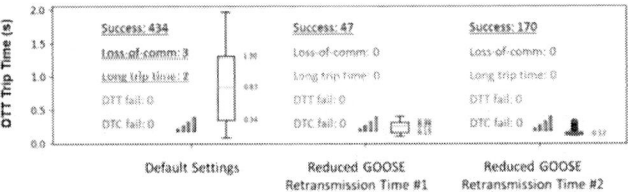

Fig. 13. Interquartile range of measured DTT trip times using default and reduced GOOSE message retransmission time.

However, this may also increase the total data transmitted on the cellular DTT system. Assuming one GOOSE message size of 250 bytes, the estimated data per month would be 33MB per DTT transmitter and receiver pair for vendor's default retransmission time. If using the reduced retransmission time, the estimated data per month can grow to 1.3GB. This may increase the monthly cost incurred.

The reason for this behavior may be related to the cellular network Idle mode discussed in Section II. B. Since it is related to the cellular network settings, the technology provider and cellular network provider may be able to establish an appropriate

set of settings to keep the communication channel active, while keeping the network traffic low. The challenge is sending data often enough to avoid idle mode, without incurring excessive data charges from the cellular operator. This is not an issue with private LTE networks, which have no data charges, but more importantly, the idle timeout can be set to whatever value is appropriate for the use case.

The same GOOSE message retransmission time settings are tested with attenuated cellular signal. The results are shown in Fig. 14. Compared with the results shown in Fig. 13, where the cellular signal is normal, it can be seen that the DTT trip time is not significantly impacted by the cellular signal strength. On the other hand, the failure rate is decreased with shorter GOOSE retransmission time.

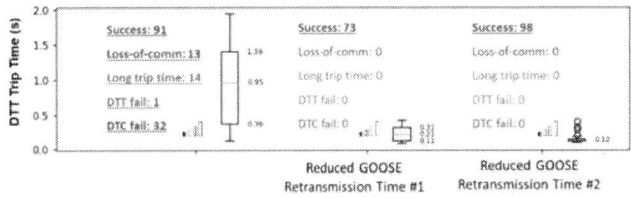

Fig. 14. Interquartile range of measured DTT trip times using default and reduced GOOSE message retransmission time, with attenuated cellular signal.

IV. UTILITY EXPERIENCE WITH CELLULAR BASED DTT

Cellular based DTT technologies are getting increasing attention by the utilities due to its lower cost and availability. A few utilities have developed internally, collaborated with equipment vendor, or acquired commercial solutions for their cellular based DTT pilot projects. Table I provides a summary list of example utility projects, with their respective network connection scheme, system architecture, models of modems, relay controllers, and cellular networks.

V. CONCLUSIONS

This paper reviews the key components and system architecture of cellular based DTT technologies. A sample commercial solution is being tested at EPRI lab. Test results have demonstrated that:

- Cellular based DTT can reliably monitor the status of feeder protection switches and trip the downstream DER.

- The median DTT trip time measured is less than 0.5s, and maximum DTT trip time is less than 2s, using the default configurations of double cellular carrier first-to-arrive logic.

- By increasing the data traffic on the cellular communication channel, DTT trip time can be reduced to around 100ms. This can be achieved by reducing the GOOSE message retransmission time.

- As cellular signal attenuation increases, the DTT trip time is not significantly impacted, but the success rate decreases.

REFERENCES

[1] IEEE 1547-2018, "IEEE Standard for Interconnection and Interoperability of Distributed Energy Resources with Associated Electric Power Systems Interfaces," IEEE standard, 2018.

978-1-7281-6118-1/22 $31.00 © 2022 IEEE

[2] *Utility Direct-Transfer-Trip Survey Results*, EPRI, Palo Alto, CA: 2019. 3002016638.

[3] *Private LTE Guidebook 2021*, EPRI, Palo Alto, CA: 2019. 3002020373.

[4] A Guide for Utilities Considering LTE for Distributed Energy Grid [Online]

[5] Mitigating the Impact of Unintentional Islanding on Electric Utility Transmission Systems from Distributed Energy Resources [Online]

[6] T. Fix, A. Smit, S. Chanda, J. Key, "New Intelligent Direct Transfer Trip Over Cellular Communication." In Proc. IEEE *Annual Conference for Protective Relay Engineers*, Mar. 2019.

[7] Intelligent Direct Transfer Trip (DTT) Leveraging Cellular Communications [Online]

TABLE II.   EXAMPLE UTILITY PROJECTS OF CELLULAR BASED DTT

Utility	Consumers Energy	Eversource [4]	Central Hudson [5]	Central Virginia Electric Cooperative (CVEC) [6]	Dominion #1 [6]	Dominion #2 [6]	Madison Gas and Electric (MGE) [7]
Cellular network connection scheme	2 Cellular carriers (first to arrive)	2 Cellular carriers (switch-over)	1 Cellular carrier	1 Cellular carrier	1 Cellular carrier + backup protection	1 Cellular carrier + fiber + backup protection	2 Cellular carriers (first to arrive)
DTT system architecture	Distributed	Centralized	Centralized	Distributed	Distributed	Distributed	Distributed
Model of Cellular Modem	Peplink Pepwave MAX HD2 Mini	Sierra Wireless AirLink MG90	Sensus RTM II	Siemens RX1400	Siemens RX1400	Siemens RX1400	Siemens RX1400
Model of DTT Relay Controller	SEL-3505-3 RTAC	SEL-3505-3 RTAC	Sensus T866 MicroRTU	Siemens 7SC80	Siemens 7SC80	Siemens 7SC80	Siemens 7SC80
Cellular Network(s)	Verizon, AT&T	Verizon, AT&T	AT&T	Unknown	Unknown	Unknown	US cellular, AT&T

Towards a Shade Tolerant Monolithically Interconnected Perovskite Module for use in Four Terminal Tandem Devices

Klaas Bakker[1], Jacopo Sala[2], Mehrdad.Najafi[1], Michaël Daenen[2] and Bart Geerligs[3]

[1]TNO partner in Solliance, High Tech Campus 21, 5656 AE Eindhoven, The Netherlands
[2]Universiteit Hasselt, Instituut voor Materiaalonderzoek, Wetenschapspark 1, B-3590 Diepenbeek, Belgium
[3]TNO Energy Transition - Solar Energy, Westerduinweg 3, 1755 LE, Petten, The Netherlands

Abstract—During partial shading of a monolithically interconnected cell string, the shaded cells can operate at a reversed voltage. This negative voltage can lead to degradation of the shaded cell. For a successful practical application of large area Perovskite modules (e.g. in four terminal tandems) a shade tolerant top module is required. In this study electrical simulations on various shading and interconnection scenarios have been performed. The simulations showed that the magnitude of reverse bias can be kept low when strings are connected in parallel, and at least one of the strings is not shaded.

Keywords—Perovskite, partial shading, electrical model, reverse bias, tandem, reliability

I. INTRODUCTION

Partial shading presents a problem for most monolithic interconnected thin film modules [1],[2]. The problem originates in the presence of a negative voltage (reverse bias) in the shaded region. Thin film solar cells generally can not tolerate reverse bias voltages without suffering permanent damage. A major disadvantage of the monolithic interconnection design is that it is almost impossible to incorporate protection against reverse bias by using bypass diodes.

Recently Perovskite – Silicon tandems have gained a lot of attention as a technology to increase the overall efficiency of photovoltaic (PV) modules. One possible way to fabricate a tandem PV module is to have a separate c-Si bottom submodule and Perovskite top submodule with each submodule having their own independent electrical terminals. In this four terminal (4T) concept the top perovskite part can be manufactured on the glass superstrate independent from the bottom c-Si part. However, in order to successfully introduce 4T tandems on an industrial scale a shade tolerant Perovskite top submodule is needed.

The most obvious strategy to mitigate reverse bias is to place strings of solar cells in parallel [1]. This is compatible with a 4T tandem module layout, as the space in between the c-Si cells would allow for current collecting tabs for the Perovskite strings that do not shade the bottom cells. In this study an LT Spice based electrical simulation model [3] is used to evaluate several interconnection and shading scenarios. The cell parameters used in this model, for both forward and reversed bias, are based on real cell measurements.

This work was partially supported by TKI Energy, through "Toeslag for TKIs", by the Dutch Ministry of Economic Affairs and Climate, project 18211.04

II. RESULTS AND DISCUSSION

A. Reverse bias parameters

The standard diode model in LT Spice offers the possibility to set reverse bias parameters in the form of two diodes with high ideality factor. The first diode sets the main reverse characteristic which is given by a point on the *JV* curve (current breakdown (*IBV*), voltage breakdown (*BV*)) and an ideality factor (*NBV*). The second diode sets the low current behavior and is not used in this study. In Fig. 1 the results of a reverse *JV* measurement of a 4×4 mm^2 cell is compared with a simulation with *IBV* = 1.5 mA/cm^2, *BV* = 4 V ,and *NBV* = 16. It can be observed that for higher currents the simulation matches the measurement nicely. At lower currents there is a mismatch, as can be seen from the inset. This mismatch is caused by the large increase in shunt resistance in the dark. For the simulation the shunt resistance value under illumination is used, as the current software version only changes the J_{sc} when a pixel is shaded. With the reverse bias parameters used the illuminated forward curve is marginally influenced, the shunt resistance obtained from the slope changes from 3426 to 3418 Ωcm^2.

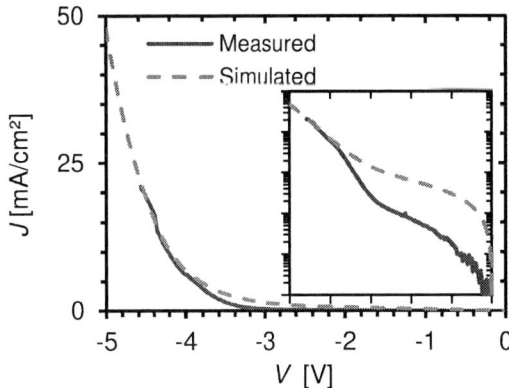

Fig. 1. Reverse *JV* curves of a measurement and a simulation of a cell of 4×4 mm^2 in the dark. The inset shows the same data on a semilog scale. The blue solid line represents the measured data and the red dashed line shows the simulation results.

During operation of a single shaded cell in a string of illuminated cells, typically current will be drawn from this string in order to get power out, and this current has to pass through the shaded cell. In order to pass the string current the shaded cell is forced to operate at a reversed (negative) voltage. For the cell used in this study the maximum power point current density at 1 sun illumination is 19.6 mA/cm^2. If this current is applied in the dark the cell will operate at a voltage of -4.52 V. This is the most negative voltage to be expected during partial shading conditions.

B. Increasingly shading one single cell in a string

In order to evaluate the impact of partial shading on a module, a model of a module consisting of a string of 10 series-connected 4×200 mm^2 cells was created. A series of simulations were performed on this module where the first cell in the string was progressively shaded from 0 to 100% with a 100% opaque shade. Fig. 2 shows the IV and PV results of a selection of these simulations. It can be seen that for higher shading fractions the maximum power point is shifting to lower voltages. The maximum power point has a large effect on the bias voltage in the shaded cells, this relation is shown in Fig. 3.

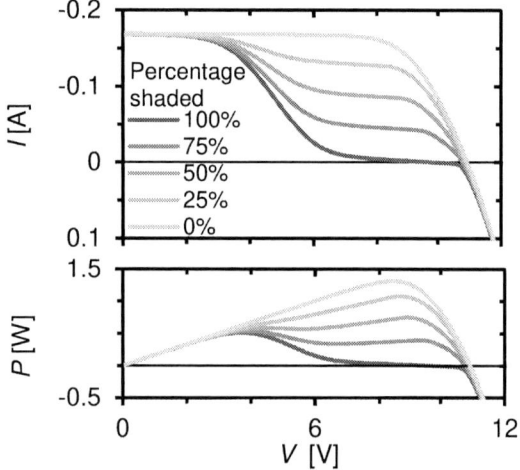

Fig. 2. IV and PV graph of simulation results of a string of 10 cells of 4 × 200 mm2 of which one cell was progressively shaded. The shades of orange represent the percentage of the cell being shaded, with the darkest color being the most shaded.

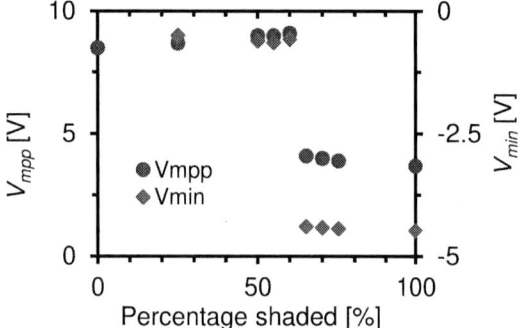

Fig. 3. Maximum power point voltage of the string (V_{mpp}) and minimum recorded bias voltage on the partially shaded cell (V_{min}) observed in simulations with 1 cell in a string being progressively shaded. On the x-axes the shading fraction is plotted. V_{mpp} is plotted in blue circles on the left axis, and V_{min} is represented in red diamonds on the right axis.

From Fig. 3 it is clear that there is a transition from a high to a low maximum power point between 60% and 65% shadow coverage. The influence of the reduced operating point on the local voltage is plotted in Fig. 4. For the high V_{mpp} (60% shaded) simulation only the shaded part is operating at a mild negative voltage, the unshaded part of the partly shaded cell is still in forward bias and thus contributing to the cell power output. The local voltage of the partly shaded cell in the module operating at a low voltage (65% shaded) is negative everywhere over the cell area. In that case, the whole of this cell, the illuminated part included, is dissipating energy.

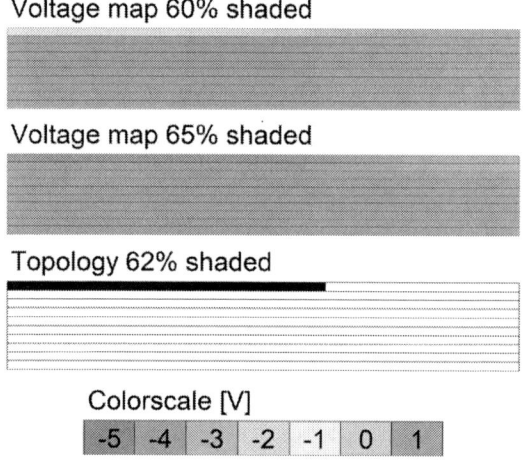

Fig. 4. Voltage maps at maximum power point of a string of 10 cells with the top cell partial shaded. The top graph shows the voltage distribution at a V_{mpp} of 9.1 V, with the top cell shaded for 60%. The graph 2nd from the top displays simulation results at V_{mpp} of 4.1 V for a string with one cell shaded for 65%. An example of the topology of the string used is drawn in the picture 3rd from the top. The black area shows the location of the shade. Both voltage maps are plotted with the same colorscale, which is given at the bottom.

C. Strings in parallel

From the previous results it is clear that the operating point of the string should be kept close to that of an unshaded string during partial shading conditions. Putting strings in parallel would potentially, as long as one of the strings is not shaded, keep the maximum power point of the array close to an unshaded string. Several simulations using multiple strings of 10 in series connected 4×200 mm^2 cells in parallel have been performed.

These simulations showed that the shaded cells can dissipate energy both at a negative voltage or a positive current. This is possible according to the relation $P = V \times I$. During normal solar cell operation the cell is generating energy (P is negative) at a positive voltage and negative current. In order for the cell to dissipate energy P has to become positive by changing the polarity of either voltage or current.

Fig. 5 shows the simulation results of three strings in parallel with one string completely shaded, one string with one cell partially shaded and one string fully illuminated. It clearly shows the features of the parallel connection. The fully shaded cells operate at a positive current and voltage. In the partly shaded string, the magnitude of the reverse bias is kept mild (-1.3V). In the power map it can be seen that the fully shaded string is dissipating energy and that in the partly shaded string most of

978-1-7281-6118-1/22 $31.00 © 2022 IEEE

the energy is dissipated in the non-shaded part of the shaded cell. This is because the total current density in this area is much higher due to the light-generated current.

D. Discussion

Monolithically interconnected strings can not easily be protect against large reverse bias voltages. A bypass diode would for the string in Fig. 4 not change the operating voltage as it operates at +4.1V. Mitigating against reverse bias can be done by controlling the operating voltage of the string. Either electronically, or as showed by the simulations, by placing strings in parallel.

A recent study [4] on the stability of perovskite solar cells at reverse bias identified three degradation mechanisms.

1. Permanent changes due to formation of shunts at the metal contacts.
2. Reversible changes, iodide is driven into the electron transport layer.
3. Permanent changes due to the formation of iodide- and bromide-rich sublayers.

Initial experiments on small cells with different layer stacks showed that the reverse curve depends on the layer stack used. The transition to the high current regime can be shifted to much more negative values than shown in the example used in this study. In our experiments we found that permanent damage occurs mostly at higher currents. As long as the cell operates at low currents (< 1 mA/cm^2) it could withstand a high reverse voltage of -6V for 30 minutes. Only small changes in *IV* parameters, mostly V_{oc}, where observed. These changes could partly be recovered by lightsoaking at *mpp*. We therefore believe that the mild reverse bias voltages as seen in the simulations with parallel strings will not be harmful.

For the simulations a limited number of cells in series was used to keep the computational time acceptable. Longer strings will probably be used in actual modules. However, we think that the simulations in this study give a good impression of the electrical conditions during partial shading events. As already identified for other thin film monolithic technologies [5], the border between light and dark in a cell is the worst case scenario with respect to energy dissipation, due to lateral current flows. A, possibly overlooked, side effect of parallel strings is the fact that partly shaded strings can operate at a reversed (positive) current. It would be easy to protect against this with a diode in series with the string, blocking the positive current (blocking diode). However this blocking diode will always carry the string current, removing output power from the string

More research is require into the long term effects of reverse bias and energy dissipation in perovskites. Not only in the dark and at reverse bias voltages but also under illumination and reverse currents.

III. SUMMARY

In this study an LT Spice based model to simulate partial shading in monolithically interconnected Perovskite modules was developed. The electrical cell parameters used were obtained from cell measurements. The reverse bias characteristic of the standard diode with high ideality factor is in good agreement with the measurement. Simulations on a string in which one cell is progressively shaded showed that the magnitude of reverse bias can be kept mild as long as the operating point of the partially shaded string is kept close to the maximum power point of the unshaded string. Further simulations proved that connecting strings in parallel will achieve this, as long as not all the strings are partially shaded. However, in a parallel configuration both shaded and non-shaded cells can dissipate power in a reverse current condition. The layout of the bottom module in a c-Si – Perovskite tandem allows some space to integrate current collecting structures (tabs) that can be used to connect strings of monolithically interconnected Perovskite cells in parallel. Supporting experiment data has shown that a module consisting of parallel connected strings of monolithically interconnected Perovskite cells can be shade tolerant.

REFERENCES

[1] T. J. Silverman, L. Mansfield, I. Repins, and S. Kurtz, "Damage in Monolithic Thin-Film Photovoltaic Modules Due to Partial Shade," *IEEE J. Photovoltaics*, vol. 6, no. 5, pp. 1333–1338, Sep. 2016

[2] K. Bakker, A. Weeber, and M. Theelen, "Reliability implications of partial shading on CIGS photovoltaic devices: A literature review," *J. Mater. Res.*, vol. 34, no. 24, pp. 3977–3987, Dec. 2019

[3] M. D. J. Carolus, Z. Purohit, T. Vandenbergh, M. Meuris, B. Tripathi, "Proposing an Electro-Thermal SPICE Model to Investigate the Effect of Partial Shading on CIGS PV Modules," in *35th European Photovoltaic Solar Energy Conference and Exhibition*, 2018, pp. 1343–1345

[4] R. A. Z. Razera *et al.*, "Instability of p–i–n perovskite solar cells under reverse bias," *J. Mater. Chem. A*, vol. 8, no. 1, pp. 242 250, 2020

[5] T. J. Silverman *et al.*, "Thermal and Electrical Effects of Partial Shade in Monolithic Thin-Film Photovoltaic Modules," *IEEE J. Photovoltaics*, vol. 5, no. 6, pp. 1742–1747, Nov. 2015

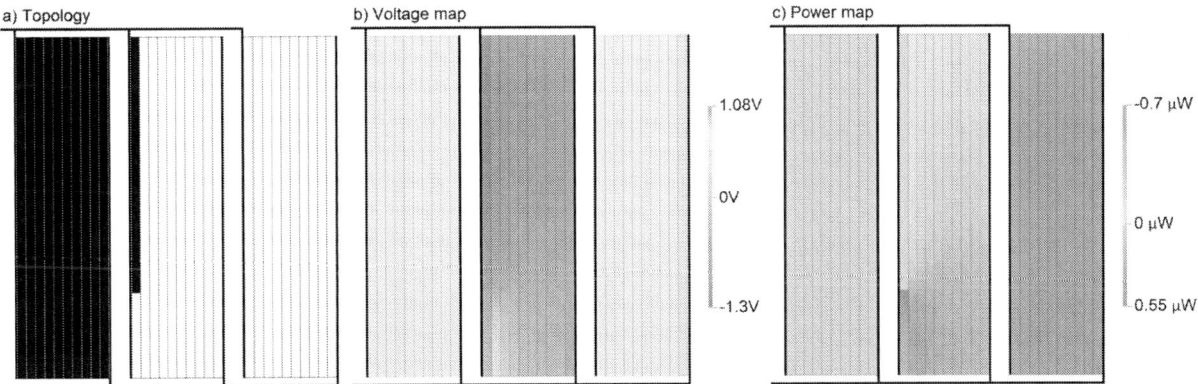

Fig. 5. Simulation results of 3 strings of 10 cells in parallel. The layout and shade topology is given in a), shades are indicated by the black areas. From the first string all cells are fully shaded, from the second string the first cell is shaded for 75% of its area, and the last string is not shaded. The local voltage map is given in b) and local power map is given in c). In the power map a negative power means the area is generating power, and a positive power means that the area is dissipating power.

978-1-7281-6118-1/22 $31.00 © 2022 IEEE

Panel Segmentation: A Python Package for Automated Solar Array Metadata Extraction using Satellite Imagery

Kirsten Perry, Christopher Campos

National Renewable Energy Laboratory (NREL), Lakewood, CO, United States

The NREL Python Panel-Segmentation package is a toolkit that automates the process of extracting accurate and valuable metadata related to solar array installations, using publicly available Google Maps satellite imagery. Our previously published work includes automated azimuth estimation for individual solar installations in satellite images. Our continued research focuses on automated detection and classification of solar installation mounting configuration (tracking, fixed-tilt) and type (rooftop, ground, carport). Specifically, a Faster-RCNN Resnet-50 feature pyramid network (FPN) model was trained and validated on over 770 manually labeled satellite images. This model was used to perform object detection on satellite imagery, locating and classifying individual solar installations' mounting configuration and type. Preliminary model results showed a combined mean average precision score (mAP) score across classes of 49.87% using an Intersection over Union (IoU) threshold of 0.5. We intend to release the complete image data set with labels on the NREL DuraMAT DataHub, to encourage further research in this area. Additionally, a pipeline for automated metadata extraction, including detection of mounting configuration and type as well as azimuth estimation, will be released via the NREL Panel-Segmentation package for public use.

Optical Determination of Carrier Concentrations in ITO, PEDOT:PSS, and (FASnI3)0.6(MAPbI3)0.4 within a PV Device

Madan K Mainali, Prakash Uprety , Zhaoning Song, Changlei Wang, Indra Subedi, Kiran Ghimire, Maxwell M Junda, Yanfa Yan, Nikolas J Podraza

The University of Toledo, Toledo, OH, United States

A low bandgap mixed tin-lead (Sn-Pb) halide perovskite based thin film solar cell is a multilayer stack consisting of indium tin oxide (ITO) as the transparent front electric contact, poly (3,4-ethylenedioxythiophene) polystyrene sulfonate (PEDOT:PSS) as the hole transport layer, low bandgap perovskite (FASnI3)0.6(MAPbI3)0.4 as the absorber layer, fullerene (C60) as the electron transport layer, bathocuproine (BCP) as the hole blocking layer, and Ag as the back contact. Determination of carrier concentration (N) of these individual layers in the solar cell stack structure using direct electrical measurements is not accessible. Here, N for some component layers in a low bandgap organic inorganic (FASnI3)0.6(MAPbI3)0.4 absorber based perovskite solar cell are determined using non-contacting magnetic field dependent terahertz (THz) range spectroscopic ellipsometry for optical Hall effect measurement to ascertain the free carrier optical absorption and corresponding N using the Drude model. From THz spectral range optical Hall effect measurements and analysis, N of the ITO, PEDOT:PSS, and low bandgap (FASnI3)0.6(MAPbI3)0.4 are determined to be $(2.8 \pm 0.6) \times 10^{20}$ cm-3, $(2.6 \pm 0.4) \times 10^{22}$ cm-3, and $(1.5 \pm 0.1) \times 10^{18}$ cm-3, respectively. All values are within expectations for this device design. These results demonstrate the capability of THz range optical Hall effect measurements to determine transport properties of layers within complete thin film polycrystalline solar cells, and the properties determined can be used as input for photovoltaic device modeling to understand device physics and optimize performance.

978-1-7281-6118-1/22 $31.00 © 2022 IEEE

Strategies to Optimize and Validate Tracking Performance of Single-Axis Trackers on Diffuse Sites

Kendra Passow, Kyumin Lee, Sanket Shah, Daniel Fusaro, Jon Sharp

Array Technologies Inc, Albuquerque, NM, United States

Tracking performance can be optimized under diffuse conditions by traveling to a flatter angle to maximize energy production of a PV power plant. Experimental results from Array Technologies' SmarTrack--TM Diffuse algorithm are presented here from a site in New York. An alternating validation strategy is described to quantify the diffuse gain over a period of roughly two months (May and June 2021). Additionally, this work includes the details of a prediction methodology used to validate the measured gains as well as estimate annual gains.

Development of Hierarchical Control for a Lunar Habitat DC Microgrid Model Using Power Hardware-in-the-Loop

Andrew R. R. Dow[1], Rachid Darbali-Zamora[1], Jack D. Flicker[1], Felipe Palacios II[1], and Jeffrey T. Csank[2]

[1]Sandia National Laboratories, Albuquerque, New Mexico, 87185, USA
[2]NASA Glenn Research Center, Cleveland, Ohio, 44135, USA

Abstract – **As interest in space exploration grows, developing a lunar habitat has become a key component of extending missions into deep space. To guarantee reliable power management of a lunar habitat's DC microgrid, control schemes are needed that can manage the different assets (batteries, photovoltaics, loads) effectively. Proposed hierarchical control schemes are further developed into hardware solutions using Opal-RT's real-time simulation software and Power Hardware-in-the-Loop platform. Experimental results of a simulated DC microgrid and physical DC/DC components can allow better realization and performance of applications such as battery discharge control.**

Keywords – photovoltaics, DC microgrids, controls, power electronics, Opal-RT, lunar habitat, PHIL, DC/DC, energy storage

I. INTRODUCTION

NASA's Artemis Plan outlines the key objectives and elements of establishing a lunar base for short and long-term habitation on Earth's moon. Lunar habitation is a steppingstone towards deeper space exploration, with missions to Mars planned in the 2030s that will require a sustainable living environment on the planet's surface [3]. A major component of a lunar habitat will be its electrical power system. To sustain a permanent presence on Earth's moon, a highly reliable and ever-evolving power infrastructure will be needed that can withstand harsh environmental conditions, be extremely resilient, and easily expandable [2]. While some existing research does focus on power architectures for spacecrafts and satellites [5-8], power systems developed specifically for lunar habitation are less common and require further investigation. Typical power systems for space applications consider photovoltaic (PV) arrays as the primary power source, providing abundant renewable energy through solar irradiance exposure and maximum power point tracking (MPPT) conversion algorithms [9]. Battery Energy Storage Systems (BESS) can provide auxiliary power when PV is not adequately available [10]. This can occur during periods of eclipse, system failure, or heavy load demand. The BESS is crucial towards providing resiliency and stability in the lunar habitat power architecture.

Hierarchical control schemes for the BESS can be employed to achieve a more resilient power architecture than localized control alone. Due to varying line impedances across a power bus, differences in BESS performance, and environmental conditions, one can expect varying discharges rates between BESS units. Lack of uniformity in battery discharge can lead to power quality and reliability issues over time [1]. Hierarchical control can help achieve uniform discharge from all BESS modules by providing coordinated control on top of each module's local control strategy (droop control, voltage regulation, etc.) [11]. One proposed approach to hierarchical control defines three levels to optimize the distributed BESS power flow [1, 4]. While these studies highlight some key aspects at the simulation level, there is a need for implementing this control with hardware and testing under realistic conditions to further develop the solution.

Herein, a hierarchical control for a DC microgrid is implemented in a real-time simulation model using Opal-RT's RT-Labs. The hierarchical control for the BESS will ensure proper operations under varying PV and system conditions. The control scheme has been tested on a hardware DC/DC converter using a Power Hardware-in-the-Loop (PHIL) platform. Experimental results demonstrate the dynamic performance of the DC/DC converter operating under a hierarchical control scheme when the real-time simulation model is subjected to varying irradiance and load conditions.

II. DC POWER HARDWARE-IN-THE-LOOP SETUP

To further investigate hierarchical control methods within the lunar habitat microgrid application, a PHIL platform consisting of a real-time simulation model and a hardware DC/DC converter are employed. The DC/DC converter functions as the battery charge-discharge unit (BCDU), regulating power between two DC source-load supplies. The current measurements of the BESS/BCDU are relayed to the real-time simulation through the Opal-RT's analog interface. Based on these values, the hierarchical control will determine how much power the DC/DC power converter should sink/source. Fig. 1 illustrates a block diagram of the experimental PHIL platform.

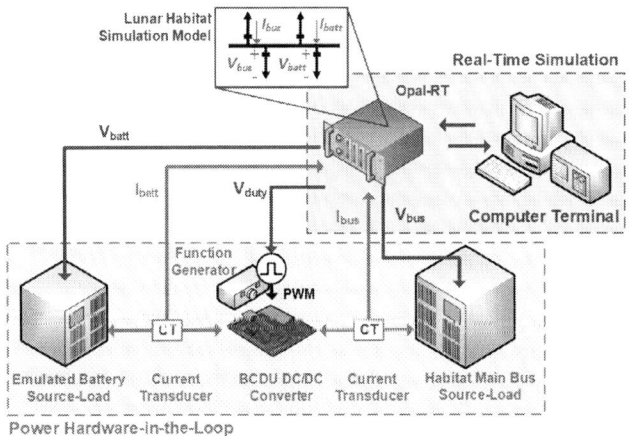

Fig. 1. Block Diagram for the Power Hardware-in-the-Loop Setup.

A. Opal-RT Real-Time Simulator

The Opal-RT OP5600 is a real-time simulation platform capable of integrating simulation models into PHIL testing. For this study, the lunar habitat model is developed in *MATLAB/Simulink* and integrated into RT-Labs for compiling and loading into the real-time processor. RT-Labs allows for control over the simulation during run-time as well as capturing and logging real-time data throughout the simulation. Furthermore, the Opal-RT provides an analog-to-digital interface for the various hardware signals. Current measurements, voltage setpoints, and control signaling can be interfaced between the real-time simulation model and hardware devices. The Opal-RT system relies on a discrete 40 μs timestep for input/output sampling time and real-time model operation.

B. BCDU Hardware Converter

The BCDU DC/DC converter consists of an evaluation module set designed and manufactured by Texas Instruments. The LMG34XX-BB-EVM and LMG3411EVM-018 operate as a synchronous half-bridge rectifier and gate driver/filter motherboard [17]. This evaluation module set is capable of bidirectional current flow (up to +/-8 A_{DC}) and step up/down voltage conversion (up to +480 V_{HI}). This DC/DC converter accepts a +5 V TTL square wave as its PWM switch waveform to control the half-bridge rectifier's gate drivers. The evaluation kit is assembled into a printed enclosure including fans for heat dissipation and insulated connectors for safety. Building the converter into an enclosure allows for a streamlined test setup and integration of this converter into a multitude of test environments.

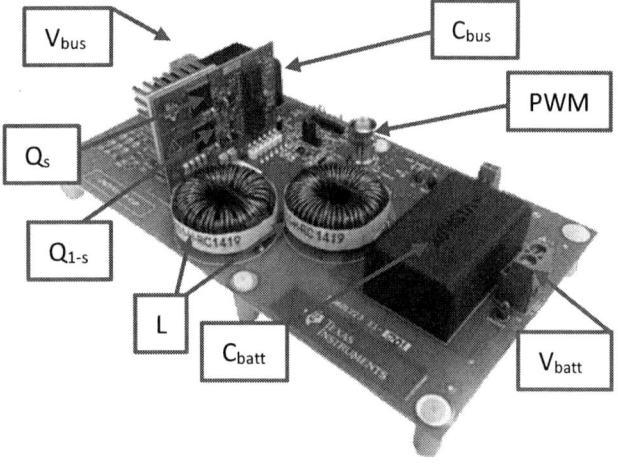

Fig. 2. TI LMG34XX Half-Bridge Bidirectional Converter [17].

C. PWM Generator

The PWM signal to the converter board is driven by a Teledyne LeCroy WaveStation 3162 function generator operating as a 100 kHz TTL square wave with a controlled, variable duty cycle. The function generator features an analog input capable of modulating the duty cycle of the PWM signal relative to the input signal's amplitude (+/-6 Vp = +/-5%). A nominal duty cycle for the BCDU is set with the real-time simulation either increasing or decreasing it based on the controller output.

D. Source-Load Power Supplies

The BCDU converter is connected to two Chroma 62180D source-load DC power supplies. These DC power supplies are capable of both sourcing and sinking current though the converter and can be voltage-controlled via analog set signals. The analog voltage set signals provided to each Chroma from the Opal-RT represent 1) an emulated battery voltage and 2) the main bus voltage for which the BCDU is connected to. Based on the control scheme of the converter and system voltages, power is transferred between the two supplies, through the BCDU, to support the main bus load and charge/discharge the battery.

E. Current Transducers

Finally, the current being fed from each source-load supply is measured using two AAC 929-50 bidirectional current transducers (CTs). The CTs provide a voltage output proportional to the current passing through the sensor. This voltage is fed into the Opal-RT's analog input and injected into the real-time simulation as a controlled current source. The hardware converter's behavior becomes integrated into the lunar habitat simulation model, contributing to the overall power source and consumption. To provide significant power from the hardware BCDU, the current signals measured in hardware are scaled up by a factor of 10x. This scaling allows the power levels from the BCDU hardware converter to be perceived as equivalent to the higher-power simulation models. To limit noise interference from the test environment and switching converter, low-pass filters are implemented in hardware and software on the current measurements.

III. LUNAR HABITAT SIMULATION MODEL

The lunar habitat simulation model consists of two subsystems: the solar array field (SAF) and the habitat. Power is sourced from the SAF's two PV arrays via solar array regulators (SAR) as well as two local BESS units (A, B). A BESS is comprised of a battery unit (BU) and BCDU capable of sourcing and sinking power based on load demand and solar array (SA) generation.

Power is sourced to the lunar habitat's DC-DC converter units (DDCU) and power distribution units (PDUs) for habitat load consumption. Two BESS units (C, D) are local to the lunar habitat alongside the power sourced from the SAF through the lunar habitat's local interconnect. A small impedance is introduced into the connection between the SAF and the lunar habitat to model the line losses present between the two subsystems. This line impedance is modeled as a lumped element, with a series resistor, inductor, and a shunt capacitor to ground.

The lunar habitat simulation model is implemented utilizing state-space average models that represent the dynamic behavior of the SAR, BCDU, and DDCU converters [9]. The simulation model is developed using *MATLAB/Simulink* with state-space average equations for each DC/DC converter model, alongside the *Simscape* Specialized Power Systems and standard library components. Fig. 3 illustrates a one-line diagram of the lunar habitat.

978-1-7281-6118-1/22 $31.00 © 2022 IEEE

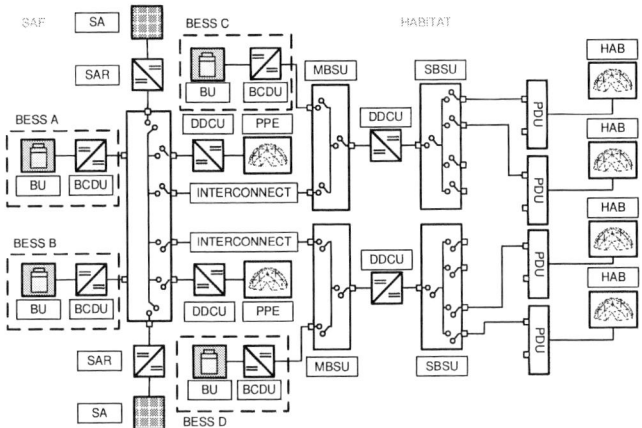

Fig. 3. One-line Diagram for the Lunar Habitat.

A. Solar Array (SA)

A dynamic mathematical PV model is used to represents the lunar habitat's SA [18]. This mathematical PV model allows varying irradiance and temperature conditions. Moreover, this model only requires the information provided from the PV manufacturer data sheet. The selected parameters provide adequate load and BESS charging power under full-irradiance conditions. Table I summarizes the parameters for the SA model used in the lunar habitat simulation model.

TABLE I:
SOLAR ARRAY SYSTEM PARAMETERS

Variable	Description	Value	Unit
N_{series}	# of Panels in Series	40	-
$N_{parallel}$	# of Panels in Parallel	230	-
V_{OC}	Panel Open-Circuit Voltage	5	V
I_{SC}	Panel Short-Circuit Current	0.35	A
$E_{e,nom}$	Nominal Irradiance	1000	Wm^{-2}
P_{NOM}	Power at Nom. Irradiance	18	kW
b	Characteristic Constant	0.0615	-

B. Solar Array Regulator (SAR)

A Single Ended Primary Inductor Converter (SEPIC) acts as the SAR for power conversion between the SA and the lunar habitat main bus voltage. The SEPIC is tasked with executing the optimal duty ratio MPPT algorithm to extract the utmost available power from the SA [16]. This algorithm provides the maximum power to the main bus from the SA by varying the output current for a given main bus voltage. The SEPIC topology is composed of two inductors, two capacitors, a diode, and a switching transistor. Fig. 4 illustrates the SEPIC topology circuit schematic.

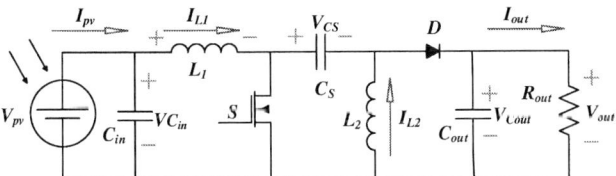

Fig. 4. The SEPIC circuit schematic. The SEPIC allows for lower, equal, or greater voltage levels at its output.

Performing an analysis on the ON and OFF switching states of the SEPIC, the dynamic equations can be obtained. The state equations for the SEPIC are shown in equations (1) through (4).

$$\frac{dI_{L_1}}{dt} = \frac{V_{C_{in}}}{L_1} + \frac{V_{C_s} + V_{C_{out}}}{L_1} \cdot (S_1 - 1) \tag{1}$$

$$\frac{dI_{L_2}}{dt} = \frac{V_{C_s}}{L_2} \cdot S_1 + \frac{V_{C_{out}}}{L_2} \cdot (S_1 - 1) \tag{2}$$

$$\frac{dV_{C_s}}{dt} = \frac{I_{L_1}}{C_s} \cdot (1 - S) - \frac{I_{L_2}}{C_s} \cdot S \tag{3}$$

$$\frac{dV_{C_{out}}}{dt} = \frac{I_{L_1} + I_{L_2}}{C_{out}} \cdot (1 - S) - \frac{V_{C_{out}}}{C_{out} \cdot R_{out}} \tag{4}$$

In these equations, S is the status of the switch (1 for ON, 0 for OFF), I_{L1} is the current through the inductor L_1, and I_{L2} is the current through the inductor L_2. V_{Cin} is the voltage across the capacitor C_{in}, V_{Cs} is the voltage across the capacitor C_s, and V_{Cout} is the voltage across the capacitor C_{out}. R_{out} represents the equivalent loads that can be connected at the output of the converter. The following design parameters are chosen for the SEPIC to operate as a SAR in the system. Inductive and capacitive elements are selected through SEPIC design equations [13] and empirical tuning within the simulation environment.

TABLE II:
SEPIC CONVERTER DESIGN PARAMETERS

Variable	Description	Value	Unit
V_{PV}	Nominal PV Array Voltage	180	V
V_{BUS}	Nominal Bus Voltage	120	V
I_{out}	Maximum Output Current	150	A
f_{sw}	Nominal Switching Frequency	100	kHz
C_{in}	Input Capacitor	470	μF
L_1, L_2	Energy Storage Inductors	100	μH
C_s	Series Capacitor	10	μF
C_{out}	Output Capacitor	100	μF

C. Battery Charge-Discharge Unit (BCDU)

The BCDU uses a bidirectional converter state-space model, realized as either a synchronous buck or boost converter, providing current flow and voltage conversion between the BESS and main bus during periods of charge and discharge. The bidirectional converters topology is composed of an inductor, a capacitor, and two switching transistors. Fig. 5 illustrates the buck-boost bidirectional converter circuit schematic.

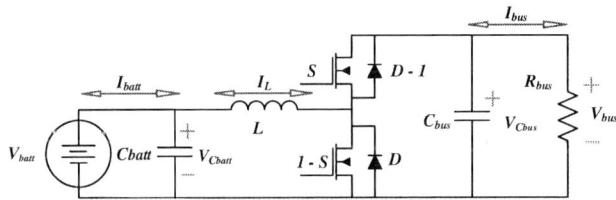

Fig. 5. The bidirectional converter circuit schematic. The converter allows charging and discharging the energy storage system.

978-1-7281-6118-1/22 $31.00 © 2022 IEEE

Performing an analysis on the ON and OFF switching states of the buck-boost bidirectional converter, the dynamic equations can be obtained and are shown in equations (5) through (7).

$$\frac{dI_L}{dt} = \frac{V_{C_{batt}}}{L} - \frac{V_{C_{bus}}}{L} \cdot (1 - S) \tag{5}$$

$$\frac{dV_{C_{batt}}}{dt} = \frac{I_{batt}}{C_{batt}} - \frac{I_L}{C_{batt}} \tag{6}$$

$$\frac{dV_{C_{bus}}}{dt} = \frac{I_L}{C_{bus}}(1 - S) - \frac{V_{C_{bus}}}{C_{bus} \cdot R_{bus}} \tag{7}$$

In these equations, S is the status of the switch (1 for ON, 0 for OFF) and I_{L3} is the current through the inductor L_3. V_{Cbatt} is the voltage across the battery capacitor C_{batt} and V_{Cbus} is the voltage across the capacitor C_{bus}. V_{batt} is the battery voltage and I_{batt} is the battery current. R_{bus} represents the equivalent loads that can be connected at the bus-side of the converter. Table III summarizes the parameters chosen for the bidirectional converter to operate as a BCDU in the system. Inductive and capacitive elements are selected through design equations [14, 15] and empirical tuning within the simulation environment.

TABLE III:
BATTERY CHARGE-DISCHARGE UNIT DESIGN PARAMETERS

Variable	Description	Value	Unit
V_{batt}	Nominal Battery Voltage	145	V
V_{bus}	Nominal Bus Voltage	120	V
f_{sw}	Nominal Switching Frequency	100	kHz
C_{batt}	Battery Capacitor	100	µF
L	Energy Storage Inductor	10	µH
C_{bus}	Bus Capacitor	100	µF

To provide charge and discharge capability through the BCDU, a droop controller with a current-controlled output is implemented within the simulation environment. This control method consists of a PI-controller regulating the converter's duty cycle based on measured current feedback and output bus voltage values. As detailed later, the current set points for the controller are modified during run-time through the hierarchical control scheme to optimize battery State-of-Charge (SoC).

D. Battery Energy Storage System (BESS)

The BESS battery model is a standard offering from the *Simscape* Specialized Power Systems toolbox. A lithium-ion battery model with nominal electrical parameters is selected. The BESS voltage, capacity, and SoC parameters are chosen to provide adequate power delivery during absence of SA-generated power, while still being small enough to charge during typical lunar habitat load and irradiance conditions.

TABLE IV:
BATTERY ENERGY STORAGE SYSTEM PARAMETERS

Variable	Description	Value	Unit
V_{nom}	Nominal Voltage	145	V
RC	Rated Capacity	62.5	Ah
SoC	Initial State-of-Charge	50	%
V_{max}	Fully charged Voltage	157.1	V
V_{min}	Cut-off Voltage	101.2	V
R_{int}	Internal Resistance	0.02	Ω

E. DC-DC Converter Unit (DDCU)

The DDCU is modeled as a buck converter, capable of regulating the voltage output and providing unidirectional current flow from the main bus to the PDU loads. The model consists of the state-space equations for the buck converter and a PI controller to regulate the output voltage. The buck converter topology is composed of a capacitor, an inductor, a switching transistor, and a diode. Fig. 6 illustrates the buck converter circuit schematic.

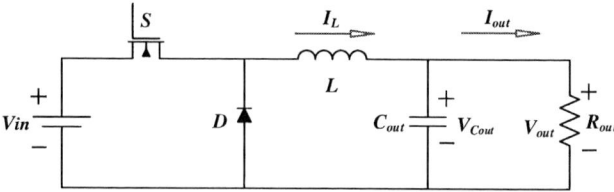

Fig. 6. The buck converter circuit schematic. The buck converter allows for lower voltage levels at its output.

To obtain the dynamic equations that model the buck converter's behavior, a simple analysis of the ON and OFF states is performed on the circuit. The dynamic equations of the converter are shown in equations (8) and (9).

$$\frac{dI_L}{dt} = \frac{V_{C_{in}}}{L} \cdot S - \frac{V_{C_{out}}}{L} \tag{8}$$

$$\frac{dV_{C_{out}}}{dt} = \frac{I_L}{C_{out}} - \frac{V_{C_{out}}}{C_{out} \cdot R_{out}} \tag{9}$$

In these equations, S is the status of the switch (1 for ON, 0 for OFF), I_L is the current through the inductor L, while V_{Cout} is the voltage across the capacitor C_{out}. R_{out} represents the equivalent loads that can be connected at the output of the converter. The following parameters are chosen for the buck converter to operate as a DDCU in the system. Inductive and capacitive elements are selected through the buck converter's design equations [15] and empirical tuning within the simulation environment.

TABLE V:
DDCU DESIGN PARAMETERS

Variable	Description	Value	Unit
V_{in}	Nominal Input Voltage	≥120	V
V_{out}	Nominal Output Voltage	120	V
I_{out}	Nominal Output Current	25	A
f_{sw}	Nominal Switching Frequency	100	kHz
L	Energy Storage Inductor	25	µH
C_{out}	Output Capacitor	10	µF

Each DDCU supplies power to the PDU and power and propulsion element (PPE) loads. There are six loads, each consuming 3 kW with an additional load step of 200 W at a constant 10 Hz rate. The load stepping is intended to test the transient power handling alongside the constant 3 kW power draw. For this simulation, main bus switching units (MBSU) and secondary bus switching units (SBSU) are held in static 'ON' states to provide maximum power to the loads without disruption.

978-1-7281-6118-1/22 $31.00 © 2022 IEEE

IV. PROPOSED HIERARCHICAL CONTROL

Hierarchical control schemes for the BESS can be employed to achieve a more resilient power architecture than localized control alone. *Level 1* (primary) is a local droop controller that regulates power and voltage to the system, while *Level 2* and *Level 3* (secondary and tertiary) use battery SoC and power flow data to provide control setpoints for each distributed BESS [1]. This research offers many practical considerations for how high-level control over BESSs will allow for longevity and resiliency within lunar power systems.

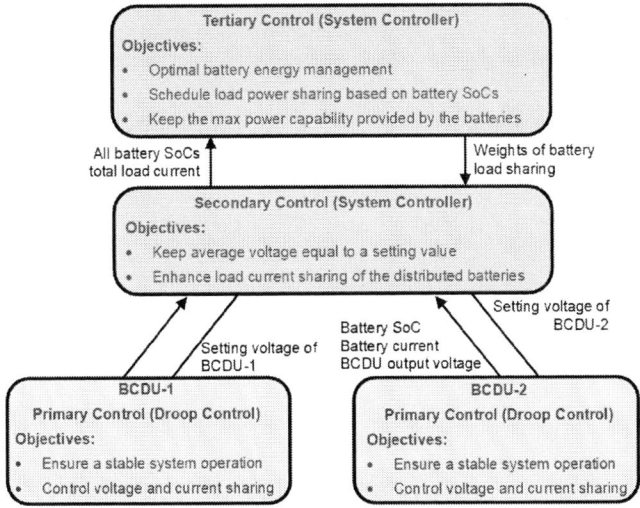

Fig. 7. Block Diagram of the Hierarchical Control Scheme.

The implementation of hierarchical control for this research relies on the battery SoC to determine the amount of current that can be charged or discharged from the battery at a given time. To preserve battery capacity, the BCDU's current controller adopts new maximum setpoints deployed from the system controller. The hierarchy to our simulation and hardware model control is as follows (highest to lowest level of control):

- **Tertiary Control:** Battery SoC is passed into system controller to determine BCDU control set points.

- **Secondary Control:** Droop controller's current set point determines average duty cycle value of the converter.

- **Primary Control:** Average duty cycle value sets PWM waveform to converter for power transfer.

Setpoints for the BCDU are defined as such for the hardware converter model to be tested. By varying the charge rate based on SoC, the hierarchical control is capable of further protection the battery and controlling overall power flow within the system.

TABLE VI:
HIERARCHICAL CONTROL BCDU SETPOINTS

SoC (%)	$I_{charge,max}$ (A)	$I_{discharge,max}$ (A)
>90	5.21	41.67
70 - 90	10.42	41.67
50 - 70	20.84	41.67
30 - 50	41.67	20.84
10 - 30	41.67	10.42
< 10	41.67	5.21

V. SIMULATION RESULTS

Developing the entire lunar habitat model in simulation provides a comprehensive platform for which the BCDU control can be tested in the PHIL environment. Fig. 8 shows the simulated power flow of the BESS units in the lunar habitat without the use of a hierarchical control for the BCDUs. The SoC and power flow from identical BESSs are disproportionate in the system during both the charge and discharge states. (an eclipse event is demonstrated at 30 sec). It is desired that battery units either have equivalent power flow or uniform SoC during their operation. This can be achieved through more advanced control methods proposed.

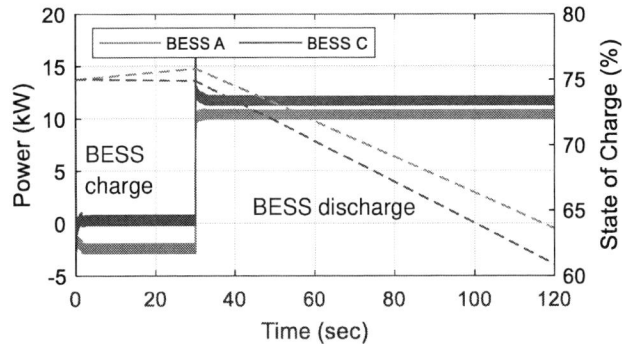

Fig. 8. BESS Power Flow and SoC Without Hierarchical Control.

By implementing new droop controller setpoints periodically during simulation runtime shows that there exists a benefit to ensuring SoC uniformity between multiple BESS systems. The hierarchical control scheme observes power flow and SoC metrics, adjusting the BCDU droop controller over time. Fig. 9 captures this behavior with a setpoint adjustment to BESS C at 60 sec. The droop controller setpoints adjustments allow variation in battery output power and guarantees uniform SoC over time.

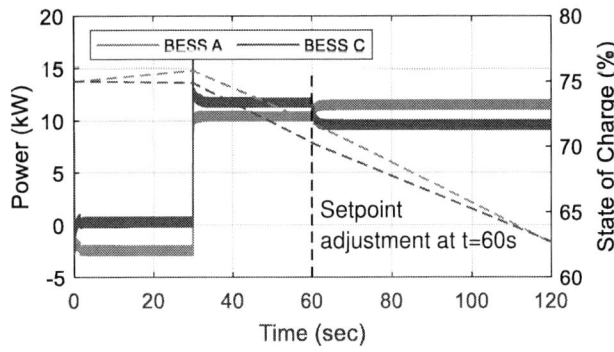

Fig. 9. BESS Power Flow and SoC with Hierarchical Control.

This control method provides a novel approach to ensuring reliable BESS power conversion within the DC microgrid (via localized droop controllers) as well as uniformity in battery power flow and discharge (via hierarchical control). The next steps towards experimental results will is to incorporate the hierarchical BCDU control model into the real-time simulation environment using the Opal-RT PHIL platform and a replacing the simulated bidirectional converter with a physical bidirectional DC/DC converter.

VI. REAL-TIME SIMULATION AND HARDWARE RESULTS

The experimental test setup, shown in Fig. 10, is prepared for the simulation and BCDU converter using a PHIL platform. To test both charge and discharge capabilities, as well as transient response of the physical BDCU converter, during source disruptions, an eclipse event is introduced halfway through the simulation time. The eclipse event causes an instantaneous loss of power from the SA to the lunar habitats main bus. This not only tests the BESS capability to provide power to the load, but also captures the ability for the BCDU controller to switch from charge to discharge modes.

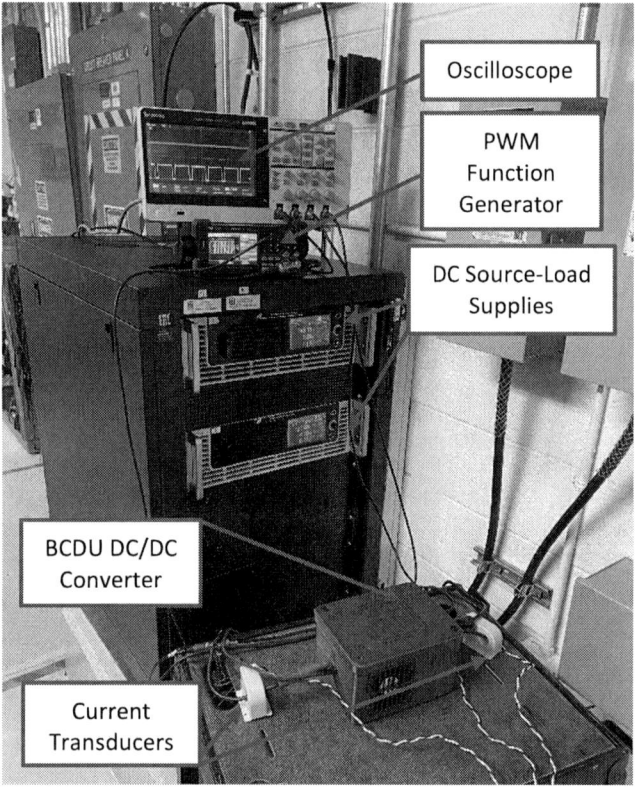

Fig. 10. Power Hardware-in-the-Loop Experimental Setup for testing the BCDU Converter. In this test setup, the lunar habitat is simulated in the Opal-RT real-time simulator to test BCDU performance.

The current set points of the hardware BCDU controller are modified by the system controller to preserve battery SoC. The primary control adjusts to implement the newly dispatched current set points throughout the simulation. These control changes are detailed as phases in the simulation (1-4) based on the power provided from the SA and system controller updates to the hardware BCDU interfacing with BESS B. Fig. 11 illustrates the power flow and SoC of the hardware-based BESS alongside the simulation models. For this test, the system controller only provides updates to the BCDU B. Omitting the hierarchical control from all other BCDUs allows for a more controlled test environment for the hardware converter. Table VII details the average power values for each power source/load during each test phase.

TABLE VII:
AVERAGE POWER CONTRIBUTIONS DURING RUN-TIME

Power Source/Load	Full Irradiance / BESS Charge (W)		Eclipse / BESS Discharge (W)	
	Phase 1 0-120s	Phase 2 120-240s	Phase 3 240-360s	Phase 4 360-480s
SA	37278	37296	0	0
BESS A	-6149	-6157	4175	4757
BESS B (PHIL)	-6141	-3052	4168	3005
BESS C	-1894	-3456	5184	5472
BESS D	-1894	-3456	5184	5472
Load	18686	18686	18533	18522

Under full irradiance conditions, BESS B matches BESS A output power until the SoC reaches 50%. At this point, the system controller dispatches a new current set point to the BCDU B controller. The charge current for the BESS reduces by half and causes other BESS units to charge at a higher rate. BESS A and B are shown to charge at a higher rate due to their proximity in the SAF (as compared to BESS C and D located across the SAF-habitat interconnect). During eclipse events, the BESS units switch to discharge to support the lunar habitat load. Phase 3 matches output power between BESS A and B and BESS C and D. BESS C and D provide additional power due to their proximity to the lunar habitat loads (shown overlapping in Fig. 11). In Phase 4, BESS B's discharge rate reduces due to the SoC of the battery dropping below 50%. The BCDU current setpoints lower to preserve the battery's capacity under the current load conditions. Phase 4 shows all other BESS units increasing their output power to support the lunar habitat load.

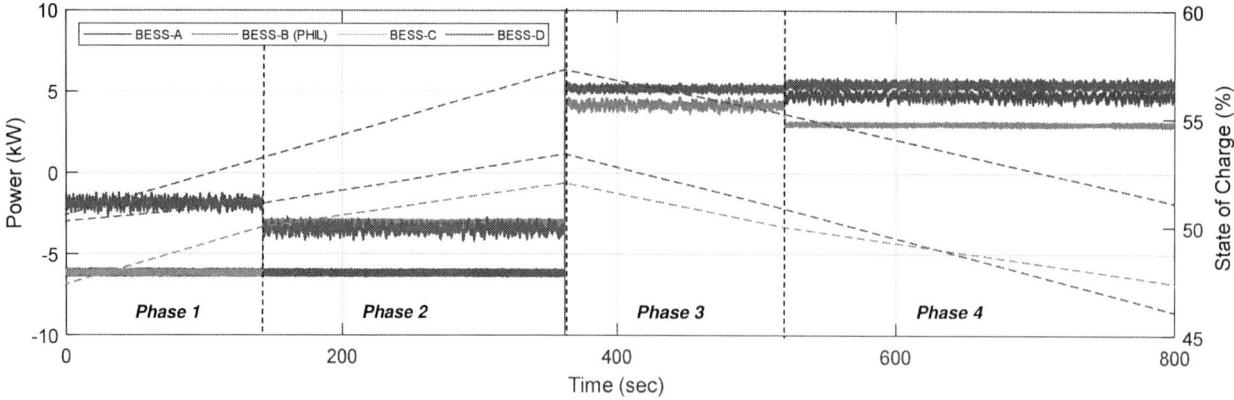

Fig 11. Results for Power Flow and State-of-Charge of BESS Units During Real-Time Simulation.

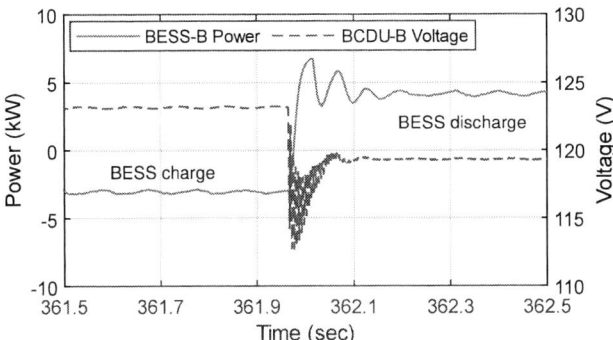

Fig. 12. Results of BESS B Transient Response During Eclipse Event.

The transient performance of the simulation shows stability throughout, though the eclipse condition is further examined. Fig. 12 shows the power and voltage performance of the BESS B unit as the BCDU controller shifts from charge to discharge modes. Some ringing is present in this transition, however the BCDU does reach a steady state within 500 msec. With further tuning of the BCDU droop controller, better transient performance could be obtained.

VII. CONCLUSION

A hierarchical control scheme for a lunar habitat's DC microgrid is proposed and implemented through a real-time simulation and PHIL environment. State-space average models are developed in *MATLAB/Simulink* that represent the lunar habitat microgrid's power converters. Environmental conditions, such as eclipses and power line impedances, are examined in simulation and through experimental testing. Developing a PHIL platform for DC hardware devices provides the ability for more critical testing. By integrating DC source-load power supplies and current transducers into the Opal-RT's real-time simulation environment, power hardware devices are tested as part of the lunar habitat with scenarios previously only tested in simulation. Furthermore, the real-time simulation introduces both primary and hierarchical control of the BCDU and tests its power sharing capability alongside all other lunar habitat models. Next steps for this work can involve more comprehensive hierarchical control methods that consider SoC control of both hardware and software BCDU models. Additionally, the DC-PHIL platform can be further expanded to include testing of more realistic power hardware devices intended for lunar and aerospace applications.

ACKNOWLEDGEMENT

Sandia National Laboratories is a multi-mission laboratory managed and operated by National Technology and Engineering Solutions of Sandia, LLC., a wholly owned subsidiary of Honeywell International, Inc., for the U.S. Department of Energy's National Nuclear Security Administration under contract DE-NA-0003525.

The NASA Glenn Research Center, and NTESS of Sandia, LLC collaborate in accordance with the National Aeronautics and Space Act (51 U.S.C. § 20113(e)) under contract SAA3-1690.

REFERENCES

[1] J. Zhang, J. T. Csank and J. F. Soeder, "Hierarchical Control of Distributed Battery Energy Storage System in a DC Microgrid," *2021 IEEE Fourth International Conference on DC Microgrids (ICDCM)*, 2021, pp. 1-8.

[2] L. Bowling, B. Horvath and C. Wohl, "Integration of Advanced Structures and Materials Technologies for a Robust Lunar Habitat," *2021 IEEE Aerospace Conference*, 2021, pp. 1-13.

[3] G. Flores, D. Harris, R. McCauley, S. Canerday, L. Ingram and N. Herrmann, "Deep Space Habitation: Establishing a Sustainable Human Presence on the Moon and Beyond," *2021 IEEE Aerospace Conference (50100)*, 2021, pp. 1-7.

[4] A. D. Bintoudi, C. Timplalexis, G. Mendes, J. M. Guerrero and C. Demoulias, "Design of Space Microgrid for Manned Lunar Base: Spinning-in Terrestrial Technologies," *2019 European Space Power Conference (ESPC)*, 2019, pp. 1-8

[5] L. Wang, D. Zhang, J. Duan and J. Li, "Design and Research of High Voltage Power Conversion System for Space Solar Power Station," *2018 IEEE International Power Electronics and Application Conference and Exposition (PEAC)*, 2018, pp. 1-5.

[6] M. D'Antonio, C. Shi, B. Wu and A. Khaligh, "Design and Optimization of a Solar Power Conversion System for Space Applications," in *IEEE Transactions on Industry Applications*, vol. 55, no. 3, pp. 2310-2319, May-June 2019.

[7] K. G. Boggs, K. Goodliff and D. Elburn, "Capabilities Development: From International Space Station and the Moon to Mars," *2020 IEEE Aerospace Conference*, 2020, pp. 1-10.

[8] K. G. Boggs and K. D. Foley, "International space station testbed for exploration," *IEEE Aerospace Conference*, 2018, pp. 1-6.

[9] R. Darbali-Zamora and E. I. Ortiz-Rivera, "A State Space Average Model for Dynamic Microgrid Based Space Station Simulations," *2017 IEEE 44th Photovoltaic Specialist Conference (PVSC)*, 2017, pp. 2957-2962.

[10] M. Kaczmarzyk and M. Musiał, "Parametric Study of a Lunar Base Power Systems," *Energies*, vol. 14, no. 4, p. 1141, Feb. 2021

[11] G. V. Somanath Reddy, V. P. Mini, N. Mayadevi and R. Hari Kumar, "Optimal Energy Sharing in Smart DC Microgrid Cluster," *2020 IEEE International Conference on Power Electronics, Smart Grid and Renewable Energy (PESGRE)*, 2020, pp. 1-6,

[12] Yu, S.-Y.; Kim, H.-J.; Kim, J.-H.; Han, B.-M. SoC-Based Output Voltage Control for BESS with a Lithium-Ion Battery in a Stand-Alone DC Microgrid. *Energies* 2016, 9, 924.

[13] D. Zhang, "AN-1484 Designing A SEPIC Converter", *Texas Instruments*, Apr. 2013. [Online]. Available: https://www.ti.com/lit/an/snva168e/snva168e.pdf [Accessed: October 2021]

[14] B. Hauke, "Basic Calculation of a Boost Converter's Power Stage", *Texas Instruments*, Jan. 2014. [Online]. Available: https://www.ti.com/lit/an/slva372c/slva372c.pdf [Accessed: October 2021]

[15] B. Hauke, "Basic Calculation of a Buck Converter's Power Stage", *Texas Instruments*, Aug. 2015. [Online]. Available: https://www.ti.com/lit/an/slva477b/slva477b.pdf [Accessed: October 2021]

[16] R. Darbali-Zamora and E. I. Ortiz-Rivera, "Optimal duty ratio maximum power point tracking technique using the SEPIC topology for photovoltaic systems applications," 2016 IEEE ANDESCON, 2016, pp. 1-4, doi: 10.1109/ANDESCON.2016.7836257.

[17] "Using the LMG3410-HB-EVM Half-Bridge and LMG34XX-BB-EVM Breakout Board EVM", *Texas Instruments*, May 2017. [Online]. Available: https://www.ti.com/lit/ug/snou140a/snou140a.pdf [Accessed: May 2022]

[18] R. Darbali-Zamora, N. Cobo-Yepes, J. E. Salazar-Duque, E. I. Ortiz-Rivera and A. A. Rincon-Charris, "Buck Converter and SEPIC Based Electronic Power Supply Design with MPPT and Voltage Regulation for Small Satellite Applications," *2017 IEEE 44th Photovoltaic Specialist Conference (PVSC)*, 2017, pp. 2963-2968.

Influence of Se Grading on the Free Carrier Profile of CdSeTe/CdTe Solar Cells

Jared D. Friedl[1], Ebin Bastola[1], Rasha A. Awni[1], Xavier Mathew[1,2], Adam B. Phillips[1], Yanfa Yan[1], and Michael J. Heben[1]

[1]Wright Center for Photovoltaics Innovation and Commercialization, Department of Physics and Astronomy, University of Toledo, Toledo, Ohio, 43606, USA

[2]Instituto de Energías Renovables, Universidad Nacional Autónoma de México, Temixco, Morelos 62580, México

Abstract—**CdTe solar cells have shown great improvements in recent years due partly to selenium alloying, leading to strong interest in CdSeTe/CdTe devices. The effects of selenium inclusion on the electronic properties of CdTe is becoming better understood, but uncertainties remain. Capacitance-voltage (CV) measurements offer a quick and accessible method for profiling the free carriers in photovoltaic devices, however little has been done to study the effect of Se alloying in CdTe via this procedure. Here we use numerical simulation of CdSeTe/CdTe photovoltaic devices to analyze the effects that Se could have on the CV behavior. We show that the charge distribution in the device changes significantly depending on how Se influences the energetic and electronic properties of the absorber, and that a poor back contact is expected to obscure the ability of CV to detect these changes. We then discuss these predictions in the context of CV profiling of devices with various Se distributions fabricated via thermal evaporation. We conclude that Se affects the carrier profile in a more complicated manner than assumed in our model and that limited spatial resolution near the front interface, likely because of back contact barriers, precludes precise extraction of information on passivation or compensation via this technique.**

Keywords—*CdSeTe, CdTe, Capacitance-Voltage, Simulation*

I. Introduction

The power conversion efficiency (PCE) of small-area CdTe devices has recently achieved 22.1% [1], largely due to the incorporation of Se at the front of the device. A graded Se distribution at the front of the device lowers the bandgap from that of pure CdTe, increasing overall absorption in the long-wavelength portion of the spectrum resulting in improved short-circuit current density (J_{SC}) [2]. In the absence of significant loss in either open-circuit voltage or fill factor, improved J_{SC} translates to an increase in PCE [3]. Other beneficial effects of Se inclusion, such as passivation of non-radiative recombination centers, have been observed [3, 4].

One potential drawback with the inclusion of Se may be the reduction in achievable p-type conductivity with Cu doping. A reduction in the apparent free carrier concentration with increasing Se concentration has been observed experimentally [5], which can be attributed to a combination of compensation by Cl-related donors and reduction of doping activation

efficiency by Se [6]. As the incorporation of both chlorine and Se have become vital to the success of CdTe and CdSeTe (CST) devices, there is no easy way to avoid these pitfalls. Nuanced optimization of the Se profile is thus required to balance these tradeoffs and produce maximally efficient devices. However, a straightforward and accessible method of assessing differences in electronic behavior due to Se has yet to be formalized.

Capacitance spectroscopic techniques are ideal for determining the spatial and energetic dependence of electronic properties in semiconductor devices. Capacitance-voltage (CV) and drive-level capacitance profiling (DLCP) measurements each provide simple estimates of the free carrier concentration of the lighter doped side of a p-n junction at room temperature but are also known to respond to bulk and interface states differently under specific conditions [7, 8]. The influence of band offsets, defect passivation, and compensation of shallow carriers in CST/CdTe devices should thus be detectable via these methods.

Here, we use numerical simulation in an effort to formalize the effect that Se may have on simple capacitance measurements of CST/CdTe devices. Several devices are then prepared via thermal co-evaporation of CdSe and CdTe source powders, whose CV carrier profiles are contrasted in the context of theoretical expectations.

II. Capacitance-Voltage Simulation

A. Numerical Model

CV behavior was modeled using SCAPS-1D [9]. The device structure was 300 nm SnO$_2$:F (FTO)/50 nm SnO$_2$/500 nm CdSe$_{0.4}$Te$_{0.6}$/1.5 μm CdSe$_x$Te$_{1-x}$/2 μm CdTe/Au. The properties used for FTO and CdTe were taken from previous work [10]. The 50 nm SnO$_2$ layer imitates the oxide layer present on NSG TEC™-12D substrates. The absorber consists of 500 nm of CdSe$_{0.4}$Te$_{0.6}$, followed by linear gradation of x from 0.4 to 0 across 1.5 μm of CdSe$_x$Te$_{1-x}$, ending at 2 μm of pure CdTe. This Se profile is consistent with EDS line scans of CST/CdTe bilayer devices [11]. The Au contact is given a work function of 5.3 eV, forming a non-Ohmic contact with CdTe that is known to have important effects on the CV profile [8].

For our model, CdSe$_x$Te$_{1-x}$ differs from CdTe only in band structure, shallow acceptor density, and defect density. Despite the well-documented bowing behavior of the bandgap of CdSe$_x$Te$_{1-x}$ alloys, the variation of the band edges is not as easily determined. Therefore, as a first-order approximation, the bandgap and electron affinity of CdSe$_{0.4}$Te$_{0.6}$ are made 1.4 eV and 4.7 eV, respectively, and progress linearly to the values of CdTe (1.5 and 4.4 eV, respectively) with x between their values for x=0.4 and 0. The net conduction band shift of 0.3 eV in this compositional range is consistent with available data [12]. The CdTe defect density and activation energy were used previously to model the effect of deep levels on CV profiling of CdTe devices [8]. Defect capture cross sections are selected such that CdTe has a 25 ns lifetime [10] and are not changed with Se addition.

To isolate the effects of passivation and compensation due to Se, the deep defect density (N_T) and shallow acceptor density (N_A), respectively, follow (1) as a function of x, such that they are reduced by a factor of 10 in CdSe$_{0.4}$Te$_{0.6}$ from their original magnitudes in CdTe.

$$N(x) = N_{CdTe} \times 10^{-x/0.4} \qquad (1)$$

In the absence of experimental determination of these quantities, this variation is a simplification meant to illustrate the potential influence of such changes.

Simulations were done without illumination at a temperature of 300 K. Bias ranged from -3 to +1 V and the frequency was 10 kHz. The most relevant electronic properties used in the simulations are summarized in Table I.

B. Simulated CV Behavior

Fig. 1(a) shows the simulated short-circuit band diagrams and corresponding absorber charge distributions for the device stack with five different absorbers: Pure CdTe, "Bands Only" CdSe$_x$Te$_{1-x}$/CdTe, "Passivation Only" CdSe$_x$Te$_{1-x}$/CdTe, "Compensation Only" CdSe$_x$Te$_{1-x}$/CdTe, and Passivation + Compensation ("P+C") CdSe$_x$Te$_{1-x}$/CdTe. The total absorber thickness is always 4 µm and CdSe$_x$Te$_{1-x}$ is graded in the described way when present. The band diagrams reveal how Se

inclusion should affect the band bending at the front interface not only due to the shift in noninteracting band positions, but also due to the way in which defect passivation and shallow carrier compensation further affect the equilibration of charge. Downward band bending is also present at the back interface due to the non-Ohmic back contact.

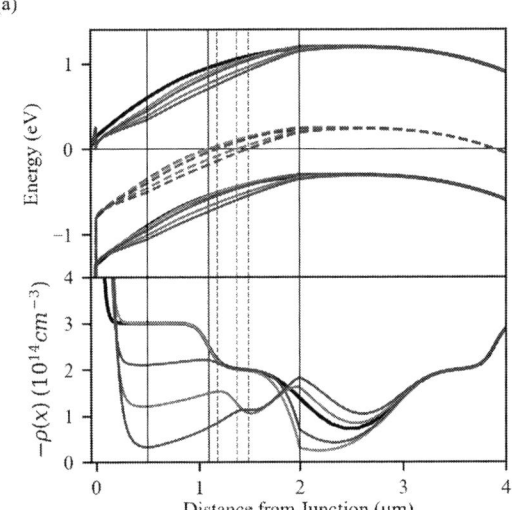

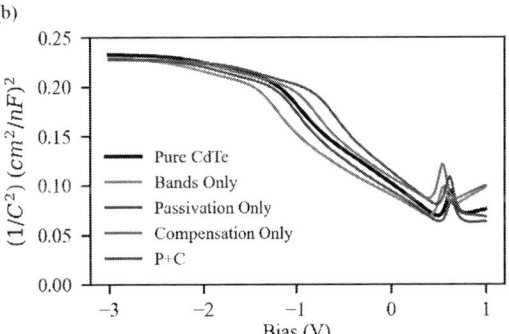

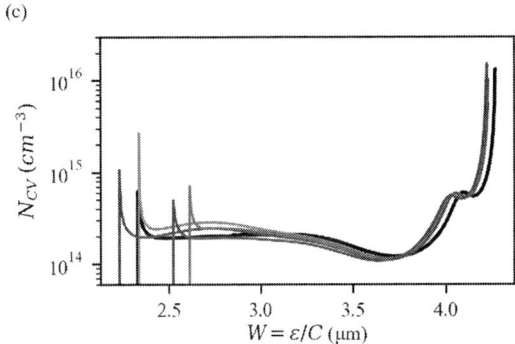

Fig. 1. (a) Simulated short-circuit band diagrams and corresponding steady-state charge distributions at short-circuit for differing absorbers as described in the text. Solid black lines denote CdSe$_{0.4}$Te$_{0.6}$/CdSe$_x$Te$_{1-x}$ and CdSe$_x$Te$_{1-x}$/CdTe interfaces at 0.5 µm and 2 µm from the main junction, respectively. Dashed colored lines denote the distance from the junction at which the Fermi level intersects the trap level. (b) Mott-Schottky plots and (c) apparent carrier profiles for absorbers.

TABLE I. ELECTRONIC PROPERTIES USED IN SIMULATION

Property	Material			
	FTO	SnO$_2$	CdSe$_{0.4}$Te$_{0.6}$	CdTe
Thickness	300 nm	50 nm	2 µm	2 µm
E$_g$ (eV)	3.6	3.6	1.4	1.5
χ (eV)	4.45	4.45	4.7	4.4
N$_D$ (cm^{-3})	5e20	1e17	0	0
N$_A$ (cm^{-3})	0	0	2e13	2e14
Defect Type	Single Donor	Single Donor	Single Acceptor	Single Acceptor
σ$_e$/σ$_h$ (cm^2)	1e-12/1e-15	1e-12/1e-15	4e-14/4e-14	4e-14/4e-14
Defect Distribution	Single	Single	Single	Single
E$_T$ (eV)	E$_V$+1.8	E$_V$+1.8	E$_V$+0.55	E$_V$+0.55
N$_T$ (cm^{-3})	1e15	1e15	1e13	1e14

978-1-7281-6118-1/22 $31.00 © 2022 IEEE

Capacitance measurements are responsive to the first moment of charge response, $\langle x \rangle$ [7]. The steady-state charge distribution as a function of bias is therefore important to understanding how, where, and when defects and carriers affect CV behavior. Because of the built-in voltage between SnO_2 and the absorber, at short-circuit the absorber has a depletion region near the interface and the net near-interface charge is approximately equal to the sum "$-(N_A+N_T)$". As shown in Fig. 1(a), the pure CdTe absorber demonstrates the expected distribution: finite depletion at the front and back interface, with gradual evolution due to the large Debye length and an additional step in charge due to band bending dragging the deep defect level below E_F close to the interface.

The vertical lines spanning Fig. 1(a) denote transitions from $CdSe_{0.4}Te_{0.6}$ to $CdSe_xTe_{1-x}$ to CdTe (solid, black) and the distance from the junction at which the trap level intersects the Fermi level (dashed, colored). Some subtlety arises in the evolution of the net charge away from the front interface due to Se grading and its chosen effects. Se affecting the Bands Only does not change the amount of net charge in the device, however the built-in voltage with SnO_2 is slightly larger. Additionally, a small built-in voltage forms between CST and CdTe, as evidenced by the difference in charge distribution 2 μm from the junction. Passivation and compensation attributed to Se each reduce the magnitude of depleted charge near the junction and throughout the graded region, requiring additional charge to deplete elsewhere in equilibrium.

Fig. 1(b) shows the simulated Mott-Schottky ($1/C^2$) behavior as a function of bias for the five absorbers. Fig. 1(c) shows the corresponding apparent carrier density profiles calculated using the slope of the Mott-Schottky plots. Despite the discussed complexity of the charge distributions of Fig. 1(a), the Mott-Schottky behavior and carrier profiles of each absorber are quite similar. As expected, they are most similar in the region corresponding to the back of the absorber, where the charge distributions differ the least. Most notably, however, rather than a decrease in the carrier density at lower depletion width, which would correspond to the CST layer, there is in fact an *increase* in N_{CV} relative to higher values of W, irrespective of absorber. The failure of the CV profile to accurately reflect reductions in charge density toward the front interface is attributed to the increased contribution of the back interface depletion charge as the share of voltage dropped at the back diode rapidly increases with forward bias. This, in effect, pins $\langle x \rangle$ to the back depletion region, preventing any precise measurement of the charge nearest to the front, where the Se concentration is highest.

C. Simulation of CV with Ohmic Back Contact

To determine how the CV behavior changes as a function of the Se, the detrimental effects of a non-Ohmic back contact were removed by implementing flat bands at the back interface. Flat band conditions guarantee that any depletion present is a result of the front junction built-in potential. Fig. 2(a) and Fig. 2(b) show the simulated Mott-Schottky behavior and carrier profiles, respectively, for the five different absorbers, now with Ohmic back contacts. The Mott-Schottky plots no longer show such significant upward deviation in forward bias due to back contact voltage sharing. This allows the measured capacitance to remain dominated by the front interface charge distribution,

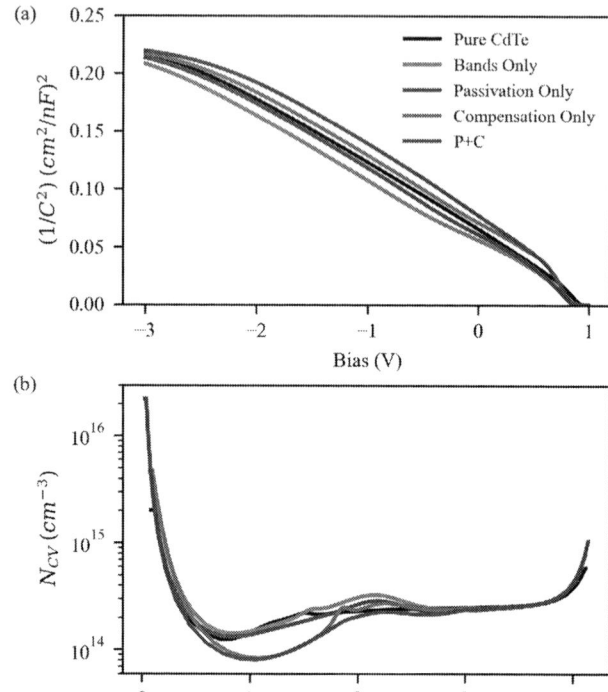

Fig. 2. (a) Simulated Mott-Schottky plots and (b) apparent carrier profiles for absorbers with differing electronic properties as described in the text.

and the carrier profiles to better reflect the true distribution of charge in the device. However, for the Passivation Only graded CST absorber, a significant reduction in charge near the front interface relative to CdTe is not detected. This is likely due to the diminished band bending in forward bias, so that fewer traps are driven below the Fermi level to contribute to the charge response. It is then only in the presence of shallow carrier compensation due to Se that capacitance-voltage is able to detect the reduction in carrier density near the front interface. The limits of this signature are a product of the chosen Se grading profile and the degree of compensation as a function of x. More Se, or more aggressive compensation by Se already present, would likely increase the ability of CV to detect such changes, and vice versa.

III. EXPERIMENTAL RESULTS

CdTe, CdSeTe, and CdSe/CdTe devices were prepared and characterized in order to apply the conclusions of our simulations to the observable effects of different Se distributions on CV carrier profiles.

A. Device Fabrication

Devices were fabricated on fluorine doped tin oxide coated glass substrates with a 50-nm thick dopant-free tin oxide layer on the surface (TEC^{TM} – 12D, NSG Pilkington). Films were deposited by thermal evaporation of source powder(s) (5N) at a substrate temperature of 400 °C and a base pressure of 2×10^{-6} Torr. One absorber, pure CdTe, was prepared with no Se. Two uniform CST absorbers were fabricated, $CdSe_{0.1}Te_{0.9}$ and $CdSe_{0.4}Te_{0.6}$, using co-evaporation of CdSe and CdTe. Lastly, the CdSe/CdTe bilayer absorber originally consisted of ~170 nm of CdSe and ~3.5 µm of CdTe, which interdiffused during CdCl$_2$ treatment leading to a graded Se profile. All the device stacks were treated with CdCl$_2$ saturated solution in methanol using the drop cast method, then annealed for 30 minutes at 400 °C in dry air. Films were rinsed with methanol twice and dried with nitrogen. Copper doping was introduced identically to all samples using the previously reported method [13], and heat treated at 250 °C for 20 mins. A 60 nm gold back electrode was thermally evaporated after Cu treatment to complete each device.

B. Capacitance-Voltage Characterization

Capacitance-voltage measurements were performed using a lock-in based measurement system consisting of a Stanford Research Systems SR865A digital lock-in amplifier and a Princeton Applied Research 273A Galvanostat/Potentiostat acting as a current-to-voltage amplifier. A high-load resistor was used to calibrate the phase shift of the lock-in. The AC signal had a fixed frequency of 10 kHz and an RMS amplitude of 30 mV. DC voltage was swept from -2 to 0.7 V. All signals were generated using the lock-in amplifier and contact to the sample was made using the electrometer leads of the potentiostat. Measurements were done in the dark at room temperature.

Fig. 3(a) and Fig. 3(b) show the Mott-Schottky behavior and CV carrier profiles, respectively, for the four devices. The Mott-Schottky curves in Fig. 3(a) exhibit broad flattening in reverse bias that is typical in CdTe devices and which our simulations would imply indicates the presence of a back barrier (see Fig. 1(c)). However, all carrier profiles (Fig. 3(b)) show limited "punch through" behavior in their left branch, initially implying that the existing back barrier may not affect the accuracy of the CV carrier profiles. The inability to measure the capacitance of pure CdTe in forward bias is likely a result of a poor interface with the thin SnO$_2$ layer, a known limitation.

The apparent carrier profiles in Fig. 3(b) show reduced carrier concentration for all Se-containing samples relative to the pure CdTe absorber. The apparent hole densities, taken from the bottom of each U-shape, are 1.5e14 cm^{-3} for CdTe, 2.9e13 cm^{-3} for CdSe$_{0.1}$Te$_{0.9}$, 4.5e13 cm^{-3} for CdSe$_{0.4}$Te$_{0.6}$, and 7.0e13 cm^{-3} for the CdSe/CdTe absorbers. Though pure CdTe exhibits the highest hole density, the hole density of the other absorbers shows no discernible trend as a simple function of Se inclusion. In fact, the lowest-Se absorber, uniform CdSe$_{0.1}$Te$_{0.9}$, exhibits the lowest apparent carrier density over much of its depth. CdSe$_{0.4}$Te$_{0.6}$ exhibits the next lowest carrier density, followed by the CdSe/CdTe absorber. The carrier profiles of all 3 Se-including absorbers overlap below ~0.8 µm of apparent depth from the junction, obscuring any comparison of the effects of Se on net charge.

These measurements are somewhat difficult to reconcile with expectations from simulation. Selenium has a clear effect on the net charge detected by CV, but of a different nature assumed in our simulation. Directly comparing the uniform CST absorbers supports the simple conclusion that any defect passivation and carrier compensation does not monotonically decrease the net hole density with increasing Se as assumed in our simple model. Their similarity in the left branch indicates limited resolution. Furthermore, the lower carrier concentration of the graded absorber toward the back interface relative to CdTe

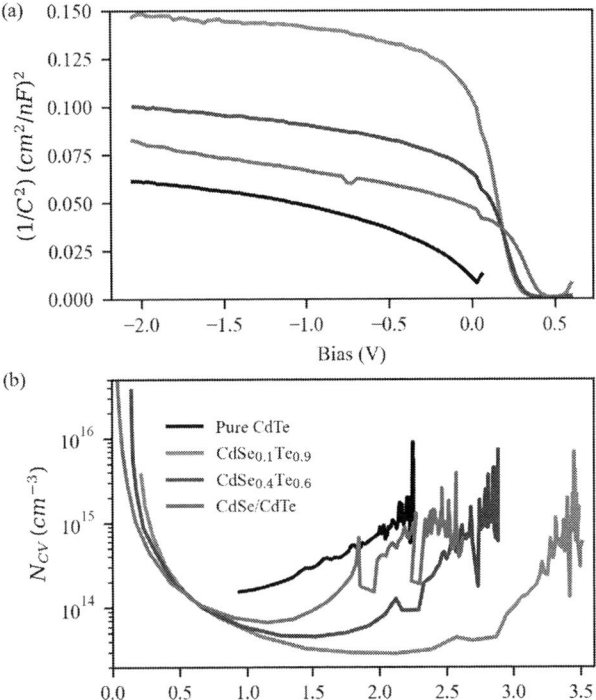

Fig. 3. (a) Mott-Schottky plots and (b) CV carrier profiles for devices of varying Se content.

implies that small amounts of Se penetrate this region and have a measurable effect.

Se clearly must have more complex and separate effects on deep defects and shallow carriers than can be expressed by an expression such as (1). Support for the passivating effect of Se in CST absorbers is strong, however the entire compositional range has yet to be scrutinized as closely as select, uniformly low Se compositions [4]. In some cases, however, increased thickness of CdSe pre-CdCl$_2$-treatment has led to an eventual increase in the concentration of some deep defects [14]. Additionally, different Cu activation processes have also been shown to affect the CV carrier concentration and overall device performance of CdSe/CdTe devices differently [15,16].

IV. CONCLUSION

Understanding the effect of Se on the electronic properties of CdTe is paramount to continuing the progress that has been made in optimizing the performance of CdTe photovoltaics as a scalable and affordable source of renewable energy. Here, we showed via numerical simulation that capacitance-voltage measurements can be used to characterize certain aspects of potential electronic changes in CdSeTe/CdTe solar cells, but that the difficulty of forming an Ohmic back contact with CdTe also introduces difficulties in this procedure.

CdTe, uniform CdSeTe, and CdSe/CdTe devices were fabricated and characterized using capacitance-voltage profiling. Measurements strongly implied that Se has a much more complex effect on the electronic properties of alloyed CdSe$_x$Te$_{1-x}$ than was afforded in our simulation. Additional characterization, such as drive-level capacitance profiling and compositional depth profiling, is needed to further probe the contributions of capacitance response in these absorbers and to better correlate the features of these profiles with Se concentration.

ACKNOWLEDGMENTS

This material is based on research sponsored by the U.S. DOE's Office of Energy Efficiency and Renewable Energy (EERE) under Solar Energy Technologies Office (SETO) Agreement DE-EE0008974 and Air Force Research Laboratory under agreement number FA9453-19-C-1002 and FA9453-21-C-0056. The U.S. government is authorized to reproduce and distribute reprints for Governmental purposes notwithstanding any copyright notation thereon. The views expressed are those of the authors and do not reflect the official guidance or position of the United States Government, the Department of Defense or of the United States Air Force. The appearance of external hyperlinks does not constitute endorsement by the United States Department of Defense (DoD) of the linked websites, or the information, products, or services contained therein. The DoD does not exercise any editorial, security, or other control over the information you may find at these locations. Approved for public release; distribution is unlimited. Public Affairs release approval #AFRL-2022-2393.

REFERENCES

[1] Green, M.A., et al., *Solar cell efficiency tables (version 52)*. Progress in Photovoltaics: Research and Applications, 2018. **26**(7): p. 427-436.

[2] Paudel, N.R. and Y. Yan, *Enhancing the photo-currents of CdTe thin-film solar cells in both short and long wavelength regions*. Applied Physics Letters, 2014. **105**(18): p. 183510.

[3] Munshi, A.H., et al., *Polycrystalline CdSeTe/CdTe absorber cells with 28 mA/cm 2 short-circuit current*. IEEE Journal of Photovoltaics, 2017. **8**(1): p. 310-314.

[4] Fiducia, T.A., et al., *Understanding the role of selenium in defect passivation for highly efficient selenium-alloyed cadmium telluride solar cells*. Nature Energy, 2019. **4**(6): p. 504-511.

[5] Lingg, M., et al., *Structural and electronic properties of CdTe1-xSex films and their application in solar cells*. Science and technology of advanced materials, 2018. **19**(1): p. 683-692.

[6] Sankin, I. and D. Krasikov, *Kinetic simulations of Cu doping in chlorinated CdSeTe PV absorbers*. physica status solidi (a), 2019. **216**(15): p. 1800887.

[7] Heath, J.T., J.D. Cohen, and W.N. Shafarman, *Bulk and metastable defects in CuIn 1− x Ga x Se 2 thin films using drive-level capacitance profiling*. Journal of Applied Physics, 2004. **95**(3): p. 1000-1010.

[8] Li, J.V., et al., *Theoretical analysis of effects of deep level, back contact, and absorber thickness on capacitance–voltage profiling of CdTe thin-film solar cells*. Solar Energy Materials and Solar Cells, 2012. **100**: p. 126-131.

[9] Burgelman, M., P. Nollet, and S. Degrave, *Modelling polycrystalline semiconductor solar cells*. Thin Solid Films, 2000. **361**: p. 527-532.

[10] Liyanage, G.K., A.B. Phillips, and M.J. Heben, *Role of band alignment at the transparent front contact/emitter interface in the performance of wide bandgap thin film solar cells*. APL Materials, 2018. **6**(10): p. 101104.

[11] Shah, A., et al., *Understanding the Role of CdTe in Polycrystalline CdSe x Te1–x/CdTe-Graded Bilayer Photovoltaic Devices*. Solar RRL, 2021. **5**(11): p. 2100523.

[12] MacDonald, B.I., et al., *Layer-by-layer assembly of sintered CDSE x te1–x nanocrystal solar cells*. ACS nano, 2012. **6**(7): p. 5995-6004.

[13] Bastola, Ebin, et al. *Doping of CdTe using CuCl2 solution for highly efficient photovoltaic devices*. 2019 IEEE 46th Photovoltaic Specialists Conference (PVSC).

[14] Hsu, Chih-An, et al. *The Effect of the CdCl2 Heat Treatment on CdSexTe1-x Solar Cells*. 2017 IEEE 44th Photovoltaic Specialist Conference (PVSC).

[15] Hsu, Chih-An, et al. *Cu-doping Effects in CdSexTe1-x/CdTe Solar Cells*. 2019 IEEE 46th Photovoltaic Specialists Conference (PVSC).

[16] Bastola, Ebin, et al. *Understanding the Interplay Between CdSe Thickness and Cu Doping Temperature in CdSe/CdTe Devices*. IEEE Journal of Photovoltaics, 2022. **12**(1): p.11-15.

978-1-7281-6118-1/22 $31.00 © 2022 IEEE

Statistical Performance Analysis on ≈ 320 Perovskite Single- and Two-junction Solar Cells and Modules from >30 Global Sources

Tao Song, Charles Mack, Rafell Williams, Josh Gallon, Allan Anderberg, Larry Ottoson, Daniel J. Friedman and Nikos Kopidakis

PV Cell and Module Performance Group, National Renewable Energy Labarotory, Golden, Colorado, 80401, USA

Abstract—As perovskite photovoltaics (PV) advance from the laboratory to commercial prototypes, their accurate and reliable performance testing is becoming increasingly important. The well documented dynamic response of perovskite solar cells to an external applied voltage has led to the development of steady-state performance measurement methods; however, these methods have not been widely adopted by the perovskite PV community. A key reason for this is that steady-state measurement methods take tens of minutes to complete, as opposed to conventional "fast" current-voltage (I-V) measurements usually lasting a few seconds. Fast I-Vs arise from a snapshot, almost always not a steady-state condition of the device; however, given their widespread use, the question arises: how do performance parameters of perovskite PV compare when measured with fast I-V and with a steady-state method? We compile results from ca. 320 perovskite PV cells and modules, including single junction, and two-terminal perovskite-perovskite and perovskite-Si tandems, and show that fast I-Vs can provide a useful measure of the open-circuit voltage of the devices, while the short-circuit current and the overall efficiency can be widely misestimated. We discuss implications of these findings on performance testing protocols and propose possible options for fast and accurate testing of perovskite PV.

Keywords—perovskite, performance testing, asymptotic, emerging PV

I. INTRODUCTION

Perovskite solar cells have made remarkable strides in efficiency and are rapidly progressing from the research laboratory to commercial prototypes. Intense research has focused on thin film perovskite cells and, more recently, small modules. In addition, tandem cell architectures of perovskites on silicon and two-junction perovskite-perovskite are very active areas of research and development. Research on these unconventional devices has also been focused on development of reliable steady-state performance measurement protocols due to the commonly seen dynamic response of perovskite devices to external applied voltage. There are essentially two types of steady-state measurements which have been adopted by all major PV calibration laboratories: 1) the Asymptotic P_{MAX} or Asymptotic I-V where a set of voltage biases around the P_{MAX} of the device under test (DUT) are chosen and the DUT is held at each voltage until the output current is stable within preset criteria [1]; 2) the maximum power point tracking (MPPT),

where a perturb-and-observe algorithm is used to keep the device at its maximum power point for a set period of time, then the average power is reported to represent the P_{MAX} of the device [2].

Despite the necessity of steady-state measurements for accurate performance testing of perovskite PV, these methods have not been widely adopted by the perovskite PV community. One key reason is testing perovskite PV via steady-state methods a) is not commonly available in a typical I-V testing station and b) is prohibitively slow in a production line testing environment. Additionally, some devices might encounter irreversible degradation during the test. A common argument in favor of fast I-V is that while it may not give an accurate power rating of a given perovskite device, it is still a useful tool for the relative comparison between devices. Hence, while the P_{MAX} extracted from the fast I-V of device A may not be accurate, if it is higher than the P_{MAX} of device B, then, the argument goes, device A will still be a better performer than device B when measured under steady-state conditions. In this contribution we test this hypothesis via a statistical performance analysis of ~320 perovskite devices from over 30 global sources. We aim to answer the following questions: what is the correlation between the performance parameters, I_{SC}, V_{OC} and efficiency (η) when measured with steady-state and with fast I-V? Hence, to what extent can we use fast I-V for perovskite characterization? And lastly, are fast I-Vs capable of providing reliable conclusions on the relative performance of perovskite devices?

II. RESULTS AND DISCUSSION

Since 2017, the PV Cell and Module Performance (CMP) group at NREL has received ca. 320 perovskite solar cells and modules, including 138 single-junction perovskite (hereafter termed 1-J PVSK), 50 perovskite/perovskite (PVSK/PVSK), 60 perovskite/Si (PVSK/Si) monolithic two-junction cells, and 68 1-J perovskite minimodules/submodules, from over 30 global sources for performance certifications. All the cells were measured with the Asymptotic P_{MAX} scan method described briefly in the previous section and in more detail in [1], as shown in the example of Figure 1. The stabilized short-circuit current, I_{SC}, and open-circuit voltage, V_{OC}, are also extracted while the DUT is held at the short-circuit and open-circuit conditions, respectively. In addition, two fast I-Vs, in the forward and

978-1-7281-6118-1/22 $31.00 © 2022 IEEE

reverse direction, are also recorded prior to the asymptotic measurement. Using this protocol, two sets of cell performance parameters (i.e., unstabilized vs. stabilized I_{SC}, V_{OC}, FF, P_{MAX}, η) can be obtained from conventional fast I-V and the asymptotic P_{MAX} scans, respectively. Then we conduct a detailed comparative analysis on the two sets of performance parameters based on the ~ 320 perovskite devices received globally and discuss how useful conventional fast I-V scans are for performance calibration of perovskite devices. We also quantify these results by linear regression analysis which can be easily used as a tool for evaluating the accuracy (and therefore the usefulness) of fast I-Vs in more specific device subsets.

All the performance results were measured at the Standard Test Conditions (STC), i.e., a device temperature of 25 °C, under the global hemispherical reference spectral irradiance, known as AM 1.5 Global with a total irradiance of 1000 W/m². This enables us to have a fair and meaningful comparative analysis between fast I-V and asymptotic measurements on different perovskite devices.

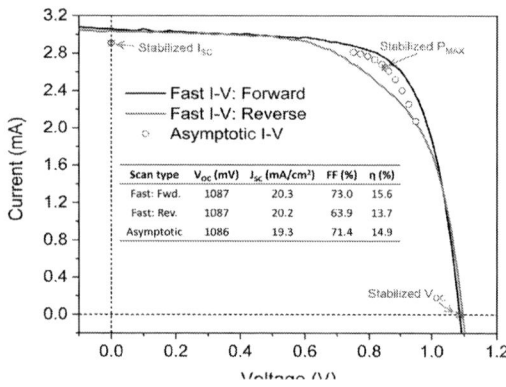

Fig. 1. Fast I-V and asymptotic I-V scans of a perovskite solar cell. The fast I-Vs are scanned in forward ($I_{SC} \rightarrow V_{OC}$) and reverse ($V_{OC} \rightarrow I_{SC}$) directions with the same scan rate of 100 mV/s. The Asymptotic I-V scan in this case lasted ca. 20 mins (detailed scan procedure can be found in [3]). The insert table lists the performance parameters obtained from these scans.

A. Performance Deviation Histograms

Figure 2 shows the percentage deviation histograms of the V_{OC}, I_{SC}, FF, and η from fast I-Vs compared to their steady-state values from asymptotic scans on these perovskite cells. The parameters from the fast I-V, hereafter denoted as "unstabilized" performance parameters, are normalized to their respective stabilized values obtained using the Asymptotic P_{MAX} method, and then subtracted from unity. The results are categorized with 1-J PVSK, 2-J PVSK/PVSK, and 2-J PVSK/Si from left to right as shown in Figure 2a, 2b, and 2c, respectively. From the histograms in Figure 2, we can observe at least three obvious performance deviation trends: 1) the deviations of all the performance parameters from the 2-J PVSK/Si cells are much smaller than those from the 1-J PVSK or the 2-J PVSK/PVSK cells, e.g., only ~20% of the 2-J PVSK/Si cells having a relative η deviation larger than 1.5% as opposed to ~ 80% of the other two types of cells and the deviation distribution range is much narrower as well; 2) for the 1-J PVSK and 2-J PVSK/PVSK cells, most of their J_{SC} deviation bins in Figure 2a and 2b distribute widely in the right side of the "0" deviation baseline

and hence the majority of the fast I-V scans generate higher J_{SC} values than asymptotic scans, i.e., ~70% of these cells giving at least 1.5% higher J_{SC}; 3) in contrast to J_{SC}, a large fraction of 1-J PVSK and 2-J PVSK/PVSK cells (i.e., ~50% of 1-J PVSK and ~90% of 2-J PVSK/PVSK) have an underestimated FF over 1.5% with fast I-V scans. Note that these performance deviation trends do show dependence on scan directions in certain parameters, and a detailed discussion on them will be given in the following linear regression analysis for each type of cell.

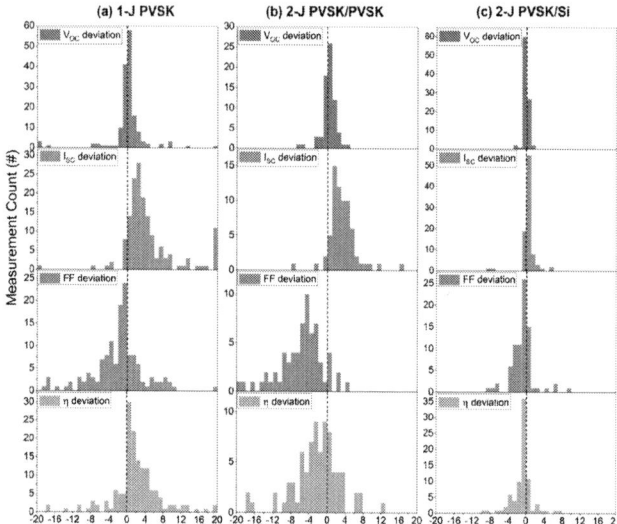

Fig. 2. Percentage deviation of V_{OC}, I_{SC}, FF, and η from fast I-Vs compared to asymptotic scans on ~320 perovskite cells. Both forward ($I_{SC} \rightarrow V_{OC}$) and reverse ($V_{OC} \rightarrow I_{SC}$) scan results are presented, but not distinguished.

B. Linear Regression Analysis

To find a more quantitative correlation between the unstabilized and stabilized parameters and make useful predictions and guidance for reliable performance measurements of perovskite PV devices, a linear regression analysis [3] is conducted on the three types of perovskite cells based on their measured fast I-V and asymptotic scan datasets, respectively. Here we set the reliable performance parameters from asymptotic scans as the best estimate of the true, steady-state value of that parameter and analyze the variation of the corresponding parameters from each cell's fast I-Vs, shown in the graphs as the dependent variable. For instance, Figure 3 shows the measured fast I-V parameters (i.e., V_{OC}, J_{SC}, η) of 1-J PVSK solar cells as a function of their corresponding asymptotic values and their fitting curves based on the linear regression analysis. The red and blue dots represent the measured data points from forward and reverse fast I-V scans respectively, and the red and blue lines are fits through the corresponding data. Also calculated are the 95% prediction bands (lighter blue and red bands), which signify a 95% probability that, based on the existing observations, a future measured parameter from fast I-V will fall within this band. The dashed curve is the assumptive steady-state baseline with a slope of 1 and intercept of 0, so any points above the dashed curve have higher cell parameters in the fast I-V than the asymptotic values or vice versa. Given the relatively large sample size of ~320 perovskite cells from over 30 global sources, the fitted curves along with their 95%

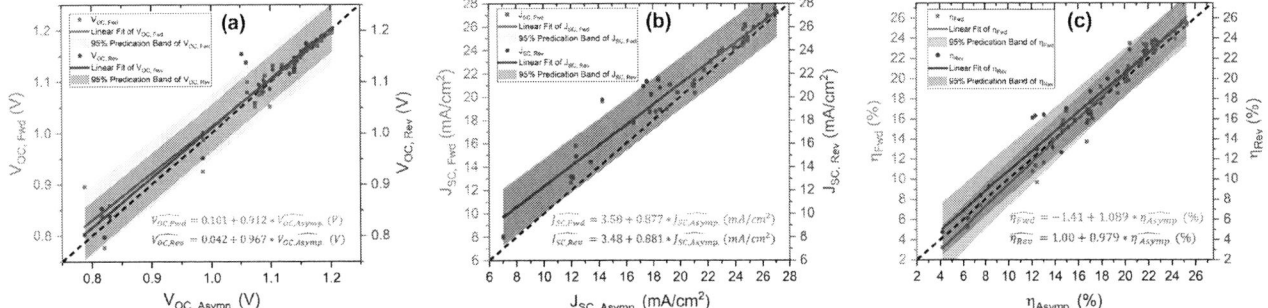

Figure 3. Measured unstabilized (a) VOC, (b) JSC, and (c) η of 1-J PVSK cells from fast I-V scans (red: forward, blue: reverse) as a function of their corresponding asymptotic values; the red and blue curves and the bands are the fitted curves and prediction intervals from the linear regression analysis. The calculated linear equations from the regression analysis are also listed in the plots

prediction interval bands can quantify the strength of the relationship between the fast I-V and steady-state performance data and forecast a predictive performance deviation range of the fast I-V parameters of an unspecified perovskite cell based on existing observations.

In Figure 3a, the calculated 95% prediction band of unstabilized V_{OC} from fast I-V has an estimated deviation width (calculated as the vertical difference between the upper and lower end of the prediction band for a given $V_{OC,Asymp}$), $\Delta V_{OC} \sim \pm 30\text{-}60$ mV, dependent on where the measured V_{OC} is, and which scan direction is chosen. The ΔV_{OC} is slightly larger for the forward scans than for the reverse scans. Overall, the fitted curves illustrate that the fast scans could give approximately same V_{OC} as the asymptotic scans but with a relatively large uncertainty ~2-8%. In Figure 3b, the calculated linear equations from the forward and reverse scans have very similar slopes and intercepts and thus the fitted J_{SC} curves overlap with each other. Meanwhile, they both lie above the assumptive steady-state baseline. As shown in the J_{SC} histogram in Figure 2a, ~ 80% 1-J PVSK cells have >1.5% (or ~50% cells >3.5%) higher J_{SC} from fast I-V scans than from asymptotic scans. In other words, the J_{SC} of 1-J PVSK devices is very likely to be overestimated when extracted from fast I-V scans. In Figure 3c, the two fitted η curves from the forward and reverse scans diverge from each other, which can be seen in their calculated fitted equations as well. This divergence trend corresponds to the commonly seen hysteresis effect in 1-J perovskite cells. Such a divergence becomes more severe with poorer cell efficiency as shown in the 95% prediction bands. For cells with higher efficiencies, it is commonly observed that the hysteresis effect is less prominent. Overall, both the measured and the predictive deviations between the unstabilized and stabilized performance parameters confirm that the fast I-V scan approach has apparent drawbacks for reliable comparison between different 1-J PVSK cells on their performance parameters , in particular on the J_{SC}. The V_{OC} parameter on these cells can probably serve as a useful performance metric to facilitate rapid-turnaround device development if a relatively large uncertainty is allowed.

Due to the 3-page abstract limit, the linear fitting results on 1-J PVSK submodules, 2-J PVSK/PVSK and PVSK/Si cells are not shown here. A detailed discussion on them will be presented in the conference. In short, for 1-J PVSK submodules and 2-J PVSK/PVSK cells, the trends are comparable to 1-J PVSK cells. In contrast to those two classes of all-PVSK devices, 2-J PVSK/Si cells show a significantly better agreement between fast and steady-state I-V parameters. This difference can be understood by the influence of the silicon bottom junction, for which there is effectively no distinction between fast and steady-state I-V, on the cell performance [3].

III. SIGNIFICANCE OF THE WORK

By comparing PVSK PV cell performance parameter values from fast vs steady-state I-V measurements for a wide range of cells, we identified and quantified trends that offer insights into the applicability and limits of fast I-V in understanding PVSK cell performance. On the basis of these results, we will present recommendations for when and how fast I-Vs may have value in both R&D and manufacturing applications, and, correspondingly, when asymptotic I-V measurements must be used. Overall, the statistical analysis here can be easily adapted as a useful tool for evaluating the accuracy of fast I-Vs by perovskite researchers and could also provide additional supporting information toward the development of fast I-V testing methods for perovskite research labs and production facilities alike.

ACKNOWLEDGMENT

This work was authored by Alliance for Sustainable Energy, LLC, the manager and operator of the National Renewable Energy Laboratory for the U.S. Department of Energy (DOE) under Contract No. DE-AC36-08GO28308. Funding provided by U.S. Department of Energy Office of Energy Efficiency and Renewable Energy Solar Energy Technologies Office (SETO) Agreement Number 38262.

REFERENCES

[1] T. Song, D. Friedman, N. Kopidakis, "Comprehensive Performance Calibration Guidance for Perovskites and Other Emerging Solar Cells," Adv. Energy Mater. 2021, 11, 2100728.

[2] N. Pellet, et al., "Hill climbing hysteresis of perovskite‐based solar cells: a maximum power point tracking investigation," Prog. Photovolt: Res. Appl. 2017; 25:942–950.

[3] T. Song, D. J. Friedman, and N. Kopidakis. "How Useful are Conventional I‐Vs for Performance Calibration of Single‐ and Two‐Junction Perovskite Solar Cells? A Statistical Analysis of Performance Data on≈ 200 Cells from 30 Global Sources." Solar RRL 6, no. 1 (2022): 2100867.

GPU-Accelerated Machine Learning for Analysis of Time-resolved Photoluminescence Data

Calvin Fai, Anthony J. C. Ladd, Charles J. Hages

University of Florida, Gainesville, FL, United States

Quantifying charge-carrier dynamics within a material or device from analysis of optoelectronic measurements is a crucial aspect of improving next-generation solar cells. However, analyses by hand are limited in information yield due to their reliance on simplified physics models. Simulation of full physics models can be computationally expensive. Here we demonstrate a GPU-accelerated machine learning approach via Bayesian parameter estimation for rapid analysis of optoelectronic data. Using time-resolved photoluminescence (TRPL) data of a perovskite absorber as a case study, we demonstrate our ability to estimate carrier mobilities, the doping level, and the radiative recombination rate. Furthermore, while most TRPL analyses are limited to determining an effective minority carrier lifetime, we reliably decompose this recombination lifetime into radiative, bulk nonradiative, and surface nonradiative components by introduction of TRPL data from multiple absorbers of different thicknesses. Our simultaneous collection of these parameters represents a significant increase in the typical information yield from TRPL measurements.

Micro-fabrication and transfer of a detachable Ge epitaxial layer grown on porous germanium

Valentin Daniel[1,2], Jeremie Chretien[1,2], Gwenaelle Hamon[1,2], Mathieu De Lafontaine[1,2], Nicolas Paupy[1,2], Zakaria Oulad El Hmaidi[1,2], Bouraoui Ilahi[1,2], Tadeàš Hanus[1,2], Maxime Darnon[1,2], Abderraouf Boucherif[1,2].

1-Institut Interdisciplinaire d'Innovation Technologique (3IT), Université de Sherbrooke,

3000 boulevard de l'Université , Sherbrooke, J1K 0A5 Québec, Canada

2-Laboratoire Nanotechnologies Nanosystèmes (LN2) - CNRS IRL-3463 Institut Interdisciplinaire

d'Innovation Technologique (3IT), Université de Sherbrooke, 3000 boulevard de l'Université,

Sherbrooke, J1K 0A5 Québec, Canada

Abstract—**Germanium (Ge) substrates are usually used for epitaxial growth of III-V materials but represent a significant part of the cell total cost. The lift-off technique using porosification by bipolar chemical etching is a promising approach to detach the active layers from the Ge substrate and allows Ge substrate reuse. However, this solution raises challenges concerning the delamination and possible structural deterioration of the membrane during the micro-fabrication of the solar cell. In this paper, we successfully apply the main micro-fabrication steps on a Ge membrane grown on porous Ge. The front side process using UV lithography and Au-Ni deposition for contacts has been successfully performed without membrane degradation. Those contacted membranes were also successfully transferred on a host substrate. X rays measurements (XRD) were also performed before and after detachment and shows no damage on the crystalline quality of Ge membrane.**

Index Terms—**Germanium, porosification, lift-off, micro-fabrication, photovoltaics**

I. INTRODUCTION

The industrial photovoltaic multi-junction (M-J) cells with the highest performances are formed from a superposition of III-V and IV elements (GaAs, GaInAs, Ge, etc.). Due to the scarcity of the constituent materials, these solar devices are often very expensive. About 50% of the cost of the cell coming from the substrate itself [1][2]. The 175 μm thick Ge substrate is mainly used as a mechanical support and only few microns are useful for light absorbtion in a Ge bottom cell for the final MJ cells. Reducing the amount of materials in M-J cell is thus mandatory to be financially competitive with other types of high-performance solar cells. In this way, concentrated photovoltaic allows to reduce lateral dimensions of the III-V materials for a given yield , using an external optical system. Another approach is to recycle the substrate after detachment of the effective solar device. The separation of the epitaxial layers from the substrate can be achieved by so-called porous lift-off. A layer of porous material made by bipolar electrochemical etching is placed in-between and used as a weak layer for detachment. This technique was firstly used on silicon [3],[4],[5] and more recently extended to other semiconductors such as Ge [6],[7] and has the advantage of being low cost, large scale and easy to set up. Here we

Study supported and funded by Umicore, UMI-LN2 and Stacelectric

report the micro-fabrication, the detachment and transfer of 1.5 μm thick epitaxial Ge membrane grown on a porosified Ge substrate.

II. METHODOLOGY

A. Sample preparation

1) Porous layer: Bipolar electrochemical etching (porosification) was used to form a sponge-like porous layer on he Germanium substrate. Fig. 1 shows a cross section SEM picture of a Ge substrate porosified on 200 nm (+/- 50 nm) with a porosity of 50% (+/- 5%). These properties were chosen enable detachment while avoiding delamination during the micro-fabrication process.

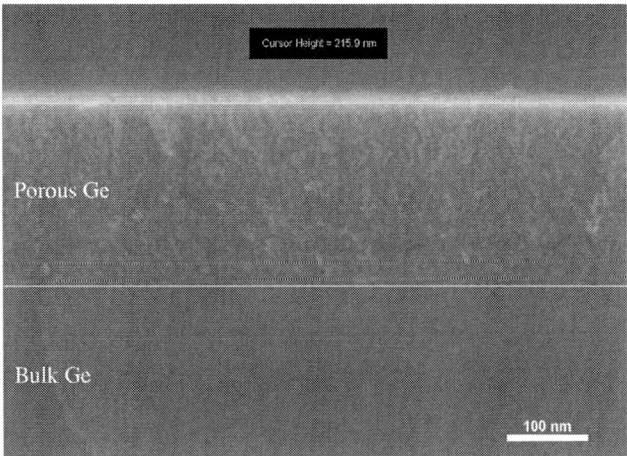

Fig. 1. Cross sectional SEM image of the porous layer after bipolar electrochemical etching (porosification).

2) Epitaxial layer: A germanium layer is then epitaxially grown on the porous germanium surface [8]. During this step, the porous Ge is annealed and transformed into a weak layer made of pillars and voids (see Fig. 2). This reduces the adhesion force at the interface between the Ge membrane and the germanium bulk. This Ge epitaxial layer is mono-crystalline with a thickness near to 1.5 μm, as showed on Fig. 2.

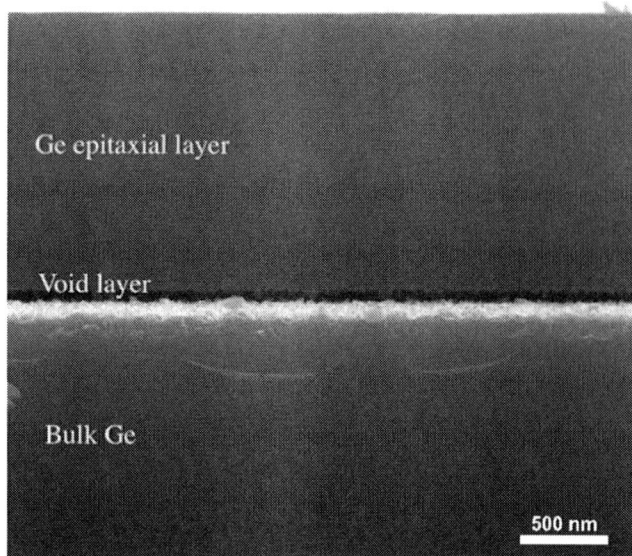

Fig. 2. Cross sectional SEM image of the germanium epitaxial layer on porous restructured after growth and annealing.

B. Micro-fabrication

As shown on the illustration Fig.3, for the front side processing, a photolithography using a 14x14mm wide mask die size, composed by different sizes of metallization patterns, is applied. After the development, a 300 nm thick Au/Ni/Ge/Ni ohmic contact is deposed by evaporation. After the resist removal, the micro-manufactured Ge epitaxial layer (membrane) is bonded to a rigid substrat (glass or silicon) coated with a sacrificial resist layer. The structure is then arranged between 2 pieces of thermal release tape and detached by applying a tensile force normal to the surface of the substrates. After detachment, the membrane transferred is fixed on a permanent substrate (glass) using an epoxy (or PDMS) and dipped in a remover solution in order to dissolve the sacrificial resist layer tand separate the epitaxial Ge thin film from the temporary superstrate.

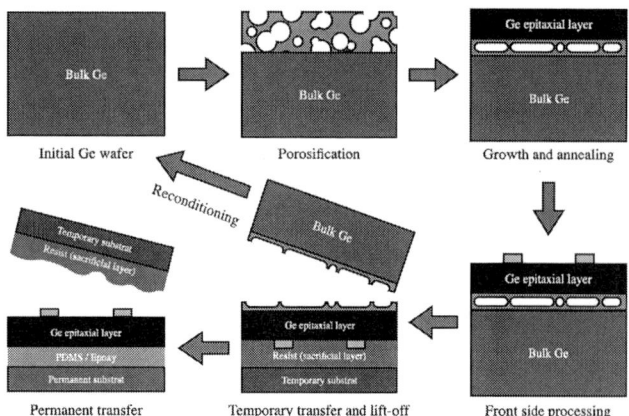

Fig. 3. Illustration of the main steps of the process.

III. PROCEDURE TEST AND RESULTS

A. Front side processing

The contacts on the top of the structure are visible in the cross section (a) and macroscopic top-view images (b) on Fig. 4. The patterns are well defined and not deteriorated. The membrane is intact, without any visible cracks or defects on the surface or at the interface with the weak layer (porous germanium after annealing) even after an immersion in resist remover during 72h. This step being a success, the characteristics of the structure and process used seems thus to be adequate to withstand the creation of front side contacts before membrane lift-off.

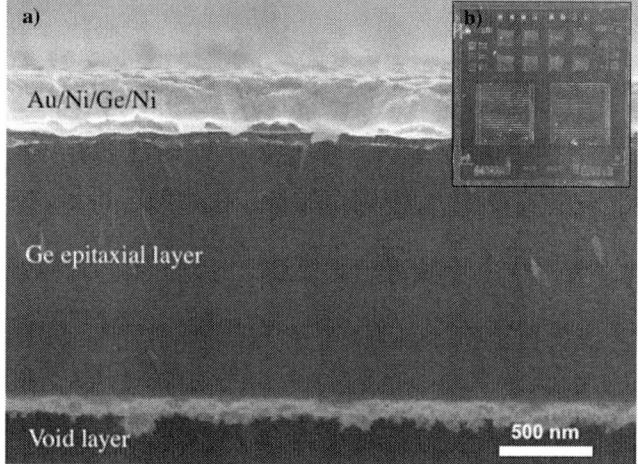

Fig. 4. a) Cross sectional SEM image and b) top view of the structure after front side lithography and metallization.

B. Lift-off and transfer

After the front side processing, the membrane must be detached. Fig. 5 shows the back side of membranes with

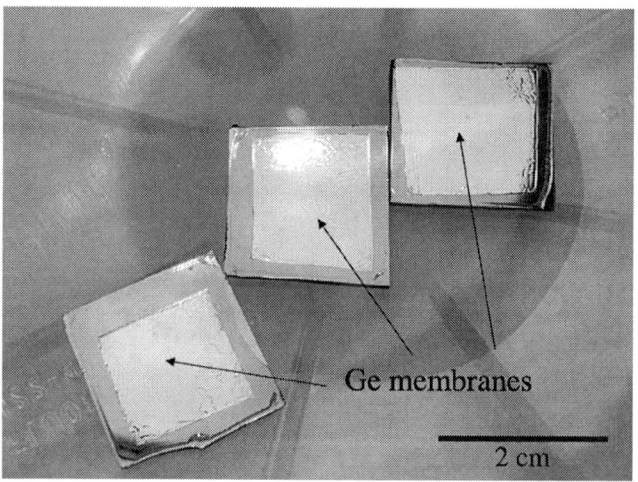

Fig. 5. Back side view of Ge thin membrane detached and transferred on temporary substrat.

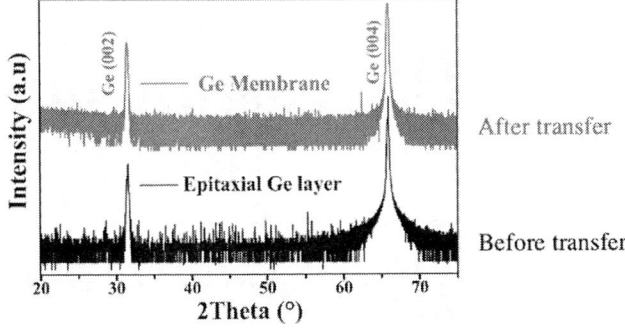

Fig. 6. 2Theta scan of germanium membrane as grown on the porous layer (non-detached) and after transfer on a glass temporary superstrate.

metallization on their front surface after bonding to a temporary superstrate and detachment. The 2 cm^2 Ge membrane are mirror-like, completely detached and transferred with no visible holes or fissures. We attribute the observable ripples (wrinkling) on the surface after the lift-off to the stress in the germanium membrane induced by the thermal expansion of the resist during the hard bake. Then, the crystallinity of the so-called transferred membranes was assessed by XRD as shown by the 2Theta scan depicted in Fig. 6. The observed pattern of the Ge membrane were associated to (002) and (004) crystalline planes similar to those of the Ge epilayer before detachment, demonstrating convincingly that the crystalline properties of the Ge membrane is preserved after all the manipulations performed.

IV. CONCLUSION

We successfully detached and transferred a Ge thin membrane grown on porous germanium which has been created by bipolar chemical etching. This study also demonstrates that the process developed maintains the integrity of a 1.5 μm-thick membrane, this process will thus be applicable to a complete solar cell structure. These results, combined with Ge substrate recycling [9] open the path towards multiple use of Ge substrates for III-V/Ge solar cell fabrication. This approach will lead to a significant cost reduction multi-junction cell.

ACKNOWLEDGMENT

The authors would like to thank NSERC, Innove, CSA, ESA, Mitacs, Umicore and Stace for financial support, and our colleagues Philippe-Olivier Provost, Chantal Simard, Caroline Roy and all the 3it.nano clean-room staff for their help and technical support.

REFERENCES

[1] J. Scott Ward,T. Remo, K. Horowitz, M. Woodhouse, B. Sopori, K. VanSant et al.,Techno-economic analysis of three different substrate removal and reuse strategies for III-V solar cells,Prog. Photovolt: Res. Appl. 2016; 24:12841292.

[2] K. A. W. Horowitz, T. Remo, B. Smith, and A. Ptak, Techno-Economic Analysis and Cost Reduction Roadmap for III-V Solar Cells. United States: N. p., 2018.

[3] I. Mizushima, T. Sato, S. Taniguchi, and Y. Tsunashima., Empty-space-in-silicon technique for fabricating a silicon-on-nothing structure, Appl. Phys. Lett. 77, (2000), 32903292.

[4] N. Milenkovic, M. Drieen, C. Weiss, and S. Janz, Porous silicon reorganization: Influence on the structure, surface roughness and strain, J. Cryst. Growth. 432 (2015) 139145.

[5] S. Kajari-Schrder, J. Ksewieter, J. Hensen, and R. Brendel, Lift-off of Free-standing Layers in the Kerfless Porous Silicon Process, Energy Procedia. 38 (2013) 919925.

[6] A. Boucherif, G. Beaudin, V. Aimez, and R. Ars, Mesoporous germanium morphology transformation for lift-off process and substrate re-use, (2013) Appl. Phys. Lett. 102, 011915.

[7] B. N. Alkurd, A. Cavalli, B.E. Ley, J. Simon, D.L. Young, A.J. Ptak, C.E. Packard, Reformed Mesoporous Ge for Substrate Reuse in III-V Solar Cells, Conf. Rec. IEEE Photovolt. Spec. Conf. (2019) 979982.

[8] N. Paupy et al., Epitaxial growth of detachable GaAs/Ge heterostructure on mesoporous Ge substrate for layer speration and substrate reuse, (2022) submitted abstract to IEEE-PVSC conference (unpublished).

[9] A. Chapotot et al., Multiple substrate reuse: the straightforward reconditioning of Ge wafers after porous separation, (2022) submitted abstract to IEEE-PVSC conference (unpublished).

Native Oxide Growth on CdSeTe for Improved Back Surface Passivation

Adam Danielson, Carey Reich, Mason Mahaffey, Arthur Onno, Zach Holman, Walajabad Sampath

Colorado State University, Fort Collins, CO, United States

Arizona State University, Tempe, AZ, United States

The use of native oxides have historically been used in numerous PV technologies to passivate surfaces and improve voltage and conversion efficiency. Using XPS, in this study we observe the growth of tellurium oxides on the back surface of CdSeTe films when stored in air for several weeks. We note considerable differences in the prevalence of tellurium oxides between as-deposited and chlorine-treated films. Finally, external radiative efficiency measurements show that ERE correlates strongly with increased tellurium oxide formation. This indicates that the oxide layer is passivating the back surface of CdSeTe, a crucial step towards improving the open-circuit voltage.

Optical Characterization of Thin Film Cu$_x$AlO$_y$ in the CdTe Device Configuration

Indra Subedi, Kamala Khanal Subedi, Prabin Dulal, Adam B. Phillips, Michael J. Heben, Randy J. Ellingson, and Nikolas J. Podraza

Department of Physics & Astronomy and Wright Center for Photovoltaics Innovation & Commercialization, The University of Toledo, Toledo, OH 43606, USA

Abstract— **Optical properties and band gap energy of solution processed p-type transparent Cu$_x$AlO$_y$ thin film deposited on sodalime glass are determined using spectroscopic ellipsometry. This is a promising material for a p-type transparent back contact and passivation layer for thin film CdTe based solar cells. The direct optical band gap obtained from Tauc plot is found to be 3.64 ± 0.01 eV. Further characterization is also done with this layer in the CdTe / CdS device stack. Cu$_x$AlO$_y$ conformally coats and smoothens the CdTe surface on the CdTe / CdS device stack indicating improved surface quality and surface passivation.**

Keywords— *Copper aluminum oxide, optical properties, thin film, CdTe, spectroscopic ellipsometry*

I. Introduction

Cadmium telluride (CdTe) based thin film solar cells are the most commercially produced photovoltaic (PV) device technology after crystalline silicon wafer-based cells. Copper aluminum oxide (Cu$_x$AlO$_y$) shows p-type conductivity and can be used as transparent conducting oxide (TCO) and passivation layers for both mono-facial and bifacial CdTe based thin film solar cells [1]. Spectroscopic ellipsometry is a robust and non-destructive characterization technique which is widely used to measure the optical properties and thicknesses of thin films, bulk crystals, glass, nanomaterials, and other component layers of PV devices. In this work, we primarily utilize spectroscopic ellipsometry to measure optical properties of thin film Cu$_x$AlO$_y$ deposited on sodalime glass as well as structural parameters in the CdTe solar cell device stack.

II. Experimental Details

Cu$_x$AlO$_y$ thin film have been deposited on sodalime glass by spin coating method described in Khanal Subedi *et al.* [1]. Ellipsometric spectra of solution processed thin film Cu$_x$AlO$_y$ deposited on sodalime glass have been collected at room temperature and 70° angle of incidence using a single rotating compensator multichannel ellipsometer [2,3] over a spectral range from 0.735 to 5.887 eV (M-2000FI, J.A. Woollam Co., Inc.). Ellipsometry measurement are also performed on the unfinished vapor transport deposited CdS / CdTe device stack before Cu$_x$AlO$_y$ layer deposition. Thin film Cu$_x$AlO$_y$ is deposited on top of the CdTe stack and the same spot on the device is again measured in the same way as the film on glass.

III. Results and Discussions

A. Cu$_x$AlO$_y$ layer on glass

Ellipsometric spectra have been analyzed by using a least square regression and an unweighted error function, σ [4]. Optical properties in the form of the complex dielectric function or complex index of refraction ($\varepsilon = \varepsilon_1 + i\varepsilon_2 = (n + ik)^2$) spectra for the Cu$_xAlO_y$ thin film and structural parameters are obtained from this analysis.

The structural model used for this analysis consists of semi-infinite sodalime glass substrate / Cu$_x$AlO$_y$ thin film / air ambient. The optical properties of the Cu$_x$AlO$_y$ film are modeled by using a combination of a constant additive term to ε_1 (ε_∞), a Sellmeier expression [5], and a Tauc-Lorentz oscillator [6-7]. The Sellmeier expression is defined as:

$$\varepsilon(E) = -\frac{A_s}{E^2} \qquad (1),$$

where A_s is the amplitude of Sellmeier expression; in this form the resonance energy is fixed at 0 eV. The Tauc-Lorentz oscillator is defined as:

$$\varepsilon_2(E) = \begin{cases} \dfrac{AE_0\Gamma\left(E - E_g\right)^2}{\left(\left(E^2 - E_0^2\right)^2 + \Gamma^2 E^2\right)E} & E > E_g \\ 0 & E \le E_g \end{cases} \qquad (2a)$$

$$\varepsilon_1(E) = \frac{2}{\pi} P \int_{E_g}^{\infty} \frac{\xi\varepsilon_2(\xi)}{\xi^2 - E^2} d\xi \qquad (2b)$$

where A is the amplitude, E_0 is the resonance energy, and Γ is the broadening parameter. E_g is an optical band gap energy corresponding to the absorption onset, and P is the Cauchy principal part of the integral. Reference optical properties for the sodalime glass substrate are used [8]. Spectra in ε as a function of photon energy is shown in Fig. 1. The Cu$_x$AlO$_y$ film thickness obtained from the analysis is found to be 18.0 ± 0.3 nm. All the fit parameters for the parametric model are reported in Table I.

U.S. Air Force Research Laboratory, agreement numbers FA9453–19–C–1002 and FA9453–21–C–0056

978-1-7281-6118-1/22 $31.00 © 2022 IEEE

TABLE I.   PARAMETERS DESCRIBING SPECTRA IN ε FOR Cu_xAlO_y FILM ON SODALIME GLASS WITH $\sigma = 3.5 \times 10^{-3}$ AND $\varepsilon_\infty = 1$

Oscillators	Fit Parameters			
	A, As	Γ (eV)	E_0 (eV)	E_g (eV)
Tauc-Lorentz	12.9 ± 0.1 eV	7.19 ± 0.08	6.27 ± 0.04	0.65 ± 0.04
Sellmeier	0.018 ± 0.008 eV²			

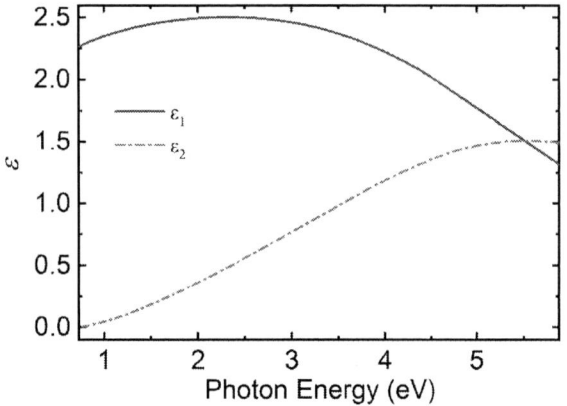

Fig. 1. Complex dielectric function ($\varepsilon = \varepsilon_1 + i\varepsilon_2$) obtained from parametric model for a Cu_xAlO_y thin film on sodalime glass substrate.

Absorption coefficient (α) spectra are determined from $\alpha = 4\pi k/\lambda$, where k is the extinction coefficient and λ is the photon wavelength. The direct optical band gap is obtained via Tauc plot by extrapolating the photon energy at which $(\alpha h\nu)^2 = 0$ as shown in Fig. 2. The linear range of the extrapolations here correspond to $\alpha = 1.59$ to 2.12×10^5 cm^{-1}. The direct band gap energy obtained from extrapolation is 3.64 ± 0.01 eV. There is also a non-zero value of α below the direct band gap which is possibility originates due to presence of defects and non-stoichiometry.

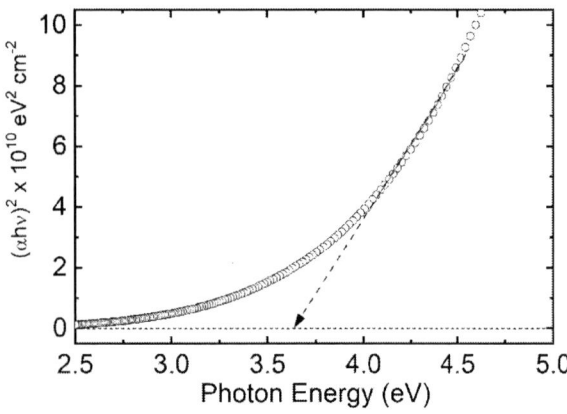

Fig. 2. Spectra in $(\alpha h\nu)^2$ as a function of photon energy used to obtain the direct band gap energy of Cu_xAlO_y film on sodalime glass.

B. Cu_xAlO_y on CdTe / CdS device stack

In the analysis of the CdTe device stack, all the optical property parameters describing the CdTe / CdS / highly resistive transparent (HRT) layer / transparent conducting oxide coated glass are used from the model based on Koirala et al. [9]. Fig. 3 shows the schematic and structural parameters of this CdTe device stack and thicknesses of individual layer obtained from the fit to measured ellipsometric spectra. CdTe surface roughness is modeled using Bruggeman effective medium approximation (EMA) layers [10] consisting of f_c fraction of the volume to be identical to the bulk CdTe material and the remaining $1-f_c$ to be void. The void fraction increases in the EMA layers towards the ambient. EMA1 is near the bulk CdTe layer and EMA3 is towards the ambient.

For the CdTe stack / Cu_xAlO_y, the Cu_xAlO_y component is also considered, and a smaller subset of photon energy range from 1.5 to 3.0 eV is chosen to simplify the model by avoiding interference fringes due to coherent multiple reflections below the band gap of CdTe (< 1.5 eV) and stronger surface scattering at higher photon energies (> 3.0 eV). Parameters describing spectra in ε for Cu_xAlO_y layer in stack are also fixed from the analysis of Cu_xAlO_y thin film on glass. In the CdTe stack / Cu_xAlO_y analysis, all the thicknesses below the EMA layers are fixed to the values from the stack without Cu_xAlO_y.

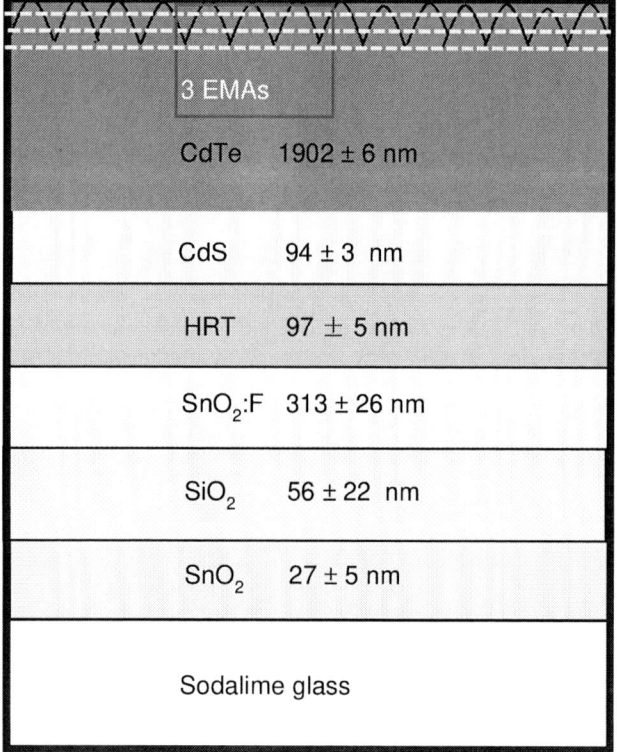

Fig. 3. Shecmatic and layer thickness for CdTe / CdS device stack. Top blue layers represent void fraction.

978-1-7281-6118-1/22 $31.00 © 2022 IEEE

TABLE II. STRUCTUARAL PARAMETERS FOR EACH EFFECTIVE MEDIUM APPROXIMATION (EMA) LAYER IN THE CdTe DEVICE STACK BEFORE AND AFTER Cu$_x$AlO$_y$ DEPOSITION

	Thickness nm	Void fraction %	Cu$_x$AlO$_y$ fraction %	Thickness nm	Void fraction %	Cu$_x$AlO$_y$ fraction %	Thickness nm	Void fraction %	Cu$_x$AlO$_y$ fraction %
	EMA1			EMA2			EMA3		
Before Cu$_x$AlO$_y$	50 ± 1	11± 1	N/A	42 ± 1	43 ± 1	N/A	76 ± 2	87 ± 1	N/A
After Cu$_x$AlO$_y$	38 ± 1	0	11	39 ± 1	0	43	51 ± 1	68	19 ± 1

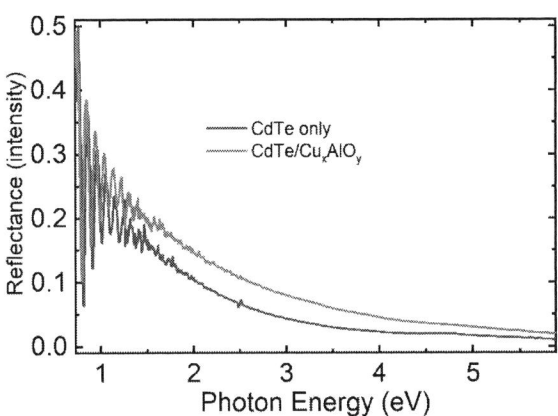

Fig. 4. Average reflected intensity of CdTe and CdTe / Cu$_x$AlO$_y$ stack measured from the film side.

The Cu$_x$AlO$_y$ layer is assumed to compeltely fills the void fractions in the bottom two EMA layers. Void is still present in EMA3 due to Cu$_x$AlO$_y$ conformally coating the CdTe. Thicknesses of the EMA layers also decrease in this fit. This decrees in EMA thickness indicates passivation of the CdTe back surface and leads to an increase in the reflected intensity due to less scattering from the surface, a greater contrast in optical properties of the top surface and void, or both. This is demonstrated by a higher specularly reflected average intensity for the CdTe / Cu$_x$AlO$_y$ stack in comparison to the CdTe device stack only at 70° angle of incidence as shown in Fig. 4. The effective thickness of Cu$_x$AlO$_y$ in the CdTe device stack is the sum of the products of each EMA layer and the respective Cu$_x$AlO$_y$ material fraction. This effective Cu$_x$AlO$_y$ thickness is 30.8 nm which is distributed over all three EMA layers.

IV. CONCULSION

Optical properties in terms of complex dielectric function of solution processed thin film Cu$_x$AlO$_y$ deposited on sodalime glass have been determined from spectroscopic ellipsometry. This material has potential application as p-type TCO and passivation layer on CdTe based solar cells. Structural parameter of Cu$_x$AlO$_y$ deposited on a CdTe / CdS device stack are determined. Cu$_x$AlO$_y$ conformally coats and smoothens the CdTe surface indicating improved surface quality of the CdTe back contact and surface passivation. This approach is also applicable to characterizing other back contact materials deposited on rough CdTe in devices.

ACKNOWLEDGMENT

This material is based on research sponsored by Air Force Research Laboratory under agreement numbers FA9453–19–C–1002 and FA9453–21–C–0056. The U.S. Government is authorized to reproduce and distribute reprints for Governmental purposes notwithstanding any copyright notation thereon. The views expressed are those of the authors and do not reflect the official guidance or position of the United States Government, the Department of Defense or of the United States Air Force. The appearance of external hyperlinks does not constitute endorsement by the United States Department of Defense (DoD) of the linked websites, or the information, products, or services contained therein. The DoD does not exercise any editorial, security, or other control over the information you may find at these locations. Approved for public release; distribution is unlimited. Public Affairs release approval #AFRL-2022-2465.

The authors would like to thank Willard and Kelsey Solar Group for providing CdS / CdTe film stack samples.

REFERENCES

[1] K. Khanal Subedi *et al.*, "Enabling bifacial thin film devices by developing a back surface field using Cu$_x$AlO$_y$," *Nano Energy*, vol. 83, p. 105827, 2021.

[2] J. Lee, P. Rovira, I. An, and R. Collins, "Rotating-compensator multichannel ellipsometry: Applications for real time Stokes vector spectroscopy of thin film growth," *Review of scientific instruments*, vol. 69, no. 4, pp. 1800-1810, 1998.

[3] B. Johs, J. A. Woollam, C. M. Herzinger, J. N. Hilfiker, R. A. Synowicki, and C. L. Bungay, "Overview of variable-angle spectroscopic ellipsometry (VASE): II. Advanced applications", presented at *Optical Metrology: A Critical Review*, pp. 29-58, 1999.

[4] B. Johs and C. Herzinger, "Quantifying the accuracy of ellipsometer systems," *physica status solidi c*, vol. 5, no. 5, pp. 1031-1035, 2008.

[5] R. W. Collins and A. S. Ferlauto, H. G. Tompkins and E. A. Irene, Eds. *Optical physics of materials* (Handbook of Ellipsometry). Norwich, NY: William Andrew Inc., 2005, pp. 93-235.

[6] G. E. Jellison, Jr. and F. A. Modine, "Parameterization of the optical functions of amorphous materials in the interband region," *Applied Physics Letters*, vol. 69, no. 3, pp. 371-373, 1996.

[7] G. E. Jellison, Jr. and F. A. Modine, "Erratum: Parameterization of the optical functions of amorphous materials in the interband region," *Applied Physics Letters, vol. 69, no. 14, pp. 2137-2137, 1996.*

[8] M. M. Junda and N. J. Podraza, "Optical properties of soda lime float glass from 3 mm to 148 nm (0.41 meV to 8.38 eV) by spectroscopic ellipsometry," *Surface Science Spectra*, vol. 25, no. 1, p. 016001, 2018.

[9] P. Koirala *et al.*, "Through-the-glass spectroscopic ellipsometry for analysis of CdTe thin-film solar cells in the superstrate configuration," *Progress in Photovoltaics: Research and Applications*, vol. 24, no. 8, pp. 1055-1067, 2016.

[10] I. Subedi, K. P. Bhandari, R. J. Ellingson, and N. J. Podraza, "Near infrared to ultraviolet optical properties of bulk single crystal and nanocrystal thin film iron pyrite," *Nanotechnology, vol. 27, no. 29 p. 295702, 2016.*

A Tool for the Simulation, Evaluation and Teaching the Operation of Low Power Microgrids

Johann A. Hernández M[1], Adolfo A. Jaramillo M[1], and Carlos A. Arredondo-Orozco[2]

[1]Grupo de investigación LIFAE. Universidad Distrital Francisco José de Caldas, Bogotá – Colombia
[2]Grupo de investigación en energía GRINEN. Universidad de Mediellín, Medellín – Colombia

Abstract— **This paper presents the development of a simulation tool to analyze the behavior of a microgrid installed at Universidad Distrital FJDC in Bogotá, Colombia. The microgrid was developed using the software DIgSILENT PowerFactory® and integrates the different elements of electrical power systems (EPS) such as electricity generation (DC and/or AC) (synchronous motors, wind turbines and photovoltaic panels), a battery storage system, a module for balancing different types of electrical loads and the input/output system to the external grid. The development allows performing different analyses such as load flows, obtaining reports for the general system or each particular subsystem and evaluating the behavior of the microgrid under different conditions including the input and/or output of the different subsystems. It was developed using specifically the DPL (DIgSILENT Programming Language) language suite, included within the DIgSILENT® software. Given the versatility of the programming language, the tool allows the inclusion of a new PES to the microgrid, which makes it a support tool for teaching (educational purposes) and research, since it allows extending its operation beyond just evaluating the existing physical microgrid.**

Keywords—microgrids, photovoltaic generator, simulations, battery storage system, DIgSILENT PowerFactory, teaching.

I. INTRODUCTION (*HEADING 1*)

The rapid growth of renewable energy sources and storage systems has allowed their increasing use in low power systems and at the same place of consumption. With this, the growth of microgrids, as an alternative to traditional large-scale energy generation and transmission systems, is increasing and is presented as an attractive alternative for users [1 – 4]. This accelerated development requires new engineers to be trained in these topics, in addition to the traditional ones of the operation of electrical systems. In this sense, this work presents a part of a larger project focused on developing training in installation, operation, operation and simulation (among others) of microgrids. At Universidad Distrital Francisco José de Caldas in Bogotá, Colombia, a microgrid was implemented, consisting of several Electrical Power Systems (EPS), which have been (and are being) developed by researchers from different areas, such as renewable energies, power electronics, power quality and electric vehicles, among others. Each of these EPS has been modeled using different simulation software, which has made it difficult to associate them in a direct way. In this work, the integration of these EPS was performed by means of their respective input/output variables, through a common simulation structure [5].

For the integration of these EPS in a simulation, the DIgSILENT® PowerFactory (DPF) software and its DPL language suite were chosen, due to its versatility in the adaptation of other software to it. Particularly because the respective representation of the behavior of the output variables can be obtained through data blocks designed internally in the software [6]. The development of this simulation tool has not only served as a study for the implemented microgrid, but has also been used as a basic tool in the connection of future elements to be developed in the microgrid, as well as a learning and research tool in this field.

For the development of the tool, the follow stages were developed: 1. Identification and characterization of the general electrical structure of the microgrid operation; 2. Characterization of each EPS component through the design of its block structure adaptable to the DPL language and the programming of the input/output variables; 3. Validation of each particular block; 4. Integration of the component blocks and validation of the simulation results of the integrated microgrid.

From this work, it can be affirmed that the development of the simulation not only serves as a case study of the implemented microgrid, but can also be used as a basic tool in the connection of future elements to be developed to it (either load or generation), or even for possible microgrids implemented in other faculties of the Universidad Distrital, thus promoting the use of renewable energies in the national electricity system.

II. IDENTIFICATION AND CHARACTERIZATION OF THE MICROGRID

This stage consisted of collecting information about the development of each component, listing the EPSs that were ready for implementation and acquiring their respective output variables. The electrical network of the building to which the microgrid is connected was taken into account, from the point of connection to the external power supply network to the common connection point. The design of the microgrid structure interconnected the EPSs that have been previously developed. This was done by accessing the output variables of each EPS and modeling their blocks within the microgrid structure in the simulation. Fig. 1 shows the electrical schematic where the node in red represents the connection point of the microgrid. The bars on Fig. 1 correspond to the interconnection points between the different components. The design contains 12 busbars, where the microgrid (MR) node is the most important as it contains the

978-1-7281-6118-1/22 $31.00 © 2022 IEEE

EPS. Thus, in this structure the MR node was implemented to represent the laboratory of the research group and interconnect all the EPS in this busbar within DPF. Two power generation blocks and three load blocks were implemented in the MR node.

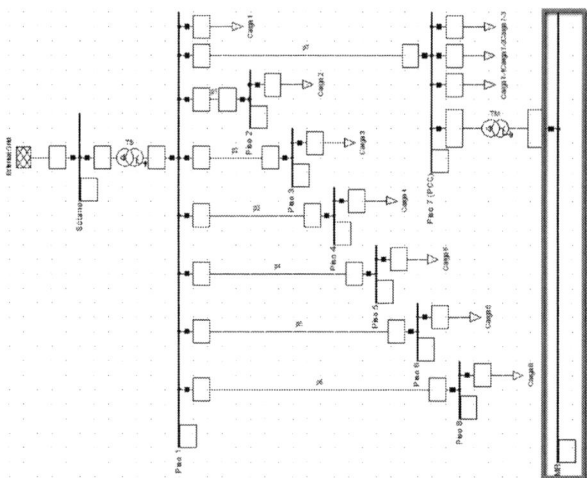

Fig. 1. Simulation diagram of the electrical structure of the building where the microgrid is connected.

A. Components Characterization

It consisted of determining the input and/or output variables of each component and their respective adaptation to the DPL language.

Photovoltaic generator: Composed of a 240 Wp system and a micro inverter for grid connection. The output variables of the PV generator were obtained directly implemented in the DPF, which made the process easier, and also allowed to perform the first tests to this block with the previously designed structure. This output profile was provided in the work done in [7].

Battery bank: Composed of 20 batteries in series of 65 Ah at 12 V each, which make up a nominal 15.6 kWh bank. The output variables of this battery bank were provided by the work done in [8].

Motors: There are two motors, one standard and one premium. The output profiles of these were obtained experimentally and characterized in an Excel® spreadsheet.

Surplus management system: It is a load that consists of a system composed of several electrical components that allow obtaining electrical energy from the photovoltaic generator and feeding certain loads of the system. Additionally, if the energy is sufficient, surpluses are delivered to the grid through an interconnection to the grid that allows the system to operate autonomously or as a grid connected one. Its characterization was obtained from [9].

Residential user load: This is a load emulation prototype that allows observing in a practical way the demand management of a user that actively participates in the grid [10].

With the exception of the photovoltaic generator (which is developed in [7]), for all the EPSs, their blocks were obtained in the simulation through the implementation of the respective output variables in DPL. The power generation EPSs (purple) were modeled with blocks made up of DPF modules (Fig. 2).

The loads are modeled with their respective module included in DPF where their characteristic variables are programmed.

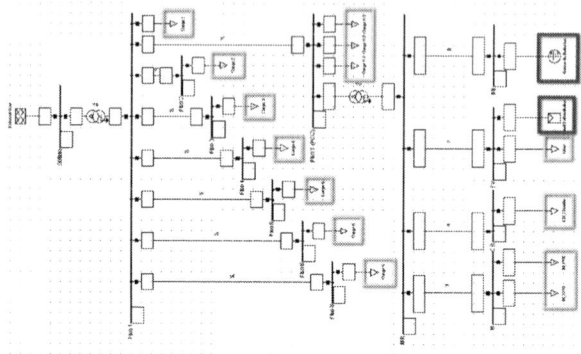

Fig. 2. Block structure of the microgrid in DPF.

To simulate the behavior of each EPS, according to the output profiles obtained, the corresponding block was determined by programming each one in DPL. This suite allows the user to generate matrices to integrate the output variables to the programming routines, which, through programming code in C, allow linking these variables to the modules implemented in the microgrid structure. The data were organized according to the parameters of each EPS. In this way, three large blocks were obtained: photovoltaic generator block, battery block and load block.

B. Implementation of Blocks for each EPS

The implementation of the simulation blocks was developed in the directory of tools and modules offered by the DPF simulation suite. In the case of power generation, the static generator modules (ElmGenStat) in DPF were used. The loads were modeled from the same load modules (ElmLodLv) in DPF and the parameters were modified according to their respective output variables.

A block is made up of two or more elements. In this way, a bar, a line and the module representing the element to be specifically modeled were obtained. Thus, the following components are obtained:

Generators: For the simulation of the microgrid, two generation components included in DPF were used as static generators. These can behave as a photovoltaic generator and as a battery bank, since they contain internal parameters that adjust to the specifications of each generation EPS.

Loads: There are two types of loads within the DPF, general loads and low voltage loads. The low voltage load was chosen for use in the simulation since its internal parameters fit better to the requirements of each EPS in the microgrid. This load required to be connected to a busbar for its implementation.

C. Programming and development of modules

For each generation and load module, the respective programming was performed in DPL. This is accessed through the Data Manager, which creates a new object within the simulation. On the options window of the DPL command, the code is developed, its internal variables, the execution of the routines, the output variables and the link with the elements of the simulation are created. All DPL blocks are integrated in a main file, which is the only executable file of the simulation.

The DPL command contains within its own library objects that allowed to integrate in a more detailed way any external variable to the program and in different ways. Thus, there are vectors (IntVec), matrices (IntMat), forms (IntForm), filters (SetFilt) and also several additional commands that allowed to execute different calculations directly in the same programming code of the block, such as load flows (ComLdf), short circuit calculation (ComShc) and harmonics (ComHmc), among others.

The DPL command offers five basic tabs for its manipulation: basic and advanced options, code, description and version (copyright). Within the basic options the input parameters to the code were created, and additionally the connection to the external objects that the code controls were generated, which in this case are the EPSs previously modeled in the microgrid structure.

The code for each generation and loading block was designed and programmed with the considerations given above, and is specific to each EPS. Fig. 3 shows an example of a script developed for one of these blocks.

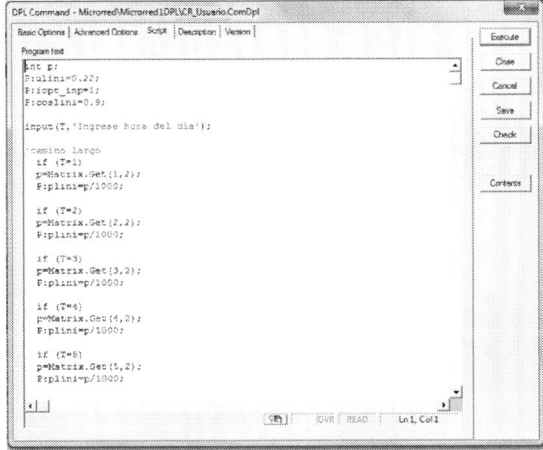

Fig. 3. DPL code (script).

After creating the input and output parameters for each block, these are included in the general code along with the objects necessary to perform the data import and any additional calculations required. The code has different restrictions determined by the behavior of each EPS. These restrictions were handled with input parameters that the user defines according to the case study to be performed, such as the time of day at which you want to simulate the power generated by the photovoltaic generator.

Each of the developed modules was validated by comparing the output variables with experimental data. In all cases there was a satisfactory error, which made it possible to determine the reliability of the developed tool.

III. Microgrid Simulation

This is represented in four basic aspects: the structure of the microgrid, its generation and load blocks, the DPL command and the export of the results.

A. Structure of the microgrid

It is formed by all the elements that were interconnected to shape the model of it within the simulation, which are located from the point of external connection of electrical energy, which supplies energy to the University building by means of a transformer, until reaching the point of common connection with the microgrid. The structure of the microgrid is composed of busbars, lines, a transformer, generators and loads, as can be seen in the Fig. 4.

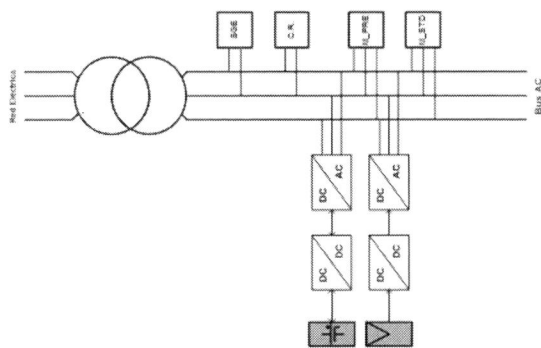

Fig. 4 Schematic of the microgrid.

B. Generation and load blocks

These blocks represent each EPS to be interconnected in the microgrid. Each of them is composed of modules that provide DPF and are programmed to emulate the behavior of each element of the microgrid.

C. DPL command

The programming of each EPS was developed in the DPL platform, which allows controlling, by means of external variables and with models based on parameters and data files, all the elements of the basic structure of the microgrid with their respective EPS, after been interconnected. The general DPL command executes the simulation with the required study cases and allows the user a more comfortable handling of the simulation, providing a graphical interface that allows the introduction of the control variables of each EPS.

D. Module of export of results

Finally, a data export module, also developed through DPL, was available to obtain output data and simulation results.

E. Simulation

A case study was developed with the following considerations, which were also validated experimentally:

Photovoltaic generator: In the case of the photovoltaic panel, it delivers power, according to the incident irradiance. Thus, an irradiance of 850 W/m^2 was chosen, which, according to statistical studies for Bogota, is a common value at 12:00 m [11]. Additionally, the operation of the solar panels depends on the ambient temperature to which they are subjected, for this, 30°C was chosen.

Battery bank: According to the output variables obtained from this EPS, the block was programmed to operate with the

978-1-7281-6118-1/22 $31.00 © 2022 IEEE

DoD (Depth of Discharge) at 20%, as it is an intermediate value of its full state of charge.

Residential load and management system: The loads have the particularity of being modeled together with the photovoltaic generator. For this reason, within its programming, the value of an hour of the day was required, which was previously linked to an irradiance value with probabilistic functions developed for this purpose. It was chosen 12:00 m according to the previously described.

Electric motors: A power quality study was performed for the motors interconnected to the microgrid. The results of this study allowed modeling their behavior in detail, since the tests were carried out for six load variations with six different speeds, based on the measurement points proposed by IEC60034-2-3 (speed variations of 25%, 50%, 75% and 100%). Based on the above and, given the average load at which the motors operate, 50% is chosen as the input parameter to the simulation for each of these [12].

IV. RESULTS

The results are shown in the respective DPF (BoxResults), where the active and reactive power values for all the loads can be observed, while for the busbars, the line-to-line voltage, the per unit voltage (p.u.) and their respective phase shift are displayed. For example, for the standard motor load a delivered active power of 347.6 W and a reactive power of 233.5 VAR were observed, as shown in Fig. 5.

From the simulation, it was found that loads (such as motors) should not be connected because they have a high-power demand that the microgrid cannot supply.

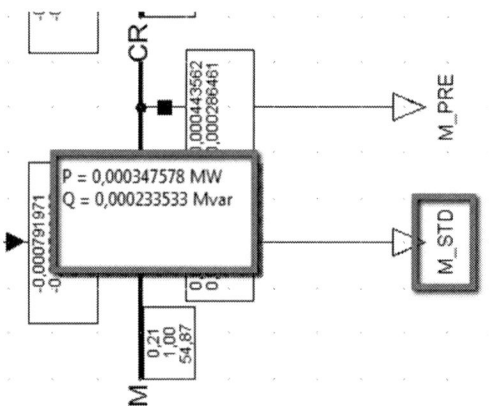

Fig. 5. P and Q standard motor values.

After executing the load flow, the behavior of the microgrid was analyzed by observing its different electrical variables: voltage, current, power and electrical losses in each of its elements. Mainly, a power analysis was made to find the power contributed by the EPS generators to the microgrid and the power consumed by the loads. The power consumed by the loads (984.3W) was added up. However, from the results of the main case study, the microgrid transformer was only supplying the microgrid with 727.6W, which indicates that the EPS generators are delivering the difference to the microgrid, as can be seen in Fig. 6.

The difference corresponds to 256.7W, of which the PV panel delivers 203.1W to the microgrid and the battery bank (for a DOD of 20%) delivers 53.6W, thus completing the total consumption of the sum of the loads connected to the microgrid as follows:

$$P_T = P_{Tr} + P_P + P_B$$

$$984.3W = 727.6W + 203.1W + 53.6W$$

The power consumed by the motors (791.97W), exceeds the value supplied by the transformer. From this it can be deduced that in this case study, the motors are the ones that consume the most power, so they cannot be connected to the microgrid if none of the SEP generators are connected supplying power.

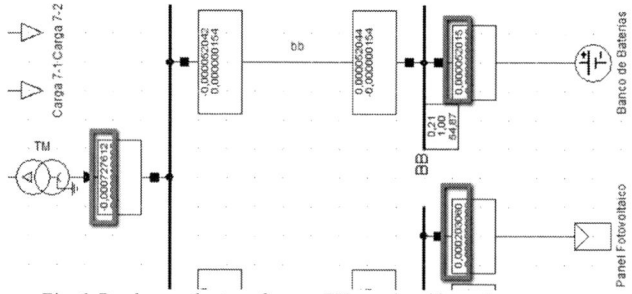

Fig. 6. P value on the transformer, PV panel and batteries.

On the other hand, the power consumed by the load (173.2W) is supplied by the power delivered by the PV generator (203.06W) and still has enough to deliver 29.7W to the microgrid.

As previously mentioned, the batteries deliver 52 W, according to the results obtained from the simulation.

From the simulations performed, the following observations can be made:

- The power of the photovoltaic panel is maintained from a minimum power value of 120 W to 230 W as it approaches its maximum irradiance value with the arrival of the zenith.

- The power delivered by the battery bank is almost constant between 50 W and 60 W and only decreases after 25% discharge (DOD).

- The surplus management system keeps its power consumption constant at around 175 W.

- The motors are the EPS with the highest consumption in the microgrid; they should not be connected until more generation EPS are added.

- The residential load maintains a power consumption between 20 W and 30 W and is the EPS that least influences the behavior of the microgrid since the loads are resistive.

A. Output profile

The output profile is a record of data that can be extracted from the simulation, which includes variables such as voltage, cur-rent, power and the loss of the elements that makes it up. The simulation has a module for exporting results for more convenient analysis, and for possible use as an output profile in other projects requiring the microgrid data. The case study is saved in a notepad file from which data can be extracted to corroborate the values provided by the DPF results table, allowing the user to have more comfortable handling of it, as shown in Fig. 7.

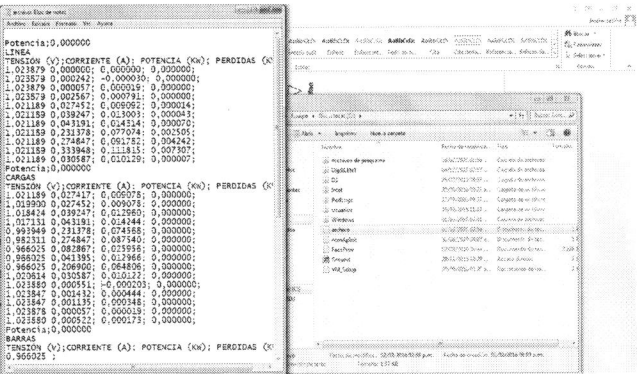

Fig. 7. Output profile in notepad

V. CONCLUSIONS

Each block of the microgrid structure that emulates the behavior of each EPS in DIgSILENT® was programmed and implemented. For this, the output profiles of the elements are used to be imported from the simulation, thus achieving the development of case studies in which load flow calculations, in steady state, are performed and output variables such as active power, voltage, current and electrical losses of the entire microgrid are obtained.

The implementation of all the EPS collected in DIgSILENT® was achieved, thus verifying that the chosen software was the right one for the present work.

DIgSILENT® is a software that allows to implement the behavior of virtually any component of an electrical system, regard-less of its size or type, such as generation, conversion, or load.

Although the software does not provide detailed models of the various components and some elements are not detailed in its libraries (such as DC/DC converters or detailed models of photovoltaic modules), these can be programmed in a DPL block, as well as their behavior (whether dynamic, static or any other type) that can be easily incorporated into any simulation. This is done effectively using matrices and programming codes, standardizing the data to be handled efficiently, which allows greater versatility of the tool. This allows its use to be extended not only to the evaluation of the implemented network, but also to more case studies and can be used in a general way in the learning and research of microgrids.

After performing the simulation and reviewing the results, it is found that is not convenient to connect motors to the microgrid, as they have a much higher power consumption than the one generated by the microgrid.

On the other hand, from the simulation results, it was found that the EPS such as the photovoltaic panel and the electric motors are the most relevant, because their input parameters significantly affect the behavior of the microgrid; while the surplus management system, the battery bank, and the residential load, have low dynamic behaviors where the output variables are almost constant for the different generation and load configurations.

Finally, the simulation is a useful and dynamic educational tool that allows the addition of modules thanks to its open and flexible programming when using the DPL.

REFERENCES

[1] S. Ullah, Ahmed M.A. Haidar, Paul Hoole, Hushairi Zena andTony Ahfock. "The current state of Distributed Renewable Generation, challenges of interconnection and opportunities for energy conversion based DC microgrids". Journal of Cleaner Production. Vol. 273, 10 November, 2020.

[2] S. Parhizi, H. Lotfi, A. Khodaei, S. Bahramirad. "State of the Art in Research on Microgrids: A Review". IEEE Access.Vol 3, 2015.

[3] K. Sarwagya, P. K. Nayak. "An extensive review on the state-of-art on microgrid protection". 2015 IEEE Power, Communication and Information Technology Conference (PCITC).

[4] E. Hossain, E. Kabalci, R. Bayindir, R. Perez. "Microgrid testbeds around the world: State of art". Energy Conversion and Management. Vol. 86, October 2014, Pages 132-153.

[5] A. Chaparro, J. D. Liscano Segura. "Diseño e Implementación de una Microrred en la Universidad Distrital Francisco José de Caldas Sede de Ingeniería". Proyecto de Grado, Universidad Distrital, Bogotá. 2017.

[6] A. Constantin, A. Ellerbrock, F. Fernandez and J. Rue, "Co-Simulation of Power Electronic Dominated Networks,". IEEE Power and Energy Magazine. Vol. 18, no. 2, pp. 84-89, March-April 2020.

[7] J. A. Hernández Mora, "Metodología para el análisis técnico de la masificación de sistemas fotovoltaicos como opción de generación distribuida en redes de baja tensión," Doctoral Thesis. Universidad Nacional de Colomnbia, 2012.

[8] A. F. Campos Fajardo, and R. A. Gómez Porras, "Metodología para incrementar los ciclos de uso de un banco de baterías de Plomo-ácido con diferentes tipos de arreglos en paralelo," Universidad Distrital Francisco José de Caldas, 2015.

[9] J. A. Hernández, C. Korez Franco, D. A. Avila, J. A. Murillo. "Design and implementation of a Management System of surplus energy generated by a distributed generation system, case study GCPVS". 2014 IEEE 40th Photovoltaic Specialist Conference (PVSC).

[10] D. A. Balaguera, A. F. Cortés, M. A. Urueña, J. A. Hernández. "Design and implementation of software to describe the behavior of a photovoltaic generator connected to the low voltage grid". 2013 ISES Solar World Congress 57, 178-187.

[11] IDEAM and UPME, "Mapas de radiacón solar global sobre una superficie plana," Colombia, 2015.

[12] F. J. Cadena Villalba and M. I. Ballesteros Camacho, "Distorsión armónica generada por accionamientos eléctricos de control de velocidad basados en motores de inducción de propósito general y eficiencia premium," Universidad Distrital Francisco José de Caldas, 2015.

Translating Material-Level Characterization of Carbon-Nanotube-Reinforced Composite Gridlines To Module-Level Degradation

Andre Chavez[1,2], Brian Rummel[1,2], April Jeffries[2], and Sang M. Han[1,2]
Nick Bosco[3]
Brian Rounsaville[4] and Ajeet Rohatgi[4]

[1]University of New Mexico, Albuquerque, NM 87131
[2]Osazda Energy, Albuquerque, NM 87102
[3]National Renewable Energy Laboratory, Golden, CO 80401
[4]Georgia Institute of Technology, Atlanta, GA 30332

Abstract — **Cell cracks in PV modules caused by poor handling during shipping and installation as well as from extreme weather events can lead to gradual or immediate power degradation. To directly address cell-crack-induced degradation, we have formulated a carbon nanotube additive for commercial screen printed silver pastes. We have shown in previous work that these metal matrix composites have little to no effect on the cell's efficiency while enhancing the metallization's fracture toughness and electrical gap-bridging capability. In this work, we focus on translating materials level characterization techniques to module level degradation. We found that we get conflicting results from two different methods of measuring the metallization's ability to electrically bridge gaps in cracked solar cells. Mini-module stress testing is currently underway to determine which materials characterization correlates well with the min-module degradation characteristics.**

Keywords—Carbon nanotubes, Cell cracks, Degradation, Metallization, Nanocomposites, Photovoltaics, Resilience, Silver.

I. INTRODUCTION

Cracks in cells can electrically isolate parts of the cell and result in current mismatch between cells in the module, which can manifest as hot spots. These hot spots can lead to severe power degradation and safety risks. When such underperforming modules are found in the field, the PV field owners and insurers are often financially responsible for replacing modules that are underperforming. To directly address cell-crack-induced module power loss, we have developed a method to incorporate a cost-effective carbon nanotube (CNT) additive into commercial, screen-printable silver pastes for terrestrial PV [1, 2]. The CNT-enhanced metal matrix composite (MMC) silver pastes match the fineness of grind (FOG) and viscosity of commercial pastes, making them an easily integratable solution within the preestablished manufacturing process flow. The screen printing and firing profile of the MMC metallization are virtually identical to the commercial paste without compromising cell performance [3].

We have previously shown that our MMC pastes provide increased fracture toughness, electrical gap-bridging, and "self-

healing" [4-6] that would enhance module lifetime against cell cracks. In this work, we focus on how materials level electromechanical characterizations to measure gap-bridging capability correlates with module level degradation. We investigate different CNT characteristics to determine the effects of the MMC at various concentrations on the overall electromechanical properties and compare them to the conventional metal lines after firing. We find that the incorporation of short multiwalled carbon nanotubes (MW-CNTs), labeled as formulation MMC-A, has a noticeable effect on the metallization's ability to electrically bridge gaps under tensile stress, whereas they show little to no improvement over the baseline when under a flexural stress. The use of long single-walled carbon nanotubes (SW-CNTs), labeled as formulation MMC-D, shows a slight improvement in gap bridging under flexural stress but does not perform as well as the short CNT formulation under tensile stress.

II. EXPERIMENTAL RESULTS AND DISCUSSION

A. Resistance Across Cleaves and Cracks

A strain failure test setup, named Resistance Across Cleaves and cracKs (RACK), is constructed in-house to evaluate the maximum gap that CNTs can electrically bridge when a gap appears in the cracked MMC lines. A voltage is applied across the metallization and it is strained at sub-micron increments, using a piezo stage in a tensile motion until the electrical continuity is lost through each gridline. The piezo stage movement has sub-nanometer translational resolution. The RACK measurements are made until the cracked and strained test lines completely fail electrically at the critical open displacement (COD), the gap is then closed, and the open- and closed-gap cycle is repeated over 20 times to mimic wearout failure.

Figure 1 shows a summary of average electrically bridgeable distance for two different MMC formulations and the baseline. Each data point represents the average electrically bridgeable distance from 16 different gridline samples. We will use the average electrically bridgeable distance and the COD

This material is based upon work supported by the U.S. Department of Energy DuraMAT (RGJ-8-82224), DOE SETO (DE-EE0009013), DOE CINT (2019AU0023), and NMSBA.

interchangeably in this work. The baseline metallization has a COD of ~35 µm upon initial fracture and levels off to ~15 µm after many cycles. For this testing method, MMC-A has a COD of ~65 µm and levels off to ~35 µm after repeated open- and closed-gap cycles, whereas MMC-D starts at ~65 µm but quickly levels off to ~15 µm similar to the baseline. In light of 4 to 20 µm crack openings observed during mini-module thermal cycle stress testing [7], the short MW-CNT formulation (MMC-A) that bridges 35 µm gaps even after repeated open- and closed-gap cycles is expected to perform well against cell cracks.

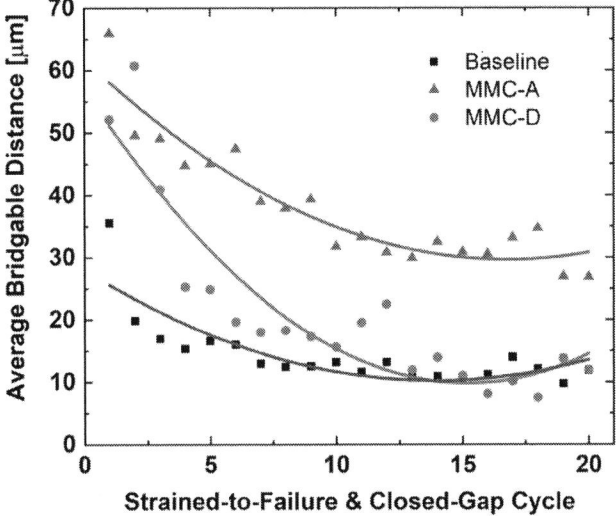

Figure 1: RACK data summary over twenty open- and closed-gap cycles. MMC-A is short MW-CNT formulation, and MMC-D is long SW-CNT formulation.

The data from the RACK testing suggests that the short MW-CNTs have a greater impact on the metallization's ability to conduct electricity across fractured cells than the long SW-CNTs. It is unclear at the moment whether this is a result of telescoping of the MW-CNTs, such that they are able to bridge gaps much larger than their initial length when the material is stressed in a tensile motion, or it is a combination of the CNT and Ag electrically bridging the gaps by creating a roughened fracture surface and more tortuous fracture path.

B. Beam Bending Resistance Measurments

We have replicated the three-point beam bending method for evaluating the fracture behavior of crystalline silicon PV cell metallization established at NREL [8]. For the beam bending method, Si PV cells with silver-based metallization, fabricated on floating-zone Si wafers, are diamond scribed into ~ 4 mm x 40 mm sections, where each contains one 2-mm-wide print. The samples are adhered to a 4.5 mm × 7.7 mm × 70 mm acrylic beams specifically designed to concentrate the lateral strain to the center of the silicon sample. Acrylic is used due to its high ratio of yield stress to elastic modulus, and a notch (1 mm *w* × 6 mm *h*) is laser cut into the midsection of the beam to encourage a hinging action when the beam is placed in the three-point bending setup. PV tabbing ribbon is soldered to both

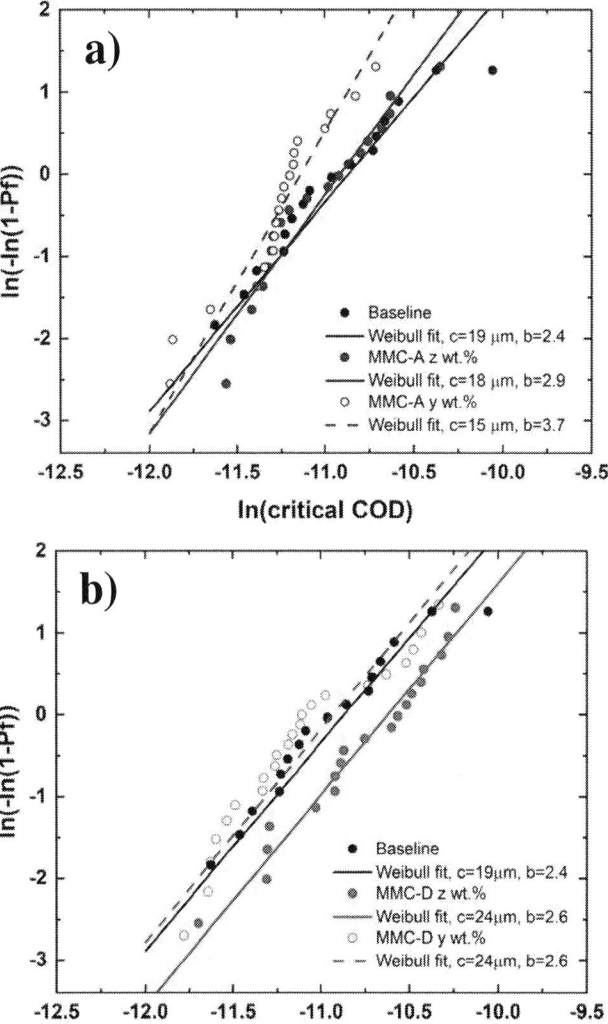

Figure 2: Weibull plot of critical COD for baseline, MMC-A, and MMC-D pastes.

ends of the metallization to provide electrical connectivity.

The silicon sample is notched with a diamond scribe to promote a singular crack, and the acrylic beam is placed in the three-point bending setup with an outer span of 50 mm. The samples are displaced to crack the silicon and characterize the COD. A 2D finite element model is used to simulate the COD to load line displacement ratio (0.096) and resulting silicon strain. The COD is the micrometer scale crack opening, and the load line displacement is the millimeter scale excursion made by the middle loading pin in the three-point beam bending setup. While the RACK testing relies on the 1:1 ratio of stage translation to COD at sub-nanometer resolution, the beam bending method also allows for a high resolution displacement of the crack opening. For every micron the beam is displaced, the COD changes by just 96 nm. Monotonic loading is applied to evaluate the critical COD. We tested 16 beams for the baseline, a total of 40 beams for the MW-CNT formulation (MMC-A) at two different CNT concentrations, and a total of 40 beams for the SW-CNT formulation (MMC-D) at two different CNT concentrations. The beams are loaded in the three-point bending setup under a constant cross head displacement ratio of 10 µm/s while the gridlines connecting the two busbars are monitored with a four-point resistance measurement. Once the gridlines become electrically open, the measurement is concluded, and the COD is recorded.

In Fig. 2 (a) the baseline is compared to two different concentrations of the MMC-A formulation containing short MW-CNTs, and no difference in characteristic COD is observed regardless of concentration. However, when the same comparison is made in Fig. 2 (b) using MMC-D containing long SW-CNTs, the characteristic critical COD is improved from 19 µm to 24 µm for one of the two CNT concentrations. This slight increase is still not as significant as the contrasting materials response under the RACK testing. At the moment, there is no clear answer as to which MMC formulation (i.e., short MW-CNTs vs. long SW-CNTs) would perform well for the module-level stress testing, so we are pursuing/testing both formulations. The RACK tensile testing method and the three-point beam bending method both have their benefits, and it will be important to understand which materials level characterization method correlates well, and to what degree, with the module-level degradation and failure associated with cell cracks. Mini-module stress testing is currently underway to shed more light on this correlation.

III. CONCLUSIONS

We have performed two different electromechanical tests that have contradicting results regarding the MMCs ability to electrically bridge cell cracks. The optimized MMC material will undergo further verification steps, including mini-module-level highly accelerated stress testing (HAST). As we complete the mini-module as well as full-size module stress testing, we will report which materials level characterization method is a good predictive indicator for the module-level degradation characteristics.

IV. ACKNOWLEDGMENT

This material is based upon work supported by the U.S. Department of Energy DuraMAT (RGJ-8-82224), DOE SETO (DE-EE0009013), DOE CINT (2019AU0023), and NMSBA.

REFERENCES

[1] O. K. Abudayyeh, A. Chavez, J. Chavez, and S. M. Han, "Development of Low-Cost, Crack-Tolerant Metallization Using Screen Printing," *Proc. 46th IEEE PVSC,* pp. 1-5, June 20 2019.

[2] O. K. Abudayyeh, A. Chavez, S. M. Han, B. Rounsaville, V. Upadhyaya, and A. Rohatgi, "Silver-carbon-nanotube composite metallization for increased durability of silicon solar cells against cell cracks," *Sol. Energy Mater. Sol. Cells,* p. under review, 2021.

[3] A. Chavez *et al.*, "Electromechanical Characterization of Crack-Tolerant, Carbon-Nanotube-Reinforced Composite Gridlines Using In Situ Scanning Electron Microscope Strain Test," in *2020 47th IEEE Photovoltaic Specialists Conference (PVSC)*, 15 June-21 Aug. 2020 2020, pp. 1689-1693, doi: 10.1109/PVSC45281.2020.9300651.

[4] O. K. Abudayyeh *et al.*, "Development of Low-Cost, Crack-Tolerant Metallization Using Screen Printing," *Proc. 46th IEEE PVSC,* 2018, doi: 10.1109/PVSC.2018.8547386.

[5] O. K. Abudayyeh, C. Nelson, G. K. Bradshaw, S. Whipple, D. M. Wilt, and S. M. Han, "Crack-Tolerant Metal Composites as

[6] Photovoltaic Gridlines," *IEEE Journal of Photovoltaics,* vol. 9, no. 6, pp. 1754-1758, 2019, doi: 10.1109/jphotov.2019.2939096.

[6] O. K. Abudayyeh, N. D. Gapp, C. Nelson, D. M. Wilt, and S. M. Han, "Silver–Carbon-Nanotube Metal Matrix Composites for Metal Contacts on Space Photovoltaic Cells," *IEEE Journal of Photovoltaics,* vol. 6, no. 1, pp. 337-342, 2016, doi: 10.1109/jphotov.2015.2480224.

[7] M. B. Timothy J. Silverman, Ali Abbas, Tom Betts, Michael Walls, and Ingrid Repins, "Movement of Cracked Silicon Solar Cells During Module Temperature Changes," *IEEE PVSC 46,* 2019.

[8] N. Bosco, A. Chavez, V. Upadhyaya, and S. M. Han, "Fatigue-Like Behavior of Silver Metallization Gridlines and Proposed Damage Mechanics Model," in *Proceedings of the 47th IEEE Photovoltaic Specialist Conference*, Calgary, Canada (Virtual), June 14-19, 2020 2020.

Testing the Abrasion Resistance of Porous SiO$_2$ Anti-reflection Coatings for Solar Cover Glass

Adam M Law, Farwa Bukhari, Luke O Jones, Ali Abbas, and John Michael Walls

CREST, Wolfson School, Loughborough University, Loughborough, LE11 3TU, United Kingdom

Abstract—The cover glass sheet on solar modules can cause reflection losses as well as soiling build-up. Reflection losses can be addressed with anti-reflection (AR) coatings, whilst soiling is removed by mechanical cleaning processes that are effective but can have adverse effects on surface coatings. In this work, multilayer broadband and commercial porous SiO$_2$ AR coatings have been subject to abrasion testing that simulates the regular cleaning of solar modules in the field, using Felt Pad and CS-10 abradant materials. The Felt Pad abrasion has no impact on the multilayer coating, but caused visible damage to the porous SiO$_2$, increasing WAR from 5.97% to 6.75% after 100 cycles. After 50 and 100 abrasion cycles of CS-10, significant scratches are visible on the porous SiO$_2$ coating, and the weighted average reflectance (WAR) of the coating increases from 5.97% to 7.08% after 100 cycles. The coating is fully removed in some abraded areas. The multilayer AR coating also experiences some damage after CS-10 abrasion, increasing WAR from 5.84% to 6.68%. Optical microscopy and Scanning Electron Microscopy (SEM) show the nature of the abrasion damage caused. Overall, the multilayer AR coating shows significantly higher abrasion resistance than the porous SiO$_2$. Significant abrasion damage to porous SiO$_2$ AR coatings is a major problem for solar asset managers resulting in long-term power losses.

Index Terms—solar, photovoltaics (PV), anti-reflection (AR), anti-soiling (AS), coatings, abrasion

I. INTRODUCTION

The glass sheet that covers solar modules provides important functions such as mechanical stability, protection of the underlying cells from physical and chemical damage, and high optical transmittance. However, two major issues still occur, front-surface reflection losses and soiling. The reflection loss is caused by a mismatch between the refractive indices of air and glass (1 and 1.5, respectively), and results in an optical loss of just over 4%. Soiling, which is the build-up of dust, dirt, and other matter on the surface of the module, can lead to much higher power losses as a result of decreased light transmission as well as accelerated degradation if left untreated. Depending on the region in which the panels are located, and other factors such as frequency of cleaning, losses can range from 3-50% [1], [2].

The reflection losses can be addressed by the use of anti-reflection (AR) coatings, and most solar modules have a single layer AR coating of porous SiO$_2$ [3]. Other materials such as MgF$_2$ are also used for single layer coatings [4], and multilayer coatings with dielectrics such as solid SiO$_2$ and ZrO$_2$ have also been developed that are effective and have superior durability [5]–[7]. The problem of soiling can also be addressed via coatings, and various anti-soiling coatings

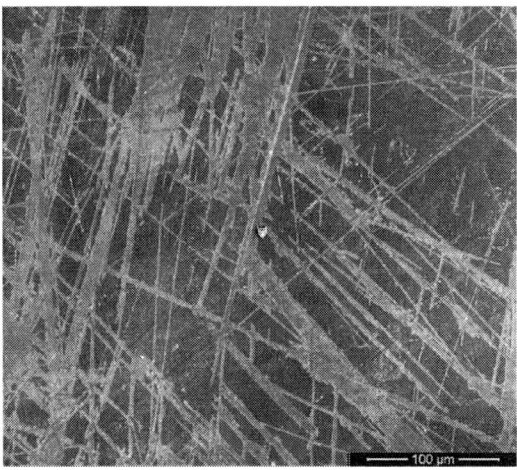

Fig. 1. SEM image showing severe abrasion damage to the AR coating on field-aged module cover glass after regular cleaning

have been developed and tested at the research level [8], [9], although at present there are no widespread commercially available options for solar modules. The ideal coating for solar modules is one that combines both anti-reflection and anti-soiling properties effectively. The porous SiO$_2$ AR coatings are hydrophilic and offer little anti-soiling effect, as anti-soiling surfaces are ideally hydrophobic [10]. Aside from natural rainfall, the current method of addressing soiling in the field is by cleaning the panels to remove accumulated dirt, and most asset managers use heavy-duty cleaning methods consisting of large brushes mounted on agricultural machinery. Cleaning frequency is highly dependent on factors such as climate and availability of water and labour but can range from once a year in milder climates to every two days in desert areas with frequent sandstorms [11], [12]. Such methods are effective at removing the accumulated dirt but can cause significant damage to the AR coatings as a result of mechanical abrasion (see Fig. 1). Damage to the AR coatings is serious as it can result in permanent power losses, with re-application of coatings not a viable solution. The typical warranty period of a solar module is 25 years, and any coatings applied to the surface should be expected to last just as long. Porous SiO$_2$ AR coatings have previously shown vulnerability to degradation as a result of abrasion [13], as well as damp heat exposure resulting in water ingress [14].

978-1-7281-6118-1/22 $31.00 © 2022 IEEE

In this work, we report results on the dry linear abrasion testing of commercially available porous SiO_2 AR-coated glass, and a comparison to a broadband multilayer AR coating comprising SiO_2 and ZrO_2, using industry standard test methods to simulate regular cleaning of PV modules. Reflection measurements, optical microscopy, and scanning electron microscopy (SEM) are used to show the effects of abrasion on the performance and morphology of the coatings.

II. EXPERIMENTAL DETAILS

5cm x 10cm samples of commercially available porous SiO_2 AR-coated glass were obtained for abrasion testing. such coatings typically have a thickness between 100nm and 150nm. Multilayer broadband anti-reflection coatings as described in [6] were deposited on 5cm x 5cm soda-lime glass substrates. Abrasion testing was performed using a Taber 5900 reciprocating abrader, with samples being subject to both 50 and 100 abrasion cycles, where 1 cycle constitutes a complete forward and backward stroke of the machine. The abrasion test method was taken from BS EN 1096-2:2012. The first abrasive material used was Felt Pad, a mild abrasive that is most similar to the type of material used in the field. A CS-10 Wearaser was the second, and more abrasive, material used, with a diameter of 3mm. A stroke length of 3.5mm was used, with a weight of 3.5N applied. Multiple passes of the abrader across adjacent areas of the sample are required to give a large enough affected area for optical measurements. Abraded samples were washed with DI water following abrasion to remove any loose debris on the surface. A photograph of the abrader is shown in Fig. 2.

Reflection measurements were taken across a wavelength range of 350-1150nm using a Varian Cary5000 UV-VIS-NIR spectrophotometer with integrating sphere attachment. The measurement area is determined by an aperture of 1.6cm diameter. A step size of 1nm was used, with a scan rate of 600nm/min. Weighted average reflectance (WAR) values are calculated from the reflectance data using equation (1). WAR provides a solar spectrum-weighted value for reflectance measurements, making them more applicable to PV as well as providing a useful figure of merit for comparison between measurements.

$$WAR(\lambda_{min}, \lambda_{max}) = \int_{\lambda_{min}}^{\lambda_{max}} \frac{\Phi \cdot R}{R} d\lambda \qquad (1)$$

Optical microscope images were taken using an Olympus CX41 microscope and InfinityAnalyze software, with magnification of up to 50X. Scanning electron microscopy (SEM) images were taken using a JEOL 7800F FEGSEM.

III. RESULTS & DISCUSSION

A. Optical Performance

1) Felt Pad: Reflection measurements for as-deposited, 50 and 100 abrasion cycles of Felt Pad on the porous SiO_2 coating are shown in Fig. 3. The measured reflectance for the MAR coating after Felt Pad abrasion is shown in Fig. 4.

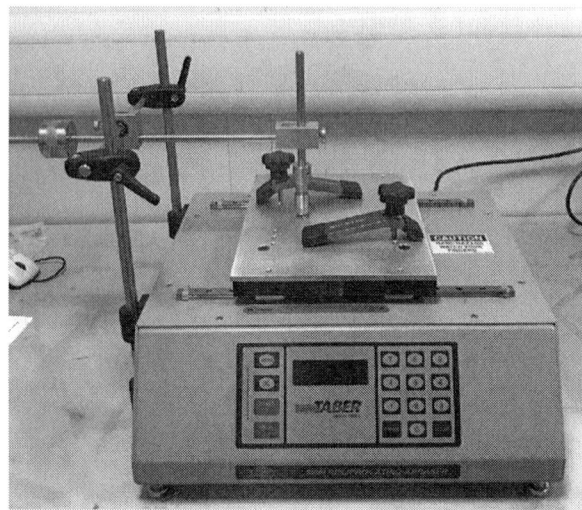

Fig. 2. The Taber 5900 reciprocating abrader used for abrasion testing

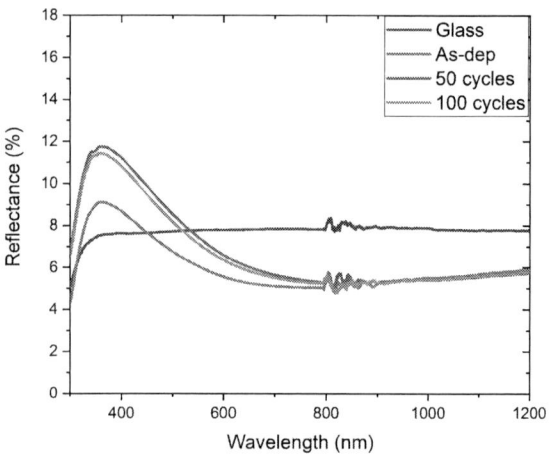

Fig. 3. Measured reflectance curves for as-deposited, 50 and 100 cycles of felt pad abrasion of the porous SiO_2 coating

Fig. 5 shows the WAR values for 0, 50 and 100 cycles of Felt Pad abrasion for both coatings. The MAR coating shows no change in WAR even after 100 cycles of abrasion, showing that the Felt Pad does not cause any damage to the coating, with WAR at 5.84% after 0 cycles, and 5.82% after 100 cycles. The porous SiO_2 coating, however, shows signs of damage and an increase in WAR from 5.97% to 6.57% and 6.75% after 50 and 100 cycles, respectively. This amounts to a relative increase of 13%. The Felt Pad material is similar to the materials used for module cleaning, so offers a more representative scenario compared to the more aggressive CS-10. These results suggest that the MAR coating would maintain performance throughout its lifetime when subject to cleaning with brushes similar to Felt Pad.

2) CS-10: Reflection measurements for the 0 (i.e., as deposited), 50, and 100 abrasion cycles of CS-10 abradant on porous SiO_2 are shown in Fig. 6. The first 50 cycles

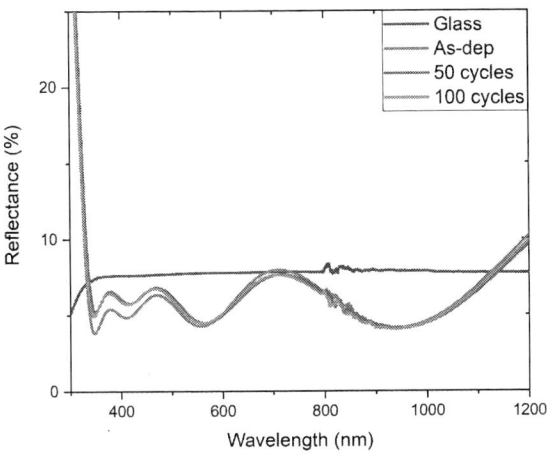

Fig. 4. Measured reflectance curves for as-deposited, 50 and 100 cycles of felt pad abrasion of the MAR coating

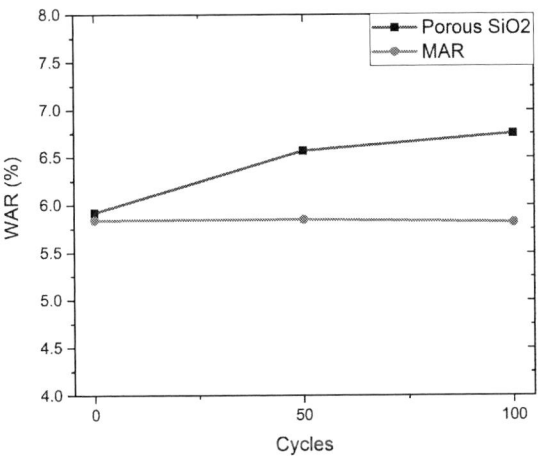

Fig. 5. WAR values against number of abrasion cycles for Felt Pad abrasion

results in a significant increase in reflection across all wavelengths, indicating serious damage to the AR coating resulting in loss of performance. There is very little change from 50 to 100 cycles, suggesting that all of the damage occurs within the first 50 cycles. This may be because the coating has been completely removed in the affected area as a result of abrasion. Localised coating removal, rather than homogeneous thinning, is further supported by the observation that the reflectance minimum stays at the same wavelength after abrasion. If the coating had undergone a reduction in thickness, then the minimum would shift to shorter wavelengths [15]. Due to the small size of the abradant contact area in comparison to the measurement area, it is possible that some unabraded sections of the coating are present in the measurement. CS 10 is used as a medium-level abrasive material, so would not constitute a 'worst-case' scenario. In milder climates such as the UK, modules are typically cleaned twice a year, so 50 cycles would amount

to a 25-year cleaning cycle. The majority of the damage occurring in the first 50 cycles is a problem because it represents 'normal service' rather than an extreme case.

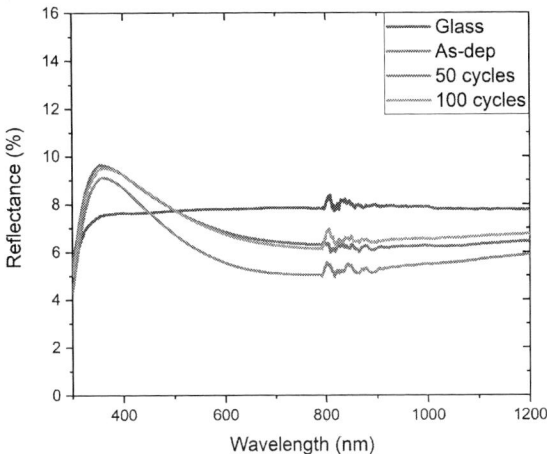

Fig. 6. Measured reflectance curves for as-deposited, 50, and 100 abrasion cycles of CS-10 abrasion for the porous SiO_2 coating.

Fig. 8 shows the weighted average reflection (WAR) values for each set of abrasion cycles. The WAR increases from 5.97% before abrasion (0 cycles) to 6.99% at 50 cycles, and to 7.08% at 100 cycles. The total relative increase is 18.6%. The significant increase in WAR after 50 cycles, followed by only a slight increase at 100 cycles, further supports the idea that the majority of the abrasion damage occurs within the first 50 cycles, as the subsequent 50 do not have a significant effect. The coating may be almost completely removed in the affected areas after 50 cycles, explaining why further abrasion has very little impact on the optical properties. The CS-10 abrasion causes a higher amount of damage and further increases reflection when compared to Felt Pad, which is to be expected as CS-10 is a more abrasive material. For significant mechanical abrasion damage to occur, the hardness of the abrasive material must be higher than that of the coating.

CS-10 abrasion also causes damage to the MAR coating, although significantly less than porous SiO_2. The WAR increases from 5.84% to 6.43% after 50 cycles, and to 6.68% after 100 cycles. This is a relative increase of 14%. Whereas almost all of the damage to the porous SiO_2 coating occurs after 50 cycles, the damage to the MAR coating is more evenly split. The nature of the change in measured reflectance also suggests that partial removal has occurred, as homogeneous thinning would shift the reflectance peaks to shorter wavelengths. The MAR coating shows significantly higher abrasion resistance compared to the porous SiO_2, and would be expected to maintain higher performance for a longer period of time in the field.

The introduction of pores into the SiO_2 coating is crucial for lowering the refractive index, and therefore achieving anti-reflection properties, but ultimately leads to poorer mechanical properties, leaving the coating vulnerable to degradation. Abrasion damage caused to AR coatings poses serious issues for PV asset managers, as cleaning methods that reduce soiling can also result in long-term power losses caused by complete or partial removal of the AR coating.

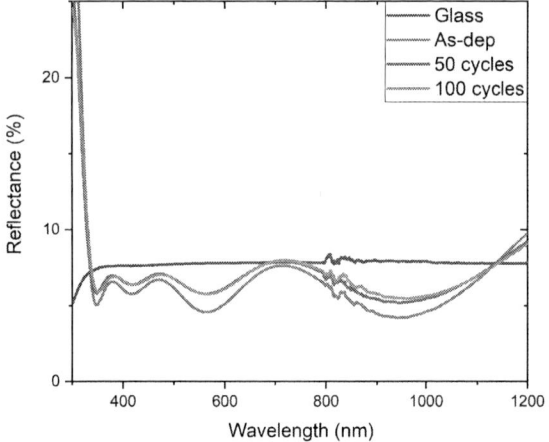

Fig. 7. Measured reflectance curves for as-deposited, 50 and 100 cycles of CS-10 abrasion for the MAR coating

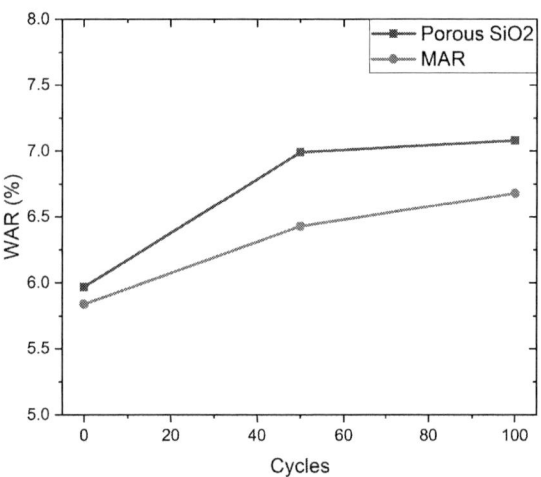

Fig. 8. WAR values against number of abrasion cycles for CS-10 abrasion

The abrasion testing in this work is carried out without the use of sand or other additional abrasives, which is a feature of many other works [16]. The addition of sand makes the abrasion testing more applicable to desert areas where soiling is typically higher, and creates additional abrasive damage which further increases reflectance. If the abrasion tests in this work were carried out with sand, the damage caused to the AR coatings is expected to be significantly higher. Factors that can potentially reduce abrasive damage caused by cleaning are the use of water i.e. wet cleaning, and rotary abrasion rather than linear. Coatings will also experience conditions such as UV exposure, thermal cycling, and high humidity in the field,

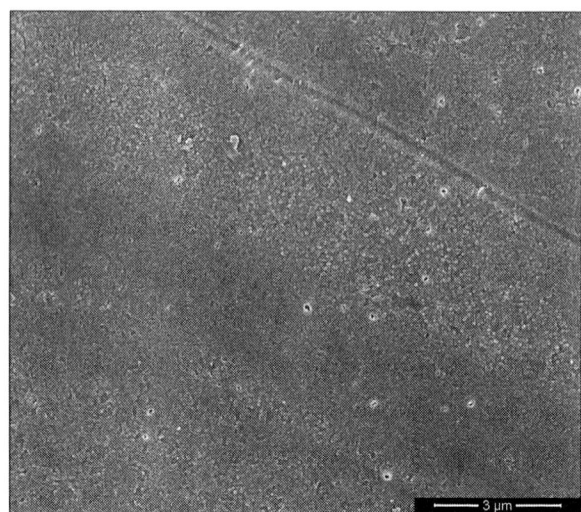

Fig. 9. Scanning electron microscope (SEM) image showing scratches after 50 abrasion cycles of Felt Pad on the porous SiO_2 coating

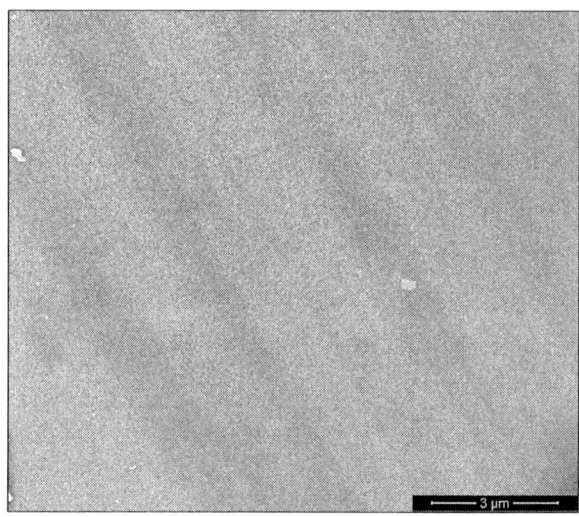

Fig. 10. Scanning electron microscope (SEM) image showing scratches after 50 abrasion cycles of Felt Pad on the MAR coating

depending on location. These factors are also expected to affect abrasion resistance, and are worth exploring in future work.

B. Microscopy

1) Felt Pad: An SEM image of the Felt Pad abraded sol-gel coating after 50 abrasion cycles is shown in Fig. 9, with a magnification of 10,000X. The directional scratches and wear caused by abrasion are visible, although the damage is not as severe as that of the CS-10 abraded samples (Fig. 12). This is to be expected, as CS-10 is a more aggressive material than Felt Pad. The surface of the MAR coating after 50 cycles of Felt Pad is shown in Fig. 10. Some directional wear can be seen, although there is no real damage to the surface, which is confirmed by the WAR values in Fig. 5.

978-1-7281-6118-1/22 $31.00 © 2022 IEEE

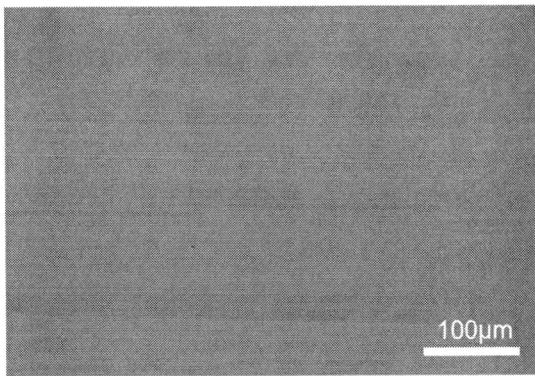

Fig. 11. Microscope image showing scratches after 50 abrasion cycles of CS-10

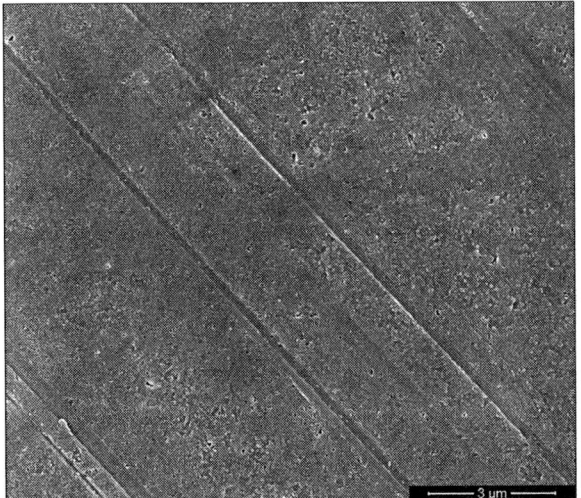

Fig. 12. Scanning electron microscope (SEM) image showing scratches after 50 abrasion cycles of CS-10

2) CS-10: A microscope image of the surface of a CS-10 abraded porous SiO_2 sample at 50X magnification is shown in Fig. 11. Clear directional scratches on the surface of the coating as a result of abrasion are visible, showing the damage caused. Fig. 12 shows an SEM image of the same sample, at 10,000X magnification. The same clear, directional scratches are visible, and the depth given in the SEM images shows the removal of the coating in affected areas. Comparing the images of Felt Pad and CS-10 abrasion for the porous SiO_2 coating, the more aggressive nature of CS-10 is clear, with deeper scratches visible.

IV. SUMMARY & CONCLUSIONS

Porous SiO_2 and multilayer broadband anti-reflection (AR) coatings have been subject to abrasion testing to determine their durability under conditions that simulate the regular cleaning of solar modules. The coatings have undergone 50 and 100 cycles of abrasion using Felt Pad and CS-10 abradants with an applied load of 3.5N. Felt Pad abrasion has no impact on the optical performance of the multilayer AR coating,

whereas the porous SiO_2 shows signs of abrasion damage and an increase in WAR from 5.97% to 6.75% after 100 cycles. Felt Pad represents a similar level of abrasion to that seen in the field, giving results that are applicable to real-world performance. The CS-10 abrasion testing results in a significant increase in the reflectance of the porous SiO_2 coating after 50 cycles resulting in loss of performance, followed by little change after 100 cycles, suggesting that no further damage occurs. This could be the result of complete removal of the coating in affected areas after 50 cycles. The porous nature of the coatings reduces their mechanical strength leaving them vulnerable to damage. The multilayer AR coating also sustains some damage after the more aggressive CS-10 abrasion, resulting in an increase in WAR from 5.84% to 6.68%. The CS-10 abrasion is seen as a worst-case scenario. Overall, the multilayer AR coating shows significantly higher durability compared to the porous SiO_2 coating under both types of abrasion, as well as improved optical performance. For both coatings, the nature of the change in reflectance suggests partial coating removal in some areas, rather than a consistent reduction in thickness across the whole affected area. Abrasion damage to AR coatings as a result of module cleaning can cause significant, and permanent, power losses for PV modules, increasing the cost of electricity produced. Greater resistance to abrasion is an important property of coatings applied to module cover glass.

REFERENCES

[1] E Klugmann-Radziemska. Degradation of electrical performance of a crystalline photovoltaic module due to dust deposition in northern Poland. *Renewable Energy*, 78:418–426, 2015.

[2] M J Adinoyi and S A M Said. Effect of dust accumulation on the power outputs of solar photovoltaic modules. *Renewable Energy*, 60:633–636, 2013.

[3] C Ballif, J Dicker, D Borchert, and T Hofmann. Solar glass with industrial porous SiO_2 antireflection coating. *Solar Energy Materials and Solar Cells*, 82(3):331–344, 2004.

[4] X Wu. High-efficiency polycrystalline CdTe thin-film solar cells. *Solar Energy*, 77(6):803–814, 2004.

[5] P M Kaminski, F Lisco, and J M Walls. Multilayer broadband antireflective coatings for more efficient thin film CdTe solar cells. *Photovoltaics, IEEE Journal of*, 4.452–456, 01 2014.

[6] A M Law, L D Wright, A Smith, P J M Isherwood, and J M Walls. 2.3% efficiency gains for silicon solar modules using a durable broadband anti-reflection coating. In *2020 47th IEEE Photovoltaic Specialists Conference (PVSC)*, pages 0973–0975. IEEE, 2020.

[7] Jimmy M Newkirk, Illya Nayshevsky, Archana Sinha, Adam M Law, QianFeng Xu, Bobby To, Paul F Ndione, Laura T Schelhas, John M Walls, Alan M Lyons, and D C Miller. Artificial linear brush abrasion of coatings for photovoltaic module first-surfaces. *Solar Energy Materials and Solar Cells*, 219:110757, 2021.

[8] M Piliougine, C Cañete, R Moreno, J Carretero, J Hirose, S Ogawa, and M Sidrach-de Cardona. Comparative analysis of energy produced by photovoltaic modules with anti-soiling coated surface in arid climates. *Applied energy*, 112:626–634, 2013.

[9] Syeda Farwah Bukhari, Fabiana Lisco, Taraneh Bozorgzad Moghim, Alan Taylor, and John Michael Walls. Development of a hydrophobic, anti-soiling coating for pv module cover glass. In *2019 IEEE 46th Photovoltaic Specialists Conference (PVSC)*, pages 2849–2853. IEEE, 2019.

[10] A M Law, F Bukhari, L O Jones, P J M Isherwood, and J M Walls. Combined anti-soiling and anti-reflection coatings for solar modules. In *2021 IEEE 48th Photovoltaic Specialists Conference (PVSC)*, pages 1379–1382, 2021.

[11] Russell K Jones, Abdulaziz Baras, Abdullah Al Saeeri, Ayman Al Qahtani, Ahmed O Al Amoudi, Yousef Al Shaya, Maher Alodan, and Shafi Ali Al-Hsaien. Optimized cleaning cost and schedule based on observed soiling conditions for photovoltaic plants in central saudi arabia. *IEEE journal of photovoltaics*, 6(3):730–738, 2016.

[12] Asad Ullah, Amir Amin, Turab Haider, Murtaza Saleem, and Nauman Zafar Butt. Investigation of soiling effects, dust chemistry and optimum cleaning schedule for pv modules in lahore, pakistan. *Renewable Energy*, 150:456–468, 2020.

[13] K Ilse, C Pfau, P Miclea, S Krause, and C Hagendorf. Quantification of abrasion-induced ARC transmission losses from reflection spectroscopy. In *2019 IEEE 46th Photovoltaic Specialists Conference (PVSC)*, pages 2883–2888. IEEE, 2019.

[14] G Womack, K Isbilir, F Lisco, G Durand, A Taylor, and J M Walls. The performance and durability of single-layer sol-gel anti-reflection coatings applied to solar module cover glass. *Surface and Coatings Technology*, 358:76–83, 2019.

[15] Muhammad Zahid Khan, Charlotte Pfau, Matthias Schak, Paul-Tiberiu Miclea, Volker Naumann, Ahmed Debess, Christian Hagendorf, and Klemens Ilse. Resilience of industrial pv module glass coatings to cleaning processes. *Journal of Renewable and Sustainable Energy*, 12(5):053504, 2020.

[16] Nicoletta Ferretti, Klemens Ilse, Aylin Sönmez, Christian Hagendorf, and Juliane Berghold. Investigation on the impact of module cleaning on the antireflection coating. *32nd EU-PVSEC*, pages 1697–1700, 2016.

Hydrothermally Deposited Antimony Sulfide Solar Cells with V_{OC} Approaching 800 mV

Dipendra Pokhrel[1], Nini Rose Mathew[1, 2], Suman Rijal[1], Ebin Bastola[1], Abasi Abudulimu[1], Tamanna Mariam[1], Xavier Mathew[1, 2], Adam B Phillips[1], Michael J Heben[1], Zhaoning Song[1], Yanfa Yan[1], and Randy J Ellingson[1]

[1] *Wright Center for Photovoltaics Innovation and Commercialization (PVIC), Department of Physics and Astronomy, University of Toledo, Ohio, 43606, USA*

[2] *Instituto de Energías Renovables, Universidad Nacional Autónoma de México, Temixco, Morelos 62580, México*

Abstract— Antimony sulfide (Sb_2S_3) represents an emerging thin-film photovoltaic light-absorber, with potential as a wide gap top cell for high-efficiency tandem devices. Here, we report the development and characterization of Sb_2S_3 absorber layers prepared by the hydrothermal method. Completed devices based on chemical bath deposited cadmium sulfide (CdS) and Spiro-OMeTAD as the electron- and hole-transport layers, respectively, have yielded promising power conversion efficiencies as high as 5.5 %. Although the typical deficit reported between the Sb_2S_3 bandgap energy and the open-circuit voltage (V_{OC}) remains high, we report high V_{OC} values approaching 800 mV.

Keywords—Antimony Sulfide, hydrothermal film deposition, solar cells

I. INTRODUCTION

The V-VI metal chalcogenides such as X_2Y_3 (X=Sb, Bi, Y= S, Se) are potential absorber layers for low-cost and environmentally benign thin-film solar cells. Among these chalcogenide materials, antimony sulfide (Sb_2S_3) is an emerging thin-film solar cell material because of the high optical absorption coefficient (1.8×10^5 cm^{-1} in the visible region), the suitable bandgap of (~1.7 eV), intrinsic p-type doping, and high material stability.[1, 2] The high bandgap of Sb_2S_3 opens the possibility of developing it as the top-cell in tandem devices.[3] Sb_2S_3 thin films can be prepared by different methods such as chemical bath deposition (CBD), hydrothermal, rapid thermal evaporation (RTE), atomic layer deposition (ALD), thermal evaporation, and spin coating.

Devices with efficiencies in the range of 4 - 7% were reported from different laboratories using Sb_2S_3 films deposited by CBD [4], thermal evaporation [5], RTE [1], and spin coating [6] [7]. The RTE process resulted in quasiepitaxial growth of vertically oriented films achieving device efficiency of 5.4 % [1]. Han et al. deposited Sb_2S_3 thin films using spin coating and to date have reported the highest PCE of 7.1 % for a planar heterojunction device. The high efficiency was a result of the 720 mV open-circuit voltage (V_{oc}) achieved by passivating the Sb_2S_3 surface with a thin layer of $SbCl_3$ [7]. The reported efficiency of planar Sb_2S_3 devices remains lower compared to the 7.5 % efficiency obtained for an Au/PCPDTBT/Sb_2S_3/mp-TiO$_2$ inorganic-organic

heterojunction solar cells in which Sb_2S_3 was used as the sensitizer on mesoporous TiO$_2$ [2].

The hydrothermal deposition method is one of the simplest and low-cost thin film deposition methods. Literature on this method as applied to Sb_2S_3 thin films remains limited. Quintero et al. deposited Sb_2S_3 thin films using the hydrothermal method and achieved an efficiency of 5.0 % with a TiO$_2$/ZnS bilayer as the electron transport layer (ETL) [8]. Similar results were reported for a device with TiO$_2$/CdS as the ETL [2]. Jin et al. observed vertically-oriented growth of Sb_2S_3 thin films during hydrothermal processing, obtaining a 6.4 % efficient cell [9]. Based on the literature, the average PCE of planar Sb_2S_3 solar cells using different deposition methods is within the range of 5-7 % and the highest V_{oc} is 770 mV [10], much lower than the thermodynamically achievable limit for material with a bandgap 1.7 eV. Here, we discuss the film deposition by hydrothermal method and the study of a 5.5% efficient Sb_2S_3 device with 795 mV V_{oc}.

II. EXPERIMENTAL METHOD

Fluorine-doped tin oxide (FTO) substrates were cleaned using a Micro-90 detergent solution and deionized (DI) water in an ultrasonic bath and dried with nitrogen. The ETL CdS was deposited on FTO using CBD as reported elsewhere [11]. Sb_2S_3 absorber layer was deposited by the hydrothermal method reported previously by Jin et al. [9]. In a typical deposition, 0.38 gm of potassium antimonyl tartrate trihydrate and 0.68 gm of sodium thiosulfate pentahydrate were dissolved in 40 ml of DI water and the solution was poured into a Teflon-lined autoclave. The substrate was placed in an autoclave facing the film side down and the autoclave was maintained at 120 °C for 17 hrs. After this the autoclave was allowed to cool naturally to room temperature, the film was cleaned with DI water, and dried in nitrogen. Three batches of CdS/Sb_2S_3 heterostructures were prepared by post-deposition annealing at 325, 350, and 375 °C for 20 mins, and subsequently the hole transport layer (HTL) Spiro-OMeTAD was deposited by spin coating, and further annealed at 100 °C for 30 min. Finally, the device was completed by depositing 60 nm of gold (Au) by thermal evaporation through a shadow mask to define multiple cells of 0.08 cm^2 on a 1.5-inch x 1-inch substrate. The current-voltage (J-V) characteristics were measured under AM1.5 illumination with an incident power of 100 mW cm^{-2}.

978-1-7281-6118-1/22 $31.00 © 2022 IEEE

III. RESULTS AND DISCUSSION

Figure 1 shows the SEM images of as-deposited and annealed Sb_2S_3 thin films. As seen in Fig. 1a, the as-deposited thin film suggests relatively small feature size with low crystallinity; the surface transforms into larger grains after annealing at 325 °C (Fig. 1b). The evolution in morphology with higher annealing temperatures are seen in Fig. 1c&d.

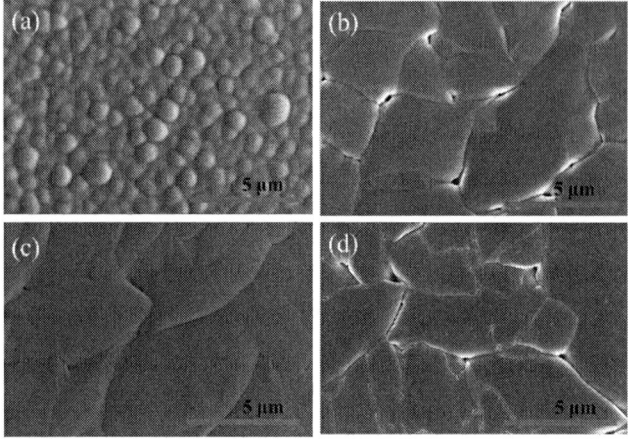

Fig. 1. SEM images of Sb_2S_3 thin films; (a) as-deposited, and annealed at (b) 325, (c) 350, and (d) 375 °C for 20 mins.

The X-ray diffraction (XRD) patterns shown in Figure 2 clearly indicate the evolution of the thin films from relatively amorphous to more highly crystalline as a result of annealing (Fig. 2). The annealed thin film of Sb_2S_3 exhibits a pure orthorhombic crystal structure in the stibnite phase. The sharp peaks in the XRD pattern confirmed the polycrystalline nature of the material.

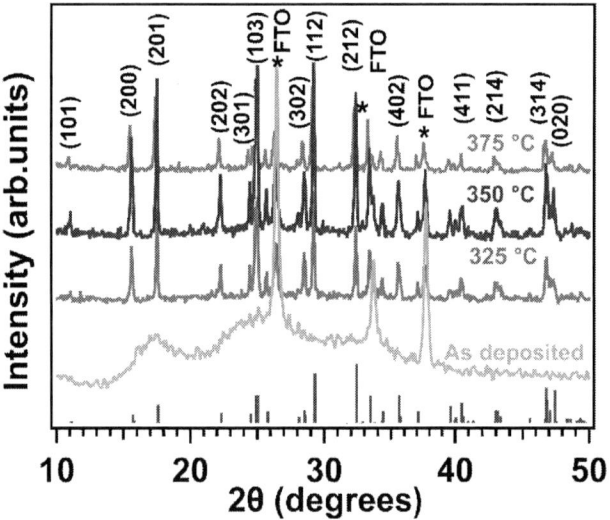

Fig.2. XRD pattern of as-deposited and annealed Sb_2S_3 thin films. The films were annealed at 325, 350, and 375 °C for 20 mins.

The J-V characteristics of the representative devices from each batch of the films annealed at different temperatures are shown in Figure 3. The dotted and solid lines represent the dark and light J-V curves respectively. The device annealed at 350 °C shows better performance, with average values of V_{OC}, J_{SC}, and PCE higher than the rest (Table 1). The champion cell has V_{oc} 795 mV, current density (J_{sc}) 14.3 mA cm^{-2}, fill factor (FF) 49.0 % and PCE of 5.5 %. The better performance of devices annealed at 350 °C is attributed to recrystallization which leads to a void-free surface with larger grains and minimal grain boundaries (Fig. 1c). The voids evident in Figs. 1b and 1c, together with higher grain boundary density, are consistent with transport barriers and higher recombination leading to deterioration in device performance. In general, the devices are clearly affected by series and shunt resistance, together with cross-over and roll-over. Further work is needed to address these issues and push the efficiency higher. Fig 3b shows the external quantum efficiency (EQE) of the devices. Parasitic absorption by the thin CdS buffer layer is negligible as indicated by the good blue response. Higher current of the 350 °C annealed device originates from the higher collection from the mid- and long-wavelength regions, a result of reduced recombination losses.

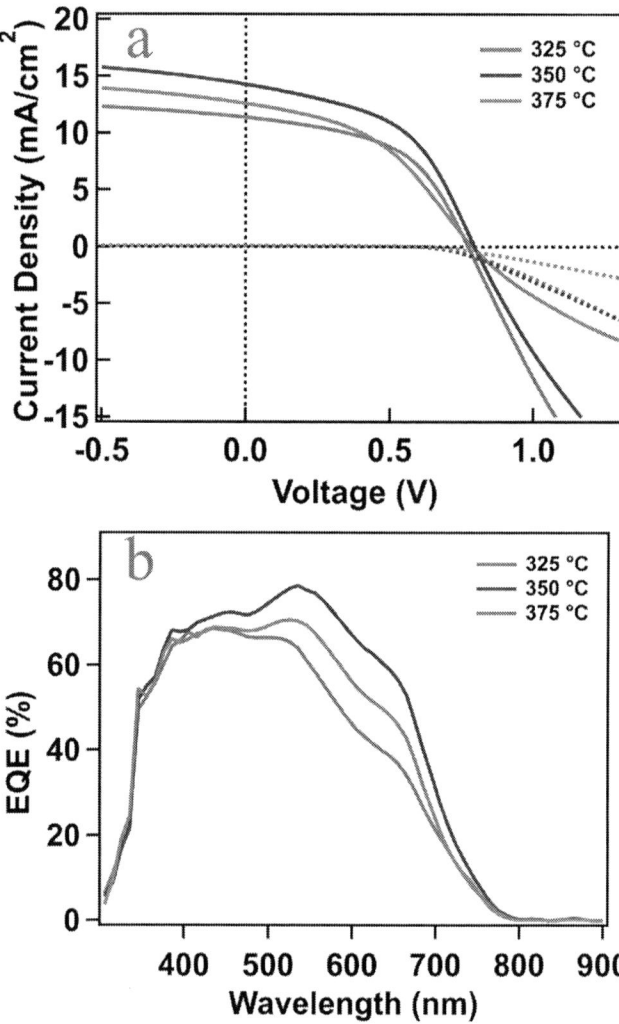

Fig. 3. (a) J-V characteristics, and (b) EQE of the best cells from the three batches of the FTO/CdS/Sb_2S_3/Spiro/Au devices in which the heterostructure CdS/Sb_2S_3 was annealed at 325, 350 and 375 °C for 20 mins.

Table 1: Photovoltaic parameters (V_{oc}, J_{sc}, FF, and efficiency) of the devices studied in this work. The average is from 8 cells. The film stack CdS/Sb_2S_3 was annealed at 325, 350, and 375 °C for 20 mins.

CdS/Sb_2S_3 anneal temperature		V_{oc} (mV)	J_{sc} (mA cm^{-2})	FF (%)	Efficiency (%)
325 °C	Best cell	772	11.3	50.6	4.4
	Average	755 ± 18	10.9 ± 0.3	47.8 ± 2.2	3.9 ± 0.3
350 °C	Best cell	795	14.2	49.0	5.5
	Average	792 ± 2	13.7 ± 0.4	46.4 ± 1.7	5.0 ± 0.6
375 °C	Best cell	788	12.5	42.6	4.2
	Average	768 ± 10	11.9 ± 0.8	42.5 ± 0.6	3.9 ± 0.2

IV. CONCLUSION

We report the hydrothermal film growth and prototype devices based on the wide band gap material Sb_2S_3. Recrystallization leading to the formation of polycrystalline films with large grains in the range of 2-5 μm were observed. The astounding grain growth at relatively low temperature, and the void-free and compact film, resulted in devices with V_{oc} = 795 mV, the highest V_{OC} reported to date for Sb_2S_3 absorber-based PV device. The unoptimized device with spiro-OMeTAD as the HTL reached a PCE of 5.5 %. The preliminary data indicates losses at the interface (V_{oc}), series and shunt resistances severally affecting the fill factor, and poor carrier collection from the bulk and the back contact as seen in the EQE. We are continuing to optimize the devices for better recrystallization and surface passivation, including alkali metal doping to improve surface morphology and grain growth. A suitable chloride treatment may help in passivating the surface and to decrease the defects and non-radiative recombination.

ACKNOWLEDGMENT

This material is based on research sponsored by Air Force Research Laboratory under agreement number FA9453-21-C-0056. The U.S. Government is authorized to reproduce and distribute reprints for Governmental purposes notwithstanding any copyright notation thereon. The views expressed are those of the authors and do not reflect the official guidance or position of the United States Government, the Department of Defense of the United States Air Force. The appearance of external hyperlinks does not constitute endorsement by the United States Department of Defense (DOD) of the linked websites, or the information, products, or services contained therein. The DoD doesn't exercise any editorial, security, or other control over the information you may find at these locations. Approved for public release; distribution is unlimited. Public Affairs release approval # AFRL-2022-2257.

REFERENCES

1. Deng, H., et al., *Quasiepitaxy strategy for efficient full-inorganic Sb2S3 solar cells.* Advanced Functional Materials, 2019. **29**(31): p. 1901720.

2. Choi, Y.C., et al., *Highly improved Sb2S3 sensitized-inorganic–organic heterojunction solar cells and quantification of traps by deep-level transient spectroscopy.* Advanced Functional Materials, 2014. **24**(23): p. 3587-3592.

3. Kondrotas, R., C. Chen, and J. Tang, *Sb2S3 solar cells.* Joule, 2018. **2**(5): p. 857-878.

4. Zimmermann, E., et al., *Toward high-efficiency solution-processed planar heterojunction Sb2S3 solar cells.* Advanced Science, 2015. **2**(5): p. 1500059.

5. Yin, Y., et al., *Composition engineering of Sb2S3 film enabling high-performance solar cells.* Science Bulletin, 2019. **64**(2): p. 136-141.

6. Jiang, C., et al., *Alkali Metals Doping for High-Performance Planar Heterojunction Sb2S3 Solar Cells.* Solar RRL, 2019. **3**(1): p. 1800272.

7. Han, J., et al., *Solution-processed Sb2S3 planar thin-film solar cells with a conversion efficiency of 6.9% at an open-circuit voltage of 0.7 V achieved via surface passivation by an SbCl3 interface layer.* ACS applied materials & interfaces, 2019. **12**(4): p. 4970-4979.

8. Jaramillo-Quintero, O.A., et al., *Cadmium-free ZnS interfacial layer for hydrothermally processed Sb2S3 solar cells.* Solar Energy, 2021. **224**: p. 697-702.

9. Jin, X., et al., *In situ growth of [hk1]-oriented Sb2S3 for solution-processed planar heterojunction solar cell with 6.4% efficiency.* Advanced Functional Materials, 2020. **30**(35): p. 2002887.

10. Savadogo, O. and K.C. Mandal, *Low-Cost Schottky Barrier Solar Cells Fabricated on CdSe and Sb2 S 3 Films Chemically Deposited with Silicotungstic Acid.* Journal of The Electrochemical Society, 1994. **141**(10): p. 2871-2877.

11. Li, D.-B., et al., *Stable and efficient CdS/Sb2Se3 solar cells prepared by scalable close space sublimation.* Nano Energy, 2018. **49**: p. 346-353.

Multiresonant Light Trapping in Ultra-thin Solar Cells with Transparent Quasi-random Structures

Eduardo Camarillo Abad, Hannah J. Joyce, Louise C. Hirst

University of Cambridge, Cambridge, United Kingdom

Ultra-thin solar cells offer advantages across different material systems such as cost reductions, flexible form factors and tolerance to absorber defects, but suffer from low absorption of incident photons. Advanced light management techniques beyond antireflection coatings and rear mirrors are needed to extend the optical path length in the absorber layer. Coupling incident illumination to waveguide modes by integrating a scattering structure can localise the field within the absorber layer and increase device performance. However, the thinnest device stacks have reduced modal structures and so introducing multiple waveguide resonances across the spectrum for maximal absorption enhancement is challenging in ultra-thin length-scales. We propose transparent (i.e. dielectric/high band gap semiconductor) quasi-random scattering structures as multiresonant light-trapping textures for ultra-thin devices, offering key advantages: i) richer device modal structure as a consequence of unhindered field propagation within the light trapping layer, effectively making the solar cell a thicker waveguide than with a metallic texture, and ii) richer scattering profile compared to ordered photonic crystals that can fully exploit the available modal structure, yield broadband absorption enhancement and be more tolerant to design variability. By performing an in-depth study of the available parameter space for QR structures, we provide design guidelines for optimal performance that can maximise these advantages across different material systems.

FAIRification, Quality Assessment, and Missingness Pattern Discovery for Spatiotemporal Photovoltaic Data

William C. Oltjen*, Yangxin Fan*, Jiqi Liu*, Liangyi Huang*, Xuanji Yu* Mengjie Li[†], Hubert Seigneur[‡],
Xusheng Xiao*, Kristopher O. Davis[†], Laura S. Bruckman*, Yinghui Wu*, Roger H. French*
*SDLE Research Center, Case Western Reserve University (CWRU), Cleveland, OH, 44106, USA
[†]University of Central Florida (UCF), Orlando, Florida, 32816, USA
[‡]Florida Solar Energy Center (FSEC), Cocoa, FL, 32922, USA

Abstract—Due to the fast growth of the photovoltaic (PV) market, more power plants have become available with data accessible for power forecasting and long-term reliability assessment. The accuracy of the modeling on this data is influenced heavily by the quality of the data and can be improved through data imputation to fill missing gaps. In this study, we introduce a FAIRification framework for ingesting data from PV power plants. This process improves the efficiency of modeling on time series data provided by different labs and companies through an automated ingestion process. We take this analysis further by investigating the use of different imputation methods for filling in large chunks of missing data. Specifically, mean interpolation, linear interpolation, and k-nearest neighbors (KNN) were used in this report to fill in missing data for module temperature and power in a PV time series. It was found that the KNN algorithm outperforms the other methods due to its ability to leverage spatial coherence from nearby systems. These results point towards the potential use of a spatio-temporal graph neural network (st-GNN) in order to impute data using spatial coherence between systems in a large data set with time series data from many PV power plants.

Index Terms—FAIRification, Spatiotemporal GNN, Missingness

I. INTRODUCTION

Photovoltaics (PV) have become a dominant force in the energy sector over the past 20 years. The total, installed solar capacity has increased 500 times since 2000 to a total of 773 GW at the end of 2020 [1]. Not only has the field expanded so much in total, but the rate of installations continues to increase as well. In 2020, the world reported a new record of solar installations by implementing 138 GW of solar energy in a year [1]. The growth of the PV market has pushed the demand for power forecasting and performance evaluation for a huge population of PV power plants which have spatiotemporal coherence that can be utilized for improving model accuracy [2]. There are many logistical challenges towards performing this kind of time series analysis. Different groups use different types of databases, different variable naming schemes,

This material is based upon work supported by the U.S. Department of Energy's Office of Energy Efficiency and Renewable Energy (EERE) under Solar Energy Technologies Office (SETO) Agreement Number DE-EE0009347 and DE-EE0009353. The views expressed herein do not necessarily represent the views of the U.S. Department of Energy or the United States Government.

different data cleaning processes, etc. The time series data itself is typically missing data, or can even have incorrect data from faulty sensors [3]. Manually addressing these kinds of issues takes time away from developing new models on the data which takes time away from producing more efficient PV modules based on the results. Through the development of an automated framework for ingesting time series data and feeding it into machine learning models, we can make our analysis methods more efficient at a larger scale. Much of the standards for our automated process are based on the FAIR principles introduced through the publication of Wilkinson et al. [4]. These principles aim to increase the ability of both humans and computers to understand data by making it Findable, Accessible, Interoperable, and Reusable (FAIR). These guiding principles have been the foundation for our automated process as we try to design a system to standardize the analysis of time series data across the whole solar field. In this paper, we propose a FAIRification framework for spatiotemporal data from PV power plants. We also propose automated methods for data quality assessment and missingness pattern classification that can be applied to time series PV data across the field. Investigating missingness patterns is essential for deciding imputation methods that can improve the model performance for degradation analysis studies. In this case, we examine missingness imputation through the application of several baseline methods including mean interpolation, linear interpolation, and KNNs.

II. METHODS

A. PV Power Plants FAIRification

The data used in this project are stored in the SDLE research center's Apache Hadoop/Hbase/Spark cluster [5], which we will henceforth refer to as CRADLE (Common Research Analytics and Data Lifecycle Environment). This environment is based on the Cloudera CDH distribution. We use a Hadoop Distributed Filesystem (HDFS) to store all of our raw data. After cleaning, the data used for analysis is moved into the Apache Hbase. Hbase takes its inspiration from Google's Big Table, a NOSQL database based on triples where each observation in the dataframe has a rowkey and a columnkey. In order to interact with CRADLE, we rely on Case Western's high performance computing cluster (HPC), an environment with over 250 compute servers, including more than 60 GPU nodes and 7000 processors. While the data used for this analysis is not made public yet, the data will be made accessible by the general PV community on OSF.io in accordance with FAIR guidelines.

We have developed a four-step data ingestion pipeline for receiving data from outside groups and ingesting it into the CRADLE ecosystem. Fig. 1 shows a visual representation of this process.

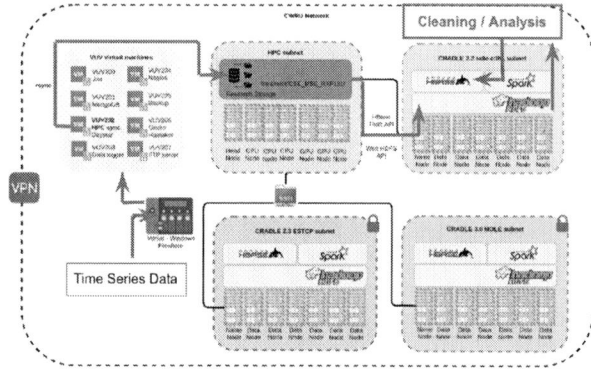

Fig. 1. Data Ingestion Pipeline

After receiving time series data, the first step is to move the data into the staging area. The staging area is stored in Case Western's V-drive, which is a Windows file sharing system hosted by the university. Every night, the contents of the V-drive get backed up automatically to the HPC for redundancy. Once the data is safe in the staging area, the next step is to move the raw data into Case's Hadoop Distributed File System (HDFS). At this point in the process, the raw data has been comfortably stored where it can be accessed should anyone ever need it again. The next step is to preprocess the data. This includes basic data cleaning, adding satellite weather data from SolarGIS, and metadata FAIRification. After the data has been processed, it is stored in an Apache Hbase table for ease of access for future analysis and modeling.

An especially important aspect of our data preprocessing step is metadata FAIRification. There are many benefits associated with FAIRifying our data. It makes our data more easily shareable with other groups because of the standards set for variable nomenclature and structure. It makes it easier for other groups to share data with us, as we can utilize our FAIRification framework to help computers understand more generally what certain variables mean. It also makes it easier to extract meaning from our modeling because of our structured, graph approach for our metadata. There has been an extensive push in the US to make metadata "FAIR" recently, as publishers, science funders, and government agencies have begun to establish requirements for the proper management of metadata. As such, we have been implementing FAIR principles into the ingestion of our data. Specifically, we use

a standardized Javascript Object Notation for Linked Data (JSON-LD) filetype to store our metadata [6]. We have defined a new structure for our JSON-LD metadata files through the creation of a solar power plant ontology. In order to create and design our solar time series ontology, we have used the Protege ontology editor [7].

An ontology is a formal dictionary of terms for a given industry or field that shows how the terms are related through densely interconnected webs. Part of the point of doing this is to standardize terms for solar time series data by defining how variables should be defined across the industry. In our model, for example, latitude is to be spelled exactly latitude (not lat, latd, etc), and it is to be measured in degrees always. This way, there is no ambiguity. An ontology not only defines terms, but it defines a structure for the metadata as well. An ontology is the blueprint for linking metadata terms together through the creation of a knowledge graph. When an ontology is filled in with real data, it becomes a knowledge graph. An ontology is made through the creation of triples, or object-relationship pairs. Fig. 2 shows an example of a triple that connects a solar power plant to a latitude by the hasLatitude property. An ontology makes use of more general terms, defining how classes of objects relate to each other. This can allow a computer to understand generally what a variable means, which can help in its understanding of data received from other groups. A knowledge graph fills the classes from an ontology with values based on the structure defined by the ontology.

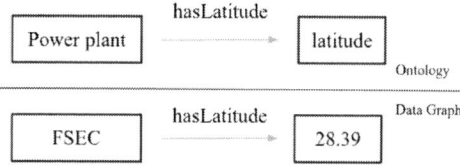

Fig. 2. Examples of Object-Relationship Pairs for the Ontology Blueprint and Resultant Graph

B. Data Quality Assessment

At SDLE, we have developed an R package for analyzing time series data called PVplr [8] that includes functions for the automated analysis of data quality. These functions include a heatmap generator and an automated data grading function. The data grading function assigns a letter grade for the data based on outliers, missingness percentage, and longest missing gap [3]. For a data set to receive an "A" in all categories, it must have outliers less than 10%, missingness less than 10%, and a longest missing gap below 15 days, for example. A full outline of the metrics for the data quality grades can be seen in Fig. 3.

III. PV POWER PLANTS DATA SETS DESCRIPTION

There are eight PV data sets that have been received from different companies and research institutions and have been ingested to the database of our research group. Table I lists

Letter grade	Outliers (%)	Missing percentage (%)	Longest gap (days)
A	Below 10	Below 10	Below 15
B	10 to 20	10 to 25	15 to 30
C	20 to 30	25 to 40	30 to 90
D	Above 30	Above 40	Above 90

Fig. 3. Standards for PV time series grading system

some basic information about each data set. The PV systems in the same data set have the same meta variables and time series variables, but the PV systems in different data sets have some differences regarding both meta and time series variables. For example, meta information about the number of strings and the number of modules in each string exists in data set 1 but not in data set 2. The irradiance data in data set 1 is global horizontal irradiance, but it is plane of array irradiance in data set 2. The PV systems in different data sets can also refer to different scales. There are individual PV modules (such as in the data set 4), inverters for a PV array (such as in the data set 1), and inverters for a PV site (such as in the data set 2).

TABLE I
LIST OF PV SYSTEMS

ID	Average Age	# of Systems	Time Interval (minute)
1	8.24	354	15
2	1.42	1088	1
3	4.24	98	5
4	5.75	8	10
5	0.95	8	30
6	3.13	8	1
7	1.72	70	1
8	2.38	28	15

The data shared from system 8 is from the Florida Solar Energy Center and is the focus of this analysis. More specifically, data from PV systems from the SunSmart Schools program were made available. From inverters that control the racks of PV modules at these schools, we have many years of time series data logged. In total, 28 sites from this program have been shared from FSEC to the SDLE lab at Case Western. This data includes 15 minute interval time series data with information about power output, ambient and reference temperature, irradiance, battery properties, and input and output current and voltage. The length of the time series varies between schools, with the longest set including about 9 years of data, and most of the data sets including data on the order of about 2 years.

IV. DISCUSSION

A. FAIRification

We have developed FAIRmaterials, both an R [9] and Python [10] package, for automating the creation of FAIRified JSON-LD files. Given a simple excel file of a user's metadata, these packages automatically generate a FAIRified JSON-LD file based off the standards that we have developed in our solar power plant ontology. We are currently going through all received data sets and collecting the variables provided with their typical names and units. This information will be provided to our collaborators, including both companies and research institutions, that own the time series PV data of multiple systems. Using feedback from these entities, we can improve our FAIRification process based off the input of real world users.

We have also developed an ontology to describe time series data for its application in solar. The steps we have taken for this process are outlined in Fig. 4.

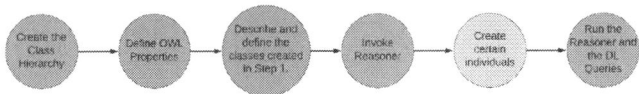

Fig. 4. Steps in Ontology Design

The first step was to create a class hierarchy that describes all of the objects that need to exist in our ontology. In this case, we need to create a general class that describes a solar power plant. Each power plant will have information about its location, array, time series, and inverters. So we have chosen these to be the main sub-classes that describe our power plants. There is a visual depiction of this design in Fig. 5.

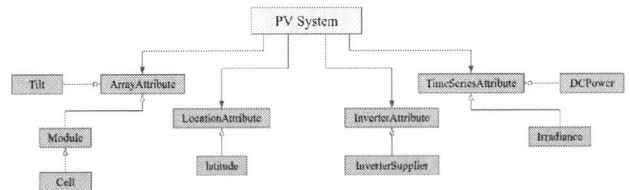

Fig. 5. Class Hierarchy for Solar Ontology

Given the structure from our class hierarchy, we can add more information to our graph by defining how our properties are connected. For example, properties can be described as functional if there is one unique value of y for each instance, x. So a PV power plant would only have one unique value for a longitude, making that a functional property. With a well defined hierarchy and correct property descriptions, we can make use of a reasoner in order to infer things about our data. As we begin to add instance level data into our ontology to create a knowledge graph, we can make use of the ontology's reasoning capabilities in order to discover important relationships in our data.

B. Data Grading

With the data from the SunSmart Schools program, we have performed an analysis of the data quality of a set of PV sites with the PVplr package [8]. In Fig. 6 we have generated a heatmap to visualize the quality of a representative data set.

We plot the time of day on the y-axis, with the date on the x-axis. The graph is then colored in by the power output. This kind of visualization is especially powerful for grading

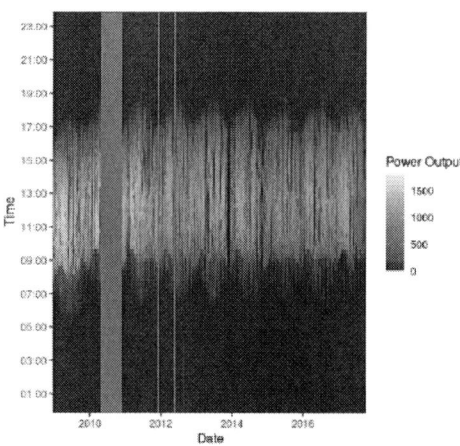

Fig. 6. Data Quality Heatmap

TABLE II
GRADED PV SYSTEMS

Site	Outlier Percentage	Missingness Percentage	Longest Missing Gap	Length Requirement
1	B	A	C	P
2	B	A	D	P
3	B	A	A	F
4	B	B	D	P
5	A	C	C	P
6	B	A	A	F
7	B	B	D	P
8	C	A	A	P
9	B	A	A	P
10	B	A	A	P
11	B	A	A	P
12	C	A	A	P
13	C	A	A	F
14	B	A	A	F
15	B	B	D	P
16	B	A	A	F
17	B	A	A	F
18	B	A	A	P
19	A	D	D	P
20	B	A	C	P
21	B	A	A	P
22	B	B	D	P
23	B	C	D	F
24	B	B	D	P
25	B	A	A	P
26	B	A	A	F
27	C	A	A	F
28	A	D	D	F

data quality, because missing chunks in the data are made especially apparent as grey bars. We can also see that the data that is not missing meets our expectations. We get high power output during the day when it is sunny, and no power output at night. Such analysis is important to perform at the beginning of a project in order to ensure that the data that we are working with is in line with reality. In Fig. 7 are heat maps plots of all the data that exist in our set. This application at scale is made easy through the use of the PVplr R package.

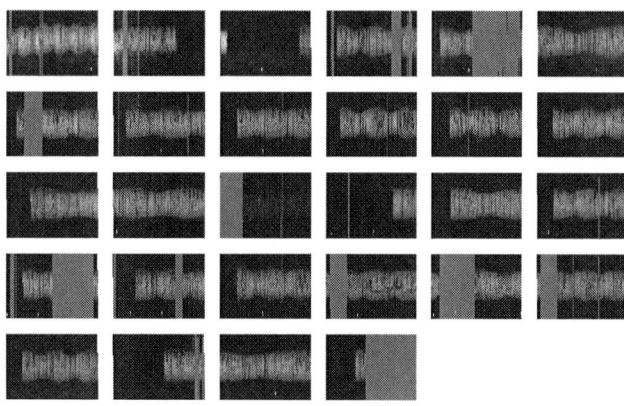

Fig. 7. Heatmaps for All of the Data in the Analysis Set

While a visual representation of the data is useful for human interpretation of the data quality, it's important to convert this into something that a computer can make sense of. We do this through assigning the data letter grades based on their percentage of outliers, missingness percentage, longest missing gap, and a pass/fail based on if the data is longer than two years or not. The grading function applied at scale on our data set is described in Table II. Using the results from our data grading process, we can easily decide which power plants provide more complete data for our analysis. This in turn allows us to focus in on the more important data sets that will allow for more in-depth analysis on our data. With an

idea of the missingness existing in the data set, we can move towards trying to impute this missing data.

C. Missingness Pattern Discovery

We can characterize the patterns of missing values between the different PV time series data sets from two aspects. From a micro perspective, missing data in a series can be categorized into single or block. Single refers to a single missing value between known values while block refers to consecutive chunks of missing values. This is why our data grading function measures both missingness percent and longest missing gap. From a macro perspective between different power plants, depending on the positions of missing values, we consider four common missingness patterns: Missing Completely at Random (MCAR), Disjoint, Overlap, and Blockout, see Fig. 8 [11]. By identifying the common missingness patterns in our PV data sets, we can possibly construct suitable missing value imputation (MVI) models whose assumptions match our data sets.

D. Missing Data Imputation

In this section, we focus on 10 specfic sites in the Sunsmart schools data set that contain data at the same time over a one year period ranging from 09/01/2014 to 08/31/2015. We aim to impute the missingness of the module temperature and power readings from these sites. We measure imputation error

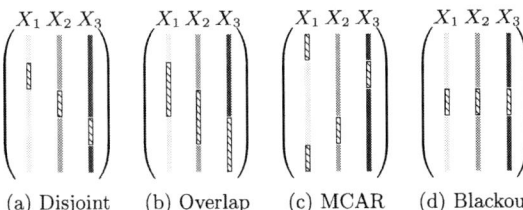

Fig. 8. Four Missingness Scenarios

by Mean Absolute Error (MAE) and Rooted Mean Squared Error (RMSE), defined as follows:

$$MAE = \frac{1}{m}\sum_{i=1}^{m}|P_i - \tilde{P}_i|; \quad RMSE = \sqrt{\frac{1}{m}\sum_{i=1}^{m}(P_i - \tilde{P}_i)^2} \tag{1}$$

where $m = card(M)$, $P_i \in P$, and $\tilde{P}_i \in \tilde{P}$, M is the set of missing data, P is set of imputed values, and $\tilde{P}$ is ground truth.

We compare the performance of three imputation methods.

(1) *Linear Interpolation (LI)* [12]: a timeseries imputation method that fits a simple linear model using two values before and after the missing data block. Each missing data point will then be estimated using the linear model between these points.

(2) *Mean Imputation (Mean)* [13]: a common approach that uses the column-wise mean to fill the missing data.

(3) *K-nearest Neighbors (KNN)* [14]: imputes data by finding and averaging the K nearest neighbors to fill in the missing value.

To evaluate the accuracy of these three methods, we inject missing values into the real-world data sets. Particularly, we are interested in how different imputation methods perform when there are large chunks of missing values (Block Missing). To achieve this, we corrupt daily time series data of each PV system by randomly injecting a 16-hours block of missing values.

Our experiments have demonstrated the superiority of KNN over the LI and Mean methods for imputing missing values in the case of Block Missingness, see Fig. 9 and Fig. 10. KNN achieves a gain from 14.18% to 63.72% in imputation accuracy compared to LI and Mean. KNN likely outperforms other methods because it leverages spatial coherence from nearby systems while LI and Mean impute each PV system separately. The rich neighboring information within PV systems can improve the accuracy of imputation and potential predictive tasks like PV degradation rate prediction.

E. Spatiotemporal GNN Autoencoders

We propose the idea of using spatiotemporal GNN autoencoders to better leverage spatial coherence of PV systems and potentially further improve the imputation accuracy over KNN.

First, we need to translate PV systems into a graph. We map the PV systems into a spatiotemporal graph $G = (V, E, X_v(t))$ where nodes V represent PV systems, edges E are assigned using similarity or spatiotemporal correlations among PV

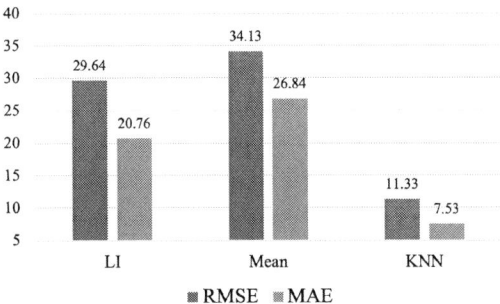

Fig. 9. Imputation Error for Module Temperature

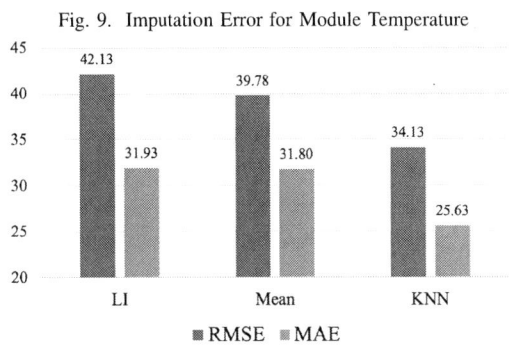

Fig. 10. Imputation Error for Power

systems, and $X_v(t)$ indicates node features. Since the locations of PV systems are fixed, the graph structure is static with time-invariant nodes and edges. However, $X_v(t)$ is time-varying. Each node consists of a set of time-series features such as power, irradiance and temperature and may have missing values in one or multiple of these features.

Spatiotemporal Graph Neural Network modeling has demonstrated its performance improvement in power forecasting for PV power systems rather than utilizing an individual PV system [2] by capturing both spatial and temporal dependencies and coherence among PV systems. We propose a new framework - St-GNN Autoencoders (STGNN-AE) to detect and impute missing values in the PV data sets. Given a set of validated and correct data, we can learn a STGNN-AE, which consists of an encoder to transform inputs into a lower dimension representation and a decoder to recover the inputs from reduced data with a small reconstruction error. STGNN-AE will detect and localize erroneous values as outliers when observing a significant reconstruction error, suggest normal values to be used for imputation with the transformed values from reconstructed embedding by the decoder for local strategy, and provide synthetic PV input for simulation analysis over PV plants and regions when and where sensors are not available. The imputation quality will be measured by the impacts on the performance of downstream learning tasks like PV performance loss rate (PLR) prediction.

V. Conclusions

We have demonstrated in this paper the FAIRification of spatiotemporal PV time series data. By creating a solar power

978-1-7281-6118-1/22 $31.00 © 2022 IEEE

plant ontology, we propose standards for the naming and structure of metadata used to describe the data from these power plants. We can also use this ontology to assist in our modeling, where the computer can infer things about our data based on the relationships we have defined. Using the structure from this ontology, we have developed both R and Python packages for the automation of the FAIRification process. Going further, we have also developed an R package that automates the analysis of the quality of a data set through letter grades and heatmaps. We have shown that imputation methods that can leverage spatial coherence (e.g.: KNN) achieve higher imputation accuracy over simple methods like Linear and Mean Interpolation. To further improve the imputation accuracy, we propose the use of St-GNN autoencoders to detect and impute missing values from a data set by utilizing the spatial coherence between the power plants in the data set.

REFERENCES

[1] J. Christiansen, "Global Market Outlook for Solar Power," SolarPower Europe, Tech. Rep., 2021.

[2] A. M. Karimi, Y. Wu, M. Koyuturk, and R. H. French, "Spatiotemporal graph neural network for performance prediction of photovoltaic power systems," in *Proceedings of the AAAI Conference on Artificial Intelligence*, vol. 35, no. 17, 2021, pp. 15 323–15 330.

[3] D. Moser, D. Bertani, A. J. Curran, R. H. French, M. Herz, and S. Lindig, "International Collaboration Framework for the Calculation of Performance Loss Rates: Data Quality, Benchmarks, and Trends," in *36th European Photovoltaic Solar Energy Conference and Exhibition*, Oct. 2019, pp. 1266–1271, citation Key Alias: moserInternationalCollaborationFramework2019a.

[4] M. D. Wilkinson, M. Dumontier, I. J. Aalbersberg, G. Appleton, M. Axton, and A. Baak, "The FAIR Guiding Principles for scientific data management and stewardship," *Scientific Data*, vol. 3, no. 1, pp. 1–9, Mar. 2016.

[5] Y. Hu, V. Y. Gunapati, P. Zhao, D. Gordon, N. R. Wheeler, M. A. Hossain, T. J. Peshek, L. S. Bruckman, G. Zhang, and R. H. French, "A Nonrelational Data Warehouse for the Analysis of Field and Laboratory Data From Multiple Heterogeneous Photovoltaic Test Sites," *IEEE Journal of Photovoltaics*, vol. 7, no. 1, pp. 230–236, Jan. 2017, tex.ids= hu2017nonrelational, huNonrelationalDataWarehouse2017a, huNonrelationalDataWarehouse2017b, huNonrelationalDataWarehouse2017c, hu_nonrelational_2017, huet.alNonrelationalDataWarehouse2017, yanghuNonrelationalDataWarehouse2017 conferenceName: IEEE Journal of Photovoltaics.

[6] A. Nihar, A. J. Curran, A. M. Karimi, J. L. Braid, L. S. Bruckman, M. Koyutürk, Y. Wu, and R. H. French, "Toward Findable, Accessible, Interoperable and Reusable (FAIR) Photovoltaic System Time Series Data," in *IEEE 48th PVSC Proceedings*, Jun. 2021.

[7] M. Horridge, R. S. Gonçalves, C. I. Nyulas, T. Tudorache, and M. A. Musen, "WebProtege: A Cloud-Based Ontology Editor," in *Companion Proceedings of The 2019 World Wide Web Conference*, ser. WWW '19. New York, NY, USA: Association for Computing Machinery, May 2019, pp. 686–689.

[8] A. Curran, T. Burleyson, S. Lindig, D. Moser, R. French, and SDLE Research Center, "PVplr: Performance Loss Rate Analysis Pipeline," Oct. 2020, tex.ids: a.j.curranPVplrSDLEPerformance2020, curran-PVplrPerformanceLoss2020. [Online]. Available: https://CRAN.R-project.org/package=PVplr

[9] Willam C. Oltjen, Liangyi Huang, and Roger H. French, "FAIRmaterials: Make Materials Data FAIR," Sep. 2021. [Online]. Available: https://CRAN-R-project.org/package=FAIRmaterials

[10] Roger H. French, Liangyi Huang, William C. Oltjen, Arafath Nihar, Jiqi Liu, Justin Glynn, and Kehley Coleman, "Fairmaterials," Oct. 2021. [Online]. Available: https://pypi.org/project/fairmaterials/

[11] M. Khayati, A. Lerner, Z. Tymchenko, and P. Cudré-Mauroux, "Mind the gap: an experimental evaluation of imputation of missing values techniques in time series," in *Proceedings of the VLDB Endowment*, vol. 13, no. 5, 2020, pp. 768–782.

[12] T. Blu, P. Thévenaz, and M. Unser, "Linear interpolation revitalized," *IEEE Transactions on Image Processing*, vol. 13, no. 5, pp. 710–719, 2004.

[13] A. R. T. Donders, G. J. Van Der Heijden, T. Stijnen, and K. G. Moons, "A gentle introduction to imputation of missing values," *Journal of clinical epidemiology*, vol. 59, no. 10, pp. 1087–1091, 2006.

[14] R. Malarvizhi and A. S. Thanamani, "K-nearest neighbor in missing data imputation," *International Journal of Engineering Research and Development*, vol. 5, no. 1, pp. 5–7, 2012.

A Sparse and Low Rank Penalized Signal Decomposition Model with Constraints: Anomaly Detection in PV Systems

Wei Yang*, Daniel Fregosi†, Michael Bolen† and Kamran Paynabar*

*Industrial and Systems Engineering, Georgia Institute of Technology, Atlanta, GA 30332, USA
†Electric Power Research Institute (EPRI), Palo Alto, CA 94304, USA

Abstract—Recently, robust PCA has seen its wide application in various industries for its ability to perform the task of anomaly detection. The essence of robust PCA approach is to break down the signal into a low rank component and sparse component. In many applications, a simple breakdown of the signal without accounting for the signs of low rank components and sparse components would violate the physical constraints of the decomposed signal. In addition, often times, the signals in the real world collected for a long duration has smooth changes within a day and between days. As an example, the power signals collected in a photovoltaic (PV) system are cyclostationary, exhibiting these characteristics. Neglecting the smoothness of signals would result in miss detection of anomalous signals which are smooth within a day but non-smooth between days and vice versa. In this paper, we developed a signal decomposition approach for the purpose of anomaly detection based on the idea of low rank and sparse decomposition taking into consideration the signs of the decomposed low rank and sparse components and the within-day and between-day smooth changes in the original signals. The proposed unsupervised approach for fault detection eliminates the need for faulty samples required by other machine learning methods. It does not require the full I-V characteristics to work. Furthermore, there is no need for complex modelling of PV systems as in the case of power loss analysis. Using Monte Carlo simulations, we demonstrate the ability of our proposed approach for detecting anomalies of different duration and severity in PV systems.

Index Terms—Anomaly detection; PV systems; signal decomposition; low rank and sparse decomposition; constrained optimization

I. Introduction

Photovoltaic systems due to their ability of converting solar energy into electricity in a clean fashion, are playing a bigger role in the evolution of the energy sector [1]. According to the projections by U.S. Energy Information Administration [2], solar generation will account for 14% of the U.S. total electricity generation in 2035 and 20% in 2050. Despite its noticeable advantages over traditional energy resources such as fossil fuel, the faults in PV systems are hindering their wide applications across industries. Without responsive identification of the faults in PV systems, undetected faults do not only impact the power output but also accelerate system aging or result in even fire hazards in worst scenario [3].

This work is funded in part by the U.S. Department of Solar Energy Technologies Office, under Award Number DE-EE-0008976.

Generally, the operation of PV systems can be categorized into three states including ideal, normal and actual operating states. The ideal operating state represents the system performing at its rated efficiency, which is almost never the case in real world scenarios. Therefore, a normal operating state depicts the systems functioning as expected despite a slight deviation from the ideal performance due to unavoidable losses such as temperature loss and inverter conversion loss. The actual operating state is when faults occur in PV systems, which lead to additional drops in the system efficiency unexpectedly [4]. The smallest unit in a PV system is the solar cells which make up the PV module, equivalently, a solar panel. When the sun strikes the solar panel, the solar energy is then converted into electricity in the form of DC current which can be converted to AC current depending on the applications. Stacking the PV modules in either parallel or series configuration transforms the individual PV modules into an array form. The array power equals to the summation of power output from individual PV modules. In field conditions, the power output of the PV arrays might not be close to the predicted power. There are numerous factors contributing to this deviation. The faults in PV systems are therefore defined as any factor reducing the power output. Based on the duration of the faults, they can be classified into two categories including temporary faults (shadows, bird droppings, and dust or snow accumulation on the surface of the PV panel) and permanent faults (electrical disconnection, wiring losses and ageing) [5].

The fault detection in PV systems is an essential task to increase system reliability, efficiency and safety. To detect the fault in a signal, it is natural to think of a detection strategy which decomposes the signal into a faulty component and non-faulty component. In our case, the faulty component is the influence of the faults which brings down the power output of the systems. The non-faulty component is the normal power output of the systems free from the impact of faults. However, directly monitoring the power signal of the system would treat the cloud influence over the system as faults which are not the actual faults of the PV systems although cloud influence reduces the power output the systems. To address this issue, the power signal used in this study is normalized by the irradiance which is measured by the irradiance sensors at the site. As the cloud influence reduces the irradiance and power output measurements simultaneously, this normalization uncovers the actual system efficiency by removing the effect of the cloud on the systems. Considering the periodic behavior of the signals across multiple days, the one-dimensional signal of the power to irradiance ratio collected at every minute is reformulated into a matrix form where the row represents the time within a day while the column represents the index of a day within a given period. Anomaly detection in the reformulated matrix can be realized through modelling the mean as a low-rank component while modelling the anomalies as the sparse matrix as the within-day and between-day correlations of the power to irradiance ratio signal implies low-rank structure of the background and the anomaly is assumed to be sparse.

Recently, low rank and sparse decomposition (LRSD) has demonstrated its successful performance in anomaly detection in various applications involving hyperspectral images [6], infrared thermal images [7] and face images [8]. Let $Y \in \mathbb{R}^{m \times n}$ denotes the original data matrix, $L \in \mathbb{R}^{m \times n}$ denotes the low-rank component and $S \in \mathbb{R}^{m \times n}$ denotes the sparse component. The problem of LRSD can be mathematically described using the following convex optimization problem:

$$\min_{L,S} \|L_*\| + \lambda \|S\|_1 \quad s.t. \quad Y = L + S \tag{1}$$

where $\|L_*\| = \sum_r \sigma_r(L)$ denotes the nuclear norm of L, $\sigma_r(L)$ $(r = 1, 2, \ldots, min(m, n))$ is the r_{th} singular value of L, $\|S\|_1$ denotes the L_1 norm of S. The well-known implementation of low rank and sparse decomposition is robust PCA [9]. Extending the concept of LRSD to detect anomalies in a smooth background, Hao et al. [10] proposed using a smooth basis matrix to extract the coefficients corresponding to the smooth background while using a predetermined basis matrix to extract the coefficients corresponding to the anomaly. A roughness matrix was used to enforce the smoothness of the extracted coefficients of the smooth component. Their proposed method demonstrated superior performance over traditional methods designed for anomaly detection in images including Sobel edge detection [11], jump regression with local polynomial kernel regression [12], the Otsu global thresholding method [13], and the Nick local thresholding method [14]. However, although their method penalizes smoothness of the coefficients corresponding to the smooth component, their method does not permit the separate control of smoothness of within-day variations and between-day variations corresponding to the row wise and column wise changes in the reformulated matrix, respectively. In addition, the low rank based method for anomaly detection in the images and the aforementioned traditional image detection methods do not take into consideration the signs of the decomposed components which carry physical meanings. In our case, the mean signal corresponds to the system efficiency defined as the ratio of power output to irradiance, which should be non-negative. The anomaly corresponds to the negative impact of the faults on system efficiency, which reduces the ratio of power output to irradiance by reducing the expected power output of the system.

In this paper, we proposed a sparse and low rank penalized signal decomposition model with constraints to decompose the signals into faulty and non-faulty components by considering within-day and between-day smoothness of the reformulated signal matrix. In addition, the proposed model takes care of the signs of decomposed components in order to meet their physical constraints. The proposed methodology in this work can be easily adapted for anomaly detection in other cyclostationary signals whose statistical properties vary cyclically with time [15]. The periodic characteristics of the cyclostationary signal is preserved through controlling the smoothness of the changes between periods.

II. RESULTS

To examine the proposed method, we performed Monte Carlo simulation. We considered nine different faulty scenarios where the power percentage drop has three levels including 30%, 60% and 90% and the duration of the faults has three levels at 3, 5 and 7 hours. 100 normal data sets were generated by randomly drawing 4 weeks out of the 20 simulated normal weeks for 100 times. These 100 normal data sets were used to determine the mean and variance of the training false rate. The training false alarm rate is calculated as the percentage of falsely flagging normal days as faulty in a normal data set. For each of the faulty scenarios, 100 test data sets were generated to calculate the mean and variance of the miss detection rate. The miss detection rate is calculated as the percentage of falsely classifying faulty days as normal in a test data set. Each test data set has a duration of 20 weeks. In each week, there is one day injected with the fault. An example of fault detection for a week in which faults with 30% power loss with a duration of 3 hours is presented in Figure 1. The mean and variance of the false alarm rate are 0.03 and 0.001 using a control limit of 0.124. The mean and variance of the miss detection rate are summarized in Table I. The results indicate

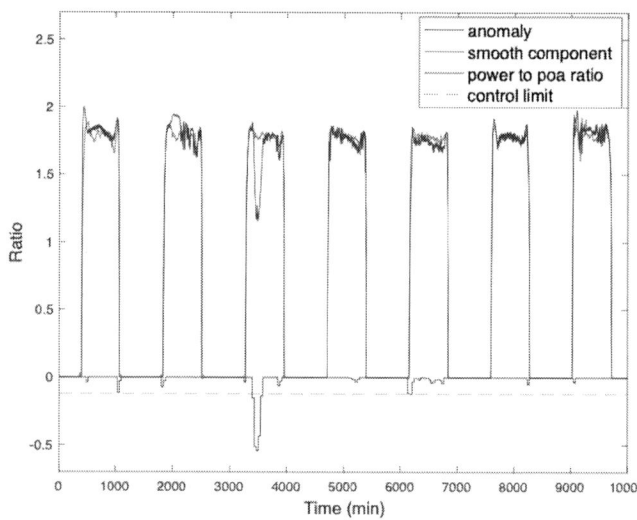

Fig. 1. An example of fault detection on the power to irradiance ratio signal with the presence of a fault leading to 30% power drop over 3 hours.

978-1-7281-6118-1/22 $31.00 © 2022 IEEE

TABLE I
MEAN AND VARIANCE OF MISS DETECTION RATE

Duration of Faults (hours)	Percentage of Power Drop		
	30%	50%	70%
3	0.0180 ± 0.00160	0.0015 ± 0.00007	0.0005 ± 0.00003
5	0.0056 ± 0.00035	0.0005 ± 0.00003	0 ± 0
7	0.0030 ± 0.00014	0 ± 0	0 ± 0

that our proposed method can detect faults with a high level of accuracy even when the fault duration is relatively short and the percentage of power loss is relatively small. As the fault severity increases, the method can detect it with higher accuracy and smaller variance.

III. DISCUSSION

The developed fault detection method taps into the power and irradiance signals collected at the site. This method does not require the full I-V characteristics to work, which is the disadvantage of the method of analyzing I-V characteristics [16], [17]. Another area of research for the fault detection in PV systems explores the use of power loss analysis technique [18], [19]. For the power loss analysis, the performance of the fault detection method is highly sensitive to the accuracy of the simulated model employed. Also, the development and calibration of the PV models for different sites are time consuming. Furthermore, the simulated model may not well represent the PV systems under complex environmental conditions. Different from the power analysis method, the proposed method does not involve complex modelling of PV systems. As recent advances in machine learning techniques, researchers also utilize machine learning tools to tackle fault detection challenges in PV systems [20], [21]. However, the machine learning based approaches require simulated or real faulty samples. On the one hand, collecting the faulty samples from the real sites is a costly and unsafe practice in some cases. On the other hand, it is impossible to include all the faulty scenarios in the training dataset. Therefore, the ability of the trained machine learning model to detect unseen faulty samples remains questionable. As an unsupervised method, this approach we developed does not require faulty samples to train the model. Our work provides a simple-to-implement yet robust tool for the detection of faults in PV systems.

IV. CONCLUSION

In this work, we developed an unsupervised fault detection algorithm using the power to irradiance ratio signal. Our proposed method was tested under various fault scenarios. Detection accuracy remains high even as fault severity decreases. Assuming smoothness both within and between periods, the developed fault detection algorithm can also be adapted to detect faults in other cyclostationary signals. As opposed to the methods in the literature, our approach does not require obtaining full I-V characteristics, simulating PV systems, or generating faulty samples.

REFERENCES

[1] S. K. Sansaniwal, V. Sharma, and J. Mathur, "Energy and exergy analyses of various typical solar energy applications: A comprehensive review," *Renewable and Sustainable Energy Reviews*, vol. 82, pp. 1576–1601, 2018.

[2] U.S. Energy Information Administration, "Annual energy outlook 2021," Feb 2021.

[3] R. Hariharan, M. Chakkarapani, G. Saravana Ilango, and C. Nagamani, "A method to detect photovoltaic array faults and partial shading in pv systems," *IEEE Journal of Photovoltaics*, vol. 6, no. 5, pp. 1278–1285, 2016.

[4] S. Firth, K. Lomas, and S. Rees, "A simple model of pv system performance and its use in fault detection," *Solar Energy*, vol. 84, no. 4, pp. 624–635, 2010, international Conference CISBAT 2007.

[5] S. R. Madeti and S. Singh, "A comprehensive study on different types of faults and detection techniques for solar photovoltaic system," *Solar Energy*, vol. 158, pp. 161–185, 2017.

[6] L. Li, W. Li, Q. Du, and R. Tao, "Low-rank and sparse decomposition with mixture of gaussian for hyperspectral anomaly detection," *IEEE Transactions on Cybernetics*, vol. 51, no. 9, pp. 4363–4372, 2021.

[7] Q. Wang, K. Paynabar, and M. Pacella, "Online automatic anomaly detection for photovoltaic systems using thermography imaging and low rank matrix decomposition," *Journal of Quality Technology*, vol. 0, no. 0, pp. 1–14, 2021.

[8] Q. Zhao, D. Meng, Z. Xu, W. Zuo, and L. Zhang, "Robust principal component analysis with complex noise," in *Proceedings of the 31st International Conference on Machine Learning*, ser. Proceedings of Machine Learning Research. Bejing, China: PMLR, 2014, pp. 55–63.

[9] E. J. Candès, X. Li, Y. Ma, and J. Wright, "Robust principal component analysis?" *Journal of the ACM*, vol. 58, no. 3, pp. 1–37, 2011. [Online]. Available: https://doi.org/10.1145/1970392.1970395

[10] H. Yan, K. Paynabar, and J. Shi, "Anomaly detection in images with smooth background via smooth-sparse decomposition," *Technometrics*, vol. 59, no. 1, pp. 102–114, 2017.

[11] Sobel and G. Feldman, "A 3x3 isotropic gradient operator for image processing," *Pattern Classification and Scene Analysis*, pp. 271–272, 1973.

[12] P. Qiu and J. Sun, "Local smoothing image segmentation for spotted microarray images," *Journal of the American Statistical Association*, vol. 102, no. 480, pp. 1129–1144, 2007.

[13] Otsu and Nobuyuki, "A threshold selection method from gray-level histograms," *IEEE Transactions on Systems, Man, and Cybernetics*, vol. 9, no. 1, pp. 62–66, 1979.

[14] D. Garcia, "Robust smoothing of gridded data in one and higher dimensions with missing values," *Computational Statistics and Data Analysis*, vol. 54, no. 4, pp. 1167–1178, 2010.

[15] W. A. Gardner, A. Napolitano, and L. Paura, "Cyclostationarity: Half a century of research," *Signal Processing*, vol. 86, no. 4, pp. 639–697, 2006. [Online]. Available: https://www.sciencedirect.com/science/article/pii/S0165168405002409

[16] A. Eskandari, J. Milimonfared, and M. Aghaei, "Line-line fault detection and classification for photovoltaic systems using ensemble learning model based on i-v characteristics," *Solar Energy*, vol. 211, pp. 354–365, 2020.

[17] M. T. Boyd, S. A. Klein, D. T. Reindl, and B. P. Dougherty, "Evaluation and Validation of Equivalent Circuit Photovoltaic Solar Cell Performance Models," *Journal of Solar Energy Engineering*, vol. 133, no. 2, p. 021005, 2011.

[18] A. Chouder and S. Silvestre, "Automatic supervision and fault detection of pv systems based on power losses analysis," *Energy Conversion and Management*, vol. 51, no. 10, pp. 1929–1937, 2010.

[19] W. Chine, A. Mellit, A. M. Pavan, and S. Kalogirou, "Fault detection method for grid-connected photovoltaic plants," *Renewable Energy*, vol. 66, pp. 99–110, 2014.

[20] Y. Zhao, L. Yang, B. Lehman, J.-F. de Palma, J. Mosesian, and R. Lyons, "Decision tree-based fault detection and classification in solar photovoltaic arrays," in *2012 Twenty-Seventh Annual IEEE Applied Power Electronics Conference and Exposition (APEC)*, 2012, pp. 93–99.

[21] S. A. Zaki, H. Zhu, M. A. Fakih, A. R. Sayed, and J. Yao, "Deep-learning-based method for faults classification of pv system," *IET Renewable Power Generation*, vol. 15, no. 1, pp. 193–205, 2021.

978-1-7281-6118-1/22 $31.00 © 2022 IEEE

Cloud Segmentation and Motion Tracking in Sky Images

Benjamin G Pierce, Joshua S Stein, Jennifer L Braid, Daniel Riley

Sandia National Laboratories, Albuquerque, NM, United States

In this work, we present two different algorithms to aid in real-time weather predictions. This information can be used to inform the movement of a tracker or short-term power predictions. Since cloud cover significantly affects the resulting insolation on a PV module, identifying and tracking cloud motion is useful to this end. This work presents a convolutional autoencoder (CAE) to identify clouds and a particle tracker to predict cloud movement. The CAE model integrates information from multiple approaches to cloud segmentation. Particle tracking is useful in areas such as Albuquerque, NM where clouds move in smaller fragments due to rapid variance in wind direction caused by nearby mountains. By combining neural networks and more classical technologies, the system becomes more robust and explainable then either image processing or pure neural network technologies, respectively.

Towards Standardization of Accelerated Stress Testing Protocols for Metal-Halide Perovskite Photovoltaic Modules

Michael Owen-Bellini, Timothy J Silverman, Michael G. Deceglie, Paul Ndione, Nikos Kopidakis, Ingrid Repins, Mickey Wilson, Dana B. Sulas-Kern, Joseph Berry, Laura T. Schelhas, Colin Sillerud, Jinsong Huang, Michael J. Heben, Yanfa Yan, Devin MacKenzie, Joshua S. Stein

National Renewable Energy Laboratory, Golden, CO, United States

CFV Labs, Albuquerque, NM, United States

University of North Caolina, Chapel Hill, NC, United States

University of Toledo, Toledo, OH, United States

University of Washington, Seattle, WA, United States

Sandia National Laboratories, Albuquerque, NM, United States

Metal-Halide Perovskite (MHP) photovoltaic (PV) modules are at the cusp of commercialization. One of the major hurdles that remains is establishing confidence in long-term durability. This work proposes an initial accelerated stress testing protocol that can be used as the foundation for the development of a standardized test procedure for qualification of commercial MHP modules. We apply the protocol to three different MHP module architectures to identify the critical degradation pathways that will inform future testing development. In addition, the modules are exposed to outdoor environments such that a cross-examination with indoor results can be used to verify the relevance of the indoor stress conditions.

THE BALANCE OF THERMODYNAMIC POTENTIALS IN SOLAR CELLS INVESTIGATED BY NUMERICAL DEVICE SIMULATIONS

Felix Komoll, Uwe Rau

This contribution presents an extension of the classical gain/loss analysis of Brendel et al. (Appl. Phys. Lett. 93, 173503, 2008) based on the balance of free energy in a solar cell. We consider the full balance of all thermodynamic potentials by separating the excess free energy into excess chemical and electrostatic potentials. A layer by layer analysis of an exemplary silicon solar cell shows that the different functionalities of different parts of the solar cell, e.g., the neutral base and the space charge regions, are well reflected in the different pictures provided by looking at the thermodynamic potentials separately.

Hybrid Functional Calculations for Antimony Doping in CdTe

Intuon Chatratin, Shagorika Mukherjee, and Anderson Janotti

Department of Materials Science & Engineering, University of Delaware, Newark, Delaware, 19716, USA

Abstract—CdTe-based solar cells are leading thin-film photovoltaic technology, with efficiencies over 22%, but still much lower than the theoretical maximum of 29%. Further improvements will rely on increasing the open-circuit voltage Voc, which, in turn, depends on carrier density and lifetime. Using hybrid density functional calculations, we investigate Sb doping of CdTe, focusing on its limitation as shallow acceptor and the formation of compensating AX center. Paying special attention to supercell size and effects of spin-orbit coupling, we predict an ionization energy of 116 meV in the dilute limit, much closer to recent experimental value of 103 meV from temperature-dependent Hall measurements in bulk single crystals, and in contrast to much larger values from previous calculations. We also find that the Sb-related AX centers are not major compensation centers in Sb-doped CdTe.

Index Terms—CdTe, group-V acceptor, defects, *p*-type

I. INTRODUCTION

Among thin-film photovoltaic technologies, cadmium telluride (CdTe) is the leading technology that is commercially available. Several factors, such as low cost, rapid industrial production, long lifetime, high absorption coefficient, and a direct band gap of 1.5 eV, have driven the research on CdTe, placing CdTe-based PV modules in direct competition to crystalline silicon (c-Si) modules [1]. A record efficiency of CdTe cells of 22.1% has been recently reported by First Solar [2]. However this efficiency is still far below the theoretical limit of $\sim$29%, mainly attributed to the low open-circuit voltage Voc which can be improved by increasing the hole concentration. From device modeling, increasing the hole concentration in the CdTe absorber from current typical values of 10^{14} to 10^{16} cm^{-3} would lead to efficiencies reaching 25% [3].

One common way to obtain high hole concentrations in semiconductors is adding impurities that act as shallow acceptors; among the most promising *p*-type dopants in CdTe, Sb stands out for the similar atomic radii as Te (1.45 vs 1.40 Å) and similar atomic electronic structures – they sit next to each other on the periodic table, with Sb on the left of Te) [4], [5]. Thus, introducing Sb on the Te sites is expected to minimize the strain energy associated with the acceptor impurity and to lead to a shallow acceptor level such that at room temperature (or operating device temperatures) a large fraction of the impurities are ionized, efficiently leaving holes in the valence. However, early theoretical studies indicated a rather high ionization energy of 230 meV for Sb substituting on the Te site [6]; more recent hybrid density

EERE Solar Energy Technologies Office, DOE grant number: DE-EE0009344

functional calculations give 150 meV [8]; but also indicates likely formation of AX compensation center, resulting in low doping efficiency and, thus, low hole concentrations [7], [8]. A recent analysis of temperature-dependent Hall measurements on CdTe single crystals doped with Sb indicates a much lower ionization energy of 103 meV [5], enabling hole concentration of $\sim$10^{16} cm^{-3}. This result indicates some compensation, but not as severe as that predicted by recent calculations [8]. To understand this apparent contradiction between theory and experiment, we revisit the properties of the Sb impurity in CdTe using hybrid density functional theory, paying special attention to the effects of spin-orbit coupling and the finite size of the supercell for dealing with shallow acceptors.

II. METHOD OF CALCULATIONS

Our calculations are based on density functional theory (DFT) and the projected augmented wave (PAW) method [9], as implemented in the Vienna ab initio simulation package (VASP) [10]. Since DFT within the standard approximations LDA and GGA severely underestimate band gaps, we use the Heyd-Scuseria-Ernzerhof (HSE) hybrid functional to correct the band gap [11]. To accurately describe not only the band gap but the ionization potential (position of the valence band with respect to vacuum level), we use a Hartree-Fock exchange mixing parameter of 33% and include the spin orbit coupling (SOC). An energy cut off of 254 eV is used for the plane-wave basis set. The calculated band gap of 1.502 eV and equilibrium lattice parameter of 6.545 Å are in good agreement with the experimental data [12], [13]. The SOC lifts the valence-band maximum (VBM) by 0.307 eV, with a spitting Δ_{SOC} of 0.940 eV. The calculated ionization potential of 5.4 eV is closer to the experimental value than standard DFT [14].

For the calculations of the Sb impurity, supercells of 64 atoms, 216 atoms and 512 atoms are used. The formation energy of a defect X (Sb$_{\text{Te}}$ in this case) in charge state q in CdTe is given by:

$$E^f(X^q) = E_{tot}(X^q) - E_{tot}(\text{bulk}) + \sum_i n_i \mu_i \quad (1)$$
$$+ q(\varepsilon_F + E_{VBM}),$$

where $E_{tot}(X^q)$ is the total energy of the supercell containing the defect in charge state q, $E_{tot}(\text{bulk})$ is the total energy of the supercell representing the perfect bulk material, μ_i is a chemical potential of atom i referenced to the total energy of the reservoir for the atom i. n_i is a number of atoms i that are removed ($n_i > 0$) or added ($n_i < 0$) to the supercell to form the defect. The Fermi level ε_f is referenced to VBM. The Fermi level position at which the formation energy of a defect in a charge state q equals to the formation of the same defect in a charge state q' defines the transition level (q/q') which is given by:

$$(q/q') = \frac{E^f(X^q; \varepsilon_F = 0) - E^f(X^{q'}; \varepsilon_F = 0)}{q' - q}, \quad (2)$$

where $E^f(X^q; \varepsilon_F = 0)$ is the formation energy of X^q taken from Fermi level at 0 eV. The transition level $(0/-)$ for the acceptor Sb_{Te} is the acceptor level or the acceptor ionization energy (E_a).

III. RESULTS AND DISCUSSION

Sb has one less electron than Te, so when it substitutes on the Te site in CdTe, it acts as an acceptor. Our calculations show that for the substitutional Sb_{Te}, we have a low-lying a_1 state buried in the valence band (well below the VBM) and three-fold degenerate t_2 state near the VBM. In the neutral charge state Sb_{Te}^0, there is one hole t_2 state, which would in principle leads to a Jahn-Teller distortion. We find, however, that this effect is quite small, which we attribute to the delocalization of the hole wavefunction. In the negative charge state Sb_{Te}^-, the t_2 state is completely filled. Due to the delocalization of the acceptor state and the interaction of the negatively charged impurity with its periodic image due to the 3D periodic boundary conditions, we expect a slow convergence of the acceptor transition level, $(0/-)$, with the supercell size, leading to an artificially large ionization energy in the finite supercell calculations.

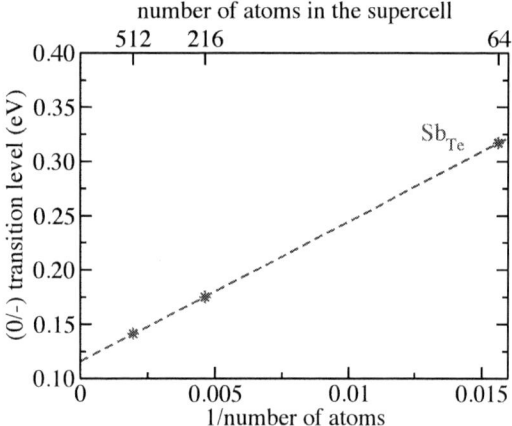

Fig. 1. Calculated $(0/-)$ acceptor transition level of Sb_{Te} in CdTe using 64-, 216-, and 512-atoms supercells as function of the inverse of the number of atoms in the supercell.

To remove the artificial interaction, we extrapolate the calculated $(0/-)$ transition level to the dilute limit. The extrapolation of the transition level as a function of $1/L$ (where L is the linear dimension of the supercell) or $1/$number of

atoms$\rightarrow 0$ is expected to give the transition level in the dilute limit regime, where the charged impurity does not interact with its periodic image. The results for Sb_{Te} using 64-, 216-, and 512-atoms supercells are shown in Fig. 1. As expected, larger supercell cell sizes lead to lower acceptor ionization energies. The extrapolation to the dilute limit gives an ionization energy of 116 meV, close to the experimental value of 103 meV obtained from temperature-dependent Hall measurements in high-quality Sb-doped CdTe bulk single crystals.

We also investigated the formation of Sb-related AX center in CdTe. By removing an electron from the charge neutral Sb_{Te}^0, leaving two hole in the t_2 state, the Sb atom undergoes a large displacement along the [110] direction, and forms a bond with a neighboring Te atom, breaking their original bonds with one of the Cd atoms. This large local lattice relaxation, that can be attributed to a Jahn-Teller distortion, splits the t_2 state into a lower energy twofold degenerate e state (occupied by four electrons) below the VBM, and an empty a_1 state (two holes) at a much higher energy, resonant in the conduction band. The resulting defect is, therefore, stable in the positive charge state, resulting in self-compensation, i.e., an acceptor center that upon ionization becomes a donor center often called an AX center [17]. The charge state associated with the AX center is quite localized around the defect and completely contained in the supercell, even for the 64-atoms supercell.

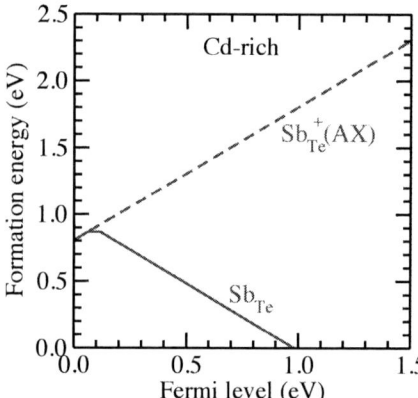

Fig. 2. Formation energies of Sb in CdTe, including the neutral (b_{Te}^0, negatively charged Sb_{Te}^-, and the donor AX center $Sb_{Te}^+(AX)$.

The calculated formation energy of Sb_{Te}^0, Sb_{Te}^-, and $Sb_{Te}^+(AX)$ are shown in Fig. 2. The results show that the AX center (dashed line) is only stable when the Fermi level is very close to the VBM, i.e., within 70 meV. This finding suggests that the AX centers are unlikely to limit hole concentration in CdTe, contrary to previous DFT and hybrid DFT calculations [7], [8]. We attribute the difference between our results and the results of previous calculations to the SOC, that raises the VBM and, consequently, places the $(+/0)$ transition level at lower energies relative to the VBM.

For comparison with experiments, we plot the calculated hole concentration as a function of temperature in Fig. 3. Taking the calculated ionization energy E_a=116 meV, and

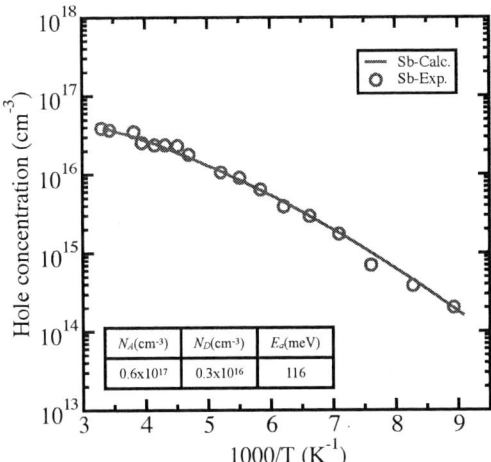

Fig. 3. Calculated hole concentration for a total Sb concentration of 6×10^{16} cm^{-3} and assuming 5% compensating donors (unknown origin), compared to temperature-dependent Hall data from [5].

assuming a total Sb concentration (N_A) of 6×10^{16} and a 5% donor concentration (N_D) as compensation centers give good agreement with temperature-dependent Hall transport measurements. We note that increasing Sb concentration by 5% would overestimate the hole concentration at high temperatures, and increasing the concentration of compensating donors by 5% would severe the agreement at low temperatures, decreasing the hole concentration. We also note that increasing the ionization energy E_a by 50 meV would significantly lower the hole concentration in the whole temperature range, and could not explain the experiment data.

Regarding the origin of the compensating centers, we first argue that the formation of AX center seems not to be a problem in the case of Sb-doped CdTe. From previous calculations that include effects of SOC [18], we would conclude that Cd interstitial (Cd_i) and Te vacancies (V_{Te}) are the most likely defects that could act as compensation centers. However, considering the rather low migration barrier of Cd_i of 0.52 eV [19], we expect these defects to be highly mobile and, therefore, not stable. We then speculate that (V_{Te}) would be the most likely compensating defects in p-type CdTe.

IV. Summary

Understanding the behavior of the Sb doping is important for designing CdTe solar cells with higher doping efficiencies. In this work, we employ hybrid density functional calculations with the inclusion of spin-orbit coupling to investigate formation energies and ionization energies of the Sb acceptor in CdTe, and explore the stability of the related AX center. The calculated ionization energy of 116 meV in the dilute limit is in good agreement with the experiment data. Furthermore, our calculations indicate that the formation of AX centers is not a problem for Sb-doped CdTe, and that other native point defects are likely acting as compensation centers. By using the calculated ionization energies, and assuming a dopant

concentration and a relatively low donor compensation, we can accurately predict the hole concentration seen in recent experiments.

Acknowledgments

This work was supported by the EERE Solar Energy Technologies Office, DOE grant number DE-EE0009344, and used resources of the National Energy Research Scientific Computing Center (NERSC), a U.S. Department of Energy Office of Science User Facility located at Lawrence Berkeley National Laboratory, operated under Contract No. DE-AC02-05CH11231, using NERSC award BES-ERCAP0021133.

References

[1] R. Schmalensee, The future of solar energy: an interdisciplinary MIT study. Energy Initiative, Massachusetts Institute of Technology, 2015.

[2] M. A. Green et al., "Solar cell efficiency tables (version 56)," Prog. Photovoltaics Res. Appl., vol. 28, no. 7, pp. 629–638, 2020.

[3] T. Ablekim, E. Colegrove, and W. K. Metzger, "Interface engineering for 25% CdTe solar cells," ACS Appl. Energy Mater., vol. 1, no. 10, pp. 5135–5139, Oct. 2018.

[4] B. E. McCandless et al., "Overcoming carrier concentration limits in polycrystalline CdTe thin films with in situ Doping," Sci. Rep., vol. 8, no. 1, p. 14519, 2018.

[5] A. Nagaoka et al., "Comparison of Sb, As, and P doping in Cd-rich CdTe single crystals: Doping properties, persistent photoconductivity, and long-term stability," Appl. Phys. Lett., vol. 116, no. 13, pp. 1–6, 2020.

[6] S.-H. Wei and S. B. Zhang, "Chemical trends of defect formation and doping limit in II-VI semiconductors: The case of CdTe," Phys. Rev. B, vol. 66, no. 15, p. 155211, Oct. 2002.

[7] J.-H. Yang et al., "Review onfirst-principles study of defect properties of CdTe as a solar cell absorber," Semicond. Sci. Technol., vol. 31, p. 083002, 2016.

[8] B. Dou, Q. Sun, and S.-H. Wei, "Optimization of doping CdTe with group-V elements: a first-principles study," Phys. Rev. Appl., vol. 15, no. 5, p. 054045, May 2021.

[9] G. Kresse and D. Joubert, "From ultrasoft pseudopotentials to the projector augmented-wave method," Phys. Rev. B, vol. 59, no. 3, p. 1758, 1999.

[10] G. Kresse and J. Furthmüller, "Efficient iterative schemes for ab initio total-energy calculations using a plane-wave basis set," Phys. Rev. B, vol. 54, no. 16, p. 11169, 1996.

[11] J. Heyd, G. E. Scuseria, and M. Ernzerhof, "Hybrid functionals based on a screened Coulomb potential," J. Chem. Phys., vol. 118, no. 18, pp. 8207–8215, May 2003.

[12] A. J. Strauss, "The physical properties of cadmium telluride," Rev. Phys. Appliquée, vol. 12, no. 2, pp. 167 184, 1977.

[13] G. Fonthal, L. Tirado-Mejia, J.I.Marin-Hurtado, H. Ariza-Calderón, and J. G. Mendoza-Alvarez, "Temperature dependence of the band gap energy of crystalline CdTe," J. Phys. Chem. Solids, vol. 61, no. 4, pp. 579–583, 2000.

[14] A. Grüneis, G. Kresse, Y. Hinuma, and F. Oba, "Ionization potentials of solids: The importance of vertex corrections," Phys. Rev. Lett., vol. 112, no. 9, pp. 1–5, 2014.

[15] C. Freysoldt, J. Neugebauer, and C. G. de Walle, "Electrostatic interactions between charged defects in supercells," Phys. status solidi, vol. 248, no. 5, pp. 1067–1076, 2011.

[16] C. Freysoldt et al., "First-principles calculations for point defects in solids," Rev. Mod. Phys., vol. 86, no. 1, p. 253, 2014.

[17] D. J. Chadi, "Predictor of p -type doping in II-VI semiconductors," Phys. Rev. B, vol. 59, no. 23, pp. 15181–15183, Jun. 1999

[18] J. Pan, W. K. Metzger, and S. Lany, "Spin-orbit coupling effects on predicting defect properties with hybrid functionals: A case study in CdTe," Phys. Rev. B, vol. 98, no. 5, pp. 1–9, 2018.

[19] J. Ma, J. Yang, S.-H. Wei, and J. L. F. Da Silva, "Correlation between the electronic structures and diffusion paths of interstitial defects in semiconductors: The case of CdTe," Phys. Rev. B, vol. 90, no. 15, p. 155208, 2014.

GaAs-Based Photovoltaic Infrared Energy Harvesting for Microscale Biomedical Implants

Y. Sun[1], J. Letner[2], J. Lee[1], N. Ahmed[3], C. Chestek[2], D. Blaauw[1], and J. Phillips[1,3]

[1]Electrical Engineering and Computer Science Department, University of Michigan, Ann Arbor, MI 48109
[2]Biomedical Engineering Department, University of Michigan, Ann Arbor, MI 48109
[3]Electrical and Computer Engineering Department, University of Delaware, Newark, DE 19716

Abstract— **Fully wireless biomedical implantable devices at the microscale are of great interest for a new generation of medical devices inlcluding precision health and brain computer interfaces. One of the greatest challenges for these devices is energy harvesting for self-powering at small dimensions. Here, we demonstrate a dual-junction (DJ) GaAs-based photovoltaic (PV) cell with an area of 200 x 200 μm^2 and explore their energy harvesting potential for near surface bio-implantables. The devices achieve a power conversion efficiency of more than 23% and open circuit voltage of greater than 1.13 V under 1.03 μW/mm^2 near-infrared irradiation at a wavelength of 850 nm. Energy harvesting through biospecimens of mouse skins are demonstrated with sufficient power density for low power CMOS electronics at irradiance levels orders of magnitude below the maximum permissible exposure defined by the American National Standards Institute (ANSI) Z136.1 standard.**

Index Terms— **optical absorption, wireless power transfer, optoelectronic devices, multi-junction photovoltaic cells**

I. INTRODUCTION

BIOMEDICAL implantable devices at the microscale have attracted great attention based on their abilities for local sensing and for large area or distributed sensor arrays. However, efficient wireless power and energy harvesting remain as a major challenge when scaling to sub-millimeter dimensions. Radio frequency (RF) [1] and ultrasound [2] have recently demonstrated success for neural implants, but both have difficulties in further scaling due to antenna coupling efficiency or size requirements for mechanical resonators. In comparison, optical devices are highly scalable at the sub-millimeter scale, where PV devices only have a weak dependence of power conversion efficiency with decreasing cell area. The near-infrared spectral region provides a key optical transparency window for tissue that enables wireless links that can provide sufficient energy to power bio-implantable devices at tissue depths ranging from millimeters to a centimeter. In addition, PV cells offer an advantage that the output voltage is relatively insensitive to device area, where only the generated current is proportional to cell area. This allows for direct power transfer without additional rectification or voltage upconversion cir-

cuitry that would degrade overall power conversion efficiency and/or size requirements of the system.

GaAs PV cells provide high power conversion efficiency that is well matched to the near-infrared high transparency window for biological tissue. The high optical absorption and conversion efficiency of GaAs can provide high power density with a small form factor for scaled bio-implantable devices, and in combination with appropriate biocompatible packaging, provides a key technology for future biomedical applications. In this work, near-infrared power conversion in tandem microscale GaAs PV cells is explored to establish near-surface bio-implantable energy harvesting capabilities.

II. EXPERIMENT

DJ GaAs PV cells were designed and optimized for 850 nm irradiation. The epitaxial structure, cell design, and fabrication technique were described previously [3]. A schematic diagram and microscope photo are shown in Fig. 1(a) and (b), respectively. The device in this work further incorporated a monolithically integrated InGaAs light emitting diode on top of the PV cell, which is intended for an optical communication link for a neural sensor [4]. The electrical characteristics of the PV cells were measured under dark and illuminated conditions using a Keithley 2400 source meter. An 850 nm near-infrared light emitting diode array was used to flood illuminate the PV cells. The optical setup is shown in Fig. 2(a), where irradiance values were measured using a calibrated Thorlabs PM120D photodetector. Two mouse biospecimens of varying thickness were used to determine optical transmission through biological tissue (Fig. 2(b) and (c)). The selected samples represent the highest loss regions in the infrared, and greatest limiting factor energy harvesting, due to reflection, absorption, and scattering in the epidermis.

III. RESULTS AND DISCUSSION

The electrical characteristics of the PV cells under 850 nm illumination at 1.03 μW/mm^2 are shown in Fig. 3(a). The PV cells demonstrate a conversion efficiency of 23.8% (without transmission through tissue) and provides at power density of over 244 nW/mm^2 at a maximum power point voltage of 0.92 V (>1.13 V open circuit voltage and >11.6 nA short

978-1-7281-6118-1/22 $31.00 © 2022 IEEE

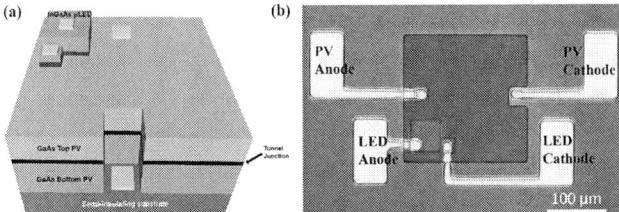

Fig. 1. (a) Schematic diagram and (b) microscope photo of the PV devices designed for biomedical implantable devices.

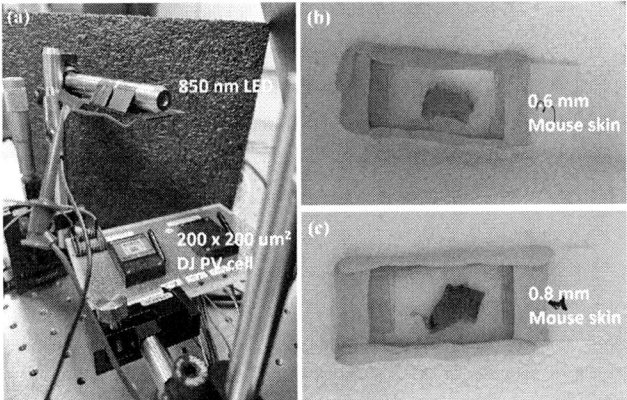

Fig. 2. (a) Optical testing setup using a 850 nm NIR LED. PV devices are placed 10 cm away from the light source. Mouse biospecimens, (b) and (c), with thickness of approximately 0.6 mm and 0.8 mm, respectively.

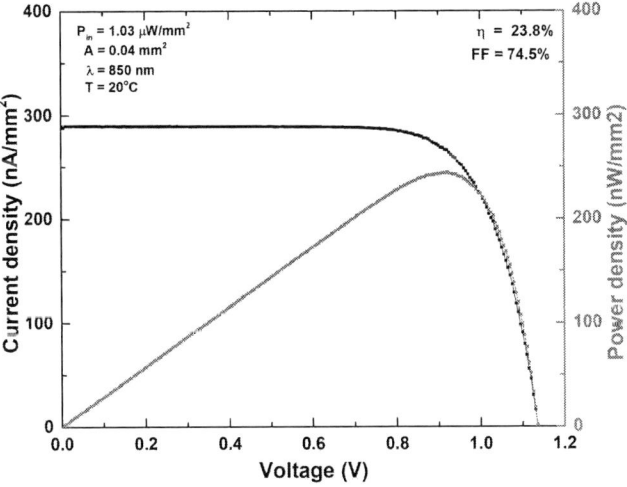

Fig. 3. Measured J-V and P-V characteristics of the DJ PV cell under 850 nm NIR LED illumination at irradiance of 1.03 μW/mm^2.

circuit current).

The electrical output power of the PV cells (at maximum power point) versus infrared irradiance is shown in Fig. 4. Comparison is shown for illumination in free space and for illumination through tissue samples. Using an estimated minimum value of 50 nW/mm^2 to achieve self-powering for a system [5], the corresponding minimum irradiance values are in the range of 100's of nW/mm^2 to 1 μW/mm^2.

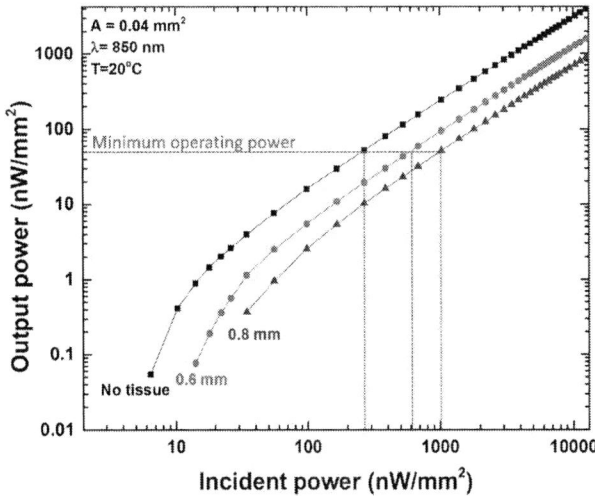

Fig. 4. Output power versus incident power density for GaAs DJ PV cells through 0.6-mm- and 0.8-mm-thick mouse skin smaples. The minimum operating power density of 50 nW/mm^2 is also plotted in the figure.

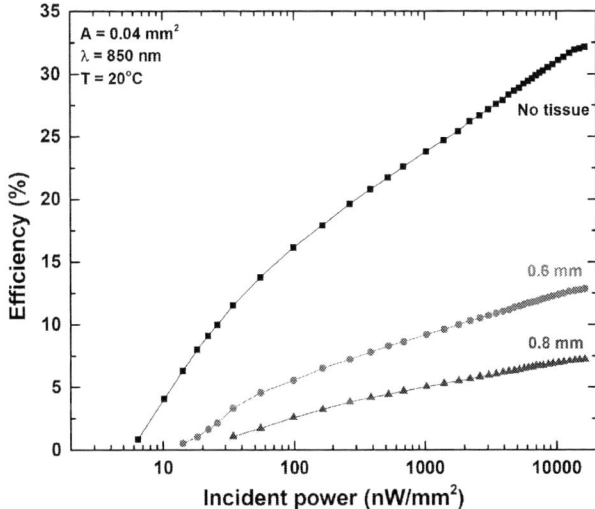

Fig. 5. PV cell power conversion efficiency versus irradiance for varying tissue samples.

The PV cell output power density shows a relatively linear behavior with respect to irradiance at the upper levels. The nonlinear behavior at lower levels are better illustrated by the power conversion efficiency values shown in Fig. 5. The efficiency values show a monotonic decrease at lower irradiance. The reduced efficiency at lower power can arise from non-ideal dark current, parasitic series resistance, and/or parasitic shunt conductance. We believe that the primary source of this behavior in our samples is due to dark current arising from perimeter recombination [5].

The irradiance dependence of short circuit current, open circuit voltage, and fill factor are further shown in Figs. 6, 7, and 8, respectively. The short circuit current and open circuit voltage behavior show relatively expected values for irradiance dependence, with steeper dropoff at very low irradiance that

978-1-7281-6118-1/22 $31.00 © 2022 IEEE

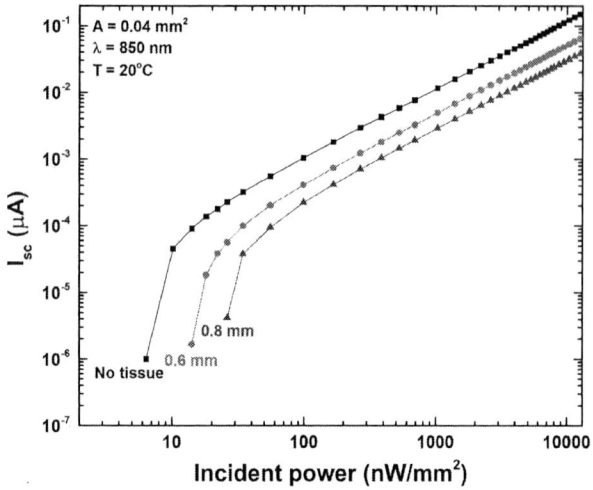

Fig. 6. Short circuit current versus incident power for varying tissue samples.

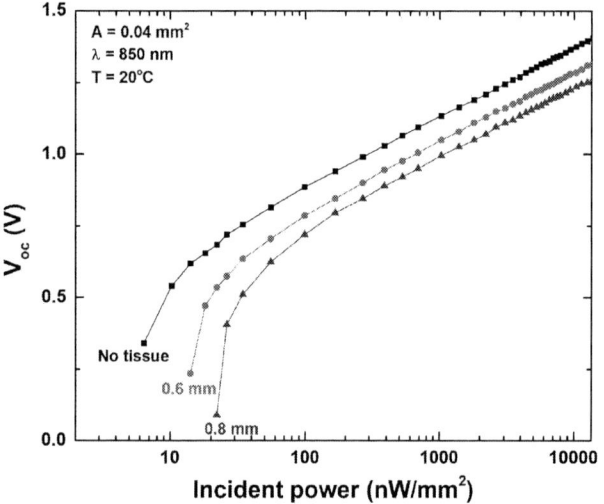

Fig. 7. Open circuit voltage versus incident power for varying tissue samples.

may be attributed to the onset of shunt conductance limitations. The open circuit voltage and voltage and the maximum power point are on the order of 1 V, providing a suitable voltage range for CMOS circuitry and showcasing an advantage of using a multi-junction approach.

IV. CONCLUSION

The PV cells in this work show that power density in excess of 50 nW/mm^2 can be generated under external irradiance conditions down to 1 μW/mm^2. This value is more than three orders of magnitude below the maximum exposure limit of 3.99 mW/mm^2 at 850 nm published in the ANSI Z136.1 standard. Under these conditions, there is substantial opportunity to provide wireless power transfer to near-surface bioimplantable devices on the order of μW for device dimensions on the order of 100 μm. The use of compound semiconductor PV cells also enables integration of optical devices for sensing and communication in complex systems

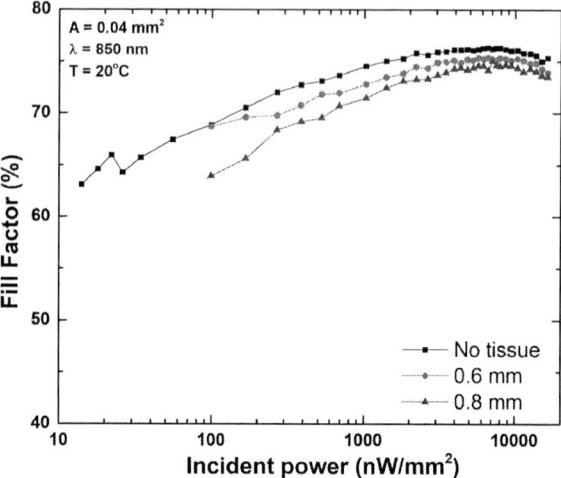

Fig. 8. Fill factor versus incident power for varying tissue samples.

such as envisioned large area arrays of wireless implantable neural probes [4], [6].

V. ACKNOWLEDGEMENT

This work was financially supported by the National Science Foundation (NSF award CBET-2129817)

REFERENCES

[1] J. Lee., V. Leung, A. H. Lee, J. Huang, P. Asbeck, P. P. Mercier, S. Shellhammer, L. Larson, F. Laiwalla, and A. Nurmikko, "Wireless ensembles of sub-mm microimplants communicating as a network near 1 GHz in a neural application," *bioRxiv*, 2020.

[2] D. K. Piech, B. C. Johnson, K. Shen, M. M. Ghanbari, K. Y. Li, R. M. Neely, J. E. Kay, J. M. Carmena, M. M. Maharbiz, and R. Muller, "A wireless millimetre-scale implantable neural stimulator with ultrasonically powered bidirectional communication," *Nature Biomed. Eng.*, vol. 4, no. 2, pp. 207–222, 2020.

[3] E. Moon, M. Barrow, J. Lim, D. Blaauw, and J. D. Phillips, "Dual-junction GaAs photovoltaics for low irradiance wireless power transfer in submillimeter-scale sensor nodes," *IEEE J. Photovolt.*, vol. 10, no. 6, pp. 1721–1726, Nov. 2020.

[4] E. Moon, M. Barrow, J. Lim, J. Lee, S. R. Nason, J. Costello, H. S. Kim, C. Chestek, T. Jang, D. Blaauw, and J. D. Phillips, "Bridging the "last millimeter" gap of brain-machine interfaces via near-infrared wireless power transfer and data communications," *ACS Photonics*, vol. 8, no. 5, pp. 1430–1438, 2021. [Online]. Available: https://doi.org/10.1021/acsphotonics.1c00160

[5] E. Moon, D. Blaauw, and J. D. Philips, "Infrared energy harvesting in millimeter-scale gaas photovoltaics," *IEEE Trans. Electron. Devices*, vol. 64, no. 11, pp. 4554–4560, 2017.

[6] J. Lim, E. Moon, M. Barrow, S. R. Nason, P. R. Patel, P. G. Patil, S. Oh, H.-S. Kim, D. Sylvester, D. Blaauw, C. A. Chestek, J. D. Phillips, and T. Jang, "A 0.19×0.17mm2 wireless neural recording IC for motor prediction with near-infrared-based power and data telemetry," in *2020 IEEE International Solid-State Circuits Conference.* San Francisco, CA: IEEE, Feb. 2020, pp. 416–418.

Evaluation of an LED simulator for single- and multi-junction PV cell performance testing

Nikos Kopidakis[1], Tao Song[1], Charles Mack[1], Rafell Williams[1], Hal Friesen[2], Justin Bertagnolli[2], John Walmsley[2]

[1]National Renewable Energy Laboratory, Golden, CO
[2]G2V Optics, Edmonton, AB

Abstract—We present the evaluation of a solar simulator comprising a Light Emitting Diode (LED) array as its light source. While multiple applications can be envisioned for such a light source, here we focus on its use for performance testing of solar cells at Standard Test Conditions. With this in mind, we present characterization of the spatial uniformity of the irradiance and its temporal stability as well as the spectral range and spectral class. For a multisource simulator, the spectral uniformity across the test plane should also be evaluated and we discuss a method for performing this evaluation that is well suited to a solar cell performance testing application. One major advantage of this simulator is the ability to easily adjust the spectral output as needed in multijunction PV cell measurements and we also discuss this application here.

Keywords—PV cells, simulator, LED, spectral irradiance

I. INTRODUCTION

Virtually all the tests of the performance of photovoltaic (PV) cells occur indoors under a solar (or sun) simulator. In general terms, the requirements that a solar simulator must fulfill, as stated in international standards for simulator classification, are: its spectral output should be similar to a reference spectrum that, in the case of terrestrial PV cells, is representative of well-defined measurement conditions under full sun illumination outdoors; its total irradiance should be similar to the total irradiance under full sun illumination as mentioned above; its total irradiance must be stable for the time period of the test; its total irradiance must be uniform within the area of the PV device that is tested. Additionally, the *spectral* output of the simulator should also be uniform across the area of the device that is tested. The latter is not a formal documented requirement for a solar simulator, however, as we discuss below, it is important for multisource solar simulators such as the one presented in this paper.

The current-voltage (IV) curve of the PV cell, measured under illumination from the solar simulator as described in ASTM and IEC standards,[1] is translated to the Standard Test Conditions (STC) using the well-documented translation equations [2]. Translation to STC enables a fair comparison of the performance of PV cells within a technology or between different technologies and is the agreed-upon standard for reporting record efficiencies of PV cells and modules [3]. The definition of STC for terrestrial PV cells is: total irradiance of

1000 W/m^2; Spectral irradiance defined in ASTM G 173 or IEC 60904-3 Ed. 4; test device temperature of 25 ^{0}C.

Most solar simulators for PV cells use a single light source, typically a Xe-arc lamp. More recently, new simulator designs use light-emitting diodes (LEDs) at different wavelengths and combine their output to achieve the required spectral output, total irradiance and uniformity as needed for accurate solar cell testing. Beyond single junction solar cells, a key advantage of multisource LED simulators is the ease of tuning the spectral output electronically, without the need for external filters, which is necessary for accurate performance testing of multijunction solar cells.[6] Considering that the spectral output is limited in the NIR (to ca. 1100 nm for a typical design), multisource LED simulators can still be used for testing cells of interest, for example all types of 2-junction 2-terminal perovskite containing tandems, such as perovskite-perovskite and perovskite-Si.

In this contribution we present the characterization of the light output of a multisource LED simulator, the Sunbrick made by G2V Optics. In addition to the established tests that a solar simulator must pass, we also discuss additional characterization needed to ensure that a *multisource* simulator can be used for accurate performance testing of single and multijunction PV cells. In the following we will briefly discuss the design that allows building the desired spectrum from distinct LED sources and then present results on the temporal stability and uniformity of the irradiance and an evaluation of the spectral class of the simulator. A common concern for multisource simulators is the uniformity of the spectral irradiance and in section II.E we present results showing minimal variation of the spectral output across a typical test plane area, which qualifies this simulator for high accuracy testing of PV cells.

II. RESULTS AND DISCUSSION

A. Description of the Sunbrick.

The Sunbrick model used in this research, show in in Figure 1, provides light in the 350-1100nm spectrum range and was reported by G2V Optics to exceed Class AAA standards for ASTM E927, IEC 60904-9 and JIS C 8904-9. The instrument incorporates 36 tunable channels to achieve 100% spectral coverage with a minimum 17% spectral deviation (as defined in the most recent IEC 60904-9:2020 standard) in the 350 nm -

978-1-7281-6118-1/22 $31.00 © 2022 IEEE

1100 nm spectral range recommendations introduced in the newest standards. The Sunbrick is a continuous-wave (CW) solid-state light emitting diode (LED) simulator, with a total illumination area of 625 cm² (25 cm x 25 cm) calibrated at the factory to Class A spatial non-uniformity for 400 cm² (20 cm x 20 cm). It can produce $0.1 - 1.1$ suns equivalent intensity, with spectral tunability controlled via software (rather than physical filters). The system is lightweight and compact - with a fully operational system weight of 26.3 kgs, and an operating footprint size of 34 cm x 44 cm x 118 cm. In the following we present the characterization of the light source after initial installation at the NREL PV Cell and Module Performance laboratory.

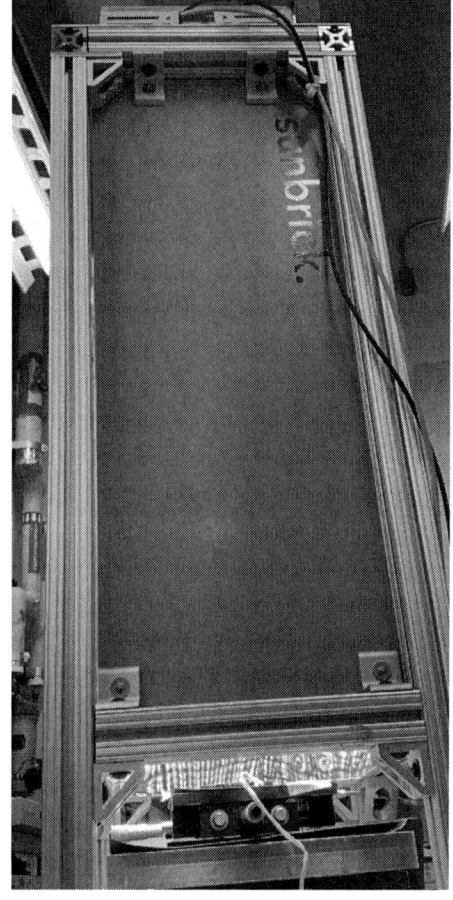

Figure 1. The Sunbrick solar simulator used in this study. The LED array is located on the top of the enclosure, and the light is guided to the test plane at the bottom of the frame. The reflection of the multisource LED array can be seen on the mirrored wall of the light guiding enclosure.

B. Spectral output

A comparison of the Spectral Irradiance defined in ASTM G 173 global [4] to the Spectral Irradiance of the Sunbrick simulator optimized to match the G 173 global spectrum is shown in Figure 2.

The calculation of the spectral class of the Sunbrick spectrum of Figure 2 with respect to the global terrestrial reference spectrum is plotted in Figure 3. The spectrum is classified as Class A in both ASTM and IEC standards [5]. We measured the spectral coverage to be 100% and the spectral

deviation to be 21% or 16% in the 350 nm - 1100 nm range, calculated according to the formulas in the ASTM and most recent IEC 60904-9:2020 standard, respectively.

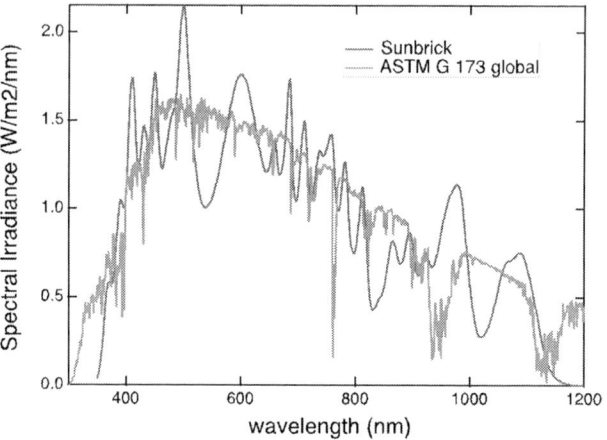

Figure 2. The Spectral Irradiance of the G2V Sunbrick LED simulator (red), measured using an ASD FieldSpec 4 spectroradiometer and the ASTM G 173 hemispherical reference spectrum (gray) as defined in [4].

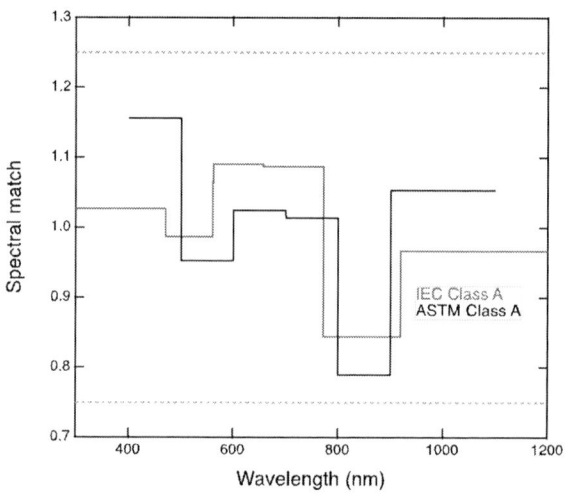

Figure 3. Spectral classification according to ASTM E0927-19 (black) and IEC 60904-9 Ed. 3 (red). The green dotted lines signify the range for Class A spectral match.

A major advantage of the LED simulator presented here is the ability to tune the spectral output easily and quickly by varying the current in each LED band. For example, while the preset spectrum shown in Figure 2 simulates the terrestrial ASTM G 173 (or similarly, IEC 60904-3 Ed. 4) standard spectrum with a total irradiance of 1000 W/m² from 200 nm - 4000 nm, with the Sunbrick producing the appropriate 791 W/m² in the 350 nm - 1100 nm range, one can easily modify the light source here to simulate an AM0 spectrum with a total irradiance of 1366.1 W/m2 from 200 nm - 4000 nm, with the Sunbrick producing the appropriate 961 W/m² in the 350 nm - 1100 nm range, as required for the calibration of PV cells for space applications. Although it is common for LED solar simulators not to produce light in the full range out to 4000 nm, many PV cells, including commercial silicon PV cells, do not respond to light above 1100 nm, so the absent irradiance at these higher wavelengths has no impact on power output or the overall accuracy of results for these PV cells. Additionally, spectral adjustability is necessary for high accuracy calibrations of the performance of multijunction cells [6]. For multijunction cell measurements our group currently uses a home-made multisource simulator based on filtered Xe and W lamps and variable apertures to adjust the irradiance of 9 total spectral bands [6]. While that multisource simulator has allowed accurate calibrations of up to 6-junction tandems [6], it is a complex and delicate instrument that few laboratories will be able to afford, let alone will have time to design and build. The goal of the initial testing and validation of the Sunbrick simulator presented here is to use it for the measurement of tandem cells that do not have spectral response beyond the 1100 nm maximum wavelength output of this LED light source. This range includes cell architectures of interest to the PV community, such as perovskite/Si and perovskite/perovskite tandems, hence the Sunbrick can provide a straightforward solution to spectral adjustability for tandem cell R&D laboratories interested in these cell types. In addition, newer versions of the Sunbrick can have extended spectral output to 1500 nm, with adjustable intensity in the IR that can be better suited for tandems with lower bandgap bottom junction.

The spectrum shown in Figure 2 was measured over an area of a few cm^2 on the test plane. One concern regarding multisource simulators is the *spectral uniformity* over smaller length scales. The active area of new solar cell designs is often quite small, 0.1 cm^2 or even smaller.[3] The concern therefore is: does the spectral output vary over small length scales? What is the *micro*-uniformity over mm^2 scales on the test plane? We address these questions in section II.E below.

C. Temporal Stability

The temporal stability of the total irradiance was tested with a Si reference cell. After a 15-minute initial stabilization of the temperature of the cell, as monitored by an RTD sensor, the variation of the I_{SC} of the cell was 0.8% over a period of 20 minutes, which gives a Class A rating for temporal stability according to both ASTM and IEC classifications [5].

D. Uniformity of Total Irradiance

The uniformity map of the total irradiance is shown in Figure 4, also tested with a Si reference cell (4 cm^2 cell area). The spatial nonuniformity of the irradiance in a 10 cm x 10 cm area is evaluated to be 1.57%, also classified as Class A [5], and within the smaller area of 4 cm x 4 cm it is 0.26%. In our group the Sunbrick simulator will be used initially for small area multijunction cell measurements, as discussed below, hence the uniformity over the larger area of the nominal test plane of the simulator (20 cm x 20 cm) is currently not of interest.

Since each LED or point source's output irradiance maps differently to the illumination field, a uniform irradiance can only be achieved by stable, individual control of each LED in the Sunbrick's array. The total irradiance is achieved in this case by tiling LED cards across the intended illumination plane.

Adjusting the LED outputs to simultaneously achieve spectral match, spatial uniformity and spectral uniformity is an iterative, non-trivial problem for which G2V has developed proprietary calibration methods. One cannot, for example, achieve uniformity in one area simply by adjusting a green LED — a balance of LEDs must be adjusted together to maintain the spectral match across the illumination plane. The NREL approach to characterization of the spectral match across the illumination plane will be discussed in the Spectral Uniformity section below.

120.0	119.3	119.4	119.5	119.2
119.6	118.6	118.3	118.0	118.0
118.7	117.5	117.0	116.6	116.6
118.6	117.3	116.6	116.4	116.3
118.6	117.6	117.0	116.8	116.9

Figure 4. Uniformity map over a 10 cm x 10 cm area. Each square corresponds to the 2 x 2 cm^2 area of the Si reference cell used, and the number shows the I_{SC} of the reference cell in mA.

E. Spectral Uniformity

A special concern for multisource simulators such as the one presented here is the *spectral* uniformity of the irradiance across the test plane. Given that the light source comprises LEDs of different color, even though the *total* uniformity does not vary significantly within a 10 x 10 cm^2 area, is a small test device of a few mm^2 area exposed to the same spectrum when placed in different locations? This is important because the spectral irradiance on the device determines the magnitude of the spectral correction needed to translate the IV measurement under the simulator to the standard spectrum for an STC calibration. The correction, quantified by the spectral mismatch factor, M [2], can be as high as a few % and therefore inaccurate knowledge of the spectrum *at the device location* can introduce several % error in measurement of the I_{SC}. Furthermore, during a tandem cell test, inaccurate knowledge of the spectrum may create false estimation of the limiting junction of the device thereby providing highly misleading information on the operation of the cell.

To evaluate the spectral uniformity of the Sunbrick simulator, we used an ASD FieldSpec 4 spectroradiometer with an aperture opening of 7 mm diameter to measure spectra at different locations on the test plane. The subtle differences observed in the Spectral Irradiance for these locations were then quantified by calculating the spectral mismatch factor, M, for a typical Si test device and a Si reference cell. We chose these devices to give a commonly observed mismatch of ca. 3%

(M≈1.03). Figure 5 shows the value of M calculated at different locations along two directions on the test plane.

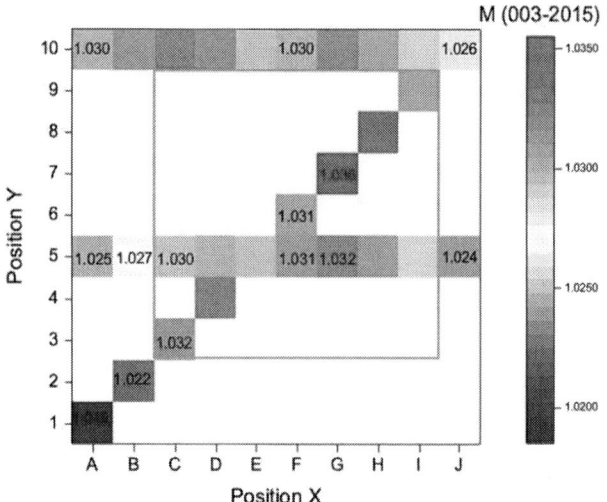

Figure 5. Variation of M for different locations on a 20 cm x 20 cm test plane. Within the smaller 14 cm x 14 cm region defined by the green square the M variation is 0.6%.

We find that M varies by 1.7% from the value in location E5, defined as the middle of the test plane in this case. Within the smaller area of 14 cm x 14 cm defined by the green square, M varies by 0.6%, which is comparable to the expected uncertainty of the spectral correction [7] for this magnitude of M. These results demonstrate that acceptable spectral uniformity can be achieved alongside a class A total uniformity and temporal stability for the Sunbrick.

To evaluate the uniformity at even smaller length scales, we used a small (<1 mm^2) area filtered photodiode mounted on an X-Y translation stage on the test plane and obtained the uniformity maps shown in Figure 6 for three wavelength bands. The maps shown on Figure 6 are centered at the center of the nominal test plane of the simulator. We find that the irradiance on the bands centered at 550 and 740 nm varies by 2.4% and 1.8% respectively, between the middle and the bottom of the 12 x 12 mm^2 test plane in this case. Using a representative spectral response of a perovskite solar cell and a typical KG-5 filtered Si reference cell, the variation of the spectral irradiance for the three bands shown in Figure 6 translates to a variation of the spectral mismatch factor from 1.0075 in the middle of the test plane to 1.011 at the bottom of the test plane. The variation of 0.0035 in this case would be considered in the uncertainty of the spectral correction, which in this case is 35% relative, an increase from the typically considered 20% relative uncertainty for M,[7] however still a small contribution to the total uncertainty of the I$_{SC}$ for a typical solar cell.

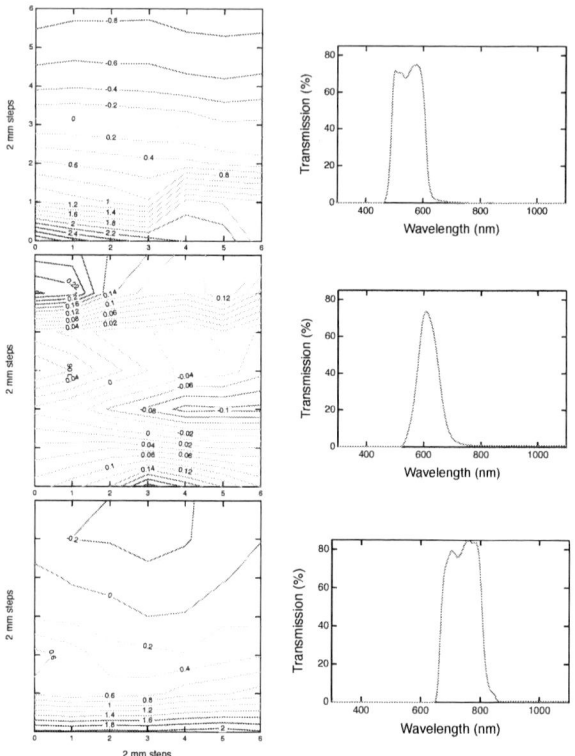

Figure 6. Uniformity of irradiance for three distinct wavelength bands, centered at 550 nm (top), 608 nm (middle) and 740 nm (bottom). The contours show % deviation from the middle of the scan plane. The transmission of the filter used in each case is shown on the right.

III.   CONCLUSIONS

Using an LED array to build a solar simulator for PV cell performance testing offers the ability to quickly and easily adjust the spectral (as well as the total) irradiance to meet the needs of tandem cell testing, or to conduct a test under essentially any spectrum as might be required in measurements for energy yield modeling. To realize this advantage however one needs to overcome the formidable challenge of maintaining both total and spectral uniformity on a several cm^2 test plane as required for high accuracy measurements. We show that the G2V Sunbrick LED simulator shows class A total uniformity and a spectral uniformity suitable for accurate performance tests of small area PV cells. The results presented here provide the validation needed for moving the LED simulator testbed to the next phase of development, that of spectrum-building algorithms for various PV cell testing scenarios.

ACKNOWLEDGMENT

This work was authored in part by Alliance for Sustainable Energy, LLC, the manager and operator of the National Renewable Energy Laboratory for the U.S. Department of Energy (DOE) under Contract No. DE-AC36-08GO28308. Funding provided by U.S. Department of Energy Office of Energy Efficiency and Renewable Energy Solar Energy Technologies Office (SETO) Agreement Number 38262. We also acknowledge funds from the Canada-NREL Cleantech Accelerator in partnership with the Trade Commissioner Service.

REFERENCES

[1] ASTM E948-15 Standard Test Method for Electrical Performance of Photovoltaic Cells Using Reference Cells Under Simulated Sunlight, 2015; IEC 60904-1 (Ed. 2) Photovoltaic devices – Part 1: Measurement of photovoltaic current-voltage characteristics, 2006

[2] C. R. Osterwald, Solar Cells, *18* (1986) *3-4*

[3] Martin A. Green, Ewan D. Dunlop, Jochen Hohl-Ebinger, Masahiro Yoshita, Nikos Kopidakis, and Xiaojing Hao. "Solar cell efficiency tables (version 57)." Prog Photovolt Res Appl. 2021; 29: 3– 15

[4] ASTM G173 Standard Tables for Reference Solar Spectral Irradiances: Direct Normal and Hemispherical on 37° Tilted Surface, 2006

[5] ASTM E927-19 Standard Classification for Solar Simulators for Electrical Performance Testing of Photovoltaic Devices, 2019; IEC 60904-9 (Ed. 3) Photovoltaic devices – Part 9: Classification of solar simulator characteristics, 2020

[6] T. Moriarty, Joe Jablonski, and Keith Emery. "Algorithm for building a spectrum for NREL's one-sun multi-source simulator." In 2012 38th IEEE Photovoltaic Specialists Conference, pp. 001291-001295. IEEE, 2012

[7] H. Field, and K. Emery. "An uncertainty analysis of the spectral correction factor." In Conference Record of the Twenty Third IEEE Photovoltaic Specialists Conference-1993 (Cat. No. 93CH3283-9), pp. 1180-1187. IEEE, 1993

978-1-7281-6118-1/22 $31.00 © 2022 IEEE

Metastability and Degradation of CdTe Solar Cells Investigated by nm-Scale Electrical Potential Imaging

Chun-Sheng Jiang, David Albin, Marco Nardone, Kassidy H. Howard, Adam Danielson, Amit Munshi, Tushar Shimpi, Walajabad Sampath, Chuanxiao Xiao, Helio R. Moutinho, Mowafak M. Al-Jassim, Glenn Teeter

National Renewable Energy Laboratory, Golden, CO, United States

Bowling Green State University, Bowling Green, OH, United States

Colorado State University, Fort Collins, CO, United States

We report on investigations of reversible metastability and irreversible degradation of MZO/CdSeTe/CdTe devices from the perspective of electric field across the device using Kelvin probe force microscopy (KPFM). The device showed reversible transitions between the light-soak state (LSS) with the best device efficiency and the dark-soak state (DSS) with an inferior efficiency. However, it showed an irreversible degradation state (DgS) driven by long-hour light soaking at an elevated temperature. The nm-scale KPFM electric field imaging on cross-sections of the devices revealed different anomalous electric field profiles. The electric field at the LSS exhibits a main peak inside the CdSeTe layer but not at the MZO/CdSeTe heterointerface, demonstrating that working junction of the device is a buried homojunction (BHJ). At the DSS, a second electric field peak was observed at the MZO/CdSeTe interface with a similar strength to the main BHJ, which probably caused the decrease in fill factor at the DSS. At the DgS, the electric field peak at the MZO/CdSeTe interface increased significantly and a third electric field was measured at the back contact of the device. Device modeling using COMSOL and in alignment with both the electric field and device current-voltage curves suggest that a slightly low n-doped CdSeTe in the region near the MZO/CdSeTe interface caused the BHJ, and that either a loss of MZO doping or increase of the conduction band offset spike due to long-term stress caused the fill factor-dominated degradation and the increased electric field near the MZO/CdSeTe interface at the DgS. The former, MZO doping decrease, is more plausible by referring the related literature.

Photovoltaic Thermal Management in Luminescent Solar Concentrators

Megan E. Phelan[1], David R. Needell[1], Maggie M. Potter[2], Haley C. Bauser[1], Catherine N. Ryczek[1], Ralph G. Nuzzo[2], Harry A. Atwater[1]

[1]California Institute of Technology, Pasadena, CA, 91125, USA
[2]University of Illinois at Urbana-Champaign, Champaign, IL, 61820, USA

Abstract—**Photovoltaic (PV) performance is dependent on device operating temperature. We show that luminescent solar concentrators (LSCs) redistribute absorbed radiation and thermalization losses away from the photovoltaic cell to the waveguide in an LSC, keeping the PV cell cooler. Here we compare PV operating temperatures in an LSC compared to a conventional planar absorber PV device. We further examine the effects of PV cell orientation and luminophore properties within the LSC waveguide on PV temperature. By modeling LSC module efficiencies, we finally analyze the effects of various LSC system configurations on PV performance, and suggest optimal luminophore designs to improve PV temperature coefficients.**

Keywords—thermalization, cooling, luminescent, concentrator

I. INTRODUCTION

Photovoltaic (PV) performance is strongly dependent on device operating temperature. Crystalline silicon solar cells conventionally exhibit temperature coefficients of approximately -0.5%/°C [1]. As solar cells are exposed to higher radiation and warm ambient air, their operating temperatures can increase significantly due to balance between heating resulting from carrier thermalization from photon absorption, and convective and radiative emission cooling.

Luminescent solar concentrators (LSCs) are notable for their *i)* ability to concentrate diffuse light, *ii)* variable degree of module transparency, useful in building integrated PV applications, and *iii)* reduced use of photovoltaic cell material in proportion to geometric concentration [2], [3]. However, another important quality of LSC modules is their ability to maintain low PV operating temperatures, which is particularly advantageous for materials such as silicon, which has a high temperature coefficient [4], [5]. Previous studies have reported reduced operating temperature for a PV cell embedded within an LSC compared to a conventional wafer PV cell [5], yet little research has been conducted to understand the thermalization pathways in an LSC and how various properties affect the PV operating temperature and overall module performance.

In an LSC module, the thermal properties of the module are influenced by the *i)* polymer waveguide, *ii)* embedded luminophore species, *iii)* photovoltaic material, and *iv)* surrounding ambient thermal environment [6], as shown in Fig. 1a. A small amount of the incident radiation is absorbed as heat by the polymer itself, depending on the polymer material,

as depicted by the glowing purple waveguide in Fig. 1a. The majority of optical absorption occurs in the luminophores, and the total absorption depends on the luminophore material and their optical density in the waveguide [7]. Photons that are not absorbed by the luminophores or polymer pass through the LSC and do not incur additional absorption. In the case of luminophore optical absorption, heating may occur via multiple pathways. For example, in the case of core/shell quantum dots (QDs), thermalization may occur due to *i)* relaxation of absorbed photons to the QD core: $Q = h\nu - E_{PL}$, and *ii)* absorbed heat due to non-unity quantum yield: $Q = E_g * (1 - \eta_{PL})$ [8]. These two phenomena are depicted schematically by the glowing red and black QDs, respectively, in Fig. 1a. This produced heat is then trapped within the polymer waveguide. Since luminophore photoluminescence (PL) emission wavelengths are typically matched to the bandgap of the edge-lined PV cell [9], carrier thermalization largely occurs at the luminophore, as opposed to at the PV cell. Finally, the ambient air surrounding the polymer waveguide acts as a cooling medium, enabling heat dissipation via convection. For each thermalization (and cooling) pathway, the majority of thermalization occurs within the LSC waveguide, thereby redistributing heat away from the PV cell [7]. By contrast, a conventional flat planar PV module incurs heat absorption from carrier thermalization uniformly across its area, with a magnitude related to the cell bandgap [10], [11], as shown by figure 1b.

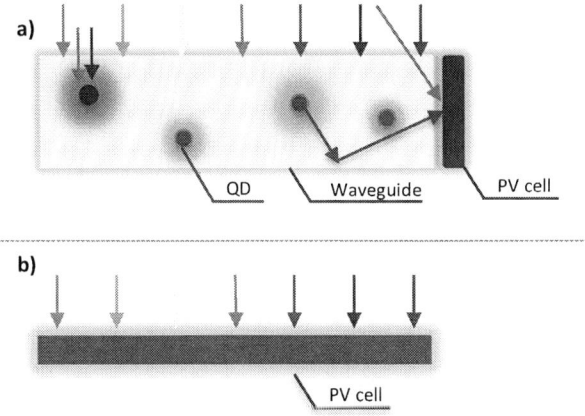

Fig. 1: Heating from incident thermal radiation is split between each the polymer waveguide, QD luminophore, and photovoltaic cell in the edge-lined LSC design (top). For a conventional wafer planar PV cell, the PV cell is subject to all incident radiation.

Here we examine the effects of PV operating temperature for PV cells embedded in an LSC waveguide compared to cells in a conventional PV module. Given their widespread use, we focus our study on Si PV cells [4]. Using simulations, we model the effects of *i)* PV cell configuration in the LSC waveguide and *ii)* luminophore optical properties – optical density, absorption spectrum, and Stokes shift – on PV operating temperature and power conversion efficiency. Using these models, we develop conclusions for optimal PV performance in an LSC module.

II. RESULTS AND DISCUSSION

To assess thermal management at the PV cell in an LSC, we compare thermal performance of a Si PV edge-lined LSC device to a flat plane Si solar cell. Thermal performance is modeled using the heat transfer module in the COMSOL Multiphysics simulation environment. Our model accounts for incident thermal irradiation in Los Angeles, CA and simulates module heating based on conduction in the polymer waveguide, thermalization at the QD luminophore, thermalization at the PV cell, and convective cooling assuming an ambient air temperature of 20°C.

The simulated edge-lined LSC consists of *i)* a 1cm x 1cm x 200μm waveguide with embedded CdSe/CdS QD luminophores (quantum yield = 99%) and *ii)* an edge-line silicon PV cell (thickness = 80μm), as shown in figure 2a. The modeled conventional wafer Si PV cell has the same dimensions as the LSC waveguide (1cm x 1cm x 200μm), as shown in figure 2b. In figure 2c, we model a waveguide optical density of 0.3 and record data for both modules after 5 min of constant AM1.5g irradiation (P = 1000 W/m^2) exposure. As shown in Fig. 2, the Si PV cell attains an operating temperature of 28.2°C when in the LSC waveguide, whereas the conventional wafer Si cell reaches a 44.6°C operating temperature. Further, the LSC keeps the Si cell nearly 3°C cooler than the waveguide center, where the majority of heat is confined, whereas the conventional Si cell is at 44.6°C across its surface, with ~0.01°C variation at the device edges.

We further study the effects of various PV cell configurations within the LSC waveguide. As shown in table 1, we model four different PV orientations – edge lined, edge lined along all 4 edges, coplanar along the bottom of the waveguide, and inverted coplanar along the waveguide top. Each orientation assumes identical PV material and cell dimensions (1cm x 200μm x 80μm). As shown, PV temperature is affected by orientation within the waveguide, with edge-lined orientations outperforming coplanar. However, all LSC cell configurations still outperform the conventional wafer PV operating temperature in similar conditions (44.6°C) by over 10°C.

PV Orientation	PV Temp (°C)	Waveguide Temp (°C)
Edge-Lined (one edge)	28.1	31
Edge-Lined (all edges)	28.2	30.6
Bottom Coplanar	30.3	30.8
Top Coplanar	32.5	32.4

Table 1: Operating temperature for varied placement orientations of a 1cm x 200μm x 80μm Si cell in a 1cm x 1cm x 200μm waveguide.

We next investigate the role of luminophore materials and properties on PV operating temperature. Figure 3 displays the spectral properties for three different quantum dot materials – CdSe/CdS, InAs/InP/ZnS, and CuInS$_2$/ZnS. We use these quantum dots to simulate three different LSC systems. In each system, we assume a single edge-lined PV cell, as this configuration was shown to have the lowest operating temperature in table 1. All other parameters between the systems are also kept constant – quantum yield = 90%, geometric gain = 50, optical density = 0.3, and waveguide dimensions = 1cm x 1cm x 200μm. Table 2 details the operating temperatures for each the Si PV device and the luminophore waveguide for each of the 3 quantum dot systems.

QD System	PV Temp (°C)	Waveguide Temp (°C)
CdSe/CdS	28.1	31
InAs/InP/ZnS	32	34.7
CuInS$_2$/ZnS	32.4	35.1

Table 2: Operating temperature for a 1cm x 200μm x 80μm edge-lined Si cell in a 1cm x 1cm x 200μm waveguide with varied quantum dot luminophores.

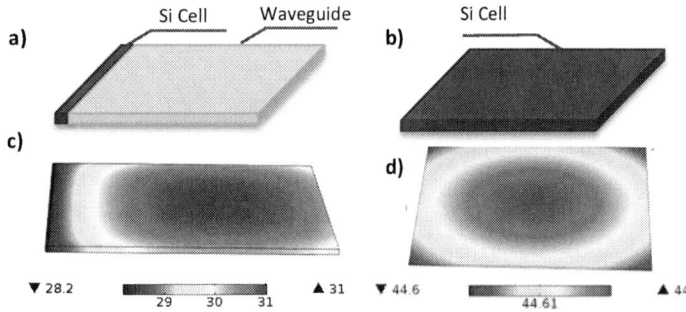

Fig. 2: Design for simulated (a) edge-lined LSC and (b) conventional wafer Si PV cell. Temperature distribution across the device after 5 min of constant 1000W/m^2 AM1.5g exposure is shown for (c) the LSC device and (d) the conventional wafer Si cell.

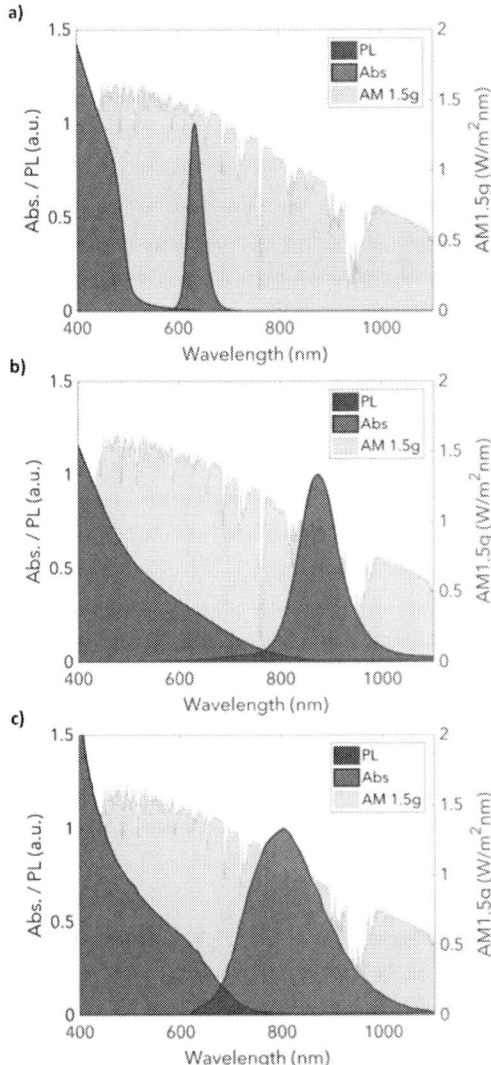

Fig. 3: Absorption and PL spectra for each quantum dot system – (a) CdSe/CdS, (b) InAs/InP/ZnSe, and (c) CuInS₂/ZnS.

To determine the effects of each luminophore property, we vary the location of each the *i)* QD PL peak, *ii)* absorption edge, and *iii)* optical density within the LSC waveguide. As shown in figure 4, optical density logarithmically affects PV operating temperature in an LSC. This logarithmic relationship is expected, given the additional photons absorbed as optical density increases in accordance with the Beer-Lambert law [12]. We further see in figure 4a that as the PL peak is red shifted, PV temperature decreases. By red-shifting towards 1000nm, the PL peak is approaching the Si solar cell bandedge. This improved performance via red-shifting towards the bandedge supports previous studies which suggest that LSC module efficiency should improve for a red-shifted luminophore system [13]. In contrast, PV temperature increases when the QD absorption edge is red-shifted (figure 4b), given the additional irradiation that is absorbed by the LSC waveguide.

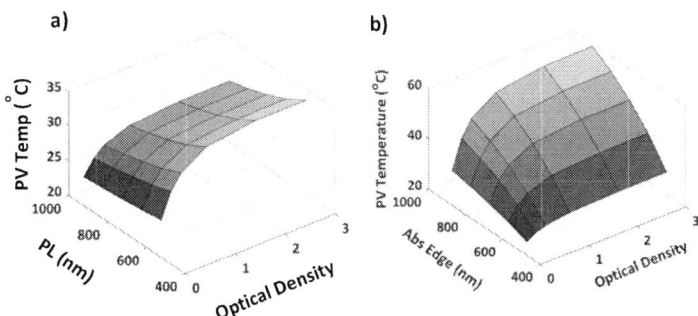

Fig. 4: PV operating temperature for (a) varied QD luminophore PL peak location and optical density, where absorption edge is held constant at 500nm, and (b) varied QD absorption edge and optical density, where PL peak is held constant at 950nm.

To understand how these results affect module performance, we use Monte Carlo ray-tracing simulations to model power conversion efficiency for the specified conditions and modeled operating PV temperature. Given the calculated power conversion efficiencies and V_{OC}, we calculate the temperature coefficient for the solar cell:

$$C_{eff} = \frac{\eta_1 - \eta_2}{T_1 - T_2}; \quad C_{Voc} = \frac{V_{OC,1} - V_{OC,2}}{T_1 - T_2}$$

In figure 5, we model the temperature coefficient – for each efficiency (blue) and V_{OC} (red) – for a Si cell in an LSC waveguide (solid lines) and conventional wafer device (dotted lines). We model two different (ideal) QD systems, keeping the Stokes shift constant (0.42meV) for both systems, with the absorption/PL spectra shown in figure 5a,c. For a 15.7% efficient Si cell (assuming ambient conditions of 20°C), the temperature coefficient for a conventional wafer device is -0.53% /°C. As shown in figure 5b, the Si temperature coefficient when embedded in the LSC, across varied optical density, is comparable to that of a conventional wafer device. However, for a red-shifted QD system (figure 4c,d), where the Stokes shift is held constant but the PL peak (950nm) is closer to the Si bandedge, we see significant improvement in the temperature coefficient. For optical densities greater than 0.3, the temperature coefficient for the LSC module is consistently higher (less negative), with efficiency coefficients of (-0.49% /°C), in comparison to that of the conventional wafer Si device (-0.53% /°C). As such, we demonstrate that an LSC device with a PL peak matched to the PV bandedge is optimal for improved module performance, with efficiency coefficient improvements up to 0.04% for every degree Celsius.

978-1-7281-6118-1/22 $31.00 © 2022 IEEE

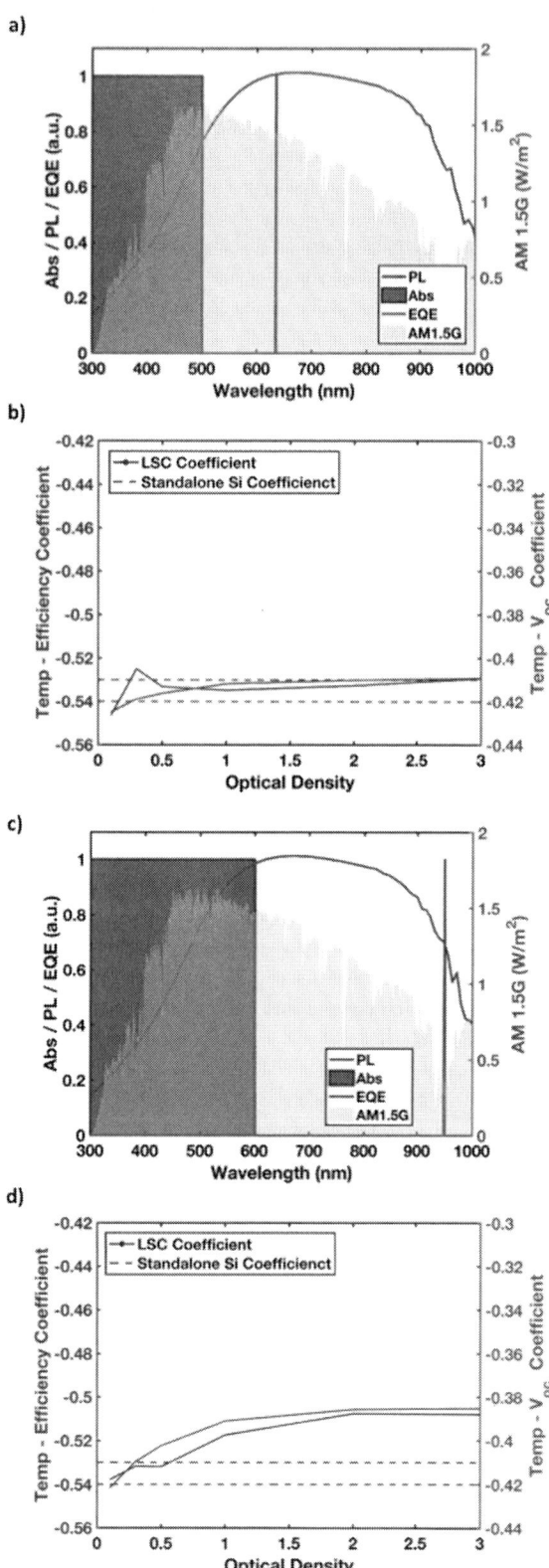

In addition to the terrestrial applications presented thus far, the ability for LSCs to maintain low photovoltaic temperatures could be particularly advantageous for space-based solar power applications. Given the lack of atmosphere (convection coefficient = 5×10^{-3}) in space, space-based photovoltaic devices are not able to rely on convective cooling for temperature control [14]. As such, photovoltaic devices employed in space could benefit from additional cooling properties. The luminophore waveguide cooling properties shown in this paper indicate that LSCs could present a promising technology to achieve lower PV operating temperatures in space. LSCs have previously been presented as potential candidates for space-based PV, given the light weight, flexibility and low-cost of the modules [15], and thus their intrinsic cooling properties could make them even stronger space-solar candidates. By employing our heat transfer simulations, we were able to show that, as with the terrestrial examples, luminescent waveguides are able to keep lower Si operating temperatures in space than a standalone silicon device. Furthermore, lower optical density QD waveguides result in a lower operating temperature for space-based silicon, thereby matching the terrestrial results in figure 4b. LSC systems could thus be designed to minimize the Si PV operating temperatures, despite the lack of convective cooling in space.

Additional methods to further cool the PV device within an LSC waveguide could include radiative cooling applications. Previous work on radiative cooling technology has demonstrated the ability to significantly cool devices by harvesting power from outgoing radiation into space, as well as through negative luminescent refrigeration [16]–[18]. By coupling these concepts – LSCs and radiative cooling – we may be able to achieve even lower PV operating temperatures than we have already shown.

III. SUMMARY AND OUTLOOK

LSCs are a promising technology to lower PV operating temperatures. We model temperature trends in LSCs for varied PV orientations and quantum dot properties to demonstrate that LSCs maintain cooler Si PV temperatures than their conventional wafer Si PV counterpart. Our models demonstrate that optimal PV performance, via temperature efficiency coefficient, occurs when the LSC luminophore PL peak is closely matched to the PV bandedge. Further studies could include research into the implementation of radiative cooling to enable additional PV cooling within the LSC.

Figure 4: Simulated temperature coefficients for ideal quantum dot systems with a Stokes shift of 0.42meV. Absorption and PL spectra are shown in a) and c), with the PV temperature coefficients for the respective LSC system in b) and d).

REFERENCES

[1] A. B. Sproul, M. A. Green, and J. Zhao, "Improved value for the silicon intrinsic carrier concentration at 300 K," *Appl. Phys. Lett.*, vol. 57, no. 3, pp. 255–257, 1990.

[2] M. G. Debije and P. P. C. Verbunt, "Thirty years of luminescent solar concentrator research: Solar energy for the built environment," *Adv. Energy Mater.*, vol. 2, no. 1, pp. 12–35, 2012.

[3] M. Phelan *et al.*, "Outdoor performance of a tandem InGaP/Si

978-1-7281-6118-1/22 $31.00 © 2022 IEEE

photovoltaic luminescent solar concentrator," *Sol. Energy Mater. Sol. Cells*, vol. 223, no. October 2020, p. 110945, 2021.

[4] T. J. Silverman, M. G. Deceglie, B. Marion, S. Cowley, B. Kayes, and S. R. Kurtz, "Outdoor performance of a thin-film gallium-arsenide photovoltaic module," *IEEE 39th Photovolt. Spec. Conf.*, pp. 103–108, 2013.

[5] V. A. Rajkumar, C. Weijers, and M. G. Debije, "Distribution of absorbed heat in luminescent solar concentrator lightguides and effect on temperatures of mounted photovoltaic cells," *Renew. Energy*, vol. 80, pp. 308–315, 2015.

[6] E. Yablonovitch, "Thermodynamics of the Fluorescent Planar Concentrator," *J. Opt. Soc. Am.*, vol. 70, no. 11, pp. 1362–1363, 1980.

[7] S. Haviv *et al.*, "Luminescent Solar Power - PV/Thermal Hybrid Electricity Generation for Cost-Effective Dispatchable Solar Energy," *ACS Appl. Mater. Interfaces*, vol. 12, no. 32, pp. 36040–36045, 2020.

[8] D. A. Hanifi *et al.*, "Redefining near-unity luminescence in quantum dots with photothermal threshold quantum yield," *Science (80-.).*, vol. 1202, no. March, pp. 1199–1202, 2019.

[9] N. D. Bronstein *et al.*, "Quantum Dot Luminescent Concentrator Cavity Exhibiting 30-fold Concentration - SI," *ACS Photonics*, vol. 2, no. 11, pp. 1576–1583, 2015.

[10] M. A. Green, *Thermal Behavior of Photovoltaic Devices*. 2017.

[11] M. A. Green, *Solar cells: Operating principles, technology, and system applications*. 1982.

[12] J. Bomm *et al.*, "Fabrication and spectroscopic studies on highly luminescent CdSe/CdS nanorod polymer composites," *Beilstein J. Nanotechnol.*, vol. 1, no. 1, pp. 94–100, 2010.

[13] D. R. Needell, C. R. Bukowsky, S. Darbe, H. Bauser, O. Ilic, and H. A. Atwater, "Spectrally Matched Quantum Dot Photoluminescence in GaAs-Si Tandem Luminescent Solar Concentrators," *IEEE J. Photovoltaics*, vol. 9, no. 2, pp. 397–401, 2019.

[14] E. E. Gdoutos *et al.*, "A lightweight tile structure integrating photovoltaic conversion and RF power transfer for space solar power applications," *AIAA Spacecr. Struct. Conf. 2018*, no. 210019, pp. 1–12, 2018.

[15] D. R. Needell *et al.*, "Ultralight Luminescent Solar Concentrators for Space Solar Power Systems," *Conf. Rec. IEEE Photovolt. Spec. Conf.*, pp. 2798–2801, 2019.

[16] K. Chen, P. Santhanam, and S. Fan, "Near-Field Enhanced Negative Luminescent Refrigeration," *Phys. Rev. Appl.*, vol. 6, no. 2, pp. 1–9, 2016.

[17] Z. Chen, L. Zhu, A. Raman, and S. Fan, "Radiative cooling to deep sub-freezing temperatures through a 24-h day-night cycle," *Nat. Commun.*, vol. 7, pp. 1–5, 2016.

[18] S. Buddhiraju, P. Santhanam, and S. Fan, "Thermodynamic limits of energy harvesting from outgoing thermal radiation," *Proc. Natl. Acad. Sci. U. S. A.*, vol. 115, no. 16, pp. E3609–E3615, 2018.

978-1-7281-6118-1/22 $31.00 © 2022 IEEE

PVRPM in Python: An overview of new capabilities

Paul Lunis, Brandon Silva, Marios Theristis, Hubert Seigneur

Florida Solar Energy Center, Cocoa, FL, United States

Sandia National Labs, Albuquerque, NM, United States

The value of investing in advanced monitoring capabilities in photovoltaic (PV) power plants is not clearly defined. Still, monitoring is essential due to the impact defective components can have on a system's energy output and hence, the levelized cost of energy (LCOE). This study introduces a new iteration of a reliablity model capable of simulating a component level analysis; a Python-based PV reliability performance model (Py-PVRPM). This tool was translated into a Python environment allowing for additional improvements on multi-component level observations and overall computational performance. With this update, real-time and reactive monitoring techniques were studied focusing on the cost-effectiveness of cross component level monitoring and aerial infrared drone imaging.

Oxygen and temperature effects on NiO buffer layers for CdTe solar cells

Nicholas Hunwick, Xiaolei Liu, Patrick J.M. Isherwood, John. M. Walls

loughborough university, Loughborough, United Kingdom

Nickel oxide has been previously shown to improve CdTe when used as a buffer layer. NiO was sputtered under varying oxygen and heating conditions. NiO films were characterised to identify changing characteristics. Increasing oxygen increases optical transmission and decreases reflectance. Post annealing at 300 °C has negligible effect on optical properties of NiO films. NiO with 5% oxygen/argon during deposition, without heating, produces the least sheet resistance of 4.51×10^5 Ω/square. These were then combined with CdCl treated CdTe and JV characteristics and carrier lifetimes measured and compared with a reference cell.

978-1-7281-6118-1/22 $31.00 © 2022 IEEE

Optimizing $CdCl_2$ Treatment on CdTe Solar Cells Using Spray Deposition Method

Prabodika N. Kaluarachchi, Shannon E. Costello, Ryan Madden, Jacob M. Gibbs, Tyler R. Brau, Aesha P. Patel, Manoj K. Jamarkattel, Jared D. Friedl, Kevin G. Schaffer, Kristof J. Nieschwitz, Ebin Bastola, Adam B. Phillips, Randy J. Ellingson, and Michael J. Heben

Wright Center for Photovoltaics Innovation and Commercialization (PVIC), Department of Physics and Astronomy, University of Toledo, Toledo, Ohio, 43606, USA

Abstract — Control over the amount of $CdCl_2$ used to process CdTe solar cells is important to achieving high-efficiency devices. Spraying allows for the deposition of a uniform film over a large area with strict management of the thickness of the film by modifying the spray parameters and/or the solution concentration. Here we report on the development of an ultrasonic spray system to deposit precisely administered amounts of $CdCl_2$ uniformly over arbitrarily sized CdTe samples up to 6"x6". We show that the deposited films are uniform and report on the optimum parameters for high-efficiency devices. Scanning electron microscopy was used to show that using the spray process reduces the amount of residual oxychlorides on the surface, likely leading to reduced recombination at the back interface.

Index Terms- $CdCl_2$ treatment, CdTe, spray coating, solar cells, Photovoltaics.

I. INTRODUCTION

With a record efficiency of 22.1% [1] and high throughput deposition techniques, CdTe-based devices have been able to achieve ~5% of the commercial photovoltaic (PV) market [2]. One of the key processes to achieving high efficiency is the $CdCl_2$ activation. During this process, Cl diffuses into the bulk and passivates defect sites at the bulk which allows carriers to be collected from the absorber during operation. This process also induces intermixing of the CdSe layer with CdTe decreasing the bandgap [3] and reducing recombination in the bulk and potentially the front interface [4].

$CdCl_2$, though, dissociates in water, leaving a toxic Cd+ ion. As a result, minimizing the amount of $CdCl_2$ used during processing is important for health and safety. In industry, where large quantities are used, precise control is employed. This may not be the case in laboratory settings where only small amounts are used for any process. Historically, we were able to achieve high-performing CdTe devices of efficiency 17% using the drop-cast method of $CdCl_2$ fabrication [5][6]. Due to the hydrophobic behavior of the CdTe surface, the use of an aqueous $CdCl_2$ solution was not possible during the

standard drop cast method. Though solubility is significantly lower in methanol (MeOH) than in deionized water, MeOH was used to make the $CdCl_2$ solution for the drop cast method because of the rapid evaporation of the solvent. Since the solubility limit is ill-defined in MeOH, $CdCl_2$ is added until significant crystallized precipitation was observed. To cover the entire surface of the sample, typically 1" x 1", ~50 µl of $CdCl_2$ solution was dispensed. Due to the poorly defined saturation limit of the solution, an unknown amount of the $CdCl_2$ compound was dispensed onto the surface. The drying process does not leave a uniform $CdCl_2$ coating on the surface due to the hydrophobic quality present in the MeOH solution which can also lead to local defects on the surface. To ensure the entire sample area is activated, it is likely that larger quantities of $CdCl_2$ are applied than necessary. This often leads to the formation of oxychlorides on the back surface of the CdTe, which can affect the device's performance [7]. More precise control of the $CdCl_2$ application that results in uniform coverage could lead to less $CdCl_2$ while achieving high performance.

Spraying offers an avenue to deposit a large area of uniform films with precise control over thickness [8]. The spray system requires the use of a computer to operate, which implies the rate of liquid delivery and motion of the substrate can both be controlled precisely. In addition, the concentration in the initial solution can also be administered specifically. With the combination of these attributes, the spray technique can fabricate very thin, uniform films on any substrate while minimizing waste and maintaining safe precautions.

Acknowledging these goals for $CdCl_2$ deposition optimization, we built a spray system to precisely control the amount of $CdCl_2$ applied to the CdTe device stack and subsequently optimized the processing parameters. Using a Sonotek Impact spray head, we control both droplet size deposited and the profile of the sprayed solution. Furthermore, by controlling the position of the substrate relative to the spray head in both the x- and y-position, we can deposit on arbitrary film sizes up to the stage size, 6" x 6". With uniform films, we have investigated and optimized the $CdCl_2$ treatment parameters, particularly the time and temperature.

II. EXPERIMENTAL

(a) $CdCl_2$ Spray System

To better control the application of $CdCl_2$ on samples, we developed a spray system. A schematic diagram is shown in Fig. 1. This system consists of four main components: (1) an x-y stepper motor controlled heated sample stage, (2) a stepper motor controlled syringe pump for solution introduction, (3) an ultrasonic spray nozzle that includes the ultrasonic power generator, and (4) a control computer.

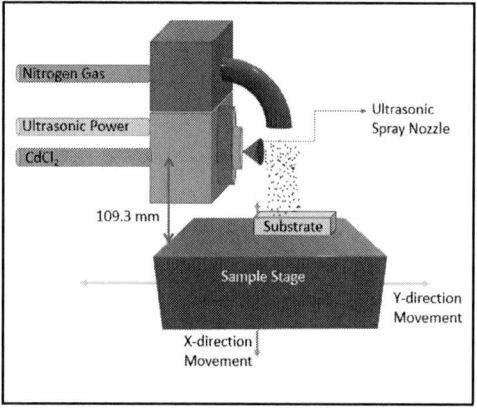

Fig.1 Schematic diagram of the ultrasonic $CdCl_2$ spray system.

The spray system was built with factors specific to the optimization of $CdCl_2$ depositions. The sample stage is a 6" x 6" copper block with cartridge heaters and an embedded thermocouple. The cartridge heaters and thermocouple are connected to a temperature controller to keep the deposition temperature consistent ensuring rapid evaporation of the solvent. The stage is mounted to x- and y- screw drives measuring 12 inches connected to 0.025 in/rev two-phase, single-shift stepper motors to control the positioning of the spray system and deposition. To achieve an uniform film, we develop an overlapping spray pattern where the stage was moved in 0.75" increments on the x-axis. A separate stepper motor-driven screw drive was built to control the solution dispersal through a syringe pump where a 30 mL syringe was used. Small diameter tubing connects the syringe to the spray head to deliver the $CdCl_2$ solution. The spray head is connected to an ultrasonic power generator (Sono-Tek ultrasonic power generator) set to 3 W. The ultrasonic power atomizes the $CdCl_2$ solution, made with deionized water, to specifically overcome the hydrophobicity of the CdTe surface onto which it is deposited. The spray nozzle also consists of an inlet for the non-reactive gas supply to direct and shape the deposited material downwards to the sample. The small droplets and heated substrate allow the solution to evaporate immediately

when in contact with the surface, resulting in a uniform deposition. The stepper motors, heating control, and the ultrasonic power generator are interfaced to a computer via an in-house build LabView virtual instrument (VI). All components were housed together in a sealed chamber with an exhaust outlet to minimize the exposure to the excess amount of $CdCl_2$.

(b) Device Fabrication and Measurement

For these samples, the device stack consisted of a F-doped SnO_2/buffer/~170 nm CdSe/~3.5 µm CdTe deposited at high temperature. Variable amounts of $CdCl_2$ (0.09, 0.27, and 0.45 mg-cm^{-2}) were sprayed on a number of device stacks. These samples were processed at 400 or 425°C for 20, 30, or 40 minutes in an N_2 environment. Fig. 2 shows an example of the thermal profile used for $CdCl_2$ activation.

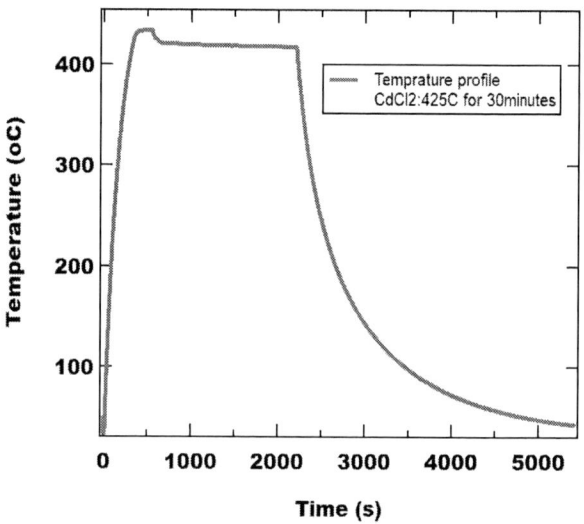

Fig.2 Temperature profile used for $CdCl_2$ heat treatment at 425°C for 30 minutes

Samples were subsequently Cu doped following our previous method[9]. Briefly, the activated device stack was dipped in a 0.1mM $CuCl_2$ in deionized water solution for 2 minutes, removed, and rinsed. The samples were heated treated at 230 °C for 20 and 30 minutes.

These samples were then finished with Au back contact by evaporating 60 nm of Au. The devices were characterized using current density-voltage (J-V), external quantum efficiency (EQE), and steady-state PL measurements.

III. RESULTS AND DISCUSSION

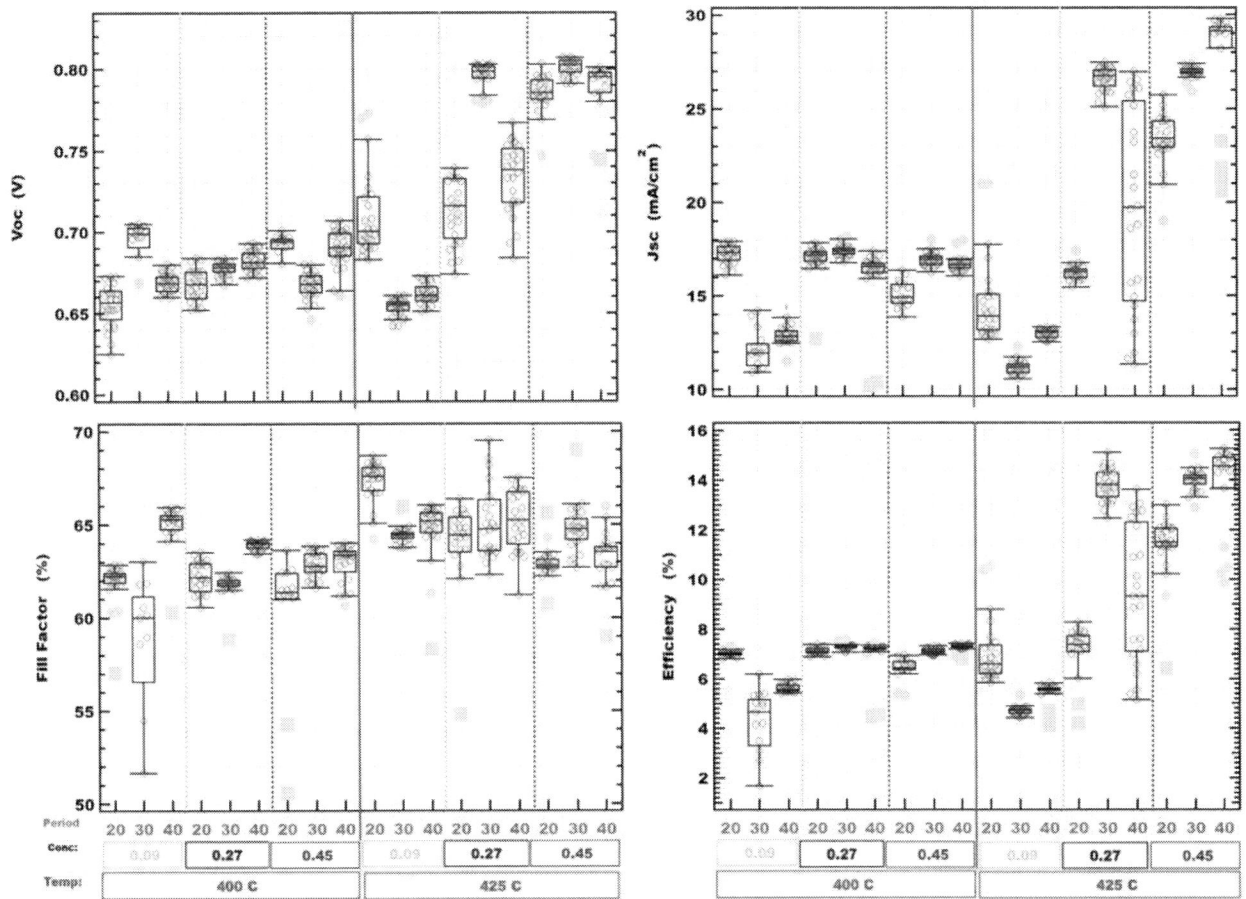

Fig.3 Statical distribution of device performance varying CdCl₂ surface concentration(0.09,0.27,0.45 mg/cm²), Heat treatment time (20,30 and 40 minutes) and temperature (400°C and 425°C)

Fig. 3 shows the statical distribution of the photovoltaic (PV) parameters for various activation processes that have undergone a CuCl₂ treated at 230°C for 30 minutes. Samples treated at 400°C show low device performance for all treatment times. The average short circuit current density (Jsc) obtained for these samples was ~16 mA/cm². These low performances were even observed when varying the CuCl₂ treatment. This suggests that insufficient Cu doping is not the main cause of Jsc loss, and it is likely the thermal energy provided at the CdCl₂ process is insufficient to result in the proper intermixing of the CdSe/CdTe layers [5].

When the CdCl₂ treatment temperature was increased to 425°C, the J_{SC}, as well as the other PV parameters, remained low for the samples with the lowest CdCl₂ surface concentration of 0.09 mg/cm². When the CdCl₂ concentration increased to 0.27 mg/cm² an increase in J_{SC} and other PV parameters is observed when CdCl₂ treatment increased from 20 to 30 minutes. The Jsc improved from 16 to 26 mA/cm²; Voc increased from 710 to 800 mV, and efficiency grew from 7% to 14%.

This higher level for Jsc, open circuit voltage (Voc), and efficiency is observed for all activation times whene the concentration of CdCl₂ is increased to 0.45 mg/cm². The 0.45 mg/cm² CdCl₂ concentration heat-treated at 425 °C for 40 minutes gave the highest performing device resulting in the following characteristics: Voc ~800mV, Jsc ~29mA/cm², Fill Factor ~64% and efficiency of 15%.

For samples activated at 400°C, the device performance was the same for both 20 and 30 minute Cu-doping times. This, though, is not the case for samples activated at 425°C. Fig. 4 shows the J-V characteristic for the highest efficiency devices activated at 425°C for 30 and 40 minutes for CdCl₂ surface concentration of 0.27 and 0.45 mg/cm² amd each CuCl₂ treatment time.

For devices fabricated with 0.27 mg/cm² CdCl₂ concentration (red and blue curves on graph), a significant improvements are observed as the CuCl₂ treatment increases from 20 minutes

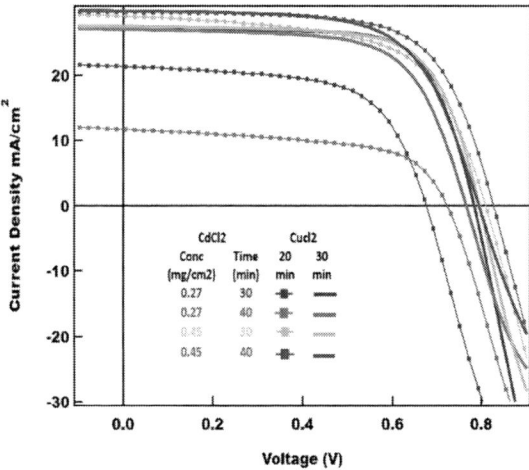

Fig.4 J-V Characteristic curves for the highest efficiency cell of devices $CdCl_2$ treated at 425°C varying $CdCl_2$ concentration, treatment time, and Cu treatment time at 230°C

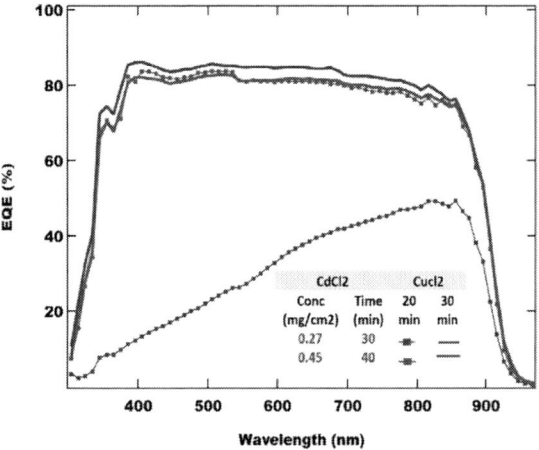

Fig.5 EQE data plot for devices $CdCl_2$ treated at 425°C varying $CdCl_2$ concentration, treatment time, and Cu treatment time at 230°C

(square markers) to 30 minutes (solid lines). For the device with $CdCl_2$ treatment of 30 minutes (blue), the Voc increased from 650 mV to 800 mV and Jsc increased from 14 mA/cm^2 to 27 mA/cm^2. Similarly, for the device with $CdCl_2$ treatment of 40 minutes (red), the Jsc increased from 12 mA/cm^2 to 27 mA/cm^2. On the other hand, for devices fabricated with 0.45 mg/cm^2 $CdCl_2$ concentration (green and yellow), the little change in the PV parameters was observed with additional $CuCl_2$ treatment time.

The external quantum efficiency for these samples can help understand these changes. Fig. 5 shows representative EQE curves for samples that resulted in low and high Jsc values. For the devices with 0.27 mg/cm^2 activated 425°C for 30 minutes (blue) and Cu doped for 20 minutes (square makers) shows a

triangle shape while devices with 30 minute doping time (solid) has a top-hat shape. The triangle shape can be due to incomplete intermixing of CdSe/CdTe or insufficient Cu doping [9]. In this case, more aggressive doping leads to a change in shape, suggesting the CdSe/CdTe are fully mixed, and the low J$_{SC}$ is due to poor doping. The samples fabricated with a 0.45 mg/cm^2 $CdCl_2$ concentration (green) demonstrate the top-hat shape for all treatment times shown here, suggesting the Cu-doping is suffient at both times.

When all the JV and EQE data is taken together, it is clear there is a fine balance between both the $CdCl_2$ and $CuCl_2$ treatment processes. When the $CdCl_2$ activation temperature is too low, the J$_{SC}$ is low, and it is likely that there is incomplete intermixing of the CdSe/CdTe at the front of the device. This leads to parasitic absorption in the photo-inactive layer [5]. The higher temperature activations demonstrate that the amount of $CdCl_2$ available can affect the device performance. Low surface concentrations of 0.09 mg/cm^2 appear to result in incomplete intermixing, though it is possible very aggressive Cu-doping treatments could improve these J$_{SC}$ values. Increasing the $CdCl_2$ concentration to 0.27 mg/cm^2 led to devices with high J$_{SC}$ for the more aggressive $CuCl_2$ treatments. This suggests the $CdCl_2$ activation led to complete intermixing of the CdSe/CdTe layer, but insufficient doping could be an issue. On the other hand, high surface concentration of 0.45 mg/cm^2 yield devices with high J$_{SC}$ for both $CuCl_2$ treatment times investigated here. This suggests the increased amount of $CdCl_2$ may have led to a higher degree of intermixing between the CdSe and CdTe, lowering the doping activation energy.

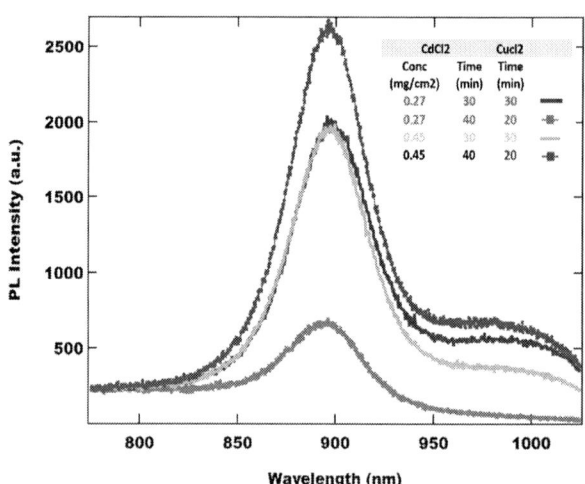

Fig.6 PL Spectrum for devices $CdCl_2$ treated at 425°C varying $CdCl_2$ concentration, treatment time, and Cu treatment time at 230°C

To investigate if the $CdCl_2$ concentration affected the intermixing we measured the steady-state photoluminescence (PL) for the devices activated 425°C for 20 minutes and 30 minutes with Cu dope for 20 minutes and 30 minutes. Fig.6 shows the PL spectrum for these samples illuminated through the glass front with excitation wavelength 633 nm and fixed intensity and integration time. For all samples, the peak intensity was observed at 875 nm. This suggests that the Se to Te ratio at the front of each of these devices is similar. Consequently, increasing the amount of $CdCl_2$ did not result in a significant migration of Se from the front of the device.

While there is not a significant difference in the peak wavelength, there is in peak amplitude. The two samples that had 0.45 mg/cm^2 $CdCl_2$ surface concentration (green and yellow) have PL responses are both high with the sample with longer $CdCl_2$ but shorter $CuCl_2$ treatment (green) approximately 25% higher than the sample with the shorter $CdCl_2$ but longer $CuCl_2$ treatment (yellow). This is consistent with the small V_{OC} increase observed in the JV curves. On the other hand, for the devices with 0.27 mg/cm^2 $CdCl_2$ concentration, the amplitude increases by a factor of four as the Cu-doping time increases from 20 (red) to 30 (blue) minutes, even though the $CdCl_2$ treatment time decreased. This significant increase in PL is consistent with Cu reaching the intermixed region at the front of the device and passivating defect states [10] and the increase in J_{SC} and V_{OC} observed in the JV response with increased doping time.

IV. SUMMARY

By developing ultrasonic Spray system we were able to control the amount of $CdCl_2$ applied to the CdTe surface prior to activation. With this ability, we systematically investigated the role of $CdCl_2$ surface concentration, activation temperature and time, and $CuCl_2$ doping time. Results obtained by varying these parameters clearly show that the $CdCl_2$ optimization process depends on all factors and that balance between each parameter is essential in the development of high performing device.

ACKNOWLEDGMENT

This material is based on research sponsored by U. S. DOE's Office of Energy Efficiency and Renewable Energy (EERE) under Solar Energy Technologies Office (SETO) Agreement DE-EE0008974 and Air Force Research Laboratory under agreement number FA9453-21-C-0056. The U.S. Government is authorized to reproduce and distribute reprints for Governmental purposes notwithstanding any copyright notation thereon. The views expressed are those of the authors and do not reflect the official guidance or position of the United States Government, the Department of Defense, or of the United States Air Force. The appearance of external hyperlinks does not constitute endorsement by the United States Department of Defense (DoD) of the linked websites, or the information, products, or services contained therein. The DoD does not exercise any editorial, security, or other control

over the information you may find at these locations. Approved for public release; distribution is unlimited. Public Affairs release approval #AFRL-2022-2391.

REFERENCES

[1] E. Artegiani, J. D. Major, H. Shiel, V. Dhanak, C. Ferrari, and A. Romeo, "How the amount of copper influences the formation and stability of defects in CdTe solar cells," *Solar Energy Materials and Solar Cells,* vol. 204, p. 110228, 2020.

[2] A. Bosio, G. Rosa, and N. Romeo, "Past, present and future of the thin film CdTe/CdS solar cells," *Solar Energy,* vol. 175, pp. 31-43, 2018.

[3] N. R. Paudel and Y. Yan, "Enhancing the photo-currents of CdTe thin-film solar cells in both short and long wavelength regions," *Applied Physics Letters,* vol. 105, no. 18, p. 183510, 2014.

[4] D. Kuciauskas, A. Kanevce, P. Dippo, S. Seyedmohammadi, and R. Malik, "Minority-carrier lifetime and surface recombination velocity in single-crystal CdTe," *IEEE Journal of Photovoltaics,* vol. 5, no. 1, pp. 366-371, 2014.

[5] E. Bastola, A. V. Bordovalos, E. LeBlanc, N. Shrestha, M. O. Reese, and R. J. Ellingson, "Doping of CdTe using CuCl2 Solution for Highly Efficient Photovoltaic Devices," *2019 IEEE 46th Photovoltaic Specialists Conference (PVSC),* pp. 1846-1850, 16-21 June 2019 2019.

[6] M. K. Jamarkattel *et al.,* "Improving CdSeTe Devices With a Back Buffer Layer of Cu_xAlO_y," *IEEE Journal of Photovoltaics,* vol. 12, no. 1, pp. 16-21, 2022.

[7] T. A. Gessert *et al.,* "Microscopic analysis of residuals on polycrystalline CdTe following wet CdCl2 treatment," *MRS Online Proceedings Library (OPL),* vol. 668, 2001.

[8] R. C. Tenent *et al.,* "Ultrasmooth, large-area, high-uniformity, conductive transparent single-walled-carbon-nanotube films for photovoltaics produced by ultrasonic spraying," *Advanced materials,* vol. 21, no. 31, pp. 3210-3216, 2009.

[9] E. Bastola *et al.,* "Understanding the Interplay Between CdSe Thickness and Cu Doping Temperature in CdSe/CdTe Devices," *IEEE Journal of Photovoltaics,* vol. 12, no. 1, pp. 11-15, 2022.

[10] I. Sankin and D. Krasikov, "Kinetic simulations of Cu doping in chlorinated CdSeTe PV absorbers," *physica status solidi (a),* vol. 216, no. 15, p. 1800887, 2019.

External Quantum Efficiency and Device Reflectance of CIGS PV for Terrestrial and Space Based Applications

Bishal Shrestha, Indra Subedi, Robert W. Collins and Nikolas J. Podraza

Department of Physics & Astronomy and Wright Center for Photovoltaic Innovation & Commercialization, University of Toledo, Toledo, OH, 43606, USA

Abstract—External quantum efficiency (EQE) of copper indium gallium diselenide (CIGS) based solar cells with different antireflection coatings (ARC) has been evaluated under the solar irradiance of airmasses (AM) 0 and 1.5G. The simulations are performed in the wavelength range of 300 - 2500 nm to investigate the absorptance and reflectance features below and above the band gap of the absorber layer. Short circuit current density is increased the most for AM 0 and 1.5G using MgF2 ARCs. However, these ARCs also reduce reflectance below the band gap energy of CIGS which will lead to absorption in other component layers, device heating, and lower operating efficiency.

Keywords—EQE, CIGS, EMA, ARC, AM 0 and 1.5G

I. INTRODUCTION

Copper indium gallium diselenide (CIGS) based photovoltaic (PV) have been established as one of the most efficient thin-film solar cell devices due to its high efficiency, low weight, high flexibility, and low-cost production [1,2]. Where its theoretical limit is 29 % [3], the current world record efficiency of a heterojunction CIGS PV cell is greater than 22% for the air mass (AM) 1.5 spectrum [4,5]. Further, a recent study shows CIGS having better stability against high energy radiation as compared to silicon wafer [6,7]. However, challenges still exist in optimizing CIGS PV and chasing its theoretical efficiency limit as most of the working module at present have the efficiencies in the range of 13 % to 15 %.

In this work, we have simulated a CIGS PV device performance operating under solar irradiance of AM 0 and 1.5G which imitates the radiation spectra of space and terrestrial conditions, respectively. Different single layered antireflection coatings (ARCs) have been incorporated into external quantum efficiency (EQE) models to assess potential increases in CIGS PV performance. Absorbance in the CIGS layer and reflectance of the complete device are calculated over the 300 - 2500 nm wavelength range. The results have been quantified in terms of optimal short-circuit current density (J_{sc}) from the active layers in each case of ARCs and has been compared to the results of the CIGS model without ARC. The reflected power calculated from the reflectance spectra has been divided below the band gap (300 - 1110 nm) and above the band gap (1111 - 2500 nm) range of the CIGS used in the simulation. This helps us to identify reflected powers and corresponding absorption in the component layers in two different regions of the spectrum. Any

absorption in other component layers of the PV devices except current generating active layers, which does not contribute to current results in heating up the device [8]. This work serves as a means of considering the effectiveness of different ARCs in increasing current generation in the absorber layer while reducing the absorption in the other component of the PV devices which is responsible for device heating and is one of the key factors reducing the device performances. The methodology can be applied to other single and multijunction photovoltaic devices which will help to identify the factors determining performance losses. Optimization can then be made to obtain enhanced performance for terrestrial and space-oriented applications.

II. SIMULATION DETAILS

e-ARC software has been used to calculate EQE of CIGS PV device with and without ARCs for the AM 0 and 1.5G solar irradiance spectra. The spectral range extends from 300 - 2500 nm which spans both above and below the absorber band gap energy and accounts for a substantial portion of the respective solar irradiance spectra. These simulations use a transfer matrix-based approach in which the complex dielectric function spectra and thickness of each component layer serve as input to simulate absorbance within each layer and reflectance from the device structure [8-10]. The structural model of the simulation consists of the bulk material layers and interfacial layers between materials represented by effective medium approximations (EMAs). For CIGS PV, this structure from illumination to substrate consists of the components as shown in Fig. 1. The optical and structural properties of each layer except the ARCs have been taken as a reference from previous literature [11].

The ARCs considered here are: MgF_2, Al_2O_3, SiO_2, and TiO_2. The thickness of each ARC has also been optimized for to give highest J_{sc} as evaluated from the simulation results. The reported optimal J_{sc} refers to the sum of current generated from the active layers which includes CIGS (EMA of CdS + CIGS, near heterojunction CIGS, and near back contact CIGS). The current contribution from the EMA of CIGS + Mo layer has been excluded assuming that the electron-hole pairs generated by the light within the CIGS component of this interface layer

978-1-7281-6118-1/22 $31.00 © 2022 IEEE

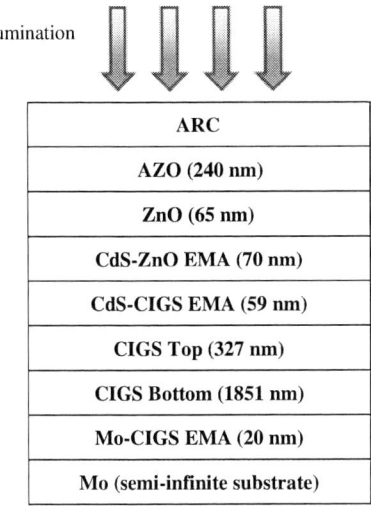

Ilumination

| ARC |
| AZO (240 nm) |
| ZnO (65 nm) |
| CdS-ZnO EMA (70 nm) |
| CdS-CIGS EMA (59 nm) |
| CIGS Top (327 nm) |
| CIGS Bottom (1851 nm) |
| Mo-CIGS EMA (20 nm) |
| Mo (semi-infinite substrate) |

Fig. 1   Schematic and nominal thickness of simulated CIGS PV device with ARC layer incorporated at the front surface.

is likely to be lost due to recombination [11]. Reflectance of the complete device has also been reported, accompanied by spectrally divided reflected power values in the wavelengths below (300 - 1110 nm) and above the band gap (1111 - 2500 nm) of the absorber layer. This helps us to track parasitic absorption at short wavelengths and IR absorption at long wavelengths in all components which are responsible for heating the device and reducing performance.

III.   RESULTS AND DISCUSSIONS

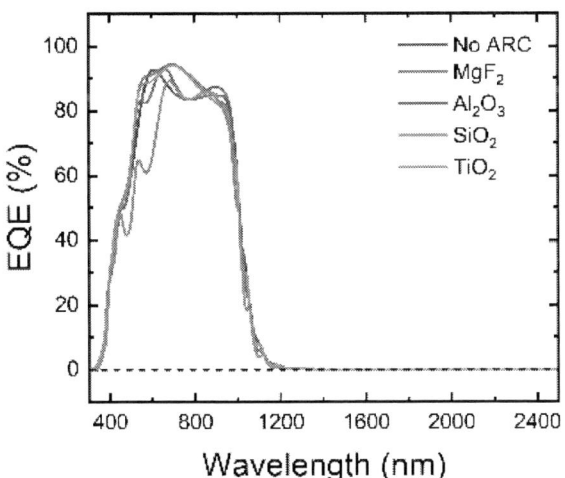

Fig. 2   External quantum efficiency (EQE) of the CIGS active layer from 300 - 2500 nm for AM 0 solar irradiance.

Fig. 2 shows the total EQE obtained from the active layers of CIGS solar cells with various ARCs for AM 0 solar irradiance. The EQE of the CIGS cell for AM 1.5G was

observed to have the similar absorbance spectra for each case of ARC as obtained for AM 0. The corresponding optimal J_{sc} and respective thickness of different ARC are shown in Table I. MgF_2 performed the best for both AM 0 and 1.5G. J_{sc} increased by 3.78% and 3.34% for AM 0 and 1.5G, respectively.

TABLE I.   THICKNESS (NM) AND OPTIMAL J_{sc} OF THE CIGS SOLAR CELL WITH DIFFERENT ARCS FROM 300 − 2500 NM FOR AM 0 AND 1.5G

AM		No ARC	MgF₂	Al₂O₃	SiO₂	TiO₂
0	Thickness (nm)	0	105	240	95	170
	Jsc (mA/cm²)	36.47	37.85	36.68	37.42	33.16
1.5G	Thickness (nm)	0	110	240	100	165
	Jsc (mA/cm²)	30.63	31.83	30.80	31.47	27.93

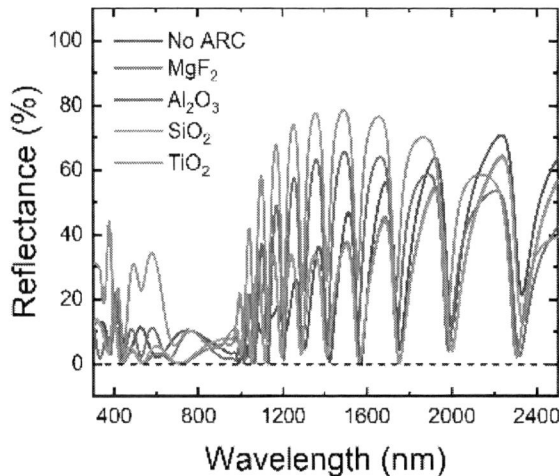

Fig. 3   Simulated reflectance of CIGS based solar cell with different ARCs from 300 - 2500 nm for AM 0.

Fig. 3 represents the total reflectance from the device for AM 0. The reflectance of the CIGS cell for AM 1.5G is similar to that for AM 0 for each ARC. Spectrally divided reflected power below and above the band gap of the absorber layer for both cases of AMs are shown in Table II. Again, CIGS with MgF_2 as the ARC exhibits the least reflectance from 300 - 1110 nm where power absorbed in the CIGS is converted to current. The lower reflectance loss in these wavelengths increases current generation in the CIGS absorber and is supported by the highest J_{sc} results obtained before in the case of MgF_2 ARC. On the contrary, TiO_2 exhibits the highest reflection throughout the measured spectrum among the other ARCs with the least values of current generated for both cases of AMs. This suggests TiO_2 would be more IR reflecting which causes less heating but at the cost of lower current generation.

978-1-7281-6118-1/22 $31.00 © 2022 IEEE   834

TABLE II. SPECTRALLY DIVIDED REFLECTED POWER BELOW AND ABOVE THE BAND OF THE CIGS ABSORBER LAYER WITH DIFFERENT ARCS

ARCs	Reflected power (mW/cm^2)			
	AM 0		AM 1.5	
	300 - 1110 nm	1111 - 2500 nm	300 - 1110 nm	1111 - 2500 nm
No ARC	77.5	83.2	60.84	50.10
MgF2	43.2	76.9	33.98	46.25
SiO2	51.7	76.9	41.35	46.27
Al2O3	72.9	116.	58.98	69.62
TiO2	187	161	147.63	97.02

We observed that the devices should be made more IR reflecting to avoid the losses at those energies and reduce the chances of device heating. Since a single layered ARC is bound to be effective in only a certain wavelength range, additional optimization may occur by adjusting the structural parameters and designs of the components such as transparent conducting oxides (TCO) and interfaces at the back contact layers of the PV devices where substantial IR absorption occur [8].

IV. CONCLUSION

EQE of CIGS PV from 300 - 2500 nm for both AM 0 and 1.5 G solar irradiances show performance improvement with the inclusion of ARC. Out of the all the implemented single layered ARCs, MgF$_2$ stands out as the most effective one while TiO$_2$ is the least effective in terms of current generation. The lower Jsc in the case of TiO$_2$ is attributed to the higher reflectance and reflected power as observed for wavelengths both, below and above the band gap of the CIGS absorber layer. The reflectance and reflected powers from each design are also obtained which shows substantial IR absorption taking place in the other component of the device which does not contribute to current generation but rather heats the device to reduce performance. Application of these modeling techniques can be very helpful in studying the behavior of PV devices when they are operated at different irradiances. These studies are crucial in identifying both optical and thermal factors which reduce device performance and is applicable to other single junction and multijunction PV technologies.

ACKNOWLEDGMENT

This material is based on research sponsored by Air Force Research Laboratory under agreement numbers FA9453-19-C1002 and FA9453-21-C-0056. The U.S. Government is authorized to reproduce and distribute reprints for Governmental purposes notwithstanding any copyright notation thereon. The views expressed are those of the authors and do not reflect the official guidance or position of the United States Government, the Department of Defense or of the United States Air Force. The appearance of external hyperlinks does not constitute endorsement by the United States Department of Defense (DoD) of the linked websites, or the information, products, or services contained therein. The DoD does not exercise any editorial, security, or other control over the information you may find at these locations. Approved for public release; distribution is unlimited. Public Affairs release approval #AFRL-2022-2394.

REFERENCES

[1] H. Zarei and R. Malekfar, "Evaluation of electrical and optical characteristics of ZnO/CdS/CIS thin film solar cell," *Chinese Physics B*, vol. 25, no. 2, p. 027103, 2015.

[2] S. Kawakita, M. Imaizumi, and H. Kusawaka, "Space Environments and Effects on CIGS Solar Cells and Modules," *MRS Online Proceedings Library*, vol. 1792, no. 1, pp. 1-7, 2015.

[3] W. Shockley and H. J. Queisser, "Detailed balance limit of efficiency of p-n junction solar cells," *Journal of applied physics*, vol. 32, no. 3, pp. 510-519, 1961.

[4] P. Jackson *et al.*, "Properties of Cu(In, Ga)Se$_2$ solar cells with new record efficiencies up to 21.7%," *physica status solidi (RRL)– Rapid Research Letters*, vol. 9, no. 1, pp. 28-31, 2015.

[5] T. M. Friedlmeier *et al.*, "High-efficiency Cu(In, Ga)Se$_2$ solar cells," *Thin Solid Films*, vol. 633, pp. 13-17, 2017.

[6] B. P. Rand, J. Genoe, P. Heremans, and J. Poortmans, "Solar cells utilizing small molecular weight organic semiconductors," *Progress in Photovoltaics: Research and Applications*, vol. 15, no. 8, pp. 659-676, 2007.

[7] A. Hamache, N. Sengouga, A. Meftah, and M. Henini, "Modeling the effect of 1 MeV electron irradiation on the performance of n+– p–p+ silicon space solar cells," *Radiation Physics and Chemistry*, vol. 123, pp. 103-108, 2016.

[8] I. Subedi, T. J. Silverman, M. Deceglie, and N. J. Podraza, "Impact of infrared optical properties on crystalline Si and thin film CdTe solar cells," in *2017 IEEE 44th Photovoltaic Specialist Conference (PVSC)*, 2017: IEEE, pp. 2771-2775.

[9] P. Nubile, "Analytical design of antireflection coatings for silicon photovoltaic devices," *Thin solid films*, vol. 342, no. 1-2, pp. 257-261, 1999.

[10] A. Nakane *et al.*, "Quantitative determination of optical and recombination losses in thin-film photovoltaic devices based on external quantum efficiency analysis," *Journal of Applied Physics*, vol. 120, no. 6, p. 064505, 2016.

[11] A.-R. A. Ibdah *et al.*, "Optical simulation of external quantum efficiency spectra of CuIn$_{1-x}$GaxSe$_2$ solar cells from spectroscopic ellipsometry inputs," *Journal of energy chemistry*, vol. 27, no. 4, pp. 1151-1169, 2018.

Radiation tolerance, high temperature stability, and self-healing of triple halide perovskite solar cells

Hadi Afshari, Sergio A Chacon, Brandon K Durant, Rose Crawford, Bibhudutta Rout, Giles E Eperon, Ian R Sellers

University of Oklahoma, Norman, OK, United States

University of North Texas, Denton, TX, United States

Swift Solar, San Carlos, CA, United States

$FA0.8Cs0.2PbI2.4Br0.6Cl0.02$ triple halide perovskite solar cells are studied for potential space power applications including exposure to high temperatures and variable radiation conditions. The radiation tolerance of these devices is investigated in response to increasing levels of proton irradiation and fluence. Parameters were chosen to investigate the relative effects of nuclear displacement and electron ionization processes upon the solar cells being assessed. The change in the photovoltaic (PV) parameters was monitored with regards to energy and fluence of irradiation at various temperatures. The experimental results indicated a considerable reduction in the PV parameters affecting Jsc and FF with minimal effect on Voc as the energy and fluence of the irradiation increased. However, a suite of complementary measurements suggests that while the irradiation negatively affects the transporting layers and interfaces in the devices: the perovskite absorber is not affected in any significant way. Moreover, these systems were observed to self-heal under ambient conditions in the dark demonstrating the unique behavior of perovskite solar cells and their potential for future space power systems. This is further substantiated by high temperature measurements that indicate that the system under investigation displays no appreciable loss up to 500 K, supporting in particular their potential as candidate systems for future lunar missions.

Hyperspectral Luminescence Imaging Analysis of Solar Cells with Localized Radiative Defects

Brianna Conrad, Behrang H. Hamadani

National Institute of Standards and Technology, Gaithersburg, MD, United States

Absolute hyperspectral electroluminescence and photoluminescence imaging is used to investigate properties of localized radiative defects in GaAs cells, including local voltage, or quasi-Fermi-level splitting, EQE and absorptivity. Previous methods to determine these quantities rely on assumptions that may not be valid in such cases, can be difficult to apply, and can produce non-physical results. We present alternative methods along with analysis of relative results. These methods are also applicable to other devices with significant heterogeneity.

Impact of In-situ Cd saturation MOCVD grown CdTe solar cells on As doping and VOC

Ochai Oklobia, Steve Jones, Giray Kartopu, Dingyuan Lu, Wes Miller, Rajni Mallick, Xiaoping Li, Gang Xiong, Vladislav Kornienko, Martin Bliss, Ali Abbas, Michael Walls, Stuart J. C. Irvine

Swansea University, Swansea, United Kingdom

First Solar Inc, California, CA, United States

Loughborough University, Leicestershire, United Kingdom

Polycrystalline CdTe thin film solar cells grown under Cd-rich conditions has been shown to be a promising strategy for further device improvement [1]. A combination of the above with group-V element doping are of interest for maximising the p-type doping concentration, without compromising minority carrier lifetime [2]. For a long time now, increasing hole concentration of >1x1016 cm-3 in thin polycrystalline CdTe films has been limiting and challenging [3]. In-situ Cd saturation growth of polycrystalline CdTe:As thin films was performed by metal organic chemical vapour deposition (MOCVD) at a low temperature of 350°C, to investigate the impact on As doping and device VOC. SIMS measurements on the Cd-rich CdTe:As layers revealed high As concentration of 1.15 - 1.20×1018 As cm-3. Device characterisation showed PCE of ~14%, a VOC of 772 mV, with a corresponding C-V derived acceptor concentration (NA) of 2.0×1016 cm-3. This is a small improvement to the baseline device with CdTe:As absorber layer grown at 390 oC and not under Cd saturation conditions [4]. A much larger improvement in PCE and Voc was achieved when the low temperature Cd saturation growth was combined with higher temperature chlorine heat treatment (CHT). This was achieved for these Cd-saturated CdTe:As devices by performing a post-growth CdCl2 activation process at an elevated temperature of 440°C for 10 mins compared with the standard 420 oC. The higher CHT temperature, however, made no significant improvement with the non-saturated baseline growth. An efficiency of ~17% was measured with a high VOC of 835 mV. Compared to the baseline device, the VOC has been enhanced by ~13%. Micro - photoluminescence (micro-PL) spectra measurements performed on these Cd-rich CdTe:As samples indicated improvement in minority carrier lifetimes, which was consistent with time resolved photoluminescence (TRPL) measurements confirming that carrier lifetime almost doubled. Optimisation of As doping efficiency as a function of Cd/Te precursor partial pressure ratio and its impact on device PV characteristics, especially VOC is also investigated and will be reported. The results that will be presented in this report further lend strength to the merit of Cd-saturated growth ambient, for higher efficiency polycrystalline CdTe:As thin film solar cells. References [1] J. Ma et al., Dependence of the Minority-Carrier Lifetime on the Stoichiometry of CdTe Using Time-Resolved Photoluminescence and First-Principles Calculations, Phys. Rev. Lett. 111, 067402 (2013) [2] B. McCandless et al., Enhanced p-Type Doping in Polycrystalline CdTe Films: Deposition and Activation, in IEEE Journal of Photovoltaics, vol. 9, no. 3, pp. 912-917, May 2019, doi: 10.1109/JPHOTOV.2019.2902356 [3] Colegrove et al., Experimental and theoretical comparison of Sb, As, and P diffusion mechanisms and doping in CdTe, J. Phys. D: Appl. Phys. 51 (2018) 075102 [4] G. Kartopu et al., Study of thin film poly-crystalline CdTe solar cells presenting high acceptor

Insertion of Photovoltaic Generation in the Planning of Electricity Distribution Systems Based on its Economic Potential

João Cardoso das Neves Neto; Miguel Edgar Morales Udaeta; Carlos Frederico Meschini Almeida; Henrique Fernandes Camilo

GEPEA//EPUSP Energy Group, Department of Energy and Automation of Electrical Engineering, Polytechnic School, University of São Paulo, São Paulo, Brasil, joaoc.neves@usp.br

Abstract — **The objective of this work is to evaluate the insertion of DG-PV (Distributed Photovoltaic Generation) and its dissemination with the support of energy storage as a value option in the electric power distribution system, aiming at generating in the form of another product available to the consumer by the distributor. Methodologically, an algorithm is developed that first involves understanding the regulated market from a specific region and calculating LCOE for photovoltaic generation and the energy storage system. The methodology will evaluate the photovoltaic system at three times: acquisition, regular operation, replacement of the inverter, and battery bank replacement, each for one year. The results are based on reducing the generation costs of DG-PV (Distributed Photovoltaic Generation) combined with its potential for savings of scales. However, for the generation of 1.5kWp, there is a savings of scales of 60.113%, and for 10000kWp, a savings of scales of 26.478%.**

Keywords— storage, distributed generation; regulation of the energy market; photovoltaic system; distribution utility.

I. INTRODUCTION

With the advancement of technology, energy distribution in the world is constantly changing and, in addition to this technological evolution, energy market regulations, attention to environmental impacts, continuity of services, including energy storage and energy quality, also contribute to this constant evolution, demanding extraordinary efforts from regulators, companies in the regulated energy market and electricity distributors. Figures 1 and 2 illustrate the distributor's responsibilities.

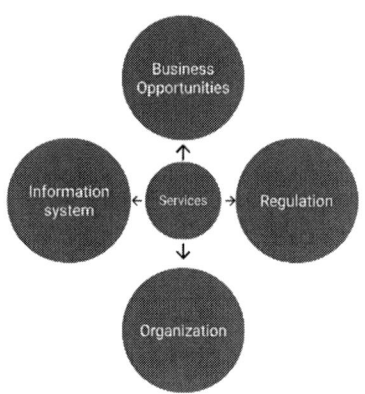

Fig 1. Distributor utility model - Adapted [1]

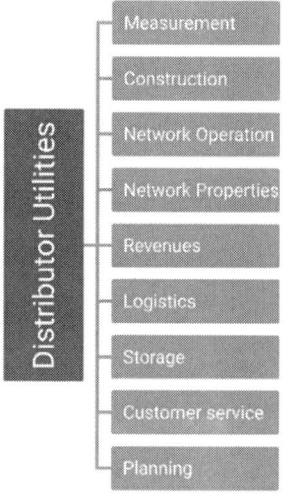

Fig 2. Distributor utility model - Adapted [1]

Thus, companies operating in the energy distribution market, meeting these new demands allied to regulatory obligations and commitments to their customers, should also be aligned with their corporate goals, thus requiring the application of new work methodologies and tools to support the challenges imposed [1].

In addition, the provision and resale of electricity to ordinary consumers is regulated in different ways, such as a limited value for consumption in MWh, for example.

While some individual differences between markets worldwide in relation to energy trade, they aim to provide the most competitive price to consumers through mechanisms that propose the lowest energy prices, better energy quality, and continuity of energy supply.

This indicates that in all business sectors, the U.S. economy is losing between $104 billion and $164 billion annually due to power system outages and another $15 billion to $24 billion due to energy quality problems [2].

Electricity prices are generally higher for residential and commercial consumers because it costs more to distribute electricity. The price of electricity for industrial consumers is generally close to the wholesale price of electricity. By 2020, the average annual electricity consumption for a residential public service customer in the U.S. was 10,715 kilowatts/hour (kWh), an average of about 893 kWh per month since the U.S. minimum wage is $7.25 an hour.

Another essential factor that reflects a lot with these high energy costs is the number of people who do not have access to electricity.

In India, the government announced that it had achieved complete access to electricity in 2019, and effective policies were implemented in several African countries.

In sub-Saharan Africa, although the number of people without access to electricity has steadily declined since 2013, it is now expected to increase by 2020, pushing many countries away from achieving the goal of universal access by 2030 [3].

Providing power at low cost and in areas that today are not met.

For this feasibility, the effective overall cost for PV projects and many other energy resources should be evaluated frequently to verify their profitability.

The study does not propose a complete replacement of the conventional generation.

II. APPLIED METHODOLOGY

As the study does not focus only on financial aspects, it is also necessary to consider other aspects of utility. Therefore, when assessing financial balance, it is necessary at the same time to understand the regulatory side to fully understand the challenges, risks, and opportunities to be addressed.

The present work proposes a study in the form of an algorithm that first goes through the understanding of the regulated market from a specific region, the calculation of LCOE for photovoltaics together with the storage system, not being considered in the study the LCOS and, analysis of the potential for savings and, in the end, composition a more concrete evaluation on the discussion of the results. In addition, it is essential to note that LCOE is a powerful tool for short-term assessment since the values for the electricity market are dynamic. Therefore, for specific long-term planning for policymakers or agencies, other robust tools can be applied together to achieve a more accurate assessment. Thus, the methodology evaluates the photovoltaic system in 5 moments, each of them for one year, according to the following considerations:

1°) Acquisition of a photovoltaic system (t=0): the beginning of the project and, therefore, no generation available, only the initial cost of implementation, if any.

2°) Operating system (t>0): a period in which the PV system is operational, considering a system degradation factor (d [%]) that will be assumed as 0.75% per year, for a gradual decrease in the generation capacity of the P V system [4]. This analysis does not consider financial incentives. The total life cycle considered for the VF and storage system is 25 years; therefore: $0 < t \leq 25$.

3°) Change of frequency inverter and battery system: similar to the second moment, although, with the need to change the frequency inverter [4], it works for approximately 12 years. In this study, half of the life of the PV system will be assumed (12 and a half years), with a price of 10% corresponding to the total initial cost [4].

4°) To evaluate the regulated market values, an annual rate of 2% (IPCA-2021) on the base value (initial investment) will be considered. You can give the following equation to compare prices on the regulated market with LCOE (1).

$$LCOE = \sum_0^t \frac{Ot+Mt}{(1+r)^t} / \sum_0^t \frac{St\,(1+d)^t}{(1+r)^t} \qquad (1)$$

C0 = Initial investment (VF and storage)
Ct = Operating Cost + Maintenance Cost = Ot + Mt
Et = Generation considering the degradation of the VF System = St (1-d)^t

5°) Potential for savings of the distributors, considering a company acquiring hundreds or even millions of VF systems, thus increasing its bargaining power and reducing the project's overall costs. To highlight this potential, the term G0 [%] (2) should be taken as the necessary savings of the scale index to acquire GD-VF systems. Like this:

$$G0\,(\%) = \frac{LCOE\text{-}MWhmr}{LCOE}\,100\% \qquad (2)$$

Thus, the algorithm below can be implemented, where it will aim to evaluate the possible financial gains of the distributor simply, choosing the distributed generation.

Phase 1: Evaluation of the general rules of the locally regulated market: Although the regulated market is already defined, each country or region has particularities related to marked relationships. Therefore, an analysis is necessarily better to understand the distributor's purchase and sale of power and, from there, understand the feasibility or not on the continuation of the algorithm analysis.

Phase 2: Comparison of PV system costs with MWhmr (3): The regulated market has known public electricity prices (MWhmr) to be purchased by the distributor. This must be recovered from the distributor itself or local agencies. It is also necessary to know the market value of photovoltaic systems in the specific region to calculate the initial investment (C0 - Vf system and storage).

$$MWhmr = \sum_0^t MWh0\,(1+R) \qquad (3)$$

MWh0 = Initial value of MWh in the regulated market for the 1st year of analysis.
R = Rate (Considered: Broad National Consumer Price Index - IPCA 2% - 2021).
t = time in years

Phase 3: Calculation for LCOE, MWhmr, and Scale Economy G0 [%]: Considering that solar systems are modular, their generation capacity must be adequate according to utility demand and their experience plan. Therefore, based on these data from the previous phase, LCOE (adapted) and MWhmr are calculated according to (1) and (3), respectively. Thus, the greater the number of solar systems to be acquired, the greater the bargaining power over manufacturers that the distributor can have - this index being G0 [%] (2).

Phase 4: Comparison of results: The values for the generation are compared with the values of the regulated market through the life cycle period of the solar system, considered at the three different times: acquisition, regular operation, replacement of the inverter, and replacement of the battery bank. The 25 years [5] shall be considered for the calculations. Due to many different applications for photovoltaic systems, this study will evaluate a comparison between the roof and medium-size systems (between 100kWp and 1MWp). The results can then be compiled into a table applied to the case study.

III. CASE STUDY

The study focuses on the metropolitan city of São Paulo (Brazil), which has one of the largest population concentrations in the world, with more than twenty million inhabitants, also taken as a large commercial and industrial center with high energy demand, responsible for more than 28% of the country's energy consumption. The present study evaluates the application of the proposed

978-1-7281-6118-1/22 $31.00 © 2022 IEEE

methodology for this region, which is served by a private company and has an annual average solar incidence of 4.58 kWh/m² [6]. Then, the application of the algorithm to this region is demonstrated.

Phase 1: Assessment of the general rules of the locally regulated market:

Based on the data released by the distributor responsible for that region, contracts registered within the ACR framework with power generation companies, an average value of $39.87/MWh [7] is considered. The photovoltaic system and storage will be considered 788.09 $/kW [8.9]. Thus, the cost of deploying the system is shown in table 1 and Figure 3 according to phase 2.

Phase 2: Comparing the cost of the PV system with the MWhmr:

TABLE 1
AVERAGE VALUES FOR PHOTOVOLTAIC SYSTEMS IN BRAZIL

P (kW)	Total Generation (MWh/year)	Average CostFV + Stor ($)
1,5	1,65	3972
2	2,20	4674
3	3,30	5947
500	550,06	789580
1000	1100,12	1436597

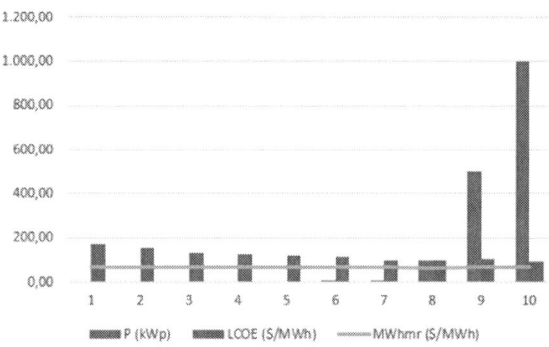

Figure. 3 Comparison of Indicators

Phase 3: Calculation for LCOE, MWhMR, and Scale Economics G0 [%]: According to data from previous table 1, and applying equations (1), (2), and (3), is presented in table 2, the compilation of data.

TABLE 2
CALCULATION OF ALGORITHM INDEXES

P (kWp)	LCOE ($/MWh)	MWhmr ($/MWh)	G0 (%)
1,5	175	70	60,113%
3	131,03	70	46,726%
500	104,37	70	33,116%
1000	94,94	70	26,478%

Phase 4: Comparing results: From this simulation, it is possible to notice that the economy of scale indices tends to decrease as the generation capacity of the solar system increases and, consequently, the decrease in LCOE, as shown in Figure 4.

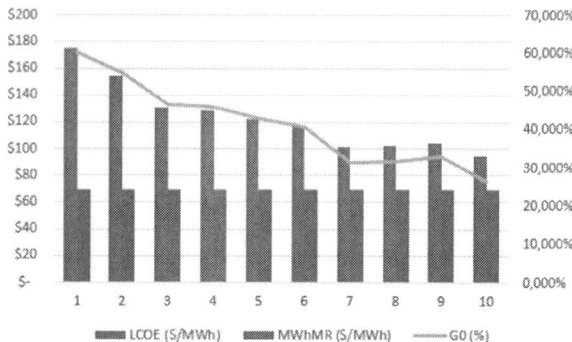

Figure. 4 Comparison of Indicators

IV. ANALYSIS OF RESULTS

The results presented here show a good opportunity for distribution companies, even within the regulated market that usually contains stricter rules for energy trading. Moreover, if distributed generation is installed in regions close to consumption, the overall marginal cost will decrease, as such large-scale application can partially avoid the need to expand large-generation plants and the respective necessary high-voltage transmission connected to it, which means projects with large budgets and environmental impact. Not only can the distribution of energy and the payment of the project be considered in this case, but also special attention of policymakers to foster the local solar industry.

In the case of the Brazilian system, the country does not yet have a local photovoltaic market developed, despite its great solar potential. In addition, the present study affirms the most realistic scenario found in Brazil today.

Other tools, such as specific credit lines for acquiring small or medium-sized renewable generation systems, could be considered. For example, the distributor delivers the PV system to the micro-generator. Both sealed a contract so that the distributor can repurchase the supplied energy price agreed upon previously and may follow the example in Figure 5.

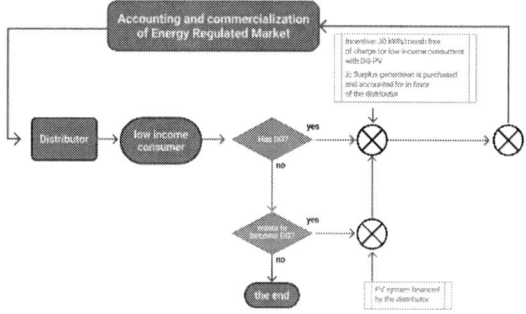

Fig 5. Diagram and a new methodology.

V. CONCLUSION

This work allows the prosumer and distributor to insert a new paradigm, exploring the potential for application from a clean, renewable, and available energy source. The results here show that, based on LCOE and the potential for savings, the application is viable even in the regulated market. The distributor can promote large-scale dissemination of DG on its network by providing a competitive energy price and making energy available to all consumers. Changes in the thoughts of policymakers should be reviewed. In addition, policymakers could provide ways to provide cheaper electricity to their population. More than that, the results also serve as the first step towards other specific research since few biographies have been found related to it.

The basis offered here can be taken as a starting point for more concrete and fair-priced searches

VI. REFERENCES

[1] CAMILO, Henrique Fernandes; UDAETA, Miguel Edgar Morales; GIMENES, André Luis Veiga; GRIMONI, Jose Aquiles Baesso. Assessment of photovoltaic distributed generation – Issues of grid-connected systems through the consumer side applied to a case study of Brazil. Renewable and Sustainable Energy Reviews, 71 (2017) 712-719

[2] Chun Sing Lai, Malcolm D. McCulloch,

[3] Levelized cost of electricity for solar photovoltaic and electrical energy storage, Applied Energy, Volume 190, 2017, Pages 191-203,

[4] ISSN 0306-2619, https://doi.org/10.1016/j.apenergy.2016.12.153.

[5] (https://www.sciencedirect.com/science/article/pii/S03062619163193 3X)

[6] SOLAR PORTAL. How much does photovoltaic solar power cost? Available in: http://www.portalsolar.com.br/quanto-custa-a-energia-solar-fotovoltaica.html. Access: May. 2017

[7] Department of Energy and Mining of the Government of the State of São Paulo

[8] Access to electricity – SDG7: Data and Projections – Analysis - IEA

[9] CCEE - Electric Energy Trading Chamber. Available in: www.ccee.org.br.

[10] IEA, Annual energy storage deployment, 2013-2019, IEA, Paris.https://www.iea.org/data-and statistics/charts/annual-energy-storage-deployment-2013-2019-2

[11] Kyriakopoulos GL, Arabatzis G, Chalikias M. Renewables exploitation for energy production and biomass use for electricity generation. A multi-parametric literature-based review. AIMS Energy 2016;4:762–80

978-1-7281-6118-1/22 $31.00 © 2022 IEEE

Physics-Guided Machine Learning Identifies 5 Optimum Test Locations to Predict Global PV Energy Yield for Arbitrary Farm Topologies

Jabir Bin Jahangir, Muhammed Tahir Patel, and Muhammad A. Alam

School of Electrical and Computer Engineering, Purdue University, West Lafayette, IN, 47907, USA

Abstract—The photovoltaics (PV) technology landscape is evolving rapidly. To gauge the relative merit of emerging PV technologies and their scalable deployability, the global performance of these systems must be understood. Historically, most experimental and computational studies have focused on PV performance in specific regional climatic conditions; however, it has been difficult to translate these isolated regional studies to a global scale. Here, we present a physics-guided machine learning (PG-ML) scheme to demonstrate that: (a) analogous to Köppen–Geiger classification, the world can be divided into just a handful of *PV-specific* climate zones, and (b) the monthly energy yield (YM) data from a few locations (only 5!) is sufficient to predict the yearly energy yield (EY) of over 250,000 locations with a high spatial resolution ($0.5° \times 0.5°$) and accuracy with root mean square error (RMSE) less than just 8 kW·h·m⁻². The map reveals that physically relevant meteorological conditions are shared across continents allowing pan-continental geographical extrapolation. Moreover, the scheme is agnostic to PV technology and farm topology, and thus, can be extended to novel PV technology/farm topology. Our results will lead to data-driven collaboration between national policymakers and research organizations to build efficient decision support systems for accelerated PV qualification and deployment across the world.

Index Terms—solar farm, global, energy, yield potential, physics-guided, machine learning, simulations

I. INTRODUCTION

The rapid evolution of module technology (e.g., PERC, PERT, bifacial, tandem, etc.) and farm topologies (e.g., vertical, tilted, tracking, floating, agro-PV) have created an urgent need to calculate the energy yield (EY) and the economic viability for various technology options and farm configurations in different regions of the world. After all, modern solar farms require significant investment, and incorrect decisions have long-term financial and public-relations implications. Currently, predicting EY of solar farms is supported either by time-intensive computational modeling involving proprietary software operated by modeling experts and/or by integrating the scattered field reports from highly-instrumented test centers located across the world. An alternative framework that allows fast model prediction and rapid experimental validation would transform the development/deployment of solar energy. In this regard, the recently proposed physics-guided machine-learning (PG-ML) models, developed by Purdue[1] and MIT [2] researchers do address the first challenge of rapid prediction (reducing the inference time from years to seconds), but

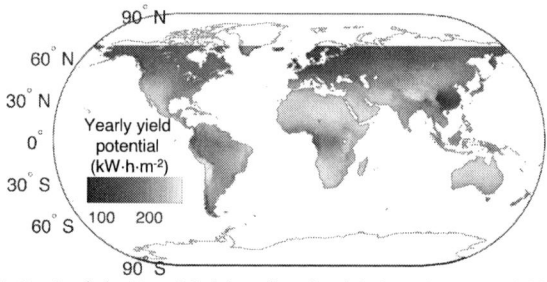

Fig. 1. Purdue Solar Farm Model predicts the global yearly energy yield (EY) potential of tilt-optimized monofacial solar farm with functional approximation approach presented in [1]. This high spatial resolution ($0.5° \times 0.5°$) map was generated based on global climate data and serves as the reference yield potential, i.e., the test data for the trained ML models in this study.

the cost-effective field-validation remains an open question of broad interest.

The PG-ML approaches are essential for PV modeling because a purely data-driven ML approach relies on voluminous data, but produces results that are often inconsistent with physical laws [3], [4]. Indeed, PG-ML approaches are being explored at all system scales ranging from cell to farm, including: material discovery[5], solar cell optimization [2], [6], solar resource prediction [7], module degradation mode identification[8], and solar farm performance prediction [1]. These PG-ML techniques could be transformative for developing decision support systems by accelerating the required time by multiple orders of magnitude. Recently, Patel et al. [1] showed a ML based functional approximation can be used to reliably predict ($R^2 \approx 1$) the global yearly EY of single-axis sun-tracking farms with a predictor set consisting of location's coordinates, average monthly global horizontal irradiance (GHI), ambient temperature (T_{amb}) and clearness index (k_t). The presented approach is agnostic to module technology and farm-topology, e.g., Fig. 1 shows the high-resolution global EY of monofacial farms. Remarkably, this worldwide EY map can be predicted with less than 10% RMSE with a relatively small random subset ($N = 100$) of yearly yield observations as training data. *Despite the impressive results, a manufacturer may find it impractical or expensive to collect EY data from 100 locations. Here, we demonstrate that we can do much better!*

In this work, we show that the EY potential of solar farms can be predicted with high accuracy and a high spatial resolution with a PG-ML model trained on a geographically

978-1-7281-6118-1/22 $31.00 © 2022 IEEE

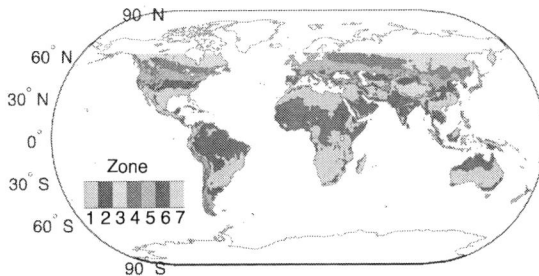

Fig. 2. Geographical distribution of physically significant features (GHI, T_{amb}, k_t) zoned with k-means clustering algorithm (k = 7). The map delineates the similarity between the features determining yield potential across continents that can be exploited to construct a training dataset with optimal set of test locations.

sparse data set. Such a data-efficient method is especially desirable given the paucity of publicly available global PV performance data representing the gamut of climatic conditions. Specifically, we will answer the following questions: (1) Is there a theoretical minimum number of test/qualifications centers that one would need to predict EY potential within a certain degree of accuracy? (2) If test centers are confined to a geographical region, what kind of predictive accuracy can be expected? (3) If one wishes to build a new test center in a given region (in the context of pre-existing centers), what is the optimal location of the new center? This advance relies on a simple observation that "West Lafayette, IN feels like Arizona during the summer, and like Alaska during the winter!"

II. METHOD

A. k-Means Clustering Defines the PV-specific Climate Zones

Inspired by the Köppen–Geiger climate classification system, we use k-means clustering of three key features identified by our previous PG-ML approach [1] (i.e., GHI, T_{amb}, and k_t) to define the geographical zones that dictate the EY of a PV technology/farm at a given geographical location. It is easy to show that other features are either correlated (e.g., module temperature, T_{mod}) or redundant (e.g., longitude/latitude).

Global gridded datasets of these meteorological features are publicly available based on surface observations from satellites. In this study, we use NASA's POWER [9] climatology data set containing the monthly averages of observations since 1984. The global dataset has a spatial resolution of $0.5° \times 0.5°$ and consists of data for more than 250,000 locations. Analysis of this high-resolution global gridded data allows us to see the geographical distribution of the meteorological factors determining yield potential. Fig. 2 shows the geographical distribution of physically significant features. It is obvious that there is symmetry in the PV-specific climate zones across the equator as clusters form in latitudinally-arranged bands and identical zones exist on either side of the equator. Moreover, some countries with large geographical areas (e.g., the United States and China) also possess large diversity of clusters. Thus, data obtained from strategically positioned test centers in these countries have the potential to predict global energy yield trends.

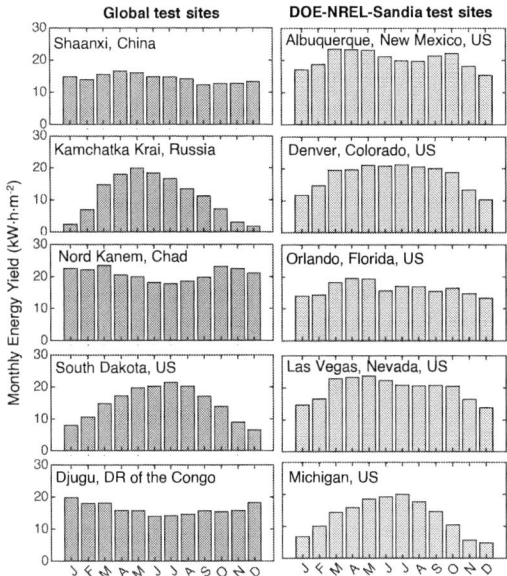

Fig. 3. Barplots showing the simulated monthly energy yields used to construct the training data sets. Dataset 1 (left panel plots) was constructed with YM from 5 randomly chosen locations from different zones in Fig. 2. Dataset 2 (right panel plots) was formed by four existing DOE-NREL-Sandia's PV test sites across US and a hypothetical site in Michigan.

B. Physics-based Machine Learning Models

With the physically significant variables identified, the problem of determining EY can be casted as a regression problem where YM potential is given by

$$\mathrm{YM} = f(GHI, T_{amb}, k_t) \qquad (1)$$

Predicted yearly EY potential is then readily calculated for each location by summing the monthly yields, i.e., $\mathrm{EY}_{ML} = \sum_{i=1}^{12} \mathrm{YM}_i$. The function in (1) may be approximated with various machine-learning algorithms. Here, we train shallow neural networks to predict the monthly energy yields using the implementations present in MATLAB's Deep Learning Toolbox.

To test and validate the machine learning model, we use the Purdue Solar Farm Model[10] to simulate the reference worldwide yearly yield potential of 24% monofacial solar farms installed in a fixed tilt configuration. The physical simulation is performed on a coarse grid with a spatial resolution of $2° \times 9°$ (1200 locations) on monthly basis for a year. Using the global simulation results and the method described in [1], a high-resolution yield potential map was obtained with a spatial resolution of $0.5° \times 0.5°$. This high-resolution map shown in Fig. 1 will serve as the actual yearly yield potential map in this study. Therefore, the test dataset will consist of the map and the features used to generate this map, namely GHI, T_{amb} and k_t from over 250,000 locations from the climatology dataset.

III. RESULTS AND DISCUSSION

A. Training Dataset: Worldwide vs. DOE-NREL-Sandia Sites

To answer the questions posed in the Introduction, two separate machine learning models are trained on two synthetic data sets. First, to understand the need for latitudinal diversity,

978-1-7281-6118-1/22 $31.00 © 2022 IEEE

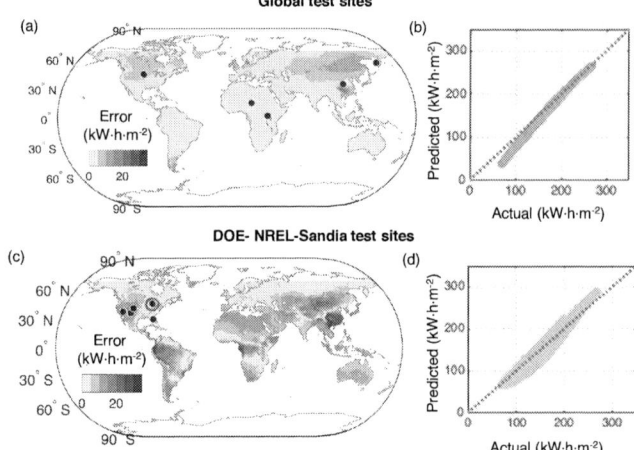

Fig. 4. Worldwide monofacial solar farm EY prediction accuracy of models trained on data from global data set (a-b) and DOE-NREL-Sandia's test site locations across US (c-d). Maps in (a) and (c) show that, with monthly yield data from 5 locations (blue dots), the models are able to predict yearly EY at most locations with less than just 16 kW·h·m^{-2} error. (b) and (d) shows that the models are able to explain most of the variabilities.

5 locations are randomly chosen from the PV-yield zones shown in Fig. 2. For each location, monthly feature data is extracted from the POWER data set. Then, corresponding monthly energy yields of monofacial solar farms are obtained through physical simulation with Purdue Solar Farm Model. The second dataset is prepared in the same manner except the five sites are the locations of the US Department of Energy's (DOE) regional PV test centers at Albuquerque, Denver, Orlando, and Las Vegas operated by Sandia National Laboratories and National Renewable Energy Laboratory (NREL), and an additional *hypothetical site* at Michigan in the US. Fig. 3 shows the results of the simulations. Each dataset contains 60 observations of monthly energy yield from 5 locations. We measure prediction accuracy by comparing the yearly yield inferences made by each model with the reference solar farm yearly yield (Fig. 1) across the globe.

B. Demonstration of High-fidelity Simulation from 5 Test-sites

Fig. 4 shows the prediction accuracy of the two models. Fig. 4(a) shows prediction error of the model trained on 5 randomly chosen global sites from zones shown in Fig. 2. For this model, we find that predictions for 95% worldwide locations are within 9 kW·h·m^{-2} of the reference values, with higher-than-normal errors are confined to locations whose PV-zone is not represented by the test-sites. Fig. 4(b) shows the model is highly accurate: The overall root mean squared error (RMSE) was only 3.85 kW·h·m^{-2}. Fig. 4(c-d) shows the prediction accuracy for a model trained on the second set of training data based on four existing DOE-NREL-Sandia sites in the US and a hypothetical test site in Michigan, US (red circled site in Fig. 4(c)). Comparing the four existing site locations with zones in Fig. 2, we find that the northernmost zones lack representation. The additional *hypothetical site* at Michigan (44.315° N) is expected to increase the representation of yield trends at the world's northernmost zones and thus increase global prediction accuracy. Indeed, Fig. 4(c-d) shows that, despite the sites being

confined to the continental US, the model is still able to predict worldwide EY potential for 95% of locations with less than 16 kW·h·m^{-2} absolute error. The overall yearly yield potential RMSE was 7.62 kW·h·m^{-2}.

IV. SUMMARY

We have presented a method to assimilate global patterns in the energy yield potential of solar farms with data from strategically chosen locations. Achieving this requires training datasets that are representative of the domain of input variables. Towards that end, we have presented the development of a map akin to the Köppen-Geiger system to facilitate the construction of a representative dataset. We have demonstrated that a relatively few strategically determined test sites (e.g., five worldwide, if possible; or 5 DOE-NREL-Sandia facilities in the USA) can predict worldwide energy yield with very high accuracy, provided that the field data is used to train the neural network. The method described herein can be extended to novel PV technologies and arbitrary farm topologies. Thus, our results define an innovative data-driven, physics-based machine learning model for creating resources to accelerate the worldwide adoption of PV technologies.

ACKNOWLEDGMENT

We acknowledge the insightful discussion with Dr. Reza Asadpour (Purdue University, USA), Professor Mohammad Ryyan Khan (East West University, Bangladesh), and Professor Peter Bermel (Purdue University, USA).

REFERENCES

[1] M. Tahir Patel, G. Thanuja Wickramaarachchi, and M. A. Alam, "Machine Learning allows Synthesis and Functional Interpolation of Computational and Field-Data for Worldwide Utility-Scale PV Systems," in *2021 IEEE 48th Photovoltaic Specialists Conference (PVSC)*, pp. 1865–1867, June 2021. ISSN: 0160-8371.

[2] S. Mann, E. Fadel, S. S. Schoenholz, E. D. Cubuk, S. G. Johnson, and G. Romano, "δPV: An end-to-end differentiable solar-cell simulator," *Computer Physics Communications*, vol. 272, p. 108232, Mar. 2022.

[3] G. E. Karniadakis, I. G. Kevrekidis, L. Lu, P. Perdikaris, S. Wang, and L. Yang, "Physics-informed machine learning," *Nature Reviews Physics*, vol. 3, pp. 422–440, June 2021.

[4] J. Willard, X. Jia, S. Xu, M. Steinbach, and V. Kumar, "Integrating Scientific Knowledge with Machine Learning for Engineering and Environmental Systems," *arXiv:2003.04919 [physics, stat]*, July 2021. arXiv: 2003.04919.

[5] Z. Ren, F. Oviedo, H. Xue, M. Thway, K. Zhang, N. Li, J. D. Perea, M. Layurova, Y. Wang, S. Tian, T. Heumueller, E. Birgersson, F. Lin, A. Aberle, S. Sun, I. M. Peters, R. Stangl, C. J. Brabec, and T. Buonassisi, "Physics-guided characterization and optimization of solar cells using surrogate machine learning model," in *2019 IEEE 46th Photovoltaic Specialists Conference (PVSC)*, pp. 3054–3058, June 2019. ISSN: 0160-8371.

[6] F. Oviedo, Z. Ren, X. Hansong, S. I. P. Tian, K. Zhang, M. Layurova, T. Heumueller, N. Li, E. Birgersson, S. Sun, B. Mayurama, I. M. Peters, C. J. Brabec, J. Fisher III, and T. Buonassisi, "Bridging the gap between photovoltaics R&D and manufacturing with data-driven optimization," *arXiv:2004.13599 [physics]*, Apr. 2020. arXiv: 2004.13599.

[7] G. Buster, M. Bannister, A. Habte, D. Hettinger, G. Maclaurin, M. Rossol, M. Sengupta, and Y. Xie, "Physics-guided machine learning for improved accuracy of the National Solar Radiation Database," *Solar Energy*, vol. 232, pp. 483–492, Jan. 2022.

[8] B. Zhang, J. Grant, L. S. Bruckman, O. Wodo, and R. Rai, "Degradation Mechanism Detection in Photovoltaic Backsheets by Fully Convolutional Neural Network," *Scientific Reports*, vol. 9, p. 16119, Dec. 2019.

[9] POWER, "Surface meteorology and Solar Energy: A renewable energy resource web site (release 6.0)," 2017.

[10] M. R. Khan, M. T. Patel, R. Asadpour, H. Imran, N. Z. Butt, and M. A. Alam, "A review of next generation bifacial solar farms: predictive modeling of energy yield, economics, and reliability," *Journal of Physics D: Applied Physics*, vol. 54, p. 323001, May 2021. Publisher: IOP Publishing.

Analyzing Effects of Solar Variability and System Location on LMP Prices

Mesude Bayrakci-Boz
Penn State Hazleton
Hazleton, PA
mzb187@psu.edu

Joseph Ranalli
Penn State Hazleton
Hazleton, PA
jar339@psu.edu

Abstract—Optimal power flow has been solved to show possible effects of solar variability and location of solar systems on electricity price using the IEEE 30 Bus Test system. Different densities of simulated solar generation plants were used, with higher-density plants exhibiting higher variability of generation. The effects of different solar variability conditions tested in this study were found to be minimal on the absolute reduction in local marginal prices (LMPs), but low-density plant distributions exhibited smaller and less frequent fluctuations in the price. In some cases, solar generation was observed to reduce the LMP to zero, resulting from congestion that limited the export of electricity. We observed that lower-density generation distributions could reduce the frequency of these rapid price fluctuations. The location of solar systems within the grid can also have a significant impact on LMPs. When solar generation is installed at a high demand bus, the LMP typically decreased at both the local and neighboring buses. When the solar systems are installed at a low demand bus, the LMPs were observed to increase or decrease depending on the demand and congestion. This work highlights the importance of the effects of solar system location on LMP.

Index Terms—solar, photovoltaics, variability, spatial aggregation, wavelet variability model, locational marginal price, LMP, demand

I. INTRODUCTION

Renewable energy generation, including solar photovoltaics (PV), naturally exhibits a higher degree of variability than traditional sources of generation and is naturally spatially distributed throughout the generation environment. The distributed nature of PV generation poses technical and economic challenges that must be overcome to enable higher levels of renewable penetration onto the electrical grid. Besides technical challenges, it is important to understand the role that distributed solar PV may play in driving electricity prices in the marketplace. Both solar variability and distribution may be expected to play a role in determining the electricity price.

An important electricity metric in the energy market is hourly wholesale price, known as the Locational Marginal Price (LMP). LMP is the marginal cost of transferring an incremental unit of energy from one network location to another location in the network. The price can change according to the balance between supply and demand, and congestion. A review of literature shows numerous studies conducted to analyze the effect of solar systems on LMP. For example, Schwabe et al. [1] made a comparison among average prices of electricity based on the combination of empirically derived energy generation from two photovoltaic systems and LMPs for two regions. Hemmati et al. demonstrated the impacts of renewable energy on flow-gate marginal pricing and LMPs using an IEEE six bus test system [2]. Albadi et al. created two scenarios; one large solar project and small geographically dispersed solar projects in Oman, and investigated the implications on transmission losses and LMPs [3]. Mohammad et al. considered the social welfare maximization problem of Independent System Operators (ISOs), and proposed a new decomposition for LMP after analyzing the effect of renewable energy systems on LMPs [4]. Jin et al. [5] created a model of LMP-based partition optimal economic dispatch with wind and photovoltaic systems that provide 42% of the total generation capacity of the test system.

However, existing studies did not account for the effects of how the distribution of generation with respect to solar variability impacts electricity pricing. This study analyzes these effects on electricity pricing by solving optimal power flow with MATPOWER for the IEEE 30 Bus Test Case. LMPs were obtained for solar systems connected to the grid considering two different generation distribution density (i.e. variability) conditions. Then, we evaluated how the solar system deployment density and its location within the test bus affects the LMPs.

II. METHODOLOGY

A. Variability Modeling

A major component of the variability inherent to solar generation is induced by cloud motion across a distributed generation facility [6]. Previous studies have described the inverse relationship between the density of the generation plant's spatial distribution and the variability of the output: that is, generation spread over a smaller spatial extent exhibits a higher degree of variability than generation spread out over a broader geographic area [7]. As variability of generation may be expected to drive short term electricity prices, we may expect that high- and low-density distribution of generation will have differing impacts on the electricity price.

A region representing a single node of the IEEE network model was approximated as covering a geographic range of 10 km x 10 km. A total of 30 MW of generation were modeled as a generator, feeding into the node. Two different generation cases were considered: one large 30 MW plant and one

978-1-7281-6118-1/22 $31.00 © 2022 IEEE

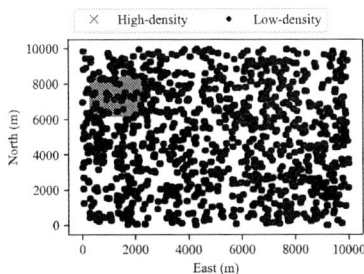

Fig. 1. Sample plant layout. In both cases, 21 MW of total rated capacity are modeled.

TABLE I
VARIABILITY METRICS FOR 2019 REF. DAYS FROM SURFRAD - PSU

Category	Date	Day of Yr	Mean k_c	VS	VI	DARR
Clear	Mar 23	082	1.07	0.07	1.20	214.7
Clear	Mar 26	085	1.11	0.07	1.04	179.2
Cloudy	Mar 01	060	0.31	0.04	1.06	152.2
Cloudy	Mar 21	080	0.16	0.04	1.03	161.9
Variable	Apr 20	110	0.88	1.45	21.46	5706.4
Variable	Jun 14	165	0.86	2.13	30.03	8093.5

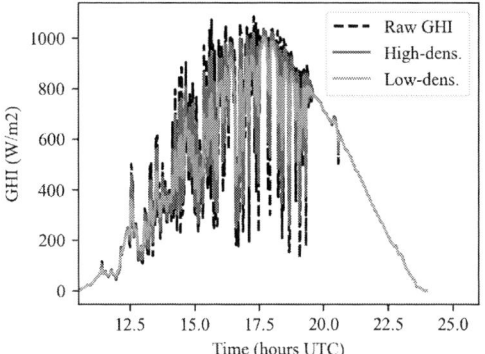

Fig. 2. Sample WVM output time series smoothed by the two different plant layouts for a variable day, Apr 20, 2019. Input in each case is a 1 minute resolution time series from SURFRAD.

thousand small 30 kW plants. A spatial density of 41 MW per acre was used to determine the geographic size of these plants. Thus, the 30 MW plant resulted in a high-density concentrated generation facility, while the 1000 smaller plants resulted in a low-density distribution within the overall area. The plants were randomly centered within the given geographic area and built out to the necessary size to match the capacity scale desired. A sample random layout is in Fig. 1.

In order to model the differences in spatiotemporal variability between the loosely- and densely-distributed PV generation, we utilized the Wavelet Variability Model (WVM) [8] as implemented in PVLIB Python [9]. The WVM represents how the variability in the irradiance time series measured at a single point is smoothed out due to a spatially distributed plant. It does so via selective reduction in the magnitude of wavelet mode amplitudes as a function of wavelet timescale. Short timescales (higher frequencies) are reduced the most, while longer (low frequency) timescales retain their amplitude. The degree to which each wavelet mode's amplitude is scaled down is determined by the modeled time series correlation between spatially disparate portions of the site; a greater spatial separation leads to more significant smoothing of the time series.

We used individual days from the SURFRAD database [10] as representative reference global horizontal irradiance (GHI) time series. These reference (i.e. point measurement) time series were taken for the SURFRAD site *PSU*, located in central Pennsylvania. Data were considered for the entire year of 2019, and several representative clear, cloudy and variable days were manually identified and were compared on the basis of common variability metrics [11]: Variability Score (VS) [12], Variability Index (VI) [13] and Daily Aggregate Ramp Rate (DARR) [14]. In the case of all of these metrics, a higher value corresponds to more variability. Values for these metrics for the reference days considered are given in Table I. As is evident, both clear and cloudy days exhibited low values of the variability as compared to the variable days. While the level of variability between the clear and cloudy days was similarly low, they are differentiated by the cloudy days exhibiting a much lower mean value clear-sky index (k_c).

In order to compute the simulated electricity generation from the high- and low-density distributed plants, the irra-

diance time series were used as an input measurement to the WVM, which provides a prediction of the smoothed irradiance over the aggregate plant. The output of the low-density plant would be expected to have lower variability (smoother time series) than the high-density plant modeled in this study. Comparisons of high- and low-density plant irradiance time series are shown in Fig. 2. It is evident that the low-density plant case exhibits reduced magnitude of fluctuations in the irradiance. The frequency dependence of the variability reduction can be seen by considering the magnitude of transfer function for each case relative to the input, as shown in Fig. 3.

As this model represents a hypothetical plant and we are principally concerned with how differences in the variability affect the electricity price, we represented the relationship between irradiance and plant electrical output as a simple scaling operation. Aggregate irradiance time series from the WVM were normalized such that a computed irradiance of $1000 \, W/m^2$ was assumed to produce electrical power at the rated capacity (30 MW) of the overall plant, as specified in Section II-A.

B. Optimization

The effects of solar variability and location of solar systems on electric prices were investigated through solving optimal power flow (OPF) using the IEEE 30 Bus Test in MAT-POWER, an open source MATLAB simulation package [15].

OPF determines the best operating levels of the generators at the lowest cost, considering any operational limits of

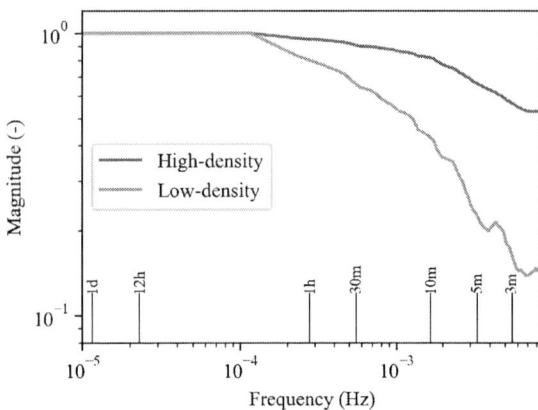

Fig. 3. Sample WVM output transfer functions for the two plant distributions demonstrating the frequency dependence of the variability reduction for a variable day, Apr 20, 2019. Several common timescales are labeled on the frequency axis.

generation and transmission facilities. The process of OPF is:

$$min[f(x)] \qquad (1)$$

Subject to the following conditions:

$$g(x) = 0 \qquad (2)$$

$$h(x) \le 0 \qquad (3)$$

$$x_{min} \le x \le x_{max} \qquad (4)$$

The objective function $f(x)$ consists of the polynomial cost of generator injections. The equality constraints $g(x)$ are the real and reactive power balance equations. The inequality constraints $h(x)$ consist of two sets of n_l branch flow limits as nonlinear functions of the bus voltage angles and magnitudes, one for the *from* end and one for the *to* end of each branch. The variable limits x_{min} and x_{max} include an equality constraint on any reference bus angle and upper and lower limits on all bus voltage magnitudes and real and reactive generator injections [16].

The IEEE 30 Bus Test Case, from December, 1961, represents data collected in the Midwestern US by the American Electric Power System [17]. The MATPOPWER package has the IEEE 30 bus test case with some modifications: the data was taken from Alsac et al. [18] with branch parameters rounded to the nearest 0.01, shunt values divided by 100, and the shunt on bus 10 moved to bus 5, with the load at bus 5 zeroed out. Generator locations, costs and limits and bus areas were taken from Ferrero at al. [19]. Generator Q limits were derived from [18], using their Pmax capacities. V limits and line limits were taken from [18]. The IEEE 30 case is shown with line limits, demand for each bus and generators in Figure 4.

The OPF was first run for the IEEE 30 bus system without any solar systems to create a LMP baseline. Then, solar systems were added to the system as a generator for each

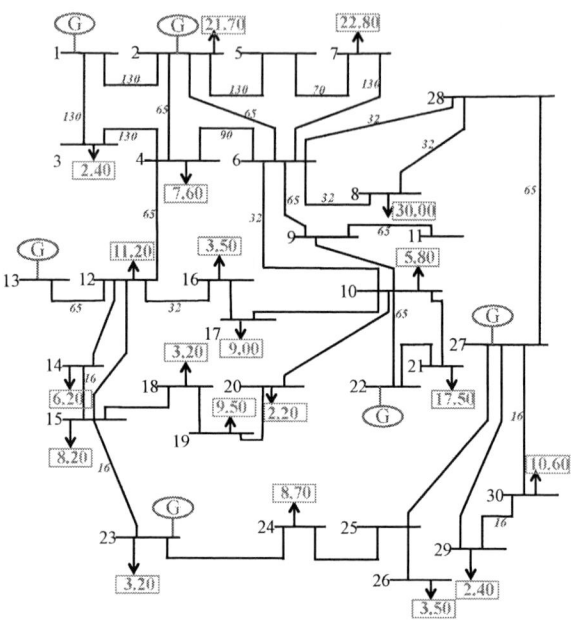

Fig. 4. IEEE 30 bus system test case. The red G in the circle shows the generators, while the grey numbers in the rectangle show the demand values, and the dark blue numbers show the line limits

TABLE II
SOLAR TEST BUS PLACEMENT DETAILS

Bus Number	Demand (MW)	Neighbour Buses
20	2.2	Bus 10, Bus 19
29	2.4	Bus 27, Bus 30
14	6.2	Bus 12, Bus 15
30	10.6	Bus 27, Bus 29
8	30	Bus 6, Bus 28

of the clear, cloudy, and variable days considering high- and low-density plant distribution scenarios. The OPF was run accordingly for each scenario. Bus demand was considered when choosing which buses to add the PV systems to. Five buses were chosen to represent different levels of demand. These five buses and the associated demand for each bus are shown in Table II. LMPs at these five buses and their neighbor buses were obtained and analyzed for each scenario.

The results were compiled on a comparative basis, relative to the baseline LMP with no solar generation. In order to do this, a percent difference range was determined for each five buses, based on the LMPs with- and without- a solar system. Equation 5 was used for this calculation.

$$LMP_{diff}(\%) = \left(\left(\frac{LMP_{with}}{LMP_{without}} \right) - 1 \right) \times 100\% \qquad (5)$$

For all these scenarios, the variables in MATPOWER were modified to include real power generation (Pg), imaginary power generation (Qg), max real power (Pmax), min real power (Pmin), max imaginary power (Qmax), min imaginary power (Qmin), and generation costs.

It was assumed that PV systems produce power proportional to irradiance, based upon their rated capacities, as stated previously. The minimum real power value for PV systems was assumed to be zero. PV systems do not produce reactive power, but inverters may be used for this purpose. In the literature, the maximum reactive power has been assumed to be 1/3 of maximum real power [20] and minimum reactive power is the negative of the maximum reactive power value. These assumptions were also used in this study. Finally, the generation costs for the PV systems were set to zero since those would be paid from the initial cost of the PV systems, with no ongoing fuel costs.

III. RESULTS AND DISCUSSION

The 30 MW solar system was first placed at bus 29 and LMP changes at this bus were analyzed with high- and low-density generation distributions for the three different days (clear, cloudy, and variable). The total generation on the bus for these days is around 190 MW. With 30 MW rated capacity from PV systems, this results in PV penetration of around 15%. On a clear day, March 26, the LMP price at bus 29 was $3.96/MWh for the base case scenario (i.e. no solar system) and decreased to a range of $3.94/MWh to $3.34/MWh when high-density solar case was added, as shown in Figure 5a. Similar results were observed for the low-density case as well. This translates into a 1 to 15% overall decrease in the range of LMPs for low- and high-density cases. On a cloudy day, March 21, the LMP decrease was very small regardless of the solar generation density, a decrease corresponding a range from 1 to 1.9%, as seen in Figure 5b. The results indicate that there is no significant difference between the high- and low-density solar cases analyzed in this study during clear and cloudy days. This makes sense, as the relatively low variability of these days leads to only small differences caused by the distribution of solar generation.

However, a different pattern of LMP changes is observed during a variable day, June 14. The LMP exhibits significant differences between the high- and low-density cases, as in Figure 5c. Overall, there is a pattern of solar causing a slight reduction in the LMP. However, multiple times the LMP fluctuates from the reduced value to zero and back in a very short time span. We attribute this to congestion on enough neighboring lines to result in no pathway for the electricity to be exported, and thus a price of zero is reached. These zero-price conditions occur rapidly, and repeatedly, and are observed more often for the high-density distribution of solar. The LMP values become zero for approximately 114 minutes of the day (8%) and 45 minutes (3%) for high- and low-density distributions, respectively. In addition, even when LMP is not zero, it experiences a higher degree of fluctuation than that seen in the low-density distribution case. This shows that accounting for the effect of solar distribution (variable day scenario) on LMPs is very important, since these sudden changes of price may give market participants misleading price signals. The low-density distribution of solar reduces the risk of these sudden changes in price and mitigates the price risk.

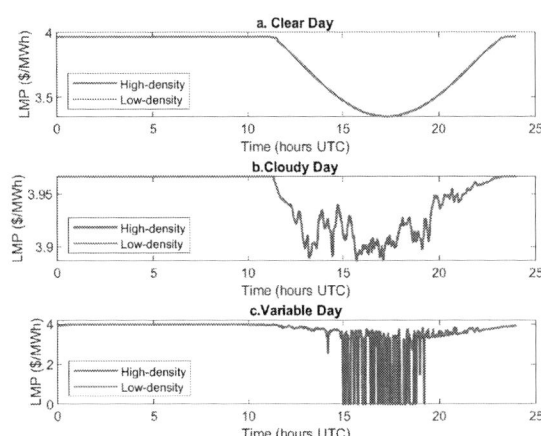

Fig. 5. LMPs at bus 29 in a clear, cloudy and variable day when solar system is placed at bus 29.

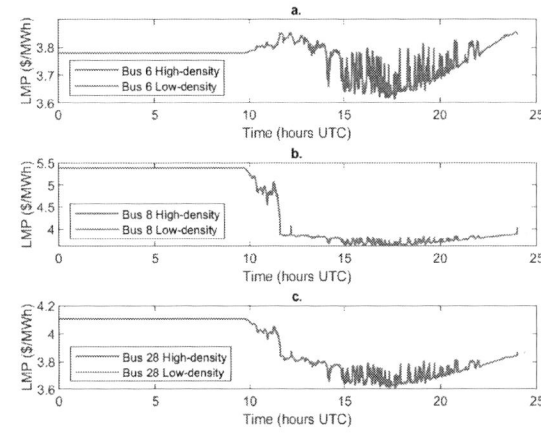

Fig. 6. Variable day LMPs at bus 6, 8, and 28 when solar system is placed at bus 8 with high-density and low-density.

In order to further investigate how LMP changes at neighboring buses to the solar system, we analyzed high- and low-density distribution plants on a variable day for each of the bus cases presented in Table II. The results are shown in Figures 6 to 10. Overall, LMP decreases at most buses for both low- and high-density distributions with a relatively lower minimum price in the high-density case. However, the price experiences a lower degree of fluctuation for the low-density case. This is consistent with the lower variability of the low-density case due to additional smoothing of the time series.

The amount of decrease in price at the bus where solar system is placed is greater than the decrease seen for the neighboring buses. This is because the solar generation would be used first by the local bus, with the surplus sent to other buses. For example, demand at bus 14 (Fig. 7) is 6.20 MW and all the demand can be supplied from the solar system, such that some is exported to neighboring buses. In this case, the LMP

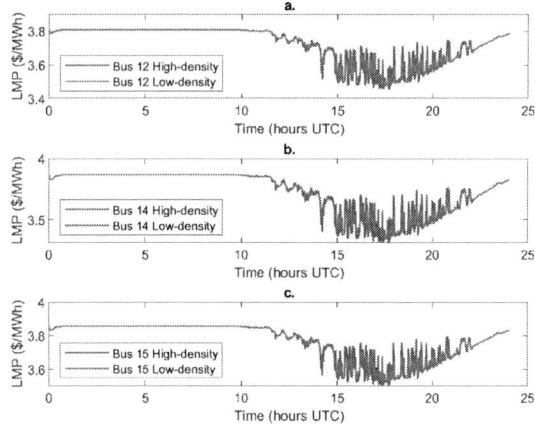

Fig. 7. Variable day LMPs at bus 12, 14, and 15 when solar system is placed at bus 14 with high-density and low-density.

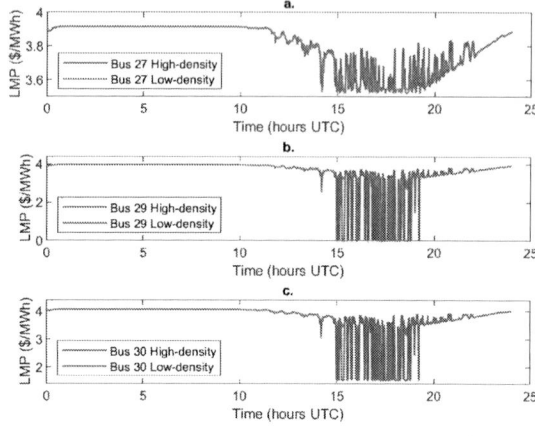

Fig. 9. Variable day LMPs at bus 27, 29, and 30 when solar system is placed at bus 29 with high-density and low-density.

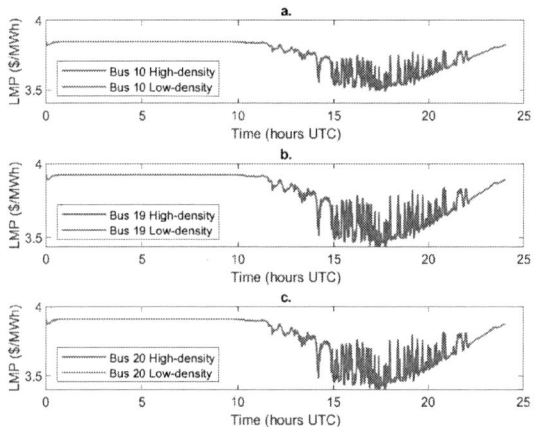

Fig. 8. Variable day LMPs at bus 10, 19, and 20 when solar system is placed at bus 20 with high-density and low-density.

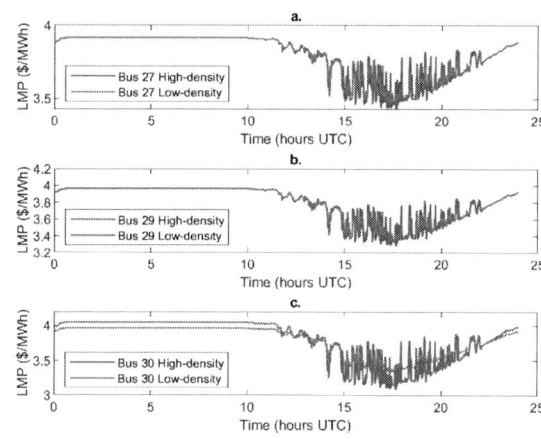

Fig. 10. Variable day LMPs at bus 27, 29, and 30 when solar system is placed at bus 30 with high-density and low-density.

decreases around 14% for bus 14 and 9% for buses 12 and 15. All the LMPs percent changes from maximum decrease to maximum increase for each case is tabulated in Table III. Negative numbers indicate a decrease in LMPs, while positive numbers an increase. Placing a solar system in the IEEE test case results in LMP decrease from 3% to 100% depending on the buses.

While installation of solar systems generally results in a reduction in LMP on neighboring buses, it can be seen that LMP may increase with solar system penetration at some buses in other parts of the grid. For example, when the solar system was added to bus 20, the LMP decreases around 13% for bus 20 with respect to the base case, but increases around 7% for bus 8, as seen in Figure 11. The demand for bus 20 is 2.2 MW and the solar system produces much more than this demand. Thus, the rest of the energy produced is sent to other buses. This causes congestion on the line between bus 6 and bus 8,

which leads to higher LMP at bus 8. The results indicate how LMPs change with respect to where solar systems are placed alongside the demand conditions and further investigation is needed to analyze the effect of solar generation on congestion with different cases.

IV. CONCLUSION

Solar generation has an impact on LMPs, and the extent of this impact is affected by many factors. In this study, the effects of solar generation distribution density (and concurrently variability of generation) and bus location of solar systems on LMPs were investigated. This objective was achieved by solving the optimal power flow using the IEEE 30 Bus Test system in MATPOWER. The results show the effects of different solar variability conditions tested in this study are similar on LMPs during clear and cloudy days. However, solar distribution density (i.e. variability) has significant effect on

TABLE III
Maximum LMP percent increase or decrease(%) for each bus location and neighbors.

		Bus 6		Bus 8		Bus 28	
		Max. ↓	Max. ↑	Max. ↓	Max. ↑	Max. ↓	Max. ↑
Solar System at Bus 8	High Density	-4.410	1.989	-32.971	0.049	-12.088	0.011
	Low Density	-3.881	1.993	-32.571	0.048	-11.619	0.011

		Bus 12		Bus 14		Bus 15	
		Max. ↓	Max. ↑	Max. ↓	Max. ↑	Max. ↓	Max. ↑
Solar System at Bus 14	High Density	-9.329	0.001	-14.414	0.002	-9.222	0.001
	Low Density	-8.596	0.001	-13.372	0.001	-8.504	0.001

		Bus 10		Bus 19		Bus 20	
		Max. ↓	Max. ↑	Max. ↓	Max. ↑	Max. ↓	Max. ↑
Solar System at Bus 20	High Density	-9.263	0.003	-12.567	0.001	-13.135	0.002
	Low Density	-8.532	0.004	-11.628	0.001	-12.160	0.002

		Bus 27		Bus 29		Bus 30	
		Max. ↓	Max. ↑	Max. ↓	Max. ↑	Max. ↓	Max. ↑
Solar System at Bus 29	High Density	-10.098	0.000	-100.000	0.000	-61.842	0.001
	Low Density	-10.092	0.000	-100.000	0.000	-61.842	0.001

		Bus 27		Bus 29		Bus 30	
		Max. ↓	Max. ↑	Max. ↓	Max. ↑	Max. ↓	Max. ↑
Solar System at Bus 30	High Density	-12.507	0.000	-17.289	0.000	-23.755	0.001
	Low Density	-11.466	0.000	-15.879	0.000	-17.631	-2.084

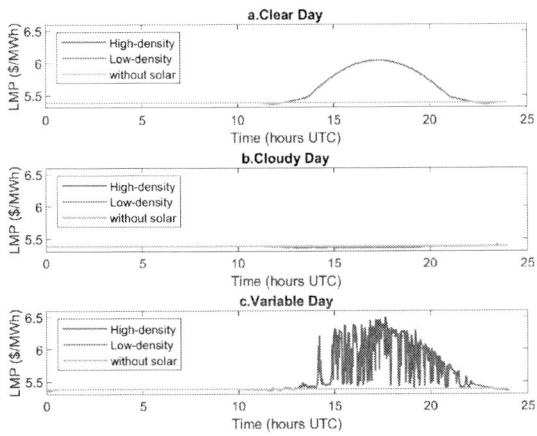

Fig. 11. LMPs at bus 8 when solar system is placed at bus 20 with high-density and low-density in a clear, cloudy and variable day.

LMPs during a variable day. The results also show that the LMP decreases for the buses in which the solar system is added and for the neighbour buses. Such decrease ranges from 3% to 100%. Cases with 100% reduction appear to result from high levels of congestion preventing the export of power. It was observed that low-density solar plant distributions reduce the occurrence of these zero-price conditions by smoothing out the variability in the generation time series.

It was also observed that when a solar system is added to a low demand bus (e.g. bus 20), the LMP decreased for the low demand bus, but it increased for high demand buses (e.g.bus 8). This results from the local surplus solar generation being sent to other buses, which can result in congestion and thus, increase in LMP. This clearly indicates the effects of solar

system location on LMPs. The results from this study can be used for informed planning and decision-making process for solar system installation. Additional work is needed to further investigate the effects of solar variability on LMPs by more comprehensively placing solar system at each bus, considering a more complete set of daily irradiance conditions days and varying solar penetration levels.

REFERENCES

[1] U. Schwabe and P. Jansson, "Utility-interconnected photovoltaic systems reaching grid parity in new jersey," in *IEEE PES General Meeting*, 2010, pp. 1–5.

[2] R. Hemmati and R.-A. Hooshmand, "Impacts of renewable energy resources and energy storage systems on the flow-gate prices under deregulated environment," *Journal of Renewable and Sustainable Energy*, vol. 9, no. 3, p. 035502, 2017. [Online]. Available: https://doi.org/10.1063/1.4984622

[3] M. H. Albadi, Y. M. El-Rayani, E. F. El-Saadany, and H. A. Al-Riyami, "Impact of solar power projects on lmp and transmission losses in oman," *Sustainable Energy Technologies and Assessments*, vol. 27, pp. 141–149, 2018. [Online]. Available: https://www.sciencedirect.com/science/article/pii/S221313881730262X

[4] M. J. Poursalimi Jaghargh and H. R. Mashhadi, "An analytical approach to estimate structural and behavioral impact of renewable energy power plants on lmp," *Renewable Energy*, vol. 163, pp. 1012–1022, 2021. [Online]. Available: https://www.sciencedirect.com/science/article/pii/S0960148120313343

[5] H. Jin, Y. Teng, T. Zhang, Z. Wang, and B. Deng, "A locational marginal price-based partition optimal economic dispatch model of multi-energy systems," *Frontiers in Energy Research*, vol. 9, 2021. [Online]. Available: https://www.frontiersin.org/article/10.3389/fenrg.2021.694983

[6] T. E. Hoff and R. Perez, "Quantifying PV power Output Variability," *Solar Energy*, vol. 84, no. 10, pp. 1782–1793, Oct. 2010. [Online]. Available: http://www.sciencedirect.com/science/article/pii/S0038092X10002380

[7] J. Marcos, L. Marroyo, E. Lorenzo, and M. García, "Smoothing of PV power fluctuations by geographical dispersion," *Progress in Photovoltaics: Research and Applications*, vol. 20, no. 2, pp. 226–237, 2012. [Online]. Available: https://onlinelibrary.wiley.com/doi/abs/10.1002/pip.1127

978-1-7281-6118-1/22 $31.00 © 2022 IEEE

[8] M. Lave, J. Kleissl, and J. S. Stein, "A Wavelet-Based Variability Model (WVM) for Solar PV Power Plants," *IEEE Transactions on Sustainable Energy*, vol. 4, no. 2, pp. 501–509, Apr. 2013.

[9] W. Holmgren, C. Hansen, and M. Mikofski, "pvlib python: a python package for modeling solar energy systems," *Journal of Open Source Software*, vol. 3, no. 29, p. 884, Sep. 2018. [Online]. Available: https://joss.theoj.org/papers/10.21105/joss.00884

[10] J. A. Augustine, J. J. DeLuisi, and C. N. Long, "SURFRAD–A National Surface Radiation Budget Network for Atmospheric Research," *Bulletin of the American Meteorological Society*, vol. 81, no. 10, pp. 2341–2358, Oct. 2000, publisher: American Meteorological Society Section: Bulletin of the American Meteorological Society.

[11] G. M. Lohmann, "Irradiance Variability Quantification and Small-Scale Averaging in Space and Time: A Short Review," *Atmosphere*, vol. 9, no. 7, p. 264, Jul. 2018. [Online]. Available: https://doaj.org

[12] M. Lave, M. J. Reno, and R. J. Broderick, "Characterizing local high-frequency solar variability and its impact to distribution studies," *Solar Energy*, vol. 118, pp. 327–337, 2015.

[13] J. S. Stein, C. W. Hansen, and M. J. Reno, "The Variability Index: A New and Novel Metric for Quantifying Irradiance and PV Output Variability," in *Proceedings of the World Renewable Energy Forum*, Denver, CO, May 2012, pp. 13–17.

[14] R. van Haaren, M. Morjaria, and V. Fthenakis, "Empirical assessment of short-term variability from utility-scale solar PV plants," *Progress in Photovoltaics: Research and Applications*, vol. 22, no. 5, pp. 548–559, 2014, _eprint: https://onlinelibrary.wiley.com/doi/pdf/10.1002/pip.2302. [Online]. Available: https://onlinelibrary.wiley.com/doi/abs/10.1002/pip.2302

[15] R. D. Zimmerman and C. E. M.-S. . M. V. . S. A. https://matpower.org.

[16] R. D. Zimmerman and C. E. Murillo-Sanchez, "Matpower user's manual, version 7.1. 2020."

[17] C.-H. Lo and N. Ansari, "Alleviating solar energy congestion in the distribution grid via smart metering communications," *IEEE Transactions on Parallel and Distributed Systems*, vol. 23, no. 9, pp. 1607–1620, 2012.

[18] O. Alsac and B. Stott, "Optimal load flow with steady-state security," *IEEE Transactions on Power Apparatus and Systems*, vol. PAS-93, no. 3, pp. 745–751, 1974.

[19] R. Ferrero, S. Shahidehpour, and V. Ramesh, "Transaction analysis in deregulated power systems using game theory," *IEEE Transactions on Power Systems*, vol. 12, no. 3, pp. 1340–1347, 1997.

Demystifying the effect of hydrogen treatment on silicon photovoltaics

Govind Nanda, Sara Almenabawy, Rajiv Prinja, Geetu Sharma, Nazir P. Kherani

Department of Electrical & Computer Engineering, University of Toronto, Toronto, ON, Canada

Interactions between hydrogen and silicon play an integral role in determining the quality of surface and bulk passivation of various device structures in silicon photovoltaics. The efficacy of a hydrogen treatment method is known to be dependent on whether the interacting hydrogen species is atomic, ionic, or molecular. Furthermore, these concentrations can be altered by controlling the substrate temperature of the silicon substrate. Moreover, an important consideration is the time and atmosphere the treatment is carried out in, as it influences the desorption of hydrogen from silicon. Hence, it is important to undertake a comprehensive investigation of both the theoretical and experimental aspects of hydrogen passivation of silicon devices, thereby developing a clear understanding of how hydrogen behaves within different solar cell structures, and its dependence on substrate properties. In the theoretical study presented, we assume that the total concentration of hydrogen in the silicon substrate does not remain constant when temperature is increased, and that longer exposures to higher temperatures may cause further loss of hydrogen. This will be augmented by experimental studies of a-Si passivation layers on silicon subjected to hydrogen treatments and follow-on stepwise annealing with resulting loss of hydrogen and examination of its effect on the minority carrier lifetime.

Bill of Materials Variation and Module Degradation in Utility-Scale PV Systems

Michael G. Deceglie, E. Ashley Gaulding, John S. Mangum, Timothy J Silverman, Steve W. Johnston, James A. Rand, Mason J. Reed, Robert Flottemesch, Ingrid L. Repins

National Renewable Energy Laboratory, Golden, CO, United States

Core Energy Works, Newark, DE, United States

Luminace, New York, NY, United States

PV modules of the same make and model are often assembled with different bills of materials (BOMs). We describe two case studies of utility-scale silicon PV systems in which these differing BOMs were associated with faster-than-expected degradation. In one of the sites we found that different metallization paste had been used for grid lines in some cells leading to loss of contact to the cell and severe series resistance degradation. In another site, we found that two different types of cell had been used, one of which suffered from light and elevated temperature induced degradation (LeTID). Our results from both sites underscore that variations in BOM, even among modules of the same make and model can lead to reliability challenges.

The thermodynamics behind the photovoltage generation and photocurrent collection in solar cells

Uwe Rau

Forschungszentrum Jülich, Jülich, Germany

The photogeneration of electron-hole pairs in solar cells leads to a non-equilibrium concentration of those charge carriers generating a non-equilibrium of chemical potentials. This contribution discusses the thermodynamic and kinetic principles behind the transformation of the initial chemical into electrical energy. It is shown that the equation for the photovoltage generation and photocurrent collection across the junction of a solar cell is a special case of the more general Butler-Volmer equation. Additionally, a model for junctions to solar cell absorbers is introduced that accounts for active photovoltaic junctions as well as for electrically passive, but selective contacts in solar cells.

Tuning the band gap of magnesium zinc oxide to enhance band alignment with CdTe based photovoltaic devices.

Kerrie M Morris, Mustafa Togay, Rachael C Greenhalgh, Jake W Bowers, John M Walls

Cadmium telluride (CdTe) based photovoltaic devices are usually coupled with a cadmium sulfide (CdS) emitter layer. Replacing the CdS with a material that has more transparency and a better band alignment with CdTe based devices could improve efficiency. The band gap of magnesium zinc oxide (MZO) has been seen to increase with increasing Mg content and sputtering at different temperatures can vary the Mg content. MZO is more transparent than CdS and the tunable band gap could help to improve band alignment between emitter and absorber layers. MZO films were deposited by radio frequency magnetron sputtering at 250°C, 300°C and 350°C and the optical properties were investigated. Cadmium selenide was then deposited by chemical bath on top of the MZO as a pre-cursor to forming a cadmium selenide telluride (CST) layer. CdSe was also deposited on Tec10 glass for comparison. Increasing the temperature of deposition of MZO from 250°C to 300°C and then 350°C saw an increase in band gap from 3.85 eV to 3.87 eV to 3.97 eV respectively. The transparency and thickness were not affected. The CdSe on MZO films had a variable reduction in transmission with the CdSe on MZO sputtered at 300°C having the largest decrease in transparency between the wavelengths 350 - 650 nm and the CdSe on MZO sputtered at 250°C the smallest decrease when compared to CdSe alone. It was determined that interactions at the interface between CdSe and MZO were responsible for the transmission shift with parasitic absorption occurring.

978-1-7281-6118-1/22 $31.00 © 2022 IEEE

Sensitivity of Sub-Hourly Modeling Error to Project Size

Christopher Hayes, Abhishek Parikh, Mark Mikofski, Rounak Kharait

DNV, Portland, OR, United States

High-frequency measurements of solar resource from the Surface Radiation Budget Network (SURFRAD) from stations in NV, MT, SD, MS, PA, IL and CO were down-sampled from 1-minute to 1-hour and used to predict energy yield and sub-hourly modeling error. A Wavelet Variability Model (WVM) incorporating an estimated solar plant layout was used to determine the sub-hourly modeling error dependency for projects ranging in size from 1 MW to 1,000 MW. Additionally, sensitivity to inverter overbuild, DC to AC ratio, average cloud speed, interannual variability and geographic location were evaluated. By incorporating the WVM to smooth the irradiance inputs we found that annual sub-hourly modeling errors exhibited a nearly logarithmic decrease as project size increased. On average, the modeling error decreases quickly for the first 200 MW and begins to asymptote for 200 - 1,000 MW. The magnitude of annual modeling errors was highly influenced by DC/AC ratio, average cloud speed and the interannual variability of the solar resource. The results of this study were implemented to develop a project size dependent sub-hourly modeling error adjustment factor for pre-construction energy assessments.

978-1-7281-6118-1/22 $31.00 © 2022 IEEE

Energy-based Soiling Loss Monitoring Approach for Solar PV System

Pavan Fuke, Shoubhik De, Narendra Shiradkar and Anil Kottantharayil

Department of Electrical Engineering, Indian Institute of Technology Bombay, Mumbai, Maharashtra 400076, India

Abstract— The energy loss caused by dust and other particles accumulating on photovoltaic (PV) modules has become one of the most important issues for PV system owners. A site-specific and dedicated soiling monitoring system may be necessary to improve yearly energy estimations and optimize cleaning schedules. Commercially available electrical soiling measurement kits compare the short circuit current (I_{SC}) or maximum power (P_{max}) of clean and naturally soiled PV modules for a specific time interval and irradiation level. In this paper, we present an energy based soiling loss monitoring approach. The I_{SC} and P_{max} based methods are benchmarked to the energy-based method for soiling data collected from a test station in Mumbai. It is found that the I_{SC} and P_{max} based methods underestimate the soiling losses by about 2.5-3%.

Keywords— *photovoltaic modules, soiling, energy loss, soiling ratio*

I. INTRODUCTION

The electric yield of solar Photovoltaic (PV) modules are dependent on factors like solar irradiance, temperature, wind speed, precipitation, humidity and dust deposition. Soiling is a commonly overlooked or underestimated external component that might be a deal-breaker for the viability of a solar project. The soiling loss rises linearly in the absence of significant rainfall and regular cleaning. It is projected that the global solar power production loss due to soiling is 3%–4%, which in turn caused 3-4 billion € of revenue losses in 2018 [1]. Various research has been published recently that presents models and methods for monitoring soiling losses [2-5].

Typically, the soiling loss of a PV system is assessed using a pair of PV modules, one of which is kept clean always time, and the other is allowed to soil naturally. The module that is allowed to soil is usually of identical make and model to the ones used for power generation, whereas the clean module can be a reference cell of the same technology or a module similar as the soiled. The output current of the PV module directly depends on light received by the solar cells, which gets reduced by soiling. So, the simplest way to calculate the soiling loss is to compare the soiled and clean modules' I_{SC}. Commercially used soiling monitoring kits such as Atonometrics RDE 300 [3], Campbell Scientific CR-PVS1 Soiling Loss Index RTU [4] and the autonomous soiling monitoring kit developed by Arizona State University [5] compute the soiling losses in terms of I_{SC}. However, dust particles tend to accumulate around the bottom edges of the module due to gravity, precipitation, and wind, resulting in non-uniform soiling. Soiling of PV panels may show significant non-uniformity within a panel, from panel to panel in a string and from string to string within a large powerplant. In the case of non-uniform soling, I_{SC} may not be a good predictor of soiling loss. As a result, some soiling monitoring equipment measures soiling loss in terms of power loss. The Atonometrics RDE 300 can also measure soiling losses in terms of power.

This study proposes an energy-based approach to analyze the soiling losses. The proposed method is applied to analyze the 2-year electrical data obtained from a rooftop PV installation in Mumbai, India. It is shown that the I_{SC} and P_{max} based soiling monitoring methods underestimate the energy yield loss due to soiling by 2.5 to 3%.

II. METHODOLOGY

To quantify the soiling loss for Mumbai, India (19.1334° N, 72.9133° E), several field experiments were undertaken. Based on four years of field data, the soiling rate in Mumbai was estimated as 0.43 percent per day during dry seasons [6]. This study analyzed the electrical data of two 327 W_P latitude-mounted PV modules installed at Photovoltaic Module Monitoring Station (PVMMS), IIT Bombay, Mumbai, as shown in Fig. 1. The bottom module was cleaned every day, while the top module was permitted to be naturally soiled, as shown in Fig. 2. The modules were connected to a multichannel I–V curve tracer (Daystar MT5 3200). I_{SC}, open circuit voltage, and P_{max} are measured with 1 minute internal and the complete I–V curves are measured once in every 10 minutes.

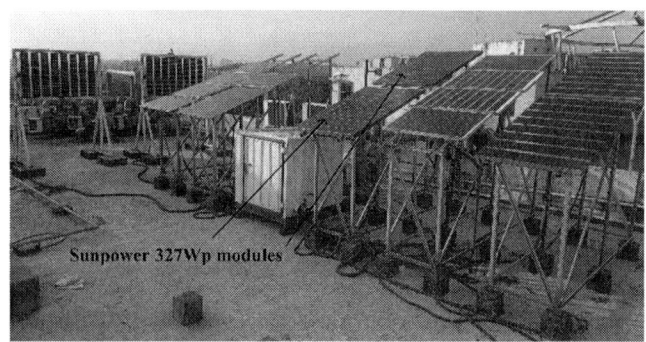

Fig. 1. PVMMS installed at IIT Bombay, Mumbai

Fig. 2. Clean (bottom) and soiled (top) modules

Gostein et. al. [2] measured the soiling ratio in terms of both I_{SC} and P_{max} using Atonometrics RDE 300 setup. The soiling ratio measurement using this method is represented by M1. Campbell Scientific CR-PVS1 Soiling Loss Index RTU and ASU autonomous soiling monitoring kit equipment are configured to measure soiling ratio using I_{SC} [4]-[5]. These soiling monitoring methods are represented by M2 and M3 respectively. Each of these methods analyzed the data only for specific time intervals and for specific solar irradiation range. Table I shows the time and irradiance filters for M1, M2 and M3.

The soiling ratios in terms of current, power, and energy are calculated using (1), (2), and (3). It may noted that measured for full day is used for the proposed energy-based method, while M1 to M3 use data collected over shorter time intervals. The soiling losses can then be calculated using the soiling ratio as given in (4). The soiling ratio difference between the proposed energy-based method and I_{SC} and P_{max} based methods are computed using SR_E as a reference for the comparative study.

$$\text{Soiling ratio in terms of current} = SR_{I_{sc}} = \frac{I_{sc\_soiled}}{I_{sc\_clean}} \quad (1)$$

$$\text{Soiling ratio in terms of power} = SR_{P_{max}} = \frac{P_{max\_soiled}}{P_{max\_clean}} \quad (2)$$

Soiling ratio in terms of energy $= SR_E$

$$= \frac{Integral\ of\ power\ generated\ by\ soiled\ module\ for\ the\ day}{Integral\ of\ power\ generated\ by\ clean\ module\ for\ the\ day} \quad (3)$$

$$\text{Soiling loss} = 1 - \text{Soiling ratio} \quad (4)$$

TABLE I. SPECIFICATIONS OF DIFFERENT SOILING MONITORING METHODS

Soiling ratio monitoring methods	Specifications		
	Soiling ratio calculation approach	*Time filter*	*Solar irradiation filter*
M1	I_{SC} and P_m	10 AM to 2 PM	Greater than 500 W/m²
M2	I_{SC}	11 AM to 1 PM	Greater than 500 W/m²
M3	I_{SC}	2 minutes at solar noon	-

III. RESULTS AND DISCUSSIONS

Fig. 3 shows the soiling ratios obtained using the 4 methods described. It is clear that if the soiled module is not cleaned, the soiling ratio drops to 0.5 or smaller during the dry seasons in Mumbai, implying more than 50% soiling losses. After heavy precipitation, the soiling ratio recovers to 1 as the module surfaces are naturally cleaned by rain. In the rainy season, the soiling ratio hovers around 1 and the data is generally noisy, more so for the I_{SC} and P_{max}. The energy-based method gives the least noisy data as it is based on the data for the entire day. After analyzing the soiling ratio by considering different approaches, it is found that the soiling ratio obtained using M3 is most noisy. This is because, it calculates the soiling ratio by considering only 2-minute intervals at solar noon, which may often lead to inaccurate measurements. The power-based and current-based soiling ratios obtained as per M1 gives similar results. The energy-based approach showed lower soiling ratio (higher losses due to soiling) values in the summer and winter seasons.

Fig. 4 depicts the differences in different soiling ratio monitoring approaches, using the energy-based soiling ratio as a reference. The current and power-based soiling ratio using M1 shows similar trends, therefore only the power-based soiling ratio is considered for soiling ratio difference calculation. During the summer and winter seasons, it is found that the current and power-based approaches overestimate the soiling ratio, or in other words, they underestimate the soiling losses.

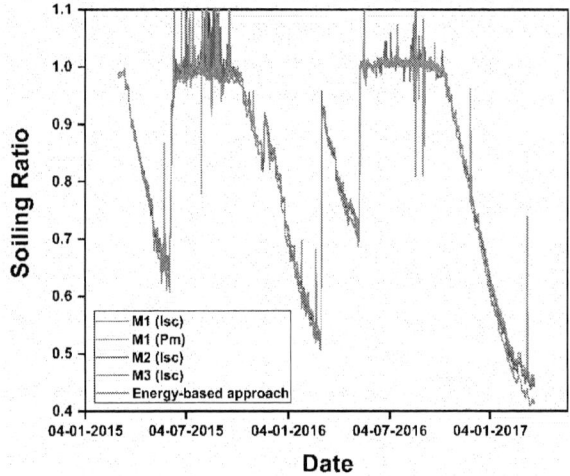

Fig. 3. Soiling ratios obtained using different soiling monitoring methods.

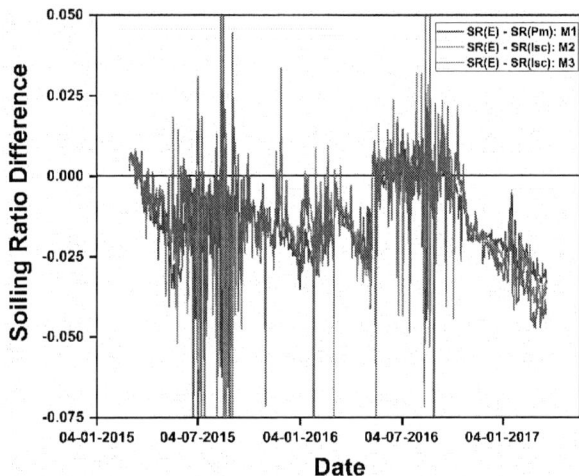

Fig. 4. Soiling ratio differences by considering energy-based soiling ratio as reference.

IV. CONCLUSION

We propose an energy-based soiling loss monitoring approach and applied it for analyzing 2 years' of data collected in Mumbai, India. It is found that soiling losses obtained in terms of I_{SC} and P_{max} are lower than the soiling loss in terms of energy by an average of 2.5-3%. So, the current and power-based soiling loss monitoring approach underestimates the soiling losses.

REFERENCES

[1] K. Ilse, L. Micheli, B. W. Figgis, K. Lange, D. Daßler, H. Hanifi, F. Wolfertstetter, V. Naumann, C. Hagendorf, R. Gottschalg, and J. Bagdahn, "Techno-economic assessment of soiling losses and mitigation strategies for solar power generation", Joule, vol. 3, no. 10, pp. 2303-2321, 2019.

[2] M. Gostein, T. Düster, and C. Thuman, "Accurately measuring PV soiling losses with soiling station employing module power measurements", In 2015 IEEE 42nd Photovoltaic Specialist Conference (PVSC), pp. 1-4, 2015.

[3] Atonometrics, "Soiling Measurement System for PV Power Plants." [Online]. Available: http://www.atonometrics.com/products/soiling-measurement-system-for-pv-modules/. [Accessed: 02- Jan-2018].

[4] Campbell Scientific, "SMP100 Solar Module Performance Monitoring System." [Online]. Available: https://www.campbellsci.com/smp100. [Accessed: 02- Jan-2018].

[5] T. Curtis, S. Tatapudi, and G. TamizhMani, "Design and operation of a waterless PV soiling monitoring station", In 2018 IEEE 7th World Conference on Photovoltaic Energy Conversion (WCPEC)(A Joint Conference of 45th IEEE PVSC, 28th PVSEC & 34th EU PVSEC), pp. 3407-3412, 2018.

[6] S. Warade, and A. Kottantharayil, "Analysis of soiling losses for different cleaning cycles", In 2018 IEEE 7th World Conference on Photovoltaic Energy Conversion (WCPEC)(A Joint Conference of 45th IEEE PVSC, 28th PVSEC & 34th EU PVSEC), pp. 3644-3647, 2018.

Predicting Materials Parameters in Colloidal Quantum Dot Photovoltaic Devices Using Machine Learning Models Trained On Experimental Data

Hoon Jeong Lee, Ariana B. Hofelmann, Yida Lin, and Susanna M. Thon

Department of Electrical and Computer Engineering, Johns Hopkins University, 3400 N. Charles Street,
Baltimore, Maryland, 21218, USA

Abstract—**Numerous characterization techniques have been developed over the last century, which have advanced progress on the development of a variety of photovoltaic technologies. However, this multitude of techniques leads to increasing experimental costs and complexity. It would be useful to have an approach that does not require the time commitment or operation costs to directly learn and implement every new measurement technique. Herein, we explore several machine learning (ML) models that output complex materials parameters, such as electronic trap state density, solely using illuminated current-voltage curves. This greatly reduces both the complexity and cost of the characterization process. Current-voltage curves were chosen as the only input to our models because this type of measurement is relatively simple to perform and most photovoltaic research labs already collect this information on all devices. We compare several different ML network architectures, all of which are trained on experimental data from PbS colloidal quantum dot thin film solar cells. We predict values for underlying materials parameters and compare them to experimentally measured results.**

Index Terms—**thin film, lead sulfide, machine learning, colloidal quantum dots**

I. INTRODUCTION

The field of photovoltaics has taken advantage of numerous experimental techniques to measure the critical underlying materials parameters which determine solar cell device performance. Techniques such as charge extraction by linearly increasing voltage (CELIV), time of flight (TOF) measurements, photocurrent transient spectroscopy, and space-charge-limited current (SCLC) measurements are used to measure charge mobility while other techniques such as the transient photovoltage method and deep level transient spectroscopy (DLTS) can be used to determine mid-gap trap state densities [1]. Determining these parameters is essential in the development cycle of solar cells as they allow researchers to compare different devices and fabrication methods, as well as identify limits to and improve device performance. However, these techniques often require specialized device architectures and an experienced researcher to determine the best underlying analytical or numerical model to fit to the data. This can not only be time consuming, but complex

This research was funded by the National Science Foundation (DMR-1807342).

as well. In addition, large upfront costs for equipment and apparatuses make some measurement techniques inaccessible to smaller labs. This sometimes leads to labs needing to choose between measuring one materials parameter versus another. One possible alternative to conventional methods is to leverage machine learning (ML) to obtain these materials parameters by taking advantage of existing datasets and correlating them with new, simpler measurements.

Herein, we test several machine learning models to aid in the solar cell development process. The key difference between our models and others found in the literature is that our algorithms are trained on experimental data rather than simulation data. The latter is usually preferred due to the high cost and complexity of fabricating numerous devices; however, data extracted from simulations is not as reliable, accurate, or comprehensive as data collected from real devices. This is due to errors such as the simulation of thermodynamically unstable, physically impossible, or idealized structures [2]. We leveraged our past work on development of a multi-modal optoelectronic scanning instrument for solution-processed solar cells [3] to generate massive training data sets on colloidal quantum dot photovoltaic devices that can be used to bolster and diversify existing experimental and computationally generated datasets.

In general, our models utilize supervised machine learning methods to predict materials parameters. The goal is to train an artificial neural network to predict an output vector $\mathbf{t} \equiv [t_1, t_2, \cdots, t_N]$ from a particular input vector $\mathbf{x} \equiv [x_1, x_2, \cdots, x_N]$. In the general case, these vectors could be of different dimensions. We train the model by providing numerous examples of input-output pairs, so that it can infer the underlying relationship between the two variables via back-propagation and the gradient descent method. Since this method uses statistical correlations instead of physical laws [4], it eliminates the need for complex user analysis and the need to encode data presumptions. There are several feed-forward network architectures that can be used to solve these classes of regression problems: the multilayer perceptron, autoencoders, and the convolutional variants of these networks. With new data being generated at an exponential rate, ML offers a time- and memory-efficient way of analyzing large

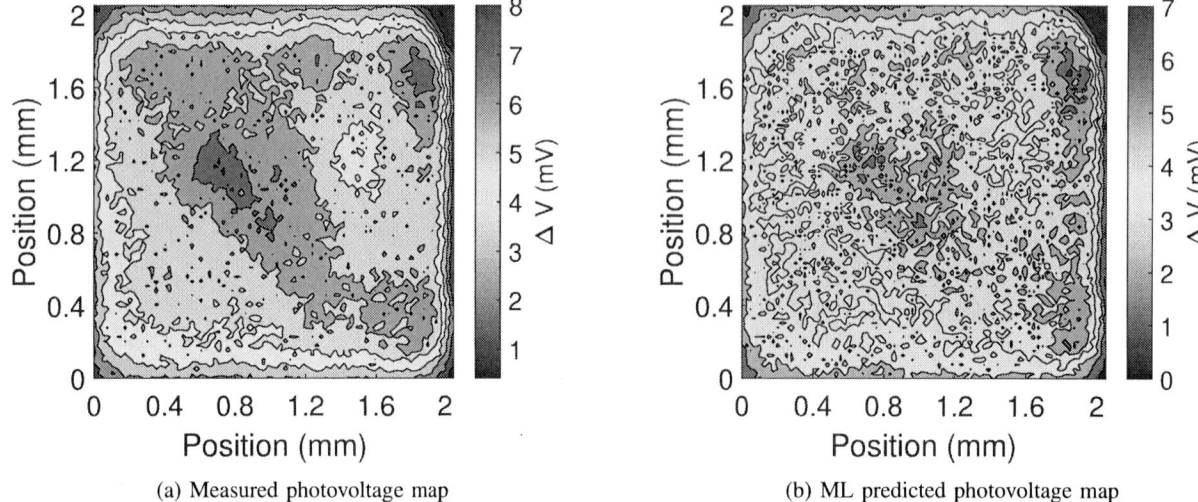

(a) Measured photovoltage map   (b) ML predicted photovoltage map

Fig. 1: a). Experimentally measured transient photovoltage map for a CQD solar cell. This device was used to validate the machine learning models. b). Predicted transient photovoltage map for the same cell by a convolutional autoencoder.

datasets and a promising alternative to traditional approaches for determining materials parameters in complex materials such as colloidal quantum dot thin films that may not behave like traditional semiconductors in all aspects.

In the following sections, we introduce our device structure and experimental setup. We explain in detail the materials parameters studied and how we measure them. Next, we discuss the inner workings of several ML models, compare their efficacy, and report their hyperparameters. We also discuss several ways in which model accuracy can be improved. Lastly, the predicted materials parameters from our models are compared to experimentally measured materials parameters.

II. Methods

A. Experimental Setup

We took measurements from several colloidal quantum dot (CQD) thin film solar cells. All the devices share the same structure: a glass substrate, a transparent electrode (fluorine-doped tin oxide), a n-type zinc oxide electron-extraction layer, a bulk absorbing layer made up of a PbS CQD film with PbX_2 (X=Br, I) ligands, a p-type hole extraction layer (PbS CQD thin film with ethanedithiol ligands), and lastly a top evaporated gold contact [5]. The absorbing layer of our devices is around $500\,\text{nm}$ thick.

A custom optoelectronic scanning setup was used to collect data on all the devices [3]. The sample is mounted on an XYZ translation stage, which allows us to create spatially-resolved materials parameter maps. In contrast to single-point measurements, this system allows us to resolve macroscopic physical phenomena such as defect regions and film inhomogeneities. We collected illuminated current-voltage curves using a Keithley 2400 Source Measurement Unit, and we used an Ocean Optics NIRQuest512 spectrometer to collect photoluminescence (PL) data. The entire system is automated

to produce parameter maps that are correlated in both space and time.

To determine the electronic trap state density n in our photovoltaic CQD thin films, we utilize the iterative transient photovoltage method [6]. A Thorlabs MCWHL5 White light emitting diode (LED) was used to provide the steady state background illumination, and a Thorlabs L520P50 Laser Diode ($\lambda = 520\,\text{nm}$) was used as a perturbation source. The pulsed laser generates excess electrons which recombine shortly after each pulse. We can calculate the excess charges generated (ΔQ) by integrating the photocurrent transient while the device is under short circuit conditions:

$$\Delta Q = \int I(t)dt$$

The photovoltage transient signal can be modeled to fit a mono-exponential of the following form:

$$\Delta V_{oc}(t) = \Delta V_{oc}(0) \exp\left(\frac{-t}{\tau_s}\right)$$

where $\Delta V_{oc}(0)$ is the maximum change in the open circuit voltage caused by the perturbation source and τ_s is the small signal lifetime [7]. We perform these measurements at different light biases corresponding to different open circuit voltages. Afterwards, a differential capacitance can be calculated using the following:

$$C = \frac{\Delta Q}{\Delta V_{oc}(0)}$$

Integrating this capacitance up to a particular V_{oc} will give us an estimation of the midgap trap state density n:

$$n = \frac{1}{Aed} \int_0^{V_{oc}} C\,dV \qquad (1)$$

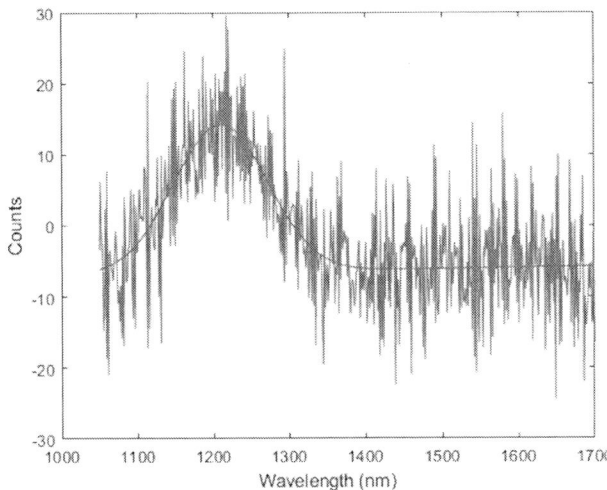

Fig. 2: Photoluminescence plot for a single point in the scan fit to a Gaussian curve.

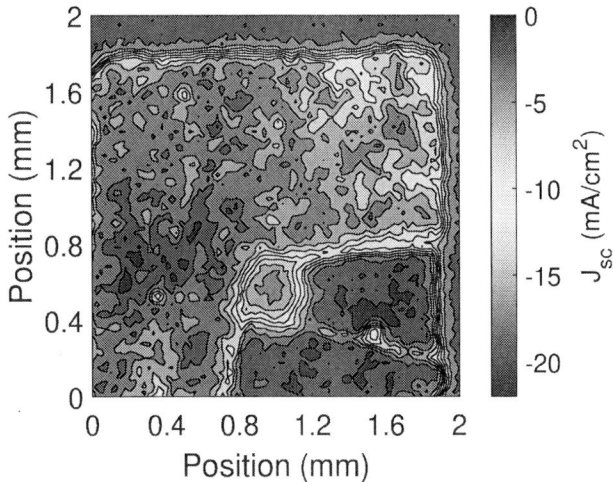

Fig. 3: Short circuit current density map of one of the CQD solar cell devices used to train the neural networks with a visible defect near the center.

where A is the area of the device, e is the electronic charge, and d is the thickness of the active layer. The spot size of the combined laser/LED beam was measured to be approximately $50\,\mu m$, but the translation stage has the capability of stepping in increments as small as $10\,\mu m$. A trade-off is made between the resolution of the transient signals and the diameter of the beam spot. Smaller spot sizes allow us to study devices in greater detail, but this may lead to longer acquisition times and lower signal-to-noise ratios.

Lastly, the collected photoluminescence data was fit to a Gaussian curve of the following form.

$$G = a \exp\left(\frac{-(x-b)}{c}\right)$$

where a, b, and c are constants. Figure 2 shows an example photoluminescence plot, from which we can obtain parameters such as the peak wavelength (λ_{peak}), peak intensity, and full width at half maximum (FWHM) of the intensity.

B. Neural Networks

We predicted materials parameters using a simple multilayer perceptron, an autoencoder (AE), and convolutional variants of each [8]. The goal of autoencoders is to learn the identity function, i.e. the desired output of the network is the input [9]. By itself, that is not useful, but if we add a bottleneck to the network, meaning if we add a layer z that has dimensions smaller than the input layer, then the autoencoder will learn a way to map the input data onto this lower dimensional space [2]. It is in this bottleneck layer that we obtain our materials parameters. We can achieve this by adding an additional term to the cost function J of the autoencoder [10]:

$$J = \frac{1}{N}\sum_{i=1}^{N}(x_i - t_i)^2 + \sum_{j=1}^{M}(k_j - z_j)^2 \qquad (2)$$

where k is the desired materials parameter that we measure and train the network with. Additional terms may be added for more materials parameters. Note that there are two parts to an autoencoder: the encoder and the decoder. The encoder performs dimensionality reduction and also learns the relationship between the input space and latent space. The decoder can be used alongside a Gaussian noise generator to create a generative adversarial network (GAN). This is one possible method of using ML to generate datasets for future use.

III. DISCUSSION/RESULTS

Four different networks were used in our study: a simple multilayer perceptron (MLP) with six hidden layers, one convolutional MLP network, one autoencoder (AE) network, and lastly a convolutional AE network [8]. All models were trained for 20 epochs with a learning rate of $\eta = 2.5 \times 10^{-3}$. The Rectified Linear Unit (ReLU) activation function was used between each layer. The input to each model was a 28×2 vector which is the illuminated current-voltage curve. For preprocessing, the input current-voltage curve and materials parameters were scaled to the range of 0 to 1 and zero-centered. We scale the data to make the problem bounded and because of the properties of the sigmoid and ReLU activation functions [11]. The data is zero centered in order to help convergence to a solution during gradient descent. In a typical solar cell device, the current is negative from zero volts all the way to the open-circuit voltage. This is undesirable because then the gradients will all have the same sign and be limited to either the negative or positive direction during training [11].

The MLP models are based on a network with six layers that have 200, 150, 150, 50, 50, and 50 neurons respectively [1]. Both convolutional MLP and AE models used a filter of size 7×2 with stride length of 1. AE networks had a hidden layer of size 1×1, which is equal to the materials parameter being

TABLE I: Mean Squared Errors for Predicted Materials Parameters from Different ML Networks

	MLP	Conv. MLP	AE	Conv. AE
Trap State Density	0.065	0.063	0.058	0.057
Peak PL Wavelength	0.026	0.022	0.019	0.018
Transient Photocurrent Decay	0.071	0.069	0.062	0.060

TABLE II: Mean Squared Errors While Predicting Multiple Materials Parameters

No. of Variables	1	2	3
Trap State Density	0.057	0.054	0.053
Peak PL Wavelength	-	0.017	0.017
Transient Photocurrent Decay	-	-	0.073

trained. We plotted these various hyperparameters against the mean squared error (MSE) for several materials parameters, and picked the values that minimized MSE across all variables.

To test our networks, we used data maps from four different devices. In total, we measured 18,124 unique points on the devices within the maps. Three of the devices were used for training, and the remaining device was used for validation. An image of one of the training devices is given in Figure 3, and an image of the validation device is given in Figure 1. Table I summarizes the results of the various ML models. Overall, we find that convolutional networks out perform their non-convolutional counterparts.

We took the best performing network (convolutional AE) and added additional hidden neurons to the bottleneck layer. The results are tabulated in Table II. Surprisingly, MSE decreased for both the peak PL wavelength as well as the electronic trap state density as we increased the number of latent variables. This result is counter-intuitive because the form of the cost function given in Equation 2 includes a trade-off between the optimization of different parameters.

From this fact, we point out that ML models can only find dependencies if they are actually provided in the dataset. For example, the bandgap of some semiconductors is a function of both interatomic spacing and temperature. If we train a model to find the bandgap, but only provide interatomic spacing data, then we would be worse off than if we combined both spacing and temperature data. We find that the increase or decrease of MSE can be used as a proxy to determine which variables are physically correlated and not just statistically correlated.

Lastly, qualitative results from our model training is shown in Figure 1. We plot both the measured and predicted values of $\Delta V_{oc}(0)$. There is strong agreement between the two values, and because the values are of the same order of magnitude,

we conclude that the system was able to properly learn the mapping function for this particular device. These preliminary results demonstrate that this method holds promise for simplifying photovoltaic materials parameter measurements.

IV. CONCLUSION

We demonstrated several simple machine learning methods to approximate key materials parameters in PbS CQD solar cells. These models not only enable faster device optimization, but also shed insight on the underlying physics and relationships between materials parameters. Compared to conventional methods, ML models are time- and cost-effective. This work is not only applicable to photovoltaic devices, but could be extended to other types of optoelectronic devices such as photodetectors and light emitting diodes. Future work will incorporate unsupervised machine learning methods (e.g. self-organizing maps) to automatically characterize different regions of devices. In addition, we plan to build a GAN in conjunction with the decoder of the AE model to allow for the creation of large and physically-motivated datasets. Because our training data is spatially resolved, we will be able to simulate non-uniform devices with features and defects such as spin-casting streaks and hairline cracks and predict their effects on device performance. We plan to eventually extend this work to other photovoltaic technologies, and encourage the field to make experimentally correlated data publicly available. This work paves the way for simplifying measurements in photovoltaics and could lead to a faster development cycle for new solar cell technologies.

REFERENCES

[1] N. Majeed, M. Saladina, M. Krompiec, S. Greedy, C. Deibel, and R. C. MacKenzie, "Using deep machine learning to understand the physical performance bottlenecks in novel thin-film solar cells," *Advanced Functional Materials*, vol. 30, no. 7, p. 1907259, 2020.

[2] J. Li, K. Lim, H. Yang, Z. Ren, S. Raghavan, P.-Y. Chen, T. Buonassisi, and X. Wang, "Ai applications through the whole life cycle of material discovery," *Matter*, vol. 3, no. 2, pp. 393–432, 2020.

[3] Y. Lin, T. Gao, X. Pan, M. Kamenetska, and S. M. Thon, "Local defects in colloidal quantum dot thin films measured via spatially resolved multimodal optoelectronic spectroscopy," *Advanced Materials*, vol. 32, no. 11, p. 1906602, 2020.

[4] F. Häse, L. M. Roch, P. Friederich, and A. Aspuru-Guzik, "Designing and understanding light-harvesting devices with machine learning," *Nature Communications*, vol. 11, no. 1, pp. 1–11, 2020.

[5] A. Chiu, C. Bambini, E. Rong, Y. Lin, and S. M. Thon, "New hole transport materials via stoichiometry-tuning for colloidal quantum dot photovoltaics," in *2020 47th IEEE Photovoltaic Specialists Conference (PVSC)*, pp. 1096–1097, IEEE, 2020.

[6] C. Shuttle, B. O'Regan, A. Ballantyne, J. Nelson, D. D. Bradley, J. De Mello, and J. Durrant, "Experimental determination of the rate law for charge carrier decay in a polythiophene: Fullerene solar cell," *Applied Physics Letters*, vol. 92, no. 9, p. 80, 2008.

[7] D. Abou-Ras, T. Kirchartz, and U. Rau, *Advanced characterization techniques for thin film solar cells.* John Wiley & Sons, 2016.

[8] C. M. Bishop and N. M. Nasrabadi, *Pattern recognition and machine learning*, vol. 4. Springer, 2006.

[9] K. P. Murphy, *Probabilistic machine learning: an introduction.* MIT press, 2022.

[10] Z. Ren, F. Oviedo, H. Xue, M. Thway, K. Zhang, N. Li, J. D. Perea, M. Layurova, Y. Wang, S. Tian, *et al.*, "Physics-guided characterization and optimization of solar cells using surrogate machine learning model," in *2019 IEEE 46th Photovoltaic Specialists Conference (PVSC)*, pp. 3054–3058, IEEE, 2019.

[11] N. Buduma and N. Locascio, "Fundamentals of deep learning: Designing next-generation machine intelligence algorithms."

Development of HVPE-Grown III-V Solar Cells Passivated with AlInP

Jacob T Boyer, Kevin L Schulte, Aaron J Ptak, John Simon

National Renewable Energy Lab, Golden, CO, United States

We present the first AlInP-passivated solar cells grown by dynamic hydride vapor phase epitaxy (D-HVPE). D-HVPE has potential to reduce the costs of III-V solar cell production, but historical challenges with the growth of high-quality Al-containing compounds placed a ceiling on photoconversion efficiencies of D-HVPE-grown solar cells. Our single junction (1J) GaAs and GaInP/GaAs (2J) tandems with AlInP passivation achieve AM1.5G efficiencies of 26.0% and 28.5%, respectively, which are the highest reported efficiencies for HVPE-grown devices of each type. 1J devices passivated with either AlInP or GaInP have an open circuit voltage of 1.06 V and similar long wavelength current collection, indicating that both windows provide a similar degree of passivation. Adding AlInP passivation to the 2J solar cell improves the VOC by ~50 mV relative to the device passivated by a GaInP emitter. AlInP windows enable short circuit current densities of >29 mA/cm2 and >14 mA/cm2 for the 1J cell and 2J GaInP top cell, respectively. These achievements remove one of the last barriers limiting parity of HVPE device efficiencies with state-of-the-art.

978-1-7281-6118-1/22 $31.00 © 2022 IEEE

Progress in PV Material Durability Test Methodologies

William J. Gambogi

Wilmington, Delaware, USA

Abstract—Accelerated test methods to assess the durability of photovoltaic modules is a crucial area of research in the PV technical community. The seminal work in this area was conducted under a US DOE program in the 1980s led by researchers at the Jet Propulsion Laboratory. Many of the industrial standards for testing PV module reliability were based on this work. As the installed PV base has grown exponentially, new materials and module designs have been introduced into the industry at an accelerated pace at the same time as price pressures have pushed for lower cost materials and designs. These cost pressures led to some notable early product field failures that were not caught by existing test standards. We will present results obtained with global PV reliability research groups on advanced targeted module and materials accelerated testing. Using field experience and assessment, we will present insights into failure mechanisms and comparison to accelerated testing results. In this paper, we will review the recent progress and direction in accelerated testing of PV materials and modules.

Keywords—materials, modules, reliability, durability, accelerated testing, field performance

I. BACKGROUND

Photovoltaic modules are unique in the electronics industry for several reasons. The expected lifetime of PV modules has grown in recent years as the demand for higher outdoor performance has grown. Warranty lifetimes have been extended to support the significant financial investment in this technology. Typically, commercial PV modules are expected to perform at a high level for 25 to 30 years and discussion of the technology needed for 50 years performance in being aggressively pursued. However, there is a significant difference between a warranty and a technical assessment of long-term performance. Warranties relate to a financial commitment on the part of the producer. Prediction of module lifetime requires a detailed understanding the of inherent degradation mechanisms and accelerated test methods that can accurately reproduce these degradation mechanisms and assess their expected outdoor performance. For commercial PV modules, this effort is further complicated by the need to have several variations in the bill of materials (BOM), so that different materials can be used in the fabrication of a given module design. Also complicating the analysis is the cost and resources needed to test a statistically significant number of PV modules. From a practical position, most PV module makers test a small number of modules against several accelerated test methods, typically those specified by industrial standards. At issue is that none of these industrial standards are predictive of product lifetime and use a threshold limit to identify materials and designs that might have a problem in the early product lifetime. PV materials testing, while also not necessarily predictive of product lifetime, has the advantage of increasing the number of test samples to a level that can be statistically significant and allow investigation of the activation energy for degradation mechanisms.

II. EXPERIMENTAL RESULTS

A. Accelerated testing and recent materials failures

The original research work by JPL on module materials and design resulted in a recommended module construction developed after testing of several module design builds under accelerated testing and outdoor exposure. The recommended structure included a glass front and a Tedlar® PVF-based backsheet rear side with EVA encapsulants used to either side of the soldered silicon cell strings. A frame was used to provide mechanical stiffness to the structure with silicone sealants at the frame and junction box. This basic design was used and improved on over many generations in the early years of the PV industry.

With this design as a starting point, new materials were developed to provide higher output power at a lower cost. Much of the early focus was applied to the encapsulant but, since ethyl vinyl acetate is a low cost encapsulant, it has continued to be used and improved and EVA is still that dominant encapsulant type. Improvements have centered on the stability of EVA and focused on improved adhesion, reduced yellowing, stable UV absorption and higher transmission. Reduced acetic acid production was also desired as early silicon cells showed high loss under damp heat (85C, 85%RH) exposure.

New backsheet materials and constructions have also been pursued. Early backsheet designs used TPT (Tedlar®/PET/Tedlar®) structures for UV protection. On the front and read side of the core PET layer. Single fluoropolymer designs (TPE) with Tedlar® only on the rear side were developed to reduce cost with the understanding that the inner layer could be protected by a non-crosslinked pigmented EVA inner backsheet layer and properly selected UV absorbing EVA encapsulant. Further changes to the backsheet design led to new materials including coextruded polyamide-based backsheets (PA). These PA backsheets were seen as having cost advantages and manufacturing simplicity as the three-layer structure was coextruded with expected improvements in mechanical integrity of the structure. While cost improvements were achieved, the mechanical properties were found to be poor under accelerated testing and cracking of backsheets in the field were observed in less than 5 years exposure.[2, 3] While these failures were a significant warranty issue for the industry, they did highlight the need for more improvements to the backsheet test methodology including the need for multistress testing where the impact of the loss in mechanical properties of the backsheet could be assessed. It was also found that sub-structures or minimodules were needed to accurately assess susceptibility to these losses in mechanical properties. Test methods were developed that included temperature, humidity, UV and thermal cycling to assess durability were developed. These methods were found to be predictive of the PA backsheet problem and identified a similar susceptibility in another fluoropolymer film (polyvinylidene fluoride (PVDF)) which has also shown field failures in the 5 to 10 year outdoor exposure timing.

The recent emergence of bifacial module technology has driven the development of bifacial PV module designs using glass/glass construction and the development of glass/backsheet constructions with transparent backsheets. The accelerated stress conditions developed under monofacial modules has been extended and modified for the specific needs of bifacial applications.

Another area of research in improved durability through materials improvements has been metallization paste formulation. Early pastes were sensitive to moisture ingress and corrosion leading to interactions with encapsulant decomposition by-products and leading to poor electrical contact resistance. Recent improvements have led to more stable performance and expected longer product lifetimes,

III. RESULTS

The seminal work at JPL led to a continuing interest in the PV module durability and PV materials durability testing by improved accelerated test sequences. Early work included extending the IEC standard test methods to multiple cycles of te established stress conditions (damp heat (DH), thermal cycling (TC), and humidity freeze (HF). While these test methods were informative on the relative stability of materials and improved the likelihood of passing IEC qualification testing, they were not as helpful in understanding long term performance. As a result, research programs were established in many PV research labs in an international effort to improve industry standards and better understand degradation mechanisms related to materials properties. NREL has been a pioneer in this area and has established many programs on the subject including an international effort to develop improved testing method under the umbrella of PVQAT (The International PV Quality Assurance Task Force) starting in 2011.[4] Under this program, the focus has been to conduct "scientific studies to assess causes of PV failures in the field and determines how best to conduct quick, inexpensive, and accurate screenings". These efforts have led to improved industrial standards through the efforts and contributions of independent government labs and industrial researchers. Examples of these efforts related to materials durability testing include the work of Miller et al. on optical

durability of encapsulants supporting the development IEC standards on extended materials testing.[5] Another PVQAT effort developed by Kempe at NREL looks to a simple coupon test with sequential stress conditions to assess the susceptibility of frontsheet and backsheet materials to cracking.[6]

Accelerated science and module materials are also important workstreams within the DuraMAT program organized and funded by the US Department of Energy.[7] This program is driven by three core national laboratories (National Renewable Energy Laboratory (NREL), Sandia National Laboratories (SNL) and Berkley Lab) and focusses on precompetitive research needs in module packaging. The goals for FY22-27 include materials and module designs for high-energy 50-year module and looks to disruptive acceleration science efforts to understand degradation mechanisms and predict product lifetime. Efforts to better understand the effects of multiple stressors in the field have led to the development of combined stress accelerated testing including the Combined Accelerated Stress Testing (CAST) developed within this program. [8].

Another effort in materials durability testing is conducted under an industrial consortium and led by NIST. It focusses on fundamental studies of materials durability of PV component materials including frontsheet and backsheets by tracking the chemical and physical properties under multiple stress conditions with a focus on service life prediction of materials.[9]

Research efforts at Case Western Reserve University SLDE Center focus on degradation mechanisms and rate to enable the design of better, long-lasting materials and accelerated more accurate accelerated testing protocols. Recent focus has included the extension of PV lifetimes beyond 50 years and the materials, test methods and module structures needed to achieve this goal.[10]

Members of the International Energy Agency Photovoltaic Power Systems Programme (IEA PVPS) have also been very active in the development and testing of new materials aimed at lowering cost and increasing performance.[11]

Another major effort in materials durability and new materials durability has been conducted at the Polymer Competence Center Leoben (PCCL) where different backsheet and encapsulant compositions and structures have been studied including failure mechanisms for PA-based co-extruded backsheets that failed in the field and newly developed co-extruded PO-based backsheets. [12] Considerable efforts have also been made at PCCL to understand the interactions between backsheets and encapsulants including a broad class of newly developed encapsulants for PV applications.[13]

The sensitivity of metallization to corrosion from encapsulant interactions is also a key degradation mechanism of interest and considerable work on this subject has been conducted by AIST.[14]. This degradation mechanism has been associated with acetic acid formation inn EVA as it degrades in the field and understanding its dependence on EVA composition and relevance for alternative encapsulants will be a key area qof interest going forward.

In the AP region, other research organizations including SERIS, Sun Yat-sen University, and others frequently work directly with module manufacturers on materials research programs and module makers and materials suppliers frequently conduct their own research which is reported at the SNEC conference and elsewhere.

IV. DISCUSSION

The evolution of PV materials and constructions and the need for longer product lifetimes has led to increased attention and efforts in materials reliability test methodology. Starting with the work at JPL, the international PV community has created several collaborative efforts to better understand degradation mechanism and the role of test methods to more accurately assess the impact of the multiple stress conditions encountered in the PV field. This reliability effort has built on materials, component and module test methods which also evaluate the synergistic effects consistent with outdoor exposure. Temperature effects including extended hot and cold exposure, localized heating, thermal cycling, direct and indirect UV/visible light exposure, thermal expansion and contraction, humidity, condensation, erosion, soiling, mechanical vibration, hail impact, contact corrosion and electromigration internal electric fields are studied individually and together under conditions relevant to the field. Recent changes to the PV module design including glass/glass module structures, transparent backsheets used in bifacial modules and new, less expensive materials have received particular attention lately. The development of sequential and simultaneous stress testing has been employed to better simulate and assess expected real world conditions. Recent changes in module design including larger module format, higher power PV cells, bifacial modules and new cell architectures have introduced new reliability challenges which must be addressed by improved test methodologies.

SUMMARY

Improvements to accelerated test methods applied to PV materials and module designs has led to a step change in the expected outdoor performance.[15] Efforts across the globe to better understand degradation mechanisms should result in a physics-of-failure approach and a more predictive model of PV outdoor performance and durability. The results of extended materials and module testing needs to be incorporated into more effective industrial standards to insure long term performance of candidate materials and module designs.[16]

REFERENCES

[1] R. Ross, "PV Reliability Development Lessons From JPL's Flat Plate Solar Array Project", IEEE J. of Photovoltaics, vol. 4, no.1, January 2014

[2] Gambogi, W.; Heta, Y.; Hashimoto, K.; Kopchick, J.; Felder, T.; MacMaster, S.; Bradley, A.; Hamzavytehrany, B.; Garreau-Iles, L.; Aoki, T.; Stika, K.; Trout, T. J.; Sample, T. A Comparison of Key PV Backsheet and Module Performance from Fielded Module Exposures and Accelerated Tests. IEEE Journal of Photovoltaics, 2014.

[3] Gambogi, W; Yu, BL; Felder, T; MacMaster, S; Choudhury, KR; Tracy, J; Phillips, N; Hu, HJ, Sequential Stress Testing to Predict Photovoltaic Module Durability, 2019 IEEE 46TH Photovoltaics Specialty Conference (PVSC), 2019.

[4] https://www.pvqat.org/index.html

[5] J. Morse, M. Thuis, D. Holsapple, R. Willis, M. Kempe, and D. Miller, "Degradation in photovoltaic encapsulant tranmittance: Results of the second PVQAT TG5 artificial weathering study", Progress in Photovoltaics, 05 April 2022.

[6] M. Kempe, "solder Bump Coupon Tetsing of Backsheets for Simplified Comprehensive Evaluation", NREL PVRW, 2019.

[7] www.duramat.org

[8] Hacke, P; Owen-Bellini, M; Kempe, M; Miller, DC; Tanahashi, T; Sakurai, K; Gambogi, WJ; Trout, JT; Felder, TC; Choudhury, KR; Philips, NH; Koehl, M; Weiss, KA; Spataru, S; Monokroussos, C; Mathiak, G, Combined and Sequential Accelerated Stress Testing for Derisking Photovoltaic Modules, Advanced Micro- and Nanomaterials for Photovoltaics, Micro & Nano Technologies, pp 279-313, 2019.

[9] A. Fairbrother, N. Phillips and X. Gu, "Degradation Processesand Mechanisms of Backsheets",Durability and Reliability of Polymers and Other Materials in Photovoltaic Modules, Elsevier, 2019.

[10] S.N. Venkat, J. Liu, N.S. Bosco, J. Dai, W.J.Gambogi, B. Brownwell, Y. Gu, J. Carter, L.S. Bruckman, J.-N. Jaubert, J.L. Braid, R.H. French, PVRW 2020.

[11] "Designing New Materials for Photovoltaics: Opportunities for Lower ing Cost and Increasing Performance through Advaned Material Innovations", Report IEA-PVPS T13-13:2021, 2021.

[12] G. Eder, Y. Voronko, G. Oreski, W. Mühleisen, M. Knausz, A. Omazic, A. Rainer, C. Hirschl, H. Sonnleitner (2019) „Error analysis of aged modules with cracked polyamide backsheets", Solar Energy Materials and Solar Cells 203, https://doi.org/10.1016/j.solmat.2019.110194

[13] Omazic, A., Oreski, G., Edler, M., Eder, GC., Hirschl, C., Pinter, G., Erceg, M. (2020), Increased reliability of modified polyolefin backsheet over commonly used polyester backsheets for crystalline PV modules. J Appl Polym Sci, 137, 48899. doi: https://doi.org/10.1002/app.48899

[14] T. Tanahashi et al., "Corrosion-induced AC impedance elevation in front electrodes of crystalline silicon photovoltaic cells within field-aged photovoltaic modules", IEEE J. Photovolt., vol. 9, no. 3, pp. 741–751, 2019.

[15] H.E. Yang, R.H. French and L.S. Bruckman (Editors), Durability and Reliability of Polymers and Other Materials in Photovoltaic Modules, Elsevier Press, 2019.

[16] N. Phillips, "IEC63209 Series: Photovoltaic Modules – Extended Stress Testing", NREL PV Reliability Workshop, 2022.

978-1-7281-6118-1/22 $31.00 © 2022 IEEE

Measuring Irradiance with Bifacial Reference Panels

Nicholas Riedel-Lyngskær, Jan Vedde, Peter B. Poulsen, Sergiu Spataru

Technical University of Denmark, Department of Photonics Engineering, Roskilde, Denmark

European Energy A/S, Søborg, Denmark

This paper describes the hardware and results from an outdoor testbed that is designed to assess rear plane-of-array (POA) irradiance measurement approaches. The measurement system contains five optical sensor types, consisting of pyranometers, reference cells, photodiodes, a spectrometer and a bifacial/monofacial reference module pair-all covering various spatial resolutions on the rear POA. The implied uncertainties associated with the different sensor types, at various positions, are assessed with the bifacial performance ratio (PRBIFI) of a 14.2 kWp string of identical bifacial technology as the reference panel. We find that a bifacial/monofacial reference module pair can be used to accurately determine PRBIFI and avoid the complications in rear POA sensor placement.

Local nm-Scale Imaging of Electrical Contact for Series Resistance Degradation of Silicon Solar Cells

C.-S. Jiang[1], S. Johnston[1], E.A. Gauding[1], M.G. Deceglie[1], R. Flottemesch[2], C. Xiao[1], R. Moutinho[1], D.B. Sulas-Kern[1], J. Mangum[1], M.M. Al-Jassim[1], and I.L. Repins[1]

[1]National Renewable Energy Laboratory, Golden, CO, USA; [2]Luminace LLC, New York, NY, USA

Abstract—**We report on an electrical conduction mechanism for series resistance (Rs) degradation observed in a utility scale solar farm by nm-scale imaging of the local resistance at the Ag/Si interface of c-Si front metallization. Scanning spreading resistance microscopy imaging revealed that the number of point or small area electrical contacts decreased in a degraded cell compared to an unaffected cell, demonstrating the direct root cause of the Rs degradation. The degraded cell shows both a morphological and chemical difference in the screen-printed finger contact compared to the unaffected cell, which likely caused the degradation during the long-term field service. The reduction in electrical contact is likely caused by a structural change: The Ag particles in contact with the Si cell aggregate into bulk Ag, and a highly resistive ceramic oxide is formed in a "belt" shape at the Ag/Si interface. This resistive belt with a thickness of ~1 μm blocks the current conduction from cell emitter to the Ag grid. Our results demonstrate an example of the multi-scale characterization approach for understanding degradation mechanisms in photovoltaics.**

Keywords—*series resistance degradation, c-Si solar cell, front metallization, scanning spreading resistance microscopy (SSRM), nm-scale imaging.*

I. INTRODUCTION

Series resistance (Rs) degradation is one of major degradation modes widely observed with long-term field service of photovoltaic (PV) modules. Rs degradation can occur in the cell or at current conduction pathways outside of the cells. In a previous paper [1], we reported significant loss of power output of a utility scale PV system in New Jersey, US with 7 years' service. Fitting current-voltage (I-V) curves of individual cells indicated exclusively Rs degradation. A combination of photoluminescence (PL) and electroluminescence (EL) imaging indicated that the degradation occurred in front Ag metal grid in the aluminum back surface field c-Si cells (Fig. 1).

The same module contained a mix of degraded and nondegraded cells (Fig. 1a and 1b). In non-degraded cells, EL is uniform along the metal grid when measured on cored pieces by injecting current through contact with a single grid (Fig. 1c). The EL contrast away from the grid increases with increasing current due to the emitter series resistance. However, in cored pieces from the central area of degraded cells, the EL shows an isolated "beads" pattern, indicating the pattern of current flow into the cell (Fig. 1d). Since the EL intensity of the beads is randomly distributed along the grid line and not weighted toward the contacted side, the degradation should be at the electrical contact

of the Ag grid with the cell emitter but not in the Ag grid itself. The cored pieces from areas along the cell perimeter area of the degraded cell show no such beads pattern with a smaller current of 10 mA. However, it appeared with a larger current of 100 mA (Fig. 1e), indicating similar degradation occurred in the perimeter area but to a lesser degree than in the central area of the degraded cells.

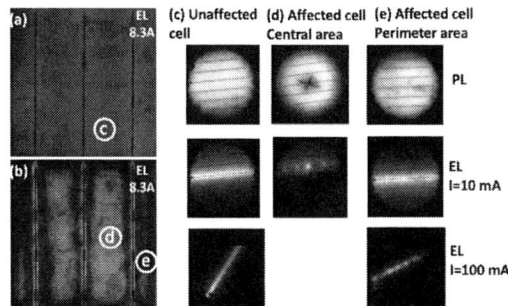

Fig. 1 EL images of (a) undegraded and (b) dergrad cells from the same module. (c)(d)(e) PL and EL images with 10 and 100 mA injection current taken on cored pieces from the (c) undegraded cell, and (d) central area and (e) perimeter area of the degraded cell.

Microscopy studies using scanning electron microscopy-based energy dispersion spectroscopy (SEM-EDS) showed that the glass frit layer between the Ag grid and Si cell has differences in the oxide components between the degraded and non-degraded cells, with the nondegraded cell being Zn-rich and the degraded cell Pb-rich [1]. In this paper, we report direct imaging of the electrical contact between the Ag grid with the Si cell using scanning spreading resistance microscopy (SSRM) [2]. We found that the number of Ag/Si point contacts decreased substantially in the degraded cell, which points directly to the Rs degradation mechanism.

II. EXPERIMENTAL

Cored pieces from the Si cells were cleaved in the direction perpendicular to the Ag grid and encapsulated into a sandwich structure with two pieces of Si wafer using Ag epoxy. The cross-

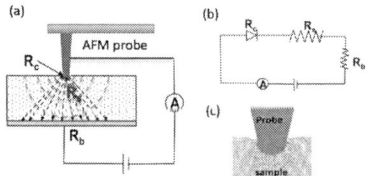

Fig. 2. Schematics of (a) SSRM setup, (b) equivalent circuit, and (c) structure around the probe/sample contact.

This work is funded by US Department of Energy, Solar Energy technology Office, with agree number 34357.

sections were mechanically polished followed by Ar-ion milling polishing.

SSRM is based on a contact mode atomic force microscopy (AFM) (Veeco D5000) using a logarithmic current amplifier that measures a wide range of resistance ($10^3 \sim 10^{15}$ Ω) (Fig. 2). The contact resistance of probe/sample was minimized by pressing the probe (diamond-coated Si tip, Bruker DDESP) to the sample in ~1 μN and applying a probe/sample bias voltage >3 V. The measured resistance is dominated by the resistance of nm-scale volume of the sample right beneath probe, because resistance contribution by other part of the sample along the current path reduces rapidly with distancing from the probe.

III. RESULTS AND DISCUSSIONS

SSRM maps the local resistivity in the nanometer scale. SSRM local resistance imaging was conducted on samples cored from a nondegraded cell, a degraded central area of a degraded cell, and a less-degraded perimeter area of the degraded cell.

Fig. 3a shows an optical image where the SSRM images were taken at the Ag/Si boundary, Fig. 3b shows SSRM resistance images, and Fig 3c shows the corresponding AFM images simultaneously taken with the SSRM images. The SSRM images were 5 μm x 5 μm each and pieced together. The small scan size is to ensure high imaging quality and adequate resolution. Examining the corresponding AFM images is necessary to exclude artifacts from rough surface morphology.

At the Ag/Si interface of this nondegraded cell, we observe low resistance Ag particles embedded in a ceramic oxide matrix. The Ag particles contact the Si emitter in an isolated points or small areas in sub-μm sizes and further connect to Ag grid bulk, forming the electrical conduction pathway [3]. The interfacial ceramics that adhere the Ag grid and Si together was formed in the firing process of screen-printed Ag slurry that contains recipe components of Ag and multiple ceramic oxides. The line profile in Fig. 3g shows the different local resistance corresponding to the Si bulk (10^9 Ω), Si emitter (10^7 Ω), ceramic

(10^{13} Ω), and Ag (10^4 Ω). We note that the resistance of ~10^4 Ω of Ag is not accurate because the low Ag resistance approaches the instrument measurement limitations, and the probe/sample contact resistance can significantly contribute to the measured value. In another example (Fig. 3h), the line profile shows the connections of Ag and Si emitter without a high resistance gap within the ~10 nm SSRM resolution [4], illustrating the intimate contact of Ag/Si in this case.

The Ag/Si contact should be examined closely in high-resolution images. Two examples of intimate electrical contact are shown in Fig. 3d and 3e. Another example shows possible contact in the larger-size mapping (Fig. 3b), but the higher resolution in Fig. 3f shows a break in contact. However, we cannot exclude that this Ag particle contacts the Si at a different point; it simply does not appear to contact on this particular cross-sectional plane cutting through the grid. We have conducted a statistical analysis on multiple polished cross-sections and found the electrical contact line density of ~340/mm in the nondegraded cell, that corresponding to an areal density of ~1.2×10^5/mm^2 (Table 1).

TABLE I.   NUMBERS OF ELECTRICAL CONTACT FOUND IN SAMPLS OF UNAFFCTED CELL, CENTRAL AND PERIMETER AREAS OF THE DEGRADED CELL

	Undegraded cell	Central area, degraded cell	Perimeter area, degraded cell
Scan length (μm)	172	287	291
Contact point numbers	58	13	24
Line contact density (1/mm)	340	69	82
Areal contact density (1/mm^2)	1.2×10^5	4.7×10^3	6.7×10^3

On the central area of the degraded cell, where the EL showed the beaded pattern with small injection current of 10 mA, the SSRM images show significantly different configurations of Ag and ceramic phases compared to the undegraded cell (Fig. 4). The Ag and Si are separated by a highly

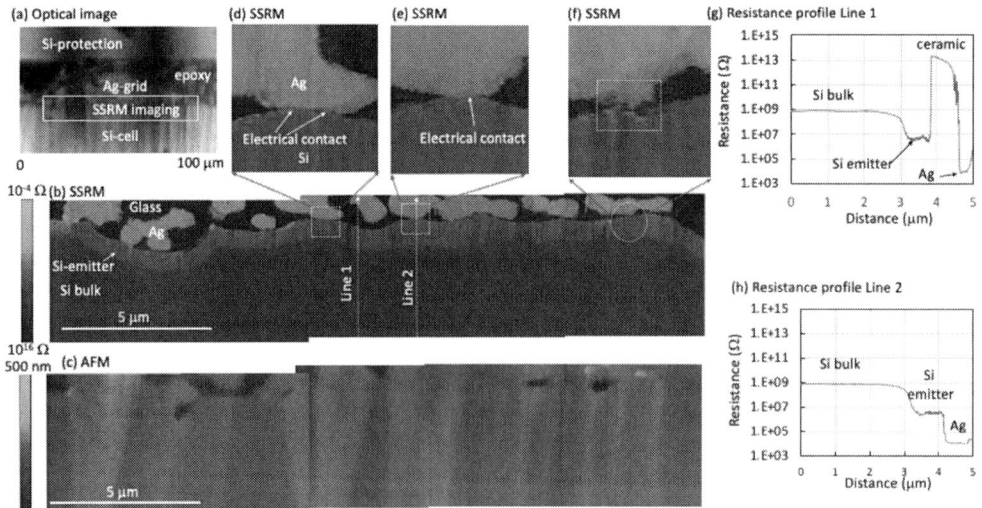

Fig. 3. SSRM resistance imaging take on the Ag/Si interface of a undegraded cell; (a) shows an optical image of the interface where SSRM was taken; (b) SSRM images and (c) the corresponding AFM images; (d)(e)(f) example zoomed-in areas with the electrical contact and without the contact; (g) and (h) line profiles of the resistance along lines in (b).

resistive "belt"-like region with a thickness of ~1 μm. There are much fewer Ag/Si point contacts along the cross section of the finger, with a line density of 69/mm (Table 1) that would translate to an areal density of 4.7×10^3/mm^2. This is about 2 orders of magnitude less than the nondegraded cell, indicating that Rs can increase 2 orders of magnitude if the Ag/Si contact resistance significantly contributes to the cell Rs. However, IV curve fits and cell level EL in our previously published work [1] on the same module show that the series resistance varied over the the affected cells and increased 2-16x, depending on location and severity of the degradation. This suggests that the statistical sampling in a length of ~300 μm may not be adequate.

The corresponding AFM image (Fig. 4b) shows a fracture at the interface with a thickness of ~300 nm. This fracture is observed in some areas but not everywhere along the interface. Fig. 4c and 4d is a continued segment of 4a and 4b, showing no such fracture, unlike Fig. 4b. Regardless, the highly resistive belt was observed throughout the area. A question arises: Is this separation of Ag/Si contact caused by (1) fracture or delamination of the Ag grid from Si, or (2) structural/compositional changes in the Ag/Si interface that further promote Ag particle detachment from Si? If it is the former, the Ag structure should not change significantly with degradation, and the high resistance belt should be accompanied with the AFM fracture everywhere, but this is not the case. In fact, our previous work showed that SEM imaging (Fig. 5) revealed a different Ag grid structure between the undegraded and degraded cells [1]. In the undegraded cell, Ag and ceramic oxides appear more mixed and with more voids. In the degraded cell, Ag bulk appears to be formed more solidly with less voids and ceramic mixtures. It is unknown whether this structural difference exists before the field service and is caused by the Ag past chemical component difference between the two cells and/or the subsequent firing process. However, because of the progressive field degradation, this separation of Ag particles

from the Si emitter occurred gradually during the module field service. The Ag particles in contact with Si could change its shape and location to separate from the Ag/Si interface and tend to segregate into the Ag bulk. In the meantime, more oxide components are left in or moved toward the Ag/Si interface forming the resistive belt. This structural change can induce a fracture or delamination of the grid from Si, but not necessarily everywhere along the interface. Ceramic material degradation at the grid/Si interface was reported in the literature [5]. The driving force for the degradation is unknown, can be thermal, UV-excitation, corrosion, or their combinations, or others. One such study suggests that acetic acid, presumably from the degradation of the ethyl-vinyl acetate (EVA) encapsulant layer, corrodes the PbO glass frit layer at the Ag/cell interface, forming lead acetate [6]. Investigation of this as the primary degradation mechanism for this particular set of modules is currently underway.

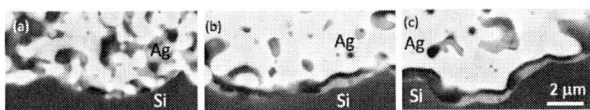

Fig. 5 SEM images take on the Ag grid of (a) undegraded cell, (b) central area and (c) perimeter area of a degraded cell.

The less degraded region in the perimeter of degraded cell shows similar SSRM and AFM images to the central region of the cell (images are not shown). The statistical line density of electrical contacts is 82/mm (Table 1), slightly more than that of the central region (69/mm). But this slight difference is not solid based on the inadequate sampling numbers. On the other hand, the beads EL pattern showed up with 100 mA versus the continuous pattern with 10 mA indicates a significant difference in Rs between the two areas, and Rs degradation in the perimeter regions is on the way to approach that of the central area.

IV. CONCLUSION

We have investigated the direct electrical conduction mechanism for a field observed Rs degradation by using SSRM. We found that the separation of electrical contact between Ag particles and Si emitter caused Rs degradation, and structural degradation of the Ag grid lines during aging likely caused the electrical contact separation.

REFERENCES

[1] E.A. Galding et al., "Differences in printeed contacts leads to suscetibility of silicon cells to sereis resistance degradation", IEEE J. Photovoltaics, (2022), in press.

[2] P. Eybens et al., "Scanning spreadng resistance microscopy and spectroscopy for routine and quantitative two-dimensional carrier mapping", J. Vac. Scie. Technol. B20, 471 (2002).

[3] C. Ballif et al., "Silver thick film contacts on highly doped n-type silicon emitters: structual and electronic properties of the interface", Appl. Phys. Lett. 82, 1879 (2003).

[4] L. Zhang et al., "High resolution characterization of ultrashallow junction by measuring in vacuum with scanning spreading resistance microscopy", Appl. Phys. Lett. 90, 192103 (2007).

[5] N. Iqbal et al., "Multiscale characterization of photovoltaic modules – case studies of contact and interconnect degradation", IEEE J. Photovoltaics 12, 62 (2022).

[6] A. Kraft et al., "Investigation of acetic acid corrosion impact on printed solar cell contacts", IEEE J. Photovoltaics 5, 736 (2015).

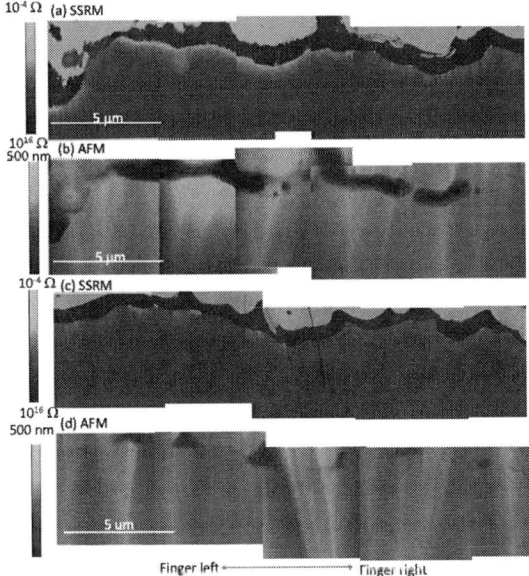

Fig. 4. (a)(c) SSRM resistance images and (b)(d) the corresponding AFM images take on the Ag/Si interface of a central areas in the degraded cell. (c)(d) is a continuous segment to the right of (a)(b) on the same cross-section.

Perimeter recombination in GaAs solar cells with different geometries

Natasha Gruginskie, Gerard Bauhuis, Peter Mulder, Elias Vlieg, John Schermer

Radboud University, Nijmegen, Netherlands

In this study, we evaluate the effects of perimeter recombination to the performance of GaAs solar cells with different geometries. Two wafers with identical epi-structures were processed into both thin-film and substrate-based devices, which allowed the precise determination of the effect of a rear mirror as well as of the thin-film processing steps to the cells output. In each wafer, a series of solar cells with varying areas was fabricated, and the different contributions to the solar cells dark currents were extracted. We observed that, aside from an expected lower $J01$, the thin-film devices also showed a larger reduction in illuminated performance with decreasing cell area. From these results, we demonstrate that, in high quality solar cells with low interface recombination velocities and high photon recycling factors, the power output is limited by $J02$, which is highly affected by perimeter recombination. Therefore, the employment of perimeter passivation techniques might become necessary for these solar cells to achieve higher efficiencies, particularly in small scale applications.

Global ranking of losses to photovoltaic power

A. Kubiniec, K. Seymour. A. Bhat, J. Hazari, T. Haley and Marc Perez.

Clean Power Research, Kirkland, WA, 98003, United States

Abstract—**Solar power is growing quickly and especially helpful in achieving decarbonization goals. With more installed solar generation capacity, understanding losses becomes increasingly important for optimizing solar development and planning. This paper will attempt to quantify and attribute solar losses globally, focusing on soiling, snow, and temperature as individual losses and how they relate to each other. This will be done by comparing simulated solar power output under a variety of different scenarios with and without the effects of soiling.**

Keywords—solar power modeling, soiling, snow, temperature, wildfires.

I. Introduction

Solar power is growing quickly in the energy sector, 23% growth from 2019 to 2020, ~156 TWh of capacity added. [1]. More and more solar is being added to the energy mix every year. Solar costs are coming down, approximately a 70% reduction of costs from 2010 to 2020 [2], and solar is an excellent energy source for meeting decarbonization goals. Many countries, states, and utilities have set aggressive renewable energy targets. This all increases the pressure to add more solar power. The same losses also affect disturbed rooftop systems, although the stakeholders may have somewhat different concerns about losses. While homeowners are more likely to be concerned about system performance, energy system managers may be more concerned about grid stability as they relate to solar system losses.

As solar power becomes a larger percentage of the energy mix, understanding losses becomes more important. Losses are defined as energy that could be generated but is not due to exogenous factors. This can result in lower revenue, or higher energy costs for the owner, making a large project less economically viable or increasing the return-on-investment period for a homeowner. Losses, depending on the cause, can also create challenges for grid management. If "behind-the-meter" power generation is lower than predicted due to soiling or snow, load balancing entities will need to procure power at potentially costlier rates to make up the difference. Losses may also mask or obscure equipment failures or malfunctions, either way, reducing power output and ultimately increasing costs.

Wildfires have been occurring more frequently [8,9]. Many wildfire-prone areas also have heavy solar development. Wildfires may have an outsized impact on solar, and in the context of this work, they result in increased soiling of panels.

This paper focuses on losses from three sources: soiling, snow, and temperature. Soiling and snow may be reduced with management strategies, such as periodic cleaning or clearing. Temperature losses are less easily managed, but may be reduced

via intelligent system design. Temperature losses are also useful in putting into context an avoidable loss vs. an unavoidable loss. The mechanism for lost power in the case of soiling or snow is reduced irradiance transmittance to the panels due to some amount of optical obstruction. Temperature losses occur because the panels lose efficiency every degree above a temperature threshold (STC, or standard test conditions, 25 C), this lower efficiency results in lower power output[3].

The benefit to understanding losses includes the ability to optimize solar plant location and design to minimize missed power generation due to soiling, temperature, and snow. Quantifying losses allows for direct comparison between fixed axis and single axis tracker systems in terms of how much power is lost. Once losses are understood well, it is possible to determine an optimized cleaning or clearing schedule based on region and loss profile. Given inputs such as the price of power and cost of cleaning, it may even be possible optimize on a cost basis.

II. Methods

The way soiling losses are assessed in this paper relies on modeling and satellite-derived irradiance data. The modeling package, pvlib [4], is used to model solar power. The input data comes from the SolarAnywhere V3.5 model [5]. Snowfall and temperature data, as well as other ancillary variables, are available through SolarAnywhere but are originally sourced from MERRA2 [6] and ERA5 [7].

Power simulations are run using the pvlib model at an hourly time resolution. This was done for the year 2020. A 1998-present assessment is pending. Simulations were run globally at 100 km spacing, from +/-60 degrees latitude. This comes to ~650,000 hourly timeseries evenly spread across the globe. For fixed tilt systems, the tilt was adjusted as a function of latitude; the tilt of the system was the determined by the latitude of the system. The azimuth of the fixed tilt systems was south facing in the northern hemisphere, and north facing in the southern hemisphere. A 50 MW DC system was simulated for fixed tilt, and single axis tracking, with a DC/AC ratio of 1.2. Simulations were run with soiling applied, and no soiling applied. The hourly simulations of power were then averaged to monthly values to simplify the analysis.

Loss assessment is listed below:

- Soiling was assessed on a percent difference basis, shown with equation 1.

- *Equation 1: Soiling Loss* $= \frac{Clean - Soiled}{Clean}$

- Snowfall was summed on a monthly interval. The total monthly solid precipitation was compared to the median total monthly snowfall for each point to have a rough snow loss percentage. This will be addressed in future work by simulating snow losses directly. Snow modeling tends to be somewhat binary and may have high uncertainty, so the sensitivity to the snow input and snow modeling configuration in pvlib will be explored in future work.

- Temperature losses were calculated using a standard temperature loss equation. As the panels get hotter, the assumption is that the panels become less efficient. The current results use monthly average temperatures, which are then scaled to correct for the nighttime values that are included in the monthly temperature averages. Future work will correct for this, as well as model temperature losses directly. The temperature loss equation is shown below, with equation 2. Temperature losses were only calculated if the temperature was above 25 C.

- *Equation* 2: $Temperature\ Loss = (Temperture\ (C) - 25) \times (0.5\ \%)$

III. RESULTS / DISCUSSION

Simulation results were plotted and assessed for each month in 2020. Several points of interest include:

-Soiling is the dominant loss in the Sahara Desert. From a global perspective, the Sahara and some locations in the Middle East experience the greatest soiling losses. Soiling losses are generally very region specific. Desert areas which generate a lot of dust, appear to be the main driver of soiling losses globally. Wildfires also contribute to soiling but are seasonally dependent.

-Temperature losses, although less avoidable, are the largest loss out of the three considered when comparing total losses globally. This is an expected result, and temperature losses may be unavoidable. It is important to consider the magnitude of the loss, as it may result in lower losses to optimize a site for better cooling, as compared to reduced soiling loss

-In Europe, snow losses are only the largest for a small fraction of northern countries and only during one to two winter months. Otherwise, soiling and temperature are much larger sources of loss.

-In The Pacific Northwest, during the months with wildfires, soiling losses are the largest loss. Soiling due to wildfires may fall outside standard cleaning schedules and may need to be accounted for separately.

Simulation results for September 2020 are plotted in Figure 1.

Comparing the impact of losses on fixed-tilt systems relative to single-axis tracking systems will require further investigation. An initial comparison of simulation results suggests that single axis trackers have slightly higher soiling losses than fixed tilt systems. This may be due to the higher power output, and therefore larger losses.

Figure 2 compares the global average soiling losses between fixed tilt systems and single axis trackers. Losses are very similar except for the first four months of the year.

Figures 1 and 2 can be found at the end of the paper, below.

IV. FUTURE WORK / CONCLUSIONS

Soiling losses appear to be very regional. Deserts and areas prone to forest fires contribute to larger soiling losses. Outside those regions, soiling losses appear to be small. Temperature losses are the largest of the three compared and this is an expected result. Temperature losses may be the most unavoidable of the three losses currently compared.

The next step for this project is to directly simulate cleaning events, either from liquid precipitation or scheduled as part of plant maintenance. Snow-clearing events will also be simulated. The goal is to quantify and better understand benefits of cleaning or clearing panels, or the costs of not doing so. Higher resolution plots may also be investigated. A possible outcome of this work would be a tool that calculates the ideal, if any, cleaning interval for a system, given inputs of soiling, energy costs, and the cost of cleaning panels, or clearing snow off panels.

REFERENCES

[1] Cherp, A., Vinichenko, V., Tosun, J. et al. National growth dynamics of wind and solar power compared to the growth required for global climate targets. Nat Energy 6, 742–754 (2021). https://doi.org/10.1038/s41560-021-00863-0J.

[2] Feldman, David, Ramasamy, Vignesh, Fu, Ran, Ramdas, Ashwin, Desai, Jal, and Margolis, Robert. U.S. Solar Photovoltaic System and Energy Storage Cost Benchmark: Q1 2020 [PowerPoint]. United States: N. p., 2021. Web. doi:10.2172/1765601.

[3] Vaillon, R., Dupré, O., Cal, R.B. et al. Pathways for mitigating thermal losses in solar photovoltaics. Sci Rep 8, 13163 (2018). https://doi.org/10.1038/s41598-018-31257-0.

[4] Holmgren, William F., Clifford W. Hansen, and Mark A. Mikofski. "pvlib python: A python package for modeling solar energy systems." Journal of Open Source Software 3.29 (2018): 884.

[5] Perez, Richard, et al. "Detecting calibration drift at ground truth stations a demonstration of satellite irradiance models' accuracy." 2017 IEEE 44th Photovoltaic Specialist Conference (PVSC). IEEE, 2017.

[6] Gelaro, Ronald, et al. "The modern-era retrospective analysis for research and applications, version 2 (MERRA-2)." Journal of climate 30.14 (2017): 5419-5454.

[7] Hersbach, Hans, et al. "The ERA5 global reanalysis." Quarterly Journal of the Royal Meteorological Society 146.730 (2020): 1999-2049.

[8] USGCRP (U.S. Global Change Research Program). 2018. Impacts, risks, and adaptation in the United States: Fourth National Climate Assessment, volume II. Reidmiller, D.R., C.W. Avery, D.R. Easterling, K.E. Kunkel, K.L.M. Lewis, T.K. Maycock, and B.C. Stewart (eds.). https://nca2018.globalchange.gov/downloads. doi:10.7930/NCA4.2018

[9] Westerling, A.L. 2016. Increasing western U.S. forest wildfire activity: Sensitivity to changes in the timing of spring. Phil. Trans. R. Soc. B. 371:20150178

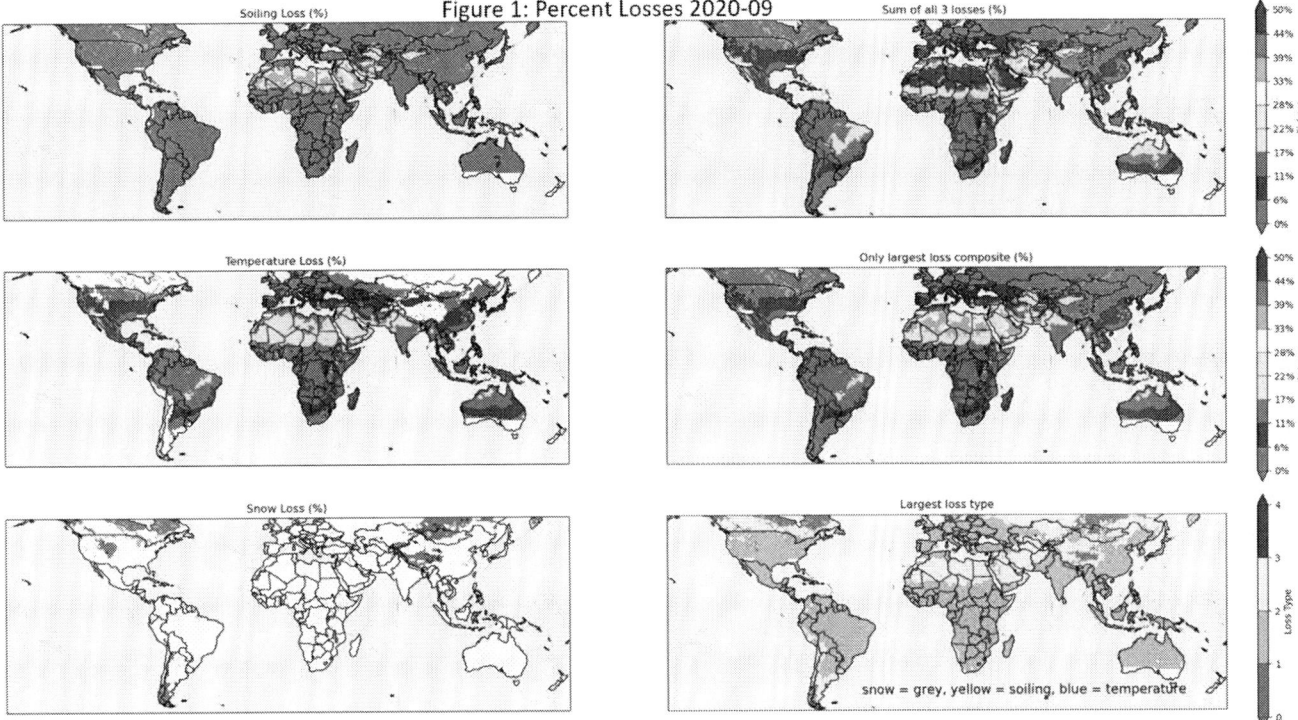

Figure 1: Global 100km monthly averages of losses. Data is based on simulations from pvlib, and input data is SolarAnywhere V3.5. Top left: Soiling Losses as a percent. Middle left: Temperature losses as a percent. Bottom left: Snow losses as a percent. Top right: Sum of all 3 losses for the given month. Middle Right: A composite map, only including the largest loss source for each point. Bottom Right: The largest loss type for each point. Note that in the bottom right figure, soiling losses are the largest loss type in the Pacific Northwest part of North America. This is likely due to enhanced soiling, mainly due to forest fires during that time. This is not evident when looking at the plot of just soiling, as the Sahara has the largest soiling losses globally, and so small deviations in soiling are less obvious in other non-Sahara regions.

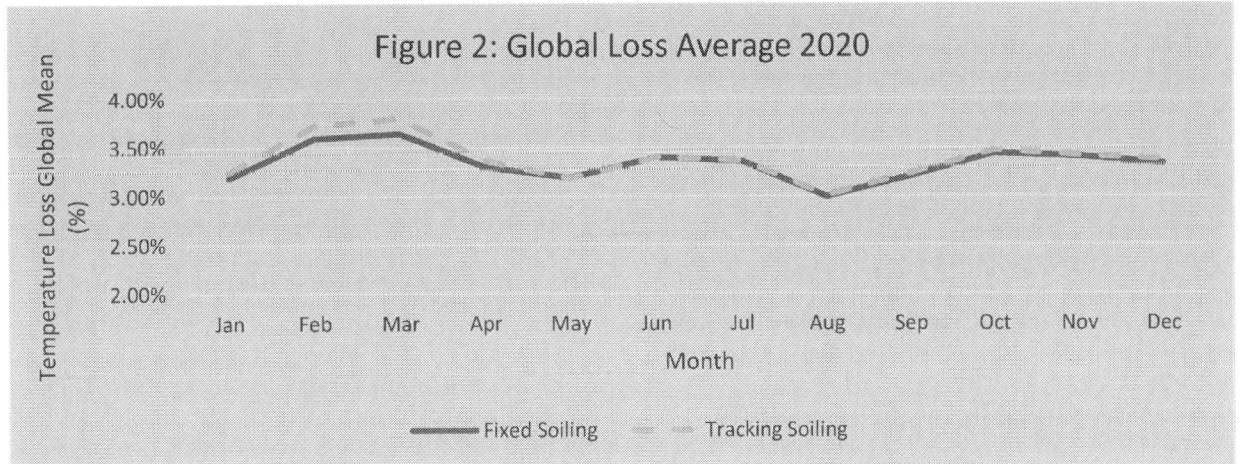

Figure 2: Global average soiling losses of fixed tilt system simulations vs single axis tracker simulations. Note that the y-axis starts at 2%. This result shows a difference during Jan-April months for soiling when comparing fixed tilt systems vs. single axis tracker systems. This needs to be investigated further and will be covered under future work. Two areas that will be investigated are, is this difference real, or related to a different loss percentage due to single axis trackers higher power output or is this a modeling related artifact.

PV Module Toxicity Testing Methods and Results: A Literature Review

F. Li[1], S. Shaw[2], C. Libby[2], B. Bicer[1], G. TamizhMani[1]

[1]Photovoltaic Reliability Laboratory, Arizona State University (ASU-PRL) Mesa, Arizona, 85212, USA
[2]Electric Power Research Institute (EPRI) Palo Alto, California, 94304, USA

Abstract—Solar photovoltaic (PV) modules may contain a variety of toxic elements in the electrical contact and/or semiconductor material that could pose environmental and health risks during end-of-life management. This paper provides an overview of the literature related to toxicity testing methodologies, standards and experimental results. It covers various topics including: number and scope of publications available in the toxicity related areas; identification of toxic elements present in silicon-based modules and thin film modules; toxicity testing standards/protocols currently required/used in various countries/jurisdictions; size and location of particles removed from modules to create samples for toxicity testing; particle cutting methodologies used by various industry stakeholders/researchers; leach testing results; and a new international standard for sampling modules for toxicity testing.

Keywords— toxicity, leaching test, PV module, TCLP, WET, lead, cadmium

I. INTRODUCTION

By 2030, the mass of solar modules at end-of-life (EoL) will be reaching up to 1 million tons in the United States, or 1% of the world's e-waste [1][2]. This amount is expected to increase to 10 million tons by 2050 as PV installation continues to grow exponentially [2]. These developments create challenges and opportunities with regards to PV waste management, material recovery and secondary markets.

About 135 publications from a variety of sources (government documents, scholarly journals, magazines, books, dissertations, or on websites) published between 1995 and 2021 were collected and reviewed. Fig. 1-2 summarizes the number of related articles published in each year and in each of six different areas. The primary focus of this paper is to provide a literature review on existing toxicity testing standards in various countries and on the test results reported by various organizations.

II. TOXIC ELEMENTS IN PV MODULES

The safe management of environmentally hazardous wastes in the United States is required by the Resource Conservation and Recovery Act (RCRA). The U.S. Environmental Protection Agency (EPA) regulates the proper disposal of certain types of waste through monitoring a list of 8 metallic elements known as RCRA 8 metals. These regulated metals are: arsenic (As), barium (Ba), cadmium (Cd), chromium (Cr), lead (Pb), mercury (Hg), selenium (Se) and silver (Ag). Commercial PV modules contain many of the RCRA 8 metals in the form of electrical

contacts and/or semiconductor materials. Table I provides the toxic elements that may be present and typical material composition (in weight percent) of crystalline silicon modules (monofacial and bifacial) and thin-film modules (a-Si, CIGS and CdTe).

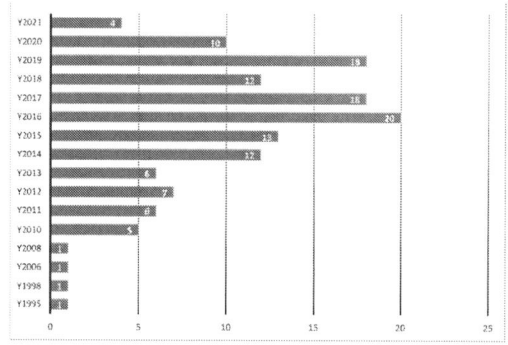

Fig. 1. Counts of papers and reports published between 1995 and 2021.

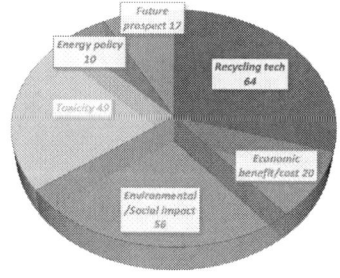

Fig. 2. Report counts by six subject types associated with PV module toxicity and recycling.

TABLE I.   COMPOSITION OF SILICON-BASED AND THIN FILM PV MODULES (WEIGHT%) [3]–[7].

Weight proportion in %	c-Si					Thin Film			
						a-Si	CIGS	CdTe	
Resource	[3] in 2019	[4] in 2010	[5] in 2017	[6] in 2017	[7] in 2021	[6] in 2017	[6] in 2017	[6] in 2017	[7] in 2021
Glass	74.16	74	74	75	76 (monofacial module) 83 (bifacial glass-glass module)	85	85	95	97
Aluminum [Al]	10.3	10	10	10	8-12	10	8	<0.01	
Polymers (plastics, films, adhesives)	10.13	~6.5	~6.5	8	5-10	5	5	3	3
Silicon [Si]	3.48	~3	~3	5	2.5-5	<0.1			
Copper [Cu]	0.57	0.6	0.6	0.8	1	0.5	0.8	0.9	
Tin [Sn]	0.12	0.12	0.12		<0.1%				
Lead [Pb]	0.07	<0.1		<0.1	<0.1%	<0.1	<0.1	<0.01	
Silver [Ag]	0.01	<0.006		0.005	<0.1%			0.01	
Zinc [Zn]			0.12	0.15		<0.1	0.12	0.02	
Indium [In]						<0.002	0.02		
Selenium [Se]							0.03		
Tellurium [Te]								0.08	<0.1
Cadmium [Cd]							0.01	0.08	<0.1
Molybdenum [Mo]							0.05	0.05	

978-1-7281-6118-1/22 $31.00 © 2022 IEEE

III. Toxicity Test

A. Sampling Location

The weight of a PV module typically ranges between 20 and 22 kg. The weight of a test sample to be supplied to toxicity test labs depends on the country/jurisdiction specific standard. For example, about 100g test sample plus 10g for the pH determination is required to conduct the TCLP (toxicity characteristic leaching procedure) test according to the US EPA's 1311 standard. Prior work by the authors determined that locations for the sample pieces within the module laminate should be representative and proportional to the individual areas of the entire module. Two potential examples of biased (inadvertently or intentionally) sample removal approaches are presented here. In the first example, the test module can be passed by removing the particles from the cell areas between the cell interconnect ribbons of crystalline silicon modules as these areas typically do not contain significant toxic materials. In the second example, the test module, on the other hand, can be failed by removing the particles from the cell interconnect ribbon areas where the modules typically contain significant amounts of toxic elements such as lead (Pb). Arizona State University in collaboration with Electric Power Research Institute developed an unbiased representative sample removal approach as presented in various scientific publications [8]–[11]. In this area-proportional representative approach, four areas have been identified for the particle extraction from a test module: cell area, cell interconnect area, string interconnect area and non-cell/non-interconnect area.

B. Particle Size

The particle size plays a major role in the leaching results as indicated in [12]: the lower the particle size, the higher the expected amount of leaching due to larger exposed surface area. The methods shown in Table II provide an upper particle size limit (for example, <1 cm for the TCLP testing per EPA 1311 standard) but they do not provide a limit for the minimum particle size. For example, a particle size of less than 1 cm and 2 mm is allowed in the regulated TCLP (US, Federal) and California's Waste Extraction Test (WET) (US, California) tests, respectively. Since there is no minimum particle size limit, it is possible that some test labs or others may crush the sample, again inadvertently or intentionally, to micrometer size particles; but, in the real world, PV modules destined for landfills do not likely contain dominant amounts of micrometer-size shattered particles. To demonstrate this issue, it was shown that when EoL CdTe modules are intentionally crushed by a heavy-duty landfill compactor, 75% of the crushed module fragments were typically larger than 1cm [13][14]. Therefore, to more closely represent a shattered module and to avoid biasing leach testing results associated with smaller, higher surface area particles, it is critical that the size of particles removed from PV modules and the number of particles is proportional to the four areas across the module without compromising the particle size requirement of the test standard/protocol. Also, some changes in the current standards for the particle size may be needed if the modules are made of annealed glass (e.g., CdTe modules) compared to tempered glass (e.g., c-Si modules) because annealed glass tends to have only a few large cracks in the landfill whereas the tempered glass tends to completely shatter into small pieces [13].

TABLE II. LEACHING STANDARDS AND LIMITS IN DIFFERENT COUNTRIES [15]–[17].

	US		Germany	Japan
	Federal	California		
Leaching method	EPA method 1311 (TCLP)	WET	Standard 12457-4:01-03	JLT-13
Sample Size (cm)	<1	<0.2	<1	<0.5
Solvent	Sodium acerate/acetic acid (PH 2.88 for alkaline waste; PH 4.93 for neutral to acidic waste)	Citric acid (PH 5.0±0.1)	Distilled water	Distilled water
Liquid-solid ratio (amount of liquid used in relation to the solid material)	20:1	10:1	10:1	10:1
Treatment method	End-over-end agitation (30 rotations per minute)	Not specified	End-over-end agitation (5 rotations per minute)	End-over-end agitation (200 rotations per minute)
Headspace	Air or nitrogen	Air or nitrogen	Air or nitrogen	Air or nitrogen
Test temperature	23±2 °C	20 °C	20 °C	20 °C
Test duration	18±2 hr	48 hr	24 hr	6 hr

C. Standards and Toxicity Limits

Various sample treatment procedures used in toxicity leaching standards, as required in various countries to classify solid waste, are summarized in Table II. Within the US, different states may use their own leaching procedures, such as WET. The leaching limits for various hazardous metals in the US are shown in Table III (e.g., Pb of 5 mg/L, Cd of 1 mg/L, Se of 1 mg/L). Many of the referenced research studies utilize either TCLP or WET. An extensive tabular collection of the test parameters used in various toxicity testing procedures around the world has been achieved during the course of this work but not provided in this abstract due to space limitations though it can be potentially provided in the full paper.

TABLE III. LEACHING LIMITS IN THE UNITED STATES FOR VARIOUS METALS PRESENT IN DIFFERENT PV MODULE TECHNOLOGIES.

	EPA 1311 Method (TCLP hazardous waste limit in mg/L)[1]	California WET Method (mg/L)[2]	US Primary Drinking Water Limit (mg/L)[3]
Lead (Pb)	5	5	<0.015
Copper (Cu)	–	25	1
Zinc (Zn)	–	250	5
Silver (Ag)	5	5	0.1
Nickel (Ni)	–	20	0.1
Cadmium (Cd)	1	1	0.005
Selenium (Se)	1	1	0.05
Tellurium (Te)	–	–	–
Molybdenum (Mo)	–	350	–
Arsenic (As)	5	5	0.01

[1,3]. Data collected from EPA
[2]. From California Environmental Protection Agency Department of Toxic Substances Control (DTSC)

D. Sample Cutting Methods

Various sample cutting methods used by the industry or researchers for commercial modules are listed in Table IV. Detailed descriptions of various factors and experimental setups used in various studies for both commercial and laboratory devices will be presented in the full paper. Common cutting methods that have been used to obtain sample pieces for leach testing include various mechanical methods and waterjet cutting. Mechanical cutting methods, which often use diamond cutting wheels, dremel tools, coring tools, hammer, ball mill or micro-shear wire cutters, have been traditionally adopted due to availability and ease of operation. The waterjet method is an

978-1-7281-6118-1/22 $31.00 © 2022 IEEE

erosive process that uses high-pressure water and grit to cut through the module laminate to obtain sample pieces for testing. An important benefit of the waterjet method compared to the mechanical method is the smooth cut with 100% glass coverage in the cut pieces and ability to cut material without separating layers of the laminate into glass, encapsulant, cell and backsheet/backglass, as there is no heat-affected zone, and glass vibration is minimized. It is critical that the selected cutting method generate repeatable pieces within a cutting facility irrespective of the operator and reproducible pieces between cutting facilities, irrespective of the machine type/model. If not, there may be unacceptable variability in TCLP results, and pass/fail determinations may differ depending on the facility or operator preparing samples. As demonstrated in [11], the mechanical method resulted in an average variability of about 30%, whereas the waterjet cutting method had an average variability of about 8%. Therefore, the waterjet cutting method is being used in the recently developed ASTM (American Society for Testing and Materials) practice standard entitled *"ASTM E3325-21, Standard Practice for Sampling of Solar Photovoltaic Modules for Toxicity Testing"*.

TABLE IV.   TEST METHOD, SAMPLING METHOD, PARTICLE SIZE AND PASS/FAIL RESULTS FOR COMMERCIAL PV MODULES OF CRYSTALLINE SILICON AND CdTe TECHNOLOGIES.

Cell Technology	Test Method	Sampling Method	Particle Size	Toxicity (Pass/Fail)	Reference
mc-Si	EPA method 1311 (TCLP)	- Mechanical Coring	9.5mm	Pass (Lab1)	[8]
				Fail (Lab2)	
		- Mechanical Cell-cut		Pass (Lab1)	
				Pass (Lab2)	
		- Mechanical Strip-cut		Pass (Lab1)	
				Fail (Lab2)	
		- Mechanical Hybrid (combined strip-cut and coring)		Pass (Lab1)	
				Pass (Lab2)	
mc-Si	EPA method 1311 (TCLP)	Water jet	9.5mm	Pass (Lab1)	[9]
				Pass (Lab2)	
mc-Si	EPA method 1311 (TCLP)	Sharp scissors to remove encapsulant and ribbon	< 1cm	Pass (encapsulated sample)	[18]
				Fail (non-encapsulated sample)	
mc-Si	EPA 1311 (TCLP), WET	Shredded or crushed	< 9.5mm	Pass	[19]
			< 2mm		
mc-Si	EPA 1311 (TCLP), WET	Crushed	< 9.5mm	Pass (regular TCLP test)	[20]
				Fail (regular WET test)	
				Fail (extended TCLP test)	
				Fail (extended WET test)	
c-Si, mc-Si	EPA 1311 (TCLP)	Not mentioned	< 9.5mm	Pass (c-Si sample)	[21]
				Fail (mc-Si sample)	
c-Si, CdTe	EPA 1311 (TCLP)	Crushed	0.5-1cm	Pass (c-Si sample)	[13]
				Pass/Fail (CdTe sample)	
CdTe	Norwegian Standard - European Norm - 12457	Crushed	< 4mm	Pass	[22]
CdTe	DIN EN 12457-4:01-03 Germany		< 1cm	Pass	
CdTe	Notice 13/JIS K0102:2013 method (JTL-13) Japan	Not mentioned	0.5cm	Pass	[23]
CdTe	EPA 1311 (TCLP)		< 1cm	Pass	

IV.   SUMMARY AND CONCLUSION

This literature review paper presents an overview of toxicity testing methodologies, standards and experimental results. It covers toxic elements present in commercial modules; toxicity testing standards/protocols currently required/used in various countries/jurisdictions; particle size and location of particles of

the removed samples from the module for toxicity testing; and particle cutting methodologies used by various industry stakeholders/researchers. This review identifies preferred approaches for sample cutting, a key step in the toxicity testing process. Finally, this paper aims to bring attention to a newly developed ASTM standard which provides an industry recognized and approved methodology to remove representative and proportional sample pieces for toxicity testing of PV modules.

REFERENCES

[1] G. Heath, "PV Modules End of Life Management Setting the Stage," *EPA Sustain. Mater. Manag. Webinar*, 2019.

[2] Stephanie Weckend, Andreas Wade, and Garvin Heath, *End of Life Management Solar PV Panels*. 2016.

[3] V. Fiandra, L. Sannino, C. Andreozzi, and G. Graditi, "End-of-life of silicon PV panels: A sustainable materials recovery process," *Waste Manag.*, vol. 84, pp. 91–101, 2019.

[4] A. Hahne and H. Gerhard, "Recycling photovoltaic modules," *Fed. Minist. Environ. Nat. Conserv. Nucl. Saf.*, vol. October, no. 02, pp. 1–4, 2010.

[5] P. Sinha, S. Raju, K. Drozdiak, and A. Wade, "Life cycle management and recycling of PV systems," *Photovoltaics Int.*, pp. 47–48, 2017.

[6] S. Kusch and M. A. T. Alsheyab, "Waste electrical and electronic equipment (WEEE): A closer look at photovoltaic panels," *Int. Multidiscip. Sci. GeoConference Surv. Geol. Min. Ecol. Manag. SGEM*, vol. 17, no. 41, pp. 317–324, 2017.

[7] Electric Power Research Institute, "Solar Photovoltaics End-of-Life Management Infographi," *Palo Alto, CA, 3002021132*, 2021.

[8] G. TamizhMani *et al.*, "Evaluating PV Module Sample Removal Methods for TCLP Testing," *2018 IEEE 7th World Conf. Photovolt. Energy Convers.*, pp. 2610–2615, 2018.

[9] G. TamizhMani, S. Shaw, C. Libby, A. Patankar, and B. Bicer, "Assessing Variability in Toxicity Testing of PV Modules," in *IEEE Photovoltaic Specialists Conference*, 2019.

[10] G. TamizhMani *et al.*, "Sampling Methods for Toxicity Testing of PV Modules for End-of-Life Decisions," in *IEEE Photovoltaic Specialists Conference*, 2021.

[11] G. TamizhMan, S. Shaw, C. Libby, and P. Sinha, "Photovoltaic Module Sampling Methods for Toxicity Testing and Drivers of Leaching Results," in *Solar Power and Wildlife/Natural Resources Symposium*, 2021.

[12] D. W. Cunningham, "Discussion about TCLP protocols presented on the Photovoltaics and Environment 1998 Workshop," 1998.

[13] P. Sinha and A. Wade, "Assessment of leaching tests for evaluating potential environmental impacts of PV module field breakage," *2015 IEEE 42nd Photovolt. Spec. Conf. PVSC 2015*, pp. 15–17, 2015.

[14] A. Rix, M. J. Rudman, U. Terblanche, and J. van Niekerk, "First Solar's CdTe module technology - performance, life cycle, health and safety impact assessment," 2015.

[15] Method-1311, "Toxicity Characteristic Leaching Procedure (TCLP)," *US Environmental Protection Agency (EPA), Washington, D.C.* 1992.

[16] California Code of Regulations, "Waste Extraction Test (WET) Pro cedures," 1991.

[17] P. Sinha and A. Wade, "Assessment of Leaching Tests for Evaluating Potential Environmental Impacts of PV Module Field Breakage," *IEEE J. Photovoltaics*, vol. 5, no. 6, pp. 1710–1714, 2015.

[18] H. B. Sharma, K. R. Vanapalli, V. K. Barnwal, B. Dubey, and J. Bhattacharya, "Evaluation of heavy metal leaching under simulated disposal conditions and formulation of strategies for handling solar panel waste," Sci. Total Environ., vol. 780, p. 146645, 2021.

[19] M. K. Collins and A. Anctil, "Implications for current regulatory waste toxicity characterisation methods from analysing metal and metalloid leaching from photovoltaic modules," Int. J. Sustain. Energy, vol. 36, no. 6, pp. 531–544, 2017.

[20] K. Collins and A. Anctil, "Photovoltaic waste characterization with

environmental considerations," 2014 IEEE 40th Photovolt. Spec. Conf. PVSC 2014, pp. 1419–1423, 2014.

[21] G. Panthi, R. Bajagain, Y. J. An, and S. W. Jeong, "Leaching potential of chemical species from real perovskite and silicon solar cells," Process Saf. Environ. Prot., vol. 149, pp. 115–122, 2021.

[22] G. Okkenhaug, "Environmental risks regarding the use and final disposal of CdTe PV modules (Document No. 20092155-00-5-R)," 2010.

[23] C. Hagendorf, M. Ebert, M. Raugei, D. Lincot, J. Bengoechea, and M. J. Rodriguez, "Assessment of performance, environmental, health and safety aspects of First Solar's CdTe PV technology," 2017.

Validation of Novel Bifacial Photovoltaic Performance Model with 3D shading for Fixed-Tilt and Single-Axis Tracked Systems

Annie Russell, Christopher E. Valdivia, Cédric Bohémier, Joan E. Haysom, Karin Hinzer

SUNLAB, Centre for Research in Photonics, Ottawa, ON, Canada

J. L. Richards & Associates Limited, Ottawa, ON, Canada

Existing bifacial photovoltaic (PV) performance models fall primarily into two categories: (1) ray tracing models that capture complex shading but lack the computational efficiency required for optimization applications; and (2) view factor (VF) models that efficiently simulate energy transfer but rely on user-defined losses which neglect temporal variation in phenomena such as shading and electrical mismatch. Hybrid VF / ray tracing models selectively employ ray tracing while balancing computational efficiency. This paper describes the validation of a novel hybrid model, DUET, which combines a 3D VF model with deterministic ray-object intersections. The software provides 2D irradiance profiles and mismatch-inclusive current-voltage curves for each scale of components: from cells to the full array. Validation against open-access data from Denmark shows that DUET predicts bifacial energy yield at 0.76% and 0.65% lower than measured yield for fixed-tilt and horizontal single-axis tracked (HSAT) rows, respectively, over 3370 and 2731 daylight hours. Monthly relative error in bifacial energy ranges from < 1% to ~4.5% for both systems. The mean absolute error (MAE) in hourly bifacial power is 18.1 mW/Wp for fixed-tilt and 18.4 mW/Wp for HSAT. These errors fall below the lowest previously reported MAE for six software at the same field site by ~0.77 mW/Wp for fixed-tilt and ~1.1 mW/Wp for HSAT. Modelled average rear insolation agrees with pyranometer data within +4.4% for fixed-tilt and -0.76% for HSAT. For both configurations, rear irradiance MAE aligns with the lowest error previously reported for other software at the site.

Characterizing the Back-Contact Interface of Bi-Facial Poly-Crystalline CdTe Devices Using Transmission Electron Microscopy

John Farrell,[1] Ebin Bastola,[2] Manoj Jamarkattel,[2] Michael Heben,[2] Robert F. Klie[1]

[1]University of Illinois at Chicago, Chicago, IL, 60607, U.S.A. [2]University of Toledo, Toledo, OH, 43606, U.S.A.

Abstract— Poly-crystalline Cd(Se)Te based thin film solar cells have shown to be competitive in terms of efficiency and cost of electricity production. Yet, the presence of hetero-interfaces in Cd(Se)Te structure and low minority carrier lifetime have limited the thin film devices from reaching their maximum theoretical efficiency of approximately 30 percent. The back-contact of CdSeTe devices has been identified as one significant limitation to increased device performance since no metal has been identified that has a sufficiently high work function to create an Ohmic contact with the CdTe absorber at the back-surface of the film stack. Here, we will explore novel back-contact film layers in an effort to overcome this energy band mismatch. Atomic-resolution imaging in a scanning transmission electron microscope (STEM) combined with electron energy-loss spectroscopy (EELS) and energy-dispersive X-ray spectroscopy (XEDS) are used to characterize these devices and to inform the production process. The goal is to identify the ideal atomic and electronic structures, as well as any interfacial diffusion of elements.

Keywords—CdTe, Thin Film, HRSTEM, XEDS, Back Contact

I. Introduction

Polycrystalline CdTe thin film solar cells prove to be highly competitive technology because of the high absorption coefficient, nearly optimum direct band gap energy and simplicity of manufacturing. However, laboratory efficiencies of CdTe solar cells are still below the theoretical efficiency limit (30%)[1]. The back contact of the CdTe Photovoltaic (PV) cell has been a serious limit to performance due to a deep valence band at 5.8 eV, much higher than the work function of many metals (e.g. Ni at 5.2 eV)[2]. The semiconductor bands go down near the interface and this creates a barrier for holes and accumulates electrons for higher recombination events. The objective is the development of major reductions in back-surface interfacial recombination. Partial solutions include Cu doping to move the Fermi level closer to the valence band edge and shrink the depletion region and a Te-rich buffer layer with a valence band maximum between CdTe and typical metal contact work functions. However, Cu diffuses rapidly in the absorber bulk and its interstitials increase recombination events. The Te-rich material needs to be heavily doped with Cu, but the Cu can be shown to be localized in the Te, however a lattice mismatch between Te and CdTe results in many interfacial defects.

II. Methodology

The poly-crystalline CdTe thin film PV devices studied were deposited on NSG TEC 10 soda lime glass coated with indium tin oxide (ITO). A CdS emitter layer approximately 150 nm thick was deposited on the FTO layer using RF sputter deposition. The CdTe layer was grown by sublimation. A $CdCl_2$ heat treatment was applied after the absorber deposition. A Cu_yAlO_x solution is then used to form a back contact layer between the absorber and the back electrode. In this study, the back contact of the device consists of a very thin (5nm) layer of suspected Cu_yAlO_x directly on the absorber followed by either another layer of ITO or by an Au contact.

The structural characterization of the CdTe devices and back-contact layers is performed using a JEOL ARM200CF aberration-corrected scanning transmission electron microscope (STEM) operated at acceleration voltage of 200 kV. The STEM images were acquired using a probe semi-convergence angle of 24 mrad and two annular detectors, a high-angle annular darkfield (HAADF) detector and low-angle annular dark-field (LAADF) detector. X-ray energy dispersive spectroscopy (XEDS) maps and line-scans were obtained using a windowless XEDS silicon drift detector X-MaxN 100 TLE from Oxford Instruments. Electron energy-loss spectroscopy is performed using a Gatan post-column spectrometer, the GIF-Continuum. The combination of these techniques allows for spatially resolved characterization of interfacial atomic structures and chemical compositions with atomic resolution. Cross-sectional TEM samples were prepared using the focused ion-beam (FIB) lift-out technique in a FEI Helios Nanolab 600 dual-beam FIB/SEM system.

III. Results and Discussion

An Au-contact device passivated with Cu_yAlO_x is characterized below. This device is a CdTe absorber and CdS emitter. The back-contact passivator was found to be a ~5nm oxide layer showing up as the dark-contrasted strip in Figure

---

This material is based upon work supported by the U.S. Department of Energy's Office of Energy Efficiency and Renewable Energy (EERE) under the Solar Energy Technologies Office Award Number DE-EE0008974.

978-1-7281-6118-1/22 $31.00 © 2022 IEEE

1B. This oxide-passivation layer is shown with sharp contrast at its interface between the absorber and the contact. This is the case intermittently for ~25% of the back contact interface (see Figure 2). The lack of visibility for the oxide layer in other regions of the interface is most likely due to surface roughness of the polycrystalline CdTe; the electron beam is not always incident in the plane of the interface. It is also possible that the oxide film is truly absent when not visible.

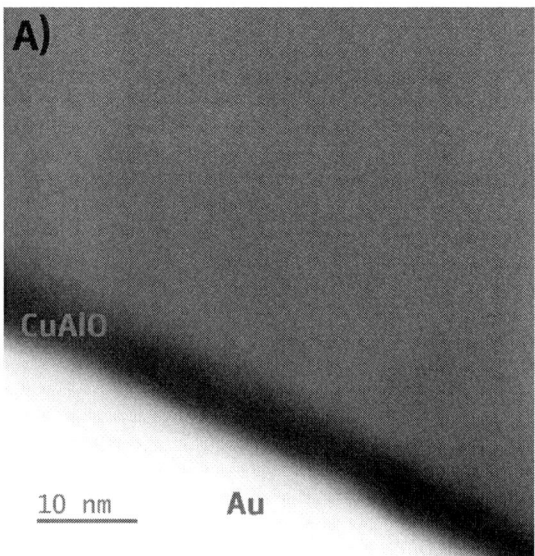

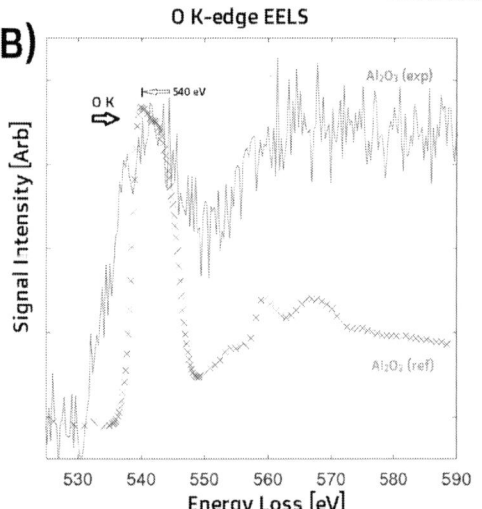

Figure 1: A) HAADF image of the back contact region with layer components labeled and B) EELS plot of the signal from dark-contrasted oxide layer and a reference signal for alumina

EELS data is acquired from the blue rectangle region shown in Figure 1A and the spectrum is plotted in Figure 1B. The EELS data collected from the region is plotted in blue against a reference spectrum for Al_2O_3 with both spectra normalized to the initial O K-edge peak. These spectra are superficially similar although the ratio of the initial O peak to the next peak in the signal is much smaller in the experimental

data (~1:1 rather than ~2:1 in the reference). The experimental data also exhibits a small peak just before the primary 540eV peak near 537eV. This could be the result of the CuO contribution in the oxide or the effect of Cu on the O fine structure. More experimentation is needed to confirm the presence of Cu in the oxide layer.

XEDS analysis in this region proves troublesome due to spurious Al signals that are exacerbated when the probe is near the Au back contact. This makes quantification of Al-signals at the back contact very challenging. The Al signals from XEDS in this region have been rejected after significant study. XEDS analysis for Cu near this interface returns no signal above the background, it is likely the Cu has diffused out of the passivation layer as Cu is highly mobile in these devices. XEDS confirms the presence of oxygen in the passivation layer. However, XEDS signal doesn't have sufficient energy resolution to make energy fine structure characterization and is substituted with EELS.

Figure 2 shows a lower magnification HAADF image of the back contact region. Along some of the Au-contact interface there is an oxide passivation layer visible, as well as some stretches where it is not. Note that both of the instances indicated in Figure 2 occur in the interface of the same grain. The grain would need to terminate with differently oriented surfaces for the layer to be obscured by the other materials where indicated. Also visible in this image are several voids within the absorber, one of which is labeled in the top right. These voids appeared throughout the entire cross-section of the specimen and accounted for ~2% image area of the full lamella.

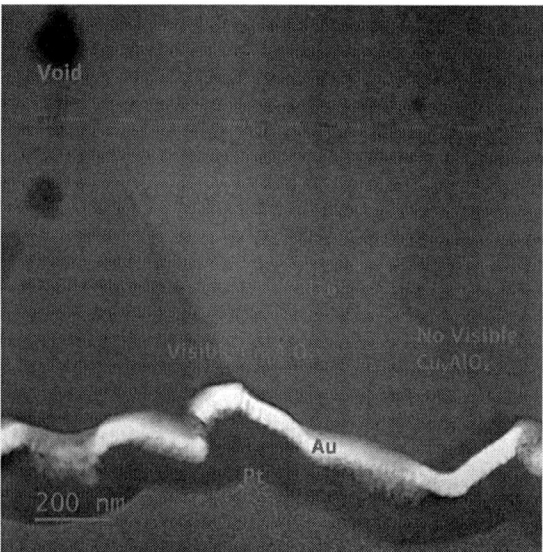

Figure 2: HAADF image of the back contact region. The absorber and contact are labeled. Note the large void in the upper right and curvature associate with the Au-contact.

An ITO-contact device passivated with Cu_yAlO_x is characterized for comparison. This device is also a CdTe absorber and CdS emitter. The device detailed in Figure 3 is an identical architecture to the Au-contact device discussed above with the exception of an ITO back-contact substituted for Au as it is a bifacial device. This makes for a meaningful comparison of the two. The back contact passivator is similarly observed to be ~5nm thick and in dark-contrast intermittently for ~25% of the interface; an example of this is shown in Figure 3A.

The ITO device shows an interesting character along the back contact interface. Intermittently between the absorber and the ITO-contact are stretches of ~200 nm appearing like a

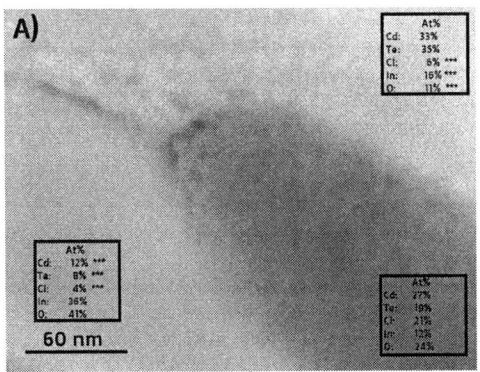

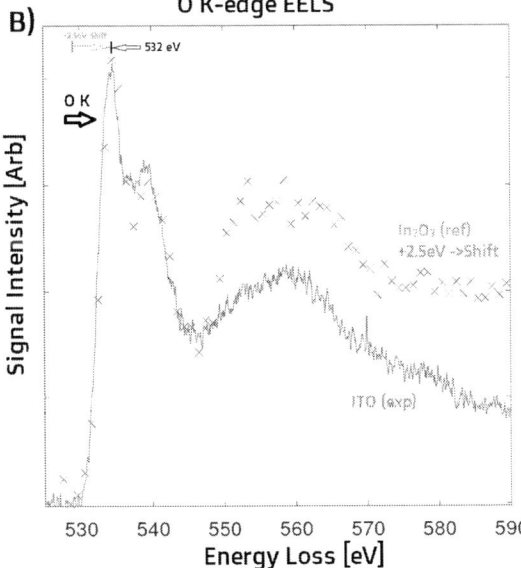

Figure 3: A) HAADF image of an interfacial anomaly between the CdTe and ITO back contact. Regions in black rectangles report atomic percentages as per XEDS measurements at that location. B) EELS plot of the O-*K* edge recorded from the Cl-rich anomaly against a reference for ITO with a 2.5 eV upshift.

'bubble' darkly contrasted against the ITO and CdTe in HAADF. One of these 'bubbles' is the subject of Figure 3.

These bubble-regions prove to be Cl-rich when XEDS data is taken. XEDS data was analyzed to give atomic percentage estimates for 3 areas marked in Figure 3A. In the two areas chosen outside the Cl-rich region, the Cl-At% estimate is ~5% which can be taken to be the noise floor for this signal. In the Cl-rich region, the Cl-At% estimate is 21% which is ~4 times higher than the background and is comparable to the Te-At% estimate in the same region (19%). Oxygen seems also to be present in higher-than-background levels in the Cl-rich region, being twice as concentrated as compared to the signal in the absorber (11% vs. 24%) .

EELS of the O *K*-edge is taken from the Cl-rich region and is plotted in Figure 3B. The EELS spectrum in Figure 3 is similar to an In_2O_3 reference. The most notable difference between the spectrum and the reference is that the experimental spectrum is energy up-shifted by 2.5 eV; the experimental O *K*-edge peak occurs at 2.5 eV higher than the reference. The reference spectrum has been plotted with a +2.5eV shift so the signal traces align at the O *K*-edge maximum for the experimental spectrum at 532 eV. This apparent energy shift is seemingly not due to zero-loss peak misalignment in the experimental data. Both of these spectra are poor fits for an Al_2O_3 reference or a CuO reference suggesting that this is more of the ITO in this region.

IV. CONCLUSION

We utilized high-resolution STEM imaging, EELS, and XEDS chemical analysis to quantify the morphological and elemental changes in the structure of the back contact materials as it relates to the fabrication process of the device. Structural defects and chemical species migration are noted when they are in contrast with expectation and when they correlate to device performance. Comparative studies of different back-contact layer configuration and layer chemistries will be conducted and correlated with the device performance. The goal is to guide the fabrication process in the development of a back contact interface that erodes the valence band barrier to encourage charge carrier transport through the device and improve efficiency.

ACKNOWLEDGMENT

This material is based upon work supported by the U.S. Department of Energy's Office of Energy Efficiency and Renewable Energy (EERE) under the Solar Energy Technologies Office Award Number DE-EE0008974.

REFERENCES

[1] W. Shockley and H. J. Queisser, "Detailed balance limit of efficiency of p-n junction solar cells," J. Appl. Phys., vol. 32, no. 3, pp. 510–519, 1961.

[2] A. R. Davies and J. R. Sites. Effects of non-uniformity on rollover phenomena in CdS/CdTe solar cells. In 33rd IEEE Photovoltaic Specialists Conference, 2008.A. R. Davies and J. R. Sites. Effects of non-uniformity on rollover phenomena in CdS/CdTe solar cells. In 33rd IEEE Photovoltaic Specialists Conference, 2008.

Photon Management in CdSeTe Absorber Solar Cells: The Case for Increased Attention to Optical Cell Design

Carey L. Reich, Arthur Onno, Adam Danielson, Zachary C. Holman, Walajabad S. Sampath

Colorado State University, Fort Collins, CO, United States

Arizona State University, Tempe, AZ, United States

It is well known in mature technologies that smart optical design in solar cells is essential to approaching the detailed balance limit. In CdTe, the SRH recombination that was limiting the radiative efficiencies (both internal and external) is slowly being reduced to levels at which the contribution to recombination from that which is radiative in nature is non-negligible. As such, it is important to start considering the effects of photon management since the re-absorption of photons becomes increasingly feasible. Here we model the implied voltage of absorbers with different back reflectance and demonstrate that real absorbers are now within the range where optical considerations are needed to achieve the highest possible voltages.

Geographic Analysis for Determining the Value of Different Photovoltaic Performance Factors

Madhuri Kumari, Marios Theristis, Joshua S. Stein

Sandia National Laboratories, Albuquerque, New Mexico, 87185, USA

Abstract—Geographic analysis of photovoltaic (PV) performance factors across large regions can help relevant stakeholders make informed, and reduced risk decisions. High temporal and spatial resolution meteorological data from the National Solar Radiation Database are used to investigate performance and cost as an effect of varying system characteristics such as the module temperature coefficients, mounting configurations and coatings. The results demonstrated the strong climatic dependence that these characteristics have on annual energy yield whereas the revenues were dominated by the electricity price.

Keywords—photovoltaics, geographic analysis, atlas, map, energy yield, interannual variability, cost

I. INTRODUCTION

Knowledge of the value of different photovoltaic (PV) performance factors and how this varies across large geographical regions is important for relevant stakeholders. Such factors include temporal and spatial variations in weather and climate (e.g., irradiance, temperature, humidity, wind, etc.) and topography (elevation, slope, and horizon, etc.). PV module and system technology and features influence how geographic factors affect energy yield and their corresponding economic value. However, these effects are typically complex involving several processes that sometimes work in opposite directions and make quantification more difficult. For example, trackers can increase the plane-of array insolation, but also can result in higher module operating temperatures. Both effects depend on geographical factors as well as module and system technology characteristics.

This work aims to create a PV performance atlas, comprised of maps that demonstrate the multitude of geographic effects on PV performance and economics across the U.S. As such, in this paper we quantify the value of a) module temperature coefficients, b) system configuration (i.e., fixed vs. tracked) and c) anti-reflective coatings.

II. METHODOLOGY

We developed a computational environment that includes input meteorological data, a PV performance model and electricity prices. Hourly solar resource and meteorological data were retrieved from the National Solar Radiation Database (NSRDB) from 2010 to 2020 [1] over a 0.5°×0.5° spatial grid encompassing the continental U.S. (> 5,000 locations). The PV performance modeling was simulated in *pvlib-python* [2, 3], and the electricity prices were sourced from the U.S. Energy Information Administration (EIA) [4]. We use this environment

to calculate energy yield across the U.S for different scenarios to investigate the effects of varying module or system characteristics such as temperature coefficients, coatings, mounting configurations, etc.

In this work, we focus on how the performance of modules varies with geographic location for cells of:

1. Different temperature coefficients:
   a. -0.4%/°C
   b. -0.2%/°C
2. Different mounting configurations:
   a. Fixed latitude tilt (Latitude)
   b. Single axis tracking (SAT)
   c. 2-axis tracking (2-axis)
   d. Fixed tilt of 10°
3. Different anti-reflective coatings
   a. Yes anti-reflective coating ("Yes ARC"): Angular loss coefficient, a_r, set to 0.12 in the Martin and Ruiz Incidence Angle Modifier (IAM) model
   b. No anti-reflective coating ("No ARC"): Angular loss coefficient, a_r, set to 0.24 in the Martin and Ruiz IAM model

A diagram depicting the above scenarios simulated is shown in Figure 1. A scenario refers to the unique combination of input parameters of temperature coefficient, tracking, and anti-reflective coating for modelling.

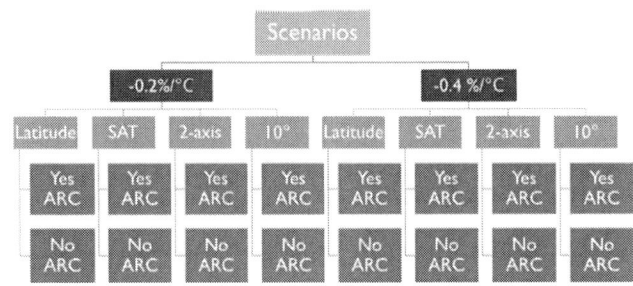

Fig. 1. A diagram of all the simulated scenarios. Temperature coefficients of -0.2%/°C and -0.4%/°C; Fixed latitude tilt mounting configuration (Latitude), single axis tracking (SAT), 2-axis tracking (2-axis), and fixed tilt of 10° for commercial rooftops (10°); and a_r = 0.12 and 0.24 for the Martin and Ruiz IAM Model.

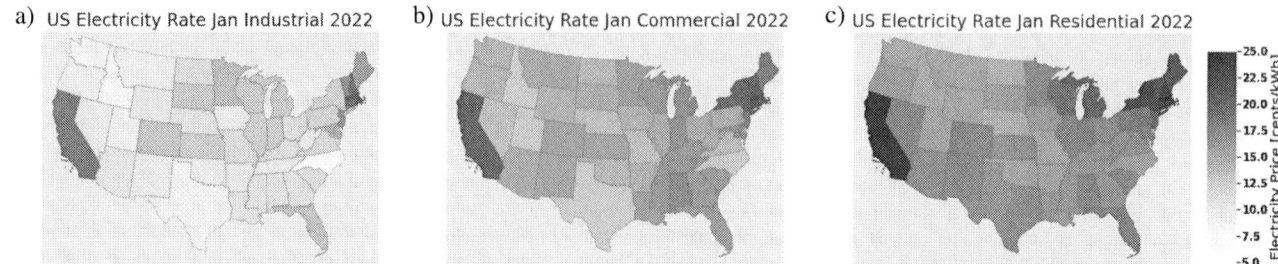

Figure 2. Electricity price for contiguous US by state for January 2022 for the a) industrial sector; b) commercial sector; c) residential sector.

The performance analysis is based on the PVWatts model [5] implemented in *pvlib-python*. This model requires as inputs the: a) STC power rating for module, b) temperature coefficient for power, c) module temperature and d) effective irradiance.

To evaluate the value of different temperature coefficients we chose two end member values from the range of temperature coefficients for commercially available modules. A value of -0.4%/°C represents typical values for aluminum back surface field (Al-BSF) modules while a value close to -0.2%/°C is more typical of modules with heterojunction with intrinsic thin layer (HIT) cells and even cadmium telluride (CdTe).

As mentioned, four types of mounting configurations are evaluated. The fixed tilt of 10° represents the tilt angle of commercial rooftops. Therefore, the revenue gains in such cases will use commercial prices of electricity, whereas the other scenarios will use the industrial price of electricity.

To evaluate the value of different anti-reflective coatings, angular loss coefficients, a_r, of 0.12 and 0.24 were selected. An $a_r = 0.12$ corresponds to lower losses at higher angles of incidence and can be thought of as the scenario where an anti-reflective coating is present, or "Yes ARC". An $a_r = 0.12$

corresponds to higher losses with increasing angles of incidence and can be thought of as a module with uncoated glass or no anti-reflective coating, i.e., "No ARC".

For each location, ten-year simulations were run, using input parameters as described out for each scenario. We used NREL Solar Position model [6], the Hay-Davies transposition model [7], the Sandia module temperature model [8].

The hourly dc power from each simulation was summed for each year and used to calculate the annual energy yield in MWh/kWp/year. The annual energy yields were also averaged to obtain the mean energy yield over the 10-year period. Differences between scenarios for mean annual energy yield, or energy gain, are reported in kWh/kWp/year. These differences are multiplied by the price of electricity by state for the industrial, commercial, and residential sectors for January 2022 from the U.S. EIA, as shown in Fig. 2, to obtain the mean annual revenue difference, or revenue gain, in $/MW/year. The energy gains and revenue gains are plotted on a map and used to evaluate economic implications.

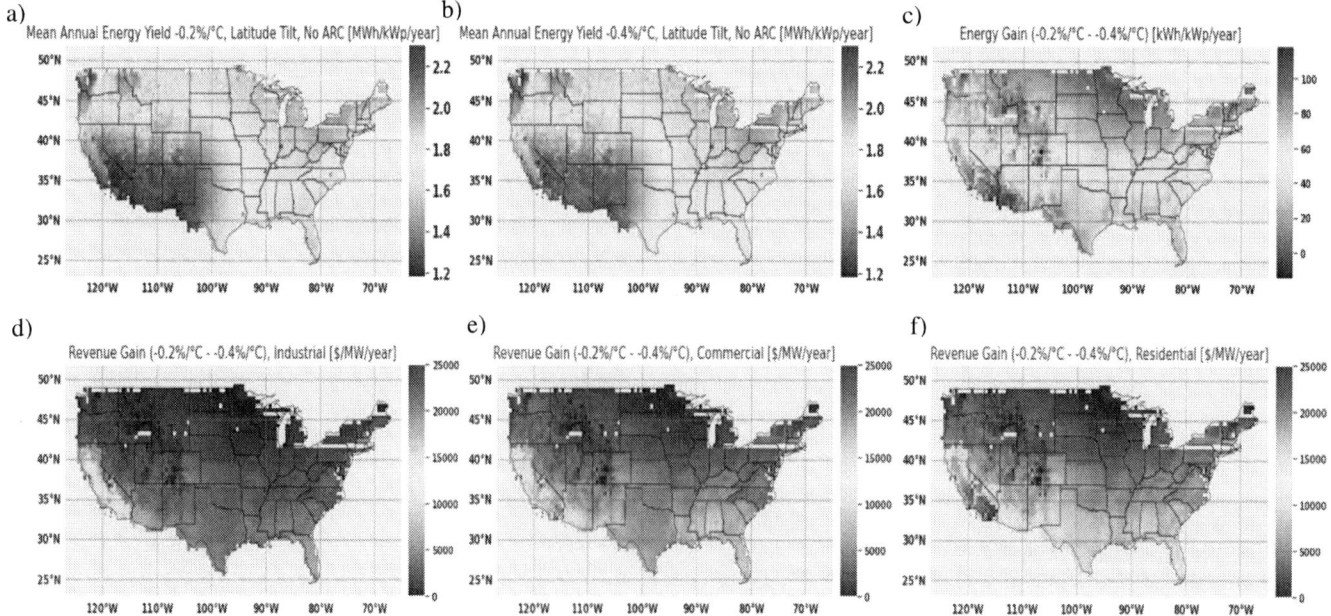

Fig. 3. Temperature coefficient analysis using 10-year mean estimates: a) energy yield for a module temperature coefficient of -0.2%/°C; b) energy yield for a module temperature coefficient of -0.4%/°C; c) difference between the energy yield in a) and b); d) revenue difference using industrial electricity prices by state; e) revenue difference using commercial electricity prices by state; f) revenue difference using residential electricity prices by state. Negative values have been clipped for uniformity purposes.

III. RESULTS

A. Temperature Coefficient Analysis

The results of the temperature coefficient analysis are shown in Fig. 3a and 3b show calculated energy yield over the contiguous U.S. for a temperature coefficient of -0.2%/°C and -0.4%/°C, respectively, with fixed latitude tilt and "No ARC". Fig. 3c shows the difference in mean annual energy yield between Fig. 3a) and Fig. 3b). Fig. 3d, 3e, and 3f take the energy gain from Fig. 3c and multiply it by the price of electricity for the industrial, commercial, and residential sectors, respectively, by state for January 2022 to obtain the revenue gains in $/MW/year. To standardize the colorbar across Fig. 3d to 3f, negative values and values past 25,000 $/MW/year have been clipped.

Fig. 3c. shows high energy gains in the south band of the USA, especially southern California and southern Arizona. The revenue gain plots, Fig. 3d, 3e, and 3f, show the highest gains in California due to the higher electricity prices.

For the industrial sector, the range in value due to changing the temperature coefficient from -0.4%/°C to -0.2%/°C ranges from -1,172 to 16,218 $/MWp/year. For the commercial sector, the revenue gains range from to -1,489 to 21,788 $/MWp/year. Finally, for the residential sector, the revenue gains range from -1,930 to 27,651 $/MWp/year. As expected, the revenue gains are higher in the residential sector because the electricity price dominates the effect. However, commercial and industrial systems are much larger in capacity and therefore, cumulative gains should be larger in those sectors.

B. Mounting Configuration Analysis

With respect to the value of different mounting configurations, Fig. 4a, 4b, and 4c show the energy gains, respectively, for single axis tracking minus fixed latitude tilt, 2-axis tracking minus fixed latitude tilt, and fixed latitude tilt minus fixed 10° tilt. Similarly, Fig. 4d, 4e, and 4f show the revenue gains, respectively, for single axis tracking minus fixed latitude tilt, 2-axis tracking minus fixed latitude tilt, and fixed latitude tilt minus fixed 10° tilt. Fig 4d and 4e use the industrial price of electricity while Fig 4f uses the commercial price of electricity.

The highest energy gains in Figs. 4a and 4b for single axis tracking minus fixed latitude tilt and for 2-axis tracking minus fixed latitude tilt occur in southern California, Arizona, and New Mexico. The highest energy gains for fixed latitude tilt minus fixed 10° tilt occur in the southwest, less so in California compared to 4a and 4b. However, the highest revenue gains for these scenario differences all occur most significantly in California.

C. Anti-reflective Coating Analysis

Fig. 5a shows the mean annual energy differences for "Yes ARC" minus "No ARC" in order to quantify the value of anti-reflective coatings. These simulations are for fixed latitude tilt and temperature coefficient equal to -0.4%/°C. Fig. 5b shows the mean annual revenue difference due to the use of ARC.

Energy gains are most prominent in southern California and the southwest while revenue gains are most prominent primarily in California.

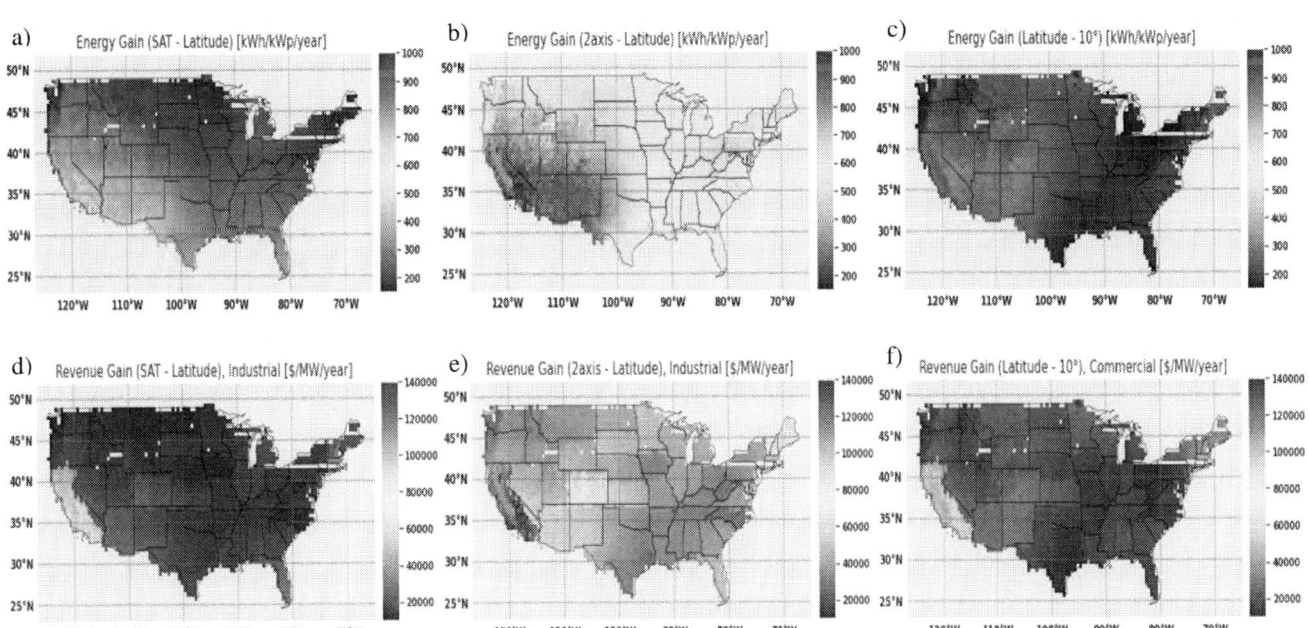

Fig. 4. Mounting configuration analysis using 10-year mean estimates: mean annual energy yield difference for a) Single-Axis Tracking – Latitude with temperature coefficient = -0.4%/°C and No ARC, b) Single-Axis Tracking – Latitude with temperature coefficient = -0.4%/°C and No ARC; c) Latitude - 10° tilt with temperature coefficient = -0.4%/°C and no ARC; 10 year mean estimates: mean annual revenue difference for d) Single-Axis Tracking – Latitude with temperature coefficient = -0.4%/°C and No ARC, e) Single-Axis Tracking – Latitude with temperature coefficient = -0.4%/°C and No ARC; f) Latitude - 10° tilt with temperature coefficient = -0.4%/°C and no ARC

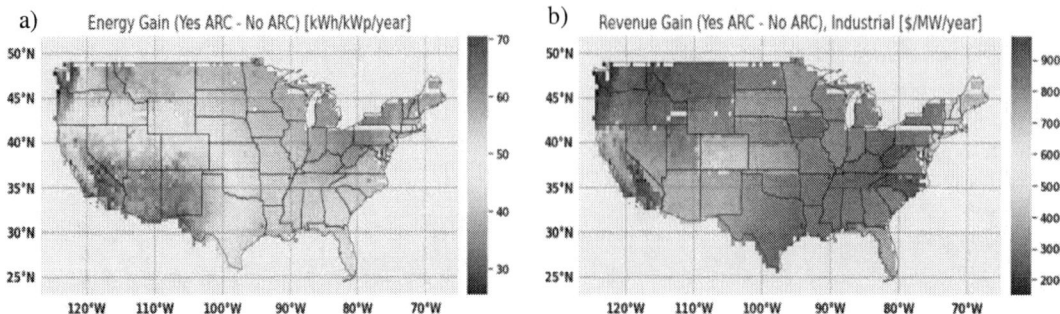

Fig. 5. Anti-reflective coating analysis using 10-year mean yield estimates for "Yes ARC" minus "No ARC" with fixed latitude tilt and temperature coefficient -0.4%/°C of a) annual energy yield difference and b) annual revenue difference.

IV. DISCUSSION

While energy gain spatial distributions are qualitatively more varied, the resulting revenue gains display lower variations across various analyses, with the highest revenue gains notably being in California. This is due to California's high price of electricity relative to the rest of the states. Electricity price is the primary driver for revenue gains.

A. Temperature Coefficient Analysis

For the temperature coefficient analysis, the value of modules with a "lower" temperature coefficient depends on both the site climate and the cost/value of electricity generated by the PV plant. Irradiance and temperature are the primary drivers for energy yield and energy gains, in this case. Our analysis shows that the most significant annual energy difference occurs in the Southwest U.S., due to the higher irradiance and temperatures in this region. These conditions result in the highest annual revenue difference.

Another interesting feature of Fig 3c is that in regions with a cold climate, it is possible to produce more energy using modules with higher temperature coefficients, as the modules operate at temperatures below 25°C for much of the year and the high temperature coefficients are beneficial to energy yields.

B. Mounting Configuration Analysis

The energy gains and revenue gains of 2-axis tracking minus fixed latitude tilt are about twice as much as those of single axis tracking minus fixed latitude tilt. For example, let us take again Los Angeles, CA and assume a 1 MW plant. The revenue gain for single axis tracking minus fixed latitude tilt is 54,065 $/MWp/year while for 2-axis tracking minus fixed latitude tilt it is 111,451 $/MWp/year.

Fig 4c shows the energy gain of a module with fixed latitude tilt minus fixed 10° tilt, conversely the energy loss of a module with fixed 10° tilt minus fixed latitude tilt. Using commercial electricity prices, the energy loss ranges from -346 to -140 kWh/kWp/year. The greatest loss occurs in the southwest and southern California in particular. The lowest losses occur in coastal Washington and Oregon and parts of the Midwest and southeast Texas.

C. Anti-reflective Coating Analysis

The lowest energy and revenue gains come from the antireflecting coating analysis compared to the energy gains from the other analyses.

As an example, the revenue gain in Los Angeles, CA (34.0, -118.0) using the industrial price of electricity would be 7,706 $/MWp/year. Assuming a 1 MW plant, a 30-year project, and a fixed electricity price, the difference in the 30-year revenue estimates would be approximately $231,180. Thus, these apparently small differences in temperature coefficients add up to significant financial differences over the life of a plant.

V. SUMMARY AND CONCLUSION

Three analyses were conducted in this work to investigate the impact of different PV performance factors on revenue: one on the value of a lower temperature coefficient, another on the value of different mounting configurations, and the last on the value of anti-reflective coating.

For the temperature coefficients, irradiance and temperature drive the energy gains. Highest energy gains are seen across the southern US. The 2-axis tracking produces approximately twice as much energy gains and revenue gains compared to the single axis tracking. For the anti-reflective coating analysis, highest energy gains are seen in the southwest and southern California. Although it provides the least energy and revenue gains compared to the other module parameter variations, these small differences amount to significant differences across the lifetime of a project. Nevertheless, all analyses showed that the electricity price is the main driver of revenue gains.

ACKNOWLEDGMENT

This work was supported by the U.S. Department of Energy's Office of Energy Efficiency and Renewable Energy (EERE) under the Solar Energy Technologies Office Award Number 38267. Sandia National Laboratories is a multimission laboratory managed and operated by National Technology & Engineering Solutions of Sandia, LLC, a wholly owned subsidiary of Honeywell International Inc., for the U.S. Department of Energy's National Nuclear Security Administration under contract DE-NA0003525. This paper describes objective technical results and analysis. Any subjective views or opinions that might be expressed in the paper do not necessarily represent the views of the U.S. Department of Energy or the United States Government.

REFERENCES

[1] Sengupta, M., Y. Xie, A. Lopez, A. Habte, G. Maclaurin, and J. Shelby. 2018. "The National Solar Radiation Data Base (NSRDB)." Renewable and Sustainable Energy Reviews 89 (June): 51-60.

[2] Stein, J. S., W. F. Holmgren, J. Forbess and C. W. Hansen (2016). PVLIB: Open Source Photovoltaic Performance Modeling Functions for Matlab and Python. 43rd IEEE Photovoltaic Specialist Conference Portland, OR.

[3] Holmgren, W. F., C. W. Hansen and M. A. Mikofski (2018). "pvlib python: a python package for modeling solar energy systems " The Journal of Open Source Software 3(29): 3.

[4] U.S. Energy Information Administration. *Electric Power Monthly,* U.S. Energy Information Administration, March, 2022. Acessed on: March, 2022. [Online]. Available: https://www.eia.gov/electricity/monthly/epm_table_grapher.php?t=epmt_5_6_a

[5] Dobos, A. P.. *PVWatts Version 5 Manual.* United States: N. p., 2014. Web. doi:10.2172/1158421.

[6] Reda, I.; Andreas, A. (2003). Solar Position Algorithm for Solar Radiation Applications. 55 pp.; NREL Report No. TP-560-34302, Revised January 2008.

[7] Hay, J. E. and J. A. Davies (1980). Calculations of the solar radiation incident on an inclined surface. First Canadian Solar Radiation Data Workshop. J. E. Hay and T. K. Won. Canada, Ministry of Supply and Services.

[8] King, D. L., E. E. Boyson and J. A. Kratochvil (2004). Photovoltaic Array Performance Model. Albuquerque, NM, Sandia National Laboratories. SAND2004-3535.

Computerized Tool for Students Training in Solar Geometry

Johjan Stiven Zea Fernández[a], Mario Luna-delRisco[b], Sebastián Villegas Moncada[b], Carlos Ernesto Arrieta González[b], Johann A. Hernández M[c], Carlos A Arredondo Orozco[b]

[a]Ingeniero en Energía, Universidad de Medellín
[b]Programa de Ingeniería en Energía, Grupo de Investigación en Energía – GRINEN, Universidad de Medellín
Medellín, Colombia
[c]Grupo LIFAE, Universidad Distrital Francisco José de Caldas
Bogotá, Colombia
stivenzea2499@gmail.com; mluna@udemedellin.edu.co; svillegas@udemedellin.edu.co;
carrieta@udemedellin.edu.co; jahernandezm@udistrital.edu.co; caarredondo@udemedellin.edu.co

Abstract— **Renewable energies have experienced significant growth in recent years. The new installed capacity from renewable sources has surpassed the new installed capacity from conventional sources. Solar energy is the fastest growing and the most pervasive. In this sense, it is essential that the sizing and design of solar energy generation systems (such as photovoltaic systems or solar thermal systems) be carried out considering the solar geometry to ensure that that the location of the generation system allows capturing the greatest amount of solar radiation on any given day or time of the year and, thus, generate more energy. This paper presents the development of a computerized tool, for educational purposes, that allows performing solar geometry calculations to be used in solar energy generation systems. The computerized tool was developed using MATLAB® AppDesigner. The computerized tool facilitates and helps the study of how solar geometry affects the performance of solar energy systems. The computerized tool was tested with students of Energy Engineering Program at Universidad de Medellín and allowed to verify that the learning process of the subject have improved substantially.**

Keywords— *renewable energies, solar energy, computerized tool, education, training, solar geometry, AppDesigner, MATLAB®.*

I. Introduction

Climate change and the predicted future decline of traditional energy resources such as oil and coal have made energy transition an important need for all countries [1], which has driven the use and growth of non-conventional sources of energy generation. New non-conventional renewable energy sources such as solar, wind and biomass are becoming more widespread and have new applications as research and new developments make them more efficient and affordable [2].

Solar technology is the most widely used and widespread form of non-conventional renewable energy generation worldwide. The extensive use of solar energy is reflected in the approximately 627 GW installed in 96 countries, 18 of which have more than 1GW of installed capacity according to the REN21 report for 2020 [3]; the growth for the solar market in this year was approximately 115 GW.

With international agreements related to climate change such as COP21 in Paris [4] and the Kyoto Protocol [5], countries are increasingly allocating more resources for research and implementation of projects related to solar energy to mitigate the effects of climate change. In Colombia, the national government has not been oblivious to these guidelines, therefore special laws and regulations have been sanctioned [6], such as CREG 030 of 2018 [7] and Law 1715 of 2014 [8] which encourage the use of clean technologies for power generation from clean sources, in addition, tax benefits are granted. These benefits have boosted the renewable industry in the country, and according to the UPME (Unidad de Planeación Minero Energética), more than 800 solar generation projects have been registered from 2007 to 2020 [9], which represents more than 50% of the projects registered before this entity for this period.

The energy sector requires professionals with comprehensive knowledge of technologies, including solar photovoltaic technology; these professionals must be able to design and size properly any solar photovoltaic installation, which requires extensive knowledge and mastery of solar geometry so that the systems can generate as much energy as possible.

Therefore, a computer tool has been developed for the teaching process of solar geometry concepts. The tool allows to evaluate, calculate, and understand, in a graphic and intuitive way, the principles of solar geometry applied to solar photovoltaic and solar thermal systems.

The tool has been tested with students of Energy Engineering at the University of Medellin (Colombia) and it was verified that helped them to easily understand the concepts of solar geometry and its application in solar power generation systems.

978-1-7281-6118-1/22 $31.00 © 2022 IEEE

This article presents the most relevant aspects for the development of the tool, as well as the description of the tool and the results obtained from its use.

II. TEORETICAL APPROACH

To estimate some of the solar parameters, the software tool is based on the solar geometry equations to perform the calculations considering the input variables presented in Table 1.

TABLE I: Input Variables of the informatic Tool

Variable	Name	Unit
Φ	Local latitude	Degrees (°)
Λ	Local longitude	Degrees (°)
LT	Local time	Hours (h)
GMT	Deviation due to UTC	Hours (h)

The solar time parameters can be calculated using equations 1 to 5 [10], [11]:

$$LSTM = \frac{15°}{h}(GMT) \quad (1)$$

$$B = \frac{360\ dias}{365\ dias}(n - 81\ dias) \quad (2)$$

$$EoT = 9.87\,\mathrm{m}\,\sin(2B) - 7.53\,\mathrm{m}\,\cos(B) - 1.5m\,\sin(B) \quad (3)$$

$$TC = 4\frac{min}{°}(\Lambda - LSTM) + EoT \quad (4)$$

$$LST = LT + \frac{TC}{60m} \quad (5)$$

Where: $LSTM$ represents the standard longitude, B is a constant or parameter for the equation of time, EoT is the equation of time, TC is the time correction, n is the number of days of the year and LST is the solar local time.

The position parameters and solar angles are calculated below using equations 6 through 9 [10], [11]:

$$HRA = \frac{15°}{h}(LST - 15h) \quad (6)$$

$$\delta = 23.45°\sin\left[\frac{360\ dias}{365\ dias}(248 + n)\right] \quad (7)$$

$$\alpha = \sin^{-1}(\sin\delta\sin\Phi + \cos\delta\cos HRA\cos\Phi) \quad (8)$$

$$\psi = \cos^{-1}\left[\frac{\sin\delta\cos\Phi - \cos\delta\sin\Phi\cos HRA}{\cos\alpha}\right] \quad (9)$$

Where: HRA is the solar hour angle, δ is the declination angle of the earth with respect to the sun, α is the solar altitude angle and ψ is the solar azimuth angle.

The daily parameters are defined by equations 10, 11 and 12 [10], [11]:

$$SR = 12h - \left(\frac{1h}{15°}\right)\cos^{-1}(-\tan\Phi\tan\delta) - \frac{TC}{60m} \quad (10)$$

$$SS = 12h + \left(\frac{1h}{15°}\right)\cos^{-1}(-\tan\Phi\tan\delta) - \frac{TC}{60m} \quad (11)$$

$$DD = SS - SR \quad (12)$$

Where: SR represents the sunrise time, SS the sunset time and DD the day length in hours.

Finally, the optimal tilt angle for photovoltaic module - βop (with positive sign if north or negative sign if south), is defined in equation 13 [10]:

$$\beta op = 3.7° + 0.69(\Phi) \quad (13)$$

III. USE OF THE INFORMATIC TOOL

The tool was designed in MATLAB®, using the AppDesigner and is described in the following sections.

A. Daily Calculations

When starting the tool, a window opens in which the latitude, longitude, time (GMT), day, month, year, hour and minute data must be entered in order to perform the calculations at that exact time, date and location (Fig. 1).

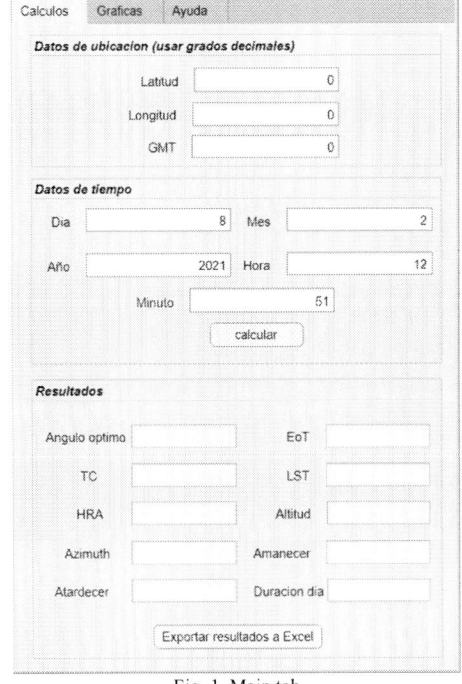

Fig. 1. Main tab

In this tab, 3 modules can be found which correspond to the location data, time, and calculated results. For the location data, the user must enter the latitude, longitude, and deviation from UTC of the area for which the analysis is to be made. For the time data, which is automatically configured at system time, the user must enter the year, month, day, hour and minute, the latter in 24 h format.

By clicking on the calculate button, the software tool uses the equations previously presented in section I to calculate the

978-1-7281-6118-1/22 $31.00 © 2022 IEEE

solar variables of interest and presents them in the same tab, in a section called results (Fig. 2). The calculated variables are exported as an Excel file (by clicking on the "Export results to Excel" button) that allows visualizing the input variables and the results obtained (Fig. 3).

B. Graphs

The second tab of the tool shows the daily or annual variation of some of the variables, which can be accessed by clicking on the "Graphs" tab.

In this tab, the user can select the variable to be analyzed, as well as the time period of the analysis for that variable; user can choose a daily period and in this option the computer tool takes the location data registered in the "calculations" tab and shows the change of the variable during the whole day from sunrise to sunset.

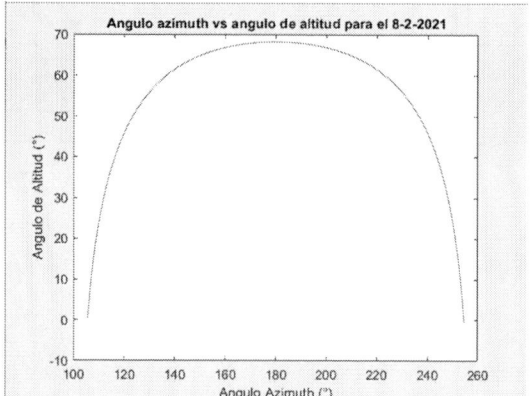
Fig. 4. Example of altitude vs. azimuth graph for a day in Medellin

The graphs are displayed as a MATLAB® tab, where it can be analyzed point by point, change the characteristics of the graph or save it either as an image or as a MATLAB® file for further analysis.

IV. RESULTS ANALYSIS AND DISCUSSION

The value of the program lies mainly in its contribution to training and education as a tool for teaching the concepts of solar geometry. The calculation module offers students the opportunity to compare results with those they can perform manually to identify errors in the handling of the formulas and to strengthen the concepts learned. The graphing tab or module also provides an intuitive approach to understanding the motion of the sun with respect to the earth, helping to visualize the variations of different solar parameters, either for a day or the annual variation, and how these are affected by the position of the sun and the effect they will have on the correct positioning of a solar power generation system.

Fig. 2. Results module for data of Medellin - Colombia

Variables de entrada		
Variable	**valor**	
Latitud	6.28	
Longitud	-75.56	
GMT		-5
Dia		8
Hora		12
Minuto		51
Año		2021
Resultados		
Variable	**Resultado**	
Angulo optimo	8.03°S	
EOT	-14.45	
TC	-16.69 min	
LST		12:34
HRA	8.58°	
Altitud	66.65°	
Azmiuth	201.27°	
Amanecer		6:23
Atardecer		18:09
Duracion del dia	11 horas y 46 minutos	

Fig. 3. Exported data to Excel

The study can also be done for a given year at a certain location by clicking on the "calculations" tab.

The variables to analyze for both a day and a year can be declination, time correction, equation of time, solar local time,

solar hour angle, solar altitude angle, solar azimuth angle, sunrise time, sunset time, and day length. In addition, the variable "altitude vs. azimuth" can be chosen to plot and analyze these. Examples of the generated and obtained graphs are illustrated in Figs. 4 and 5.

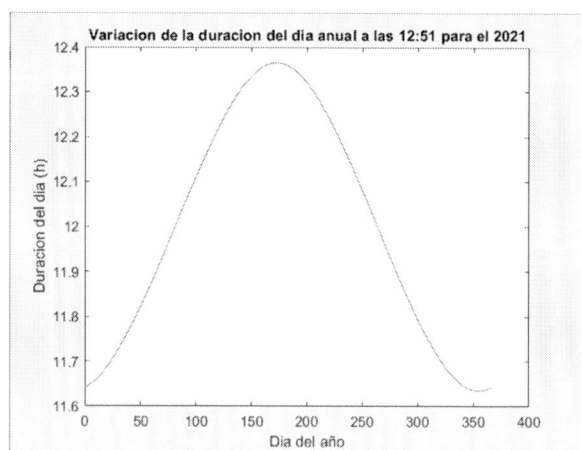
Fig. 5. Example of an annual graph for the duration of the day in the city of Medellin

An example of how the computer tool helps students to understand concepts in a straightforward way is shown in Fig. 6, where a graph of local weather as a function of the day of the year can be visualized. This allows to understand how the sun changes or varies its trajectory or "does not travel along the same path" during the year. For the city of Medellin, the sun "may transit" to the south or to the north depending on the day of the year; this may be counterintuitive, since the optimal tilt angle for a photovoltaic module in Medellin is approximately 8° facing south.

Observing the graph for the azimuth angle generated with the computer tool (Fig. 6) we can see that the number of days that the sun spends in the north corresponds to only one third of the total days per year. With these graphs and the comprehension of the concepts, it is possible to properly design and size solar photovoltaic or solar thermal systems in such a way that a greater amount of solar radiation can be captured for energy generation, which allows the systems to be more efficient from the design point of view.

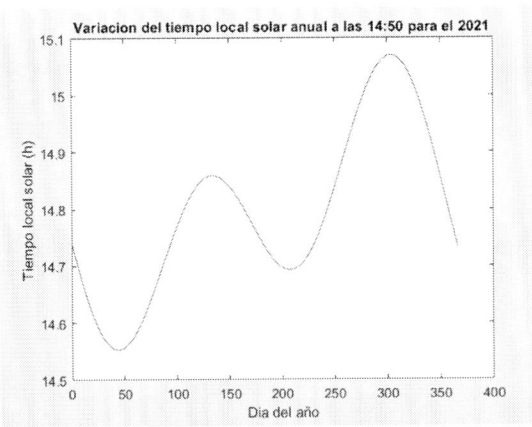

Fig. 6. Example of the annual variation in local solar time

The value of the computer tool developed is not limited to students, but also to trained professionals who need to apply the concepts previously learned, offering additional possibilities to apply these concepts in their field of work. For example, the developed tool can be used to check how a solar generation system with a tracking system is performing to maximize efficiency or to reposition fixed systems considering the solar geometry parameters allowing to enhance the performance and increase the energy generated.

Likewise, professionals who have learned geometry concepts and need to apply them anywhere in the world can use the computerized tool to familiarize themselves with the characteristics of the sun's position and adapt them quickly to carry on their work efficiently (although there are websites such as the National Oceanic and Atmospheric Administration - NOAA [12], but with less functionality or more basic compared to the one presented in this paper).

V. CONCLUSIONS

A computer tool was developed for educational purposes to calculate solar geometry variables that can be used in solar energy generation systems. Even though the computer tool has an educational focus, it can also be used by professionals who need to use solar geometry in the work environment.

The tool allows to strengthen the concepts of solar geometry and use them where required. Although online programs that calculate the position of the sun can be found, these are very general, they are not focused on education and do not have additional functions, such as the graphs for a specific time of day or throughout the year and the export of data to an Excel sheet.

The software tool developed has been tested with students of the Energy Engineering program of the University of Medellin in the subject "Renewable Energy Seminar 2: solar energy" and has proven to be a tool that helps to significantly improve the understanding of the concepts of solar geometry, facilitating the understanding and proper use of solar geometry in the design and sizing of photovoltaic and solar thermal systems. The software tool is currently in the registration process.

REFERENCES

[1] X. Labandeira, P. Linares and K. Würzburg "Energías renovables y cambio clímatico," Cuad. Económicos ICE, 2012.

[2] D. Gielen, F. Boshell, D. Saygin, M. D. Bazilian, N. Wagner, and R. Gorini, "The role of renewable energy in the global energy transformation," Energy Strateg. Rev, vol. 24, 2019.

[3] D. Henner and REN21, Ren21. 2020.

[4] S. Di Pietro, "Acuerdo de París," Coop. Desarro., vol. 25, no. 111, 2017.

[5] T. Gerden, "The adoption of the kyoto protocol of the United Nations framework convention on climate change," Prisp. za Novejso Zgodovino, vol. 58, no. 2, 2018.

[6] Ministerio de Ambiente, El Acuerdo de Paris, así actuará Colombia frente al Cambio Climático. 2009.

[7] Comision de Regulación de Energía y Gas CREG, "Resolución No. 30 de mayo de 2018," Mme. p. 13, 2018.

[8] Congreso de Colombia, Ley 1715 de 2014. Diario oficial 2014.

[9] UPME, "Registro de proyectos de generación," 2022. Disponible online en: https://app.powerbi.com/view?r=eyJrIjoiNzBhN2Q4YmMt N2IxMy00Mjg2LWJhZTctMjRkNWE2NDdlMzI0IiwidCI 6IjgxNTAwZjZkLWJjZTktNDgzNC1iNDQ2LTc0YjVmY jljZjEwZSIsImMiOjh9

[10] S. V. Szokolay, Solar geometry, vol. 2, no. 9. 2007.

[11] Soteris A. Kalogirou, Solar energy engineering, processes and systems. 2014.

[12] NOAA, "Solar position calculator." Available online: https://www.esrl.noaa.gov/gmd/grad/solcalc/azel.html

Anisotropy-induced fluctuations in Cu(In,Ga)Se2

Diego Colombara

Università degli Studi di Genova, Genova, Italy

The attractiveness of Cu(In,Ga)Se2 (CIGS) is still curtailed by the R&D gap that separates it from silicon. Overcoming the gap requires strategic approaches, leaving plenty of room for research in industry and lab. Yet, the progress of this technology hinges on our understanding of the diffusion phenomena that occur during and after the material growth, particularly in combination with alkali metal doping. This contribution introduces a sponge cake simplified model of atomic diffusion in CIGS, based on insights drawn from recent and older (but crucial) literature. The concept of anisotropy-induced fluctuations emerges, with the ambition to stir the community and unlock the full potential of the technology. Keywords-Alkali metal doping, PDT, atomic diffusion, texture

Development of Spatial Mapping and Degradation Monitoring for Perovskite Films

Emily J Miller, Biwas Subedi, Jaehoon Chung, Chongwen Li, Yanfa Yan, Nikolas J Podraza

The University of Toledo, Toledo, OH, United States

Hybrid organic-inorganic metal halide based perovskite films prepared in device relevant structures are examined by mapping spectroscopic ellipsometry measurements to determine spatial variations in structural and complex optical properties of the perovskite film. The measurements are repeated several times over the course of 48 hours in order to track changes in these properties as well as the spatial dependence in how the film degrades. Preliminary analysis of mapping and time dependent measurements of a $(FAPbI3)0.95(MAPbBr3)0.05$ perovskite film in a device-like structure consisting of soda lime glass superstrate / indium tin oxide top contact / PEDOT:PSS hole transport layer / perovskite film suggests that degradation is indicated by decreases in relative density of the perovskite nucleation layers, changes in perovskite effective film thickness, and reduction in the quality of fit indicated by increases in thee the mean square error (MSE) over time when fixing the complex optical properties of the perovskite to those obtained prior to degradation. The increase in the MSE and reduced quality of fit implies that the initial perovskite optical properties are insufficient to describe changes in the opto-electronic response during degradation occurring simultaneous to changes in the film structure. These changes vary with position from the center of the sample to the corner. This approach develops the methodology in analyzing spatially dependent variations in perovskite materials which will be necessary during scale up for large area depositions and industrial fabrication.

PV+ Storage Operation and Maintenance

Natalie Gayoso, Nicole D Jackson, Thushara Gunda, Jal Desai, Andy Walker

Sandia National Laboratories, Albuquerque, NM, United States

National Renewable Energy Laboratory, Golden, CO, United States

Photovoltaic (PV) technology is a rapidly developing technology in response to supply-demand balancing needs. Although there is some understanding of costs associated with PV O&M, costs associated with emerging technologies such as PV plus storage lack details about the specific systems and/or activities that contribute to the cost values. This study aims to address this gap by exploring the specific factors and drivers contributing to utility-scale PV plus storage (UPVS) systems O&M costs, how particular storage technologies were selected, O&M data collection, and ongoing challenges in this space. Here, we present an initial analysis of data collected from 10 semi-structured interviews and questionnaires representing 50 MW of installed battery storage capacity. More detailed analysis of collected results and insights into additional cost drivers will be presented in the full conference presentation.

Effect of Microstructure on the Photoactivity of Thin Film CdSe

Rachael Greenhalgh, Kerrie Morris , Vladislav Kornienko, Martin Bliss, Ali Abbas, Jake Bowers and Michael Walls

Centre of Renwable Energy Systems Technology, CREST, Loughborough University, Loughborough, LE11 3TU, UK

Abstract—CdSe is increasingly being used as a Se source in CdSeTe/CdTe photovoltaic device fabrication. In this paper, a CdSe thin film has been deposited by sputtering These films have then been subject to a $CdCl_2$ activation treatment. The as deposited and treated films have then been characterized with XRD, EBSD, TEM, EDX and PL to analyse the relationship between photoactivity and the microstructural changes that occur during $CdCl_2$ activation. The as deposited CdSe is photo-inactive, the structure is columnar and is a heavily faulted mix of cubic and hexagonal. Cross-sectional TEM reveals a high density of stacking faults. After the $CdCl_2$ treatment, CdSe is photoactive and the microstructure is clearly hexagonal and indexes well in EBSD. The stacking faults are removed. It is important that thin film CdSe at the front of the device is photoactive. It must be fully exposed to the cadmium chloride process and recrystallized if its use as a source of Se in devices is to be successful. Untreated CdSe will have a detrimental effect on device performance.

Keywords—CdSe, Solar Photovoltaics, CdTe, Thin Film, Photoactive, Microstructure, CdSeTe

I. Introduction

Including a $CdSe_{(x)}Te_{(1-x)}$ alloy at the front of the absorber improves CdTe device efficiency by increasing the current density by grading the band gap through the device. It also increases carrier lifetime by reducing defects in the bulk and assisting the passivation of grain boundaries [1]–[4]. The Se can be incorporated at the front of the device with the deposition of a CdSeTe alloy [5], or as a CdSe thin film followed by CdTe and then interdiffused during the $CdCl_2$ activation process [6]. This thin film CdSe based architecture is shown schematically in Fig. 1a).

CdSe has been deposited previously by chemical bath [7], RF sputtering [6] DC sputtering [8] vapor transport deposition [9] and evaporation [10]. Deposition of CdSe with magnetron sputtering causes the Ar working gas to be trapped in the film, this leads to gas bubbles and surface blisters during post-processing anneals [8]. This effect has also been seen in sputtered CdTe after post processing [11]. CdSe has also been used in quantum dot (QD) form to produce Light Emitting Diodes (LEDs). These LEDS rely on ligands to passivate the surfaces of the CdSe quantum dots [12]. Hexagonal CdSe quantum dots have been found to be photoactive with a bandgap of up to 2.29 eV depending on the size. This bandgap is higher than commonly reported bulk CdSe values at 1.7 eV, this

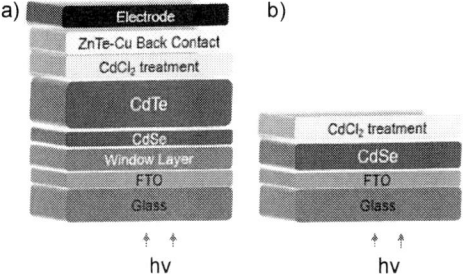

Fig. 1. a) Architcture of CdSeTe/CdTe device produced using a CdSe thin film layer. Se diffuses into the CdTe during $CdCl_2$ activation to form CdSeTe b) the structure of the treated CdSe films in this research.

difference is due to quantum effects of the small size of the CdSe QDs [13]. This shows that high quality hexagonal CdSe is photoactive with particular ligands. A previous study by Poplawsky et al. [14] showed that untreated hexagonal CdSe is not photoactive [15]. When a thick layer of CdSe is incorporated into CdTe, it is possible that some CdSe remains undiffused at the front of the device causing a non-photoactive parasitic layer that is detrimental to device performance. In this paper we examine the effect of microstructure on CdSe photoactivity before and after $CdCl_2$ activation. The thin film structure used in this work is shown in Fig. 1b.

II. Experimental

A. Thin Film Preparation

Deposition of the sputtered CdSe films was performed using a PV Solar (PowerVision Ltd) load locked sputtering system using a pulsed dc power supply. This system is fitted with four 6-inch vertically mounted, circular magnetrons. The substrate holder carousel can hold up to four 5 cm x 5 cm substrates and rotates in the center of the chamber at 100 rpm around IR heating lamps. The CdSe was deposited at 440°C on a stationary (non-rotated) TEC12D™ (Pilkington, NSG) substrate (Fig. 1b), using 500 W deposition power. The pulse frequency used was 200 kHz with a 2 μs reverse time and 15 second ramp. Ar was used as the working gas at a pressure of 2.7 mTorr. The treatment was performed by evaporating a layer of $CdCl_2$ onto the CdSe film, followed by an anneal on a hotplate for 10 minutes at 440 °C.

978-1-7281-6118-1/22 $31.00 © 2022 IEEE

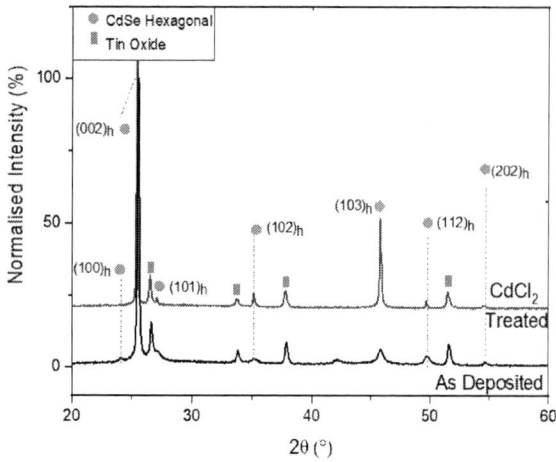

Fig. 2. X-Ray Diffraction patterns of as deposited sputtered CdSe and CdCl$_2$ treated CdSe

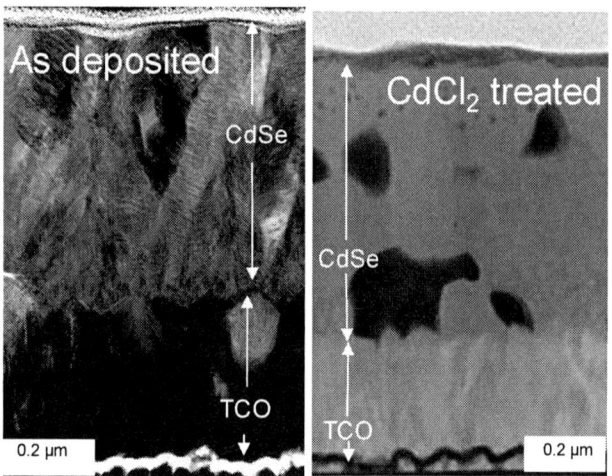

Fig. 3. TEM cross-section of as deposited sputtered CdSe on the left, and a CdCl$_2$ treated sputtered CdSe on the right

B. Characterisation

X-ray diffraction (XRD) patterns were obtained using a Bruker benchtop D2, using Cu Kα X-rays. The Transmission Electron Microscope (TEM) images were obtained on a FEI Tecnai 20, equipped with an Oxford Instruments C-max N8 TLESDD for compositional EDX measurements and maps. A Ga Focussed Ion Beam (FIB) was used to prepare TEM lamellar specimens. Electron Backscatter Diffraction (EBSD) was performed using a G4 Xenon Plasma Focused Ion Beam (PFIB) equipped with an Oxford symmetry CMOS detector. Scan parameters were in immersion mode at 3 kV with a 15 nm step size. Photoluminescence (PL) measurements were performed on an in house-built system [15] between the wavelengths of 500 – 1500nm with a 1 nm step size with a 640 nm laser.

III. RESULTS

A. As Deposited CdSe

CdSe deposited by sputtering on a substrate held at 450°C is a mixed phase material of hexagonal and cubic, with hexagonal

the dominant phase. The XRD spectra in Fig. 2 indicates a strongly (001)$_h$ and (111)$_c$ preferentially oriented film. The TEM image in Figure 3 shows a columnar grain structure with a high density of stacking faults present in the 750nm as deposited film. An EBSD analyis of the thin film cross section in Fig. 4 confirms a mostly hexagonal film with a (001)$_h$ texture. The EBSD cross section of the as deposited CdSe material is not well indexed because the thin film is heavily faulted and mixed cubic and hexagonal phases. As mentioned previously [8], Ar gas is incorporated during the sputter deposition of CdSe. This Ar in the grain boundaries can be observed in the EDX in Figure 5, and areas where the Ar is present there is reduced Cd and Se signal.

B. CdCl$_2$ Treated CdSe

The XRD spectra in Figure 2 shows that sputtered CdSe treated with a CdCl$_2$ annealing treatment is hexagonal with a change in the preferential orientation. The (101)$_h$ and (110)$_h$ Miller planes are much more intense after the CdCl$_2$ treatment, corresponding to a hexagonal, polycrystalline film with increased randomization of grain orientations. The peaks are also narrower, indicating larger grain sizes. Figure 3 shows the fully re-crystallized rounded grains with rounded voids in the TEM cross-section. image. Figure 4 c) and d) shows CdSe after CdCl$_2$. an increase in grain size and improvement of pattern quality can be observed, despite the significant number of voids. The grains lose their columnar structure after recrystallisation with a weaker texture but still shows a (100)$_h$

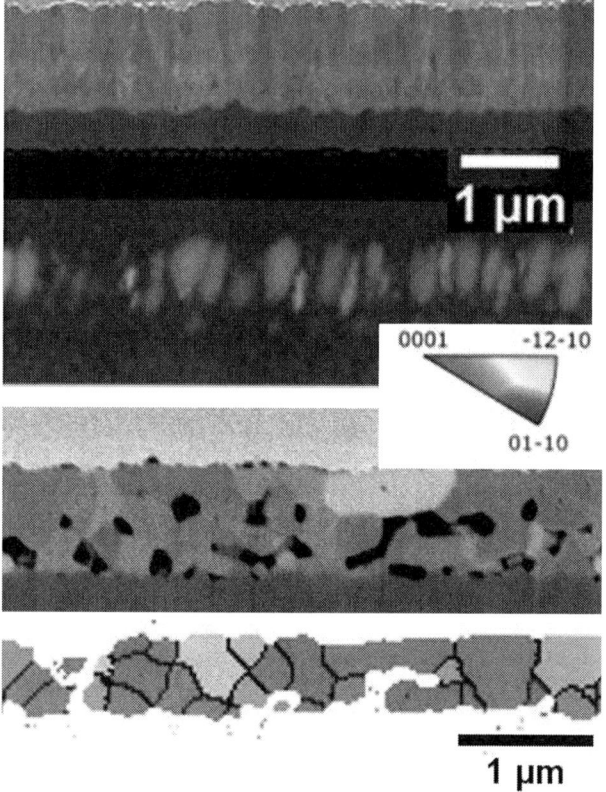

Fig. 4. EBSD cross-sections of CdSe film as deposited (a) electron backscatter image and (b) phase orientation map, and after CdCl$_2$ treatment (c) electron image and (d) phase orientation map.

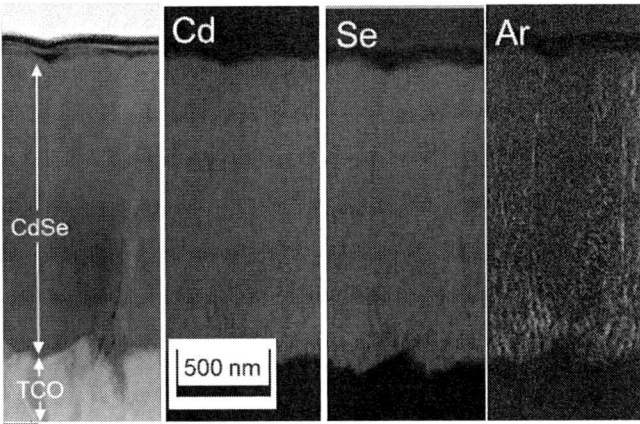

Fig. 5. Compositional EDX maps of a CdSe film as deposited, with Ar present along the columnar grain boundaries.

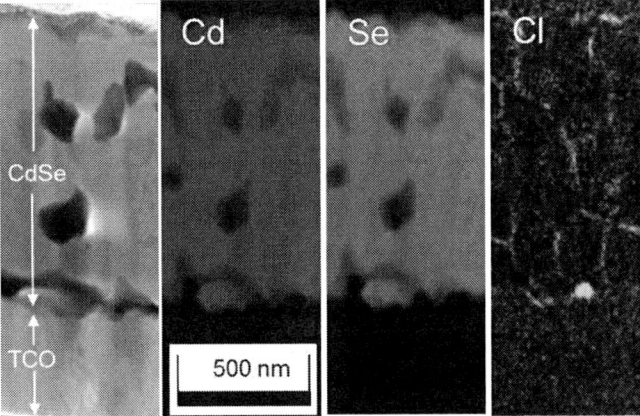

Fig. 6. Compositional EDX maps of a CdSe film after CdCl$_2$ treatment with Cl decorating the grain boundaries.

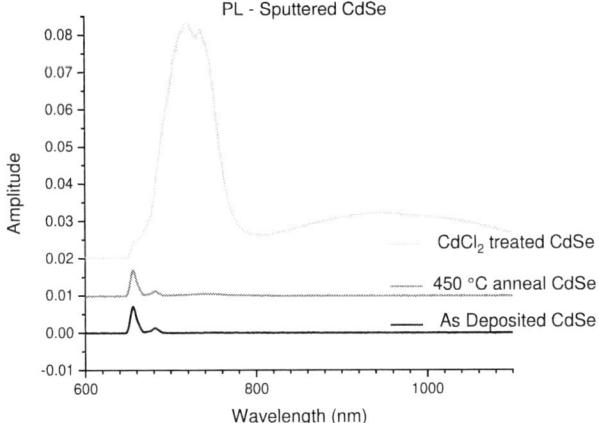

Fig. 7. Photoluminesence spectra of sputtered CdSe as deposited, after a 450°C anneal and after CdCl$_2$ treatment.

preference along the growth direction. The EBSD indexes well indicating that the thin film is now much less faulted and hexagonal. These effects of CdCl$_2$ treatment are similar to those previously observed to occur in CdTe [16] and have been confirmed using cross-sectional TEM.

Chlorine can clearly be observed decorating the grain boundaries of CdSe after CdCl$_2$ treatment (Figure 6), this shows that CdSe and CdTe interact with Cl and CdCl$_2$ treatment in the same way.

C. Photoluminescense

The photoluminescence in Fig. 7 of as deposited, mixed-phase CdSe shows no peak at the bandgap of ~730 nm (1.7 eV). However, an intense peak at 719-734 nm (1.72-1.68 eV) is present after the CdCl$_2$ treatment. As seen in the EBSD and XRD, the CdCl$_2$ treatment has removed defects such as stacking faults and caused the thin film to recrystallise into a pure hexagonal phase. This hexagonal CdSe produces a strong photoluminescent intensity.

IV. DISCUSSION

The as-deposited sputtered thin film CdSe is columnar, heavily faulted and is not photoactive. XRD and EBSD show that it is mixed cubic and hexagonal. Previously, sputtered CdSe films were interpreted to be amorphous and cubic, however it is more likely these are highly faulted hexagonal films [17] with a preferent growth direction in the (002)$_h$ direction .The TEM images show clear crystallinity and supports the interpretation of mixed cubic and hexagonal structure as deposited.

Thin film CdSe following the CdCl$_2$ activation treatment undergoes recrystallisation similar to that observed with CdTe. Stacking faults are removed and the structure is hexagonal as confirmed with well indexed EBSD images. Grain growth and re-orientation is also observed in EBSD, TEM and XRD results. Accompanying these microstructural changes after activation, the hexagonal CdSe is photoactive at its bandgap of 1.71 eV (726 nm wavelength). It is likely that Cl decorates and passivates the CdSe grain boundaries as observed in CdTe after the CdCl$_2$ treatment [16]. Passivation of the grain boundaries and other defects reduces non-radiative recombination and allows radiative recombination across the band gap, as seen in the luminescence observed in the PL spectra.

Photoactivity of CdSe after CdCl$_2$ treatment is an important and useful effect, this could be beneficial in the production of L.E.Ds. However, CdSe at the front of a CdSeTe device could cause negative effects. If the diffusion of Se into the CdTe layer is incomplete or if untreated non-photoactive CdSe remains at the front of the device [15], [18], this photo-inactive layer would cause parasitic absorption at the front of the device. CdSe is also hexagonal and n-type, so may lead to a buried junction or lattice mismatch at the CdSe/CdSeTe interface.

Further work includes full device manufacture with sputtered CdSe to observe how detrimental these voids are in a full device. Further CdCl$_2$ treatments of sputtered, chemical bath and evaporated CdSe films and analysis by optical and electrical characterisation will be important to fully analyse this material's properties.

ACKNOWLEDGMENTS

The authors are grateful to the EPSRC for funding this project through grants EPW00092X/1.

REFERENCES

[1] A. H. Munshi et al., "Polycrystalline CdTe photovoltaics with efficiency over 18% through improved absorber passivation and current collection," Sol. Energy Mater. Sol. Cells, vol. 176, pp. 9–18, 2018.

[2] N. R. Paudel and Y. Yan, "Enhancing the photo-currents of CdTe thin-film solar cells in both short and long wavelength regions," Appl. Phys. Lett., vol. 105, no. 18, p. 183510, Nov. 2014.

[3] X. Zheng et al., "Recombination and bandgap engineering in CdSeTe/CdTe solar cells," APL Mater., vol. 7, no. 7, p. 71112, Jul. 2019.

[4] T. A. M. Fiducia et al., "Understanding the role of selenium in defect passivation for highly efficient selenium-alloyed cadmium telluride solar cells," Nat. Energy, vol. 4, no. 6, pp. 504–511, 2019.

[5] D. E. Swanson, J. R. Sites, and W. S. Sampath, "Co-sublimation of CdSe x Te 1 À x layers for CdTe solar cells," Sol. Energy Mater. Sol. Cells, vol. 159, pp. 389–394, 2016.

[6] T. Baines et al., "Incorporation of CdSe layers into CdTe thin film solar cells," Sol. Energy Mater. Sol. Cells, vol. 180, no. February, pp. 196–204, Jun. 2018.

[7] K. M. Morris, C. Potamialis, F. Bittau, J. W. Bowers, and J. M. Walls, "Chemical bath deposition of thin film CdSe layers for use in Se alloyed CdTe solar cells," in 2019 IEEE 46th Photovoltaic Specialists Conference (PVSC), 2019, pp. 1857–1862.

[8] R. Greenhalgh et al., "The Origins of Void formation in Sputtered CdSe," in 2021 IEEE 48th Photovoltaic Specialists Conference (PVSC), 2021, pp. 886–889.

[9] T. Hussain, M. F. Al-Kuhaili, S. M. A. A. Durrani, and H. A. Qayyum, "Effect of collision during vapor transport between Cd and X (X = Te2, Se2, or S2) molecules on the properties of thermally evaporated CdTe, CdSe, and CdS thin films," Results Phys., vol. 8, pp. 988–1000, 2018.

[10] S. Mathuri, K. Ramamurthi, and R. Ramesh Babu, "Effect of Sb incorporation on the structural, optical, morphological and electrical properties of CdSe thin films deposited by electron beam evaporation technique," Thin Solid Films, vol. 660, pp. 23–30, Aug. 2018.

[11] P. M. M. Kaminski et al., "Blistering of magnetron sputtered thin film CdTe devices," IEEE J. Photovoltaics, pp. 3430–3434, Jun. 2017.

[12] Z. Liu, C. Chang, W. Zhang, M. Yang, and Q. Zhang, "Research on Ligand Properties of CdSe Quantum Dots," IOP Conf. Ser. Mater. Sci. Eng., vol. 562, no. 1, 2019.

[13] V. A. Amin, K. O. Aruda, B. Lau, A. M. Rasmussen, K. Edme, and E. A. Weiss, "Dependence of the Band Gap of CdSe Quantum Dots on the Surface Coverage and Binding Mode of an Exciton-Delocalizing Ligand, Methylthiophenolate," J. Phys. Chem. C, vol. 119, no. 33, pp. 19423–19429, Aug. 2015.

[14] K. B. Subila, G. Kishore Kumar, S. M. Shivaprasad, and K. George Thomas, "Luminescence properties of CdSe quantum dots: Role of crystal structure and surface composition," J. Phys. Chem. Lett., vol. 4, no. 16, pp. 2774–2779, 2013.

[15] J. D. Poplawsky et al., "Structural and compositional dependence of the CdTe x Se 1-x alloy layer photoactivity in CdTe-based solar cells," Nat. Commun., vol. 7, no. August, pp. 1–10, Nov. 2016.

[16] A. Abbas et al., "The effect of a post-activation annealing treatment on thin film CdTe device performance," in 2015 IEEE 42nd Photovoltaic Specialist Conference (PVSC), 2015, pp. 1–6.

[17] C. Li et al., "Characterization of sputtered CdSe thin films as the window layer for CdTe solar cells," Mater. Sci. Semicond. Process., vol. 83, no. January, pp. 89–95, 2018.

[18] E. Bastola et al., "Understanding the Interplay between CdSe Thickness and Cu Doping Temperature in CdSe/CdTe Devices," IEEE J. Photovoltaics, vol. 12, no. 1, pp. 11–15, Jan. 2022.

AgriPV Citizen Science Lab: A Collaborative Model for Engineers, Youth Scholars and Communities

Stuart Bowden, Jazmine Cordon, Myla Dykes, Michael Hernandez, Michelle Jordan, Alex Killam, Jasmine Martinez Castillo, Alex Park, Alondra Pita, Maryan Robledo, Steve Zuiker

Arizona State University, QESST Solar Energy Engineering Research Center

Abstract—**This paper presents a model for PV education that brings together photovoltaics engineers with 6th-12ᵗʰ-grade students and teachers to test agrivoltaics applications using citizen science. A summer research experience program engaged high school students in building a novel agrivoltaics school garden monitoring system, facilitated by PV engineering researchers and education scholars. The high schoolers are mentoring middle grade students as they design and install school gardens and collect data on garden conditions with and without solar panels covering crops. Ideas for multi-generational collaboration are explored.**

Keywords—agrivoltaics, K-12 education, citizen science

I. INTRODUCTION

This paper provides an overview of an interdisciplinary partnership that sought to develop a collaborative model for photovoltaic (PV) education that brings together PV researchers and industry members, educational scholars and teachers, and middle and high school youth in multi-generational and cross-institutional PV research. As a pilot of this model, a team of engineering and education scholars designed a series of summer research experience programs to support an agrivoltaics (agriPV) citizen science lab. This paper seeks to introduce our educational model. Future reports will focus on findings from applied agriPV research being piloted on middle school campuses in the geographic location of the Sonoran Desert region of the U.S. In keeping with our commitment to multi-generational collaboration, the authors consist of university researchers and high school scholars who contributed to the project. The voices of youth authors are privileged in this paper.

The purpose of the project is to introduce young scholars to a new and innovative PV technology and provide opportunities for them to contribute to PV innovation through citizen science applied projects at their local school campuses. Middle and high school students are the newest generation of scientists and researchers. Introducing them to agriPV at such an early stage gives them the chance to develop a passion for sustainability and engineering which is integral to the future of our earth. Usually, new technologies are not shared with youth let alone youth who come from an economically disadvantaged community. Moreover, youth are rarely afforded a seat at the table for contributing to designing, testing, or championing new technologies. In this program, underserved youth are given opportunities to dive into the engineering field and understand the future of solar technology and farming. They are introduced to the idea of combining new and old practices to enhance their communities' and help them become more sustainable.

In the summer of 2021, eight high school students participated in a six-week program focused on agriPV citizen science research. The Urban Energy Engineering (UEE) Youth Scholars worked with PV engineering faculty and graduate students on an applied project to pilot test small-scale a*griPV* installations on school campuses. Their challenge was two-fold: cconduct solar energy research with the nation's top solar energy scientists, learn how to support energy transitions in their local community. The UEE team built and tested (a) a novel solar-powered garden monitoring device and (b) a PV system for use in school-based agriPV citizen science garden beds. This work built on research conducted the previous summer by a team that included K-12 teachers who built the first generation of the garden monitoring system, co-designed a mobile PV racking system for safe installation in school contexts, and created agriPV classroom curricula [1].

Educational research suggests that learning is facilitated when students engage in real engineering work with real consequences [2], when they are provided a rightful presence in science and engineering [3] and see an authentic purpose for their work. Thus, UEE fostered STEM learning while inspiring youth to enlist energy science and engineering as tools for navigating opportunities and challenges in their communities as well as to see the ramifications of past technologies for the earth and real social consequences of innovation. University scholars are continuing to partner with the UEE scholars as they support middle-grade students and teachers in implementing agriPV citizen science on their school campuses.

II. AGRIVOLTAICS BACKGROUND

Agrivoltaics is a relatively new technology that combines agriculture and PV. Researchers are exploring the effects of agriPV in different areas of the world [4]. The goal is to create a more reliable energy source where it's needed. Solar panels positioned overhead provide shading so the plants receive less intense dappled light and extreme temperatures in places like Arizona where the high temperatures and solar irradiance can have a negative effect on crop plants. So far, AgriPV shows a promising future for more sustainable farming practices and reliable energy sources. In addition, it creates multi-use for land because solar cells and agricultural crops form a symbiotic relationship. The crops keep the solar panels cool so that they are more efficient and, in return, the panels protect the plants from harsh environmental conditions and reduce water evaporation. Overall, agriPV improves growth of some crops, especially and reduces water use in desert agriculture [5].

978-1-7281-6118-1/22 $31.00 © 2022 IEEE

AgriPV dates back to the 1980's, starting out as a concept for efficient use of land [6]. AgriPV systems have been built all around the world since then, most often on pre-existing agricultural land. One early example of a successful agriPV project was implemented in Japan by Akira Nagashimaby in 2004. Nagashimaby was a retired agricultural machinery engineer whose interest in biology led him to explore the light saturation point and to the idea of combining PV energy on farmland [7]. He implemented 348 PV panels on stilts over crops such as peanut, yam, taros and into a community that saw benefit from it in the form of profit to support their families and an increase in interest in farming. Another way agriPV systems are created is with the solar first, through the employment of large-scale solar farms, to which agricultural aspects are then added (e.g., planting native wildflowers to support pollinators, grazing sheep around panels to reduce foliage while decreasing animal heat stress). For instance, a recent study in upstate New York, found that the sheep eat the grass so that the solar panels don't get covered, and the panels boost the ecosystem to the point that some wildflowers and rare pollinators returned to the area [8]. Both methods are being experimented with and different models are created with environmental situations in mind.

Agrivoltaics may be particularly beneficial in areas such as the Sonoran Desert region, known for its hot dry climate. Native plants include various species of cacti, palo verde, mesquite, yucca, gourds, legumes, and the creosote bush, which gives Arizona its signature smell of rain. Long before modern civilization came to the Sonoran Desert region, several indigenous tribes, including the Yaqui and Tohono O'odham, inhabited the land and learned from it, creating innovative technologies such as trincheras, lithic mulch, Ak-chin cultivation, and companion planting. While Arizona has some native crops, farmers also cultivate plants not native to the dry climate and require excess watering, shade, or cooler temperatures. Results from regional studies suggest that when a solar panel is above crops, more water is retained in the soil and soil moisture remains higher than in a control bed where there is no solar panel. Thus, less water needed for crops because the solar panel helps prevent evaporation and retain moisture [9].

III. AGRIVOLAICS RESEARCH FOR YOUTH SCHOLARS

Our interdisciplinary team is interested in furthering applied research in agrivoltaics, as well as how such research can benefit communities and PV learners. Thus, the Youth Scholars program sought both to educate youth and to learn from youth by engaging them as members of an intergenerational citizen science research team. Building on previous work of the lab, UEE youth started off this past summer by working on imagining a better future using PV technologies, some even submitting their solar futures narratives to the IEEE PVSC conference high school competition. Throughout the summer, the youth interacted with a broad range of PV and sustainability mentors as well as teachers and industry members to grow their understanding of PV science and understand multiple aspects of the project. They learned, for example, about the Food-Energy-Water Nexus, and the importance of considering sustainability in energy engineering research projects. To help them prepare to facilitate middle school students who would be implementing agriPV in the fall, the UEE youth explored the use of community ethnography, surveying the local and social geography using GIS and surveying local community members to gain perspectives on their research and to make sure they knew how to use and care for the projects' PV systems.

The authorship team imagined a multi-generational, citizen science approach to agriPV research. The AgriPV Citizen Science lab is therefore composed of diverse collaborators, each assuming different roles in a socio-technical system. Furthermore, the AgriPV Citizen Science Lab team has cultivated partnership with community members, energy users, maintenance providers, researchers, financial sponsors, equipment/resource suppliers and general sponsors. When approaching the social aspects of agriPV, we first recognized the network of social actors present in the designing technical system (Figure 1), how each supporter/actor contributed, challenges supporters might face when interacting with the system or helping in its design, and connections among actors. For instance, the high school members have played a significant role in building system elements, recognizing social factors of the community, and working alongside middle students to design a system that fulfills the community's needs.

Figure 1. AgriPV Citizen Science Lab Roles & Partners

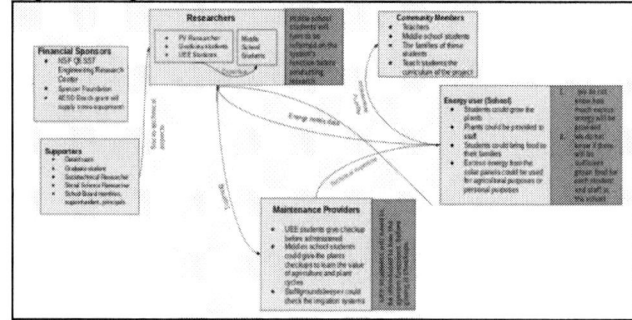

IV. AGRIVOLTAICS GARDEN MONITORING SYSTEM

The agrivoltaics school garden monitoring system is a device that uses 4 DHT22 sensors, two soil moisture sensors, two solar irradiance sensors, and a PCB that connects them together. The purpose of these sensors being used in conjunction is to be able to monitor a garden bed's soil moisture, air humidity, and the level it gets. The importance of these sensors is to be able to adjust the agriPV garden beds on each middle school partner's campus based on the data received, as well as to use that data to monitor differences in experimental (beds with solar panel) and control (beds without solar panels). Over the summer, these garden monitoring systems were built by high schoolers via soldering sensors to wiring, as well as soldering circuitry to the PCB which was designed by Stuart Bowden and Alex Killam.

The ESP32 contains Bluetooth, Wi-Fi, and voltage measurement in addition to the primary microcontroller. To make the ESP32 interact with the sensors and Wi-Fi network, we wrote the code to program it. The ESP32 has specific connections, called pins, to connect to the sensors, and we use specific code to talk to each sensor. Steps for programming and wiring needed to complete our device include to following: (a) install the programming environment: install the Arduino IDE, test the installation, prepare to test the ESP32, (b)

978-1-7281-6118-1/22 $31.00 © 2022 IEEE

assemble and test the components on the PCB: wires, resistor for DHT22, terminal blocks, ESP32 board, (c) attach the sensors: DHT22 sensor and cable, soil monitor, solar sensor cell, (d) write the programs: We use the C/C++ language the Arduino IDE to write code to tell the ESP32 how and when to take measurements and what to do with the data, (e) troubleshoot and test the system; work with campus and district leaders to determine data collection protocols. The finished systems are able to measure and send data to a web dashboard.

The UEE members also built a PV system that connected to the solar panels and provided the middle school users electricity to power tools of their choice. This system is made up of a battery, an inverter, and a charge controller housed with the garden monitoring circuit board inside a locked storage box. A solar panel is first connected to the charge controller that controls the amount of energy the battery gets, and then to the inverter which converts the DC energy coming from the solar panels to AC. With the energy converted, the middle grade partners can use the energy to benefit their own campus community. On the outside of the box there is a water gallon and a hose that serves as the water pump to water the garden. All of these things combined create the PV system (Figure 2).

Figure 2. AgriPV Garden Monitoring Device, PV System, and School Garden Installation

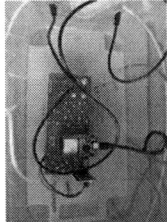

V. AGRIVOLTAICS SCHOOL GARDEN CITIZEN SCIENCE

K-12 teachers who built the first version of the agriPV garden monitoring system are partnering with university and high school members of the agriPV citizen science team to implement the applied project. Students and teachers at each partner campus are installing garden beds with solar panels (experimental) and without solar panels (control). In addition to installing the garden monitoring device and PV system, middle school students at each partner campus are building and adapting a mobile racking system for 305-watt donated panels (Figure 2). They are also collecting and analyzing data to compare the environmental conditions and crops. They also decide on uses for the energy generated (e.g., irrigation pump, lighting, charging laptops, mixers for food prep).

Following the summer program, the UEE youth continued to support the project, meeting frequently to develop and deliver presentations to middle school student and teacher partners who started installing the agriPV system and collecting and reporting preliminary data on their campuses in the fall. They provided feedback to their middle grade partners on citizen science lab reports and on their ideas, sketches, and pitches for improving the initial design of the PV racking

system. They also transformed data collected by the middle graders, creating graphs so they could interpret changes over time in garden conditions for control and experimental beds. Our team is also working with a local high school to implement an agriPV citizen science research site, continuing to modify the system constructed in the summer of 2021, to then enhance systems implemented on middle grade campuses.

The purpose of this project was to combine cutting edge technology with local communities. AgriPV is a new and up and coming technology that not many people know about. By educating young people in one local community shared by the researchers, and showing community members the importance of it, the agriPV lab members were able to cultivate passion and understanding of the project. We are conducting this citizen science study in order to get the community involved in work that directly affects them and their well-being. We want to find out how people interact with this technology and gain their perspectives on its viability and value in their community.

VI. ACKNOWLEDGEMENT

This material is based upon work supported in part by the National Science Foundation under award No. 1560031 and by the DOE under NSF Cooperative Agreement No. EEC-1041895. Additional support was provided by the IEEE Electron Devices Society and the Arizona State University Knowledge Exchange for Resilience Fellowship.

REFERENCES

[1] S. Bowden, M. Jordan, A. Killam... and Cortez, A. (2021). "Agrivoltaics Citizen Science: A Model for Collaboration between Engineers and K-12 Schools." Proceedings of the 2021 IEEE PVSC Conference, virtual.

[2] M. E. Jordan, S. Zuiker, W. Wakefield, and M. DeLaRosa. "Real Work with Real Consequences: Enlisting Community Energy Engineering as an Approach to Envisioning Engineering in Context," *Journal of Pre-College Engineering Education Research (J-PEER)*, *11*(1), Article 13. https://doi.org/10.7771/2157-9288.1294

[3] Angela Calabrese Barton & Edna Tan (2019) Designing for Rightful Presence in STEM: Journal of the Learning Sciences, 28:4-5, 616-658, DOI: 10.1080/10508406.2019.1591411

[4] Weselek, A., Ehmann, A., Zikeli, S. *et al.* Agrophotovoltaic systems: applications, challenges, and opportunities. A review. *Agron. Sustain. Dev.* **39,** 35 (2019). https://doi.org/10.1007/s13593-019-0581-3

[5] G.A. Barron-Gafford, M.A. Pavao-Zuckerman; R.L. Minor, et al. "Agrivoltaics provide mutual benefits across the food-energy-water nextus in drylands," *Nat. Sustain, vol. 2.,* pp. 848-855, 2019. https://doi.org/10.1038/s41893-019-0364-5

[6] Goetzberger, A. and Zastrow, A. (1982-01-01). "On the Coexistence of Solar-Energy Conversion and Plant Cultivation". *International Journal of Solar Energy.* **1** (1): 55–69. doi:10.1080/01425918208909875.

[7] Flesher, J. and Webber, T. "Bees, sheep, crops: Solar developers tout mulitple benefits. Panasonic, 2021. https://na.panasonic.com/us/green-living/bees-sheep-crops-solar-developers-tout-multiple-benefits

[8] D. Majumdar, and M.J. Pasqualetti. "Dual use of agricultural land: Introducing 'agrivoltaics' in Phoenix Metropolitan Statistical Area, USA," *Landscape and Urban Planning, Vol 170*, pp. 150-168, 2018. ISSN 0169-2046, https://doi.org/10.1016/j.landurbplan.2017.10.011.

[9] Cottrell, C. "Reimagining the Three Sisters: Finding a New Symbiosis in Tribal Food and Energy Sovereignty" Native Business, Sept, 24, 2019. https://www.nativebusinessmag.com/reimagining-the-three-sisters-finding-a-new-symbiosis-in-tribal-food-and-energy-sovereignty/

[10] Movellan J (2013) Japan next-generation farmers cultivate crops and solar energy. http://www.renewableenergyworld.com/articles/2013/10/japan-next-generation-farmers-cultivate-agriculture-and-solar-energy.html

Behavioral and Population Data-Driven Distribution System Load Modeling

Isaac Bromley-Dulfano, Xiangqi Zhu, and Barry Mather
Power Systems Engineering Center
National Renewable Energy Laboratory
Golden, CO, USA
isaac.bromleydulfano@nrel.gov , xiangqi.zhu@nrel.gov, barry.mather@nrel.gov

Abstract—Distribution system residential load modeling and analysis for different geographic areas within a utility or an independent system operator territory are critical for enabling small-scale, aggregated distributed energy resources to participate in grid services under Federal Energy Regulatory Commission Order No. 2222 [1]. In this study, we develop a methodology of modeling residential load profiles in different geographic areas with a focus on human behavior impact. First, we construct a behavior-based load profile model leveraging state-of-the-art appliance models. We simulate human activity and occupancy using Markov chain Monte Carlo methods calibrated with the American Time Use Survey data set. Second, we link our model with cleaned Current Population Survey data from the U.S. Census Bureau. Finally, we populate two sets of 500 households using California and Texas census data, respectively, to perform an initial analysis of the load in different geographic areas with various group features (e.g., different income levels). To distinguish the effect of population behavior differences on aggregated load, we simulate load profiles for both sets assuming fixed physical household parameters and weather data. Analysis shows that average daily load profiles vary significantly by income and income dependency varies by locality.

Keywords—Behavior-based Load Profile Modeling, U.S. Census, American Time Use Survey, Markov chain Monte Carlo

Nomenclature

age	ATUS respondent age
ATUS	American Time Use Survey
cat	ATUS respondent occupation category
C_h	HVAC equivalent heat capacity (J/°C)
C_p	Specific heat of water (J/kg °C)
CS	Lighting calibration scalar
i	Current activity state
I	Irradiance (W/m²)
I_{max}	Irradiance lighting threshold (W/m²)
j	Transitioned activity state
l	Lightbulb index
N	Number of ATUS respondents
occ	ATUS respondent occupation status
O_{eff}	Lighting effective occupancy
par	ATUS respondent parental status
P	Activity transition probability
Q_h	HVAC equivalent heat rate (W)
Q_w	Water heater power input (W)
RF	Lighting relative use factor
R_h	HVAC equivalent thermal resistance (°C/W)
R_w	Water heater tank thermal resistance (m² °C/W)
SA	Water heater tank surface area (m²)
t	Time
T_a	Ambient air temperature (°C)
T_h	Water heater tank temperature (°C)
T_{inc}	Water heater incoming water temperature (°C)
T_{int}	Interior air temperature (°C)
V	Water heater tank volume (m³)
W_D	Hot water demand (L/s)
Δt	Time step

I. Introduction

To address the ongoing climate crisis, major shifts are taking place in both the bulk power and distribution systems. Investments in utility-scale renewable energy continue to soar [2], and distributed energy resources (DER) are increasingly considered essential for decarbonized grid operations [3]. In 2020, the Federal Energy Regulatory Commission (FERC) released Order No. 2222 [1], which enables small-scale, aggregated DERs to engage in grid services. One small-scale DER service of interest is demand response, in which aggregators reduce or shift flexible residential loads during peak load hours, typically through consumer incentives or direct appliance control. With the expansion of demand response programs under FERC 2222, planners require advanced load profile models and analysis to predict outcomes for individual residential consumers and system reliability at large [4]–[5].

Existing literature has taken several approaches to modeling loads in distribution systems, particularly residential household loads, with both "top-down" [6]–[8] and "bottom-up" [9]–[19] methods. Although top-down methods require relatively simple data inputs, they typically do not distinguish load contributions of individual components. This is undesirable, especially under FERC 2222. Without the knowledge of small group or individual load profiles, it is difficult to predict and evaluate the performance of demand response programs.

Conversely, bottom-up studies model subcomponents within a complex load and aggregate to obtain the expected demand profile. To provide visibility to load aggregators and grid operators for demand response potential analysis, we focus on the bottom-up approach in this paper.

The state of the art offers a range of modeling options (e.g., different physical appliance models and population clustering methods), upon which we can build our model. A popular bottom-up strategy in the state of the art [9]–[19] is to model human activity using Markov Chain processes calibrated to behavioral time use surveys (TUS) and derive the associated residential load profiles. These studies associate each simulated human activity (cooking, laundry, etc.) with residential appliances (oven, washer, dryer, etc.). They use physical models or measured data to represent the load profiles of active appliances. Studies have applied this method using public TUS data in the United Kingdom [9], [12], [18], [20], Sweden [10], [16], France [21], the Netherlands [17], and the United States [11], [13], [22], [23]; and several have validated their behavioral load models against metered household data [9]–[11], [17], [19].

Although behavioral load modeling is growing in the realm of research, only a limited number of studies have considered behavior-defined load modeling across geographic and socioeconomic dimensions. Diao et al. [22] take a necessary first step in clustering population activity patterns, but the question remains how these activity patterns—and their resulting load profiles—will aggregate for distinct geographically distributed populations. For example, to gauge demand response potential, aggregators need to know how load profiles differ among small areas within a larger city. If available census data exist to capture population distributions within the city, a behavior-based model can provide insights that would otherwise require detailed metered data.

To address this question, we propose linking U.S. census data with a behavior-based household load model to capture geographic variations in load profiles at the aggregator level or census block level. This paper presents a proof of concept. First, we construct a behavior-based household load profile model by integrating the model approaches in the literature. Second, by using a generic set of appliances and fixed weather data, we model the load profiles for two sets of 500 households populated with occupants from the California and Texas census subsets, respectively. By fixing the household characteristics and weather profiles, we demonstrate the sensitivity of the aggregated household load profiles to the population characteristics and associated behavior pattern differences.

The rest of the paper is organized as follows: Section II introduces the methodology for behavior-based load profile modeling. Section III presents an initial analysis of the aggregated load profile for a representative socioeconomic dimension—income level. Section IV concludes the paper and discusses potential future work.

II. MODELING METHODOLOGY

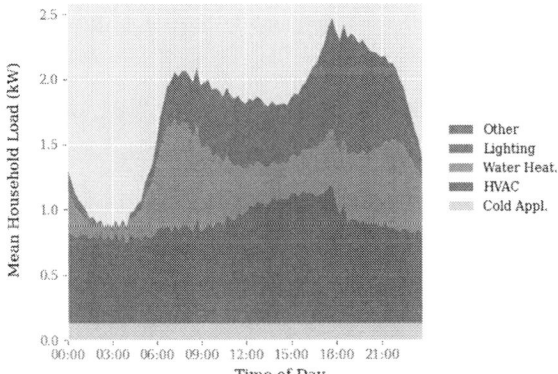

Fig. 1 Mean household load profile sample.

We construct a household load profile model following three main steps: (1) We model occupancy for household members based on their personal characteristics, i.e., age, work status, work type, and parental status; (2) We translate household member activities into appliance-use profiles, such as hot water consumption, heating, ventilating, and air-conditioning (HVAC) needs, etc.; and (3) We use appliance load profile models or power conversion factors in conjunction with the output from Step 2 to derive the final load profile.

The model outputs minute-resolution load profiles by end use (a sample is shown in Fig. 1). To connect the model with U.S.

census data, we directly populate households with members from the 2019 Current Population Survey (CPS).

A. Occupancy Modeling

To model the activity and occupancy of household members, we use a Markov chain Monte Carlo simulation calibrated with 2016–2019 American Time Use Survey (ATUS) data [24]. The ATUS contains detailed, 24-hour activity logs for 50,000 respondents on both weekdays and weekends. To clean the ATUS data, we classify each activity into one of nine activity states, as shown in Table. I. Additionally, we label each respondent by their age, work status, work type, and parental status because these indicators have been shown to affect activity patterns [25], [26]. Our model follows the literature [11], [16], [22], [23], [26] and uses hourly transition probability matrices for the weekday and weekend to sample transitions between activities. We populate the weekday or weekend transition probability matrices according to Equation (1), which describes the probability of an arbitrary household member transitioning between two activities at a given time. Finally, we sample transitions between activities every 10 minutes.

$$P_{age,occ,cat,par}^{t,i,j} = \frac{N_{age,occ,cat,par}^{t,i,j}}{N_{age,occ,cat,par}^{t,i}} \quad (1)$$

Table I. Classified ATUS Activity States.

#	Activity State	#	Activity State
0	Away	5	Cleaning
1	Sleeping	6	Laundry
2	Grooming	7	Leisure
3	Cooking	8	Other
4	Dishwashing	-	-

B. Appliance Modeling

1) HVAC

To simulate HVAC load profiles, we use a thermal model described by Lu [27]. The model characterizes interior temperature as a function of ambient temperature and three effective parameters tuned to the behavior of a typical household (Equation (2)).

$$T_{int}^{t+1} = T_a^{t+1} + Q_h R_h - (T_a^{t+1} + Q_h R_h - T_{int}^t) \cdot e^{\frac{\Delta t}{R_h C_h}} \quad (2)$$

The time variant equivalent heat rate, Q_h, indicates the active HVAC state (i.e., heating, cooling, neither). We select the HVAC state to maintain interior temperatures within the set point deadband. Although this simplified model does not capture differences between households (i.e., physical house size, level of insulation), it is well suited for our central experiment, which uses a fixed household for the entire population (Section III). To integrate occupancy in the HVAC model, we adjust the set point when all household members are away.

2) Water Heater

To account for water heater loads, we use a thermal model presented by Dolan et al. [28]. The model characterizes tank temperature behavior as a function of ambient temperature, tank size, incoming water temperature, and hot water demand (Equations (3)–(7)).

$$T_h^{t+1} = T_h^t \cdot e^{-\frac{\Delta t}{R'C_w}} + [G \cdot T_a + B^t \cdot T_{in} + Q_w] \cdot R' \cdot [1 - e^{-\frac{\Delta t}{R'C_w}}] \quad (3)$$

$$R' = \frac{1}{G + B^t} \quad (4)$$

$$G = \frac{SA}{R_w} \quad (5)$$

$$B = W_D^t * C_p \quad (6)$$

$$C_w = C_p * V \quad (7)$$

To integrate human behavior in the water heater model, we sample hot water consumption events with predetermined consumption profiles associated with occupant activity states (Table. II).

Table. II: Hot Water Consumption Profiles.

	Activity State	Hot Water Consumption Activity	Flow Rate (L/min.)	Duration (min.)
0	Away	-	-	-
1	Sleeping	-	-	-
2	Grooming	Showering	8.0	8
3	Cooking	Washing hands, filling pots	.1	10
4	Dishwashing	Filling washer	4.0	5
5	Cleaning	Mopping, washing counters	1.2	5
6	Laundry	Filling washer	2.5	30
7	Leisure	-	-	-
8	Other	-	-	-

3) Lighting

To model lighting loads, we use an agent-based method presented by Richardson et al. [29]. Following their method, we sequentially sample turn-on events for individual lightbulbs based on outdoor global irradiance and household occupancy.

We convert absolute occupancy into effective occupancy to account for shared-use lighting. Additionally, we assign relative use factors to individual lightbulbs and include a calibration scalar given in the literature [29], [30]. In each time step, we use Equation (8) to find the probability of turning on each lightbulb. Following turn-on events, we sample the duration for which lights remain active from a distribution provided by Richardson et al. [29].

$$P_{on,l}^t = (I^t > I_{max}) \cdot O_{eff}^t \cdot RF_l \cdot CS \quad (8)$$

4) Cold Appliances

To account for cold appliance loads—namely, refrigerators and freezers—we follow a simple procedure described by Muratori et al. [11]. For each appliance, we sample operation in 10-minute intervals assuming a Bernoulli distribution. The exact distribution is calculated such that the expected annual cold appliance consumption is consistent with that reported in the Residential Energy Consumption Survey [31].

5) Other Appliances

For all other appliances—including ovens, televisions, computers, etc.—we use power conversion factors and predetermined appliance load profiles [10], [11]. When at least one occupant is engaged in an activity state, we include the power conversion factor associated with that state in the load profile. For laundry and dishwashing events, we assume a predetermined load profile because the associated appliances typically remain active after the occupant has transitioned to another state [10]. The conversion factors and predetermined load profiles are shown in Table III.

Table. III: Power Conversion Factors and Profiles.

	Activity State	Appliances	Load (W)	Duration (min.)
0	Away	-	-	-
1	Sleeping	-	-	-
2	Grooming	Hair dryer	100	-
3	Cooking	Oven, toaster, microwave	3500	-
4	Dishwashing	Dishwasher	1800	60
5	Cleaning	Vacuum cleaner	1500	-
6	Laundry	Washing machine, dryer	425/3400	30/90
7	Leisure	Television, computer	120	-
8	Other	-	-	-

III. Load Profile Analysis

To demonstrate the benefit of linking census data with behavior-based load modeling at the aggregator level, we conduct an analysis to isolate the effect of population behavior on average and aggregated load profiles. First, we build a generic fixed household. We assign the model parameters in Table IV, and we use 2019 temperature and irradiance data from Austin, Texas [31]. Next, we populate two sets of 500 households using 2019 CPS census data. Because the household characteristics and weather data are fixed, the only difference between households is the number of occupants and their respective labels (i.e., age, work status, work type, parental status). For each respective set, we assign the number of occupants and occupant labels from random households in the Texas and California CPS subsets.

In our analysis, we compare the overall load profiles and load profiles by income for the two populations. Finally, because the California and Texas populations differ in household size distributions, we repeat the income analysis for households with only two members.

Table IV: Assumed Household Parameters.

Parameter	Value	Units	Model(s)
C_h	40,000	J/°C	HVAC
R_h	.18	°C/W	HVAC
Q_h (heating)	450	W	HVAC
Q_h (cooling)	-150	W	HVAC
HVAC set point (home)	21	°C	HVAC
HVAC set point (away)	21 ± 5	°C	HVAC
HVAC deadband	2	°C	HVAC
Heater rating	6000	W	HVAC
AC rating	4500	W	HVAC
WH set point	55	°C	Water heater
WH deadband	4	°C	Water heater
T_{inc}	10	°C	Water heater
V	190	L	Water heater
SA	2	m²	Water heater
C_p	4186	J/kg °C	Water heater
R_w	1.2	m² °C / W	Water heater
Q_w	3000	W	Water heater

Lightbulbs	30	Quantity	Lighting
CS	.008	-	Lighting
Refrigerator rating	200	W	Cold appliances
Freezer rating	50	W	Cold appliances

A. Aggregated Household Load

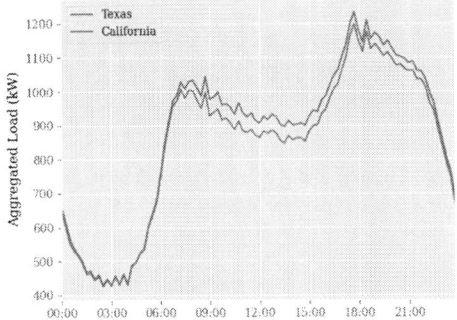

Fig. 2: Aggregated load for 500 Texas (red) and California (blue) households.

The results from the 500-household analysis indicate that the purely behavior-driven aggregated load profiles of the Texas and California samples are similar in shape and comparable in magnitude. The shape is primarily driven by the HVAC load profile (as shown in Fig. 1 in Section II); however, because the weather data are fixed, differences in HVAC consumption caused by occupancy are difficult to identify in the aggregate load profiles. Overall, the Texas sample consumes 3% more electricity annually and demands up to 6% more power during the middle hours of the day (Fig. 2). We attribute this difference to a larger average household size (2.6 occupants) relative to the California sample (2.4 occupants).

B. Load Profiles by Income

Households with similar characteristics, such as income, are likely to cluster in the same geographic areas; therefore, distinguishing load profiles by income bracket can capture the load differences among geographic areas that aggregators might observe within a city. Hence, we further investigate the load profile differences between income brackets.

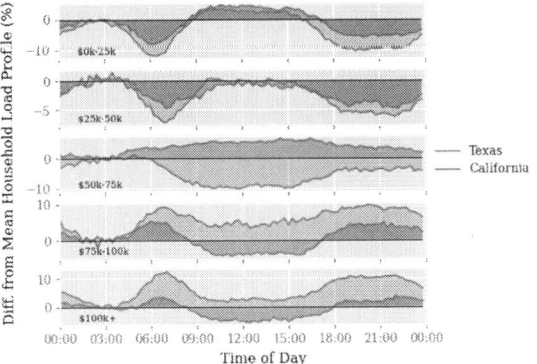

Fig. 3: Difference from mean household load profile by income bracket.

Fig. 3 shows how the mean load profile for each income bracket differs from the overall mean load profile throughout the day by state. In California, loads in the lowest income bracket are up to 11% less than the mean California loads during peak load hours. Conversely, loads in the highest income bracket are up to 12% greater than the mean loads during peak

load hours. For households making $50K–75K, the Texas load profile is up to 7% greater than the mean Texas loads, whereas the California load profile is up to 10% less than the mean California loads.

We largely attribute differences between states to their respective income-dependent household size distributions. Generally, average household size grows with income, however, the largest average household size occurs in the $50K–75K bracket and the +$100K bracket for the Texas and California samples, respectively.

C. Load Profiles by Income with Fixed Household Size

To remove the effect of house size on load profiles, we repeat the analysis in Part III.B, but we only include census households with two occupants. Fig. 4 shows that the fixed-size California and Texas samples yield generally consistent results across incomes. Loads in lower-income households are up to 6% less than mean loads during peak residential load periods and up to 10% more during the middle hours of the day. Loads in high-income households are up to 5% greater than mean loads during the morning peak and up to 10% less during the middle hours of the day. To be clear, we can attribute the results in Fig. 4 purely to behavioral differences across income brackets.

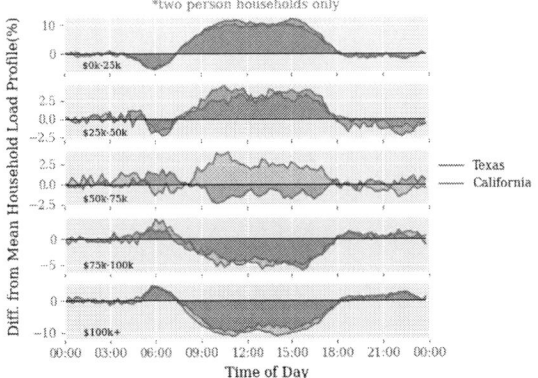

Fig. 4: Difference from mean household load profile by income bracket two-person households.

This analysis can indicate the demand response potential of small areas (i.e., neighborhood or census block level) for which there are available Census data. For example, Fig. 4 indicates that *considering only family size and behavior*, the $75K–100K households show the most potential for load shifting in the mid-to-late afternoon hours in California, whereas in Texas the $50–75K households show more potential during the same period.

IV. CONCLUSIONS AND FUTURE WORK

This paper demonstrates the potential of distinguishing load profiles and demand response potential for different geographic regions using behavior-based load models in conjunction with aggregated population characteristics. Based on previous reported work, we construct an integrated behavior-based load profile model. Then we link data from the U.S. Census Bureau CPS to the load profile model, and we compare the load profiles for two sets of 500 households populated by Texas and California census data, respectively, assuming a fixed set of appliances and weather data. From our analysis, we find that the mean load profiles differ significantly according to income in both California and Texas, which can be used by aggregators to estimate demand response potential in geographically distributed income clusters. In our future work, we will perform further analysis from more socioeconomic dimensions for different geographic areas.

ACKNOWLEDGMENTS

This work was authored by the National Renewable Energy Laboratory, operated by Alliance for Sustainable Energy, LLC, for the U.S. Department of Energy (DOE) under Contract No. DE-AC36-08GO28308. Funding provided by U.S. Department of Energy SULI program. The views expressed in the article do not necessarily represent the views of the DOE or the U.S. Government. The U.S. Government retains and the publisher, by accepting the article for publication, acknowledges that the U.S. Government retains a nonexclusive, paid-up, irrevocable, worldwide license to publish or reproduce the published form of this work, or allow others to do so, for U.S. Government purposes.

REFERENCES

[1] Federal Energy Regulatory Commission, Order No. 2222. 2020.

[2] L. Perea, Austin (Wood Mackenzie et al., "Solar Market Insight Report - 2018 Q2 - Executive Summary," 2018.

[3] P. de Martini and L. Kristov, "Distribution Systems in a High Distributed Energy Resources Future," 2015.

[4] M. Mohammed, A. Abdulkarim, A. S. Abubakar, A. B. Kunya, and Y. Jibril, "Load modeling techniques in distribution networks: a review," Journal of Applied Materials and Technology, vol. 1, no. 2, 2020, doi: 10.31258/jamt.1.2.63-70.

[5] A. J. Collin, G. Tsagarakis, A. E. Kiprakis, and S. McLaughlin, "Development of low-voltage load models for the residential load sector," IEEE Transactions on Power Systems, vol. 29, no. 5, 2014, doi: 10.1109/TPWRS.2014.2301949.

[6] E. Hirst, "A model of residential energy use," Simulation, vol. 30, no. 3, 1978, doi: 10.1177/003754977803000301.

[7] M. Saviozzi, S. Massucco, and F. Silvestro, "Implementation of advanced functionalities for Distribution Management Systems: Load forecasting and modeling through Artificial Neural Networks ensembles," Electric Power Systems Research, vol. 167, 2019, doi: 10.1016/j.epsr.2018.10.036.

[8] X. Zhu and B. Mather, "Data-Driven Distribution System Load Modeling for Quasi-Static Time-Series Simulation," IEEE Transactions on Smart Grid, vol. 11, no. 2, 2020, doi: 10.1109/TSG.2019.2940084.

[9] I. Richardson, M. Thomson, D. Infield, and C. Clifford, "Domestic electricity use: A high-resolution energy demand model," Energy and Buildings, vol. 42, no. 10, 2010, doi: 10.1016/j.enbuild.2010.05.023.

[10] J. Widén and E. Wäckelgård, "A high-resolution stochastic model of domestic activity patterns and electricity demand," Applied Energy, vol. 87, no. 6, 2010, doi: 10.1016/j.apenergy.2009.11.006.

[11] M. Muratori, M. C. Roberts, R. Sioshansi, V. Marano, and G. Rizzoni, "A highly resolved modeling technique to simulate residential power demand," Applied Energy, vol. 107, 2013, doi: 10.1016/j.apenergy.2013.02.057.

[12] E. Lampaditou and M. Leach, "Evaluating participation of residential customers in demand response programs in the UK," eceee 2005 summer study - What works & who delivers?, 2005.

[13] J. Zhu, Q. Liao, Y. Lin, W. Lei, and R. Cui, "Residential high-resolution electricity demand optimization with a cooperative PSO algorithm," Energy Reports, vol. 7, 2021, doi: 10.1016/j.egyr.2021.02.031.

[14] A. Mammoli, M. Robinson, V. Ayon, M. Martínez-Ramón, C. fei Chen, and J. M. Abreu, "A behavior-centered framework for real-time control and load-shedding using aggregated residential energy resources in distribution microgrids," Energy and Buildings, vol. 198, 2019, doi: 10.1016/j.enbuild.2019.06.021.

[15] S. Ge, J. Li, H. Liu, X. Liu, Y. Wang, and H. Zhou, "Domestic energy consumption modeling per physical characteristics and behavioral factors," in Energy Procedia, 2019, vol. 158. doi: 10.1016/j.egypro.2019.01.399.

[16] J. Widén, A. Molin, and K. Ellegård, "Models of domestic occupancy, activities and energy use based on time-use data: Deterministic and stochastic approaches with application to various building-related simulations," Journal of Building Performance Simulation, vol. 5, no. 1, 2012, doi: 10.1080/19401493.2010.532569.

[17] M. Nijhuis, M. Gibescu, and J. F. G. Cobben, "Bottom-up Markov Chain Monte Carlo approach for scenario based residential load modelling with publicly available data," Energy and Buildings, vol. 112, 2016, doi: 10.1016/j.enbuild.2015.12.004.

[18] G. Tsagarakis, A. J. Collin, and A. E. Kiprakis, "Modelling the electrical loads of UK residential energy users," 2012. doi: 10.1109/UPEC.2012.6398593.

[19] K. McKenna and A. Keane, "Residential Load Modeling of Price-Based Demand Response for Network Impact Studies," IEEE Transactions on Smart Grid, vol. 7, no. 5, 2016, doi: 10.1109/TSG.2015.2437451.

[20] I. Richardson, M. Thomson, and D. Infield, "A high-resolution domestic building occupancy model for energy demand simulations," Energy and Buildings, vol. 40, no. 8, 2008, doi: 10.1016/j.enbuild.2008.02.006.

[21] U. Wilke, F. Haldi, J. L. Scartezzini, and D. Robinson, "A bottom-up stochastic model to predict building occupants' time-dependent activities," Building and Environment, vol. 60, 2013, doi: 10.1016/j.buildenv.2012.10.021.

[22] L. Diao, Y. Sun, Z. Chen, and J. Chen, "Modeling energy consumption in residential buildings: A bottom-up analysis based on occupant behavior pattern clustering and stochastic simulation," Energy and Buildings, vol. 147, 2017, doi: 10.1016/j.enbuild.2017.04.072.

[23] Y. S. Chiou, K. M. Carley, C. I. Davidson, and M. P. Johnson, "A high spatial resolution residential energy model based on American Time Use Survey data and the bootstrap sampling method," Energy and Buildings, vol. 43, no. 12, 2011, doi: 10.1016/j.enbuild.2011.09.020.

[24] U.S. Bureau of Labor Statistics, "American time use survey," Washington, D.C., 2019.

[25] D. S. Hamermesh, H. Frazis, and J. Stewart, "Data watch the American time use survey," Journal of Economic Perspectives, vol. 19, no. 1, 2005, doi: 10.1257/0895330053148029.

[26] F. Farzan, M. A. Jafari, J. Gong, F. Farzan, and A. Stryker, "A multi-scale adaptive model of residential energy demand," Applied Energy, vol. 150, 2015, doi: 10.1016/j.apenergy.2015.04.008.

[27] N. Lu, "An evaluation of the HVAC load potential for providing load balancing service," IEEE Transactions on Smart Grid, vol. 3, no. 3, 2012, doi: 10.1109/TSG.2012.2183649.

[28] P. S. Dolan, M. H. Nehrir, and V. Gerez, "Development of a Monte Carlo based aggregate model for residential electric water heater loads," Electric Power Systems Research, vol. 36, no. 1, 1996, doi: 10.1016/0378-7796(95)01011-4.

[29] I. Richardson, M. Thomson, D. Infield, and A. Delahunty, "Domestic lighting: A high-resolution energy demand model," Energy and Buildings, vol. 41, no. 7, 2009, doi: 10.1016/j.enbuild.2009.02.010.

[30] E. J. Palacios-Garcia, A. Chen, I. Santiago, F. J. Bellido-Outeiriño, J. M. Flores-Arias, and A. Moreno-Munoz, "Stochastic model for lighting's electricity consumption in the residential sector. Impact of energy saving actions," Energy and Buildings, vol. 89, 2015, doi: 10.1016/j.enbuild.2014.12.028.

[31] U.S. Energy Information Administration, "Residential Energy Consumption Survey (RECS)," 2015.

This page intentionally left blank.

Thin-Film Multijunction Inverted Metamorphic Solar Cells with Light Management for Space Applications

Julia D'Rozario, Steve Polly, Rao Tatavarti, Seth Hubbard

Rochester Institute of Technology, Rochester, NY, United States

MicroLink Devices, Niles, IL, United States

Inverted metamorphic (IMM) solar cells are an attractive power generation source for space-related applications due to their high power conversion efficiency (PCE), however, the beginning-of-life (BOL) PCE declines after radiation exposure to highly energized particles in the space environment as crystalline defects impact carrier transport. Thinning the bottom subcell, which suffers most from crystalline defects, reduces the distance carriers must travel to support power, making the thin-film design tolerant to radiation damage compared to their optically thick counterparts. Light management is essential in these devices as transmission loss occurs after one optical pass through the thinned solar cell. This work focuses on light trapping structures in the form of back surface reflectors (BSR) that increase the optical path length in the thin-film bottom InGaAs subcell to meet the current output realized in their optically thick counterparts. The textured BSR utilizes a facile in situ texturing method to reduce the time and cost of the light-trapping development, suitable for integration into already-established space PV manufacturing. The light current-voltage results on the in situ BSR device show no degradation in the VOC while exhibiting an increased JSC due to the photon scattering. The triple-junction IMM results also show improved current output using the in situ textured BSR compared to a planar mirror geometry. The prediction model in the remaining factor shows the 700 nm-thick devices with the in situ BSR surpasses the optically thick control beyond 1 MeV electrons at a fluence of 2x1014 e-/cm2, which supports the use of the in situ texturing in photovoltaic manufacturing.

978-1-7281-6118-1/22 $31.00 © 2022 IEEE

Probing Dynamic Influence of Moisture Ingress on Cell Deflection in Photovoltaic Modules

Ian M Slauch, Rishi E Kumar, Tala Sidawi, Jared Tracy, William Gambogi, Rico Meier, David P Fenning, Mariana I Bertoni

Fulton School of Engineering, Arizona State University, Tempe, AZ, United States

Department of Nanoengineering, University of California San Diego, La Jolla, CA, United States

DuPont Photovoltaic and Advanced Materials, Wilmington, DE, United States

Investigations into photovoltaic (PV) modules are driven by the desire to increase their reliability and lifetime, thereby decreasing the levelized cost of electricity (LCOE) and improving the bankability of PV installation projects. Critical to both the reliability and lifetime of a module is the mechanical stability of the cell since cell cracking accounts for up to 8% annual degradation in power output. This study uses Water Reflectometric Detection (WaRD) and X-Ray Topography (XRT) to investigate the effects of water incorporation on the encapsulated cell deflection in-situ. Repeated XRT imaging of the cell as it dries from pre-conditioning at 85 °C and 85% relative humidity show changes in the obtained diffraction signal over time, corresponding to changes in cell deflection. Measurements of local water concentration over time show that water diffuses out of the rear of module after saturation through the backsheet, while moisture remains in the encapsulant on the front side of the module. We show that drying out from a humid condition can increase the curvature of the cell near its edges, which we hypothesize is due to changes in encapsulant strain or modulus as water diffuses out of the rear EVA. We have shown that the cell responds mechanically to changes in environmental humidity, raising the question of the role of humidity in fatigue, mechanical cycling, and cell cracking as the module ages.

Potovoltaic Module R&D Considerations for Soiling Mitigation

Lin J. Simpson, Matthew Brantl, Ryo Huntamer

National Renewable Energy Laboratory, Golden, CO, United States

Stevens Institute of Technology, Hoboken, NJ, United States

University of California, Riverside, Riverside, CA, United States

Photovoltaic (PV) modules work best in the sunniest environments. Unfortunately, often the sunniest places also have substantial amounts of airborne "dust" that deposits on the front surface of the modules and blocks the sunlight; reducing energy output. In fact, natural soiling has reduced the energy output of PV systems since the technology was first used, and viable mitigation strategies have remained elusive ever since. With the ever-increasing deployments around the world, especially in dusty environments, soiling is becoming a billion-dollar problem, worldwide. While substantial work has been done to examine and resolve some of the issues with PV soiling, often mitigation comes down to physically cleaning the modules. However, a more systematic evaluation of the different module properties correlations to soiling mitigation needs to be done. In many instances, the causal connections between module properties and soiling are simply not known. This lack of knowledge results in a substantial increase in time and effort to evaluate and qualify appropriate soiling mitigation protocols based on site specific issues and the intrinsic module properties that are typically not optimized for mitigating soiling in a given environment. Thus, module property protocols and/or standards are needed to more quickly help identify appropriate module and site-specific mitigation. Thus, in this paper, we will present a review of the different issues between module properties and their relationship to soiling mitigation, and then outline a roadmap of the issues that still need to be resolved with additional research and development. Issues from frameless modules to anti-fungal glass compositions will be discussed.

Seven-Level Cascaded H-Bridge Multilevel Single-Phase Inverter Implemented with an ATMEGA Microprocessor

Edgardo Desarden-Carrero[1], Rachid Darbali Zamora[2], Erick Aponte-Bezares[1], and Eduardo I. Ortiz-Rivera[1]

[1]University of Puerto Rico-Mayagüez, Mayagüez, Puerto Rico 00682, USA
[2]Sandia National Laboratories, Albuquerque, New Mexico, 87185, USA

Abstract – An inverter is a power electronics circuit that can convert DC power into AC power. Inverters have a wide range of applications, from battery-powered devices to renewable energy systems such as photovoltaics. Typically, inverters rely on switching devices to create an AC signal. In most cases, these switching devices are arranged in the form of an H-bridge, the circuit topology taking its name from the arrangement of the switching devices. Although most inverters rely on utilizing a single H-bridge to achieve either two or three voltage levels, the use of multiple H-bridges increases levels as well as voltage. This proves to be advantageous when trying to reduce total harmonic distortion (THD). A low-cost multilevel (ML) single-phase inverter based on a cascade of H-bridge connections and PWM voltages was developed using an ATMEGA microprocessor. The ML inverter can supply up to 30 W with an RMS output voltage of 120 V, 60 Hz, and a THD less than 30% without a lowpass filter. An 18 V battery per level provided the primary DC voltage. This document presents simulation and experimental results when the ATMEGA-based inverter prototype was tested under various lighting load technologies, including linear loads such as filament and nonlinear loads such as compact fluorescent lighting (CFL) and light-emitting diode (LED).

Index Terms – inverter, multilevel (ML) inverter, microprocessor, total harmonics distortion (THD), pulse width modulation, ATMEGA

I. INTRODUCTION

The use of DC to AC conversion plays a critical role in power systems applications [1]. Applications range from small everyday battery-powered household appliances to larger power devices such as uninterruptable power supplies, photovoltaic systems, and even electric vehicles. Typically, inverters consist of four switching devices arranged in the form of an H-bridge, allowing three voltage level options. An H-bridge alone yields significant total harmonic distortion (THD), requiring an output filter. More complex inverter topologies use multilevel (ML) configurations, which enable more voltage levels and help reduce the THD. Although the ML inverter's effectiveness in reducing the THD increases with the number of voltage levels, the switching frequency and the microprocessor of choice might limit the performance of the ML inverter.

A seven-level cascaded H-bridge ML single-phase inverter was developed based on the ATMEGA microprocessor's maximum switching frequency. The inverter is designed to operate at an RMS voltage of 120 V, a frequency of 60 Hz, and supply up to 30 W of power. An 18 V rechargeable battery per level provided the primary DC voltage to supply the load including losses. The inverter was tested under different lighting load technologies, including incandescent, compact fluorescent lighting (CFL) and light-emitting diode (LED).

II. MULTILEVEL INVERTER DESIGN

A. Inverter Design Description

The developed inverter is a ML inverter operating at a pulse width modulation (PWM) frequency of 1.8 KHz. As the number of voltage levels increase, the THD decreases, the required amount of trigger signals increases, and the ATMEGA microprocessor velocity to switch its outputs decreases. The trade-off point between the lowest possible THD and an acceptable voltage PWM pattern was found with a seven-level ML inverter working at a switching frequency of 1.8 KHz. The seven-level ML inverter is based on three H-bridges connected in a cascade [2]. Each level provides 18 V through a rechargeable battery connected to it. Next, the reference sinewave signal at 60 Hz, and the six PWM signals at 1.8 kHz are implemented with the ATMEGA microcontroller. In addition, at the final stage, a 1:4 step-up transformer was used to increase the RMS voltage from 30 V to 120 V. Fig. 1 shows the proposed seven-level cascaded H-bridge ML single-phase inverter.

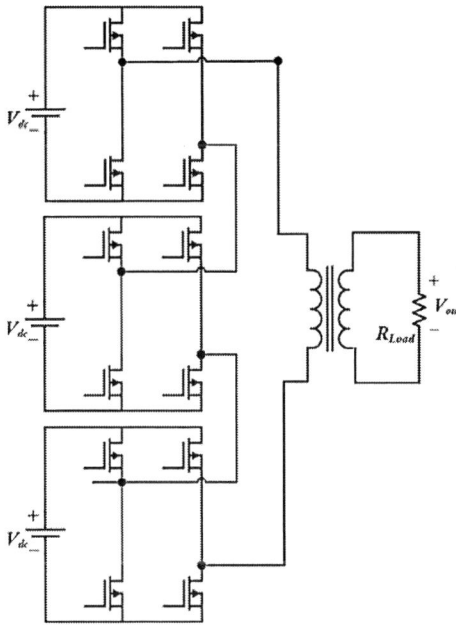

Fig. 1. Circuit diagram of the seven-level cascaded H-bridge multi-level single-phase inverter with no lowpass filter.

978-1-7281-6118-1/22 $31.00 © 2022 IEEE

B. Inverter H-Bridge Configuration

The H-bridge contains four N-type MOSFETs, a DC Source, and load terminals, as shown in Fig. 2. Depending on the desired output voltage polarity, a signal will be sent to the switching devices [3], [4]. There are three H-bridge switching patterns that prevent a short circuit event from happening, and they are shown in Fig. 3. The H-bridge experimental implementation is demonstrated in Fig. 4.

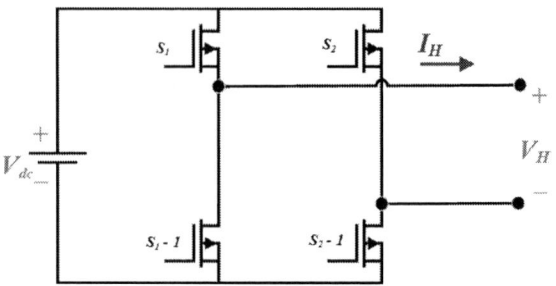

Fig. 2. H-bridge configuration with DC voltage source and load terminals.

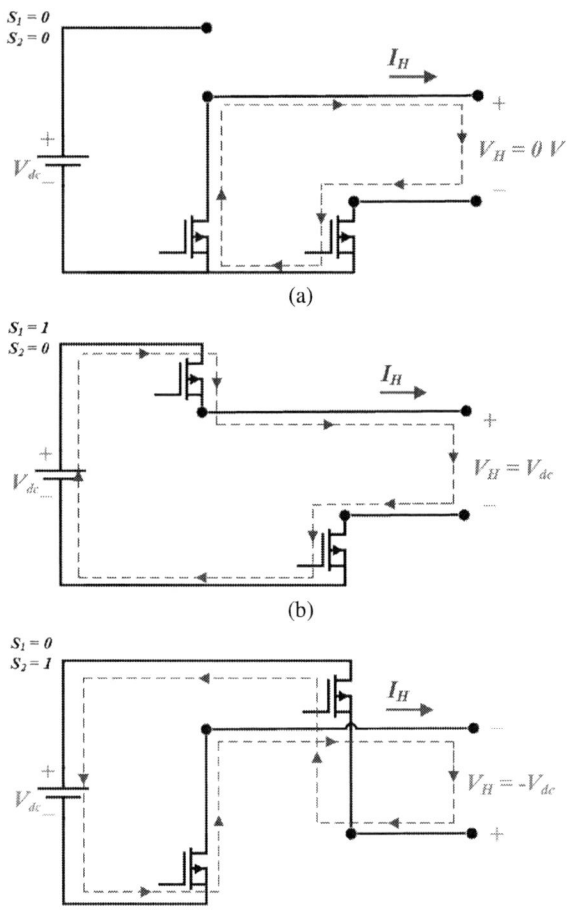

Fig. 3. The different ON and OFF switching states of the H-bridge inverter. (a) With switches S_1 and S_2 open, the switches S_1-1 and S_2-1 yield 0 V at the output of the inverter. (b) With switches S_1 closed, and S_2 open, the inverter yields V_{dc} at the output. (c) With switches S_1 open, and S_2 closed, the inverter yields -V_{dc} at the output.

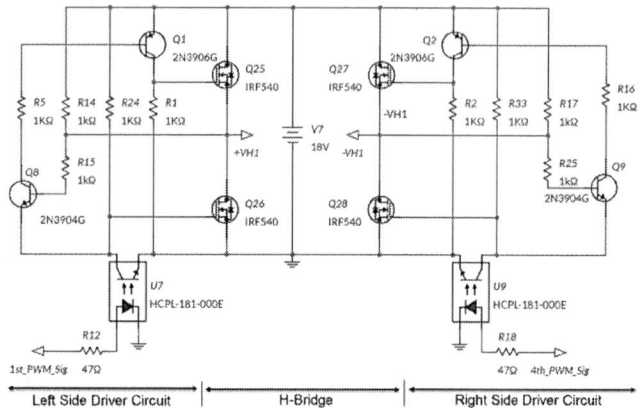

Fig. 4. H-bridge implemented with N-type MOSFET and their gate driver circuits.

C. H-Bridge Implementation

The H-bridge blocks implemented in the ML inverter were constructed using four IRF540 N-type MOSFETs, as seen in Fig. 4. The MOSFETs were connected in the H-bridge configuration to provide positive or negative voltage depending on the signal received. Therefore, the timing of the gate drive signals is crucial to make this complex ML inverter work properly. The gate driver circuit and the ATMEGA simplify this complex control of the MOSFETs in the H-bridge with relatively low power and provide isolation to protect the ATMEGA's outputs in case of circuit malfunction or overvoltage. The isolation in the gate driving circuit is provided by two optocouplers that interface between the ATMEGA's outputs and the H-bridge. The H-bridge has two driving circuits. When the H-bridge receives no signal, Q26 and Q28 are activated, providing an output voltage of zero between the terminals +VH1 and -VH1. This condition is presented in Fig. 3 (a). A voltage signal should be sent to optocoupler U7 (1st_PWM_Sig) when a positive voltage is desired. This voltage will turn on the optocoupler, and the optocoupler will ground the gate of MOSFET Q26 and turn it off. The same ground that turns off Q26 will turn on Q1 (PNP transistor) when Q8 is activated. Q8 provides an interlock between MOSFETs Q25 and Q26 to prevent a short circuit. Only when the ground is not reflected at the terminal +VH1, transistor Q8 is allowed to turn on, providing an activation path for transistor Q1 and MOSFET Q25. This positive voltage condition is shown in Fig. 3 (b). Finally, when a negative voltage is desired, a voltage signal should be sent to optocoupler U9 (4th_PWM_Sig). The driving circuit on the right side works in the same manner as the left, with the only difference that the resulting voltage between terminals +VH1 and -VH1 is negative, as shown in Fig. 3 (c).

D. ATMEGA Microcontroller Considerations

An ATMEGA microcontroller was selected to implement the reference sinewave signal and the six PWM signals because of its versatility, low cost, and extensive online support. However, despite those advantages, its most significant disadvantage is its speed. A considerable reduction in the ATMEGA's speed will be manifested when implementing I/O instructions like, pinMode( ), digitalRead( ), digitalWrite( ), and serialPrint( ). I/O instructions included in the ATMEGA program reduce the processing speed [5]. For a seven-level ML inverter application, six digital outputs will be needed. Those outputs signal will decrease the ATMEGA's velocity below 400 Hz. Two PWM commutations per voltage level require at least a frequency of 30 times higher than nominal. In other words, the ATMEGA should switch its outputs at 1.8 kHz or faster to generate an acceptable PWM output voltage pattern. Direct PORT manipulation should be used instead of the classical I/O instructions to achieve the 1.8 kHz minimum velocity in the ATMEGA. Direct PORT manipulation will increase the velocity up to 25 times depending on the number of the program's instructions. The port instructions are described as follows:

DDR = Determines whether the pin is IN (0) or OUT (1)
PIN = Reads the state of the INPUT pins
PORT = Controls whether the pin is HIGH (1) or LOW (0)

Fig. 5 shows the ATMEGA pin mapping necessary to implement direct PORT control. PORT B digital outputs eight to thirteen were used for this ML inverter.

Atmega168 Pin Mapping

Arduino function			Arduino function
reset	(PCINT14/RESET) PC6	1 28	PC5 (ADC5/SCL/PCINT13) analog input 5
digital pin 0 (RX)	(PCINT16/RXD) PD0	2 27	PC4 (ADC4/SDA/PCINT12) analog input 4
digital pin 1 (TX)	(PCINT17/TXD) PD1	3 26	PC3 (ADC3/PCINT11) analog input 3
digital pin 2	(PCINT18/INT0) PD2	4 25	PC2 (ADC2/PCINT10) analog input 2
digital pin 3 (PWM)	(PCINT19/OC2B/INT1) PD3	5 24	PC1 (ADC1/PCINT9) analog input 1
digital pin 4	(PCINT20/XCK/T0) PD4	6 23	PC0 (ADC0/PCINT8) analog input 0
VCC	VCC	7 22	GND GND
GND	GND	8 21	AREF analog reference
crystal	(PCINT6/XTAL1/TOSC1) PB6	9 20	AVCC VCC
crystal	(PCINT7/XTAL2/TOSC2) PB7	10 19	PB5 (SCK/PCINT5) digital pin 13
digital pin 5 (PWM)	(PCINT21/OC0B/T1) PD5	11 18	PB4 (MISO/PCINT4) digital pin 12
digital pin 6 (PWM)	(PCINT22/OC0A/AIN0) PD6	12 17	PB3 (MOSI/OC2A/PCINT3) digital pin 11(PWM)
digital pin 7	(PCINT23/AIN1) PD7	13 16	PB2 (SS/OC1B/PCINT2) digital pin 10 (PWM)
digital pin 8	(PCINT0/CLKO/ICP1) PB0	14 15	PB1 (OC1A/PCINT1) digital pin 9 (PWM)

Fig. 5. Diagram of the ATMEGA microprocessor pin mapping.

E. Inverter H-Bridge Pulse Width Modulation Control

The 60 Hz sinewave reference signal and the six 1.8 kHz PWM triangular signals are implemented in the ATMEGA microprocessor, as shown in Fig. 6. With the velocity enhanced by the direct port manipulation, the reference sinewave signal of 60 Hz and 0.9 magnitude was implemented in the ATMEGA using equation (1) where the variable t is time.

$$ys = 0.9 \cdot sin(2 \cdot \pi * 60 \cdot t) \tag{1}$$

The PWM triangular signal at 1.8 kHz is implemented with a mathematical series. It has a magnitude of 1/3 to create the three PWM positive references, three PWM negative references, and the zero reference for a total of seven-signal references. The triangular reference signal corresponding to the PWM1 is implemented using equation (2). In this equation, variable f_s is the switching frequency, set to 1.8 kHz.

$$yt_1 = \frac{1}{3} \cdot abs\big[2 * [t \cdot f_s - floor(t \cdot f_s + 0.5)]\big] \tag{2}$$

The reference signals PWM2 to PWM6 are shown in equation (3) through equation (7), and they are derived from equation (2).

$$yt_2 = \frac{1}{3} + yt_1 \tag{3}$$

$$yt_3 = \frac{2}{3} + yt_1 \tag{4}$$

$$yt_4 = -\frac{1}{3} + yt_1 \tag{5}$$

$$yt_5 = -\frac{2}{3} + yt_1 \tag{6}$$

$$yt_6 = -1 + yt_1 \tag{7}$$

Fig. 6 shows the generated 60 Hz sinewave reference signal, the six triangular 1.8 kHz PWM signals, and the switching pattern (black) that controls the microcontroller's outputs.

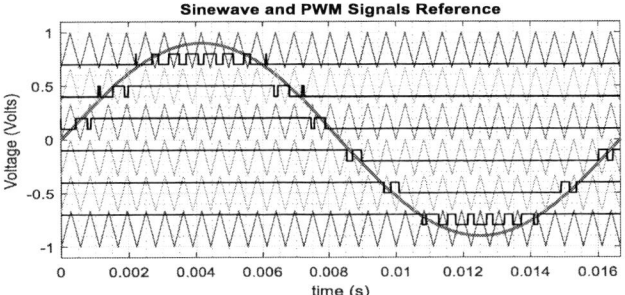

Fig. 6. Sinewave, PWM and trigger signals references for the microcontroller.

F. Lowpass LC Filter

An LC Lowpass Filter (LPF) could be implemented to eliminate undesired harmonics, including the 1.8 kHz switching harmonics at the voltage signals [6]. The main objective in this work was to develop a low cost ML inverter, because one of its advantages is reducing the THD as levels increases. Therefore, when operating nonsensitive loads like lights, increasing the number of levels in the ML inverter until we get an output voltage with low THD could eliminate the use of an LC LPF. The LC LPF tends to be the heaviest part in an inverter circuit, and its volume increases with the inverter capacity. The possibility of eliminating this component for nonsensitive load applications reduces cost, weight, components, and complexity.

G. Inverter Step-Up Transformer

The final stage of the seven-level cascaded H-bridge ML single-phase inverter is a step-up transformer with a turn ratio of 1:4 that will increase the peak AC voltage from ~42 V (with losses) in the primary winding to a peak AC voltage of ~170 V in the secondary winding. The inductance in the transformer's winding affects the current and provokes changes in the voltage PWM shape and pattern, as shown in Fig. 7 [7].

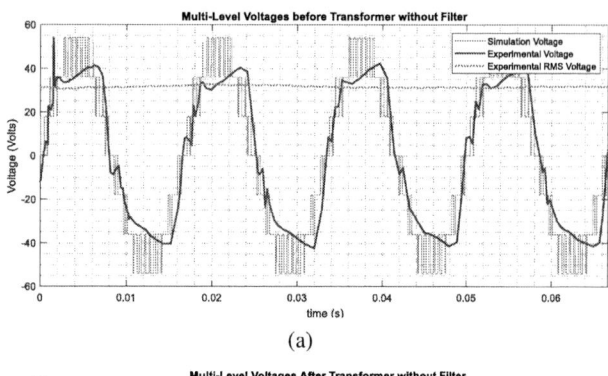

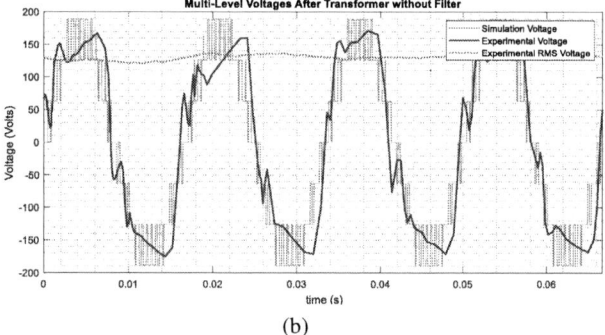

Fig. 7. Voltages before and after the multilevel circuit without LPF. (a) Voltages at multilevel output - before transformer without LPF (b) Voltages at load - after transformer without LPF.

III. EXPERIMENTAL SETUP

An experimental seven-level cascaded H-bridge ML single-phase inverter was developed. Power is supplied to the three cascaded H-bridges through 18 V batteries. Fig. 8 illustrates the experimental inverter, consisting of three H-bridge circuits and an ATMEGA microcontroller. The seven-level cascaded H-bridge ML single-phase inverter was tested with different lighting technologies.

Fig. 8. Seven-level cascaded H-bridge multi-level single-phase inverter implemented with discrete electronics components and an ATMEGA Microprocessor.

IV. EXPERIMENTAL RESULTS

Filament, CFL, and LED lights are considered nonsensitive loads, and they could be powered with a seven-level ML inverter if the THD is around 30%. The THD is a measurement of the harmonic distortion present in the voltage signal, and it's the ratio of the sum of the powers of all harmonic components to the power of the fundamental frequency. It is expected that a significant amount of harmonic distortion to the signal comes from the fundamental and the switching frequency. Therefore, as the THD is lowered, the voltage signal becomes sharper. The ML inverter configuration provides THD reduction with increasing levels. Still, transformer and passive energy storage elements in CFL and LED could affect the anticipated THD. Therefore, to know how much those elements affect the voltage THD, a Fast Fourier Transformation (FFT) is performed on every experimental voltage.

In this prototype, working with a switching frequency of 1.8 kHz, a ML inverter with a single H-bridge configuration provides three voltage levels and a THD of 62.33%. Two H-bridges provide five voltage levels and a THD of 32.41%. Finally, as shown in Fig. 9, three H-bridges ML circuit provides seven voltage levels with a THD of 21.63%.

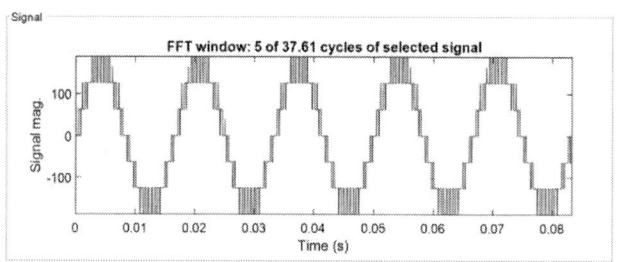

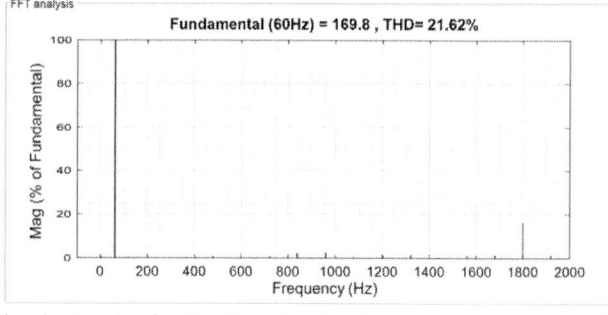

Fig. 9. Results for the Fast Fourier Transform analysis of the simulation voltage in the seven-level ML inverter after transformer with no load and no LPF.

A. Filament Light

The simulation model used for the filament light was a 14 Ω resistance. The results shown in Fig. 10 reveal that the expected experimental and simulation voltage differs in filament light. However, as shown in Fig. 12 the experimental value has an acceptable THD for this application even with the difference. Fig. 11 reveals that the experimental current contains high-frequency oscillations because of the current change in direction caused by switching in the H-bridge. This condition is reflected in the light as a random flickering. This effect could be minimized by connecting snubber circuits to MOSFETS.

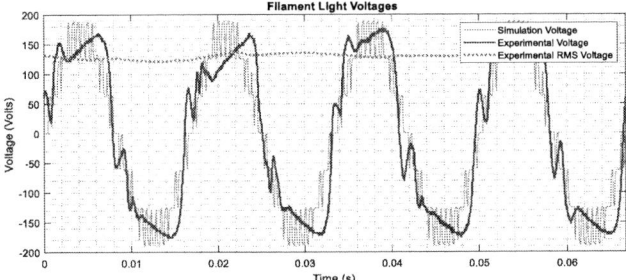

Fig. 10. Results for the voltages in filament light.

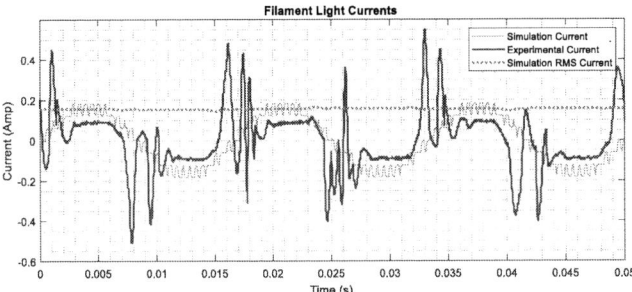

Fig. 11. Results for the currents in filament light.

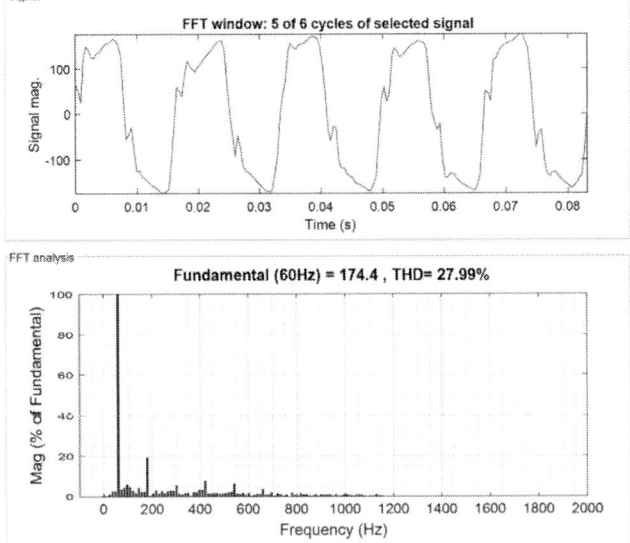

Fig. 12. Results for the Fast Fourier Transform analysis of the experimental voltage in a seven-level ML inverter after transformer with no LPF and filament light as load.

Fig. 9 shows the simulated THD for a pure Seven-level ML inverter without LPF, which yields 22%. Fig. 12 shows an experimental THD of 28% for the voltage applied to the filament light. Even though the voltage patterns look slightly different, the difference in THD between the simulation and experimental voltage is only 6%. This slight difference in THD is acceptable for operating a nonsensitive load like a filament light.

B. Compact Fluorescent Light (CFL)

An RL parallel circuit with a power factor of 0.6 was used to model the compact fluorescent light. The resistor was set to an active power of 14W, and the inductor to a reactive power of 18.67 vars [8]–[11]. As a result, the experimental and simulation voltage shown in Fig. 13 looks to be quite different, but the difference in THD is about only 5%, as shown in Fig. 15. Fig. 14 illustrates the currents in a CFL light. A power factor close to 0.6 in CFL light introduces an inductive load where the effects can be seen in the current. The inductive load is opposed to sudden changes in the current's direction, providing a light-filtering effect to the current. This condition could be observed when comparing Fig. 11 and Fig. 14.

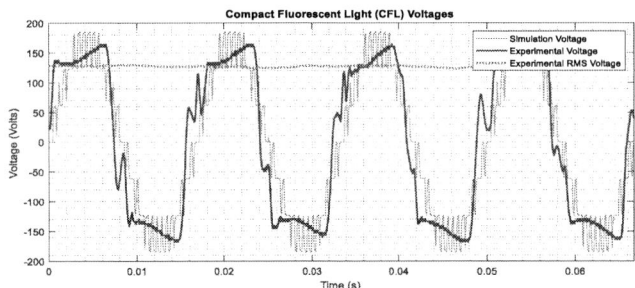

Fig. 13. Results for the voltages in compact fluorescent light.

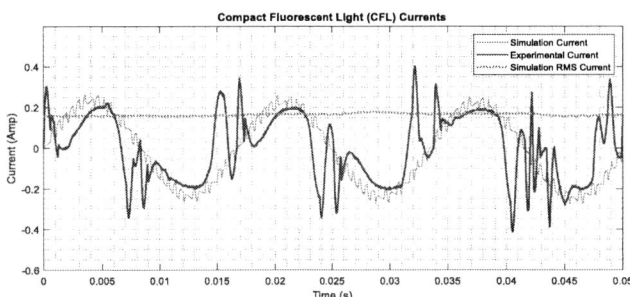

Fig. 14. Results for the currents in compact fluorescents light.

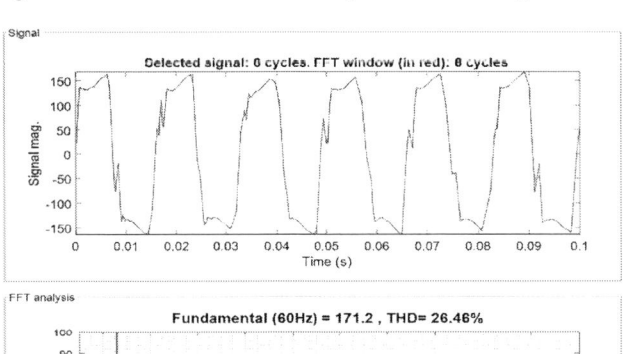

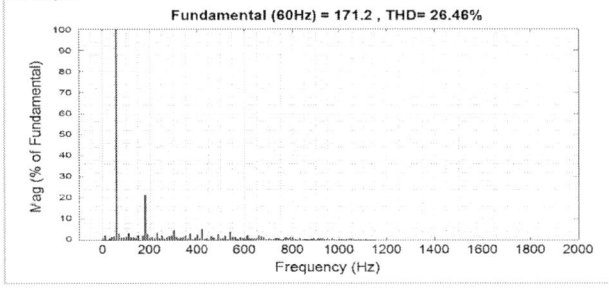

Fig. 15. Results for the Fast Fourier Transform analysis of the experimental voltage in a seven-level ML inverter after transformer with no LPF a and CFL light as load.

C. Light Emitting Diode (LED) Light

The nonlinear model used in the simulation for the LED light is shown in Fig. 16. The ballast circuit used to turn on the LED is composed of passive energy-storing elements that distort the LED voltage and current and increase THD [12].

The LED simulation and experimental voltages and currents are plotted in Fig. 17 and Fig. 18, respectively. The LED's experimental voltage looks very different from what was expected. The large difference in expected voltage can be attributed to the high THD. Interestingly, LEDs create more distortions to voltage signals than the CFLs. The LED light used in this experiment has a power factor close to 0.2 (which is not good). In addition, the low power factor makes it difficult for the ML inverter to supply the LED with active power. More expensive LEDs include power factor correction, which increases its power factor up to 0.85. Fig. 19 reveals that LED experimental voltage THD is about 39%. The high THD on LED makes it difficult for it to turn on and triggers more frequent light flickering [13], [14].

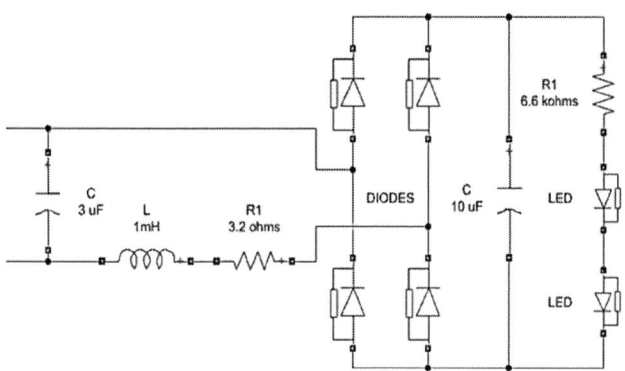

Fig. 16. Diagram of Light Emitting Diode model for simulation results.

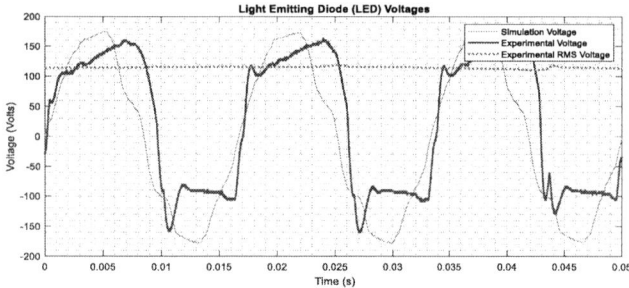

Fig. 17. Results for the voltages in Light Emitting Diode (LED) light.

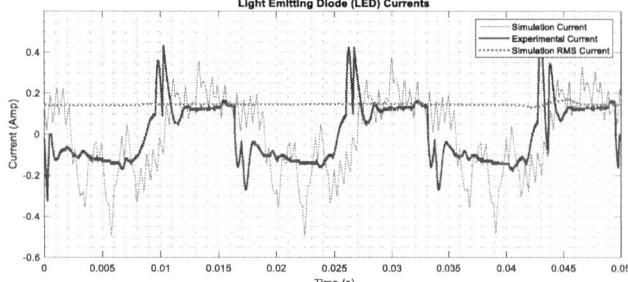

Fig. 18. Results for the currents in Light Emitting Diode (LED) light.

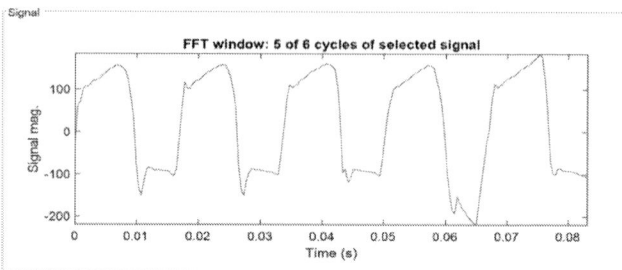

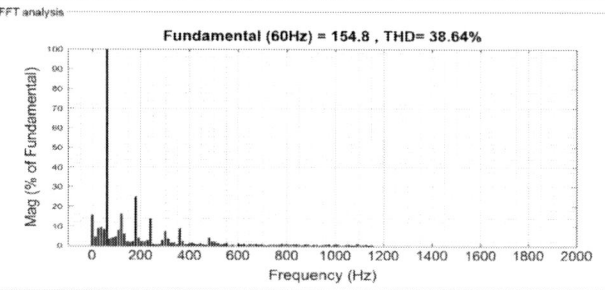

Fig. 19. Results for the Fast Fourier Transform analysis of the experimental voltage in a seven-level ML inverter after transformer without LPF and a LED light as load.

V. CONCLUSION AND FUTURE WORK

A seven-level ML cascaded H-bridge inverter was developed using an ATMEGA microprocessor in this work. Simulation and experimental results proved that a low-cost ML inverter could be designed using an ATMEGA microprocessor. Results demonstrate that each ML stage can provide multiple voltage levels and reduce THD [2]. The circuit topology keeps the system inexpensive, simple, and allows for the operation of nonsensitive loads like filament, CFL, and LED lights. Future work includes snubbers circuits installation on MOSFETs to improve signal quality and a faster microprocessor and one that allows for the control of more H-bridge levels.

ACKNOWLEDGMENT

Sandia National Laboratories is a multi-mission laboratory managed and operated by National Technology and Engineering Solutions of Sandia, LLC., a wholly owned subsidiary of Honeywell International, Inc., for the U.S. Department of Energy's National Nuclear Security Administration under contract DE-NA-0003525.

This work was sponsored in part by the Consortium for Hybrid Resilient Energy Systems (CHRES) under grant number DE-NA0003982 from the National Nuclear Security Administration part of the U.S. Department of Energy.

REFERENCES

[1] J. N. C. Sekhar and A. Tejasree, "Hardware Implementation of 31-Level Inverter using Arduino Uno Controller," vol. 10, no. 03, pp. 61–67, 2020, doi: 10.9790/9622-1003016167.

[2] M. Kumar Sahu, M. Biswal, and J. Mohana Rao Malla, "THD Analysis of a Seven, Nine, and Eleven Level Cascaded H-Bridge Multilevel Inverter for Different Loads," *Teh. Glas.*, vol. 14, no. 4, pp. 514–523, 2020, doi: 10.31803/tg-20180206150332.

[3] M. J. Scott *et al.*, "Inverter based on Switched-Capacitor Cells," pp. 3057–3061, 2013.

[4] M. J. Scott, R. D. Zamora, A. Long, C. Li, F. Zhang, and J. Wang, "Inverter for Utility Applications," pp. 3930–3933, 2014.

[5] S. Robinson, "Speeding Up Arduino," *stackabuse.com*, 2016. https://stackabuse.com/speeding-up-arduino/ (accessed Dec. 09, 2021).

[6] A. E. W. H. Kahlane, L. Hassaine, and M. Kherchi, "LCL filter design for photovoltaic grid connected systems," *Third Int. Semin. new Renew. energies*, vol. 8, no. 2, pp. 227–232, 2014.

[7] A. S. K. Chowdhury, M. S. Shehab, M. A. Awal, and M. A. Razzak, "Design and implementation of a highly efficient pure sine-wave inverter for photovoltaic applications," 2013, doi: 10.1109/ICIEV.2013.6572634.

[8] M. A. Adelabu, A. L. Imoize, and G. U. Ughegbe, "Analysis of the electronic circuits of 11 W and 15 W compact fluorescent lamps," *Niger. J. Technol.*, vol. 40, no. 3, pp. 501–517, 2021, doi: 10.4314/njt.v40i3.16.

[9] G. Malagon-Carvajal, J. Bello-Peña, G. Ordóñez Plata, and C. Duarte Gualdrón, "Analytical and experimental discussion of a circuit-based model for compact fluorescent lamps in a 60 Hz power grid," *Ingenieria e Investigacion*, vol. 35, no. 1. pp. 89–97, 2015, doi: 10.15446/ing.investig.v35n1Sup.53618.

[10] M. Hanan, X. Al, S. Salman, A. Masood, and K. Hashmi, "A Shunt Active Filter for Extenuation of Harmonics by Compact Fluorescent Lamp Drives," *2nd IEEE Conf. Energy Internet Energy Syst. Integr. EI2 2018 - Proc.*, pp. 3–9, 2018, doi: 10.1109/EI2.2018.8581908.

[11] A. Kumar, U. Prasad, and R. P. Gupta, "Analysis and Simulation of CFL Ballast circuit With MOSFET &IGBT based Inverters," vol. 2, no. 9, pp. 684–689, 2013.

[12] L. Kukacka, P. Dupuis, R. Simanjuntak, and G. Zissis, "Simplified models of LED ballasts for spice," *2014 IEEE Ind. Appl. Soc. Annu. Meet. IAS 2014*, pp. 1–5, 2014, doi: 10.1109/IAS.2014.6978426.

[13] F. Ion, "A Model of a LED for Street Illumin Simulat tions and Measuremen nts," pp. 662–667, 2015.

[14] Y. Hassan, M. Orabi, M. Ismeil, and A. Alshreef, "Study the effect of series and parallel LEDs connections on the output current ripple for LED driver of solar street lighting," *2017 19th Int. Middle-East Power Syst. Conf. MEPCON 2017 - Proc.*, vol. 2018-Febru, no. December, pp. 1492–1499, 2018, doi: 10.1109/MEPCON.2017.8301380.

Annual Energy Production Uncertainty of Bifacial PV Plants Caused by Inaccuracies in Albedo Data: Case Studies Using SAM

Vicente Lara Fanego

Solargis s.r.o., Bratislava, Slovakia

Bifacial PV technology has a rapidly growing presence in the solar industry. Although several studies have tried to quantify bifacial gains, there is limited information about the associated uncertainty-an essential quantity in solar PV projects. In this regard, a sensitivity analysis of the uncertainty in annual energy production of a bifacial PV plant with respect to the ground albedo uncertainty is presented here. A bifacial PV plant is configured using NREL' public-domain SAM simulation model and realistic scenarios at three U.S. sites. Results show that the uncertainty in the annual energy production directly depends on the albedo uncertainty and, in a non-linear way, on the albedo value itself.

Arsenic doped CdSeTe solar cells: Charge collection and Defects

Niranjana Mohan Kumar, Srisuda Rojsatien, Trumann Walker, Tara Nietzold, Barry Lai, Arun K.M. Kanakkithodi, Maria Chan, Dan Mao, Mariana Bertoni

Arizona State University, Tempe, AZ, United States

Argonne National Lab, Lemont, IL, United States

Purdue University, West Lafayette, IN, United States

First Solar, Perrysburg, OH, United States

The selection of arsenic as a dopant has been an integral part of achieving high performing, 20.8%, polycrystalline CdTe solar cells. Although arsenic has significantly improved the long-term performance of the cells and the p-type doping levels reached inside the absorber, the activation ratio of dopant remains quite low, ~1%. Understanding the origins of this activation would aid in further bettering device performance. Herein, devices of two different activation levels but identical arsenic concentration are studied using nanoscale correlative X-ray microscopy in cross-section. Charge collection in cross-section shows a distinctive change between activation levels and the local environment around the As atom, as measured by X-ray absorption, shows the signature of several defects and phases that could be the culprit.

Observations On A Colorado Electric-Utility Resource Plan For Increasing Renewables From 55% to 80% By 2030

Ronald A. Sinton

Sinton Instruments, Boulder, CO, 80305

Abstract—**This paper presents observations of a PV specialist concerning the actual implementation of renewables into an electric utility resource plan. In 2021, Public Service Company of Colorado introduced a resource plan to increase renewables from 55% to 80% by 2030. The actual process is a regulatory proceeding that involves the technical capability of resources, but also balances the interest of many stakeholders including the regulated monopoly utility, state legislation, ratepayers, independent power producers, environmentalists and local economies impacted by power-plant retirements. This paper will discuss many ways in which this process may result in different choices than a purely technical optimization might find. A particular focus will be on the subtle assumptions that reduce the choice of low-cost PV as the preferred clean energy addition.**

Keywords—Photovoltaics, wind, utility-resource planning

I. INTRODUCTION

In 2019, the U.S. state of Colorado legislature passed a bill to require an 80% reduction in CO_2 by 2030 from the major investor-owned utilities relative to a 2005 baseline. In 2021, one utility proposed a resource plan to accomplish this goal. This paper will present observations on this ongoing process using the treasure trove of technical documentation provided by the regulatory utility proceeding [1]. Central to the plan are the negotiation of retirement dates for coal plants; and the cost and reliability evaluation of new resources including: PV, wind, and batteries. The legislature mandated that a social cost of carbon of $68/U.S. ton ($0.0748/kg CO_2) be utilized in the cost evaluations for new resources including a net present value calculation of the social cost of CO_2 emissions. Although the utility uses a capacity-expansion model to evaluate the potential options, the results are seen to be highly constrained by a set of input assumptions for future hourly-demand profiles; coal retirement dates; capacity factors of the coal plants until retirement; and reliability metrics for the variable renewables and storage. These assumptions became the focus of negotiations during the regulatory process. The areas of particular interest to PV specialists that will be discussed include the pricing that the ratepayers see concerning the various "kinds" of PV, given the different laws and subsidies prescribed for each. For example, utility PV obtained with a power purchase agreement through an all-source bid is projected to cost $25/MWH or $35/MWH including transmission. On the other hand, residential rooftop is compensated at $110/MWH through a net metering approach. It is noted that hourly demand projections assumed that the greatest source of new load growth, EVs, would be at night. This has the effect of limiting the penetration of PV despite the excellent resource available in Colorado.

II. COST OF PV

The utility resource plan includes additions to only the utility-scale PV. However, the distributed PV, determined by another process, is comparable in scope with prices and volumes determined in other regulatory proceedings. The result is that the two are not co-optimized with respect to price or performance in the utility. A table of projected costs for the different types of PV in the utility area is shown in Fig. 1. Since the local PV and utility PV produce power during the same times of day, the two trade off almost 1:1 in a resource plan. The assumption of how much distributed PV will be installed that occurs through those separate programs effectively limits the amount of utility PV that is determined from capacity-expansion modeling. In the resource plan proposal, modeling projects 1550 MW of utility PV to be installed in the period 2023 - 2030, in contrast to 1011 MW of distributed PV. The blended cost of distributed PV and utility PV is therefore significantly higher than the cost of utility PV. This negates the potential for PV to lower the cost of electricity rates despite the expected low prices of utility PV power-purchase agreements shown in Fig. 1. For

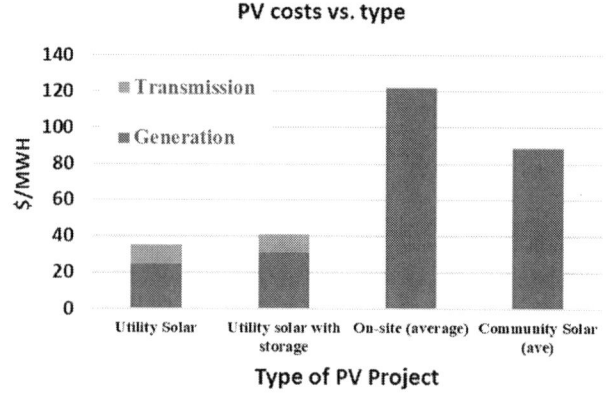

Fig. 1. The projected price for various "types" of PV in the resource plan. The values shown for on-site and community solar are weighted averages encompassing many different programs [2,3].

the quantities of each in this resource plan, we estimate that the total PV, including transmission, costs an estimated USD $2.7B more than the same total quantity of PV installed as utility PV only. The cost was estimated by the author for the following quantities: 25 years at 1600, 2000, and 2200 hours per year of production in behind the meter, community gardens, and utility scale power plants using the prices shown in Fig. 1.

Residential rooftop PV is projected to cost $111/MWH, while utility PV, including the costs of a new transmission project, is projected to cost $35/MWH. As a specific example, the latest price for community PV, approved by the PUC, was $117/MWH and was designed with a social equity focus in contrast to lowest cost [4].

III. EVOLUTION OF THE HOURLY FOSSIL FUEL CONTENT OF ELECTRICITY

The proceeding includes projections from the utility for the percentage of wind and PV relative to total generation for various years within the proceeding. By using publicly available data from the Energy Information Administration [5] for the balancing area for the utility (of which the utility comprises

85%), the fossil fuel intensity of the electricity supply in future years can be projected using the percentages reported by Public Service Company of Colorado. This is the total generation minus the available renewables. The CO_2 content of this electricity can't be estimated in this way, because the fossil fuel used to meet the difference between renewable generation and load could be coal or gas from various types of plants. The results from this study, comparing the historical data from 2019 with the projected results for the renewable fractions for 2023 and 2030 in the resource plan, are shown in Fig. 2.

In 2019 (top graphic), the historical data was 25% wind and 4% PV. Due to large installations of PV in 2022; 2023 (middle) will have 39% wind and 15% PV; while 2030 (bottom) is projected to have 55% wind with 25% PV. The difference between 2019 and 2023 is striking. The effect of 15% PV generation is to make daytime quite clean, while night can be dirty or clean depending on whether the wind is blowing.

The utility planning process has been to use historical data for demand projections. Based on this strategy, the main requirement to minimize capacity of conventional power plants

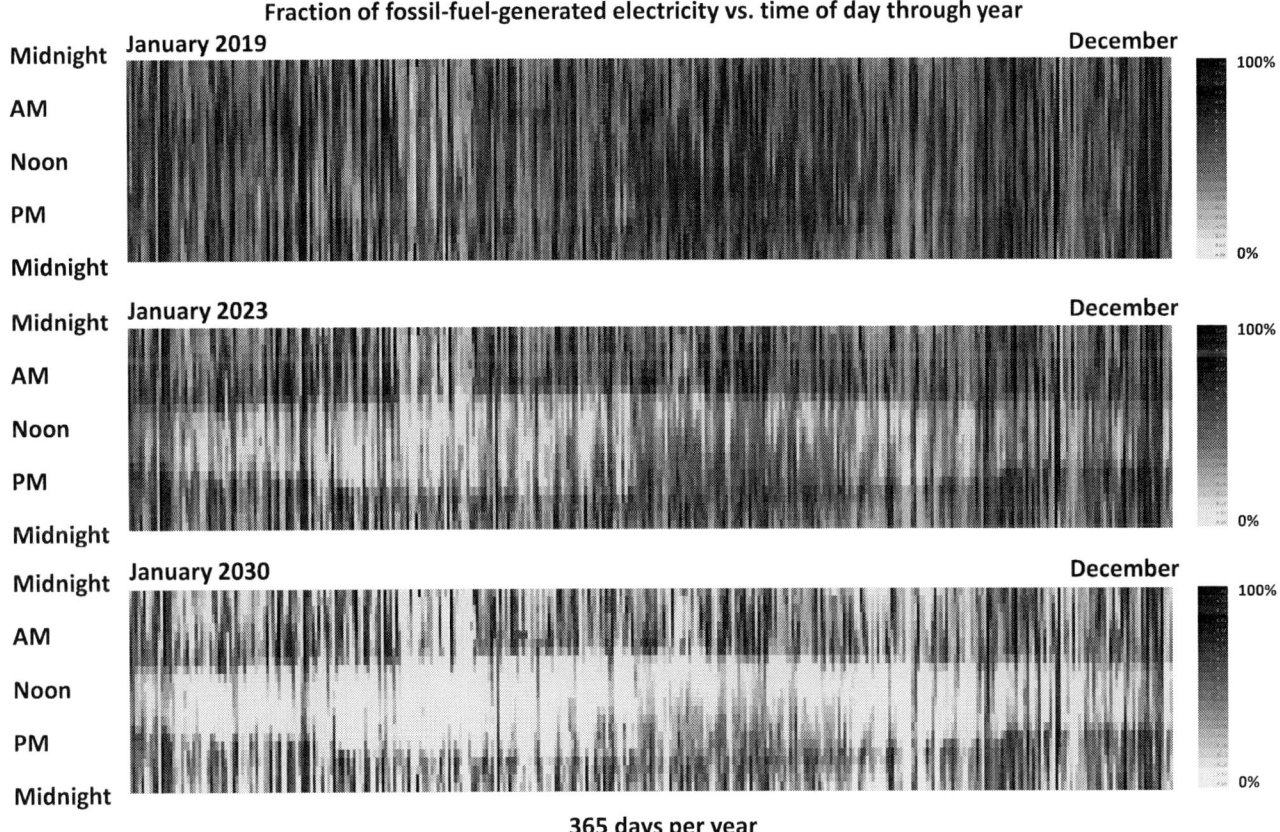

Fig. 2. Visualization of the fraction of the power generated by fossil fuel on the grid vs. hour of day and day of year. The scale is at right, with black being 100% fossil-fuel-generated power and green being 0%. Hours of the day are at left reading down from 1AM to midnight. The top graph is historical data from the Energy Information Agency (EIA) with the total generation and renewables data plotted as percent fossil fuel (total – renewables) for 2019 in the Public Service Company of Colorado balancing area. This was with 25% wind and 4% PV. In 2023 (middle) the renewable fraction will increase to 39% wind and 15% PV. By 2030 (bottom) there will be 55% wind and 25% PV. The plots for 2023 and 2030 simply scale the wind and PV up to the appropriate projected fractions with the same underlying data from 2019 for the demand and weather patterns.

978-1-7281-6118-1/22 $31.00 © 2022 IEEE

was to minimize power use during summer demand peaks which is the main feature prior to 2023. As a result, a "time-of-use" tariff was implemented in 2021 (to be phased in during 2022-2025) that was designed primarily to reduce power consumption in the afternoon and evening during summer, but was implemented year-around. In the resource plan, this time-of-use rate structure, with high rates during daytime after 1PM in summer, results in the assumption that EVs will charge primarily at night. Essentially the characteristics of the fossil grid prior to 2023, with excess capacity at night, were projected to be true through for the entire planning period.

Figure 2 illustrates several key features of both the demand profile and the renewables resources for this utility. First, daytime will become much cleaner than night after 2023. Because Colorado has negligible hydro generation, the renewable resource at night is wind. Nights can be clean (on windy nights) but within each season, there may be very calm nights when the demand is met by fossil fuel. Daytime is more consistently clean, due to the sunny climate in Colorado, as well as the fact that both wind and PV are available during daytime. Winter is challenging with both short days and calm periods without wind.

However, for purposes of hourly demand projection, these characteristics for EV charging were assumed to be constant throughout the planning period and based on a period prior to the installation of large amounts of PV that will occur in 2022. Since the effects of large amounts of PV only become prevalent starting in 2023, the opportunity to use demand flexibility to integrate renewables was excluded from the capacity-expansion models due to the fixed assumptions for the hourly demand profiles based on historical data from before there was

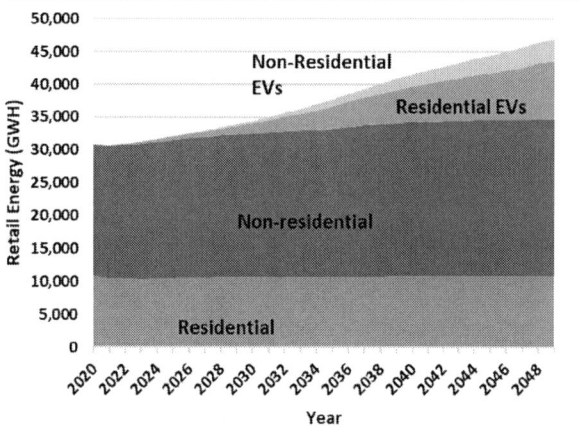

Fig. 3. Demand growth projected by Public Service Company of Coloardo to 2040 indicating that EVs will be the largest source of demand growth[1].

significant PV.

IV. DEMAND PROJECTIONS AND EV CHARGING

The EV charging profile projected by Public Service Company of Colorado as a fraction of total demand is shown in Fig. 3. It consists of ~75% charging during night, and ~25% during daytime. A comparison of this assumption for 2040, and the evolution of the grid by 2030 in Fig. 2 indicates that the opportunity to use EV demand flexibility to integrate renewables into the grid was not recognized in this planning

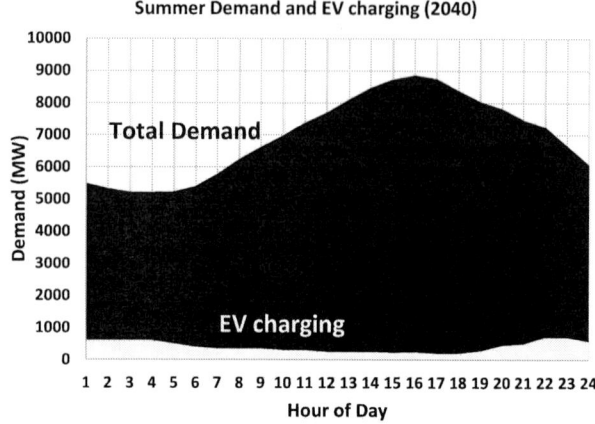

Fig. 4. The demand profile projected for 2040 in the resource plan[1], replotted to illustrate the EV charging profile relative to the total load as a function of the 24 hours in the day.

process. This oversight was very significant with respect to PV. Nearly all demand growth in this resource plan, through 2040 was assumed to be due to light-vehicle EV charging. This new EV-charging load was projected to be 23% of the 2020 total load by 2040 (Fig. 3). By assuming that this charging would be primarily at night, Fig. 4, much of the peak demand shifts into the night, reducing the capacity value of PV (which quantifies the ability of a resource to meet peak loads) as well as limiting the potential for PV to meet the load growth.

V. CAPACITY VAULES FOR STORAGE

This section discusses the Effective Load Carrying Capacity (ELCC) for storage to illustrate the level of detail that can affect the choice of resources in a plan. ELCC is a measure of how valuable a resource is for meeting peak loads, to ensure that a portfolio of resources can meet load reliably. In this resource plan, many of the intervenors noted that the ELCC for batteries was found in the Public Service Company of Colorado study [6] to be 55% of the ELCC from a combustion turbine for the first tranche of added storage, then falling to 38% for subsequent additions. This is significantly lower than the ELCC used in many other states with good PV resources, and much lower than the 90% ELCC for PV + storage in the 2016 Public Service Company of Colorado resource plan. It is not possible to do ELCC calculations from publicly available data here, but the trends can be illustrated using examples and a graphical method as previously applied [7, 8].

Figure 5 shows the total generation for August 13, 2020, a day where demand response was called due to high air-conditioning loads [9]. This data was downloaded from the EIA for the Public Service Company of Colorado balancing area in Colorado, where 85% of the total power is determined by the resource plan. The curves shown are the total generation (top), the total generation minus the available renewables on that day (middle curve) and the total generation minus renewables that are anticipated using the PV and wind that will be available in 2023. The 2023 wind and PV resources are scaled from the 2020 data using an annual basis to agree with the totals projected by Public Service Company of Colorado.

978-1-7281-6118-1/22 $31.00 © 2022 IEEE

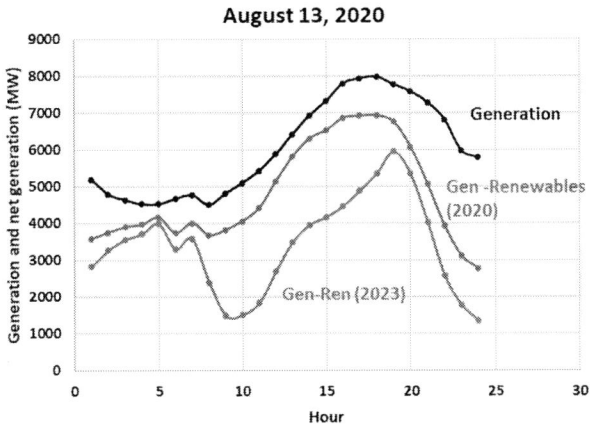

Fig. 5. An example of a summer demand peak in 2020. Shown are the total generation, the generation minus renewables, and the generation minus renewables that would be present in 2023 based on the projections in this resource plan.

The total generation curve, and the generation minus renewables curve for 2023 are reproduced in Fig. 6 to show the implications of the different curve shapes. The addition of more than 1 GW of PV in 2022 in addition to more wind changes the shapes of the net demand peaks. PV during daytime meets a significant amount of demand from 7AM to 7PM, but when the sun sets there is now a sharp peak of demand not met by renewables. There is an increase in the wind in the evening on this day, which reduces net demand after 8PM. The result is a very sharp peak at 7PM. If storage were used to meet the original demand peak, without renewables, it would require 5GWH of storage to reduce the peak by 1GW, from 2PM to 10PM. However, with the renewables, the peak is so sharp and abrupt that only 1.6GWH is required to reduce the peak net demand by 1GW (red curve large triangle). The addition of PV and wind, especially PV, has made the storage much more valuable as a capacity resource. However, another observation is that if you split this 1.6 GWH into two equal 0.8 GWH additions of storage, the first addition (shown as an inset triangle) already reduces the peak by 0.71GW. The second 0.8GWH only reduces the peak by an additional 0.29 GW. This is an example of how dependent the expected usability calculation for storage is on the order in which the storage is evaluated for capacity, because of the nuances previously mentioned. Because of this it subtly played a significant role in the PUC proceeding and power mix determination.

For calculations of new storage capacity value, 500 MW of load-reduction demand response was considered ahead of assessing the capacity value of new batteries. This demand response is largely air conditioners, which reduces load during summer peaks. This had a similar effect as the 0.8 GWH of storage shown here, reducing both the ELCC of batteries and therefore the ELCC of PV + storage by blunting the peak on the generation minus renewables curve. In addition, the synergistic effects between PV and storage are damped by this assumption. This is another indication of the very significant effects that subtle assumptions can have on the results of a resource procurement. In the resource plan, the assumptions applied were those that would result in the lowest bound on the capacity value for batteries. In the settlement agreement currently being

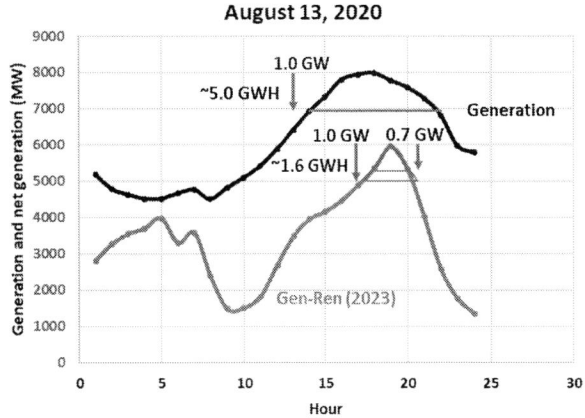

Fig. 6. Two of the curves from Fig. 5. The use of storage to reduce the peak by 1GW is shown for the generation curve without renewables, and then for the curve after 2023 renewables reduce the load. 0.8 GWH of storage reduces the red-curve peak by 0.7GW. Another 0.8GWH completes 1GW reduction.

considered by the commissioners, the ELCC for storage will be recalculated. However, the specific issue discussed in this section was not called out.

VI. OVERALL RESULT IN CARBON REDUCTION

This resource plan, while still in progress, is very significant in reaching the legislative mandate of 80% CO_2

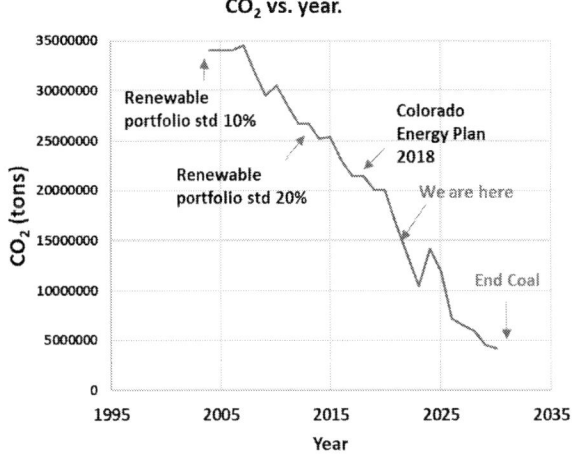

Fig. 7. The utility reduction in CO_2 emission since 2005 with historical data and projections from this resource plan [10,11].

reduction from a 2005 baseline, a point when the utility had mostly coal-fired power. The trend in CO_2 is shown in Fig. 7. In the details of the plan, there are many factors that have biased the result against a maximum deployment of PV that could contribute to even more reduction in CO_2. This may be an opportunity, as it indicates that it may not be difficult to make further progress towards a fully clean grid on this utility system.

978-1-7281-6118-1/22 $31.00 © 2022 IEEE

VII. CONCLUSIONS

Phase I of the 2021 resource plan for the utility Public Service Company of Colorado is currently being considered by the Colorado public utility commission. This Phase I considered the modeling assumptions with which the all-source bid for resources in Phase II will be evaluated. These assumptions, to be used in capacity expansion models applying the prices and technical capabilities of the bids that are submitted, constrain the probable results into a narrow range. Notable assumptions that potentially reduce the amount of PV expansion include the retirement dates of coal plants and the capacity factors at which they will run until retirement. The assumption that a majority of EV charging which is assumed accounts for most load growth, will be at night limits both the capacity value (in MW) and the energy value (MWH) of PV to meet the load. Assumptions that determine the capacity value of battery storage can vary through a large range, from 38% to 90% within the discussion in this proceeding. Low values limit storage and the PV that would likely charge the storage. Assumptions that favor PV in the proceeding are the legislative requirement for a minimum of 80% CO_2 reduction by 2030 relative to 2005 and a legislative requirement to use a social cost of carbon of $68 USD/US ton of CO_2 in the evaluation of new resources.

All of this is quite abstract, because the topic of this Phase I is to define the parameters of the modeling that will be used to assess the resources from an all-source bid to meet the grid requirements. The actual results in GW of PV to be installed in the 2023-2030 interval depend on both the bids and the modeling using these assumptions.

Seeing the extent of assumptions that constrain the choice of resources, I would very much like to test a few of my own. More speculatively and simply stated than the known facts in the proceeding, I wonder if:

1. Much of the EV charging demand could be shifted into daytime. Since it was nominally assumed to be at night, could this 23% increase in electricity demand by 2040 in Fig. 3 translate into a ~20% increase in PV penetration on the grid, increasing the nominal 25% PV (Fig. 2) towards ~40+%? This would result in much cleaner nights than in the modeling presented.

2. Traditionally, pumped storage was charged with coal at night, and discharged into the summer late-afternoon demand peak, for example. Currently, it is likely that it is charged at night by wind, gas, and coal. If this 350MW pumping load were matched with ~400-500MW of new PV, would this increase the penetration of PV by another 500MW and result in significant CO_2 reduction?

3. The demand projections relied on a time-of-use rate encouraging power use at night. This rate was designed for a nominally fossil-fueled grid with excess capacity at night. How much demand will shift into daytime after the TOU rate shifts to accommodate the net demand resulting from 15% PV in 2023 and 25% PV in 2030? These TOU rates are being re-evaluated in April 2025.

4. This proceeding assumed that behind-the-meter PV was counted on the supply side in evaluating the ELCC for PV additions. As a result, behind-the-meter PV contributed to a rapidly declining ELCC (capacity value) with increasing PV deployment. Was this also taken into account on the demand projections, adding this PV back into the hourly demand profiles used by the capacity expansion models?

Finally, it is notable to an engineer that building an 80% renewable grid by 2030, as proposed in this resource plan, is an ambitious engineering project. There was no discussion of the cost or technical difficulty of balancing the proposed grid. For example, there was no discussion that I can find on grid-forming inverters, synchronous condensers, or synthetic inertia that might be useful in an 80% renewable grid that has relatively little transmission to adjacent balancing areas. It will be interesting to see when and how these topics come up.

REFERENCES

[1] Direct testimony of Alice K. Jackson, Proceeding number 21A-0141E of the Colorado Public Utilities Commission, Plan overview and appendix 2. Pg. 47-48 and 296, March 2021 (Unpublished)

[2] Direct testimony of Jack W. Ihle, Proceeding number 21A-0141E fo the Colorado Public Utilities Commision, page 59. March 2021 (Unpublished)

[3] Direct testimony of Kerry R. Klemm, Proceeding number 21A-0625EG of the Colorado Public Utilities Commission, pg 20. Dec. 2021 (Unpublished)

[4] Supplemental Testimony of Jack Ihle, Proceeding number 21A0625EG of the Colorado Public Utilities Commission, pg. 30-31, May 2022 (Unpublished).

[5] https://www.eia.gov/electricity/gridmonitor/dashboard/electric_overview/balancing_authority/PSCO

[6] Alice K Jackson Appendix E_ELCC Study Report, Colorado PUC Proceeding, May 2021 (Unpublished).

[7] Keith Parks, Declining Capacity Credit for Energy Storage and Demand Response With Increased Penetration, IEEE Transactions on Power Systems, Volume: 34, Issue: 6, Nov. 2019

[8] N. Schlag, Z. Ming, A. Olson, L. Alagappan, B. Carron, K. Steinberger, and H. Jiang, "Capacity and Reliability Planning in the Era of Decarbonization: Practical Application of Effective Load Carrying Capability in Resource Adequacy," Energy and Environmental Economics, Inc., Aug. 2020 Available at: https://www.ethree.com/elcc-resource-adequacy/

[9] AC Rewards Program Evaluation, Prepared for Xcel Energy, Submitted by Guidehouse, Boulder, CO 80302 pg. 14, May 2022 (Unpublished).

[10] Public Service 2017 All Source Solicitation 120-day Report, Colorado PUC Proceeding 16A-0396E page 10, June, 2018 (Unpublished)

[11] Attachment D – Settlement Agreement Modeling Results, Colorado PUC Proceeding No. 21A-0141E, May 2022 (Unpublished)

Estimation of Soiling Losses in Unlabeled PV Data

Bennet Meyers[1,2]

[1] SLAC National Accelerator Laboratory, Menlo Park, CA, 94025, USA
[2] Stanford University, Stanford, CA, 94305, USA

Abstract—**We provide a methodology for estimating the losses due to soiling for photovoltaic (PV) systems. We focus this work on estimating the losses from historical power production data that are unlabeled, *i.e.* power measurements with time stamps, but no other information such as site configuration or meteorological data. We present a validation of this approach on a small fleet of typical rooftop PV systems. The proposed method differs from prior work in that the construction of a performance index is not required to analyze soiling loss. This approach is appropriate for analyzing the soiling losses in field production data from fleets of distributed rooftop systems and is highly automatic, allowing for scaling to large fleets of heterogeneous PV systems.**

Index Terms—**photovoltaic systems, solar energy, distributed power generation, energy informatics, machine learning, statistical learning, unsupervised learning, soiling, unlabeled data**

I. INTRODUCTION

Soiling can cause significant energy yield reduction in photovoltaic (PV) systems, as high as -1%/day, but the effects are quite variable and impacted by many factors from system geometry to local climate conditions to nearby industry and agriculture [1]. Quantifying the losses due to soiling is important for understanding and mitigating this effect, which in turn improves the overall reliability and dependability of PV as an energy generation source. This importance is seen in the dedicated subarea for soiling at this conference, as well as the strong emphasis on soiling at other events such as the yearly PV Reliability Workshop hosted by NREL. Largely absent from this conversation, however, is quantification of the impacts of soiling in large fleets of heterogenous, distributed PV systems, which comprised over 40% of the installed capacity in 2020 [2]. The reason for this is a technical one: it is very difficult to analyze soiling trends without a reliable reference, and the unlabeled nature of distributed PV data make generating this reference difficult or impossible.

In this work, we present a methodology, based on recent research on machine learning for signal processing, that extracts estimates of system soiling losses from *unlabeled* production data. This work eliminates the requirement of constructing a performance index to analyze soiling loss and is highly automated, thus enabling the large-scale analysis of fleets of thousands of heterogenous systems. The proposed approach is based on an implementation of the *signal decomposition* (SD) framework [3].

This material is based on work supported by the U.S. Department of Energy's Office of Energy Efficiency and Renewable Energy (EERE) under the Solar Energy Technologies Award Number 38529.

Our approach takes unlabeled PV power generation measurements as an input and returns an estimate of the soiling loss over time, given as a percent loss relative to the unsoiled performance. This trend may be used to calculate secondary statistics such as the total energy loss or seasonal loss patterns. We validate this method on synthetic data, labeled data from a soiling test site, and on representative unlabeled data. The algorithm is available as a module in the Solar Data Tools package [4], [5]. This approach is uniquely suited to the analysis of fleet-scale PV systems, where it can be difficult or impossible to get suitable reference data for normalization.

II. RELATED WORK AND CONTRIBUTIONS

We are not the first to propose a method for estimating soiling losses from PV production data [6]–[9]. As described in [8], it has been determined that a combined model of degradation, soiling, and seasonal bias outperforms estimates of each loss term separately.

In this paper, we build upon previous work in a few ways. First, we propose a clear unified signal model (implemented as an SD problem) to describe the underlying components, rather than invoking iterative heuristics. Second, the use of the extensible SD framework allows for the expression of component classes that are designed specifically for this application Third, our approach estimates soiling losses from unlabeled system power generation time series data, without the need to construct a performance index (PI). As discussed in the introduction, this unlocks the analysis of around 40% of the installed PV capacity in the United States. An example of a PI constructed from data labels and an unlabeled daily energy signal are shown in figure 1. The proposed method allows for the analysis of *both* unlabeled and labeled data. Labeled data will provide more accurate results, when available, but reasonable estimates may be obtained when such labels are unavailable.

III. METHODS

We construct an SD problem [3] that models the decomposition of measured daily PV system energy or normalized energy (*i.e.*, a performance index, "PI") into a number of components, one of which represents the soiling loss in the system. This approach can be thought of as an *unsupervised* machine learning (ML) method for finding structure in time series data, similar to model-based clustering methods like Gaussian mixture models [10, §14.3.7]. Unlike supervised ML, there is no "training" of the method; we simply design the mathematical optimization problem, input the data for

978-1-7281-6118-1/22 $31.00 © 2022 IEEE

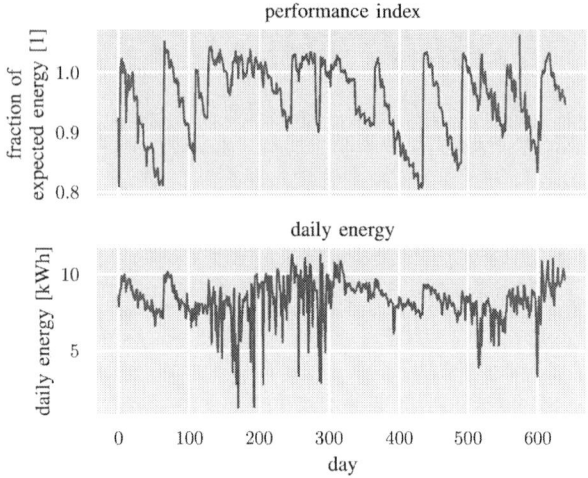

Fig. 1. A comparison of labeled data (top) and unlabeled data (bottom) from a single PV system in a desert environment.

analysis, and receive the soiling estimate. In this section, we describe the data preparation, SD formulation, and validation procedure. Finally, we briefly describe a procedure for utilizing the estimate of system soiling to "correct" the measured power data.

A. Data preparation

The proposed method of signal decomposition operates on a discrete daily time series representing raw or normalized daily system energy production, which we refer to as "unlabeled" and "labeled" data respectively. The purpose of generating a performance index is to remove known sources or variation in data, particularly due to available irradiance and operating temperature, and is a typical analytical approach for PV performance analysis, but requires additional knowledge about the system and its operating environment beyond real power production.

The raw data may be any measurements of PV system power or energy production indexed in time. This may be 1-minute measurements of instantaneous power, 15-minute interval averaged power, or daily energy production. If starting from high-frequency power measurements, one simply integrates to get the daily energy production. If one wishes to construct a PI, you then normalize by expected daily energy at this time.

When starting from sub-daily measurements, any prefiltering step may be applied, and rejected days can have their values replaced with `NaN` values (which we represent as ? in our notation). The SD framework optimally handles missing data points, so there is no need to use corrupted or untrustworthy data nor replace such data with interpolated values.

In this paper, we validate on both synthetic and real data. As described in §III-C0a, the synthetic data is already a daily time-series (a normalized PI), so no preprocessing is required. For the real data, which has a 5-minute measurement interval,

we use the data cleaning and filtering tools provided in Solar Data Tools [4], [5], and replace days that do not pass the quality check with ? values. After obtaining a representation of daily energy production (possibly with missing values), the signal is scaled so the 95th percentile is equal to 1. This is our input to the *signal decomposition problem* (SD problem). We do not construct a performance index on the real data and instead analyze the raw energy signal.

B. SD problem formulation

Utilizing the notation of signal decomposition defined in [3], we say that our data for the SD problem is a signal $y \in (\mathbf{R} \cup \{?\})^{T \times p}$ with length T equal to the number of days in the data set and measurement dimension $p = 1$. In this case, because the measurement dimension is equal to 1, y may also be thought of as a column vector of length T. We model the signal y as the composition of $K = 4$ components, x^1 to x^4, the sum of which must be equal to the signal y at the entries that do not contain missing values, or in other words,

$$y_t = x_t^1 + x_t^2 + x_t^3 + x_t^4, \text{ for } t \in \mathcal{K},$$

where $\mathcal{K}$ is the set of time indices that do not contain missing values (the "known" set). The four components are defined in the SD model by their cost functions $\phi_k(x^k)$ for $k = 1, \ldots, 4$. (We drop the superscript k on x to keep the notation lighter when not distinguishing between particular components.) As we will see, the last component, x^4, will represent the soiling signal which we wish to estimate from the data. The other three components represent other processes which impact the energy production of the system.

a) Component definitions: The first component represents the residual of the model, and it is taken to be the quantile cost function [11], [12],

$$\phi_1(x) = \mathbf{quant}_\tau(x) = \sum_{t=1}^{T} (1/2) |x_t| + (\tau - 1/2)x_t,$$

where $\tau \in (0,1)$ is a parameter. When $\tau < 0.5$, positive residuals are prefered to negative residuals, and vice versa when $\tau > 0.5$. When working with raw energy data, we set $\tau = 0.85$, which strongly prefers negative residuals, accounting for the fact that we have not normalized for weather effects, and clouds tend to reduce rather than increase the system energy product. When operating on normalized, PI data (such as the synthetic data set in this paper), we set $\tau = 0.5$ since we expect the deviations from the expected output to be symmetric, or at the very least more symmetric then when no normalization is performed.

The second component is a seasonal term, which is smooth and periodic each year,

$$\phi_2(x) = \begin{cases} \lambda_2 \|D_2 x\|_2^2 & x_t = x_{t+Y}, \text{ for } t = 1, \ldots T - Y \\ \infty & \text{otherwise,} \end{cases}$$

where $D_2 \in \mathbf{R}^{(T-2) \times T}$ is the second-order discrete difference operator. (See, for example, [13, §6.4] for information on difference matrices.) λ_2 is a weighting parameter, and $Y = 365$ is

978-1-7281-6118-1/22 $31.00 © 2022 IEEE

the period of the component. The normalization of the energy signal affects the expected amplitude of this component. That is, we would expect raw energy data to have a larger seasonal component than normalized data. The component definition presented here covers both cases well and is not sensitive to the amplitude of the component.

The third component represents the bulk, long-term degradation rate, and is given by

$$\phi_3(x) = \begin{cases} 0 & x_0 = 0 \text{ and } x_t = mt + b \text{ for } t = 1, \ldots, T \\ \infty & \text{otherwise,} \end{cases}$$

for some values of m and b. This just constraints the component to be linear with respect to time with an initial value equal to zero. We have chosen a linear degradation model due to its popularity in the literature, but we note that other trend models could be employed within the SD framework, e.g., a smooth, monotonically decreasing signal. Note that the third component does not include a weight parameter, as the penalty function only takes on values of zero or infinity.

The fourth and final component represents what we are interested in measuring, the soiling losses in the system. This cost is defined as,

$$\phi_4(x) = \begin{cases} \ell_{4a}(x) + \ell_{4b}(x) + \ell_{4c}(x) & x \preceq 0 \\ \infty & \text{otherwise,} \end{cases}$$

where the summed functions are

$$\begin{aligned} \ell_{4a} &= \lambda_{4a}\|D_2 x\|_1 \\ \ell_{4b} &= \lambda_{4b}\sum_{t=0}^{T}(-x_t) \\ \ell_{4c} &= \lambda_{4c}\mathbf{quant}_\tau(D_1 x) \end{aligned}$$

with parameters λ_{4a}, λ_{4b}, and λ_{4c}. D_2 is again the second-order discrete difference operator, and D_1, similarly, is the first-order difference. The quantile cost parameter τ is taken to be 0.9 here. This cost is a composite of simpler functions, which combine to select for signals with the following characteristics:

- non-positive (soiling can only reduce the system power)
- sparse in second-differences (i.e., piecewise linear)
- with values "close" to zero (is sum-absolute sense)
- a preference for more values with a negative slope than a positive one (i.e., a tendency towards slow degradation and quick recovery).

Component cost ϕ_4 demonstrates the *extensibility* of the SD framework. We are able to build up a complex cost function from smaller units and design it in a way to capture domain knowledge about the component.

b) SD parameters: The SD problem formulating includes four parameters, λ_2, λ_{4a}, λ_{4b}, and λ_{4c}. These parameters are tunable, and different values can greatly effect the characteristics and quality of the resulting decomposition. A deep discussion on the role of parameters in SD problems can be found in [3, §2.6]. While a method is provided in [3, §2.7] for selecting optimal parameter values, we find that in this context it makes more sense to rely on the practical experience of the

TABLE I
SD PROBLEM PARAMETERS

param.	value	description
λ_2	5×10^2	stiffness of seasonal baseline
λ_{4a}	2	effects number of soiling component breakpoints
λ_{4b}	3×10^{-2}	penalizes large values of soiling component
λ_{4c}	2×10^{-1}	encourages asymmetric rates in soiling component

analyst. In other words, we have found values that work well in many cases, and a small amount of hand-tuning is accepted in other cases. A description of these parameters and their default values are given in table I.

c) Solution method: The SD problem is convex (inequality-constrained quadratic program [14, §4.4]) and of modest size (around 2.1k variables for 3-year data set to around 15k variables for a 10-year data set), so we simply use CVXPY [15], [16] and the commercial Mosek solver [17], which is sufficient for research purposes. An implementation of the algorithm described in [3] would allow for the removal of the dependence on Mosek and is an area for future work.

C. Validation

The most desirable method of validation for an unsupervised machine learning algorithm such as the methods described in this paper would be access to real PV system production data that has been hand-labeled with soiling trends. Because this is difficult to obtain or generate, we take a multi-modal approach to validation in this paper, with three different approaches to validation, described below, ordered by how well labeled the data source is.

a) Synthetic data: We follow the methods published in [8] to evaluate the performance of the algorithm on synthetic data that represents normalized energy productions. This approach generates random realizations of PI signals, with components drawn from pre-defined statistical models. The synthetic data model includes a 'seasonal' term which represents the seasonal variation in performance. Normalization with a performance index is typically expected to lower this seasonal variation, but it does not fully remove it. The noise term in the model is Gaussian white noise, representing the assumption that a PI signal normalized for weather phenomenon. We therefore set $\tau = 0.5$ in ϕ_1 of the SD formulation to reflect the expectation of symmetric residuals.

Because the synthetic model explicitly models the system soiling losses, we are able to directly assess the ability of the SD soiling algorithm to estimate the hidden soiling signal. We select *mean-absolute error* (MAE) as a summary error metric for comparing the known synthetic soiling loss to the SD estimate, which is preferred over *root-mean-square error* when the errors are not expected to be normally distributed [18, §6.1.2]. Because analysts often want to be able to estimate soiling *rates* on a PV system, in addition to understanding the total energy loss, we calculate the MAE on both the soiling loss component and the first-order difference [19] of the loss, which we call the soiling rate. Finally, we note a small error in estimate of a cleaning event (*i.e.*, one day before or after the true event) is of small consequence to the analyst but will result in very large error values, especially for the analysis of soiling rates. Therefore, we introduce a third summary statistic which is the MAE of *filtered* soiling rate, which simply selects for time periods when both the synthetic soiling component and the SD estimate agree that the instantanous soiling rate is negative, *i.e.*, neither component is currently in a cleaning event. This final metric provides useful insight into the ability of the algorithm to accurately estimate the rate of soiling loss between cleaning events.

b) Labeled production data: Labeled soiling data was presented in [20] and analyzed for soiling trends. We leverage these published results to validate the SD soiling algorithm on the *unlabeled* power production data. For this power, we ignore the reference system, treating the test system as an unlabeled data source, and estimate the soiling losses of the test system using the SD formulation. Then we compare results of the proposed method to the results from [20]. We calculate the same three error metrics for this data set, as described previously for the synthetic data.

c) Unlabeled production data: We demonstrate the application of the SD soiling method on a selection of 50 PV systems, with only access to measured real power. There is no known soiling trend to compare to in this case. Instead, this represents what we see as a typical use case, and we show how outlier sites may be identified.

D. Soiling correction for downstream analysis

We briefly note that the estimate of soiling loss may be used to "correct" for soiling for other analysis. For example, we note in two other papers submitted to this conference [21], [22], that it is beneficial to account for soiling in production PV data prior to analyzing shade losses. We provide a simple procedure for doing this correction here.

The soiling loss component is a daily signal representing the fractional loss in daily energy from soiling, typically between 0 and 1. Therefore, the expected power output of the system in the absence of soiling would be the measured power divided by the instantaneous soiling loss. The Solar Data Tools implementation of the methods described in this paper includes a feature for automatically performing this correction.

IV. RESULTS

A. Synthetic data

Following the scenario generation procedure defined in [8, §II-D], we define six generative model configurations, with different levels of soiling, seasonality, and noise. These scenarios are briefly described in II. Two characteristic examples of the synthetic PI signals are shown in figure 2, and the SD estimates of the soiling trends are compared to the true values in figure 3. The error metric for these two examples is given in table III. The soiling rate were on average around -0.001 for the first example and -0.0005 for the second example.

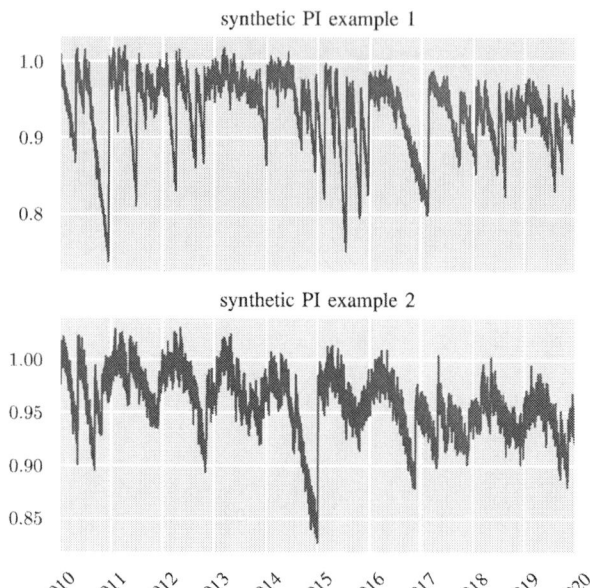

Fig. 2. Two typical synthetic soiling signals generated by NREL software. The top signal is drawn from scenario 1 and the bottom from scenario 2.

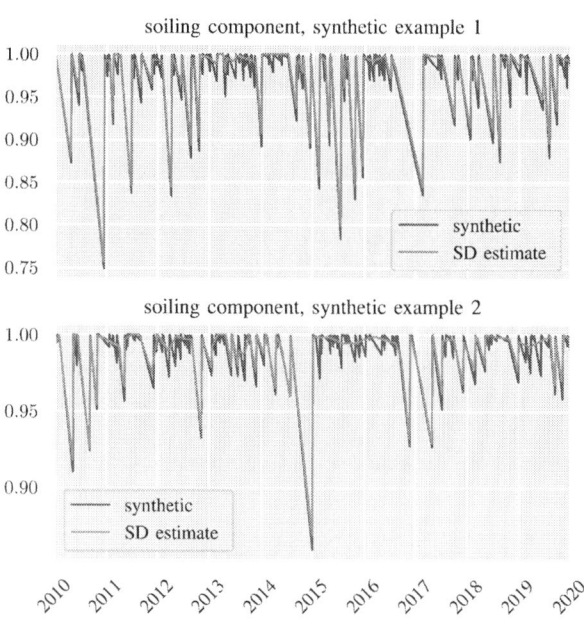

Fig. 3. Comparison of the actual and estimated soiling components in the two synthetic examples shown in figure 2.

TABLE II
SYNTHETIC DATA SCENARIOS

number	name	description
1	normal	soiling rates (sr) uniform in $[0, 0.003]$
2	M soil, H season	$sr \in [0, 0.001]$, double seasonal amplitude
3	M Soil, H noise	$sr \in [0, 0.001]$, double noise amplitude
4	seasonal cleaning	$sr \in [0, 0.005]$, cleaned seasonally
5	M soil	$sr \in [0, 0.005]$
6	L soil	$sr \in [0, 0.001]$

TABLE III
ERROR METRICS FOR TWO SYNTHETIC SOILING EXAMPLES

	loss MAE	rate MAE	filtered rate MAE
Ex. 1	0.008698	0.002257	0.000379
Ex. 2	0.005366	0.000919	0.000202

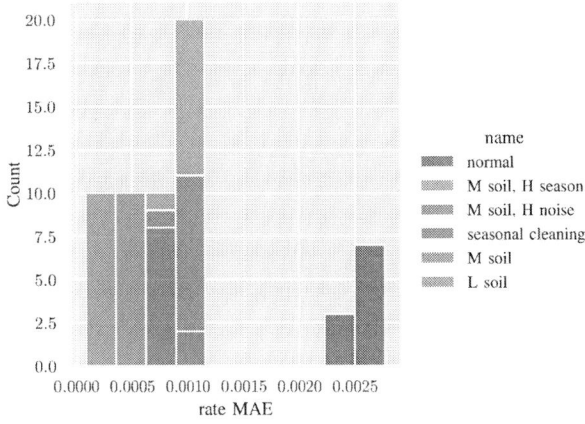

Fig. 5. Distribution of rate MAE for the 60 realizations, labeled by scenario.

We sample the 6 scenarios 10 times each, and we solve the associated SD problem for each of the 60 realizations. Finally, we calculate the three error metrics for each realization. The distribution of loss MAE is given in figure 4; the distribution of rate MAE is given in figure 5, and the distribution of the filtered rate MAE is given in figure 6.

B. Labeled data

The labeled performance index and corresponding unlabeled energy signal was previously shown in figure 1. In figure 7, these signals are overlaid with the "denoised" SD signal estimate, *i.e.*, the sum of the estimated components excluding the first residual term. Note how the SD model for the PI assumes symmetric residuals ($\tau = 0.5$) while the model for the energy signal assumes highly asymmetric residuals ($\tau = 0.85$). Figure 8 shows the comparison between the soiling loss components estimated from the PI and from the unlabeled energy signal. Taking the loss component derived from the PI signal as groundtruth, we then calculate the three error metrics for the energy-derived soiling estimate. The loss MAE

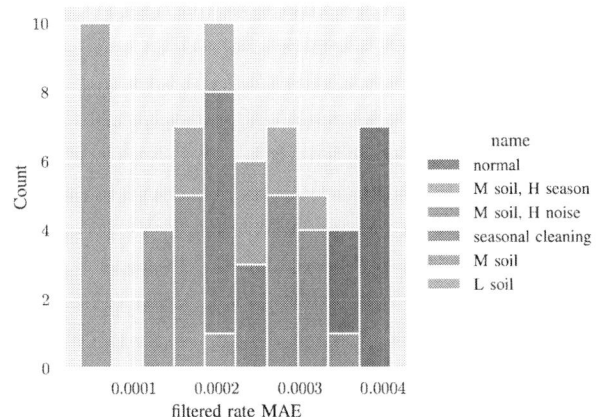

Fig. 6. Distribution of the filtered rate MAE for the 60 realizations, labeled by scenario. This metric accurately captures how closely the SD method estimated the loss rates during soiling periods.

is 0.042558; the rate MAE is 0.004734, and the filtered rate MAE is 0.001756.

We find that the analysis of the unlabeled energy produces an estimate of soiling losses that agrees well with the PI-derived estimate. We observe that the unlabeled analysis does worst around days 125–200, which was particularly rainy and cloudy. This lack of clear sky baseline during this period is seen to negatively impact the soiling estimate. However, with additional years of data, this may be improved due to the seasonal structure in the second component.

C. Unlabeled data

In [21], we present a preliminary shade loss analysis of an unlabeled rooftop PV data set. As described in that manuscript, the shade analysis depends on first estimating and correcting the soiling losses in the signal. A view of the soiling signal decomposition for one system in this data set is shown in figure 9, and the soiling component in isolation is shown in figure 10.

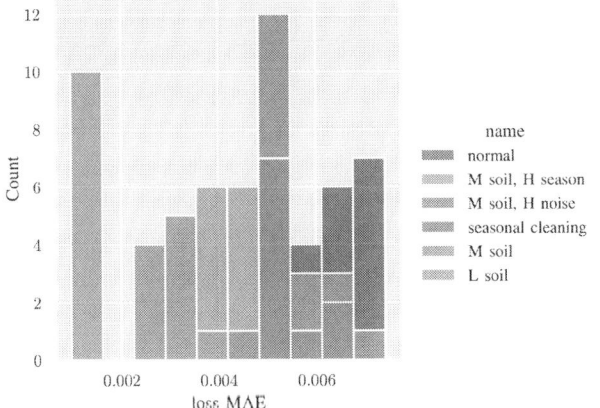

Fig. 4. Distribution of loss MAE for the 60 realizations, labeled by scenario.

978-1-7281-6118-1/22 $31.00 © 2022 IEEE

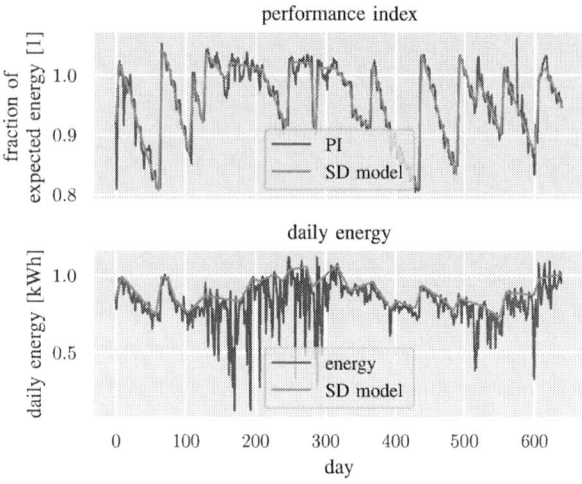

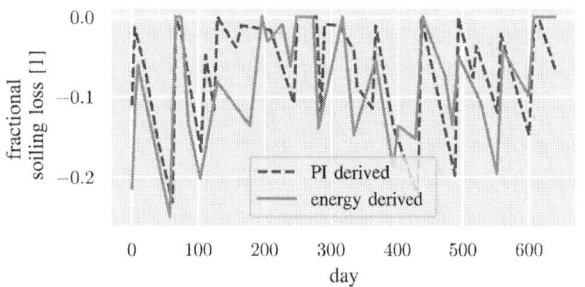

Fig. 7. PI and energy signals for the labeled soiling test system, with the denoised SD estimates overlaid in orange.

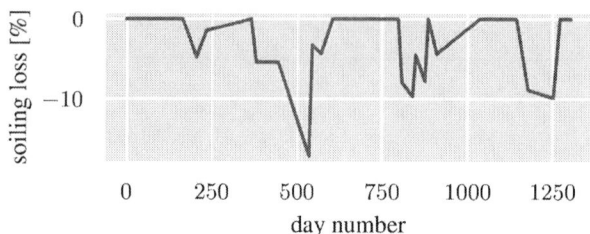

Fig. 10. The isolated soiling trend for the unlabeled data analysis.

V. CONCLUSIONS

We present a methodology, based on the signal decomposition (SD) framework, for estimating soiling losses PV system production data, *i.e.*, time series measurements of generated real power or energy, typically over multiple years. Unique to this work is the capability to estimate soiling losses in raw, unlabeled power/energy data, rather than requiring a performance index. By utilizing the extensibility of the SD, we are able to design a signal decomposition model that is bespoke to the problem of estimating soiling losses. This results in a robust model for the soiling loss component, x^4, as well as adjustable residual component, x^1, that is able to model both unlabeled ($\tau = 0.85$) and labeled ($\tau = 0.5$) data. The ability to analyze unlabeled PV data potentially unlocks huge potential in the form of fleet-scale datasets of hetergenous, distributed PV systems, which typically have internet-connected power electronics which generate time series of real power, but lack correlated meteorological measurements and possibly accurate system models. A software implementation is available in the Solar Data Tools package [4], [5] and a demonstration notebook of the code usage is available online [23].

ACKNOWLEDGMENTS

The author would like to thank Stephen Boyd, Justin Luke, Elsa Kam-Lum, Mayank Malik, and the entire GISMo Team at SLAC National Accelerator Laboratory for their input and feedback on this work. I also recognize the Seaborn plotting package for Python, which made the figures possible [24].

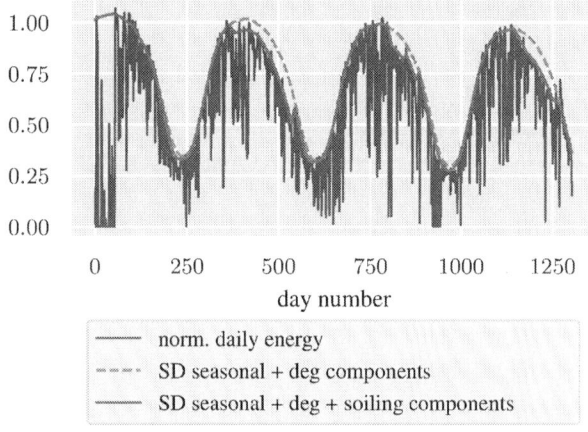

Fig. 9. An illustration of the soiling decomposition results for the unlabeled rooftop PV data discussed in [21]. The soiling trend is estimated to be the difference between the orange and green trends.

Fig. 8. Comparison of the estimated soiling loss component from the PI signal and the unlabeled energy signal.

REFERENCES

[1] K. Ilse, L. Micheli, B. W. Figgis, K. Lange, D. Daler, H. Hanifi, F. Wolfertstetter, V. Naumann, C. Hagendorf, R. Gottschalg, and J. Bagdahn, "Techno-economic assessment of soiling losses and mitigation strategies for solar power generation," *Joule*, vol. 3, no. 10, pp. 2303–2321, 2019. [Online]. Available: https://www.sciencedirect.com/science/article/pii/S2542435119304222

[2] M. Davis, C. Smith, B. White, R. Goldstein, X. Sun, M. Cox, G. Curtin, R. Manghani, S. Rumery, C. Silver, and J. Baca, *U.S. Solar market insight executive summary, 2020 year in review*. Wood Mackenzie and SEIA, 2021.

[3] B. Meyers and S. Boyd, "Signal decomposition using masked proximal operators," pp. 1–60, feb 2022. [Online]. Available: http://arxiv.org/abs/2202.09338

[4] B. Meyers, E. Apostolaki-Iosifidou, and L. Schelhas, "Solar data tools: Automatic solar data processing pipeline," in *2020 47th IEEE Photovoltaic Specialists Conference (PVSC)*, 2020, pp. 0655–0656.

[5] B. Meyers, "solar-data-tools," may 2022. [Online]. Available: http://dx.doi.org/10.5281/zenodo.6450368

[6] M. Deceglie, L. Micheli, and M. Muller, "Quantifying soiling loss directly from pv yield," *IEEE Journal of Photovoltaics*, vol. 8, no. 2, pp. 547–551, 2018.

[7] Å. Skomedal, H. Haug, and E. S. Marstein, "Endogenous soiling rate determination and detection of cleaning events in utility-scale pv plants," *IEEE Journal of Photovoltaics*, vol. 9, no. 3, pp. 858–863, 2019.

[8] A. Skomedal and M. Deceglie, "Combined Estimation of Degradation and Soiling Losses in Photovoltaic Systems," *IEEE Journal of Photovoltaics*, vol. 10, no. 6, pp. 1788–1796, nov 2020. [Online]. Available: https://ieeexplore.ieee.org/document/9186286/

[9] L. Micheli, M. Theristis, A. Livera, J. Stein, G. Georghiou, M. Muller, F. Almonacid, and E. Fernandez, "Improved PV soiling extraction through the detection of cleanings and change points," *IEEE Journal of Photovoltaics*, vol. 11, no. 2, pp. 519–526, 2021.

[10] T. Hastie, R. Tibshirani, and J. Friedman, *The Elements of Statistical Learning*, ser. Springer Series in Statistics. New York, NY: Springer New York, dec 2009. [Online]. Available: http://ieeexplore.ieee.org/document/6727256/ http://link.springer.com/10.1007/978-0-387-84858-7

[11] R. Koenker and G. Bassett, "Regression quantiles," *Econometrica*, vol. 46, no. 1, p. 33, jan 1978. [Online]. Available: https://www.jstor.org/stable/1913643 https://www.jstor.org/stable/1913643?origin=crossref

[12] R. Koenker and K. F. Hallock, "Quantile regression," *Journal of Economic Perspectives*, vol. 15, no. 4, pp. 143–156, nov 2001. [Online]. Available: https://pubs.aeaweb.org/doi/10.1257/jep.15.4.143

[13] S. Boyd and L. Vandenberghe, *Introduction to Applied Linear Algebra*, 2018.

[14] ——, *Convex optimization*. Cambridge University Press, 2009.

[15] S. Diamond and S. Boyd, "CVXPY: A Python-embedded modeling language for convex optimization," *Journal of Machine Learning Research*, vol. 17, no. 83, pp. 1–5, 2016.

[16] A. Agrawal, R. Verschueren, S. Diamond, and S. Boyd, "A rewriting system for convex optimization problems," *Journal of Control and Decision*, vol. 5, no. 1, pp. 42–60, 2018.

[17] E. D. Andersen and K. D. Andersen, "The Mosek Interior Point Optimizer for Linear Programming: An Implementation of the Homogeneous Algorithm," in *High performance optimization*, 2000, pp. 197–232. [Online]. Available: http://link.springer.com/10.1007/978-1-4757-3216-0_8

[18] S. Boyd and L. Vandenberghe, "Convex Optimization," in *Cambridge University Press*, 2004.

[19] "Numpy.diff," *NumPy v1.22 Manual*. [Online]. Available: https://numpy.org/doc/stable/reference/generated/numpy.diff.html

[20] E. Kam-Lum, B. E. Meyers, D. Cosme, B. Aissa, and G. Scabbia, "Soiling Rate Determination from Referenced Systems in Desert Climate using PVInsight Soiling Algorithm," in *2021 IEEE 48th Photovoltaic Specialists Conference (PVSC)*, no. July 2019. IEEE, jun 2021, pp. 2552–2554. [Online]. Available: https://ieeexplore.ieee.org/document/9518459/

[21] B. Meyers and D. F. Florez Rodriguez, "Estimation of shade losses in unlabeled PV data," *Submitted to PVSC49*, June 2022.

[22] D. F. Florez Rodriguez and B. Meyers, "Solar panel power simulation for shade detection," *Submitted to PVSC49*, June 2022.

[23] B. Meyers, "Soiling analysis demonstration," *solar-data-tool GitHub repository*, May 2022. [Online]. Available: https://github.com/slacgismo/solar-data-tools/blob/shade-dev/notebooks/Soiling_analysis_demonstration.ipynb

[24] M. L. Waskom, "seaborn: statistical data visualization," *Journal of Open Source Software*, vol. 6, no. 60, p. 3021, 2021. [Online]. Available: https://doi.org/10.21105/joss.03021

978-1-7281-6118-1/22 $31.00 © 2022 IEEE

Understanding the Solar Cell Contacts With Atmospheric Screen-printed Copper

Sandra Huneycutt[1], Abasifreke Ebong[1], Krishnamraju Ankireddy[2], Ruvini Dharmadasa[2], and Thad Druffel[2]

1) The University of North Carolina at Charlotte, Department of Electrical and Computer Engineering, Charlotte, North Carolina, 28223
2) Bert Thin Films, LLC., Louisville, Kentucky, 40208

Abstract—**Although Cu is very close to Ag in conductivity, there are still some concerns with its high diffusivity into Si. There are two Cu ions of focus; interstitial (Cu_i^+) and substitutional (Cu_s^+), whereby, with regard to Si the fast-diffusing impurity is the interstitial, Cu_i^+. An isolated Cu_i^+ acts as a shallow donor, it reacts with impurities and defects to alter the electrical properties of the material. However, Cu passivates shallow acceptors, forms pairs with various impurities, including itself, and precipitates at defects. Thus, these Cu precipitates become strong electron-hole recombination centers. With regard to n-type Si, the Cu impurity precipitates much easier than in p-type Si, provided that several of the Cu_i^+'s precipitate without trapping an electron. The diffusivity of species in the semiconductor generally depends on the time and temperature, thus, a diffusivity of $7x10^{15}$ cm^{-3} can be inferred for Cu at 600°C for 20 minutes. For the atmospheric screen-printed Cu solar cell contacts, the sintering is performed on a PERC wafer with a finger width of 83 μm fired at a peak a temperature of 593°C at 325 ipm for approximately 2 seconds; thus, the measured diffusion coefficient would be different. More so, since the paste consists of glass frits and Cu powder, the glass must react first with the SiN_x to produce the molten glass which would then react with Cu. Since the reaction time is very short, the Cu will not have enough time to diffuse into the Si before cooling down and subsequent sequestration by the reformed glass. STEM will be used to understand the mechanisms which enable or disable the sequestration of Cu and the associated challenges will be discussed. Additionally, the solar cell electrical output parameters comparing the results of sequestered Cu on PERC Si wafers will be presented.**

Keywords—**diffusivity, interstitial, substitutional, precipitate**

I. INTRODUCTION

For many decades Ag has been the metal of choice for Si based solar cells because of its high conductivity and compatibility with the n-type emitter. Unfortunately, the economic market volatility controls the price, which can unpredictably change and render the cost of manufacturing almost prohibitive. In 2011, the cost of Ag rose to approximately $1562 per kg [1] where the offset in cost was eventually passed along to the consumer and in turn makes solar a less attractive option. To overcome this barrier, there needs to be a more cost-effective option for metallization of the solar cell than Ag. It is here that Cu seems to be the best contender because Cu is close in conductivity to Ag and about 100 times lower cost [1, 2]. However, there are several challenges that

arise when using Cu such as oxidation in air and diffusion at high temperatures leading to degradation of cells over time, delamination, and fast precipitation through diffusion into Si. These challenges often make Cu the villain with regards to PV and its practical implementation in the global consumer market. To fully consider Cu as a viable candidate to replace the traditional, yet prohibitively expensive, Ag, an understanding of the physics regarding Cu's associated challenges in the presence of Si is imperative.

II. THE ANATOMY OF CU DIFFUSION

A. *Understanding the physics of Cu*

Cu typically precipitates at nano defect regions due to its hexavacancy [3]. It was reported [3] that the Cu_i^+ does not distribute its outmost orbital electrons in the same configuration as would be expected for a free Cu ion (i.e. $3d^{10}$ $4sp^0$), rather, it actually transfers some of its electrons from the 3d shell to the 4 sp shell and borrows electron density from its nearest neighbors. This seemingly odd distribution is responsible for the various covalent interactions between the host crystal and the Cu impurity [3]. Cu_i exists almost entirely as a positive ion and is thought to be a shallow donor in the presence of Si [3]. The Cu_i^+ reacts with many other impurities, including itself, however, its diffusion is trap limited [3]. Previous measurements report [4] that the diffusivity of Cu in Si was found to be activated at 0.43 eV which is close to the predicted value of 0.24 eV by ab initio Hartree-Fock calculations [5] for the interstitial tetrahedral-hexagonal-tetrahedral path of Cu_i diffusion. Through interpolation of reported values [4] a diffusivity value of $7x10^{15}$ cm^{-3} for Cu at 600°C over 20 minutes can be obtained.

In addition to diffusion, (i) Cu also forms pairs [6,7] with itself (Cu_s and Cu_i) [8], (ii) reacts with transition metals like Pt and Au [7, 8] and (iii) interacts with common impurities such as C, P, As [3]. If there is a presence of interstitial O (typically introduced during the growth of the Si ingot) Cu will weakly trap within the vicinity of it since the intrinsic diffusivity of the positive interstitial Cu is higher in FZ grown Si than in CZ below room temperature [10]. Additionally, Cu also has a strong tendency to precipitate (i) at defects [11, 12], (ii) grain boundaries [13], (iii) radiation damaged regions [14], and (iv) stacking faults [15]. In the (111) plane Cu can take the form of star shaped etch pits and platelets [16]. Precipitates such as these can cause a reduction in the charge carrier lifetimes [17, 18] and form effective recombination centers for electron-hole pairs.

978-1-7281-6118-1/22 $31.00 © 2022 IEEE

With regard to Cu_s^+, it presents itself as an electron trap [3] and the estimated amount of relaxation around Cu_s using an empirical valance force potential was reported [19] to have a symmetric outward relaxation of the nearest lattice neighbors valued at 0.24 angstrom. The four nearest lattice neighbors to Cu_i was reported to relax outward by less than 0.05 angstrom. In order to insert a free Cu ion into Si it would require an energy $\cong$ 1.67 eV [3]. The act of promoting electrons from the 3d to 4sp orbitals is largely the reason it requires energy to insert Cu into the crystal lattice [3]. This value is very near the experimentally determined activation energy of solubility, whereby, the difference of energy is related to the free Cu ion [20, 21].

The ability of Cu to interact with itself occurs when the Cu_i^+ and a neutral vacancy get within close proximity of each other and then the Cu_i^+ transitions and becomes a Cu_s^+ [3], whereby, the reaction has an energy gain of 2.71 eV. The energy gained is not very large especially when considering there is an interstitial impurity and a vacancy reaction corresponding to about 0.6 eV for every Cu-Si bond. To compare, it was reported that a gain of 3 eV occurs when a single H atom is inserted into a single vacancy [22], whereby, the vacancy self-interstitial recombination releases 8.2 eV [23]. It was stated [3] that there is a possibility that a single interstitial H could expel Cu from its substitutional site and that the use of two H's certainly would.

B. Cu behavior while annealing at high temperatures

The device properties can be sustained using gettering techniques to thwart the impact of contamination by metallic impurities [24]. However, the deep levels with which have been associated with Cu diffusion into Si are not indicative of the Cu concentration in total; it depends largely on the thermal history of the sample [25]. It was reported [4] that positively charged Cu atoms are dissolved interstitially during the process of annealing. When fast cooling (or quenching) proceeds a high temperature anneal of 1000°C [26] only a fraction of the interstitially dissolved Cu atoms will remain. If the Cu_i^+ is allowed to accumulate near the junction the capacitance of the Schottky barrier will be affected [24].

Heiser et. al. [24] reportedly used two sample types, (i) 100 nm thick Cu layer was deposited by sputtering and thermally annealed at 800°C for 30 minutes (to allow complete diffusion of Cu) and (ii) the samples were Cu implanted at 150 keV with a high dose (10^{17} cm^{-2}) in order to saturate the implantation induced defects with Cu atoms [21]. They [24] studied the low-level copper contamination in Si by transient ion drift detection (TIDD) method. For the two sets of samples; (i) annealed at temperature 800°C for 30 minutes (to avoid the diffusion enabling properties brought by the presence of native oxide) followed by a 30-minute cooling to allow the Cu concentration to reach the limit of solubility at T (450°C – 800°C) and then quenched in ethylene glycol. (ii) The second sample set was annealed at a temperature, T (450°C - 600°C), then quenched to room temperature. Their results showed that Cu solubility in Si depends strongly on temperature.

Solubility tends to increase linearly with increasing temperature (>600°C) and it decreases exponentially for lower temperatures. Additionally, the maximum concentration of Cu_i following quench was reported [24] to have increased linearly with acceptor concentration. However, the precipitation of Cu,

triggered by supersaturation, may be influenced by the cool down as a function of decreased fermi energy related to the energy gain [24].

This work reports on the use of an atmospheric Cu paste which can operate as an Ag counterpart. By noting the temperature regime that could exacerbate the diffusion of Cu into silicon, the paste was formulated to fire at a lower temperature. Additionally, the reformed glass used successfully blocked the diffusion of the atmospheric Cu and sequestered the particles to prevent degradation. This necessitates a sequential annealing of the metal-contacts and optimized peak temperature profile for the PERC structure reported in this work.

III. CELL FABRICATION

The PERC cells used were fabricated with front metal contacts using a Cu paste that was printed, dried at 200°C for 2 minutes, and then fired with fast belt-speeds (300 ipm and 325 ipm) with some modification to the gas flow rate. The electrical output parameters of some of the cells are shown in Table 1 for peak temperatures measuring 576°C and 593°C, respectively. One of the cells, which showed good results, was cut for interface study as depicted in Fig. 1 and visually aids in the understanding of the physics regarding the screen-printed atmospheric Cu paste.

IV. RESULTS AND DISCUSSION

A. Practical applications of atmospheric Cu

As shown in Fig. 1, there is an obvious Si signature underneath the metal contact, but not throughout the contact bulk. Additionally, there is small trace layer of N underneath the bulk paste, coupled with the irregularity of the pyramids it seems as though the SiN_x is removed from the surface of the cell and deposited on top of the metal contact. As the metal/glass layers reform during cool down there are small signatures of N that remain in the bulk of the Cu/reformed oxide contact.

Fig. 1. STEM taken of cross section for an atmospheric Cu contacted cell

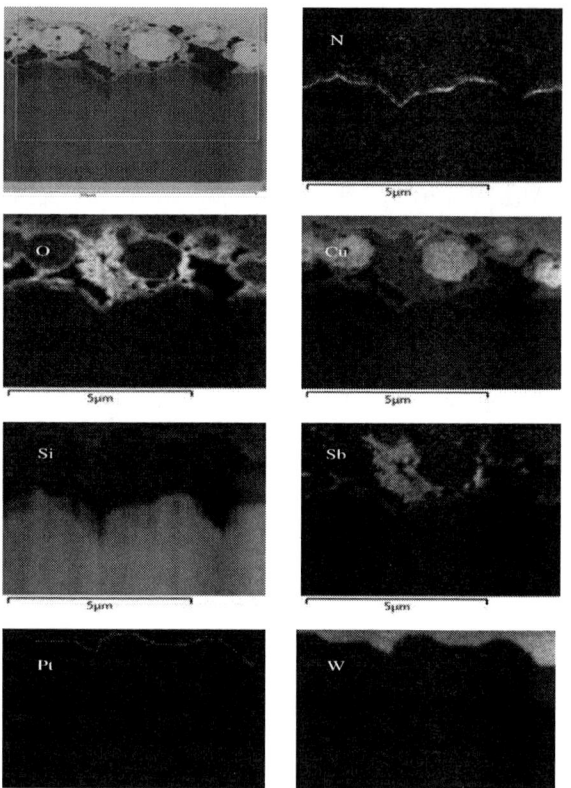

During the sintering process when the organics in the paste reach their critical temperatures and are violently outgassed it is possible that the surrounding topmost layer of the Si wafer blows off and deposits on top of the metal contact. The STEM images above give further credence to this observation because there was not a trace of Si within the bulk of the metal contact; however, there is on top. This signature leads to believe that the Si did not simply diffuse through the metal during contact formation. Additionally, Cu did not diffuse into the Si bulk as shown by the sequestering of Cu using the reformed oxide which aids in the blocking of diffusion [2]. More work will be needed to understand why the N is brought into the bulk during the sintering process. Through Ramon spectroscopy, XRD, and other characterization techniques the particular form and geometry of the CuSi compounds (i.e. Cu_3Si or Cu_4Si) can be identified.

TABLE I.   COMPARED IV OUTPUT CHARACTERISTICS

Cell ID	VOC (V)	FF (%)	Eff. (%)
37-7	665.96	72.1	18.5
14-10	661.3	77.3	19.2

TABLE II.   IV OUTPUT CHARACTERISTICS CONTINUED

Cell ID	RS Ω-cm^2	N-factor	Jo2 (A/cm^2)
37-7	1.8	1.1	1.19E-8
14-10	1.0	1.0	7.59E-9

The above tables show the compared IV characteristics for two cells that were fired at different times and conditions. 37-7 was fired at 576 °C at 300 ipm and sample 14-10 was fired at 593°C at 325 ipm. As shown, a higher temperature coupled with a faster belt speed is desired so that the Cu does not have the time or heighted temperature it needs to precipitate into the Si [27].

V. THE CONCLUSION

As shown in this work, Cu in the presence of Si is not the villain after all. The STEM images allow for a visual vindication of the IV characteristics that the Cu has not, in fact, diffused and/or shunted the p-n junction. This is buttressed by the understanding of the physics regarding Cu's tendency to diffuse into Si, the conditions of which contribute, and how to circumvent such challenges in an effective manner. When all attributes of Cu have been considered there is no need for the use of forced inert gas, expensive equipment aside from current abilities, or over complicated metallization techniques. This particular Cu formulation was printed using a normal screen printing method as would be used for Ag and fired in ambient air.

Acknowledgment

We would like to thank the Energy Production and Infrastructure Center (EPIC) at The University of North Carolina at Charlotte for collaborating with Bert Thin Film, LLC. to formulation and optimization of the atmospheric Cu paste.

REFERENCES

[1] T. Druffel, R. Dharmadasa, K. Ankireddy, K. Elmer, A. Ebong and S. Huneycutt, "Copper based front side metalization contacts screen printed and fired in air demonstrating durability," 2020 47th IEEE Photovoltaic Specialists Conference (PVSC), pp. 2609-2611, 2020.

[2] A. Ebong, S. Huneycutt, S. Grempels, K. Ankireddy, R. Dharmadasa and T. Druffel, "Progress of Atmospheric Screen-printable Cu Paste for High Efficiency PERC Solar Cells," 2021 IEEE 48th Photovoltaic Specialists Conference (PVSC), pp. 1417-1420, 2021.

[3] S. Estreicher, "Rich chemistry of copper in crystalline silicon," The American Phy. Soc., PRB 60, pp. 5375-5382, Jan. 1999.

[4] R. N. Hall and J. H. Racette, "Diffusion and solubility of copper in extrinsic and intrinsic germanium, silicon, and gallium arsenide," Journal of Applied Phy. 35, 379-397, 1964.

[5] D. Woon, D. Marynick, and S. Estreicher, "Titanium and copper in Si: Barriers for diffusion and interactions with hydrogen," Phys. Rev. B 45, 13 383, Jun. 1992.

[6] T. Prescha and J. Weber, "Interaction of a copper-induced defect with shallow acceptors and deep centers in silicon," Mater. Sci. Forum 83–87, 167–172, 1992.

[7] A. Mesli and T. Heiser, Phys. Rev. B45 11 632, 1992.

[8] J. Weber, H. Bauch, and R. Sauer, "Optical properties of copper in silicon: excitons bound to isoelectronic copper pairs," Phys Rev. B 25, 7688, 1982.

[9] A. Istratov, H. Hieslmair, T. Heiser, C. Flink, and E. Weber, "The dissociation energy and the charge state of a copper-pair center in silicon," Appl. Phys. Lett. 72, 474, 1998.

[10] T. Heiser, A.A. Istratov, C. Flink, E.R. Weber, "Electrical characterization of copper related defect reactions in silicon," Mat. Sci. and Eng., B Vol 58, Iss. 1–2, pp. 149-154, 1999.

[11] J. weber, solid state phenom 37-38, 13, 1994.

[12] Istratov, A., Weber, E., "Electrical properties and recombination activity of copper, nickel and cobalt in silicon," Appl Phys A 66, 123–136, 1998.

[13] J.-L. Maurice, C. Colliex, "Fast diffusers Cu and Ni as the origin of electrical activity in a silicon grain boundary," Appl. Phys. Lett. 55, 241-243, 1989.

[14] Scott A. McHugo, E. R. Weber, S. M. Myers and G. A. Petersen, "Competitive gettering of copper in Czochralski silicon by implantation-induced cavities and internal gettering sites," Appl. Phys. Lett. 69, 3060-3062, 1996.

[15] M. Kaniewska, J. Kaniewski, A. Peaker, "Deep States Associated with Copper Decorated Oxidation Induced Stacking Faults in Silicon," Mat. Sci. Forum, 83-87, 1457-1462, 1992.

[16] Istratov, A., Weber, E., "Electrical properties and recombination activity of copper, nickel and cobalt in silicon," Appl Phys A 66, 123–136, 1998.

[17] A. Rohatgi, J.R. Davis, R.H. Hopkins, P. Rai-Choudhury, P.G. McMullin, J.R. McCormick,"Effect of titanium, copper and iron on silicon solar cells," Solid-State Electronics, Volume 23, Issue 5, Pages 415-422, 1980.

[18] A. A. Istratov, C. Flink, H. Hieslmair, T. Heiser, and E. R. Weber, "Influence of interstitial copper on diffusion length and lifetime of minority carriers in p-type silicon," Appl. Phys. Lett. 71, 2121-2123, 1997.

[19] U. Lindefelt, "Symmetric lattice distortions around deep-level impurities in semiconductors: Vacancy and substitutional Cu in silicon," Phys. Rev. B 28, 4510, 1983.

[20] E. Weber, "Transition metals in silicon," Appl. Phys. A, 30, 1–22, 1983.

[21] S. Meyers, D. Follstaedt, Jou. of App. Phys, 1996.

[22] Y. Park, S Estreicher, C. Myles, and P. Fedders, "Molecular-dynamics study of the vacancy and vacancy-hydrogen interactions in silicon," Phys. Rev. B 52, 1718, 1995.

[23] S. Estreicher, J. Hastings, P. Fedders, "Radiation-Induced Formation of H_2^* in Silicon," Phys. Rev. Lett. 82, 815, 25, 1999.

[24] T. Heiser, S. McHugo, H. Hieslmair, and E. R. Weber, "Transient ion drift detection of low level copper contamination in silicon," Appl. Phys. Lett. 70, 3576-3578, 1997.

[25] S. Brotherton, J. Ayres, A. Gill, H. van Kesteren, F. Greidanus, "Deep levels of copper in silicon," Journal of Applied Physics 62, 1826-1832, 1987.

[26] Heiser, T., Mesli, A., "Determination of the copper diffusion coefficient in silicon from transient ion-drift," Appl. Phys. A 57, 325–328, 1993.

[27] S. Grempels, S. Huneycutt, A. Ebong, R. Dharmadasa, K. Ankireddy and T. Druffel, "Rapid Thermal Annealing of Screen-printable Atmospheric Cu Pastes for PERC Solar Cell," 2020 IEEE 17th International Conference on Smart Communities: Improving Quality of Life Using ICT, IoT and AI (HONET), pp. 244-248, 2020.

978-1-7281-6118-1/22 $31.00 © 2022 IEEE

Na Diffusion and Device Performance of AgBr Treated CuGaSe₂ Thin Films

Elizabeth Palmiotti[a], Polyxeni Tsoulka[b], Thomas Lepetit[b], Nicolas Barreau[b], and Angus Rockett[a]

[a]Colorado School of Mines, Department of Metallurgical and Materials Engineering, Golden, CO, U.S.A.

[b]Université de Nantes, CNRS, Institut des Matériaux Jean Rouxel, IMN, F-44000, Nantes, France

Abstract—**Previous work demonstrated that uniform CuGaSe₂ (CGS) thin films with large grains could be grown using a short AgBr vapor treatment during growth. Devices made with this treated CGS showed better performance compared to devices made with standard material. Here, it is shown that AgBr treated CGS device performance worsens over time and is attributed to the suppression of Na diffusion. A NaF post-deposition treatment is shown to effectively introduce Na into the AgBr treated CGS film and prevent device degradation.**

Keywords— *Copper Gallium Diselenide, Post-Deposition Treatment, Co-Evaporation, Thin Film*

I. INTRODUCTION

The development of tandem photovoltaics with a silicon bottom cell and wide band gap top cell is a path to quickly increase efficiency while utilizing existing manufacturing lines. CuGaSe₂ (CGS) is a top cell candidate due to its band gap (1.7 eV) which is well matched to silicon. Typical co-evaporation deposition procedures optimized for $CuIn_{1-x}Ga_xSe_2$ (CIGS) are often used for CGS deposition, resulting in devices with limited efficiencies due to $Cu_\delta Se$ phase formation and grain size non-uniformities [1], [2]. This is likely attributed to the slower rate of formation of CuGaSe₂ compared to CuInSe₂ [3], [4], [5], meaning CIGS deposition procedures are not suitable for CGS.

This was first addressed by a modified co-evaporation procedure which introduced two, thirty-minute anneals at a high substrate temperature [1]. Although the modified procedure improved CGS material properties and device performances, the long process is not suitable for manufacturers. In our previous work we demonstrated that the long anneals in the modified procedure could be replaced by a short AgBr vapor treatment after the second stage of the co-evaporation process [6]. The AgBr treatment resulted in large grains, removal of the $Cu_\delta Se$ phase, and enhanced device properties driven by V_{oc} in half the time of the previous modified procedure [6].

Typically, laboratory-scale CIGS devices improve with ageing. Characterization of the AgBr treated devices after many weeks showed a decay of performance and significant efficiency losses. In this work it is noted that AgBr treated films had suppressed Na diffusion from the soda-lime glass substrate. The ageing and Na diffusion of AgBr treated CGS devices were studied and a NaF post-deposition treatment tested to solve this problem.

II. EXPERIMENTAL

A. Sample and Device Fabrication

CGS films were deposited from elemental Cu and Ga sources by co-evaporation under a Se overpressure onto molybdenum-coated soda-lime glass (SLG) substrates held at 575°C. The co-evaporation deposition followed standard and modified versions of the Cu-poor/Cu-rich/Cu-off (CuPRO) [7] procedure. Reference samples followed a standard CuPRO procedure and were labelled as 'CuPRO.' The modified procedure included two thirty-minute anneals in Se overpressure as described in [1] and were labelled 'CuPRO(M).' Samples undergoing AgBr treatment were introduced to AgBr vapor in a Se overpressure as described in [6]. AgBr treated samples were labelled as 'CuPRO+AgBr' and 'CuPRO(M)+AgBr', respectively. Some films received a NaF post-deposition treatment (PDT).

Devices were completed with the following architecture: SLG/Mo/CGS/CdS/ZnO/ZnO:Al. After the CGS deposition, the samples were dipped in a 0.05 M KCN solution for two minutes. The CdS buffer layer was grown by chemical bath deposition. The ZnO/ZnO:Al window layer was deposited using rf-sputtering. The 0.5 cm² devices were completed with Ni-Al-Ni metallic grids and separated by mechanical scribing.

B. Characterization

Elemental depth profiling was studied by time-of-flight secondary ion mass spectroscopy (TOF-SIMS) with an Ion TOF 5 SIMS. Depth profiling was performed using a 1 kV Cs⁺ ion beam and analysis using a 30 kV Bi_3 beam. Current density voltage (J-V) data was collected under standard testing conditions (STC). The sample was placed on a metal plate which was held at 25°C using a thermoelectric cooler and illuminated under AM1.5 using a solar simulator at 1000 W/m².

III. RESULTS

A. TOF-SIMS

The elemental distribution through the CGS films was studied by positive ion TOF-SIMS. To compensate for minor drift in the analyzer beam intensity, elemental intensity was normalized to the total ion yield. Cu, Ga, and Se normalized intensities were mostly constant through all films and devices in this study though the Na signal varied per sample; a representative example is shown in Appendix A. This work investigates the influence of Na on CGS device performance, thus, only the Na+ signal is plotted.

Figure 1 shows the Na+ ion signal on a logarithmic scale for the standard CuPRO with and without AgBr treatment (Figure 1(a)) and CuPRO(M) with and without AgBr treatment (b). The low sputter time region of all samples shows the Na+ signal in the CGS film. The increase at ~5000 s sputter time in Figure 1(a) is at the CGS/Mo interface and the second step at ~7000 s shows the Mo/SLG interface. Note that the same sputtering parameters were used for all samples.

The high intensity Na+ signals for the CuPRO and CuPRO(M) samples are typical for films grown on SLG at elevated temperatures. Both films that received a AgBr treatment show significantly less Na+ in the film. The Na+ signal for CuPRO+AgBr is approximately three orders of magnitude less than that for CuPRO. The CuPRO(M)+AgBr Na+ signal is approximately one order of magnitude less than CuPRO(M). Although a AgBr treatment to a CuPRO(M) sample is not standard, it shows that introducing two, thirty-minute anneals does not promote Na diffusion as seen in the samples without AgBr treatment.

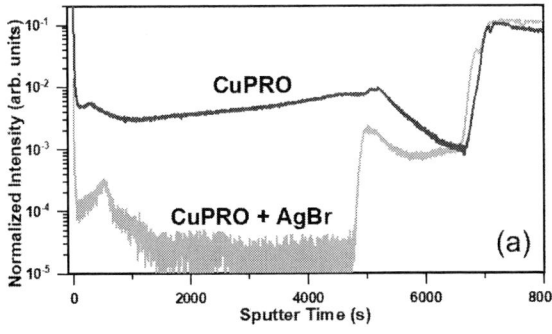

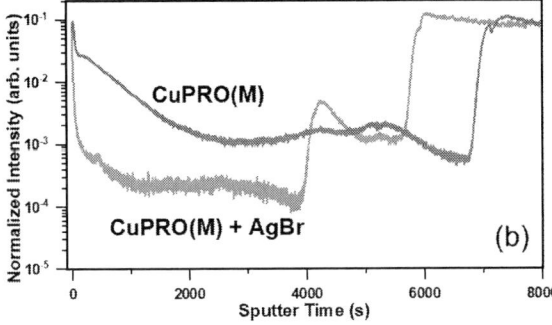

Fig. 1. Positive ion Na+ TOF-SIMS scans collected for (a) CuPRO with (light blue) and without (blue) AgBr treatment and (b) CuPRO(M) with (light red) and without (red) AgBr treatment.

Na diffusion in CIGS films is dominated by grain boundary diffusion as shown using atom probe tomography [8], [9]. The AgBr vapor treatments were inspired by the CdCl$_2$ treatment of CdTe films, which acts through Cl segregation into grain boundaries [10]. If it is assumed that the AgBr treatment of CGS films similarly works through grain boundaries, we propose that the AgBr displaces Na in the grain boundaries. The increased Na+ signal at the CGS/Mo interface for both AgBr treated samples suggests the Na may be displaced to and collects at this interface.

A NaF post-deposition treatment was performed on a CuPRO+AgBr CGS film. The Na+ signal for this CuPRO+AgBr+NaF CGS films was overlaid on Figure 1(a) and is shown in orange in Figure 2. CuPRO+AgBr+NaF has two orders of magnitude more Na compared to CuPRO+AgBr and slightly less than CuPRO. Notably, the Na barrier at the CGS/Mo interface of CuPRO+AgBr disappears.

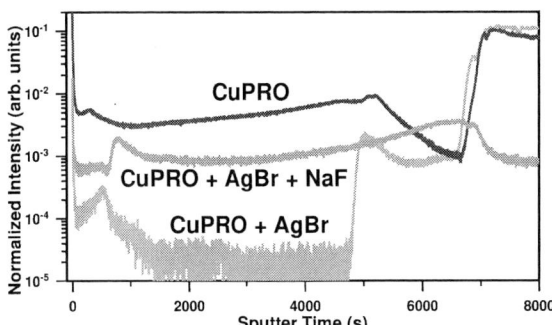

Fig. 2. Positive ion Na+ TOF-SIMS scans collected for CuPRO with (light blue) and without (blue) AgBr treatment. The scan in orange is the CuPRO sample prepared with AgBr treatment and NaF post-deposition treatment.

B. Device Results

Devices were completed using the CuPRO+AgBr and CuPRO+AgBr+NaF films studied in Figure 2. The highest efficiency device per sample was studied using J-V and device parameters were extracted (Figure 3). Device parameters were collected when the devices were completed (Week No. 0) and weekly thereafter. The initial CuPRO+AgBr+NaF device has a higher efficiency, as expected [11], [12], due to a high J_{sc} and V_{oc}.

All device parameters decrease on a weekly basis for CuPRO+AgBr. The decrease in efficiency is largely driven by decreases in J_{sc} and Fill Factor. The CuPRO+AgBr+NaF device parameters fluctuate, but generally do not change over time.

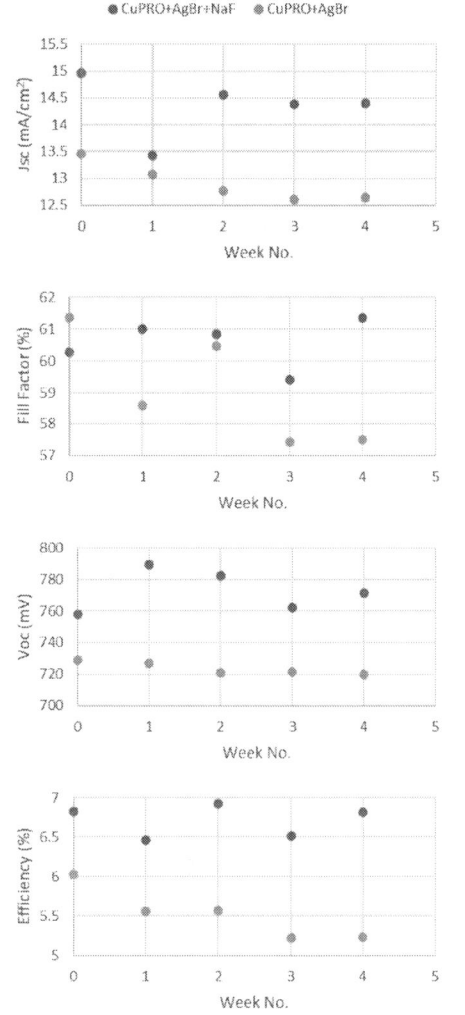

Fig. 3. Device parameters extracted from current density-voltage measurements for CuPRO+AgBr and CuPRO+AgBr+NaF collected weekly for four months.

IV. CONCLUSION

It is hypothesized that the Na barrier at the CGS/Mo interface caused by the AgBr treatment causes the device to degrade rapidly with time. In this work we have demonstrated that a NaF PDT can re-introduce Na into the AgBr treated CGS film, remove this Na barrier, and prevent device degradation. Weekly EQE data was also collected and will be reported at the conference. This work shows that further deposition and PDT optimization is necessary for metal halide treated CGS films and devices.

ACKNOWLEDGMENT

This material makes use of the TOF-SIMS system at the Colorado School of Mines, which was supported by the National Science Foundation under Grant No.1726898.

This material is based upon work supported by the U.S. Department of Energy's Office of Energy Efficiency and Renewable Energy (EERE) under Solar Energy Technologies Office (SETO) Award Number DE-EE0007551.

REFERENCES

[1] P. Tsoulka, A. Rivalland, L. Arzel, and N. Barreau, "Improved CuGaSe$_2$ absorber properties through a modified co-evaporation process", Thin Solid Films, 709, 2020. DOI: 10.1016/j.tsf.2020.138224.

[2] P. Tsoulka, N. Barreau, I. Braems, L. Arzel, and S. Harel, "Detrimental copper-selenide bulk precipitation in CuIn$_{1-x}$Ga$_x$Se$_2$ thin-film solar cells. A possible reason for the limited performance at large x?", Thin Solid Films, 712, 2020. DOI: 10.1016/j.tsf.2020.138297.

[3] M. Purwins, A. Weber, P. Berwian, G. Müller, F. Hergert, S. Jost, and R. Hock, "Kinetics of the reactive crystallization of CuInSe$_2$ and CuGaSe$_2$ chalcopyrite films for solar cell applications", Journal of Crystal Growth, 287, 408-413, 2006. DOI: 10.1016/j.jcrysgro.2005.11.054.

[4] S. Kim, W.K. Kim, R.M. Kaczynski, R.D. Acher, S. Yoon, T.J. Anderson, and O.D. Crisalle, "Reaction kinetics of CuInSe$_2$ thin films grown from bilayer InSe/CuSe precursors", Journal of Vacuum Science & Technology A: Vacuum, Surfaces, and Films, 23, 310-315, 2005. DOI: 10.1116/1.1861051.

[5] W.K. Kim, E.A. Payzant, S. Kim, S.A. Speakman, O.D. Crisalle, and T.J. Anderson, "Reaction kinetics of CuGaSe$_2$ formation from a GaSe/CuSe bilayer precursor film", Journal of Crystal Growth, 310, 2987-2994, 2008. DOI: 10.1016/j.jcrysgro.2008.01.034.

[6] E. Palmiotti, P. Tsoulka, D. Poudel, S. Marsillac, N. Barreau, A. Rockett, and T. Lepetit, "Homogeneous CuGaSe$_2$ Growth by the CuPRO Process with In-Situ AgBr Treatment", unpublished.

[7] J. Kessler, C. Chityuttakan, J. Lu, J. Schöldström, and L. Stolt, "Cu(In,Ga)Se$_2$ Thin Films Grown with a Cu-Poor/Rich/Poor Sequence: Growth Model and Structural Considerations", Progress in Photovoltaics: Research and Applications, 11, 319-331, 2003. DOI: 10.1002/pip.495.

[8] F. Couzinie-Devy, E. Cadel, N. Barreau, L. Arzel, and P. Pareige, "Atom probe study of Cu-poor to Cu-rich transition during Cu(In,Ga)Se$_2$ growth", Applied Physics Letters, 99, 2011. DOI:10.1063/1.3665948.

[9] A. Stokes, M. Al-Jassim, D. Diercks, A. Clarke, and B. Gorman, "Impact of Wide-Ranging Nanoscale Chemistry on Band Structure at Cu(In,Ga)Se$_2$ Grain Boundaries", Scientific Reports, 7, 1, 1-11, 2017. DOI: 10.1038/s41598-017-14215-0.

[10] M. Kim, S. Sohn, and S. Lee, "Reaction kinetics study of CdTe thin films during CdCl2 heat treatment", Solar Energy Materials and Solar Cells, 95, 8, 2295-2301, 2011. DOI:10.1016/j.solmat.2011.03.044.

[11] D. Rudmann, A. da Cunha, M. Kaelin, F. Kurdesau, H. Zogg, A. Tiwari, and G. Bilger, "Efficiency enhancement of Cu(In,Ga)Se$_2$ solar cells due to post-deposition Na incorporation", Appl. Phys. Lett., 84, 7, 1129-1131, 2004. DOI: 10.1063/1.1646758.

[12] F. Pianezzi, P. Reinhard, A. Chirilˇa, B. Bissig, S. Nishiwaki, S. Buecheler, and A. Tiwari, "Unveiling the effects of post-deposition treatment with different alkaline elements on the electronic properties of CIGS thin film solar cells", Phys. Chem. Chem. Phys., 16, 8843-8851, 2014. DOI:10.1039/c4cp00614c.

APPENDIX A

Figure A.1 shows an example of the Cu, Ga, Se, versus Na elemental distributions for CGS films prepared by CuPRO(M)+AgBr and CuPRO(M). The Cu, Ga, and Se signals are mostly constant through these films which is representative of all films and devices in this study.

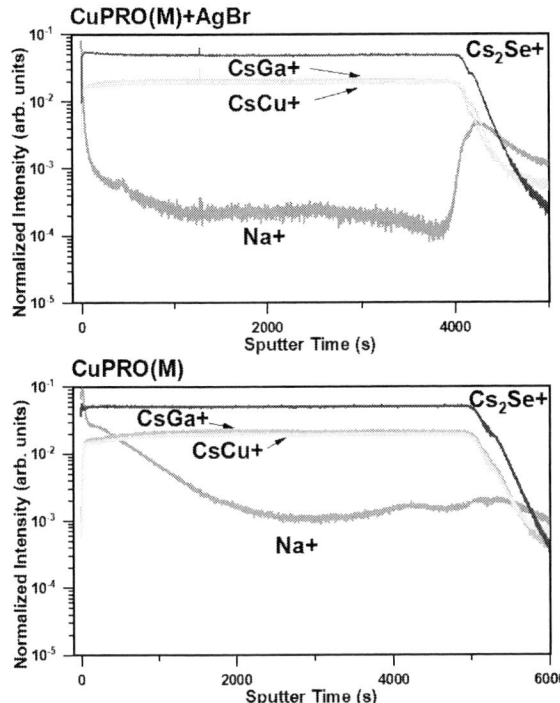

Fig. A1. Positive ion TOF-SIMS Cu, Ga, Se, and Na signals for CGS films prepared by CuPRO(M)+AgBr and CuPRO(M).

Extensive Evaluation and Uncertainty Estimation of Albedo Data Sources

Vicente Lara-Fanego[1], Christian A. Gueymard[2], Jose A. Ruiz-Arias[1,3], Tomas Cebecauer[1], and Juraj Betak[1]

[1] Solargis s.r.o., Bratislava (Slovakia), [2] Solar Consulting Services, Colebrook, NH (USA), [3] Applied Physics Department I, University of Malaga (Spain)

Abstract—Surface albedo has been constantly used in solar energy modeling. In recent times, the growing development of bifacial PV modules has ignited a surge of interest for this variable. However, there is a lack of knowledge in the energy industry about several key aspects concerning albedo. In this work, we present an evaluation of 8 different sources of gridded global albedo data against 5 years of data at 29 ground observing stations in North America. Based on that, we discuss and estimate the uncertainty associated with each source.

Keywords—*Albedo, uncertainty, bifacial PV.*

I. INTRODUCTION

Albedo is a quantity characterizing the fraction of radiation incident on a given surface that it reflects under specific illumination conditions. From a broadband (as opposed to spectral) radiation perspective, the albedo of a surface is simply defined as the ratio of the total irradiance reflected to the total hemispherical irradiance received, i.e.,

$$\alpha = \text{RHI} / \text{GHI} \tag{1}$$

where α denotes the albedo, GHI is the global (total) horizontal irradiance, and RHI is the total horizontal irradiance reflected by the surface. Albedo is dimensionless and varies over the range 0–1. Most land areas not covered by ice or snow have an albedo in the approximate range 0.1–0.4. In contrast, water bodies have a relatively low albedo, typically around 0.05. At the other extreme, regions covered with fresh snow or clean ice can have a very high albedo, which may exceed 0.8.

The simple definition above has relatively complex physical implications. Even though the reflectiveness of the surface depends on its physical characteristics (roughness, opacity, type of material, etc.), albedo is not merely an intrinsic property of a surface but depends also on the spectral and angular distributions of the incident light, which in turn are governed by atmospheric composition and sun position. In that sense, strictly speaking, albedo is rather a characteristic of the coupled atmosphere-surface system. For that reason, in general, albedo presents a high variability both in space (at scales of only a few meters in the worst cases) and time (at scales from minutes to daily, seasonal, and even inter-annual).

On the other hand, GHI can be decomposed as the sum of two components: direct and diffuse irradiances. This results in a convenient definition of two conceptual cases of albedo under extreme illumination conditions, sometimes called "albedo components", namely: white-sky albedo (WSA, i.e., reflectance under isotropic diffuse illumination) and black-sky albedo (BSA, i.e., reflectance under direct-beam illumination) [1]. Hence, the actual albedo under real conditions is sometimes referred to as blue-sky albedo. In that framework, RHI can be expressed in terms of these individual components as:

$$\text{RHI} = \text{BSA} \cdot \text{DNI} \cdot \cos(\theta) + \text{WSA} \cdot \text{DIF} \tag{2}$$

where DNI and DIF denote the direct normal irradiance and the diffuse horizontal irradiance, respectively, and θ is the solar zenith angle. For solar energy applications, two alternate approaches (assuming isotropic or "Lambertian" reflectance) can be considered to estimate the albedo in practice:

$$\alpha \approx \text{WSA} \cdot K + \text{BSA} \cdot (1-K) \tag{3}$$

or

$$\alpha \approx \text{WSA} \tag{4}$$

where $K = \text{DIF/GHI}$ is the diffuse fraction.

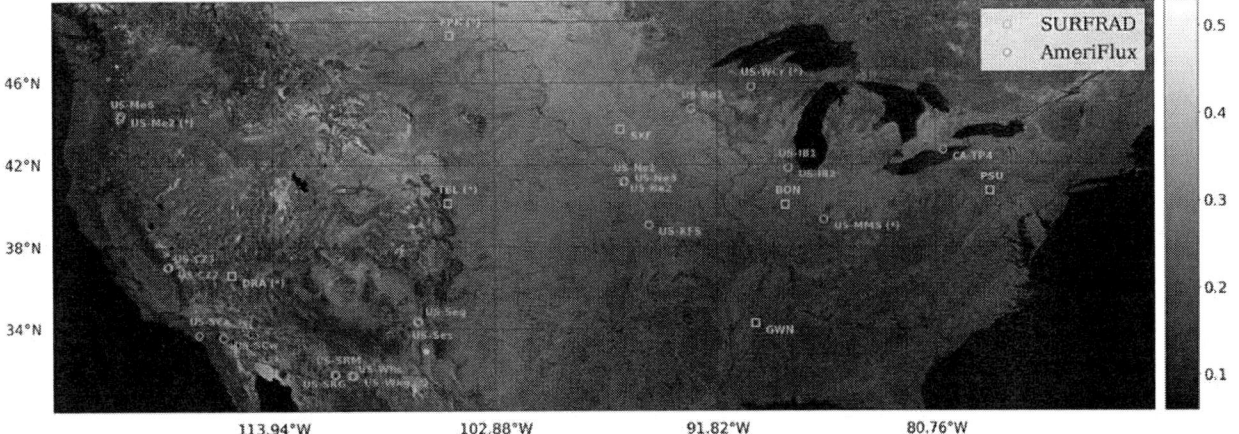

Fig. 1. Geographical distribution of the stations superimposed on the mean-annaul albedo map from the 1-km Solargis climatology. Sites highlighted with (*) are considered representative of a larger, 500-m MODIS pixel area.

RHI is of particular importance for bifacial PV technology, and as a matter of consequence, the energy production of PV plants using that technology is very sensitive to temporal variations in albedo. Moreover, the design, bankability and performance of such bifacial-PV plants depend on the assumptions made regarding the site's albedo conditions, which must be obtained as accurately as possible from existing databases. Thus, albedo has become more relevant for the solar energy industry, now driven by the growing interest in bifacial modules. The annual power production of a bifacial PV project can be increased by up to ≈15% compared to conventional monofacial installations (depending on the particular conditions of the project), which is sizeable [2]. One of the most practical and difficult constraints is the determination of the uncertainty associated with albedo data. In that sense, the concept of representativeness of the albedo at utility scales plays a key role. In this work, we discuss this notion from an applied perspective. To that end, an extensive evaluation of 8 sources of gridded global albedo data is carried out at daily and monthly resolutions. The evaluation is conducted over 5 complete years at 29 locations in North America. The results pertaining to the monthly scale are presented here.

II. METHODOLOGY

The evaluation described below is carried out for a 5-year period, 2011–2015, and consists in comparing modeled data to actual ground albedo measurements. The results are presented in terms of the usual metrics, namely: mean bias deviation (MBD), mean absolute deviation (MAD), and relative root mean square deviation (RMSD).

A. Ground measurements

Data from 29 ground radiometric stations located in North America have been used for the analysis. These stations belong to the NOAA-SURFRAD and AmeriFlux networks. Fig. 1 shows their geographical distribution. The dataset covers a wide range of climatic conditions, from hot and dry environments to cold and humid ones. Many of them have snowy periods of variable duration. Measurements were collected by means of pyranometers, one facing up to measure GHI and the other facing down to measure RHI. In all cases, these measurements are quality-checked first to improve data quality, and used to derive the albedo from Eq. (1). They are installed at different heights—from a few meters up to more than 30 meters with high towers—depending on location. The height of installation of the instrument directly determines the surface area covered. For an installation at 1.5 m from the surface (an ideal setup for operation and maintenance of the instrument), the circle-shaped area covered by 99% of the measured signal is approximately 700 times smaller than the area covered by the best possible resolution (500-m) in existing MODIS-derived data. In contrast, for an installation on a high tower of 30 m, the area covered is still approximately 1.8 times smaller than the same MODIS pixel. This difference of representativeness of the area covered is a critical element to consider in the experimental validation of albedo data. Seven of the 29 stations are located in zones whose surface characteristics are considered homogeneous and, therefore, representative of a large area surrounding the station, and comparable to a 500-m MODIS pixel [3].

TABLE I. SOURCES OF ALBEDO DATA AND THEIR CHARACTERISTICS

Source	Type	Temporal res.	Spatial res.	Variable
CMSAF	Satellite CLARA-A2-SAL	5 days	25 km	BSA
GLASS	Satellite AVHRR	8 days	5.6 km	WSA
MCD43A3	Satellite MODIS	1 day	5.6 km	WSA
Mines ParisTech [2004–2011]	Satellite MODIS	Monthly averages	5.6 km	WSA
NREL NSRDB	Satellite MODIS	1 day	4 km	WSA
SGClim [2006–2015]	Satellite MODIS+Solargis	Monthly averages	1 km	WSA
ERA5	ECMWF Reanalysis	1 hour	≈30 km	α
MERRA2	NASA Reanalysis	1 hour	≈50 km	α

B. Sources of albedo data

An exhaustive catalogue of currently available sources of data of albedo of this type is presented in [1]. A small subset of five sources of gridded global albedo data is selected here to carry out the analysis and evaluate their quality. These sources are based on either satellite-based observations or reanalysis models, and have a wide variety of spatiotemporal resolution (Table I).

To obtain mutually consistent results, the monthly averages are calculated for the same period of evaluation, i.e., 2011–2015 for all sources, except for Mines ParisTech (MPT) and Solargis (SGclim), which are monthly climatologies derived from different periods, as detailed in Table I.

III. RESULTS

Fig. 2 shows the scatter plots comparing the gridded albedo values from each source to their measured counterpart at all 29 stations combined. The worst results are obtained in winter, mostly because of snow, whose albedo may vary widely due to melting and soiling. The large NSRDB errors are mostly due to its reliance on an excessive albedo value in the presence of snow [1]. In general, sources with a coarser resolution, like ERA5 or MERRA2, are less precise, as could be expected. A notable exception is CMSAF, whose results are reasonably good. In general, the best results are obtained by the sources based on MODIS observations (Table I). This is particularly the case with SGclim, despite the difference in averaging period. Fig. 3 shows specific results at one of the stations with "homogeneous" area conditions (Willow Creek, US-WCr).

The determination of uncertainty is quite complex because of the many factors involved. For simplification, it is estimated here from the model uncertainty only, relatively to measurements [4], and obtained statistically from the residuals. The standard and expanded uncertainties are presented in Table II, with best results shown in boldface.

TABLE II. UNCERTAINTY ESTIMATIONS FOR ALL SOURCES

Source	Standard U [68%]	Expanded U [95%]
CMSAF	0.0320	0.1144
GLASS	0.0417	0.1273
MCD43C3	**0.0284**	**0.0942**
MPT	0.0351	0.1263
NSRDB	0.0504	0.2546
SGClim	0.0305	0.1019
ERA5	0.0389	0.1120
MERRA2	0.0407	0.1786

IV. CONCLUSIONS

In practice, estimating the albedo value that best represents the area of a PV plant is a challenge. Here, we have presented an extensive evaluation of different albedo data sources, and have discussed the difficult determination of the associated uncertainty. One of the most important factors is the representativeness of the surface area covered by the measurements in comparison with that of the data source's grid cell. This can be

particularly significant at PV utility scales, when using bifacial PV technologies.

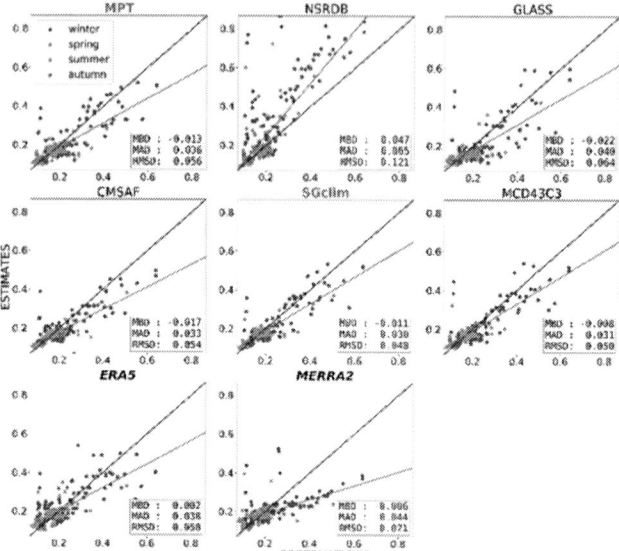

Fig. 2. Scatter plots of monthly-mean estimated and measured albedo values for all data sources and all 29 stations combined.

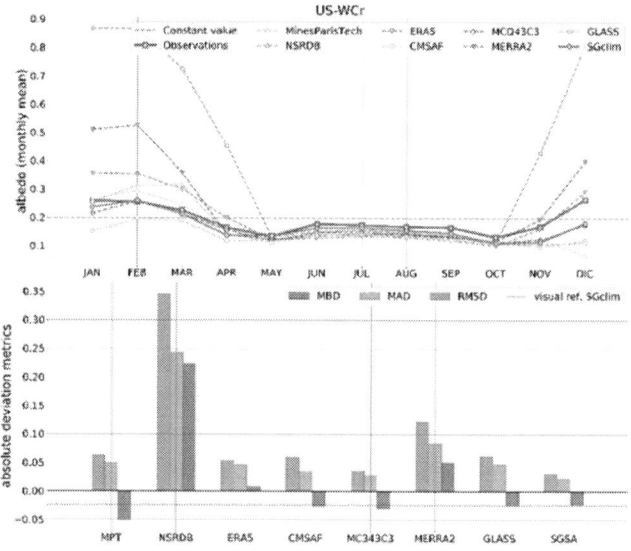

Fig. 3. Time series plot (top) and deviation metrics (bottom) for all sources at the US-WCr station, which is representative of "homogeneous" conditions.

REFERENCES

[1] C. A. Gueymard, V. Lara-Fanego, M. Sengupta, Y. Xie. 2019. Solar Energy, 182, 194-212.

[2] Reise C, Schmid A. Realistic yield expectations for bifacial PV systems – an assessment of announced, predicted and observed benefits. Proc. 31th EUPVSEC Conf., 2015. pp.1775-1779.

[3] Wang D, Liang S L, He T, Yu Y, Schaaf C, Wang Z, 2015. Estimating daily mean land surface albedo from MODIS data J. Geophys. Res. Atmos., 120, 4825-4841.

[4] M. Sury and T. Cebecauer. Satellite-based solar resource data: model validation statistics versus user's uncertainty. ASES SOLAR Conference, San Francisco, 7-9 July 2014.

Sampling Solar Irradiance with Copula

Mesude Bayrakci-Boz
Penn State Hazleton
Hazleton, PA
mzb187@psu.edu

Abstract—Solar radiation data at the place of interest is not available all the time mainly due to the cost of measuring instruments. In these cases, it is possible to get reasonable accurate estimation using the different models such as copula method. This paper presents a model ultimately provides a framework for generating simulated solar irradiation data using a copula with beta distributions. Based on normalized global irradiance and normalized beam irradiance, a model is developed to create synthetic solar irradiation data for use in solar energy systems and planning. The model is applied to the city of Philadelphia using hourly measurements of solar irradiation from 2003-2012 using the Solar Anywhere website. The results are compared with the real data and it is shown that the copula approach performs good.

Index Terms—solar irradiation, normalized global irradiance, normalized beam irradiance, copula, beta distribution

I. INTRODUCTION

Understanding and quantifying solar irradiation is important since the sun is the driving force of the environment. A precise model of solar irradiation can be used in many fields, including agriculture, architecture, and engineering planning [1], and it is especially useful for solar energy systems and planners. Seasonal and daily solar data is usually used for achieving day lighting and glazing energy balance, whereas hourly and sub-hourly data is used for system simulation modeling and rating, as well as economic analyses and determinations of systems' lifetimes [2]. Generators and system operators are the two main groups that use solar irradiation forecasts. An accurate and reliable solar forecast allows grid operators to predict generation and balance the supply and demand chain [3]. Such a forecast also helps promote better solar penetration to mitigate resource uncertainty and reduces the need for the scheduling of ancillary generation[4]. Finally, an effective solar forecast provides information for the energy market to predict electricity prices [5]; establishes better control of energy storage dispatch [6, 7]; helps reduce penalty charges in CSP systems;and promotes grid stability and unit commitment [4].

Solar irradiation data at the place of interest is not always possible. In these cases, it is possible to obtain reasonably accurate estimations using different models such as copula models. Copulas have been used to examine the relationship between solar irradiation and other factors. Abbedi et al. presented the correlation of the solar irradiation patterns between the two PV farms [8]. Haghi et al. used the Archimedean copula to examine the integration of wind and PV in a distribution network, presenting a comprehensive case study of Iran [9].

Some studies have generated synthetic data using copulas. Campo et al. provided a statistical approach using copulas in order to define a criterion for accepting or rejecting ground sensors' observations [10]. They generated a dataset that could be used to correct Italian solar radiation maps. Munkhammar et al. proposed a method of creating correlated data and estimating the clear-sky index for any set of locations [11]. They used clear-sky index data from 14 different locations and calculated the correlation matrices for the clear-sky index data for all location pairs. They then created synthetic irradiance data using copulas.At the end, the original and synthetic data was compared, and the copula model had better goodness-of-fit than the other models.

Copulas have also been used to examine the relationship between solar irradiation and other factors. Naksrisuk and Kulyos Audomvongseree proposed a dependable capacity evaluation method considering load uncertainty [12]. During their solar power calculation, the researchers used the Frank, Gumbel, and Clayton's copulas to describe the relationship between ambient temperature and solar irradiation. Bazrafshan et al. used five different copulas: (1) normal, (2) student's t, (3) Clayton, (4) Frank, and (5) Gumbel. This allowed them to model joint behavior between monthly mean solar radiation and sunshine duration data at nine stations in Iran[13.

In this paper a framework is provided for generating simulated solar irradiation data using a copula with beta distributions.

II. DATA AND METHODOLOGY

A. Copula Method

A copula is a function that joins or "couples" a multivariate distribution function to its one-dimensional marginal distribution functions. Sklar introduced the copula in order to decompose an n-dimensional distribution function into two parts, the marginal distribution functions and the copula, describing the dependence part of the distribution [14].

The main theorem is called Sklar's theorem [14]. This theorem shows that a multivariate cumulative distribution function (CDF) can be expressed in terms of a multivariate uniform distribution function with marginal density functions $U(0,1)$. Sklar's theorem was subsequently modified by Nelson [15]. In Nelson's formulation, the CDF of two random variables(X and Y) was expressed as:

$$H(x,y) = C[F(x), G(y)] = C(u,v) \qquad (1)$$

$$x, y \in R \text{and} F(x), G(y) \in [0,1] \qquad (2)$$

where $F(x) = u = $ marginal distribution of x; $G(y) = v =$ marginal distribution of y; $H(x,y) =$ joint distribution of x and y; and $C[F(x), G(y)] =$ copula function of the marginal distributions of $F(x)$ and $G(y)$ [13].

Various copula functions have since been introduced. These functions are generally classified into explicit types, such as the Gaussian copula and the t-Student copula, and implicit types, such as the Clayton copula and the Gumbel copula.

B. Solar Irradiation

Solar irradiation can be broken down into its various components, including beam, circumsolar diffuse, sky diffuse, horizon diffuse, and ground diffuse [16]. Beam irradiance is the type of irradiance received from the sun without being scattered by the atmosphere. Diffuse irradiance is the irradiance after scattering has changed its direction by the atmosphere. The total solar irradiance on a surface is called global irradiation.

This study is based on modeling the k_t (normalized global irradiance) and k_b (normalized beam irradiance) for every hour during the year as probability distributions. A copula is then used to obtain correlated samples from the distributions. The steps to be taken for the presented methodology are as follows:

- A. Solar Irradiation Data Process
    1) Obtain the publicly available hourly measurements of solar irradiation for Philadelphia, PA, from 2003-2012 using the Solar Anywhere website[17].
    2) Normalize global and beam irradiance using average-top-of-atmosphere radiation. Create the variables k_t (normalized global irradiance) and k_b (normalized beam irradiance).
    3) Reorganize the data to show each hour in a day over a ten-year period from January 1:00 am to December 11:00 pm. In this case, each hour consists of 300 or 310 data points, depending on the month. (For example, there are 31 days in January, so 31 instances of 9.00 am data over 10 years results in 310 data points.)
    4) Create the histograms for the hourly observations of k_t and k_b.
    5) Apply Hartigan's dip test in order to measure multimodality [18].
    6) Apply the Curren and Flay approach to decide which distribution best fits the CDFs of k_t and k_b.
- B. The Sampling Procedure:
    1) Find the dependence structure of the data by calculating the correlation matrices of:
        - hourly k_t's and hourly k_b's
        - between k_t and k_b for each hour
    2) Select the Copula family according to AIC and BIC.
        - All available copulas should be fitted using maximum likelihood.

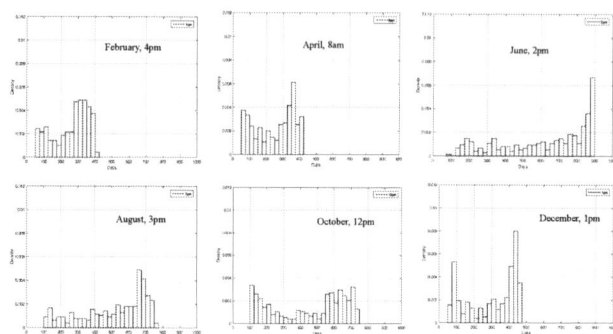

Fig. 1. Histograms of k_t.

- The criteria should then be computed for all copula families.
    3) Generate the synthetic data using Copula
    4) Create the plots
    5) Apply the GOF and K-S test in order to validate the results
- C. Model simulation:
    All processes for the model are implemented in R [19]. In order to reorganize data and calculate k_t and k_b, a function is created in R (Step A. 1– 3). Next, the histograms are created (Step A. 4). Hartigan's dip test and the Curren and Flay approach are applied in R using the diptest and fitdistrplus libraries, respectively (Step A. 5-6). Finally, the copula library is used in R for the sampling procedure.

III. RESULTS AND DISCUSSION

Based on the datasets k_t and k_b for each hour, histograms were created and the Hartigan dip test was applied to multimodality. Some of the histograms are shown in Figure 1. As can be seen from the figures, most of the summer months' hours had one peak each, whereas the other months had at least two peaks each. This resulted from the cloudy weather in these months.

When the Cullen and Flay approach was applied, it showed the beta distributions' most suitable functions. One beta distribution was used for single peaks, while a mixed-beta distribution was used for the multiple peaks of the k_t and k_b data. Figure 2 shows the k_t data for June 4:00 pm, and Figure 3 shows the k_t for January 11:00 am as examples of the distributions.

The correlation matrices were calculated for k_t and k_b. Figure 4 shows the correlation matrix of k_t for July. The rows and columns correspond to the k_t values of each hour. A high correlation between every two consecutive hours was observed, ranging from 0.75 to 0.92. A moderate correlation

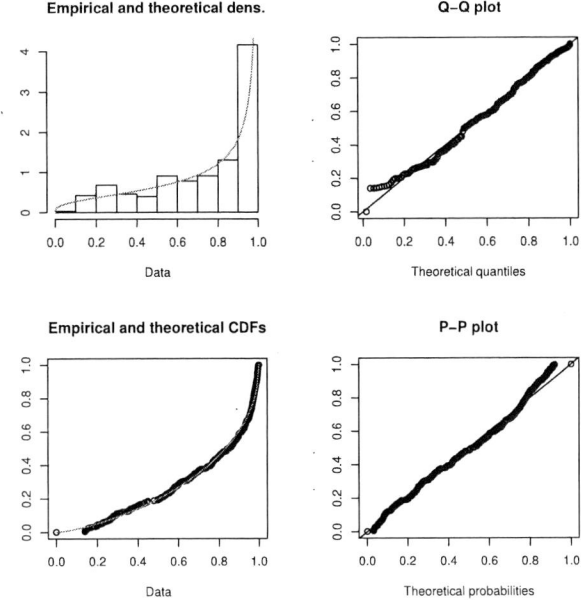

Fig. 2. Beta distribution and k_t data at 4.00 pm in July

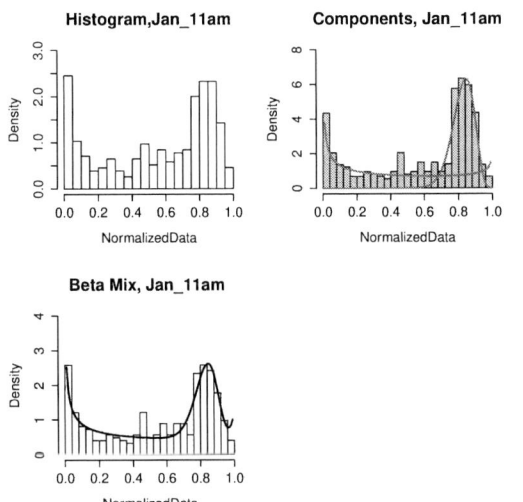

Fig. 3. Mixed-beta distribution and k_t data at 11.00 am in January

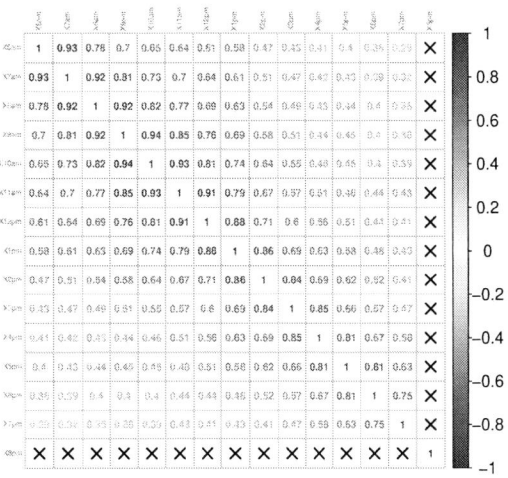

Fig. 4. Correlation matrix of k_t's in July

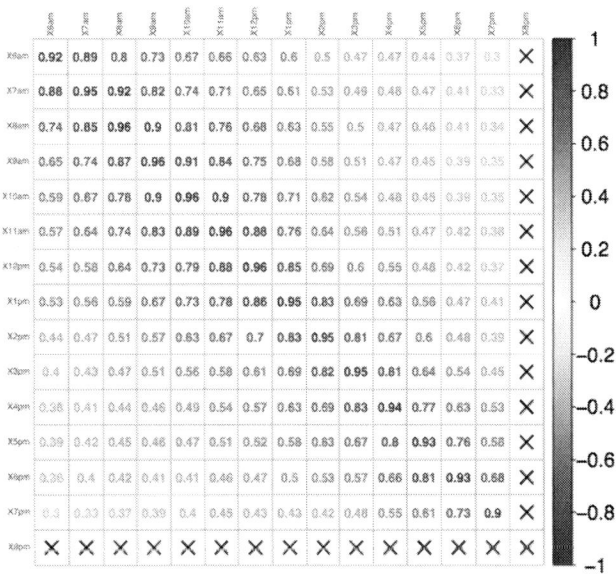

Fig. 5. Correlation matrix between k_t and k_b in June

was observed between 3-4 hours later or before, ranging from 0.61 to 0.81.

Figure 5 shows the correlation between k_t and k_b for Jun4. The rows correspond to k_t, and the columns correspond to k_b. There was a high correlation between k_t and k_b for each hour, ranging from 0.9 to 0.96. There was a moderate correlation observed for the other hours, around 0.5.

A sampling of the solar irradiation data was performed using a copula based on the observed data. In Figure 6, the observed data is compared to the simulated data obtained

from the model for certain hours. The simulated distributions were very accurate approximations of the observed ones. The results showed that the Frank, Gumbel, and t-copulas were the most suitable functions. Finally, the K–S test values were low, under 0.5, meaning that they were a good match. All of these results indicated that a copula is a good method for generating synthetic solar irradiation data.

IV. SUMMARY AND CONCLUSION

This study used copula functions to generate synthetic solar irradiation data. First, global and beam solar irradiation was normalized using average-top-of-atmosphere radiation, k_t and k_b, respectively. It was determined that k_t and k_b had two

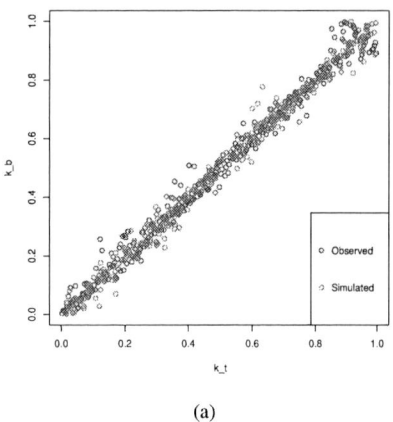

(a)

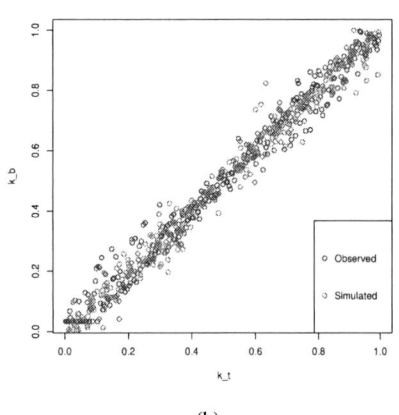

(b)

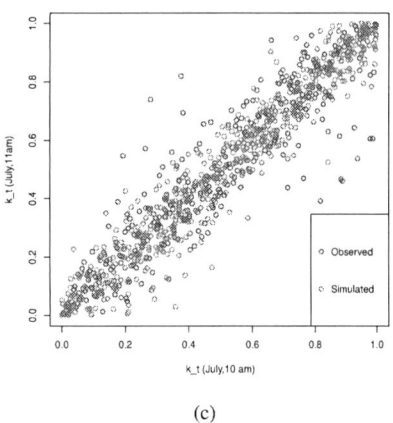

(c)

Fig. 6. Copula modeling a)July, 1pm (Frank Copula, param=54.01) b) July, 5pm (Frank Copula, param=40.91) c) July, between 10 am and 11 am (t Copula, rho=0.92, df=3.11)*(Colored version is available in online print)*

different patterns based on months and hours. Almost all summer months had one-peak distributions, whereas other months had two-peak distributions. The correlation values between consecutive hours were high, ranging from 0.75 to 0.98. Moreover, k_t and k_b had a different correlation, 0.98.

Copula functions were used to construct the joint distribution of the dependent variables; all the marginal distributions were fixed as beta. Finally, the results were tested graphically, indicating that the model appropriately fit the reanalysis data. The Kolmogorov–Smirnov test was applied, and it showed a good fit. An accurate and reliable solar forecast can be very useful to generator and system operators.

This study can be further developed in a number of ways. In this model, we did not take into consideration that the weather may be cloudy, partly cloudy, or clear after one hour, and it may depend on the next hours, too. These different types of weather states and possibilities could be incorporated by fitting both distributions and correlations to different subsets of data. Other meteorological variables such as temperature and precipitation could also be added into the study.

V. ACKNOWLEDGMENT

The author acknowledges the review and feedback by Jeffrey Brownson and Mort Webster of The Pennsylavaia State University.

REFERENCES

[1] C. K. Pandey and A. K. Katiyar, "Solar Radiation: Models and Measurement Techniques," J. Energy, vol. 2013, pp. 1–8, 2013, doi: 10.1155/2013/305207.

[2] D. R. Myers, "Solar radiation modeling and measurements for renewable energy applications: data and model quality," Energy, vol. 30, no. 9, pp. 1517–1531, Jul. 2005, doi: 10.1016/j.energy.2004.04.034.

[3] Skip Dise Clean Power Research. What is the value of accurate solar forecasting for utility-scale pv plants?, 2017.

[4] D. P. Larson, L. Nonnenmacher, and C. F. M. Coimbra, "Day-ahead forecasting of solar power output from photovoltaic plants in the American Southwest," Renew. Energy, vol. 91, pp. 11–20, Jun. 2016, doi: 10.1016/j.renene.2016.01.039

[5] R. Weron, "Electricity price forecasting: A review of the state-of-the-art with a look into the future," Int. J. Forecast., vol. 30, no. 4, pp. 1030–1081, Oct. 2014, doi: 10.1016/j.ijforecast.2014.08.008.

[6] R. Hanna, J. Kleissl, A. Nottrott, and M. Ferry, "Energy dispatch schedule optimization for demand charge reduction using a photovoltaic-battery storage system with solar forecasting," Sol. Energy, vol. 103, pp. 269–287, May 2014, doi: 10.1016/j.solener.2014.02.020.

[7] S. Abedi, G. Hossein, S. Hossein, and M. Farhadkhani, "Improved Stochastic Modeling: An Essential Tool for Power System Scheduling in the Presence of Uncertain Renewables," in New Developments in Renewable Energy, H. Arman, Ed. InTech, 2013. doi: 10.5772/52161.

[8] H. Valizadeh Haghi, M. Tavakoli Bina, M. A. Golkar, and S. M. Moghaddas-Tafreshi, "Using Copulas for analysis of large datasets in renewable distributed generation: PV and wind power integration in Iran," Renew. Energy, vol. 35, no. 9, pp. 1991–2000, Sep. 2010, doi: 10.1016/j.renene.2010.01.031.

[9] L. Campo and F. Castelli, "A statistical model for the selection of ground observations of solar radiation: an application in producing a five-year dataset of radiation maps on Italian territory through correction of MSG-derived data," Prague, Czech Republic, Oct. 2011, p. 81741T. doi: 10.1117/12.897922.

[10] J. Munkhammar, J. Widén, and L. M. Hinkelman, "A copula method for simulating correlated instantaneous solar irradiance in spatial networks," Sol. Energy, vol. 143, pp. 10–21, Feb. 2017, doi: 10.1016/j.solener.2016.12.022.

[11] C. Naksrisuk and K. Audomvongseree, "Dependable Capacity Evaluation of Wind Power and Solar Power Generation Systems", ECTI-EEC, vol. 11, no. 2, pp. 58–66, Sep. 2013.

[12] J. Bazrafshan, N. Heidari, I. Moradi, and Z. Aghashariatmadary, "Simultaneous Stochastic Simulation of Monthly Mean Daily Global Solar Radiation and Sunshine Duration Hours Using Copulas," J. Hydrol. Eng., vol. 20, no. 4, p. 04014061, Apr. 2015, doi: 10.1061/(ASCE)HE.1943-5584.0001051.

[13] B. Schweizer and A. Sklar, Probabilistic metric spaces. Mineola (New York): Dover Publications, 2005.

[14] R. B. Nelsen, An introduction to copulas, 2nd ed. New York: Springer, 2006.

[15] J. R. S. Brownson, Solar energy conversion systems. 2014. Accessed: May 28, 2022. [Online]. Available: https://www.sciencedirect.com/science/book/9780123970213

[16] Scalable Real Time and Forecast Data. https://www.solaranywhere.com/.

[17] Hartigan, J and Hartigan, PM, "The dip test of unimodality," Ann. Stat., vol. 13, pp. 70–84, 1985.

[18] R Core Team (2013). R: A language and environment for statistical computing. R Foundation for Statistical Computing, Vienna, Austria. URL http://www.R-project.org/

Analysis of Temperature Dependence of Solar Cell Performance Through Light Soaking

Samuel Seibert, Aesha P. Patel, Manoj Rajakaruna, Sandip S. Bista, Lei Chen, Randy J. Ellingson, Yanfa Yan, Zhaoning Song

Department of Physics and Astronomy, and Wright Center for Photovoltaics Innovation and Commercialization (PVIC), University of Toledo, Toledo, OH 43606, USA

Abstract — Solar cells working at elevated temperatures generally exhibit inferior device performance than that characterized under standard test conditions due to increased internal carrier recombination. This work seeks to investigate the temperature coefficient of solar cells composed of crystalline and thin-film semiconductor materials, including Si, GaAs, CdTe, metal halide perovskite under AM1.5G and AM0 irradiance spectra. The increasing temperature through continuous irradiance results in decreases in almost all the photovoltaic parameters, including short-circuit current density (J_{sc}), open-circuit voltage (V_{oc}), fill factor (FF), and power conversion efficiency (η), with some exceptions for specific devices. The results in the cell underperforming overall as the temperature increases are used to determine the temperature coefficient of cell performance for each type of solar cell.

Keywords—temperature coefficient, Si, GaAs, CdTe, perovskite

I. INTRODUCTION

Solar cells are devices designed to generate electrical energy by harvesting solar radiation. Solar cells in real-world conditions, whether they are terrestrial or in space, will heat up under solar irradiance and see a decrease in their performance [1]. Being able to study and model the temperature dependence of cell performance is of the utmost importance to those using the energy as it will allow them to predict the energy yield of a photovoltaic system and determine the required installation capacity. This is critical for emerging photovoltaic (PV) technologies such as perovskite solar cells because their behaviors at operating temperatures and outdoor performance are yet to be understood [2].

The decrease in PV performance will come in terms of a lower efficiency rating of a solar panel, given by the temperature coefficient (TC) of a particular photovoltaic technology. Under continuous solar irradiance, the temperature of a solar cell will stabilize at a maximum value, depending on the generation of heat on the surface of the cell and surrounding environment, and its power conversion efficiency (η), open-circuit voltage (V_{oc}), and fill factor (FF) will begin to stabilize around a minimum, and short-circuit current density (J_{sc}) will begin to increase slightly from its minimum and stabilize around a new point as the temperature remains stable [3].

To determine the TCs of different solar cells based on single crystalline semiconductors (c-Si and GaAs) and polycrystalline thin films (metal halide perovskite), we performed temperature-dependent solar cell performance measurements under continuous AM1.5G or AM0 solar irradiance conditions without additional temperature control. This allows us to discover a generalized formula for the change in photovoltaic parameters of different types of solar cells as a factor of the temperature when under AM1.5G and AM0 solar spectral conditions.

II. EXPERIMENTAL DETAILS

The light source chosen for this study is the Sunbrick solar simulator manufactured by G2V Optics. This product replicates a solar spectrum using a programmable array of light emitting diodes (LEDs) which allows the testing cell to experience a real-world simulation of AM1.5G and AM0 spectral conditions as well as the associated temperature increase. In order to record as many pieces of data and while also controlling the Sunbrick, an in-house LabVIEW program was created to include the Sunbrick spectrum control as well as a thermocouple to monitor the device temperature. This program allows recording of current density-voltage (J-V) scans and calculations of J_{sc}, V_{oc}, η, and FF, as well as the series resistance (R_s), shunt resistance (R_{sh}), and max power (P_{max}) of a device through a solar simulation. This allows a user of the program to develop a model of performance for the cell over the entirety of a light soaking cycle.

For the purposes of this work, we measured six types of solar cells, including commercial Si and GaAs reference cells, thin-film CdTe [4], wide-bandgap (WBG) perovskite [5], low-bandgap (LBG) perovskite [6], and tandem perovskite solar cells [7, 8] fabricated in our lab. The temperature-dependent J-V

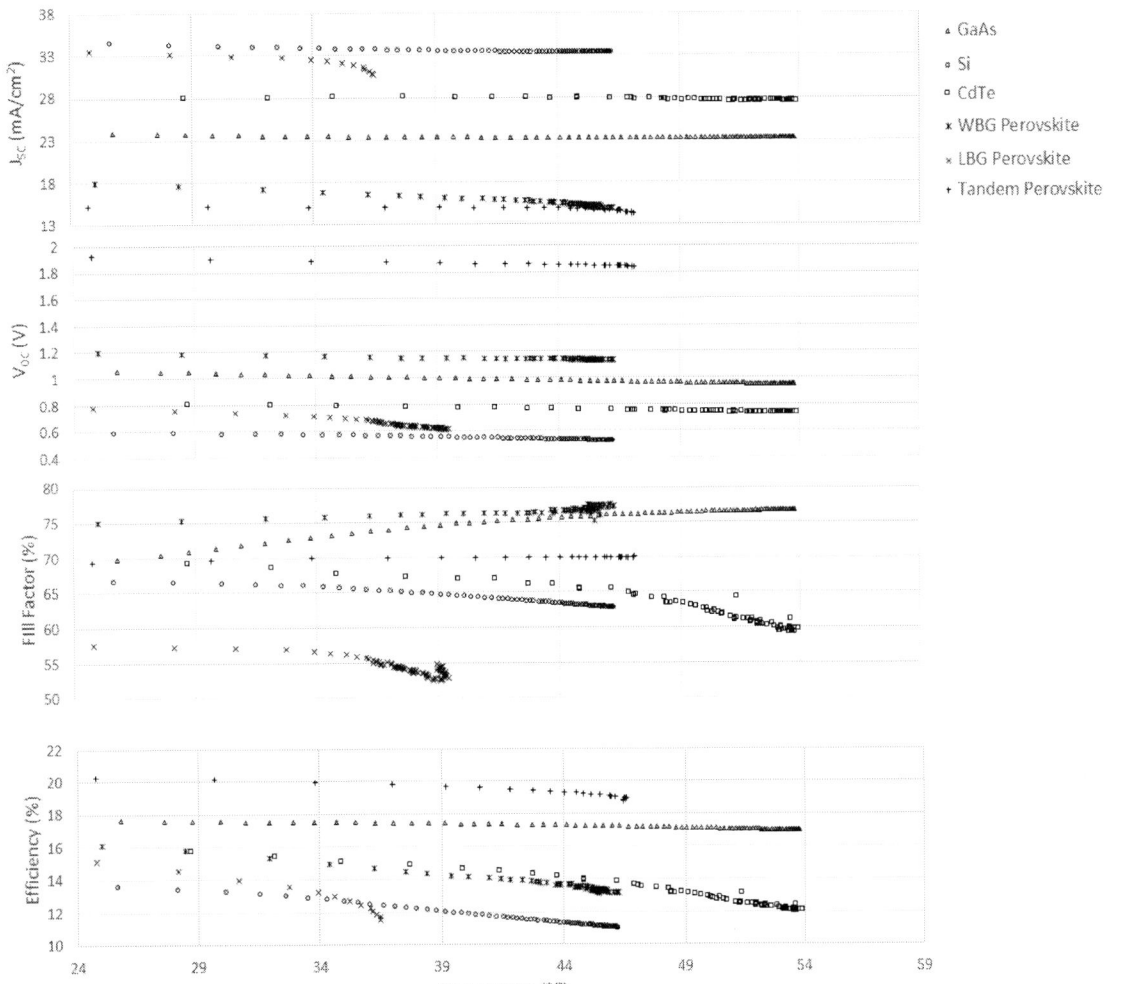

Fig. 1. Variations in PV parameters of Si, GaAs, CdTe, and WBG, LBG, and tandem perovskite solar cells as a function of temperature under continuous AM1.5G solar spectrum.

measurements were run for 60 minutes, and an analysis of the cell's performance was taken 86 individual times. A 1-hour cycle was chosen because the increase in temperature for the individual cells had drastically slowed down around the 40-minute mark, or the 50th test. Beyond that point, most gains were within a 10th of a degree, and the change in the cell parameters had stagnated immensely alongside the decreasing rate of temperature change. Table I lists the approximate stabilization temperature for each type of solar cell.

III. RESULTS AND DISCUSSION

A. Theoretical Considerations

The equations for analyzing the temperature dependence of solar cell performance have already been well established in literature [1, 3]. In general, the following equations can be used to describe the temperature-dependent solar cell performance.

$$J = J_o \left(e^{\frac{qV}{nKT}} - 1 \right) - J_{ph}. \qquad (1)$$

where J simply refers to the total current density in the circuit of the cell, J_{ph} is the photogenerated current density and can be assumed $J_{ph} \approx J_{sc}$, J_0 is the reverse saturation current density, n is the ideality factor, K is the Boltzmann constant, T is the temperature of the cell, q is the unit charge, V is the voltage of the cell. This equation allows us to then define the open circuit voltage of a cell as it changes depending on the temperature, and is shown to be inversely proportional to J_o, is given by

$$V_{oc} = \frac{kT}{q} ln \left(\frac{J_{ph}}{J_o} + 1 \right). \qquad (2)$$

978-1-7281-6118-1/22 $31.00 © 2022 IEEE

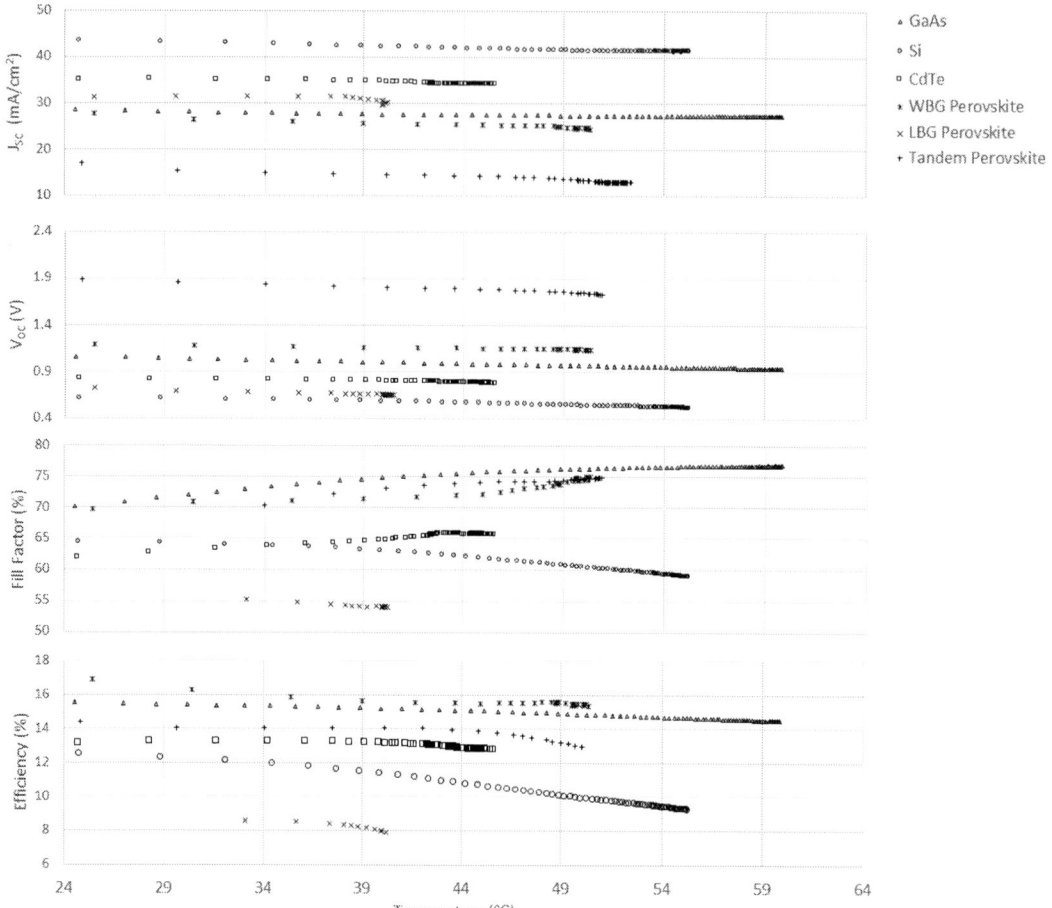

Fig. 2. Variations in PV parameters of Si, GaAs, CdTe, and wide-bandgap, low-bandgap, and tandem perovskite solar cells as a function of temperature under continuous AM0 solar spectrum.

J_{SC} can now be calculated as a resultant of the spectral irradiance of light that is interacting with the cell given by,

$$J_{sc} = q \int_{hv=E_g}^{\infty} \frac{dN_{ph}}{dhv} d(hv). \qquad (3)$$

where E_g is the bandgap of the semiconduction, hv is photon energy, and N_{ph} is the flux of photons with an energy of hv, given by AM1.5G or AM0 spectrum.

With respect to J_o, this value also has a level of dependence on the temperature of the material, and it is given by

$$J_o = CT^3 e^{-E_g/kT}. \qquad (4)$$

where C is a constant determined by a series of diffusion constants, diffusion lengths, carrier densities, and electron and hole masses [1, 3].

FF can be determined by an empirical formula:

$$FF = \frac{v_{oc} - ln\,(v_{oc} + 0.72)}{v_{oc} + 1}. \qquad (5)$$

where

$$v_{oc} = \frac{qV_{oc}}{kT}. \qquad (6)$$

Based on the equations above, for a typical solar cell, J_{SC} increases slightly with temperature because the bandgap of an absorber material decreases, allowing absorption of more photons. V_{OC} and FF, however, decreases with increasing temperature due to increased internal carrier recombination and resistance losses.

B. Experimental Results

Figs. 1 and 2 compare the variations in PV parameters of Si, GaAs, CdTe, WBG perovskite, LBG perovskite, and tandem perovskite solar cells as a function of temperature under continuous AM1.5G

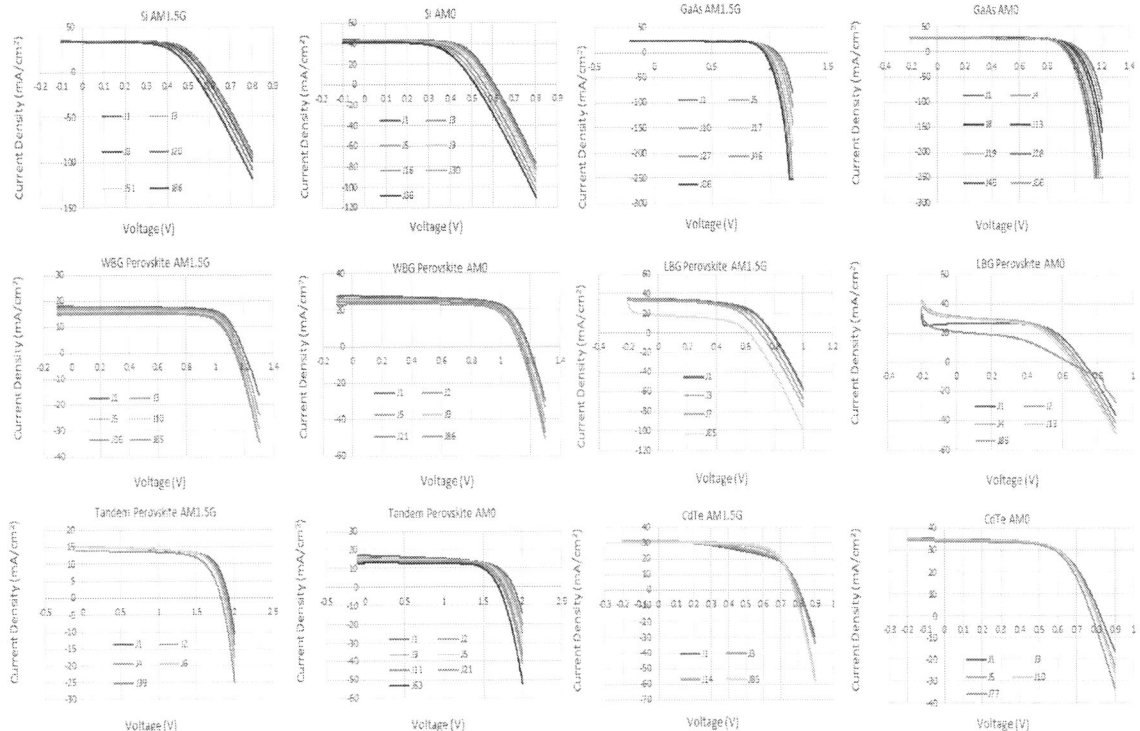

Fig. 3. Selected J-V curves of Si, GaAs, CdTe, and wide-bandgap, low-bandgap, and tandem perovskite solar cells under continuous AM1.5G and AM0 solar spectra.

and AM0 solar spectra, respectively. The selected J-V curves of corresponding devices are plotted in Fig. 3.

As discussed in theoretical considerations, the maximum possible J_{SC} is dependent upon the bandgap energy, where the bandgap energy defines the lowest possible photon energy that can be absorbed by the solar cell. In testing, there also appeared to be a dependence on the J_{SC} on other nonideal factors, such as cell design, materials properties, and possible light-induced degradation. For instance, we observed a decrease in the J_{SC} of the perovskite solar cells with temperature. This is partially due to the abnormal material characteristic of metal halide perovskites [9], which show an increased bandgap and with temperature. The degradation of perovskite due to photodecomposition [10] and moisture-induced corrosion [11] also contributed to the loss of J_{SC} at elevated temperatures after a long time exposure to the ambient air.

For other cells, we observed a stabilization of J_{SC} that coincided with η, at which point an increase began for the GaAs, and a slight increase began for Si and CdTe. The overall drop in all parameters for the materials are recorded in Table II.

The changes in V_{oc}, η, and FF were linearly fit for all of the data before the temperature stabilized, and this allowed us to create a linear equation to model the change in all three parameters as the temperature increased. Formulae are modelled around the proper V_{oc} at 25°C, for standard test conditions (S.T.C.), and were created using linear fitting.

The change in η can be modelled using the formulae:

Si: y = -0.1422 x + 17.537
GaAs: y = -0.0263 x + 18.4
CdTe: y = -0.1690 x + 21.234
WBG Perovskite: y = -0.1360 x + 19.623
LBG Perovskite: y = -0.2589 x + 21.774
Tandem Perovskite: y = -0.0623 x + 22.011

The change in V_{OC} can be modelled using the formulae:

Si: y = -0.0038x + 0.703
GaAs: y = -0.0037x + 1.1492
CdTe: y = -0.0032x + 0.9021
WBG Perovskite: y = -0.0029x + 1.2689
LBG Perovskite: y = -0.0134x + 1.1524
Tandem Perovskite: y = -0.0035x + 2.0067

The change in FF can be modelled with the formulae:

Si: y = -0.2308x + 73.512
GaAs: y = 0.2111x + 65.808
CdTe: y = -0.4658x + 85.291
WBG Perovskite: y = 0.1116x + 71.867
LBG Perovskite: y = -0.4121x + 69.684

Tandem Perovskite: $y = 0.0138x + 69.34$

The changes in AM0 were calculated using an identical method as was done in the AM1.5G tests. While all of the same phenomena were studied for the parameters it is important to note that as would be expected, the surface of the cell did become significantly hotter during the AM0 tests than in the AM1.5G tests, despite starting around the same temperature of 25°C. The resulting formulae were also gathered using the same Excel method of finding a line of best fit.

The change in η can be modelled using the formulae:
Si: $y = -0.1255x + 16.258$
GaAs: $y = -0.0327x + 16.478$
CdTe: $y = -0.0305x + 14.322$
WBG Perovskite: $y = -0.0451x + 17.677$
LBG Perovskite: $y = -0.0592x + 10.496$
Tandem Perovskite: $y = -0.0495x + 15.815$

The change in V_{OC} can be modelled using the formulae:
Si: $y = -0.0036x + 0.7163$
GaAs: $y = -0.0033x + 1.1351$
CdTe: $y = -0.0027x + 0.9011$

WBG Perovskite: $y = -0.0019x + 1.2354$
LBG Perovskite: $y = -0.0042x + 0.8197$
Tandem Perovskite: $y = -0.0057x + 2.0284$

Finally, the change in FF under AM0 light can be modelled using the formulae:
Si: $y = -0.2276x + 71.732$
WBG Perovskite: $y = 0.2119x + 63.688$
LBG Perovskite: $y = -0.1689x + 60.774$
Tandem Perovskite: $y = 0.2101x + 64.288$
GaAs: $y = 0.1495x + 68.35$
CdTe: $y = 0.1904x + 57.341$

Based on the linear fittings, TCs for the PV parameters of each type of solar cells are calculated, the results are tabulated in Table III. Among all types of solar cells, GaAs shows the lowest TCs of less than -0.2%/°C. Thin-film CdTe and perovskite-based devices show similar or lower average TCs (-0.2 to -0.8%/°C) than crystalline Si (~-0.8%/°C), consistent with other reports in the literature [12, 13]. However, we observed significant photodegradation in perovskite cells (Fig. 3), particularly LBG perovskite cells, under continuous illumination, resulting in some uncertainties in determining TCs. To accurately assess the TCs for perovskite solar cells, better device

TABLE I.   APPROXIMATE STABILIZATION TEMPERATURE (°C)

Material	AM1.5G	AM0
Si	46	55
WBG PK	46	51
LBG PK	39	42
Tandem PK	48	52
GaAs	54	60
CdTe	53	57

TABLE II.   OVERALL CHANGE IN PARAMETERS

Material	AM1.5G				AM0			
	J_{SC}	V_{OC}	η	FF	J_{SC}	V_{OC}	η	FF
Si	-2.2%	-2.6%	-18.3%	-5.7%	-5.1%	-8.5%	-26.2%	-8.5 %
WBG PK	-16.8%	-5.5%	-18.6%	3.5%	-12%	-4.1%	-9.2%	7.7%
LBG PK	-12.2%	-14.1%	-28.3%	-4.7%	-5.4%	-4%	-11%	-2%
Tandem PK	5.9%	-4.4%	-9.3%	0.8%	-11.4%	-5.5%	-10.8%	6.4%
GaAs	-2.2%	-9.7%	-3.7%	9.1%	-3.9%	-11.3%	-6.7%	9.4%
CdTe	-1.8%	-9.8%	-23.9%	-14.2%	-2.5%	-4.4%	-1%	6.3%

TABLE III.   TEMPERATURE COEFFICIENTS(%/°C)

Material	AM1.5G			AM0		
	V_{OC}	η	FF	V_{OC}	η	FF
Si	-0.54	-0.81	-0.31	-0.50	-0.77	-0.32
WBG PK	-0.23	-0.69	-0.16	-0.15	-0.25	0.33
LBG PK	-1.1	-1.12	-0.59	-0.51	-0.56	-0.28
Tandem PK	-0.17	-0.28	0.02	-5.7	-0.31	0.33
GaAs	-0.32	-0.14	0.32	-0.29	-0.20	0.22
CdTe	-0.35	-0.80	-0.55	-0.30	-0.21	0.33

encapsulation and stable materials and device structure designs are needed. It is worth noting that tandem perovskite solar cells show relatively low TCs (-0.3%/°C) comparable to that of GaAs. If long-term stability issues can be addressed, these new tandem cells will become a promising PV technology.

IV. CONCLUSION

We built a solar cell measurement setup that allows us to monitor the solar cell performance variation under different solar irradiance spectra. Combining the experimental data and the theory allows us to model, very simply, the expected trajectory of each major PV parameter as the temperature increases on the solar cell under continuous illumination. Having this predictive model can allow large solar fields to use weather modelling data to find an accurate amount of energy that will be produced, especially by emerging PV technologies.

The established solar cells based on GaAs and Si retained much of their power production while not risking degradation, and therefore, would be suggested PV materials for applications. The CdTe cell tested in this study showed some photo-induced metastability, likely caused by the Cu doping. Meanwhile, the WBG perovskite and tandem perovskite tended to produce more power than the LBG perovskite, but all three failed to produce consistently high power through testing. Perovskite tended to react relatively well to begin, however increasing temperature appeared to increase the speed of the degradation of the device. More study will be required to make a conclusive statement on the effect of the temperature on the perovskite cells.

ACKNOWLEDGEMENTS

This material is based upon work supported by the U.S. Department of Energy's Office of Energy Efficiency and Renewable Energy (EERE) under the Solar Energy Technologies Office Award Numbers DE-EE0008753 and DE-EE0008790, and by the U.S. Air Force Research Laboratory under agreement number FA9453–19-C-1002. The U.S. Government is authorized to reproduce and distribute reprints for Governmental purposes notwithstanding any copyright notation thereon. Disclaimer: The views and conclusions contained herein are those of the authors and should not be interpreted as necessarily representing the official policies or endorsements, either expressed or implied, of the U.S. Air Force Research Laboratory or the U.S. Government.

REFERENCES

[1] Singh, P., and Ravindra, N.M.: 'Temperature dependence of solar cell performance—an analysis', Sol. Energy Mater. Sol. Cells, 2012, 101, pp. 36-45.

[2] Song, Z., Li, C., Chen, L., and Yan, Y.: 'Perovskite Solar Cells Go Bifacial—Mutual Benefits for Efficiency and Durability', Adv. Mater., 2022, 34, (4), pp. 2106805.

[3] Green, M.A.: 'General temperature dependence of solar cell performance and implications for device modelling', Prog. Photovoltaics, 2003, 11, (5), pp. 333-340.

[4] Bista, S.S., Li, D.-B., Awni, R.A., Song, Z., Subedi, K.K., Shrestha, N., Rijal, S., Neupane, S., Grice, C.R., Phillips, A.B., Ellingson, R.J., Heben, M., Li, J.V., and Yan, Y.: 'Effects of Cu Precursor on the Performance of Efficient CdTe Solar Cells', ACS Appl. Energy Mater., 2021, 13, (32), pp. 38432-38440.

[5] Chen, C., Song, Z., Xiao, C., Awni, R.A., Yao, C., Shrestha, N., Li, C., Bista, S.S., Zhang, Y., and Chen, L.: 'Arylammonium-Assisted Reduction of the Open-Circuit Voltage Deficit in Wide-Bandgap Perovskite Solar Cells: The Role of Suppressed Ion Migration', ACS Energy Lett., 2020, 5, (8), pp. 2560-2568.

[6] Li, C., Song, Z., Zhao, D., Xiao, C., Subedi, B., Shrestha, N., Junda, M.M., Wang, C., Jiang, C.-S., Al-Jassim, M., Ellingson, R.J., Podraza, N.J., Zhu, K., and Yan, Y.: 'Reducing Saturation-Current Density to Realize High-Efficiency Low-Bandgap Mixed Tin–Lead Halide Perovskite Solar Cells', Adv. Energy Mater., 2019, 9, (3), pp. 1803135.

[7] Li, C., Song, Z., Chen, C., Xiao, C., Subedi, B., Harvey, S.P., Shrestha, N., Subedi, K.K., Chen, L., Liu, D., Li, Y., Kim, Y.-W., Jiang, C.-s., Heben, M.J., Zhao, D., Ellingson, R.J., Podraza, N.J., Al-Jassim, M., and Yan, Y.: 'Low-bandgap mixed tin–lead iodide perovskites with reduced methylammonium for simultaneous enhancement of solar cell efficiency and stability', Nat. Energy, 2020, 5, (10), pp. 768-776.

[8] Song, Z., Zhao, D., Chen, C., Ahangharnejhad, R.H., Li, C., Ghimire, K., Podraza, N.J., Heben, M.J., Zhu, K., and Yan, Y.: 'Monolithic Two-Terminal All-Perovskite Tandem Solar Cells with Power Conversion Efficiency Exceeding 21%', in Editor (Ed.)^(Eds.): 'Book Monolithic Two-Terminal All-Perovskite Tandem Solar Cells with Power Conversion Efficiency Exceeding 21%' (2019, edn.), pp. 0743-0746

[9] Moot, T., Patel, J.B., McAndrews, G., Wolf, E.J., Morales, D., Gould, I.E., Rosales, B.A., Boyd, C.C., Wheeler, L.M., Parilla, P.A., Johnston, S.W., Schelhas, L.T., McGehee, M.D., and Luther, J.M.: 'Temperature Coefficients of Perovskite Photovoltaics for Energy Yield Calculations', ACS Energy Lett., 2021, 6, (5), pp. 2038-2047.

[10] Song, Z., Wang, C., Phillips, A.B., Grice, C.R., Zhao, D., Yu, Y., Chen, C., Li, C., Yin, X., Ellingson, R., Heben, M., and Yan, Y.: 'Probing the Origins of Photodegradation in Organic-Inorganic Metal Halide Perovskites with Time-Resolved Mass Spectrometry', Sustainable Energy Fuels, 2018, 2, (11), pp. 2460-2467.

[11] Song, Z., Shrestha, N., Watthage, S.C., Liyanage, G.K., Almutawah, Z.S., Ahangharnejhad, R.H., Phillips, A.B., Ellingson, R.J., and Heben, M.J.: 'Impact of Moisture on Photoexcited Charge Carrier Dynamics in Methylammonium Lead Halide Perovskites', J. Phys. Chem. Lett., 2018, 9, (21), pp. 6312-6320.

[12] Jošt, M., Lipovšek, B., Glažar, B., Al-Ashouri, A., Brecl, K., Matič, G., Magomedov, A., Getautis, V., Topič, M., and Albrecht, S.: 'Perovskite Solar Cells go Outdoors: Field Testing and Temperature Effects on Energy Yield', Adv. Energy Mater., 2020, 10, (25), pp. 2000454.

[13] Lamb, D.A., Irvine, S.J.C., Baker, M.A., Underwood, C.I., and Mardhani, S.: 'Thin film cadmium telluride solar cells on ultra-thin glass in low earth orbit—3 years of performance data on the AlSat-1N CubeSat mission', Prog. Photovoltaics, 2021, 29, (9), pp. 1000-1007.

Demonstrating the Thermoradiative Diode: Generating Electrical Power Through Radiative Emission

Nicholas J Ekins-Daukes, Michael P Nielsen, Andreas Pusch, Muhammad H Sazzad, Phoebe M Pearce, Peter J Reece

UNSW Sydney, Kensington, Australia

Thermoradiative power generation is achieved through the emission of light from a warm ambient into a cold surroundings representing a thermodynamically symmetric counterpart to photovoltaic solar power generation. The thermoradiative diode provides a semiconductor implementation of this process whereby radiative emission from a warm diode into a cold environment expels more entropy than supplied by the flow of heat to the diode, hence permitting work to be performed. Under these conditions a reverse bias spontaneously forms across the diode allowing an electrical current to flow and hence delivering electrical power. We report the full IV characteristics from a 0.3eV HgCdTe diode was held at 20.5C and exposed to a radiant surface of different temperatures. Thermoradiative power is generated when the diode faces a cold (< 20.5C surface) delivering a positive photocurrent and negative voltage. In the radiative limit, where all parasitic processes are eliminated, a thermoradiative diode exposed to the cold night sky could deliver electrical power densities of the order of tens of W/m2, offering the tantalizing prospect of generating useful quantities of electrical power. The linear behaviour of our IV curves is consistent with Auger mediated generation and recombination processes that reduce the power density for currently available commercial diodes to mW/cm2.

978-1-7281-6118-1/22 $31.00 © 2022 IEEE

Vertical Bifacial Solar Panels as a Candidate for Solar Canal Design

Jeremiah B Reagan, Sarah Kurtz

UC Merced, Merced, CA, United States

A vertical bifacial + reflector configuration is presented as a candidate for solar canal design. Simulations show output to be competitive with fixed 20° tilt systems, with South-facing vertical orientation showing 117% and 87% of annual output of South-facing 20° systems with and without a reflector, respectively. South-facing vertical orientations have better performance in non-summer months relative to other systems, resulting in a flatter seasonal curve, with useful implications for load balancing and energy storage. East- and West-facing vertical orientations outperform their fixed tilt defaults, even without a reflector, and tolerate higher DC/AC inverter ratios than similar South-facing vertical orientations before appreciable clipping effects are seen.

Arrhenius Analysis of the Degradation Modes in Emerging Photovoltaic Backsheets

Naila M. Al Hasan[1], Rachael Arnold[1], David C. Miller[1*], Jimmy Newkirk[1], Emily Rago[1], Michael Thuis[1], Bruce H. King[2], Laura T. Schelhas[1], Archana Sinha[3], Kent Terwilliger[1], Soňa Uličná[3], Peter Pasmans[4], Christopher Thellen[5]

[1]National Renewable Energy Laboratory, Golden, CO 80401 USA

[2]Sandia National Laboratories, Albuquerque, NM 87123, USA

[3]SLAC National Accelerator Laboratory, Menlo Park, CA 94025, USA

[4]Endurans Solar Solutions B.V., Urmond 6129, NL

[5]Endurans Solar, Nashua, NH 03063, USA

Abstract— With increasingly large-scale deployments of photovoltaic (PV) modules to meet global energy demands, interest in better establishing modules' lifespans is growing. We report on the performance of co-extruded polyolefin- (PO)-based backsheets as environmentally friendly alternatives to fluoropolymer-reinforced polyethylene-terephthalate (PET)-based backsheets using three hygrometric accelerated test conditions. After completing cumulative 4000 hours of aging, we analyzed data from electrical performance (I-V), surface roughness (gloss), and appearance (L, a*, b* color) characterizations to quantify degradation rates, quantify the corresponding activation energy, and cross-correlate between the characteristics examined, thereby providing insights into the relationship between physical characteristics and operating performance.

Keywords—BackFLIP, photovoltaic (PV), backsheet, damp heat testing, durability, DuraMAT, IEC TS 62788-7-2, PET, polyolefin, polyamide, UV weathering

I. BACKGROUND AND INTRODUCTION

Photovoltaic (PV) backsheets play an important role in the durability of PV modules. They are cost effective packaging materials with enhanced mechanical strength that also protect the electrical components of the modules against physical and chemical stresses such as UV radiation, temperature variation, humidity, electrical shunting, moisture, and fires. Backsheets themselves must withstand different environmental conditions during their service life outdoors [1]–[3]. Thus, their weathering stability is also critical.

Backsheets are often three-layer laminates consisting of an adhesive inner layer, an electrically insulating core layer, and a weather-resistant outer layer [4]. They have been observed in field inspections to be vulnerable to cracking and delamination, thereby compromising module insulation and performance, in addition to presenting unsafe working conditions for personnel maintaining PV systems [5]. Furthermore, hydrolysis of the core layer can generate acid species that may corrode metals present in interconnects as well as catalyze degradation of other

packaging layers like poly(ethylene-co-vinyl acetate) (EVA) [4].

Traditionally, backsheets have been made with an electrically insulating polyethylene-terephthalate (PET) core that is further layered with adhesives and fluoropolymers for improved durability. A common backsheet construction is laminated layers of PET film with poly-vinyl fluoride (PVF, named "Tedlar") in a three-layer PVF/PET/PVF configuration, or "TPT". Due to the difficulty of decomposing and recycling fluoropolymers, efforts have been made to develop fluorine-free backsheets. Polyolefins (POs) have been shown to be environmentally friendly alternatives to PET [6], where when POs are used, backsheets can be co-extruded, reducing the risk of delamination and the cost of fabrication.

Using steady state aging experiments, we investigate the durability of developmental PO-based backsheets relative to the traditional PET- and polyamide (PA)-based backsheets on the market. To evaluate the degradation modes and rates of degradation of commercial and novel backsheets, we performed accelerated stress experiments in a double-blind study. Seven different backsheets in coupon and mini-module (MiMo) form were subjected to three hygrometric and four photolytic aging conditions for 4000 hours[7]. Degradation models were developed by correlating accelerated aging with changes in electrical performance, appearance, mechanical performance, polymer chemical structure, crystal structure, and degree of crystallinity. Accelerated testing will be compared with natural weathering in Albuquerque, NM, and Cocoa, FL, to identify characteristics common to artificial and natural aging so we can predict degradation. For conciseness, this abstract focuses on the Arrhenius modeling of thermal degradation of backsheets induced in the three hygrometric tests.

II. EXPERIMENTAL

A. Test Specimens

Seven different backsheets from three groups of polymers for the core layer were used to fabricate coupon and MiMo samples. **Error! Reference source not found.** lists the b acksheets studied: three traditional ("TPT," "PPE," and "KPf"); one known bad ("AAA"); and three novel PO-based backsheets (PO-1, PO-2, and APO). PPE is a laminate of pigmented PET/unpigmented PET/EVA; KPf is a laminate of polyvinylidene fluoride (PVDF)/PET/fluorinated coating; and AAA is a coextruded backsheet composed of PA/blended PA and polypropylene (PP)/PA layers.

TABLE 1: POLYMER BACKSHEETS INVESTIGATED IN THIS STUDY.

Arbitrary Index	Backsheet	Construction	Comment
BS-1	PO-1	Coextruded	New
BS-2	PO-2	Coextruded	New
BS-3	TPT	Laminate	Traditional (reference)
BS-4	APO	Coextruded	New
BS-5	PPE	Laminate	Traditional (contemporary)
BS-6	AAA	Coextruded	Known bad
BS-7	KPf	Laminate	Traditional (contemporary)

MiMos were prepared using non-tempered float glass (Planibel Clearvision, AGC Inc.) with no antireflective coating. A single 156-mm, stabilized Si-Cz, p-PERC solar cell (diced into four equal pieces) was connected with ribbon to an edge-mounted junction box. EVA encapsulant of a UV-transparent front layer and a UV-blocking rear layer was laminated to attach the MiMo components. They were first outdoor light-soaked and then verified using electroluminescence (EL) imaging and electrical performance (I-V) measurements.

Coupon specimens consisted of polymer backsheet that was run through the same lamination process to provide the same thermal history as the MiMos.

B. Accelerated Testing

In this abstract, we focus on samples weathered in the three hygrometric accelerated test conditions listed in Table 2. Combined temperature/humidity chambers (e.g., BTX-475, ESPEC North America Inc.) were used, and weathering was performed at read points of 0, 1000, 2000, 3000, and 4000 hours of cumulative duration.

TABLE 1. HYGROMETRIC TEST CONDITIONS EVALUATED IN THIS STUDY.

Arbitrary Experiment Index	UV Irradiance (Wm^{-2} at 340 nm)	Temperature (°C)	Relative Humidity (%)
1	0	85	85
2	0	65	85
3	0	45	85

C. Characterization Methods

As described in Ref. [7], MiMo and coupon samples were characterized for: color (L, a*, and b*), gloss (at 20°, 60°, and 85°), visual appearance, optical microscopy, wide-angle X-ray scattering (WAXS) of the polymer crystalline structure, and Fourier transform infrared (FTIR) polymer chemical structure. Electrical performance of the MiMos was characterized using I-V flash testing (at 1000 W·m^{-2}) followed by EL imaging. Destructive characterization, including mechanical tensile testing, DC breakdown voltage, and differential scanning calorimetry (DSC) of the crystalline content, were conducted after cutting the coupons into smaller specimens. For this presentation, we will focus on the I-V characteristics.

D. Arrhenius Modeling of Degradation

The degradation of certain characteristics is thermally activated, and the effective rate-limiting reaction of a degradation mode can be modeled with an Arrhenius fit:

$$k = A e^{\left[\frac{-E_a}{RT}\right]} \quad (1)$$

where k is the degradation rate (units s^{-1}), A is the frequency factor (units s^{-1}), E_a is the activation energy (kJ mol^{-1}), R is the gas constant (8.3145 J mol^{-1}K^{-1}), and T is the temperature (K). Where applicable, the change in maximum power will be evaluated to determine the activation energy of degradation induced by the accelerated tests.

III. RESULTS AND DISCUSSION

The electrical performance, as change in maximum power ΔP_{max}, averaged for all seven backsheets over 4000 hours in the three hygrometric test conditions, is shown in Fig. 1. P_{max} decreases with increased exposure time with the most damage observed in test condition 1 at 85 °C and 85% relative humidity. P_{max} degradation is similar up to 2000 hours, beyond which PET-based backsheets 3, 5, and 7 degrade significantly more (~3x), indicated by a steeper inflection. From analysis of the

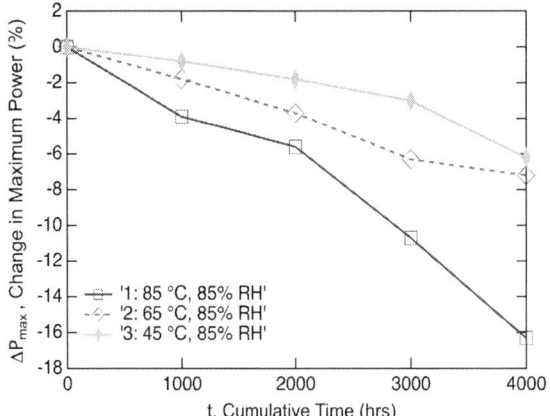

Fig. 1. Performance, ΔP_{max}, averaged for all seven backsheets, in the three hygrometric aging experiments. It decreases with increased exposure time for all hygrometric tests, with test 1 at 85 °C and 85% relative humidity being the most damaging. Modified from Ref [7].

series and shunt resistance, performance degradation (ΔP_{max}) results from corrosion of the busbars, ribbons, and solder in addition to damage to the cell [7].

Hygrometric degradation of P_{max} is thermally activated. Thus, the rate-limiting reaction can be modeled with an Arrhenius fit (Equation 1). Using backsheet 4 (coextruded APO) MiMo as an example, we consider the linear relationship of change in P_{max} with time (Fig. 2). From Equation 1, the slopes of these lines are then analyzed to determine their relationship with inverse temperature (of the three hygrometric test conditions, as shown in the Fig. 2 inset). The slope in the Fig. 2 inset is multiplied by R to obtain E_a for the P_{max} degradation. The E_a values for the MiMos of each backsheet are listed in Table 3. It is important to note that as ΔP_{max} of backsheets 3, 5, and 7 degraded more beyond 2000 hours, a different rate and, therefore, a different mode of degradation was assumed. Hence, data points beyond 2000 hours were not

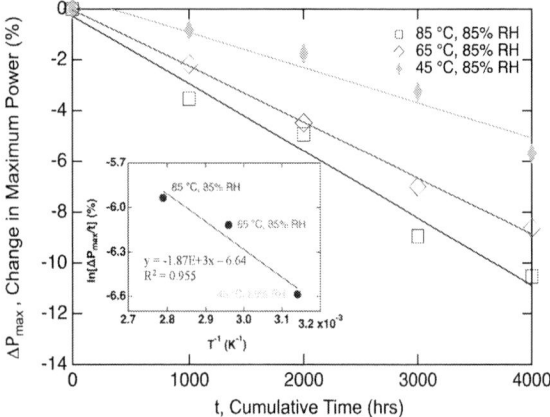

Fig. 2. Change in maximum power, ΔP_{max}, of the recent PO-based backsheet 4 over time in the three hygrometric tests. Inset: Change in natural log of the slopes from Fig. 2 with inverse temperature of the three hygrometric tests. Activation energy, E_a, for P_{max} is obtained from the slope of this graph.

considered for these backsheets.

From the table, E_a ranges from 15 kJ mol^{-1} to 28 kJ mol^{-1}, being similar within the range of variation for two standard deviations. The activation energy of diffusion of water has been reported to be 18 kJ mol^{-1} [8] and 129 kJ mol^{-1} for hydrolysis of PET [9]. For all backsheets, E_a is consistent with the diffusion of water. This suggests ΔP_{max} in Fig. 1 results from the mass transport of water through the backsheets, rather than the degradation of the backsheet materials. Minor variation in E_a in Table 3 may result from differences in the backsheets (diffusivity of water, porosity, etc).

IV. SUMMARY

To extend the lifespan of PV modules, it is important to understand the long-term effects of operating conditions on physical characteristics and thus on, PV module performance. Here, we have described initial calculations of E_a for P_{max} degradation of new and traditional backsheets after 4000 hours

of accelerated hygrometric aging. The BackFLIP study will compare E_a for each of the characteristics examined (gloss, color, etc) so that the effect of each degradation mode can be predicted and compared relative to MiMo performance. The most significantly affected characteristics and test conditions will be identified and correlated to provide a comprehensive picture of backsheet degradation.

TABLE 2. E_a FROM P_{MAX} IN FIG. 1.

	Backsheet	E_a, effective activation energy [kJ mol^{-1}]
1	PO-1	15.3 ± 15
2	PO-2	19.7 ± 15
3	TPT	23.9 ± 13
4	APO	15.6 ± 15
5	PPE	20.5 ± 14
6	AAA	28.4 ± 15
7	KPf	10.1 ± 14

REFERENCES

[1] W. Gambogi, S. Kurian, B. Hamzavytehrany, A. Bradley, and J. Trout, "The Role of Backsheet in PV Module Performance and Durability," 2011.

[2] Z. Xia, J. H. Wohlgemuth, and D. W. Cunningham, "A semi-empirical method of predicting the lifetime of EVA encapsulant and polyester based backsheet materials," in *Reliability of Photovoltaic Cells, Modules, Components, and Systems II*, 2009, vol. 7412. doi: 10.1117/12.825472.

[3] A. W. Czanderna and F. J. Pern, "Encapsulation of PV modules using ethylene vinyl acetate copolymer as a pottant: A critical review," *Solar Energy Materials and Solar Cells*, vol. 43, no. 2, 1996, doi: 10.1016/0927-0248(95)00150-6.

[4] Y. Lyu *et al.*, "Drivers for the cracking of multilayer polyamide-based backsheets in field photovoltaic modules: In-depth degradation mapping analysis," *Progress in Photovoltaics: Research and Applications*, vol. 28, no. 7, pp. 704–716, Jul. 2020, doi: 10.1002/pip.3260.

[5] Y. Voronko, G. C. Eder, M. Knausz, G. Oreski, T. Koch, and K. A. Berger, "Correlation of the loss in photovoltaic module performance with the ageing behaviour of the backsheets used," *Progress in Photovoltaics: Research and Applications*, vol. 23, no. 11, 2015, doi: 10.1002/pip.2580.

[6] A. Omazic *et al.*, "Increased reliability of modified polyolefin backsheet over commonly used polyester backsheets for crystalline PV modules," *Journal of Applied Polymer Science*, vol. 137, no. 30, 2020, doi: 10.1002/app.48899.

[7] M. Thuis *et al.*, "A Comparison of Emerging Nonfluoropolymer-Based Coextruded PV Backsheets to Industry-Benchmark Technologies," *IEEE Journal of Photovoltaics*, pp. 1–9, 2021, doi: 10.1109/JPHOTOV.2021.3117915.

[8] J. Konya and N. M. Nagy, "Physicochemical Application of Radiotracer Methods," *Nuclear and Radiochemistry: Second Edition*, pp. 247–286, 2018, doi: 10.1016/B978-0-12-813643-0.00009-3.

[9] J. E. Pickett and D. J. Coyle, "Hydrolysis kinetics of condensation polymers under humidity aging conditions," *Polymer Degradation and Stability*, vol. 98, no. 7, pp. 1311–1320, Jul. 2013, doi: 10.1016/J.POLYMDEGRADSTAB.2013.04.001.

Comparison of Measured and Modeled Snow Losses for Photovoltaic Systems in Colorado

Owen W. Westbrook[1], Sara M. MacAlpine[2], and David A. Bowersox[2]

[1]Sunrise Technologies LLC, Louisville, CO, 80027, USA
[2]juwi Inc., Boulder, CO, 80301, USA

Abstract—We quantify measured and modeled snow losses at four utility-scale single-axis tracking photovoltaic (PV) power plants in Colorado. Across 50 site-months of data collected over three winters, the Marion and Townsend snow loss models exhibited similar absolute bias errors, although the Townsend model had lower monthly root mean square error. Based on our results, we recommend that, for PV systems in Colorado and similar climates, the Townsend and Marion model predictions be averaged together to generate the most accurate snow loss predictions for monofacial tracking PV facilities, and that solely the Townsend model be used for bifacial tracking PV facilities.

Keywords— photovoltaic systems, power system modeling, solar energy, solar power generation, system performance.

I. INTRODUCTION

Snow losses are a major contributor of uncertainty when forecasting energy production for large-scale solar photovoltaic ("PV") systems in cold climates. Widely used existing models for snow loss prediction, such as the Marion model [1] and the Townsend model [2], can provide acceptable annual results but offer poor monthly accuracy and have not been thoroughly validated for single-axis tracking systems. We employed Sunrise Technologies' proprietary PRISM® software [3] to detect and quantify snow losses for four utility-scale PV facilities in Colorado using a combination of historical operating data and National Oceanic and Atmospheric Administration ("NOAA") daily weather data. We then compared the detected snow losses to the predictions of the Marion and the Townsend models.

II. DATA AND ANALYSIS METHODS

A. Photovoltaic Facilities

The four PV facilities ("projects") in this study all feature horizontal single-axis trackers with a ±60° range of motion and monofacial crystalline silicon ("c-Si") modules in a one-high portrait mounting configuration. The four projects cover a diverse geographic range within the state. One project (Project D) is located in a mountain community within the Rocky Mountains, and three are located along the Front Range. The Front Range projects span 250 miles from north of Fort Collins to south of Pueblo. Table I summarizes the projects.

B. Data

The analysis data set consisted of ten-minute interpolated performance data collected by the projects' SCADA systems between 2018 and 2021 (the "Measured Data"). For each project, the Measured Data included average plane-of-array ("POA") irradiance and global horizontal irradiance ("GHI") measured by one or more Class A pyranometer(s), average module temperature measured by three or more RTD sensors, ambient temperature, wind speed, and relative humidity measured by on-site weather station(s), and revenue meter and inverter AC power.

Project B contains a test bed with six monofacial and four bifacial c-Si module strings installed side by side and connected to string-level DC optimizers collecting one-minute DC input power measurements. These data were used to compare monofacial and bifacial snow losses. The monofacial and bifacial modules were produced by different manufacturers but shared the same front-side nameplate power rating, 72-cell design, and form factor.

In addition, we utilized daily NOAA precipitation, snowfall, and snow depth data from nearby weather stations. To run the PRISM snow detection algorithm, we linearly interpolated the NOAA snow depth values to the same ten-minute interval of the Measured Data and distributed the daily precipitation and snowfall totals evenly at ten-minute intervals over the preceding 24-hour period. For the Marion model, snow depth was interpolated by backfilling NOAA snow depth values over the preceding 24-hour period, in alignment with the methods described in [1].

TABLE I. SUMMARY OF PHOTOVOLTAIC FACILITIES STUDIED

Project Designation	Closest Municipality	DC Capacity (MW)	Commercial Operations Date
Project A	Wellington	36.2	Oct 2016
Project B	Fountain	81.7	Apr 2020
Project C	Trinidad	37.9	Dec 2016
Project D	Buena Vista	2.7	Jan 2019

C. PRISM Snow Loss Detection

Sunrise Technologies has developed a suite of software tools, collectively known as PRISM®, built from both open-source PV modeling algorithms in the pvlib library [4] as well as proprietary algorithms. Designed to work with measured PV performance data, PRISM has been used to analyze more than 60 PV plants with an aggregate capacity of more than 1.5 GW. PRISM's performance modeling engine can incorporate a PV facility's original model assumptions to produce a weather-

978-1-7281-6118-1/22 $31.00 © 2022 IEEE

adjusted performance benchmark based on sub-hourly measurements of irradiance, temperature, and wind speed.

PRISM compares measured power production to a modeled power prediction to quantify facility energy losses. PRISM classifies these losses into multiple categories, including transposition and module temperature model error, inverter availability, curtailment, and snow/soiling loss. After loss classification, the remaining measured versus model energy delta is designated "uncategorized." PRISM's algorithms incorporate snowfall and snow depth data from NOAA to identify snow-impacted periods. Snow losses are then quantified according to the delta between measured and modeled AC power after accounting for other loss factors. We processed the Measured Data with PRISM in monthly increments to detect and quantify snow losses for each month.

D. Modeled Snow Losses

We generated snow loss predictions within PRISM using pvlib's Marion model implementation, run at the ten-minute interval of the Measured Data. Inputs to the Marion model included POA irradiance and ambient temperature from the Measured Data and snow depth data from the nearby NOAA weather stations processed as described above. We used the default coefficients of the Marion model as implemented in pvlib, except that the sliding coefficient, which determines the fraction of snow that slide off in an hour, was set to 0.6 per the reported value for ground-mounted systems in [1].

We ran the Townsend model at monthly intervals using the following inputs, summed or averaged over the month as appropriate: NOAA daily snowfall totals, NOAA days with snow events >0.01 inch, and POA irradiance, ambient temperature, and relative humidity from the Measured Data. We used the coefficient values specified in [2] for all Townsend model parameters.

III. RESULTS AND DISCUSSION

A. Monofacial PV System Summary Results

We completed snow loss detections with PRISM and snow loss predictions with the Marion and Townsend models for a total of 50 site-months of data across the four projects. One month for Project A and two months for Project D during which plant-wide availability issues coincided with snowfall were excluded from this final data set. The final data set included at least six months of data and at least one full winter season for each facility.

Table II summarizes the POA irradiance-weighted monthly average snow losses detected with PRISM and predicted with the Marion and Townsend models for these four Colorado projects. On an irradiance-weighted basis, detected snow losses ranged from 1.5% to 2.8% at each plant. Across all projects, the average of the irradiance-weighted monthly average snow losses was 2.2%, within 1% of the Townsend model's prediction of 1.3% and the Marion model's prediction of 3.1%. The Townsend model had the lowest mean bias error for the three Front Range projects (facilities A, B, and C), while the Marion model had the lowest mean bias error for Project D.

Figures 1 and 2 show the dispersion in detected and modeled losses for Marion and Townsend, respectively. The Townsend

loss predictions cluster more closely to the 1:1 line relative to detected losses than do the Marion loss predictions. Monthly snow loss root mean square error ("RMSE") was 2.7% for Marion compared to 2.1% for Townsend.

TABLE II. COLORADO MEASURED VERSUS MODELED SNOW LOSSES

Project Designation	PRISM Detected Loss (%)	Marion Predicted Loss (%)	Townsend Predicted Loss (%)
Project A	1.5%	2.0%	1.3%
Project B	2.8%	4.5%	1.4%
Project C	1.8%	2.6%	1.1%
Project D	2.6%	3.2%	1.2%
All Site Avg.	**2.2%**	**3.1%**	**1.3%**

Fig. 1. Detected snow losses vs. Marion model predictions, all site-months.

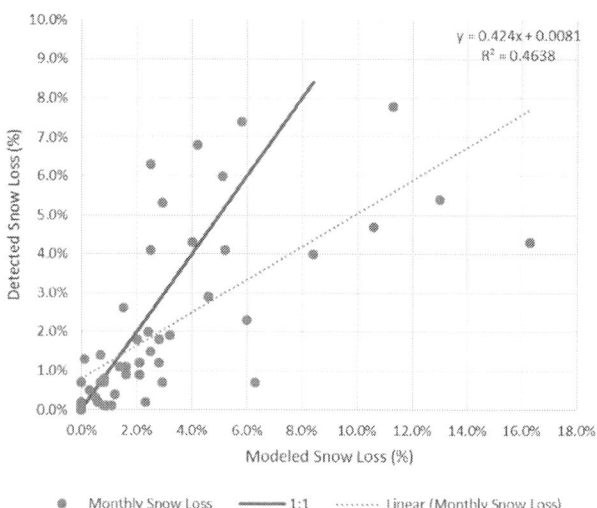

Fig. 2. Detected snow losses vs. Townsend model predictions, all site-months.

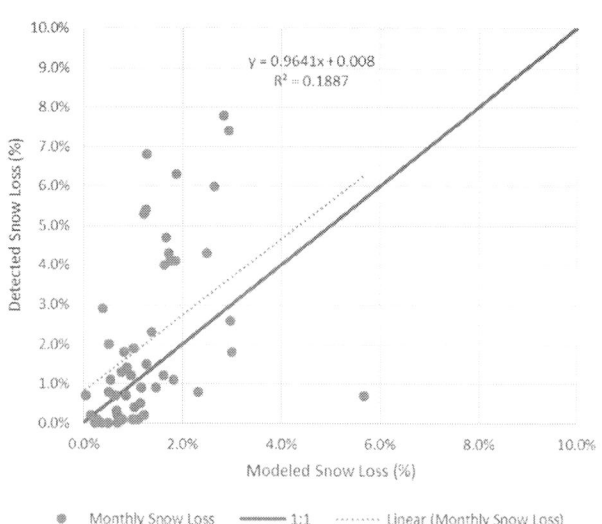

B. Monofacial Versus Bifacial Snow Losses

To examine differences in snow losses for bifacial versus monofacial modules, we analyzed the side-by-side monofacial and bifacial string-level data collected at Project B using an adaptation of the snow loss detection methodology employed at the facility level. To quantify losses, the PRISM software was used to generate expected module-level power values based on facility environmental data. No adjustments were made to the original model monthly average albedo values when calculating the expected bifacial module energy.

For each month, detected monofacial snow losses were within 0.6% of the losses found for the entire project, demonstrating that the snow detection methodology returns consistent loss values on both the string and facility level. Bifacial strings on both interior and exterior tracker rows experienced significantly lower production losses than the monofacial strings (Fig. 3), and snow losses for the bifacial strings were far lower than the losses predicted by either the Marion or Townsend models. Over the entire six-month measurement period, the irradiance-weighted losses for the interior bifacial strings were only 0.8%, compared to 2.8% for the facility as a whole, 4.5% predicted by the Marion model, and 1.4% predicted by Townsend. An inspection of daily power trends suggests that bifacial snow losses were lower than monofacial losses at Project B because the snow physically cleared earlier from the bifacial strings, allowing for faster power recovery after a snow event.

Fig. 3. Detected snow losses for the monofacial/bifacial test bed at Project B.

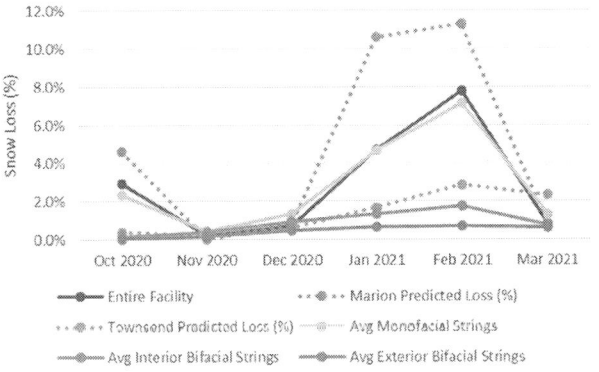

C. PRISM Snow Detection Accuracy

We performed a detailed review of the PRISM results for each month to verify that snow losses were detected accurately and completely. Any facility performance shortfall that PRISM cannot classify is marked as "uncategorized." Should PRISM underestimate losses from snow, then significant uncategorized losses would appear in snow-impacted periods. By visual inspection of PRISM's daily performance breakdown charts for each site-month studied, we estimate that PRISM's monthly snow loss detections were biased low by at most approximately 0.1% on average. In the worst-case month, potential undetected snow losses reached 1.1%, but 39 of 50 site-months had no more than 0.2% undetected losses.

D. Limitations of the Marion Model

The Marion model offers several advantages over the Townsend model, including validation in a greater variety of climates and the capability of predicting time series snow losses rather than only monthly average losses. However, we consistently observed the Marion model to over predict snow losses in this study. This discrepancy can most likely be traced to differences in system design between the tracking facilities analyzed in this work and the fixed-tilt systems studied in [1]. The parameters of the Marion model related to snow shedding were derived from three rooftop systems in Wisconsin, two rooftop systems in Colorado, and one 30° fixed-tilt ground-mount system in Colorado. The Marion model "is expected to give good results…for [tilt] values of 10° to 45°" [1] and may not capture effects for trackers with a ±60° range of motion.

The Marion model has also not been validated against bifacial system performance data. As the data presented above suggest, bifacial modules may experience significantly lower snow losses not only due to their rear-side power production capabilities but also from increased snow clearing efficiency.

Overall, we consider the basic principles behind the Marion model to be sound. Further tuning and validation of the Marion model parameters might improve loss prediction accuracy for one-high portrait tracking facilities and for bifacial systems.

IV. CONCLUSIONS

We compared measured and modeled snow losses at four Colorado utility-scale single-axis tracking PV facilities built with monofacial c-Si modules. The Townsend model's snow loss predictions had a lower monthly RMSE than did the Marion model, but both models had similar absolute bias errors. The Townsend model's loss predictions were biased low for every site, while the Marion model's losses were biased high. For the data examined in this study, averaging the predictions of the Townsend and Marion models would reduce the average monthly bias error to less than 0.1%. We therefore recommend that the Townsend and Marion model predictions be averaged together to generate the most accurate snow loss predictions for monofacial tracking PV facilities in Colorado and similar sunny, dry, high-altitude climates. We also found that at a test bed of mixed bifacial and monofacial strings within one of the projects, the bifacial strings experienced significantly lower snow losses than the monofacial strings. For bifacial tracking PV facilities in climates similar to Colorado's, we recommend solely utilizing the Townsend model to predict snow losses.

REFERENCES

[1] B. Marion, R. Schaefer, H. Caine, and G. Sanchez, "Measured and modeled photovoltaic system energy losses from snow for Colorado and Wisconsin locations," Solar Energy, Vol. 97, pp.112-121, 2013.

[2] T. Townsend and L. Powers, "Photovoltaics and snow: An update from two winters of measurements in the Sierra," in 37th IEEE Photovoltaic Specialists Conference, 2011.

[3] "PRISM | Sunrise Technologies," Retrieved from https://www.sunrise-tech-llc.com/prism on January 13, 2021.

[4] W. Holmgren, C. Hansen, and M. Mikofski, "pvlib Python: A python package for modeling solar energy systems." Journal of Open Source Software, vol. 3(29), p. 884, 2018.

PV Hosting Capacity Estimation: Experiences with Scalable Framework

Wenbo Wang, Daniel Thom, Kwami Senam Sedzro, Sherin Ann Abraham, Yiyun Yao, Jianli Gu, Shibani Ghosh

Energy Systems Integration
National Renewable Energy Laboratory
Golden, CO, USA
{Wenbo.Wang, Daniel.Thom, Kwami.Sedzro, SherinAnn.Abraham, Yiyun.Yao, Jianli.Gu, Shibani.Ghosh}@nrel.gov

Abstract—Hosting capacity is an indication of the amount of solar photovoltaics (PV) that can be hosted in a distribution system without additional changes to infrastructure or operations. This paper presents a framework for estimating the PV hosting capacity at scale. First, we analyze computational, modeling and other key challenges of performing relevant, large-scale simulations, provided along with the experiences and lessons learned. Then, we develop two open-source Python-based software tools to conduct repeatable distribution analyses: the Distribution Integration Solution Cost Options (DISCO) for configuring and analyzing simulations and the Job Automation and Deployment Engine (JADE) for parallelizing jobs on high-performance computing clusters. A case study of hosting capacity estimation for the SMART-DS San Francisco (SFO) 2000+ synthetic feeders, is used to demonstrate the capability of the developed DISCO+JADE framework and tools. The framework and tools can help utilities assess the overall hosting capacity of their service territory, which can help them better plan for the overall upgrade costs to integrate more PV in the future. The experiences are shared to aid the tool users and researchers to conduct relevant studies and research.

Index Terms—Distributed energy resources, Hosting capacity, Large-scale simulation

I. INTRODUCTION

The fast deployment of distributed solar photovoltaics (PV) stretches the electric grid toward limitations and creates operational concerns for utilities. The grid's ability to accommodate PV is typically estimated through hosting capacity. In the field, PV interconnection screening processes are often evaluated based on the understanding of the feeder hosting capacity [1]. The concept of hosting capacity is defined as the total PV capacity that can be accommodated on a given feeder without violating operational constraints [2]. Note that in this paper, we focus on distributed PV (DPV); thus PV and DPV are used interchangeably.

Estimating hosting capacity normally involves simulating many scenarios of different locations and sizes of PV to evaluate the impact and identify boundary scenarios from operational violations. This can be done through steady-state or time-series simulations, and the corresponding results are often called static or dynamic hosting capacity (SHC or DHC), respectively. Using steady-state analysis, [3] considers a limited number of scenarios, including PV distributed evenly, aggregated near the beginning, and aggregated near the end of feeders; [4] proposes to estimate hosting capacity

more comprehensively based on stochastic analysis, generating many scenarios and penetration levels for PV deployments. Reference [5] proposes using year-long time-series simulations to estimate hosting capacity, where the duration of violations and the movement of the legacy controllers can be captured. However, because of the large number of scenarios that need to be considered, past research mostly focused on a limited number of feeders. It would be beneficial to have a flexible and scalable framework for estimating feeder hosting capacity, e.g., with thousands of feeders and potentially millions of scenarios. This will help utilities better plan for the overall upgrade costs to integrate more PV [6], and it will facilitate data analytics of the interconnection process [7]. To estimate hosting capacity at scale, one can (but is not limited to) use a smaller number of scenarios for more feeders [7], speed up each simulation [8], optimize the computing execution, and use more computing power [9].

Setting this work apart from previous hosting capacity research, this paper develops a scalable hosting capacity solution through optimizing models and the computing execution. The main contributions can be summarized as follows:

- A framework is developed for scalable hosting capacity estimation. As a byproduct, two open-source Python-based software tools are developed for conducting repeatable distribution analyses and simulation job submission: namely, the Distribution Integration Solution Cost Options (DISCO) [10] and the Job Automation and Deployment Engine (JADE) [11], respectively.
- The challenges and experiences of scalable hosting capacity estimation are analyzed and discussed, including computational and modeling challenges.
- The capability of the developed framework and tools are demonstrated through the SMART-DS San Francisco (SFO) region 2000+ feeders [12].

II. HOSTING CAPACITY ESTIMATION

A. Methodology

This paper uses a Monte Carlo-based stochastic approach [4], [13] to estimate hosting capacity. Fig. 1 shows the analysis flow; both SHC and DHC [5] are shown. Starting with the feeder models and weather data, the stochastic approach is used to generate the PV deployment scenarios at different

TABLE I
SHC METRICS

Metric	Threshold
Voltage	± 5% deviation from the nominal value
Thermal	100% asset loading
Power quality	Voltage unbalance 3%, etc.
Protections	Coordination, set points (false/miss detection)

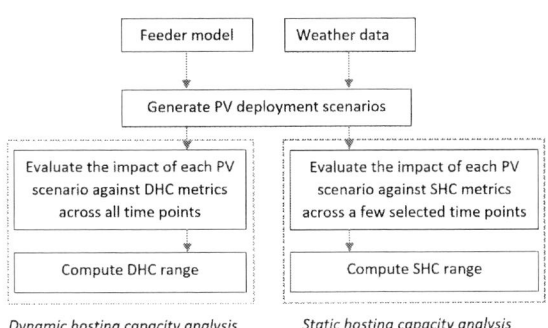

Fig. 1. Hosting capacity analysis flowchart

penetration levels and for a diverse spatial distribution according to the technique introduced in [14]. Next, the impact of each PV scenario is independently assessed with regard to operational metrics and thresholds. Examples of SHC metrics and thresholds are shown in Table I [2], [13]. Based on the impact assessment, the hosting capacities of the system under study are determined.

In addition, because of the stochastic nature of the PV deployment (location, size), the analysis typically results in a range of hosting capacities for each system, which are characterized by minimum and maximum hosting capacities [4].

1) Generating PV Deployments: Because it is nearly impossible to perfectly predict the adoption pattern of PV in terms of location and size distribution, we develop several adoption patterns or deployment samples to capture diversified and realistic PV scenarios [14]. The developed process considers three spatial placement types: close, random, and far. In each spatial placement type, the deployment of PV is incremental from one penetration level to next; therefore, each PV scenario is uniquely identified by its placement, sample, and penetration level.

2) Multi-Time-Point SHC Analysis: Instead of a single snapshot, the multi-time-point SHC analysis considers several grid conditions that are often used in grid planning studies. The most common four conditions include minimum daytime load, maximum PV output, minimum net load, and maximum load, which are extracted from load and solar irradiance profiles. The thermal and voltage impacts of integrating each PV scenario into the distribution grid are assessed for these conditions. In the end, the worst-case results from the four conditions are selected.

3) DHC Analysis: Unlike the multi-time-point SHC analysis, which evaluates the grid impact at a few selected time points, the DHC analysis assesses the impact of each PV scenario across year-long time series. DHC analysis allows violations for a short duration, and it can track moving averages and device operation counts. More details on the DHC metrics and suggested thresholds can be found in [5].

B. Challenges and experiences of Estimating Hosting Capacity at Scale

This subsection lists the challenges and considerations of estimating hosting capacity at scale. In each listed challenge or consideration, it is also provided alongside with a description of how the developed framework (i.e. DISCO, JADE tools) approaches the them.

1) Computational Challenge: The main challenge is how to use computational resources efficiently to manage large numbers of jobs under the constraints of central processing unit (CPU) cores, memory, and storage space. Here we consider the use of high-performance computing, with access to multiple compute nodes simultaneously. Ideally, one can use as many as compute nodes as are available; however, not only are the resources limited, but also, in many cases, the benefit of using more nodes is outpaced by the burden of the communication among nodes [9]. Conveniently, in the case of estimating hosting capacity, the PV deployment scenarios can be run independently (naturally partitioned) with little communication required (data dependence and synchronization) among nodes.

2) Modeling Challenge: This is related to the standardization of the data and models. Models can take many forms, and it is impractical to support all models. DISCO [10] defines standard models and then provides transformations for specific formats. For example, DISCO can run simulation both at the feeder level and the substation transformer level. Also, note that the actual power flow is conducted through OpenDSS [15]; DISCO leverages PyDSS [16], an OpenDSS Python wrapper that provides PV control functions with enhanced convergence (i.e., volt-var, volt-watt) and many other functions.

3) Other Practical Considerations:

- Computational burden load balancing refers to the practice of distributing approximately equal amounts of work among processors so that all processors are kept busy all the time [17]; otherwise, the slowest task will determine the overall run time. In the case of estimating hosting capacity, different feeders with different numbers of circuit elements that create challenges to computational load balancing (not to be confused with the load in customer demand in kilowatts). In DISCO, a linear regression model is developed to predict the job run times with the predictor variables that include the numbers of PV units and circuit elements. Based on the predicted run time, the jobs can be batched and allocated roughly evenly to the processors. This linear regression model builds its estimates based on a training dataset created by dry run jobs.

- Often, not all jobs will successfully run the first time due to issues including model errors, computational limits, or convergence challenges. The capability to test run, debug, and rerun failed or missing jobs is critical for managing large numbers of jobs. Both DISCO and JADE have comprehensive logging functionalities for each steps of the simulation providing meaningful debug information to the users and developers. In addition, JADE records status for each simulation job (i.e. pass or fail), and has a function that directly re-submit the failed jobs.

- Job monitoring and reporting are important because the execution information can be used for tuning the simulation parameters such as required computational nodes on HPC, simulation wall time, job batch size per computational node, etc. The reported metrics for job execution in JADE include individual job status, errors and events, job execution times, and compute resource utilization statistics such as CPU and memory usage, and networked communication related metrics[1] (e.g. time consumed transmitting packets from CPU to hard drives, hard drives to CPU).

- Care is required in data architecture and formats when working with large quantities of input and output data. Data storing, query, sharing are critical to data management and analytics. DISCO has build-in function to ingest raw output of the simulation results into a SQLite database. The current database schema are designed for distribution impact analysis, e.g. hosting capacity.

III. DEVELOPED OPEN-SOURCE TOOLS

A. Hosting Capacity Estimation with DISCO

DISCO—an NREL-developed, open-source tool—is a collection of integrated functions that can be used to automate a wide range of electric distribution analyses at scale. For instance in the LA100 distribution analysis effort, it was used to conduct impact analysis and estimate upgrade costs for thousands of feeders with hundreds of scenarios each [18]. Here we focus on its use for distributed PV hosting capacity estimates. Fig. 2 shows the flowchart for the main steps to run the hosting capacity estimation. These blocks are briefly described as follows, and more details about the implementation and examples can be found in [10]:

- Prepare the OpenDSS models and directory structures according to the data sources defined by DISCO, then provide the input path to DISCO. Four types of data sources are currently supported.

- Transform the source of the OpenDSS models into DISCO models. In the case of hosting capacity, the transformed DISCO models include PV deployment scenarios and OpenDSS instances through PyDSS with PV control enabled as well as functions such as selectively saving simulation results. These functions and the files

are described by a JSON file, which is the output of this step.

- Configure the JADE jobs based on the DISCO models with customized execution requirements. Execution on a high-performance computer (HPC) is highly configurable depending on the job resource requirements, e.g., the number of computational nodes to use, the number of jobs to run in parallel on each node. The output is an updated JSON file from the previous step with the added entries including all the job configuration information.

- Submit the jobs with JADE based on the JSON file. Underneath, JADE uses subprocess management [19] to parallelize the execution of the jobs on either HPC clusters (with Slurm [20]) or stand-alone computers. The submitted jobs will be run once the requested resources become available.

- After the jobs are complete, JADE can assist with the execution analysis by showing summaries of the individual job status, errors and events, job execution times, and compute resource utilization statistics. DISCO provides simulation results analysis and certain visualizations.

B. JADE for Submitting Jobs

JADE [11] automates the parallelized execution of jobs. It has specific support for distributing work on HPC compute nodes, but it can also be executed on stand-alone computers. Some important features are described as follows; for more information, see [11]:

- Maximizing the number of jobs that can be completed on a given node in a specific time duration is critical to optimize jobs on HPC systems, even more so if the HPC systems are managed such that the computational nodes are typically allocated for a limited period of time and are not always available. JADE constructs per-node job batches by accounting for job duration, number of required and available CPUs, and allocation time to maximize the use of each node. JADE allows customization of all parameters.

- For job monitoring and reporting, for example, after the simulation jobs are submitted, JADE provides ways to monitor the simulation status and results, find failed jobs, and restart them.

- For pipeline capability, JADE allows users to specify inter-job dependencies and pipeline stages to submit all work in one step. JADE implements a distributed submitter protocol[2] whereby a node can submit new jobs once dependent jobs are complete. This obviates the need to monitor jobs from a software application that must remain running for the duration of the work, which can take multiple days or weeks, depending on node availability.

IV. CASE STUDIES

The capability of the developed framework is demonstrated on the SMART-DS synthetic SFO 2000+ feeders [12]. This

---

[1]https://nrel.github.io/jade/tutorial.html#resource-monitoring

[2]https://nrel.github.io/jade/distributed_submission.html

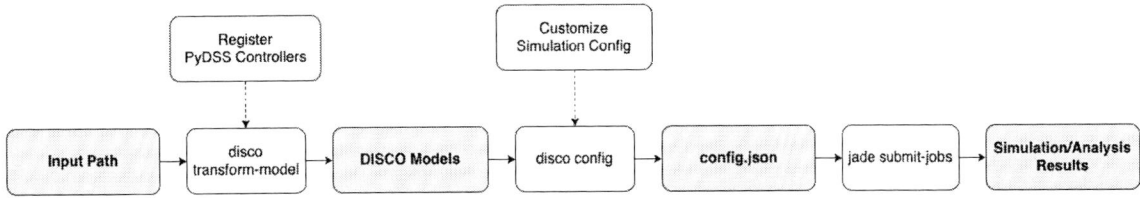

Fig. 2. DISCO workflow: bold fonts with grey boxes indicate data input and output; and dashed arrows indicate customized configurations and features.

section analyzes the large-scale simulations and results. We ran simulations on NREL's Eagle HPC. The HPC contains 2500+ compute nodes, where each node has 36 cores and at least 90 GB of memory. Only a fraction of these nodes were used at any given time for this analysis. The synthetic SFO system is built in the geographic area of the extended San Francisco Bay Area, California. The data set contains 40 subregions that span both urban and rural geographies. It covers a total of 2,236 feeders, 4.3 million consumers, 9.8 million electrical nodes, 632 primary substations, and 559,151 distribution transformers and is publicly available at [21].

A. Large-Scale Hosting Capacity Results

The large-scale hosting capacity estimation result is visualized as the hosting capacity map shown in Fig. 3. The map color codes feeders based on the percentage of PV hosting capacity to peak loads. It shows a diverse hosting capacity results for all 2000+ feeders. Fig. 4 provides a zoom-in example results for 3 feeder near San Mateo area, which shows the example feeders can host relatively high PV penetration. Fig. 5 gives the distribution of the hosting capacity results in terms of the number the feeders, it roughly follows a normal distribution except the extreme 0 and 200 percent results.

B. Computational Efficiency

In addition to a large number of feeders under study, the stochastic approach for estimating hosting capacity requires running many power flows with different PV scenarios for each feeder. In this study, there are a total of 849,719 jobs to run, and each job contains 8 snapshot power flows, including 2 control modes for PV (unity power factor and volt-var) and 4 time points (see Section II.A.2). All the jobs are packed in batches, and each batch is assigned to compute nodes based on an estimated run time (see Section II.B.3). Fig. 6 gives the job simulation time distribution. The average job simulation time is 12 minutes, with a standard deviation of 8 minutes. The total simulation time is 10,012,654 minutes, which is equivalent to 19 years. This is the amount of time needed if all the simulations were run in a serial program. Using HPC with the developed framework, we required 1000 computational nodes, and the total simulation was done in approximately 35 hours, plus 5 more hours to post-process the results.

V. CONCLUSIONS

This paper has described our experiences estimating PV hosting capacity at scale for distribution systems. First, we analyze the key challenges of performing relevant, large-scale simulation, including computational and modeling challenges,

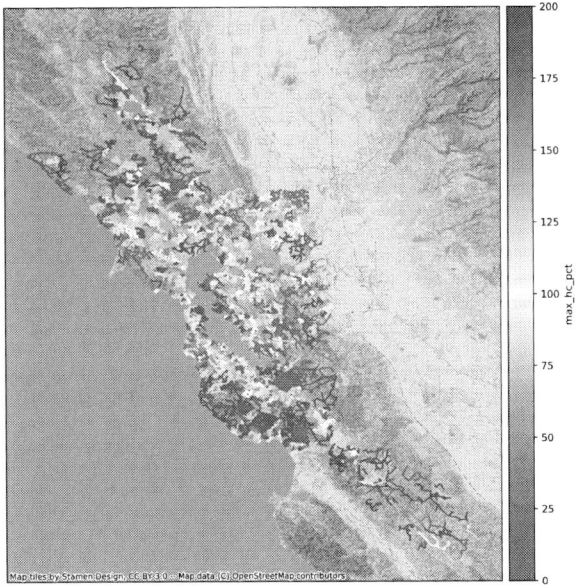

Fig. 3. Hosting capacity map for synthetic SFO feeders, where the maximum hosting capacity percentage relative to peak loads are displayed. Note that this figure serves as a demonstration only, the values of the hosting capacity might be different when the analysis assumptions are different, e.g. PV DC-AC ratio, load and PV values, legacy controls, etc.

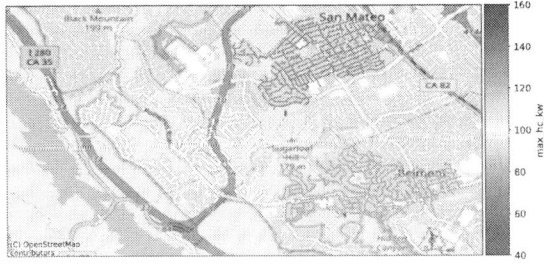

Fig. 4. Zoomed-in hosting capacity map for 3 feeders, other adjacent feeders are not displayed for better readability.

our experiences and lesson learned are also provided. Then, two Python-based, open-source tools are created for modeling and running the simulations. The case study using the SMART-DS SFO 2000+ feeders demonstrates the capability of the developed framework and tools. Our experience shows that estimating hosting capacity at scale requires large number of power flow simulations, it is critical to efficiently manage the limited computational resources.

The outcomes from this research can help utilities better plan for the overall costs of integration, and it can help enable

978-1-7281-6118-1/22 $31.00 © 2022 IEEE

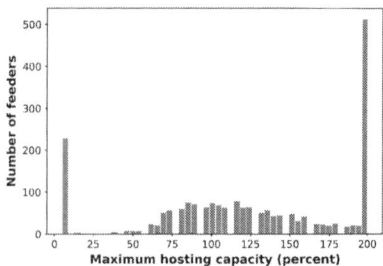

Fig. 5. Maximum hosting capacity (percent) distribution. Note that the maximum PV deployment tested was 200% of peak load, such that the right most histogram bar indicates 200% or higher hosting capacity.

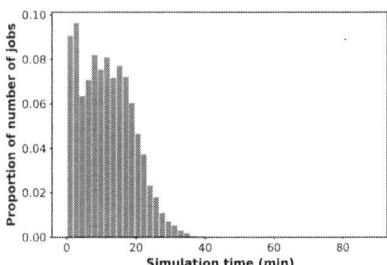

Fig. 6. Simulation time histogram

data analytics of the interconnection process. Future work will include large-scale simulations of DHC and cost-benefits analysis of traditional upgrade and non-wire alternatives.

ACKNOWLEDGMENTS

This work was authored by the National Renewable Energy Laboratory, operated by Alliance for Sustainable Energy, LLC, for the U.S. Department of Energy (DOE) under Contract No. DE-AC36-08GO28308. Funding provided by the U.S. Department of Energy Office of Energy Efficiency and Renewable Energy Solar Energy Technologies Office. The views expressed in the article do not necessarily represent the views of the DOE or the U.S. Government. The U.S. Government retains and the publisher, by accepting the article for publication, acknowledges that the U.S. Government retains a nonexclusive, paid-up, irrevocable, worldwide license to publish or reproduce the published form of this work, or allow others to do so, for U.S. Government purposes.

This research was performed using computational resources sponsored by the Department of Energy's Office of Energy Efficiency and Renewable Energy and located at the National Renewable Energy Laboratory.

REFERENCES

[1] W. Wang, J. Keen, J. Giraldez, K. Baranko, B. Lunghino, D. Morris, F. Bell, A. Shumavon, K. Levine, T. Ward, A. Dave, and J. Bank, "Supervised learning for distribution secondary systems modeling: Improving solar interconnection processes," *IEEE Transactions on Sustainable Energy*, 2022, early access.

[2] S. M. Ismael, S. H. Abdel Aleem, A. Y. Abdelaziz, and A. F. Zobaa, "State-of-the-art of hosting capacity in modern power systems with distributed generation," *Renewable Energy*, vol. 130, pp. 1002–1020, 2019.

[3] R. A. Kordkheili, B. Bak-Jensen, J. R-Pillai, and P. Mahat, "Determining maximum photovoltaic penetration in a distribution grid considering grid operation limits," in *2014 IEEE PES General Meeting — Conference Exposition*, 2014, pp. 1–5.

[4] J. Smith, "Stochastic analysis to determine feeder hosting capacity for distributed solar pv," Elect. Power Res. Inst., Palo Alto, CA, USA, Tech. Rep. 1026640, 2012.

[5] A. K. Jain, K. Horowitz, F. Ding, K. S. Sedzro, B. Palmintier, B. Mather, and H. Jain, "Dynamic hosting capacity analysis for distributed photovoltaic resources—framework and case study," *Applied Energy*, vol. 280, p. 115633, 2020.

[6] K. A. Horowitz, B. Palmintier, B. Mather, and P. Denholm, "Distribution system costs associated with the deployment of photovoltaic systems," *Renewable and Sustainable Energy Reviews*, vol. 90, pp. 420–433, 2018.

[7] M. J. Reno and R. J. Broderick, "Statistical analysis of feeder and locational pv hosting capacity for 216 feeders," in *2016 IEEE Power and Energy Society General Meeting (PESGM)*, 2016, pp. 1–5.

[8] R. J. Broderick *et al.*, "Rapid QSTS simulations for high-resolution comprehensive assessment of distributed PV." Sandia National Lab, Albuquerque, NM, USA, Tech. Rep. 2021-2660, 2021.

[9] Q. Mu, L. Niu, and Y. Cheng, "The communication methodology on the large-scale power system realtime simulation," in *The 16th IET International Conference on AC and DC Power Transmission (ACDC 2020)*, vol. 2020, 2020, pp. 497–504.

[10] D. Thom. DISCO. Accessed: Dec. 10, 2021. [Online]. Available: https://github.com/NREL/disco

[11] ——. JADE. Accessed: Dec. 10, 2021. [Online]. Available: https://github.com/NREL/jade

[12] C. Mateo, F. Postigo, F. D. Cuadra, T. G. S. Roman, T. Elgindy, P. Duenas, B. M. Hodge, V. Krishnan, and B. Palmintier, "Building large-scale u.s. synthetic electric distribution system models," *IEEE Transactions on Smart Grid*, vol. 11, pp. 5301–5313, 11 2020.

[13] F. Ding and B. Mather, "On distributed PV hosting capacity estimation, sensitivity study, and improvement," *IEEE Transactions on Sustainable Energy*, vol. 8, no. 3, pp. 1010–1020, 2017.

[14] K. S. A. Sedzro, M. Emmanuel, and S. A. Abraham, "Generating sequential PV deployment scenarios for high renewable distribution grid planning ," *submitted for publication*, 2022.

[15] R. C. Dugan and D. Montenegro, *The Open Distribution System Simulator (OpenDSS Manual)*, 2020. [Online]. Available: https://smartgrid.epri.com/SimulationTool.aspx

[16] A. Latif, M. Ikechi, I. Hawaiian Electric Company, and S. E. Industries, "NREL/PyDSS," 2018-12. [Online]. Available: https://www.osti.gov/biblio/1512458

[17] B. Barney. Introduction to parallel computing. [Online]. Available: https://hpc.llnl.gov/training/tutorials/introduction-parallel-computing-tutorial

[18] B. Palmintier, K. Horowitz, S. Abraham, T. Elgindy, K. S. Sedzro, B. Sigrin, J. Lockshin, B. Cowiestoll, and P. Denholm, "Chapter 7: Distribution System Analysis," in *The Los Angeles 100% Renewable Energy Study (LA100)*, J. Cochran and P. Denholm, Eds. Golden, CO: National Renewable Energy Lab, Mar. 2021. [Online]. Available: https://www.nrel.gov/docs/fy21osti/79444-7.pdf

[19] Subprocess management. Accessed: Dec. 10, 2021. [Online]. Available: https://docs.python.org/3/library/subprocess.html

[20] SchedMD. Slurm workload manager. [Online]. Available: https://slurm.schedmd.com/

[21] OEDI. SMART-DS. Accessed: Dec. 10, 2021. [Online]. Available: https://data.openei.org/submissions/2981

Properties of Co-Sputtered $(In_xGa_{(1-x)})_2O_3$ Layers Used in CdTe Solar Cells

Manoj K. Jamarkattel, Adam B. Phillips, Indra Subedi, Abasi Abudulimu, Ebin Bastola, Deng-Bing Li, Zhaoning Song, Xavier Matthew, Yanfa Yan, Randy J. Ellingson, Nikolas J. Podraza, and Michael J. Heben.

Wright Center for Photovoltaics Innovation and Commercialization, Department of Physics and Astronomy, University of Toledo, Toledo, OH, 43606, USA

Abstract—We recently demonstrated that $(In_xGa_{1-x})_2O_3$ (IGO) alloys have the potential to be high-performing emitters in CdTe based photovoltaic devices, readily producing devices with efficiencies in excess of 16%. Here we present characterization data for the $(In_xGa_{1-x})_2O_3$ (IGO) films as x was varied from 0 to 1. As grown IGO films exhibited band gaps ranging from 3.3 eV to 4.77 eV and were amorphous and highly resistive. After heating through a temperature profile that would be experienced during CdTe deposition, Hall effect measurements found n-type conductivity and carrier concentrations ranging from 10^{19} to 10^{20} cm^{-3}. The best performing solar cell was fabricated with x = 0.36, which showed a bandgap of 4.02 eV, a carrier concentration of 2.5×10^{19} cm^{-3}, and a mobility of 9.1 cm^2/V.s. PL measurements showed the brightest emission for this same composition.

Keywords—Emitter, Indium Gallium oxide, CdTe

I. INTRODUCTION

CdTe based solar cells have a record efficiency of 22.1 % [1]. Recent modeling has shown that a longer carrier lifetime, higher absorber doping density, and properly aligned front and back interfaces are needed to achieve higher efficiencies [2, 3]. Promising results ascribed to improved back interface layer performance have been reported [4-8]. However, beyond the undoped SnO_2 layers that are included as "HRT" layers deposited by the manufacturer in NSG TEC products [9], only magnesium zinc oxide (MZO) has been demonstrated as an effective front emitter material for CdTe based solar cells [10, 11]. While devices with MZO have exhibited efficiencies exceeding 19 % [11], the so-called "s-kink" that decreases FF and, therefore efficiency, is often seen in the J-V characteristics[10, 12].

Alloyed layers of $(In_xGa_{(1-x)})_2O_3$ (IGO) have been recently proposed as a potential emitter layer that would form a positive conduction band offset (CBO) with CdTe[13]. We recently demonstrated experimentally that, indeed, IGO could be used as an alternative to MZO [14]. Here, we present data relating to the optical and electrical properties of IGO films with band gaps ranging from 3.3 to 4.8 eV. We were not able to probe the electrical properties through Hall measurements for as-deposited samples. So, the samples went through CdTe deposition temperature profile in high vacuum close-space

sublimation (CSS) system, where source and substrate graphites (without CdTe source material) were heated up to 560 °C and 450 °C respectively under high vacuum (lower 10^{-6} torr) and naturally cool down to room temperature. This will correspond to heat treatment that would experience by IGO film stacks during CdTe deposition. High carrier concentration within the range of $10^{19} - 10^{20}$ cm^{-3} is measured for IGO layers. From photoluminescence (PL) measurements of CdSe/CdTe devices with IGO as the front emitter showed the highest PL intensity for films with x = 0.36 when the bandgap was 4.02 eV, which is the same composition that showed the best PV performance.

II. EXPERIMENTAL DETAILS

Samples were deposited on bare soda lime glass (SLG) for characterization purposes. 75 nm of IGO was deposited on SLG glass by co-sputtering using 4" diameter In_2O_3 and Ga_2O_3 sputtering targets, both from Plasmaterials. The composition of IGO was adjusted by varying the sputtering powers from 0 W-250 W [14]. Spectroscopy ellipsometry was used to calculate the optical bandgap. The composition of the IGO was determined by energy dispersive X-ray spectroscopy (EDX). Surface images of IGO films were acquired using a Hitachi S-4800 UHR-Scanning electron microscope (SEM). XRD diffraction patterns were obtained using a Rigaku Ultima III X-ray diffractometer equipped with a Cu k-alpha source. The carrier concentrations, resistivities, and Hall mobilities were measured by Lake Shore FastHall measurements system (M91 FastHall™). Photoluminescence (PL) measurements were performed using 633 nm pulsed laser with photon fluence density of 1.2×10^{11} photons/sec.cm^2. To investigate the electrical properties of IGO film stacks that would experience during a heat treatment during CdTe deposition, SLG/IGO samples went through CdTe deposition temperature (source and substrate temperature at 560 °C and 450 °C respectively, under high vacuum of lower 10^{-6} torr) but without CdTe source material. This will correspond to heat treatment that IGO would experience during CdTe deposition. For photoluminescence (PL) measurements, IGO samples were deposited on TEC10 glass substrate, 150 nm of CdSe was sputtered through RF sputtering and 3.5-micron CdTe was deposited through a high vacuum CSS system at 560 °C source temperature $CdCl_2$ treatments were performed in air ambient at $400\,°C$ for 30 min.

III. RESULTS AND DISCUSSIONS

The band gaps of IGO film stacks were calculated using ellipsometry data and normal incidence transmittance measurements from the same ellipsometer. First, the imaginary part of the refractive index (k) was calculated from combined data analysis. Then k was converted into the absorption coefficient using the relation $\alpha = 4\pi k/\lambda$. The optical band gap was determined to be direct and was

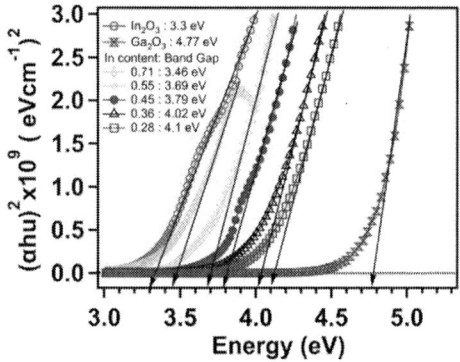

Fig .1. Band gap measurements from spectroscopy ellipsometry measurements.

evaluated with an extrapolation of the linear portion of $(\alpha E)^2$ to zero using a Tauc plot as shown in Fig. 1. The optical band gap was measured to be 3.3 eV for In_2O_3 and 4.77 eV for Ga_2O_3 and varied from 3.46 eV to 4.1 eV as a function of the different indium fractions in $(In_xGa_{1-x})_2O_3$. To investigate the crystallography, low angle x-ray diffraction (XRD) was used to determine the crystal structure of the film deposited on soda lime glass, and the results are shown in Fig. 2(a) XRD data shows weak diffraction from the polycrystalline structure with preferred orientation along (222) direction for the In_2O_3 film. However, for other film compositions, there were no clear XRD peaks. Hence, the films exhibit

insufficient long-range order. Further, surface SEM imaging was performed to study the surface morphology of the film stacks. Fig. 2(b) shows the grains of In_2O_3 which are irregular in shape. In contrast, the surface of Ga_2O_3 (Fig. 2(g)) was difficult to image, presumable because of high resistivity. From SEM, all the IGO images shown here have small round features with some larger features appearing and increasing in size and density in the SEM images as x increases. Interestingly, these features appear larger but with lower density for x equal to 0.36 with the density again increasing for x equal to 0.28. Since we did not observe clear XRD patterns for IGO films stacks (Fig. 2(a)), further investigation is needed to identify whether those round features were related to phase transition between the bixbyite-hexagonal-monoclinic to the monoclinic phase occurring at x equal ~0.3 [15] or could be due to local changes in composition that may be related to sputtering parameters that were varied to deposit films with the different compositions.

For investigating electrical properties, samples were prepared as described in the experimental section. As shown in Fig. 3., the resistivity of film stacks measured from Hall measurements and four point probe was increased with a decrease in indium fraction in the IGO layer. The carrier concentration was measured to be $4.2 \times 10^{20}\,cm^{-3}$ for the IGO sample with x = 0.71 and $2.5 \times 10^{19}\,cm^{-3}$ for x = 0.36. Respective samples have Hall mobility of 18.0 cm^2/V and 9.1 cm^2/V. Recent numerical simulations [13] suggest that IGO with a bandgap of ~ 4 eV is needed for efficient CdTe devices to create the positive CBO with CdTe at the front interface. In our case, 0.36 Indium content in IGO is very close with the bandgap of 4.02 eV. For this layer, we measured $2.5 \times 10^{19}\,cm^{-3}$ carrier concentration with hall mobility of 9.1 cm^2/V. The carrier concentration that we measured here is in agreement with the numerical work which shows higher efficiencies of above 17% CdTe devices using a highly doped n-type layer of IGO ($1 \times 10^{19}\,cm^{-3}$) as an emitter layer but mobility is lowered than their estimated

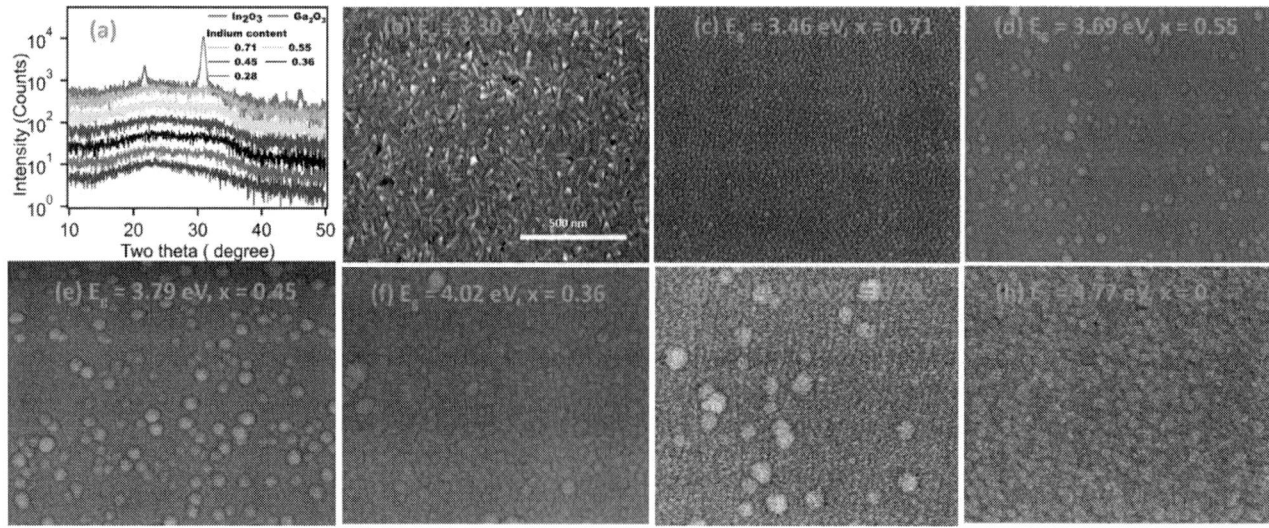

Fig .2 Surface SEM of different film stacks and XRD pattern.

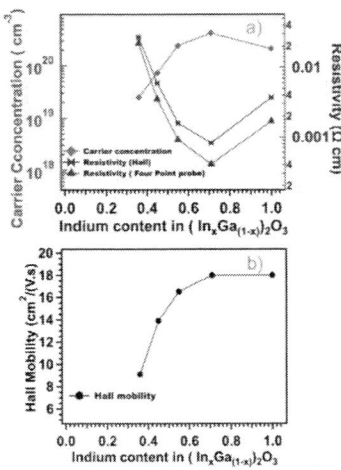

Fig .3. Carrier concentration, resistivity, and Hall mobility measurements

value (25 cm^2/V) [13]. This further illustrates the importance of the IGO layer as a TCO layer too. For indium fraction of x = 0.28 in (In$_x$Ga$_{1-x}$)$_2$O$_3$ and for pure Ga$_2$O$_3$ samples, we were not able to perform Hall measurement due to the high resistivity, likely greater than the detection limit for the instrument. To further investigate the recombination mechanism that could happen with these layers in actual solar devices, we fabricated TEC10/IGO/CdSe/CdTe devices with different values of the indium content. After CdCl$_2$ treatment in flowing dry air, PL intensities were measured from the

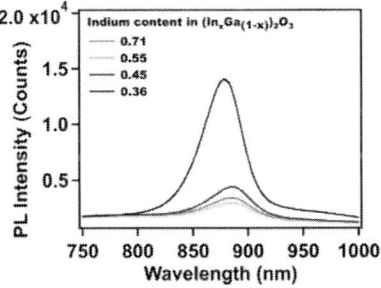

Fig. 4. PL measurements of different IGO samples

glass side of the device stacks. The variation that we applied here was only at the front emitter layer with different band gaps. Hence, any changes in Pl intensities can be related due to different front interfaces. As seen from Fig. 4., PL intensity for the sample with x = 0.36 indium content in IGO is an order of magnitude more intense than other compositions. This suggests that the minority carrier recombination rate is lower for the sample with x = 0.36 indium content with band gap of 4.02 eV, most probably due to proper band alignment at the front interface which is consistent with the best device performance measured for the same composition [14].

IV. CONCLUSIONS

In this work, we studied the optical and electrical properties of the co-sputtered IGO film layer. With adjusting sputtering power, we demonstrate that a wide bandgap of the IGO layer (3.46 eV – 4.1 eV) can be fabricated. High n-type carrier

concentration in a range of $10^{19} – 10^{20}$ cm^{-3} has been measured for IGO films. With carrier concentration of 2.5 x 10^{19} cm^{-3} and brightest PL intensity, we believe, IGO with the bandgap of 4.02 eV could be a favorable emitter layer among other compositions for CdTe based solar cells.

ACKNOWLEDGMENT

This report is based on research sponsored by Air Force Research Laboratory under agreement numbers FA9453 - 19C-1002 and FA9453-21-C-0056, and by the U.S. DOE's Office of Energy Efficiency and Renewable Energy (EERE) under Solar Energy Technologies Office (SETO) Agreement DE-EE0008974. The U.S. Government is authorized to reproduce and distribute reprints for Governmental purposes notwithstanding any copyright notation thereon.

REFERENCES

[1] M. Green, E. Dunlop, J. Hohl - Ebinger, M. Yoshita, N. Kopidakis, and X. Hao, "Solar cell efficiency tables (version 57)," *Progress in Photovoltaics: Research and Applications,* vol. 29, no. 1, pp. 3-15, 2020.

[2] G. K. Liyanage, A. B. Phillips, F. K. Alfadhili, R. J. Ellingson, and M. J. Heben, "The Role of Back Buffer Layers and Absorber Properties for >25% Efficient CdTe Solar Cells," *ACS Applied Energy Materials,* vol. 2, no. 8, pp. 5419-5426, 2019/08/26 2019.

[3] T. Ablekim, E. Colegrove, and W. K. Metzger, "Interface Engineering for 25% CdTe Solar Cells," *ACS Applied Energy Materials,* vol. 1, no. 10, pp. 5135-5139, 2018/10/22 2018.

[4] R. S. Hall, D. Lamb, and S. J. C. Irvine, "Back contacts materials used in thin film CdTe solar cells—A review," *Energy Science & Engineering,* vol. 9, no. 5, pp. 606-632, 2021/05/01 2021.

[5] F. K. Alfadhili *et al.*, "Back-Surface Passivation of CdTe Solar Cells Using Solution-Processed Oxidized Aluminum," *ACS Applied Materials & Interfaces,* vol. 12, no. 46, pp. 51337-51343, 2020/11/18 2020.

[6] K. K. Subedi *et al.*, "Semi-transparent p-type barium copper sulfide as a back contact interface layer for cadmium telluride solar cells," *Solar Energy Materials and Solar Cells,* vol. 218, p. 110764, 2020/12/01/ 2020.

[7] D. Pokhrel *et al.*, "Copper iodide nanoparticles as a hole transport layer to CdTe photovoltaics: 5.5 % efficient back-illuminated bifacial CdTe solar cells," *Solar Energy Materials and Solar Cells,* vol. 235, p. 111451, 2022/01/01/ 2022.

[8] M. K. Jamarkattel *et al.*, "Improving CdSeTe Devices With a Back Buffer Layer of Cu_xAlO_y," *IEEE Journal of Photovoltaics,* vol. 12, no. 1, pp. 16-21, 2022.

[9] Pilkington, "SolarEnergy brochure." [Online]. Available: https://www.pilkington.com/en/us/products.

[10] R. Pandey *et al.*, "Mitigation of J–V distortion in CdTe solar cells by Ga-doping of MgZnO emitter," *Solar Energy Materials and Solar Cells,* vol. 232, 2021, doi: 10.1016/j.solmat.2021.111324.

[11] T. Ablekim *et al.*, "Thin-Film Solar Cells with 19% Efficiency by Thermal Evaporation of CdSe and CdTe," *ACS Energy Letters,* vol. 5, no. 3, pp. 892-896, 2020/03/13 2020, doi: 10.1021/acsenergylett.9b02836.

[12] D.-B. Li *et al.*, "Eliminating S-Kink To Maximize the Performance of MgZnO/CdTe Solar Cells," *ACS Applied Energy Materials,* vol. 2, no. 4, pp. 2896-2903, 2019.

[13] A. Dive, J. Varley, and S. Banerjee, "In2O3−Ga2O3 Alloys as Potential Buffer Layers in CdTe Thin-Film Solar Cells," *Physical Review Applied,* vol. 15, no. 3, 2021.

[14] M. K. Jamarkattel *et al.*, "Indium Gallium Oxide Emitters for High Efficiency CdTe-based Solar Cells," *Under Review*.

[15] J. E. N. Swallow *et al.*, "Indium Gallium Oxide Alloys: Electronic Structure, Optical Gap, Surface Space Charge, and Chemical Trends within Common-Cation Semiconductors," *ACS Appl Mater Interfaces,* vol. 13, no. 2, pp. 2807-2819, Jan 20 2021.

978-1-7281-6118-1/22 $31.00 © 2022 IEEE

High efficiency solar cells grown on spalled germanium without polishing

John S. Mangum, Anthony D. Rice, Jie Chen, Jason Chenenko, Evan Wong, Anna K. Braun, Steve Johnston, John F. Geisz, Aaron J. Ptak, Corinne E. Packard

National Renewable Energy Laboratory, Golden, CO, United States

Colorado School of Mines, Golden, CO, United States

Implementing device removal and substrate reuse provides an opportunity for substrate cost reduction. Controlled spalling allows removal of devices; however, the fracture-based process generates surfaces with significant morphological changes compared to polished wafers. We create small single junction devices across full 2" spalled germanium wafers without polishing before epitaxial growth. We identify device defects related to surface morphology and their impact on cell performance using physical and functional characterization techniques. A single junction efficiency above 23% and VOC of 1.01 V is achieved, demonstrating that spalled germanium does not need to be returned to a pristine, polished state to achieve high quality device performance.

The Effect of CdSe$_X$Te$_{1-X}$ Thickness on the CdSe$_X$Te$_{1-X}$/CdTe Solar Cell Performance

Md Zahangir Alom, Sheikh Tawsif Elahi, Vasilios Palekis, Wei Wang, Chris Ferekides

University of South Florida, Tampa, FL, 33620, USA

Abstract—The effect of the CdSe$_X$Te$_{1-X}$ (CST) thickness on the performance of CdSe$_X$Te$_{1-X}$/CdTe solar cells at different Se (x) compositions and annealed at various CdCl$_2$ temperatures has been studied. The cell configuration was superstrate as ITO/MZO/CdSexTe$_{1-X}$/CdTe/Back Contact. Both CST and CdTe were deposited by the close-spaced sublimation (CSS) process. The CST thickness and Se composition varied from 0.25-1 μm and 6-29% respectively. The open circuit voltage (V$_{OC}$) was found to decrease with increasing CST thickness and Se composition. V$_{OC}$ decreases, and J$_{SC}$ increases due to the decrease in the bandgap of CST. A CST thickness of 0.50 μm produced the optimum V$_{OC}$ and fill factor, thus giving the optimum efficiency. Also, 430^{0}C CdCl$_2$ annealing temperature appears to result in a higher minority carrier lifetime, which is responsible for the improved cell performance observed for devices annealed at this temperature.

Keywords—Thickness, Annealing, Se Composition, Bandgap, Lifetime

I. INTRODUCTION

CdTe thin-film solar cell has a maximum efficiency of 22.1% [1]. This value is still substantially lower than the theoretical limit. The main limitation of the efficiency has been V$_{OC}$ and J$_{SC}$. Due to the use of CST in recent years, the J$_{SC}$ reached values close to its theoretical limit (31.7 mA/cm^2) for the CdTe bandgap. At present, the difference between the theoretical and experimental efficiency is primarily due to the V$_{OC}$. For polycrystalline p-type CdTe solar cells, V$_{OC}$ is limited to approx. 900 mV. Only for monocrystalline p-type solar cells, V$_{OC}$ above 1000 mV has been reported [2]. So, there still exists an opportunity to increase the efficiency of p CdTe solar cells by achieving V$_{OC}$ above 1000 mV. To achieve this goal, a net p-type carrier concentration of more than 10^{16} cm^{-3} with a lifetime of several to tens of nanoseconds is required [3].

Using CST as part of the absorber structure (i.e. CST/CdTe bilayers) has led to increases in the J$_{SC}$ due to the lower bandgap of the absorber (CST). Moreover, CST is easier to dope compared to CdTe [4]-[5]. It can be doped by group V elements (As, P) to increase the net p-doping concentration for achieving higher V$_{OC}$ [6]. Also, the introduction of Se leads to a higher minority carrier lifetime, which can further boost V$_{OC}$. This can be seen in the smaller decrease in V$_{OC}$ relative to the change in the CST bandgap as will be discussed below.

The CdCl$_2$ heat treatment is another important process step for high-efficiency CST/CdTe solar cells; because it controls the interdiffusion between CdTe and CST, thus controlling the bandgap of the absorber, and therefore the J$_{SC}$. It also helps to increase the minority carrier lifetime, thus increasing V$_{OC}$.

The interdiffusion between CST and CdTe at a certain CdCl$_2$ annealing temperature impacts the final absorber bandgap near the junction. The quality and the bandgap of the absorber near the junction also depend on the CST thickness. For this study, the CST thickness was varied from 0.25 to 1 μm by varying the deposition time.

In this paper, the effect of CST thickness at different Se compositions, and processed at various CdCl$_2$ annealing temperatures is presented. The effect of CST thickness on the solar cell parameters, especially open-circuit voltage (V$_{OC}$), short circuit current (J$_{SC}$), fill factor (FF), bandgap (E$_g$), doping concentration have been studied. Also, using I-V, and spectral response measurements, the effect of CdCl$_2$ annealing temperature, and CST thickness on the cell performance has been investigated in order to identify the optimum CST thickness and annealing temperature.

II. EXPERIMENTAL

Our superstrate cell configuration is: ITO/MZO/CST/CdTe/Back Contact. The glass substrate used is the corning EagleXG. The glass substrate is cleaned with dilute HF solution and rinsed with DI water. Indium Tin Oxide (ITO) and Magnesium Zinc Oxide (MZO) were deposited by rf sputtering at 250^{0}C and at room temperature respectively. The CST was deposited by CSS from CdTe (99.999%), and CdSe (99.999%) mixtures at source and substrate temperature of 680 and 580^{0}C respectively. For depositing As doped CST, Cd$_3$As$_2$ (99.999%) powder was introduced during source preparation. The Se(x) composition of the CST alloys was controlled by varying the source making temperature; the thickness was controlled by varying the deposition time. Following the CST deposition, the CdTe layer was deposited at the same condition as CST. The CdCl$_2$ annealing was performed at 410 and 430^{0}C temperatures following the evaporation of approx. 1 μm CdCl$_2$ film onto the cell structure. The sample was then bromine etched, and Cu-doped graphite paste was applied as the back contact. The samples were characterized by Current-Voltage (J-V), Spectral Response (SR), and Capacitance-Voltage (C-V) measurements using optical filters with 20 nm bandwidth measurements. Hitachi SU800 was used to analyze the Electron Dispersive Spectroscopy (EDS) to estimate the Se composition

978-1-7281-6118-1/22 $31.00 © 2022 IEEE

of the CST source. The X-ray diffraction peak was measured using a panalytical X'Pert MRT.

III. RESULTS AND DISCUSSION

The addition of the CST alloy in the CdTe absorber leads to improve J_{SC} by decreasing the bandgap; Se alloys also exhibit improved minority carrier lifetimes. The final bandgap of the absorber depends upon the interdiffusion of CST and CdTe, which is controlled by the CST thickness and the subsequent annealing conditions. The CST thickness was varied by changing the deposition time with keeping other conditions identical.

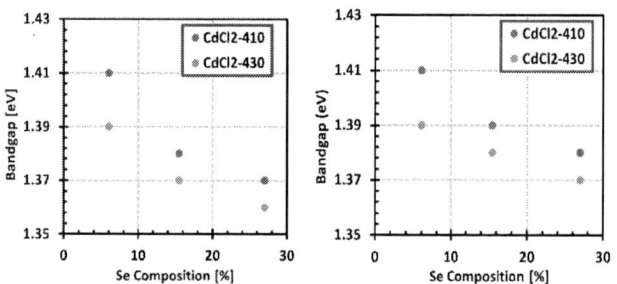

Fig. 1. Bandgap Vs Se composition for 1 (left) and 0.50 (right) μm As doped CST cells at 410 and 430⁰C annealing temperature.

From Fig. 1, it can be seen that with the increase of Se composition, bandgap decreases that is due to the bandgap bowing effect of CST. The bandgap is estimated from the absorption edge of the spectral response of the devices. For 1 μm CST thickness, the overall bandgap is 10-20 meV lower compared to 0.50 μm of CST cells. Thicker CST makes more interdiffusion between CST and CdTe, which can make the bandgap even lower. Also, higher CdCl₂ annealing temperature reduces the bandgap more by enhancing the interdiffusion between the CST and CdTe.

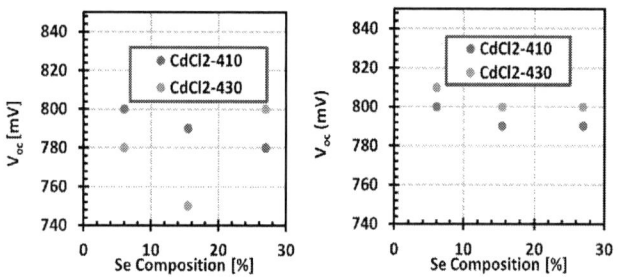

Fig. 2. V_{OC} Vs Se composition for 1 (left) and 0.50 (right) μm As doped CST cells at CdCl₂ 410 and 430⁰C annealing temperature.

Figure 2 shows the effect of 1 and 0.50 μm of CST thickness on the V_{OC}; the CST films were doped with As and CdCl₂ annealing temperatures were 410 and 430⁰C. Cells fabricated with CST films with 0.50 μm thickness produce 10-20 mV higher V_{OC} compared to 1 μm. This is primarily because of the higher bandgap of 0.50 μm cells.

For both 1 and 0.50 μm CST cells, V_{OC} decreases with increasing Se composition due to the bandgap reduction of CST ; within this composition range, the CST bandgap initially decreases with increasing composition due to the bowing effect.

The CdCl₂ annealing temperature also affects the V_{OC}. V_{OC} increases with the increase of CdCl₂ annealing temperature. 430⁰C CdCl₂ annealing temperature shows comparatively higher V_{OC} because of the higher minority carrier lifetime.

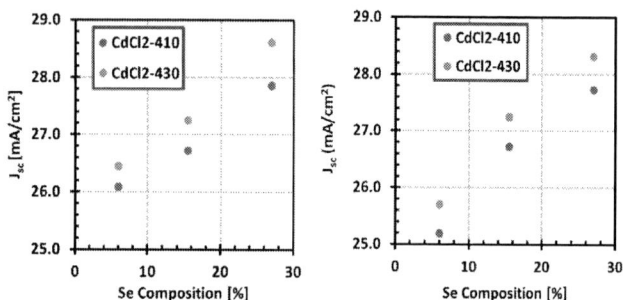

Fig. 3. J_{SC} Vs Se composition for 1 (left) and 0.50 (right) μm As doped CST cells at CdCl₂ 410 and 430⁰C annealing temperature.

Figure 3 shows the variation of J_{SC} with 1 and 0.50 μm CST thickness at different Se compositions and CdCl₂ annealing temperatures. For both cases, J_{SC} increases with the increase of Se composition due to the lowering bandgap. For 1 μm of CST, J_{SC} appears to be slightly higher than 0.50 μm CST cells, most likely due to higher absorption in the CST region for the thicker CST devices. J_{SC} is also at 430⁰C CdCl₂ annealing temperature, because higher temperature increases the interdiffusion between CST and CdTe, thus reducing the bandgap and increasing the carrier collection.

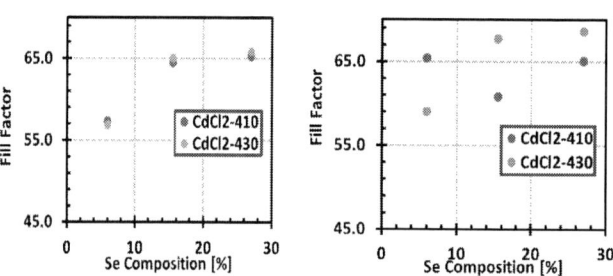

Fig. 4. Fill Factor Vs Se composition for 1 (left) and 0.50 (right) μm n As doped CST cells at 410 and 430⁰C annealing temperature.

Figure 4 depicts the fill factor variation with CST thickness, Se composition, and CdCl₂ annealing temperatures. The fill factor increases for both thicknesses with increasing of Se composition. This can be the effect of minority carrier lifetimes as it varies with Se composition. Also, for 0.50 μm of CST cells, the fill factor is higher than 1 μm. Moreover, Higher CdCl₂ annealing temperature has a positive impact on the fill factor, which means cells processed under 430⁰C CdCl₂ annealing temperature have a greater fill factor, maybe because of higher minority carrier lifetime at 430⁰C.

978-1-7281-6118-1/22 $31.00 © 2022 IEEE

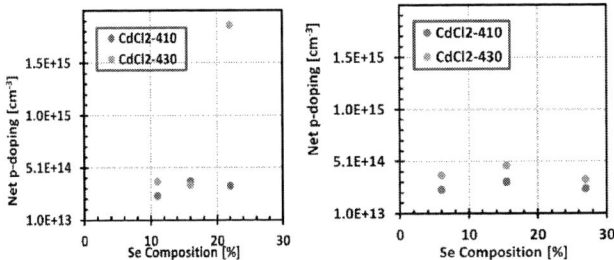

Fig. 5. Net p-doping Vs Se composition for 1 (left) and 0.50 (right) μm As doped CST cells at CdCl₂ 410 and 430⁰C annealing temperature.

Figure 5 shows the doping concentration variation with the Se composition and $CdCl_2$ annealing temperature for 1 and 0.50 μm of CST cells. There is no noticeable difference in net p-doping for 1 and 0.5 μm of CST cells. Also, higher $CdCl_2$ annealing temperatures tend to lead to an increase in net p-doping.

TABLE 1. V_{OC}, J_{SC}, and FF for 0.25 and 0.50 μm CST Cells at 410 and 430⁰C $CdCl_2$ Annealing Temperature for 29% Se composition.

Condition	V_{OC} [mV]	FF [%]	J_{SC} [mA/cm²]
CST-0.25 μm, CdCl₂-410⁰C	770	61.2	26.75
CST-0.25 μm, CdCl₂-430⁰C	780	51.7	26.24
CST-0.50 μm, CdCl₂-410⁰C	790	57.00	27.72
CST-0.50 μm, CdCl₂-430⁰C	810	68.60	28.35

Table 1 shows the performance variation (V_{OC}, J_{SC}, and FF) of the 0.25, and 0.50 μm CST cells at different $CdCl_2$ annealing temperatures. For both thicknesses, V_{OC} increases with the increase of $CdCl_2$ annealing temperature because of the higher minority carrier lifetime at 430⁰C. It is clear from the high absorption near the bandgap. For 0.50 μm, FF increases with the increase of $CdCl_2$ annealing temperature because of the higher minority lifetime. A reverse trend is found for 0.25 μm because of more chlorine in the front interface. For 0.50 μm, with the increase of $CdCl_2$ annealing temperature, J_{SC} increases because of higher carrier collection. But for 0.25 μm, J_{SC} decreases with the increase of the $CdCl_2$ annealing temperature due to lower carrier collection. So, from these results, we can conclude that a certain CST thickness is needed for collecting optimized current at different $CdCl_2$ annealing temperatures.

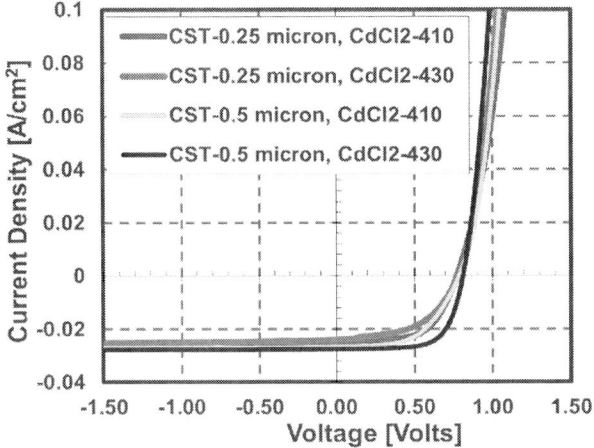

Fig. 6. I-V curve for 0.25 and 0.5 μm of As doped CST cells at CdCl₂ 410 and 430⁰C annealing temperature.

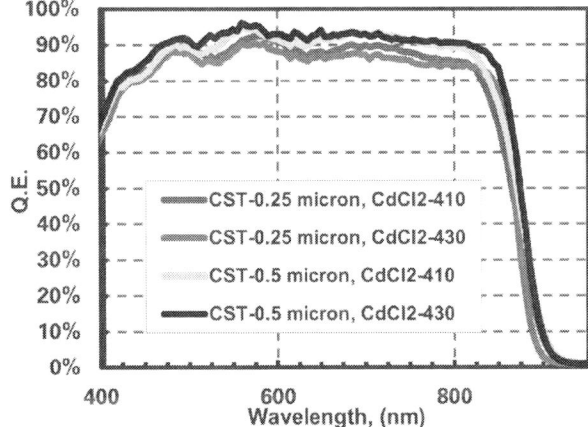

Fig. 7. Spectral Response for 0.25 and 0.5 μm of As doped CST cells at CdCl₂ 410 and 430⁰C annealing temperature.

From the I-V and spectral response, it is seen that 0.50 μm CST cells result in higher V_{OC}, fill factor, and J_{SC}. The spectral response shows that 0.50 μm CST cells exhibit enhanced collection near the absorption. This is due to the smaller bandgap for the 0.50 μm CST cells as shown in Fig. 1. That's why 0.50 μm CST devices have higher J_{SC}. According to the bandgap, V_{OC} should be higher for 0.25 μm CST cells. However, it is the opposite. The reason is not clear at this time. Maybe the change in the bandgap of the CST affects the interface between CST and MZO, thus decreasing the V_{OC}.

The same trend of V_{OC}, J_{SC}, fill factor, bandgap, and doping concentration was also seen for undoped CST cells.

IV. CONCLUSION

The effect of the CST thickness on the performance of $CdSe_XTe_{1-X}/CdTe$ solar cells at different Se (x) compositions and annealed at various $CdCl_2$ temperatures has been studied. With the increase of CST thickness, V_{OC} decreases, and J_{SC} increases due to the bandgap of CST. 0.50 µm CST gives the optimum V_{OC} and fill factor compared to other thicknesses. Also, higher Se composition increases the J_{SC} more as expected. Moreover, the Se in the CST helps to improve the minority carrier lifetime. Due to this reason, the change in V_{OC} is smaller compared to the reduction in the bandgap of CST, and the fill factor is also higher. $CdCl_2$ annealing also affects the performance by changing the interdiffusion between CST and CdTe. Besides that, it affects the minority carrier lifetime. 430^0C $CdCl_2$ annealing shows a higher minority lifetime. That is why V_{OC} is higher at this temperature compared to 410^0C. Cells fabricated with 29% Se composition at 0.50 µm CST thickness and 430^0C $CdCl_2$ annealing temperature produce the optimum efficiency.

ACKNOWLEDGMENT

This research work was funded by the National Science Foundation (NSF) (grant no. EPMD-1711716).

REFERENCES

[1] Green, Martin A., et al. "Solar cell efficiency tables (version 49)." *Progress in photovoltaics: research and applications* 25.1 (2017): 3-13.

[2] Burst, J. M., et al. "CdTe solar cells with open-circuit voltage greater than 1 V. Nat." *Energy* 1 (2016).

[3] Duenow, Joel N., et al. "Relationship of open-circuit voltage to CdTe hole concentration and lifetime." *IEEE Journal of Photovoltaics* 6.6 (2016): 1641-1644.

[4] Wei, Su-Huai. "Overcoming the doping bottleneck in semiconductors." *Computational Materials Science* 30.3-4 (2004): 337-348.

[5] Ablekim, Tursun, et al. "Self-compensation in arsenic doping of CdTe." *Scientific reports* 7.1 (2017): 1-9.

[6] Alom, Md Zahangir, et al. "The Effect of Arsenic Doping on the Performance of CdSe x Te 1-x/CdTe Solar Cells." 2021 IEEE 48th Photovoltaic Specialists Conference (PVSC). IEEE, 2021.

[7] Hsu, Chih An, et al. "The Effect of the CdCl2 Heat Treatment on CdSexTe1-x Solar Cells." 2017 IEEE 44th Photovoltaic Specialist Conference (PVSC). IEEE, 2017.

[8] Hsu, Chih-An, et al. "Cu-doping Effects in CdSe x Te 1-x/CdTe Solar Cells." 2019 IEEE 46th Photovoltaic Specialists Conference (PVSC). IEEE, 2019.

[9] Collins, Shamara, et al. "Se profiles in CST films formed by annealing CdTe/CdSe bi-layers." 2018 IEEE 7th World Conference on Photovoltaic Energy Conversion (WCPEC)(A Joint Conference of 45th IEEE PVSC, 28th PVSEC & 34th EU PVSEC). IEEE, 2018.

[10] Swanson, Drew E., James R. Sites, and Walajabad S. Sampath. "Co-sublimation of CdSexTe− x layers for CdTe solar cells." *Solar Energy Materials and Solar Cells* 159 (2017): 389-394.

[11] Çiriş, Ali, et al. "Effect of ultra-thin CdSexTe− x interface layer on parameters of CdTe solar cells." *Solar Energy* 234 (2022): 128-136.

[12] Munshi, Amit H., et al. "Polycrystalline CdSeTe/CdTe absorber cells with 28 mA/cm 2 short-circuit current." *IEEE Journal of Photovoltaics* 8.1 (2017): 310-314.

[13] Artegiani, Elisa, et al. "A new method for CdSexTe1-x band grading for high efficiency thin-absorber CdTe solar cells." *Solar Energy Materials and Solar Cells* 226 (2021): 111081.

[14] Artegiani, Elisa, et al. "Effects of CdTe selenization on the electrical properties of the absorber for the fabrication of CdSexTe1-x/CdTe based solar cells." *Solar Energy* 227 (2021): 8-12.

[15] Zheng, Xin, et al. "Recombination and bandgap engineering in CdSeTe/CdTe solar cells." *APL Materials* 7.7 (2019): 071112.

Overall Performance Losses and Activated Mechanisms in Double Glass and Glass-backsheet Photovoltaic Modules with Monofacial and Bifacial PERC Cells, under Accelerated Exposures

Jiqi Liu*, Sameera Nalin Venkat*, Jennifer L. Braid*†, Xuanji Yu*, Brenton Brownell§, Xinjun Li§,
Jean-Nicolas Jaubert‡, Kaushik Roy Choudhury ¶, Laura S. Bruckman*, Roger H. French*
*SDLE Research Center, Materials Science Department, Case Western Reserve University, Cleveland, OH, 44106, USA
†Sandia National Laboratories, Albuquerque, NM, 87123, USA
‡CSI Solar Co. Ltd., Suzhou, Jiangsu, China
§Cybrid Technologies Inc., Suzhou, Jiangsu, China
¶DuPont Photovoltaic Solutions, Wilmington DE, 19803, USA

Abstract—Commercial PV modules have various packaging choices nowadays, which influence their long-term reliability. This study compared the degradation behaviors of sixteen module variants from two brands with varying encapsulant materials (EVA or POE), encapsulant types, module architectures (GB or DG), and cell types (monofacial or bifacial) using null hypothesis testing to determine statistical significant findings. The modules were exposed for 2,520 hours under two accelerated exposures: modified damp heat (mDH) and modified damp heat with full-spectrum light (mDH+FSL). For both brands, two DG module variants with UV-Cutoff rear encapsulant are found to have significantly lower average power loss than the module variants of EVA+GB with opaque rear encapsulant after each accelerated exposure. Metallization interconnect corrosion is identified as the primary degradation mechanism. Unsupervised hierarchical clustering finds that the degradation behaviors of modules from one brand with a more strict manufacturing quality control depends on module architectures only.

Index Terms—double glass, POE, EVA, damp heat, statistical significance

I. INTRODUCTION

Annual installed solar energy systems in 2020 have larger power generation capacity than any other renewable energy sources [1]. Since the degradation of PV modules depends on the design of the module's packaging and exposure conditions, various indoor accelerated exposures have been designed to study degradation behaviors of PV modules under specific environmental stressors. Commercial PV modules can be categorized into two different module architectures: double glass

This material is based upon work supported by the U.S. Department of Energy's Office of Energy Efficiency and Renewable Energy (EERE) under Solar Energy Technologies Office (SETO) Agreement Number DE-EE-0008550. The views expressed herein do not necessarily represent the views of the U.S. Department of Energy or the United States Government. Sandia National Laboratories is a multimission laboratory managed and operated by National Technology & Engineering Solutions of Sandia, LLC, a wholly owned subsidiary of Honeywell International Inc., for the U.S. Department of Energy's National Nuclear Security Administration under contract DE-NA0003525. The authors acknowledge Alan Curran, Raymond Wieser, and William Oltjen for I-V measurements.

(DG) and glass-backsheet (GB). With the rising popularity of bifacial PV modules, DG modules have an increased market share [1] due to their bifacial nature. At the same time, a new polyvinyl fluoride (PVF)-based transparent backsheet has been commercialized to compete with glass used as the rear "sheet" or "cover" of the PV module, and is found to be very durable. W. Gambogi et al. studied PVF-based transparent backsheet using 500 hours of UV exposure [2]. The UV absorption decreased by 18%, the elongation at break decreased by 30% [2].

The market growth of DG modules motivates the replacement of EVA encapsulants. EVA has been the dominant encapsulant for nearly four decades with over 80% market share due to its balance of properties and its low cost. However, its degradation products contain acetic acid, which can diffuse outward through the polymeric backsheet, but in a DG module the acetic acid would be sealed inside the module [3]. Recently, another encapsulant material POE, which does not generate acetic acid upon degradation, has risen as a competitor of EVA. POE is a copolymer of polyethylene and octene [4] and the significant advantage of POE is the absence of acetic acid produced during EVA degradation due to the replacement of vinyl acetate side group with alkanes [5]. Multiple studies have shown that PV modules using POE have a better resistance to potential induced degradation (PID) than modules using EVA encapsulants [6]–[8]. In addition, Barretta et al.found that the cross-linked film of EVA and POE had very similar stability under damp heat and UV accelerated exposures [9]. The changes to packaging materials and module architectures, in recent years, have brought challenges to reliability studies of PV modules. A few recent studies have compared DG and GB modules using different encapsulant materials. In outdoor studies, there are fewer DG modules than GB modules, and their reliability performance differs from that of recent DG modules due to changes in manufacturing and materials [10], [11]. Also recent laboratory-based accelerated studies are

978-1-7281-6118-1/22 $31.00 © 2022 IEEE

generally limited to being observational, and therefore lack the ability to provide statistically significant findings. Therefore, no firm conclusion could be made on whether the degradation performance has a significant difference for modules with different encapsulant materials or module architectures [5], [12]–[15].

The degradation behaviors of sixteen variants of DG and GB modules using different types of EVA or POE encapsulants were investigated using the following study protocol. The "study object" in these accelerated exposure studies are photovoltaic minimodules with four solar cells connected in series. The interconnect ribbons are configured with five single wire junction boxes such that the electrical properties of each cell can be measured separately, in addition to the 4 cells in series. The minimodules are exposed, in a step-wise manner, to two indoor accelerated exposure conditions, mDH and mDH+FSL. The evaluation methods used in the study include current-voltage (I-V) curves, $Suns$-V_{oc}, electroluminescence (EL), and photoluminescence (PL). The statistical significance of our results was determined from null hypothesis testing [16] and confidence intervals were used for two-sample t-tests, so as to compare average power loss and degradation mechanism features across different modules variants. This statistical analysis enables us to identify stable and relatively unstable module variants. Furthermore, the dependency of degradation performance on the choice of materials, module architectures, cell types, and manufacturing process was explored through unsupervised hierarchical clustering.

II. STUDY PROTOCOL: STUDY OBJECT, EXPOSURES, EVALUATIONS AND ANALYSIS

This section introduces study protocol, including the study object, accelerated exposure conditions, evaluation methods with their corresponding feature extraction, and the data analysis methods used, including principal component analysis (PCA) and unsupervised hierarchical clustering.

A. Four-cell Minimodules

The study has sixteen module variants from two brands: A and B. Brand A minimodules are fabricated by Canadian Solar Inc. using its research facilities, while Brand B minimodules are fabricated in the CWRU SDLE Research Center. Two minimodules from each brand are put under each accelerated exposure, contributing to measurements from a total of eight cells at each measurement step, thus providing the benefit that the standard error of the results is reduced by the $\sqrt{8}$. The sixteen module variants with differences, as shown in Fig. 1, are divided into four sets. Each set includes four different combinations: EVA+GB, EVA+DG, POE+GB, and POE+DG. Across the sets, cell type (monofacial or bifacial) and rear encapsulant type (transparent, UV-Cutoff, and opaque) vary. The first three sets are monofacial modules with the polymeric backsheet of the GB minimodules being a KPf backsheet, which is a polyvinylidene fluoride-based backsheet. Set #4 modules are bifacial modules with the backsheet for GB minimodules as PVF transparent backsheet. The glass is made of

3.2 mm tempered glass or 2.5 mm heat-strengthened glass for GB and DG minimodules, respectively. Detailed specifications and quantities for the minimodules under indoor accelerated exposures are listed in Table I. Fig. 2 shows the front and backside of a DG minimodule. The use of minimodules tend to amplify the degradation compared to commercial full-size modules due to size reduction.

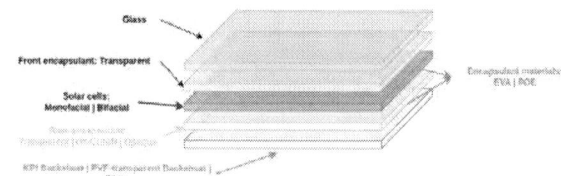

Fig. 1: Possible material choices for each layer within the sixteen module variants.

TABLE I: Specifications of sixteen module types

Set	Module Variant ID	Cell Type	Encap.	Rear Encap.	Architecture	Module Type
1	1	mono-facial	EVA	UV-Cutoff	GB	mono-facial
1	2	mono-facial	EVA	UV-Cutoff	DG	mono-facial
1	3	mono-facial	POE	UV-Cutoff	GB	mono-facial
1	4	mono-facial	POE	UV-Cutoff	DG	mono-facial
2	5	bi-facial	EVA	Opaque	GB	mono-facial
2	6	bi-facial	EVA	Opaque	DG	mono-facial
2	7	bi-facial	POE	Opaque	GB	mono-facial
2	8	bi-facial	POE	Opaque	DG	mono-facial
3	9	mono-facial	EVA	Opaque	GB	mono-facial
3	10	mono-facial	EVA	Opaque	DG	mono-facial
3	11	mono-facial	POE	Opaque	GB	mono-facial
3	12	mono-facial	POE	Opaque	DG	mono-facial
4	13	bi-facial	EVA	UV-Cutoff	GB	bi-facial
4	14	bi-facial	EVA	Transparent	DG	bi-facial
4	15	bi-facial	POE	UV-Cutoff	GB	bi-facial
4	16	bi-facial	POE	Transparent	DG	bi-facial

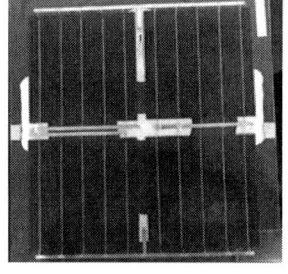 

(a) Front side. (b) Backside.

Fig. 2: The front and back side appearance of one DG minimodule.

B. Indoor Accelerated Exposures

We conducted two kinds of step-wise indoor accelerated exposures; mDH and mDH+FSL. For the modified damp heat (mDH), the temperature was adjusted to be 80 °C with a relative humidity (RH) of 85%. Compared to the conditions of standard damp heat exposure, we lower the temperature by 5 °C to avoid excessive stress on the module such as acceleration of the hydrolysis of PET in the backsheet [3], therefore bringing a more realistic comparison between the DG and GB modules. The total exposure time was up to 2,520 hours with 504 hours as one step of the exposure, with evaluation measurements performed at "baseline" (step 0) and each of the 5 exposure steps. In the mDH+FSL, the module spent 2/3 of the time (336 hours) under mDH, and the rest 1/3 of the time under FSL (168 hours) for each step. The average irradiance intensities were 420.4 W/m^2 and 85.1 W/m^2 for the front and back sides of the exposed modules under FSL. Each module was connected to a 0.5 Ω load resistor to make the module operate around its maximum power point. The main purpose of the FSL exposure is to have the PV module be exposed under fully operational conditions, with current flowing through the module. The module temperature was lower than 70 °C during FSL exposure.

C. Evaluations: Characterization and Analysis Methods

After each exposure step, current-voltage (I-V) curves of three irradiance levels (1000 W/m^2, 500 W/m^2, and 250 W/m^2), $Suns$-V_{oc} curve, electroluminescence (EL) images at three current levels (9.4 A (I_o), 4.7 A (0.5 I_o), and 2.4 A (0.25 I_o)), and open-circuit (OC) and short-circuit (SC) photoluminescence (PL) images were measured. EL and PL images were taken with a Tau Science PixEL system, using a 20.2 megapixel (5496 x 3672 pixels) ZWO ASI183MM Pro monochrome camera with a Peltier cooled Sony IMX183CLK J back-illuminated CMOSsensor, and green LED illuminators for PL measurements [17]. Another PL image denoted by $PL@OC-SC$ was obtained by subtracting the short-circuit PL image from the open-circuit PL image. The corresponding dark images were subtracted from each EL image, and the cell extraction Python pipeline [18] was applied to extract individual cell images from minimodule images, producing EL and PL cell images at about 1485 $\times$ 1485 pixels spatial resolution. The series resistance (R_s) was extracted using all three I-V curves following IEC 60891. The maximum power (P_{mp}) and short-circuit current I_{sc} were extracted from the 1000 W/m^2 I-V curve using the ddiv package [19]. Another P_{mp} was obtained from the Pseudo I-V curve (PIV) converted from the $Suns$-V_{oc} curve [20] using a constant I_{sc} as 9.465 A, which was the nameplate value for the cell. In addition, the standard deviation (F_{sd}) from the three EL images at different current levels ($EL@I_o$, $EL@0.5I_o$, and $EL@0.25I_o$), the open-circuit PL image ($PL@OC$), and the $PL@OC-SC$ was calculated. Features extracted from multiple characterization methods were further normalized by the value measured at baseline from the same cell. Outliers were removed separately by checking the observations of each module variant from each brand at each measurement step under each exposure using (1), where k was set as 3, and Q_1 and Q_3 were the lower and upper quartiles, respectively.

$$[Q_1 - k(Q_3 - Q_1), Q_3 + k(Q_3 - Q_1)] \quad (1)$$

D. Unsupervised Hierarchical Clustering

Principal component analysis (PCA) is a process of computing the principal components to perform a basis transformation on the data while retaining as much information as possible in a lower dimension [21], [22]. Its computation relies on the singular value decomposition, and the resulting principal components are orthogonal to each other. Nine normalized features for each brand of minimodules after exposures were selected for PCA, including $^nI_{sc,IV}$, $^nP_{mp,IV}$, $^nR_{s,IV}$, $^nP_{mp,PIV}$, $^nF_{sd,EL@I_o}$, $^nF_{sd,EL@0.5I_o}$, $^nF_{sd,EL@0.25I_o}$, $^nF_{sd,PL@OC}$, and $^nF_{sd,PL@OC-SC}$. The data were scaled and centered for input for PCA. The first three principal components were taken for the agglomerative hierarchical clustering using Ward's linkage method. Ward's method, also called the minimal increase of sum-of-squares (MISSQ) method, evaluates the proximity between two clusters as the quantity by which the summed square in their joint cluster will be greater than the combined summed square in the two clusters [23]. Intuitively, this method aims at finding compact, spherical clusters. It was chosen due to its better resistance to making outliers as individual clusters.

III. RESULTS

A. Performance and Degradation Mechanisms for Module Variants under Two Exposures

The normalized maximum power extracted from I-V curves ($^nP_{mp,IV}$) after each accelerated exposure is shown in for brands A and B are shown in Fig. 3 and Fig. 4, respectively. The x-axis is the $^nP_{mp,IV}$ at the last exposure step, and the y-axis is the module variant. The two blue dashed lines mark 1 ± 0.005. The red circles mark the average value for each module variant and the purple bar is the 95% confidence interval (CI). The black bar indicates the 83.4% confidence interval (CI) and is used for the two sample t-test, and this 83.4% CI also corresponds for two samples to a 0.05 significance level [24]. If two of such CIs of two samples have no overlap, then their corresponding averages have a statistically significant difference at the significance level of 0.05. Comparing these black bars provides a way to visualize the null hypothesis two sample t-test results of comparing changes of two module variants groups [25], [26].

The results of $^nR_{s,IV}$, $^nI_{sc,IV}$, and $^nP_{mp,PIV}$ for brand A minimodules after each accelerated exposure are presented in Fig. 5, Fig. 6, and Fig. 7, respectively. Although these three electrical features are overall electrical parameters, they are also closely related to specific degradation mechanisms [27]. The increase in $^nR_{s,IV}$ indicates interconnection corrosion, the decrease in $^nI_{sc,IV}$ correlates to encapsulant discoloration, and the reduction of $^nP_{mp,PIV}$ reveals easier recombination after exposure. The association of these features to the degradation mechanisms are also based on understanding on how PV module should degrade under our specific accelerated exposure conditions. While the amount of change in each mechanism feature is not linearly proportional to its contributed power loss, a significant change, generally, indicates

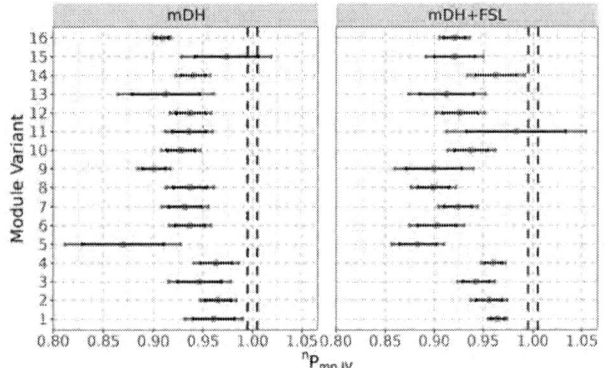

Fig. 3: $^{n}P_{mp,IV}$ for brand A after each accelerated exposure.

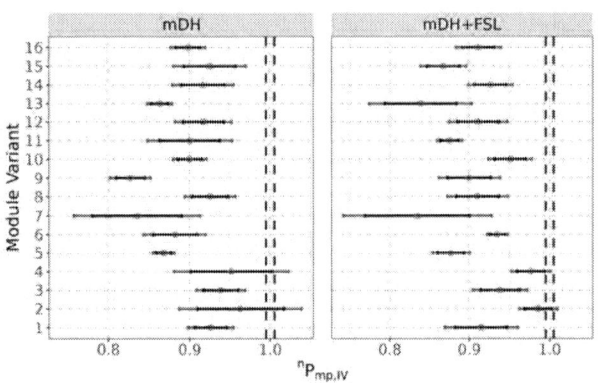

Fig. 4: $^{n}P_{mp,IV}$ for brand B after each accelerated exposure.

the activated degradation. The average and standard error (SE) of the change in each normalized electrical feature for both brands under each accelerated exposure are listed in Table II.

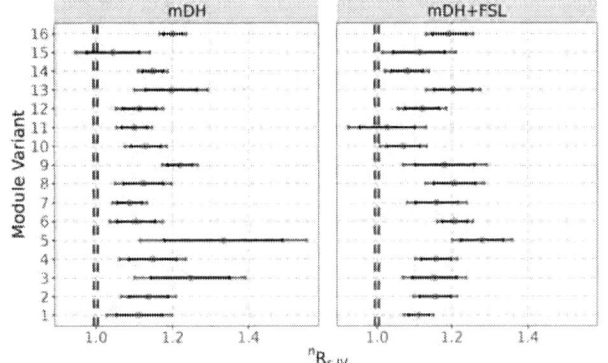

Fig. 5: $^{n}R_{s,IV}$ for brand A after each accelerated exposure.

B. Hierarchical Clustering of Principal Components to Rank Order Module Variants and Exposures

For each of the two brands A and B, the first three principal components explain 80.0% data variance, and they were used as input for the hierarchical clustering algorithm. The resulting

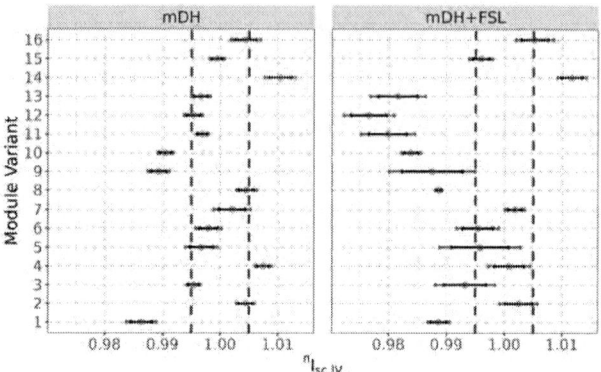

Fig. 6: $^{n}I_{sc,IV}$ for brand A after each accelerated exposure.

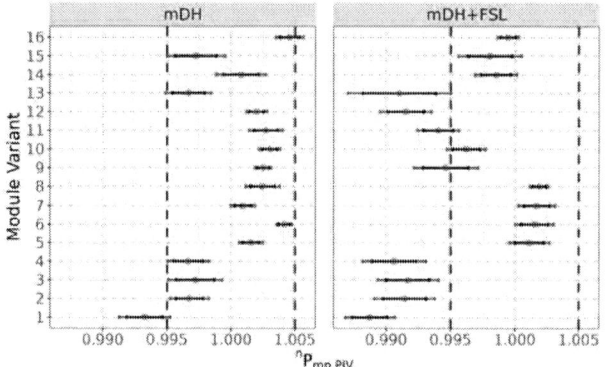

Fig. 7: $^{n}P_{mp,PIV}$ for brand A after each accelerated exposure.

dendrograms for brands A and B are shown in Fig. 8. Cutting the dendrogram at the height of 19.5 for both brands resulted in three clusters.

For the result of brand A minimodules, Fig. 9 shows the clustered points using the principal component basis with the variable vector. Table III and Table IV lists the number of observations in each cluster under different subsetting conditions for brands A and B, respectively. Table V lists the medium value for $^{n}F_{sd,EL@I_o}$, $^{n}P_{mp,IV}$, and $^{n}I_{sc,IV}$ for each cluster of each brand to associate the degree of degradation to each cluster. These three features are selected based on the importance of the variable to each principal component.

TABLE II: The average and standard error (SE) of the change in the four normalized electrical features, including $^{n}P_{mp,IV}$, $^{n}I_{sc,IV}$, $^{n}R_{s,IV}$, and $^{n}P_{mp,PIV}$, for brands A and B under each accelerated exposure.

Brand	Exposure	Average ± SE (%)			
		$^{n}P_{mp,IV}$	$^{n}I_{sc,IV}$	$^{n}R_{s,IV}$	$^{n}P_{mp,PIV}$
A	mDH	-6.59 ± 0.42	-0.14 ± 0.06	+15.46 ± 1.30	+0.02 ± 0.03
A	mDH+FSL	-6.93 ± 0.47	-0.72 ± 0.10	+14.82 ± 1.10	-0.41 ± 0.05
B	mDH	-9.78 ± 0.64	-0.42 ± 0.11	+24.49 ± 1.85	-0.23 ± 0.07
B	mDH+FSL	-8.87 ± 0.59	-0.60 ± 0.08	+21.32 ± 1.87	-0.58 ± 0.07

TABLE III: The number of observations for each cluster under different subsetting conditions, including encapsulant materials, module architectures, and cell types after each accelerated exposure for brand A.

Cluster ID	mDH						mDH+FSL					
	Encapsulant		Architecture		Cell Type		Encapsulant		Architecture		Cell Type	
	EVA	POE	DG	GB	Monofacial	Bifacial	EVA	POE	DG	GB	Monofacial	Bifacial
1	28	31	28	31	24	35	29	25	17	37	30	24
2	15	19	25	9	20	14	26	26	38	14	23	29
3	9	3	3	9	7	5	0	0	0	0	0	0

TABLE IV: The number of observations for each cluster under different subsetting conditions, including encapsulant materials, module architectures, and cell types after each accelerated exposure for brand B.

Cluster ID	mDH						mDH+FSL					
	Encapsulant		Architecture		Cell Type		Encapsulant		Architecture		Cell Type	
	EVA	POE	DG	GB	Monofacial	Bifacial	EVA	POE	DG	GB	Monofacial	Bifacial
1	22	20	16	26	29	13	18	16	15	19	26	8
2	14	33	30	17	18	29	23	36	38	21	24	35
3	8	1	1	8	6	3	7	1	0	8	4	4

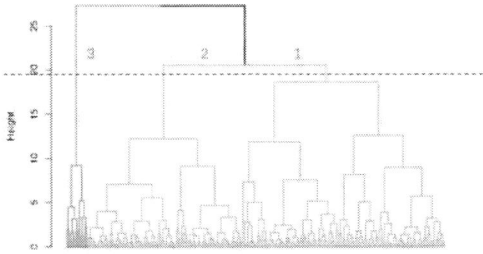

(a) Brand A.

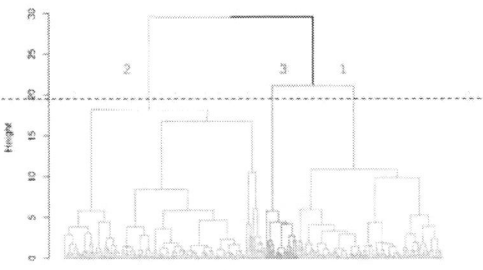

(b) Brand B.

Fig. 8: The hierarchical clustering dendrograms for brands A and B. Three clusters are obtained for each by cutting at the height of 19.5.

IV. DISCUSSION

A. Study Protocol for Parametric Variations across Module Variants and Exposures

Solar technology and products have changed rapidly in recent years, bringing significant challenges to reliability stud-

TABLE V: The medium value of selected features of each cluster for brands A and B.

Cluster ID	Brand	$^{n}F_{sd,EL@I_o}$	$^{n}P_{mp,IV}$	$^{n}I_{sc,IV}$
1	A	1.05	0.940	0.997
2	A	0.979	0.926	0.998
3	A	1.28	0.918	0.997
1	B	1.19	0.900	0.992
2	B	1.03	0.927	1.00
3	B	1.65	0.859	0.986

ies. In most studies on comparing the PV module reliability performance, only average characterization results have been used, ignoring confidence intervals caused by sample and measurement uncertainties. Most studies evaluated only the P_{mp} and paid less attention to features related to specific degradation mechanisms. A sufficient number of samples are essential for evaluating the statistical significance of the results. In our study, eight cells laminated in two minimodules were used to compare each module variant of each brand under each accelerated exposure. Such sample size allows us to use the 83.4% confidence interval (CI) to visualize the t-test result for comparing two module variants. Our study evaluated sixteen module variants of two brands, A and B, under two different accelerated exposures: mDH and mDH+FSL. The sixteen module variants take different encapsulant materials (EVA and POE), encapsulant types, cell types, and module architectures (GB and DG) into account. The detailed specifications of the sixteen module variants are listed in Table I. The two accelerated exposures were chosen to evaluate the module reliability against high temperature and humid conditions, with or without operating solar cells. It is worth noting that the minimodules tend to amplify the degree of degradation due to the reduced size compared to the commercial PV module, which usually contains about 60 cells or more.

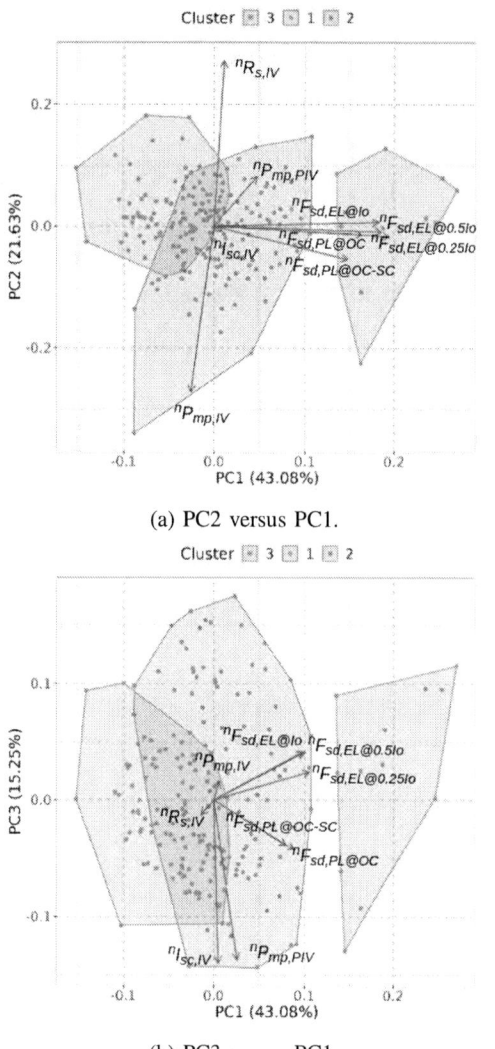

(a) PC2 versus PC1.

(b) PC3 versus PC1.

Fig. 9: Principal component scores for brand A of each input observations, colored by the clusters identified by the hierarchical clustering result. The arrow displays the loading of each variable, of which the projected length can be understood as the weight for each original variable when calculating the principal component.

B. Performance and Degradation Mechanism Features

$^{n}P_{mp,IV}$ is chosen as an indicator for the overall degradation since the power output decides the performance of PV modules. In addition, $^{n}R_{s,IV}$, $^{n}I_{sc,IV}$ and $^{n}P_{mp,PIV}$ were selected as indicators for different degradation mechanisms. It is worth mentioning that the features selected in this study can also be obtained from the outdoor timeseries I-V curves through modeling methods [27], [28]. Thus, it provides common variables for data-driven methods to compare the degradation behaviors of PV modules under indoor accelerated and outdoor exposures. EL reveals the spatial information of the solar cell through light emission under current excitation,

while PL reveals the spatial information through light emission under light excitation. The standard deviation is an indicator for measuring signal uniformity. If local defects are presented in the solar cell and detectable under the excitation mechanism, the standard deviation of the corresponding image will increase.

C. Rank Ordering of Variant Factors on Degradation and Performance

This section discusses the differences in the reliability performance of the sixteen module variants of brand A modules under each accelerated exposure. From Fig. 3, under mDH, the average $^{n}P_{mp,IV}$ drop for module variants 1, 2, 4, and 15 is lower than 5%, and these module variants are considered to be stable. Module variants 5, 9, and 16 under mDH are observed to have a significantly greater average power loss than these stable module variants. A similar analysis is conducted for module variants under mDH+FSL exposure. Module variants 1, 2, 4, 11, 14 are found to have an average $^{n}P_{mp,IV}$ drop that is less than 5%. Then module variants 5, 6, 8, 9, and 16 are identified to have a significantly greater average power loss. Therefore, module variants 5, 9, and 16 perform relatively poorly under both accelerated exposures. Both module variants 5 and 9 are EVA+GB modules with opaque rear encapsulant, while module variant 16 is for the bifacial POE+DG modules. Module variants 1, 2 and 4 are identified to be stable under both exposures and they are EVA+GB, EVA+DG and POE+DG modules with UV-Cutoff rear encapsulant, respectively.

From the results of $^{n}R_{s,IV}$, $^{n}I_{sc,IV}$, and $^{n}P_{mp,PIV}$, shown in Fig. 5, Fig. 6, and Fig. 7, respectively, $^{n}R_{s,IV}$ has changed much more significantly than the other two features. The average change in $^{n}R_{s,IV}$ is much higher than that of $^{n}I_{sc,IV}$ and $^{n}P_{pm,PIV}$ for brand A modules under both accelerated exposures as shown in Table II. While more changes occur in $^{n}I_{sc,IV}$ and $^{n}P_{mp,PIV}$ for modules under mDH+FSL than mDH, they are still much smaller than the change in $^{n}R_{s,IV}$. Several module variants have a relatively higher $^{n}R_{s,IV}$ shown in Fig. 5. Under mDH, module variants 3, 5, 9, 13, and 16 have an average $^{n}R_{s,IV}$ increase of over 19%, and under mDH+FSL, module variants 5, 6, 8, 9, 13, and 16 have an average $^{n}R_{s,IV}$ increase of over 18%. The module variants that have a more significant increase in the average of $^{n}R_{s,IV}$ include all module variants identified as relatively unstable. Furthermore, none of the module variants with an average power loss of less than 5% are identified in the module variants with a relatively greater $^{n}R_{s,IV}$ increase, which is another evidence that corrosion dominates power loss for modules under both defined exposures. The largest $^{n}I_{sc,IV}$ reduction is at 1.37% and 2.34% for mDH and mDH+FSL, respectively. The change in $P_{mp,PIV}$ is even minor with all cases expect module variant 1 under mDH+FSL having an average change less than 1%.

The PCA result of brand A modules using the centered and unit variance scaled data as the input is shown in Fig. 9. From Fig. 9, the first principal component (PC_1) is found

978-1-7281-6118-1/22 $31.00 © 2022 IEEE

to represent most image features. PC_2 is mainly influenced by $^nR_{s,IV}$ and $^nP_{mp,IV}$. The opposite directions of these two feature loadings is another evidence that the increased R_s dominates the power reduction. Moreover, PC_3 is mainly influenced by $^nI_{sc,IV}$ and $^nP_{mp,PIV}$ with some ability to separate features obtained from EL and PL. Using the first three principal components obtained from the scaled and centered data as the input to the clustering algorithm, the resulting clusters avoid influences from the variance difference of selected features and the selection of correlated features. Based on the dendrogram shown in Fig. 8 for brand A, the two large clusters are more similar to each other than the tiny cluster in the red color. Table V lists the median value for $^nF_{sd,EL@I_o}$, $^nP_{mp,IV}$, and $^nI_{sc,IV}$ to associate the degree of degradation to each cluster. Cluster 3 has the most significant power loss and $^nF_{sd,EL@I_o}$ increase. Therefore, it has the highest degree of degradation. Cluster 1 has the least power loss with a slightly increased $^nF_{sd,EL@I_o}$. Cluster 2 has the smallest $^nF_{sd,EL@I_o}$ but a slightly greater power loss than that of cluster 1.

Table III reveals the dependency of these clusters on the module specifications. Cluster 3 is only made of samples under mDH, and it has 50% more EVA samples than POE, and 50% more GB samples than DG. However, cluster 3 has a smaller number of observations. Therefore, the revealed dependency does not indicate average performance differences between EVA and POE or DG and GB. A better interpretation is that a few samples under the category of EVA or GB have a higher risk of performing like outliers than modules of its counterpart. In addition, cluster 2 is found to have 47.1% more DG samples than GB samples under mDH. Under mDH+FSL, cluster 1 has 37.0% more GB samples, and cluster 2 has 46.2% more DG samples. Therefore, unsupervised hierarchical clustering result finds that the identified clusters of brand A minimodules mainly depend on the module architecture rather than the encapsulant material or the cell type.

D. Impact of Manufacturing Variability on Degradation and Performance: Two Brands

This section discusses the degradation performance of brand B minimodules and compares the results to brand A. From the $^nP_{mp,IV}$ result shown in Fig. 4, module variants 2 and 4 have an average $^nP_{mp,IV}$ drop of less than 5% under mDH, which is also significantly less than that of module variants 5, 7, 9, and 13. These four module variants are found to have an average $^nR_{s,IV}$ increased by 25%. Under mDH+FSL, module variants 2, 4, and 10 have an average $^nP_{mp,IV}$ drop of less than 5%, which is also significantly lower than that of module variants 5, 7, 9, 11, 13, 15, and 16. Among them, module variants 5, 7, 11, and 13 have an average $^nR_{s,IV}$ increase of over 25%, and the rest have an average increase of over 18.5%. Therefore, module variants 2 and 4, two DG module variants with UV-Cutoff rear encapsulant are identified as the stable module variants, and module variants 5 and 9, two EVA+GB with opaque rear encapsulant module variants are identified as

the relatively unstable module variants in both brands under both accelerated exposures.

The average amount of change in $^nP_{mp,IV}$ and $^nR_{s,IV}$ is greater for brand B modules than for brand A as shown in Table II, indicating reliability performances differences due to manufacturing (lamination, soldering, and cell storage) since the packaging materials used are the same. Interconnection corrosion is again identified as the dominant degradation mechanism contributing to the power loss of brand B modules. The influence of different features on each principal component for brand B modules is similar to that of brand A. In the dendrogram shown in Fig. 8, cutting at the same height at 19.5 results in three clusters again, in which two clusters are relatively large. However, the small cluster is more similar to one large cluster colored in blue than the other large cluster. In brand A, the two large clusters are from the same branch and more similar to each other. Therefore, the reliability performance of most brand A minimodules is more consistent than that of brand B. Based on the median feature value listed in Table V for brand B modules, cluster 3 is found to have the highest level of degradation indicated by the pronounced $^nP_{mp,IV}$ drop and $^nF_{sd,EL@I_o}$ increase, and cluster 2 has the slightest degree of degradation.

From Table IV, brand B results have more complex dependencies for the cluster separation than that of brand A. Under mDH, most observations under cluster 3 are EVA samples or GB samples, which is similar to that of brand A. However cluster 3 also has 25% more samples that use monofacial cells than that use bifacial cells. Under mDH, the most significant differences in the number of samples for cluster 2 occurs between encapsulant materials at 40.4%. In addition, apparent differences are also presented in different module architectures and cell types for cluster 2 under mDH. Differences between module architectures and cell types are also shown in cluster 1 under mDH, and the most prominent difference occurs between cell types at 38.1%. Under mDH+FSL, we find cluster 3 has more EVA samples and more GB samples than their counterparts. Cluster 1 has 52.9% more samples with monofacial cells than bifacial cells. The most noticeable difference in cluster 2 under mDH+FSL occurs between module architectures, but it is only 28.8% due to many observations in cluster 2. Comparing Table III and Table IV, the difference in reliability performance caused by cell types and encapsulant materials could be reduced by a more strict manufacturing quality control. The difference between cell types shown in brand B modules could be related to the storage duration and the compatibility with the manual soldering process.

V. CONCLUSIONS

By comparing the characterization results of sixteen module variants from two brands under the accelerated exposure of mDH and mDH+FSL of up to 2,520 hours, two DG with UV-Cutoff rear encapsulant module variants are identified to experience an average power loss of less than 5%, which is significantly less than that of two EVA+GB with opaque rear encapsulant module variants. The power loss under both accelerated exposures is mainly due to interconnection corrosion as indicated by the increase in series resistance. With the same exposed hours, mDH+FSL leads to more changes in $^{n}I_{sc,IV}$ and $^{n}P_{mp,PIV}$ than mDH but similar changes in $^{n}P_{mp,IV}$ and $^{n}R_{s,IV}$ on average for both brands. The manufacturing process is found to influence the reliability performance. Brand A modules, which have a more strict quality control in manufacturing, experience less change than brand B in power and series resistance. The unsupervised hierarchical clustering result shows that the reliability performance of most brand A modules only has a dependency on the module architecture, which is GB or DG. However, the cluster of brand B modules shows differences in the number of observations between encapsulant materials, module architectures, and cell types. Therefore, different fabrication process could lead to the differences in the reliability performance caused by encapsulant materials and cell types under our accelerated exposure conditions.

References

[1] Walburga Hemetsberger, Micheal SchMela, and Gianni Chianetta, "Global Market Outlook 2021-2025 – SolarPower Europe," Solar Power Europe, Tech. Rep., Jul. 2021.

[2] W. Gambogi, M. Demko, B.-L. Yu, S. Kurian, S. MacMaster, K. R. Choudhury, J. Tracy, D. Hu, and H. Hu, "Transparent Backsheets for Bifacial Photovoltaic Modules," in *2020 47th IEEE Photovoltaic Specialists Conference (PVSC)*, Jun. 2020, pp. 1651–1657.

[3] A. Omazic, G. Oreski, M. Halwachs, G. C. Eder, C. Hirschl, L. Neumaier, G. Pinter, and M. Erceg, "Relation between degradation of polymeric components in crystalline silicon PV module and climatic conditions: A literature review," *Solar Energy Materials and Solar Cells*, vol. 192, pp. 123–133, Apr. 2019.

[4] B. Lin, C. Zheng, Q. Zhu, and F. Xie, "A polyolefin encapsulant material designed for photovoltaic modules: From perspectives of peel strength and transmittance," *Journal of Thermal Analysis and Calorimetry*, vol. 140, no. 5, pp. 2259–2265, Jun. 2020.

[5] G. Oreski, A. Omazic, G. C. Eder, Y. Voronko, L. Neumaier, W. Mühleisen, C. Hirschl, G. Ujvari, R. Ebner, and M. Edler, "Properties and degradation behaviour of polyolefin encapsulants for photovoltaic modules," *Progress in Photovoltaics: Research and Applications*, vol. 28, no. 12, pp. 1277–1288, 2020.

[6] Q. Wang, "Research on the effect of encapsulation material on anti-PID performance of 1500,V solar module," *Optik*, vol. 202, p. 163540, Feb. 2020.

[7] D. B. Sulas-Kern, M. Owen-Bellini, P. Ndione, L. Spinella, A. Sinha, S. Uličná, S. Johnston, and L. T. Schelhas, "Electrochemical degradation modes in bifacial silicon photovoltaic modules," *Progress in Photovoltaics: Research and Applications*, vol. n/a, no. n/a, Dec. 2021.

[8] B. M. Habersberger and P. Hacke, "Impact of illumination and encapsulant resistivity on polarization-type potential-induced degradation on n-PERT cells," *Progress in Photovoltaics: Research and Applications*, vol. n/a, no. n/a, Oct. 2021.

[9] C. Barretta, G. Oreski, S. Feldbacher, K. Resch-Fauster, and R. Pantani, "Comparison of Degradation Behavior of Newly Developed Encapsulation Materials for Photovoltaic Applications under Different Artificial Ageing Tests," *Polymers*, vol. 13, no. 2, p. 271, Jan. 2021.

[10] A. Skoczek, T. Sample, and E. D. Dunlop, "The results of performance measurements of field-aged crystalline silicon photovoltaic modules," *Progress in Photovoltaics: Research and Applications*, vol. 17, no. 4, pp. 227–240, 2009.

[11] A. P. Patel, A. Sinha, and G. Tamizhmani, "Field-Aged Glass/Backsheet and Glass/Glass PV Modules: Encapsulant Degradation Comparison," *IEEE Journal of Photovoltaics*, vol. 10, no. 2, pp. 607–615, Mar. 2020.

[12] R. E. Kumar, G. Von Gastrow, N. Theut, A. M. Jeffries, T. Sidawi, A. Ha, F. DePlachett, H. Moctezuma-Andraca, S. Donaldson, M. I. Bertoni, and D. P. Fenning, "Glass vs. Backsheet: Deconvoluting the Role of Moisture in Power Loss in Silicon Photovoltaics With Correlated Imaging During Accelerated Testing," *IEEE Journal of Photovoltaics*, vol. 12, no. 1, pp. 285–292, Jan. 2022.

[13] Y. Zhang, J. Xu, J. Mao, J. Tao, H. Shen, Y. Chen, Z. Feng, P. J. Verlinden, P. Yang, and J. Chu, "Long-term reliability of silicon wafer-based traditional backsheet modules and double glass modules," *RSC Advances*, vol. 5, no. 81, pp. 65 768–65 774, Jul. 2015.

[14] T. C. Felder, W. Gambogi, H. Hu, T. J. Trout, L. Garreau-Iles, S. MacMaster, and K. R. Choudhury, "Analysis of glass-glass modules," in *New Concepts in Solar and Thermal Radiation Conversion and Reliability*, J. N. Munday, P. Bermel, and M. D. Kempe, Eds. San Diego, United States: SPIE, Sep. 2018, p. 1.

[15] J. Tang, C. Ju, R. Lv, X. Zeng, J. Chen, D. Fu, J.-N. Jaubert, and T. Xu, "The Performance of Double Glass Photovoltaic Modules under Composite Test Conditions," *Energy Procedia*, vol. 130, pp. 87–93, Sep. 2017.

[16] R. Fisher, "Statistical Methods and Scientific Induction," *Journal of the Royal Statistical Society. Series B (Methodological)*, vol. 17, no. 1, pp. 69–78, 1955.

[17] Greg Horner, "PixEL: Versatile Electroluminescence and Photoluminescence System," May 2022. [Online]. Available: http://tauscience.com/pixel

[18] Jennifer L. Braid, Ahmad Maroof Karimi, Benjamin G. Pierce, Justin S. Fada, Nicholas A. Parrilla, and Roger H. French, "Pvimage: Package for pv image analysis and machine learning modeling," May 2020.

[19] W.-H. Huang, Xuan Ma, Jiqi Liu, Menghong Wang, Alan J. Curran, J. S. Fada, Jean-Nicolas Jaubert, Jing Sun, Jennifer L. Braid, Jenny Brynjarsdottir, and Roger H. French, "Ddiv: Data Driven I-v Feature Extraction," Comprehensive R Archive Network (CRAN), Apr. 2021.

[20] R. A. Sinton and A. Cuevas, "A quasi-steady-state open-circuit voltage method for solar cell characterization," in *16th European Photovoltaic Solar Energy Conference*, Glasgow, UK, May 2000.

[21] K. Pearson, "LIII. On lines and planes of closest fit to systems of points in space," *The London, Edinburgh, and Dublin Philosophical Magazine and Journal of Science*, vol. 2, no. 11, pp. 559–572, Nov. 1901.

[22] G. James, D. Witten, T. Hastie, and R. Tibshirani, *An Introduction to Statistical Learning: 2nd Ed., with Applications in R,*, 2nd ed. New York: Springer, Jul. 2021.

[23] F. Murtagh and P. Legendre, "Ward's Hierarchical Agglomerative Clustering Method: Which Algorithms Implement Ward's Criterion?" *Journal of Classification*, vol. 31, no. 3, pp. 274–295, Oct. 2014.

[24] G. Cumming and S. Finch, "Inference by eye: Confidence intervals and how to read pictures of data," *American Psychologist*, pp. 170–180, 2005.

[25] A. Hazra, "Using the confidence interval confidently," *Journal of Thoracic Disease*, vol. 9, no. 10, pp. 4124–4129, Oct. 2017. [Online]. Available: http://jtd.amegroups.com/article/view/16406/13455

[26] J. Leppink, "A Pragmatic Approach to Statistical Testing and Estimation (PASTE)," *Health Professions Education*, vol. 4, no. 4, pp. 329–339, Dec. 2018. [Online]. Available: https://www.sciencedirect.com/science/article/pii/S2452301117301487

[27] J. Liu, M. Wang, A. J. Curran, E. Schnabel, M. Köhl, J. L. Braid, and R. H. French, "Degradation mechanisms and partial shading of glass-backsheet and double-glass photovoltaic modules in three climate zones determined by remote monitoring of time-series current–voltage and power datastreams," *Solar Energy*, vol. 224, pp. 1291–1301, Aug. 2021.

[28] M. Wang, Tyler J. Burleyson, Jiqi Liu, Alan J. Curran, Abdulkerim Gok, Eric J. Schneller, Kristopher O. Davis, Jennifer L. Braid, and Roger H. French, "SunsVoc: Constructing Suns-Voc from Outdoor Time-Series I-V Curves," Comprehensive R Archive Network (CRAN), Apr. 2021.

Chlorine Doped n-Type CdTe Solar Cells

Wei Wang, Vasilios Palekis, Md Zahangir Alom, Sheikh Elahi Tawsif, and Chris Ferekides

Electrical Engineering, University of South Florida, Tampa, FL 33620, USA

Abstract— **This paper investigated the effect of chlorine (Cl) doping on CdTe thin film solar cells. Polycrystalline CdTe films are grown by the elemental vapor transport (EVT) process with various Cd/Te gas phase ratios and Cl vapor concentrations. Solar cell devices of the configuration glass/TCO/CdS/n-CdTe/p-ZnTe/BC have been fabricated and characterized. Increasing the Cl vapor concentration results in higher n-type doping in CdTe. Films deposited at higher Cd/Te vapor ratios (i.e. under Cd-rich vapor composition) exhibited higher n-type doping. The higher Cd/Te ratios facilitate the creation of Te-vacancies, which are required for substitutional Cl doping, resulting in the net doping increases at higher Cd/Te ratios. Minority carrier lifetimes ~ 7 ns have been measured for Cl doped CdTe absorbers by TRPL measurements.**

Keywords—CdTe solar cell, chloride, n-type doping, minority carrier lifetime

I. INTRODUCTION

Cadmium telluride (CdTe) is one of the most promising materials for the fabrication of low-cost, high-efficiency thin-film solar cells. This is due to its near-ideal band gap (~1.5eV) and high optical absorption coefficient ($>10^4$/cm) in the range of visible wavelengths. Recently, the efficiency of CdTe solar cells has improved significantly, with a world record of 22.1%, as demonstrated by First Solar [1].

CdTe can be intrinsically or extrinsically doped to achieve p- or n-type conductivity. Due to the presence of intrinsic defects, p-type or n-type CdTe films can be achieved by controlling the film stoichiometry. Under Cd-rich (Cd/Te ratio >1) conditions, the intrinsic CdTe films will be n-type due to the creation of Te vacancies, and on the other hand, under Te rich (Cd/Te ratio <1) conditions, the intrinsic CdTe films will be p-type. As for the extrinsic doping, p-type CdTe films can be achieved by replacing Cd and Te sites with groups-I (Cu, Au) and -V (As, P, Sb) elements [2], respectively. Similarly, n-type CdTe films can be achieved by replacing Cd and Te sites with groups-III (In, Al) and -VII (Cl, I) elements, respectively [3].

One of the most challenging problems that limits CdTe devices is the difficulty to achieve high hole concentration doping in p-CdTe. Moreover, it is even harder to achieve high majority carrier concentration and long minority carrier lifetime at the same time. Recently, it has been shown that high p-type doping (10^{16}-10^{17} cm^{-3}) was achieved by in situ doping with As [4].

On the other hand, highly doped n-type CdTe is easier to obtain compared to p-type. Several studies have shown high concentration of n-type doping (10^{16}-10^{19} cm-3) in CdTe using Bridgman or molecular beam epitaxy (MBE) methods [5] [6] [7]. Recent results have shown that n-type single-crystal CdTe solar cells doped with In have a V_{OC} of more than 1 volt and a lifetime of more than 2 µs [8].

This paper presents experimental results on the performance of polycrystalline thin film devices fabricated at different Cd/Te ratios and different Cl doping concentrations.

II. EXPERIMENTAL

The device configuration of the n-CdTe/p-ZnTe solar cells is shown in *Fig. 1* The device structure includes Glass/TCO/CdS/CdTe/ZnTe/Cu/ITO.

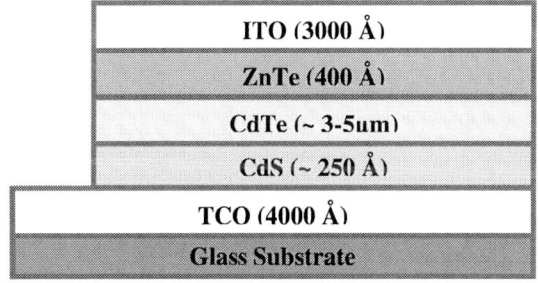

Fig. 1. Baseline device configuration of n-CdTe/p-ZnTe

The cells are fabricated on corning EagleXG glass. The transparent contact (TC) is indium tin oxide (ITO) which is deposited by RF sputtering on the glass substate. Cadmium sulfide (CdS) is then deposited by close-spaced sublimation (CSS). Polycrystalline CdTe films are grown by the elemental vapor transport (EVT) process. The EVT process is used to achieve stoichiometric control of the Cd/Te ratio and various dopant concentrations. Cadmium chloride is used as the n-type dopant source. Zinc telluride (ZnTe) is used as the p-type window layer and is deposited by sputtering at a substrate temperature of 300 °C. Finally, a 3000Å ITO is used as the top n-type electrode contact to complete the device.

Current-voltage (J-V), spectral response (SR), capacitance-voltage(C-V) and time-resolved photoluminescence (TRPL) measurements were used to characterize CdTe devices and films. J-V measurements of the devices were performed with a four-terminal connection and a Keithley 2410 source meter. SR measurements were performed using an Oriel monochromator

978-1-7281-6118-1/22 $31.00 © 2022 IEEE

(model 74100) with a light source whose intensity was calibrated using a standard calibrated silicon reference solar cell. C-V measurements were done with a HP 4194A impedance/gain-phase analyzer.

We are thankful to Steve Johnston and Mowafak Al-Jassim of the National Renewable Energy Laboratory (NREL) and to Charles Hages of the University of Florida for the TRPL measurements.

III. RESULT AND DISCUSSION

In our previous work, we were using In as our n-type dopant. Indium occupying a Cd vacancy (In_{Cd}) is a shallow donor defect that can increase the n-type conductivity of CdTe. It was found that CdTe films deposited under Cd rich conditions resulted in higher lifetimes [9]. Under Cd-rich conditions it would be more suitable to employ a group VII dopant and replace Te. Cl occupying a Te vacancy (ClTe) is also a shallow donor defect that can increase the n-type doping of CdTe. Cd-rich conditions will favor the creation of Te vacancies needed for the formation of the Cl_{Te} defect. CdTe films and solar cells have been prepared using the EVT process. N-type CdTe films were deposited under various Cd/Te vapor ratios and under various Cl vapor concentrations. The results are presented below.

A. Cl as n-type Dopant for CdTe

Two Cl vapor concentrations were used: 500 ppm and 1Kppm. The Cd/Te ratio was varied from intrinsic to Cd-rich conditions to facilitate Cl incorporation into Te vacancies. Cd/Te ratios used: 1.0, 1.5, 2.0, and 3.0. *Fig. 2* shows the spectral response data for devices deposited at various Cd/Te ratios and two Cl vapor concentrations; (top) 500 ppm, and (bottom) 1K ppm. Low carrier collection is observed on the top figure for the devices deposited near stoichiometric ratios; 1.0, and 1.5. These devices behave *like intrinsic* solar cells based on previous results for undoped films. As the Cd/Te ratio becomes more Cd-rich carrier collection increases. Devices shown to the bottom in *Fig. 2* have similar spectral response except the one deposited at the highest Cd/Te ratio.

TABLE 1 lists the net n-type doping calculated from C-V measurements. These results indicate that as the ratio shifts from intrinsic to Cd-rich the net doping increases. Increasing the amount of Cl vapor concentration also results in higher n-type doping in CdTe. These results are consistent with the hypothesis that Cd-rich conditions should favor the creation of V_{Te} and at higher $CdCl_2$ vapor concentration there is more Cl_{Te} donor defects formed. The highest doping level achieved was $6.57E+16$ cm^{-3}. *Fig. 3* shows the net doping of devices deposited at Cd/Te ratio of 1.5 with various Cl vapor concentrations. There is a clear trend of net doping increase with an increase in Cl vapor concentration. A similar trend was observed for devices deposited at Cd/Te ratio 2.0.

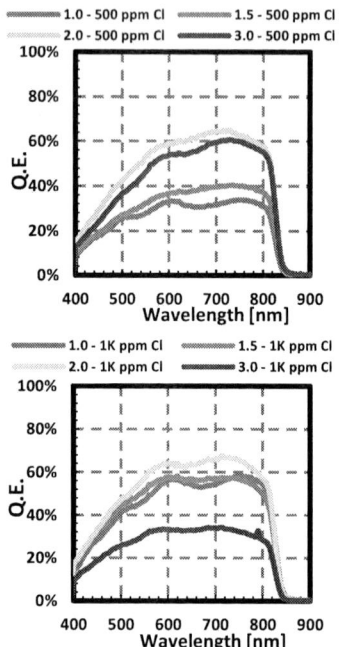

Fig. 2. SR data for cells deposited at various Cd/Te ratios and two Cl vapor concentrations: (top) 500 ppm, and (bottom) 1K

TABLE 1. NET DOPING VS. CD/TE RATIO AND 2 CL VAPOR CONCENTRATIONS

Cd/Te Ratio – Cl Vapor Concentration [ppm]	Net Doping [cm^{-3}]
1.0 - 500	Depleted
1.5 – 500	Depleted
2.0 – 500	4.32E+15
3.0 - 500	6.58E+15
1.0 – 1K	7.57E+15
1.5 – 1K	1.38E+16
2.0 – 1K	1.02E+16
3.0 – 1K	6.57E+16

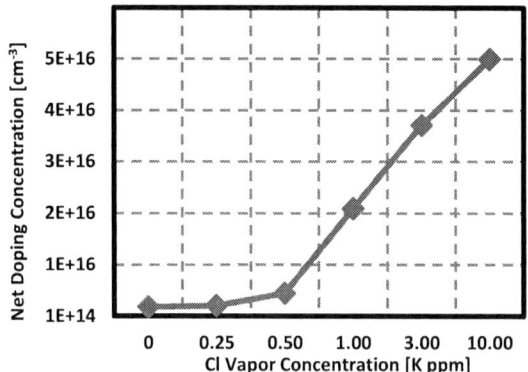

Fig. 3. Net doping versus Cl vapor concentration

B. Lifetime Measurements for Cl Doped CdTe Absorbers and Solar Cell Devices

A series of n-CdTe absorber layers were fabricated for lifetime measurements to investigate the effects of Cd/Te ratio and Cl concentration on the quality of the absorber layer. The structure configuration was: glass/TCO/CdS/CdTe. Four Cd/Te ratios and four Cl vapor concentrations for a total of 16 deposition conditions were used for this study.

- Cd/Te vapor ratios: 1.5, 2.0, 2.5, and 3.0

- Chlorine vapor concentrations: 0, 1, 10, and 20K ppm

Fig 4 shows the lifetime measurements for the n-CdTe absorber layers (glass/TCO/CdS/CdTe). These films were deposited at Cd/Te 1.5, 2.0, 2.5, and 3.0 ratios doped with 10K ppm Cl. The longest minority-carrier lifetime, ~ 7 ns, was achieved for the film deposited at the lowest Cd/Te ratio (1.5). Absorbers deposited under higher Cd/Te ratios had similar lifetimes around ~ 1.2 ns. It's not clear at this time why lifetimes decrease at higher Cd/Te ratios. From *TABLE 1* there is an increase in doping concentration at higher Cd/Te ratios. Cd-rich conditions create more V_{Te} which result in the formation of more Cl_{Te} donor defects. It is possible that the formation of these excess defects affects the lifetime of the absorber. TRPL measurements for the rest of the films is under way and will be presented in the future.

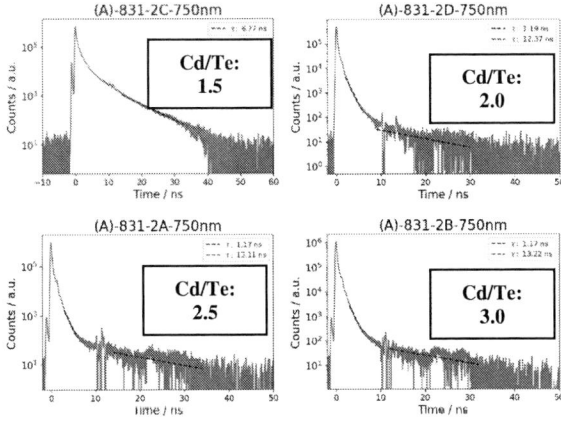

Fig 4. TRPL decays for CdTe absorbers deposited at various Cd/Te ratios doped with 10K ppm Cl

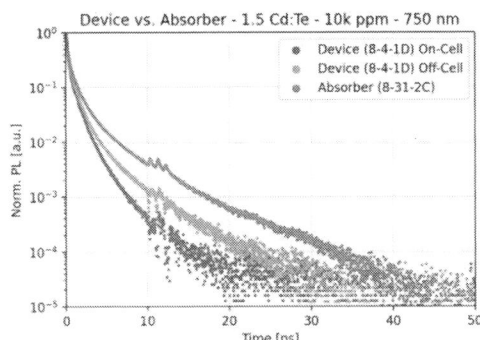

Fig. 5. TRPL decays of the absorber Vs a complete device for CdTe deposited at the same conditions

Fig. 5 shows the TRPL decays for the CdTe absorber versus a complete solar cell. In both cases the CdTe was deposited under the same conditions. The structure of these devices is:

- Absorber: glass/TCO/CdS/CdTe

- Complete device: glass/TCO/CdS/CdTe/ZnTe/Cu/ITO

From the decays the absorber in unfinished cells has a higher lifetime when compared to the completed device. Possible explanation for the lower lifetimes in completed devices are (1) interface recombination at the junction of CdTe/ZnTe, or (2) Cu diffusing into the n-CdTe causing compensation.

IV. CONCLUSION

The elemental vapor transport process was used to grow n-type CdTe films using various Cl gas phase concentrations. Net doping concentration increases with an increase in Cl concentration. Higher Cd/Te ratios (Cd-rich conditions) favor the creation of V_{Te}, which is needed for the formation of the n-type Cl_{Te} defect, result in higher net n-type doping. TRPL measurements indicate that minority carrier lifetimes ~ 7 ns have been measured for Cl doped CdTe absorbers. Lifetimes decrease for solar cell devices.

ACKNOWLEDGMENT

This work was supported in full by the Department of Energy under award DE-EE0008745. TRPL measurements of n-CdTe films were performed by professor's Charles Hages group at the University of Florida.

REFERENCES

[1] M. A. Green, E. D. Dunlop, J. Hohl-Ebinger, M. Yoshita, N. Kopidakis and X. Hao, "Solar cell efficiency tables (version 59)," *Progress in Photovoltaics: Research and Applications*, pp. 3-12, 2022.

[2] I. S. Khan, V. K. Evani, V. Palekis and C. Ferekides, "Effect of Stoichiometry on the Lifetime and Doping Concentration of Polycrystalline CdTe," in *IEEE Journal of Photovoltaics*, 2017.

[3] Y. Marfaing, "Impurity doping and compensation mechanisms in CdTe," *Thin Solid Films*, vol. 387, no. 1-2, pp. 123-128, 2001.

[4] W. K. Metzger, S. Grover, D. Lu, E. Colegrove, J. Moseley, C. L. Perkins, X. Li and R. Mallick, "Exceeding 20% efficiency with in situ group V doping in polycrystalline CdTe solar cells," *Nature Energy*, vol. 4, no. 10, pp. 837-845, 2019.

[5] M. Turker, J. Kronenberg, M. Deicher, H. Wolf and T. Wichert, "Formation of DX-centers in indium doped CdTe," in *HFI/NQI*, 2007.

[6] S. Seto, K. Suzuki, J. Abastillas, N. V and K. Inabe, "Compensating related defects in In-doped bulk CdTe," *Journal of crystal growth*, 2000.

[7] F. Bassani, S. Tatarenko, K. Saminadayar, J. Bleuse, N. Magnea and J. L. Pautrat, "Luminescence characterization of CdTe: In grown by molecular beam epitaxy," *Applied physics letters*, 1991.

[8] Y. Zhao, M. Boccard, S. Liu, J. Becker, X. H. Zhao, C. M. Czmpbell, E. Suarez, M. B. Lassise, Z. Holman and Y. H. Zhang, "Monocrystalline CdTe solar cells with open-circuit voltage over 1 V and efficiency of 17%," *Nature Energy*, 2016.

[9] V. Palekis, W. Wang, S. E. Tawsif, Z. A. Md and C. Ferekides, "Thin Film Solar Cells with n-type CdTe Absorber and p-type ZnTe Window Layers," in *2021 IEEE 48th Photovoltaic Specialists Conference (PVSC)*, 2021.

26.7% AM0, 30.2% AM1.5G dual junction solar cell with 50x InGaAs quantum wells, GaAsP strain compensation, and distributed Bragg reflector

Stephen J Polly, Brandon Bogner, Anastasiia Fedorenko, Subhra Chowdhury, Dhrubes Biswas, Seth M Hubbard

Rochester Institute of Technology, Rochester, NY, United States

Magnolia Optical Technologies, Inc., Woburn, MA, United States

Dual junction upright solar cells composed of a homojunction InGaP top cell and a heterojunction InGaP/GaAs bottom cell incorporating 50 layers of $In_{0.1}Ga_{0.9}As$ with $GaAs_{0.9}P_{0.1}$ and a 12-pair $Al_{0.9}Ga_{0.1}As/Al_{0.1}Ga_{0.9}As$ distributed Bragg reflector were designed for operation under either AM0 or AM1.5G and grown by MOVPE. The QWs allowed an increase in current density of 1.98 mA/cm2 under AM0 (1.07 mA/cm2, AM1.5G design), with an exhibited Voc reduction of only 10 mV as compared to a control tandem without QWs or a DBR. This lead to a efficiency increase of 2.55% absolute under AM0 (1.79% absolute, AM1.5G design). Details on growth considerations for this system of strain balanced QWs, and development of single and dual junction device designs, will be presented.

Spectral rear irradiance testing and modeling for degradation and performance of solar fields

Silvana Ovaitt, Matthew Brown, Chris Deline, Michael D. Kempe

National Renewable Energy Laboratory, Golden CO, US 80401, UC Boulder

Abstract—This work investigates how the spectrum of irradiance incident on the rear of solar modules impacts the degradation and performance of backsheets. We model the spectral irradiance incident on the rear of modules through raytrace simulations and validate with measured field data collected from a 75kW single axis-tracked bifacial test site. A generic equation to estimate relative degradation is proposed, and we show that current acceleration factors for UV damage in chambers can be sub-estimate up to 4.5% absolute from the usually assumed 10% dosage on the rear surfaces.

Keywords—solar PV, rear irradiance, UV degradation, bifacial performance

I. INTRODUCTION

Recently, ~90% of the photovoltaic (PV) modules in the field are less than ten years old. There are significant unknowns regarding the durability of new technologies and materials being deployed without established performance hisotories. While accelerated testing attempts to capture degradation modes, the fast roll-out of new materials poses a challenge resulting in high probabilities of false-negaitve or false-positive results. In particular, the aging behavior and failure modes of backsheets is of interest to ensure PV modules' performance. Current degradation testing for backsheets commonly assumes 10% of UV dosage on the back relative to the front. This study explores that assumption.

Light on a module's front side comes primarily from direct solar irradiation, some diffuse scattering from the sky, and small contributions from light scattered from ground sources. In contrast, light incident on the rear of PV modules may have no direct irradiation and only a smaller portion of diffuse light from the sky. Here, a significant amount of light comes from reflection from the ground, mounting structures, and other nearby objects. However, most surfaces have lower reflection in the UV range relative to visible light. This causes the intensity and spectrum to be significantly modified relative to the front side exposure (Fig. 1). However for special cases such as snow, measurements not only have a stronger spectral content but also appear 'blue-shifted' as there is less absorption in the NIR and VIS spectrum. As a result, UV-irradiance levels and resulting degradation experienced by PV backsheets is dependent on the mounting structure of the array, the location of the modules within the array, the environmental albedo and weather conditions including temperature, and spectral direct normal irradiation (DNI) and diffuse horizontal irradiance (DHI) characteristics. In the context of bifacial PV, it has been shown that spectra reflected from different ground materials may cause

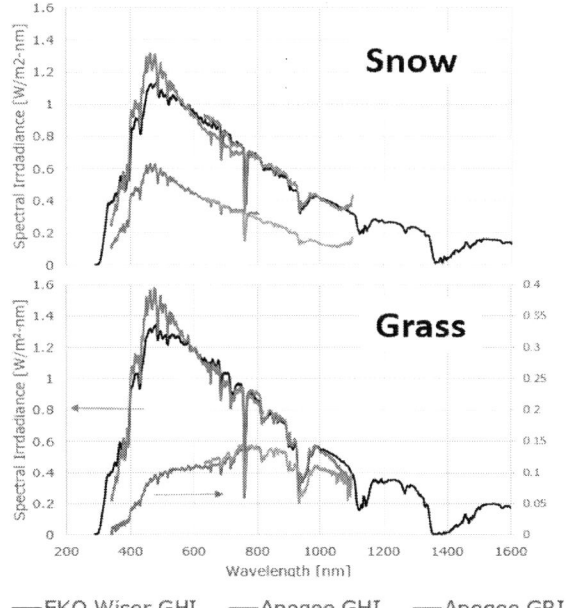

Fig. 1 GHI and Ground Reflected Irradiance spectral irradiance. Apogee instruments were taken in the field at noon with horizontal tilt, and the EKO wiser was located nearby at the SRRL. The ground snow measurements not only have a stronger spectral content but also appear 'blue-shifted' as there is less absorption in the NIR and VIS spectrum.

variations between power output predictions and measurments of up to 3.1% for grass or 5.2% for sand [1].

UV induced degradation modes typically occur more readily at lower wavelengths where the photon energy is higher and absorption is stronger. To understand the impacts of the albedo on degradation, the reflectivity of materials in the PV field and the spectral distribution of direct and diffuse light in the UV-spectra are essential. The raytracing tool used for modeling front and rear irradiance in PV modules bifacial_radiance has been modified for spectral calculations as part of this project. Raytracing results are compared to field degradation in a PV array in Maryland, MD [3]. Field measurements by different narrow- and broadband irradiance sensors and spectrophotometers taken at the 75-kW single-axis tracked bifacial field at the National Renewable Energy Laboratory (NREL) (Golden, CO) are also used to validate the methodology described below [4].

978-1-7281-6118-1/22 $31.00 © 2022 IEEE

II. METHODOLOGY

A. Degradation Calculation

The degradation (D) experienced by the backsheet material is often modeled as a function of time t and wavelength λ such that with typical functional forms of:

$$D = D_o \int_0^t RH(t)^n \cdot e^{\frac{-E_a}{RT(t)}} \int_\lambda \left[e^{-C_2\lambda} \cdot G(\lambda, t) \right]^x d\lambda dt, \quad (1)$$

where $T(t)$ is temperature in Kelvin at a specific hour, λ is the wavelength, E_a is the activation energy, R is the universal gas constant, RH is the relative humidity, n is a coefficient denoting the sensitivity of the material to humidity, C_2 is the coefficient for the exponential degradation of the material as a function of wavelength, $G(\lambda)$ is the irradiance in W/m^2 at the specific wavelength, and x is the scaling of the degradation effect due to irradiance intensity. Eq 1 is just one of many different functional forms for degradation in response to UV light. Furthermore, this can model only one degradation mechanism and is an empirical function. For C_2 and x, both the functional form and the values are empirical. However, from surveying many different degradation processes, we know that values between $x = 0.64 \pm 0.22$ [5] are common and that C_2 is commonly around 0.07 (1/nm) [6]. The activation energy for UV processes is typically around 40±15 kJ/mol. For humidity, the dependence can be nonexistent ($n=0$), negative or positive, and is set to n=1.0 for this paper calculations. This provides a basis to begin to characterize the relative degradation expectations of the backsheets in different environments and mounting configurations, and in comparison to the frontsheet degradation as,

$$D_{ratio} = D_{rear}/D_{front} \quad (2)$$

To describe the degradation on the backsheet, one must know the spectral distribution and intensity of light on the backside of the module as a function of time with corresponding meteorological data. By considering the relative degradation one can minimize the issues with temperature and humidity dependence, and can ignore the prefactor D_o of Eq. 1. It is far easier to make comparative assessments than to predict actual degradation rates.

B. Rear Irradiance Modeling

The Simple Model of the Atmospheric Radiative Transfer of Sunshine (SMARTS2) [7] was used to generate spectral DNI, DHI, and global horizontal irradiance (GHI) and spectral ground reflectance for various ground types. This tool generates clear-sky irradiance spectra. The spectrum between 300 and 2500 nm was considered for the simulations. A method to correlate field measurements of DNI, DHI and GHI with the spectra generated by SMARTS was implemented so that

$$E_{scaled}^*(\lambda) = \frac{E_{meas}}{\int E^*(\lambda)\, d\lambda} \times E^*(\lambda) \quad (3)$$

In this equation, $E_{scaled}^*(\lambda)$ is the resulting scaled irradiance spectrum, E_{meas} is the field measured irradiance value, and $E^*(\lambda)$ is the modeled irradiance spectrum from SMARTS. This methodology is valid only for mostly clear skies, as the scaling does not alter the relative spectral distribution of the components. Ground reflectance spectrum is generated depending on the type of ground cover most common in the season (dry or green gas), and high sustained albedo values (>0.5) are generated as snow albedo. Ground-reflected spectra are likewise scaled to match field-measured albedo values with eq. (3).

C. Rear-side Irradiance and Degradation Measurements in Gaithersburg, MD (NIST Site)

The degradation patterns and gradient irradiance conditions in the rear of the modules were published in [3] for a PV array in the National Institute of Standards and Technology (NIST) campus in Gaithersburg, Maryland (USA). In the NIST site, the edge module degradation effect observed for yellowing and gloss appears similar to irradiance's measured patterns. The conditions and geometry of the site were modeled for April 28th 2017, reproducing Fig. 5 from this paper and the degradation for a typical model year was calculated.

D. NREL Spectral Irradiance Measurements

Meteorological and spectroradiometer data were measured for one day with calibrated sensors 0.5 km distant from the bifacial horizontal single-axis tracker field (HSAT) at NREL's Solar Radiation Research Laboratory [8]. Ground-Horizontal Spectral irradiance is measured at the Solar Radiation Research Laboratory (SRRL) at NREL using an EKO WISER spectroradiometer, with a frequency response between 290-1650nm. Plane of Array spectral measurements were taken in the bifacial field with two co-located Apogee field spectroradiometers – the SS-110 with sensitivity from 340 nm to 820 nm and the SS-120 with sensitivity from 635 nm to 1100 nm. Both instruments were calibrated against the higher-accuracy EKO unit at SRRL. Front and rear POA were also measured with Silicon reference cells (IMT).

III. RESULTS

A. Gaithersburg, MD (NIST Site)

As expected, there are PV array edge effects (Fig. 2a) seen through modeling the NIST PV array in addition to shading effects that vary the irradiance across the collector width. Comparing the modeled to measured irradiance, the absolute root mean square error is less than 10 W/m^2. This is within the range of uncertainty for a photodiode measurement used by Fairbrother et al. [3]. Uncertainty in the measurements is further exacerbated by the position they were taken during the field survey. As shown in Figure 2b, the beams contribute to a shading factor of 13.7%. However, depending on where the measurement is taken, the irradiance level can be half to 3x higher than the average.

Table 1 shows the relative degradation of the backsheet versus the front of the module as computed using Eq. 4. For the hour modeled, the top of the edge modules in the center row experience 75% more degradation than the middle of the center module of the array, or 4.5% absolute more degradation than the 10% assumed on back sheets. Edge effects of the module itself also represent a 20% increase in degradation in the top of the modules relative to the center. For the full proceedings, full year degradation will be computed.

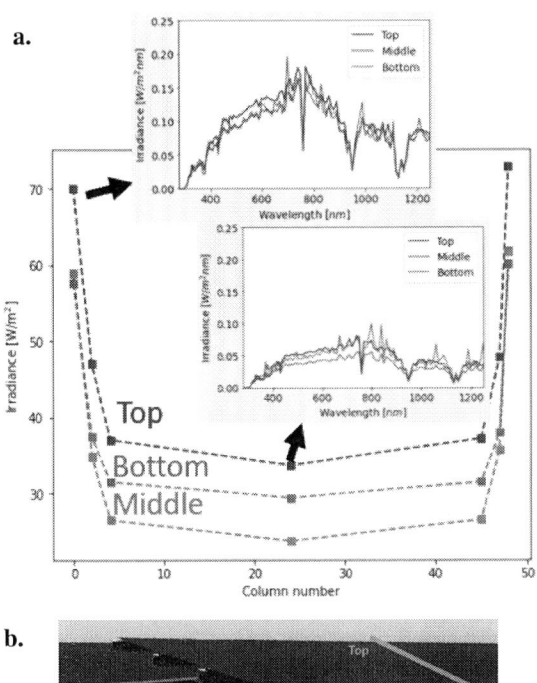

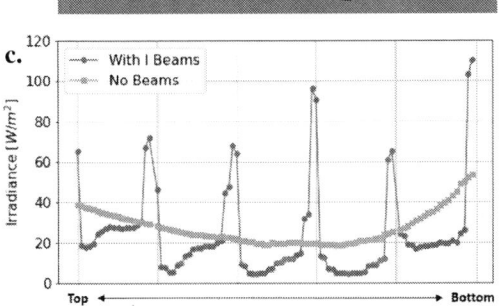

Fig. 2. a) Modeled spectra and the integrated irradiances for the rear of the central row of NIST array for 04/28/17 at 12 PM. b) Computer simulation model of the array. c) Rear irradiance across the central module, showing specular reflections and shading from the I Beams (shading factor of 13.7).

TABLE I. DEGRADATION RATIO

| Sensor Location | Degradation Ratio (back/front * 100%) | | | |
	Center Row, West Edge	Center Row, Center	Center Row, East Edge	Southern Row, Center
2 (top)	14.5	9.9	14.4	10.0
1 (middle)	13.3	8.3	13.1	8.4
0 (bottom)	13.0	8.6	12.8	8.9

Spectral effects of albedo have been studied in [9], pointing to how different sensors affect measured ground reflected irradiance. The absorption effects of ground are difficult to isolate in spectral measurements because there can be a competing enhancement of irradiance resulting from multiple reflections between the sky, clouds, structures in the PV field, and the ground surface itself. Fig. 4 shows a rear-spectral irradiance measurement and model in the tracker field. There is good agreement on wavelengths of interest for degradation (< 450 nm). Modeled and measured integrated values of the rear-POA spectra fall in the uncertainty range of the IMT-reference cell.

IV. CONCLUSIONS

We propose a method to calculate relative backsheet degradation. This is enabled by raytracig, showing that while rear irradiance can be below 10% of the front irradiance, spectral effects lead to higher degradation ratios than the 10% currently used in accelerated testings. The full manuscript will expand on this proceeding results.

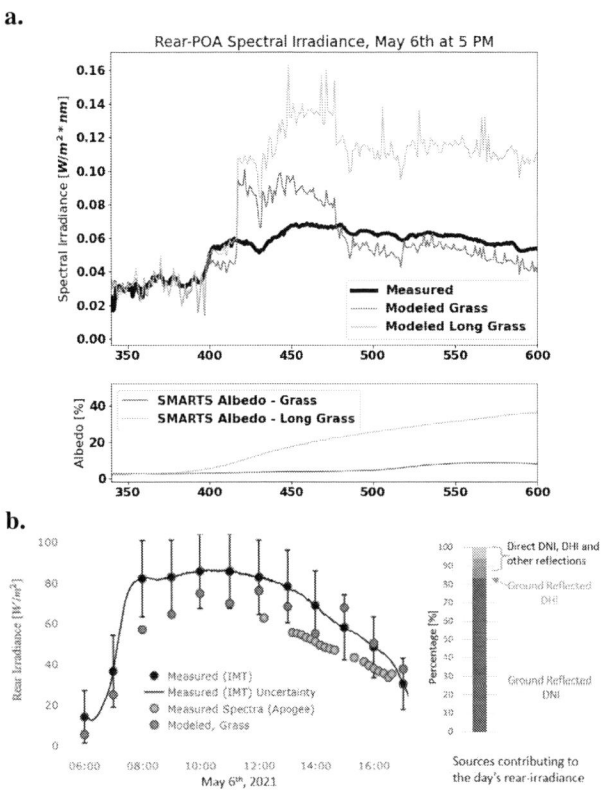

Fig. 3 a) Spectral rear irradiance measured and modeled on the rear-side of the HSAT, east-most edge at 5 PM. b) Modeled and measured spectral integrated spectral irradiance, compared to the measured IMT irradiance and it's uncertainty. (right) Shows the contribution of ground reflected DNI and DHI, versus direct DNI, DHI and other secondary reflections.

REFERENCES

[1] Russell (2017); [2] Kim (2018); [3] Fairbrother (2018); [4] BEST Field Data 2020; [5] Fischer [2004]; [6] Miller (2009 [7] Gueymard (1995); [8] Stoffer (1981); [9] Gostein (2021)

Effects of Satellite Sampling on Subhourly Modeling Errors

Mark A. Mikofski, William F. Holmgren, Jeff Newmiller, Rounak Kharait

DNV, Oakland, CA, United States

Modeling errors due to hourly inputs averaged from high frequency measurements can be significant where solar resource variability and inverter loading ratios are both high. However, satellite data sources average low frequency measurements. This paper studies the effects of satellite sampling frequency on subhourly modeling errors. We simulate satellite data from high frequency measurements by selecting instantaneous measurements at a lower sampling rate then averaging selected instantaneous measurements for the hour. With simulated satellite data sampled every 30-minutes or shorter we observe modeling errors, but for sampling rates longer than 30-minutes, modeling errors appear to cancel out annually possibly due to random errors in the sampled data. Similar effects have been observed by other researchers, and the results of this study seem to confirm their findings. However, the modeling error is not completely canceled out. Even with input sampled every 30-minutes, we still see some modeling error, and the modeling error increases as sampling rates get shorter. Based on our observations, we recommend that a modeling error correction be applied whenever hourly inputs are used, especially at sites with high solar variability and DC/AC ratios greater than one.

Room Temperature, Dip Coating Organic Passivation for c-Si surface

Kejun Chen, Abigail. R Meyer, Harvey Guthrey, William Nemeth, San Theingi, Matthew Page, Sumit Agarwal, David. L Young, Paul Stradins

Colorado School of Mines, Golden, CO, United States

bNational Renewable Energy Laboratory, Golden, CO, United States

Effective surface passivation of crystalline silicon (c-Si) surface by reducing the carrier recombination rate has led to modern c-Si solar cells with efficiencies > 25% in both laboratory and industrial settings. Typical surface passivation techniques include high-temperature silicon oxide (SiOx), amorphous silicon (a-Si:H), hydrogen-rich silicon nitride (SiNx), and aluminum oxide (Al2O3). However, almost all passivation techniques require high-temperature processes that are > 300°C, which poses challenges to certain PV applications that are sensitive to temperature. For example, the above-mentioned passivation methods are not compatible with the degradation study of cells or field modules. Ideally, a room-temperature surface passivation technique that can passivate the edge of a laser-cut cell fragment is necessary for advanced characterization techniques, such as EPR, EDMR, DLTS, local cell J-V and Suns-Voc tests. In this work, we show the passivation results of Nafion polymer on bare nCz and doped n- and p-type poly-Si/SiOxpassivating contact structures We demonstrated that the best Nafion passivated nCz sample has an iVoc of 716 mV, with a J0 of 7 fA/cm2. Compared with Al2O3 passivation, Nafion has slightly lower performance, but it can be extended to wider applications, such as processes sensitive to high temperature. This feature is important, especially when studying the degradation mechanisms of field modules to not destroy the degraded states. Our EPR confirmed that the decrease in Si dangling bond signal is observed after Nafion treatment. We also brought insights into the different morphologies of samples before Nafion.

978-1-7281-6118-1/22 $31.00 © 2022 IEEE

Investigation of Underperformance in Fielded N-type Monocrystalline Silicon Photovoltaic Modules

E. Ashley Gaulding, Steve W. Johnston, Dana B. Sulas-Kern, Mason J. Reed, James A. Rand, Robert Flottemesch, Timothy J Silverman, Michael G. Deceglie

National Renewable Energy Lab, Golden, CO, United States

Core Energy Works, Newark, DE, United States

Luminance, New York, NY, United States

As photovoltaic (PV) modules continue to evolve, it is important to catch and understand the causes behind new failure modes. Herein, we study n-type monocrystalline silicon PV modules that have been fielded at a utility scale power plant for 5 years, all of which have already degraded to < 90% of the nameplate max power (Pmp). High resolution electroluminescence (EL) and photoluminescence (PL) imaging suggests multiple possible factors contributing to the modules' underperformance, including series resistance issues (Rs) and wafer non-uniformities. Dark lock-in thermography (DLIT) measurements on a selected module suggests two specific module strings have high Rs. We then tabbed out all 60 cells of the same module. Suns-Voc measurements confirm relatively higher Rs values for the cells in these two strings. Multi-irradiance IV scans show the largest underperformance at the cell level for these same cells. This implicates Rs, rather than wafer non-uniformity, to be the largest contributor to the cell and therefore module underperformance.

Fill Factor Loss in Perovskite Solar Cells Using Fullerene ETLs Caused by Air Exposure

Austin G Kuba, Alexander J Harding, Raphael Richardson, Ujjwal K Das, Kevin D Dobson, William N Shafarman

Institute of Energy Conversion, University of Delaware, Newark, DE, United States

The effect of air exposure on perovskite solar cells using fullerene electron transport layers (ETLs) was investigated. Coplanar conductivity measurements showed a loss of conductivity in C60 thin films, decreasing by 1 order of magnitude in 5 minutes, and 2 orders of magnitude in 35 minutes. N-i-p solar cells using C60 ETLs and MAPbI3 processed by 2-step close space vapor transport showed a progressive loss of FF and onset of s-shaped J-V curve over 15 minutes of air exposure. This effect was not observed for over 2 hours of air exposure for solar cells with SnO2 ETLs, indicating that the C60 is the source of the degradation. SCAPS1D was used to simulate perovskite solar cells with varying C60 ETL carrier concentration and mobility. A loss of FF and onset of s-shaped curve was predicted as the conductivity drops below 2×10^{-6} S/cm, in accordance with the experimental J-V curves. This work shows that air exposure to fullerene ETLs can cause rapid performance degradation in perovskite solar cells due to a corresponding increase in C60 resistivity.

The Effect of Residual PbI2 on 2-Step Vapor-Processed p-i-n and n-i-p MAPbI3 Solar Cells

Austin G Kuba, Alexander J Harding, Chaiwarut Santiwipharat, Ujjwal K Das, Kevin D Dobson, William N Shafarman

Institute of Energy Conversion, University of Delaware, Newark, DE, United States

Solar cells were made with MAPbI3 deposited by a two-step close space vapor transport (CSVT) deposition process and identical materials and deposition techniques in n-i-p and p-i-n architectures. Evaporated C60 ETL and copper phthalocyanine (CuPC) HTLs were used as contact layers. P-i-n solar cell J-V curves are well-behaved while n-i-p solar cells initially have s-shaped J-V curves that become well-behaved upon light soaking, possibly due to the substrate work function or integer charge transfer effects from the substrates. The CSVT reaction proceeds by conversion of a PbI2 layer via reaction with methylammonium iodide (MAI), where the diffracted intensity of the residual PbI2 reflects the degree of conversion of the film. N-i-p solar cells lose performance when there is no residual PbI2 in the film, which may be explained by excess mobile ions diffusing through the film into the C60 layer and reducing the carrier concentration. This is tested by treatment of C60 with KI/IPA solution, which produces a stronger s-shaped J-V curve. In contrast, p-i-n solar cells perform well without residual PbI2 due to the improved chemical stability of the CuPC HTL with the perovskite and excess MAI. This study highlights the flexibility of the 2-step close space vapor process for producing multiple solar cell architectures and the need to choose contact layers that are not degraded by the perovskite and excess MAI in 2-step vapor processes.

Characterization of DER Momentary Cessation and Rate-of-Change-of-Frequency Response

Rasel Mahmud*, Li Yu†, and Andy Hoke*

*Power Systems Engineering Center, National Renewable Energy Laboratory, Golden, CO 80401, USA
†Hawaiian Electric
email: Rasel.Mahmud@nrel.gov, Li.Yu@hawaiianelectric.com, Andy.Hoke@nrel.gov

Abstract—Momentary cessation (MC) and response to rate-of-change-of-frequency (ROCOF) are inverter responses that can have serious impact on the stability of the grid during abnormal conditions. Though IEEE Std 1547-2018 provides fairly well defined expected responses from inverter-based DERs during abnormal grid conditions, a significant portion of currently installed DERs in the distribution network do not have a well defined response to abnormal grid conditions. Any analysis of the grid involving inverter MC and ROCOF response must consider the specific characteristics of these responses for the installed DERs to have better confidence in the analysis results, especially for grids with high levels of DERs. However, there is lack of information on MC and ROCOF response for the inverters already installed in the field. In this paper, we examine a large data set of installed inverters to map the dominant population of installed inverters. This information is then used to characterize the most dominant MC and ROCOF responses of installed inverters using experimental data.

Index Terms—DER, momentary cessation, rate-of-change-of-frequency, IEEE Std 1547

I. INTRODUCTION

For power systems with very high levels of DERs, the behavior of the DER inverters during transient events can have crucial impacts on power system stability. For IEEE Standard 1547-2018 Category III compliant inverters, the standard gives a pretty good boundary on the undervoltage and overvoltage momentary cessation (MC) limit as well as current recovery ramp rate [1]. Similarly, 1547-2018 requires well defined rate-of-change-of-frequency (ROCOF) ride-through behavior. But the response of existing (pre-IEEE Std 1547-2018) inverters to voltage or frequency disturbances in terms of MC and high ROCOF events is largely unstandardized and undocumented, so studies have made justifiably conservative assumptions, sometimes based on data for utility-scale inverters, many of which are known to use MC just outside the normal voltage operating region [2]. The majority of the distribution-connected inverters installed today are not compliant to IEEE 1547-2018; these existing DERs comprise a significant portion of generation during midday in areas such as California and Hawaii. Therefore it is important to understand and better model the behavior of these inverters, especially their MC and ROCOF responses.

This paper presents an experiment-based characterization of MC and response to high ROCOF for installed inverter models

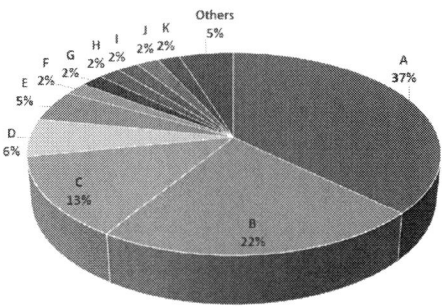

Fig. 1. Percentage distribution of installed DER inverter capacity

using DER interconnection application data from the Hawaiian Electric service area. The DER inverter data used here list all installed and approved distribution-connected inverter systems on Oahu. The authors processed the this data to estimate the most prevalent models of DER inverters. The authors then examined available laboratory test data from the National Renewable Energy Laboratory (NREL) [3]–[5] to attempt to characterize MC and ROCOF behavior. While the past tests were not designed to answer these questions, they contain partial answers for several prevalent inverter models.

II. SUMMARY OF INSTALLED INVERTERS

Hawaiian Electric provided the list of approved DER systems in an Excel file. This dataset provides the percentage contribution of each inverter manufacturer in the total installed capacity, as shown in Fig. 1. The actual names of the inverter manufacturers have been masked with alphabetic names. It can be seen in Fig. 1 that five manufacturers dominate this market: Manufacturer A with (37%), Manufacturer B (22%), Manufacturer C (13%), Manufacturer D (6%), and Manufacturer E (5%), followed by other manufacturers with smaller portions. When total inverter capacity for Manufacturer A is broken down by inverter models, the percentage share of the inverter models is as shown in Fig. 2, which indicates that there is one dominant inverter model (A1) from Manufacturer A. The full paper will include additional information on the prevalence of inverter models and how it has changed over time.

978-1-7281-6118-1/22 $31.00 © 2022 IEEE

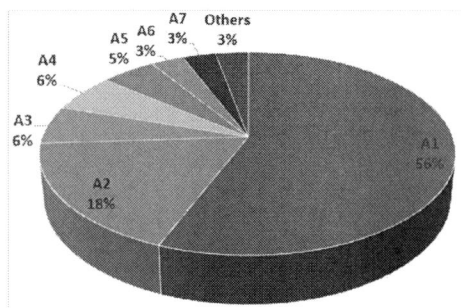

Fig. 2. Capacity share of inverter models from Manufacturer A

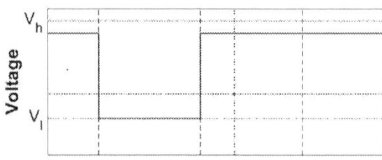

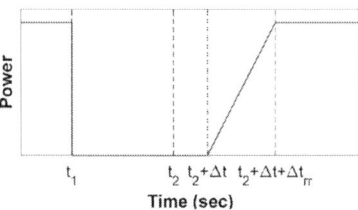

Fig. 3. A simple characterization of momentary cessation [2]

TABLE I
INVERTER RESPONSES TO VOLTAGE DISTURBANCES

Inverter Model	V_h (pu)	V_l (pu)	t_1 (sec)	t_2 (sec)	Δt (sec)	Δt_{rr} (sec)	MC/ trip
A2		0.48	12.6	14.6	304	13.9	trip
C1	1.1		97.9				trip
F1		0.46	19.1	27.5	338	30.3	trip

TABLE II
INVERTER RESPONSES TO FREQUENCY DISTURBANCES

Inverter Model	f_h (pu)	f_l (pu)	t_1 (sec)	t_2 (sec)	Δt (sec)	Δt_{rr} (sec)	MC/ trip
A2	64.17		11.9	13.85	311.35	14.2	trip
B1	62		24.24	63.12	0.94	2.58	trip
C1		55.85	10.85	11.85	3.7		trip
F1	64.14		17.2	19.7	41.3		trip

value after two cycles even though the voltage has not returned to within nominal range. Thus this response is not a classical momentary cessation. However, similar responses were not observed for other inverters, as shown in Fig. 5.

A common inverter response often observed in the laboratory experiments, also evident from Fig. 5, is that an inverter keeps injecting current to the grid even when there is a significant voltage or frequency disturbance. A recent publication [5] also reveals such a response from inverters.

IV. RATE OF CHANGE OF FREQUENCY

We also reviewed the past experimental data to characterize the inverter response to high ROCOF events; a few examples are shown in Table III; the full paper will include a more extensive version of this table. In Table III, ROCOF was calculated as $ROCOF = \frac{f_f - f_i}{\delta t}$ where δt is the time to change the frequency from from f_i to f_f. There was no record of inverter tripping on ROCOF found from the available test data. The recorded test data shows that the inverters were subjected to up to 2 Hz/s ROCOF and the inverters continued to export current without tripping, as shown in Table III. This is an encouraging result because there are many inverters from these model families in the field, and there were no requirements

III. CHARACTERIZATION TEST RESULTS

Momentary cessation in response to a low voltage event can be characterized in a simplified fashion as shown Fig. 3 following the NERC modeling guideline [2]. This modeling guideline is a piecewise linear approximation of real inverter behavior, which can vary significantly within some bounds. This approximation is believed to be sufficient to capture the effects of MC on bulk power system stability. After analyzing the existing laboratory test results for multiple inverters, we extracted potential MC responses. Table I lists a few of the potential MC responses for voltage disturbances, while Table II lists a few of the potential MC responses to frequency disturbances. The full paper will include more extensive versions of these tables. As can be seen in Table I and Table II, in most cases the observed delay time, Δt, is too long for momentary cessation, indicating the inverters tripped. A MC event would have a Δt of a few milliseconds or seconds. The relatively low delay time Δt observed for inverter B1 in Table II is not a MC response, but rather the power curtailment response of the frequency-watt function.

There are inverter behaviors observed from the test data which are closely related to MC. One such behavior is shown in Fig. 4a and Fig. 4b where inverter A2 reduced current to close to zero for about two cycles in response to a voltage disturbance. However, the current returns to its pre-disturbance

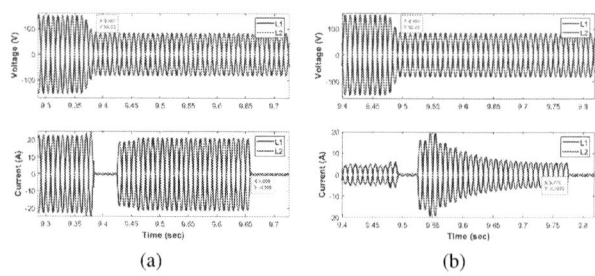

Fig. 4. Momentary current reduction during voltage disturbance observed for inverter A2 a) pre-disturbance power level at 100%, and a) pre-disturbance power level at 20%

978-1-7281-6118-1/22 $31.00 © 2022 IEEE

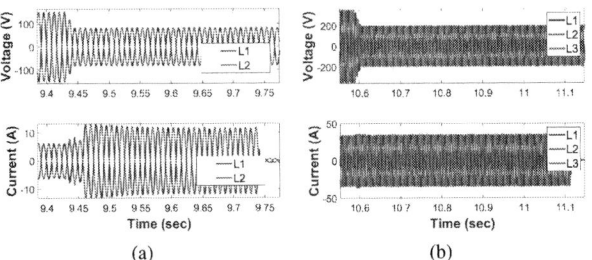

Fig. 5. Response of inverters for sudden voltage drop (a) inverter A7 with pre-disturbance power level at 20%, and (b) inverter B1 with pre-disturbance power level at 100%

TABLE III
INVERTER RESPONSE TO ROCOF

Inverter Model	f_i (Hz)	f_f H(z)	δt (s)	ROCOF (Hz/s)	Inverter Tripped?
A2	60.02	62.85	1.8	1.57	No
B1	59.98	57.05	2.95	0.99	No
C1	59.6	57.16	1.45	1.68	No
F1	60.02	62.88	3.5	0.82	No

for ROCOF ride through at the time these inverters were being installed. In contrast, since the advent of UL 1741 Supplement SA testing, DER inverters have been subject to a de facto requirement to ride through ROCOFs of at least 1.0 Hz/s [6]. No data was available for ROCOFs beyond 2.28 Hz/s.

V. CONCLUSION

This report presents an analysis of some prevalent inverter models on Oahu and a meta-analysis of past lab test data with a goal of making a rough characterization of the Hawaii DER inverter fleet's response to voltage and frequency transients. Specifically, test results were examined in an effort to estimate the DER fleet's use of momentary cessation and its ability to ride through high ROCOF events. The findings can be summarized as follows:

- The tested inverters include at least one model from the three most prevalent manufacturers in Hawaii. Together, these manufacturers make up about 72% of the Oahu DER generation fleet. However, only one or two models from each manufacturer were tested, so this data is not representative of that full 72% of Hawaii's DER fleet.
- For voltages above 0.5 p.u., MC was found only in extremely limited circumstances and for very short durations (i.e. only for 2 line cycles, and only in one inverter model). Instead, the tested inverters tend to continue exporting current until they reach a trip setting designated by the interconnection standard. (In contrast, IEEE 1547-2018 Category III compliant inverters, which will begin to be installed in 2022, are *required* to use momentary cessation in specific circumstances [1].) Therefore it would be reasonable to model existing inverters as not performing MC at all, and simply model the inverters as tripping at

the voltage and frequency trip settings from Rule 14H at the time they entered the interconnection queue. If a more conservative approach is desired, a portion of the inverter fleet could be modeled using MC, representing the inverter models for which no test data was found. This is in sharp contrast to existing *central* inverters, many of which have been identified by NERC as using MC at voltages just outside the continuous operating region [2]. Because no significant MC was found, we provide no data on the parameters to be used to model any portion of the DER fleet presumed to use MC.

- For voltages below 0.5 p.u., no data was found indicating whether MC is used. In the absence of such data, it would be reasonable to make the conservative assumption that MC is used given the high risk to power system stability.
- The inverters tested appear to ride through up to about 2 Hz/s ROCOF at least. No data was found for higher ROCOFs. Therefore a uniform and somewhat conservative approach would be to model existing DERs as tripping for ROCOF over 2 Hz/s. A more conservative approach would be to model some DERs as tripping at 2 Hz/s and others tripping at 1 Hz/s, representing inverters for which no relevant test data was found.
- Given that 1547-2018 allows DERs to trip for ROCOF beyond 3 Hz/s, modeling should be conducted to evaluate the system risk of massive DER trips for ROCOFs between 2 and 3 Hz/s.
- A significant unanswered question is how long a ROCOF of 2 Hz/s must persist for DERs to potentially trip. Because DERs in the lab typically do not trip on voltage transients (which often show up as brief high ROCOF spikes in calculated frequency measurements), it can be estimated that a high-ROCOF event must last at least a few cycles to cause a trip (if it causes a trip at all).
- It may be possible to increase the confidence in these findings by examining a wider swath of test data. Targeted testing could also greatly increase the confidence in these findings, thereby reducing the risk of future large-scale DER trip events and reducing the operating costs incurred due to necessarily conservative modeling assumptions.

REFERENCES

[1] IEEE, "1547-2018: IEEE Standard for Interconnection and Interoperability of Distributed Energy Resources with Associated Electric Power Systems Interfaces," 2018.

[2] NERC, "Modeling Notification: Recommended Practices for Modeling Momentary Cessation Initial Distribution," 2018.

[3] A. Nelson, A. Nagarajan, K. Prabakar, V. Gevorgian, B. Lundstrom, S. Nepal, A. Hoke, M. Asano, R. Ueda, J. Shindo, K. Kubojiri, R. Ceria, and E. Ifuku, "Hawaiian electric advanced inverter grid support function laboratory validation and analysis," 12 2016.

[4] A. Hoke, A. Nelson, J. Tan, R. Mahmud, V. Gevorgian, M. Elkhatib, C. Antonio, D. Arakawa, and K. Fong, "The Frequency-Watt Function: Simulation and Testing for the Hawaiian Electric Companies," National Renewable Energy Laboratory (NREL), Golden, CO, Tech. Rep., 2017.

[5] R. Mahmud, D. Narang, and A. Hoke, "Reduced-Order Parameterized Short-Circuit Current Model of Inverter-Interfaced Distributed Generators," *IEEE Transactions on Power Delivery*, 2021.

[6] UL1741, "UL 1741 Supplement SA: Grid Support Utility Interactive Inverters and Converters," *Underwriters Laboratories*, 2016.

978-1-7281-6118-1/22 $31.00 © 2022 IEEE

No Time to Waste: Quickly Optimizing Perovskite Composition with Off-the-Shelf Active Learning Methods

Rishi E Kumar, Moses Kodur, Arun Kumar Mannodi Kannakithodi, David P Fenning

University of California San Diego, La Jolla, CA, United States

Purdue University, West Lafayette, IN, United States

The compositional flexibility of halide perovskite presents both opportunity for discovery of tailored materials and risk of extraneous effort. Active learning - a class of machine learning methods that alternates between modeling a response surface and suggesting interesting points to be tested next - enables optimization of a system while testing just a small fraction of the total search space, and can alleviate the cost of large compositional searches. We demonstrate that a single generic active learning algorithm is effective across five typical tasks in halide perovskite development under the three research paradigms of hand-done, high-throughput robotic, and computational experiments. These tasks involve tuning the composition at one, two, or all three of the A,B, and X sites. Moving beyond generic active learning, representing ABX_3 compositions by their physical properties at each site -- which, in practice, entails simply looking up and averaging tabulated values -- further improves the rate of composition optimization by 1-1.4x. The experimental budget of a typical successful active learning optimization was about 15 samples -- well within the budget of tedious manual experiments, and a drop in the bucket for high-throughput automated or computational studies. Our findings resonate with the growing body of active learning demonstrations across the scientific literature, and advocate for the integration of active learning into composition optimization efforts in halide perovskites.

Reactive Silver Inks as Front Electrodes for TCO Coated Solar Cells

Michael W. Martinez-Szewczyk[1], Steven DiGregorio[2], Owen Hildreth[2], Mariana I. Bertoni[1]

[1]Arizona State University, Tempe, Arizona 85281, USA

[2]Colorado School of Mines, Golden, Colorado, 80401, USA

Abstract— **Silicon heterojunction (SHJ) cells have shown record efficiencies of 23-25% in recent years and continue to show a promising pathway towards high-efficiency solar cells [1-3]. The efficiencies of these cells are typically limited by their series resistance, which results from the high resistivity of the low-temperature silver paste front-grid metallization that are required in SHJ cells. Using reactive silver ink (RSI) and dispense printing offers an alternative to fabricate solar cell front-grids with a low-resistivity compatible with low-temperature processes and all while using 97% less silver. Here we present latest results of our RSI metallization for use on TCO surfaces. We will show how these perfectly metallic inks allow us to redefine the form factor needed in advanced metallization.**

Keywords— Contacts, low-temperature metallization, silver

I. INTRODUCTION

Current research in photovoltaics is strongly driven by a need to reduce the cost of manufacturing and increase the efficiency of solar cells. An already promising next step for higher efficiency solar cells is silicon heterojunction (SHJ) solar cells. These devices use amorphous silicon (a-Si) to passivate the emitters and achieve efficiencies above 25% [1]. A consequence of this technology is its sensitivity to temperature during the fabrication. Temperatures over 200 °C severely degrade the devices, resulting in an alternative metallization paste that can form solid contacts below 200 °C. This low-temperature (LT) silver paste is screen-printed (SP) onto the cell and contains more silver than normal solar cells and as such is more expensive and susceptible to the volatility of the price of silver. Reactive silver ink (RSI) is a high-performance and low-cost alternative to low-temperature silver paste that uses considerably less silver and can form solid contacts below 60 °C. RSI contains no metal particles in solution or binding agents like with LT SP paste, but rather metal ions in solution that can be easily dispensed through nozzles, which sizes are mostly limited by their fabrication. In the past we have shown the performance of a baseline RSI formula, herein a new formulation with improved drying properties is characterized and compared to the baseline.

With this work, we look to redefine not only what a good transparent conductive oxide (TCO) contact looks and performs

like but how little silver one actually needs in order to achieve good performance

II. EXPERIMENTAL

The baseline formulation of RSI was proposed by Walker and Lewis [4] and was thoroughly tested in our previous publication [5], herein noted as Walker's ink. Our advanced formulation, called Gaitan's ink, was synthesized using 0.66 g silver acetate, 7.84 mL of ethylamine (CH3CH2NH2 66-72% in H2O) instead of ammonium hydroxide, and 0.15 mL of formic acid. This formula of ink does not need to be diluted and results in 8 mL of 0.77 M ink.

The system used to dispense the RSI utilizes a 3- axis Nordson printer robot, syringe pump ink delivery system, and a flow sensor. Using various flow rates and printer head speeds, the ink is dispensed onto a heated stage equipped with a vacuum chuck. Lines with higher silver contents are achieved by adding subsequent layers of ink onto the original print.

Media resistivity samples were created on polished Si wafers coated with 75 nm of indium tin oxide (ITO), and DC sputtered 1 x 1 mm Ag pads spaced 5 cm apart. Contact resistivity samples were prepared by plasma-enhanced chemical vapor deposition (PECVD) of an intrinsic a-Si layer (50nm), 75 nm ITO, and DC sputtered 1 x 1 mm Ag pads spaced 8 mm apart.

SHJ solar cells were fabricated from 156 mm^2 180 µm thick n-type CZ Si wafers. Following texturing, samples had various a-Si layers deposited using PECVD immediately followed by DC sputtering of the front ITO, back ITO, and Ag back contact. The solar cell stack is described by the following: ITO 75 nm | (p) a-Si 17 nm | (i) a-Si 6 nm | (n) c-Si 180 µm | (i) a-Si 6 nm | (n) a-Si 5 nm | ITO 75 nm | Ag 200 nm.

All electrical measurements on the media and contact resistivity samples were measured using the four-point probe method. Solar cell characterization was performed at ASU's Solar Power Lab using various techniques such as IV/Suns Voc, electroluminescence (EL) imaging, and photoluminescence (PL) imaging.

III. RESULTS AND DISCUSSION

Figure 1 shows the media resistivity of both formulas of RSI as a function of silver content versus the media resistivity of the LT SP paste and of bulk silver. Media resistivity was measured following the general effective media model for a composite

---

U.S. Department of Energy, Energy Efficiency and Renewable Energy Program, Award Number DE-EE0008166

978-1-7281-6118-1/22 $31.00 © 2022 IEEE

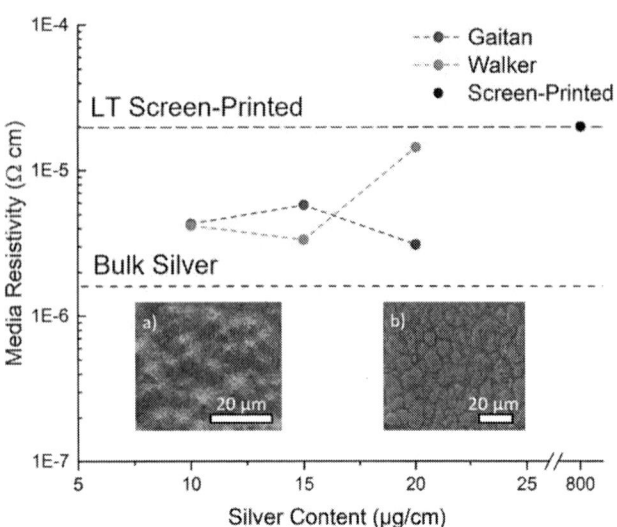

Figure 1: Media resistivity vs silver content for Walker's and Gaitan's ink. SEM images below show the top-down morphology of the a) RSI and b) LT SP paste

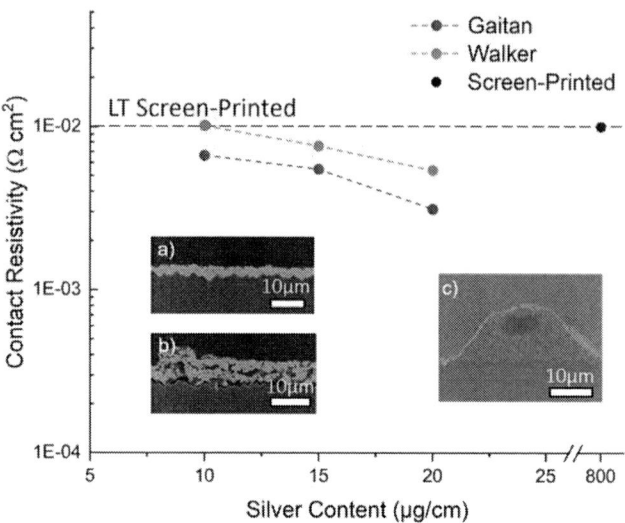

Figure 2: Contact resistivity vs silver content for Walker's and Gaitan's ink. SEM images below show cross-section of a) Gaitan's ink b) Walker's ink, and c) LT SP paste

material and using a best fit percolation exponent of to identify the type and distribution of pores in our samples. A percolation exponent of 3/2 resulted in the best fit and is indicative of having mostly spherical pores within the RSI [6]. Porosity data was used as the experimental data for fitting and was measured using a combination of X-ray transmission and profilometry measurements. These results show that both RSI formulations have a lower media resistivity than the LT SP ink. This can be explained by the difference in the pre-print composition of the RSI and LT SP systems. The RSI precursor contains metal ions in solution that result in a pure silver metallization after deposition, while the LT SP contains binding agents. Once annealed, these binding agents remain in the metallization and decrease the electrical performance.

It is important to note that the large increase in the media resistivity of Walker's ink above 15 μg/cm is believed to be a direct result of the differences in precipitation of these two inks which are described in detail elsewhere [7]. This phenomenon results in "puckering" of subsequent printed layers of Walker's ink and results in a more porous and therefore resistive metallization. These results motivated the subsequent development of Gaitan's ink.

Figure 2 shows the contact resistivity of Gaitan's ink is lower than Walker's ink at all silver contents as well as below the contact resistivity of the LT SP paste. All contact resistivity samples were printed using identical printing parameters as the media resistivity samples and were measured according to the transfer length method. The differences between the two RSI formulations can be attributed to the precipitation methods described previously [7]. As is shown in the SEM images in Figure 2, Gaitan's ink has a much more conformal and non-porous contact with the TCO surface which helps both validate the difference in precipitation mechanism and explain the improved contact resistivity of Gaitan's ink. As a result of these experiments and previous work [5], Gaitan's ink was the only

RSI formulation chosen to metallize the solar cells described in the next section.

Full-scale 156 mm^2 SHJ solar cells were fabricated and subsequently metallized by either RSI or LT SP paste and characterized as described in the experimental section. All cells had an identical design of the front-grid. Figure 3 shows the IV performance of the RSI metallized cells with varying silver content and the LT SP cell. The 10 μg/cm cell shows an extremely low fill factor compared to the other samples. This

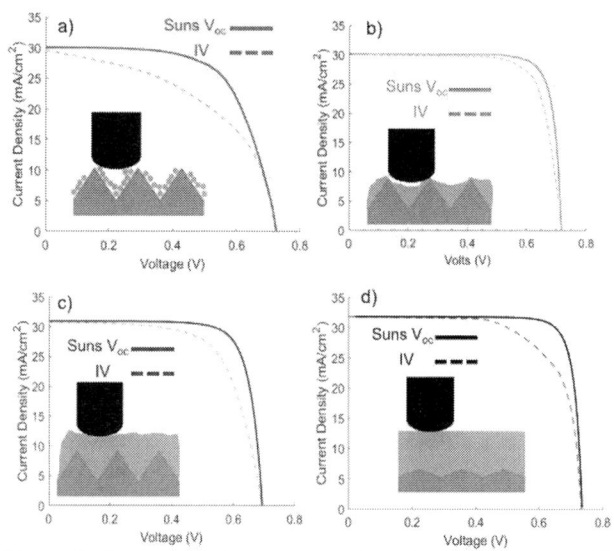

Figure 3: Current density vs voltage IV and Suns V_{oc} curves of RSI solar cells with a) 10 μg/cm, b) 15 μg/cm, and c) 20 μg/cm of silver and LT SP solar cell with d) 800 μg/cm of silver. The diagrams below each curve demonstrate the ability of the probe tip to contact the silver metallization.

can be explained by the diagrams depicted in Figure 3. The 10 μg/cm

Table 1 Solar Cell Characteristics

	V_{oc} (mV)	J_{sc} (mA/cm^2)	pFF (%)	FF (%)	R_s (Ω•cm^2)	η (%)
10 μg/cm	723	29.7	83.2	48.6	8.8	10.4
15 μg/cm	681	29.8	78.4	77.4	2.6	13.8
20 μg/cm	685	28.2	78.1	73.1	1.4	14.2
LT SP paste	684	31.7	78.6	57.0	3.3	12.3

RSI cell has 98% less silver than the LT SP paste cell and as a result has issues with the probes contacting the silver that remains largely in the valleys of the textured silicon pyramids. This issue can be avoided with higher silver contents of the front-grid. When comparing the performance of the RSI and LT SP cells in , the main difference is in the short circuit current density (Jsc), namely the consistently lower Jsc of the RSI cells. This can be attributed directly to the difference in form factors of these two types of metallization. The LT SP cell finger dimensions are 70 μm wide and 30 μm in height, while the RSI finger dimensions are 100 μm wide by about 3-5 μm in height depending on the silver content. This increase in finger width of the RSI contributes directly to shading of the solar cell and subsequently results in a reduction of the Jsc. Further reduction of the line width and improvement of the overall cell quality is currently underway.

This stark difference in the form factor of the metallization begs the question of exactly how the shading and subsequently the power generation of the cell changes at varying angles of incidence. We have shown the difference in performance when the incident light is normal to the surface of the solar cell, but how much of that performance difference is still exhibited when the light comes in at a glancing angle? The shadow cast by a metallization that is 30 μm tall is very different than a metallization that is only 3-5 μm tall. At the time of presentation, we will show optical and electrical modeling that look to answer this question. In addition, preliminary corrosion experiments of the RSI have been performed and indicate that this metallization undergoes a different mechanism of corrosion than standard LT SP paste. At the time of presentation, we will also show results from the reliability studies of RSI.

IV. CONCLUSION

Samples of RSI and LT SP paste metallization were fabricated and characterized using four-point probe and profilometry measurements. As a result of these measurements, we chose to down select the Gaitan formula for its improved media and contact resistivity. The subsequent 156 mm^2 solar cells metallized with Gaitan's ink and LT SP sister cells had shown comparable, if not improved, performance of the RSI cells at higher silver content without any grid optimization and around 97% less silver per cell. Further reduction of the RSI finger width along with grid optimization would undoubtedly prove beneficial to the overall cell performance and further prove the relevancy of RSI as a novel metallization for the temperature sensitive next generation of solar cell technology.

ACKNOWLEDGMENT

The authors would like to thank Dr. Tara Nietzold for the X-ray transmission measurements.

REFERENCES

[1] K. Masuko et al., "Achievement of More Than 25% Conversion Efficiency With Crystalline Silicon Heterojunction Solar Cell," *IEEE Journal of Photovoltaics*, vol. 4, no. 6, pp. 1433–1435, Nov. 2014, doi: 10.1109/JPHOTOV.2014.2352151.

[2] X. Ru et al., "25.11% efficiency silicon heterojunction solar cell with low deposition rate intrinsic amorphous silicon buffer layers," *Solar Energy Materials and Solar Cells*, vol. 215, Sep. 2020, doi: 10.1016/j.solmat.2020.110643.

[3] J. Dréon et al., "23.5%-efficient silicon heterojunction silicon solar cell using molybdenum oxide as hole-selective contact," *Nano Energy*, vol. 70, Apr. 2020, doi: 10.1016/j.nanoen.2020.104495.

[4] S. B. Walker and J. A. Lewis, "Reactive silver inks for patterning high-conductivity features at mild temperatures," *Journal of the American Chemical Society*, vol. 134, no. 3, pp. 1419–1421, Jan. 2012, doi: 10.1021/ja209267c.

[5] A. M. Jeffries, A. Mamidanna, L. Ding, O. J. Hildreth, and M. I. Bertoni, "Low-Temperature Drop-on-Demand Reactive Silver Inks for Solar Cell Front-Grid Metallization," *IEEE Journal of Photovoltaics*, vol. 7, no. 1, pp. 37–43, Jan. 2017, doi: 10.1109/JPHOTOV.2016.2621351.

[6] D. S. Mclachlan, M. Blaszkiewicz, and R. E. Newnham, "Electrical Resistivity of Composites."

[7] A. Mamidanna, "MORPHOLOGY PREDICTION OF REACTIVE SILVER INK SYSTEMS by Avinash Mamidanna."

Flexible CdTe/MgCdTe Double-Heterostructure Solar Cells Made from Epitaxial Lift-off Thin Films

Xin Qi[1], Jia Ding[1], Zheng Ju[2], Stephen Schaefer[1], and Yong-Hang Zhang[1]

[1] School of Electrical, Computer and Energy Engineering, Arizona State University, Tempe, AZ, 85287, USA

[2] Department of Physics, Arizona State University, Tempe, AZ, 85287, USA

Abstract—**An improved epitaxial lift-off (ELO) technique is developed and monocrystalline CdTe/MgCdTe double-heterostructure (DH) thin films on conductive flexible superstrates are demonstrated. The post-ELO DH shows strong photoluminescence and uniform surface morphology comparable to that prior to the lift-off process. Solar cells fabricated with the post-ELO thin films exhibit an open circuit voltage of 0.79 V and an efficiency of 9.8%, showing that the ELO technique is practical for fabrication of flexible, light-weight monocrystalline CdTe solar cells for space and terrestrial applications.**

Keywords—CdTe, solar cell, double heterostructure, epitaxial lift-off, flexible

I. INTRODUCTION

Highly efficient, flexible thin-film solar cells are desirable for both terrestrial and space applications. Epitaxial lift-off (ELO) techniques enable the development of these thin-film solar cells and have already been demonstrated for ZnTe-based II-VI single-crystal thin films and GaAs flexible solar cells by introducing sacrificial layers and using highly selective etchants. [1–4] Polycrystalline CdTe-based thin films were obtained by similar methods using water-soluble sacrificial layers and superstrates to achieve a lift-off process. [5, 6] However, due to the large lattice-mismatch between CdTe and those substrates or sacrificial layers, the polycrystalline thin films obtained using these methods have poor carrier mobility and lifetime and fail to demonstrate high performance devices.

An ELO technique for monocrystalline CdTe-based thin films using a water-soluble sacrificial MgTe layer was proposed and successfully used to lift off single crystal CdTe/MgCdTe double-heterostructures (DHs) with few extended defects. [7–9] Both CdTe/MgCdTe DH and the MgTe layers were grown on nearly lattice-matched InSb substrates using molecular-beam epitaxy (MBE). Deionized (DI) water was used as a highly selective etchant to dissolve only the MgTe layer and complete the lift-off process. Previous studies used Kapton tape and hard-baked photoresist as superstrates to support the post-ELO thin films, and the resulting thin films remained highly crystalline with smooth surface morphology. [8, 9] However, due to the poor conductance of those superstrates, photogenerated carriers from the thin films could not be extracted. Further study of ELO using conductive superstrates is needed to enable fabrication of practical devices.

In this study, an improved ELO technique is developed to demonstrate monocrystalline CdTe/MgCdTe DH thin films on conductive flexible superstrates. Solar cells using the thin films were successfully fabricated.

II. EXPERIMENTAL DESIGN

A. Sample growth

The CdTe/MgCdTe DH and the MgTe layer were grown on InSb substrates using MBE as reported previously. [8–10] The layer structure used in this study is shown in Fig. 1. The thickness of the MgTe layer was optimized at 32 nm to ensure both high etching rate and high thin-film quality. Te and In were used as n-type dopants in III-V and II-VI layers, respectively.

B. ELO process

Either polyimide films or silicon wafers coated with Ti/Pt/Au were used as conductive superstrates to support the post-ELO thin film. Indium was used as bonding medium to attach the superstrates to the as-grown ELO sample as demonstrated in previous studies. [11] As shown in Fig. 2, indium was deposited onto both the ELO sample and conductive superstrates, followed by bonding the two surfaces using either (i) mechanical pressure or (ii) silver paste. The bonded samples were annealed at 150 °C for 10 min. After bonding and annealing, the samples were immersed into 65 °C DI water for approximately 3 minutes to complete the lift-off by dissolving the MgTe layer.

C. Device processing

After the ELO process, a 50 nm thick indium tin oxide (ITO) layer was sputtered onto the post-ELO thin films to form a front contact. The active area of the devices was 0.005 cm^2 defined by the shadow mask during ITO sputtering. The complete process flow of the device, including ELO, is shown in Fig. 2.

III. RESULTS AND DISCUSSION

A. Thin film characterization

Optical images of the post-ELO thin film using the polyimide superstrate and indium bonding technique is shown in Fig. 3. The thin film is mirror-like in the center with smooth surface morphology and few extended defects such as cracks or pinholes. Although the post-ELO thin film was also successfully bonded to the polyimide superstrate with silver paste, the surface

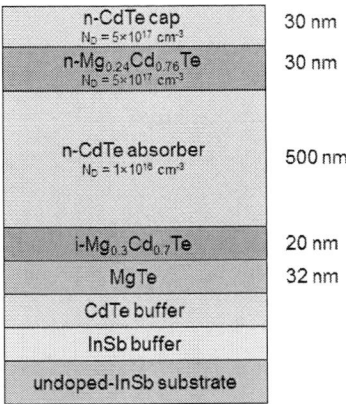

Fig. 1. Layer structure of the MBE-grown ELO sample.

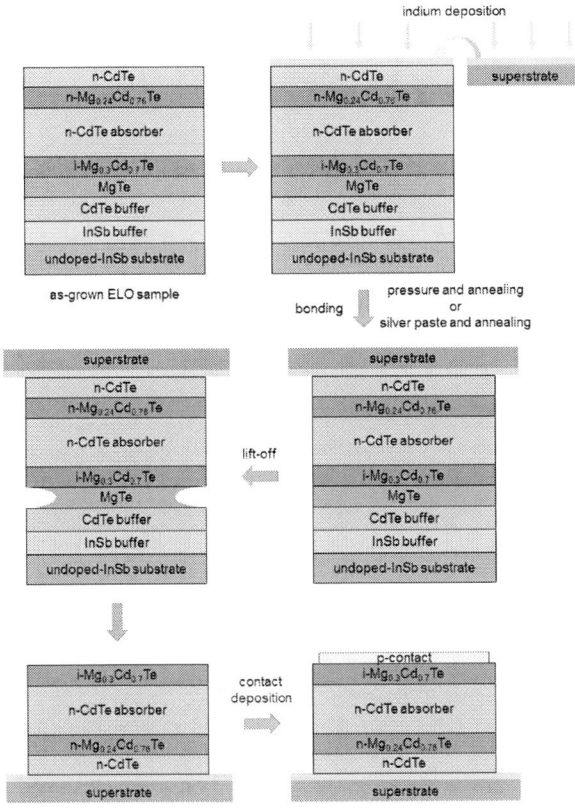

Fig. 2. Schematic process flow of the ELO and solar cell device fabrication processes.

is rough, indicating defects inside the thin film. The post-ELO thin film using a rigid silicon wafer as the superstrate exhibits poor morphology, likely due to poor attachment between the superstrate and the thin film. As a reference, ELO was repeated using Kapton tape as the superstrate. The post-ELO thin film is mirror-like in most areas but exhibits numerous cracks, consistent with previous results. [8]

Fig. 3. Optical images of a post-ELO thin film using the polyimide superstrate and indium bonding technique.

The optical properties of the post-ELO thin films are characterized by room temperature steady state photoluminescence (PL) as shown in Fig. 4a. The post-ELO thin film using the Kapton tape shows enhanced PL compared to the as-grown ELO sample, which is attributed to the reflection of the superstrate. However, a noticeable red-shift of the peak wavelength is observed, indicating strain relaxation of the CdTe layer as confirmed by the cracks on its surface. The thin film using the polyimide superstrate and the indium bonding technique shows the same PL peak wavelength as the as-grown ELO sample, indicating a negligible deformation of the thin film. The PL peak intensity of this post-ELO thin film is close to that of the as-grown ELO sample, confirming the ELO process does not significantly degrade the quantum efficiency of the thin film. The slight decrease of the PL intensity is probably due to extended defects such as dislocations generated during ELO process resulting in an increase of non-radiative recombination. Thin films using the silicon superstrate or the silver paste technique show both decreased PL intensity and red-shift of the peak wavelength, indicating an increase in defect density and partial strain relaxation of the DH.

The average PL peak intensity of the post-ELO thin films using different lift-off techniques is shown in Fig. 4b to illustrate the lateral uniformity of the thin films. Although some points on the thin film using Kapton tape as superstrate show stronger PL, the average PL intensity is not enhanced due to poor uniformity of the thin film. The thin film obtained using the polyimide superstrate and the indium bonding technique shows improved PL uniformity and average intensity compared to the other two films, consistent with the observed thin-film morphology.

B. Photovoltaic performance

Solar cell devices fabricated with the post-ELO CdTe/MgCdTe thin film using the polyimide superstrate and indium bonding technique exhibit a photovoltaic current density-voltage (J-V) curve under AM1.5G spectra as shown in Fig. 5. The V_{OC} of the device is 0.79 V. The short circuit current density (J_{SC}) is calculated using the designed mask area, which can be larger than the actual value due to current spreading. The

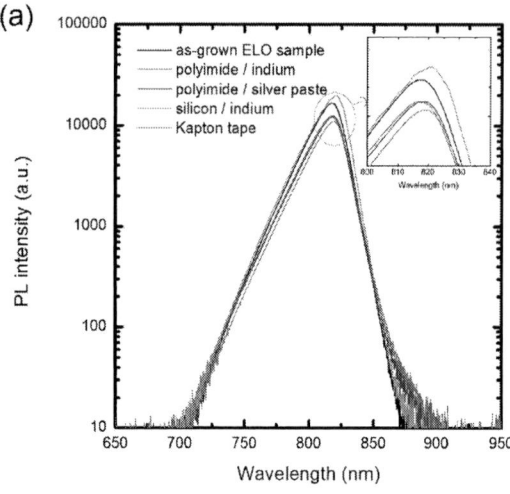

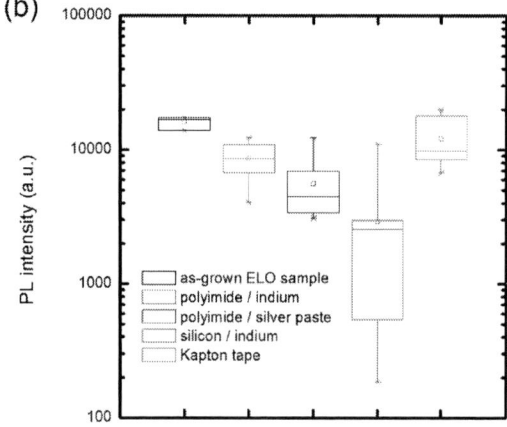

Fig. 4. (a) The PL spectra and (b) the average PL peak intensity of the as-grown ELO sample and the post-ELO thin films.

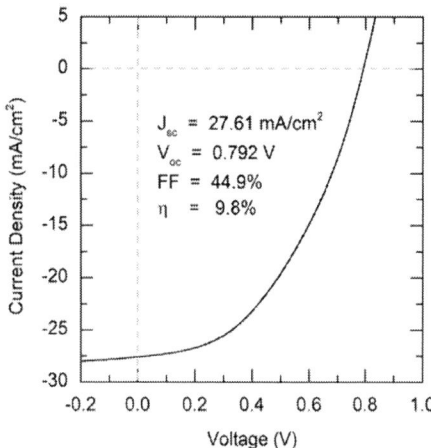

Fig. 5. J-V curve of the solar cell device fabricated with the post-ELO CdTe/MgCdTe thin film using the polyimide superstrate and indium bonding technique.

low fill factor (FF) of 44.9% is mainly attributed to the poor contact between the probe and the ITO layer, which can be improved by depositing Ag contact on top of the ITO. The calculated efficiency is 9.8%.

IV. SUMMARY

An ELO technique for monocrystalline CdTe/MgCdTe DHs using conductive superstrates is studied and photovoltaic devices fabricated with this technique are demonstrated. The post-ELO thin film using polyimide as superstrate and indium-bonding technique shows strong PL with good lateral uniformity. The device fabricated with the post-ELO thin film exhibits photovoltaic performance with a V_{OC} of 0.79 V. These findings demonstrate a method to fabricate lightweight monocrystalline CdTe solar cells, and the ELO technique enables the fabrication of tandem CdTe thin film/Si solar cell devices with high power conversion efficiency.

ACKNOWLEDGMENT

This work was partially supported by AFRL (Grant No. FA9453-20-2-0011) and the ARO program with contract number W911NF2010225. The authors acknowledge ASU NanoFab for the fabrication of the devices.

REFERENCES

[1] M. Konagai, M. Sugimoto, and K. Takahashi, "High efficiency GaAs thin film solar cells by peeled film technology", Journal of Crystal Growth 45, 277–280 (1978).

[2] E. Yablonovitch, T. Gmitter, J. P. Harbison, and R. Bhat, "Extreme selectivity in the lift‐off of epitaxial GaAs films", Applied Physics Letters 51, 2222‐2224 (1987).

[3] T. Kenning, "Alta Devices sets GaAs solar cell efficiency record at 29.1%, joins NASA space station testing", https://www.pv-tech.org/alta-devices-sets-gaas-solar-cell-efficiency-record-at-29-1-joins-nasa-spac/.

[4] A. Balocchi, A. Curran, T. C. M. Graham, C. Bradford, K. A. Prior, and R. J. Warburton, "Epitaxial liftoff of ZnSe-based heterostructures using a II-VI release layer", Appl. Phys. Lett. 86, 011915 (2005).

[5] A. N. Tiwari, A. Romeo, D. Baetzner, and H. Zogg, "Flexible CdTe solar cells on polymer films", Prog. Photovoltaics Res. Appl. 9, 211–215 (2001).

[6] D. J. Magginetti, J. A. Aguiar, J. R. Winger, M. A. Scarpulla, E. Pourshaban and H. P. Yoon, "Water-assisted liftoff of polycrystalline CdS/CdTe thin films using heterogeneous interfacial engineering", Adv. Mater. Interfaces. 6, 1900300 (2019).

[7] B. Seredyński et al., "(Cd,Zn,Mg)Te-based microcavity on MgTe sacrificial buffer: Growth lift-off and transmission studies of polaritons", Phys. Rev. Mat. 2, 043406 (2018).

[8] C. M. Campbell, C.-Y. Tsai, J. Ding, and Y.-H. Zhang, "Epitaxial lift off of II-VI thin films using water-soluble MgTe", IEEE J. Photovoltaics 9, 1834–1838 (2019).

[9] J. Ding, C.-Y. Tsai, Z. Ju, and Y.-H. Zhang, "Epitaxial lift-off CdTe/MgCdTe double heterostructures for thin-film and flexible solar cells applications", Appl. Phys. Lett. 118, 181101 (2021).

[10] Y. Zhao, et al, "Monocrystalline CdTe solar cells with open-circuit voltage over 1 V and efficiency of 17%", Nat Energy 1, 16067 (2016).

[11] W. Yang, et al., "Ultra-thin GaAs single-junction solar cells integrated with a reflective back scattering layer", J. Appl. Phys. 115, 203105 (2014).

Machine Learning Driven Studies of Performance Degradation in a-Si:H/c-Si Heterojunction Solar Cells

Davis Unruh[1], Reza Vatan Meidanshahi[2], Zitong Zhao[1], Stephen M. Goodnick[2], Gergely T. Zimanyi[1]

[1]Physics Department, University of California Davis, Davis, CA, 95616, USA
[2]School of Electrical, Computer and Energy Engineering, Arizona State University, Tempe, AZ, 85287, USA

Abstract— **a-Si:H/c-Si heterojunction solar cells hold the efficiency world record around 27%, yet their market penetration is delayed. One concern is the migration of passivating hydrogen away from the interface, that some suspect may speed up the degradation of their performance. Mitigating the performance degradation necessitates the understanding of the structural evolution of a-Si:H/c-Si structures, with a focus on hydrogen migration. To this end, we have developed the SolDeg structural simulation platform that is capable of capturing extremely slow degradation processes. SolDeg integrates molecular dynamics methods that optimize the Si structure with femtosecond time steps, with the nudged elastic band method that captures the defect generation on time scales extending to gigaseconds. The molecular dynamics layer of SolDeg requires a high quality Si-H interatomic potential. While classical parametric interatomic potentials have been used extensively, the recent development of machine-learning driven interatomic potentials ignited the ambition of achieving DFT-level accuracy with classical molecular dynamics simulations. In this paper we report the development of the first machine-learning driven Gaussian Approximation Potential (GAP) to describe Si-H interactions. This potential will be used in the SolDeg platform to determine the performance degradation of a-Si:H/c-Si heterojunction solar cells.**

Keywords—silicon heterojunctions, degradation, molecular dynamics, machine learning

I. INTRODUCTION

Si heterojunction (HJ) solar cells have world record efficiencies around 27%, due to the excellent surface passivation provided by the a-Si layer, which leads to low surface recombination velocity (SRV) and high open circuit voltage (V_{OC}). In spite of the impressive efficiency records, Si-HJ cells have not yet been widely adopted in the market because of various perceived challenges, including the cost of n-wafers, the low temperature deposition of Ag leads, and the higher capital expense costs. Several of these are getting rapidly resolved, making Si-HJ technology competitive. It has been estimated by the International Technology Roadmap for Photovoltaic (ITRPV) that the market share of Si-HJ cells will be 12% by 2026, and 15% by 2029 [1].

One of the remaining widely held reservation concerning the adoption of Si-HJ cells is that they may exhibit faster than usual performance degradation, possibly related to the a-Si layer. Traditional c-Si modules typically exhibit about a 0.5%/yr efficiency degradation, primarily via I_{sc}, typically attributed to moisture ingress and contacts. In contrast, the degradation of fielded Si-HJ modules was studied over 5-10 years by research groups at NREL [2], and AIST in Japan [3]. Both reported degradation rates close to 1%/yr, about twice the rate of the

traditional cells, and attributed it to a new degradation channel, the decay of V_{OC}, at a rate of about 0.5%/yr. The decay of V_{OC} suggests degradation processes internal to the cells, possibly due to an increased recombination at the a-Si/c-Si interface, or in the a-Si layer. Either way, the degradation is suspected to be caused by defects induced by mobile hydrogen. A recent study by Bertoni et al. measured the surface recombination velocity at the c-Si/a-Si interface in HJ stacks, and provided further evidence for such internal degradation processes, probably related to hydrogen effusion, increasing the density of defect states [4]. This is consistent with hydrogen incorporation during deposition impacting the implied V_{OC} for tunneling silicon oxides [5].

The interest in solar cell performance degradation was further intensified recently by another development. The ITPRV industry roadmap shows that in the next couple of years, the fraction of advanced Passivated Emitter and Rear Contact (PERC) modules among all newly installed modules will rapidly rise, exceeding 70% in 3 years [1]. In these advanced PERC cells extra hydrogen is used for passivation. Alarmingly, at WCPEC 2018 and PVSC 2019 several talks reported that these advanced PERC cells also exhibited unexpectedly large V_{OC} degradation [6]. Some talks related this degradation to the presence of excess hydrogen, analogous to processes in HJ cells. The fact that the V_{OC} degradation is also seen in PERC cells suggests that its physics is tied to the interface, rather than to the bulk a-Si.

To summarize, the accelerated degradation of V_{OC} in Si-HJ modules is well established but poorly understood. Whether the degradation is driven by defects or other structural processes is an open question. The new discovery of V_{OC}-linked degradation of H-passivated PERC solar cells has lent more weight to the idea that the degradation is a result of interfacial defects induced by mobile hydrogen diffusion, but the exact root cause remains unknown. The possibility of higher degradation rates in Si HJ cells, and the lack of understanding of its causes continue to fuel the anxiety about the Si HJ technology and limits its market acceptance in spite of the record efficiencies, since higher degradation rates adversely impact the levelized cost of energy (LCOE), the module lifetime, and warranty obligations.

II. RESULTS AND DISCUSSION

Given the above, we decided to comprehensively explore the device loss processes at the interfaces of Si-HJ stacks, and their impact on long-term device degradation by focusing on (1) light- and temperature-induced defect formation as a result of H diffusion; (2) the surface recombination velocity; (3) the ratio of interface to bulk hydrogen over time; (4) the impact of the

Funded by the DOE SETO grant DE-EE0009835

degree of crystallinity in the amorphous structure. Point (2) was inspired by some experiments suggesting that the electron lifetime is greatly affected by the surface recombination velocity. While there are good models for understanding the surface recombination itself [7, 8], it is not understood how the surface recombination velocity changes over time. Our goal is to develop a model to describe how the surface recombination velocity, and thus electron lifetime, changes over time. This paper presents our analysis of point (1), progress on (2-4) is underway as future work.

Our model, the SolDeg (**Sol**ar cell **Deg**radation) platform starts with creating a-Si:H/c-Si heterojunction structures, or stacks using molecular dynamics (MD) simulations. First, we create a-Si:H structures, with input parameters and conditions chosen to produce structures which match lab-grown a-Si:H as closely as possible. The matching is tracked by the radial distribution function, angle distribution function and excess energy relative to crystalline Si. Second, we place these optimized a-Si:H structures on top of slabs of c-Si, and then anneal the interface region. This approach was chosen in order to create the most realistic a-Si:H atomic structures possible, while still yielding a reasonable a-Si:H/c-Si interface regions.

The choice of interatomic potential used for these MD simulations has a substantial impact on the results. Reproducing structural properties of lab-grown structures, and accurately measuring the energetics of interactions within these structures, requires an interatomic potential with accuracy on the level of density functional theory (DFT). However, a small number of parametric potentials which describe Si-H interactions have been developed previously, but none of them come close to matching the accuracy of DFT.

To improve the accuracy of our MD simulations compared to those using existing interatomic potentials, we developed the first machine-learning (ML) driven Si-H interatomic potential. Our ML driven approach uses the framework of the Gaussian Approximation Potential (GAP) with a SOAP (Smooth Overlap of Atomic Positions) kernel, developed by the Csanyi group to yield DFT-level accuracy with an increase of efficiency of 10x or more, thus enabling the fast simulation of large systems [9]. To enhance the efficiency of the training of the potential, we supplemented the SOAP kernel with a predefined core pair potential, chosen to be purely repulsive, for all relevant atomic pairs: Si-Si, H-H, and Si-H. For details of the GAP method with a SOAP kernel and a core potential, see Ref. 10.

We trained this Si-H GAP by adaptive benchmarking relative to DFT measurements of energies, forces and virial stresses on a wide variety of structures. "Adaptive training" refers to enlarging the training database over time from which GAP learns in an iterative process. In this process, the GAP is fitted to an initial training set, and then is used to perform calculations such as structure optimizations or basic MD simulations. The results of these calculations are then validated against DFT calculations, and ill-fitting results are added to the training database. Using this procedure, it was not necessary to add thousands of structures indiscriminately to the training set, as the adaptive training procedure on its own preferentially

gravitated only towards structures that needed to be included into the training set without any external input, thereby avoiding the problem of over-fitting.

We conducted 27 full rounds of iterative training of the Si-H GAP. The initial training database was assembled by adding H to pure Si structures in various phases. These initial pure Si structures were taken both from our previous work on a-Si/c-Si interface degradation [11], as well as from the reference database on which the published Si-only GAP was trained [10]. The representative phases of Si were: 1) Amorphous Silicon; 2) Liquid Silicon; 3) Diamond Silicon with a vacancy; 4) Diamond Silicon with a divacancy; 5) Diamond Silicon with an interstitial Si. The atomic concentration of H added was around 12% for the liquid and amorphous phases. and between 4% and 8% for the diamond phases, as sufficient H was added to passivate all dangling or strained bonds. Since we specifically want to model hydrogen-related defects, structures with plus or minus a H ad-atom were also added into the training database. Late iterations included optimized and annealed structures, and structures in the intermediate temperature range between liquid and amorphous silicon. Then interface structures were added to the training set, followed by revisiting the phases already in the training database for a final benchmarking.

We validated our Si-H GAP by conducting MD calculations of experimentally observable quantities, and comparing the results to those obtained by DFT and by experiments. The Si-H GAP results were significantly closer to DFT results than any existing interatomic potentials in all tests, including the partial pair correlation functions, bond angle distributions, and

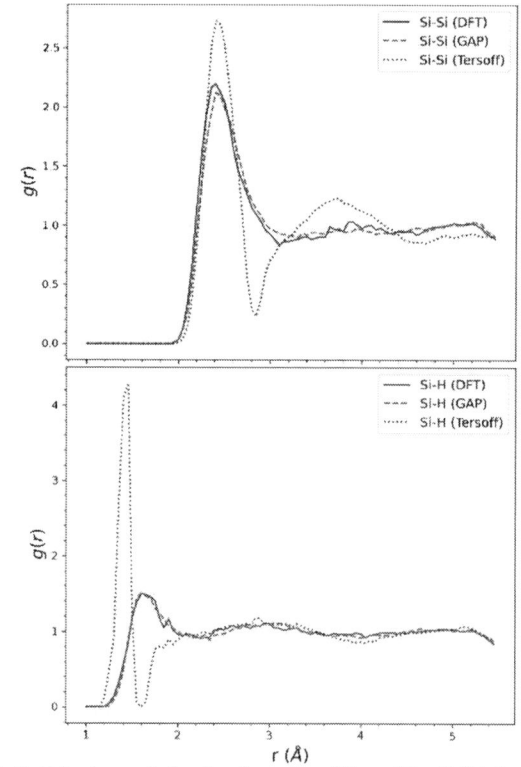

Fig. 1: Partial pair correlation functions of equilibrated liquid Si infused with H, with a density of 2.58 g/cm³, and T=2000K. Top: Si-Si; bottom: Si-H. Comparison provided between the GAP, a Si-H Tersoff potential, and DFT.

coordination statistics. Fig. 1 shows an example: the partial pair correlation functions of liquid Si:H at 2000K.

To demonstrate the scalability of our Si-H GAP, we created a large a-Si:H supercell with 4096 Si and 558 H atoms: see Fig. 2. This system size is prohibitively expensive for DFT-BOMD, while comfortably attainable by our Si-H GAP MD. Our simulations showed the formation of vacancies and nano voids in numbers large enough for statistical evaluation. [12]

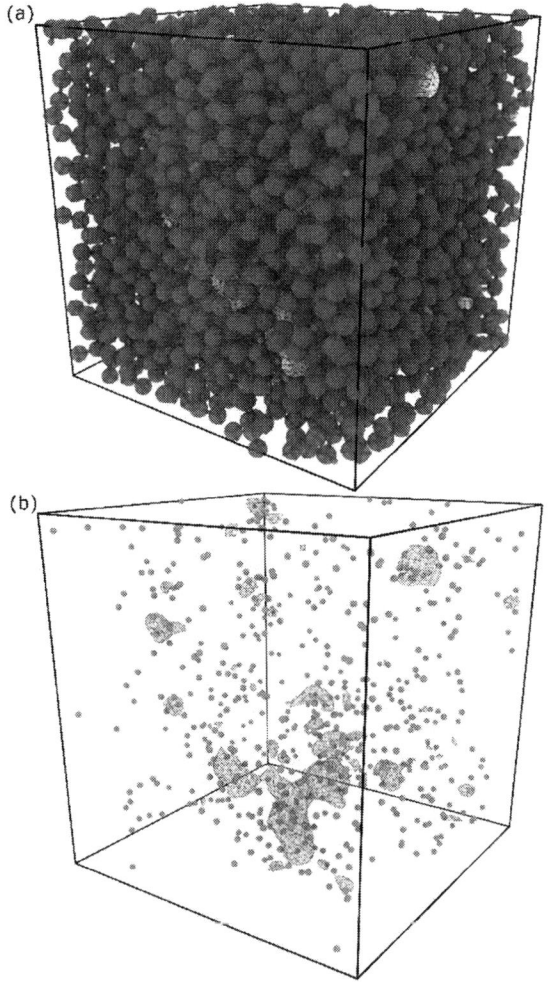

Fig. 2: (a) Ball-and-stick rendering of an a-Si:H slab, containing 4096 Si atoms (blue) and 558 H atoms (red). (b) A visualization of the H atoms and voids, obtained by suppressing Si. Voids are highlighted with grey surface meshes.

With the GAP successfully developed, the project will proceed with a modified version of our previously developed SolDeg project [11]. We will analyze the defect generation induced by H migration. The procedure will be as follows. After creating a-Si:H/c-Si interface structures, we will create electronic defect candidates by breaking Si-H bonds at the interface, moving the H atoms into the bulk a-Si region far enough to prevent spontaneous reformation of the Si-H bond. Then, the structure will be re-annealed. In the next step, DFT calculations of orbital localization of the electronic states will be used to measure the inverse participation ratio (IPR) per atom in order to determine if the H migration has resulted in the creation, or annihilation, of an electronic defect at the interface.

We will also endeavor to develop an ML-trained defect-locator that will identify localized electronic defect states exclusively from MD structural simulations, without the need of expensive DFT calculations. If a carefully trained ML model is able to determine electronic defects with only atomic descriptors, we will be able to study dynamics of defect generation and degradation combined with GAP MD.

III. Conclusions

The central message of this paper is that we believe that the degradation of Si-HJ solar cells is a result of interfacial defect formation due to the migration of mobile hydrogen away from the interface. We have made substantial progress toward modelling this defect formation process by training a new machine-learning driven Gaussian Approximation Potential to model Si-H interactions, and by developing the SolDeg platform to comprehensively simulate H migration in Si-HJs by (1) generating crystalline-amorphous interfaces; (2) characterizing the pathways of H migration from the interface into the bulk; (3) measuring energy barriers via the NEB method; and (4) using the resulting energetic barrier distribution to model the H dynamics with an accelerated kinetic Monte Carlo method to reach extended time scales up to gigaseconds, i.e. years.

Acknowledgment

We thank M. Bertoni and S. Manzoor for great discussions.

References

[1] International Technology Roadmap for Photovoltaic. [2020]. *Results for 2019 including maturity report 2020.* https://itrpv.vdma.org/ueber-uns.

[2] D. Jordan, C. Deline, S. Johnston, S. Rummel, W. Sekulic, P. Hacke, S. Kurtz, K.O. Davis, E.J. Schneller, X. Sun, M.A. Alam, and R. Sinton, "Silicon Heterojunction System Field Performance," IEEE J. of Photovoltaics, vol. 8, pp. 177-182, 2018.

[3] T. Ishii, S. Choi, R. Sato, Y. Chiba, A. Masuda, "Annual degradation rates of recent c-Si PV modules," Prog. Photovoltaic DOI: 10.1002/pip.2903 (2017); and DOI: 10.1109/ PVSC.2018.8548088.

[4] S. Bernardini, T. U. Naerland, A. L. Blum, G. Coletti, and M. Bertoni, "Unraveling bulk defects in high-quality c-Si material via TIDLS," Prog. Photovolt: Res. Appl. Vol. 25, pp. 209–217, 2016.

[5] B. Nemeth et al., "Effect of the SiO_2 interlayer properties with solid-source hydrogenation on passivated contact performance and surface passivation," Energy Procedia, vol. 124, pp. 295–301, 2017.

[6] S. Wenham plenary talk, WCPEC 2018, unpublished.

[7] S. Olibet, E. Vallat-Sauvain and C. Ballif, "Model for a-Si:H/c-Si interface recombination based on the amphoteric nature of silicon dangling bonds," Phy. Rev. B, vol. 76, pp. 1-14, 2007.

[8] R. Vasudevan, I. Poli, D. Deligiannis, M. Zeman, A. Smets, "Temperature dependency of the silicon heterojunction lifetime model based on the amphoteric nature of dangling bonds," AIP Advances, vol. 6, 2016.

[9] A. Bartók, M. Payne, R. Kondor, and G. Csányi, "Gaussian approximation potentials: The accuracy of quantum mechanics, without the electrons," Phys. Rev. Lett., vol. 104, pp. 136403, 2010; A. Bartók, R. Kondor, and G. Csányi, "On representing chemical environments," Phys. Rev. B, vol. 87, pp. 184115, 2013.

[10] A. P. Bartók, J. Kermode, N. Bernstein, and G. Csányi, "Machine learning a general-purpose interatomic potential for silicon," Phys. Rev. X, vol. 8, pp. 041048, 2018.

[11] D. Unruh et al., "From femtoseconds to gigaseconds: performance degradation in silicon heterojunction solar cells," arXiv: 2012.01703, unpublished.

[12] A. Smets, W.M. Kessels, M.C. van de Sanden, "Vacancies and voids in hydrogenated amorphous Silicon," Appl. Phys. Lett. Vol 82, pp. 1547.

Stability Analysis and Volt-Watt Control Setting Guideline for Distributed Energy Resources

Wenzong Wang[1], Wei Ren[1], Aminul Huque[1], Devin Van Zandt[1] and Reigh Walling[2]

[1]Electric Power Research Institute, Knoxville, TN, USA

[2]Walling Energy Systems Consulting, LLC, Clifton Park, NY, USA

Abstract—**Smart inverter grid support functions such as volt-var and volt-watt have the potential to alleviate over-voltage and voltage variation issues caused by high penetration of distributed energy resources, especially solar photovoltaic (PV), on the distribution circuit. However, because of the dynamic nature of volt-var and volt-watt control functions, they may cause voltage oscillation, especially when aggressive control settings are used. This paper investigates the stability of volt-watt control function through lab testing, analytical analysis, and simulation studies. Key factors affecting the stability of volt-watt control are identified. Moreover, a guideline to identify the volt-watt control setting range to avoid the voltage instability is developed and is verified via dynamic simulation on a simple test system.**

Keywords— Control stability, smart inverter, voltage oscillation, volt-watt.

I. INTRODUCTION

With increasing penetration of solar photovoltaic (PV) in power distribution systems, unacceptable voltage variations caused by intermittent PV generation challenges the secure operation of distribution grids. To address these issues, distributed energy resources (DERs) are required by standards, such as IEEE Std 1547™-2018, to have voltage support capability. Among others, smart inverter volt-var and volt-watt functions have been proven effective in mitigating voltage issues and have been used by some distribution utilities [1].

However, there is still relatively limited field experience with these types of advanced active and reactive power control functions on distribution systems and distribution engineers have raised concerns about potential voltage oscillation caused by the interaction of these controls with the distribution system, with each other, and with other existing voltage regulating equipment on the circuit.

Moreover, even though the stability of volt-var control has been analyzed in the literature [2]-[7], the stability of volt-watt function has been rarely studied. A stability vulnerability when volt-watt and volt-var functions are activated simultaneously is discussed in [8]. However, the vulnerability is associated with active power priority and does not exist with reactive power priority specified in IEEE Std 1547™-2018.

Previous research [5] has investigated the oscillatory behavior caused by inverter volt-var control function and developed guidelines to identify the setting range that ensures control stability. As a continuation of the work in [5], this paper investigates the stability of inverter volt-watt control and addresses two main research questions:

1) Can aggressive volt-watt control function cause unstable voltage oscillations?

2) What's the setting guideline to avoid unstable voltage oscillations with volt-watt control?

Lab testing of commercial inverters, analytical state-space analysis and simulation studies have been conducted to identify the conditions for volt-watt function induced voltage oscillations and to develop a setting guideline to avoid control instability. The setting guideline takes into account the location and operating condition of the DER and can be obtained analytically through power flow and eigenvalue calculations without simulation studies. Furthermore, the setting guideline is demonstrated and verified through dynamic simulation studies on a simple test system.

II. LAB TESTING OF COMMERCIAL INVERTERS

A. Test Setup

To investigate the possibility of volt-watt induced oscillations with commercial inverter, a three-phase 33kW PV inverter was tested in the lab. The inverter was connected to a programmable grid simulator through an impedance module and a transformer in series, as shown in Fig. 1. The impedance module was set up such that the short circuit ratio (SCR) at the inverter terminal is 5, which represents a weak system connection scenario. The X/R ratio of the inverter external Thevenin equivalent impedance is 3.

During the lab testing, the inverter would output 1 pu active power without volt-watt control. Three scenarios were tested: 1) volt-watt control is disabled, 2) volt-watt control is enabled with

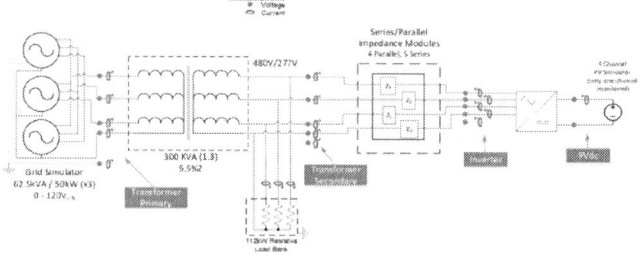

Fig. 1. Lab setup for volt-watt function testing of a commercial three-phase inverter.

a gentle slope (setting 1 in Fig. 3) and 3) volt-watt control is enabled with an aggressive slope (setting 2 in Fig. 3). For all the testing scenarios, the inverter reactive power control was set to unity power factor.

B. Lab Testing Results

The inverter current output and terminal voltage were recorded with a sampling rate of 10kHz. The active and reactive power output calculated based on the measurements are shown in Fig. 3 together with the three-phase terminal voltage.

As can be seen from the lab testing results, the inverter active power output is around 1 pu without volt-watt control and the inverter terminal voltage is around 1.06 pu. When volt-watt control is enabled with the gentle slope (setting 1), the active power is reduced to 0.9 pu which brings the terminal voltage down slightly to 1.055 pu, without causing any voltage oscillations. When the volt-watt slope is increased (setting 2), the active power is further reduced which lowers the terminal voltage. However, sustained voltage and active power oscillations occur with this aggressive slope setting.

The lab testing results indicate that aggressive volt-watt settings can cause sustained oscillations when the inverter is connected to a weak system location. It is worth pointing out that the impact of volt-watt open loop response time was not tested because for this inverter, which is not IEEE 1547-2018 certified, the response time is not user configurable.

III. INVERTER STEADY STATE OPERATION WITH VOLT-WATT CONTROL

The effectiveness of inverter volt-watt control in dealing with voltage violations varies with system conditions. To illustrate this and to prepare for the discussion on volt-watt induced oscillations, this section introduces the steady state operation characteristics for inverters with volt-watt control.

A. Inverter Steady State Operation with Volt-Watt Control

To illustrate the impact of volt-watt control on inverter steady state operation, a single inverter is considered with its external grid modeled as a Thevenin equivalent circuit, which is shown in Fig. 4. The inverter terminal voltage magnitude and angle are denoted as V and θ. P and Q represent the inverter output active and reactive power. Z and φ are the magnitude and angle for the external impedance and E denotes the voltage magnitude of the equivalent voltage source.

The power flow equations of the single inverter system can be derived as

$$\begin{cases} P = \dfrac{V^2}{Z}\cos\varphi - \dfrac{VE}{Z}\cos(\theta + \varphi) \\ Q = \dfrac{V^2}{Z}\sin\varphi - \dfrac{VE}{Z}\sin(\theta + \varphi) \end{cases} \quad (1)$$

With the following algebraic manipulation,

$$\begin{cases} sin(\theta + \varphi) = \dfrac{V}{E}sin\varphi - \dfrac{QZ}{VE} \\ cos(\theta + \varphi) = \pm\sqrt{1 - sin^2(\theta + \varphi)} = \pm\sqrt{1 - \left(\dfrac{V}{E}sin\varphi - \dfrac{QZ}{VE}\right)^2} \end{cases} \quad (2)$$

the relationship between the inverter terminal voltage and its active power output can be expressed as

$$P = \frac{V^2}{Z}cos\varphi \pm \frac{VE}{Z}\sqrt{1 - \left(\frac{V}{E}sin\varphi - \frac{QZ}{VE}\right)^2} \quad (3)$$

Equation (3) is the physical constraint of the system and it holds at any time instant. This equation is shown in Fig. 5 as the $f(V, P)$ curve. The plus minus sign in the equation indicates that

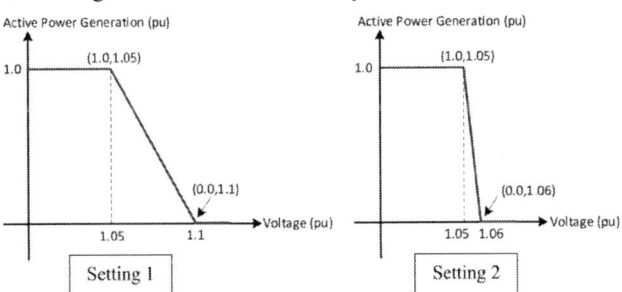

Fig. 3. Different volt-watt settings considered in lab testing.

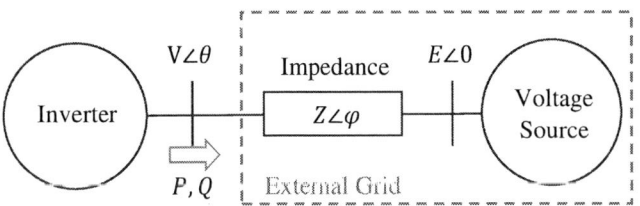

Fig. 4. Single inverter connected to the grid.

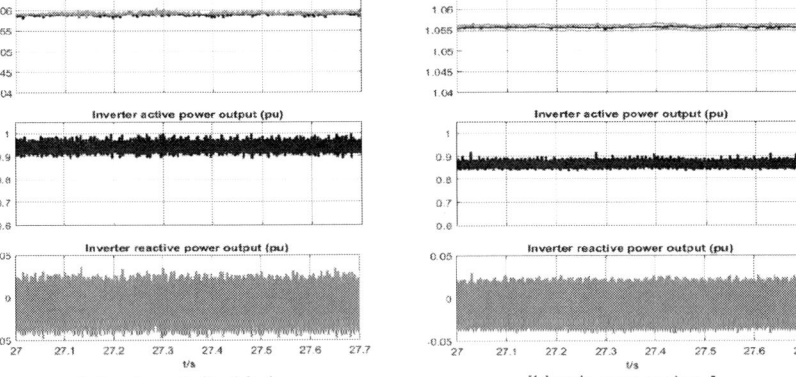

Fig. 2. Lab testing results with different volt-watt settings.

978-1-7281-6118-1/22 $31.00 © 2022 IEEE

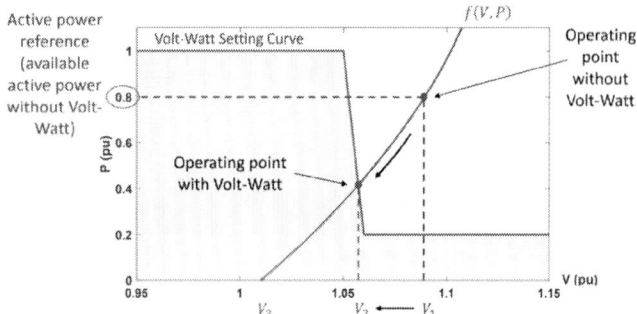

Fig. 5. Illustration of inverter steady state operating point with volt-watt control.

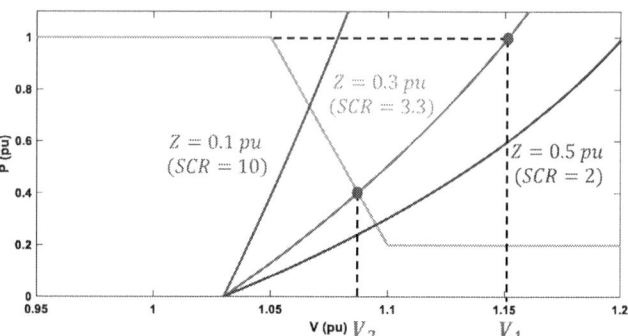

Fig. 6. Effectiveness of volt-watt control with different SCR (X/R=1.5).

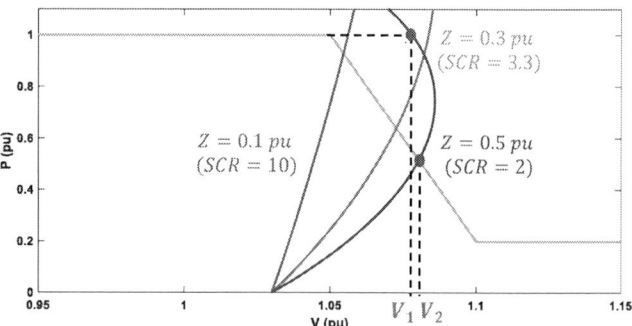

Fig. 7. Effectiveness of volt-watt control with different SCR (X/R=3).

the curve can be highly nonlinear (i.e., a given voltage may correspond to two possible power levels) in certain conditions and may affect the effectiveness of volt-watt control, which will be illustrated in the next subsection.

Meanwhile, the volt-watt curve determines a region in the voltage-active power plane that the steady state operating point needs to fall within, which is shown as the shaded area in Fig. 5. If the available active power is high such that without volt-watt control the steady state operating point is outside the region, the volt-watt function, if activated, will reduce the inverter active power output and the final steady state operating point is the intersection of the $f(V, P)$ curve and the volt-watt curve. This process is also illustrated in Fig. 5.

In Fig. 5, V_3 is the inverter terminal voltage when the inverter has no power output. When the inverter is generating maximum available power, V_1 and V_2 are the voltages without and with volt-watt control, respectively. A "degree of effectiveness" index can be defined as $(V_1 - V_2)/(V_1 - V_3)$ to characterize the effectiveness of volt-watt control. When the index is positive, it indicates that the volt-watt control is mitigating the overvoltage condition. The closer the index is to 1, the more effective the volt-watt control is. As can be seen from Fig. 5, the flatter the $f(V, P)$ curve (higher voltage sensitivity to inverter active power output) is, the more effective the volt-watt control (more overvoltage reduction) can be.

B. Effectiveness of Volt-Watt Control Function

Based on sensitivity studies, it is found that the slope and nonlinearity of the $f(V, P)$ curve mainly depend on the magnitude and X/R ratio of the inverter external impedance. To illustrate the impact on volt-watt control effectiveness, two cases with different X/R ratios (1.5 and 3) are compared. For each case the magnitude of the impedance (or equivalently the SCR) is also varied.

The case with X/R=1.5 is shown in Fig. 6. The volt-watt control is effective in reducing the inverter terminal voltage for the range of SCR considered. For example, with SCR=3.3, the voltage is reduced from V_1 to V_2. It can be seen that the amount of voltage reduction is increased, and the volt-watt is more effective when the SCR is lower.

In comparison, Fig. 7 illustrates the results when X/R is increased to 3. In this case, if the SCR is 2, the $f(V, P)$ curve becomes very nonlinear and the voltage sensitivity to active

power output can become negative, which indicates that by reducing the inverter active power output, the terminal voltage will increase rather than decrease. As a result, the volt-watt control will further elevate the overvoltage (increase the voltage from V_1 to V_2), instead of mitigating it.

In summary, when the X/R ratio of the external impedance is low, volt-watt control is effective across a wide range of SCR. The effectiveness increases with lower SCR. However, with higher X/R ratio, the effectiveness of volt-watt control reduces when SCR is low and may even become negative. Note that the system considered in this study (Fig. 4) is highly simplified and the aim is to show the trend of the influence of X/R ratio and SCR. In practice, the loading and generation from other DERs on the same feeder may also affect the effectiveness of volt-watt control. Therefore, the exact values from this study (e.g., X/R=1.5 SCR=2) should not be used to determine whether volt-watt control is effective. Instead, it is recommended that the voltage sensitivity to active power output be checked and make sure it is positive before activating volt-watt control for a DER plant, especially in the case with high X/R ratio and low SCR.

IV. STABILITY ANALYSIS FOR INVERTER VOLT-WATT CONTROL AND A SETTING GUIDELINE

In this section, stability analysis for a DER operating with volt-watt function is conducted through analytical state-space analysis and dynamic simulation. A setting guideline to avoid control instability is also developed.

A. State-space Stability Analysis

Small-signal state-space analysis has been utilized for volt-var stability analysis in [2]. Here the same approach is applied to analyze the stability of volt-watt control. As a first step, the state-space block diagram of volt-watt control is constructed as in Fig. 8, where m_p is the slope of the volt-watt curve and dV/dP is the voltage sensitivity to inverter active power output at the DER reference point of applicability (RPA).

The block $1/(1 + s\tau)$ represents in Laplacian form a first-order lag, or low-pass filter, characteristic and is used to model the various lags or delays in the control loop. Time constants τ_1, τ_2 and τ_3 are associated with the open loop response time (OLRT) of volt-watt control, the time lag/delay in active power output caused by inverter inner control and the time lag/delay introduced by voltage measurement. In the remainder of this paper, the total time lag/delay caused by voltage measurement and inverter inner control is termed measurement and processing delay, which is different from the configurable control lag that is adjusted to obtain the desired OLRT.

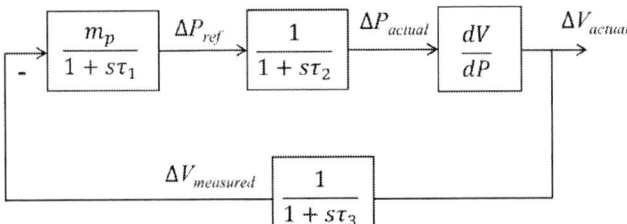

Fig. 8. State-space block diagram of volt-watt control.

Eigenvalues of the state-space model can then be calculated to evaluate whether volt-watt induced oscillation occurs and if so, what its damping ratio is. A negative damping ratio indicates the oscillation is unstable. If the damping ratio is positive, the oscillation is stable. The higher the damping ratio, the faster the oscillation magnitude goes to zero. If the damping ratio is greater than 0.6, the oscillations will be well damped with an overshoot of less than 10%. This can be used as a conservative limit for acceptable dynamic performance.

To analyze the stability of inverter volt-watt control, the state-space analysis is applied on a simple system (Fig. 9) where a PV plant with volt-watt control is connected to the grid (modeled as a Thevenin equivalent voltage source) through a step-up transformer. The grid impedance is chosen to be 0.3 pu based on the PV plant kVA rating, which renders the SCR at the plant RPA to be 3.3, representing a weak grid. The results are illustrated in Fig. 10 and Fig. 11.

The slope and the OLRT of the volt-watt function are required by IEEE Std 1547™-2018 to be configurable within a defined range. When they are set within the green stability region shown in Fig. 10 and Fig. 11, the voltage response will be stable. The blue line corresponds to zero damping ratio and is the stability boundary or limit. When the settings are below the magenta line, the damping ratio is above 0.6 and the oscillation, if occurs, will be well damped.

From Fig. 10 and Fig. 11, the chance for instability increases with larger slope or shorter OLRT setting. Longer OLRT allows larger slope to be used without causing instability. Comparing Fig. 10 and Fig. 11, it can be seen that longer measurement and processing delay reduces the stability region. Besides these factors associated with inverter control, as can be seen from the

control diagram in Fig. 8, the RPA voltage sensitivity to inverter active power output is another key factor that affects the volt-watt stability in a similar way as the volt-watt slope. In other words, higher voltage sensitivity reduces the stability region.

It is worth pointing out that the default volt-watt settings in IEEE Std 1547™-2018 has a slope of 25 and an OLRT of 10s, which is within the stability region in both Fig. 10 and Fig. 11 with a large margin to the stability limit, indicating that the default settings are quite stable even in this weak system condition considered.

B. Simulation Verification

To verify the analytical results, a detailed PV inverter model was set up in MATLAB-Simulink and it includes modeling of the outer and inner control loops for real and reactive power control, the phase locked loop (PLL) and the volt-watt control function. The PV inverter is connected to the same system in Fig. 9 and the dynamic simulation results are shown in Fig. 12 and Fig. 13. The volt-watt control is activated at t=3s. With a slope of 50, the volt-watt control successfully reduces the overvoltage without causing instability, whereas with a slope of 100, sustained oscillations in voltage and active power occur, indicating volt-watt control instability. The results correspond well with the two red dots shown in Fig. 11 and therefore verifies the analytical analysis.

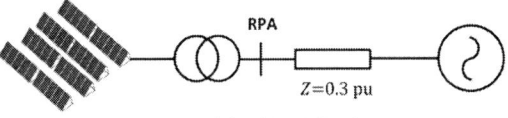

$SCR = 3.3, X/R = 1.5$ at the RPA

Fig. 9. System considered to analyze volt-watt control stability.

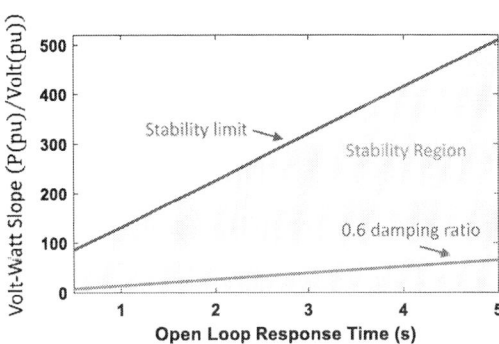

Fig. 10. Volt-watt stable setting region with measurement and processing delay $\tau_2 + \tau_3 = 0.15s$

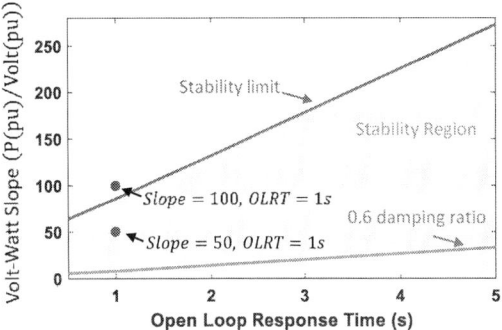

Fig. 11. Volt-watt stable setting region with measurement and processing delay $\tau_2 + \tau_3 = 0.3s$

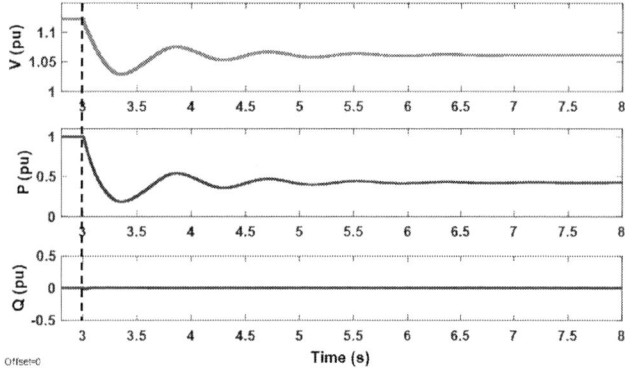

Fig. 12. Dynamic simulation results with $slope = 50, OLRT = 1s, \tau_2 + \tau_3 = 0.3s$

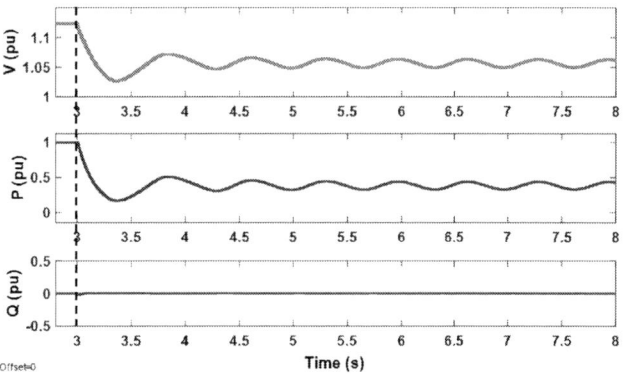

Fig. 13. Dynamic simulation results with $slope = 100, OLRT = 1s, \tau_2 + \tau_3 = 0.3s$

C. A Setting Guideline to Avoid Instability

As discussed in section IV part A, there are mainly four factors that affect the stability of inverter volt-watt control, namely the slope of the volt-watt setting curve, the OLRT, the voltage sensitivity to inverter active power output, and the measurement and processing delay. Of the four factors, the volt-watt slope and the OLRT are defined in IEEE Std 1547™-2018 to be settable by a distribution operator.

Therefore, the goal of the setting guideline is to identify the range of those two settings that ensures voltage stability. In other words, given a feeder model and the location of the DER, the stability regions similar to those shown in Fig. 10 and Fig. 11 should be identified and the volt-watt settings should be inside the region to avoid instability. Note that this setting guideline is to avoid control instability rather than identifying the optimal settings for feeder voltage regulation. Once the stability region is identified, the settings can then be optimized inside the stability region as needed, based on other criteria which is outside the scope of this paper.

To identify the stability region, the first step is to obtain the voltage sensitivity to active power output (dV/dP) at the RPA of the DER plant. This can be done by running two power flow calculations following the procedure below

1) In the power flow solver where the feeder is modeled, set the active power output of the DER under consideration to $P_1 = 0$ and run power flow. Obtain the RPA voltage of the DER plant (V_1)

2) Change the power generation by a small amount (e.g. 0.05 pu) by setting the DER output active power to $P_2 = 0.05\ pu$ (based on the plant kW rating) and run power flow again. Obtain the updated RPA voltage of the DER (V_2)

3) Calculate $dV/dP = \frac{V_2 - V_1}{P_2 - P_1}$

Notice here the active power generation is perturbed around 0. The reason to choose 0 is that based on Fig. 6 and Fig. 7, the voltage sensitivity is higher at lower active power generation level, so the value at 0 pu can be used as a conservative estimate to account for all the possible operating conditions of the DER plant.

The second step is to determine the measurement and processing delay. In practice, this delay is not user configurable and could be difficult to obtain from plant developers or inverter manufacturers. If the delay information is unavailable, it is recommended to use 0.5s as a conservative estimate. Since IEEE Std 1547™-2018 requires the open loop response time of volt-watt function to be adjustable between 0.5s and 60s, the measurement and processing delay should be less than 0.5s such that the fasted open loop response can be achieved.

After dV/dP and the delay are obtained, eigenvalues of the system shown in Fig. 8 can be calculated with different OLRT settings. For a given OLRT, the volt-watt slope should be gradually increased until it causes the damping ratio to drop to 0.6. The last value of the volt-watt slope then defines the limit of the slope associated with the OLRT under consideration and it corresponds to one point on the 0.6 damping ratio curve in Fig. 10 and Fig. 11.

The same procedure can be repeated for other OLRT settings to obtain the entire 0.6 damping ratio curve. Note that based on results shown in Fig. 10 and Fig. 11 and other study results, this curve is close to a straight line, which means the entire curve can be extrapolated after calculating the slope limit for at least two OLRT settings. Moreover, this process can be automated by developing a script in software such as MATLAB.

Once the 0.6 damping ratio curve is obtained, the volt-watt slope and OLRT should be set in the region below this curve to ensure voltage stability, as shown in Fig. 10 and Fig. 11. Note that the damping ratio of 0.6 is used as a conservative limit for acceptable dynamic performance. As discussed in the previous subsection, if the damping ratio is greater than 0.6, the oscillations will be well damped with less than 10% overshoot.

V. CONCLUSION

This paper investigates the volt-watt control induced voltage oscillations in distribution systems. Results indicate that the default volt-watt settings in IEEE Std 1547™-2018 are unlikely to cause unstable voltage oscillations in practice. However, voltage instability can occur if a DER plant connected at a weak feeder location employs aggressive volt-watt settings. Moreover, the instability is more likely to happen with higher volt-watt control slope, shorter open loop response time, longer inverter voltage measurement and inner control delay, and higher RPA voltage sensitivity to active power output. A volt-watt control setting guideline is provided to avoid unstable voltage oscillations, which does not require dynamic simulations and therefore is relatively easy to be applied by distribution engineers.

REFERENCES

[1] *Application Guidelines for DER Advanced Functions & Settings*. EPRI, Palo Alto, CA: 2019. 3002015738.

[2] *Oscillation Mechanism and Setting Guideline for Inverter Volt-Var Control*, EPRI, Palo Alto, CA: 2020. 3002019776.

[3] *Feeder Closed Loop Control Interactions on High Penetration DER Systems*, EPRI, Palo Alto, CA: 2020. 3002018657.

[4] *Stability Analysis and Volt-Watt and Volt-Var Control Setting Guideline for DERs*, EPRI, Palo Alto, CA: 2021. 3002021716.

[5] W. Wang, X. Shi, C. Brewster and A. Huque, "Oscillation Mechanism and Setting Guideline for Inverter Volt-Var Control," *2020 47th IEEE Photovoltaic Specialists Conference (PVSC)*, 2020, pp. 2032-2039.

[6] A. Singhal, V. Ajjarapu, J. Fuller and J. Hansen, "Real-Time Local Volt/Var Control Under External Disturbances With High PV Penetration," in *IEEE Transactions on Smart Grid*, vol. 10, no. 4, pp. 3849-3859, July 2019.

[7] J. Schoene *et al.*, "Investigation of oscillations caused by voltage control from smart PV on a secondary system," *2017 IEEE Power & Energy Society General Meeting*, Chicago, IL, 2017, pp. 1-5.

[8] J. H. Braslavsky, L. D. Collins and J. K. Ward, "Voltage Stability in a Grid-Connected Inverter With Automatic Volt-Watt and Volt-VAR Functions," in *IEEE Transactions on Smart Grid*, vol. 10, no. 1, pp. 84-94, Jan. 2019.

Moisture Ingress and Distribution in Bifacial Silicon Photovoltaics

Rishi E Kumar, Tala Sidawi, Ian M Slauch, Rico Meier, Mariana I Bertoni, David P Fenning

University of California San Diego, La Jolla, CA, United States

Arizona State University, Tempe, AZ, United States

Water drives various modes of degradation in photovoltaic (PV) modules, ranging from encapsulant yellowing and delamination to contact corrosion. In silicon PV modules, the primary route of moisture ingress is diffusion through the polymeric module components (encapsulant and, when present, backsheet). Understanding the behavior of moisture in modern encapsulants and architectures is crucial for projection and, ultimately, mitigation of moisture-induced degradation. To this effect, we quantify the solubility and diffusivity of water in four state-of-the-art encapsulants for bifacial silicon PV modules: ethylene vinyl acetate (EVA) and polyolefin (POE), each with and without UV-blocking additives. Using water reflectometry detection (WaRD), we track the "breathing" rate of water indiffusion and outdiffusion in miniature bifacial modules of both glass-glass and glass-backsheet configurations. Crucially, we separate module moisture content from in front and back of the silicon cell. Overall, we present a quantitative picture of how moisture moves within modern bifacial silicon PV modules.

Evaluation of PV Module Packaging Strategies of Monofacial and Bifacial PERC Using Degradation Pathway Network Modeling

Sameera Nalin Venkat*, Jiqi Liu*, Xuanji Yu*, Jakob Wegmueller*, Kunal Rath*, Xinjun Li[†],
Jean-Nicolas Jaubert[‡], Jennifer L. Braid[§]*, Roger H. French*, Laura S. Bruckman*

*SDLE Research Center, Department of Materials Science and Engineering, Case Western Reserve University (CWRU),
Cleveland, OH, 44106, USA
[†]Cybrid Technologies Inc., Suzhou, Jiangsu, China
[‡]CSI Solar Co. Ltd., 199 Lushan Road, SND, Suzhou, Jiangsu, China, 215129
[§]Sandia National Laboratories, Albuquerque, NM, USA

Abstract—As the PV industry is rapidly expanding, it is important to thoroughly investigate the long-term impact of packaging strategies on the performance of PV modules. In this study, the variants in sets differ on the basis of manufacturer (A/B), encapsulant (EVA/POE), rear encapsulant (UV-cutoff/opaque/transparent), module architecture (GB/DG) and cell type (monofacial/bifacial). The minimodules were exposed for 2520 hours in modified damp heat, with or without full spectrum light. Every 504 hours, stepwise electrical characterization techniques were employed to track changes in minimodules. Degradation pathway modeling using network structural equation modeling was employed to study pairwise relationships between variables and service lifetime prediction in minimodules. Through this study, differences in quality control are identified in minimodules made by different manufacturers. Minimodules with UV-cutoff rear encapsulant show relatively better stability, whereas the ones with opaque rear encapsulant show greater power loss. In addition, GB having UV-cutoff rear encapsulation and GB with POE having opaque rear encapsulation were identified to be stable as they lack a best model fit. The primary power loss mechanism in degrading variants is interconnect corrosion.

Index Terms—pathway modeling, accelerated exposures, PV degradation, electrical measurements, bifacial, monofacial

I. INTRODUCTION

The PV industry continues to experience a massive growth over the past few years. While silicon PV is still dominating the market, the impact of packaging strategies on module degradation is being actively explored by researchers. There are emerging polymeric materials like thermoplastic olefin (TPO) and polyolefin elastomer (POE) were designed to

This material is based upon work supported by the U.S. Department of Energy's Office of Energy Efficiency and Renewable Energy (EERE) under Solar Energy Technologies Office (SETO) Agreement Number DE-EE-0008550. Alan Curran, Raymond Wieser, Leean Jo, Hein Htet Aung, Max Atkinson are acknowledged for electrical measurements. The views expressed herein do not necessarily represent the views of the U.S. Department of Energy or the United States Government. Sandia National Laboratories is a multimission laboratory managed and operated by National Technology & Engineering Solutions of Sandia, LLC, a wholly owned subsidiary of Honeywell International Inc., for the U.S. Department of Energy's National Nuclear Security Administration under contract DE-NA0003525.

avoid problems seen in ethylene vinyl acetate (EVA) such as chemical instability due to the formation of acetic acid, potential induced degradation,.

Bifacial PV is gaining traction in the market due to its ability of absorbing photons from front and rear sides, leading to more electrical power. Double glass module architecture has been used exclusively for bifacial PV. In order to offset the issues with extra weight, additional source of sodium ions that lead to PID and elevated cost [1], PV manufacturers developed transparent backsheets aimed for bifacial PV [2]. The long-term degradation of monofacial versus bifacial PV, the impact of different packaging strategies and rank-ordering of durable/degrading variants are not fully explored.

The current methods of degradation modeling in the PV field aim at studying degradation rates based on a single mechanism. Arrhenius model, a popular method for estimating degradation rate in the PV field, does not apply for situations where there are multiple competing reactions. Hence, there is a great opportunity to explore multiple degradation modes impacting power loss in PV modules using network structural equation modeling (netSEM).

netSEM was developed at the SDLE Center as part of the lifetime and degradation science framework to capture the complex behavior of real-life degradation in materials systems [3], [4], [5]. In the netSEM R package, it is possible to map the stressor (S), the mechanistic/tracking variables (M) and the response (R) using two types of principles. Principle 1, based on the Markov property, allows for investigating relationships between variables in a pairwise manner while keeping the rest of the variables constant. Principle 2 is multiple regression by using step Akaike information criterion (step-AIC) (non-Markov property): it considers multivariate relationships between variables simultaneously.

In this article, we use netSEM to investigate causes of power loss in 4 cell PV minimodule with a variety of packaging strategies and cell types in indoor accelerated conditions. Pairwise relations between variables using predictive modeling, rank-ordering using predictive confidence intervals using

Principle 1 and service lifetime prediction using Principle 2 are discussed.

II. MINIMODULE SPECIFICATIONS

The samples in this study are 4-cell PV minimodules which have passivated rear and emitter cells (PERC) connected in series. There are four sets of minimodules with varying packaging combinations, which we will refer to as sets #1-#4. Within each set, the minimodules differ in terms of the manufacturer (A/B), encapsulant (EVA/POE) and architecture (GB/DG). Between sets, the minimodules have different PERC type (monofacial/bifacial) and rear encapsulant (UV-cutoff/opaque/transparent). The minimodule specifications are detailed in Table I.

TABLE I: Specifications of sets #1-#4 4-cell PV minimodules.

Set #	PERC Cell Type	Encapsulant	Rear Encapsulant	Architecture
1	Monofacial	EVA	UV-cutoff	DG
				GB
		POE		DG
				GB
2	Bifacial	EVA	Opaque	DG
				GB
		POE		DG
				GB
3	Monofacial	EVA	Opaque	DG
				GB
		POE		DG
				GB
4	Bifacial	EVA	Transparent UV-cutoff	DG
				GB
		POE	Transparent UV-cutoff	DG
				GB

III. STUDY PROTOCOL AND METHODS

The study protocol is organized in four components: fabrication of 4-cell PV minimodules, indoor accelerated exposures, data collection using electrical characterization techniques, and data-driven degradation pathway modeling. These components are highlighted in detail in our previous publication for minimodules with monofacial multicrystalline PERC, and UV-cutoff rear encapsulant [6].

The fabrication process consists of two important steps: soldering of solar cells to enable series connectivity, and lamination to fuse polymeric layers, glass, and soldered solar cells. Both the manufacturers utilized different methods of soldering, and slightly different lamination recipes, which led to differences in the production quality. It is to be noted that minimodules by manufacturer A were fabricated in a reliable manner.

Once the minimodules were fabricated, they were preconditioned to verify stability. Then, the minimodules were exposed in modified damp heat (mDH), with or without full spectrum light (FSL). mDH refers to 80°C and 85% relative humidity, and FSL makes use of 420 Wm^{-2} intensity light from high-intensity discharge lamp. In each set, half of the minimodules were exposed in mDH, and the other half in mDH+FSL. The total exposure time for each cycle is 2520 hours (105 days or about 3.5 months).

At the end of 504 hours (21 days), changes in minimodules were monitored using characterization techniques. The data that was collected using electrical characterization techniques, namely I-V, and $Suns$-V_{oc}, was subsequently employed for data-driven modeling.

For degradation pathway modeling, network structural equation modeling (netSEM) was used. netSEM was developed at the SDLE Research Center to expand from linear modeling in structural equation modeling (SEM) to include nonlinear cases like quadratic, logarithmic, change point, etc. It is available as an R package on CRAN (the version used to generate the results is 0.6.0). For extracting electrical features, ddiv R package was employed. Some of the data points were found to be missing due to issues with junction box connection and/or broken minimodules. No additional data imputation was done to address this issue.

IV. <STRESSOR|MECHANISM|RESPONSE> MODELING USING PRINCIPLE 1

The variables selected for constructing degradation pathway models were obtained from I-V, and $Suns$-V_{oc}. The stressor is exposure time in decimal year (dy); decimal year indicates that the value is not a whole number but a value with a decimal place (ranging from 0-0.3). The mechanistic variables track degradation modes occurring in minimodules; short circuit current obtained from I-V measurements ($I_{sc,IV}$) monitors optical transmission losses [7], series resistance from I-V measurements ($R_{s,IV}$) tracks interconnect corrosion [8], and voltage at maximum power from $Suns$-V_{oc} measurements ($V_{mp,PIV}$) helps in identifying recombination and shunting losses [9]. The subscript 'PIV' indicates pseudo I-V; from the $Suns$-V_{oc} curve, one can do a mathematical transformation to obtain an I-V curve without the effect of series resistance. The response is power from I-V measurements ($P_{mp,IV}$). In order to reduce noise, normalized mechanistic variables, and response were used. So, the normalized mechanistic variables are $^nI_{sc,IV}$, $^nR_{s,IV}$, and $^nV_{mp,PIV}$, whereas the normalized response is $^nP_{mp,IV}$.

<Stressor|Mechanism|Response> model maps the stressor, mechanistic variables, and response; it is also referred to as the <S|M|R> model for short. Each <S|M|R> model obtained from Principle 1 has several pathways connecting variables in a pairwise fashion. The pathways represent the best fitting model/functional form (also referred to as best model) between variables and the corresponding adjusted R-squared (R_{adj}^2). A higher value of R_{adj}^2 indicates that the best model is a good fit for the pair of variables and that the variables are strongly correlated to each other.

Figure 1 shows an <S|M|R> model for GB with EVA in mDH+FSL exposure fabricated by manufacturer B (set #3) from Principle 1. It can be noted that most of the pathways show that the best fits are nonlinear. The direct pathway or <S|R> has exponential best fit with an R_{adj}^2 of 0.26. This shows that $^nP_{mp,IV}$ is not as strongly affected by dy. The <S|M| pathways relating dy to $^nR_{s,IV}$, and $^nV_{mp,PIV}$ and the |M|R> pathway relating $^nR_{s,IV}$ to

978-1-7281-6118-1/22 $31.00 © 2022 IEEE

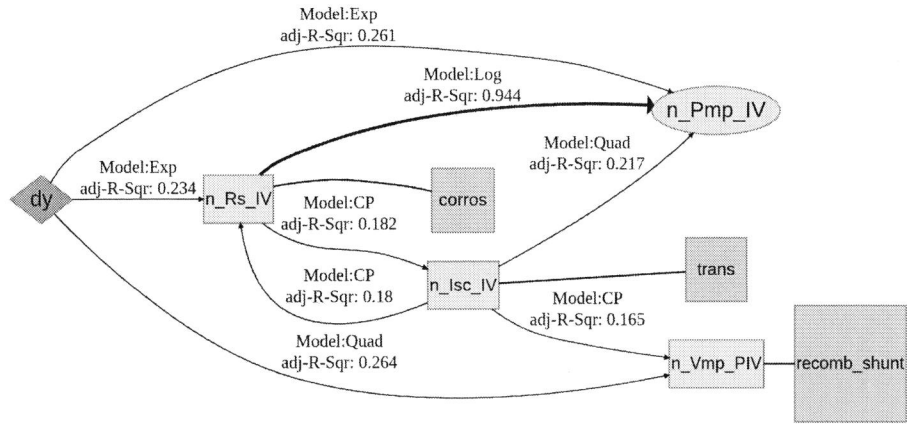

Fig. 1: $<$S$|$M$|$R$>$ model of GB with EVA in mDH+FSL by manufacturer B (set #3) using Principle 1. The stressor (exposure time in decimal year, represented as dy) is in red diamond, the mechanistic variables ($^{n}I_{sc,IV}$, $^{n}R_{s,IV}$, and $^{n}V_{mp,PIV}$) are in yellow boxes, and the response ($^{n}P_{mp,IV}$) is in green oval. The blue boxes represent short-hand descriptions of degradation modes corresponding to mechanistic variables. Each pathway between variables has best fit and corresponding adjusted R-squared (R^2_{adj}) information. The paths with $R^2_{adj} > 0.1$ are shown in the figure.

$^{n}P_{mp,IV}$ show that there is an active degradation pathway, namely $<dy|^{n}R_{s,IV}|^{n}P_{mp,IV}>$. A more detailed version of $<$S$|$M$|$R$>$ modeling can be found in our previous proceedings article [6].

V. PREDICTIVE PAIRWISE MODELING ($<$STRESSOR$|$RESPONSE$>$, $<$STRESSOR$|$MECHANISM$|$ AND $|$MECHANISM$|$RESPONSE$>$) FROM PRINCIPLE 1

Predictive pairwise modeling allows for comparison of different variants by looking at two variables at a time. In this section, $<$S$|$R$>$ results for sets #1-#4 will be included along with examples of $<$S$|$M$|$ and $|$M$|$R$>$ using $^{n}R_{s,IV}$ as the mechanistic variable.

Predictive pairwise modeling results are represented using facet plots which comprise a matrix of smaller panels/grids. Each grid pertains to a variant (DG/GB, EVA/POE, and A/B) in a particular exposure type (mDH/mDH+FSL). Within each grid, there is normalized data (indicated by blue data points), the best model equation from netSEM Principle 1 (shown as a curve/line), 83.4% confidence intervals (CIs) with estimated mean (red hollow circle) at exposure step 5 (end of exposure cycle), best model type. It is possible for some variants to not have a best model fit due to R^2_{adj} being lower than 0.01. Such variants are considered to be stable. For $|$M$|$R$>$, 83.4% CIs are not shown as dy is not an independent variable (X-axis) in the plot.

As highlighted in the study protocol, minimodules by manufacturer A were reliably made. Hence, generalizations in trends in this section are based on manufacturer A; such points are indicated by an asterisk (*). Significant observations corresponding to minimodules B will be highlighted as well.

A. $<$Stressor$|$Response$>$ Results from Principle 1

Set #1 minimodules were fabricated using monofacial PERC, UV-cutoff rear encapsulant. For GB type, the backsheet used was KPf. Figure 2 shows a facet plot between dy and $^{n}P_{mp,IV}$ (or $<$S$|$R$>$ results).

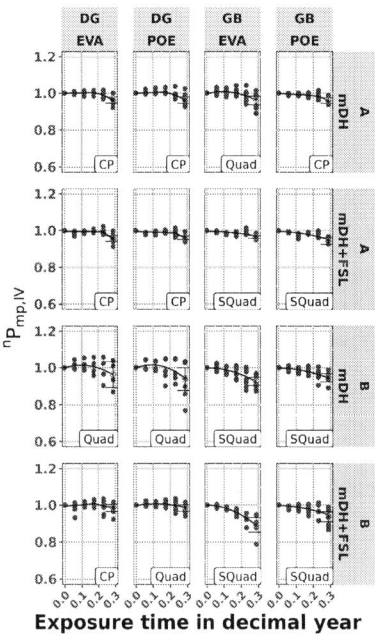

Fig. 2: Set #1 minimodule facet plot with dy as the stressor and $^{n}P_{mp,IV}$ as the response. Each facet grid has normalized data points and best fit information along with 83.4% confidence intervals at exposure step 5. The estimated mean is indicated by red hollow circle.

It can be seen that DG minimodules are relatively stable*. In the case of DG by manufacturer B in mDH exposure, there is data scatter: this could be likely due to sample/measurement variance. GB minimodules fabricated by manufacturer B show greater power loss in comparison to those by manufacturer A. This reflects differences in the manufacturing quality control in both the facilities.

Set #2 minimodules utilized bifacial PERC and opaque rear encapsulant. The backsheet used for GB minimodules was KPf. The $<S|R>$ results for set #2 minimodules are included in Figure 3. Even though all the set #2 variants undergo significant power loss, DG minimodules are relatively stable*. GB with POE is nearly as stable as DG minimodules*. It can also be seen that GB with EVA degrades more than GB with POE in both the exposures*.

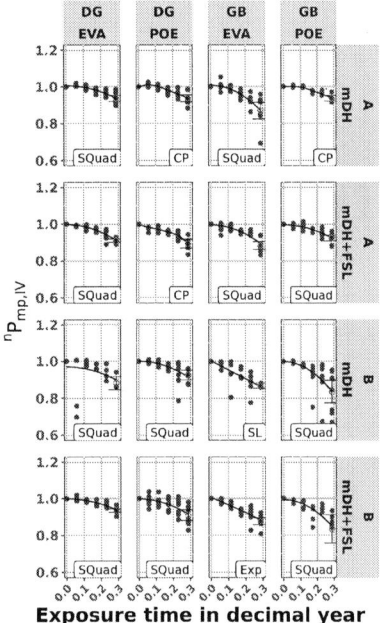

Fig. 3: Set #2 minimodule facet plot with dy as the stressor and $^nP_{mp,IV}$ as the response. Each facet grid has normalized data points and best fit information along with 83.4% confidence intervals at exposure step 5. The estimated mean is indicated by red hollow circle.

Set #3 minimodules were made using monofacial PERC and opaque encapsulant. The only difference between set #1 and set #3 minimodules is the rear encapsulant: set #1 minimodules implemented UV-cutoff, whereas set #3 utilized opaque encapsulant. The backsheet used for GB minimodules was KPf. Figure 4 corresponds to set #3 $<S|R>$ results

In comparison to set #1 $<S|R>$ results, most set #3 variants undergo greater power loss. GB minimodules by manufacturer B exhibit greater power loss compared to the ones by manufacturer A. GB with POE ($^nP_{mp,IV}$ estimated mean at step 5 = 0.96) is seen to have better relative stability than GB with EVA ($^nP_{mp,IV}$ estimated mean at step 5 = 0.9)*. It can be seen that GB with POE by mDH+FSL by manufacturer A does not have a best model fit, so it is considered to be a stable variant. It is also interesting to observe that for some of the variants, the best model fit stabilizes towards the end of exposure. DG minimodules from set #3 undergo more power

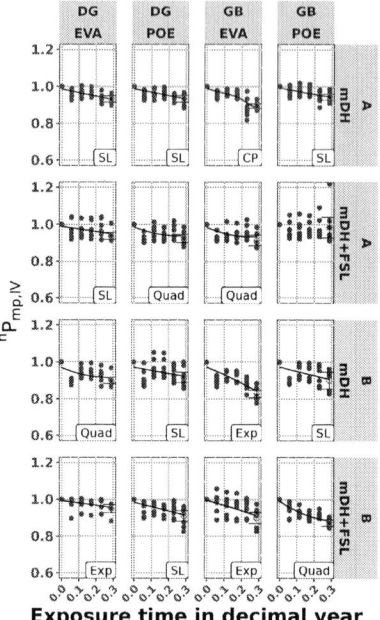

Fig. 4: Set #3 minimodule facet plot with dy as the stressor and $^nP_{mp,IV}$ as the response. Each facet grid has normalized data points and best fit information along with 83.4% confidence intervals at exposure step 5. The estimated mean is indicated by red hollow circle.

loss ($^nP_{mp,IV}$ estimated mean at step 5 = 0.91) compared to set #1 counterparts ($^nP_{mp,IV}$ estimated mean at step 5 = 0.95). This points to evidence that the opaque encapsulant may be playing a role in causing higher power loss in set #3, but no firm conclusion should be made without further studies.

The only variation between set #2 and set #3 minimodules is the cell type: set #2 used bifacial PERC, whereas set #3 used monofacial PERC. Currently, from our analysis, we have not identified a clear trend as to how the cell type impacts power loss. Further investigation is required.

Set #4 minimodules were manufactured using bifacial PERC. The rear encapsulant used for GB minimodules was UV-cutoff and the backsheet was transparent Tedlar backsheet with white grid. For DG minimodules, the rear encapsulant was transparent. Figure 5 highlights set #4 $<S|R>$ results.

Among all the set #4 minimodules, GB with POE in mDH+FSL by manufacturer A is stable as it lacks a best model fit. The best model fits flatten at step 5 for some variants*.

The only distinction between set #2 and set #4 DG minimodules is the rear encapsulant: set #2 DG made use of opaque encapsulant and set #4 DG utilized transparent encapsulant. Both use the same cell type (bifacial). We see that there is more power loss in set #2 DG minimodules ($^nP_{mp,IV}$ estimated mean at step 5 = 0.92) in comparison to set #4 ($^nP_{mp,IV}$ estimated mean at step 5 = 0.94)*.

Between set #2 and set #4 GB minimodules, the difference is again in the rear encapsulant: set #2 GB used opaque, whereas set #4 GB employed UV-cutoff as the rear encap-

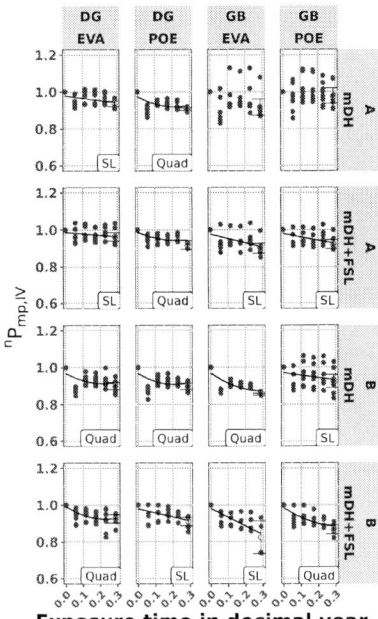

Fig. 5: Set #4 minimodule facet plot with dy as the stressor and $^nP_{mp,IV}$ as the response. Each facet grid has normalized data points and best fit information along with 83.4% confidence intervals at exposure step 5. The estimated mean is indicated by red hollow circle.

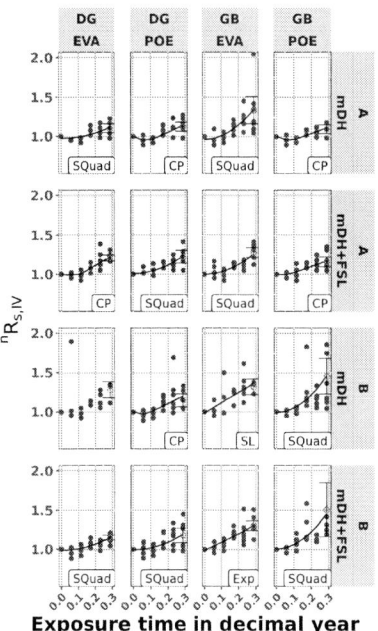

Fig. 6: Set #2 minimodule facet plot with dy as the stressor and $^nR_{s,IV}$ as the mechanistic variable. Each facet grid has normalized data points and best fit information along with 83.4% confidence intervals at exposure step 5. The estimated mean is indicated by red hollow circle.

sulant. Comparing GB minimodules, set #2 GB undergoes greater power loss ($^nP_{mp,IV}$ estimated mean at step 5 = 0.9) when compared to set #4 GB ($^nP_{mp,IV}$ estimated mean at step 5 = 0.93).

All these results point to the significant differences in minimodule variants under indoor exposure and different trends in power loss. Further insights can be realized when $<S|M|$ and $|M|R>$ are closely investigated for every mechanistic variable.

B. $<Stressor|Mechanism|$ and $|Mechanism|Response>$ Results from Principle 1

In order to further explore the cause of power loss, $<S|M|$ and $|M|R>$ can be taken into consideration. Here, set #2 minimodule results will be shown as an example. The results for the rest of the sets have been examined.

From Figures 6 and 7, some observations can be captured. With increasing dy, $^nR_{s,IV}$ increases: this is a trend exhibited by all variants. With increasing $^nR_{s,IV}$, $^nP_{mp,IV}$ decreases: this is shown by all set #2 variants.

Among all the mechanistic variables in this study, $^nR_{s,IV}$ shows a greater change. It is to be kept in mind that $^nR_{s,IV}$ has a different range of values (ranging from 1 to infinity) than $^nI_{sc,IV}$ and $^nV_{mp,PIV}$ (ranging from 0-1). In order to look at a similar range, we have also inspected the inverse of $^nR_{s,IV}$: the results still indicate that $^nR_{s,IV}$ has the highest contribution of power loss in minimodules under indoor accelerated exposures.

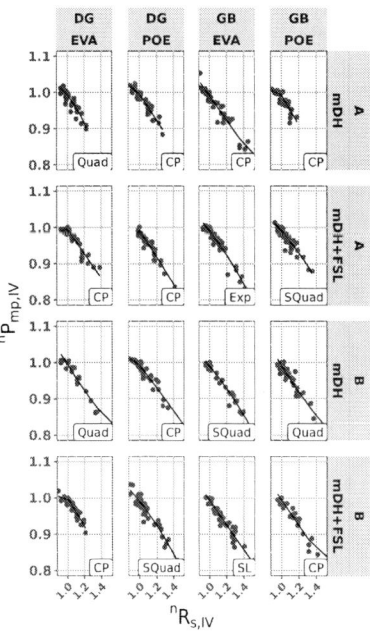

Fig. 7: Set #2 minimodule facet plot with $^nR_{s,IV}$ as the mechanistic variable and $^nP_{mp,IV}$ as the response. Each facet grid has normalized data points and best fit information.

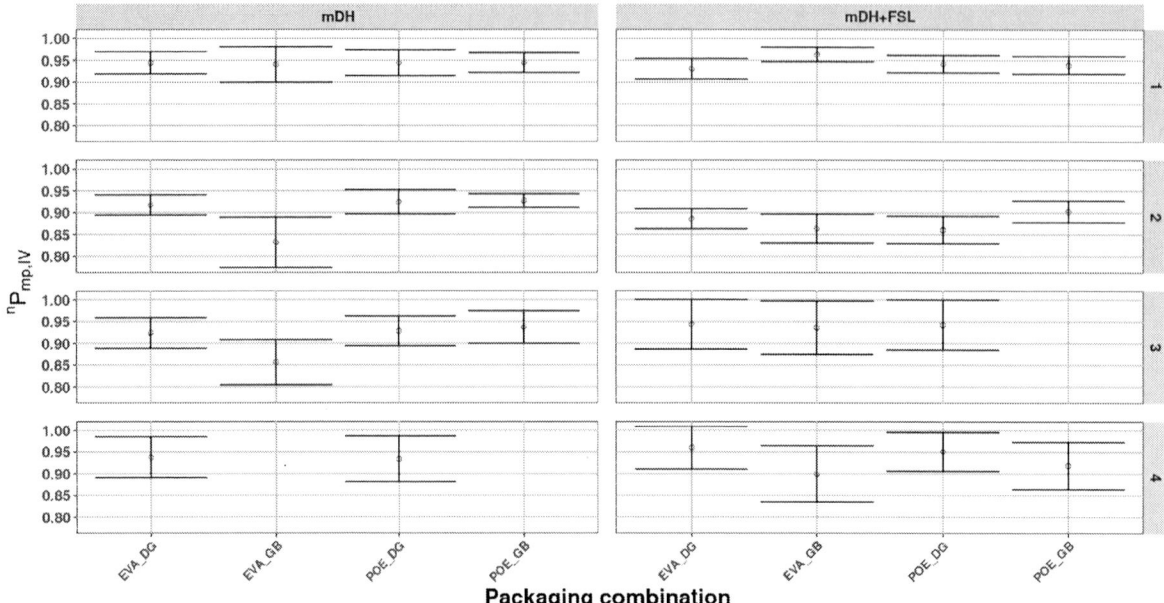

Fig. 8: Predictive 83.4% CIs for minimodule variants. The estimated mean is included as a red hollow circle in each of the CIs. Some of the CIs are missing due to the absence of best fit model for those variants

VI. Rank Ordering Minimodule Variants Based on Predictive Confidence Intervals Using Principle 1

We have been able to rank-order minimodule variants by manufacturer A using Principle 1 <S|R> results. Predictive 83.4% CIs from Principle 1 best models were employed for this purpose. The reason we have used predictive CIs is because it captures how degradation occurs in all the steps (and not just step 5) whilst providing uncertainty in estimated mean.

Figure 8 shows the predictive 83.4% CIs used for rank ordering minimodule variants. The estimated mean is included as a red hollow circle. The facet plot is constructed based on exposure types (mDH/mDH+FSL) and set number (1/2/3/4).

The list of durable and degrading variants are highlighted in Table II.

TABLE II: Rank-ordering durable variants fabricated by manufacturer A based on predictive confidence intervals from best models/functional forms.

Set #	Variant	Category	Comments
1	POE_DG_mDH	Durable	
1	EVA_GB_mDH+FSL	Durable	
1	POE_GB_mDH+FSL	Durable	
1	EVA_DG_mDH	Durable	
3	POE_GB_mDH+FSL	Durable	No best model fit
4	POE_GB_mDH	Durable	No best model fit
4	EVA_GB_mDH	Durable	No best model fit
2	EVA_DG_mDH+FSL	Degrading	
2	EVA_GB_mDH+FSL	Degrading	
2	POE_DG_mDH+FSL	Degrading	

VII. <Stressor|Mechanism|Response> Model Using Principle 2

Using Principle 2, we can obtain <S|M|R> models that specify each variable as a function of the rest of the variables. For example, $^{n}P_{mp,IV}$ could be a function of dy, $^{n}I_{sc,IV}$, $^{n}R_{s,IV}$ and/or $^{n}V_{mp,PIV}$.

Figure 9 shows GB with POE in mDH made by manufacturer A (set #4). It can be seen that there are mathematical functions defined for different variables. $^{n}P_{mp,IV}$ is a function of dy, $^{n}I_{sc,IV}$, $^{n}R_{s,IV}$ and $^{n}V_{mp,PIV}$ in this case. By using mathematical equations obtained from Principle 2 output, it is possible to visualize how power degrades for minimodule variants.

VIII. Service Lifetime Prediction and Identifying Causes of Power Loss Using Principle 2

Using Principle 1, we have been able to explore trends between variables in a pairwise manner. This approach does not capture the complexity of degradation. In the real world, PV module degradation is a phenomenon in which multiple stressors and mechanisms act simultaneously. In order to capture the complexity, we present our ongoing work on service lifetime prediction and visualizing trends using surface plots using Principle 2. In this section, set #1 minimodule results are used as an example.

From Figure 10, we see the service lifetime prediction plot in which set #1 minimodule variant, GB with EVA, in mDH is compared on the basis of manufacturer (A versus B). We are able to see that the variant by manufacturer B undergoes greater power loss than that of manufacturer A. In addition, we are able to see that the highest contribution is from $^{n}R_{s,IV}$

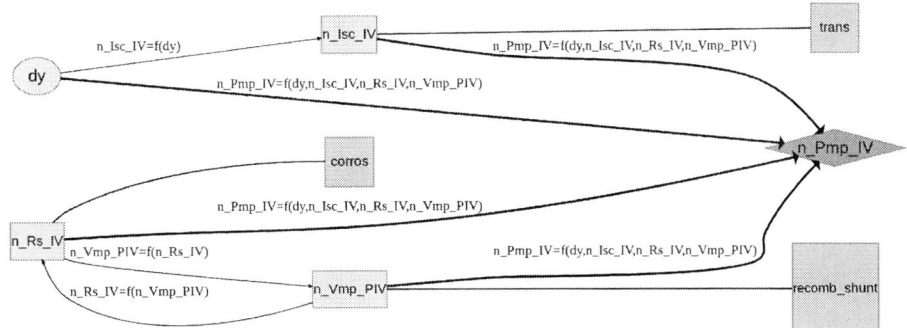

Fig. 9: <S|M|R> model of GB with POE in mDH by manufacturer A (set #4) using Principle 2. The stressor (exposure time in decimal year, represented as dy) is in green oval, the mechanistic variables ($^nI_{sc,IV}$, $^nR_{s,IV}$, and $^nV_{mp,PIV}$) are in yellow boxes, and the response ($^nP_{mp,IV}$) is in red diamond. The blue boxes represent short-hand descriptions of degradation modes corresponding to mechanistic variables. Each pathway has information about mathematical relationship between variables.

(and the inverse of $^nR_{s,IV}$ closely follows the $^nP_{mp,IV}$ plot) from Figure 11.

Equation 1 provides the Principle 2 general equation for degradation in GB with EVA in mDH fabricated by both manufacturers A and B Table III shows the corresponding change points and coefficient values. By comparing each coefficient between the manufacturers, we can quantify the degradation difference using the values in Equation III.

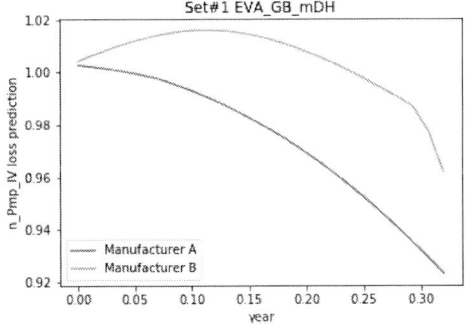

Fig. 10: Service lifetime prediction plot comparing set #1 GB with EVA in mDH manufactured by A and B.

$$^nP_{mp,IV} = \eta + \gamma_{dy,1}dy + \gamma_{dy,2}dy^2 + \delta_{I_{sc}}(^nI_{sc,IV}) \quad (1)$$
$$+ \delta_{R_s}(^nR_{s,IV}) + \delta_{V_{mp}}(^nV_{mp,PIV})$$

In addition to service lifetime prediction, surface plots were generated using the equation relating $^nP_{mp,IV}$ to mechanistic variables and stressor, namely, dy, $^nI_{sc,IV}$ and $^nR_{s,IV}$ for GB with EVA in mDH by manufacturer B before the change point. The plot is shown in Figure 12. It can be seen that with increasing exposure time, there is a decrease in power and some fluctuations at $dy > 0.2$. $^nR_{s,IV}$ increases above 1 (the value is about 1.2) at $dy > 0.2$. The change in $^nI_{sc,IV}$ is $< 1\%$. This shows that the minimodule variant is impacted primarily by interconnect corrosion. Future work will

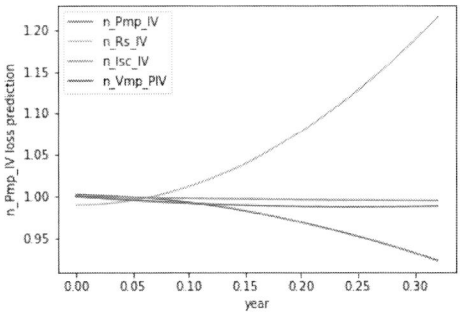

Fig. 11: Contributions of variables compared against power loss with year (dy).

be directed towards combining the results of before and after change point.

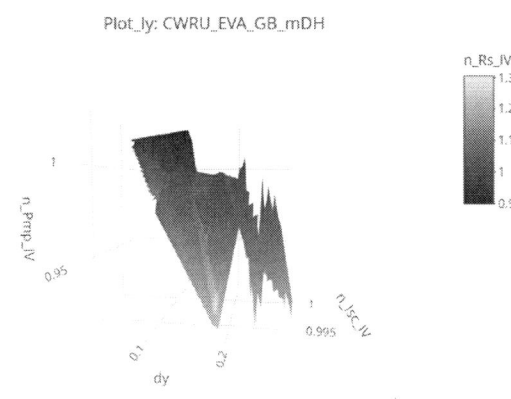

Fig. 12: Surface plot depicting change in $^nP_{mp,IV}$ based on dy, $^nI_{sc,IV}$ and $^nR_{s,IV}$.

978-1-7281-6118-1/22 $31.00 © 2022 IEEE

TABLE III: Equations obtained from Principle 2 for GB with EVA in mDH (set #1) for manufacturers A and B.

Manufacturer	Change point	η	$\gamma_{dy,1}$	$\gamma_{dy,2}$	$\delta_{I_{sc}}$	$\delta_{V_{mp}}$	δ_{R_s}
A	$dy \leq 0.0674$	0.28	0	0	1.05	0	-0.33
	$dy > 0.0674$	0.19	0	0	1.05	0	-0.26
B	$dy \leq 0.3066$	-0.74	0.07	-0.19	2.79	-0.87	-0.18
	$dy > 0.3066$	-0.45	0.07	-0.19	2.79	-0.87	-0.48

IX. CONCLUSIONS AND PATH FORWARD

In the previous sections, Principle 1 and 2 results have been shown for different minimodule variants in two types of indoor accelerated exposures. One of the important aspects we can identify through these results is the differences in manufacturing process: manufacturer A made minimodules in a very reliable manner from the very beginning. The results from Principle 1 and 2 validate this point. Overall, set #1 minimodules with UV-cutoff encapsulant have relatively better stability. 3 variants of minimodules were identified to have stability due to lack of best model fit: GB having UV-cutoff rear encapsulation and GB with POE having opaque rear encapsulation. In most minimodules having opaque rear encapsulation, there are signs of higher power loss in comparison to the other options. Except GB with POE, rest of the variants exhibit power loss among set #2 minimodules with opaque encapsulant. It can also be seen that in the minimodule variants undergoing greater power loss, interconnect corrosion (tracked by series resistance) plays an important role. This is also further confirmed by Principle 2.

We plan on rank-ordering minimodule variants using Principle 2 in the near future and correlate the results to what we have obtained from Principle 1. In addition, service lifetime prediction would be visualized and quantified for variants within and across different sets of minimodules.

REFERENCES

[1] M. K. da Silva, M. S. Gul, and H. Chaudhry, "Review on the Sources of Power Loss in Monofacial and Bifacial Photovoltaic Technologies," *Energies*, vol. 14, no. 23, p. 7935, Jan. 2021.

[2] W. Gambogi, M. Demko, B.-L. Yu, S. Kurian, S. MacMaster, K. R. Choudhury, J. Tracy, D. Hu, and H. Hu, "Transparent Backsheets for Bifacial Photovoltaic Modules," in *2020 47th IEEE Photovoltaic Specialists Conference (PVSC)*, Jun. 2020, pp. 1651–1657.

[3] L. S. Bruckman, N. R. Wheeler, J. Ma, E. Wang, C. K. Wang, I. Chou, J. Sun, and R. H. French, "Statistical and Domain Analytics Applied to PV Module Lifetime and Degradation Science," *IEEE access : practical innovations, open solutions*, vol. 1, pp. 384–403, 2013.

[4] W.-H. Huang, N. Wheeler, A. Klinke, Y. Xu, W. Du, A. K. Verma, A. Gok, D. Gordon, Y. Wang, J. Liu, A. Curran, J. Fada, X. Ma, J. Braid, J. Carter, L. Bruckman, and R. French, "netSEM: Network Structural Equation Modeling," Nov. 2018.

[5] Abdulkerim Gok, Cara L. Fagerholm, Roger H. French, and Laura S. Bruckman, "Temporal evolution and pathway models of Poly(Ethylene-Terephthalate) degradation under multi-factor accelerated weathering exposures," *PLOS ONE*, vol. 14, no. 2, p. e0212258, Feb. 2019.

[6] S. Nalin Venkat, J. Liu, J. Wegmueller, B. Yu, B. Gould, X. Li, J.-N. Jaubert, J. L. Braid, L. S. Bruckman, and R. H. French, "Degradation Pathway Modeling of PV Minimodule Variants with Different Packaging Materials Under Indoor Accelerated Exposures," in *2021 IEEE 48th Photovoltaic Specialists Conference (PVSC)*, Jun. 2021, pp. 1725–1731.

[7] J. Ahmad, A. Ciocia, S. Fichera, A. F. Murtaza, and F. Spertino, "Detection of Typical Defects in Silicon Photovoltaic Modules and Application for Plants with Distributed MPPT Configuration," *Energies*, vol. 12, no. 23, p. 4547, Jan. 2019.

[8] T.H.Kim, N.C.Park, and D.H.Kim, "The effect of moisture on the degradation mechanism of multi-crystalline silicon photovoltaic module," *Microelectronics Reliability*, vol. 53, no. 9-11, pp. 1823–1827, Sep. 2013.

[9] Mohammad Jobayer Hossain, Geoffrey Gregory, Eric John Schneller, Andrew M. Gabor, Adrienne L. Blum, Zhihao Yang, Dana Sulas, Steve Johnston, and Kristopher Olan Davis, "A Comprehensive Methodology to Evaluate Losses and Process Variations in Silicon Solar Cell Manufacturing," *IEEE Journal of Photovoltaics*, vol. 9, no. 5, pp. 1350–1359, Sep. 2019.

978-1-7281-6118-1/22 $31.00 © 2022 IEEE

Material Use and Life Cycle Impact of Crystalline Silicon PV Modules Over Time

Luyao Yuan, Annick Anctil

Department of Civil and Environmental Engineering, Michigan State University, East Lansing, Michigan 48824, USA

Abstract—Most life cycle assessment (LCA) of crystalline silicon photovoltaics (c-Si PV) modules are based on public life cycle inventory (LCI) datasets with limited use of actual manufacturing data. We collect and calculate the amount of material used for production of different PV modules installed in the U.S. to analyze the trend in material intensity over and compare the numbers among various tier manufacturers and module reliability. Furthermore, results of LCA models using the public LCI data and the actual manufacturing material (specifically aluminum) data are compared to investigate the impact of material use on the life-cycle impact assessment of c-Si PV modules. Results show a trend of material use decrease over time and indicate a potential connection between material usage and the manufacturer tier – better manufacturers tend to use more materials for modules production which may lead to higher quality performance. Additional work will complete the life cycle assessment, explore more materials, and fill the data gap of PV modules produced by different manufacturer tiers in different years.

Keywords—c-Si PV, Aluminum, Life cycle assessment, Trend analysis, Manufacturer tier comparison

I. INTRODUCTION

As the end of 2020, the cumulative photovoltaics (PV) capacity installed worldwide reached nearly 714 GW, and 127 GW was installed within the year [1]. In the U.S., the cumulative PV capacity exceeded 113 GW in 2021 and the number keeps increasing – 300 GW of new PV capacity will be installed in the next ten years [2]. Among the installed PV capacity in the U.S. in Q3, 2021, residential PV accounts for almost 25%, and the rest is commercial or utility-scale [3]. The most common type of PV modules is crystalline silicon (c-Si), which account for 96.4% of the global PV production in 2020 [4]. The application of PV systems can reduce environmental impacts compared to conventional fuels. However, the production phase of PV modules can impose negative impacts on the environment.

Life cycle assessment (LCA) is a common tool or method to analyze the environmental impacts of PV systems from a holistic perspective. Most LCA study of PV systems has been based on two life cycle inventory (LCI) datasets. The first one is Ecoinvent PV dataset [5], which reflects the status of c-Si PV technology production in 2005. The second one is IEA PVPS 2015 dataset [6], reflecting the status of c-Si PV technology in year 2011. In 2020, PVPS published the up-to-date LCI data describing the status for c-Si PV technology in year 2018 (except for some manufacturing data in 2011) [7]. However, the

site-specific inventory data are often not available to compare the performance of various PV modules.

Among the life cycle inventory data, materials used for PV module production are an important category and a small difference in material usage amount can lead to a significant difference in environmental impacts. As the technology mature it is expected that PV modules will use less and less materials such as aluminum and glass. For example, ITRPV forecasted a mass reduction for module frame and a thickness reduction of the front side glass from 3mm to 2mm between 2013 to 2029 [8][9]. From the 2015 to 2020 IEA PVPS LCI reports, the material intensity for aluminum and glass has remained constant [6][7], which may not really reflect technology status of the PV industry.

This study collects and calculates the amount of material used for production of different PV modules to get some insights of the trend of material intensity over time and to compare the numbers among various tier manufacturers and module reliability. The hypothesis we are testing are: 1) the material intensity of PV modules has been decreasing over time, 2) PV modules produced by Tier 1 manufacturers have higher material intensity than non-Tier 1, and 3) modules with better quality and reliability performance have higher material intensity than those with lower performance. Based on the material data collected, LCA models of PV modules are built. The LCA results are compared between using the LCI datasets mentioned above and using the actual module data. We choose aluminum since it contributes to 8% of the total mass of framed c-Si PV module [10] but 10% of the carbon footprint. Quantifying the amount of aluminum used in frame is also important to plan for PV module recycling since it is one of the most materials recycled in a module [10].

II. METHODS

The scope of this study is limited to silicon PV modules commonly installed in residential applications in the U.S. The material investigated is aluminum but additional data on glass will be collected using a similar method. Most modules selected for this study were commonly installed in California in 2019. To make sure representative samples are collected for the analysis, modules produced by certain manufacturer tiers or with certain years stated in the specification sheet were added.

A. Material data collection and calculation

The amount of aluminum and other materials used for PV modules are rarely disclosed by manufacturers or suppliers.

978-1-7281-6118-1/22 $31.00 © 2022 IEEE

Since aluminum is mainly used for the frame, we calculate the module frame weight using frame dimensions from the specification data sheet published by the manufacturer of the PV module product. With the assistance of an area calculator tool called "SketchAndCalc" [11], the frame cross section area is measured, which then multiplies the depth or height of the frame to calculate the total volume. The density is kept constant at $2.7 g/m^3$. Normalized material weight is calculated as weight per square meter of the module:

$$\text{Normalized weight} = \text{Weight} / (\text{length} \times \text{width}) \quad (1)$$

Examples of PV frame cross section are shown in Fig. 1. In addition, publication year of the specification sheet is collected to better understand the trend of material use for PV modules.

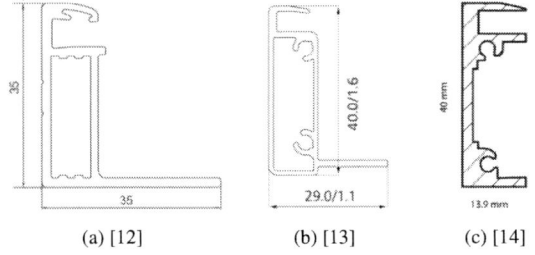

(a) [12]   (b) [13]   (c) [14]

Fig. 1.   Examples of PV module frame cross section.

B. Manufacturer tiers

This study refers to two Tier 1 lists to further analyze and compare the material use and environmental impacts of PV module products among different manufacturers and tiers. The first list is Bloomberg PV Module Tier 1 List, Q3 2021 [15], which divides the PV market into tiers based on manufacturers' financial situations. And the second one is PVEL 2021 Top Performers [16], which is based on the product quality and reliability through the PVEL's Production Qualification Program.

C. Life cycle assessment

For the selected PV module, two different LCA models are built for comparison. One uses a commonly used LCI dataset (Ecoinvent PV dataset, IEA PVPS 2015 dataset, or IEA PVPS 2020 dataset) depending on the year stated in the specification sheet. Another one uses the actual production data which is the aluminum amount calculated by the author. Results of these two models are interpreted and compared.

III. RESULTS AND DISCUSSIONS

In total, specification sheets of 64 PV modules produced by 14 different manufacturers are collected. A summary of the manufactures' tiers is listed in TABLE I.

TABLE I.   SUMMARY OF MODULES BY MANUFACTURER TIERS

Bloomberg Tiers, Q3 2021	PVEL Tiers, 2021		Total
	Tier 1	Non-Tier 1	
Tier 1	34	7	41
Non-Tier 1	7	16	23

National Science Foundation #2044886.

Bloomberg Tiers, Q3 2021	PVEL Tiers, 2021		Total
	Tier 1	Non-Tier 1	
Total	41	23	64

A. Trend of aluminum amount used for PV modules over time

Based on the 64 modules collected, trend of material weight over time is presented in Fig. 2. The aluminum weight for PV module frames is decreasing over time which is what was expected [8][9]. In IEA PVPS 2015 and 2020 datasets, the inventory data of aluminum used for PV module production is 2.13 kg/m^2, which is higher than the numbers presented in Fig. 2. Additional analysis will be performed to understand the increase seen for 2013 modules and 2016 which is probably associated with change in module design. In 2013 and 2016, certain modules have frames with more complicated structure or larger dimensions.

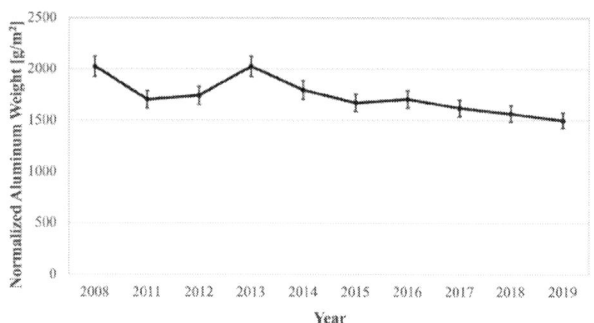

Fig. 2.   Average normalized aluminum weight by year.

When looking at specific manufacturers, as Fig. 3 shows, the trend is not obvious. However, it can still be noted that some manufacturers, e.g., Suniva, Hyundai, and JA Solar, are using less and less aluminum for the module frames. While some manufacturers, e.g., Astronergy and Trina Solar, are using more materials for their frames than the previous year.

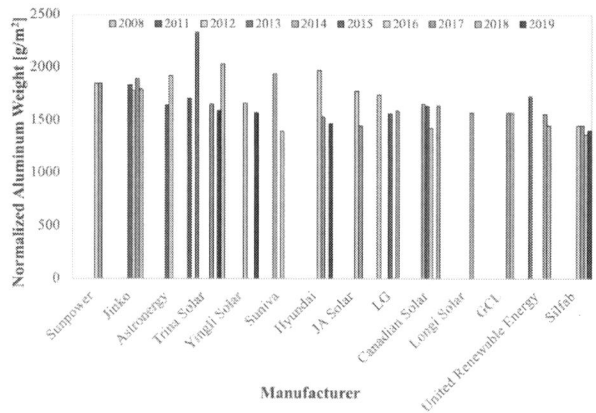

Fig. 3.   Average normalized aluminum weight by manufacturer and year.

B. Compare aluminum amount used for PV modules between different manufacturers

Average normalized aluminum weight used for PV modules by different manufacturer tiers is presented in Fig. 4. It indicates that tier 1 manufacturers are using more aluminum than non-tier 1 on average. The average amount of all these 64 modules is 1653.87 g/m². Fig. 5 shows more specific data by year. In year 2013, 2016, 2018, and 2019, both Bloomberg and PVEL Tier 1 manufacturers have a higher aluminum intensity than non-Tier 1 manufacturers. The results indicate a possible connection between material usage and the manufacturer tier – better manufacturers tend to use more materials for modules production which may lead to higher quality performance. Additional data will be collected for certain years and tiers. There is a data gap that tiers in some years are missing, which will be added in the future work. For example, in Fig. 5, Bloomberg Tier 1 and PVEL Tier 1 are missing in 2008, and Bloomberg Non-Tier 1 and PVEL Non-Tier 1 are missing in 2011.

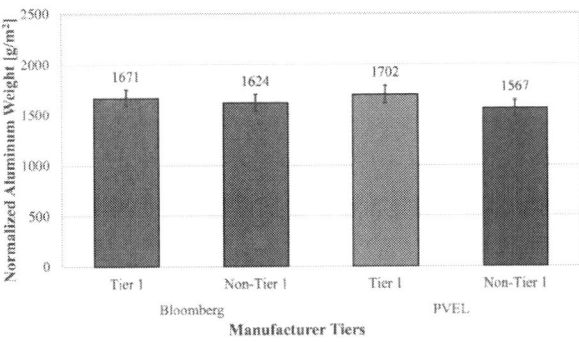

Fig. 4. Average normalized aluminum weight by manufacturer tiers.

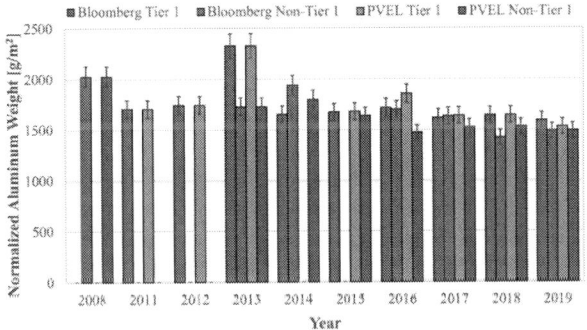

Fig. 5. Average normalized aluminum weight by manufacturer tiers and year.

C. Compare LCA results between using LCI dataset and actual production data

Life cycle assessment is in progress and will be presented at the conference.

IV. CONCLUSION AND FUTURE WORK

Preliminary results show a trend of material use decrease over time for the PV modules installed in the U.S. and indicate a potential connection between material usage and the manufacturer tier – better manufacturers tend to use more materials for modules production which may lead to higher quality performance. Overall, this study demonstrates how the sustainability impacts of PV modules production has been

changing over time and the variability within manufacturers that is currently not considered in life cycle assessment.

Future work will consider additional materials such as glass and plastic. With regards to life cycle inventory data, location-variable data will be considered to build models that consider the manufacturing location (including electricity intensity) and transportation for module installation in the U.S. Finally, we will calculate the life cycle error associated with the use of constant material rather than specific per module data.

ACKNOWLEDGMENT

This work was supported by the National Science Foundation grant number NSF-2-44886: *CAREER: Environmental Sustainability of Photovoltaics in the US.*

REFERENCES

[1] International Renewable Energy Agency (IRENA), "Renewable capacity statistics 2021," 2021. Available: https://www.irena.org/-/media/Files/IRENA/Agency/Publication/2021/Apr/IRENA_RE_Capacity_Statistics_2021.pdf

[2] Solar Energy Industries Association (SEIA), "Solar data cheat sheet," 2021. Available: https://www.seia.org/research-resources/solar-data-cheat-sheet

[3] Solar Energy Industries Association (SEIA), "U.S. solar market insight," 2021. Available: https://www.seia.org/us-solar-market-insight

[4] International Energy Agency (IEA), "Trends in photovoltaic applications 2021," 2021. Available: https://iea-pvps.org/wp-content/uploads/2022/01/IEA-PVPS-Trends-report-2021-1.pdf

[5] Ecoinvent database v.3.7., www.ecoinvent.org, 2020.

[6] R. Frischknecht, R. Itten, P. Sinha, M. de Wild-Scholten, J. Zhang, V. Fthenakis, H. C. Kim, M. Raugei, M. Stucki, 2015, Life Cycle Inventories and Life Cycle Assessment of Photovoltaic Systems, International Energy Agency (IEA) PVPS Task 12, Report T12-04:2015.

[7] R. Frischknecht, P. Stolz, L. Krebs, M. de Wild-Scholten, P. Sinha, V. Fthenakis, H.C. Kim, M. Raugei, M. Stucki, 2020, Life Cycle Inventories and Life Cycle Assessment of Photovoltaic Systems, International Energy Agency (IEA) PVPS Task 12, Report T12-19:2020.

[8] International Technology Roadmap for Photovoltaics (ITRPV), "International Technology Roadmap for Photovoltaics (ITRPV), Results 2013," 2014. Available: https://www.semi.org/sites/semi.org/files/docs/ITRPV_2014_Roadmap_Revision1_140324.pdf

[9] International Technology Roadmap for Photovoltaics (ITRPV), "International Technology Roadmap for Photovoltaics (ITRPV), Results 2018," 2019. Available: https://pv-manufacturing.org/wp-content/uploads/2019/03/ITRPV-2019.pdf

[10] International Renewable Energy Agency (IRENA), "End-of-life management: Solar Photovoltaics Panel," 2016. Available: https://www.irena.org/-/media/Files/IRENA/Agency/Publication/2016/IRENA_IEAPVPS_End-of-Life_Solar_PV_Panels_2016.pdf

[11] SketchAndCalc. https://www.sketchandcalc.com/

[12] Jinko Solar, "Specification sheet of Eagle Perc 60 280-300 Watt," 2018. Available: https://si-datastore.s3.us-west-2.amazonaws.com/documents/ePrryWZm0K6lBjEadjaS2Wg73ylHLE6bSUG1uauX.pdf

[13] LG, "Specifiaction sheet of LG350Q1C-A5," 2017. Available: https://www.lg.com/us/business/download/resources/BT00002151/lg-business-solar-spec-neon-r-350q1c-A5-051118_V2.pdf

[14] Astronergy, "Specification sheet of ASM6610P Series," 2015. Available: https://si-datastore.s3.us-west-2.amazonaws.com/documents/Ch46aCkbeKOv865rtR3t2TTdQso1D5do15WdsMaC.pdf

[15] Bloomberg, "BloombergNEF PV module Tier 1 list, Q3 2021", 2021.

[16] PVEL., "Introducing the 2021 top performers," 2021. Available: https://modulescorecard.pvel.com/top-performers/

Drift-Diffusion Modelling of Four-Junction InGaP/InGaAs/SiGeSn/Ge Solar Cells

Laurier S. Baribeau, Robert F.H. Hunter, Christopher E. Valdivia, Karin Hinzer

SUNLAB, Centre for Research in Photonics, University of Ottawa, Ottawa, ON, Canada

The ternary alloy silicon germanium tin is a versatile candidate to extend the industry standard lattice matched InGaP/InGaAs/Ge multijunction solar cell to four junctions. Here, the SiGeSn composition space is discussed and its bandgap trend is visualized. Then, InGaP/InGaAs/SiGeSn/Ge solar cells are simulated using drift-diffusion modelling to ascertain SiGeSn quality limits, and the design challenges in the four-junction material system. Power conversion efficiencies of 42.6% and 41.6% at 1000 suns AM1.5D are determined for designs implementing surface recombination velocities of 103 cm/s and 5×104 cm/s, respectively, at important interfaces in the device. These signify absolute efficiency gains of 1.3% and 0.4% with respect to like-modelled InGaP/InGaAs/Ge designs. The obtained power conversion efficiencies assume a Shockley-Read-Hall recombination lifetime of 1 μs in the SiGeSn material, however, lifetimes of 100 ns drop the efficiency by only ~1% (absolute). The external quantum efficiency of the four-junction devices is near 90% across most of the solar spectrum. A plot of the fraction of incident light lost to various physical mechanisms in the solar cell is given and has been used to optimize the surface field layers to reduce minority charge carrier loss currents. Designs have been optimized for output power by thinning the top three subcells to ensure that the germanium does not limit the device' 11.25 A/cm2 operating current at its maximum power point. This result indicates that current-matching limited by inefficient absorption and Auger recombination in the Ge subcell is one of the main design challenges of this material system and suggests avenues for possible improvements to the design, such as improved light trapping, refined bandgap engineering, and subcell segmentation.

Impact of Humidity, Temperature, and Oxygen on the Stability of FA0.7MA0.3Sn0.5Pb0.5I3 Perovskites

Alex Bordovalos , Marie S Tumusange, Biwas Subedi , Lei Chen, Zhaoning Song, Yanfa Yan, Nikolas J Podraza

Department of Physics and Astronomy and The Wright Center for Photovoltaics Innovation and Commercialization, Toledo, OH, United States

Narrow-bandgap hybrid organic-inorganic mixed tin (Sn) + lead (Pb) halide perovskites have optimal bandgap for use as the absorber layer in the bottom subcell of tandem photovoltaics. However, mixed Sn + Pb perovskites are susceptible to degradation via oxidation and other environmental factors. Therefore, it is important to be able to understand the mechanism and extent of degradation under those considerations. A set of narrow-bandgap FA0.7MA0.3Sn0.5Pb0.5I3 (FA = formamidinium, MA = methylammonium, I = iodine) perovskite films are exposed to 21% oxygen and 85% relative humidity in nitrogen in a climate-controlled chamber and measured using real time spectroscopic ellipsometry. Additionally, films of the same composition are exposed to either laboratory ambient air, ambient air with increased relative humidity (85% RH), or N2. Each of these three environments has a film heated to 25, 50, and 85 °C. Results show that the films exposed to oxygen in nitrogen show more substantial degradation than films exposed to either high temperature (85 °C) or high relative humidity (85%) in nitrogen; however, films exposed to oxygen and high humidity, oxygen and high temperature, or combination of all degrade the most severely.

The influence of wind and module tilt on the operating temperature of single-axis trackers

Keith R. McIntosh,[1] Malcolm D. Abbott,[1] Ben A. Sudbury,[1]
Saurabh Aneja,[2] Mitch Bowman,[2] Lance Brown,[2] Ben Kahane,[2] Norm Nicholas[2] and Kristian Nolde[2]

[1] PV Lighthouse, Coledale, NSW 2515, Australia
[2] FTC Solar, 9020 N Capital of Texas Hwy, Building 1, Suite 260, Austin, Texas 78759, USA

Abstract—We measure the module temperature T_m in 1P and 2P single-axis trackers, analyzing how T_m depends on wind speed and direction, as well as on irradiance, ambient temperature and module tilt β. On a clear day, we find that the typical temperature variation within a tracker is 1−4 °C for 1P and 2−6 °C for 2P trackers, where the coolest region tends to be nearest the torque tube. Whether for 1P or 2P, we find that when $\beta <$ ~25°, the wind cools the windward side of the tracker by 1−1.5 °C more than the leeward side; but when $\beta >$ ~25°, the upper side is 0.5−1.5 °C cooler than the lower side, irrespective of wind direction (for wind speeds < 5 m/s). We also find that the commonly used NOCT and Faiman temperature models overpredict T_m by, on average, 7.4 °C and 3.3 °C. Even after calibrating these models to our trackers, they only predict T_m at any given time to ±6.6 °C with 95% confidence. Without adding any free variables, the modelling accuracy is improved to ±3.8 °C by accounting for radiative loss to the sky and transient effects; the accuracy is improved further to ±2.8 °C by accounting for module tilt, wind direction and ground temperature. This study expands upon the PV industry's understanding of how single-axis trackers are influenced by wind speed, wind direction, and tilt, and it refines our ability to accurately predict T_m of FTC's Voyager 2P tracker.

Keywords—module temperature, modeling, simulation, wind, tracking.

I. INTRODUCTION

There are many models that predict the module temperature T_m of PV systems: the NOCT [1], Sandia [2], and Faiman [3] models are the best known, and there are other more sophisticated models (e.g., [4]–[10]) that distinguish between radiative, convective and conductive heat flow, and incorporate transient effects. All effectively treat convection with one or two parameters, U_c and $U_v \times w$, where U_c and U_v are constants and w is wind speed. Studies find U_c and U_v vary strongly from site to site [11]–[14] and depend on the height that w was measured [11], [13], [15]. Whether simple or complex, models become significantly more accurate after being calibrated to a particular system layout (e.g., to determine U_c and U_v) due to the complexities of forced convection. We provide more detail in an expanded version of this paper that will be submitted to the IEEE Journal of Photovoltaics following this conference.

We therefore gather a large set of experimental data from FTC's test facility to assess the thermal behavior of horizontal single-axis trackers (HSATs), determining how T_m depends on w, wind direction, and tilt. We evaluate the variation of temperature within modules and between the eastern and western modules of a 2P configuration; and we test a variety of thermal models to quantify the trade-off between complexity and accuracy. The described findings will assist the industry in making more accurate predictions of T_m and hence electrical yield, and in improving layout and module selection for HSATs.

II. EXPERIMENTAL FACILITY

Fig. 1 shows FTC's experimental facility. It is located at the Solar Technology Acceleration Center (SolarTAC), 20 miles east of downtown Denver. The facility contains several rows of HSATs, some with a 1P configuration (one module in portrait) and others with FTC's 2P configuration (two module in portrait). Over 2.5 GW of the 2P configuration has been deployed worldwide under the trade name, Voyager.

The temperature of three modules is monitored using four thermocouples attached to their rear. The modules are Longi LR6-72BP-360M bifacial glass-glass modules, with an efficiency under standard testing conditions of 18.3%, a thermal coefficient of −0.37%/K, and a nominal operating temperature (NOCT) of 45 ± 2 °C.

We measured T_m in five-minute intervals throughout a 19-day period in late summer 2021: from August 20th (Day 232) to September 7th (Day 250). Concurrent weather measurements were taken from the SolarTAC weather station operated by NREL; it measured w and wind direction (at 10 m), the ambient temperature T_a, and GHI, DHI and DNI.

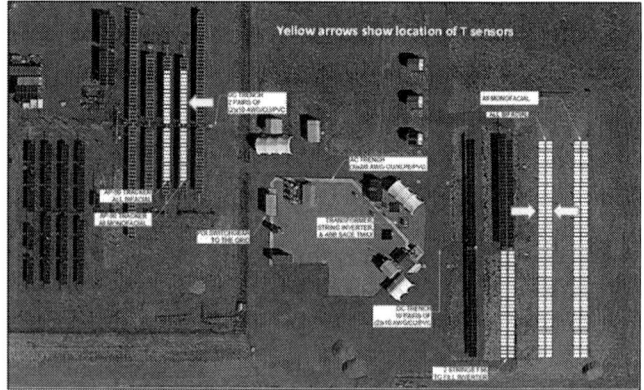

Fig. 1. Birds-eye view of the FTC test facility (39.756 N 104.621 W).

III. Weather

The evaluation period was mostly sunny with light winds and almost no rain. Fig. 2 plots the irradiance and direct fraction (defined as $f_D = 1 - DHI/GHI$). More figures of GHI, T_a, w, and wind direction is provided in the expanded paper.

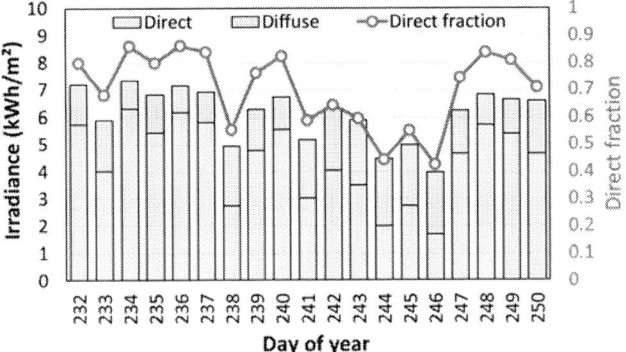

Fig. 2. Direct irradiance, diffuse irradiance and direct fraction on a horizontal surface during the evaluation period.

IV. Experimental Observations

A. Within-module temperature variation

We first evaluate the temperature variation ΔT within the modules. Table I gives the typical range in ΔT over the course of all 19 days. It shows that there is a wide range in ΔT; that is, the spatial variation in cell temperature is sometimes small (~1 °C) and sometimes large (~6 °C).

We find that ΔT tends to be lower when w is higher, when it is cloudier, and when the modules are more tilted. The first two of these three trends are evident in Fig. 3, which plots every daytime datapoint between 14:30 and 16.30 when $f_D > 0.6$ (top graphs) and $f_D < 0.6$ (bottom graphs). This result agrees with [12] that found the within-array ΔT decreases with increasing w, consistent with [16], [17].

We further find that the cells nearer the torque tube tend to be cooler than the other cells. Table II gives results for the average of all outer thermocouples T_O minus the average of the inner thermocouples T_I. In the expanded paper, we provide additional figures and describe these trends, concluding that ΔT is only partially explained by the proximity of the thermocouples to the torque tube.

TABLE I.   TYPICAL RANGE OF ΔT (°C) OVER ALL 19 DAYS.

Module	Nighttime	Daytime, cloudy ($f_D < 0.6$)	Daytime, clear ($f_D > 0.6$)
1P	0–1	1–2	1–4
2P-West	0.5–1.5	1–3	1–5
2P-East	0–1	1–3	2–6
2P-Both	1–2	1–4	2–6

TABLE II.   TYPICAL RANGE OF $T_O - T_I$ (°C).

Configuration	Nighttime	Daytime, cloudy ($f_D < 0.6$)	Daytime, clear ($f_D > 0.6$)
1P	0.0 – 0.4	0–1	0.3–1.5
2P-Both	−0.2 – 0.4	0–2	0.3–3.5

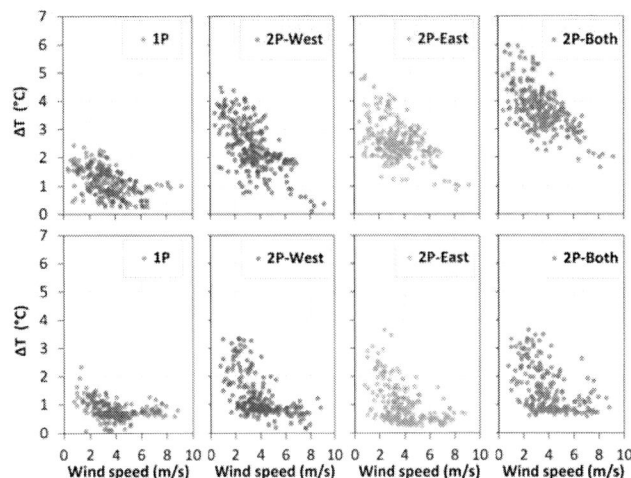

Fig. 3. ΔT between 14:30 and 16:30 when $f_D > 0.6$ (top) and $f_D < 0.6$ (bottom).

B. East vs west, upper vs lower, windward vs leeward

We next address some open questions relating to module temperatures within 2P configurations: Which is hotter: the upper or lower module, and by how much? Does wind direction affect these trends? And do the same trends apply within 1P?

We address these questions by taking the average of all thermocouples on the western module T_W and the eastern module T_E (i.e., four readings each), and plotting their difference, $\Delta T = T_W - T_E$, in terms of wind direction, module tilt and w in Fig. 4 and Fig. 5. The expanded paper explains these figures in more detail, but for now we distil the major findings for the 2P configuration. (The findings were almost identical in relation to which side is hotter in the 1P config.)

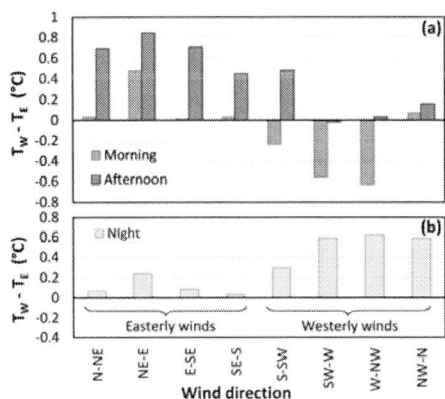

Fig. 4.   Average $T_W - T_E$ vs wind direction for the 2P configuration.

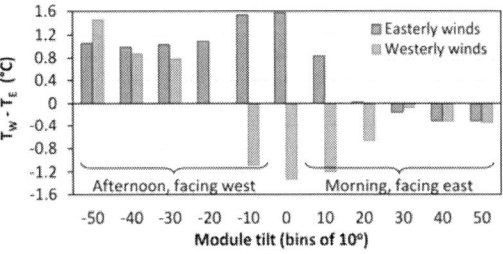

Fig. 5.   Average $T_W - T_E$ vs module tilt for the 2P configuration. When module tilt is positive, the eastern module is closest to the ground.

Firstly, when the modules are horizontal, the windward side is more affected by forced convection. At night, more heat is transferred into the windward side, increasing its temperature by ~0.6 °C more than the leeward side. In the day, more heat is removed from the windward side, decreasing its temperature by 1–1.5 °C more than the leeward side.

Secondly, when the modules have a high tilt (40°–55°), the lower modules are 0.5–1.5 °C hotter than the upper modules, and this is not affected by wind direction.

Finally, the impact of wind direction on the difference in windward and leeward module temperatures tends to increase as wind speed increases from 0 to about 4 or 5 m/s, but probably decreases at higher speeds (there's insufficient data to be sure).

V. MODELLING

We now quantify how accurately we can predict T_m with thermal models of increasing complexity, including expansions to include wind speed, wind direction and module tilt. In the process, we illustrate how simple thermal models can be greatly improved without incorporating additional free variables. The models are defined in the appendix.

Table III presents the results. It lists their equations and pertinent inputs, and gives their accuracy in terms of the mean-bias error (MBE) and the root-mean-square error (RMSE).

We find that the three simplest models—all commonly used by the PV industry—have an MBE above +3 °C. That is, these models overestimate T_m by, on average, more than 3 °C.

We find that the calibrated Faiman model without wind (Model 4) and with wind (Model 5) have an RMSE of 4.0 and 3.3. Thus, with 95% confidence (2 × RMSE), they predict T_m to within ±8.0 °C and ±6.6 °C, respectively. Thus, accounting for w provides a large improvement, consistent with [13], [18].

We reduce RMSE much further by making two other refinements, neither of which introduce a free variable. They are, the inclusion of radiative loss to the sky (Model 6), and transient effects (Model 7). Similarly large improvements were observed by Luketa-Hamlin and Stein [19].

We then expand the convection model to account for module tilt and wind direction, which are convoluted, as we saw in Section IV. Combined, they reduce the RMSE to 1.5 °C. When we allow all variables in Table III to be free, we predict T_m of the Voyager 2P tracker to within ± 2.8 °C at any time of day with 95% confidence (an RMSE of 1.39 °C).

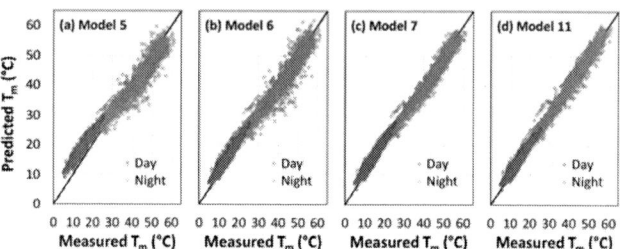

Fig. 6.   Predicted vs measured T_m for various thermal models.

VI. APPENDIX

The appendix of the expanded paper describes the models of Table III. In this version, we simply list the more interesting equations, since they pertain to tilt and wind direction.

We observed a roughly proportional dependence of U_c on the absolute tilt β,

$$U_c = U_{c0} + U_{c\beta}|\beta|, \tag{9}$$

where U_{c0} is U_c for horizontal modules ($\beta = 0$), and $U_{c\beta}$ is the constant of proportionality that defines how much additional heat loss occurs due to free convection when the modules are tilted. (The effect of tilt is already incorporated into the radiative model.)

We incorporate wind direction δ into the forced-convection term U_v via the equation,

$$U_v = U_{v0}[1 + f_v \cos\delta], \tag{10}$$

where the best-fit $f_v(\beta)$ is plotted as symbols in Fig. 7, where parameterization of those points is the dampened hyperbolic tangent, plotted as lines in Fig. 7:

$$f_v = f_{v0} \tanh\left(\frac{\beta - \beta_0}{\beta_v}\right) \exp\left(-\left|\frac{\beta}{\beta_e}\right|\right), \tag{11}$$

where the best-fit constants are $f_{v0} = -7$, $\beta_v = 100°$, $\beta_e = 14°$, and $\beta_0 = -4°$ and $+4°$ for the western and eastern modules.

TABLE III.   MODEL EQUATIONS, INPUTS AND FIT METRICS FOR T_M OF THE WESTERN 2P MODULE. (SEE APPENDIX FOR MODELS.)

M	Eqs	U_c	U_v	ε	C	U_g	$U_{c\beta}$	f_{v0}	MBE	RMSE
1	2,3,6	22.9	—	—	—	—	—	—	+7.4	[8.6]
2	2,3,6	29	0	—	—	—	—	—	+3.3	[4.8]
3	2,3,6	25	1.2	—	—	—	—	—	+3.3	[4.5]
4	2,3,6	36.6	0	—	—	—	—	0	4.0	
5	2,3,6	26.8	3.2	—	—	—	—	0	3.3	
6	2,3,6,7	17.0	2.8	0.9	—	—	—	0	2.7	
7	1,3,6,7	15.0	3.4	0.9	833	—	—	0	1.9	
8	1,3,6-8	10.5	3.2	0.9	833	2.5	—	0	1.8	
9	1,3,6-9	12.9	2.8	0.9	833	2.5	−5.0	0	1.6	
10	1,3,6-11	11.9	2.7	0.9	833	2.5	−5.0	−7	0	1.5
11	1,3,6-11	12.4	2.7	0.88	1200	2.5	−5.0	−7	0	1.4

The units are U_c, U_g in W·m⁻²·K⁻¹; U_v in W·s·m⁻³·K⁻¹; $U_{c\beta}$ in W·m⁻²·K⁻¹rad⁻¹, ε and f_{v0} are unitless; MBE, RMSE in °C; c in J·kg⁻¹·K⁻¹.

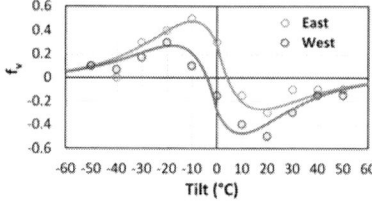

Fig. 7.   Best-fit f_v for eastern and western modules in the 2P configuration

REFERENCES

[1] R. G. Ross Jr and M. I. Smokler, "Electricity from photovoltaic solar cells: Flat-Plate Solar Array Project final report. Volume VI: Engineering sciences and reliability," NASA, 1986.

[2] D. L. King, "Photovoltaic Module and Array Performance Characterization Methods for All System Operating Conditions," *Rev. Lit. Arts Am.*, pp. 1–22, 1997.

[3] D. Faiman, "Assessing the outdoor operating temperature of photovoltaic modules," *Prog. Photovoltaics Res. Appl.*, vol. 16, no. 4, pp. 307–315, 2008.

[4] M. W. Davis, A. H. Fanney, and B. P. Dougherty, "Prediction of building integrated photovoltaic cell temperatures," *J. Sol. Energy Eng.*, vol. 123, no. 3, pp. 200–210, 2001.

[5] G. Notton, C. Cristofari, M. Mattei, and P. Poggi, "Modelling of a double-glass photovoltaic module using finite differences," *Appl. Therm. Eng.*, vol. 25, no. 17–18, pp. 2854–2877, 2005.

[6] I. Haedrich, D. C. Jordan, and M. Ernst, "Methodology to predict annual yield losses and gains caused by solar module design and materials under field exposure," *Sol. Energy Mater. Sol. Cells*, vol. 202, p. 110069, 2019.

[7] M. R. Vogt, H. Holst, M. Winter, R. Brendel, and P. P. Altermatt, "Numerical Modeling of c-Si PV Modules by Coupling the Semiconductor with the Thermal Conduction, Convection and Radiation Equations," *Energy Procedia*, vol. 77, pp. 215–224, 2015.

[8] S. Armstrong and W. G. Hurley, "A thermal model for photovoltaic panels under varying atmospheric conditions," *Appl. Therm. Eng.*, vol. 30, no. 11–12, pp. 1488–1495, 2010.

[9] M. K. Fuentes, "A Simplified Thermal Model," 1987.

[10] L. Weiss, M. Amara, and C. Ménézo, "Impact of radiative-heat transfer on photovoltaic module temperature," *Prog. Photovoltaics Res. Appl.*, vol. 24, no. 1, pp. 12–27, 2016.

[11] C. Chaudhari and B. Bourne, "Deriving Thermal Response Coefficients for PVSyst," in *PV Performance Symposium (PVPMC)*, 2019.

[12] L. Ghabuzyan, K. Pan, A. Fatahi, J. Kuo, and C. Baldus-Jeursen, "Thermal effects on photovoltaic array performance: Experimentation, modeling, and simulation," *Appl. Sci.*, vol. 11, no. 4, pp. 1–15, 2021.

[13] C. Schwingshackl *et al.*, "Wind effect on PV module temperature: Analysis of different techniques for an accurate estimation," *Energy Procedia*, vol. 40, pp. 77–86, 2013.

[14] H. Goverde *et al.*, "Spatial and temporal analysis of wind effects on PV modules: Consequences for electrical power evaluation," *Sol. Energy*, vol. 147, pp. 292–299, 2017.

[15] M. V Oliphant, "Measurement of wind speed distributions across a solar collector," *Sol. Energy*, vol. 24, no. 4, pp. 403–405, 1980.

[16] H. Goverde *et al.*, "Spatial and temporal analysis of wind effects on PV module temperature and performance," *Sustain. Energy Technol. Assessments*, vol. 11, pp. 36–41, 2015.

[17] H. Goverde *et al.*, "Impact of wind on intra-module energy yield variations," in *32nd European Photovoltaic Solar Energy Conference and Exhibition: proceedings of the international conference held in Munich, Germany, 20 June-24 June 2016*, 2016, pp. 1665–1668.

[18] P. Mora Segado, J. Carretero, and M. Sidrach-de-Cardona, "Models to predict the operating temperature of different photovoltaic modules in outdoor conditions," *Prog. Photovoltaics Res. Appl.*, vol. 23, no. 10, pp. 1267–1282, 2015.

[19] A. Luketa-Hanlin and J. Stein, "Improvement and validation of a transient model to predict photovoltaic module temperature.," 2012.

Impacts of Nonuniform Soiling on Photovoltiac Production

Lin J. Simpson, Ian K. Teague, Jody Ford, Nathan Shih, Mahfujur Rahman, Jorge I. T. Marchand, Kirsten Perry, Chris Deline

National Renewable Energy Laboratory, Golden, CO, United States

Monroe Community College, Rochester, NY, United States

Red Rocks Community College, Lakewood, CO, United States

Bellevue Community College, Bellevue, WA, United States

City University of New York Queensborough, New York, NY, United States

Even though soiling of photovoltaic (PV) modules can substantially reduce power production, cost effective mitigation remains a challenge. The differences in soiling type, environmental conditions, and module properties all contribute to substantial soiling variability between sites and even within sites. Here, we develop tools to evaluate the production data from individual modules and observe specific incidences where nonuniform soiling produces an imbalance of light throughput resulting in a "hot cell" which induces a module string diode to shut down power delivery from that cell string. Thus, demonstrating that even a small amount of nonuniform soiling can substantially reduce power production.

Operability of a Power System with Synchronous Condensers and Grid-Following Inverters

Marena Trujillo[*†‡], Rick Wallace Kenyon[*†‡], Gemini Yau[§], Li Yu[§], Andy Hoke[‡], and Bri-Mathias Hodge[*†‡]

[*]Department of Electrical, Computer, and Energy Engineering, University of Colorado of Boulder, Boulder, Colorado 80309
marena.trujillo, richard.kenyonjr, brimathias.hodge@colorado.edu; andy.hoke@nrel.gov
[†]Renewable and Sustainable Energy Institute, University of Colorado of Boulder, Boulder, Colorado 80309
[‡]National Renewable Energy Laboratory, Golden, Colorado 80401
[§]Hawaiian Electric Company, Honolulu, Hawaii 96820

Abstract—Growing shares of inverter-based resources generally correlate with a reduction in inertia as synchronous generators are displaced. Along the path to high shares of inverter-based resources, in particular with only conventional grid-following inverter controls, a proffered solution is the use of synchronous condensers as a source of inertia to maintain the frequency–power balance relationship. An outstanding question is whether only grid-following inverters and synchronous condensers yield a viable power system; i.e., all frequency response is derived from inverters. A validated electromagnetic transient model of the Maui system, with many nonlinear elements such as load-shedding, line tripping, and inverter ride through criteria disabled, is used to investigate the stability of such a system. Two types of perturbations were applied, a 15% generation loss and a fault event, and it was found that the system remains stable.

Index Terms—inverter-based resources, grid-following inverters, inertia, PSCAD, synchronous condensers

I. INTRODUCTION

As the large-scale deployment of renewable energy sources continues, challenges relating to the stability and control of power systems with reduced relative quantities of synchronous machines (SMs) have emerged [1]. For instance, reduced inertia can lead to faster frequency dynamics, while a lack of SMs generally yields a reduced short circuit current. Grid-following (GFL) inverters are the contemporary standard integration technology for renewable sources, where the primary assumption is the presence of a stable voltage waveform at the point of interconnection. An additional assumption is that the power system frequency continues to evolve according to machine swing dynamics [2], [3], such that GFL grid support mechanisms can be designed around these characteristics to participate in frequency regulation [4]. A synchronous condenser (SC) is a synchronous generator operating without a primer mover/governor, with the standard primary objective to provide reactive power compensation for voltage support [5]. Recently, these devices have been identified as a potential solution to low inertia challenges [3], [6]. While the presence of SCs will contribute inertia to the system at large, of question is whether the presence of only SCs and GFL inverters yields a viable power system, as presented in [7]; i.e., can a power system function with all frequency response (FR) derived solely from GFL devices?

The frequency response of a SM resulting from a power imbalance is described by the classical swing equation (1) [2]:

$$\dot{f} = \frac{f}{2HS_B}(P_m - P_e) \qquad (1)$$

where S_B is the rated power of the generator, H is the time for which rated power can be supplied +by the machine with stored kinetic energy, f is the rotational frequency of the generator, P_m and P_e is the supplied mechanical power and electric power demand, respectively [2]. In SM dominated systems, (1) will approximate the center of inertia frequency trajectory with aggregate H, S_B, P_m, and P_e values, where P_m represents all real power delivered to the system. The operational philosophy behind a GFL and SC only system is that the aggregate inertial characteristics of the SCs on the system will yield a change in frequency when $P_m - P_e \neq 0$. With frequency response control based on the dynamics of (1), the GFL devices will arrest changes in frequency by matching the aggregate injected power, P_m, to the electrical demand, P_e.

In [7], the feasibility of maintaining system stability with GFLs and SCs alone was investigated with PSCAD simulations of a two-bus system. Island systems typically have far higher annual and instantaneous renewable shares than larger power systems on account of more expensive generation sources. As a result, these island systems are at the fringe of operational thresholds, and the applicability of a GFL/SC only operating scenario is in the near future. In [8], a highly-detailed electromagnetic transient (EMT) model of the Maui Hawaiian island power system was developed and validated in PSCAD. This research utilizes the Maui EMT model with high order dynamical models of the GFLs to simulate a scenario in which only GFLs provide frequency response.

II. THE MAUI MODEL

The PSCAD model of the Maui power system is operated on 32 cores with a rough partition into six areas as shown in Fig. 1. The map depicts aggressive aggregation for visual purposes; the actual system model contains over 200 buses and nearly 200 distinct sources of generation. The system dispatch for the scenario investigated is outlined in Table I. While the majority of real power is delivered by distributed generation, only three hybrid power plants (HPPs), plants with coupled solar photovoltaic and energy storage systems, and a single energy storage system (ESS) provide frequency response. There are no synchronous generators online.

To focus on the oscillatory stability of the system, nonlinearities including underfrequency load shedding, inverter trip and momentary cessation, and line tripping were disabled for this study.

Reduced 2023 Maui Transmission Map:

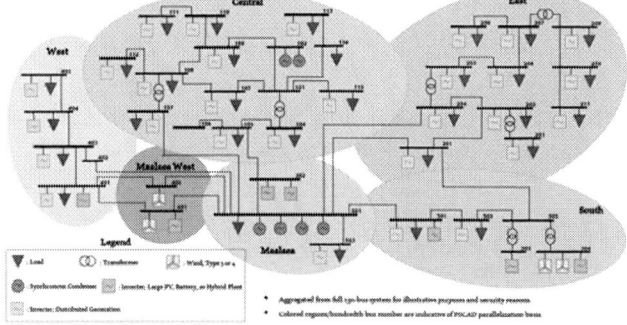

Fig. 1: The Maui network as partitioned into regions for the system resource allocation description [9].

TABLE I: System Dispatch

Device	Bus	Quantity*	Rating (MVA)	Dispatch (MW)
Hybrid Power Plant**	502	2	60.0	12.6
Hybrid Power Plant**	305	1	19.1	0.2
Synchronous Condenser	501	4	107.2	0.0
Synchronous Condenser	102	2	29.1	0.0
Utility Solar	301 & 401	2	5.8	5.3
Wind (Type 3)	601 & 602	2	56.7	3.9
Wind (Type 4)	304	2	24.0	21.0
Energy Storage System***	304	1	11.0	0.0
Distributed Generation	see Fig. 1	179	131.1	105.7
Total Generation	–	–	–	148.7
Total Load	see Fig. 1	90	–	144.7

* Ratings and dispatch apply to the aggregate quantity.
** The hybrid power plants are the only devices with an active frequency response on the system.
*** The objective of the energy storage system at bus 304 is to mitigate fluctuations in wind output, with an auxiliary frequency response with available headroom.

GFL inverters interface with a power system by tracking an external voltage signal as reference for synchronization with the larger power system [1]. A current waveform is injected in conjunction with this tracked voltage waveform to achieve power set points. An open-source GFL model for use in PSCAD is provided in [10]; this model is used for all distributed generation and HPP instances in this study. The 15th order control scheme consists of nested current and power controllers, with an LCL filter, and an averaged voltage source implementation. To participate in frequency response, the HPP GFL models employ a frequency droop control, which alters the real power set points based on phase-locked loop (PLL)

measured frequency values [10], at a 5% droop relation. The GFL control diagram is shown in Fig. 3.

Two types of perturbations are applied to assess the large signal stability of the Maui power system.

- Generation Loss: the two type 4 wind plants in the South region at bus 304 are disconnected, for a loss of generation of 21 MW (approximately, 14% of total generation).
- Fault: a three phase, zero impedance fault is applied for 5 cycles (0.083 s) in the Maalaea West region at bus 602.

III. RESULTS

The results of the two simulations are presented, highlighting the dynamic responses of the SCs and HPPs.

A. Load-Generation Imbalance

The PLL measured frequency and SC 501a rotation speed at the Maalaea region generation station (bus 501) following the trip of both type 4 wind plants at bus 304 for a 21 MW generation loss is shown in Fig. 2a. Frequency stabilization is due to the response of the elements in the system, including the ESS. The output of the ESS at bus 304 is shown in Fig. 2b, where the real power ramps to the maximum output to mitigate the loss in power from the two tripped wind plants and restore generation/load balance.

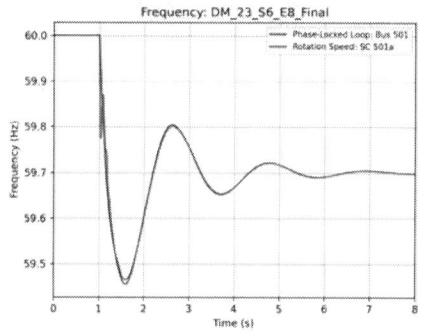

(a) Frequency

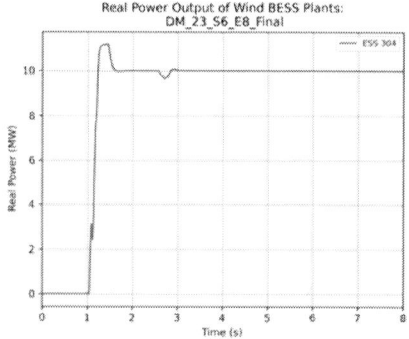

(b) Energy storage system

Fig. 2: Frequency response and wind energy storage system real power output following a 21 MW generation loss.

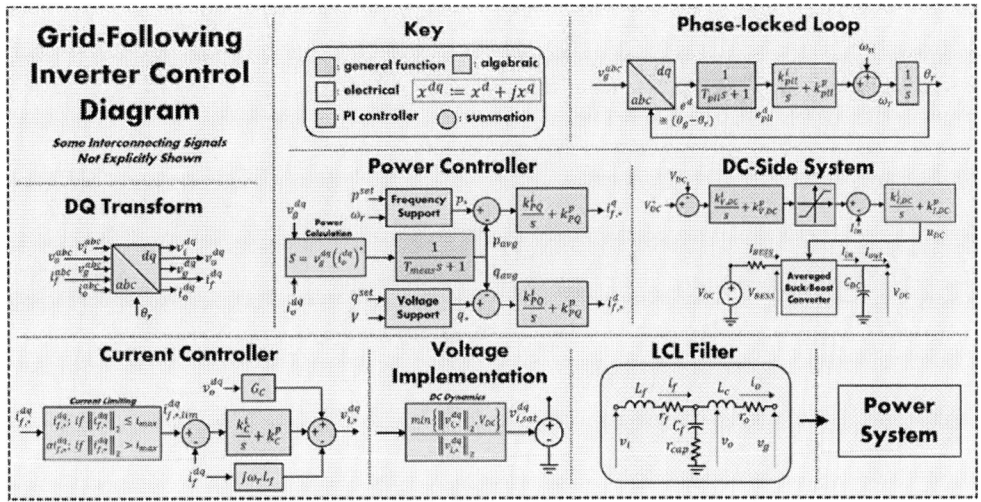

Fig. 3: The control diagram of the GFL model used for all distributed generation and HPP instances in this study.

Of the six synchronous condensers online, four are located in the Maalaea region at bus 501, and two located in the Central region at bus 102. The real power output of each synchronous condenser following the 21MW loss in generation is shown in Fig. 4.

power over time. The real power output of the SCs decreases to zero as other elements respond.

In Fig. 5, the real power output of each HPP directly following the 21MW generation loss is presented. The HPPs are the only devices in the system with a 5% frequency–droop response, excepting the ESS at bus 304 with a more aggressive response comparable to a step output in real power.

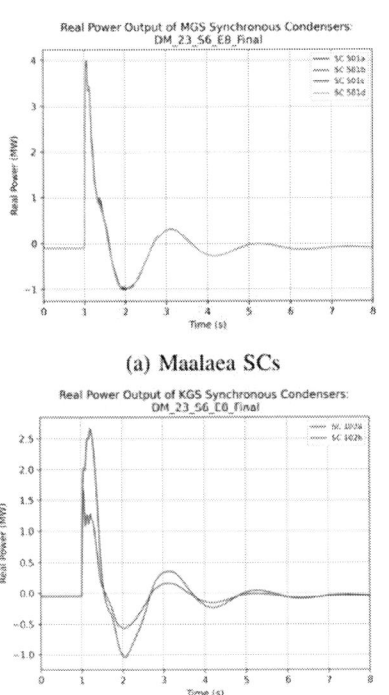

(a) Maalaea SCs

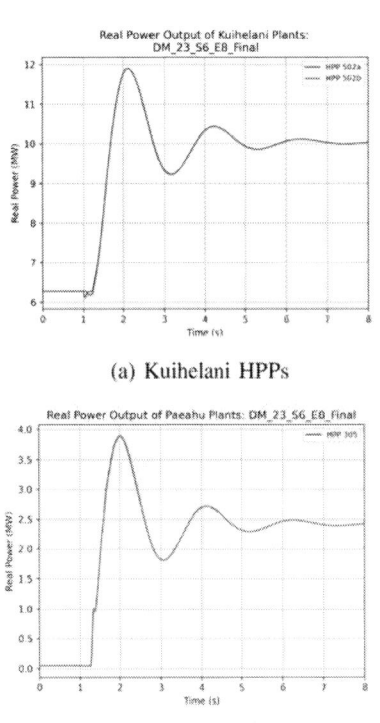

(a) Kuihelani HPPs

(b) Kahului SCs

Fig. 4: Real power output of the six online synchronous condensers following a 21 MW generation loss.

(b) Paeahu HPPs

Fig. 5: Real power output of the Kuihelani and Paeahu hybrid power plants following a 21 MW generation loss.

Although co-located, the responses of SC 102a/b differ on account of different machine ratings and exciter models. Kinetic energy is extracted from the SCs and delivered as real

B. Fault

The fault at bus 601 causes significant voltage drops at most buses in the system, as indicated by the select profiles presented in Fig. 6a. The PLL and SC 501a rotational velocity presented in Fig. 6 show a frequency perturbation, but no substantial settling deviation, as expected.

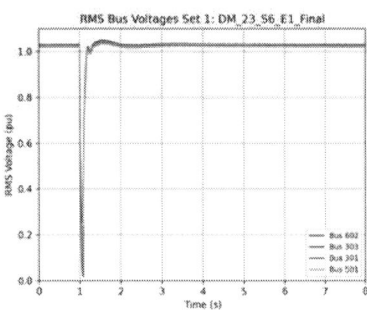

(a) Voltage

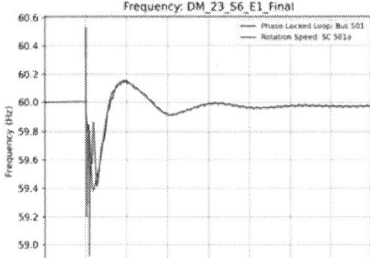

(b) Frequency

Fig. 6: Voltage at select buses, and frequency at bus 501, following at fault at bus 601.

The behavior of the Maalaea SCs in response to the fault event is shown in Fig. 7. As expected, a large quantity of reactive power is delivered during the fault. The spike in real power output is in part due to the large decrease in power exported by the IBRs on the system when the voltage drops at the IBR terminals.

The real and reactive power output of the Kuihelani HPP is shown in Fig. 8. The system returns to a stable steady state. The real power response is due to the change in frequency after the fault, which occurs because of the real power delivered by the SCs, which is essentially drawn out of the devices because the IBRs on the system cannot export power under low voltage conditions. Furthermore, the reactive power output is the response of the HPP to the low voltage conditions. The response from the Kuihelani HPP is not as abrupt as that of the SCs, which is because the HPP is driven by a relatively slow controller, whereas the SC is delivering the reactive power due to the much faster exciter action.

IV. CONCLUSION

This paper presented the results of an investigation into the operation of the island of Maui EMT model in which

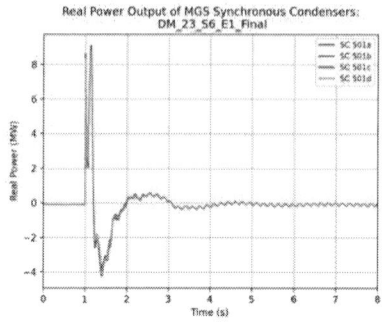

(a) Real Power

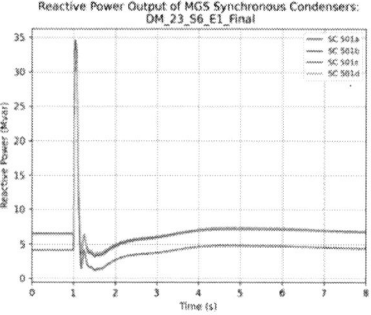

(b) Reactive Power

Fig. 7: Real and reactive power output of Maalaea synchronous condensers following a fault at bus 601.

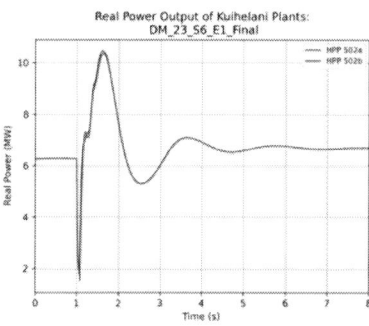

(a) Real Power

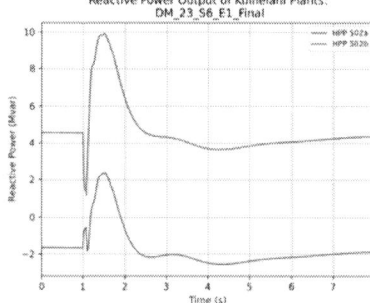

(b) Reactive Power

Fig. 8: Real and reactive power output of the Kuihelani hybrid power plant following a fault at bus 601.

only grid-following inverters and synchronous condensers are present on the system (i.e. no synchronous generation). The only primary frequency response on the system is derived from the grid-support functionality of three hybrid power plants, with a separate step-like response from a single energy storage device, while the actual sinusoidal voltage profiles are maintained by the synchronous condensers. A 21 MW generation loss and a fault were simulated to assess the large signal stability of the system, with both simulations yielding acceptable responses, pointing to the potential viability of such an operational scenario. This finding highlights the benefit of synchronous condensers in stabilizing highly renewable power systems. Further analyses may be required to confirm stability of the system with these resources under a wider range of fault and operating conditions. This study did not examine other important aspects of the operation of such a power system, such as protection system operation and resource adequacy.

REFERENCES

[1] R. W. Kenyon, M. Bossart, M. Marković, K. Doubleday, R. Matsuda-Dunn, S. Mitova, S. A. Julien, E. T. Hale, and B.-M. Hodge, "Stability and control of power systems with high penetrations of inverter-based resources: An accessible review of current knowledge and open questions," *Solar Energy*, 2020. [Online]. Available: http://www.sciencedirect.com/science/article/pii/S0038092X20305442

[2] A. Ulbig, T. S. Borsche, and G. Andersson, "Impact of Low Rotational Inertia on Power System Stability and Operation," *IFAC Proceedings Volumes*, vol. 47, no. 3, pp. 7290–7297, Jan. 2014.

[3] M. Nedd, C. Booth, and K. Bell, "Potential Solutions to the Challenges of Low Inertia Power Systems with a Case Study Concerning Synchronous Condensers," in *2017 52nd International Universities Power Engineering Conference (UPEC)*, Aug. 2017, pp. 1–6.

[4] Inverter-Based Resource Performance Task Force, "Fast Frequency Response Concepts and Bulk Power System Reliability Needs," NERC, Atlanta, GA, Tech. Rep., Mar. 2020.

[5] Y. Liu, S. Yang, S. Zhang, and F. Z. Peng, "Comparison of synchronous condenser and STATCOM for inertial response support," in *2014 IEEE Energy Conversion Congress and Exposition (ECCE)*, Sep. 2014, pp. 2684–2690.

[6] Ha Thi Nguyen, Guangya Yang, A. H. Nielsen, and P. H. Jensen, "Frequency stability improvement of low inertia systems using synchronous condensers," in *2016 IEEE International Conference on Smart Grid Communications (SmartGridComm)*, Nov. 2016, pp. 650–655.

[7] R. W. Kenyon, A. Hoke, J. Tan, and B. Hodge, "Grid-Following Inverters and Synchronous Condensers: A Grid-Forming Pair?" in *2020 Clemson University Power Systems Conference (PSC)*, 2020, pp. 1–7.

[8] R. W. Kenyon, B. Wang, A. Hoke, J. Tan, C. Antonio, and B.-M. Hodge, "Validation of Maui PSCAD Model: Motivation, Methodology, and Lessons Learned," in *2020 52nd North American Power Symposium (NAPS)*, Apr. 2021, pp. 1–6.

[9] R. W. Kenyon, A. Sajadi, A. Hoke, and B.-M. Hodge, "Criticality of Inverter Controller Order in Power System Dynamic Studies – Case Study: Maui Island," 2022, in Review.

[10] ——, "Open-Source PSCAD Grid-Following and Grid-Forming Inverters and A Benchmark for Zero-Inertia Power System Simulations," in *2021 IEEE Kansas Power and Energy Conference (KPEC)*, Apr. 2021, pp. 1–6.

Vapor Treatment for Growth of High-Quality Oxide Barriers within P-I-N Perovskite Solar Cells and Tandems

Samuel A. Johnson, Michael D. McGehee, Joseph J. Berry, Axel F. Palmstrom

University of Colorado, Boulder, CO, United States

National Renewable Energy Laboratory, Golden, CO, United States

We investigate oxide growth on fullerene (C60) by atomic layer deposition (ALD) for C60/oxide bilayer electron selective contacts in P-I-N perovskite solar cells. A key inhibitor to the formation of high-quality oxide barriers on C60, despite the dense, conformal nature of ALD-grown materials, is the poor reactivity of the fullerene surface. This leads to sub-surface diffusion and exothermic growth and, in turn, low-quality oxide barriers. We demonstrate an in-situ vapor-phase approach to enhance the activity of the fullerene surface and fill pinholes within ultra-thin oxide films for high-quality oxide barriers on top of perovskite absorbers in P-I-N solar cells.

Long-term UV Durability of Laminated Glass/Transparent Backsheet Coupons for Bifacial Photovoltaics: Backsheet Side Exposure

Soshana Smith, Stephanie Moffitt, Stefan Mitterhofer, Song-Syun Jhang, Stephanie Watson, Li-Piin Sung, LaKesha Perry, Deborah Jacobs , Xiaohong Gu

Engineering Laboratory, National Institute of Standards and technology, Gaithersburg, MD, United States

Bifacial modules with glass/transparent backsheet (GB) structure offer many advantages over traditional glass/glass (GG) modules, such as lighter weight, smaller heat capacity, and higher hail resistance. However, research on long-term durability of transparent backsheets and their usage in bifacial modules is lacking. In this study, durability of GB laminated coupons constructed with three fluoropolymer-based transparent backsheets were investigated, along with GG coupons laminated with either polyolefin elastomer (POE) or ethylene vinyl acetate (EVA) encapsulants. Laboratory accelerated weathering was performed using the NIST SPHERE (Simulated Photodegradation via High Energy Radiant Exposure) under UV/65 °C/50 % relative humidity (RH), followed by thermal cycling. Optical and chemical properties of the specimens were characterized by UV-visible spectroscopy and attenuated total reflectance Fourier transform infrared spectroscopy with microscope (micro-ATR-FTIR) periodically during UV exposure. The non-destructive depth profiling of yellowing in GB coupons was performed by laser scanning confocal microscope-based fluorescence mapping. Results indicated that the long-term performances of the GB coupons strongly depended on the materials and structures of the transparent backsheets. The backsheet cracking was observed in a fluoroethylene vinyl ether (FEVE)/ polyethylene terephthalate (PET)/ EVA-based transparent GB coupons after UV exposure and subsequential thermal cycling. The UV durability of PET core layer appeared to be critical to the application of GB structure for bifacial modules. Additional characterizations on aged GB and GG coupons are on-going. This study has provided a scientific basis for material choice and product development for a more reliable bifacial PV technology.

Silicon Heterojunction Solar Cells with High Bulk Resistivities Over 1,000 Ω·cm in Relevant Field Conditions of Illumination and Temperature

Anh Huy Tuan Le, Apoorva Srinivasa, Stuart G. Bowden, Ziv Hameiri, André Augusto

School of Photovoltaic and Renewable Energy Engineering, University of New South Wales, Sydney, Australia

School of Electrical, Computer and Energy Engineering, Arizona State University, Tempe, AZ, United States

As we design solar cells with better surface passivation, it is important to revisit the bulk properties. The use of lightly doped wafers provides a promising way to mitigate Auger recombination and increase the breakdown voltage of solar cells, which could lead to new module and system designs. Thus, studying the performance of silicon (Si) solar cells and modules using such wafers in relevant field conditions is of significant interest. In this study, we experimentally investigate the impact of the bulk resistivity (up to >15,000 Ω·cm) on the properties of Si heterojunction solar cells under different illuminations (0.1-1 suns) and temperatures (25-70 °C). We also study the dependency between the breakdown voltage and the bulk resistivity. The results indicate that for very low illuminations intensities down to 0.1 suns, cells with very high bulk resistivities, over 15,000 Ω·cm, have comparable performances to cells with much lower bulk resistivities. The temperature coefficients measured on these cells are also comparable with values previously reported for cells using wafers with standard resistivities. The cells with bulk resistivities over 1,000 Ω·cm show breakdown voltages larger than -1,000 V, almost two orders of magnitude higher than in typical Si solar cells. Our simulations indicate that in the absence of bypass diodes, shaded solar cells with larger breakdown voltages still operate in forward-bias, even under extreme shading conditions, protecting the integrity of the cell and module. Together, these results highlight the large potential of using high-resistivity wafers to manufacture high-efficiency Si solar cells suitable to operate under relevant field conditions, and with the prospect of more robust and cost-effective module designs.

Electroluminescence Analysis and Grading of Hail Damaged Solar Panels

Andrew M. Gabor, Phillip J. Knodle, Maurice Covino, Dylan J. Colvin, Kristopher O. Davis

BrightSpot Automation LLC, Westford, MA, United States

University of Central Florida, Orlando, FL, United States

We analyzed over 4000 electroluminescence images of hail damaged solar panels from a cluster of houses in Texas. We enhanced the images for ease of analysis and classified the defects within each solar cell into categories of glass breakage, installer damage, inactive substrings, crack severity, interconnect wire problems, and whether the damage was likely caused by hail. From these statistics, we quantified each panel into 5 levels of hail damage for insurance claims, and 4 levels of overall quality for potential resale pricing. We share here some statistics regarding the defects with the hope that the data is useful to others attempting to predict the invisible damage to systems based just on the easily observable glass breakage statistics.

WHAT IS THE ROLE OF RECYCLING IN THE SOLAR TERAWATT FUTURE?

Pablo R Dias, Moonyong Kim, Alison Lennon, Brett Hallam

UNSW, Sydney, Australia

UFRGS, Porto Alegre, Brazil

SunDrive, Sydney, Australia

Recycling can play an important role in sourcing the materials required for deploying TW-scale photovoltaics while also assisting with reducing carbon emissions associated with the production of key materials. Secondary silver can supply anywhere from single digits to 100% of the demand for the solar industry under certain scenarios, depending on the silver learning rate, the PV deployment rate, and the lifespan of the modules. Likewise, secondary aluminum can supply about 10% of the required material at peak PV manufacturing and as much as 50% closer to 2050, as the available aluminum pool increases. Aluminum recycling is crucial due to the large carbon footprint associated with the production of primary aluminum.

Solar Panel Power Simulation for Shade Detection

David Jose Florez Rodriguez[1] and Bennet E. Meyers[1,2]

[1] Stanford University, Stanford, CA, 94305, USA,

[2] SLAC National Accelerator Laboratory, Menlo Park, CA, 94025, USA

Abstract—We construct models of the AC power production from real residential solar systems, with the goal of estimating the losses in the systems due to the shading from nearby objects. The purpose of this paper is to document the procedure used to model the systems. Our goal is to generate modeled PV power signals that are a reasonable match to the shape and scale of the measured data, and to use that model to estimate the losses experienced by these systems due to shade. In this paper, we describe the details of how this model is implemented and how the necessary model parameters are derived, and we present a summary of shade losses in 25 residential rooftop PV systems.

Index Terms—photovoltaic systems, solar energy, distributed power generation, energy informatics, PV digital operations and management, modeling, clear sky, irradiance, solar cell temperature, soiling

I. INTRODUCTION

We present a method for modeling real-world, rooftop photovoltaic (PV) systems using open source software tools. We construct these models based on very little external data, validating a basic geometric model from satellite imagery and using a clear sky irradiance model. The goal of this work is to generate a small data set—consisting of 25 unique PV systems, with good quality historical power production data and a basic digital twin model—that can be labeled with estimates of shade losses in the systems.

It is well known that losses due to nearby object shading are a major source of under-performance in real-world PV systems, particularly for distributed PV in urban environments [1]–[3]. Previous work on analyzing shade losses from production data have focused on generating high-accuracy models of mismatch losses [4]–[6]; whereas, we focus this work on developing a method for practical shade analysis based on minimal external data. This method is not automatic, requiring some amount of bespoke model design, but it nonetheless is practical to scale to a fleet of a couple dozen PV systems.

This labeled data set is intended to be a validation data set for a proposed algorithm for estimating shade losses in *unlabeled* PV data (*i.e.*, power data and nothing else). This application is described in [7]. The data set is not available to be made public at this time, but we are exploring options for making this data set open source. The relevant code is publicly available in the author's GitHub [8].

This material is based on work supported by the U.S. Department of Energy's Office of Energy Efficiency and Renewable Energy (EERE) under the Solar Energy Technologies Award Number 38529.

II. METHODS

Making use of the open-source PV modeling software pvlib-python [9], we develop models of system power production for 25 residential rooftop PV systems in Southern California. We model the expected power generation under a typical clear sky year, estimating the physical system geometry from satellite images of the system locations. To analyze shade losses, we filter the measured data for cloudy days using Solar Data Tools [10], [11], and then we compare the measured output of the systems on sunny days to the pvlib-python models. The remainder of this section describes the details of how the models were identified and implemented, and how the analysis of yearly clear sky energy loss was carried out.

It's important to note that the models below are not strict upper bounds for the data at hand. We allow for measured power to seldom surpass modelled power, as factors like cloud lensing or data-collection errors could generate data spikes.

A. Model identification and implementation

Site coordinates, system specifications, altitude, azimuth angle, and surface tilt angle are all essential for building an accurate model. We originally have the coordinates and the inverter type, and infer the remaining values. Google Earth Pro (a free product) facilitated obtaining these missing parameters from the given latitude and longitude. Altitude, for example, is immediately available for rendered structures (including roofs) in the zoomed-in globe view.

Many models exist for predicting the AC power of PV systems with given ambient conditions. This work uses an implementation of Sandia National Laboratories PV Array Performance Model [12] (SAPM) to produce a clear sky power signal. The SAPM relies on irradiance data and solar cell temperature on top of the geometric parameters discussed above; unfortunately, irradiance and cell temperature data are absent from most residential PV systems, as previously specified.

We also use an implementation of SAPM for the cell temperature, and the Simplified Solis [13] model for generating irradiance values (all implemented via pvlib in python). This work chooses the aforementioned models due to their strong documentation and reasonable output given our scarce meteorological data. A lot of the variance between different irradiance and cell temperature models comes from their employing of different functions on unavailable data like humidity and wind speed. Lacking this data, the choice to switch from one model to another would not likely reduce the error in our shade estimation.

978-1-7281-6118-1/22 $31.00 © 2022 IEEE

Panel type and the system's wiring are also essential inputs to SAPM. Our method for obtaining of these two parameters, along with all other essential data mentioned above, follows.

1) System geometry: We must develop a reliable method for extracting surface tilt and azimuth angles from a rendered roof on Google Earth. These parameters' role in a PV system is defined in section 1.6 of Solar Engineering of Thermal Processes [14] (which refers to surface tilt and azimuth as slope and surface azimuth angle, respectively); however, one must note that the azimuth definition employed in this work is $180°$ greater than that in the Solar Engineering textbook. Trigonometry applied to 3D polygons created in Google Earth will provide these angles.

Google's desktop software can generate triangles on the surface of a roof containing the relevant solar panels. These triangles share the desired orientations in space of our panels, and their coordinates invite trigonometric analysis. Google Earth can save these triangles' coordinates in a .kml file. Originally, the points are in the coordinate space $\begin{bmatrix} \alpha & \omega & h \end{bmatrix}$ defined by degrees of longitude α, degrees of latitude ω and altitude above sea level h in meters. A transformation is necessary to extract the tilt and azimuth, however, as coordinates must be in the same units.

If we approximate the Earth as a sphere with radius $r_e = 6371$ km [15], each degree of latitude has constant length $k = r_e(\frac{\pi}{180})$ m, and each degree of longitude measures $k \cos \omega$ meters. These conversions yield longitude, $k(\cos \omega)\alpha$, and latitude, $k\omega$, in meters, while we keep the previous altitude values h.

With each triangle's coordinates defined in meters, displacement vectors can be defined between the points. The cross product of two such displacement vectors within a triangle provides the normal vector to the plane containing the triangle; which is also normal to the panels at the site. Let the normal vector, $\begin{bmatrix} x & y & z \end{bmatrix}$, to triangle (choosing the normal vector with the positive z coordinate). The surface tilt angle,

$$\tau = \arctan\left(\frac{\sqrt{x^2 + y^2}}{z} \right),$$

comes directly from said normal vector.

SAPM will also need the azimuth angle. The function `arctan2()` from the NumPy library [16] can provide an azimuth angle from the x and y coordinates of the panels' normal vector. This function takes these two values in the plane and generates an angle $\sigma \in [-\pi/2, \pi/2]$, which increases counterclockwise from the positive x-axis. In contrast, the desired azimuth angle $\theta \in [0, 2\pi]$ increases clockwise from the positive y-axis. We derive our desired azimuth angle

$$\theta = \left\{ \begin{array}{ll} 90° - \sigma & \sigma \leq 90° \\ 450° - \sigma & \sigma > 90° \end{array} \right\}.$$

In practice, the above method computes multiple triangles' tilts and azimuths, and the median angles in degrees become the inputs to the `PVSystem` object from the pvlib library [9].

These methods and additional 3D plotting tools make up a python class written for this project, which is publicly available [8].

2) Irradiance: Irradiance values inform the final model of the sunlight vectors hitting the panels. Simplified Solis [13] in pvlib generates irradiance values via assuming an atmospheric aerosol type, ozone content and aerosol optical depth common to urban areas. This model only requires the relevant dates and coordinates. It models global horizontal irradiance (the total irradiance reaching a horizontal surface), beam normal irradiance (irradiance received by a plane normal to the sun), and diffuse horizontal irradiance (irradiance received by a surface that was scattered by the atmosphere and thus originates from all directions).

The authors briefly considered a constant irradiance model. The resulting modelled PV signal didn't resemble regular PV data at all, even generating negative power in some hours, so the above Simplified Solis model was used instead.

3) Cell temperature: The SAPM [12] cell temperature model takes in air temperature, wind speed, mounting type, and panel backing. From the mounting type and panel backing, the model chooses from internal, empirically derived parameters relating the wind speed and air temperature to cell temperature.

All models herein use the 'insulated back glass polymer' parameter when implementing a cell temperature model in pvlib. The mounting type is absent from the data, so we make our selection on the basis of 'glass polymer' installation, which is more common in residential PV than 'glass glass' installations due to their reduced mass. Additionally, the choices for backing were 'insulated back' and 'open rack', the latter of which refers to sites installed in an open space like those in a desert utility scale system. Given the residential focus of the modelling, 'insulated back' seemed most appropriate.

Given mounting type, air temperature and wind speed, the SAPM first computes back-surface temperature from air temperature and irradiance. The model then gets cell temperature from back-surface temperature. We set air temperature at a fixed $25°$C (room temperature, where manufacturers determine nameplate voltages), and wind speed at 0 mph.

It's worth noting that early models attempted to include air temperature data interpolated from lowest and highest daily values for the site's zip code; however, this increased PV power signal variance without improving model accuracy. For this reason, a constant air temperature model at $25°$C is the input to the SAPM cell temperature model.

4) Panels and panel wiring: We select a panel type from the SAPM PV module model database provided by pvlib by first limiting our selection to a single panel manufacturer. Then, we select from available PV panel types by determining the cell count of the panels at the site. This particular manufacturer produces panels in 72-cell and 96-cell layouts, which have aspect ratios of $2 : 1$ and $3 : 2$, respectively. Google Earth's satellite view of this length-to-width ratio of the panels thus informs the choice of panel type.

978-1-7281-6118-1/22 $31.00 © 2022 IEEE

The panel size then informs our choice of the system wiring: given n total panels (counted in satellite view), there must be s parallel strings with $\frac{n}{s}$ panels connected in series, where s divides n. The number of panels connected in series affects the DC voltage going into the inverter and thus the AC voltage and power present in our data. We note this positive correlation and tune s accordingly to generate models whose maximum daily powers match those in the data.

B. Shade energy loss analysis

1) Soiling correction: Soiling describes the accumulation of opaque matter (dust, bird droppings, leaves, etc.) on PV panels and the resulting power losses. Soiling is most prevalent in regions with arid climate, where rains don't frequently wash panels; and it's often a seasonal phenomenon, as the dry season results in increasing soiling until the rains return. The data's soiling losses are best removed before quantifying shading losses by the model. We employ SLAC's soiling detection algorithm [17]. The soiling detection algorithm provides an estimate for the percentage of daily maximum power that is available to a panel on a day with soiling. Higher soiling will cause a lower percentage and a lower amplitude signal for that day's PV power signal. While the original data is in T-minute interval power readings, it's readily converted into daily energy values as described in the following section. Element-wise division of this daily energy by their respective percentages of available power after soiling provides soiling-corrected data. This produces an approximation of what the measured PV values would 'look' like if there were no soiling losses.

Starting with clear sky days—and thereby removing days with weather and malfunctioning losses—the remaining soiling-corrected signal should only contain shading-losses.

2) Average daily loss: We compute the energy obtained from a site on clear sky days and average the daily energy over the given years. First, we sum over the power readings for a day and scale by the appropriate coefficient to yield that day's energy in kilowatt-hours. Said coefficient depends on the intervals of time, T on which measurements are available. Assuming that measurements are in kilowatts, $\frac{60}{T}$ is the number of measurements per hour and $\frac{T}{60}$ is the multiplication factor providing kilowatt-hours. With these daily energy values, we compute the average energy for each calendar day of the year over the years available. This averaging may reduce variance and does not hurt the shading loss estimate as shade should not substantially change over various years, as it pertains to the physical environment around a panel (buildings, trees, public infrastructure, etc.).

The same procedure on pvlib model data (generated in the same T-minute intervals) returns the model's estimate for the energy obtained in a shading-free world. The difference between the two signals is our shading loss estimate in the form of a 365-length vector.

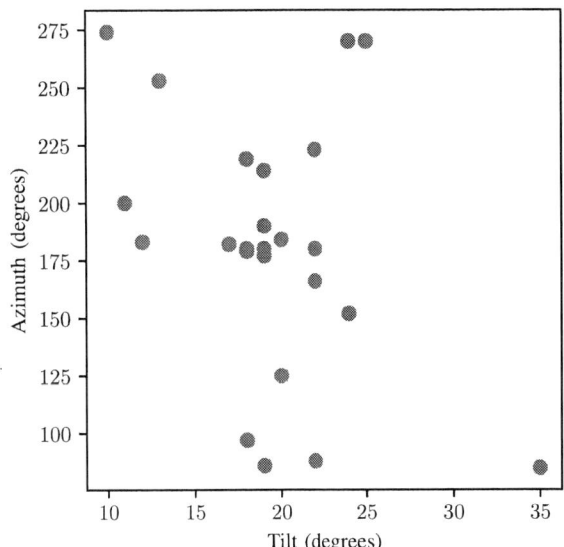

Fig. 1. Sites' azimuthal and surface tilt angle distribution

III. RESULTS

A. Data

A solar distributor provided data sets with the power readings of residential installations. Most sites have several years of power signals collected at regular intervals—with a few days or weeks of malfunctions in most cases—the site coordinates, and the inverter type. There are outliers, of course: some sites have less than two years of data, with malfunctions for near half of available days. In total, we studied 25 sites in the Northern Hemisphere.

These sites include a variety of spatial configurations: figure 1 shares their azimuthal and surface tilt angles' scatter. As expected in the Northern Hemisphere, many sites are pointing south; although, there are also a pair of sites pointing east and west. The surface tilt distribution is much more limited, never leaving the 10 to 35 degree interval.

These diverse orientations inevitably affect the sunlight available at each site, and the resulting shading losses. Some sites had multiple orientations, as residents with limited space couldn't fit all panels on the same section of their roof. For these sites, each orientation gets its own pvlib model, and then the sum of all modelled signals at a site produce its clear sky signal.

B. Heatmap analyses and shade categorization

Solar power matrices facilitate visualizing the data within a site. These are matrices where a day's power readings go down a column and the next day's power readings go down the adjacent column to the right. Time of day progresses down a column, and days progress from left to right. The visualization consists of a matrix' heat plot where brighter colors are higher

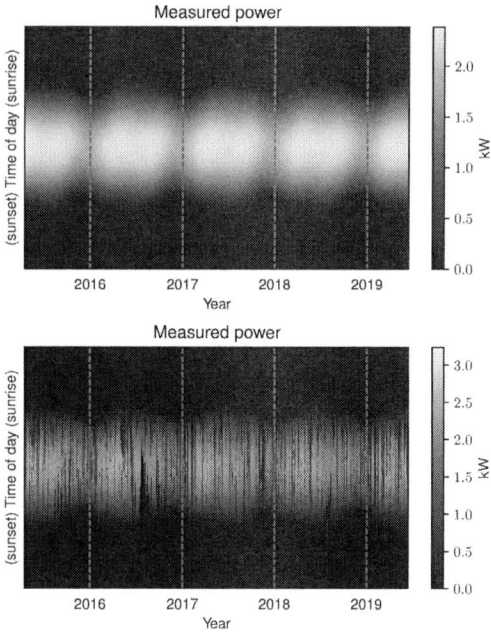

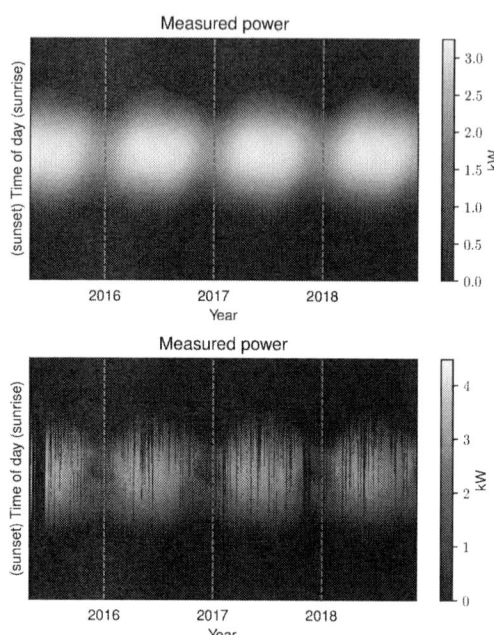

Fig. 2. Low shade solar power matrices from model (above) and site (below)

Fig. 3. Medium shade solar power matrices from model (above) and site (below)

power readings and nighttime values, even if absent in the raw data, are all. With d days of T-minute readings, Solar Data Tools [10] produces a $k \times d$ matrix, where $k = (24)(60)\frac{1}{T}$ is the number of daily measurements.

From these heat plots, one can get a rough idea of the shade present; we thereby categorize our 25 sites into Low, Medium, and High shade. One site from each category will be discussed in depth. Sample solar power matrices for our sites are in figures 2, 3, and 4.

C. Case studies

The low shade site has clear malfunctions starting mid-2016 and around November 2017, but is otherwise relatively a clean signal and its shape match that of its model. Most importantly, there are barely any visible shading losses.

A similar case is present in our medium site (figure 3): the data's shape matches that of the model closely. This case, unlike the low shade site, has clear shading losses every winter around noon. This site also has a visibly malfunctioning group of days next to its shading losses in late 2017, which allows a visual comparison of the two phenomena. Malfunctions are often sharp and aren't usually cyclical, unlike shading losses, which repeat seasonally across the years.

The visually striking losses in figure 4 show how high shading can get in a high shade site. Between the data malfunctions and the extreme shade, it seems that the site never sees a single shade-less day. The comparison to its relevant model's heat plot communicates that vast amounts of energy were lost. We aim to quantify how much.

The method for generating the average daily loss in §II-B2 also builds 365-long vectors for the average daily measured

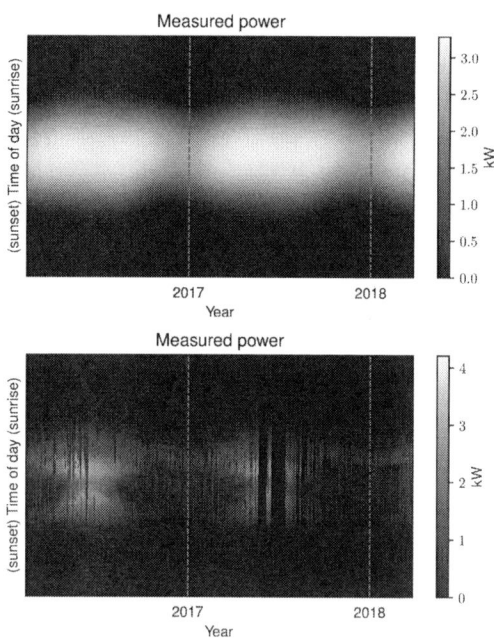

Fig. 4. High shade solar power matrices from model (above) and site (below)

978-1-7281-6118-1/22 $31.00 © 2022 IEEE

energy and the average daily modelled energy. Figure 5 plots these three signals together to underscore how the model matches the data and allow for an analysis of shade as it varies through the seasons.

Each plot has the clear-sky average energy for each day of the year in the 'measured' curve, the corresponding modelled value in the 'modelled' curve, and the difference between the two in the 'difference' curve.

The low shade site is essentially shade-less most of the year, and the average daily measure energy often surpasses the model. Whatever shading is present in this case is prevalent in the fall.

The medium shade site has much more noticeable shading, especially in the winter. The connecting of the 'measured' and 'difference' curves around calendar day 320 means that half of the energy available according to the model was lost due to shade. Though the data nearly fulfills its potential for most of the year, these winter losses are far from negligible. Shading losses, in most panels studied, also cluster around the winter. The sun is lower on the panel's horizon, and its rays are therefore more likely to hit an object before reaching PV cells.

The heavy shade site follows this fall and winter shade pattern. For nearly half of the year, the shading losses encompass most of the energy available. Briefly, in the summer, the energy absorbed by the panels surpasses that which is lost. There is likely a large object obscuring a lot of the panels at the site.

Now with modelled PV power, we can extract days from the model and clear sky days in the data for a higher resolution comparison. In figure 6, the model overestimates the power of a late October day with no losses at our low shade site.

At the medium shade site, figure 7 reveals slight shading on Christmas morning, 2015. Shading losses are the heaviest around noon and taper out in the afternoon. This agrees with the yearly patterns previously seen in the heat plot for the same site in figure 3. Winter is a particularly shady time for this site.

The heavy shade site's measured and modelled days in figure 8 are in February, which we know to be heavily shaded at this site from figure 4. Indeed, there is very heavy shade throughout the whole day. In fact, the shade at this site is so high that one can't verify that the general envelope of the model and the data agree. This envelope agreement in figure 7 and figure 6 validates the respective models. In this case, however, the validity of the model can't be verified with daily plots. There is a slight underlying curve at the lower edge of the real data in this site. This implies that some sunlight does get to the panel, but studying plots at other scales and ensuring proper estimation of inputs to the SAPM are the only methods for validating the model when such shade is present.

D. Fleet results

Finally, estimating shade loss gives a scalar result (the total energy lost due to shade over an average year) by summing the difference between the data and the model on clear sky days. The quotient of this lost energy divided by the area under the

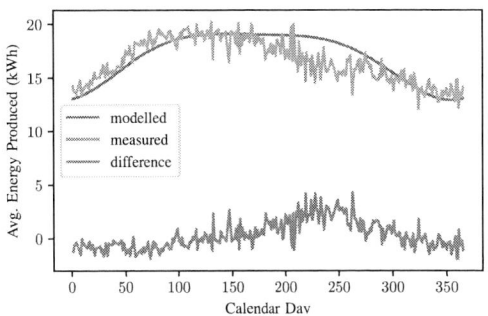
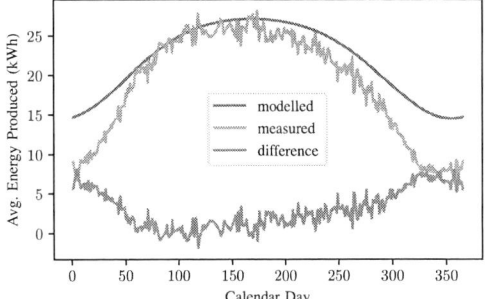
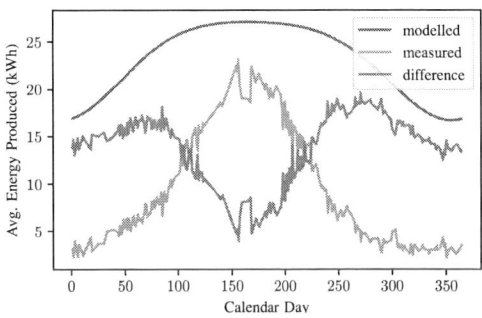

Fig. 5. Average daily energy comparisons between data and model. From top to bottom: low shade, medium shade, and high shade

modeled energy curve (one of 'modelled' in figure 5) is the estimated percent of all energy available lost to shade on an average year.

A summary of modeled energy and shade losses for each site in the fleet is available in table I. The distribution of lost energy as a percent is skewed. Most are well under 20%, with multiple below 10%, but there is also one site with a 42% loss and another with a 59% loss. Figure 9 has a histogram of the percentages of total modelled energy lost to shade. The plot is organized by the low, medium, and high shade categories that sites were split into internally. These results agree with the original ranking of sites based on their heat map's apparent shading loss: the high shade models tend to have higher percentages than all others, and most of the medium shade models have higher percentages than the low shade models.

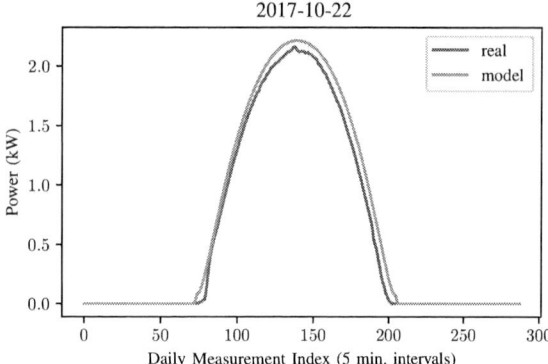

Fig. 6. Low shade model and data

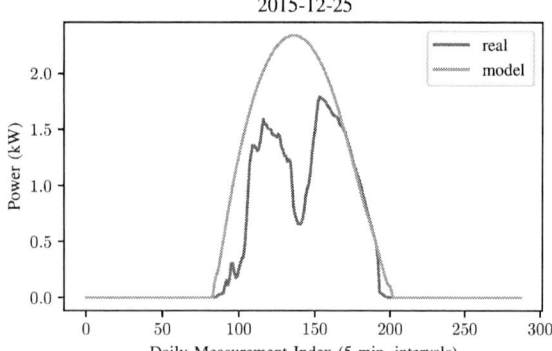

Fig. 7. Medium shade model and data

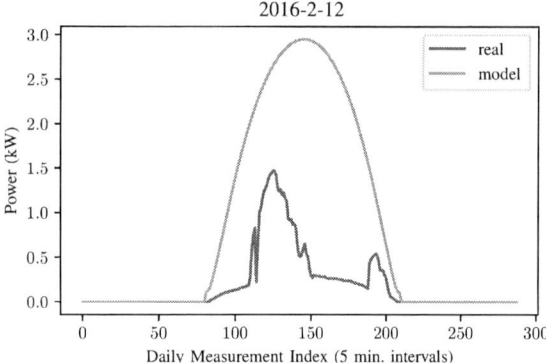

Fig. 8. High shade model and data

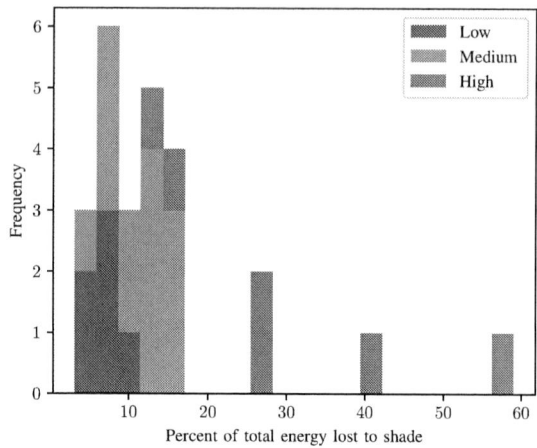

Fig. 9. Histogram of shading losses as a percent, sorted by shade category

TABLE I
ENERGY, LOSSES, AND COST FOR ALL SITES

Shading	Energy (kWh)	Loss (kWh)	Loss (%)	Costs ($US)
L	4903.0	415.0	8.5	60.2
L	14858.0	1106.0	7.4	160.3
L	9772.0	668.0	6.8	96.8
L	7592.0	268.0	3.5	38.8
L	5186.0	93.0	1.8	13.5
L	4248.0	510.0	12.0	74.0
M	15822.0	1469.0	9.3	213.0
M	7026.0	1144.0	16.3	165.9
M	8560.0	1110.0	13.0	160.9
M	6897.0	1014.0	14.7	147.0
M	4008.0	458.0	11.4	66.4
M	6351.0	591.0	9.3	85.7
M	10004.0	735.0	7.3	106.6
M	10407.0	1530.0	14.7	221.9
M	10431.0	837.0	8.0	121.4
M	13157.0	1605.0	12.2	232.8
M	10827.0	2008.0	18.6	291.2
M	6481.0	990.0	15.3	143.6
M	5366.0	517.0	9.6	75.0
H	4429.0	1220.0	27.5	176.9
H	9213.0	3872.0	42.0	561.4
H	9243.0	1369.0	14.8	198.6
H	5549.0	3272.0	59.0	474.5
H	23456.0	6430.0	27.4	932.3
H	16800.0	2594.0	15.4	376.1

Looking at energy, most sites lost a few hundred kilowatt-hours, and one site lost over 6000 kWh—more than was available in the pvlib model of several other smaller sites. Converting kilowatt-hours to dollars at the average 14.5 cent price of energy published by the U.S. Energy Information Administration [18] reveals the annual financial impact of shading losses for each site in the fleet. Many sites lost under $100, but 4 sites lost over $300 with one site losing over $900. With some of these larger costs, the importance of looking into shading losses for both planning and feedback becomes clear.

Ongoing work on these models could expand the number of sites studied, to better understand the shortcomings and strengths of the shading loss approximation. With a larger dataset, researchers could study connections between a site's configuration and its propensity for shading losses. A denser distribution in figure 1 could support such studies, for example.

Given publicly available data, it may also be possible to provide the model more detailed information on air temperature, wind speed and rain patterns without greatly increasing model variance, as mentioned in §II-A3. Finally, this work has largely ignored the effects of degradation, although many sites have over 4 years of data; future work may improve model accuracy by accounting for losses due to degradation.

IV. CONCLUSION

We constructed models of the clear sky behavior of 25 real-world, rooftop PV systems. These models were implemented in pvlib-python, a popular open source software package [9]. These models appear to be a good fit to the observed data and provide reasonable estimates of the energy lost to shade in these systems. Constructing these models was a manual, time intensive process, and not easily scalable to large fleets of PV systems. As such this work is intended to support the development of a method, based on unsupervised machine learning and signal processing [19], for estimating shade losses automatically in PV system power generation signals, as described in [7].

V. ACKNOWLEDGMENTS

This work was completed under the guidance and mentorship of Ph.D. Candidate Bennet E. Meyers-Im and Professor Stephen Boyd. The pvlib library was instrumental to this work and their documentation was very helpful in building the first models.

This material is based upon work supported by the U.S. Department of Energy's Office of Energy Efficiency and Renewable Energy (EERE) under Solar Energy Technologies Office Award Number 38529.

REFERENCES

[1] A. Woyte, J. Nijs, and R. Belmans, "Partial Shadowing of Photovoltaic Arrays with Different System Configurations: Literature Review and Field Test Results," *Solar Energy*, vol. 74, no. 3, pp. 217–233, mar 2003. [Online]. Available: https://linkinghub.elsevier.com/retrieve/pii/S0038092X03001555

[2] C. Deline, "Partially Shaded Operation of a Grid-tied PV System," in *2009 34th IEEE Photovoltaic Specialists Conference (PVSC)*. IEEE, jun 2009, pp. 001 268–001 273. [Online]. Available: http://ieeexplore.ieee.org/document/5411246/

[3] B. Meyers and M. Mikofski, "Accurate Modeling of Partially Shaded PV Arrays," in *2017 IEEE 44th Photovoltaic Specialist Conference (PVSC)*. IEEE, jun 2017, pp. 3354–3359. [Online]. Available: https://ieeexplore.ieee.org/abstract/document/8521559

[4] J. P. N. Torres, S. K. Nashih, C. A. Fernandes, and J. C. Leite, "The effect of shading on photovoltaic solar panels," *Energy Systems*, vol. 9, 2018.

[5] S. MacAlpine, C. Deline, and A. Dobos, "Measured and Estimated Performance of a Fleet of Shaded Photovoltaic Systems with String and Module-level Inverters," *Progress in Photovoltaics: Research and Applications*, vol. 25, no. 8, pp. 714–726, aug 2017. [Online]. Available: http://dx.doi.org/10.1002/pip.1160 https://onlinelibrary.wiley.com/doi/10.1002/pip.2884

[6] A. Fairbrother, H. Quest, E. Özkalay, P. Wälchli, G. Friesen, C. Ballif, and A. Virtuani, "Long-Term Performance and Shade Detection in Building Integrated Photovoltaic Systems," *Solar RRL*, vol. 6, no. 5, p. 2100583, may 2022. [Online]. Available: https://onlinelibrary.wiley.com/doi/10.1002/solr.202100583

[7] B. Meyers and D. Florez Rodriguez, "Estimation of Shade Losses in Unlabeled PV Data," *Submitted to 2022 49th IEEE Photovoltaic Specialists Conference (PVSC)*, June 2022.

[8] D. J. F. Rodriguez, "Pv system modelling public," https://github.com/elsirdavid/PV_system_modelling_public, accessed: 2022 05 30.

[9] W. F. Holmgren, C. W. Hansen, and M. A. Mikofski, "pvlib python: a python package for modeling solar energy systems." *Journal of Open Source Software*, 2018.

[10] B. Meyers, E. Apostolaki-Iosifidou, and L. Schelhas, "Solar data tools: Automatic solar data processing pipeline," in *2020 47th IEEE Photovoltaic Specialists Conference (PVSC)*, 2020, pp. 0655–0656.

[11] B. Meyers, "solar-data-tools," may 2022. [Online]. Available: http://dx.doi.org/10.5281/zenodo.6450368

[12] J. A. Kratochvil, W. E. Boyson, and D. L. King, "Photovoltaic Array Performance Model." 8 2004. [Online]. Available: https://www.osti.gov/biblio/919131

[13] P. Ineichen, "A Broadband Simplified Version of the Solis Clear Sky Model," *Solar Energy*, vol. 82, no. 8, pp. 758–762, 2008. [Online]. Available: https://www.sciencedirect.com/science/article/pii/S0038092X08000406

[14] J. Duffie and W. Beckman, *Solar Engineering of Thermal Processes*, 4th ed. John Wiley, 2013.

[15] J. D. R. Lide, *CRC handbook of chemistry and physics, 2000-2001*, 2000.

[16] C. R. Harris, K. J. Millman, S. J. van der Walt, R. Gommers, P. Virtanen, D. Cournapeau, E. Wieser, J. Taylor, S. Berg, N. J. Smith, R. Kern, M. Picus, S. Hoyer, M. H. van Kerkwijk, M. Brett, A. Haldane, J. F. del Río, M. Wiebe, P. Peterson, P. Gérard-Marchant, K. Sheppard, T. Reddy, W. Weckesser, H. Abbasi, C. Gohlke, and T. E. Oliphant, "Array programming with NumPy," *Nature*, vol. 585, no. 7825, pp. 357–362, Sep. 2020. [Online]. Available: https://doi.org/10.1038/s41586-020-2649-2

[17] B. Meyers, "Estimation of Soiling Losses in Unlabeled PV Data," *Submitted to PVSC49*, June 2022.

[18] U. E. I. Administration, "Table 5.6.A. Average Price of Electricity to Ultimate Customers by End-Use Sector," https://www.eia.gov/electricity/monthly/epm_table_grapher.php?t=epmt_5_6_a, 2022, [Online; accessed 30-May-2022].

[19] B. Meyers and S. Boyd, "Signal Decomposition Using Masked Proximal Operators," pp. 1–60, feb 2022. [Online]. Available: http://arxiv.org/abs/2202.09338

978-1-7281-6118-1/22 $31.00 © 2022 IEEE

Evaluation and Demonstration of Slot-Die Coating for Perovskite Thin Film Mini-Modules for Space Photovoltaics

Manoj Rajakaruna, Amir Hossein Ghahremani, Tao Zhu, Jaehoon Chung, Tamanna Mariam, Tyler Brau, Adam Phillips, Michael J. Heben, Zhaoning Song, Randy J. Ellingson, Yanfa Yan

Wright Center for Photovoltaics Innovation and Commercialization (PVIC), Department of Physics and Astronomy, University of Toledo, Toledo, Ohio, 43606, USA

Abstract— **Using the slot-die coating technique, we demonstrate a scalable production process for perovskite solar minimodules (PSMs). To enable uniform coating of perovskite films on FTO glass substrates in an ambient environment, we carried out an investigation on the perovskite precursor solution with a focus on chemistry and processing optimization. Perovskite solar modules (PSMs) prepared by scalable slot-die coating deliver power conversion efficiencies of up to 14% with an active area of 39.2 cm^2 under AM1.5G solar irradiation and ~10% PCE under AM0 spectrum. Current density vs. voltage characterization under AM0 and AM1.5G simulated solar spectra, acquired within an irradiance and temperature control chamber.**

Keywords— *Perovskite solar mini-modules, slot die coating, AM0, AM1.5G.*

I. INTRODUCTION

Due to their high power conversion efficiency (PCE) and ability to be produced on a large scale with printing processes, metal halide perovskite solar cells (PSCs) are identified as one of the most promising emerging photovoltaic technologies over the past few years [1]. To date, small-area (< 1 cm^2) perovskite solar cells have achieved PCEs > 25% under the AM1.5G solar spectrum and remain on an upward trend [2]. Desirable properties -- such as long carrier lifetimes (> 1 μs), long diffusion lengths (> 1 μm), large absorption coefficients, and suitable bandgap (~ 1.5 eV) – enable perovskite solar cells with high PCEs. In addition to the abovementioned qualities, PSCs also can be considered potential candidates for space applications due to compatibility with production on flexible substrates, and promising radiation resistance [3] [4].

High-efficiency PSCs have commonly been fabricated using anti-solvent assisted spin-coating methods, which are not suitable for large scale manufacturing. Thus, one needs to consider other scalable printing techniques, such as blade coating, injection printing, slot-die coating [5], and spray coating [6]. Among these, slot-die coating is a potential low-cost scalable deposition, capable of high-rate coating of various solution precursors over large areas [7][8]. In this study, we demonstrate regular structure perovskite solar modules (PSMs) based on metal oxide electron transport layers (ETLs), perovskite absorber layer and hole transport layer deposited using a commercial FOM alphaSC slot die coater. We investigate the need to adjust the precursor properties and deposition parameters to achieve efficient and stable PSCs produced by the slot-die method. Together with proper stoichiometric incorporation of the precursor materials and solvent properties, deposition parameters include the coating speed, precursor injection rate, and slot-die head to substrate gap during the coating process. Optimization of these process parameters leads to a uniform wet film that results in a smooth and homogenous morphology with desired thickness upon post-deposition processing. Furthermore, an irradiance and temperature-controlled test chamber has been developed to simulate current density vs. voltage (J-V) measurements at varying irradiance and temperatures ranging from -15 to 80 °C, to simulate the temperature-dependent operation of perovskite mini-modules under AM0 conditions present in space[9]. This irradiance chamber is capable to produce AM0, AM1.5G, and AM1.5D solar spectra within 25% of tolerance.

II. EXPERIMENTAL

To fabricate PSMs, 3-inch squared FTO glass substrates were consecutively cleaned by sonicating in detergent (Micro 90), DI water, Acetone, Isopropanol alcohol, and were then treated under UV-ozone environment for 30 min to remove contaminants and improve the surface hydrophilicity. Later, a 40 nm thick layer of SnO$_2$ was deposited on the substrates using the slot-die machine (FOM alphaSC) with a coating speed of 2.5 mm/s, a flow rate of 40 μL/min, and a slot-die to substrate coating gap of 150 μm, using a stock 15% H$_2$O colloidal SnO$_2$ solution (Alfa Aesar). The SnO$_2$ coated slides were annealed at 150 °C for 30 min. Upon annealing and cool down, the substrates were treated under UV illumination for 15 min followed by the perovskite film deposition. The perovskite film was developed from a precursor solution of Cs$_{0.17}$FA$_{0.83}$PbI$_3$ containing 10% PbCl$_2$ with the addition of L-α-Phosphatidylcholine (LP) and diphenyl sulfoxide (DPSO) additives in a mixed solvent of N,N-Dimethylformamide (DMF) and N-methyl-2-pyrrolidone (NMP). Perovskite films were prepared by both spin-coating and by slot-die coating. Spun films used a 2 M perovskite solution without additives, coated in a nitrogen-filled glove box for 5 s at 3000 rpm followed by nitrogen blowing to convert the yellow perovskite

978-1-7281-6118-1/22 $31.00 © 2022 IEEE

into the black phase in the ambient environment. The substrates were then annealed in the glovebox for 15 min at 150 °C. For slot-die coating, different concentrations of the perovskite films were deposited in the ambient with a relative humidity of 20%. It was found that 1.3M of concentration can lead to a smooth and uniform perovskite layer along with the following slot die coating parameters. Coating and pumping speeds were 0.2 m/min and 100 μl/min, respectively. In addition, 0.5 mg/mL L-α-Phosphatidylcholine (LP) surfactant and 40% Diphenyl sulfoxide (DPSO) was added to allow for tuning the solution surface tension and improving the stability of the wet film after the deposition. Upon slot-die deposition, the wet perovskite film was converted to black phase using a dry air knife, blowing air at the flow of 1.5 SCFM. The substrates were then annealed on a hotplate for 10 min at 150 °C in a 20% humid cleanroom. Upon cool down, a solution of Spiro-OMeTAD was spin-coated for 30 s at 2000 rpm in a cleanroom with 30% humidity. The Spiro-OMeTAD solution was made by dissolving 0.1 g spiro-OMeTAD in 1.1 mL chlorobenzene and mixing 39 μL 4-tert-butyl pyridine, 23 μL of 540 mg/mL Li-TFSI in anhydrous acetonitrile, and 10 μL of 300 mg/mL cobalt dopant FK209 in anhydrous acetonitrile.

III. RESULTS AND DISCUSSION

In the first stage, n-i-p PSCs with the glass/FTO/SnO$_2$/perovskite/spiro-OMeTAD/Au structure were fabricated, where the SnO$_2$ ETL was slot-die coated on FTO-glass and the perovskite film was processed in the glovebox to determine the functionality of SnO$_2$ films deposited using the FOM alphaSC slot die coater. Figure 1(A) shows the top surface SEM image of a perovskite film prepared by spin-coating of the perovskite precursor solution in 1 mL DMF and 96 μL NMP. Figure 1(B) shows the cross-sectional SEM image, indicating a perovskite layer thickness of 860 nm.

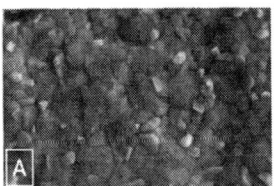 

Figure 1 Characterization of perovskite film. (A) Surface SEM image of a perovskite film. (B) Cross-sectional SEM images of a perovskite solar cell prepared by spin coating.

According to the SEM images above, compact pinhole-free uniform perovskite film was obtained even without using antisolvent during the process of depositing the perovskite layer. Furthermore, it is evident that a considerable amount of PbI$_2$ formed during the annealing of Cs$_{0.17}$FA$_{0.83}$PbI$_3$ ink (Figure 3). The PCEs measured by reverse and forward scanning were 18.16% and 14.25%, respectively, indicating 27.4% hysteresis between forward and reverse scans for PSCs prepared with the slot-die coated ETL. Figure 2 shows the JV curves of best performing small area cell.

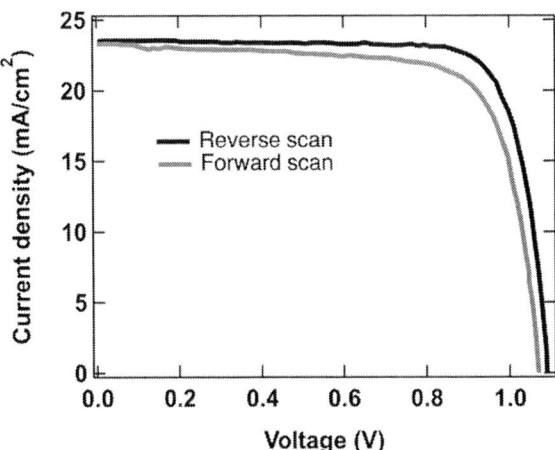

Figure 2 Current density Vs Voltage curves of best unit cell

The obtained results from spin-coating indicate the potential to utilize the perovskite precursor solution to fabricate efficient PSCs without the use of an antisolvent, which makes it suitable for upscaling to slot-die coating. However, precursor and processing optimization were carried out to achieve uniform and smooth slot-die-coated films at the end of post-process annealing.

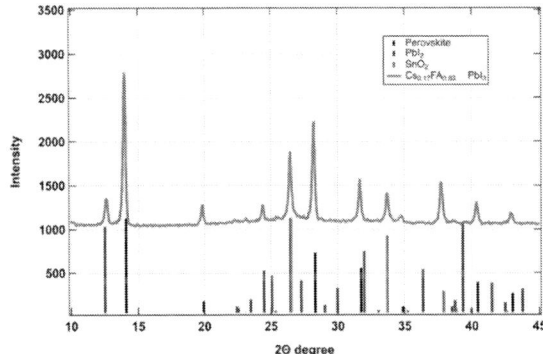

Figure 3 X-Ray diffraction pattern of a perovskite film prepared by slot-die coating.

Figure 4 (A) depicts the slot-die coated 3" × 2" perovskite film using Cs$_{0.17}$FA$_{0.83}$PbI$_3$ ink, which resulted in a uniform and smooth morphology with a 600 nm of thickness. This optical image was taken by placing the substrate on an optical stage. Figure 4 (B) shows the reflection and transmission spectra and the calculated optical absorption of the perovskite film. The absorption band edge of this perovskite film is estimated to be 1.55 eV.

978-1-7281-6118-1/22 $31.00 © 2022 IEEE

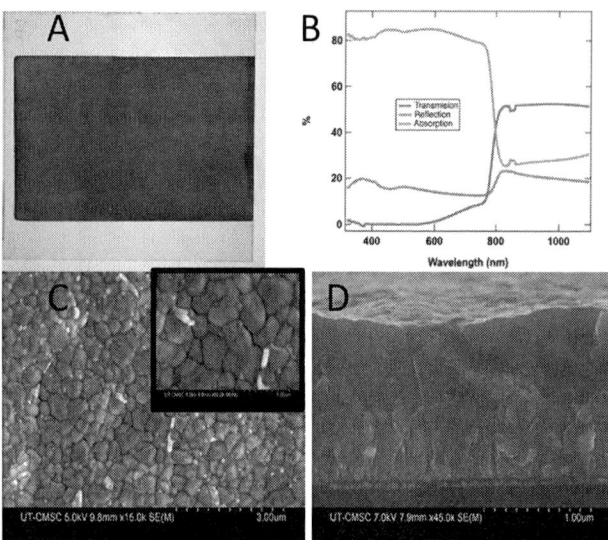

Figure 4 Slot-die coated perovskite film characterization. (A) Optical image of a perovskite film on an optical stage (B) UV-Vis transmittance, reflectance, and absorption spectra. (C) Surface SEM images. (D) Cross-sectional SEM image of a perovskite film deposited on an FTO glass substrate.

According to Figure 4 (C) and Figure 4 (D) it is evident that we obtained large compact perovskite grains and pin hole free surface using above mentioned precursor and slot die coating parameters. Furthermore, as SEM images and XRD patten illustrated, this ink still leaves a significant amount of PbI_2 residuals in the dry film upon annealing.

Finally, to deposit spiro-OMeTAD as the hole transport layer, a few parameters have been changed in slot die coater without air quenching. Since chlorobenzene has a relatively low boiling point (132 °C), the solvent starts to evaporate quickly, hence leading to an inhomogeneous and very thin spiro-OMeTAD layer. To mitigate inhomogeneity and obtain thick enough layer, coating speed, pumping rate, and coating height were changed to 0.4 m/min, 200 μL/min, and 200 μm, respectively.

Figure 5 Optical image of a perovskite mini-module fabricated by slot-die coating. The device active area is 39 cm².

Figure 5 shows an optical image of a slot-die coated perovskite mini-module with an active area of 39 cm². There are nine strip cells connected in series using laser patterned monolithic integration. For the module integration, P1 lines were laser patterned prior to cleaning the substrates to isolate front contact. P2 scribes were laser etched after depositing the SnO_2 ETL, perovskite absorber layer, and the spiro-OMeTAD HTL, to open the channels for metal connection. Finally, the P3 lines were scribed after the deposition of Au back contact layer to separate individual cells. The distance between each line was approximately 300 μm. Figure 6 shows current density vs voltage (J-V) curves under the forward and reverse scans for two mini-modules. Table 1 summarizes the PV parameters of these mini modules under AM1.5G spectrum.

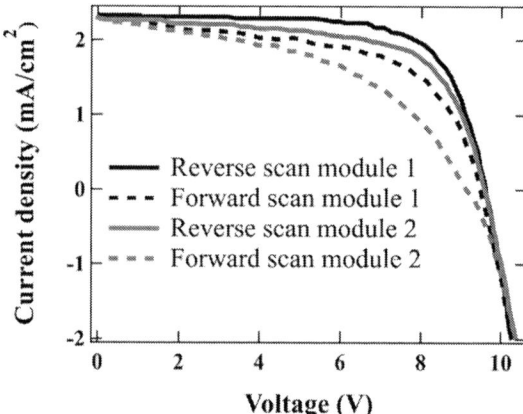

Figure 6 J-V curves of two mini-modules prepared by slot die coating, under AM1.5G spectrum.

Table 1 J-V parameters of two mini modules under AM1.5G solar spectrum.

Module	Scan	Voc V	Jsc mA/cm²	Fill Factor	η %
1	Forward	9.45	2.32	56.7	12.46%
	Reverse	9.59	2.33	69.9	15.68%
2	Forward	9.11	2.28	47.1	9.98%
	Reverse	9.57	2.27	65.3	14.23%

Finally, Figure 7 shows the calibrated AM0, AM1.5D, and AM1.5G solar spectrum in a temperature – irradiance control chamber. The light source of the JVT chamber consists of a 3x3 array of halogen bulbs, and nine led bulbs surround each halogen bulb. Each LED and halogen bulb emits a range of frequencies. The peak emission frequency of each LED and the halogen has been controlled by a LabView program with ten controllers, and each controller consists of 1024 levels of intensities. AM0, AM1.5D, and AM1.5G solar spectrums were obtained by adjusting each light bulb's intensity levels. It is possible to measure current-voltage measurements in this chamber at temperatures ranging from -15 °C to 80 °C.

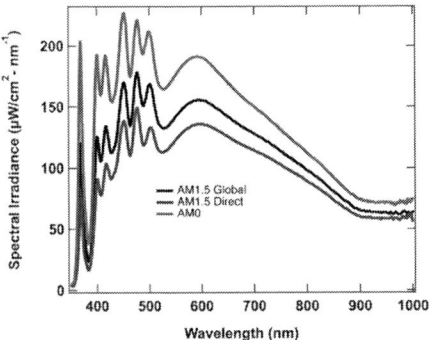

Figure 7 Simulated solar spectrums (AM0, AM1.5G and AM1.5D) in temperature – irradiance control chamber.

Furthermore, J-V curves of modules 1 and 2 were measured under AM0 solar spectrum using the irradiance-temperature dependent chamber. Both modules showed significant degradation before the measurement. The results are plotted in Figure 8 and summarized in Table 2.

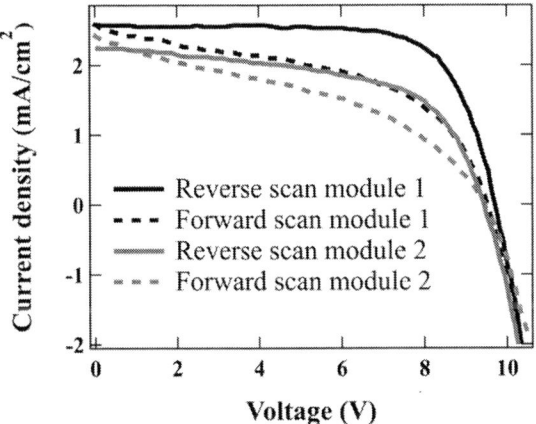

Figure 8 Current density Vs Voltage curves under AM0 spectrum for two mini modules prepared by slot die coating.

Table 2 J-V parameters of two mini modules under AM0 solar spectrum.

Module	Scan	Voc V	Jsc mA/cm²	Fill Factor	η %
1	Forward	9.54	2.59	48.8	8.93%
	Reverse	9.7	2.59	72.2	13.46%
2	Forward	9.44	2.42	40.2	6.82%
	Reverse	9.43	2.27	57.7	9.16%

IV. CONCLUSION

This study reports our progress in tuning the perovskite precursor solution to enable the scaling-up slot-die deposition of PSMs. Our preliminary results show successful preparation of the SnO_2 electron transport layer, perovskite absorber layer and hole transport layer by slot-die coating. Perovskite solar cells with the n-i-p structure fabricated by slot-die coating show PCEs of 16.9% for small area solar cells, and PSMs show PCEs of 14% under AM1.5G solar spectrum. Optimizing P2 laser scribing, and surface treatments will lead to further improvement of mini modules by slot die coating. Also, the irradiance-temperature dependent JV measurement chamber developed in house will be used to simulate space-like environments to test PSMs under different conditions.

ACKNOWLEDGMENT

This work is based on research sponsored by the U.S. Department of Energy's Office of Energy Efficiency and Renewable Energy (EERE) under the Solar Energy Technologies Office Award Number DE-EE0008790 and U.S. Air Force Research Laboratory under agreement numbers FA9453-19-1002 and FA9453-21-C0056. The U.S. Government is authorized to reproduce and distribute reprints for Governmental purposes notwithstanding any copyright notation thereon. Approved for public release; distribution is unlimited. Public Affairs release approval # AFRL-2022-2364.

REFERENCES

[1] Y. Wang et al., "Printing strategies for scaling-up perovskite solar cells," National Science Review, vol. 8, no. 8, 2021, doi: 10.1093/nsr/nwab075.

[2] G. Kim, H. Min, K. S. Lee, D. Y. Lee, S. M. Yoon, and S. I. Seok, "Impact of strain relaxation on performance of α-formamidinium lead iodide perovskite solar cells," Science, vol. 370, no. 6512, pp. 108-112, 2020.

[3] Y. Tu et al., "Perovskite Solar Cells for Space Applications: Progress and Challenges," Advanced Materials, vol. 33, no. 21, p. 2006545, 2021.

[4] Z. Song et al., "High Remaining Factors in the Photovoltaic Performance of Perovskite Solar Cells after High-Fluence Electron Beam Irradiations," The Journal of Physical Chemistry C, vol. 124, no. 2, pp. 1330-1336, 2019.

[5] I. Zimmermann et al., "Sequentially Slot‐Die‐Coated Perovskite for Efficient and Scalable Solar Cells," Advanced Materials Interfaces, vol. 8, no. 18, p. 2100743, 2021.

[6] R. Wu et al., "Progress in blade-coating method for perovskite solar cells toward commercialization," Journal of Renewable and Sustainable Energy, vol. 13, no. 1, p. 012701, 2021.

[7] Z. Yang et al., "Slot-die coating large-area formamidinium-cesium perovskite film for efficient and stable parallel solar module," Science Advances, vol. 7, no. 18, p. eabg3749, 2021, doi: doi:10.1126/sciadv.abg3749.

[8] T. Bu et al., "Lead halide–templated crystallization of methylamine-free perovskite for efficient photovoltaic modules," Science, vol. 372, no. 6548, pp. 1327-1332, 2021.

[9] N. Katakumbura et al., "Irradiance and Temperature Control Chamber for Testing Solar Cell Performance," in 2021 IEEE 48th Photovoltaic Specialists Conference (PVSC), 2021: IEEE, pp. 1813-1820.

Deleterious Effect of Light Trapping on the Temperatures of Solar Modules

Nicholas P. Irvin, D. Martínez Escobar, Aaron Wheeler, Tomas Leijtens, Hyunjong Lee, Annikki Santala, Richard R. King, Christiana B. Honsberg, Sarah R. Kurtz

Arizona State University, Tempe, AZ, United States

University of California Merced, Merced, CA, United States

Swift Solar, San Carlos, CA, United States

Increased temperatures generally reduce the efficiencies and life spans of photovoltaic modules. Experimental measurements show that thin-film GaAs modules operate more than 10°C cooler than Si modules. This study identifies the main thermal advantage of the GaAs modules as their high sub-bandgap reflection of 77%, which dwarfs the 15-26% measured on various Si architectures. This paper proves that the sub-bandgap reflection in modules with textured Si cells is fundamentally limited compared to reflection for cells without light trapping, due to the amplification of parasitic absorption that occurs with light trapping. Now, this finding is being tested on perovskite-silicon tandems. It is expected that light trapping will increase the tandems' temperatures by several degrees.

978-1-7281-6118-1/22 $31.00 © 2022 IEEE

Evaluating the Environmental Benefit of Residential Photovoltaic Modules Early Retirement in California

Mallika Kothari, Annick Anctil

Michigan State University, East Lansing, MI, United States

The state of California is the foremost leader in solar photovoltaics (PV) installations in the United States. With 1,390,240 installations and 24.76% of the state's energy coming from solar, the demand for PV modules is steadily increasing. Most PV modules have an expected lifetime of 25-30 years. However, due to repowering or early module failure, module lifetime can often be shorter than anticipated. Current studies calculate the environmental impact of PV systems based on ideal installation conditions and a full 25-year module lifetime. This study considers the impact on the life cycle of PV systems from early PV module retirement and actual system installation in California. Using the life cycle cumulative energy demand, electricity data from the Energy Information Administration (EIA), and greenhouse gases, carbon payback time (CPBT) was evaluated. Data from various PV module rooftop residential installations in 2019 were collected from the California NEM database. Information on the system design (tilt, azimuth, module model) and module specification sheets were used to calculate the cumulative electricity generated in kilowatt-hours (kWh) over the system' lifetime. The calculated average CPBT was 2.8 years, shorter than most of the system lifetimes, and the mean number of zero carbon years experienced by earlier retired systems was about 5 years. Although the rapid movement towards solar energy is promising and essential as reliance on greener energy increases, attention must be paid to the diverse lifespans of PV modules, system design, and performance to substantiate or reject the assumption that PV always have a positive impact on the environment.

From Femtoseconds to Gigaseconds: The SolDeg Project to Analyze Si Heterojunction Cell Degradation with Machine Learning

Gergely Zimanyi, Davis Unruh, Reza Vatan, Zitong Zhao, Andrew Diggs, Stephen Goodnick

University of California, Davis, Davis, CA, United States

Arizona State University, Tempe, AZ, United States

We are reporting the results of the SolDeg project for analyzing performance degradation in Si heterojunction solar cells. First, femtosecond molecular dynamics (MD) simulations were performed to create a-Si/c-Si stacks, using a Machine-Learning-based Si-Si Gaussian Approximation Potential GAP. The silicon- and hydrogen-related defects were determined next by combining MD and DFT methods. The defect generation energies were determined by the Nudged Elastic Band method. Finally, an accelerated Monte Carlo method was developed to simulate the thermally activated time dependent defect generation across the barriers, out to gigaseconds. We have shown that a stretched exponential analytical form can successfully describe the defect generation $N(t)$ over at least ten orders of magnitude in time. We also developed the Time Correspondence Curve to calibrate and validate the accelerated testing of solar cells. We found a compellingly simple scaling relationship between accelerated and normal testing times: $t(normal) \sim t(accel)^{\wedge}(T(accel)/T(normal))$. - Second, we used Machine Learning to develop our own Si-H GAP to reach unparalelled, DFT-level precision with computation times 10-100 times faster than DFT. We showed that in typical c-Si/a-Si:H HJ cells the hydrogen atoms experienced a potential gradient that sloped away from the interface, making the hydrogen atoms drift away from the interface and thus generating defect states at the interface. This degradation of the passivation is quite likely a key driver of the cell performance degradation. Finally, we discovered that the hydrogen potential gradient was caused by the crystallinity gradient of a-Si:H. Thus, the hydrogen potential gradient can be reversed to slope toward the interface by reversing the a-Si crystallinity gradient. In such reversed-gradient stacks, the hydrogen does not drift away from the interface. This is a key message of the Soldeg project: HJ cell degradation can be stopped by deposition protocols that have a crystallinity minimum at the HJ interface.

978-1-7281-6118-1/22 $31.00 © 2022 IEEE

Critical transport behavior in quantum dot solids

Michael Kovtun, Zachary Crawford, Adam Goga, Gergely T. Zimanyi

Physics Department, University of California Davis, Davis, CA, 95616, USA

Abstract— Due to recent advances, silicon solar cells are rapidly approaching the Shockley-Queisser limit of 33% efficiency. Quantum Dot (QD) solar cells have the potential to surpass this limit and enable a new generation of photovoltaic technologies beyond the capabilities of any existing solar energy modalities. The creation of the first epitaxially-fused quantum dot solids showing broad phase coherence and metallicity necessary for solar implementation has not yet been achieved, and the metal-insulator transition in these materials needs to be explored. We have created a new model of electron transport through QD solids informed by 3D-tomography. We used the transfer matrix method, finite-size scaling to create a dynamic metallic-insulator transition phase diagram. For a large portion of the parameter space, our model shows a critical exponent distinct from previous studies.

Keywords—metal-insulator transition, kinetic disorder, quantum dot, nanoparticle

I. INTRODUCTION

In recent years, quantum dot photovoltaics (QD-PV) had the fastest rising efficiency of any PV, having broken the 10% level around 2015, reaching 18% in 2021 [1,2]. They key factors controlling this efficiency are absorption and transport. The industry workhorse PbSe QDs have highly localized electron wavefunctions, and thus layers made of these PbSe QDs are in the insulating phase with carrier mobilities orders of magnitude lower than in c-Si. The most promising direction to improve transport is to drive these QD solids through the Metal-Insulator Transition (MIT). Metallicity can be reached by establishing phase coherence and mini-band formation. The field of MIT is decades-old and extensive. However, most studies focused on simplified, paradigmatic models.

Techniques for the fabrication and characterization of (epi) QD solids are rapidly advancing [3-6]. In particular, projects by Law and Moulé have generated high-quality 7-layer QD solids, and imaged them with 3D near-atomistic tomography. They determined detailed features of individual QDs, as well as couplings between the QDs [7]. Their samples are approaching metallicity, but so far did not cross the MIT. Thus, charting a roadmap for the fabrication processes and parameters to actually drive the QD solids to a phase-coherent metallic phase through the MIT remains elusive. To achieve this goal, more realistic simulations of the transport properties of QD solids are needed that capture both the on-site and the kinetic disorders.

Countless numerical studies have been conducted exploring the MIT with either exclusively on-site disorder, or kinetic (hopping) disorder [8,9]. However, real materials exhibit disorder of both types. Samples by Law, characterized by Moulé have shown epitaxial couplings between 72% of the QD-QD pairs [7], inducing a much-enhanced kinetic coupling

between them. Thus, the kinetic disorder can be best captured by a binary distribution. Increasing the uniformity and quality of the epitaxial QD couplings will be the key to drive these QD solids across the MIT, and thus creating the first QD solid showing high enough electron mobility for photovoltaic use.

II. RESULTS AND DISCUSSION

We created a realistic model of electron transport in QD solids that has all the realistic essentials to capture the metal-insulator transition from the metallic phase. Our model tight-binding Hamiltonian (1) has an on-site disorder, as well as kinetic disorder in its nearest-neighbor hopping term:

$$H = \Sigma_i \, \varepsilon_i \, c_i^{\dagger} c_i + \Sigma_{<i,j>} \, t_{ij} \, c_j^{\dagger} c_i \qquad (1)$$

Several studies have explored the second-order phase transition that occurs when increasing the disorder W of the onsite energy ε_i, while assuming a uniform hopping parameter [8-10]. However, real systems have disorder in both on-site energy and in the hopping integral t_{ij}, so this must be also accounted for to accurately describe the transition in QD solids. Our present study has investigated kinetic disorder combined with onsite disorder. We draw ε_i from a box distribution [-W/2, W/2]. t_{ij} is drawn with probability c to be t_{hi} to represent epi-coupled QD pairs, and with probability (1-c) to be t_{lo} to represent QD pairs with no epitaxial coupling. To fix the energy scale, we set t_{hi}=1. We will explore the MIT on the (W,c) plane by keeping W constant with a value such that W/t_{hi}<16.53 <W/t_{lo}, because W_c/t=16.53 is the known critical value for the MIT in the purely diagonally-disordered Anderson model with a constant hopping t. We will drive the system across the MIT by keeping W fixed and varying the parameter c [9]. We have set t_{lo}=0.3 throughout this paper. The maximum kinetic disorder is achieved when c=0.5. We also performed scans of the same (W,c) plane by holding c fixed and sweeping with W.

A. Transfer matrix method

We used the transfer matrix method (TMM) together with finite-size scaling (FSS) developed by Slevin to determine the mobility edge c_{crit} and the critical exponents of the localization length and the conductivity [8-10]. We considered the transmission of electrons with an energy E relative to the Fermi level (at E=0) through long disordered wires with a uniform square cross section: $L_z \gg L_x = L_y = L$. The transfer matrix method begins with a reformulation of the Schrodinger equation,

$$H_{n,n+1} \mathbf{A_{n+1}} = (E_- H_n) A_n - H_{n,n-1} \qquad (1)$$

978-1-7281-6118-1/22 $31.00 © 2022 IEEE

where $\mathbf{A}_n$ is a vector representing the wave function amplitudes at sites i in the n_{th} slice of the wire. H_n is the intra-slice Hamiltonian and $H_{n,n+1}$ is the inter-slice Hamiltonian, both are $L^2 \times L^2$. This was rearranged into a transfer matrix:

$$\begin{pmatrix} A_{n+1} \\ A_n \end{pmatrix} = T_n \begin{pmatrix} A_n \\ A_{n-1} \end{pmatrix} \tag{3}$$

where the transfer matrix was defined by:

$$T_n = \begin{pmatrix} H_{n,n+1}^{-1}(E\mathbb{I} - H_n) & -H_{n,n+1}^{-1} H_{n,n-1} \\ \mathbb{I} & 0 \end{pmatrix} \tag{4}$$

The transfer matrix (TM) propagated the wavefunction amplitudes from slice-to-slice. By iterating the TM equation (4) recursively, it was possible to calculate the amplitudes out to a large length L_z. As the disorder was increased, the system crossed from its initially metallic phase across the MIT into the (Anderson-) localized phase. Here, the amplitude of the wavefunction decayed exponentially with a quasi-one-dimensional localization length, which the method captured in the $L_z \gg \xi$ regime. The TM method estimated ξ by computing the Lyapunov exponents of the product matrix M, generated by multiplying the transfer matrix TM L_z times [10]. The Lyapunov exponents are defined as:

$$\gamma_i = \lim_{L_z \to \infty} \nu_i / 2L_z \tag{5}$$

where ν_i were the eigenvalues of $\ln(MM)$. In practice, after a handful of transfer matrix multiplications the eigenvalues of interest exponentially shrunk to numerical precision limits, so a QR decomposition was performed at regular intervals, obtaining the eigenvalues in the process. The Lyapunov exponents could be estimated by truncating the above equation at a large but finite length L_z. Finally, the quasi-one-dimensional localization length could be related to the smallest positive Lyapunov exponent γ_+ by $\xi = 1/\gamma_+$.

B. Finite-size scaling

Once data for ξ had been obtained for a range of disorder values around the critical disorder for as wide a range of system sizes as practical, the finite size scaling method was used to extract information about the critical phenomena. This allowed us to estimate universal properties such as the critical exponent. The first step was to assume that the disorder and system size dependences of the dimensionless quantity $\Gamma = L/\xi$ were described by a scaling function of the form $\Gamma = F(\chi L^{1/\nu}, \psi L^y)$, where χ/ψ were the relevant/irrelevant scaling variables. F was Taylor-expanded in both relevant and irrelevant scaling variables up to orders n_R, and n_I respectively. The scaling variables $\chi(c)$ and $\psi(c)$ were Taylor-expanded as well to account for nonlinearities, up to orders m_R, m_I respectively. The contribution of the irrelevant scaling variable vanished in the large L limit, such that the exponent y turned negative, and ν remained positive. The quality of the fit was controlled by minimizing the relevant χ^2 statistic:

$$\chi^2 = \Sigma_i^N (F_i - \Gamma_i)^2 / \Gamma_i \tag{6}$$

where F_i was the value of the finite-size scaling function evaluated at the parameters used at the i_{th} run of the TMM; Γ_i was the computed value of Γ in the TMM, and N was the total number of datapoints. We used the subsampling (a.k.a. bootstrap) method to determine 95% confidence intervals.

TMM calculations were performed with $L_z > 6*10^6$, extending until the desired accuracy of 1% was achieved. Periodic boundary conditions were used. The FSS was performed using $n_R=3$, $n_I=1$, $m_R=2$, $m_I=1$ and the minimization of the χ^2 was done by the CMA-ES algorithm implemented in pagmo [11]. Fig. 1 shows the critical (MIT) concentration values c_c for the t_{hi} couplings as a function of the diagonal disorder W, for a set of energies E. For $c > c_c$, the system is metallic, for $c < c_c$, the system is in its insulating phase.

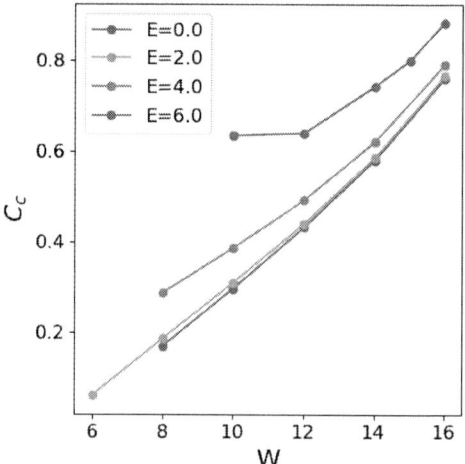

Fig. 1. Critical c-values vs. W for various energies E. Localized states exist for $c < c_c$, extended/metallic/phase coherent states exist $c > c_c$. Increasing E shrinks the region of extended states.

Next, we determined the universal critical exponents of the just-identified MIT. For a review of known exponents and their universality classes, see [12]. Figs. 2-3 show that for most of the $c_c(W,E)$ space of our model, the correlation length exponent $\nu = \sim 1.15$. As such, ν appears not to converge to any of the values of the known universality classes identified to present [13]. The reliability of our work was well-validated by the fact that we did reproduce the known exponent of $\nu=1.57$ for the only- diagonal disorder MIT. Based on this, our results are quite suggestive that the inclusion of the kinetic disorder into the diagonal disorder Hamiltonian changed the universality class of the MIT.

Supported by NSF-DMR-2005210

978-1-7281-6118-1/22 $31.00 © 2022 IEEE

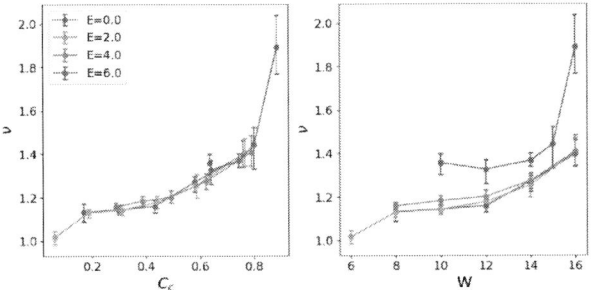

Fig. 2. Exploration of critical exponent ν across C_c (left), W (right), and E parameter space. Error bars are the 95% confidence interval determined by the bootstrap method. Large values of W or E cause the estimates of ν and the width of the error bars to increase.

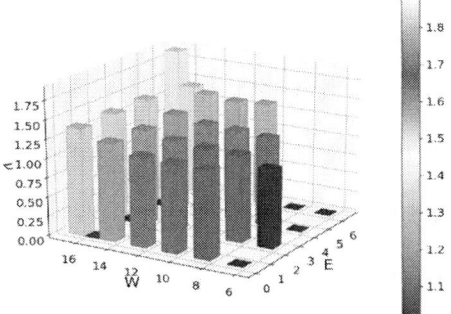

Fig. 3. Critical exponent vs. W and E. is stably ~1.15-1.25 for much of the W-E plane but diverges at extreme values. Critical exponents of 0 mark locations in E,W space not yet considered.

We note that for extreme values of W or E that brought the system close to the band edge at E=6, the exponent ν deviated from the (approximately) parameter-independent universal value of 1.2. Our confidence level in this result is lower, as in this (W, E) region the precision of the TMM decreased, and the method exhibited stronger finite-size effects. When we increased the finite sizes to $L \geq 8$ and $L_z > 10^7$, the finite size effects decreased, but did not disappear. Thus, our results do not provide robust evidence for a possible distinct scaling behavior at the band's edge. It is possible that at these extreme values of W and E the single-parameter scaling hypothesis, critical to FSS, fails. Or, that we are picking up an extended crossover regime. We end by reminding that in 3D the localization length exponent ν equals the conductivity exponent s, and thus should be an experimentally directly observable quantity.

III. CONCLUSION

This work investigated transport in realistic experimental epitaxially-fused quantum dot solids, created and imaged by our collaborators. Our goal was to identify pathways to cross the MIT, to achieve delocalization of electron wavefunctions, and to see mini-band formation. To this end, we have mapped out the dynamic phase diagram of these QD solids. Our work is strongly suggestive that the introduction of kinetic disorder into the Anderson diagonal disorder Hamiltonian changes the critical behavior around the MIT, at least for a binary distribution. For a large portion of the (W, E, c) parameter space, the critical exponent ν ≈ 1.15 was found to markedly differ from the purely diagonal-disordered case of ν ≈ 1.57, although it does fall within error of the Chiral Orthogonal Class. Additional runs averaging 10 realizations of disorder yielded a critical exponent of ν ≈ 1.10±0.05 However, the limitations of the numerical procedures used must be kept in mind. The largest cross-section width that we could practically compute was L=20, which put a limit on the longest fluctuations the TMM could capture. This effect could have influenced the estimation of ν in the FSS procedure. Future TMM computations with a wider L range may yield a more accurate estimation of ν for our model.

ACKNOWLEDGMENT

We thank our invaluable collaborators Matt Law and Adam Moulé for their insight, analysis, and characterization efforts.

REFERENCES

[1] M. Hao, et al., "Ligand-assisted cation-exchange engineering for high-efficiency colloidal Cs1−xFAxPbI3 quantum dot solar cells with reduced phase segregation," Nat. Energy, vol. 5, pp. 79–88, January 2020.

[2] E. M. Sanehira et al., "Enhanced mobility CsPbI3 quantum dot arrays for record-efficiency, high-voltage photovoltaic cells," Science Advances, vol. 3, no. 10, p. eaao4204, 2017, doi: 10.1126/sciadv.aao4204.

[3] K. Whitham, J. Yang, B. Savitzky, L. Kourkoutis, F. Wise, T. Hanrath, "Charge transport and localization in atomically coherent quantum dot solids," Nat. Mater., vol. 15, no. 5, pp. 557-563, February 2016

[4] W. Evers, et al., "High charge mobility in two-dimensional percolative networks of PbSe quantum dots connected by atomic bonds," Nat. Commun., vol. 6, pp. 1-8, September 2015.

[5] B. T. Diroll, N. J. Greybush, C. R. Kagan, C. B. Murray, "Smectic Nanorod Superlattices Assembled on Liquid Subphases: Structure, Orientation, Defects, and Optical Polarization", Chem. Mater., vol 27, no 8, bll 2998–3008, Apr 2015.

[6] Altamura D, Sibillano T, Siliqi D, Caro LD, Giannini C, "Assembled Nanostructured Architectures Studied by Grazing Incidence X-Ray Scattering", Nanomaterials and Nanotechnology, December 2012.

[7] X. Chu et al., "Structural characterization of a polycrystalline epitaxially-fused colloidal quantum dot superlattice by electron tomography", J. Mater. Chem. A, vol 8, bll 18254–18265, 2020.

[8] A. MacKinnon, B. Kramer, "The scaling theory of electrons in disordered solids: Additional numerical results", Zeitschrift für Physik B Condensed Matter, vol 53, no 1, bll 1–13, Mrt 1983.

[9] K. Slevin, T. Ohtsuki, "Critical exponent for the Anderson transition in the three-dimensional orthogonal universality class", New Journal of Physics, vol 16, no 1, bl 015012, Jan 2014.

[10] K. Slevin, Y. Asada, L. I. Deych, "Fluctuations of the Lyapunov exponent in two-dimensional disordered systems", Phys. Rev. B, vol 70, bl 054201, Aug 2004.

[11] F. Biscani, D. Izzo, "A parallel global multiobjective framework for optimization: pagmo", Journal of Open Source Software, vol 5, bl 2338, 09 2020.

[12] F. Evers, A. D. Mirlin, "Anderson transitions", Rev. Mod. Phys., vol 80, bll 1355–1417, Okt 2008.

[13] Altland, A., Zirnbauer, M. R., "Nonstandard symmetry classes in mesoscopic normal-superconducting hybrid structures", Physical Review B, vol. 55, no. 2, pp. 1142–1161, 1997. doi:10.1103/PhysRevB.55.1142.

Extended Accelerated Stress Testing (EAST) of Glass/Glass, Glass/Backsheet and Glass/Transparent Backsheet PV Modules: Influence of EVA and POE Encapsulants

Akash Kumar[1], Ashwini Pavgi[1], Peter Hacke[2], Kaushik Roy Choudhury[3] and GovindaSamy TamizhMani[1]

[1]Photovoltaic Reliablity Laboratory, Arizona State University, Mesa, AZ, United States [2]National Renewable Energy Laboratory,Golden, CO, United States [3]DuPont, Central Research and Development, Wilmington, USA

Abstract— This paper presents the indoor extended accelerated stress testing (EAST) results of glass/glass (GG), glass/backsheet (GB) and glass/transparent backsheet (GT) modules having identical cells and two different encapsulant types, ethyl-vinyl-acetate (EVA) and polyolefin-elastomer (POE). Six 4-cell modules having the above-mentioned construction combinations were subjected to extended ultraviolet (UV; 600 kWh/m^2), damp-heat (DH; 2000 hours) and thermal-cycling (TC; 600 cycles) tests. The post-stress UV fluorescent imaging, electroluminescent imaging, reflectance spectrophotometry and colorimetry results indicated that the grid finger degradation and encapsulant browning are slightly higher in the GG modules compared to the GB modules. The post-stress IV test results indicated, in general, that the GG/EVA modules tend to perform inferior to the GG/POE modules with the EAST evaluation.

Keywords- indoor accelerated stress testing, glass/glass, glass/backsheet, EVA, POE, reliability

I. INTRODUCTION

Glass/Glass (GG) photovoltaic modules are claimed to offer several advantages compared to the conventional glass/backsheet (GG) modules and the percentage of market share for GG modules has also steadily increased over the recent years. By having glass as both superstrate and substrate, the modules are reported to offer better mechanical support and humidity/temperature tolerance [1]. Further, more than 90 % bifacial modules (GG and glass/transparent backsheet, GT) employ ethylene vinyl acetate (EVA) and polyolefin-elastomer (POE) encapsulants. Considering an estimated market share of over 35% by 2030 [2], it becomes critical for the industry to understand the reliability of GG modules compared to the well-established glass/backsheet (GB) modules with common encapsulants such as EVA and POE .

We recently reported two reliability studies related to field-aged (10-35 years) GG modules with different cell technologies/manufacturers and two different encapsulant types of EVA and ionomer [3], [4]. The primary degradation modes observed in the field-aged GG/EVA modules were encapsulant browning and delamination. The GG/ionomer modules did not suffer from the browning issue, but they experienced significant corrosion and delamination issues.

Other studies indicated some limitations of GG modules in terms of weight and operating temperature.

This paper focuses on an indoor extended accelerated stress testing (EAST) per IEC 63209 [5] of GG, GB and GT modules having identical cells and two dominant encapsulant types, EVA and POE. In this study, six 4-cell modules having the above-mentioned construction combinations were subjected to extended UV, damp-heat and thermal-cycling tests to determine the degradation modes and degradation magnitude for each of the module construction combination.

II. METHODOLOGY

A. Module Construction

In this work, 6 monofacial 4-cell crystalline silicon (PERC) modules with three different substrate types and two different encapsulant types (two types of encapsulant per substrate type) were used for the extended accelerated stress (EAST) testing. In addition, we also constructed control modules for each of the three combinations. The three different substrate types are: glass-glass (GG – 3.2 mm substrate), glass- polymer backsheet (GB – TPT backsheet) and glass-transparent backsheet (GT – Tedlar PVF clear backsheet). The two encapsulant types are: EVA and POE. To be consistent with the current industry practice, we used UV pass (UVp) encapsulant above the cell and UV cut (UVc) encapsulant below the cell in all module constructions. All the six modules were subjected to the following three extended stress tests in parallel with 2 modules (one with EVA and the other with POE) per test: UV (600 kWh/m^2 at 60 °C; beyond Qualification Plus testing per NREL protocol), thermal cycling (TC) with 600 cycles between 40 °C and +85 °C per IEC 63209 standard, and damp heat for 2000 hours at 85° C/ 85 % RH per IEC 63209 standard. All the test and control modules were characterized before and after the stress tests, and they include: EL, IR, UVF, spectral reflectance and colorimetry.

III. RESULTS AND DISCUSSIONS

Due to space limitation of this extended abstract, only EL, UVF, reflectance and colorimetry results are analyzed and presented. Detailed I-V performance results will be presented in the full paper.

1) UV Extended (UV600)

Fig.1 shows the UVF images indicating four GB and GT modules showed the presence of fluorescence rings close to the cell edge but without covering the cell edge and of two GG modules showed the presence of fluorescence at the cell edges without any ring formation. The fluorescence pattern in all the modules is thought to originate from the bottom UV cutting encapsulant, where the additives could have migrated from backside of the cell to the front edge of the cell during the lamination process at about 150 °C. During the accelerated

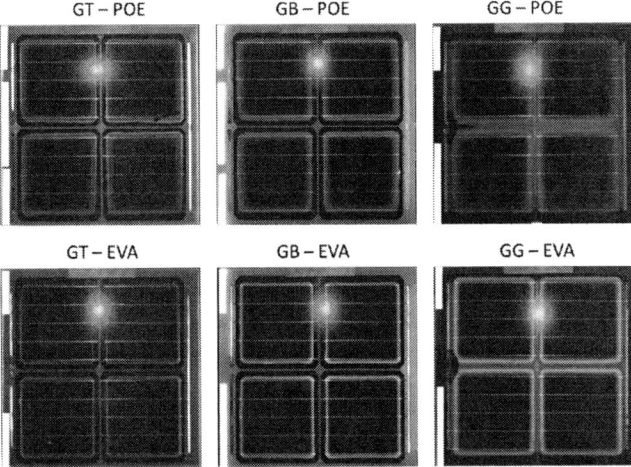

Fig 1. UVF images of 6 test modules after UV600 kWh/m²

testing, the migrated additives from back to front of the cells are exposed to UV irradiation causing the encapsulant discoloration. The presence of oxygen bleaching of chromophores leads to the lack of fluorescence right at the cell edge in GB and GT modules. However, in the GG modules, the glass substrate doesn't allow oxygen transportation for the bleaching reaction to occur, leading to fluorescence all the way to the cell edge absent of oxygen bleaching. The fluorescence pattern in POE and EVA encapsulant were marginally different in all the modules. The POE modules exhibited a wider fluorescence ring width than the EVA modules. However, the fluorescence intensity was higher in EVA modules.

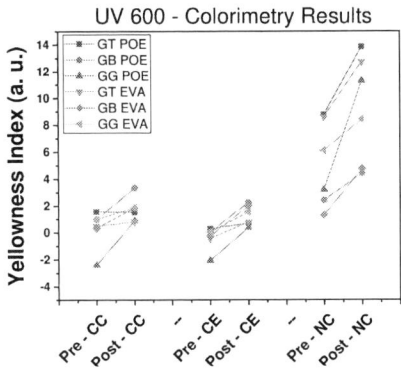

Fig 2. Pre- and post-UV600 colorimetry results of 6 modules performed at three different locations i) CC – Cell center ii) CE – Cell edge and iii) NC – non-cell location

Evidence of yellowing/browning after UV600 was also observed from colorimetry results as shown in Fig. 2. The test

was performed at three different locations in the module i) CC- cell center ii) CE- cell edge iii) NC- non-cell. As shown in Fig. 2, the non-cell area displayed the highest increase in YI and the cell-center and cell-edge areas showed only small increase in YI. Additional analysis on the colorimetry results will be presented in the full paper.

2) DH Extended (DH2000)

During intermittent characterization, one module, GT/POE, shattered due to handling issue and hence the results of GT modules are not presented for the DH2000 test. The reflectance spectrophotometry was performed at same location as colorimetry. The results, as shown in Fig. 3 (inset pictures of UVF images after DH2000), indicate little or no reflectance change on the cell-center areas (bandgap absorption only) and a significant decrease in reflectance at non-cell area and cell-edge

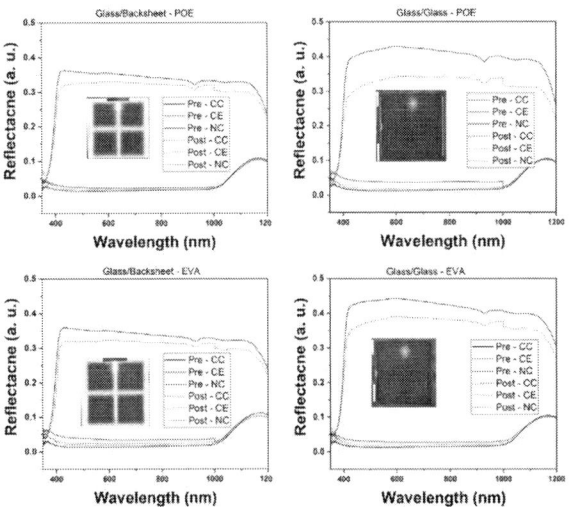

Fig 3. Reflectance results of GB and GG modules from three different locations i) CC- cell center ii) CE- cell edge iii) NC- non-cell for pre and post DH 2000 h (85 °C, 85 % RH). Inset showing the UVF images after DH2000.

area in all the modules. The decrease in reflectance could be due to ingress of water vapor into the polymeric encapsulant. Moisture ingress through backsheet and/or the edges of the

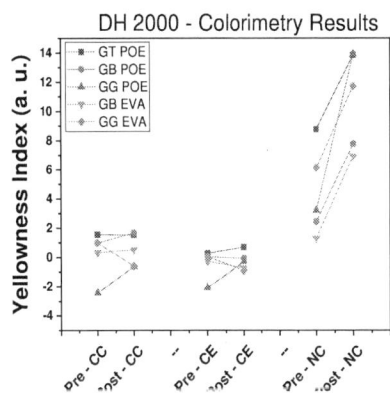

Fig 4. Pre- and post-DH2000 colorimetry results of 6 modules performed at three different locations i) CC – Cell center ii) CE – Cell edge and iii) NC – non-cell location

laminate could be associated with reflection reduction during the DH2000 test

As shown in Fig. 4, the non-cell area displayed the highest increase in YI and the cell-center and cell-edge areas showed only small increase in YI. Additional analysis on the colorimetry results will be presented in the full paper.

3) TC Extended (TC600)

The EL images of TC600 test modules are presented in Fig. 5. Severe grid finger degradation was observed in the GG/EVA module. The GT/POE module also displayed minor grid finger detachment or breakage in cells below the junction box. The failures in metal fingers might lead to increase in the series resistance (R_S), and hotspots if the degradation becomes severe [6].

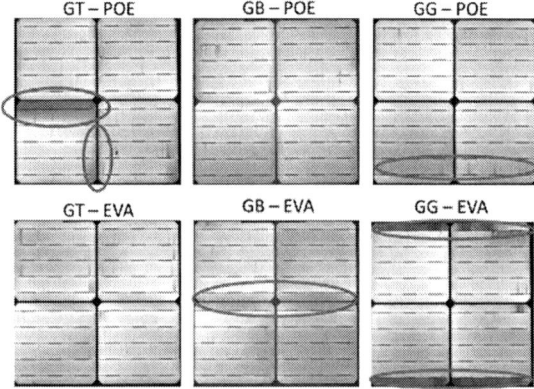

Fig 5. EL images taken at 100% Isc with 60s exposure of 6 modules after TC600

IV. CONCLUSION

An indoor extended accelerated stress testing (EAST) of UV600, DH2000 and TC600 was performed on 4-cell modules with EVA and POE encapsulants. Severity ranking color codes (red color>amber color>green color) for each degradation mode in GG and GB/GT is summarized in Fig. 6. The key degradation modes observed in GG are listed below:

- Encapsulant yellowness index increase in all constructions in UV600 and DH2000: Ring browning in GB and GT is attributed to oxygen bleaching. Edge browning in GG modules is attributed to absence of oxygen bleaching reaction. Even though UV transmitting encapsulant was on the top layer of cell, the additives from the bottom UV cutting encapsulant are suspected to have migrated during the lamination process causing ring/edge browning.

- Grid finger degradation in GG/EVA construction in TC600: Could possibly be attributed to the glass/glass rigidity.

- POE encapsulant appears to be better than EVA as the module displayed lesser fluorescence, sustained EL emission and lower changes in reflectance and colorimetry.

ACKNOWLEDGMENT

This material is based upon work supported by the Department of Energy, Office of Energy Efficiency and Renewable Energy (EERE), under Award Number DE-EE-0008565. We would like to sincerely thank our research colleagues at ASU-PRL who partly helped performing characterizations presented in this paper.

REFERENCES

[1] A. Sinha, D. Sulas-Kern, M. Owen-Bellini, L. Spinella, S. Ulicna, S. Pelaez, S. Johnston and L. Schelhas, "Glass/glass photovoltaic module reliability and degradation: a review," *J. Phys. D. Appl. Phys.*, vol. 54, no. 41, 2021.

[2] P. Klemchuk, M. Ezrin, G. Lavigne, W. Halley, J. Galid, and S. Agro, "Investigation of the degradation and stabilization of EVA-based encapsulant in field-aged solar energy modules," *Polym. Degrad. Stablity*, vol. 55, pp. 347–365, 1997.

[3] A. P. Patel, A. Sinha, and G. Tamizhmani, "Field-Aged Glass / Backsheet and Glass / Glass PV Modules : Encapsulant Degradation Comparison," *IEEE J. Photovoltaics*, vol. PP, pp. 1–9, 2019, doi: 10.1109/JPHOTOV.2019.2958516.

[4] P. M. Thorat, S. P. Waghmare, A. Sinha, A. Kumar, and G. TamizhMani, "Reliability Analysis of Field-aged Glass / Glass PV Modules : Influence of Different Encapsulant Types," 2020.

[5] IECTS 63209-1:2021, " Photovoltaic modules - Extended-stress testing - Part 1: Modules", 2021

[6] I. M. Slauch *et al.*, "Manufacturing Induced Bending Stresses: Glass-Glass vs. Glass-Backsheet," *Conf. Rec. IEEE Photovolt. Spec. Conf.*, pp. 1943–1948, 2021, doi: 10.1109/PVSC43889.2021.9518938.

Construction	Encapsulant Type	Degradation modes after UV 600 kWh/m^2, TC 600 cycles and DH 1000 hours			
		Encapsulant Browning	Encapsulant Delamination	Glass Breakage	Grid Finger Degradation
Glass/Glass	EVA	Edge browning	No significant sign	No breakage	Severe after TC600
	POE	Edge browning	No significant sign	No breakage	Minor after TC600
Glass/Backsheet*	EVA	Ring browing	No significant sign	No breakage	Minor after TC600
	POE	Ring browing	No significant sign	No breakage**	Minor after TC600
Severity	High	Medium	Low		

* Both white and transparent backsheets

**One of the two modules broke due to handling issue

Fig 6. Severity ranking color codes (red color>amber color>green color) for each degradation mode observed in GG and GB/GT

Development of a co-anneal process for double-side TOPCon precursor fabricated by ex-situ POCl3 and APCVD boron diffusion

Wook-Jin Choi, Young-Woo Ok, Keeya Madani, Shubham Duttagupta, Ajeet Rohatgi

Georgia Institute of Technology, Atlanta, GA, United States

Solar Energy Research Institute of Singapore, 7 Engineering Drive 1, Singapore

The aim of this study was to develop a simple and industrially attractive co-anneal process to fabricate a high-quality DS-TOPCon precursor with textured n-TOPCon on front and planar p- TOPCon on rear by ex-situ POCl3 and APCVD boron diffusion. This requires only one high temperature anneal with no additional masking steps. Excellent iVOC of 733mV and iFF of ~86% were achieved after SiNX passivation on both sides, prior to contact firing. Our device modeling projects that this precursor in combination with a manufacturing-friendly poly-Si patterning technique on front can enable > 25% DS-TOPCon cells at low-cost.

Complex Refractive Index and Complex Dielectric Function Modeling of Film Stack in Perovskite Solar Cells using Spectroscopic Ellipsometry

Maria Fernanda Villa Bracamonte, Jose Raul Montes Bojorquez, Arturo Ayon

The University of Texas at San Antonio, San Antonio, TX, United States

We report a comprehensive single layer modeling approach to investigate the complex refractive index of ITO, PEDOT:PSS, MAPbI3 perovskite film stack deposited on a glass substrate. The optical constants such as refractive index and extinction coefficient as well as the complex dielectric function are studied by spectroscopy ellipsometry, We propose that spectroscopic ellipsometry characterization can be used at the different stages of the fabrication process of each layer to study the mechanisms that impact the final performance of a photovoltaic device.

Spatially-Resolved X-Ray Excited Optical Luminescence of Metal Halide Perovskites

Connor Dolan, Deniz N. Cakan, Rishi E. Kumar, Moses Kodur, Yanqi Luo, Barry Lai, David P. Fenning

University of California, San Diego, La Jolla, CA, United States

Argonne National Laboratory, Lemont, IL, United States

X-Ray Excited Optical Luminescence (XEOL) is a photon-in, photon-out technique that enables probing of electronic structure and optoelectronic quality of materials using synchrotron-generated X-rays. Coupled with nanoprobe X-ray microscopy such as X-Ray Fluorescence (XRF), XEOL can provide valuable information in helping draw correlations between local compositional and electronic structure without the need to metallize or complete a device. Using an off-axis parabolic mirror to improve luminescence collection, we demonstrate spatially resolved XEOL using the X-Ray Nanoprobe at Advanced Photon Source Sector 2-ID-D on luminescent halide perovskites. We observe a decay in luminescence intensity after several seconds of continuous X-ray irradiation of hybrid halide perovskites, indicating a degradation of the perovskite over long times relative to typical scanning conditions due to X-ray beam damage, consistent with previous findings in synchrotron nanoprobe diffraction experiments. To optimize for spatially resolved measurements, we evaluate the XEOL signal to noise ratio as a function of of different single point dwell time and point-to-point spacing to achieve a balance of strong luminescence signal and fine spatial resolution. Because of the minimal experimental hardware required, XEOL holds promise to benefit correlative microscopy experiments of halide perovskites and other luminescent materials, offering in situ measurement of the key luminescence optoelectronic figure of merit.

978-1-7281-6118-1/22 $31.00 © 2022 IEEE

Exploring the Composition Space of Wide Band-gap Absorbers for Silicon-Perovskite Tandems

Moses Kodur, Rishi E. Kumar, Deniz N. Cakan, Connor Dolan, Yanqi Luo, Barry Lai, David P. Fenning

University of California, San Diego, La Jolla, CA, United States

Argonne National Laboratory, Lemont, IL, United States

Herein high throughput synthesis and characterization is used to analyze 60 unique perovskite compositions that could be incorporated in silicon-perovskite tandem devices. By analogy to advances made for single junction application, we focus our efforts on FAPbI3 based systems with MA/Cs substitution on the A-site and Br/Cl incorporation on the X-site. The impact of these changes on the formation of the perovskite, halide incorporation, optoelectronic properties, photostability, and photovoltaic properties is investigated. These efforts unveil absorbers with promise for tandem partnering with silicon and provide a template for tailoring of individual perovskite compositions, with co-optimization of several relevant parameters, for high-performance perovskite-based tandem devices.

A machine vision tool for facilitating the optimization of large-area perovskite photovoltaics

Mathilde Fievez, Nina Taherimakhsousi, Benjamin P. MacLeod, Edward P. Booker, Muriel Matheron, Matthieu Manceau, Stéphane Cros, Solenn Berson, Curtis P. Berlinguette

CEA, Le Bourget du lac, France

UBC, Vancouver, DC, Canada

A bottleneck to deposit homogeneous large-area perovskite films is the inability to quickly quantify the homogeneity of these films. Standard stylus profilometry measurement is destructive, and the acquisition time scales with device area and thus goes up dramatically when working on large samples. Once perovskite films are integrated into devices, techniques such as electroluminescence and light-beam-induced current can provide spatially resolved information. However, device preparation is time-consuming, and the performance of a full device may be limited by other layers inhomogeneities. Therefore, researchers often evaluate the perovskite film homogeneity prior to device fabrication by either cutting large-area substrates into smaller pieces for individual characterization, or by relying on visual inspection alone. Here, we combine fast optical imaging (~ 10 s / sample) with machine vision to obtain a reliable and non-destructive method for quantifying the homogeneity of perovskite films. We adapt existing algorithms to spatially quantify multiple perovskite film properties (substrate coverage, film thickness, defect density) with 10 μm x 10 μm pixel resolution from pictures of 25 cm2 samples. Our machine vision tool - called PerovskiteVision - can be combined with an optical model to predict photovoltaic cell and module current density from the perovskite film thickness. We use the extracted film properties and predicted device current density to identify a posteriori the process conditions that simultaneously maximize the device performance and the manufacturing throughput for a large-area perovskite deposition process (gas-knife assisted slot-die coating). PerovskiteVision thus facilitates the transfer of a new deposition process to large-scale photovoltaic module manufacturing. This work shows how machine vision can accelerate slow characterization steps essential for the multi-objective optimization of thin film deposition processes.

978-1-7281-6118-1/22 $31.00 © 2022 IEEE

Effective Irradiance Monitoring Using Reference Modules

Jennifer L. Braid[1], Joshua S. Stein[1], Bruce H. King[1], Christopher Raupp[2], Jaya Mallineni[2], Justin Robinson[3], Steve Knapp[3]

[1]Sandia National Laboratories, Albuquerque, NM, 87123, USA

[2]SOLV Energy, San Diego, CA, 92127, USA

[2]GroundWork Renewables, Holladay, UT, 84117, USA

Abstract— We evaluate the use of reference modules for monitoring effective irradiance in PV power plants, as compared with traditional plane-of-array (POA) irradiance sensors, for PV monitoring and capacity tests. Common POA sensors such as pyranometers and reference cells are unable to capture module-level irradiance nonuniformity and require several correction factors to accurately represent the conditions for fielded modules. These problems are compounded for bifacial systems, where the power loss due to rear side shading and rear-side plane-of-array (RPOA) irradiance gradients are greater and more difficult to quantify. The resulting inaccuracy can have costly real-world consequences, particularly when the data are used to perform power ratings and capacity tests. Here we analyze data from a bifacial single-axis tracking PV power plant, (175.6 MW$_{dc}$) using 5 meteorological (MET) stations, located on corresponding inverter blocks with capacities over 4 MW$_{dc}$. Each MET station consists of bifacial reference modules as well pyranometers mounted in traditional POA and RPOA installations across the PV power plant. Short circuit current measurements of the reference modules are converted to effective irradiance with temperature correction and scaling based on flash test or nameplate short circuit values. Our work shows that bifacial effective irradiance measured by pyranometers averages 3.6% higher than the effective irradiance measured by bifacial reference modules, even when accounting for spectral, angle of incidence, and irradiance nonuniformity. We also performed capacity tests using effective irradiance measured by pyranometers and reference modules for each of the 5 bifacial single-axis tracking inverter blocks mentioned above. These capacity tests evaluated bifacial plant performance at ~3.9% lower when using bifacial effective irradiance from pyranometers as compared to the same calculation performed with reference modules.

Keywords—irradiance monitoring, reference module, capacity test, performance modeling, bifacial

I. INTRODUCTION

Irradiance monitoring has a few applications in real photovoltaic (PV) power plants. Most commonly, measured irradiance is used to perform long-term performance analysis and short-term capacity tests, wherein system energy yield is evaluated compared to performance guarantees from the developer or engineering, procurement, and construction company (EPC). Long-term irradiance measurements are used for operations and maintenance (O&M) performance monitoring to diagnose plant issues and schedule maintenance activities (e.g., cleaning). The irradiance measurement accuracy and system representativeness are incredibly important for these applications, to both system owners, EPCs, and O&M providers. However, the common use of point sensors to estimate the irradiance for the PV array causes inherent discrepancy in calculating the system energy yield.

Traditionally, plane-of-array (POA) irradiance sensors fall into two categories: pyranometers and reference cells. Pyranometers (photodiode or more often thermopile) are expensive but offer a flat absorption profile. However, they have a different spectral and thermal response than PV cells or modules. Reference cells consisting of an encapsulated silicon cell are better spectrally matched and offer lower measurement uncertainty compared to pyranometers [1], [2]. But as point sensors, both pyranometers and reference cells have inherent disadvantages for representing array-level irradiance. Point sensors can have installation differences from the modules in the array, such as different locations (e.g., on a weather station), tilt angle, and/or field of view. There can also be variability in the measured irradiance due to the device scale: a point sensor cannot capture the effects of irradiance gradients or partial shading on a module. This is especially problematic for bifacial PV arrays, where rear facing POA irradiance sensors cannot capture the spatial variation of irradiance that commonly occur on the back of the module [3]. Furthermore, spectral albedo can be of great consequence when coupled with the differences in spectral response of rear-facing pyranometers vs. reference modules [4].

To combat these disadvantages, we explore the use of reference modules for PV array irradiance monitoring. Broadly, this method uses I-V curves measured *in situ* on a module installed and located within the array. Lab characterization data of the same module (or nameplate values), along with module temperature, are used to convert the measured short circuit current (I_{SC}) to effective irradiance. We hypothesize that this method alleviates issues with traditional irradiance sensors including spectral and temperature mismatch, installation and fielding differences, irradiance variability related to scale, and issues with measuring effective irradiance for bifacial systems.

II. METHODS

A. Dataset

Data for this study were provided by SOLV Energy and consist of 8 days of data from the utility-scale bifacial single-axis tracking PV power plant, recorded during the capacity testing phase shortly after commissioning. For this study, we examine the 5 inverters and their corresponding co-located weather stations. Each of these 5 stations has a monofacial and a bifacial reference module, 1 forward facing and 1 rear facing SR30 pyranometer, as well as windspeed and ambient temperature sensors. The rear facing pyranometers are mounted on the underside of the torque tube, 3 modules interior from the north end of the row. Reference module electrical and back-of-module temperature measurements were taken with an Atonometrics RDE300. All weather parameters are measured by LUFFT WS500.

B. Reference module irradiance monitoring approach

Previous studies using reference modules have used continuous I_{SC} measurements [5]. For the reference module method to accurately represent the conditions in the field, our approach uses a module physically located within the array (electrically isolated from the production modules within the array), connected to a RDE300 device which measures I_{SC} and an I-V curve at regular intervals. While nameplate coefficients may be used, the reference module should be characterized in a laboratory to determine the temperature coefficient for current (α) and 1-sun I_{SC}. Then outdoor I_{SC} measurements can be corrected with concurrent module temperatures and used to determine the effective irradiance for the array by:

$$E_e = 1000W/m^2 * \frac{I_{SC-m}}{I_{SC-1sun}} / \left(1 + \alpha\left(T_{ref} - T_m\right)\right)$$

where I_{SC-m} is the measured I_{SC}, T_{ref} is the reference temperature, usually 25 °C, and T_m is the measured module temperature. Temperature correction could also be performed using V_{OC} determined module temperature with necessary calibration.

C. Pyranometer irradiance monitoring approach

The use of pyranometers to monitor POA irradiance is well established for monofacial systems. Thermopile pyranometers have a flat absorption profile, so the measured irradiance is ~3% higher than that of a PV device and does not require temperature correction. In this work, our front side pyranometer measurements were spectrally corrected using the airmass spectral correction implemented through *pvlib-python* [6], and the angle of incidence response measured on the specific module model in the field (Fig. 1).

The current standard for irradiance monitoring of bifacial systems lacks detail on positioning and corrections necessary for measurement accuracy [7]. Waters et al. [8] suggest scaling the measured rear side POA irradiance by the module bifaciality constant and adding the front side POA to obtain the total irradiance. Gostein et al. [3] suggested that the bifaciality constant could be expanded to account for rear side shading and irradiance non-uniformity, and that additional factors could be used for spectral, angular, and mismatch losses. For the analysis

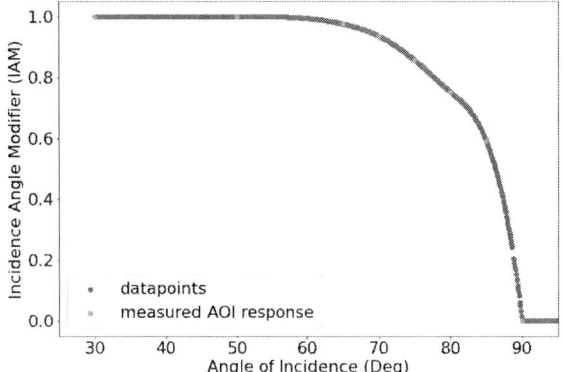

Fig. 1: Measured and interpolated front side angle of incidence response for the fielded modules in this study.

presented here, we scaled the front side POA according to the incidence angle modifier [9] to account for module reflective losses.

For the rear facing pyranometers in this work, the rear side POA was scaled using the tracker torque tube manufacturer's reported shading factor (0.123), and the modules' bifaciality constant (0.7).

D. Capacity test comparison

To compare pyranometer and reference module irradiance measurements for use in a real-world application, we performed capacity test regression according to ASTM E2848 [10], implemented with the python package *pvcaptest* [11], using each effective irradiance measurement method. This procedure filters the power and irradiance data for range (200 to 2000 W/m²) and outliers, then uses multilinear regression to model measured system power as a function of irradiance, ambient temperature, and windspeed. Then the parameters determined via multilinear regression are used to predict the system output at a reference condition determined based on the range of available data, also according to the ASTM E2848 standard. After performing this regression and evaluation on measured data, it is repeated on modeled power data, here using PVWatts [12]. The last step of the capacity test is to calculate the ratio of the measured power and modeled power regressions evaluated at the reference condition, which is here referred to as the capacity ratio.

III. RESULTS

Here we compare effective irradiance measurements made with pyranometers and reference modules and evaluate their effects on capacity test results for 5 bifacial PV systems.

A. Pyranometer vs. reference module effective irradiance

First, we directly compare effective irradiance measured by thermopile pyranometer and reference module for a monofacial configuration. Because this site has monofacial reference modules in addition to bifacial reference modules for all 5 inverters, we can directly compare these front side POA irradiance measurements. The linear regression for one such

---

Sandia National Laboratories is a multimission laboratory managed and operated by National Technology & Engineering Solutions of Sandia, LLC, a wholly owned subsidiary of Honeywell International Inc., for the U.S. Department of Energy's National Nuclear Security Administration under contract DE-NA0003525. This work is funded in part by the U.S. Department of Energy Solar Energy Technologies Office, under Award Number 38268.

978-1-7281-6118-1/22 $31.00 © 2022 IEEE

comparison is shown in Fig. 2. For all 5 weather stations, the pyranometer measured effective POA irradiance is 2.6-3.7% higher than that measured by monofacial reference module (measured by linear regression), after accounting for spectral and angle of incidence corrections.

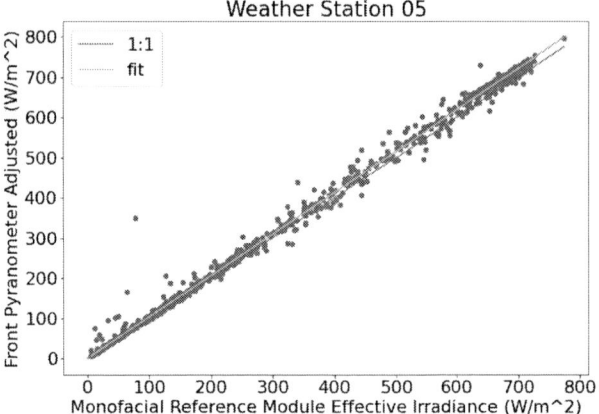

Fig. 2: Linear regression of effective irradiance calculated from monofacial reference module vs. POA irradiance measured by front-facing pyranometer. $R2=0.995$, slope = 1.031, with intercept fixed at the origin.

Next we compare bifacial effective irradiance as measured by forward and rear facing pyranometers to that measured by bifacial reference module. Rear facing pyranometer measurements were adjusted based on module bifaciality and torque tube shading factor, in addition to spectral corrections also applied to the forward facing pyranometer. Only the forward facing pyranometer had angle of incidence correction. For all 5 weather stations, the bifacial effective irradiance averaged 3.6% higher when measured with a pair of pyranometers than when measured by bifacial reference module. The pyranometers irradiance overreporting percentages are given in TABLE II.

TABLE I. BIFACIAL EFFECTIVE IRRADIANCE PERCENT DIFFERENCE FROM REFERENCE MODULE TO PYRANOMETERS, MEASURED BY LINEAR REGRESSION. PYRANOMETERS EFFECTIVE IRRADIANCE WAS CORRECTED FOR ANGLE OF INCIDENCE, SPECTRAL RESPONSE, AND REAR SHADING.

Station	Pyranometers Effective Irradiance Overreporting %
1	3.478
2	3.650
3	3.615
4	2.967
5	4.421
Average	**3.626**

B. Capacity test comparison

Here we compare the regressions for measured and modeled power using reference module and pyranometers measured bifacial effective irradiance, evaluated at a) common (averaged) reference condition values for irradiance windspeed, and ambient temperature, and b) individual reference condition values for each irradiance measurement type. We report the

evaluated DC power regression values for each case in the sections below.

1) Measured DC Power Regression Evaluation:
We first evaluate the regressions on measured DC power at a common reference condition, that is the same values for effective irradiance, windspeed, and ambient temperature. A scatterplot for the values used in one such pair of regressions is shown in Fig. 3. The lower effective irradiance values measured by reference module result in a higher slope for the power regression on this variable as compared to the pyranometer effective irradiance.

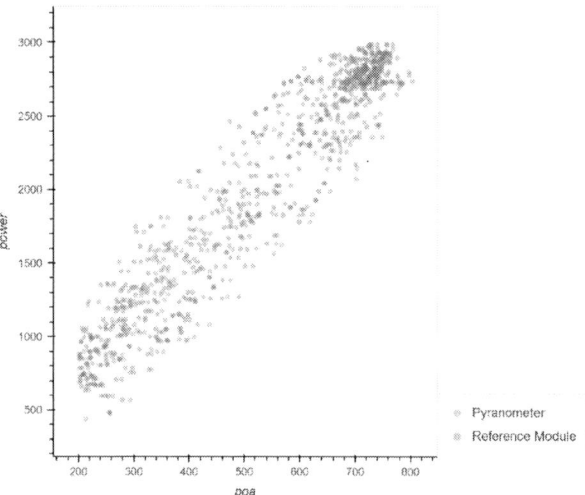

Fig. 3: Filtered measured DC power vs. bifacial effective irradiance measured by pyranometers (blue) or reference module (red), used for multilinear regression analysis.

TABLE II. REPORTED DC POWER % DIFFERENCE FROM REFERENCE MODULE TO PYRANOMETERS, EVALUATED FROM CAPACITY TEST REGRESSIONS USING A COMMON REFERENCE CONDITION.

Station	Pyranometers Measured Power Underreporting %
1	4.334
2	3.998
3	4.184
4	2.830
5	4.727
Average	**4.014**

The difference in measured power regression evaluations for each irradiance measurement type at a common (averaged) reference condition are given in TABLE II. The regression evaluated measured DC power is consistently higher when using the reference module data when using common reference condition values between the two effective irradiance measurement methods, due to the lower measured irradiance values for the reference modules. When using individually determined reference condition values (based on the distribution of filtered irradiance and weather data values), there is little to no difference between effective irradiance methods for the regression evaluated measured DC power values.

2) Estimated DC Power Calculation and Regression Evaluation:

We use the PVWatts method to predict system DC power, which uses the effective irradiance, module temperature, and maximum power point temperature coefficient. We employed this method to predict the DC power for each data point across all 5 inverters using both pyranometers and reference module measured irradiance.

We then performed multilinear regressions on the modeled DC power with effective irradiance, windspeed, and ambient temperature (as for the measured DC power). The modeled power regression for one pair of irradiance sensors is shown in Fig. 4.

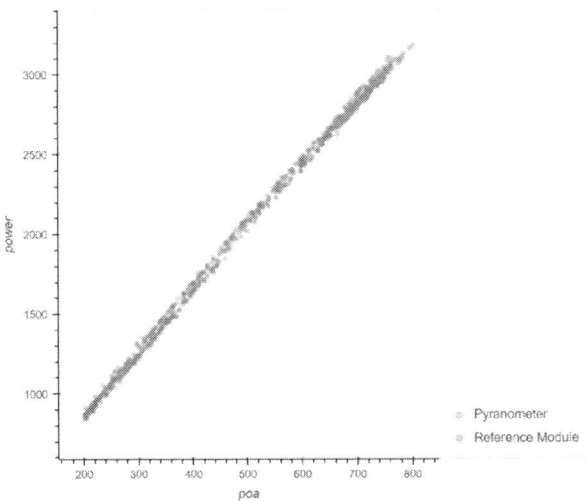

Fig. 4: Filtered expected DC power data modeled with PVWatts vs. effective irradiance as measured by pyranometers (blue) and reference module (red), as used for capacity test multilinear regressions.

As can be seen in the figure above, the expected DC power and regression lines align well between the two effective irradiance measurement methods. Therefore, at a common reference condition, the expected DC power regression evaluations are very similar (as these are essentially multilinear regressions of the PVWatts model). However, when evaluated at individual reference conditions, based individually on the distributions of effective irradiance measurements of each type, the expected DC power averages 1.4% lower when determined with the reference module compared to pyranometers.

Finally, we compare the capacity ratios calculated using each effective irradiance measurement method. The capacity ratios evaluated at common reference conditions between the two methods, as well as the percent differences, are shown in TABLE III. On average, the capacity ratio at a common reference condition is 4.02% higher when regressions are performed with reference module measured effective irradiance versus pyranometers.

TABLE III. CAPACITY RATIOS EVALUATED AT COMMON REFERENCE CONDITIONS ON REGRESSIONS PERFORMED ON REFERENCE MODULE AND PYRANOMETERS EFFECTIVE IRRADIANCE MEASUREMENTS.

Station	Pyranometers Capacity Ratio	Reference Module Capacity Ratio	% Difference
1	0.918	0.960	4.34
2	0.939	0.978	4.01
3	0.922	0.962	4.17
4	0.936	0.962	2.73
5	0.958	1.006	4.82
Average	**0.934**	**0.974**	**4.02**

We also evaluated the capacity ratios at individually determined reference conditions for each effective irradiance measurement type at each inverter/weather station. The capacity ratios at individual reference conditions average 3.84% higher when calculated with reference module effective irradiance vs. pyranometers. The results for each station are shown in TABLE IV.

TABLE IV. CAPACITY RATIOS EVALUATED AT INDIVIDUAL REFERENCE CONDITIONS ON REGRESSIONS PERFORMED ON REFERENCE MODULE AND PYRANOMETERS EFFECTIVE IRRADIANCE MEASUREMENTS.

Station	Pyranometers Capacity Ratio	Reference Module Capacity Ratio	% Difference
1	0.925	0.965	4.12
2	0.945	0.982	3.78
3	0.928	0.965	3.84
4	0.937	0.965	2.88
5	0.963	1.009	4.61
Average	**0.940**	**0.977**	**3.84**

3) Inverter Expected DC Power vs. Inverter Measured DC Power

To further evaluate the difference being seen in with the use of pyranometers vs reference modules for system performance analysis, a perfectly clear sky day from this site was evaluated. On the clear sky day, the timeseries expected DC power, per station, was calculated by:

$$DC\ Power_{exp}\ (kW) = DC\ Capacity\ (kW) * TA * IA * (1 - WL)$$

where the DC Capacity is the rated STC dc capacity of the inverter on the corresponding station, TA is the temperature adjustment factor calculated by:

$$TA = 1 + \gamma(T_{ref} - T_m)$$

where γ is the maximum power (P_{max}) temperature coefficient of installed PV modules within the array, T_{ref} is the reference temperature, 25 °C, and T_m is the measured module temperature, IA is the irradiance adjustment factor calculated by:

$$IA = \frac{POA_{total}}{POA_{ref}}$$

where, POA_{total} is the effective total irradiance measured by pyranometers (using the same adjustment factors as outlined in Section II part C, of this paper) and POA_{ref} is the reference irradiance condition, 1000 W/m², WL is the overall DC ohmic wire loss, of the corresponding station, calculated based on the cable gauge size and total installed wire length. As seen in Fig.

5, the expected inverter DC power calculated using the bifacial reference module irradiance is observed to be very close in alignment to the total measured inverter DC power. As can be seen, the expected inverter DC power as calculated using the pyranometers deviates significantly from the measured power.

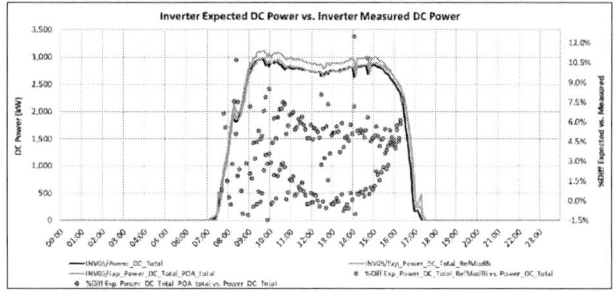

Fig. 5: Inverter Expected DC Power derived using the effective irradiance as measured by pyranometer (orange) and reference module (blue) vs. Inverter Measured DC Power (black)

As previously mentioned, the datasets used were from the site's capacity test period, and thus any potential DC health or other field related issues are minimal, if at all present. On an average the expected inverter DC power calculated using the reference module is 0.9% higher than the measured DC power whereas the inverter DC power calculated using the pyranometers effective irradiance is 5.3% higher than the measured inverter DC power.

4) Reference Module Expected DC Power vs Reference Module Measured DC Power

As seen in *Fig. 6* the total effective irradiance from the front side facing POA and rear side facing POA, from the same weather station used to derive the expected inverter DC power in *Fig. 5*, were also used to derive the expected power of the corresponding bifacial reference module to further validate the results. As previously described, the reference modules are measuered with use of RDE300 units that perform full I-V curve sweeps. This thus allowed for measured P_{max} of the reference module to measured and collected throughout the testing perioed. The expected reference module power was calculated by:

$$DC\ Power_{exp}\ (W) = STC\ Rated\ Pmax\ (W) * TA * IA$$

where the STC Rated Wattage is the flashtested Pmax of the reference module, TA is the temperature adjustment factor using the back of module temperature sensor installed on the reference module, and IA is th total effect irradiance of the pyranometers (adjusted per Section II part *C* of this paper). As can be seen in *Fig. 6*, the calculated expected power of the reference module is deviating significantly compared to actual measured P_{max} of the bifacial reference module.

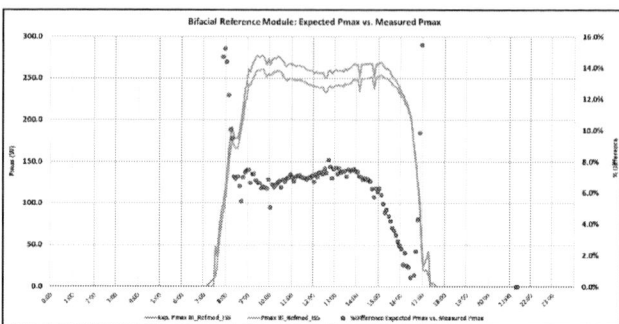

Fig. 6: Bifacial Reference Module Measured Pmax (Blue) vs Expected Pmax (Orange) using Pyranometers effective total irradiance

The expected P_{max} is on an average of 6.4% higher than that of measured P_{max} of the bifacial reference module. This result is in alignment with the results described in the previous section where the expected inverter DC power, when calculated with pyranometers, averaged 5.3% higher than the measured inverter DC power. With the bifacial reference modules having been previsouly flashed and newly installed, the modules are known to not be underperforming, thus the difference in the expected vs. measured P_{max} is directly attributeable to the differnece in total irradaince measured by pyranometers (even with adjustments) as compared to the usable irradiance "seen" by the reference module.

IV. DISCUSSION

We have shown that effective irradiance measured with reference modules averages 3.1% lower for monofacial POA irradiance and 3.6% lower for bifacial POA irradiance when compared to equivalent pyranometer measurements adjusted for spectral, non-uniformity, and angle of incidence effects. While additional measurements could be made to further adjust pyranometer measurements to reflect the effective irradiance more accurately for a PV array, we have demonstrated that reference modules offer a simple method for representative irradiance measurement without these corrections. We have also shown that effective irradiance monitoring with reference modules yields ~3.9% higher capacity ratios in standard ASTM E2848 capacity tests as compared with pyranometers, across 5 bifacial PV arrays at a large scale utility PV site. This result demonstrates that the calculated expected DC power more closely aligns with the measured DC power when reference modules are used in place of pyranometers for bifacial effective irradiance measurement.

We evaluated the sensitivity of the capacity test results by running the capacity tests again using nameplate (rather than flash test) I_{SC} values. The results showed that the capacity tests performed with reference module effective irradiance calculated on nameplate values varied from the flash test equivalent by the same percentage and direction as the difference between the nameplate and flash test I_{SC} values. That is, a nameplate I_{SC} value 2.5% lower than the flash test value results in a capacity ratio ~2.5% lower than the flash test equivalent capacity ratio. This both emphasizes the need for accurate flash test I_{SC} values for reference modules and provides a basis for uncertainty determination.

The reference module data collected and used in this abstract was obtained with an Atonometrics RDE300, which holds the modules at short circuit or open circuit condition (the latter used here) when not performing measurements. This means that the reference module is not at the same operating point or temperature as the rest of the array, and that the reference module is not an active power producer within the system. GroundWork Renewables is addressing these issues by developing a device that can be attached to a module in series with the rest of the array, but which electrically disconnects the module from the array for short time periods to measure I_{SC} and an I-V curve at a specified frequency. This will allow for the reference module to participate in power production with the rest of array and remove the need for adjustments in array string wiring and additional independent modules for measurement.

Reference modules also provide additional advantages over pyranometers and reference cells for irradiance monitoring. Generally, reference modules and associated hardware are lower cost and can serve dual purposes – performance monitoring and soiling monitoring. Alongside irradiance monitoring, the timeseries I-V curves measured on reference modules can be used for more advanced power loss analysis, such as outdoor Suns-V_{OC} [13]. This type of performance loss monitoring can be used to inform O&M activities, as well as diagnose mechanisms of long-term degradation in fielded modules, but is not possible without time series I-V characterization.

V. Conclusion

In this work, we have directly compared effective irradiance measured by reference PV modules to standard pyranometers. We showed that in the case of both monofacial and bifacial effective irradiance, pyranometers overestimate the irradiance reaching a PV array by 2.5-4.5%, even after adjusting for PV angle of incidence, spectral response, and irradiance nonuniformity effects. Additionally, we showed that when using effective irradiance measurements to perform capacity tests, this difference in measured irradiance results in a ~3.9% lower capacity ratio when using pyranometers instead of reference modules to evaluate the performance of bifacial single axis tracking PV arrays. These results are consistent when comparing the expected DC power calculated by use of pyranometer at both the system level as well as module level. In addition to more accurately predicting and evaluating the performance of a PV system, the reference module approach to effective irradiance monitoring has several inherent advantages over the use of pyranometers or reference cells, including cost, reduction or elimination of correction factors, and ease of implementation, particularly for bifacial systems. Inaccuracies in effective irradiance measurements have real world consequences for system developers and owners, so it is critical that new methods are explored and developed to keep up with advances in PV system technologies.

References

[1] L. Dunn, M. Gostein, and K. Emery, "Comparison of pyranometers vs. PV reference cells for evaluation of PV array performance," in *Photovoltaic Specialists Conference (PVSC), 2012 38th IEEE*, 2012, pp. 002899–002904.

[2] C. Reise and M. J. Rivera Aguilar, "Silicon Sensors vs. Pyranometers – Review of Deviations and Conversion of Measured Values," in *37th European Photovoltaic Solar Energy Conference and Exhibition*, Oct. 2020, pp. 1449–1454. doi: 10.4229/EUPVSEC20202020-5BV.3.3.

[3] M. Gostein *et al.*, "Measuring Irradiance for Bifacial PV Systems," in *2021 IEEE 48th Photovoltaic Specialists Conference (PVSC)*, Jun. 2021, pp. 0896–0903. doi: 10.1109/PVSC43889.2021.9518601.

[4] N. Riedel-Lyngskær *et al.*, "The effect of spectral albedo in bifacial photovoltaic performance," *Sol. Energy*, vol. 231, pp. 921–935, Jan. 2022, doi: 10.1016/j.solener.2021.12.023.

[5] J. Polo, W. G. Fernandez-Neira, and M. C. Alonso-García, "On the use of reference modules as irradiance sensor for monitoring and modelling rooftop PV systems," *Renew. Energy*, vol. 106, pp. 186–191, Jun. 2017, doi: 10.1016/j.renene.2017.01.026.

[6] W. F. Holmgren and D. G. Groenendyk, "An open source solar power forecasting tool using PVLib-Python," in *2016 IEEE 43rd Photovoltaic Specialists Conference (PVSC)*, Jun. 2016, pp. 0972–0975. doi: 10.1109/PVSC.2016.7749755.

[7] "IEC 61724-1 2021 | Photovoltaic system performance - Part 1: Monitoring." Accessed: Feb. 08, 2022. [Online]. Available: https://webstore.iec.ch/publication/65561

[8] M. Waters, C. Deline, J. Kemnitz, and J. Webber, "Suggested Modifications for Bifacial Capacity Testing," in *2019 IEEE 46th Photovoltaic Specialists Conference (PVSC)*, Jun. 2019, vol. 2, pp. 1–6. doi: 10.1109/PVSC40753.2019.9198974.

[9] A. F. Souka and H. H. Safwat, "Determination of the optimum orientations for the double-exposure, flat-plate collector and its reflectors," *Sol. Energy*, vol. 10, no. 4, pp. 170–174, Oct. 1966, doi: 10.1016/0038-092X(66)90004-1.

[10] "ASTM-E2848: Standard Test Method for Reporting Photovoltaic Non-Concentrator System Performance," American Society for Testing and Materials. Accessed: Jan. 17, 2022. [Online]. Available: https://www.astm.org/e2848-13r18.html

[11] *pvcaptest*. pvcaptest, 2022. Accessed: Jun. 06, 2022. [Online]. Available: https://github.com/pvcaptest/pvcaptest

[12] A. P. Dobos, "PVWatts Version 5 Manual," National Renewable Energy Lab. (NREL), Golden, CO (United States), NREL/TP-6A20-62641, Sep. 2014. doi: 10.2172/1158421.

[13] M. Wang *et al.*, "Analytic Isc–Voc Method and Power Loss Modes From Outdoor Time-Series I–V Curves," *IEEE J. Photovolt.*, vol. 10, no. 5, pp. 1379–1388, Sep. 2020, doi: 10.1109/JPHOTOV.2020.2993100.

Designing a Multi-Quantum-Dot Array for Efficient Light-Harvesting in Solar Cells

Jose Raul Montes-Bojorquez, Maria Fernanda Villa-Bracamonte, and Arturo A. Ayon
The University of Texas at San Antonio, San Antonio, Texas, 78249, USA

Abstract—One of the major loss mechanisms leading to low power conversion efficiencies in photovoltaic (PV) devices arises from the limited spectral response of solar cells to the wideband solar spectrum. The capture of photons that otherwise would contribute to thermalization of charge carriers, could be targeted by a luminescent down-shifting (LDS) layer capable to absorb high energy photons and re-emit them at lower energies where the PV material exhibit a significantly better response. In addition, when the LDS layer is placed at the surface, it modifies the reflectance of the solar cell, and consequently, its thickness and refractive index (RI) are key factors in the performance of the resulted PV structure. Thus, for PV applications, the LDS layer must be optimized for the best trade off between the antireflective and LDS capabilities. In this study, we present a new approach to increase the spectral response of a single-junction solar cell by exploiting the size-dependent refractive index and down-shifting capabilities of quantum dots. Compared to a conventional LDS layer, our design allows for thicker films and thus higher optical absorptions, while the reflectance is reduced.

Index Terms—LDS, quantum dot, CdTe, ARC, FDTD

I. INTRODUCTION

Luminescent down-shifting (LDS) is an approach that addresses the thermalization losses due to high-energy photons by modifying the incident solar spectrum to better suit the cell's optimal absorption regions [1]. It involves the utilization of luminescent species that absorb high energy photons before they reach the solar cell to re-emit them at lower energies. Contrary to other methods addressing spectral losses, LDS is a passive approach, meaning that carrier collection is still performed via the single pn junction, thus the application of LDS in photovoltaics (PV) does not require modification of the existing solar cell [4].

In this regard, quantum dots (QDs) are known to exhibit a number of favorable characteristics as LDS materials, such as their reported ability to reach efficiencies in excess of 90% for wavelengths in the near-infrared and visible range, large stokes shift and broad absorption spectra. Moreover, QDs can be synthesized from different materials and are known for their size-dependent opto-electronic properties. In this sense, a variety of different materials and sizes could be employed to address different regions of the electromagnetic spectrum in order to achieve a better absorption by the active material. In addition to the potential efficiency enhancement due to LDS effects, the reported size-dependent refractive index of QDs opens the possibility of designing structures that resemble graded-index layers. Herein, we propose a new approach to

Consejo Nacional de Ciencia y Tecnologia (CONACYT).

increase the spectral response of a solar cell by exploiting the size-dependent refractive index and down-shifting capabilities of QDs.

II. EXPERIMENTAL DETAILS

A. Quantum dot synthesis

Cadmium tellurite (CdTe) QDs are particularly attractive for the proposed studies due to their high PLQY, absorption and emission spectra strongly dependent on particle size, and synthesis by relatively affordable chemical methods. Moreover, CdTe QDs can be synthesized with different surface ligands and have proved to be compatible with layer-by-layer deposition which allows the precise control of the composition and size needed to fabricate the proposed designs [3]. Thioglycolic acid (TGA) capped CdTe QDs were synthesized employing a hydrothermal synthesis method [2], where the size of the resulting TGA capped CdTe QDs is determined by the reflux time.

B. Characterization

The UV–Vis absorption spectra and the photoluminescent effects of QDs in solution were recorded using an Ocean Optics Flame-S-UV-VIS spectrometer. A Zetasizer Nano Series in dynamic light scattering (DLS) mode was employed to measure particle size of QDs in colloidal solution.

C. Optical Modelling

The multilayered structure was optically simulated by the finite-difference time-domain (FDTD) model of Lumerical FDTD solutions 8.25.2621. The auto non-uniformmesh was applied in the simulation region which has a dimension of 0.5 μm (width) $\times$ 3 μm (length). The upper and lower boundaries at the y-direction were perfectly matched layers which allow radiation to propagate out of the computational area without disturbing the fields inside. The surface was modelled as an infinite plane by setting periodic boundaries at the x-direction, which allows to save computational memory. A plane wave with a Gaussian distribution of frequencies between 450 nm and 1000 nm providing uniform lateral illumination and normal incidence was used as the source. A frequency domain power monitor (FDPM) was used to record the reflection at the surface of the thin film.

III. RESULTS AND DISCUSSION

The DLS measurements showed that the particle mean size increases monotonically from 2.60 to 4.12 nm when the reaction time is varied from 15 min to 11 h, and is expected to further increase for longer reaction times. Fig. 1 shows the absorption spectra of the synthesized QDs in aqueous solution for different reaction times. For all cases, the absorption spectra increases exponentially towards the UV, while a red-shift of the first absorption peak is observed when increasing the reaction time. On the other hand, as shown in Fig. 2, the emission can be tailored over the entire visible spectrum.

The refractive index of a material is known to decrease with band gap [5], and since the size-dependent band gap of QDs is a well-known and widely studied quantum confinement effect, it is expected that the size and RI of QDs will have certain correlation. Reference [6] calculated the RI for spherical CdTe QDs using 5 different models, showing for all models a RI that decreases quickly as the size is reduced. For the QD sizes obtained in the present work, a RI ranging from 1.8 to 3 (bulk value) is expected.

Theoretically, an antireflective coating where the RI decreases gradually from the substrate to ambient, would give zero reflection. Unfortunately, this is not feasible in practice because the high-optical-index materials are generally absorbent. Inspired by these incentives, a design of LDS film with a QD size gradually decreasing towards the air interface will be optimum in terms of antireflective properties, which are known to play a dominant role in LDS layer for solar cell applications. Since the goal of the present study is to design a structure for future fabrication with QDs, a discrete number of layers with a given RI was the preferred approach rather than continuum profiles. Therefore, to simulate the size-gradient nanostructures, a film with a QD size gradient containing 1 to 7 different layers with RI scaling from 1.8 to 3 on top of a silicon substrate was designed. Each layer of the thin film

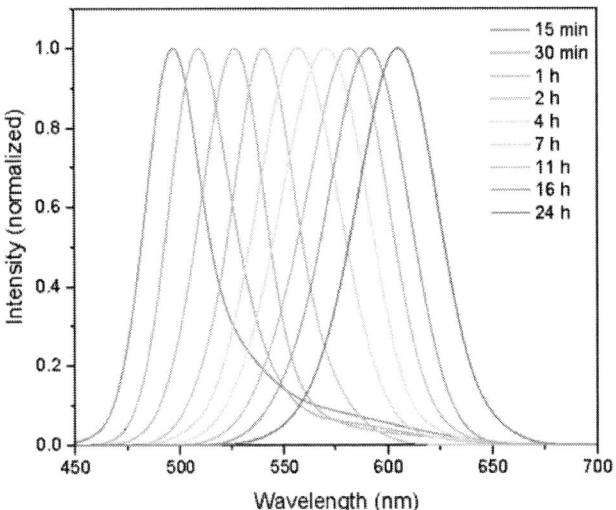

Fig. 2. Photoluminescent spectra of CdTe QDs in colloidal solution for different reaction times.

was modelled as a bulk material for which the thickness and refractive index were set as parameters to be optimized. For all the cases, particle swarm optimization was used to find the parameters producing the minimum average reflectance (R_{avg}).

In the FDTD model of a single layer, the optimized thickness and RI is obtained as 71 nm and 1.98 respectively. Since the optimum RI is very close to the average RI reported for high-QY QDs films (~1.80) [3], a layer with RI=1.80 was included in all the designs. With the optimized thickness, the R_{avg} for a single layer reaches 5.85%. However, in the 7 layers model the optimum thickness for 5 of the 7 layers was found to be zero, so that the optimum design was equivalent to a film composed of only two layers (RI=1.8, 3). This constitutes an important design rule in the fabrication of multilayered antireflection films, since although the lowest reflectance is obtained when the RI gradually decreases from the substrate to the environment, for a discrete number of layers the RI of each layer must be distributed over the gap between the substrate and the environment.

To further decrease the reflectance, a three layers film was designed where the layer with RI=1.8 was placed at the middle of two layers with RI to be optimized. The RI of the layer on top was set to vary between 1 and 1.8, while the RI of the layer at the interface with the substrate was set to vary between 1.8 and 3.2. With this design, R_{avg} decreases to 0.86%. The optimized parameters of each design are summarized in Table 1, and the reflectance of each design is showed in Fig. 3.

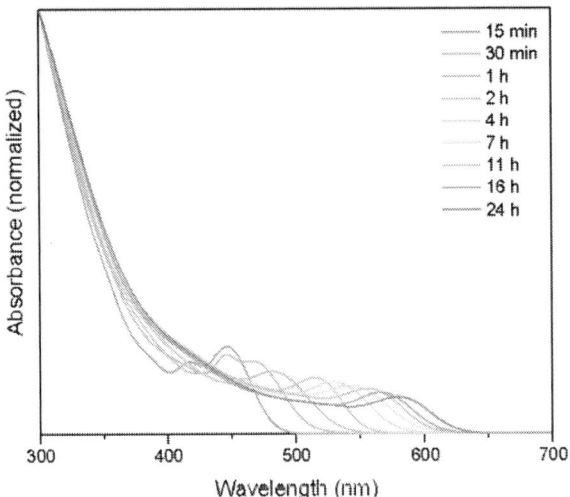

Fig. 1. Absorption spectra of CdTe QDs in colloidal solution for different reaction times.

Table 1. Optimized parameters for each film design.

Design	RI	Thickness (nm)	R_{avg} (%)
1 Layer	1.8	82	5.85
2 Layers	1.8, 3	87, 51	2.63
3 Layers	1.15, 1.8, 3.2	113, 84, 45	0.86

Although the lowest reflectance is obtained with the 3 layers design, the difficulty of finding a material with very low refractive index must be consider when evaluating the practicality of such design. On the other hand, the 2 layers design shows that compared to the one obtained in a single-size LDS, an important decrease in the reflectance is possible by adding two different sizes, and that a R_{avg} as low as 2.63% can be achieved with QD layers.

Since QDs are high absorbing materials, a more complete study using the measured complex index of refraction is needed to fully understand the antireflection properties of the resulting films. Moreover, the proposed structures present important characteristics for the study of the LDS effects of the QDs in the film. For example, since the total thickness of the film increases with the number of layers (see Table 1), the optical absorption of LDS later would benefit from the proposed structures. Furthermore, light harvesting finds further improvement when different band gap chromophores communicate with each other, for example Förster resonant energy transfer (FRET). Non-radiative transductions of the excited state energy funnel photons unidirectionally from the large band gap chromophores to the small band gap chromophores [7]. In this regard, the close proximity of two monolayers with QDs of different sizes enable exciton migration from the layer of smaller size to the layer of larger size, and it has shown to extend to span several layers when the size is changed monotonically between layers [8] (see Fig. 4). This has already been demostrated in graded multilayered TGA capped CdTe QDs films [9]. If FRET is present across the different single-size films, it is expected that the larger QDs will dominate the emission of the multi-size structure. We hypothesize that upon hybdrization with the designed multi-QD structures, solar cells will benefit from funneling high-energy photons towards the Si interface. This will enable us to simultaneously utilize LDS, FRET, and antireflection effects, while explore for first time

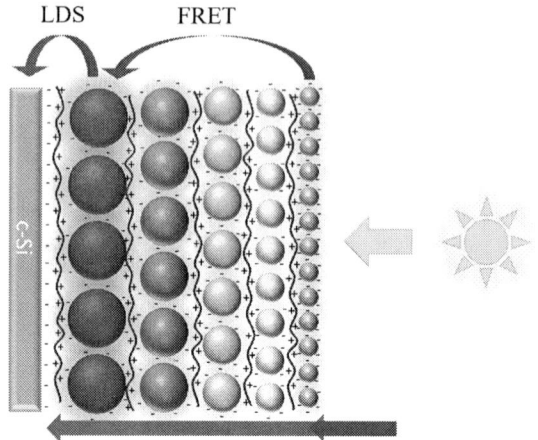

Fig. 4. Schematics of multi-size QD film on top of a silicon substrate.

the potential of FRET cascades in the context of down-shifting enhanced photovoltaics. down-shifting

CONCLUSION

In this work, a multilayered structure that harvest both UV and visible photons was designed. The computational optical studies showed that a film composed of two different sizes of downshifting quantum dots resulted in a reduction in the average reflectance from 35% to 2.63%, and to 0.86% when an extra layer is added. The RI and thickness of each layer of the thin film were deduced by the FDTD model. The application of the presented approach allows the use of downshifting materials to fabricate graded-index antireflective thin films with superior antireflection properties compared to single layers while increasing the QD load.

REFERENCES

[1] J. Joseph, S. Senthilarasu, and T. Mallick, "Improving spectral modification for applications in solar cells: A review," Renew. Energy, vol. 132, pp. 186–205, September 2019.

[2] S. Wu, J. Dou, J. Zhang, and S. Zhang, "A simple and economical one-pot method to synthesize high-quality water soluble CdTe QDs," J. Matter. Sci., vol. 22, pp. 14573–14578, August 2012.

[3] R. Montes and A. Ayon, "Enhanced performance of a solar cell based on polyelectrolyte-quantum dot multilayered films," IEEE 48th PVSC, pp. 264–268, June 2021.

[4] E. Klampaftis, D. Ross, K. McIntosh, and B. Richards, "Enhancing the performance of solar cells via luminescent down-shifting of the incident spectrum: A review," Sol. Energy Mater Sol. Cells, Vol. 93, pp. 1182–1194, August 2009.

[5] S. Tripathy, "Refractive indices of semiconductors from energy gaps," Sol. Energy Mater Sol. Cells, Vol. 46, pp. 240–246, August 2015.

[6] A. Kaddouri, A. Kouzou, A. Hafaifa, and A. Khadir, "Optimization of anti-reflective coatings using a graded index based on silicon oxynitride," J. Comput. Electron., Vol. 18, pp. 971–981, September 2019.

[7] A. Ruland, C. Schulz-Drost, V. Sgobba, and D. Guldi, "Enhancing photocurrent efficiencies by resonance energy transfer in CdTe quantum dot multilayers: towards rainbow solar cells," Adv. Mater., Vol. 23, pp. 4573–4577, October 2011.

[8] T. Franzl, T. Klar, S. Schietinger, A. Rogach, and J. Feldmann, "Exciton recycling in graded gap nanocrystal structures," Nano Letters, Vol. 9, pp. 1599-1603, July 2004.

[9] T. Klar, T. Franzl, A. Rogach, and J. Feldmann, "Super-efficient exciton funneling in layer-by-layer semiconductor nanocrystal structures," Adv. Mater., Vol. 17, pp. 769–773, March 2005.

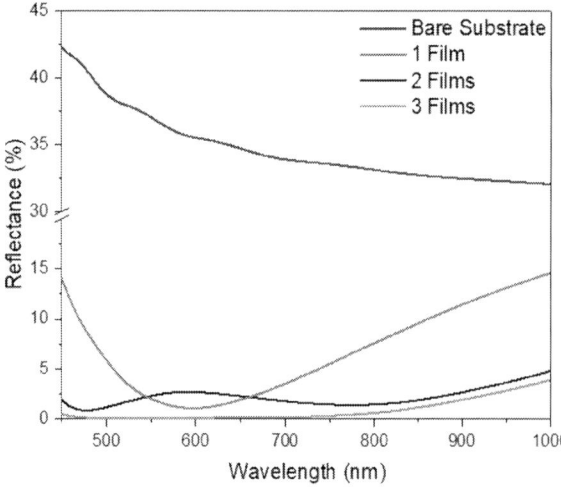

Fig. 3. Modeled reflectance for each film design.

Chemomechanics of Halide Perovskites: Linking Mechanical Behavior with Reliability

Nicholas J Rolston

Arizona State University, Tempe, AZ, United States

Organic-inorganic metal halide perovskite semiconductors have demonstrated real promise as a next-generation thin film photovoltaic technology based on unprecedented gains in device efficiencies and electronic tunability; however, the lack of device reliability remains a key drawback. Fracture properties of thin film materials play a key role in determining reliability but have been largely overlooked in halide perovskites. All of the fracture studies to date for halide perovskites were performed in the absence of chemical, thermal, optical, or electrical stresses, where crack growth through a thin film does not occur below Gc and there is no external driving force for diffusion or reaction. Real-world operational conditions are far from equilibrium, however, and photovoltaic materials are especially likely to endure environments where conditions are subject to rapid change. When debonding of device interfaces is initiated under these non-ideal conditions, external species (e.g., water, UV radiation) near the crack tip may strain or break neighboring bonds, resulting in subcritical fracture processes even if the mechanical driving force on the material is below the fracture energy (i.e., G

Planar and Nanowire InP Thin Solar Cells for Ultralight Space Power Applications

Sara Anjum, Pilar Espinet Gonzalez, and Harry A. Atwater

California Institute of Technology, Pasadena, California, 91125, USA

Abstract—**Indium phosphide (InP) thin film solar cells have considerable potential for low-cost space photovoltaic applications due to their efficiency, ultralight weight form factor, favorable surface recombination properties, optimal bandgap, and innately high radiation resistance compared to silicon and gallium arsenide (GaAs). However, InP cells have received less attention than their GaAs and GaInP/GaAs counterparts for space photovoltaic application. However, future ultralight space photovoltaics with specific power greater than 1kW/kg will require innovative designs for flexible radiation-hard cells without conventional cover glass radiation shielding. We investigate here designs for ultralight InP space photovoltaics, including nanowire array cells, planar thin-film cells with TiO$_2$/ITO optimized antireflection layers, and planar thin-film cells with pyramidally-textured TiO$_2$ antireflection layers, and assess these designs with respect to efficiency, specific power, and scalable fabrication.**

Index Terms—**Space, photovoltaic, solar cell, nanowires, thin-film, pyramid**

I. Introduction

To operate well in space for a long time, solar cells for space applications need to be lightweight, highly-efficient, and radiation resistant. Indium phosphide (InP) as a solar cell material is promising for space applications because of its bandgap and its lower surface recombination and higher innate resistance to radiation damage relative to gallium arsenide (GaAs) and silicon (Si). However, InP is less well-understood as a material in general compared to Si and GaAs, and limited studies with the constraints of space photovoltaics in mind.

Previous work has suggested the potential of both InP thin-film planar and nanowire (NW) solar cells as alternatives with better radiation resistance than their GaAs counterparts, protecting both short-circuit current (J_{SC}) and open-circuit voltage (V_{OC}) from degradation [1] [2]. However, the beginning-of-life efficiencies of InP NW cells are lower than for the GaAs NW cells in [2], so further investigation is needed to compare the performance and end-of-life efficiency in state-of-the-art NW cells of both materials, which currently are lower than those for planar bulk cells. In addition, beyond absolute device efficiency, fabrication complexity, deployability, and device robustness are important factors to consider for space photovoltaic applications.

In this work, we simulated the AM0 spectral absorption of InP nanowire solar cells with a geometry like those previously prototyped in [3], and compared it to the spectral absorption in planar devices of equivalent thickness. Our planar cell

We acknowledge financial support from Space Solar Power Project.

features a 10 nm titanium dioxide (TiO$_2$) layer as an electron-selective contact along with a 60 nm indium tin oxide (ITO) layer as a transparent top contact, which also serves as an antireflection (AR) coating, while the pyramidally-textured cell combines a TiO$_2$ electron selective contact layer with TiO$_2$ pyramids, previously studied as an AR layer for silicon cells [4]. We report the carrier photogeneration profiles and spectral absorption in all three geometries.

II. NW and Planar InP Cell Geometries

For the NW geometry, we simulated an array similar to the one reported in [3], the current record for InP NW solar cells. The top radius of the tapered NW was $r_{top} = 75$ nm, the bottom radius was $r_{bot} = 175$ nm, the height was $h = 1.6\,\mu$m, and the distance between adjacent wires was set to $a = 500$ nm. Such a wire has an equivalent planar thickness of $t_{eq} \approx 330$ nm, calculated as:

$$t_{eq} = \frac{V}{a^2} = \frac{\pi h(r_{top}^2 + r_{top}r_{bot} + r_{bot}^2)}{3a^2} \quad (1)$$

so for purposes of comparison, we investigated planar and pyramidally-textured InP thin cells with similar thicknesses. The NW solar cell in [3] relied on absorption enhancement due to ITO hemispherical nanoparticles that formed during ITO deposition, so we simulated a similar structure with a planar ITO layer of thickness $t_{ITO} = 300$ nm and a nanoparticle with a radius of $r_{ITO} = 175$ nm on top.

For the planar InP thin film cell, we employed a 10 nm TiO$_2$ layer on InP followed by a 60 nm ITO layer as an AR coating, as reported in [5]. The 10 nm TiO$_2$ serves as an electron-selective contact so that the thin 330-nm InP layer does not require an internal pn junction for carrier collection.

For the pyramidally-textured InP cell, the 10 nm TiO$_2$ layer was used as an electron-selective contact, and antireflection was enhanced by formation of an array of TiO$_2$ square pyramids. The base of each pyramid was $b = 450$ nm, while the height of each was $h = 540$ nm. All cells had 50 nm of Ag and 50 nm of Zn as a bottom ohmic contact to p-type InP. Fig. 1 illustrate the 2D cross-sectional schematics of the three solar cells discussed.

III. Absorption and Photocurrent Density Calculations

The spectral absorption and photocurrent density calculations employed finite difference time domain (FDTD) full electromagnetic simulations, which used polynomial fitting of

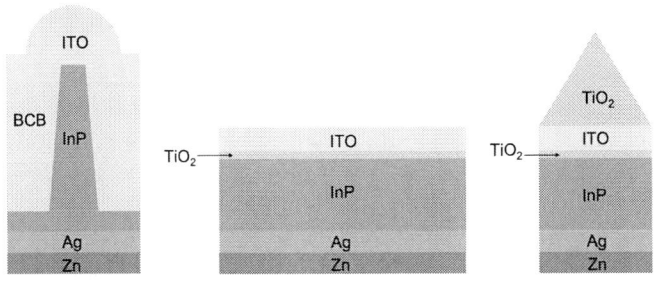

Fig. 1. Schematics of the solar cell architectures evaluated.

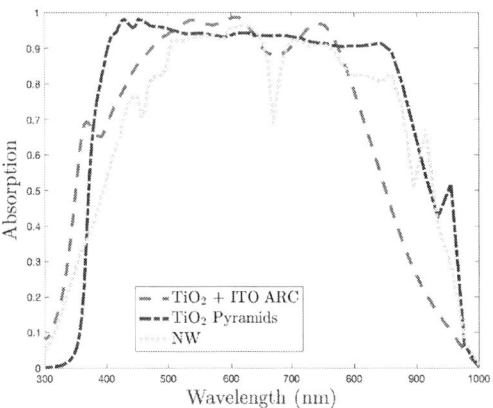

Fig. 2. Absorption in the structures illustrated in Fig. 1.

experimentally-measured complex refractive indices for each material to interpolate the indices of refraction at intermediate wavelengths and accordingly model light absorption. The carrier photo generation analysis calculates an absorption spectrum and a predicted J_{SC} with the assumption that each absorbed photon leads to one collected charge carrier, i.e., unity internal quantum efficiency. A single period of each structure was simulated, as illustrated in the schematics above, with appropriate symmetric and anti-symmetric periodic boundary conditions used to simulate an entire cell. The simulations employ the AM0 spectrum provided in [6]. Due to the polynomial fitting of the material data, care must be taken with the solar spectrum range so as not to simulate unphysical sub-bandgap absorption. This is why we limited the wavelength range from 300-1000 nm.

IV. RESULTS AND DISCUSSIONS

Table I summarizes the predicted J_{SC} for the different geometries. We can see that the different methods of absorption enhancement can theoretically achieve similar J_{SC}. Fig. 2 plots the simulated absorption spectra of the cells to illustrate the differences between the geometries.

The planar InP cell with the AR coating enhances absorption at shorter wavelengths more than the other geometries, while the other two geometries better enhance absorption at longer wavelengths. All three structures absorb well in the 500-750 nm wavelength range. The NWs show a dip in absorption at around 675 nm, largely due to the presence of the ITO structure, as illustrated by Fig. 4. The predicted J_{SC} of the pyramidally-structured cells is the highest. Figure 2 illustrates that the pyramidally-structured cell generally outperforms the other two structures at the longer and shorter wavelengths while exhibiting comparable performance throughout the middle of the wavelength range.

However, one caveat is that the currently-simulated pyramidally-structured cell represents a limiting case, since a transparent conductor is needed on top of the InP to transport charges to the metal ohmic contacts with minimal absorption loss. The planar InP cell with TiO_2 and ITO needs the ITO layer to create the 70 nm thick AR coating since previous work showed that 10 nm TiO_2 was the optimal thickness for maximizing V_{OC}. However, the TiO_2 pyramids already serve an antireflection function, so the ITO layer will simply hurt its performance. Fig. 3 illustrates this by comparing the absorption spectra of the pyramidally-textured cell with one containing a 40-nm ITO layer as shown in the inset. Table II summarizes the different J_{SC} values for the different ITO configurations across the cell geometries.

The ITO layer hurts absorption throughout most of the wavelength range as compared to the structure without ITO since ITO is not completely transparent in the wavelength range of interest. The thinner the ITO used, the closer the performance of such a cell can be to the limiting case shown here. The TiO_2 pyramids themselves can be fabricated using nanoimprinting, which makes them potentially easier and cheaper to fabricate than NWs.

Similarly, the performance of the NW geometry suffers due to the ITO. Fig. 4 plots the absorption of the NW array described above along with one with only a 50-nm thick planar ITO layer and one with no ITO layer, with the latter representing another limiting but unrealistic situation.

TABLE I
J_{SC} OF DIFFERENT GEOMETRIES

Geometry	J_{SC} (mA/cm^2)
NW	35.0852
Planar (no pyramids)	33.9873
Pyramids	38.0511

TABLE II
J_{SC} WITH DIFFERENT ITO CONFIGURATIONS

Geometry	J_{SC} (mA/cm^2)
NW (Planar + Hemispherical ITO)	35.0852
NW (No ITO)	38.4783
NW (Planar ITO Only)	33.8494
Pyramids (No ITO)	38.0511
Pyramids + Planar ITO	35.8569

978-1-7281-6118-1/22 $31.00 © 2022 IEEE

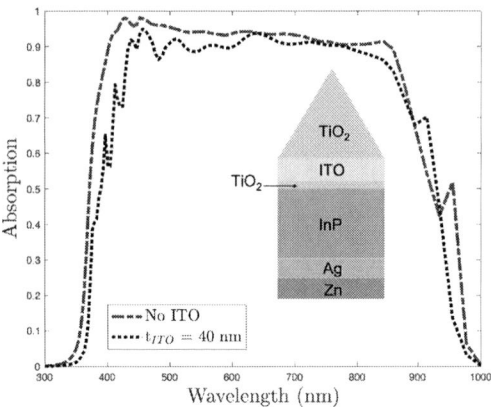

Fig. 3. Absorption of the pyramidal structure with and without ITO below the TiO$_2$ pyramids, along with a schematic of the ITO-containing structure.

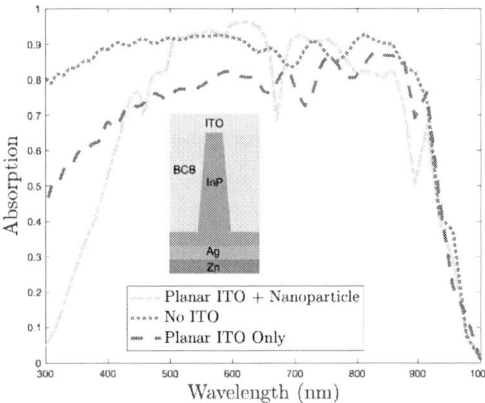

Fig. 4. Absorption of the NW structure with different ITO configurations and a schematic of the structure with only planar ITO of thickness t$_{ITO}$ = 50nm.

The absorption plot shows that the geometry with the ITO hemisphere, which results in Mie scattering of the light into the wire, hurts short-wavelength absorption relative to the structures with and without just planar ITO. This is in part due to the larger quantity of ITO used in the structure containing the hemispherical nanoparticle, in which the planar ITO layer was 300 nm thick as opposed to the 50 nm layer we used in our simulation. Despite the lower amount of ITO used in the geometry shown in the inset of Fig. 4, the NW array with planar ITO alone does not absorb as much light in the middle of the wavelength (500-800 nm) range. Coupling a thinner planar ITO layer and an optimized nanoparticle radius could still further improve the device performance. However, given that the nanoparticles were a byproduct of depositing ITO onto a NW array with the wire top poking through the benzocyclobutene (BCB) infill surrounding it, more work or new fabrication methodologies would be needed to optimize the ITO dimensions for improved performance. Additionally, the creation of the nanoparticles has been difficult to replicate, which currently means we cannot consistently rely on the nanoparticles for absorption enhancement in real devices.

When taking all of these factors into consideration, as well as the fabrication challenges of each method, it becomes clear that a simple heterojunction thin-film planar InP cell with an optimized AR coating warrants further study and consideration for space solar cells alongside structures with more complex means of absorption enhancement. Recent experimental work on fabricating thin films of InP includes the spalling technique demonstrated in [7]. While the fabricated films demonstrated thus far are currently 15 μm thick, further work, both on this technique and other fabrication methods, could help achieve thin InP films.

V. CONCLUSIONS

This paper compared different potential InP solar cell geometries for efficient solar cells suited for space. They all had the same equivalent planar thickness in order to isolate the effects of geometry device on device absorption. The planar and NW cells exhibited similar overall absorption, while the pyramidally-textured cell exhibited stronger performance at the edges of the spectrum examined, resulting in higher predicted J$_{SC}$. However, the properties and configuration of the ITO used in the devices has a heavy impact on device absorption, to the point that thin-film heterojunction planar InP photovoltaics with simple AR coatings still offer much promise.

Further work will involve an analysis of the electrical properties of such devices to understand device recombination and final device performance, simulations of the radiation resistance of some of these devices, fabrication, and testing both device performance and radiation resistance.

REFERENCES

[1] C. J. Keavney, R. Walters, and P. J. Drevinsky, "Optimizing the radiation resistance of InP solar cells: Effect of dopant density and cell thickness," *Journal of Applied Physics*, vol. 73, no. 1, pp. 60–70, 1993. [Online]. Available: https://doi.org/10.1063/1.353830

[2] P. Espinet-Gonzalez, E. Barrigón, G. Otnes, G. Vescovi, C. Mann, R. M. France, A. J. Welch, M. S. Hunt, D. Walker, M. D. Kelzenberg, I. Åberg, M. T. Borgström, L. Samuelson, and H. A. Atwater, "Radiation tolerant nanowire array solar cells," *ACS Nano*, vol. 13, no. 11, pp. 12 860–12 869, 11 2019. [Online]. Available: https://doi.org/10.1021/acsnano.9b05213

[3] G. Otnes, E. Barrigón, C. Sundvall, K. E. Svensson, M. Heurlin, G. Siefer, L. Samuelson, I. Åberg, and M. T. Borgström, "Understanding InP nanowire array solar cell performance by nanoprobe-enabled single nanowire measurements," *Nano Letters*, vol. 18, no. 5, pp. 3038–3046, 05 2018. [Online]. Available: https://doi.org/10.1021/acs.nanolett.8b00494

[4] T. K. Chong, J. Wilson, S. Mokkapati, and K. R. Catchpole, "Optimal wavelength scale diffraction gratings for light trapping in solar cells," *Journal of Optics*, vol. 14, no. 2, p. 024012, jan 2012. [Online]. Available: https://doi.org/10.1088/2040-8978/14/2/024012

[5] X. Yin, C. Battaglia, Y. Lin, K. Chen, M. Hettick, M. Zheng, C.-Y. Chen, D. Kiriya, and A. Javey, "19.2% efficient InP heterojunction solar cell with electron-selective tio2 contact," *ACS Photonics*, vol. 1, no. 12, pp. 1245–1250, 12 2014. [Online]. Available: https://doi.org/10.1021/ph500153c

[6] "2000 ASTM Standard Extraterrestrial Spectrum Reference E-490-00," 2000. [Online]. Available: https://www.nrel.gov/grid/solar-resource/spectra-astm-e490.html

[7] Y. Lee, I. Yang, H. H. Tan, C. Jagadish, and S. K. Karuturi, "Monocrystalline inp thin films with tunable surface morphology and energy band gap," *ACS Applied Materials & Interfaces*, vol. 12, no. 32, pp. 36 380–36 388, 08 2020. [Online]. Available: https://doi.org/10.1021/acsami.0c10370

Effective Passivation of CdTe Rear Interface via Thin Selenium Interface Layer Indicated by Surface Photovoltage Spectroscopy

Michael A Scarpulla, Nathan D Rock, Amit Munshi

University of Utah, Salt Lake City, UT, United States

Colorado State University, Fort Collins, CO, United States

We present evidence of the passivation of the CdTe back interface by the application of a thin arsenic doped selenium layer. Surface photovoltage spectroscopy measurements are presented indicating the reduction of band bending on the rear surface of completed cadmium telluride photovoltaic cells.

Study of Perovskite Solar Cells under High-Fluence, Low-Energy Proton Radiation

Michael D Kelzenberg, Ahmad R. Kirmani, Kaitlyn T. VanSant, Joseph M. Luther, Harry A. Atwater

California Institute of Technology, Pasadena, CA, United States

National Renewable Energy Laboratory, Golden, CO, United States

Perovskite photovoltaics are of interest for power generation in space applications owing to their low mass, ease of fabrication, and tolerance to radiation. Here, we study the performance of perovskite solar cells irradiated by low-energy (30 and 75 keV) protons, with fluences ranging from 4.3E13 cm-2 to 1.7E14 cm-2. These fluences are relatively high, intended to study the viability of unshielded perovskite solar cells for long-duration missions to harsh radiation environments. We find that the cells are considerably degraded at these fluences, suggesting that some amount of radiation shielding may be necessary for such missions. The degradation varies considerably with cell architecture.

Photophysical Properties of CdSe/CdTe Bilayer Solar Cells: A Confocal Raman and Photoluminescence Microscopy Study

Abasi Abudulimu[1]*, Jaroslav Kulicek[2], Ebin Bastola[1], Adam B Phillips[1], Aesha Patel[1], Dipendra Pokhrel[1], Manoj K. Jamarkattel[1], Michael J Heben[1], Bohuslav Rezek[2]*, and Randy J Ellingson[1]*

[1] Wright Center for Photovoltaics Innovation and Commercialization (PVIC), Department of Physics and Astronomy, The University of Toledo, Toledo, OH 43606 USA

[2] Faculty of Electrical Engineering, Czech Technical University in Prague, 166 27 Prague, Czech Republic

Abstract—**Understanding and controlling the optical and electrical properties of the solar cells, from the absorber layer to the complete devices, is one of the key elements for engineering high-efficiency devices. Such an understanding, especially the correlation between device performance and optical-structural-morphological properties of film stack, is still lacking in the field of cadmium selenide/telluride alloy-based solar cells. Here, we report confocal Raman and photoluminescence microscopy study results obtained through exciting both film and glass sides of cadmium selenide and cadmium telluride bilayer device stacks treated with cadmium chloride. We show that the device stack, especially the glass side, losses significant charge carries to the recombination arising from uniformity issues related to material composition, energetics, and defects. Furthermore, there is a high energy tail emission peak originating from CdTe, and CdSe can suppress it significantly under CdCl₂ treatment.**

Keywords— CdTe/CdSe bilayer, Confocal Raman, Photoluminescence, Charge Recombination

I. INTRODUCTION

Polycrystalline CdS/CdTe-based thin film solar cells have been considered a promising technology for solar energy conversion [1]. Recently, replacing CdS with CdSe was found to improve device performance significantly by regulating the bandgap of the absorber and reducing the parasitic absorption in the front side of the device stack, which fueled the CdTe research field further [2]. It is understood that, under CdCl₂ treatment, CdSe and CdTe form CdSeTe alloy with a bandgap lower than that of the CdTe (depending on the Se concentration), leading to a significant increase in carrier lifetime and device power conversion efficiency [3]. The attempt of reducing carrier recombination at the back of the device stack by repelling electrons back into the absorber also succeeded in CdSe/CdTe bilayer solar cells and a relatively high device efficiency has been achieved [4]. Bothwell et al. reported an increased Photoluminescence (PL) intensity along with a PL peak shift (from 1.5 eV to 1.42 eV) for the glass-side illuminated CdSeTe sample [5]. Similar results have been reported in several other studies, and the PL peak shift, from 1.5 to 1.42 eV, is assigned to emission from direct band-to-band charge recombination [6-

7]. However, understanding the optoelectronic properties of the device stack in correlation to the morphology at each layer and along with the interaction they have with the next neighboring layer is still lacking. Thus, following our previous work [8], here we present the photophysical properties of the CdSe/CdTe bilayer film stack studied with correlative confocal PL and Raman microscopy.

II. EXPERIMENTAL

A typical device structure for the data presented in this report is shown in the schematics 1. 100 nm thick Al2O3 layer was deposited on soda lime glass by reactive RF sputtering of Al target at a power of 150 W, maintaining 10 mT pressure of argon gas containing 5% oxygen followed by deposition of CdSe and CdTe bilayers, as reported previously [8], by using a multi-source deposition (MSD) system. During the deposition of these bilayers CdSe/CdTe, the base pressure was maintained at 2.6x10-6 Torr, and the substrate temperature of 400 ℃. After the deposition of CdSe/CdTe layers, CdCl₂ treatment (saturated CdCl2 solution in methanol) of the device stack was carried out at 400 ℃ for 30 mins in a dry air environment. Later, the device stack was rinsed with methanol twice to remove the excess CdCl2 from the CdTe surface. For device completion, 50 nm of gold (Au) back contact was thermally evaporated at 1 Å/s through a mask with a dot cell area of 0.08 cm².

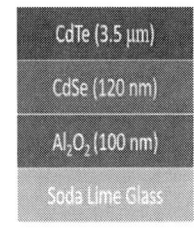

Schematics 1. Device Structure

WItec Alpha 300ARS (AFM-SNOM-Raman combined microscope system equipped with CCD-based spectrometer) is used to map the confocal PL and confocal Raman images. A fiber-coupled 532 nm laser (Raman specific, supplied by WItec) is applied through 20x and 100x objectives as the excitation source for Raman mapping, while a filter-based supercontinuum laser from NKT Photonics (at 532 nm FWHM=10nm) is used as the excitation source for PL mapping. In both measurements, the signal is collected in reflection mode. Sample positioning (X–Y piezo stage), signal/image optimization, and data recording & processing are achieved via WItec Control/Project-Pro software

III. RESULTS AND DISCUSSION

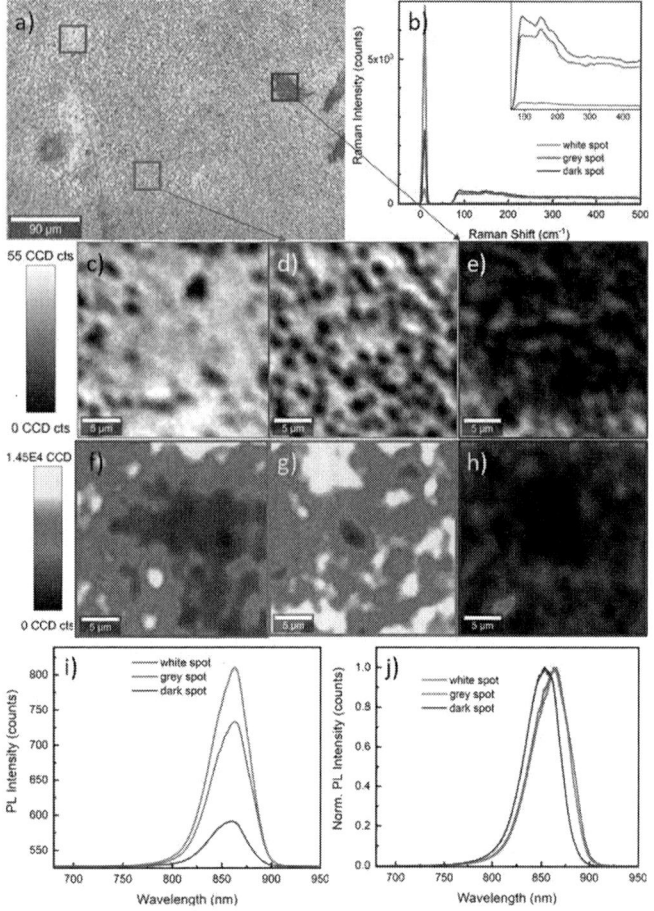

Fig. 1. Glass side of the sample: (a) optical image; (b) averaged Raman spectra over the marked area in (a); (c-e) Raman images (filtered at the laser signal) of the marked areas in (a); (f-h) PL images of the marked areas in (a); (i) PL spectra, (j) normalized, resulted from spatially averaging f), g), and h).

Fig. 1 shows an exemplary optical image of a CdCl$_2$ treated CdSe/CdTe sample measured from the glass side. The sample surface exhibits different areas in terms of color (white, grey, and dark, respectively). To see if there are any differences in optical properties, we first acquired confocal Raman images of those spots (marked with blue, red, and black rectangular) to better visualize them. It is worth mentioning that the thick substrate (3.3 mm thick soda lime glass) jeopardizes the quality of Raman images/spectrum we can acquire without burning the sample with high laser intensity with a 20x objective. Therefore, the Raman images are filtered for the reflected laser signal to visualize the morphological differences among those different areas. Fig. 1(c-d) and (b) show the corresponding Raman images of the three marked regions and the averaged spectra, respectively. They indicate that the dark area is relatively amorphous.

Fig. 1(f-h) and (i-j) display the PL images and spectra corresponding to the three marked spots in Fig 1. They show that the dark spot emits significantly less, and its emission peak also blue shifted (see the normalized PL spectra) in contrast to the grey spot (which is the dominant feature of the film). The lower PL intensity correlates with defects (or less-activated CdTe) while the emission peak shift correlates with the CdTe/CdSe alloy composition (in terms of selenium concentration). In other words, the dark spots are the charge recombination centers. The difference in the emission peaks between the spots can be understood as the presence of inconsistent bandgap in the film, which could act as energy barriers for charge transport. Both PL peak position and the intensity variations suggest that the film uniformity (energetic or structural) is one of the performance limiting factors that must be resolved in order to improve the device efficiency. Resolving such a problem means seeing just the grey spot (marked red in Fig.1a) everywhere on the glass side of the film.

The optical and confocal Raman/PL images of the same sample measured from the film side are given in Fig. 2. Interestingly, the film side of the sample is relatively uniform (see Fig. 2(a-b, g)). Taking advantage of the high-resolution confocal microscope, we compared the PL spectra of three different spots, dark, blue, and bright spots in the PL map (Fig. 2c, marked with circles). Fig. 2e shows that there is a noticeable difference in the PL intensities but spectral shapes and peak positions. It implies that although the material composition on the film side is very uniform (which is expected as the film side is dominated by CdTe, compare the PL peak positions for the glass side and film side), there is still significant carrier loss from film non-uniformity.

Another important point to mention is that the high energy tail observed in the PL spectra of the film side (around 775nm, see Fig. 2e-f) does not appear (or suppressed significantly) on the glass side of the sample. This tail emission is also present on both sides of the CdTe-only sample (neither CdCl$_2$ treated nor CdSe incorporated). Thus, it is tempting to assign the higher energy tail in the PL spectra of the film side to the emission peak arising from defects in CdTe (rich) film. And it is also convenient to interpret the reduction of this high energy tail PL intensity on the glass side as a result of passivating the High energy defects in CdTe by CdSe under CdCl$_2$ treatment. However, its origin and impact on the device performance should be further investigated, as tail or defect emission often occurs on the lower energy (longer wavelength) side of the main emission peak causing device performance loss.

To have a better resolution we imaged the film side of the sample with a 100x objective, which was not possible for the glass side measurement due to the thick glass substrate. While the optical image again shows the film to be very uniform (Fig. 2g), the Raman and PL images demonstrate that there are still noticeable variations in the material uniformity and consequently the PL emission properties within a few micrometers range. And the correlative Raman and PL images reveal that the less PL emitting spots on the film correlate with a stronger transverse optical mode signal of the tellurium, see the regions with a red circle in Fig. 2h-I (note that the Raman image, Fig. 2h, is filtered for the Tellurium transverse optical signal at 126 cm^{-1}, see also the inset on Fig. 2d).

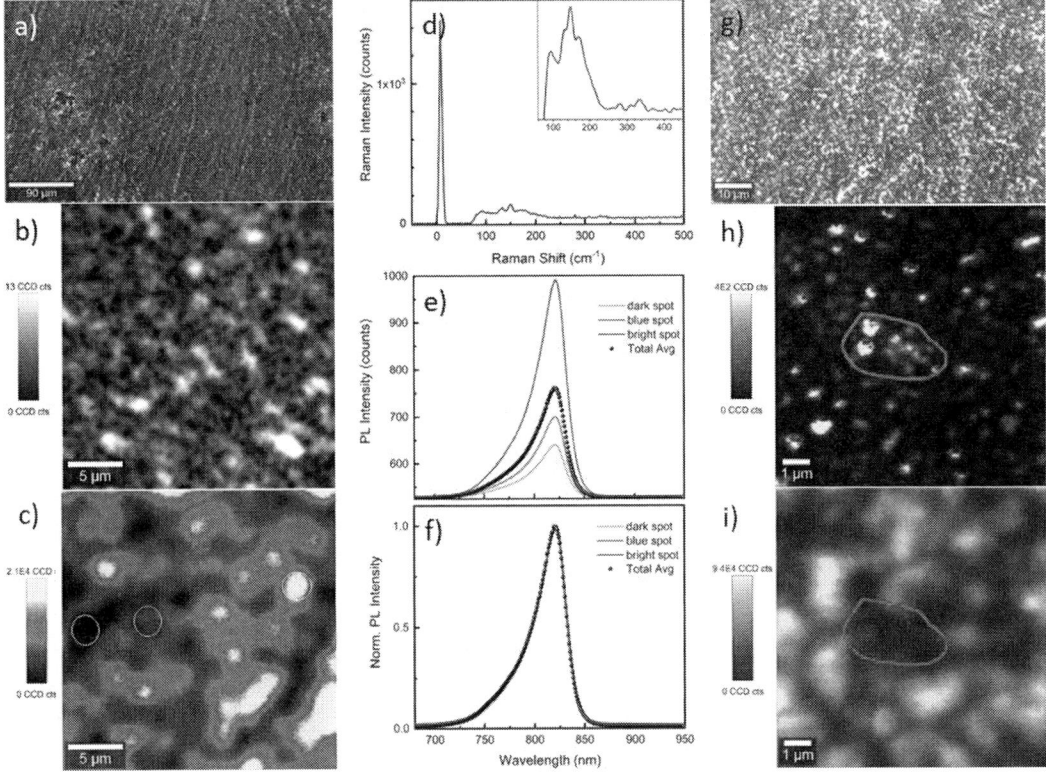

Fig. 2. Film side of the sample: (a) optical image; (b) raman image (filtered at laser signal); (c) PL image; (d) raman spectra; (e) recorded at the marked area in(a); (e-f) as-measured and normalized PL spectra recorded at dark, blue, bright bright spots marked with circles in (c); (g) optical image taken with 100x objective; (h) correlative Raman (filtered at 126 cm^{-1}, the transverse optical mode of tellurium) and PL (i) images taken with 100x objective.

IV. CONCLUSIONS

We have studied the photophysical properties of the CdSe/CdTe bilayer device stack using confocal Raman and PL microscopy. We found that sample uniformity, in terms of material composition/energetics and defects, is one of the main problems for quenching charge carries (up to 80%), especially illumination from the glass side is concerned. Such a film uniformity issue is less critical on the film side of the device stack. However, the film side suffers from both inefficient CdCl$_2$ activation and high energy defects/barriers. The former causes fast carrier recombination (due to reduced grain size) and less carrier generation (due to less photon absorption in the near-infrared region of the solar spectrum). The latter causes an increase in carrier recombination and a reduction in carrier extraction. Moreover, we noticed that there is a high energy tail emission peak on the PL spectrum of the sample originating from CdTe, and CdSe can suppress it significantly under CdCl$_2$ treatment.

ACKNOWLEDGMENT

Authors from the UT-PVIC acknowledge the support of Air Force Research Laboratory under agreement number FA9453-19-C-1002. The U.S. Government is authorized to reproduce and distribute reprints for Governmental purposes notwithstanding any copyright notation thereon. The views and conclusions contained herein are those of the authors and should not be interpreted as necessarily representing the official policies or endorsements, either expressed or implied, of Air Force Research Laboratory or the U.S. Government. Authors from CTU in Prague acknowledge the support of ERDF and MEYS through the Centre of Advanced Photovoltaics project (CZ.02.1.01/0.0/15_003/0000464).

REFERENCES

[1] M. Green, E. Dunlop, J. Hohl-Ebinger, M. Yoshita, N. Kopidakis, and X. Hao, "Solar cell efficiency tables (version 57)," Progress in Photovoltaics: Research and Applications, vol. 29, pp. 3–15, 2021.

[2] Xiong, G. and Metzger, W., "Thin Film Cadmium Telluride Photovoltaics," Comprehensive Renewable Energy, pp.362-387, 2022.

[3] D. Kuciauskas et al., "Microsecond carrier lifetimes in polycrystalline CdSeTe heterostructures and in CdSeTe thin film solar cells," PVSC, pp. 82–84, 2020.

[4] M. K. Jamarkattel et al., "Improving CdSeTe Devices with a Back Buffer Layer of CuxAlOy," in IEEE Journal of Photovoltaics, vol. 12, no. 1, pp. 16-21, 2022.

[5] Bothwell, A.M., Drayton, J.A., Jundt, P.M., Sites, J.R. "Close-Space Sublimation-Deposited Ultra-Thin CdSeTe/CdTe Solar Cells for Enhanced Short-Circuit Current Density and PL," J. Vis. Exp. (157), e60937, 2020.

[6] D. Kuciauskas, J. Moseley, P. Š'cajev, and D. Albin, "Radiative efficiency and charge-carrier lifetimes and diffusion length in polycrystallineCdSeTe heterostructures," Physica Status Solidi Rapid Res. Lett., vol. 14, 1900606, 2019

[7] Mia M.D et al., "Electrical and optical characterization of CdTe solar cells with CdS and CdSe buffers—A comparative study," J. Vac. Sci. Technol., B, 5, 36, 052904, 2018.

[8] Bastola E, Phillips AB, Barros-King G, Jamarkattel MK, Li DB, Quader A, Pokhrel D, Friedl J, Gibbs JM, Mathew X, Yan Y. Understanding the Interplay Between CdSe Thickness and Cu Doping Temperature in CdSe/CdTe Devices. IEEE JPV, 20;12(1):11-5, 2021.

The Capability of a Grid-Forming Inverter to Support Dynamic Microgrids with High Penetrations of Photovoltaics Systems

Rachid Darbali-Zamora[1], C. Birk Jones[1] Matthew S. Lave[1], and Erick E. Aponte-Bezares[2]

[1]Sandia National Laboratories, Albuquerque, New Mexico, 87185, USA
[2]University of Puerto Rico-Mayagüez, Mayagüez, Puerto Rico 00682, USA

Abstract – **Grid-forming (GFM) inverters that support photovoltaic (PV) and dynamic microgrids, that expand and contract depending on available resources, must adapt. Unlike common microgrid approaches that require significant financial investments to maintain operations of a system of a set size, the approach presented here provides necessary resilience for a critical load at a reduced cost. To work, a GFM inverter must maintain appropriate voltage and frequency as the magnitude of PV output and loads change quickly during a switching event that acts to reduce the difference between generation and consumption. The simulation effort, described here, studies the potential capabilities of the GFM to support variable PV generation and fast changes in the microgrid's size that either increase or decrease the system's size. Simulations in *MATLAB/Simulink* showed that a GFM inverter can maintain a frequency around 60 Hz and voltage between 0.995 pu and 1.005 pu during variable PV generation output if microgrids change their size by opening and closing switches in order to maintain balanced operations.**

Index Terms – *photovoltaic inverter, grid-forming, distribution systems, circuit analysis, simulation, stability*

I. INTRODUCTION

Unlike traditional photovoltaic (PV) inverters that operate as grid-following (GFL) devices, grid-forming (GFM) inverters control voltage and frequency at their point of common coupling (PCC) [1], [2]. GFM inverters support electric grids like synchronous generators (SG) but can operate using carbon free resources [3], such as battery energy storage systems. GFM inverters are anticipated to play a large role in ensuring stable operations of microgrids by providing frequency and voltage support. These GFM inverters allow for underserved communities, such as some areas in Puerto Rico, to install and operate low-cost microgrids.

Development and testing of low-cost microgrid solutions is important since many vulnerable communities in Puerto Rico (and other regions of the U.S.) have suffered significantly in recent history. Two of the most notable natural disasters were Hurricane Irma and Maria, which struck the island of Puerto Rico in 2017 [4]. The combination of these two natural disasters caused loss of power resulting in the largest power outage in U.S. history [5], [6]. Even after the power was restored, the electrical systems throughout the island continued to experience outages, especially in rural and remote regions like Corcovada, a small community in western Puerto Rico.

After the devastating 2017 hurricanes, Corcovada's water system was restored in a matter of hours because of an existing diesel generator that powered the 3.5 kW water pump. However, other critical and non-critical loads could not be served without receiving power from the main grid. In 2015, a PV system was installed to operate the community's water pump. In the future, the critical water pump and other loads in the community can be powered during an outage without significant upgrades to the existing system using a GFM inverter.

To understand the GFM inverter's potential, this paper studies the operations of a GFM inverter inside a low-cost microgrid powering the water pump continuously as well as providing power to non-critical loads when distributed PV is able to support it. The administration of such a microgrid requires switching capabilities that will allow for the expansion and reduction of the system that optimally balance the load demand with power generation as conditions change throughout the day. An initial study found that this approach is possible using a Particle Swarm Optimization (PSO) [7]. To power the essential services, a GFM inverter, battery storage, and PV array are located at the critical load per the set up described in [8].

Initial studies, such as the PSO assessment of a "breathable" microgrid that expands and reduces its size as conditions change [7] or the inclusion of a distant PV array into a microgrid [8], assumed that the GFM inverter could maintain the microgrid's frequency and voltage during switching events. But a GFM inverter's specific capabilities for this type of dynamic microgrid switching operation has yet to be explored. This work, therefore, performs an advanced simulation of a potential microgrid in the rural community of Corcovada, Puerto Rico. The simulation effort emulated the GFM inverter, dynamic loads, GFL PV inverters, and electrical lines to provide an initial understanding of a GFM inverter's ability to support transitions in system size and PV availability.

This paper evaluates the integration of a GFM inverter in an isolated microgrid with a large penetration of GFL inverters scattered throughout the system. A dynamic *MATLAB/Simulink* model of an actual power distribution network based on this community in Puerto Rico was created using local load data provided by the community and nearby irradiance profiles to emulate realistic operations. This meant that the irradiance profile includes variability, which is typical for Puerto Rico as rain clouds move into and through the mountainous regions in the afternoons.

The focus of this evaluation was on the GFM inverter's ability to support switching transitions intent on providing a more balanced system, and significant changes in PV power generation. Transient results for this application, which are unreported in existing literature, describe the frequency and voltage dynamics of the microgrid under varying configurations and conditions.

978-1-7281-6118-1/22 $31.00 © 2022 IEEE

II. Distribution Feeder Model Description

Corcovada is a small community in the mountainous region of the municipality of Añasco, located to the West of Puerto Rico, with a population of 627 [2010]. The community has been operating their aqueduct system for 42 years.

In order to evaluate the dynamic switching microgrid, a simulation model of a distribution system representing the community was developed in *MATLAB/Simulink*. Fig. 1 illustrates the distribution feeders and the different switching group locations. The distribution feeder model consists of 50 buses, 48 distribution lines, 43 single-phase transformers, 14 single-phase GFL inverters, a three phase GFM inverter, 36 residential loads and a critical load representing the community water pump. Table I summarizes the switching group generation and load demand.

TABLE I:
MICROGRID SWITCHING GROUP DESCRIPTION

Switching Group	Photovoltaic Inverter			Total Load Demand	
	ID	Phases	Rating (kVA)	Active Power (kW)	Reactive Power (kVAR)
1	-	-	-	28.60	0.00
2	PV 1	AB	5.00	25.80	0.00
3	-	-	-	19.50	0.00
4	PV 3	BC	26.00	44.00	0.00
5	PV 4	BC	23.00	99.35	0.00
	PV 5	BC	5.00		
	PV 6	BC	10.00		
	PV 7	BC	44.00		
6	PV 2	AB	12.00	100.50	0.00
7	-	-	-	2.60	0.00
8	PV 11	AB	32.00	66.55	4.00
	PV 12	AB	5.00		
	PV 13	AB	8.00		
	PV 14	AB	21.00		
9	PV 8	BC	10.00	53.00	0.00
	PV 9	BC	6.50		
	PV 10	BC	4.50		
10	-	-	-	47.60	0.00
Total			212.00	487.50	4.00

The system is a portion of a larger feeder. Hence, the loads are only distributed throughout phases AB and BC, with no loads connected to phases AC. Moreover, there are no lines or transformers for phase AC. This caused a voltage imbalance in the system. The total peak load consumption of the microgrid is 487.50 kW. In addition to the residential loads, the community's 5 hp (3.7 kW) water pump serves as the critical load of the system, connected to phase AB. The total PV generation of the microgrid model is 191 kVA.

In addition to the household PV systems, the water pump has a dedicated PV system with a rated capacity of 21 kVA. The dynamic microgrid is divided into ten switching groups. Notice that some switching groups do not have PV generation. PV systems are located in switching groups 2, 4, 5, 6, 8 and 9. Switching group 8 is the nucleus of the dynamic microgrid, the critical as well as a the GFM inverter. The 100 kVA GFM inverter is a three-phase inverter connected to switching group 8. Switching group 8 has a total load consumption of 66.5 kW and 4 kVAR, while there is a total of 66 kVA in GFL inverter power generation. During the dynamic microgrid group switching transitions, the total load consumption within the dynamic microgrid is 206.62 kW, while the total PV generation within the dynamic microgrid is 221.62 kW.

III. Distributed Energy Resource Modelling

Both GFL and GFM inverters models are tasked with generating power for the microgrid locally but also to safeguard coordinated continuous operations [10]. GFL inverters operate as current sources and following the grid voltage [11]. Multiple GFL inverter models are distributed throughout the microgrid system, as indicated in Fig. 1. GFM inverters define the reference voltage and frequency to ensure continuous operation of the community [12].

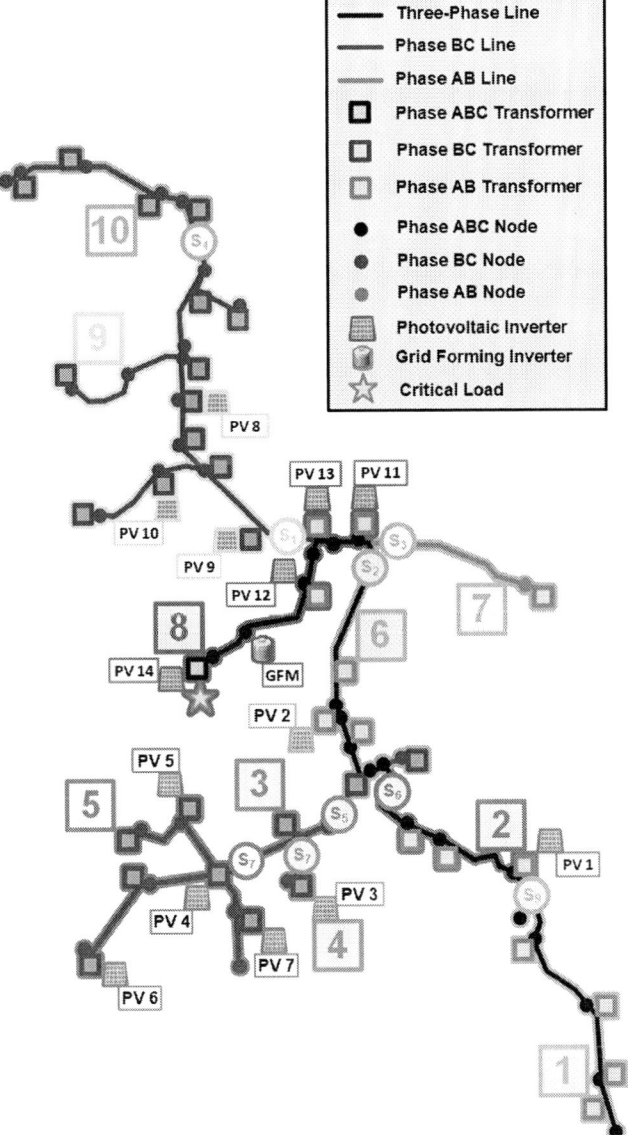

Fig. 1. Diagram of the Corcovada Distribution Feeder shows the lines phases, switching groups, PV locations, loads, and GFM inverter.

A. Grid-Following Inverter Model

To emulate the GFL inverter dynamics, a current controlled single-phase PV inverter model is implemented [13]. The GFL inverter model utilizes a Phase-Locked Loop (PLL) to synchronize to the grid [14]. Fig. 2 illustrates the block diagram for the GFL inverter model. In this diagram, the variable P_{ref} and Q_{ref} are the active and reactive power reference values. The active power is derived from a linear relationship to solar irradiance. The control is represented using the direct (d) and quadrature (q) axis reference frame, and a dq-current controller. Adjusting the d and q axis currents allows controlling both the active and reactive powers, respectively. The resulting equations for the total inverter active and reactive power injected to the grid using the dq-frame are shown in equation (1) and equation (2), respectively.

$$P_{ref} = \frac{3}{2} \cdot \left(V_d \cdot I_d + V_q \cdot I_q \right) \qquad (1)$$

$$Q_{ref} = \frac{3}{2} \cdot \left(V_q \cdot I_d - V_d \cdot I_q \right) \qquad (2)$$

In these equations, the variables V_d and V_q are the d and q voltages in the dq-frame, respectively. The variables I_d and I_q are the d and q currents in the dq-frame, respectively. Table II summarizes the GFL inverter control parameters.

TABLE II:
GRID-FOLLOWING INVERTER PARAMETERS

Variable	Description	Value	Unit
Q_{ref}	Reference Reactive Power	0.00	kVAR
P_{ref}	Reference Active Power	5.00	kW
k_p	Proportional Gain	15.00	-
k_i	Integral Gain	200.00	-
f_o	Grid Frequency	60.00	Hz

A scaling method is used to calculate the controller and filter parameters for parallel-connected GFL inverters [15]. This scaling method is used to represent GFL inverters with different power ratings.

B. Grid-Forming Inverter Model

The dynamic behavior of a three-phase GFM inverter is modeled using the Electric Reliability Technology Solutions (CERTS) control [16], [17], [18], [19]. Fig. 3 illustrates the block diagram for the CERTS GFM inverter model. The active power, reactive power, and voltage magnitude are calculated using the instantaneous values of the three-phase voltages and currents in the $\alpha\beta$-frame. Equation (3) through equation (5) are used to calculate P_o, Q_o, and V_m, respectively.

$$P_o = \frac{3}{2} \cdot \left(V_{g\alpha} \cdot I_{g\alpha} + V_{g\beta} \cdot I_{g\beta} \right) \qquad (3)$$

$$Q_o = \frac{3}{2} \cdot \left(V_{g\beta} \cdot I_{g\alpha} - V_{g\alpha} \cdot I_{g\beta} \right) \qquad (4)$$

$$V_m = \sqrt{v_{g\alpha}^2 + v_{g\beta}^2} \qquad (5)$$

In these equations, variables $v_{g\alpha}$, $v_{g\beta}$, $i_{g\alpha}$, and $i_{g\beta}$ are the $\alpha\beta$-frame voltages and currents, respectively. The active power/frequency droop control adjusts the frequency ω for the GFM inverter voltage. The variable m_p is the active power/frequency droop gain, P_{set} is the power setpoint, and ω_o represents the grid frequency. The reactive power/voltage droop control regulates the voltage of the GFM inverter. The variable V_{set} is the voltage setpoint. The modulation is calculated using a PI controller, where the variables k_{pv} and k_{iv} are the proportional and integral gains, respectively. Table III summarizes the GFM inverter control parameters.

TABLE III:
GRID-FORMING INVERTER PARAMETERS

Variable	Description	Value	Unit
V_{set}	Voltage Setpoint	1.00	pu
P_{set}	Power Setpoint	0.00	pu
m_p	Active Power/Frequency Droop Gain	3.77	pu
m_q	Reactive Power/Voltage Droop Gain	0.05	pu
k_{pv}	Proportional Gain	0.00	-
k_{iv}	Integral Gain	5.86	-
f_o	Grid Frequency	60.00	Hz

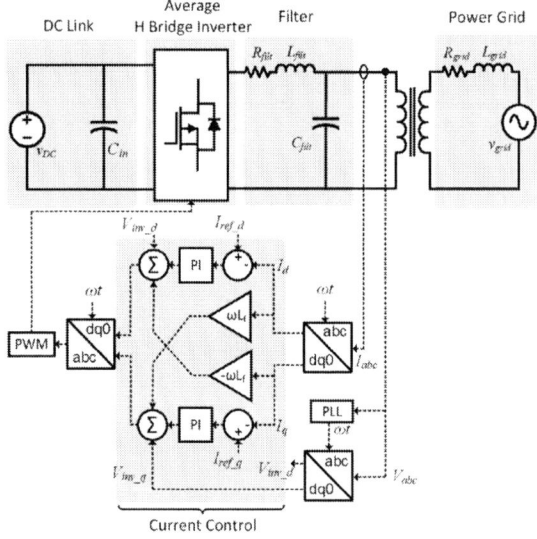

Fig. 2. Block Diagram Describing the dq-current controlled Single-Phase Grid-Following PV Inverter Control Scheme.

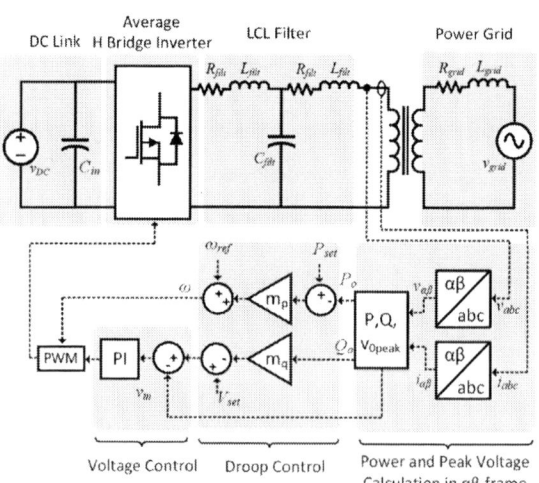

Fig. 3. Block Diagram Describing the CERTS Three-Phase Grid-Forming Inverter Control Scheme.

978-1-7281-6118-1/22 $31.00 © 2022 IEEE

IV. Microgrid Switching Methodology

The simulations involved the operations of a changing microgrid supported by a GFM inverter. Throughout a single day of operations the microgrid expanded and reduced its size depending on the load demand and available PV power generation.

A. Microgrid Reconfiguration Approach

The switching control methodology considers an assessment and reconfiguration that includes monitoring of the system's performance as well as the modification of the switching states. Both the load consumption as well as the PV power generation from the microgrid and the aggregation of the net power for each switching group is monitored. The net power for each switching group and the system topology provides necessary inputs data for the PSO algorithm. The PSO algorithm processes the net power data and considers the topology of the microgrid to determine the new switching states. To determine the switching states, the PSO algorithm considers the overall topology and the location of the critical load. The solution obtained by the PSO algorithm is constrained to include the critical load switching group.

The reconfiguration optimization (that utilizes the PSO algorithm) determined the states for each switch, which defined the size of the microgrid, as described in [7]. The size of the microgrid depended on the optimization's ability to balance the system at different levels of PV power generation and load consumption. When a system was properly balanced, appropriate grid voltage level could be maintained. This will require that both the load consumption and PV power generation be as close as possible. The PSO algorithm minimizes the difference between load demand and PV power generation within the microgrid, as shown in equation (6).

$$\min_{P_l, P_g} \sum_{i=1}^{N} |P_l - P_g| \qquad (6)$$

In this equation, the variable N is the number of iterations and the variables P_l and P_g are microgrid's load demand and power generation, respectively.

The objective of the PSO algorithm was constrained to only consider the switching group conditions that included the critical load. At the very least the microgrid included the critical load's switching group. At most, the optimization results could form a microgrid that included all the switching groups.

The optimization determined the best set of binary switching states that minimized this objective function within predetermined constraints. In this case, the number of particles was equal the number of switches, and the optimal switching configuration produced the smallest difference between load demand and PV power generation. This approach implemented the standard binary PSO algorithm to define the binary state for each switch. Initially, the algorithm set the particles to a random state; then at each iteration the particle state are updated based on two improved values. Finally, the velocity and position equations updated based on the two new values. The PSO algorithm considered all the neighboring switching groups to obtain the global minimum solution.

B. Expansion of the Microgrid

The expansion of the microgrid occurred when there was enough PV power generation to support a larger portion of the feeder's loads. During this type of event, the load increased from 94 kW to 203 kW, as depicted in Fig. 4. With the increase in switching groups, the PV power generation also increased to a maximum of 190 kW. However, because of the startup delay of the GFL inverters, it took 0.3 s after the expansion of the microgrid to reach the maximum PV power generation for the available irradiance.

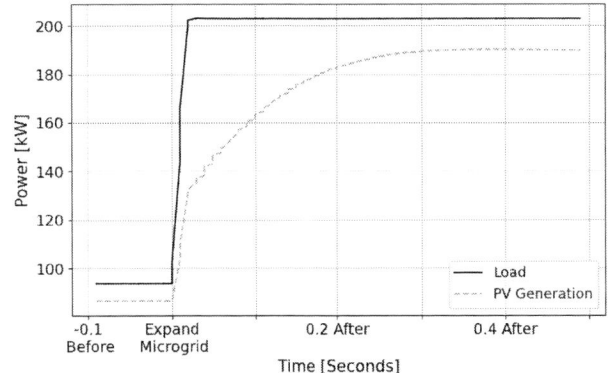

Fig. 4. To improve the balance between load consumption and PV power generation, the system expanded its boundaries to include more non-critical loads. Because of the start of delay of PV inverters, it took approximately 0.3 s for the generation to match closely with the load.

C. Reduction of the Microgrid

The reduction of the microgrid involved the opening of switches that excluded one or more switching groups from the microgrid. This reduced the load demand and the amount of PV power generation, as shown in Fig. 5. Notice from Fig. 5 that the load demand and PV power generation took 0.4 s after the reduction of the microgrid to reach a steady state. The PV power generation took a sharp drop in output when the system was reduced to 95 kW, which matched well with the load demand of 98 kW. However, the solar irradiance starts to decrease, which limited the output of the connected PV arrays.

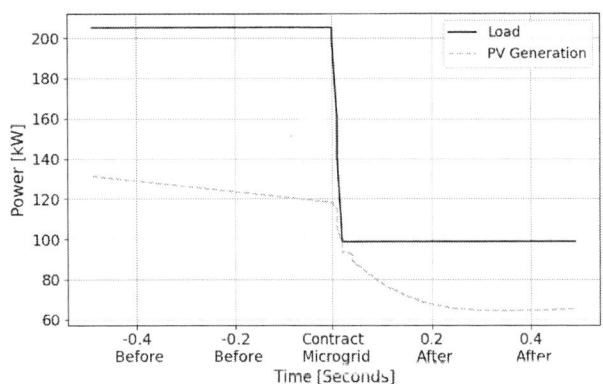

Fig. 5. A reduction in the system size occurs in order to improve the growing difference between load demand and PV power generation. In this case, a few seconds after the switching the active PV power generation.

978-1-7281-6118-1/22 $31.00 © 2022 IEEE

V.   MICROGRID SIMULATION

The *MATLAB/Simulink* model of Corcovada's electric power system was run for a single day of operations. The model considered the PSO switch states for each instance in time. It also used realistic irradiance and load profiles for the area, as shown in Fig. 6 and Fig. 7. Fig. 6 illustrates an example of an irradiance curve collected at the western region of Puerto Rico. Fig. 7 illustrates the total load demand profile used to represent the community residential household power consumption. Fig. 8 illustrates the active and reactive power demand of the community's critical load.

VI.   SIMULATION RESULTS

The expansion and reduction of the dynamic microgrid, via optimized switching control, were supported by the GFM inverter. Sample maps of the dynamic microgrid around 11:00 AM and 4:00 PM are shown in Fig. 8. Between 11 AM and 11:10 AM the isolated electric grid experienced an expansion of its boundaries from four switch groups to eight. A second example shows the reduction in the microgrid's size from three switch groups at 3:40 PM to two at 4:00 PM. In both cases, the GFM inverter was able to supply the necessary active power and reactive power to support the microgrid's frequency and voltage. It also injected or absorbed reactive power needed to maintain proper voltage. The GFM inverter responded well to the change in system size and easily maintained the voltage and frequency around 1.0 pu and 60 Hz, respectively.

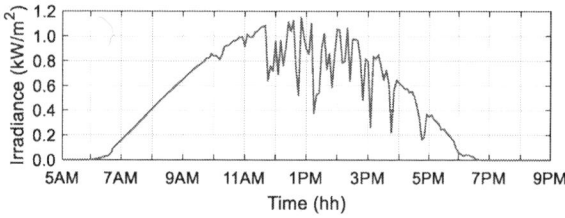

Fig. 6. Irradiance Profile from the Western Region of Puerto Rico.

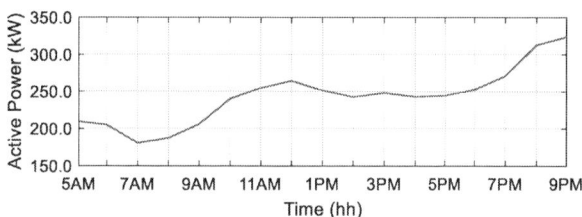

Fig. 7. Total Residential Load Demand Profile for the Community.

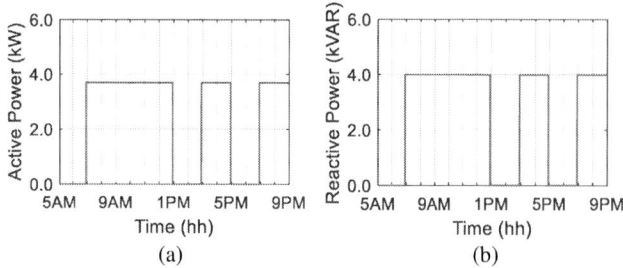

Fig. 8. Water Pump Critical Load Demand Profile. (a) Active Power. (b) Reactive Power.

Fig 9 illustrates an example of how the dynamic microgrid can change its switching states to either expand or reduce the size of the system.

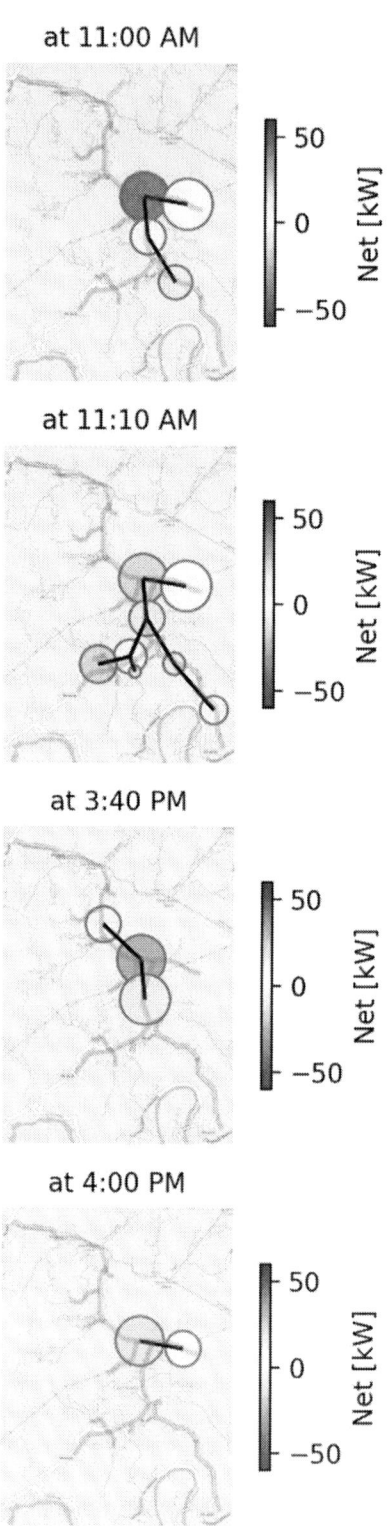

Fig. 9. Example operations of the dynamic microgrid show the reduction in the system size between 11:00 AM to 11:10 AM. At 3:40 PM the system powered three switch groups, which changed at 4:00 PM to include only two groups.

Fig. 10 illustrates the simulation results obtained for the total load demand and PV generation. Fig.11 illustrates the results for the per-phase voltage measured at the GFM inverter PCC. At the point, where the microgrid altered its boundaries by either increasing or decreasing its size, the voltage had noticeable deviations shown in plots in Fig. 11. However, in both situations, the GFM inverter's voltage did not change by more than 0.5%. Fig. 12 and Fig. 13 illustrate the power generation and load demand, RMS voltage and frequency during the microgrid expansion and reduction, respectively. When expanding the microgrid there was a frequency overshoot of less than 0.07% at most.

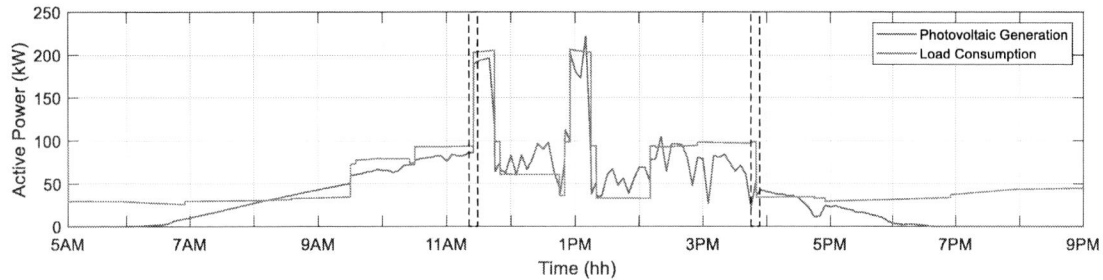

Fig. 10. Simulation Results for the Total Load demand and Generation within the Main Microgrid.

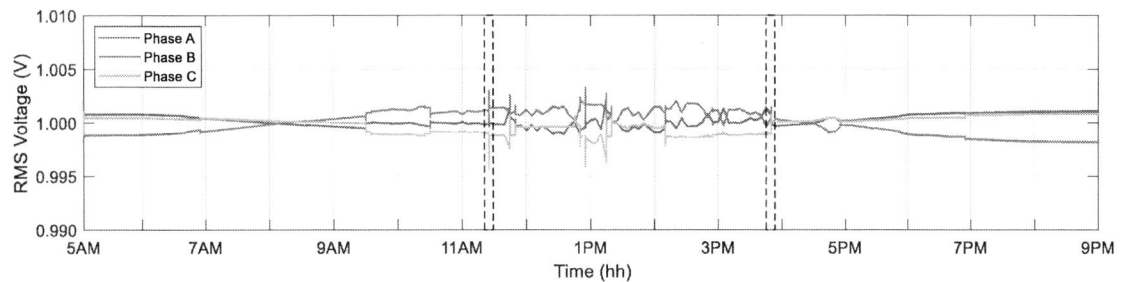

Fig. 11. Simulation Results for the GFM Inverters Phase Voltages During Switching Transitions.

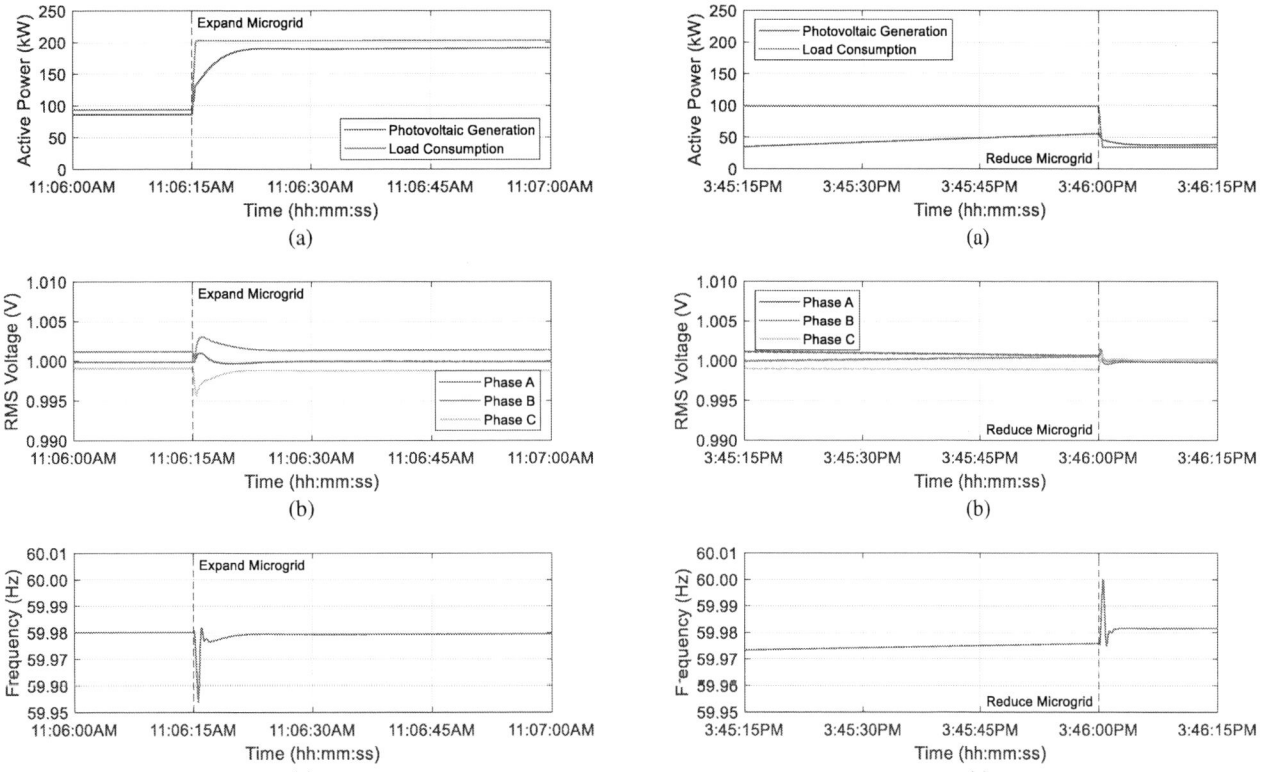

Fig. 12. Simulation results for the microgrid during an Expansion. (a) Generation Load. (b) Voltage. (c) Frequency.

Fig. 13. Simulation results for the microgrid during a Reduction. (a) Generation Load. (b) Voltage. (c) Frequency.

978-1-7281-6118-1/22 $31.00 © 2022 IEEE

Within the first seconds after the event, the load had reached its maximum. The PV on the other hand, took about 0.4 s to reach its maximum value. As a result, the voltage on each phase changed differently. Phase A voltage experienced a slight jump that reached a steady state about 0.2 s after the event. The other two phases took longer to reach steady state due to a more drastic change in power. The frequency had a noticeable dip in Fig. 12 of 0.05% before returning to the original value about 0.1 s since event. The reduction of the microgrid, intent on improving the load and generation balance, also resulted in minimal changes to the system voltage. Phase C experienced the largest change in voltage, as shown in Fig. 13 (b), of 0.063%. The other two phases had a decrease in voltage that

returned to a steady state in 0.2 s after the event. The reduction in system size caused the frequency to increase by 0.28%.

Fig. 14 and Fig. 15 illustrate the simulation results for the per-phase active and reactive power of the GFM inverter, respectively. To manage the changes in load and PV generation, the GFM inverter altered its active and reactive power output. This is evident in Fig. 16 and Fig. 17 for the expansion and reduction cases, respectively. The microgrid is located at the very end of a very long feeder and the loads are connected not evenly distributed among the three phases. This caused an unbalance in the system. This imbalance is evident in the GFM inverter simulation results; each of the three phases have different active and reactive power behaviors.

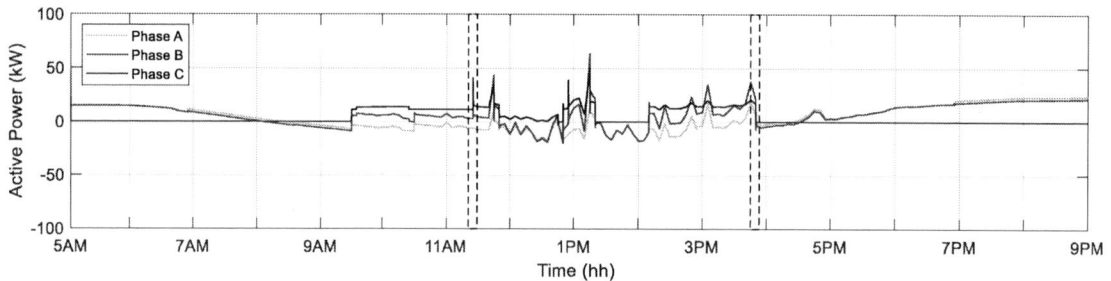

Fig. 14. Simulation Results for the GFM Inverters Active Power During Switching Transitions.

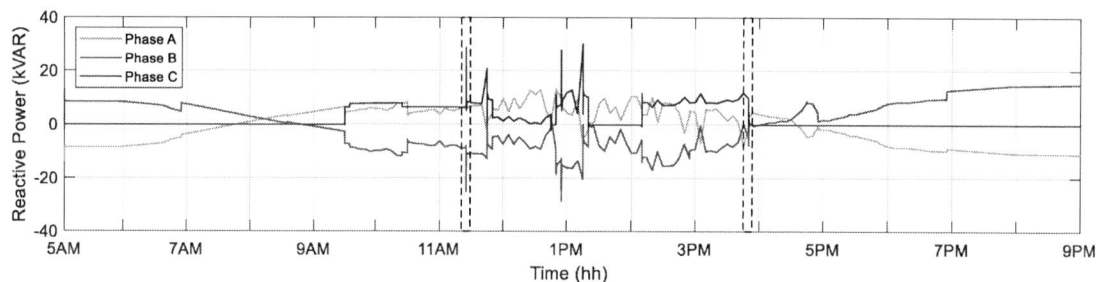

Fig. 15. Simulation Results for the GFM Inverters Reactive Power During Switching Transitions.

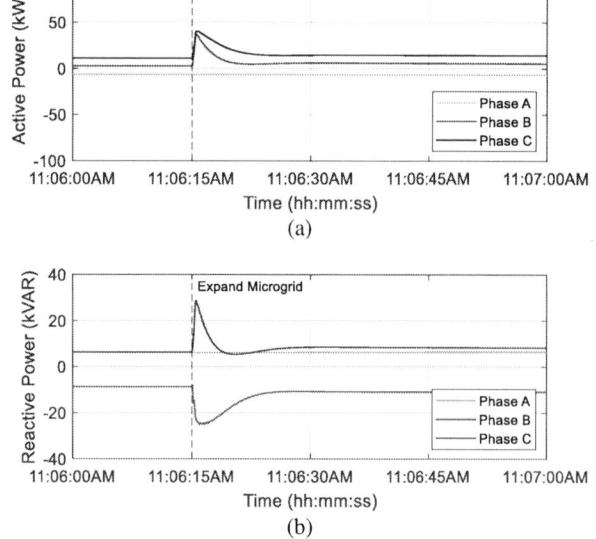

Fig. 16. Simulation results for the Microgrid During an Expansion. (a) Active Power. (b) Reactive Power.

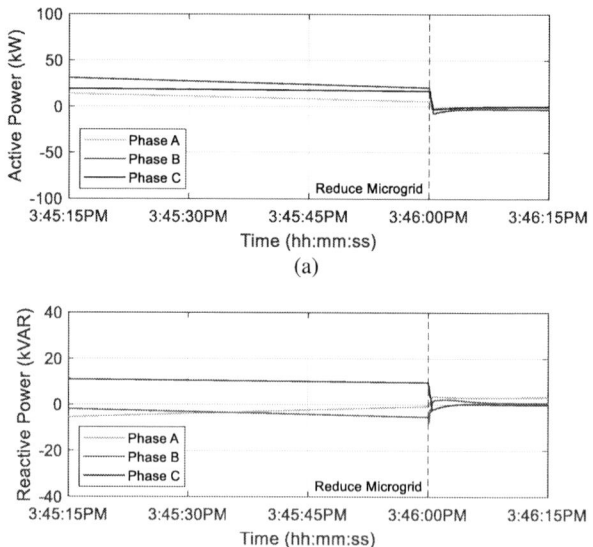

Fig. 17. Simulation results for the Microgrid During a Reduction. (a) Active Power. (b) Reactive Power.

978-1-7281-6118-1/22 $31.00 © 2022 IEEE

Results from Fig. 16 illustrate that when the system expanded, the active power provided by the GFM inverter increased for Phases B and C. The active power provided by the GFM inverter did not change for Phase A. At the same time, reactive power was injected on Phase C, absorbed on Phase B, and remained a constant value on Phase A.

Fig. 17 shows that the active power for Phase B and C has similar behavior before and after the microgrid reduction. In this situation, Phase A active power did not remain constant and experienced a sudden drop of almost 10 kW. Prior to the event, The GFM inverter generated reactive power to Phase A, while absorbing reactive power from Phase B and C. Once the change in size occurred, reactive power was reduced to zero. The GFM inverter generated reactive power since the voltage was slightly below 1.0 pu. At the same time, the GFM inverter's reactive power on Phase B was negative in order to decrease the voltage.

VII. CONCLUSION

This paper presents the integration of a GFM inverter in a hypothetical, isolated dynamic microgrid with a large penetration of GFL inverters. A dynamic *MATLAB/Simulink* model based on an existing power distribution feeder is used to study the system performance. The simulation results were obtained for the entire microgrid, consisting of 10 switching groups. Dynamic simulation results are obtained for the GFM inverter operating under microgrid reconfiguration conditions. Varying irradiance data, collected from the West of Puerto Rico, was used for the GFL PV inverters. Varying load profiles were used to add more dynamic conditions. Microgrid reconfiguration controls were developed and successfully used to determine when and what microgrids should be switched. The GFM inverter was able to support stable operations as the system changed its boundaries to either include or exclude loads and PV generation. Thus showing that dynamic operations is a viable solution for providing a low-cost, PV-based microgrid to remote communities.

ACKNOWLEDGEMENT

Sandia National Laboratories is a multi-mission laboratory managed and operated by National Technology and Engineering Solutions of Sandia, LLC., a wholly owned subsidiary of Honeywell International, Inc., for the U.S. Department of Energy's National Nuclear Security Administration under contract DE-NA-0003525.

This material is based upon work supported by the U.S. Department of Energy's Office of Electricity under agreement with the FEMA. The authors acknowledge the leadership of the Corcovada Communal Committee Inc. NPO. They wholeheartedly collected data for this work and welcomed the authors into their community.

REFERENCES

[1] P. P. Beires, C. L. Moreira and J. P. Lopes, "Grid-forming inverters replacing Diesel generators in small-scale islanded power systems", *IEEE Milan PowerTech*, 2019, pp. 1-6.

[2] B. K. Poolla, D. Groß and F. Dörfler, "Placement and Implementation of Grid-Forming and Grid-Following Virtual Inertia and Fast Frequency Response", *IEEE Transactions on Power Systems*, vol. 34, no. 4, pp. 3035-3046, July 2019.

[3] R. H. Lasseter, Z. Chen and D. Pattabiraman, "Grid-Forming Inverters: A Critical Asset for the Power Grid", *IEEE Journal of Emerging and Selected Topics in Power Electronics*, vol. 8, no. 2, pp. 925-935, June 2020.

[4] A. Kwasinski, "Effects of Hurricane Maria on Renewable Energy Systems in Puerto Rico", *7th International Conference on Renewable Energy Research and Applications (ICRERA)*, 2018, pp. 383-390.

[5] M. Gallucci, "Rebuilding Puerto Rico's Grid", *IEEE Spectrum*, vol. 55, no. 5, pp. 30-38, May 2018.

[6] C. Keerthisinghe, M. Ahumada-Paras, L. D. Pozzo, D. S. Kirschen, H. Pontes, W. K. Tatum, M. A. Matos, "PV-Battery Systems for Critical Loads During Emergencies: A Case Study from Puerto Rico After Hurricane Maria", *IEEE Power and Energy Magazine*, vol. 17, no. 1, pp. 82-92, Jan.-Feb. 2019.

[7] C. B. Jones, M.E. Ropp, J.H. Alvidrez, and R. Darbali-Zamora, "Optimized Control of Distribution Switches to Balance a Low Cost Photovoltaic Microgrid", *IEEE 48th Photovoltaic Specialists Conference (PVSC)*, 2021.

[8] C. B. Jones, M. Theristis, R. Darbali-Zamora, M. E. Ropp, M. J. Reno and M. Lave, "Switch Location Identification for Integrating a Distant Photovoltaic Array into a Microgrid", *IEEE Access*, 2022, pp. 1-12.

[9] M. Eriksson, M. Armendariz, O. O. Vasilenko, A. Saleem, and L. Nordström, "Multiagent-based distribution automation solution for self-healing grids", *IEEE Trans. Ind. Electron.*, vol. 62, no. 4, pp. 2620–2628, Apr. 2015.

[10] A. Singhal, T. L. Vu and W. Du, "Consensus Control for Coordinating Grid-Forming and Grid-Following Inverters in Microgrids", *IEEE Transactions on Smart Grid*, 2022, pp. 1-11.

[11] R. Darbali-Zamora, J. Johnson, N. S. Gurule, M. J. Reno, N. Ninad and E. Apablaza-Arancibia, "Evaluation of Photovoltaic Inverters Under Balanced and Unbalanced Voltage Phase Angle Jump Conditions", *47th IEEE Photovoltaic Specialists Conference (PVSC)*, 2020, pp. 1562-1569.

[12] R. Darbali-Zamora, N. S. Gurule, J. Hernandez-Alvidrez, S. Gonzalez and M. J. Reno, "Performance of a Grid-Forming Inverter Under Balanced and Unbalanced Voltage Phase Angle Jump Conditions", *IEEE 48th Photovoltaic Specialists Conference (PVSC)*, 2021.

[13] A. Yazdani and R. Iravani, "Voltage-Sourced Converters in Power Systems: Modeling, Control, and Applications", *Wiley-IEEE Press*, 2010, pp.1-541.

[14] M. Ebrahimi, S. A. Khajehoddin and M. Karimi-Ghartemani, "Fast and Robust Single-Phase DQ Current Controller for Smart Inverter Applications", *IEEE Transactions on Power Electronics*, vol. 31, no. 5, pp. 3968-3976, May 2016.

[15] V. Purba, S. V. Dhople, S. Jafarpour, F. Bullo and B. B. Johnson, "Reduced-order structure-preserving model for parallel-connected three-phase grid-tied inverters", *IEEE 18th Workshop on Control and Modeling for Power Electronics (COMPEL)*, 2017.

[16] W. Du, R. H. Lasseter and A. S. Khalsa, "Survivability of Autonomous Microgrid During Overload Events", *IEEE Transactions on Smart Grid*, vol. 10, no. 4, pp. 3515-3524, July 2019.

[17] W. Du and R. H. Lasseter, "Overload mitigation control of droop-controlled grid-forming sources in a microgrid", *IEEE Power & Energy Society General Meeting*, 2017, pp. 1-5.

[18] P. Piagi and R. H. Lasseter, "Autonomous control of microgrids", *2006 IEEE Power Engineering Society General Meeting*, 2006, pp. 1-8.

[19] M. E. Elkhatib, W. Du and R. H. Lasseter, "Evaluation of Inverter-based Grid Frequency Support using Frequency-Watt and Grid-Forming PV Inverters", *IEEE Power & Energy Society General Meeting (PESGM)*, 2018, pp. 1-5.

Parametric Analysis of Capacitance-Voltage Data for In-Situ Heat and Light Soaking Behavior of CIGS Solar Cells

Shubhra Bansal, Mohsen Jahandardoost

University of Nevada Las Vegas, Las Vegas, NV, United States

Parametric studies of Cu(In,Ga)Se2 defect properties are examined using SCAPS-1D device model for the purpose of understanding heat and light soaking behavior. The variables and interdependencies modeled in this study include: (1) two different buffer layers: CdS vs. Zn(O,S); (2) conduction band offset between buffer and absorber; (3) buffer shallow donor density (0.05 eV below conduction band); (4) CIGS shallow acceptor density (0.05 eV above valence band); (5) CIGS deep acceptor density (0.55 eV above valence band); (6) CIGS shallow donor density ( ); (7) ionized acceptor concentration in the ordered vacancy compound near CIGS/buffer interface ( ); and (8) back contact work-function. The device metrics simulated include open-circuit voltage (VOC), apparent hole density of CIGS and space charge width at reverse bias, 0V and forward bias. While an increase in VOC is seen by increasing the shallow donor and acceptor densities, or decreasing the buffer concentration, the behavior of the forward bias space charge width W(V>0) behaves differently between the two buffer types based on the conduction band offset. Inflection in the apparent doping density (in reverse bias) is also observed by an increase in the bulk deep acceptor density and shallow donor density near the back contact. An increase in the space charge width is also observed by formation of a Schottky barrier at the back contact.

MoS2 solar cell with 120 nm-absorber and 3.8% AM1.5G efficiency

Elisa Antolin, Simon A. Svatek, Carlos Bueno-Blanco, Antonio Marti, Der-Yuh Lin, Micaela Rodriguez-Peña, Monica Luna

Universidad Politécnica de Madrid , Madrid, Spain

National Changua University of Education, Changua, Taiwan

Consejo Superior de Investigaciones Científicas, Madrid, Spain

Recent research on semiconductor materials with layered crystalline structure, like the transition metal dichalcogenides (TMDCs) MoS2 and WeS2, has opened the path to the realization of ultrathin, ultralight microelectronic devices. TMDCs enable the exfoliation or growth of ultrathin laminae - from a monolayer to hundreds of nm thick - that have self-passivated surfaces and are mechanically and chemically stable. Furthermore, applying just mechanical methods several laminae can be assembled at room temperature to form a device (a van der Waals structure). Among the potential applications of TMDCs, the development of ultrathin solar cells is attracting attention because of the possible reduction in cost and in semiconductor material consumption, and because of the suitability of these materials for flexible and ultralight photovoltaics. In this work, a 120 nm thick MoS2 p-n junction is presented. We have developed a straightforward method to fabricate ohmic contacts to both n and p MoS2 and we have added an h-BN layer on top of the semiconductor to minimize the reflectance of the front surface. The homojunction device exhibits (3.8 ± 0.2)% efficiency and 57% fill factor under AM1.5G illumination. This work contributes to the maturity of the emerging technology of TMDC-based ultrathin solar cells.

978-1-7281-6118-1/22 $31.00 © 2022 IEEE

Effects of Arsenic Doping on $CdSe_xTe_{1-x}$/CdTe Solar cells

Sheikh Tawsif Elahi, Md Zahangir Alom, Wei Wang, Vasilios Palekis, and Chris Ferekides

Electrical Engineering, University of South Florida, Tampa, FL 33620, USA

Abstract—**The objective of this study was to investigate the effects of Arsenic (As) doping in CdTe/CST solar cells. The absorber layer was deposited by Closed Space Sublimation (CSS) process under the presence of ambient He and O_2. Both CdTe and CST were deposited with a substrate and source temperature of 580^0C and 680^0C respectively. Devices were investigated with various Se compositions, with only CST doped or both CdTe & CST doped with As. Devices with only CST doped exhibited improved open circuit Voltage (V_{OC}) 810 mV and Fill Factor (FF) of 64.5% compared to tcompletely undoped and both CdTe/CST doped.**

Keywords— Heat Treatment, Cadmium Telluride, Cadmium Selenium Telluride, Poly Crystalline

I. INTRODUCTION

Thin film CdTe solar cells have shown significant improvement in terms of efficiency in recent years. For polycrystalline CdTe (PX-CdTe) solar cells, First Solar demonstrated record breaking cell efficiency of 22.1% and for module 18.6% [1], [2]. This has mainly been possible due to the enhancement of short circuit current density (J_{SC}) 31.69 mA/cm^2, a gain of more than 5 mA/cm^2 [3]. Introducing Selenium (Se) into CdTe, an alloy of $CdSe_xTe_{1-x}$ (CST) is created, which causes the absorber layer to be graded and the bandgap to decrease, hence the absorption in the long wavelength region increases, which leads to improved J_{SC} and overall efficiency of the cell. As the short circuit current density (J_{SC}) has reached near the theoretical limit with the record efficiency cell (31.7 mA/cm^2), to further improve the efficiency of the cell, research interest has focused on improving the V_{OC} with carrier lifetime and net p-doping.

PX-CdTe can be doped extrinsically with both n and p type dopants. Depending on the stoichiometry and native defects CdTe can be n or p type intrinsically. Under Te rich conditions with the presence of Cd vacancies (V_{Cd}) CdTe behaves like p type material and under Cd rich conditions with the presence of Te vacancies (V_{Te}) and Cd interstitials (Cd_i) CdTe shows n type conductivity. Extrinsically p type conductivity can be achieved by substituting Cd and Te with group I (Cu, Au) and group V (As, P, Sb) elements respectively [4]. Similarly, by substituting Cd and Te sites with group III (In, Al) and group VII (Cl, I) can result n type conductivity.

In this study we are investigating the effects of Arsenic (As) a p-type dopant, on the absorber layer when only the CST layer is doped and when both the CST & CdTe layers are doped.. Arsenic has higher solubility than Phosphorus (P) and it is safer to use for high volume manufacturing; however, As doped cells

show the formation of Ax-centers, a self-compensating mechanism which affects the performance of the cell severely by decreasing the Fill Factor (FF) and Open Circuit Voltage (V_{OC}) [5], [6].

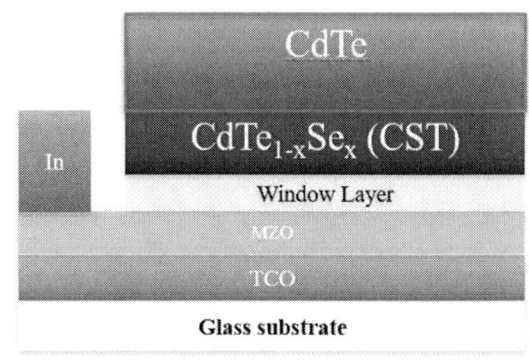

Fig 1. Device structure of CST/CdTe Cell

II. EXPERIMENTAL

All the samples fabricated were in superstrate configuration and Corning EagleXG glass substrates were used to fabricate the devices. A layer of Transparent Conducting Oxide (TCO) was deposited by sputtering with a thickness of 6000 Å and sheet resistance of 8 Ω/□ approximately. Magnesium Zinc Oxide (MZO) was deposited as emitter layer by sputtering as well. A very thin layer of CdS was deposited on top of MZO using Closed Spaced Sublimation (CSS). The CST alloy was deposited also using CSS deposition technique with a thickness of 1 µm and on top of that CdTe was deposited approximately with a thickness of 3-4 µm. Both for CST and CdTe the substrate temperature was 580°C and for source it was 680°C. For both of these depositions Oxygen and Helium ambient were used with the pressure of 10 Torr and 15 Torr respectively. Figure 1 shows the structure of the studied device. For this study two sets of samples were fabricated one is with only CST doped and another set is doped with both CST and CdTe. All the samples were synthesized in house with CdTe (99.999%), CdSe (99.999%) and Cd_3As_2 (99.999%) powders. After the CdTe deposition all the samples were $CdCl_2$ heat treated (HT) with the ambient of He:O_2 with four different temperatures 390°C, 410°C, 420°C, and 430°C. Then the samples were etched with a bromine methanol solution, and a Cu doped graphite paste was applied back contact. Then the samples were HT with temperature of 285°C and 300°C. The samples were characterized with Current

978-1-7281-6118-1/22 $31.00 © 2022 IEEE

Voltage (J-V), Spectral response (SR), and Capacitance Voltage (C-V) measurement. The J-V measurement was done using four-point connectors with Keithley 2410 source meter. For SR measurements an Oriel 74100 monochromator was used, and the intensity of light was adjusted using a calibrated Silicon solar cell.

III. RESULTS AND DISCUSSION

As mentioned earlier for this study two sets of samples were prepared one with only CST doped and another with both CST and CdTe doped. All the samples are processed with CdCl$_s$ post annealing heat treatment. In TABLE I different parameters for only CST doped samples are shown. From TABLE I we can see that as the Se composition increases the short circuit current density also increases as a result of the bandgap also decreasing [7]. Also, as the Se composition got higher V$_{OC}$ did

TABLE I SOLAR CELL PARAMETERS FOR ONLY CST DOPED SAMPLES

Se Comp. [%]	Voc [mV]	FF [%]	Jsc [mA/cm²]	Dop. Conc. [cm⁻³]
14	780	67.10	25.11	2.7×10^{14}
16	790	57.30	26.51	1.3×10^{14}
18	790	50.80	27.98	1.5×10^{14}
29	810	64.50	29.37	1.7×10^{15}
43	820	50.60	28.50	2.3×10^{14}

TABLE II SOLAR CELL PERFORMANCE PARAMETERS FOR BOTH CST & CDTE DOPED SAMPLES

Se Comp. [%]	Voc [mV]	FF [%]	Jsc [mA/cm²]	Dop. Conc. [cm⁻³]
14	810	61.70	26.75	2.4×10^{14}
16	810	65.70	25.54	1.5×10^{14}
18	820	44.80	25.98	1.3×10^{14}
29	830	48.80	26.61	2.1×10^{14}
43	830	56.20	24.13	1.8×10^{14}

not change significantly indicating improved grain structure near the interface [8]. Very high Se composition (i.e. 29% and 43%) samples were back contact annealed at 300°C, which could decrease the FF due to excessive Cu doping. If we turn our attention to TABLE II, we can observe that FF of the devices remained poor for both doped cases, which is an indication of recombination in the device This is caused by the formation of self compensating Ax centers, which is an acceptor induced defects working as a donor to compensate the acceptor itself. Se composition did not have much effect on the net doping concentration on both sets of samples, meaning Se does not

control the doping concentration of the device. If we closely

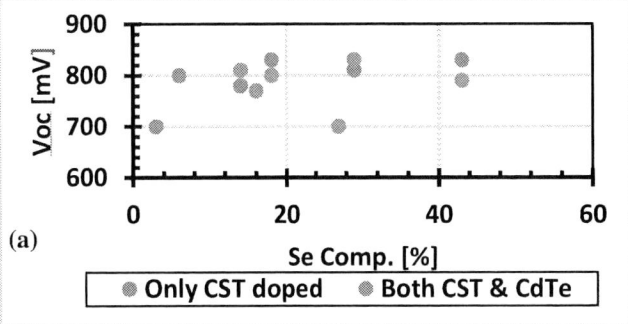

(a)

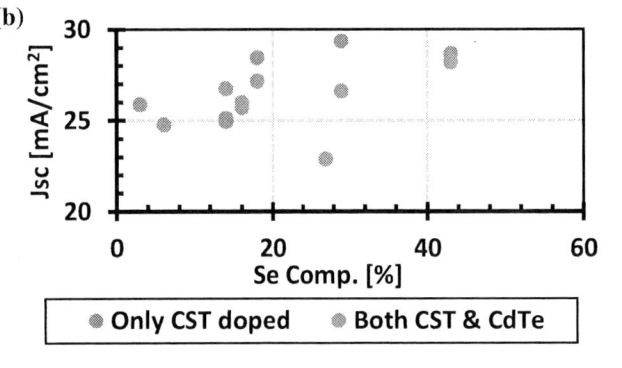

(b)

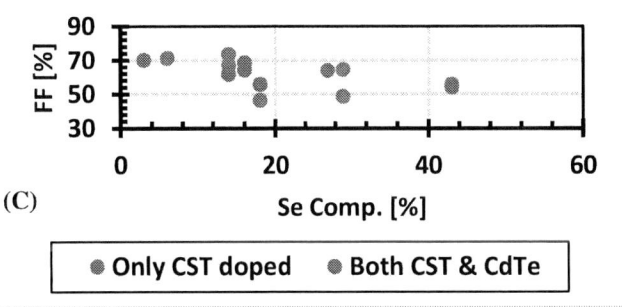

(C)

Fig 2. (a) V$_{OC}$Vs Se Comp. for only CST doped and both doped cells (b) Jsc vs Se comp. for only CST doped and both doped cells (C) FF vs Se composition

look at the effects of Se on the device in Figure 2(a), for only CST doped and both doped samples V$_{OC}$ remained in the 800 mV region, showing that Se did not have any detrimental effects on open circuit voltage. In Figure 2(b) we can observe that as the Se composition gets higher so does the J$_{SC}$, clearly indicating graded bandgap which enhances absorption and therefore carrier generation/collection at long wavelengths. In figure 2(c) as the Se composition increases the FF decreases, this can be the effects of Se reaching to the CdS layer causing front recombination of the device, however, this does not happen to all the cells because throughout the bulk of the cell Se profile is not uniform for processing variation [9], affecting the overall performance of the devices. Only CST doped with 29% Se showed best performance with an efficiency of 15.34%, for same Se composition both doped samples showed better V$_{OC}$, but very poor FF and J$_{SC}$, showing doping of CdTe is much difficult than CST, due to lower dopant activation, self-

978-1-7281-6118-1/22 $31.00 © 2022 IEEE

compensation of dopants as a by product of the creation of Ax centers.

Another very important step for CdTe solar cell is CdCl₂ HT process. In figure 3(a) for low annealing temperature (390°C) the FF is more than 70%, however, in figure 3(b) it also reveals that V_{OC} is less than 800 mV for only CST doped samples. As the CdCl₂ annealing temperatures got higher V_{OC} improved suggesting for low temperatures there is not enough Cl to passivate the gain boundaries [10]. Though there is no clear indication that CdCl₂ annealing temperatures increase the net doping concentration, however, from figure 3(C) we can see that as the temperature gets higher carrier concentrations increase

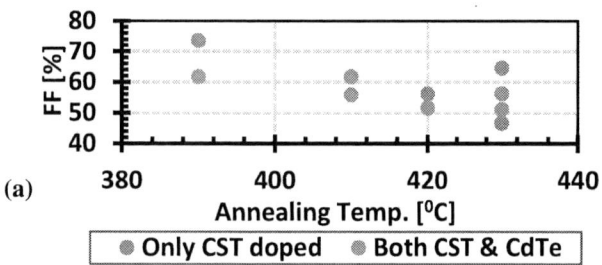

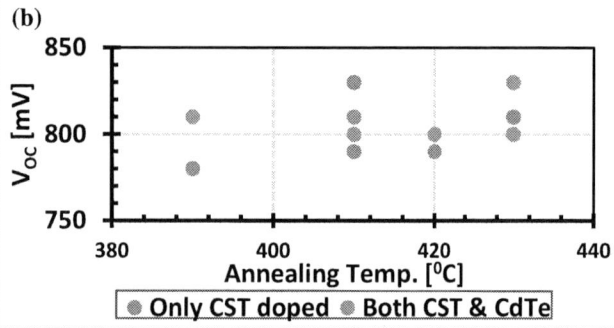

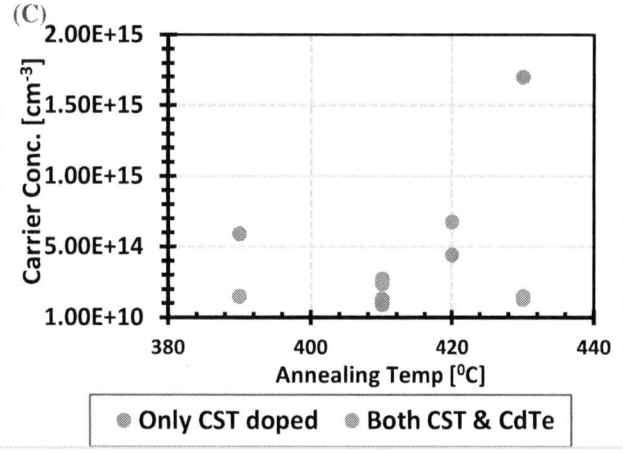

Figure 3. Effects of CdCl₂ annealing temperatures on (a) FF (b) V_{OC} (c) Carrier concentration

above 10^{14} cm⁻³, suggesting CdCl₂ does help the doping of the cells. Again, for CST only doped sample the carrier concentration reached to $1.7×10^{15}$ cm⁻³ indicating it is easier to dope CST than CdTe.

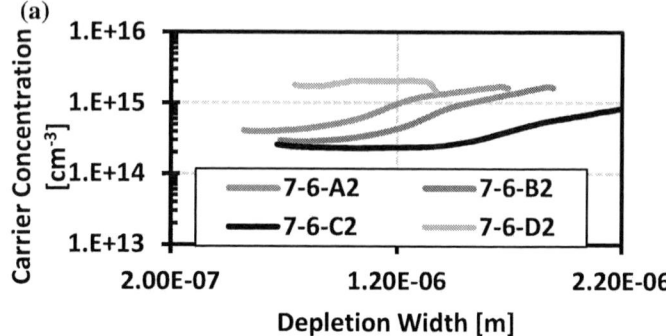

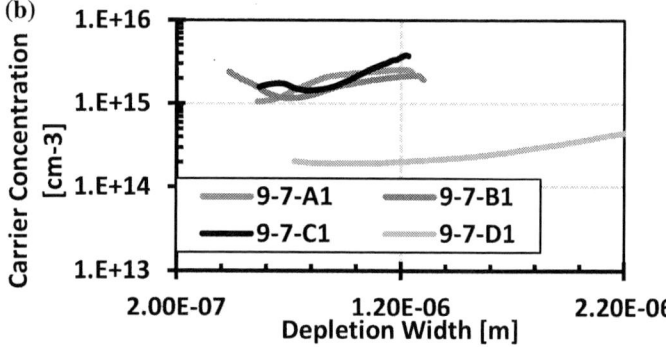

Figure 4. CV measurement with Se 29% (a) Only CST doped (b) Both CST and CdTe

In Fig. 4 CV responses of Se 29% with (a) only CST doped and (b) both CST and CdTe doped are shown. Samples from A to D are CdCl₂ annealed from 390°C, 410°C, 420°C, and 430°C respectively. As none of the samples is showing flat CV curve suggesting they are not uniformly doped [11]. This also means inside the bulk of the device not all the dopants are activated uniformly, hence, the overall net doping of the device in the range of 10^{14} cm⁻³. . Both doped samples are showing higher carrier concentration compared to the only CST doped one, meaning both CST and CdTe doping do help to improve the net doping. One interesting point is for 430°C annealing temperature the cell performance for only CST doped with 29% Se showed the best performance and highest carrier concentration $1.7×10^{15}$ cm⁻³ , but for both doped samples with the same condition it got the lowest doping concentration. This can be better understood if we look at the light IV of both doped samples for high CdCl₂ annealing temperature. In figure 5 the samples processed at 410°C and 430°C CdCl₂ annealing temperatures and back contact annealing at 285°C and 300°C are listed. All the samples consistently showed V_{OC} over 800mV, reached maximum V_{OC} of 850 mV, not shown in this paper. However, the FF of the samples were really poor in the mid- fifties. One reason could be that too much Cu getting into the device which is detrimental to the cell performance. For 300°C back contact anneal the "S" kink starts to appear meaning a barrier a discontinuity in the conduction band [8].

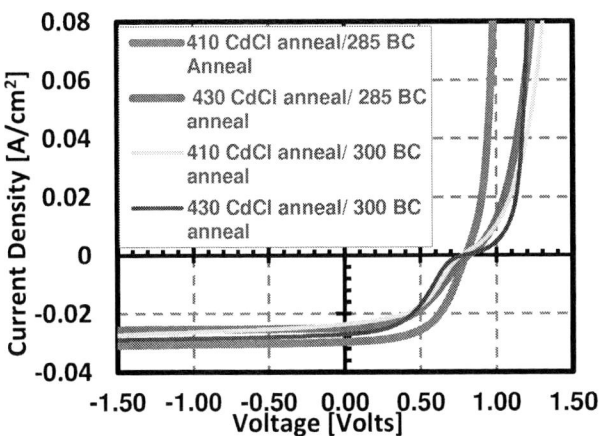

Figure 5. Light IV for both doped samples

Table III shows bandgap calculation for only CST doped and both doped samples. Bandgap of the samples were calculated using the formula:

$$Eg = \frac{hc}{\lambda}$$

Where E_g is the bandgap of the device, h is Plank's constant, λ is wavelength and c is the speed of light. These calculations were made by fitting the Spectral Response (SR) curve's edge of the samples shown above. From the table we can see that for any

TABLE III BANDGAP FOR ONLY CST AND BOTH DOPED DEVICES

Se Comp. (%)	Bandgap for Only CST doped (eV)	Bandgap for both doped (eV)
14	1.41	1.42
18	1.41	1.40
29	1.39	1.39
43	1.37	1.38

device Se does change the bandgap of that device from 0.04 eV to 0.08 eV. As the Se composition increases the bandgap decreases, which is usual for CdTe cells with Se, hence the short circuit current density of the samples are above 25 mA/cm². We can also suggest that Se does not affect the carrier concentration if we look at the TABLE II and TABLE III as well.

IV. . CONCLUSION

In this study the effects Arsenic doping on only CdSexTe1-x alloy and both CdSexTe1-x and CdTe bilayer has been investigated. It shows that only CST doped samples exhibit the best performance with an efficiency of 15.34% with 29% Se. For both doped samples the open circuit voltage was always more than 800 mV but the low FF affected the overall performance of the cell. Better VOC for both doped cells do suggest doping both

absorber layer does help to improve the net doping, however, dopant solubility and formation of Ax centers have always been a pathway for self-compensation of the dopants. Composition of Se does not have any effect on the net doping of the cells, but it does help to improve the JSC by reducing the bandgap of the device. Furthermore, post CdCl2 helps the VOC by improving the grain structure of the cell. Further study is required to understand the role of Arsenic as a dopant with CST and CdTe at high temperature processing if cooling down the samples after deposition can be controlled.

ACKNOWLEDGMENT

This work was supported by the National Science Foundation under grant EPMD-1711716 to study the effects of doping on CdTe solar cells.

REFERENCES

[1] M. Gloeckler, "Realization of the potential of CdTe thin-film PV", Proc. 43rd IEEE PVSC, 2016, pp. 1292-1292.

[2] M. A. Green, et al. "Solar cell efficiency tables (version 49): Solar cell efficiency tables (version 49)." Progress in Photovoltaics 25.NREL/JA--5J00-67687 (2016).

[3] S. Krum, S. Haymore, First Solar Achieves yet Another Cell Conversion Efficiency World Record, First Sol, 2016

[4] I. S. Khan, V. K. Evani, V. Palekis, and C. Ferekides, "Effect of stoichiometry on the lifetime and doping concentration of polycrystalline CdTe", IEEE Journal of Photovoltaics, vol. 7, no. 5, pp. 1450-1455, 2017

[5] Burst, J. M. et al. CdTe solar cells with open-circuit voltage breaking the 1 V barrier. Nature Energy 1, 16015, doi:10.1038/nenergy.2016.15 (Ablekim, T., Swain, S. K., Kuciauskas, D., Parmar, N. S. & Lynn, K. G. Fabrication of single-crystal solar cells from phosphorousdoped

[6] CdTe waferh. Photovoltaic Specialist Conference (PVSC), 2015 IEEE 42nd 1-4, doi:10.1109/PVSC.2015.7356372 (2015)2016).

[7] Burst, J. M. et al. CdTe solar cells with open-circuit voltage breaking the 1 V barrier. Nature Energy 1, 16015, doi:10.1038/nenergy.2016.15 (Ablekim, T., Swain, S. K., Kuciauskas, D., Parmar, N. S. & Lynn, K. G. Fabrication of single-crystal solar cells from phosphorousdop[8]Shamara Collins; Chih An Hsu; Vasilios Palekis ; Ali Abbas ; Michael Walls ; Chris Ferekides, "Se Profiles in CST Films Formed by Annealing CdTe/CdSe Bi-Layers", IEEE 7th World Conference on Photovoltaic Energy Conversion (WCPEC), 2018, pp. 0114-0118Ali Çiris, a,b, Bülent M. Bas¸ol c, Yavuz Atasoy a,d, Tayfur Küçük¨omero¨glu e, Abdullah Karaca e, Murat Tomt takin f, Emin Bacaksız, "Effect of CdS and CdSe pre-treatment on interdiffusion with CdTe in CdS/ CdTe and CdSe/CdTe heterostructures" https://doi.org/10.1016/j.mssp.2021.105750

[8] Chih An Hsu; Vasilios Palekis ; Imran Khan ; Shamara Collins ; Don Morel ; Chris Ferekides "The Effect of the CdCl2 Heat Treatment on CdSexTe1-x Solar Cells", IEEE 44th Photovoltaic Specialist Conference (PVSC), pp. 3413-3416

[9] Sachit Grover, Xiaoping Li, Dingyuan Lu, Rajni Mallick, Gang Xiong, Markus Gloeckler, "Comparison of P, As & Sb doped Polycrystalline CdTe Solar Cells", IEEE explore, 2019

[10] C.S. Ferekides, D. Marinskiy, V. Viswanathan, B. Tetali, V. Palekis, P. Selvaraj, D.L. Morel, "High efficiency CSS CdTe solar cells", Thin Solid Films, 2000, Vol. 361-362, pp. 520-526 520-526.

What is the Optimal Electricity Share for Very Inexpensive Solar PV?

Adam Dvorak[a], Marta Victoria[a,b,*]

[a]*Department of Mechanical and Production Engineering, Aarhus University, Inge Lehmanns Gade 10, 8000 Aarhus, Denmark*
[b]*iCLIMATE Interdisciplinary Centre for Climate Change, Aarhus University*

Abstract

Solar electricity is already cost-optimal in many world regions. Moreover, by 2050, the price of utility solar could decrease by over an order of magnitude from today's costs with present-day learning rates and projected solar capacities. In this work, we focus on investigating what is the optimal energy mix in a framework of very inexpensive solar PV and what are the main drivers that foster solar penetration. We show the effects of cheap solar on four contrasting "cold" and "hot" regions across the United States and Europe using an open, hourly resolved, copper plate model. In addition to varying by region, we find that the optimal mix of renewable resources varies with stricter CO_2 constraints. Finally, we identify the temperature-dependency of the electricity demand of these regions, show the effects of global-warming induced temperature increase on demand and discuss the implications for optimal solar penetration.

---

1. Introduction

Decarbonization by 2050 would require large and fast growth of renewable technologies, fueled by cost reductions through research as well as improvements in the fabrication and implementation process. Solar is well positioned to become the driver of this growth, given the ubiquity of the resource across the globe as well as having already become the cheapest source of electricity in history [1]. The cost of solar has decreased rapidly, dropping two orders of magnitude in the past 40 years [2], but we have every reason to believe that the price will drop even further between now and 2050. We show the plausibility of the price of solar dropping another order of magnitude, as well as the consequences of this to a given energy system.

The optimal electricity share of solar PV in decarbonized future energy systems is impacted by the available resource of solar and other renewable energy sources, as well as assumptions on costs and modelling of strategies to balance renewable fluctuations. The latter includes different types of storage technologies, interconnection and coupling the electricity with other sectors such as heating, transport, and industry. There are indicators of a potential for very high penetration of solar in our future grids. On the one hand, several models predict high solar contribution by 2050 [3–5]. On the other hand, solar contribution in real power systems keeps breaking records with California supplying about 20% of its electricity demand with solar PV in 2020 [6] and South Australia covering 100% of its electricity demand in one hour with rooftop solar [7].

In an environment with very inexpensive solar PV, we investigate the following research questions: What is the optimal share of electricity supplied by solar PV? What are the main factors impacting the optimal share? In particular, for extremely low-cost solar PV, are cost and prevention of curtailment significant parameters?

One of the potential limitations of solar penetration in high latitudes is that it shows a high seasonality which is anti-correlated with heating. The opposite is true for regions closer to the equator. We select four representative regions to understand better the sensitivity to location and weather. Specifically, over the course of this paper, we investigate contrasting "cold" and "hot" regions between the United States and Europe, selecting the states of Colorado and California and the countries of Denmark and Spain. See the main seasonal patterns in Fig. 1.

The rise in solar capacity through the next century might begin to mitigate the rise in global temperature due to global warming, but until there are zero fossil fuel emissions we can expect this increase in temperature to continue. An increase in temperature would result in a change in electricity demand, which could go up or down depending on whether a region's temperature-dependent electricity demand is cooling dominated or heating dominated, or both. We also include this factor in our analysis.

2. Methods

In our model, we determine the optimal capacity and dispatch of wind and solar PV generators that minimize the system cost while supplying uninterrupted hourly demand for a full year. Intentionally, we use a very simple model, including only the minimum required elements to represent the main fluctuations in wind, solar and electricity demand. The simplified model allows us to unveil fundamental dynamics by running parameter sweeps.

---

*Corresponding author
  Email address: mvp@mpe.au.dk (Marta Victoria)

Preprint submitted to   *January 25, 2022*

Our model uses Python for Power System Analysis (PYPSA) [8] to model electricity generation. For each region, we assume a copper plate, with no interconnections. We assume the addition of storage.

As stated in the introduction, we use the four regions of Colorado (CO), California (CA), Denmark (DNK), and Spain (ESP) as the method of comparison between "cold" and "hot' regions between the United States and Europe. We assume generators of solar, wind, gas (OCGT), hydrogen storage and batteries, and assume zero CO_2. Baseline prices in 2020 are shown in Table 1.

Technology	Unit	Cost
Utility Solar	€/kW	529
Onshore wind	€/kW	1118
OCGT	€/kW	453
Battery storage	€/kWh	232
Battery inverter	€/kW	270
Hydrogen storage	€/kWh	3.0
Fuel cell	€/kW	1300
Electrolysis	€/kW	650

Table 1: Table of costs for 2020, given in 2015 money. Costs taken from[9]

To complete the model, we use electricity demand from 2011 from [10] for CO and CA as well as electricity demand from 2015 from ENTSOE[11]. Capacity factors for solar and wind are obtained from the following repositories for DNK and ESP[12, 13], and from www.renewables.ninja/ for CO and CA.

The electricity demand and capacity factors are plotted in Figure 1. Seasonal patterns are strong in some regions (DNK) but not as dominant in others.

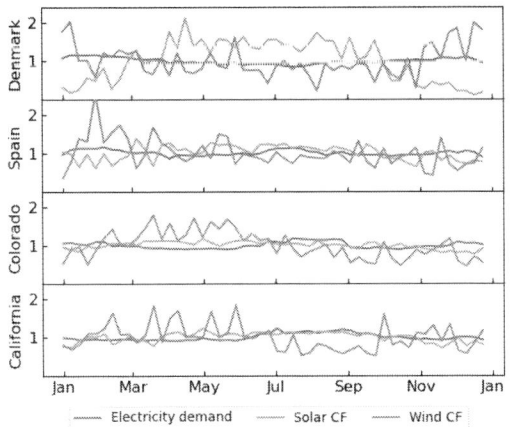

Figure 1: Plot of weekly solar and wind resources and electricity demand over the course of the year for all countries, normalized to the average.

3. Results

To estimate the price of solar in 2050, we use the learning rate, which is the percent decrease in cost of a technology per doubling of cumulative capacity produced. The learning rate of solar has historically been 23%, but has more recently reached 40% [6]. Haegel et al. projects cumulative global capacity of solar to reach 30-70TW by 2050. The global capacity of solar for 2020 was 770GW [14], meaning that by 2050, the capacity of solar will have doubled between 5.3 and 6.5 times. The price of utility scale solar in 2020 can be estimated to be around 0.529 EUR/Wp [14], but can also reach up to 1.3 EUR/Wp[15]. The cost of solar is known to vary widely by region [16].

We can extrapolate the cost of utility solar under different scenarios. If we take the least optimistic scenario, with a learning rate of 23% and 30TW installed (doubling 5.3 times), that would correspond with a price of 0.132 EUR/Wp for a given solar installation, or about four times cheaper than the current price of 0.529EUR/Wp. If we were to take the most optimistic scenario, with 70TW installed and a learning rate of 40%, we would expect a price of 0.019 EUR/Wp, which is more than an order of magnitude cheaper than the current price. We then have "less optimistic" and "optimistic" scenarios from which to compare. [1]

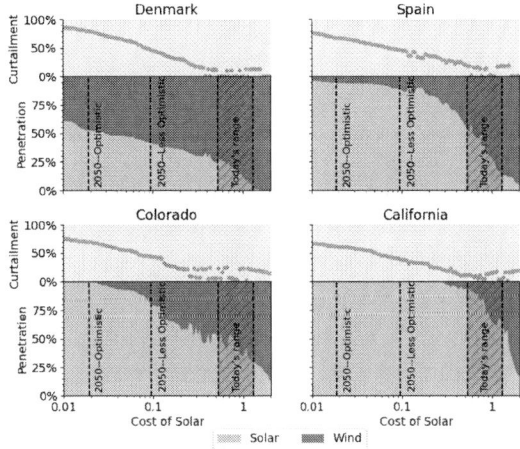

Figure 2: Curtailment and solar penetration, with cost in €/Wp on the x axis

In Figure 2, we plot electricity mix on the y axis and vary solar cost on the x-axis on a logarithmic scale, marking the "less optimistic" and "optimistic" price points in

---

[1]We acknowledge that the "optimistic" scenario may be more optimistic than realistic, considering that we consider the learning rate of solar modules to be that of the entire PV system. In addition, even with a high learning rate, there may be costs associated with grid and land usage that are non-negligible. Nevertheless, we plot values up to 0.01 EUR/Wp as an academic exercise to better understand what would happen in a universe with extremely cheap solar.

2050, and the range of solar that is present today. In addition, we plot the percent of solar energy that is curtailed. The ideal proportion of solar varies by country. For DNK, there is never a scenario in which 100% solar energy is realized, whereas CA experiences 100% solar energy at a cost similar to today's. For CO and ESP, we see that there is a large difference between current cost points and the least optimistic 2050 scenario, but that there is a smaller difference between the "less optimistic" and "optimistic" scenarios. In our model, it is more important to get to the first 30 TW than the next 40 TW. We also see that solar curtailment increases significantly as the price of solar decreases, approaching near-100% of produced energy curtailed. This illustrates how we can expect high levels of solar curtailment accompanied in regions with high solar penetration.

We are also interested in the pressures of a decreasing CO2 allowance on an energy system. In Figure 3, we plot resource fraction on the y axis against the fraction of energy provided by a flexible, potentially CO_2 emitting source, assuming baseline 2020 costs. In our model, this is the fraction of energy provided by gas, but this could also represent a dispatchable energy source, e.g. reservoir hydropower.

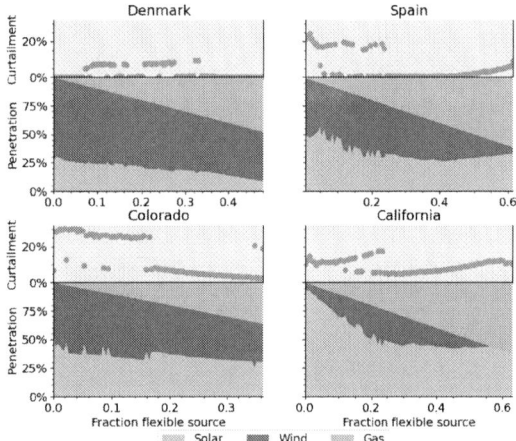

Figure 3: This plots solar penetration and curtailment against gas penetration, which decreases due to a CO_2 limit. Values are with present day costs.

The noise in the solar curtailment plot can be explained by a flat optimization landscape. The algorithm chooses to stop searching for optimal solutions when the change in objective (system price) due to a change in resource fails to reach a minimum threshold ϵ, leading our model to choose alternatively between a scenario with more solar and a scenario with less solar.

We see that each region has different proportions of solar and wind under no CO2 restriction, as is expected, but we also see that different regions favor different resources as CO2 limits are tightened. In ESP, additional wind is

favored over adding solar. In CA, whether additional wind or additional solar is favored depends on how much capacity of solar or wind already exists. In DNK and CO, solar and wind are added proportionally. This shows that when CO2 restrictions are added, one cannot expect the demand for solar and wind to increase similarly.

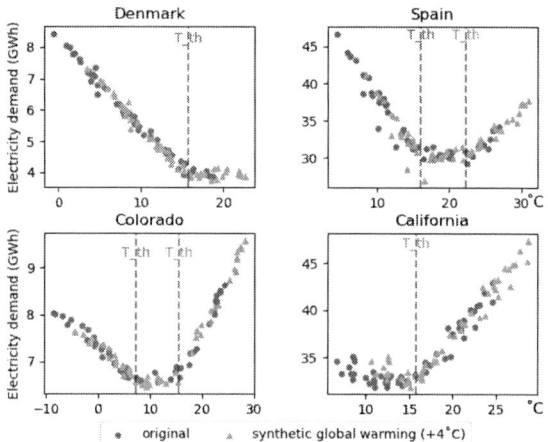

Figure 4: Weekly average of electricity demand by temperature is plotted for a historical weather year and with a synthetic temperature increase of 4C, estimating the effect of global warming on future electricity demand. Electrification of heating is assumed where we have heating data (DNK, ESP).

It is clear that there is a relationship between a country's electricity demand and its temperature. We observe that a profile of a given country's electricity demand vs temperature can have three distinct regions–a negative correlation between electricity demand and temperature, a positive correlation, or a correlation with a slope of 0. We find that these three regions begin or end at a temperature threshold T_{th}, which varies by country.

If we assume that these temperature thresholds and correlations remain the same, then we can extrapolate what would happen to the electricity demand given a temperature change due to global warming. We map all data points from our year of study to a year with temperature change by adding the degree change to each point, and plotting the corresponding change in electricity demand. This is the result of Figrue 4.

We can use this method on all datapoints to model energy demand of the future due to global warming. In a full paper, we will investigate how the electrical demand changes due to temperature shifts would impact the model, and what it means for the electricity mix of a future electricity system. We will also investigate plausible prices for wind in 2050 given present learning rates, the effect of the CO2 constraint on the total system cost, as well as their combined effect with extremely cheap solar.

References

[1] International Energy Agency (IEA) (2020). Renewables 2020. URL https://www.iea.org/reports/renewables-2020/solar-pv.

[2] Haegel, N. M. *et al.* Terawatt-scale photovoltaics: Transform global energy **364**, 836–838. URL https://www.science.org/doi/10.1126/science.aaw1845.

[3] Frew, B. *et al.* Sunny with a chance of curtailment: Operating the US grid with very high levels of solar photovoltaics **21**, 436–447. URL https://linkinghub.elsevier.com/retrieve/pii/S2589004219303967.

[4] Breyer, C. *et al.* On the role of solar photovoltaics in global energy transition scenarios: On the role of solar photovoltaics in global energy transition scenarios **25**, 727–745. URL https://onlinelibrary.wiley.com/doi/10.1002/pip.2885.

[5] Victoria, M., Zhu, K., Brown, T., Andresen, G. B. & Greiner, M. Early decarbonisation of the european energy system pays off **11**, 6223. URL http://arxiv.org/abs/2004.11009. 2004.11009.

[6] Victoria, M. *et al.* Solar photovoltaics is ready to power a sustainable future **5**, 1041–1056. URL https://linkinghub.elsevier.com/retrieve/pii/S2542435121001008.

[7] Parkinson, G. Rooftop solar helps send south australia grid to zero demand in world first. URL https://reneweconomy.com.au/rooftop-solar-helps-send-south-australia-grid-to-zero-demand-in-world-first/.

[8] Brown, T., Hörsch, J. & Schlachtberger, D. PyPSA: Python for power system analysis **6**, 4. URL http://openresearchsoftware.metajnl.com/articles/10.5334/jors.188/.

[9] Technology data for generation of electricity and district heating, update november 2019. Tech. Rep. URL https://ens.dk/en/our-services/projections-and-models/technology-data/technology-data-generation-electricity-and.

[10] Welty, E., Selvans, Z. & Kumar, Y. Pudl us hourly electricity demand by state (2021). URL https://doi.org/10.5281/zenodo.5348396.

[11] Muehlenpfordt, J. URL https://doi.org/10.25832/time_series/2020-10-06.

[12] Victoria, M. & Andresen, G. B. Photovoltaic time series for European countries and different system configurations (2019). URL https://doi.org/10.5281/zenodo.2613651.

[13] Victoria, M. & Andresen, G. B. Validated onshore and offshore time series for European countries (1979-2017) (2019). URL https://doi.org/10.5281/zenodo.3253876.

[14] International Energy Agency. Snapshot of global PV markets - 2020. 21. URL https://iea-pvps.org/wp-content/uploads/2021/04/IEA_PVPS_Snapshot_2021-V3.pdf.

[15] Vimmerstedt, L. 2021 annual technology baseline (ATB) cost and performance data for electricity generation technologies. URL https://doi.org/10.25984/1807473.

[16] Renewable power generation costs 2020 77.

Transparent Oxides as a Protective Encapsulant for Perovskite Solar Cells in Low Earth Orbit

Kyle M Crowley, Kaitlyn VanSant, Timothy J Peshek, Lyndsey B McMillon-Brown

NASA Glenn Research Center, Cleveland, OH, United States

Perovskite Solar Cells (PSCs) are a low-cost, light-weight emerging technology for photovoltaic energy generation in space. Low earth orbit (LEO), the region home to most artificial satellites, contains atomic oxygen (AO), which presents a challenge for PSC durability. In this study, various metal oxide films are deposited onto triple cation PSCs as encapsulation layers. Cells are then exposed to AO via a plasma etching system, in order to simulate the LEO environment up to a period of 5 years. PSC electrical performance is then benchmarked at various exposure intervals. This study successfully illustrates the effectiveness of metal oxides as AO barrier layers and explores the practicality of utilizing PSCs for future space photovoltaic systems.

Barriers to Solar Photovoltaic (PV) Adoption on a National Scale in the United States

Casey Corrado, Emily Holt, and Dr. Lauren Schambach

The MITRE Corporation, Bedford, MA, 01730, U.S.

Abstract—**The U.S. Government is facing immense pressure to reduce carbon emissions and shift towards renewable energy sources due to the pressing issue of climate change. Solar Photovoltaic (PV) technology, when implemented at the levels of residential, commercial businesses, and government agencies, has the potential to slow the rate of global warming, one step in addressing the climate crisis. Solar PVs provide an alternative to fossil fuels, with both lower carbon emissions and cost. The falling price of solar energy has made solar PVs increasingly cost effective compared to traditional, non-renewable energy sources. There are, however, a myriad of technological, environmental, political, economic, and social hurdles that prevent wider-spread solar adoption. While several solar-focused government policies and incentives have already been put in place, the government lacks a full understanding of the constraint space preventing widespread solar adoption. Identifying these barriers is a crucial step in developing effective and impactful plans and policy to expedite nation-wide implementation. This work evaluates current barriers to solar PV adoption within the U.S. and provides potential mitigation steps to address them. A list of recommendations for the U.S. federal government are also provided.**

Keywords— photovoltaics, renewable energy, policy

I. INTRODUCTION

With concerns ranging from rising sea levels to increased natural disasters, climate change poses a risk to the safety and security of our nation. Confronted with these concerns, the U.S. government is facing pressure to reduce carbon emissions and shift towards renewable energy sources. Solar Photovoltaic (PV) technology, when implemented at the levels of residential, commercial businesses, and government agencies, has the potential to slow the rate of global warming, one step in addressing the climate crisis. To this end, following Presidential Executive Order 14008, several federal agencies mentioned solar PVs in their Federal Climate Adaptation Plans as a method of reducing their carbon footprint and investing in sustainable energy sources (e.g., USDA, DOC, DOE, CDC, HUD, DOI plans) [1]. Solar PVs provide an alternative to fossil fuels, with both lower carbon emissions and cost, the falling price making PVs increasingly effective compared to traditional sources [2].

There are, however, a myriad of technological, environmental, political, economic, and social hurdles that prevent wider-spread solar adoption. While several solar-focused government policies and incentives have already been put in place, the government lacks a full understanding of the constraint space preventing widespread solar adoption. Identifying these barriers is a crucial step in developing effective and impactful plans and policy to expedite nation-wide

implementation. Once identified, mitigation steps can be evaluated, and the best path forward can be determined.

II. BARRIERS TO NATIONWIDE ADOPTION

In 2006, the National Renewable Energy Lab (NREL) outlined major non-technical barriers to solar [3]. The following were the most frequently identified barriers:

- Lack of government policy supporting energy efficiency (EE) and renewable energy (RE)
- Lack of information dissemination and consumer awareness about energy and EE/RE
- High cost of solar and other EE/RE technologies compared with conventional energy
- Difficulty overcoming established energy systems
- Inadequate financing options for EE/RE projects
- Failure to account for all costs/benefits of energy choices
- Inadequate workforce skills and training
- Lack of adequate codes, standards, and interconnection and net-metering guidelines
- Poor public perception public of renewable aesthetics
- Lack of stakeholder/community participation in energy choices and EE/RE projects.

This work expands on these previously identified barriers and provides an update to their status, since the technology, economics, policies, and costs of solar have changed dramatically over the last decade [3]. In addition, the U.S. social perception of solar will also be discussed.

III. TECHNOLOGICAL BARRIERS

While widescale adoption of solar PVs shows a great potential to impact climate change, there are several technological limitations that must be addressed for the effort to be successful. These include issues associated with the manufacturing process, efficiency, and energy storage. Investing in the development of PVs, their manufacturing, as well as energy storage can decrease cost and improve energy output, increasing the feasibility of nationwide adoption.

A. Manufacturing

Cost is a major limiting factor for solar PVs. From a technological perspective, there are many barriers that keep the cost of this technology high. For example, most current solar panels utilize monocrystalline or polycrystalline silicon for the semiconductor material. Monocrystalline panels are more efficient but are also more expensive to manufacture than polycrystalline. Because of this, most panels on the market utilize polycrystalline to decrease cost [2]. By investing in monocrystalline development and manufacturing, the production cost can be decreased.

B. Efficiency

Continuing to invest in the research and development of PV technology can create a new generation of solar panels with increased electricity output, making both residential and powerplant level solar energy a feasible nationwide energy source. For existing panels, the most critical factor in determining electricity output is sun exposure. This is dependent on many environmental factors, such as location on the earth's surface. Environmental barriers will be further addressed in Section 2.2, while this section will provide a focus on how the efficiency of the technology itself can be impacted. For example, many environmental factors, like temperature and weather, can affect efficiency. A comprehensive list of factors impacting the efficiency of a solar panel are shown below in Table 1, with additional information on mitigation strategies.

TABLE I.   FACTORS IMPACTING SOLAR PV PANEL EFFICIENCY

Factor	Summary	Mitigation
Temperature	Increased efficiency at low temps. [4]	Improved temperature regulation (water cooling, etc.) [4]
Reflection	Increased efficiency by minimizing light reflected away	Anti-reflection coatings, textured surfaces, etc. [5]
Installation	Increased efficiency with proper installation [6]	Proper mounting position (tilt, orientation, etc.) and maintenance
Time	Efficiency decreases over time [5]	Invest in research to increase lifespan
Weather	Large weather events damage panels	Proper weather protection and maintenance
Soiling	Sediment buildup decreases efficiency	Regular cleanings and maintenance

C. Energy Storage

As we continue to improve solar energy output, increased emphasis is placed on energy storage. Without implementing storage solutions, energy must be generated and consumed simultaneously, which can impact grid reliability as generation must periodically go offline to avoid overgeneration issues. For solar energy specifically, electricity generation is dependent on sun exposure, which does not often align with power usage. For example, summer afternoons and evenings are often when energy usage is at its peak, the same time of day when solar energy generation decreases. Implementing energy storage into the grid can help even out this power imbalance, creating more reliable energy supply. There are many different types of energy storage that can be considered, including pumped-storage hydropower, electrochemical batteries, etc. [7]. Storage methods can be customized for a specific location or application, such as using excess solar generation to heat buildings. Both residential and commercial solar can be improved using solar-plus-storage systems, and investment in energy storage is critical to the success of solar energy.

IV. ENVIRONMENTAL BARRIERS

In their review of environmental impacts from the installation and operation of large-scale solar power plants, Turney and Fthenakis identify and appraise 32 impacts, covering the themes of land use intensity, human health and wellbeing, plant and animal life, geohydrological resources, and climate change [8]. Most impacts considered were found to be beneficial, and there were no negative effects when compared to traditional power generation. In addition, when investigating adverse environmental impacts of the entire solar PV lifecycle, including manufacturing and end of life, Tawalbeh et al. identified six areas of interest including land usage, water usage, hazardous materials, air pollution, noise, and visual [9]. Air pollution, hazardous materials, and noise were found to be greatest during the manufacturing phase, and while hazardous materials need to be considered [10], air pollution and noise were minimal to moderate when compared to standing up traditional energy production facilities. Visual impacts will be discussed under social barriers in a following section.

Environmental barriers to adoption of solar PV technology can be split into two categories, large scale facilities with the intent of distributing the generated solar power, as well as small scale, on the order of individual homes and businesses. Here, we discuss potential environmental barriers with a focus on siting location considerations and recycling of PV materials to reduce overall environmental impact of the technology.

A. Insolation, Extreme Weather, and Land Use Intensity

Insolation is defined as the measure of the amount of solar radiation that reaches the surface of the earth and is often used to determine appropriate siting for solar PV installations. Latitude, climate, weather patterns, and time of day are major factors that affect insolation [11]. Other factors can include cloud cover, dust, volcanic ash, and pollution. Seasonal variation in insolation is greatest in higher latitudes. In general, it's desirable to install solar PV technology in areas with high insolation, which are commonly areas in lower latitudes with arid climates [11]. However, the lower latitudes and arid climate are usually hot which can degrade the panels, and a lack of water resources in these areas can pose an issue for cooling and cleaning off dust. With respect to water usage, Macknick and Cohen found that even with water requirements for cleaning and cooling, solar PVs use dramatically less water than classic forms of energy production [12].

Solar PVs are designed to have maximum exposure to sunlight, which also makes them vulnerable to extreme weather. While the panels are waterproof and able to withstand heavy rain, snow, and even hail, areas that experience these weather patterns need ensure the selected panels are appropriately rated. Specifically in regions that experience a lot of snowfall, costs of clearing the snow off the solar panels may need to be considered. Solar PV panels in regions that often experience hurricane force winds have shown the most damage from debris smashing into the panels rather than the panels being directly affected by the wind or rain [13]. The types of severe weather conditions that can be expected in a specific area should inform the design of solar PV systems selected for that location, which could increase costs in some locations where extreme weather is common.

Land use intensity can be defined as a measure of the amount of land area required, multiplied by the amount of time the land is in use, per unit power generated [8]. Turney and Fthenakis compared the land use intensity for the lifecycles of solar PV power and coal power, taking into account the power plant lifetime, areas used for gathering and transporting fuel, areas used for the generating facilities, land and energy required for manufacturing components, and finally the recovery time for the land that had been transformed. They found that a 30-year-old PV plant has 15% less land use intensity than a coal power plant of the same age, and as the age of the solar PV plant increases, its land use intensity becomes significantly less because additional land for mining isn't needed [8]. Even though land use intensity for solar PV has been found to be less than that of coal power for example, the significant amount of open space needed for large scale solar PV facilities can still pose an issue. A mitigation strategy for the land use intensity is to prioritize building solar PV fields in land that is already in use (e.g. installing in parking structures or on top of buildings) or land that has been degraded (e.g. used previously for mining, undesirable for crop growth, areas with minimal biodiversity).

B. Impact on Wildlife and Ecology

There is limited published literature on the impact of solar PV on wildlife, especially large scale commercial solar PV facilities [14]. Due to this lack of research, it is currently unclear if the development of solar PV farms will have major detrimental impacts on wildlife. Chock et al. performed a survey of experts and professionals in the fields of ecology, conservation, and energy to identify key questions to understand the impacts of solar PV facilities on wildlife [14]. This includes investigation on what wildlife is being killed or negatively impacted by solar fields and to what extent this is happening. For example, some species of birds are known to mistake solar PV panels as water and fly into them. Chock et al. concluded that more investigation is needed to determine the wildlife's perception of solar facilities, if they can generally be considered a natural attraction or deterrence, as well as habitat use in and around facilities in resident and migratory species [14].

Tanner et al. recently studied the effect of the microhabitat created by the solar fields in desert environment and found that panel shading and water runoff on the panel configuration had impacts on the richness and diversity of the ecological habitats [15]. They suggest that further study will be required to gain a full understanding of how large-scale solar fields will change the biodiversity and landscape of the land they are situated on.

Because there is limited research in this area, funding should be provided to continue investigation on our understanding of the impacts of solar PV on the natural world. An example of a way that large scale solar PV fields can enhance the land habitat on which they exist is by becoming pollinator friendly. Fresh Energy, a nationally recognized source of expert knowledge on solar sites planted with deep-rooted native flowers and grasses, have helped communities develop scorecards to help them become pollinator-friendly [16]. The flowers and grasses can capture and filter storm water, build topsoil, and provide abundant and healthy food for bees and other insects.

C. Hazardous Materials

In Bakhiyi et al.'s review of the environmental and occupational health issues associated with the solar PV industry, they assert that manufacturers should cooperate with workers, researchers, and government agencies toward improved and transparent research, the adoption of specific and stricter regulations, and the implementation of preventive risk management of occupational health and safety [10]. The hazardous materials in the panels also create challenges for disposal or recycling at the end of their lifecycle [17].

As we approach the end of the lifecycle for solar panels installed in the early 2000's, the recycling or disposing of solar PV technology will become an increasingly important issue that needs to be addressed. In an NREL report released in March 2021, it was indicated that Government-funded research and analysis is needed to study and inform: 1) the value of and the markets for recovered materials, 2) the volume and composition of end-of-life PV modules, 3) module recycling technology and infrastructure needs, 4) permitting requirements and liabilities, and 5) costs associated with PV module recycling [18]. It was also found that no federal statutes or regulations expressly speak to recycling-based recovery of PV modules in the United States, however state- and industry-led policies have started to emerge to address end of life PV module management concerns, such as that by the State of Washington. Funding for research and education on best practices for recycling of these systems should be given a high priority in the coming years.

V. POLITICAL BARRIERS

A variety of political barriers currently hinder widespread solar adoption. These barriers include existing policies and regulations, as well as political challenges in passing new legislation and policies to accelerate the process. Existing policies and regulations present a barrier to nationwide adoption of solar PVs, from investment regulations, permitting regulations, and policies favoring traditional sources of energy and centralized power grids.

A. Investment Regulation

Shared solar, or community solar, is an increasingly popular business model for deploying distributed solar technology. These installations allow customers that do not have sufficient solar resources, rent their homes, or are otherwise unable or unwilling to install solar on their residences to buy or lease a portion of a shared solar system. The participant's share of the electricity generated is credited to their electricity bill, as if the solar system were located at their home [19]. This model for solar participation broadens the potential customer base and offers the benefits of economies of scale to project developers.

However, current policy and regulatory barriers inhibit share solar projects. Uncertainty exists about the applicability of Securities and Exchange Commission (SEC) requirements for registration and disclosure of these programs. Due the infancy of the shared solar market, there is a lack of legal precedent, market research, and data on project successes. One of the top concerns raised by shared solar stakeholders is the applicability of SEC requirements for registration and disclosure of shared solar projects. Central to this issue is whether an interest in a shared solar project is a 'security' [20].

In community solar project, solar is not used 'on-site' by residents but is instead externally installed and shared among subscribers. Virtual net metering (VNM) is a bill crediting system for community solar where customers receive credits on their electric bill for excess energy produced by their share of the community solar project [21]. Some authors claim that all community solar projects involve the sale or offer of securities, while others claim that the electricity bill-crediting process relieves a community solar project any securities liability [21]. The electricity bill-crediting process ensures that subscribers are not viewed as 'investors' and that energy bill savings are not considered income, which treats customers as paying for the energy preemptively, then subsequently receiving a credit [22].

Depending on the model that is used, selling shares in community solar projects may implicate federal and state securities laws. Although the federal statutes do not clearly define what a security is, in S.E.C. v. W.J. Howey Co., the Supreme Court articulated that a security is any transaction that involves an investment of capital with an expectation of profits without effort on the part of the investor [23]. Whether a community solar project would be a "security" is under this definition is still not clear [22], demonstrating the need for state action on authorizing virtual net metering to remove the barrier regarding uncertainty about S.E.C. requirements.

In addition to the lack of clarity in existing policies and the need for additional policies, there also exists a high churn of current policies, resulting in a lack of stability of incentives for the adoption of PVs. Examples are inconsistencies between policy measures and socioeconomic factors, or the sudden removal of existing subsidies. While most countries have policy measures to support renewable energies, the market loses trust when policy decisions are reversed, such as the recent retrospective reduction of feed-in tariffs in Italy and Spain [22]. Failure to involve all the relevant stakeholders in energy policy planning and regulatory issues, such as difficulties acquiring building permits, constituted further barriers.

B. Permitting and Zoning Regulations

Permitting processes for commercial or utility-scale solar PV installations can be complicated and present additional barriers, particularly in urban areas. Specifically, large rooftop projects are more difficult to get permitted than ground-mounted projects because of zoning and contractual issues, and in some jurisdictions, there is no clear right of appeal for permitting decisions [24]. Permitting issues can be addressed by putting statues in place that give solar the right of way, but some states have forbidden local land-use restrictions. In these instances, the local authority cannot deny a permit for a solar system for many superficial reasons including visual aesthetics.

C. Political Divide and Difficulty Passing Legislation

As discussed above, current policies limit the deployment of solar PV technologies. There are, however, further barriers preventing future legislation from being passed. Researchers have found that there has been significant partisan and ideological polarization on support for environmental spending since 1992—consistent with the expectations of party sorting theory. Such polarization likely will inhibit the further development and implementation of environmental policy and the diffusion of environmentally friendly behaviors [25].

While polarization of environmental spending is present, surveys demonstrate that there exists majority support for government action on climate change. Pew Research Center studies have shown that about two-thirds of U.S. adults (67%) say the federal government is doing too little to reduce the effects of climate change [26]. Additionally, Pew Research Center studies found that 65% of Americans give priority to developing alternative energy sources, compared with 27% who would emphasize expanded production of fossil fuel sources [27]. While these studies demonstrate the barriers in the political divide to passing future climate change-focused legislation, studies show that three-quarters of Americans (77%) agree that the more important energy priority should be developing alternative energy sources such as wind and solar power and hydrogen technology rather than increasing U.S. production of fossil fuels [26], demonstrating the public's interest in investing in alternative energy sources such as solar.

Global support for renewables is widespread [28] and support for these technologies is growing across the U.S. Prior research demonstrates that most Americans have positive attitudes toward renewable energy sources and toward policies for increasing the use of renewable energy [29] [30] [31]. This body of research indicates that support for renewable energy tends to be motivated primarily by people's perceptions that it creates economic benefits and reduces environmental harms. However, while most Americans want energy that is cheap and clean, the relative priority of these considerations vary across political divides. Support for renewable energy is driven by two main camps: major focus on considerations of economic costs/benefits vs. concern about global warming [32]. Even with strong public support for alternative energy sources, however, interview data has found that the language surrounding climate resiliency-based legislation has a major impact on how it is received [33] [34] [35]. For example, documents that focus on environmental benefits have had more difficulty being passed than those that address energy sufficiency and resiliency or the economic benefits of these practices [36].

To make lasting, impactful legislation on nationwide implementation of energy sources like solar PVs, a middle ground needs to be found across political parties. This can be accomplished by focusing on not only environmental benefits and energy sufficiency, but also universally supported bipartisan issues such as the key role renewable energy has in the Great Power Competition. China, for example, has been heavily investing in renewable energy technology research and development. While China is the number one producer of the world's greenhouse gas emissions, it also generates the most wind energy (twice as much as the U.S., the second largest wind generator) in addition to generating one third of the world's solar energy [37]. The Chinese government has also indicated an investment in sustainability, making a pledge to reach carbon neutrality by 2060 [38]. Without establishing a nationwide stance, the U.S. will not be able to take the steps necessary to be a key player in this space and maintain international influence.

VI. ECONOMIC BARRIERS

Historically one of the driving limitations of solar panels was cost. However, since 2014 the average cost of solar PV panels has dropped nearly 70% [39]. Markets for solar energy are maturing rapidly around the country since solar electricity is now economically competitive with conventional energy sources in most states [40]. With the cost of solar becoming more reasonable on a national scale, factors like homeowner return on investments and the soft costs of solar must be considered for successful implementation.

A. Homeowner Return on Investments

A standard warranty for residential solar PV panels is 90% efficiency after 10 years and 80% after 25 years whereby the manufacturer will guarantee 80% or better performance compared to the rated efficiency [41]. Solar PV panel warranties will depend on manufacturer, installer, as well as location (i.e. varies from state to state, and even from locality to locality). The return on investments for homeowners varies from an average of 5 years to 16+ years due to tax incentives and the cost of electricity from other sources [42]. The main mitigation strategy to tackle this issue is to institute both tax incentives for different types of energy. For example, a higher tax would be placed on "dirty" methods (oil, coal, etc.) and tax incentives would be established to lower the return on investment for going solar.

B. Unequally Funded Subsidies

Existing policies and subsidies to traditional energy sources such as fossil fuels create an unequal playing field for renewable energy sources such as solar PV. Studies have demonstrated the necessity of financial support and subsidies to promote the spread in adoption of solar energy [43], yet researchers from the International Renewable Energy Agency (IRENA) estimates that the United States totaled $23 billion in renewable energy subsidies while researchers from the International Monetary Fund (IMF) estimate that the United States funded $649 billion fossil fuel subsidies in 2017 [44]. Globally, IMF researchers estimates that global fossil-fuel subsidies to be at least totaled to $447 billion in 2017, with an estimated total of $220 billion going to petroleum products alone, while subsidies to renewables was estimated to only be $166 billion [45]. This drastic difference in estimated subsidies creates an unequal playing field for solar PV and inhibits wider spread adoption of solar energy. Researchers have demonstrated that "state solar financial incentives systematically encouraged market deployment of solar photovoltaic (PV) technology", and that "states offering cash incentives such as rebates and grants experienced more extensive and rapid deployment of grid-tied PV technology than states without cash incentives over the study period" [46].

C. Installation Costs

The relatively high fraction of transaction cost is inhibiting the growth of the midscale solar market, specifically for projects larger than single residential instillations but smaller than utility-scale solar. Transaction costs, which is like raw cost compared to larger systems, can represent 5%–20% of the total for a midscale project [24]. Further, the interconnection process for midsized can be similar to that for larger utility-scale systems, even though much of the electricity is consumed on site rather than exported entirely to the grid, resulting in a disproportionally large cost burden for mid-sized solar installations [24]. Key challenges driving these high installation costs include a lack of interconnection cost certainty and relatively long processing and approval times. Project delays are more common for the midscale market than smaller distributed PV projects (250 kW and under), often requiring a detailed impact study for any project greater than 1 MW [47]. As a result, PV installers often downsize projects because long duration and high costs can place financing at risk or render a project economically unfeasible [24].

D. Soft Costs of Solar

Solar hardware costs have continued to decrease in the past decades, but other economic barriers persist. "Soft costs" of solar, representing the non-hardware costs related to solar installations (permitting, financing, and customer acquisition) are becoming an increasingly larger fraction of the total cost of solar, now constituting up to 65% of the cost of a residential PV system [48], which includes transaction costs, indirect corporate costs, installer/developer profit, supply chain costs, and sales tax [49]. Soft costs are driven up when processes for going solar are slow or inefficient, often a result of the many jurisdictions, utilities, and differing state and local laws involved, resulting in inefficiencies in permitting, inspection, and grid interconnection. As a result, customers often experience a significant lag time between the purchase of a solar system and the installation of that system [50]. The lag time not only creates a negative customer experience, but also potentially adds costs.

Several factors limit certain customers from adopting solar including: the high cost and up-front expense of solar systems (residential solar installation prices can range from $26,000 - $38,000 [51]); the lack of competitive interest rates for personal loans in order to finance an installation; the need for high credit scores in order to securing loans to finance solar installations; and the inability of tax-exempt businesses and certain populations to use the Solar Investment Tax Credit [52] [50].

Soft costs limit utility-scale solar installations in addition to residential installations. Though the soft costs of large-scale PV projects (which include permitting, inspection, and interconnection; sales tax; engineering, procurement, and construction; and developer overhead and profit) have decreased, they have decreased less drastically compared to the soft costs of residential installations. Costs to cover these activities have remained at about the same proportion of total costs, between 32 to 44 percent, for utility-scale PV [53].

Improved educational resources and training across the solar product cycle is one way to address the high cost of solar. Technical assistance programs can help to increase efficiency and decrease these costs by engaging experienced solar professionals to provide governing bodies the knowledge and tools they need to start their own programs [50]. In addition, solar also impacts professionals working in neighboring industries, such as real estate agents, code officials, and firefighters, who need to understand how solar energy affects their jobs. Educating these professionals lowers costs by improving solar sales and speeding up installations [50].

VII. SOCIAL BARRIERS

Societal barriers for solar adoption exist across both residential and large-scale generation. Consistent pushback has been seen for the aesthetic look of solar panels, particularly for residential installations. In addition, concern has been raised that visual pollution from solar farms has the potential to impact wildlife. Beyond this, there is lack of trained workers to perform installation, maintenance, etc.

A. Austhetics

One of the biggest drawbacks for consumer adoption of solar panels is the aesthetics. Bao et al. investigated the role of visual appeal in consumer preference for residential solar panels and found that consumers were willing to pay higher prices for better looking rooftop solar PV systems [54]. In recent years, the advancement of solar technology and especially aesthetic design considerations have made rooftop solar PV more desirable. While different colors of solar PV panels are now available to consumers, Bao et al. found that in general, black was the most preferred color because it was assumed to absorb more sunlight and thus be more efficient, however, when the panels were shown to survey participants on a colored roof, the responses indicated that panels that matched the roof color and blended in were more desirable. In the context of reliability, efficiency, and price, solar panels that were rated highly in appearance scored the highest even if the remaining factors were not as high.

To address aesthetic concerns, several new solar options are coming to market. While improvements like this aim to solve customer concerns, they are currently much too costly for the average consumer or for many government funded buildings. As it stands, residential installations are dependent on the homeowner, with limitations like cost, visual appeal, etc. Even with proper government incentives, public opinion on the aesthetics of solar needs to change to change to see widespread adoption on a residential level. Proper advertising and media coverage surrounding the benefits of solar is one solution.

B. Workforce Education

Adoption of solar is hindered by a lack in trained solar workers to handle the increasing demand for solar installations. In 2019, the US Energy Employment Report found that 87% of construction employers engaged in the solar industry — who employ the majority of the solar workforce — reported that hiring was either somewhat difficult or very difficult, while 76% of professional services and 81% of manufacturing employers reported that hiring was somewhat difficult or very difficult in 2019 [55]. As the solar industry continues to grow, there is more demand and opportunity for qualified solar workers. Additionally, the Department of Energy has stated that "as solar becomes more ubiquitous, it will continue to impact people working in other industries. Professionals need access to high-quality, local, accessible training in solar energy system design, installation, sales, and inspection, as well as power systems engineering and related professions like building safety officials and first responders. Solar training and energy education play a crucial role in securing the future of solar adoption" [56].

VIII. MITIGATIONS

The sections above discussed the technological, environmental, political, economic, and social barriers that prevent the adoption of solar energy in the U.S. Various mitigation steps and strategies must be taken to address these challenges, tax incentives to technology development. A full summary including mitigation steps is included in Table 2.

TABLE II.   SOLAR BARRIERS MITIFATION STEPS

Category	Barrier	Summary	Mitigation Steps
Technical Barriers	Existing PV Inefficiencies	Energy output of PVs is limited by both environmental factors as well as the operational limit of the technology	Proper cooling, installation, maintenance, etc. to ensure efficiency of existing units; improve baseline efficiency
	Manufacturing Setbacks	Highly efficient semiconductors used in PVs are challenging and costly to produce	Improve manufacturing process to decrease cost
	Energy Storage Issues	Solar energy generation does not typically align with power usage, resulting in grid reliability issues	Integrate next generation energy storage solutions (ex. advanced batteries) into the grid
Environmental Barriers	Location Limitations	Energy output highly dependent on location, weather, etc., and not all locations are ideal for solar power	Integrate solar energy generation on existing buildings or structures; invest in solar at good candidate locations
	Hazardous Materials	Solar PVs panels contain hazardous materials that are challenging to recycle	Continue research to determine the best processes and regulations for recycling PVs
	Concern over Wildlife Impact	Potential negative impacts to wildlife and ecology are not well understood by the solar community	Continue research to fully evaluate risk to wildlife
Political Barriers	Political Divide	No unified stance on renewables within the U.S. government, making passing legislation challenging	Increased focus on universally accepted benefits, such as the role solar will play in the great power competition
	Investment Regulations	U.S. government does not have a unified stance on the status of community solar as a "security"	Establish guidelines and legal frameworks for community solar projects that work within the SEC guidance
Economic Barriers	High Cost	The high cost of solar, particularly "soft costs," like permitting/financing, limit the widespread feasibility	Increased education and training for all involved in the solar process to speed up installation, decrease cost, etc.
	Lack of Subsidies	U.S. is currently funding subsidies for fossil fuels at higher rates than subsidies for renewable energy	Reduce investments in fossil fuel subsidies, and invest in renewable energy sources, like solar, at higher rates
	Return on Investments	Can take upwards of 16+ years for the return on investment of solar, limiting implementation	Increased tax incentives for implementing solar, increased taxes on "dirty" energy sources
Social Barriers	Aesthetic Public Opinion	Public approval of the aesthetics of solar is low across both residential and solar farm installations	Increased advertising/media coverage on the benefits of solar
	Workforce	Lack of skilled, trained workers within the solar sector	Increased solar energy education and training

978-1-7281-6118-1/22 $31.00 © 2022 IEEE   1115

IX. RECOMMENDATIONS

The desired outcome of this investigation was to develop a set of guiding recommendations to provide the U.S. government when developing policy and legislation focused on widespread implementation of solar PV technology. Using the identified mitigation steps as a baseline, the following central recommendations were made:

A. Improved Renewable Energy Incentive Programs

Recommend that the U.S. government reinvigorate existing tax incentive programs for residential and commercial solar installations to encourage solar implementation, in addition to stronger disincentive programs aimed toward decreasing the use of "dirty" energy nationwide; in addition, recommend that the U.S. government reduce investments in fossil fuel subsidies and invest in renewable energy sources (solar PV in particular) at higher rates.

B. Increased Government Research Funding

Recommend that the U.S. government increase funding for the following research areas regarding solar PVs: improving solar technologies, their manufacturing, and energy storage; evaluating the impact of large-scale solar installations on wildlife; and establishing best practices for recycling the parts and materials used for this technology.

C. Shift in Focus of Renewable Energy Legislation

Recommend that moving forward, all federal policies, legislation, etc. regarding renewable energy, such as solar, focus on universally agreed benefits, such as the key role renewables will play in the Great Power Competition.

D. Increased Educational Resources

Recommend that the U.S. government invest in developing the educational system around solar energy, including government funded educational assistance for workers to be trained in solar installation and maintenance, as well as increased awareness for costumers on topics like: solar PV options; the installation and adoption process; available incentives and assistance programs; and legal regulations surrounding solar.

E. Solar Benefits Campaign

Recommend that the U.S. government launch a campaign on the benefits of solar in order to increase the public awareness on the benefits of solar beyond just environmental impact (decreased cost, job creation, etc.) as well as educate the public on the process of switching to solar energy.

X. CONCLUSIONS

Solar energy, specifically solar PVs, have great potential to make lasting impact on the U.S. energy sector. While great strides have been made to further develop and effectively utilize this technology, nationwide adoption has been slowed by a variety of technological, environmental, political, economic, and social barriers. This work provided a detailed look into each of these limitation areas, providing potential mitigations steps as well as a list of recommendations for the U.S. government to increase PV adoption on a nationwide scale.

ACKNOWLEDGMENT

We would like to acknowledge Dr. Laura Leets, the Agile Connected Government Innovation Area Lead for the MITRE Internal Research Program, who funded this research.

REFERENCES

[1] Office of the Federal Chief Sustainability Officer, "Federal Climate Adaption Plans," Council on Environmental Quality, [Online]. Available: https://www.sustainability.gov/adaptation/.

[2] International Energy Agency, *World Energy Outlook,* 2020.

[3] R. Margolis and J. Zuboy, "Nontechnical Barriers to Solar Energy Use: Review of Recent Literature," September 2006. [Online]. Available: https://www.nrel.gov/docs/fy07osti/40116.pdf.

[4] National Renewable Energy Laboratory, "Lifetime of PV Panels," 23 April 2018. [Online]. Available: https://www.nrel.gov/state-local-tribal/blog/posts/stat-faqs-part2-lifetime-of-pv-panels.html.

[5] Freedom Solar Power, "How Long Do Solar Panels Last? The Average Lifespan of Solar Panels," 2021. [Online]. Available: https://freedomsolarpower.com/blog/how-long-do-solar-panels-last.

[6] U.S. Office of Energy Efficiency & Renewable Energy, "Solar Photovoltaic System Design Basics," 2021. [Online]. Available: https://www.energy.gov/eere/solar/solar-photovoltaic-system-design-basics.

[7] U.S. Office of Energy Efficiency & Renewable Energy, "Solar Integration: Solar Energy and Storage Basics," 2021. [Online]. Available: https://www.energy.gov/eere/solar/solar-integration-solar-energy-and-storage-basics.

[8] D. Turney and V. Fthenakis, "Environmental impacts from the installation and operation of large-scale solar power plants," *Renewable and Sustainable Energy Reviews,* pp. 3261-3270, 2011.

[9] M. Tawalbeh, A. Al-Othman, F. Kafiah, E. Abdelsalam, F. Almomani and M. Alkasrawi, "Environmental impacts of solar photovoltaic systems: A critical review of recent progress and future outlook," *Science of teh Total Environment,* 2021.

[10] B. Bakhiyi, F. Labreche and J. Zayed, "The photovoltaic industry on the path to a sustainable future - Environmental and occupational health issues," *Environment International,* pp. 224-234, 2014.

[11] US Energy Information Administration, "Solar explained: Where solar is found and used," [Online]. Available: https://www.eia.gov/energyexplained/solar/where-solar-is-found.php.

[12] J. Macknick and S. Cohen, "Water Impacts of High Solar PV Electricity Penetration," *NREL Technical Report TP-6A20-63011,* p. 21 pp, 2015.

[13] SolarReviews, "What happens to solar panels in a hurricane?," 10 6 2021. [Online]. Available: https://www.solarreviews.com/blog/what-happens-to-solar-panels-in-a-hurricane.

[14] R. Y. Chock, B. Clucas, E. Peterson, B. Blackwell, D. T. Blumstein, K. Church, E. Fernandez-Juricic, G. Francescoli, A. L. Greggor, P. Kemp, G. M. Pinho, P. M. Sanzenbacher and B. Schulte, "Evaluating potential effects of solar power facilities on wildlife from an animal behavior perspective," *Conservation Science and Practice,* 2020.

[15] K. E. Tanner, K. A. Moore-O'Leary, I. M. Parker, B. M. Pavlik, S. Haji and R. R. Hernandez, "Microhabitats associated with solar energy development alter demography of two desert annuals," *Ecological Applications,* 2021.

[16] Fresh Energy, "Fresh Energy," [Online]. Available: https://fresh-energy.org/beeslovesolar/pollinator-friendly-solar-scorecards.

[17] M. S. Chowdhury, K. S. Rahman, T. Chowdhury, N. Nuthammachot, K. Techato, M. Akhtaruzzaman, S. K. Tiong, K. Sopian and N. Amin, "An overview of solar photovoltaic panels' end-of-life material recycling," *Energy Strategy Reviews,* 2020.

[18] T. L. Curtis, H. Buchanan, G. Heath, L. Smith and S. Shaw, "Solar Photovoltaic Module Recycling: A Survey of U.S. Policies and Initiatives," *National Renewable Energy Laboratory Technical Report,* 2021.

[19] National Renewable Energy Lab, "Community Shared Solar Policy and Regulatory Considerations," Golden, CO, 2014.

[20] D. Feldman, A. Brockway, E. Ulrich and R. Margolis1, *Shared Solar: Current Landscape, Market Potential, and the Impact of Federal Securities Regulation,* Golden, CO: National Renewable Energy Laboratory, April 2015.

[21] L. Richardson, "Virtual net metering: What is It? How does it Work?," EnergySage, May 2019. [Online]. Available: https://news.energysage.com/virtual-net-metering-what-is-it-how-does-it-work/.

[22] J. Meltzer, "Community Solar: A Solution to Solar Problems with Significant Legal Barriers," Georgetown Environmental Law Review, 2 April 2020. [Online]. Available: https://www.law.georgetown.edu/environmental-law-review/blog/community-solar-a-solution-to-solar-problems-with-significant-legal-barriers/#_ftn4.

[23] U. .S. Supreme Court, "SECURITIES AND EXCHANGE COMMISSION v. W. J. HOWEY CO. et al.," May 1946. [Online]. Available: https://www.law.cornell.edu/supremecourt/text/328/293.

[24] L. Bird, P. Gagnon and J. Heeter, "Expanding Midscale Solar: Examining the Economic Potential, Barriers, and Opportunities at Offices, Hotels, Warehouses, and Universities," National Renewable Energy Lab, Golden, CO, 2016.

[25] A. M. McCright, C. Xiao and E. R. Dunlap, "Political polarization on support for government spending on environmental protection in the USA, 1974–2012," *Social Science Research*, vol. 48, pp. 251-260, 2014.

[26] C. Funk and M. Hefferon, *U.S. Public Views on Climate and Energy*, Pew Research Center, November 2019.

[27] B. Kennedy, *Two-thirds of Americans give priority to developing alternative energy over fossil fuels,* Pew Research Center, 2017.

[28] R. Wüstenhagen, M. Wolsink and M. Bürrer, "Social acceptance of renewable energy innovation," *Energy Policy*, vol. 35, p. 2683–2691, 2007.

[29] M. Ballew, A. Leiserowitz, C. Roser-Renouf, S. Rosenthal, J. Kotcher, J. Marlon, E. Lyon, M. Goldberg and E. Maibach, "Climate Change in the American Mind: Data, Tools, and Trends," *Environment: Science and Policy for Sustainable Development*, vol. 61, no. 3, pp. 4-18, 2019.

[30] J. McCarthy, "Most Americans Support Reducing Fossil Fuel Use," Gallup, 2019.

[31] S. Mills, B. Rabe and C. Borick, "Widespread Public Support for Renewable Energy Mandates Despite Proposed Rollbacks," *Energy and Environmental Policy*, no. 22, 2015.

[32] A. Gustafson , M. H. Goldberg, J. E. Kotcher, S. A. Rosenthal, E. W. Maibach, M. T. Ballew and A. Leiserowitz, "Republicans and Democrats differ in why they support renewable energy," *Energy Policy*, vol. 141, 2020.

[33] K. Schulze, "Do parties matter for international environmental cooperation? An analysis of environmental treaty participation by advanced industrialised democracies," *Environmental Politics*, vol. 23, no. 1, pp. 115-139, 2014.

[34] F. M. Farstad, "What explains variation in parties' climate change salience?," *Party Politics*, vol. 24, no. 6, pp. 698-707, 2018.

[35] B. Tranter and K. Booth, "Scepticism in a changing climate," *Global Environmental Change*, vol. 33, p. 154–164, 2015.

[36] C. Horne and E. Huddart Kennedy, "Explaining support for renewable energy: commitments to self-sufficiency and communion," *Environmental Politics*, vol. 28, no. 5, pp. 929-949, 2019.

[37] S. O'Meara, "China's plan to cut coal and boost green growth," *Nature,* 20 August 2020.

[38] S. L. Myers, "China's Pledge to Be Carbon Neutral by 2060: What It Means," *New York Times,* 23 September 2020.

[39] Solar Energy Industries Association, *Solar Means Business*, 2019.

[40] International Energy Agency, *World Energy Outlook*, 2020.

[41] EnergySage, "What to know about solar panel warranty," [Online]. Available: https://news.energysage.com/shopping-solar-panels-pay-attention-to-solar-panels-warranty/.

[42] SolarReviews, "How long does it take for solar panels to pay for themselves?," 24 09 2021. [Online]. Available: https://www.solarreviews.com/ blog/how-to-calculate-your-solar-payback-period.

[43] Y. Heng, C.-L. Lu, L. Yu and Z. Gao, "The heterogeneous preferences for solar energy policies among US households," *Energy Policy*, vol. 137, 2020.

[44] D. Coady, I. Parry, N.-P. Le and B. Shang, *Global Fossil Fuel Subsidies Remain Large: An Update Based on Country-Level Estimates*, International Monetary Fund, 2019.

[45] M. Taylor, *Energy subsidies: Evolution in the global energy transformation to 2050*, Abu Dhabi: International Renewable Energy Agency, 2020.

[46] A. Sarzynski, J. Larrieu and G. Shrimali, "The impact of state financial incentives on market deployment of solar technology," *Energy Policy*, vol. 46, pp. 550-557, 2012.

[47] K. Ardani, C. Davidson, R. Margolis and E. Nobler, "A State-Level Comparison of Processes and Timelines for Distributed Photovoltaic Interconnection in the United States," National Renewable Energy Lab, Golden, CO, 2015.

[48] Solar Energy Industries Association, "Solar Soft Costs," June 2019. [Online]. Available: https://www.seia.org/sites/default/files/2019-07/Solar-Soft-Costs-Factsheet.pdf.

[49] B. Friedman, K. Ardani, D. Feldman, R. Citron, R. Margolis and J. Zuboy, *Benchmarking Non-Hardware Balance-of-System (Soft) Costs for U.S. Photovoltaic Systems, Using a Bottom-Up Approach and Installer Survey – Second Edition,* National Renewable Energy Lab, 2013.

[50] Department of Energy, *Solar Futures Study*, 2021.

[51] D. Feldman, V. Ramasamy, R. Fu, A. Ramdas, J. Desai and R. Margolis, "U.S. Solar Photovoltaic System and Energy Storage Cost Benchmark: Q1 2020," National Renewable Energy Lab, Golden, CO, 2021.

[52] M. Vargas, J. Heeter, K. Fricker and K. Laymon, "National Community Solar Partnership," Department of Energy, National Renewable Energy Lab, 2020.

[53] D. Feldman, V. Ramasamy, R. Fu, A. Ramdas, J. Desai and R. Margolis, "U.S. Solar Photovoltaic System and Energy Storage Cost Benchmark: Q1 2020," National Renewable Energy Lab, Golden, CO, 2021.

[54] Q. Bao, T. Honda, S. E. Ferik, M. M. Shaukat and M. C. Yang, "Understanding the role of visual appeal in consumer preference for residential solar panels," *Renewable Energy*, pp. 1569-1579, 2017.

[55] Energy Futures Initiative with the National Association of State Energy Officials , "The 2020 U.S. Energy & Employment Report," Energy Futures Initiative, 2020.

[56] Department of Energy, "Solar Workforce Development," [Online]. Available: https://www.energy.gov/eere/solar/solar-workforce-development.

Laser-weld qualification methods for Al foil interconnection of back-contacted cells to predict module reliability

Barry B. Hartweg, Kathryn C. Fisher, Zhengshan J. Yu, Zachary C. Holman

Arizona State University, Tempe, AZ, United States

SunFlex Solar, LLC, Tempe, AZ, United States

Conductive backsheet modules have been fabricated using inexpensive Al foil, where the bond between foil and cells are made using a laser-welding process. The durability of this interconnection has been proven via thermocycling (TC) on a millisecond laser in the past. However, this laser is too slow for large-scale manufacturing, so a faster nanosecond laser needs to be proven still. A laser-shunted module demonstrated a durable interconnection that can pass TC, but it resulted in an upfront power loss of 34% from the laser-processing. Microscopy demonstrated that the laser-welding ablated the Ag contact down to the Si surface, causing a shunt of the n-type contact. Quality welds only interact with the surface of the Ag to promote adhesion without damaging the cell performance.

Highly Efficient Perovskite-on-Silicon Tandem Solar Cells on Planar and Textured Silicon

Christian M. Wolff, Xin Yu Chin, Deniz Türkay, Kerem Artuk, Mohammadreza Golobostanfard, Florent Sahli, Daniel Jacobs, Quentin Guesnay, Peter Fiala, Mostafa Othman, Bosky Sharma, Brett Kamino, Aïcha Hessler-Wyser, Mathieu Boccard, Quentin Jeangros, Christophe Ballif

EPFL, STI IEM PVLAB, Neuchatel, Switzerland

CSEM, PV-Center, Neuchatel, Switzerland

Beyond the limitations of single-junction solar cells, multi-junction devices offer the possibility to harness the sun' light more efficiently. In particular perovskite-on-silicon (Pk/Si) tandems hold the great promise of high efficiencies >30 % while maintaining low cost. We report on our latest progress in the development of Pk/Si tandems comparing our efforts on single-side and double-side textured Pk/Si tandems, reaching a VOC up to 1.95 V, summed short-circuit current densities above 41 mA/cm2, and certified efficiencies >29 %, on an active area of 1cm2. We achieved these results by dedicated electrical and optical optimizations of all layers within the stack. Specifically, we improved the transparency of the front stack electrodes and contacts through simulation-guided optimizations of the front grid and layer thicknesses, and reduced recombination and transport losses in the Pk absorbers through process and additive engineering for both solution-processed one-step and hybrid two-step deposited Pks. Furthermore, we investigated the stability of single-junction Pk and tandem devices under reverse-bias and standardized accelerated aging conditions.

Corrosion testing of solar cells: Insights to wear-out mechanisms

Andrew Fairbrother, Luca Gnocchi, Christophe Ballif, Alessandro Virtuani

EPFL, Neuchatel, Switzerland

CSEM, Neuchatel, Switzerland

Corrosion is a major end-of-life degradation mode in photovoltaic modules. Herein, an accelerated corrosion test for screening new cell, metallization, and interconnection technologies is presented. The top glass and encapsulation layers were removed from modules to expose the solar cells. These "opened" modules were then placed in acetic acid baths under varying concentration, temperature, and bias. Three cell technologies were tested, including Al-BSF, PERC, and SHJ. The corrosion test can be optimized to match wear-out corrosion behavior of field modules with greater precision and shorter times than standard tests, and can be applied to corrosion sensitive components to assess their durability.

NRG-X-Change and Cooperative Game Strategies as an Alternative to Net-Metering for Solar Generation

Hector Lopez, Ali Zilouchian

Florida Atlantic University, Boca Raton, FL, United States

In this paper, the net metering payment approach and a novel trading paradigm for buying and selling local produced solar energy scheme known as NRG-X- Change are compared. The NRG-X-change does not relay on energy market or matching order ahead of the time. The scheme provides incentives to customer to balance their production. The payment are carried out using NRGcoin, a digital currency similar to Bitcoin. Therefore, the collaborating individuals may maximize their profit using various approaches including the cooperative game theory as presented in this summary. The simulation of both strategies were conducted using actual data sets. The simulation results shows the effective of forming a coalition and utilization of NRG-X change. Such strategy is an attractive alternative to net metering to incentivize the adaptation of renewal technology.

Nanostructured ZnO electron transporting materials for hysteresis-free perovskite solar cell

Vilko Mandić, Ivana Panžić, Floren Radovanović-Perić, Thomas Rath

Faculty of Chemical Engineering and Technology, Zagreb, Croatia

Institute for Chemistry and Technology of Materials, Graz, Austria

An effective perovskite solar cell (PSC) usually reposes on perovskite absorber interfaced between an electron (ETL) and a hole transport layer (HTL), and outer electrodes. Planar zincite is viable for preparing PSCs with high power conversion efficiency (PCE) at low-temperatures due to appropriate electronic structure and physical properties. Recently, semiconductors with advanced geometries have attracted interest for ETLs. To rise up the specific surface area and thereof the charge transfer, zincite in nanorods configuration (ZNR) was tailored, but ZNR-PSCs show only limited increase of PCE due to considerable recombination in 1D ZNR. Methyl ammonium & formamidinium iodide (MAI, FAI) PSCs show highest PCE in ZNR-based devices, but still suffer from hysteresis and degradation. Finally, by eliminating the MAI from the composition, overall reach out to better fill factors and lower rate of hysteresis were achieved. Here we derived ZNR-(chemical-bath) PSCs without MAI (spin coated; assembled in protective atmosphere), to understand the preparing effects and influence on the overall PSC efficiency. Characterization using electric (J/V, EQE), structural (XRD) and morphologic (SEM) methods took place immediately upon PSC closure. Reliable interfacing of the layers can be confirmed, pointing out to importance of such configurations for the development of PSCs performance.

Variable Renewable Energy Participation in U.S. Ancillary Services Markets: Economic Evaluation and Key Issues

James Hyungkwan Kim,* Fredrich Kahrl, Andrew Mills, Ryan Wiser, Cristina Crespo Montañés, and Will Gorman

Lawrence Berkeley National Laboratory, 1 Cyclotron Road, MS 90-4000, Berkeley, CA 94720, USA

Abstract— This research estimates the economic value of standalone and hybrid (battery-paired) variable renewable energy (VRE) participation in ancillary services (AS) markets, from resource owner and electricity system perspectives, in each of the seven U.S. independent system operator and regional transmission organization (ISO/RTO) markets. Across ISO/RTO markets, average (2015-2019) simulated incremental revenues from regulation market participation were $0.0-2.9/MWh (+0-15% of revenue without participation) for standalone VRE owners and $1-33/MWh (+1-69%) for hybrid VRE owners. In most markets, standalone and hybrid VRE were able to provide regulation reserves during periods with high regulation prices, suggesting that VRE participation in AS markets could have high system value. The analysis highlights the value of separate upward and downward regulation products and suggests that ISOs/RTOs might consider initially focusing on enabling hybrid VRE provision of AS.

Keywords—electricity market, renewables, energy storage, ancillary services

I. INTRODUCTION

In the United States, however, VRE participation in organized AS markets is currently low or nonexistent and many questions around the economic value of VRE participation in these markets remain unanswered. For instance, how would the economic value of AS market participation to resource owners and to the electricity system as a whole compare between solar and wind generation, between standalone and hybrid VRE, across the seven organized electricity markets, and between different AS products? How might the economic value change with higher VRE and storage penetrations? What changes in market rules would be needed to allow VRE to participate in AS markets?

This paper examines the economic value of VRE participation in AS markets from resource owner and electricity system perspectives across the seven U.S. electricity markets. The analysis uses a price-taker dispatch model with simple, consistent assumptions that facilitate comparisons across technologies, VRE configurations, and markets over time. It considers two kinds of VRE configurations: (1) standalone VRE facilities, with a standalone solar or wind facility; and (2) hybrid VRE facilities, with a solar or wind facility paired with battery storage.

In a base case, the analysis focuses on VRE participation in regulation markets using historical market prices, with interconnection capacity limits sized to the VRE facility's nameplate capacity. It also examines sensitivities in which VRE participates in spinning reserve markets, VRE participates in future regulation markets in electricity systems with higher renewable penetration, and where interconnection capacity limits are sized to the maximum output of the combined generator and battery capacity (for hybrids). The paper closes with a discussion of three key issues for the results: barriers to VRE participation in AS markets, the potential impacts of higher VRE and storage penetrations on the results, and other emerging AS opportunities for VRE not considered in the analysis.

II. METHODS

A. Metrics

The analysis examines the value of VRE AS market participation from a VRE resource owner's and an electricity system perspective in each of the seven ISO/RTO markets. In both cases, we compare a scenario in which the VRE resource does not participate in AS markets to one in which it does.

For resource owners, we measure value to the resource owner in terms of incremental unit revenues (Δr, \$/MWhPC) from participating in AS markets, where

$$\Delta r = \frac{EN_1 + AS_1 - EN_0}{G_{PC}}$$

- EN is the VRE facility's annual energy market revenues, in \$/yr
- AS is the VRE facility's annual AS market revenues, in \$/yr
- G_{PC} is the pre-curtailment (PC) amount of annual generation from the VRE facility, in MWh$_{PC}$/yr
- Subscript 1 is the scenario in which the facility provides both energy and AS (energy + AS)
- Subscript 0 is the scenario in which the facility only provides energy (energy only)

B. Modeling Framework and Assumptions

To estimate Δr, v, and AR, we use a linear optimization model that maximizes wholesale market revenues against zonal energy and AS market prices for standalone and hybrid VRE resources in each ISO/RTO market, with consistent assumptions across markets to allow for comparability.

The analysis considers four resource types:

1) a 20-MW standalone solar PV plant
2) a 20-MW standalone onshore wind plant
3) a 20-MW hybrid solar PV plant paired with 10 MW/40 MWh of battery storage
4) a 20-MW hybrid onshore wind plant paired with 10 MW/40 MWh of battery storage

The hybrid results should not be compared against the standalone results to assess whether storage would be cost-effective for VRE owners. There are other potential benefits to hybridization, such as interconnection cost savings and capacity value, that are not considered in this analysis.

III. RESULTS

A. Value to Standalone VRE Owners

Figure 1 and 2 show the base case results for standalone VRE owners, by year and ISO/RTO. The tables below each figure show simple average incremental unit revenues (Δr) across 2015-2019 and the percentage change in 2015-2019 average revenues from providing regulation reserves and energy, relative to only providing energy.

Fig. 1. Incremental Unit Revenue ($/MWh$_{PC}$) to Standalone Solar Owner

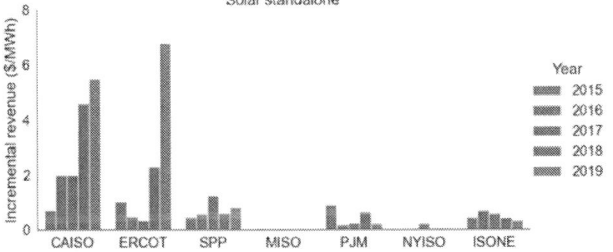

Fig. 2. Incremental Unit Revenue ($/MWh$_{PC}$) to Standalone Wind Owner

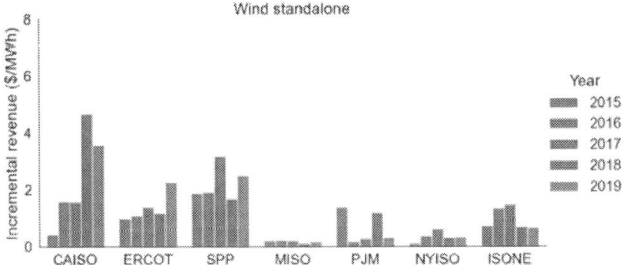

As the figures show, incremental value to standalone VRE owners varies significantly among ISOs/RTOs, across years, and between wind and solar resources. Differences among ISOs/RTOs stem from different regulation products, price levels, and the relationship between energy and regulation prices. The incremental value for resource owners is generally higher in ISOs/RTOs with separate upward and downward regulation products (CAISO, ERCOT, SPP) than in ISOs/RTOs with bidirectional regulation (MISO, PJM, NYISO, ISO-NE).

The main reason for this result is that, with separate products, VRE can provide downward regulation in most hours, whereas with bidirectional regulation products VRE will only provide downward regulation in a limited number of hours in which regulation prices are higher than energy prices.

B. Value to Hybrid VRE Owners

As Figure 3 and Figure 4 show, incremental value for hybrid VRE owners is significantly higher than for standalone VRE owners in most markets. This result shows batteries will tend to provide reserves unless energy price differences are high or reserve prices are low, whereas standalone VRE will tend to provide energy unless reserve prices are high relative to energy prices.

Fig. 3. Incremental Unit Revenue ($/MWh$_{PC}$) to Hybrid Solar Owner

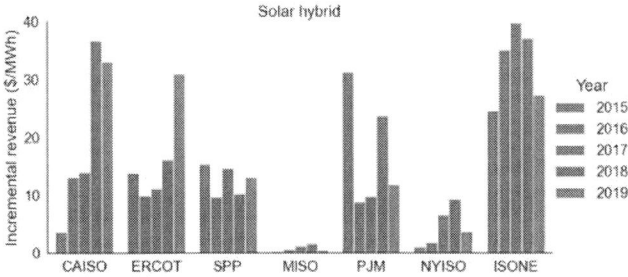

Fig. 4. Incremental Unit Revenue ($/MWh$_{PC}$) to Hybrid Wind Owner

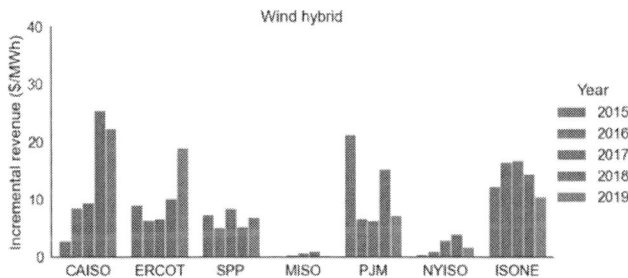

Differences in value among ISOs/RTOs and years are mainly driven by differences in market design and regulation price levels. For instance, MISO and SPP had similar real-time energy and regulation price levels and price variance during 2015-2019, but hybrid batteries provide significantly more incremental revenue in SPP than in MISO because SPP has separate upward and downward regulation products, which allows batteries to provide regulation more efficiently. For PJM and ISO-NE, higher regulation prices

IV. CONCLUSION

Standalone and hybrid VRE resources are not currently participating at meaningful levels in U.S. ISO/RTO markets for frequency regulation and spinning reserves. This paper examined the value of regulation and spinning reserve market participation from a VRE owner and an electricity system perspective.

For standalone VRE owners, the results suggest that the incremental revenues from providing regulation and spinning reserves would vary significantly across ISO/RTO markets, across years, and between solar and wind. For some resources in some markets, the average incremental value may be non-trivial. For instance, average (2015-2019 market prices) incremental revenues for providing regulation services in CAISO (solar/wind), ERCOT (solar/wind), and SPP (wind) were $1.4-3/MWhPC (+6-15%). In other markets and for solar in SPP, incremental revenues were $1.0/MWhPC (+3%) or less. Regulation markets are, however, relatively thin (< 800 MW in each direction), and even in ISOs/RTOs with higher incremental value expanding market participation to VRE and energy storage may lead to market saturation and a decline in AS prices.

Participating in spinning reserve markets added little incremental value for standalone VRE owners, outside of ERCOT and, to a lesser extent, CAISO. This result underscores that, in most markets, most of the reserve market value for standalone VRE owners would be in providing regulation reserves, though differences between ERCOT and other markets suggest that this result is sensitive to differences in market design and AS procurement practices. The high VRE penetration sensitivity showed significant increases in the incremental value of regulation market participation for standalone VRE, due to higher regulation prices and a higher frequency of hours in which regulation prices exceed energy prices.

At current market prices, revenues from regulation and spinning reserve markets are not large enough to meaningfully offset declines in solar and wind resources' energy and capacity value as their penetrations increase. The price forecasts on which the high VRE penetration sensitivity are based did not include higher levels of energy storage, which would tend to depress regulation prices. Relying on high future AS prices to fill revenue gaps will present risks for VRE developers.

For hybrid VRE owners, incremental revenues were, as expected, several-fold higher than for standalone owners, though variation across markets highlights differences in storage value due to different market designs and resource mixes. In the near term, the results suggest that AS revenues could be a significant part of hybrid VRE business models, with the POI sensitivity showing that most of the regulation value of hybrids could be captured with POI capacity limited to the VRE facility's nameplate capacity when storage is sized to 50% of VRE capacity. However, hybrid VRE faces the same uncertainty around AS market prices that standalone VRE does.

In most ISOs/RTOs, standalone and hybrid VRE participation in regulation markets could provide significant value to the electricity system as a whole, as measured by the difference between VRE resources' average regulation value and average regulation market prices. In other words, VRE could provide regulation during periods with high market prices, which would put downward pressure on average market prices and provide ISOs/RTOs with a larger toolset to resolve emerging, higher-cost system constraints. The results show that, in general, VRE provision of regulation services in ISOs/RTOs with separate upward and downward regulation products was higher than in ISOs/RTOs with bidirectional products. Hybrid VRE provided more regulation service and often, but not always, had higher regulation value than standalone VRE.

The results provide insights on two priority areas for considering VRE participation in ISO/RTO reserve markets. First, developing separate upward and downward regulation products, for ISOs/RTOs that do not have them, will enable more efficient use of VRE and storage resources in regulation markets by taking advantage of the fact that these resources have very different opportunity costs for upward and downward reserves and that prices for upward and downward regulation tend to be poorly correlated. Second, focusing initially on VRE hybrid participation in AS markets may be a more efficient first step toward expanding market participation, given that hybrids will provide more reserves than standalone VRE and will generally have higher AS value. That being said, ultimately it may be beneficial to enable both kinds of resources to participate in AS markets.

ACKNOWLEDGMENT

The work described in this study was conducted at Lawrence Berkeley National Laboratory and supported by the U.S. Department of Energy's Office of Energy Efficiency and Renewable Energy Strategic Analysis, Wind Energy Technologies Office, and Solar Energy Technologies Office through the DOE SunShot National Laboratory Multiyear Partnership (SuNLaMP) under Lawrence Berkeley National Laboratory under Contract No. DE-AC02-05CH11231. We would like to especially thank Patrick Gilman, Ammar Qusaibaty, and Paul Spitsen of DOE for their support of this work. The authors thank the following experts for reviewing this report (affiliations do not imply that those organizations support or endorse this work): Clyde Loutan (CAISO), Vinod Syberry (DOE), Jian Fu (DOE), Guohui Yuan (DOE), Adria Brooks (DOE), Paul Denholm (NREL), Elina Spyrou (NREL), Trieu Mai (NREL), Raymond Byrne (Sandia)

Amorphous Manganese Sulfide Enables Efficient and Stable All-Inorganic Antimony Selenosulfide Solar Cells

Chen Qian, Jianjun Li, Kaiwen Sun, Chenhui Jiang, Jialiang Huang, Rongfeng Tang, Martin Green, Bram Hoex, Tao Chen, Xiaojing Hao

University of New South Wales, Sydeny, Australia

University of Science and Technology of China, Hefei, China

Antimony selenosulfide, $Sb_2(S,Se)_3$, has emerged as a promising light-harvesting material for its high absorption coefficient, suitable bandgap, low-toxic and low-cost constituents. An n-i-p device architecture has to be adopted in $Sb_2(S,Se)_3$ based solar cell to accommodate its anisotropic property and low carrier mobility. It has been realized that the high-efficiency antimony selenosulfide solar cells are obtained exclusively using Spiro-OMeTAD as the hole-transporting material. However, the poor stability and high cost of Spiro-OMeTAD may restrict its potential in practical applications in solar cells. Here, we report an all-inorganic $Sb_2(S,Se)_3$ solar cell enabled by using an evaporated inorganic manganese sulfide (MnS) hole-transporting layer. We identify the critical factors that influence the device performance: the carrier concentration and work function of MnS layer, and the junction-quality of $MnS/Sb_2(S,Se)_3$ interface, thus obtaining the highest efficiency of 9.7% in all-inorganic $Sb_2(S,Se)_3$ solar cells. In addition, the unencapsulated $Sb_2(S,Se)_3$ solar cell with MnS demonstrates remarkably enhanced stability than those fabricated using conventional Spiro-OMeTAD as the hole-transporting layer. Our findings provide a new understanding and practical material fabrication strategy regarding how to obtain high-efficiency solar cells when using MnS as a hole-transporting layer. This low-cost, efficient, stable, and up-scalable MnS hole-transporting layer may also be applicable to other emerging solar cells, rendering a better pathway toward commercialization.

A Data-Driven Feeder Selection Method for Distribution System Planning Studies

Alexandre B. Nassif, *SM'IEEE*, and Fernanda C. L. Trindade, *SM'IEEE*

LUMA Energy, San Juan, PR, 00907, United States, and University of Campinas, Campinas, SP, Brazil

Abstract—Distribution system operators across all jurisdictions depend on simulation models to analyze multiple scenarios and derive investment strategies. Relying on century old antiquated assessments can lead to incorrect and onerous decisions, handicapping electric utilities' strategies. Even though load flow and related modeling and associated analysis is today's industry adopted practice, there are many electric utilities that lack such models for their distribution feeders and can benefit from a balance that entails modeling a strategically defined portion of their systems. Additionally, there are niche studies that do not require running individual models of every single distribution feeder and can also rely on sample analysis and subsequent extrapolation. This paper presents a clustering method to derive representative samples of distribution feeders for common distribution planning studies. This work was driven by the needs of a Caribbean electric utility that operates about 1,400 distribution feeders concentrated in an island, but only about 3% of these feeders have a certified load flow model.

Index Terms—Clustering methods, distribution planning, hosting capacity, k-means, technical losses, unsupervised learning.

I. INTRODUCTION

Grid modernization initiatives require effective and efficient distribution system analysis techniques to estimate emerging parameters such as hosting capacity, inverter-based Distributed Energy Resources (DER) grid support requirements, and distribution technical losses. The most effective means of obtaining these parameters is through a detailed study, analysis and customization for each individual distribution feeder, adequately determining feeder-specific impacts [1]. This detailed analysis requires not only time and effort, but also an accurate and verified model that allows a reliable calculation of load flow, short-circuit, and in some cases dynamic response of a feeder. On the other hand, table-top analyses are the most efficient assessment method but are not detailed enough, yielding approximate results at best. In this context, the balance between effectiveness and efficiency will depend on how developed an electric utility's distribution planning resources are. Oftentimes utilities do not have good models for their entire distribution system, and creating them represents a significant effort, especially to those utilities that cannot rely on a decent Geographic Information System (GIS).

Feeder sampling needs have been identified in [2]-[3] for DER interconnection screening and in [4]-[5] for system-wide technical loss estimation. These publications provided great insight for feeders within their jurisdiction and illustrated the effectiveness of unsupervised learning-based sampling. This paper leverages these past contributions and derives a selection method to obtain a small set of distribution feeders of an electric utility that currently has less than 3% of their feeders certified in their GIS, which is the basis for the creation of power flow models. The proposed method uses a k-means based technique to group similar feeders into clusters. Within each cluster, an optimal representative feeder is selected for analysis. The optimization method attempts to bias the search towards a feeder that is part of the pool of 3% modeled feeders, in order to further reduce the amount of effort toward creating new models. The adopted practice conceives a detailed analysis on the representative feeders and subsequent extrapolation to the other feeders that have not been modeled.

The main contribution of this paper is the development of a method to derive a sample of representative feeders, allowing making quick and accurate decisions for three applications:

1. Hosting Capacity, to streamline the DER interconnection process;

2. Smart inverter grid support requirements;

3. Technical losses to report to the energy board.

This technique was adopted by a Caribbean utility that operates about 1,400 feeders but only about 3% of them are certified.

II. DISTRIBUTION PLANNING STUDIES

A. The Need for Sampling

Obtaining a representative sample of an entire distribution system is an effective way of increasing efficiency in studies that are intended to provide system-wide information about a system and where performing system wide studies is either not possible or not practical. It is possible to obtain insight and even make investment decisions based on the direction provided by these studies.

This can be the strategic direction under two scenarios.

978-1-7281-6118-1/22 $31.00 © 2022 IEEE

First, an electric utility may not have all their systems modeled. In this case, barring specific needs, modeling can be directed at a representative sample of systems. Second, there are studies that are, by nature, intended to provide system-wide indices or parameters. One of such examples is the calculation of technical losses. This section addresses the motivation for the three applications.

B. Hosting Capacity Studies

Hosting capacity can be stated as the total DER capacity that can be accommodated on a given feeder without violating operational constraints. Usually these constraints are voltage, ampacity, protection coordination, or short circuit levels, with the voltage rise often being the first bottleneck [6]. Electric utilities have adopted many screening methods to assess the hosting capacity. Among many, the 15% threshold gained widespread acceptance as a somewhat conservative means of quickly screening feeders that may require a detailed impact study [3].

Nowadays, most North American electric utilities have created hosting capacity maps and published them in their website. These maps typically contain all feeders operated by the utility and either show the amount of DERs that can be interconnected without requiring system upgrades. It becomes a challenge to derive an accurate, load flow-based results when a utility does not have models for all feeders or is not able to run them all in batch [7]. Finding a method to classify the feeders in a utility service territory and determine the sensitivity of particular groups of distribution feeders to the impacts of high DER deployment levels is an effective way to circumvent this limitation.

C. Smart Inverter Setting Creation

As discussed earlier, the increased adoption of inverter-based DERs can cause adverse impacts to a distribution network. These impacts can be reduced or mitigated by prescribing these inverters to export and absorb reactive power, or to curtail active power in response to voltage deviations. Furthermore, if properly chosen, the settings can improve distribution operational metrics that include technical losses, reduce the wear and tear of inline voltage regulator tap changer drivers, and reduce voltage variability [8].

It would be impractical to perform studies intended to develop the best inverter settings for a very large number of feeders, as the amount of effort for each feeder is considerable. These studies include placing inverters in different locations along a feeder and testing the many grid support functions like Volt-VAR, Volt-Watt, etc., under multiple settings. Like in the previous case, a practical solution is to properly sample the feeders for study and extrapolate their result to similar feeders.

D. Technical Loss Calculations

The proper estimation of technical losses is fundamental to the operation, planning and economics of electric distribution networks. Current industry practice requires that some, if not most of the system losses, be calculated using load flow

software. Conversely, there is not a single best practice that can be followed. Difference in adopted methodologies vary according to individual utility practices [10].

This exercise can also benefit from the clustering approach. While there are atypical feeders, a representative sample from each voltage class can be selected that are similar enough to others on the system to allow reasonable extrapolation.

III. APPLYING CLUSTERING ANALYSIS TO DISTRIBUTION FEEDER DATA

K-Means is a popular clustering method used in many machine learning applications as an unsupervised learning algorithm. It is adopted very widely when dealing with large datasets due to its efficiency and simplicity [2]-[5]. In this paper, this method was selected to process the operation dataset containing feeder operational data. This section provides the background on the application of this method into distribution feeder data for each of the three applications described in the previous section.

A. Description of Operational Dataset

For each of the three specific studies, the dataset contains many important parameters defined as follows.

- Voltage level: the electric utility driving this study currently operates distribution systems of four different primary voltage levels: 4.16kV, 7.2kV, 8.32kV, and 13.2kV.

- Presence of step transformers (step-down between two distribution voltage levels): Since the utility operates four different voltage levels and has been slowly converting voltages to 13.2kV to unify its supply, it commonly adopts step transformers to supply portions of a feeder at different voltage levels.

- Number of overhead miles of a feeder: hosting capacity reduces the farther a point of interconnection is from the source [9], and typically longer feeder have higher losses.

- Number of underground miles of a feeder: like the overhead case.

- Total feeder miles: the addition of both parameters above.

- Presence of fixed capacitor banks: the electric utility driving this project only adopts fixed (non-controllable) capacitor banks. These have an impact on hosting capacity [9] as well as on technical losses [10].

- Presence of voltage regulators and voltage boosters: the need for regulating voltage is directly related to the tendency of experiencing low or high voltage on a distribution system.

- Critical or priority customers and associated customer criticality index (CCI): the presence of critical customers is a parameter that can often be used for reliability analysis and is considered as a sensitivity parameter.

- SAIDI: system average interruption duration index, another parameter often used in reliability studies and considered in this paper as a sensitivity index.

SAIFI: system average interruption frequency index,

considered in the same manner as SAIDI.

- Customer count and feeder peak MVA, two related parameters that affect the studies in this paper as well.

B. Definition of Variables to be Used for the Clustering Analysis

It is necessary to determine which variables to use from the dataset by measuring their impact to the project goals as well as by how they help attain the optimum number of clusters. The optimum choice of variables is conductive to achieving the optimum number of clusters. Ideally, these variables should be as independent as possible, to reduce bias. A decision was made to decouple the voltage level from the clustering approach by conducting the clustering analysis for each voltage level group. Then, the independence of the variables is analyzed by analyzing their correlation. This is shown, for each voltage level, in the surface plots of Fig. 1.

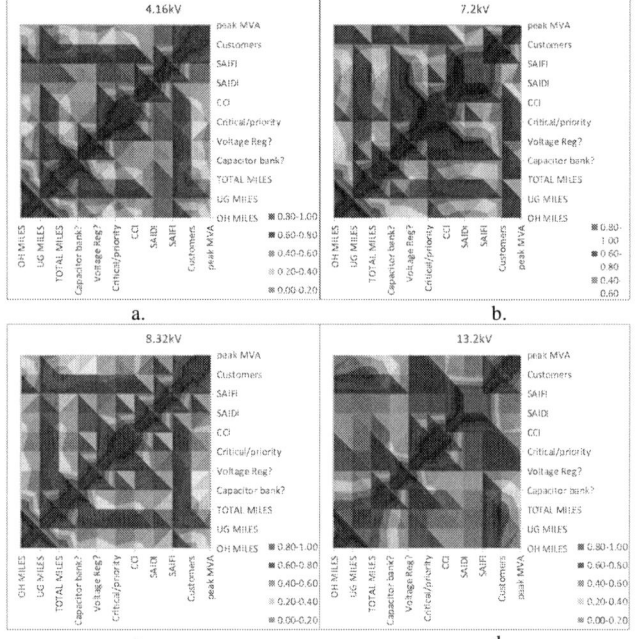

Figure 1. Correlation analysis of all variables used for clustering, (a) 4.16kV, (b) 7.2kV, (c) 8.32kV, (d) 13.2kV.

These figures allow us to conclude that some variables are closely correlated with others and can be condensed. This analysis suggested that it is doable to simplify the number of groups and merge them into five as follows:

- Voltage level group: there are four primary voltage levels, and we believe this is the most important variable to be considered; this first group includes the presence of step transformers. No 4.16kV feeders have step transformers as those are already the lowest voltage operated by the utility. For other voltage levels, the presence of step transformers is primordially related to voltage levels.

- Total feeder length group (overhead, underground, and total): The total feeder length is directly correlated with overhead, underground and total miles.

- Feeder loading group: it can be quantified by the peak demand as well as customer count.

- Voltage regulation devices (capacitor banks, voltage boosters and voltage regulators) group: these can be condensed and more importantly, incorporated into the feeder length group, as in most cases they were highly correlated with the feeder total length; hence, this variable is removed from the analysis.

- Reliability data (critical customers, SAIDI and SAIFI) group: although this group can provide great insight, we believe they are affected by factors that are not related to the clustering analysis. Hence, we decided to remove the reliability group from the analysis, as it does not have significance to the three tasks addressed in this paper.

As a result, only three variables are being used: voltage level, feeder length, and loading, as summarized in Table I. Interestingly, this grouping can be used for the three studies conducted in this research.

TABLE I. GROUPING OF HIGHLY CORRELATED VARIABLES

Voltage	Length	Voltage Regulation	Loading	Reliability
Voltage level Step transformer	OH length UG length Total length	Regulators Capacitors	MVA_{peak} Customer count	SAIDI SAIFI CCI

C. Background on K-Means Clustering

K-Means clustering is a popular unsupervised learning method to process data that may be categorized in multiple groups where components are similar to each other. In order to run the algorithm, the number of expected clusters (k) needs to be provided, representing an interactive process that often involves trial and error. The description of K-Means clustering is described as follows. For a given a set of observations (x_n), where each observation is a d-dimensional real vector, K-Means clustering has as the main objective to segment these observations into k sets, where each set is represented as $S = \{S_1, S_2, ..., S_k\}$. A minimization based on least squares regression is run to minimize the within-cluster sum of squares. Formally, the objective is to find:

$$\arg_S \min \sum_{i=1}^{k} \sum_{x \in S_i} \|x - \mu_i\|^2 = \arg_S min \sum_{i=1}^{k} |S_i| Var S_i,$$

(1)

where μ_i is the mean of S_i. For the three studies datasets, an example of a feature vector (FV) is given as below.

$$FV = (V_{class}, l_{feeder}, MVA_{peak}),$$

(2)

In this example, the voltage level, total feeder length, and feeder peak demand are recorded into the feature vector. In practice, the selection of features depends on feature data availability, the variation of feature data and the sensitivity of feature data to the clustering results [11]. In summary, the steps followed in this paper are:

1. Randomly initialize k centroids within the data set;

2. Associate all data points to their nearest centroids, creating *k* clusters;
3. Update each cluster centroid with all its members;
4. Repeat Steps 2 and 3 until the algorithm converges.

D. Normalization and Weighting

The proposed method requires all features to be normalized to a numerical range (i.e., [0,1]) because the raw operation data use different units and their resulting magnitude can also vary dramatically. This paper uses the Min-Max normalization to achieve this purpose, as described by:

$$x_{norm} = \frac{x_{raw} - Min}{Max - Min}, \qquad (3)$$

where *Max* and *Min* are the maximum and minimum observed values of a feature. Furthermore, weighting factors are assigned as shown below:

$$d(X,Y) = \sum_{j=1}^{p} w_j \left(x_j - y_j \right)^2 + \sum_{j=p+1}^{m} w_j (x_j, y_j), \quad (4)$$

where w_j is the empirical weighting factor of a feature *j*. These techniques allow grouping data points using peak consumption, feeder length and voltage levels, which are very different parameters. Following the steps presented above, distribution planning engineers can:

- Determine how data are distributed and how they belong within each cluster;

- Determine the centroids, which are the averaged feature vector of all members of each cluster and the most representative component of each cluster.

IV. APPLICATION TO A DISTRIBUTION SYSTEM

This study was driven by an island distribution system in the Caribbean, which operates the electricity infrastructure in three islands. It supplies electricity to about 2.5M residents and has a 230kV backbone ring and a meshed 115kV networked transmission system. It contains a highly dispersed subtransmission system operating at 38kV, supplying almost all industrial and commercial customers in the island. As mentioned earlier in this paper, the utility also operates about 1,400 distribution feeders, currently supplied at four different voltage levels: 13.2kV, 8.32kV, 7.2kV, and 4.16kV, which are about 25%, 15%, 3%, and 55% of the feeders, respectively.

A. Determination of Parameters and Centroids for each Application

The dataset described in the previous section was processed to extract the three main features for all the 1,400 feeders using their most current data, including peak demand for 2019. Out of these 1,400 feeders, about 450 had to be removed from the analysis because they are either spare breakers, were out of service, or, for a small number of them, not enough data were available, e.g., demand data were not captured due to failed monitoring equipment. This resulted in 917 feeders which

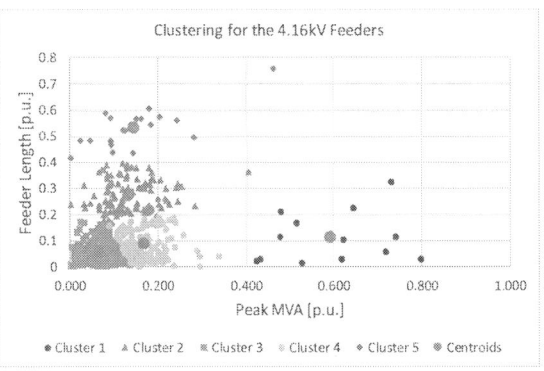

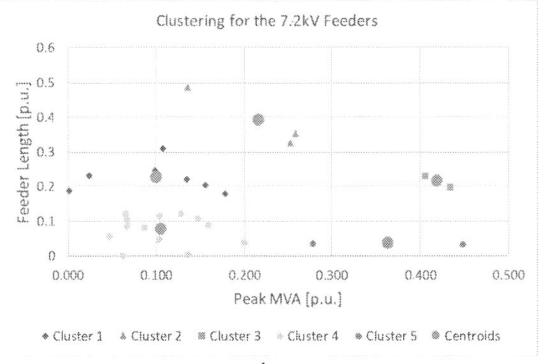

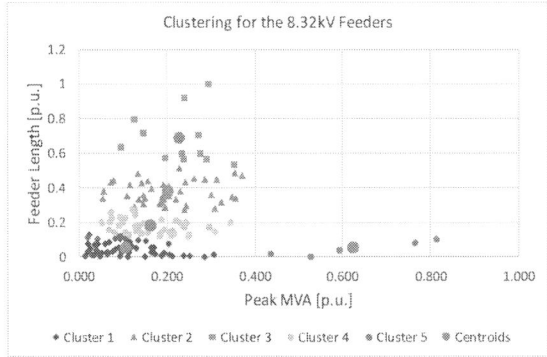

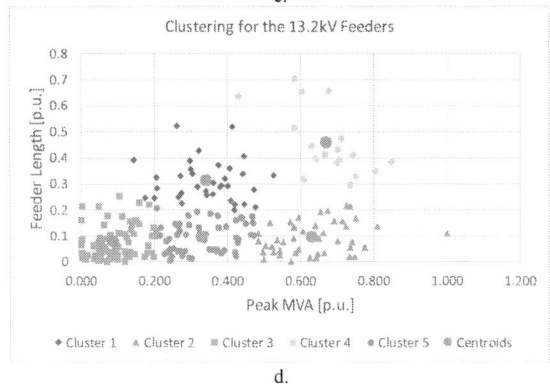

Figure 2. Clustering results for the 917 feeders, (a) 4.16kV, 7.2kV, 8.32kV, and 13.2kV.

could be included in the analysis. After condensing the variables as described in Section III, we arrived at about 1,834 records (two records per feeder). These non-categorical features were normalized using (3). Subsequently, these features were taken into the steps described in III.C for K-means clustering. Following this procedure, we arrived at k = 5 per voltage level, totaling 20 feeders. Fig. 2 shows the clustering results for all feeders and Table II summarizes the produced cluster composition.

TABLE II.   CLUSTER COMPOSITION RESULTS

Voltage Level [kV]	Cluster ID	Number of Members	Average peak load [MVA]	Average length [miles]
4.16	1	13	1.99	55.62
4.16	2	111	4.87	12.60
4.16	3	235	0.89	6.23
4.16	4	126	1.51	15.88
4.16	5	19	9.55	13.75
7.2	1	7	4.05	9.39
7.2	2	3	7.00	20.26
7.2	3	2	3.85	39.29
7.2	4	13	1.34	9.86
7.2	5	2	0.61	34.05
8.32	1	54	0.89	10.18
8.32	2	40	6.76	19.09
8.32	3	12	12.32	21.55
8.32	4	44	3.18	15.32
8.32	5	5	0.86	58.69
13.2	1	38	5.60	32.19
13.2	2	44	1.65	59.43
13.2	3	78	1.32	8.41
13.2	4	18	8.20	63.14
13.2	5	53	1.70	30.95

V. CONCLUSIONS

This paper presented a data-driven method for distribution planning studies. Compared to previous research, the proposed method has the following advantages:

- It is reliant on loading data, such as that from SCADA, and GIS data and does not require other monitoring devices.

- It provides insights of feeder characteristics and allows further analysis that goes beyond these three applications.

- Many types of studies can be conducted with the results of this exercise, since it has shown condensing categorical data can be done for multiple applications.

One of the most important findings of this paper, obtained from the correlation analysis, is that the same data categorization can be used for the three applications, namely hosting capacity, smart inverter grid-supporting functions, and technical losses. The proposed method was adopted by a Caribbean utility and the exercise created great value to its distribution planning engineers.

VI. REFERENCES

[1] CEATI Report T104700-5095, Distribution Planner's Manual vol. one: Planning of a System, August 2015.

[2] R. J. Broderick, J. R. Williams, K. Muñoz-Ramos, Clustering method and representative feeder selection for the California solar initiative SANDIA Technical Report (2014).

[3] R. J. Broderick and J. R. Williams, "Clustering methodology for classifying distribution feeders," 2013 *IEEE 39th Photovoltaic Specialists Conference* (PVSC), 2013, pp. 1706-1710, doi: 10.1109/PVSC.2013.6744473.

[4] G. Gheorghe, G. Cartina, F. Rotaru, "Using K-Means Clustering Method in Determination of the Energy Losses Levels from Electric Distribution Systems" 2010 *International Conference on Mathematical Methods and Computational Techniques in Electrical Engineering*.

[5] S. Wang, P. Dong, Y. Tian, "A Novel Method of Statistical Line Loss Estimation for Distribution Feeders Based on Feeder Cluster and Modified XGBoost". Energies 2017, 10, 2067. https://doi.org/10.3390/en10122067

[6] A. Nassif and X. Long, "Mitigating overvoltage scenarios caused by large penetration of distributed energy resources," 2016 *IEEE Electrical Power and Energy Conference* (EPEC), 2016, pp. 1-5, doi: 10.1109/EPEC.2016.7771699.

[7] F. Ding and B. Mather, "On Distributed PV Hosting Capacity Estimation, Sensitivity Study, and Improvement," in *IEEE Transactions on Sustainable Energy*, vol. 8, no. 3, pp. 1010-1020, July 2017, doi: 10.1109/TSTE.2016.2640239.

[8] S. R. Abate, T. E. McDermott, M. Rylander and J. Smith, "Smart inverter settings for improving distribution feeder performance," 2015 *IEEE Power & Energy Society General Meeting*, 2015, pp. 1-5, doi: 10.1109/PESGM.2015.7286560.

[9] M. Rylander, J. Smith and W. Sunderman, "Streamlined Method for Determining Distribution System Hosting Capacity," in *IEEE Transactions on Industry Applications*, vol. 52, no. 1, pp. 105-111, Jan.-Feb. 2016, doi: 10.1109/TIA.2015.2472357.

[10] D. Ćetković, S. Vlahinić, D. Franković and V. Komen, "Analysis of justification for using capacitor banks in distribution networks with low power demand," 2020 *43rd International Convention on Information, Communication and Electronic Technology* (MIPRO), 2020, pp. 923-927, doi: 10.23919/MIPRO48935.2020.9245272.

[11] M. Dong, A. B. Nassif and B. Li, "A Data-Driven Residential Transformer Overloading Risk Assessment Method," in *IEEE Transactions on Power Delivery*, vol. 34, no. 1, pp. 387-396, Feb. 2019, doi: 10.1109/TPWRD.2018.2882215.

Fireable passivating tunnel oxide contacts for crystalline silicon solar cell

Franz-Josef Haug, Mario Lehmann, Sofia Libraro, Ezgi Genç, Audrey Morisset, Christophe Ballif

EPFL, PV-Lab, Neuchatel, Switzerland

We present fireable n- and p-type passivating contacts for silicon solar cells, using a tunnelling oxide and a doped SiCx layer. Rapid annealing is used to crystallize the layer and to activate its dopants. Subsequent hydrogenation reveals a fast passivation of interfacial defects and a slow passivation of bulk defects created by the rapid annealing in our floating zone silicon. We apply the contacts in proof-of-concept solar cells with ITO/Ag contacts, resulting in efficiencies above 20%. Towards an industrially more relevant process using a fireable metallization, we present preliminary results on contacting through the nitride layer with aluminium.

Antisolvent effect on Acetamidinium Substituted 2D Ruddlesden-Popper Perovskite Solar Cells

Vani Pawar, Anuj Kumar Palariya, Nisheka Anadkat, Sandeep Kumar, and Sushobhan Avasthi*

Centre for Nano Science and Engineering, Indian Institute of Science, Bangalore, Karnataka – 560012, India

*Email: savasthi@iisc.ac.in

Abstract— **2D Ruddlesden-Popper (RP) metal-halide perovskite solar cells (PSCs) have attracted growing attention for their fascinating optoelectronic properties and enhanced stability as compared to 3D counterparts. Here, we report the fabrication and characterization of acetamidinium substituted 2D RP $(PEA)_2(MA_{(1-x)}AA_x)_{n-1}Pb_nI_{3n+1}$ (PEA: phenylethylammonium, MA: methylammonium, AA: acetamidinium, x = 0 and 0.1, and n = 2) perovskites. We have investigated the effect of antisolvent, chlorobenzene (CB) on the structural, optical, and photovoltaic properties of 2D PSCs. Perovskite films displayed the uniform polycrystalline nature with high coverage and blended texture after the application of antisolvent. The structural and optical investigations suggest that n = 2 ($(PEA)_2MAPb_2I_7$ and $(PEA)_2(MA_{0.9}AA_{0.1})Pb_2I_7$) films manifest the triclinic crystal structure with tunable optical bandgap ranges from 2.08 to 2.13 eV. The champion PSC based on $(PEA)_2(MA_{0.9}AA_{0.1})Pb_2I_7$ with CB addition exhibits a power conversion efficiency of $\approx 0.66\%$ and a short-circuit current density (J_{sc}) of 3.80 mA/cm^2. The open-circuit voltage (V_{oc}) and fill-factor (FF) for this device are 0.56 V and 30%, respectively. This work is currently in progress to achieve better performing photovoltaic devices.**

Keywords—Two-dimensional Perovskite solar cell, Ruddlesden Popper perovskite, Antisolvent

I. INTRODUCTION

The world record for power conversion efficiency (PCE) of perovskite solar cells (PSCs) has been broken many times over the decade and now it is > 25 %; very close to conventional silicon solar cells [1]. Despite excellent optoelectronic properties, 3D perovskite still suffers from long-term stability issues. The low phase stability under elevated temperatures and the hydrophilic nature of methylammonium and formamidinium cation contribute to the poor stability of PSCs [2]. Two-dimensional (2D) or quasi-2D Ruddlesden Popper (RP) perovskites appeared as a strong candidate to improve the long-term stability, and hence, recently gained an extensive research interest. These perovskites have a general chemical formula of $R_2A_{n-1}M_nX_{3n+1}$, where R refers to a long-chain organic group or bulky organic group, A stands for small organic cation (FA$^+$, MA$^+$ or Cs$^+$), M corresponds to metal cation (Pb^{2+} or Sn^{2+}), X stands for halide anion (Cl, Br, and I), and n refers to number of octahedrons in each individual perovskite layer that defines the number of 2D perovskites. The bandgap and binding energy of perovskites can be tuned by increasing the dimensionality by a change in n values. In order to achieve an appropriate bandgap

for PSCs application while improving the moisture resistance, the mixed-dimensionality approach has become one of the solutions [3].

In this work, we report acetamidinium (AA$^+$) substituted phenylethylammonium based 2D RP $(PEA)_2(MA_{(1-x)}AA_x)_{n-1}Pb_nI_{3n+1}$ (x = 0 and 0.1) perovskite matrix for $n = 2$ (the number of Pb-I sheets in each inorganic slab). Thin films of the perovskite compounds are readily formed by the one-step spin coating method and the effect of chlorobenzene (CB) as an antisolvent on the structure, morphology, bandgap and photovoltaic properties are investigated. Photovoltaic devices using $PEA_2MAPb_2I_7$ (without AA) and $PEA_2MA_{0.9}AA_{0.1}Pb_2I_7$ (with 10% AA) as the absorber material were fabricated. For AA substituted perovskites, an improved PCE of $\approx 0.66\%$ with CB treatment is achieved compared to $\approx 0.39\%$ without any antisolvent incorporation.

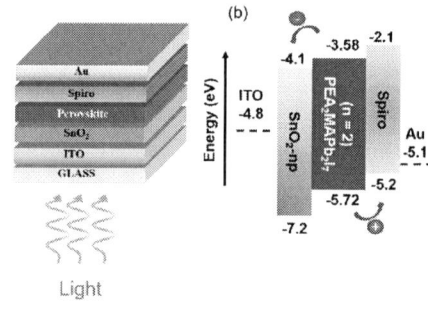

Fig. 1. (a) Device schematic and (b) energy-levels of various constituent layers.

II. EXPERIMENTAL DETAIL

The schematic of fabricated device structure and energy levels of various constituents [4] are represented in Fig. 1. The 2D perovskite precursor solutions of matrix $PEA_2MAPb_2I_7$ (without AA0) and $PEA_2MA_{0.9}AA_{0.1}Pb_2I_7$ (with 10% AA) were prepared by dissolving in a specific stoichiometric quantity of PEAI, MAI, AAI, and PbI$_2$ (2:1:2 molar ratio) in DMF and DMSO (9:1 v/v) with vigorous stirring. For material characterizations, the freshly prepared perovskite solutions were spin-coated on the cleaned glass substrate at 5000 rpm for 40s followed by two-step heating at 65 °C for 1 minute and subsequently at 100 °C for 5 minutes. During the spin coating process, 50 µl CB (antisolvent) was pipetted and dynamically poured on the substrate at 10s before the last spin-off. The as-formed perovskite film without any antisolvent (CB) dripping is abbreviated as AA0 Bare and AA10 Bare. Further, the 2D perovskite films with CB as antisolvent will be referred hereafter as AA0 CB and AA10 CB.

For fabricating PSCs, the ITO coated glass substrates were cleaned sequentially in a soap solution, distilled water, acetone, and isopropyl alcohol for 15 minutes each. The substrates were dried using a nitrogen gun and heated at 100 °C on the hotplate for 15 minutes. After UV-treatment for 30 minutes, commercially purchased SnO_2 (ETL) nano-particle solution was spin-coated at 4000 rpm for 60 seconds. The film was annealed at 130 °C on a hotplate for 40 minutes in an ambient atmosphere. After annealing, SnO_2 coated ITO substrates were placed into an ozonization chamber for 30 minutes and then transferred to a N_2-ambient glove box having oxygen and moisture levels < 0.1 ppm for further processing. The 2D perovskite solution was spin-coated at 5000 rpm for 40 seconds. The antisolvent CB was dynamically spin-coated at the 10th second before the spin-off. ITO/SnO$_2$/Perovskite deposited samples were annealed at 65°C for 1 min, and 5 mins at 100 °C on the hot plate. For HTL, Spiro-OMeTAD solution was prepared by dissolving 73 mg Spiro-OMeTAD in 1 ml of CB by adding 18 µl of Li-TFSI solution (prepared by dissolving 156 mg in 300 µl acetonitrile) and 29 µl of 4-tert-butylpyridine. This solution was dynamically spin-coated on the perovskite layer at 4000 rpm for 35 sec. Finally, 100 nm Au top electrode was deposited in a thermal evaporator outside the glove box at a base pressure of ~1×10^{-6} mbar. The fabricated cells exhibit a device area of 4.5 mm^2.

The X-ray diffraction pattern (XRD) was recorded by using Rigaku smartlab X-ray diffractometer at room temperature using Cu-kα radiation. Scanning electron microscope (SEM) (Zeiss Ultra-55 FESEM) measurements were conducted to obtain the morphology of the perovskite films. The absorption spectrum was recorded by using the Perkin Elmer Lambda UV-vis-NIR spectrophotometer. The current-density versus voltage (J-V) characteristics of devices were measured using Keithley 2450 source-meter. For the light J-V measurements, Oriel Instrument ΛΛΛ solar simulator with AM 1.5G having intensity of 1 sun (100 mW/cm^2) is used.

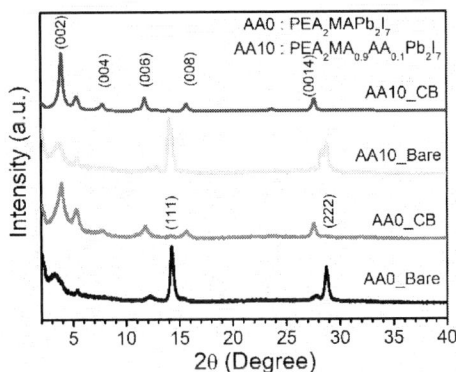

Fig. 2. X-ray diffraction pattern of AA0 (PEA$_2$MAPb$_2$I$_7$) and AA10 (PEA$_2$MA$_{0.9}$AA$_{0.1}$Pb$_2$I$_7$) perovskite without or with chlorobenzene as antisolvent.

III. RESULTS & DISCUSSION

The crystallinity of the perovskite films was investigated using the XRD (Figure 2). The diffraction patterns were corroborated with the standard data corresponding to n = 2 (PEA$_2$MAPb$_2$I$_7$) perovskites [5]. The perovskite compounds AA0 and AA10 manifest triclinic crystal structure of the 2D RP phase. The peaks around 14.2° and 28.7° are identified as the scattering from (111) and (222) crystal planes, majorly dominating in AA0 Bare and AA10 Bare samples (without antisolvent). In contrast, (002), (004), and (006) planes at 2θ ≈ 3.9°, 7.8°, 11.7° are strongly observed in AA0 CB and AA10 CB samples (with antisolvent) [4, 6]. As 2D perovskites consist the tendency to form well-defined films with the layers oriented parallel to the substrates [7, 8]. This trend turns out to be true for the perovskite films AA0 CB and AA10 CB, as revealed from Fig. 2. The evolution of sharp (002) peak indicates the good crystallinity and high degree of preferred orientation along <001> direction [3]. In addition, the occurrence of (111) peak indicates that besides the parallel growth of the 2D perovskite, some part of the perovskite slabs grows vertically with respect to the glass substrates. This vertical growth is more dominating in AA0 and AA10 bare samples having no antisolvent treatment.

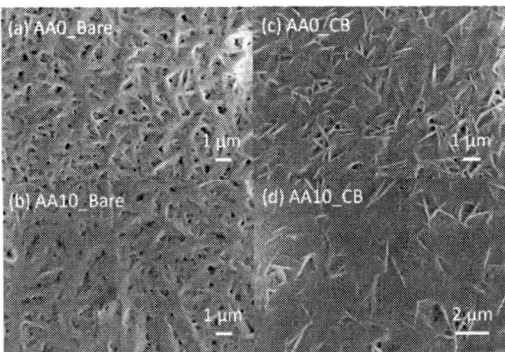

Fig. 3. Scanning electron microscopy image of RP perovskites on glass substrates without [(a) AA0 Bare, (b) AA10 Bare] or with [(c) AA0 CB, (d) AA10 CB] antisolvent addition.

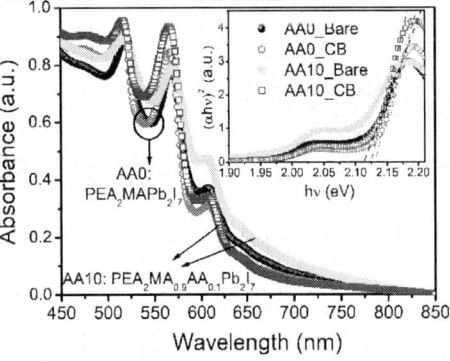

Fig. 4. UV-vis absorption spectra of 2D RP perovskite films with or without antisolvent addition. Inset shows the optical bandgap evaluated using Tauc's relation.

Figure 3 represents the SEM micrographs of perovskite films with and without CB treatment of AA0 and AA10 samples. AA0 bare and AA10 bare samples (w/o CB) exhibit the dendritic kind of morphology with pinholes. However, minute observation suggests that acetamidinium substituted AA10 bare sample have less pinholes as compared to AA0 Bare sample, shown in Fig. 3 (a and b). The application of CB as an antisolvent leads to high coverage and blended texture of rod-like morphology in both type of perovskite films, displayed in Fig. 3 (c and d).

Figure 4 shows the room temperature UV-visible absorption spectra of spin-coated 2D RP perovskite films on glass substrates recorded between 450-850 nm range. The absorbance spectrum of AA0 and AA10 films displays absorption peaks around ~ 518 nm, ~ 568 nm, and ~ 608 nm. The absorption peak at ~ 568 nm can be assigned to $PEA_2MAPb_2I_7$ (n = 2) perovskite system, while the peak at ~ 518 nm attributes to PEA_2PbI_4 (n = 1) 2D perovskite. It suggests that trace amounts of n = 1 perovskite phase other than n = 2 coexist in the corresponding AA0 and AA10 compounds [4]. Although, this trace amount of PEA_2PbI_4 (n = 1) perovskite phase is not visible in the XRD pattern of the films. Further, the direct optical bandgap of the films was evaluated from the Tauc's plot $(\alpha h v)^2$ vs hv [9] (inset of Fig.4) and it ranges from 2.08 - 2.13 eV tabulated in Table I.

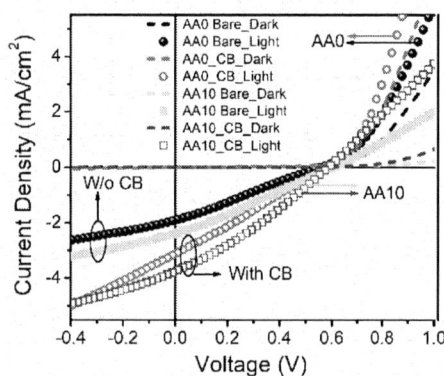

Fig. 5. Current density versus voltage characteristics of 2D perovskite solar cell.

Table I. Various photovoltaic parameters of the devices and optical bandgap of perovskite films.

System	J_{sc} (mA/cm^2)	V_{oc} (V)	Fill Factor (%)	η (%)	E_g (eV)
AA0 Bare	1.93	0.56	24.92	0.27	2.11
AA0 CB	3.12	0.59	25.49	0.47	2.13
AA10 Bare	2.44	0.61	25.98	0.39	2.08
AA10 CB	3.80	0.59	30.05	0.66	2.12
Reported [4] **n = 2**	2.38	0.77	65	1.19	2.12

The J-V characteristics of the 2D RP PSCs are demonstrated in Fig. 5. The photovoltaic parameters of the devices fabricated with and without CB antisolvent application are summarized in Table I below. The best performing device (AA10 CB) with antisolvent incorporation exhibits an open-circuit voltage (V_{oc}) of 0.59 V, a short-circuit current density (J_{sc}) of 3.80 mA/cm^2, a fill-factor of 30 %, and a PCE of 0.66%. This work shows an

improvement in J_{sc} as compared to the literature reported value [4]. The low device performance could be the resultant of the energetic mismatch between the absorber layer and the hole acceptor layer.

IV. SUMMARY

In summary, we have fabricated the 2D RP perovskite matrix of $(PEA)_2(MA_{(1-x)}AA_x)_{n-1}Pb_nI_{3n+1}$ by using the one-step spin coating method, with and without incorporation of CB as an antisolvent. Our results reveal that the antisolvent assistance approach supports the 2D perovskite growth and crystallinity along with high coverage. Overall, we observed a significant improvement in structural, morphological, optical, and photovoltaic properties with the incorporation of antisolvent. Further attempts are being made to improve the device performance. This work demonstrates the great potential of 2D RP perovskites as a promising photovoltaic material.

ACKNOWLEDGMENT

The authors would like to acknowledge the support of the National Nano Fabrication Center (NNfC) and Micro-Nano Characterization Facility (MNCF) for providing access to fabrication and characterization facilities. NNfC and MNCF are funded by a grant from the Ministry of Electronics and Information technology, Government of India.

REFERENCES

[1] M. Green, E. Dunlop, J. Ebinger, M. Yoshita, N. Kopidakis, and X. Hao, "Solar cell efficiency tables (version 57)," *Progress in Photovoltaics: Research and Applications,* vol. 29, pp. 3-15, 2020.

[2] F. Zhang, H. Lu, J. Tong, J. Berry, M. C. Beard, and K. Zhu, "Advances in two-dimensional organic–inorganic hybrid perovskites", *Energy Environ. Sci.*, 13, 1154, 2020.

[3] A. Krishna, S. Gottis, M. K. Nazeeruddin, and F. Sauvage, " Mixed Dimensional 2D/3D Hybrid Perovskite Absorbers: The Future of Perovskite Solar Cells?, *Adv. Funct. Mater.* 29, 1806482, 2019.

[4] X. Gan, O. Wang, K, Liu, X. Du, L. Guo, and H, Liu, "2D homologous organic-inorganic hybrids as light-absorbers for planer and nanorod-based perovskite solar cells", *Solar Energy Materials and Solar Cells,* 162, 93-102, 2017.

[5] M. H. Jung, "Exploration of two-dimensional perovskites incorporating methylammonium for high performance solar cells", *CrystEngComm*, 23, 1181, 2021

[6] A. D. Taylor, Q. Sun, K. P. Goetz, Q. An, T. Schramm, Y. Hofstetter, M. Litterst, F. Paulus, and Y. Vaynzof, "A general approach to high-efficiency perovskite solar cells by any antisolvent", *Nature Communications,* 12, 1878, 2021

[7] C. O. Cervantes, P. C. Monroy, ans D. S. Ibarra, "Two-Dimensional Halide Perovskites in Solar Cells: 2D or not 2D?" *ChemSusChem*, 12, 8, 1560-1575, 2019.

[8] P. Singh, R. Mukherjee, and S. Avasthi, "Acetamidinium-Substituted Methylammonium Lead Iodide Perovskite Solar Cells with Higher Open-Circuit Voltage and Improved Intrinsic Stability," *ACS Appl Mater Interfaces*, vol. 12, pp. 13982-13987, Mar 25 2020.

[9] V. Pawar, M. Kumar, P.A. Jha, S. K. Gupta, P. K. Jha, and P . Singh, "Cs/MAPbI3 composite formation and its influence on optical properties", *Journal of Alloys and Compounds*, 783, 935-942, 2019.

[10] V. Pawar, M. Kumar, P.A. Jha, S. K. Gupta, A. S. K.Sinha, P. K. Jha, and P . Singh, "Ambient atmospheric temperature processed lead halide perovskites", *Journal of Thermal Analysis and Calorimetry*, 139, 3073-3078, 2020

Performance and stability of electrodeposited mixed perovskites MAPbI$_{3-x}$Cl$_x$ and MA$_{1-y}$FA$_y$PbI$_{3-x}$Br$_x$

Mirella Al Katrib, Lara Perrin, and Emilie Planes

Univ. Grenoble Alpes, Univ. Savoie Mont Blanc, CNRS, Grenoble INP, LEPMI, Grenoble, 38000, France

Abstract— **Electrodeposition was investigated in this work as a substitute method to develop large area perovskite active layer for solar device application. Along with the simple MAPbI$_3$ perovskite, the deposition of mixed MAPbI$_{3-x}$Cl$_x$ and MA$_{1-y}$FA$_y$PbI$_{3-x}$Br$_x$ perovskites was studied. This present study is one of its kind, since these mixed perovskite were never developed using electrodeposition before. It was detected that using these mixed perovskites in a solar device enhances its photovoltaic activity. It also enhances its stability when evaluated in mild ageing conditions (40°C, under vacuum or ambient atmosphere) during 500h. The different perovskites fabricated using electrodeposition experience a maturation phenomenon.**

Keywords—mixed perovskites, solar cells, electrodeposition

I. Introduction

Hybrid organic / inorganic perovskites (PK) have considerably arisen as efficient active materials in the solar community[1], because of their low cost and performance's fascinating enhancement in the last decade[2]. Nowadays, the current techniques for depositing the perovskite are essentially based on spin coating in glove box. This method showed good results[3], but operates with small active areas, limiting the industrialization of perovskite solar cells (PSC). In our previous works (under review), electrodeposition was explored as an efficient alternative for CH$_3$NH$_3$PbI$_3$ (MAPbI$_3$) perovskite fabrication. This innovative process could plausibly be the best for low cost industrial development, especially for large active surfaces with an homogeneous performance[4]. Due to electrodeposition process simplicity, not only simple perovskites could now be developed but also mixed cations and mixed halides perovskites, such as chlorine or bromine doped perovskites. In the case of perovskite deposited by spin-coating, incorporating Cl in the MAPbI$_3$ lattice have somehow a profound and beneficial effect on the film morphology, carrier diffusion length, and stability[5]. The study of Cl incorporation effects is even still a very up to date center of interest for the PSC community[6], [7]. Also, the methyl ammonium (MA) cation have been replaced by formamidinium (FA) in many studies which showed good durability improvements[8], [9]. Furthermore, the bromide incorporation in the MAPbI$_3$ perovskite lattice has led to an increase in the optical band gap, tunable according to the ratio of added Br, which raises the open circuit voltage V$_{oc}$ of the corresponding PSCs[10]. High power conversion efficiencies (PCEs) were achieved when using composition engineering of MA$_{1-y}$FA$_y$PbI$_{3-x}$Br$_x$ perovskite[11]. Given the opposing effects of FA and Br substitutions on the perovskite in terms of band gap change and photoluminescence shift, the role of their incorporation together is not fully understood yet. In addition, a fundamental question remains regarding the function of Cl atoms during the formation process of the perovskite.

In this work, a PbO$_2$ layer is electrodeposited as a starting material and then perovskite is obtained using a two-step conversion route where PbO$_2$ is first converted to PbI$_2$. The second conversion was conducted in a solution containing a mixture of MAI and MACl to obtain MAPbI$_{3-x}$Cl$_x$ or MAI and FABr to obtain MA$_{1-y}$FA$_y$PbI$_{3-x}$Br$_x$, while variating the ratio of Cl and Br (Fig. 1). This present study is one of its kind, since these mixed perovskite were never developed using electrodeposition before.

II. Results and Discussion

MCl30, MCl40, and MCl50 are the IPA-based solutions used to convert PbI$_2$ into the mixed perovskite MAPbI$_{3-x}$Cl$_x$, and FB30, FB40, FB50 into MA$_{1-y}$FA$_y$PbI$_{3-x}$Br$_x$. The number in their nomenclature determines the percentage of Cl or Br in the conversion solution. Before testing these active layers in solar devices, their morphology, microstructure, chemical composition and properties were investigated using a set of characterization methods.

At the microscopic scale, the in-plane morphology of the different converted PKs was examined by SEM. The MAPbI$_3$ film converted from PbI$_2$ has a fully covered surface of small cubic-shaped grains with a non-uniform orientation, and a grain size ranging between 150-200 nm. When converting PbI$_2$ into the mixed perovskites, we notice similar surface morphologies for the six films. It consists of cubic grains, slightly larger than MAPbI$_3$ cubes, more entangled and less spaced. This can limit the presence of gap areas and counter a partial short-circuit across the junction when we use the perovskite in a solar device.

UV-visible absorption spectroscopy and XRD analysis were conducted on the fabricated films to identify the obtained

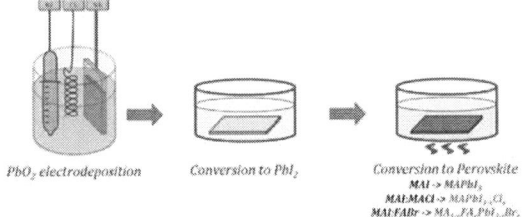

Fig. 1. Schematic representation of the deposition process

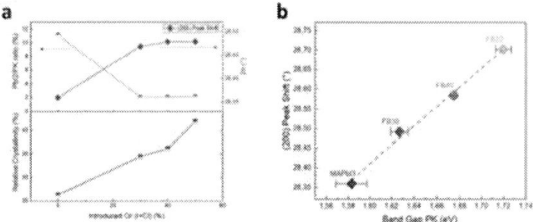

Fig. 2. a) the PbI_2/ PK ratio determined from XRD analyses of the MCl samples, the XRD peak shift at 28° and the relative crystallinity as a function of the Cl percentage present in the conversion solutions ($MAPbI_3$ being the one with 0% Cl), b) the XRD peak shift at 28° as a function of the Band Gap of the FB perovskite films

materials, and better understand their optical properties and their phase composition. For the different samples, two optical transitions can be identified. The first transition at around 770 nm corresponds to the one of $MAPbI_3$ used to determine the band gap of the perovskite. The second transition at 500 nm is attributed to the lead iodide octahedral present in the $MAPbI_3$ perovskite lattice. The band gaps (BGs) of the developed films were calculated using the Tauc plots. We notice that $MAPbI_{3-x}Cl_x$ films exhibit a rather constant band gap of 1.6 eV, with slight variations, whereas $MAPbI_3$ has a band gap of 1.58 eV. This band gap increase confirms here the infiltration of Cl in the $MAPbI_3$ lattice, due to the presence of the Cl ions that modify the perovskite lattice, thereby widening the band gap. However, even for a high content of Cl in the perovskite, the band gap cannot be tuned widely as observed for our $MAPbI_{3-x}Cl_x$ films. $MA_{1-y}FA_yPbI_{3-x}Br_x$ converted via FB30, FB40 and FB50 have band gaps of 1.63 eV, 1.68 eV, and 1.72 eV successively. Apparently, we notice a band gap increase when the Br content of the conversion solution is increased from 0% to 50%. According to the literature, increasing the Br ratio leads to an increase in the BG while increasing the FA ratio leads to a decrease is the band gap, which means that here the Br incorporation has a greater effect than FA on the optical properties of the perovskite film.

In XRD, the $MAPbI_{3-x}Cl_x$ samples show a typical XRD pattern of tetragonal $MAPbI_3$, with a peak shift towards higher angles, which validates the incorporation of Cl in the lattice. The crystalline PbI_2 ratio in these samples was calculated, indicating that $MAPbI_{3-x}Cl_x$ samples have lower PbI_2 content (Fig. 2.a). This also validates the incorporation of Cl in the lattice, which favors the conversion reaction and thus reduces the presence of PbI_2 in the mixed perovskites. Furthermore, increasing the Cl percentage in the conversion bath increases the relative crystallinity of the films (Fig. 2.a). Moving to the $MA_{1-y}FA_yPbI_{3-x}Br_x$ samples, they present a cubic pattern, which is a sign of the incorporation of Br. Furthermore, the XRD pattern encounters a shift toward higher theta for mixed perovskites, and continues to shift in the same direction when we increase the Br ratio in the conversion solution. A linear relation could be then detected between the band gap increase and the peak shift (Fig. 2.b).

Photoluminescence (PL) spectroscopy was also conducted to investigate the emission properties of the developed perovskite films (Fig. 3), coupled with PL imaging to extract irradiance levels overall the surface of the layers and compare

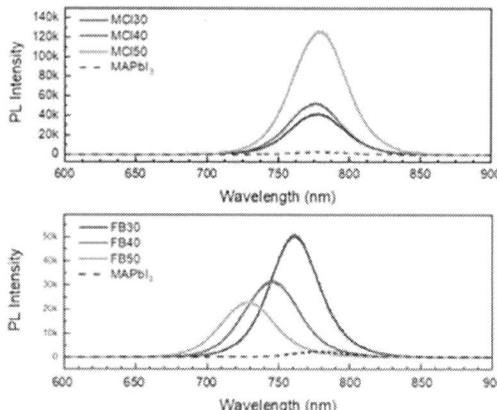

Fig. 3. PL spectra of the seven developed perovskite films

the homogeneity of the different films. The average radiant flux intensity is less homogenous overall the substrate for the mixed perovskite films, comparing to the $MAPbI_3$ film. When drawing the extracted λ_{max} as a function of the Cl%, we also detect a blue shift for the $MAPbI_{3-x}Cl_x$ layers comparing to $MAPbI_3$, confirming the incorporation of Cl, but no specific shift between the $MAPbI_{3-x}Cl_x$ samples. However, for the $MA_{1-y}FA_yPbI_{3-x}Br_x$ samples, a blue shift is detected when incorporating and increasing the Br ratio, and a linear relation could be thus detected between λ_{max} and the Br ratio in the conversion solution. Such a behavior was expected since prior studies already proved that the PL peak exhibits a blue shift when using Br anions in the perovskite and a less intense red shift when using FA cations.

To evaluate the photovoltaic performance of the developed electrodeposited perovskite active layers, we measured the J(V) curves of Glass/ITO/c-TiO_2/mp-TiO_2/PK/P3HT/C solar cells. The chlorine doping of the perovskite does not always enhance the performance of the solar cell. It only works when it is implemented in a small dose, which is the case here (Fig. 4.a). Concerning the $MA_{1-y}FA_yPbI_{3-x}Br_x$ PKs, FA-based perovskites were proven to provide better photovoltaic activity than MA ones because of their broader absorption of the solar spectrum. Nevertheless, doping with Br usually lowers the efficiency of the PSCs, because of the consequent increase in the band gap. In this study, we notice an increase in the J_{sc} when increasing the Br ratio, but similar V_{oc} (Fig. 4.b). Hence, the mean PCE value increases when increasing the Br ratio, being slightly higher than

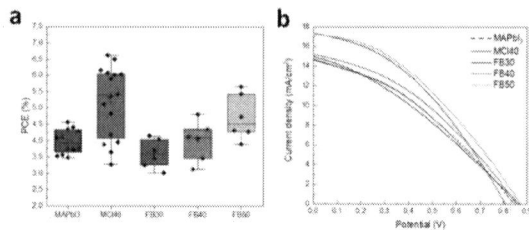

Fig. 4. a) PCE distribution and b) J(V) curves of the 5 developed solar devices

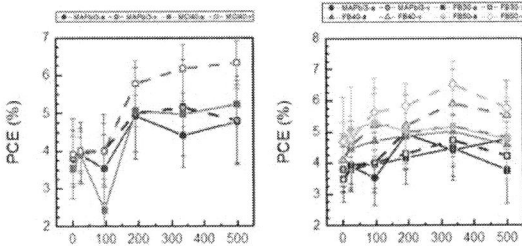

Fig. 5. PCE variation with time for the five electrodeposited solar cells aged at 40°C either at ambient atmosphere or under vacuum

that of MAPbI$_3$ for FB40 and much higher for FB50. We deduce that increasing the FABr percentage in the conversion bath engenders an increase in the V$_{oc}$, the J$_{sc}$ and the FF, leading to a PCE enhancement. This is probably due to an increase of the FA content in the perovskite, as previously explained.

In addition to the PCE enhancement that some mixed samples deliver comparing to electrodeposited MAPbI$_3$, it seems also relevant to verify the impact on stability. It was proven in the literature that incorporating Cl in the MAPbI$_3$ lattice have somehow a profound and beneficial effect on the stability. It was also proven that a substitution by Br increases the stability of the conventional perovskite by forming a pseudo-cubic phase. For that, we studied the aging of our different perovskites types in solar cells. The architecture of the latter was as follows: Glass/ITO/c-TiO$_2$/mp-TiO$_2$/PK/P3HT/C. Applied conditions were: 40°C for 500h, either under ambient atmosphere (R.H. = 20%, noted "a") or under vacuum (noted "v") and presented respectively with plain lines and symbols or with dash lines and empty symbols (Fig. 5).

We notice an increase of the PCE with time, for MAPbI$_3$ or MCl40, whatever the ageing conditions, which validates a maturation process. This increase stops at 350 h for the MAPbI$_3$ and the MA$_{1-y}$FA$_y$PbI$_{3-x}$Br$_x$ based PSCs but continues for the MAPbI$_{3-x}$Cl$_x$-based PSCs (Fig. 5). This maturation process seems to be more pronounced for the treatment under vacuum than in ambient atmosphere, since the devices do not probably endure the same concurrent degradation mechanisms. PCE enhancement under vacuum reached 60% for MCl40, 40% for FB50 and FB40, 30% for FB30, and 35% for MAPbI$_3$, comparing to their initial PCE value. We also notice that the PCE variates at the same pace as the short-circuit current density J$_{sc}$ and the fill factor FF. We deduce that the developed perovskites are rather stable, and even after 500 h of ageing, the maturation had still a greater impact than the degradation.

III. CONCLUSION

In this study, the electrodeposition method was investigated to deposit the perovskite layer. This method is easy to manipulate, and the electrodeposited films could grow on surfaces as large as desired, inside an electrolyte, allowing its application in ambient atmosphere, with no need of a glove box.

This work investigates the possibility of developing a mixed perovskite using electrodeposition. After studying the electrodeposition of MAPbI$_3$ perovskite layers, the same process

was applied on mixed perovskites such as MAPbI$_{3-x}$Cl$_x$ and MA$_{1-y}$FA$_y$PbI$_{3-x}$Br$_x$ while variating the ratio of Cl and Br. This successful attempt is one of its kind, since these mixed perovskites were never developed using electrodeposition before. The photovoltaic performances of the developed MAPbI$_3$ and mixed perovskites were then compared in a solar device. The electrodeposited mixed perovskites offer a PCE enhancement, and are subject to maturation when aged in specific conditions, with a PCE improvement reaching 60% for the chlorine doped perovskites after 500 h of ageing under vacuum at 40°C. The obtained stability is impressive, since the developed solar devices are not encapsulated, and such ageing usually leads to a direct degradation and decrease in the photovoltaic performance under such conditions.

ACKNOWLEDGMENT

This work was financed by Grenoble INP and University Savoie Mont Blanc. This work has been supported by the French National Research Agency, through Investments for Future Program (ref. ANR-18-EURE-0016 -Solar Academy). The research unit LEPMI is member of the INES Solar Academy Research Center. The authors gratefully acknowledge the CMTC (Consortium des Moyens Technologiques Communs) more particularly Francine Roussel-Dherbey et Frédéric Charlot for the SEM observations and Thierry Encinas and Stéphane Coindeau for XRD analyses.

REFERENCES

[1] W.-J. Yin, J.-H. Yang, J. Kang, Y. Yan, and S.-H. Wei, "Halide perovskite materials for solar cells: a theoretical review," *J. Mater. Chem. A*, vol. 3, no. 17, pp. 8926–8942, 2015.

[2] D. Zhao *et al.*, "Four-Terminal All-Perovskite Tandem Solar Cells Achieving Power Conversion Efficiencies Exceeding 23%," *ACS Energy Lett.*, vol. 3, no. 2, pp. 305–306, 2018.

[3] Z. Xiao *et al.*, "Efficient, high yield perovskite photovoltaic devices grown by interdiffusion of solution-processed precursor stacking layers," *Energy Environ. Sci.*, vol. 7, no. 8, pp. 2619–2623, 2014.

[4] G. Popov, M. Mattinen, M. L. Kemell, M. Ritala, and M. Leskelä, "Scalable Route to the Fabrication of CH3NH3PbI3 Perovskite Thin Films by Electrodeposition and Vapor Conversion," *ACS Omega*, vol. 1, no. 6, pp. 1296–1306, 2016.

[5] J. Chae, Q. Dong, J. Huang, and A. Centrone, "Chloride incorporation process in CH3NH3PbI3−x Cl x perovskites via nanoscale bandgap maps," *Nano Lett.*, vol. 15, no. 12, pp. 8114–8121, 2015.

[6] A. Jamshaid *et al.*, "Atomic-scale Insight into Enhanced Surface Stability of Methylammonium Lead Iodide Perovskite by Controlled Deposition of Lead Chloride," *Energy Environ. Sci.*, vol. 14, pp. 4541–4554, 2021.

[7] J. Hieulle *et al.*, "Unraveling the impact of halide mixing on perovskite stability," *J. Am. Chem. Soc.*, vol. 141, no. 8, pp. 3515–3523, 2019.

[8] F. C. Hanusch *et al.*, "Efficient planar heterojunction perovskite solar cells based on formamidinium lead bromide," *J. Phys. Chem. Lett.*, vol. 5, no. 16, pp. 2791–2795, 2014.

[9] T. J. Jacobsson *et al.*, "Exploration of the compositional space for mixed lead halogen perovskites for high efficiency solar cells," *Energy Environ. Sci.*, vol. 9, no. 5, pp. 1706–1724, 2016.

[10] R. Singh, M. Parashar, S. Sandhu, K. Yoo, and J.-J. Lee, "The effects of crystal structure on the photovoltaic performance of perovskite solar cells under ambient indoor illumination," *Sol. Energy*, vol. 220, pp. 43–50, 2021.

[11] Z. Yang *et al.*, "Effects of formamidinium and bromide ion substitution in methylammonium lead triiodide toward high-performance perovskite solar cells," *Nano Energy*, vol. 22, pp. 328–337, 2016.

978-1-7281-6118-1/22 $31.00 ©2022 IEEE

Performance Assessment of A Residential Building Integrated Photovoltaic (BIPV) System in Dhaka City

Md. Mahbub Ali, Nur Jahan Beanta Sorower, Abu Niem Seum, Md. Shifain Mahathir Alvi,
Rawnak Reza Raka, Mohaimenul Islam, Md. Mosaddequr Rahman

Brac University, Dhaka, 66-Mohakhali, 1212, Bangladesh

Abstract— Residential buildings, being an unavoidable part of day-to-day life, can have an impactful subsidy to the Renewable Energy Source (RES). Therefore, in this work, the energy-focused outcome of integrating the BIPV (Building Integrated Photovoltaic) system into a pre-built residential building in Dhaka city has been assessed. To begin with, segmentation of the building into subsystems and incident energy estimation have been considered. The building encounters a total of 391.66 MWh of incident energy over the course of one year. Taking 17% panel efficiency and 32% system losses into account, the system generates 45.80 MWh of electrical output energy per year without considering cloud impact. It is around 53% higher than the energy requirement (21 MWh) of the building. The system remained grid independent for more than half a year. It can reduce the emission of carbon by about 20,751 kg. Assuming the longevity of the system to be 20 years, the financial analysis shows that the initial cost can be retrieved within 4 years. It demonstrates that the BIPV system can meet the energy demand of the studied building while reducing grid dependency and carbon emissions. .

Keywords—BIPV, Incident Energy, Tilt Angle, Irradiation, Dhaka.

I. INTRODUCTION

As the world's demand for renewable and sustainable energy is increasing, zero-emission buildings are gaining much attention nowadays. To achieve zero-emission, a building needs to harvest energy from the sun, which is one of the most prominent sources of renewable energy. A Building Integrated Photovoltaic (BIPV) system integrates with the conventional building elements such as roof tiles, façade elements, and shading elements, replacing them with photovoltaic (PV) modules that should produce electrical energy. It has some advantages, such as low installation cost, shortened energy payback time, and a pleasant architectural view. At present, the BIPV system is one of the most popular and cost-effective applications of solar energy [1].

The number of high-rise buildings has been seen to increase day by day. Statics says that buildings consume around 40% of global energy, which is one of the main sources of consumption around the world [2]. Today, BIPV is being considered by building planners and designers as the new option for the future design of buildings to generate clean energy and decrease greenhouse gases [3].

Numerous research on the impact of BIPV systems on the energy performance of buildings has been conducted. Sun and Yang have assessed the optimal tilt angle for a mounted BIPV system in Hong Kong [4]. Based on observations at eight different places in Bangladesh, including Bogura, Chittagong, Cox's Bazar, and Dhaka, NREL has given the optimal tilt angle as 23.81°, which is the same as Bangladesh's latitude [1, 5]. The ecological criteria to optimize the solar architecture and BIPV system have been described in a paper [6]. BIPV systems can be more cost-effective simply because their composition and location replace a number of conventional components and thus provide multiple gains, which are reviewed in detail in the paper [7].

However, not much study has been done in Bangladesh on the prospects of BIPV. In this paper, a BIPV system has been considered to assess the feasibility and viability of a BIPV system for a residential building in Dhaka city.

The remainder of the paper is coordinated as follows: Section II contains the proposed system description. Section III provides the necessary understanding of the theoretical framework, and results and analysis are clarified in Section IV. Finally, Section V sums up the effectiveness of the proposed system.

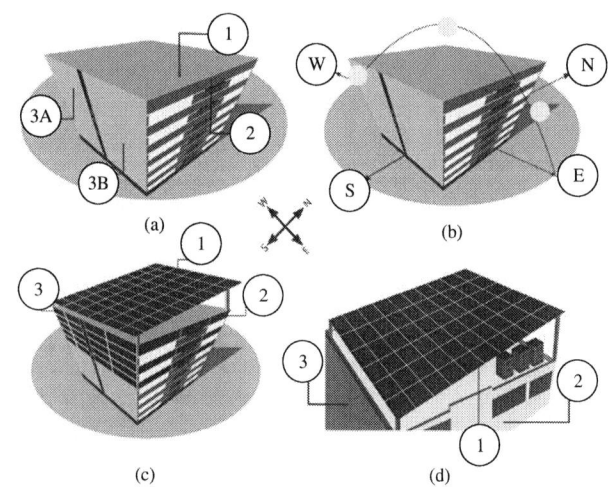

Fig. 1 (a) Subsystem allocation of BIPV construction: Subsystem 1 (Rooftop), Subsystem 2 (Front wall), Subsystem 3A (Left_South Wall), Subsystem 3B Right_South Wall) ; (b) Cardinal direction ; (c) PV module integrated subsystem; (d) PV module integrated Subsystem 1 (Rooftop) with tilt angle 23.81°

TABLE I: FUNCTIONAL AREA AND NUMBER OF PANELS OF THE SUBSYSTEMS

System No.	Position	Area (m²)	Number of Panels	Panel Wattage (W)
Subsystem-1	Rooftop	100.3	60	380
Subsystem-2	Front Wall(East)	24	9	280
Subsystem-3a	Left_South Wall	44.5	10	280
Subsystem-3b	Right_South Wall	35.3	10	280
Total		**204.1**	**89**	

II. SYSTEM DESCRIPTION

A. Proposed System Details

The building considered in order to design the system is located in Mirpur, Dhaka city (lat. 23.81°, long. 90.35°). Following that, the PV installation area has been selected based on the optimal solar intensity level.

Following that, the PV installation area has been selected based on the optimal solar intensity level. The three-prime subsystems: Subsystem-01, Subsystem-02, and Subsystem-03, respectively, represent the roof, front wall, and south wall of the proposed BIPV system, as shown in Fig. 1(a-d). Here, the front wall and the south wall take up the least amount of space due to the shadow effect of the other residential buildings during solar window time, while the rooftop area takes up the most. So the top three floors have been selected for the PV installation on the front wall and south wall. Therefore, the amount of harnessed incident solar energy varies depending on the particular usable area. The measured functional area and the number of calculated PV panels are shown in Table I. A total of 89 solar panels have

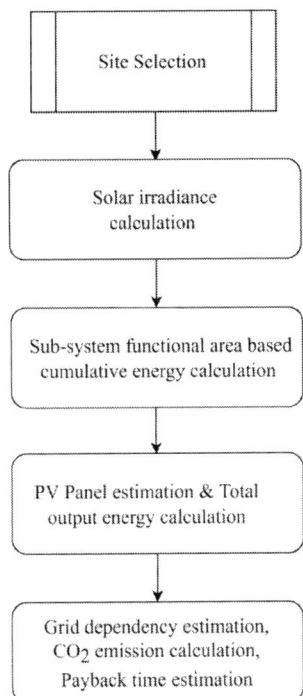

Fig. 2. Analysis approach of the BIPV system of this study in Block diagram.

been considered for this system, with 60 panels on the roof rated at 380Watt and 29 panels on the front and south wall rated at 280Watt. In addition, considering the evening hours, 70 (12V, 150 Ah) batteries have been considered in this study.

B. Methodology:

In order to assess the overall amount of energy harnessed by the BIPV, first a site survey is necessary, as solar intensity levels are impacted by latitude and cardinal direction. Then, solar irradiance in that particular area has been calculated. Furthermore, functional area-based cumulative solar energy has been calculated considering a clear day, which led to the

TABLE II: DIFFERENT TYPES OF LOAD QUANTITY OF THE BUILDING

Equipment's	Power(W)	QTY
AC	2100	4
Fan	70	30
Energy Saving Light	22	80
Tube Light	38	32
Exhaust fan	55	32
Refrigerator	150	12
Miscellaneous	1550	-

determination of the overall system energy of the selected site. Also, the required PV panel on the particular usable area has been calculated. On the other hand, the total amount of output energy has been calculated utilizing the efficiency factors that take the cloud impact into account. The grid dependence and carbon emissions have been determined based on the load demand. Table II shows the number of different types of loads that exist in the studied building. Lastly, the payback period has been calculated using the most updated global price list of the equipment needed for the system. Fig. 2 shows the methodology in a block diagram.

III. THEORETICAL BACKGROUND

A. Solar Irradiance

The rate at which solar energy strikes the unit space of a surface is known as solar irradiance. The total solar irradiance (I) for a tilt surface is comprised of three types of solar radiation: direct-beam (I_B), diffuse (I_D), and ground reflected (I_R) solar radiation, which can be written as [8]:

$$I = I_B + I_D + I_R \qquad (1)$$

978-1-7281-6118-1/22 $31.00 © 2022 IEEE

TABLE III: TILT AND SURFACE AZIMUTH ANGLE OF THREE SUBSYSTEMS

Subsystems	Position's Name	Angles
1	Roof Top	Surface Azimuthal 0°, tilt angle 23.8°,
2	Front Wall	Surface Azimuthal +90°, tilt angle 90°,
3	South Wall	Surface Azimuthal 0°, tilt angle 90°,

B. Beam Irradiance

Direct-beam irradiation (I_B) incident on the PV panel can be calculated from equation (2) [8]:

$$I_B = I_{Bo} cos\theta \qquad (2)$$

Beam solar irradiation (I_{Bo}) flows in a straight line through the atmosphere and strikes the PV panel, whereas direct-beam solar irradiation (I_B) is a simple function of the angle of incidence (θ) between a line drawn normal to the PV module face and the incident beam [8].

The incident angle (θ) can be calculated by using surface azimuth angle (γ_s), solar azimuth angle (γ), solar altitude angle (α_s) and tilt angle (β) [8].

$$cos\theta = cos\alpha_s cos(\gamma - \gamma_s) + sin\alpha_s cos\beta \qquad (3)$$

In order to determine the total solar energy collected by the BIPV, three subsystems were evaluated separately. The different tilt angles along with surface azimuth angle for the three subsystems listed in Table III.

C. Diffuse & Reflected Irradiance:

Diffuse irradiance can be calculated according to the equation (4) [8].

$$I_D = CI_B \frac{1+cos\beta}{2} \qquad (4)$$

Where C is a sky diffuse factor. In this study, reflected irradiance (I_R) has been ignored due to so modest irradiance value.

D. Energy calculation:

The total solar incident energy (E) can be calculated using the following equation [9]:

$$E = A_p \int_{TR}^{TS} I \, dt \qquad (5)$$

Where A_p is the area of the panel, TR and TS are the sunrise and sunset times respectively, and E is the incident solar

TABLE IV: SYSTEM EFFICIENCY FACTORS [9].

Factors Name	Percentage
Battery utilization efficiency, η_B	90%
Wiring efficiency, η_W	98%
Inverter efficiency, η_{Inv}	90%
PV array degradation factor, DF_p	90%

energy. The total output energy from the system has been estimated considering some significant factors that have been listed in Table IV.

Total output energy (E_o) of the system for each day can be accumulated from the PV panel maximum output voltage (V_{PO}), Number of solar panels connected in series (N_{series}), Solar Insolation hour (H), Inverter efficiency (η_{Inv}) and current generated from PV panel array (I_{PV}) [9].

$$E_o = V_{PO} * H * \eta_{Inv} * I_{PV} * N_{series} \qquad (6)$$

From the datasheet published by PV industrial manufacturer of the selected panel, the maximum rated current (I_{PO}) and voltage (V_{PO}) have been taken.

An insolation hour (H) is just the length of time in hours required at an irradiance level of 1 KW/m^2 to produce the daily irradiation obtained by integrating irradiance over all daylight hours [9]. Here, Insolation hour (H) can be calculated from this equation (7)

$$H = \frac{E}{1 kW/m^2} * \frac{1}{A_p} \qquad (7)$$

The effective array current, I_{PV} of the system, can be calculated using the equation (8) [9].

$$I_{PV} = I_{PO} * N_{parallel} * DF_p * \eta_W * \eta_B \qquad (8)$$

Where I_{PO} is the PV panel maximum output current, $N_{parallel}$ represents number of solar panels connected in parallel, DF_p is the PV array degradation factor, η_B is the battery utilization efficiency and η_W is the Wiring efficiency.

E. Cloud Impact

The presence of clouds reduces the efficiency of photovoltaic devices. Using the following set of mathematical calculations, Fig.3 illustrates the monthly percentage of sunny and cloudy days in Bangladesh throughout the year. [11]:

$$E_{sunny} = \int_{T_{SR}}^{T_{SS}} (Isin\alpha + 0.1I) \, dx \qquad (9)$$

$$E_{Cloudy} = \int_{T_{SR}}^{T_{SS}} 0.2I \, dx \qquad (10)$$

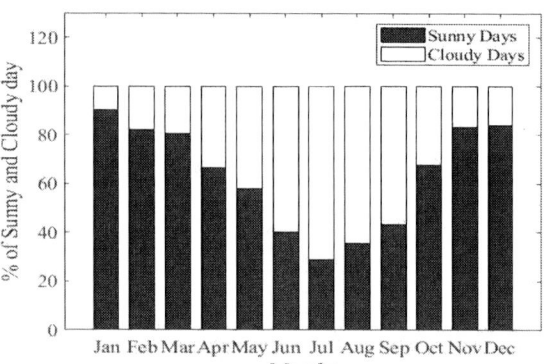

Fig.3. Percentage of sunny and cloudy days per month in Bangladesh throughout the year [11]

F. Panel Calculation:

In solar panel system design or installation, the number of panels must be estimated. Therefore, several elements and sets of equations need to be considered for the energy calculation, such as the area (A) of the portion of the subjected building, selected PV panel size (S_{PZ}), PV panel maximum output voltage (V_{PO}) and, System battery voltage (V_{batt}). The highest possible integer (floor) number of Panel setup (P_{nos}), can be calculated with the following equations [9]:

$$P_{nos} = \frac{A}{S_{PZ}} = N_{parallel} * N_{series} \qquad (11)$$

where the number of Parallel and Series PV panels denoted by $N_{parallel}$ and N_{series} respectively.

Required integer (ceiling) number of Panel in series, N_{series} [9]:

$$N_{series} = \frac{V_{batt}}{V_{PO}} \qquad (12)$$

G. Payback Time

The amount of time required for a PV system to achieve an electricity balance is referred to as the energy payback time (*EPBT*) [12]. In order to figure out the economic benefits of a BIPV system, the net present values (*NPV*), internal rate of return (*IRR*), and energy payback time (*EPBT*) of a BIPV system has been calculated over the building's lifetime.

The *NPV* a of an investment is given by equation (13)[14],

$$NPV = \sum_{n=1}^{y}(CI - CO)(1 + DR)^{-n} \qquad (13)$$

where *CI, CO, DR, y,* and *n* have been initialized with cash inflows and cash outflows, discount rate, BIPV lifespan, and the number of the year, respectively. *CI* is the revenue generated by the BIPV system, such as revenue from electricity production. *CO* denotes the amount of money spent on the system, such as the investment, inverter replacement, and battery replacement. The discount rate (*DR*) is the interest rate that banks charge on their loans. The discount rate applied to this study is 3 %.

Finally, *EPBT*, and the *IRR* can be calculated from the following equations,

$$\sum_{n=1}^{y} NPV = Q \qquad (14)$$

$$-Q + \sum_{n=1}^{y}(CI - CO)(1 + IRR)^{-n} = 0 \qquad (15)$$

Where (*y*) represents the payback year and *Q* denoted as initial investment of this system. The internal rate of return (*IRR*) [13] is the expected annual rate of growth for an investment. It is estimated in the same way as *NPV*, except that the *NPV* is set to zero. [12].

IV. RESULT ANALYSIS

Fig.4 shows the variance in daily solar irradiance (top) and cumulative incident solar energy (bottom) of roof mounted, south facing and the front facing (east side) system. The Important parameters of the subsystems are the tilt angle from the surface of the subsystems and the shifting position of the sun

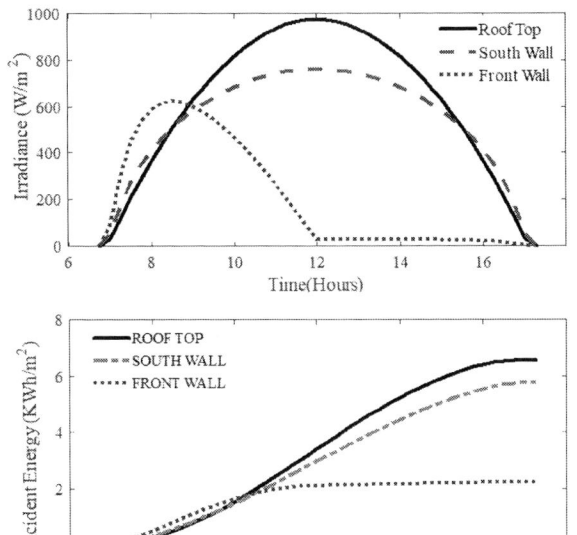

Fig. 4. Comparison of Daily Irradiance (Top) and Daily cumulative Incident Energy (Bottom) of three BIPV Subsystems calculated for 15th December.

from East to West with increasing time of day. Because of the optimal tilt angle and superior surface azimuthal direction based on the BIPV installation plan, Subsystem-I accumulates a substantial quantity of incident energy, 6.55 KWh/m² on daily basis.

Due to its acute disposition to the East, the (front-mounted) Subsystem-II does not get much direct-beam radiation and mostly receives diffused irradiance after mid-day (12.00 solar hour). As a result, Subsystem-II can only accumulate 2.22 KWh/m² incident energy per day. Though Subsystem-III maintains the same disposition as Subsystem-I, it gathers 5.74 KWh/m² incident energy, which is slightly less than Subsystem-I due to tilt angle disparity.

Fig. 5 shows the monthly incident energy per unit area. It is seen from Fig.5 that, Subsystem-I can accumulate maximum energy in May (239.66 KWh/m²) and minimum energy in

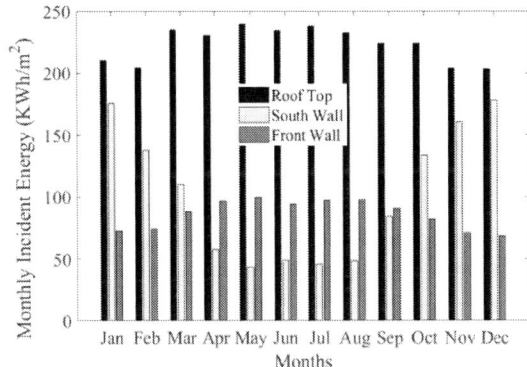

Fig 5: Distribution of total incident energy per unit area per month calculated for the three sub systems without considering cloud impact.

978-1-7281-6118-1/22 $31.00 © 2022 IEEE

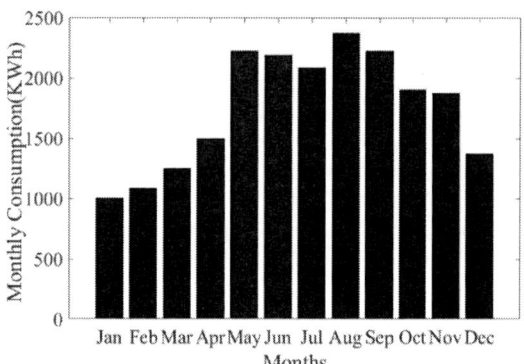

Fig.6: Monthly consumption of the studied building for the year 2021.

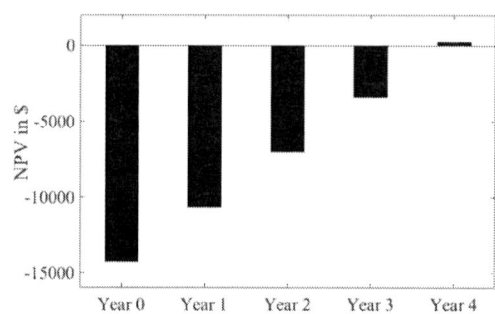

Fig.8: Capital investment and payback time.

December (203.23 KWh/m²). Likewise, accumulated monthly incident energy of Subsystem-II is 99.83 KWh/m² in May and 68.70 KWh/m² in December. On the contrary, Subsystem-III follows the opposite curve form, where it can gather minimum energy in May (43.56 KWh/m²) and maximum energy (178.14 KWh/m²) in December.

Based on the given load, total load consumption of the building has been depicted in Fig. 6. The data has been collected from the energy meter of the building for the year 2021. The bar chart shows that the seasonal variation of the load consumption throughout the year. The maximum consumption occurs in August and the minimum is in the January. Considering the data, calculation shows that the building consumes 21 MWh of electrical energy yearly whereas the system generates 45.80 MWh of output energy without considering the cloud impact.

Fig.7 shows a substantial graphical comparison between BIPV System Output and Load Consumptions with the considerations of the cloud impact. The system generates 32.4 MWh of total output energy due to the cloud impact. The figure shows that BIPV system can be able to supply to the grid for the seven months with its excess energy, and for the other five months, the system requires to consume from the grid. The building must depend on the utility grid for five months. In January, a maximum of 59.13% energy can be supplied to the grid, while in August, a maximum of 83% energy must be consumed from the grid to meet the building's demand. The graph also illustrates that during the summer and rainy seasons,

the demand for grid dependency evolves due to the BIPV system's low energy output.

In order to realize the energy payback period of the proposed BIPV System, Fig.8 needs to be observed. In this work, the achieved payback period is 3.99 years after initial investment. A higher internal rate of return (IRR) is necessary in order to establish a profitable investment. The IRR algorithm [14] in this system has been assessed at a 25% rate of return, indicating that the project is justified and acceptable..

Bangladesh contributes very little to global CO_2 emissions where the carbon emission factor (0.64 Kg/KWh) of electricity production in local power plants been utilized in this study [10]. The estimated equation of the reduced carbon emission ($Kg\ CO_2$) for a BIPV system is usually presented as follows:

$$Carbon\ Emission\ (Kg\ CO_2) = Generated\ Energy\ (KWh) \times CO_2\ emission\ factor(Kg/KWh) \qquad (16)$$

On that account, from equation (16) it can be estimated that total CO_2 emission reduction will be 20,751 Kg in a year whereas total carbon emission will be 3,604 Kg, which is depicted in Fig.9.

V. CONCLUSIONS

The building consumes 21 MWh of electrical energy whereas the building encounters a total of 391.66 MWh incident energy and output energy 45.80 MWh can be retrieved

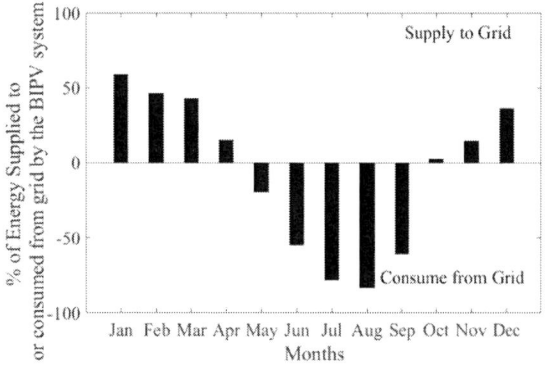

Fig.7: Percentage of Energy Supplied to and Consumption from Grid.

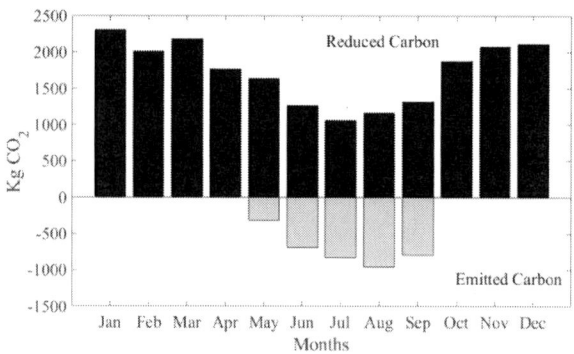

Fig.9: Carbon emission throughout the year.

according to the proposed design scheme. Therefore, it can be estimated that the BIPV system is providing 53% more electrical energy as per this study. After considering cloud impact, the total output energy (32.4 MWh) can not only meet the existing load demand of the building but can also be supplied to the grid, and thus building would remain grid independent for more than half a year. The system can reduce the emission of carbon by about 20751 Kg. Assuming the longevity of the system to be 20 years, the financial analysis yields that the initial costing can be retrieved within 3.99 years. It is noted so far, the assessment has destined that, implementing BIPV system in the subject would successfully address the challenge of environment, sustainable source of Renewable Energy Despite the fact that BIPV is optimally suited for high-rise buildings, visualizing and simulating photovoltaic generation for this pre-built, partial shading conditioned non-high-rise building yielded a positive result.

REFERENCES

[1] A. H. Khan, M. Islam, A. Islam, and M. S. Rahman, "A systematic approach to find the optimum tilt angle for meeting the maximum energy demand of an isolated area," in 2015 International Conference on Electrical Engineering and Information Communication Technology (ICEEICT), 2015.

[2] Hasan, M. Rakibul, Akter and Jhumana, "Energy Performance Analysis of a Residential Building: A Case Study on a Typical Residential Building at Mohammadpur in Dhaka, Bangladesh," 2019.

[3] A. K. Shukla, K. Sudhakar, P. Baredar, and R. Mamat, "BIPV based sustainable building in South Asian countries," Sol. Energy, vol. 170, pp. 1162–1170, 2018.

[4] L.L. Sun, H.X. Yang, "Impacts of the shading-type building-integrated photovoltaic claddings on electricity generation and cooling load component through shaded windows," Energy and Buildings, vol. 42, pp. 455-460, 2010.

[5] Nadim, Mohammad et al. "Estimation of optimum tilt angle for PV cell: A study in perspective of Bangladesh." *2016 9th International Conference on Electrical and Computer Engineering (ICECE)* (2016): 271-274.

[6] S.-H. Yoo, "Optimization of a BIPV system to mitigate greenhouse gas and indoor environment," Sol. Energy, vol. 188, pp. 875–882, 2019.

[7] D. E. Attoye, A. Hassan, and K. A. T. Aoul, "A review on building integrated photovoltaic façade customization potentials," Sustainability, vol. 9, no. 12, p. 2287, 2017.

[8] M. M. Ali, Rokonuzzaman, M. D. Z. Shuvo, A. Das, M. Islam and M. M. Rahman, "Performance Investigation of Bifacial Module Based Time Varying Multilevel Solar Panel System," 2021 IEEE 48th Photovoltaic Specialists Conference (PVSC), 2021, pp. 0564-0568.

[9] R. A. Messenger and J. Ventre, "Photovoltaic Systems Engineering," 2nd Edition, CRC Press, Boca Raton, 2004.

[10] Das, B. K., Hoque, N., Mandal, S., Pal, T. K., & Raihan, M. A. (2017). A techno-economic feasibility of a stand-alone hybrid power generation for remote area application in Bangladesh. Energy (Oxford, England), 134, 775–788.

[11] H. M. Fahad, A. Islam, M. Islam, M. F. Hasan, W. F. Brishty and M. M. Rahman, "Comparative Analysis of Dual and Single Axis Solar Tracking System Considering Cloud Cover," *2019 International Conference on Energy and Power Engineering (ICEPE)*, 2019, pp. 1-5.

[12] Tripathy, M., Joshi, H., Panda and S.K., Energy payback time and life-cycle cost analysis of building integrated photovoltaic thermal system influenced by adverse effect of shadow, vol. 208, Applied Energy, 2017.

[13] Ganti, A. (2011) *Internal rate of return (IRR) rule*, Investopedia. Available at: https://www.investopedia.com/terms/i/internal-rate-of-return-rule.asp (Accessed: May 30, 2022).

[14] H. Gholami, H. N. Røstvik, N. M. Kumar, S. S. and Chopra, Lifecycle cost analysis (LCCA) of tailor-made building integrated photovoltaics (BIPV) façade: Solsmaragden case study in Norway, vol. 211, Solar Energy, 2020, pp. 488-502.

Elucidating Materials Paradigm of CIGS by Structure--Composition--Performance Correlations

Niklas Pyrlik, Christina Ossig, Giovanni Fevola, Svenja Patjens, Jan Hense, Catharina Ziska, Martin Seyrich, Frank Seiboth, Andreas Schropp, Jan Garrevoet, Gerald Falkenberg, Christian G. Schroer, Romain Carron, Michael E. Stuckelberger

Deutsches Elektronen-Synchrotron DESY, Hamburg, Germany

Universität Hamburg, Hamburg, Germany

Empa, Dübendorf, Switzerland

Recent developments in focusing hard X-rays to nanoscale beams have enabled scanning X-ray microscopy modalities and their simultaneous exploitation in multi-modal measurement campaigns. Specifically, X-ray beam induced current and X-ray fluorescence measurements have been established for the correlation of the electrical performance with the distribution of absorber and trace elements for thin-film solar cells with absorbers from CIGS to CdTe and perovskites. For CIGS, the composition is in an especially complex interplay with the synthesis conditions and the crystallographic structure due to the tetragonal lattice distortions, steep vertical In/Ga gradients, and lateral inhomogeneities that introduce lattice strain and structural defects. For this contribution, we have added scanning X-ray nano-diffraction to the multi-modal envelope of scanning X-ray microscopy to assess crystallographic properties of a solar-cell series with a varying In/Ga ratio. For the first time, this combination has been used to characterize a statistically significant number of CIGS grains embedded in as-deposited solar cells: mapping out the real and reciprocal space, we have isolated nearly 500 individual grains. This enabled us to elucidate Materials Paradigm of CIGS, by (1) correlating the lateral Cd and In/Ga distribution with the local performance and lattice spacing with unprecedented sensitivity, (2) differentiating voids in the absorber layer that appear (not) to be filled with CdS, and (3) evaluating the crystallographic properties including the grain orientation and grain-boundary classification with sub-grain resolution and powerful statistics in fully assembled devices. In the full presentation, we will elaborate on our methodological advances and unveil performance-relevant findings from the CdS coverage to the strain distribution at small- and large-angle grain boundaries. Beyond applications to CIGS, our work highlights the latest developments in the field of X-ray imaging and paves the way for advanced correlative nanoscopy at diffraction-limited storage rings that will become operational within the next few years.

An experimental comparison between view factor and ray tracing models for energy estimation of bifacial modules

Hugo Sánchez*[†] ⓘ, Sebastian Dittmann*[‡] ⓘ , Carlos Meza*[†] ⓘ, Ralph Gottschalg*[‡] ⓘ

*Anhalt University of Applied Sciences, Bernburger Str. 55, 06366 Köthen, Germany
E-mail: hugo.sanchez@hs-anhalt.de, Tel.: +49 3496 67 2354
[†]Costa Rica Institute of Technology, Cartago, Costa Rica
[‡]Fraunhofer-Center for Silicon Photovoltaics CSP, Halle, Germany

Abstract—Albedo, or the ground reflectance irradiance, is one of the crucial variables in the energy estimation for bifacial photovoltaic modules due to the modules' ability to collect electricity from both front and rear sides. There are several modeling tools for the energy estimation of these modules. Our paper reports on the comparison of on two available energy estimation models with view factor and ray tracing and compares the estimated data with actual measured values. For a one-year data gathered at the Anhalt Photovoltaic Performance and Lifetime Laboratory (APOLLO) in Bernburg, Germany, the view factor model-based algorithm for a bifacial module is more accurate when predicting PV energy yield. Furthermore, an albedo sensitivity analysis concludes that the tested Ray-tracing algorithm overestimates bifacial PV energy for large albedo values. However, the tested modeling methods follow the measured data trend with a deviation between -8 % to 6 %.

Index Terms—bifacial modules, energy estimation, simulation

I. INTRODUCTION

In the past decade, there has been an increasing interest in bifacial photovoltaic (PV) modules. With a minimal increase in production costs, bifacial modules can increase the energy output by up to 30 % concerning monofacial PV modules. Also, bifacial PV modules allow exploring new ways to use PV technology, such as agrivoltaics [1], building architecture [2], and highway noise barriers [3].

Estimating the energy yield of PV applications using bifacial PV modules is more challenging and less studied than for monofacial-based PV modules or plants. The reason is that several factors affect the energy production of a bifacial PV module, such as clearance height (module elevation), azimuth angle, tilt angle, and rear-side irradiance characteristics (magnitude, distributions, spectrum), [4], [5], [6].

Ground reflectance is one of the most critical input for challenging topics in estimating the energy yield of bifacial PV modules. Two of the most widely used and promising modeling techniques are the View Factor and Ray Tracing models.

View Factor models, also known as Configuration Factor, estimate, through geometrical considerations, the fraction of irradiance scattered or reflected from adjacent surfaces to the collection on a bifacial PV module [7]–[10]. Optimizing of

these models can be found in [7]. The most known commercial tool that uses view factor models is PVSyst. This tool was initially developed at the University of Geneva in 1992 and has integrated bifacial PV technology in their simulation library since 2017.

On the other hand, Ray Trace Models allow the possibility to create complex scenes to evaluate the bifacial PV systems. These scenes can also consider the mounting system and the effects of self-shadowing. Besides, it is possible to consider variations in the solar module, such as the distance between the cell matrix, the characteristics of the glass, and other optical effects. Nevertheless, these methods request higher computational processing than view factor models. Some authors offer complete studies regarding the Ray Trace Models for Bifacial PV [8], [9], [11], [12]. The software Radiance is a rendering system developed at the Lawrence Berkeley Laboratory (LBL) in California and Switzerland's Ecole Polytechnique Federal de Lausanne (EFPL) [13]. Initially, they developed the software to study the problem of lighting design. It helps designers, planning builders, engineers, and architects create a more efficient lighting design by optimizing ray-tracing algorithms. Ray Tracing can be defined as a technique by tracing the path of light as pixels in an image plane and simulating the effects of its encounters with a virtual object [14]. The National Renewable Energy Laboratory (NREL) developed an open-source wrapper that integrates the Radiance Software to estimate energy for a bifacial PV module. The software is based on Python's programming language and allows the user to configure site details such as latitude, longitude, and ground albedo [15]. It is also possible to configure the specific ground clearance height, row spacing, tilt, and azimuth angle. As a result, the front- and rear-side irradiance are created for annual and hourly values.

The present paper analyzes and compares the PV bifacial energy estimation results of two simulation algorithms using i) tthe view factor model of PVSyst and ii) the ray-tracing model of NREL's Radiance. These simulated results will be compared with experimental data for a commercial bifacial PV module measured at the Anhalt Photovoltaic Performance and Lifetime Laboratory (APOLLO) at Anhalt University of

978-1-7281-6118-1/22 $31.00 © 2022 IEEE

Applied Sciences in Bernburg, Germany.

II. METHODOLOGY

The bifacial View Factor based energy estimation (VFEE) and Ray Tracing based energy estimation (RTEE) are analyzed and compared using simulations and measurement data. Temperature and front- and rear-side irradiance were measured and fed into the VFEE and RTEE algorithms. Then, the energy estimation results were compared with measured energy. In the following subsections, a more detailed description of this procedure is explained.

A. Front- and Rear-side Irradiance Measurements

We designed an experiment to measure the rear- and front-side irradiance variation. For this purpose, an albedometer has been installed at APOLLO's outdoor test field. Fig.1 shows the outdoor test facility at the Anhalt University of Applied Sciences (HSA) located in Bernburg, Germany.

Fig. 1: Outdoor test field at Anhalt Photovoltaic Performance and Lifetime Laboratory (APOLLO)

The albedometer consists of a combination of two pyranometers (both Kipp & Zonen SMP10). One is mounted facing the sky, the other one inverted to measure the ground-reflected irradiance. Figure 2 shows the measurement set-up with the horizontal measurement and the measurement of the rear-side plan-of-array (POA) at 35°.

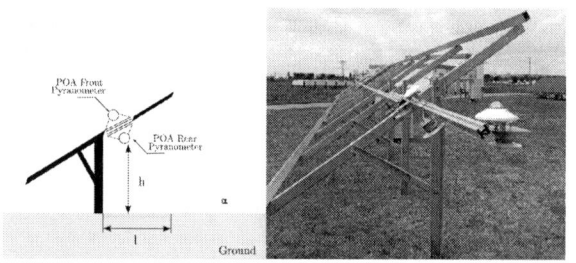

Fig. 2: Albedometer installed at APOLLO, Bernburg Germany.

The sky-facing pyranometer is equipped with the standard sunscreen; meanwhile, the inverted pyranometer is fitted with a glare screen as requested by Kipp & Zonen. According to the manual, the glare screen has an angle of 5° and is fitted to the pyranometer to prevent direct irradiance at high sun azimuth angels.In addition to the meteorological data current

voltage (IV) curves are measured in 10 seconds intervals. The measurement period started in October 2019 and finished in September 2020. During the measurement period, the substrate changed from green grass to yellow due to an arid summer.

The ratio of the rear- and front-side irradiance is denoted by the letter α and it is given by:

$$\alpha = \frac{G_{\text{rear}}}{G_{\text{front}}} \tag{1}$$

where G_{rear} and G_{front} are the rear and front irradiance, respectively.

A preprocessing, analysis and quality check were performed following the recommendations and the previous work presented in [16]–[19]. First, the rear-/front-side ratio, the ambient temperature, and module performance data are normalized using a classical seasonal decomposition (CSD) method, which is applied to consider only the relevant data of the front- and rear-sides. [20]. After the normalization, a cumulative moving average (CMA) is used to group the daily data.

Fig. 3 shows the monthly values of α and its distribution. The seasonal variation of α is visible, ranging from 0.08 to 0.25 with a mean value of 0.15 and a standard deviation of 0.038.

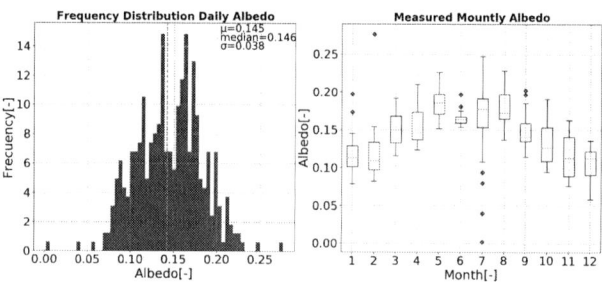

Fig. 3: Histogram and monthly distribution of α measured at APOLLO's test side in Bernburg

B. Simulations

The calculated α based on the measurements was used as input for both PVSyst v6.8.6 and NREL Radiance based on the on-site measurements. The site-specific parameters such as longitude, latitude, POA irradiance, azimuth angle, tilt angle, and height clearance were used in both simulation models. The input parameters for both simulations are presented in Table I. The hourly averages of the global horizontal irradiance (GHI) and ambient temperature (Tamb) were used for all the modeling tools.

III. RESULTS

Figure 4 shows the estimated energy yield of both simulation models and the real measured energy yield of the used bifacial PV module.

The deviations of the estimated energy from the measured values are calculated using the *Root Mean Square Error*

TABLE I: Main simulation values for the validation

Parameter	Value	Unit
Latitude	51.773	°
Longitude	11.763	°
Tilt Angle	35	°
Commercial PV Module	LG300N1T-G4	m
Cleareance height	1.5	m
Azimut Angle	180	°
Bifacial System Definition	Unlimited Sheds 2D Model	-
Structure Shading Factor	5	%
Pitch	6.60	m
Shed total width	3.04	m

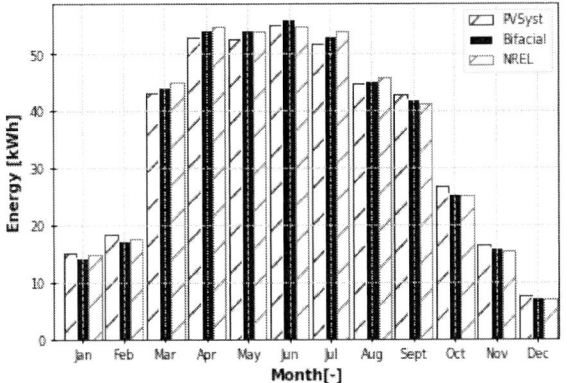

Fig. 4: Comparison of the simulated energy yield versus real measured values.

(RMSE) and *Mean Bias Error (MBE)*. Figures 5 and 6 show both models' RMSE and MBE concerning the real measurement.

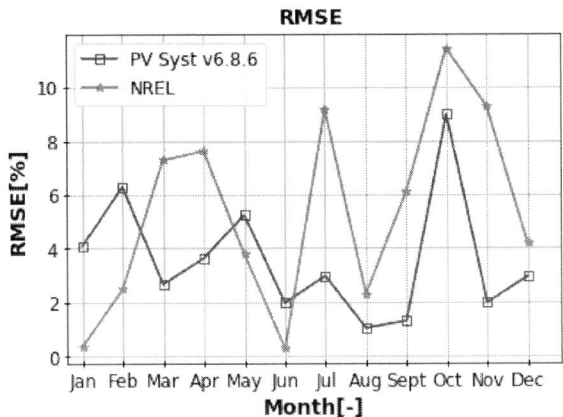

Fig. 5: RMSE of the simulated energy yield of PVSYST and the NREL model concerning the real measured values.

Statistical analysis was performed to rank the overall predictions of the methods. For the present study, the most effective ways that correspond to the parametric ranking are Chi-Squared ($\tilde{\chi}^2$) and Distance Correlation (d), which can be described in detail in [21], [22]. Other non-parametric methods

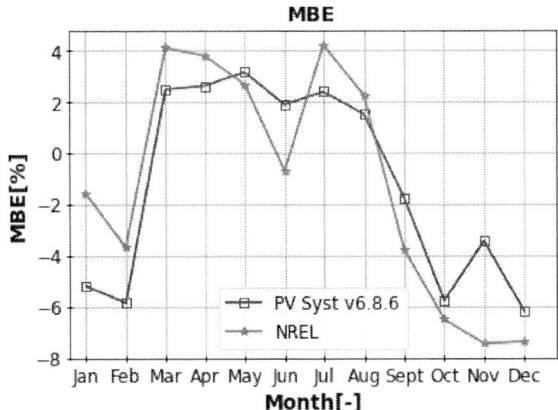

Fig. 6: MBE of the simulated energy yield of PVSYST and the NREL model concerning the real measured values.

were tested; however, it was not possible to rank the methods. Figure 7 shows the obtained results when the *Chi-Squared* and *correlation distance* are estimated for every method of prediction.

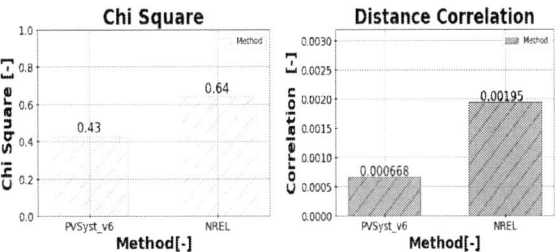

Fig. 7: Chi-Squared and Correlation Distance for each model

IV. DISCUSSION OF RESULTS

Figure 4 shows the values corresponding to the total monthly amount of energy harvested in the solar module. Roughly 70 % of the total energy is produced between April and August, as is expected in this latitude. The monthly energy goes from 7.66 kWh in December to 55.04 kWh in June. It is possible to see how both modeling tools follow the trend of the measured values. However, as shown in Fig.5, there is a significant deviation margin between the simulations and the actual values at specific months, e.g., October. From March to August, according to Fig.6 both algorithms overestimate the energy production with a lower error compared to the other months, i.e., $0\% < \text{MBE} < 4\%$. On the other hand, during the period from September to February, both models underestimate the energy with significantly higher errors, i.e., $-8\% < \text{MBE} < -2\%$.

The accuracy of the methods is shown in Figure 7, where it is clear that the most accurate method for the used data corresponds to *PVSystV6* with a $\tilde{\chi}^2 = 0.43$, $d = 0.000668$. The

NREL algorithm achieved a $\tilde{\chi}^2 = 0.64$, $d = 0.00195$. There is a normalized difference of 4.5% between both models.

A. Albedo Sensibility Analysis

An additional simulation study was performed to understand the impact of different albedo values. The albedo value was increased from 10% to 70% for each simulation software in 5% steps.Other parameters have been set constant and equal for both algorithms. Figure 8 show the results.

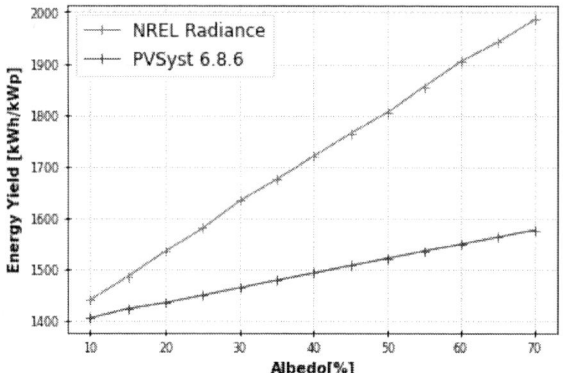

Fig. 8: Albedo sensibility simulation comparison

There is a significant difference between both algorithms. The NREL Radiance algorithm overestimates the energy concerning the PVSyst estimation for higher albedo values. The overestimation increases when the albedo increases. The reason might be the assumptions related to clear-sky conditions and/or uniformity in the backside of the module taken by NREL Radiance.

B. Challenges in the modeling of bifacial modules

The simulation process of bifacial modules is complex. Most PV simulation tools are initially designed considering only monofacial modules. The challenge for simulating bifacial modules is the estimation of the rear-side irradiance for any time of the year. In addition, the irradiance is affected by several factors: the clearance height, albedo, tilt angle, azimuth, and seasonal variation of the specific location where the system is installed. For this reason, these main barriers turn highly important to accelerate the optimization of modeling tools and, with that, the adoption of the technology.

Concerning the lack the standards, the main progress occurs for the monofacial modules. International standards such as IEC 61724 [23] and IEC 61853 for energy rating [24] are well adopted in the industrial and scientific community for outdoor evaluation and monitoring. However, since they are designed for single-face systems, they do not allow the consideration of the energy yield potential on the rear-side. On the other hand, the indoor measurement of bifacial modules presents more progress in standardization. The IEC 60904-1-2 [25], [26] includes an indoor evaluation procedure for bifacial modules and cells.

The creation and enhancement of standards for evaluating, simulation, and monitoring bifacial systems is only possible with collaboration. The industrial and scientific communities must keep working on the research and proof of concepts for the bifacial technology, improving the accuracy of simulation tools and the adoption of the technology. Bifacial technology can become an ally for the transition to a green economy. The present work pretends to contribute in this direction. Efforts should be continued to reduce errors in bifacial PV simulation tools.

V. Conclusion and outlook

This paper compares two simulation models for energy estimation of bifacial PV modules using view factor and ray tracing. As inputs for the VFEE and RTEE, temperature and albedo measurements were used. The obtained estimation values are compared with actual energy measurements. Both VFEE and RTEE algorithms follow the measured data trend with an -8% to 4% error margin. The annual energy estimation presents higher errors for the low irradiance months, i.e., September and March. The most accurate method in this validation corresponds to the VFEE algorithm (*PVSystV6*) with a $\tilde{\chi}^2 = 0.43$ and $d = 0.000668$. An albedo sensitivity analysis determines that the RTEE algorithm tested overestimates the energy production for large albedo values. In this regard, we infer that the albedo overestimation is a factor that makes the RTEE algorithm present higher errors for the energy estimation.

VI. Acknowledgment

This work was funded by the Federal Ministry fo Economic Affairs and Climate Action within the project Anomalous (03TN0025D).

References

[1] M. H. Riaz, H. Imran, and N. Z. Butt, "Optimization of pv array density for fixed tilt bifacial solar panels for efficient agrivoltaic systems," in *2020 47th IEEE Photovoltaic Specialists Conference (PVSC)*, pp. 1349–1352, 2020.

[2] A. Elnosh, H. O. Al-Ali, J. J. John, A. Alnuaimi, E. R. Ubinas, M. Stefancich, and P. Banda, "Field study of factors influencing performance of pv modules in buildings (bipv/bapv) installed in uae," in *2018 IEEE 7th World Conference on Photovoltaic Energy Conversion (WCPEC) (A Joint Conference of 45th IEEE PVSC, 28th PVSEC 34th EU PVSEC)*, pp. 565–568, 2018.

[3] T. Nordmann, T. Vontobel, and L. Clavadetscher, "15 years of practical experience in development and improvement of bifacial photovoltaic noise barriers along highways and railways lines in switzerland," in *27th PV Conference*, 2012.

[4] H. Sánchez, C. Meza, S. Dittmann, and R. Gottschalg, "The effect of clearance height, albedo, tilt and azimuth angle in bifacial pv energy estimation using different existing algorithms," in *Proceedings of the III Iberoamerican Conference on Smart Cities (ICSC-2020)*, pp. 315–331, 2020.

[5] L. Kreinin, A. Karsenty, D. Grobgeld, and N. Eisenberg, "Pv systems based on bifacial modules: Performance simulation vs. design factors," in *2016 IEEE 43rd Photovoltaic Specialists Conference (PVSC)*, pp. 2688–2691, 2016.

[6] C. Deline, S. MacAlpine, B. Marion, F. Toor, A. Asgharzadeh, and J. S. Stein, "Evaluation and field assessment of bifacial photovoltaic module power rating methodologies," in *2016 IEEE 43rd Photovoltaic Specialists Conference (PVSC)*, pp. 3698–3703, 2016.

[7] C. W. Hansen, J. S. Stein, C. Deline, S. MacAlpine, B. Marion, A. Asgharzadeh, and F. Toor, "Analysis of irradiance models for bifacial pv modules," in *2016 IEEE 43rd Photovoltaic Specialists Conference (PVSC)*, pp. 0138–0143, IEEE, 2016.

[8] S. A. Pelaez, C. Deline, S. M. MacAlpine, B. Marion, J. S. Stein, and R. K. Kostuk, "Comparison of bifacial solar irradiance model predictions with field validation," *IEEE Journal of Photovoltaics*, vol. 9, no. 1, pp. 82–88, 2018.

[9] M. Chiodetti, J. Kang, C. Reise, and A. Lindsay, "Predicting yields of bifacial pv power plants–what accuracy is possible?," *System*, vol. 2, p. 1, 2018.

[10] D. Berrian and J. Libal, "A comparison of ray tracing and view factor simulations of locally resolved rear irradiance with the experimental values," *Progress in Photovoltaics: Research and Applications*, vol. 28, no. 6, pp. 609–620, 2020.

[11] C. A. Deline, S. Ayala Pelaez, W. F. Marion, W. R. Sekulic, M. A. Woodhouse, and J. Stein, "Bifacial pv system performance: Separating fact from fiction," tech. rep., National Renewable Energy Lab.(NREL), Golden, CO (United States), 2019.

[12] C. Reise and A. Schmid, "Realistic yield expectations for bifacial pv systems—an assessment of announced predicted and observed benefits," in *Proc. 31st Eur. Photovolt. Sol. Energy Conf. Exhib*, pp. 1775–1779, 2015.

[13] G. J. Ward, "The radiance lighting simulation and rendering system," in *Proceedings of the 21st annual conference on Computer graphics and interactive techniques*, pp. 459–472, 1994.

[14] G. R. Hofmann, "Who invented ray tracing?," *The Visual Computer*, vol. 6, no. 3, pp. 120–124, 1990.

[15] C. Deline and S. Ayala, "Bifacial radiance." [Computer Software], 12 2017.

[16] N. Riedel, D. Berrian, D. Alvarez Mira, A. Protti, P. Poulsen, J. Libal, and J. Vedde, "Validation of bifacial photovoltaic simulation software against monitoring data from large-scale single-axis trackers and fixed tilt systems in denmark," *Applied Sciences*, vol. 10, p. 8487, 11 2020.

[17] A. Driemel, J. Augustine, K. Behrens, S. Colle, C. Cox, E. Cuevas-Agulló, F. M. Denn, T. Duprat, M. Fukuda, H. Grobe, *et al.*, "Baseline surface radiation network (bsrn): structure and data description (1992-2017)," *Earth System Science Data*, vol. 10, no. 3, pp. 1491–1501, 2018.

[18] J. Palacios, "Assessment of Uncertainty and Data Quality of PV Systems for Intercomparison of PV Module Technologies.," Master's thesis, Anhalt University of Applied Sciences, Kothen, Germany, 2017.

[19] S. Dittmann, H. Sanchez, L. Burnham, R. Gottschalg, S.-Y. Oh, A. Ben-larabi, B. Figgis, A. Abdallah, C. Rodriguez, R. Rüther, *et al.*, "Comparative analysis of albedo measurements(plan-of-array and horizontal at multiple sites worldwide," in *36th European Photovoltaic Solar Energy Conference (EU PVSEC)*, pp. 1388–1393, 2019.

[20] S. Wheelwright, S. Makridakis, and R. J. Hyndman, *Forecasting: methods and applications*. John Wiley & Sons, 1998.

[21] W. H. Press, S. A. Teukolsky, W. T. Vetterling, and B. P. Flannery, *Numerical recipes 3rd edition: The art of scientific computing*. Cambridge university press, 2007.

[22] P. R. Bevington and D. K. Robinson, "Data reduction and error analysis," *McGrawâ€"Hill, New York*, 2003.

[23] I. Commission *et al.*, "International standard iec 61724: Photovoltaic system performance monitoring—guidelines for measurements, data exchange and analysis," *IEC*, 1998.

[24] I. E. Commission *et al.*, "Iec 61853-1," *Photovoltaic (PV) module*, 2011.

[25] I. E. Commission *et al.*, "Standard iec 60904-1: photovoltaic devices," *Part I: Measurement of photovoltaic current-voltage characteristics*, 2006.

[26] T. S. Liang, M. Pravettoni, C. Deline, J. S. Stein, R. Kopecek, J. P. Singh, W. Luo, Y. Wang, A. G. Aberle, and Y. S. Khoo, "A review of crystalline silicon bifacial photovoltaic performance characterisation and simulation," *Energy & Environmental Science*, vol. 12, no. 1, pp. 116–148, 2019.

Field assessment of Transparent Conductive Oxides Stability Under Outdoor Conditions

Brahim Aïssa (SM IEEE), Amir A. Abdallah and Juan Lopez Garcia

Qatar Environment and Energy Research Institute (QEERI), Hamad bin Khalifa University (HBKU), Qatar Foundation, P.O. Box 5825, Doha, Qatar

Abstract— Large variety of optoelectronic and photosensitive devices are using essentially transparent conductive oxides (TCOs) as electrodes, such as panel displays, light emitting diodes, electrochromic devices and silicon heterojunction (SHJ) solar cells. The Sn doped In_2O_3 (ITO) is the most commonly used transparent electrode-contact in SHJ technology. However, alternative TCO coatings have shown low parasitic absorption losses and higher electron mobility (μ_n), and could be suitable alternatives enabling high sheet conductivities with low free-carrier absorption pending to show long term stability especially in outdoor real-world conditions. We report here on the electronic and optical properties of different TCOs, namely, ITO, IZO and IO:H deposited by DC or RF magnetron sputtering on glass substrates. TCOs were furthermore integrated into SHJ solar cell precursors and their carrier lifetimes were systematically investigated together with Hall-effect and spectroscopic analysis. The long-term stability of these different TCO coatings was evaluated at QEERI outdoor Test Facility site under real world outdoor conditions during 15 weeks characterized by heat and humidity, and their optoelectronic properties were systematically assessed.

Keywords— Transparent Conductive Oxides, Stability, Harsh environment, Silicon Heterojunction Solar Cell, Magnetron Sputtering

I. INTRODUCTION

Most of optoelectronic devices needing transparent electrical contacts or electrodes for charge collection and transport are using transparent conductive oxides (TCOs). This includes touch screens, thin film solar cells, electrochromic devices and high-efficiency silicon heterojunction (SHJ) solar cells. The latter conjugate crystalline silicon (c-Si) as absorber and thin-film hydrogenated amorphous silicon (a-Si:H) as a light window. SHJs solar cells are a very propitious configuration that has already shown records open circuit voltages values leading to power conversion efficiencies beyond 26 % [1]. However, in this precise case, the low electrical conductivity of the a-Si:H necessitates the use of TCOs as a front contact layer on top of it, to be able to transport the photocurrent. Yet, TCOs must gather at the same time many characteristics, including a low contact resistance with juxtaposed layers, high electrical conductivity, high optical transmittance and an adapted refractive index [2].

Several deposition techniques have already been explored to fabricate high-mobility TCOs at low temperatures, including sputter deposition [3], chemical vapor deposition [5] and atomic layer deposition [5].

Among this variety of methods, magnetron sputtering deposition is the most established technique, despite the fact that it can lead to damage of underlying layers. Among all the employed TCOs, ITO (i.e. Sn doped In_2O_3) is definitely the most used coating as a transparent front-contact in SHJ configuration. However, its charge carrier mobility is rather modest ranging between 20 to 40 cm^2/V s. In addition, ITO might be characterized by serious IR parasitic absorption losses in the scenario where a high electrical conductivity is needed. To deal with such a challenge, lowering the parasitic absorption losses has been found to be possible through the use of hydrogen doped In_2O_3 (IO:H) which places itself as a serious and suitable alternative. Indeed, IO:H demonstrates a much higher electron mobility than its peers thereby offering high sheet electrical conductivities with low free-carrier absorption, leading thus to a low IR parasitic absorption losses. Moreover, when applied to SHJ cells, compared to ITO, the IO:H with a high electron mobility has shown an improvement of the short-circuit current density (Jsc), with a drawback of lowering the fill factor (FF) due to an increase of the interface IO:H/ silver front grid contact resistance [6]. Morales and coworkers reported on amorphous Zn-doped In_2O_3 (IZO) as high mobility front contact TCOs in SHJ [5].

IZO combines high conductivity and high mobility allowing the improvement in Jsc as compared to ITO. These front TCO films coat the functional solar cell and are subjected to harsh environmental conditions in the photovoltaic module (temperature gradient, humidity, etc.) and their long-term stability is hence a critical factor for their large-scale deployment).

Here, we studied the optoelectronic characteristics of three types of–doped In_2O_3 based TCOs, namely the classical Sndoped In_2O_3 (ITO), the H-doped In_2O_3 (IO:H), and the Zn-doped In_2O_3 (IZO), employed as front contact electrodes under real world outdoor conditions, characterizing our Outdoor Test Facility (OTF) located in Doha, state of Qatar. The OTF is a 35,000 m^2 field station, operated by Qatar Environment & Energy Research Institute (QEERI) that measures the real-world performance of solar related technologies. Its core mission is to determine which technologies work best in

978-1-7281-6118-1/22 $31.00 © 2022 IEEE

desert climate, and how to maximize their energy production and durability while lowering operational costs.

ITO and IZO show high stability where the sheet resistance change was found to be within 15% from the initial measured value (i.e. as-deposited), while for IO:H, the relative variation is significant and goes beyond 38% after 15 weeks in OTF. This Rs degradation is likely due to losses in electron Hall mobility (μHall) and might probably originate from chemisorption of OH- and/or H_2O-related species in the material as pointed out with the indoor damp heat (DH) conditions testing performed by Morales et al. [5]. Moreover, while μ_{Hall} clearly decreases during the outdoor exposure, the variation in the optical characteristics was somehow negligible, indicating that the degradation mainly occurs at the structural and/or morphological state.

II. EXPERIMENTAL

ITO, IO:H and IZO films were grown on glass substrates by dc- and rf-magnetron sputtering, under argon and oxygen gases flow (see details in Table 1). ITO thin films were DC-sputtered using an ITO target (90 wt.% In_2O_3 and 10 wt.% SnO_2) at room temperature. The DC power density was 1.9 W/cm^2, and the oxygen to total flow ratio $r(O_2) = O_2/(Ar+O_2)$, introduced during the deposition, was varied from 1% to 8%.

For IO:H coatings, water vapor was used as source of H, and was introduced into the reactor during the deposition process [7,8]. The deposited TCOs were subsequently thermally annealed in air at 190 °C for 20 min [9,10]. Prior to the SHJs solar cell fabrication, we have optimized the sputtering parameters of each TCO film to standardize the thicknesses at 100 nm and the carrier concentrations (Ne) between 1 and 2.5×10^{20} cm^{-3} after the thermal annealing treatment. These particular parameters are optimized in such a way to decrease as much as possible the optical reflectance in the visible range of the silicon substrates and to increase the optical transmittance in the IR.

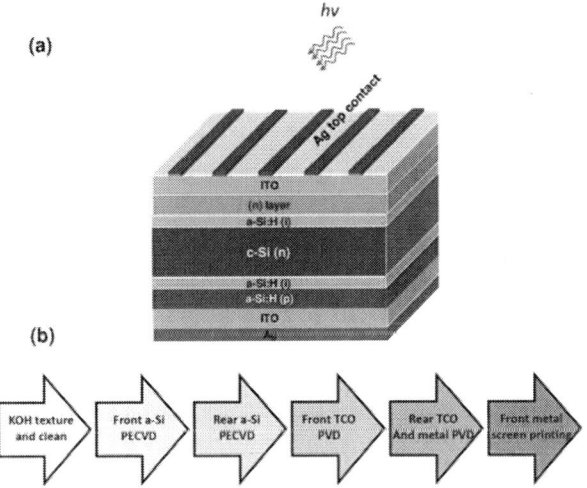

Fig. 1: (a) Schematic structure of the fabricated SHJ solar cell studied in this work. (b) Details of the wafer processing.

Stability of the annealed films was tested by placing the samples in OTF for 15 weeks of time under real world conditions (3 of May-15 of August 2021). The electrical conductivity (σ), carrier concentration (Ne), and Hall mobility (μHall) were performed with a HMS-5000 system using the Van Der Pauw method. The optical transmittance (T) and reflectance (R) of the films were measured by an UV-Vis-NIR spectrometer with an integrating sphere, and the absorptance was determined from 100%–T–R. Samples were carefully cleaned by dry nitrogen flux to remove the accumulated dusts prior to any measurements. Depth profiling was carried out using IONTOF TOF-SIMS5 model to verify the quality of the deposition throughout it thickness. Grazing-incidence X-ray diffraction was used to investigate the structural properties of the annealed films. [9-12]

III. RESULTS

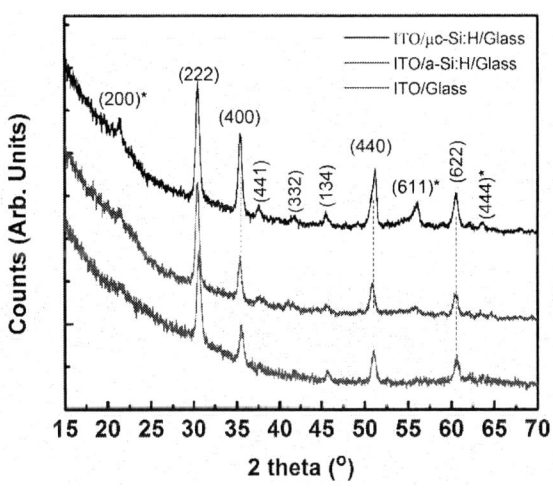

Fig. 2.: GI-XRD diagram of ITO films deposited onto (n)μc-Si:H/(i)a-Si:H/glass, (n)a-Si:H/(i)a-Si:H/ /glass , and naked glass substrates, at $r(O_2) = 0.08$, showing multiples polycrystalline planes.

The rear junction SHJ solar cell schematic is depicted in Fig. 1a. SHJ cells were fabricated on high-quality n-type float-zone c-Si wafers ($\langle 100 \rangle$; 180 μm; 1–5 Ω cm). As detailed in Fig. 1b, the wafers were first random-pyramidal textured in an alkaline solution, wet chemically cleaned, and dipped in hydrofluoric acid prior to plasma-enhanced chemical vapor deposition of intrinsic and doped hydrogenated amorphous silicon layers (a-Si:H). ITO films were used as the front electrode and deposited through a shadow mask (2×2 cm^2), defining the cell size. On the rear side of the wafer, an ITO layer was used in all

978-1-7281-6118-1/22 $31.00 © 2022 IEEE

devices, followed by a silver back reflector also sputtered immediately after the back ITO. A silver front grid was screen-printed on the front of the five ITO pads, and the solar cells were cured for 30 min at 210 °C. Finally, the complete solar cells were characterized at STC conditions by current–voltage (J–V) measurements on a AAA sun simulator under Air Mass 1.5 global illumination.

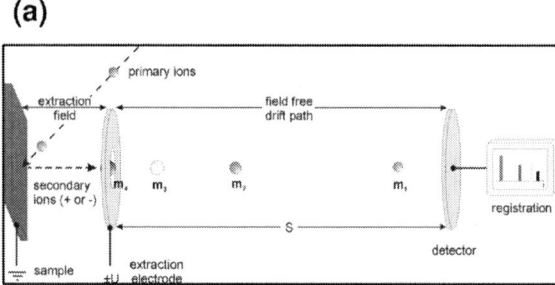

(a)

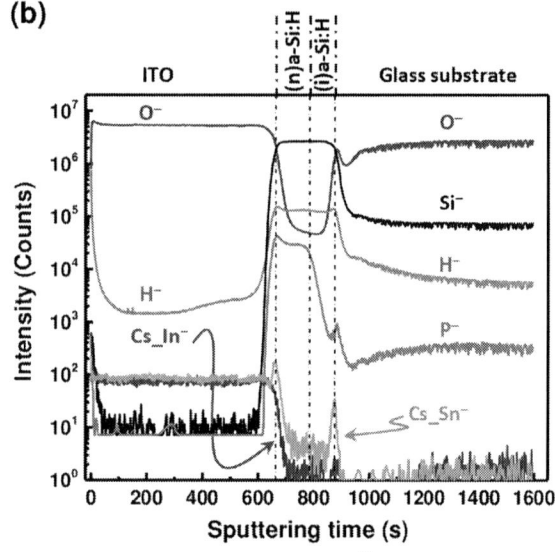

(b)

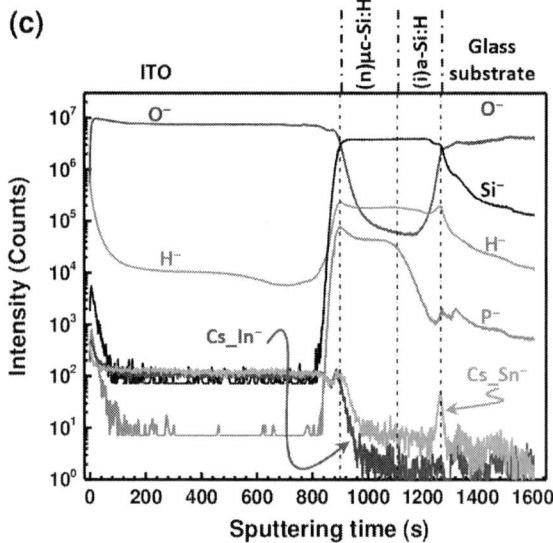

(c)

Fig. 3: Schematic illustrating the TOF SIMS measurement principle. (b) Typical TOF-SIMS depth profiling of the ITO deposited onto (n)a-Si:H(n)/(i)a-Si:H/glass and (c) onto (n)μc-Si:H(n)/(i)a-Si:H/glass, at r(O₂) = 0.01 (Sn and In are probed using Cs_Sn and Cs_In to avoid overlapping with other compounds).

Furthermore, the time of flight-secondary ion mass spectrometry ToF-SIMS technique was used to determine the depth profiling of indium, oxygen, tin, silicon, phosphorous and hydrogen throughout the various thicknesses and interfaces of ITO/a-Si:H(i or n)/Glass, simulating the TCO properties after full SHJ solar cell fabrication. The SIMS data show clear and distinctive steps, witnessed by the abrupt variation of the species profiles occurring at the different interfaces [13].

First, regardless of the oxygen content (patterns are shown here for r(O₂) = 8%), all ITO films show a polycrystalline nature, where diffraction peaks corresponding to orientations along (222), (400), (440) planes are the strongest ones observed, in addition to the presence of (441), (332), (134) and (622) phases [14,15].

No clear preferential growth is noticed. However, comparatively to the other underlying substrates, ITO deposited on (n)μc-Si:H/(i)a-Si:H/glass shows additional peaks (indicated by stars in the Fig. 2), namely (200), (611) and (444), in addition to display better defined and more intense bands with somehow a narrower FWHM. In fact, for ITO deposited on (n)μc-Si:H/(i)a-Si:H/glass, the main (222), (400), (440) and (622) peaks show respective FWHM of 0.58°, 0.55°, 0.77° and 0.66°, while they are of 0.65°, 0.59°, 0.70° and 0.66° for films deposited on (n)a-Si:H/(i)a-Si:H/glass and of 0.65°, 0.60°, 0.69° and 0.67° for those deposited on glass, which may suggest that the underlying (n)μc-Si:H could improve the crystallinity of the ITO films [16-19].

Furthermore, time-of-flight secondary ion mass spectrometry (TOF-SIMS) was employed to analyze deposited thin films and interfaces and detect associated species (functioning principle is schematized in Fig. 4a).

Figures 3 (b-c) show representative TOF-SIMS spectra performed on ITO films, at r(O₂) = 1%, for the two main configurations mentioned in the experimental section, namely ITO/(n)a-Si:H/(i)a-Si:H/glass, and ITO/(n)μc-Si:H/(i)a-Si:H/glass. Here, TOF-SIMS was used to determine the depth profiling of indium, oxygen, tin, silicon, phosphorous and hydrogen throughout the various thicknesses and interfaces of ITO/silicon layers.

All profiles are found quite similar in shape, in the sense that all signals related to ITO layer decrease when reaching the interface with n-silicon windows, while all those related to the n-windows increase, whether the r(O₂) changed from 1 to 8 % (therefore only the results for the sample of r(O₂)= 0.01 are shown). Although the SIMS measurements are not employed as quantitative analyses, the intensities of the hydrogen, phosphorous, and silicon depth profiles are found to clearly increase by orders of magnitude in the (n)a-Si:H/(i)a-Si:H and/or (n)μc-Si:H/(i)a-Si:H zones, then decreasing when reaching the

978-1-7281-6118-1/22 $31.00 © 2022 IEEE   1153

glass substrate interface [20-36]. Conversely, oxygen, indium and tin depth profiles decrease considerably when reaching the ITO end-edge interface. Moreover, by comparing the depth profiling of Figs. 3b and 3c, the following remarks can be summarized:

(i)   by observing the phosphorous profile in the (i) zone, P seems to diffuse into the intrinsic silicon layer when both (n)a-Si:H and/or (n)μc-Si:H layers analysis, this diffusion is found to be enhanced when the multicrystalline silicon is used, for which the grains boundaries may favour the diffusibility. Indeed, the P diffusion rate is even doubling and ranges from ~120 counts/s to 260 counts/s, for the (n)a-Si:H and (n)μc-Si:H layers, respectively,

(ii)   the hydrogen specie is well contained within the intrinsic and the (n)-layers zone, with a certain diffusion into glass substrate. In fact, it was already reported in the relevant literature the diffusion of hydrogen into fused silica and borosilicate glass, where most of the hydrogen permeation has taken place in the form of hydrogen molecules passing through the large holes induced in the structure with random arrangements of SiO_4 tetrahedra [37], and finally,

(iii)   no oxygen atoms were diffused from ITO into the intrinsic and/or the (n)-layers.

TABLE I. Comparison of the deposition parameters and electrical properties before and after annealing of the studied TCOs.

Deposition method			Electrical properties before and after annealing			
Magnetron Sputtering						
Sample	T°C	Target	Rs (Ω/□)	μHALL (cm²/V s)	Ne (10²⁰ cm⁻³)	
ITO	dc	RT	In_2O_3 with 10 wt.% SnO	120/105	35/28	1.08/2.03
IO:H	rf	RT	In_2O_3	33/38	48/104	3.4/1.3
IZO	rf	60	In_2O_3 with 10 wt.% ZnO	78/55	49/61	1.4/1.8

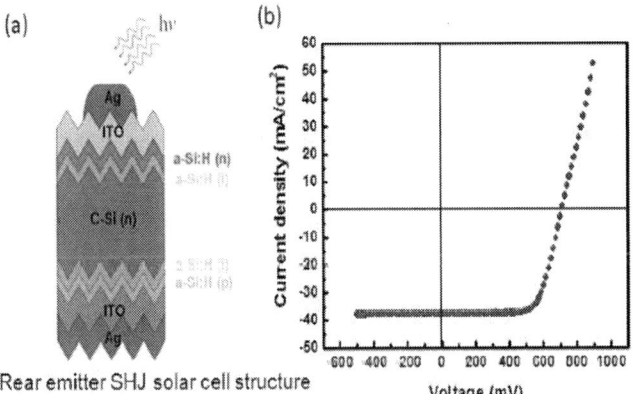

(a) Rear emitter SHJ solar cell structure

(b)

Fig. 4: (a) Schematic the measured SHJ cell. (b) Typical example of the corresponding I-V measurement.

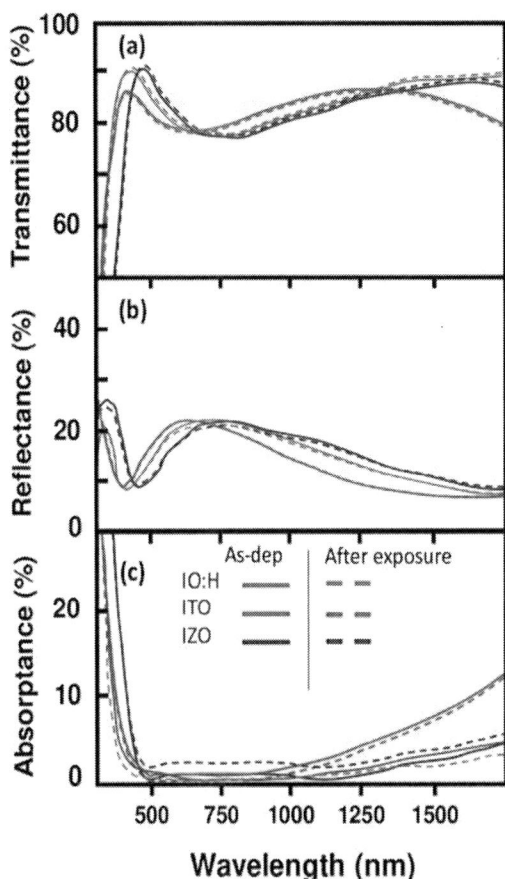

Fig. 5: Optical (a) transmittance, (b) reflectance and (c) absorptance of ITO, IO:H, IZO, and IO:H/ITO bilayers before (full lines) and after Outdoor exposure in OTF for 15 weeks (dashed lines).

Table 1 summarizes the electrical properties for the as-dep (i.e. before annealing) and the thermally annealed films (ITO, IO:H and IZO). The as-deposited ITO was found to be polycrystalline, the IO:H and IZO were both amorphous. After the thermal annealing process, IO:H becomes crystalline and the IZO remain unchanged and amorphous (results not shown in this short version of the paper). [38-45] The results of our measurements summarized in Table 1 show different scenarios, where for these TCO films high electron Hall mobility (μHall) (e.g. IO:H: 104 cm²/V s), low mobility (i.e. ITO: 28 cm²/V s), and intermediate μHall (obtained for IZO: 61 cm²/V s) are obtained after the thermal annealing process. Charge carrier density Ne decreases in the case of IO:H but augment in both ITO and IZO films after annealing. As mentioned briefly above, the recorded drop in the Ne of IO:H films might be attributed to the OH radicals and H_2O molecules desorption triggered by the annealing process, as well as a possible oxidation of the films. In contrary, the crystallization, the decrease in Ne, and the probable hydrogen passivation at the grain boundaries could result in a frank improvement of the μ_{Hall}. Resistance (Rs), carrier concentration (Ne), and electron Hall mobility (μ_{Hall}) of annealed ITO, IO:H and IZO as a function of their exposure time at OTF. For ITO and IZO, Rs changes is within 15% from the initial value while for IO:H, the relative change is more than 38% after 15 weeks in outdoor conditions (Fig. 5(a)). For Ne, all films show less than 13% relative change (Fig. 5(b)). In terms of μ_{Hall} (Fig. 5(c)), we observe only 5% decrease after 15 weeks outdoor for ITO, 8 % in case of IZO and about 25% for IO:H/ITO. The increase of the carrier density of ITO results in a further stability improvement, with small change in hall mobility even after 15 weeks exposure.

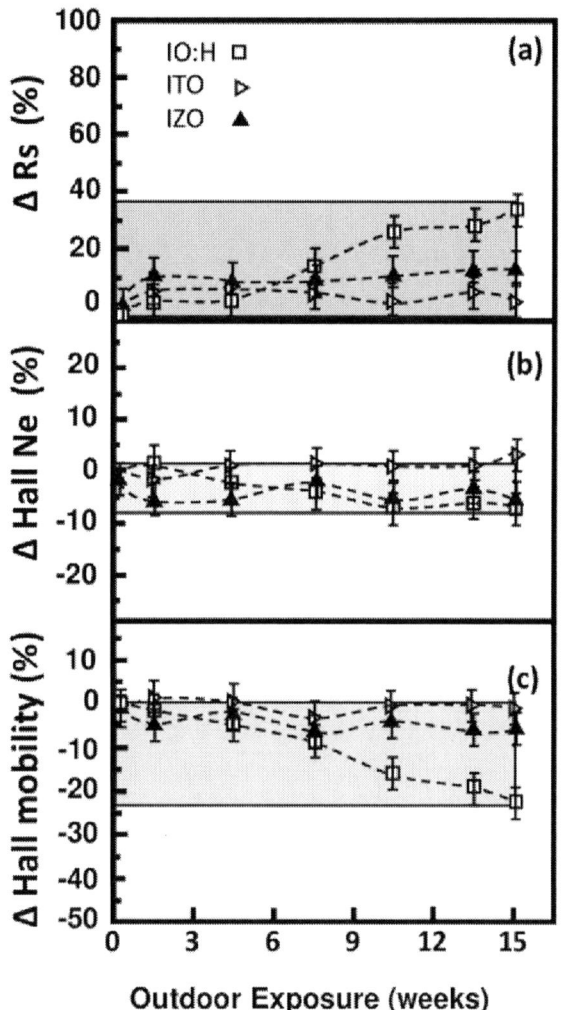

Fig. 6: Relative change in electrical properties of annealed TCOs w/r to exposure time (a) Rs, (b) Ne, and (c) μ_{Hall}.

For IO:H, we witness rather a marked reduction of μ_{Hall} (25% in relative change) after exposure to outdoor conditions, idem for the Rs degradation. Figure 6 displays the relative change (%) in sheet

IV. CONCLUSIONS

In summary, we have shown the influence of the oxygen content applied during DC magnetron sputtering deposition on the optical and carrier transport properties of the ITO thin films, and on the PV performance of the associated SHJ devices. We first explored a possible effect of the underlying substrates on the ITO structural properties by growing the films on naked glass substrate, and on n-doped amorphous and multicrystalline silicon substrates. The latter was found to favour the crystallinity of the ITO films, which was corroborated by the Hall effect measurement that showed a higher mobility of the ITO when deposited on multicrystalline silicon. All ITO films with $r(O_2) > 1\%$ presented a very low absorptance in the visible and NIR region of the spectra and the optical absorptance of the films in the NIR was found to strongly increase with decreasing oxygen, which could be associated to an increase of the carrier concentration as confirmed by Hall effect measurement. The lowest electrical resistivity was achieved for the lowest $r(O_2)$. With increasing the oxygen content, the Hall mobility increased and the highest mobility achieved for the ITO films was about 38 cm^2/Vs. Moreover, when applied as front contact in SHJ solar cells, the lower $r(O_2)$ into the ITO was found to yield to the best PV performance, which is attributed to lower resistive losses. The low temperature deposition and excellent optoelectronic properties of ITO confirm this TCO as one of the best front electrodes candidates not only for SHJ solar cells but for a large range of temperature-sensitive devices like organic LEDs and perovskite solar cells. Furthermore, we have studied the impact of outdoor conditions (sun, heat and humidity) characterizing the OTF site located in Doha (the state of Qatar) on the stability of the grown ITO, IO:H and IZO thin films deposited by magnetron sputtering and used as front contacts in solar cells. Our findings show that polycrystalline ITO and amorphous IZO films demonstrate rather a descent stability against the outdoor conditions, and are promising for large-scale deployment of optoelectronic devices, while polycrystalline IO:H films showed a clear degradation of its electronic properties over outdoor exposure. The degradation of IO:H films might be attributed to the H removal at the grain boundaries by attachment with chemisorbed OH- groups to create H_2O molecules, thereby increasing the potential barrier for electron transport and hence lowering the μ_{Hall}.

ACKNOWLEDGMENT

The authors thank PV-LAB (EPFL) in Neuchatel (Switzerland) for the PECVD deposition and Core labs (QEERI) in Qatar for the materials characterizations.

REFERENCES

[1] Best Research-Cell Efficiency Chart at: https://www.nrel.gov/pv/cellefficiency.html

[2] S. De Wolf, A. Descoeudres, Z. C. Holman, and C. Ballif, Green 2, 7 (2012).

[3] Z. C. Holman, M. Filipic, A. Descoeudres, S. DeWolf, F. Smole, M. Topic, and C. Ballif, J. Appl. Phys. 113, 013107 (2013).

[4] T. Koida, H. Fujiwara, and M. Kondo, Sol. Energy Mater. Sol. Cells 93, 851 (2009).

[5] M. Morales-Masis, S. Martin de Nicolas, J. Holovsky, S. De Wolf, and C. Ballif, IEEE J. Photovoltaics 5, 1340 (2015).

[6] L. Barraud, Z. C. Holman, N. Badel, P. Reiss, A. Descoeudres, C. Battaglia, S. De Wolf, and C. Ballif, Sol. Energy Mater. Sol. Cells 115, 151 (2013).

[7] T.Koida, H. Shibata, M.Kondo, K. Tsutsumi, A. Sakaguchi, M. Suzuki, and H. Fujiwara, J. Appl. Phys. 111, 063721 (2012).

[8] T. Koida, M. Kondo, K. Tsutsumi, A. Sakaguchi, M. Suzuki, and H. Fujiwara, J. Appl. Phys. 107, 033514 (2010).

[9] B. Macco, H. C. M. Knoops, and W. M. M. Kessels, ACS Appl. Mater. Interfaces 7(30), 16723 (2015).

[10] H. F. Wardenga, M. V. Frischbier, M. Morales-Masis, and A. Klein, Materials 8, 561 (2015).

[11] H. Fujiwara and M. Kondo, "Effects of carrier concentration on the dielectric function of ZnO:Ga and In$_2$O$_3$:Sn studied by spectroscopic ellipsometry: analysis of free-carrier and band-edge absorption," *Physical Review B,* vol. 71, pp. 075109–1–075109-10, 2005.

[12] S. DeWolf, A. Descoeudres, Z. C. Holman, and C. Ballif, "High-efficiency silicon heterojunction solar cells: a review," *Green,* vol. 2, 7–24, 2012.

[13] A.K. Gnosh, C. Fishman, and T. Feng, "SnO$_2$/Si solar cells—heterostructure or Schottky-barrier or MIS-type device," *Journal of Applied Physics,* vol. 49, pp. 3490, 1978.

[14] T. Feng, A.K. Gnosh, and C. Fishman, "Efficient electron beam deposited ITO/n-Si solar cells," *Journal of Applied Physics,* vol. 50, pp. 4972, 1979.

[15] T. Feng, A.K. Gnosh, and C. Fishman, "Structure, photovoltaic properties, and angle-of-incidence correlations of electron-beam-deposited SnO$_2$/n-Si solar cells," *Journal of Applied Physics,* vol. 50, pp. 8070, 1979.

[16] H. Kobayashi, T. Ishida, Y. Nakato, and H. Tsubomura, "Mechanism of carrier transport in highly efficient solar cells having indium tin oxide/Si junctions," *Journal of Applied Physics,* vol. 69, pp. 1736, 1991.

[17] M. Tanaka, M. Taguchi, T. Matsuyama, T. Sawada, S. Tsuda, S. Nakano, H. Hanafuza, and Y. Kuwano, "Development of new a-Si/c-Si heterojunction solar cells: ACJ-HIT (artificially constructed junction-heterojunction with intrinsic thin-Layer)," *Japanese Journal of Applied Phyics,* vol. 31, pp. 3518, 1992.

[18] M. Moreno, M. Labrune, and P. Rocai Cabarrocas, "Dry fabrication process for heterojunction solar cells through in-situ plasma cleaning and passivation," *Solar Energy Materials and Solar Cells,* vol. 94, pp. 402-405, 2010.

[19] A.G. Ulyashin, R. Job, M. Scherff, M. Gao, W.R. Fahrner, D. Lyebyedyev, N. Roos, and H.C. Scheer, "The influence of the amorphous silicon deposition temperature on the efficiency of the ITO/A-Si:H/C-Si heterojunction (HJ) solar cells and properties of interfaces," *Thin Solid Films,* vol. 403, pp. 359- 362, 2002.

[20] M.P. Houng, K.C. Lai, J.H. Wang, C.H. Lu, F.J. Tsai, and C.H. Yeh, "Plasma-induced TCO texture of ZnO:Ga back contacts on silicon thin film solar cells," *Solar Energy Materials and Solar Cells,* vol. 95, pp. 415-418, 2011.

[21] A. Klein, C. Korber, A. Wachau, F. Sauberlich, Y. Gassenbauer, R. Schafranek, S.P. Harvey, and T.O. Mason, "Surface potentials of magnetron sputtered transparent conducting oxides," *Thin Solid Films,* vol. 518, 1197-1203, 2009.

[22] A. Chen and K. Zhu, "Computer simulation of a-Si/c-Si heterojunction solar cell with high conversion efficiency," *Solar Energy,* vol. 86, pp. 393-397, 2012.

[23] L. Zhao, C.L. Zhou, H.L. Li, H.W. Diao, and W.J. Wang, "Role of the work function of transparent conductive oxide on the performance of amorphous/crystalline silicon heterojunction solar cells studied by computer simulation," *Physica Status Solidi A,* vol. 205, pp. 1215-1221, 2008.

[24] E. Centurioni and D. Iencinella, "Role of front contact work function on amorphous silicon/crystalline silicon heterojunction solar cell performance," *IEEE Electron Device Letters,* vol. 24, pp. 177-179, 2003.

[25] Y.J. Kim, S.B. Jin, S.I. Kim, Y.S. Choi, I.S. Choi, and J.G. Han, "Effect of oxygen flow rate on ITO thin films deposited by facing targets sputtering," *Thin Solid Films,* vol. 518, pp. 6241-6244, 2010.

[26] M. Bender, W. Seelig, C. Daube, H. Frankenberger, B. Ocker, and J. Stollenwerk, "Dependence of oxygen flow on optical and electrical properties of DC-magnetron sputtered ITO films," *Thin Solid Films,* vol. 326, pp. 72-77, 1998.

[27] B. Zhang, X. Dong, X. Xu, P. Zhao, and J. Wu, "Characteristics of zirconium-doped indium tin oxide thin films deposited by magnetron sputtering," *Solar Energy Materials and Solar Cells,* vol. 92, pp. 1224-1229, 2008.

[28] A. Thøgersen, M. Rein, E. Monakhov, J. Mayandi, and S. Diplas, "Elemental distribution and oxygen deficiency of magnetron sputtered indium tin oxide films," *Journal of Applied Physics,* vol. 109, pp. 113532, 2011.

[29] S. Li, X. Qiao, and J. Chen, "Effects of oxygen flow on the properties of indium tin oxide films," *Materials Chemistry and Physics,* vol. 98, pp. 144-147, 2006.

[30] M.-C. Chen and S.-A. Chen, "Influence of oxygen deficiency in indium tin oxide on the performance of polymer light-emitting diodes," *Thin Solid Films* 517, 2708–2711 (2009).

[31] C.G. Choi, K. No, W.-J. Lee, H.-G. Kim, S.O. Jung, W.J. Lee, W.S. Kim, S.J. Kim, and C. Yoon, "Effects of oxygen partial pressure on the microstructure and electrical properties of indium tin oxide film prepared by d.c. magnetron sputtering," *Thin Solid Films,* vol. 258, pp. 274-278 ,1995.

[32] J.S. Kim, P.K.H. Ho, D.S. Thomas, R.H. Friend, F. Cacialli, G.W. Bao, and S.F.Y. Li, "X-ray photoelectron spectroscopy of surface-treated indium-tin oxide thin films," *Chemical Physics Letters,* vol. 315, pp. 307-312, 1999.

[33] J.C.C. Fan and J.B. Goodenough, "X-ray photoemission spectroscopy studies of Sn-doped indium-oxide films," *Journal of Applied Physics,* vol. 48, pp. 3524 ,1977.

[34] R.X. Wang, C.D. Beling, A.B. Djurisic, S. Li, and S. Fung, "Properties of ITO thin films deposited on amorphous and crystalline substrates with e-beam evaporation," *Semiconductor Science and Technology,* vol. 19, pp. 695—698, 2004.

[35] A. Thøgersen, M. Rein, E. Monakhov, J. Mayandi, and S. Diplas, "Elemental distribution and oxygen deficiency of

magnetron sputtered indium tin oxide films," *Journal of Applied Physics*, vol. 109, pp. 113532, 2011.

[36] J.P. Perdew, K. Burke, and M. Ernzerhof, "Generalized gradient approximation made simple," *Physical Review Letters,* vol. 77, pp. 3865-3868, 1996.

[37] N. Kurita, N. Fukatsu, H. Otsuka, and T. Ohashi, "Measurements of hydrogen permeation through fused silica and borosilicate glass by electrochemical pumping using oxide protonic conductor," *Solid State Ionics*, vol. 146, 101-111, 2002.

[38] R. Martins, P. Almeida, P. Barquinha, L. Pereira, A. Pimentel, I. Ferreira, and E. Fortunato, "Electron transport and optical characteristics in amorphous indium zinc oxide films," *Journal of Non-Crystalline Solids,* vol. 352, pp. 1471-1474, 2006.

[39] L. Ding, S. Nicolay, J. Steinhauser, U. Kroll, and C. Ballif, "Relaxing the conductivity/transparency trade-off in MOCVD ZnO thin films by hydrogen plasma," *Advanced Functional Materials,* vol. 23, pp. 5177-5182, 2013.

[40] M. Iqbal, M.M. Nauman, F.U. Khan, P.E. Abas, Q. Cheok, A. Iqbal, B. Aissa, "Vibration-based piezoelectric, electromagnetic, and hybrid energy harvesters for microsystems applications: A contributed review, International journal of energy research 45 (1), 65-102, https://doi.org/10.1002/er.5643

[41] H. Gavi, Balla D. Ngom, A.C. Beye, A.M. Strydom, B. Aissa, V.V. Srinivasu, M. Chaker, N. Manyala, "Low-field microwave absorption in pulse laser deposited FeSi thin film, Journal of Magnetism and Magnetic Materials 324 (6), 1172-1176, https://doi.org/10.1016/j.jmmm.2011.11.003

[42] J. Haschke, R. Lemerle, B. Aïssa, A.A. Abdallah, M.M. Kivambe, M. Boccard, C. Ballif, "Annealing of silicon heterojunction solar cells: Interplay of solar cell and indium tin oxide properties, " IEEE Journal of Photovoltaics 9 (5), 1202-1207, DOI: 10.1109/JPHOTOV.2019.2924389

[43] B. Aïssa, M.A. El Khakani, "The channel length effect on the electrical performance of suspended-single-wall-carbon-nanotube-based field effect transistors," Nanotechnology 20 (17), 175203, https://doi.org/10.1088/0957-4484/20/17/175203

[44] A. Bentouaf, R. Mebsout, H. Rached, S. Amari, A.H. Reshak, B. Aïssa "Theoretical investigation of the structural, electronic, magnetic and elastic properties of binary cubic C15-Laves phases TbX2 (X = Co and Fe)," Journal of alloys and compounds 689 (25), 885-893, https://doi.org/10.1016/j.jallcom.2016.08.046

[45] R. Nechache, M. Nicklaus, N. Diffalah, A. Ruediger, F. Rosei, "Pulsed laser deposition growth of rutile TiO2 nanowires on Silicon substrates," Applied Surface Science 313, 48-52

Influence of Business Models on PV-Battery Dispatch Decisions and Market Value: A Pilot Study of Operating Plants

Joachim Seel, Cody Warner, Andrew Mills

Lawrence Berkeley Lab, Berkeley, CA, United States

University of California, Berkeley, Berkeley, CA, United States

PV-battery hybrid projects dominate interconnection queues in some regions in the United States. But few large-scale projects have been in use long enough to assess how the hybrid capabilities may be used in practice and the existing literature rarely discusses observed operational strategies. We interview plant operators and analyze empirical dispatch data for eleven large-scale PV-battery hybrids in three organized wholesale markets in the United States. We find the empirical increase in market value of a PV-battery hybrid relative to a standalone PV plant varies by project and ranges from $1 to $48 per MWh, often aided by a large boost in capacity value. This premium is driven by market, location, technical characteristics of the PV and battery asset, and battery dispatch strategies. In contrast to the PV-battery hybrid modeling literature, only three of the eleven project operators optimize battery usage for wholesale market revenue. Instead, load-serving entities target peak load reductions, incentive program participants focus on compliance with program requirements, and large energy consumers prioritize resiliency and utility bill minimization. These alternative business models can result in high revenues for the project operators but do not optimize the storage dispatch from a grid perspective. Understanding real-world dispatch signals and aligning them with grid needs will be critical for electric grid operators and system planners.

Wide Bandgap AlGaInP-based Photovoltaic Cell for Indoor Ambient Energy Harvesting

Aditya Prabaswara [†], Jack Browne[†], Yongjie Zou[‡], Richard King[‡], Stephen Goodnick[‡], Brian Corbett[†]

[†]Tyndall National Institute, University College Cork, Cork T12 R5CP, Ireland

[‡]School of Electrical, Computer, and Energy Engineering, Arizona State University, Tempe, Arizona 85287-5706, USA

Abstract—**Wide bandgap III-V materials are suitable as an efficient absorber for indoor photovoltaic (IPV) cell as they can cover the 2.0 eV bandgap required for maximum efficiency. In this work, we present our progress on solving the challenge associated with the development of III-V IPV cell, namely (i) design of efficient IPV cell structure, (ii) nanosphere lithography-based surface roughening to enhance light trapping, and (iii) chemical passivation to suppress nonradiative sidewall recombination. Our result highlights the cell design and treatment required to realize efficient wide bandgap III-V indoor photovoltaic cell.**

Index Terms—**indoor photovoltaics, wide bandgap, AlGaInP**

I. INTRODUCTION

The internet of things (IoT) is a broad term that define a vast interconnected network of devices, enabling the combination of thing-based physical functions and IT-based digital services. As more and more devices are connected into the network, the availability of a reliable power source becomes crucial. Currently, these devices rely on battery power, which places a constraint on their functionality due to limited battery life.

Indoor photovoltaics (IPV) cells has the potential to solve this issue by performing indoor light harvesting. cm^2-sized devices would be able to power individual sensor nodes, acting as a reliable power source under artificial light sources such as compact fluorescent (CFL), halogen, and light emitting diode (LED) bulbs. The market size for IPV is predicted to reach US $850 million by 2023 [1].

To realize highly efficient cells, the IPV cell should be based on wide bandgap material, close to the 2.0 eV optimum. The III-V material system, specifically AlGaInP, is suitable for this purpose. AlGaInP is widely used in red LEDs, and can be grown lattice-matched to GaAs substrates, ensuring high material quality. Furthermore, III-V materials can be processed to fabricate bio-compatible devices for powering miniature-sized sensors [2]. Currently, there is a push towards a universally accepted method for IPV cell measurement in order to have a better comparison between IPV cell performance [3], [4].

Despite their use as a component for red LEDs, not much work has been done for PV cells based on wide bandgap AlGaInP material. Furthermore, several limitations must be overcome in order to realize highly-efficient AlGaInP based IPV cells. The first limitation is the large (30%) Fresnel reflection at the interface between the cell surface and air, due to the large refractive index of III-V material [5]. The second limitation is the large surface recombination at the cell sidewalls, which becomes more severe as the cell size is reduced.

In this work, we report on the solution for the technological limitations related to AlGaInP-based IPV cells. The IPV cell structure was designed and simulated using a commercially available Silvaco TCAD software suite. A commercial multiple quantum well (MQW) red LED wafer (Kingsoon, China) was used to study the effect of surface roughening on the cell reflectivity and the effect of chemical passivation on surface recombination. Our work shows the design considerations and device processing required to realize a highly-efficient AlGaInP-based IPV cell.

II. DESIGN OF ALGAINP-BASED INDOOR PHOTOVOLTAIC CELL

To determine the best configuration for the AlGAInP-based cell, we performed a device simulation using a comercially available Silvaco TCAD simulation package. The cells considered here are n on p structures with two different configurations, which are the standard upright cell and the rear junction cell. The rear junction cell has been shown to exhibit better fill factor and higher V_{OC} [6], making it suitable for powering IoT devices.

The cell configurations are based on InGaP solar cells developed by the National Renewable Energy Laboratory (NREL) [7]. No antireflective coating is used. The structure is optimized as an upright cell, and the rear junction cell is designed using the same parameters for the window and BSF layers. The cell designs are shown in Fig. 1. The structures are simulated using a 3000K white light spectrum provided by the National Institute of Standards and Technology (NIST), as the input [3]. The input power is set at 2.9267 W/m^2.

TABLE I
COMPARISON OF IPV CELL SIMULATION RESULTS UNDER 3000K INDOOR WHITE LIGHT ILLUMINATION

	Upright	Rear Junction
Jsc (mA/cm^2)	0.0418	0.04126
Voc (V)	1.4925	1.505
Fill factor (%)	91.26	91.25
Efficiency (%)	19.43	19.37

The simulation result shows that both device configuration can reach about 19.4% efficiency under ambient lighting. The rear junction device exhibits higher V_{OC} when compared to the upright device configuration, but worse overall efficiency due to reduced J_{SC}. Further optimization of the layer thickness

(a)

| GaAs contact 5e18n |
| AlInP LM window 25 nm 2e18n |
| AlInGaP (23% Al) emitter 90 nm 9e17n |
| AlInGaP (23% Al) base 900 nm 5e16p |
| AlInGaP (54% Al) BSF 200nm 2e18p |
| AlGaAs lower clad layer 500 nm |
| GaAs buffer 200 nm |
| P-doped GaAs substrate |

Upright

(b)

| GaAs contact 5e18n |
| AlInP LM window 80 nm 2e18n |
| AlInGaP (23% Al) emitter 1000 nm 1e16n |
| AlInGaP (23% Al) base 27 nm 5e16p |
| AlInGaP (54% Al) BSF 200nm 2e18p |
| AlGaAs lower clad layer 500 nm |
| GaAs buffer 200 nm |
| P-doped GaAs substrate |

Rear junction

Fig. 1. Indoor photovoltaic cell designs used in the simulation. (a) upright cell and (b) rear junction cell.

and doping level should be done to improve the performance of the rear junction cell.

From our simulation and optimization process, we find that using lower doping level on the absorber helps increase the efficiency by increasing carrier mobility. The window layer should be thick enough and highly doped to contain the Fermi pinning at the surface, but without causing parasitic absorbance at shorter wavelength.

III. NANOSPHERE LITHOGRAPHY-BASED SURFACE ROUGHENING FOR ENHANCED LIGHT EXTRACTION

To study the effect of reflection at the interface and non-radiative recombination at the perimeter, we fabricated IPV cells using a commercial MQW LED epi structure (Kingsoon, China). Due to its high refractive index in the visible wavelength (n=3.5), large Fresnel reflection occurs at the interface between the III-V IPV cell surface and air. For PV cells with high internal luminescence efficiency such as GaAs, enhancing the luminescence extraction efficiency has been shown to be crucial to improve the overall efficiency of the IPV cell [8]. Under light trapping condition, the maximum

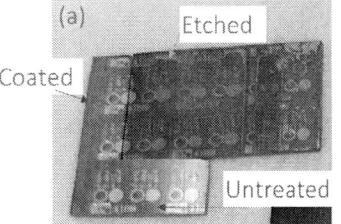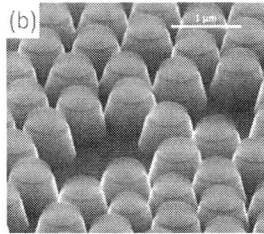

Fig. 2. (a) Photograph of fabricated samples with bare, coated, and etched surface. (b) SEM micrograph of etched sample surface, showing conical shapes under where the nanospheres were.

amount of voltage boost ΔV is expected fall between $\ln\{n^2\} < (q\Delta V/kT) < \ln\{4n^2\}$, where n is the material refractive index, k is the Boltzmann constant, and T is the temperature.

By performing surface roughening, the probability of photons generated by radiative recombination escaping the cell surface increases, resulting in increased extraction efficiency. Surface roughening was done on the IPV cell surface using nanosphere lithography as a mask [9].

Three IPV cells are fabricated to compare the effect of surface roughening on device performance. The first sample is fabricated without any surface roughening. The second sample (coated) has nanosphere coated on its surface, where silica nanospheres with 750 nm diameter were deposited on the top surface of coated sample using dip-coating technique. The third sample (etched) is coated with nanosphere, followed by surface roughening using inductive coupled plasma (ICP) dry etching. Following dry etching, any remaining nanosphere is removed using chemical etching. The sample photograph is shown in Fig. 2 (a), where the etched sample shows darker surface indicating decreased reflectivity. SEM micrograph of the etched sample surface is shown in 2 (b), where cone-shaped structures are formed underneath where the nanospheres were.

The EQE of the samples are shown in Fig. 3. In general, the EQE of the AlGaInP-based cells increases following nanosphere coating and surface roughening and shifts to longer wavelength. The absorption edge on the long wavelength side is due to the absorption in the MQW region, while the absorption on the short wavelength side is due to absorption within the LED's AlInP cladding layer.

IV. SUPPRESSION OF SIDEWALL RECOMBINATION USING CHEMICAL PASSIVATION

One particular issue with sub-mm IPV cells is the large nonradiative recombination at the cell perimeter. After surface roughening, the dark current in 1000 μm diameter device increases from < 100 pA to 3 nA. This increase is caused by increased nonradiative recombination on the cell sidewalls. As the cell size is reduced, the effect of nonradiative surface recombination becomes more severe, preventing cell miniaturization.

To quantify the effect of surface recombination, we fabricated cells with different diameters. We performed electrical characterization and fit the measured current density vs voltage

978-1-7281-6118-1/22 $31.00 © 2022 IEEE

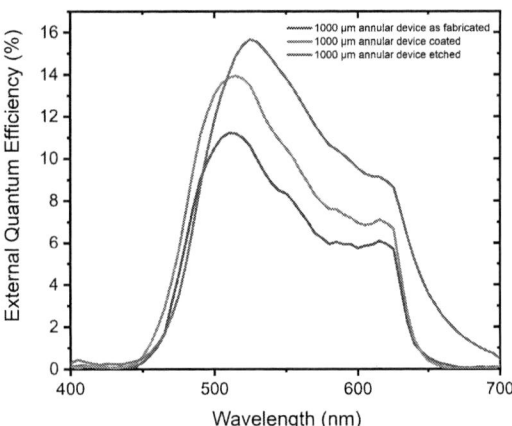

Fig. 3. External quantum efficiency comparison between bare sample, coated sample, and etched sample.

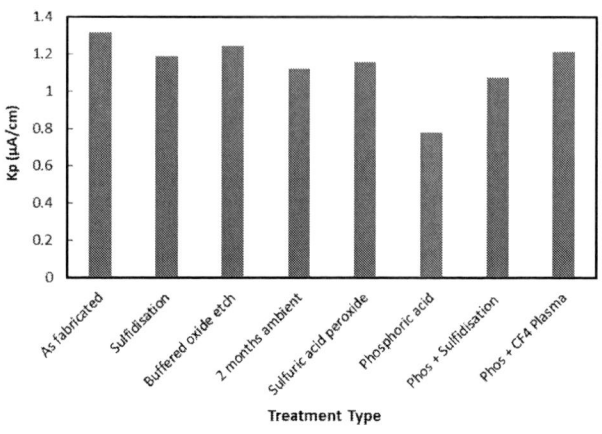

Fig. 4. Comparison of K_p obtained after passivating the cell with different chemical treatments. The graphs include as-fabricated cells , treatment by ammonium sulphide (sulfidisation), buffered oxide etchant (BOE), sample left in ambient condition for two months, sulfuric acid and peroxide mixture, phosphoric acid, phosphoric acid followed by sulfidisation, and phosphoric acid followed by CF$_4$ plasma treatment (Phos CF4).

to a two-diode model with ideality factor n=1 and n=2. We then used the following equation to evaluate the surface recombination:

$$J_{total} = J_{bulk} + K_P \frac{P}{A} \qquad (1)$$

Where J_{total} is the current density obtained by using a fit to the n=2 component, J_{bulk} is current density due to recombination at the material bulk, P is the cell perimeter length, A is the total cell area. The influence of the surface is described experimentally by the term K_P with the unit A cm^{-1}, where higher K_P indicates higher surface recombination.

Currently, there is no effective sidewall passivation scheme for AlGAInP-based devices. We performed several passivation experiments using different chemicals and calculate K_P to assess which method would give the best result. The results are summarized in Fig. 4. From our experiment, phosphoric acid treatment gives the most promising result for sidewall treatment, and can help increase the efficiency of IPV cells.

V. CONCLUSIONS

We have investigated the technological challenges of using wide band gap III-V materials as an absorber for IPV cell. We investigated two cell configurations using upright cell design and rear junction cell designs, with the rear junction configuration giving better fill factor and open circuit voltage at the cost of lower short circuit current. Two main limitations associated with using AlGAInP as IPV cells were identified, namely large reflection at the front surface and surface recombination at the cell perimeters. Nanosphere lithography was used to perform surface roughening, resulting in decreased reflectivity and increased EQE, while chemical passivation treatment was performed to suppress nonradiative recombination at the perimeter. Quantum well-based IPV cells are promising for the realization of IPV cells. Our work highlights the design

considerations and device treatment required to realize highly efficient III-V wide bandgap IPV cell.

ACKNOWLEDGMENT

The project was funded by Science Foundation Ireland under 19/US-C2C/3579 and 12/RC/2276_P2. This work was partially supported through the Center to Center Program through National Science Foundation (NSF) and the Department of Energy (DOE) under NSF CA No. EEC-1041895.

REFERENCES

[1] I. Mathews, S. N. Kantareddy, T. Buonassisi, and I. M. Peters, "Technology and Market Perspective for Indoor Photovoltaic Cells," *Joule*, vol. 3, no. 6, pp. 1415–1426, 2019.

[2] E. Moon, D. Blaauw, and J. D. Phillips, "Subcutaneous Photovoltaic Infrared Energy Harvesting for Bio-implantable Devices," *IEEE Transactions on Electron Devices*, vol. 64, no. 5, pp. 2432–2437, may 2017.

[3] B. H. Hamadani and M. B. Campanelli, "Photovoltaic Characterization under Artificial Low Irradiance Conditions Using Reference Solar Cells," *IEEE Journal of Photovoltaics*, vol. 10, no. 4, pp. 1119–1125, 2020.

[4] B. H. Hamadani, Y.-s. Long, M.-a. Tsai, and T.-c. Wu, "Interlaboratory Comparison of Solar Cell Measurements Under Low Indoor Lighting Conditions," *IEEE Journal of Photovoltaics*, pp. 1–6, 2021.

[5] Y. M. Song, E. S. Choi, J. S. Yu, and Y. T. Lee, "Light-extraction enhancement of red AlGaInP light-emitting diodes with antireflective subwavelength structures," *Optics Express*, vol. 17, no. 23, p. 20991, nov 2009.

[6] G. Bauhuis, P. Mulder, Y.-Y. Hu, and J. Schermer, "Deep junction III-V solar cells with enhanced performance," *physica status solidi (a)*, vol. 213, no. 8, pp. 2216–2222, aug 2016.

[7] E. E. Perl, J. Simon, J. F. Geisz, W. Olavarria, M. Young, A. Duda, D. J. Friedman, and M. A. Steiner, "Development of High-Bandgap AlGaInP Solar Cells Grown by Organometallic Vapor-Phase Epitaxy," *IEEE Journal of Photovoltaics*, vol. 6, no. 3, pp. 770–776, may 2016.

[8] V. Ganapati, M. A. Steiner, and E. Yablonovitch, "The Voltage Boost Enabled by Luminescence Extraction in Solar Cells," *IEEE Journal of Photovoltaics*, vol. 6, no. 4, pp. 801–809, 2016.

[9] C. Zhang, S. Cvetanovic, and J. M. Pearce, "Fabricating ordered 2-D nano-structured arrays using nanosphere lithography," *MethodsX*, vol. 4, pp. 229–242, 2017.

A comparison study of the performance of vertical vs single axis tracking bifacial agrivoltaic systems in Belgium

Brecht Willockx, Jan Cappelle

KU Leuven, Ghent, Belgium

In this paper, the performances of the two bifacial agrivoltaic systems is compared in Grembergen, Belgium: a fixed vertical set-up and a dynamic single axis tracker. A novel tracking algorithm is presented, including crop specific critical periods and time dependent economic electricity values. The vertical set-up has the lowest operations costs; However, the tracker exceeds drastically (+20% in winter months and +50% in perfect tracking summer months) the energy production return and consequently the lowest LCOE . For places with a moderate climate, and light competition between PV and crop growth, the light availability is considered as key indicator for the crop production. Both the fixed vertical and dynamic tracker system limited the ground light reductions, with exception to the close area around the structures itself. This can be solved with additional shade loving flower strips.

Power Factors 2022 PV System Efficiency Benchmarks

Stephen Lightfoote, Samantha Wilson Ph.D, Steve Voss

Power Factors Inc., Brossard, QC, J4Z 1A7, Canada

Abstract— **This report presents benchmark analysis of inverter DC to AC conversion efficiency and AC side collection system efficiency characterized from field measurements in Power Factors database. It represents an update and extension to our 2021 study [10] with similar objectives which aimed to benchmark inverter efficiency across multiple dimensions (time, age, manufacturer, model), validate manufacturer specifications and inform energy modelers tasked with predicting inverter performance from manufacturer specifications. Major updates include incorporating additional field measurements from more devices over a longer period of time (including 2021 data), summarizing results of AC side collection system efficiency, updating methodology to use the Sandia inverter model [8] to characterize inverter efficiency and some additional data validation quantifying the impact of measurement bias. It was found that the majority of the most common inverter models perform within 1.0% of specifications, with a mean across all inverters of 97.5%. Additional analysis of AC side collection efficiency indicates that these losses are negligible with a mean of 99.5%.**

Keywords—

I. INTRODUCTION

Inverters are key components of solar energy production, but most solar developers do not require independent verification of inverter performance [1]. Module performance is often verified by independent test labs, but inverter specifications are usually accepted with no additional testing. Additionally, although there has been some lab testing of inverter performance, reports on inverters in commercial operation are sparse [2-5]. Power Factors endeavors to correct the lack of information on asset field performance through insights gleaned from our large database of commercial photovoltaic plants. Through our Drive asset performance management software, we provide customers with industry leading analytics and with our database we can compare predicted versus actual performance across the world [6]. This report will focus on how one aspect of this database was developed and some preliminary results and applications of the data.

Inverters are well known for requiring more service calls and having a smaller mean-time-to-failure than modules [2]. This is unsurprising given that they are made of hundreds more components and perform many more functions than a PV module [3]. Due to the complexity of their operation, inverter specifications are multifaceted and difficult to test. They cover inverter efficiency versus input power, inverter AC capacity versus temperature and elevation, operating voltage windows etc. [7]. This paper will focus on energy conversion efficiency of inverters in the field and will compare measured efficiency against manufacturer specifications. A recent report by PV Evolution Labs found that a significant number of inverters are deficient in some aspect of their specified performance, including 20% of inverters that deviated from their published efficiencies by greater than 1% [1]. Previous, smaller reviews of inverter field performance have also consistently found some inverters are underperforming efficiency values [5].

This paper represents an update to our 2021 study [10] and presents detailed findings from >200,000 months of inverter operation from > 10,000 inverters within ~1000 solar power plants distributed all over the world. Results indicate that most inverters perform within 1.0%% of specification, newer inverter models with higher specifications tend to perform better and that performance as a function of time in the field appears to be stable.

II. METHODS

AC and DC data were collected for each qualifying inverter in the Power Factors database at data resolutions of 5-15 minutes using measurements taken from the period of January 2018 – December 2021. Monthly inverter efficiency for a given inverter was characterized using the Sandia Inverter Performance Model[8] implemented in pvlib [9]. An example model fit is shown in Figure 1**Error! Reference source not found.**.

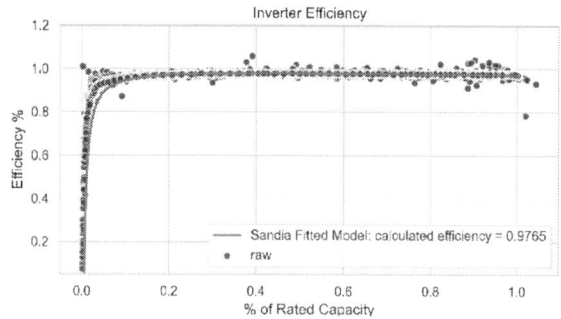

Figure 1: Example Characterized Inverter Efficiency using Sandia Inverter Performance Model

Additional post-processing filters were applied to the aggregated statistics to control for inverter x months with data quality issues (i.e., poor regression fit, non-physical results) or insufficient samples.

AC side collection system efficiency was characterized by the slope of the linear model between the plant's revenue meter energy and the sum of the inverters' energy, filtered for periods where all inverters are in communication. An example model fit is shown in Figure 2.

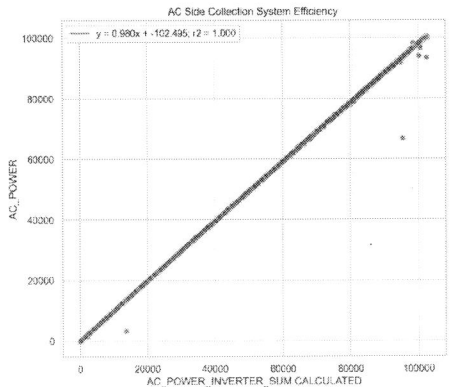

Figure 2: Example Characterized AC side collection efficiency

Inverter manufacturer and model are based on the plant schematics provided by clients [6]. Manufacturer specified efficiency is based on the California Energy Commission (CEC) database [7].

III. RESULTS AND DISCUSSION

Overall characterized distribution statistics for inverter x months and plant x months are shown in Figure 3 and Figure 4 respectively.

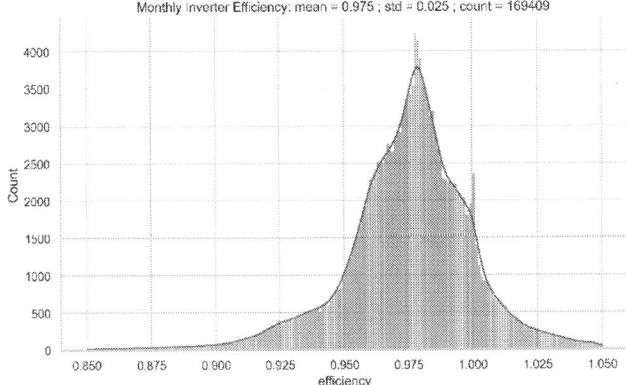

Figure 3: Global Monthly Inverter Efficiency Benchmark

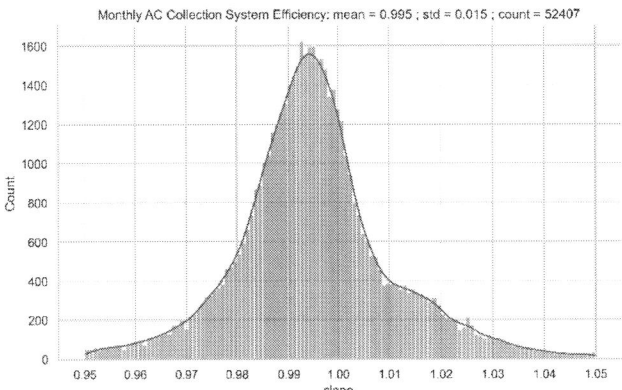

Figure 4: Global Monthly AC side Collection System Efficiency Benchmark

It was found that there is a weak correlation between the characterized inverter efficiency and the resulting AC collection system efficiency (Figure 5). A possible inference here is that the measured inverter efficiency is partially a function of bias in the inverter's measurements where a higher(lower) characterized inverter efficiency equates to a slightly lower(higher) characterized AC collection system efficiency.

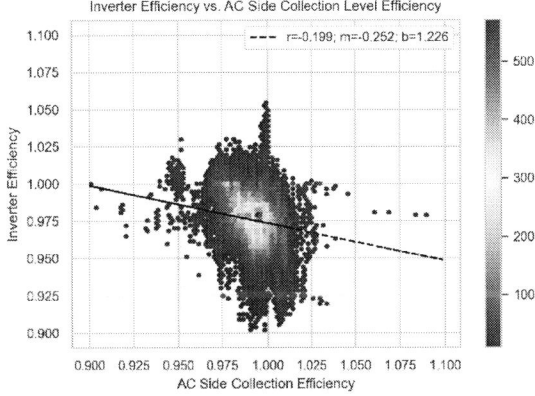

Figure 5: Inverter Efficiency vs. AC side collection system efficiency

Despite this correlation, it was found that applying this linear model bias to the inverter efficiency had a negligible impact on the resulting inverter efficiency distribution, reducing the standard deviation from 0.24% to 0.23%. As such, while inverter measurement bias can impact individual inverter months (as evidenced by efficiencies > 100%), it is unlikely that systematic inverter measurement bias is impacting the central tendency metrics.

We also updated our previous results of comparing inverter efficiency against age of the system where we concluded, perhaps prematurely, that inverter efficiency of newer models was increasing. Updated results continue to show a strong increase in efficiency from 2015 – 2019 vintage inverters (which constitute the majority of inverters in the distribution) but the trend is does not appear to hold for years prior to 2015 nor for 2020. It is of note that these years constitute substantially lower sample counts, so perhaps these years are not representative of the industry as a whole.

978-1-7281-6118-1/22 $31.00 © 2022 IEEE

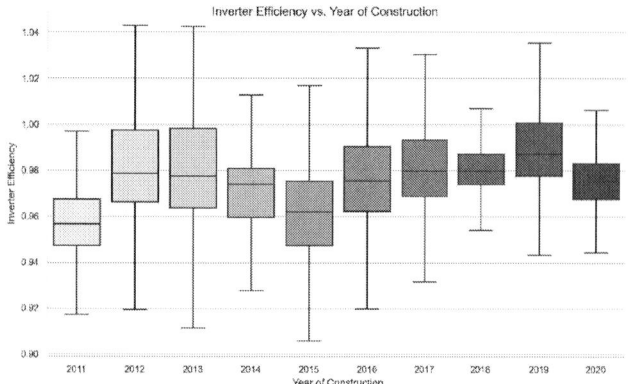

Figure 6: Inverter Efficiency vs. Year of Construction (COD)

Lastly, characterized inverter efficiency distributions and their deviations from CEC Spec Sheet Efficiencies were broken down by Original Equipment Manufacturer (OEM) in Figure 7 and Figure 8 and model in Figure 9 and Figure 10. The make and models have been anonymized and only manufacturers (models) with >1000 (500) inverter months are shown.

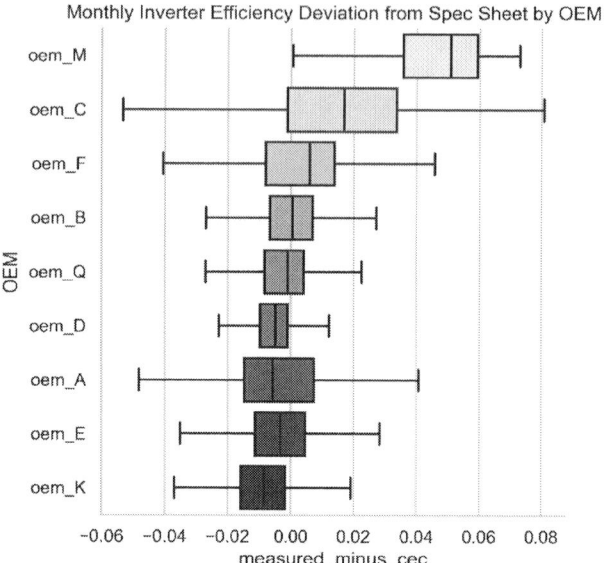

Figure 8: Monthly Inverter Efficiency Deviation from CEC by OEM

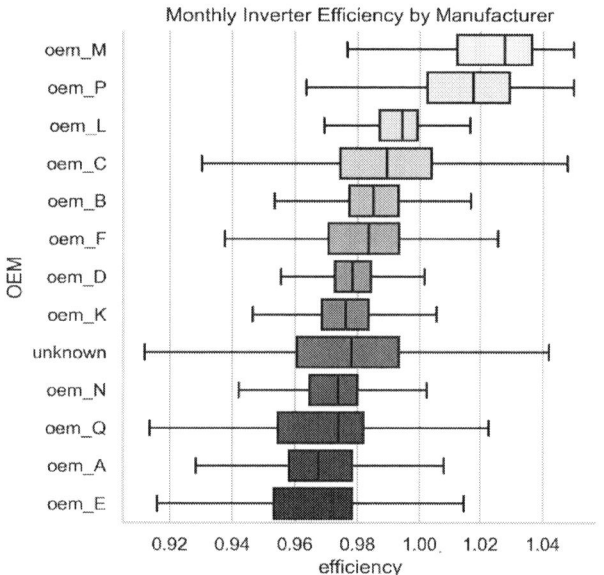

Figure 7: Monthly Inverter Efficiency by Manufacturer

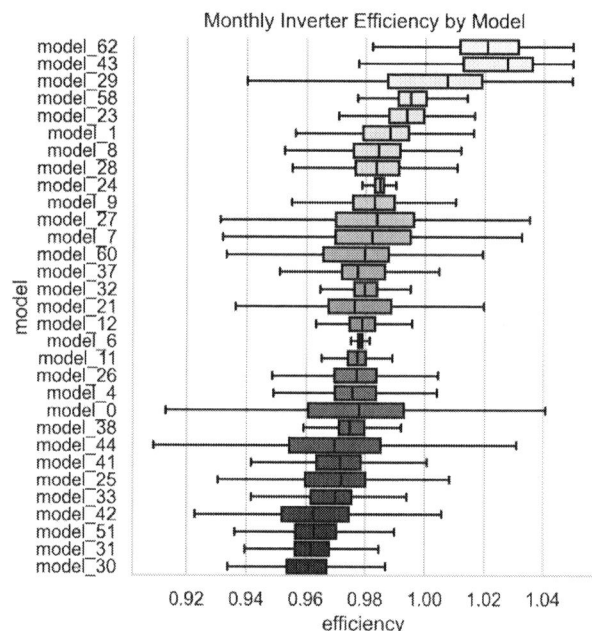

Figure 9: Monthly Inverter Efficiency by OEM Model

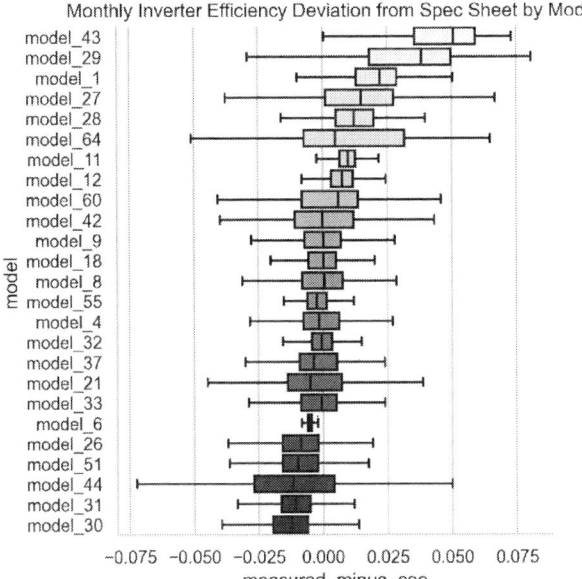

Figure 10: Monthly Inverter Efficiency Deviation from CEC by OEM Model

From these figures we can discern that the majority of inverter manufacturer and models perform in line with their spec sheet efficiencies, with a few exceptions at both ends of the spectrum. It is assumed that inverter models with efficiencies systematically > 100% are subject to measurement bias.

IV. CONCLUSIONS

Inverter energy conversion efficiency was characterized from >200,000 months of inverter operation from > 10,000 inverters within ~1000 solar power plants distributed all over the world in the Power Factors database. It was found that the majority of the most common inverter models perform within 1% of specifications, with a mean across all inverters of 97.5%. Additional analysis of AC side collection efficiency indicates that these losses are negligible on average with a mean of 99.5%. All of these conclusions are facilitated by the size of the Power Factors photovoltaic field performance database, which allows analysis at scales rarely attempted. Future planned work consists of additional data validation and applications of this dataset operationally, including, but not limited to OEM and O&M oversight, vendor selection, fleet performance metrics and feedback for energy modeling assumptions.

REFERENCES

[1] T. Doyle, R. Desharnais, and M. Mills-Price, "2019 PV Inverter Scorecard," https://www.pvel.com/inverter-scorecard/

[2] A. Nargarajan, R. Thiagarajan, I. Repins, and P. Hacke, "Photovoltaic Inverter Reliability Assessment" Golden, CO: National Renewable Energy Laboratory

[3] J. Flicker, and S. Gonzalez, "Performance and reliability of PV inverter component and systems due to advanced inverter functionality" Proceedings of 42nd IEEE PVSC

[4] P. Hacke, S. Lokanath, P. Williams, A. Vasan, P. Sochor, G. TamizhMani, H. Shinohara, and S. Kurtz, "A status review of photovoltaic power conversion equipment reliability, safety, and quality assurance protocols" Renewable and Sustainable Energy Reviews, Vol. 82, pp. 1097-1112, Feb. 2018.

[5] N. Allet, F. Baumgartner, J. Sutterlueti, S. Sellner, and M. Pezzotti, "Inverer performance under field conditions" Proceddings from the 27th PVSEC

[6] Power Factors Inc. https://pfdrive.com/

[7] California Energy Commision, "Grid Support Inverter List Simplified Data ADA.xlsx" https://www.energy.ca.gov/media/2

[8] D. King, S. Gonzalez, G. Galbraith, W. Boyson, "Performance Model for Grid-Connected Photovoltaic Inverters", SAND2007-5036, Sandia National Laboratories.

[9] William F. Holmgren, Clifford W. Hansen, and Mark A. Mikofski. "pvlib python: a python package for modeling solar energy systems." Journal of Open Source Software, 3(29), 884, (2018). https://doi.org/10.21105/joss.00884

[10] S. Lightfoote, S. Wilson and S. Voss, "Investigation of More Than 100,000 Months of Inverter Power Conversion Efficiency Data Using the Power Factors Database," 2021 IEEE 48th Photovoltaic Specialists Conference (PVSC), 2021, pp. 2360-2362, doi: 10.1109/PVSC43889.2021.9518706.

Thin, Radiation-Resilient III-V PV Devices Utilizing Quantum Structures and Epitaxial Light Reflectors

Brandon M. Bogner, Stephen J. Polly, Seth M. Hubbard, Roger E. Welser

Rochester Institute of Technology, Rochester, NY, United States

Magnolia Optical Technologies, Inc., Woburn, MA, United States

To improve BOL performance with EOL degradation in mind, nip InGaP-GaAs Heterojunction PV devices were grown and tested investigating a variety of epitaxial designs. Various numbers of quantum wells were added i-region, ranging from 6-45x in combination with a distributed Bragg reflector to promote additional current production at and beyond the GaAs band-edge. Utilizing the extra current production from the wells, the GaAs base was thinned from an optically thick 2,680 nm to 1,500 nm to prepare the devices for radiation-induced minority carrier lifetime degradation. Under AM0, the 45xSBQWs with a 1,500 nm base increased Jsc by 1.3mA/cm2 compared to the optically thick GaAs device containing 0xQW, as well as maintaining a Voc within 8mV.

Rapid Thermal Annealing (RTA) of Hydrogenated Poly-Si under Air and Nitrogen and Blister Formation

Arpan Sinha, Sagnik Dasgupta, Ajeet Rohatgi, Mool Gupta

University of Virginia, Charlottesville, VA, United States

Georgia Inst. of Tech., Atlanta, GA, United States

The high-efficiency poly-Si-based TOPCon cells are getting significant attention in the global PV market. Thermal annealing is an important fabrication step for poly-Si crystallization, passivation, and dopant activation. We investigated rapid methods of annealing using RTA and laser at different temperature regimes under N2 and air atmosphere. The surface morphology, degree of crystallization, and passivation were studied using SEM, Raman, and photoluminescence methods. Extensive formations of blisters under N2 conditions were observed while very low blister density under air annealing, indicating atmosphere dependence. High photoluminescence was observed after air annealing and forming gas treatment for boron-doped Si.

Investigation of High Open Circuit Voltage in CdTe-based Solar Cells Using Oxide Back Buffer Layers

Abdul Quader, Manoj K. Jamarkattel, Ebin Bastola, Kamala Khanal Subedi, Dipendra Pokhrel, Indra Subedi, Adam B. Phillips, Nikolas J. Podraza, Randy J. Ellingson and Michael J. Heben

Wright Center for Photovoltaics Innovation and Commercialization (PVIC), Department of Physics and Astronomy, University of Toledo, Toledo, Ohio, 43606, USA

Abstract — **Polycrystalline CdTe-based photovoltaic (PV) devices currently account for ~5-10% of the PV market. This is despite the fact that the open circuit voltage (V_{oc}) of these devices is only ~71% of its detailed-balanced limit. Increasing the V_{OC} will lead to further efficiency gains and potentially reduce the already low cost per Watt production. To improve the V_{OC} will require reducing recombination, likely at the back interface for the current generation of Cu-doped devices. To do this, we employ sputtered oxide buffer layers, specifically, Al_2O_3 and $CuAlO_2$. The high bandgap and valence band positions deeper than that of CdTe suggest that these materials have the potential to reduce the recombination at the back interface. When thick oxide layers are used, we observe and s-kink in the current density-voltage curve, but V_{oc} values above 900 mV are observed for both materials. As the thickness of the oxide buffers decreases the V_{OC} decreases with an increase in the fill factor. These results point to a pathway to achieve high V_{OC} while maintaining a high fill factor.**

Keywords—CdTe, Al_2O_3, $CuAlO_2$, Back contact, Photovoltaics

ACKNOWLEDGEMENTS

This material is based on research sponsored by Air Force Research Laboratory under agreement number FA9453-21-C-0056. The U.S. Government is authorized to reproduce and distribute reprints for Governmental purposes notwithstanding any copyright notation thereon. Approved for public release; distribution is unlimited. Public Affairs release approval #AFRL-2022-2493.

Solution-Processed Copper Selenium Oxide (CuSeO₃) as Hole Transport Layer for CdS/CdTe Solar Cells

Sandip S Bista, Deng-Bing Li, Suman Rijal, Sabin Neupane, Manoj K Jamarkattel, Rasha A Awni, Zhaoning Song, Adam Phillips, Michael Heben, Randy J. Ellingson, and Yanfa Yan

Department of Physics and Astronomy, and the Wright Center for photovoltaics Innovation and Commercialization, The University of Toledo, Toledo, OH, 43606 USA

Abstract— Solution-processed Cu doping with Cu contained hole transport layer (HTL) in CdTe solar cells has attracted great attention due to its low cost, ease of process, low-temperature activation, and better Cu control advantages. Here, we introduced solution-processed copper selenate (CuSeO₃) as a Cu source and hole transport layer. Steady-state photoluminescence (PL) quenching effect and back-barrier height reduction together indicate efficient hole extraction from the bulk and suppressed carrier recombination at the back surface. As a result, a cell efficiency of 16.7% was demonstrated with a V_{OC} of 0.861 V and a fill factor of 76.65%.

Keywords— Cadmium telluride (CdTe), copper selenium oxide (CdSeO₃), copper selenite, hole transport material.

I. INTRODUCTION

CdTe is a leading thin-film technology, which produced a record power conversion efficiency (PCE) of 22.10 % for small cells and around 18 % for module production[1]. These remarkable PCEs were achieved through introducing effective passivation and doping of the CdTe absorber layer. Traditionally, cadmium chloride (CdCl₂) is used for grain boundary passivation and Cu is used for p-type doping in CdTe. It is also observed that Cu forms Cu_xTe alloy to reduce the back-barrier height. However, the low solubility and fast diffusivity of Cu in CdTe indicate that an optimum Cu incorporation is required to reduce the formation of undesired Cu related trap states and to improve the device stability[2, 3].

Recently, solution-processed Cu doping and Cu contained hole transfer layer (HTL) have attracted attention due to their low cost, easy process, low-temperature activation, and better Cu control characteristics. Our previous work suggests that ionic Cu (using CuCl and CuCl₂ solution) can give better Cu control in terms of reducing the Cu dosage and the formation of gradient Cu distribution[4, 5]. Nonetheless, a suitable hole transport material with positive conduction band offset (CBO) and higher work function are still required to further reduce the carrier recombination at the back surface. Therefore, it is believed that inserting a back buffer layer between CdTe and metal electrode is expected to enhance carrier collection and device performance.

CuSeO₃ has a large bandgap of 3.9 eV[6]. It can act as an alternative Cu source for p-type doping of CdTe with controllable Cu dosage and a hole transport material in CdTe solar cells. Here, we report on fabrication of efficient CdTe thin film solar cells using solution-processed CuSeO₃ as the hole transport material as well as Cu source for p-type doping. Steady-state photoluminescence (PL) and temperature-dependent J-V measurement indicate an improved carrier collection efficiency between the CdTe and CuSeO₃, reduced back-barrier height in comparison to Cu-metal treated devices. We demonstrate CdS/CdTe solar cells with PCEs up to 16.7%.

II. EXPERIMENTAL DETAILS

All devices were fabricated on commercial T12D soda-lime glass provided by Pilkington, North America. First, the oxygenated cadmium sulfide (CdS:O) window layer was RF sputtered at room temperature under 10 mT pressure with a gas flow of 2% oxygen and 98% argon. A 4 µm cadmium telluride (CdTe) absorbing layer was then deposited at 10 Torr pressure using the closed space sublimation (CSS) method at a source temperature of 660 °C and a substrate temperature of 590 °C. Cadmium chloride (CdCl₂) treatment was carried out at 390 °C for 30 minutes in dry air ambient. The sample was removed from the chamber after cooling down and rinsed with methanol to remove chlorine and oxide residue. Subsequently, a CuSeO₃ was deposited by solution process, with the solution prepared by dissolving CuSeO₃•2H₂O powder in ammonium hydroxide (28-30 wt.%) with a 2 mg/ml concentration. 100 µL solution was pipetted and spun at 6000 rpm for the 30 s on a 1.5"×1.5" sample surface. After spin coating, samples were post-annealed at different temperatures. Finally, devices were completed by thermally evaporating a 40 nm gold (Au) back metal contact. Under a similar condition, control devices without CuSeO3 were fabricated with a thermally evaporated 4 nm Cu and 40 nm Au bilayer and annealed at 200 °C for 20 minutes.

AM1.5 illumination current-voltage(J-V) measurement and 0V bias external quantum efficiency (EQE) measurement were conducted to evaluate the device performance. Steady-state photoluminescence (PL) measurement using a 532 nm laser excitation from the film side and temperature-dependent current-voltage (J-V-T) measurement (performed in a closed-cycle helium cryostat, where the temperature varied from 190 to 310 K, with a step size of 10 K) were conducted to understand the differences of device performance.

III. RESULTS AND DISCUSSIONS

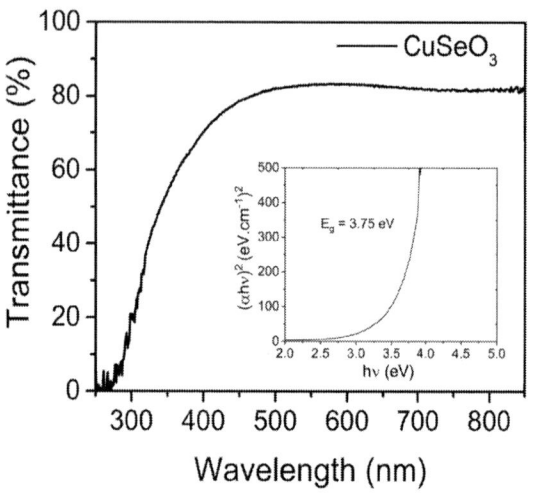

Figure 1. UV-Vis spectrum and Tauc plot of a CuSeO₃ thin film.

Figure 1 shows a transmittance spectrum of the CuSeO₃ thin film deposited on soda lime glass using UV-Vis spectroscopy. From the Tauc plot, an energy band gap E_g = 3.75 eV is measured which is consistent with the literature report.

We optimized the thermal activation of CuSeO₃ based on the device performance. According to our previous experience on CdTe solar cells using CuSCN as a back contact, we chose low temperature and shorter annealing time to activate CuSeO₃. We annealed CuSeO₃ processed samples in a CSS chamber at three different temperatures: 190 ℃, 210 ℃, and 230 ℃, with no holding time. Figure 2 shows the statistical distribution of photovoltaic parameters under different annealing conditions and the comparison with standard Cu metal treated cells. CuSeO₃ processed cells annealed at 210 ℃ give the highest V_{OC} and FF. The high V_{OC}s might be due to a better doping profile. The enhancements in FF are due to reduced series resistance (R_S) and increased shunt resistance (R_{SH}). The poorer photovoltaic parameters for cells annealed at the lower temperature could be due to insufficient Cu diffusion, while for the cells annealed at a higher temperature (230 ℃), the lower device performance might be due to excessive diffusion of Cu.

Figures 3 (a) and (b) show the JV curve and external quantum efficiency of the best cell. Table 1 summarizes the photovoltaic parameters of the best cells at different annealing temperatures. EQE spectral responses clearly show the effect of annealing temperatures. The device annealed at 210 ℃ has a better spectral response at the longer wavelength regions (600 – 840 nm), suggesting reduced back-barrier height. All CdTe-CuSeO₃ devices show higher spectral response at short wavelength regions between 350 to 600 nm than Cu metal doped devices, suggesting CuSeO₃ offers a better control on Cu diffusion, leading to better front junction quality.

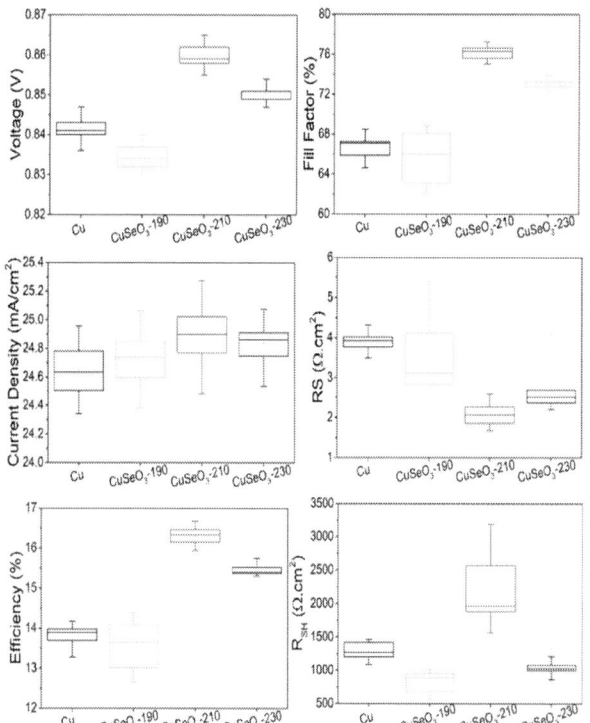

Figure 2. Box plots of (a) open-circuit voltage (V_OC) (b) short circuit current density (J_SC) (c) efficiency (d) fill factor (FF), (e) series resistance (R_S), (f) shunt resistance (R_SH) of CdTe solar cells with Cu metal and CuSeO₃ back contact annealed at different temperatures.

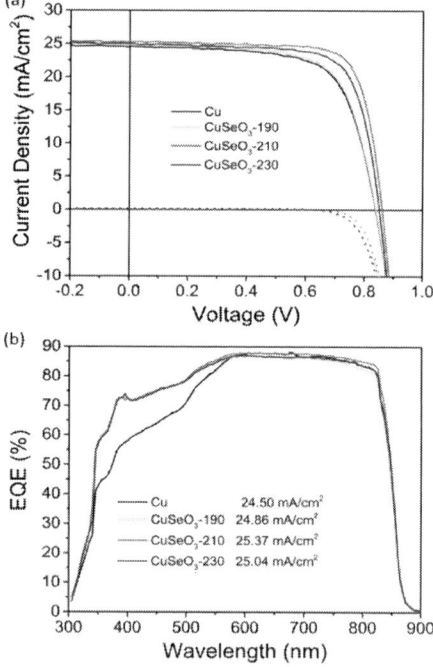

Figure 3. (a) Current density-voltage (J-V) and (b) external quantum efficiency (EQE) curves for best devices annealed different temperatures for CdTe-CdSeO₃ cells and the reference cell.

978-1-7281-6118-1/22 $31.00 © 2022 IEEE

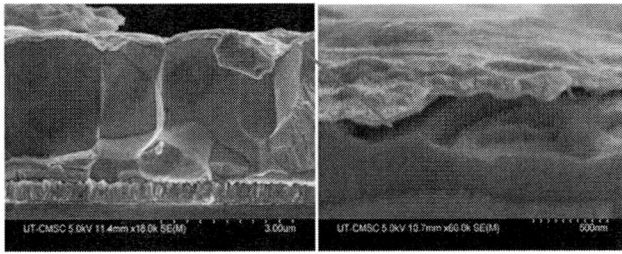

Figure 4 (a) SEM image of a complete CdTe-CdSeO₃ device (b) thick CuSeO₃ Voc layer.

To understand the effect of $CuSeO_3$ thickness, we carried out a few experiments with a thicker $CuSeO_3$ back-buffer layer as shown in Figure 4. The thickness was controlled by increasing the concentration of $CuSeO_3.2H_2O$ powder in NH_4OH solvent and reducing the spin-coating speeds. Thicker $CuSeO_3$ was highly resistive, resulting in poorer device performance.

Table 1. Photovoltaic parameters of cells annealed at different temperatures.

Process	V_{OC} (mV)	J_{SC} (mA/cm²)	FF (%)	Eff. (%)
Cu	841	24.60	68.47	14.17
$CuSeO_3$-190 °C	840	24.85	68.84	14.37
$CuSeO_3$-210 °C	861	25.27	76.65	16.70
$CuSeO_3$-230 °C	851	25.01	74.0	15.75

Steady-state PL measurement and ttemperature depended dark J-V measurement were carried out Tto understand the effects of $CuSeO_3$ on the device performance. As shown in Figure 5 (a), the CdTe-$CuSeO_3$ sample has significantly lower PL intensity than the CdTe-Cu sample. The PL quenching measured in CdTe-$CuSeO_3$ suggests that $CdSeO_3$ acts as a hole transfer layer, and it extracts holes from CdTe into the $CuSeO_3$. Temperature depended dark J-V measurements were used to extract the back-barrier height. The Cu/Au device shows a severer roll-over at the lower temperatures compared to $CuSeO_3$/Au devices, as shown in Figures 5 (b) and (c). The presence of roll-over is directly related to the back-barrier height. Figure 5 (d) shows the Arrhenius plots that is used calculate the back-barrier height. The calculated back-barrier height for Cu/Au and $CuSeO_3$/Au devices is 0.325 eV and 0.266 eV, respectively. The reduced back-barrier is consistent with the improvement of device performance.

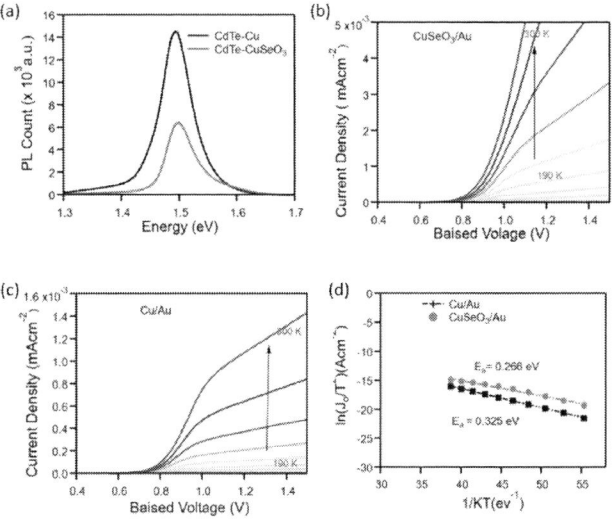

Figure 5. (a) steady-state photoluminescence (PL) curves, (b) (c) temperature-dependent dark J-V curves for device contain CuSeO₃ and Cu metal back contact respectively. (d) Arrhenius plots for the calculation of back-barrier heights for CuSeO₃ and processed cells.

IV. CONCLUSION

A solution processed Cu contained hole transfer layer $CuSeO_3$ has been successfully applied to CdTe solar cells. The best performing cell achieved a 16.70 % power conversion efficiency with a VOC of 861 mV, a JSC 25.27 mA/cm² and a FF of 76.65 %. T-J-V and PL quenching measurement show that $CuSeO_3$ processed cells exhibit improved hole extraction than the cells using copper metal as the doping source.

ACKNOWLEDGMENT

This material is based on research sponsored by Air Force Research Laboratory under agreement number FA9453-21-C-0056. The U.S. Government is authorized to reproduce and distribute reprints for Governmental purposes notwithstanding any copyright notation thereon. Approved for public release; distribution is unlimited. Public Affairs release approval #AFRL-2022-2620.

REFERENCES

1. Green, M., et al., *Solar cell efficiency tables (version 57).* Progress in Photovoltaics: Research and Applications, 2021. **29**(1): p. 3-15.
2. Woodbury, H. and M. Aven, *Some diffusion and solubility measurements of Cu in CdTe.* Journal of Applied Physics, 1968. **39**(12): p. 5485-5488.
3. E.D. Jones, N.M.S., J.B. Mullin, *The diffusion of copper in cadmium telluride.* Journal of Crystal Growth, 1992. **117**(1992)(244—248).
4. Bista, S.S., et al., *Effects of Cu Precursor on the Performance of Efficient CdTe Solar Cells.* ACS Applied Materials & Interfaces, 2021.
5. Li, D.-B., et al., *Maximize CdTe solar cell performance through copper activation engineering.* Nano Energy, 2020. **73**: p. 104835.
6. Tomar, R., et al., *Multiple helimagnetic phases in triclinic CuSeO3.* Journal of Magnetism and Magnetic Materials, 2020. **497**.

Prediction of Electron Band Gap of A_2XY_6 Perovskite Compounds using Machine Learning

1st Jatin Chaudhary
Department of Computing
University of Turku
Turku, Finland
jatin.chaudhary@utu.fi

1st Swastik Bhattacharya
Department of Electrical, Computer and Energy Engineering
University of Colorado
Boulder, USA
Swastik.Bhattacharya@colorado.edu

2nd Jukka Heikkonen
Department of Computing
University of Turku
Turku, Finland

3rd Rajeev Kanth
School of Information Technology
Savonia University of Applied Sciences
Kuopio, Finland

Abstract—Increasing population and industrialization haveled to an uptick in energy requirements. Many traditional energy sources are not anymore attractive due to climate change, instead, the interest has turned to power generation from renewable sources, such as wind energy, hydro-power, and solar energy. The wide availability of sunlight and simplicity in converting sunlight to electricity has led to the search for synthesized semiconductors that give high efficiency in this conversion. A family of such semiconductors attains the perovskite structure, the most established being Methyl Ammonium Lead Iodide. The shortcomings of this compound include lead poisoning, motivating the search for perovskite structures that have low electron band-gap and are stable. A family of such perovskite structures is compounds that attain an A_2XY_6 type structure. This paper demonstrates some methods that can be used to calculate the electron band-gap of such compounds. The metrics found from Support Vector Machine Regression and Random Forest Regression are compared and analyzed to propose a scalable model for predicting electron band-gap.

Index Terms—A_2XY_6 Perovskite Compounds, Band Gap Calculation, Support Vector Machine, Random Forest

I. INTRODUCTION

The shift of energy dependencies from traditional origins to renewables has been prominently taking place in the past decade. Among all, solar Energy has been one of the most promising source of renewable energy [1]. In 2020-2021, solar energy has been second most eminent renewable source [2]. Researches to improve the efficiency of photovoltaic(PV) cells has been an imperative task looking over the potential PVs possess [3]. Over the years, the perovskite technology has accelerated the optimization of PV cell due to its stable nature and high potential of increased power conversion efficiency [4]. Perovskite Cells have definitive crystal structure as ABX_3 (where, X= oxygen, hallide). A_2XY_6 (A= K, Cs, Rb, Tl; X= tetravalent cation, Y= F, Cl, Br, I) Perovskite compunds have been identified to have incredible semiconductor properties along with exclusive optical properties making it suitable for opto-electronic devices like PV cells [5].

Designing of highly efficient PV cells have been a notable state-of-the-art research in the field and the inter-linkage to the perovskite technology with PV have been very assuring [6] [7]. The increasing investigation of perovskite compounds are bringing an evolution to the PV industry [8]. Band gap energies are a crucial consideration when the PV cells are designed as a wide band gap energies may make the compound unsuitable for PV applications [9]. Calculation of the bandgaps for different compounds is an cumbersome process as it requires optical diffuse reflectance measurements followed by computations using Kubelka–Munk equation [10]. Studies have shown that direct band gaps can vary with the experimental structure as band gap is susceptible to functional group employed for structure optimization [11].

Keeping Volonakis et al., 2017's study in consideration, the authors of this paper decided to study the dependence of bandgap upon the experimental structure of the compound and further, a method to predict the bandgap based on the experimental structure. This has been done so as to predict the bandgap of newly simulated or developed A_2XY_6 Perovskite Compounds. In this paper, we have presented an efficient method for prediction of band gap of A_2XY_6 Pervoksite Compounds using Support Vector Machine (with Radial basis Kernel and Linear Kernel) and Random Forest [12] [13]. This study has been performed in order to present a solution to the band gap energy for prediction of upcoming compounds which are in the primitive stage of designing or manufacturing. Physical parameters including ionic radii, electronegativities, Miller Indices, formation energy, lattice index and lattice constants are used for the training of the model. The data used for the training of the model is taken from the literature available upon the experimentation.

The paper consists of methodology in section II, and results of the model trained and it's discussion in section III, and conclusion and future scope of the work in section IV.

978-1-7281-6118-1/22 $31.00 © 2022 IEEE

II. METHODOLOGY

A. Datasets

Data used for the training of models consists of experimental output of eighty nine A_2XY_6 perovskite compounds with lattice constant of 8.109 to 11.790 Å [14]. Seventy one compounds as data points are taken from the [14], whereas the rest are the repeated compounds which exist in more than one structure, hence having different Lattice Constants and Miller Indices. The value of lattice constants has been taken in normal room temperature and pressure conditions. Compounds which qualified the condition A_2XY_6 perovskite compounds with lattice constant of 8.109 Å to 11.790 Å but had unstable structural values have not been considered as a part of the dataset.

B. Models and model performance evaluation

For the purpose of this study, due to a simpler complexity of the dataset, Random Forest (RF) Regression and Support Vector Machines (SVMs) with linear and radial-basis function kernels. The models were evaluated using Leave-One Out (LOO) cross-validation approach, having divided the data into 5 batches and cross-validating the results after leaving one of the batches.

A Random Forest (RF), as defined by Breiman et al. [15] is a classifier consisting of a collection of tree-structured classifiers $\{h(\mathbf{x}, \Theta_k), k = 1, ...\}$ where the Θ_k are independent identically distributed (i.i.d.) random vectors and each tree casts a unit vote for the most popular class at input $\mathbf{x}$. RF for regression problems are created by growing trees depending on a random vector, such that the tree estimator $h(\mathbf{x}, \Theta)$ takes on numerical values, and not class labels [15]. The output values are numerical and it is assumed that the training set is independently drawn from the distribution of the random vector Y, $\mathbf{X}$. The mean-squared generalization error for any numerical predictor $h(\mathbf{x})$ is:

$$E_{\mathbf{X}, Y}(Y - h(\mathbf{x}))^2 \quad (1)$$

SVMs are used in both classification and regression problems. The goal of SVMs in a regression problem is to find a function $f(x)$ that can provide predictions under a margin of error ε. The function $f(x)$ can be expressed as:

$$f(\mathbf{x}) = \mathbf{w}^T \Phi(\mathbf{x}) + b \quad (2)$$

where $\Phi(\mathbf{x})$ is the mapping result into the input space, $\mathbf{w}$ is the weight matrix, and b is the bias vector. The weight and bias are trained by minimizing the risk function:

$$R = min\frac{1}{2}||\mathbf{w}||_2^2 + C\frac{1}{l}L_\varepsilon(y, f(\mathbf{x})) \quad (3)$$

Regression problems that are of non-linear nature can be solved by using non-linear functions. For any input vectors $\mathbf{x}$ and $\mathbf{z}$, a kernel function must satisfy the condition:

$$k(\mathbf{x}, \mathbf{z}) = \Phi(\mathbf{x})^T \Phi(\mathbf{z}) \quad (4)$$

In this study, the regression problem has been attempted by using linear and radial-basis function kernel. A linear kernel is expressed as:

$$k(\mathbf{x}, \mathbf{z}) = \mathbf{x}^T \mathbf{z} \quad (5)$$

A radial-basis function kernel is defined as:

$$k(\mathbf{x}, \mathbf{z}) = exp(\frac{1}{2\sigma^2}||\mathbf{z} - \mathbf{x}||_2^2) \quad (6)$$

For the purpose of this study, the number of tree estimators in the RF regressor is assigned as 100. For the SVRs, for both kernels, the value of C in equation (3) for risk minimization is set to 100, and the error margin ε is set to 0.1. Rest of the hyper-parameters are assigned the default values as per the scikit-learn package of Python [16]. The model was trained using the experimental data discussed above.

TABLE I
RESULT MATRIX OF THE MODELS TRAINED.

	RMSE	Relative RMSE
SVM (Linear Kernel)	2.20 eV	0.31
SVM (RBF Kernel)	2.30 eV	0.32
Random Forest	2.33 eV	0.32

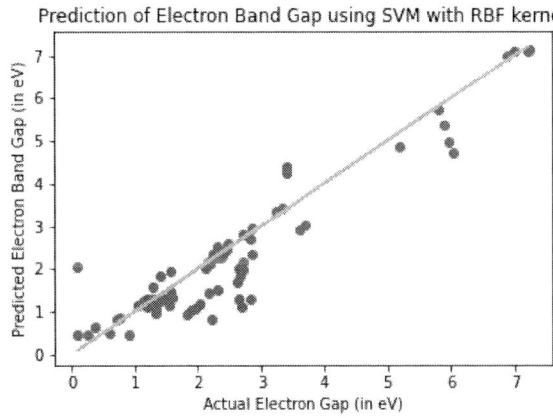

Fig. 1. Predicted vs Actual Bandgap graph of Radial Basis Function SVM.

III. RESULTS AND DISCUSSION

The three models trained presented promising results. Table 1, shows the result matrix consisting of Root Mean Square Error(RMSE), and Relative RMSE for all the three models. Figure 1, depicts the predicted vs the actual band gap graph of the RBF SVM model trained. Similarly, figure 2 and 3 also depicts the predicted vs the actual band gap graph of the linear kernel SVM and random forest algorithm respectively. These graphs are an indication to the accuracy of the model. Furthermore, the error curves for the cross-validation of the SVM models were analyzed. These were obtained by performing a 5-fold Leave-One-Out cross validation over the models. For both the kernels in case of SVM, as shown in figures 4 and 5, the MSE converged to 1.25eV in case of linear

978-1-7281-6118-1/22 $31.00 © 2022 IEEE

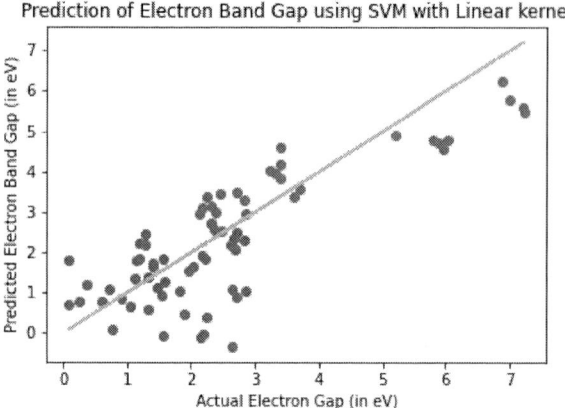

Fig. 2. Predicted vs Actual Bandgap graph of Linear SVM.

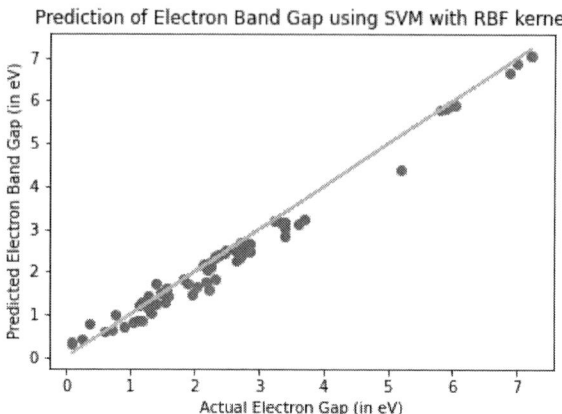

Fig. 3. Predicted vs Actual Bandgap graph of Random Forest.

kernel SVM, and 1.5eV in case of the RBF kernel SVM. The convergence of all the models affirm the optimization of the model. Figures 6, 7 and 8 depict scatter plots between the ground-truth and cross-validation predictions. It is observed that there is a better fitting on testing ground-truth in case of RF regression upon cross-validation as compared to both the SVM models. However, as stated in table 1, it is also seen that the overall accuracy in terms of RRMSE is better in case of both the SVM models.

The calculation of the bandgap of A_2XY_6 perovskite compounds requires the computation over first principles and experimental results [11]. This study has been focused upon developing a model which can predict the band gap value using the physical properties (input features) of stable double perovskite material (A_2XY_6). Authors' focus has been towards developing a model which can be used to predict the bandgap of the newly simulated A_2XY_6 Perovskite Compounds. This method of bandgap prediction has been developed to bypass the computationally expensive calculations of bandgap, for further estimation of the compound's practical viability. Our focus has been towards building a model which can predict bandgap values which can be made into a scalable solution to avoid the current computationally intensive models meant for this problem.

IV. CONCLUSION AND FUTURE SCOPES

Our paper presents novel output in terms of prediction of band gap in a computationally economically way towards bandgap prediction of A_2XY_6 Perovskite Compounds. Among the models we have trained, the random forest model has given promising results towards the predictor function. The support vector machine with radial basis function and linear function has been trained, and Support Vector Machine with Radial Basis Function has given better results considering the potential intervention of noise signals while training. This paper can be extended with the addition of latest compounds to the dataset and optimizing the existing model with the feedback of more validation methods. This study has presented

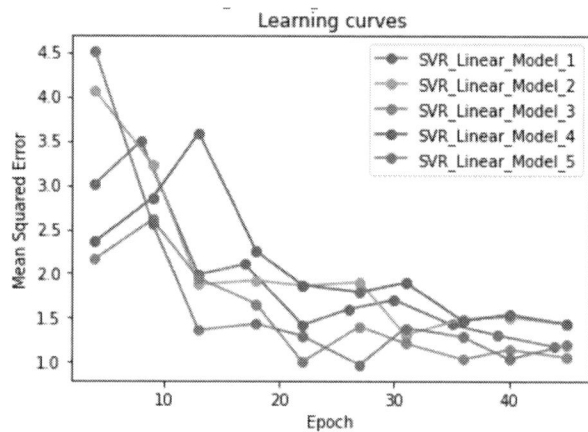

Fig. 4. Learning Curve for SVM Linear Kernel. Mean Square Error was calculated to be 1.25eV.

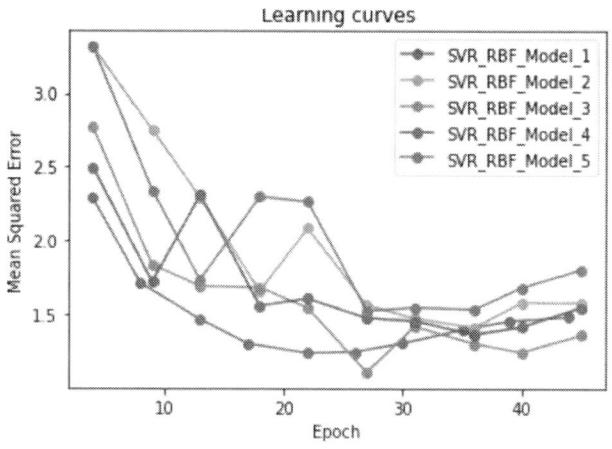

Fig. 5. Learning Curve for SVM RBF Kernel. Mean Square Error was calculated to be 1.5eV.

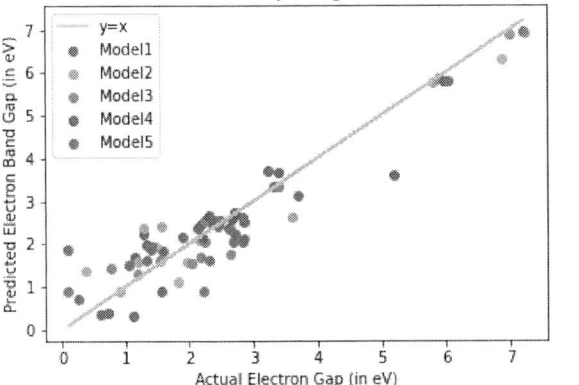

Fig. 6. Cross Validation graph of Random Forest.

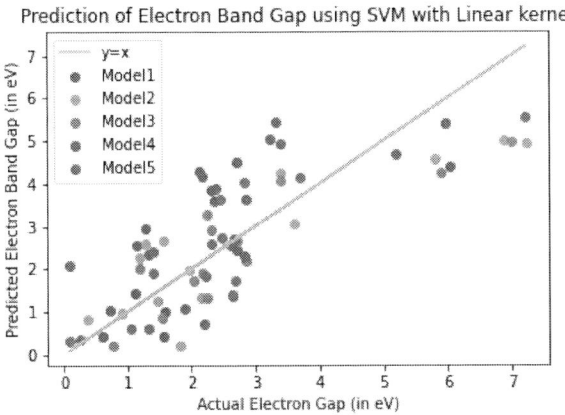

Fig. 7. Cross Validation graph of Support Vector Machine with Linear Kernel.

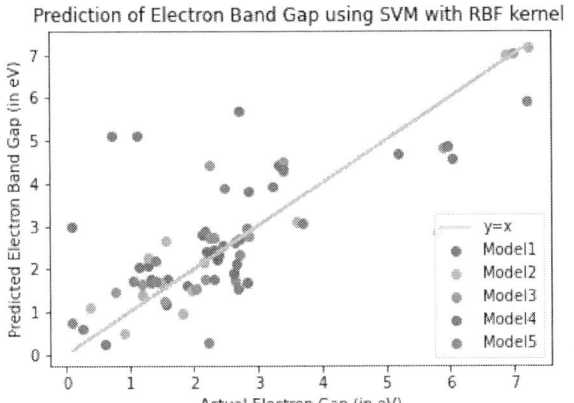

Fig. 8. Cross Validation graph of Support Vector Machine with RBF Kernel.

a preliminary study in the field of bandgap prediction of A_2XY_6 Perovskite Compounds and hence, possesses huge opportunities towards extending this study. Such efforts can also be applied towards predicting bandgap for perovskite compounds that have a different stoichiometry and elements present in their structure which may not necessarily form a perovskite of A_2XY_6 nature.

REFERENCES

[1] N. Kannan and D. Vakeesan, "Solar energy for future world:-a review," *Renewable and Sustainable Energy Reviews*, vol. 62, pp. 1092–1105, 2016.

[2] Iea, "Renewable electricity generation increase by technology, 2019-2020 and 2020-2021 –nbsp;charts – data amp; statistics." [Online]. Available: https://www.iea.org/data-and-statistics/charts/renewable-electricity-generation-increase-by-technology-2019-2020-and-2020-2021

[3] J. K. Chaudhary, R. Kanth, J.-P. Skön, and J. Heikkonen, "Analysis and enhancement of quantum efficiency for multi-junction solar cell," in *2019 IEEE 46th Photovoltaic Specialists Conference (PVSC)*, 2019, pp. 0210–0214.

[4] N.-G. Park, "Perovskite solar cells: an emerging photovoltaic technology," *Materials Today*, vol. 18, no. 2, pp. 65–72, 2015. [Online]. Available: https://www.sciencedirect.com/science/article/pii/S1369702114002570

[5] S. Chadli, A. Bekhti Siad, M. Baira, M. Siad, A. Allouche, and A. Reguig, "Physical properties of double perovskites rb2xcl6 (x= sn, te, zr): Competitive candidates for renewable energy devices," *Solid State Communications*, vol. 342, p. 114633, 2022. [Online]. Available: https://www.sciencedirect.com/science/article/pii/S0038109821004105

[6] J. Chaudhary, R. Kanth, and J. Heikkonen, "Performance analysis of back surface field (bsf) effects in multijunction photovoltaic cell," in *2020 47th IEEE Photovoltaic Specialists Conference (PVSC)*, 2020, pp. 1207–1211.

[7] M. I. H. Ansari, A. Qurashi, and M. K. Nazeeruddin, "Frontiers, opportunities, and challenges in perovskite solar cells: A critical review," *Journal of Photochemistry and Photobiology C: Photochemistry Reviews*, vol. 35, pp. 1–24, 2018. [Online]. Available: https://www.sciencedirect.com/science/article/pii/S1389556717301144

[8] A. Mutalikdesai and S. K. Ramasesha, "Emerging solar technologies: Perovskite solar cell," *Resonance*, vol. 22, no. 11, pp. 1061–1083, 2017.

[9] R. Nechache, C. Harnagea, S. Li, L. Cardenas, W. Huang, J. Chakrabartty, and F. Rosei, "Bandgap tuning of multiferroic oxide solar cells," *Nature Photonics*, vol. 9, no. 1, pp. 61–67, 2015.

[10] I. Chung, J.-H. Song, J. Im, J. Androulakis, C. D. Malliakas, H. Li, A. J. Freeman, J. T. Kenney, and M. G. Kanatzidis, "Cssni3: semiconductor or metal? high electrical conductivity and strong near-infrared photoluminescence from a single material. high hole mobility and phase-transitions," *Journal of the American Chemical Society*, vol. 134, no. 20, pp. 8579–8587, 2012.

[11] G. Volonakis, A. A. Haghighirad, R. L. Milot, W. H. Sio, M. R. Filip, B. Wenger, M. B. Johnston, L. M. Herz, H. J. Snaith, and F. Giustino, "Cs2inagcl6: a new lead-free halide double perovskite with direct band gap," *The journal of physical chemistry letters*, vol. 8, no. 4, pp. 772–778, 2017.

[12] G. L. Prajapati and A. Patle, "On performing classification using svm with radial basis and polynomial kernel functions," in *2010 3rd International Conference on Emerging Trends in Engineering and Technology*, 2010, pp. 512–515.

[13] G. Biau and E. Scornet, "A random forest guided tour," *Test*, vol. 25, no. 2, pp. 197–227, 2016.

[14] Y. Zhang and X. Xu, "Machine learning lattice constants from ionic radii and electronegativities for cubic perovskite a2xy 6 compounds," *Physics and Chemistry of Minerals*, vol. 47, no. 9, pp. 1–15, 2020.

[15] L. Breiman, "Random forests," *Machine learning*, vol. 45, no. 1, pp. 5–32, 2001.

[16] F. Pedregosa, G. Varoquaux, A. Gramfort, V. Michel, B. Thirion, O. Grisel, M. Blondel, P. Prettenhofer, R. Weiss, V. Dubourg *et al.*, "Scikit-learn: Machine learning in python," *the Journal of machine Learning research*, vol. 12, pp. 2825–2830, 2011.

A Silicon learning curve and polysilicon requirements for broad-electrification with photovoltaics by 2050

Brett Hallam, Moonyong Kim, Robert Underwood, Storm Drury, Li Wang, Pablo R Dias

School of Photovoltaic and Renewable Energy Engineering, UNSW Sydney, Kensington, Australia

UFRGS, Porto Alegre, Brazil

This paper investigates the current and future projected silicon demand for the photovoltaics industry towards broad electrification scenarios with over 60 TW of PV installed by 2050. The current silicon consumption contained in cells/modules is 1510-1900 tonnes/GW. However, this does not account for silicon losses during purification, ingot growth and wafering. The global polysilicon demand by the PV industry in 2020 of 452 kt equates to a silicon consumption of approximately 3150 tonnes/GW, suggesting a current utilization factor of 48-60%. Depending on physical constraints determining the lower limit for future silicon consumption, (eg. 1550 tonnes/GW for 30% tandems made on 100 μm thick wafers, with 50% silicon utilization), the cumulative silicon demand to 2050 could be in the range of 45-123 Mt, with an annual demand of 2-9 Mt in 2050. To reduce the environmental impact of silicon wafers, we must increase efficiencies, use thinner wafers, reduced kerf-loss and explore alternative purification methods with low emissions intensities.

Photovoltaic Surfaces to Reverse Global Warming

Christiana B. Honsberg, Stuart G. Bowden, Ian R. Sellers, Richard R. King, Stephen M. Goodnick

Arizona State University, Tempe, AZ, United States

University of Oklahoma, Norman, OK, United States

Climate changes and its many associated impacts are one of the most critical global challenges. Photovoltaics has been instrumental in mitigation of CO_2 through the generation of electricity. However, the goal of limiting global warming to 1.5 °C increasingly requires additional approaches. The paper presents how PV surfaces can be designed to reverse the Earth' radiative imbalance from increased greenhouse gasses that lead to higher global temperatures. The new PV surface generate electricity, reflect sub-band gap radiation, minimize their temperature, generate thermal radiation and emit additional IR through the atmospheric, with these processes totaling 650 W/m². This is realized by: (1) PV system efficiency at operating temperature > 20% and sub-band gap reflection of 150 W/m² for a total of 350 W/m²; (2) Thermally emitted radiation (radiative cooling) of 150 W/m²; and (3) Active IR emission through an atmospheric window at 1.5 mm of 150 W/m². With such PV surfaces, we show that 10 TW of installed PV can reverse global warming. Using PV to balance global temperatures introduces additional considerations for PV, focusing on high efficiency, particularly high efficiency at operating temperatures, radiative cooling, and new processes for 1.5 mm emission. We find that depending on their design, PV panels can increase or decrease global temperatures.

Strategies for implementing of Very Large Scale Solar and Wind power plants in the Gobi desert for the Northeast Asia regional energy market

Enebish Namjil[1], Keiichi Komoto[2]

[1]Institute of Physics and Technology, Academy of Sciences of Mongolia

[2]Mizuho Information & Research Institute, Japan, keiichi.komoto@mizuho-ir.co.jp

*Corresponding author:enebishnamjil@gmail.com

Abstract-Northeast Asian countries are cooperating on development of cost-effective regional energy supply through renewable energy market integration and advanced technologies of ultra-high capacity DC and AC transmission network and innovative grid integration solutions of renewable energy. Results of the more than 20 years study of the IEA PVPS Task 8 by international research team consisted of experts from industries, international organisations and non-government organisations of IEA member and non-member countries, including Mongolia has showed that the Gobi desert area of Mongolia is one of the most suitable places for establishment of the Very Large Scale Photovoltaic (VLS-PV) power generation plants. Gobi desert of Mongolia is characterized with abundant solar and wind energy resources and has vast arid and semi-arid plate land territory with very low population, low temperature, dry climate and high wind speed, which are additional big advantages for development of the VLS-PV, as well as Very Large Scale Wind Farms (VLS-PV&WF). The results of the latest studies on the strategies for practical realization of Northeast Asian Super grid by utilizing Mongolia's vast renewable energy resources are discussed in this paper.

Keywords: VLS-PV, Gobi desert, Gobitec, Northeast Asian supergrid

I. INTRODUCTION

Growing concerns for air pollution in Mongolia and China, problem of climate change and power shortage have caused the Northeast Asian countries to focus on increased use of renewable energy and on the possibility to improve energy supply through energy market integration and energy system interconnection between the countries in the Northeast Asian region. Northeast Asian countries are seeking to improve energy market efficiency and secure cost-effective energy supply through energy market integration and system interconnection. Extensive research has been performed on the development and construction of the VLS-PV power generation plants and Wind farms in Gobi desert [1-6]. With increasing share of renewable energy in the total energy mix, it may create certain issues due to intermittency of solar and wind power generation, weather instability, and low energy density, etc. In order to address the issues of intermittency of solar and wind power sources and instability, large scale battery energy storage system is typically added into VLS-PV &W power generation complex. Greater attention has been paid to the optimization of such complex and synchronization operation of power plants in different locations. Thus hybridization of VLS-PV &W power generation with large scale pumped storage hydropower plants and large scale battery storage complex is becoming an important research topic.

As part of the Asian Development Bank (ADB) country operations, the Government of Mongolia sought ADB technical assistance (TA) to prepare a strategy for Northeast Asia power system interconnection (NAPSI) using Mongolia's abundant solar and wind energy resources and advanced technologies of grid network and interconnection solutions of renewable energy, specially focusing on the high voltage DC transmission (HVDC) system and renewable energy technologies will be discussed in this paper. The preliminary results of the study on possible options for interconnection of the power system in Northeast Asia by utilizing Mongolia's vast renewable energy resources are described in this paper.

II. FIRST STEPS TOWARDS THE GOBITEC AND NORTHEAST ASIAN SUPERGRID

The studies on Very Large Scale Photovoltaic Power Generation (VLS-PV) systems first began under the umbrella of the IEA PVPS Task8 in 1998. After that, the new Task8 – Study on Very Large Scale Photovoltaic power generation (VLS-PV) systems in deserts was established during the European PVSEC conference in Glasgov, UK, in 1999 under leadership of Prof.Kosuke Kurokave [1]. IEA PVPS Task 8 consisted of experts from industries, international organisations and non-government organisations of IEA member and non-member countries, including Mongolia, China, Japan, Korea, Isreal, USA, Germany, Italy, Holland, Canada and USA.

The scope of Task 8 was to examine and evaluate the potential of VLS-PV systems, which have a capacity ranging from several megawatts to gigawatts, and to develop practical project proposals for realising VLS-PV systems in the future. At the time when Task 8 established, it was just a dream of scientists, but establishment of gigawatt scale solar plants has become reality and many of the gigawatt scale PV plants are operating around the world.

Since Task 8 established, it has been published extensive research reports as a series of 'Energy from the Desert', focusing on VLS-PV systems. Our first extensive report entitled "Energy from the Desert-Feasibility of VLS-PV Power Generation Systems" was published in 2003 [1]. In this study a fundamental concept of the VLS-PV systems and key factors to enable implementation of the VLS-PV in deserts were studied, and Mid- and Long term development scenario options for implementation of the VLS-PV systems feasible in some desert areas were also proposed. In the second book, published in 2007, entitled "Energy from the Desert-Practical Proposals for Very Large Scale Photovoltaic Systems", a realistic and practical next step, demonstrative approaches toward the realization of VLS-PV in different deserts around the world, e.g., the Mediterranean region, the Middle East, Northeast and Central Asia were studied in-depth. This report contained new knowledge about how to realise VLS-PV systems in the desert in sustainable way [2]. The third book, published in 2009, entitled "Energy from the Desert-Very Large Scale Photovoltaic Systems: Socio-Economic, Financial, Technical and Environmental Aspects" contains comprehensive analysis of economic, social and environmental impact of VLS-PV projects. This report focused on various subjects on VLS-PV and proposed a VLS-PV roadmap toward 2100 [3].

These studies on VLS-PV systems are well-known all over the world, especially in academic society. Some countries and regions, in particular Northeast Asian countries use the results of these studies as a important reference for large scale deployment of PV plants in Gobi desert.

V. UNIQUENESS OF GOBI DESERT OF MONGOLIA WITH REGARD TO THE DEVELOPMENT VLS-PV

The Gobi Desert is the fifth largest desert in the world and covers southern Mongolia and northwest part of China with an approximate size of around 1.3×10^6 km^2 in size and is located between 40°N and 45°N. Within IEA Task 8 study comprehensive studies on optimal site selection for the VLS-PV in world deserts using NASA satellite images and geographic information system technology it has been carried out. The results of those studies has shown that the Gobi Desert is an ideal place the installation site of VLS-PV systems both in terms of potential solar irridance, and faviourable weather conditions (windy, dry, cooler climate and plate land covering large area) for the construction of the VLS-PV and has some advantages. Because of very low population, Gobi desert of Mongolia has large unused land area and solar irradiation is much higher than other areas. In the Gobi desert alone, the total 2357.84×10^3 TWh (8486.67×10^{21} Jule) which equals almost 20 times the total world primary energy supply 10,723 MTOE (IEA, 2003) (3479.51×10^{20} Jule). Secondly, Gobi desert of Mongolia is a stone desert rather than sand and has faviourable weather conditions (windy, dry, cooler climate and plate land covering large area) for the effective operation of the VLS-PV with high efficiency.

Fig.3 Results of Task8 studies on the suitability land area and solar resources of the world deserts for deployment of the VLS-PV [1.2]

III. THE GOBITEC AND SUPERGRID FOR NORTHEAST ASIA

There are two different initiatives for power interconnection for Northeast Asia: Gobitec and Asian Supergrid. The "Gobitec" is result of idea for large scale deployment of clean renewable energy power from the Mongolian Gobi desert and in basic concept, it corresponds to the "Desertec" or "Seatec" initiatives. Desertec initiative was developed by the scientists and industrial experts from across Europe, the Middle East and North Africa (EU-MENA), and it is an industrial proposal to deploy clean electricity generated by the renewable energy plants in the Middle East and North African countries to meet domestic electricity demand and to export surplus electricity to meet rising power demand Europe. "Seatec" is the name of a project proposal for interconnection North Sea offshore wind farms to Northern European grid through undersea power cables.

GOBITEC is a visionary initiative that aims to deploy large-scale electrical power from solar and wind energy plants in the Gobi desert of Mongolia and to deliver this clean energy to regions with high energy demand in Northeast Asia.

The concept of "ASIAN SUPERGRID" is the result of an idea by Masayoshi Son, founder of the Telecom and Internet giant SoftBank Group to establish an electric power systems for Asian countries, enabling mutual benefits by exchanging abundant renewable energy resources, such as wind, solar and hydropower.

In the opening speech of the international conference on "Grid integration and renewable energy cooperation in Northeast Asia" held in Sept. 2012, in Ulaanbaatar, Mongolia, the Gobitec initiative was officially announced for the first time by Mr.Ts.Elbegdorj, President of Mongolia.

V. TOWARDS THE REALIZATION OF THE GOBITEC AND NORTHEAST ASIAN SUPERGRID

During the last two decades, Northeast Asia countries has suggested numerous initiatives on joint development of the Northeast Asia Supergrid initiatives in reality, unfortunately they had failed due to the lack of effective cooperative mechanism on government level between China, Russia, Japan, South Korea, and Mongolia. None of the initiatives were ever successfully realized as each country has its own vision and interest for implementation of the Northeast Asian supergrid. China has promoted the Belt and Road Initiative (BRI) as a mechanism which could incorporate the Asian Super Grid and give China a leadership position in Northeast Asian energy cooperation. However, it is indefinite if other Northeast Asian countries would support that idea. Thus, it is crucial to understand how such effective, mutually beneficial collaborations should be organized. The Asian Super Grid will make progress if Governments of the Northeast Asian countries can agree on the form of a multilateral mechanism. This matter is currently under serious discussions and consultations, supported by UNESCAP, ADB and IRENA, as well as Northeast Asian participating countries. Mongolian Government has made a clear commitment to jointly implement the Northeast Asia Super Grid Project and at the Eastern Economic Forum held in April, 2019 in Vladivostok, Mr.Khaltmaa Battulga, former President of Mongolia has called the leaders of Russia, China, Japan, South Korea to support the establishment of the organization in Mongolia and jointly implement and coordinate the Northeast Asia Super Grid Project.

V. THE MAIN OUTCOMES OF THE NORTHEAST ASIAN SUPERGRID INTERCONNECTION STUDY (NAPSI)

The main goal of the Gobitec initiative and Supergrid for Northeast Asia proposal is to build a sustainable energy infrastructure based on renewable energy sources, where electricity produced by the vast potential of solar and wind energy in the Gobi desert of Mongolia and hydropower plants in Siberia of Russia will be transmitted through HVDC supergrid to all countries in the Northeast Asia.

For the Northeast Asian regional supergrid projects, it is proposed that Gobi desert of Mongolia can serve as the 1st pilot project place to generate and supply electricity to the Chinese grid at first using the UHVDC transmission supergrid. Figure 3 shows the one of the proposed topology of the Northeast Asian regional supergrid, which includes the main direction of the supergrid between Russia and China across Mongolia (Gobi desert) and further connection within China before leading to Japan and Korea.

As part of the Asian Development Bank (ADB) country operations business plan for 2015, the Government of Mongolia has requested ADB technical assistance to prepare a strategy for Northeast Asia power system interconnection (NAPSI) using abundant solar and wind energy resources in Gobi desert of Mongolia. In accordance of the development scenarios carried out within NAPSI study, it can be suggested that Northeast Asian regional supergrid can be implemented in following three phases focusing establishment of the ground mounted VLSPV and Wind power plants in Gobi desert of Mongolia and over ±800kV HVDC transmission lines:

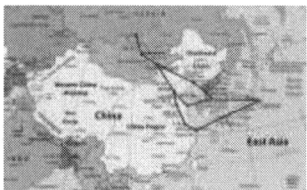 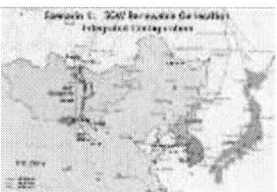

a) Author's proposal   *b) NAPSI Proposal for Phase 1*
Fig.2 The proposed topology of the North East Asian supergrid

Phase 1: Construction of the 5GW of VLSPV and Wind power plants in Gobi desert of Mongolia, mainly for exportation to China

Phase 2: Establishment of the additional 5GW of VLSPV and Wind power plants in Gobi desert of Mongolia, mainly for exportation to China

Phase 3: Increasing capacity up to 100GW of VLSPV and Wind power plants in Gobi desert of Mongolia for exportation to China and other neighbouring countries.

The other option would be in the first phase to establish 5GW of VLSPV and Wind power plants in Gobi desert with construction of the HVDC supergrid connecting Hydropower plants of Far East Russia as a green battery for planned North East Asian supergrid.

According to the ADB supported NAPSI study, under Phase 2 period the development of three new 500kV of HVDC interconnectors has been planned to be constructed, one between China and Mongolia, another between Mongolia and Russia, and the last between China and South Korea

In the Phase 2, additional 5GW of VLSPV and Wind power plants will be developed in the Gobi desert by 2036, bringing total renewable generation capacity to 10GW. Within Phase 2 period a 500kV AC substation would be expanded to double the transformation capacity. The HVDC interconnection between VLSPV and Wind power plants in Gobi desert and Chinese interconnection substation would be uprated to 800kV. In addition, the interconnection between China and South Korea would be doubled to 4GW, still operating at 500kV.

The phase 3 is a long term development scenario of 100GW VLSPV and Wind power generation capacity in Gobi desert. Under this scenario, there are planned to construct eight ±800kV of HVDC interconnectors to China with a total capacity of 70GW. Interconnection to Buryatia, Russia would be increased to 10GW. In this phase, development interconnection capacity between China and South Korea would be increased to 30GW with 3 ±800kV HVDC projects. Interconnection between South Korea and Japan would be increased to 20GW with 2 multi-terminal VSC-HVDC operating at ±800kV. Russia Far East would be interconnected with South Korea with a ±800kV HVDC project of 10GW, with Japan with a ±800KV HVDC link with the capacity of 6GW, and with Harbin of China with a ±800kV HVDC interconnection of 7GW. In addition, the Russia Siberia and

IV. CONCLUDING REMARKS

-All countries of Northeast Asia have recognized that implementation of the Gobitec and Northeast Asian Supergrid initiatives will open bright opportunities for accelerating energy transition forward to shifting renewable energy supply. -Successful implementation of the Gobitec and Northeast Asian Supergrid initiatives will enable increase share of renewable energies which can lead in further reduction of the

cost of electricity for Northeast Asian countries because of high renewable energy potential in the Gobi Desert and dramatic cost reductions of the PV, wind and other renewable energy technologies every year.

The power generation capacity from Solar power plants in Gobi desert area will have a maximum output during the summer, while peak consumption in China, Japan and Korea are usually during the day time and has maximum consumption in the summer. This factor favor that GW capacity generated by solar and wind power plants from Gobi desert could be used potentially in high energy demand regions via regional HVDC transmission supergrid.

Regional energy cooperation, especially establishment of regional HVDC transmission supergrid to harness the full potential of renewable energy sources from the Gobi desert, and other parts of the region is not only economic cooperation and environmental cooperation; it is a high level political collaboration of the Governments to maximize the benefit from use of large scale renewable energy integration.

In order to harness the full potential of renewable energy sources from the Gobi desert, and other parts of the region, all the governments need to have high level political commitment for long term energy cooperation, a better policy and regulatory framework for cross-border energy trade, as well as consolidated regional and national transmission planning on energy integration.

Government of Mongolia have recently approved a 'New revival policy", which highly encourages the possibility of exporting clean electricity using vast solar and wind energy resources of the Gobi desert and has called the leaders of Russia, China, Japan, South Korea to cooperate closely for the implementation of Gobitec initiative and supergrid for the Northeast Asian region. Successful implementation of the Gobitec initiative and Northeast Asia Super Grid should become a real regional project for clean energy supply and might act as a blueprint for other regions in the world.

V. REFERENCES

[1] Energy from the Desert: Feasibility of Very Large Scale Photovoltaic Power Generation (VLS-PV) Systems, Edited by Kosuke Kurokawa, James & James, 2003

[2] Energy from the Desert: Practical Proposals for Very Large Scale Photovoltaic Systems, Edited by Kosuke Kurokawa, Keiichi Komoto, Peter Van Der Vleuten, David Faiman, Earthscan, 2007

[3] Energy from the Desert: Very Large Scale Photovoltaic Systems: Socio-economic, Financial, Technical and Environmental Aspects, Edited by Keiichi Komoto, Masakazu Ito, Peter van der Vleuten, David Faiman, Kosuke Kurokawa, Earthscan, 2009

[4] Energy from the Desert: Very Large Scale Photovoltaic Power – State of the Art and Into the Future, Earthscan, Edited by Keiichi Komoto, Tomoki Ehara, Christian Breyer, Sicheng Wang, Edwin Cunow, Earthscan from Routage, London, 2013

[5] Energy from the Desert: Very Large Scale PV Power Plants for Shifting to Renewable Energy Future, 5th Edition, IEA PVPS Task8, External Final Report IEA-PVPS, February 2015, NEDO, Japan

[6] Komoto K., Enebish N., Song J., 2013. Very Large Scale PV Systems for North-East Asia: Preliminary project proposals for VLS-PV in the Mongolian Gobi desert, 39th PVSC, Tampa, June 16-21 2013

Interposed versus Juxtaposed Solar Array Configurations for Agrivoltaics

M. Sojib Ahmed[1], M. Rezwan Khan[2], Anisul Haque[1], Muhammad A. Alam[3] and M. Ryyan Khan[1, a)]

[1]East West University, Dhaka, Bangladesh
[2]United International University, Dhaka, Bangladesh
[3]Purdue University, West Lafayette, Indiana, USA

Abstract— The agrivoltaics (AV) farm are traditionally configured as interposed photovoltaic (IP-AV) panel arrays on crops, however IP-PV requies heightened panel fixtures to provide clearance for cropping machines. This complicates the architecture and increases the price. One wonders if one could achieve the same integrated output (with same crop-loss) by subdividing the land for panels and crops for a juxtaposed (JP-AV) setup. If viable, the JP-AV configuration would simplify AV design considerably. Here we have used state-of-art PV and crop modeling tools (i.e. Purdue Solar Farm Simulator and APSIM) to numerically explore the design considerations of these AV setups in China and India. We find that for a finite land constraint, IP-AV will always yield higher output and the design can be optimized to produce 4 times higher energy than JP-AV. In both cases, output per panels are comparable. IP-AV essentially allows for more panels to be installed on the finite land while maintaining the predefined or allowable crop-loss. These results have important implcations for policymaker in deciding on renewable energy readmaps for a country with limited lands and PV equipment manufacturers in deciding the creating innovative trackers for AV systems.

Keywords— *Agrivoltaics, solar panel, food-energy nexus, bifacial, crop yield.*

I. INTRODUCTION

The energy production costs from large-scale photovoltaics (PV) farms have become competitive with energy production from other renewable/non-renewable energy sources. The world-wide expansion of PV will eventually create a land-use conflict between agriculture and solar farms for many countries in the world. Among the strategies proposed, agrivoltaics (AV) provides a viable approach of co-harvesting crops and solar energy from single land use to balance both food and energy security. Farmer may find the approach attractive because it would stabilize their income and increase economic security.

Many of the previous work on AV has been aimed to evaluate the relative change in microclimate and crop yield under the solar panel considering a certain panel density [1], [2]. Studies on performance of AV farms for different PV array configuratrions primarily focus on leafy crops and vegetables [3]. The relative potential of AV with major crops (e.g., corn, rice, wheat, maize, etc.) and its optimization remains unclear. Since the energy and crop yield both are highly dependent on the received sunlight, the yield of these individual components will compete in an AV farm. For example, a denser PV

arrangement will significantly affect the crop yield. That is why AV farm's layout design and optimization should be based on ensuring a minimum threshold of crop production. In fact, some countries set policies on minimum crop production requirement to adopt agrivoltaics [4].

In this paper, we will focus on AV farms on major crops— this is of special interest due to the vast arable lands with the possibility of nationwide PV expansion. The common concept of AV farms involves sparsely placed solar panel arrays *interposed (IP)* on a crop land, see Fig.1 (a). Thus a land and the incident insolation is shared by the panels and the crops. This tradiational design, however, introduces added complexity in the array architecture. For example, the panel fixtures need to be high enough for the crops and the machines underneath, the wiring need to go underground or overhead with proper safety— all such details adds to initial and maintenance costs, and difficulty in logistics. Once the AV farm is operational, we would expect a fractional loss in crop (f_A) due to the sparse shadows from the panels.

Keeping this crop-loss and structural challenges in mind, we can imagine an alternative design of an AV farm, where f_A-fraction of the land could be given exclusively for densely packed panel array. The rest of the $(1 - f_A)$ fraction of land would have unhindered crop production resulting in the same crop yield as in the *interposed (IP)* AV farm. In a such *juxtaposed (JP)* AV farm shown in Fig. 1(b), the panels can have conventional PV-farm-like packing and architecture within its allotted sub-land. We would now need to answer: (i) how much difference in energy production would we see in IP-AV versus JP-AV farms? (ii) Is the energy difference small enough to justify the simpler architecture of JP-AV?

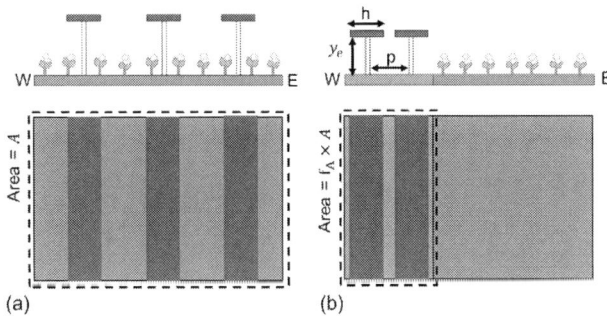

Fig. 1. Schematic of (a) interposed (IP) AV, and (b) juxtaposed (JP) AV.

---

a) email: ryyan@ewubd.edu , ryyan.khan.eee@gmail.com

978-1-7281-6118-1/22 $31.00 © 2022 IEEE

In this study, we investigate the performance of AV farms consisting of horizontal East-West (h-E/W) bifacial and monofacial panel arrays on rice (*Oryza sativa*). In the h-E/W panel array configuration, the shadow on the ground moves throughout the day, resulting in a overall uniform, daily-integraed light distribution. Through detailed and coupeld PV-array and crop modeling we compare IP-AV and JP-AV farms for various crop-loss thresholds f_A. We analyze the systems for China (Jiangsu) and India (Haryana), the two largest rice-producing countries.

II. Modeling

To analyze the entire AV system, we have used several sub-model (e.g., light collection model, temperature-dependent PV efficiency model, PV output electric model, rice yield model). All these sub-models are integrated with a common framework that predicts crop and energy production for any weather and system design considerations.

A. PV farm modeling

The highly sophisticated Purdue Solar Farm Simulator (PSFS) uses solar position model [5] implemented in PVLIB [6] to estimate time-varying and location-specific solar positions and corresponding solar angles. Location-specific ambient data (e.g., Global Horizontal Irradiance (GHI), ambient temperature) are collected from NASA's POWER database [7]. Then we decompose the GHI data into direct (DNI) and diffuse (DHI) components using Orgill-Hollands and Perez models [8], [9]. We use the View-Factor Model to calculate direct, diffuse, and albedo light collection by PV array and non-uniform light distribution on the ground. The amount of albedo light collection by the panel depends on the ground albedo R_A value. For this study, we assume that the albedo value of the crop is $R_A = 0.23$. The total light collection is then used in an electrical model for panel and integrated over time to get daily or yearly energy. Besides the local ambient conditions, the PV yield and ground light distribution depend on PV design parameters, e.g., the panel tilt β, row spacing or pitch p, panel elevation from the ground y_e, panel width h, and array azimuth angle γ_A, albedo R_A. Therefore, the crop and energy yield can be optimized based on these parameters.

B. Rice yield modeling

For rice yield modeling, we couple our PV farm light collection model with APSIM-Oryza [10]. The spatial distribution of daily insolation on ground calculated from our location-aware PV-model is used as input in the APSIM simulator to calculate non-uniform rice yield distribution on the ground. Of course, we also include the location-specific weather data (e.g., maximum temperature, minimum temperature, and rainfall) and soil and management information.

C. Performance metric

Consider an AV-farm of area A (m^2). For a given pitch to panel-width ratio ($p_h = p/h$), we can find the yearly energy yield per land area, E_{PV} (kWh/m^2).

For an allowable crop-loss f_A, optimum energy generation from an AV farm depends on the panel density. Accordingly, the (interposed) IP-AV farm will have panels sparsed at pitch p_{IP}

encompassing all of area A. Then total harvested energy from the IP-AV farm is given by

$$E_{AV-IP}^t = A \times E_{PV}(p_{IP}) \tag{1}$$

Now, for a (juxtaposed) JP-AV farm, by definition, the panel arrays cover an area of $f_A \times A$ out of the area A of the AV farm. In this case, the panels can be much densely packed with pitch $p_{JP} < p_{IP}$. The total energy production from the JP-AV farm will be,

$$E_{AV-JP}^t = (f_A \times A) \times E_{PV}(p_{JP}) \tag{2}$$

Typically, for a practical solar farm, after leaving row spacing for maintenance, p_{JP}/h ratio is ~2.5 to 3. For both AV farm configurations, the crop output is the same but the net energy contribution would be different. The relative performance can be compared through the following energy ratio:

$$f = \frac{E_{PV}(p_{IP})}{E_{PV}(p_{JP}) \times f_A} \tag{3}$$

Here, $f > 1$ would indicate higher energy yield from the more commonly known IP-AV configuration.

III. Results

A. Rice yield in IP-AV

Fig. 2 shows the single season rice yield production (Y_C) of an IP-AV farm as a function of PV array density. The 'open farm' rice yield ($Y_{C(open)}$) shown by the dashed lines represent conventional agriculture (no panel array). The rice production increases with decreasing panel density (or increasing p/h). This relationship is linear up to $p/h = 4$ and starts to saturate afterwards. The crop-loss fraction, i.e.,

$$f_A = 1 - \frac{Y_C(p_{IP})}{Y_{C(open)}},$$

is also shown on the right hand axis in Fig. 2. This gives us the mapping between p_{IP} and f_A.

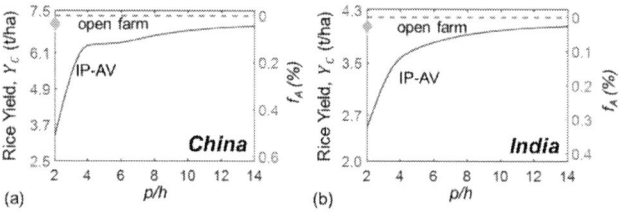

Fig. 2. The rice yield from IP-AV farm as a function of the panel array pitch shown for (a) China, and (b) India. The crop-loss factor is also shown on the right axis. The diamond symbols represent recent open farm rice production data [11].

B. Energy output at different panel pitch

Fig. 3 shows the yearly energy output (E_{PV}) of a horizontal E/W oriented panel array as a function of pitch. As E_{PV} is the output per land area, this is expected to decrease as we space out the panels (i.e., increasing p). For example, yield from each of the monofacial panel will not vary with p. Therefore, E_{PV} for the monofacial array decreases as $1/p$. Bifacial panels have higher output compared to monofacials due to the back-face light collection. The %-gain is more prominent at higher p, which is of interest while desgning IP-AV farms. In fact, implementing a bifacial AV farm would be the more economical choice over monofacial.

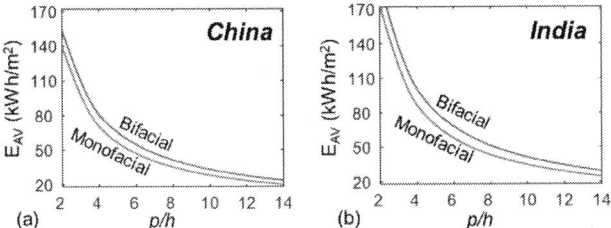

Fig. 3. The yearly energy yield (per land area) for horizontal E/W oriented bifacial and monofacial panel array shown for (a) China, and (b) India.

C. Comparison between IP-AV and JP-AV farms

Fig. 4 shows the energy ratio (IP-AV to JP-AV), f as defined in Eq. (3). For all cases, $f > 1$ indicating IP-AV will always yield higher energy compared to JP-AV, for a given land. For these calculations, we have chosen $p_{JP}/h = 3$ as the typical panel array period. In both China and India, for a reasonable crop-loss of $f_A = 10\text{-}15\%$, we see that $f > 4$, i.e., IP-AV will give more than 4-folds energy compared to JP-AV. Although, the analysis is done for horizontal panels, the south facing optimally tiled bifacial panels would result in only ~5% additional output E_{PV} at $p/h = 3$. This will have insignificant change in the predictions of energy ratio f shown in Fig. 4.

Finally, we can explain the decreasing trends in f vs. f_A as follows. In the limit (albeit impractical) when f_A is large (i.e., crop production is very low), we would be using almost all of the cropland for dense PV-array in JP-AV. Therefore, the optimum IP-AV energy output approaches JP-AV, and f is almost unity. As we increase p/h beyond 2 or 3, crop yield Y_C in IP-AV increases (i.e., f_A decreases) steeply, see Fig. 2. As Y_C increases (i.e., f_A decreases) faster than the p/h-ratio, IP-AV becomes favourable over JP-AV. Essentially, given a fixed land area, we will be able to install more panels in IP-AV compared to JP-AV, and hence the large energy ratio f.

IV. DISCUSSIONS AND CONCLUSIONS

Common agrivoltaics farm design involve higher panel fixtures for cropping machines to move freely. This complicates the architecture and increases the relevant costs. Such interposed panel arrays on crops (IP-AV) will reduce the crop yield. We could in principle, partition the land for exclusive panel-array and crop in each subdivision. The juxtaposed setup (JP-AV) will simplify the panel array fixtures and reduce its costs. Both these configurations may be under consideration for nationwide expansion of AV to scale-up renewable energy capacity. In this work, we have numerically modeled PV yield and crop output in IP-AV for China and India. Ideally, we would want the crop-loss in AV to be less than 15%. In such cases, given a fixed land, IP-AV may produce 4 times higher energy than JP-AV. Between the two configurations, each panel would produce the same energy—the underlying difference is that, IP-AV allows us to place more panels on the same land while maintaining the same, predefined crop loss.

Finally, IP-AV is expected to have marginally higher levelized cost of energy (LCOE) compared to PV-farm or JP-AV. It may be a matter of policy and the limitation of land for progressing through PV-scaling roadmap in a country which will dictate the choice of IP-AV over JP-AV. Like other industries, as we implement more AV farms throughout the world, the IP-AV architecture may become standardized and simplified.

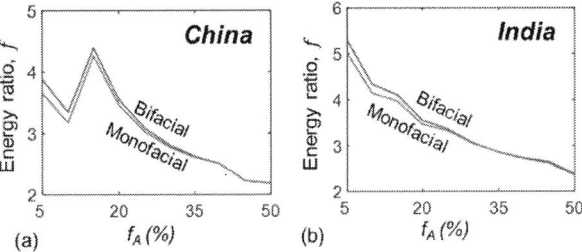

Fig. 4. The energy ratio (IP-AV to JP-AV) are shown for (a) China, and (b) India. The sudden jump in (a) near $f_A = 10\%$ is due to the jump in crop yield (and f_A) around $p/h = 6$ seen in Fig. 2(a).

ACKNOWLEDGMENT

The work was funded by UIU-IAR, ref. UIU/IAR/02/2019-20/SE/09. We also thank United International University (UIU) and BdREN for providing computational support.

REFERENCES

[1] G. A. Barron-Gafford *et al.*, "Agrivoltaics provide mutual benefits across the food–energy–water nexus in drylands," *Nat. Sustain.*, vol. 2, no. 9, Art. no. 9, Sep. 2019, doi: 10.1038/s41893-019-0364-5.

[2] H. Marrou, J. Wery, L. Dufour, and C. Dupraz, "Productivity and radiation use efficiency of lettuces grown in the partial shade of photovoltaic panels," *Eur. J. Agron.*, vol. 44, pp. 54–66, Jan. 2013, doi: 10.1016/j.eja.2012.08.003.

[3] M. H. Riaz, H. Imran, R. Younas, M. A. Alam, and N. Z. Butt, "Module Technology for Agrivoltaics: Vertical Bifacial Versus Tilted Monofacial Farms," *IEEE J. Photovolt.*, vol. 11, no. 2, pp. 469–477, Mar. 2021, doi: 10.1109/JPHOTOV.2020.3048225.

[4] S. Schindele *et al.*, "Implementation of agrophotovoltaics: Techno-economic analysis of the price-performance ratio and its policy implications," *Appl. Energy*, vol. 265, p. 114737, May 2020, doi: 10.1016/j.apenergy.2020.114737.

[5] M. R. Khan, M. T. Patel, R. Asadpour, H. Imran, N. Z. Butt, and M. A. Alam, "A review of next generation bifacial solar farms: predictive modeling of energy yield, economics, and reliability," *J. Phys. D: Appl. Phys.*, vol. 54, no. 32, p. 323001, May 2021, doi: 10.1088/1361-6463/abfce5.

[6] J. S. Stein, W. F. Holmgren, J. Forbess, and C. W. Hansen, "PVLIB: Open source photovoltaic performance modeling functions for Matlab and Python," in *2016 IEEE 43rd Photovoltaic Specialists Conference (PVSC)*, Jun. 2016, pp. 3425–3430. doi: 10.1109/PVSC.2016.7750303.

[7] "NASA POWER | Prediction Of Worldwide Energy Resources." https://power.larc.nasa.gov/ (accessed Oct. 07, 2021).

[8] J. F. Orgill and K. G. T. Hollands, "Correlation equation for hourly diffuse radiation on a horizontal surface," *Sol. Energy*, vol. 19, no. 4, pp. 357–359, Jan. 1977, doi: 10.1016/0038-092X(77)90006-8.

[9] R. Perez, R. Seals, P. Ineichen, R. Stewart, and D. Menicucci, "A new simplified version of the perez diffuse irradiance model for tilted surfaces," *Sol. Energy*, vol. 39, no. 3, pp. 221–231, Jan. 1987, doi: 10.1016/S0038-092X(87)80031-2.

[10] D. P. Holzworth *et al.*, "APSIM – Evolution towards a new generation of agricultural systems simulation," *Environ. Model. Softw.*, vol. 62, pp. 327–350, Dec. 2014, doi: 10.1016/j.envsoft.2014.07.009.

[11] "FAOSTAT." http://www.fao.org/faostat/en/#data/QC (accessed Apr. 21, 2021).

Fabrication of Microscale Back-Contact Arrays for Local Charge Transport Measurements

Kaden M. Powell[1], Yu-Lin Hsu[1], Etee Kawna Roy[1], David J. Magginetti[2], and Heayoung P. Yoon[1, 2*]

[1]Electrical and Computer Engineering, [2]Materials Science and Engineering, University of Utah, Salt Lake City, UT 84112, USA

Abstract — **Remarkable progress has been achieved in CdTe thin-film solar cells by maximizing open-circuit voltage (V_{oc}) and short-circuit current (J_{sc}) via bandgap optimization and interfacial engineering. Fundamental knowledge of the local carrier dynamics at microstructures in advanced PV architectures is required for further improvement of the performance. Scanning probe techniques have been extensively used for investigating inhomogeneous microstructural properties of CdTe solar cells. While extremely useful, the tip-based platforms are often limited to surface roughness, slow scan speed, and narrow field of view (< 10's of µm). In this work, we report the fabrication of microscale back-contact array to measure the optoelectronic properties of individual grains and grain boundaries. This platform allows the intergranular (across a grain boundary) and intragranular (within a grain) local transport measurements. The laser-beam lithography developed in this work enables the parallel production of many identical microscale contacts, providing sufficient datasets for statistical analysis. We discuss the critical lithography conditions (e.g., exposure power) to optimize the fabrication processes for rough surface CdTe solar cells. As proof of concept, we design and fabricate micro-contacts on individual grains of CST ($CdSe_{0.1}Te_{0.9}$) grown by colossal grain growth. We show intragranular charge transport of this sample is promoted compared to intergranular transport under 1-sun illumination. Our results confirm that innate grain boundaries of CGG-grown CST film act as an active barrier, often observed in conventional polycrystalline semiconductors. The fabrication versatility of microscale contact arrays offers a robust measurement platform for studying inhomogeneous local carrier transport in CdTe-based solar cells and other types of thin-film PVs (e.g., CIGS, CZTS, perovskites).**

Keywords — *CdTe solar cells, microstructures, CdSeTe, local transport, microcontacts*

I. INTRODUCTION

Thin-film solar cells are promising photovoltaic (PV) technologies owing to high optical absorption/conversion properties and cost-effective manufacturing processes. Researchers optimized the front contact that leads the J_{sc} over 31 mA/cm² using (Zn, Mg)O buffer layer and/or alloying CdTe with Se [1, 3]. Reese *et al.* reported a V_{oc} over 1 V with group-V doped single crystal [4]. Recent studies proposed the V_{oc} improvement in polycrystalline CdTe PVs requires well-passivated back contact [5, 8]. Fundamental knowledge of the local carrier dynamics at microstructures in advanced PV architectures is required for further improvement. [9].

Numerous techniques have been applied to measure the spatially inhomogeneous optoelectronic properties of thin-film solar cells. Scanning probe-based methods have revealed significant electronic fluctuations at CdTe grain boundaries compared to adjacent grain interiors. Examples include Kelvin probe force microscopy (KPFM) [10], conductive atomic force microscopy (c-AFM) [11], and near-field scanning optical microscopy (NSOM) [12]. While extremely useful, these methods are often limited to surface roughness, thermal drift, and narrow field of view (< 10's of µm), which together are not insufficient for statistical analysis. Other studies have attempted to directly measure the local carrier transport using small contacts that were defined on individual CdTe microstructures [Fig. 1(a)]. Kephart *et al.* demonstrated the fabrication of point contacts for Al_2O_3 passivated CdTe solar cells [13]. While they successfully fabricated point contacts in the oxide layer with diameters of about 3 µm, the devices suffered from poor fill factor and insignificant improvement in V_{oc}. In the past, we reported a local probing method using platinum (Pt) nanocontacts fabricated by a focused-ion beam [FIB; Fig. 1(b)]. Nanomanipulators were positioned on the contacts, while the excess carriers were generated via electron-beam excitation in a scanning electron microscope (SEM). We observed about 2X higher carrier collection at grain boundaries than adjacent grain interiors, which may be attributed to the local band-bending [14]. However, the FIB-produced contacts significantly reduced the conductivity, limiting the local cell operation (i.e., poor V_{oc} and *FF*).

In this work, we report the fabrication of microscale back-contact array to measure the optoelectronic properties of individual grains and grain boundaries. By optimizing the laser-beam dose, we pattern microscale back-contact arrays on the rough CdTe surface (peak-to-valley roughness ≈ 1 µm). The current-voltage (*I-V*) characteristics at different doses are

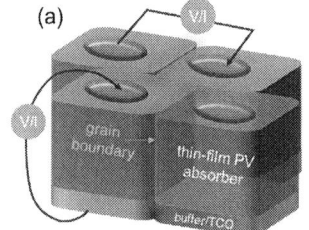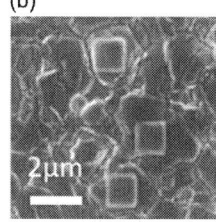

Fig. 1. (a) Schematic illustrating a small contact platform. (b) SEM image of Pt-contacts on CdTe produced by a focused-ion beam (FIB) system.

compared to control devices fabricated using a shadow mask. As proof of concept, we design and fabricate micro-contacts on individual grains of CST ($CdSe_{0.1}Te_{0.9}$) grown by colossal grain growth (CGG). We show intragranular charge transport of this sample is promoted compared to intergranular transport under 1-sun illumination.

II. EXPERIMENTAL METHODS

The CdS/CdTe solar cells used in this work were cleaved from a commercially available CdTe panel. The size of individual samples was a few centimeters in length and width. A similar extraction process was described in our previous work [15]. The back contacts on the samples were fabricated using laser beam photolithography. Fig. 2 illustrates each step of the fabrication process.

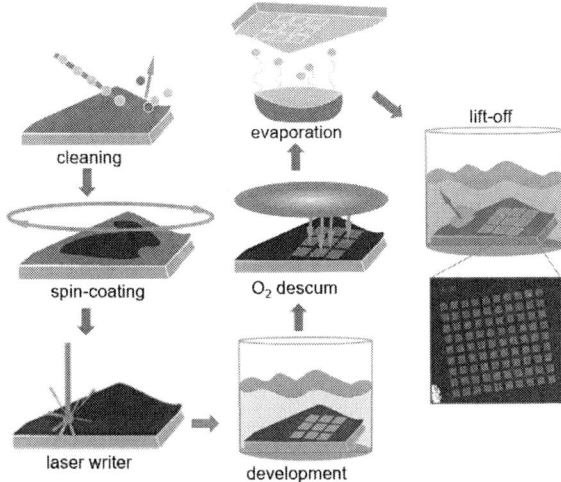

Fig. 2. Schematics illustrating small contact fabrication. An optimized exposure condition is required to fabricate well-defined microscale contacts on a rough surface CdTe.

Prior to lithography, all samples were sequentially cleaned with acetone, isopropanol, deionized water, and N_2 blown dry. A photoresist (Shipley, S1813) was spun on the sample at 3000 rpm for 1 min followed by a 90 s soft-bake at 110 °C. While the final photoresist thickness was at approximately 1.5 µm on a smooth and planar Si control sample, the thickness of the photoresist on CdTe can be in the range of 1 µm to 3 µm due to its surface roughness. The sample was then exposed using a laser writer (405 nm) at various exposure powers from a pixel pulse of 60 % duration at 10 mW to 75 % of 20 mW. The photoresist was then developed in the developing solution (AZ 1:1) at room temperature for 60 s to 90 s. An O_2 plasma descum (base pressure of 100 mTorr) was then performed at 50 W for 60 s to remove the residual photoresist. In preparation for metal deposition, the edges of the CdS/CdTe sample were sealed using a photoresist with an additional soft bake at 85 °C for 30 s to prevent shorting of the junction. Au was evaporated onto the sample with thicknesses ranging from 50 nm to 150 nm. Lastly, a liftoff process was performed in acetone to remove the photoresist and the metal flakes, leaving Au small contacts. An optical microscope image of representative contacts fabricated with this method is shown at the far right of Fig. 2. As a control, we used a shadow mask consisting of an array of square holes (90 µm × 90 µm). The

shadow mask was placed on the clean CdTe surface, and the same metallization conditions of the lithography were used.

III. EXPERIMENTAL RESULTS AND ANALYSIS

Our lithography approach to fabricating microscale back-contact arrays is versatile, and it can be applicable for other types of inorganic solar cells. The individual contact size can be fabricated in a range of a few centimeters to 0.7 µm (diffraction limit of our laser beam source). We note that the optimizing laser dose for patterning/developing photoresist for the ***rough CdTe surface*** is critical to producing well-defined contact arrays. In this work, we designed an array of squares (90 µm × 90 µm) to form microscale back-contacts on CdS/CdTe solar cells at different doses. To assess the role of photoresist residue, we also used a commercially available shadow mask having the same size of square arrays. The measured dark and light *I-V*s using ten contacts of each set are shown in Fig. 3.

The exposure conditions of 10 mW at 60 % (S1) , 18 mW at 45 % (S2), and 18 mW at 75 % (S3) are corresponding to the net power of 6.0 mW, 8.1 mW, and 13.5 mW, respectively. We observed the improved *I-V* characteristics with the contacts fabricated under high exposure power. As seen in Fig. 3, the overall V_{oc} is increased from 0.77 V to 0.82 V with the

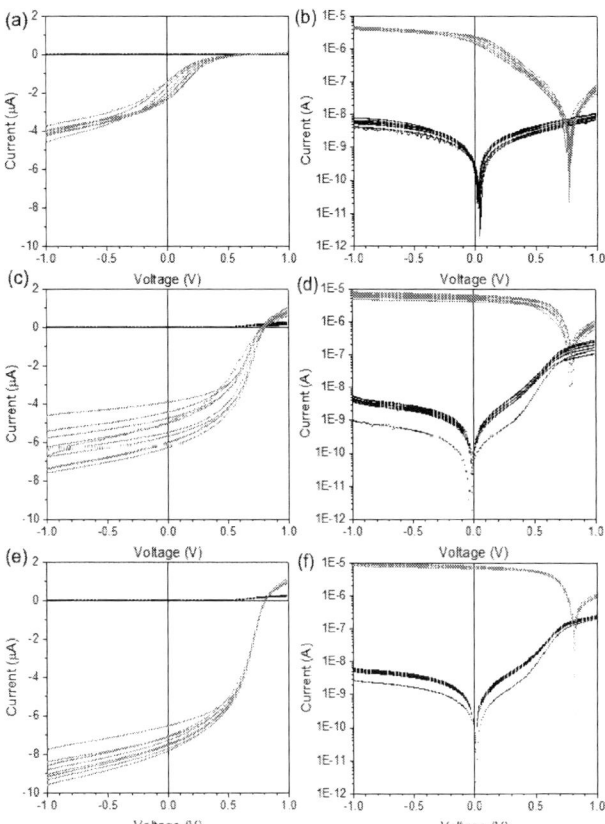

Fig. 3. *I-V* characteristics of CdS/CdTe solar cells with lithographically patterned contacts (90 µm × 90 µm). The dark (black) and the light (red) *I-V*s were measured with the contacts fabricated at an exposure power of (a, b) 10 mW at 60 %, (c, d) 18 mW at 45 %, and (e, f) 18 mW at 75 %.

exposure power from 6 mW to 13.5 mW. The short-circuit current (I_{sc}) values of S3 devices are approximately 2X higher than those of S1 devices. The wide distribution of I_{sc} of S2 is notably reduced. The calculated Jsc of the small contacts is higher than the estimated value. We speculate the stray light penetration to the thick glass substrate under 1-sun illumination and the variation of the completed contact area. The fill-factor of 0.11 of S3 is significantly improved from 0.11 (S1) to 0.46 (S3) with an increase of the exposure power. The poor electrical behaviors (Si, S2) are likely attributed to photoresist residue, particularly on the valleys of CdTe grains. Almost identical I-Vs measured with the small back contacts with a shadow mask (i.e., no photoresist residue; S4) support our observation.

The fill factor values of 0.46 measured with S3 (optimized dose) and S4 (shadow mask) were far below than that of our control CdS/CdTe solar cells extracted from the same panel (V_{oc} = 0.82 V, J_{sc} = 24 mA/cm², FF = 0.64, efficiency ≈ 12 %). The contact area of the control devices was 500 μm × 500 μm, approximately 30X larger than the small contacts used in this work. An external bias was applied to the entire layer of n-CdS/FTO (e.g., 2 cm × 1cm), while limited to the small contact on p-CdTe (90 μm × 90 μm). Presumably, the asymmetric contact geometry increases the series resistance and shunt resistance, decreasing the FF value of the micro-contacts. Isolation of individual small contacts (e.g., local laser scribing) may help the improvement of the FF.

As proof of concept, we designed and fabricated micro-contacts on individual grains of CGG-grown CST ($CdSe_{0.1}Te_{0.9}$) on an Al_2O_3/glass substrate [16]. An optimized lithography condition developed above was applied to produce an array of contacts (exposure power of 13.5 mW, 70 nm thick Au contacts, 90 μm × 90 μm square arrays). Fig. 4 displays the measured dark and light I-V characteristics. Both

intergranular and intragranular dark I-Vs display slightly non-linear behaviors with the applied voltages (Fig. 4 a, b). The current at ±1 V is in the range of 5 pA to 25 pA. This non-linear characteristic can be attributed to the intrinsically high resistivity of the CST film. Under 1-sun illumination, the light I-Vs became linear with both intergranular and intragranular contact pairs. Interestingly, the magnitude of the current with the intergranular contact pairs increases to 150 nA, whereas 100 nA with the intragranular contact pairs. We speculate that innate grain boundaries of CGG-grown CST film act as an active barrier, often observed in conventional polycrystalline semiconductors.

IV. SUMMARY

In summary, we have demonstrated the fabrication of microscale back-contact arrays to measure the local electronic properties of individual grains and grain boundaries. Optimizing exposure power in lithography is critical to producing high-quality microscale contacts on thin-film PVs. We have shown that the intragranular charge transport of CGG-grown CST film is promoted compared to intergranular transport under 1-sun illumination. Our results confirm that innate grain boundaries of this CST sample act as an active barrier, often observed in conventional polycrystalline semiconductors.

ACKNOWLEDGMENT

The authors thank O. Lam, B. Baker, G. Xiong, and C. Lee for experimental assistance and valuable discussions in this work. We thank D. Albin at NREL for providing CGG-grown CST samples studied in this work. We acknowledge support, in part, by NSF (#1711885) and DOE (DE-EE0008983).

REFERENCES

[1] A. H. Munshi *et al.*, "Polycrystalline CdSeTe/CdTe Absorber Cells With 28 mA/cm2 Short-Circuit Current," *IEEE Journal of Photovoltaics*, vol. 8, no. 1, pp. 310–314, Jan. 2018, doi: 10.1109/JPHOTOV.2017.2775139.

[2] D. E. Swanson, J. R. Sites, and W. S. Sampath, "Co-sublimation of CdSexTe1−x layers for CdTe solar cells," *Solar Energy Materials and Solar Cells*, vol. 159, pp. 389–394, Jan. 2017, doi: 10.1016/j.solmat.2016.09.025.

[3] M. A. Green *et al.*, "Solar cell efficiency tables (Version 53)," *Progress in Photovoltaics: Research and Applications*, vol. 27, no. 1, pp. 3–12, 2019, doi: 10.1002/pip.3102.

[4] M. O. Reese *et al.*, "Intrinsic surface passivation of CdTe," *J. Appl. Phys.*, vol. 118, no. 15, p. 155305, Oct. 2015, doi: 10.1063/1.4933186.

[5] J. N. Duenow and W. K. Metzger, "Back-surface recombination, electron reflectors, and paths to 28% efficiency for thin-film photovoltaics: A CdTe case study," *Journal of Applied Physics*, vol. 125, no. 5, p. 053101, Feb. 2019, doi: 10.1063/1.5063799.

[6] J. Sites and J. Pan, "Strategies to increase CdTe solar-cell voltage," *Thin Solid Films*, vol. 515, no. 15, pp. 6099–6102, May 2007, doi: 10.1016/j.tsf.2006.12.147.

[7] C. A. Wolden *et al.*, "Photovoltaic manufacturing: Present status, future prospects, and research needs," *Journal of Vacuum Science & Technology A*, vol. 29, no. 3, p. 030801, May 2011, doi: 10.1116/1.3569757.

[8] J. Liang *et al.*, "Rectification and tunneling effects enabled by Al2O3 atomic layer deposited on back contact of CdTe solar cells," *Appl. Phys. Lett.*, vol. 107, no. 1, p. 013907, Jul. 2015, doi: 10.1063/1.4926601.

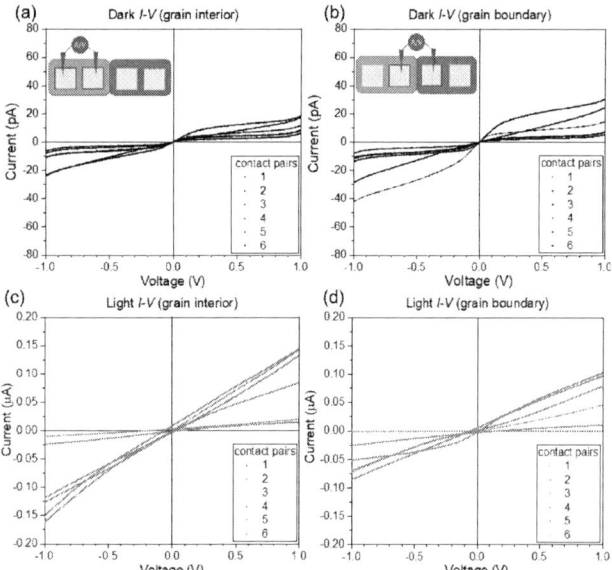

Fig. 4. Dark (black) and light (red) *I-V*s (1-sun illumination) measured with the microscale contact pairs on CST. (a) and (c): intragranular transport (within a grain); (b) and (d): intergranular transport (across a grain boundary).

[9] M. M. Taheri, T. M. Truong, S. Li, W. N. Shafarman, B. E. McCandless, and J. B. Baxter, "Distinguishing bulk and surface recombination in CdTe thin films and solar cells using time-resolved terahertz and photoluminescence spectroscopies," *Journal of Applied Physics*, vol. 130, no. 16, p. 163104, Oct. 2021, doi: 10.1063/5.0064730.

[10] B. E. McCandless and S. Rykov, "Cross-section potential analysis of CdTe/CdS solar cells by kelvin probe force microscopy," in *2008 33rd IEEE Photovoltaic Specialists Conference*, May 2008, pp. 1–4. doi: 10.1109/PVSC.2008.4922532.

[11] H. R. Moutinho and R. G. Dhere, "Conductive Atomic Force Microscopy Applied to CdTe/CdS Solar Cells," p. 7.

[12] L. Marina, M. Abashin, H. Lezec, A. Gianfrancesco, A. Talin, and N. Zhitenev, "Nanoscale Imaging of Photocurrent and Efficiency in CdTe Solar Cells," vol. 8, no. 11, pp. 11883–11890, Oct. 2014.

[13] J. M. Kephart et al., "Sputter-Deposited Oxides for Interface Passivation of CdTe Photovoltaics," in IEEE Journal of Photovoltaics, vol. 8, no. 2, pp. 587-593, March 2018, doi: 10.1109/JPHOTOV.2017.2787021.

[14] H. P. Yoon *et al.*, "Local electrical characterization of cadmium telluride solar cells using low-energy electron beam," *Solar Energy Materials and Solar Cells*, vol. 117, pp. 499–504, Oct. 2013, doi: 10.1016/j.solmat.2013.07.024

[15] K. Powell, Y.-L. Hsu, D. Magginetti, and H. Yoon, "Local Photovoltaic Measurements of CdTe Solar Cells Using Microscale Point Back-Contacts," Univ. of Utah, Salt Lake City, UT (United States), Jun. 2021. Accessed: Jan. 24, 2022. [Online]. Available: https://www.osti.gov/biblio/1810572

[16] D. S. Albin, M. Amarasinghe, M. O. Reese, J. Moseley, H. Moutinho, and W. K. Metzger, "Colossal grain growth in Cd(Se,Te) thin films and their subsequent use in CdTe epitaxy by close-spaced sublimation," *J. Phys. Energy*, vol. 3, no. 2, p. 024003, Jan. 2021, doi: 10.1088/2515-7655/abd297.

978-1-7281-6118-1/22 $31.00 © 2022 IEEE

Agrivoltaic Modules Optimizing Light for Crops in Dryland Regions

Christiana B. Honsberg, Greg Barron-Gafford, Stuart G. Bowden, Robert Sampson

Arizona State University, Tempe, AZ, United States

University of Arizona, Tucson, AZ, United States

Sunesta Solar, Peoria, AZ, United States

The paper presents experimental results on agrivoltaics and an optimized optical design. We show: (1) The water utilization benefit of an agrivoltaic module on crops could be quantified, which is the first time this is demonstrated in agrivoltaic literature; (2) Simultaneous benefits were observed across a range of different crops and impacts were compared on these different crops, which is also the first time this is demonstrated in agrivoltaic literature; (3) Optimal light for plants requires full sun early in the day and shade later in the day; (4) An optical design for the module achieves a light profile which also produces electrical generation that is comparable with conventional PV solar modules.

Understanding Device Performance Limiting Factors by Reproducing The Current-Voltage Characteristics from Transient Optoelectrical Measurements

Abasi Abudulimu, Klaus Eckstein, Mirella El Gemayel, Imge Namal, Adam B Phillips, Michael J Heben, Tobias Hertel, Sebastian B Meier, Larry Lüer, Randy J Ellingson

Wright Center for Photovoltaics Innovation and Commercialization (PVIC), Department of Physics and Astronomy, The University of Toledo, Toledo, OH, United States

IMDEA Nanociencia, C/ Faraday 9, 28049 Cantoblanco, MADRID, Spain

Institute of Physical and Theoretical Chemistry, Julius-Maximilian University Würzburg, Würzburg, Germany

Belectric OPV GmbH, Landgrabenstr. 94, 90943 Nürnberg, Nürnberg, Germany

Institute of Materials for Electronics and Energy Technology (i-MEET), Friedrich-Alexander University Erlangen-Nürnberg, Erlangen, Germany

Pushing the performance of the solar cell to the theoretical limit is the matter of better understanding the charge dynamics from the absorber to the complete device and laying the direction for device/material engineering. Here we study the charge dynamics on a set of bulk heterojunction organic solar cells with/o incorporating semiconducting single-walled carbon nanotubes into the absorber layer. Employing nano-microsecond transient optoelectrical characterization techniques along with femtosecond transient absorption spectroscopy and reproducing the steady-state current-voltage characteristics of the device from the parameter extracted from the transient optoelectrical measurements, we demonstrate that the cost-effective optoelectrical techniques capable of determining the performance limiting factors of the devices under operational conditions.

Impact of the 2019-2020 Australian Black Summer Wildfires on Photovoltaic Energy Production

Ethan Ford[1], Bram Hoex[1], and Ian Marius Peters[2]

[1]UNSW, Sydney, NSW, 2052, Australia
[2]Forschungszentrum Jülich GmbH, 52425 Jülich, Germany

Abstract—**Air pollution produced by the Australian Black Summer wildfires caused extreme haze events across New South Wales (NSW). We analyzed 30-minute resolution energy data from 160 residential PV systems in NSW from 6 November 2019 to 15 January 2020. A percentile data analysis technique was adapted to derive a mean reduction rate for PV energy generation with PM2.5. The mean power reduction rate for PV systems was approximately -12.5% ± 2.2% per 100 μg/m³ of PM2.5 for airmass 1.0. The energy loss for residential PV systems was estimated as 39.8 ± 7.9 GWh, equating to a worst-case financial impact of 9.27 ± 1.85 million AUD. This work aims to help inform PV system planning and energy storage options of new PV systems; and raise awareness of the impact of wildfires and air pollution on solar PV.**

Keywords—*photovoltaic power generation, air quality, wildfires, particulate matter, PM2.5, solar photovoltaic generation*

I. INTRODUCTION

The 2019-20 Australian wildfires were a natural disaster with state- and nation-wide impacts. Areas not directly affected by fire were often subject to extreme smoke, haze and poor air quality, with strong winds carrying heavy smoke hundreds of kilometres (Fig. 1). In January 2020, smoke from the wildfires traversed the Pacific Ocean and was observed in the stratosphere over Punta Arenas, Chile [2]. Such circumstances induced by the wildfires likely led to significant reductions in power output of residential, commercial and utility photovoltaic (PV) systems, particularly in New South Wales (NSW). Consisting predominantly of fine aerosol particulate matter less than 2.5 μm in diameter (PM2.5) [3], wildfire smoke attenuates solar irradiance and leads to soiling via the deposition of particles on the solar modules' surfaces. This reduction in sunlight decreases the electric energy yield of photovoltaic (PV) systems and thus revenue for PV system owners and investors. PV capacity in Australia is projected to increase almost 3.5-fold to 80 GW by 2030 [4] and extreme wildfires are a growing issue due to climate change. Hence, a thorough and accurate understanding of the behaviour of PV output during wildfire events is essential.

Prior work investigating the correlation between air quality and insolation or solar PV performance predominantly entails experimental field data, computational models or field testing. Anthropogenic air pollution in Delhi was observed to reduce sunlight intensity by 12.5% for every 100 μg/m³ of PM2.5 particle concentration [5]. Poor air quality and haze originating from wildfires yielded PV system losses of 15-25% in Singapore [6]. A fire burn event on a clear sky afternoon in Canberra in 2014 resulted in peak reductions in PV generation of 27% [7].

Finally, the first of two recent studies focusing on wildfires in California found a reduction in normalized PV generation of 9.4-37.8% when PM2.5 ranged from 50-200 μg/m³ [8]. The second found a PV capacity derate of 9-49% for aerosol optical depth (AOD) of 0.5-4.5, and geospatial derates in PV capacity of up to 15% for 12:00-13:00 for the Californian fire season [9].

This work aims to quantify the relative PV energy generation loss from wildfire-induced ambient particulate matter; and the resulting financial impacts of the 2019-20 Australian wildfires. The below sections detail the datasets, analysis processes and statistical results regarding the effects of wildfire-induced PM2.5 on residential PV system production.

Fig. 1. NASA satellite image of NSW showing wildfires (red) and smoke across the North-East of the state on 8 November 2019 [1].

II. METHODOLOGY

A. Data Filtering

The dataset consisted of 30-minute cumulative energy data from 710 residential PV systems, and hourly average PM2.5 concentrations from 50 meteorological sites across NSW from 6 Nov 2019 to 15 Jan 2020. A data filtering process was created in Python to detect curtailment and non-recordings (subpar performance) in the PV generation data and exclude PV sites with such issues. Finally, the analysis only investigated PV sites located within 5 km of a PM2.5 monitoring site with air pollution data of at least 95% completeness.

B. Geospatial Analysis

Sites that passed the above filtering process are plotted in Fig. 2 and Fig. 3. Geographical coordinates were available for

978-1-7281-6118-1/22 $31.00 © 2022 IEEE

the PM2.5 sites, but PV site locations were only known accurate to the mean coordinates of their respective postal codes [10].

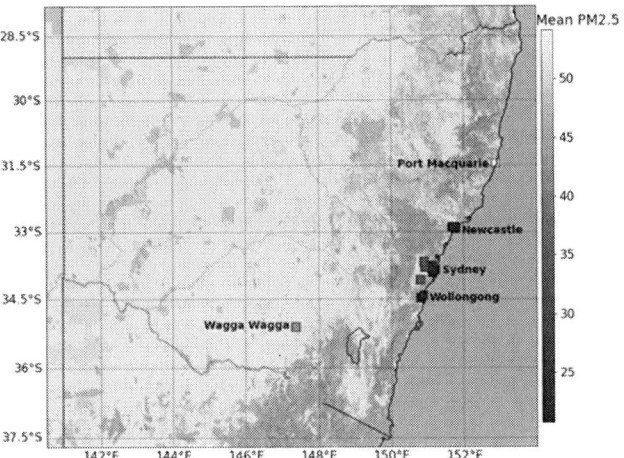

Fig. 2. PM2.5 monitoring sites used in the analysis with mean daytime PM2.5 concentrations for the wildfire period in μg/m³.

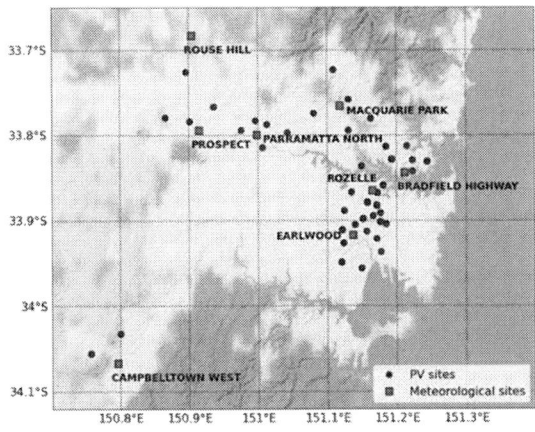

Fig. 3. Distribution of analyzed PV and PM2.5/meteorological sites across Sydney. Blue dots indicate postal codes that contain at least one PV site.

C. 80-Percentile Clear Sky Filter

PV generation data was sorted into a two-dimensional array of bins according to hour of the day and PM2.5 concentration. The 80-percentile of each bin represents clear sky (sunny) conditions for each PM2.5 range as in Fig. 4.

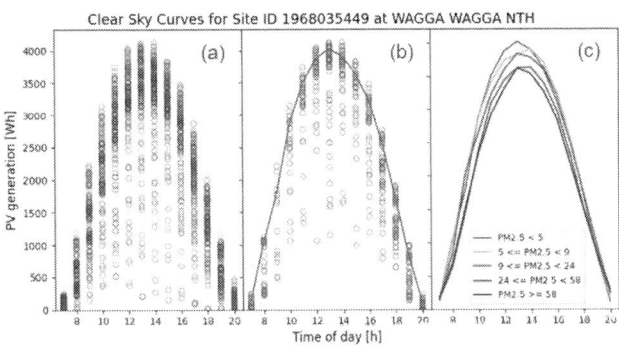

Fig. 4. Creation of clear sky curves. (a) Raw binned data; (b) taking the 80-percentile; (c) repeating step 'b' for all PM2.5 ranges (units μg/m³).

III. RESULTS AND DISCUSSION

A. Relative Power Reduction Rates

PV generation was normalized to negligible air pollution (i.e., the green curve of Fig. 4) for each hour and corrected for atmospheric optical pathlength (airmass (AM)). The results for a single PV site are plotted in Fig. 5 showing the trend in normalized generation for each hour of the day (color bar). This process was repeated for all PV sites and the results are plotted in Fig. 6 in the form of a 2D kernel density of normalized PV data points for all PV sites. Beer Lambert's Law was fit to the data to find the mean relative reduction rate as in equation (1).

$$PV(PM2.5) / PV_0 = \exp(-R * AM * PM2.5) \qquad (1)$$

Where:

- PV(PM2.5) is the measured PV energy production;
- PV_0 is the energy production for zero air pollution;
- R is the relative reduction rate for the PV system;
- AM is the mean airmass for the given hour; and
- PM2.5 is the mean PM2.5 concentration for the given hour.

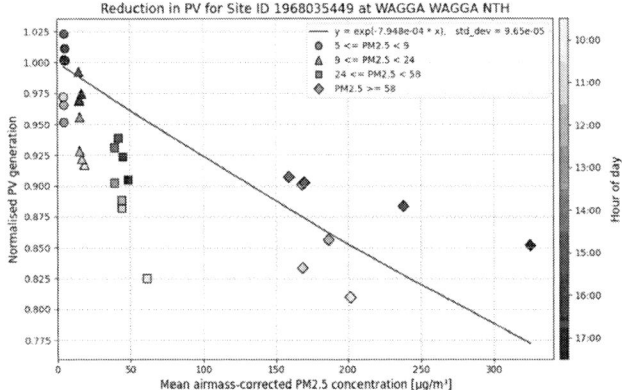

Fig. 5. Normalized PV generation by hour (from 80-percentile filter) and PM2.5 range (μg/m³) plotted against the mean product of airmass and PM2.5.

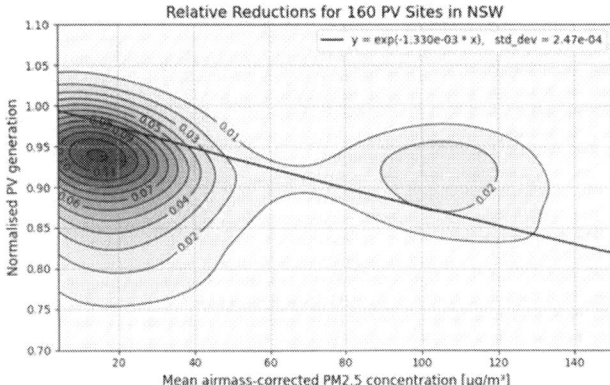

Fig. 6. 2D kernel density plot of normalized PV generation for all PV sites with clean data showing the mono-exponential decay best fit of Beer-Lambert's Law (blue line) and the corresponding equation in the top-right of the chart.

B. Energy and Financial Impact

The relative reduction rate from equation (1) was used to correct PV generation for no air pollution (0 µg/m³ PM2.5) and thus estimate the effective PV energy loss from wildfire smoke using equation (2). This involved finding the relative difference between the sum of the measured and corrected PV energy generation time series data of each PV system.

$$PV_{loss} = [\Sigma(PV_0) - \Sigma(PV(PM2.5))] / \Sigma(PV_0) \qquad (2)$$

The worst-case financial loss (R_{loss}) assumed self-consumption of all PV generation and was calculated as per equation (3) with a residential electricity price (C) of 0.2329 AUD/kWh [11].

$$R_{loss} = C * PV_{loss} \qquad (3)$$

This process was also carried out for a single PV site in Wagga Wagga Nth for a clear sky but smokey day with an average daytime PM2.5 concentration of 111 µg/m³ on 23 Dec 2019. The greatest relative energy loss of 42.4% was observed from 7:30am-8am local time and the mean energy loss for the day was 16.8%. A summary of the findings discussed in this section are presented in Table I.

It should be noted this analysis utilized point measurements of PM2.5 near the ground as a proxy for the integral of wildfire smoke through the entire atmosphere. It also did not directly correct for environmental factors that affect PV performance like ambient air temperature, wind speed and soiling.

TABLE I.   IMPLICATIONS FOR 582 000 RESIDENTIAL PV SYSTEMS IN NSW WITH STD. DEVIATIONS AS ERROR BOUNDS [12].

State-wide loss scenario	Energy loss [%]			Energy loss [GWh]			Financial loss [million AUD]		
	Lower	Mean	Upper	Lower	Mean	Upper	Lower	Mean	Upper
71-day wildfire season	3.42	4.22	5.03	32.0	39.8	47.8	7.45	9.27	11.1
Smokey day [111 µg/m³]	13.9	16.8	19.7	1.83	2.23	2.63	0.426	0.520	0.613

IV. CONCLUSIONS AND FUTURE WORK

A Python data analysis program was developed to analyze historic PV system energy data for 160 PV systems, and PM2.5 concentration data for 17 meteorological stations across NSW. A self-referencing 80-percentile data analysis technique developed by Peters et al. [5] was employed to describe the correlation between PV system energy production and ambient PM2.5 concentration for the 71-day period from 6 Nov 2019 to 15 Jan 2020. A novel technique for airmass correction was developed to account for the optical path length of sunlight through the atmosphere, which builds upon the previous analysis of Peters et al..

The reduction in PV generation due to wildfire smoke was -12.5% ± 2.2% per 100 µg/m³ of PM2.5. The corresponding statewide energy loss for residential PV systems was approximately 39.8 ± 7.9 GWh for the 71-day wildfire period, giving a worst-case economic loss of approximately 9.27 ± 1.85 million AUD for rooftop PV system owners and investors.

Wildfire smoke reduced PV power output by 16.8% over the course of a moderately hazy day. However, mean long-term energy losses from the wildfires were relatively small at just 4.22% and may have been exceeded by soiling or curtailment.

Recommended future work could include:

- Correcting PV generation for environmental variables;
- Accounting for soiling losses;
- Running the analysis on a different PM2.5 dataset;
- Investigating a PV dataset with voltage and current measurements;
- Deriving alternative percentile values for the 80-percentile filter to suit different climates; and
- Exploring long term damages caused by wildfires and wildfire smoke on the reliability and longevity of PV modules in the field.

ACKNOWLEDGMENT

This work has been inspired by the devastating natural disaster of the 2019-20 Australian wildfires, of which the author's family and friends were directly affected. The authors acknowledge the work of UNSW and Solar Analytics to allow access to photovoltaic system data, and to the NSW Department of Planning, Industry and Environment for access to meteorological data. This work describes objective scientific results and data analysis. Any subjective views that might be expressed in this work are not necessarily representative of the views of UNSW or Forschungszentrum Jülich GmbH.

REFERENCES

[1] "EOSDIS Worldview." [online] Available at: https://worldview.earthdata.nasa.gov/

[2] K. Ohneiser et al., "Smoke of extreme Australian bushfires observed in the stratosphere over Punta Arenas, Chile, in January 2020: optical thickness, lidar ratios, and depolarization ratios at 355 and 532 nm," Atmos. Chem. Phys. Discuss., no. December 2019, pp. 1–16, 2020, doi: 10.5194/acp-2020-96.

[3] E. Ward, "Smoke from wildland fires," Heal. Guidel. Veg. FIRE EVENTS Backgr. Pap., no. October 1998, pp. 70–85., 1999, [online]. Available at: http://www.preventionweb.net/files/1905_VL206106.pdf%5Cnpapers2://publication/uuid/87FA08A8-ECDD-41CD-952E-16982F7699A6.

[4] "Australia's solar capacity could hit 80 GW by 2030, says GlobalData – pv magazine International." [online] Available at: https://www.pv-magazine.com/2021/05/11/australias-solar-capacity-could-hit-80-gw-by-2030-says-globaldata/

[5] I. M. Peters, S. Karthik, H. Liu, T. Buonassisi, and A. Nobre, "Urban haze and photovoltaics," Energy Environ. Sci., vol. 11, no. 10, pp. 3043–3054, 2018, doi: 10.1039/c8ee01100a.

[6] A. M. Nobre et al., "On the impact of haze on the yield of photovoltaic systems in Singapore," Renew. Energy, vol. 89, pp. 389–400, 2016, doi: 10.1016/j.renene.2015.11.079.

[7] M. Perry and A. Troccoli, "Impact of a fire burn on solar irradiance and PV power," Sol. Energy, vol. 114, pp. 167–173, Apr. 2015, doi: 10.1016/J.SOLENER.2015.01.005.

[8] S. D. Gilletly, N. D. Jackson, and A. Staid, "Quantifying Wildfire-Induced Impacts to Photovoltaic Energy Production in the western United States."

[9] D. L. Donaldson, D. M. Piper, and D. Jayaweera, "Temporal Solar Photovoltaic Generation Capacity Reduction from Wildfire Smoke," IEEE Access, vol. 9, pp. 79841–79852, 2021, doi: 10.1109/ACCESS.2021.3084528.

[10] "Australian Post Codes - Matthew Proctor." [online] Available at: https://www.matthewproctor.com/australian_postcodes

[11] "Electricity Costs Per kWh | QLD, SA, VIC, NSW Rates – Canstar Blue." [online] Available at: https://www.canstarblue.com.au/energy/electricity/electricity-costs-kwh/

[12] "Solar energy | Energy NSW." [online] Available at: https://www.energy.nsw.gov.au/renewables/renewable-generation/solar-energy

978-1-7281-6118-1/22 $31.00 © 2022 IEEE

Combining nanoscale 3D printing with spark ablation to achieve novel nanostructured surfaces for photovoltaic applications

Ivana Panzic, Alexander Jelinek, Floren Radovanovic-Peric, Daniel Kiener, Vilko Mandic

Faculty of Chemical Engineering and Technology, Zagreb, Croatia

Department of Materials Physics, Montanuniversität Leoben, Leoben, Austria

Laser polymerization has emerged as a direct writing technique allowing the fabrication of complex 3D structures with microscale resolution. The technique provides rapid prototyping capabilities for a broad range of applications, but to meet the growing interest in 3D nanoscale structures the resolution limits need to be pushed beyond the 100 nm benchmark, which is challenging in practical implementations. By using a two-photon polymerization process precise structures in the range of 40 to 50 nm can be achieved. Subsequent post-processing of the printed nanostructures by means of plasma etching or pyrolysis opens the possibilities to obtain even smaller 3D structures, only limited by the mechanical properties of the polymerize resist and the geometry. On the other hand, spark ablation recently emerged as a technique capable of preparing reproducibly sized and clean nanoparticles in a cost-effective manner. Here we employ the outcome of combining the abovementioned processes. Spark ablation process was used to decorate the printed 3D surface to yield specific surfaces with metal/metal oxide core-shell nanoparticles. Broad characterization was applied using microscopy (SEM, AFM), mechanical testing (in situ SEM mechanical testing), diffraction analysis (XRD), and electrical characterization (J/V)) before and after the assembly of complete solar cells. Namely, such formations were found to be prosperous for electron transport layers in perovskite solar cells.

978-1-7281-6118-1/22 $31.00 © 2022 IEEE

Understanding configuration of geopolymer materials for application in solar-cells

Arijeta Bafti, Filip Brleković, Vilko Mandić, Luka Pavić, Ivana Panžić, Andraž Krajnc

University of Zagreb, Faculty of Chemical Engineering and Technology, Zagreb, Croatia

Ruđer Bošković Institute, Zagreb, Croatia

National Institute of Chemistry, Ljubljana, Slovenia

Geopolymers are ceramic-like inorganic polymers, that nowadays push through as alternatives to concrete materials, which is important for civil engineering as the geopolymers do not contribute to energy and environmental issues. Specifically, a much lower environmental footprint is present which significantly reduces CO2 emission, and opens up the space for energy conservation by applying a completely new approach. A particularly neat contribution in the overlapping areas of construction and energy materials could arise from preparing geopolymers in the form of a paste that can further be developed into thin-films (TF). Such conductive and transparent films open op the possibility for application in solar-cells. In this work, we focused on the chemical and (micro)structural changes that occur during the (pre)mullite-based geopolymer curing. Various factors influenced changes, such as type of ions in activation solution and additives that improve targeted properties of material. Having this in mind, we examined the conditions behind the (geo)polymerization to obtain optimized samples which were characterized by DTA-TGA, FTIR, XRD, SEM and IS. Particular attention was devoted to ssNMR study of the evolution course of the geopolymer system. Obtained results enabled a better understanding of the influence of chemical composition and homogeneity of constituents on the resulting (micro)structural and electrical characteristics of studied samples. We demonstrated a challenging development process including shift from relatively porous bulk to thin-film configuration, to broaden the applicability of geopolymers to vertical facade photovoltaic application.

978-1-7281-6118-1/22 $31.00 © 2022 IEEE

The potential use of spark ablation in development of AgNP decorated copper oxide thin films for photodetection applications

Floren Radovanović-Perić, Vilko Mandić, Ivana Panžić

University of Zagreb, Faculty of Chemical Engineering and Technology, Zagreb, Croatia

Recently, CuO (p-type) semiconductor thin films have been investigated for sensing applications due to their excellent optical properties and narrow bandgap. It has been proposed that the performance of CuO thin films in photosensing applications depends strongly on the grain size, morphology and nanostructure which introduces the possibility of fabricating these materials by spark ablation, a novel, low cost and efficient method capable of producing controlled and clean nanoparticles with various subsequent deposition methods that further broaden the synthesis possibilities. Here we investigated the potential of this method for fabricating photosensing devices through the ability to control the properties of both the deposited copper oxide thin-films as well as nanoparticles of gold. Copper oxide films were obtained on Si wafers by; i) vacuum jet deposition of either Cu or CuxOy layers that were thermally treated to obtain pure CuO phase, ii) spin coating of the Cu or CuxOy nanoparticles (NPs) solution produced by spark ablation which were collected in 2-methoxyethanol. After the CuO nanofilms were obtained, they were decorated with AuNPs by vacuum jet deposition. Phase purity, morphology and particle size were investigated by Grazing Incidence X-ray Diffraction (GIXRD), Atomic Force Microscopy (AFM) and Scanning Electron Microscopy (SEM), while optical absorption was determined by UV/Vis spectrometry and photoluminescence spectroscopy (PL). To determine the photocurrent, I/V characteristics were performed both in light and dark conditions. It was determined that produced films show comparable properties with competitive commercial devices.

Impact of Anti-soiling Coating on Potential Induced Degradation of Silicon PV modules

Farrukh ibne Mahmood, GovindaSamy TamizhMani

Photovoltaic Reliability Laboratory, Arizona State University (ASU-PRL), Mesa, AZ, USA

Abstract—**Potential induced degradation of the shunting type (PID-s) can adversely affect module performance. Most of the current solutions to mitigate PID require changes in the manufacturing process. This paper presents an approach to reduce PID-s in crystalline silicon (c-Si) photovoltaic (PV) modules after module manufacturing by applying an off-the-shelf anti-soiling (AS) coating on the front glass surface. Two identical one-cell modules, one with AS coating and the other without the AS layer, were stressed in an indoor environmental chamber for PID-s. The results indicate that the AS coating application reduces power (Pmax) degradation by nearly half (53%) compared to the module stressed without the AS coating. The outcome of this study can help reduce PID-s of modules after manufacturing.**

Index Terms— **Potential induced degradation (PID), anti-soiling coating, reliability, PV modules, solar cell characterization.**

I. INTRODUCTION

Modules are connected in series in large commercial PV systems to improve the overall system efficiency. Depending on the system voltage, this can result in a voltage difference of 1000 to 1500 V, which is a cause of leakage currents between the cells and the grounded module frame. These leakage currents are a pre-cursor to PID, drastically reducing module performance [1]. There are three main types of PID, i.e., PID-s, PID-polarization (PID-p), and PID-corrosion (PID-c). The scope of this paper is limited to PID-s. PID-s mainly occurs when the cells are at a negative voltage bias to the frame. This causes sodium ions in the glass to travel to the cell leading to sodium shunts in the junction.

PID-s is characterized by a reduction in the shunt resistance (R_{sh}), leading to a decline in the Fill Factor (FF) and Pmax of the module [2]. In most cases, PID-s is nearly reversible by applying a reverse voltage [3]; the ideal approach would be to prevent PID-s from happening. Numerous methods are examined in the literature to avoid PID. These include appropriate grounding methods, sodium-free glass, application of better encapsulants, alteration of the emitter, and antireflection (ARC) stack [1]. But all these methods require changes in the cell or module manufacturing process.

This paper presents an approach that can be implemented after the module is manufactured by applying an off-the-shelf AS coating to reduce PID-s. In this paper, two one-cell modules with identical cells are stressed for PID-s using the humidity method in an indoor environmental chamber with and without AS coating. Pre- and post-PID characterization techniques are used to determine the changes in the performance parameters due to the application of the coating.

II. EXPERIMENTAL SETUP

Two identical one-cell modules were used for this study. The single-cell module construction mirrors the commercial modules. i.e., glass / encapsulant / cell / encapsulant / backsheet.

Solar glass with low iron was used, measuring 203mm x 280mm x 3.2mm; the encapsulant was fast-cure EVA with a thickness of 0.46 mm. The backsheet used had a thickness of 0.39 mm. A monocrystalline silicon aluminum back-surface field solar cell measuring 156mm x 156mm was used. Soldering 60-Sn/40-Pb tabbing ribbon onto the busbars of the cells with a semiautomatic tabbing machine was used to make the cell connections. Lamination was done at 150 °C. At the rear of the cell, single-pole junction boxes were connected using silicone PV-804 sealant [4]. 3M Aluminum tape was used to cover all four sides of the one-cell modules to construct a mock frame for PID stress. The front and rear of the single-cell modules are shown in Fig. 1(a) and (b), respectively. One of the two modules was uniformly coated by applying an anti-soiling coating solution purchased from a commercial vendor.

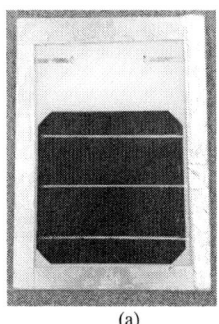 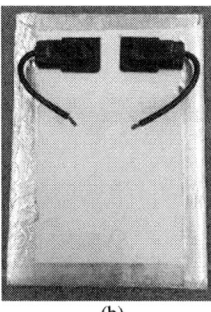

(a)   (b)

Fig. 1 (a) Module front. (b) module rear.

To monitor the loss of performance parameters, pre- and post-PID characterization tests including indoor light IV, electroluminescence (EL), and dark IV were carried out on both modules. The modules were subjected to PID stress in an indoor environmental chamber at -1000 V, 85 °C, and 85% relative humidity (RH) per IEC standard 62804-1-1 for 96 hours. Leakage current (LC) was monitored throughout the experiment. The results indicated that after the first 96 hours, the modules did not experience any significant degradation due to PID. Therefore, the single-cell coupons were subjected to another round of PID stress in the chamber for 192 hours to determine the effect of PID on both modules.

III. RESULTS AND DISCUSSION

The percent change between pre-characterization at 0 h and post-characterization at 96h plus 192h of stress is discussed in this section.

Fig. 2 and Fig. 3 show the % Pmax degradation and % FF degradation obtained using pre- and post-light IV data. The % degradation for all parameters is obtained using (1).

978-1-7281-6118-1/22 $31.00 © 2022 IEEE

$$\% \ degradation = \frac{Post \ data - Pre \ data}{Pre \ data} * 100 \qquad (1)$$

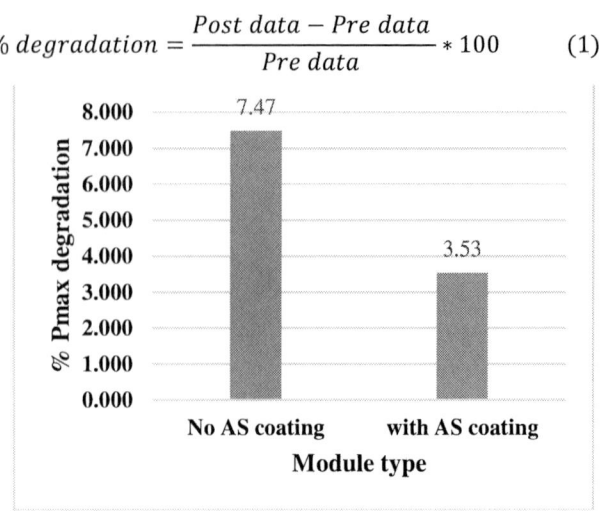

Fig. 2: % Pmax degradation for the coated and non-coated module.

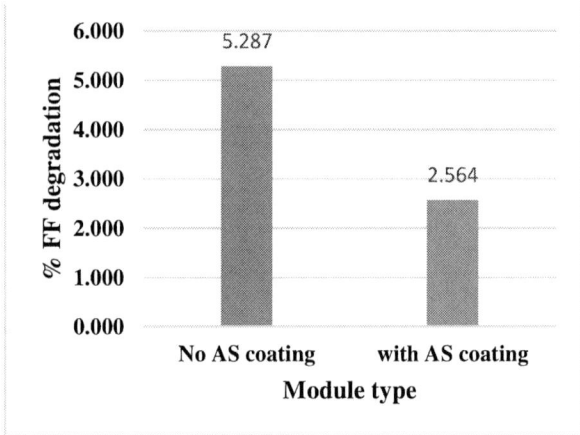

Fig. 3: % FF degradation for the coated and non-coated module.

It is evident from Fig. 2 and Fig. 3 that the uncoated module undergoes more degradation in Pmax and FF than the coated module. The % Pmax degradation is approximately 53% more in the uncoated module.

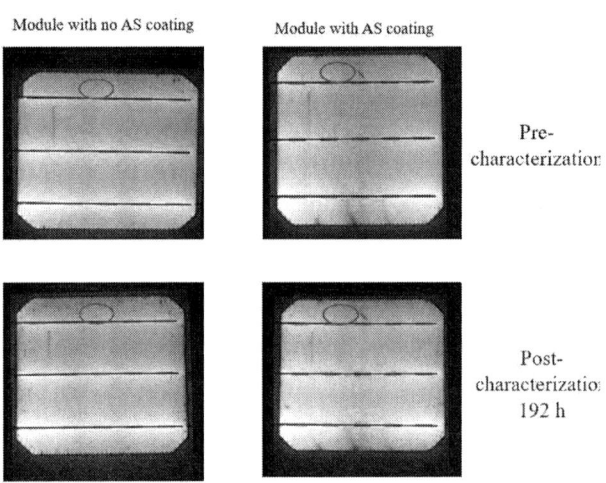

Fig. 4: pre-and post-EL images at 100 % short circuit current (Isc) for the coated and uncoated module.

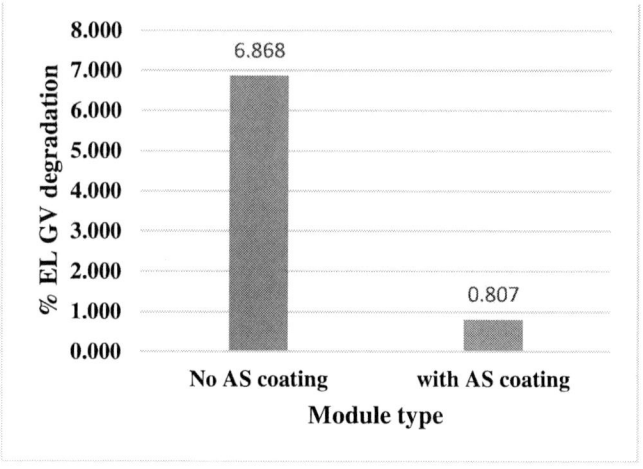

Fig. 5: % Grey value (GV) degradation analysis for the pre-and post-EL images at 100% Isc for the coated and uncoated module.

Fig. 4 shows the EL images for pre-and post-characterization for the coated and uncoated module. Using these EL images, the gray value (GV) analysis is performed at the marked red circle, and the % change in the value between pre-and post-characterization is shown in the plot in Fig.5. It is noticeable in Fig 4 and Fig 5 that the uncoated module experiences more PID than its counterpart.

Fig. 6 shows the % R_{sh} degradation acquired using pre-and post-dark IV tests. Dark IV also indicates that the coated module experiences less degradation in the $R_{sh,}$ leading to lower loss due to PID than the uncoated module.

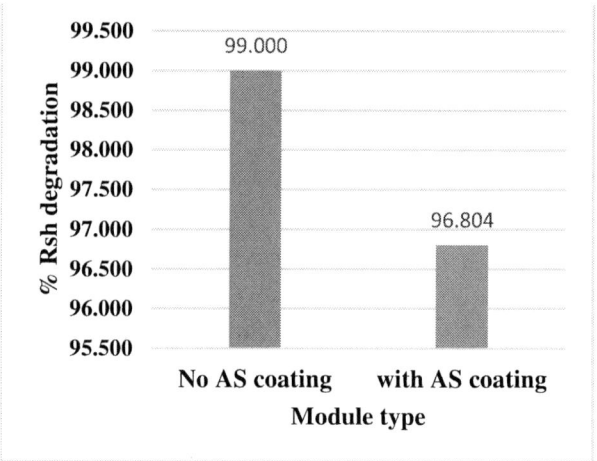

Fig. 6: % Rsh degradation for the coasted and non-coated module

978-1-7281-6118-1/22 $31.00 © 2022 IEEE

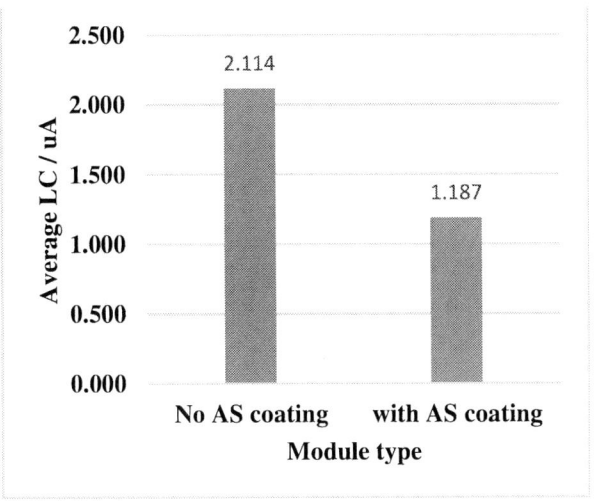

Fig. 7: Average LC for the coated and uncoated module during 192 h of PID stress

Fig.7 shows the average LC from the coated and uncoated module during 96 h plus 192 h of PID stress. The coated module has a lower LC than the uncoated module.

As the LC level is typically a good indicator for PID, the results clearly show that the AS coating application results in a lower LC generation, leading to a lesser degradation in the Pmax of the coated module than the uncoated module. This result is further supported by lower degradation seen in FF, EL GV, and R_{sh} of the coated module. The main reason for this reduced degradation is attributed to the hydrophobic properties of the AS coating, which minimizes a continuous moisture accumulation on the glass surface. Consequently, the AS coating disrupts the continuity of glass surface conductivity, which is a previously proven method by ASU-PRL to lower the effect of PID [5], [6]. As explained in a previous publication [7], the main voltage drop of the applied voltage (1000V) during PID stress test occurs on the surface of the glass, bulk of glass, and bulk of encapsulant, leaving only a tiny fraction of the voltage (called activation potential) available for sodium deposition reaction on the cell surface. In the presence of moisture on the glass surface, the surface conductivity becomes high, and hence the voltage drop becomes low on the glass surface, leaving a higher fraction of voltage available for the sodium deposition reaction. If the voltage drop on the glass surface is increased by removing moisture from the glass surface or applying hydrophobic AS coating on the glass surface, then the voltage drop is increased on the glass surface, leaving lower voltage available for the sodium deposition reaction on the cell surface and hence the PID loss is reduced.

IV. CONCLUSION AND FUTURE WORK

The PID issue is traditionally reduced or eliminated at the manufacturing stage itself by adjusting the cell processing method (e.g., silicon-rich AR coating composition) or changing module encapsulant material (e.g., from EVA to POE). In this paper, we have presented a method to reduce the PID issue of the modules which are already manufactured and are susceptible to the PID problem. In our method, we use a hydrophobic anti-soiling (AS) coating on the glass surface to reduce PID. It is known that the accumulated soil layer on the glass surface can enhance the PID issue [8]. It is, therefore,

envisioned that hydrophobic AS coating on the installed/soiled modules in the field would have three benefits: reducing the soiling issue; reducing the PID issue due to lower soiling level; reducing the PID issue due to lower surface conductivity level. The AS coating, we applied did not fully eliminate the PID issue though it dramatically reduced the issue. It is possible that our manually coated layer contains some pinholes leading to some leakage current through water-filled pinholes. We plan to determine if the coated layer contains some pinholes using microscopic techniques.

REFERENCES

[1] G. J. M. Janssen *et al.*, "Minimizing the Polarization-Type Potential-Induced Degradation in PV Modules by Modification of the Dielectric Antireflection and Passivation Stack," *IEEE J. Photovoltaics*, vol. 9, no. 3, pp. 608–614, 2019.

[2] V. Naumann *et al.*, "Explanation of potential-induced degradation of the shunting type by Na decoration of stacking faults in Si solar cells," *Sol. Energy Mater. Sol. Cells*, vol. 120, no. PART A, pp. 383–389, 2014.

[3] J. Oh, S. Bowden, and G. S. TamizhMani, "Potential-Induced Degradation (PID): Incomplete Recovery of Shunt Resistance and Quantum Efficiency Losses," *IEEE J. Photovoltaics*, vol. 5, no. 6, pp. 1540–1548, 2015.

[4] J. Oh *et al.*, "Reduction of PV module temperature using thermally conductive backsheets," *IEEE J. Photovoltaics*, vol. 8, no. 5, pp. 1160–1167, 2018.

[5] S. R. V. Tatapudi, "Potential Induced Degradation (PID) of Pre-Stressed Photovoltaic Modules: Effect of Glass Surface Conductivity Disruption," Arizona State University, 2012.

[6] B. Bora *et al.*, "Mitigation of PID in commercial PV modules using current interruption method," in *Reliability of Photovoltaic Cells, Modules, Components, and Systems X*, 2017, no. 10370, pp. 89–94.

[7] G. TamizhMani and J. Kuitche, "Accelerated Lifetime Testing of Photovoltaic Modules Solar America Board for Codes and Standards," 2013.

[8] M. Koehl and S. Hoffmann, "Impact of rain and soiling on potential induced degradation," *Prog. Photovoltaics Res. Appl.*, vol. 20, no. 1, pp. 6–11, 2015.

Extreme Solar: Towards 24-7 Renewable Energy

Sijo Augustine[*], Sathishkumar Ranade[*], Valerio De Angelis[#], Gabriel Cowles[†], Jinchao Huang[†],
Olga Lavrova[*], and Stanley Atcitty[#]

[*] New Mexico State University, Las Cruces, NM , USA
[†] Urban Electric Power, Pearl River, NY, USA
[#] Sandia National Laboratories, Albuquerque, NM, USA

Abstract—**This paper describes a method of connection and control algorithm for photovoltaic panel-integrated storage that can be interfaced to a grid using commercial dc-ac microinverters. Emulation of an appropriate i-v characteristic allows the integrated panel to operate with any inverter, e.g., peak-power tracking inverter, droop mode or stand-alone. or a battery inverter in droop mode. Rechargeable zinc-manganese dioxide (ZnMnO₂) battery cells developed by Urban Electric Power (UEP) are used as a source of energy storage. The batteries are charged with excess solar and discharged in the evening or when additional power is required. A battery management algorithm is developed, charge/discharge data is collected from each cell, and a proof of concept is demonstrated.**

Keywords—dc-dc bidirectional converter, Rechargeable ZnMnO₂ battery cells, battery, solar-PV, microinverter, bidirectional power flow, dc coupling, energy storage.

I. INTRODUCTION

Microinverters represent a mature technology and the high efficiency and reliability of microinverter enhances the power generation from individual PV panel and improve the system performance, thus, providing economic benefits [1, 2, 3]. Energy storage systems are crucial while considering standalone or grid-connected solar-PV systems. The most feasible and common configurations to couple the energy storage devices with PV systems are; (a) ac coupled PV energy storage systems (b) dc coupled PV energy storage systems, and (c) reverse dc coupled PV energy storage systems [5]. Reference [4] discusses a few other connection methods to integrate the energy storage system with PV. In dc coupling, popular method is the direct series connection of the battery in which, a diode is used to isolate the PV panel and the battery.

Most of the string inverters available in the market can work either in maximum power point tracking (MPPT) mode or the grid storage mode, and the control algorithm controls the transition between the two modes. But, the microinverters available in the market are always operating on MPPT mode to maximize power generation. Therefore, most of the microinverter-based solar-PV systems follow the ac coupled connection method as shown in Fig. 1. The main disadvantage of this connection is the poor efficiency due to multiple power electronic conversions for charge-discharge cycles [5]. To overcome this and improve the system reliability, this paper proposes a dc-coupled energy storage system for microinverter-based solar-PV power generation. The proposed method of

connection and battery charge-discharge control algorithm is used to manage the PV and battery power effectively.

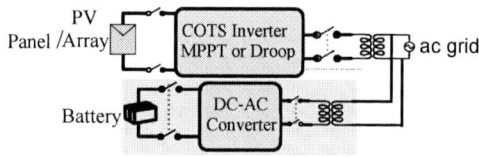

Fig. 1. Existing ac coupled system for energy management.

II. SYSTEM CONFIGURATION AND CONTROL ALGORITHM

The proposed dc coupled configuration is shown in Fig. 2. It consists of a PV panel connected to a grid-tied microinverter, and a bidirectional dc-dc converter with energy storage. A bidirectional converter based on buck and boost topologies is shown for illustration, but other topologies can be used. The dc-dc converter output is connected in parallel with the PV panel. The proposed control algorithm controls the dc-dc converter output voltage-current characteristic in a manner that allows the battery to directly charge from the PV panel and discharge through the microinverter. For example, if the microinverter operates in a MPPT mode, the controller synthesizes a peak power point corresponding to the desired battery discharge current.

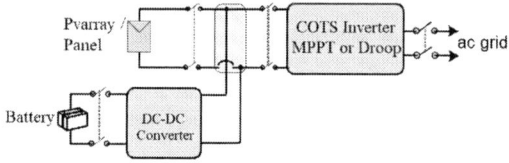

Fig. 2. Proposed dc coupled energy storage configuration with microinverter.

A. Battery Cells

The rechargeable ZnMnO₂ battery cells developed by UEP are used as energy storage. Fig. 3(a) shows a schematic diagram of the battery in a large cylindrical form factor. The battery cells are manufactured in fully charged state which eliminates the need of initial charge. The battery cell's open circuit voltage (OCV) is around 1.55V and the self-discharge rate is very slow at about 0.2% per month. The nominal discharge voltage is about 1.3 V. The ZnMnO₂ battery cell can be charged in 4 hours or less in a wide range of power levels, temperatures (between -10 to 50 °C) and applications. Fig. 3(b) is a plot demonstrating the performance of a ZnMnO₂ cell tested under the IEC 61427-1 protocol for secondary cells for photovoltaic off-grid application. This solar cycling protocol has been developed so that 45 days of testing at 40°C is equivalent to 1 year of daily solar cycles at the specified nameplate capacity. In 45 days of testing, periods of low solar insolation are captured by cycling at low SoC (state of charge). Periods with excess insolation are captured by cycling at high SoC. These conditions on the average try to capture a realistic range of scenarios to which the battery systems are exposed. The UEP ZnMnO₂ battery cell demonstrated a cycle life equivalent to 7 years under this extremely demanding protocol.

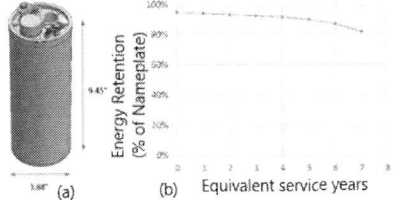

Fig. 3(a) Schematic diagram of the a UEP ZnMnO₂ battery cell. (b) Energy retention of a UEP ZnMnO₂ cell.

B. Control Algorithm

The proposed battery charge-discharge control algorithm combines a bi-directional converter current control and a

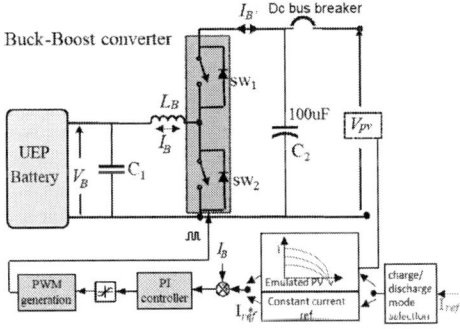

Fig. 4. Proposed algorithm for battery management.

current reference generation method for emulating the PV curve as shown in Fig. 4. The algorithm handles charging mode and discharging mode separately based on the operating conditions. The battery charges directly from PV through the dc-dc bidirectional converter with a constant current reference. In this mode, the charging current is, of course, less than the PV current at the peak-power point, thus allowing the microinverter to peak power track. But during the discharging mode, the converter emulates a PV characteristic to accommodate the microinverter MPPT Mode. The discharging mode has two different operating scenarios; (a) with PV power, and (b) without PV power. In case (a), the microinverter extracts power from both the PV panel and from the battery. In case (b), PV power is zero and battery is the only power source to the microinverter.

III. EXPERIMENTAL ANALYSIS

A proof-of-concept has been developed and consists of rooftop PV panels (Fig. 5(a)), 16 battery cells in series (Fig. 5(b)), an Enphase IQ 7A microinverter, and a dc-dc bidirectional converter (Fig. 5(c)). Texas Instruments LAUNCHXL-F28379D development kit is used as the controller, and Orion Jr. Battery Management System (BMS) is used to record the individual cell voltage, total current, and battery pack temperature. The BMS logs the data at a sampling rate of 5 samples/sec. Additionally, a novel data acquisition system is developed to collect data from the PV panel and microinverter at a sampling rate of 12 samples/min. The system parameters are given in Table I.

The proposed system is tested under different weather conditions to validate the performance. Fig. 6 shows the experimental result on a sunny day. The microinverter turns on automatically when the solar irradiation is enough to produce PV power. The red color curve indicates the PV panel output current, and blue curve represents the microinverter DC side current. When the battery discharges, then the DC side current of the inverter will be more than the PV current (blue curve above the red curve). When the battery charges from the PV current, then the inverter DC side current will be less than the PV current (blue curve below the red curve).

TABLE I. NOMINAL SYSTEM PARAMETERS

Parameters	Value
PV panel power	345W
PV panel V_{MPP}	27-35V
DC-DC bidirectional converter power	200W
DC-DC bidirectional converter inductor	270µH
DC-DC bidirectional converter capacitor	100µF
Nominal cell voltage	1-2V
Battery pack voltage	16-30V
Nominal switching frequency	15kHz

Fig. 5. Experimental setup in NMSU.

As demonstrated in Fig. 6, around 10:30 AM, the battery pack begins charging with 5A current. The PV panel current (red curve) remains same and there is change in the microinverter current (blue curve). After half an hour, the battery charging current reference is reduced to 4A and after one hour the reference changed to 3A. During the transition from 4A to 3A the converter went to unstable mode due to the microinverter ac voltage variation. After charging the battery with 3A, the converter is kept idle for few hours. During battery discharge, the PV panel breaker is turned off to test the dc-dc converter

978-1-7281-6118-1/22 $31.00 © 2022 IEEE

performance. A 6A discharging current reference is activated for a duration of two hours. From Fig. 6, it is clear that, both charging and discharging is seamless and the proposed method of connection and control algorithm is effective.

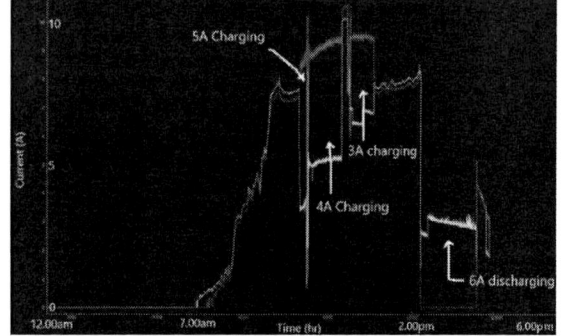

Fig. 6. Battery charge-discharge sequence on a sunny day.

The individual cell voltages from the BMS are shown in Fig. 7. As mentioned earlier, there are 16 cells in series with an upper voltage limit of 2V (each cell) and lower limit of 1V. From time 10:30am -12.30pm, the cells are in charging mode until the cell reaches a voltage of 1.8V. From 12.30pm to 2.30pm, the cells kept idle. After 2.30pm, the cells started discharging at a rate of 6A. The corresponding battery pack current is shown in Fig. 8.

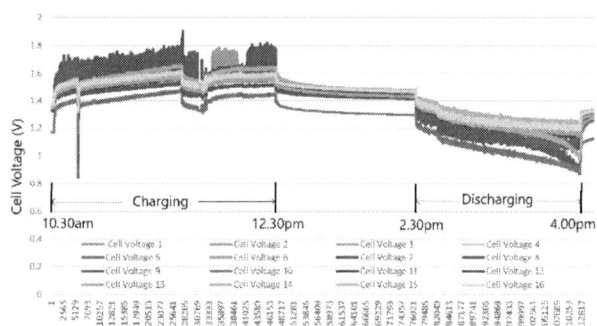

Fig. 7. Individual cell voltages.

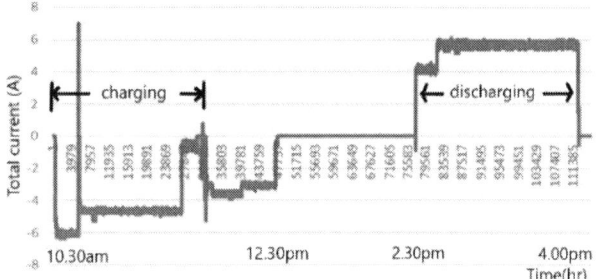

Fig. 8. Total battery pack current.

Fig. 9 shows system performance on a cloudy day. Around noon, a 2A charging current reference is given to the dc-dc converter. It can be observed that, the converter is operates in a stable during sudden change in the solar radiance due to clouds. After few hours, the PV panel circuit breaker is turned off and the battery started discharging at 5A. The above experimental results show the proof concept system is stable during the charging/discharging process.

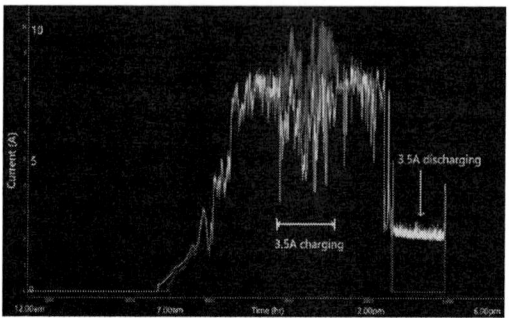

Fig. 9. Battery charge/discharge sequence on a cloudy day.

IV. CONCLUSION

A dc coupled Panel-Converter-Battery system with a novel connection method and control algorithm is proposed to achieve the seamless charging-discharging of energy storage. The dc-dc converter acts as a 'PV emulator' so that, the combination can be operated with any inverter that has a peak power tracker. As a unit, the battery and dc-dc converter can be simply connected in parallel with an existing PV system at the panel or array level. The unit can also be used for grid connection of battery energy storage using off-the-shelf PV inverters. Future work includes converter optimization and further testing including battery management and efficiency measurement. This kind of fully integrated, low-cost solar-storage system holds significant commercial potential in markets across the globe.

ACKNOWLEDGMENT

The authors gratefully acknowledge support for this work from the U.S. Department of Energy, Office of Electricity, ES program and Dr. Imre Gyuk, and Sandia National Laboratories (SNL), Albuquerque (SAND2022-3809 C). SNL is a multi-mission laboratory, managed and operated by National Technology and Engineering Solutions of Sandia, LLC., a wholly owned subsidiary of Honeywell International, Inc., for the U.S. Department of Energy's National Nuclear Security Administration under contract DE-NA0003525. The views expressed in the article do not necessarily represent the views of the U.S. Department of Energy or the United States Government. This work was partially supported by the NSF Grants #OIA-1757207(NM EPSCoR), HRD-1345232 & HRD-1914635.

REFERENCES

[1] T. Lodh and V. Agarwal, "Single stage multi-port Flyback type solar PV module integrated micro-inverter with battery backup," IEEE Int. Conf. on Power Elect., Drives and Energy Syst. (PEDES), 2016,

[2] Valerio De Angelis, Nataraj Pragallapati, Satish J Ranade, et al., "Extreme Distributed Storage for Photovoltaic Systems", DOE ESS Peer Review in Santa Fe, 2018.

[3] J Tabarez, N Pragallapati, et al., "Extreme Solar: Towards 24-7 Renewable Energy" DOE ESS Peer Review in Albuquerque, NM, 2019.

[4] H. Wang and D. Zhang, "The Stand-alone PV Generation System with Parallel Battery Charger," Int. Conf. on Electrical and Control Eng., 2010,

[5] Corrie Austin, "Solar Plus Energy Storage: Comparing System Options Dynapower.com.

[6] John C. Palombini, Apurva Somani "Energy storage system for photovoltaic energy and method of storing photovoltaic energy" U.S. Patent WO2018213157A1, Nov 22, 2018.

Impact of Indium Chloride Treatment on the Properties of CuInSe$_2$ Thin Films

Deewakar Poudel[1], Adam Masters[1], Benjamin Belfore[1], Elizabeth Palmiotti[2], Angus Rockett[2] and Sylvain Marsillac[1]

[1]Virginia Institute of Photovoltaics, Old Dominion University, Norfolk, VA 23529, USA

[2]Dept. of Metallurgical and Materials Engineering, Colorado School of Mines, Golden, CO 80401, USA

Abstract— Copper Indium diselenide (CIS) semiconductor thin films were deposited by single-stage process. Following the deposition, annealing and recrystallization of CIS was carried out in the presence of InCl$_3$ at different temperatures. Increase in grain size was observed by SEM in all cases. Increase in peak intensity was observed by XRD after treatment, correlating well with the SEM results. Measurements of the composition by both XRF and EDS indicate significant changes, notably a decrease in copper content. This will likely be an issue for device performance and will need to be addressed.

Keywords— *CuInSe$_2$, recrystallization, annealing, InCl$_3$*

I. INTRODUCTION

The chalcopyrite semiconductor CuInSe$_2$ has long be known as a highly attractive material for photovoltaic applications, due to its characteristic advantages of variable bandgap, high absorption coefficient, stability under high energy radiation and excellent thermal stability [1]. In 1976, Kazmerski et al. first reported the fabrication of CuInSe$_2$ based solar cells with energy conversion up to 5% [2]. Since then, a lot of progress has been made and CuInSe$_2$ has reached record efficiency of more than 16%, with good long-term stability and device performance [3]. In terms of fabrication process, the conventional three-stage deposition process used often for Cu(In,Ga)Se$_2$ usually involves high temperature deposition, is time consuming and can lead to high cost. On the other hand, the single-stage process is faster, lower cost, and has a high output beneficial for industrial manufacturing process [4]. This will permit to improve the manufacturing supply chain and logistics of the fabrication [5-12]. Finally, enhancement in the grain size for polycrystalline chalcopyrite is known to improve the device performance. To achieve that, post-deposition treatments by metal halides seem to be one of the most promising approaches [13-28].

In this work, we therefore studied the effect of InCl$_3$ vapor treatment on the recrystallization of CuInSe$_2$ films deposited by single-stage process at 450°C.

II. EXPERIMENTAL METHODS

CIS films were deposited on Mo coated soda lime glass (SLG) substrates by single-stage co-evaporation method in which a constant flux of Cu, In, and Se was supplied during the film growth at a substrate temperature of 450°C in a high vacuum chamber with a base pressure of less than 2E-7 Torr. Fabrication of Mo bilayer was done by dc magnetron sputtering [29-31].

Annealing was performed in a custom-built annealing chamber (Figure 1). The annealing chamber consisted of a stainless-steel nipple with a blank flange on one end which was machined to connect to a stainless-steel tube fitted with a valve. The annealing chamber was connected to a dry roughing pump with a detachable K-type flange. The samples were loaded in a quartz cylinder (with 5 mg of InCl$_3$ and 50 mg of Se) and placed into the annealing chamber. The open end of the annealing chamber was then closed with a C-type blank stainless-steel flange. The chamber was evacuated to a pressure below 70 mT and loaded in a cylindrical quartz tube furnace, which was heated to the desired temperature (either 400°C, 450°C or 500°C) for 30 minutes. After annealing, the annealing chamber was allowed to cool down to room temperature by switching off the tube furnace power.

Fig. 1. Custom built annealing chamber used for ex-situ recrystallization experiments.

The surface morphology of the annealed samples was studied using scanning electron microscopy (SEM). The composition was measured using energy dispersive spectroscopy (EDS). The complex dielectric functions were extracted using a M-2000 ellipsometer and calibrated samples [32-48]. The crystallographic structure was determined by symmetric Θ-2Θ x-ray diffraction (XRD) and analyzed using the International Center for Diffraction Data (ICDD) database.

III. RESULTS AND DISCUSSIONS

The surface and cross-section SEM of as-deposited and recrystallized samples are shown in Figure 2. Significant changes were observed as a function of temperature and

This research was supported by the Department of Energy Contract No. DE-EE0007551.

recrystallization. One can see that the as deposited films (substrate temperature of 450°C) appear to have relatively small grains. With an InCl₃ at 400°C, large grains appear at the surface while the bulk seems relatively unchanged. At 450°C, no more grains appear clearly on the surface, while the bulk of the film tends to have larger grain. Finally, at 500°C, some grains appear to be as large as the thickness of the films. Figure 3 are reported the EDS results for all the films with the corresponding values shown in Table I.

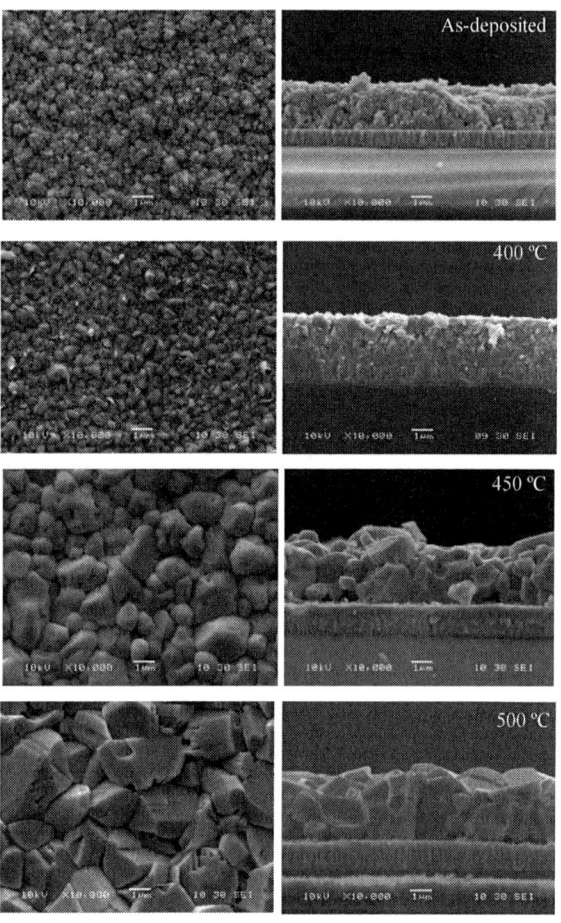

Fig. 2. Surface and cross-section Scanning Electron Microscopy micrographs of CIS films: as-deposited and recrystallized at 400°C, 450°C and 500°C by InCl₃ vapor treatment

TABLE I. EDS COMPOSITION OF AS-DEPOSITED AND INCL₃ RECRYSTALLIZATED FILM

	Element	As-dep.	400°C	450°C	500°C
EDS	Cu (%)	36	32	30	27
	In (%)	19	21	22	24
	Se (%)	44	47	48	49

As the compositional readings show, the film started out intentionally copper rich, to intentionally compensate for the effect of InCl₃ vapor treatment. One can see that all the films become slowly more copper poor with increase treatment temperature. Such process could potentially be used also to grow alternative buffer layers to CdS, such as In_xS_y or In_xSe_y [49-62]. One can see that there is still an excess of Cu even for the 500°C treatment, but this should be easily taken into account by modifying the composition of the as-deposited films.

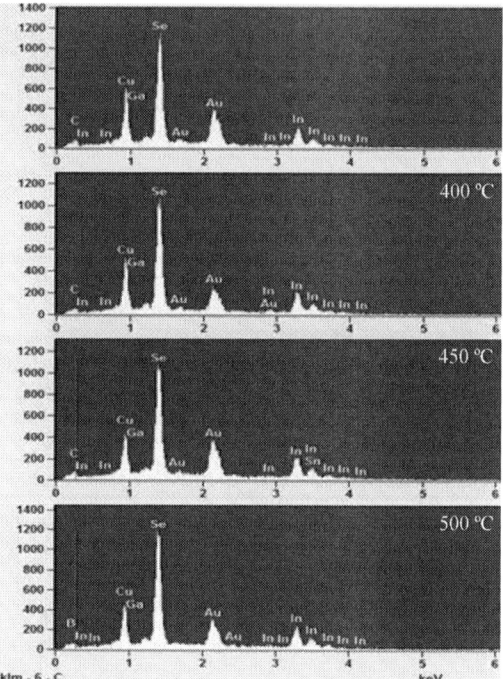

Fig. 3. Energy Dispersive X-ray spectrocsopy (EDS) of CIS films: as-deposited and recrystallized at 400°C, 450°C and 500°C by InCl₃ vapor treatment.

To understand how successful the recrystallization process was in terms of grain growth, a 3-stage CIS deposition was performed at 500 °C and compared to the films recrystallized. As Figure 4 shows, the grain size between the films is nearly indistinguishable. This is remarkable because for the 3-stage process film, the deposition was more than twice as long, required a higher temperature, and had a more complex deposition process.

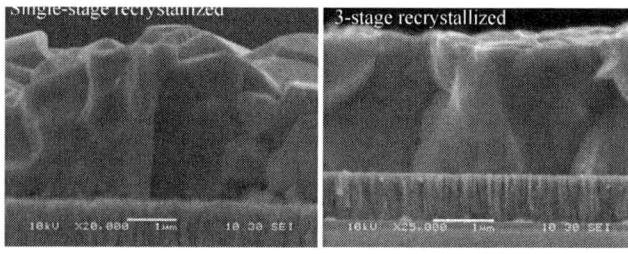

Fig. 4. Comparison of SEM cross-section micrographs for a single stage film recrystallized at 500°C by InCl₃ vapor treatment (left) and a 3-stage CIS deposition at 500°C (right)

XRD was performed to further analyze the effect of recrystallization on crystallinity. Figure 5 shows the XRD measurements on the as-deposited and recrystallized samples, while the results are reported Table II. An increase in peak intensity was observed after the treatment, with higher values at high temperature.

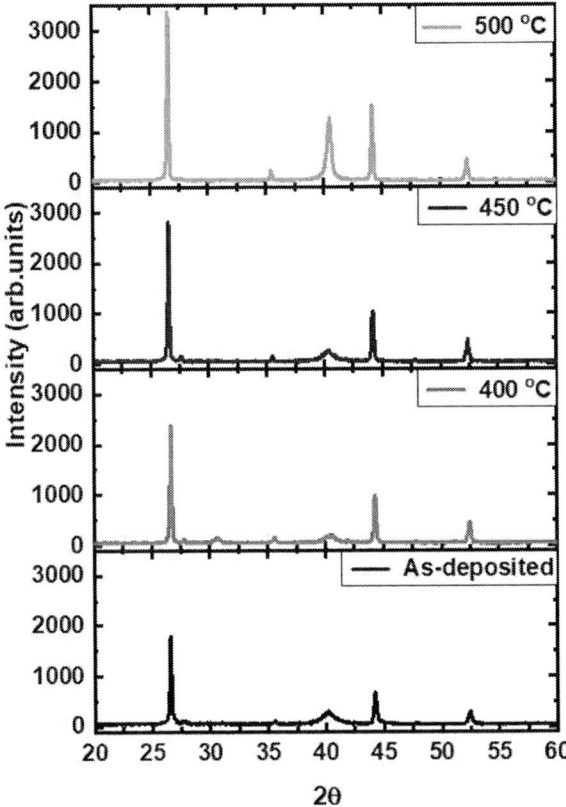

Fig. 5. XRD plots (focusing on (112), (220)/(204) and (312) peaks) for as-deposited and InCl₃ treated CIS samples.

TABLE II. XRD RESULTS FOR AS-DEPOSITED AND INCL₃ TREATED SAMPLES AT 400 °C, 450 °C AND 500 °C

Sample	Peak	Angle (deg)	FWHM (deg)	Int.I (cps deg)
As-deposited	(112)	26.6	0.20	1785
	(220)/(204)	44.2	0.23	668
	(312)	52.4	0.28	293
400°C	(112)	26.6	0.15	2381
	(220)/(204)	44.2	0.19	980
	(312)	52.4	0.17	439
450°C	(112)	26.6	0.13	2812
	(220)/(204)	44.1	0.18	1026
	(312)	52.3	0.13	484
500°C	(112)	26.5	0.12	3371
	(220)/(204)	44.2	0.16	1527
	(312)	52.3	0.12	456

Furthermore, the full width at half maximum (FWHM) decreases with temperature after annealing. These results are consistent with the SEM results, indicating an increase in grain size after treatment. No shift in the peak position was observed as expected, while a small change in preferential orientation was observed, with a shift towards a (220)/(204) orientation versus a (112) orientation.

IV. CONCLUSION

Single stage CIS deposition was performed on molybdenum coated soda lime glass at a substrate temperature of 450°C. The annealing was done at various temperature of 400°C, 450°C and 500°C under InCl₃ vapor. Changes in morphology were observed by SEM images, with grain sizes increasing with temperature and becoming more uniform and compact. An increase in peak intensity and decrease in FWHM was also observed by XRD with increasing annealing temperature, in good agreement with SEM results. The composition of the films was measured by EDS and indicated a decrease in copper content with increase in annealing temperatures. Even though the composition is currently off stoichiometry, it will be easy to compensate for that by modifying the composition of the as-deposited films, while paying attention to the sodium content, which is known to influence device characteristics [63-70]. More characterizations will be done, notably electrical and compositional depth profile, to fully understand the potential viability of this process for device fabrication.

ACKNOWLEDGMENT

This research was supported by the Department of Energy Contract No. DE-EE0007551.

REFERENCES

[1] S. Ruffenach, Y. Robin, M. Moret, R.-L. Aulombard, and O. Briot, "(112) and (220)/(204)-oriented CuInSe2 thin films grown by co-evaporation under vacuum," *Thin solid films,* vol. 535, pp. 143-147, 2013.

[2] L. Kazmerski, F. White, and G. Morgan, "Thin-film CuInSe2/CdS heterojunction solar cells," *Applied Physics Letters,* vol. 29, no. 4, pp. 268-270, 1976.

[3] S. Zhang, S.-H. Wei, A. Zunger, and H. Katayama-Yoshida, "Defect physics of the CuInSe 2 chalcopyrite semiconductor," *Physical Review B,* vol. 57, no. 16, p. 9642, 1998.

[4] H. Wang *et al.*, "Effect of substrate temperature on the structural and electrical properties of CIGS films based on the one-stage co-evaporation process," *Semiconductor science and technology,* vol. 25, no. 5, p. 055007, 2010.

[5] D. M. Nelson, E. Marsillac, and S. S. Rao, "Antecedents and evolution of the green supply chain," *Journal of Operations and Supply Chain Management,* no. Special Issue, 2012.

[6] K. Liao, E. Marsillac, E. Johnson, and Y. Liao, "Global supply chain adaptations to improve financial performance: supply base establishment and logistics integration," *Journal of Manufacturing Technology Management,* 2011.

[7] R. Romero-Silva, E. Marsillac, S. Shaaban, and M. Hurtado-Hernández, "Serial production line performance under random variation: dealing with the 'Law of Variability'," *Journal of Manufacturing Systems,* vol. 50, pp. 278-289, 2019.

[8] E. Marsillac, "Management of the photovoltaic supply chain," *International Journal of Technology, Policy and Management,* vol. 12, no. 2-3, pp. 195-211, 2012.

[9] Y. Liao and E. Marsillac, "External knowledge acquisition and innovation: the role of supply chain network-oriented flexibility and

organisational awareness," *International Journal of Production Research,* vol. 53, no. 18, pp. 5437-5455, 2015.

[10] R. Romero-Silva, S. Shaaban, E. Marsillac, and M. Hurtado, "Exploiting the characteristics of serial queues to reduce the mean and variance of flow time using combined priority rules," *International Journal of Production Economics,* vol. 196, pp. 211-225, 2018.

[11] E. L. Marsillac, "Environmental impacts on reverse logistics and green supply chains: similarities and integration," *International Journal of Logistics Systems and Management,* vol. 4, no. 4, pp. 411-422, 2008.

[12] E. Marsillac, "Closing the Photovoltaic Supply Chain Loop-Invest Now for Future Returns," in *2018 IEEE 7th World Conference on Photovoltaic Energy Conversion (WCPEC)(A Joint Conference of 45th IEEE PVSC, 28th PVSEC & 34th EU PVSEC),* 2018: IEEE, pp. 2498-2500.

[13] G. Rajan *et al.,* "Impact of Post-Deposition Recrystallization by Alkali Fluorides on Cu (In, Ga) Se2Thin-Film Materials and Solar Cells," *Thin Solid Films,* vol. 690, p. 137526, 2019.

[14] E. Palmiotti, B. Belfore, D. Poudel, S. Marsillac, and A. Rockett, "In-Situ Study of the Crystallization of Amorphous CuInSe2 Thin Films and the Effect of InCl3 Treatment," *Thin Solid Films,* p. 139095, 2022.

[15] B. Belfore *et al.,* "Recrystallization of Cu (In, Ga) Se2 Semiconductor Thin Films via InCl3 Treatment," *Thin Solid Films,* vol. 735, p. 138897, 2021.

[16] S. Marsillac, M. Zouaghi, J. Bernede, T. B. Nasrallah, and S. Belgacem, "Evolution of the properties of spray-deposited CuInS2 thin films with post-annealing treatment," *Solar energy materials and solar cells,* vol. 76, no. 2, pp. 125-134, 2003.

[17] D. Poudel *et al.,* "Analysis of Post-Deposition Recrystallization Processing via Indium Bromide of Cu (In, Ga) Se2 Thin Films," *Materials,* vol. 14, no. 13, p. 3596, 2021.

[18] S. Karki *et al.,* "Analysis of recombination mechanisms in RbF-treated CIGS solar cells," *IEEE Journal of Photovoltaics,* vol. 9, no. 1, pp. 313-318, 2018.

[19] D. Poudel *et al.,* "In Situ Recrystallization of Co-Evaporated Cu (In, Ga) Se2 Thin Films by Copper Chloride Vapor Treatment towards Solar Cell Applications," *Energies,* vol. 14, no. 13, p. 3938, 2021.

[20] D. Poudel *et al.,* "Studying the Recrystallization of Cu (InGa) Se 2 Semiconductor Thin Films by Silver Bromide In-situ Treatment," in *2021 IEEE 48th Photovoltaic Specialists Conference (PVSC),* 2021: IEEE, pp. 2307-2311.

[21] D. Poudel *et al.,* "Effect of Indium Bromide Treatments Post-Deposition Recrystallization Temperature on Cu (In, Ga) Se 2 Thin Films," in *2021 IEEE 48th Photovoltaic Specialists Conference (PVSC),* 2021: IEEE, pp. 0429-0432.

[22] E. Palmiotti, B. Belfore, D. Poudel, S. Marsillac, and A. Rockett, "CuIn (1− x) Ga x Se 2 Recrystallization by Metal Halide Treatments," in *2021 IEEE 48th Photovoltaic Specialists Conference (PVSC),* 2021: IEEE, pp. 0669-0671.

[23] B. Belfore *et al.,* "Study of Indium Chloride Vapor Treatment on Cu (In, Ga) Se 2 Semiconductor Thin Films," in *2021 IEEE 48th Photovoltaic Specialists Conference (PVSC),* 2021: IEEE, pp. 2320-2323.

[24] B. Belfore *et al.,* "Vapor Treatment and In-situ Recrystallization by Copper Chloride on Cu (In, Ga) Se 2 Thin Film," in *2021 IEEE 48th Photovoltaic Specialists Conference (PVSC),* 2021: IEEE, pp. 2316-2319.

[25] B. Belfore, D. Poudel, T. Ashrafee, S. Karki, G. Rajan, and S. Marsillac, "Morphological Study of Indium Chloride Post Deposition Treated CuInSe 2 Thin Films," in *2021 IEEE 48th Photovoltaic Specialists Conference (PVSC),* 2021: IEEE, pp. 2312-2315.

[26] B. Belfore *et al.,* "In-Situ Recrystallization of CIGS via Metal Halides," in *2020 47th IEEE Photovoltaic Specialists Conference (PVSC),* 2020: IEEE, pp. 1131-1133.

[27] B. Belfore *et al.,* "Ex-Situ Recrystallization of CIGS via Metal Halides," in *2020 47th IEEE Photovoltaic Specialists Conference (PVSC),* 2020: IEEE, pp. 1102-1104.

[28] B. Belfore, G. Rajan, S. Karki, D. Poudel, A. Rockett, and S. Marsillac, "The Impact of Deposition Temperature on Sodium Fluoride Recrystallization in Cu (In, Ga) Se 2 Solar Cells," in *2019 IEEE 46th Photovoltaic Specialists Conference (PVSC),* 2019: IEEE, pp. 1851-1853.

[29] E. Gourmelon, J. Bernede, J. Pouzet, and S. Marsillac, "Textured MoS 2 thin films obtained on tungsten: electrical properties of the W/MoS 2 contact," *Journal of Applied Physics,* vol. 87, no. 3, pp. 1182-1186, 2000.

[30] K. Aryal, H. Khatri, R. Collins, and S. Marsillac, "In situ and ex situ studies of molybdenum thin films deposited by rf and dc magnetron sputtering as a back contact for CIGS solar cells," *International Journal of Photoenergy,* vol. 2012, 2012.

[31] O. Ayala *et al.,* "Theoretical Analysis of Experimental Data of Sodium Diffusion in Oxidized Molybdenum Thin Films," *Energies,* vol. 14, no. 9, p. 2479, 2021.

[32] S. Marsillac, S. Little, and R. Collins, "A broadband analysis of the optical properties of silver nanoparticle films by in situ real time spectroscopic ellipsometry," *Thin Solid Films,* vol. 519, no. 9, pp. 2936-2940, 2011.

[33] P. Koirala *et al.,* "Real time spectroscopic ellipsometry for analysis and control of thin film polycrystalline semiconductor deposition in photovoltaics," *Thin solid films,* vol. 571, pp. 442-446, 2014.

[34] L. R. Dahal, D. Sainju, N. Podraza, S. Marsillac, and R. Collins, "Real time spectroscopic ellipsometry of Ag/ZnO and Al/ZnO interfaces for back-reflectors in thin film Si: H photovoltaics," *Thin Solid Films,* vol. 519, no. 9, pp. 2682-2687, 2011.

[35] P. Koirala *et al.,* "Through-the-glass spectroscopic ellipsometry for analysis of CdTe thin-film solar cells in the superstrate configuration," *Progress in Photovoltaics: Research and Applications,* vol. 24, no. 8, pp. 1055-1067, 2016.

[36] P. Aryal *et al.,* "Parameterized complex dielectric functions of CuIn1− xGaxSe2: applications in optical characterization of compositional non-uniformities and depth profiles in materials and solar cells," *Progress in Photovoltaics: Research and Applications,* vol. 24, no. 9, pp. 1200-1213, 2016.

[37] S. Marsillac *et al.,* "Spectroscopic ellipsometry studies of In2S3 top window and Mo back contacts in chalcopyrite photovoltaics technology," *physica status solidi c,* vol. 5, no. 5, pp. 1244-1248, 2008.

[38] V. Ranjan, R. Collins, and S. Marsillac, "Real-time analysis of the microstructural evolution and optical properties of Cu (In, Ga) Se2 thin films as a function of Cu content," *physica status solidi (RRL)– Rapid Research Letters,* vol. 6, no. 1, pp. 10-12, 2012.

[39] T. Begou, J. D. Walker, D. Attygalle, V. Ranjan, R. Collins, and S. Marsillac, "Real time spectroscopic ellipsometry of CuInSe2: growth dynamics, dielectric function, and its dependence on temperature," *physica status solidi (RRL)–Rapid Research Letters,* vol. 5, no. 7, pp. 217-219, 2011.

[40] S. Karki *et al.,* "In situ and ex situ investigations of KF postdeposition treatment effects on CIGS solar cells," *IEEE Journal of Photovoltaics,* vol. 7, no. 2, pp. 665-669, 2016.

[41] L. R. Dahal *et al.,* "Correlations between mapping spectroscopic ellipsometry results and solar cell performance for evaluations of nonuniformity in thin-film silicon photovoltaics," *IEEE Journal of Photovoltaics,* vol. 3, no. 1, pp. 387-393, 2012.

[42] P. Aryal, D. Attygalle, P. Pradhan, N. J. Podraza, S. Marsillac, and R. W. Collins, "Large-area compositional mapping of Cu (In $ _ {1-x} $ Ga $ _ {x} $) Se $ _ {2} $ materials and devices with spectroscopic ellipsometry," *IEEE Journal of Photovoltaics,* vol. 3, no. 1, pp. 359-363, 2012.

[43] A.-R. Ibdah *et al.,* "Spectroscopic ellipsometry for analysis of polycrystalline thin-film photovoltaic devices and prediction of external quantum efficiency," *Applied Surface Science,* vol. 421, pp. 601-607, 2017.

[44] J. Walker, H. Khatri, V. Ranjan, J. Li, R. Collins, and S. Marsillac, "Electronic and structural properties of molybdenum thin films as determined by real-time spectroscopic ellipsometry," *Applied Physics Letters,* vol. 94, no. 14, p. 141908, 2009.

[45] S. Little, T. Begou, R. Collins, and S. Marsillac, "Optical detection of melting point depression for silver nanoparticles via in situ real time spectroscopic ellipsometry," *Applied Physics Letters,* vol. 100, no. 5, p. 051107, 2012.

978-1-7281-6118-1/22 $31.00 © 2022 IEEE

[46] P. Pradhan *et al.*, "Real time spectroscopic ellipsometry analysis of the three-stages of CuIn 1− x Ga x Se 2 co-evaporation," in *2014 IEEE 40th Photovoltaic Specialist Conference (PVSC)*, 2014: IEEE, pp. 2060-2065.

[47] D. Attygalle *et al.*, "Optical monitoring and control of three-stage coevaporated Cu (In 1− x Ga x) Se 2 by real-time spectroscopic ellipsometry," in *2012 IEEE 38th Photovoltaic Specialists Conference (PVSC) PART 2*, 2012: IEEE, pp. 1-6.

[48] H. Fujiwara and R. W. Collins, *Spectroscopic Ellipsometry for Photovoltaics: Volume 2: Applications and Optical Data of Solar Cell Materials*. Springer, 2019.

[49] S. Marsillac, J. Bernede, R. Ny, and A. Conan, "A new simple technique to obtain In2Se3 polycrystalline thin films," *Vacuum*, vol. 46, no. 11, pp. 1315-1323, 1995.

[50] N. Barreau, S. Marsillac, and J. Bernede, "Physico-chemical characterization of β-In2S3 thin films synthesized by solid-state reaction, induced by annealing, of the constituents sequentially deposited in thin layers," *Vacuum*, vol. 56, no. 2, pp. 101-106, 2000.

[51] N. Barreau, J. Bernede, S. Marsillac, C. Amory, and W. Shafarman, "New Cd-free buffer layer deposited by PVD: In2S3 containing Na compounds," *Thin Solid Films*, vol. 431, pp. 326-329, 2003.

[52] R. Robles, N. Barreau, A. Vega, S. Marsillac, J. Bernede, and A. Mokrani, "Optical properties of large band gap β-In2S3− 3xO3x compounds obtained by physical vapour deposition," *Optical Materials*, vol. 27, no. 4, pp. 647-653, 2005.

[53] J. Bernede, S. Marsillac, and A. Conan, "Electrical properties of γ-In2Se3 layers synthesized by solid state reaction between In and Se thin films," *Materials chemistry and physics*, vol. 48, no. 1, pp. 5-9, 1997.

[54] S. Marsillac, J. Bernede, and A. Conan, "Change in the type of majority carriers in disordered ln x Se 100− x thin-film alloys," *Journal of materials science*, vol. 31, no. 3, pp. 581-587, 1996.

[55] N. Barreau, J. Bernede, and S. Marsillac, "Study of the new β-In2S3 containing Na thin films. Part II: Optical and electrical characterization of thin films," *Journal of crystal growth*, vol. 241, no. 1-2, pp. 51-56, 2002.

[56] N. Barreau, J. Bernede, C. Deudon, L. Brohan, and S. Marsillac, "Study of the new β-In2S3 containing Na thin films Part I: Synthesis and structural characterization of the material," *Journal of crystal growth*, vol. 241, no. 1-2, pp. 4-14, 2002.

[57] K. D'Almeida, J. Bernede, F. Ragot, A. Godoy, F. Diaz, and S. Lefrant, "Carbazole-based electroluminescent devices obtained by vacuum evaporation," *Journal of applied polymer science*, vol. 82, no. 8, pp. 2042-2055, 2001.

[58] N. Barreau, S. Marsillac, J. Bernede, and L. Assmann, "Evolution of the band structure of β-In 2 S 3− 3x O 3x buffer layer with its oxygen content," *Journal of applied physics*, vol. 93, no. 9, pp. 5456-5459, 2003.

[59] C. Amory, J. Bernede, and S. Marsillac, "Study of a growth instability of γ-In 2 Se 3," *Journal of applied physics*, vol. 94, no. 10, pp. 6945-6948, 2003.

[60] K. Benchouk *et al.*, "New buffer layers, large band gap ternary compounds: CuAlTe2," *The European Physical Journal Applied Physics*, vol. 10, no. 1, pp. 9-14, 2000.

[61] J. Bernede, N. Barreau, S. Marsillac, and L. Assmann, "Band alignment at β-In2S3/TCO interface," *Applied surface science*, vol. 195, no. 1-4, pp. 222-228, 2002.

[62] N. Barreau, S. Marsillac, J. Bernede, and A. Barreau, "Investigation of β-In2S3 growth on different transparent conductive oxides," *Applied surface science*, vol. 161, no. 1-2, pp. 20-26, 2000.

[63] D. Poudel *et al.*, "Assessment of Cu (In, Ga) Se2 Solar Cells Degradation due to Water Ingress Effect on the CdS Buffer Layer," *Journal of Energy and Power Technology*, vol. 3, no. 1, 2021.

[64] P. Paul *et al.*, "Direct nm-scale spatial mapping of traps in CIGS," *IEEE Journal of Photovoltaics*, vol. 5, no. 5, pp. 1482-1486, 2015.

[65] S. Karki *et al.*, "Degradation mechanism in Cu (In, Ga) Se 2 material and solar cells due to moisture and heat treatment of the absorber layer," *IEEE Journal of Photovoltaics*, vol. 9, no. 4, pp. 1138-1143, 2019.

[66] S. Karki *et al.*, "Impact of water ingress on molybdenum thin films and its effect on Cu (In, Ga) Se 2 Solar Cells," *IEEE Journal of Photovoltaics*, vol. 10, no. 2, pp. 696-702, 2019.

[67] D. Poudel *et al.*, "Degradation Mechanism Due to Water Ingress Effect on the Top Contact of Cu (In, Ga) Se2 Solar Cells," *Energies*, vol. 13, no. 17, p. 4545, 2020.

[68] D. Poudel *et al.*, "Numerical Analysis of Water Ingress Effect on the Window Layer of Cu (In, Ga) Se 2 Solar Cells using SCAPS-1D," in *2021 IEEE 48th Photovoltaic Specialists Conference (PVSC)*, 2021: IEEE, pp. 0438-0442.

[69] D. Poudel, B. Belfore, S. Karki, G. Rajan, A. Rockett, and S. Marsillac, "Process Dependent Instabilities In Cu (In, Ga) Se 2 Solar Cells Under Water Ingress," in *2021 IEEE 48th Photovoltaic Specialists Conference (PVSC)*, 2021: IEEE, pp. 0433-0437.

[70] G. Rajan *et al.*, "Study of Instabilities and Degradation due to Moisture Ingress in the Molybdenum back contact of Cu (In, Ga) Se 2 Solar Cells," in *2018 IEEE 7th World Conference on Photovoltaic Energy Conversion (WCPEC)(A Joint Conference of 45th IEEE PVSC, 28th PVSEC & 34th EU PVSEC)*, 2018: IEEE, pp. 3037-3039.

On the Effect of Indium Chloride Dose on the Recrystallization of Cu(In,Ga)Se$_2$ Thin Films and associated Devices

Deewakar Poudel[1], Adam Masters[1], Benjamin Belfore[1], Angus Rockett[2] and Sylvain Marsillac[1]

[1]Virginia Institute of Photovoltaics, Old Dominion University, Norfolk, VA 23529, USA

[2]Dept. of Metallurgical and Materials Engineering, Colorado School of Mines, Golden, CO 80401, USA

Abstract— **Cu(In,Ga)Se$_2$ thin films deposited by a single-stage co-evaporation process at 350 °C on molybdenum coated soda lime glass substrate were annealed post-deposition in InCl$_3$ vapor. The amount of InCl$_3$ and Se was varied. The annealing treatment was done at 450 °C for 30 minutes. Increase in grain size was observed after the treatment in all cases by X-ray diffraction. Device performance was low, but improved slightly after KCN etching.**

Keywords— *Cu(In,Ga)Se$_2$, recrystallization, annealing, InCl$_3$*

I. INTRODUCTION

High-throughput production process make thin film photovoltaics a potential alternate to silicon based solar cells. Such a process is the fabrication of CdTe photovoltaic modules. The high-rate deposition of CdTe layer includes the annealing in CuCl$_2$ vapor, recrystallizing the CdTe and resulting in enhanced cells performance [1, 2]. Similarly, the application of metal halides recrystallization in the case of Cu(In,Ga)Se$_2$ or CIGS could potentially produce high efficiency devices. Along with enhancement in its supply chain management, this might allow this technology to compete with more conventional ones such as crystalline silicon [3-10]. One has to keep in mind that an optimum compositional profile, including proper gallium grading and ideal Cu content, is likely required for high device performance, while only large grain size does not always yield higher efficiency devices [11].

We have already established that several metal halides post-deposition treatment can act as exceptional transport agents in CIGS, yielding larger grain and uniform layers [12-27]. In this work, we explore the ex-situ recrystallization of CIGS using indium chloride vapor treatment at various doses and its effect on both thin film properties and device performance.

II. EXPERIMENTAL METHODS

CIGS thin films were deposited at substrate temperature of 350 °C by a single-stage co-evaporation process onto molybdenum coated soda-lime glass. Molybdenum was fabricated by dc magnetron sputtering [28-30]. The composition of the samples determined by x-ray fluorescence (XRF). After deposition, half of the samples were placed in a quartz tube with different amount of InCl$_3$ for recrystallization (along with an amount of elemental Se to ensure overpressure of Se). The quartz tubes were then sealed and placed into a

furnace. The annealing treatment was performed at a constant temperature of 450 °C for 30 minutes. After annealing samples were cool to room temperature and were rinsed with deionized water to remove any deposit on the surface. The other half of the samples were kept as reference. Some of these samples were exposed to KCN etching for surface treatment.

The resulting samples were characterized by several methods. The crystallographic structures analysis was done by symmetric θ-2θ X-ray diffraction and analyzed using the International Center for Diffraction Data (ICDD) database. Cross-section morphological analysis were performed by scanning electron microscope (SEM). Using a Woollam M-2000 rotating compensator multichannel system, we characterized samples and their complex dielectric functions in the spectral range of 0.75 eV to 6.5 eV [31-48]. The photovoltaic characteristics were evaluated by external quantum efficiency (QE) measurements (QEX7, PV measurements Inc.) and current density-voltage (J-V) measurements (IV5, PV measurements Inc.) done under AM 1.5G with a light intensity of 100 mW/cm^2 at 25°C.

A summary of the various types of samples is presented in Table I. Both the dose of InCl$_3$ and Se were changed giving rise to 3 types of samples. The duration of annealing was kept constant at 30 minutes.

TABLE I. PROCESS PARAMETERS FOR THE SAMPLES RECRYSTALLIZED IN THE ANNEALING CHAMBER

Sample	Se (g)	InCl$_3$ (g)	Time (min)
I	0.25	0.25	30
II	0.10	0.10	30
III	0.03	0.03	30

III. RESULTS AND DISCUSSIONS

The amount of Se and InCl$_3$ was varied to optimize the vapor during recrystallization. The details of the amount included are indicated in Table I. The composition measured by XRF indicated a condensation of Se and In on the samples, when an excess quantity was used (as in sample I). The amount

---

This research was supported by the Department of Energy Contract No. DE-EE0007551.

of Se and InCl₃ was then optimized (for sample II), but excess Se still appeared. Sample III had a more uniform composition. As can be seen in Figure 1, the cross section SEM taken on all samples revealed an evolution of the grain size, with samples annealed having larger crystallite sizes as compared to the reference samples.

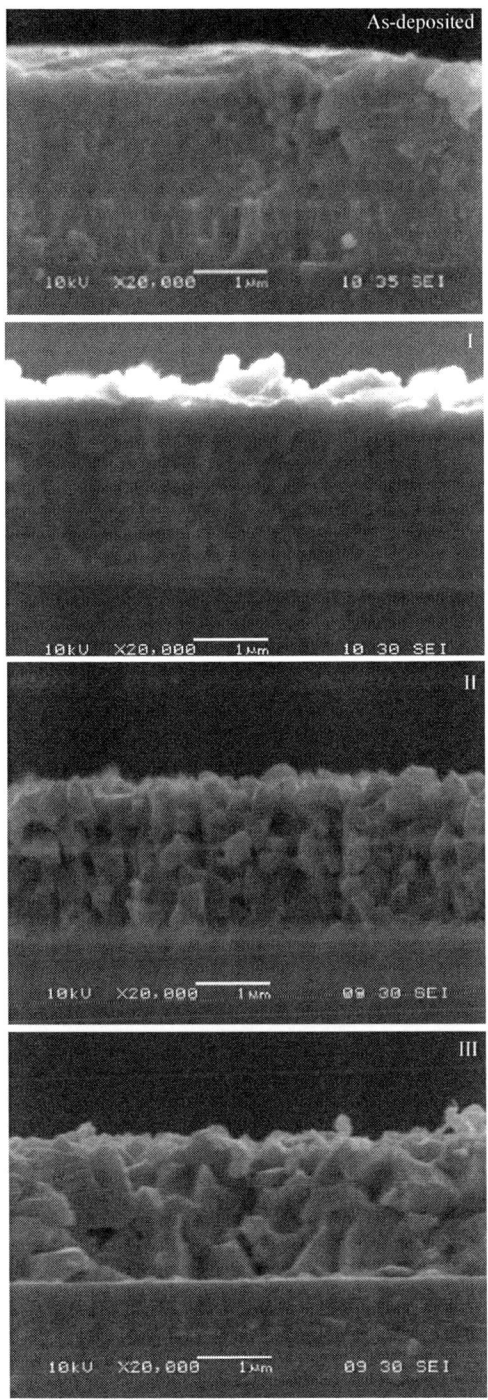

Fig. 1. Cross-section Scanning Electron Microscopy micrographs of CIGS films: as-deposited and recrystallized at 450 °C by InCl₃ vapor treatment at various doses (see Table I).

XRD characterization was performed on the films. XRD profiles of the samples annealed with InCl₃ with different doses are shown in Figure 2 and the peak properties are shown in Table II. XRD profiles shows that the samples have a strong (112) orientation and no change of preferential orientation occurs due to the recrystallization. All samples after recrystallization have larger peak intensities and smaller full width a half maximum (FWHM) indicating larger cystallite sizes for these samples.

TABLE II. XRD RESULTS FOR AS-DEPOSITED AND INCL₃ TREATED SAMPLES AT 450 °C

Sample	Peak (112)/(220)	Int.I (cps deg)	FWHM (deg)
As-deposited	26.7/44.6	395/73	0.41/0.73
I	27.0/44.9	703/128	0.22/0.51
II	26.9/44.7	1183/392	0.26/0.42
III	27.0/44.7	1188/216	0.33/0.35

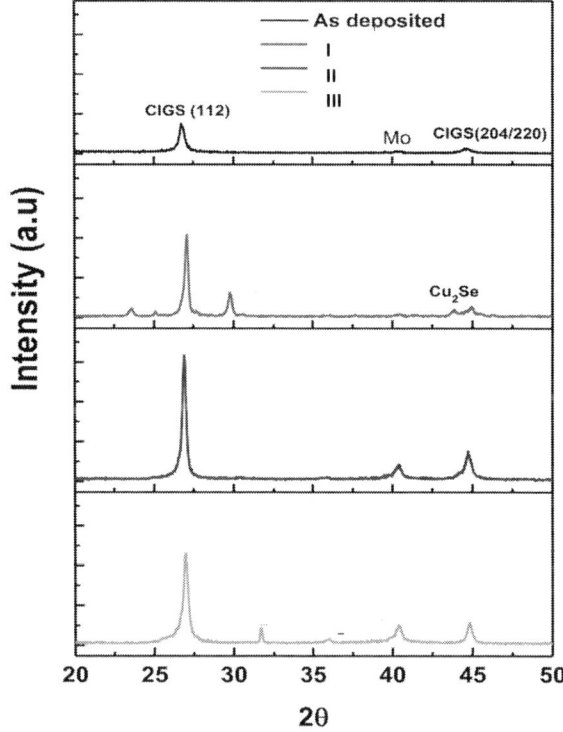

Fig. 2. XRD plots ((112) and (220)/(204) peaks) for as-deposited and InCl₃ treated CIGS samples.

A small shift in the peak position for all samples after recrystallization might indicate a change in gallium distribution or a change in the film strain. This will be confirmed by secondary ion mass spectrometry measurements.

Raman Spectra were measured at RT before and after annealing at 450 °C for samples of type III. The Raman spectra shows a clear intense peak ascribed to A1 mode of CIGS compound with a lower FWHM compared to the as-deposited samples due to better crystallinity. The A1 peak of the annealed sample shifted towards CuInSe$_2$. The peak at 252 cm^{-1} for the as-deposited sample might be related to Cu$_{2-x}$Se. The spectrum also shows a broader peak or the as-deposited samples due to the contribution of E and B$_2$ modes of CIGS, implying the presence of In-Se and Ga-Se phases.

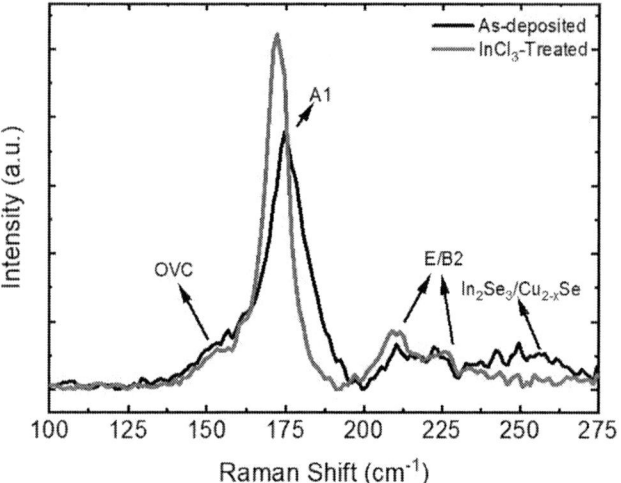

Fig. 3. Raman spectra of as-deposited and InCl$_3$ treated CIGS samples III at 450 °C.

All types of samples were processed into full devices. However, most of the devices resulted in shunted devices, except for devices of Type III. The representative current-voltage and external quantum efficiency curve for devices of Type III are shown in Figure 4. Table III shows the corresponding photovoltaic parameters. The device performance was lower than expected. Because of its influence on device performance, it would be of interest to study the connection between recrystallization and alkali content [49-56]. One also need to assess whether the InCl$_3$ treatment modify the main element profiles, and would require a different n-type layer, other than CdS [57-70]. To improve the device performance, surface treatment with KCN was performed. KCN is known to remove metallic rich phase at interface, improve junction and mitigate shunting. While there can be a significant removal of alternate phases from the film's surface, the actual CIGS material remains generally untouched. The overall device performance increases after the KCN treatment as can be seen from Figure 4 and Table III, mostly due to an increase in short circuit current density.

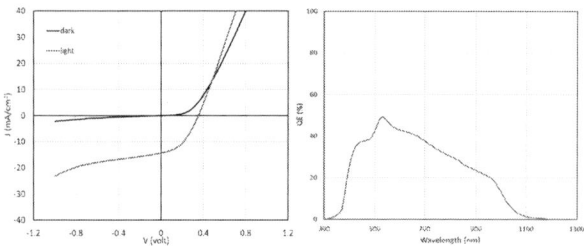

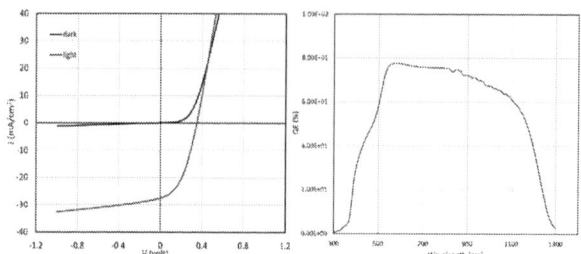

Fig. 4. Representative I-V and QE curves for samples of Type III with and without KCN treated CIGS samples.

TABLE III. XRD RESULTS FOR AS-DEPOSITED AND INCL$_3$ TREATED SAMPLES AT 450 °C

Sample	V$_{OC}$ (V)	J$_{SC}$ (mA/cm^2)	FF (%)	η (%)
InCl$_3$	0.35	14.4	42	2.1
KCN	0.34	27.5	42.6	4.0

IV. CONCLUSION

The recrystallization of CIGS thin films by InCl$_3$ vapor treatment at various doses was studied by morphological and structural analysis. Grain enhancement was observed by SEM in treated samples in all cases. A decrease in FWHM and an increase in peak intensity were observed by XRD, suggesting a change in crystallinity. The performance of most of the devices was poor, with a lot of shunting. The performance was slightly improved after KCN treatment, which tends to remove metallic phases and reduce surface roughness. Further improvement in the precise dose, annealing temperature and optimum composition is still needed to improve the grain quality and device performance.

ACKNOWLEDGMENT

This research was supported by the Department of Energy Contract No. DE-EE0007551.

REFERENCES

[1] B. McCandless, L. Moulton, and R. Birkmire, "Recrystallization and sulfur diffusion in CdCl2-treated CdTe/CdS thin films," *Progress in photovoltaics: Research and Applications*, vol. 5, no. 4, pp. 249-260, 1997.

[2] C. Ferekides, U. Balasubramanian, R. Mamazza, V. Viswanathan, H. Zhao, and D. Morel, "CdTe thin film solar cells: device and technology issues," *Solar energy*, vol. 77, no. 6, pp. 823-830, 2004.

[3] D. M. Nelson, E. Marsillac, and S. S. Rao, "Antecedents and evolution of the green supply chain," *Journal of Operations and Supply Chain Management*, no. Special Issue, 2012.

[4] K. Liao, E. Marsillac, E. Johnson, and Y. Liao, "Global supply chain adaptations to improve financial performance: supply base establishment and logistics integration," *Journal of Manufacturing Technology Management*, 2011.

[5] R. Romero-Silva, E. Marsillac, S. Shaaban, and M. Hurtado-Hernández, "Serial production line performance under random variation: dealing with the 'Law of Variability'," *Journal of Manufacturing Systems*, vol. 50, pp. 278-289, 2019.

[6] E. Marsillac, "Management of the photovoltaic supply chain," *International Journal of Technology, Policy and Management*, vol. 12, no. 2-3, pp. 195-211, 2012.

[7] Y. Liao and E. Marsillac, "External knowledge acquisition and innovation: the role of supply chain network-oriented flexibility and

organisational awareness," *International Journal of Production Research,* vol. 53, no. 18, pp. 5437-5455, 2015.

[8] R. Romero-Silva, S. Shaaban, E. Marsillac, and M. Hurtado, "Exploiting the characteristics of serial queues to reduce the mean and variance of flow time using combined priority rules," *International Journal of Production Economics,* vol. 196, pp. 211-225, 2018.

[9] E. L. Marsillac, "Environmental impacts on reverse logistics and green supply chains: similarities and integration," *International Journal of Logistics Systems and Management,* vol. 4, no. 4, pp. 411-422, 2008.

[10] E. Marsillac, "Closing the Photovoltaic Supply Chain Loop-Invest Now for Future Returns," in *2018 IEEE 7th World Conference on Photovoltaic Energy Conversion (WCPEC)(A Joint Conference of 45th IEEE PVSC, 28th PVSEC & 34th EU PVSEC),* 2018: IEEE, pp. 2498-2500.

[11] D. Poudel *et al.,* "In Situ Recrystallization of Co-Evaporated Cu (In, Ga) Se2 Thin Films by Copper Chloride Vapor Treatment towards Solar Cell Applications. Energies 2021, 14, 3938," ed: s Note: MDPI stays neutral with regard to jurisdictional claims in published ..., 2021.

[12] G. Rajan *et al.,* "Impact of Post-Deposition Recrystallization by Alkali Fluorides on Cu (In, Ga) Se2Thin-Film Materials and Solar Cells," *Thin Solid Films,* vol. 690, p. 137526, 2019.

[13] E. Palmiotti, B. Belfore, D. Poudel, S. Marsillac, and A. Rockett, "In-Situ Study of the Crystallization of Amorphous CuInSe2 Thin Films and the Effect of InCl3 Treatment," *Thin Solid Films,* p. 139095, 2022.

[14] B. Belfore *et al.,* "Recrystallization of Cu (In, Ga) Se2 Semiconductor Thin Films via InCl3 Treatment," *Thin Solid Films,* vol. 735, p. 138897, 2021.

[15] S. Marsillac, M. Zouaghi, J. Bernede, T. B. Nasrallah, and S. Belgacem, "Evolution of the properties of spray-deposited CuInS2 thin films with post-annealing treatment," *Solar energy materials and solar cells,* vol. 76, no. 2, pp. 125-134, 2003.

[16] D. Poudel *et al.,* "Analysis of Post-Deposition Recrystallization Processing via Indium Bromide of Cu (In, Ga) Se2 Thin Films," *Materials,* vol. 14, no. 13, p. 3596, 2021.

[17] S. Karki *et al.,* "Analysis of recombination mechanisms in RbF-treated CIGS solar cells," *IEEE Journal of Photovoltaics,* vol. 9, no. 1, pp. 313-318, 2018.

[18] D. Poudel *et al.,* "In Situ Recrystallization of Co-Evaporated Cu (In, Ga) Se2 Thin Films by Copper Chloride Vapor Treatment towards Solar Cell Applications," *Energies,* vol. 14, no. 13, p. 3938, 2021.

[19] D. Poudel *et al.,* "Studying the Recrystallization of Cu (InGa) Se 2 Semiconductor Thin Films by Silver Bromide In-situ Treatment," in *2021 IEEE 48th Photovoltaic Specialists Conference (PVSC),* 2021: IEEE, pp. 2307-2311.

[20] D. Poudel *et al.,* "Effect of Indium Bromide Treatments Post-Deposition Recrystallization Temperature on Cu (In, Ga) Se 2 Thin Films," in *2021 IEEE 48th Photovoltaic Specialists Conference (PVSC),* 2021: IEEE, pp. 0429-0432.

[21] E. Palmiotti, B. Belfore, D. Poudel, S. Marsillac, and A. Rockett, "CuIn (1− x) Ga x Se 2 Recrystallization by Metal Halide Treatments," in *2021 IEEE 48th Photovoltaic Specialists Conference (PVSC),* 2021: IEEE, pp. 0669-0671.

[22] B. Belfore *et al.,* "Study of Indium Chloride Vapor Treatment on Cu (In, Ga) Se 2 Semiconductor Thin Films," in *2021 IEEE 48th Photovoltaic Specialists Conference (PVSC),* 2021: IEEE, pp. 2320-2323.

[23] B. Belfore *et al.,* "Vapor Treatment and In-situ Recrystallization by Copper Chloride on Cu (In, Ga) Se 2 Thin Film," in *2021 IEEE 48th Photovoltaic Specialists Conference (PVSC),* 2021: IEEE, pp. 2316-2319.

[24] B. Belfore, D. Poudel, T. Ashrafee, S. Karki, G. Rajan, and S. Marsillac, "Morphological Study of Indium Chloride Post Deposition Treated CuInSe 2 Thin Films," in *2021 IEEE 48th Photovoltaic Specialists Conference (PVSC),* 2021: IEEE, pp. 2312-2315.

[25] B. Belfore *et al.,* "In-Situ Recrystallization of CIGS via Metal Halides," in *2020 47th IEEE Photovoltaic Specialists Conference (PVSC),* 2020: IEEE, pp. 1131-1133.

[26] B. Belfore *et al.,* "Ex-Situ Recrystallization of CIGS via Metal Halides," in *2020 47th IEEE Photovoltaic Specialists Conference (PVSC),* 2020: IEEE, pp. 1102-1104.

[27] B. Belfore, G. Rajan, S. Karki, D. Poudel, A. Rockett, and S. Marsillac, "The Impact of Deposition Temperature on Sodium Fluoride Recrystallization in Cu (In, Ga) Se 2 Solar Cells," in *2019 IEEE 46th Photovoltaic Specialists Conference (PVSC),* 2019: IEEE, pp. 1851-1853.

[28] E. Gourmelon, J. Bernede, J. Pouzet, and S. Marsillac, "Textured MoS 2 thin films obtained on tungsten: electrical properties of the W/MoS 2 contact," *Journal of Applied Physics,* vol. 87, no. 3, pp. 1182-1186, 2000.

[29] K. Aryal, H. Khatri, R. Collins, and S. Marsillac, "In situ and ex situ studies of molybdenum thin films deposited by rf and dc magnetron sputtering as a back contact for CIGS solar cells," *International Journal of Photoenergy,* vol. 2012, 2012.

[30] O. Ayala *et al.,* "Theoretical Analysis of Experimental Data of Sodium Diffusion in Oxidized Molybdenum Thin Films," *Energies,* vol. 14, no. 9, p. 2479, 2021.

[31] S. Marsillac, S. Little, and R. Collins, "A broadband analysis of the optical properties of silver nanoparticle films by in situ real time spectroscopic ellipsometry," *Thin Solid Films,* vol. 519, no. 9, pp. 2936-2940, 2011.

[32] P. Koirala *et al.,* "Real time spectroscopic ellipsometry for analysis and control of thin film polycrystalline semiconductor deposition in photovoltaics," *Thin solid films,* vol. 571, pp. 442-446, 2014.

[33] L. R. Dahal, D. Sainju, N. Podraza, S. Marsillac, and R. Collins, "Real time spectroscopic ellipsometry of Ag/ZnO and Al/ZnO interfaces for back-reflectors in thin film Si: H photovoltaics," *Thin Solid Films,* vol. 519, no. 9, pp. 2682-2687, 2011.

[34] P. Koirala *et al.,* "Through-the-glass spectroscopic ellipsometry for analysis of CdTe thin-film solar cells in the superstrate configuration," *Progress in Photovoltaics: Research and Applications,* vol. 24, no. 8, pp. 1055-1067, 2016.

[35] P. Aryal *et al.,* "Parameterized complex dielectric functions of CuIn1− xGaxSe2: applications in optical characterization of compositional non-uniformities and depth profiles in materials and solar cells," *Progress in Photovoltaics: Research and Applications,* vol. 24, no. 9, pp. 1200-1213, 2016.

[36] S. Marsillac *et al.,* "Spectroscopic ellipsometry studies of In2S3 top window and Mo back contacts in chalcopyrite photovoltaics technology," *physica status solidi c,* vol. 5, no. 5, pp. 1244-1248, 2008.

[37] V. Ranjan, R. Collins, and S. Marsillac, "Real-time analysis of the microstructural evolution and optical properties of Cu (In, Ga) Se2 thin films as a function of Cu content," *physica status solidi (RRL)– Rapid Research Letters,* vol. 6, no. 1, pp. 10-12, 2012.

[38] T. Begou, J. D. Walker, D. Attygalle, V. Ranjan, R. Collins, and S. Marsillac, "Real time spectroscopic ellipsometry of CuInSe2: growth dynamics, dielectric function, and its dependence on temperature," *physica status solidi (RRL)–Rapid Research Letters,* vol. 5, no. 7, pp. 217-219, 2011.

[39] S. Karki *et al.,* "In situ and ex situ investigations of KF postdeposition treatment effects on CIGS solar cells," *IEEE Journal of Photovoltaics,* vol. 7, no. 2, pp. 665-669, 2016.

[40] L. R. Dahal *et al.,* "Correlations between mapping spectroscopic ellipsometry results and solar cell performance for evaluations of nonuniformity in thin-film silicon photovoltaics," *IEEE Journal of Photovoltaics,* vol. 3, no. 1, pp. 387-393, 2012.

[41] P. Aryal, D. Attygalle, P. Pradhan, N. J. Podraza, S. Marsillac, and R. W. Collins, "Large-area compositional mapping of Cu (In $ _ {1-x} $ Ga $ _ {x} $) Se $ _ {2} $ materials and devices with spectroscopic ellipsometry," *IEEE Journal of Photovoltaics,* vol. 3, no. 1, pp. 359-363, 2012.

[42] A.-R. Ibdah *et al.,* "Spectroscopic ellipsometry for analysis of polycrystalline thin-film photovoltaic devices and prediction of external quantum efficiency," *Applied Surface Science,* vol. 421, pp. 601-607, 2017.

[43] J. Walker, H. Khatri, V. Ranjan, J. Li, R. Collins, and S. Marsillac, "Electronic and structural properties of molybdenum thin films as determined by real-time spectroscopic ellipsometry," *Applied Physics Letters,* vol. 94, no. 14, p. 141908, 2009.

978-1-7281-6118-1/22 $31.00 © 2022 IEEE

[44] S. Little, T. Begou, R. Collins, and S. Marsillac, "Optical detection of melting point depression for silver nanoparticles via in situ real time spectroscopic ellipsometry," *Applied Physics Letters,* vol. 100, no. 5, p. 051107, 2012.

[45] D. R. Sapkota *et al.*, "Evaluation of CuInSe 2 Materials and Solar Cells Co-evaporated at Different Rates Based on Real Time Spectroscopic Ellipsometry Calibrations," in *2021 IEEE 48th Photovoltaic Specialists Conference (PVSC)*, 2021: IEEE, pp. 0451-0458.

[46] P. Pradhan *et al.*, "Real time spectroscopic ellipsometry analysis of the three-stages of CuIn 1− x Ga x Se 2 co-evaporation," in *2014 IEEE 40th Photovoltaic Specialist Conference (PVSC)*, 2014: IEEE, pp. 2060-2065.

[47] D. Attygalle *et al.*, "Optical monitoring and control of three-stage coevaporated Cu (In 1− x Ga x) Se 2 by real-time spectroscopic ellipsometry," in *2012 IEEE 38th Photovoltaic Specialists Conference (PVSC) PART 2*, 2012: IEEE, pp. 1-6.

[48] H. Fujiwara and R. W. Collins, *Spectroscopic Ellipsometry for Photovoltaics: Volume 2: Applications and Optical Data of Solar Cell Materials.* Springer, 2019.

[49] D. Poudel *et al.*, "Assessment of Cu (In, Ga) Se2 Solar Cells Degradation due to Water Ingress Effect on the CdS Buffer Layer," *Journal of Energy and Power Technology,* vol. 3, no. 1, 2021.

[50] P. Paul *et al.*, "Direct nm-scale spatial mapping of traps in CIGS," *IEEE Journal of Photovoltaics,* vol. 5, no. 5, pp. 1482-1486, 2015.

[51] S. Karki *et al.*, "Degradation mechanism in Cu (In, Ga) Se 2 material and solar cells due to moisture and heat treatment of the absorber layer," *IEEE Journal of Photovoltaics,* vol. 9, no. 4, pp. 1138-1143, 2019.

[52] S. Karki *et al.*, "Impact of water ingress on molybdenum thin films and its effect on Cu (In, Ga) Se 2 Solar Cells," *IEEE Journal of Photovoltaics,* vol. 10, no. 2, pp. 696-702, 2019.

[53] D. Poudel *et al.*, "Degradation Mechanism Due to Water Ingress Effect on the Top Contact of Cu (In, Ga) Se2 Solar Cells," *Energies,* vol. 13, no. 17, p. 4545, 2020.

[54] D. Poudel *et al.*, "Numerical Analysis of Water Ingress Effect on the Window Layer of Cu (In, Ga) Se 2 Solar Cells using SCAPS-1D," in *2021 IEEE 48th Photovoltaic Specialists Conference (PVSC)*, 2021: IEEE, pp. 0438-0442.

[55] D. Poudel, B. Belfore, S. Karki, G. Rajan, A. Rockett, and S. Marsillac, "Process Dependent Instabilities In Cu (In, Ga) Se 2 Solar Cells Under Water Ingress," in *2021 IEEE 48th Photovoltaic Specialists Conference (PVSC)*, 2021: IEEE, pp. 0433-0437.

[56] G. Rajan *et al.*, "Study of Instabilities and Degradation due to Moisture Ingress in the Molybdenum back contact of Cu (In, Ga) Se 2 Solar Cells," in *2018 IEEE 7th World Conference on Photovoltaic Energy Conversion (WCPEC)(A Joint Conference of 45th IEEE PVSC, 28th PVSEC & 34th EU PVSEC)*, 2018: IEEE, pp. 3037-3039.

[57] S. Marsillac, J. Bernede, R. Ny, and A. Conan, "A new simple technique to obtain In2Se3 polycrystalline thin films," *Vacuum,* vol. 46, no. 11, pp. 1315-1323, 1995.

[58] N. Barreau, S. Marsillac, and J. Bernede, "Physico-chemical characterization of β-In2S3 thin films synthesized by solid-state reaction, induced by annealing, of the constituents sequentially deposited in thin layers," *Vacuum,* vol. 56, no. 2, pp. 101-106, 2000.

[59] N. Barreau, J. Bernede, S. Marsillac, C. Amory, and W. Shafarman, "New Cd-free buffer layer deposited by PVD: In2S3 containing Na compounds," *Thin Solid Films,* vol. 431, pp. 326-329, 2003.

[60] R. Robles, N. Barreau, A. Vega, S. Marsillac, J. Bernede, and A. Mokrani, "Optical properties of large band gap β-In2S3− 3xO3x compounds obtained by physical vapour deposition," *Optical Materials,* vol. 27, no. 4, pp. 647-653, 2005.

[61] J. Bernede, S. Marsillac, and A. Conan, "Electrical properties of γ-In2Se3 layers synthesized by solid state reaction between In and Se thin films," *Materials chemistry and physics,* vol. 48, no. 1, pp. 5-9, 1997.

[62] S. Marsillac, J. Bernede, and A. Conan, "Change in the type of majority carriers in disordered ln x Se 100− x thin-film alloys," *Journal of materials science,* vol. 31, no. 3, pp. 581-587, 1996.

[63] N. Barreau, J. Bernede, and S. Marsillac, "Study of the new β-In2S3 containing Na thin films. Part II: Optical and electrical characterization of thin films," *Journal of crystal growth,* vol. 241, no. 1-2, pp. 51-56, 2002.

[64] N. Barreau, J. Bernede, C. Deudon, L. Brohan, and S. Marsillac, "Study of the new β-In2S3 containing Na thin films Part I: Synthesis and structural characterization of the material," *Journal of crystal growth,* vol. 241, no. 1-2, pp. 4-14, 2002.

[65] K. D'Almeida, J. Bernede, F. Ragot, A. Godoy, F. Diaz, and S. Lefrant, "Carbazole-based electroluminescent devices obtained by vacuum evaporation," *Journal of applied polymer science,* vol. 82, no. 8, pp. 2042-2055, 2001.

[66] N. Barreau, S. Marsillac, J. Bernede, and L. Assmann, "Evolution of the band structure of β-In 2 S 3− 3x O 3x buffer layer with its oxygen content," *Journal of applied physics,* vol. 93, no. 9, pp. 5456-5459, 2003.

[67] C. Amory, J. Bernede, and S. Marsillac, "Study of a growth instability of γ-In 2 Se 3," *Journal of applied physics,* vol. 94, no. 10, pp. 6945-6948, 2003.

[68] K. Benchouk *et al.*, "New buffer layers, large band gap ternary compounds: CuAlTe2," *The European Physical Journal Applied Physics,* vol. 10, no. 1, pp. 9-14, 2000.

[69] J. Bernede, N. Barreau, S. Marsillac, and L. Assmann, "Band alignment at β-In2S3/TCO interface," *Applied surface science,* vol. 195, no. 1-4, pp. 222-228, 2002.

[70] N. Barreau, S. Marsillac, J. Bernede, and A. Barreau, "Investigation of β-In2S3 growth on different transparent conductive oxides," *Applied surface science,* vol. 161, no. 1-2, pp. 20-26, 2000.

Effect of Metal Halides Treatment on High Throughput Low Temperature CIGS Solar Cells

Deewakar Poudel[1], Benjamin Belfore[1], Adam Masters[1], Angus Rockett[2] and Sylvain Marsillac[1]

[1]Virginia Institute of Photovoltaics, Old Dominion University, Norfolk, VA 23529, USA

[2]Dept. of Metallurgical and Materials Engineering, Colorado School of Mines, Golden, CO 80401, USA

Abstract— Copper indium gallium diselenide (CIGS) semiconductor thin films were deposited at high rate and low temperature using single-stage thermal co-evaporation process on molybdenum back contact. A post deposition treatment was done by flashing AgBr at 350 ºC to induce recrystallization. Changes in morphology were confirmed by SEM, with an observed increase in grain size, as well as by XRD measurements, with a decrease in FWHM. Device results show an improvement of the performance after the AgBr vapor treatment, as all the photovoltaic parameters enhanced. Overall, AgBr seems to be a suitable transport agent and beneficial for device fabrication.

Keywords— Cu(In,Ga)Se$_2$, recrystallization, AgBr

I. INTRODUCTION

CIGS high absorption coefficient and tunable band gap makes it a good candidate in the field of thin film photovoltaic technology [1]. Conventional deposition processes for this technology often require high substrates temperature (up to 600 ºC) to get large grain size and are also time consuming. The economic viability of CIGS solar modules is affected by this duration and high temperature process. A single-stage process (where all source temperatures are kept constant) is simple to operate and faster as compared to the traditional three stage process, potentially providing high throughput for industrial applications [2]. This could also improve the manufacturing supply chain and logistics of the fabrication [3-10]. Furthermore, using a recrystallization process, enhancing grain growth, might be a suitable approach to get high quality film and better device performance [11]. This recrystallization and grain growth could be achieved by the post deposition treatment of CIGS semiconductor thin films by metal halides [12-27].

In this work, we studied the post deposition treatment by silver bromide vapor of CIGS semiconductor thin films, deposited by single-stage process at a high rate and low temperature.

II. EXPERIMENTAL METHODS

A single stage co-evaporation process was used to fabricate the CIGS semiconductor thin films. The final device structure is the following: SLG/Mo/CIGS/CdS/i-ZnO/ITO/Grids. The back contact molybdenum bilayer was deposited by dc sputtering magnetron [28-30]. The Cu-poor films were grown at substrate temperature of 350 ºC for both as-deposited and AgBr treated samples. The substrate and source temperatures were kept constant throughout the process. The silver bromide post deposition treatment was done by flashing 60 mg in 2 minutes. Half of the samples were set aside for characterization, while the other half were converted to completed devices.

The crystallographic structure analysis was done by symmetric θ-2θ X-ray diffraction and analyzed using the International Center for Diffraction Data (ICDD) database. Cross-section morphological analyses were performed by scanning electron microscope (SEM). The surface roughness was measured by atomic force microscopy (AFM). The samples were characterized by spectroscopic ellipsometry in the spectral range of 0.75 eV to 6.5 eV. Using previously measured results, the complex dielectric functions were calculated [31-48]. The photovoltaic characteristics were evaluated by external quantum efficiency (QE) measurements (QEX7, PV measurements Inc.) and current density-voltage (J-V) measurements (IV5, PV measurements Inc.) done under AM 1.5G with a light intensity of 100 mW/cm^2 at 25°C.

III. RESULTS AND DISCUSSIONS

The CIGS samples were recrystallized at 350 °C in an AgBr environment by flashing 60 mg for 2 minutes after CIGS deposition. Samples with no treatment, named as-deposited, were also prepared separately.

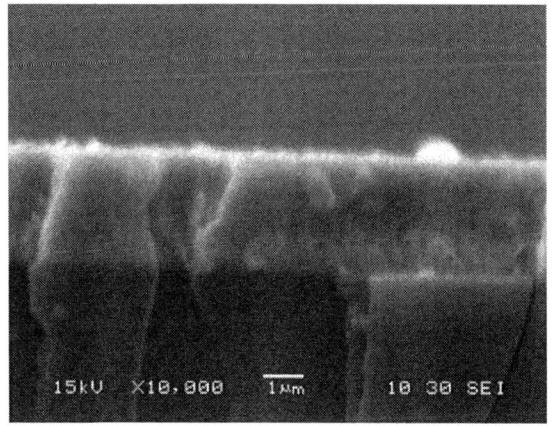

This research was supported by the Department of Energy Contract No. DE-EE0007551.

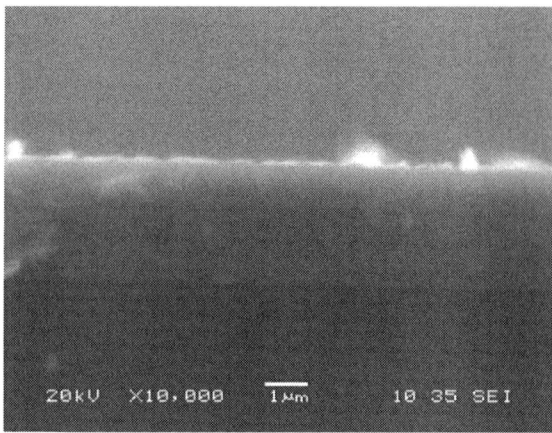

Fig. 1. Cross-section scanning electron microscopy micrographs of CIGS films: as-deposited (top) and recrystallized (bottom) at 350 °C by AgBr vapor treatment.

Figure 1 shows the cross-sectional SEM images of as-deposited and recrystallized samples. A change in grain structure can be observed as small grain changes into larger ones after the treatment. The grains for the recrystallized films seemed to be compact and uniform, as compared to the as-deposited samples. The AFM images of the as-deposited and recrystallized samples are shown in Figure 2. The roughness of the surface increases slightly after the treatment as the rms roughness increases from 20.1 nm (as-deposited) to 26.7 nm (AgBr treated).

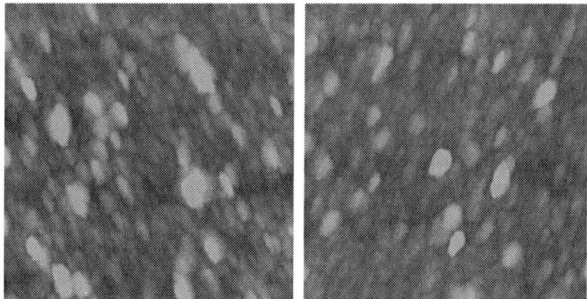

Fig. 2. Atomic Force Microscope images of CIGS films: as-deposited (top) and recrystallized (bottom) at 500 °C by AgBr vapor treatment.

Figure 3 shows the XRD measurements for the as-deposited and recrystallized films. As one can see Table 1, an increase in peak intensity and decrease in full width half maxima (FWHM) was observed for all the major peaks ((112), (220)/(204) and (312)) suggesting an increase in crystallinity which is consistent with the SEM results. One can also see Figure 3 that for the (112) peak, the as-deposited films seem to have two peaks, while only one peak is observed for the AgBr treated samples, indicating potentially a redistribution of the gallium content in the films. This phenomena was observed previously, whereby the rerystallization process leading to an increase in grain size, is also associated with a redistribution of the elements in the films. This will be further studied by SIMS. We should also study how recrystallization affects alkali content, and the

influence on device performance [49-56]. Also, if elemental profiles are changed after the AgBr treatment, it could modify the conduction band offset, and require different junction partners [57-70].

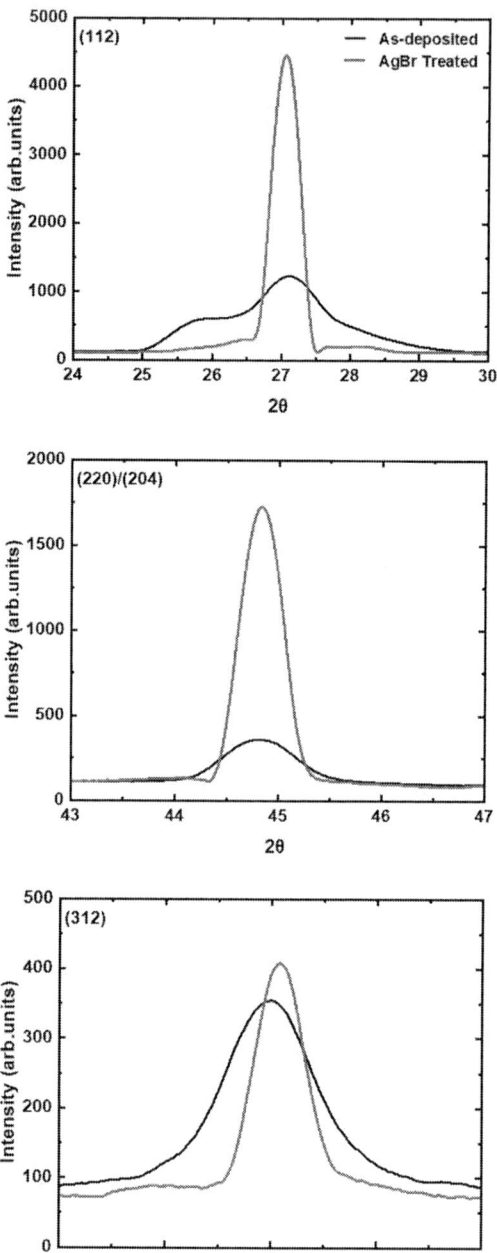

Fig. 3. XRD plots (full spectrum, (112), (204) and (312) peaks) for as-deposited (black) and AgBr treated (red) CIGS samples.

TABLE I. XRD RESULTS FOR AS-DEPOSITED AND AGBR TREATED SAMPLES AT 350 °C

	As-deposited			AgBr-Treated		
Peaks	(112)	(204)	(312)	(112)	(204)	(312)
Angles (deg)	26.9	44.7	52.8	27.2	44.7	53.2
FWHM (deg)	1.27	0.63	0.85	0.23	0.29	0.16
Int. I (cps deg)	1284	357	355	4454	1725	408

Representative current-voltage and external quantum efficiency curves for the devices are shown Figure 4. The solar cell devices were completed for as-deposited and AgBr treated samples. After the post deposition treatment by AgBr, the device shows significant improvements in device performance. All the photovoltaic parameters, V_{OC}, J_{SC} and FF increases from 0.35 V to 0.53 V, 28.5 mA/cm^2 to 30.3 mA/cm^2 and 40.2 % to 56.1 % respectively. The overall efficiency increases from 4 % to 8.9 %. The diode parameters were extracted by fitting a single diode model to the dark J-V curves.

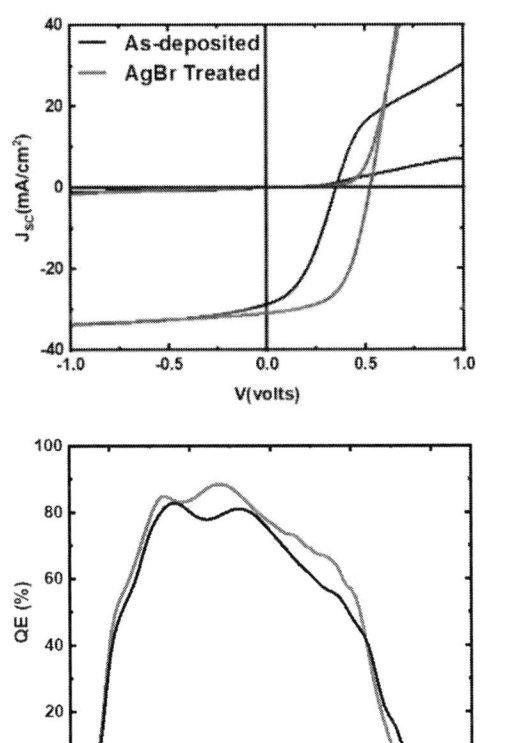

Fig. 4. Representative I-V and QE curves for the as-deposited and AgBr treated CIGS samples.

The reverse saturation current density decreased (from 5.2E-04 mA/cm^2 to 4.4E-05 mA/cm^2), the series resistance decreased (from 5.7 Ω.cm^2 to 2.5 Ω.cm^2), the diode ideality factor decreased (from >2 to 1.88) whereas the shunt resistance increased (from 800 Ω/cm^2 to 1100 Ω/cm^2) after the AgBr treatment. All these changes in parameters correlate well with an increase in performance for the device after AgBr treatment. The QE curve confirm a small increase in current collection at all wavelengths.

IV. CONCLUSION

The post-deposition treatment and recrystallization by AgBr of CIGS thin films deposited at low temperature and high rate by single stage process was studied. Changes in morphology were observed by SEM and confirmed by XRD measurements, with a decrease in FWHM and an increase in peak intensity. The surface roughness does not seem much affected by the treatment. Device performance increases after the AgBr treatment as V_{OC}, J_{SC} and FF all increase. These results indicate that AgBr acts as a good transport agent even in case of low temperature CIGS deposition. Further characterization will be needed to understand fully the effect of AgBr, including secondary ion mass spectrometry to assess how the elements are redistributed in the films, and Hall effect measurements to assess the impact of the treatment on electrical properties.

ACKNOWLEDGMENT

This research was supported by the Department of Energy Contract No. DE-EE0007551.

REFERENCES

[1] T. Kato, "Cu (In, Ga)(Se, S) 2 solar cell research in Solar Frontier: Progress and current status," *Japanese Journal of Applied Physics,* vol. 56, no. 4S, p. 04CA02, 2017.

[2] H. Wang *et al.*, "Effect of substrate temperature on the structural and electrical properties of CIGS films based on the one-stage co-evaporation process," *Semiconductor science and technology,* vol. 25, no. 5, p. 055007, 2010.

[3] D. M. Nelson, E. Marsillac, and S. S. Rao, "Antecedents and evolution of the green supply chain," *Journal of Operations and Supply Chain Management,* no. Special Issue, 2012.

[4] K. Liao, E. Marsillac, E. Johnson, and Y. Liao, "Global supply chain adaptations to improve financial performance: supply base establishment and logistics integration," *Journal of Manufacturing Technology Management,* 2011.

[5] R. Romero-Silva, E. Marsillac, S. Shaaban, and M. Hurtado-Hernández, "Serial production line performance under random variation: dealing with the 'Law of Variability'," *Journal of Manufacturing Systems,* vol. 50, pp. 278-289, 2019.

[6] E. Marsillac, "Management of the photovoltaic supply chain," *International Journal of Technology, Policy and Management,* vol. 12, no. 2-3, pp. 195-211, 2012.

[7] Y. Liao and E. Marsillac, "External knowledge acquisition and innovation: the role of supply chain network-oriented flexibility and organisational awareness," *International Journal of Production Research,* vol. 53, no. 18, pp. 5437-5455, 2015.

[8] R. Romero-Silva, S. Shaaban, E. Marsillac, and M. Hurtado, "Exploiting the characteristics of serial queues to reduce the mean and variance of flow time using combined priority rules," *International Journal of Production Economics,* vol. 196, pp. 211-225, 2018.

[9] E. L. Marsillac, "Environmental impacts on reverse logistics and green supply chains: similarities and integration," *International*

Journal of Logistics Systems and Management, vol. 4, no. 4, pp. 411-422, 2008.

[10] E. Marsillac, "Closing the Photovoltaic Supply Chain Loop-Invest Now for Future Returns," in *2018 IEEE 7th World Conference on Photovoltaic Energy Conversion (WCPEC)(A Joint Conference of 45th IEEE PVSC, 28th PVSEC & 34th EU PVSEC),* 2018: IEEE, pp. 2498-2500.

[11] H. Rodriguez-Alvarez *et al.,* "Recrystallization of Cu (In, Ga) Se2 thin films studied by X-ray diffraction," *Acta Materialia,* vol. 61, no. 12, pp. 4347-4353, 2013.

[12] G. Rajan *et al.,* "Impact of Post-Deposition Recrystallization by Alkali Fluorides on Cu (In, Ga) Se2 Thin-Film Materials and Solar Cells," *Thin Solid Films,* vol. 690, p. 137526, 2019.

[13] E. Palmiotti, B. Belfore, D. Poudel, S. Marsillac, and A. Rockett, "In-Situ Study of the Crystallization of Amorphous CuInSe2 Thin Films and the Effect of InCl3 Treatment," *Thin Solid Films,* p. 139095, 2022.

[14] B. Belfore *et al.,* "Recrystallization of Cu (In, Ga) Se2 Semiconductor Thin Films via InCl3 Treatment," *Thin Solid Films,* vol. 735, p. 138897, 2021.

[15] S. Marsillac, M. Zouaghi, J. Bernede, T. B. Nasrallah, and S. Belgacem, "Evolution of the properties of spray-deposited CuInS2 thin films with post-annealing treatment," *Solar energy materials and solar cells,* vol. 76, no. 2, pp. 125-134, 2003.

[16] D. Poudel *et al.,* "Analysis of Post-Deposition Recrystallization Processing via Indium Bromide of Cu (In, Ga) Se2 Thin Films," *Materials,* vol. 14, no. 13, p. 3596, 2021.

[17] S. Karki *et al.,* "Analysis of recombination mechanisms in RbF-treated CIGS solar cells," *IEEE Journal of Photovoltaics,* vol. 9, no. 1, pp. 313-318, 2018.

[18] D. Poudel *et al.,* "In Situ Recrystallization of Co-Evaporated Cu (In, Ga) Se2 Thin Films by Copper Chloride Vapor Treatment towards Solar Cell Applications," *Energies,* vol. 14, no. 13, p. 3938, 2021.

[19] D. Poudel *et al.,* "Studying the Recrystallization of Cu (InGa) Se 2 Semiconductor Thin Films by Silver Bromide In-situ Treatment," in *2021 IEEE 48th Photovoltaic Specialists Conference (PVSC),* 2021: IEEE, pp. 2307-2311.

[20] D. Poudel *et al.,* "Effect of Indium Bromide Treatments Post-Deposition Recrystallization Temperature on Cu (In, Ga) Se 2 Thin Films," in *2021 IEEE 48th Photovoltaic Specialists Conference (PVSC),* 2021: IEEE, pp. 0429-0432.

[21] E. Palmiotti, B. Belfore, D. Poudel, S. Marsillac, and A. Rockett, "CuIn (1− x) Ga x Se 2 Recrystallization by Metal Halide Treatments," in *2021 IEEE 48th Photovoltaic Specialists Conference (PVSC),* 2021: IEEE, pp. 0669-0671.

[22] B. Belfore *et al.,* "Study of Indium Chloride Vapor Treatment on Cu (In, Ga) Se 2 Semiconductor Thin Films," in *2021 IEEE 48th Photovoltaic Specialists Conference (PVSC),* 2021: IEEE, pp. 2320-2323.

[23] B. Belfore *et al.,* "Vapor Treatment and In-situ Recrystallization by Copper Chloride on Cu (In, Ga) Se 2 Thin Film," in *2021 IEEE 48th Photovoltaic Specialists Conference (PVSC),* 2021: IEEE, pp. 2316-2319.

[24] B. Belfore, D. Poudel, T. Ashrafee, S. Karki, G. Rajan, and S. Marsillac, "Morphological Study of Indium Chloride Post Deposition Treated CuInSe 2 Thin Films," in *2021 IEEE 48th Photovoltaic Specialists Conference (PVSC),* 2021: IEEE, pp. 2312-2315.

[25] B. Belfore *et al.,* "In-Situ Recrystallization of CIGS via Metal Halides," in *2020 47th IEEE Photovoltaic Specialists Conference (PVSC),* 2020: IEEE, pp. 1131-1133.

[26] B. Belfore *et al.,* "Ex-Situ Recrystallization of CIGS via Metal Halides," in *2020 47th IEEE Photovoltaic Specialists Conference (PVSC),* 2020: IEEE, pp. 1102-1104.

[27] B. Belfore, G. Rajan, S. Karki, D. Poudel, A. Rockett, and S. Marsillac, "The Impact of Deposition Temperature on Sodium Fluoride Recrystallization in Cu (In, Ga) Se 2 Solar Cells," in *2019 IEEE 46th Photovoltaic Specialists Conference (PVSC),* 2019: IEEE, pp. 1851-1853.

[28] E. Gourmelon, J. Bernede, J. Pouzet, and S. Marsillac, "Textured MoS 2 thin films obtained on tungsten: electrical properties of the

W/MoS 2 contact," *Journal of Applied Physics,* vol. 87, no. 3, pp. 1182-1186, 2000.

[29] K. Aryal, H. Khatri, R. Collins, and S. Marsillac, "In situ and ex situ studies of molybdenum thin films deposited by rf and dc magnetron sputtering as a back contact for CIGS solar cells," *International Journal of Photoenergy,* vol. 2012, 2012.

[30] O. Ayala *et al.,* "Theoretical Analysis of Experimental Data of Sodium Diffusion in Oxidized Molybdenum Thin Films," *Energies,* vol. 14, no. 9, p. 2479, 2021.

[31] S. Marsillac, S. Little, and R. Collins, "A broadband analysis of the optical properties of silver nanoparticle films by in situ real time spectroscopic ellipsometry," *Thin Solid Films,* vol. 519, no. 9, pp. 2936-2940, 2011.

[32] P. Koirala *et al.,* "Real time spectroscopic ellipsometry for analysis and control of thin film polycrystalline semiconductor deposition in photovoltaics," *Thin solid films,* vol. 571, pp. 442-446, 2014.

[33] L. R. Dahal, D. Sainju, N. Podraza, S. Marsillac, and R. Collins, "Real time spectroscopic ellipsometry of Ag/ZnO and Al/ZnO interfaces for back-reflectors in thin film Si: H photovoltaics," *Thin Solid Films,* vol. 519, no. 9, pp. 2682-2687, 2011.

[34] P. Koirala *et al.,* "Through-the-glass spectroscopic ellipsometry for analysis of CdTe thin-film solar cells in the superstrate configuration," *Progress in Photovoltaics: Research and Applications,* vol. 24, no. 8, pp. 1055-1067, 2016.

[35] P. Aryal *et al.,* "Parameterized complex dielectric functions of CuIn1− xGaxSe2: applications in optical characterization of compositional non-uniformities and depth profiles in materials and solar cells," *Progress in Photovoltaics: Research and Applications,* vol. 24, no. 9, pp. 1200-1213, 2016.

[36] S. Marsillac *et al.,* "Spectroscopic ellipsometry studies of In2S3 top window and Mo back contacts in chalcopyrite photovoltaics technology," *physica status solidi c,* vol. 5, no. 5, pp. 1244-1248, 2008.

[37] V. Ranjan, R. Collins, and S. Marsillac, "Real-time analysis of the microstructural evolution and optical properties of Cu (In, Ga) Se2 thin films as a function of Cu content," *physica status solidi (RRL)– Rapid Research Letters,* vol. 6, no. 1, pp. 10-12, 2012.

[38] T. Begou, J. D. Walker, D. Attygalle, V. Ranjan, R. Collins, and S. Marsillac, "Real time spectroscopic ellipsometry of CuInSe2: growth dynamics, dielectric function, and its dependence on temperature," *physica status solidi (RRL)–Rapid Research Letters,* vol. 5, no. 7, pp. 217-219, 2011.

[39] S. Karki *et al.,* "In situ and ex situ investigations of KF postdeposition treatment effects on CIGS solar cells," *IEEE Journal of Photovoltaics,* vol. 7, no. 2, pp. 665-669, 2016.

[40] L. R. Dahal *et al.,* "Correlations between mapping spectroscopic ellipsometry results and solar cell performance for evaluations of nonuniformity in thin-film silicon photovoltaics," *IEEE Journal of Photovoltaics,* vol. 3, no. 1, pp. 387-393, 2012.

[41] P. Aryal, D. Attygalle, P. Pradhan, N. J. Podraza, S. Marsillac, and R. W. Collins, "Large-area compositional mapping of Cu (In $ _ $ {1-x} $ Ga $ _ $ {x} $) Se $ _ $ {2} $ materials and devices with spectroscopic ellipsometry," *IEEE Journal of Photovoltaics,* vol. 3, no. 1, pp. 359-363, 2012.

[42] A.-R. Ibdah *et al.,* "Spectroscopic ellipsometry for analysis of polycrystalline thin-film photovoltaic devices and prediction of external quantum efficiency," *Applied Surface Science,* vol. 421, pp. 601-607, 2017.

[43] J. Walker, H. Khatri, V. Ranjan, J. Li, R. Collins, and S. Marsillac, "Electronic and structural properties of molybdenum thin films as determined by real-time spectroscopic ellipsometry," *Applied Physics Letters,* vol. 94, no. 14, p. 141908, 2009.

[44] S. Little, T. Begou, R. Collins, and S. Marsillac, "Optical detection of melting point depression for silver nanoparticles via in situ real time spectroscopic ellipsometry," *Applied Physics Letters,* vol. 100, no. 5, p. 051107, 2012.

[45] D. R. Sapkota *et al.,* "Evaluation of CuInSe 2 Materials and Solar Cells Co-evaporated at Different Rates Based on Real Time Spectroscopic Ellipsometry Calibrations," in *2021 IEEE 48th Photovoltaic Specialists Conference (PVSC),* 2021: IEEE, pp. 0451-0458.

[46] P. Pradhan *et al.*, "Real time spectroscopic ellipsometry analysis of the three-stages of CuIn 1− x Ga x Se 2 co-evaporation," in *2014 IEEE 40th Photovoltaic Specialist Conference (PVSC)*, 2014: IEEE, pp. 2060-2065.

[47] D. Attygalle *et al.*, "Optical monitoring and control of three-stage coevaporated Cu (In 1− x Ga x) Se 2 by real-time spectroscopic ellipsometry," in *2012 IEEE 38th Photovoltaic Specialists Conference (PVSC) PART 2*, 2012: IEEE, pp. 1-6.

[48] H. Fujiwara and R. W. Collins, *Spectroscopic Ellipsometry for Photovoltaics: Volume 2: Applications and Optical Data of Solar Cell Materials*. Springer, 2019.

[49] D. Poudel *et al.*, "Assessment of Cu (In, Ga) Se2 Solar Cells Degradation due to Water Ingress Effect on the CdS Buffer Layer," *Journal of Energy and Power Technology*, vol. 3, no. 1, 2021.

[50] P. Paul *et al.*, "Direct nm-scale spatial mapping of traps in CIGS," *IEEE Journal of Photovoltaics*, vol. 5, no. 5, pp. 1482-1486, 2015.

[51] S. Karki *et al.*, "Degradation mechanism in Cu (In, Ga) Se 2 material and solar cells due to moisture and heat treatment of the absorber layer," *IEEE Journal of Photovoltaics*, vol. 9, no. 4, pp. 1138-1143, 2019.

[52] S. Karki *et al.*, "Impact of water ingress on molybdenum thin films and its effect on Cu (In, Ga) Se 2 Solar Cells," *IEEE Journal of Photovoltaics*, vol. 10, no. 2, pp. 696-702, 2019.

[53] D. Poudel *et al.*, "Degradation Mechanism Due to Water Ingress Effect on the Top Contact of Cu (In, Ga) Se2 Solar Cells," *Energies*, vol. 13, no. 17, p. 4545, 2020.

[54] D. Poudel *et al.*, "Numerical Analysis of Water Ingress Effect on the Window Layer of Cu (In, Ga) Se 2 Solar Cells using SCAPS-1D," in *2021 IEEE 48th Photovoltaic Specialists Conference (PVSC)*, 2021: IEEE, pp. 0438-0442.

[55] D. Poudel, B. Belfore, S. Karki, G. Rajan, A. Rockett, and S. Marsillac, "Process Dependent Instabilities In Cu (In, Ga) Se 2 Solar Cells Under Water Ingress," in *2021 IEEE 48th Photovoltaic Specialists Conference (PVSC)*, 2021: IEEE, pp. 0433-0437.

[56] G. Rajan *et al.*, "Study of Instabilities and Degradation due to Moisture Ingress in the Molybdenum back contact of Cu (In, Ga) Se 2 Solar Cells," in *2018 IEEE 7th World Conference on Photovoltaic Energy Conversion (WCPEC)(A Joint Conference of 45th IEEE PVSC, 28th PVSEC & 34th EU PVSEC)*, 2018: IEEE, pp. 3037-3039.

[57] S. Marsillac, J. Bernede, R. Ny, and A. Conan, "A new simple technique to obtain In2Se3 polycrystalline thin films," *Vacuum*, vol. 46, no. 11, pp. 1315-1323, 1995.

[58] N. Barreau, S. Marsillac, and J. Bernede, "Physico-chemical characterization of β-In2S3 thin films synthesized by solid-state reaction, induced by annealing, of the constituents sequentially deposited in thin layers," *Vacuum*, vol. 56, no. 2, pp. 101-106, 2000.

[59] N. Barreau, J. Bernede, S. Marsillac, C. Amory, and W. Shafarman, "New Cd-free buffer layer deposited by PVD: In3S3 containing Na compounds," *Thin Solid Films*, vol. 431, pp. 326-329, 2003.

[60] R. Robles, N. Barreau, A. Vega, S. Marsillac, J. Bernede, and A. Mokrani, "Optical properties of large band gap β-In2S3− 3xO3x compounds obtained by physical vapour deposition," *Optical Materials*, vol. 27, no. 4, pp. 647-653, 2005.

[61] J. Bernede, S. Marsillac, and A. Conan, "Electrical properties of γ-In2Se3 layers synthesized by solid state reaction between In and Se thin films," *Materials chemistry and physics*, vol. 48, no. 1, pp. 5-9, 1997.

[62] S. Marsillac, J. Bernede, and A. Conan, "Change in the type of majority carriers in disordered In x Se 100− x thin-film alloys," *Journal of materials science*, vol. 31, no. 3, pp. 581-587, 1996.

[63] N. Barreau, J. Bernede, and S. Marsillac, "Study of the new β-In2S3 containing Na thin films. Part II: Optical and electrical characterization of thin films," *Journal of crystal growth*, vol. 241, no. 1-2, pp. 51-56, 2002.

[64] N. Barreau, J. Bernede, C. Deudon, L. Brohan, and S. Marsillac, "Study of the new β-In2S3 containing Na thin films Part I: Synthesis and structural characterization of the material," *Journal of crystal growth*, vol. 241, no. 1-2, pp. 4-14, 2002.

[65] K. D'Almeida, J. Bernede, F. Ragot, A. Godoy, F. Diaz, and S. Lefrant, "Carbazole-based electroluminescent devices obtained by vacuum evaporation," *Journal of applied polymer science*, vol. 82, no. 8, pp. 2042-2055, 2001.

[66] N. Barreau, S. Marsillac, J. Bernede, and L. Assmann, "Evolution of the band structure of β-In 2 S 3− 3x O 3x buffer layer with its oxygen content," *Journal of applied physics*, vol. 93, no. 9, pp. 5456-5459, 2003.

[67] C. Amory, J. Bernede, and S. Marsillac, "Study of a growth instability of γ-In 2 Se 3," *Journal of applied physics*, vol. 94, no. 10, pp. 6945-6948, 2003.

[68] K. Benchouk *et al.*, "New buffer layers, large band gap ternary compounds: CuAlTe2," *The European Physical Journal Applied Physics*, vol. 10, no. 1, pp. 9-14, 2000.

[69] J. Bernede, N. Barreau, S. Marsillac, and L. Assmann, "Band alignment at β-In2S3/TCO interface," *Applied surface science*, vol. 195, no. 1-4, pp. 222-228, 2002.

[70] N. Barreau, S. Marsillac, J. Bernede, and A. Barreau, "Investigation of β-In2S3 growth on different transparent conductive oxides," *Applied surface science*, vol. 161, no. 1-2, pp. 20-26, 2000.

Grain Enhancement in Polycrystalline CuGaSe$_2$ by AgBr Vapor Treatment

Deewakar Poudel[1], Benjamin Belfore[1], Adam Masters[1] Elizabeth Palmiotti[2], Angus Rockett[2] and Sylvain Marsillac[1]

[1]Virginia Institute of Photovoltaics, Old Dominion University, Norfolk, VA 23529, USA

[2]Dept. of Metallurgical and Materials Engineering, Colorado School of Mines, Golden, CO 80401, USA

Abstract— **Copper Gallium diselenide (CuGaSe$_2$ or CGS) thin film were deposited using a three-stage thermal co-evaporation process on molybdenum coated soda lime glass. Recrystallization was carried after the second stage by flashing AgBr for 2 mins. The change in morphology, structure and depth profile were studied after the treatment. SEM and XRD showed an increase in grain size and enhanced crystallinity. A decrease in sodium profile after the treatment was observed through SIMS measurements. Overall, AgBr treatment of CGS seems to be promising for the improvement of film quality, which could help with enhanced device fabrication in the future.**

Keywords— CuGaSe$_2$, recrystallization, AgBr

I. INTRODUCTION

The use of wide bandgap chalcopyrite material devices could potentially allow the fabrication of high efficiency photovoltaic devices in tandem structure. Due its bandgap values, comparatively high absorption coefficient and cost, CGS with a direct bandgap of 1.7 eV is thought to be one of the most important contenders for top cell fabrication in thin films solar cells, as its with low band gap CuInSe$_2$ in tandem devices promises high efficiency. The highest efficiency reported for CGS solar cells to date remain somewhat low, reaching value of 11.9 % and thus additional development in enhancing material properties and device performance are necessary to achieve potential application. The control of defects present at the surface, interfaces, and bulk with alkali-metal doping along with Cu-deficient phases at grain boundaries is essential to improve this photovoltaic device performance [1, 2]. Developing a process that allows a low levelized cost of energy is one of the key stages, along with the development of proper logistics and supply chain, to allow thin film technologies to be viable [3-10]. It is important to note that the microstructure of these chalcopyrite thin films can change significantly after recrystallization due to change in orientation and grain growth [11].

We have previously shown that metal halides can act as transport agents to recrystallize CIGS and enhance grain growth [12-27]. High-rate depositions were achieved by post-deposition vapor treatment by alkali and metal halides, and recrystallization of CIGS thin films. In this work, we studied various experimental conditions allowing post deposition

recrystallization with AgBr vapor treatment of CGS films fabricated at 500 °C.

II. EXPERIMENTAL METHODS

A bilayer of molybdenum film was deposited by dc magnetron sputtering on soda lime glass (SLG) substrates, with a first layer at low pressure for adhesion and a second layer at high argon pressure for conductivity [28-30]. A three stage CGS deposition was performed on these SLG/Mo substrates at a high deposition rate (around 10 µm/hr). The CGS grown was about 1.5 µm thick. The first, second and third stage temperatures were kept at 350 °C, 500 °C and 500 °C respectively. The substrate temperature was observed using a thermocouple positioned at the back of the substrate and a pyrometer targeted at the growing film. The recrystallization was performed between the second and third stage. 40 mg of AgBr was flashed over 2 minutes. The flux was varied by changing the AgBr source temperature. As-deposited samples were also fabricated without AgBr treatment.

Time of flight secondary ion mass spectrometry (SIMS) was used to measure the compositional variation as a function of depth in the device. The samples were characterized by spectroscopic ellipsometry in the spectral range of 0.75 eV to 6.5 eV, using a Woollam M-2000 rotating compensator multichannel system. Using calibrated results, we extracted the complex dielectric functions [31-48]. The crystallographic structures analysis was done by symmetric θ-2θ X-ray diffraction and analyzed using the International Center for Diffraction Data (ICDD) database. Cross-section morphological analysis were performed by scanning electron microscope (SEM). The surface roughness was measured by atomic force microscopy (AFM). The composition was measured by X-ray fluorescence (XRF).

III. RESULTS AND DISCUSSIONS

The CGS samples were recrystallized at 500 °C, by flashing AgBr for 2 minutes at the end of the 2nd stage. The Ag halide was introduced in between the 2nd and 3rd stages to control the composition of group I, and avoid having group I rich films. The cross-section of the films as-deposited and after recrystallization was observed by SEM as shown in Figure 1. Significant changes were observed after the AgBr vapor treatment. The small grains observed for the as-deposited films

This research was supported by the Department of Energy Contract No. DE-EE0007551.

transformed into larger grains of about a micrometer size, indicating that the AgBr acts as a suitable transport agent to generate large grains. Looking at the AFM (Figure 2), one can see that the surface roughness increases slightly after the treatment, as the rms roughness increases from 35.4 nm (as-deposited) to 39.3 nm (AgBr treated). This is in good agreement with the SEM images, where the as-deposited films appear quite smooth while the recrystallized films have some non-uniformities associated with some larger grains.

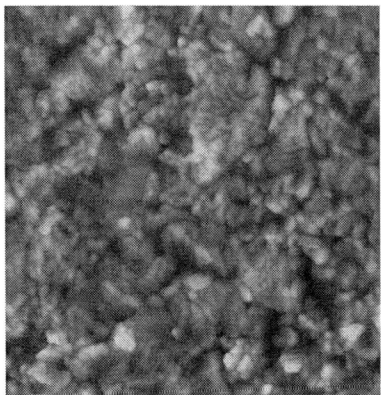

Fig. 2. Atomic Force Microscope images of CGS films: as-deposited (top) and recrystallized (bottom) at 500 °C by AgBr vapor treatment.

XRD measurements were completed on the as-deposited and recrystallized films to understand how the crystalline structure of the films changed with AgBr treatment. For the four main peaks observed ((112), (220)/(204) and (312)), one can observe a decrease in full width at half maxima (FWHM) in all the peaks, which is consistent with an increase in grain size observed with SEM. One can also notice the increase in peak intensity in the case of the (220)/(204) peak, with a decrease in the other directions, which indicates a change in preferred orientation towards the (220)/(204) peak after recrystallization. There is a small shift in peak position between the as-deposited and the recrystallized films, with slightly higher angles for the recrystallized films. Such behavior was previously observed for CGS films with slightly different gallium content [2]. The authors mentioned also that as the gallium content increases, the (220)/(204) doublet becomes less resolved. This seems to be the case for our work too, as a secondary peak is not clearly seen. Composition measurements by X-ray fluorescence did not indicate a modification of the composition between the as-deposited and the recrystallized films, and a Cu/Ga ratio between 0.92 and 1. Higher resolution scans will be perform to try to resolve the two peaks. Interestingly, higher efficiency devices are often obtained in CIGS solar cells when the films are oriented along the (220)/(204) direction as is the case here.

Fig. 1. Cross-section Scanning Electron Microscopy micrographs of CGS films: as-deposited (top) and recrystallized (bottom) at 500 °C by AgBr vapor treatment.

TABLE I. XRD RESULTS FOR AS-DEPOSITED AND AGBR TREATED SAMPLES AT 500 °C

	As-deposited			AgBr Treated		
Peaks	(112)	(220)/(204)	(312)	(112)	(220)/(204)	(312)
Angle (deg)	27.7	45.7	54.3	27.8	45.8	54.3
FWHM (deg)	0.29	0.25	0.32	0.17	0.13	0.13
Int.I (cps deg)	1190	6402	508	396	9815	470

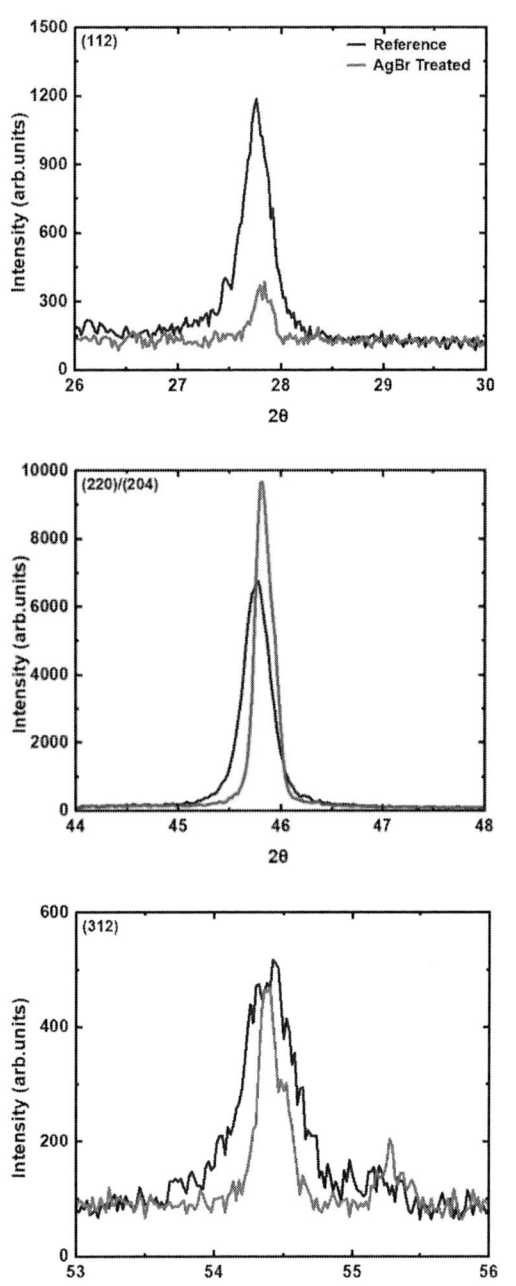

Fig. 3. XRD plots (full spectrum, (112), (204) and (312) peaks) for as-deposited (blue) and AgBr treated (red) CIGS samples.

To assess the effect of AgBr treatment on the elemental depth profile, the samples were investigated by SIMS. The main elements profiles (Cu, Ga and Se) remain the same after the AgBr treatment, as can be seen in Figure 4. This might have to be taken into account when fabricating good devices, as the band offset could be affected, and would require alternative buffer layers [49-62]. On the other hand, the sodium profile was modified after the treatment, with a lower concentration for the recrystallized film compared to the as-deposited [63-70]. It is possible that the increase in grain size for the recrystallized films decreases the grain boundaries density. Since sodium has

been hypothesized to mostly resides at the grain boundaries, this could be correlated with a decrease in the sodium content.

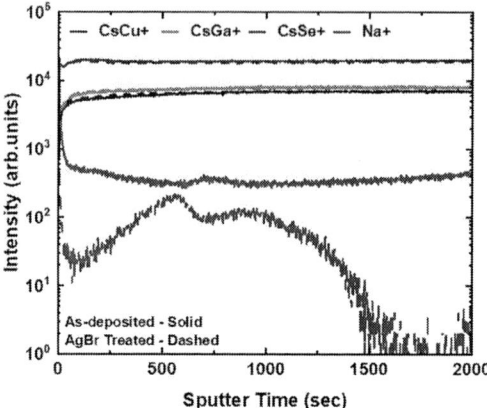

Fig. 4. Secondary Ions Mass Spectroscopy (SIMS) depth profile (positive ions) of the main elements for the as-deposited (solid) and AgBr treated (dashed) CGS samples.

IV. CONCLUSION

A three-stage thermal co-evaporation process was used to deposit $CuGaSe_2$ thin films and the effects of AgBr vapor treatment on the CGS films were studied. The structural and elemental depth profile properties were analyzed, indicating that the AgBr treatment results in larger grain throughout the film, as confirmed by SEM and XRD. No change in the main elements distribution was observed by SIMS, while the sodium profile changed drastically with a decrease in content after AgBr treatment. Electrical measurements will be performed to assess fully the effect of the treatment on the films. Additional research is still required in terms of AgBr doses and substrate temperature for potential application in solar cells device fabrications

ACKNOWLEDGMENT

This research was supported by the Department of Energy Contract No. DE-EE0007551.

REFERENCES

[1] S. Ishizuka and P. J. Fons, "Polycrystalline CuGaSe2 thin film growth and photovoltaic devices fabricated on alkali-free and alkali-containing substrates," *Journal of Crystal Growth,* vol. 532, p. 125407, 2020.

[2] G. Orsal, N. Romain, M.-C. Artaud, and S. Duchemin, "Characterization of CuGaSe/sub 2/thin films grown by MOCVD," *IEEE Transactions on Electron Devices,* vol. 46, no. 10, pp. 2098-2102, 1999.

[3] D. M. Nelson, E. Marsillac, and S. S. Rao, "Antecedents and evolution of the green supply chain," *Journal of Operations and Supply Chain Management,* no. Special Issue, 2012.

[4] K. Liao, E. Marsillac, E. Johnson, and Y. Liao, "Global supply chain adaptations to improve financial performance: supply base establishment and logistics integration," *Journal of Manufacturing Technology Management,* 2011.

[5] R. Romero-Silva, E. Marsillac, S. Shaaban, and M. Hurtado-Hernández, "Serial production line performance under random variation: dealing with the 'Law of Variability'," *Journal of Manufacturing Systems,* vol. 50, pp. 278-289, 2019.

[6] E. Marsillac, "Management of the photovoltaic supply chain," *International Journal of Technology, Policy and Management,* vol. 12, no. 2-3, pp. 195-211, 2012.

[7] Y. Liao and E. Marsillac, "External knowledge acquisition and innovation: the role of supply chain network-oriented flexibility and organisational awareness," *International Journal of Production Research,* vol. 53, no. 18, pp. 5437-5455, 2015.

[8] R. Romero-Silva, S. Shaaban, E. Marsillac, and M. Hurtado, "Exploiting the characteristics of serial queues to reduce the mean and variance of flow time using combined priority rules," *International Journal of Production Economics,* vol. 196, pp. 211-225, 2018.

[9] E. L. Marsillac, "Environmental impacts on reverse logistics and green supply chains: similarities and integration," *International Journal of Logistics Systems and Management,* vol. 4, no. 4, pp. 411-422, 2008.

[10] E. Marsillac, "Closing the Photovoltaic Supply Chain Loop-Invest Now for Future Returns," in *2018 IEEE 7th World Conference on Photovoltaic Energy Conversion (WCPEC)(A Joint Conference of 45th IEEE PVSC, 28th PVSEC & 34th EU PVSEC),* 2018: IEEE, pp. 2498-2500.

[11] H. Rodriguez-Alvarez et al., "Recrystallization of Cu (In, Ga) Se2 thin films studied by X-ray diffraction," *Acta materialia,* vol. 61, no. 12, pp. 4347-4353, 2013.

[12] G. Rajan et al., "Impact of Post-Deposition Recrystallization by Alkali Fluorides on Cu (In, Ga) Se2Thin-Film Materials and Solar Cells," *Thin Solid Films,* vol. 690, p. 137526, 2019.

[13] E. Palmiotti, B. Belfore, D. Poudel, S. Marsillac, and A. Rockett, "In-Situ Study of the Crystallization of Amorphous CuInSe2 Thin Films and the Effect of InCl3 Treatment," *Thin Solid Films,* p. 139095, 2022.

[14] B. Belfore et al., "Recrystallization of Cu (In, Ga) Se2 Semiconductor Thin Films via InCl3 Treatment," *Thin Solid Films,* vol. 735, p. 138897, 2021.

[15] S. Marsillac, M. Zouaghi, J. Bernede, T. B. Nasrallah, and S. Belgacem, "Evolution of the properties of spray-deposited CuInS2 thin films with post-annealing treatment," *Solar energy materials and solar cells,* vol. 76, no. 2, pp. 125-134, 2003.

[16] D. Poudel et al., "Analysis of Post-Deposition Recrystallization Processing via Indium Bromide of Cu (In, Ga) Se2 Thin Films," *Materials,* vol. 14, no. 13, p. 3596, 2021.

[17] S. Karki et al., "Analysis of recombination mechanisms in RbF-treated CIGS solar cells," *IEEE Journal of Photovoltaics,* vol. 9, no. 1, pp. 313-318, 2018.

[18] D. Poudel et al., "In Situ Recrystallization of Co-Evaporated Cu (In, Ga) Se2 Thin Films by Copper Chloride Vapor Treatment towards Solar Cell Applications," *Energies,* vol. 14, no. 13, p. 3938, 2021.

[19] D. Poudel et al., "Studying the Recrystallization of Cu (InGa) Se 2 Semiconductor Thin Films by Silver Bromide In-situ Treatment," in *2021 IEEE 48th Photovoltaic Specialists Conference (PVSC),* 2021: IEEE, pp. 2307-2311.

[20] D. Poudel et al., "Effect of Indium Bromide Treatments Post-Deposition Recrystallization Temperature on Cu (In, Ga) Se 2 Thin Films," in *2021 IEEE 48th Photovoltaic Specialists Conference (PVSC),* 2021: IEEE, pp. 0429-0432.

[21] E. Palmiotti, B. Belfore, D. Poudel, S. Marsillac, and A. Rockett, "CuIn (1− x) Ga x Se 2 Recrystallization by Metal Halide Treatments," in *2021 IEEE 48th Photovoltaic Specialists Conference (PVSC),* 2021: IEEE, pp. 0669-0671.

[22] B. Belfore et al., "Study of Indium Chloride Vapor Treatment on Cu (In, Ga) Se 2 Semiconductor Thin Films," in *2021 IEEE 48th Photovoltaic Specialists Conference (PVSC),* 2021: IEEE, pp. 2320-2323.

[23] B. Belfore et al., "Vapor Treatment and In-situ Recrystallization by Copper Chloride on Cu (In, Ga) Se 2 Thin Film," in *2021 IEEE 48th Photovoltaic Specialists Conference (PVSC),* 2021: IEEE, pp. 2316-2319.

[24] B. Belfore, D. Poudel, T. Ashrafee, S. Karki, G. Rajan, and S. Marsillac, "Morphological Study of Indium Chloride Post Deposition Treated CuInSe 2 Thin Films," in *2021 IEEE 48th Photovoltaic Specialists Conference (PVSC),* 2021: IEEE, pp. 2312-2315.

[25] B. Belfore et al., "In-Situ Recrystallization of CIGS via Metal Halides," in *2020 47th IEEE Photovoltaic Specialists Conference (PVSC),* 2020: IEEE, pp. 1131-1133.

[26] B. Belfore et al., "Ex-Situ Recrystallization of CIGS via Metal Halides," in *2020 47th IEEE Photovoltaic Specialists Conference (PVSC),* 2020: IEEE, pp. 1102-1104.

[27] B. Belfore, G. Rajan, S. Karki, D. Poudel, A. Rockett, and S. Marsillac, "The Impact of Deposition Temperature on Sodium Fluoride Recrystallization in Cu (In, Ga) Se 2 Solar Cells," in *2019 IEEE 46th Photovoltaic Specialists Conference (PVSC),* 2019: IEEE, pp. 1851-1853.

[28] E. Gourmelon, J. Bernede, J. Pouzet, and S. Marsillac, "Textured MoS 2 thin films obtained on tungsten: electrical properties of the W/MoS 2 contact," *Journal of Applied Physics,* vol. 87, no. 3, pp. 1182-1186, 2000.

[29] K. Aryal, H. Khatri, R. Collins, and S. Marsillac, "In situ and ex situ studies of molybdenum thin films deposited by rf and dc magnetron sputtering as a back contact for CIGS solar cells," *International Journal of Photoenergy,* vol. 2012, 2012.

[30] O. Ayala et al., "Theoretical Analysis of Experimental Data of Sodium Diffusion in Oxidized Molybdenum Thin Films," *Energies,* vol. 14, no. 9, p. 2479, 2021.

[31] S. Marsillac, S. Little, and R. Collins, "A broadband analysis of the optical properties of silver nanoparticle films by in situ real time spectroscopic ellipsometry," *Thin Solid Films,* vol. 519, no. 9, pp. 2936-2940, 2011.

[32] P. Koirala et al., "Real time spectroscopic ellipsometry for analysis and control of thin film polycrystalline semiconductor deposition in photovoltaics," *Thin solid films,* vol. 571, pp. 442-446, 2014.

[33] L. R. Dahal, D. Sainju, N. Podraza, S. Marsillac, and R. Collins, "Real time spectroscopic ellipsometry of Ag/ZnO and Al/ZnO interfaces for back-reflectors in thin film Si: H photovoltaics," *Thin Solid Films,* vol. 519, no. 9, pp. 2682-2687, 2011.

[34] P. Koirala et al., "Through-the-glass spectroscopic ellipsometry for analysis of CdTe thin-film solar cells in the superstrate configuration," *Progress in Photovoltaics: Research and Applications,* vol. 24, no. 8, pp. 1055-1067, 2016.

[35] P. Aryal et al., "Parameterized complex dielectric functions of CuIn1− xGaxSe2: applications in optical characterization of compositional non-uniformities and depth profiles in materials and solar cells," *Progress in Photovoltaics: Research and Applications,* vol. 24, no. 9, pp. 1200-1213, 2016.

[36] S. Marsillac et al., "Spectroscopic ellipsometry studies of In2S3 top window and Mo back contacts in chalcopyrite photovoltaics technology," *physica status solidi c,* vol. 5, no. 5, pp. 1244-1248, 2008.

[37] V. Ranjan, R. Collins, and S. Marsillac, "Real-time analysis of the microstructural evolution and optical properties of Cu (In, Ga) Se2 thin films as a function of Cu content," *physica status solidi (RRL)–Rapid Research Letters,* vol. 6, no. 1, pp. 10-12, 2012.

[38] T. Begou, J. D. Walker, D. Attygalle, V. Ranjan, R. Collins, and S. Marsillac, "Real time spectroscopic ellipsometry of CuInSe2: growth dynamics, dielectric function, and its dependence on temperature," *physica status solidi (RRL)–Rapid Research Letters,* vol. 5, no. 7, pp. 217-219, 2011.

[39] S. Karki et al., "In situ and ex situ investigations of KF postdeposition treatment effects on CIGS solar cells," *IEEE Journal of Photovoltaics,* vol. 7, no. 2, pp. 665-669, 2016.

[40] L. R. Dahal et al., "Correlations between mapping spectroscopic ellipsometry results and solar cell performance for evaluations of nonuniformity in thin-film silicon photovoltaics," *IEEE Journal of Photovoltaics,* vol. 3, no. 1, pp. 387-393, 2012.

[41] P. Aryal, D. Attygalle, P. Pradhan, N. J. Podraza, S. Marsillac, and R. W. Collins, "Large-area compositional mapping of Cu (In $ _ {1-x} $ Ga $ _ {x} $) Se $ _ {2} $ materials and devices with spectroscopic ellipsometry," *IEEE Journal of Photovoltaics,* vol. 3, no. 1, pp. 359-363, 2012.

[42] A.-R. Ibdah et al., "Spectroscopic ellipsometry for analysis of polycrystalline thin-film photovoltaic devices and prediction of external quantum efficiency," *Applied Surface Science,* vol. 421, pp. 601-607, 2017.

978-1-7281-6118-1/22 $31.00 © 2022 IEEE

[43] J. Walker, H. Khatri, V. Ranjan, J. Li, R. Collins, and S. Marsillac, "Electronic and structural properties of molybdenum thin films as determined by real-time spectroscopic ellipsometry," *Applied Physics Letters,* vol. 94, no. 14, p. 141908, 2009.

[44] S. Little, T. Begou, R. Collins, and S. Marsillac, "Optical detection of melting point depression for silver nanoparticles via in situ real time spectroscopic ellipsometry," *Applied Physics Letters,* vol. 100, no. 5, p. 051107, 2012.

[45] D. R. Sapkota *et al.*, "Evaluation of CuInSe 2 Materials and Solar Cells Co-evaporated at Different Rates Based on Real Time Spectroscopic Ellipsometry Calibrations," in *2021 IEEE 48th Photovoltaic Specialists Conference (PVSC),* 2021: IEEE, pp. 0451-0458.

[46] P. Pradhan *et al.*, "Real time spectroscopic ellipsometry analysis of the three-stages of CuIn 1− x Ga x Se 2 co-evaporation," in *2014 IEEE 40th Photovoltaic Specialist Conference (PVSC),* 2014: IEEE, pp. 2060-2065.

[47] D. Attygalle *et al.*, "Optical monitoring and control of three-stage coevaporated Cu (In 1− x Ga x) Se 2 by real-time spectroscopic ellipsometry," in *2012 IEEE 38th Photovoltaic Specialists Conference (PVSC) PART 2,* 2012: IEEE, pp. 1-6.

[48] H. Fujiwara and R. W. Collins, *Spectroscopic Ellipsometry for Photovoltaics: Volume 2: Applications and Optical Data of Solar Cell Materials.* Springer, 2019.

[49] S. Marsillac, J. Bernede, R. Ny, and A. Conan, "A new simple technique to obtain In2Se3 polycrystalline thin films," *Vacuum,* vol. 46, no. 11, pp. 1315-1323, 1995.

[50] N. Barreau, S. Marsillac, and J. Bernede, "Physico-chemical characterization of β-In2S3 thin films synthesized by solid-state reaction, induced by annealing, of the constituents sequentially deposited in thin layers," *Vacuum,* vol. 56, no. 2, pp. 101-106, 2000.

[51] N. Barreau, J. Bernede, S. Marsillac, C. Amory, and W. Shafarman, "New Cd-free buffer layer deposited by PVD: In2S3 containing Na compounds," *Thin Solid Films,* vol. 431, pp. 326-329, 2003.

[52] R. Robles, N. Barreau, A. Vega, S. Marsillac, J. Bernede, and A. Mokrani, "Optical properties of large band gap β-In2S3− 3xO3x compounds obtained by physical vapour deposition," *Optical Materials,* vol. 27, no. 4, pp. 647-653, 2005.

[53] J. Bernede, S. Marsillac, and A. Conan, "Electrical properties of γ-In2Se3 layers synthesized by solid state reaction between In and Se thin films," *Materials chemistry and physics,* vol. 48, no. 1, pp. 5-9, 1997.

[54] S. Marsillac, J. Bernede, and A. Conan, "Change in the type of majority carriers in disordered In x Se 100− x thin-film alloys," *Journal of materials science,* vol. 31, no. 3, pp. 581-587, 1996.

[55] N. Barreau, J. Bernede, and S. Marsillac, "Study of the new β-In2S3 containing Na thin films. Part II: Optical and electrical characterization of thin films," *Journal of crystal growth,* vol. 241, no. 1-2, pp. 51-56, 2002.

[56] N. Barreau, J. Bernede, C. Deudon, L. Brohan, and S. Marsillac, "Study of the new β-In2S3 containing Na thin films Part I: Synthesis

and structural characterization of the material," *Journal of crystal growth,* vol. 241, no. 1-2, pp. 4-14, 2002.

[57] K. D'Almeida, J. Bernede, F. Ragot, A. Godoy, F. Diaz, and S. Lefrant, "Carbazole-based electroluminescent devices obtained by vacuum evaporation," *Journal of applied polymer science,* vol. 82, no. 8, pp. 2042-2055, 2001.

[58] N. Barreau, S. Marsillac, J. Bernede, and L. Assmann, "Evolution of the band structure of β-In 2 S 3− 3x O 3x buffer layer with its oxygen content," *Journal of applied physics,* vol. 93, no. 9, pp. 5456-5459, 2003.

[59] C. Amory, J. Bernede, and S. Marsillac, "Study of a growth instability of γ-In 2 Se 3," *Journal of applied physics,* vol. 94, no. 10, pp. 6945-6948, 2003.

[60] K. Benchouk *et al.*, "New buffer layers, large band gap ternary compounds: CuAlTe2," *The European Physical Journal Applied Physics,* vol. 10, no. 1, pp. 9-14, 2000.

[61] J. Bernede, N. Barreau, S. Marsillac, and L. Assmann, "Band alignment at β-In2S3/TCO interface," *Applied surface science,* vol. 195, no. 1-4, pp. 222-228, 2002.

[62] N. Barreau, S. Marsillac, J. Bernede, and A. Barreau, "Investigation of β-In2S3 growth on different transparent conductive oxides," *Applied surface science,* vol. 161, no. 1-2, pp. 20-26, 2000.

[63] D. Poudel *et al.*, "Assessment of Cu (In, Ga) Se2 Solar Cells Degradation due to Water Ingress Effect on the CdS Buffer Layer," *Journal of Energy and Power Technology,* vol. 3, no. 1, 2021.

[64] P. Paul *et al.*, "Direct nm-scale spatial mapping of traps in CIGS," *IEEE Journal of Photovoltaics,* vol. 5, no. 5, pp. 1482-1486, 2015.

[65] S. Karki *et al.*, "Degradation mechanism in Cu (In, Ga) Se 2 material and solar cells due to moisture and heat treatment of the absorber layer," *IEEE Journal of Photovoltaics,* vol. 9, no. 4, pp. 1138-1143, 2019.

[66] S. Karki *et al.*, "Impact of water ingress on molybdenum thin films and its effect on Cu (In, Ga) Se 2 Solar Cells," *IEEE Journal of Photovoltaics,* vol. 10, no. 2, pp. 696-702, 2019.

[67] D. Poudel *et al.*, "Degradation Mechanism Due to Water Ingress Effect on the Top Contact of Cu (In, Ga) Se2 Solar Cells," *Energies,* vol. 13, no. 17, p. 4545, 2020.

[68] D. Poudel *et al.*, "Numerical Analysis of Water Ingress Effect on the Window Layer of Cu (In, Ga) Se 2 Solar Cells using SCAPS-1D," in *2021 IEEE 48th Photovoltaic Specialists Conference (PVSC),* 2021: IEEE, pp. 0438-0442.

[69] D. Poudel, B. Belfore, S. Karki, G. Rajan, A. Rockett, and S. Marsillac, "Process Dependent Instabilities In Cu (In, Ga) Se 2 Solar Cells Under Water Ingress," in *2021 IEEE 48th Photovoltaic Specialists Conference (PVSC),* 2021: IEEE, pp. 0433-0437.

[70] G. Rajan *et al.*, "Study of Instabilities and Degradation due to Moisture Ingress in the Molybdenum back contact of Cu (In, Ga) Se 2 Solar Cells," in *2018 IEEE 7th World Conference on Photovoltaic Energy Conversion (WCPEC)(A Joint Conference of 45th IEEE PVSC, 28th PVSEC & 34th EU PVSEC),* 2018: IEEE, pp. 3037-3039.

978-1-7281-6118-1/22 $31.00 © 2022 IEEE

Post-deposition Metal Halide Treatment of CuGaSe$_2$ for Photovoltaic Application

Deewakar Poudel[1], Benjamin Belfore[1], Adam Masters[1], Elizabeth Palmiotti[2], Angus Rockett[2] and Sylvain Marsillac[1]

[1]Virginia Institute of Photovoltaics, Old Dominion University, Norfolk, VA 23529, USA

[2]Dept. of Metallurgical and Materials Engineering, Colorado School of Mines, Golden, CO 80401, USA

Abstract— Copper gallium diselenide (CGS) semiconductor thin films were deposited by three-stage thermal co-evaporation process. Post-deposition treatments and recrystallization were performed at various AgBr doses of 40 mg, 60 mg, and 80 mg after the 2nd stage. The changes in surface morphology were confirmed by SEM. The electrical properties were also modified, as the resistivity of the film decreases with increasing doses. The device performance after the treatment did not change as expected, as the overall device performance only slightly increases in the case of 60 mg of AgBr. Substantial changes in the fabrication process will therefore be required for better device results.

Keywords— CuGaSe$_2$, recrystallization, AgBr

I. INTRODUCTION

Chalcopyrite materials are critical materials for achieving high efficiency and low-cost polycrystalline thin film solar cells, thanks to their attractive properties such as their high absorption coefficient, appropriate bandgap energy and tunable band edges [1]. As compared to Cu(In,Ga)Se$_2$ and CuInSe$_2$, the reports on CuGaSe$_2$ are fewer and the maximum recorded efficiency is only 11.9 % [2]. One of the constraining aspects of CGS is its low conductivity. To overcome this limitation, it is important to investigate the materials properties in detail to attain the optimum device performance and is a required step to enhance the PV manufacturing capacity and its supply chain [3-10]. One reason for such research is that tandem solar cells with wide bandgap top cell and high efficiency bottom cell have drawn substantial interests. Based on the required properties, CGS with a band gap of 1.67 eV and its high absorption coefficient is considered as an excellent candidate for the top cell of the tandem solar cell [11]. To fabricate this suitable wide bandgap absorber layer, a recrystallization process might be required to enhance the grain growth and provide a high-quality absorber layer.

Previously, we have demonstrated that various metal halides post-deposition treatment can act as excellent transport agents in CIGS, generating larger grain and uniform layers [12-27]. In this work, we studied silver bromide as a transport agent for the recrystallization of CGS thin film deposited at temperature of 500 °C using various doses and investigated its effect on photovoltaic devices.

II. EXPERIMENTAL METHODS

DC magnetron sputtering was used to deposit the molybdenum back contact on soda lime glass substrate. The molybdenum layer is a bilayer with the bottom layer deposited at low Ar pressure resulting in a tensile stress, whereas the top layer was deposited at high Ar pressure resulting in compressive stress [28-30]. Traditional, three-stage thermal co-evaporation deposition process was used to deposit the CGS layers. The 1st stage temperature was 300 °C, whereas the 2nd and 3rd stages temperature were maintained at 500 °C. The vapor treatment and thereafter recrystallization process was carried out in between the 2nd and 3rd stage of the CGS deposition. Different AgBr doses were used, ranging from 40 mg, 60 mg to 80 mg. The AgBr was flashed over 2 minutes. As-deposited samples were also fabricated. After completion of the CGS deposition, half of the samples were set aside for characterization, while the other half were converted into completed devices using our standard process, resulting in a SLG/Mo/CGS/CdS/ZnO/ITO/grids structure.

Surface morphological analysis was performed by scanning electron microscope (SEM). Electrical properties of the film were measured by Hall effect measurements performed on films deposited on glass. Spectroscopic ellipsometry measurements were performed to further analyze the optical properties of the films. The photovoltaic characteristics were evaluated by external quantum efficiency (QE) measurements (QEX7, PV measurements Inc.) and current density-voltage (J-V) measurements (IV5, PV measurements Inc.) done under AM 1.5G with a light intensity of 100 mW/cm^2 at 25°C.

III. RESULTS AND DISCUSSIONS

CGS samples were recrystallized at 500 °C in an AgBr environment by flashing either 40 mg, 60 mg or 80 mg of AgBr for 2 minutes at the end of the 2nd stage. The surface morphology of the films was observed by SEM before and after recrystallization as shown in Figure 1. Slight variations in the surface morphology were observed after the 40 mg and 60 mg AgBr treatments as compared to the reference, while clear and significant changes on the surface were observed in the case of the 80 mg dose.

This research was supported by the Department of Energy Contract No. DE-EE0007551.

fitting procedure revealed that there is no significant shift in the $E_0(A,B)$ or $E_0(C)$ CP energies with the AgBr treatment.

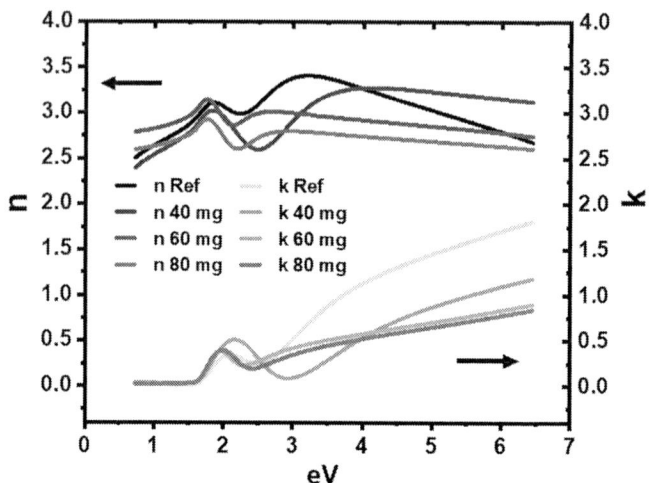

Fig. 2. Dielectric functions for the as-deposited and AgBr treated CGS samples.

Hall effect measurements were also performed to assess the electrical properties, such as carrier concentration, mobility, and resistivity, as shown in Table I. This shows that the electrical properties of the film are enhanced with an increase in the amount of AgBr dose, due both to a modification of the mobility and of the carrier concentration, with the higher dose yielding the best properties.

TABLE I. ELECTRICAL PROPERTIES OF CGS FILMS FOR AS-DEPOSITED AND AGBR TREATED SAMPLES AT 500 °C WITH 40 MG, 60 MG AND 80 MG DOSES.

Samples	Carrier concentration (cm^{-3})	Mobility (cm^2/s)	Resistivity (Ω cm)
As-deposited	6.9E +13	4.4	2.0E +4
AgBr 40 mg	2.7E +15	5.6	4.1E +2
AgBr 60 mg	5.5E +15	6.7	1.7E +2
AgBr 80 mg	7.8E +15	7.7	1.0E +2

After the CGS fabrication process, devices were completed by depositing ~50 nm of CdS by chemical bath deposition, followed by i-ZnO (~50 nm) and ITO (~250 nm) by r.f. sputtering, and Ni/Al/Ni grids by e-beam evaporation. Figure 3 shows representative current-voltage and external quantum efficiency curve for the devices as a function of the AgBr doses, while Table II shows the corresponding photovoltaic parameters. All devices have a voltage dependent current collection, indicative of a high level of trap density. This could be coming from the effect of sodium or other alkali [49-56] or could be due to band offsets between the absorber and the buffer layers, and mandate an alternative choice to CdS [57-70]. The major problem observed with the treatment is that some of the samples are shunted (40 mg and 80 mg samples). This result is not correlated with a specific change in morphology, optical or

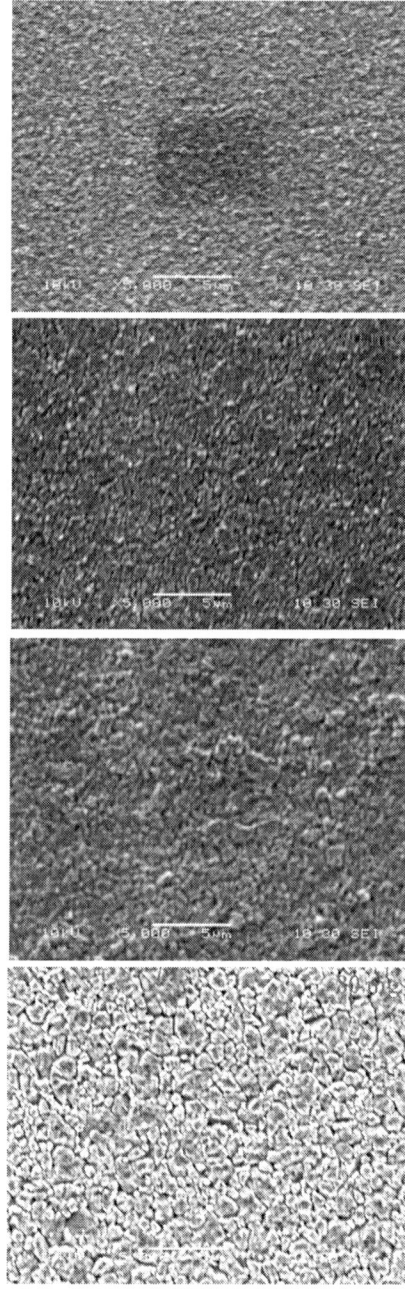

Fig. 1. Surface Scanning Electron Microscopy micrographs of CGS films: as-deposited and recrystallized at 500 °C by 40 mg, 60 mg, and 80 mg AgBr vapor treatment.

The film optical properties of the CGS film were analyzed by spectroscopic ellipsometry (Figure 2) using previously reported methods [31-48]. The observed features in (n,k) are associated with interband transitions that appear at the van Hove singularities or critical points (CPs) of the joint density of states. These features were fitted assuming parabolic bands (PBs), yielding CPPB oscillators. Here the fundamental transitions were fitted with excitonic CPs and the higher energy transition points were fitted with 2-dimensional CPs. This

electrical properties of the films so is likely to be due to a problem in processing or scribing the device rather than the AgBr effect.

Interestingly, for the devices that were not shunted (60 mg dose), we can observe an enhancement for all parameters (open circuit voltage, fill factor, short circuit current density). The diode parameters were extracted from the dark J-V curves for the two most efficient devices (as-deposited and 60 mg samples). The reverse saturation current density changed from $3.1E-3$ mA/cm^2 to $5.3E-4$ mA/cm^2, the series resistance from 9.7 $\Omega.cm^2$ to 5.5 $\Omega.cm^2$, the shunt resistance from 500 Ω/cm^2 to 900 Ω/cm^2 and the diode ideality factor from 2.4 to 2.1 after the AgBr treatment. All these parameters indicate a slight enhancement of the overall device after AgBr treatment.

TABLE II. PHOTOVOLTAIC PARAMETER FOR AS-DEPOSITED AND AGBR TREATED SAMPLES AT 500 °C FOR 40 MG, 60 MG AND 80 MG DOSES.

Samples	V_{OC} (volt)	J_{SC} (mA/cm^2)	FF (%)	η (%)
As-deposited	0.58	7.2	50.8	2.3
AgBr 40 mg	0.03	3.4	24.3	0.03
AgBr 60 mg	0.66	10.0	54.9	4.23
AgBr 80 mg	0.18	9.6	43.2	0.86

The quantum efficiency correlates well with the spectroscopic ellipsometry measurements, where no change in the bandgap was observed. One can see a clear enhancement of QE at all wavelengths for the 60 mg and 80 mg devices.

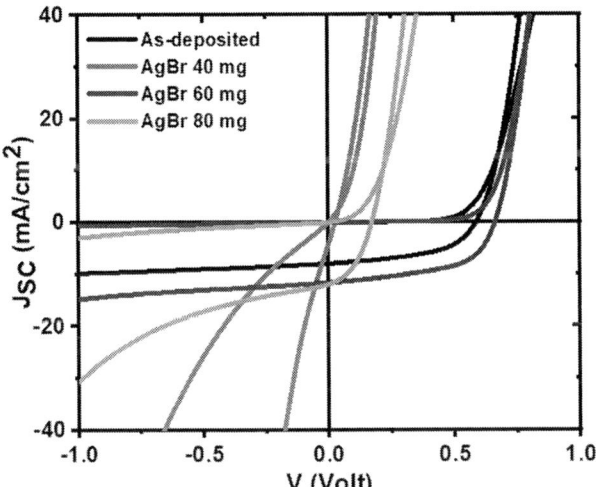

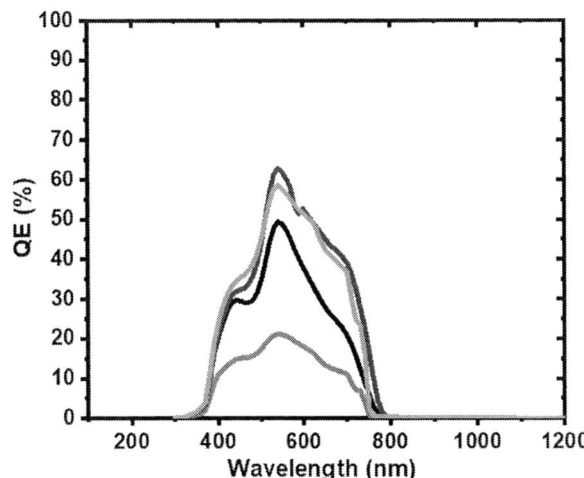

Fig. 3. Representative I-V and QE curves for the as-deposited and AgBr treated CGS samples.

IV. CONCLUSION

The effects of various doses of silver bromide treatment on CGS semiconductor thin films deposited by 3-stage were analyzed. The recrystallization was done by flashing three different doses of 40 mg, 60 mg, and 80 mg of AgBr for 2 minutes. The change in surface morphology was observed with increased surface roughness with increasing doses. The increase in conductivity was also observed after the treatment with higher conductivity at higher doses. In the case of devices, several issues were observed both before and after treatment. The reference sample efficiency was very low, indicating a fundamental problem in the way we fabricated these devices. This will need to be addressed to properly study the effect of AgBr treatment. Nevertheless, one can observe that potential benefits of this AgBr treatment are present, notably an enhancement of the electrical properties. An increase in efficiency was observed for the films deposited with a 60 mg dose, with an improvement of all the device parameters. Additional characterizations will be performed on the samples, including X-ray diffraction for structural analysis, and secondary ion mass spectrometry for elemental depth profile.

ACKNOWLEDGMENT

This research was supported by the Department of Energy Contract No. DE-EE0007551.

REFERENCES

[1] T. Kato, "Cu (In, Ga)(Se, S) 2 solar cell research in Solar Frontier: Progress and current status," *Japanese Journal of Applied Physics,* vol. 56, no. 4S, p. 04CA02, 2017.

[2] S. Ishizuka and P. J. Fons, "Polycrystalline CuGaSe2 thin film growth and photovoltaic devices fabricated on alkali-free and alkali-containing substrates," *Journal of Crystal Growth,* vol. 532, p. 125407, 2020.

[3] D. M. Nelson, E. Marsillac, and S. S. Rao, "Antecedents and evolution of the green supply chain," *Journal of Operations and Supply Chain Management,* no. Special Issue, 2012.

[4] K. Liao, E. Marsillac, E. Johnson, and Y. Liao, "Global supply chain adaptations to improve financial performance: supply base

establishment and logistics integration," *Journal of Manufacturing Technology Management,* 2011.

[5] R. Romero-Silva, E. Marsillac, S. Shaaban, and M. Hurtado-Hernández, "Serial production line performance under random variation: dealing with the 'Law of Variability'," *Journal of Manufacturing Systems,* vol. 50, pp. 278-289, 2019.

[6] E. Marsillac, "Management of the photovoltaic supply chain," *International Journal of Technology, Policy and Management,* vol. 12, no. 2-3, pp. 195-211, 2012.

[7] Y. Liao and E. Marsillac, "External knowledge acquisition and innovation: the role of supply chain network-oriented flexibility and organisational awareness," *International Journal of Production Research,* vol. 53, no. 18, pp. 5437-5455, 2015.

[8] R. Romero-Silva, S. Shaaban, E. Marsillac, and M. Hurtado, "Exploiting the characteristics of serial queues to reduce the mean and variance of flow time using combined priority rules," *International Journal of Production Economics,* vol. 196, pp. 211-225, 2018.

[9] E. L. Marsillac, "Environmental impacts on reverse logistics and green supply chains: similarities and integration," *International Journal of Logistics Systems and Management,* vol. 4, no. 4, pp. 411-422, 2008.

[10] E. Marsillac, "Closing the Photovoltaic Supply Chain Loop-Invest Now for Future Returns," in *2018 IEEE 7th World Conference on Photovoltaic Energy Conversion (WCPEC)(A Joint Conference of 45th IEEE PVSC, 28th PVSEC & 34th EU PVSEC),* 2018: IEEE, pp. 2498-2500.

[11] A. Popp and C. Pettenkofer, "Epitaxial growth of CuGaSe2 thin-films by MBE—Influence of the Cu/Ga ratio," *Applied Surface Science,* vol. 416, pp. 815-823, 2017.

[12] G. Rajan *et al.,* "Impact of Post-Deposition Recrystallization by Alkali Fluorides on Cu (In, Ga) Se2Thin-Film Materials and Solar Cells," *Thin Solid Films,* vol. 690, p. 137526, 2019.

[13] E. Palmiotti, B. Belfore, D. Poudel, S. Marsillac, and A. Rockett, "In-Situ Study of the Crystallization of Amorphous CuInSe2 Thin Films and the Effect of InCl3 Treatment," *Thin Solid Films,* p. 139095, 2022.

[14] B. Belfore *et al.,* "Recrystallization of Cu (In, Ga) Se2 Semiconductor Thin Films via InCl3 Treatment," *Thin Solid Films,* vol. 735, p. 138897, 2021.

[15] S. Marsillac, M. Zouaghi, J. Bernede, T. B. Nasrallah, and S. Belgacem, "Evolution of the properties of spray-deposited CuInS2 thin films with post-annealing treatment," *Solar energy materials and solar cells,* vol. 76, no. 2, pp. 125-134, 2003.

[16] D. Poudel *et al.,* "Analysis of Post-Deposition Recrystallization Processing via Indium Bromide of Cu (In, Ga) Se2 Thin Films," *Materials,* vol. 14, no. 13, p. 3596, 2021.

[17] S. Karki *et al.,* "Analysis of recombination mechanisms in RbF-treated CIGS solar cells," *IEEE Journal of Photovoltaics,* vol. 9, no. 1, pp. 313-318, 2018.

[18] D. Poudel *et al.,* "In Situ Recrystallization of Co-Evaporated Cu (In, Ga) Se2 Thin Films by Copper Chloride Vapor Treatment towards Solar Cell Applications," *Energies,* vol. 14, no. 13, p. 3938, 2021.

[19] D. Poudel *et al.,* "Studying the Recrystallization of Cu (InGa) Se 2 Semiconductor Thin Films by Silver Bromide In-situ Treatment," in *2021 IEEE 48th Photovoltaic Specialists Conference (PVSC),* 2021: IEEE, pp. 2307-2311.

[20] D. Poudel *et al.,* "Effect of Indium Bromide Treatments Post-Deposition Recrystallization Temperature on Cu (In, Ga) Se 2 Thin Films," in *2021 IEEE 48th Photovoltaic Specialists Conference (PVSC),* 2021: IEEE, pp. 0429-0432.

[21] E. Palmiotti, B. Belfore, D. Poudel, S. Marsillac, and A. Rockett, "CuIn (1− x) Ga x Se 2 Recrystallization by Metal Halide Treatments," in *2021 IEEE 48th Photovoltaic Specialists Conference (PVSC),* 2021: IEEE, pp. 0669-0671.

[22] B. Belfore *et al.,* "Study of Indium Chloride Vapor Treatment on Cu (In, Ga) Se 2 Semiconductor Thin Films," in *2021 IEEE 48th Photovoltaic Specialists Conference (PVSC),* 2021: IEEE, pp. 2320-2323.

[23] B. Belfore *et al.,* "Vapor Treatment and In-situ Recrystallization by Copper Chloride on Cu (In, Ga) Se 2 Thin Film," in *2021 IEEE 48th*

Photovoltaic Specialists Conference (PVSC), 2021: IEEE, pp. 2316-2319.

[24] B. Belfore, D. Poudel, T. Ashrafee, S. Karki, G. Rajan, and S. Marsillac, "Morphological Study of Indium Chloride Post Deposition Treated CuInSe 2 Thin Films," in *2021 IEEE 48th Photovoltaic Specialists Conference (PVSC),* 2021: IEEE, pp. 2312-2315.

[25] B. Belfore *et al.,* "In-Situ Recrystallization of CIGS via Metal Halides," in *2020 47th IEEE Photovoltaic Specialists Conference (PVSC),* 2020: IEEE, pp. 1131-1133.

[26] B. Belfore *et al.,* "Ex-Situ Recrystallization of CIGS via Metal Halides," in *2020 47th IEEE Photovoltaic Specialists Conference (PVSC),* 2020: IEEE, pp. 1102-1104.

[27] B. Belfore, G. Rajan, S. Karki, D. Poudel, A. Rockett, and S. Marsillac, "The Impact of Deposition Temperature on Sodium Fluoride Recrystallization in Cu (In, Ga) Se 2 Solar Cells," in *2019 IEEE 46th Photovoltaic Specialists Conference (PVSC),* 2019: IEEE, pp. 1851-1853.

[28] E. Gourmelon, J. Bernede, J. Pouzet, and S. Marsillac, "Textured MoS 2 thin films obtained on tungsten: electrical properties of the W/MoS 2 contact," *Journal of Applied Physics,* vol. 87, no. 3, pp. 1182-1186, 2000.

[29] K. Aryal, H. Khatri, R. Collins, and S. Marsillac, "In situ and ex situ studies of molybdenum thin films deposited by rf and dc magnetron sputtering as a back contact for CIGS solar cells," *International Journal of Photoenergy,* vol. 2012, 2012.

[30] O. Ayala *et al.,* "Theoretical Analysis of Experimental Data of Sodium Diffusion in Oxidized Molybdenum Thin Films," *Energies,* vol. 14, no. 9, p. 2479, 2021.

[31] S. Marsillac, S. Little, and R. Collins, "A broadband analysis of the optical properties of silver nanoparticle films by in situ real time spectroscopic ellipsometry," *Thin Solid Films,* vol. 519, no. 9, pp. 2936-2940, 2011.

[32] P. Koirala *et al.,* "Real time spectroscopic ellipsometry for analysis and control of thin film polycrystalline semiconductor deposition in photovoltaics," *Thin solid films,* vol. 571, pp. 442-446, 2014.

[33] L. R. Dahal, D. Sainju, N. Podraza, S. Marsillac, and R. Collins, "Real time spectroscopic ellipsometry of Ag/ZnO and Al/ZnO interfaces for back-reflectors in thin film Si: H photovoltaics," *Thin Solid Films,* vol. 519, no. 9, pp. 2682-2687, 2011.

[34] P. Koirala *et al.,* "Through-the-glass spectroscopic ellipsometry for analysis of CdTe thin-film solar cells in the superstrate configuration," *Progress in Photovoltaics: Research and Applications,* vol. 24, no. 8, pp. 1055-1067, 2016.

[35] P. Aryal *et al.,* "Parameterized complex dielectric functions of CuIn1− xGaxSe2: applications in optical characterization of compositional non-uniformities and depth profiles in materials and solar cells," *Progress in Photovoltaics: Research and Applications,* vol. 24, no. 9, pp. 1200-1213, 2016.

[36] S. Marsillac *et al.,* "Spectroscopic ellipsometry studies of In2S3 top window and Mo back contacts in chalcopyrite photovoltaics technology," *physica status solidi c,* vol. 5, no. 5, pp. 1244-1248, 2008.

[37] V. Ranjan, R. Collins, and S. Marsillac, "Real-time analysis of the microstructural evolution and optical properties of Cu (In, Ga) Se2 thin films as a function of Cu content," *physica status solidi (RRL)– Rapid Research Letters,* vol. 6, no. 1, pp. 10-12, 2012.

[38] T. Begou, J. D. Walker, D. Attygalle, V. Ranjan, R. Collins, and S. Marsillac, "Real time spectroscopic ellipsometry of CuInSe2: growth dynamics, dielectric function, and its dependence on temperature," *physica status solidi (RRL)–Rapid Research Letters,* vol. 5, no. 7, pp. 217-219, 2011.

[39] S. Karki *et al.,* "In situ and ex situ investigations of KF postdeposition treatment effects on CIGS solar cells," *IEEE Journal of Photovoltaics,* vol. 7, no. 2, pp. 665-669, 2016.

[40] L. R. Dahal *et al.,* "Correlations between mapping spectroscopic ellipsometry results and solar cell performance for evaluations of nonuniformity in thin-film silicon photovoltaics," *IEEE Journal of Photovoltaics,* vol. 3, no. 1, pp. 387-393, 2012.

[41] P. Aryal, D. Attygalle, P. Pradhan, N. J. Podraza, S. Marsillac, and R. W. Collins, "Large-area compositional mapping of Cu (In $ _ {1- x} $ Ga $ _ {x} $) Se $ _ {2} $ materials and devices with

spectroscopic ellipsometry," *IEEE Journal of Photovoltaics,* vol. 3, no. 1, pp. 359-363, 2012.

[42] A.-R. Ibdah *et al.*, "Spectroscopic ellipsometry for analysis of polycrystalline thin-film photovoltaic devices and prediction of external quantum efficiency," *Applied Surface Science,* vol. 421, pp. 601-607, 2017.

[43] J. Walker, H. Khatri, V. Ranjan, J. Li, R. Collins, and S. Marsillac, "Electronic and structural properties of molybdenum thin films as determined by real-time spectroscopic ellipsometry," *Applied Physics Letters,* vol. 94, no. 14, p. 141908, 2009.

[44] S. Little, T. Begou, R. Collins, and S. Marsillac, "Optical detection of melting point depression for silver nanoparticles via in situ real time spectroscopic ellipsometry," *Applied Physics Letters,* vol. 100, no. 5, p. 051107, 2012.

[45] D. R. Sapkota *et al.*, "Evaluation of CuInSe 2 Materials and Solar Cells Co-evaporated at Different Rates Based on Real Time Spectroscopic Ellipsometry Calibrations," in *2021 IEEE 48th Photovoltaic Specialists Conference (PVSC),* 2021: IEEE, pp. 0451-0458.

[46] P. Pradhan *et al.*, "Real time spectroscopic ellipsometry analysis of the three-stages of CuIn 1− x Ga x Se 2 co-evaporation," in *2014 IEEE 40th Photovoltaic Specialist Conference (PVSC),* 2014: IEEE, pp. 2060-2065.

[47] D. Attygalle *et al.*, "Optical monitoring and control of three-stage coevaporated Cu (In 1− x Ga x) Se 2 by real-time spectroscopic ellipsometry," in *2012 IEEE 38th Photovoltaic Specialists Conference (PVSC) PART 2,* 2012: IEEE, pp. 1-6.

[48] H. Fujiwara and R. W. Collins, *Spectroscopic Ellipsometry for Photovoltaics: Volume 2: Applications and Optical Data of Solar Cell Materials.* Springer, 2019.

[49] D. Poudel *et al.*, "Assessment of Cu (In, Ga) Se2 Solar Cells Degradation due to Water Ingress Effect on the CdS Buffer Layer," *Journal of Energy and Power Technology,* vol. 3, no. 1, 2021.

[50] P. Paul *et al.*, "Direct nm-scale spatial mapping of traps in CIGS," *IEEE Journal of Photovoltaics,* vol. 5, no. 5, pp. 1482-1486, 2015.

[51] S. Karki *et al.*, "Degradation mechanism in Cu (In, Ga) Se 2 material and solar cells due to moisture and heat treatment of the absorber layer," *IEEE Journal of Photovoltaics,* vol. 9, no. 4, pp. 1138-1143, 2019.

[52] S. Karki *et al.*, "Impact of water ingress on molybdenum thin films and its effect on Cu (In, Ga) Se 2 Solar Cells," *IEEE Journal of Photovoltaics,* vol. 10, no. 2, pp. 696-702, 2019.

[53] D. Poudel *et al.*, "Degradation Mechanism Due to Water Ingress Effect on the Top Contact of Cu (In, Ga) Se2 Solar Cells," *Energies,* vol. 13, no. 17, p. 4545, 2020.

[54] D. Poudel *et al.*, "Numerical Analysis of Water Ingress Effect on the Window Layer of Cu (In, Ga) Se 2 Solar Cells using SCAPS-1D," in *2021 IEEE 48th Photovoltaic Specialists Conference (PVSC),* 2021: IEEE, pp. 0438-0442.

[55] D. Poudel, B. Belfore, S. Karki, G. Rajan, A. Rockett, and S. Marsillac, "Process Dependent Instabilities In Cu (In, Ga) Se 2 Solar Cells Under Water Ingress," in *2021 IEEE 48th Photovoltaic Specialists Conference (PVSC),* 2021: IEEE, pp. 0433-0437.

[56] G. Rajan *et al.*, "Study of Instabilities and Degradation due to Moisture Ingress in the Molybdenum back contact of Cu (In, Ga) Se 2 Solar Cells," in *2018 IEEE 7th World Conference on Photovoltaic Energy Conversion (WCPEC)(A Joint Conference of 45th IEEE PVSC, 28th PVSEC & 34th EU PVSEC),* 2018: IEEE, pp. 3037-3039.

[57] S. Marsillac, J. Bernede, R. Ny, and A. Conan, "A new simple technique to obtain In2Se3 polycrystalline thin films," *Vacuum,* vol. 46, no. 11, pp. 1315-1323, 1995.

[58] N. Barreau, S. Marsillac, and J. Bernede, "Physico-chemical characterization of β-In2S3 thin films synthesized by solid-state reaction, induced by annealing, of the constituents sequentially deposited in thin layers," *Vacuum,* vol. 56, no. 2, pp. 101-106, 2000.

[59] N. Barreau, J. Bernede, S. Marsillac, C. Amory, and W. Shafarman, "New Cd-free buffer layer deposited by PVD: In2S3 containing Na compounds," *Thin Solid Films,* vol. 431, pp. 326-329, 2003.

[60] R. Robles, N. Barreau, A. Vega, S. Marsillac, J. Bernede, and A. Mokrani, "Optical properties of large band gap β-In2S3− 3xO3x compounds obtained by physical vapour deposition," *Optical Materials,* vol. 27, no. 4, pp. 647-653, 2005.

[61] J. Bernede, S. Marsillac, and A. Conan, "Electrical properties of γ-In2Se3 layers synthesized by solid state reaction between In and Se thin films," *Materials chemistry and physics,* vol. 48, no. 1, pp. 5-9, 1997.

[62] S. Marsillac, J. Bernede, and A. Conan, "Change in the type of majority carriers in disordered In x Se 100− x thin-film alloys," *Journal of materials science,* vol. 31, no. 3, pp. 581-587, 1996.

[63] N. Barreau, J. Bernede, and S. Marsillac, "Study of the new β-In2S3 containing Na thin films. Part II: Optical and electrical characterization of thin films," *Journal of crystal growth,* vol. 241, no. 1-2, pp. 51-56, 2002.

[64] N. Barreau, J. Bernede, C. Deudon, L. Brohan, and S. Marsillac, "Study of the new β-In2S3 containing Na thin films Part I: Synthesis and structural characterization of the material," *Journal of crystal growth,* vol. 241, no. 1-2, pp. 4-14, 2002.

[65] K. D'Almeida, J. Bernede, F. Ragot, A. Godoy, F. Diaz, and S. Lefrant, "Carbazole-based electroluminescent devices obtained by vacuum evaporation," *Journal of applied polymer science,* vol. 82, no. 8, pp. 2042-2055, 2001.

[66] N. Barreau, S. Marsillac, J. Bernede, and L. Assmann, "Evolution of the band structure of β-In 2 S 3− 3x O 3x buffer layer with its oxygen content," *Journal of applied physics,* vol. 93, no. 9, pp. 5456-5459, 2003.

[67] C. Amory, J. Bernede, and S. Marsillac, "Study of a growth instability of γ-In 2 Se 3," *Journal of applied physics,* vol. 94, no. 10, pp. 6945-6948, 2003.

[68] K. Benchouk *et al.*, "New buffer layers, large band gap ternary compounds: CuAlTe2," *The European Physical Journal Applied Physics,* vol. 10, no. 1, pp. 9-14, 2000.

[69] J. Bernede, N. Barreau, S. Marsillac, and L. Assmann, "Band alignment at β-In2S3/TCO interface," *Applied surface science,* vol. 195, no. 1-4, pp. 222-228, 2002.

[70] N. Barreau, S. Marsillac, J. Bernede, and A. Barreau, "Investigation of β-In2S3 growth on different transparent conductive oxides," *Applied surface science,* vol. 161, no. 1-2, pp. 20-26, 2000.

Current or power matching? A third option for monolithic all-perovskite tandem solar cells

Yuan Gao, Renxing Lin, Ke Xiao, Xin Luo, Jin Wen, Xu Yue, Hairen Tan

Lawrence Berkeley National Laboratory, Berkeley, CA, United States

National Laboratory of Solid State Microstructures, Jiangsu Key Laboratory of Artificial Functional Materials, College of Engineering and Applied Sciences, Nanjing University, Nanjing, China

Jiangsu Key Laboratory of Atmospheric Environment Monitoring and Pollution Control, Collaborative Innovation Center of Atmospheric Environment and Equipment Technology, School of Environmental Science and Engineering, Nanjing University of Information Sci, Nanjing, China

Multijunction solar cells offer high potential to improve the efficiency beyond the single-junction limit. The record power conversion efficiency of monolithic all-perovskite tandem solar cells (26.4%) has now surpassed that of the single-junction counterparts (25.7%). The two-terminal tandem architectures, in most cases, require a "matched" current between the top and bottom subcells for an optimal tandem performance. In some cases, where the two subcells have distinct fill factors, a "power matching" strategy shows better performance than "current matching". In this study, we prove that optimal performance of monolithic all-perovskite tandem solar cells is obtained when the top subcell has a higher short-circuit current than the bottom one. This is attributed to both optical and electrical properties of the all-perovskite tandem cells. We also investigate the optimal tandem configurations under real-world conditions, where solar spectra, angle of incidence, and cell temperature are different from the standard test conditions. We find that the top subcell shifts to thinner optimal thickness under blue-rich solar spectra. We also observe that the bottom subcell shifts to thicker optimal thickness due to larger angle of incidence under real-world conditions. Since two perovskite subcells have similar temperature coefficients, the real-world cell temperatures have little impact the optimal tandem configurations, but will affect the annual energy yield. We further obtain the real-world optimized tandem configurations, which can increase the annual energy yield compared with the one optimized under standard test conditions. The current-mismatching losses under real-world conditions can be reduced by optimized configurations, and even be turned into energy gains in certain cases. This finding enhances the confidence in promoting high-efficient, customizable tandem solar cells for different regions and installation methods.

Simulation of High open-circuit voltage Perovskite/CIGS-GeTe tandem cell

Mohamed Mousa[1], Mostafa M. Salah[1], A. Zekry[2], Mohamed Abouelatta[2], Ahmed Shaker[2], Fathy Z. Amer[3], Roaa I. Mubarak[3], Ahmed Saeed[1]

[1]Electrical Engineering Department, Faculty of Engineering and Technology, Future University in Egypt, Cairo, 11835, Egypt.

[2]Electronics and Communication Engineering Department, Faculty of Engineering, Ain Shams University, Cairo, 11535, Egypt.

[3]Electronics and Communication Engineering Department, Faculty of Engineering, Helwan University, Cairo, 11795, Egypt.

Abstract— This research work aims to improve the performance of the tandem solar cells based on a novel design that uses perovskite material as an absorber layer of the top sub-cell and Copper indium gallium selenide (CIGS) and Germanium telluride (GeTe) materials as absorber layers of the bottom sub-cell. Using two different materials with different energy gaps and distinct doping levels as absorbers of the bottom sub-cells leads to improvement of the sub-cell performance. These two absorbers, of the bottom sub-cell layers, act electrically series-connected, which improves the sub-cell voltage but at the same time limits the current density of the sub-cell to the minimum one of the used materials. After studying the performance of both sub-cells with AM 1.5 incident spectrum at room temperature, both are used to configure Perovskite/CIGS-GeTe tandem cells. The performance of the designed tandem cell was studied to find the optimum thickness with the variation of the top sub-cell absorber layer thickness. The results show that PVK/CIGS-GeTe tandem cell has a high power conversion efficiency of 33.48 %, open-circuit voltage of 1.972 V, short-circuit current of 20.50 mA/cm^2, and fill factor of 82.81 %.

Keywords— *CIGS; High performance; High voltage; GeTe; Perovskite; Tandem-solar cell; Two absorbers; SCAPS-1D.*

I. INTRODUCTION

Nowadays, the need for energy increases at a very fast rate. Clean energy, especially solar cells, shows a promising solution [1]. Solar energy meets this need and could be used in remote regions where the grids are not accessible [2]. Currently, the crystallized silicon solar cells in multi-crystalline and monocrystalline versions dominate the photovoltaic market. So far, the efficiency of crystalline silicon solar cells has surpassed 25% [3], [4]. Copper indium gallium selenide (CIGS) solar cells are very competitive thin-film solar cells and have achieved a power conversion efficiency (PCE) of 23.35% [5]. Perovskite solar cells (PSCs) have also recently developed, demonstrating rapid progress, and providing new goals in photovoltaics. PSCs have a record efficiency of over 22% [6], and the development of PSC in the recent few years shows simulation results up to 30% [7]. Lower recombination rates, wide absorption spectrum,

long diffusion lengths, high open-circuit voltage (Voc), and bandwidth adjustment are all factors that contribute to perovskite cells' improved performance. There is a maximum PCE for a single-junction solar cell. For a bandgap of 1.34 eV, Shockley and Queisser drove a maximum PCE of 33.7 % in 1961 [8]. This type of solar cell has a limited performance because it can only absorb photons with energies equal to or greater than the energy gap of the used material. At the same time, the rest of the incident spectrum is lost, and even photons with higher energies lose the energy difference due to thermalization loss. A simple perovskite/crystalline silicon monolithic tandem cell with an efficiency greater than 21% was recorded by Jérémie Werner, et al in 2016 [9]. The tandem cells based on silicon as a bottom sub-cell are reported with a PCE of up to 28%. A novel structure for a monolithic PSC/silicon tandem cell with a mesoscopic perovskite and a homojunction silicon bottom sub-cells that can withstand high temperatures is introduced by YiLiang Wu, et al with 22.5% efficiency [10]. Florent Sahli, et al improved a recombination junction using nanocrystalline silicon to overcome the losses. This leads to increase the bottom cell photocurrent, which improves the tandem cell current density and efficiency up to 22.7% [11]. This team also developed a deposition process for the top sub-cell that achieved a controlled optoelectronic property of the sub-cell. Controlling the optical properties of the perovskite/silicon sub-cells of the tandem cell increases its efficiency up to 25.2%. Coupling an infrared-tuned silicon heterojunction bottom cell with the cesium formamidinium lead halide perovskite in a two-terminal tandem cell shows 23.6% efficiency and high stability [11]. Kevin A. Bush, et al. demonstrated reducing parasitic absorption and reflection loss based on optical optimization to improve the incident spectrum harvesting, improving the tandem efficiency to 25% [12]. Increasing the perovskite grain size, and lowering its defect density to improve the photocurrent and voltage by using combined additives is tested by BoChen, et al and shows a 25.4% efficiency for a perovskite/silicon tandem cell [12]. Marko Jošt, et al. used a textured light management foil on the tandem cell front side processed on a wafer with a textured backside and planar front side to make

978-1-7281-6118-1/22 $31.00 © 2022 IEEE

silicon wafers compatible with perovskite solution, which improves the tandem cell efficiency to 25.5% [13]. Using methyl-substituted carbazole as a hole material in the PSC to enhance hole's extraction in the perovskite/silicon tandem cell was recorded by Amran Al-Ashour with an efficiency of 29.15% [14]. A new tandem cell using selenium as a top sub-cell (instead of always used perovskite sub-cell) and a copper zinc tin sulfide bottom sub-cell was introduced by H.Ferhati, et al. with an efficiency over 30% [15]. M. Benaicha, et al demonstrate a gallium indium phosphide/silicon tandem cell with an efficiency of 31.1% [16]. A two terminal and three junction GaInP/GaAs/Si tandem cell shows a high efficiency over 33% as demonstrated by Romain Cariou, et al [17]. Stephanie Essig, et al. used the same GaInP/GaAs/Si materials but using GaInP/GaAs 2 junction solar cell stacked on silicon bottom sub-cell with a thicker front TCO, this cell records 35.9% efficiency [18]. A six junctions monolithic tandem cell designed by John F. Geisz, et al. recorded a very high efficiency up to 39.2% [19]. The remainder of this paper is organized as follows; the detailed structure of the sub-cells is introduced in section II, section III shows the results of Perovskite/CIGS-GeTe tandem cell; and finally the conclusion in section IV.

II. SUB-CELLS

A. Perovskite top sub-cell

The MAPbI₃ perovskite (PVK) sub-cell and the energy diagrams of the constituting layers are shown in Fig.1. It consists of copper oxide (Cu₂O) used as a Hole Transport Layer (HTL), PCBM as an Electron Transport Layer (ETL), and the MAPbI₃ as an active layer. MAPbI₃ perovskite sub-cell simulation is carried out in this section. The Cu₂O layer's maximum value is greater than the MAPbI₃ layer's highest occupied molecular orbital energy level (5.5 eV), and the holes at the MAPbI₃/Cu₂O interface are extracted. There is also the fact that the Cu₂O layer conduction edge is 2.60 eV, which is significantly higher than the Lowest Unoccupied Molecular Orbital (LUMO) of MAPbI₃ (−3.9 eV). This successfully prevents electrons from MAPbI3 from reaching the Cu₂O layer. In contrast, the HOMO level of the PCBM layer is lower than the MAPbI₃ layer, and PCBM has a deeper LUMO level compared to MAPbI₃. The holes and electrons can drift to the contacts and make contact because of this arrangement of the used materials. Table I gives the used simulation parameters for the perovskite (MAPbI₃) sub-cell. Different active layer (MAPbI₃) thickness ranges of 50 nm to 1000 nm are used to study the perovskite sub-cell's performance. Figure 2 shows the PCE, V_{OC}, Short Circuit Current Density (J_{SC}), and Fill Factor (FF) with the changes in MAPbI₃ layer thickness.

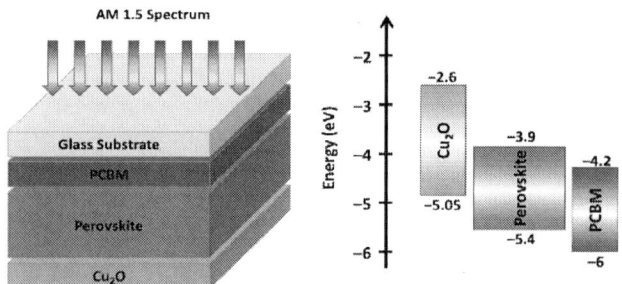

(a)   (b)

Fig. 1. (a) Perovskite sub-cell and (b) the energy diagrams of the constituting layers.

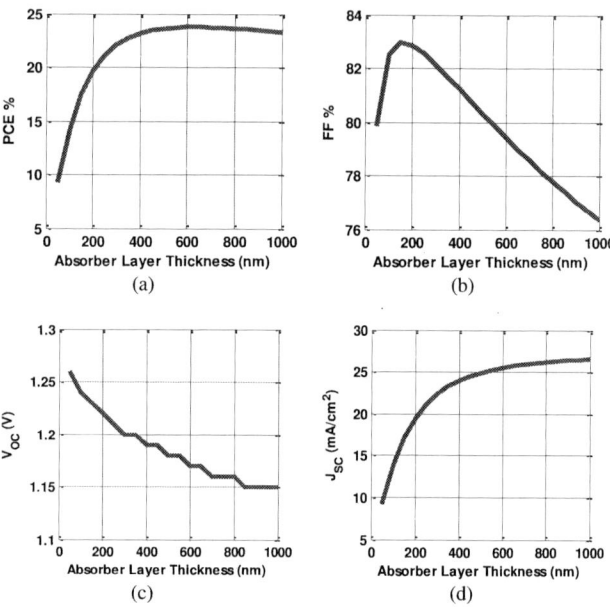

Fig. 2. Perovskite characteristics vs absorber thickness: (a) PCE, (b) FF, (c) V_{OC}, and (d) $J_{SC.}$

TABLE I. SIMULATION PARAMETERS OF PEROVSKITE SUB-CELL

Parameters	Unit	PCBM	MAPbI₃	Cu₂O
E_g	(eV)	1.80	1.50	2.45
Electron Affinity	(eV)	4.2	3.9	2.6
Relative permittivity		3.90	6.50	7.11
Thickness	(nm)	25	(variable)	20
CB Effective density of states	(cm⁻³)	10^{21}	2.75×10^{18}	2.02×10^{17}
μ_e	(cm⁻² V⁻¹s⁻¹)	10^{-3}	10	200
μ_p	(cm⁻² V⁻¹ s⁻¹)	2×10^{-3}	10	80
N_D	(cm⁻³)	10^{18}	10^{13}	0
VB Effective density of states	(cm⁻³)	10^{21}	3.9×10^{18}	1.1×10^{19}
N_A	(cm⁻³)	0	10^{13}	10^{18}

B. CIGS-GeTe bottom sub-cell

One of the possible methods to improve cell efficiency is increasing V_{OC} by increasing the doping level. One of the materials that can be highly doped is the GeTe. On the other hand, this limits the current as the increase in the doping affects the lifetime of the holes and electrons, which in turn decreases the diffusion length. The used doping of GeTe is N_D 7.5×10^{15} cm^{-3} and N_A 7.5×10^{20} cm^{-3}; the other parameters are given in Table II. A proposed solution to this problem is by decreasing the thickness of the GeTe layer and adding a CIGS layer with a thinner thickness before the GeTe layer. This allows increasing the voltage of sub-cells as they work as a series connection. The negative side of this connection is limiting the sub-cell current to the minimum of one of the used sub-layers materials.

Despite GeTe small energy gap that decreases V_{OC}, its highly doped level than CIGS can explain this improvement, where GeTe optimal doping level is in order of 10^{20}. The V_{OC} depends on the difference of the energy between the acceptor material LUMO (E^A_{LUMO}) and the donor material HOMO (E^D_{HOMO}). This difference is called the donor-acceptor effective energy gap (E_{DA}). An empirical value indicating the energy losses incurred while transferring charge carriers to electrodes is the ΔE_{loss}, which has many reasons such as Coulombic interactions, bimolecular recombination, and energetic disorder.

Also, the voltage loss was minimized by increasing the doping level of GeTe layer.

TABLE II.   SIMULATION PARAMETERS OF CIGS-GETE SUB-CELL

Parameters	Unit	CIGS	GeTe	ZnO	CdS
E_g	(eV)	1.10	0.8	3.30	2.45
Electron Affinity	(eV)	4.5	4.8	4.6	4.4
Relative permittivity		13.6	36	9.0	10.0
Thickness	(nm)	500	1500	20	50
CB Effective density of states	(cm^{-3})	2.2 ×10^{18}	10^{16}	2.2×10^{18}	2.2×10^{18}
VB Effective density of states	(cm^{-3})	1.8 ×10^{19}	10^{17}	1.8×10^{19}	1.8×10^{19}
μ_e	(cm^{-2}V^{-1}s^{-1})	100	100	100	100
μ_p	(cm^{-2}V^{-1}s^{-1})	25	20	25	25
N_D	(cm^{-3})	0	7.5×10^{15}	10^{20}	10^{20}
N_A (cm^{-3})		2×10^{16}	7.5×10^{20}	0	0

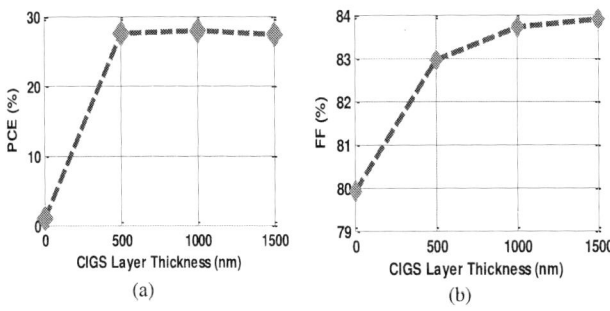

(a)

(b)

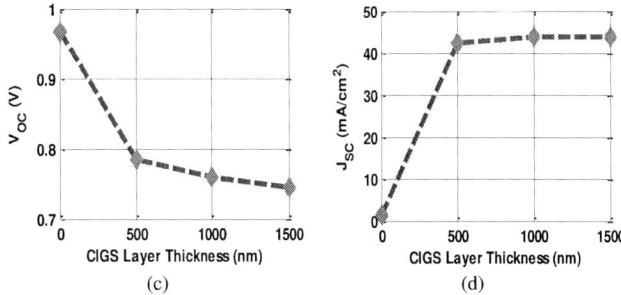

(c)

(d)

Fig. 3. CIGS-GeTe sub-cell performance parameters with different CIGS layer thickness: (a) *PCE*, (b) *FF*, (c) V_{OC}, and (d) $J_{SC.}$

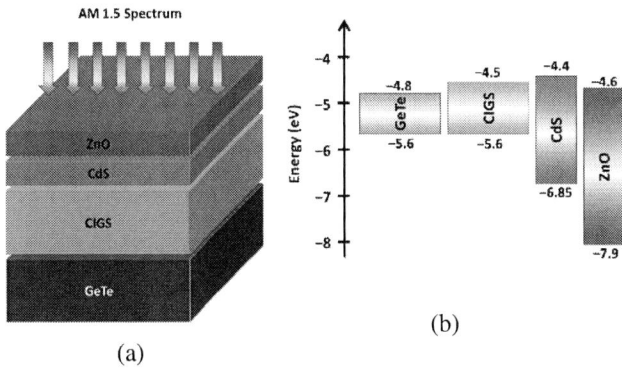

(a)

(b)

Fig. 4. (a) CIGS-GeTe sub-cell and (b) the energy diagrams of the constituting layers.

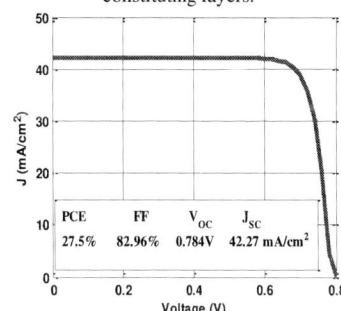

Fig. 5. CIGS-GeTe characteristics *JV* curve

III. PEROVSKITE/CIGS-GETE TANDEM CELL

To find the optimum performance of the designed tandem cell, an algorithm for determining the top sub-cell absorber layer thickness is used [20]. It is found that the optimum thickness of the top sub-cell (perovskite (MAPbI$_3$)) is 230 nm, which gives a PVK/CIGS-GeTe tandem cell efficiency of 33.48% as shown in Fig.6. The tandem cell with incident spectrum is AM 1.5 to the top sub-cell and transmitted spectrum (filtered AM 1.5) from top to bottom sub-cell are shown in Fig.7. Also, the *J/V* curves of the PVK, CIGS-GeTe sub-cells, and PVK/CIGS GeTe tandem cell are shown in Fig.8. The tandem PVK/CIGS-GeTe cell has *PCE* 33.48%, V_{OC} 1.9721 V, J_{SC} 20.50 mA/cm^2,

978-1-7281-6118-1/22 $31.00 © 2022 IEEE   1232

and *FF* 82.81 %. The Quantum Efficiency (QE) of top and bottom sub-cells and tandem cell is shown in Fig.9.

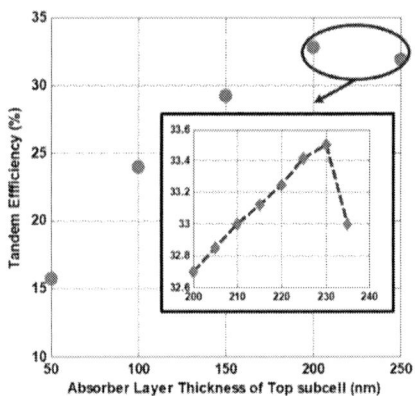

Fig. 6. Tandem cell efficiency of PVK/CIGS-GeTe with a top sub-cell absorber layer thickness.

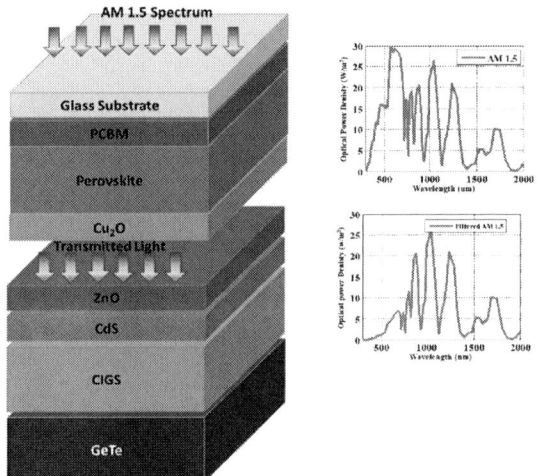

Fig. 7. PVK/CIGS-GeTe tandem solar cell as AM1.5 illuminated on the top sub-cell and filtered AM 1.5 is illuminated on the bottom sub-cell.

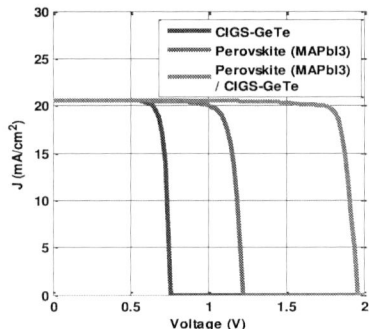

Fig. 8. *JV* curve of PVK/ CIGS-GeTe sub-cells and tandem cell.

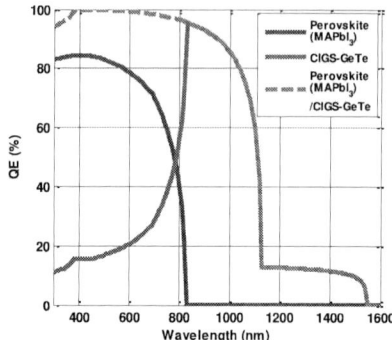

Fig. 9. QE of PVK/ CIGS-GeTe sub-cells and tandem cell

IV. CONCLUSION

The main idea of using CIGS with GeTe as absorbers is to improve the bottom sub-cell's open-circuit voltage as the CIGS and GeTe are equivalent to a series connection, which in turn enhances the tandem cell voltage. Besides the increase in the open-circuit voltage of the sub-cell, the CIGS layer in this equivalent series connection limits the current of the GeTe layer and the overall bottom sub-cell's current. This explains the unusual results of the QE at wavelengths higher than the cutoff wavelength of CIGS. The formation of tandem cell using perovskite (MAPbI$_3$) as a top sub-cell and the CIGS-GeTe as a bottom sub-cell gives an efficiency of 33.48%

REFERENCES

[1] A. Zekry, "A road map for transformation from conventional to photovoltaic energy generation and its challenges." *Journal of King Saud University - Engineering Sciences*, vol. 32, no. 7, pp. 407-410, 2020.

[2] Sahbel, A., Hassan, N., Abdelhameed, M. and Zekry, A., 2013. Experimental Performance Characterization of Photovoltaic Modules Using DAQ. *Energy Procedia*, 36, pp.323-332.

[3] Okil, M., Salem, M., Abdolkader, T. and Shaker, A., 2021. From Crystalline to Low-cost Silicon-based Solar Cells: a Review. Silicon, 14(5), pp.1895-1911.

[4] Yoshikawa, K., Kawasaki, H., Yoshida, W., Irie, T., Konishi, K., Nakano, K., Uto, T., Adachi, D., Kanematsu, M., Uzu, H. and Yamamoto, K., 2017. Silicon heterojunction solar cell with interdigitated back contacts for a photoconversion efficiency over 26%. *Nature Energy*, 2(5).

[5] M. Nakamura, K. Yamaguchi, Y. Kimoto, Y. Yasaki, T. Kato and H. Sugimoto, "Cd-Free Cu(In,Ga)(Se,S)$_2$ Thin-Film Solar Cell With Record Efficiency of 23.35%," in *IEEE Journal of Photovoltaics*, vol. 9, no. 6, pp. 1863-1867, Nov. 2019, doi: 10.1109/JPHOTOV.2019.2937218.

[6] Yang, W., Park, B., Jung, E., Jeon, N., Kim, Y., Lee, D., Shin, S., Seo, J., Kim, E., Noh, J. and Seok, S., 2017. Iodide management in formamidinium-lead-halide–based perovskite layers for efficient solar cells. *Science*, 356(6345), pp.1376-1379.

[7] Salah, M., Abouelatta, M., Shaker, A., Hassan, K. and Saeed, A., 2019. A comprehensive simulation study of hybrid halide perovskite solar cell with copper oxide as HTM. *Semiconductor Science and Technology*, 34(11), p.115009.

[8] W. Shockley and H. J. Queisser, "Detailed balance limit of efficiency of p-n junction solar cells," *J. Appl. Phys.*, vol. 32, no. 3, pp. 510–519, 1961.

[9] J. Werner, "Efficient Monolithic Perovskite/Silicon Tandem Solar Cell with Cell Area >1 cm2," *J. Phys. Chem. Lett*, vol. 7, no. 1, pp. 161–166, 2016

[10] F. Sahli *et al.*, "Improved optics in monolithic perovskite/silicon tandem solar cells with a nanocrystalline silicon recombination junction," *Adv. Energy Mater.*, vol. 8, no. 6, p. 1701609, 2018.

[11] Y. Wu *et al.*, "Monolithic perovskite/silicon-homojunction tandem solar cell with over 22% efficiency," *Energy & Environmental Science*, vol. 10, no. 11, pp. 2472–2479, Nov. 2017, doi: 10.1039/C7EE02288C.

[12] F. Sahli *et al.*, "Fully textured monolithic perovskite/silicon tandem solar cells with 25.2% power conversion efficiency," *Nat. Mater.*, vol. 17, no. 9, pp. 820–826, 2018.

[13] M. Jošt, E. Köhnen, A. B. Morales-Vilches, B. Lipovšek, K. Jäger, B. Macco, A. Al-Ashouri, J. Krč, L. Korte, B. Rech, R. Schlatmann, M. Topič, B. Stannowski, and S. Albrecht, "Textured interfaces in monolithic perovskite/silicon tandem solar cells: Advanced light management for improved efficiency and energy yield," *Energy & Environmental Science*, vol. 11, no. 12, pp. 3511–3523, 2018.

[14] A. Al-Ashouri *et al.*, "Monolithic perovskite/silicon tandem solar cell with> 29% efficiency by enhanced hole extraction," *Science*, vol.

370, no. 6522, pp. 1300–1309, 2020.

[15] H. Ferhati and F. Djeffal, "Exceeding 30 % efficiency for an environment-friendly tandem solar cell based on earth-abundant Se/CZTS materials," *Physica E Low Dimens. Syst. Nanostruct.*, vol. 109, pp. 52–58, 2019.

[16] M. Benaicha, L. Dehimi, F. Pezzimenti, and F. Bouzid, "Simulation analysis of a high efficiency GaInP/Si multijunction solar cell," *J. Semicond.*, vol. 41, no. 3, p. 032701, 2020.

[17] R. Cariou, "III–V-on-silicon solar cells reaching 33% photoconversion efficiency in two-terminal configuration." *Nature Energy*, vol. 3, no. 4, pp. 326-333, 2018, doi: 10.1038/s41560-018-0125-0.

[18] S. Essig, "Raising the one-sun conversion efficiency of III–V/Si solar cells to 32.8% for two junctions and 35.9% for three junctions." *Nature Energy*, vol. 2, no. 9, 2017, doi: 10.1038/nenergy.2017.144.

[19] J. F. Geisz, "Building a Six-Junction Inverted Metamorphic Concentrator Solar Cell." *IEEE Journal of Photovoltaics*, vol. 8, no. 2, pp. 626-632, 2018, doi: 10.1109/jphotov.2017.2778567.

[20] M. Mousa, F. Z. Amer, R. I. Mubarak and A. Saeed, "Simulation of Optimized High-Current Tandem Solar-Cells With Efficiency Beyond 41%," in *IEEE Access*, vol. 9, pp. 49724-49737, 2021, doi: 10.1109/ACCESS.2021.3069281.

Unlocking 1550 nm Laser Power Conversion by InGaAs Single- and Multi-Junction PV Cells

Henning Helmers, Oliver Höhn, Thomas Tibbits, Meike Schauerte, H. M. Noman Amin, David Lackner

Fraunhofer Institute for Solar Energy Systems ISE, Freiburg, Germany

Photovoltaic cells based on In0.53Ga0.47As absorber material grown lattice-matched on InP substrates are developed, targeting laser power conversion around 1550 nm in the optical C-band. Single- and multi-junction devices are fabricated and characterized. In addition, a thin film route which allows for the implementation of a back surface reflector is developed based on wet-chemical etching of the InP substrate. The influence of the junction type (homo-, front-hetero-, rear-heterojunction) on luminescence coupling in 2-junction devices is explored. Also, a first 10-junction device is investigated. First experimental results will be presented in this conference contribution.

Estimation and Degradation Analysis of Physics-based Circuit Parameters for PV Systems Using Only DC Operation and Weather Data

Baojie LI, Xin CHEN, Todd KARIN, Anubhav JAIN

Energy Technologies Area, Lawrence Berkeley National Laboratory, Berkeley, CA, United States

PV Evolution Labs (PVEL), Napa, CA, United States

Physics-based circuit parameters like series and shunt resistance are also essential to provide insights into the degradation modes of PV arrays. However, the calculation of these parameters typically relies on I-V curves, which require specific measurement devices and may impede the regular operation of PV systems. Although a reference module equipped with I-V tracers may be installed next to the array, the recorded I-V curves cannot fully represent the array condition. Therefore, I-V curves of the entire array are not always available to obtain the circuit parameters, especially for large-scale PV power plants. Therefore, this paper proposes a methodology (PVPRO) to estimate these parameters for degradation analysis without the need for I-V tracers. Rather, it estimates these parameters using only operation (DC voltage and current) and weather data (irradiance and module temperature). PVPRO first performs a multi-stage data cleaning and filtering. Next, the time-series DC data are clipped into windows to fit the equivalent single-diode model to estimate the circuit parameters on minimizing the loss between real and estimated values. Finally, the estimated parameters are plotted as function of time to quantitatively analyze the degradation trends. PVPRO is evaluated on synthetic datasets in the presence of noise and is applied to a well-qualified field system (2015-2019). The degradation trend by PVPRO is in reasonable agreement with the ground truth, especially for the seasonal variation (correlation coefficient of 0.71). Future work will apply PVPRO to more large-scale PV systems and deconvolve the degradation pathways. The current version of PVPRO is available on Github: https://github.com/DuraMAT/pvpro

The effect of moisture ingress on titania antireflection coatings in field-aged photovoltaic modules

Oscar Kwame Segbefia[1], Naureen Akhtar[1], Tor Oskar Sætre[1]

[1]Department of Engineering Sciences, University of Agder, 4879 Grimstad, Norway

Abstract—Titanium dioxide (TiO_2) or titania antireflection coating (ARC) enhances photovoltaic (PV) module efficiency. Yet, degraded TiO_2 can affect the performance reliability of PV modules. In the present work, the effect of moisture ingress on the degradation of TiO_2 ARC in a field-aged multicrystalline silicon PV module is investigated. Scanning electron microscopy (SEM) and energy dispersive X-ray spectroscopy (EDS) analyses show degradation of the TiO_2 ARC. Disintegration of the TiO_2 nanoparticles (NPs) were also observed. The assumed Ti-O stoichiometry of the degraded TiO_2 ARC in the field-aged PV module was found to be higher than 1:2. It turned out that moisture ingress strongly influences the surface morphology and defects, crystallinity, and stoichiometry of TiO_2 ARC in the PV module during field operation. Silver and aluminium NPs migrate to and aggregate on the surfaces of the TiO_2 NPs which might likely lead to the formation of titania-metal complexes such as titania-alumina and silver-titania complexes. These degradation mechanisms affect the opto-electrical properties of the TiO_2 ARC in the PV module.

Keywords— titanium dioxide, moisture ingress, degradation, morphology, titania-metal complex

I. INTRODUCTION

Crystalline silicon (c-Si) photovoltaic (PV) modules are made up of a framework of solar cells, front glass, antireflection coating, polymeric front encapsulation (e.g., ethylene vinyl acetate, EVA), backsheet (polymeric/glass), aluminum frame, and junction boxes [1, 2]. The solar cells convert sunlight to clean electricity via the photovoltaic effect. The other components play complementary roles to enhance performance reliability, and hence, optimize the PV module's efficiency [2, 3]. The presence of moisture within the PV module initiates different electrochemical reaction pathways [3]. This leads to different defects and fault mechanisms [4]. Some of these defects and fault modes are cracks, corrosion, optical degradation, and potential induced degradation [5, 6]. This leads to drop in power output of the PV modules [4-6].

In spite of the various degradation mechanisms, a major problem for low efficiency in c-Si solar cells is the high surface reflectivity of the polished silicon substrate [7-10]. Antireflection coating (ARC) improves the fraction of incident light reflection and coupling of the absorbed photon within the PV module. Hence, ARC is used in commercial PV modules to improve the device efficiency [11, 12]. Titanium dioxide or titania (TiO_2) ARC, in particular, is a well-established technology in the PV industry [7] and the most popular ARC used in photovoltaic devices [9, 13]. Stoichiometric TiO_2 films exhibit optimum refractive index, low absorption coefficient, excellent thermal and chemical stability, and is cost-effective [7]. However, it is known that the properties of the TiO_2 film depend on the sintering and deposition conditions [7, 12, 14, 15]. For instance, diffused moisture in porous TiO_2 films during deposition influences the refractive index [13, 16, 17].

It is also believed that TiO_2 possesses superior photocatalytic self-cleaning property under ultraviolet (UV) light ($\lambda < 400$ nm). This property could be enhanced by doping TiO_2 with noble metals with its associated cost constraints [18]. The excellent photoactivity of TiO_2 can also lead to the formation of free radicals within the PV module bulk, hence, accelerated degradation of the polymeric encapsulation [19]. Moreover, the main limitation of TiO_2 ARC is that it possesses poor passivation characteristics [13]. Deficient passivation characteristics means degradation of TiO_2 ARC under UV radiation [20]. Besides, the morphology, surface defects, and presence of moisture can affect the efficiency of the TiO_2 ARC [21]. When moisture enters the PV module during operation in the field, degradation of the TiO_2 film is likely. In this case, the opto-electrical efficiency could also be compromised. Yet, we have not found any publication on the effect of moisture on the TiO_2 ARC in PV modules in the field.

In the present work, the effect of moisture on the degradation of the TiO_2 ARC in a field-aged multicrystalline silicon (mc-Si) PV module is investigated using scanning electron microscopy (SEM) and energy dispersive spectroscopy (EDS) techniques. Section 2 presents the experimental methods used for the investigation. This is followed by the discussion of the results on the effects of moisture on the TiO_2 ARC in Section 3.

II. MATERIAL AND METHODS

A field-aged PV module from a group of PV modules which were installed outdoors in Dømmesmoen, Grimstad (58.3° N, 8.59° E), Norway between the year 2000 and 2011 was chosen for this investigation [22]. The manufacture's data sheet and the measured average electrical data of the chosen field-aged PV module is presented in Table 1. The parameters measured were the maximum power (P_{max}), open circuit and maximum power point voltage (V_{oc}, V_{mpp}), short circuit and maximum power point current (I_{sc}, I_{mpp}), and fill factor (FF), The values in Table 1 are normalized to Standard Test Conditions (STC).

The PV module was manufactured using anodized aluminium (Al-) frame, low iron tempered front glass, ethylene vinyl acetate (EVA) encapsulation, white multi-layered Tedlar®/Polyester/Tedlar® (TPT) backsheet, and 2 weatherproof plastic casing junction boxes (accommodating a bypass diode each). The screen-printed multicrystalline solar cells with dimensions of 100 x 100 mm^2 were made using a TiO_2 ARC and tinned copper (Cu) busbars. The PV module is an assembly of (12 x 2) series connected solar cells in 3 substrings. The cells were cut out from the panel using a water jet cutting machine and the cells were extracted by treating them with toluene.

A. I-V measurements

The selected field-aged PV module was taken through current-voltage (I-V) measurements using a handheld I-V 500w I-V Curve Tracer, following the IEC 60904- 1 standard. These measurements provided information on the electrical characteristics, irradiance, and module temperature characteristics at Standard Test Conditions (STC). STC specifies cell temperature of 25 °C, an irradiance of 1000 W/m^2 and air mass 1.5 (AM1. 5) spectrum for commercial PV modules. Measurements were done under clear sky in-plane irradiance (GI) (960 - 1060 W/m^2) conditions, and the I-V tracer used converted all measurements to STC automatically. This

TABLE I. AVERAGE ELECTRICAL PARAMETERS OF THE PV MODULE

Year	P_{max} (W)	V_{oc} (V)	V_{mpp} (V)	I_{mpp} (A)	I_{sc} (A)	FF (%)	η (%)
2000	100	21.6	16.7	6.0	6.7	70	13
2020	76.0	19.8	14.4	5.3	6.0	64	10

means the operating conditions were optimally resolved by the device to minimize errors in measuring and recording data.

B. Microstructural Investigation

The as-cut solar cells (which comprised of the front glass, EVA, and backsheet) was immersed in toluene to extract the solar cells. The reclamation of the solar cells took 14 days under room temperature conditions. Finally, the extracted solar cell samples were investigated using SEM-EDS techniques to identify the effect of moisture on the TiO_2 ARC. The SEM-EDS analysis was done utilizing a field emission scanning electron microscope (SEM) (JEOL 7200F) equipped with an energy dispersive X-ray spectrometer (Octane Elect EDS system from EDAX®-AMETEK®). An overview of the experimental method is shown in Fig. 1.

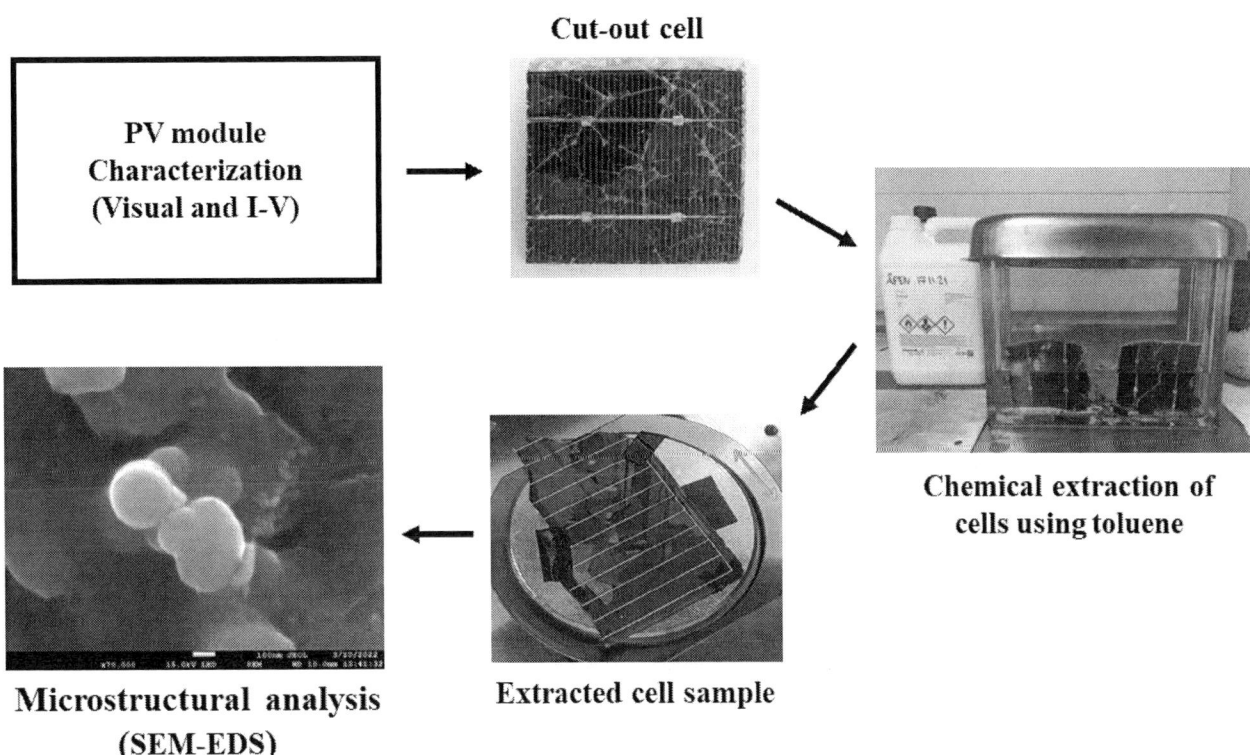

Fig. 1. Overview of the experimental method.

III. RESULTS AND DISCUSSION

A. Visual inspection

Fig. 2 shows the results from the visual inspection of the field-aged PV module. Figs. 2a and 2b show signs of moisture ingress from the edges of the modules. Aside from influencing the optical and electrical efficiency of the ARC, the ingressed moisture can cause the production of acetic acid in the presence of ethylene vinyl acetate (EVA) encapsulant , and the presence of acidic species catalyzes TiO_2 degradation [19]. This can initiate several degradation mechanisms within the PV module [2, 3].

In Figs. 2a and 2b, there are clear indications of encapsulant discoloration, metal grid corrosion, and solar cell degradation. The front and the rear sides of the extracted cell after toluene treatments are shown in Fig. 2c and Fig. 2d, respectively. Some parts of the Cu busbars and the silver fingers appear to be darker than other parts of the same solar cell. It appears that the Cu busbars and the silver fingers have undergone some degradation. However, the focus of this paper is on the TiO_2 ARC. The degradation of Cu and other components of the solar cells is the subject of future investigation.

B. SEM-EDS characteristics

To investigate the effect of moisture on the degradation of the ARC, the extracted solar cells were taken through SEM-EDS analysis. Fig. 3 shows the SEM-EDS analyses from the region about 10 mm from the edge of a solar cell extracted from the field-aged PV module. The SEM micrograph in Fig. 3a and its corresponding EDS analyses is shown in Figs. 3b - d. The elemental mappings for oxygen (O) and titanium (Ti) of the SEM micrograph as displayed in Fig. 3a are shown in Figs. 3c and 3d, respectively. Highlighted in yellow circle, nanoparticles (NPs) of TiO_2 ARC, that appear bright in the SEM micrographs are randomly distributed on the surface of the solar cells, see Fig. 3a. The assumed titanium-oxygen (Ti-O) stoichiometry of TiO_2 is 1:2.

Fig. 3b represents the EDS elemental composition of the TiO_2 ARC in the field-aged PV module investigated. Obviously, silicon (Si) assumed the greatest abundance, as the TiO_2 ARC were deposited on multicrystalline silicon solar cells. The amount of oxygen in Fig. 3b suggests that the effect of moisture ingress on the solar cells is likely. The stoichiometry of Ti and O also confirms that the stoichiometric TiO_2 ARC has undergone oxidation. This observation agrees with other reports on the effect of moisture on TiO_2 [14, 23]. The presence of phosphorus (P) is thought to be from the degradation of the EVA encapsulant which contains phosphate stabilizing antioxidants [24].

The Ti-O elemental maps of oxygen (Fig. 3c) and titanium (Fig.3d) confirm that O and Ti are located along the same areas on the solar cells. It also shows the relative positions of O and TiO_2 NPs around the edges of the solar cell's microcrystals. This suggests that moisture induced degradation starts from the edges of the solar cells' crystals.

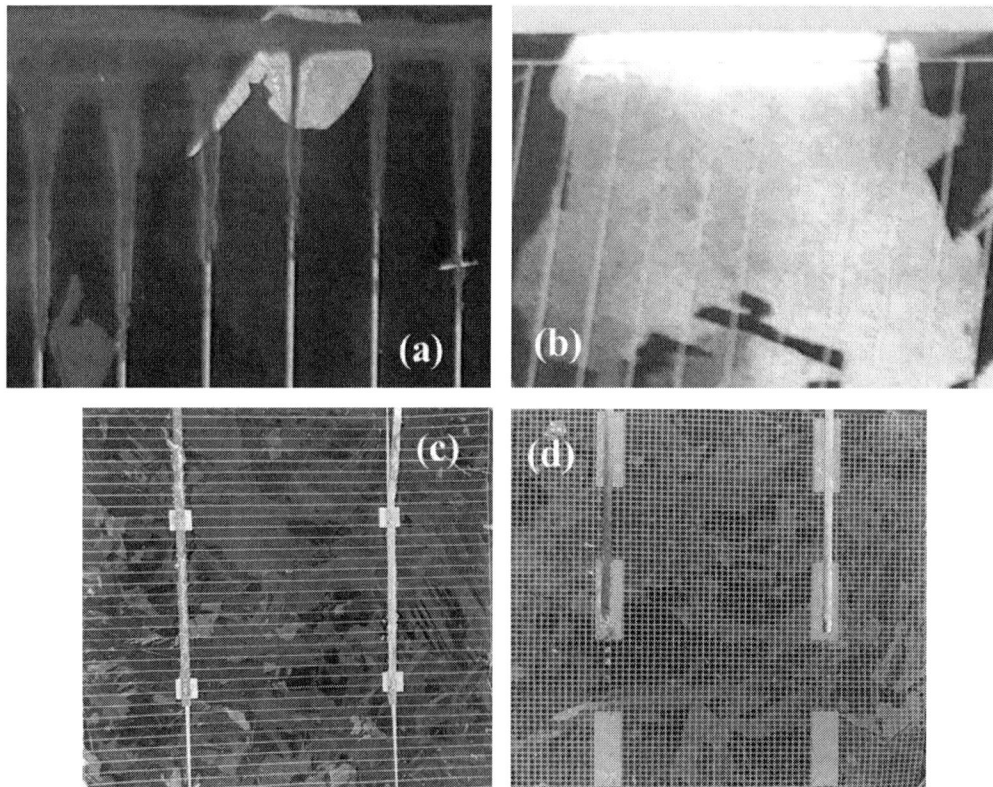

Fig. 2. Visual images of the field-aged PV module showing (a) - (b) signs of moisture ingress from the edges and the (c) front and (d) rear sides of the extracted solar cell.

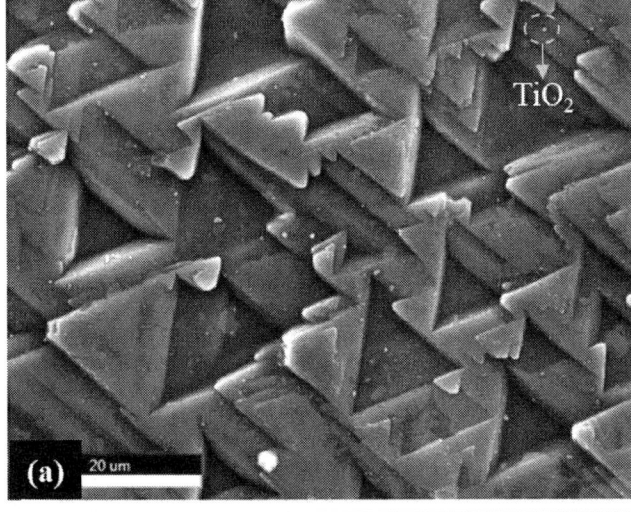

Element	Atomic %
O	11.6
Si	85.8
P	0.4
Ti	2.1

(b)

Fig. 3. (a) SEM micrograph and (b) EDS analysis and elemental mapping of (c) oxygen and (d) titanium of a solar cell extracted from the field-aged PV module. The data were taken from the middle of the solar cell, about 10 mm from the edge of the solar cell.

Fig. 4 shows the SEM-EDS analyses of a solar cell extracted from the edge of the field-aged PV module. Fig. 4a is the SEM micrograph with the corresponding EDS analyses in Figs. 4b - 4d. Figs. 4b and 4c show the relative positions of the TiO_2 NPs on the solar cell. It is evident that most of the oxygen content is associated with these particles. Also, it is clear from Fig. 4d that the dark spots correspond to the relative positions of the oxidized TiO_2 ARC NPs. This further suggests that these NPs are just not oxidized, but it is likely that the observed NPs form titania complexes with other elements. The EDS spectra and analysis of the SEM micrograph in Fig. 4a is shown in Figs. 5a and 5b, respectively. The Ti to O ratio as deduced from the quantification results shown in Fig. 5b is approximately 1:6, which is similar to what was observed in Fig. 3b. This suggests that the degradation of TiO_2 ARC appears to follow a general trend across the solar cells.

Leaching of aluminium (Al) from rear contacts of the solar cells under the influence of moisture ingress is also evident, see Figs. 5a and 5b. This agrees with other reports [25]. In Fig. 5a, the vertical and horizontal axes represent X-ray counts and electron energy (keV), respectively. Si shows the highest intensity. The EDS spectra in Fig. 5a corroborates with the elemental composition analysis in Fig. 5b. The higher amount of oxygen observed on the solar cells extracted from the edge of the PV module suggests that the influence of moisture ingress near the edges of the PV module is greater. Fig. 6 shows the high magnification SEM micrographs of some of the NPs of the TiO_2 ARC in their degraded forms. The surfaces of these NPs have been modified significantly. Fig. 7 shows the SEM-EDS analyses of some of the NPs of the TiO_2 ARC.

The presence of moisture can influence the Ti-O bond by introducing extra oxygen vacancies in the TiO_2 ARC [14]. In acetic acid environments, the morphology of TiO_2 NPs can change and become more porous, as observed elsewhere [26]. That is, degradation of the EVA encapsulant in the presence of moisture can further accelerate the degradation of the TiO_2 ARC. Another possible explanation for the observed degradation of the ARC might be the synergetic effect of UV radiation and moisture [23].

978-1-7281-6118-1/22 $31.00 © 2022 IEEE

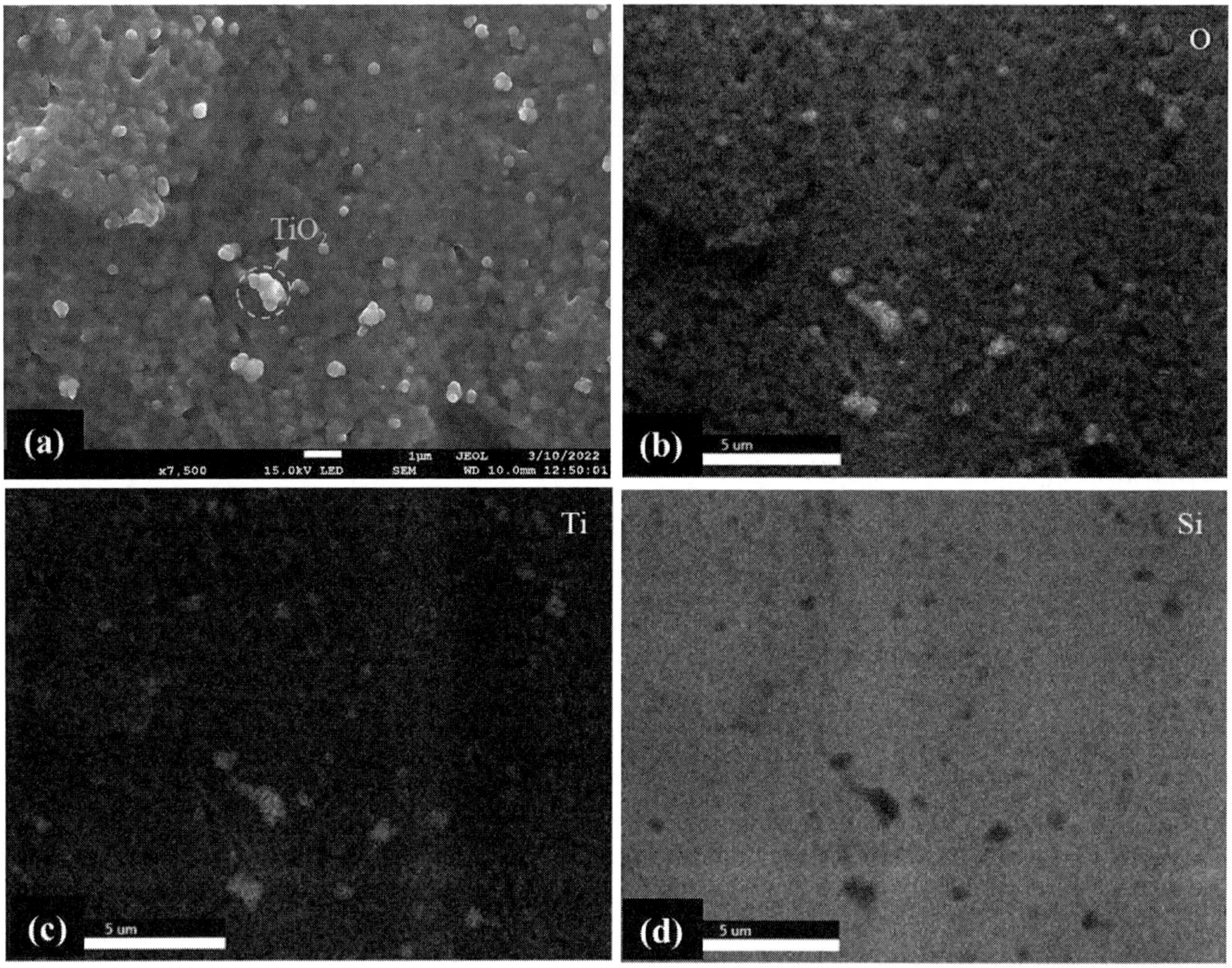

Fig. 4. (a) SEM micrograph and the corresponding EDS elemental mappings of (b) oxygen, (c) titanium, and (d) silicon of a solar cell extracted from the edge of the field-aged PV module.

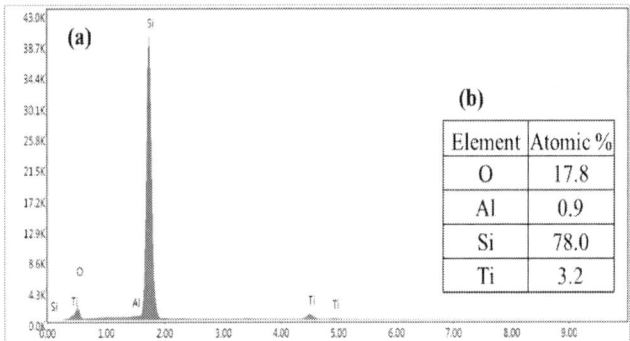

Element	Atomic %
O	17.8
Al	0.9
Si	78.0
Ti	3.2

Fig. 5. (a) EDS spectra and (b) EDS analysis of the SEM micrograph in Fig. 4a.

Fig. 7c displays the EDS analysis of Fig. 7a and the EDS point analysis at Point 1 in Fig. 7b is shown in Fig. 7d. As expected, the observed Ti-O stoichiometry of TiO_2 ARC is significantly higher, which strongly suggests moisture induced degradation. Moreover, the morphologies of the two main

phases of stoichiometric TiO_2: anatase and rutile; that are used in PV applications are different from what was observed in Figs. 6 and 7. Both phases of TiO_2 are more efficient in the crystalline phase [15, 23, 26]. The structural features of the TiO_2 ARC NPs are also modified substantially, see Fig. 7. The surface morphology and defects, crystallinity, and stoichiometry are known to affect the efficiency of TiO_2 NPs [15, 21]. Figs. 6 and 7 suggest that moisture ingress is the underlying factor for the observed disintegration of the TiO_2 NPs. In Fig. 7b, NPs of Ag (from the silver paste) were found to migrate to and aggregate on the surfaces of the TiO_2 ARC NPs under the influence of moisture ingress. This can lead to the formation of silver-titania (Ag-TiO_2) metal complexes. Other metal ions (e.g., lead, tin, sodium, and zinc) from the silver paste and solder are also capable of migrating to the surface of the TiO_2 ARC NPs under the influence of moisture ingress [27-29]. Moisture assisted migration of metal ions to the surface of the TiO_2 ARC can cause potential induced degradation [30].

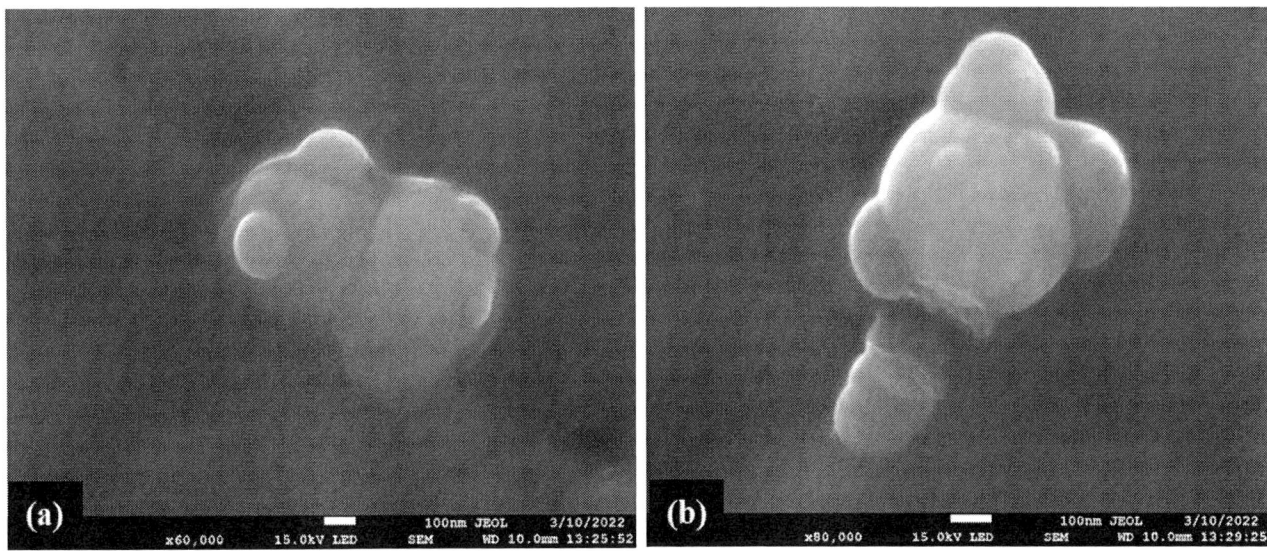

Fig. 6. Degraded TiO$_2$ ARC nanoparticles in the field-aged PV module.

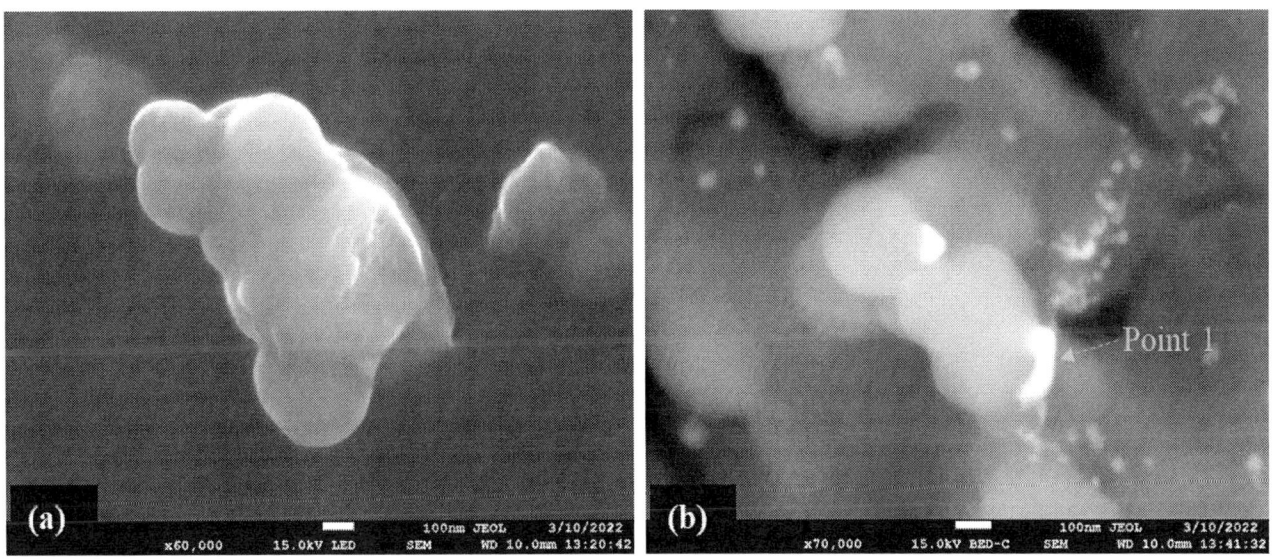

(c)

Element	Atomic %
O	58.2
Si	30.1
Ti	11.6

(d)

Element	Atomic %
O	35.2
Al	0.9
Si	55.2
Ag	2.1
Ti	6.6

Fig. 7. (a) - (b) High magnification SEM micrographs of TiO$_2$ ARC NPs in Fig. 4a and corresponding EDS (c) full area analysis of Fig. 7a and (d) point analysis at Point 1 in Fig. 7b.

One interesting observation from Figs. 5b and 7d is that the amount of Al (0.9 %) is the same. This suggests the formation of titania-alumina (Ti-O-Al) complexes under the influence of moisture ingress. The formation of these metal complexes can affect opto-electrical properties of the TiO_2 ARC [31]. That is, refractive index, scattering properties, and extinction coefficient of the TiO_2 ARC during field operation, are strongly influenced by moisture ingress.

IV. CONCLUSION

Titania (TiO_2) antireflection coatings play a vital role in enhancing PV module efficiency. However, in degraded form, TiO_2 can affect the performance reliability of PV modules. In the present work, the effect of moisture ingress on the degradation of TiO_2 ARC in a 20-year-old field-aged PV module is investigated. Visual inspection suggests the incidence of moisture ingress in the field-aged PV module. The SEM-EDS analyses show that moisture ingress leads to the degradation of the TiO_2 ARC. The assumed Ti-O stoichiometry of the oxidized TiO_2 ARC in the module was found to be higher than 1:2. Also, the surface morphology of the TiO_2 ARC NPs appeared to be modified under the influence of moisture. Finally, Ag and Al-NPs were observed to migrate to and aggregate on the surfaces of the TiO_2 ARC NPs which can induce the formation of titania-metal complexes e.g., titania-alumina and silver-titania complexes. These metal complexes can influence the opto-electrical properties of the TiO_2 ARC, hence, the performance reliability of PV modules. Specifically, moisture ingress strongly influences the refractive index, scattering properties, and extinction coefficient of the TiO_2 ARC in PV modules during field operation. This could be an important cause of the observed 1.2 %/year degradation in the P_{max} of the field-aged PV module, refer to Table I.

REFERENCES

[1] M. C. C. de Oliveira, A. S. A. D. Cardoso, M. M. Viana, and V. d. F. C. Lins, "The causes and effects of degradation of encapsulant ethylene vinyl acetate copolymer (EVA) in crystalline silicon photovoltaic modules: A review," Renew. Sust. Energ. Rev., vol. 81, pp. 2299-2317, 2018, doi: https://doi.org/10.1016/j.rser.2017.06.039.

[2] M. Köntges et al., "Review of failures of photovoltaic modules," 2014.

[3] O. K. Segbefia, A. G. Imenes, and T. O. Saetre, "Moisture ingress in photovoltaic modules: A review," (in English), Sol. Energy, vol. 224, pp. 889-906, Aug 2021, doi: 10.1016/j.solener.2021.06.055.

[4] O. K. Segbefia, A. G. Imenes, and T. O. Saetre, "Outdoor Fault Diagnosis of Field-Aged Multicrystalline Silicon Solar Modules," in 37th EU PVSEC, 2020, 2020.

[5] O. K. Segbefia, A. G. Imenes, I. Burud, and T. O. Sætre, "Temperature profiles of field-aged multicrystalline silicon photovoltaic modules affected by microcracks," in 2021 IEEE 48th Photovoltaic Specialists Conference (PVSC), 2021: IEEE, pp. 0001-0006, doi: 10.1109/PVSC43889.2021.9518939.

[6] O. K. Segbefia and T. O. Sætre, "Investigation of the Temperature Sensitivity of 20-Years Old Field-Aged Photovoltaic Panels Affected by Potential Induced Degradation," Energies, vol. 15, no. 11, p. 3865, 2022. [Online]. Available: https://www.mdpi.com/1996-1073/15/11/3865.

[7] B. S. Richards, "Single-material TiO2 double-layer antireflection coatings," (in English), Sol. Energy Mater. Sol. Cells, vol. 79, no. 3, pp. 369-390, Sep 15 2003, doi: 10.1016/S0927-0248(02)00473-7.

[8] D. Bouhafs, A. Moussi, A. Chikouche, and J. M. Ruiz, "Design and simulation of antireflection coating systems for optoelectronic devices: Application to silicon solar cells," (in English), Sol. Energy Mater. Sol. Cells, vol. 52, no. 1-2, pp. 79-93, Mar 16 1998, doi: Doi 10.1016/S0927-0248(97)00273-0.

[9] S. Chhajed, M. F. Schubert, J. K. Kim, and E. F. Schubert, "Nanostructured multilayer graded-index antireflection coating for Si solar cells with broadband and omnidirectional characteristics," (in English), Appl. Phys. Lett., vol. 93, no. 25, p. 251108, Dec 22 2008, doi: Artn 25110810.1063/1.3050463.

[10] W. Kern and E. Tracy, "Titanium-Dioxide Antireflection Coating for Silicon Solar-Cells by Spray Deposition," (in English), Rca Review, vol. 41, no. 2, pp. 133-180, 1980.

[11] A. S. Sarkın, N. Ekren, and Ş. Sağlam, "A review of anti-reflection and self-cleaning coatings on photovoltaic panels," Sol. Energy, vol. 199, pp. 63-73, 2020, doi: https://doi.org/10.1016/j.solener.2020.01.084.

[12] M. Murozono, S. Kitamura, T. Ohmura, K. Kusao, and Y. Umeo, "Titanium dioxide antireflective coating for silicon solar cells by spinning technique," Jpn. J. Appl. Phys., vol. 21, no. S2, p. 137, 1982, doi: https://doi.org/10.7567/JJAPS.21S2.137.

[13] N. Shanmugam, R. Pugazhendhi, R. M. Elavarasan, P. Kasiviswanathan, and N. Das, "Anti-Reflective Coating Materials: A Holistic Review from PV Perspective," (in English), Energies, vol. 13, no. 10, p. 2631, May 2020, doi: https://doi.org/10.3390/en13102631.

[14] T. Fuyuki, T. Kobayashi, and H. Matsunami, "Effects of Small Amount of Water on Physical and Electrical-Properties of Tio2 Films Deposited by Cvd Method," (in English), J. Electrochem. Soc., vol. 135, no. 1, pp. 248-250, Jan 1988, doi: Doi 10.1149/1.2095566.

[15] Y. Leprince-Wang and K. Yu-Zhang, "Study of the growth morphology of TiO2 thin films by AFM and TEM," (in English), Surface & Coatings Technology, vol. 140, no. 2, pp. 155-160, May 30 2001, doi: Doi 10.1016/S0257-8972(01)01029-5.

[16] V. N. Van et al., "Growth of Low and High Refractive-Index Dielectric Layers as Studied by in-Situ Ellipsometry," (in English), Thin Solid Films, vol. 253, no. 1-2, pp. 257-261, Dec 15 1994, doi: https://doi.org/10.1016/0040-6090(94)90331-X.

[17] J. P. Borgogno et al., "Refractive index and inhomogeneity of thin films," Appl Opt, vol. 23, no. 20, p. 3567, Oct 15 1984, doi: 10.1364/ao.23.003567.

[18] M. S. Mozumder, A. H. I. Mourad, H. Pervez, and R. Surkatti, "Recent developments in multifunctional coatings for solar panel applications: A review," (in English), Sol. Energy Mater. Sol. Cells, vol. 189, pp. 75-102, Jan 2019, doi: 10.1016/j.solmat.2018.09.015.

[19] S. L. Cashmore, A. J. Robinson, and D. A. Worsley, "The Effect Of Humidity On Titanium Dioxide Photocatalysed PVC Degradation," (in English), Coatings for Corrosion Protection, vol. 25, no. 29, pp. 95-104, 2010, doi: 10.1149/1.3327228.

[20] A. G. Aberle, "Surface passivation of crystalline silicon solar cells: A review," (in English), Prog Photovoltaics, vol. 8, no. 5, pp. 473-487, Sep-Oct 2000, doi: Doi 10.1002/1099-159x(200009/10)8:5<473::Aid-Pip337>3.3.Co;2-4.

[21] T. Luttrell, S. Halpegamage, J. Tao, A. Kramer, E. Sutter, and M. Batzill, "Why is anatase a better photocatalyst than rutile?-Model studies on epitaxial TiO2 films," Scientific reports, vol. 4, no. 1, pp. 1-8, 2014, doi: https://doi.org/10.1038/srep04043.

[22] O. K. Segbefia, A. G. Imenes, I. Burud, and T. O. Sætre, "Temperature profiles of field-aged photovoltaic modules affected by optical degradation. Manuscript submitted for publication. Department of Engineering Sciences, University of Agder.," Sol. Energy Mater. Sol. Cells, 2022.

[23] P. C. Ricci et al., "Anatase-to-rutile phase transition in TiO2 nanoparticles irradiated by visible light," (in English), J. Phys. Chem. C., vol. 117, no. 15, pp. 7850-7857, Apr 18 2013, doi: 10.1021/jp312325h.

[24] I. Duerr, J. Bierbaum, J. Metzger, J. Richter, and D. Philipp, "Silver grid finger corrosion on snail track affected PV modules–investigation on degradation products and mechanisms," Energy Procedia, vol. 98, pp. 74-85, 2016.

[25] N. Kyranaki, J. Zhu, R. Gottschalg, and T. Betts, "Investigating the degradation of front and rear sides of c-Si PV cells exposed to acetic acid," 2018.

[26] X. Zhang, X. Ge, and C. Wang, "Synthesis of titania in ethanol/acetic acid mixture solvents: phase and morphology variations," Crystal growth & design, vol. 9, no. 10, pp. 4301-4307, 2009.

[27] S. Kumar, R. Meena, and R. Gupta, "Imaging and micro-structural characterization of moisture induced degradation in crystalline silicon photovoltaic modules," (in English), Sol. Energy, vol. 194, pp. 903-912, Dec 2019, doi: 10.1016/j.solener.2019.11.037.

[28] T. H. Kim, N. C. Park, and D. H. Kim, "The effect of moisture on the degradation mechanism of multi-crystalline silicon photovoltaic module," (in English), Microelectronics Reliability, vol. 53, no. 9-11, pp. 1823-1827, Sep-Nov 2013, doi: 10.1016/j.microrel.2013.07.047.

[29] N. Park, C. Han, and D. Kim, "Effect of moisture condensation on long-term reliability of crystalline silicon photovoltaic modules," (in English), Microelectronics Reliability, vol. 53, no. 12, pp. 1922-1926, Dec 2013, doi: 10.1016/j.microrel.2013.05.004.

[30] J. Bauer, V. Naumann, S. Großer, C. Hagendorf, M. Schütze, and O. Breitenstein, "On the mechanism of potential‐induced degradation in crystalline silicon solar cells," physica status solidi (RRL)‐Rapid Research Letters, vol. 6, no. 8, pp. 331-333, 2012.

[31] M. Waleczek et al., "Influence of Alumina Addition on the Optical Properties and the Thermal Stability of Titania Thin Films and Inverse Opals Produced by Atomic Layer Deposition," Nanomaterials (Basel), vol. 11, no. 4, p. 1053, Apr 20 2021, doi: 10.3390/nano11041053.

Improved STC and energy yield performance of bifacial modules with white-grid rear reflectors

Robert Witteck[1], Michael Siebert[1], Tobias Wietler[1], Marc Köntges[1], Paulius Laurikenas[2], and Julius Denafas[2]

1. Institute for Solar Energy Research Hamelin (ISFH), Am Ohrberg 1, 31860 Emmerthal, Germany
2. SoliTek R&D, Mokslininkų st. 6A, LT-08412 Vilnius, Lithuania

Abstract— **We investigate the performance of bifacial white-grid solar modules using a white reflector on the module rear glass along the cell gaps. Considering the optical and long-term reliability properties of different colors, we determine the optimal geometry of the white reflector in ray tracing simulation. Based on these results we build test-modules and compare their performance under standard testing conditions as well as the annual energy yield with bifacial and monofacial references. White-grid modules outperform the bifacial references under single side indoor-measurements achieving a 1.6% higher module power output. Moreover, they improve the annual energy yield by 1.3% for a location in Hamelin. Comparing the performance gain for varying irradiance conditions indicates that white-grid modules are especially advantageous for locations with high fractions of direct sun light or the application on trackers.**

Keywords—energy yield, bifacial module performance

I. INTRODUCTION

Bifacial modules usually achieve higher energy yields compared to monofacial modules due to the additional use of rear light [1]–[3]. Yet, bifacial modules exhibit higher cell-to-module losses compared to monofacial modules with a highly reflective white backsheet when measured under standard testing conditions (STC). In a monofacial module, light impinging on the white backsheet in the cell gaps may be reflected onto the cell due to total internal reflection. The application of a white reflector in the cell gaps can combine both advantages of mono and bifacial modules, as it maintains the bifaciality of the module while reflecting light incident on the cell gaps. In this work, we refer to this type of module as a white-grid module.

Van Aken *et al.* reported a 2.2% higher power due to an increased module current and an enhanced energy yield of 10% on a summer day for white-grid modules compared to typical bifacial modules without a reflector in the cell gaps [4]. Saw *et al.* evaluated various coatings for the cell gap reflector, resulting in an optical gain of 3% for a white-grid module [5]. Sng *et al.* developed an analytical model to simulate the effect of varying angle of incidents and an annual energy yield gain of 2.2% for white-grid modules [6], [7]. Jang *et al.* developed a detailed analytical model and simulated the power gain for various reflector sizes, bifacial factors and angles of incidents resulting in 1.6% to 1.8% power gain for white-grid modules employing cells with 65% to 100% bifaciality [8].

The listed works usually neglect the influence of different optical color properties and the long-term stability of the colors. Furthermore, experimental annual energy yield studies for modules with white-grid compared to typical bifacial and monofacial modules are missing.

In this work we investigate the optical properties and reliability of various white colors for the application in white-grid solar modules. With ray tracing computer simulations we determine the optimal geometry for the white reflector in the cell gaps and build test modules to evaluate the power gain for white-grid compared to typical bifacial and monofacial modules. Moreover, we investigate the outdoor performance for these modules over a period of one year. At the conference we will also demonstrate the mass manufacturing of white-grid modules and their advantage over typical bifacial modules in terms of STC measurements.

II. EXPERIMENTAL SECTION

A. Glass color samples

We investigate 26 white colors applied to glass color samples based on typical solar module rear glass and characterize their optical performance as well as long-term reliability. The glass color samples differ in that they are based on organic as well as ceramic color particles and are applied using different screen-printing processes. For the evaluation of the optical performance we measure the hemispherical reflectance of the glass color samples employing a Varian Cary 5000 two-channel spectrophotometer with an integrating sphere. Besides the glass color samples, we also measure the reflectance of 7 commercially available white polymer backsheets for a comparison.

We test the reliability of the glass color samples in UV irradiance and humidity freeze (HF) tests. All samples are exposed to a UV light source for 96 kWh m^{-2}. During the test we cover the samples with solar front glass. For the HF test, we laminate the glass color samples with a polyolefin elastomer (POE) and a cover glass. The samples are stressed in the HF test for 30 cycles in accordance to the IEC 61215:2019 MQT 12 standard.

B. Ray tracing simulations

We perform Monte-Carlo based ray tracing simulations employing our in-house ray tracing software *Daidalos*. The simulations are based on our previous publications [9], [10] and simulates a unit cell of one solar module encapsulated with a

978-1-7281-6118-1/22 $31.00 © 2022 IEEE

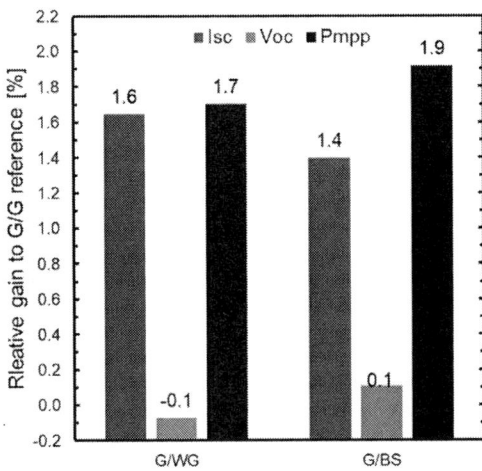

Fig. 1 Gain in I_{sc}, V_{oc} and P_{mpp} for the glass-white-grid and glass-backsheet (G/BS) relative to the glass-glass (G/G) reference module. The data is based on STC IV flasher measurements.

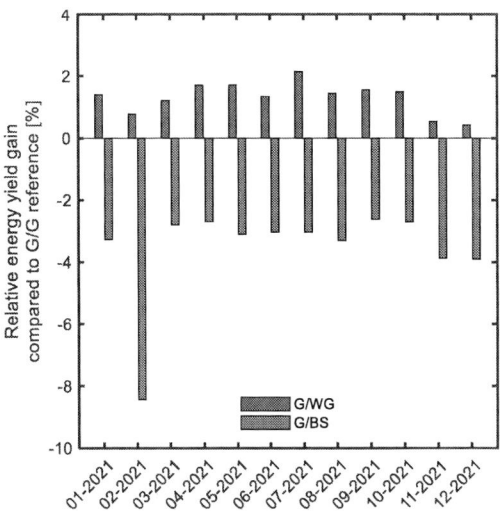

Fig. 2 Monthly energy yield gain for the G/WG and G/BS module relative to the bifacial G/G reference for the year 2021.

UV-transparent polymer and a low-iron front and rear glass. We implement the white-grid rear glass as a reflection interface on the inner boundary of the rear glass. The optical properties are based on our reflectance measurements of the glass color samples assuming a Lambertian factor of 0.95. In the simulations we vary the reflector width and ground albedo to determine the optimal geometry for the application in a white-grid solar module. The distance between the cells is 3 mm in all simulations as in the experimental test-modules.

C. Test module fabrication at ISFH

We build four test-modules at *ISFH* each featuring 9 bifacial n-type based passivated emitter, rear totally diffused (PERT) solar cells provided by *SoliTek*. The cells are binned according to their maximum power point (MPP) to obtain modules with similar cell performances. In each module the cells are interconnected via 5 cell interconnect ribbons and a cell distance of 3 mm in all directions. All modules employ POE encapsulation polymers on the cells' front and rear side.

(i) The reference module named G/G is a glass-glass bifacial module featuring a solar module front and rear glass. (ii) Module G/BS is a monofacial glass-backsheet module with a solar module front glass and white polymer backsheet. (iii) Module G/WG is a white-grid module using a solar module glass on the front and rear side with the rear glass having a white colored grid at the positions of the cell gaps. After building the modules we measure their current-voltage (IV) characteristics using a halm cetisPV-Moduletest3 system [11].

D. Outdoor testing at ISFH test site

The test-modules are installed at the *ISFH* test site with a module tilt of 35° and azimuth angle of 180° facing south. All modules are equipped with a temperature sensor. An IV tracer measures the module's IV characteristics every 5 min. The modules are connected to a load resistor to operate the modules close to the maximum power point (MPP) between IV measurements. Additionally, we monitor the global horizontal G_{GHI}, diffuse horizontal G_{DHI}, and beam horizontal G_{BHI} irradiance with thermopile pyranometers [12]. For the plane of array irradiance G_{POA} we employ a silicon sensor [13].

III. RESULTS AND DISCUSSION

The evaluation of the reflectance measurements of the colored glass samples reveal that the colors with the highest average reflectance perform similar as an average backsheet. Weighted with the AM1.5g spectrum [14], the highest integrated reflectance of the glass color sample is 8% lower than our backsheet with the highest integrated reflectance. Due to the screen-printing process the color film is partially transparent, which results in a lower reflectance. This could be reduced by drying and printing the colors several times. However, this would significantly increase the printing costs. The reliability results suggest that organic colors are unsuitable since the integrated reflectance tend to degrade after UV aging.

The ray tracing results indicate that the current gain rises fastest from 0 mm to the cell spacing of 3 mm, increasing the current by 1.6% for the white-grid module compared to the bifacial reference. This is in agreement with findings from Jang *et al.* [8]. When the reflector width becomes larger than the cell gap the additional current gain reduces from about 0.54%/mm to 0.02%/mm. Finally, we chose a reflector width of 5 mm for our test modules with a cell spacing of 3 mm to account for cell positioning tolerances and cell movement during the lamination process.

Figure 1 shows the relative gain in short circuit current I_{sc}, open circuit voltage V_{oc}, and maximum module power P_{mpp} for the G/WG and G/BS test-modules. The bifacial G/G is used as reference. We measure a gain in I_{sc} of 1.6% for the white-grid module compared to the bifacial reference. This manifests in an additional power gain of 1.7%. The I_{sc} gain of the monofacial G/BS module is 1.4% compared to the bifacial reference. The measured gain in power is 1.9% for G/BS, which is 0.2% higher than for the G/WG module. The open circuit voltage indicates that this might be due to a higher V_{oc} of the individual cells in the module.

Figure 2 shows the monthly energy yield gain of the G/WG and G/BS modules relative to the bifacial G/G reference. The monofacial G/BS module has the lowest energy yield, which is 2.6% lower than the G/G reference module on an annual average. We attribute the higher yield of the reference to the

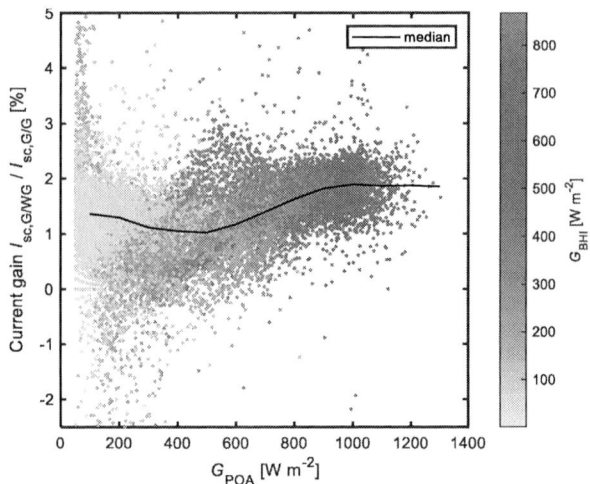

Fig. 3 I_{sc} gain for the G/WG relative to the G/G module as function of plane of array irradiance. Color coding indicates the beam horizontal irradiance. Solid line is the moving median for a 100 W m^{-2} window.

additional usage of rear light. Especially in the winter months, the advantage of the G/G reference is higher. We measure the highest energy yield for the G/WG module. On an annual average the G/WG module produces 1.3% more energy than the G/G reference. The advantage is lower in the winter months and highest in July with a maximum energy yield gain of 2.1%.

We infer the higher energy yield of the G/WG module from a higher module current due to the white-grid reflector. Figure 3 shows the gain in I_{sc} for the G/WG module relative to the G/G reference module. Above 800 Wm^{-2} the I_{sc} of the G/WG module is 1.9% higher compared to the G/G module. The colormap indicating the beam irradiance suggests that a higher direct irradiance is beneficial for the G/WG module. With lower irradiance and increasing diffuse irradiance the advantage reduces to about 1%.

CONCLUSION AND OUTLOOK

The application of white-grid rear glasses improves the STC performance of standard bifacial modules by 1.6%. Considering different cell IV characteristics and measurement tolerances of our flasher system the white-grid module shows a similar STC performance as the monofacial glass-white backsheet module. Investigating the outdoor performance shows that white-grid modules also improve the annual energy yield by 1.3% and 5.1% compared to a bifacial and monofacial reference, respectively. We observe the highest advantage predominantly in the summer months and when the fraction of direct irradiance is high. Therefore, white-grid modules are advantageous for regions with high amounts of direct sun light or the installation on trackers.

At the conference we will present further results on the mass production of these modules at the SoliTek manufacturing site. Furthermore, we will report on the outdoor performance of bifacial and white-grid modules for various test sites across Europe with varying irradiance conditions.

ACKNOWLEDGMENT

The results were generated in the SUPERPV project (grant agreement no. 792245) funded by the European Union's Horizon 2020 Research and Innovation Program.

REFERENCES

[1] A. Cuevas, A. Luque, J. Eguren, and J. del Alamo, "50 Per cent more output power from an albedo-collecting flat panel using bifacial solar cells," *Solar Energy*, vol. 29, no. 5, pp. 419-420-419–420, Jan. 1982, doi: 10.1016/0038-092X(82)90078-0.

[2] G. Sala *et al.*, *Albedo collecting photovoltaic bifacial panels*. 1984, pp. 565–569. [Online]. Available: https://ui.adsabs.harvard.edu/abs/1984pvse.conf..565S

[3] R. Hezel, "Novel applications of bifacial solar cells," *Progress in Photovoltaics: Research and Applications*, vol. 11, no. 8, pp. 549-556-549–556, 2003, doi: 10.1002/pip.510.

[4] B. B. Van Aken, L. A. G. Okel, J. Liu, S. L. Luxembourg, and J. A. M. Van Roosmalen, "White Bifacial Modules – Improved STC Performance Combined with Bifacial Energy Yield," 2016. doi: 10.4229/eupvsec20162016-1bo.12.2.

[5] M. H. Saw, Y. S. Khoo, J. P. Singh, and Y. Wang, "Enhancing optical performance of bifacial PV modules," *Energy Procedia*, vol. 124, pp. 484–494, 2017, doi: 10.1016/j.egypro.2017.09.285.

[6] E. Sng, C. X. Ang, and I. L. H. Lim, "Investigation and Analysis of Bifacial Photovoltaics Modules with Reflective Layer," 2018. doi: 10.4229/35THEUPVSEC20182018-5CV.1.45.

[7] E. Sng *et al.*, "Optimisation of Bifacial Photovoltics Module with Reflective Layer in Outdoor Performance," 2019. doi: 10.4229/EUPVSEC20192019-4AV.1.12.

[8] J. Jang, A. Pfreundt, M. Mittag, and K. Lee, "Performance Analysis of Bifacial PV Modules with Transparent Mesh Backsheet," *Energies*, vol. 14, no. 5, pp. 1399–1399, Jan. 2021, doi: 10.3390/en14051399.

[9] M. R. Vogt, T. Gewohn, K. Bothe, C. Schinke, and R. Brendel, "Impact of Using Spectrally Resolved Ground Albedo Data for Performance Simulations of Bifacial Modules," in *35th EUPVSEC*, 2018, pp. 1011–1016. doi: 10.4229/35thEUPVSEC20182018-5BO.9.5.

[10] M. R. Vogt *et al.*, "Ray Tracing of Complete Solar Cell Modules," 2019. doi: 10.1364/pvled.2019.pw2c.1.

[11] h.a.l.m. elektronik gmbh, "cetisPV-Moduletest3." https://www.halm.de/products/module-lab (accessed Mar. 09, 2022).

[12] "CMP11 Standard-Pyranometer bei PV- und CSP-Anlagen - Kipp & Zonen." https://www.kippzonen.de/Product/476/CMP11-Pyranometer (accessed Mar. 09, 2022).

[13] Ingenieurbüro Mencke & Tegtmeyer, "Si Sensor," *Ingenieurbüro Mencke & Tegtmeyer GmbH*. https://www.imt-solar.com/solar-irradiance-sensors/si-sensor/ (accessed Mar. 09, 2022).

[14] Technical Committee ISO/TC 180, "ISO 9845-1:1992(en), Solar energy — Reference solar spectral irradiance at the ground at different receiving conditions — Part 1: Direct normal and hemispherical solar irradiance for air mass 1,5," 1992.

4D-Printed Shape Memory Polymer Based Solar Tracker

Serhii Tytov,[1,2] Fhad Al-Modaf,[2] Shicheng Su[2,3] and Nazek El-Atab[2]

[1]National Technical University of Ukraine "Igor Sikorsky Kyiv Polytechnic Institute" 37, prospect Peremohy, 03056 Kyiv, Ukraine

[2]Electrical and Computer Engineering, Computer Electrical Mathematical Science and Engineering Division, King Abdullah University of Science and Technology (KAUST), Thuwal 23955-6900, Kingdom of Saudi Arabia

[3]University of Electronic Science and Technology of China, 610056, China

Abstract—**In this work, soft actuators made of shape memory polymers (SMPs) are used to actuate and control a solar tracker in response to heat generated by sunlight. To achieve this, a thermo-mechanical design of a solar tracker is 4D printed. The results show that the black-colored elliptical-shaped solar tracker can shrink and tilt by up to 30° when exposed to sunlight with a response time of 30 s, enabling the solar cell to remain exposed to the highest light intensity and therefore, allowing for continuously optimizing the solar cell power output.**

Keywords—*solar tracker, shape memory polymer, solar energy, 4D printing.*

I. INTRODUCTION

Solar energy has been drawing an increasing amount of interest as a result of being renewable, clean and reliable. The efficiency of solar panels depend on a number of factors, and one way to enhance the power output is by using a solar tracker which ensures that the incident sunlight is always normal to the surface of the solar panel [1-5]. Nevertheless, currently used solar tracker systems in photovoltaic plants include sensors, microcontrollers and motors, and thus are power-hungry, require regular maintenance and are heavy which limits their range of potential applications. Therefore, in order to maximize the world's benefit from solar energy, it is necessary to develop smart and lightweight solar trackers which do not consume any energy.

The use of smart materials which respond to the sunlight stimulus and change their shape accordingly can be promising for this purpose. More specifically, shape memory polymers (SMPs) have been previously programmed to deform and recover to a memorized shape in response to different stimuli such as light and heat [6]. In addition, because of their ability to deform and create different physical movement such as twisting, contraction and bending, their application into different applications as artificial muscles and micromachines was possible [7]. Taking the physical movement of sunflowers as an example, we can see how it tracks the sun by relying on the sun rays. Inspired by this natural phenomenon, SMPs are sensitive to sunlight, and they can deform from the

programmable shape to the initial (permanent) shape upon being incident to sunlight.

In this work, we introduce a solar tracker that can tilt and track the sun by titling in either one-axis or two-axes. This is achieved by utilizing actuators made of smart materials to improve the performance of the solar tracker. The main idea that stands behind this concept is placing solar cell on a platform with legs printed from SMP filament in some compressed shape and then expanded under some external pressure as shown in Fig. 1. After heating under sunlight, the structure would compress and return to its permanent shape. The solar tracker is developed using a 4D printing process.

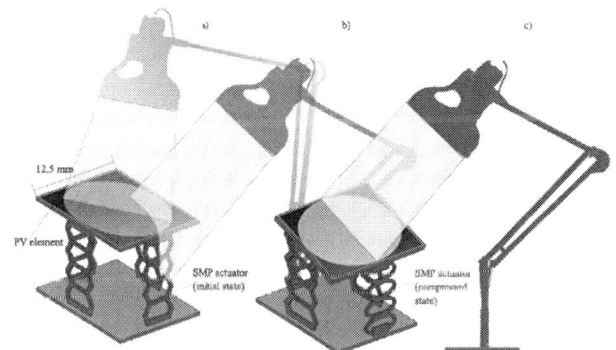

Fig. 1. Concept of solar tracker with shape memmory polymer actuators: changing incidence angle of light from a) to b) causes heating of SMP actuator and thus, it returns into its compressed state c) the resulting tilting of the photovoltaic cell towards the light source.

II. MATERIALS AND METHODS

4D printing is used to develop the solar tracker. 4D printing stands for 3D printing using smart materials that can react to and respond to surrounding external stimuli such as light, heat, humidity, etc. Structures are affected even during the printing itself so it is necessary to take into account combination of different effective parameters such as platform temperature, printing speed, liquefier temperature, among others [8], otherwise properties of the resulting specimen would severely differ from desired ones [9].

978-1-7281-6118-1/22 $31.00 © 2022 IEEE

The design of the solar tracker includes 4 legs which can be compressed when exposed to the light stimulus. Each leg is based on multiple SMP based ellipses as shown in Fig. 1. Moreover, the different ellipses can be actuated separately which enhances the sensitivity of the solar tracker. The actuators were printed using Creality Ender-3 3D printer based on the fused filament fabrication process. The material used in the experiment is polyurethane SMP filament, which is developed by SMP technologies Inc. and KYORAKU Co., Ltd., with a diameter of 1.75 mm. The glass transition temperature Tg of the SMP filament was 328 K. Extruder head temperature was set to 473K, while the build plate temperature was set to 323K and the used print speed was 50 mm/s.

For further consideration, we define the relative elongation ratio of the actuator (δ) as a relation between the differences of the length of the actuator in extended ($h_{extended}$) and compressed states ($h_{compressed}$) to the length in compressed

$$\delta = \frac{h_{extended} - h_{compressed}}{h_{compressed}}. \qquad (1)$$

The final tested system includes a 5 inch by 5 inch SunPower C60 monocrystalline solar cell with maximum voltage and power output of around 0.6 V and 3.4 W, respectively, mounted on the developed smart solar tracker.

III. RESULTS AND DISCUSSION

Fig. 2 shows the large scale commercial grade solar cell mounted on the SMP based smart solar tracker. Since the SMP legs were added on the edges of the solar cell, as a result, the cell can be tilted towards the direction of the sunlight in either uniaxial or biaxial directions (Fig. 2). The main goal, that was standing behind the conducted study, was to test different designs suitable for this application. The initial idea of using just a helix shaped actuator coiled around a telescopic tube [10] was discarded as inappropriate because it does not meet the key requirements of this study. The helix spring shows high elongation ratio, but it is not stable without the use of internal support with moving parts, which requires further maintenance in addition to severely losing its initial shape after actuation.

The final design is selected based on an optimum combination of relative elongation ratio and ability to maintain the original form of the actuator. One of the main parameters that further influences the performance of the actuator is the ratio

of horizontal radius a of the ellipse to the vertical radius b. The key features and advantages of the 4D printed system are the following: 1) it is easy to print because all parts are printed in a planar fashion and do not requiere any support layers; 2) the different parts (i.e. different ellipses) contract separately because they are exposed to the the sun gradually which enhances the sensitivity of the system; and 3) the relative elongation ratio of the initial and actuated states can be controlled by varying the horizontal radius of the ellipse as shown in Fig. 3. The main drawback, however, is that the smart leg does not return to its original state upon removal of sunlight/heat, instead manual elongation/programming is required. This can be mitigated in future designs where materials with the capability to reverse to the initial state is possible, such liquid crystal elastomers.

After fabricating several samples with different ellipse radii ratios, it is possible to generate the relationship between the elongation and dimensions of the ellipse (a and b), which allows the printing of the actuator with a predefined relative elongation ratio, as shown in Fig. 4. The approximation of the curve was made according to the definition of arc length of the ellipse $l \sim \sqrt{\left(\frac{a}{b}\right)^2 + 1}$.

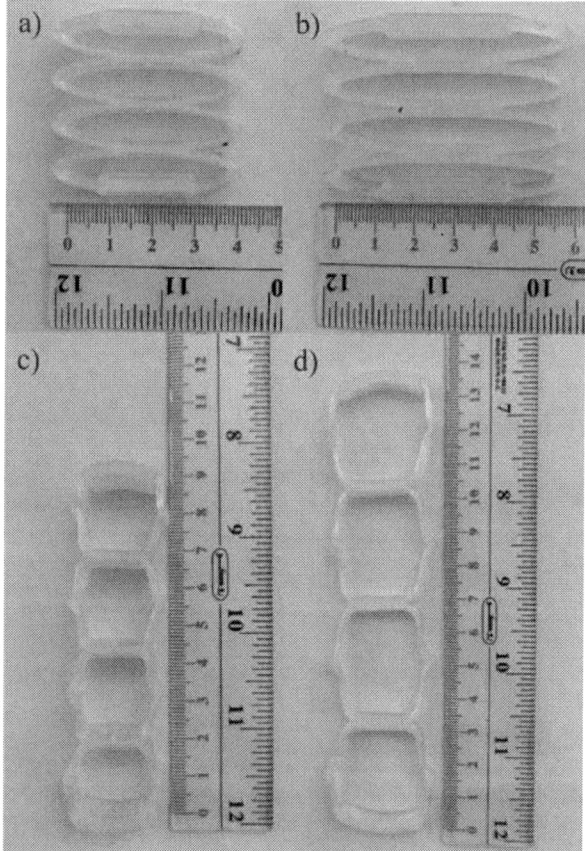

Fig. 3. Comparison of actuators with a), c) 3.81 and b),d) 5.45 ellipse radius ratio in contracted and elongated states respectively.

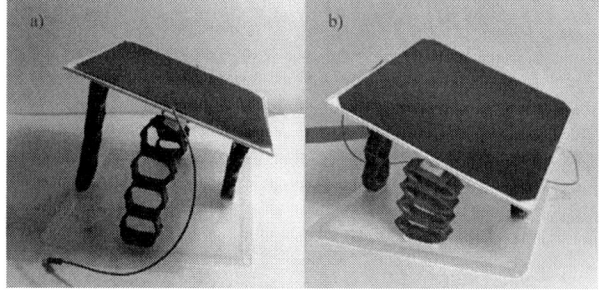

Fig. 2. Fabricated solar tracker tilted towards direct light along a) 1 axis, b) 2 axes.

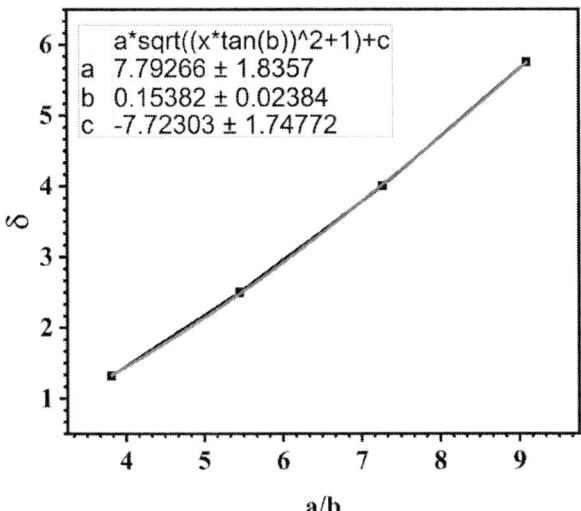

Fig. 4. Relative elongation ratio of actuator as a function of ellipse radius ratio.

It is worth to note that the used SMP requires a temperature of at least $55°C$ to react and compress. Such a temperature can be reached in many regions of the world including Saudi Arabia. However, one way to further enhance the ability of the actuator to absorb the light radiation is by simply painting it in a black color as shown in Fig. 5. As a result, the black actuator can fully respond to the sunlight and be completely compressed in 30 s (40% compression) while the original white colored actuator got only compressed by 10% of the original length. This approach helps in decreasing the exposure time to the stimulus by a factor of more than half.

The effectiveness of solar tracker was checked by measuring the open circuit voltage of the solar cell at different

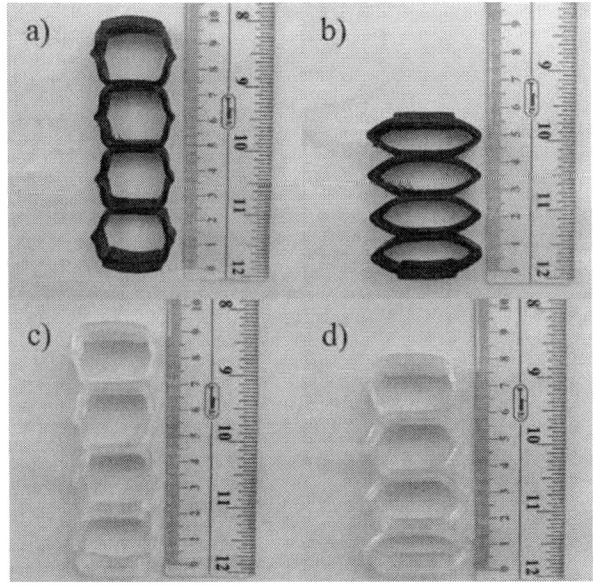

Fig. 5. Comparison of contraction speed of a), b) black actuator *vs.* c), d) original color of filament actuator. For the black actuator, 30 s was enough to return to its initial state, while the white actuator is only starting to compress.

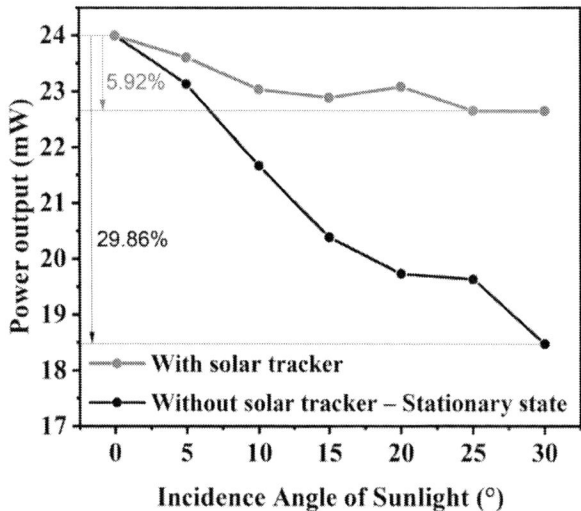

Fig. 6. omparison of the power output of the photovoltaic cells with and wihout solar tracker, when placed at a distance of 60 cm from the light source.

angles of light incidence as shown in Fig. 6. The experimental setup is a light source placed at 60 cm distance from the solar cell, which was connected to a multimeter. The stationary state curve shows a decreasing perfomance of the photovoltaic cell as the light source is deviated from the normal direction. Using the SMP based solar tracker enables the solar cell to reamin always exposed to a highest light intensity, as a result, its power output degrades slightly compared to the fixed cell.

While the improvement in the solar cell power output due to the solar tracker is obvious, it is worth to note that during the day and thorughout the year, some incidence angles prevail ver others. For instance, in the Ma'an area, Jordan, adjusting the tilting angle of the solar cell 4 times per year to the incidance angle of 28-30 degrees can encrease the power output by up to 45% [11].

IV. CONCLUSIONS

In this work, a novel thermo-mechanical design of a smart and lightweight solar tracker is developed. 4D printing is employed for the simple fabrication of the actuators using shape memory polymers. The elliptical design of the actuator was optimized to achieve a stable and fast response to the sunlight. The dependency of the relative elongation ratio of the actuator on the ellipse radius ratio was also obtained. In addition, we have shown that one of the most effective ways to speed up the actuation time was to paint it in black color, which allows the actuator to respond to the light stimulus almost twice faster. The final results show that using the solar tracker, the power output of the solar cell show a negligible degradation as the incidence angle of the light increases. Future work will focus on enabling the automated recovery of the elongated actuator state.

ACKNOWLEDGMENT

The authors acknowledge generous funding from the King Abdullah University of Science and Technology (KAUST) and the KAUST Climate and Livability Initiative.

REFERENCES

[1]   T. Salameh, C. Ghenai, A. Merabet, and M. Alkasrawi, "Techno-economical optimization of an integrated stand-alone hybrid solar PV tracking and diesel generator power system in Khorfakkan, United Arab Emirates," Energy, 2020, doi: 10.1016/j.energy.2019.116475.

[2]   C. Jamroen, P. Komkum, S. Kohsri, W. Himananto, S. Panupintu, and S. Unkat, "A low-cost dual-axis solar tracking system based on digital logic design: Design and implementation," Sustain. Energy Technol. Assessments, 2020, doi: 10.1016/j.seta.2019.100618.

[3]   N. El-Atab, et al., "Ultraflexible Corrugated Monocrystalline Silicon Solar Cells with High Efficiency (19%), Improved Thermal Performance, and Reliability Using Low-Cost Laser Patterning", *ACS Applied Materials & Interfaces*, 2019. Available: 10.1021/acsami.9b15175.

[4]   N. El-Atab, Net al., "Nature-inspired spherical silicon solar cell for three-dimensional light harvesting, improved dust and thermal management", *MRS Communications*, vol. 10, no. 3, pp. 391-397, 2020. Available: 10.1557/mrc.2020.44.

[5]   N. El-Atab and M. Hussain, "Flexible and stretchable inorganic solar cells: Progress, challenges, and opportunities", *MRS Energy & Sustainability*, vol. 7, 2020. Available: 10.1557/mre.2020.22.

[6]   A. Lendlein and O. E. C. Gould, "Reprogrammable recovery and actuation behaviour of shape-memory polymers," Nature Reviews Materials. 2019, doi: 10.1038/s41578-018-0078-8.

[7]   H. M. Chen, L. Wang, and S. B. Zhou, "Recent Progress in Shape Memory Polymers for Biomedical Applications," Chinese Journal of Polymer Science (English Edition). 2018, doi: 10.1007/s10118-018-2118-7.

[8]   M. Bodaghi, A.R. Damanpack, W.H. Liao, Adaptive metamaterials by functionally graded 4D printing, Materials & Design, vol. 135, 2017, pp.26-36.

[9]   K. Takeda, S. Hayashi, K. Hayashi, Deformation Properties of 3D Printed Shape Memory Polymer. KEM, vol.725, 2016, pp.378–382.

[10]   Amine Riad, Mouna Ben Zohra, Abdelilah Alhamany, Mohamed Mansouri, Bio-sun tracker engineering self-driven by thermo-mechanical actuator for photovoltaic solar systems, Case Studies in Thermal Engineering, vol. 21, 2020, 100709.

[11]   Ibrahem S. Altarawneh et.al, "Optimal tilt angle trajectory for maximizing solar energy potential in Ma'an area in Jordan", Journal of Renewable and Sustainable Energy, vol.8, 033701, 2016

Study of Degradation of Cu(In,Ga)Se$_2$ Solar Cell Parameters Due to Temperature.

Rabee B. Alkhayat[1*] and Dhurba R. Sapkota[2*]

[1]Department of Physics, College of Education for Pure Science, Univ of Mosul, 41002 Iraq; [2] Wright Center for Photovoltaics Innovation & Commercialization, Univ, Toledo, Toledo, Ohio, 43606, USA

Abstract —Copper indium gallium diselenide, Cu(In,Ga)Se$_2$, thin-film has been completed by using thermal co-evaporation of Cu, In, Ga, and Se sources at a substrate temperature of 570°C. A CIGS solar cell with a 12% efficiency and a 1.5 μm absorber thickness has been completed. The thin film is characterized using UVVS, XRD, and SEM/EDS. The effects of temperature on the degradation of CIGS solar cells are also investigated. Furthermore, the degradation of cell characteristics such like open-circuit voltage (V_{oc}), fill factor (FF), current density (J_{sc}), and efficiency (η) with temperature is examined. The value of cell characteristics like V_{oc} and FF decreases significantly as the temperature increases. Although there is no clear trend in the J_{sc} value, overall cell efficiency is found to decrease with temperature. The ultimate purpose of this work is the optimization of narrow band-gap CIGS-related solar cells for the bottom layer in tandem devices.

Keywords— CIGS thin films, Degradation, Composition, Bandgap, Cell parameters, and Temperature.

I. INTRODUCTION

Cu(In,Ga)Se$_2$ (CIGS) thin-film photovoltaic cells are widely considered to be among the most efficient thin-film solar cell technology. More than 37 years ago, the first effective cells of this sort were produced [1]. In recent years, significant advances have been made in the understanding of electronic transport measurements on CIGS-based heterostructures. Admittance spectroscopy has shown to be a very valuable method for investigating bulk and surface flaws in these devices. As a result of this investigation, we can immediately draw conclusions about the effects of chemical treatments on the heterostructure's band diagram [2-5]. Undoubtedly, they are now considered to be among the highest-performing devices, alongside crystalline silicon solar cells. High conversion efficiencies of CIGS devices are strongly associated to the use of soda lime glass (SLG) as the substrate [6].The role of the SLG has been discussed in terms of thermal expansion matching to the CIGS layer as well as leakage of various impurities into the CIGS layer during deposition, such as Na. It's been proposed that the presence of Na causes a higher concentration of melted Cu$_2$Se during deposition, increasing atomic mobilities. It has also been demonstrated that adding Na to the CIGS during growth results in large grains [7]. The amount of Na from the SLG that can reach the growing CIGS is determined by the diffusivity of Na through the Mo layer. Then, the Mo back contact layer plays a significant role as well [8]. CIGS has gained increased in popularity as an absorber material in thin-film PV devices over time [4, 9-15]. The efficiency of a CIGS film with a three-stage thermal co-evaporation absorber layer is unprecedented among chalcopyrite thin films. In 2015 and 2016, Jackson *et al.* produced CIGS cells with an efficiency of 21.7% and 22.6%, respectively[14, 16]. 22.6% efficiency CIGS was obtained with RF post-deposition treatment. Controllable p-type conductivity with copper stoichiometry and adjustable bandgap are the main two features due to which CIGS is being widely used. Cadmium free recorded efficiency of Cu(In,Ga)(Se, S)$_2$ is 23.35 % as reported in [17]. Magnesium zinc oxide (MZO) with alloys, and indium gallium oxide (IGO) would be the potential alternative of Cd-free window layers whose detail study and properties are presented in [18, 19].

The current focus of research for the CIGS-based system is the development of tandem or multijunction solar cells, with the goal of surpassing single-junction limits. Shockley-Queisser limits for single-junction solar cells are approaching [20]. The CI(G)S/perovskite tandems combination make good photovoltaic properties for both absorbers with the proper pairing of their bandgaps. Tandem cell can be completed by depositing perovskite on the top of CI(G)S without annealing window layer and TCO layer in CI(G)S device. The goal of this study is to figure out how the temperature factor affects the characteristics of CIGS. The ultimate intent of this work is to optimize CIGS and reduce Ga in a systematic order to fabricate/optimize thin-film CIS devices. Optical properties, bandgap, and Urbach tail energy for the different compositional ratios y = [Cu]/[In], for CIS material were studied previously [21, 22]. Photoluminescence experiments of CIS thin-films at various y parameters were given earlier, which revealed disorder, trapped levels, and transitions from donor to acceptor [23, 24]. Hole concentration and mobility were determined by optical Hall influence measured value of CIS film for y = 0.9 [25, 26]. Optimal tandem solar cell

978-1-7281-6118-1/22 $31.00 © 2022 IEEE

designs were also reported in the prior work based on the obtained optical characteristics [24, 27, 28]. Structural-optical properties of two-stage CIS and material properties of CIS deposited at different rates were studied by Real-time Spectroscopic Ellipsometry (RTSE) [29, 30]. Structural evolution and optical properties of CIS thin films are studied by real time spectroscopic ellipsometry and presented in the previous work in [21, 22, 31]. CIGS devise of ~16% efficiency by three-stage thermal co-evaporation was reported previously [32]. Here, the effect of temperature factors on different parameters of the CIGS solar cell has been explored and studied its effect on open circuit voltage, external quantum efficiency and short-circuit current density.

II. EXPERIMENTAL DETAILS

A. CIS Solar Cell Fabrication

Soda-lime glass (SLG) substrates measuring 2.5 cm x 7.5 cm x 1.5 mm are used for the CIGS solar cells. Molybdenum is the most common material reportedly used in the solar cell contacts of $Cu(In,Ga)Se_2$ (CIGS). Several characteristics are required, particularly chemical and mechanical inertness during the other mechanisms of deposition, high conductivity, low resistance to CIGS layer, and appropriate thermal expansion coefficient. However, it does not by itself guarantee a highly efficient solar cell to deposit molybdenum as back contact. The deposition and process parameters play a major role in achieving the appropriate properties for a layer [33]. Mo was deposited via a two-step sputtering technique at a substrate temperature of 250°C. A 0.1 μm layer is first deposited on the SLG at a pressure of 15 mTorr to confirm adhesion. To achieve an intended thickness of 0.8 μm with a low sheet resistance of < 0.25 Ω/sq, a thicker second layer is deposited at the lowest system pressure of 4 mTorr. The CIGS absorber is deposited on the Mo surface by four source co-evaporations at a substrate temperature of 570°C and at deposition rate ~ 7.0 Å/s to effective thicknesses ~1.5 μm. A one-stage thermal co-evaporation of CIGS deposition procedure was carried out with a measured composition of y = 0.90 and rate ~7Å/s range. Also, cadmium sulphide (CdS) is a common II–VI compound semiconductor material used in a variety of heterojunction photovoltaic systems [34]. CBD is a popular method for producing ultra-thin CdS thin films for solar cells, while RF sputtering is utilized to produce smooth CdS thin films [35]. Chemical bath deposition (CBD) is used to deposit CdS on the SLG/Mo/CIGS structures to an intended thickness of ~ 50 nm using an aqueous solution of cadmium acetate and thiourea with a molar ratio of [Cd]/[S] = 0.0292, as well as an ammonium hydroxide catalyst at 1.1 M in the solution. Moreover, due to its high transparency in the visible part of the spectrum and low electrical resistivity of 10^{-4} Ω cm, indium tin oxide (ITO) is utilized as the front contact in solar cells [36]. The transparent conducting top contact, a ZnO/ITO bilayer, was formed by sputtering at room temperature with thicknesses of 50/250 nm, respectively.

Finally, a Ni/Al/Ni tri-layer with thicknesses of (50nm)/(2μm)/(50nm) was deposited by electron beam evaporation through masks to serve as the electrical contact grids.

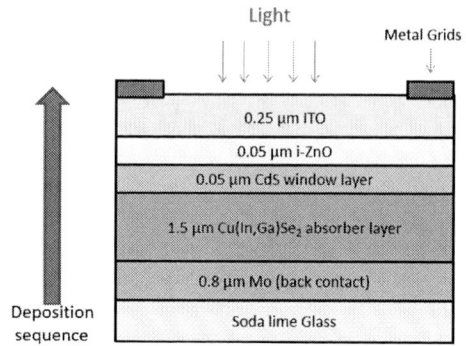

Fig.1. Schematic of the CIGS solar cell device in substrate configuration used in this study.

B. CIGS Device and Characterization

The CIGS absorber layer was examined for morphological features, crystallite size, composition, and stoichiometry using scanning electron microscopy (SEM), X-ray diffraction (XRD), and energy-dispersive X-ray spectroscopy (EDS), respectively. The EDS data for the CIGS layer shows the p-type CIGS with copper stichometry y = 0.90 ± 0.03 and Ga content x ≈ 0.3. The morphology of the CIGS absorber was examined using SEM as shown in Fig. 2. Also, the cross-sectional SEM of the whole device was performed, and the thickness of each component layer was obtained as presented above in Fig.1. Using the Scherrer equation, the average crystallite size of this CIGS absorber layer was calculated as 293 Å from the (112) and (204) diffractions from the XRD spectra presented in Fig.3. UVV's spectrometer was used to measure the light irradiance and transmission through the CIGS film over the range of 300-1050 nm as shown in Fig. 4(a). The absorption coefficient α was calculated by using the transmission and reflections information. By using the value of absorption bandgap of CIGS thin-film was calculated as 1.19 eV as presented in Fig. 4 (b). Commercial instrument (Model IVEQE8-C, PV Measurements, Inc.) was used for the J-V measurement. The solar cell exhibited the following performance parameters: V_{oc} = 0.569 V, J_{sc} = 30.87 mA/cm², R_{sh} = 537.94 Ω cm² R_S = 2.58 Ω cm², FF = 62.97 %, and efficiency, η = 12.13 % from the illumination area of 0.5 cm² cell. The solar cell parameters are presented in Fig. 5 along with the J-V curve and external quantum efficiency.

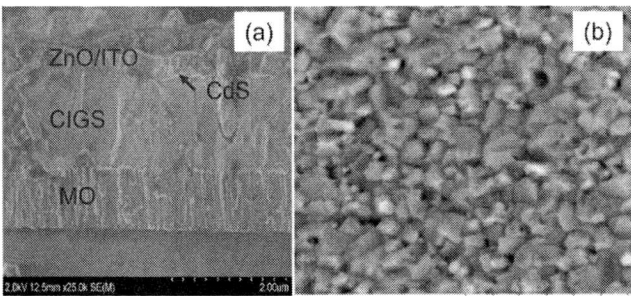

Fig.2. (a) Cross-Sectional SEM of CIGS device, (b) surface SEM of CIGS absorber layer.

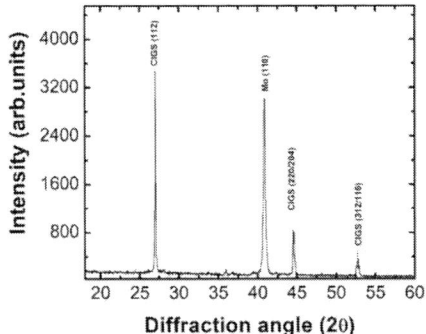

Fig. 3. XRD diffractions of CIGS absorber layer

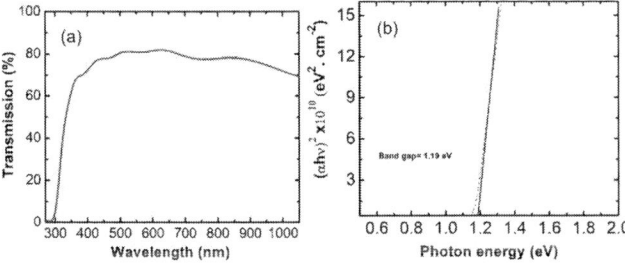

Fig.4.(a) Transmission spectrum and (b) bandgap determination of CIGS absorber layer using absorption coefficient.

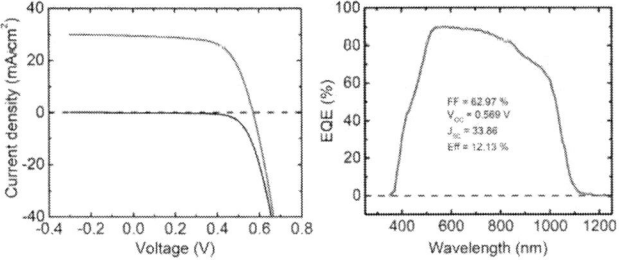

Fig. 5. (left) J-V characteristics of CIGS solar cells in the dark and under AM 1.5 illumination. (right) an experimental quantum efficiency spectrum.

III. *TEMPERATURE DEPENDENCE OF CIGS SOLAR CELL PARAMETERS.*

Most commercial CIGS solar cells are currently manufactured on SLG substrates, which naturally provide the absorber with a sufficient supply of Na (alkali metal) and can withstand high temperatures during deposition and annealing processes[37].Alkali diffusion from the substrate to the absorbers is facilitated by high substrate temperatures, as well as enhancing the growth of the absorber layer. The majority of high-efficiency CIGS modules were made at a substrate temperature (T_{sub}) of above 773 K, which is extremely close to the softening temperature of SLG substrates [38]. Fewer studies have focused on heat treatment of entire solar cells, that could provide more understanding into how well CIGS can withstand harsh environmental conditions. Furthermore, exposure to high temperatures is required for tandem cells that use CIGS as the bottom cell, additionally to the placement of module packaging, such as the encapsulating layer. The sensitivity of solar cells with temperature (T) is crucial because they are subjected to temperatures varying from 288 to 323 K in working conditions [39], and even higher temperatures (370–380 K) in space and concentrator systems. Photovoltaic (PV) parameters such as short circuit current density (J_{sc}), open circuit voltage (V_{oc}), and fill factor (FF) determine the solar cell's performance. These parameters are temperature-dependent and thus have an impact on solar cell performance. Furthermore, V_{oc} decreases as temperature increases, while J_{sc} increases only slightly. The slight decrease in J_{sc} is due to reflection losses (series and shunt resistance) and recombination losses, as indicated. The FF and efficiency both decrease as the temperature increases as a result of this degradation V_{oc} [39-41]. The change in R_s and R_{sh} has a minor effect on efficiency [42]. As a result, most PV variables (e.g., FF, V_{oc}, and J_{sc}) follow a similar pattern., which are mostly negative. Electron-hole couples may recombine because of the aluminum in the absorber layer. The extraction of electrons from solar cells may be hampered as a result of these effects, resulting in a reduction in performance. The IV barrier could be caused by diffusion from the front contact to the CIGS absorber layer, whereas diffusion from the front contact to the CIGS absorber layer might have a secondary effect: This could reduce the conductivity of the front contact layer, affecting the series resistance of the solar cells [12]. The entire CIGS device was exposed to various conditions of temperature in air environment for a heat treatment of at 15 minutes. The heat was accumulated in the solar cell as a result of this thermal process after each cycle, exposing the same cells to the next temperature, and so on. The effect of temperature on the operational characteristics of CIGS-based solar cells in a wide temperature range of 120 to 180 ˚C is investigated in this study. This study will assist in the planning and implementation of future studies on the performance of single junction and tandem solar cells in regard to temperature.

978-1-7281-6118-1/22 $31.00 © 2022 IEEE

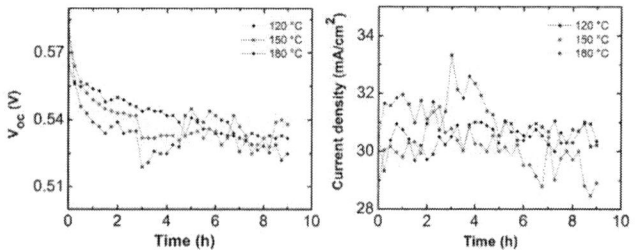

Fig. 6. Degradation of open-circuit voltage (left panel) and current density (right panel) of CIGS solar cell with time at different temperatures.

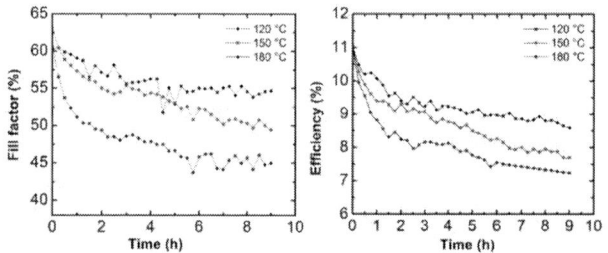

Fig.7. Degradation of fill factor (left panel) and efficiency (right panel) of CIGS solar cell with time at different temperatures.

IV. CONCLUSIONS AND FUTURE WORK

Significant decrease in the value of cell parameters; open circuit voltage (V_{oc}) and fill factor (FF); is observed with increase in temperature. Although clear trend of decrease in J_{sc} value is not observed the overall cell efficiency is found to decrease with temperature. Future work will continue to optimize the CIGS solar cells by increasing thickness, utilizing two and three-stage thermal co-evaporation procedure [43], higher substrate temperature, and post alkali metal treatments. Recrystallization of CIGS absorber layer prepared by multi-stage thermal co-evaporation by using fluxing agent InBr$_3$ and copper chloride vapor treatment in three-stage co-evaporation process may enhance its quality [44, 45]. Ultimate goal is to reduce the Ga from optimized CIGS and make CIS solar cell which is lower band gap material ∽ 1.0 eV. Use the optimized CIGS/CIS solar cells in the perovskite tandem cells as a bottom cell.

Two authors contributed equally

ACKNOWLEDGEMENTS

Department of Physics, College of Education for Pure Science, Univ of Mosul, 41002 Iraq. Wright Center for Photovoltaics Innovation & Commercialization, University of Toledo, Toledo, Ohio, 43606, USA.

REFERENCES

[1] R. Mickelsen, W. S. Chen, Y. R. Hsiao, and V. Lowe, "Polycrystalline thin-film CuInSe$_2$/CdZnS solar cells," *IEEE Transactions on Electron Devices,* vol. 31, no. 5, pp. 542-546, 1984.

[2] U. Rau *et al.*, "Oxygenation and air-annealing effects on the electronic properties of Cu(In,Ga)Se$_2$ films and devices," *Journal of Applied Physics,* vol. 86, no. 1, pp. 497-505, 1999.

[3] T. Walter, R. Herberholz, C. Müller, and H. Schock, "Determination of defect distributions from admittance measurements and application to Cu(In,Ga)Se$_2$ based heterojunctions," *Journal of applied physics,* vol. 80, no. 8, pp. 4411-4420, 1996.

[4] I. Repins *et al.*, "19· 9%-efficient ZnO/CdS/CuInGaSe$_2$ solar cell with 81· 2% fill factor," *Progress in Photovoltaics: Research and applications,* vol. 16, no. 3, pp. 235-239, 2008.

[5] R. Herberholz, M. Igalson, and H. Schock, "Distinction between bulk and interface states in CuInSe$_2$/CdS/ZnO by space charge spectroscopy," *Journal of Applied Physics,* vol. 83, no. 1, pp. 318-325, 1998.

[6] L. Stolt, J. Hedström, J. Kessler, M. Ruckh, K. O. Velthaus, and H. W. Schock, "ZnO/CdS/CuInSe$_2$ thin-film solar cells with improved performance," *Applied Physics Letters,* vol. 62, no. 6, pp. 597-599, 1993.

[7] S.-H. Wei, S. Zhang, and A. Zunger, "Effects of Na on the electrical and structural properties of CuInSe$_2$," *Journal of Applied Physics,* vol. 85, no. 10, pp. 7214-7218, 1999.

[8] T. Walter and H. Schock, "Crystal growth and diffusion in Cu(In,Ga)Se$_2$ chalcopyrite thin films," *Thin Solid Films,* vol. 224, no. 1, pp. 74-81, 1993.

[9] W. N. Shafarman, S. Siebentritt, and L. Stolt, "Cu(In,Ga)Se$_2$ Solar Cells," *Handbook of photovoltaic science and engineering,* pp. 546-599, 2010.

[10] W. N. Shafarman, R. Klenk, and B. E. McCandless, "Device and material characterization of Cu(InGa)Se$_2$ solar cells with increasing band gap," *Journal of Applied Physics,* vol. 79, no. 9, pp. 7324-7328, 1996.

[11] P. Jackson *et al.*, "New world record efficiency for Cu(In,Ga)Se$_2$ thin-film solar cells beyond 20%," *Progress in Photovoltaics: Research and Applications,* vol. 19, no. 7, pp. 894-897, 2011.

[12] A. Chirilă *et al.*, "Highly efficient Cu(In,Ga)Se$_2$ solar cells grown on flexible polymer films," *Nature materials,* vol. 10, no. 11, pp. 857-861, 2011.

[13] P. Jackson, D. Hariskos, R. Wuerz, W. Wischmann, and M. Powalla, "Compositional investigation of potassium doped Cu(In,Ga)Se$_2$ solar cells with efficiencies up to 20.8%," *physica status solidi (RRL)–Rapid Research Letters,* vol. 8, no. 3, pp. 219-222, 2014.

[14] P. Jackson *et al.*, "Properties of Cu(In,Ga)Se$_2$ solar cells with new record efficiencies up to 21.7%," *physica status solidi (RRL)–Rapid Research Letters,* vol. 9, no. 1, pp. 28-31, 2015.

[15] P. Jackson, R. Wuerz, D. Hariskos, E. Lotter, W. Witte, and M. Powalla, "Effects of heavy alkali elements in Cu(In,Ga)Se$_2$ solar cells with efficiencies up to 22.6%," *physica status solidi (RRL)–Rapid Research Letters,* vol. 10, no. 8, pp. 583-586, 2016.

[16] P. Jackson, R. Wuerz, D. Hariskos, E. Lotter, W. Witte, and M. Powalla, "Effects of heavy alkali elements in Cu (In, Ga) Se2 solar cells with efficiencies up to 22.6%," *physica status solidi (RRL)–Rapid Research Letters,* vol. 10, no. 8, pp. 583-586, 2016.

[17] M. Nakamura, K. Yamaguchi, Y. Kimoto, Y. Yasaki, T. Kato, and H. Sugimoto, "Cd-free Cu(In,Ga)(Se,S)$_2$ thin-film solar cell with record efficiency of 23.35%," *IEEE Journal of Photovoltaics,* vol. 9, no. 6, pp. 1863-1867, 2019.

[18] M. A. R. Alaani *et al.*, "Optical Properties of Magnesium-Zinc Oxide for Thin Film Photovoltaics," *Materials,* vol. 14, no. 19, p. 5649, 2021.

[19] M. K. Jamarkattel *et al.*, "Indium Gallium Oxide Emitters for High-Efficiency CdTe-Based Solar Cells," *ACS Applied Energy Materials,* 2022.

[20] W. Shockley and H. J. Queisser, "Detailed balance limit of efficiency of p-n junction solar cells," *Journal of applied physics,* vol. 32, no. 3, pp. 510-519, 1961.

[21] D. Sapkota, P. Koirala, P. Pradhan, and R. Collins, "Real Time Spectroscopic Ellipsometry Analysis of the Structural Evolution and Optical Properties of $CuInSe_2$," *Bulletin of the American Physical Society,* vol. 63, 2018.

[22] D. R. Sapkota *et al.*, "Spectroscopic Ellipsometry Investigation of $CuInSe_2$ as a Narrow Bandgap Component of Thin Film Tandem Solar Cells," in *2018 IEEE 7th World Conference on Photovoltaic Energy Conversion (WCPEC)(A Joint Conference of 45th IEEE PVSC, 28th PVSEC & 34th EU PVSEC)*, 2018: IEEE, pp. 1943-1948.

[23] N. Shrestha *et al.*, "Identification of defect levels in copper indium diselenide ($CuInSe_2$) thin films via photoluminescence studies," *MRS advances,* vol. 3, no. 52, pp. 3135-3141, 2018.

[24] P. Uprety *et al.*, "Transport properties of photovoltaic devices via optical Hall effect."

[25] P. Uprety *et al.*, "Optical Hall effect of PV device materials," *IEEE Journal of Photovoltaics,* vol. 8, no. 6, pp. 1793-1799, 2018.

[26] P. Uprety *et al.*, "Application of Optical Hall Effect to PV Relevant Materials," *Bulletin of the American Physical Society,* vol. 63, 2018.

[27] R. H. Ahangharnejhad *et al.*, "Optical design of perovskite solar cells for applications in monolithic tandem configuration with $CuInSe_2$ bottom cells," *MRS Advances,* vol. 3, no. 52, pp. 3111-3119, 2018.

[28] R. Hosseinian Ahangharnejhad *et al.*, "Optical design of perovskite solar cells for applications in tandem configuration with $CuInSe_2$ bottom cells," *Bulletin of the American Physical Society,* vol. 63, 2018.

[29] D. R. Sapkota *et al.*, "Evaluation of $CuInSe_2$ Materials and Solar Cells Co-evaporated at Different Rates Based on Real Time Spectroscopic Ellipsometry Calibrations," in *2021 IEEE 48th Photovoltaic Specialists Conference (PVSC)*, 2021: IEEE, pp. 0451-0458.

[30] D. R. Sapkota *et al.*, "Structural and Optical Properties of Two-Stage $CuInSe_2$ Thin Films Studied by Real Time Spectroscopic Ellipsometry," in *2019 IEEE 46th Photovoltaic Specialists Conference (PVSC)*, 16-21 June 2019 2019, pp. 0943-0948, doi: 10.1109/PVSC40753.2019.8980671.

[31] D. R. Sapkota *et al.*, "Optimization of the $CuInSe_2$ Absorber for the Bottom Cell of a Polycrystalline Thin Film Tandem Solar Cell," *Bulletin of the American Physical Society,* vol. 65, 2020.

[32] V. Ranjan, T. Begou, S. Little, R. W. Collins, and S. Marsillac, "Non-destructive optical analysis of band gap profile, crystalline phase, and grain size for Cu(In,Ga)Se2 solar cells deposited by 1-stage, 2-stage, and 3-stage co-evaporation," *Progress in Photovoltaics: Research and Applications,* vol. 22, no. 1, pp. 77-82, 2014, doi: https://doi.org/10.1002/pip.2350.

[33] K. Orgassa, H. W. Schock, and J. Werner, "Alternative back contact materials for thin film Cu (In,Ga)Se2 solar cells," *Thin Solid Films,* vol. 431, pp. 387-391, 2003.

[34] I. Repins, S. Glynn, J. Duenow, T. J. Coutts, W. K. Metzger, and M. A. Contreras, "Required material properties for high-efficiency CIGS modules," in *Thin Film Solar Technology*, 2009, vol. 7409: SPIE, pp. 156-169.

[35] F. Ouachtari, A. Rmili, B. Elidrissi, A. Bouaoud, H. Erguig, and P. Elies, "Influence of bath temperature, deposition time and S/Cd ratio on the structure, surface morphology, chemical composition and optical properties of CdS thin films elaborated by chemical bath deposition," *Journal of modern physics,* vol. 2011, 2011.

[36] H. Kim *et al.*, "Indium tin oxide thin films for organic light-emitting devices," *Applied physics letters,* vol. 74, no. 23, pp. 3444-3446, 1999.

[37] W. Li, X. Yan, A. G. Aberle, and S. Venkataraj, "Effect of a TiN alkali diffusion barrier layer on the physical properties of Mo back electrodes for CIGS solar cell applications," *Current Applied Physics,* vol. 17, no. 12, pp. 1747-1753, 2017.

[38] B. Dimmler, M. Powalla, and R. Schaeffler, "CIS solar modules: pilot production at Wuerth Solar," in *Conference Record of the Thirty-first IEEE Photovoltaic Specialists Conference, 2005.*, 2005: IEEE, pp. 189-194.

[39] P. Singh, S. Singh, M. Lal, and M. Husain, "Temperature dependence of I–V characteristics and performance parameters of silicon solar cell," *Solar Energy Materials and Solar Cells,* vol. 92, no. 12, pp. 1611-1616, 2008.

[40] P. Singh and N. M. Ravindra, "Temperature dependence of solar cell performance—an analysis," *Solar energy materials and solar cells,* vol. 101, pp. 36-45, 2012.

[41] A. Mahfoud, F. Mohamed, S. Mekhilef, and F. Djahli, "Effect of temperature on the GaInP/GaAs tandem solar cell performances," *International Journal of Renewable Energy Research,* vol. 5, no. 2, pp. 629-634, 2015.

[42] M.-J. Jeng, Y.-L. Lee, and L.-B. Chang, "Temperature dependences of $In_xGa_{1-x}N$ multiple quantum well solar cells," *Journal of Physics D: Applied Physics,* vol. 42, no. 10, p. 105101, 2009.

[43] P. Pradhan *et al.*, "Real time spectroscopic ellipsometry analysis of the three-stages of $CuIn_{1-x}Ga_xSe_2$ co-evaporation," in *2014 IEEE 40th Photovoltaic Specialist Conference (PVSC)*, 2014: IEEE, pp. 2060-2065.

[44] D. Poudel *et al.*, "Analysis of Post-Deposition Recrystallization Processing via Indium Bromide of Cu(In,Ga)Se2 Thin Films," *Materials,* vol. 14, no. 13, p. 3596, 2021.

[45] D. Poudel *et al.*, "In Situ Recrystallization of Co-Evaporated Cu(In,Ga)Se2 Thin Films by Copper Chloride Vapor Treatment towards Solar Cell Applications," *Energies,* vol. 14, no. 13, p. 3938, 2021.

978-1-7281-6118-1/22 $31.00 © 2022 IEEE

Flight Demonstration Test of State-of-the-Art Photovoltaic Devices on JAXA’s New ISS Transfer Vehicle HTV-X

Mitsuru Imaizumi, Teppei Okumura, Tetsuya Nakamura, Shusaku Kanaya, Taishi Sumita

JAXA, Tsukuba, Japan

We plan to perform an on-orbit demonstration test of a variety of state-of-the-art photovoltaic devices by using the test flight module of JAXA' new ISS transfer vehicle HTV-X. The demonstration mission is called SDX, and it includes InGaP/GaAs//CIGS PHOENIX triple-junction, perovskite, CIGS, micro-concentrator module and #503 InGaP/GaAs/Ge triple-junction solar cells. The devices for SDX demonstration test have suffered environmental tests of UV, proton/electron irradiation, thermal cycle, and vibration. The photovoltaic device exposure board and the measurement unit are now under assembling. The expected launch date is early 2023.

Mechanical Degradation Studies on flexible CIGS cells and modules for floating PV

Wim Soppe, Aldo Kingma and Dorrit Roosen

TNO-Solliance, Eindhoven, The Netherlands

Abstract—**This paper describes mechanical degradation experiments carried out to simulate possible bending effects of flexible CIGS modules when mounted on flexible floaters on sea. In an offshore environment the PV modules are subjected to millions of deformations per year. It is concluded that the strain induced by the bending can lead to serious performance loss of the PV modules and measures are advised to reduce the strain in the modules.**

Keywords—CIGS, degradation, floating PV

I. INTRODUCTION

A large part of the human population lives in densely crowded areas in the vicinity of sea or ocean. Floating PV is therefore considered as an important option to supply the world with renewable energy. The wave conditions vary largely between these locations but in this paper we use the conditions of one of the more difficult locations: the North Sea, as a starting point.

Within the framework of a Dutch research project Solar@Sea, a concept for offshore floating PV is developed, based on flexible floaters and flexible PV modules.

Figure 1: Solar@Sea floater with PV modules.

The flexible floaters are made of double wall fabric, are filled with air and contain water ballast bags underneath for stability. The flexibility of the floaters makes them excellent wave riders, resulting in low energy absorption from the waves. This, in turn, results in small mooring forces for which relatively simple (and low cost) mooring systems are required.

The drawback of the excellent wave riding properties is that the PV modules will undergo very frequent deformations. For the North Sea we have to take into account tens of millions of deformations per year.

In this paper we will describe the nature of these deformations in more detail, explain how we simulate these wave induced deformations and report on test results that we obtained

II. NORTH SEA WAVE CONDITIONS

The North Sea is a challenging environment for floating PV with a significant wave height of 7.3 m for a 1 year recurrence period, (that is a storm that can be expected once per year) [1]. More important than these extreme wave heights are the wave heights, lengths and periods under "normal" conditions. The most dominant waves on the North Sea can be divided into 10 categories, that are shown in the table below.

Table I: dominant wave types at the North Sea

Wave type	Period (s)	Length (m)	Amplitude (m)
1	4.5	31	0.8
2	5.2	43	0.9
3	5.4	44	1.2
4	5.3	44	2.4
5	6.3	64	4.2
6	8.2	99	7.2
7	10.7	156	9.8
8	15.8	260	11.0
9	31.6	540	10.6
10	58.2	1030	13.0

The most frequent occurring waves are types 1-4, with short periods: in the order of 5 seconds. This implies that the floaters and the modules are subjected to millions of deformations per year (an interval time of 5 s corresponds to 6.3 million waves/yr). Another observation is that the dominant higher waves have longer wavelengths such that the maximum deformations occur for the shorter waves. We will elaborate on that in the next paragraph.

III. EXPERIMENTAL SETUP

To simulate the frequent deformations that PV modules will be subjected to in the Solar@Sea concept, we have built a dedicated setup that is shown in the figure below.

This work was funded by the Topsector Energy of the Dutch Ministry of Economic Affairs (project no. TEUE119003).

Figure 2: Test setup for mechanical degradation studies of flexible PV modules.

In the test setup small modules of typically 50×40 cm² are fixated on six rods which can move independently of each other in a direction perpendicular to the plane of the PV module. The amplitude can be varied in the range 0.5-5 cm and the frequency can be varied in the range of 1-5 Hz. The setup can be combined with a climate chamber, allowing the tests to be carried out between -45 °C and + 85 °C.

The modules that we used in this study were home made using CIGS cells from Miasole [2]. These cells are made on steel foil and have typical dimensions of 30×5 cm and a nameplate efficiency of 15%.

The full stack of the modules consisted of a front sheet (ETFE) 0.20 mm; an encapsulant (POE) 0.45 mm; a solar cell (CIGS) 0.33 mm; an encapsulant (POE) 0.45 mm; and a back sheet (PET+Al) 0.35 mm.

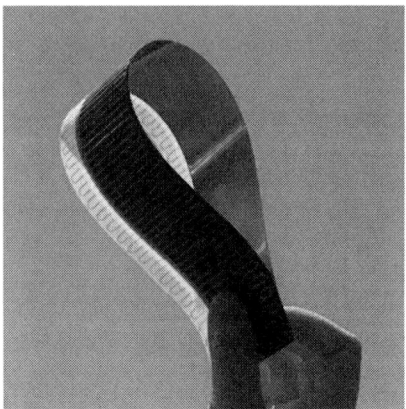

Figure 3: CIGS cell from Miasole

In this study we assume that the floater will bend such that the wave profile is completely followed. From wave basin test we know that this is a conservative assumption. In order to determine the amplitudes to be applied in the test setup we had to calculate the maximum bending heights over a length of 40 cm for the different wave categories shown in Table I. The resulting amplitudes are shown in Table II. In this table also the maximum theoretical strain is shown, that would result if the module would be rigidly fixated to the rods. The minimum amplitude of 5 mm of our setup is much larger than the expected bending heights and would lead to unrealistically high strains if the module would be rigidly fixated. In our setup therefore the strain is relieved by springs between the PV module and the rods. Consequently, the experiments we describe here focus rather on bending effects than on strain effects. In practice the strain of the modules will also be mitigated by a certain elasticity of the adhesion material that is applied between the floater and the modules.

Table II: Maximum bending heights over a length of 40 cm and corresponding strains for the dominant North Sea wave types.

Wave type	Maximum bending height (mm)	Maximum theoretical strain (%)
1	0.96	1.24
2	0.61	0.79
3	0.76	0.98
4	1.55	2.00
5	1.26	1.63
6	0.91	1.17
7	0.49	0.64
8	0.20	0.26
9	0.04	0.06
10	0.02	0.02

Two types of modules were made: one type with cells in horizontal orientation and one type with cells in vertical orientation.

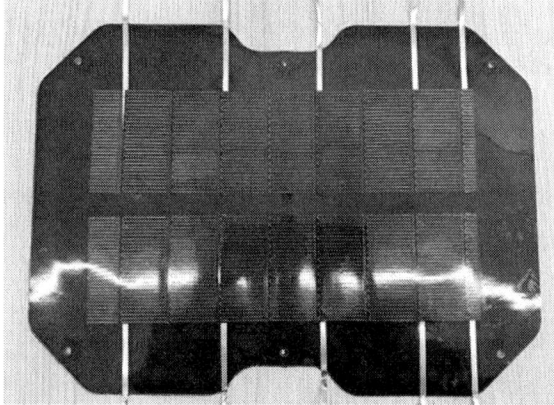

Figure 4: test module with cells in vertical orientation.

The testing procedure was as follows: after fabrication the performance of the individual cells was measured and EL measurements were performed. Then the module was subjected to 1,600,000 sinusoidal deformations, at a frequency of 2 Hz and an amplitude of 1 cm, with waves traveling in the horizontal or oblique direction. The IV and EL measurements were repeated and so on until the module failed or more than one million cycles were reached.

IV. Results and Discussion

A. Cells oriented inline with the wave direction

The modules with cells in horizontal orientation degraded significantly. All IV parameters deteriorated but there was significant spread over the different cells in each module. The IV characteristic of the best performing cell is shown in Figure 5. After 1.6 million cycles the I_{sc} has dropped by 7%; the V_{oc} by 3% and the FF by 2%, resulting in a power loss of 3.5%. Other cells in that module lost more performance after 1.6 million cycles, namely up to 18%. For the worst performing cells V_{oc} and FF had degraded by about 10% and the I_{sc} degraded by about 5%.

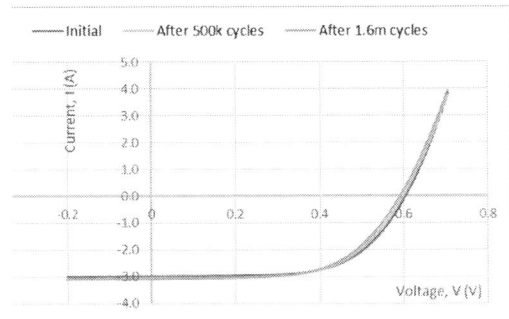

Figure 5: IV results for one cell in a module with cells oriented horizontally. Cycling at room temperature.

We assume that the cells, when oriented in line with the wave direction, will be strained significantly, despite the strain mitigation by the springs. This assumption is supported by EL measurements as shown in Figure 6 where many shunted areas have occurred after cycling.

According to [3], the yield stress of CIGS is between 640 and 1100 MPa. We could not measure the real strain of the samples during the test. But, assuming an elastic modulus of 83 GPa [4], a strain of 1 % would already lead to a tensile stress of 830 MPa: i.e. to a plastic deformation of the CIGS absorber layer. This is confirmed by a study of Yang et al. [5] who conducted a degradation study on similar CIGS cells strained in the same horizontal direction. They observed a severe drop in performance for strains larger than 1.5%. They attribute the performance loss to either plastic deformation of the CIGS layer (grain boundary cracking) or to delamination of the CIGS/Mo interface. They also report a complete failure of the cells when strained once by more than 2.5%. Since we did not observe such a complete loss of performance, we can assume that in our experiments the strain was well below 2.5%.

Figure 6: EL image of two cells after 1.6 million cycles at room temperature.

B. Cells oriented perpendicular to the wave direction

Tests with cells oriented vertically in the module have started only recently and have not been finalized yet. First tests were carried out at -45 °C and reached 173.000 cycles by now. After this amount of cycles the cells showed negligible loss in performance. The relative efficiency loss was on average less than 1%. The test however needs to be extended to many more cycles before we can draw conclusions for this orientation of the cells.

C. Cells oriented oblique to the wave direction

For these tests we used modules with cells oriented vertically but modified the phase of the individual rod movements to simulate oblique incidence of waves. The modules were subjected to 1 million cycles at -45, +25 and +85 °C. For temperatures of -45 and +25 °C a performance loss of 4-5% was observed. A much larger performance loss of 20% was observed after cycling at +85 °C. This performance loss is mainly caused by delamination of the metal grid on the front side of the cells, due to softening of the encapsulants in combination with the deformation.

V. Conclusions and Discussion

The experiments described in this paper confirm that the performance of CIGS cells can degrade when strained. We investigated the effects of very high numbers of strain inducing deformations that flexible CIGS PV modules would be subjected to when installed on flexible floaters at the North Sea. The results indicate that under these circumstances the efficiency of modules may deteriorate significantly. It is recommended to adopt the design of the total construction such that the strain of the floaters is not fully transferred to the PV modules.

Acknowledgment

Debanjan Paul and James Ho are gratefully acknowledged for their significant contributions to the experiments carried out in this study.

References

[1] Erik Asp "MetOcean Investigations for the Wind Farm Zone Hollandse Kust (noord) ". Netherlands Enterprise Agency report CR-SC-DNVGL-SE-0190-02453-4 (2019).

[2] https://miasole.com/products/

[3] Shi Luo, Jiun-Haw Lee, Chee-Wee Liu, Jia-Min Shieh, Chang-Hong Shen, Tsung-Ta Wu, Dongchan Jang, and Julia R. Greer, "Strength, stiffness, and microstructure of Cu(In,Ga)Se2 thin films deposited via sputtering and co-evaporation" Applied Physics Letters 105, 011907 (2014)

[4] Tang-Yu Lai,, Yu-JenHsiao and Te-Hua Fang, "Mechanical properties of CIGS film with different metallic composition by co-evaporation method" Mater. Res. Express 4 (2017) 115006.

[5] Chen Yang, Kai Song, XinLiang Xu, Gang Yao, ZhenYu Wu, "Strain dependent effect on power degradation of CIGS thin film solar cell" Solar Energy 195 (2020) 121–128.

978-1-7281-6118-1/22 $31.00 © 2022 IEEE 1261

Influence of temperature and magnetic field on the transient voltage decay of a silicon solar cell with parallel vertical junction in open circuit.

pape Diop, Papa Touty Traore, Papa Monzon Samake, Babou Dione , Fatimata Ba, Modou Pilor

Dakar, Senegal

A study of the effect of magnetic field temperature on a parallel vertical junction silicon solar cell in open circuit transient is presented in this work. The transitory density of excess minority carriers appears as the sum of infinite terms. The decay time between the density of minority carriers and the different harmonic states is studied. Optimizing the thickness of the base allows us to determine optimal temperatures that will constantly increase the transient decay time.

Progression in Grain Size of Novel Photoferroic Absorber Bournonite (CuPbSbS3)

Oliver M Rigby, Budhika G Mendis, Marek Szablewski

Durham University, Durham, United Kingdom

Bournonite (CuPbSbS3) is an emerging absorber material for solar cells. While initial solar cell devices with bournonite as an absorber layer are promising, the grain size is a limiting factor restricting current power conversion efficiencies. In this work we synthesise bournonite thin-films through a thiol-amine dissolution of bulk oxides and investigate the effects of temperature on controlling the grain sizes, morphology and phase purity of bournonite thin-films. Higher annealing temperatures of 500°C result in ~3μm grain sizes, without the formation of PbS or CuSbS2 secondary phases. This is a significant improvement over the sub-micrometer grain sizes reported previously, although the larger grain size is achieved at the expense of some surface coverage.

978-1-7281-6118-1/22 $31.00 © 2022 IEEE

Direct observation of an atomic thin inversion layer at the native oxide/ n-Si interface

Yibo Zhang, Joel Y. Y. Loh, Andrew G. Flood, Chengliang Mao, Geetu Sharma, Nazir P. Kherani

University of Toronto , Toronto, ON, Canada

A recombination-free high-quality interface is desired for all semiconductor solar cells. An induced inversion layer and the resulting concentrated surface electric field are critical factors for the reduction of photocarrier recombination. The surface inversion layer has been widely predicted at the interface of two materials with different work functions leading to large energy band bending. Herein, we present the direct observation of an atomic thin hole inversion layer between cubic-phase indium tin oxide (c-ITO)/native oxide/n-Si interface using transmission electron microscopy. Excellent lattice matching, atomic oxidation of pristine Si substrate and interfacial charge engineering enable this high-quality interface and its revelation. A facile process of air-annealing and commensurate adjacent thin film phase transition is explored. The device exhibits an ultra-low recombination rate at the interface and internal quantum efficiency (IQE) of over 97% for a broad range of wavelengths. This presentation will provide an understanding of the silicon - native oxide semiconductor interface as developed via facile processing and electron microscopy.

978-1-7281-6118-1/22 $31.00 © 2022 IEEE

Ultrathin III-V solar cells with light-trapping structures fabricated in situ using an HVPE reactor

Allison N. Perna, Anna K. Braun, Kevin L. Schulte, John Simon, Corinne E. Packard, Aaron J. Ptak

Colorado School of Mines, Golden, CO, United States

National Renewable Energy Laboratory, Golden, CO, United States

We developed a back surface scattering technique for inverted III-V solar cells by texturing GaInP with HCl and PH3 in situ within a hydride vapor phase epitaxy (HVPE) reactor. Back surface texturing is a promising light management method for ultrathin solar cells to enhance photocurrent collection, however most methods demonstrated to date are expensive. This fully in situ method is a potentially low-cost and high-throughput technique, in which the rear GaInP contact layer is textured via vapor phase etching and redeposition of Ga-rich GaInP. We demonstrate single junction GaAs solar cells with a relative 5% boost in short circuit current density for in situ back surface textured devices without any loss in open circuit voltage or fill factor. This work supports the development of low-cost ultrathin III-V photovoltaics.

New substituted small A cation(Acetamidinium) based Tin perovskite solar cell

Soumen Kundu, Sushobhan Avasthi*

Centre for Nano Science and Engineering, Indian Institute of Science, Bangalore, Karnataka – 560012, India

*Email: savasthi@iisc.ac.in

Abstract— There is a need to search for high-efficiency Pb-free metal halide perovskites. Here we report a novel organic cation (Acetamidinium) for substituted formamidimium tin iodide perovskite. Using the acetamidinium-substituted formamidinium tin perovkite as the absorber, we demonstrate solar cells in the p-i-n configuration. The best device exhibited a power conversion efficiency $\approx$ 1%; with a short-circuit current density (J_{sc}) of 8.53 mA/cm^2, open circuit-voltage (V_{oc}) of 0.3V, and fill-factor (FF) of 39%. Poor V_{oc} of the device is due to conduction band offset of absorber layer with PCBM and interfacial recombination.

Keywords—Perovskite solar cell, steel-substrate, semi-transparent-metal contact

I. INTRODUCTION

Hybrid organic-metal halide perovskite is a class of cost-effective absorber that is easy to process at low temperatures [1]. Lead-based perovskite solar cells have achieved peak efficiency of 25.6 % [2]. However, lead in the lead-perovskite is highly toxic and soluble in water, posing a significant environmental hazard [3] [4]. The research community has been evaluating Pb-free alternatives, like Sn, Ge, Sb and Bi [5]. However, the most promising continue to Sn-based perovskite[6], [7]. Unfortunately, the performance of Sn-based perovskites is low due to bulk recombination, surface recombination and band alignment with existing ETL. Bulk recombination is largely caused by defects due to the tendency of Sn^{+2} to oxidize to Sn^{+4} [8]. The same tendency also explains the poor ambient stability of Sn-perovkite

Cation engineering has emerged as a promising method to improve stability of perovskite thin-films. Large cation like GA, can stabilize the perovskite matrix. We have previously reported that Acetamidinium (AA) cation forms more number of hydrogen bonds than MA$^+$ cation inside octahedral cages PbI$_6$^{4}. AA-substituted MAPbI3 solar cells show higher V_{OC} and improved stability [9]. In this report, we report acetamidinium-substituted formamidinium Sn-iodide. Devices with p-i-n configuration were fabricated with PEDOT:PSS as the hole-transport layer (HTL) and [6,6]-phenyl C$_{61}$ butyric acid methyl ester (PCBM) and Bathocuporine (BCP) as the electron transport layers (ETL). The devices shows a short-circuit current density (J_{sc}) of 8.53 mA/cm^2, open circuit voltage of 0.3V, fill factor(FF) of 47 %, and efficiency of 1%.

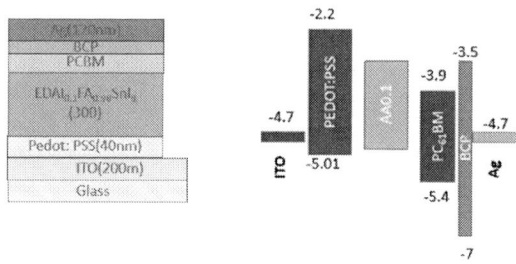

Fig.1 (a) Device stack (b) energy level of different layers

II. EXPERIMENTAL DETAIL

The device structure and band-energy alignment of the corresponding layers are shown in figure 1(a) and 1(b), respectively. Devices were fabricated on glass coated with 200 nm ITO, which was precleaned with deionized water, acetone and IPA for 10 minutes each. After 30 minutes ozone exposure, PEDOT: PSS is spin coated at 4000 rpm followed annealed at 140 °C for 20 minutes. The perovskite precursor solution was spin coated at 4000 rpm with chlorobenzene as antisolvent. The antisolvent is dripped at 30th second in a spin duration of 40s and then annealed at 65°C for 10 minutes. The perovskite precursor solution was prepared by dissolving tin iodide (SnI$_2$): formamidinium iodide (FAI): acetamidinium iodide (AAI) salts in 1:0.9:0.1 molar ratio at 1 ml of DMF and DMSO (8:2) solvent. For ETL layer, 20 mg of PCBM was dissolved in 1 ml of chlorobenzene (CB). The PCBM solution was spin coated at 2000 rpm. for 45 second. After that, 0.5 mg BCP dissolved in isopropanol spin coated at 6000 rpm. for 45 seconds. For top electrode, silver (Ag) was evaporated 100 nm at a base pressure of 5×10^{-7} mbar. Device area of each cell is 0.045 cm^2.

To investigate absorber layer's morphology scanning electron microscope (Zeiss Ultra-55 FESEM) was used. UV visible (Perkin Elmer Lambda UV-vis-NIR) was used to study absorption spectra of absorber layer. Solar cell measurement was conducted using Keithley 2420 source meter and solar simulator (Oriel Instrumentation). Under illumination solar simulator was calibrated to 1 sun (or 100mW/cm^2). XRD spectra was scanned using a Rigaku smart-lab X-ray diffractometer at room temperature using Cu-Kα radiation. Parallel beam optics with medium resolution was taken for theta/2theta scan. 2° /min scan speed and 0.01° step size were selected for the XRD scan.

III. RESULTS AND DISCUSSION

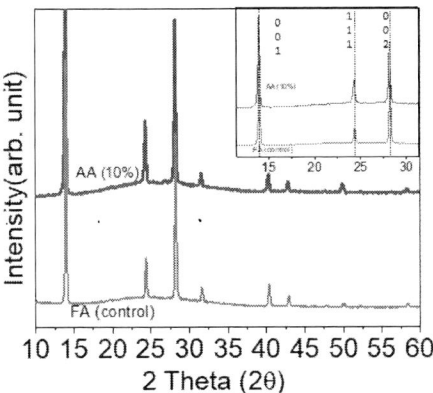

Fig. 2 X-ray diffraction pattern: red curve shows FASnI$_3$ (control) perovskite and Black curve shows 10 % AA substituted AA$_{0.1}$FA$_{0.9}$SnI$_3$ perovskite, inset graph shows the zoom view of the XRD pattern.

Figure 2 shows the X-ray diffraction patterns of FASnI$_3$ perovskite thin-film. Incorporation of Acetamidinium ion (AA$^+$) in place of formamidinium ion (FA$^+$) does not change the host crystal structure as both spectra show peaks of an orthorhombic crystal. However, we can see inset of figure 2 (111) and (002) plane's peaks in AA substituted sample shifts 0.04° and 0.06° respectively in leftward with respect to control sample. This peak shift attributes to expansion lattice spacing of those corresponding planes. In addition, we show in figure 3 size-strain plot. It can segregate peak broadening effect into size effect and strain effect. From this analysis, it is observed that AA incorporation increases strain little higher than control sample. Therefore, this little residual strain can affect charge carrier dynamics.

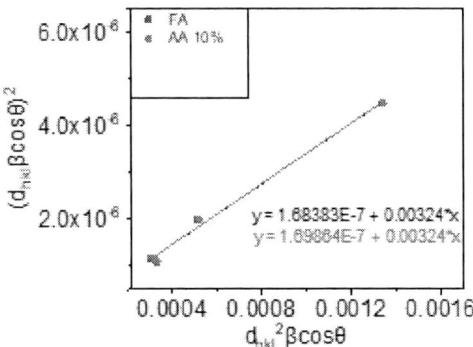

Fig. 3 SSP plot FASnI3 perovskite (control) sample and AA$_{0.1}$FA$_{0.9}$SnI$_3$ perovskite sample.

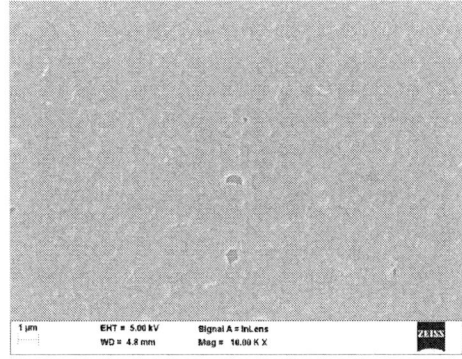

Fig. 4 Scanning electron microscope image of AA$_{0.1}$FA$_{0.9}$SnI$_3$ perovskite

The figure 4 shows absorption spectra of control and 10% AA substituted sample. With addition of AA cation perovskite band edge shifts leftward in wavelength scale from 892 nm to 867 nm. That means, this AA substituted sample's band gap increase 0.4 ev with respect to control sample. This band gap range is suitable for photovoltaic application. The figure 5 exhibits scanning electron microscope image AA 10 % substituted perovskite material on ITO. This image shows average grain size of 500-700 nm. and the film coverage is good. The figure 5(a) shows J-V characteristics of AA$_{0.1}$FA$_{0.9}$SnI$_3$ perovskite based single junction p-i-n solar cell. V$_{oc}$ of the best was obtained 0.3 v, J$_{sc}$ was 8.53 mA/cm^2 and FF 39 % and efficiency 1%. Device performance is poor due to loss of Voc and fill factor, at low bias high dark (J$_o$) current

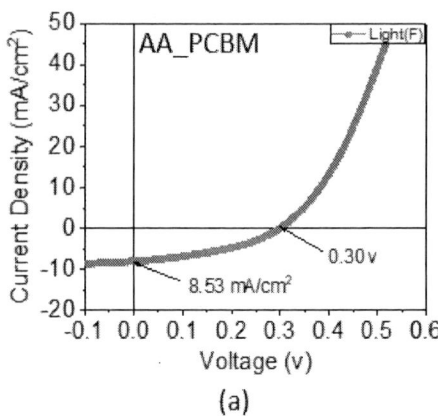
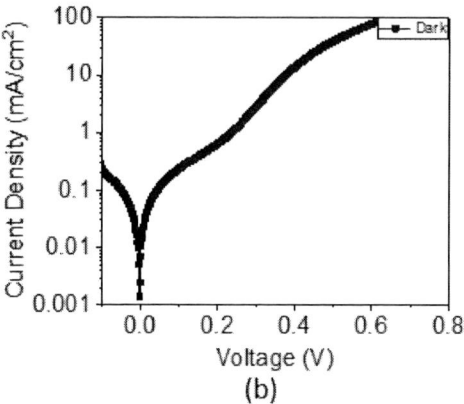

(a) (b)

Fig. 5 J-V characteristics of the best measured $AA_{0.1}FA_{0.9}SnI_3$ perovskite p-i-n solar cell a) J-V plot of AA 10% substituted solar cell under illumination of 1sun(100 mW/cm^2) b) J-V plot under dark condition

[2]

in figure 5(b) attributes generation current in depletion region by bulk traps. Next, if we further look at little higher bias region of the dark J-V curve we still find high dark current, because of surface recombination. Fill factor is also poor because of poor shunt resistance.

IV. SUMMARY

This is a new substituted A cation(AA) attempted in Sn based perovskite in p-i-n solar cell. In first attempt 1% solar cell was obtained. Further improvement is in progress.

ACKNOWLEDGE

The authors acknowledge the Ministry of Education, Government of India, for financial support under Scheme for Transformational and Advanced Research in Sciences (STARS) project number MoE-STARS/STARS1/135. This work was conducted at NNfC and MNCF with generous support from MeitY under grant MeitY 5(3)/2017-NANO, by the DST under grant DST/NM/NNetRA/2018(G)-IISc and MHRD.

REFERENCE

[1]   Q. Jiang *et al.*, "Surface passivation of perovskite film for efficient solar cells," *Nat. Photonics*, vol. 13, no. 7, pp. 460–466, 2019.

J. Jeong *et al.*, "Pseudo-halide anion engineering for α-FAPbI3 perovskite solar cells," *Nature*, vol. 592, no. 7854, pp. 381–385, 2021.

[3]   W. Ke and M. G. Kanatzidis, "Prospects for low-toxicity lead-free perovskite solar cells," *Nat. Commun.*, vol. 10, no. 1, p. 965, 2019.

[4]   G. Flora, D. Gupta, and A. Tiwari, "Toxicity of lead: A review with recent updates," *Interdiscip. Toxicol.*, vol. 5, no. 2, pp. 47–58, Jun. 2012.

[5]   R. Nie, R. R. Sumukam, S. H. Reddy, M. Banavoth, and S. Il Seok, "Lead-free perovskite solar cells enabled by hetero-valent substitutes," *Energy Environ. Sci.*, vol. 13, no. 8, pp. 2363–2385, 2020.

[6]   A. Toshniwal and V. Kheraj, "Development of organic-inorganic tin halide perovskites: A review," *Sol. Energy*, vol. 149, pp. 54–59, 2017.

[7]   F. Giustino and H. J. Snaith, "Toward Lead-Free Perovskite Solar Cells," *ACS Energy Lett.*, vol. 1, no. 6, pp. 1233–1240, Dec. 2016.

[8]   D. Meggiolaro, D. Ricciarelli, A. A. Alasmari, F. A. S. Alasmary, and F. De Angelis, "Tin versus Lead Redox Chemistry Modulates Charge Trapping and Self-Doping in Tin/Lead Iodide Perovskites," *J. Phys. Chem. Lett.*, vol. 11, no. 9, pp. 3546–3556,

2020.

[9] P. Singh, R. Mukherjee, and S. Avasthi, "Acetamidinium-Substituted Methylammonium Lead Iodide Perovskite Solar Cells with Higher Open-Circuit Voltage and Improved Intrinsic Stability," *ACS Appl. Mater. Interfaces*, vol. 12, no. 12, pp. 13982–13987, 2020.

Significance of Power and Energy Ratings of Modules in Large-scale PV Plants

Bijaya Paudyal and Donny Campos Paniagua

Main Equipment and Solar Center of Excellence, Enel Green Power North America, Andover, MA, 01810, USA

Abstract— **Power and energy rating measurement processes of PV modules are described in IEC standards 60904-3 and IEC61853. PV modules are purchased, and revenue forecasts are being made using the power and energy rating for large-scale PV plants. The energy rating of a PV module is based on the laboratory tests, which provide data required for a PAN file. PAN files are widely used for energy yield and financial simulations of PV plants. This work investigates the sensitivity of the PAN files on energy yield and LCOE of large-scale PV plants operating in a different part of the United States.**

Keywords— *photovoltaic, power, energy, pan file*

I. INTRODUCTION

Photovoltaic (PV) modules are being purchased using the cost ($/Wp) based on the power rating (Wp), which is also a basis for the design and revenue forecast of a large-scale PV plant. The measuring techniques for the power rating of PV modules are described in the standard, IEC 60904-3:2019 [1], and provide the maximum output power of PV modules (P_{max}) at the standard test condition (STC, the irradiance at 1000W/m2, module temperature at 25 °C, and air-mass at 1.5 global spectrum). The expected energy output of the plant is often calculated using the power rating of the module [2]:

$$E_{Expected} = \sum P_{Array} \left(\frac{G_{POA}}{G_{STC}} \right) \left(1 + C_t (T_{Mod} - T_{STC}) \right) \quad (1)$$

where G_{POA} is global solar irradiance at the plane of the array (POA), G_{STC} is global solar irradiance at STC (1000 W/m²), $(1+C_t (T_{Mod} - T_{STC}))$ is the temperature normalizing factor, and C_t is the power temperature coefficient of a PV module. T_{Mod} is the module temperature, and T_{STC} is the temperature at STC. Equation (1) utilizes the power rating of the PV array and normalizes for the expected irradiance and temperature in the field linearly. However, various site-specific operational parameters, i.e., temperatures, irradiances, and sunshine spectrum, affect the actual energy generation of the field-installed PV array, and the linear relationship on one parameter is no longer retained [2].

IEC 61853 [4] illustrates the technique for evaluating the energy efficiency of PV modules, allowing for the estimation of energy yield for specific environmental conditions. The energy rating process entails taking a comprehensive set of measurements on modules under a variety of operational scenarios. The test results are used with a model to predict energy production for a set of standardized reference climates. The energy rating provides comprehensive and reliable information required to estimate the energy yield of a PV plant. However, it should be noted that the energy rating process is limited to module-specific factors. It does not address other site-specific factors such as soiling, shading, inverter clipping, mismatch, and so on, which can affect the energy yield of the PV plant.

PVSyst [5], a modeling software, initiated the PAN format and has been accepted by the solar industry to generate the energy yield forecast for a PV plant. The file extension PAN comes from the French word "PANeau Solaire," which means "solar panel." A PAN file contains detailed information about a module beyond the datasheet and can be plug-in with the site-specific weather data for performance simulation. Fig. 1 depicts a generic process flow for the energy yield simulation of a PV plant.

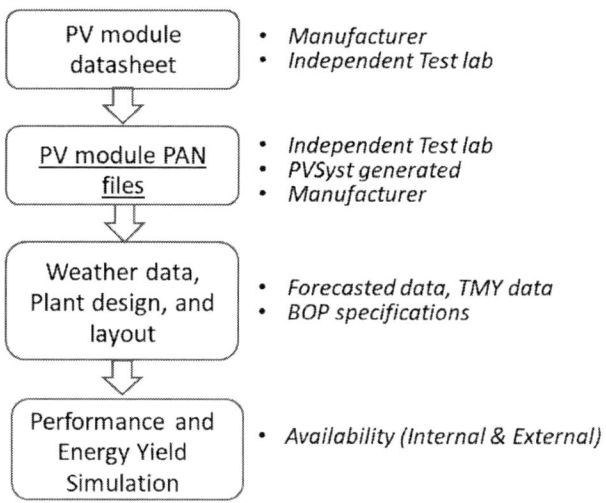

Fig. 1. A generic process flow for the energy yield simulation of a large-scale PV plant.

The information required to construct PAN files is provided by the data supplied by tests outlined in the standards IEC 61853 (1 and 2). However, there is a lack of consensus on the methodology for transferring the measured data of a PV module to a PAN file. Further exacerbating the situation, the PAN files have been created using synthetic data generated by different models that may not be relevant to a specific PV cell or module type they have been applied to. The inappropriate PAN files of a PV module led to an inaccurate performance forecast of a PV plant.

This work investigates the sensitivity of the PAN file parameters in place of the power and energy rating of PV modules in estimating equivalent operating hours (EOH), capital expenses (CapEx), and levelized cost of electricity (LCOE) for two large-scale PV plants. The PAN files provided by a Tier-1 manufacturer and the independent test laboratory for the same PV module are used for the study. The PV plants selected for this work are in two different weather zone in the USA. This paper describes the study approach in the methodology section (Section II), provides results and discussion in Section III, and concludes with remarks in the last section (Section IV).

II. METHODOLOGY

This work selected a PV module from a tier-1 manufacturer that provided six different PAN files for a PV module with the same power rating. The PAN files are based on the third-party laboratory tests according to IEC 61853 and from the manufacturer, created by using the generic models in modeling software. Table I shows the deviation (%) of crucial PAN file parameters (in symbols) from the average value. The normalized values for the PV module parameters are used throughout this paper to maintain the anonymity of the manufacturer.

TABLE I. KEY PAN FILE PARAMETERS (DEVIATION, % FROM THE AVERAGE)

Parameters	PAN-1	PAN-2	PAN-3	PAN-4	PAN-5	PAN-6
I_{sc}	0.85%	0.85%	-0.70%	-0.70%	-0.16%	-0.16%
V_{oc}	0.75%	0.75%	1.97%	1.97%	-2.72%	-2.72%
I_{mp}	0.59%	0.59%	-0.87%	-0.87%	0.30%	0.26%
V_{mp}	-0.58%	-0.58%	0.85%	0.85%	-0.27%	-0.27%
μ_{ISC}	12.86%	12.86%	4.86%	4.86%	-17.72%	-17.72%
$\mu_{VocSpec}$	-0.17%	-0.17%	-0.17%	-0.17%	0.33%	0.33%
μ_{PmpReq}	-2.95%	-2.95%	-11.46%	-0.57%	8.97%	8.97%
R_{Shunt}	-4.39%	-4.39%	-0.63%	-14.07%	11.74%	11.74%
R_{p_0}	76.05%	76.05%	-28.10%	-40.69%	-41.65%	-41.65%
R_{p_Exp}	-21.74%	-21.74%	-21.74%	-21.74%	43.48%	43.48%
R_{Seris}	4.47%	4.47%	6.72%	20.53%	-18.09%	-18.09%
Gamma	5.33%	5.33%	3.27%	5.10%	-9.52%	-9.52%

Parameters: - Isc: Short circuit current, Voc: Open circuit voltage, Imp: Current at maximum power, Vmp: Voltage at maximum power, μIsc: Temperature coefficient of Isc, μVoc: Temperature coefficient of open-circuit voltage, Rshunt: Shunt resistance, Rp_0: parallel resistance at STC, Rseris: Series resistance, and Gamma: Diode ideality factor.

Two large-scale PV plants situated in the continental united states' Northeast and Southwest climate regions are selected for this study. The satellite-based climate data from SolarGIS [6] and Meteonorm [7] are used for analysis that includes the multiyear time series and the typical metrological year (TMY). The simulations of the selected PV plants are performed using the commercially available software, PVSyst [5], and proprietary software, Conceptual Design 2.0 [7].

The PV plants in this study are 70 and 310 Mega Watt (DC) in size. Both plants are designed with a central inverter, and PV modules are mounted in the horizontal single-axis tracker

(HSAT). Hence the sensitivity analysis is performed based on normalized values for each PAN file. The EOH with exceedance probability, P50, is calculated using individual PAN files for the first year and the lifetime average (30-years). The sensitivity analysis is performed for the CapEx and LCOE, considering the average value of the PAN file parameter as the base value.

III. RESULTS AND DISCUSSIONS

A. Northeast PV Plant Study

A PV plant in the Northeast climate region shows a maximum deviation of 2.77% and 2.89% on EOH value for the first year and the plant's lifetime average (30-years), respectively, with SolarGIS TMY data. Fig. 2 shows the first-year EOH of the PV plant at the inverter and grid level.

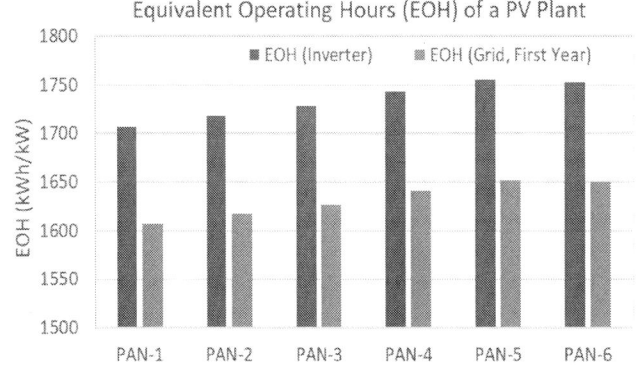

Fig. 2. First-year EOH of a PV plant in the Northeast climate region of the US at the inverter and interconnection level for the same power-rated PV module with six different PAN files, using climate data from SolarGIS.

Fig. 3 shows the first year and thirty-year average EOH numbers for the PV plant in the Northeast climate zone. The plant shows the variability of 3.01% for the first year and 3.02% for the lifetime average on EOH while using [7] Meteonorm TMY data. The variation in EOH numbers due to the utilization of TMY data from a different source and uncertainty level is insignificant compared to the variation resulting from the PAN files provided.

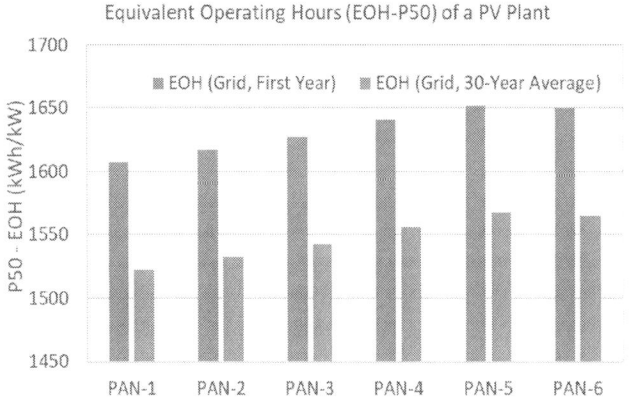

Fig. 3. First-year and thirty-year (lifetime) average EOH at a PV plant in the Northeast climate region of the US for the same power-rated PV module with six different PAN files, using climate data from SolarGIS.

The CapEx and LCOE sensitivity of a PV plant in the Northeast region is shown in Fig. 4 and 5. The variation in CapEx is the result of a change in a number of modules and tracking structures (tracker tables) to maintain the optimum performance of the central inverter for the interconnection limit. The base value is the average value calculated from six PAN files.

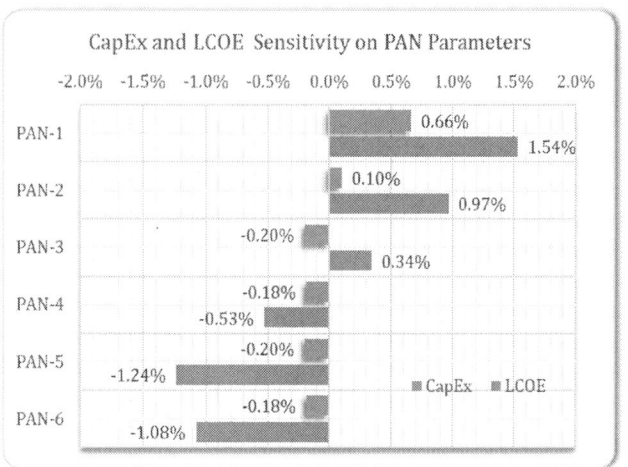

Fig. 4. Sensitivity of CapEx and LCOE of a PV plant in the Northeast region of the US against the six different PAN files a PV module with the same power rating, using climate data from SolarGIS.

The CapEx for the PV plant in the Northeast region with SolarGIS data shows a maximum of 0.86% variation while the LCOE varies by 2.77%. The same plant depicts the CapEx varies by 1.7% and LCOE by 2.8% for [7] Meteonorm data.

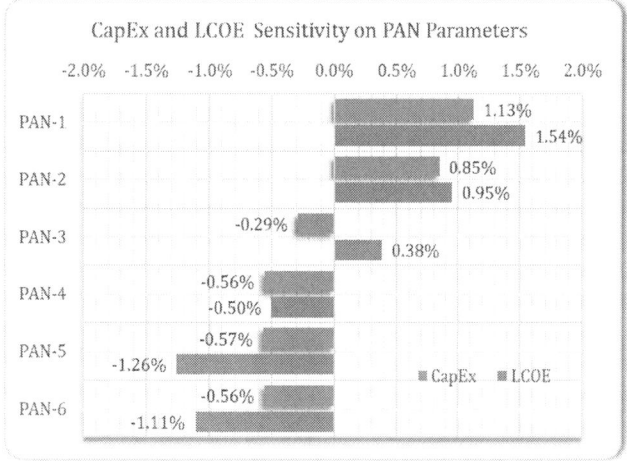

Fig. 5. Sensitivity of CapEx and LCOE of a PV plant in the Northeast region of the US against the six different PAN files a PV module with the same power rating, using climate data from Meteonorm.

B. Southwest PV Plant Study

Fig. 6 shows the actual EOH for the first year and the lifetime average. The deviations in the first year and lifetime average EOH for a PV plant in the Southwest are estimated as 1.77% and

2.16%, respectively, for the six PAN files. The deviations are slightly less than the variations seen in the Northeast PV plant.

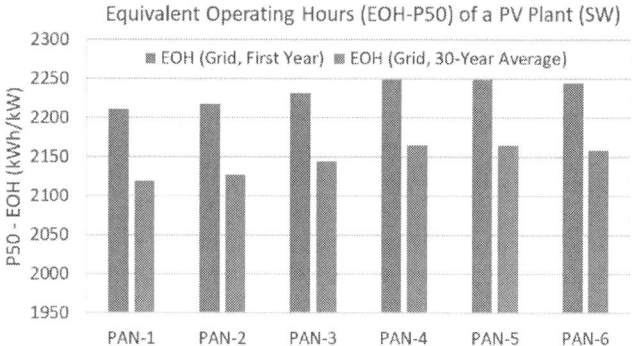

Fig. 6. First-year and thirty-year (lifetime) average EOH at a PV plant in the Southwest region of the US for the same power-rated PV module with six different PAN files, using climate data from SolarGIS.

Fig. 7 shows the sensitivity of CapEx and LCOE of a PV plant in the Southwest climate zone of the US. The CapEx and LCOE variations concerning the different PAN files are 0.23% and 2.24% for the selected PV PV plant in the Southwest. These variations are less than the selected PV plant in the Northeast.

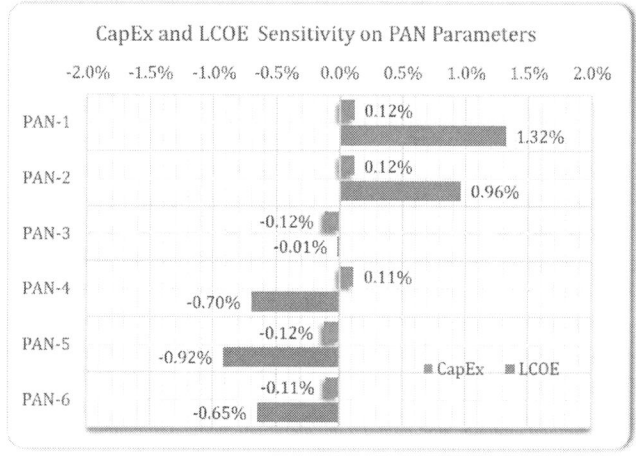

Fig. 7. Sensitivity of CapEx and LCOE of a PV plant in the Southwest region of the US against the six different PAN files a PV module with the same power rating, using climate data from SolarGIS.

A poor or no correlation between CapEx and LCOE variations in Fig.4, 5, and 7 rationalizes the significance of each PAN file parameter (Table I) on the estimated energy yield, CapEx, and LCOE of a PV plant.

C. PV Plant Performance Prediction and Business Decisions

The improvement in PAN file parameters has a significant impact on the estimated performance of a PV plant, hence having the financial value. A 2.8% variation in LCOE, as seen for the Northeast PV plant in this paper, is sufficient to influence an investment decision for a PV power plant. Furthermore, that may lead to incorrect performance predictions and uncertainty in meeting the key performance index (KPI) during the entire operational life of the PV plant [9].

D. Underrating PV module: Not a Viable Solution

Some manufacturers have been underrating PV modules to maintain goodwill with the customer or avoid potential litigation. However, the underrated PV module specifications from the manufacturer further exacerbate the uncertainty in the energy yield forecast of a PV plant. Table II shows the difference in independent laboratory-measured data versus manufacturer claimed data for Climate Specific Energy Rating (CSER) parameters and the angle of the incident (AOI) measurement data.

TABLE II. THE DIFFERENCE IN MANUFACTURER'S CLAIM AND LAB-TESTED VALUES ON THE ENERGY RATING OF PV MODULES

CSER Locations	Difference (Claim–Test)	AOI (Angel, Deg.)	Difference (Claim–Test)
Desert	0.00%	0	0.00%
High Elevation	-0.80%	30	0.00%
Mediterranean	-0.30%	55	-0.51%
Temperate Continental	-0.90%	60	-1.03%
Temperate Coastal	-1.50%	65	0.00%
Tropical Humid	-0.60%	70	-1.11%
Tropical Semi-Arid	-0.40%	75	0.00%

Normalized values are listed in the table to maintain the anonymity of the supplier.

E. Discussions and Takeaways

Independent laboratories now produce most PAN files, and they are associated with the IEC61853 test data, which helps define the parameters. However, a lack of industry-wide consensus on the standard procedure to transfer the measured data into the PAN file parameters creates confusion in the industry.

An industry-wide awareness is necessary to discourage the improper creation and modification of PAN file parameters. A PAN file provides the information on the exact behavior of PV modules and could be part of the qualification certificate. Thus, validation of the origination of the PAN files and the measurement uncertainties, and the confidence interval of the test data that derives the PAN file parameters are critical.

It is also essential for the end-user to cross-validate the PAN file provided by the supplier before using it for simulation to minimize inaccuracy in the energy yield forecast and associated financial risks.

IV. CONCLUSION

The sensitivity of PV modules' PAN parameters on the energy yield and other financial parameters are illustrated in two different large-scale PV plants at different climate zones in the continental USA. PAN files created by a different process of a PV module with the same power rating may result in different energy forecasts and the cost of a PV plant. An industry-wide initiative is needed to develop and promulgate a standard methodology to transfer energy rating test data to the PAN file. A procedure to track the changes and cross-validation of the PAN file parameter is needed to uphold the confidence in the energy yield forecast of a large-scale PV plant.

REFERENCES

[1] International Electrotechnical Commission, "IEC 60904-3 Photovoltaic devices - Part 3: Measurement principles for terrestrial photovoltaic (PV) solar devices with reference spectral irradiance data," Geneva, 2016.

[2] International Electrotechnical Commission [IEC], "IEC 61724 Photovoltaic system performance - Part 3: Energy evaluation method," IEC, Geneva, 2016.

[3] M. Schweiger, W. Herrmann, G. Andreas, and U. Rau, "Understanding the energy yield of photovoltaic modules in different climates by linear performance loss analysis of the module performance ratio," IET Renewable Power Generation, pp. 558-565, 2017.

[4] International Electrotechnical Commission, "IEC 61853 Photovoltaic (PV) module performance testing and energy rating," Geneva, 2018.

[5] PVSYST SA, "PVsyst Photovoltaic Software," PVSYST SA, 8 4 2022. [Online]. Available: https://www.pvsyst.com/. [Accessed 8 4 2022].

[6] Solargis, "Data and software architects for bankable solar investments," Solargis, 8 4 2022. [Online]. Available: https://solargis.com/. [Accessed 8 4 2022].

[7] Meteonorm, "Meteonorm Software Worldwide irradiation data," Meteonorm, 29 May 2022. [Online]. Available: https://meteonorm.com/en/. [Accessed 29 May 2022].

[8] Exprivia, "Automating the conceptual design of photovoltaic systems: now a reality for Enel Green Power," Exprivia, 29 May 2022. [Online]. Available: https://www.exprivia.it/en-tile-7927-automating-the-conceptual-design-of-photovoltaic-systems-now-a-reality-for-enel-green-power/. [Accessed 29 May 2022].

[9] kWh Analytics, "Solar Risk Assessment: 2021, Quantitative Insights from the Industry Experts," kWh Analytics, San Francisco, California, 2021.

FEM based thermal model of an agrivoltaic system

Karan Rane[1], Navni Verma[2], Ardeshir Contractor[2], and Narendra Shiradkar[1]

[1]National Centre for Photovoltaic Research and Education (NCPRE), Department of Electrical Engineering,
Indian Institute of Technology Bombay, Mumbai, Maharashtra, 400076, India
[2]Department of Mechanical and Aerospace Engineering, The Ohio State University,
Columbus (OH), 43210, USA

Abstract—**This study aims at building a finite element model of an agrivoltaic system for thermal simulation. The system under consideration consists of an array of PV modules over a shadenet covering an array of plant crops. The plant crops are modelled as porous mediums for computing air flow. The solar radiative energy recieved by the various components of the system is computed using the surface to surface radiation equations. Heat loss in PV modules and plants by convection is modelled. This model also accounts for heat loss in plants through evaporative cooling. The physical properties of plants are assumed from literature. Properties that need experimental determination are treated as variable parameters.**

Index Terms—**agrivoltaics, fem, thermal, model, shadenet**

I. INTRODUCTION

An agrivoltaic (AV) systems are PV installations which provide space for growing food crops, thus sourcing electrical energy and food from the same location. The configuration of an agrivoltaic system alters the air flow and irradiance received by the plants thus affecting the temperature and the growth of food crops, as noted in [1] [2]. Estimation of impacts of an agrivoltaic configuration on plant temperatures needs parameters that are determined empirically through experimental data. This paper describes a 2D physics based model for the thermal assessment of plants under photovoltaic modules. Such a model will help in prediction of PV and plant temperatures, which are required for the calculation of crop yield and energy yield for a given configuration. The configuration under consideration includes a shade-net to prevent the detrimental overheating of plants. Comsol multiphysics was used for defining and simulating the model.

II. GEOMETRY AND MATERIALS

The agrivoltaic configuration, in this study, consists of PV modules and shade nets at 3 meters and 2 meters above the ground respectively. The PV module is represented by domain layers of $3\,mm$ glass, $140\,\mu m$ EVA, $180\,\mu m$ silicon and $170\,\mu m$ backsheet [3]. The assembly is tilted at 20° with respect to ground level. The shade-net is represented by a $2\,mm$ thick layer. Under the shadenet, $1\,m$ high and $0.5\,m$ wide rectangles represent plant bodies, considering full grown tomato plants [4]. The air domain is $15\,m$ high and $40\,m$ wide. The location of the components in the geometry are illustrated in Fig. 2.

The thermal properties of the PV modules and plants were obtained from [3] and [5] respectively. The properties of air and moist air were inbuilt in the software.

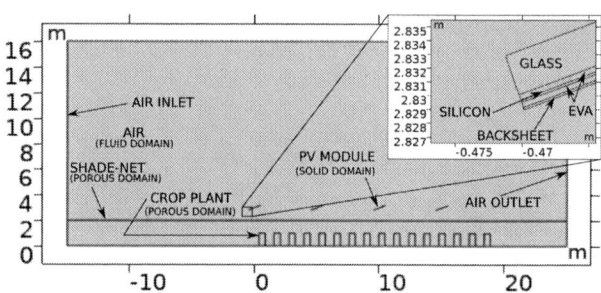

Fig. 1. The domains and the boundaries defined for the simulation. The left boundary of the air domain is the inlet for air at ambient temperature and the right boundary is the outlet

III. SIMULATION MODEL

The external irradiance, ambient temperature, moisture source and the wind speed at the inlet form the inputs to the simulation model, which outputs humidity, air velocity, irradiance and temperature distributions. The sub-models associated with the steps in the simulation are shown in fig.

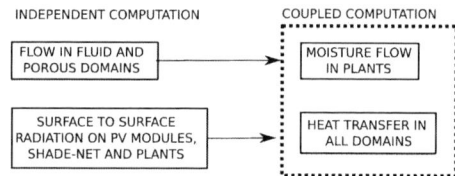

Fig. 2. The irradiance and the air velocity distributions generated by the surface to surface and the CFD model are fed to the heat transfer with moisture flow coupled model for generation of moisture and temperature distributions

A. Flow in fluid and porous domains

The free and porous medium flow physics node was used for flow field computation. The free flow in the fluid domain used the equations of $k-\epsilon$ turbulent model. The flow in porous domain, representing the plants, used the Darcy-Forchheimer equations, (1). It calculates the velocity field **u** considering the porous medium flow parameters, ϵ representing porosity, c_F representing Forchheimer parameter, and κ representing the permeability. This simulation assigned the values 0.9,

978-1-7281-6118-1/22 $31.00 © 2022 IEEE

0.245, and 0.017 1/m for porosity, Forchheimer parameter, and permeability respectively obtained from [4].

$$\frac{1}{\epsilon_p}\rho(\mathbf{u}\cdot\nabla)\mathbf{u}\frac{1}{\epsilon_p} = \nabla\cdot[-p2\mathbf{I}+\mathbf{K}]$$

$$-\left[\mu\mathbf{K}^{-1}+\beta\rho|\mathbf{u}|+\frac{Q_m}{\epsilon_p^2}\right]\mathbf{u}+\mathbf{F}$$

$$\rho\nabla\cdot\mathbf{u} = Q_m \qquad (1)$$

$$\mathbf{K} = \frac{\mu}{\epsilon_p}[\nabla\mathbf{u}+(\nabla\mathbf{u})^T]-\frac{2\mu}{3\epsilon_p}(\nabla\mathbf{u})$$

$$\beta = \frac{c_F}{\sqrt{\kappa}}$$

The boundary conditions included air velocity of 1 m/s at the inlet and pressure at the outlet of 1 atm. The lower wall of the air domain was assigned a no-slip condition. Air resistance due to shade-net was included using the Ergun model. We assumed a porosity of 0.5 and particle size of 1 mm in the present study.

B. Surface to surface radiation

The surface to surface radiation node calculates the heat flux at the opaque or transparent surfaces for a given geometric configuration. An $1000W/m^2$ external radiation source irradiating the system at right angle to the ground level was added to the model. The silicon domain layer of the PV module was checked to be opaque to the radiation. The external surfaces of the shade-net and the plant domains were considered to be semitransparent with reflectivity of 0.1 and transmittivity of 0.9. The heat flux obtained using this model was modified to exclude the energy lost as electrical output by the PV modules. We assumed the efficiency to be 0.17. The irradiance received by the plant domain was used for calculation of heating power for heat transfer computation.

C. Moisture transport in plants

The heat loss through evaporation is counted through the coupling of moisture transport with heat transfer physics nodes. The equations governing the transport of moisture, (2), are formulated in [6] and [7]. It uses the velocity field of moist air, $\mathbf{u_g}$ (derived from III-A) and liquid water, $\mathbf{u_l}$, where subscripts g and l stand for gas and liquid phases respectively, to compute relative humidity, ϕ_w considering the porous media properties described in III-A, and parameters related to liquid - gas phase equilibrium.

$$d_z\rho_g\mathbf{u_g}\cdot\nabla\omega_v+\nabla\cdot\mathbf{g_w}+d_z\mathbf{u_l}\cdot\nabla\rho_g+d_z\nabla\cdot\mathbf{g_{lc}} = d_zG$$

$$w(\phi_w) = \epsilon_p[\rho_l s_l+\rho_g\omega_v(1-s_l)]$$

$$\omega_v = \frac{M_v\phi_w c_{sat}}{\rho_g}$$

$$g_w = -d_z\rho_g D_{eff}\nabla\omega_v$$

$$\mathbf{u_l} = -\frac{\kappa_{rl}\kappa}{\mu_l}\nabla p_a$$

$$(2)$$

The boundary conditions in this node are the moisture source, G and the relative humidity of the inflowing air. This study assumes moisture source of $100\,\mu kg/(m^3\cdot s)$ and an inflow of moist air at relative humidity of 0.75.

D. Heat transfer

This physics node uses the modified heat flux (III-B) as the heat source in the PV modules. Since heat loss through convection is dominant in an PV module, we neglet radiative the heat loss, in the current study. The irradiance received at the upper surface of the plant domain (top of the canopy), Rg_0 is used for the evaluation of heating rate using equation for Beer's law [8]. It relates the irradiance, Rg(z) at height z to the plant of height H_0 and an extinction coefficient, k'.

$$Rg(z) = Rg_0 exp(-k'LAD(H_0-z)) \qquad (3)$$

The other boundary condition is the temperature of the air at the inlet of the air domain which is 25^oC in this study.

IV. SIMULATION RESULTS

The results of the model settings in III are presented in this section. The air velocity field for the settings in III-A is shown in Fig. 3. The result exhibits division of the fast air stream into two due to presence of shadenet . The air speed profile shows a boundary layer, with the front modules and front plants experiencing highest momentum followed by the trailing modules and plants. Fig. 4. shows the radiation incident on the surfaces of opaque or semitransparent domains. The heat flux at the PV module surface is adjusted to exclude the portion of the energy flux that is converted to electrical energy. The radiation on the canopy top is used for calculating the heat source for the plant domain as described in III-B.

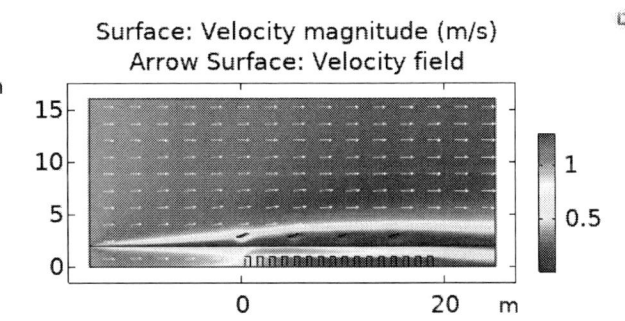

Fig. 3. Results for flow in fluid and porous domains.

The results of moisture flow and heat transfer are shown in Fig. 5 and Fig. 7. The temperature distribution in the plant domain shows larger temperature variation over the height. This variation in plants at the back is further amplified by the cooling effect from the evaporative cooling. The area average of the plant temperatures were found to vary from 25.06^oC to 28.2^oC, from inlet to outlet. The simulation was run excluding the PV modules in the set up. This resulted in the area average of the plant temperatures to vary from 25.07^oC to 28.5^oC, from inlet to outlet.

V. DISCUSSION

A plant responds to higher irradiance by allowing higher exchange of water and carbon dioxide. Its' leaves also generates heat during respiration. The random arrangement of leaves,

978-1-7281-6118-1/22 $31.00 © 2022 IEEE

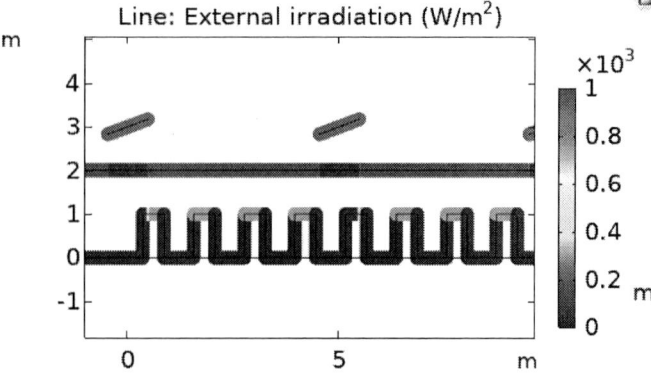

Fig. 4. Results for surface to surface radiation (magnified to a selecte portion) on PV modules, shade-net and plant canopies.

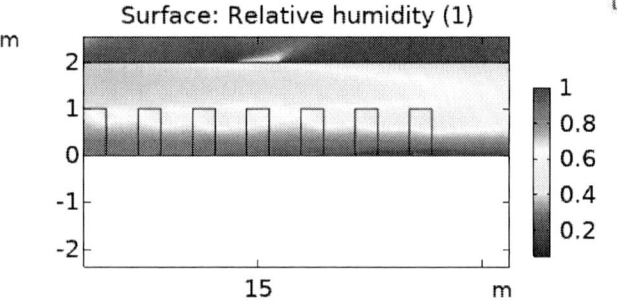

Fig. 5. Relative humidity adds up down the flow as air passes through multiple plants

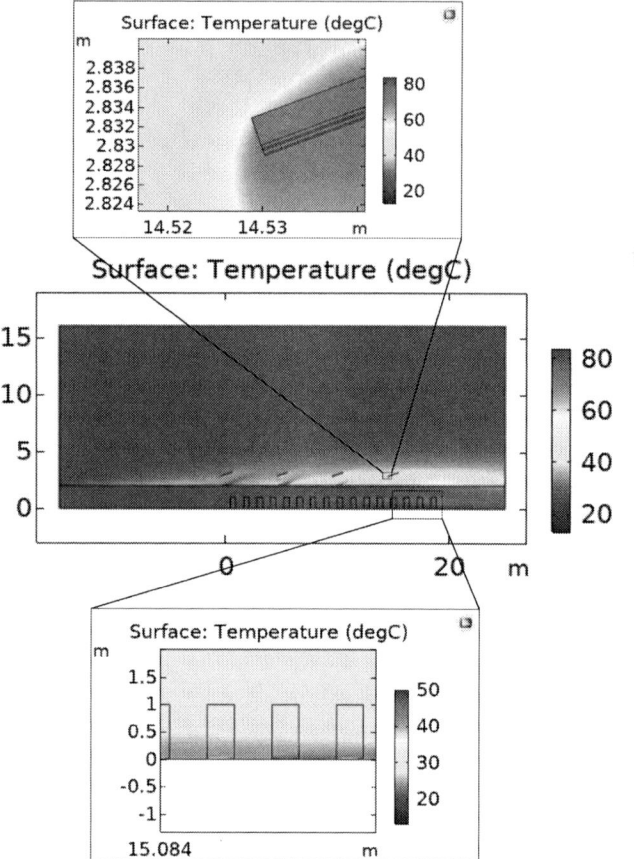

Fig. 6. Temperature distribution in all domains with portion of PV module and plant magnified to capture the range.

also calls for a statistical approach for the development of the model. Thermal IR image of leaves of a well watered crepe jasmine plant were captured for measuring difference of plant temperature with the ambient. We found that the leaf temperatures varied from 31^oC to 34^oC for the ambient of 33^oC. As expected, the shaded leaves exhibited lower temperatures compared to irradiated ones.

VI. CONCLUSION

FEM based model was built and simulated for thermal assessment of an agrivoltaic configuration. Plant specific parameters for tuning the model for flow, moisture and thermal computation were identified. The average model-plant temperatures for the setup with and with out PV modules were found to vary less than 1^oC, which indicates a dominant cooling rate through transpiration, in the studied configuration.

REFERENCES

[1] H. Marrou, L. Guilioni, L. Dufour, L., Dupraz, C., and J. Wery, J. " Microclimate under agrivoltaic systems: Is crop growth rate affected in the partial shade of solar panels?." Agricultural and Forest Meteorology, vol. 177, pp. 117-132.

[2] J. Hatfield and J. Prueger , "Temperature extremes: Effect on plant growth and development." Weather and climate extremes, 10, pp.4-10

[3] Vogt, Malte R., Hendrik Holst, Matthias Winter, Rolf Brendel, and Pietro P. Altermatt. "Numerical modeling of c-Si PV modules by coupling the semiconductor with the thermal conduction, convection and radiation equations." Energy Procedia 77 (2015): 215-224.

Fig. 7. Thermal IR image of leaves of a Crepe Jasmine plant in the afternoon, with ambient temperature of 33^oC

[4] S. Sase, M. Kacira, T. Boulard, and L. Okushima, "Wind tunnel measurement of aerodynamic properties of a tomato canopy". Transactions of the ASABE, vol. 55(5), pp.1921-1927.

[5] J.C. Vieccelli , D.L.D. Siqueira, W. Bispo and L. Lemos, 2016. "Characterization of leaves and fruits of mango (Mangifera indica L.)" cv. Imbu. Revista Brasileira de Fruticultura, vol. 38.

[6] A.K. Datta, " Porous media approaches to studying simultaneous heat and mass transfer in food processes. I: Problem formulations." Journal of food engineering, vol. 80(1), pp.80-95.

[7] A.K. Datta, "Porous media approaches to studying simultaneous heat and mass transfer in food processes. II: Property data and representative results." Journal of food engineering, vol. 80(1), pp.96-110.

[8] A. Kichah, P. Bournet, C. Migeon, and T. Boulard. "Measurement and CFD simulation of microclimate characteristics and transpiration of an Impatiens pot plant crop in a greenhouse." Biosystems Engineering, vol. 112(1), pp.22-34.

Towards Understanding of Cementation of Particulate Soils on PV Cover Glass Materials

Mohamed Adawi, Adedoyin Abe, Min Zou, Robert A. Fleming

Arkansas State University, Jonesboro, AR, United States

University of Arkansas, Fayetteville, AR, United States

Accumulation of soils and other particulate matter on the front cover glass of solar photovoltaic (PV) modules results in transmission losses that detrimentally affect the power output of PV installations. Cementation reactions, which occurs due to interactions between the dust and the glass surface in the presence of temperature, humidity, and pH, results in the dust becoming rigidly attached and potentially difficult to remove with conventional cleaning methods. In this study, accelerated soiling and cementation tests on glass coupons have been performed using a custom instrumented soiling chamber and several standardized soils (Arizona Test Dust, ARAMCO Test Dust, and China Test Dust) to assess soil adhesion and cementation behaviors. Micromechanical scratch testing, along with supporting water contact angle (WCA) and X-ray photoelectron spectroscopy (XPS) measurements are used to characterize the surface chemistry of the soiled coupons and surface energy evolution of the deposited soils before and after cementation. The end result of this study is a better understanding of the surface properties of cemented soils, which can potentially lead to the development of novel soiling mitigation technologies for PV applications.

Thin-film Solar Cells with MgF$_2$/Ag back mirror patterning for improved near-IR reflectance

Lara Barros Rebouças [1*], Gerard J. Bauhuis [1], Jens Olhmann [2], Jeroen Maasen [1],
Elias Vlieg [1] and John J. Schermer [1]

1 Institute for Molecules and Materials, Radboud University, Nijmegen, the Netherlands
2 Institute for Solar Energy Systems ISE, Freiburg, 79110, Germany
*l.reboucas@science.ru.nl

Abstract— **Combining epitaxial lift-off with a dielectric-metal back mirror boosts III-V solar cells efficiencies without sacrificing the costly growth wafer. In this work, GaAs and GaInP/GaInAs solar cells are produced with a patterned MgF$_2$/Ag mirror. The sub-bandgap reflectance increases by 3.0% and 2.6%, respectively, compared to the devices with a full Ag back mirror. Initial results indicate that, during operation, the temperature-induced open-circuit voltage degradation decreases due to the enhanced reflection of unused near-infrared photons.**

Keywords— epitaxial lift-off, back mirror, MgF$_2$, infrared reflectance

I. Introduction

Producing thin-film solar cells by epitaxial lift-off (ELO) allows the reuse of the wafer lowering the production costs of high-efficiency III-V devices [1]. ELO solar cells further benefit from photon recycling because the lifted thin-films are commonly mounted on a metal foil that works not only as a mechanical support and back electrical contact but also as a mirror to the incident light [2].

The back mirror reflects emitted photons back into the absorber layers, increasing the probability of internal photon recycling and avoiding parasitic absorption losses to the substrate. As a result, the device's open-circuit voltage (V$_{OC}$) increases [3]. More recent studies show that the back mirror design can be further optimized with the addition of a low refractive index dielectric layer between the semiconductor and back metal contact. To allow electric contacting of the rear side, the dielectric layer is typically applied in a pattern to allow metal deposition directly onto the contact layer. ZnS [2], SiO$_2$ [4], SU-8 photoresist [5], MgF$_2$ [6], and a MgF$_2$/AlO$_x$ stack [7] are shown to improve the standard performance of III-V solar cells. More importantly, in a quasi-planar configuration, this design could reduce the operating temperature of single and multi-junction solar cells by reflecting the unused near-infrared photons.

However, combining ELO processing with a dielectric-metal reflector is challenging due to the interaction of HF with dielectric materials. Oxides such as silica are readily dissolved, whereas photoresists such as SU-8 are found to easily peel off when immersed in HF 48% solution [8]. In this work, we demonstrate the production of thin-film solar cells with a patterned MgF$_2$/Ag back mirror by standard substrate etching

and ELO. Underetching of the MgF$_2$ in HF is prevented by maintaining a silver/gold protective edge rim. Further, we investigate the influence of the patterned dielectric/metal back-side mirror on the operating temperature of solar cells by simultaneously measuring the V$_{OC}$ of devices with a patterned MgF$_2$/Ag back mirror and a full Ag mirror a during a 30 min simulated operation.

II. Methods

A. Suitability for ELO processing

To test the stability of MgF$_2$ during ELO, a 100 nm thick layer was evaporated onto the AlGa(In)As p-contact and exposed to concentrated HF 48%. Half of the sample was masked during evaporation to allow surface profile step height measurements. Optical microscopy images were obtained using a Reichert-Jung Polyvar MET microscope.

B. Growth and processing of solar cells

The solar cell structures, shown in figure 1, were grown in an inverse order to facilitate the wafer removal. The single-junction GaAs solar cells were grown using a Aixtron 200 reactor on 2 inch (100) GaAs substrates with 2° off to (110) orientation. The GaInP/GaInAs solar cell structures were grown in an AIX2800 G4 TM MOVPE reactor on 4 inch (100) p-Ge wafers with a 6° offcut towards <111> direction.

For both cell structures, the back mirror pattern was defined using standard photolithography consisting of regularly spaced contact points with a 127 µm diameter covering 10% of the solar cell area. A 100 nm MgF$_2$ layer was deposited followed by 1 nm Pd to improve adhesion dielectric/metal adhesion. Note that this was only performed on half of the wafer to allow comparison with 100% Ag mirror cells produced from the same wafer. After resist removal of the masked contact points, a 3 µm Ag layer was evaporated in the whole wafer. The back mirror was capped by 50 nm Au to prevent oxidation.

Subsequently, wafer removal was performed by etching in [1:3] NH$_4$OH:H$_2$O$_2$ for the single junction GaAs. For the dual-junction GaInP/GaInAs, ELO was performed by etching a 10 nm AlAs sacrificial layer in [1:1] 25% HF:Acetone.

The released thin-films were processed in a similar procedure; after the removal of the buffer and etch stop layers, a 200 nm Au layer was evaporated in a photo-lithographically defined front grid (16.6% metal coverage). Metal lift-off was performed in acetone, followed by selective etching of the cap layer between the grid fingers using [1:1:10] $NaOH(1M):H_2O_2:H_2O$. The cell areas are defined by MESA etching with HCl (37%) for the phosphorous-containing layers and [1:1:8] $H_3PO_4:H_2O_2:H_2O$ for the GaAs-based layers. 44 nm ZnS and 94 nm MgF_2 are applied as a top anti-reflection coating (ARC).

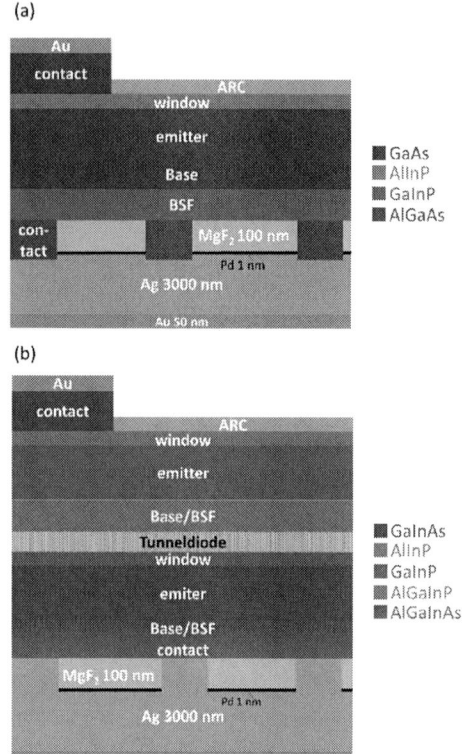

Fig. 1. Schematic representation of the (a) single-junction GaAs and (b) dual junction GaInP/GaInAs solar cells with a patterned MgF_2/Ag back mirror and a reference Ag mirror processed on the same wafer.

C. Solar Cells Characterization

Wavelength-dependent reflectance spectra from the top of the processed device was measured using a ReRa SpeQuest system with ReRa Photor 3.1 software using a step size of 5 nm coupled with an integrating sphere (Bentham DTR6).

Standard AM1.5G JV characteristics of the solar cells were measured at 25 °C with an ABET Technologies Sun 2000 Class AAA solar simulator, which provides illumination with a power density of 1000 W m^2 over a 10 x 10 cm^2 area, a Keithley 2401 source meter, and ReRa Tracer 3.0 software. The same setup was equipped with a second source meter, Keithley 2601B, to allow simultaneous measurement of two solar cells over a 30 min operating cycle. Data acquisition was performed by an in-house developed software. During the measurement, the back side of solar cells is kept at 25 °C using a heating/cooling water thermostat and Pt100 temperature sensing. The ambient temperature under illumination was monitored with a type K thermocouple placed between to two probed devices.

This work was supported by the Regeneration project (grant agreement 17043), funded by EIT Raw Materials from the European Institute of Innovation and Technology (EIT), an independent body supported by the European Union's Horizon 2020 program.

III. RESULTS AND DISCUSSION

A. MgF_2 in HF

The 100 nm MgF_2 layer exposed to HF for 30, 60, and 120 seconds showed no significant step height variation in the contact profilometer measurements. However, as shown in figure 2, the dielectric layer is under etched and detaches from the semiconductor surface in smaller pieces. The observed under etching seems to occur initially in the edge of the deposited layer and readily propagates through the structure. Therefore, a 5 mm mask was applied to cover the edge rim during MgF_2 and Pd deposition to process an epitaxial lift-off solar cell with a patterned mirror. This protective edge is then fully covered with silver during the back contact evaporation. Subsequently, the ELO of a GaInP/GaInAs demonstrated that Subsequently, ELO of a GaInP/GaInAs solar cell demonstrate that this scheme prevented structural damage of the released thin-film.

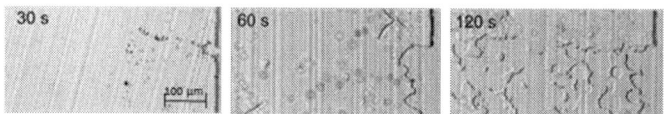

Fig. 2. Polarized optical microscopy of a 100 nm MgF_2 layer on AlGa(In)As after exposure to 48% HF for 30, 60 and 10 seconds.

B. Sub-bandgap Reflectance

Figure 3 shows the total reflectance measured from the top of the fully processed solar cell structures.

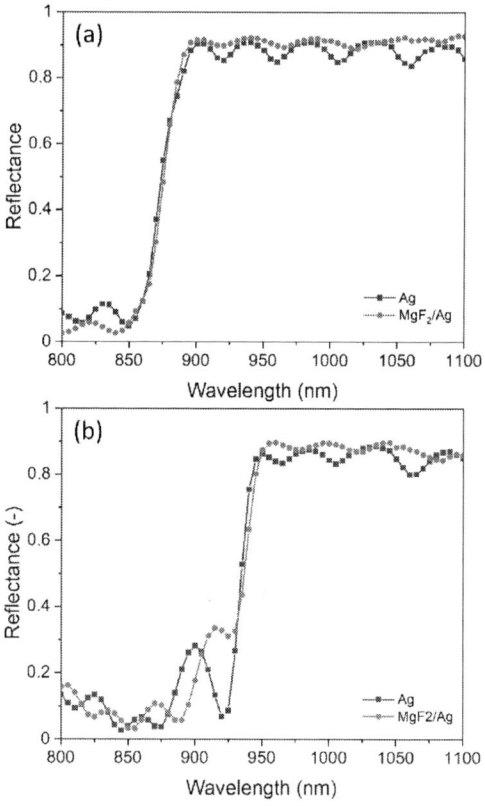

Fig. 3. Wavelength dependent reflectance of a (a) single-junction GaAs and (b) dual junction GaInP/GaInAs solar cells with a patterned MgF_2/Ag back mirror and a reference Ag back mirror.

As expected, a strong absorption is observed for wavelengths below the GaAs and the GaInAs bandgap. In the longer wavelengths (lower energy photons) the reflectance increases when a MgF$_2$/Ag mirror is used instead of a 100% Ag mirror. An increase of 3.0% in the range of 895-1100 nm is observed for the single junction GaAs and 2.6% in the range of 945-1100 nm for the GaInP/GaInAs dual junction.

C. Operating Temperature and V_{OC}

JV-characteristics of the GaAs and the GaInP/GaInAs solar cells are given in table I. The parameters are given by the average of the three best-performing cells. The similar performances indicate that no deterioration occurs from the added MgF$_2$ layer processed by either substrate etching or epitaxial lift-off, although no significant improvement is observed for those initial cells in standard 25 °C measurements.

TABLE I.   AM1.5G JV-CHARACTERISTICS.

Solar Cell – Back Mirror	J_{sc}	V_{OC}	FF	η
	$mA.cm^{-2}$	V	%	%
Sub. Etch GaAs - Ag	23.2	1.04	73.8	17.9
Sub. Etch GaAs - MgF$_2$/Ag	22.4	1.04	76.9	17.9
ELO GaInP/GaInAs - Ag	10.2	2.31	83.6	19.8
ELO GaInP/GaInAs - MgF$_2$/Ag	10.2	2.31	84.2	19.8

However, III-V solar cells rarely operate in a controlled room temperature environment. Elevated temperatures have a negative impact in the efficiency of the solar cells. This degradation is mainly apparent in the V_{OC}. Figure 4 shows the variation in V_{OC} during a 30 min cycle under simulated AM1.5G.

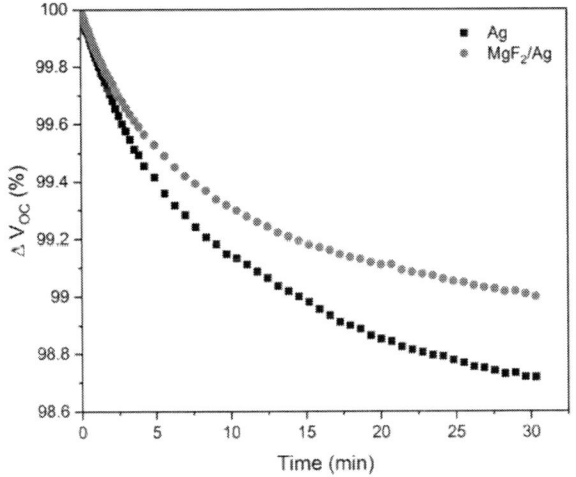

Fig. 4. VOC under simulated AM1.5G ilumination over 30 min for a sinlgle junction GaAs with patterned MgF$_2$/Ag back mirror and a reference full Ag mirror.

While the bottom of the solar cells was kept at 25 °C, the air temperature under illumination increased 5 °C. This result indicates that the enhanced sub-bandgap reflection of the MgF$_2$/Ag mirror plays a role in reducing the heat load of the solar cell. This effect is expected to have a higher impact in space applications, where waste heat is only removed through thermal radiation.

IV. CONCLUSION

Thin film solar cells with a patterned MgF$_2$/Ag back mirror were produced using substrate etch and epitaxial lift-off. Since MgF$_2$ detaches from a semiconductor surface when exposed to hydrofluoric acid, a protective edge rim was employed to prevent the delamination of the thin-film from its carrier during ELO. The initial solar cells produced with a MgF$_2$/Ag mirror showed enhanced sub-bandgap reflection, which leads to more efficient heat dissipation, reducing the operating temperature of the device, as observed in the V_{OC} measurement during a 30 min operating cycle. Further solar cell optimization should allow for an increased V_{OC} also in standard 25 °C measurements by enhanced photon recycling, which is not yet observed in the current cells.

ACKNOWLEDGMENT

The authors would like to thank Daan van der Woude for his support with the epitaxial lift-off processing and the technical assistance of Wil Corbeek and Peter Mulder form Radboud University.

REFERENCES

[1] G. J. Bauhuis, P. Mulder, and J. J. Schermer, Thin-film III-V solar cells using epitaxial lift-off, in: High-efficiency solar cells-physics, Springer, Chap 203 pp 623-643

[2] N. Gruginskie, S.C.W. van Laar, G.J. Bauhuis, et al., Increased performance of thin-film GaAs solar cells by rear contact/mirror patterning. Thin Solid Films. 2018;660:10-18.

[3] O.D. Miller, E.Yablonovitc, and S.R. Kurtz, Strong internal and external luminescence as solar cellsapproach the Shokley-Queisser limit. IEEE J. Photovolt. 2012;2(3):303-311.

[4] D.Alonso-Álvareza, C.Weiss, J.Fernandez et al., Assessing the operating temperature of multi-junction solar cells with novel rear side layer stack and local electrical contacts. Solar Energy Materials and Solar Cells. 2019;200:110025.

[5] M.K. Arulanandam, M.A. Steiner, E.J. Tervo et al., GaAs thermophotovoltaic patterned dielectric back contact deviced wit improved sub-bandgap reflectance. Solar Energy Materials and Solar Cells. 2022;238:111545.

[6] C.L. Schilling, O Höhn, D.N, Micha, et al., Combining Photon Recycling and Concentrated Illumination in a GaAs Heterojunction Solar Cell, IEEE J. Photovolt. 2018;8(1):348–354.

[7] H. Helmers, E. Lopez. O. Höhn, ct al. 68.9% efficient GaAs-based photonic power conversion by photon recycling and optical resonance. Phys. Status Solidi. 2021;15:2100113.

[8] D.C.S. Bien, P.V. Rainey, S.J.N. Mitchell, et al., Characterization of masking materials for deep glass micromachining. Journal of Micromechanics and Microengineering. 2003;13(4):34-40

InAs Thermophotovoltaic Cells with Low Reverse Saturation Current

Eric J. Tervo, Andrew J. Ferguson, Myles A. Steiner, Ryan M. France

National Renewable Energy Laboratory, Golden, CO, United States

To efficiently convert heat from sources < 1000 °C to electricity with thermophotovoltaic cells, low-bandgap devices < 0.7 eV with good electrical characteristics are required. III-V semiconductors are the best material system for these applications due to their high quality and compatibility with a variety of cell architectures. However, low-bandgap III-V cells operating at ambient temperatures suffer from challenging nonradiative losses, including Auger recombination and diffusion current from the contacts. We report the modeling, fabrication, and characterization of low-bandgap InAs (0.35 eV) thermophotovoltaic cells with good electrical characteristics as evidenced by low reverse saturation currents < 20 mA/cm2. Auger losses are mitigated with a double-heterojunction p-i-n architecture that minimizes minority carrier densities in the central intrinsic InAs layer. Our results should provide strategies to design efficient thermophotovoltaic systems for solar thermal energy, waste heat recovery, and other low-temperature heat sources.

Toxicity assessment of lead and other metals used in perovskite solar panels

Gonzalo Rodriguez-Garcia, Jon J. Kellar, Zhengtao Zhu, and Ilke Celik

South Dakota School of Mines and Technology, Rapid City, SD, 501 E. St. Joseph St., 57701, USA

Abstract—. We evaluated the potential life cycle toxicity impacts of Pb, and five other metals found in perovskite solar panels—Al, Ag, Cu, In, and Sn. We focused on their use in integrated applications—urban, agrivoltaic, buildings, and floating solar, but also included their mining and recycling. Results indicated only the mining of silver is more ecotoxic than that of lead. During a catastrophic break, aluminum emissions in general and silver for floating photovoltaics, are more ecotoxic than those of lead. In all other cases, metals evaluated are potentially as toxic as lead. Finally, the use of virgin materials for the manufacture of the panel has similar impacts as recycling those materials. However, the recovery of the bottom glass and cell is environmentally beneficial due to its silver content.

Keywords—End-of-Life, Life Cycle Assessment, Raw Material Extraction, Use, USEtox

I. INTRODUCTION

From very early in the development of perovskite solar cells (PSCs), the presence of lead—a well-known toxicant—has received substantial attention. Opinions range from significant concern [1] to consider the complete leaching of lead from a solar farm a low level of contamination [2]. As a holistic method for the evaluation of the environmental profile of products and processes, life cycle assessment has been applied extensively to PSCs [3]. It has also been used to offer an additional perspective on the lead debate. Literature thus far indicates Pb has a limited contribution to the PSCs' environmental burden. The electricity produced by PSCs is about four times less lead intense than from US electricity mixes, and their lead emissions are about 20 times less toxic [4]. PSCs with alternative metals like tin do not only suffer from lower efficiency and stability, but also entail larger environmental impacts through their entire lifetime [5]. That said, of the six metals used in emerging photovoltaics (PV) in [6], lead and copper are the two whose emissions during use might cause a larger environmental impact than their extraction.

As perovskite solar panels (PSP) are getting ready for commercialization, it becomes necessary to offer a broader insight into their toxicity. To that end, we evaluated metals beyond those used as part of the active layer, and acknowledged the wide uncertainty present by available life cycle toxicity assessment methods (LCTA). We calculated the freshwater ecotoxicity and human toxicity potential of mining six metals present in PSP—Ag, Al, Cu, In, Pb, Sn—as well as the toxicity impacts that could result from a catastrophic loss of metal under four different integrated applications. In addition, we propose a treatment train for the recycling of PSC, focusing on the recovery of metals, and present its toxicity impacts.

II. METHODOLOGY

Our purpose is to offer PSP developers a deeper insight into the potential risks the materials they use entail, and ways to reduce their usage. To that end, we conducted an LCTA of metals used in the manufacturing of PSP [7] using USEtox 2.0 [8]. This LCTA includes the extraction of raw materials based on [9], their use as part of a PSP, and a recycling process based on [10], [11].Our functional unit is 1 m^2 PSP, whose metal content is presented on Table I.

Here we modeled a catastrophic loss of metal as a worst-case scenario for the applications included in Table II. Pervious scenarios are those for which the emission compartment is the soil: natural soil for urban—where sealed areas of a city like parking lots are used—and agricultural soil for agrivoltaics —where agricultural activities are performed in the vicinity of solar panels. In the impervious scenarios metals leak directly to water. In the building integrated scenario, any leakage would be collected by the pipes and end up in a river or other freshwater body. At the same time, in floating PV, the emission compartment would be the sea, assuming that is where the panels would be located [12].

USEtox is a UNEP-SETAC endorsed LCTA method developed for organic pollutants. As such, it presets three important limitations affecting our evaluation. 1) there is no current characterization factor for In. Thus, we did not it included in the catastrophic release scenarios. 2) Al is considered non-toxic for humans, Sn is regarded as non-carcinogenic, and there is a lack of characterization factors for its non-carcinogenic impacts. Therefore, it does not appear in the human toxicity assessment of catastrophic release scenarios. 3) The characterization factor of all metals is based on the EC$_{50}$—the concentration that affects 50% of a population—of truly dissolved metals. As seen on Table I, most metals assessed are not present in ionic form, and therefore not likely to become immediately dissolved and released to the environment during an accident. Thus, it is expected we are overestimating the impact of non-ionic metals.

TABLE I METAL CONTENT OF A PSP

	Ag	Al	Cu	Pb		In	Sn	
Oxidation state	0	0	0	0	+2	+3	0	+2
g/m^2 [7]	11.4	2130	103	0.73	1.6	7.5	13	7.5

This work was funded by the US Department of Energy (DE-EE0009836), National Science Foundation (ERI: CAS: 2138293) and South Dakota School of Mines and Technology (Nelson Grant, and Start-up Grant)

TABLE II APPLICATIONS FOR PEROVSKITE SOLAR PANELS AND USEtox
COMPARTMENT WHERE THEIR EMISSIONS WOULD BE RELEASED

	Applications	USEtox compartment
Pervious	Urban	Natural soil
	Agrivoltaic	Agricultural soil
Impervious	Building-integrated	Continental freshwater
	Floating	Continental seawater

III. RESULTS AND DISCUSSION

A. Extraction impacts

We present the ecotoxicity potential of extracting and refining metals present in a PSP in Fig. 1 a). Since the uncertainty of USEtox is of three orders of magnitude, we used lead as reference substance, and considered metals within ±500% of its impact as essentially equal [8]. Therefore, on a per kg basis only the extraction of silver is more harmful than that of lead (data not shown), while the extraction of other metals can be seen as very similar. In addition to silver, when we account for the metal that is present in 1 m² PSP, the extraction of aluminum and copper are more ecotoxic than that of lead. In terms of human toxicity, the extraction of all six metals can be seen as roughly equally impactful per mass (data not shown). However, the large mass of aluminum and copper makes them more toxic than lead on an area basis (Fig. 1 b).

B. Use impacts

A similar picture emerges when we compare the potential emissions of metals during the use of the PSP (Fig. 2 a). On a per kg of metal emitted (data not shown), most metals would be similarly harmful to freshwater aquatic life. The exceptions would be silver and aluminum in the floating solar scenario, where they would be significantly more toxic than lead. In that scenario, Ag, Al, and Cu would also be more toxic than Pb during the breakage of 1 m² panel. In addition, a complete release of aluminum in the other three scenarios would also be more ecotoxic than that of lead, due to its large mass. When comparing the four different scenarios, urban and agrivoltaic present identical profiles. This is because in USEtox there is no

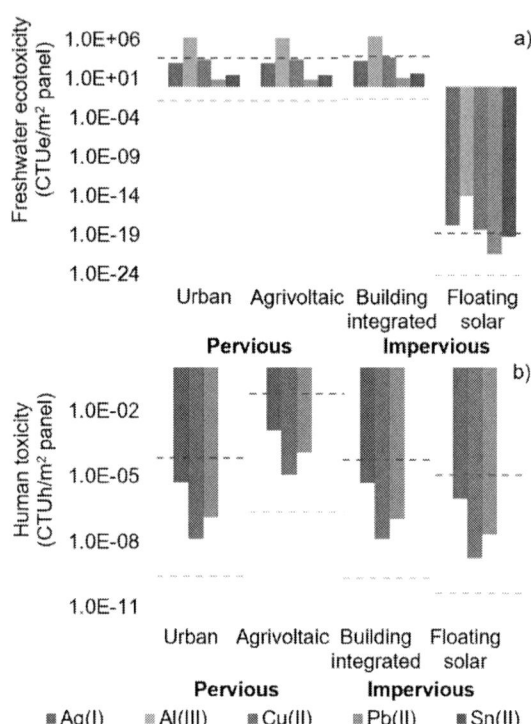

Fig. 1. a) Freshwater ecotoxicity and b) combined ecotoxicity impacts for the complete released for all the metals in a 1 m² panel. The dashed lines indicate the metals that would be within 1000 times the impact of lead, and therefore could be considered equally toxic to it.

distinction between kinds of soils—natural and target compartment—freshwater. Because the latter is the emission compartment for building integrated PV, it is only natural that its impacts are the largest when it comes to ecotoxicity. Finally, the freshwater ecotoxicity potential of floating solar would be negligible for all metals, as the fraction of them released to sea returning to freshwater would be very close to non-existent.

In terms of human toxicity, the differences between scenarios are not so marked. Emissions to seawater caused by the breaking of floating solar are more relevant than for ecotoxicity due to the presence of fish in our diets (Fig. 2 b). For that same reason, if the failure of an agrivoltaic system were to release all metals to the environment, its impacts would be higher than if that failure were to happen in any other application. Regardless of where the PSP is installed, all three metals evaluated have a similar human toxicity potential on a per m² basis. On a per mass basis, the release of Cu would be clearly less toxic than that of lead in the floating solar scenario.

C. End-of-life impacts

In Figure 3, we show the freshwater ecotoxicity and human toxicity impacts of recycling the PSP, and the impacts of obtaining the components of the PSP from virgin materials. In addition, we also show the impacts of recovering individual components vs. those same virgin materials, allocating them by mass. For both categories, the recycling of PSPs has an overall impact similar to that of the manufacturing of the panel using virgin materials, (Fig 3.1). The contributions, however, are clearly distinct. For both freshwater ecotoxicity and human

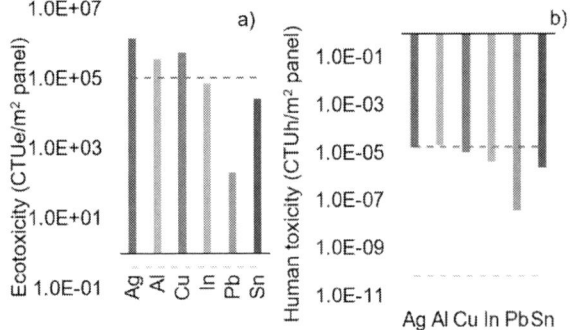

Fig. 2. a) Freshwater ecotoxicity and b) combined human toxicity impacts during the extraction of metals required for 1 m² panel. The dashed lines indicate the metals that would be within 1000 times the impact of lead, and therefore could be considered equally toxic to it.

toxicity, the recovery of the top glass is the largest source of impact, followed by the recovery of the aluminum frame (Fig 3.2). For the manufacturing, however, the bottom glass and cell is the largest source of impact—due to its silver content—followed by copper and aluminum.

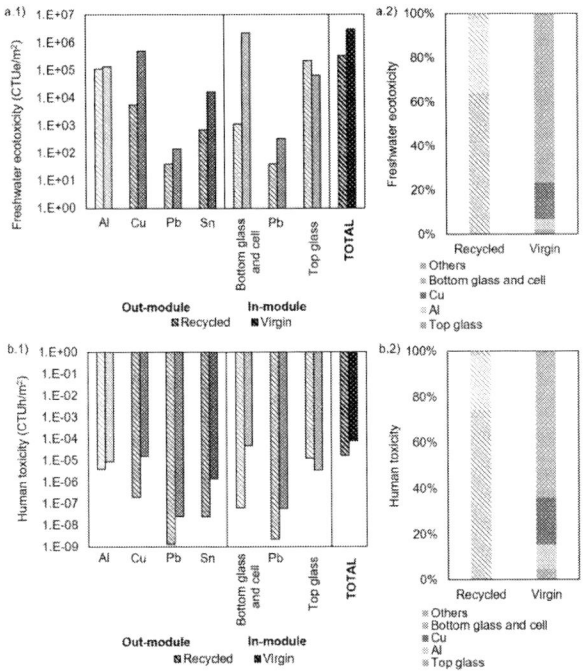

Fig. 3. a) Freshwater ecotoxicity and b) combined human toxicity impacts for the recycling of 1 m² panel, and the mining and manufacturing of the same products using virgin materials. 1) per metal/ layer, 2) fractions of the total impact.

When we assess the PSP components, only recovering the bottom glass and cell is clearly more beneficial than manufacturing using virgin materials. The recycling of the top glass however is nominally more impactful than its manufacturing, but this difference is not significant. Similarly, the recovery of all metals is slightly less impactful than their manufacturing, although not in a conclusive way. The recycling of lead for example, is between 4 and 26 times less toxic than its production. Due to the small amount of lead recovered in-module—when compared to the weight of recovered top glass—this recycling is more efficient than the out-module one. This is, more toxicity is avoided in Pb in-module recycling than in out-module, but none of them is efficient enough to conclude that Pb recovery is less toxic than Pb mining and refining.

IV. SUMMARY

There is a legitimate concern for the presence of lead in perovskite solar panels. However, our preliminary results suggested that from a life cycle toxicity perspective, neither its extraction, nor its complete release to the environment is significantly more harmful than those of other metals present in these PVs. Of the four integrated PV scenarios, floating PV is safer in terms of freshwater ecotoxicity, while agrivoltaics is the most harmful one in terms of human toxicity. Lead recycling, although desirable, is less of a priority than the recovery of the bottom glass and cell.

REFERENCES

[1] A. Babayigit, H.-G. Boyen, and B. Conings, "Environment versus sustainable energy: The case of lead halide perovskite-based solar cells," *MRS Energy Sustain.*, vol. 5, no. 1, May 2018, doi: 10.1557/mre.2017.17.

[2] D. Fabini, "Quantifying the Potential for Lead Pollution from Halide Perovskite Photovoltaics," *Journal of Physical Chemistry Letters*, vol. 6, no. 18. American Chemical Society, pp. 3546–3548, Sep. 17, 2015, doi: 10.1021/acs.jpclett.5b01747.

[3] E. Leccisi and V. Fthenakis, "Life-cycle environmental impacts of single-junction and tandem perovskite PVs: a critical review and future perspectives," *Prog. Energy*, vol. 2, no. 3, p. 032002, Jul. 2020, doi: 10.1088/2516-1083/ab7e84.

[4] P. Billen *et al.*, "Comparative evaluation of lead emissions and toxicity potential in the life cycle of lead halide perovskite photovoltaics," *Energy*, vol. 166, pp. 1089–1096, 2019, doi: 10.1016/j.energy.2018.10.141.

[5] G. Schileo and G. Grancini, "Lead or no lead? Availability, toxicity, sustainability and environmental impact of lead-free perovskite solar cells," *J. Mater. Chem. C*, vol. 9, no. 1, pp. 67–76, 2021, doi: 10.1039/d0tc04552g.

[6] I. Celik, Z. Song, A. B. Phillips, M. J. Heben, and D. Apul, "Life cycle analysis of metals in emerging photovoltaic (PV) technologies: A modeling approach to estimate use phase leaching," *J. Clean. Prod.*, vol. 186, pp. 632–639, Jun. 2018, doi: 10.1016/J.JCLEPRO.2018.03.063.

[7] R. Frischknecht *et al.*, *Life Cycle Inventories and Life Cycle Assessments of Photovoltaic Systems*. IEA PVPS Task 12, International Energy Agency Photovoltaic Power Systems Programme, 2020.

[8] P. Fantke *et al.*, "USEtox® 2.0 Documentation (Version 1.00).," 2017.

[9] G. Wernet, C. Bauer, B. Steubing, J. Reinhard, E. Moreno-Ruiz, and B. Weidema, "The ecoinvent database version 3 (part I): overview and methodology," *Int. J. Life Cycle Assess.*, vol. 21, no. 9, pp. 1218–1230, Sep. 2016, doi: 10.1007/s11367-016-1087-8.

[10] F. Ardente, C. E. L. Latunussa, and G. A. Blengini, "Resource efficient recovery of critical and precious metals from waste silicon PV panel recycling," *Waste Manag.*, vol. 91, no. 2007, pp. 156–167, 2019, doi: 10.1016/j.wasman.2019.04.059.

[11] G. Rodriguez-Garcia, E. Aydin, S. De Wolf, B. Carlson, J. Kellar, and I. Celik, "Life Cycle Assessment of Coated-Glass Recovery from Perovskite Solar Cells," *ACS Sustain. Chem. Eng.*, vol. 9, no. 45, pp. 15239–15248, Nov. 2021, doi: 10.1021/ACSSUSCHEMENG.1C05029.

[12] Fraunhofer ISE, "Integrated Photovoltaics – Areas for the Energy Transformation," 2022. https://www.ise.fraunhofer.de/en/key-topics/integrated-photovoltaics.html (accessed Apr. 06, 2022).

978-1-7281-6118-1/22 $31.00 © 2022 IEEE

Molecular Beam Epitaxy Growth of CdSe for Si-based Tandem Cell Application

Stephen Schaefer[1], Zheng Ju[2], Allison McMinn[1], Xin Qi,[1] and Yong-Hang Zhang[1]

[1] School of Electrical, Computer and Energy Engineering, Arizona State University, Tempe, AZ 85287, USA

[2] Department of Physics, Arizona State University, Tempe, AZ, 85287 USA

Abstract—The II-VI compound semiconductor CdSe has a bandgap energy of 1.71 eV [1] and 1.68 eV [2] in the wurtzite (hexagonal) and zincblende (cubic) crystal structures, respectively, making it an ideal candidate material for the top cell in tandem application with a Si bottom cell. However, the growth of monocrystalline CdSe and control of its phase, wurtzite or zincblende, remains a challenge. Molecular beam epitaxy (MBE) growth of CdSe thin films nearly lattice matched to InAs (100), (111)A, and (111)B oriented substrates is investigated. Growth temperature ranges from 250 to 350 °C, Cd/Se flux ratio ranges from 0.74 to 1.35, and the growth rate ranges from 0.14 to 0.84 monolayers per second. Single crystal zincblende material luminescing at 1.668 eV is demonstrated on the (100) substrates, while polycrystalline mixed-phase material luminescing from 1.589 to 1.726 eV is demonstrated on the (111) substrates. *In-situ* reflection high energy electron diffraction (RHEED) patterns show a clear transition from zincblende 1×1 surface reconstructions with four-fold symmetry to wurtzite 1×1 reconstructions with six-fold symmetry. The results indicate that CdSe crystal phase and thin film morphology is highly sensitive to growth temperature, Cd/Se flux ratio, and polar (111) surface preparation.

Keywords—CdSe, II-VI, thin film, tandem cell, photovoltaic, molecular beam epitaxy

I. Introduction

One of the most effective approaches to reducing solar energy systems' total installed cost is by increasing PV solar cell efficiency and reducing materials use. Currently, the most cost-effective solar cells are Si solar cells and CdTe thin-film solar cells. A low-cost solar cell composed of polycrystalline II-VI alloys (CdSe, MgCdTe, or ZnCdTe) for the top cell in tandem with a crystalline Si bottom cell has the potential to achieve power conversion efficiency greater than 30% at concentrations below 50 suns. The optimal bandgap for the top II-VI cell is 1.73 eV [3]. The II-VI compound CdSe crystallizing in the wurtzite (hexagonal) crystal structure has a bandgap energy of 1.714 eV [2] at 300 K, making it an ideal candidate material for the top cell. A CdSe / Si tandem solar cell has a theoretical power conversion efficiency as high as 40% [3]. The tandem solar cell efficiency exceeds 30% for minority carrier lifetimes as short as 1 ns in the upper CdSe cell, a readily achievable figure even in polycrystalline thin-film II-VI materials [4]. The absorption coefficient of wurtzite CdSe at 1.72 eV is 4.19×10^4 cm^{-1} [5]

yielding an absorption length of 239 nm. To function effectively as absorbers, high quality CdSe films up to 1 μm thick are desired.

Previous works on CdSe for thin film solar cells primarily investigated non-epitaxial growth techniques such as closed-space sublimation, thermal evaporation, and electrodeposition [6-8]. The resulting CdSe material is polycrystalline with poor carrier lifetimes, mobility, and open-circuit voltage (V_{oc}). Additionally, p-type doping of polycrystalline CdSe remains challenging. PlantPV [7] studied the feasibility of CdSe thin film solar cells and p-type doping techniques. Polycrystalline CdSe was grown on sapphire wafers and ZnTe on Si wafers by closed-space sublimation, and p-type doping $\geq 10^{16}$ cm^{-3} using arsenic was achieved, but complete solar cell devices were not demonstrated. Bagheri [6] investigated CdSe thin films grown by thermal evaporation. CdCl$_2$ and Se flux treatments and thermal annealing yielded modest improvements in CdSe grain size. PN junction solar cells with $V_{oc} = 0.8$ V and short circuit current $J_{sc} = 8$ mA/cm^2 were demonstrated. Shaikh *et al.* [8] investigated electrodeposition of n-CdSe/p-Cu$_2$Se heterojunction solar cells, however photovoltaic performance was limited with poor $V_{oc} = 0.35$ V and $J_{sc} = 0.4$ mA/cm^2.

This work investigates the molecular beam epitaxy (MBE) growth of CdSe thin films nearly lattice matched to InAs (100), (111)A, and (111)B oriented substrates. InAs is an attractive substrate for epitaxial growth of monocrystalline CdSe due to the small lattice mismatch of -0.31% for the zincblende (ZB) phase [9] and -0.35% for the wurtzite (WZ) phase [10]. Additionally, growth on the polar InAs (111)A indium-terminated or (111)B arsenic-terminated crystal planes offers an additional degree of freedom for control over the CdSe crystal phase. Precise control over the CdSe crystalline properties and phase, wurtzite or zincblende, is required to achieve the target tandem solar cell efficiency.

II. Experiment

A. Molecular beam epitaxy growth

Bulk CdSe layers are grown using a dual chamber VG V80H solid source molecular beam epitaxy system on cleaved pieces of 50 mm (100), (111)A, and (111)B oriented InAs substrates indium mounted to Si wafers. InAs substrate preparation, including oxide desorption and growth of a 100 –

500 nm thick InAs buffer layer, occurs in a separate III-V chamber followed by transfer under vacuum to the II-VI chamber for CdSe epitaxy. The InAs buffer growth proceeds at a temperature of 450 °C, As/In flux ratio of 2, and growth rate of 0.83 monolayer/s (ML/s) on (100) oriented substrates, and at 500 °C, As/In = 20, and 0.12 ML/s on (111) oriented substrates. The substrate temperature ranges from 250 °C to 350 °C for CdSe growth and is calibrated using an Ircon Modline 3 (model 3G-10C05) pyrometer. Effusion cells containing elemental Cd and Se enable independent control of the fluxes. Cd/Se flux ratio ranges from 0.74 for Se rich (Cd limited) growth to 6.75 for Cd rich (Se limited) growth. The CdSe growth rate ranges from 0.14 to 0.84 ML/s. Reflection high energy electron diffraction (RHEED) monitors the surface reconstructions during the growth of InAs buffer layers and bulk CdSe layers.

The prepared InAs surface is soaked with either Cd or Se for 3 minutes prior to introduction of the other species to initiate CdSe growth. The InAs polar (111)A surface is terminated by the cation In, while the (111)B surface is terminated by the anion As. Therefore, there are four possible initial conditions for CdSe growth on the polar (111) InAs surface, with significant impact on the crystalline properties of the CdSe thin film.

B. Physical and optical characterization

High-resolution coupled ω-2θ X-ray diffraction (XRD) scans of the (400) and (111) crystal planes are acquired using a Panalytical X'Pert Pro MRD triple-axis diffractometer and are simulated using the Panalytical X'Pert Epitaxy dynamical diffraction modeling program. A four-bounce hybrid monochromator provides Cu K-α illumination with wavelength λ = 1.5406 Å. The CdSe material parameters including lattice constant, Poisson ratio, Debye-Waller factor, and X-ray scattering factors are obtained from literature [9-11]. Scanning electron microscopy (SEM) images are acquired using a Hitachi S-4700 field emission microscope at a beam accelerating voltage of 15 kV. Photoluminescence (PL) spectroscopy measurements are acquired using a Horiba iHR 550 monochromator equipped with a CCD detector and InGaAs array detector. A 532 nm CW diode laser is used for sample excitation.

III. RESULTS AND DISCUSSION

A. Reflection high energy electron diffraction

Prior to initiation of CdSe growth the prepared InAs substrate exhibits a streaky 1×1 surface reconstruction with four-fold symmetry, characteristic of an atomically smooth zincblende crystal surface. The 1×1 reconstruction is initially present during CdSe layer growth, followed by the gradual appearance of doubled spots on the 1× streaks connected by Kikuchi lines, and finally a complete transition to a spotty/streaky 1×1 surface reconstruction with six-fold symmetry, characteristic of a roughened wurtzite (0001) crystal surface. Fig. 1 shows the initial streaky 1×1 and transitional spotty/streaky 1×1 reconstruction after 30 minutes of CdSe growth on InAs (111)B at 300°C.

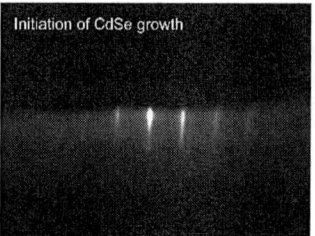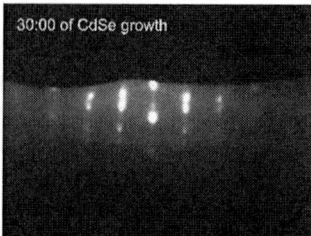

Fig 1. RHEED patterns for CdSe grown on InAs (111)B at 300 °C, Cd/Se flux ratio of 6.75, and growth rate of 0.16 ML/s.

B. X-ray diffraction

Coupled ω-2θ XRD scans of the (111) plane for CdSe layers grown on InAs (111) oriented substrates, shown in Fig. 2, indicate primarily compressively strained material in agreement with reported values for zincblende and wurtzite lattice constants [8, 9]. The sample grown on InAs (111)B at 350 °C shows anomalously large compressive strain and peak broadening, attributed to the highly non-uniform film morphology and presence of nanometer-scale crystallites on the surface (see Fig. 4). In contrast, the sample grown on InAs (111)A at 300 °C lacks a clearly defined layer peak with some evidence of both compressively and tensilely strained material, attributed to the presence of mixed phase CdSe and the formation of nanocolumns in the film (see Fig. 5). The origin of tensile strain in the sample is unclear but may result from the large distribution of crystallite size and orientation in the film.

C. Scanning electron microscopy

Figs. 3-5 show SEM images in cross-section (left) and angled/plan views (right) for a series of CdSe bulk layers on InAs (111)B and (111)A substrates. The film morphology is strongly dependent on the choice of polar surface, growth temperature, and Cd/Se flux ratios. Cd rich growth at 250 °C (Fig. 3) yields a relatively uniform film with an apparent phase transition after about 100 nm indicated by the change in cleavage plane. In contrast, Cd rich growth at 350 °C (Fig. 4)

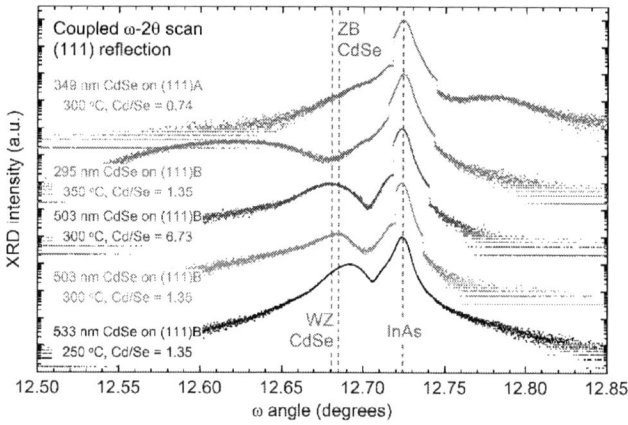

Fig. 2. Coupled ω-2θ scans of the (111) reflection for bulk CdSe layers grown on InAs (111) oriented substrates. Dashed vertical lines indicate the Bragg angles for ZB InAs, ZB, CdSe, and WZ CdSe. Growth conditions are labeled on each curve. (The discontinuities near the InAs peaks are machine artifacts)

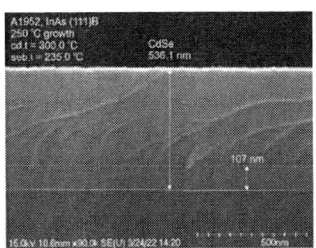
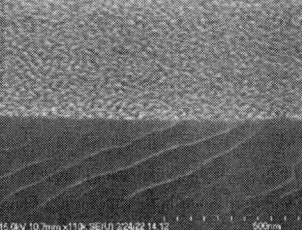

Fig. 3. SEM cross-section (left) and surface (right) images of bulk CdSe grown on InAs (111)B at 250 °C, Cd/Se flux ratio = 1.35, and growth rate = 0.84 ML/s.

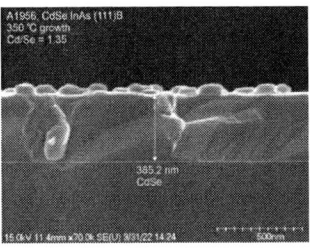
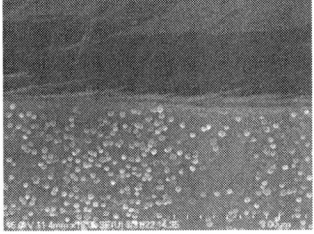

Fig. 4. SEM cross-section (left) and surface (right) images of bulk CdSe grown on InAs (111)B at 350 °C, Cd/Se flux ratio = 1.35, and growth rate = 0.83 ML/s.

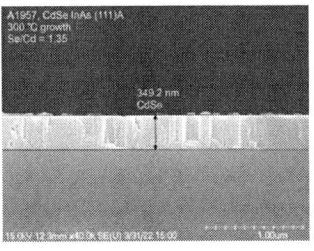

Fig. 5. SEM cross-section (left) and surface (right) images of bulk CdSe grown on InAs (111)A at 300 °C, Cd/Se flux ratio = 0.74, and growth rate = 0.55 ML/s.

yields a rough film with numerous polygonal crystallites on the surface. Finally, Se rich growth on the (111)A plane at 300 °C results in nanocolumns with a roughly hexagonal cross-section, in agreement with the rapid RHEED transition to the six-fold 1×1 reconstruction during growth suggesting wurtzite CdSe.

D. Photoluminescence spectroscopy

Fig. 6 shows the room temperature (300 K) PL spectra for bulk CdSe layers grown on InAs (100), (111)A, and (111)B substrates. Two distinct sets of PL peaks are observed at 743.3 nm (1.668 eV) and 718.5 nm (1.726 eV) attributed to zincblende (ZB) and wurtzite (WZ) phase CdSe, respectively. The sample grown under Se-rich conditions on InAs (111)A exhibits bright but very broad luminescence centered around 780 nm (1.589 eV). The PL results indicate that both phases of CdSe are present in the MBE grown material. The extreme spectral broadening of the sample grown on the (111)A polar surface suggests a large distribution of crystal grain sizes in the nanocolumns shown in Fig. 5.

IV. SUMMARY

MBE growth of CdSe indicates the physical and optical properties of the material is strongly dependent on the growth temperature, Cd/Se fluxes, growth rate, and InAs polar (111)A or (111)B surface. Both zincblende and wurtzite phases

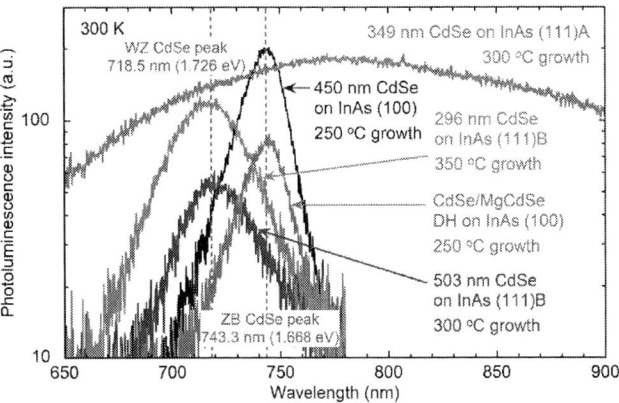

Fig. 6. Room temperature PL spectra for bulk CdSe layers grown on InAs (100), (111)A, and (111)B substrates. Growth conditions are indicated directly on the figure. Dashed lines indicate the PL peaks attributed to zincblende (ZB) and wurtzite (WZ) CdSe at 743.3 nm and 718.5 nm, respectively.

are present in the epitaxially grown material as shown by XRD, SEM, and PL measurements. There remains considerable scope for optimization of the MBE growth conditions to obtain monocrystalline wurtzite CdSe for thin-film solar cells. Preliminary results indicate that growth at higher temperatures yields superior photoluminescence intensity. Future experiments will explore variations in the Cd/Se fluxes, as well as initial preparation of the polar InAs (111) substrate surface.

ACKNOWLEDGMENT

This work is partly supported by the NSF I/UCRC SPF2050 (Award number: IIP-2052814) and First Solar. The authors acknowledge the use of facilities within the Eyring Materials Center at Arizona State University supported in part by NNCI-ECCS-2025490.

REFERENCES

[1] D W Palmer, www.semiconductors.co.uk, 2008.03, retrieved from http://www.semiconductors.co.uk/propiivi5410.htm.

[2] Gutowski J, Sebald K and Voss T, Semiconductors, vol. 44B (Berlin: Springer) p. 75 (2009).

[3] D. Ding, S. R. Johnson, S.-Q. Yu, S.-N. Wu, and Y.-H. Zhang, "A semi-analytical model for semiconductor solar cells", J. Appl. Phys. **110**, 123104 (2011).

[4] J. Ma, D. Kuciauskas, D. Albin, R. Bhattacharya, M. Reese, T. Barnes, J. V. Li, T. Gessert, and S.-H. Wei, "Dependence of the Minority-Carrier Lifetime on the Stoichiometry of CdTe Using Time-Resolved Photoluminescence and First-Principles Calculations", Phys. Rev. Lett. **111**, 067402 (2013).

[5] S. Ninomiya and S. Adachi, "Optical properties of cubic and hexagonal CdSe", J. Appl. Phys. **78**, 4681-4689 (1995).

[6] B. Bagheri, "Research project to study cadmium selenide (CdSe) solar cells", Iowa State University 2020.

[7] Plant PV Inc., "Low Cost, Epitaxial Growth of II-VI Materials for Multijunction Photovoltaic Cells" (2014).

[8] A. V. Shaikh, S. G. Sayyed, S. Naeem, S. F. Shaikh, and R. S. Mane, "Electrodeposition of n-CdSe/p-Cu$_2$Se Heterojunction Solar Cells", Eng. Sci. **14**, 14-26 (2021).

[9] P. D. Lao, Y. Guo, G. G. Siu and S. C. Shen, Phys. Rev. B **48**, 11701 (1993).

[10] A. W. Stevenson and Z. Barnea, Acta Cryst. **B40**, 530 (1984).

[11] C. H. Macgillavry and G. D. Rieck, International Tables for Crystallography (The Kynoch Press, Birmingham, 1968).

Fabrication of ultrathin Ge template for growth of multijunction solar cells based on wafer-scale porous Ge

Tadeáš Hanuš, Javier Arias-Zapata, Bouraoui Ilahi, Philippe-Olivier Provost, Alexandre Chapotot, Abderraouf Boucherif

Institut Interdisciplinaire d'Innovation Technologique (3IT), Université de Sherbrooke, 3000 Boulevard de l'Université, Sherbrooke, J1K OA5 Québec, Canada , Sherbrooke, QC, Canada

Laboratoire Nanotechnologies Nanosystèmes (LN2) – CNRS IRL-3463 Institut Interdisciplinaire d'Innovation Technologique (3IT), Université de Sherbrooke, 3000 Boulevard Université, Sherbrooke, J1K OA5 Québec, Canada, Sherbrooke, QC, Canada

Multijunction solar cells (MJSC) currently hold the highest efficiency on the market. However, their widespread in terrestrial applications is getting held back by the high devices cost. A considerable part of the cost mainly comes from the substrate materials such as Ge and GaAs making them nonviable for terrestrial application compared to much cheaper silicon-based solar cells. Consequently, the Ge based MJSC deployment is restrained to niche domains such as spatial applications. Accordingly, the development of nanostructured substrates allowing MJSC detachment and wafer reuse stands out as a promising approach to overcoming these limitations. In this work, we demonstrate the formation of homogenous edge-to-edge porous Ge (PGe) layers on an industry-standard 100 mm wafer-scale produced by bipolar electrochemical etching. The produced nanostructured substrates' properties are easily assessable by production line compatible, fast, and nondestructive techniques such as ellipsometry. The PGe layers have been found to exhibit excellent uniformity over the wafer' surface with a relative variation of 1% in porosity and 2% in thickness. Furthermore, we show that the PGe structural properties can be finely tuned to create on-demand characteristics including the suitability for epitaxial growth. Accordingly, low-temperature growth of ultrathin crystalline Ge layer on top of PGe structure is demonstrated. The fabricated structure has been shown to be compatible with III-V heterostructures growth drawing the way for wafer-scale detachable MJSC and substrate reuse.

Prediction of Novel Phosphors using Machine Learning for Efficiency Enhancement of Silicon Solar Cells

Tae-Gwan Kim, Eun-gyeong Kim, M.Shaheer Akhtar, O-Bong Yang

Graduate School of Integrated Energy-AI, Jeonbuk National University, Jeonju, South Korea

School of Semiconductor and Chemical Engineering, Jeonbuk National University, Jeonju, South Korea

New and Renewable Energy Materials Development Center (NewREC), Jeonbuk National University, Jeonbuk, South Korea

In order to achieve the theoretical efficiency over 30%, the mismatch between solar spectrum and spectral response of Si solar cells is needed to improve. In general, the thermalization of charge carriers, that usually generated by the absorption of photons having higher energy greater than a bandgap of semiconductor is the most common loss mechanism in Si solar cells. To control the thermalization of charge carriers, the introduction of sensitizer ions (such as Eu^{2+}) into phosphor fluorescence materials can enhance the absorption of photons with lower energy. In this work, $BaMgAl10O17:Eu+2$, $SrMgAl10O17:Eu+2$ nano-phosphors were synthesized by solution combustion method. The synthesized phosphors were extensively analyzed in terms of morphology, crystal structure, compositional, photoluminescence, and electrical properties by various analytical tools like X-ray Diffraction (XRD), Field Emission Scanning Electron Microscope-Energy Dispersive X-ray Spectrometer (FESEM-EDX), PL Spectroscopy, and quantum efficiency analyzer, etc. The database were extracted from raw data to enable the application of algorithms for machine learning training. From optimized regression model like Random Forest Regressor, Gradient Boosting, we can predict appropriate crystal structure and desirable wavelength of phosphors. The establishment of predicted models is efficient to work, so researchers can consider to find the potential materials such as $ZnMgAl10O17$ and $CaMgAl10O17$ in advance and save the huge demand for energy resources. It is hope that this work can facilitate the development of desirable synthesis methods under various conditions for phosphors with minimal light scattering, and good luminous efficiency which may helpful for improving the performance of Si solar cells.

978-1-7281-6118-1/22 $31.00 © 2022 IEEE

Predictive Modeling of Cracks within Flexible Perovskite Thin Films

Melissa A Davis, Rebekah Sweat, Zhibin Yu

Florida State University, Tallahassee, FL, United States

High Performance Materials Institute, Tallahassee, FL, United States

Cracking behaviors within perovskite thin films were modeled through finite element analysis. Coupons of polyethylene terephthalate or PET, Indium Tin Oxide or ITO, and perovskite layers were placed in a three-point bend set-up and deformed. Modeling gave insight into ranges for physical properties of each layer, when absent or greatly varied in literature. The model proved the relationship between perovskite crystal size and its mechanical tolerance. Models with smaller crystals withstood more than 30% further displacement than those with crystals twice its size. A material replacement of ITO, the driving force causing cracking in films, was also modeled and cracking within the perovskite layer occurred significantly later, showing an improved resilience to bending.

Results of Environmental-Based PV Soiling Models after Extreme Dust Events: The Case of Saharan Dust Intrusions in Southern Spain

João Gabriel Bessa, Álvaro F Solas, Florencia A Cruz, Eduardo F Fernández , Leonardo Micheli

Advances in PV technology Research Group (AdPVTech), CEACTEMA, University of Jaén, Jaén , Spain

Department of Astronautical, Electrical and Energy Engineering (DIAEE), Sapienza University of Rome, Rome, Italy

Soiling is a major issue that greatly determines the operation and maintenance cost of PV systems. The occurrence of extreme dust events in sites where typically the losses due to soiling are not high can significantly alter the cleaning schedule. In this work, two different environmental-based models are applied to estimate the soiling losses that two recent Saharan dust intrusions caused in a location in southern Spain. During the first one, a peak of PM10 equal to 904 µg/m³ was reached, and it became the most intense dust event registered in 10 years for both its intensity and its length, as it lasted almost three days. Due to these two factors that led to very low irradiance values during the episode, no measurements from a soiling station installed in the site were available. Therefore, the only approach to quantify the impact of this event was through environmental-based soiling extraction models. The results of these models showed both relatively high differences between them and a strong dependence with the cleaning threshold value. These issues are expected to impact the decision-making about the cleaning of PV modules. For example, non-negligible differences were found when considering a cleaning threshold of 1 mm/day or a cleaning threshold of 10 mm/day. In the first case, only 1 day with significant soiling losses (8.3% using Coello' model) is detected; whereas in the second case, the PV modules are supposed to be notably soiled (losses >= 8.3%) during 5 days.

Optical Design Considerations for Thin Photonic Power Converters with Textured Back Reflector

Neda Nouri, Christopher E. Valdivia, Meghan N.Beattie, Jacob J.Krich and Karin Hinzer

SUNLAB, Centre for Research in Photonics, University of Ottawa, Ottawa, Ontario, Canada

Abstract— **Photonic power converters (PPC) convert narrow-band light into electricity via the photovoltaic effect. In this study, we discussed optical design considerations of thin InAlGaAs PPCs with integrated pyramidal nano-textured back reflectors (BRs) under 1310 nm illumination. Simulation results using the finite difference time domain method revealed that multiple combinations of design space parameters yield similar total absorptance at 1310 nm but with different absorptance spectra. A tolerance study performed using one optimized design showed that a 70 nm variation in device thickness shifted BR-induced resonances from constructive to destructive interference and a 50 nm variation of the height and base width of nano-pyramids dropped the absorptance by more than 25% (absolute). Among all simulated designs, those with overlapping resonances around the target wavelength that incorporated a thin-film antireflection coating were the least sensitive to variations in the illumination wavelength.**

Keywords— *photonic power, photovoltaic device, light trapping, textured back reflector, antireflection coating, optimization, numerical simulation.*

I. Introduction

Photonic power converters (PPCs) are a main component of power over fiber and power beaming systems, generating electrical power from incident narrow-band illumination via the photovoltaic effect. Ultra-thin PPCs reduce material use, enable flexible devices, and exceed the efficiency of their optically thick counterparts when high absorptance is maintained through efficient light trapping. In our recent works, we showed through simulations that back reflector (BR) integration in thin and ultra-thin PPCs enables high light absorptance (by increasing the optical path length and minimizing reflection loss), while textured BRs exhibit superior light trapping capability than planar BRs due to guided mode resonances [1-2]. In those designs, aligning BR-induced resonances at the wavelength of interest was crucial to achieve high absorptance.

Spectral resonances in monochromatic PPCs with textured BRs are sensitive to the PPCs thickness and nanotexture geometries due to interference effects. Aligning the resonance with the desired wavelength requires fine tunning and precise optimization of PPC and nanotexture dimensions. Also, deviation of the incident light angle and operational wavelength due to manufacturing tolerances and temperature fluctuations can lead to resonance misalignment during operation. In this work, we use optical modeling to investigate the design considerations including optimization of the design space parameters and the impact of the thin-film antireflection coating

NSERC Award Number 497981

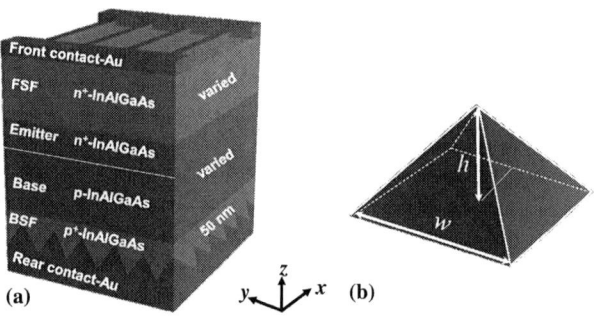

Fig. 1. (a) Schematic structure of ultra-thin PPCs with pyramidal textured back reflector. (b) Dimensions of pyramidal back reflector with height h and base width w indicated.

(ARC) on the resonances for an ultra-thin PPC with an integrated square pyramidal textured BR under 1310 nm illumination. In particular, this paper quantifies the spectral variation for several designs and the tolerance of absorptance to a wide variation in the geometrical design parameters.

II. Optical Modeling

Our modeled thin PPCs, shown in Fig. 1(a), are comprised of InAlGaAs absorber and transparent layers grown lattice-matched to an InP substrate. The absorber region (emitter and base) is comprised of $In_{0.532}Al_{0.097}Ga_{0.371}As$ with a bandgap energy of 0.88 eV, and is surrounded by wider bandgap (1.25 eV) front and back surface field (FSF and BSF) layers with the composition $In_{0.527}Al_{0.356}Ga_{0.117}As$. Gold acts as both rear contact and pyramidal nanotextured BR, shown in Fig. 1(b). 3D optical simulations over a single unit cell were conducted under x-polarized normal-incidence 1310 nm plane wave illumination with a spectral linewidth of 10 nm and an intensity of 100 mW/cm², using the FDTD Solutions (finite-difference time-domain based software) by Lumerical. Wavelength-dependent refractive indices for the absorber and transparent layers were extracted from ellipsometry measurements of samples with the same composition [3]. Periodic and perfectly matched boundary condition were applied in the lateral (x and y) directions, while boundary regions on the top and bottom of modeled PPCs assumed perfectly matched layers. The modeled geometry of the pyramidal BR employed a staircase approximation using a series of 2.5-nm thick stacked squares of varying lateral size. To avoid the design dependence on (x or y) polarized light, pyramid dimensions were kept equal in the x and y directions.

III. OPTIMIZATION

Optical cavity formation between the BR and PPC layers leads to resonances in the absorptance (or reflectance) spectrum. The first design consideration is to tune the resonance peak in the absorptance spectrum (or resonance dip in the reflectance spectrum) to coincide with the target wavelength. Variations of design space parameters, including the absorber and FSF layer thicknesses and the height h and base width w of the pyramidal BR, can align resonances at the target wavelength but not necessarily with the highest achievable absorptance. Due to the large number of design space parameters, we employed a particle swarm algorithm [4] to determine optimum parameters that minimize the absorber layer thickness while obtaining a target absorptance of 90%, corresponding to a photocurrent J_{ph} of ~95 mA/cm^2. Optical simulation results show that guided mode resonances induced by the textured BR yield up to an 8-fold reduction in absorber layer thickness compared to an optically thick PPC on an InP substrate with the same absorptance (Fig. 2). To allow comparison of currents between thick and ultra-thin designs, a double-layer SiO$_2$/TiO$_2$ antireflection coating (ARC) was necessarily applied to the thick designs to match the near-zero reflection possible in ultra-thin designs via optimization of their internal layer structures. The optimized parameters for both ultra-thin and thick PPCs are listed in the inset table of Fig. 2.

IV. TOLERANCE STUDY

Fig. 3 shows absorptance maps of the optimized PPC as a function of (a) absorber and FSF layer thickness, and (b) width and height of nanotextured BR pyramids. In each case, the remaining design parameters were fixed to the values specified

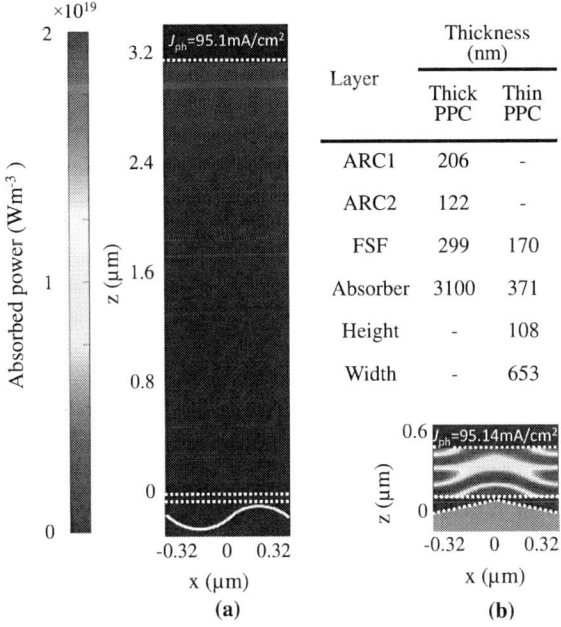

Fig. 2 Cross-sectional absorbed power profiles through the center of optimized: (a) thick PPC on InP substrate; (b) thin PPC with pyramidal nanotextured BR. Structure dimensions are listed in the inset table. Note that the region with wavy line in (a) represents the substrate, which was simulated, but its thickness is not fully depicted here.

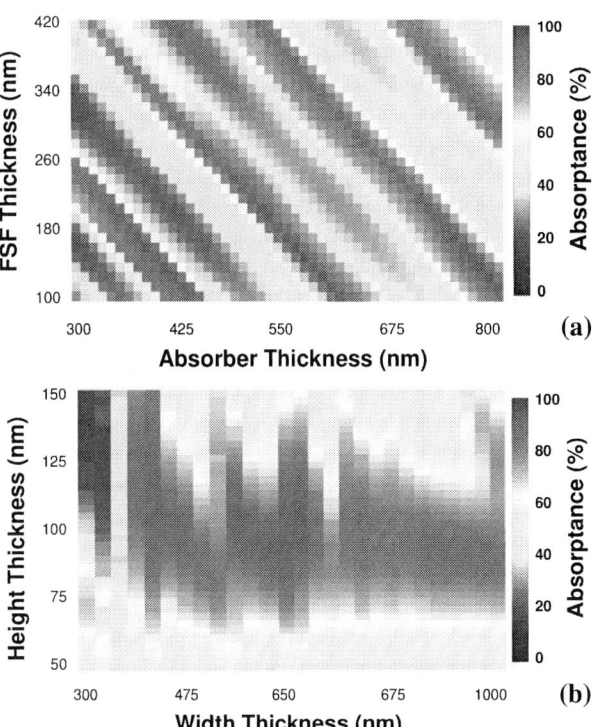

Fig. 3 Absorptance as a function of: (a) absorber and FSF layers thicknesses; (b) height and width of the back reflector nanotextured square pyramids.

in the inset table of Fig. 2. These results show that an inversion from constructive to destructive interference occurs with variations of 60-70 nm and 70-80 nm for the absorber and FSF layer thicknesses, respectively. For the BR design, a 50 nm variation of width or height can drop absorptance by >25%. We note that a thick absorber layer does not universally guarantee high absorption, and therefore the absorber and FSF layer thicknesses and nanopyramid width and height must all be simultaneously and precisely tuned to produce a resonance peak at the desired wavelength.

As evident from Fig. 3, multiple combinations of design space parameters result in high absorptance, enabling considerable design freedom. To compare differing high-absorptance designs, we investigated the absorptance spectra of four combinations of absorber and FSF layer thicknesses, each with an absorptance of 90.25±0.05% at 1310 nm. The absorptance, reflectance, and BR loss at 1310 nm for these

TABLE 1. LAYER THICKNESS AND LOSSES AT 1310 NM, AND ABSORPTANCE AS A FUNCTION OF LASER WAVELENGTH FOR FOUR DESIGNS WITH SIMILAR TOTAL ABSORPTANCE

		Design			
		A	B	C	D
Absorber (nm)		540	610	540	610
FSF (nm)		398	314	175	100
Reflectance (%)		1.5	1.9	1.6	2.2
Back reflector loss (%)		8.3	7.9	8.1	7.5
Absorptance (%)	1310 nm	90.2	90.2	90.3	90.3
	1320 nm	85	86.4	86.7	85.3
	1330 nm	81	83	78.0	74.5
	1340 nm	84.2	83.8	68.4	63.6
	1350 nm	79.0	78.1	61.2	55.2

978-1-7281-6118-1/22 $31.00 © 2022 IEEE

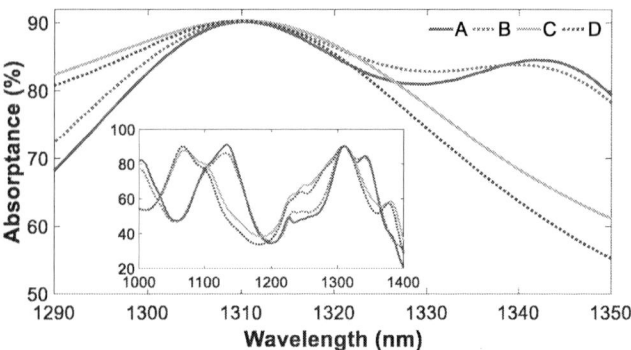

Fig.4. Absorptance spectra of four combinations of absorber and FSF layer thicknesses, yielding similar absorptance at 1310 nm but varying peak widths. Inset shows the absorptance spectra over a broadened wavelength range of 1000-1400 nm.

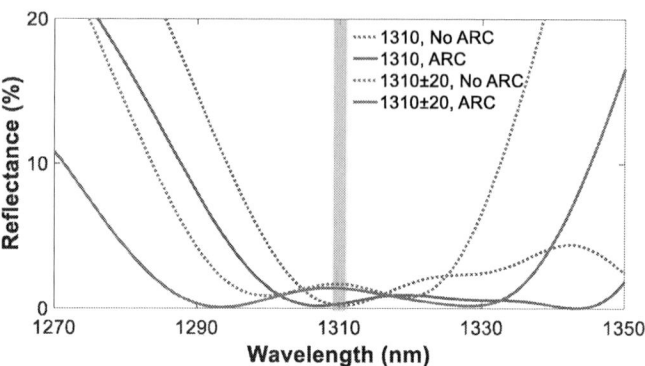

Fig.5. Reflectance spectra of optimized ultra-thin PPCs with and without ARC (solid and dashed lines) for single (1310 nm) and wide (over 1290 to 1330 nm) optimization (blue and red lines). The vertical green line indicates laser bandwidth.

selected designs (A-D) are listed in Table 1. Parasitic BR loss is dependent upon the number of round trips at the resonance wavelength, yielding a greater interaction with the back mirror where scattering and absorption can occur. This effect is demonstrated by comparing designs A and C (540-nm thick absorbers) to designs B and D (610 nm), for which the thinner designs result in higher BR loss. Comparing the absorptance spectra for all four designs over a spectral range of 1000-1400 nm (inset of Fig. 4), we notice that the absorptance spectra of design A closely matches the spectra for design B, exhibiting overlapping resonances around 1310 nm, and the spectra of design C matches design D. In the case of laser-illuminated PPCs, where spectral linewidth of the incident light is ~10 nm, these variations on resonance widths can be negligible under ideal conditions. However, wider resonances are more tolerant of wavelength deviations due to laser fabrication variations and temperature instabilities [5], or deviation of incidence angle from the normal. Resonance shapes for designs A-D are shown in Fig. 4, demonstrating the variation in peak widths across designs. Table 1 reports the absorptance of each design for wavelength deviations of 0 to +40 nm from the design wavelength of 1310 nm. Among these designs, A and B are more wavelength-tolerant with overlapping resonances near 1310 nm and absorptance drops of ~ 5.2-11.2% and 3.8-12.1% (abs.), respectively, for 10-40 nm wavelength deviations. Design D is least tolerant with 5-35.1% (abs.) absorptance drop under the same conditions.

V. ARC EFFECT

In broadband photovoltaics (e.g. solar cells), it is common to add a single or double layer ARC to minimize reflection losses across the broad solar spectrum. In the case of PPCs under narrow-band illumination, reflection can be minimized by aligning BR-induced resonances at the wavelength of interest without an ARC. For our modeled ultra-thin PPC shown in Section III (Fig. 2 (b)), we achieved near-zero reflection loss of ~0.04% for a wavelength of 1310 nm without incorporating an ARC. The spectral absorptance of this baseline design is shown as the blue dotted line in Fig. 5. By adding a single layer SiO_2 ARC, optimization further minimizes the reflectance in the wavelengths around 1310 nm, as indicated by the blue solid line

of Fig. 5. Due to high simulation time and memory requirement both optimizations were performed for a single wavelength at 1310 nm. As can be seen in Fig. 5, the resulting reflectance spectra were not symmetric around 1310 nm, making these designs more tolerant to longer wavelength variations.

To form a more symmetrical and wider tolerance, we extended optimizations over a wider bandwidth (1310 nm±20 nm) both with and without ARC incorporation. The parameters of four optimized designs which are illustrated in Fig. 5 are listed in Table. 2. For fair comparison of these optimizations, the absorber layer thickness was selected to be the same as the baseline design (371 nm) ±10 nm for minimum and maximum limits.

Compared to the optimizations for the single wavelength, these results show reflectance spectra that are centered on the design wavelength, although at the cost of a increased reflectance at 1310 nm. While the optimization without ARC does not exhibit satisfactory performance (Fig. 5 dashed red line), the opimized design with ARC shows a relatively flat absorption across the optimized wavelength range (Fig. 5 solid red line).

VI. CONCLUSION

We discussed design considerations for thin PPCs with pyramidal textured BRs, using numerical simulations. The effects of variations in illumination wavelength and geometrical properties including layer thickness and nano-pyramid dimensions were quantified. A variation of about 70 nm in FSF or absorber layer thickness can shift BR induced resonances from constructive to destructive interference and a 50 nm variation of the nano-pyramid dimensions drops absorptance by more than 25% (absolute), which illustrates the precision required in the fabrication of high efficiency ultra-thin PPCs.

TABLE 2. OPTIMIZED PARAMETERS OF PPCs WITH AND WITHOUT ARC

Parameters thicknesses	Optimization			
	1310		1310±20	
ARC (nm)	0	101	0	143
FSF (nm)	170	155	143	140
Absorber (nm)	371	372	366	375
Width (nm)	653	652	456	648
Height (nm)	108	135	163	113

Our results showed that multiple combinations of design space parameters yield similar absorptance at 1310 nm but with different absorptance spectra and resonance widths. Designs with overlapping resonances around the target wavelength combined with an ARC layer are less sensitive to variation in the illumination wavelength. We expect that the results shown in this study will be useful for the development of BR-integrated PPCs with improved tolerance to wavelength deviations, both near 1310 nm and for other wavelengths.

REFERENCES

[1] N. Nouri, C. E. Valdivia, Meghan. N. Beattie, M. S. Zamiri, and K. Hinzer, "Thin Photonic Power Converters Designs with Back Reflector Operating at a 1310 nm Wavelength," in *2020 47th IEEE Photovoltaic Specialists Conference (PVSC)*, Jun. 2020, pp. 2359–2362. doi: 10.1109/PVSC45281.2020.9300435.

[2] N. Nouri, C. E. Valdivia, M. N. Beattie, M. S. Zamiri, J. J. Krich, and K. Hinzer, "Ultrathin monochromatic photonic power converters with nanostructured back mirror for light trapping of 1310-nm laser illumination," in *Physics, Simulation, and Photonic Engineering of Photovoltaic Devices X*, Mar. 2021, vol. 11681, p. 116810X. doi: 10.1117/12.2584689.

[3] M. Beattie, "Semiconductor Materials and Devices for High Efficiency Broadband and Monochromatic Photovoltaic Energy Conversion," Thesis, Université d'Ottawa / University of Ottawa, 2021. doi: 10.20381/ruor-26695.

[4] D. Wang, D. Tan, and L. Liu, "Particle swarm optimization algorithm: an overview," *Soft Comput.*, vol. 22, no. 2, pp. 387–408, 2018.

[5] N. I. Khan, S. H. Choudhury, and A. A. Roni, "A comparative study of the temperature dependence of lasing wavelength of conventional edge emitting stripe laser and vertical cavity surface emitting laser," in *Proceedings of the International Conference on Data Communication Networking and Optical Communication System*, Seville, Spain, 2011, pp. 1–5. doi: 10.5220/0003512101410145.

Interrelation of CdTe grain size, post-growth processing and window layer selection on solar cell performance

Thomas P Shalvey, Heath Bagshaw, Jonathan D Major

Stephenson Insitute for Renewable Energy, Department of Physics, University of Liverpool, Liverpool, United Kingdom

SEM Shared Research Facility, School of Engineering, University of Liverpool, Liverpool, United Kingdom

This work studies three different device architectures in parallel, allowing for an in-depth comparison of processing conditions for CdTe solar cells grown on CdS, SnO2 and CdSe coated substrates. Direct replacement of the CdS window layer with a wider band gap SnO2 layer is hindered by poor growth of the absorber, producing highly strained CdTe films and a weak junction. This is alleviated by inserting a CdSe layer between the SnO2 and CdTe, which improves the growth of CdTe and results in a graded CdSexTe1-x absorber layer. For each substrate, the CdTe deposition rate and post growth chloride treatment is systematically varied, highlighting the distinct processing requirements of each device structure.

DIrect Sunlight into CO conversion

Thierry de Vrijer, Arno Smets

Delft University of Technology, 2628 CD Delft, Netherlands

Abstract—in this abstract an overview is presented of research performed in the DISCO project, on the development of a silicon-based high voltage multijunction device for autonomous solar to fuel applications. "

Keywords— solar-to-fuel, multijunction PV, silicon, germanium, high voltage

I. INTRODUCTION

In this abstract, the highlights of the DISCO project are discussed. In the DISCO project the processing of wireless silicon-based high-voltage 2-terminal multijunction (MJ) photovoltaic (PV) devices, is investigated. Such devices can be used in autonomous solar-to-fuel synthesis systems, as well as other innovative approaches in which the MJ solar cell is used not only as a photovoltaic current-voltage generator, but also as an ion-exchange membrane, electrochemical catalysts and/or optical transmittance filter. A prerequisite for the development of stand-alone solar-to-fuel synthesis devices is a wireless PV component that generates sufficient voltage to drive the desired electrochemical reduction reaction. The aim of DISCO is therefore to provide a framework that can be used for the flexible application of earth abundant and chemically inert Group IV alloys to realize a range of V_{oc}'s, facilitating electrochemical (EC) reactions ranging from water splitting to CO_2 reduction, in different wireless MJ architectures.

The framework is developed through the investigation of I. PV materials and single junction solar cells, including a plasma enhanced chemical vapour deposition (PECVD) processed group IV low bandgap alloy based on Ge(Sn):H, and II textures on c-Si. In these first two areas, research was focused on developing a better understanding of the fundamental relations between: i) processing parameters ↔ ii) material structure ↔ iii) opto-electrical properties & chemical stability ↔ iv) PV device performance. Here i) includes parameters like the deposition technique, substrate type/morphology, and conditions such as temperature, power pressure and ii) includes characteristics like the elemental composition/stoichiometry, void fraction, crystalline phase fraction.

Finally, in part III of DISCO, the additional challenges introduced in MJ architectures are considered. These include the structural investigation of the tunnel recombination junction (TRJ) as well as the J_{SC} vs $V_{OC}*FF$ trade-offs as a function of absorber characteristics, such as elemental composition and thickness, and those introduced by using an intermediate reflective layer. Additionally, in DISCO, the design considerations for developing a Photoelectrochemical (PEC) device from a PV device are investigated.

II. DISCO RESULTS

A. Photovoltaic materials

The influence of individual deposition conditions is characterized in DISCO for a range of materials. These include hydrogenated (:H) doped Si and siliconoxide (SiO_X) [1], [2] as well as intrinsic SiGe:H [3], Ge:H [4]–[6] and GeCSn:H [7]. These detailed works do not only provide tools for achieving certain structural properties in the different materials, but also high-level relations between structural properties and the chemical stability and opto-electrical nature of the materials.

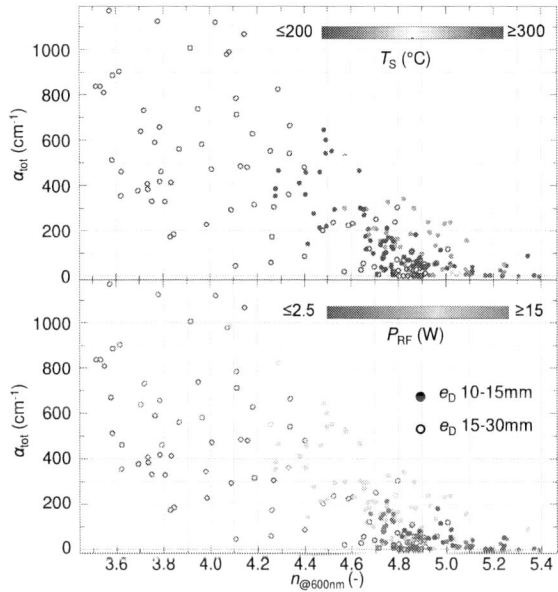

Fig.1: α_{tot} (metric for chemical instability) as a function of $n_{@600nm}$ for all Ge:H films processed in the DISCO project. Closed icons and open icons indicate samples processed with $e_D \leq 15mm$ and $e_D > 15mm$, respectively. Colour indicates T_S (top) and P_{RF} (bottom). From [8].

An example of the ii) material structure ↔ iii) chemical stability relation is shown in Fig.1, where the α_{tot}, a film-thickness independent metric for chemical stability [8] based on infrared oxidation and carbisation signatures [9], is plotted as a function of the refractive index at a wavelength of 600nm ($n_{@600nm}$), which can be considered a metric for the material density.

Fig.1 not only shows that the densest films do not suffer from post-deposition oxidation ($\alpha_{tot}=0$) and that the degree of oxidation increases ($\alpha_{tot}\uparrow$) with decreasing $n_{@600nm}$, it also demonstrated that the densest films are only realized in a specific processing window. This window involves relatively high

978-1-7281-6118-1/22 $31.00 © 2022 IEEE

substrate temperature (T_S), low RF power (P_{RF}) and a small electrode distance (e_D).

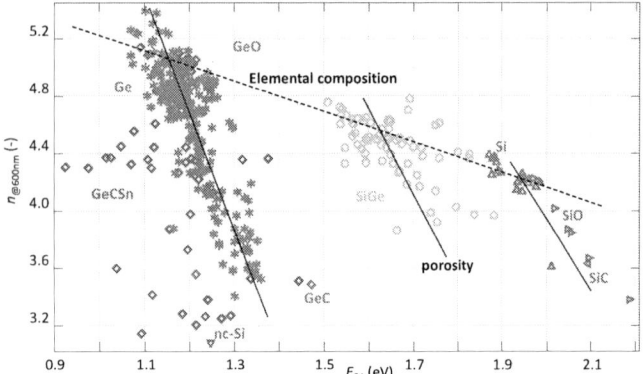

Fig. 2: $n_{@600nm}$ as a function of the E_{04} optical bandgap energy of a range of hydrogenated group IV alloys processed in the DISCO project. Trendlines indicating the effect of elemental composition and porosity are referenced in text. From [8].

An example of the ii) material structure ↔ iii) opto-electrical properties relation is shown in Fig.2, where the optical bandgap energy (E_{04}) is plotted as a function of $n_{@600nm}$ for of over 400 films consisting of a wide range of group IV alloys. The influence of structural properties such as the elemental composition and porosity/void fraction, are indicated by the dashed lines.

Naturally, a lack of chemical stability on a film level is not desirable, as this will result either in potential trade-offs between performance and stability, or more complex and expensive encapsulation on a device level. The former trade-off is shown in Fig.3, which demonstrates that increasing the deposition pressure of the p- or n-doped SiO_X:H layer results in improved chemical stability of single junction p-i-n a-Si:H solar cells. However, this improvement comes at the cost of initial conversion efficiency, as the opto-electrical qualities of the doped SiO_X:H films are reduced when the chemical stability is improved.

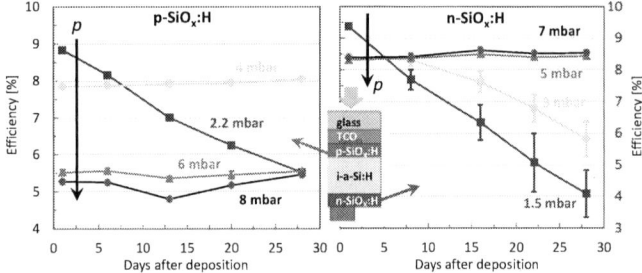

Fig. 3: Evolution over time of the conversion efficiency of single junction a-Si:H p-i-n superstrate PV devices. Coloured curves indicate depostion pressure of the p-SiO_X:H layer (left) and n-SiO_X:H layer (right). From [1].

B. Textures on c-Si

Surface textures that result in high optical yields are crucial for high efficiency PV devices. Some of the architectures investigated in DISCO combine a crystalline (c-) silicon heterojunction with a thin film silicon junction with a nanocrystalline (nc-) absorber. This nano-crystalline material is

incompatible with the relatively steep slopes of the conventional <111> crystal orientation. For that reason, several texturing approaches were explored for developing smooth concave surface features on mono-c-Si. Four approaches were explored, resulting in widely different surface morphologies, as shown in Fig.4. The figure demonstrates that three of the 4 textures could sustain the growth of crack-free device-quality nc-Si:H. For the periodic honeycomb texture (right) this requires some tuning of periodicity and radius.

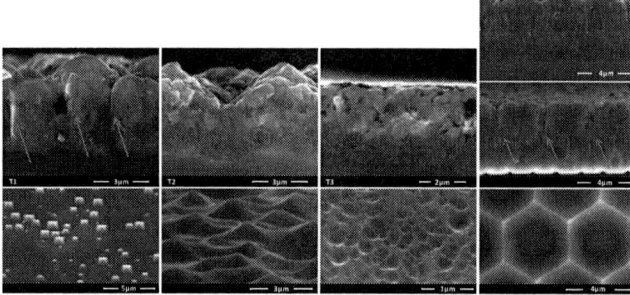

Fig. 4: SEM images of surface features (bottom) and cross sectional images of a 3-5um nc-Si:H layer grown on top of the textured silicon (top) of the different textures on mono-c-Si developed in the DISCO project. Red arrows indicate cracks fromed in the nc-Si:H layer in the focal point of steep features. From [10] and [11].

The influence of the various processing steps in each of the different approaches was extensively characterized. This characterization revealed the relation between the surface morphology and optical behavior, both in terms of overall light in-coupling and light-scattering, as well as the tunability of spectrally dependent scattering of the periodic surface features. Additionally, the performance of devices grown on the different textures was compared both through optical simulations, as demonstrated in Fig.5, as well as experimentally. The latter revealed the trade-offs between surface roughness related optical gains and $V_{oc}*FF$ losses.

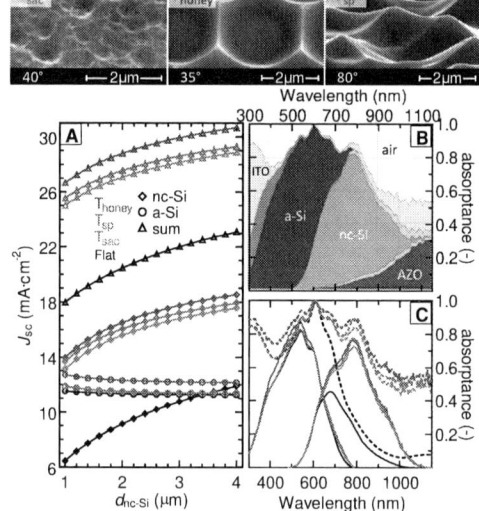

Fig. 5: Simulated performance of a-Si/nc-Si tandem devices on flat substrates and substrates with T_{sac} (red), T_{sp} (green) and T_{honey} (blue) textures (textures indicated by SEM images). **A** shows the J_{SC} of the a-Si junction, nc-Si junction and the sum of both junctions, for different textures, as a function of nc-Si absorber thickness (d_{nc-Si}). **B** shows the spectral absorptance in each of the layer of the tandem device on T_{honey}. **C** shows the a-Si and nc-Si absorptance curves

and 1-R curves for the different substrate types, for tandem device with $d_{a\text{-}Si}$=300nm and $d_{nc\text{-}Si}$=1.2μm. From [11].

C. Multijunction devices

Having developed a better understanding of the materials on a film and single junction device level, part III of DISCO focused on MJ devices. In [12], the qualitative fundamental working principles of TRJ's based on p- and n-doped Si(O$_X$) alloys are revealed using both electrical modelling and experiments based on a unique set of tandem lab cells (four types based on four different PV materials) combined with structural variations in TRJs architectures. The study resulted in design rules for the integration of silicon-oxide based TRJs and provides fundamental insights into the sensitivity of the electrical performance of the TRJ's to doping concentrations, to alignment of the conduction and valence bands of consecutive sub-cells, to the nature of interface defects, to the growth of amorphous and crystalline phases and its dependence on substrate or seed layers and to the nanoscale thicknesses of the TRJ layers.

Then, in [13], optimal current matching in MJ devices is investigated. Specifically, the influence was studied of variations in absorber thickness as well as thickness variations of different intermediate reflective layers based on SiO$_X$, various transparent conductive oxides and metallic layers on all-silicon MJ PV devices. Using the design rules from this study, the SHJ/nc-Si:H(/a-Si:H) devices, shown in Fig.6, are processed with a V$_{oc}$≈2V and conversion efficiencies close 15%, the highest reported conversion efficiency for an all-silicon solar cell that generates at least 1V.

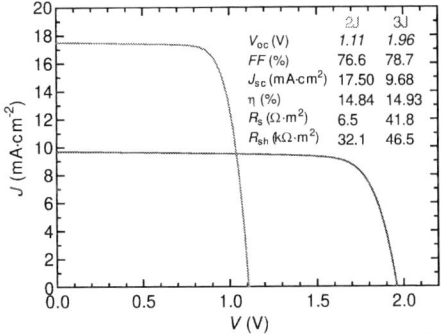

Fig. 6: The J-V and EQE curves of the champion 2J and 3J devices. An EQE diagram of the champion device, including all three subcells as well as the sum of three subcells and 1-R curves (dashed line) is presented. From [13]

In addition to the optimization of this particular device architecture, the flexible application of earth abundant and chemically inert silicon and silicon-germanium alloys is demonstrated in two distinct device architectures, as shown in Fig.7, combining up to 4 different junctions, yielding a relatively continuous V$_{oc}$ range of 0.5V to 2.8V. The figure and the V$_{oc}$'s presented therein, can be used as a framework for selecting a suitable device architecture for facilitating a range of wireless and autonomous photo-electrochemical devices by fulfilling the voltage requirement for a range of electrochemical reduction reactions and electrocatalysts.

Finally, converting a multijunction PV device into a PEC device capable of continuous, autonomous stand-alone operation, requires the use of electrocatalytically active contacts and the development of micropores through the PV device to prevent large pH-gradient related overpotentials. Therefore, in DISCO, the influence of the size and distribution of micropores on the photovoltaic performance and electrochemical performance of a PEC device is simulated.

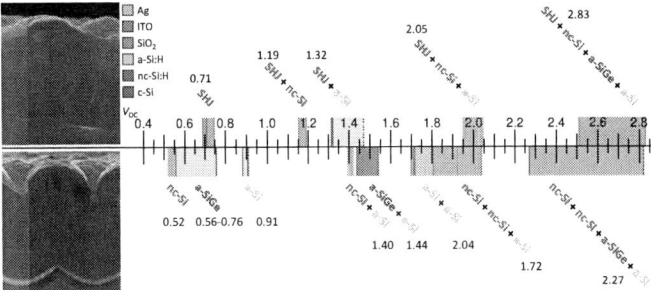

Fig. 7: V_{OC} range of different single- and multijuction PV device architectures. Top row indicates SHJ and hybrid SHJ and thin film silicon (SHJ+) multijunction devices. Bottom row indicates thin film silicon devices. Cross-sectional SEM images indicate the device architectures of the SHJ+ and thin film silicon multijunction devices. Colours added in the SEM images represent different materials, as indicated by the legend. In the V_{OC} range, 3 lines are present for each device architecture. Of these, the two solid lines indicate the highest measured V_{OC} and the V_{OC} of the device with the highest V_{OC}*FF product processed in our lab. Highest measured V_{OC}'s are also indicated next to the device architectures. The dashed line indicates the V_{OC} of devices with the highest reported conversion efficiency, for which references can be found in [14].

It was shown that the influence on PV and EC performance is opposite in trend and that current density losses can be limited to less than 20% for a range of pore diameter-period combinations Additionally, as shown in Fig.8, the successful processing of micropores using deep reactive ion etching, and the processing of Pt microdots with a 2μm diameter and 5% surface coverage is demonstrated

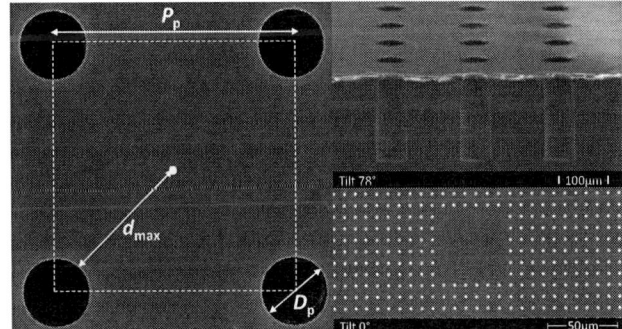

Fig. 8: SEM image of the surface of a porous substrate with D_p, P_p and d_{max} schematically indicated (left). Cross-sectional SEM image of pores with D_p=60μm processed partly through the wafer (right, top). SEM images of the substrate surface following successful Pt microdot application (right, bottom). From [14]

III. CONCLUSION

In this abstract an overview is presented of the DISCO project, where a framework is developed for the processing of high-voltage wireless silicon-based monolithically integrated 2-terminal multijunction photovoltaic devices. That can be used for stand-alone solar-to-fuel synthesis devices and other innovative PV applications.

ACKNOWLEDGMENT

The authors would like to gratefully acknowledge the financial support from the Netherlands Organization for Scientific Research (NWO) Solar to Products grant awarded to Arno Smets and the support provided by Shell International Exploration & Production Dense Energy Carriers Program.

REFERENCES

[1] T. de Vrijer, F. T. Si, H. Tan, and A. H. M. Smets, "Chemical Stability and Performance of Doped Silicon Oxide Layers for Use in Thin-Film Silicon Solar Cells," *IEEE J. Photovoltaics*, vol. 9, no. 1, pp. 3–11, 2019, doi: 10.1109/JPHOTOV.2018.2882650.

[2] T. de Vrijer and A. H. M. Smets, "The Relation Between Precursor Gas Flows, Thickness Dependent Material Phases, and Opto-Electrical Properties of Doped a/nc-SiO$_{x\geq0}$:H Films," *IEEE J. Photovoltaics*, vol. 11, no. 3, pp. 591–599, 2021, doi: 10.1109/JPHOTOV.2021.3059940.

[3] T. de Vrijer, H. Parasramka, S. J. Roerink, and A. H. M. Smets, "An expedient semi-empirical modelling approach for optimal bandgap profiling of stoichiometric absorbers: A case study of thin film amorphous silicon germanium for use in multijunction photovoltaic devices," *Sol. Energy Mater. Sol. Cells*, vol. 225, p. 111051, 2021, doi: 10.1016/j.solmat.2021.111051.

[4] T. de Vrijer, A. Ravichandran, B. Bouazzata, and A. H. M. Smets, "The impact of processing conditions and post-deposition oxidation on the opto-electrical properties of hydrogenated amorphous and nano-crystalline Germanium films," *J. Non. Cryst. Solids*, vol. 553, p. 120507, 2021, doi: 10.1016/j.jnoncrysol.2020.120507.

[5] T. de Vrijer, J. E. C. van Dingen, P. J. Roelandschap, K. Roodenburg, and A. H. M. Smets, "Improved PECVD processed hydrogenated germanium films through temperature induced densification," *Mater. Sci. Semicond. Process.*, vol. 138, p. 106285, 2022, doi: 10.1016/j.mssp.2021.106285.

[6] T. de Vrijer, B. Bouazzata, and A. H. M. Smets, "Spectroscopic review of hydrogenated , carbonated and oxygenated group IV alloys," *Vib. Spectrosc.*, under review.

[7] T. de Vrijer, K. Roodenburg, F. Saitta, T. Blackstone, G. Limodio, and A. H. M. Smets, "PECVD Processing of low bandgap-energy amorphous hydrogenated germanium-tin (a-GeSn:H) films for opto-electronic applications," *Appl. Mater. Today*, vol. 27, p. 101450, 2022, doi: 10.1016/j.apmt.2022.101450.

[8] T. de Vrijer *et al.*, "Opto-electrical properties of group IV alloys : the inherent challenges of processing hydrogenated germanium," *Adv. Sci.*, accepted for publication, 2022.

[9] T. de Vrijer and A. H. M. Smets, "Infrared analysis of catalytic CO2 reduction in hydrogenated germanium," *Phys. Chem. Chem. Phys.*, 2022, doi: 10.1039/D2CP01054B.

[10] T. de Vrijer and A. H. M. Smets, "Advanced textured monocrystalline silicon substrates with high optical scattering yields and low electrical recombination losses for supporting crack-free nano- to poly-crystalline film growth," *Energy Sci. Eng.*, vol. 9, no. 8, pp. 1080–1089, 2021, doi: 10.1002/ese3.873.

[11] T. de Vrijer *et al.*, "The optical behavior of random and periodic textured crystalline silicon surfaces for photovoltaic applications," *Prog. Photovoltaics Res. Appl.*, under review

[12] T. de Vrijer *et al.*, "The fundamental operation mechanisms of nc-SiO$_{x\geq0}$:H based tunnel recombination junctions revealed," *Sol. Energy Mater. Sol. Cells*, vol. 236, p. 111501, 2022, doi: 10.1016/j.solmat.2021.111501.

[13] T. de Vrijer, S. Miedema, T. Blackstone, D. van Nijen, C. Han, and A. H. M. Smets, "Application of metal, metal-oxide and silicon-oxide based intermediate reflective layers for current matching in autonomous high voltage multijunction photovoltaic devices," *Prog. Photovoltaics Res. Appl.*, under review

[14] T. de Vrijer, *High Voltage Photovoltaic Devices for Autonomous Solar-to-Fuel Applications*. Delft University of Technology, 2022.

Novel 1D van der Waals SbSeI micro-columnar solar cells by a self-catalyzed high pressure process

Ivan Caño-Prades, Alejandro Navarro-Güell, Sergio Giraldo, Joaquim Puigdollers, Marcel Placidi, Edgardo Saucedo

Universitat Politècnica de Catalunya (UPC), Barcelona, Spain

Emerging 1D van der Waals chalco-halide semiconductors are attracting a lot of interest as photovoltaic absorbers, due to their unique structural, electrical and optical properties. In particular, the mixed chalco-halide compound SbSeI, that tends to easily form highly crystalline nanowires, have demonstrated efficiencies exceeding 4%, with a bandgap of 1.80 eV, and a synthesis temperature below 300ºC. These properties are very interesting for its possible future application in tandem solar cells concepts among others. In this work, SbSeI micro-columnar solar cells are obtained by an innovative process based on the selective iodination of Sb2Se3 layers at high pressures, by using SbI3 as iodine source. Annealing parameters such as temperature, time and pressure are investigated, and a complete morphological, structural and compositional characterization of the absorbers is performed. We observe a self-catalyzed solid-liquid-vapor transformation process of Sb2Se3 thin films into micro-columnar SbSeI at annealing temperatures above 250ºC. Highly crystalline micro-columnar structures are obtained as it is demonstrated by XRD and TEM analysis, which sizes strongly depends on the annealing temperature, time and pressure. The extracted activation energy of the processes suggests that the SbSeI micro-columns are formed very fast thanks to the release of liquid Se during the decomposition of Sb2Se3 under iodine atmosphere. First solar cell prototypes were fabricated using standard thin film solar cell substrate configuration, demonstrating devices with an open-circuit-voltage higher than 550 mV and conversion efficiencies close to 1%, which are currently limited by the low short-circuit current and fill factor, due to the deficient coverage of the front contact. Finally, a complete analysis of the SbSeI materials and solar cell devices will be presented, together with strategies to control de size and orientation of the micro-columns, and the improvement of the front contact interfaces.

978-1-7281-6118-1/22 $31.00 © 2022 IEEE

Developement of phosphors by magnetron sputtering for solar cells improvement

Eduardo Salas , Miguel Modesto Tardio, Elisa García-Tabares, Gracia Belén Perea, Rosa de la Cruz, Stavros Athanasopoulos, Clement Kanyinda-Malu, Juan Enrique Muñoz-Santiuste, Beatriz Galiana

Universidad Carlos III de Madrid, Leganes, Spain

Universidad Rey Juan Carlos, Móstoles, Spain

Phosphors based on LaNbO doped with Nd3+ on Si substrates have been achieved by means of magnetron sputtering together with ex-situ annealing at 1200 ºC for 12 hours. NIR luminescence shows emission peaks at 1060 nm when a 520 nm laser source is applied. This response is equivalent to those detected in samples fabricated by solid state reaction method. According to the structural analysis carried out, it seems that two phases (LaNbO4 and La3NbO7) coexist, being LaNbO4 dominant. This result opens the possible of develop conventional substrates, such as Si, covered by luminescent materials as a previous stage for solar cells improvement based on photon recycling mechanisms.

High efficiency Silicon Heterojunction Metal Wrap Through produced in industrial pilot line

Marina Foti(1)*, Nicolas Guillevin(2), Eric Kossen(2), Lars Okel(2), Eelko Hoek(2), Anna Carr(2), Bas van Aken(2), Petra Manshanden(2), Francesco Rametta(1), Marcello Sciuto(1), Antonio Spampinato(1), Alfredo Di Matteo(1), Antonino Ragonesi(1), Gianluca Coletti(2,3), Cosimo Gerardi(1)

*Corresponding author: marina.foti@enel.com

1. Enel Green Power S.p.A, 3SUN, Contrada Blocco Torrazze Zona Industriale 95121, Catania, Italy
2. TNO Energy Transition, Westerduingweg 3, 1755LE Petten, the Netherlands
3. School of Photovoltaic and Renewable Energy Engineering, University of New South Wales, Sydney, NSW, Australia

Abstract— In this work, we describe the development of metal wrap through (MWT) silicon heterojunction (SHJ) solar cells and modules using an industrial pilot line. We believe that MWT technology applied to SHJ cell can pave the way to a larger use of SHJ for single side PV module for residential and BIPV applications. The up to 5% power gain demonstrated is extremely important for the future of PV in such market segment and beyond and show once more the high potential of this technology.

Keywords— *heterojunction, silicon solar cells, MWT interconnection*

I. Introduction

SHJ solar cells, based on crystalline silicon (c-Si) absorber and thin hydrogenated amorphous silicon (a-Si:H) passivation and selective contacts, have been demonstrated to reach efficiencies higher than 25% at industrial scale. The SHJ production process presents a reduced number of steps for cells production with respect to conventional technologies like PERC and the possibility to use high level of automation for module assembly line. The market share of SHJ technology is rapidly increasing and further growth is forecasted in the near future. SHJ solar cells, shows high bifacial ratio that can reach 95%, thanks to their very symmetric structure. For this reason, SHJ is particularly suitable for application on utility scale using glass-to-glass module architecture.

However, different module architectures can be used to optimize the use of SHJ for single side PV module for commercial and residential PV for buildings.

MWT solar cells have the same architecture as a conventional solar cell with the addition that the front metal contact is wrapped through the wafer by metallized via-holes, providing both emitter and base contacts on the rear side (Fig. 1). Compared to conventional H-pattern cells and modules, MWT-SHJ modules do not require front and rear tabbing and are based on TNO's cell and interconnection with conductive back-foil technologies [1, 2, 3, 4] . Consequently, this interconnection method allows to significantly slimming down the front busbars of the cell metallization enhancing the light capture. In addition the front and rear metal finger can be extremely reduced in width and aspect ratio when a design with multiple vias is employed and therefore enabling a Ag cost reduction of up to 50% [3]. This is especially relevant for SHJ technology where the low temperature silver paste is about three times less conductive than conventional firing through silver paste used in PERC solar cells. Thanks to the concept of unit cells, which is the area surrounding each vias, MWT architecture can be easily scaled to any wafer size. The conductive back-foil is typically combined with very low temperature cell interconnection, which is advantageous for SHJ which have lower temperature tolerances than PERC cells.

MWT-SHJ cells and modules combine the positive benefits of both underlying technologies: namely high Voc from the SHJ, and higher Jsc and FF maintained before and after module encapsulation from the MWT technology.

At the present time, there are no commercial modules using both technologies except the ground-breaking work by TNO and partners in the Dutch Whooper projects and early research project by TNO [3, 4]. We believe that MWT technology applied to SHJ cell can pave the way to a larger use of SHJ for single side PV module for residential and BIPV applications.

II. Experimental Set-up and Results

In this work we aim to demonstrate the combination of MWT architecture with SHJ technology in industrially proven tools. The MWT-SHJ solar cells are fabricated in the industrial production line of ENEL while the laser drilling and isolation take place in the pilot plant at TNO.

MWT solar cells have the same architecture as a conventional solar cell with the addition that the front metal contact is wrapped through the wafer by metallized via-holes, providing both emitter and base contacts on the rear side (Fig. 1) as developed by TNO [3]. Compared to conventional H-pattern cells, interconnection of MWT cells will be based on TNO's conductive back-foil technology [1, 2], which does not require tabbing. Consequently this interconnection methods allows to significantly slimming down the front busbars of the cell metallization enhancing the light capture. MWT-SHJ cells and modules combine the positive benefits of both underlying technologies: namely high Voc, higher Jsc and higher FF maintained before and after module encapsulation.

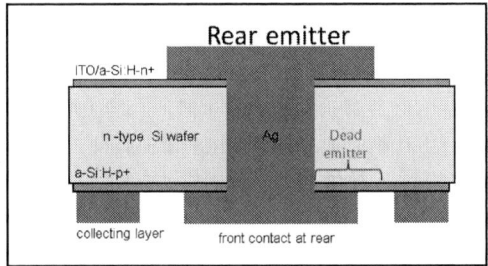

Figure 1: Scheme of MWT architecture

We compared PV single cell laminates based on SHJ cells using H-pattern ECA busbar interconnection with single cell laminates based on MWT-SHJ and with back contact interconnection. Reference H-pattern cells have been processed in EGP 3SUN factory in Catania. The factory consists in fully automated production line for cells and modules manufacturing with annual capacity of 200MWp/year. Reference modules are bifacial glass-glass with 72 cells monocrystalline n-type SHJ cells [5, 6].

The solar cell were realized using n-type M2 silicon substrates, which were inspected in terms of contamination, mechanical defects and electrical characteristics. They were processed in inline wet benches where a sequence of chemical processes took care of surface texturing and cleaning. The a-Si:H intrinsic and doped layers were deposited on both side of the textured wafer using a Plasma Enhanced Chemical Vapor Deposition reactor (PECVD), to form the emitter (on the rear side) and the Front Surface Field (FSF) (on the front side) of the solar cell. To complete the electrical contact Transparent Conductive Oxide (TCO) ITO was deposited through a sputtering tool on both wafer sides to enhance light transmission and collect the electrical charges produced inside the cell. Metal grid especially designed for MWT-SHJ by TNO has been deposited by screen printing process. The use of wafers with holes did not compromise the quality of the process nor the use of automation at EGP 3Sun factory.

A first set of MWT-SHJ cells were manufactured at ENEL to evaluate the performance and reliability of the technology at laminate level. With interconnection pads for both polarity integrated to the rear metallization grid, single MWT-SHJ cells were successfully connected and laminated to a conductive back-sheet at TNO. The MWT-SHJ laminates were compared to single-cell H-pattern-SHJ laminates processed in the same run on ENEL's pilot line. Fig. 2 shows pictures of the samples. Two

MWT-SHJ configurations have been compared for a preliminary optimization of the electrical/optical performance for the future pilot production.

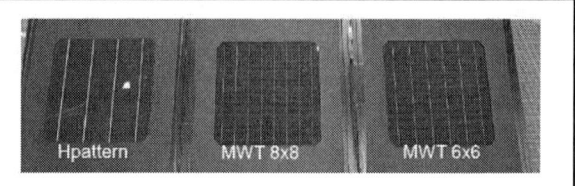

Figure 2: First set of one-cell module laminates

The H-pattern SHJ laminates are glass-glass with transparent encapsulant while the MWT laminates employ a black encapsulant and a conductive back sheet on glass. To minimize the difference in light coupling of the two configurations, the laminates have been masked on the front (see Fig. 2) while the bifacial laminates are measured with a black rear to mimic the black encapsulant. We still expect optical reflection on the rear side at the cell/transparent encapsulant/rear glass interfaces of the SHJ bifacial module but we expect the contribution to the extra J_{sc} to be <1%.

The results of this first experiment (see Tab.1) reveal that the back-contact MWT-SHJ laminates outperform the front-to-back contact H-pattern SHJ laminates. The Voc of the laminates is identical demonstrating that the excellent passivation of the SHJ is retained in the MWT cell architecture. Up to $5\%_{rel}$ higher power, or equivalently up to $1\%_{abs}$ higher encapsulated cell efficiency, was measured on the MWT-SHJ laminates compared to the H-pattern SHJ laminates due to an increased FF. These excellent initial results show that the MWT-SHJ cells do not suffer from passivation losses related to the rear-contact cell architecture and demonstrate the high potential of the technology. The absence of Isc gain expected for the MWT-SHJ is still under investigation. Preliminary characterization indicates that the difference in laminate structure resulting in different light coupling between the MWT and H-pattern laminates is the cause of this absence of Isc gain. In addition we expect further improvement of the laminate performance once the metallization grid and finger geometry are fully optimized.

Single cell laminate	Pmax (W)	Isc (A)	Voc (V)	FF (%)	Power gain to SHJ
MWT-SHJ_6x6	4.86	8.62	0.732	77.0	+3.8%$_{rel}$
MWT-SHJ_8x8	4.90	8.61	0.732	77.9	+4.7%$_{rel}$
Hpattern_Mod5	4.68	8.62	0.732	74.1	-

TABLE I.   ELECTRICAL RESULTS

External Quantum Efficiency (EQE) measurements were also carried out on the laminates reported in Table I. the EQE response is very comparable between both H-pattern and MWT-SHJ cell technology as expected from the similar Isc measured. Furthermore there is no difference between the 6x6 and 8x8 MWT-SHJ vias configuration despite the large difference number in vias and therefore larger dead area that effectively does not play a significant role in the cell performance.

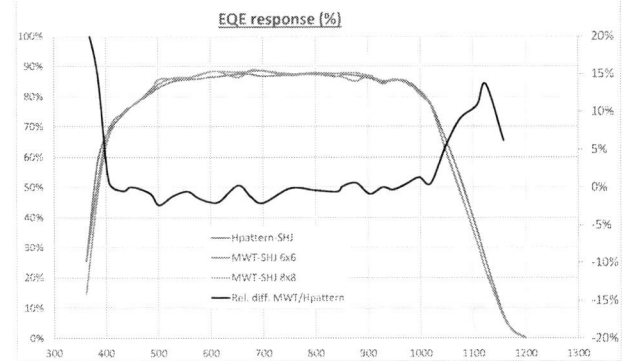

Figure 3: External Quantum Efficeincy of the the three different laminates of Fig 2 (Colour curves, left axis). Relative difference of the MWT/H-Pattern (black, right axis).

ElectroLuminescence (EL) measurements were carried out on the single cell laminates and presented in Fig. 4. The EL images are rather uniform and do not show obvious surface passivation or resistive losses issues.

Figure 4: Electroluminescence of the MWT-SHJ 6x6 vias configuration.

A series of laminates from the same batch of the MWT-SHJ 6x6 underwent accelerated aging processes of damp heat (85 degrees C and 85% humidity) and thermal cycle (-40 degrees C to 85 degrees C, under load) following the IEC standards (See Fig. 5). It shows the relative power for each of the individual laminates, designated by the Axxx numbers, after each cycle of testing. The control samples were not placed in the climate chambers but used as a baseline reference for the tests.

The tests took four months as each cycle (equivalent to ½ IEC) takes 4 weeks, 3 weeks in the climate chamber and 1 week in characterization. Among the preliminary bill of materials tested, one combination based on commercially available material (BOM2) passed all four runs (2 x IEC), except for one laminate in damp heat A7053. A7053 and as well as BOM1 sample A7062 both employ slightly different laminate build up compared to the other laminates and this made them vulnerable in damp heat, which is seen back in the results as a larger drop in performance.

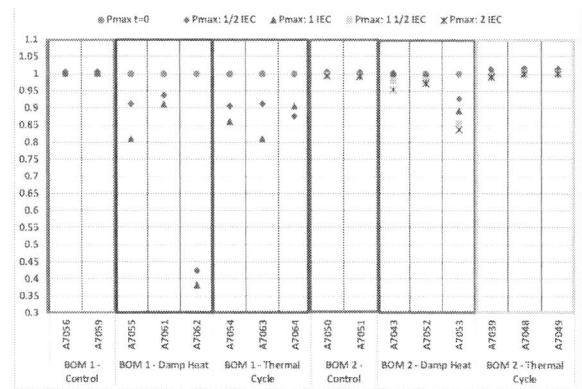

Figure 5: Damp heat and thermal cycle relative power results of two Bill of Materials. BOM1 (experimental) left and BOM 2 (employing commercial materials) right. The bill of material employing commercial material passed two times IEC test.

In order to study the behavior of the MWT-SHJ laminates in condition of low illumination we measured their power at different illumination levels. We measured both the SHJ H-pattern and MWT-SHJ solar cells as shown in Figure 6.

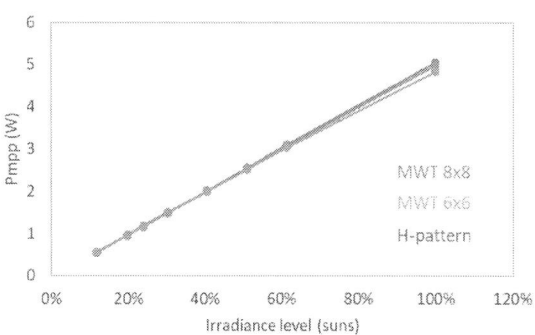

Figure 6: Power as function of irradiance level of the three different laminates of Fig 2.

Confirming previous results [4], no significant difference in the illumination response is observed for MWT-SHJ solar cells with respect to H-pattern silicon heterojunction solar cells. This indicates that no particular shunt resistance impact is visible in MWT-SHJ or at least comparable to that of conventional front and rear contacts H-pattern SHJ laminates.

More recently, about a thousand MWT-SHJ cells were processed on ENEL's pilot line with the objective to manufacture full size MWT-SHJ modules (60 cells). The module manufacturing is currently in preparation. The MWT-SHJ cells have been sorted and the selection of the required materials (conductive back-sheet, encapsulants, glass) is ongoing. The MWT-SHJ modules will then be placed in an outdoor test facility together with other commercial modules for comparison.

III. CONCLUSIONS

In this work, we describe the development of metal wrap through (MWT) silicon heterojunction (SHJ) solar cells and modules using an industrial pilot line. The power gain demonstrated is extremely important for the future of PV in such

978-1-7281-6118-1/22 $31.00 © 2022 IEEE   1308

market segment and beyond and show once more the high potential of this technology. We believe that there is still significant room to further improvement of the performance once the metallization grid and finger geometry are fully optimized.

ACKNOWLEDGEMENT:

The Whooper HER+ project is being carried out with a Topsector Energy subsidy of the Ministry of Economic Affairs and Climate Policy, implemented by Netherlands Enterprise Agency (RVO) with number TEHE119008.

REFERENCES

[1] N. Guillevin et al., "High Power n-Type Metal-Wrap-through Cells and Modules Using Industrial Processes", 28th European Photovoltaic Solar Energy Conference and Exhibition, pp. 1304-1310, 2013.

[2] Newman, B., Kroon, J., Guillevin, N., Okel L., Sommeling, P., Goris, M., Eerenstein, W., Gonzalez. J. 2018, "Materials Development and Increased Module Efficiency for 15% Cost Reduction of Back Contact Modules". Submitted to WCPEC-7: 45th IEEE PVSC, 28th PVSEC, 34th EU PVSEC, Hawaii, 2018.

[3] Coletti, G., Wu, Y., Janssen, G., Loffler, J., Van Aken, B.B., Li, F., Shen, Y., Yang, W., Shi, J., Li, G. and Hu, Z., 2014. 20.3% MWT Silicon Heterojunction Solar Cell—A Novel Heterojunction Integrated Concept Embedding Low Ag Consumption and High Module Efficiency. IEEE Journal of Photovoltaics, 5(1), pp.55-60.

[4] Coletti, G., Ishimura, F., Wu, Y., Bende, E.E., Janssen, G.J.M., van Aken, B.B., Hashimoto, K. and Watabe, Y., 2016, June. 23% Efficiency metal wrap through silicon heterojunction solar cells. In 2016 IEEE 43rd Photovoltaic Specialists Conference (PVSC) (pp. 2417-2420). IEEE.

[5] G. Condorelli et al., "High Efficiency Hetero-Junction: From Pilot Line To Industrial Production," 2018 IEEE 7th World Conference on Photovoltaic Energy Conversion (WCPEC) (A Joint Conference of 45th IEEE PVSC, 28th PVSEC & 34th EU PVSEC), Waikoloa Village, HI, 2018, pp. 1970-1973, doi: 10.1109/PVSC.2018.8548197.

[6] G. Condorelli et al., "Initial Results of Enel Green Power Silicon Heterojunction Factory and Strategies for Improvements," 2020 47th IEEE Photovoltaic Specialists Conference (PVSC), Calgary, OR, 2020, pp. 1702-1705, doi: 10.1109/PVSC45281.2020.9300806

DERConnect – A Distributed Energy Resources Testbed for Solar Power Integration

Jan Kleissl, Adil Khurram, Keaton Chia, Scott Brown, Aditya Mishra, Jorge Cortes, Raymond de Callafon, Rajesh Gupta, Sonia Martinez, David Victor

University of California, San Diego, La Jolla, CA, United States

Distributed Energy Resources Connect (DERConnect) is a National Science Foundation user facility to facilitate testing of distributed communication and controls algorithms at scale. DERConnect caters to industry and academic users in the electric power sector. DERConnect will provide testing capabilities of 1,000s of real DERs and millions of simulated DERs. DERConnect is designed to test intelligence on the grid edge by configuring the DERs in any communication architecture such as peer-to-peer, hierarchical, and centralized. DERConnect also enables cybersecurity test, social science tests, and advanced building controls. DERConnect will open to the research community in 2025. This paper describes DERConnect use cases and instructs potential future users of when and how to engage.

Slot-die Fabrication of Solution-processed Kesterite Solar Cells for Product Integrated Photovoltaics

Xinya Xu, Matthew C Naylor, Michael Jones, Bethan Ford, Stephen Campbell, Yongtao Qu, Vincent Barrioz, Guillaume Zoppi, Neil S Beattie

Northumbria University, Newcastle upon Tyne, United Kingdom

Traditional photovoltaics (PV) has made excellent progress over the last decade with decreasing installation costs and a levelised cost of electricity that is competitve with fossil fuel based sources of electricity. This has been achieved using economies of scale manufacturing at a relatively small number of mega-factories that are mostly based in a single geographical region. One consequence of this approach to PV manufacturing is that PV modules are highly standardised and the ubiquitous and low-cost deployment of PV on any surface is still not a commercial reality. There is therefore clear scope to perform the manufacturing research that enables this goal. In this work, slot-die deposition of an inorganic nanoparticle ink is used to create proof-of-concept photovoltaic devices in any geometry and pattern using a low-cost masking technique. The results are discussed in the context of a new design-led approach to photovoltaics manufacturing that has the potential to provide complementary new global electricity capacity for distributed applications in the built-environment.

N-type CdTe Thin Films via In-Situ Indium Doping

Theodore D C Hobson, Luke Thomas, Laurie J Phillips, Leanne A H Jones, Matthew J Smiles, Christopher H Don, Pardeep K Thakur, Vinod R Dhanak, Tim D Veal, Jonathan D Major, Ken Durose

Stephenson Institute for Renewable Energy, University of Liverpool, Liverpool, United Kingdom

Diamond Light Source Ltd, Didcot, United Kingdom

Indium-doping of CdTe thin films to produce n-type conductivity for solar cell device applications was investigated. Films were deposited via close spaced sublimation (CSS) from a 1 at% in-situ In-doped CdTe source. Films analyzed with SIMS exhibited a difference in indium incorporation of several orders of magnitude depending on whether a reducing or oxidizing ambient was used during CSS deposition. X-ray photoemission measurements using both soft and hard x-rays further indicated that for samples rapidly cooled after deposition under a high N_2 gas flow, a Cd-rich surface layer developed, which was not observed for more slowly cooled films. Photoemission measurements also allowed the conductivity type of the films to be inferred from the valence band positions in hard x-ray spectra, with n-type conductivity confirmed for In-doped films as-deposited, as well as an undoped CdTe film, with the latter thought to be due to a small deviation from stoichiometry. The Fermi levels of In-doped films varied with deposition conditions, with In-doping, a reducing ambient, and/or slow cooling tending to produce higher Fermi levels. Chloride activation treatments using $MgCl_2$ tended to lower the Fermi level, with more significant impacts on less highly doped films, in the case of the lowest doped film causing a switch from n-type to p-type conductivity. These results offer insights expected to be of great utility to the production of n-CdTe based solar cells, especially those fabricated via CSS.

Tellurium Availability for the PV Industry Using a System Dynamics Approach

Francis Hanna, Annick Anctil

Department of Civil and Environmental Engineering, Michigan State University, East Lansing, MI, United States

Tellurium is one of the rarest elements on earth. With the increased deployment of solar energy, the CdTe PV market is expected to grow substantially, exacerbating resource depletion. This study uses system dynamics to assess the availability of tellurium for the PV sector by 2050. Historical data analysis on tellurium production, demand and price shows a negative correlation between tellurium price and annual tellurium surplus. The accumulated tellurium surplus was estimated to be 4,030 metric tons by 2018. Preliminary results from the system dynamics model show a potential production gap, which is when demand surpass production between 2027-2034 and 2038-2043. The peak production gap is 443 kg in 2030 and 404 kg in 2042. However, considering historical production and accumulated tellurium stocks, no supply gap is expected. Finally, global tellurium stocks are expected to decrease by 21.6%, from 6960 metric tons in 2023 to 5454 metric tons in 2050.

978-1-7281-6118-1/22 $31.00 © 2022 IEEE

Progress on Substrate Reuse Using Sonic Lift-Off for GaAs-Based Photovoltaics

Andrew B. Sindermann, Stephen J. Polly, Pablo Guimera Coll, Elijah J. Sacchitella, Brandon M. Bogner, Mariana I. Bertoni, Seth M. Hubbard

Rochester Institute of Technology, Rochester, NY, United States

Crystal Sonic Inc., Phoenix, AZ, United States

Arizona State University, Tempe, AZ, United States

Sonic lift-off is able to reduce average surface facet amplitude without degrading bulk material quality and is thus a promising technology for enabling repeated substrate reuse with GaAs-based photovoltaics. 1-sun AM1.5G illuminated current density-voltage data and spectral response data from four devices at different stages of the spalling process have been presented. Devices grown on a commercial substrate and then acoustically-spalled as well as devices grown on a previously acoustically-spalled substrate with minimal surface treatment did not show degradation in device performance compared to the control grown on a commercial substrate, which was 17% efficient without anti-reflection coatings. Devices grown on a previously acoustically-spalled substrate and then spalled exhibited degradation in both short-circuit current density and open circuit voltage for a final 8% efficiency, indicating further process improvements are necessary to realize efficient substrate reuse.

Bandgap Dependence of Near-Conduction Band State in $(Ag_yCu_{1-y})(In_xGa_{1-x})Se_2$ Solar Cells

Michael F. Miller[1], Alexandra M. Bothwell[2], Nicholas Valdes[3,*], Stefan Paetel[4], Rouin Farshchi[5,*], Ana Kanevce[4], William Shafarman[3], Darius Kuciauskas[2], and Aaron R. Arehart[1]

[1]Department of Electrical & Computer Engineering, The Ohio State University, Columbus, OH 43210, USA
[2]National Renewable Energy Laboratory, Golden, CO 80401, USA
[3]Department of Materials Science & Engineering, University of Delaware, Newark, DE 19716, USA
[4]Zentrum für Sonnenenergie-und Wasserstoff-Forschung (ZSW), 70563 Stuttgart, Germany
[5]MiaSolé Hi-Tech Corp., Santa Clara, CA 95051, USA
[*]Currently at First Solar, Inc., Santa Clara, CA 95050, USA

Abstract—**(Ag,Cu)(In,Ga)Se₂-based solar cells have achieved high collection efficiencies, but defects still limit efficiencies well below the theoretical limit. The near-conduction band defect, typically observed at Ev+0.98 eV, has been ubiquitous across (Ag,Cu)(In,Ga)Se₂ samples from multiple vendors. The current work explores a wider range of composition and demonstrates the trap energy varies relative to the valence band but is approximately constant relative to the conduction band (~Ec-0.13 eV). There is also no definitive dependence of the trap concentration on composition.**

Keywords—ACIGS, CIS, ACIS, CIGS, traps, defects, DLOS, deep level optical spectroscopy

I. INTRODUCTION

Cu(In,Ga)Se₂ (CIGS) absorbers have long been established as a promising thin-film solar cell technology due to their low cost and high absorption [1]. By changing the Ga ratio and/or adding Ag, a wide range of bandgaps can be attained [2]. While Ag and Ga incorporation has been well studied, few studies have been done on the effect of composition on material defects, which are known to limit efficiencies in CIGS. In particular, a near-conduction band defect at Ev+0.98 eV has been observed in all CIGS and (Ag,Cu)(In,Ga)Se₂ (ACIGS) samples, though the range of composition has been limited [3]–[5]. This trap has been attributed to the V_{Se}-V_{Cu} divacancy defect, and has been known to cause light and heat-based metastabilities and limit device performance [5], [6].

Here, we explore Ga/(Ga+In) and Ag/(Ag+Cu) compositions from 0 to 50%. To properly predict and model the impact of this trap on solar cell performance, the precise energy dependence of this trap must be obtained. In this study, we use deep level optical spectroscopy (DLOS) to quantitatively characterize trap energies and concentrations.

II. APPROACH

In this study, CIS, ACIS, CIGS, and ACIGS samples from five suppliers (ZSW, MiaSolé, University of Delaware (UDel), Old Dominion University (ODU), and Solibro) were investigated. Although alkali treatments are common for (Ag,Cu)(In,Ga)Se₂ solar cells, all samples in this study were baseline samples without post-deposition treatment. Samples from these suppliers had varying Ga/(Ga+In) and Ag/(Ag+Cu) ratios, and the individual structures and growth processes for each supplier are summarized in Table I. All samples were full solar cells, which were physically scribed to isolate devices with an area of approximately 2 mm².

To characterize the near-conduction band trap, DLOS was performed by shining monochromatic light, and recording the resulting capacitance transient using a 1 MHz test signal for increasing photon energies. More details of the equipment and measurement setup can be found in Ref. [3] and Ref. [7]. The steady-state photocapacitance (SSPC) was measured during the experiment to determine the trap onset energies and concentrations. The trap concentrations were calculated from the step height of the SSPC using [8]

$$N_T = 2N_A \frac{C_\infty(hv) - C_0}{C_\infty(hv)} \tag{1}$$

where N_A is the sample doping, C_∞ is the steady-state capacitance at the photon energy corresponding to the full step height of the SSPC for that trap, hv is the photon energy, and C_0 is the initial capacitance in the dark with the traps filled. More accurate trap energy was determined by fitting the DLOS optical cross section using the Lucovsky model, which is [9]

$$\sigma_p^{op}(hv) \propto \frac{(E_T)^{\frac{1}{2}}(hv - E_T)^{\frac{3}{2}}}{(hv)^3} \tag{2}$$

where σ_p^{op} is the optical cross section of the trap, E_T is the trap energy. Sample bandgaps were determined by fitting the external quantum efficiency (EQE) to models from [10]. We note that this method was used for samples from ZSW, Delaware, and MiaSolé while the Solibro and ODU results are from previous studies [11], [12]. EQE has been reported to slightly underestimate the bandgap, which is evidenced by the slightly lower than theoretical CIS and ACIS bandgaps as shown in Table I [13]. The Ga/(Ga+In) and Ag/(Ag+Cu) ratios, shown in Table I, are given for the depletion depth measured by DLOS, and were measured using glow discharge optical emission spectroscopy (GDOES) for ZSW and Solibro [11], secondary-ion mass spectrometry (SIMS) for MiaSolé [14] and ODU [4],

978-1-7281-6118-1/22 $31.00 © 2022 IEEE

Table I: Sample description from each supplier

Supplier	Sample	Bandgap (eV)*	Ga/(Ga+In)†	Ag/(Ag+Cu)†	Absorber Width	Buffer	Window Layer 1	Window Layer 2	Source	Near-Conduction Trap	
										Conc. (cm⁻³)	Energy (eV)
ZSW	CIGS (3 samples)	1.13	~0.25	0	2-2.5 µm	CdS (50 nm)	i-ZnO (80nm)	Al-ZnO (400 nm)	[16]	3.5×10^{15} 2.7×10^{15} 6.9×10^{14}	E_V+1.03 eV E_V+1.0 eV E_V+1.02 eV
UDel	ACIS	1.00	0	0.18	2.2 µm	CdS (50 nm)	i-ZnO (50 nm)	ITO (150 nm)	[15]	1.5×10^{14}	E_V+0.87 eV
	CIS	0.99	0	0	1.9 µm	CdS (50 nm)	i-ZnO (50 nm)	ITO (150 nm)	[15]	1.8×10^{15}	E_V+0.87 eV
	CIGS	0.99	0.05	0	1.9 µm	CdS (50 nm)	i-ZnO (50 nm)	ITO (150 nm)	[15]	2.0×10^{15}	E_V+0.88 eV
MiaSolé	ACIGS	1.21	~0.4	< 0.5	1.5 µm	CdS (30 nm)	i-ZnO (100 nm)	ZnO (300 nm)	[3]	3.9×10^{15}	E_V+1.07 eV
Solibro	CIGS (3 samples)	1.18	~0.4	0	2-3 µm	CdS (50 nm)	i-ZnO (50 nm)	ITO (250 nm)	[5], [11]	2.4×10^{15} 5.1×10^{15} 6.0×10^{15}	E_V+1.05 eV E_V+1.05 eV E_V+1.05 eV
ODU	CIGS (2 samples)	1.12	~0.3	0	2 µm	CdS (50 nm)	i-ZnO (50 nm)	Al-ZnO (1 µm)	[4], [17]	2.9×10^{14} 5.2×10^{15}	E_V+0.99 eV E_V+0.99 eV

*The average bandgaps in the depletion region for the ZSW, UDel, MiaSolé, and Solibro samples were extracted from EQE The bandgap of the ODU CIGS was estimated using the composition measured using scanning transmission electron microscopy based energy dispersive X-ray spectroscopy [12].

†Ga/(Ga+In) and Ag/(Ag+Cu) ratios were measured using GDOES for ZSW and Solibro [11], SIMS for ODU[4], and MiaSolé [14], and XRF for UDel [15].

and x-ray fluorescence (XRF) for UDel [15]. Although composition and bandgap are measured using different methods, small errors from measurement variation will not impact the conclusions.

III.   RESULTS AND DISCUSSION

Representative SSPC plots for MiaSolé, ZSW, and UDel are shown in Fig. 1a, with the trap concentrations for all samples summarized in Table I. In SSPC, the trap onset energy is where the SSPC slope changes indicating trap emission above this energy. In Fig. 1a, the SSPC plots have similar shapes, but markedly different onset energies of the near conduction band trap, which indicates the samples have different trap energies. Additional traps further from the conduction band exist in some

samples, and the concentrations are significantly lower than the near conduction band trap concentration. These traps are interesting, but not the subject of the current study. The DLOS optical cross sections were fit using the Lucovsky model that assumes negligible lattice relaxation energy (Franck-Condon energy (d_{FC})), and representative fits are shown in Fig. 1b, with trap energies for all samples listed in Table I. The SSPC and DLOS onset energies are similar indicating consistent results, and the good fits indicate d_{FC} is indeed minimal, which is consistent with theory for a V_{Se}-V_{Cu} defect [6]. These trap energies are consistent among diodes from the same sample but differ among samples from the same distributor. Previously, this trap has been labeled as E_V+0.98 eV because it was observed over the small range of compositions studied. Now, however,

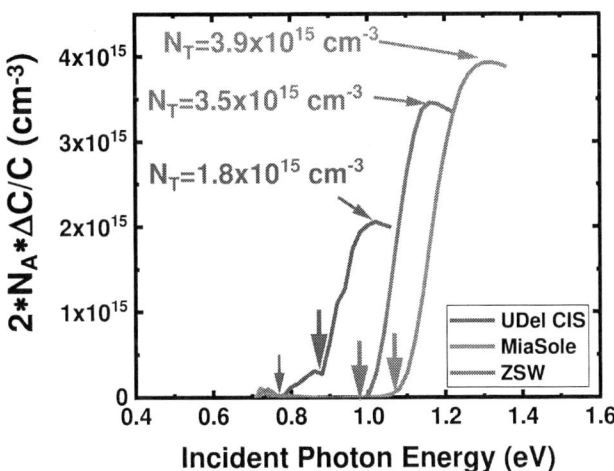 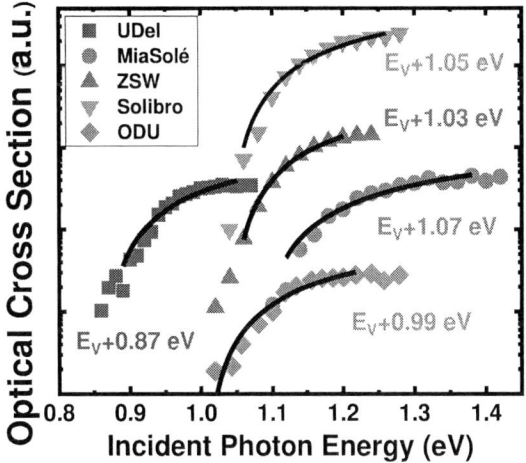

Fig. 1. a) SSPC of the UDel CIS, ZSW CIGS, and MiaSolé ACIGS. All SSPC have similar shapes but markedly different onset energies (arrow) and concentrations. (b) Specific trap energies are found by fitting the optical cross section, and a wide range of energies are observed among suppliers. The energy of the UDel CIGS is 0.16 eV lower than ZSW CIGS and 0.20 eV lower than the MiaSolé ACIGS. All cross sections fit well to the Lucovsky model indicating minimal Franck-Condon energy.

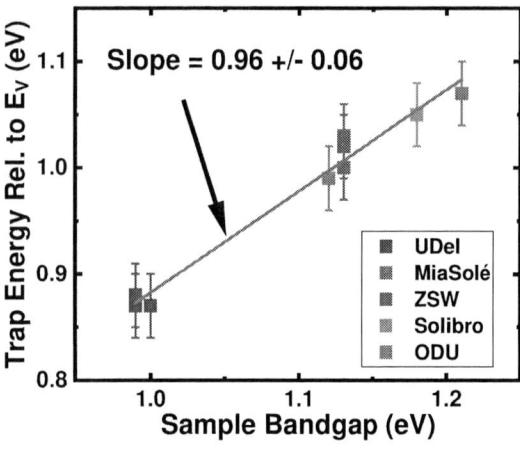 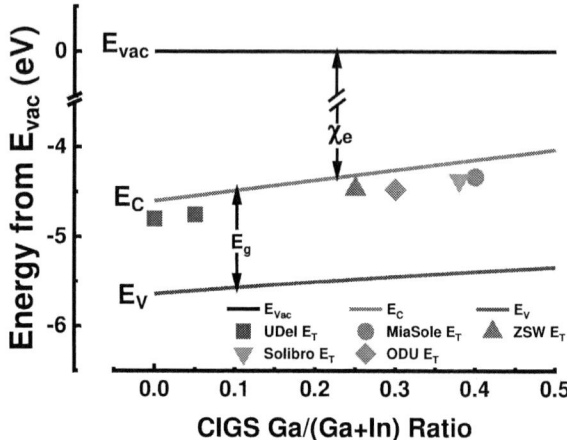

Fig 2. a) Sample bandgap vs. near-conduction band trap energy relative to E_V, including error bars from fitting and measurement error. A linear correlation is observed with a slope of 0.96, which suggests that the trap energy is almost constant with respect to the conduction band with a trap energy of ~E_C-0.13 eV. (b) Band diagram vs CIGS Ga/(Ga+In) ratio with trap energies plotted. Band diagram data from [18] and [19].

after studying a wider range of compositions, it is clear that the energy variation is above the experimental error. Comparing the trap energies to the bandgap in Fig. 2a, one observes that the measured trap energy increases with the $(Ag,Cu)(In,Ga)Se_2$ bandgap. A similar relationship is predicted by Lany and Zunger for the V_{Se}-V_{Cu} divacancy defect; the V_{Se}-V_{Cu} defect optical absorption energy of $CuGaSe_2$ is predicted to be 0.14 eV higher than that of $CuInSe_2$, indicating that it increases as the bandgap increases [6]. Thus, the trap energy increasing with the sample bandgap observed in Fig. 2a is consistent with the relationship predicted for the V_{Se}-V_{Cu} defect, providing further evidence that this is the likely source for the near-conduction band trap observed by DLOS in $(Ag,Cu)(In,Ga)Se_2$.

An approximately linear relation is observed in Fig. 2a with a slope of 0.96 ± 0.06, which fits well within the trap energy error associated with optical cross section fitting and measurement error. This quite close to unity slope indicates the trap energy is approximately constant with respect to the conduction band within the bandgap range here, with a trap energy with respect to the conduction band of ~E_C-0.13 eV. This relationship is illustrated in Fig. 2b. Although this diagram is shown for CIGS only and does not include the Ag-containing samples. However, this relation is consistent for the limited set of Ag-containing samples as well.

In addition to trap energy, the impact of the bandgap on the near-conduction band trap concentration was also investigated for the samples in this study. Previous studies have shown low solar cell collection efficiency for Ga to Ga+In ratios above 40% mainly due to open circuit voltage (V_{OC}) losses [20]. Hence, any dependence of the near-conduction band trap concentration was investigated. As shown in Fig. 3, there is possibly a weak trend of increasing trap energy with bandgap, but the fit error is very large. The fit suggests a near-conduction band trap concentration increase of 1.4×10^{16} cm^{-3}/eV bandgap, but the error is $\pm 6.6 \times 10^{15}$ cm^{-3}/eV. Thus, statistically it cannot be concluded that the trap concentration has any dependence on bandgap in this bandgap/composition range. However, there are clearly other variables not accounted for that matter more. Consider the

ZSW data where the bandgap is constant, but the trap concentration changes ~8X. Indeed, trap energy is specific to the atomic configuration of the defect, but the trap concentration depends more strongly on the growth process and, more specifically, the treatments performed on the samples. We have previously shown that alkali treatments can have a large impact on the near-conduction band trap concentrations, for example [4], [17]. As shown in Table I, there is a large variation of sample structure and growth, which likely obscured the impact of bandgap on the near-conduction band trap concentration. Additionally, this lack of dependence on bandgap is not entirely unexpected: V_{OC} is the primary efficiency degradation mechanism in CIGS with high Ga to Ga+In ratios, but the near conduction band trap is not an efficient recombination center unlike the mid-gap trap [3]. Thus, more work is needed at higher Ga content to see if this behavior changes.

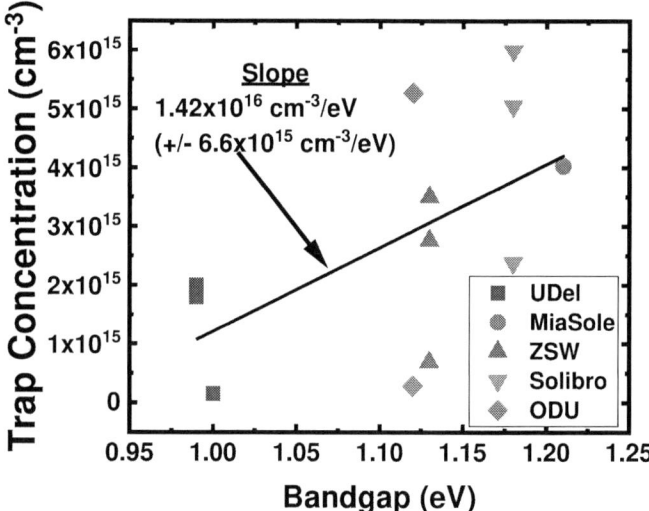

Fig. 3. a) Sample bandgap vs. near-conduction band trap concentration. There is possibly a weak correlation between near-conduction band trap concentration and bandgap, but statistically no correlation can be established at this point.

978-1-7281-6118-1/22 $31.00 © 2022 IEEE

IV. CONCLUSIONS

In this study, DLOS was performed on CIS, ACIS, CIGS, and ACIGS samples, which enabled characterization of the near-conduction band trap on materials to determine its dependence on composition and bandgap. By fitting the trap's optical cross section to obtain the trap energy via DLOS, it was found that the trap energy of the near-conduction band trap relative to the valence band edge depends approximately linearly on bandgap and appears to have no other dependence on composition within the experimental error. This indicates the previously labeled "E_V+0.98 eV" defect level varies with respect to the valence band but is constant at ~E_C-0.13 eV relative to the conduction band. Additionally, there was a possible weak trend between the sample bandgap and near-conduction band trap concentration but clearly other variables dominate given this relationship is largely obscured by the different growth parameters of the different suppliers, and further study is required to investigate this trend. A more precise energy for this trap has been established, which will aid in the solar cell modeling on the impacts of defects.

ACKNOWLEDGMENT

This material is based upon work supported by the U.S. Department of Energy's Office of Energy Efficiency and Renewable Energy (EERE) under the Solar Energy Technology Office (SETO) Award Number DE-EE0008755. This work was authored in part by the National Renewable Energy Laboratory, operated by Alliance for Sustainable Energy, LLC, for the U.S. Department of Energy (DOE) under Contract No. DE-AC36-08GO28308. The authors acknowledge German Federal Ministry for Economic Affairs and Energy under Project Number 03EE1078 (ODINCIGS).

REFERENCES

[1] K. L. Chopra, P. D. Paulson, and V. Dutta, "Thin-film solar cells: an overview," *Progress in Photovoltaics: Research and Applications*, vol. 12, no. 2–3, pp. 69–92, 2004.

[2] P. T. Erslev, J. Lee, G. M. Hanket, W. N. Shafarman, and J. D. Cohen, "The electronic structure of Cu(In1−xGax)Se2 alloyed with silver," *Thin Solid Films*, vol. 519, no. 21, pp. 7296–7299, Aug. 2011.

[3] A. J. Ferguson, R. Farshchi, P. K. Paul, P. Dippo, J. Bailey, D. Poplavskyy, A. Khanam, F. Tuomisto, A. R. Arehart, and D. Kuciauskas, "Defect-mediated metastability and carrier lifetimes in polycrystalline (Ag,Cu)(In,Ga)Se₂ absorber materials," *Journal of Applied Physics*, vol. 127, no. 21, p. 215702, Jun. 2020.

[4] S. Karki, P. K. Paul, G. Rajan, T. Ashrafee, K. Aryal, P. Pradhan, R. W. Collins, A. Rockett, T. J. Grassman, S. A. Ringel, A. R. Arehart, and S. Marsillac, "In Situ and Ex Situ Investigations of KF Postdeposition Treatment Effects on CIGS Solar Cells," *IEEE Journal of Photovoltaics*, vol. 7, no. 2, pp. 665–669, Mar. 2017.

[5] P. K. Paul, T. Jarmar, L. Stolt, A. Rockett, and A. R. Arehart, "Role of Ev+0.98 Ev trap in light soaking-induced short circuit current

instability in CIGS solar cells," in *2017 IEEE 44th Photovoltaic Specialist Conference (PVSC)*, 2017, pp. 30–32.

[6] S. Lany and A. Zunger, "Light- and bias-induced metastabilities in Cu„In, Ga…Se2 based solar cells caused by the „VSe-VCu… vacancy complex," *J. Appl. Phys.*, p. 15.

[7] A. R. Arehart, A. A. Allerman, and S. A. Ringel, "Electrical characterization of n-type Al0.30Ga0.70N Schottky diodes," *Journal of Applied Physics*, vol. 109, no. 11, p. 114506, Jun. 2011.

[8] P. Blood and J. W. Orton, *The Electrical characterization of semiconductors: Majority carriers and electron states*. London: Academic Press, 1992.

[9] G. Lucovsky, "On the photoionization of deep impurity centers in semiconductors," *Solid State Communications*, vol. 3, no. 9, pp. 299–302, Sep. 1965.

[10] U. Rau, B. Blank, T. C. M. Müller, and T. Kirchartz, "Efficiency Potential of Photovoltaic Materials and Devices Unveiled by Detailed-Balance Analysis," *Phys. Rev. Applied*, vol. 7, no. 4, p. 044016, Apr. 2017.

[11] V. Gusak, O. Lundberg, E. Wallin, S.-O. Katterwe, U. Malm, and L. Stolt, "Optimization of alkali supply and Ga/(Ga+In) evaporation profile for thin (0.5 μm) CIGS solar cells," in *2018 IEEE 7th World Conference on Photovoltaic Energy Conversion (WCPEC) (A Joint Conference of 45th IEEE PVSC, 28th PVSEC 34th EU PVSEC)*, 2018, pp. 1641–1644.

[12] J. I. Deitz, S. Karki, S. X. Marsillac, T. J. Grassman, and D. W. McComb, "Bandgap profiling in CIGS solar cells via valence electron energy-loss spectroscopy," *Journal of Applied Physics*, vol. 123, no. 11, p. 115703, Mar. 2018.

[13] M. Turcu, I. M. Kötschau, and U. Rau, "Composition dependence of defect energies and band alignments in the Cu(In1−xGax)(Se1−ySy)2 alloy system," *Journal of Applied Physics*, vol. 91, no. 3, pp. 1391–1399, Feb. 2002.

[14] A. M. Bothwell, S. Li, R. Farshchi, M. F. Miller, J. Wands, C. L. Perkins, A. Rockett, A. R. Arehart, and D. Kuciauskas, "Large-Area (Ag,Cu)(In,Ga)Se2 Thin-Film Solar Cells with Increased Bandgap and Reduced Voltage Losses Realized with Bulk Defect Reduction and Front-Grading of the Absorber Bandgap," *Solar RRL*, vol. n/a, no. n/a, p. 2200230.

[15] N. Valdes, J. Lee, and W. Shafarman, "Comparison of Ag and Ga alloying in low bandgap CuInSe2-based solar cells," *Solar Energy Materials and Solar Cells*, vol. 195, pp. 155–159, Jun. 2019.

[16] A. Kanevce, S. Paetel, D. Hariskos, and T. M. Friedlmeier, "Impact of RbF-PDT on Cu(In,Ga)Se2 solar cells with CdS and Zn(O,S) buffer layers," *EPJ Photovolt.*, vol. 11, p. 8, 2020.

[17] S. Karki, P. Paul, G. Rajan, B. Belfore, D. Poudel, A. Rockett, E. Danilov, F. Castellano, A. Arehart, and S. Marsillac, "Analysis of Recombination Mechanisms in RbF-Treated CIGS Solar Cells," *IEEE Journal of Photovoltaics*, vol. 9, no. 1, pp. 313–318, Jan. 2019.

[18] S. Wei and A. Zunger, "Band offsets and optical bowings of chalcopyrites and Zn-based II-VI alloys," *Journal of Applied Physics*, vol. 78, no. 6, pp. 3846–3856, Sep. 1995.

[19] N. Khoshsirat, N. A. Md Yunus, M. N. Hamidon, S. Shafie, and N. Amin, "Analysis of absorber layer properties effect on CIGS solar cell performance using SCAPS," *Optik*, vol. 126, no. 7, pp. 681–686, Apr. 2015.

[20] S. Jung, S. Ahn, J. H. Yun, J. Gwak, D. Kim, and K. Yoon, "Effects of Ga contents on properties of CIGS thin films and solar cells fabricated by co-evaporation technique," *Current Applied Physics*, vol. 10, no. 4, pp. 990–996, Jul. 2010.

Improvement in PV Plant Performance for Convection Heat Transfer Changes from Altered Plant Layout

Matthew Prilliman, Sarah Smith, Brooke Stanislawski, Marc Calaf, Raul Bayoan Cal, Tim Silverman, Janine M.F. Keith

National Renewable Energy Laboratory, Golden, CO, United States

Portland State University, Portland, OR, United States

University of Utah, Salt Lake City, UT, United States

Heat transfer modeling that accounts for how convective cooling changes with PV array layout has been found to improve system LCOE in certain climates conditions. Analysis of fixed tilt systems performed using the System Advisor Model reveals that reducing system ground coverage ratio from 0.46 to 0.35 can lead to as much as a 1.7% increase in module annual energy output in Phoenix. Depending on climate conditions, these energy increases due to changing convective cooling flow can lead to LCOE improvements for systems with increased row spacing despite the increased wiring and land costs associated with increased module row spacing. While the energy gain from decreasing system ground coverage ratio can be largely attributed to increased plane of array irradiance, the convection cooling considerations presented here can have a non-negligible impact on PV power plant energy output and economic viability depending on climate conditions and array spacing parameters.

High Altitude Flight Results using Selenium, A PV Measurement Ecosystem

Don Walker, Colin J. Mann, John Nocerino, Kevin Lopez, Alexandra Pettengill, Jonathan Ortiz, Katrina Baumgarten, Misha Dowd, Yao Lao, Simon H. Liu

The Aerospace Corporation, El Segundo, CA, United States

We demonstrate the reproducibility and variability of 8 high altitude, near space solar cell characterization flights using the Selenium Ecosystem to be equal to prior JPL solar cell calibration flights. The Selenium Ecosystem consists of hardware and software that can measure and process high altitude solar cell characterization data. The modular nature of Selenium enables it to piggyback on larger balloon platforms as a secondary payload and light enough to be flown as primary payload on small weather balloons (< 2kg). The Selenium software is free to download.

Time-Evolving Electroluminescence Imaging in Perovskite Solar Cells

Jackson W. Schall, Hsinhan Tsai, Harvey Guthrey, Chun-Sheng Jiang, Steve Johnston, Dana Kern, Andrew Norman, Mowafak Al-Jassim

National Renewable Energy Laboratory, Golden, CO, United States

Colorado School of Mines, Golden, CO, United States

First Solar, Perrysburg, OH, United States

We present an approach to evaluate perovskite solar cell reliability and metastability during 100-hour stress/rest cycling under electrical bias with in-situ electroluminescence (EL) imaging. We show an example of time-evolving EL and associated voltage transients under constant current bias for triple-cation mixed-halide devices with defects of lead-iodide-rich wrinkles. We observe the greatest degradation in the regions of the wrinkles. Our results demonstrate how time-evolving luminescence imaging can be applied to better understand impacts of spatial imperfections on perovskite photovoltaic device stability.

The Nuts and Bolts of PV: Maturing Solar PV Racking and Module Mounting Critical Bolted Joint Technologies for LCOE Reductions and Increased Reliability

James Elsworth, Gerald Robinson, Jon Ness, Joe Cain, PE

National Renewable Energy Laboratory, Golden, CO, United States

Lawrence Berkeley National Laboratory, Berkeley, CA, United States

Matrix Engineering, Chicago, IL, United States

Solar Energy Industries Association, Washington, DC, United States

Industry stakeholders have to date largely overlooked both the critical role and uniqueness of bolted joints found in solar PV systems. Bolted joints seen in solar PV racking and module mounting lack the technological maturity exhibited in comparable industries to deliver low cost and high reliability solutions critically needed for further advancement of the industry. As a result they have regularly failed, which can have results ranging from unexpected system maintenance to entire PV system failures--in severe weather events and in normal operating conditions. This paper will overview and categorize the current state of PV bolted joint technologies, provide an engineering analysis of failure modes, identify codes and standards gaps leading to inconsistent bolted joint application in the field, summarize and publicize data gathered from a surveying and interviewing effort on field failures, share results of structural lab physical testing of common PV system bolted joint typets and the CFD flow models they inform, and develop a lifetime cost accounting model and tool for determining the cost impacts of various bolted joint and hardening options. The paper format is that of a technical guidance document, backed by empirical evidence aimed at near and long-term influence; product engineers for near-term and codes and standards committees for long-term advancement.

978-1-7281-6118-1/22 $31.00 © 2022 IEEE

Atomic Layer Deposited Bilayers and the Influence on Metal-Insulator-Semiconductor Schottky Barriers

Benjamin E. Davis and Nicholas C. Strandwitz

Lehigh University, Bethlehem, PA, 18015, United States

Abstract—**Atomic layer deposited oxide bilayers have been investigated as a means to tune the Schottky barrier height of metal-insulator-semiconductor contacts. Inserting LaO$_x$ between AlO$_x$ and a hydrogen-terminated silicon substrate increased the average Schottky barrier height of both n- and p-type silicon contacts by up to 0.14 eV. When the substrate was terminated with a chemical oxide, or the LaO$_x$ was inserted above the AlO$_x$ instead, the direction of the barrier height shifts was insensitive to the deposition order of the high-κ oxides. LaO$_x$ insertion then increased the p-type barrier height by 0.04-0.05 eV and decreased the n-type barrier height by 0.20-0.47 eV. Possible mechanisms for the shifts are discussed, including oxygen areal density differences, Fermi level depinning, and oxide charges. The data presented demonstrate new and more complex dielectric layer stacks to tune contact properties in metal-insulator-semiconductor structures.**

I. Introduction

It is desirable to control and predict the Schottky barrier height (Φ_B) at metal-insulator semiconductor (MIS) tunneling contacts for a variety of electronic applications. According to the Schottky model of M-S interfaces, Φ_B should be defined by the relationship between the vacuum work function of the pure metal $\Phi_{M, vac}$ and the electron affinity of semiconductor X_S so that

$$\Phi_B = \Phi_{M, vac} - X_S \qquad (1)$$

for n-type Si [1]. This is rarely the case for experimental data. Φ_B is often observed to be independent of the pure metal work function and the Fermi level is considered "pinned." Thus, Φ_B is often expressed as

$$\Phi_B = \Phi_{CNL} + S(\Phi_{M, vac} - \Phi_{CNL}) - X_S \qquad (2)$$

where Φ_{CNL} is the charge neutrality level determined by the transition energy between donor-like and acceptor-like interfacial energy states and S is a pinning factor that determines the sensitivity of Φ_B to $\Phi_{M, vac}$ (S = 1 indicates no Fermi level pinning, and S = 0 indicates strong pinning) [2].

In the case of transistors, reducing Φ_B decreases the contact resistance, allowing more current flow and decreasing Joule heating. MIS photovoltaics (PVs), on the other hand, require a large Φ_B to induce band bending in the semiconductor absorber and generate a large photovoltage.

MIS PVs were a subject of significant research interest in the 1970s and 80s [3], [4]. While the large Φ_B between the metal and semiconductor is what drives charge carrier

separation [5], the insulating tunnel layer can increase the open-circuit voltage for a given Φ_B by suppressing majority carrier tunneling compared to a cell with no tunnel insulator [6], [7]. The MIS type PV is perhaps a simpler, potentially lower-cost device than typically-used p-n junctions, but has lower efficiencies than champion diffused junction and a-Si:H heterojunction technologies [8], [9].

The widespread adoption of atomic layer deposition (ALD) in recent decades presents an opportunity to improve device figures of merit and increase the understanding of MIS tunnel structures, as ALD allows excellent control of insulator thickness and the deposition of many different compositions [10]. It is known that the insulating layer in a tunneling-based MIS stack can have various effects on Φ_B. For example, the effective work function of the metal contact has been shown to change, depending on the composition of the insulating layer [2]. The introduction of certain insulating layers "depins" the Fermi level, corresponding to a larger S-factor in equation (2) [11]. Some dielectric thin films, such as alumina, may also contain fixed charges [12]. However, results have suggested that any fixed charges in ultra-thin alumina do not significantly affect Φ_B of certain MIS devices [13].

When a dielectric-dielectric heterostructure exists in a MIS insulator, it has been observed that diploes form between the different dielectric layers (Fig. 1) [13]–[17]. One model attributes these dipoles to differences in the oxygen areal density (OAD) between oxide-based dielectrics [18]. According to this model, the OAD difference drives the oxygen ions to displace from the high-OAD side of the interface to the low-OAD side, taking its negative charge with

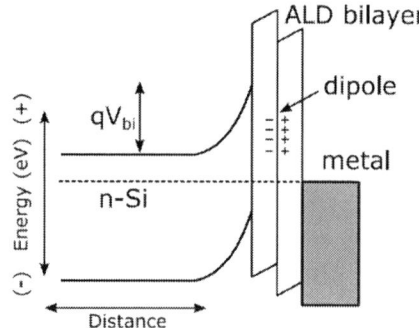

Fig. 1. Proposed band diagram of MIS device (not drawn to scale) demonstrating increased band bending with the introduction of an interfacial dipole.

it and leaving a positive charge behind. These dipoles will have an electrostatic effect such that

$$\Phi_B = \Phi_{CNL} + S(\Phi_{M, vac} - \Phi_{CNL}) - X_S + \Delta\Phi_{B,d}. \quad (3)$$

Previous studies have examined this dipolar effect by separately employing lanthanum oxide (LaO_x), which has a lower OAD than SiO_x, and AlO_x, which has a higher OAD than SiO_x [17], [18]. On p-type substrates, stacking LaO_x onto SiO_x was found to increase Φ_B relative to SiO_x-only insulators, while use of AlO_x instead decreased the barrier height [13], [17]. To the authors' current knowledge, however, this effect on Φ_B has only been studied stacking a single high-κ oxide on SiO_x. Thus, the present work tested the hypothesis that stacking two high-κ oxides (LaO_x and AlO_x) that have a greater OAD difference than either high-κ oxide has with SiO_x will yield a larger dipole, maximizing the Φ_B shift.

II. EXPERIMENT

MIS structures were prepared on both (100) p-type (8.6 Ω·cm) and n-type Si (2.0 Ω·cm) substrates (Virginia Semiconductor). Samples were cleaned with the standard RCA process comprising a 5-minute sonication in IPA, followed by sequential soaking in 1:1:5 (by volume) NH_4OH:H_2O_2:H_2O and 1:1:6 HCl:H_2O_2:H_2O solutions for 10 minutes each at temperatures between 70 and 80 °C [21]. Half of the samples were hydrogen-terminated (H-terminated) using a 1-minute soak in 5 wt% HF (Transene, Inc). Samples were rinsed with 18.2 MΩ water after every step. The samples that were not treated with HF retained an interfacial chemical oxide are hereafter referred to as RCA-terminated.

After substrate cleaning, the p-type samples received full-area aluminum back contacts via thermal evaporation and were subsequently annealed at 500 °C in N_2. Prior to ALD, the same half of samples that were previously H-terminated had HF applied to the front surface until the surface was observed to be hydrophobic.

ALD was performed at 200 °C in a Cambridge Nanotech Savannah S100 reactor. LaO_x was deposited using Tris(N,N'-di-i-propylformamidinato)lanthanum (La(^{i}PrFMD)$_3$) as a precursor, supplied by Dow Chemicals and later supplied by Strem Chemicals. The co-reactant was ozone (~3.5 wt%) generated from >99.99% oxygen gas in an A2Z Ozone Lab Benchtop Generator. The La(^{i}PrFMD)$_3$ was held at 160 °C during deposition and delivered to the reaction chamber via three consecutive pulses with a nitrogen boost. The growth rate of ~2 Å/cycle across the chamber was confirmed by a J.A. Woollam VASE spectroscopic ellipsometer. After the La(^{i}PrFMD)$_3$ was refilled by Strem, the observed growth rate decreased to 1.4 Å/cycle. AlO_x was deposited using trimethylaluminum (TMA, Strem Chemicals) and ozone precursors, with a growth rate of 1.1 Å/cycle.

P-type substrates were used to fabricate both single-high-κ oxide diodes (for comparison with Coss *et al.* [17]) and high-κ bilayers. For the single-high-κ layers, samples received either 0, 5, 9, or 14 ALD cycles of AlO_x, or 0, 4, 8, or 12 cycles LaO_x. Control samples for the bilayers consisted

only of 14 cycles AlO_x. The bilayers themselves received sequential depositions of either 4 cycles LaO_x followed by 7 cycles AlO_x, or 7 cycles AlO_x followed by 4 cycles LaO_x to achieve a similar total thickness to the 14-cycle AlO_x layers. Finally, circular Al front contacts with ~0.005 cm² area were deposited onto the p-type samples by thermal evaporation through shadow masks. Samples with LaO_x as the outermost oxide were transferred to the evaporation chamber as quickly as possible (within 1 hour) after ALD to minimize hygroscopic reaction with the atmosphere.

The preparation of n-type samples was identical to that of p-type samples, with the following exceptions. The n-$Si/AlO_x/LaO_x$ stacks (i.e., the right half of Fig. 3(b) below) received 5 cycles of LaO_x instead of 4 to compensate for a change in our measured growth rate using La(^{i}PrFMD)$_3$. After ALD, Ni front contacts were electron beam evaporated onto the n-type samples. The top contact metals were chosen to provide a Φ_B large enough to be measured unambiguously for each substrate polarity. Finally, a diamond-tipped scribe was used to scratch through the SiO_x on the rear side of the n-type samples before applying silver paste and an aluminum plate back contact.

Four samples were prepared for XPS. Two of these samples consisted of 12-cycles (~2 nm) LaO_x on H- and RCA-terminated p-type Si. The others comprised 77-cycles (~10 nm) LaO_x on RCA- and H-terminated n-Si.

Φ_B values were measured by Mott-Schottky analysis of capacitance-voltage curves obtained on an HP 4194a impedance spectrometer [22]. C-V measurements were conducted at a frequency of 100 kHz for p-type samples and 1 MHz for n-type samples, chosen to consistently obtain an average phase angle ≥89 °.

III. RESULTS & DISCUSSION

The Φ_B of diodes with single-high-κ oxide insulators was measured for comparison with previous research and to form a basis for comparison to samples with two high-κ oxide insulators within a single stack [17]. Φ_B decreased with increasing AlO_x thickness (Fig. 2(a)) on RCA-terminated p-type substrates, with a smaller thickness dependence of Φ_B observed on H-terminated substrates. Both observations are consistent with previous literature [13], [17], [23]. LaO_x (Fig. 2(b)), on the other hand, was expected to increase Φ_B with increasing thickness as in previous work [17]. Instead, most samples with LaO_x produced a similar Φ_B to SiO_x-only samples, with 12 cycles of LaO_x on H-terminated Si exhibiting significantly lower Φ_B than the other LaO_x samples. Three differences exist between the present work and reference [17] that may account for the discrepancy. First, reference [17] deposited LaO_x via physical vapor deposition where the present work used ALD, which may result in different material properties. Second, reference [17] utilized sputtered TaN top contacts while the present work used thermally evaporated Al top contacts. Third, reference [17] performed a post-metallization anneal in forming gas while all samples in the present work remained as deposited. It is also noted that the LaO_x synthesized in the present work contained significant amounts of lanthanum hydroxide (discussed further below). Further, Φ_B of the SiO_x-only

978-1-7281-6118-1/22 $31.00 © 2022 IEEE

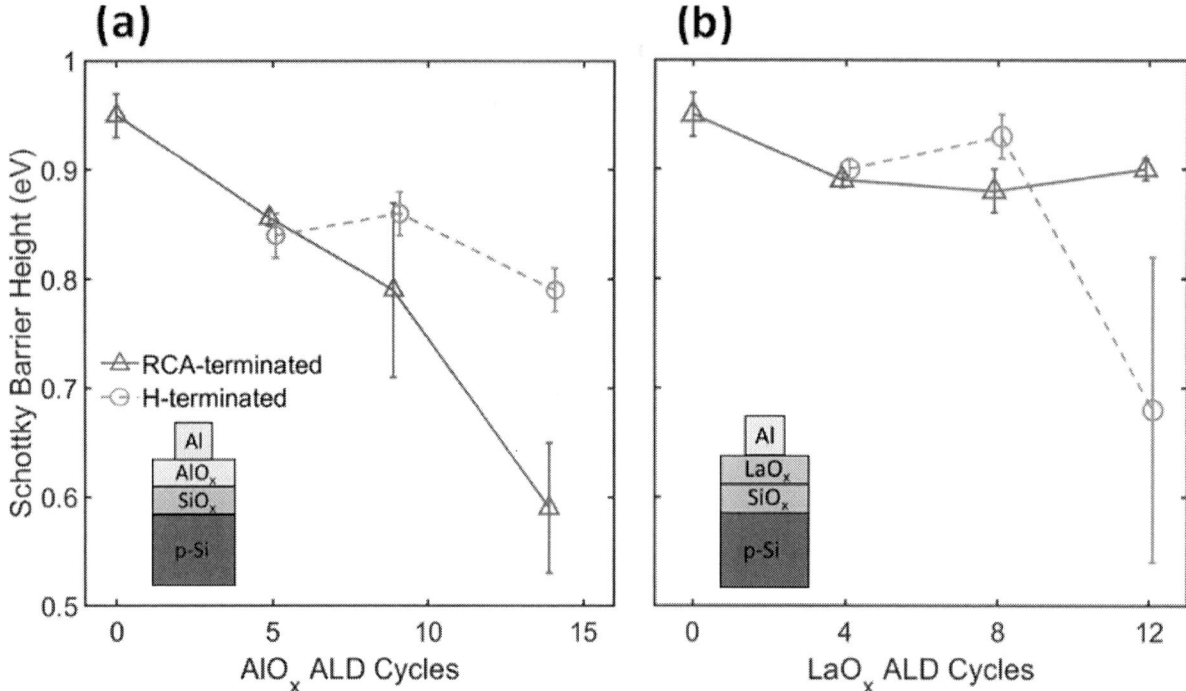

Fig. 2. Φ_B of diodes using a single ALD oxide as a tunnel layer, comprising (a) aluminum oxide and (b) lanthanum oxide. The error bars represent standard deviations of the two samples fabricated for each condition. Insets: layer schematics of the test structures utilized.

samples in Fig. 2 was already near the Si band gap, leaving little opportunity for increasing Φ_B.

Fig 3(a) shows Φ_B data for bilayer structures with LaO_x between Si and AlO_x (i.e., $Si/LaO_x/AlO_x$). On H-terminated substrates, this LaO_x introduction increased the average Φ_B for both n- and p- type substrates relative to samples with only AlO_x. This observation is inconsistent with the dipole model, which would predict an identical dipole forming at the LaO_x/AlO_x interface in both cases and hence an opposite Φ_B shift for opposite substrate majority carrier types. These results are better explained by the LaO_x layer depinning the Fermi level, i.e., a larger S-factor in equations (2) and (3). LaO_x insulators have been shown to depin the Fermi level in MIS capacitors more effectively than HfO_2 [24]. Such depinning has been attributed to solubility of Si in LaO_x and the formation of ionic lanthanum silicate during deposition rather than covalent SiO_x, satisfying charge neutrality at the interface [25]. La-silicate formation has previously been observed for ALD LaO_x using the same precursors as in the present study and is further addressed below [26]. With LaO_x insertion in the present study, the Φ_B values approach the Schottky model predictions (equation (1)), which are estimated to be 1.1 eV for p-Si/Al [27] and 1.0 eV for n-Si/Ni [28]. Such a Φ_B increase with LaO_x deposition on p-type Si was likely not observed in Fig. 2(b) because Φ_B was already near the Schottky-Mott predicted value for p-Si/Al prior to LaO_x introduction.

The incorporation of an RCA oxide in the AlO_x-only samples decreased Φ_B for p-type samples and increased Φ_B for n-type samples (Fig. 3(a), 3(b)). Such results are consistent with an OAD-induced dipole at the SiO_x/AlO_x interface [17], [18], [23]. Introducing LaO_x between RCA-

terminated Si and AlO_x resulted in an average Φ_B increase of 0.04 eV for p-type Si and decrease of 0.47 on n-type Si (Fig. 3(a)). Because oxygen displacement from AlO_x to LaO_x would induce a positive charge in the AlO_x and a negative one in the LaO_x, these Φ_B shifts were of the opposite directions of those hypothesized from an OAD-induced dipole at the LaO_x/AlO_x interface [18]. A consistent dipole is also expected to yield shifts of similar magnitude (though opposite direction) for each substrate type, which were not observed here.

When the deposition order of AlO_x and LaO_x was reversed, LaO_x introduction between AlO_x and the top contact metal increased Φ_B by 0.05 eV on p-type Si of both terminations and decreased Φ_B by 0.20 and 0.37 eV on H-terminated and RCA-terminated n-type Si, respectively (Fig. 3(b)). It Is noted that these shifts are similar to those observed for the RCA-terminated samples in Fig. 3(a). The fact that the direction of the Φ_B shift with LaO_x introduction is independent of the deposition order is further evidence against a significant dipole at the LaO_x/AlO_x interface. Such a dipole would result in equal and opposite shifts when the deposition order is reversed. LEIS spectra collected on LaO_x films capped with 7 ALD cycles of AlO_x (not shown) indicated that the AlO_x layer was continuous. Therefore, interactions between SiO_x and LaO_x are unlikely to be a source of dipoles or barrier height shifts.

Previous experiments have observed a negative flat band voltage shift when incorporating LaO_x into dielectric stacks, and the authors at least partially attributed the effect to positive charges in the LaO_x [29]–[32]. Positive charges would potentially explain an increased Φ_B for p-type Si and decreased Φ_B for n-type Si. In the present work, ~10 nm LaO_x

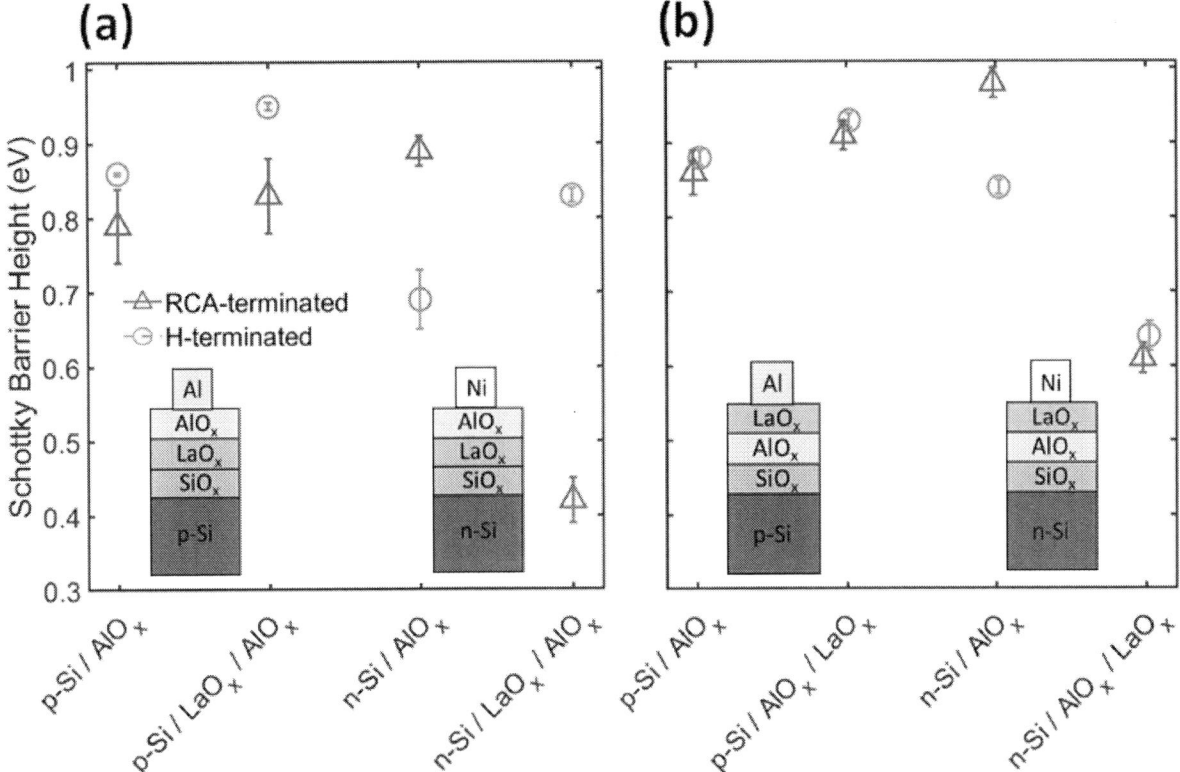

Fig. 3. Effects on Φ_B of introducing lanthanum oxide between (a) aluminum oxide and the Si substrate and (b) aluminum oxide and the metal top contact (Ni for n-type, Al for p-type). The error bars represent sample standard deviations. Insets: layer schematics of the test structures utilized.

films were deposited for C-V measurements. The average flatband voltage value extracted indicated that positive charges may have been present in the films. Fixed charge has been shown not to significantly impact Φ_B in tunnel MIS systems using only AlO_x [13], but this may not be true of systems incorporating LaO_x. The effects of these positive charges may compete with those of a SiO_x/AlO_x dipole, resulting in the Φ_B modulation observed.

It was noted that the AlO_x-only controls in Fig. 3(b), though prepared identically, had Φ_B values slightly different from their counterparts in Fig. 3(a). We attribute this to sample variation, however, the present work focused on changes in Φ_B within each set of concurrently processed samples.

To attain a better understanding of the mechanisms of Φ_B modulation, the chemical properties of LaO_x and its interactions with the substrate were studied via X-ray photoelectron spectroscopy (XPS). "Bulk" 10 nm LaO_x films were compared with ultrathin 2 nm films, each on RCA and H-terminated substrates.

The dominant O1s peak (Fig. 4) for all samples at ~531.5 eV was attributed to $La(OH)_3$ [33]. The peak was shifted towards 532 eV for ultrathin samples, likely due to contributions from the underlying SiO_x [34]. $La(OH)_3$ was also observed in samples capped with 7 cycles of AlO_x (not shown), indicating that the AlO_x was not sufficient to protect the underlying LaO_x layer from hygroscopic reaction with the atmosphere. Thus, it is likely that the samples used to

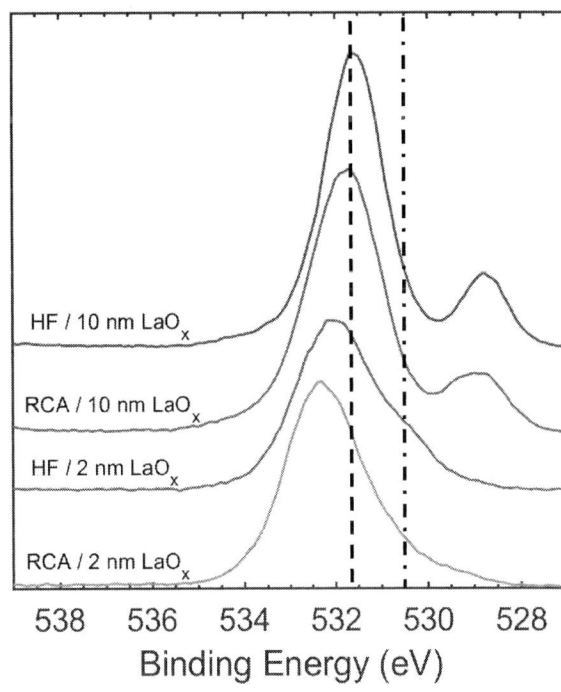

Fig. 4. O1s core-level XP spectra of LaO_x films grown with different thicknesses and substrate terminations. The dashed line indicates the position of the $La(OH)_3$ peak and the dotted and dashed line indicates the position of the La-silicate shoulder.

978-1-7281-6118-1/22 $31.00 © 2022 IEEE

measure Φ_B also contained significant La(OH)$_3$. Hydroxide presence was corroborated by the low dielectric constant of the lanthanum species: 9.8 as measured from ~10 nm LaO$_x$ films in the present work. The literature value of the dielectric constant for LaO$_x$ films grown using the same ALD chemistry is 29 [35], and La(OH)$_3$ content has separately been shown to reduce the dielectric constant to values as low as 7 [36].

The peaks near 529 eV (or tails in the ultrathin case) were attributed to stoichiometric La$_2$O$_3$ buried under a La hydroxide layer [37]. The 2 nm sample on the H-terminated substrate also exhibited a shoulder at ~530 eV, which may be attributed to lanthanum silicate formed at the interface [26]. Because this shoulder appeared for LaO$_x$ grown on the same termination for which depinning was observed above, and because Fermi level depinning has been attributed to lanthanum silicate formation in the past, silicate formation may explain the depinning observed in the present work.[25]

IV. CONCLUSION

The use of high-k dielectric stacks as insulators in MIS diodes has been demonstrated to have various effects on the Schottky barrier height. Increases in Φ_B up to 0.14 eV and decreases up to 0.47 eV were observed when introducing LaO$_x$ into AlO$_x$ tunnel layers, depending on the substrate termination and polarity. The Φ_B shifts observed were consistent with oxygen areal density-induced dipoles at SiO$_x$/AlO$_x$ interfaces, Fermi level depinning when introducing LaO$_x$ between H-terminated Si and AlO$_x$, and effects from positive oxide charges when introducing LaO$_x$ in other cases. XPS data indicated the presence of La-silicate at the interface between LaO$_x$ and H-terminated Si, which may be responsible for Fermi level depinning. Thus, while a direct interaction between the high-κ oxides could not be confirmed, Φ_B modulation arises from competing effects of the individual high-κ oxides. The data presented provide insight into the interactions between different oxide thin films and demonstrate the promise of using such tunnel stacks to tune Φ_B of future MIS devices.

ACKNOWLEDGMENT

This work was supported by the National Science Foundation CBET program under Grant No. 1605129. The authors acknowledge Dr. Ryan Thorpe for collection of XPS data and guidance on interpretation.

REFERENCES

[1] D. K. Schroder, *Semiconductor Material and Device Characterization*, 3rd ed., vol. 44, no. 4. John Wiley & Sons, Inc., 2006.

[2] Y. C. Yeo, P. Ranade, T. J. King, and C. Hu, "Effects of high-κ gate dielectric materials on metal and silicon gate workfunctions," *IEEE Electron Device Lett.*, vol. 23, no. 6, pp. 342–344, 2002.

[3] D. L. Pulfrey, "MIS Solar Cells: A Review," *IEEE Trans. Electron Devices*, vol. 25, no. 11, pp. 1308–1317, 1978.

[4] Y. W. Lam, "min m.i.s. Schottky barrier solar cells—a review," *Radio Electron. Eng.*, vol. 51, no. 9, pp. 447–454, 1981.

[5] D. L. Pulfrey and R. F. McOuat, "Schottky-barrier solar-cell calculations," *Appl. Phys. Lett.*, vol. 24, no. 4, pp. 167–169, 1974.

[6] H. C. Card and E. H. Rhoderick, "Studies of tunnel MOS diodes I. Interface effects in silicon Schottky diodes," *J. Phys. D. Appl. Phys.*, vol. 4, no. 10, pp. 1589–1601, 1971.

[7] H. C. Card and E. S. Yang, "MIS-Schottky theory under conditions of optical carrier generation in solar cells," *Appl. Phys. Lett.*, vol. 29, no. 1, pp. 51–53, 1976.

[8] R. Hezel, R. Meyer, and A. Metz, "New generation of crystalline silicon solar cells: simple processing and record efficiencies for industrial-size devices," *Sol. Energy Mater. Sol. Cells*, vol. 65, no. 1, pp. 311–316, 2001.

[9] M. Green, E. Dunlop, J. Hohl-Ebinger, M. Yoshita, N. Kopidakis, and X. Hao, "Solar cell efficiency tables (version 57)," *Prog. Photovoltaics Res. Appl.*, vol. 29, no. 1, pp. 3–15, 2021.

[10] S. M. George, "Atomic Layer Deposition : An Overview," *Chem. Rev.*, pp. 111–131, 2010.

[11] D. Connelly, C. Faulkner, P. A. Clifton, and D. E. Grupp, "Fermi-level depinning for low-barrier Schottky source/drain transistors," *Appl. Phys. Lett.*, vol. 88, p. 012105, 2006.

[12] B. Hoex, J. J. H. Gielis, M. C. M. Van De Sanden, and W. M. M. Kessels, "On the c-Si surface passivation mechanism by the negative-charge-dielectric Al2 O3," *J. Appl. Phys.*, vol. 104, no. 11, 2008.

[13] R. J. Marstell, A. Pugliese, and N. C. Strandwitz, "Absence of Evidence for Fixed Charge in Metal–Aluminum Oxide–Silicon Tunnel Diodes," *Phys. Status Solidi Basic Res.*, vol. 256, no. 3, pp. 1–6, 2019.

[14] Y. Yamamoto, K. Kita, K. Kyuno, and A. Toriumi, "Study of La-induced flat band voltage shift in metal/HfLaO x/SiO2/Si capacitors," *Japanese J. Appl. Physics, Part 1 Regul. Pap. Short Notes Rev. Pap.*, vol. 46, no. 11, pp. 7251–7255, 2007.

[15] K. Iwamoto *et al.*, "Experimental evidence for the flatband voltage shift of high- k metal-oxide-semiconductor devices due to the dipole formation at the high- kSi O 2 interface," *Appl. Phys. Lett.*, vol. 92, no. 13, pp. 1–4, 2008.

[16] P. D. Kirsch *et al.*, "Dipole model explaining high-k /metal gate field effect transistor threshold voltage tuning," *Appl. Phys. Lett.*, vol. 92, no. 9, pp. 1–4, 2008.

[17] B. E. Coss, W. Y. Loh, R. M. Wallace, J. Kim, P. Majhi, and R. Jammy, "Near band edge Schottky barrier height modulation using high- κ dielectric dipole tuning mechanism," *Appl. Phys. Lett.*, vol. 95, no. 22, p. 222109, 2009.

[18] K. Kita and A. Toriumi, "Origin of electric dipoles formed at high-k/ SiO2interface," *Appl. Phys. Lett.*, vol. 94, no. 13, pp. 92–95, 2009.

[19] J. Fei and K. Kita, "Opportunity of dipole layer formation at non-SiO2 dielectric interfaces in two cases: Multi-cation systems and multi-anion systems," *Microelectron. Eng.*, vol. 178, pp. 225–229, 2017.

[20] L. Lin and J. Robertson, "Atomic mechanism of electric dipole formed at high-K: SiO2 interface," *J. Appl. Phys.*, vol. 109, no. 9, p. 094502, 2011.

[21] W. Kern, "RCA Critical Cleaning Process," *MT Syst.*, pp. 1–7, 2007.

[22] K. Gelderman, L. Lee, and S. W. Donne, "Flat-band potential of a semiconductor: Using the Mott-Schottky equation," *J. Chem. Educ.*, vol. 84, no. 4, pp. 685–688, 2007.

[23] B. E. Coss *et al.*, "Dielectric dipole mitigated Schottky barrier height tuning using atomic layer deposited aluminum oxide for contact resistance reduction," *Appl. Phys. Lett.*, vol. 99, no. 10, p. 102108, 2011.

[24] K. Ohmori *et al.*, "Wide Controllability of Flatband Voltage in La2O3 Gate Stack Structures – Remarkable Advantages of La2O3 over HfO2 –," *ECS Trans.*, vol. 3, no. 3, pp. 351–362, 2006.

[25] N. Umuzawa *et al.*, "Relation between Solubility of Silicon in High-k Oxides and the Effect of Fermi Level Pinning," *ECS Trans.*, vol. 13, no. 2, pp. 15–20, 2008.

[26] T. J. Park, Y. C. Byun, R. M. Wallace, and J. Kim, "Impurity and silicate formation dependence on O3 pulse time and the growth temperature in atomic-layer-deposited La2O3 thin films," *J. Chem. Phys.*, vol. 146, no. 5, pp. 1–6, 2017.

[27] R. M. Eastment and C. H. B. Mee, "Work function measurements on (100), (110) and (111) surfaces of aluminum," *J. Phys. F Met. Phys.*, vol. 3, pp. 1738–1745, 1973.

[28] B. G. Baker, B. B. Johnson, and G. L. C. Maire, "Photoelectric work function measurements on nickel crystals and films," *Surf. Sci.*, vol. 24, no. 2, pp. 572–586, 1971.

[29] H. N. Alshareef *et al.*, "Work function engineering using

lanthanum oxide interfacial layers," *Appl. Phys. Lett.*, vol. 89, p. 232103, 2006.

[30] M. Kouda *et al.*, "Charged defects reduction in gate insulator with multivalent materials," *Dig. Tech. Pap. - Symp. VLSI Technol.*, pp. 200–201, 2009.

[31] B. L. Yang, H. Wong, K. Kakushima, and H. Iwai, "Improving the electrical characteristics of MOS transistors with CeO 2 / La 2 O 3 stacked gate dielectric," *Microelectron. Reliab.*, vol. 52, pp. 1613–1616, 2012.

[32] C. K. Chiang *et al.*, "Effects of La2O3 Capping Layers Prepared by Different ALD Lanthanum Precursors on Flatband Voltage Tuning and EOT Scaling in TiN/HfO2/SiO2/Si MOS Structures," *J. Electrochem. Soc.*, vol. 158, no. 4, pp. H447–H451, 2011.

[33] J. P. H. Li *et al.*, "Understanding of binding energy calibration in XPS of lanthanum oxide by: In situ treatment," *Phys. Chem. Chem. Phys.*, vol. 21, no. 40, pp. 22351–22358, 2019.

[34] J. Finster, E. D. Klinkenberg, J. Heeg, and W. Braun, "ESCA and SEXAFS investigations of insulating materials for ULSI microelectronics," *Vacuum*, vol. 41, no. 7–9, pp. 1586–1589, 1990.

[35] B. Lee *et al.*, "Electrical properties of atomic-layer-deposited La 2 O 3 films using a novel La formamidinate precursor and ozone," *Microelectron. Eng.*, vol. 86, no. 7–9, pp. 1658–1661, 2009.

[36] Y. Zhao, M. Toyama, K. Kita, K. Kyuno, and A. Toriumi, "Moisture-absorption-induced permittivity deterioration and surface roughness enhancement of lanthanum oxide films on silicon," *Appl. Phys. Lett.*, vol. 88, no. 7, pp. 10–13, 2006.

[37] T. Jiang *et al.*, "La2O3 catalysts with diverse spatial dimensionality for oxidative coupling of methane to produce ethylene and ethane," *RSC Adv.*, vol. 6, no. 41, pp. 34872–34876, 2016.

978-1-7281-6118-1/22 $31.00 © 2022 IEEE

Optimizing Perovskite Solar Cells by Understanding the Bulk Properties of Contact Layers

Mason Mahaffey, Zhengshan (Jason) Yu, Vidya Krishnan, David Quispe, Arthur Onno, Zachary Holman

School of Electrical, Computer and Energy Engineering, Arizona State University, Tempe, AZ 85287, USA

Abstract—Along with optimizing perovskite-contact interfaces is the need to optimize the bulk contact properties in relation to passivation, selectivity, and conductivity. In this work, we investigate the thickness-dependence of the implied-JV, pseudo-JV, and JV curves for devices made with different perovskite contact layers. We are able to extract from these curves selectivity (V_{oc}/iV_{oc}), series resistance, efficiencies, and fill factors in order to demonstrate the reasons for optimal device performance and to offer paths forward for improving each contact. Additionally, we show the spatial variance of iV_{oc} of perovskite grown on top of hole contact layers to demonstrate the effects of this variation on final device performance.

Keywords—*perovskite, PTAA, implied voltage, ERE, selectivity*

I. INTRODUCTION

Optimal electrical contacts for photovoltaic solar cells need to be "carrier-selective"; this is defined as being *passivating*, *conductive*, and *selective* [1]. *Passivating* refers to a high implied open-circuit voltage (iV_{oc}), *conductive* refers to low series resistance at the maximum power point (R_s, FF), and *selective* refers to low voltage losses within the contact leading to a high open-circuit voltage (V_{oc}).

Perovskite-based photovoltaic solar cells currently tend to suffer from significant performance loss related to the electrical contacts [2]. Many efforts have been taken to reduce these contact losses through the engineering of a well-passivated absorber-contact interface [2,8]. For instance, Zhu et. al. demonstrated a 1.5% absolute efficiency improvement through the deposition of a layer of 2-phenylethylammonium iodide (PEAI) in-between perovskite and Spiro-OMeTAD [8]. While these layers have been shown to improve the iV_{oc} (improve *passivation*), they do not necessarily lower the R_s (improve *conductivity*) or lower the open-circuit voltage losses within the bulk of the contact (improve *selectivity*) [8]. To be "carrier-selective", a contact needs to have both an optimized bulk and optimized surfaces.

In this work, we show how a simple single-factor experiment – varying the contact thickness – can identify the causes of non-optimized carrier selectivity. To ensure that we are varying contact thickness (rather than a process parameter), we utilize ellipsometry to determine the thickness of each layer. For each finished cell, we measure the implied-JV (injection-dependent QFLS measurement), pseudo-JV (Suns-V_{oc}), and JV curves. We then utilize these curves to extract selectivity S (V_{oc}/iV_{oc}), series resistance R_s, as well as the implied-, pseudo-, and device fill factors and efficiencies. We plan to utilize this information in

the future to improve the bulk contact performance through the optimization of doping.

II. MEASUREMENT THEORY

Important to this work is our ability to construct implied-JV curves from injection-dependent quasi-Fermi level splitting (QFLS) measurements (which we refer to as Suns-ERE). These curves have been previously demonstrated [4]. First, we define $V_{oc,ideal}$ as:

$$V_{oc,ideal} = \frac{k_B T}{q} \ln \left(\frac{\int a(\lambda)\phi_{exc}(\lambda)d\lambda}{\int a(\lambda)\phi_{BB}(\lambda,T)d\lambda} + 1 \right) \quad (1)$$

where k_B is the Boltzmann constant, T is the temperature of the sample Kelvin, q is the elementary charge, λ is the wavelength, $a(\lambda)$ is the sample's absorptance, $\phi_{BB}(\lambda,T)$ is the blackbody photon current, and $\phi_{exc}(\lambda)$ is the excitation photon current. The QFLS of the device is limited by non-radiative recombination to less than the theoretical limit of $q \times V_{oc,ideal}$. We call this the *implied* or *internal* voltage, defined as:

$$iV = \frac{QFLS}{q} \quad (2)$$

As the non-radiative recombination, and therefore QFLS limitation, can be quantified by ERE, we further define the iV to be equal to:

$$iV = V_{oc,ideal} - \frac{k_B T}{q} |\ln(ERE)| \quad (3)$$

Unity is perfect radiative efficiency and therefore maximum implied voltage.

During a Suns-ERE measurement the ERE is measured while the excitation is varied, producing different implied voltages; changes in excitation change the $V_{oc,ideal}$ via the $\phi_{exc}(\lambda)$ and can also change the ERE. We assume that the corresponding current at each implied voltage is defined as:

$$iJ = J_{sc}(1 - Suns) \quad (4)$$

where J_{sc} refers to the short circuit current of the cell and *Suns* refers to the excitation of the cell as a function of the 1-sun excitation. At 1-sun intensity, the implied current is equal to zero and therefore the implied open-circuit voltage (iV_{oc}) is equal to:

$$iV_{oc} = V_{oc,ideal} - \frac{k_B T}{q} |\ln(ERE_{1-Sun})| \quad (5)$$

where ERE_{1-Sun} is the external radiative efficiency at 1-sun intensity.

Beyond implied-JV curves, we also show common pseudo-JV (Suns-V_{oc}) and JV curves, and from these curves we can

978-1-7281-6118-1/22 $31.00 © 2022 IEEE

extract series resistance (R_s). The method we use for measuring R_s has been described previously by Pysch, Mette, and Glunz [5].

III. EXPERIMENTAL

A. Substrate and Solution Preparation

Patterned glass/ITO substrates (Xin-Yan Technology) were cleaned through 20 minutes of sonication in a solution of DI water with 5% Extran (Extran MA 02, Sigma-Aldrich), a rinse in DI water, two 20-minute rounds of sonication in acetone and isopropyl alcohol respectively, and a 10-minute UV-Ozone cleaning.

Poly[bis(4-phenyl)(2,4,6-trimethylphenyl)amine] (PTAA, Sigma-Aldrich, 702471), the hole contact layer, was dissolved in various concentrations in chlorobenzene (Sigma-Aldrich, 284513). Poly(9,9-bis(3'-(N,N-dimethyl)-N-ethylammoinium-propyl-2,7-fluorene)-alt-2,7-(9,9-dioctylfluorene))dibromide (PFN-Br, Sigma-Aldrich, 906980) was dissolved in methanol at a concentration of 0.5mg/mL.

The perovskite selected to be made was $Cs_{0.22}FA_{0.78}Pb(I_{0.85}Br_{0.15})_3$, plus 3 mol% $MAPbCl_3$ which is a wide-bandgap perovskite recipe with a bandgap near 1.67eV [9]. The precursor solution was made stoichiometrically with lead iodide (TCI, L0279), lead bromide (Sigma-Aldrich, 398853), cesium iodide (Sigma-Aldrich, 203033), formamidinium iodide (Sigma-Aldrich, 901436), methylammonium chloride (Sigma-Aldrich, 806020) and lead chloride (Sigma-Aldrich, 203572) dissolved in a 4:1 ratio of DMF:DMSO at a 1M concentration.

B. Sample Preparation

Layers of PTAA were deposited by spincoating 150 microliters of solution at 4000 rpm for 60s; the samples were then annealed for 10 minutes at 100C. After the samples were finished annealing, representative samples were removed for ellipsometry measurements.

The remaining samples were spincoated with 100 microliters of PFN-Br solution at 4000 rpm for 60s with no annealing.

The perovskite layer was formed using the anti-solvent quench method. 300 microliters of perovskite precursor solution was placed onto the samples and spun at 5000 rpm for 60s. At the 30s mark, 100 microliters of methyl acetate (Sigma-Aldrich) was dropped onto the substrate. After spincoating, the sample was annealed for 30 minutes at 100C. At this stage, representative samples were removed for use in later measurements.

Finally, for completed cells 30nm of C60 (Sigma-Aldrich, 572500) and Ag (Sigma-Aldrich) was evaporated onto the samples.

IV. RESULTS AND INITIAL CONCLUSIONS

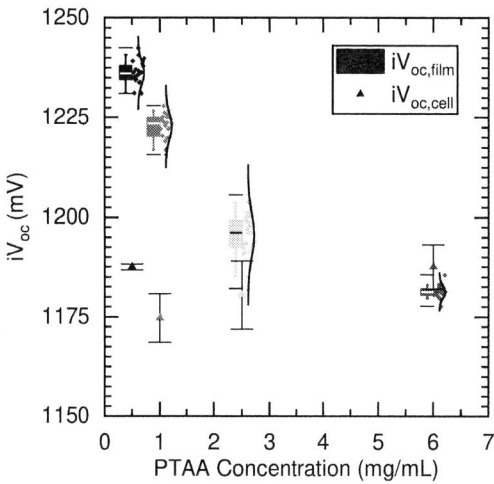

Fig. 1. Distribution of iV_{oc} values measured at different 4-mm² points across perovskite films made on ITO/PTAA/PFN-Br contact stacks of varying PTAA thickness (rectangle), as well as the distribution of iV_{oc} values for completed cells made with the same hole contact (triangle).

The data shown in Figure 1 demonstrates the varying passivation of perovskite films (made on hole contact stacks) and cells as a function of PTAA solution concentration (and therefore PTAA thickness). As the thickness increases, the iV_{oc} of the perovskite films decreases, signifying a lower passivation and lower limit to V_{oc}. However, the iV_{oc} values of the completed cells do not follow this pattern; they have an indeterminate trend, with the means falling within a 15-mV range. This suggests that the passivation of the completed cells is not being limited by PTAA and instead is being limited by non-radiative recombination from the addition of the C60/Ag electron contact. Because of this limitation, the benefit of passivation from a lower PTAA thickness is lost; these results are consistent with our current understanding of the role C60 has in limiting perovskite device performance, as it is known that the absorber/C60 interface can be a source of non-radiative recombination [6-7].

TABLE I.   SELECTIVITY DATA

PTAA Concentration (mg/mL)	$V_{oc,ideal}$ (mV)	iV_{oc} (mV)	V_{oc} (mV)	iV_{oc}-V_{oc} (mV)	S (V_{oc} /iV_{oc})
6	1381	1188±6	1114±4	73±4	0.94
2.5[a.]	1381	1180±9	1067±117	113±120	0.90
1	1381	1175±6	1029±29	146±24	0.88
0.5	1381	1187±1	937±12	251±12	0.79

[a.] Lower sample count for this concentration led to higher variance in V_{oc}.

Table 1. Data for calculating selectivity, with averages and standard deviations shown for iV_{oc}, V_{oc}, and iV_{oc}-V_{oc}.

While there is no trend in passivation for cells with different PTAA thicknesses, there is a trend in selectivity. Table 1 shows how the selectivity emerges from the characteristics of each batch of cells. The $V_{oc,ideal}$, calculated from external quantum efficiency of a representative cell, is assumed to be the same for each cell. As the PTAA thickness decreases, the V_{oc} decreases. Because the passivation (iV_{oc}) is invariant, each decrease in V_{oc} leads to a greater difference between iV_{oc} and V_{oc}. This leads to the thickest (6 mg/mL) contact producing an external voltage which is 94% of the internal voltage, versus the thinnest (0.5 mg/mL) producing 79%. Figures 2 and 3 show the implied-, pseudo-, and JV curves of representative samples with the thinnest and thickest PTAA contacts. The increase in selectivity due to thickness can be seen in how the distance between the implied and psuedo-JV curves lowers from Figure 2 to Figure 3.

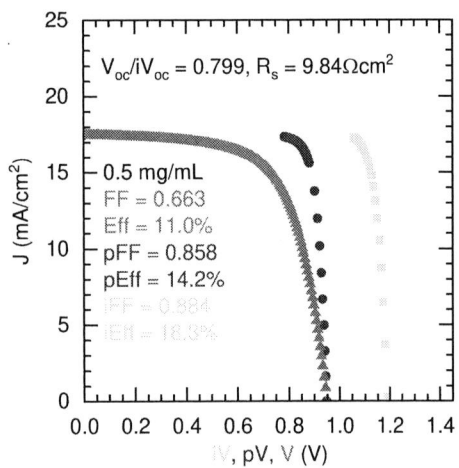

Fig. 2. Implied-, pseudo-, and JV curves for a sample made with the 0.5 mg/mL contact layer, along with summary data. The over 200mV gap between the iV_{oc} and V_{oc} shows the relatively low selectivity.

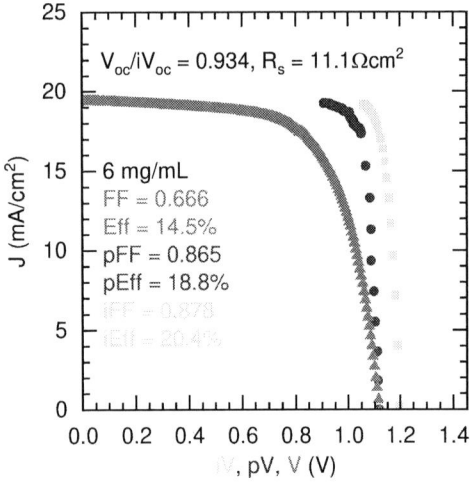

Fig. 3. Implied-, pseudo-, and JV curves for a sample made with the 6 mg/mL contact layer, along with summary data. The gap between the iV_{oc} and V_{oc} is less than 100mV.

TABLE II. DEVICE OUTPUT DATA

PTAA Concentration (mg/mL)	V_{oc} (mV)	J_{sc} (mA/cm^2)	FF (%)	Eff. (%)	R_s (Ω-cm^2)
6	1114±4	19.5±0.2	66±2	14.3±0.5	12.9±2
2.5[a.]	1067±117	18.6±1.3	64±6	12.8±2.5	11.5±6
1	1029±29	19.6±0.3	70±0.4	14.1±0.4	8.2±2
0.5	937±12	16.6±1.4	62±5	9.7±1.54	13.8±5

a. Lower sample count for this concentration led to higher variance in V_{oc}.

Table 2. Cell output data, with averages and standard deviations shown for V_{oc}, J_{sc}, FF, efficiency (Eff.), and R_s. There were no clear trends in R_s, and therefore no clear trends in conductivity.

V. DISCUSSION

In this work, we have shown how variation in passivation and selectivity due to PTAA contact thickness emerge in perovskite solar cells and have identified limitations to the cell's performance. This technique can be applied to optimize any contact layer that is able to be varied in thickness.

The conductivity, while measured, did not produce clear trends; overall low sample yield contributed to higher variance in some measured values, and future work will aim to reduce this uncertainty. Work is ongoing to produce ellipsometry models to determine the thickness of our hole contact layers. Finally, future work will demonstrate this analysis for a wider range of contact layers.

REFERENCES

[1] A. Onno, C. Reich, S. Li, A. Danielson, W. Weigand, A. Bothwell, S. Grover, J. Bailey, G. Xiong, D. Kuciauskas, W. Sampath, and Z. C. Holman, "Understanding what limits the voltage of polycrystalline CdSeTe solar cells," *Nature Energy*, 2022.

[2] E. Aydin, M. Bastiani, and S. Wolf, "Defect and contact passivation for perovskite solar cells," *Advanced Materials*, vol. 31, no. 25, p. 1900428, 2019.

[3] N. Mundhaas, Z. J. Yu, K. A. Bush, H. P. Wang, J. Häusele, S. Kavadiya, M. D. McGehee, and Z. C. Holman, "Series resistance measurements of perovskite solar cells using $J sc - V oc$ measurements," *Solar RRL*, vol. 3, no. 4, p. 1800378, 2019.

[4] F. Lang, E. Köhnen, J. Warby, K. Xu, M. Grischek, P. Wagner, D. Neher, L. Korte, S. Albrecht, and M. Stolterfoht, "Revealing fundamental efficiency limits of monolithic perovskite/Silicon tandem photovoltaics through Subcell characterization," *ACS Energy Letters*, vol. 6, no. 11, pp. 3982–3991, 2021.

[5] D. Pysch, A. Mette, and S. W. Glunz, "A review and comparison of different methods to determine the series resistance of solar cells," *Solar Energy Materials and Solar Cells*, vol. 91, no. 18, pp. 1698–1706, 2007.

[6] J. Benduhn, K. Tvingstedt, F. Piersimoni, S. Ullbrich, Y. Fan, M. Tropiano, K. A. McGarry, O. Zeika, M. K. Riede, C. J. Douglas, S. Barlow, S. R. Marder, D. Neher, D. Spoltore, and K. Vandewal, "Intrinsic non-radiative voltage losses in fullerene-based organic solar cells," *Nature Energy*, vol. 2, no. 6, 2017.

[7] M. Stolterfoht, C. M. Wolff, J. A. Márquez, S. Zhang, C. J. Hages, D. Rothhardt, S. Albrecht, P. L. Burn, P. Meredith, T. Unold, and D. Neher, "Visualization and suppression of interfacial recombination for high-efficiency large-area pin perovskite solar cells," *Nature Energy*, vol. 3, no. 10, pp. 847–854, 2018.

[8] T. Zhu, D. Zheng, J. Liu, L. Coolen, and T. Pauporté, "Peai-based interfacial layer for high-efficiency and stable solar cells based on a macl-mediated grown FA0.94ma0.06pbi3 perovskite," *ACS Applied Materials & Interfaces*, vol. 12, no. 33, pp. 37197–37207, 2020.

[9] J. Xu, C. C. Boyd, Z. J. Yu, A. F. Palmstrom, D. J. Witter, B. W. Larson, R. M. France, J. Werner, S. P. Harvey, E. J. Wolf, W. Weigand, S. Manzoor, M. F. van Hest, J. J. Berry, J. M. Luther, Z. C. Holman, and M. D. McGehee, "Triple-halide wide–band gap perovskites with suppressed phase segregation for efficient tandems," *Science*, vol. 367, no. 6482, pp. 1097–1104, 2020

Development and Qualification of IMMβ and Z4J+, Radiation Hard III-V Solar cells

John T Hart, Dan Aiken, Zac Bittner, Ben Cho, Daniel Derkacs, Khalid Emshadi, Andrew Espenlaub, Frank Fencl, Jeremy Leshin, Ahmad Mansoori, Nate Miller, Pravin Patel, Albert Perry, Janine Walker

Solaero by Rocket Lab, Albuquerque, NM, United States

Solaero has begun AIAA-S111 qualification of its new, radiation-hard, solar cells: the 33.3% IMMβ (AM0 1353 W/m2) and 31.3% Z4J+ (AM0 1353 W/m2). These solar cells are designed targeting ideal performance in real operating environments including charged particle irradiation and elevated temperatures. This presentation outlines the performance of the cells, the advancements in material quality and device design, as well as unique features, such as the reduction in solar absorptance of IMMβ.

The Profound Influence of Substrate Thermal Resistance on the Photovoltaic Properties of Solution-Processed Cu(In,Ga)Se2

Kyle G Weideman, Rakesh Agrawal

Purdue University, West Lafayette, IN, United States

Decreased material usage and manufacturing costs associated with the solution-processing of thin film PV materials such as Cu(In,Ga)Se2 (CIGS) offer a route towards larger scale PV production and deployment. Unfortunately, recent progression in the power conversion efficiency (PCE) of solution-processed CIGS has been slow, and the normal buildup of understanding through worldwide collaborative research has been hampered by batch to bath and lab to lab irreproducibility. This work presents a potential solution to these issues by identifying the previously understudied variable of substrate thermal resistance during the high temperature growth step of CIGS fabrication. Subtle changes in transient heat transfer are shown to have a large impact on the final device performance, with identical films processed under nominally the same conditions controllably displaying final device PCE' from < 5% to >14% through variation of only thermal resistances within the furnace system. A model demonstrating how transient heat transfer determines the interface quality and current collection is proposed, and use of these ideas is shown to reliably manipulate all relevant PV parameters. This improved understanding of the high temperature growth step then eliminates a previously unaccounted for variable that hampered globally cohesive property improvement and disguised the true impact of other parameters studied in previous works. This will then open up a new avenue of PV property control that is not only limited to CIGS but should also translate to many other material systems such as Cu(Zn,Sn)Se2 where similar high temperature growth steps involving a liquid flux agent are used.

Soiling Measurement Based on Checkered Pattern Image Analysis

Bing Guo, Wasim Javed

Texas A&M University at Qatar, Doha, Qatar

Experiments were carried out in laboratory conditions using various types of dust samples, various lighting conditions and two checkered patterns. The checkered pattern, at various soiling levels, was imaged using an industrial camera with a linear profile. A black-to-white ratio (BWR) was determined using the checkered pattern image's histogram. The soiling level (SL) was quantified in terms of dust surface coverage through analysis of micrographs of the soiled surface or in terms of transmission loss measured with a PV reference cell. In addition, a 5-week field test was carried out using one of the checkered patterns, with checked pattern photographs taken using the same industrial camera and soiling level determined based on short-circuit current measurements of PV module receiving the same soiling. The results show that, with a solar simulator as the illumination source there is a very strong, reproducible linear correlation between the BWR and the soiling level with camera altitude angle correction. The data points from three different dust samples nearly fell onto a straight line. The BWR-SL correlation was reproducible using a larger checkered pattern using a consumer "ring light" as the illumination source, with a lower R-squared value for the linear fitting. The BWR-SL correlation for the field test was still significant but with an even lower R-square value. The findings from this study suggest that using checkered pattern image analysis to measure soiling is a potentially feasible method, but addition work is needed to account for the various factors affecting the BWR-SL relationship.

978-1-7281-6118-1/22 $31.00 © 2022 IEEE

Modeling Efficiency of Inverters with Multiple Inputs

Clifford Hansen, Jay Johnson, Rachid Darbali-Zamora, Nicholas Gurule

Sandia National Laboratories, Albuquerque, NM, USA

Abstract—**Inverters convert DC power to AC power that can be injected into the grid. Many inverters offer multiple, independent maximum power point trackers (MPPTs) to accommodate photovoltaic arrays with different orientations or capacities. No validated model for overall DC-to-AC power conversion efficiency is available for such inverters. Herein, we propose a mathematical model that describes the efficiency of a multi-MPPT inverter and present validation using a commercial inverter with six MPPT inputs.**

Keywords—inverter, modeling, efficiency, power electronics

I. INTRODUCTION

Modeling photovoltaic (PV) system performance requires a model for the DC-to-AC power conversion efficiency of the system's inverters. Available inverter models [1], [2], [3] describe inverter conversion efficiency as a function of input DC power and DC voltage. These models were developed and validated for inverters with a single maximum power point tracker input. Each model employs parameters fitted to observed conversion efficiency curves to predict conversion efficiency at any condition.

Many inverters now offer multiple, independent maximum power point trackers (MPPTs). Multiple MPPTs allow an inverter to maximize energy conversion from PV arrays with different orientation, capacity and shading, and thus with DC power and DC voltage varying among the arrays. To our knowledge, no model specific to multi-MPPT inverters has been published and validated, although some PV simulation software (e.g. [5]) include models for PV systems with multiple arrays.

Here, we extend the single-input model in [1] to a form applicable to multi-input inverters. The procedure is not specific to the model form in [1] and thus may also indicate how the models in [2] and [3] could also be extended. A python implementation of the resulting model is available in pvlib-python [4] as the `pvlib.inverter.sandia_multi` function.

Bower et al. [5] published a procedure for measuring inverter efficiency over a range of test conditions. This procedure produces data that can be used to fit the inverter model in [1]. Test results for many inverters are recorded in the California Energy Commission (CEC) Equipment List[1]. The test procedure is also being applied to multi-input inverters, but in a limited manner: inverter efficiency measurements are made with equal DC voltage and DC power applied to each input (according to private communications).

For fitting and validation of the extended model, inverter AC power is measured for a commercial device with six MPPT inputs, with the DC voltage and DC power at each input varied over a matrix of test conditions. We calibrate the extended inverter model using only this "equal input" data and show that the model accurately predicts inverter conversion efficiency when unequal DC voltage and/or DC power are supplied on different inputs.

We compare our model with the multi-input inverter model implemented in the System Advisor Model (SAM) [6]. Our new model predicts conversion efficiency without a bias toward underprediction at low input power that is observed in the output of SAM.

II. DERIVATION OF THE MULTI-INPUT INVERTER MODEL

Available PV inverter models (e.g., [1]) are of the form:

$$P_{AC} = min\{f(P_{DC}, V_{DC}), P_{AC,max}\} \tag{1}$$

where P_{DC} is the total input DC power, V_{DC} is input DC voltage (assumed to be the same at each DC input) and $P_{AC,max}$ is the PV inverter's AC power limit. The function form f accounts for the dependence of conversion efficiency on input power and voltage, as well as factors such as input or output power limiting, minimum start-up power, self-consumption by the inverter, and voltage limits.

An inverter with several MPPTs comprises two functional stages in sequence:

1. A DC-DC converter on each input, which holds the connected array at the array's MPP, and converts the input DC voltage to a DC bus at a common DC voltage.

2. A DC-AC inverter stage which produces AC power from the DC power on the DC bus.

Fig. 1 illustrates a block diagram of a PV inverter with multiple MPPTs and assigning variables to DC voltage and power on each input and on the bus. Not every input needs to be connected to a PV array.

---

[1] https://www.energy.ca.gov/programs-and-topics/programs/solar-equipment-lists

978-1-7281-6118-1/22 $31.00 © 2022 IEEE

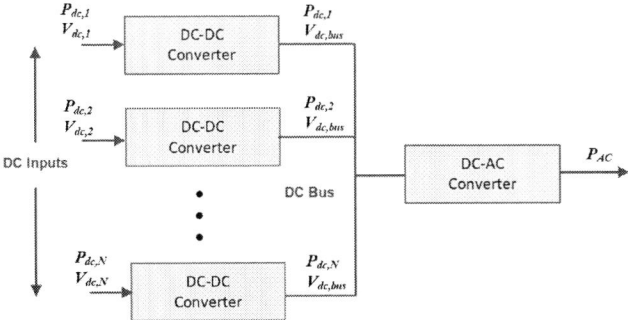

Fig. 1. Block Diagram of a Multi-Channel PV Inverter.

We derive an extension of the model represented by Eq. 1 to multi-MPPT PV inverters that can be calibrated using only the "equal input" data collected according to [6]. Denote the AC power that results from DC input to one MPPT as

$$P_{AC,i} = g_i(P_{dc,i}, V_{dc,i}) \qquad (2)$$

where i indexes the MPPT inputs. Assume that all DC-DC converters are equally efficient, i.e., the function g_i is the same for every MPPT input and we can drop the subscript from g. Assume that each DC-DC converter acts independently, i.e., $P_{AC,i}$ is independent of $P_{AC,j}$ for $i \neq j$. Then the sum over all MPPT inputs results in the total output AC power:

$$P_{AC} = f(P_{DC}, V_{DC}) = \sum_{i=1}^{N} P_{ac,i} = \sum_{i=1}^{N} g(P_{dc,i}, V_{dc,i}) \qquad (3)$$

When multi-input inverters are tested using the "equal power" method, then at each test point (P_{DC}, V_{DC}) the conditions for each MPPT input are the same: $V_{dc,1} = V_{dc,2} = \cdots V_{dc,N} = V_{DC}$ and $P_{dc,1} = P_{dc,2} = \cdots P_{dc,N} = P_{dc} = \frac{1}{N} P_{DC}$. In these conditions $f(P_{DC}, V_{DC})$ is the single-input model, i.e., [1]. It follows that:

$$f(P_{DC}, V_{DC}) = \sum_{i=1}^{N} g(P_{dc,i}, V_{dc,i})$$
$$= N \times g\left(\frac{1}{N} P_{dc,i}, V_{dc,i}\right) \qquad (4)$$

Applying the change of variables $P_{dc} = \frac{P_{DC}}{N}$ yields:

$$g(P_{dc}, V_{DC}) = \frac{1}{N} f(P_{DC}, V_{DC}) \qquad (5)$$

We take Eq. 5 to define the form of the function g, i.e., congruent to the function f evaluated at the total DC power P_{DC} but scaled in amplitude. We generalize from Eq. 5 to define the function g at any set of conditions $P_{dc,i}, V_{dc,i}$ to be:

$$g(P_{dc,i}, V_{dc,i}) = \frac{P_{dc,i}}{P_{DC}} f(P_{DC}, V_{DC}) \qquad (6)$$

With this definition of g, the extended model for the AC output of an inverter with multiple MPPTs is a weighted sum of the output of the single-input inverter model, applied at each

MPPT input to total DC power and the DC voltage at the MPPT input:

$$P_{AC} = min\left\{ \sum_{i=1}^{N} \frac{P_{dc,i}}{P_{DC}} f(P_{DC}, V_{dc,i}), P_{AC,max} \right\} \qquad (7)$$

$$P_{DC} = \sum_{i=1}^{N} P_{dc,i} \qquad (8)$$

III. MEASUREMENTS

AC power is measured for a SMA Tripower Core1 inverter with a power rating of 33kVA, an operating voltage of 480 V_{AC} and six independent MPPT inputs. Tests were conducted using the open-source System Validation Platform (SVP) and the power hardware-in-the-loop architecture in Fig. 2. Scaled analog voltage signals were sent to a 180 kVA, 480 VAC AMETEK AC power amplifier to provide an AC voltage signal to the PV inverter. DC power for each input to the inverter was provided by a 200 kW, 1000 VDC AMETEK TerraSAS programmable PV simulator. The current and voltage responses from the PV inverter were recorded using MATLAB/Simulink.

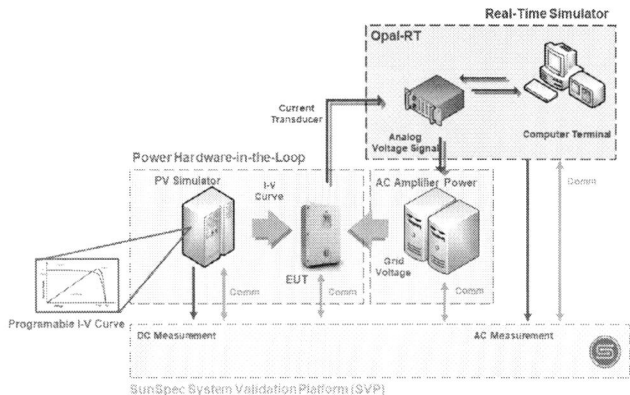

Fig. 2. Testbed for measuring inverter efficiency.

IV. RESULTS

Parameters for the inverter model [1] were determined by applying the 'fit_sandia' function in pvlib-python [3] to the "equal power and voltage" subset of the test results. The fitted model is used to predict AC power at all test conditions and predicted power is compared to measurements. Fig. 3 illustrates the predicted inverter efficiency (top) and the relative error at each test point (bottom). The results demonstrate that the proposed model is generally unbiased with prediction accuracy between ±0.5%. Variance in inverter efficiency is observed at each level of DC input power and DC voltage. This variance appears to arise from variability in the laboratory measurements or from the dynamics of the inverter's MPPT algorithm. The variance does not correlate with the DC power level on any specific MPPT input (Fig. 4).

978-1-7281-6118-1/22 $31.00 © 2022 IEEE

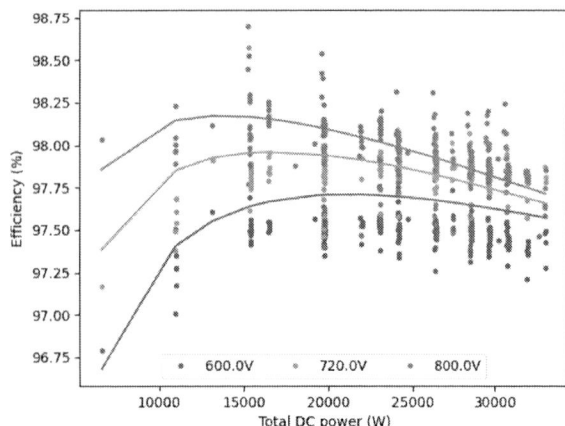

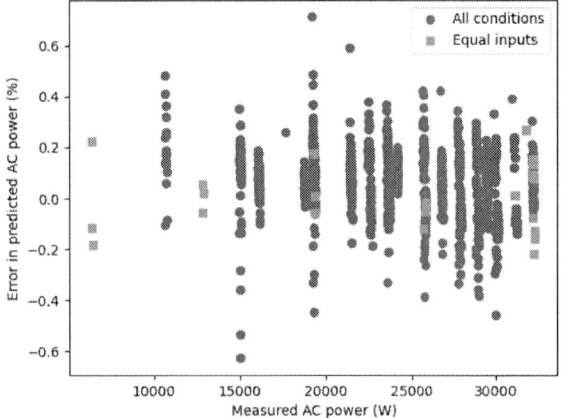

Fig. 3. Measured and predicted efficiency (top), and error in predicted AC power (bottom).

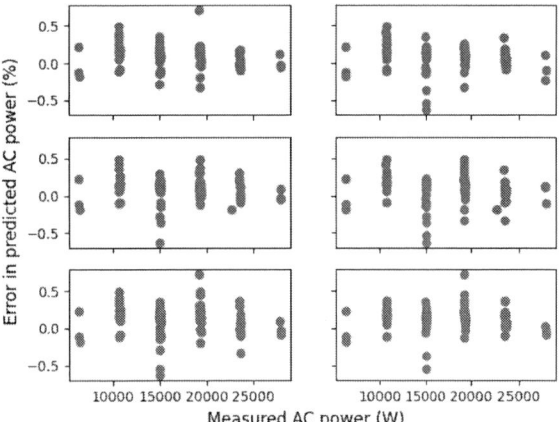

Fig. 4. Error in predicted AC power separated by input.

V. COMPARISON WITH THE SAM INVERTER MODEL

The multi-inverter model implemented in SAM version 2021.12.02 is shown in Eq. 9. The function f represents the model described in [1] fit to data measured with equal power and voltage on each input. This model omits the weighting of AC

power produced from the DC power at each MPPT input and applies the function f to the DC power on each input $P_{dc,i}$, rather than to the total DC power P_{DC}. Consequently, the model uses only on the lower range of the curve relating input DC power to efficiency (Fig. 3 (top)) as consequently, underestimates AC power at all power levels (Fig. 5). The SAM development team plans to update the multiple-input inverter model to be consistent with the model described in Eq. 7 and Eq. 8 (private communication).

$$P_{AC} = min\left\{\sum_{i=1}^{N} f\left(P_{dc,i}, V_{dc,i}\right), P_{AC,max}\right\} \quad (9)$$

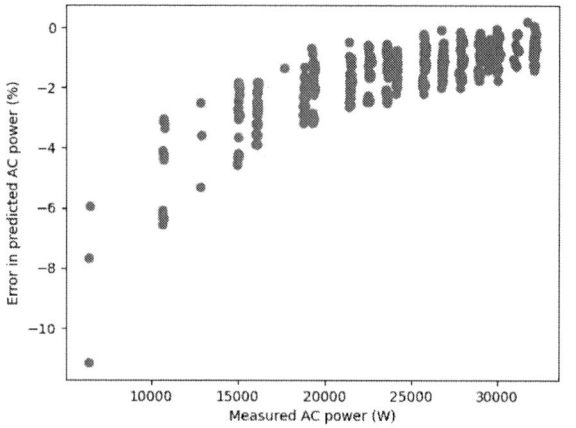

Fig. 5. Error in predicted AC power using the SAM model.

ACKNOWLEDGMENT

Sandia National Laboratories is a multimission laboratory managed and operated by National Technology and Engineering Solutions of Sandia, LLC., a wholly owned subsidiary of Honeywell International, Inc., for the U.S. Department of Energy's National Nuclear Security Administration under contract DE-NA-0003525.

REFERENCES

[1] D. L. King, S. Gonzalez, G. M. Galbraith, and W. E. Boyson. "Performance Model for Grid-Connected Photovoltaic Inverters". Sandia National Laboratories Report SAND2007-5036, September 2007.

[2] A. Driesse, P. Jain and S. Harrison, "Beyond the curves: Modeling the electrical efficiency of photovoltaic inverters," 2008 33rd IEEE Photovoltaic Specialists Conference, 2008, pp. 1-6, doi: 10.1109/PVSC.2008.4922827.

[3] A. Dobos. "PVWatts Version 5 Manual". National Renewable Energy Laboratory Report NREL/TP-6A20-62641. September 2014.

[4] William F. Holmgren, Clifford W. Hansen, and Mark A. Mikofski. "pvlib python: a python package for modeling solar energy systems." Journal of Open Source Software, 3(29), 884, (2018). https://doi.org/10.21105/joss.00884

[5] W. Bower, C. Whitaker, W. Erdman, M. Behnke, M. Fitzgerald. "Performance Test Protocol for Evaluating Inverters Used in Grid-Connected Photovoltaic Systems". Available at https://www.energy.ca.gov/sites/default/files/2020-06/2004-11-22_Sandia_Test_Protocol_ada.pdf

[6] System Advisor Model (SAM), available at https://sam.nrel.gov/.

Pathways to High Efficiency Perovskite Monolithic Solar Modules

Xuezeng Dai, Shangshang Chen, Yehao Deng, Allen Wood, Guang Yang, Chengbin Fei, Jinsong Huang

Department of Applied Physical Sciences, University of North Carolina at Chapel Hill , CHAPEL HILL, NC, United States

Department of Chemistry, University of North Carolina at Chapel Hill, CHAPEL HILL, NC, United States

With the rapidly improving efficiency and stability of perovskite solar cells, the transition of small area device fabrication innovations into modules is becoming increasingly important for the commercialization of this technology. The record efficiencies of small perovskite cells are already approaching that of the best silicon crystal solar cells, but the module efficiencies are still far behind. Understanding the factors causing the cell-to-module (CTM) efficiency loss is critical for large area perovskite module development. Here, we experimentally validated a comprehensive model that analyzes the CTM efficiency loss with a precision better than 97%. Using the model, we deciphered the impact of the critical module components and fabrication variables, including perovskite bandgap, transparent electrodes, scribing lines, and film uniformity, on module aperture efficiency. Our analysis provides pathways toward the aperture efficiency ceiling of 25.8% for single-junction perovskite solar modules with a bandgap of 1.49 eV. Enlightening by the model, we found that the tandem structures have intrinsic merit to achieve high-efficiency perovskite modules of 28.4% with much lower CTM derate due to the smaller photocurrent but larger photovoltage.

978-1-7281-6118-1/22 $31.00 © 2022 IEEE

Simulation of hot-carrier filtering in InAs-InP nanowire heterostructures

Urs Aeberhard

ETH Zurich, Zurich, Switzerland

Fluxim AG, Winterthur, Switzerland

A microscopic simulation of hot carrier filtering across potential barriers formed by InP-InAs heterostructures in InAs nanowires with localized illumination is discussed. Qualitative agreement with the experimentally observed dependence of the extracted current on photon energy and localized source position is demonstrated. Additionally, the impact on the hot carrier extraction efficiency of carrier relaxation due to electron-phonon interaction is analyzed.

A simple approach to ohmic contacts for transition metal dichalcogenide solar cells

Mario Martinez, Simon A. Svatek, Carlos Bueno-Blanco, Der-Yuh Lin, Ines Duran, Antonio Marti, Elisa Antolin

Instituto de Energia Solar, Universidad Politecnica de Madrid, Madrid, Spain

Department of Electronics Engineering, National Changua University of Education, Changua, Taiwan

Transition metal dichalcogenide (TMDC) semiconductors are promising materials for the manufacture of ultrathin solar cells due to their optoelectronic properties and their potential for low-cost fabrication. However, they still present several technological challenges, such as the development of ohmic contacts. The most common contact technology is based on the deposition of metals on the TMDC and subsequent annealing. It is known that this process damages the crystalline structure of the TMDC, leading to Fermi level pinning at the contact interface (Schottky barrier). In this work we explore an easy-to-implement ohmic contact for TMDC solar cells, in which a very flat metal surface has been prepatterned on the substrate and the TMDC laminae are transferred onto it. The TMDC atomic layers remain intact, and they are joined to the metal surface only by van der Waals forces. If a metal of suitable working function is chosen, an ohmic contact is produced without the need of thermal annealing. Using the transfer length method (TLM) we demonstrate that it is possible to obtain contact resistances in the order of $1 \cdot 10^{-3} \ \Omega \cdot cm^2$ for n and p doped MoS_2, which means that this simple fabrication method for van der Waals metal/TMDC contacts produces sufficiently low series resistance for one-sun applications.

978-1-7281-6118-1/22 $31.00 © 2022 IEEE

Luminescent Solar Concentrators for Building Integrated Photovoltaic Devices

Liam J. Halloran

Trinity College Dublin, Dublin, Ireland

Luminescent Solar Concentrators are used to collect sunlight across a large area, convert it into luminescent radiation, and concentrate it onto a small photovoltaic cell for electricity generation. The main advantage of this type of Luminescent Solar device is that it works effectively with both direct and diffuse solar radiation, and therefore enables increased solar electricity generation in areas or climates with high levels of diffuse sunlight. This research used computer modelling software to validate existing designs for Luminescent Solar Concentrators and determine expected behaviours of a LSC of a specific size and luminescent dye concentration. Following that, LSC layers were fabricated in a lab using a luminescent red perylene dye and an optical encapsulant that were then experimentally tested for performance comparison using artificial sunlight in a solar simulator and using real-world conditions in an outdoor experimental installation. The results of this research include a modelled optical efficiency of about 5.7% and an experimentally tested optical efficiency in outdoor conditions of only 1.34%. From these results, it is clear that there are significant areas of growth and potential for improvement when it comes to the fabrication and physical design of these Luminescent Solar devices. Further analysis and discussion of the various results from the assorted testing methodologies also inspired recommendations for improvements and future work that will be essential to strengthen these optical efficiencies if LSCs are to be used in Building Integrated Photovoltaic Devices in the coming years.

978-1-7281-6118-1/22 $31.00 © 2022 IEEE

LETID in legacy and modern PV modules: accelerated testing and field deployment

Joseph Karas, Ingrid Repins

National Renewable Energy Laboratory, Golden, CO, United States

The kinetics of light- and elevated temperatureinduced degradation (LETID) in silicon solar cells depend on the precise operating excess carrier density of the device. This dependency causes differences in the way LETID manifests in modern, higher-efficiency devices compared to lower-efficiency, legacy devices that might have been deployed in the field in previous years. In this work we model how different vintages of devices are expected to behave in both accelerated laboratory testing, as well as field deployment. The differing excess carrier densities encountered in various module vintages has implications both for interpreting accelerated test data, as well as identifying, diagnosing, and potentially treating LETID in the field.

Measurement of Snow Loading on a Tilted PV Module in Northern Michigan

Daniel Riley[1], Laurie Burnham[1], William Snyder[1], Bruce King[1], and Paul Dice[2]

Sandia National Laboratories[1], Albuquerque, NM, USA
Michigan Technological University[2], Houghton, MI, USA

Abstract— As PV systems are proliferating at latitudes, there is additional need to understand the effects of snow accumulation on PV modules. Sandia National Laboratories has installed an experimental platform in Calumet, MI to measure both static and dynamic snow loading on a photovoltaic (PV) module throughout the course of a winter characterized by periods of persistent snow, in which the module remains under continuous load, as well as repeated snowfalls of varying load. This paper details the magnitude of snow load, the rate of load change, the spatial distribution of load on a PV module, and the affect of wind on the measurement of snow load over the 2021-2022 winter.

Keywords—PV System, Snow, Loading

I. INTRODUCTION

With the rapid growth of solar across northern regions, the impact of snow shading on modules is a growing concern[1]. Published estimates of energy losses range from 1 to 12 percent annually, with monthly losses as high as 100 percent, depending on location and weather conditions; in addition, snow creates excessive and uneven stress on modules, cells and systems. That stress can threaten the reliability of PV systems in at least two ways: 1) cell cracking and 2) racking deformation. Yet to be determined is how the combined stressors of snow, cold and wind loading affect overall module integrity, especially when the different thermal properties of the layers are taken into account.

Sandia has developed a patent-pending platform for measuring loading on a tilted module that captures both static load and dynamic load created by the rapid shedding of snow from the upper areas of the module. Depending on prevailing weather conditions, the momentum of sliding snow is often impeded by the module frame, resulting in non-uniform loading of the module surface, with most weight along the bottom edge of the module. In addition, the force of the sliding snow can loosen fasteners and deform the module frame.

Late in the winter of 2021, Sandia National Laboratories installed the measurement platform (*see Figure 1*) at the Michigan Regional Test Center (RTC) site in Calumet MI, at a latitude of 47°N. The site, which allows for cross-technological comparisons under identical and relatively predictable conditions, provides ideal field conditions for evaluating solar technologies in winter: snow and low temperatures here are both predictable and persistent: the average annual snowfall is 202 inches and can be as much 300 inches or more. [1]

II. METHODOLOGY

Data collected from the snow load measurement station during the winter of 2021-2022 is correlated here with onsite, high-fidelity weather data. These data show the range of loads that can occur on a PV module during a single snow event, as well as the cumulative impact over the course of a winter. The data also show the rate of snow load accumulation during a single snowfall, and the rate of load release during a snow shedding event. Furthermore, the load data can be correlated with wind speed to determine the extent that high wind speed conditions contribute to PV module loading.

The snow load station was installed in late fall of 2021 and equipped with a 60-cell photovoltaic (PV) module that is 1.675 m by 1.001 m in a glass/cell/backsheet construction with an aluminum frame. The PV module is held to a rigid structural frame at an angle of 35 degrees relative to horizontal, and the downward load is measured once per second. Note that the load measurement is downward only and does not measure shear forces (i.e., forces that attempt to slide the PV downward). Every 15 seconds, the minimum, maximum, and average of all load measurements is recorded (i.e., 1 second measurements are summarized every 15 seconds). The PV module, structural frame and data logging enclosure are shown in Fig. 1.

The onsite weather station similarly measures data at a 1-second rate, but then averages the measurements into a 1-minute average that is recorded. The precipitation gauge of the weather station is of the tipping-bucket style with a heated inner cone to melt snow into a rain-water equivalent.

Fig. 1. The structural frame supporting the PV module at an angle of 35 degrees

III. ANALYSIS OF SNOW-LOADING DATA

A. Maximum loading

The maximum average load observed over the winter of 2021-2022 occurred on March 23, 2022 with a peak load of 25.68 kg (56.6 lb), following a winter storm that dropped the snow equivalent to 29 mm of rain. This load corresponds to a pressure of 15.33 kg/m^2 or 150.3 Pa. The average load above the unloaded (snow-free) state is shown in Fig 2, where it can also be seen that snow loading of greater than 10 kilograms (22 lbs, 58.5 Pa) occurred five times over the course of the winter.

B. Maximum load change rate

In addition to maximum load values, it is also important to determine the maximum rates at which snow loads change as snow either accumulates or slides from PV modules. Accumulation of snow on a PV module is typically a slow process, one that is affected by the rate od snowfall rate, wind speed and direction (which can remove snow from modules), the module's tilt angle, etc. In our data set, the maximum rate of snow accumulation led to load changes of less than 1.5 kg per hour. A particularly quick deposit of snow is shown in Fig 3, where the snow accumulated and added approximately 16 kg in 12 hours.

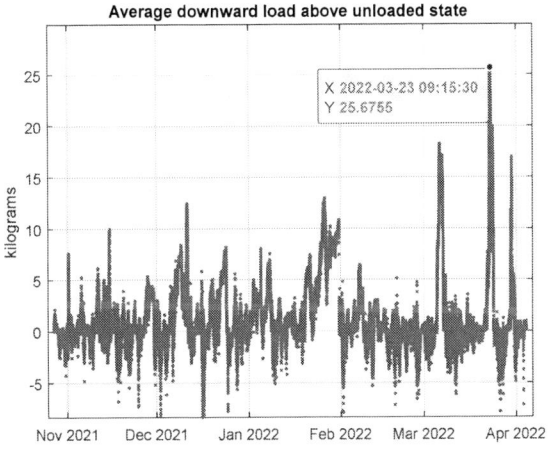

Fig. 2. Maximum average load observed on March 23 is 25.6755 kg.

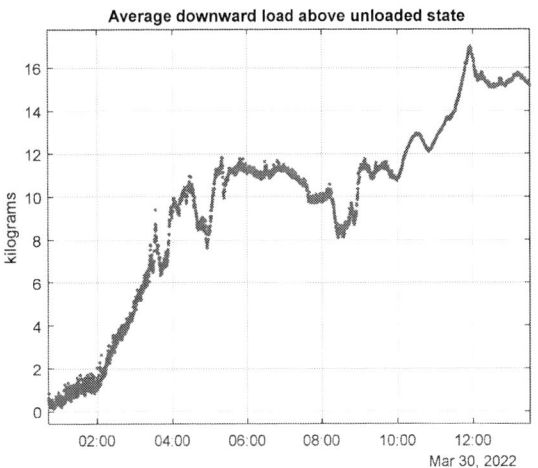

Fig. 3. Snow accumulation added 16 kg of snow in 12 hours.

Depending on weather conditions, the morphology of snow, tilt angle and module architecture, snow can adhere to a PV module for days, with the possibility of a new snow event adding new load. But snow can also slide from a PV module very quickly under situations that are conducive to snow sliding. This sliding effect will dominate the "dynamic" changes in snow load across a PV system [3]. Fig. 4 shows the most dynamic sliding condition observed during the winter of 202122. In this event, the maximum rate of total load change (total of all 4 load cells) was 4.23 kg over a single 15-second measurement period, equal to almost 17 kg/min at the maximum rate. The entire sliding event was slower and removed 8.5 kg of snow in 1 minute.

While the total load change of this sliding event reached as high as 17 kg/min, the effect is not evenly distributed across the PV module. Fig. 5 shows the average absolute load of each individual load cell supporting the PV module. Note that the northern two load cells only lose 2-3 kg, but the southern two load cells drop by approximately 7 kg each over the 1-minute sliding event. Thus, the change in load is primarily affecting the southern (lower) edge of the PV module.

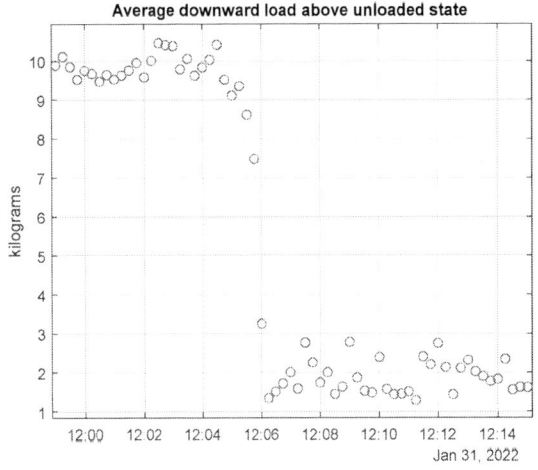

Fig. 4. The fastest observed sliding event where the PV module shed 8.5 kg of snow in 1 minute.

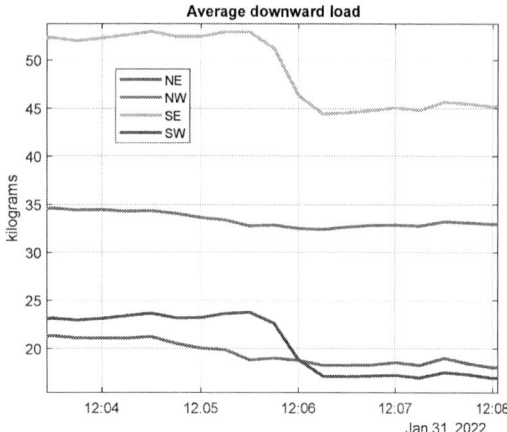

Fig. 5. Load measurements at each of the four load cells during the fast sliding event on January 31. Load change primarily occurs at the module's lower edge.

IV. WIND SPEED AS A FACTOR AFFECTING LOAD

In an effort to better understand data from the snow load measurement station, the load data was correlated with wind speed as measured at a 10-meter height. We expect that wind blowing toward the face of the module (i.e., from the South) will increase the load measured, and wind blowing toward the back of the module (from the North) will decrease the load measured as the module lifts but wind can shift direction quicky and create lateral as well as vertical loads. Because of its complexity, the effect of wind direction has not yet been studied from this data set, and the following analysis will focus only on wind speed.

As noted earlier, 15 measurements were made (one per second) for each 15-second recording period, including minimum, maximum, and average measurements. For this analysis, the difference between the maximum and minimum measurement within each recording period is compared against the wind speed. As shown in Fig. 6, as wind speed increases, so too does the difference between the maximum and minimum measurement. This indicates that higher wind speeds create loads that are highly variable in time and may affect the ability to isolate snow load from wind loading. Future analyses with higher data sampling rates may illuminate these effects further.

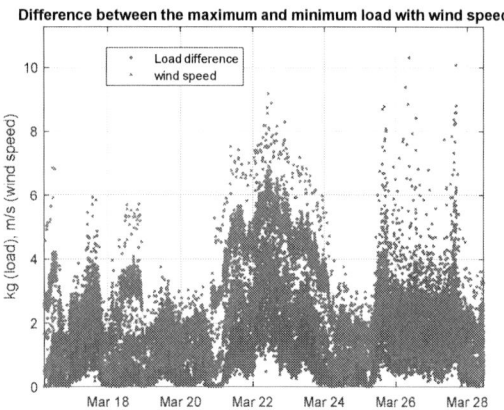

Fig. 6. Difference between the maximum and minimum recorded total load in each recording period and the 10-meter wind speed

V. CONCLUSIONS AND FUTURE WORK

Sandia's snow load measurement station in Calumet, MI represents the first known attempt to collect accurate field data to quantify patterns of snow loading on tilted solar panels and to capture the magnitude of snow loading on PV modules in northern climates. The correlation of weather and load data from the 2021-2022 winter has shown that on a single 1.675 m^2 PV module at a 35° tilt:

- Downward load often exceeds 10 kg or 58 Pa, with maximum loads approaching 26 kg or 150 Pa.

- Loads typically accumulate relatively slowly at rates up to 1.5 kg/hr.

- Loads can be removed very quickly due to sliding of snowpack from a module at rates up to 17 kg/min.

- The load reduction due to sliding snow is not evenly distributed around the PV module, most of the reduction occurs on the lower edge of the PV module.

- Wind speed can affect the measurement of snow load. Further studies may better quantify this effect.

The effects of these rapid load changes may be present in electroluminescence images which may show additional cell cracking due to high snow loads at low module temperatures. Future work by Sandia and Michigan Technological University will entail the collection of time-series image to determine the prevalence and future formation of module cracks, as result of both single high-load snow events and repeated events that result in punctuated as well persistent loading.

This measurement station remains operational at Sandia's RTC in Calumet year-round and is configured for rapid sampling during the warmer months to measure the effect of wind on module loading.

Future measurement systems are being designed to measure additional effects of snow loading on PV such as shear forces, DC power losses, PV tilt angle, and PV module size. These systems will be installed prior to the 2022-2023 winter.

ACKNOWLEDGMENTS

Sandia National Laboratories is a multimission laboratory managed and operated by National Technology & Engineering Solutions of Sandia, LLC, a wholly owned subsidiary of Honeywell International Inc., for the U.S. Department of Energy's National Nuclear Security Administration under contract DE-NA0003525. This material is based upon work supported by the U.S. Department of Energy's Office of Energy Efficiency and Renewable Energy (EERE) under the Solar Energy Technologies Office Award Number 34367.

REFERENCES

[1] R.E. Pawluk, Y. Chen, Y.She, "Photovoltaic electricity generation loss due to snow—a literature review on influence factors, estimation and mitigation,",2019, Renewable and Sustaiunable Energy Reviews 107:171-182.

[2] L. Burnham, D. Riley, B. King, J. Braid, P. Dice, A. Dyreson, W. Snyder, C. Pike. Dedicated cold-climate field laboratory for photovoltaic system and component studies: the Michigan Regional Test Center as a case study, 49th Photovoltaic Specialists Conference, 2022, Philadelphia, PA. [Accepted; *in prep.]*

[3] J. Bogenrieder, C. Camu, M.Huttner, P. Offermann, J. Hauch, C. Brabec, 2018. "Technology-dependent analysis of the snow melting and sliding behavior on photovoltaic modules," Journal of Renewable and Sustainable Energy 10, 021005.

Probabilistic Assessment of Narrowband vs Broadband Solar Irradiance Temporal Variability in Ottawa

Nick Anderson, Viktar Tatsiankou, Karin Hinzer, Richard Beal, Henry Schriemer

University of Ottawa, Ottawa, ON, Canada

Spectrafy, Ottawa, ON, Canada

Using a recently-created database for Ottawa, Canada, a 9-month longitudinal study of the solar irradiance, measured with a custom spectral pyranometer every 250ms, was conducted with a specific focus on comparing the narrowband and broadband response to the temporal variability. Deterministic diurnal and orbital dependencies of the spectral irradiance were removed by clear sky normalization, and the resulting clear sky index was forward differenced across time steps ranging from the sub-second to ~30 minutes. The stochastic behavior of this spectral clear sky index increment was assessed by kernel density estimates of the probability distributions for data from each of the nine narrow wavelength channels of the spectral pyranometer and the derived broadband global horizontal irradiance (GHI). Scaling analyses of their peak densities and their full widths at half maximum (FWHM) with increment time step revealed power law scaling with consistent stationary breaks at very-short and short times. Broadband scaling was consistent with some narrowband dependencies, but not with others, which may reflect the wavelength dependence of different sky conditions. The existence of three distinct scaling regimes, each of which coincides with an operationally significant time period (primary/secondary control, automatic generation control, and real time market dispatch), has implications for short term probabilistic forecasting.

Evaluation of Solar Capacity Factor of ~2000 Solar Plants Across the United States Using Multilayer Perceptron Regressor Models

Samantha S. Wilson, Stephen Lightfoote, and Stephen Voss

Power Factors Inc., Brossard, QC, J4Z 1A7, Canada

Abstract— **The measured and budgeted solar capacity factor for ~2000 fixed tilt and HSAT sites were collected from the Power Factors database. Both budgeted and measured capacity factors were fit to multilayer perceptron regressor models using the age of the plant and 192 monthly climate features. The model for the measured data shows several regions in the United States where capacity factor is predicted to be higher, particularly the Southwest. Comparison between the models for measured and budgeted data shows budgets for the eastern United States may be more inaccurate than budgets for the Southwestern United States.**

Keywords—capacity factor, United States, neural networks

I. INTRODUCTION

Capacity factor is the ratio of a solar plant's theoretical maximum energy output to its actual energy production over the course of some measure of time [1, 2]. Typical measures of time are either one year or the plant's lifetime. For solar plants, the maximum output is fixed by the AC capacity of the inverters. Solar capacity factor is also fundamentally limited by the availability of solar power. Since plants cannot produce power at night it is generally impossible for them to have capacity factors exceeding 35%. This also means capacity factor is limited by anything that impacts a plant's solar resource including location and plant design. For example, fixed tilt plants have fundamentally lower capacity factors than horizontal single-axis tracking plants since modules in fixed tilt plants experience lower incident solar radiation in the morning and afternoon. Previous surveys of the capacity factor of existing utility plants have found median capacity factors are higher in the southwest due to increased availability of direct normal irradiance in the Southwestern United States [3].

Before a plant is built its capacity factor is assessed through pre-build analysis based on projected understandings of local weather conditions and plant design. However, pre-build modeling is based on our understanding of basic climate data and the physics of specific plant design. A detailed understanding of the local micro-climate is difficult to capture in modeling. For example, soiling is usually assessed as a flat monthly loss with little consideration of local ground conditions or rain patterns that may impact soiling accumulation.

Due to the limitations of physical models for assessing microclimate, we have attempted to generate a model for capacity factor entirely based on measured capacity factor for existing sites and climatology data from the NASA POWER API which is sourced from a collection of open data initiatives including MERRA.[4] We have limited our analysis to the continental United States and have attempted to generate a performance prediction algorithm using machine learning which only takes widely available satellite data and limited information on plant construction.

II. METHODS

Energy data were collected for each qualifying plant in the Power Factors database at data resolutions of 5-15 minutes using measurements taken from the period of January 2018 – March 2022. Capacity factor was characterized as a simple ratio between the actual energy produced in the interval examined and the maximum possible production over the same time period. The maximum possible production was set simply as the AC capacity times the number of hours in the interval of every site based on inverter information provided to us by clients. Additional post-processing filters were applied to the aggregated statistics to control for plants with data quality issues (i.e., non-physical capacity factor) or insufficient samples. Plants were required to have at least one year of available energy data.

Multilayer perceptron regressor models were used to fit the data using the open-source python package sci-kit learn [5]. Models were trained with 3 layers of 193, 17, and 193 nodes.

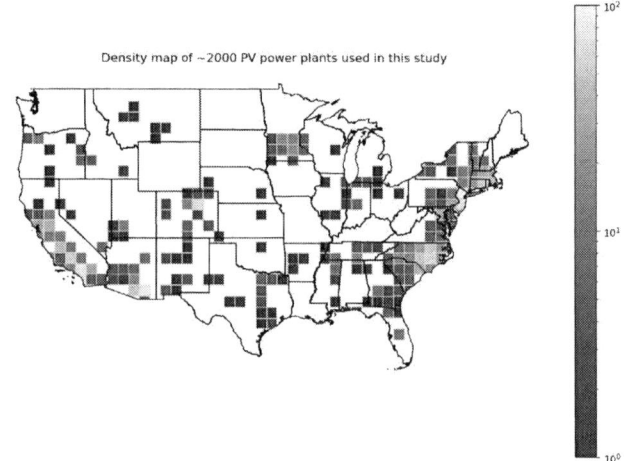

Fig. 1. Geographic density of solar plants used in this analysis. Most plants are in the Southwestern United States as well as on the Eastern Seaboard.

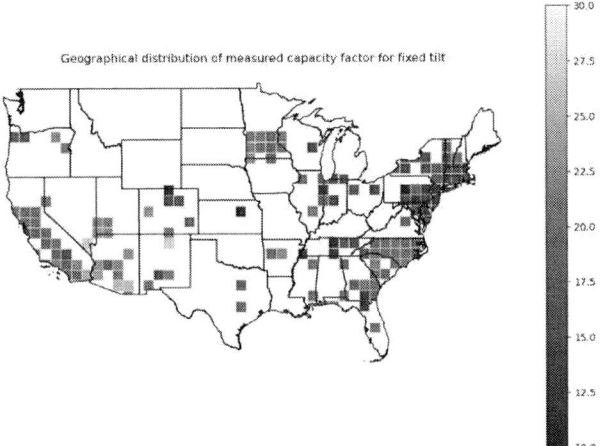

Fig. 2. Measured capacity factor for 1600 fixed tilt plants. The median capacity factor for fixed tilt plants is 17.2% and the median plant age in the data set is nearly 6 years old. Local averages are shown.

Goodness of model fit was assessed by calculating the root means square error of a reserved set of test data. Variables fit were the age of the plants as well as monthly climatology values for horizontal insolation, temperature at 2 meters, dew point temperature at 2 meters, wet bulb temperature at 2 meters, minimum temperature at 2 meters, wind speed, temperature range at 2 meters, horizontal clear sky insolation, insolation on a horizontal surface at midday, total precipitation, relative humidity at 2 meters, surface albedo, clearness index, direct normal insolation, and diffuse insolation. Resolution of the database is 0.5 x 0.5 degrees longitude and latitude.

III. RESULTS AND DISCUSSION

Approximately 2000 plants were used for this study. A map of the location density is seen in Fig 1. Most plants used in this study are in the Southwestern United States, the Eastern Seaboard, Colorado, and Minnesota. This represents a broad

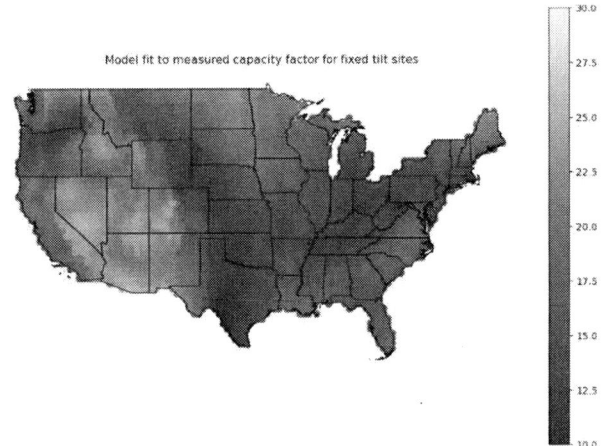

Fig. 3. The measured capacity factor for fixed tilt plants was then fit to a type of neural network called a multilayer perceptron regressor model.

distribution of climates, although the interior is more sparsely populated than the coasts.

Climatology data for this study was retrieved from the NASA POWER API [4]. This is an open-source collection of high-quality climate data based on measurements taken between 1990 and 2019. Annual insolation for this dataset follows expected trends and is higher in the Southwestern United States.

For this analysis we separated plants into two main categories, fixed-tilt and horizontal single-axis trackers. The geographic distribution of measured capacity factor for fixed tilt plants is shown below. Fixed tilt plants had a median capacity factor of 17.2% and the median age of the fixed tilt plants was nearly 6 years old.

The measured results were matched with 192 monthly climate parameters from the NASA POWER API database as well as the age of the plant in years. results are shown below in Fig 3. The model fit was good with RMSE = 0.59 for the reserved test data set. The plot indicates the southwest is indeed

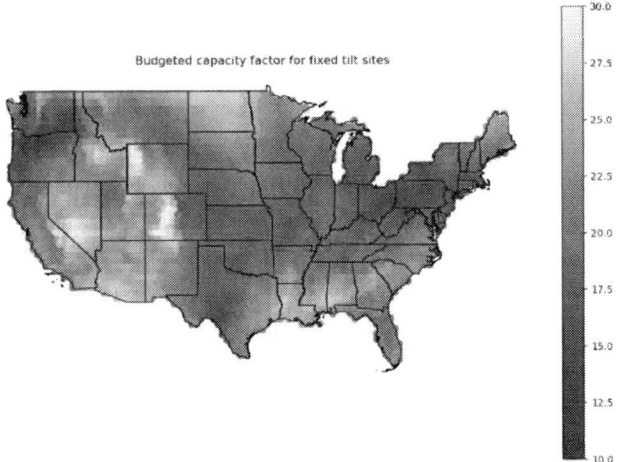

Fig. 4. The budgeted capacity factors for the plants above were also fit to a multilayer perceptron regressor model. Fit quality was better than for the measured data with RMSE = 0.72.

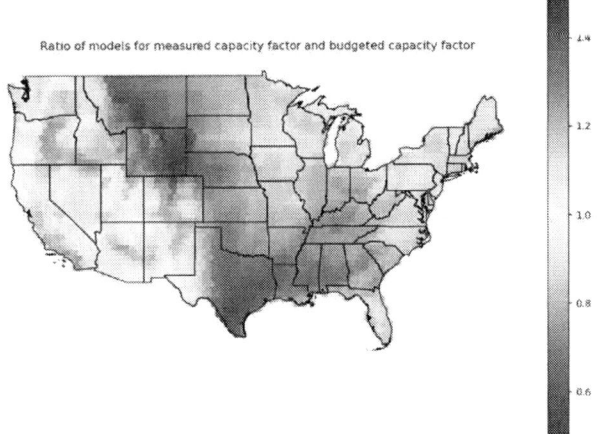

Fig. 5. The ratio of the models for budgeted and measured capacity factor. In the plot above 1 (represented by the white area) is where the budget model and measured model are equal. Most plants underperformed versus their budget and thus most regions are red.

978-1-7281-6118-1/22 $31.00 © 2022 IEEE

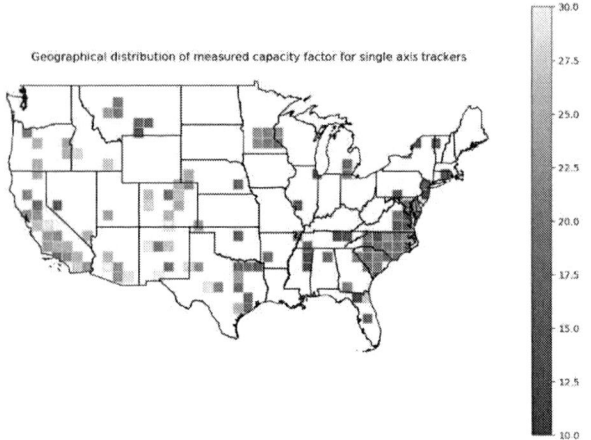

Fig. 6.   Measured capacity factors for 400 horizontal single axis tracking plants. Local averages are shown.

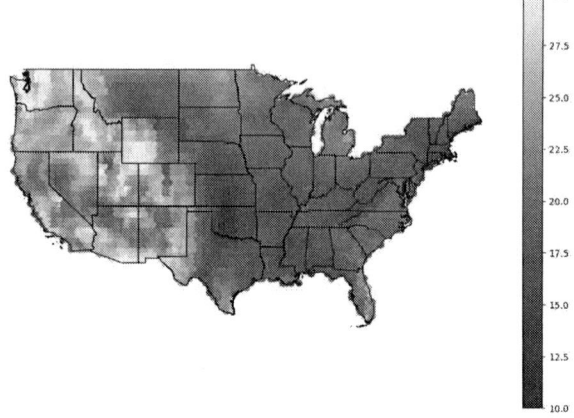

Fig. 7.   Modeled capacity factor for trackers. As expected, capacity factors are predicted to be much higher in the Southwestern United States.

seeing significantly higher capacity factors. However, there are some interesting details that emerge. The California central valley has a significantly lower expected capacity figure than Arizona, due mostly to increased soiling and slightly lower insolation. The California central valley is a heavily agricultural area that sees higher soiling than most parts of the United States. [6]

There is also a large area of low predicted capacity factor in the interior of the country running from Texas along the Rocky Mountains and through Montana. This could be due to model error given that there is very low coverage of plants in that region. Additionally, the few plants that are present in the region are poorly performing and that could be biasing the dataset. We will continue to monitor that area as we collect more data. The south Atlantic coast also is expected to be a very favorable area for solar as well as indicated. There is also a favorable region near Minnesota where there is concentration of high capacity factor plants.

We compared these results with the expected capacity factors from budget information provided by clients. The budgets were taken from the plants featured in Fig. 2 and fit to the same model. Fit for the budgeted data was even better, with a RMSE = 0.72 for the reserved test data.

Once we had model fits for both the measured and budgeted data we compared the ratio of the two models. The results are shown in Fig 5. In this plot a ratio of 1 is colored white, and it is where the budget model and measured model are equal. However, most plants underperformed versus their budget and thus most regions are red. The median ratio of budget to actual capacity factor is 0.88. For red areas in the interior of the country, the modeled budget capacity factor greatly exceeds the modeled measured capacity factor. However, the density of actual measured data is low in the reddest regions so those may be outliers.

The Eastern United States, and in particular the Southeast seems to perform the worst based on these metrics. Greater care may be needed when estimating budgets in this region.

We also looked briefly at horizontal single trackers, but due to the lower number of plants the model fitting was more inconclusive. The RMSE for the horizontal tracker dataset was only 0.22 for this dataset.

IV. CONCLUSION

The measured and budgeted solar capacity factor for ~2000 fixed tilt and HSAT sites were collected from the Power Factors database. Both budgeted and measured capacity factors were fit to multilayer perceptron regressor models using the age of the plant and 192 monthly climate features. The model for the measured data shows several regions in the United States where capacity factor is predicted to be higher, particularly the Southwest. Comparison between the models for measured and budgeted data shows budgets for the eastern United States may be more inaccurate than budgets for the Southwest.

REFERENCES

[1]   "Solar Energy and Capacity Value," Golden, CO: National Renewable Energy Laboratory. Accessed April 2022: https://www.nrel.gov/docs/fy13osti/57582.pdf

[2]   S. H. Madaeni, R. Sioshansi, and P. Denholm. (2012). "Comparison of Capacity Value Methods for Photovoltaics in the Western United States." Golden, CO: National Renewable Energy Laboratory. Accessed April 2022: www.nrel.gov/docs/fy12osti/54704.pdf.

[3]   "Southwestern states have better solar resources and higher solar PV capacity factors." US Energy Information Asscociation. Accessed April 2022: https://www.eia.gov/todayinenergy/detail.php?id=39832

[4]   "NASA POWER Docs" NASA. Accessed April 2022: https://power.larc.nasa.gov/docs/services/api/

[5]   "Scikit Learn" Accessed April 2022: https://scikit-learn.org/stable/modules/generated/sklearn.neural_network.MLPRegressor.html

[6]   "Photovoltaic Module Soiling Map" Golden, CO: NREL. Accessed April 2022: https://www.nrel.gov/pv/soiling.html

Radiant/non-Radiant Lifetime Switching in Chlorophyll and Application to Energy Storing Photovoltaic Cells

Charles M. Fortmann[1] Julie B. Liu[2], Nahian Rahman[1], Aaron Song[3], Elizabeth Nazginov[4], Mia Pancari[4], Amina Exilhomme[4], and Charles M. Fortmann[2]

[1]St. John's University, Physics Dept., Queens NY 11439 U.S.A., [2]St. John's University, Chemistry Dept., Queens NY 11439 U.S.A, [3]Ossining High School, Ossining, NY, [4]St. John's University, Biology Dept., Queens NyY11439 U.S.A

Abstract—The absorption, light scattering, and anti-Stoke's fluorescence of chlorophyll and closely related sodium copper chlorophyllin were examined in various solutions with a wide range of refractive index and permittivity. No evidence of anti-Stoke's fluorescence was found in any of the solvents used. On the other hand, it was evident that radiative photo electron decay corresponded to low permittivity solvents, but not correlated to solvent refractive index. Conversely, non-radiative decay was dominant in high permittivity solvents. The properties suggest evolution has provided plants with an ability to redirect light to chlorophyll in the high permittivity (water) photosynthesis center. Furthermore, exploratory solar cells exploiting lifetime switching were developed and tested. These cells have the unique property of charge storage owing to an insulating transport layer.

Keywords— chlorophyll, permittivity gradient solar cells, battery

I. INTRODUCTION

Previously, one of us (Fortmann) described a molecular-based solar cell wherein photo-excited molecules drift in a permittivity (dielectric) gradient due to a photo-induced increase in polarity [1].

The permittivity or dielectric gradient solar requires a mobile molecule, which although remaining neutral, is excited charge carrier. Charge collection may occur by two different mechanisms. One, a molecule becomes polar (or more polar) upon photo excitation. This now polar molecule drifts towards higher polarity media where charge collection occurs via transfer to a highly polar contact with and appropriate conduction band and work function energies. Second, a molecule may be engineered to have a photon capture cross section that increases when in or near polar media thereby only absorbing photon energy when in or near an appropriately polar contact or contact region.

In this work we explore the possibility of basing solar cells on *chlorophyll a* (ChAa) and closely related synthetic Sodium Copper Chlorophyllin, SCC. Its noteworthy that ChAa is hydrophobic while SCC is hydrophilic. Before embarking on a solar cell design, the photo absorption and scattering was measured.

Almost every aspect of chlorophyll in plant physiology is well known and documented for example see [2]. Nonetheless, the universally accepted ChA photo excitation model involving antenna like oscillations has no semiconductor analog [3]. Here the ChA and SCC absorption and scattering were studied as a function of solvent refractive index and permittivity.

II. EXPERIMENT

A. Molecular Aborption and Scattering

Commercial purified ChlA (from spinach) was obtained from Sigma Aldrich Company and measured immediately upon delivery. SCC powder was obtained from various commercial sources as it's a commonly used nutritional supplement.

ChA and SCC absorption and light scattering were measured in various solvents having reported refractive indices and permittivity (e.g., see [4],[5]) having a wide range of solvent refractive index and permittivity (polarity). The ChA absorption systematically decreased with decreasing permittivity as seen in Fig.1. SCC exhibited similar behavior.

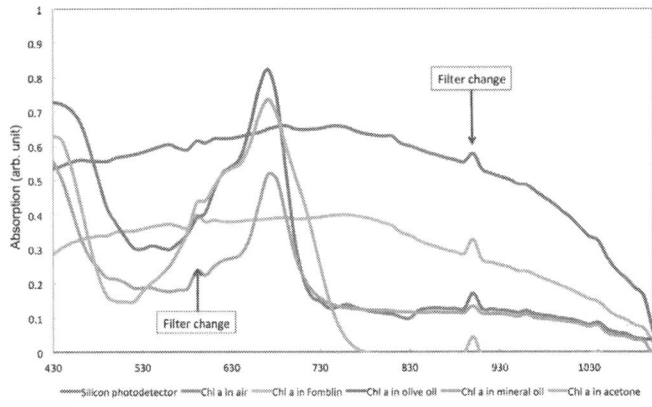

Figure 1. ChlA absorption as a function of wavelength and solvent permittivity as indicated. Note the lack of absorption in the lowest permittivity cases.

978-1-7281-6118-1/22 $31.00 © 2022 IEEE

SCC absorption also decreased with decreasing solvent permittivity. Fig. 2 shows the scattered light in SCC as a function of solvent permittivity.

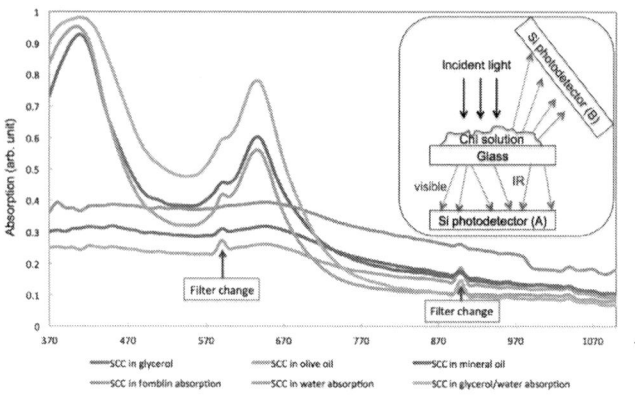

Figure 2. SCC absorption as a function of wavelength and solvent polarity as indicated (silicon photodetector is at position A). In the lowest polarity cases (perfluoropolyether, olive, and mineral oils), absorption is completely absent. Inset is the photo-detector configuration used for absorption and scattered light experiments

Light scattering broadly increased across the entire spectrum measured with decreasing solvent permittivity. This effect is readily visible by illuminating SCC solutions with various visible light laser diodes.

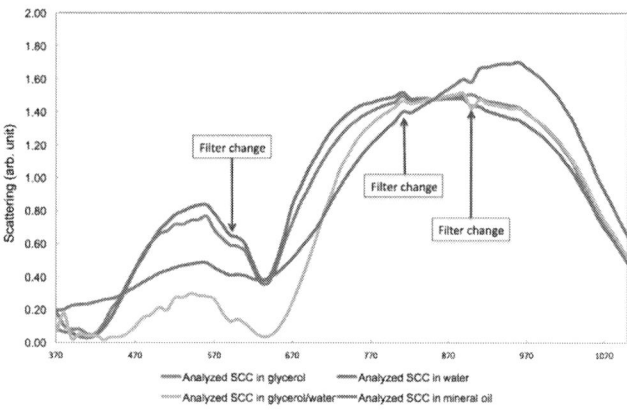

Figure 3. SCC scattered light as a function of wavelength and solvent polarity as indicated (silicon photodetector is at position B). Broadening is evident in the lower solvent polarity case (mineral oil)

The width (Full width at half maximum, FWHM) of ChA and SCC were also measured as a function of solvent permittivity and shown in Tables 1 and Table II. Importantly, there is no systematic change in absorption with changing refractive index.

These observed optical properties suggest that a radiant photo-excited electron lifetime dominates at low solvent permittivity ($\varepsilon < \sim 2\varepsilon_0$) and a non-radiant lifetime involving the ambient dominates at larger permittivity. This property may have evolved to improve photosynthesis efficiency.

Table 1: Chlorophyll *a* 640 nm absorption peak width vs. permittivity

Solvent	Permittivity	Refractive index	FWHM (cm^{-1})
Methanol[1]	32.66	1.329	520
Ethanol[1]	24.30	1.361	488
Pyridine[1]	13.06	1.509	428
Tetrahydrofuran[1]	7.39	1.407	408
Acetone	20.7	1.359	2230
Olive oil	3	1.467-1.471	1315
Mineral oil	2.1-2.3	1.457-1.487	771
Perfluoropolyether	2.1	1.295-1.304	no absorption
Air	1	1	no absorption

Table 2 Sodium-copper-chlorophyllin absorption peak widths as a function of solvent polarity

Solvent	Dielectric constant	Refractive index	FWHM (cm^{-1})
Water	80.4	1.33	1462
Water/glycerol	65.6	1.398	1640
Glycerol	42	1.473	1291
Olive oil	3	1.467-1.471	No absorption
Mineral oil	2.1-2.3	1.457-1.487	No absorption
Perfluoropolyether	2.1	1.295-1.304	No absorption

III. DISCUSSION

In plants membrane proteins, have a relatively small local permittivity, $\sim 4\varepsilon_0$[6] ChA would therefore more likely scatter light in this unfavorable environment for photosynthesis. Chamorovsky et al. [7] noted that the permittivity varied over the photosynthetic reaction center. While the effective permittivity at the lipid–water interface was found to be relatively high ($\sim 20\ to\ 30\varepsilon_0$)[8] and that of water, $\varepsilon \approx 84\ \varepsilon_0$, is greater. On the other hand, chlorophyll not in favorable positions for photosynthesis such as within a membrane absorbs less photon energy and scatter at higher probability. Thereby poorly positioned chlorophyll would pass photons onto more favorably positioned chlorophyll while also reduce the thermal load by reflection (emitted scatting light) across a broad spectrum.

A solar cell design was developed to exploit the lifetime switching found in ChA and SCC. Fig. 4 shows a solar cell design wherein the bottom contact is comprised of a high permittivity, thin, water film deposited on a titanium oxide (TiO₂) coating glass substrate. The water layer is thin to ensure that only ChA (or SCC) positioned near to the TiO₂ contact have high photo absorption probability owing to the high permittivity environment. Thus, the electron transfer to the TiO₂ layer is increased (as further detailed in Section IV below).

SCC is likely to accumulate in the near TiO₂ water layer because it is hydrophilic. This is readily observable. Although, ChA is hydrophobic some molecules diffuse into the near TiO₂ water film as well. A film of insulating oil is deposited onto water films, this film acts as a ChA reservoir and prevents front contacts from shorting to the back contact.

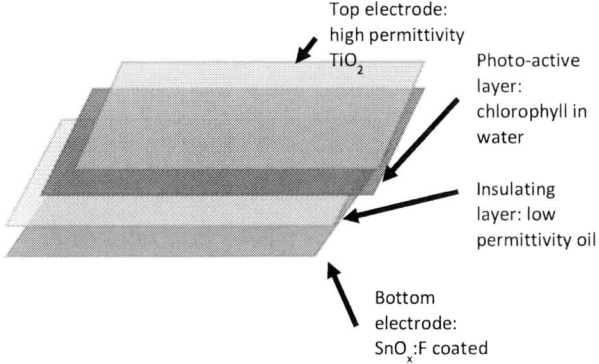

Figure 4. A solar cell designed employs a reservoir of ChA (or SCC) dispersed in layers of oil and water. Only the water solvated, near TiO₂ interface molecules non-radiatively absorb photons and thereby transfer electrons to the TiO₂ interface.

Completing the circuit is the most challenging component of the design. Once an electron is transferred at the front contact the now ionized molecule needs transport across the insulating layer via diffusion. When layers are appropriately thin the diffusion is current is sufficient.

ChA and SCC solar cells prepared where prepared on TiO₂ coated tin oxide (SnOₓ:F) coated glass substrates. The counter electrode was SnOₓ:F without TiO₂. Non-conductive 7 μm particles where used to provide spacing between the electrodes to reduce the possibility of direct contact shorting.

High purity ChA was sourced from Sigma-Aldrich and UHP SSC was obtained commercially. The ChA and SSC were maintained below $0^0 C$ until use.

Exploratory solar cells of this configuration obtained ~ 240 mV V_{oc}'s under ~ 100mW/cm² illumination at room temperature. Surprisingly, these solar retain V_{oc} in the dark for several minutes after the illumination has ended. Evidently there is no dark current mechanism capable of discharging the cell as this would require the back contact to inject an electron

to a neutral SCC or (ChA) molecule. In the case of SCC this is more unlikely as it aggregates in the water film due to its hydrophilicity.

IV. THEORETICAL IMPLICATIONS

A more detailed characterization of photovoltaic performance at this time is made difficult by charge related hysteresis and the lack of optimized structures.

Nonetheless, progress requires the proposal and testing of concepts describing the underlying photovoltaic mechanisms. The mechanisms of charge generation and charge separation withing mobile molecules (especially natural biological molecules) moving within solvents of varying polarity (permittivity) is complicated and unfamiliar territory to the solid-state photovoltaic perspective.

For example, the absorption process of ChA is most descried by a resonant phenomenon wherein photon energy is absorbed by setting mobile electrons oscillating. While this description explains many biological processes it fails to explain the lifetime switching process observed here.

Since oscillation and antenna-like excitation (absorption) are predicated upon an aperture or photon capture cross section (A_{eff}) that is predicated upon wavelength as described by Eq.1.

$$A_{eff} = \frac{\lambda^2}{4\pi} \xrightarrow[\lambda=c/n\upsilon]{} \frac{c^2}{4\pi n^2 \upsilon^2} \qquad (1)$$

Where c is the speed of light in vacuum, λ is the wavelength of light, υ is the light frequency (invariant with light energy) and n is the refractive index. Therefore, antenna theory predicts decreasing absorption with increasing refractive index. No systematic relation between solvent refractive index and absorption (or light scattering) were observed. Since the molecules studied are small relative to the wavelengths of light examined and since the wavelength of light corresponds to that of the solvent it is necessary to conclude that the present observations are not antenna related phenomena.

There are several solvent dependent mechanisms that may explain the observed lifetimes. Fig. 5 illustrates photo-induced charge redistribution within a ChA (or SCC) molecule. The radiant lifetime sensitivity to solvent permittivity suggests that either an electron cloud (or orbit) of a photo excited electron and/or the subsequent migration of photo-excited electrons relative to the co-generated positive atom are stabilized by a polar environment. Since, in both cases the polarization of the environment would lower the energy of the photo-excited state.

Without delving further into the underlying photo chemistry of ChA and SSC absorption we can elucidate design principles that exploit the absorbed lifetime switching for photovoltaic energy production and storage.

978-1-7281-6118-1/22 $31.00 © 2022 IEEE

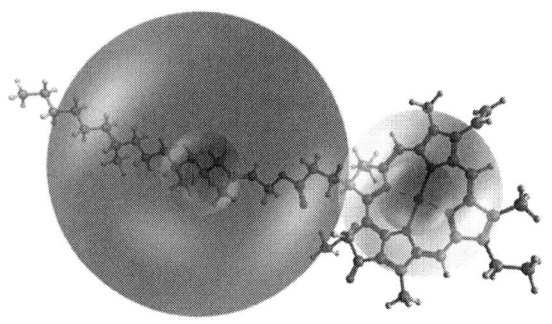

Figure 5. A Cha molecule showing photo generated charge separation within the molecule and a polarized solvent shell about each charged region. He solvent shell both reduces the energy and stabilizes the excited state(s).

Figure 6 illustrates an ideal photovoltaic system with two different permittivity solvents.

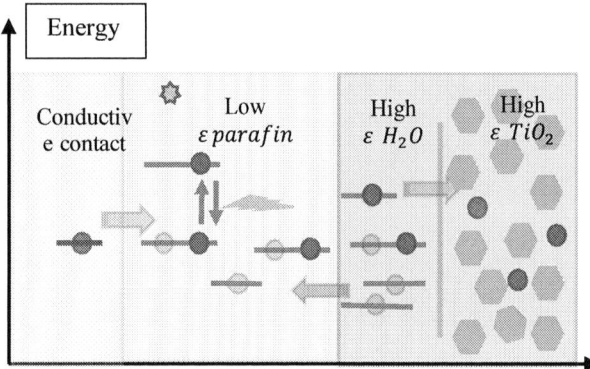

Figure 6. Illustrated is the design of ChA or SSC solar cell wherein a low permittivity solvent (e.g., paraffin) provides electrical isolation between positive (left) and negative contacts. Photo-excited electrons in this region are likely to recombine radiatively. Absorption within the high permittivity solvent is stabilized by solvent and by the high permittivity TiO₂ contact polarization. Excited electron transfer to the contact can be aided by tailoring the various permittivity of the solvent relative to that of the contact. Flow of discharged molecules is by diffusion.

Photoexcitation of the molecules within the low permittivity solvent do not necessarily result in energy loss as photo emission (scattering) is the likely outcome. One the other hand photo-absorption within the high permittivity solvent will likely result in non-radiant recombination which in a well-designed structure result in electron transfer to the TiO₂ contact.

Transfer of the photo-generated electron to the TiO₂ contact is increased through ensuring the water layer (film) is thin $\lesssim 20$ Å. Also, it is important to recognize that my mixing

lower permittivity solvents (e.g., methanol) in water the permittivity can be reduced with respect to TiO₂.

One of the more challenging aspects of the design is the need for discharged molecules to diffuse the conductive contact. Two design elements can promote the diffusion. First, the low permittivity (and insulating) layer thickness must be thin. Second, the conducting contact provides near proximity attractive force through the image charge. Both these measures require smooth flat surface contact and TiO₂ layers as well as much thinner liquid films than used thus far.

Solar cells were measured under a xenon AM 1.5 white light solar simulator and under a 780 nm red light provided by a commercial 5 mW laser diode. Since ChA absorbs only narrow bands of the white spectrum as seen in Fig. 1 low efficiency under white (AM 1.5) light made quantum efficiency estimation difficult and overall efficiency less than 1%.

Under 780 nm illumination corresponding to the lower of the two prominent energy absorption bands of ChA and SCC the measured current was $3 \times 10^{-6} A$ which was approximately 0.3% of that produced by a standard silicon reference cell a standard refence cell has near 100% internal quantum efficiency at this wavelength.

Under Xenon solar simulated AM 1.5 illumination the current increased by an order of magnitude but was of course much less than 1% of that of a standard silicon cell. We look forward to our next set of testing under $\sim 425\ nm$ illumination corresponding to the higher energy ChA and SCC absorption peaks.

An important caveat is that the photo-excited ChA or SSC may non-radiatively absorb photons by water dissociation. Afterall, chlorophyll-based water dissociation is the basis for all plant life. The probability of water dissociation can be decreased through limiting water layer thickness. Whereby all ChA (or SSC) are positioned within close proximity to the TiO₂ contact. Interestingly, water dissociation components (H^+, OH^- are charged and therefore might act as intermediaries, or alternative charge collection and transport pathways.

V. CONCLUSIONS

The optical properties of SCC and ChA were measured and a solvent dependent switching between radiative and non-radiative lifetime found. In low permittivity solvents the non-radiative dominates. In-turn this property was exploited to develop charge storing photovoltaic devices. Significantly, high voltages and efficiencies are expected from improved materials and more optimized designs. While many questions remain, and many avenues are yet to be explored the design principles provide many opportunities for optimization. Further device performance and designs details will be discussed separately.

ACKNOWLEDGMENT

We thank the Brookhaven National Laboratory and the DOE of use of the Center Functional Nanomaterials, Julie B. Liu Thanks the St. John's University S-STEM Scholars, and The Clare Boothe Luce Foundation for funding.

REFERENCES

[1] C.M. Fortmann, Patent, US 8,501,332 B2 2013

[2] R. S. Ochs, Biochemistry, Jones&Bartlett Learning, Burlington, MA, 2pp. 197-198, 2014

[3] A. Freer, et al., Pigment-pigment interactions and energy transfer in the antenna complex of the photosynthetic becterium Rhodopseudomonas acidophila, Structure 4 (4), 449 (1996).

[4] R. Vladkove, Photochemistry and Photobiology 71 (1), 71 (2000).

[5] I. Renge and K. Mauring, Spectrochimica acta. Part A, Molecular and biomolecular spectroscopy 102, 301 (2013).

[6] K. Sharp and B. Honig, Annual review of biophysics and biophysical chemistry 19, 301 (1990).

[7] S. K. Chamorovsky, C.S. Chamorovsky, and A. Yu. Semenov, Biochemistry 70 (2), 257 (2005).

[8] J. L. Tichadou Lakhdar-Ghazal F., and J. F. Tocanne, European Journal of Biochem 134, 531 (1983).

Measuring Carrier Concentration on the Back Side of Thin Film Solar Cells

Nathan Rosenblatt[1,2], Alex Polizzotti[1], Sachit Grover[1], Xiaoping Li[1], Wyatt K. Metzger[1]

[1]First Solar Inc. California Technology Center, Santa Clara, CA 95050, USA

[2]School of Electrical, Computer, and Energy Engineering, Arizona State University Tempe, AZ 85281, USA

Abstract— **The ability to profile and understand carrier concentration throughout thin film absorbers is important to advance solar cell technology. Here, the historical impediments of grain boundary potentials, lateral resistance, and high work functions to such measurements on polycrystalline solar cells are overcome by applying electrochemical capacitance-voltage profiling across the solar cell p-n junction and an electrolyte/semiconductor junction made at the back. Despite the presence of two junctions, modeling indicates that accurate carrier concentrations at the rear of the device can be measured under certain conditions. This is validated by experiments on CdTe solar cells.**

Keywords—solar cells, semiconductor physics

I. INTRODUCTION

Standard capacitance-voltage (C-V) measurements are widely used to probe carrier concentrations in semiconductor devices, including solar cells [1]. These measurements probe the carrier concentration at the depletion width edge of the primary p-n junction. The depletion width of many absorbers limits the measurement to the region near this junction. Yet, for many solar cell technologies, the carrier concentration near the back of the device is important to create doping gradients to enhance collection, back surface fields to reduce recombination, and appropriate band alignment for back contact carrier transport. Consequently, being able to measure the carrier concentration in the back region is also critical.

For example, CdTe solar technology provided nearly 30% of the market in the past decade. Efficiency progressed significantly by making a compositionally graded absorber of CdSeTe to CdTe from the front to back. CdSeTe tends to shift from p-type to n-type doping with increasing Se composition, so carrier concentration can naturally shift across the absorber. Advances in group V doping have pushed carrier concentrations higher in CdTe solar cells while avoiding device degradation associated with Cu incorporation. CdTe has a high work function, so positioning the Fermi level at the back is important with these different dopants to avoid hole barriers and engineer recombination. Yet, there have not been well established methods to measure carrier concentration near the back contact for CdTe and other thin film solar cell technologies.

Here, we implement a distinct approach to electrochemical capacitance-voltage (ECV) measurements using CdTe solar cells as an example. The standard approach is to probe the capacitance across a single junction formed by electrolyte against the back surface of the film, with one contact in the electrolyte and the second on the back surface of the film. However, this configuration has provided unreliable results due to the intrinsic absorber resistance, impediments to transport across many grain boundaries, and the high work function of the back surface. Here, the front contact is instead made to the transparent conducting oxide, which is straightforward, and the electrolyte serves as a tunable back contact introduced without subjecting the device to additional processing or deposition. This distinct configuration poses the challenge of extracting out carrier concentration from a capacitance-voltage measurement across a) the primary solar cell p-n junction and b) the CdTe/electrolyte junction. Modeling indicates the conditions required to determine carrier concentration in this configuration, and experiments validate the results.

II. MODELING

To interpret ECV results in a structure with two junctions, it is necessary to understand and model the current and voltage distributions of the back semiconductor-electrolyte junction and front solar cell p-n junction.

The semiconductor-electrolyte junction is similar in nature to a Schottky junction [2]. The current generally can be well described by the thermionic emission (TE) model, where reverse current mechanisms include thermal generation in the space charge region and barrier lowering. The current characteristic can be written [3]:

$$J_{Sch} = J_{0,Sch}\left(\exp\left(\frac{qV}{kT}\right)-1\right) - \frac{qn_i\sqrt{\frac{2\varepsilon}{qN_A}(V_{bi}-V)}}{\tau_p+\tau_n} \quad (1)$$

where $J_{0,Sch}$ is:

$$J_{0,Sch} = A^*T^2\exp\left(\frac{q\phi_{B0}}{kT}\right)\exp\left(\frac{q\left(\frac{q^3N_A(V_{bi}-V)}{8\pi^2\varepsilon^3}\right)^{\frac{1}{4}}}{kT}\right). \quad (2)$$

Here V_{bi} defines the potential variation across the space charge region at equilibrium, and the equilibrium barrier height ϕ_{B0} is defined as:

$$\phi_{B0}=V_{bi}+E_F-E_V. \quad (3)$$

The reverse conductance is:

$$G_{Sch} = \frac{qn_i\sqrt{\frac{2\varepsilon}{qN_A}}}{2(\tau_p+\tau_n)\sqrt{(V_{bi}-V)}} + \frac{d}{dV}\left(J_{0,Sch}\right)\left(\exp\left(\frac{qV}{kT}\right)-1\right) + \frac{q}{kT}J_{0,Sch}\exp\left(\frac{qV}{kT}\right). \tag{4}$$

This treatment specifically assumes that charge transfer is not rate limited and any non-equilibrium potential variation drops entirely across the semiconductor, rather than the Helmholtz layer. In short, it assumes the junction behaves like a traditional Schottky barrier dominated by TE transport. It is intended as an upper bound on reverse conductance and leakage current for carefully prepared semiconductor-electrolyte surfaces. Experimental results in Si-electrolyte junctions demonstrate voltage ranges where barrier lowering accurately describes current-voltage characteristics [4].

The front semiconductor-semiconductor junction, consisting of TCO/buffer/CdTe, has a forward current characteristic which can be approximated by [2]:

$$J_{pn} = q\left[\frac{n_i^2 D_n}{N_A L_n}\left(\exp\left(\frac{qV}{kT}\right)-1\right) + \frac{qW_D n_i}{2\tau_n}\left(\exp\left(\frac{qV}{2kT}\right)-1\right)\right] \tag{5}$$

or:

$$J_{pn} = J_{01}\left(\exp\left(\frac{qV}{kT}\right)-1\right) + J_{02}\left(\exp\left(\frac{qV}{2kT}\right)-1\right). \tag{6}$$

Only electron diffusion current must be considered due to the relatively higher doping of the TCO and large valence band offset. The conductance is found by:

$$G_{pn} = \frac{q}{kT}J_{01}\exp\left(\frac{qV}{kT}\right) + \frac{q}{2kT}J_{02}\exp\left(\frac{qV}{2kT}\right). \tag{7}$$

The appropriate model consists of two series diode, each with a voltage-dependent capacitance, voltage-dependent resistance, and constant shunt resistance in parallel. Similar models have been used to explore admittance spectroscopy results in CIGS solar cells and justify the extraction of the back contact barrier height via thermal admittance spectroscopy [7]. Here, this model is implemented to solve for the total current, the voltage across each diode, their associated capacitance and conductance, and the total C-V response of the circuit as a function of the back contact barrier height, CdTe absorber doping, and measurement frequency. Appropriate values for shunt resistance, series resistance, J_{01}, and J_{02} in high quality CdTe devices are used. The capacitances of each are calculated from the single sided junction depletion region approximation [5]:

$$C_{dep} = \left[\frac{qN_A\varepsilon\varepsilon_0}{2(V_{bi}-V)}\right]^{1/2}. \tag{8}$$

Equations (4) and (7) are solved to determine the total current-voltage characteristic and voltage drop across each diode as a function of external applied voltage. A barrier of 0.7eV, results in the current-voltage characteristics given in Fig. 1. These characteristics display well-known rollover phenomenon as the back junction becomes more resistive. The voltage across each diode as a function of total external voltage is given in Fig. 2. Beyond a certain threshold, the back contact is totally limiting, and any additional voltage drops across it. It

is this regime that allows for the measurement of the carrier concentration at the back of the device.

The capacitance of this circuit is given by [5]:

$$C_{tot} = \frac{C_{Sch}G_{pn}^2 + C_{pn}G_{Sch}^2 + \omega^2 C_{Sch}C_{pn}(C_{Sch}+C_{pn})}{\left(G_{pn}+G_{Sch}\right)^2 + \omega^2\left(C_{pn}+C_{Sch}\right)^2}. \tag{9}$$

The frequency ω is a crucial parameter. It can easily be seen that in the low frequency limit, the total capacitance reduces to:

$$C_{tot} = \frac{C_{Sch}G_{pn}^2 + C_{Sch}G_{pn}^2}{\left(G_{Sch}+G_{pn}\right)^2}. \tag{10}$$

In this low frequency limit, the conductance of each diode determines the extracted capacitance. If the front junction has a much higher conductance G_{pn} than the semiconductor-electrolyte G_{Sch}, the total measured capacitance reduces to C_{Sch}. Fig. 3 illustrates the C-V response of the individual diodes and the total circuit, generated by applying (8) and (9) to solved voltage drops across each diode. The "s curve" in the observed capacitance across voltage has been observed in older CdTe devices and explained via a two-diode model previously [6]. The frequency dependence in accurately extracting C_{Sch} is illustrated. While the model behavior with a barrier of 0.7eV helps to demonstrate a more general two-diode behavior, our aim is to introduce as large a barrier as possible. For a larger barrier height of 0.9eV, presuming it is not significantly shunted, all applied potential immediately drops across the rear junction, and the total measured capacitance reflects the space charge only at the back.

The choice of redox system is critical to introduce a large barrier to probe carrier concentrations accurately at the rear of the device.

III. METHODS AND EXPERIMENT

To validate ECV as an approach, samples with spatially uniform chemical composition were characterized both with traditional C-V and ECV. C-V curves were collected on CdTe samples with finished back contact structures showing minimal evidence of back contact barriers. These samples were then etched and polished to prepare for ECV. ECV measurements were carried out on a Dage wafer profiler CVP 21.

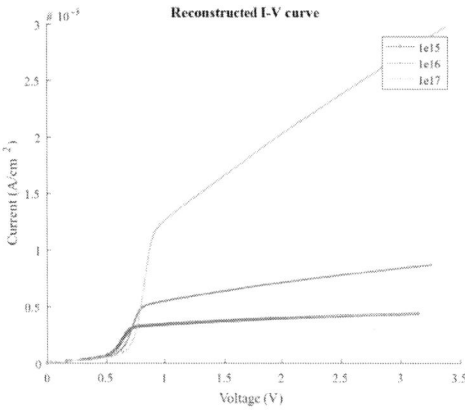

Fig. 1. I-V curves from two-diode model, 0.7eV barrier

978-1-7281-6118-1/22 $31.00 © 2022 IEEE   1356

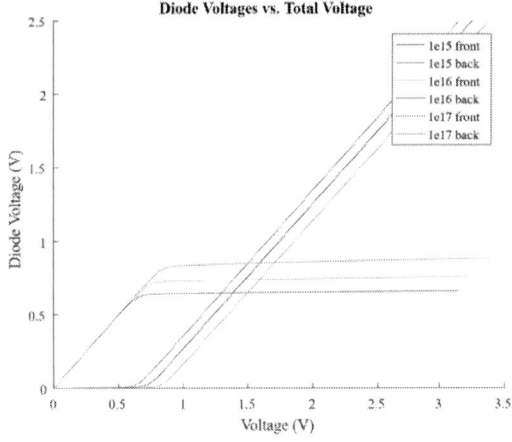

Fig. 2. V-V curves from two-diode model, 0.7eV barrier

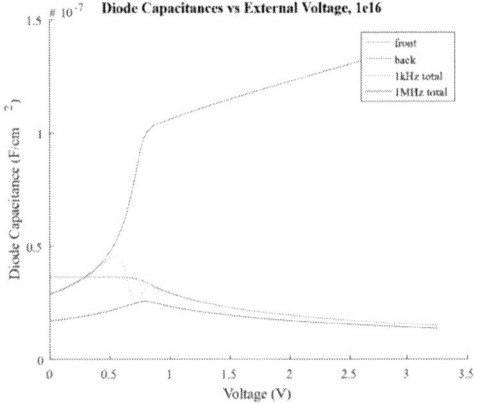

Fig. 3. Capacitance response of two-diode model at low and high frequencies, 0.7eV barrier

The redox system used was an aqueous solution of HCl in IPA at 0.3 %/vol. concentration, giving an expected redox potential close to the SHE potential of 4.44V [7]. This is close to the conduction band of CdTe with an affinity of 4.5eV and therefore is expected to introduce a very large hole barrier on the order of the CdTe bandgap. CV and ECV data are presented in Fig. 4. The capacitance-voltage characteristic from ECV is presented in Fig. 5.

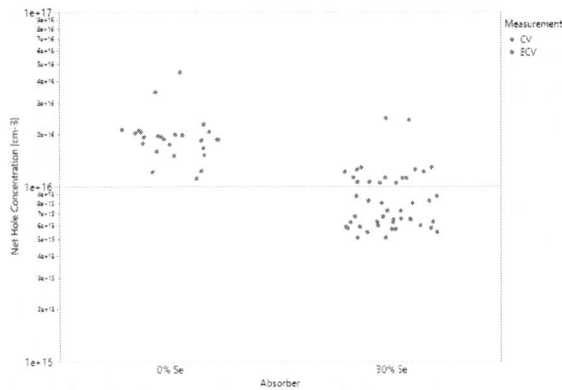

Figure 4. C-V and ECV data for uniformly doped samples

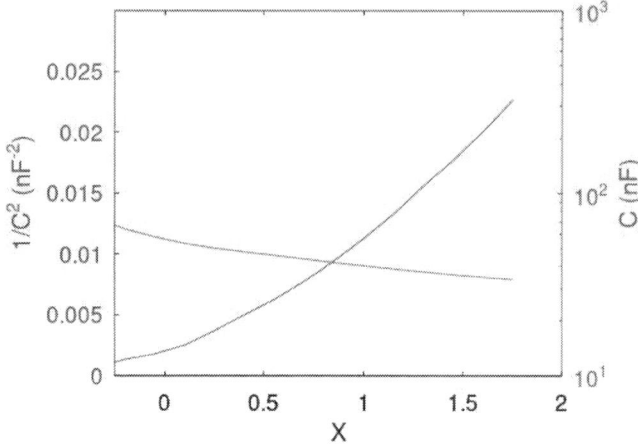

Figure 5. Capacitance and $1/C^2$ measured by ECV on uniform sample

DISCUSSION

The linear $1/C^2$ characteristic present in the ECV data extending over a large voltage range is indicative of uniform doping accurately measured at a single junction, as can be seen from (8). The carrier concentrations measured by ECV at the back are similar in magnitude to those at the front, typically within a factor of two. This is encouraging given that the group V dopant is expected to dope uniformly. Measurements on more than 40 devices (not shown) demonstrate reasonable values in agreement with expected outcomes. Consequently, this ECV approach can help determine carrier concentrations at the rear of semiconductor devices, and in particular overcomes previous barriers to such measurements on polycrystalline thin film solar cells.

REFERENCES

[1] Jian V. Li et al., "Theoretical analysis of effects of deep level, back contact, and absorber thickness on capacitance–voltage profiling of CdTe thin-film solar cells," in Solar Energy Materials and Solar Cells, Vol. 100, 2012, pp. 126-131

[2] R. Memming, "Semiconductor Electrochemistry," John Wiley & Sons, 2015, Ch. 7 pp. 202-203

[3] Sze, Simon M., Yiming Li, and Kwok K. Ng. "Physics of semiconductor devices". John wiley & sons, 2021, Ch. 2

[4] J-N. Chazalviel, "Schottky barrier height and reverse current of the n-Si-electrolyte junction." Surface Science 88.1 (1979): 204-220.

[5] Eisenbarth, T., Unold, T., Caballero, R., Kaufmann, C. A., & Schock, H. W. (2010). "Interpretation of admittance, capacitance-voltage, and current-voltage signatures in Cu(In, Ga)Se$_2$ thin film solar cells." Journal of Applied Physics, 107(3), 034509.

[6] A. Niemegeers, M. Burgelman, "Effects of the Au/CdTe back contact on IV and CV characteristics of Au/CdTe/CdS/TCO solar cells." Journal of applied physics, 81(6), 2881-2886.1997

[7] Trasatti S. "The absolute electrode potential: an explanatory note." Pure App. Chem., 58(7), 955–966.

Reproducibility and Photostability of High-Efficiency Perovskite Solar Cells in Scalable Manufacturing

Rohit Prasanna

Swift Solar, San Carlos, CA, United States

Metal halide perovskite solar cells represent an exciting new generation of high efficiency solar cells including in single junction and tandem configurations. Having made a rapid progress in academic research, they now need to prove their ability to be manufactured using high-throughput and inexpensive methods while maintaining high efficiency. In addition, ensuring long term intrinsic stability of the active light-absorbing perovskite layer under stressors of light and elevated temperature is crucial to enabling long solar cell lifetimes in commercial products. This study designs a deposition system for high-throughput manufacturing of efficient perovskite solar cells. It identifies factors that lead to good reproducibility of perovskite layers deposited using scalable manufacturing techniques capable of high throughput. By identifying a key mechanism of photodegradation, we identify strategies to achieve intrinsic stability of the perovskite active layer under heat and light. On this latter point, we also identify specific research directions that will be important to identifying and further developing perovskite compositions that are likely to be most stable against the degradation mechanisms that we identify.

Optimized Near-Field Thermophotovoltaic Cell using InAs and InAsSbP

Gavin P Forcade, Christopher E Valdivia, Sean Molesky, Shengyuan Lu, Alejandro W Rodriguez, Jacob J Krich, Raphael St-Gelais, Karin Hinzer

SUNLAB, Center for Research in Photonics, University of Ottawa, Ottawa, ON, Canada

Department of Electrical and Computer Engineering, Princeton University, Princeton, NJ, United States

Department of Engineering Physics, Polytechnique Montreal, Montreal, QC, Canada

Department of Physics, University of Ottawa, Ottawa, ON, Canada

Department of Mechanical Engineering, University of Ottawa, Ottawa, ON, Canada

Industrial waste heat is a free and abundant energy source, a quarter of which exists at medium grade temperatures of 600-900 K. For this temperature range, near-field thermophotovoltaics (NFTPVs) are theorized to be the most effective solid-state technology to recycle the waste heat into electrical power. NFTPV devices rely on the enhanced radiation transfer between a radiator and a thermophotovoltaic (TPV) cell separated < 0.2 μm, which can improve power density and efficiency. Unoptimized NFTPV devices and/or large bandgap TPV cells have limited experimental efficiencies of $< 1\%$ for 600-900 K radiators. In this work, we employ a validated 2D drift-diffusion model to optimize an NFTPV device composed of a lattice matched double-heterostructured InAsSbP/InAs/InAsSbP TPV cell positioned 0.1 μm away from a 700 K radiator. Our optimized device shows a $5.6\times$ higher above-bandgap energy transfer compared to the blackbody limit, which produces enhanced power density. Within the 600-900 K temperature range, we locate a maximum efficiency of 13.1% and maximum power output of 1.56 W/cm2 at 900 K, although for different radiator-TPV gaps of 0.1 μm and 0.01 μm, respectively. The maximum efficiency was reached for a larger radiator-TPV gap because of the high parasitic sub-bandgap energy transfer for radiator-TPV gaps less than 0.1 μm. At all temperatures, device efficiency is higher with near-field than far-field illumination.

22% efficiency module combining Silicon Heterojunction Solar and Shingle interconnection

Marina Foti[(1)]*, Marco Galiazzo[(2)], Enrico Sovernigo[(2)], Nicola Frasson[(2)], Cosimo Gerardi[(1)], Alfredo Guglielmino[(1)], Grazia Litrico[(1)], Marcello Sciuto[(1)], Antonio Spampinato[(1)], Antonino Ragonesi[(1)], Francesco Rametta[(1)], Andrea Canino[(3)], Agata Carbonaro[(3)], Fabrizio Coco[(3)], Agnese Di Stefano[(3)], Fabrizio Bizzarri[(4)]

1. Enel Green Power S.p.A, 3SUN, Contrada Blocco Torrazze Zona Industriale 95121, Catania, Italy
2. Applied Materials Italia Srl, Via Postumia Ovest 244, Olmi di San Biagio di Callalta (TV) Italy
3. Enel Green Power S.p.A, Innovation Lab Contrada Passo Martino Zona Industriale 95121, Catania, Italy
4. Enel Green Power S.p.A, Solar Innovation, V.le Regina Margherita 125, Roma, Italy

*Corresponding author: marina.foti@enel.com

Abstract—Efficiency remains the key index among manufacturers' advancements towards an increased module power output. Silicon heterojunction (SHJ) and shingling interconnection are the best in class technologies to obtain the highest efficiency from cells and modules respectively, and in this work, we demonstrate their combination in 22% efficiency full size modules.

Keywords— shingle interconnection, heterojunction, silicon solar cells

I. INTRODUCTION

Shingled interconnection concept applied to c-Si solar cells has been discovered in 1960 and it has been extensively investigated for more than 26 years [1].

In shingled modules solar cells are cut into strips and overlapped to form the so called "strings", with the characteristic roof-tiles shape. Shingling allows the elimination of the space between the cells; increases the active surface per module; removes exposed busbars, reducing the overall resistive losses and promoting a highly aesthetic appearance.

The rise of high-efficiency and highly bifacial solar cell technologies has brought shingle technology developments back to the forefront, with ITRPV forecasting a market share around 10% by 2030 [2]. Originally applied to residential and commercial applications, recently Sunpower and Tongwei deployed this technology in the utility-scale market [3].

It is known that standard (i.e. soldering based) module processing are not completely suitable for Silicon Heterojunction (SHJ) cells [4] due to the high soldering temperatures (>280°C) and thermo-mechanical stress present in the interconnection process. Shingling instead provides industrial feasibility thanks to the use of Electrically Conductive Adhesive (ECA) to join the portions of cell, establishing a permanent bond after curing at less than 200 °C. Due to this low-temperature interconnection, shingling is currently the only industrial solution compatible with thinner wafers (<120 um) and larger wafer formats.

On the other hand, SHJ combining crystalline silicon (c-Si) absorber and thin hydrogenated amorphous silicon (a-Si:H) passiving allow to reach efficiencies higher than 25% [5].

Previous works combining shingle and SHJ showed good results, in terms of both power output as well as reliability [6,7]

II. EXPERIMENTAL SET-UP AND RESULTS

Reference modules based on ECA- busbar interconnections have been processed in EGP 3SUN factory in Catania. The factory consists in fully automated production line for cells and modules manufacturing with annual capacity of 200MWp/year. Modules are bifacial glass-glass with 72 cells monocrystalline n-type SHJ cells [8,9].

Most of the process steps to fabricate shingled modules (cells precursor up to ITO deposition and strings assembly in modules) have been carried out in 3SUN fab for a better result comparison.

For shingle cell, shingling cell metallization, testing and sorting and stringing process have been carried out in Applied Materials laboratory.

Modules assembly and module I-V measurement were performed at EGP.

Experimental set-up was extensively described into our previous work [7].

A. Module layout

Shingled module layout, including string size and connection, has been studied to have module output electrical parameters in line with the standard ones.

Results are compared with standard EGP production module, obtained by cells of similar efficiency. We used same module size and materials type for assembly. Standard modules have been processed in a full automated way, instead the shingled module were processed manually in more steps, including shipment of strings to the different sites. Despite the

978-1-7281-6118-1/22 $31.00 © 2022 IEEE

manual process, we did not observe significant defectivity in shingled modules.

We explored two different layouts for shingled modules to maximize the silicon active area on module and increase efficiency. "Shingle" and "Shingle LA" modules present an increment of active area with respect to the standard 72 cells module of 4.8% and 6.2% respectively. Modules are glass-glass assembled.

We also considered modules in bifacial and monofacial configuration. Bifacial configuration is particularly useful for utility scale application where 15-20% gain in power generated is typically observed with respect to an equivalent monofacial module [10-12]. Monofacial configuration is instead suitable for residential application. Monofacial modules are obtained through a white back sheet.

B. Module electrcal results

Comparison of electrical parameters is reported in Tab. 1. for (A) bifacial configuration and (B) monofacial configuration; percentage of gain due to back-sheet is reported in (C).

The results are in line with the increased active area (resulting in higher Isc) and reduced resistive loss (resulting in improved FF). Shingling technology allows to reach a Cell-To-Module (CTM) factor equal to 1 or higher (CTM is 0.98 for the standard module). For bifacial shingled, we measured a bifaciality factor value comparable with the standard (>90%).

Picture of shingled module is showed in Fig. 1.

Figure 1: Two shingled module in outdoor testing.

III. CONCLUSIONS

In this work, we showed the results of the combination of two highly efficient PV technologies: silicon heterojunction cells (SHJ) and shingling interconnection on full size modules (72 cells equivalent).

The results of shingled modules have been compared with results of reference ECA-busbar interconnections modules, both for monofacial and bifacial modules.

Shingled modules present Cell-To-Module (CTM) factor equal to 1 or higher and

- 22% efficiency on monofacial module (+10% with respect to reference)

- 21.6% efficiency on bifacial module (+11% with respect to reference), with a bifaciality factor >90%.

These results demonstrate the high potential of shingle interconnection applied to SHJ.

TABLE 1   ELECTRICAL RESULTS

(A) Measured values for bifacial modules (measure is referred to front only)

	Isc [A]	Voc [V]	FF [%]	Pmpp [W]	PCE [%]	PCE gain [%]	Active Area Gain [%]	PCE of cells [%]	CTM
72 Cells (Ref)	9.417	52.522	78.0	385.886	19.499	-------	-------	22.832	0.96
Shingle	9.820	54.871	78.8	424.343	21.442	10.0	4.8	23.006	0.999
Shingle LA	9.804	55.540	78.7	428.395	21.647	11.0	6.2	22.832	1.004

(B) Measured values for monofacial modules with Back-sheet

	Isc [A]	Voc [V]	FF [%]	Pmpp [W]	PCE [%]	PCE gain [%]	Active Area Gain [%]	PCE of cells [%]	CTM
72 Cells (Ref)	9.671	52.534	78.0	396.222	20.021	-------	-------	22.832	0.986
Shingle	9.990	54.840	79.0	433.057	21.882	9.3	4.8	23.006	1.020
Shingle LA	9.980	55.510	78.8	436.600	22.061	10.2	6.2	22.832	1.024

(C) Gain due to back-sheet [%]

	Isc [%]	Voc [%]	FF [%]	Pmpp [%]	PCE [%]
72 Cells (Ref)	2.697	0.023	-0.041	2.679	2.679
Shingle	1.729	-0.056	0.254	2.053	2.053
Shingle LA	1.797	-0.054	0.127	1.915	1.915

ACKNOWLEDGMENT

The authors want to acknowledge Applied Materials laboratory for processing the cells and CEA-INES (Armand Bettinelli and Samuel Harrison) for providing the metallization layout.

We observe that module output power of shingled modules is higher with respect to the standard, corresponding to an increase of 10% for "Shingle" and 11% for "Shingle LA" in case of bifacial configuration. For monofacial configuration, the gain is 9% and 10%.

REFERENCES

[1] D. M. Chapin, C. S. Fuller and G. L. Pearson, "A New Silicon p-n Junction Photocell for Converting Solar Radiation into Electrical Power," Journal of Applied Physics 25, p. 676, 1954.

[2] International Technology Roadmap for Photovoltaic (ITRPV), "Results 2019 including maturity report 2020," Eleventh Edision, Oct 2020.

[3] PV, "SNEC 2020 Wrap-Up II: Large formats, high module power output," in SNEC 2020, 2020.

[4] S. K. Chunduri and M. Schmela, "Heterojunction solar Technology," TaiyangNews, 2020.

[5] W. Favre et al, "25% efficient large area silicon solar cell: paving the way for premium PV manufacturing in Europe" in EU PVSEC, 11 September 2020

[6] C. Carrière, V. Barth, A. Bettinelli, S. Harrison, A. Derrier, L. Cerasti and M. Galiazzo, "Toward shingling interconnection with SHJ solar cells," in EU PVSEC, 11 September 2020.

[7] M. Foti et al., "Silicon Heterojunction Solar Module using Shingle interconnection," 2021 IEEE 48th Photovoltaic Specialists Conference (PVSC), 2021, pp. 1092-1095, doi: 10.1109/PVSC43889.2021.9518670.

[8] G. Condorelli et al., "High Efficiency Hetero-Junction: From Pilot Line To Industrial Production," 2018 IEEE 7th World Conference on Photovoltaic Energy Conversion (WCPEC) (A Joint Conference of 45th IEEE PVSC, 28th PVSEC & 34th EU PVSEC), Waikoloa Village, HI, 2018, pp. 1970-1973, doi: 10.1109/PVSC.2018.8548197.

[9] G. Condorelli et al., "Initial Results of Enel Green Power Silicon Heterojunction Factory and Strategies for Improvements," 2020 47th IEEE Photovoltaic Specialists Conference (PVSC), Calgary, OR, 2020, pp. 1702-1705, doi: 10.1109/PVSC45281.2020.9300806

[10] G. Razongles, L. Sicot, M. Joanny, E. Gerritsen, P. Lefillastre, S. Schroder, andP. Lay, "Bifacial photovoltaic modules: Measurement challenges",Energy Procedia, vol. 92, pp. 188–198, 2016.

[11] J. S. Stein, D. Riley, M. Lave, C. Hansen, C. Deline, F. Toor, "Outdoor Field Performance from Bifacial Photovoltaic Modules and Systems," 2017 IEEE 44th Photovoltaic Specialist Conference (PVSC), 2017, pp. 3184-3189, doi: 10.1109/PVSC.2017.8366042.

[12] H. Park, S. Chang, S. Park, W. K. Kim, "Outdoor performance test of bifacial n-type silicon photovoltaic modules," Sustainability,vol. 11, no. 22, 2019, Art. no. 6234, doi: 10.3390/su11226234.

978-1-7281-6118-1/22 $31.00 © 2022 IEEE

Polyethienimine interface dipole tuning for electron selective contacts

Eloi Ros Costals[1], Thomas Tom[2], Gerard Masmitjà[1], Benjamin Pusay[1], Estefania Almache[1], Maykel Jimenez[1], Julià Lopez[1], Edgardo Saucedo[1], Pablo Ortega[1], Joan Bertomeu[2], Joaquim Puigdollers[1], Cristobal Voz[1]

[1]Universitat Politècnica de Catalunya, Barcelona, Catalunya, 08034, España
[2]Universitat de Barcelona, Barcelona, Catalunya, 08007, España

Abstract— **This work studies the use of thin layers of polyethylenimine (PEI) as an interface film to produce electron selective contacts for photovoltaic applications in crystalline silicon. Generally, in conjugated polyelectrolytes such as PEI with a high Lewis basicity, charge is accumulated along the chain of the polymer and counter anions from the solvent create an intense dipole array. In this work, part of the amine groups in PEI are protonated by the solvent that behaves as a weak Brønsted acid during the process. The PEI band modification is able to eliminate Fermi level pinning at metal/semiconductor junctions as it shifts the work function of the metallic electrode by more than 1 eV. As a consequence, induced charge transport between the metal and the semiconductor forms an electron accumulation region and promotes enhanced selectivity.**

Keywords—, Silicon Photovoltaics, Dipoles, Selective contacts, Conjugated Polyelectrolites

I. INTRODUCTION

There are many factors involved in the transport properties of a Metal/Semiconductor (MS) junction. From a simple point of view, the difference in work function between the metal and the semiconductor leads to charge transfer and barrier formation. This effect can be captured in essence by the Schottky-Mott Rule for MS junctions and the Anderson Rule for Heterojunctions. Recently, it has been observed that dipole thin films are able to tune the workfunction of a given electrode [1–4], therefore promoting particular types of charge migration as known from standard knowledge of MS junctions.

The consequence of this workfunction modification at the interface results in an enhanced selectivity of one type of carrier with respect to the other, primarily related to this indirect interface doping. In particular, conjugated polyelectrolyte's such as PFN, PEI, PEIE [5–7]... have been explored throughout the literature mostly to improve contact resistivity and carrier injection in different kind of solar cells (OSCs, Perovskites, Silicon...[8–11]). The advantage of using these types of polymers is that one can fine tune the dipole strength by properly choosing their counter anion [9]. The straightforward mechanism by which these contacts operate allows for a direct route to improve selectivity.

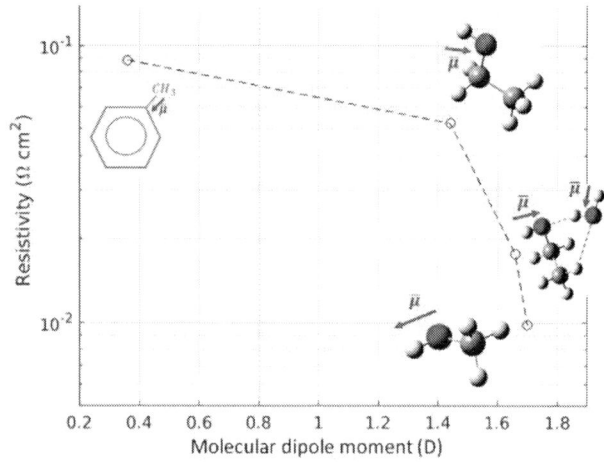

Fig. 1. Specicfic contact resistance of c-Si/PEI$_{(X)}$/Al structure as a function of different solvents and their dipole moment, from left to right: Toluene, Ethanol, Ethanol-Water 50:50 mix, Methanol.

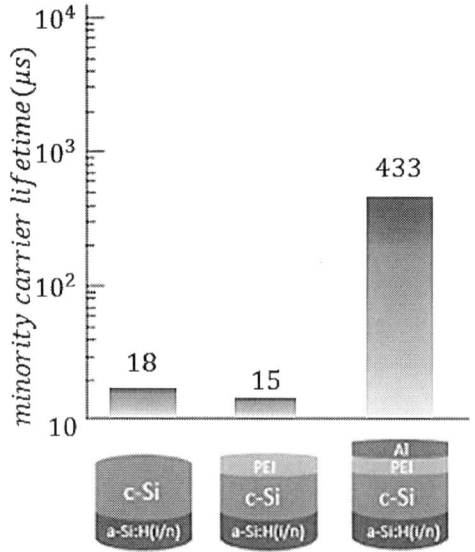

Fig. 2. Minority carrier lifetime measured on a sinton consulting QSSPC setup with a pseudo-ideal rear passivation a-Si:H(i/n) and the effect of PEI$_{et}$ and its notorious emnhancement by adding the semi-transparent metal.

978-1-7281-6118-1/22 $31.00 © 2022 IEEE

In this work we focus on the physical effect produced by the dipole and how chemically modifying the dipoles of the PEI film translate into different physical effects. This is mainly done through the analysis of the contact resistance by means of Transfer Length Method (TLM), the passivation of the silicon surface by means of Quasi Steady State Photo Conductance (QSSPC) as well as interface dipole analysis via Ultraviolet Photon-electron Spectroscopy (UPS).

II. RESULTS

PEI solution was studied with four solvents (i.e. Toluene, Ethanol, Ethanol-Water 50:50 mix, Methanol) and for simplicity they will be referred with the notation $PEI_{(Solvent)}$. The polymer was deposited on pristine c-Si cleaned with HF. Wettability was good in all the presented cases while in stronger polar solvents such as water was not properly spreading as one could expect from the highly hydrophobic Silicon surface. One can see in Fig.1 how the use of stronger polar solvents has positive effect on the contact resistance.

The monotonous trend could be understood from the chemical implications of an increase in polarity of the solvent. A stronger polar solvent is also likely to form an ionic form, which is also related to the pK_a of the solution. In practically all of the used solvents except for toluene, the hydroxide groups can act as proton donors to the amino groups found in the polyelectrolyte. The increased Brønsted acidity (i.e. capacity to give up a proton [H⁺]), is one of the limiting factors of the dipole formation along the polymer chain. This effect is combined with the counter ion condensation of the leftover ionic solvent which also reinforces the dipole strength. Toluene on the other hand has some negative charge density in the aromatic ring that may also interact with the amino groups.

Fig. 3. UPS spectra of reference silicon and their effect by PEI films in ethanol and methanol and PEI in ethanol with a capping of Aluminium.

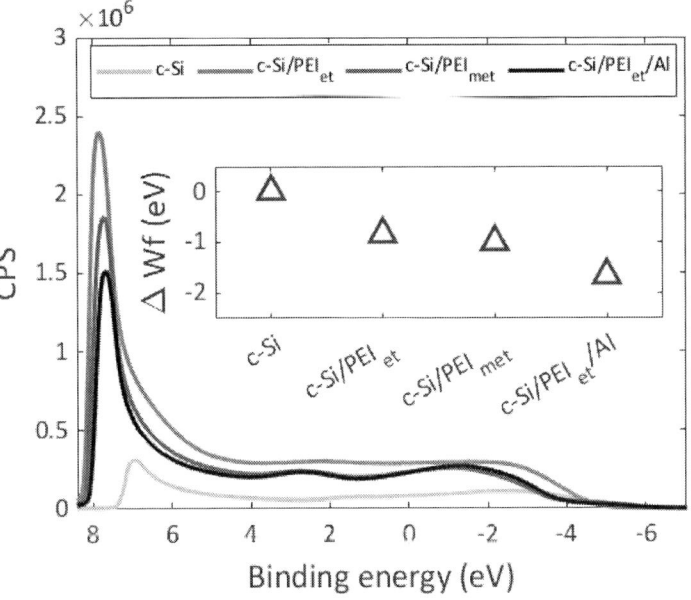

Regarding the physical band structure and the effect of the polymer, it's a reasonable assumption that the effect correlates to the presence of charge in the PEI_{et} film. However, empirical data indicates otherwise. QSSPC results show that after dipole deposition surface passivation remains unchanged. This means that the polymer itself does not provide any kind of surface doping and therefore field effect passivation. On the contrary, the dipole is merely a driving force by which carriers from the electrode migrate to the semiconductor.

One can see how by adding this electron reservoir in the form of a semitransparent aluminum electrode, the dipole effect becomes prominently observable, and the field effect passivation increases lifetime of minority carriers by an order of magnitude up to 400 microseconds.

Finally, by means of Ultraviolet photon-electron spectroscopy the magnitude of the dipole was measured. In this experiment the work function can be calculated from the obtained spectrums with $\phi_s = 21.2 - E_{onset} - E_{Offset} - 10eV$. Comparing with respect to the silicon reference, a first observation might be that the PEI_{Et} film has an overall surface dipole value of 0.9eV while the PEI_{Met} has an overall dipole value of 1.05 eV. The increased workfunction displacement measured by the UPS technique agrees with the first TLM experiments and therefore allows for a direct correlation between the strength and density of the formed dipoles in the films with the dipole moment of the solvent and its pK_A. Furthermore, the PEI_{Et} sample with a small capping of aluminium exhibited a more pronounced displacement of the workfunction. Reported workfunction of aluminium actually lies closer to the reference silicon. Therefore, evidence seems to indicate that a secondary mechanism is allowing for a better electron extraction by reducing the workfunction even further at the interface. This seems to be in agreement with the QSSPC results and enhanced passivation obtained from charge transfer. Indirect doping from the presence of the metallic film allows for a band bending effect which modifies workfunction. This would seem to indicate that the presence of the metal is fundamental to the operation of this type of polymers as selective contacts.

Fig. 4. IV curve of a dopant free ITO/V2O5/c-Si (n)/PEI$_{et}$/Al and its reference without dipole.

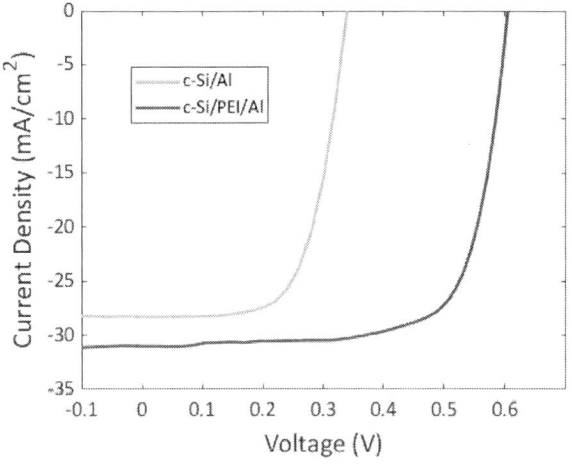

978-1-7281-6118-1/22 $31.00 © 2022 IEEE

A test structure photovoltaic device was fabricated with and without dipole (i.e. ITO/V2O5/c-Si(n)/PEI$_{et}$/Al and ITO/V2O5/c-Si(n)/Al) to measure its effect on efficiency. The solar cell without dipole exhibited a low efficiency produced by a massive loss of V$_{oc}$ in the ETL contact at 0.33V, this can be mostly due to the strong pinning of the semiconductor surface. On the other hand, the deposition of the thin dipole layer produces an increased V$_{oc}$ of more than 200 mV with an obtained value of 0.6V. This seems to be evidence of the elimination of Fermi Level Pinning. FF was largely improved from 64% up to 72% also in line with the decreasing contact resistance seen from TLMs. Finally obtained efficiencies ranged from 6% PCE for the reference cell and 13.7% for the device with the polymer.

III. CONCLUSIONS

This work consists on an investigation between the physical and chemical effects of Metal/Dipole/Semiconductor heterojunctions using an Aluminium/PEI/c-Si(n) structure. The increased dipole moment-pK_a of the solvent exhibited a better contact resistance of the stack measured by TLM. Furthermore, the presence of the metal seems to be fundamentally related to the surface passivation observed by QSSPC. This phenomenon is exhibited via band bending and consequent field effect passivation. UPS experiments seems to agree that PEI$_{Et}$ films have an overall dipole value of 0.15 eV lower than PEI$_{Met.}$. Finally, the presence of the aluminium metallic electrode also seems to be a positive reinforcement of the dipole effect as indicated by the UPS results. Photovoltaic devices indicate that the dipole potential is preventing the surface to be pinned by the metal while also providing some passivation and improved contact resistance.

ACKNOWLEDGMENT

This research has been supported by Spanish government through Grants PID2019-109215RB-C41 (SCALED), PID2019-109215RB-C43, PID2020-116719RB-C41 (MATER ONE) and PID2020-115719RB-C21 (GETPV) and funded by MCIN/AEI/ 10.13039/501100011033. Besides this the work is also supported by the international Grants SENESCYT-2018 funded by Ecuadorian government.

REFERENCES

1. Lin X, Jumabekov AN, Lal NN, et al (2017) Dipole-field-assisted charge extraction in metal-perovskite-metal back-contact solar cells. Nature Communications 8:613. https://doi.org/10.1038/s41467-017-00588-3
2. Reichel C, Würfel U, Winkler K, et al (2018) Electron-selective contacts via ultra-thin organic interface dipoles for silicon organic heterojunction solar cells. Journal of Applied Physics 123:. https://doi.org/10.1063/1.5010937
3. Ford WE, Gao D, Knorr N, et al (2014) Organic Dipole Layers for Ultralow Work Function Electrodes. ACS Nano 8:9173–9180. https://doi.org/10.1021/nn502794z
4. Chen L, Chen Q, Wang C, Li Y (2020) Interfacial dipole in organic and perovskite solar cells. J Am Chem Soc 142:18281–18292. https://doi.org/10.1021/jacs.0c07439
5. Min X, Jiang F, Qin F, et al (2014) Polyethylenimine aqueous solution: a low-cost and environmentally friendly formulation to produce low-work-function electrodes for efficient easy-to-fabricate organic solar cells. ACS Appl Mater Interfaces 6 24:22628–33
6. Ros E, Barquera Z, Ortega PR, et al (2019) Improved Electron Selectivity in Silicon Solar Cells by Cathode Modification with a Dipolar Conjugated Polyelectrolyte Interlayer. ACS Applied Energy Materials 2:5954–5959. https://doi.org/10.1021/acsaem.9b01055
7. Kim S, Lee J, Dao VA, et al (2013) Effects of LiF/Al back electrode on the amorphous/crystalline silicon heterojunction solar cells. Materials Science and Engineering B: Solid-State Materials for Advanced Technology 178:660–664. https://doi.org/10.1016/j.mseb.2012.10.029
8. Würfel U, Seßler M, Unmüssig M, et al (2016) How Molecules with Dipole Moments Enhance the Selectivity of Electrodes in Organic Solar Cells – A Combined Experimental and Theoretical Approach. Advanced Energy Materials 6:. https://doi.org/10.1002/aenm.201600594
9. Wang C, Luo Y, Zheng J, et al (2018) Spontaneous Interfacial Dipole Orientation Effect of Acetic Acid Solubilized PFN. ACS Applied Materials and Interfaces 10:10270–10279. https://doi.org/10.1021/acsami.8b00975
10. Chen L, Chen Q, Wang C, Li Y (2020) Interfacial dipole in organic and perovskite solar cells. J Am Chem Soc 142:18281–18292. https://doi.org/10.1021/jacs.0c07439
11. Shih YC, Lan YB, Li CS, et al (2017) Amino-Acid-Induced Preferential Orientation of Perovskite Crystals for Enhancing Interfacial Charge Transfer and Photovoltaic Performance. Small 13:. https://doi.org/10.1002/smll.201604305

Impact of Snow Depth on Single-Axis Tracked Bifacial Photovoltaic System Performance

Annie C. J. Russell[1], Christopher E. Valdivia[1], Joan E. Haysom[1,2], and Karin Hinzer[1]

[1] SUNLAB, Centre for Research in Photonics, University of Ottawa, Ottawa, Ontario, Canada
[2] J. L. Richards & Associates Limited, Ottawa, Ontario, Canada

Abstract— **Photovoltaic (PV) capacity is rapidly expanding in mid-to-high latitude jurisdictions due to drastic decreases in PV system cost and global decarbonization efforts. However, uncertainty around performance impacts of latitude-specific conditions, such as ground-accumulated snow, contribute to investment risk. In this work, we employed DUET, our custom bifacial PV modelling tool, to study variable ground clearance resulting from snow accumulation. We model the impact of snow depth on a module in four generic, 2-in-portait single-axis tracked (SAT) systems with baseline ground clearances ranging from 1.6-2.8 m. Over the snowy season in Ottawa, Ontario (45°N) and Cambridge Bay, Nunavut (69°N), Canada, rear insolation and energy yield decrease by 3.4-9.6% and 0.36-0.69%, respectively – comparable to annual structure shading and electrical mismatch loss factors. The average daily energy yield loss in both locations is 0.034% per centimeter of accumulated snow. When hourly energy yield loss throughout the snowy season is binned by hour of day, hourly averages peak at 1.2-1.4% loss, suggesting implications for real-time and short-term forecasting.**

Keywords—bifacial, albedo, model, ground clearance, energy yield, rear irradiance, derate, forecasting, high latitude, Northern

I. INTRODUCTION

Competitive photovoltaic (PV) system costs and global electricity decarbonization efforts are driving PV installation at increasingly higher latitudes [1,2]. In Canada, for example, the population – and thus, the demand for affordable, renewable energy – spans latitudes of 41.7-82.3°N. PV systems in such regions, where extended summer daylight hours offset low winter light conditions, benefit from colder average yearly temperatures. In snowy months, *bifacial* PV systems in particular benefit from both high albedo and preferential snow shedding as compared to monofacial modules [3]. However, the dominant factors influencing mid-to-high-latitude PV performance remain uncertain, since performance models have historically been designed and validated for low-latitude regions. For example, rear irradiance of a single bifacial module has been shown to increase with ground clearance [4], but the impact of varying snow depth on multi-row systems in high albedo snowy months is still unknown. Understanding performance under mid-to-high latitude conditions reduces uncertainty in energy yield predictions for this region and, consequently, reduces financial risk for companies and communities looking to invest in bifacial PV [5].

In this work, we used the University of Ottawa SUNLAB's bifacial PV modelling tool DUET [6, 7] which has been validated against field data from 2-in-portrait (2P) fixed-tilt and single-axis tracked (SAT) systems at 55°N [8]. DUET was customized to enable variable ground clearance at each timestamp. We paired this feature with snow depth data to model a time-varying distance between the top surface of accumulated snow and the modules. We compare modelled rear irradiance and energy yield for this snow-dependent ground clearance method against the traditional fixed ground clearance method for a SAT system at 69°N and 45°N during the snowy season. Since forecasting, performance monitoring, and project financing require energy yield predictions at varied timescales, we present a discussion of the hourly, daily, and seasonal impacts of snow depth.

II. METHODOLOGY

DUET's 3D optical model calculates the irradiance incident on the front and rear of each PV cell in a module based on the geometry of the array. This includes the ground clearance, defined here for a SAT system as the distance between the ground and the center of rotation of the torque tube. Eq. 1 and Fig. 1 demonstrate the *snow-dependent ground clearance* ($C_s(t)$) as the difference between a fixed ground clearance (C_f) and snow depth ($S(t)$) at a given time, t:

$$C_s(t) = C_f - S(t) \tag{1}$$

We investigated snow depth impacts at two Canadian locations: Cambridge Bay, Nunavut at 69°N and Ottawa, Ontario at 45°N. The daily snow depth data for each location was retrieved from [9] for all available years in 1990-2020. The average snow depth for each day of the year is found from this 30-year dataset, and the *snowy season* is defined as the period

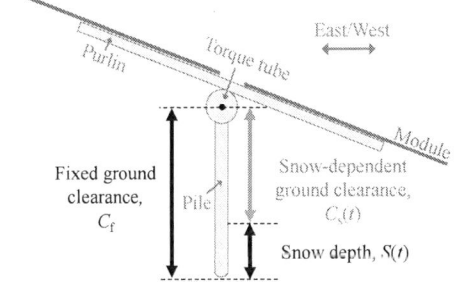

Fig. 1: Cross-section of a 2-in-portrait single-axis tracker with snow-dependent ground clearance as the difference between fixed ground clearance and snow depth for time, t.

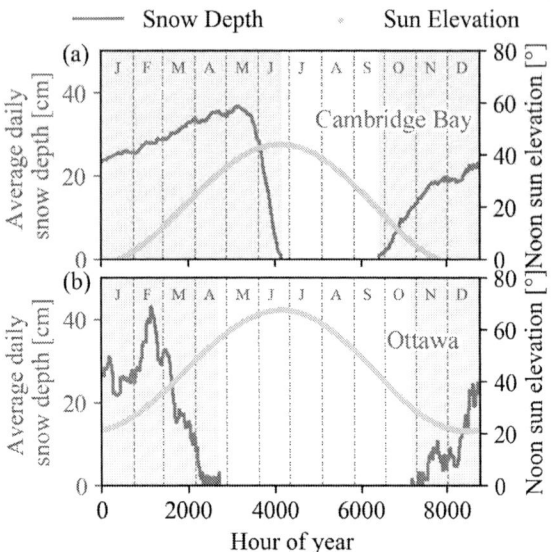

Fig. 2: Hourly snow depth and noon sun elevation for (a) Cambridge Bay, Nunavut and (b) Ottawa, Ontario. The gray region shows the snowy season.

between the first and last snowfall, including days with 0 cm snow depth. Fig. 2 shows that the resulting snowy season runs from September 25 - June 22 (271 days) in Cambridge Bay and October 27 - April 23 (179 days) in Ottawa. The average daily snow depth ranges from 0-37 cm and 0-43 cm for each location, respectively. For the following analysis, the snow depth of a given hour is set equal to the average snow depth on that day.

Hourly irradiance and temperature are taken from the Typical Meteorological Year in each location. In both ground clearance methods, the albedo for each hour is based on [10] with the following values:

(1) 0.8 (fresh snow) for $S(t) \geq 2$ cm;
(2) 0.5 for 2 cm $> S(t) > 0$ cm; and
(3) 0.2 (grass / brown dirt) for $S(t) = 0$.

We simulated the eastern and western modules in the center row of a 5-row, 2-in-portrait (2P) SAT system with ±60° tracking angle and backtracking. The morning performance of

a western string is approximately equivalent to the afternoon performance of an eastern string, and vice versa. The full system performance can be approximated by the average of irradiance and the sum of power for eastern and western modules.

The simulations are executed for all hours in the snowy season with both the fixed ground clearance and the snow-dependent ground clearance methods for four C_f: 1.6, 2.0, 2.4, and 2.8 m. While a true snow surface may be shaped by ground elevation profile, blowing snow, site operation and maintenance, and snow shedding from the front of the panels, these simulations assume a flat snow surface.

This analysis does not account for snow cover on the front of modules. Ground-accumulated snow will have a significant impact on energy yield during these periods since rear-incident light is the primary remaining irradiance source; therefore, the following analysis provides a conservative baseline for performance impacts across the snowy season.

III. RESULTS & DISCUSSION

A. Hourly impact

Since hourly energy yield estimates are required for near real-time forecasting, we investigate the hourly performance impact of snowy conditions. We define the hourly performance impact of snow depth as:

$$\Delta(t) \ [\text{absolute}] = X_s(t) - X_f(t) \quad (2)$$

$$\Delta(t) \ [\text{relative}] = \frac{X_s(t)}{X_f(t)} - 1 \quad (3)$$

where $X_s(t)$ is either the hourly irradiance or the power with a snow-dependent ground clearance at time, t; $X_f(t)$ is the same quantity with a fixed ground clearance.

To demonstrate the impact of snow depth across the day, we bin the hourly performance impacts throughout the snowy season by hour of day for the $C_f = 2$ m case. Figure 3 shows the average of each bin as a solid line and provides the minimum and maximum value in each bin as the bounds of the shaded region. This figure demonstrates eastern module performance,

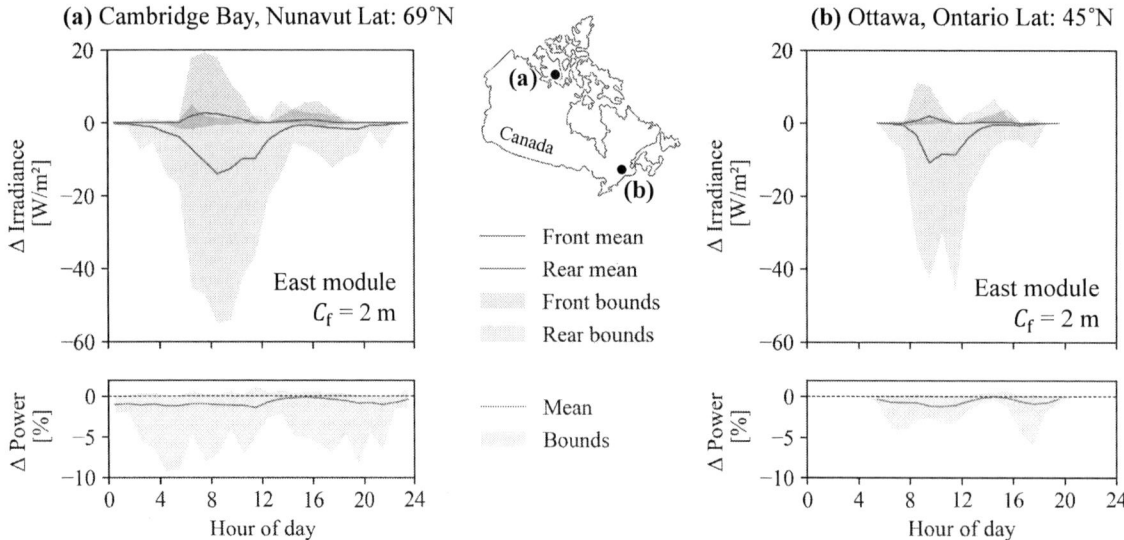

Fig. 3: Change (Δ) in front irradiance, rear irradiance, and energy yield induced by snow-dependent ground clearance by hour of day for an eastern module in a 2P bifacial SAT system in (a) Cambridge Bay, Nunavut, and (b) Ottawa, Ontario. Bounds indicate the highest and lowest change in each metric for a given hour.

978-1-7281-6118-1/22 $31.00 © 2022 IEEE

but the western module performance can be approximated as the mirror image of Fig. 3, flipped about solar noon. The absolute increase in front irradiance and absolute decrease in rear irradiance both peak mid-morning when the module is closest to the ground, which suggests that low-lying fixed-tilt modules would be subject to these peak changes at most hours of the day. The average change in rear irradiance peaks at -14.0 W/m^2 for Cambridge Bay and -10.8 W/m^2 for Ottawa. The average change in electrical power peaks at -1.4% and -1.2%, respectively.

In the days with deepest snow, the maximum absolute hourly change in rear irradiance for the eastern module is -55.1 W/m^2 in Cambridge Bay and -46.0 W/m^2 in Ottawa, corresponding to 44.7% and 22.5% rear irradiance loss and 2.0% and 2.8% power loss, respectively. The maximum relative module power losses – 9.2% in Cambridge Bay and 5.6% in Ottawa – are found in a different set of highly-diffuse evening hours in which rear irradiance contributes 20-25% of total irradiance in the fixed ground clearance case. Since snow depths in this analysis are 30-year averages, larger-than-average snowfalls would see higher hourly impacts than those reported here.

As seen in Fig. 3, higher snow depths can also *increase* module power under some conditions. In the afternoon, for example, reducing the ground clearance of a multi-row array shifts the ground shading pattern in a direction that can increase both front and rear irradiance on the eastern module. However, any afternoon power gain for the higher eastern modules coincides with a more significant power loss for the lower western modules, resulting in an overall negative performance impact to the array during most of these hours. When the eastern and western module performance are combined to represent the full system behavior, the mean impact of snow-dependent ground clearance is more stable throughout the day. Figure 4 shows this full-system relative energy yield loss with a slight peak in the mean at -1.1%.

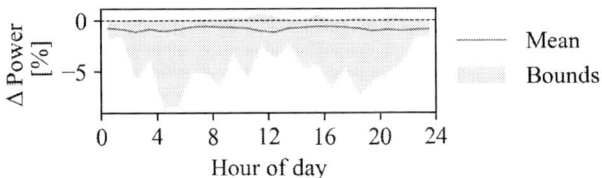

Fig. 4: Relative change in energy yield induced by snow-dependent ground clearance by hour of day for a 2P bifacial SAT system in (a) Cambridge Bay, Nunavut.

B. Daily impact

The snow-dependent ground clearance decreases daily energy yield of the full system by an average of 0.79% in Cambridge Bay and 0.57% in Ottawa across the snowy season. Therefore, adjusting for ground accumulated snow in day-ahead forecasting could avoid mild overestimation of supply. Such a procedure would require an energy yield derate based on accumulated snow depth. Interestingly, daily energy yield losses amount to an average 0.034% per cm of accumulated snow for both locations. Figure 5 shows the spread of daily relative energy yield loss around the loss calculated using the average loss/cm/day derate (dark grey line) in Ottawa. Since Cambridge Bay sees between 0 and 24 hours of daylight

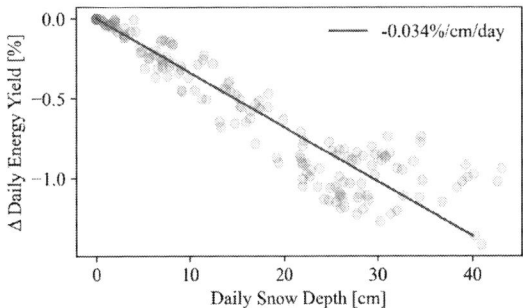

Fig. 5: Daily relative change in energy yield by snow depth in Ottawa, Ontario. Dark gray line shows loss calculated using the -0.034%/cm average derate.

throughout the year, the daily losses are spread more widely around this line than in Ottawa, suggesting that this simple derate may be less useful for forecasting applications at higher latitudes.

C. Seasonal impact

The overall impact of ground-accumulated snow on full-system snowy season insolation is shown in Table 1 for the range of fixed ground clearances, C_f. Rear insolation over the snowy season is 3.4-9.6% lower depending on location and C_f. This irradiance reduction is comparable to the 0-9% shading factors estimated for torque tubes in SAT bifacial PV systems [11]. Figure 6 shows that the absolute decrease in rear irradiance is offset slightly by an increase in front insolation, resulting in 0.37-0.69% energy yield loss in Cambridge Bay and 0.36-0.52% energy yield loss in Ottawa across the snowy season – on-par with annual electrical mismatch loss for SAT systems [12]. Since snow-dependent ground clearance affects energy yield and rear irradiance to a similar degree as widely adopted derate factors like structure shading and electrical mismatch loss, snow depth may be an equally important consideration in modelling applications that rely on accurate yield estimates in snowy months such as storage capacity planning for microgrids.

IV. CONCLUSION

Over the 179- and 271-day snowy seasons in two Canadian locations at 69°N and 45°N, modelling snow-dependent ground clearance for a 2P SAT bifacial system increases front insolation by 0.02-0.37% and decreases rear insolation by 3.4-9.6%, as compared to a fixed ground clearance. The hourly impact of snow depth on front and rear irradiance of a 2P

Table 1: Change (Δ) in front/rear insolation and energy yield for snow-dependent ground clearance with respect to fixed ground clearance over each snowy season.

	C_f	Δ Front Insolation		Δ Rear Insolation		Δ Energy Yield	
	m	%	kWh/m^2	%	kWh/m^2	%	Wh/W$_p$
Cambridge Bay	1.6	0.37	3.2	-8.8	-11.5	-0.37	-3.6
	2	0.20	1.7	-9.6	-14.3	-0.69	-6.7
	2.4	0.14	1.1	-7.0	-11.6	-0.58	-5.7
	2.8	0.04	0.3	-5.5	-9.9	-0.54	-5.3
Ottawa	1.6	0.19	1.0	-6.5	-6.1	-0.40	-2.5
	2	0.11	0.6	-6.2	-6.5	-0.52	-3.2
	2.4	0.07	0.4	-4.1	-4.8	-0.38	-2.4
	2.8	0.02	0.1	-3.4	-4.3	-0.36	-2.3

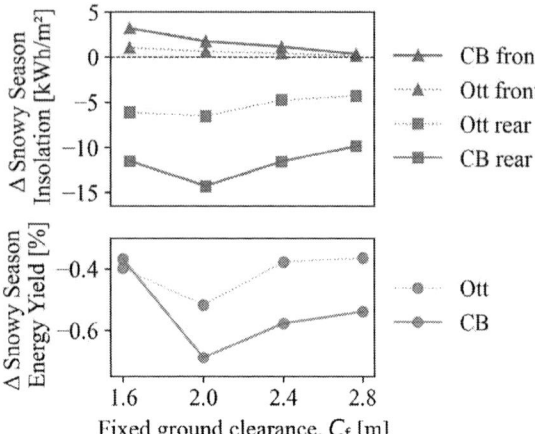

Fig. 6: (a) Absolute change (Δ) in front/rear insolation and (b) relative change (Δ) in energy yield for snow-dependent ground clearance with respect to fixed ground clearance over the full snowy season in each location. CB = Cambridge Bay. Ott = Ottawa.

tracked module is most significant whenever the module is closest to the ground, suggesting that deep snow would affect the lowest lying modules in a fixed tilt system throughout the day. When power loss is binned by hour of day, the average per-bin loss peaks at 1.2-1.4%, suggesting implications for real-time and short-term forecasting.

The snow depth results in an average daily energy yield loss of 0.034% per cm of accumulated snow for both sites. The impact on energy yield over the snowy season ranges from a 0.36-0.69% loss which is comparable to typical annual electrical mismatch and structure shading losses. Thus, a snow depth derate may be warranted for snowy season energy yield

predictions. Further research is required to understand impact of snow depth on fixed-tilt and 1-in-portrait SAT systems, and knowledge of site operation, maintenance, and snow-drift patterns would be beneficial for a full analysis of a given site.

REFERENCES

[1]. International Renewable Energy Agency, "Renewable Capacity Statistics", March 2020. ISBN: 978-92-9260-239-0

[2]. International Renewable Energy Agency, "Renewable power generation costs in 2020", 2021, ISBN: 978-92-9260-348-9

[3]. K. S. Hayibo, A. Petsiuk, P. Mayville, L. Brown, and J. M. Pearce, "Monofacial vs Bifacial Solar Photovoltaic Systems in Snowy Environments," *SSRN Electron. J.*, vol. 193, pp. 657–668, 2022.

[4]. C. Deline et al., "Assessment of Bifacial Photovoltaic Module Power Rating Methodologies—Inside and Out," *IEEE JPV*, Mar. 2017.

[5]. R. Kopecek and J. Libal, "Towards large-scale deployment of bifacial photovoltaics," *Nature Energy*, 2018.

[6]. C. E. Valdivia et al., "Bifacial Photovoltaic Module Energy Yield Calculation and Analysis," in *44th IEEE PVSC*, 2017.

[7]. J. S. Stein et al., *Bifacial PV modules & systems: Experience and Results from International Research and Pilot Applications*. IEA PVPS, 2021.

[8]. A. Russell et al., "DUET: A novel energy yield model with 3D shading for bifacial photovoltaic systems", *under review*.

[9]. Historical Weather Data, Environment & Climate Change Canada, 2021. [Online]. Available: https://climate.weather.gc.ca/historical_data/search_historic_data_e.html

[10]. Perovich, D. "Light reflection and transmission by a temperate snow cover," *Journal of Glaciology*, 53(181), 2007, 201-210.

[11]. S. A. Pelaez, et al., "Effect of torque-tube parameters on rear-irradiance and rear-shading loss for bifacial PV performance on single-axis tracking systems," *46th IEEE PVSC*, 2019.

[12]. C. Deline et al., "Estimating and parameterizing mismatch power loss in bifacial photovoltaic systems," *Prog. PV. Res. Appl.*, no. Dec 2019, pp. 1–13, 2020.

AUTHOR INDEX

Aarnio-Winterhof, Minna ... 485
Abad, Eduardo Camarillo ... 795
Abbas, A. ... 705
Abbas, Ali ... 63, 414, 786, 838, 900
Abbas, Muhammad A. ... 208
Abbott, Malcolm D. ... 1033
Abdallah, Amir A. ... 52, 1151
Abdullah-Vetter, Zubair ... 472, 476
Abe, Adedoyin ... 1279
Abido, Mahmoud Y. ... 212
Ablinger, Ron ... 725
Abouelatta, M. ... 336
Abouelatta, Mohamed ... 1230
Abraham, Sherin Ann ... 967
Abudulimu, Abasi ... 414, 792, 972, 1088, 1190
Abzieher, Tobias ... 351, 565
Ackermann, Mathieu ... 413, 629
Adawi, Mohamed ... 1279
Adua, Habeebullah ... 354
Aeberhard, Urs ... 479, 1339
Afshari, Hadi ... 836
Agarwal, Anusha ... 614
Agarwal, Sumit ... 996
Agrawal, Rakesh ... 1333
Agresti, Antonio ... 576
Aguirre, Aranzazu ... 217
Ahangharnejhad, Ramez Hosseinian ... 701
Ahmad, Rukhsar ... 532
Ahmed, M. Sojib ... 1182
Ahmed, N. ... 811
Aiken, Dan ... 1332
Aimé, Jérémie ... 169
Aimez, Vincent ... 254
Aïssa, Brahim ... 52, 98, 1151
Akhtar, M. Shaheer ... 1292
Akhtar, Naureen ... 1237
Akiyama, Hidefumi ... 468
Akopian, Arkadi ... 244
Al Hasan, Naila M. ... 961
Al Katrib, Mirella ... 1136
Al-Jassim, M. M. ... 872
Al-Jassim, Mowafak M. ... 819
Al-Jassim, Mowafak ... 75, 1321
Al-Modaf, Fhad ... 1248
Al-Shidhani, Mazin ... 395, 396
Alam, Muhammad A. ... 843, 1182
Albadwawi, Omar ... 390, 452, 457
Alberts, Vivian ... 390, 452, 457

Albin, David ... 722, 819
Albrecht, Steve ... 529
Alfadhili, Fadhil K. ... 348, 667
Alhamadani, Hebatalla ... 390
Alhammadi, Aisha ... 12
Ali, Adnan ... 98
Ali, Md. Mahbub ... 1139
Alkhayat, Rabee B. ... 1252
Allami, Hassan ... 575
Almache, Estefania ... 1363
Almeida, Carlos Frederico Meschini ... 839
Almenabawy, Sara M. ... 74
Almenabawy, Sara ... 854
Almutawah, Zahrah S. ... 667
Alnajideen, Mohammad ... 395, 396
Alnaqbi, Wafa ... 12
Alom, Md Zahangir ... 976, 988, 1101
Alshehhi, Badreyya ... 452
Alvarez, Genesis ... 693
Alvi, Md. Shifain Mahathir ... 1139
Alvidrez, Javier Hernandez ... 190, 578
Aly, Shahzada Pamir ... 457
Amer, Fathy Z. ... 1230
Amin, H. M. Noman ... 1235
Anadkat, Nisheka ... 1133
Ananthanarayanan, Divya ... 80
Anctil, Annick ... 144, 1028, 1060, 1313
Anderberg, Allan ... 766
Anderson, Kevin S. ... 714
Anderson, Kevin ... 733
Anderson, Nick ... 1346
Andreas, Afshin ... 146
Aneja, Saurabh ... 733, 1033
Anitat, Remi ... 740
Anjum, Sara ... 1083
Ankireddy, Krishnamraju ... 937
Antolin, Elisa ... 538, 1100, 1340
Antón, Ignacio ... 413, 629, 739
Anwar, Bhuiyan M. ... 667
Aponte-Bezares, Erick E. ... 1091
Aponte-Bezares, Erick ... 398, 916
Araki, Kenji ... 58
Arbaretaz, Sébastien ... 562
Arcebal, John Derek ... 251
Arehart, Aaron R. ... 1315
Arès, Richard ... 430, 550
Arias-Zapata, Javier ... 530, 1291
Armour, Eric ... 164

Arnold, Rachael...961
Arredondo-Orozco, Carlos A.............................778
Artuk, Kerem..1119
Arvinte, Roxana.....................................430, 439
Ascencio-Vásquez, Julián..............................680
Askins, Steve A..629
Askins, Steve.....................................413, 739
Assmann, Nicole...525
Atcitty, Stanley...1201
Athanasopoulos, Stavros...............................1305
Atwater, Harry A...................247, 820, 1083, 1087
Augustine, Sijo..1201
Augusto, André...1045
Ault, David J..419
Avasthi, Sushobhan..............................1133, 1266
Awni, Rasha A....................................464, 761, 1170
Awni, Rasha...126
Ayala, Silvana...255
Ayari, Ahmed......................................439, 627
Ayon, Arturo A..1079
Ayon, Arturo...1069
Azzolini, Joseph A................62, 183, 204, 431
Ba, Fatimata..1262
Babcock, Sean J...460
Babin, Markus...544
Bae, Soohyun.......................................481, 527
Baerwaldt, Daniel.......................................121
Bafti, Arijeta..1196
Bagshaw, Heath...1299
Bai, Jing..345
Bailey, Jeff..425
Bakker, Klaas.....................................381, 740, 748
Ballif, Christophe.........604, 629, 1119, 1120, 1132
Baloch, Ahmer A. B......................................452
Bannister, Mike...480
Bansal, Shubhra..1099
Baranek, Philippe.......................................625
Baribeau, Laurier S....................................1031
Barnes, Teresa...299
Barreau, Nicolas..................................381, 415, 941
Barretta, Chiara..................................485, 633, 680
Barrioz, Vincent...................73, 516, 577, 1311
Barron-Gafford, Greg..................................1189
Barth, Kurt L...658
Barth, Kurt.......................................63, 387
Barthel, Armin...528
Bartholomäus, Martin...................................608
Bashardoust, Sattar......................................18
Bastola, Ebin.........348, 414, 701, 761, 792, 828, 884,
....972, 1088, 1169
Bauhuis, Gerard J......................................1280
Bauhuis, Gerard..875

Baumgarten, Katrina...................................1320
Bauser, Haley C..820
Baxter, Jason B..43
Bayrakci-Boz, Mesude..............................847, 948
Beal, Richard...1346
Beattie, Meghan N................................107, 1295
Beattie, Neil S..................................516, 1311
Beattie, Neil..73
Bedilion, Robin...111
Belfore, Benjamin.........1204, 1209, 1214, 1219, 1224
Bellani, Sebastiano....................................576
Belledin, Udo..18
Bendfeld, Jörg..539
Bengasi, Giuseppe......................................576
Berlinguette, Curtis P.................................1072
Berry, Joseph J..1043
Berry, Joseph..806
Berson, Solenn...1072
Bertagnolli, Justin.....................................814
Bertomeu, Joan..1363
Bertoni, Mariana I.........8, 199, 327, 914, 1004, 1019, 1314
Bertoni, Mariana...................425, 429, 708, 924
Bessa, João Gabriel....................................1294
Betak, Juraj..945
Bhat, A...876
Bhat, Akanksha...314
Bhattacharya, Swastik.................................1173
Bicer, B..879
Bista, Sandip S....................108, 464, 953, 1170
Bista, Sandip Singh.....................................154
Bista, Sandip...126
Biswas, Dhrubes..991
Bittner, Zac...1332
Bivour, Martin...109
Bizzarri, F...554
Bizzarri, Fabrizio..............214, 567, 576, 1360
Blaauw, D...811
Blakely, Logan...204
Bläsi, Benedikt..38
Bliss, Martin.....................................838, 900
Boccard, Mathieu......................................1119
Bogachuk, Dmitry.......................................114
Bogner, Brandon M................................1167, 1314
Bogner, Brandon..991
Bohémier, Cédric.......................................883
Bojorquez, Jose Raul Montes...........................1069
Bolen, Michael...................111, 116, 307, 802
Bonaccorso, Francesco..................................576
Bonilla, Ruy Sebastian.................................443
Booker, Edward P.....................................1072
Bordovalos, Alex.................................653, 1032
Borland, John O..127

Borojevic, Nino 443
Bosco, Nick 106, 298, 783
Bosman, Johan 381
Bothwell, Alexandra M 440, 1315
Boucher, Evan 255
Boucherif, Abderraouf 430, 439, 530, 550, 627,
.... 770, 1291
Bowden, Stuart G 1045, 1178, 1189
Bowden, Stuart 904
Bowers, Jake W. 658, 857
Bowers, Jake 63, 900
Bowersox, David A. 964
Bowman, Mitch 1033
Boyce, Kenneth P. 255
Boyer, Jacob T 867
Brabec, Christoph 31
Bracamonte, Maria Fernanda Villa 1069
Braid, Jennifer L. 805, 980, 1020, 1073
Braid, Jennifer 333
Brantl, Matthew 915
Brau, Tyler R. 828
Brau, Tyler 1055
Braun, Anna K. 198, 975, 1265
Brendel, Rolf 529
Brewster, Charles 741
Brlekovic, Filip 1196
Bromley-Dulfano, Isaac 907
Brown, Lance 1033
Brown, Matthew 255, 992
Brown, Scott 1310
Browne, Jack 1159
Brownell, Brenton 980
Bruckman, Laura S. 255, 668, 796, 980, 1020
Bruhat, Elise 497
Bründlinger, Roland 725
Bu, Xiaotong 384
Buencuerpo, Jeronimo 625
Bueno-Blanco, Carlos 538, 1100, 1340
Bukhari, F. 705
Bukhari, Farwa 604, 786
Buratti, Yoann 353, 476
Burgos, Rolando 693
Burnham, Laurie 205, 333, 1343
Bush, Meghan E 208, 232
Buster, Grant 480
Cabal, Raphaël 607
Cabarrocas, Pere Roca I 109
Cacciato, Mario 214
Cai, Zhonghou 526
Cain, Joe 1322
Cakan, Deniz N. 526, 1070, 1071
Cai, Raul Bayoan 1319

Calaf, Marc 1319
Calvo-Barrio, Lorenzo 284
Camilo, Henrique Fernandes 839
Campagna, Rory M 283
Campbell, Stephen 73, 516, 577, 1311
Campos, Christopher 751
Canino, A. 554
Canino, Andrea 214, 1360
Caño-Prades, Ivan 1304
Cao, Liqi 484
Cappelle, Jan 1162
Carbonaro, Agata 1360
Carr, Anna 1306
Carron, Romain 1145
Case, Chris 328
Caselles, Jaime 739
Cassini, Dênio Alves 534
Castillo, Jasmine Martinez 904
Cattoni, Andrea 625, 722
Cebecauer, Tomas 945
Celik, Ilke 1284
Cha, Seung I. 209
Chacon, Sergio A 836
Chan, Maria K. Y. 425, 429
Chan, Maria 924
Chapotot, Alexandre 530, 1291
Chard, Julie 684
Chatratin, Intuon 808
Chaudhary, Jatin 1173
Chavez, Andre 783
Cheetham, Kieran J 577
Chen, Cong 626, 653
Chen, Jie 975
Chen, Kejun 996
Chen, Lei 126, 626, 653, 953, 1032
Chen, Shangshang 1338
Chen, Shaoqiang 468
Chen, Tao 1126
Chen, Xin 213, 1236
Chenenko, Jason 975
Cheng, Kai 7
Chestek, C. 811
Chia, Keaton 1310
Chin, Robert A Lee 368
Chin, Robert Lee 329, 472
Chin, Xin Yu 1119
Cho, Ben 1332
Cho, Eun-Chel 507
Cho, Jinyoun 235
Cho, Yasuo 461
Chockalingam, Nitin K. 668
Choi, Wook-Jin 251, 366, 1068

Choudhury, Kaushik Roy 980, 1065
Choudhury, Kausik R. 255
Chowdhury, Subhra 991
Chrétien, Jérémie 530
Chretien, Jeremie 627, 770
Christians, Jeffrey A 283, 605
Chung, Jaehoon 108, 898, 1055
Clenney, Jacob A. 327
Click, Natalie 248
Coathup, Trevor J 406
Coco, Fabrizio 1360
Colegrove, Eric 295
Coleman, Kehley A. 668
Coletti, Gianluca 1306
Coll, Pablo Guimera 1314
Colletti, C. 554
Collin, Stéphane 625, 722
Collins, Robert W. 405, 407, 833
Colombara, Diego 897
Colvin, Dylan J. 426, 1046
Compaan, Alvin D. 348
Connelli, Carmelo 567, 576
Conrad, Brianna 164, 837
Contractor, Ardeshir 1275
Cook, Tim 307
Corbett, Brian 1159
Cordon, Jazmine 904
Corrado, Casey 1110
Corso, Roberto 551, 567
Cortes, Bruno 517
Cortes, Jorge 1310
Costa, Suellen C. Silva 534
Costals, Eloi Ros 1363
Costello, Shannon E 828
Couderc, Romain 169
Courtois, Guillaume 235
Covino, Maurice 1046
Cowles, Gabriel 1201
Crawford, Rose 836
Crawford, Zachary 1062
Critchlow, Gary 683
Cros, Stéphane 239, 497, 1072
Crowley, Kyle M 1109
Cruz, Florencia A 1294
Csank, Jeffrey T 754
Curcija, Charlie 155
Cutlip, Elizabeth V 605
D'Rozario, Julia 913
Daenen, Michaël 748
Dai, Xuezeng 1338
Daiber, Benjamin 328
Dalal, Vikram 244
Dancza, Viktor 7
Daniel, Valentin 430, 530, 770
Danielson, Adam 773, 819, 887
Darbali-Zamora, Rachid 398, 431, 754, 1091, 1335
Darnon, Maxime 254, 530, 770
Das, Sandip 424
Das, Ujjwal K 570, 592, 998, 999
Das, Ujjwal 76, 643
Dasgupta, Sagnik 366, 1168
Dauskardt, Reinhold H 106
Davis, Benjamin E. 1323
Davis, Kristopher O. 426, 668, 796, 1046
Davis, Melissa A 1293
De Angelis, Valerio 1201
De Callafon, Raymond 1310
De La Cruz, Rosa 1305
De Lafontaine, Mathieu 254, 770
De Luna, Gabby 251
De Rose, Angela 231
De Vito, Eric 497
De Vrijer, Thierry 1300
De, Shoubhik. 859
Deans, Gordon 77, 80
Debije, Michael G. 89
Deceglie, M. G. 872
Deceglie, Michael G 298, 806, 855, 997
Deline, Chris 992, 1037
Deminico, Mathew R 145
Denafas, Julius 1245
Deng, Yehao 1338
Denis, Christine. 607
Denz, Cornelia 69
Depauw, Valérie 235
Derkacs, Daniel 1332
Desai, Jal 899
Desai, Umang 512, 574
Desarden-Carrero, Edgardo 398, 916
Despeisse, Matthieu 629
Desrues, Thibaut 607
Dessein, Kristof 235
Dhanak, Vin 577
Dhanak, Vinod R 1312
Dharmadasa, Ruvini 937
Di Carlo, Aldo 576
Di Matteo, Alfredo 1306
Di Stefano, A 554
Di Stefano, Agnese 1360
Diallo, Thierno Mamoudou 430
Dias, Pablo R 1047, 1177
Dice, Paul 333, 1343
Diederich, Marvin 529
Diggs, Andrew 1061

Digregorio, Steven 1004
Dimroth, Frank 625
Ding, Jia 1007
Diniz, Antonia Sonia A. C. 534
Dione, Babou 1262
Diop, Pape 1262
Dittmann, Sebastian 1146
Dixon, Richard 413
Dobson, Kevin D 998, 999
Dolan, Connor 526, 1070, 1071
Dominguez, César 413, 629
Don, Christopher H 1312
Don, Eric 658
Dorman, Kyle R 732
Dow, Andrew R. R. 754
Dowd, Misha 1320
Driesse, Anton 172
Druffel, Thad 937
Drury, Storm 1177
Du, Wei 190
Duan, Xiaomeng 154
Dubois, Sébastien 607
Duchamp, Martial 73
Duchemin, Mathilde 629
Duenow, Joel 722
Dulal, Prabin 774
Duran, Ines 1340
Durant, Brandon K 836
Durose, Ken 233, 577, 1312
Duttagupta, Shubham 251, 366, 1068
Dvorak, Adam 1105
Dwivedi, Priya 329, 353, 442, 472, 476
Dykes, Myla 904
Dyreson, Ana 333
Dziechciarz, Mikolaj 740
Ebner, Rita 217
Ebong, Abasifreke 937
Eckstein, Klaus 1190
Ecoffey, Serge 254
Eder, Gabriele 485
Edwards, Paul 480
Eggink, Wouter 731
Eidlisz, Jordan 588
Ekins-Daukes, Nicholas J 959
El Gemayel, Mirella 1190
El Hmaidi, Zakaria Oulad 530, 770
El-Atab, Nazek 1248
Elahi, Sheikh Tawsif 976, 1101
Elhmaidi, Zakaria Oulad 430
Ellingson, Randall J 154
Ellingson, Randy J. 348, 407, 414, 464, 667, 701,
............774, 792, 828, 953, 972, 1055, 1088, 1169, 1170, 1190

Ellinguson, Randy 108
Elshamy, Mohamed 114
Elsworth, James 1322
Emshadi, Khalid 1332
Eperon, Giles E 836
Erickson, Joel 345
Erion-Lorico, Tristan 200
Escarra, Matthew D 208
Escobar, D. Martínez 1059
Esmaielpour, Hamidreza 732
Espenlaub, Andrew 1332
Estrada, Joseph 741
Exilhomme, Amina 1350
Fafard, Simon 254
Fafarman, Aaron T. 43
Fai, Calvin 769
Fairbrother, Andrew 1120
Falkenberg, Gerald 1145
Fan, Yangxin 796
Fanego, Vicente Lara 923
Fardin, Laura R. 545
Farha, Cynthia 59, 239
Farrell, John 884
Farshchi, Rouin 1315
Fassl, Paul 368
Faustini, Marco 625
Favela, Carlos A 115
Fedorenko, Anastasiia 991
Fei, Chengbin 1338
Feichtner, Markus 485
Feldmann, Frank 369
Fencl, Frank 1332
Fenning, David P 327, 526, 914, 1003, 1019, 1070, 1071
Ferekides, Chris 976, 988, 1101
Ferguson, Andrew J. 1283
Fernandes, Paulo A. 630
Fernández, Eduardo F 1294
Fernández, Johjan Stiven Zea 893
Ferreira, Jonathan G. 354
Ferry, David K. 732
Feurer, Thomas 533
Fevola, Giovanni 1145
Fiala, Peter 1119
Fievez, Mathilde 1072
Filho, E. A. Sarquis 223
Fina, Jeffrie 668
Fisher, Kathryn C. 1118
Flandin, Lionel 59, 239
Fleming, Robert A. 1279
Flicker, Jack D. 754
Flicker, Jack 183, 578
Flood, Andrew G. 1264

Florea, Ileana 109
Florides, Michalis 131
Flottemesch, R. 872
Flottemesch, Robert 855, 997
Fonoll-Rubio, Robert 591
Forcade, Gavin P. 107, 1359
Ford, Bethan 1311
Ford, Ethan 1191
Ford, Jody 1037
Fortmann, Charles M. 1350
Foti, M. 554
Foti, Marina 567, 576, 1306, 1360
France, Ryan M. 1283
Frank-Rotsch, Christiane 247
Frasson, Nicola 1360
Fregosi, Daniel 111, 116, 307, 802
Freitas, Walmir 83, 517
French, Roger H. 255, 796, 980, 1020
Friedl, Jared D. 414, 701, 761, 828
Friedman, Daniel J. 766
Friedrich, Lorenz 566
Friesen, Hal 814
Frouin, Bérengère 722
Fthenakis, Vasilis M. 43
Fthenakis, Vasilis 447
Fuke, Pavan 859
Fusaro, Daniel 753
Gabor, Andrew M. 1046
Gabor, Andrew 426
Galiana, Beatriz 1305
Galiazzo, Marco 1360
Gallon, Josh 766
Galstyan, Eduard 115
Gambogi, William J. 255, 868
Gambogi, William 914
Gangemi, W. 554
Gao, Munan 620
Gao, Yuan 155, 1229
Garcia, Juan Lopez 52, 1151
García-Tabares, Elisa 1305
Gardner, Mathew 693
Garrevoet, Jan 1145
Gauding, E. A. 872
Gaulding, Ashley 299
Gaulding, E. Ashley 855, 997
Gay, Guillaume 254
Gayoso, Natalie 899
Gayot, Félix 497
Geerligs, Bart 748
Gehrke, Kai 228, 557
Geisz, John F. 975
Genç, Ezgi 1132

Georghiou, George E. 131, 156, 217, 267
Gerardi, C. 554
Gerardi, Cosimo 567, 576, 1306, 1360
Gevorgian, Vahan 419
Ghahremani, Amir Hossein 1055
Gharabeiki, Sevan 532
Ghimire, Kiran 752
Ghosh, Shibani 967
Gibbs, Jacob M. 828
Gibson, Elizabeth A. 516
Giraldo, Sergio 284, 1304
Glunz, Stefan W. 114
Gnocchi, Luca 1120
Godfrey, Tim 741
Goga, Adam 1062
Goldschmidt, Jan Christoph 566
Golobostanfard, Mohammadreza 1119
Golubev, Timofey 362
González, Carlos Ernesto Arrieta 893
Gonzalez, Pilar Espinet 1083
Good, Brian 295
Goodnick, Stephen M. 1010, 1178
Goodnick, Stephen 1061, 1159
Gorman, Will 1123
Gostein, Michael 285, 291
Gotseff, Peter 146
Gottschalg, Ralph 1146
Götz-Köhler, Maximilian 557
Goumenos, Panagiotis 131
Grätzel, Michael 114
Green, Martin 1126
Greenhalgh, Rachael C. 857
Greenhalgh, Rachael 900
Gregory, Christopher T. 460
Grijalva, Santiago 204
Grover, Sachit 1355
Gruenhagen, Philip 197
Gruginskie, Natasha 875
Gu, Jianli 967
Gu, Xiaohong 255, 1044
Guc, Maxim 591
Guerrero-Perez, Javier 406
Guesnay, Quentin 1119
Gueymard, Christian A. 72, 945
Guglielmino, Alfredo 1360
Guillevin, Nicolas 1306
Gunda, Thushara 899
Guo, Bing 1334
Gupta, Mool 1168
Gupta, Rajesh 1310
Gurian, Patrick 43
Gurule, Nicholas S. 190, 431, 578

Gurule, Nicholas 1335
Guthrey, Harvey 996, 1321
Gutierrez, Carlos M. 208
Habte, Aron 146, 480
Hacke, Peter...................................... 15, 183, 1065
Hackett, Sean 111
Hadipour, Afshin 217
Hadjipanayi, Maria 217
Hages, Charles J...769
Haines, Thad..276
Haley, T. 876
Hallam, Brett 1047, 1177
Halloran, Liam J. 1341
Hamadani, Behrang H.................................. 164, 837
Hameiri, Ziv...................... 329, 352, 353, 367, 368, 442, 443,
.... 472, 476, 1045
Hamers, Edward 606
Hammann, Benjamin 525
Hammann, Liv 95
Hamon, Gwenaëlle 530, 770
Han, Can 484
Han, Sang M. 783
Hanna, Francis 1313
Hansen, Clifford W...22
Hansen, Clifford 1335
Hansen, Ole 461
Hanuš, Tadeáš........................... 430, 439, 530, 627, 770, 1291
Hao, Xiaojing.. 352, 1126
Haque, Anisul 1182
Harding, Alexander J 998, 999
Hariskos, Dimitrios..370
Hart, John T 1332
Härtel, Marlene S..529
Hartley, James Y...199
Hartweg, Barry B.. 1118
Harvey, Steven 643
Hassan, Shaikha...390
Hauch, Jens..31
Haug, Franz-Josef... 1132
Hauschild, Dirk.. 76
Hauser, Adam W. 255
Hauser, Hubert... 38
Hayes, Christopher 858
Haysom, Joan E. 883, 1366
Hazari, J. 876
He, Guojun...352
Heben, Michael J. 348, 407, 414, 667, 701, 761,
..................774, 792, 806, 828, 972, 1055, 1088, 1169, 1190
Heben, Michael................................ 154, 464, 884, 1170
Hegedus, Steven S. 570
Hegedus, Steven 418, 441, 643
Heikkonen, Jukka 1173

Heintz, Alexandre 430, 550
Helfer, Eric 633
Helmers, Henning 38, 107, 1235
Henriques, Jonathan 550
Hense, Jan 1145
Hernández, M Johann A...........................778, 893
Hernandez, Michael 904
Hertel, Tobias..1190
Herterich, Jan Philipp....................................... 114
Heske, Clemens... 76
Hessler-Wyser, Aïcha 1119
Hieulle, Jeremy 532
Hildreth, Owen 1004
Hinsch, Andreas 114
Hinzer, Karin...................... 8, 107, 143, 406, 883, 1031, 1295,
....1346, 1359, 1366
Hirst, Louise C. 528, 795
Ho-Baillie, Anita 368
Höahn, Oliver.. 107
Hobaon, Theo DC 577
Hobbs, William B. 419, 714
Hobson, Theodore D C 1312
Hodge, Bri-Mathias..1038
Hoehn, Oliver... 625
Hoek, Eelko..1306
Hoelscher, Torsten 591
Hoex, Bram 1126, 1191
Hofelmann, Ariana B. 862
Hoff, Thomas E... 197
Hoff, Tom 654
Hoffman, Adam 285
Hoffman, Anthony J... 208
Hoffmann, Stephan 231
Hoheisel, Raymond 624
Höhn, Oliver 38, 1235
Hoke, Andy 1000, 1038
Holland, Nicolas... 223
Holman, Zach 773
Holman, Zachary C. 887, 1118
Holman, Zachary 1329
Holmgren, William F. 714, 995
Holt, Emily...1110
Holt, Martin.. 526
Honsberg, Christiana B. 143, 1059, 1178, 1189
Hour, Socheata 121
Howard, Kassidy H. 819
Hsu, Yu-Lin 1185
Hu, Xiaobo 468
Hua, Amandee... 76
Huang, Jialiang..1126
Huang, Jinchao..1201
Huang, Jing 314, 654

Huang, Jinsong 806, 1338
Huang, Liangyi 796
Huang, Ying-Yuan 366
Hubbard, Seth M 624, 991, 1167, 1314
Hubbard, Seth 913
Huddy, Julia E 28
Huneycutt, Sandra 937
Huntamer, Ryo 915
Hunter, Robert F. H. 1031
Hunwick, Nicholas 387, 827
Huque, Aminul 725, 741, 1013
Hussin, Shahzada Qamar 507
Hutter, Oliver S. 516
Hyndman, David W 144
Iandolo, Beniamino 461
Ilahi, Bouraoui 430, 439, 530, 550, 627, 770, 1291
Imaizumi, Mitsuru 1257
Imbrock, Jörg 69
Infante-Ortega, Luis 387
Ingrish, George B 208
Iqbal, Nafis 668
Irvin, Nicholas P. 143, 1059
Irvine, Stuart J. C. 838
Irvine, Stuart 63, 387
Irving, Richard 407
Isaacs, Mark 577
Isabella, Olindo 484, 623
Isherwood, Patrick J. M. 827
Ishikawa, Yasuaki 44
Islam, Kazi M. 208
Islam, Mohaimenul 1139
Ismael, Timothy 208
Izquierdo-Roca, Victor 591
Jackson, Nicole D 899
Jacobs, Daniel 1119
Jacobs, Deborah 1044
Jäger, Klaus 529
Jäger-Waldau, Arnulf 508
Jahandardoost, Mohsen 1099
Jahangir, Jabir Bin 843
Jahelka, Phillip R 247
Jain, Anubhav 213, 1236
Jakobsen, Michael L. 544
Jamarkattel, Manoj K 154, 348, 414, 667, 828,
.... 972, 1088, 1169, 1170
Jamarkattel, Manoj 884
Janotti, Anderson 808
Jansson, P. M. 330
Jaouad, Abdelatif 254
Jaramillo, M Adolfo A 778
Jaubert, Jean-Nicolas 980, 1020
Javed, Wasim 1334

Javier, Gaia Maria N 353
Jay, Frédéric 607
Jayswal, Niva K. 405
Jeangros, Quentin 1119
Jeffries, April 783
Jelinek, Alexander 1195
Jhang, Song-Syun 1044
Ji, Liang 255
Jia, Yun 468
Jiang, C.-S. 872
Jiang, Chenhui 1126
Jiang, Chun-Sheng 15, 819, 1321
Jiang, Nan 76
Jimenez, Alex 284
Jimenez, Maykel 1363
Jin, Shuangshuang 178
John, Jim Joseph 457
Johnson, Andrew 528
Johnson, Jay 183, 1335
Johnson, Samuel A. 1043
Johnston, S. 872
Johnston, Steve W. 855, 997
Johnston, Steve 200, 643, 649, 975, 1321
Jones, C. Birk 276, 1091
Jones, Derek 646
Jones, Kevin 693
Jones, L. 705
Jones, Leanne A H 1312
Jones, Luke O. 683, 786
Jones, Luke 387, 604
Jones, Michael 73, 1311
Jones, Russell K 27
Jones, Stephen 387
Jones, Steve 63, 838
Jonsson, Jacob 155
Jordan, Michelle 904
Jost, Norman 413
Joyce, Hannah J 795
Ju, Zheng 1007, 1288
Junda, Maxwell M 752
Jundt, Pascal 47
Kaewnukultorn, Thunchanok 418, 441
Kahane, Ben 1033
Kahrl, Fredrich 1123
Kajari-Schröder, Sarah 529
Kakoulaki, Georgia 508
Kaluarachchi, Prabodika N. 414, 667, 828
Kamino, Brett 1119
Kaminski, Anne 607
Kanakkithodi, Arun K. M. 924
Kanaya, Shusaku 1257
Kanevce, Ana 1315

Kang, Yoonmook...481, 527
Kannakithodi, Arun Kumar Mannodi.................................1003
Kanth, Rajeev...1173
Kanyinda-Malu, Clement..1305
Karas, Joseph...1342
Karin, Todd...213, 1236
Kartopu, Giray..63, 838
Kasik, Camden L..702
Kasik, Camden...264
Kau, Wylie..351
Kaufmann, Kai...231
Kazmerski, Lawrence L...534
Keelin, Patrick..314, 654
Keith, Janine M. F...1319
Keith, Janine M. Freeman..302
Kellar, Jon J..1284
Kelzenberg, Michael D..1087
Kelzenberg, Michael...247
Kempe, Michael D...255, 992
Kendall, Anthony D..144
Kenyon, Rick Wallace...1038
Kern, Dana...1321
Kessler, Rich...146
Kessler-Lewis, Emily..624
Khadka, Dhruba B..1, 4
Khalifa, Sherif A..43
Khan, M. Rezwan..1182
Khan, M. Ryyan...1182
Kharait, Rounak A...714
Kharait, Rounak..858, 995
Kherani, Nazir P..74, 854, 1264
Khokhar, Muhammad Quddamah..456, 507
Khurram, Adil..1310
Kiefer, Klaus...223
Kiener, Daniel...1195
Kiessling, Frank..247
Killam, Alex..904
Kim, Donghwan..481, 527
Kim, Eun-Gyeong..1292
Kim, James Hyungkwan...1123
Kim, Moonyong...1047, 1177
Kim, Tae-Gwan..1292
Kim, Youngkuk..456
King, Bruce H......................................285, 291, 333, 961, 1073
King, Bruce..1343
King, Richard R..........................143, 460, 1059, 1178
King, Richard..1159
Kingma, Aldo...740, 1258
Kirmani, Ahmad R...........................320, 351, 1087
Kizilkaya, Orhan..208
Klawitter, Markus...531
Kleissl, Jan...1310

Klie, Robert F..884
Knapp, Steve...1073
Knodle, Phillip J..1046
Ko, Jongwon...481
Kodur, Moses...........................526, 1003, 1070, 1071
Köhnen, Eike..529
Kollosch, Bernd...223
Komoll, Felix...807
Komoto, Keiichi..1179
Köntges, Marc..680, 1245
Kopidakis, Nikos...766, 806, 814
Kornienko, Vlad...414
Kornienko, Vladislav.............................63, 838, 900
Korte, Lars...529
Kossen, Eric...1306
Kothari, Mallika...1060
Kottantharayil, Anil..859
Kottokkaran, Ranjith..244
Kovtun, Michael..1062
Krajne, Andraž...1196
Krasowski, Michael J..145
Krauter, Stefan...539, 650
Krich, Jacob J..........................575, 1295, 1359
Krishnan, Vidya..1329
Kroupa, Daniel M..351
Krügener, Jan...369
Krzywicki, Alfred...476
Kuba, Austin G..........................275, 998, 999
Kubiniec, A...876
Kubiniec, Alex..314
Kuciauskas, Darius...............................440, 1315
Kujovic, Luksa..387
Kulicek, Jaroslav..1088
Kumar, Akash...1065
Kumar, Niranjana Mohan...924
Kumar, Niranjana...425, 429
Kumar, Rishi E............526, 914, 1003, 1019, 1070, 1071
Kumar, Sandeep...1133
Kumar, Vibhor...620
Kumari, Madhuri...888
Kundu, Soumen..1266
Kurstjens, Rufi...235
Kurtz, Sarah R..212, 1059
Kurtz, Sarah...27, 121, 960
Kwapil, Wolfram...525
Kyranaki, Nikoleta..239
Lackner, David..........................38, 107, 625, 1235
Ladd, Anthony J. C..769
Lahti, Gabriella D..320
Lahti, Gabriella..351
Lai, Barry.....................425, 429, 526, 924, 1070, 1071
Lam, Isaac K..275

Lam, Issac	76
Lancia, Adrian A. Santamaria	544
Landis, Geoffrey A.	321
Lanterne, Adeline	607
Lao, Yao	1320
Lara-Fanego, Vicente	945
Laurikenas, Paulius	1245
Lave, Matthew S.	1091
Lavrova, Olga	1201
Law, A.	705
Law, Adam M.	683, 786
Law, Adam	604
Le, Anh Huy Tuan	443, 1045
Le, Tien T.	369
Leaver, Jacob F	233
Leccisi, Enrica	447
Lee, Changhyun	527
Lee, Dong Yoon	209
Lee, Hae-Seok	481, 527
Lee, Hoon Jeong	862
Lee, Hyunjong	1059
Lee, Hyunju	527
Lee, J.	811
Lee, Kyumin	753
Lee, Ross	721
Lehmann, Mario	1132
Leijtens, Tomas	1059
Lennon, Alison	1047
Leonardi, Marco	551, 567
Lepetit, Thomas	415, 941
Leshin, Jeremy	1332
Letner, J.	811
Levrat, Jacques	629
Lewis, Mandy R.	406
Lewis, Timothy	354
Li, Baojie	1236
Li, Bor	529
Li, Chongwen	126, 626, 898
Li, Deng-Bing	108, 154, 464, 667, 701, 972, 1170
Li, Dinica	77
Li, F.	879
Li, Fang	426
Li, Jianjun	1126
Li, Liying	468
Li, Mengjie	796
Li, Wayne	675
Li, Xiaoping	838, 1355
Li, Xinjun	980, 1020
Li, Zelin Zack	255
Li-Kao, Zacharie Jehl	284, 591
Libby, C.	879
Libby, Cara	298, 675
Libraro, Sofia	1132
Lightfoote, Stephen	1163, 1347
Limodio, Gianluca	606
Lin, Der-Yuh	1100, 1340
Lin, Renxing	1229
Lin, Yida	862
Linse, Michael	531
Lisco, Fabiana	604
Litrico, Grazia	1360
Liu, Anyao	369
Liu, Grace	442
Liu, Jiqi	796, 980, 1020
Liu, Julie B.	1350
Liu, Simon H.	1320
Liu, Xiaolei	387, 827
Liu, Xing-Quan	584
Liu, Yuhang	114
Livera, Andreas	131, 267
Liyanage, Geethika K.	348
Loh, Joel Y. Y.	1264
Lombardo, Salvatore A.	551, 567
Loo, Roger	235
Lopes, Miguel	630
Lopez, Hector	1121
Lopez, Julià	1363
Lopez, Kevin	1320
Lopez-Lorente, Javier	156, 267
Loran, Erick Martinez	327
Lorenz, Adam	447
Lorenz, Andreas	531
Lu, Dingyuan	838
Lu, Shengyuan	1359
Lüer, Larry	1190
Lumb, Matthew P.	164
Luna, Monica	1100
Luna-Delrisco, Mario	893
Lunis, Paul	826
Luo, Xianjia	468
Luo, Xin	1229
Luo, Yanqi	526, 1070, 1071
Lustig, Zachary F.	475
Luther, Joseph M.	320, 351, 1087
Luthy, Claire E.	208
Lyons, Alan M.	588
Lyra, Christiano	545
Ma, Yiwei	725, 741
Maasen, Jeroen	1280
Macalpine, Sara M.	964
Macdonald, Daniel	75, 369
Machado, Joana A. F.	532
Macher, Astrid E.	633, 680
Macher, Astrid	485

Machon, Denis 439
Mack, Charles 766, 814
Mack, Sebastian 18
Mack, Shawn 240
Mackenzie, Devin 806
Maclaurin, Galen 480
Macleod, Benjamin P. 1072
Madani, Keeya 1068
Madden, Ryan 828
Magginetti, David J. 1185
Magliano, Erica 576
Mahaffey, Mason 773, 1329
Mahmood, Farrukh Ibne 1198
Mahmud, Rasel 1000
Mainali, Madan K 752
Major, Jon D 577
Major, Jonathan D 233, 1299, 1312
Makhlouf, Seba 606
Makrides, George 156, 267
Mallajosyula, Arun Tej 504
Mallick, Rajni 838
Mallineni, Jaya 1073
Manceau, Matthieu 497, 1072
Mandic, Vilko 1122, 1195, 1196, 1197
Manganiello, Patrizio 623
Mangum, J. 872
Mangum, John S. 855, 975
Mann, Colin J. 1320
Mannino, Gaetano 214
Mannodi-Kanakkithodi, Arun 425, 429
Manshanden, Petra 1306
Mansoori, Ahmad 1332
Manzoor, Salman 708
Mao, Chengliang 1264
Mao, Dan 429, 924
Marchand, Jorge I. T. 1037
Mariam, Tamanna 792, 1055
Mariolle, Denis 497
Mariotti, Silvia 529
Marquis, Audrey 285, 291
Marsillac, Sylvain 407, 1204, 1209, 1214, 1219, 1224
Marti, Antonio 1100, 1340
Martí, David 739
Martin, Ina T. 668
Martineau, David 59, 239
Martinez, Juan F 629
Martinez, Mario 1340
Martinez, Sonia 1310
Martinez-Szewczyk, Michael W 1004
Martinez-Szewczyk, Michael 8
Masmitjà, Gerard 1363
Masters, Adam 1204, 1209, 1214, 1219, 1224

Mastroianni, Simone 114
Masuda, Taizo 58, 467
Matam, Manjunath 596, 692
Matas, Joaquin Coll 381
Mather, Barry 907
Matheron, Muriel 1072
Mathew, Nini Rose 792
Mathew, Xavier 414, 761, 792
Mathiak, Gerhard 390, 457
Matteocci, Fabio 576
Matthew, Xavier 972
Mavromatakis, Fotis 146
Mazzarella, Luana 484
Mbeunmi, Alex Brice Poungoué 430
McDermott, Daniel 235
McGehee, Michael D. 320, 1043
McHann, Stanley 204
McIntosh, Keith R. 1033
McMahon, William E. 198
McMillon-Brown, Lyndsey B. 320, 1109
McMinn, Allison 1288
McNatt, Jeremiah S. 145, 208
Meddeb, Hosni 228, 557
Meidanshahi, Reza Vatan 1010
Meier, Rico 199, 327, 914, 1019
Meier, Sebastian B 1190
Meira, Paulo 83
Mendis, Budhika G 1263
Merkel, Milena 69
Mesropian, Shoghig 584
Meßmer, Marius 18
Metzger, Wyatt K 1355
Meyer, Abigail. R. 996
Meyers, Bennet E. 326, 1048
Meyers, Bennet 930
Meza, Carlos 1146
Micheli, Leonardo 110, 1294
Mikofski, Mark A. 714, 995
Mikofski, Mark 858
Milazzo, Gabriella 567
Miller, David C. 319, 961
Miller, David 15
Miller, Emily J 898
Miller, Michael F. 1315
Miller, Nate 1332
Miller, Wes 838
Mills, Andrew 1123, 1158
Min, Gao 395, 396
Minz, Manoranjan 504
Mira, Daniel Alvarez 608
Mirletz, Heather M. 299
Mirzokarimov, Mirzo 351

Mishima, Tetsuya D.	732
Mishra, Aditya	1310
Mitra, Anirban	98
Mitterhofer, Stefan	1044
Miyano, Kenjiro	1, 4
Moffitt, Stephanie L.	255
Moffitt, Stephanie	1044
Molesky, Sean	1359
Möller, Marius	539, 650
Monakhov, Eduard V.	525
Moncada, Sebastián Villegas	893
Montañés, Cristina Crespo	1123
Montbach, Erica N.	232
Montes-Bojorquez, Jose Raul	1079
Montes-Romero, Jesús	267
Moore, David T.	351
Morais, Modesto	630
Morales-Masis, Monica	109
Morisset, Audrey	1132
Morris, Kerrie M.	857
Morris, Kerrie	900
Morse, Joshua	15
Moseley, John	722
Mouchel, Laurie	439
Mouri, Tasnim K.	76, 592
Mousa, Mohamed	1230
Moutinho, Helio R.	819
Moutinho, Helio	200
Moutinho, R.	872
Mubarak, Roaa I.	1230
Mughal, Maqsood Ali	354
Mukherjee, Shagorika	808
Mulder, Peter	875
Müller, Björn	223
Muller, Matthew T.	628
Muller, Matthew	709
Mundt, Laura E.	320
Muñoz-Rojas, David	722
Muñoz-Santiuste, Juan Enrique	1305
Munshi, Amit	819, 1086
Muttillo, Mirco	623
Muzzillo, Christopher P.	351
Myers, S. M.	330
Najafi, Mehrdad	748
Nakado, Takashi	58, 467
Nakamura, Tetsuya	1257
Nakarmi, Upama	197
Namal, Imge	1190
Namjil, Enebish	1179
Nanda, Govind	854
Nardin, Gaël	629
Nardone, Marco	819
Nassif, Alexandre B.	1127
Natale, Stephen	354
Navarro, Alejandro	284
Navarro-Güell, Alejandro	1304
Nayfeh, Ammar	12
Naylor, Matthew C.	1311
Nayshevsky, Illya	588
Nazginov, Elizabeth	1350
Ndione, Paul	806
Needell, David R.	820
Nejand, Bahram Abdollahi	565
Nemeth, William	996
Ness, Jon	1322
Neto, João Cardoso Das Neves	839
Neugebohrn, Nils	228, 557
Neumaier, Lukas	485
Neupane, Sabin	108, 464, 1170
Newberry, M. G.	330
Newkirk, Jimmy	961
Newmiller, Jeff	995
Nguyen, Dong C.	44
Nguyen, Hieu	75
Nguyen, Phuong-Linh	625
Nicholas, Norm	1033
Nielsen, Michael P	959
Nieschwitz, Kristof J.	828
Nietzold, Tara	425, 924
Niewelt, Tim	525
Niquille, Xavier	629
Nishinaga, Jiro	533
Nishioka, Kensuke	58, 467
Nobre, André M.	614
Nocerino, John	1320
Nolde, Jill A.	240
Nolde, Kristian	1033
Nonni, Elisa	576
Norman, Andrew	1321
Norton, Matthew	217
Nouri, Neda	143, 1295
Nuzzo, Ralph G.	820
O'Brien, Colleen	255
O'Brien, Greg S.	255
O'Neill, Mark	137
Obikoya, Gbenga D.	570
Obikoya, Gbenga	643
Ochoa, Eduardo Prieto	443
Ok, Young-Woo	251, 366, 1068
Okel, Lars	1306
Oklobia, Ochai	63, 387, 838
Okumura, Kenichi	58
Okumura, Teppei	1257
Olhmann, Jens	1280

Oltjen, William C. 796
Onno, Arthur 773, 887, 1329
Oreski, Gernot 485, 633, 680
Orozco, Carlos A Arredondo 893
Ortega, Pablo 1363
Ortiz, Jonathan 1320
Ortiz-Rivera, Eduardo I. 916
Ossig, Christina 1145
Osterthun, Norbert 228, 557
Ostrowski, David P. 320
Ota, Yasuyuki 58
Othman, Mostafa 1119
Ottanà, A. 554
Ottoson, Larry 766
Ourinson, Daniel 531
Outen, Jonathan 283
Ovaitt, Silvana 299, 992
Owen-Bellini, Michael 298, 319, 649, 806
Packard, Corinne E. 198, 975, 1265
Padhamnath, Pradeep 251
Paetel, Stefan 533, 1315
Paetzold, Ulrich Wilhelm 565
Paetzold, Ulrich 368
Page, Matthew 996
Pal, Shweta 134
Palacios, Felipe 754
Palariya, Anuj Kumar 1133
Palekis, Vasilios 976, 988, 1101
Palmiotti, Elizabeth C. 199
Palmiotti, Elizabeth 941, 1204, 1219, 1224
Palmstrom, Axel F. 1043
Pancari, Mia 1350
Pandey, Ramesh 264
Paniagua, Donny Campos 1271
Panžic, Ivana 1122, 1195, 1196, 1197
Papadopoulos, Ioannis 267
Papaioannou, Estefania 721
Paphitis, George 267
Parada, Gabor 658
Paraskeva, Vasiliki 217
Pargon, Erwine 254
Parikh, Abhishek 858
Park, Alex 904
Parquette, William J. 62, 339
Pasmans, Peter 961
Passow, Kendra 753
Patel, Aesha P. 828, 953
Patel, Aesha 1088
Patel, Jay B. 320
Patel, Muhammed Tahir 843
Patel, Pravin 1332
Patjens, Svenja 1145

Pato, Pedro A. V. 83
Paudyal, Bijaya 1271
Paupy, Nicolas 430, 530, 627, 770
Pavgi, Ashwini 1065
Pavic, Luka 1196
Pawar, Vani 1133
Paynabar, Kamran 802
Pazniak, Hanna 576
Pearce, Phoebe M. 959
Peibst, Robby 529
Pelland, Sophie 72
Perea, Gracia Belén 1305
Pereira, Rui N. 98
Perez, Marc 197, 314, 654, 661, 876
Perez, Richard 314, 654, 661
Pérez-Rodríguez, Alejandro 591
Perez-Wurfl, Ivan 352
Perna, Allison N. 198, 1265
Perrin, Greg 15
Perrin, Lara 59, 239, 1136
Perry, Albert 1332
Perry, Kirsten R. 714
Perry, Kirsten 709, 751, 1037
Perry, Lakesha 1044
Perullo, Christopher 307
Pescetelli, Sara 576
Peshek, Timothy J. 145, 232, 320, 1109
Peters, Ian Marius 31, 1191
Peterson, Josh 146, 684
Petit-Etienne, Camille 254
Petri, Delphine 629
Pettengill, Alexandra 1320
Pham, Duy Phong 483, 507
Phelan, Megan E. 820
Phillips, Adam 154, 464, 1055, 1170
Phillips, Adam B. 348, 407, 414, 667, 701, 761,
.... 774, 792, 828, 972, 1088, 1169, 1190
Phillips, J. 811
Phillips, Jamie D. 9
Phillips, Laurie J 577, 1312
Phirke, Himanshu 532
Pierce, Benjamin G 805
Pietzcker, Robert 566
Pike, Christopher 205, 333
Pikolos, Loucas 267
Pilor, Modou 1262
Pilot, Nicholas 111
Pinney, David 204
Pita, Alondra 904
Placidi, Marcel 284, 591, 1304
Planès, Emilie 59, 239, 1136
Plotnikov, Victor V. 348

Podraza, Nikolas J.	405, 407, 653, 752, 774, 833, 898, 972, 1032, 1169
Poessl, Sebastian	608
Pokhrel, Dipendra	108, 792, 1088, 1169
Polizzotti, Alex	1355
Polly, Stephen J.	624, 991, 1167, 1314
Polly, Steve	913
Polo, Christian R.	62, 339
Polzin, Jana-Isabelle	369
Pomares, Luis	457
Porret, Clément	235
Potter, Maggie M.	820
Poudel, Deewakar	1204, 1209, 1214, 1219, 1224
Poulsen, Peter B.	608, 871
Powell, Kaden M.	1185
Powicki, Chris	675
Prabaswara, Aditya	1159
Pradhan, Puja	407
Pranav, Sai	614
Prasanna, Rohit	1358
Prilliman, Matthew J.	302
Prilliman, Matthew	1319
Prinja, Rajiv	74, 854
Privitera, Stefania M. S.	567
Procel, Paul	484
Prokop, Norman F	145
Provost, Philippe-Olivier	1291
Ptak, Aaron J.	198, 867, 975, 1265
Ptak, Aaron	247
Puigdollers, Joaquim	284, 1304, 1363
Purkayastha, Atanu	504
Pusay, Benjamin	1363
Pusch, Andreas	959
Pyrlik, Niklas	1145
Qi, Xin	1007, 1288
Qian, Chen	1126
Qu, Yongtao	73, 1311
Quader, Abdul	154, 414, 667, 1169
Quaegebeur, Nicolas	627
Quispe, David	1329
Rabbani, Zulkifl H.	414, 701
Radovanovic-Peric, Floren	1122, 1195, 1197
Rafhay, Quentin	607
Rago, Emily	961
Ragonesi, Antonino	1306, 1360
Rahman, Farhan	199
Rahman, Mahfujur	1037
Rahman, Md. Mosaddequr	1139
Rahman, Nahian	1350
Rajakaruna, Manoj	953, 1055
Raka, Rawnak Reza	1139
Rakotoniaina, Jean Patrice	169
Ramanujam, Balaji	407
Rametta, F.	554
Rametta, Francesco	1306, 1360
Ramirez, Omar	533
Ranade, Sathishkumar	1201
Ranalli, Joseph	32, 847
Rand, James A.	855, 997
Rane, Karan	1275
Ransome, Steve	375
Rashwan, Ola	444
Rath, Kunal	1020
Rath, Thomas	1122
Rau, Uwe	807, 856
Raupp, Christopher	1073
Raza, Hamza Ahmad	261
Reagan, Jeremiah B.	960
Rebouças, Lara Barros	1280
Redinger, Alex	532
Reece, Peter J	959
Reed, Mason J.	855, 997
Reese, Matthew O.	295, 344
Reese, Samantha B	344
Rehder, Eric M.	584
Reich, Carey L.	887
Reich, Carey	773
Reinders, Angèle H. M. E.	89
Reinders, Angele	489, 731
Reise, Christian	223
Remund, Jan	661
Ren, Wei	1013
Reno, Matthew J.	183, 190, 204, 431, 578
Repecaud, Pierre-Alexis	109
Repins, I. L.	872
Repins, Ingrid L.	855
Repins, Ingrid	806, 1342
Rezek, Bohuslav	1088
Rezk, Ayman	12
Rhee, Kurt	29
Riboldi, Victor B.	517
Ricciardi, Tiago R.	83, 517
Rice, Anthony D.	975
Richardson, Raphael	998
Riedel-Lyngskær, Nicholas	871
Rigby, Oliver M	1263
Rijal, Suman	108, 126, 405, 464, 792, 1170
Riley, Daniel	205, 333, 805, 1343
Ritzer, David Benedikt	565
Roberts, Billy	480
Robinson, Gerald	1322
Robinson, Justin	684, 1073
Robledo, Maryan	904
Rocha, Daniel	630

Rock, Nathan D 1086
Rockett, Angus 415, 941, 1204, 1209, 1214, 1219, 1224
Rodriguez, Alejandro W 1359
Rodriguez, David J. F. 326
Rodriguez, David Jose Florez........................... 1048
Rodriguez-Garcia, Gonzalo 1284
Rodriguez-Peña, Micaela........................ 1100
Rohatgi, Ajeet.................. 76, 251, 366, 592, 783, 1068, 1168
Rojsatien, Srisuda 425, 429, 924
Rollins, Nathan J...................................275
Rolston, Nicholas J 1082
Roosen, Dorrit 1258
Rosenblatt, Nathan........................ 1355
Rosenlieb, Evan 480
Rossol, Michael 480
Rounsaville, Brian 251, 783
Rout, Bibhudutta......................836
Roy, Etee Kawna 1185
Ruffolo, Peter........................283
Ruhstaller, Beat........................479
Ruiz-Arias, Jose A. 945
Ruiz-Preciado, Marco Alejandro 565
Rummel, Brian........................783
Russell, Annie C. J. 406, 1366
Russell, Annie........................883
Ryczek, Catherine N. 820
Saavedra-Peña, Nelson E. 398
Sabin, Neupane 154
Sacchitella, Elijah J. 1314
Sacchitella, Elijah 624
Saddedine, Karima........................239
Saeed, Ahmed....................336, 1230
Sætre, Tor Oskar 1237
Sahli, Florent 1119
Saif, Omar M. 336
Saive, Rebecca........................ 134, 374
Sala, Jacopo 748
Salah, Mostafa M. 1230
Salas, Eduardo 1305
Salome, Pedro M. P. 630
Samake, Papa Monzon........................ 1262
Sampath, Walajabad S. 475, 887
Sampath, Walajabad 387, 773, 819
Sampson, Robert........................ 1189
Sánchez, Hugo 1146
Santala, Annikki 1059
Santiwipharat, Chaiwarut........................999
Santos, Michael B. 732
Sapkota, Dhurba R. 407, 1252
Saraswat, Govind........................419
Sasala, Chase 444
Satou, Akinori........................58

Saucedo, Edgardo.....................284, 1304, 1363
Sauter, Evan........................ 354
Sayre, Larkin........................ 528
Sazzad, Muhammad H 959
Scarpulla, Michael A........................ 1086
Schaefer, Stephen.....................1007, 1288
Schaffer, Kevin G........................ 828
Schall, Jackson W. 1321
Schambach, Lauren........................ 1110
Schauerte, Meike........................38, 1235
Scheer, Roland 591
Scheideler, William J 28
Scheiner, Aaron........................ 22
Schelhas, Laura T......................319, 320, 649, 806, 961
Schermer, John J. 1280
Schermer, John 875
Schindler, Florian 525
Schlemmer, James 314
Schmieder, Kenneth J.....................164, 240
Schneider, Kevin 190
Schriemer, Henry 1346
Schroer, Christian G.....................1145
Schropp, Andreas........................ 1145
Schubert, Martin C. 525
Schulte, Kevin L. 198, 867, 1265
Schygulla, Patrick 38, 114
Sciuto, Marcello 1306, 1360
Scuto, Andrea.....................551, 567
Sedzro, Kwami Senam 967
Seel, Joachim 1158
Segbefia, Oscar Kwame........................ 1237
Seibert, Samuel 953
Seiboth, Frank........................ 1145
Seigneur, Hubert 596, 692, 796, 826
Sekkat, Abderrahime........................ 722
Sellers, Ian R 732, 836, 1178
Selvamanickam, Venkat 115
Semichaevsky, Andrey........................ 92
Sengupta, Manajit.....................146, 480
Sepúlveda-Mora, Sergio........................ 441
Seren, Sven 18
Seron, Charles 607
Serrano-Escalante, Valentina 533
Seum, Abu Niem........................1139
Seymour, K. 876
Seyrich, Martin........................ 1145
Shafarman, William N. 275, 592, 998, 999
Shafarman, William........................ 1315
Shah, Akash 264, 475
Shah, Sanket........................ 753
Shaker, Ahmed.....................336, 1230
Shalvey, Thomas P 1299

Shan, Ambalanath 407
Sharikadze, Saba 244
Sharma, Bhuwanesh Kumar 574
Sharma, Bosky 1119
Sharma, Geetu 854, 1264
Sharma, Sahil 115
Sharp, Jon 753
Shaw, Daniel Z. 702
Shaw, S. 879
Sheil, Huw 577
Sheppard, Scott 307
Shi, Xiaojie 725
Shih, Nathan 1037
Shimpi, Tushar M. 475
Shimpi, Tushar 387, 658, 819
Shiradkar, Narendra 859, 1275
Shirai, Yasuhiro 1, 4
Shirazi, Eli 731
Shrestha, Bishal 833
Shukla, Siddharth 144
Sidawi, Tala 914, 1019
Siebentritt, Susanne 532, 533
Siebert, Michael 1245
Siefer, Gerald 629
Sillerud, Colin 806
Silva, Brandon 826
Silverman, Tim 1319
Silverman, Timothy J. 298, 806, 855, 997
Sim, Yeon Hyang 209
Simon, John 867, 1265
Simpson, Lin J. 915, 1037
Sindermann, Andrew B. 1314
Singh, Ajay 532, 646
Singh, Aparna 512, 574
Singh, Pritpal 721
Sinha, Archana 319, 649, 961
Sinha, Arpan 1168
Sinha, Parikhit 95
Sinton, Ronald A. 925
Sites, James R. 702
Sites, James 47, 264
Sky, Haiku 480
Slauch, Ian M. 199, 914, 1019
Smets, Arno H. M. 484
Smets, Arno 606, 1300
Smiles, Matthew J. 1312
Smirnov, Yury 109
Smith, Ryan M. 596, 692
Smith, Sarah 1319
Smith, Soshana 1044
Snider, Sean 146
Snyder, William 333, 1343

Solas, Álvaro F. 1294
Soman, Anishkumar 570, 643
Somasundaram, Sakthi Guhan 489
Søndenå, Rune 525
Song, Aaron 1350
Song, Tao 766, 814
Song, Zhaoning 108, 126, 405, 464, 626, 653, 701,
.... 752, 792, 953, 972, 1032, 1055, 1170
Sonkar, Ramesh Kumar 504
Soppe, Wim 1258
Sorower, Nur Jahan Beanta 1139
Soufiani, Arman 368
Sovernigo, Enrico 1360
Sowmya, Arcot 476
Sozzi, Giovanna 370
Spampinato, Antonio 1306, 1360
Spatari, Sabrina 43
Spataru, Sergiu V. 544, 608
Spataru, Sergiu 871
Spinella, Laura 649
Sridhar, Seetharaman 299
Srinivasa, Apoorva 1045
St-Gelais, Raphael 1359
Stalker, Amy R. 145
Stanbery, Billy J. 168
Stanislawski, Brooke 1319
Stein, Joshua S. 172, 805, 806, 888, 1073
Steiner, Marc 629
Steiner, Myles A. 1283
Steinhauser, Bernd 369
Stekli, Joseph 111
Stepec, Murielle 562
Stevens, Margaret A. 164, 240
Stid, Jacob T. 144
Stiff-Roberts, Adrienne D. 234
Stradins, Paul 996
Strandwitz, Nicholas C. 1323
Stricher, Romain 254
Stuckelberger, Josua 75
Stuckelberger, Michael E. 1145
Su, Shicheng 1248
Subedi, Biwas 405, 653, 898, 1032
Subedi, Indra 405, 752, 774, 833, 972, 1169
Subedi, Kamala Khanal 774, 1169
Sudbury, Ben A. 1033
Sulas-Kern, D. B. 872
Sulas-Kern, Dana B. 200, 649, 806, 997
Sultana, Nadera 588
Sumita, Taishi 1257
Sun, Kaiwen 352, 1126
Sun, Y. 811
Sung, Li-Piin 1044

Surel, Josephine I.	605
Sutterlueti, Juergen	267
Svatek, Simon A.	1100, 1340
Svatek, Simon Aurel	538
Sveinbjörnsson, Kári	529
Swatton, Nicole	92
Sweat, Rebekah	1293
Szablewski, Marek	1263
Szábo, Sandor	508
Taherimakhsousi, Nina	1072
Takahashi, Takuji	482
Takamoto, Tatsuya	467
Talkington, Samuel	204
Tamizhmani, G.	879
Tamizhmani, Govindasamy	261, 426, 1065, 1198
Tan, Hairen	1229
Tan, Kelvin	62, 339
Tang, Rongfeng	1126
Tao, Meng	62, 248, 339
Tarbuck, Neil	577
Tardio, Miguel Modesto	1305
Tatavarti, Rao	913
Tatsiankou, Viktar	1346
Tawsif, Sheikh Elahi	988
Taylor, Nigel	508
Teague, Ian K.	1037
Teeter, Glenn	819
Teixeira, Jennifer P.	630
Terlier, Tanguy	327
Tervo, Eric J.	1283
Terwilliger, Kent	15, 961
Thaikattil, Greeta J	145
Thakur, Pardeep K	1312
Theelen, Mirjam	381, 740
Theingi, San	996
Thellen, Christopher	961
Theocharides, Spyros	156
Theristis, Marios	172, 826, 888
Thiagarajan, Ramanathan	178, 183
Thom, Daniel	967
Thomas, Luke	577, 1312
Thomere, Angelica	591
Thon, Susanna M.	862
Thornton, Patrick	106
Thorsteinsson, Sune	544
Thuis, Michael	961
Tibbits, Thomas	1235
Till, Micah J.	693
Tina, Giuseppe Marco	214
Tiwari, Ayodhya N.	533
Tiwari, Kunal Jogendra	284
To, Bobby	15

Todaro, Lorenzo	214
Togay, Mustafa	658, 857
Tom, Thomas	1363
Tomko, Brian J.	145
Tonita, Erin M.	8
Topic, Marko	680
Torquato, Ricardo	517
Tracy, Jared	199, 255, 914
Traore, Papa Touty	1262
Treglia, Andrew C.	702
Trindade, Fernanda C. L.	83, 517, 545, 1127
Trujillo, Marena	1038
Truong, Thien	75
Trupke, Thorsten	329, 353, 367, 368, 442, 472, 476
Tsai, Hsinhan	1321
Tse, Yau Yau	63
Tsoulka, Polyxeni	415, 941
Tumusange, Marie S	1032
Turala, Artur	254
Türkay, Deniz	1119
Tutsch, Leonard	109
Tytov, Serhii	1248
Udaeta, Miguel Edgar Morales	839
Uddin, Muhammad Hammad	354
Ulicná, Sona	319, 649, 961
Underwood, Robert	1177
Unruh, Davis	1010, 1061
Upadhyaya, Ajay D.	251
Upadhyaya, Ajay	76, 592
Upadhyaya, Vijaykumar D.	251, 366
Uprety, Prakash	752
Valdes, Nicholas	1315
Valdivia, Christopher E.	8, 107, 143, 406, 883, 1031, 1295, 1359, 1366
Vallerotto, Guido	629, 739
Van Aken, Bas	1306
Van De Lagemaat, Jao	168
Van De Voorde, Mathis	374
Van Den Berg, Joran	740
Van Herpt, Javier	739
Van Nijen, David A.	623
Van Zandt, Devin	1013
Vansant, Kaitlyn T.	320, 1087
Vansant, Kaitlyn	1109
Varonides, AC	637
Vatan, Reza	1061
Veal, Tim D	577, 1312
Vedde, Jan	871
Vehse, Martin	228, 557
Venkat, Sameera Nalin	980, 1020
Verma, Navni	1275
Victor, David	1310

Victoria, Marta 1105
Vignola, Frank 146
Villa, Simona 740
Villa-Bracamonte, Maria Fernanda 1079
Vining, William F. 276
Virtuani, Alessandro 1120
Vlieg, Elias 875, 1280
Volatier, Maïté 254
Von Gastrow, Guillaume 327
Voroshazi, Eszter 169, 562
Voss, Stephen 1347
Voss, Steve 1163
Voyce, Ryan 516
Voz, Cristobal 1363
Wagner, Lukas 114, 239, 566
Walajabad, Sampath S. 658
Walker, Andy 899
Walker, Don 1320
Walker, Janine 1332
Walker, Trumann 429, 924
Walling, Reigh 1013
Walls, J. M. 705
Walls, J. Michael 658
Walls, John M. 683, 827, 857
Walls, John Michael 604, 786
Walls, Michael 63, 387, 414, 838, 900
Walmsley, John 814
Wands, Jake 415
Wang, Biqi 693
Wang, Changlei 752
Wang, Haomiao 384
Wang, Li 1177
Wang, Liwei 178
Wang, Taowen 533
Wang, Wei 976, 988, 1101
Wang, Wenbo 967
Wang, Wenzong 1013
Wang, Xiaoming 126
Wang, Youyang 468
Warner, Cody 1158
Watson, Stephanie 1044
Weber, Julian 231
Weeber, Arthur 381, 484
Wegmueller, Jakob 1020
Weideman, Kyle G 1333
Weinhardt, Lothar 76
Weiser, Philip M. 525
Weiss, Thomas P. 533
Welser, Roger E. 1167
Wen, Bo 693
Wen, Jin 1229
Weng, Guoen 468

Went, Cora 247
Westbrook, Owen W. 964
Wheeler, Aaron 1059
Whiteside, Vincent R. 732
Wieliczka, Brian M. 320
Wieser, Raymond J. 255
Wietler, Tobias 529, 1245
Wikoff, Hope 344
Williams, Benjamin 328
Williams, Henry J. 384
Williams, Rafell 766, 814
Willockx, Brecht 1162
Wilson, Mickey 806
Wilson, Samantha S. 1347
Wilson, Samantha 1163
Wiser, Ryan 1123
Witte, Wolfram 370, 533
Witteck, Robert 1245
Wolf, Andreas 18
Wolff, Christian M. 1119
Wong, Evan 975
Wong, Johnson 77, 80
Wood, Allen 1338
Wright, Brendan 472
Wright, Niara E. 234
Wu, Kunlin 517
Wu, Yinghui 796
Würfel, Uli 114
Wylie, Zachery R 283
Xia, Daixi 575
Xiao, C. 872
Xiao, Chuanxiao 15, 819
Xiao, Ke 1229
Xiao, Xusheng 796
Xie, Yi Hao 121
Xie, Yu 480
Xiong, Gang 838
Xu, Qianfeng 588
Xu, Xinya 1311
Yamada, Ayaka 482
Yamada, Kazumi 58, 467
Yamaguchi, Masafumi 58, 467
Yan, Chang 352
Yan, Yanfa 108, 126, 154, 405, 464, 626, 653, 667,
.... 701, 752, 761, 792, 806, 898, 953, 972, 1032, 1055, 1170
Yanagida, Masatoshi 1, 4
Yang, Dazhi 654
Yang, Guang 1338
Yang, Guangtao 484
Yang, Jaemo 480
Yang, O-Bong 1292
Yang, Wanli 76

Yang, Wei 802
Yang, Zhongshu 369
Yao, Yiyun 967
Yao, Zhirong 484
Yau, Gemini 1038
Ye, Youxiong 28
Yi, Junsin 456, 483, 507
Yilmaz, Pelin 740
Yoon, Heayoung P. 1185
Young, David. L 996
Young, Matthew 75
Youtsey, Chris 137
Yu, Bo 115
Yu, Li 1000, 1038
Yu, Xuanji 255, 796, 980, 1020
Yu, Zhengshan J. 1118
Yu, Zhengshan Jason 1329
Yu, Zhibin 1293
Yuan, Luyao 1028
Yue, Xu 1229
Yun, Min Ju 209
Zabalza, Ruben 255
Zahid, Muhammad Aleem 456, 507
Zakeeruddin, Shaik M 114
Zamora, Rachid Darbali 916
Zareafifi, Farzan 121
Zeder, Simon J. 479
Zekry, A 1230
Zekry, Abdelhalim 336
Zeman, Miro 484, 623
Zhang, Chaomin 143
Zhang, K. Max 384
Zhang, Simon M. F. 352
Zhang, Yibo 74, 1264
Zhang, Yong-Hang 1007, 1288
Zhang, Zheyu 178
Zhao, Yifeng 484
Zhao, Zitong 1010, 1061
Zheng, Jianghui 368
Zhou, Tao 526
Zhu, Junhao 244
Zhu, Tao 126, 1055
Zhu, Xiangqi 907
Zhu, Xitong 89, 489, 731
Zhu, Yan 367
Zhu, Zhengtao 1284
Zilouchian, Ali 1121
Zimanyi, Gergely T. 1010, 1062
Zimanyi, Gergely 1061
Zin, Ngwe 620
Ziska, Catharina 1145
Zoppi, Guillaume 73, 516, 1311

Zou, Min 1279
Zou, Yongjie 1159
Zouaghi, Firas 627
Zouhair, Salma 114
Zuiker, Steve 904
Zunft, Heiko 18

IEEE
445 Hoes Lane
Piscataway, NJ 08854-4141

ISBN 978-1-7281-6118-1